Springer Handbook of Medical Technology

Springer Handbook provides a concise compilation of approved key information on methods of research, general principles, and functional relationships in physics and engineering. The world's leading experts in the fields of physics and engineering will be assigned by one or several renowned editors to write the chapters comprising each volume. The content is selected by these experts from Springer sources (books, journals, online content) and other systematic and approved recent publications of physical and technical information.

The volumes will be designed to be useful as readable desk reference book to give a fast and comprehensive overview and easy retrieval of essential reliable key information, including tables, graphs, and bibliographies. References to extensive sources are provided.

Springer Handbook of Medical Technology

Rüdiger Kramme, Klaus-Peter Hoffmann,
Robert S. Pozos (Eds.)

With 1008 Figures and 139 Tables

Springer

Editors
Rüdiger Kramme
Titisee, Germany

Klaus-Peter Hoffmann
Fraunhofer Institute for Biomedical Engineering
Medical Engineering and Neuroprosthetics
St. Ingbert, Germany

Robert S. Pozos
San Diego State University
Department of Biology
San Diego, CA, USA

ISBN: 978-3-540-74657-7 e-ISBN: 978-3-540-74658-4
DOI 10.1007/978-3-540-74658-4
Springer Heidelberg Dordrecht London New York

Library of Congress Control Number: 2011933994

Production and typesetting: le-tex publishing services GmbH, Leipzig
Senior Manager Springer Handbook: Dr. W. Skolaut, Heidelberg
Translation of Chaps. 1, 3–7, 17, 28, 35, 47–51, 54, 58, 60 from German by Grace Hughes on behalf of Translearning GbR, Mannheim
Typography and layout: schreiberVIS, Seeheim
Illustrations: le-tex publishing services GmbH, Leipzig, Hippmann GbR, Schwarzenbruck
Cover design: eStudio Calamar Steinen, Barcelona
Cover production: WMXDesign GmbH, Heidelberg
Printing and binding: Stürtz GmbH, Würzburg

Printed on acid free paper

Springer is part of Springer Science+Business Media (www.springer.com)

89/3180/YL 5 4 3 2 1 0

Preface

This Springer Handbook is an overview of the expanding and exciting field of medical technology in which the reader will find a modern presentation of the relevant aspects of research, design, manufacturing, and application of different medical devices. The following components: Basics, Functional Diagnostics Devices, Monitoring, Medical Imaging, Therapeutic Devices, Rehabilitation, Medical Information Processing, Telemedicine, Equipment and Tools cover the major aspects of this field.

The handbook was compiled to be an indispensable resource for professionals working directly or indirectly with medical systems and appliances. Just as importantly, it was organized for graduate and postgraduate students in hospital management, medical engineering, and medical physics.

Medical technology has a long and productive tradition of developing medical devices, and innovative approaches to solve critical problems in medicine, biology, and environmental sciences. Hence, biomedical engineering is a rapidly developing field, which exemplifies multidisciplinary approaches such as biotechnology, microsystems technology and telematics. Biomedical engineers develop devices and systems which ultimately contribute to the identification, treatment, abatement and monitoring of diseases and to the compensation of disabilities. The use of modern diagnostic methods enables the early and safe identification of numerous diseases and improving therapeutic outcomes. Using engineering methods to address medical problems will foster additional breakthroughs in clinical treatment and management.

The transfer of ideas from basic research and prototyping to the final medical product, including the methodological questions of application requires continued cooperation between teams. The result of the interaction of basic and clinical medical sciences, information technology, engineering, materials science, and cell biology will open up undreamed-of possibilities in diagnostics and therapy. Challenges include 4-D imaging, e.g. for beating heart diagnostics, coupling of microsystems to neurons, e.g. neural prostheses, the application of new biomaterials with surface modifications at the nanoscale, e.g. for the fabrication of a lifelong stable joining of hip prostheses, and the computer modeling of a virtual patient for the verification of diagnosis and direction of therapy.

Protection of patients, cost reduction and the consideration of progress in medicine as well as the technological state-of-the-art are significant challenges for the development of medical products. Hopefully this Springer Handbook will assist in the continued development of new medical products that will enhance the well being of patients.

The editors would like to thank the authors for their fruitful, successful and collegial cooperation. It was a pleasure for us to collect views from the different fields of medical technology and bring them together in the handbook. Special thanks to Dr. Werner Skolaut, Senior Manager Springer Handbooks, and Dr. Thomas Ditzinger, Senior Editor Engineering/Applied Sciences, from Springer publishing for their time, help and kind support.

May 2011

Rüdiger Kramme — Titisee
Klaus-Peter Hoffmann — St. Ingbert
Robert S. Pozos — San Diego

About the Editors

Rüdiger Kramme, Graduate Engineer, studied Biomedical and Hospital Engineering as well as Economics in Gießen and Freiburg, Germany. After his studies, he acquired several years of professional experience in sales, marketing and human resources development in the medical device industry for consumer and investment capital goods. Since 1993 he is working for the German Federal State of Baden-Württemberg and the German Federal Armed Forces, where he is involved in planning and project development of University Hospitals and Military Medical Facilities in Germany and abroad. He is a lecturer for Medical Engineering at the University of Applied Sciences in Gießen, Germany. Rüdiger Kramme has authored numerous scientific publications in magazines and books, and is the founder and editor of the Springer standard work *Medizintechnik*, which will be published in its fourth edition in 2011, as well as the Springer Dictionary *Technische Medizin*.

Klaus-Peter Hoffmann received his Doctorate degree in Biomedical Engineering from the University of Technology in Ilmenau, Germany, in 1987. He is currently Professor of Biomedical Engineering at the University of Applied Sciences in Saarbrueck, and Head of the Department of Medical Engineering and Neuroprosthetics at the Fraunhofer Institute for Biomedical Engineering in St. Ingbert. His main research interests include methods and devices of Clinical Neurophysiology especially Neuromonitoring and saccadic eye movement as well as the use of Microsystems in medicine especially sensors and actuators for Neuroprostheses. He is also active in the field of cognitive technical systems. He has coordinated various European and national research projects, has published more than 200 journal papers, conference papers and book chapters. He is member of several scientific societies as well as in the advisory board of the journal *Das Neurophysiologie-Labor*. Since 2004, Professor Hoffmann has led the Expert Group Neuroprosthetics in the Initiative Micromedicine of of the German Association for Electrical, Electronic & Information Technologies (VDE). He is also member of the Technical Committee on Cardiopulmonary Systems of the IEEE Engineering in Medicine and Biology Society.

Robert Pozos, Professor of Biology at San Diego State University, has been actively engaged in numerous studies utilizing various biomedical technologies to study human performance in extreme environments. His expertise extends to motor control studies in which he has two patents dealing with quantification of fatigue generated during typing using dynamic finger force measurements. His current studies deal with the quantification of surface electromyographic signals during various athletic events utilizing wifi technology. In addition he is continuing his studies that combine the use of NIRS measurements with EMG signals during movement.

List of Authors

Dino Accoto
Università Campus Bio-Medico di Roma
CIR – Center for Integrated Research
Via A. Del Portillo
00128 Rome, Italy
e-mail: *d.accoto@unicampus.it*

Albert J. Augustin
Städtisches Klinikum Karlsruhe
Augenklinik
Moltkestr. 90
76133 Karlsruhe, Germany
e-mail: *albertjaugustin@googlemail.com*

Natasha Avila
e-medicis
26 rue George Sand
75016 Paris, France
e-mail: *natasha.avila@emedicis.com*

Michael Bateman
University of Minnesota
Department of Biomedical Engineering and Surgery
420 Delaware St. SE
Minneapolis, MN 55455, USA
e-mail: *batem034@umn.edu*

Andreas Bermann
Siemens Healthcare
Imaging and Therapy Division, SYNGO
Henkestr. 127
91052 Erlangen, Germany
e-mail: *andreas.bermann@siemens.com*

Ulrich Böckler
Sorin Group Deutschland GmbH
Lindberghstr. 25
80939 München, Germany
e-mail: *ulrich.boeckler@sorin.com*

Armin Bolz
Karlsruhe Institute of Technology
Institute for Biomedical Engineering
Kaiserstr. 12
76128 Karlsruhe, Germany
e-mail: *armin.bolz@kit.edu*

Jörn Borgert
Philips Technologie GmbH Forschungslaboratorien
Tomographic Imaging Group
Röntgenstr. 24–26
22335 Hamburg, Germany
e-mail: *joern.borgert@philips.com*

Michael Born
San Diego State University
Department of Biology
5500 Campanile Drive
San Diego, CA 92182, USA
e-mail: *Michael.j.born@gmail.com*

Guenther Braun
University Medical Center Freiburg
Paediatrics and Adolescent Medicine
Mathildenstr. 1
79106 Freiburg, Germany
e-mail: *guenther.braun@uniklinik-freiburg.de*

Dirk Büchel
Medical Faculty of Tübingen
Ernst-Simon-Str. 16
72072 Tübingen, Germany
e-mail: *buechel.dirk@t-online.de*

Thorsten M. Buzug
University of Lübeck
Institute of Medical Engineering
Ratzeburger Allee 160
23538 Lübeck, Germany
e-mail: *buzug@imt.uni-luebeck.de*

Marco Capogrosso
Scuola Superiore Sant Anna
ARTS lab
Piazza Martiri della Liberta, 33
56100 Pisa, Italy
e-mail: *m.capogrosso@sssup.it*

Peter H. Cossmann
Med Tech Consulting
Säntisstr. 10
5430 Wettingen, Switzerland
e-mail: *peter.cossmann@medtechconsulting.ch*

Andreas Demosthenous
University College London
Electronic and Electrical Engineering
Torrington Place
London, WC1E 7JE, UK
e-mail: *a.demosthenous@ee.ucl.ac.uk*

Thomas M. Deserno
RWTH Aachen University
Pauwelsstr. 30
52057 Aachen, Germany
e-mail: *Deserno@ieee.org*

Nick Donaldson
University College London
Medical Physics & Bioengineering
Gower St.
London, WC1E 6BT, UK
e-mail: *nickd@medphys.ucl.ac.uk*

William K. Durfee
University of Minnesota
Department of Mechanical Engineering
111 Church St SE
Minneapolis, MN 55455, USA
e-mail: *wkdurfee@umn.edu*

Günter Edlinger
g.tec medical engineering GmbH
Herbersteinstr. 60
8020 Graz, Austria
e-mail: *edlinger@gtec.at*

Amir Eftekhar
Imperial College London,
South Kensington Campus
Centre for Bio-Inspired Technology
London, SW7 2AZ, UK
e-mail: *a.eftekhar@imperial.ac.uk*

Mariana Fernandes
University of Minho
Department of Industrial Electronics,
DEI – Campus de Azurém
4800-058 Guimarães, Portugal
e-mail: *mfernandes@dei.uminho.pt*

Eduardo Fernandez
Universidad Miguel Hernández
Bioengineering Institute
Avenida de la Universidad, s/n
03202 Elche, Spain
e-mail: *e.fernandez@umh.es*

Fabiola Fernandez-Gutierrez
University of Dundee
Institute for Medical Science and Technology
1 Wurzburg Loan, Dundee Medipark
Dundee, DD2 1FD, UK
e-mail: *f.fernandezgutierrez@dundee.ac.uk*

Celso P. Figueiredo
University of Minho, Campus de Azurém
Department of Industrial Electronics
4800-058 Guimarães, Portugal
e-mail: *celso.figueiredo@dei.uminho.pt*

Harald Fischer
Creamedix GmbH
Königsberger Str. 2
76356 Weingarten, Germany
e-mail: *hafi@creamedix.eu*

Martin R. Fischer
Private University Witten/Herdecke gGmbH
Fakultät für Gesundheit
Alfred-Herrhausen-Str. 50
58448 Witten, Germany
e-mail: *martin.fischer@uni-wh.de*

Óscar S. Gama
Universidade do Minho, Campus de Gualtar
Departamento de Informática
4710-057 Braga, Portugal
e-mail: *osg@di.uminho.pt*

Stephan Garbe
University Hospital of Bonn
Radiology Department
Sigmund-Freud-Str. 25
53105 Bonn, Germany
e-mail: *garbe@uni-bonn.de*

Belinda Garner
Imperial College London,
South Kensington Campus
Institute of Biomedical Engineering
London, SW7 2AZ, UK
e-mail: *b.garner@imperial.ac.uk*

Armin Gärtner
Ingenieurbüro für Medizintechnik
Edith-Stein-Weg 8
40699 Erkrath, Germany
e-mail: *armin.gaertner@t-online.de*

Bernhard Gleich
Philips Technologie GmbH
Tomographic Imaging Group
Röntgenstr. 24–26
22335 Hamburg, Germany
e-mail: *bernhard.gleich@philips.com*

Rudolf Götz
Aloka GmbH
Halskestr. 25
47877 Willich, Germany
e-mail: *R.Goetz@aloka-europe.com*

Christoph Guger
g.tec medical engineering GmbH, Guger Technologies OG
Sierningstr. 14
4521 Schiedlberg, Austria
e-mail: *guger@gtec.at*

Eugenio Guglielmelli
Universita' Campus Bio-Medico
CIR – Center For Integrated Research
Via Alvaro Del Portillo 21
00128 Rome, Italy
e-mail: *E.Guglielmelli@Unicampus.It*

Martin Haag
Heilbronn University
Medical Informatics
Max-Planck-Str. 39
74081 Heilbronn, Germany
e-mail: *martin.haag@hs-heilbronn.de*

Reiner Haag
Lawton GmbH & Co. KG
Württemberger Str. 23
78567 Fridingen, Germany
e-mail: *reiner.haag@lawton.de*

Peter Haas
University of Applied Sciences and Arts Dortmund
Medical Informatics
Emil-Figge-Str. 42
44147 Dortmund, Germany
e-mail: *haas@fh-dortmund.de*

Andreas Hahn
Sorin Group Deutschland GmbH
General Management
Lindberghstr. 25
80939 München, Germany
e-mail: *andreas.hahn@sorin.com*

Sibylle Hanus
Textile Research Institute Thuringia-Vogtland
Textile Structures and Materials
Zeulenrodaer Str. 42
07973 Greiz, Germany
e-mail: *s.hanus@titv-greiz.de*

Michael Heinlein
Mednovo Medical Software Solutions GmbH
Hohenzollerndamm 150
14199 Berlin, Germany
e-mail: *info@mednovo.de*

Ewald Hennig
Deutsches Herzzentrum Berlin
Augustenburger Platz 1
13353 Berlin, Germany
e-mail: *Hennig@dhzb.de*

Roland Hentschel
Universitätsklinikum Freiburg
Neonatologie/Intensivmedizin, Zentrum für Kinder- und Jugendmedizin
Mathildenstr. 1
79106 Freiburg, Germany
e-mail: *roland.hentschel@uniklinik-freiburg.de*

Roland Hetzer
Deutsches Herzzentrum Berlin
Augustenburger Platz 1
13353 Berlin, Germany
e-mail: *hetzer@dhzb.de*

Lothar Heuser
Klinikum der Ruhr-Universität Bochum
Institut für Diagnostische & Interventionelle Radiologie, Knappschaftskrankenhaus
In der Schornau 23–25
44892 Bochum, Germany
e-mail: *lothar.heuser@rub.de*

Ullrich Hieronymi
Dräger Medical Deutschland GmbH
Moislinger Allee 53–55
23558 Lübeck, Germany
e-mail: *ullrich.hieronymi@draeger.com*

Robert Hitchcock
University of Utah
Bioengineering
Salt Lake City, UT 84112, USA
e-mail: *r.hitchcock@utah.edu*

Gerhard Hoffmann
Orangedental GmbH & Co KG
Aspachstr. 11
88400 Biberach, Germany
e-mail: *gh@orangedental.de*

Klaus-Peter Hoffmann
Fraunhofer Institute for Biomedical Engineering
Medical Engineering and Neuroprosthetics
Emsheimer Str. 48
66386 St. Ingbert, Germany
e-mail: *klaus.hoffmann@ibmt.fraunhofer.de*

Gerald C. Holst
JCD Publishing Company
Winter Park, FL 32789, USA
e-mail: *Jerry@JCDPublishing.com*

Sebastian Hoth
University of Heidelberg
ENT Hospital – Audiology
Im Neuenheimer Feld 400
69120 Heidelberg, Germany
e-mail: *sebastian.hoth@med.uni-heidelberg.de*

Stephen A. Howard
University of Minnesota
Department of Biomedical Engineering and Surgery
420 Delaware St. SE
Minneapolis, MN 55455, USA
e-mail: *howa0255@umn.edu*

Bernhard Hug
KLS Martin GmbH + Co. KG
Am Gansacker 1b
79224 Umkirch, Germany
e-mail: *Bernhard.Hug@klsmartin.com*

Paul A. Iaizzo
University of Minnesota
Department of Surgery
420 Delaware St. SE
Minneapolis, MN 55455, USA
e-mail: *iaizz001@umn.edu*

Erwin Immel
University of Dundee
Institute for Medical Science and Technology
Wurzburg Loan
Dundee, DD2 1FD, UK
e-mail: *e.z.immel@dundee.ac.uk*

Klaus-Martin Irion
Karl Storz GmbH & Co. KG
Mittelstr. 8
78532 Tuttlingen, Germany
e-mail: *K.Irion@karlstorz.de*

Heinz-Michael Just
Klinikum Nürnberg
Institute of Hospital Epidemiology, Medical Microbiology and Infectious Diseases
Prof.-Ernst-Nathan-Str. 1
90419 Nürnberg, Germany
e-mail: *just@klinikum-nuernberg.de*

Daniel W. Kauff
University Medicine of the Johannes Gutenberg University Mainz
Department of General and Abdominal Surgery
Langenbeckstr. 1
55131 Mainz, Germany
e-mail: *Kauff@ach.klinik.uni-mainz.de*

Rüdiger Klar
Universitätsklinik Freiburg
Institut für Medizinische Biometrie und Medizinische Informatik
Stefan-Meier-Str. 26
79104 Freiburg, Germany
e-mail: *ruediger.klar@web.de*

Werner Kneist
University Medicine of the Johannes Gutenberg University Mainz
Department of General and Abdominal Surgery
Langenbeckstr. 1
55131 Mainz, Germany
e-mail: *werner.kneist@unimedizin-mainz.de*

Klaus P. Koch
University of Applied Sciences Trier
Department of Engineering
54293 Schneidershof, Trier, Germany
e-mail: *koch@fh-trier.de*

Heike Kramme
Fuchsweg 14
79822 Titisee, Germany
e-mail: *kramme-titisee@t-online.de*

Rüdiger Kramme
Fuchsweg 14
79822 Titisee, Germany
e-mail: *kramme-titisee@t-online.de*

Ursula Krechel
University Clinic Freiburg
IT-Department
Robert-Koch-Str. 1
79106 Freiburg, Germany
e-mail: *Ursula.krechel@uniklinik-freiburg.de*

Harald Kronberg
Am Hohen Rain 6
64720 Michelstadt, Germany
e-mail: *haraldkronberg@freenet.de*

Klaus A. Kuhn
Technische Universität München (TUM)
Institute of Medical Statistics and Epidemiology
München, Germany
e-mail: *klaus.kuhn@tum.de*

Bernhard Kulik
Maquet GmbH & Co. KG
Kehler Str. 31
76437 Rastatt, Germany
e-mail: *b.kulik@maquet.de*

Timothy G. Laske
Medtronic, Inc.
8200 Coral Sea St. NE.
Mounds View, MN 55112, USA
e-mail: *tim.g.laske@medtronic.com*

Martin Leonhard
Karl Storz GmbH & Co. KG
Research and Technology
Mittelstr. 8
78532 Tuttlingen, Germany
e-mail: *m.leonhard@karlstorz.de*

Xiao Liu
University College London
Department of Electronic and Electrical Engineering
Torrington Place
London, WC1E 7JE, UK
e-mail: *x.liu@ee.ucl.ac.uk*

Martin Maier
Philips Healthcare
Hewlett-Packard-Str. 2
71034 Böblingen, Germany
e-mail: *martin.maier@philips.com*

Giovanni Maio
University of Freiburg
Institute of Bioethics and History of Medicine
Stefan-Meier-Str. 26
79104 Freiburg, Germany
e-mail: *maio@ethik.uni-freiburg.de*

Albrecht Malkmus
GE Medical Systems Information Technologies
Munzinger Str. 5
79111 Freiburg, Germany
e-mail: *albrecht.malkmus@ge.com*

Kei Masani
Toronto Rehabilitation Institute
Lyndhurst Centre
520 Sutherland Drive
Toronto, Ontario M4G 3V9, Canada
e-mail: *k.masani@utoronto.ca*

Ulrich Matern
wwH-c GmbH
Ernst-Simon-Str. 16
72072 Tübingen, Germany
e-mail: *ulrich.matern@wwH-c.com*

John McNulty
San Diego State University
Department of Biology
5500 Campanile Drive
San Diego, CA 92182, USA
e-mail: *jmcnulty24@gmail.com*

Andreas Melzer
University Dundee
IMSaT
1 Würzburg Loan
Dundee, Scottland DD2 1FD, UK
e-mail: *a.melzer@dundee.ac.uk*

Paulo M. Mendes
University of Minho, Campus de Azurém
Department of Industrial Electronics
4800-058 Guimarães, Portugal
e-mail: *paulo.mendes@dei.uminho.pt*

Silvestro Micera
ETH Zürich, ETL K 10.1
Institut für Automatik
Physikstr. 3
8092 Zürich, Switzerland
e-mail: *micera@control.ee.ethz.ch*

Kostis Michelakis
Imperial College London,
South Kensington Campus
Institute of Biomedical Engineering
London, SW7 2AZ, UK
e-mail: *k.michelakis@imperial.ac.uk*

Uwe Möhring
Textile Research Institute Thuringia-Vogtland
Zeulenrodaer Str. 42–44
07973 Greiz, Germany
e-mail: *u.moehring@titv-greiz.de*

Wolfgang Müller-Wittig
Nanyang Technological University (NTU)
Fraunhofer IDM@NTU
Nanyang Avenue
639798 Singapore
e-mail: *wolfgang.mueller-wittig@fraunhofer.sg*

Andreas Neudeck
Textile Research Institute Thuringia-Vogtland
Zeulenrodaer Str. 44
07973 Greiz, Germany
e-mail: *a.neudeck@titv-greiz.de*

Wolfgang R. Nitz
Siemens AG Healthcare Sector
Allee am Röthelheimpark 2
91052 Erlangen, Germany
e-mail: *wolfgang.nitz@siemens.com*

Heike Oschatz
Textile Research Institute Thuringia-Vogtland
Textile Structures and Materials
Zeulenrodaer Str. 42
07973 Greiz, Germany
e-mail: *h.oschatz@titv-greiz.de*

Ernst Pelikan
University Medical Center Freiburg
Hospital Computer Department
Agnesenstr. 6–8
79106 Freiburg, Germany
e-mail: *ernst.pelikan@uniklinik-freiburg.de*

Thomas Peyn
Dräger Medical AG & Co. KG
Department for Respiratory Care
Lübeck, Germany
e-mail: *thomas.peyn@draeger.com*

Doris Pommi
Siemens Healthcare
Siemensstr. 1
91301 Forchheim, Germany
e-mail: *doris.pommi@siemens.com*

Milos R. Popovic
University of Toronto
Rehabilitation Engineering Laboratory, Institute of Biomaterials and Biomedical Engineering
164 College St.
Toronto, Ontario M5S 3G9, Canada
e-mail: *milos.popovic@utoronto.ca*

Robert S. Pozos
San Diego State University
Department of Biology
5500 Campanile Dr.
San Diego, CA 92182-4616, USA
e-mail: *bpozos@gmail.com*

Anna Radomska-Botelho Moniz
Imperial College London,
South Kensington Campus
Institute of Biomedical Engineering
London, SW7 2AZ, UK
e-mail: *a.radomska@imperial.ac.uk*

Stanisa Raspopovic
Scuola Superiore Sant Anna
ARTS lab
Piazza Martiri della Liberta, 33
56100 Pisa, Italy
e-mail: *s.raspopovic@sssup.it*

Annette Reinhardt
Klinikum Nürnberg
Institute of Hospital Epidemiology
Prof.-Ernst-Nathan-Str. 1
90419 Nürnberg, Germany
e-mail:
annette.reinhardt@klinikum-nuernberg.de

Cristiano Rizzo
Micromed S.p.A.
Via Giotto, 2
31021 Mogliano Veneto (TV), Italy
e-mail: *cristiano.rizzo@micromed-it.com*

Eckhard Roggenkamp
Klinikum Nuremberg
Institute of Hospital Epidemiology
Prof.-Ernst-Nathan-Str. 1
90419 Nürnberg, Germany
e-mail:
eckhard.roggenkamp@klinikum-nuernberg.de

Christopher Rolfes
University of Minnesota
Department of Biomedical Engineering and Surgery
420 Delaware St. SE
Minneapolis, MN 55455, USA
e-mail: *rolf0053@umn.edu*

Christian Rotsch
Fraunhofer Institute for Machine Tools and Forming Technology IWU
Department Adaptronics and Acoustics
Nöthnitzer Str. 44
01187 Dresden, Germany
e-mail: *christian.rotsch@iwu.fraunhofer.de*

Georg-Friedemann Rust
Rendoscopy AG
Grubmühlerfeldstr. 54
82131 Gauting, Germany
e-mail: *gfr@rendoscopy.de*

Christina Sampogna
University of Cambridge
Newmarket, Suffolk C88 7YY, UK
e-mail: *Christina.Sampogna@cantab.net*

Gregor Schaefers
MR:comp GmbH
Buschgrundstr. 33
45894 Gelsenkirchen, Germany
e-mail: *schaefers@mrcomp.com*

Rolf M. Schlegelmilch
SMT medical GmbH & Co.
Im Kreuz 9
97076 Würzburg, Germany
e-mail: *r.schlegelmilch@smt-medical.com*

Oliver Scholz
HTW Saarland,
University of Applied Sciences
Goebenstr. 40
66117 Saarbrücken, Germany
e-mail: *oliver.scholz@htw-saarland.de*

Frank Schön
Aloka Holding Europe AG
Steinhauserstr. 74
6300 Zug, Switzerland
e-mail: *F.Schoen@aloka-europe.com*

Arthur Schultz
Hannover Medical School
EEG Monitoring Research Group
Carl-Neuberg-Str. 1
30625 Hannover, Germany
e-mail: *ab.schultz@t-online.de*

Barbara Schultz
Hannover Medical School
Informatik/Biometrie
Podbielskistr. 380
30659 Hannover, Germany
e-mail: *ab.schultz@t-online.de*

Dirk Schulze
DDZ Breisgau
Kaiser-Joseph-Str. 263
79098 Freiburg, Germany
e-mail: *dirk.schulze@ddz-breisgau.de*

Wilhelm Schütte
Kreiskrankenhaus Gummersbach, Academic
Teaching Hospital of the University of Cologne
Department of Medical Radiation Physics
Wilhelm-Breckow-Allee 20
51643 Gummersbach, Germany
e-mail: *wilhelm.schuette@kkh-gummersbach.de*

Danny Schwabe
Textile Research Institute Thuringia-Vogtland
Zeulenrodaer Str. 44
07973 Greiz, Germany
e-mail: *d.schwabe@titv-greiz.de*

Sanjiv Sharma
Imperial College London,
South Kensington Campus
Institute of Biomedical Engineering
London, SW7 2AZ, UK
e-mail: *sanjiv.sharma@imperial.ac.uk*

Erich Siegel
Dräger Medical AG & Co. KG
Moislinger Allee 53–55
23558 Lübeck, Germany
e-mail: *erich.siegel@draeger.com*

Florian Solzbacher
University of Utah
Electrical and Computer Engineering
Salt Lake City, UT 84112, USA
e-mail: *florian.solzbacher@utah.edu*

Silvia Sterzi
Università Campus Bio-Medico
Physical Medicine and Rehabilitation
Via Alvaro del Portillo 21
00128 Rome, Italy
e-mail: *s.sterzi@unicampus.it*

Wilfried Storz
Gebrüder Martin GmbH & Co. KG
Ludwigstaler Str. 132
78532 Tuttlingen, Germany
e-mail: *wilfried.storz@klsmartin.com*

Dirk Sunderbrink
Siemens AG – Healthcare Sector
Imaging and Therapy Department
Hartmannstr. 16
91301 Forchheim, Germany
e-mail: *dirk.sunderbrink@siemens.com*

Hajo Tanck
Mednovo Medical Software Solutions GmbH
Hohenzollerndamm 150
14199 Berlin, Germany
e-mail: *info@mednovo.de*

Prashant Tathireddy
University of Utah
Electrical and Computer Engineering
Salt Lake City, UT 84112, USA
e-mail: *p.tathireddy@utah.edu*

Rachel Toomey
University of Dundee
Institute for Medical Science and Technology
1 Wurzburg Loan, Dundee Medipark
Dundee, DD2 1FD, UK
e-mail: *r.toomey@dundee.ac.uk*

Iasonas F. Triantis
Sensors Systems and Circuits Research Group
Department of Electronic and Electrical Engineering
University College London
London, WC1E 7JE, UK
e-mail: *i.triantis@ee.ucl.ac.uk*

Jakub Trzebinski
Imperial College London,
South Kensington Campus
Institute of Biomedical Engineering
London, SW7 2AZ, UK
e-mail: *jakub.trzebinski09@imperial.ac.uk*

Friedrich Ueberle
Hamburg University of Applied Sciences
Department of Life Sciences/Biomedical Technology
Lohbrügger Kirchstr. 65
21033 Hamburg, Germany
e-mail: *friedrich.ueberle@haw-hamburg.de*

Hans-Peter Uhlig
Österreicher Str. 69a
01279 Dresden, Germany
e-mail: *cu-uhlig@t-online.de*

Jörg Vienken
Fresenius Medical Care
Else Kroener Str. 1
61342 Bad Homburg, Germany
e-mail: *joerg.vienken@fmc-ag.com*

Udo Voges
Karlsruher Institut für Technologie
Institut für Angewandte Informatik
Hermann-von-Helmholtz-Platz 1
76344 Eggenstein-Leopoldshafen, Germany
e-mail: *udo.voges@kit.edu*

Birgit Wacker
Philips Healthcare
Hewlett-Packard Str. 2
71034 Böblingen, Germany
e-mail: *birgit.wacker@philips.com*

Golam Abu Zakaria
Hospital of the University of Cologne
Department of Medical Radiation Physics
Wilhelm-Breckow-Allee 20
51643 Gummersbach, Germany
e-mail: *zakaria@kkh-gummersbach.de*

Christian Zapf
Siemens AG – Healthcare Sector
Imaging and Therapy – SYNGO
Hartmannstr. 16
91052 Erlangen, Germany
e-mail: *christian.zapf@siemens.com*

Frank Zgoda
Laser- und Medizin-Technologie GmbH
Fabeckstr. 60–62
14195 Berlin, Germany
e-mail: *f.zgoda@LMTB.de*

Loredana Zollo
Università Campus Bio-Medico di Roma
Laboratory of Biomedical Robotics and Biomicrosystems
Via Álvaro Del Portillo, 21
00128 Rome, Italy
e-mail: *l.zollo@unicampus.it*

Contents

List of Abbreviations XXXVII

Part A Medical Technology Basics

1 Technology in Medicine: Its Role and Significance in Terms of Health Policy
Rüdiger Kramme, Heike Kramme 3
1.1 A Short History 3
1.2 Early Breakthroughs of Medical Technology 3
1.3 Analog to Digital 4
1.4 Health Policy 5
1.5 New Key Areas 5
1.6 Innovation Versus Financial Resources 6

2 Medicine Is More Than Applied Technology for Human Beings
Giovanni Maio 7
2.1 Technology Suggests Feasibility and Controllability 7
2.2 Technology Knows No Bounds 8
2.3 Technology Is Unable to Answer the Question of Meaning 9
2.4 Technology Alone Does Not Make Medicine Humane 9
References 10

3 Hygiene in Medical Technology
Heinz-Michael Just, Eckhard Roggenkamp, Annette Reinhardt 11
3.1 Background 12
3.2 Causes of Infection 13
3.3 Vaccinations 13
3.4 Disinfection Methods 14
3.5 Sterilization Methods 21
3.6 Hygiene of Noninvasive Technology Equipment 25
3.7 Hygiene of Invasive Technology Equipment 26
3.8 Practical Examples 26
3.9 Regulations 31
References 33

4 Technical Safety of Electrical Medical Technology Equipment and Systems
Rüdiger Kramme, Hans-Peter Uhlig 35
4.1 General Information Regarding the Safety of Technical Systems 36
4.2 Attaining Safety in Medical Institutions 36
4.3 Minimum Requirements for ME Equipment 37

4.4 Areas Used for Medical Purposes ... 40
4.5 Electrical Systems According to the Nature of the Connection to Earth ... 42
4.6 Protection Against Shock Currents ... 42
4.7 Power Supply ... 44
4.8 Power Sources for Safety Purposes with Accumulators ... 44
4.9 Final Circuits and Plug Sockets ... 45
4.10 Static Electricity ... 45
4.11 Electromagnetic Compatibility ... 46
4.12 Conclusions ... 47
References ... 47

5 **Quality Management in Medical Technology**
Albrecht Malkmus ... 49
5.1 Objectives of a Quality Management System ... 49
5.2 Elements of a Quality Management System ... 54
5.3 Organization of a Quality Management System ... 54
5.4 Implementation of a QMS ... 56
5.5 Product Quality ... 57
5.6 Concluding Remarks ... 57
References ... 58

6 **Usability of Medical Devices**
Ulrich Matern, Dirk Büchel ... 59
6.1 What Is Usability? ... 59
6.2 Usability in Medical Technology – Obligation or Opportunity? ... 60
6.3 Usability in Medical Technology – Why? ... 61
6.4 Development of Usable Devices – How Is this Done? ... 62
6.5 Testing of Usable Devices – How Is this Done? ... 64
6.6 Assessment of Usability ... 67
6.7 Usability Development, Testing, and Assessment – An Example ... 67
References ... 70

Part B Functional Diagnostics Devices

7 **Basic Diagnostics in Cardiology**
Rüdiger Kramme ... 75
7.1 Electrocardiography ... 75
7.2 Electrocardiograph Equipment Technology and PC ECG ... 76
7.3 ECG Methods ... 79
7.4 Lead Systems ... 80
7.5 Methodological Notes ... 83
7.6 The Diagnostic Value of the ECG ... 83
7.7 Complications ... 84
7.8 Technical Safety Aspects of ECG Systems ... 84
7.9 Long-Term ECG ... 84

7.10 Long-Term ECG Systems ... 85
7.11 Computer-Based Assessment ... 85
7.12 Heart Rate Variability and Heart Rate Turbulence ... 87
7.13 Indications for Long-Term Electrocardiography ... 87
7.14 The Significance of the Long-Term ECG ... 87
7.15 The Exercise ECG ... 88
7.16 Equipment Technology ... 88
7.17 Reduced Exercise ECG Leads ... 89
7.18 Automatic ST Measuring Programs ... 90
7.19 Exercise Test ... 90
7.20 Methodological Notes ... 93
7.21 The Diagnostic Value of Ergometry ... 93
7.22 Indications ... 93
7.23 Abort Criteria and Safety Measures ... 93
7.24 Technical Safety Aspects ... 94
7.25 Notes on Planning ... 94
Further Reading ... 94

8 **Pulmonary Function Testing**
Rolf M. Schlegelmilch, Rüdiger Kramme ... 95
8.1 Spirometry ... 95
8.2 Advanced Cardiopulmonary Function Testing ... 105
References ... 116

9 **Devices and Methods in Clinical Neurophysiology**
Klaus-Peter Hoffmann, Ursula Krechel ... 119
9.1 Basics ... 119
9.2 Electroencephalograph ... 129
9.3 Electromyograph ... 138
Further Reading ... 157

10 **Sleep Diagnostic Systems**
Klaus-Peter Hoffmann, Robert S. Pozos ... 159
10.1 Function and Application ... 159
10.2 Sleep Diagnostics, Sleep Laboratories, and Sleep Apneas ... 160
10.3 Technology ... 162
10.4 Sleep Diagnostic Procedures ... 165
10.5 Signal Recording and Signal Processing ... 166
10.6 Fields of Application ... 171
10.7 Methodical Instructions ... 172
10.8 Medical Significance of Sleep Diagnostics ... 173
10.9 Therapy ... 175
10.10 Safety Aspects ... 176
10.11 Planning Advice ... 176
Further Reading ... 176

11 **Nystagmography**
Klaus-Peter Hoffmann, Eduardo Fernandez ... 179
11.1 Application ... 179
11.2 Eye Movements ... 180
11.3 Technology and Methods ... 181
11.4 Methods ... 184
11.5 Signal Recording and Signal Processing ... 187
11.6 Medical Significance ... 187
11.7 Safety Aspects ... 188
11.8 Spatial Planning ... 188
References ... 189

12 **Audiometry**
Sebastian Hoth ... 191
12.1 Physical, Technical and Physiological Bases of Audiometry ... 191
12.2 Behavioral Audiometric Assessment ... 200
12.3 Objective Audiometric Assessment ... 212
12.4 Technical Hearing Devices ... 228
References ... 242

13 **Measurement Techniques in Ophthalmology**
Albert J. Augustin ... 245
13.1 Measurement of Intraocular Pressure ... 246
13.2 Optical Coherence Tomography (OCT) ... 247
13.3 Laser-Scanning Tomography with the Heidelberg Retina Tomograph (HRT) ... 249
13.4 Nerve Fiber Polarimetry with GDx ... 251
13.5 The Rostock Cornea Module (Confocal Laser Microscope) ... 252
13.6 Automatic Refractometry ... 252
13.7 Visually Evoked Potential (VEP) ... 253
13.8 The Ganzfeld ERG (Ganzfeld Electroretinogram) ... 254
13.9 Pattern Electroretinography (Pattern ERG, PERG) ... 257
13.10 Multifocal ERG (mfERG) ... 259
13.11 Electrooculograms (EOG) ... 260
13.12 Adaptometry ... 261
13.13 Aberrometry (Wavefront Analysis) ... 261
13.14 Keratometry ... 263
13.15 Retinoscopy or Skiascopy ... 263
13.16 Ultrasound ... 263
13.17 Corneal Topography ... 265
13.18 The Orbscan ... 265
13.19 Scheimpflug Examination ... 267
13.20 Fluorescence Angiography of the Retina (Sodium-Fluorescein) ... 267
13.21 Fluorescence Angiography of the Retina (Indocyanine Green) ... 269
13.22 Visual Field Measurement (Perimetry) ... 269
13.23 Exophthalmometry ... 271
References ... 271

14 **Functional Force Assessment of Skeletal Muscles**
Paul A. Iaizzo, William K. Durfee ... 273
14.1 The Need for Skeletal Muscle Force Assessment ... 273
14.2 Manual Muscle Strength Testing ... 275
14.3 Advanced Muscle Assessment Methods ... 278
14.4 Stimulated Muscle Force Assessment ... 279
14.5 Stimulated Muscle Force Assessment in Animal Models ... 283
14.6 Conclusion ... 285
References ... 285

Part C Medical Imaging

15 **Digital Radiography**
Lothar Heuser ... 291
15.1 Historical Background ... 291
15.2 From Analog to Digital Image ... 292
15.3 Digital Imaging Systems in Radiology ... 296
15.4 Digital Image Processing ... 305
15.5 Image Communication and Archiving ... 307
References ... 309

16 **Computed Tomography**
Thorsten M. Buzug ... 311
16.1 Background ... 311
16.2 Milestones of Computed Tomography ... 313
16.3 Computed Tomography Technology ... 318
16.4 Image Reconstruction ... 323
16.5 Scan Planning and Applications ... 329
16.6 Dose ... 335
16.7 Special System Designs ... 338
References ... 341

17 **Ultrasound Diagnostics**
Rudolf Götz, Frank Schön ... 343
17.1 Basic Physical Principles ... 344
17.2 Visualization of the Blood Flow and Vascular System ... 348
17.3 Equipment Technology ... 350
17.4 Three-Dimensional Ultrasound (3-D, Real-Time 3-D) ... 360
17.5 Operation of an Ultrasound Unit ... 367
Further Reading ... 367

18 **Medical Infrared Imaging**
Gerald C. Holst, Thorsten M. Buzug ... 369
18.1 Background ... 369
18.2 Infrared System Design ... 370
18.3 Infrared Physics ... 372

18.4 IR Imaging in Medical Applications 373
18.5 Specific Applications 374
18.6 Limitations of IR Imaging in Medical Applications 377
18.7 Summary 377
References 378

19 **Endoscopy**
Martin Leonhard, Klaus-Martin Irion 379
19.1 Basics 380
19.2 Endoscopes and Endoscopic Accessories 382
19.3 Integrated Operating Theaters 391
19.4 Medical Applications 393
19.5 Tissue Differentiation 394
19.6 Further Future Developments 398
References 402

20 **Cone-Beam Computed Tomography and Navigation**
Dirk Schulze, Gerhard Hoffmann 405
20.1 Technical Background of Dental Digital Volume Tomography 405
20.2 Areas of Application of Dental CBCT 408
References 413

21 **Interventional Radiology – Angiography**
Doris Pommi 417
21.1 Definition of Digital Subtraction Angiography 417
21.2 Application Range for Angiography 419
21.3 Advantages of Interventional Radiology Procedures 420
21.4 Trends of Development 420
Further Reading 421

22 **Near-Infrared Spectroscopy (NIRS)**
John McNulty, Michael Born, Robert S. Pozos 423
22.1 NIRS – Technical 423
22.2 NIRS Technology: Engineering Aspects 425
22.3 Instrumentation and Equipment 427
22.4 New Developments: Multidepth Differential Approach 430
22.5 Clinical Application and Study of NIRS 431
22.6 Does Skin Blood Flow Affect NIRS Measurements? 433
22.7 Future of NIRS 436
References 437

23 **Magnetic Resonance Imaging**
Wolfgang R. Nitz 439
23.1 History of MRI 439
23.2 MRI – System Components 441

23.3 MRI – Basic Principles and Applications 446
23.4 MRI – Safety-Relevant Aspects 453
23.5 MRI – Pictures of the Future 456
References 458

24 **Magnetic Particle Imaging**
Jörn Borgert, Bernhard Gleich, Thorsten M. Buzug 461
24.1 Introduction 461
24.2 A Brief History of Magnetic Particle Imaging 462
24.3 How Magnetic Particle Imaging Works 463
24.4 From Data to Images – Reconstruction 468
24.5 Beyond General Purpose Systems – Special Geometry 470
24.6 Putting MPI to Use – Applications 472
References 474

25 **MR-Guided Interventions and Surgery**
Andreas Melzer, Erwin Immel, Rachel Toomey,
Fabiola Fernandez-Gutierrez 477
25.1 MRI Basics 478
25.2 MRI Image Guidance for Interventions and Surgery in Comparison with CT and Ultrasound 479
25.3 MR Systems Design and Setup for Interventions and Surgery 479
25.4 Instruments for Interventional and Intraoperative MRI 482
25.5 MR-Applicable Endoscopic Instrument Systems 482
25.6 Instrument Representation and Tracking in MRI 483
25.7 MR-Guided Robotics and Navigation 485
25.8 Hybrid Multimodal Imaging for MR-Guided Diagnosis and Therapy .. 491
25.9 Therapeutic MR-Guided Imaging 493
25.10 MR-Guided Delivery of Implants 494
25.11 Conclusions 498
References 498

26 **Devices and Materials in MRI**
Gregor Schaefers, Andreas Melzer 503
26.1 MR Safety 504
26.2 Interactions in the MR Environment 504
26.3 Examples of MR Artifacts Caused by Medical Devices 510
26.4 Evaluation of MRI Artifacts of Implants 512
26.5 MR Safety Labeling 514
26.6 Interpretation of MR Labeling 517
26.7 Discussion 518
References 519

Part D Therapeutic Devices

27 **Long-Term Ventilators for Intensive Therapy**
Thomas Peyn 525
27.1 Tasks of the Ventilator 525
27.2 Function and Components of a Long-Term Ventilator 526
27.3 Technical Implementation 529
27.4 Controlling the Ventilator 530
27.5 Ventilation Procedures 531
27.6 Ventilation Extras and Special Functions 540
27.7 Patient Monitoring and Alarm Limits 543
27.8 Weaning Strategy and SmartCare/PS 543

28 **Defibrillators and ICD Systems**
Rüdiger Kramme 545
28.1 Defibrillator Technology 546
28.2 Therapeutic Intervention 549
28.3 Methodological Notes 550
28.4 Complications 551
28.5 Technical Safety Aspects 551
28.6 Implantable Cardioverter-Defibrillators 551
References 556

29 **Laser Systems**
Frank Zgoda 557
29.1 History of the Laser 558
29.2 Physics and Technology 558
29.3 Application Methods 564
29.4 Biophysical Effects on Tissue 564
29.5 Laser Types in Medicine 567
29.6 Fields of Use 571
29.7 Safety Aspects 574
29.8 Future Prospects 577
References 577

30 **Inhalational Anesthesia Devices**
Erich Siegel 579
30.1 Anesthesia Devices in General Anesthesia 579
30.2 Functional Principle, Medical Aspects 580
30.3 Functional Principle of the Main Components 582
30.4 Safe Operation Prerequisites 594
Further Reading 595

31 **Extracorporeal Blood Purification Systems**
Jörg Vienken 597
31.1 Historical Perspective 598
31.2 Blood Purification for the Therapy of Renal Failure 600

31.3 Dialysis Machines and Additional Equipment: Use and Conditions ... 611
31.4 Blood Purification in Liver Replacement Therapy ... 614
References ... 617

32 **Heart–Lung Machines**
Ulrich Böckler, Andreas Hahn ... 621
32.1 Historical Development of Extracorporeal Circulation ... 621
32.2 Extracorporeal Circulation ... 623
32.3 Structure and Function of the Heart–Lung Machine ... 623
32.4 Components of the Heart–Lung Machine ... 627
32.5 Extracorporeal Circulation ... 629
32.6 Differentiation of Heart–Lung Machines ... 635
32.7 Aspects of Technical Safety ... 636
32.8 Prospects for Further Development ... 637
References ... 638

33 **Application of Shock Waves and Pressure Pulses in Medicine**
Friedrich Ueberle ... 641
33.1 Introduction – Historical Development ... 642
33.2 Definitions of Physical Terms: Acoustics – Sound Waves – Pressure Pulses – Shock Waves ... 643
33.3 The Acoustic Field of a Lithotripter – Basics of Measurement Technology ... 649
33.4 Generation of Pressure Pulses for Extracorporeal Lithotripsy (ESWL) and Extracorporeal Shock Wave Therapy (ESWT) ... 660
33.5 Extracorporeal Lithotripsy (ESWL) and Extracorporeal Shock Wave Therapy (ESWT) in Practice ... 664
33.6 The Patient ... 667
33.7 Assessment of the Clinical Efficiency of Lithotripters ... 668
33.8 Associations and Societies for Lithotripsy and Pressure Pulse Therapy ... 670
References ... 671

34 **High-Frequency Surgery**
Bernhard Hug, Reiner Haag ... 677
34.1 Development of High-Frequency Surgery ... 678
34.2 Physical and Technical Principles ... 681
34.3 Technology and Techniques ... 684
34.4 Types of Current and Their Application ... 687
39.5 Methodical Instructions for Application and Safety ... 697
34.6 Outlook ... 700
References ... 701

35 **Medical Radiation Therapy**
Peter H. Cossmann ... 703
35.1 X-Radiation ... 704
35.2 Historical Development of Radiation Therapy ... 704

35.3 Physical and Technical Principles of Radiation Physics 705
35.4 Forms of Therapy 708
35.5 Equipment Technology for the Generation of Radiation 710
35.6 Special Techniques and Newer Developments in Teletherapy 716
References 721

36 **Mechanical Circulatory Support Systems**
Roland Hetzer, Ewald Hennig 723
36.1 Introduction – History 724
36.2 Indications for Application of MCSS 725
36.3 Classification of MCSS 727
36.4 Today's Systems 728
36.5 Complications 739
36.6 Technical Follow-Up and Care 742
36.7 Psychosomatic Syndromes and Quality of Life During Treatment with MCS 742
36.8 Overview and Outlook 744
References 747

37 **Neural Interfaces for Implanted Stimulators**
Xiao Liu, Andreas Demosthenous, Nick Donaldson 749
37.1 Stimulating Electrodes 751
37.2 Implantable Cable Management 753
37.3 Design of Stimulator Output Stage 756
37.4 Conclusions 763
References 764

38 **Cardiac Pacemaker Systems**
Armin Bolz 767
38.1 Structure of a Pacemaker System 768
38.2 The Functionality of a Cardiac Pacemaker 771
38.3 Stimulation Modes 774
References 782

39 **Introduction to Neuroprosthetics**
Klaus-Peter Hoffmann, Silvestro Micera 785
39.1 Neuroprostheses 785
39.2 Application of Neural Prostheses 787
39.3 Specific Technological Features 788
39.4 Biological–Technical Interface 790
39.5 Future Developments 798
References 799

40 **Implantable Microsystems**
Prashant Tathireddy, Florian Solzbacher, Robert Hitchcock,
Klaus-Peter Hoffmann 801
40.1 Market, Applications, and Common Requirements 801
40.2 Sensors 805
40.3 In vitro and in vivo Testing 814
References 816

41 **Visual Prostheses**
Eduardo Fernandez, Klaus-Peter Hoffmann 821
41.1 The Case for Artificial Vision 822
41.2 Visual Pathways: From Real Vision to Visual Neuroprostheses 822
41.3 Current Approaches to Visual Prostheses 824
41.4 Engineering Visual Neuroprostheses 827
41.5 Safe and Effective Stimulation of Visual Pathways
Through Multiple Microelectrodes 829
41.6 Selection of Suitable Subjects for a Visual Prosthesis 830
41.7 Challenges and Future Perspectives 831
References 832

42 **Rehabilitation and Therapeutic Robotics**
Loredana Zollo, Dino Accoto, Silvia Sterzi, Eugenio Guglielmelli 835
42.1 Background 835
42.2 Human-Centered Approach to Rehabilitation Robot Design 837
42.3 Robot-Based Measure of Patient's Performance 844
42.4 Conclusions and Further Readings 850
References 851

43 **Cardiac Devices and Testing**
Michael Bateman, Stephen A. Howard, Christopher Rolfes,
Timothy G. Laske, Paul A. Iaizzo 855
43.1 Background 856
43.2 Selected Landmark Events in Cardiac Devices and Surgery 856
43.3 Market Released Cardiac Devices 857
43.4 Device Development 857
43.5 The Anatomy of a Device 867
43.6 Emerging Cardiac Device Technology 871
43.7 Conclusions 874
References 874

44 **Functional Electrical Stimulation in Rehabilitation and Neurorehabilitation**
Kei Masani, Milos R. Popovic 877
44.1 The Basis of Electrical Stimulation 878
44.2 Neuroprosthetic Use of FES 883

44.3 FES Therapy ... 889
44.4 Other Uses of Electrical Stimulation ... 890
44.5 Concluding Remarks ... 890
References ... 890

45 **Treatment Planning and Patient Treatment**
Golam Abu Zakaria, Wilhelm Schütte, Stephan Garbe ... 897
45.1 Principles of Radiotherapy and Treatment Planning ... 898
45.2 Imaging in Treatment Planning ... 901
45.3 Basic Techniques in External Beam Therapy ... 902
45.4 Target Volumes and Organ at Risk ... 909
45.5 Modern Treatment Planning Systems ... 910
45.6 Simulation of the Patient and the First Treatment ... 915
45.7 Quality Control in Radiation Therapy ... 917
References ... 919

Part E Monitoring

46 **Recording and Processing of Biosignals**
Klaus-Peter Hoffmann, Florian Solzbacher ... 923
46.1 Measuring in Medicine ... 923
46.2 Registration of Biological Signals ... 934
46.3 Measurement and Signal Analysis from a Metrological Point of View 940
46.4 Test Planning and Clinical Studies ... 944
Further Reading ... 945

47 **Monitoring Systems**
Ullrich Hieronymi, Rüdiger Kramme ... 947
47.1 Fields of Use for Patient Monitoring Systems ... 948
47.2 Types of Monitors ... 949
47.3 Monitor Screen Content ... 951
47.4 Handling ... 952
47.5 Alarms and Events ... 952
47.6 Trend Display ... 952
47.7 Automatic Calculations ... 953
47.8 Advanced System Properties ... 953
47.9 Central Monitoring and Documentation ... 953
Further Reading ... 954

48 **Cardiovascular Monitoring**
Ullrich Hieronymi, Rüdiger Kramme ... 955
48.1 Monitoring the Cardiac Function ... 955
48.2 Monitoring the Circulatory Function (Hemodynamic Monitoring) ... 957
References ... 969

49 **Respiratory Monitoring and Pulse Oximetry**
Ullrich Hieronymi, Rüdiger Kramme, Harald Kronberg ... 971
49.1 Respiratory Mechanics ... 971
49.2 Gas Exchange ... 973
References ... 985

50 **Temperature Monitoring**
Rüdiger Kramme, Ullrich Hieronymi ... 987
50.1 Hyperthermia and Hypothermia ... 987
50.2 Measuring Sites for Temperature Measurement ... 987
50.3 Temperature Sensors and Probes ... 987
50.4 Methodological Notes ... 989
Further Reading ... 990

51 **Cerebral Monitoring**
Barbara Schultz, Arthur Schultz, Harald Kronberg ... 991
51.1 EEG Monitoring ... 991
51.2 Intracranial Pressure ... 995
References ... 1000

52 **Brain Computer Interface**
Günter Edlinger, Cristiano Rizzo, Christoph Guger ... 1003
52.1 Introduction to BCI ... 1003
52.2 Measuring Brain Activity ... 1004
52.3 BCI System Structure ... 1005
52.4 Conclusions ... 1014
References ... 1015

53 **Fetal Monitoring**
Birgit Wacker, Martin Maier ... 1019
53.1 Cardiotocography (CTG) ... 1019
53.2 Obstetric Monitoring Systems ... 1022
Further Reading ... 1030

54 **Neonatal Monitoring**
Roland Hentschel ... 1031
54.1 Electrocardiogram ... 1032
54.2 Impedance Pneumography ... 1033
54.3 Combined Cardiorespiratory Analysis ... 1034
54.4 Pulse Oximetry ... 1035
54.5 Transcutaneous Measurement of the Partial Pressure ... 1037
54.6 Measurement of the $PtcCO_2$ (Transcapnode) ... 1038
54.7 Measurement of the $PtcO_2$ (Transoxode) ... 1038
54.8 Monitoring the Oxygenation – Which Method? ... 1039
54.9 Setting Alert Limits and Limit Values ... 1040
References ... 1041

55 **Intraoperative Neuromonitoring**
Werner Kneist, Daniel W. Kauff ... 1043
55.1 General Principles ... 1043
55.2 Neuromonitoring Signals ... 1046
55.3 Scope of Application ... 1048
55.4 Quality Management ... 1056
55.5 Guidelines and Legal Aspects ... 1056
Further Reading ... 1057

56 **Ionic Neural Sensing**
Iasonas F. Triantis, Anna Radomska-Botelho Moniz, Kostis Michelakis, Sanjiv Sharma, Jakub Trzebinski, Belinda Garner, Amir Eftekhar ... 1059
56.1 Central and Peripheral Nervous System Monitoring ... 1060
56.2 Chemistry of Neural Activity ... 1064
56.3 Chemical Neural Sensing Technology and Challenges ... 1065
56.4 Conclusion ... 1069
References ... 1070

Part F Medical Information Processing and Communication

57 **Fusing Medical Engineering and Information Technology – Structure, Integration and Process Optimization**
Hajo Tanck, Michael Heinlein ... 1075
57.1 Standards of Interfaces ... 1076
57.2 Data Structure ... 1077
57.3 Integrating the Healthcare Enterprise ... 1078
57.4 Integration of Medical Devices ... 1078
57.5 Sample Integration – From Findings to Medical Documentation ... 1082
57.6 Résumé ... 1082
Further Reading ... 1083

58 **Communicating Medical Systems and Networks**
Armin Gärtner ... 1085
58.1 Medical Networks ... 1085
58.2 Requirements for Medical Networks ... 1086
58.3 Interconnected Medical Networks ... 1087
58.4 Risk Management, DIN EN ISO 14971 ... 1088
58.5 Shared Networks ... 1089
58.6 Security Aspects of Medical Networks from a Regulatory Viewpoint .. 1092
58.7 Future Standard IEC 80001-1 ... 1093
References ... 1093

59 **Hospital Information Systems**
Peter Haas, Klaus A. Kuhn ... 1095
59.1 Background ... 1095
59.2 Necessity, Objectives, and Benefits of Comprehensive HIS ... 1096

59.3 Dimensions of IT Support ... 1098
59.4 Case Study ... 1099
59.5 Architecture and Components of HIS ... 1103
59.6 Current Trends and Prospects ... 1108
59.7 Selection and Implementation of HIS ... 1110
59.8 Conclusion ... 1117
References ... 1118

60 **Telemedicine in Germany**
Rüdiger Klar, Ernst Pelikan ... 1119
60.1 The Peculiar Features of German Telemedicine ... 1120
60.2 Consequences of the Peculiarities of the German System for Telemedicine ... 1125
References ... 1126

61 **Telemedicine Using Active Implants**
Klaus P. Koch, Oliver Scholz ... 1129
61.1 Telemedicine in the Operating Theater ... 1129
61.2 Telemedicine in Domestic Care ... 1130
61.3 Implant Telemetry ... 1131
61.4 Inclusion of Active Medical Implants in Telemedicine Systems ... 1136
References ... 1136

62 **Fundamentals of Medical Image Processing**
Thomas M. Deserno ... 1139
62.1 Background ... 1139
62.2 Remarks on the Terminology ... 1141
62.3 Image Enhancement ... 1142
62.4 Feature Extraction ... 1147
62.5 Segmentation ... 1148
62.6 Classification ... 1154
62.7 Quantitative Measurements ... 1157
62.8 Interpretation ... 1158
62.9 Image Data Visualization ... 1158
62.10 Image Management ... 1161
62.11 Conclusion and Outlook ... 1163
References ... 1165

63 **Virtual Reality in Medicine**
Wolfgang Müller-Wittig ... 1167
63.1 Virtual Reality ... 1168
63.2 Medical Applications ... 1168
63.3 VR-Based Medical Simulation ... 1174
63.4 Model Generation – Virtual Anatomy ... 1175
63.5 Manipulations – Surgical Interventions ... 1178
63.6 Outlook ... 1182
References ... 1184

64 **Computer-Supported Teaching and Learning Systems in Medicine**
Martin Haag, Martin R. Fischer ... 1187
64.1 Historical Development ... 1187
64.2 Moves Towards the Reform of Medical Studies ... 1189
64.3 Developing Learning and Teaching Systems ... 1189
64.4 Learning Environments ... 1193
64.5 Application Scenarios for Learning and Teaching Systems ... 1195
64.6 Status of and Outlook for e-Learning in Medicine ... 1195
References ... 1197

65 **PACS and RIS**
Christian Zapf, Andreas Bermann, Dirk Sunderbrink ... 1199
65.1 Radiological Workflow ... 1200
65.2 Integrating PACS/RIS into the Hospital Environment ... 1205
65.3 State-of-the-Art IT Infrastructure ... 1206
65.4 Summary ... 1208
References ... 1208

66 **3-D Postprocessing in Virtual Endoscopy**
Georg-Friedemann Rust ... 1209
66.1 What Is Virtual Reality? ... 1210
66.2 Why Virtual Reality? ... 1210
66.3 Advantages of 3-D Visualization ... 1210
66.4 Conclusions ... 1216
References ... 1216

67 **e-Health – Ambient Assisted Living and Personal Health Systems**
Natasha Avila, Christina Sampogna ... 1217
67.1 Background ... 1218
67.2 AAL and PHS Approaches ... 1221
67.3 Benefits and Challenges Ahead ... 1234
67.4 Conclusion ... 1241
References ... 1243

68 **Electrical Stimulation of the Nervous System**
Stanisa Raspopovic, Marco Capogrosso, Silvestro Micera ... 1247
68.1 Background ... 1247
68.2 Biophysics Models of Neuronal Response to External Fields ... 1248
68.3 Finite Element (FE) Models ... 1252
68.4 Conclusion ... 1256
References ... 1256

Part G Equipment and Tools

69 **Operating Tables – the Surgeon's Workplace**
Bernhard Kulik ... 1261
69.1 The History of the Operating Table ... 1262
69.2 The OR Table System ... 1263
69.3 Technology of Operating Room Table Systems ... 1264
69.4 Safe Patient Positioning ... 1268
69.5 Preparation: Care, Maintenance, and Hygiene ... 1271
Further Reading ... 1272

70 **Medical Robotics**
Harald Fischer, Udo Voges ... 1273
70.1 Fundamentals ... 1273
70.2 Development of Medical Robots ... 1274
70.3 Overview of Systems ... 1275
70.4 Medical Applications ... 1280
70.5 Technical Aspects ... 1281
70.6 Outlook ... 1282
References ... 1283

71 **Incubators**
Guenther Braun, Roland Hentschel ... 1285
71.1 Historical Background ... 1285
71.2 Construction and Function of an Incubator ... 1286
71.3 Incubator Models ... 1287
71.4 Risks of Incubator Therapy ... 1289
71.5 Hygiene ... 1289
71.6 Unsolved Problems ... 1290
References ... 1290

72 **Surgical Scissors**
Reiner Haag, Wilfried Storz ... 1291
72.1 The History of Scissors ... 1292
72.2 The Function and Design of Scissors ... 1293
72.3 Materials ... 1294
72.4 Manufacture of Surgical Scissors ... 1294
72.5 Diversification Overview ... 1297
72.6 Handling and Care ... 1310
72.7 Inspection, Testing, and Care ... 1314
72.8 Packaging ... 1314
72.9 Current Terminology ... 1315
72.10 Steam Sterilization with Saturated Steam ... 1315
72.11 Quality Characteristics ... 1316
72.12 Future Developments ... 1317
72.13 Bipolar Scissors ... 1318
References ... 1319

73 **Intelligent Textiles and Trends**
Christian Rotsch, Sibylle Hanus, Danny Schwabe, Heike Oschatz, Andreas Neudeck, Uwe Möhring 1321
73.1 Textile Manufacturing Technologies and Applications 1321
73.2 Sensory Applications of Textiles 1326
73.3 Active Textiles – Therapeutical Applications 1330
73.4 Passive Medical Textiles for Therapy 1333
References 1335

74 **Electronics in Medicine**
Paulo M. Mendes, Celso P. Figueiredo, Mariana Fernandes, Óscar S. Gama . 1337
74.1 Basics 1338
74.2 Electronic Sensing 1341
74.3 Electronics for Wireless Health Monitoring 1344
74.4 Power Supply 1348
74.5 Wearable Medical Electronics 1353
74.6 Electronics in Medicine at Work 1358
References 1373

Appendix 1377
Acknowledgements 1391
About the Authors 1393
Detailed Contents 1415
Subject Index 1455

List of Abbreviations

μTAS	micro total analytical system
1-D	one-dimensional
2-D	two-dimensional
3-D	three-dimensional
3-D CSI	three-dimensional chemical shift imaging
4-D	four-dimensional

A

A-mode	amplitude mode
A/D	analogue/digital
AAA	abdominal aortic aneurysm
AAL	ambient assisted living
AAMI	Association for the Advancement of Medical Instrumentation
ABI	auditory brainstem implant
ABLB	alternate binaural loudness balance
ABR	auditory brainstem responses
AC	air conduction
AC	alternating current
ACC	American College of Cardiology
ACD	absolute claudication distance
ACT	activated clotting time
ADC	analog-to-digital converter
ADHD	attention-deficit/hyperactivity disorder
ADL	advanced distributed learning
ADSL	asymmetric digital subscriber line
ADT	admission, discharge, and transfer
AEC	automatic exposure control
AED	automated external defibrillator
AEP	auditory evoked potential
AEP	acoustic evoked potential
AGC	automatic gain control
AGC/i	input-controlled automatic gain control
AGC/o	output-controlled automatic gain control
AGIT	Arbeitsgemeinschaft Informationstechnik
AHA	American Heart Association
AI	artificial intelligence
AIx	augmentation index
ACF	autocorrelation function
ALL	acute lymphatic leukemia
ALS	amyotrophic lateral sclerosis
AMFR	amplitude modulation following response
AMI	alternate mark inversion
AMI	auditory midbrain implant
AMIGO	advanced multimodality image-guided operating room
AML	acute myeloic leukaemia
ANSD	auditory neuropathy spectrum disorder
AP	anaesthetic proof
APD	auditory processing disorder
APOD	adaptive probe off detection
APm	mean arterial pressure
AR	augmented reality
AR	automated refractometer
ARM	Aspen return monitor
ART	algebraic reconstruction technique
ASB	assisted spontaneous breathing
ASD	atrial septal defect
ASHA	American Speech and Hearing Association
ASICS	application-specific integrated circuit
ASL	arterial spin labeling
ASSR	auditory steady-state response
ASTM	American Society for Testing and Materials
AT	adaptive tripole
ATC	automatic tube compensation
ATMP	advanced therapy medicinal product
ATP	antitachycardic pacing
ATP	adenosine triphosphate
ATPD	ambient temperature and pressure, dry
ATPS	ambient temperature and pressure, saturated
ATS	American Thoracic Society
AV	atrioventricular
AVC	automatic volume control
AWIGS	advanced workplace for image guided surgery

B

B-mode	brightness mode
B2B	business-to-business
BAHA	bone-anchored hearing aid
BANG	bis acrylamide nitrogen gelatin
BAR	billing and accounting request
BART	breathing-adapted radiotherapy
BC	bone conduction
BCI	brain–computer interface
BCM	body composition monitor
BER	bit error ratio
BERA	brainstem electric response audiometry
BERA	brainstem evoked response audiometry
BF	body floating
BGA	blood gas analysis
BGO	bismuth germanate
BILD	binaural intelligibility level difference
BIS	bispectral index
bit	binary digit
BL	blended learning
BMLD	binaural masking level difference

BOLD	blood oxygenation-dependent imaging
BPEG	British Pacing and Electrophysiology Group
BPH	benign prostate tissue
BPS	battery-supported power supply
BSE	bovine spongiform encephalopathy
BSN	body sensor network
BTB	bridge to bridge
BTD	bridge to decision
BTE	behind the ear
BTPS	body temperature and pressure, saturated
BTR	bridge to recovery
BTT	bridge a patient to transplantation
BVAD	biventricular assist device
BfArM	Bundesinstitut für Arzneimittel und Medizinprodukte
Bipol-TUR	bipolar transurethral resection

C

CABG	coronary artery bypass grafting
CAD	computer-aided diagnosis
CAD	coronary artery disease
CAL	computer-assisted learning
CAM	computer-aided manufacture
CAP	contention access period
CAPD	central auditory processing disorder
CAS	compressed analog stimulation strategy
CAS	computer-aided surgery
CAT	computerized axial tomography
CBCT	cone-beam computed tomography
CBF	cerebral blood flow
CBI	computer-based instruction
CBIR	content-based image retrieval
CBT	computer-based training
CBV	cerebral blood volume
CCC	China compulsory certificate
CCD	charge-coupled device
CCITT	Comité Consultatif International Téléphonique et Télégraphique
CCR	continuity of care record
CCU	camera control unit
CCU	coronary care unit
CCU	critical care unit
CD	compact disc
CD-ROM	compact disc read-only memory
CDA	clinical document architecture
CDC	Center for Disease Control
CDM	central display and control module
CE	contractile element
CEN	Comité Européen de Normalisation
CENELEC	European Committee for Electrotechnical Standardization
CERA	cortical electric response audiometry
CF	cardiac floating
CFP	contention-free period
CGM	continuous glucose monitoring
CGMS	continuous glucose monitoring system
CGS	centimeter-gram-second
CI	cochlear implant
CI	confidence interval
CIC	completely in the canal
CIM	ceramic injection molding
cIONM	continuous intraoperative neuromonitoring
CIRS	computer integrated radiology system
CJD	Creutzfeldt–Jakob disease
CL	comfortable stimulation level
CLT	color lookup table
CMCT	central motor conduction time
CML	chronic myeloic leukemia
CMOS	complementary metal–oxide–semiconductor
CMRR	common mode rejection ratio
CNAP	continuous noninvasive arterial pressure
CNC	computer numerically controlled
CNCA	Certification and Accreditation Administration of the Peoples Republic of China
CNS	central nervous system
CNT	carbon nanotube
CNV	choroidal neovascularizations
CNV	contingent negative variation
CO	cardiac output
CPA	continuous positive airway
CPAP	continuous positive airway pressure
CPK	creatine phosphokinease
CPOE	computerized physician order entry
CPP	cerebral perfusion pressure
CPU	central processing unit
CQC	China Quality Certification Centre
CQM	contact quality monitor
CR	computed radiography
CROS	contralateral routing of signals
CRT	cardiac resynchronization therapy
CRT	cathode-ray tube
CSCL	computer-supported cooperative/collaborative learning
CSCN	customer support clinical network
CSCW	computer-supported cooperative work
CSF	cerebrospinal fluid
CSMA-CA	carrier sense multiple access with collision avoidance
CSSD	central sterilization supply department
CT	computer tomograph(y)
CTA	CT-angiography
CTG	cardiotocography
CTI	computed tomography imaging
CTV	clinical target volume
CUNY	City University of New York
CVC	central venous catheter
CVD	cardiovascular disease

CVD	congenital vascular disorder
CVP	central venous pressure
CW	continuous wave
CWRUVA	Case Western Reserve University/Department of Veterans Affairs

D

D2D	doctor-to-doctor
DAC	digital-to-analog converter
DART	dynamic adaptive radiotherapy
DAS	data-acquisition system
DAS	detector angular subtense
DBS	deep brain stimulation
DC	direct current
DCT	dynamic contour tonometry
DECT	digital enhanced cordless telecommunication
DFM	design for manufacture
DFT	defibrillation threshold
DFT	detail financial transaction
DGSL	Deutsche Gesellschaft für Stosswellenlithotripsie
DGSM	German Sleep Society
DHZB	Deutsches Herzzentrum Berlin
DICOM	digital imaging and communications in medicine
DIHK	Deutscher Industrie- und Handelskammertag
DIMDI	Deutsches Institut für Medizinische Dokumentation und Information
DIT	differential infrared thermography
DMD	Duchenne muscular dystrophy
DNA	deoxyribonucleic acid
DNEP	descending neurogenic evoked potential
DOF	degree of freedom
DPF	differential path length
DPOAE	distortion product otoacoustic emission
DQE	dose quantum efficiency
DR	direct radiography
DR	dual rate
DRC	dynamic range compression
DRG	diagnosis related group
DRL	driven right leg
DRR	digitally reconstructed radiograph
DRR	dynamic range reduction
DSA	digital subtraction angiography
DSL	desired speech level
DSO	distribution system operator
DSP	digital signal processor
DT	destination therapy
DTI	diffusion tensor imaging
DTL	Dawson–Trick–Litzkow
DVD	digital versatile disc
DVH	dose volume histogram
DWI	diffusion-weighted imaging
DoD	Department of Defense
DoF	degree of freedom

E

E-ABR	evoked auditory brainstem response
e-HC	electronic health card
EAEP	early auditory evoked potential
EAS	electric and acoustic stimulation
EBCT	electron-beam computed tomography
EBUS	endobronchial ultrasonography
ECC	enhanced cornea compensation
ECC	extracorporeal circulation
ECG	electrocardiogram
ECG	electrocardiograph(y)
ECMO	extracorporeal circulatory support
ECW	extracellular water
ECoG	electrocortical grid
ECoG	electrocorticography
ED	energy flux density
EDG	electrodermography
EDP	electronic data processing
EDTA	ethylenediaminetetraacetic acid
EEG	electroencephalogram
EEG	electroencephalograph(y)
eFA	electronic case record
EFOV	extended field of view
EFT	electrical fast transient
EG	electrogram
EGFET	extended gate field effect transistor
EGG	electrogastrography
EHG	electrohysterography
EHR	electronic health record
EHT	electrohydrothermosation
EIRP	effective isotropically radiated power
EIS	electrochemical impedance spectroscopy
EL	electroluminescent
EM	electromagnetic
EMC	electromagnetic compatibility
EMEA	European Medicine Agency
EMG	electromyogram
EMG	electromyograph(y)
EMI	Electric and Musical Industries Ltd.
EMI	electromagnetic interference
EMS	electrical muscle stimulation
EMSE	electromagnetic source
ENG	electronystagmography
ENG	electroneurogram
ENG	electroneurography
ENT	ear, nose, throat
EO	electrooptic
EO	ethylene oxide
EOAE	evoked otoacoustic emission
EOG	electrooculogram
EOG	electrooculography

EP	electropolished
EP	evoked potential
EPR	electronic patient record
EPSP	excitatory postsynaptic potential
ER	enhanced reality
ERA	electric response audiometry
ERB	equivalent rectangular bandwidth
ERD	event-related desynchronization
ERG	electroretinography
ERG	electroretinogram
ERP	early receptor potential
ERP	event related potential
ERS	European Respiratory Society
ERS	event-related synchronization
ERV	expiratory reserve volume
ESC	European Society of Cardiology
ESD	electrostatic discharge
ESWL	extracorporeal shock wave lithotripsy
ESWT	extracorporeal shock wave treatment
ESWT	extracorporeal shock wave therapy
ETL	echo train length
ETO	ethylene oxide
ETSI	European Telecommunications Standards Institute
EUG	ectopic pregnancy (Extrauteringravidität)
EUS	endosonographic
EUTox	European Uremic Toxin Working Group
EVC	expiration volume capacity

F

FA	flip angle
FAEP	brainstem auditory evoked potentials (frühe akustisch evozierte Potentiale)
FBP	filtered back-projection
FCC	US Federal Communications Commission
FDA	US Food and Drug Administration
FDI	Fédération Dentaire Internationale
FDL	flashlamp pumped dye laser
FDRC	full dynamic range compression
FEL	free-electron laser
FEM	finite element method
FES	functional electrical stimulation
FET	FES therapy
FFE	fast field echo
FFP	field free point
ffs	form/fill/seal
FFT	fast Fourier transformation
FHN	Fitzhugh–Nagumo
FINE	flat interface nerve electrode
FIR	far-infrared
FIV1	forced expiratory volume in 1 s
FLAIR	fluid-attenuated inversion recovery
FLASH	fast low-angle shot
FMD	fibromuscular dysplasia
FMEA	failure modes and effects analysis
fMRI	functional magnetic resonance imaging
FMT	floating mass transducer
FO	formaldehyde
FOV	field of view
FP	Framework Programme
FRC	functional residual capacity
FSE	fast spin echo
FSK	frequency shift keying
FSM	Frederick S. Mikelberg
FSP	fine structure processing
FT	Fourier transformation
FTIR	Fourier transform infrared
FVC	forced vital capacity
FWHM	full width at half maximum
FoV	field of view

G

GA	genetic algorithm
GCP	good clinical practice
GEDV	global end-diastolic volume
GEF	global ejection fraction
GIF	graphics interchange format
GM	gray matter
GMDS	Society for Medical Informatics, Biometry and Epidemiology
GMG	Law on the Modernization of Healthcare
GOx	glucose oxidase
GPU	graphics processing unit
GRAPPA	generalized autocalibrating partially parallel acquisition
GRE	gradient echo
GS	general supply
GSDOM	German Society of Dental, Oral and Craniomandibular Sciences
GSM	Global System for Mobile Communications
GSR	galvanic skin response
GTV	gross tumor volume
GTWM	Georgia Tech Wearable Motherboard
GUI	graphical user interface

H

hcg	human chorionic gonadotropic
HCV	hepatitis C virus
HD	hemodialysis
HDD	hard-disc drive
HDF	hemodiafiltration
HDR	high-dose rate
HEMO	hemodialysis
HF	hemofiltration
HF	high frequency
HFCS	high-frequency current-switching
HFITT	HF-induced interstitial tumor therapy

HH	Hodgkin and Huxley
HI	hypopnea index
HINT	hearing in noise test
HIS	hospital information system
HL	hearing level
HL	hearing loss
HL7	Health Level 7
HLD	high level disinfection
HLM	heart–lung machine
HMD	head-mounted display
HME	heat and moisture exchanger
HMEF	heat and moisture exchange filter
HPC	health professional card
HR	heart rate
HRT	Heidelberg Retina Tomograph
HTX	heart transplant
HU	Hounsfield unit
HV	Vickers hardness
Hb	hemoglobin
Hbt	hemoglobin concentration
HpD	hematoporphyrin derivative

I

IC	inspiratory capacity
IC	integrated circuit
ICD	implantable cardioverter-defibrillator
ICD	International Classification of Diseases
ICD	implantable cardioverter–defibrillator
ICD	initial claudication distance
ICG	impedance cardiogram
ICG	indocyanine green
ICP	intracranial pressure
ICSD	International Classification of Sleep Disorders
ICSPE	International Commission for the Standardization of Ergometry Application
ICT	information and communication technology
ICT	information and computer technology
ICU	intensive care unit
ICW	intracellular water
ID	identification number
IDE	investigational device exemption
IDEFIX	identification of dental fixtures
IEC	International Electrotechnical Commission
IEGM	intracardial electrogram
IFOV	instantaneous field of view
IG	insertion gain
IGRT	image guided radiotherapy
IGV	intra-thoracic gas volume
IHC	inner hair cell
IHE	integrating healthcare enterprises
IHS	inspiratory help system
iIONM	intermittent intraoperative neuromonitoring
ILR	implantable loop recorder
ILV	independent lung ventilation
IMAT	intensity-modulated arc therapy
IMRT	intensity-modulated radiotherapy
INA	instrumentation amplifier
IOERT	intraoperative electron radiation therapy
IOL	intraocular lens
IOM	Institute of Medicine
IONM	intraoperative neuromonitoring
IOP	intraocular pressure
IP	intellectual property
IPPV	intermittent positive pressure ventilation
IPSP	inhibitory postsynaptic potential
IQ	installation qualification
IR	infrared
IRFI	infrared functional imaging
IRMA	image retrieval in medical applications
IROG	infrared oculography
IRV	inspiratory reserve volume
ISDN	integrated services digital network
ISE	ion-selective electrode
ISF	interstitial fluid
ISFET	ion-sensitive field effect transistor
ISM	industrial, scientific, and medical
ISO	International Standardization Organization
IT	information technology
ITBV	intrathoracic blood volume
ITGV	intra-thoracic gas volume
ITS	initial transmission slot
ITU	International Telecommunication Union
IVC	inspiration vital capacity

J

jnd	just noticeable difference

K

KIT	Karlsruhe Institute of Technology
KV	health insurance (Krankenversicherung)

L

LAN	local area network
LAP	left atrial pressure
LASER	light amplification by stimulated emission of radiation
LCA	Leber's amaurosis
LCD	liquid-crystal display
LDA	linear discriminant analyzer
LDH	lactate dehydrogenase

LDR	low-dose rate
LED	light-emitting diode
LF	low frequency
LFP	local field potential
LGN	lateral geniculate nucleus
LI	laser iridotomy
LITT	laser-induced thermotherapy
LLLT	low-level laser therapy
LLT	laser lithotripsy
LMS	learning management system
LOINC	logical observation identifiers names and codes
LPRT	low-power, real-time protocol
LSHD	light spot hydrophone
LTP	low-temperature plasma sterilization
LUT	look-up table
LVAD	left ventricular assist device
LWIR	long-wavelength infrared
LiDCO	lithium ion dilution

M

M-mode	motion mode
MAC	media access control
MAEP	middle latency auditory evoked potential
MAP	mean arterial pressure
MARS	adsorbent recirculating system
MC	microcontroller
MCS	mechanical circulatory support
MCSS	mechanical circulatory support system
MDCT	multidetector computed tomography
MDD	medical device directive
MDM	manage document message
MDR	medium-dose rate
MDS	move during scan
MDS	myelodysplastic syndrome
MDX	dystrophic mice
ME	medical electrical
ME	medical engineering
MEA	multielectrode array
MEDARPA	medical augmented reality for patients
MEDDEV	medical device
MEG	magnetoencephalography
MEMS	microelectromechanical system
MEP	motor evoked potential
MEQ	modified essay question
mfERG	multifocal ERG
MFI	micro flex interconnection technique
MFT	multifrequency tympanometry
MHLW	Ministry of Health, Labor and Welfare
MIB	medical information bus
MICS	medical implant communication service
MIP	maximum-intensity projection
MIS	minimally invasive surgery
MITOS	multimodality image-guided diagnosis and therapy setup
MLC	multileaf collimator
MLEM	maximum-likelihood expectation maximization
MLRA	middle latency response audiometry
MMN	mismatch negativity
MMS	multimedia message service
MMT	manual muscle testing
MOD	magnetooptical disc
MOG	magnetooculography
MPDA	microphotodiode array
MPE	maximum permissible exposure
MPI	magnetic particle imaging
MPI	master patient index
MPPS	modality performed procedure step
MPR	multiplanar reconstruction
MPS	magnetic particle spectrometer
MPS	maximum physical frame size
MR	magnetic resonance
MRI	magnetic resonance imaging
MRS	MR spectroscopy
MRSA	methicillin-resistant staphylococcus aureus
MRT	magnetic resonance tomography
MRT	minimum resolvable temperature
MRgFUS	MR-guided focused ultrasound
MS	multiple sclerosis
MSCT	multislice computer tomography
MSLT	multiple sleep latency test
MT	movement time
MTF	modulation transfer function
MTT	mean transit time
MUAP	motor unit action potential
MUP	motor unit potential
MV	minute volume
MVCT	megavoltage computer tomography
MVP	MedBiquitous virtual patient
MWIR	mid-wavelength infrared
MWT	maintenance of wakefulness test
Mb	myoglobin
MetHb	methemoglobin
Mi	mechanical index

N

NA	numerical aperture
NAA	*n*-acetyl-aspartate
NAL	National Acoustics Laboratories
NAS	network attached storage
NASPE	North American Society of Pacing and Electrophysiology
NBG	NASPE/BPEG generic pacemaker code
NBI	narrowband imaging
NBP	noninvasive blood pressure

NCIGT	National Center for Image Guided Therapy
nCPAP	nasal continuous positive airway pressure
NCV	nerve conduction velocity
NDIR	nondispersive infrared
NDT	nondestructive testing
NE	neutral electrode
NEDT	noise equivalent differential temperature
NET	noise equivalent temperature
NETD	noise equivalent temperature difference
NFD	nephrogenic fibrosing dermopathy
NFI	nerve fiber indicator
NIBP	noninvasive blood pressure
NICU	neonatal intensive care unit
NIHL	noise-induced hearing loss
NIP	needle image plate
NIR	near-infrared
NIRS	near-infrared spectroscopy
NIV	noninvasive ventilation
NMES	neuromuscular electrical stimulation
NMR	nuclear magnetic resonance
NN	nearest neighbor
NOTES	natural orifice translumenal endoscopic surgery
NP	neural pathway
NRZ	no return to zero
NSF	nephrogenic systemic fibrosis
NTC	negative-temperature coefficient
NTP	normal transmission period
NTSC	National Television System Committee
NYHA	New York Heart Association

O

O_2Hb	oxyhemoglobin
O2C	oxygen to see
OAE	otoacoustic emission
OAR	organ at risk
OCT	optical coherence tomography
ODI	oxygen desaturation index
OEG	open ear gain
OHC	outer hair cell
OID	object identifier
OKN	optokinetic nystagmus
OL-HDF	online hemodiafiltration
OME	otitis media with effusion
ONH	optic nerve head
OP	operative field
OPCAB	off-pump coronary artery bypass
OPS	classification of operational procedures (Germany)
OQ	operational qualification
OR	operating room
ORC	oxygen ratio controller
ORM	order message
ORU	observation results unsolicited
OSA	obstructive sleep apnoea
OSCE	objective structured clinical examination
OSI	open system interconnection
OT	operating theater
OXI	oximetry

P

PA	polyamide
PAAM	polyacrylamide
PACC	patient contact control
PACS	picture archiving and communication system
PAD	peripheral arterial disease
PAL	pharmaceutical affairs law
PAN	polyacrylonitrile
PAP	pulmonary artery pressure
PAPm	mean pulmonary artery pressure
PAT	parallel acquisition technique
PAV	proportional assist ventilation
PAW	airway pressure
PBL	problem-based learning
PC	peak clipping
PC	personal computer
PC	pressure controlled
PC-AC	pressure control–assist control
PC-APRV	pressure control airway pressure release ventilation
PC-BIPAP	pressure control, biphasic positive airway pressure
PC-CMV	pressure control-continuous mandatory ventilation
PC-SIMV+	pressure control-synchronized intermittent mandatory ventilation plus
PCB	printed circuit board
PCR	principal component regression
PCS	patient control system
PCV	pressure-controlled ventilation
PCWP	pulmonary capillary wedge pressure
PDA	patent ductus arteriosus
PDA	personal digital assistant
PDCA	plan–do–check–act
PDD	percentage depth dose
PDF	portable document format
PDM	permanent dynamic monitoring
PDMA	Pharmaceuticals and Medical Devices Agency
PDR	pulsed-dose rate
PDT	photodynamic therapy
PDw	proton density weighting
PE	parallel element
PE	protective earth
PEEP	positive end-expiratory pressure
PEF	peak expiratory flow

PEG	polyethylene glycol
PELV	protective extra low voltage
PEMS	programmable electrical medical system
PENG	photoelectronystagmography
PERG	pattern ERG
PET	positron emission tomography
PF	pulmonary function
PFO	patent foramen ovale
PFT	pulmonary function test
PHS	personal health system
PI	perfusion index
PID	proportional–integral–derivative
PII	pulse intensity integral
PIN	personal identification number
PLS	partial least-squares
PLV	pressure-limited ventilation
PMMA	polymethylmethacrylate
PMS	periodic leg movement
PMV	pump minute volume
PNS	peripheral nerve stimulation
PNS	peripheral nervous system
PPG	photoplethysmogram
ppm	parts per million
PPS	proportional pressure support
PQ	performance qualification
PRF	proton resonance frequency
PRF	pulse repetition frequency
PRVC	pressure regulated volume controlled
PS	power supply
PS	pressure support
PSN	pelvic splanchnic nerve
PSu	polysulfone
PT	perception threshold
PT	programmed teaching
PTA	percutaneous transluminal angioplasty
PTCA	percutaneous transluminal coronary angioplasty
PTS	permanent threshold shift
PTT	pulse transit time
PTV	planned target volume
pTX	parallel transmit
PUK	personal unblocking key
PUR	polyurethane
PUVA	psoralen and ultraviolet A
PVAD	paracorporeal ventricular assist device
PVARP	postventricular atrial refractory period
PVC	polyvinyl chloride
PVDF	polyvinyl difluoride
PVDF	polyvinylidene fluoride
PVI	pleth variability index
PVP	polyvinylpyrrolidone
PVR	pulmonary vascular resistance
PW	pulsed wave
PW	pulse width
PWC	pulse working capacity
PWV	pulse wave velocity

Q

QBE	query by example
QM	quality management
QMS	quality management system
QSRL	Q-switched ruby laser
QT	quasi-tripole
QWIP	quantum-well infrared photodetector
QoS	quality of service

R

RAID	redundant array of independent discs
RARE	rapid acquisition with relaxation enhancement
RBC	red blood cell
RCD	residual current protective device
RCTX	radiochemotherapy
RDI	respiratory disturbance index
RECD	real ear to coupler difference
REFET	reference FET
REM	rapid eye movement
RF	radiofrequency
RFID	radiofrequency identification
RGB	red–green–blue
RGRT	respiratory guided radiotherapy
RIM	reference information model
RIS	radiology information system
RKI	Robert Koch Institute
RLN	recurrent laryngeal nerve
RMS	root mean square
ROI	region of interest
ROM	range of motion
RPE	retinal pigment epithelium
RR	respiration rate
RSNA	Radiological Society of North America
RST	rotation, scale, and translation
RT	radiation therapy
RV	residual volume
R&TTE	radio and telecommunications terminal equipment

S

S-MAC	sensor MAC
SA	sinoatrial
SAN	storage area network
SAP	systemic arterial pressure
SAR	specific absorption rate
SARS	severe acute respiratory syndrome
SB	spontaneous breathing
SBRT	stereotactic body radiation therapy
SCI	steered compound imaging
SCI	spinal cord injury
SCORM	sharable courseware object reference model

SCP	slow cortical potential
SD	standard deviation
SE	spin echo
SE	series element
SEM	scanning electron microscopy
sEMG	spontaneous electromyography
SENSE	sensitivity encoding
SEP	somatosensory evoked potential
SET	signal extraction technology
SF	standard flash
SFOAE	stimulus frequency otoacoustic emission
SIDS	sudden infant death syndrome
SIMV	synchronized intermittent mandatory ventilation
SISI	short increment sensitivity index
SLARSI	sacro-lumbar anterior root stimulator implant
SMD	surface mount device
SME	small and medium-sized enterprises
SMR	sensorimotor rhythmic
SMS	short message service
SNOMED	Systematized Nomenclature of Medicine
SNR	signal-to-noise ratio
SOA	service oriented architecture
SOAE	spontaneous otoacoustic emission
SOC	system-on-a-chip
SPECT	single photon emission computed tomography
SPGR	spoiled gradient recalled
SPIN	speech in noise
SPIO	superparamagnetic iron oxide
SPL	sound pressure level
SPN-CPAP	spontaneous-continuous positive airway pressure
SPN-CPAP/PS	spontaneous continuous positive airway pressure/pressure support
SPN-PPS	spontaneous proportional pressure support
SQUID	superconductive quantum interference device
SRS	stereotactic radiosurgery
SRT	stereotactic radiotherapy
SRT	speech perception threshold
SSEP	somatosensory evoked potential
SSVEP	steady-state visually evoked potential
STAN	ST waveform analysis
STIR	short-tau inversion recovery
STPD	standard temperature and pressure, dry
STR	scotopic threshold response
SV	stroke volume
SVC	superior vena cava
SVES	ventricular super-extrasystole
SVR	systemic vascular resistance
SWI	susceptibility-weighted imaging
SWIR	short-wavelength infrared
SkBF	skin blood flow

T

TA	technical assistant
TAH	total artificial heart
TASP	téléassistance en soins de plaies
TBI	total body irradiation
TC	tube compensation
TCI	target-controlled infusion
TCP/IP	transmission control protocol/Internet protocol
TDI	tissue Doppler imaging
TDMA	time-division multiple access
TE	time of echo
TEB	thoracic electrical bioimpedance
TEC	thermoelectric cooler
TEE	transesophageal echocardiography
tEMG	triggered electromyography
TENS	transcutaneous electrical nerve stimulation
TEOAE	transient evoked otoacoustic emission
tf-LIFE	thin-film longitudinal intrafascicular electrode
TFT	thin-film technology
TFT	thin-film transistor
THC	tissue hemoglobin measurement
THI	tissue hemoglobin index
TIFF	tagged image file format
TIVA	total intravenous anesthesia
TLC	total lung capacity
TLD	thermoluminescent dosimeter
TMP	transmembrane pressure
TMS	transcranial magnetic stimulation
TN-S	terre neutre séparé
TOI	tissue oxygenation index
TPR	total peripheral resistance
TPS	treatment planning system
TPU	thermoplastic polyurethane
TR	time of repetition
TRICKS	time-resolved imaging of contrast kinetics
TSE	turbo spin echo
TSEBT	total skin electron beam therapy
TT	true-tripole
TTDT	threshold tone decay test
TTP	time to peak
TTS	temporary threshold shift
TTS	transdermal therapeutical system
TUR	transurethral resection
TUR-B	transurethral resection, bladder
TUR-P	transurethral resection, prostate
TV	television
TWIST	time-resolved imaging with interleaved stochastic trajectory
TimCT	total imaging matrix with continuous table movement
ToF-MRA	time-of-flight MR angiography

U

UCUM	Unified Code for Units of Measure
UDP	user datagram protocol
UEA	Utah electrode array
UITDD	ultrasound-induced targeted drug delivery
ULF	ultralow frequency
UPG	ultrasound pneumography
UPS	uninterruptible power supply
URR	urea reduction rate
US	ultrasound
USB	universal serial bus
USRDS	United States Renal Data System
UV	ultraviolet

V

VAD	ventricular assist device
VAH	Verbund für angewandte Hygiene
VAP	ventilator-associated pneumonia
VC	vital capacity
VC	volume controlled
VC-AC	volume control-assist control
VC-MMV	volume control-mandatory minute volume
VC-SIMV	volume control-synchronized intermittent mandatory ventilation
vCJD	variant Creutzfeldt–Jakob disease
VCT	volume computer tomography
VDE	Verband der Elektrotechnik, Elektronik und Informationstechnik
VEP	visual evoked potential
VES	ventricular extrasystole
VG	volume guarantee
VLAN	virtual local area network
VLF	very low frequency
VLWIR	very longwave infrared
VNA	vendor-neutral archive
VOG	videooculography
VOR	vestibulo-ocular reflex
VOT	vascular occlusion test
VR	ventricular rate
VR	virtual reality
VRE	vancomycin-resistant enterococcus faecium
VRT	volume rendering technique
VSV	vacuum–steam–vacuum
VT	tidal volume
VTC	videoteleconference

W

W3C	World Wide Web Consortium
WBT	web-based training
WD	washer-disinfector
WDRC	wide dynamic range compression
WHMS	wireless health monitoring system
WIM	wireless interface module
WLAN	wireless local area network
WM	white matter
WOB	work of breathing
WPW	Wolff–Parkinson–White-syndrome
WSN	wireless sensor network

X

XDS	cross-enterprise document sharing
XML	extensible markup language

Z

ZVEI	Zentralverband der Elektrotechnischen Industrie

Medical

Part A

Part A Medical Technology Basics

1 Technology in Medicine: Its Role and Significance in Terms of Health Policy
Rüdiger Kramme, Titisee, Germany
Heike Kramme, Titisee, Germany

2 Medicine Is More Than Applied Technology for Human Beings
Giovanni Maio, Freiburg, Germany

3 Hygiene in Medical Technology
Heinz-Michael Just, Nürnberg, Germany
Eckhard Roggenkamp, Nürnberg, Germany
Annette Reinhardt, Nürnberg, Germany

4 Technical Safety of Electrical Medical Technology Equipment and Systems
Rüdiger Kramme, Titisee, Germany
Hans-Peter Uhlig, Dresden, Germany

5 Quality Management in Medical Technology
Albrecht Malkmus, Freiburg, Germany

6 Usability of Medical Devices
Ulrich Matern, Tübingen, Germany
Dirk Büchel, Tübingen, Germany

1. Technology in Medicine: Its Role and Significance in Terms of Health Policy

Rüdiger Kramme, Heike Kramme

New avenues in diagnosis and therapy are today increasingly being opened up as a result of sophisticated and advanced technology, and at the forefront of this are evolutionary developments in existing technology. Many medical devices and pieces of equipment are developing at lightning speed as a result of digital technologies, which enable new medical concepts, strategies, and visions to be implemented faster than ever before. This means that developments which previously took a decade to implement are now being introduced at a rate of one a year. Technology thus not only has a dynamic interrelationship with medicine; it influences and shapes modern medical science on the basis of new technical possibilities. First-class health care would be inconceivable without progress and innovation in the field of medical technology.

1.1 A Short History 3
1.2 Early Breakthroughs of Medical Technology 3
1.3 Analog to Digital 4
1.4 Health Policy 5
1.5 New Key Areas 5
1.6 Innovation Versus Financial Resources 6

1.1 A Short History

Medicine (from the Latin *ars medicīna*, *the art of healing*) and technology (from the Greek, meaning *skill, craft*) have inspired and fascinated mankind since its early beginnings. Technical instruments and devices have always had their place in medicine. Acupuncture needles are known to have been used in Far Eastern medicine since approximately 2500 BC. Hippocrates (460–370 BC), the founder of scientific medicine in the Western world and a prominent doctor of his time, was already using a proctoscope to inspect his patients' intestines. He also gave descriptions of a variety of instruments and apparatuses for the treatment of wounds. These included, for example, apparatuses with weights and straps which, in the case of an arm fracture, positioned the broken bones in relation to one another, straightened them, and simultaneously immobilized them. As striking evidence from archeological digs in the buried town of Pompeii has shown, sophisticated instruments and devices for surgical interventions were already being used in the Roman Empire (from 63 BC onwards). The vision aids known as glasses, on which many of us rely, are not an achievement of the 20th century but had already been invented by a craftsman at the end of the 13th century.

1.2 Early Breakthroughs of Medical Technology

The first major breakthrough in medical technology and boom in modern medicine took place around the turn of the 20th century with Röntgen's discovery of x-rays in 1895. Although the nomenclature of the electrocardiograph (ECG) – which is still in use today – had already been decided by Einthoven in 1895, use of the first clinically viable ECG was not possible until 1903. In 1896, Riva-Rocci introduced the method of noninvasive palpatory measurement for determining blood pressure. The electroencephalogram (EEG) was first recorded in 1924 by Berger using a string galvanometer. Other milestones in medical technology were the invention and introduction of the artificial kidney (1942), the heart–lung machine (1953), hip-joint prostheses (1960), artificial cardiac valves (1961), and the first clinical patient monitoring devices (around 1965). Criteria which had already been used for classification in the USA were developed for measurement and standardization of the ECG according to the Minnesota Code around 1960. In the early 1940s, the construction of the first electronic computer ushered in a new era, and a new technology was born which was to revolutionize medical technology once more: data processing and information technology. This new technology overshadowed all the technological developments which went before it. If a modern calculator were equipped with electronic components (e.g., transistors) from 40 years ago, that calculator would require a power of 6000 W, provided by an electricity supply and emitted to the surroundings as heat. A weight of 50 kg and cube edges approximately 1 m in length would more likely suggest an oven than a calculator.

1.3 Analog to Digital

The radical change in technology from analogue to digital opened up new dimensions in medical technology: the computer tomograph (CT), which generates cross-sectional images of the body, was developed by Hounsfield and Cormack, and a prototype was installed and tested in a hospital in 1971. In 1977, Mansfield found success with a breakthrough for medical applications of magnetic resonance tomography using the magnetic resonance method, and the human thorax was imaged for the first time without the use of x-rays. Unique and sophisticated possibilities in diagnosis were introduced by a large-scale medical technology system which is used in nuclear medicine: the positron emission tomograph (PET). As an imaging system, the PET enhances the diagnostic range because it enables representations of physiological and metabolic processes in the human body to be determined both quantitatively and on a location-dependent basis. Molecular imaging with hybrid PET/CT scanners offers a view of things which had previously not been visible. However, other hybrids such as ultrasound and magnetic resonance imaging also not only have the advantage that they offer an image quality which is much more precise and accurate in every detail when compared with other imaging methods, but they can also be used without any exposure to radiation. It has so far been possible to reduce the radiation dose for a full body scan to as little as 40% compared with older systems.

As a result of the increasing integration of computer-based systems in x-ray technology, imaging methods are being redeveloped in ever shorter time cycles. The rapid growth of the spectrum of clinical applications and the continuous further development and implementation of new technologies have not only led to an altered and extended range of indications for these methods. Furthermore, imaging technologies are increasingly being developed as a complete solution, such as hybrid systems for interventional radiology or integrated IT solutions (picture archiving and communication system (PACS), radiology information system (RIS), etc.) which aim to optimize processes and thus increase efficiency in hospitals. The increasing interconnectedness of technology will change the health system.

To outline the progress and development of all the devices and achievements in medical technology would be to go beyond the scope of this book. Although medical technology is in most cases not original but rather adopts technological developments from fields such as electronics, optics, precision engineering, and plastics technology among others, and these developments are only thought of as being part of medical technology when applied to living creatures, medical technology has nevertheless established itself as a field, and medical care today would be unthinkable without it. This fact reveals the real significance of medical technology:

Medical technology devices and equipment (including in the laboratory and research field) are individual or interlinked instruments, apparatuses, machines, appliances, and auxiliary devices, and any necessary equipment which is used because of its function for the identification (diagnosis), treatment (therapy), observation (monitoring), and prevention (prophylaxis) of illness in humans.

1.4 Health Policy

The aim of health policy must be to provide human, modern, high-performance, efficient, and people-orientated medical care both in hospitals and on an ambulatory basis, with the focus on the patient. In the future, diagnosis and therapy will be adjusted according to the genetics of the patient, and technical solutions will be orientated towards the interaction between diagnosis and therapy. Another trend is that of orientation towards disease patterns. This aspect is even more important, because the risks of acute illnesses will increase as a result of ageing society. Investment in health care should provide benefits, not only in terms of administration at the level of individual hospitals and clinics but also in terms of national economics.

The development of medical technology as an essential part of health care is in permanent interaction with the changes in social lifestyles. The significance of medical technology in terms of health policy is therefore essentially based on the following points.

- The quality and security of medical care as a result of continuous modifications and improvements to diagnostic and therapeutic options and promotion of medical and technological research, and furthermore with broad application and extension to large population and patient groups using equipment-based mass screening (e.g., within the scope of illness prevention).
- Shortening the duration of illness or the length of hospital stay, which will reduce costs and therefore bring about associated benefits in terms of national economics.
- Relieving staff from time-consuming routine jobs.
- Meeting the expectations and demand level of the population in terms of the quality of the processes and of the results in health care.

Future developments in technical medicine must be geared towards the additional demands of health care as a result of limited resources.

- Medical technological diagnosis and therapy with high cost-savings potential, using environmentally friendly equipment and systems.
- Further development of minimally invasive procedures with the aim of reducing morbidity rates and convalescence times.
- Miniaturized compact systems, which are less time and cost intensive in terms of installation and servicing.
- User-friendly and operationally reliable design, which substantially avoids faulty operation. Invasive techniques will increasingly be replaced by less invasive and/or noninvasive techniques, such as disintegration of kidney and gall stones using a lithotripter instead of surgical intervention, endoscopic minimally invasive interventions instead of conventional surgery, three-dimensional (3-D) echocardiography to show complex malformations of the heart, pathomorphological changes in the mitral or tricuspid valve, and atrial or ventricular septal defects instead of the complex and high-risk procedure of cardiac catheterization, and imaging the coronary vessels using magnetic resonance imaging instead of contrast angiography or diagnosis by cardiac catheterization.

It is becoming apparent that the boundaries between diagnosis and therapy are becoming increasingly blurred by the use of current technological solutions such as interventional radiological or endoscopic procedures, for example.

Where operative interventions in traditional surgery were performed using a scalpel and surgical instruments, in the foreseeable future these will to a large extent be replaced by the *light and sound* of noninvasive surgery. Successful high-energy ultrasound operations on the brain have already been achieved in the field of neurosurgery, meaning not only that there are new methods of treatment opening up but that there is even talk of a paradigm shift in neurosurgical therapy. Neurosurgical treatment by means of ultrasound in the case of psychiatric disorders, such as affective psychoses, for example, should also be possible in the future.

1.5 New Key Areas

A key area of technology in health care of the 21st century is telematics, which has the potential to bring enormous advantages to all those involved in health care but will also mean that health care organizations are faced with many new organizational, technological, and legal requirements. In the future, hospitals will be centers of telemedicine applications. Telemedical communication and systems – that is to say all IT applications in the health care system which are provided via public or long-distance communications networks – enable large amounts of data to be transferred quickly, meaning that physical distance is no longer an obstacle. This is also a reason for the fact that increasingly great importance is placed on telemedicine internationally. These endeavors are aimed at developing a uniform platform for telematics, so that use of modern telecommunications and information technology will improve the quality of care and economic efficiency in the future.

1.6 Innovation Versus Financial Resources

Today, limited financial resources in hospitals mean that it is only rarely possible to introduce and exploit every technical innovation and possibility. It is therefore imperative for the user to evaluate any investment decisions on a commercial and performance-related basis (e.g., by process-orientated technology assessment, which takes account primarily of criteria such as performance, effectiveness, and efficiency). Particularly with respect to the advantages of a real investment, it is important that it is not emotional but rather rational criteria which are at the fore in the decision-making process. One of the key questions is whether there are limits to technical progress, and where these limits might lie. Assessment of technological possibilities with respect to their benefits for patients requires an understanding of modern technology and its limits. Frequently, the aim of medical technological manufacturers and suppliers is to provide medical technological products and medical data-processing systems which are better and more technically perfect every time. The result is that, these days, the functions of many medical technological products go far beyond the needs and possible uses for them. Users – who are usually not technophiles – will pay for something extra that they cannot use. Numerous sophisticated products are perhaps technically perfect but are rarely tailored to suit a need. There is a lot which is feasible technologically, but equally it is obvious that humans can barely control this technology, as in the case of the complexity of various software interfaces, which are no longer completely understood even by highly qualified technicians. This means that the technical possibilities are frequently beyond the ability of many users to use them. Uncritical enthusiasm for technology can therefore very quickly turn into technophobia.

However, to do nothing or to make do without innovations and to cling to outdated technical products is not a solution either. In the future, the service provided to customers in hospitals or in a doctor's practice will itself become a product with greater potential for differentiation than the quality and technical performance of medical technological products. The innovations in medical technology which will also be indispensable in the future must have a *human dimension* and be tailored to suit a need. This will inevitably be embedded in the area of tension between technical and scientific knowhow, market orientation, and orientation towards individual customers. From the point of view of the user, virtually all products are becoming increasingly similar. Good customer service will be another factor in the success of medical products: the focus must increasingly be on the demand for products and not just supplying them *on offer*.

2. Medicine Is More Than Applied Technology for Human Beings

Giovanni Maio

Medicine owes many of its undeniable achievements to technological developments; without the elaboration of technologies, medicine would have been unable to devise or apply many methods of treatment which are, indisputably, a blessing for mankind. And yet, curative science has sometimes been dazzled by the alliance between medicine and technology. Medicine has been taken in by technology to such an extent that it has lost sight of what characterizes it as curative science and what constitutes its actual essence. Technology is not just a method to be chosen, but also a programme. Ethical reflections on the relationship between medicine and technology are presented in this chapter.

2.1 Technology Suggests Feasibility and Controllability ... 7
2.2 Technology Knows No Bounds ... 8
2.3 Technology Is Unable to Answer the Question of Meaning ... 9
2.4 Technology Alone Does Not Make Medicine Humane ... 9
References ... 10

From the very moment when medicine presents a purely technical solution to the crisis of falling ill, medicine has not only chosen a method, but has devoted itself to a certain view of the world and humankind. If the relationship between medicine and technology is to be understood adequately, these basic preconceptions about humanity must be contemplated.

2.1 Technology Suggests Feasibility and Controllability

Despite the uncertainty that still remains, a system that backs technology alone assumes a high degree of controllability of the system. In this case, that which is incalculable merely represents a challenge to the developer to perfect technology to make it controllable. If technology is now used to heal the ill, and as a central instrument at that, this engineer's way of thinking also infiltrates how physicians think. In other words, physicians who rely fully on technology are under the tacit assumption that the problems faced by medicine are generally problems that can be solved using technology. If a problem remains, however, technology is to blame for not being sophisticated enough, according to this credo. This way, it is assumed that basically everything is feasible and that all problems faced by mankind can be solved using technology. The reproductive-medicine complex is one example of coupling treatment with technology. Reproductive medicine, in particular, which increasingly regards itself as a market-oriented service industry, equates remedy with the application of technical instruments; it responds to many persons' crisis of meaning by offering technical solutions. What is more, reproductive medicine implicitly declares the technical solution to be the only possible response to the challenge arising from involuntary childlessness. Medicine that regards itself in this way does not only create an offer, but also establishes standards that those who are confronted with involuntary childlessness are virtually unable to avoid [2.1]. In particular, however, reproductive medicine comprehended in this way fails to make use of the opportunity of making couples aware of the potential of alternative life concepts at a sufficiently early stage. Such an absolutization of the technical solution leads to a continuation of infertile couples being dependent on the technical solutions offered by medicine, rather than making them aware that a crisis of meaning can also be overcome by giving life a new meaning, a meaning that may arise by opening up new prospects of life [2.2]. It is the example of reproductive medicine, in particular, that highlights the fact that humane medicine is more than about just being technically adept. It is equally important to invest in good consultation, in a good conversation which should, in particular cases, also include touching upon the potential failure of technology. Technology suggests feasibility – as can be clearly seen using reproductive medicine as an example – and puts the *victims* of technology in a situation where they are unable to escape from the resulting technical imperative. Thanks to technology, every woman has the opportunity to fall pregnant, and if they still remain infertile, they just have not invested enough or have not had a sufficient number of attempts. It hardly occurs to the majority of physicians, nor, secondarily, to many couples, that the technical solution may not be the adequate one. Thus, technology creates a maelstrom, a feasibility maelstrom, which can only be resisted with difficulty. This aspect must always be taken into consideration with every new technology.

2.2 Technology Knows No Bounds

Technology has no boundaries; it progresses into unchartered territory, it never shies away from the new, nor does it spare the essence of being; technology is always oriented towards change and dynamism. With such a basic concept, however, technology ensures that there is no longer such a thing as a reasonable boundary within medicine, and that there is no state that might not undergo technical optimization or change. Research on embryos is one example of this. Here, in particular, it can be seen that technical methods are considered almost blindly without thinking about morally tenable boundaries. There are grounds here for criticizing the basic approach of using an inherently problematic technical means for an undeniably good cause. The fact that the destruction of embryos was at all taken into account as an option is the actual core of the ethical problem, which is that, in their basic approach, the natural sciences and technology grasp blindly at methods simply because they are technically feasible or simply because they are required to make promises come true. Offering such options alone creates a problematic denial of any boundaries whatsoever. It is like asking a counterpart a certain question that simply should not be asked, no matter what the situation is, because it is, for example, an unreasonable demand. In the same way, there should be certain methods in technical research and development that one simply does not select, because they are bound to offend the feelings of far too many people, and should therefore be regarded as unacceptable. Consequently, the basic problem that research on embryos has to face only arose because research methods were chosen blindly from the start, and, from the very beginning, no respect was shown for the fact that merely the request of sacrificing embryos is an unreasonable demand for many people – and this is just one example representing many more. We need only think of cloning techniques, the creation of chimera, and so on. In this case, technicians cannot retract and deny any responsibility. Moreover, simply by selecting their methods, technicians assume responsibility, and this responsibility must become apparent prior to the development so that no *blind* mechanization occurs, but a mechanization on a humane scale. And this humane scale must also take boundaries into consideration, and must bear in mind that certain processes are inherently problematic. If they are still selected, despite being aware of these problems, presenting the whole of society with a fait accompli, this could signify a heavy burden in certain constellations.

And yet, it is not only the moral boundary that developers of technology defy without further reflection – the denial of the boundary itself is the basic problem of technical development. Technology knows no point at which it could be said that it is perfect now as it is; technology, i. e., those who work on technical products, are always anxious to surpass existing technology with one that is even faster, even more sophisticated, even smaller or even more comprehensive. This attitude towards endless development is based on the lack of a notion about the ideal state and on a glorification of efficiency. The credo is: the faster, the better; the more, the better, and so on. Admittedly, this credo may be useful when dealing with certain utensils. In the context of medicine, however, the credo cannot be generalized in all cases.

What about the discussion on enhancement? This is precisely where the esteem for *more and more* reaches its limits and where *less is more* makes sense [2.3]. The human being is not a machine that operates better the faster it runs; humans need efficiency as much as they crave leisure, they need to achieve their goals quickly, but also need changes in their lives, and resistance to be able to mature at all. Accelerating everything – human beings included – does not automatically mean that human beings are doing themselves a favor. With regard to humanity, offering less, decelerating or back-pedalling is often more conducive to reaching the goal, provided, of course, that humanity finds fulfilment is the desirable goal rather than rapid production.

2.3 Technology Is Unable to Answer the Question of Meaning

Technology knows only purpose-rational thinking, and if technology is proclaimed as the solution for mankind, the human being will primarily be regarded as a body machine within such thought. Technology does not ask about meaning, about the superior sense; indeed, it cannot ask what is meaningful, because it lacks the tools to assess the answer to this question – since the question of meaning cannot be expressed in figures and values. With regard to technology in the context of medicine, this aspect is particularly precarious. As a result of technology's triumph, medicine has sometimes fallen victim to an absolutization of the natural sciences and technology, with the grave consequence that medicine is time and again inclined to define not only what is proper but also what is good via the natural sciences. That which takes place as an applied natural science and technology in the course of the self-image of medicine is equating functionality with the good, equating natural scientific properness with what is good for mankind. In this context, organic functionality is occasionally seen as a value in itself; in other words, functionality should always be restored. Secondly, the loss of functionality is per se interpreted as something negative or even as a failure on the part of medicine under this natural scientific dictum. Both conclusions, however, are inherently problematic. First of all, trying to restore functionality by all means may be problematic if sight of the whole picture is lost. Concerning the superior sense, restoring functionality cannot be equated with the creation of meaningfulness; for example, organs can be restored but the treatment may be senseless regardless. The more medicine becomes specialized and regards itself as a natural science, the more it sometimes treats organs and x-ray photographs, blood gases, and laboratory values, but by doing this does not automatically treat the human being. Precisely this, then, becomes a serious problem if medicine sees itself only as a technically oriented specialist medicine.

2.4 Technology Alone Does Not Make Medicine Humane

Bearing all this in mind, it should become apparent that it is not a matter of demonizing technology. Technology per se is not the problem of modern medicine. The problem starts when the importance of technology is overestimated, i. e., when it is assumed that existential questions arising from a patient falling ill can only be solved by technological means. The central criticism is therefore not directed at technology itself, but at the thought that technology is the one and only solution. Medicine that categorically rejects technology is doomed to failure, because in this case, it often fails to exploit the potential of being able to assist. Medicine, however, that relies solely on technology and ignores everything else will equally fail. For this reason, humane medicine has to focus on implementing technology, whilst always bearing in mind that human beings need more than effective technology to recover. The art of healing is to precisely recognize where technology would be a good solution for human beings and where it is merely an apparent solution. This art of healing requires a pronounced practical power of judgement in a more comprehensive sense, and not only technical expertise.

The technical credo in modern medicine overlooks the fact that physicians often successfully bring about a cure by a good relationship in which technology has to be embedded. Technology without a relationship will generally achieve little. Great importance is attached to the technical aspect, whereas the human relationship is often completely ignored. This change in priorities represents the greatest challenge mechanized medicine has to face. The more technology is used as a substitute for a relationship, the more this will lead to the loss of a curative culture. Consequently, as long as we do not expect more from technology than it is actually able to solve, and avoid using it as a substitute for everything, the sick will rightly hope that they will regain their health if possible or find comfort and meaning when there is no further chance of a cure. As long as medicine wishes to see itself as a curative science and not as a repair service, it will have to invest the same amount of energy in comforting the suffering and giving their life new meaning as in technical development.

Further Reading

- M.L. Eaton, D. Kennedy: *Innovation in Medical Technology: Ethical Issues and Challenges* (Johns Hopkins Univ. Press, Baltimore 2007)
- R. Chadwick (Ed.): *The Concise Encyclopedia of the Ethics of New Technologies* (Academic, New York 2000)

References

2.1 G. Maio: Auf dem Weg zum Kind als erkauftes Dienstleistungsprodukt? Eine ethische Kritik der modernen Reproduktionsmedizin, Z. Evang. Ethik **52**(3), 194–205 (2010)

2.2 G. Maio: Zur Hilflosigkeit der modernen Medizin im Hinblick auf die Frage nach dem Sinn, Ethica **18**(1), 3–9 (2010)

2.3 J. Boldt, G. Maio: Neuroenhancement. Vom technizistischen Missverständnis geistiger Leistungsfähigkeit. In: *Das technisierte Gehirn. Neurotechnologien als Herausforderung für Ethik und Anthropologie*, ed. by O. Müller, J. Clausen, G. Maio (Mentis, Paderborn 2009) pp. 381–395

3. Hygiene in Medical Technology

Heinz-Michael Just, Eckhard Roggenkamp, Annette Reinhardt

The application of new technologies in medicine leads to therapeutic and diagnostic advancements, yet also causes risks for patients to aquire health-care associated infections. In this chapter precautions to prevent the transmission of infectious agents from inanimate medicotechnical sources are shown.

Disinfection and sterilization processes are described in detail aside with requirements for cleaning equipement used for noninvasisve and invasive technology on the patient (Sects. 3.4–3.7). Targeted measures with focus on technical means for preventing the four most important device-related infections are pointed out in practical examples (Sect. 3.8). Furthermore special attention is given to dialysis departments because of high risk of infection both for patients and staff and to the special processing of medical devices that have been used on patients with proven or strongly suspected Creutzfeld–Jacob disease (CJD) respectively its new variant (vCJD).

Finally technical regulations and standards focused on german and european circumstances give an overview of what must be observed by manufacturers and users of medical devices (Sect. 3.9).

3.1 **Background** ... 12
3.1.1 Employee Protection ... 12
3.1.2 Patient Protection ... 12

3.2 **Causes of Infection** ... 13

3.3 **Vaccinations** ... 13

3.4 **Disinfection Methods** ... 14
3.4.1 Basics of Disinfection ... 14
3.4.2 Disinfection Processes ... 14
3.4.3 Chemical Disinfecting Agents ... 15
3.4.4 Carrying out Manual Disinfection ... 15
3.4.5 Physical Disinfection Processes ... 16
3.4.6 Application Times and Ranges of Action ... 19
3.4.7 Comparison of Chemical and Physical Disinfection Processes 20

3.5 **Sterilization Methods** ... 21
3.5.1 Sterilization Processes ... 21

3.6 **Hygiene of Noninvasive Technology Equipment** ... 25
3.6.1 Equipment Used on the Patient ... 25
3.6.2 Equipment Not Used on the Patient 26
3.6.3 Repair and Maintenance ... 26

3.7 **Hygiene of Invasive Technology Equipment** ... 26

3.8 **Practical Examples** ... 26
3.8.1 Postoperative Wound Infections ... 28
3.8.2 Ventilator-Associated Pneumonias (VAP) ... 28
3.8.3 Catheter-Related Septicemia ... 29
3.8.4 Catheter-Related Urinary Tract Infection ... 29
3.8.5 Dialysis ... 29
3.8.6 Creutzfeldt–Jakob Disease ... 31

3.9 **Regulations** ... 31
3.9.1 Technical Regulations for Hazardous Substances ... 31
3.9.2 Standards ... 32

References ... 33

Technology is increasingly finding its way into medicine. Many diagnostic and therapeutic advancements have only become possible as a result of corresponding technical processes and further developments.

3.1 Background

The significance of hygiene becomes clear when we realize that even many years ago almost half of all infections contracted by patients in hospital were associated with medicotechnical measures or were (partly) caused by them [3.1]. Medical progress uses increasingly complicated and sophisticated technical facilities and equipment, the use and preparation of which also endangers employees.

It is difficult to obtain reliable data which is more or less representative from Germany, because to date there is no central collection point for this purpose. An indication can be gained by drawing on observations which are based on data from the Trade Association for Health and Welfare in Hamburg [3.2]. According to these observations, infectious diseases represent the second largest group after dermatosis with a frequency of 7.3%, but their numbers increase to one-third in the case of those occupational diseases for which compensation is awarded for the first time. The data do not show what percentage of the cases of dermatosis can also be attributed to technical use in the widest sense, such as handling of detergents and disinfectants.

Hygiene measures in the context of medical technology devices must therefore pursue the goals of

1. Protection of employees during handling,
2. Protection of patients during use of these devices *against the transmission of germs*, which can lead to
   a) Contamination,
   b) Colonization, or
   c) Infection.

The measures that are necessary in individual cases to achieve these goals depend on several factors.

3.1.1 Employee Protection

When using these devices on patients, the rule is to act such that the risk of coming into contact with the patient's germs is kept to a minimum. This is achieved by providing appropriate briefing regarding correct handling before a medical technology device is used for the first time. Hygiene guidelines govern which protective measures are necessary and when, but these protective measures are also dependent on the illness of the patient, the suspected bacterial colonization, and the possible transmission path. When dealing with medical technology devices in the course of reprocessing, maintenance, and repair, the employee can himself monitor whether the equipment is already visibly contaminated on the outside or components are dirty, for example. He/she must in particular have been instructed by the operator regarding whether the device has been used immediately beforehand for a patient with a contagious disease or with certain germs.

In such cases, disinfectant preliminary cleaning must be carried out before maintenance or repair is begun. Disinfection as a first step is also always necessary when handling of the device is linked with an increased risk of injury. Where reprocessing work is concerned, processes should be used in which the devices are cleaned and disinfected by machine, and this should be done with the application of heat and in one process. Under some circumstances, certain protective clothing (e.g., gloves) is sensible or even compulsory.

3.1.2 Patient Protection

How the medical technology device is used on the patient is crucial to the necessary measures. A pacemaker implanted in the patient must be and remain sterile and pyrogen-free during insertion. Disinfectant pretreatment is sufficient for medical equipment with only external (skin) contact, and in the case of equipment which stands at the bedside next to the patient, cleaning is generally sufficient. However, if parts of a piece of equipment which is situated remotely from the patient come into contact with sterile areas of the patient (e.g., tube systems which convey blood in a dialysis machine or in cardiac surgery equipment), then this system component must of course satisfy the same criteria as an implanted device. The same also applies when equipment is used to introduce fluids or medication into sensitive (e.g., lungs in the case of machine-assisted artificial respiration) or sterile regions of the body (e.g., infusion apparatus) [3.3–5].

The text which follows explains the principles of targeted hygiene measures and demonstrates, with reference to examples, how the risks can be recognized and which risk-based measures are necessary. Reference is made to sets of regulations which must be observed, although it is the duty of the person responsible for the area in question to adapt the catalog of measures to new scientific findings and recommendations in the course of regular training. General guidelines which are not orientated towards practical use, the specific risk of infection, and the path of infection are often expensive, as they require a lot of personnel and time, but are rarely effective.

3.2 Causes of Infection

Requirements for an infection to develop are an infectious agent, a person susceptible to infection, and contact which enables the germs to colonize the individual such that an infection can develop. This multitude of requirements makes it clear that there is no reason to live in general fear of microorganisms, be they bacteria, viruses or fungi. Bacteria colonize our skin and mucous membranes and are an important part of our body's defenses. Over 40 different species can be isolated from our nasopharyngeal cavity, and there are up to 10^{12} germs living in every gram of feces.

The natural bacterial colonization present in the skin of every human being can be divided into *permanent* and *temporary*. Permanent germs are always present, whereas temporary germs are acquired and therefore change according to what the person has been handling or what work he/she has been carrying out. Washing the hands eliminates the majority (> 90%) of this acquired *contamination* but leaves the permanent bacterial colonization undisturbed. Disinfecting the hands or skin should completely eliminate acquired germs, but it also has an adverse effect on the permanent skin colony.

The skin and mucous membranes are mechanical barriers which, when they are not damaged, prevent microorganisms from penetrating into our bodies. This explains why damage to the skin and mucous membranes is always accompanied by an increased risk of infection, whether that be in the form of a local, superficial infection (pustule, abscess) or whether it be in the form of a widespread infection – usually occurring in immunocompromised patients – of the soft tissue (ulceration, gangrene), which can also result in sepsis with high fever.

In many fields, our body has also developed further defense mechanisms, such as the acid mantle of the skin, microorganism-killing enzymes in secretions and excretions (e.g., tear fluid), and special structures in our blood whose primary role is to eliminate intruders. These include the white blood cells which *eat up* and digest bacteria (phagocytose), and so-called antibodies which help blood cells identify structures in the body which they should destroy. These specific antibodies are formed in the lymphatic tissue of our body following such *stimulation*. Stimulation of this kind may be due to contact with the infectious agent itself (*natural immunization*) or may be as a result of vaccination (*artificial immunization*, see later).

If a germ nonetheless manages to attach to skin or mucous membranes, then the first important step has been successful. If this *colonization* persists, although it does not result in illness, the patient or member of staff would become an (undetected) source of further transmission, in case the germ in question is a problematic germ (infectious agent, multiresistant bacterium). However, if in the second step the germ is able to deploy its pathogenic properties and the person affected is not immune, then this would lead to an *infection* which, depending on the state of health of the affected individual, can result in an illness which varies in its severity.

3.3 Vaccinations

One of the most important measures for protecting against infections is vaccination. The vaccinations which are recommended and constantly updated by the Standing Committee on Vaccination (Ständige Impfkommission – STIKO) at the Robert Koch Institute (RKI) [3.6] are of particular importance for employees in the health care system. The vaccinations in category S (standard vaccinations with general application = standard vaccinations) are the vaccinations for infants and children and should be given to all employees in the health care system and, if necessary, should be regularly boosted. These include important vaccinations against such as tetanus, poliomyelitis, and diphtheria. Depending on the field of activity, so-called *indicated vaccinations* (category I) may also be added, such as vaccinations against hepatitis A and B, influenza, and varicella. The point of contact for questions regarding personal vaccine protection and work-related requirements is generally the occupational health officer.

3.4 Disinfection Methods

3.4.1 Basics of Disinfection

Like sterilization (Sect. 3.5), disinfection also has the aim of preventing transmission of pathogens. Complete freedom from germs (sterility) is not guaranteed, however. Disinfection of equipment and materials is always sufficient if, although the aim is to prevent transmission of microorganisms which are capable of multiplying, the body physiologically speaking has a certain level of self-protection in these areas as well as in other areas of the human body which have colonies of germs (e.g., the gastrointestinal tract). Obligate pathogens (those which always cause illness) must not be present on disinfected items, however.

When using disinfectants and disinfection processes, both the respective microbiological spectrum of activity and the field of use must be taken into consideration. Thermal disinfection processes, where they can be used, must always be given precedence over chemical disinfectants and disinfection processes. Provided they do not contain any other special directions, chemical disinfectants are usually only suitable for killing vegetative bacteria and fungi.

Before they come onto the market, disinfectants are tested for their antimicrobial action by means of microbiological analysis. There are standardized methods for this, whose results also determine whether a substance will be included in the list of substances permitted by the Robert Koch Institute in accordance with the German Infection Protection Act [3.7].

Part A | 3.4

3.4.2 Disinfection Processes

Thermal processes are only suitable for thermostable objects, while *chemical* processes are also suitable for thermolabile objects and surfaces.

A distinction is made between the fields of use for chemical disinfection, as follows:

- Disinfection of hands, skin, and mucous membranes
- Disinfection of surfaces
- Disinfection of instruments.

Disinfection of Hands, Skin, and Mucous Membranes and Disinfection of Surfaces

Disinfection of hands, skin, and mucous membranes and disinfection of surfaces can only be carried out in the form of chemical disinfection, with various germicidal substance groups being used (Table 3.1). When selecting a substance, the purpose for which it will be used and the required strength of its effect as well as the required scope of its effect are crucial in this selection. Appropriate definitions should be regulated in area-specific or process-specific *hygiene plans* based on the corresponding KRINKO (Kommission für Krankenhaushygiene und Infektionsprävention) recommendation [3.8].

Table 3.1 Advantages and disadvantages and fields of use of the most common active substances in disinfectants

Active substance	Advantages	Disadvantages	Field of use
Alcohols	Fast-acting, no residues, low toxicity, pleasant odor	Not sporocidal, combustible/explosive, expensive	Hand disinfection, skin disinfection, small surfaces
Iodine/iodophosphorus compounds	Does not irritate mucous membranes, fast-acting	Allergies possible, naturally colored, (side-effects on thyroid?)	Skin disinfection, mucous membrane disinfection, hand disinfection
Formaldehyde/aldehyde	Broad spectrum of activity, biodegradable	Irritant, allergenic, moderately toxic, (carcinogenic?)	Surfaces, instruments, disinfection of rooms
Quaternary ammonium compounds	Good detergent action, low odor, low toxicity	Gaps in effectiveness, inactivated by soap and protein	Disinfection of surfaces in special areas (kitchen)
Peracids/peroxides	Broad spectrum of activity, fast-acting	Inactivated by protein, corrosive, irritant, unstable	Surfaces, instruments
Phenols	Low impact because of environment	Gaps in effectiveness, barely biodegradable	Disinfection of excretions, otherwise obsolete

Disinfection of Instruments

Instruments and equipment can be disinfected thermally, thermochemically, or else purely chemically. The choice of the process is dependent on the suitability of the material for certain types of disinfection, on the local conditions (infrastructure), and if applicable, on certain requirements. The most reliable option is mechanical thermal disinfection in special washer-disinfectors, because it is only when exposed to appropriate temperatures that the desired reduction in germs is guaranteed with sufficient certainty. The machines report faults in the program sequence, which means that it is not possible to remove items inadvertently before the disinfection process is complete, and errors are for the most part ruled out. Purely chemical processes, such as immersion in solutions, etc., are in contrast susceptible to errors in processing and require a high degree of reliability on the part of the staff performing the process.

3.4.3 Chemical Disinfecting Agents

The most common active substances in disinfectants, their advantages and disadvantages, and their fields of use are reproduced in Table 3.1. In the rarest cases, commercial disinfectants contain only one active substance, but they frequently consist of mixtures of active substances in order to achieve the most optimum antimicrobial action possible.

3.4.4 Carrying out Manual Disinfection

Selection of Disinfectants

Disinfectants are usually selected on the basis of the Network for Applied Hygiene (Verbund für angewandte Hygiene, VAH) list. However, in so doing it is necessary to bear in mind that, in addition to the field of use, the concentration, and the application time (which are dependent on one another), the necessary scope of activity is also ensured. Where there is any doubt, reference must be made to appropriate expert advice. A further important source of information, in particular from the point of view of occupational health, is the material safety data sheets according to 91/155/EEC – amended by 2001/58/EEC – for the disinfectants in question. With regard to material compatibility and effectiveness, particular care must be taken with materials which contain rubber and plastic.

Exposure Time

The maximum period during which the substance has its intended effect can be gathered from the relevant data sheets. If the solution becomes visibly dirty, however, then it should be replaced immediately. If a combination of detergents and disinfectants is used, the exposure time is generally only 24 h.

Sequence: Disinfection and Cleaning

In the case of instruments where there is a risk of injury, disinfection must be carried out *prior* to cleaning. In other cases, disinfection is performed *with or after* cleaning.

Procedures

Disinfectants for treating surfaces and instruments are provided by manufacturers in the form of a concentrate in various packaging sizes, from sachets to large packs, and must be made up to the appropriate usage concentration by the user by adding water. To avoid foam formation when preparing a disinfectant solution, water is added first and then the disinfectant. The solution is prepared by hand by means of dosing aids or using mixing equipment. The advantage of using mixing equipment is the automatic dosing of the disinfectant.

Disinfectants must only be used for the stated purpose and must not be mixed with detergents without prior testing, because this can result in a loss of effectiveness of the disinfectant (follow the information from the manufacturer). It becomes necessary to change the disinfectant solution when the exposure time stated by the manufacturer has been reached or when the solution becomes visibly dirty.

Effectiveness may be impaired in tubes and pipes with narrow lumen, e.g., as a result of air bubbles or impurities. It is therefore necessary to ensure that the item to be disinfected is completely submerged, that there are no bubbles, and that all surfaces are completely and thoroughly wetted.

All items should be disassembled as far as possible. Instruments should be immersed in the solution with care to prevent damage.

Cost-Saving Hints

The following options can be used for cost savings:

- It is often sufficient to carry out cleaning instead of disinfection prior to subsequent sterilization. Exceptions: Only pointed, sharp items must be disinfected prior to cleaning.
- Where feasible from a time perspective, low concentrations should be selected with a longer application time.

- If possible, make do without additional detergents, because a disinfectant solution with added detergent must be changed daily.
- Contaminate the solution as little as possible with organic materials, so that it is not necessary to change the solution before the end of the exposure time.
- Wipe down items instead of immersing them.

Pouring Away

To protect drainage systems against corrosion, care must be taken to ensure that the solution is diluted sufficiently before it is poured away. Disinfectants for treating instruments are generally provided with corrosion inhibitors. High concentrations can nevertheless be problematic for drains. The municipal wastewater bylaws must also be taken into account. Disinfectant concentrates are considered hazardous substances, and their disposal requires special supervision.

Personal Protection

When preparing the solution, immersing/removing items, and emptying and cleaning the bowl, gloves (disposable or household gloves) and protective clothing (waterproof apron) must be worn. If there is a danger of splashing, protective goggles and a face mask must be worn. Bowls filled with disinfectant must be covered to minimize evaporation into the surrounding air. Disinfectant solutions must always be made up with cold water (no warmer than lukewarm) for the same reason. Disinfectant concentrates must not be stored above eye level. The workplace-specific regulations to protect staff should take into account the relevant technical regulations (the German Technical Regulations for Biological Agents, TRBA, and the Technical Regulations for Hazardous Substances, TRGS) (Bundesanstalt für Arbeitsschutz und Arbeitsmedizin, BAuA).

Special Circumstances

There may be undesired effects on materials if unsuitable disinfectants are used. The chemical resistance of the items must therefore always be taken into account. Where there is any doubt, enquiries should be made to the instrument manufacturer.

3.4.5 Physical Disinfection Processes

With physical disinfection processes, a distinction is made between thermal and thermochemical disinfection processes.

Thermal Disinfection Processes

In thermal disinfection processes, pathogens are rendered harmless as a result of the influence of heat. The higher the temperature and the longer the application time, the more effective the process is. In practical applications, a distinction is made between *dry heat* and *damp heat*, depending on the presence or absence of free water. Only *damp heat* is of significance for combating hospital infections. When using a treatment with *damp heat*, a distinction is made between two processes:

1. Rinsing with hot water (washer-disinfectors) and
2. Treating with steam (steam disinfection process).

Washer-disinfectors

Washer-disinfectors are devices in which instruments, anesthesia accessories, laboratory materials (glassware and the like), and other thermostable items are processed by machine.

The series of standards DIN EN ISO 15883 specifies the performance and device requirements for washer-disinfectors.

Depending on the design, a distinction is made between washer-disinfectors with one processing chamber and devices with multiple processing chambers, so-called batch washer systems.

Washer-disinfectors are available in a front-loading design (loaded and unloaded in the same area) or in a through-loading design with two doors (separated into a *clean* and *dirty* side).

Batch washer systems (multichamber systems) consist of multiple washing chambers and drying chambers through which the items to be treated are passed on loading trolleys. The loading trolleys differ according to the items to be treated. Sensors on the loading trolleys allow the control unit in the system to identify what items are on the trolley and to automatically select the correct processing program, which means that operating errors as a result of incorrectly selected programs, temperatures, and times are ruled out.

The disinfection process which runs in each case is usually thermochemical or thermal. An ultrasonic cleaning tank can be added to the system as an additional feature for precleaning of heavily soiled items.

Different processing steps are performed in each chamber. Special tanks with the appropriate detergents are assigned to each chamber. The detergent solution is collected in the tanks and reused for the next processing batch. Because not all of the detergent solution is recaptured, some of the solution must be supplemented.

The exact dosing of the detergent is done using dosing pumps which are controlled via contact water meters. The contents of the processing tank must be emptied and refilled every working day.

Owing to their relatively high throughput rate, batch washer systems are predominantly used in central sterile supply departments.

Carrying out Machine Cleaning and Disinfection

To achieve a good cleaning and disinfection result, the way in which the machine is loaded is of crucial importance. It is important to ensure that all items are disassembled as far as is possible and that hollow parts are inserted with the opening pointing downwards. Fluid must also flow through the interior of instruments with long or narrow cavities, such as metal catheters, metal suction apparatus, special needles, etc. Special loading trolleys must be used for this.

In the case of use of detergents or combined disinfectants and detergents, the information provided by the manufacturer (application time, concentration, and temperature) must be followed precisely.

Only the correct dosage will ensure a perfect disinfection and cleaning result while providing the greatest possible protection of the material. Underdosing of alkaline detergents involves the risk that pitting may occur, as this is avoided at pH values of above 10.5. When using acidic detergents, corrosion may occur as a result of chlorides in the water, and this can only be precluded by using demineralized water.

In the case of machine cleaning, all residues from the cleaning phase must be reliably removed in the rinsing phase, as otherwise stains and discolorations may appear on the surgical instruments. Additional use of a suitable neutralizing agent can support this process and improve the result of rinsing.

Documentation

In the course of quality assurance, when processing medical equipment it is necessary for the process-relevant processing steps of the individual batches to be documented with direct assignment to the relevant items to be treated.

Inspections and Maintenance

Disinfection measures in cleaning equipment are only effective if maintenance and inspection of these machines are not neglected. The necessary inspection and maintenance are specified in the operating instruction, which should be issued by the manufacturer. Maintenance should be carried out at least once a year by trained specialists.

Tests

The ordinance regarding the installation, operation, and use of medical devices (German Medical Devices Operator Ordinance – Medizinprodukte-Betreiberverordnung, MPBetreibV) states the requirement that medical devices must be processed using suitable, validated processes such that reproducible success is ensured and the safety and health of patients, users, and third parties are not endangered. The KRINKO recommendation [3.9] likewise requires validated processing of medical devices.

Specific information about carrying out the validation and the subsequent periodic tests is set out in DIN EN ISO 15883-1 – Washer-disinfectors – General Requirements, Definitions, and Tests, and in the guideline from the German Society for Hospital Hygiene, the German Society for Sterile Supply, and the Working Group Instrument Preparation for routine monitoring of machine washer-disinfectors for thermolabile medical devices.

Validation is a documented process for providing, recording, and interpreting the results necessary to show that a process constantly produces the desired quality which conforms with the given specifications. For washer-disinfectors (WD), the validation consists of installation qualification, operating qualification, and performance qualification, performed for devices which have documented proof from the manufacturer of compliance with the requirements of DIN EN ISO 15883.

The installation qualification is performed to ensure that the WD and accessories have been properly supplied and installed and that the supply of operating media satisfies the special requirements. The tests and inspections which are to be carried out for the installation qualification must be defined and performed, and the results documented.

Tests and inspections which must be carried out are:

- Testing the scope of supply and delivery (in the case of existing installations, testing the stock)
- Loading trolleys/baskets, cartridges, and also plugs/adapters
- Installation plan, instructions for use
- Testing the connections and supply of media, and comparing them with the installation plan
- Electricity
- Water (cold/warm/demineralized)
- Steam

- Wastewater
- Exhaust air/ventilation.

The operating qualification is carried out to ensure that the WD and the supply of media comply with the manufacturer's specifications and the requirements set out in DIN EN ISO 15883. The tests and inspections which are to be carried out for the operating qualification must be defined and performed, and the results documented. In the performance testing, the specified washer-disinfector programs are tested for reference loading, and the results are documented. When observing the regulations, it should be ensured that results are obtained which can be reproduced at any time. Any reference loading must cover instruments with contamination which is typical of operation as well as critical design features. Reference loading is always operator specific and must be documented. A requirement for the performance qualification is specification and documentation of the necessary programs with the corresponding process flows. The description of the process must also include the preconditions for cleaning. The process description must be documented in detail, including precise information about the chemicals. The following tests are performed as part of the performance qualification.

Testing the Cleaning. The cleaning is tested using two different methods. Test instruments (hemostats after Crile) with defined test soiling in accordance with DIN EN ISO 15883 and instruments which are actually soiled following use are used. Every program used must be tested. The test instruments are removed from the WD using gloves following the cleaning phase and prior to the disinfection phase.

The results of the cleaning process are first evaluated visually, and these findings are documented. The test instruments must be visibly clean. The test instruments must then be tested for protein residues using a protein detection method which is at least semiquantitative. In practice, detection of protein residues using the Biuret method has proven successful. Testing of the instruments which are actually soiled is carried out in the same way.

Assessment. The limit value is that all test instruments must neither reach nor exceed the protein content of 100 µg protein per ml eluate. If this limit is exceeded, then the WD is immediately shut down. The guide value is a maximum of 50 µg protein per ml eluate for a test instrument, at which no measures are necessary.

Testing the Disinfection. The disinfection is tested using thermoelements which are distributed in the disinfection chamber at critical locations and which record the temperature profile during processing. The resulting temperature graphs show whether the temperature necessary for killing microorganisms was present at all locations in the chamber. The A_0 value can be calculated from the temperature graph in accordance with DIN EN ISO 15883. The A_0 value of a disinfection process with damp heat is a measure of the rate at which microorganisms are killed, given as a time in seconds at a temperature of 80 °C applied to the medical item by the process.

The A_0 value which must be achieved depends on the nature and quantity of the microorganisms on the contaminated medical device and on the subsequent use. In the case of crucial medical devices and medical devices which are or may be contaminated with heat-resistant viruses such as hepatitis B, the A_0 value must reach 3000. This corresponds to an application time of 5 min at a temperature of 90 °C or an application time of 50 min at a temperature of 80 °C.

An A_0 value of 600 is used in the case of noncritical medical devices which can only come into contact with undamaged skin. This corresponds to an application time of 1 min at a temperature of 90 °C or an application time of 10 min at a temperature of 80 °C.

In addition to the thermoelectric measurements (A_0 concept), biological indicators can also be used to make statements about the killing of microorganisms. Biological indicators are germ carriers which are contaminated with a blood/germ mixture which has a defined resistance to the disinfection process in question. The Robert Koch Institute stipulates the use of contaminated screws and tubes for testing thermal disinfection processes in washer-disinfector machines. Meanwhile, equally good biological indicators are available which allow testing which is easier for the user.

The performance qualification must be repeated annually. When the programs or process chemicals are changed or new medical devices are introduced which have to be processed differently, the performance qualification must be carried out once more.

Decontamination Systems

Decontamination systems have in the past been used first and foremost for cleaning and disinfection (decontamination) of bed frames and accessories. Requirements regarding hygiene, economic considerations, and occupational safety requirements (TRBA 401 hazard due to skin contact) have led to decontamination systems increasingly being used for other articles in the field of medicine as well, such as transport trolleys, containers (e.g., for medications, medical devices, sterile articles, and meals), operating theater shoes, transporting containers for small conveyor systems, and similar items.

A decontamination system consists of a decontamination chamber, which receives the items to be treated, and an apparatus compartment, which contains the units and components necessary for operation. The systems are generally constructed with a two-door design.

The items to be treated are pushed into the decontamination chamber on the *dirty side* (in some cases using special loading trolleys). Combined cleaning and disinfection of the items to be treated is carried out in the first phase using a separate nozzle system by means of a recirculation pump. The temperature of the decontamination agent solution and the decontamination period can be set and changed on the control panel, making it possible to optimally adapt the process quickly and simply to the items being treated. The decontamination agent solution is supplied from a heated storage tank.

Following the decontamination process, the items being treated are sprayed with a rinsing agent solution to remove residues of the decontamination agent solution and ensure fast and spotless drying. During the drying time, a ventilator extracts the damp warm air from the interior of the compartment, while at the same time sucking in fresh air from the clean side. The decontamination agent solution is pumped back into the storage tank via a recirculation pump, so that only about 20 l of water is required for each batch.

With regard to the requirements, operation, and testing of the effectiveness of decontamination systems, reference should also be made to DIN standards 58955 section 1–7.

Steam Disinfection Processes

Steam disinfection processes are preferably used to disinfect bedding (mattresses, laundry, and textiles) but also for waste which needs to be disinfected. Simultaneous use of these systems for disinfecting waste as well can be considered to be problematic because of the offensive smell and risk of contamination of the apparatus. With appropriate separation, however, joint use of the apparatus to disinfect both bedding and waste is also perfectly feasible.

The items to be disinfected are subjected to the effect of saturated steam in the steam disinfection apparatus. To ensure that all surfaces to be disinfected are exposed to unobstructed steam, the air must be removed from the disinfection chamber and from the items.

A distinction of steam disinfection processes can be made depending on the procedure:

1. Steam flow process, and
2. Fractionated vacuum process (vacuum–steam–vacuum (VSV) process).

Steam Flow Process (Range of Action, ABC: Sect. 3.4.6). In the steam flow process the air is forced out of the chamber and the items to be disinfected using saturated steam. The disinfection temperature is 100–105 °C, with an application time of at least 15 min. For porous items, the application time may be more than 1 h. The steam flow process is suitable for disinfecting waste which contains sufficient water, e.g., microbiological cultures.

Fractionated Vacuum Process. The process (Fig. 3.1) is characterized by:

1. Removal of the air from the chamber and the items to be disinfected by repeated evacuation alternated with influx of saturated steam
2. Disinfection with saturated steam
3. Drying of the disinfected items by evacuation.

To perform this process, steam which is largely free of air and foreign gases is necessary (cf. DIN EN 285). The disinfection chamber must be vacuumtight. The fractionated vacuum process is mainly used to disinfect porous items such as mattresses, blankets, and waste.

3.4.6 Application Times and Ranges of Action

The application times and ranges of action are presented in Table 3.2. In the list of disinfectants and disinfection processes tested and approved by the Robert Koch Institute [3.7], the ranges of action are identified by letters; these are:

- A – suitable for killing vegetative bacteria, including mycobacteria, as well as fungi, including fungal spores
- B – suitable for inactivating viruses
- C – suitable for killing spores of the anthrax pathogen.

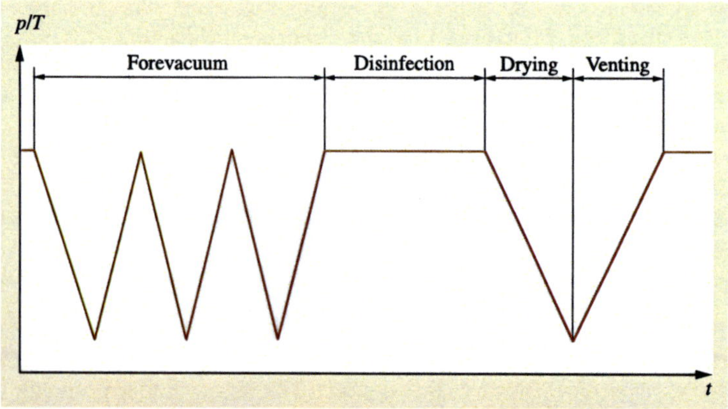

Fig. 3.1 Diagram of the fractionated vacuum process

Table 3.2 Application times and ranges of action of disinfection processes

Temperature (°C)	Duration (min)	Range of action
75	20	A, B (except viral hepatitis)
105	1	A, B
105	5	A, B, C

Higher temperatures and longer application times are sometimes used for disinfecting waste. Approved processes can be found in the RKI list. The requirements, operation, and testing of the effectiveness of steam disinfection apparatus are laid down in the DIN standards 58949 section 1–7.

3.4.7 Comparison of Chemical and Physical Disinfection Processes

Disadvantages of Chemical Disinfection

- Gaps in effectiveness, contamination
- (Primary) bacterial resistance
- Adaptation (biofilm formation)
- Possible distribution of germs in the hospital (central units)
- Dependence on concentration, temperature, and pH
- Decomposability, loss of effectiveness
- Inactivation by soap and protein
- Limited ability to penetrate organic material
- Risk of decontamination
- Disinfectant residues in the material (e.g., rubber)
- Material corrosion
- Health effects for staff and patients
- Pollution of the workplace and environmental damage
- High costs
- Increase in the volume of refuse.

Advantages of Physical Disinfection Processes

- Lower costs
- Lower impact on the environment
- Higher degree of reliability
- Automated operation possible
- Cleaning, disinfection, and drying in one process
- No toxicity and no allergization
- Testing for effectiveness.

3.5 Sterilization Methods

3.5.1 Sterilization Processes

- Physical processes
- Steam sterilization
- Hot air sterilization
- Physicochemical processes
- Ethylene oxide gas sterilization
- Formaldehyde gas sterilization
- H_2O_2 low-temperature plasma sterilization.

Physical Processes

Steam Sterilization. Sterilization with the aid of saturated and compressed steam, also sometimes referred to as damp heat, is the most reliable sterilization process and, because of its simple handling, is the most important process for sterilizing medical devices.

The principle of steam sterilization is based on the transfer of thermal energy to the contaminated surfaces as a result of condensation of compressed steam. Energy is released through condensation of the steam on the items to be sterilized, which causes irreversible damage to microorganisms.

The pressure and temperature of steam are dependent on one another; for example, compressed, saturated steam at a temperature of 121 °C has a pressure of 2 bar (1 bar = 105 Pa), while at a temperature of 134 °C it has a pressure of 3.2 bar. In practice, two standard conditions are used:

- 121 °C with application time of 15 min
- 134 °C with application time of 3 min (only for correspondingly heat-resistant items).

Pathogen Resistance to Damp Heat. The resistance of germs to damp heat is classified into four levels (Table 3.3).

A complete effect of the steam on the items to be sterilized is only possible if the air has been removed from the chamber and from the items which are to be sterilized. Processes for removing the air from the items to be sterilized include:

1. Fore-vacuum process. In the fore-vacuum process (Fig. 3.2), the air is removed from the sterilizer chamber using a vacuum pump. The process features the following operating phases:
   - Single evacuation of the sterilizer chamber to pressure of 20–70 mbar
   - Admission of steam until the operating pressure has been reached.

   The fore-vacuum process is not suitable for sterilizing porous items (e.g., laundry) in sterilization containers with a filter or valve in the lid of the sterilization container.
2. Fractionated vacuum process. The fractionated vacuum process (Fig. 3.3) features operating phases such as:
   - Evacuation to pressure < 130 mbar (1 mbar = 100 Pa), repeated several times
   - Alternated with influx of steam to a pressure which is below or above atmospheric pressure
   - Admission of steam until the operating pressure has been reached.

The fractionated vacuum process is suitable for all sterilization items in packaging approved for steam sterilization.

According to the German Medical Devices Operator Ordinance § 4, medical devices must be sterilized us-

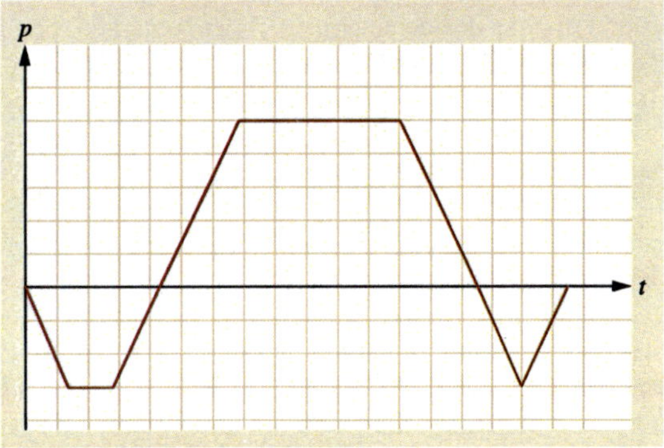

Fig. 3.2 Fore-vacuum process

Table 3.3 The four levels of resistance of germs to damp heat

Level of resistance	Temperature (°C)	Application time	Pathogens recorded
I	100	Seconds to minutes	Vegetative bacteria, fungi including fungal spores, viruses, protozoa
II	105	5 min	Bacterial spores with a lower level of resistance, e.g., anthrax spores
III	121 or 134	15 min or 3 min	Bacterial spores with a higher level of resistance
IV	134	Up to 6 h	Bacterial spores with a high level of resistance

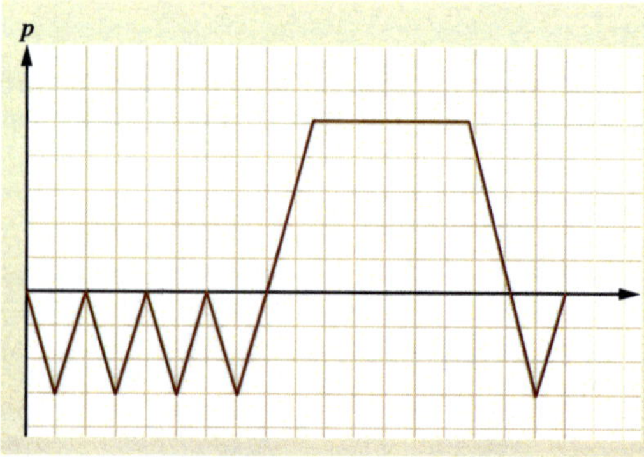

Fig. 3.3 Fractionated vacuum process

ing suitable, validated processes, such that reproducible success of these processes is ensured and the safety and health of patients, users, and third parties are not endangered. Validation serves to prove the effectiveness of the sterilization process under the operating conditions present in the location where the equipment is installed, with the items which are to be sterilized in routine operation, in the appropriate packaging, and with the loading model used. Validation consists of commissioning and performance qualification. DIN EN 554 and DIN 58946-6 have been replaced by DIN EN ISO 17665-1. Validation according to DIN EN ISO 17665-1 consists of installation qualification (IQ), operational qualification (OQ), and performance qualification (PQ).

In the installation qualification, evidence must be found that the device equipment, documentation, operating media, and installation comply with the standard. It must also be demonstrated that the device is in good working order, and that there is no leakage of the operating media or in the equipment.

In the operational qualification, evidence must be provided that the sterilizer is able to perform the specified sterilization programs.

In the performance qualification, all sterilization items and the types of packaging used must be recorded and commissioned. The sterilization parameters of the resulting commissioned items are then measured and assessed in the sterilizer using physical test methods.

In addition to the physical test methods, DIN EN ISO 17665 also approves microbiological testing. Thermoelectric tests cannot be performed on medical devices (e.g., instruments for minimally invasive surgery). In this case, there is the option of contaminating the instrument directly with test germs or using suitable medical device simulators as per DIN 58921. The results must be documented in the validation report.

Documentation. As part of the quality assurance in the central sterilization supply department (CSSD), the sterilization batches must be documented. This is done firstly by recording the process-relevant sterilization parameters (pressure, temperature, and time) and secondly by testing each batch using a suitable chemical indicator in a special test specimen. Using a label with the batch number on ensures that the item to be sterilized is assigned to the batch. Once sterilization has taken place, the parameters are checked, and the batch is released and documented.

Hot Air Sterilization. Hot air sterilization is sterilization by dry heat. Because dry air is a poor thermal conductor, relatively high temperatures and long application times (e.g., 180 °C for 30 min sterilization time) are necessary to ensure reliable sterilization. Table 3.4 presents the sterilization times for hot air sterilization.

Germs, spores, and viruses are killed as a result of protein coagulation. The sterilization effect is strongly dependent on the preparation of the items to be sterilized. The items to be sterilized must be clean and dry (evaporatively cooled).

The following must be taken into consideration when loading the sterilizer.

- It must be possible for air to flow unhindered around all of the items.
- The direction of the air flow must be taken into account.
- Larger items can create a slipstream.
- The items to be sterilized must not be stacked in blocks.

Larger sterilizers must be equipped with forced air circulation equipment (ventilator). Because of their unreliable operation, hot air sterilizers should only be used to a very limited degree. Their fields of use are:

- Glass (laboratory)
- Metal
- Porcelain.

Suitable packaging materials are:

- Metal cases
- Glass bowls
- Aluminium foil.

Note that textile and paper packaging are not suitable.

Physicochemical Processes

Ethylene Oxide Sterilization. Sterilization with ethylene oxide must only be used if the item to be sterilized cannot be sterilized using any other process [3.10].

Table 3.4 Sterilization times for hot air sterilization

Sterilization temperature (°C)	Sterilization time (min)
160	120
170	60
180	30

According to the German Ordinance on Hazardous Substances and the TRGS 513, since 1 January 1995, ethylene oxide (EO) may only be used in validated, fully automatic sterilizers, in which an automated degassing program follows the sterilization program, with the sterilizer remaining locked until the program is finished.

Substance Properties. Ethylene oxide is a colorless, sweet-smelling, highly flammable gas, which can form explosive mixtures with air. It is toxic and can cause cancer and inheritable damage. A maximum allowable concentration (MAC value), in the sense of harmless concentration, can therefore not be given. The technical reference concentration (TRC value) for the breathing air in the workplace is 1 ppm-vol. EO irritates the eyes, the respiratory organs, and the skin. It is classified as hazardous to water (water hazard class WHC 2).

How It Works. Its good ability to penetrate cells makes it possible for various vital biochemical components in the metabolism of microorganisms, such as DNA, proteins, vitamins, and enzymes, to be exposed to EO. The alkylation reaction of proteins with EO kills the microorganisms.

The effectiveness of EO is influenced by various parameters. The relative humidity of the gas mixture is optimally 33% at a temperature of $55\pm3\,°C$. Because the materials to be sterilized and their packaging absorb water to different degrees depending on how the sterilization chamber is loaded, relative humidity of 100% is in practice aimed for at the beginning of the sterilization process. Water loss due to vacuum and absorption then results in a reduction in the relative humidity. This initially high humidification should in practice prevent the relative humidity from dropping below the limit value of 33% which is necessary for reliable EO sterilization.

Adsorption and Desorption. Ethylene oxide binds to surfaces of solids (is adsorbed) depending on the material. With regard to EO sterilization this means that residues adhere to the materials following EO exposure. A relatively long degassing or desorption time is therefore necessary for the treated items following sterilization. These may contain a maximum residual concentration of 1 ppm EO before use on patients [3.11]. As already mentioned above, in accordance with TRGS 513, the desorption process must be performed in the sterilization chamber, which is automatically locked, after sterilization has ended. Reloading of the sterile items directly after sterilization into so-called ventilation cabinets is not permitted.

Exhaust Air from EO Sterilization Systems. A mass flow of 2.5 ppm EO (according to the German Technical Instructions on Air Quality Control, TALuft) must not be exceeded in the exhaust air from the system. Methods used for reducing the EO concentration are listed here.

- Combustion. This must be supported using auxiliary gas firing. The consumption of fuel gas is $\approx 0.5\,m^3/h$. In the process, EO is fully converted into carbon dioxide and water.
- Catalytic reaction. The necessary temperature of the catalyst is reached by supplying energy in the form of steam or electrical energy. It is virtually impossible to achieve complete decomposition of the EO.
- Gas wet scrubbing process (only for pure EO). The firm VIG provides a process in which the EO is bonded with a washing agent. Using diluted sulfuric acid in water, the EO is converted to ethylene glycol. The EO is disposed of completely, and there is no need to supply exhaust air.

Sterilization Processes

Vacuum Process. This process involves working with 100% EO in the vacuum range. After evacuation to below absolute pressure of 55 hPa and humidification of the sterilization chamber, the EO flows into the chamber, and the items to be sterilized are treated at an operating temperature of 50–60 °C for between 1 h and 6 h. The application time is essentially determined by the process parameters of EO concentration, pressure, humidity, and temperature. The pressure range is kept below the lower explosive limit of EO. The sterilization chamber is then purged and ventilated multiple times.

Overpressure Process. Following a pre-vacuum and humidification, sterilization is performed in the overpressure range using a gas mixture of 6% EO and 94% CO_2. As in the vacuum pressure process, the application time is essentially determined by the process parameters of EO concentration, pressure, humidity, and temperature. The inert gas (mostly carbon dioxide) helps to avoid explosion. After the sterilization phase, multiple vacuum and ventilation phases take place alternately to achieve desorption of the EO from the sterile items. The desorption time is generally 8–10 h.

Requirements for Sterilizers. The requirements for reliable sterilization are described in detail in DIN 58948 and DIN EN 1422. According to the German ordinance for minimizing hazardous substances, care should be taken to ensure that sterilizers work with gas mixtures of 6% EO and 94% CO_2 and that the risks for staff, patients, and the environment are therefore reduced.

Requirements for Operating Staff. According to TRGS 513, operating staff must demonstrate knowledge of the subject through an approved course.

Formaldehyde Gas Sterilization

Substance Properties. Formaldehyde (FO) is a colorless, pungent-smelling gas with a broad spectrum of biocidal activity. It is toxic, allergenic, suspected to cause cancer, combustible, and can form explosive mixtures with air. With the exception of its allergenicity, the potential for danger in the areas just mentioned is lower than in the case of ethylene oxide, but it is also less effective. The maximum allowable concentration (MAC value) is 0.5 ppm. Formaldehyde is commercially available as a 30–50 weight percent solution (formalin).

Adsorption and Desorption. Formaldehyde is adsorbed by the surfaces of solids. With regard to FO sterilization, this means that residues remain on materials following FO exposure. After the sterilization phase, the FO residues adsorbed on the sterile items are removed by purging with air and steam (desorption). The extent of adsorption and desorption is dependent, among other things, on the type of material of the solids. Experiments looking at its desorption behavior have shown that the residues on sterilized products can often be removed relatively easily, but that the residues in the sterilization packaging are many times higher in comparison. FO-sterilized items stored in poorly ventilated spaces can lead to the MAC value for formaldehyde being reached in the air of the room. Storage locations in which sufficient dilution and aeration are guaranteed must accordingly be selected.

Process Sequence

1. Ventilation and humidification, often using a fractionated vacuum process. The FO sterilization process is not able to penetrate to the same extent as the EO sterilization process. To achieve sufficient penetration in the process, the FO sterilization process may only be performed using the fractionated vacuum process.
2. Sterilization, at a constant, low vacuum and high atmospheric humidity (at least 60%); i. e., formaldehyde and steam are used for sterilization in combination at temperature of 60–75 °C.
3. Desorption: purging and ventilation using steam/air scrubbing; i. e., the chamber and the items being sterilized are purged in 15–20 changes in pressure with air or steam.

Requirements for Sterilizers. Sterilizers must comply with the requirements of DIN 58948.

Requirements for Operating Staff. According to TRGS 513, operating staff must demonstrate knowledge of the subject through an approved course.

H_2O_2 Low-Temperature Plasma Sterilization (LTP)

Field of Use. Because of their well-known problems in respect of residues, there are considerable restrictions on operation of conventional gas sterilizers (ethylene oxide/formaldehyde) (German Ordinance on Hazardous Substances, TRGS 525). No harmful residues of the active substance are to be expected in the case of H_2O_2 LTP sterilization, which is based on the active substance H_2O_2, i. e., hydrogen peroxide plasma, and whose chamber temperature is 45 °C.

Active Principle. The active substance used is hydrogen peroxide (H_2O_2). In a vacuum it is evaporated, diffused through the sterilization packaging, and then excited by radiofrequency to form hydrogen peroxide plasma. Hydroxyl and hydroperoxy radicals form in the plasma, which inactivate the microorganisms. After the radiofrequency field is switched off, the radicals lose their high levels of energy and recombine to form water and oxygen [3.12, 13].

Process Sequence

First Phase: Vacuum. The sterilization chamber is evacuated to residual pressure of ≈ 1 mbar (1 mbar = 100 Pa). The radiofrequency generator is then switched on, and an air plasma is generated. The chamber is then ventilated and evacuated once more to prepare for the injection. The air plasma allows any residual moisture to be dried for better preparation for loading.

Second Phase: Injection. At room temperature, 1.8 ml H_2O_2 is injected into the chamber and evaporated at pressure of ≈ 11 mbar. A short period of ventilation then follows. This ensures that the active substance penetrates quickly into the lumina.

Third Phase: Diffusion. The H_2O_2 diffuses into the items being sterilized. A vacuum is generated once again before the plasma phase.

Fourth Phase: Plasma. The pressure in the chamber is reduced to 0.7 mbar, and the plasma phase is begun by means of radiofrequency in the MHz range. The hydrogen peroxide vapor is thereby ionized, i. e., converted into gas plasma. This consists, among other things, of highly reactive hydroxy and hydroxyl radicals, which bond with functional building blocks of microorganisms, thus causing them irreversible damage. Phases 2–4 are carried out twice, the so-called first and second halves of the cycle.

Fifth Phase: Ventilation/Pressure Equalizing. Following completion of the 10 min plasma phase, the gas residues in the chamber and in the sterile items are removed harmlessly by fractionated air purging and active carbon filtering. The vacuum is equalized. When atmospheric pressure is reached, the cycle is ended and the door can be opened.

Process Duration. When fully loaded, the process lasts for 75 min. Smaller loads sometimes mean shorter process times.

Information for Use. Because of the way in which it works, this process can be used for sterilization of thermolabile items, but with certain restrictions [3.14]. Users of this process should draw up a list of all the thermolabile items which must be sterilized and then decide which items this process can be used for. A so-called positive list of articles which are approved for sterilization is issued by the manufacturer.

Based on comprehensive tests, open plastic tubes with internal lumen > 3 mm and up to 200 mm in length can currently be sterilized. Tubes with internal lumen < 3 mm must be provided with a diffusion amplifier before sterilization. Catheters which are closed at one end cannot be reliably sterilized.

The disadvantages of this process should be discussed very specifically: hydrogen peroxide vapor is strongly adsorbed by materials containing cellulose. Plasma sterilization therefore cannot be used for items to be sterilized which contain cellulose or for instruments in packaging containing cellulose. All packaging materials must be free from cellulose and are currently only available from the operating company. Packaging bags, for example, are also on average three times more expensive than similar products for steam sterilization. Containers for instruments, such as those which are available for steam sterilization, are only in the development stages. The instruments must be completely dry before loading the plasma sterilizer. In the case of organic contamination of the surface, the effect of plasma sterilization is restricted considerably, but this is also the case with all surface sterilization processes.

Inspection and Testing for Effectiveness. Test systems with biological indicators in accordance with ISO 14937 must be used to test effectiveness.

3.6 Hygiene of Noninvasive Technology Equipment

This category includes all equipment and technical measures which are used for diagnosis and therapy but which are not inserted into the patient. These may be electrocardiogram (ECG) electrodes, ultrasonic transducers or monitors by the patient's bed or in function-testing departments, and they may equally be small conveyor systems, ventilation technology or processing machines.

3.6.1 Equipment Used on the Patient

These must be included in regular cleaning. If the equipment does *not* come into contact with the patient, then cleaning – often only of the accessible surfaces – is usually sufficient. Environmentally friendly detergents should be given preference for this, and it is also important to bear in mind material compatibility and area-specific regulations (e.g., explosion protection).

Where such equipment is in areas with increased requirements in terms of sterility (e.g., operating theaters) or in isolation rooms (infectious patients, patients with multiresistant colonized bacteria, immunosuppressed patients), disinfectant cleaning must be stipulated. The disinfectant used must be suitable for the field of use (disinfectants for treating surfaces or instruments) and must comply with requirements in the scope of its effect. When used in areas with increased requirements in terms of sterility, *all* potential pathogens should be covered, and when used in isolation areas, substances are preferred which act very quickly *against the known pathogens* with a long-lasting effect.

Equipment or equipment parts which come *into contact with the patient* must be cleaned carefully prior to use in accordance with the guidelines and must always be disinfected if there are corresponding regulations (e.g., after use on patients with multiresistant germs) or if, despite proper cleaning, there is still a risk of infection, e.g., in the case of severely immunocompromised patients for whom even harmless environmental germs can become a threat.

3.6.2 Equipment Not Used on the Patient

Initially, the same requirements apply for this equipment as for use outside a hospital. It must be maintained according to its intended purpose and regularly cleaned. Different requirements may be necessary when the equipment is used in *areas with increased requirements in terms of sterility*, and, when necessary, these increased requirements should be governed by in-house guidelines. If, as a result of use, the equipment/unit is contaminated with material from the patient, then disinfectant cleaning is recommended, and if the material is from an infectious patient then disinfectant cleaning is a requirement. For certain equipment (e.g., air-conditioning units, cleanroom benches), special regulations may need to be taken into account (see below).

3.6.3 Repair and Maintenance

Equipment which is handed in for repair and maintenance to relevant departments should have been given at least basic cleaning beforehand. Equipment parts which have been in contact with patients/patient material must only be disinfected before being worked on if they are visibly dirty, or if it can be assumed that it is contaminated with germs which should be prevented from spreading in the hospital (multiresistant germs), which can lead to infections in the maintenance staff, or which are subject to special regulations by law (German Infection Protection Act). These also include germs such as hepatitis and tuberculosis pathogens, and other similar pathogens.

Whether this pretreatment is carried out by the user, somebody in another position (processing unit) or the maintenance department itself may depend on the structure of the organization and also on the size of the hospital. It is therefore recommended that the procedure and the respective responsibility are set down in writing and that *hygiene plans* are used to regulate who has to do what, when, and in particular precisely how, and in what sequence. This should in any case be agreed upon together with the person responsible for hygiene (hygiene officer/infection control nurse) and possibly also with the member of staff responsible for occupational health and safety.

3.7 Hygiene of Invasive Technology Equipment

This includes all equipment, equipment parts, and technical measures which are used for diagnosis or therapy and which in the process are inserted *into* the patient. These may either be instruments or equipment which are used in the patient *without* penetrating the skin or mucosa (e.g., bronchoscope, oesophageal arteriography, suction aspirator), in which case there may be intentional (biopsy) or unintentional damage to the mucosa, or they may be instruments or equipment which are designed for use *with* or *after* penetration of the barrier of the skin/mucosa (e.g., biopsy forceps, arthroscope, and vascular catheter).

Disinfectant measures are usually sufficient for the first group, whereas a sterilization process is obligatory for the second group (see above).

Here, too, the following general *rule* applies: Primarily use thermal disinfection (washing) processes, and only consider thermochemical or even purely chemical processes when the materials are not compatible with thermal disinfection processes.

3.8 Practical Examples

Targeted measures to prevent transmission of germs and infections must be adapted to the pathogen, the group of people, and the hazard potential. Vaccination alone may be sufficient, or specific disinfection or sterilization measures or even isolation of the patient may be necessary. The route of transmission plays a crucial role in this (Table 3.5).

The German Infection Protection Act, which has been in force since 1 January 2001 and has replaced the old Federal Contagious Diseases Act, takes into consideration these findings but also takes into account the effects of technology in the changing world of medicine by stipulating in § 23 that infections acquired in hospitals must be continuously recorded, even those which

Table 3.5 Types of infection transmission and their features and protective measures (after [3.15])

Type of transmission	Features	Examples	Protective measures
Airborne transmission	Microorganisms attached to particles in the air with size of <5 μm, movement over a relatively long period of time therefore possible	1. Reasonable suspicion of or confirmed tuberculosis 2. Measles 3. Varicella/disseminated herpes zoster 4. HIV patients with cough, fever, and opaque pulmonary infiltrates, provided TB cannot be ruled out	1. Isolation in a single room (door and windows closed), cohort isolation potentially possible 2. Respiratory protection when entering the room if open-lung TB is identified or there is strong clinical suspicion 3. In the case of certain diseases (measles, varicella) nonimmune people should not enter the room; if unavoidable, only with respiratory protection
Droplet transmission	Microorganisms attached to particles >5 μm (these droplets are created when speaking, coughing, and sneezing)	1. Bacterial diseases: *H. influenzae* (type B) infections, meningococcal infections, multiresistant pneumococcal infections, diphtheria, pertussis, mycoplasma pneumonia infections 2. Viral diseases: influenza, mumps, rubella, parvovirus infections	1. Single room, cohort isolation if necessary; if not possible a distance of at least 1 m should be kept between the infectious patient and other patients or visitors 2. Mouth and nose protection required when working close to the patient (<1 m distance)
Contact transmission	Direct contact (touching) or indirect contact (secondary, e.g., via contaminated surfaces) with epidemiologically important pathogens in the case of infected or colonized patients	1. Infectious diarrheal diseases 2. *C. difficile* enteritis 3. Respiratory infections in children (bronchiolitis, croup) 4. Multiresistant pathogens such as Methicillin-resistant *Staphylococcus aureus* (MRSA), Vancomycin-resistant *Enterococcus faecium* (VRE) (except multiresistant TB) 5. Abscess or secreting wounds which cannot be covered	1. If possible single room; cohort isolation if necessary 2. Gloves and gowns depending on the pathogen and site of the infection (follow infection control recommendations) 3. Disinfect hands on leaving the room

are particularly influenced by technical factors (*device related*):

- *Postoperative* wound infections
- *Ventilator associated* pneumonias
- *Catheter-related* septicemias
- *Catheter-related* urinary tract infections.

Because these are also the four most common health-care acquired infections in hospitals and the measures required by law make an important contribution to sensible quality management, the following practical examples are limited to these infections and demonstrate the vital contribution that technology can make here.

3.8.1 Postoperative Wound Infections

The pathogens for postoperative wound infections either originate from the patient himself (germs in the skin or mucous membranes) or are introduced into the patient from outside during surgery as a result of a lack of hygiene. Before many surgical procedures, hair removal up to now is still carried out using a razor (medicotechnical measure). If this is done on the evening before the procedure, then the bacteria from skin flora have time overnight to migrate deep into the skin via the microscopic cuts in the skin which are unavoidably caused during shaving to cause inflammation there. The result is a significantly higher risk of developing surgical site infection. It is therefore recommended either not to shave at all and to simply cut the hair short using hair trimmers, or to shave immediately before the surgery and thus immediately before the skin is disinfected (see also KRINKO recommendation for the prevention of postoperative infections in the operating field [3.9]).

Further important measures for the avoidance of postoperative wound infections which are proven in their effectiveness and which relate to technical means are: sterile instruments, sterile implants, reliably sterilized equipment parts which are required for surgery (suction apparatus, counter with connection cable, etc.), correctly functioning air-conditioning units, sufficient vacuum in the case of reprocessed reusable drainage equipment, etc. [3.16].

3.8.2 Ventilator-Associated Pneumonias (VAP)

Pneumonia is the second most common overall infection in hospitals and is the most important in the case of intensive care patients. It is seen as the primary or secondary cause in 30–50% of deaths. With respect to diagnosis related groups (DRGs), it is also significant that health-care associated pneumonia increases the length of hospital stay by an average of 11.5 days. The main risk factor for developing health-care associated pneumonia is mechanical ventilation.

According to the KRINKO evidence criteria, orientated towards the criteria of the Centers for Disease Control and Prevention (CDC) [3.17], the following technically relevant points are important for targeted prevention [3.17–20].

1. Oral intubation is better than nasal (development of maxillary sinusitis); in the case of long-term artificial respiration prefer tracheotomy.
2. Use sterile or disinfected tracheal tubes.
3. Clean all equipment and tools thoroughly before disinfection and sterilization.
4. Do not carry out routine sterilization or disinfection of the circulation system of ventilators and anesthetic apparatus.
5. Do not change ventilator breathing circuits more frequently than every 48 h, including tubes and expiratory valves and also nebulizers and steam humidifiers, provided the equipment is only used for one patient (according to recent studies an interval of 7 days between changing is even possible).
6. Do not use atmospheric humidifiers, which form aerosols (= atomizer), if sterilization/disinfection and sterile water are not used on at least a daily basis.
7. Use sterile (not distilled or nonsterile) water for rinsing the processed equipment and tools which are used on the respiratory tract after they have been chemically disinfected.
8. Do not reprocess equipment and tools which have been manufactured for single use, unless there is data to show that reprocessing does not pose any threat to the patient and is cost-effective and that the functionality of the equipment and tools is not altered.
9. Sterilize or disinfect ventilation breathing circuits and humidifiers between use on different patients.
10. Do not use bacterial filters between the humidifier reservoir and the inspiratory tube.
11. Do not change the respiratory tube routinely if the system is connected to an heat and moisture exchange (HME) or heat and moisture exchange filter (HMEF), provided it is only used on one patient.
12. Change tubes between patients, including nose clamps or masks, which are used to supply oxygen from a wall outlet.
13. Sterilize atmospheric humidifiers which are used in inhalation therapy, e.g., for tracheostomy patients,

or disinfect between patients and every 24 h when used on the same patient.

14. Sterilize or disinfect portable spirometers, oxygen probes, and other respiratory tools which are used on various patients between uses.
15. Anesthetic equipment: clean and then sterilize or thermally/chemically disinfect the reprocessable parts of the respiratory circuit (such as the endotracheal tube or mask, inspiratory and expiratory tube, Y-piece, bag valve mask, humidifier, and tube) between use on different patients and observe the relevant manufacturer's instructions.
16. Pulmonary function testing: sterilize or disinfect reusable mouthpieces and tubes between different patients or in accordance with the manufacturer's instructions.

3.8.3 Catheter-Related Septicemia

Most cases of septicemia acquired in hospitals are the result of using a vascular catheter. The most important points to avoid resulting infections are checking of the indication for access, selection of the correct catheter and the correct access site, aseptic placement of the catheter, and aseptic dressing change. The technical component is comparatively low here and covers the following points [3.21]:

1. Change IV tubes including the three-way valves only every 72 h (every 24 h when blood/blood products or lipid solutions are administered), except where there are signs of infection.
2. When choosing transducers (pressure sensors), preference should if possible be given to disposable items (as opposed to reusable equipment).
3. The transducer, the tube system, and the rinse solution must be changed at least every 96 h.
4. All components of the blood pressure monitoring system must be sterile (including the calibration apparatus and the rinsing fluid). The entire pressure system (tube lines, transducers, and rinse solution) must be handled aseptically.
5. Reusable pressure systems must be processed and sterilized taking into account the manufacturer's instructions.
6. If preparation of mixed infusions in areas close to patients is unavoidable, then it must be done under controlled aseptic conditions.

In past years it has been increasingly common to reprocess certain expensive intravascular catheters within the hospital or to have them reprocessed by external providers (see later). This practice is currently the subject of controversial debate for very many different reasons (German Medical Devices Act, costs, borine spongiform enphalopaty (BSE)). In this context, reference should in particular be made to the recommendations of the German Commission for Hospital Hygiene and Infection Prevention at the RKI [3.21] and [3.9] with the explanations from the RKI and the final report of the vCJD taskforce: variant Creutzfeldt–Jakob disease (vCJD), epidemiology, identification, diagnosis, and prevention, with particular consideration of minimizing the risk of iatrogenic transmission via medical devices, particularly surgical instruments [3.22]. Reference is further made to specific literature [3.23–26].

3.8.4 Catheter-Related Urinary Tract Infection

Urinary tract infections make up more than 40% of all infections acquired in hospital and, as the starting point for urosepsis, are responsible for up to 15% of cases of septicemia. Up to 80% of these urinary tract infections are found in patients with urinary catheters, which emphasizes their significance. Technically relevant aspects of targeted prevention are as follows:

- Urinary catheters should only be placed if medically necessary and only for as long as is absolutely necessary; an indication as part of nursing care must be rejected.
- Only sterile, permanently sealed urine drainage systems with an antireflux valve should be used (i. e., without disconnection to empty the bag) [3.27, 28].

3.8.5 Dialysis

Because of the high risk of infection both for patients and also for staff, dialysis departments deserve special attention. The risks of infection are:

1. For the patient:
   a) Infections via the vascular access
   b) Bloodborne infections
   c) Contamination of the dialysate and dialyzer.
2. For the staff:
   a) Through infected dialysis systems
   b) Infections via blood and dialysate.

In comparison with peritoneal dialysis, hemodialysis is more significant from a technical viewpoint and is the basis for the following explanations. However, some of the requirements also apply to peritoneal dialysis.

If drinking water is used for dialysis, it must undergo additional processing because it contains bacteria and pyrogens, even if chlorinated. The processes used for this are ion exchange (water softening), active carbon filtering, distillation, and reverse osmosis. However, it must be borne in mind that particularly the first process mentioned can provide waterborne bacteria (mainly *Pseudomonas* spp. and other Gram-negative bacteria such as acinetobacter and enterobacter) but also mycobacteria, which are present in water and are described as *atypical*, with good opportunities to multiply. Subsequent ultrafiltration to remove bacteria and bacterial toxins is therefore considered essential [3.29].

Although reverse osmosis is currently the most optimum processing method, it is still necessary to take into account that, even with this method, there may be microbial contamination of the membrane, and germs may find their way in if there are leaks. Once the dialysate is added, this forms a mixture which, because of its composition, is a good culture medium for waterborne bacteria. Various countries have therefore suggested guidelines for the assessment of dialysis water (Table 3.6).

To prevent contamination of hemodialysis equipment and supply apparatus, the following technical requirements are deemed necessary [3.30]:

- No open reservoirs for water and processed dialysis fluid
- No open reservoirs for concentrates
- Small line cross-sections in supply lines
- Route lines as a closed circular pipeline; avoid dead spaces (only for clean water)
- Complete disinfectability of the line system
- Pipe disconnection when disposing of the dialysis fluid to prevent retrograde microbial contamination.

In accordance with the hygiene guideline which is an appendix to the 2006 Dialysis Standard [3.31], dialysis equipment (hemo- and peritoneal dialysis) used must comply with the regulations of the German Medical Devices Act and must be maintained, operated, cleaned, and disinfected in accordance with the manufacturer's instructions (instructions for use, technical manual) [3.30]. Due to this required disinfection after each patient, the formerly required separation into a so-called *yellow* (for infectious patients) and *white* (for noninfectious patients) region is now obsolete. The operator is responsible for ensuring that all parts which come into contact with the used dialysate or even with the patient's blood are treated as potentially infectious and that the equipment is disinfected after each dialysis treatment.

This can be done by [3.29]:

- Sterilization with steam at a temperature of 121 °C, provided this is technically possible from a materials point of view (equipment with stainless-steel tanks)
- Disinfection using hot water (90–95 °C for 20 min); citric acid is automatically added during the process to prevent deposits in the equipment
- Thermochemical disinfection (for ecological reasons, preferably with peracetic acid, possibly also with formaldehyde or glutaraldehyde or with sodium hypochlorite).

Dialyzers have been regularly reprocessed in the past. Figures for the USA show that the proportion of dialysis centers reprocessing their equipment rose from 18% in 1976 to 82% in 1997 [3.32], and corresponding processing guidelines issued by the Association for the Advancement of Medical Instrumentation (AAMI) were adopted by official authorities [3.33]. The percentage then dropped to 62% by 2002. The literature describes infectious complications through to outbreaks in the course of reprocessing dialyzers, without being able to prove causality beyond doubt [3.34]. For quality assurance purposes it is therefore necessary to ensure that effective processes are used, which are currently based on thermochemical disinfection. The hygiene guideline which is an appendix to the 2006 Dialysis Standard – written by the German Commit-

Table 3.6 Guidelines for assessment of dialysis water in various countries

	Dialysis water (generally permeate) (CFU/ml)	Endotoxin	Dialysis fluid (CFU/ml)	Endotoxin
AAMI (USA 2004)	≤ 200	2 EU/ml	≤ 2000	2 EU/ml
European Pharmacopoeia (2008)	≤ 100	≤ 0.25 IU/ml	No information	No information
Swedish Pharmacopoeia (1997)	< 100	< 0.25 IU/ml	< 100	
Japanese Society for Dialysis Therapy (2008)	< 100	< 0.05 EU/ml	< 100	0.05 EU/ml

tee for Clinical Nephrology in collaboration with the Association of German Nephrology Centers run by the German Society of Nephrologists in Private Practice and also with the German Society of Pediatric Nephrology, in agreement with the German Commission for Hospital Hygiene and Infection Prevention – likewise points to the requirements for a processing method with appropriate validation and stresses that the German Committee for Nephrology currently does not favor reuse of dialyzers and tube systems, despite the financial ramifications [3.30]. Tests to monitor clean water and dialysis fluid must be performed and documented at least twice a year [3.30, 35]. The relevant water-conducting systems (e.g., closed circular pipelines) must be fitted with suitable sampling points.

The following procedure is considered necessary to test the microbiological quality of the water [3.35]:

- Disinfect hands before each sample is taken
- Use sterile glass bottles with a screw closure for collecting the sample; to test for endotoxins use pyrogen-free containers made from polystyrene
- Draw off ≈ 100 ml in each case
- *Demineralized water*
  - From the closed circular pipeline at the bedside
  - Aseptic sampling (disinfected adapter, at least twice a year)
- *Basic bicarbonate*
  - Only if it is taken as a concentrate from a closed circular pipeline at the bedside (not from canisters or cartridges)
  - Aseptic sampling (disinfected adapter)
  - Once a month
- *Dialysis fluid*
  - Sample taken from the dialyzer
  - Before the start and after the end of dialysis
  - Every 6 months
- *The assessment is performed according to the following guidelines*
  - In the processed water and in the dialysis fluid before the beginning of dialysis: 100 CFU/ml.

In an ISO standard [3.36] which appeared in 2009 and in which the quality of fluids for hemodialysis is formulated as an international standard, requirements, and limit values for the chemical constituents and impurities in water, concentrate, and dialysis fluid are listed in addition to microbiological quality standards. The operator of specialist dialysis departments is responsible for monitoring of microbiological water quality and also for chemical monitoring.

3.8.6 Creutzfeldt–Jakob Disease

After medical devices have been used on patients with a proven or strongly suspected case of Creutzfeldt–Jakob disease (CJD) or its new variant (vCJD), special processing procedures are necessary [3.22, 25]. According to the currently favored prion theory, particular significance is attributed to the cleaning which takes place during processing, because proteins must primarily be removed. For disinfectant processing, 1–2 M sodium hydroxide (NaOH), 2.5–5% sodium hypochlorite (NaOCl), or 4 M guanidinium thiocyanate (GdnSCN) is currently recommended [3.22].

Instruments which cannot be steam-sterilized are then subsequently processed with aldehyde disinfectant and finally rinsed with 70% alcohol (e.g., endoscope) and gas-sterilized.

Instruments which can be steam-sterilized undergo chemical decontamination before then being subjected to machine processing at 93 °C, and are finally autoclaved at 134 °C for 1 h.

The relevant appendices to the RKI guideline must be observed [3.22].

3.9 Regulations

3.9.1 Technical Regulations for Hazardous Substances

- TRGS 513. Fumigations with ethylene oxide and formaldehyde in sterilization and disinfection systems (edition June 2008).
- TRGS 525. Handling of hazardous substances in facilities for human medical care (as of May 1998 BArbBl. no. 5/1998, p. 58, currently undergoing revision).
- TRGS 401. Hazards due to skin contact (as of June 2008).
- TRBA/TRGS 406. Sensitizing substances for the airways (as of June 2008).
- TRBA 250. Technical regulations for biological agents (as of February 2008).

3.9.2 Standards

The importance of standards has been re-evaluated as a result of Directive 93/42/EEC about medical devices and the correspondingly harmonized German Medical

Table 3.7 Standards for the (minimum) standard for medical devices ▶

Type of standard	Source
DIN	German standard
E DIN	Published German draft standard
EN	European standard
DIN EN	Harmonized (European) standard
prEN, E	Published European draft standard
DIN EN/ISO	International standard

Table 3.8 Overview of major standards in the field of sterilization technology for the health care system

Standard	Steam sterilization	As of
DIN EN 285	Sterilization – steam sterilizers – large sterilizers	8/2009
DIN EN/ISO 17665-1	Requirements for the development, validation, and routine control of a sterilization process for medical devices	11/2006
DIN 58948	Sterilization – low-temperature steam formaldehyde sterilizers	
Section 17	Requirements for the installation and operation of low-temperature steam formaldehyde and formaldehyde sterilizers and their supply sources	3/2009
DIN EN 14180	Sterilizers for medical purposes Low-temperature steam and formaldehyde sterilizers – requirements and testing	1/2010
Standard	**Ethylene oxide sterilization**	**As of**
DIN EN 1422	Sterilizers for medical purposes – ethylene oxide sterilizers – requirements and test methods	8/2009
DIN 58948-7	Requirements on the installation and requirements on the service supply for ethylene oxide sterilizers	1/2010
DIN 58949	**Steam disinfection apparatus**	**As of**
Section 1	Terminology	1/2001
Section 2	Requirements	1/2001
Section 3	Efficiency testing	2/2004
Section 4	Biological indicators for efficacy tests	10/2006
Section 6	Operating of steam disinfection apparatus	2/2004
Section 7	Structural requirements and requirements on service supply	1/2001
DIN 58955	**Decontamination equipment for medical use**	**As of**
Section 1	Terminology	1/2003
Section 2	Requirements	7/2005
Section 3	Efficiency testing	9/1998
Section 4	Biological indicators for efficacy tests	3/2006
Section 6	Operation	3/2001
Section 7	Structural requirements and requirements on service supply	3/2001
DIN EN/ISO 11138	Sterilization of health care products – biological indicators	
Section 1	Requirements	8/2008
Section 2	Biological indicators for ethylene oxide sterilization processes	9/2009
Section 3	Biological indicators for moist heat sterilization processes	9/2009
Section 4	Biological indicators for dry heat sterilization processes	9/2006
Section 5	Biological indicators for low-temperature steam and formaldehyde sterilization processes	9/2007
DIN 58921	Draft standard: Test method to demonstrate the suitability of a medical device simulator during steam sterilization – medical device simulator testing	12/2008
DIN EN/ISO 11140	Sterilization of health care products – chemical indicators	
Section 1	General requirements	9/2009
Section 3	Class 2 indicator systems for use in the Bowie and Dick-type steam penetration test	9/2009
Section 4	Class 2 indicators as an alternative to the Bowie and Dick-type test for detection of steam penetration	7/2007

Devices Act. They specify the (minimum) state of the art which the manufacturer must observe during the design, manufacture, operation, and use of medical devices (Table 3.7). Table 3.8 gives an overview of major standards in the field of sterilization technology for the health care system.

References

3.1 W.E. Stamm: Infections due to medical devices, Part 2, Annu. Intern. Med. **89**, 764–769 (1978)

3.2 F. Hofmann: Arbeitsmedizin und Gesundheitsschutz im Krankenhaus. In: *Praktische Krankenhaushygiene und Umweltschutz*, ed. by F. Daschner (Springer, Berlin, Heidelberg 1997)

3.3 A.R. Aitkenhead, J.F. Dhainaut: International standards for safety in the intensive care unit, Intensive Care Med. **19**, 178–181 (1993)

3.4 P. Gastmeier, K. Weist, O. Weigt, H. Rüden: Prävention nosokomialer Infektionen in der Intensivstation und im OP, Anästhesist **48**, 575–590 (1999)

3.5 M.A. Martin: Nosocomial infections related to patient care support services: Dietetic services, central services department, laundry, respiratory care, dialysis, and enoscopy. In: *Prevention and Control of Nosocomial Infections*, ed. by R.P. Wenzel (Williams, Philadelphia 1997)

3.6 Robert-Koch-Institut: Impfempfehlungen der Ständigen Impfkommission (STIKO) am Robert-Koch-Institut/Stand Juli 2009 (2000) http://www.rki.de/ (last accessed July 2011)

3.7 Robert-Koch-Institut: Liste der vom Robert-Koch-Institut geprüften und anerkannten Desinfektionsmittel und -verfahren, Stand vom 31.05.2003, 14. Ausgabe, Bundesgesundheitsblatt **46**, 72–95 (2003)

3.8 KRINKO: Anforderungen an die Hygiene bei der Reinigung und Desinfektion von Flächen, Mitteilung der Kommission für Krankenhaushygiene und Infektionsprävention am Robert Koch-Institut, Bundesgesundheitsblatt **47**, 51–61 (2004)

3.9 KRINKO: Anforderungen an die Hygiene bei der Aufbereitung von Medizinprodukten, Mitteilung der Kommission für Krankenhaushygiene und Infektionsprävention am Robert Koch-Institut, Bundesgesundheitsblatt **44**, 1115–1126 (2001)

3.10 C. Ruef, S. Harbarth, A. Henry, D. Pittet: Sterilisation mit Ethylenoxid: Anwendungen und Grenzen, Swiss-NOSO **4**, 3–6 (1997)

3.11 V. Herzog: Desorption von Ethylenoxid (EO) zur Einhaltung der Grenzwerte, Zentralsterilisation **8**, 204–209 (2000)

3.12 P. Mecke: Wasserstoffperoxid-Plasma, Ein interessantes mikrobizides Prinzip, Hyg. Med. **17**, 537–542 (1992)

3.13 A.F. Widmer, H. Siegrist: Plasmasterilisation: Eine revolutionäre Technik für thermolabile Instrumente, Swiss-NOSO **1**, 7 (1994)

3.14 M. Borneff, U. Färber, H. Getreuer, P. Heeg, C. Höller, U. Junghannß, H. Martiny, R. Mechmerth, P. Mecke, J. Peters: Zur Wirksamkeit und Prüfung von H_2O_2-Plasma-Sterilisatoren, Hyg. Med. **18**, 557–558 (1993)

3.15 H.M. Just, R. Ziegler: Isolierungsmaßnahmen. In: *Praktische Krankenhaushygiene und Umweltschutz*, ed. by F. Daschner (Springer, Berlin, Heidelberg 2006)

3.16 KRINKO: Prävention postoperativer Infektionen im Operationsgebiet, Mitteilung der Kommission für Krankenhaushygiene und Infektionsprävention am Robert Koch-Institut, Bundesgesundheitsblatt **50**, 377–393 (2007)

3.17 KRINKO: Prävention der Nosokomialen Pneumonie, Mitteilung der Kommission für Krankenhaushygiene und Infektionsprävention am Robert Koch-Institut, Bundesgesundheitsblatt **43**, 302–309 (2000)

3.18 D. Craven, K.A. Steger: Nosocomial pneumonia in mechanically ventilated adult patients: Epidemiology and prevention in 1996, Semin. Respir. Infect. **11**, 32–53 (1996)

3.19 M.H. Kolleff: The prevention of ventilator-associated pneumonia, Curr. Concepts **340**, 627–634 (1999)

3.20 M. Lacour, P. Gastmeier, H. Rüden, F. Daschner: Prävention der nosokomialen Pneumonie, Empfehlungen des Nationalen Referenzzentrums (NRZ) für Krankenhaushygiene, 1. Prävention der Beatmungspneumonie, Intensivmedizin **35**, 87–94 (1998)

3.21 KRINKO: Prävention Gefäßkatheter-assoziierter Infektionen, Mitteilung der Kommission für Krankenhaushygiene und Infektionsprävention am Robert Koch-Institut, Bundesgesundheitsblatt **45**, 907–924 (2002)

3.22 Verhütung und Kontrolle von TSE: Die Variante der Creutzfeldt-Jakob-Krankheit (vCJK). Epidemiologie, Erkennung, Diagnostik und Prävention unter besonderer Berücksichtigung der Risikominimierung einer iatrogenen Übertragung durch Medizinprodukte, insbesondere chirurgische Instrumente – Abschlussbericht der Task Force vCJK zu diesem Thema, Mitteilung der Robert Koch Instituts, Bundesgesundheitsblatt **45**, 376–394 (2002)

3.23 H. Baron, J. Safar, D. Groth, S.J.. DeArmond: Prions. In: *Disinfection, Sterilization, and Preservation*, ed. by S.S. Block (Lippincott, Philadelphia 2001)

3.24 P. Gastmeier: Device-associated nosocomial infection surveillance in neonatal intensive care using specified criteria for neonates, J. Hosp. Infect. **38**, 51–60 (1998)

3.25 B. Hörnlimann, D. Riesner, H. Kretzschmar (Eds.): *Prionen und Prionkrankheiten* (De Gruyter, Berlin 2001)

3.26 M. Lacour: Prävention von Infektionen durch intravasale Katheter, Intensivmedizin **35**, 582–592 (1998)
3.27 D.H. Forster, P. Gastmeier, H. Rüden, F. Daschner: Prävention nosokomialer Harnweginfektionen, Intensivmedizin **36**, 15–26 (1999)
3.28 J. Martius, P. Brühl, M. Dettenkofer, U. Hartenauer, S. Niklas, H.-J. Piechota: KRINKO Prävention und Kontrolle Katheter-assoziierter Harnwegsinfektionen. Mitteilung der Kommission für Krankenhaushygiene und Infektionsprävention am Robert Koch-Institut, Bundesgesundheitsblatt **42**, 806–809 (1999)
3.29 H. Gartmann, M. Dettenkofer: Dialyse. In: *Praktische Krankenhaushygiene und Umweltschutz*, ed. by F. Daschner (Springer, Berlin, Heidelberg 2006)
3.30 Dialyseeinheiten (2008) Hygieneleitlinie als Ergänzung zum Dialysestandard 2006 der Deutschen Arbeitsgemeinschaft für Klinische Nephrologie e.V. in Zusammenarbeit mit dem Verband deutsche Nierenzentren der DD nÄ e.V. sowie der Gesellschaft für pädiatrische Nephrologie (GPN) in Abstimmung mit der KRINKO www.nephrologie.de/Leitlinie.pdf bzw. www.rki.de (last accessed July 2011)
3.31 Dialysestandard 2006 der Deutschen Arbeitsgemeinschaft für klinische Nephrologie, Mitt. Arb. Klin. Nephro. **35**, 121–184 (2006), http://www.nephrologie.de/Richtlinien.html (last accessed July 2011)
3.32 J.I. Tokars, E.R. Miller, M.J. Alter, M.J. Arduino: National surveillance of dialysis-associated diseases in the United States, Semin. Dial. **13**, 75–85 (2000)
3.33 AAMI: *American National Standard: Reuse of Hemodialyzers ANSI/AAMI RD47-2002/A1:2003* (Association Advancement of Medical Instrumentation, Arlington 2003)
3.34 J.M. Arduino: Dialysis-associated complications and their control. In: *Bennett and Brachmann's Hospital Infections*, ed. by W.R. Jarvis (Lippincott, Philadelphia 2007)
3.35 A. Podbielski, M. Herrmann, E. Kniehl, H. Mauch: *MiQ: Qualitätsstandards in der mikrobiologisch-infektiologischen Diagnostik, Teil 1* (Elsevier, Munich 2005) pp. 22–27
3.36 ISO 11663: *Quality of Dialysis Fluid for Hemodialysis and Related Therapies* (International Organization for Standardization, Geneva 2009), http://www.iso.org/ (last accessed July 2011)

4. Technical Safety of Electrical Medical Technology Equipment and Systems

Rüdiger Kramme, Hans-Peter Uhlig

Medical electrical (ME) equipment and systems are used primarily in medical institutions such as doctors' surgeries, dental surgeries, medical centers, outpatient clinics, rehabilitation clinics, sanatoriums, health clinics, nursing homes for the elderly, homes for the disabled and those with learning difficulties, and hospitals and clinics. According to the regulations and standards for electrical systems, suitable rooms – areas used for medical purposes – must be present in these institutions in which *mains-operated* ME equipment can be operated. Many questions emerge concerning the use of this equipment, on the one hand relating to the safety of patients and those operating the equipment, and on the other hand relating to the smooth operation of this apparatus. The main risks posed by these pieces of ME equipment and ME systems may be of electrical, mechanical or thermal nature. Other possible potential risks are ionizing radiation, explosion or fire as outlined in this chapter.

4.1 **General Information Regarding the Safety of Technical Systems** 36

4.2 **Attaining Safety in Medical Institutions** .. 36

4.3 **Minimum Requirements for ME Equipment** 37

4.4 **Areas Used for Medical Purposes** 40

4.5 **Electrical Systems According to the Nature of the Connection to Earth** 42

4.6 **Protection Against Shock Currents** 42

4.7 **Power Supply** 44

4.8 **Power Sources for Safety Purposes with Accumulators** 44

4.9 **Final Circuits and Plug Sockets** 45

4.10 **Static Electricity** 45

4.11 **Electromagnetic Compatibility** 46

4.12 **Conclusions** 47

References 47

Part A | 4

What is meant by risks and hazard? *Risks* in an operational process are components of the working system which can damage the life or health of operating staff or patients. These risks are energy sources or are connected to them. When this energy is greater than the electrical resistance of the human body or an affected body part, the energy is of a damaging nature. The damaging force is derived from the ratio of energy to body resistance. A *hazard* is described as the possibility of humans encountering risks in space and time. To provide safe medical workplaces, it is therefore necessary to clarify how and under what operating conditions humans encounter risks and the risk of injury occurs, and what successful safety measures (protection objectives) are or should be in use to guard against this [4.1]. The largest proportion of possible hazards result from ME equipment and ME systems for diagnosis, therapy, and monitoring. The commonest causes of an electrical accident can be attributed to malfunctioning protective devices, defective insulation on cables and wires, defective or missing protective covers on equipment and plug connectors, and missing, disconnected or interchanged protective earths.

4.1 General Information Regarding the Safety of Technical Systems

The current state of science and technology which has been achieved in principle makes it possible to develop and build technical products and systems which are very safe. Absolute safety nevertheless does not exist, either in life or in technology. The important thing is therefore to ascertain the justifiable residual risk in each individual case, then to consciously live with this risk [4.2].

DIN 31000 general guidelines for the safety design of technical products (*Allgemeine Leitsätze für das sicherheitsgerechte Gestalten technischer Erzeugnisse*), dated 03/1979, forms the basis for the content of safety standards and VDE (Verband der Elektrotechnik, Elektronik und Informationstechnik, Association for Electrical, Electronic and Information Technology) specifications. In DIN 31000 part 2, dated 12/1987, which was unfortunately withdrawn in 2005, *borderline risk* is described as the greatest risk relating to a certain technical process or condition that is still considered justifiable. Safety is defined according to DIN VDE 31000 part 2 as a circumstance in which the residual risk remains below a defined borderline risk. The level of risk is formed as an abstract variable from the product of the extent of damage and probability of damage occurring [4.3]. The temporal and content-based limit on safety resulting from this is achieved through the description of appropriate requirements or measures.

The measures which ensure safety are divided into three classes according to DIN 31000 part 1 (three-class theory).

1. *Direct safety engineering* includes measures which provide safety mainly through constructive, planning-related engineering of products and systems and their parts.
2. *Indirect safety engineering* is understood as meaning the application of safety-related means, e.g., locking mechanisms.
3. *Indicative safety engineering* is produced by guidelines regarding the conditions for risk-free use – e.g., in operating instructions and on signs – when direct and indirect safety engineering do not, or do not completely, achieve their goal.

In the case of all of the safety-engineering measures mentioned, use of the products and systems according to their intended purpose is a precondition. The *first-fault philosophy*, described in DIN VDE 0752 supplementary sheet 1, dated 03/2003, forms an important foundation when assessing hazard situations. The findings of the international standardization on the subject of safety, such as the contents of IEC 300-3-9, dated 12/1995 (Dependability Management; Part 3: Application Guide; Section 9: Risk Analysis of Technological Systems), can also be used [4.2]. The possible consequences of first faults are virtually ruled out by protective measures and by scheduled testing and maintenance. Suitable monitoring devices must immediately discover any *first faults* which might occur.

4.2 Attaining Safety in Medical Institutions

In Germany there are very comprehensive regulations, provisions, standards, and guidelines which apply to the planning, installation, refurbishment, and testing of electrical systems in medical institutions. The same applies to the development, manufacture, and operation of ME equipment and ME systems. The crossovers or points of intersection between these documents are unfortunately not defined but are instead fluid. So, for example, different provisions and standards for ME equipment contain important requirements regarding the planning and installation of electrical systems.

Provisions and standards in principle contain only the minimum requirements, and there is not always conformity between them. In addition, they do not always reflect the state of the art, because in many cases technological developments move at a faster rate than revision of these documents is carried out. Each individual project must therefore be checked both with regard to observance of the statutory and normative minimum requirements and with regard to additional requirements. The required safety can in principle only be achieved if the overall system, consisting of the physical structure and technical systems as well as the ME equipment, is safe.

The safety of the system consisting of electrical system/ME equipment is essentially described by the protection objectives [4.2].

- Protection of patients and staff from shock currents and from the risks of power failure or of poor-quality voltage or frequency
- Safety of the interface between the ME equipment and the electrical system, in particular when using equipment for vital purposes
- Safety of the escape and emergency routes
- Protection from fires and their effect on the system of electrical installations and ME equipment over a required period of time
- Safety as a result of using reliable technology
- Safety as a result of using all of the system components according to their intended purpose
- Safety as a result of operation in accordance with regulations and regular maintenance.

4.3 Minimum Requirements for ME Equipment

DIN EN 60601-1 (classified as VDE 0750-1) Medical electrical equipment: General requirements for basic safety and essential performance (*Medizinische elektrische Geräte; Allgemeine Festlegungen für die Sicherheit einschließlich der wesentlichen Leistungsmerkmale*), dated 2007 (which replaces the 1996 edition), is the basic standard in this series and describes the general requirements for ME equipment. All approved ME equipment in Germany must be built and tested according to the standards of the DIN EN 60601 series. The basic standard DIN EN 60601-1 is evaluated here based mainly in connection to the electrical system. The protection classes I and II are approved for mains-operated ME equipment.

In *protection class I*, all nonactive parts of the ME equipment are connected to the protective earth (PE). The equipment housing is composed of metal and is likewise connected to the PE. In the event of a fault to frame or earth fault internally in the equipment (e.g., insulation fault) between an active conductor and the PE, a protective device is triggered in the equipment and/or in the electrical system and the electrical circuit affected is disconnected. During a fault situation, a contact voltage occurs between the housing of the equipment and other earthed components.

The definition of ME equipment in protection class I which is given in DIN 60601-1.

The equipment in *protection class II* is double insulated (all-insulated). A fault can only occur between active conductors or an active conductor and the neutral conductor (N). No contact voltage to earth occurs during a first fault to frame or earth fault.

The definition of ME equipment in protection class II which is given in DIN EN 60601-1.

Another category is made up mostly by *battery-operated ME equipment* which does not rely on mains electricity. As in mains-operated ME equipment of protection class II, with regard to protection against shock currents, this equipment is not dependent on the installed protective measures.

Equipment in class AP (anaesthetic proof (low risk)) must comply with the requirements regarding construction, labeling, and accompanying documentation to avoid ignition sources in explosive mixtures of inhalational anesthetic agents with air.

Equipment in class APG (anaesthetic proof, Cat. G (high risk)) must comply with the requirements regarding construction, labeling, and accompanying documentation to avoid ignition sources in explosive mixtures of inhalational anesthetic agents with oxygen or nitrous oxide.

A distinction is also made between ME equipment according to the *insulation classes of the applied parts*.

- Insulation class B: The applied part is connected to earth. External application to the patient. Provides protection against electric shock by taking into account the fault current. Not suitable for direct application to the heart.
- Insulation class BF: The applied part is insulated from the ME equipment (no connection to earth). External application to the patient, higher-quality protection against electric shock than in insulation class B. Not suitable for direct application to the heart.
- Insulation class CF: The applied part is insulated to a high level from the ME equipment (no connection to earth). External and intracardiac application

Ser. no.	Consumables	Applied for	Medical area of group	Protection against electric shock					Power supply			
				Disconnection	IT system	ΔU ≤ 25 V	ΔU ≤ 10 mV	Protective extra low voltage (PELV) 25 V	For safety purposes with permissible interruption time			Via general power supply
									≤ 15 s	≤ 0.5 s	0.0 s	
1	ME equipment protection class 1 or 2	Monitoring and therapy in intensive care, not directly on the heart and not repeatable	2	Prohibited	Prescribed (VDE 0100-710)	Prescribed (VDE 0100-710)	Not applicable	Allowed in case of protection through insulation and coating or coverage	Only when ME equipment is suitable	Only when ME equipment is suitable	If required for safe operation of the ME equipment	Prohibited (VDE 0100-710)
2	ME equipment protection class 1 or 2	Investigations and/or therapies directly on the heart and not repeatable	2	Prohibited	Prescribed (VDE 0100-710)	Not permissible in patient environment	Mandatory	Possible in protection class II	Only when ME equipment is suitable	Only when ME equipment is suitable	If required for safe operation of the ME equipment	Prohibited (VDE 0100-710)
3	ME equipment protection class 1 and 2	Investigations and/or therapies not directly on the heart and not repeatable	2	Prohibited	Prescribed (VDE 0100-710)	Prescribed (VDE 0100-710)	Not applicable	Allowed in case of protection through insulation and coating or coverage	Only when ME equipment is suitable	Only when ME equipment is suitable	If required for safe operation of the ME equipment	Prohibited (VDE 0100-710)
4	ME equipment protection class 1 and 2	Investigations and/or therapies not directly on the heart and repeatable	1	Allowed (VDE 0100-710)	Allowed, but not prescribed	Prescribed (VDE 0100-710)	Not applicable	Allowed in case of protection through insulation and coating or coverage	Allowed, but not prescribed	Allowed, but not prescribed	Allowed, but not prescribed	Allowed
5	Operation theatre (OT) lights	Illumination of work area	2	Prohibited	Prescribed at $V_n > 25\,V_{AC}$ (VDE 0100-710)	Prescribed (VDE 0100-710)	For therapies: to be tested directly on the heart	Prescribed at $V_n \leq 25\,V_{AC}$ (VDE 0100-710)	Prohibited (VDE 0100-710)	Prescribed (VDE 0100-710)	Best solution	Prohibited (VDE 0100-710)

Table 4.1 Requirements of the electrical system for ME equipment

Ser. no.	Consumables	Applied for	Medical area of group	Protection against electric shock					Power supply			
				Disconnection	IT system	$\Delta U \leq 25\,V$	$\Delta U \leq 10\,mV$	Protective extra low voltage (PELV) 25 V	For safety purposes with permissible interruption time			Via general power supply
									$\leq 15\,s$	$\leq 0.5\,s$	0.0 s	
6	OT microscopes	Illumination of work area	2	Prohibited	Prescribed at $V_n > 25\,V_{AC}$ (VDE 0100-710)	Prescribed (VDE 0100-710)	For therapies: to be tested directly on the heart	Prescribed at $V_n \leq 25\,V_{AC}$ (VDE 0100-710)	Prohibited (VDE 0100-710)	Prescribed (VDE 0100-710)	Best solution	Prohibited (VDE 0100-710)
7	Examination lights	Illumination of work area for simple examinations	1	Allowed (VDE 0100-710)	Allowed, but not prescribed	Prescribed (VDE 0100-710)	Not applicable	Not applicable	According to VDE 0100-710 not required, to be tested in individual cases	Allowed, but not prescribed	Allowed, but not prescribed	Allowed, but not sensible in all cases
8	Examination lights	Illumination of work area for simple examinations	0	Allowed (VDE 0100-710)	Not applicable	Not applicable	Not applicable	Not applicable	According to VDE 0100-710 not required, to be tested in individual cases	Not applicable	Not applicable	Allowed, but not sensible in all cases

to the patient, higher-quality protection against electric shock than in insulation class BF. Suitable for direct application to the heart.

DIN EN 60601-1 applies accordingly to the safety of ME equipment.

Prerequisites for this are performance of risk management processes, use of the equipment according to its intended purpose, and compliance with the instructions from the manufacturer. The following are considered to be first faults for ME equipment.

1. Break in the protective earth.
2. Break in a power supply cable.
3. Occurrence of an external voltage at applied parts with code letter "F".
4. Occurrence of an external voltage at signal input or output parts.
5. Failure of an electrical component.

The rated voltage may be a maximum of 250 V for handheld ME equipment. For ME equipment and ME systems which are operated on direct current or single-phase alternating current, the maximum value is 250 V at a power limitation of 4 kV A. For ME equipment and ME systems operated on polyphase current, the upper limit of the rated voltage is 500 V.

The tolerance of the nominal voltage for ME equipment is ±10% and that of the mains frequency is ≤ 1 Hz. According to the standards which apply to electrical systems, other values may also occur in this case (Sect. 4.7).

The protective earth of the equipment must be checked regularly. A limit value of 0.1 Ω is permitted between the PE of the plug and exposed metal parts (housing) for equipment with a permanently connected power supply cord, and of 0.2 Ω for equipment with a removable power supply cord.

The collateral standard DIN EN 60601-1-1 stipulates that, when using portable multiple sockets, it must be possible to connect the ME equipment of an ME system to a multiple socket only with the aid of a tool. Multiple sockets of this kind are available factory-built in accordance with DIN EN 60601-1-1. Alternatively, the portable multiple socket must be electrically isolated (e.g., isolating transformer).

It is worth noting that electrically wired toroidal transformers are frequently used as isolating transformers. There is no proof of safety from first faults for this arrangement. The same applies to the supply of entire ME systems from one power socket. Complete power failure is possible at any time.

The conclusion is that the safest kind of power supply is the supply of ME systems from fixed single power sockets, especially for vital purposes, being connected to an information technology (IT) system.

Some of the requirements for ME equipment in DIN EN 60601-1 which have been mentioned have a direct or indirect bearing on the electrical system. Some additional requirements for the electrical system can be found there [4.2].

Table 4.1 summarizes the various applications of ME equipment such that the most important requirements can be recognized at the interface with the electrical system and additional requirements can be defined.

4.4 Areas Used for Medical Purposes

Medical examinations and treatments must be carried out in appropriate rooms, areas used for medical purposes, regardless of where these rooms are. It may be a hospital, medical center or health center, an outpatient clinic or a private doctor's practice. The use of medical procedures requires the areas used for medical purposes to be divided into the groups 0, 1 or 2 (DIN VDE 0100-710).

There are specific hazards caused by the possible condition of patients, such as unconsciousness, or the nature of the medical examination or treatment, e.g., an operation. A patient's skin resistance can be compromised as a result of medical interventions, and a patient's cognitive ability or the body's defenses can be impaired or limited as a result of medical treatment. It is mandatory that the required classification and division of the areas used for medical purposes into groups be carried out – generally by the representative of the purchaser (= planner) together with the medical and technical staff (= operators). An essential basis for the decision regarding which group a room is allocated to (Table 4.2) is the nature of the contact between the applied parts of the ME equipment and the body of the patient. In areas belonging to group 1, invasive examinations and therapies can even be performed, but no examinations or therapies can be performed directly on the heart. A power failure with a maximum duration of 15 s can be dangerous

Table 4.2 Normative criteria for the classification of rooms in groups

Group 0	Group 1	Group 2
Consulting rooms	Medical and dental practice rooms	Surgery preparation rooms
Surgery consulting rooms	Surgical outpatient clinics	Operating theaters
Dressing rooms	Rooms for home dialysis	Recovery rooms
Wards	Wards	Intensive care units
Massage rooms	Rooms for hydrotherapy	Special baby care units
Fitness rooms	Endoscopy rooms	Endoscopy rooms
OT adjoining rooms	Rooms for minimally invasive surgery	Rooms for minimally invasive surgery
Respiratory chambers	Angiography rooms	Angiography rooms
	Cardiac catheterization laboratories (investigation)	Cardiac catheterization laboratories (investigation and treatment)
	CT rooms	
	Magnetic resonance imaging (MRI) rooms	
	Electrocardiography (ECG) rooms	
	Electroencephalography (EEG) rooms	
	Delivery room	

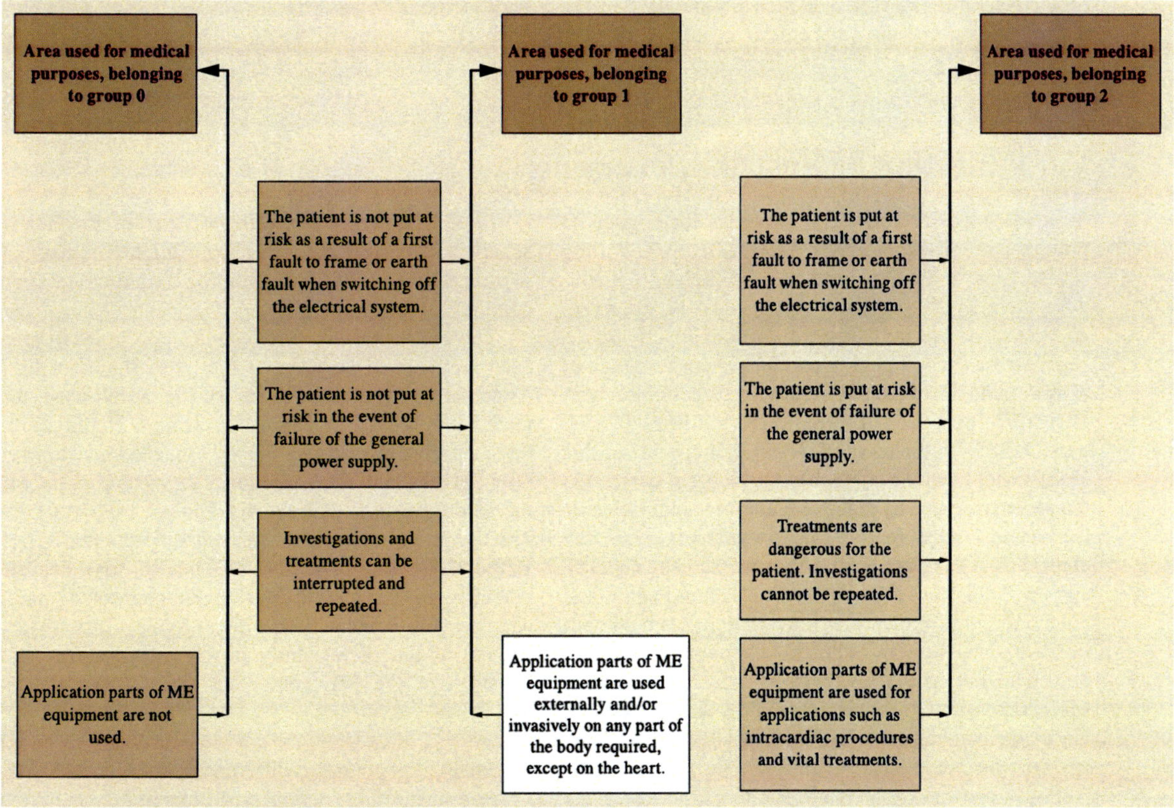

Fig. 4.1 Examples of the allocation of rooms to groups

in this situation, however. Classification in group 2 and the use of a battery-supported power supply for safety purposes (BPS, previously PS) should therefore be considered.

In group 2, the criteria applied are *intracardiac or open-heart treatments or examinations* (in addition, consider DIN 57753-2) and *essential examinations and treatments*. The additional description of the electrical conditions in DIN VDE 0100-710 for classification into groups should help in making the decision (Fig. 4.1). All criteria which are important for the safety of the patient must always be consulted when selecting the group. Where there is any doubt, the decision should be made to choose the higher level of safety [4.2].

4.5 Electrical Systems According to the Nature of the Connection to Earth

Consistent application of the terre neutre séparé (TN-S) system in accordance with DIN VDE 0100-100 and DIN VDE 0100-710 within a building is of fundamental importance for the effectiveness and the proper application of protective measures in the electrical system, in particular of measures to protect against shock currents and against electromagnetic interference (external voltages). In this system, the neutral conductor and the protective earth (PE) are conducted separately in the entire building and are only connected to one another at one point, e.g., in the low-voltage main distribution board. As a result, only fault currents but no operating currents can continue to flow in the protective earth. In normal operation, small, unavoidable fault currents occur in the protective earth of the final circuits, in particular small inductively and capacitively induced fault currents and those induced by insulation resistances. These currents can be several amps in the protective earth of the building's main distributor. In the case of fault to frame or earth faults, relatively high fault currents of up to several kA may briefly occur in final circuits until the protective device is triggered [4.2].

4.6 Protection Against Shock Currents

In areas used for medical purposes belonging to groups 1 and 2, according to DIN VDE 0100-710 an agreed limit of the *contact voltage* (DIN VDE 0100-200) of $U_L \leq 25\,V$ applies, which must be observed. *Examinations and treatments performed directly on the heart* are an exception. A maximum contact voltage of 10 mV is applicable in this case, in accordance with DIN 57753 part 2 of 02/1983 (VDE guideline 0753-2 part 2/02.83). In the event of a fault, a low value of this kind cannot be achieved purely by physical means via the protective earth by means of measures in the electrical system. For this reason, ME equipment in protection class I can in this case only be operated at the medical IT system (DIN VDE 0100-710). The flexible additional potential equalization cables must be connected without fail. With interventions of this kind, measures must also be carried out on the medical side, such as the use of ME equipment with applied parts of insulation class CF. Under the conditions mentioned, risks caused by leakage currents are extremely unlikely. Leakage currents (Fig. 4.2) and their effects are limited in terms of the electrical system by the application of the medical IT system and by the additional potential equalization.

Protection against the effect of leakage currents is achieved in terms of equipment by using ME equipment in protection class II and by limiting the patient leakage current (Table 4.3).

The *additional potential equalization* in every area used for medical purposes is the most important fundamental protective measure in the areas used for medical purposes belonging to groups 1 and 2. In the areas belonging to group 2, in accordance with DIN VDE 0100-710, additional connector bolts for the additional potential equalization according to DIN 42801 are installed in the patient environment (DIN VDE 0100-710, DIN EN 60601-1-1). The flexible potential equalization cables to be connected there, which establish an additional direct connection between the PE of the permanently installed system and the housings of the ME equipment, serve the purposes of

- equalizing potential differences between the ME equipment in the patient environment,
- preventing failure of the protective earth in equipment in protection class I (first-fault safety), and

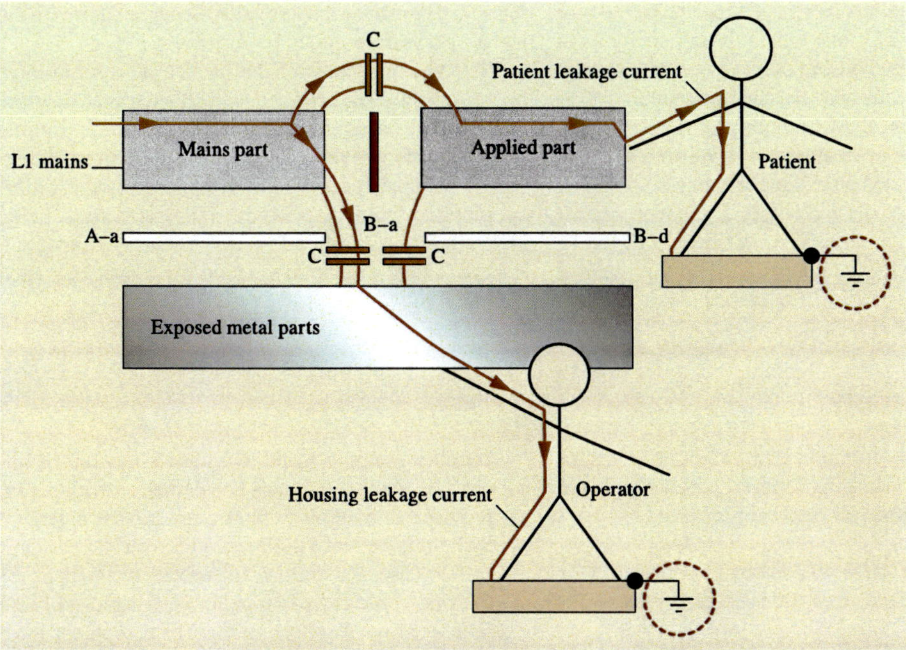

Fig. 4.2 Patient and operator connected to ME equipment (after [4.4]). The various equipment parts form capacitors C with the safety-related insulation which is situated between them, and the patient leakage current and housing leakage current flow via these capacitors. In equipment belonging to protection class I, the earth leakage current also flows via the protective earth. (A–a Insulation mains part/exposed metal parts, B–a Insulation mains part/applied part, B–d Insulation applied part/mains part)

- ensuring that the required contact voltage at the housing of equipment in protection class I is maintained.

These cables are of great importance for protecting against shock currents and must always be connected when equipment in protection class I is used [4.2].

Further *important protective measures* against shock currents include

1. *Protection by automatic disconnection according to DIN VDE 0100-410*
   The electrical circuits of power sockets must be disconnected by the associated protective device within 0.3 s in the event of a fault (zero-impedance short circuit). In accordance with DIN VDE 0100-410 of 2007, this protective measure may only be used for power sockets of up to 20 A when the power socket is monitored (in this regard see the German government safety organization regulation BGV A3). In all other situations, 2 shall apply.
2. *Protection by automatic disconnection of the power supply using residual current protective devices (RCD) release current $I_{\Delta N} = 30\,\mathrm{mA}$*
   This protective measure is prescribed for power sockets in areas in groups 1 and 2 if there is no risk to the patient. ME equipment for vital purposes must not be connected here.
3. *Protection by notification of insulation faults in the medical IT system*
   Where ME equipment is used for vital purposes or for investigations which cannot be repeated, the equipment must be supplied from a medical IT system. In accordance with DIN VDE 0100-710, power sockets of this type are only installed in areas belonging to group 2. Here, the medical IT system constitutes the preferred protective measure. Owing to the limitation of power in the IT isolating transformers to a maximum of 8 kV A, however, it cannot always be used universally.

Table 4.3 Maximum permissible leakage currents from ME equipment according to DIN EN 60601-1

Type	Type B	Type BF	Type CF
Earth leakage current (mA)	0.5	0.5	0.5
Housing leakage current (mA)	0.1	0.1	0.1
Patient leakage current (mA)	0.1	0.1	0.01

The *medical IT system* forms an isolated and locally restricted network. The conductive housing of the ME equipment in protection class I is connected to the protective earth. When only one first fault to frame or earth fault occurs, only the housing leakage current emerges as a fault current and there is no disconnection. This ensures that a piece of ME equipment which is vital for the examination or treatment will not fail and can continue to be operated until the treatment is complete. The medical staff are notified of this. Protection against overload is not permissible. The overloading of the IT isolating transformer is communicated to the medical staff via a monitoring device. The causes of the overload must be determined and remedied because otherwise the isolating transformer can be destroyed.

The power sockets of the medical IT system must be accordingly labeled or designed such that they are unmistakable when several systems are present in one room according to the type of earth connection. The intention of this is to prevent ME equipment which must be supplied from the medical IT system from being inadvertently connected to power sockets of the TN-S system. There are no normative guidelines for such color coding.

4.7 Power Supply

A safe power supply is another important criterion for safe operation. The following forms of power supply are available for areas used for medical purposes.

- *General supply* (GS) via a house connection or a high-voltage feed from a distribution system operator (DSO);
- *Power supply for safety purposes* (PS) via power generation units with internal combustion reciprocating piston engines;
- *Battery-supported central power supply for safety purposes* (BPS).

When supplying safety-relevant consumers in areas belonging to group 2, it is necessary to switch between these power supplies (power sources and mains networks). The requirements for this switching can be found in DIN VDE 0100-710 under several sections. Considering these normative sources, the following switching devices are necessary in the subdistributors for the supply to areas in group 2.

- Switching between a PS and GS feed.
  In this case, the voltage of the outer conductor of the first cable (PS) is monitored. If the first cable fails, continued supply to the consumers mentioned under 710.564.4 and 5 must be guaranteed within 15 s, provided this supply is fed from a subdistributor.
- Switching between a PS and BPS feed.
  In this case, the voltage of the outer conductor of the first cable (PS) is monitored. If the first cable fails, continued supply to the ME equipment must be guaranteed within 0.5 s. If the BPS is used in continuous operation and the first cable is supplied with power from the BPS, the voltage provided is uninterruptible.

The stability and quality of the supply voltage is an important criterion for a safe power supply. When the standards for the power socket are observed, in individual cases the nominal voltage of the mains electricity can deviate by 14% below the nominal voltage. When supplying the mains from a power generation unit, in the case of certain load changes deviations of ±1 Hz from the permissible tolerance band for the ME equipment are possible. On account of the lower impedance of the generator compared with mains electricity, harmonic currents can also affect the voltage and lead to interference. The deviations mentioned can be compensated by suitable planning measures [4.2].

4.8 Power Sources for Safety Purposes with Accumulators

In accordance with DIN VDE 0100-710, supply from a BPS with a maximum permissible interruption time of 0.5 s and a minimum supply period of 1 or 3 h is required for surgical lights. A minimum supply period of 3 h is required when only one power generation unit is present.

Central BPS units for the supply of ME equipment are unfortunately not covered by DIN VDE 0100-710. DIN EN 60601-1 requires that ME equipment be designed such that an interruption in the power supply does not present a hazard before it is restored. On this basis, the function of the BPS is in many cases adopted by what are known as internal power sources built into the ME equipment. There is still demand for central BPS units, however, as is evidenced by many examples. Some reasons for this are

- a central BPS generates lower costs, in particular in large medical institutions, when the high testing and maintenance costs for the internal power sources (small accumulators) are taken into consideration in addition to the investment costs,
- other indispensable lighting for the operating theater lighting (e.g., OT microscopes, endoscopes) are operated with a 230 V alternating current,
- the internal power sources of ME equipment are not sufficiently reliable,
- the internal power sources of ME equipment have no standby system,
- there is ME equipment without internal power sources in circulation,
- the central BPS units are required under building law.

For battery-supported central power supply systems (BPS) for safety purposes for supplying areas used for medical purposes, the German national standard DIN VDE 0558-507 has been applicable since October 2008.

A uninterruptible power supply (UPS) must not be used to supply ME equipment because it does not meet the necessary requirements.

4.9 Final Circuits and Plug Sockets

Important requirements for the circuits of plug sockets are mentioned in DIN VDE 0100-710. The plug sockets in the patient space must therefore be split between at least two circuits. The note includes the requirement that two medical IT systems must be set up where there are more than two electrical circuits per patient space. It is therefore virtually recommended by the standards to set up *two medical IT systems per patient space*.

When using *portable multiple sockets*, individually protected electrical circuits are prescribed. For safety reasons, multiple sockets must not be used, with the exception of wheeled equipment trolleys (Sect. 4.3).

DIN VDE 0100-710 furthermore stipulates that all power sockets which supply ME equipment must be equipped with an optical voltage indicator. As a result, the medical personnel will immediately be able to recognize which power sockets are still carrying voltage following a fault.

4.10 Static Electricity

Static charges are dependent on the conductivity and separation speed of the materials involved. Conductive substances are those solid or liquid substances whose specific resistance is up to $10^4\ \Omega/\text{m}$. Nonconductive substances on the other hand have specific resistance of more than $10^4\ \Omega/\text{m}$. Particularly in the field of surgery, where there is a large amount of medical technology equipment, and anesthetic gases as well as combustible liquids such as disinfectants are used, an electrostatic discharge could ignite the *explosive OT atmosphere*, with devastating consequences. The following protective measures are therefore necessary in rooms used for medical purposes which have areas with potentially explosive atmospheres [4.5].

- Work clothes and also sheets and blankets must be made of a blended fabric consisting of at least 30% natural cotton or viscose (without a resin finish). Rubber blankets, mattresses, and pillows must be made or covered with conductive rubber.
- The leakage resistance of the flooring may be $10^7 - 10^8\ \Omega$. To achieve a certain standard insulation, the flooring leakage resistance should not fall below $5 \times 10^4\ \Omega$.
- Equipment and furnishings which are exposed and are conductive must be conductively connected both to one another and to the floor. The flooring should be electrostatically conductive and at the same time connected to the potential equalization. Because explosive, volatile anesthetics such as halothane or enflurane are no longer available, a controversial discussion has arisen about the need to equip operating theaters with expensive static-dissipative flooring. Nevertheless, according

to experts, the problem of static charge remains and can negatively influence or even damage medical technology equipment (e.g., monitoring), which means that it is still sensible to keep the floor dissipative.
- Operating table systems must be connected to the potential equalization conductor or protective earth. The tabletop pad must be electrically conductive so that any electricity generated as a result of friction can be discharged via the earthed table without sparking.
- Conductive footwear (including overshoes) must have a leakage resistance of at least $5 \times 10^4\ \Omega$.
- Breathing bags of inhalational anesthetic equipment and oxygen ventilation equipment must only be manufactured from conductive material. Electrically conductive latex mixtures or polypropylene are particularly suitable for this purpose.

4.11 Electromagnetic Compatibility

A very high level of electromagnetic compatibility (EMC) is required in ME equipment and ME systems to ensure the proper functioning of other electrical devices. IEC 50-161 gives the definition for EMC [4.7].

Magnetic and electrical interference fields can impair or prevent the safe operation of medical technology equipment and systems. Interference radiated or conducted from the environment can on the one hand affect medical technology electrical devices, but can also originate from such devices. One example of a radiated interference variable is electrosurgery or radiofrequency (RF) surgery. The construction of an RF generator (frequency spectrum 0.4–5 MHz) is in principle comparable to a transmitter; all that is missing are electronic components for the transmission of information and an antenna. A high proportion of the radiofrequency energy of the RF generator is radiated wirelessly from the cables into the environment, especially in the case of relatively old equipment. Further interference occurs when electrical cables are conducted parallel to the cables of the RF generator, because stray currents arise in the electrical supply cables of neighboring equipment as a result of capacitive and inductive coupling. To ensure the smooth operation of electromedical equipment and systems it is also necessary for the required mains voltage to be electromagnetically compatible, i. e., for the effective value to be stable within prescribed limits and the level of interference in the voltage to be less than the level of compatibility of the equipment being operated (EMC environment class 1 and/or 2). Major conduction-related parasitic inductions are caused by what are known as interference phenomena, such as

- fluctuations in the voltage and periodic voltage fluctuations as a result of quick load changes in the supply network (flicker),
- harmonics and interharmonics,
- voltage drops and short interruptions,
- transient overvoltages,
- imbalances,
- fluctuations in the mains frequency.

As the number and use of pieces of ME equipment increase, there is also a proportional increase in the problems with electromagnetic compatibility, which is ultimately a synonym for concepts including overvoltage, RF noise, 50 Hz mains hum, earth loops, and circuit feedback (Table 4.4).

The following EMC features are of particular interest for electromedical equipment [4.8].

- In the case of life-sustaining and life-supporting medical technology equipment and systems, correct operation without any impairment must be ensured.
- In the case of medical technology equipment and systems which are not for life-sustaining purposes, impairment may occur which is not relevant to safety and can be recognized by the operator.

Table 4.4 Reference values for interference parameters (after [4.6])

Interference variable		Unit	Reference value
Frequency range		[Hz]	$0–10^{10}$
Voltage		[V]	$10^{-6}–10^{6}$
Voltage variation		[V/s]	up to 10^{11}
Current		[A]	$10^{-9}–10^{5}$
Current variation		[A/s]	up to 10^{11}
Field strength, electrical		[V/m]	up to 10^{5}
Field strength, magnetic		[A/m]	$10^{-6}–10^{8}$
Power		[W]	$10^{-9}–10^{9}$
Pulse energy		[J]	$10^{-9}–10^{7}$
Pulses	Rise time	[s]	$10^{-10}–10^{-2}$
	Duration	[s]	$10^{-8}–10$

- Equipment with various functions which are directly connected to the patient must operate reliably, without any mutual interference.
- Specific requirements apply for the operating theater area.
- There are special interference-suppression concepts regarding leakage currents (e.g., housing leakage current).

To guarantee electromagnetic compatibility in medical technology, the following technical and organizational measures are available, among others [4.9].

1. For *direct safety*, grounding, potential equalization, shielding, filtering, protection against overvoltage, selection of cables, and circuit layouts suitable for EMC are all possibilities.
2. *Indirect safety* is achieved by spatial measures, signal conversion of the usage variables, use of interference-free software, and installation of alert technologies.
3. *Administrative safety* is achieved by measures such as a code of conduct, EMC planning, repair, maintenance, and servicing, among others.

4.12 Conclusions

The primary requirements for ME equipment and the electrical system in areas used for medical purposes are based on a safe power supply, protection against shock currents and magnetic or electrical interference, and protection against the risk of explosion and fire.

When dealing with ME equipment in the clinical environment, it is inevitable that each operator will become familiar with the functioning and intended purpose of the electromedical equipment and with the possible combinations with ME systems. In addition, monitoring of the condition of the equipment, the accessories, and the cable lines and function checking of the ME equipment and ME systems are also crucial for safe operation.

References

4.1 B. Schneider: Aufgabe und Arbeitsweise der Fachkräfte für Arbeitssicherheit. In: *Grundlehrgang A für Sicherheitsfachkräfte*, Vol. II, ed. by Bundesanstalt für Arbeitsschutz (TÜV, Cologne 1976)

4.2 H.-P. Uhlig, N. Sudkamp: *Elektrische Anlagen in medizinischen Einrichtungen* (Hüthig Pflaum, Heidelberg 2004)

4.3 ISO Guide 51: *Guidelines for inclusion of safety aspects in standards* (Beuth, Berlin 1990)

4.4 T. Flügel, W. Linke, E. Möller, H.-J. Slischka, K. Tillmanns: *Starkstromanlagen in medizinisch genutzten Gebäuden mit stationären oder ambulanten Bereichen*, 3rd edn. (VDE, Berlin 2004)

4.5 W. Twachtmann: Sichere Stromversorgung im Krankenhaus, Krankenh. Tech. **1**, 42–44 (1992)

4.6 H.G. Meyer: Elektromagnetische Verträglichkeit. In: *Biomedizinische Technik*, Medizinische Sondergebiete, Vol. 4, ed. by H. Hutten (Springer/TÜV, Berlin Heidelberg 1991) pp. 301–339

4.7 IEC 50-161: *International Electrotechnical Vocabulary (IEV)*, Electromagnetic Compatibility (Beuth, Berlin 1990), Chap. 161

4.8 R. Sitzmann: Elektromagnetische Verträglichkeit in der Medizintechnik, Electromedica **2**, 84–86 (1998)

4.9 C. Hartung: Szenario elektromagnetischer Störungen am Beispiel der Medizintechnik im Krankenhaus, mt **5**, 172–177 (1999)

5. Quality Management in Medical Technology

Albrecht Malkmus

This chapter aims to provide a basic understanding about the objectives, elements, and structure of a quality management system in the field of medical technology. It will show the elements and the organization of a quality management system. Practical guidance is given for setting up a quality management system.

5.1 **Objectives of a Quality Management System** 49
5.1.1 Concepts 49
5.1.2 Objectives and Clientele of a QMS ... 50
5.1.3 Regulatory Requirements 50

5.2 **Elements of a Quality Management System** 54

5.3 **Organization of a Quality Management System** 54

5.4 **Implementation of a QMS** 56
5.4.1 The QMS and the (End) Customers... 56
5.4.2 The QMS and the Management....... 56
5.4.3 The QMS and the Employees 56
5.4.4 The QMS and Satisfaction of the Regulatory Requirements..... 56

5.5 **Product Quality** 57

5.6 **Concluding Remarks** 57

References 58

5.1 Objectives of a Quality Management System

5.1.1 Concepts

What exactly is a quality management system, and what does it achieve?

The standard EN ISO 9000:2008 gives the following answers to this question:

- A quality management system is a management system to direct and control an organization with regard to quality.
- Quality is the degree to which a set of inherent characteristics fulfils requirements.

Expressed in rather more everyday language, this means that, with the help of a quality management system (QMS), the relevant processes in an organization are designed and controlled such that the finished products or services rendered satisfy the requirements of the customers to the highest possible degree.

According to this understanding, other management activities such as finance management or environmental management are *not* part of quality management (QM) – unless the quality of the products and services would be directly affected by these management systems. All of a company's customer-based and value-adding processes are accordingly the subject of QM:

- Marketing and sales
- Development, procurement, and production
- Delivery, installation, and maintenance.

A particular feature of medical technology is the comprehensive regulatory guidelines, that is to say the legal requirements which are attached to the marketing and operation of medical devices. Since these requirements have an impact on virtually all product realization processes, it makes sense in the field of medical technology to regard the systematic fulfilment of the regulatory requirements (regulatory affairs) as part of QM.

Another feature is the critical nature of medical devices: The consequent minimization of risks for the patient, user, and third parties during development, production, installation, and use of a medical device must be pursued with a dedicated risk management process.

Because the requirement for safe medical devices is part of the essential customer requirements, the technical risk management must also be an integral part of QM.

Quality Costs

Another concept which is important in understanding QM is the idea of quality costs. Quality costs are understood as being the sum of the following cost groups:

- *Prevention costs*
  All endeavors which are undertaken to prevent faults from emerging: implementation of a QMS, internal audits, staff training, identifying customer requirements, designing the infrastructure and working environment, etc.
- *Appraisal costs*
  All endeavors which are undertaken to detect faults: verification and validation of products and processes.
- *Nonconformity costs*
  All endeavors which are undertaken to remedy faults and their consequences: reworking, warranty costs, etc.
- *External quality assurance costs*
  All endeavors which are undertaken to demonstrate to third parties the existence of an effective QMS: certifications and product approvals, demonstration of the QMS to customers, etc.

5.1.2 Objectives and Clientele of a QMS

Why do businesses with high costs implement a QMS? The increase in the competitiveness of the company is the *intrinsic motivation* for implementing a QMS. This means:

- Realizing the products and services such that they fulfil the requirements of the customers as best possible
- Introducing new products onto the desired markets in as short a time as possible, while fulfilling all applicable regulatory requirements
- Minimizing the quality costs while maintaining a given level of quality.

The numerous regulatory requirements and the requirements of the market are the extrinsic motivation:

- An effective QMS is required by law as a prerequisite for marketing medical devices. The existence of an effective QMS is checked by independent certification authorities (*notified bodies*).
- The customers in the business-to-business sector (B2B) and also large end customers will only accept suppliers which have a certified QMS.
- The liability risk of the manufacturer within the scope of product liability can be reduced by an appropriate design of the QMS. A checklist covering these points can be found in a handbook provided by the DIHK (Deutscher Industrie- und Handelskammertag, German Association of Chambers of Industry and Commerce) [5.1].

As a result of these different objectives for a QMS, a QMS must also serve different customer groups.

The diagram in Fig. 5.1 shows the most important customer groups for a QMS in medical technology.

5.1.3 Regulatory Requirements

The numerous regulatory requirements in the field of medical technology relate to:

- The product realization processes and the manufacturer itself
- The products manufactured and the services rendered, as well as
- The marketing, installation, operation, and maintenance of medical devices.

These regulatory requirements therefore have a direct impact on the processes of the company which fall within the domain of QM and must therefore be taken into consideration when setting up a QMS. Table 5.1 gives an overview of the applicable requirements in the most important medical technology markets.

In addition to the directives, numerous guidelines have also been created for the European Union (EU) in a consensus between authorities and manufacturers of medical devices, so-called MEDDEVs, which represent a general interpretation of legal requirements. These guidelines deal with different issues, from definitions of concepts, through general and specific production requirements, to market observation. A complete list of the MEDDEVs can be found on the Internet pages of the European Commission [5.2].

In the case of the European Union, the national laws must also be taken into account, which arise as a result of the translation of European law into national law.

Those which are applicable for Germany are those shown in Table 5.2.

In addition to the European directives and guidelines and the national laws, the approval of medical

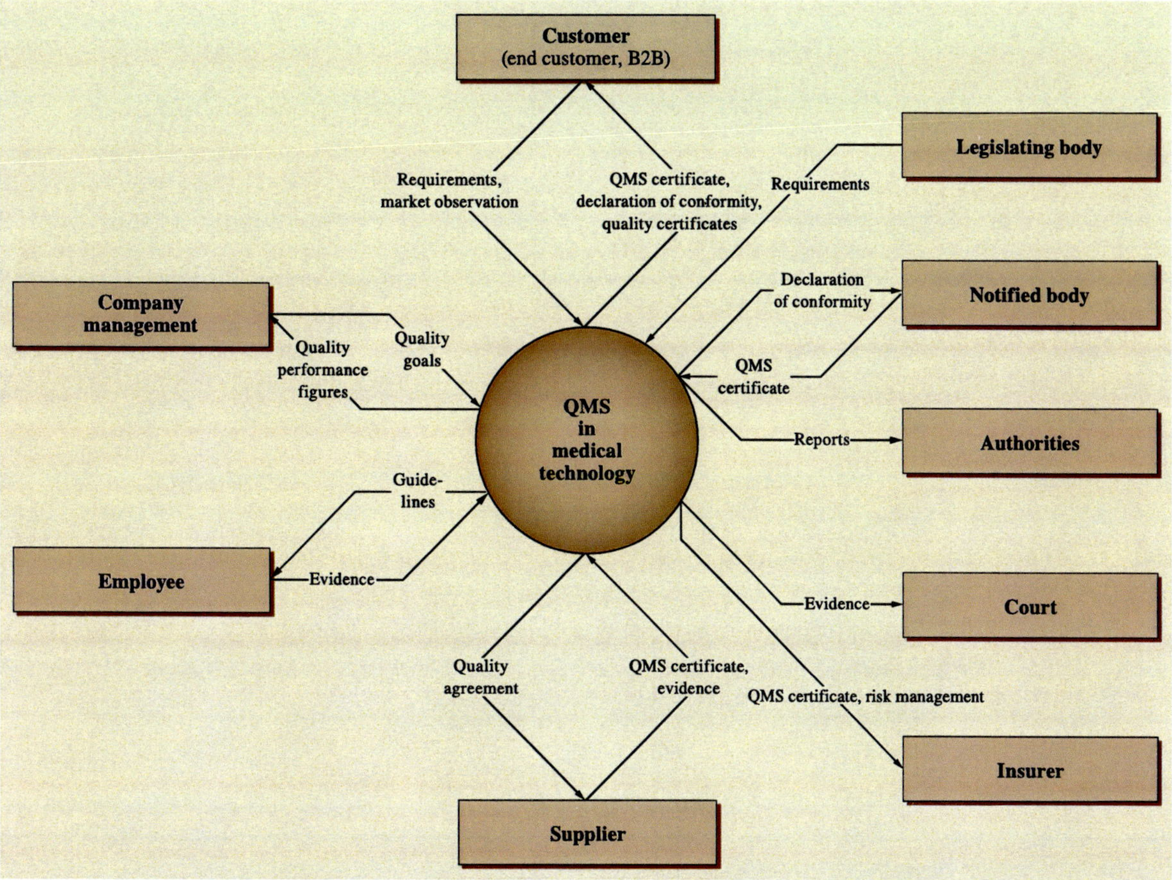

Fig. 5.1 Customer groups for a QMS in medical technology

devices in the European economic zone is predominantly based on harmonized standards:

- Harmonized standards are developed by the European standardization bodies CEN (Comite Europeen de Normalisation, Europeaan Committee for Standardization), CENELEC (European Committee for Electrotechnical Standardization), and ETSI (European Telecommunications Standards Institute) with the aim of specifying the requirements of the CE directives.
- Harmonized standards adapt the requirements to the EU member states. When a harmonized standard is satisfied, it is assumed that the requirements of the corresponding CE directive on which it is based will be satisfied.
- The implementation of (harmonized) standards is not compulsory. However, the manufacturer must then demonstrate in some other way that the underlying requirements are satisfied.
- The harmonized standards include standards for quality management models such as EN ISO 13485, standards regarding general safety, and standards regarding the safety and effectiveness of specific product groups or products.

A complete list of the harmonized standards in the field of medical devices is provided by the European Commission on the Internet (http://www.newapproach.org/Directives/DirectiveList.asp). The European Directives for medical devices distinguish between different development stages of a quality management system:

- Complete quality management system.
  The QMS includes the areas of design, manufacture, and final checking of the products in question.

Table 5.1 International regulatory requirements in medical technology

Economic zone	Applicable requirements	Authority/link
European Union directives	93/42/EEC Medical device directive (1993) 90/385/EEC Active implantable medical devices (1990) 98/79/EC In vitro diagnostic medical devices (1998)	National authorities in Germany Bundesinstitut für Arzneimittel und Medizinprodukte (BfArM): http://www.bfarm.de/ Deutsches Institut für Medizinische Dokumentation und Information (DIMDI): http://www.dimdi.de/
USA	21 CFR Part 820 Quality system regulation 21 CFR Part 11 Electronic records 21 CFR Parts 800–1299 Medical devices	Federal Drug Administration (FDA) http://www.fda.gov/
Canada	Medical devices regulations	Health Canada: http://www.hc-sc.gc.ca/
Japan	New Japanese pharmaceutical affairs law (PAL; 2005)	Ministry of Health, Labor and Welfare (MHLW), Pharmaceuticals and Medical Devices Agency (PDMA): http://www.mhlw.go.jp/
China	China compulsory certificate (CCC)	Certification and Accreditation Administration of the Peoples Republic of China (CNCA), China Quality Certification Centre (CQC): http://www.cnca.gov.cn/

Table 5.2 Further regulatory requirements in Germany in the field of medical technology

	Law	Applies to
MPG	Medical Devices Act; 2002/2007/2009 (*Medizinproduktegesetz*)	Manufacturer and product
MPV	Ordinance on Medical Devices; 2001/2004 (*Medizinprodukteverordnung*)	Manufacturer and product
MPSV	Ordinance on Medical Devices Safety Planning; 2003 (*Medizinprodukte-Sicherheitsplanverordnung*)	Manufacturer, operator
MPBetreibV	Medical Devices Operator Ordinance; 2003 (*Medizinprodukte-Betreiberverordnung*)	Manufacturer, operator
MPVertrV	Ordinance on Medical Devices Marketing Channels; 1997/2001/2003 (*Medizinprodukte-Vertriebswegeverordnung*)	Manufacturer
MPVerschrV	Ordinance on the Prescription Requirement for Medical Devices; 2002 (*Verschreibungspflicht von Medizinprodukten*)	Manufacturer, operator
DIMDIV	Ordinance on Database-Assisted Information Systems Concerning Medical Devices (*Verordnung Datenbankgestütztes Informationssystem über Medizinprodukte*)	Manufacturer, operator

- Quality management system production.
  The QMS only includes the areas of manufacture and final checking of the products in question.
- Quality management system product.
  The QMS only includes the final checking of the products in question.

With the lower development stages of the QMS, depending on the critical nature of the product, product-related tests must be carried out by a notified body – process steps which under certain circumstances can lead to sensitive delays in market introduction. Because the international requirements moreover do not make these distinctions, it can only be recommended that a complete QMS be implemented.

The requirements for a QMS in the field of medical technology are specified in the following standards.

- EN ISO 13485:2009. Medical devices – Quality management systems – Requirements for regulatory purposes.
- EN ISO 14971:2008. Medical devices – Application of risk management to medical devices.

Whoever is certified according to ISO 13485:2009 satisfies the European legal requirements for a QMS. In addition, compliance with ISO 13485 and ISO 14971 also for the most part means satisfaction of the international requirements (USA, Japan, Canada, etc.) for a QMS.

Table 5.3 The most important QM elements according to ISO 13485

Responsibility of the management and quality management officer	Quality management is the responsibility of the company management. It is the responsibility of the company management to ensure the implementation and maintenance of a QMS and to regularly check its effectiveness. A member of the company management (*management representative*) must be nominated who has responsibility and authorization for the implementation and maintenance of the QMS and who informs the management about quality matters in the company.
Quality policy and quality goals	The responsibilities of the company management also include the obligation of the company to document quality in a quality policy and to form concrete quality goals from this policy.
Document management	Document management, which is to say the creation, release, distribution, and updating of specifications and quality certificates, must be provided in written form. The standard also stipulates what content must be documented.
Human resources	The employees whose work can influence the product quality must be adequately qualified for their jobs.
Infrastructure and working environment	The company must design the infrastructure and the working environment such that the required product properties are achieved. Examples of this include suitable premises, machines, and tools or environmental conditions, such as the atmospheric humidity, temperature, etc.
Product realization Planning of all the necessary processes within the scope of product realization (including risk management) Ascertainment of customer requirements Define product development processes Define procurement processes Define production processes Define processes for delivery, installation, and maintenance	All product realization processes must be planned and documented. The process chain under consideration ranges from marketing and sales (ascertaining customer requirements) to service in the field (maintenance and market observation). The standard stipulates at what points in the process chain checks must be performed. The (technical) risk management must be integrated in this process chain in the same way as the verifiable fulfilment of all regulatory requirements (conformity to standards, product approvals).
Measurement, analysis, and improvement	Setting up meaningful and valid performance measurement systems which provide information about the conformity of the product and the conformity and effectiveness of the QMS. The spectrum of the performance figures covers both technical process parameters and performance figures concerning customer satisfaction. Setting up effective improvement processes which make the necessary corrective and preventive measures possible.

Note: ISO 13485:2009 is based on ISO 9001:2008, which is not industry specific. Allowance must be made for the fact that ISO 13485 has *not* adopted all of the requirements of ISO 9001. Organizations which are certified according to ISO 13485 therefore do not automatically meet the requirements of ISO 9001.

5.2 Elements of a Quality Management System

The requirements for a QMS in the field of medical technology are defined in detail in ISO 13845. In case of any questions, the following standards can be used:

- DIN EN ISO 9000:2005 (Concepts)
- DIN EN ISO 9004:2000
- ISO/TR 14969 (Guidelines).

Table 5.3 aims to give a brief overview of the essential elements of a QMS according to ISO 13485: The last element "Measurement, analysis, and improvement" in particular represents a principle which forms the basis of all QM models, the plan–do–check–act (PDCA) cycle by *Deming* [5.3], a cycle for continuous improvements (Fig. 5.2):

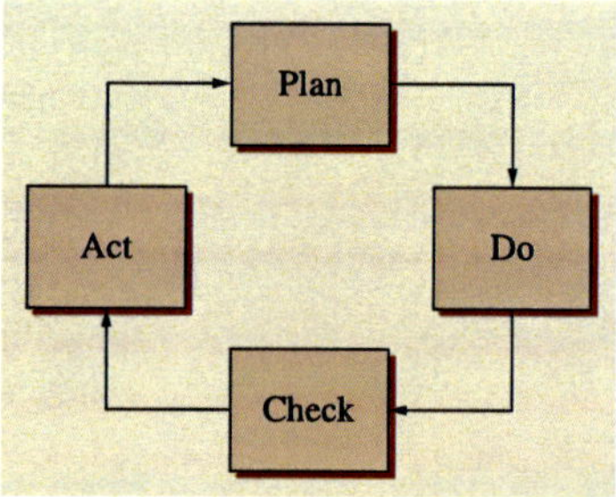

Fig. 5.2 The PDCA cycle by *Deming* [5.3], a cycle for continuous improvements

Plan: Analysis of the current situation, determination of the new quality goals, specification of the new processes

Do: Implementation of the new processes

Check: Checking to ascertain whether the planned goals are being achieved

Act: Transfer the new processes into the routine and implement countermeasures in the event of deviations.

This cycle is followed continuously for the purpose of constant improvement and is a closed control loop.

The challenge when setting up a QMS is to integrate the principle of the PDCA cycle into all elements of the QMS and thus to create an active and self-adapting QMS.

5.3 Organization of a Quality Management System

If the field of application and the elements of the QMS have already been prescribed in detail by means of the European directives and ISO 13485, then there is a great deal of room for maneuver in the organization of the QMS. The spectrum of the possible forms of organization ranges from a large, central QM department, which is responsible for all QM duties, to delegation of the QM duties to all management staff and employees and to complete abandonment of a separate QM organization.

Three principles have proved to be of value in practice and are consistent with the applicable requirements:

1. Direct feeding or integration of QM into company management
2. Preferably process-oriented responsibility for the individual elements of the QMS
3. Guarantee the independence of all testing authorities.

These principles can lead to the following organizational model:

- A separate organizational unit is formed which is responsible and authorized for all superior (company-wide) QM issues.
- This central QM position is a member of the corporate management or is directly below the corporate management.
- In all functional areas of the company which are the subject of QM, one or more employees are assigned QM tasks. Depending on the size of the functional area, they take on the QM tasks either in addition to other assignments or on a full-time basis. Within the scope of their QM tasks, they report professionally to the superior QM position.
- All employees who perform checks do not belong to the organizational unit whose processes or products they check.

Table 5.4 Organization of quality management

Superior QM	QM representative of a functional area
Formulation of the company-wide quality goals in coordination with the company management	Formulation of the area-specific quality goals in coordination with the area management and the superior QM position
Implementation and adjustment of a company-wide performance measurement system	Determination of the performance figures for the functional area
Preparation of the annual management appraisal	Provision of the necessary information
Identification of the applicable legal and normative requirements and communication of these requirements in the company	Adaptation of the processes in the functional area to the applicable requirements
Coordination/performance of QM training programmes	Performance of QM training in the functional area
Implementation and maintenance of the company-wide QM handbook	Implementation and maintenance of all sections of the QM handbook which concern the functional area
Implementation and maintenance of a superior procedure model for product realization	Implementation and maintenance of all steps of the procedure model which concern the functional area
Coordination of external quality audits and certifications	Preparation and provision of the necessary documents and points of contact
Coordination/performance of internal quality audits	Preparation and provision of the necessary documents and points of contact; if necessary, involvement in the audit
Monitoring of superior corrective and preventive measures	Monitoring of area-specific corrective and preventive measures
Performance of product approvals	Provision of the necessary documents
Contact with authorities as part of the notification obligation	Provision of the necessary information

In the model above, the tasks would be distributed as shown in Table 5.4. The representation of Table 5.4 does not show the processing of customer complaints, investigation into customer satisfaction or product observation in the field. These activities can be sensibly integrated into the superior QM position and into service or sales. Demonstration of conformity of the products with technical standards is likewise not included. According to the principle of process-oriented responsibility, it is recommended that these tasks be integrated into product development. The number of employees who are assigned QM tasks is of course dependent on the size of the company and the complexity of the products. In very small companies, all of the activities described above could also be undertaken by one employee in simultaneously held positions.

Last but not least, the principles of effective management should of course also be observed in the structuring of a QMS [5.4].

5.4 Implementation of a QMS

As in the organizational implementation, there is also a large degree of freedom in the technical implementation of a QMS. When implementing a QMS, the various customer groups of the QMS should be considered again (Sect. 5.1.2).

5.4.1 The QMS and the (End) Customers

The (end) customer is the reason for the existence of the company and the QMS. It is important that the customer is actually also integrated into the processes of the QMS as a real person. This should be done not merely reactively as part of the processing of complaints but also proactively, for example, through cooperation in the development of new products and services.

5.4.2 The QMS and the Management

The management of the company is the key customer group of the QMS. For the management, the QMS must represent a reliable instrument for managing the company. As mentioned in *Concept of a QMS* (Sect. 5.1.1) to the management the QMS is just one of many management systems. To create an easily understandable and manageable basis for decision-making for the company management, the most important performance figures from the QMS should be integrated with the most important performance figures from the other management systems to form a common instrument.

One suitable approach to this is the balanced scorecard performance and management system [5.5].

A balanced scorecard combines at least the following four perspectives of a company:

1. The financial perspective
2. The customer perspective
3. The internal processes perspective
4. The learning and development perspective.

With the help of these four perspectives, both retrospective and prospective performance figures are used to manage the company. The performance figures for the QMS fall under the customer perspective in particular, but also the internal processes perspective and the learning and development perspective.

5.4.3 The QMS and the Employees

The employees are numerically the largest customer group of a QMS. If their requirements are not met, the guidelines of a QMS do not become a valued resource but rather an unpopular burden. The regulations of the QMS should not only satisfy the regulatory requirements but should also represent valuable tools for the employee.

This results in the following requirements for the QMS.

- The structure of the quality management manual should be orientated towards the processes in the company, and not towards the structure of a standard.
- Specifications (process descriptions, work instructions, etc.) should be written for the respective target group in an easily comprehensible form and should contain implementable instructions. For each process step, the required results should be defined in content and structure.
- All specifications should be integrated into the respective working environment, and it should be possible to find them quickly and intuitively.
- It is recommended that a structured procedure model be defined for product realization and that the necessary results be reported for each process step.

This structured procedure model is ideally implemented as an electronic workflow. This workflow is used for implementation of the QMS as well as for project management. *Note*: In the case of a paperless workflow, the international regulatory requirements for electronic signatures must be observed (see, for example, [5.6]):

- All process steps of the QMS should regularly be checked for their effectiveness, their efficiency, and the reason for their existence.
- Checking of the acceptance of the QMS by the employees should be incorporated in the QMS as a fixed component.

5.4.4 The QMS and Satisfaction of the Regulatory Requirements

The implementation of a QMS should take into consideration the regulatory requirements of all countries in which the products will be marketed. A decision must also be made regarding the language in which the documents of the QMS and the product-related documents, such as specifications, design documents or test certificates, will be written. To be able to present the QMS externally with little effort or cost, it is a good idea to create associations between the QMS documents and the product-related documents on the one hand and the various national and international requirements on the other hand.

Ideally, these associations are managed in a database and are integrated directly into the electronic document management system and into the electronic workflow of product realization. All necessary regulatory perspectives can then be generated from the database on demand, and the proof of conformity for the QMS and for the products can thus be provided with little effort or cost. Conversely, the electronic workflow can automatically compile the respective process steps necessary and product-related documents on the basis of the information for the planned markets.

5.5 Product Quality

Table 5.5 Product requirements of various parties in different lifecycle phases

Lifecycle phase	Distributor	Buyer/ operator	User	Patient	Technical service	Health insurance	Notified body	Legislating body/ authority
Product realization		×			×		×	×
Marketing	×	×					×	×
Commissioning	×	×	×		×			×
Operation		×	×	×	×	×		×
Maintenance		×	×		×			×
Reprocessing		×			×			×
Decommissioning		×						
Scrapping		×						×

Following the discussion of the objectives and the implementation of quality management systems, the actual subject of a QMS should still be taken into consideration: the quality of the products manufactured. What is important here is that consideration should not only be given to the requirements of patient and user during operation of the product, but also that the manufacturer takes into account the requirements of all relevant parties over the entire lifecycle of the product. Table 5.5 illustrates which parties place requirements on the product in which phase of its lifecycle.

The scope of the requirements and the significance of the individual lifecycle phases are dependent on the nature and complexity of the medical device: nonactive or active medical device, instrument or computer-assisted device, individual product or system component, disposable product or reusable equipment, etc.

The QMS of the manufacturer of the medical device must therefore be designed such that the requirements of *all* parties for a product in the various lifecycle phases are determined at the beginning of the product realization phase and are adequately satisfied. Only in this way is it possible to guarantee that the customer receives a product which complies with their requirements over its entire lifecycle.

5.6 Concluding Remarks

The principle of QM – to systematically ensure that the requirements of the customers are met – must always also be adapted to the QMS itself. In a continuous process of reflection, or rather in a continuous plan–do–check–act cycle, checks must continuously be carried out to determine:

- Whether the QMS effectively establishes the relevant requirements for the product and ensures these requirements are met
- Whether the balance is maintained between (a) helpful guidelines and unnecessary regimentation, (b) important evidence and superflu-

ous paper, and (c) customer-orientated optimization of products and mere conformity with the QMS.

The instrument of management appraisal, as required in Sect. 5.6 of ISO 13485, can perform this reflection process and maintain the balance by carefully testing and evaluating to what degree the QMS satisfies the requirements of its many customers.

In the face of strong regulatory pressure, many companies tend to forget the intrinsic motivation for the QMS and neglect the employees and the company management as customers of the QMS. This would mean throwing away valuable potential, however a QMS should never be pursued solely to meet external requirements but should always be used to increase the competitiveness of a company.

Further Reading

- ISO: DIN EN ISO 9000:2005-12 Quality management systems – Fundamentals and vocabulary (ISO 2005)
- ISO: DIN EN ISO 9001:2000-09 Quality management systems – Requirements (ISO 2000)
- ISO: DIN EN ISO 9004:2000-12 Quality management systems – Managing for the sustained success of an organization (ISO 2000)
- ISO: DIN EN ISO 13485:2009-07 Medical devices – Quality management systems – Requirements for regulatory purposes (ISO 2003)
- ISO: DIN EN ISO 14971:2007-07 Medical devices – Application of risk management to medical devices (ISO 2007)
- ISO: ISO/TR 14969:2005-10 Medical devices – Quality management systems – Guidance on the application of ISO 13485:2003 (ISO/TR 2005)

References

5.1 DIHK: *Produkthaftung – Ein Leitfaden für Hersteller, Zulieferer, Importeure und Händler* (DIHK, Bonn 2008)

5.2 European Commission, Enterprise and Industries, Healthcare Industries, Referenced Documents: http://ec.europa.eu/enterprise/sectors/medical-devices/documents/guidelines/ (European Commission, Brussels 2010)

5.3 W.E. Deming: *Out of the Crisis* (Massachusetts Institute of Technology, Cambridge 1982) p. 88

5.4 F. Malik: *Führen, Leisten, Leben. Wirksames Management für eine neue Zeit* (Heyne, München 2001)

5.5 R.S. Kaplan, D.P. Norton: *Balanced Scorecard – Strategien erfolgreich umsetzen* (Schäffer-Poeschel, Stuttgart 1997)

5.6 US Government: Code of Federal Regulations, Title 21 Food and Drugs, Chapter I, Part 11 Electronic Records, Electronic Signatures (US Government Printing Office, Washington 2005)

6. Usability of Medical Devices

Ulrich Matern, Dirk Büchel

This chapter gives an overview on how to design medical equipment according to international standards (IEC EN DIN 62366 and IEC EN DIN 60601-1-6) with the user in mind, no matter if physician, nurse or patient. After the definition of the term usability (Sect. 6.1) and widening it to include questions of safety and user satisfaction. In Sects. 6.2 and 6.3 the general questions of usability in medical technology are positively discussed to lead to Sects. 6.4–6.6 outlining in detail the development, testing and assessment of usable devices. The chapter ends with a an illustrative example (Sect. 6.7).

6.1 **What Is Usability?** 59
6.2 **Usability in Medical Technology – Obligation or Opportunity?** 60
6.3 **Usability in Medical Technology – Why?** .. 61
6.4 **Development of Usable Devices – How Is this Done?** 62
6.5 **Testing of Usable Devices – How Is this Done?** 64
6.6 **Assessment of Usability** 67
6.7 **Usability Development, Testing, and Assessment – An Example** 67
References 70

Usability engineering (also called human factors engineering), which is the development of devices which are fit for use, is a subarea of ergonomics. Ergonomics is an interdisciplinary science which concerns itself with adapting work to people. Until the middle of the last century, it was the primary aim of engineers, medical scientists, ergonomists, anthropologists, physiologists, psychologists, and others to reduce the physical exertion of the worker and simultaneously increase production.

6.1 What Is Usability?

As a result also of the increasing digitalization of the working world and of leisure and consumer goods, in recent years the focus has shifted from the reduction of physical exertion to the reduction of mental exertion of the user when using devices, and it could also be said that the devices should not distract the operator's attention away from the actual task as a result of inexplicable operating concepts. Devices which are particularly simple and intuitive to operate are today described as *usable*.

Usability is a qualitative attribute which indicates how easy devices are to use. The term *usability* includes the following properties of a device:

- Effectiveness: Can the goal of the user be fully achieved with the device?
- Efficiency: What is the cost of achieving this goal?
- Satisfaction: What reaction does the device prompt in the operator: does it disturb, or provide assistance?

Another aspect is increasingly gaining in importance in connection with usability in the field of medicine itself, i.e., the aspect of safety. This is important because it has been recognized that risks to the operator or to other people and objects are possible both as a result of potential faulty operation and also as a result of correct use of a device. This recognition gained acceptance early on in the field of aviation and the power plant industry, and in these sectors there is a lot of investment in the development and testing of ergonomic and usable properties.

In other consumer sectors, however, it often becomes apparent that the terms *ergonomic* and *usable* are used as marketing instruments, rather than the corresponding product properties being based on verifiable criteria with a scientific basis.

In contrast, products which are truly intuitive and therefore simple to use have an indisputable competitive advantage. They are accepted by the customer as being *top of the line* and so become bestsellers. The iPod and iPhone have already become almost proverbial examples of these kinds of products which are extremely successful in the market.

6.2 Usability in Medical Technology – Obligation or Opportunity?

To provide all groups involved in health care with a guideline about how usability can be achieved and tested, an internationally valid standard was published in 2006 as a collateral standard for the general provisions for the safety of medical devices. This safety standard for *electrical* medical devices was authoritatively written by manufacturers of medical devices, scientists, testing institutes, and experts in usability, and has been validated in Germany as a harmonized standard since 1 February 2008 [6.1].

Another international standard for the application of usability for *all* medical devices gained validity on 1 September 2008 [6.2]: IEC 62366 is increasingly replacing the older IEC 60601-1-6, and for this reason reference will primarily be made below to the content of IEC 62366.

It is first necessary to establish how international standardization defines usability. According to international standardization, usability is the [6.2]:

> *character of the user–product interface, which includes the effectiveness, efficiency and learnability and the satisfaction of the user.*

Through this definition alone, the device is accredited with essential features, and the user, in contrast to the manufacturer, is to a certain extent relieved of his responsibility; the device must support the user in operating the device correctly and learning how to do so.

The introduction to this process standard already includes clarification that safe usability is not an unambiguous definition, such as, the paper size according to DIN A, but rather is a challenging task, a method, a process, which demands *entirely different capabilities than the (purely) technical realization of this interface*. Ergonomics and considerations of usability constitute an interdisciplinary science which primarily involves designing, evaluating, and finally testing the configuration of the operating elements and the man–machine interface, with an external but expert eye, both with impartiality and without blinkers. The usability-orientated developing process should therefore minimize operational errors and risks associated with use. It is closely associated with the established risk management process of ISO 14971 [6.3].

The manufacturer must establish how it specifies the usability of its medical device [6.2] and therefore minimizes risks associated with use. Which risks it accepts is left to the manufacturer according to ISO 14971 [6.3] and is thus the *policy of the manufacturer*. This provides the manufacturer with the opportunity to position itself in the market according to its concepts, goals, and possibilities. It can determine for itself how it minimizes hazards which it will not accept for its product. It can invest, for example, in an intuitive man–machine interface [6.2]:

> *If training is necessary for the medical device in question in order to ensure safe and effective use of the main operating functions by the designated user, the manufacturer must take at least one of the measures*
> - *provide the necessary training material itself;*
> - *ensure that this material is available; or*
> - *provide the training itself.*
>
> *If training of this nature is necessary, the accompanying documentation must describe the available training possibilities and give details of the approximate duration and frequency of training.*

It remains to be seen whether usability will in the future be seen by manufacturers as an inconvenient obligation which must be fulfilled, or as an opportunity to set themselves apart from competitors as a premium manufacturer.

Another advantage for the manufacturers is the result of usability tests which are to be carried out in accordance with IEC EN DIN 62366 and 60601-1-6 during the design and development process and documented in what is known as the usability engineering file. Doing so, the manufacturer can increase the ease of use of his product and can avoid usability issues and risks associated with usability. This reduces the cost and the legal risks for the manufacturers and the participating clinics in the subsequent phase of clinical testing according to MEDDEV 2.7.1 [6.4] and DIN 14155 [6.5]. Application and insurance regarding these clinical studies will become easier, because reference can then be made to the results of the usability tests which are documented in the ergonomics record. The staffing costs associated with safeguarding correct operation of the testing equipment in the clinic will be reduced if it is possible to ensure in advance that the product can be operated easily and safely even in difficult situations.

The new standards for usability also provide the operators and users of medical devices with obligations and opportunities. There is therefore a sufficient basis for a comparative cost–benefit analysis of medical devices which, if used correctly, can furthermore make an economical and medical contribution to patient safety.

6.3 Usability in Medical Technology – Why?

The possible importance of this contribution to patient safety is just as well known as the economic benefits which it will bring to health care worldwide.

National surveys of German surgeons and operating theater (OT) nursing staff have confirmed that 70% of surgeons and 50% of OT nursing staff have difficulties in correctly operating the medical devices found in the operating theater. They do not consider themselves to be sufficiently well trained in dealing with the devices, do not attend briefings, and do not read the operating instructions. The users themselves estimate the resulting risk to the patients and employees to be considerable [6.6].

An assessment of the reports of these incidents involving medical devices, which are required by law and presented to the Bundesinstitut für Arzneimittel und Medizinprodukte (BfArM, the German Federal Institute for Drugs and Medical Devices), demonstrates that this observation is correct. Approximately half of all incidents are based on a *misunderstanding between the user and the device*, which is to say they are based on a usability problem. The dangers posed by these incidents are greater and the consequences for the patients more serious than those resulting from purely technical causes [6.7].

The Institute of Medicine (IOM) proceeds from the assumption that more people are killed every year in the USA as a result of usability incidents than are killed by acquired immune-deficiency syndrome (AIDS) or traffic accidents, and that figure is between 44 000 and 98 000 patients per year [6.8].

During a Dutch study, problems were found with the technical equipment in 87% of the operations observed [6.9].

Of the working time of medical technology service centers in hospitals, 50–70% is spent ensuring that equipment which has been *adjusted* by doctors and nursing staff but which is reported as faulty is returned to full working order [6.10].

The economic impact is colossal. In German intensive care units alone, usability problems are the cause of additional costs of approximately €396 million per year for the health insurance funds [6.11]. In England, the number of incidents per year as a result of *user errors* with medical devices is taken to be 850 000, with costs of £400 million. Between 1984 and 1991, approximately 130 000 incidents were reported in the USA, and in 60% of these incidents incorrect operation was at least partly responsible [6.12]. The same conclusions were reached by *Bleyer* [6.13] and *von der Mosel* [6.14].

Manufacturers know these data and confirm the situation from their own experience. Although their medical devices have ever more functionality because of the increasing requests from doctors and nursing staff, it is the users who then complain about the complexity and demand repeated briefings and retraining from the manufacturers – at the expense of the manufacturers, of course, leading to hugely increasing costs.

It is of course right, as a manufacturer, operator, and solicitor, to refer to the duty of care of the user, and the user should read the manuals and participate in the equipment briefings. However, the clinics and those running practices must then provide their employees with the time necessary for this. At a time when we have a worldwide shortage of nurses and doctors and ever shorter hospital stays, this is an unrealistic requirement.

The approach of DIN EN IEC 62366 can be helpful here in specifying the expenditure on training for the medical device in question. All those involved can therefore calculate the value of the investment in a usable medical device.

6.4 Development of Usable Devices – How Is this Done?

An interdisciplinary approach is not all that is necessary for the development of usable medical devices. As soon as the decision has been made to develop a new device, consideration must also be given to how the device is to be operated. This means that the man–machine interface must be developed in parallel with the development of the technology, with the focus on the future users. Every operating element which has to be changed at a later date because it has not proved to be a success causes unnecessary costs. The fundamental requirements of usability thus belong just as much in the functional specifications as do the fundamental technical requirements.

When designing a usable device, it is essential to understand the context of use, the user requirements, and the organizational requirements. This involves taking into consideration not only the medical application but also the cleaning – and, if appropriate, recycling – as well as the necessary maintenance work. An essential principle here is that the users should not only be questioned regarding what they expect from a new device, but rather they should be observed in their daily work

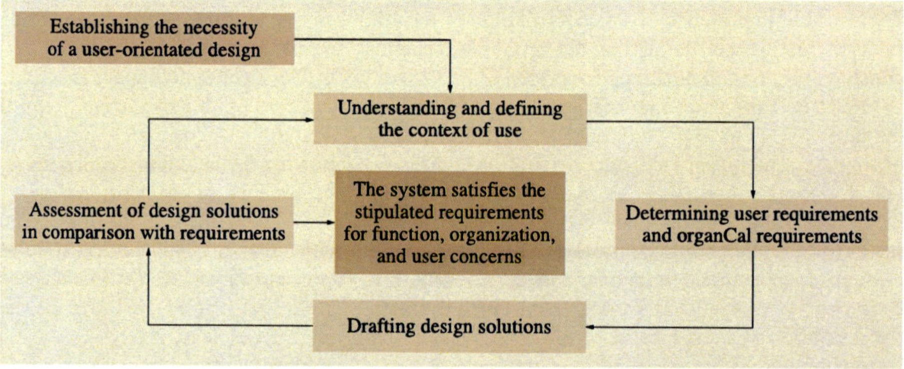

Fig. 6.1 User-oriented development process from DIN EN ISO 13407 (after [6.15])

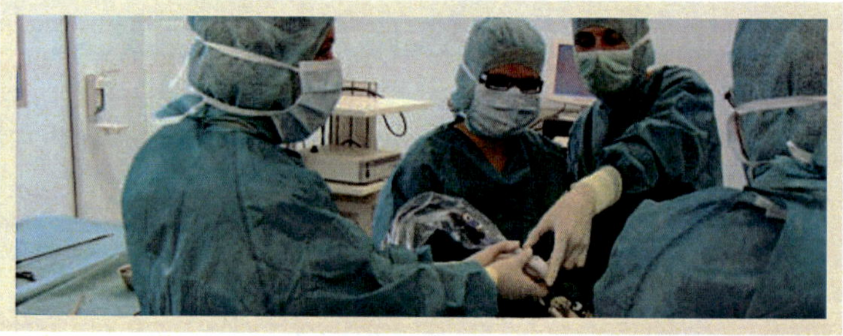

Fig. 6.2 Practical exercises as part of the training on the subject of ergonomics and usability at wwH-c GmbH

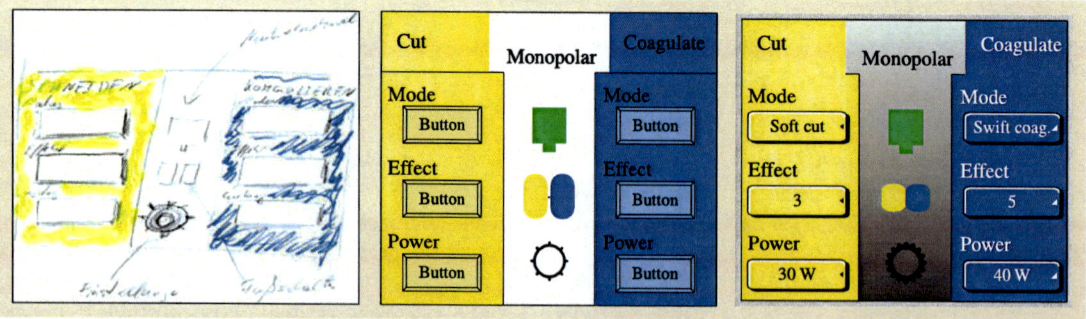

Fig. 6.3 Different stages of development in the course of the usability engineering process (after [6.16])

in a targeted and structured manner in order to identify which functions they really need. The reason for this is that users generally find it difficult to articulate their true requirements. It is only through doing this that the excessive functionality, and therefore the complexity, of the equipment can be avoided. The many unused functions of a device make it not only complex to operate but also susceptible to faults and, above all, expensive. With fewer functions, less can be significantly more in the sense of effectiveness, efficiency, and safety.

To successfully turn *less* into *more*, usability engineering uses various proven methods and fundamental technologies which are used in an iterative development process (Fig. 6.1).

The developers must have a distinct understanding of the users and the culture of usability. If this is not yet the case, training on the topic of usability should be contemplated, where possible with practical exercises in the field of the respective context of use (Fig. 6.2). At the beginning of the work, it is crucial that the developers are familiar with the concerns, needs, and working environment of their users and regard them as the focal point of the development process.

First and foremost, the context of use, including the requirements of the user and organizational requirements, must be understood and defined by the engineers: Who uses the device, in what situations and what environment, for what purpose, and on whom? Focus groups, field observations, and cognitive walkthroughs are used to answer these and other questions:

- *Focus groups*: Moderated group meetings with users and developers to ascertain the requirements and goals of the users.
- *Field observations*: Participatory observation of users in their working environment.

Table 6.1 Typical development errors

• No primary analysis of the needs and expectations of users
• No or too little understanding of the needs of the users
• Focus on using design features which are esthetic and showy
• No interdisciplinary usability engineering team
• Poor communication between the members of the development team
• No iterative development
• No or only little use of design prototypes
• No usability evaluation

- *Cognitive walkthroughs*: Usability experts run through tasks within a user interface. Here the focus is less on the device and more on the thought processes of the user during operation.

Equipped with this knowledge, the functional specifications are checked once more and refined again, in particular in terms of ergonomics and usability. What is particularly important here is that checks are continually carried out to ascertain (Fig. 6.3):

- What is really needed in the medical application; this provides a list of the main or primary functions of the devices, and
- Which supporting functions are necessary for this; the list of secondary functions should be as short as possible.

The development of the man–machine interface then begins in parallel with the development of the technical functions. To this end, the first design proposals are then drafted and checked using the simplest possible means:

- *Paper prototyping*: Operating elements, interaction concepts, and information displays are checked with users with the aid of drawings of an early operating concept.
- *Card sorting*: Interactive cards test the operational logic of menu structures and interaction concepts.

These design proposals are compared with the established requirements and tested in stages:

- *Heuristic evaluation*: Neutral experts evaluate an interface with the aid of recognized principles in software ergonomics and give suggestions for improvement.
- *User test (also usability test)*: Users tackle tasks under observation using a user interface. Experts analyze the man–machine interaction and find usability problems.

These tests usually feed into an iterative process with recurrent design and evaluation phases, as no-one succeeds in producing a perfect design at the first attempt. The planning must include the time and means to avoid the typical stumbling blocks listed in Table 6.1, which are encountered in particular when the development team is a closed group in which results are no longer sufficiently discussed. Careless mistakes of this kind mean that development quickly becomes an extremely expensive and lengthy procedure, which can be avoided if the development team obtains suggestions from external experts at an early stage which help to overcome the typical *tunnel vision* (Table 6.1).

The design phase may extend over many cycles in which functions must be repeatedly tested, optimized, and checked. At the end of this process, there should be a usable and safe device.

The technical solution has proved its functionality and reliability in the animal model. For this purpose, long-term tests and survival tests have been performed, where applicable using histological and biochemical methods:

- The recycling and/or disposal of the device have been tested.
- The experiences with the device have been incorporated in the manual and any possible training material.
- The design prototype has passed the normatively prescribed usability tests.

6.5 Testing of Usable Devices – How Is this Done?

When testing usability, IEC EN DIN 60601-1-6 and 62366 distinguish between the concepts of verification and validation.

Verification is understood as meaning the evaluation of the individual parts of a device during development. Various methods can be used to verify the individual functions of the man–machine interface, and these methods include heuristic evaluation, cognitive walkthrough, pluralistic walkthrough, feature inspection, and user test. Usability experts and typical users examine and test the individual functions, according to typical usability criteria in accordance with structured study protocols and guidelines, in the development stages which are available at the particular point in time (Fig. 6.4).

The validation of usability, in contrast, concludes the usability-orientated development process. During validation, all functions are scrutinized and not just individual functions, as in the previous verification processes. Finally, it ensures that a suitable, operable device is built in which there should be no unexpected and hazardous interactions between the user and the device.

A user test (also called a usability test or laboratory experiment) is performed for validation. In this method, the strengths and weaknesses of a device are systematically evaluated with the involvement of representatives of the user group, in an environment which is as representative as possible, with representative tasks. For this purpose, the context of use is represented as realistically as possible in a laboratory.

Usability laboratories are very expensive and are fundamentally divided into two areas. The actual use situation is simulated in the test area. This area ideally corresponds 1 : 1 to the real usage environment, to give the test person the feeling of being in his or her

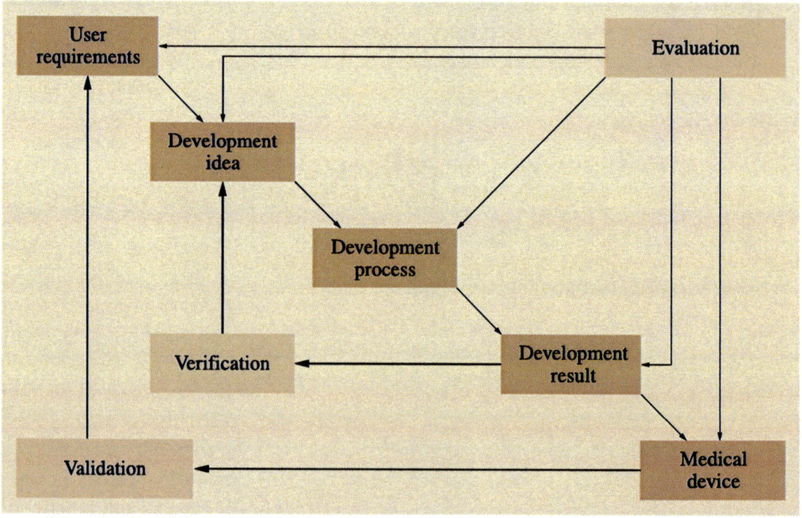

Fig. 6.4 Verification and validation in the usability-orientated development process

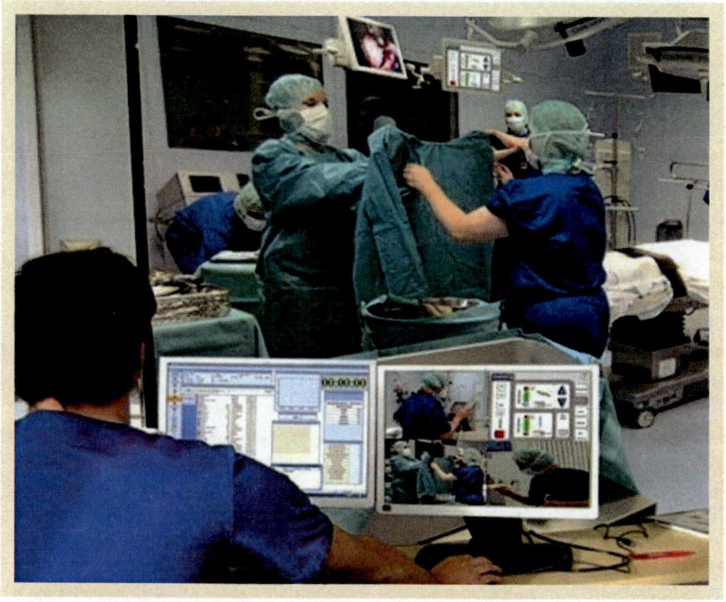

Fig. 6.5 Typical user test in the laboratory (wwH-c GmbH, Tübingen)

usual workplace. Recording equipment such as video cameras and microphones should be positioned as inconspicuously as possible here, so that the test person can work undisturbed with the equipment which is to be examined. In the monitoring area there is recording equipment and an intercom system. From here, test supervisors, minute-takers, and any other interested parties can observe the test participant, usually behind a semitransparent mirror or a window (Fig. 6.5). In this way, the test person is disturbed as little as possible, but assistance and instructions can nevertheless be given.

User tests are among the most widely used and most reliable methods in the world for determining the usability of a device. It takes only a few test people to find a relatively large number of problems. In the consumer sector, just six test users are generally sufficient to find 80% of usability problems (Fig. 6.6). In the safety-critical context of medicine, 10 test people from one occupational group are usually recommended to test the safety-critical functions. To ensure this, the test people must be selected with a great deal of care.

The particular advantage of this method is that, in the simulation environment, critical situations in which the device to be tested must be operated safely under nerve-wracking tension can be repeated with various test people in a standardized manner, without any risks to patients. Precise tasks which correspond to the evaluation goal are set for this purpose. Problems which were not noticed in the actual situation can be subsequently detected using the video and audio documentation and demonstrated to the development team. Quality factors such as objectivity, reliability, and validity can be maintained in the simulation environment in contrast to clinical testing on a patient.

It is a disadvantage that a completely real context of use can never be simulated under laboratory conditions, and these results are thus not 100% transferable to the real context of use. However, in a laboratory, those interferences and stress situations can be created which, for ethical and actuarial reasons, cannot be simulated in routine clinical operation and even in an animal labora-

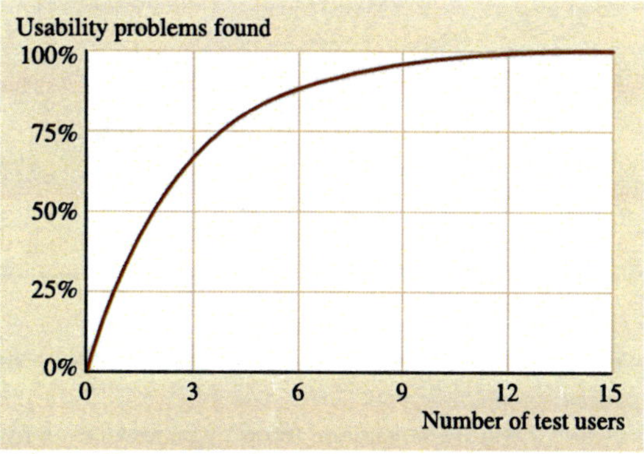

Fig. 6.6 Dependence of the usability problems found on the number of test users in a user test in the consumer sector (after [6.17])

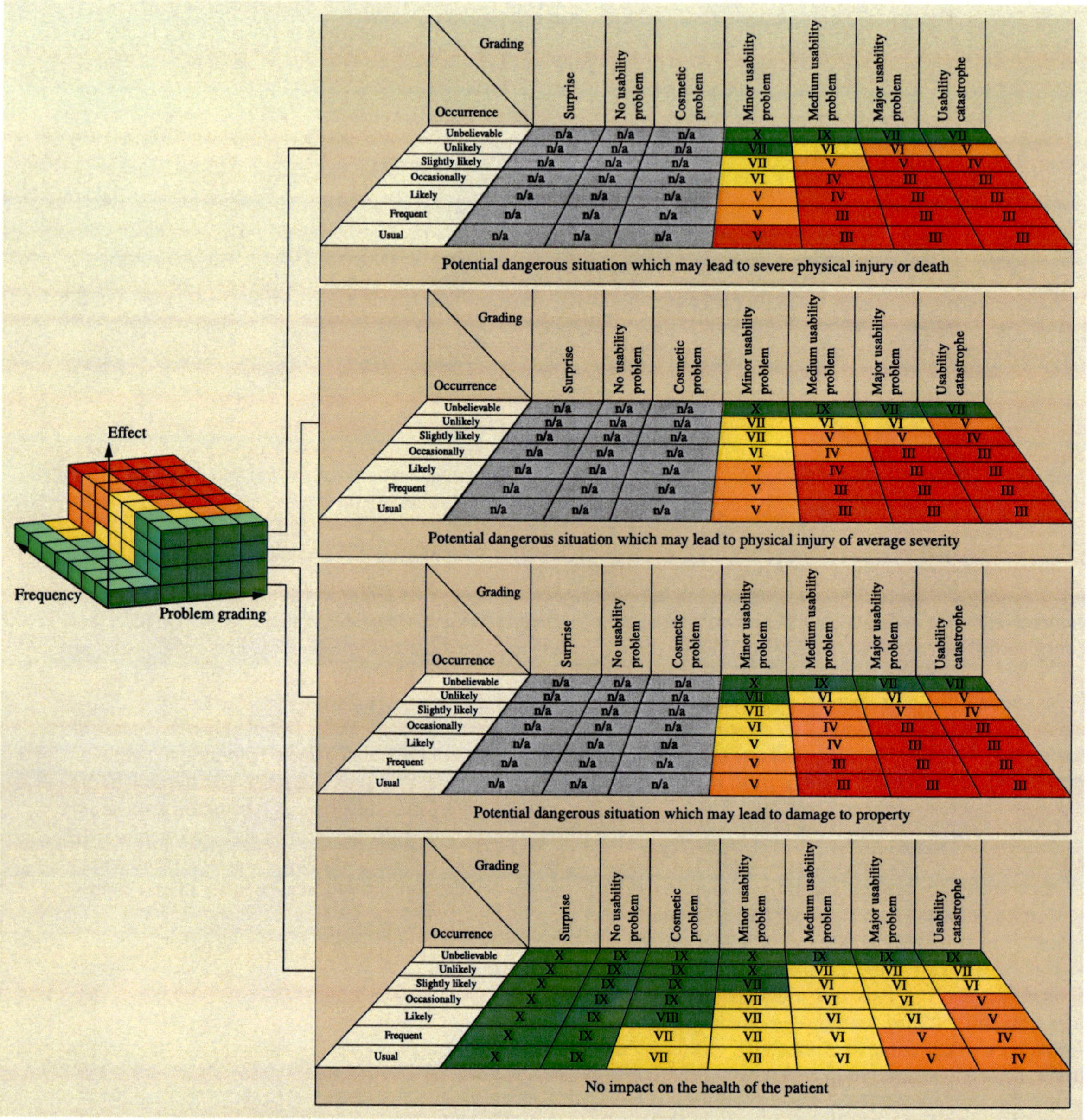

Fig. 6.7 Three-dimensional matrix for the assessment of the usability of medical devices (after [6.18])

tory, but which are the situations which are relevant to patient safety.

It is therefore sensible in terms of patient safety for a medical device to undergo such tests for the verification and validation of usability before it is used clinically. As already mentioned, this procedure reduces the cost and the legal risks for the manufacturer and the participating clinics in the subsequent phase of clinical testing according to MEDDEV 2.7.1 [6.4] and DIN 14155 [6.5].

During the user test, the test people are also urged to perform these typical work tasks using the equipment. Conclusions regarding the usability of the equipment can be drawn from the observations and testimonies during the performance of the tasks. For this purpose, the criteria of effectiveness, efficiency, and satisfaction are operationalized such that qualitative statements are possible. The results are listed and displayed, sorted according to functions and problems.

6.6 Assessment of Usability

The results of usability studies are typically documented in long accounts. However, no priorities generally emerge from these accounts for the development team. The assessment matrix developed for the field of medical technology *UseProb* provides transparency here for the manufacturer and user.

The degree of severity of the problem, the probability of occurrence, and the extent of damage are taken into consideration. Each criterion is given a grade. For the computing-based correlation of the three influencing variables, a three-dimensional matrix was selected which brings together the individual combinations to form a 10-level assessment (Fig. 6.7), which are divided into four color-coded areas. The red area (grading scale 1–3) represents safety-critical functions, that is to say those which should have been avoided by the usability engineering process prompted by the two standards. Functions whose assessment is in the orange, yellow or green areas conform to the standards, with the green area representing very simple and safe operation.

In this way, the party commissioning the usability study receives an *overview of grades* for the product which enables them to recognize immediately which functions are problematic and also which have been excellently solved (Fig. 6.8). Thus, there is an objective instrument available which the manufacturer can use not only to further develop his product but which they can also use for marketing purposes. Those buying medical devices can likewise base their purchase on these assessments.

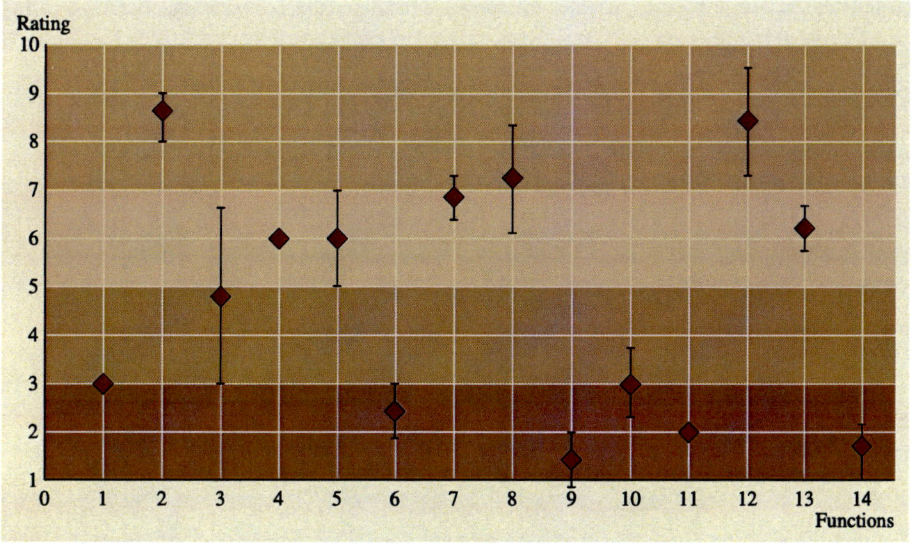

Fig. 6.8 Example of an overview of grades for a medical device which must be classified as safety-critical and therefore one which does not conform to standards. The operation of the functions 1, 6, 9, 10, 11, and 14 should be revised before the device can come onto the market (after [6.18])

6.7 Usability Development, Testing, and Assessment – An Example

In their laboratory, the *experimental OT*, wwH-c GmbH developed and tested the operator interface of a cardiac catheter pump in accordance with the normatively prescribed design process described.

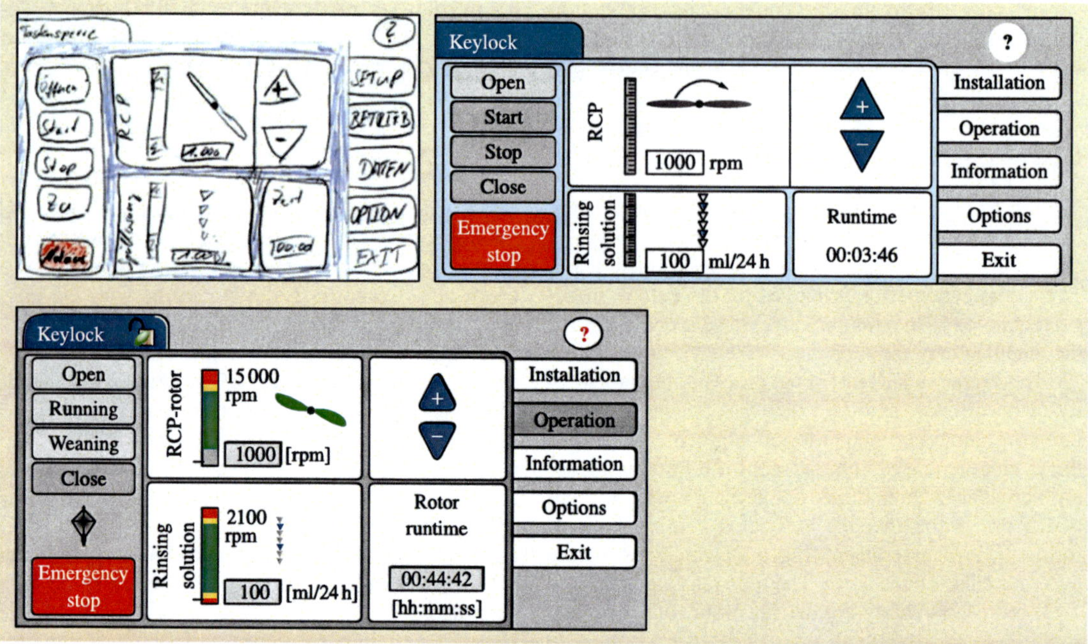

Fig. 6.9 From design drafts to the flash simulation in which the operation of all functions could be checked (after [6.18])

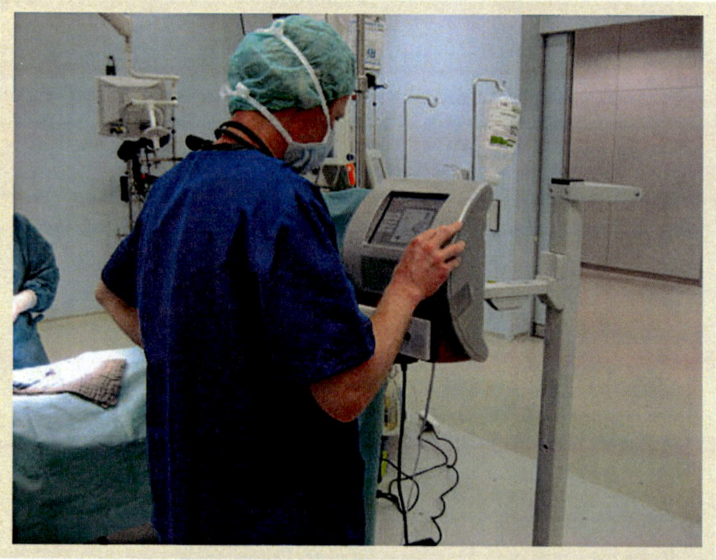

Fig. 6.10 The flash simulation was integrated into the hardware such that the test users did not notice the simulation

To determine the usability requirements for the equipment, a user profile was first created, and a task and environment analysis was performed. This process involved users in the development. In interviews they provided initial information about the requirements in practice and about problems with comparable devices to be found on the market. In accordance with the principle of *look at what your users do, but never ask them what they want* [6.19], field observations then provided the essential information about the requirements, which were recorded in the functional specifications.

The iterative design process, consisting of repeating design and evaluation phases, could then begin. Outlines of hardware and operating concepts were drawn up, discussed with users, and tested in a first expert evaluation.

The interaction concept and the operating logic were put into concrete terms and presented in detailed design drafts (Fig. 6.9). Various experts in ergonomics tested the logic of the preset interaction steps with the aid of the verification method of cognitive walkthrough, and revealed potential usability problems. The results of this evaluation were simulated in a flash simulation fully functioning in all menu structures. This simulation depicted the interaction steps to be carried out, and was integrated into the hardware of the catheter pump (Fig. 6.10). Using this animation, a realistic operator interface could be simulated for test users before any software code was written. Users were unable to detect that they were in reality not controlling the device.

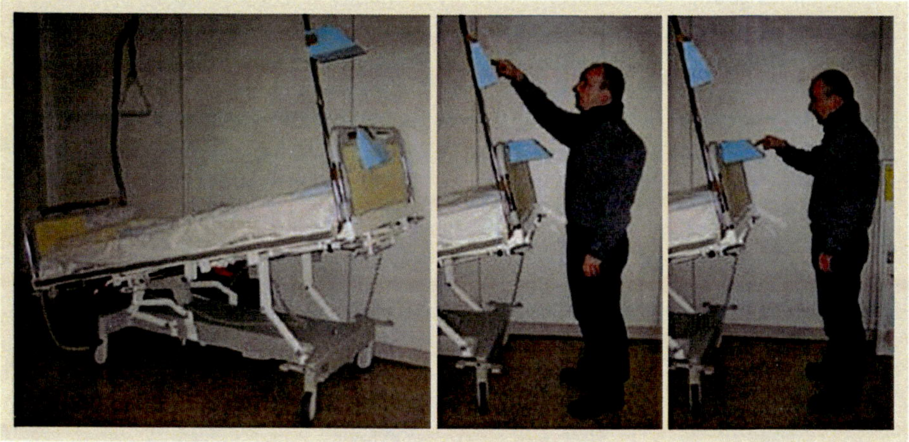

Fig. 6.11 With the aid of a rigid foam prototype, the ergonomically expedient installation on the patient bed is tested (after [6.20])

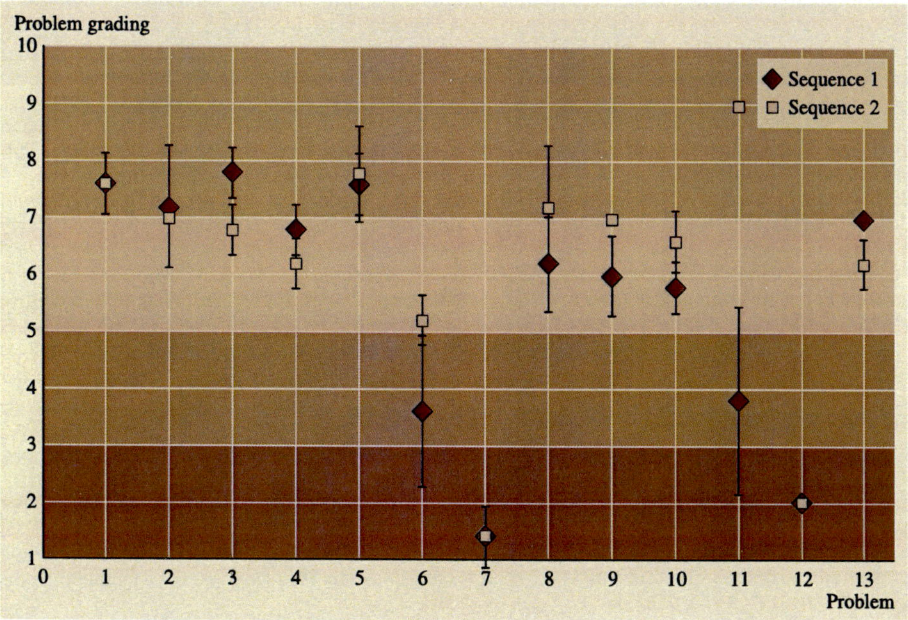

Fig. 6.12 Results from the first user test; assessment by means of *UseProb* before the second development iteration

The hardware models were also subjected to various function checks. Figure 6.11 shows a test which verifies the installation of the equipment on a patient bed.

To prepare the final user test, ergonomics experts conducted a heuristic evaluation. The evaluation method provides initial results about potential problems, which are verified in the following user test. The tasks involved in this user test consisted of the primary tasks identified in the task analysis and some secondary tasks, as well as the potentially problematic tasks. Nine doctors and nurses were used as test people. The test tasks were completed in the form of a role play in which an employee from wwH-c GmbH gave instructions to the test people in the role of the operator. Additional recorded OT noises increased the level of realism for the test people. The test was recorded and documented using four cameras and a microphone.

Even in this first usability evaluation with real users, the positive effects of the user-orientated design process were evident. The users were able to complete the tasks set relatively quickly and were satisfied with the operation of the equipment (Table 6.2).

Nevertheless, several problems were identified in this usability evaluation. These problems were assessed with the aid of the *UseProb* problem evaluation method developed by wwH-c GmbH. This method also looks at usability problems from the point of view of safety. The results of the evaluation are shown in Fig. 6.12. They show the relevance of the problems found.

Table 6.2 Statements from the test people, typical of the result for a usability engineering process performed with such consistency

When I saw the equipment, I realized for the first time how complicated our pump is. (specialist nurse)
It was easy to operate. Considering that I was doing it for the first time, it was fantastic. (trainee assistant surgical technician)
The equipment is self-explanatory. (specialist nurse)
It makes a good impression and is simple and intuitive. (perfusionist)
Compared with other equipment, the menu sequence is positive and very simple. (cardiologist)

Following the evaluation, the problems found were analyzed and the animation optimized on the basis of the proposals for improvement acquired.

Another user test was conducted with the improved use concept. It became apparent that the usability had further improved in respect of the factors of effectiveness, efficiency, satisfaction, and safety. Until this point, all development steps of the man–machine interface were simulated. Only now, once safe and efficient operation has been ensured by the users, does the programming of the software begin.

References

6.1 DIN EN IEC 60601-1-6: Medizinische elektrische Geräte – Teil 1 – 6: Allgemeine Festlegungen für die Sicherheit – Ergänzungsnorm: Gebrauchstauglichkeit, 1st edn. (Beuth, Berlin 2008)

6.2 DIN EN IEC 62366: Medizinprodukte – Anwendung der Gebrauchstauglichkeit auf Medizinprodukte (Beuth, Berlin 2008)

6.3 ISO 14971: Medizinprodukte – Anwendung des Risikomanagements auf Medizinprodukte (Beuth, Berlin 2003)

6.4 MEDDEV 2.7.1 Rev.3: Guidelines on Medical Devices, Clinical Evaluation: A guide for manufacturers and notified bodies (Global Harmonization Task Force 2009) http://www.ghtf.org/

6.5 DIN 14155: Clinical investigation of medical devices for human subjects – Good clinical practice, Draft (International Standard Organization, Geneva 2009)

6.6 U. Matern, S. Koneczny, M. Scherrer, T. Gerlings: Arbeitsbedingungen und Sicherheit am Arbeitsplatz OP, Dtsch. Ärztebl. **103**(47), 3187–3192 (2006)

6.7 K. Montag, T. Rölleke, U. Matern: Gebrauchstauglichkeitsuntersuchung von Medizinprodukten im Anwendungsbereich OP, Kongressbd. 2. Eur. Konf. Krankenhaustech., Wien (2007) pp. 60–62

6.8 L. Kohn, J. Corrigan, M. Donaldson: *To Err Is Human. Building a Safer Health System* (National Academy Press, Washington 1999)

6.9 E. Verdaasdonk, L. Stassen, M. van der Elst, T. Karsten, J. Dankelman: Problems with technical equipment during laparoscopic surgery – An observational study, Surg. Endosc. **21**, 275–279 (2007)

6.10 DGBMT Fachausschuß Ergonomie: *Ergonomie in der Medizintechnik* (VDE, Berlin 2004)

6.11 C. Backhaus: Entwicklung einer Methodik zur Analyse und Bewertung der Gebrauchstauglichkeit von Medizintechnik. PROMEDIKS – Prozessorientierte Medizintechnik in klinischen Systemen. Ph.D. Thesis (Technische Universität Berlin, Berlin 2004)

6.12 W. Laurig, U. Hölscher: *Ergonomie-Leitfaden für die Gestaltung von Medizinprodukten* (Mignon, Bonn 2004)

6.13 S. Bleyer: *Medizinisch-technische Zwischenfälle in Krankenhäusern und ihre Verhinderung*, Mitteilungen des Instituts für Biomedizinische Technik und Krankenhaustechnik der Medizinischen Hochschule Hannover (Fachverlag für Krankenhaustechnik, Hannover 1992)

6.14 H.A. von der Mosel: Der klinisch-biomedizinische Ingenieur, Schweiz. Ärzteztg. **52** (1971)

6.15 DIN EN ISO 13407: Benutzerorientierte Gestaltung interaktiver Systeme (Beuth, Berlin 1999)

6.16 F. Kleinert: Entwicklung von Interaction Design Pattern für medizinische Geräte und Überprüfung des Nutzens für die Industrie und Anwender. Bachelor Thesis (Hochschule Furtwangen/wwH-c GmbH, Furtwangen/Tübingen 2010)

6.17 J. Nielsen, T. Landauer: A mathematical model of the finding of usability problems, Proc. Conf. Hum. Factors Comput. Syst. CHI99 (1993) pp. 83–84

6.18 D. Büchel: Entwicklung einer allgemeingültigen Standardprozedur zur Überprüfung der Gebrauchstauglichkeit medizintechnischer Produkte. Ph.D. Thesis (Universität Tübingen, Tübingen 2009)

6.19 J. Raskin: *The Humane Interface* (Addison-Wesley Longman, Reading 2000)

6.20 A. Hültner: Gebrauchstaugliche und nutzerorientierte Neugestaltung der Bedienoberfläche einer Herzkatheterpumpe. Bachelor Thesis (Hochschule für Gestaltung Schwäbisch Gmünd/wwH-c GmbH, Schwäbisch Gmünd/Tübingen 2009)

Part B Function

Part B Functional Diagnostics Devices

7 Basic Diagnostics in Cardiology
Rüdiger Kramme, Titisee, Germany

8 Pulmonary Function Testing
Rolf M. Schlegelmilch, Würzburg, Germany
Rüdiger Kramme, Titisee, Germany

9 Devices and Methods in Clinical Neurophysiology
Klaus-Peter Hoffmann, St. Ingbert, Germany
Ursula Krechel, Freiburg, Germany

10 Sleep Diagnostic Systems
Klaus-Peter Hoffmann, St. Ingbert, Germany
Robert S. Pozos, San Diego, USA

11 Nystagmography
Klaus-Peter Hoffmann, St. Ingbert, Germany
Eduardo Fernandez, Elche, Spain

12 Audiometry
Sebastian Hoth, Heidelberg, Germany

13 Measurement Techniques in Ophthalmology
Albert J. Augustin, Karlsruhe, Germany

14 Functional Force Assessment of Skeletal Muscles
Paul A. Iaizzo, Minneapolis, USA
William K. Durfee, Minneapolis, USA

7. Basic Diagnostics in Cardiology

Rüdiger Kramme

The term basic cardiology diagnostics refers to the noninvasive measurement of the cardiac electrical action potential at rest and under stress, in order to assess heart function. Due to technical developments in ECG systems, the informative value of (basic) diagnostic assessments of the cardiovascular system has improved enormously with ever-increasing accuracy. This chapter introduces equipment-based diagnostic methods: Sect. 7.1 ECG, Sect. 7.9 Holter Monitoring and Sect. 7.19 Exercise ECG.

7.1 **Electrocardiography** 75

7.2 **Electrocardiograph Equipment Technology and PC ECG** 76
- 7.2.1 Physical and Technological Principles 76
- 7.2.2 Equipment Classification 77
- 7.2.3 Recording Systems 78
- 7.2.4 Electrode Technology 78
- 7.2.5 System Properties 79
- 7.2.6 Operating Modes 79

7.3 **ECG Methods** 79

7.4 **Lead Systems** 80
- 7.4.1 Measurements from the Surface of the Body 80
- 7.4.2 Standard Leads 80
- 7.4.3 Augmented and Reduced Leads from the Surface of the Body 82
- 7.4.4 Invasive Leads 83

7.5 **Methodological Notes** 83

7.6 **The Diagnostic Value of the ECG** 83

7.7 **Complications** 84

7.8 **Technical Safety Aspects of ECG Systems** 84

7.9 **Long-Term ECG** 84
- 7.9.1 Leads 84

7.10 **Long-Term ECG Systems** 85

7.11 **Computer-Based Assessment** 85

7.12 **Heart Rate Variability and Heart Rate Turbulence** 87

7.13 **Indications for Long-Term Electrocardiography** 87

7.14 **The Significance of the Long-Term ECG** 87

7.15 **The Exercise ECG** 88

7.16 **Equipment Technology** 88
- 7.16.1 Physical and Technological Principles 88
- 7.16.2 Ergometry Measuring Station 88
- 7.16.3 Types of Ergometers 88

7.17 **Reduced Exercise ECG Leads** 89

7.18 **Automatic ST Measuring Programs** 90

7.19 **Exercise Test** 90
- 7.19.1 Stress Intensity 90

7.20 **Methodological Notes** 93

7.21 **The Diagnostic Value of Ergometry** 93

7.22 **Indications** 93

7.23 **Abort Criteria and Safety Measures** 93

7.24 **Technical Safety Aspects** 94

7.25 **Notes on Planning** 94

7.1 Electrocardiography

Electrocardiography (ECG) is a method of recording electrocardiograms – which record the temporal and spatial profiles of the electrical excitation processes in the myocardium in the form of waves, peaks and lines

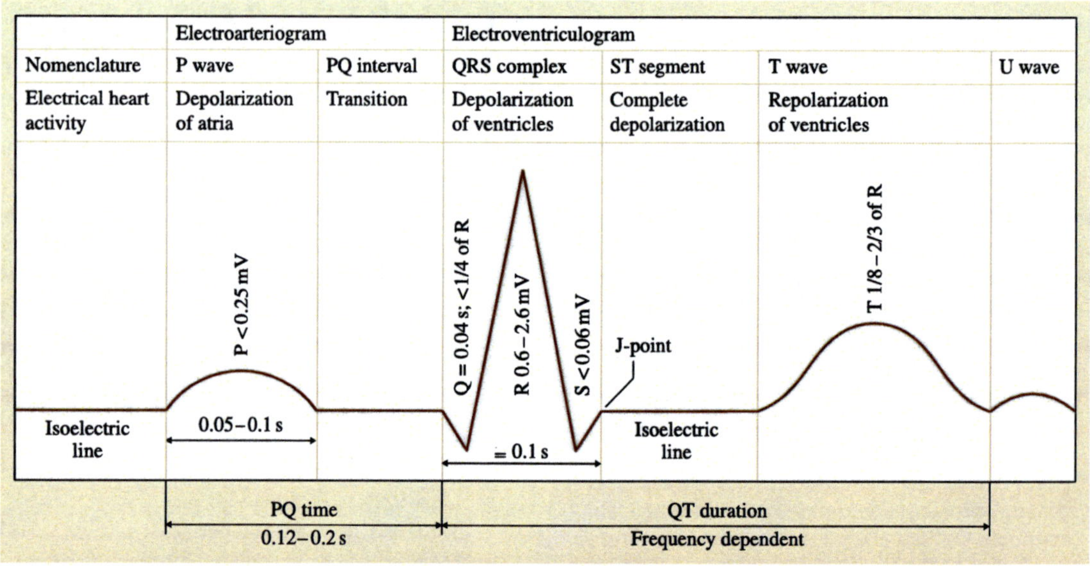

Fig. 7.1 Nomenclature of the electrocardiogram (ECG)

(Fig. 7.1) – and performing diagnostic analyses of these electrocardiograms.

Every instance of depolarization through cardiac fibres is the source of an electric potential. These noninvasive measurements generally measure potential differences that occur across an electric field on the surface of the body, which make up only a fraction of the potential generated by the heart. The heart is thus interpreted as a source of potential, so the ECG ultimately provides an image of the generation of electricity (potential shift) and reflects the excitation processes in the measurements selected.

7.2 Electrocardiograph Equipment Technology and PC ECG

Electrocardiographs (ECG devices for short) and PC ECG modules are diagnostic devices for recording, amplifying, storing, processing, analysing and documenting (registering) an electrocardiogram. Noninvasive electrocardiography has been a standard procedure for a number of decades, and ECG equipment is therefore as good as is mandatory in both hospitals and small private practices.

7.2.1 Physical and Technological Principles

ECG systems are low-noise differential amplifiers; in other words, they consist of a strongly coupled DC amplifier with a high amplification factor and inverting (reversed) and noninverting inputs. The output voltage is a multiple of the voltage present at the input terminals. It is the differential voltage of the two voltages present at the same pole. Particularly high requirements are set for an ECG preamplifier when compared to usual amplifier technology: interfering high-frequency AC voltages are attenuated by means of a high-frequency filter upstream of the preamplifier input, so that no overamplification or self-excitation can occur. Extreme interference voltages are likewise blocked by means of discharge sections and antiparallel diodes. A relatively high input resistance (generally $10\,M\Omega$ and above) keeps the input currents very low. Capacitively coupled mains-frequency AC voltages are eliminated through high common mode rejection. Following a 20- to 30-fold preamplification, the ECG signal is separated from the direct current via a high-pass filter. The time constant of this filter is 1.5 or 3.2 s. Whereas the processed ECG signal is fed to the respective channel amplifiers (1 ... > 12) via an lead selector in analogue ECG equipment, this step is omitted in digital ECG equipment. The recorded signals are instead connected to 12 leads, amplified and fed to a processor system (control computer – CPU) via multiplexers and A/D converters. The system software processes all of the digitized input signals, interpolates the individual measurement readings to give continuous curves, controls equipment processes, and communicates with the operator via the keypad, display and printout; it prepares text outputs, displays the time, produces a QRS trigger signal, and evaluates the remote start ergometer input. Output occurs via a digital recording system, the thermal comb.

In analogue ECG equipment, following further amplification to ≈1–2 V using the respective channel amplifier, the ECG signal is output via a power amplifier stage (final amplifier) and finally through a mechanical recording system (e.g. lever recorder, inkjet recorder, etc.).

ECG measurement, the signal path and signal processing resemble a measurement chain, from the test object (patient) via transducers or sensors (electrodes) and the transfer of measurement readings (electrode leads and patient cables) to signal processing, signal evaluation and documentation (ECG equipment). The recording electrodes or the intracardiac electrode (signal source) on the surface of the body receive the ECG signal (useful signal) and transmit it via a cable to the input amplifier of the cardiograph. There is a galvanic potential between the skin and the metal section of the electrode, which is substantially influenced by the electrode material, the composition of the electrolytes, and by the condition of the electrode/electrolyte interface. This electrochemical contact potential can be up to 300 mV. The transmitted input signal from the amplifier consists of four different fractions:

1. The useful signal (ECG), with an amplitude of 50 μV to approximately 1 mV, and
2. A DC voltage component of up to 300 mV that is superimposed on the useful signal; however
3. Interfering signals of up to 100 mV (50 Hz *ripple voltage*) and
4. Extreme interference voltages (resulting from defibrillators and RF surgery equipment, among other sources) of up to 3000 V can also occur.

In order to ensure that the patient is galvanically isolated from the mains voltage, the power supply and the output signal are galvanically decoupled.

In order to promote reproducibility, the ECG signal processing, recording and measuring process is standardized:

- Frequency response:
  - Lower frequency limit $f_{limit} = 0.05\,Hz$
  - Upper frequency limit $f_{limit} = >100\,Hz$
- ECG amplitudes:
  2.5, 5, 10 and 20 mm/mV
- Times:
  10, 25, 50 and 100 mm/s

7.2.2 Equipment Classification

ECG recorders are mainly classified according to the number of channels of the device and the equipment technology used (Figs. 7.2, 7.3).

From analogue to digital ECG systems with integrated recording components, a new ECG generation has developed: the PC ECG. The recorded analogue signals are amplified, digitized and supplied to data transmission equipment in an external module. The data can be transmitted to a connected computer (PC or laptop) using either a USB microcontroller or wireless data transmission technology (bluetooth technology)

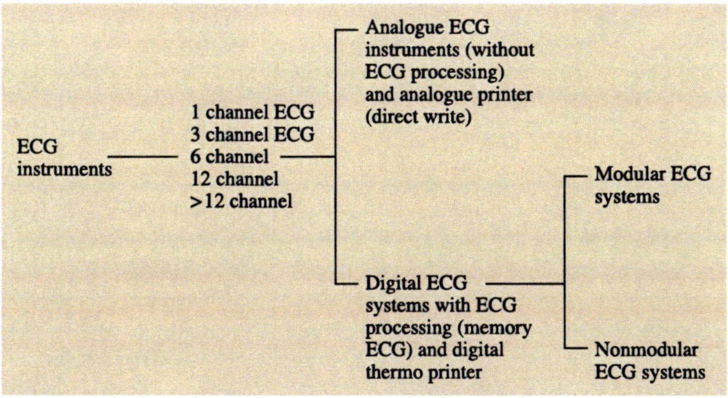

Fig. 7.2 Overview and differentiation of ECG recording equipment and PC ECG modules

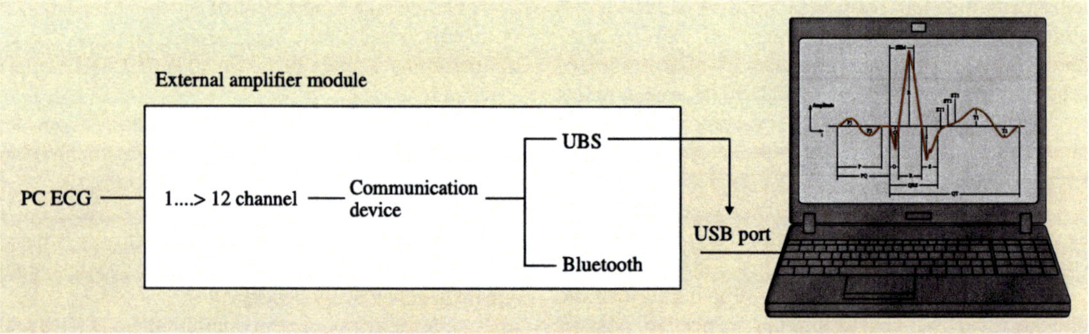

Fig. 7.3 PC ECG module

(Fig. 7.3). Whereas the USB port is also used to supply power to the module, the Bluetooth method makes mains-independent operation (10–12 h) possible with an accumulator battery. The transmitted data is processed and prepared on the computer using installed ECG application programs. In addition to the visual image on the computer display or external monitor, ECG graphs and interpretations can be printed out on a connected printer and plain paper for documentation purposes.

The PC ECG has practical advantages over conventional ECG equipment: the ability to save any desired ECG sequences at any time, the comprehensive availability of the data, the considerable reduction in the consumption of expensive ECG paper and therefore financial savings, the lack of mechanical equipment components, the ability to back up data on common storage media, the capacity for mobile use with mains-independent operation, as well as the possibility of interactive data exchange via the Internet. A particular disadvantage is the restricted data exchange possible via data carriers or over the Internet, as data can only be exchanged with an application program which is compatible with the PC ECG software. An artefact that is characteristic of PC ECG equipment, and is probably a result of computer electronics, leads to interference in the sensitive ECG analysis.

7.2.3 Recording Systems

Recording systems can be considered the various technical possibilities of the recording components (Fig. 7.3). The digital thermal array recording method is the current state of the art. Thermal comb recordings have high resolution and a low susceptibility to interference from the recording system, and allow optimum reproduction of the recorded and processed ECG signal.

7.2.4 Electrode Technology

The recording and transmission of the ECG signal from the surface of the body and the quality of this signal are fundamentally influenced by the electrodes. They repre-

Table 7.1 Overview of electrode types

Type of electrode	Electrode material	Recording region	Short application time	Long application time	Observation	Emergency application	Comments
Plate	V_2	Limbs	×				Without electrode gel
Suction	Silver/silver chloride	Chest	×				Low polarization voltage
Button	Silver/silver chloride	Limbs	×	×	×	×	Use foam plaster
Self-adhesive	Silver/silver chloride	Limbs					Comparatively low artefacts in restless patients
Single-use	Silver/silver chloride	Limbs, chest	×		×	×	Higher costs
Triple	Silver/silver chloride	Chest			×	×	Fast application possible

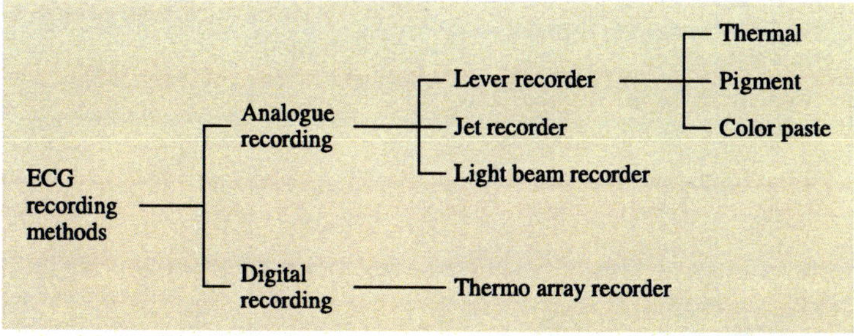

Fig. 7.4 ECG recording methods

sent a junction between the ionic conduction in the body and the metallic electronic conduction of the electrode surface and of the transmitting cable. There is therefore an electrical potential between the body and electrode, which is dependent on three factors in particular:

1. The electrode material
2. The composition of the electrolytes
3. The condition of the electrode–electrolyte interface.

Electrochemical contact potentials of up to 300 mV can occur at these interfaces. The electrode impedance (also known as electrode transfer impedance) – the sum of the electrode and skin impedances and also the resistance of the contact medium, which is dependent on the current density – is the result of the current state of the skin, body, contacting agent and electrode material. For this reason, electrode pastes or gels are used to reduce the transfer impedance.

Particularly in the case of polarizable electrode material (e.g. high-grade steel), unwanted polarization voltages may occur. These spontaneous voltage fluctuations are superimposed on the ECG signal and lead to signal instabilities in the recording (drifting leads). Nonpolarizable electrodes from Ag/AgCl (silver/silver chloride) have proven to be particularly advantageous (Table 7.1).

7.2.5 System Properties

The following properties represent the performance spectrum and at the same time the requirements for modern ECG systems:

- Simultaneous display of the 12 standard leads in a time frame of 15 s
- Various alphanumerical recording options
- High operational convenience and programmable
- Digital signal processing
- Automated control together with programs
- Interfaces for communication with other systems
- A high-resolution recording system without mechanical parts.

An upgradable system has the advantage of being able to perform more functions than just one ECG registration. The ECG device has these so-called multifunctional system properties when it is equipped with additional functions such as spirometry, long-term ECG monitoring, arrhythmia monitoring, Doppler ultrasound, pulse oximetry, capnography and late potentials, etc. Data communication with a PC and networking with other systems is also standard. The conditions for the use of medical EDP statistics, ECG data management and ECG software libraries are therefore satisfied.

7.2.6 Operating Modes

A fundamental distinction is drawn between manual and automatic operating modes. Automatic programs, such as standard ECGs with 12 leads, ergometry and arrhythmia, are typical of modern cardiographs which can be programmed by the user. The lead mode, formatting, baseline, calibration and sensitivity are for the most part automated. Alternatively, registration can be performed manually at any time.

7.3 ECG Methods

Figure 7.5 shows an overview of noninvasive and invasive ECG methods.

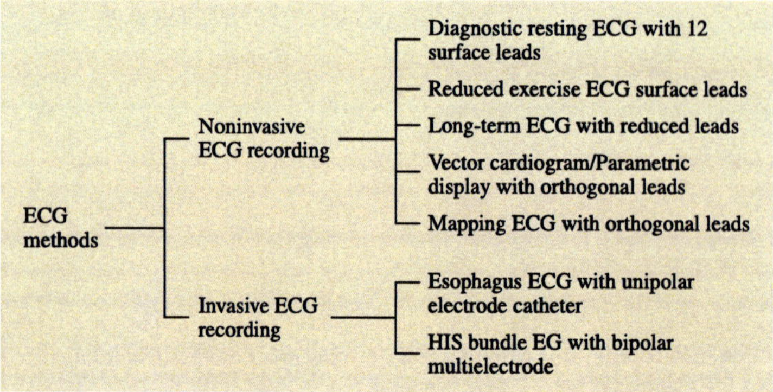

Fig. 7.5 Noninvasive and invasive ECG methods

7.4 Lead Systems

7.4.1 Measurements from the Surface of the Body

The action currents produced by biopotential differences that vary with time are measured as electrical potentials (the magnitude of the useful signal is around 1 mV) between two points on the surface of the body. Theoretically, any point on the surface of the body could be used for this purpose, but specific electrode sites have developed in practice, and the respective leads are defined according to these sites. The leads differ substantially in terms of recording technique and spatial arrangement.

Leads between two points are described as bipolar leads. In unipolar leads, in contrast, the different electrodes are connected to what is known as a zero electrode (collector electrode). This collector electrode is created by connecting together limb leads via high-impedance resistors. The limb leads provide a spatial view of the frontal plane, and the chest leads provide a spatial view of the horizontal plane.

7.4.2 Standard Leads

The standard leads are the three bipolar limb leads known as Einthoven's leads (I, II, III), the three unipo-

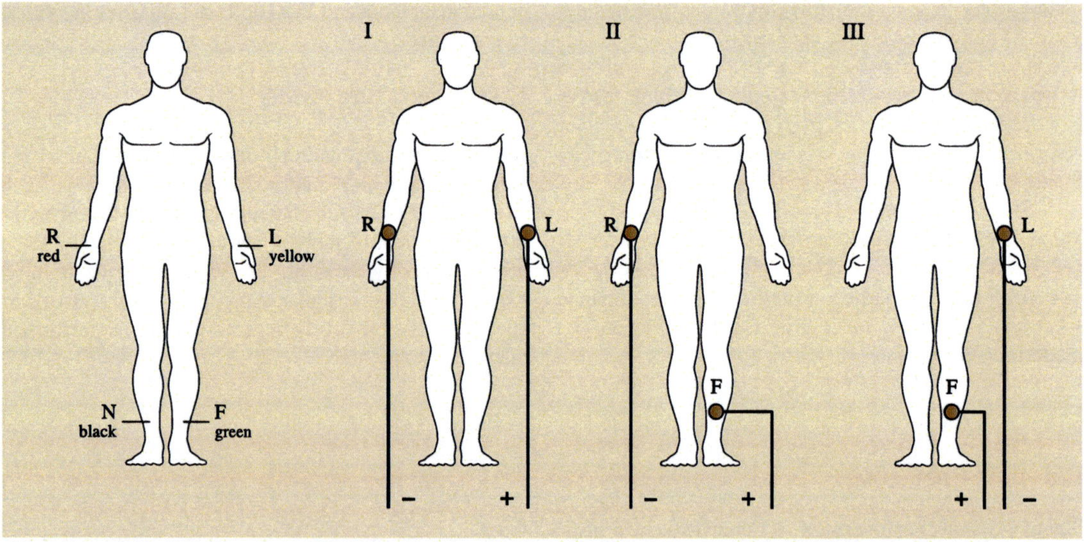

Fig. 7.6 Lead positions, connector colors and schematic diagram of the bipolar limb leads according to Einthoven

Table 7.2 Overview of the 12 standard leads of an ECG

	Limb leads		Chest leads
Description	According to Einthoven	According to Goldberger	According to Wilson
Electrodes	I, II, III	aVR, aVL, aVF	$V_1, V_2, V_3, V_4, V_5, V_6$
Measurement technique	Bipolar	Unipolar	Unipolar
Electrode sites	Upper and lower limbs	Constructed from I, II, III	Chest wall $C_1 \ldots C_6$
Projection	Distal potentials of the frontal plane		Proximal potentials of the horizontal plane

lar limb leads constructed from these three bipolar limb leads (known as Goldberger's leads: aVR, aVL, aVF), and six unipolar chest leads known as Wilson's chest leads (V1...V6) (Table 7.2).

Einthoven's Bipolar Limb Leads (Fig. 7.6)

For these, the potential difference is measured between two electrodes:

R: right arm = red connector
L: left arm = yellow connector
F: left foot = green connector
N: right foot = black connector (earth)

The three leads lie in the frontal plane and form what is known as Einthoven's triangle (Fig. 7.7).

Goldberger's Unipolar Limb Leads

For these, the potential difference is determined between each electrode on the limb and an *electrical zero* which lies at the central potential between the two other limbs (Fig. 7.8).

aVR: $R-L+F/2$

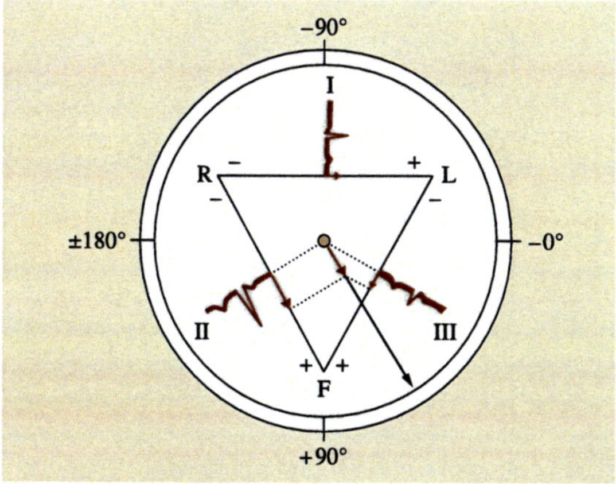

Fig. 7.7 Einthoven's triangle

aVL: $L-F+R/2$
aVF: $F-L+R/2$

Wilson's Unipolar Chest Leads

The chest electrodes ($V_1 \ldots V_6$, V = voltage) are placed at six defined points on the thorax ($C_1 \ldots C_6$, C = chest). The three limb leads are combined via

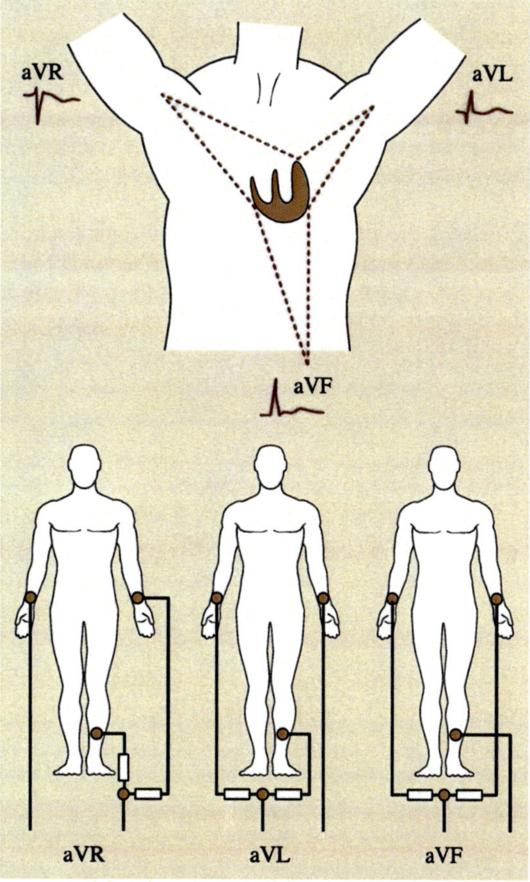

Fig. 7.8 Unipolar leads according to Goldberger, and schematic diagram

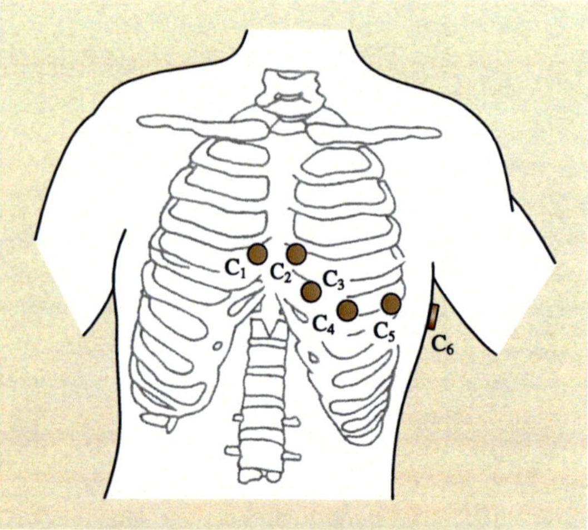

Fig. 7.9 Unipolar chest leads according to Wilson: C_1 – Fourth intercostal space at the right border of the sternum; C_2 – Fourth intercostal space at the left border of the sternum; C_3 – Fourth intercostal space at the fifth rib, midway between C_2 and C_4; C_4 – At the mid-clavicular line in the fifth intercostal space; C_5 – At the left anterior axillary line on the same horizontal level as C_4; C_6 – At the left mid-axillary line on the same horizontal level as C_4

a high-impedance resistor to form a collector electrode (indifferent electrode). Wilson's central terminal forms the electrical zero point. As a result of the proximity to the heart, the amplitudes of the chest leads are greater than those of the limb leads (Figs. 7.9, 7.10)

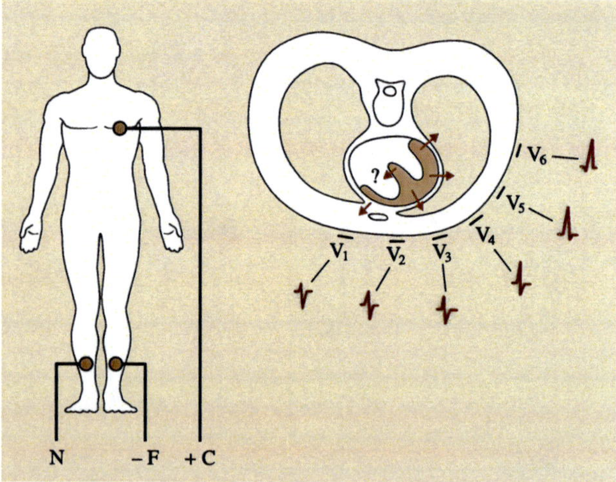

Fig. 7.10 Lead positions after Wilson, and illustration in the horizontal plane

Cabrera's Lead Sequence

Cabrera's circle is a hexaxial system that can be used, among other things, as an aid when determining the axis deviation. The circle includes leads of the frontal plane that are each rotated through 30° with respect to one another. The leads from Einthoven's triangle are shifted parallel such that lead I assumes an angle of zero, lead II +60° and lead III +120°. The Goldberger leads are also included in this system, and have angles of: aVR −150°, aVL −30° and aVF +90° (Fig. 7.11).

By flipping the aVR lead through 180°, we get −aVR at 30°. It is possible to determine the axis deviation simply by determining the highest R amplitude from the leads. The angular space allocated to the leads determines the deviation of the axis (for example, the highest R amplitude is found in lead II; i.e. the electrical axis of the heart falls at +60°. This corresponds to the normal position of the heart in a healthy adult).

7.4.3 Augmented and Reduced Leads from the Surface of the Body

Augmented Leads

Augmented leads include the corrected orthogonal leads of Frank and the additional unipolar chest leads of Wilson.

The Frank leads are based on a theoretical model in which the heart is at the centre of a square three-dimensional coordinate system consisting of a lateral axis x, a longitudinal axis y and a sagittal axis z. The excitation processes can therefore be registered in the form of vector loops (vectorcardiography).

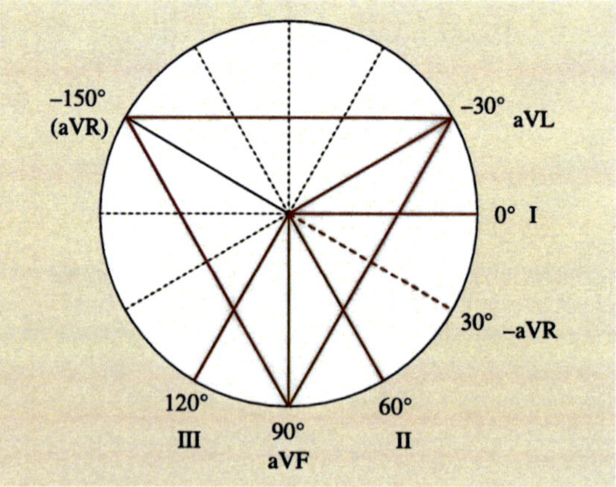

Fig. 7.11 Cabrera's circle

In addition to Wilson's standard leads on the chest, it is also possible to include other electrode sites. These can be extended precordially to the right (V_{r3}, V_{r4}, V_{r5}, V_{r6}) or to the left (V_7, V_8, V_9 or V_8 two ICS lower).

Reduced Leads

Reduced leads include leads for ECG monitoring, leads for long-term ECG and ergometry leads (see the individual chapter on this topic).

7.4.4 Invasive Leads

Unipolar Oesophageal Leads

When using oesophageal leads, an electrode catheter is inserted via the oesophagus to just above the cardia. The aim is to obtain a precise assessment of proximal potentials of the posterior myocardial wall and the left atrium.

HIS Bundle Electrocardiography

Here an intracardiac electrogram (EG) is obtained using a three-pole electrode catheter. This investigation is performed with radiological and electrocardiographic monitoring (this procedure is part of routine diagnostics in cardiac centres). The particular advantage of this lead is that it is possible to precisely investigate the proximal potentials of the right atrium, the right ventricle and the conductive system, and in this case the HIS-Purkinje system in particular (Fig. 7.12).

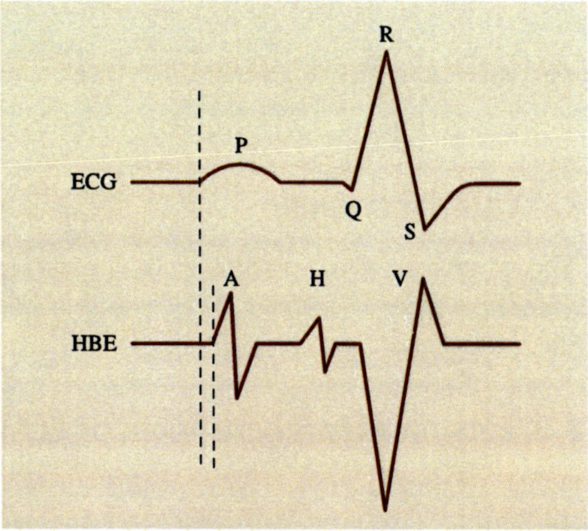

Fig. 7.12 Surface ECG compared with the intracardiac HIS bundle EG. *HBE* HIS bundle electrogram, *A* depolarization in the atria, *V* depolarization in the ventricles, *H* depolarization in the HIS bundle

7.5 Methodological Notes

Faults that occur during ECG registration can generally be differentiated into patient-, environment-, user- or equipment-related faults. Most faults (baseline shifts, irregular and regular superpositions of AC voltage) that occur during ECG registration can be traced back to external influences.

7.6 The Diagnostic Value of the ECG

As it illustrates the electrical excitation processes occurring in the heart, the ECG provides information about the origin and rhythm of excitation, the heart position and rate, pulse propagation, as well as repolarization and repolarization arrhythmias, which in turn can be caused by anatomical, mechanical, metabolic or circulatory problems. However, the ECG has no direct informative value regarding the contraction and pumping capacity of the heart (mechanical cardiac function). The medical significance of the ECG is undisputed, although its capabilities should not be overestimated. Within the field of cardiological investigation, the ECG is essential, and in routine internal investigations it is a valuable, possibly even crucial, investigation tool. Caution is advised regarding the informative power of the ECG in relation to the aetiology and pathogenesis of a cardiac disease, as well as in relation to the indications for and success of therapeutic measures. Anatomical damage and functional faults of the myocardium are not necessarily reflected in the ECG. It must be observed that the informative value of the ECG is dependent on the knowledge and experience of the individual providing the assessment. Whereas inexperienced users will be inclined to interpret the ECG too broadly, the experienced user will give a more reserved assessment and where possible avoid schematization and generalizations. It must be noted that, particularly with computerized evaluation programs (e.g. diagnostic hints), an individual's own assessment is essential.

7.7 Complications

Whereas incidents during noninvasive ECG recording are a complete novelty, serious arrhythmias (e.g. flutter, fibrillation, etc.) can occur during intracardiac ECG registration.

7.8 Technical Safety Aspects of ECG Systems

Electrocardiographs are power-operated technical medical devices, with associated provisions, regulations and standards that are applied to protect patients and operating personnel.

ECG devices that measure intracardially, or are provided for that purpose, are subject to Annex 1 of the German Medical Devices Operator Ordinance (Medizinproduktebetreiberverordnung – MPBetreibV). The protection classes that apply to ECG equipment are primarily protection class 1 (with a protective earth) and protection class 2 (without a protective earth). VDE 0750 allows devices to be equipped with a protection class switch so that operation in either protection class is possible. In order to maintain functional and operational safety, the equipment must be subjected to technical safety checks and tests at regular intervals.

7.9 Long-Term ECG

Long-term electrocardiography – long-term ECG for short, and also known as Holter monitoring, *ambulant* ECG or ambulatory monitoring – is a noninvasive routine ECG procedure that is central to primary diagnosis and the therapeutic monitoring of cardiac arrhythmias. As its name suggests, in this procedure, the ECG is recorded and analysed over a relatively long period of time (generally 24 h), under everyday stress.

Whereas other electrocardiographic investigation procedures, such as resting or exercise ECGs, only allow a diagnostic statement to be made over a limited period of time under specific conditions (*laboratory conditions*), the long-term ECG follows the various physical and emotional stresses that occur over the course of at least one day/night period.

7.9.1 Leads

To date there is no uniform agreement on the best electrode positions for long-term electrocardiography. Efforts are being made towards standardization, however, in order to ensure better comparability of the results obtained. The electrode sites must be selected such that they satisfy the following criteria:

1. The electrodes must be placed on the thorax at sites where there is little muscle, since movements of the skeletal muscles during long-term monitoring can lead to undesirable fluctuations and tremors in the ECG (artefacts).
2. The patient's freedom of movement must not be restricted.
3. The amplitude of the R wave in relation to the P and T waves must be sufficient for computer-based ECG analysis of the recordings.

As a general rule, right and left precordial leads are chosen. Preference is given to the following bipolar chest leads:

MC5: This lead is applied parallel to the electrical axis of the heart; the different electrode level with the fifth ICS, and the indifferent electrode above the sternum (manubrium sterni). The position of the electrode is varied depending on the deviation of the axis.

MX: The indifferent electrode lies above the sternum, whereas the different electrode is applied above the xiphoid process (xiphoideus) of the sternum. This lead has the advantage of having the greatest freedom from interference from muscle tremors.

CC5: The electrodes are applied to both sides of the anterior axillary line, level with the fifth ICS, with the different electrode to the right and the indifferent electrode to the left.

In addition to the leads mentioned here, modified leads are recommended by various authors in order to optimize the sensitivity of detection.

7.10 Long-Term ECG Systems

The fundamental work of Norman J. Holter (physiologist) on the development of wireless telemetry – in 1947 he transmitted an ECG signal by radio, and then in 1949 he transmitted ECG signals from ambulant patients performing physical work – had a great influence on modern long-term ECG registration, and indeed made it possible in the first place.

Long-term ECG systems consist of a portable recording device (recorder) with 2–3 channels that registers and stores the ECG signals, a computerized playback device, and a high-performance analysis unit. By means of fast Fourier transformation, application programs enable frequency analysis for artefact recognition, automatic evaluation according to various criteria, accelerated analysis, or batch processing. If we consider the recording technology, a distinction can be drawn between continuously and discontinuously recording long-term electrocardiographic systems in which the data carrier of the recording device is either a tape cartridge or solid-state memory. The tape cassette, which records in analogue, has today been virtually completely superseded by digital solid-state memory. In continuous recording, the ECG potentials are stored in their entirety over a period of between 24 and 72 h. In addition to real-time analysis, high-quality devices also make it possible to evaluate a highly amplified ECG, long-term blood pressure, heart rate variability, and other functions on the unit in parallel.

Discontinuous system-based ECGs are plotted according to the limited storage capacity on the recording device. Following a continuous recording analysis, only time-limited ECG information and results are stored and documented. With these so-called event recorders, the patient has the option of activating the recording via an activator or remote control unit if symptoms occur. Loop recorders have memory on a loop; in this case, ECG information remains stored for a limited period of time. Only ECG sequences that have been marked by the patient using the activator remain stored. A further development of the event recorder is the subcutaneous implantable loop recorder (ILR), which is used particularly in the case of symptomatic cardiac arrhythmias that occur infrequently. The device, which generally has an operating time of up to two years, is implanted subcutaneously in the region of the left sternum using local anaesthetic. The continuously recording bipolar electrogram is evaluated in the implant. ECG episodes that deviate from programmed ECG criteria are recorded. Here, too, the patient has the option of marking and storing symptoms or events that occur using a remote control.

With so-called simple sampling recorders, ECG sections are recorded only in certain time segments, without the recording being triggered by cardiac arrhythmias and without being influenced by the patient. This intermittently recording system, which stores information only randomly and nonspecifically, leads to recordings with only very limited informative value regarding the complete day/night rhythm.

The transmission of stored ECGs and complete long-term ECG data records via telephone to a computer in the hospital or in a private practice is increasingly gaining in significance. Developments in telecommunications are enabling ECGs to be transmitted using mobile phones, thus permitting *online diagnosis*.

7.11 Computer-Based Assessment

The nature and number of events (e.g. SVES, VES, bigeminy, couplets, salvos, R-on-T phenomena, etc.) within a certain time interval provide insight into cardiac arrhythmia. At the beginning of the analysis, there is

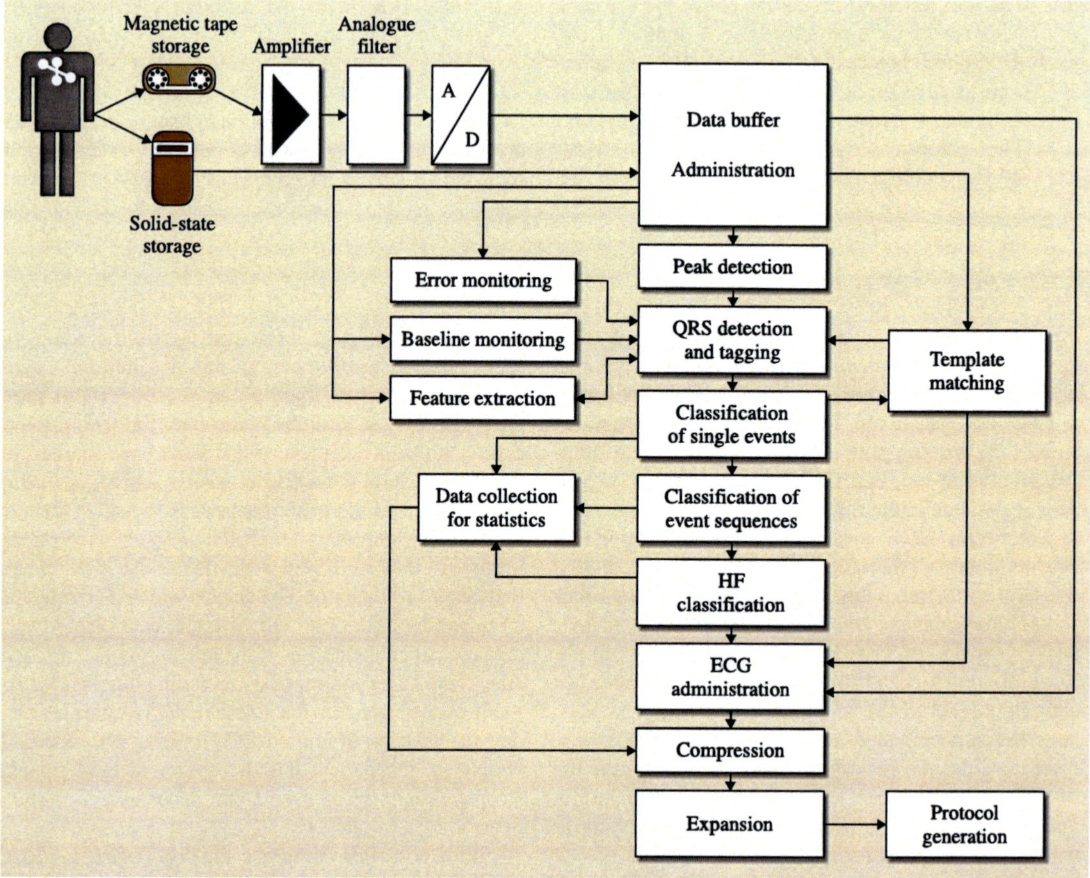

Fig. 7.13 Flow diagram of a computer-based assessment of the long-term ECG

a learning phase for the program: the QRS complex that is typical of this particular patient is calculated, and the QRS morphology is *learnt* and considered the normal curve shape, in order to be able to distinguish the QRS morphology from deviating QRS complexes and from artefacts. The arrhythmia program uses detection criteria (Fig. 7.13) for the QRS complex (a parameter comparison, also known as feature extraction or characteristics recognition) to draw a rhythmological distinction between physiological and pathological cardiac actions. The polarity of the R wave, the amplitude, the QRS area and position, the offset of the QRS from the baseline, and the slope of the R wave are calculated and determined as particular features for each QRS complex. The individual parameters are used at various points in the analysis for decision-making purposes. Thus, for example, the height of the complex provides an additional criterion when validating a QRS complex. A morphological (normal case or not) and rhythmological correlation (premature, delayed, later, etc.), as well as an individual complex assessment and ultimately a conclusion of arrhythmia, are possible using this feature determination. Continuous comparison between the current QRS complex and the learnt QRS is termed the cross-correlation method (or the template-matching method). The classification of each QRS complex forms the basis of the arrhythmia monitoring. The shape and time of occurrence of the complex are taken into account for this purpose. If the QRS shape that is typical of this particular patient is found after a learning phase, the next complex will be compared with the stored reference complex by *superimposing* the two complex shapes and checking point-for-point for a match. Where there is a low percentage of deviation, the complexes are defined as being identical in shape and are used as the reference complex for the next shape comparison. If there is no match (if the complexes have different shapes), this is stored in what is known as a QRS class memory.

In order to qualitatively assess diagnostic effectiveness and analytical accuracy, it is necessary to test analytical programs that have equipment-specific algorithms against recognized and evaluated ECG databases, and to compare the results (validation). The quality of the analysis, and in particular the reliability of the system, can be described statistically by the sensitivity, positive correctness and specificity:

- True positive = an event is correctly recognized and evaluated
- True negative = a normal complex is correctly recognized and evaluated
- False positive = an event is incorrectly recognized and evaluated
- False negative = an event is overlooked
- True positive and true negative = number of complexes
- False positive and false negative = number of errors

$$\text{Sensitivity} = \frac{\text{true positive}}{\text{true positive} + \text{false negative}} \times 100(\%)\,,$$

$$\text{Positive correctness} = \frac{\text{true positive}}{\text{true positive} + \text{false positive}} \times 100(\%)\,,$$

$$\text{Specificity} = \frac{\text{true negative}}{\text{true negative} + \text{false positive}} \times 100(\%)\,.$$

7.12 Heart Rate Variability and Heart Rate Turbulence

Prognostic indicators such as heart rate variability (HRV) and heart rate turbulence can be calculated from the recording of a long-term ECG. Whereas heart rate variability provides information on the variations in heart rate over a relatively long period of time, heart rate turbulence refers to the variations in the interval between two heart beats (RR interval, where R denotes the peak in the QRS complex) following the occurrence of a ventricular extrasystole. In order to assess the heart rate variability, the standard deviation of the RR intervals recorded over a period of 24 h is ascertained (SDNN). A spectral analysis of heart rate variability can be performed by most analytical devices from approximately 300 recorded sinus beats. Various frequency ranges can be discerned by means of fast Fourier transformation:

- ULF (ultra low frequency power): <0.0034 Hz
- VLF (very low frequency power): 0.0034–0.04 Hz
- LF (low frequency power): 0.04–0.15 Hz
- HF (high frequency power): 0.15–0.4 Hz.

7.13 Indications for Long-Term Electrocardiography

The range of indications for performing a long-term ECG has changed over the years. The most important indication is an investigation where there is suspicion of underlying cardiac arrhythmias (Class I indication). Other indications are risk stratification achieved by recording ventricular arrhythmias, and the monitoring of an anti-arrhythmic therapy. The guidelines of the American College of Cardiology (ACC), the American Heart Association (AHA) and the European Society of Cardiology (ESC) provide further possible applications.

7.14 The Significance of the Long-Term ECG

The long-term ECG is of great significance for primary cardiological diagnosis and the monitoring of therapy for arrhythmias, as well as for the diagnosis of ischaemia. The resting and exercise ECG usually do not allow a reliable judgement to be made on the frequency and nature of cardiac arrhythmias, since they record only a small period of time. A significant advantage of the long-term ECG is the continuous recording of the

ECG over a relatively long period of time (at least 24 h). In addition, the results of the analysis can be quantitatively measured, documented, or passed on for further data processing.

7.15 The Exercise ECG

Ergometry, the most commonly used exercise tolerance test in clinical and outpatient cardiology, is a noninvasive routine procedure in cardiovascular diagnostics.

By providing additional information to the resting ECG, ergometry yields essential insights into current cardiopulmonary performance and into limitations on performance which may result from coronary heart conditions.

The Difference Between Ergometry and Exercise ECG

Ergometry (from the Greek *ergon*, meaning work or performance) refers to the generation of a defined and reproducible stress with an ergometer per unit time, and the subsequent monitoring of the ECG, heart rate, blood pressure and oxygen consumption. On the other hand, in an exercise ECG, the focus is less on the blood pressure, heart rate and oxygen consumption and more on recording electrocardiographic changes during and after the period of exertion. Conventional ergometry generally does not include measurements of the oxygen consumption and saturation, whereas such measurements are absolutely essential in ergospirometry (see more Chap. 8, Sect. 8.2.5).

7.16 Equipment Technology

7.16.1 Physical and Technological Principles

The work performed by the test subjects during the exercise period at various stress stages is the product of the force (in newtons, N) and the distance (in metres, m). Including the time component, the output generated can then be measured in newton metres per second = watts = joules per second (Nm/s = W = J/s).

If we now consider a bicycle ergometer, when the bicycle peddle is rotated, a force must be expended over a distance. The speed at which the bicycle peddle is rotated is equal to the time in which this output is generated. As the bicycle pedal rotates, the output is thus calculated from $2\pi \times$ rotational speed × torque. The torque is calculated from the force with which the pedals are rotated multiplied by the length of the pedal cranks. In the process, it becomes clear that the performance changes depending on the force or rotational speed applied.

7.16.2 Ergometry Measuring Station

In order to avoid standalone systems and restrictions on the range of equipment that can be used, complete systems known as fully automatic ergometry measuring stations are currently available. These measuring stations consist of a 3-, 6- or 12-channel ECG device, an ergometer or treadmill, a 1-, 3-, 6- or 12-channel cardioscope, a defibrillator, a noninvasive blood pressure meter, an electrode application system and a portable equipment trolley that carries this equipment and thus enables flexible transportation of it (Fig. 7.14). When new equipment is acquired, conventional ECG devices are usually replaced with PC ECGs. The advantage of these systems is that they allow data storage and data transfer.

7.16.3 Types of Ergometers

In terms of technology and apparatus, there are three main types of ergometer: bicycle ergometers that are used in the sitting position, recumbent bicycle ergometers, and *treadmill ergometers (primarily in Anglo-Saxon countries)*.

Essential requirements for an ergometer – primarily bicycle ergometers – are:

- Ability to adjust the stress accuracy (max. ±5%)
- Accurate monitoring as a result of standardization and comparability
- 5 W should be the smallest stress that can be set
- Ability to display the stress and rotational speed
- Optimized adjustment options for different body sizes

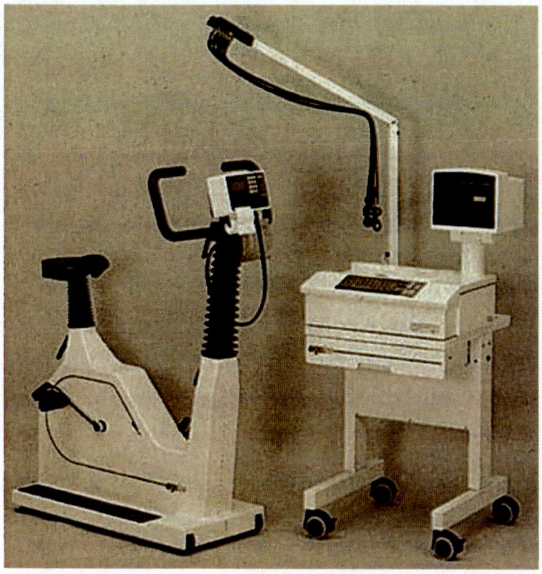

Fig. 7.14 Ergometry measuring station with bicycle ergometer (courtesy of ergoline Inc., Germany)

- Low requirements in terms of space, and high stability at the same time.

Rotational Speed Dependence of Bicycle Ergometers

In bicycle ergometers that are dependent on rotational speed, the output can be increased by both increasing the braking resistance and increasing the pedalling frequency. If a particular braking action (belts, brake shoes, weights or an electromechanical eddy-current brake) that predefines the force is set, a specific rotational speed must be maintained in order to be able to precisely determine the output. In practice, this situation is rather difficult to implement. It is for this reason (as well as the inaccuracies in the coefficients of friction for different braking systems) that bicycle ergometers which depend on rotational speed are unsuitable for ergometry. The use of these devices for medical investigations has therefore been banned. Bicycle ergometers that do not depend on rotational speed are predominantly used today. These are usually equipped with a computer-controlled eddy-current brake, and only the output to be attained is predefined; if the pedalling frequency is increased, the resistance decreases proportionally. The braking action is constantly compared with the rotational speed, the force expended and the predefined stress. Deviations are corrected electronically by increasing or decreasing the braking force. The accuracy is specific to the measuring system, and dependent on the consistency and the control rate. In good systems, deviations of between 1 and 3% in the stress set are observed; the standard allows deviations of up to 5%.

Characteristics and Criteria for Bicycle and Recumbent Ergometers

The following characteristics are important when assessing a bicycle or recumbent ergometer:

1. *Braking principle.* Computer-controlled eddy-current brake, independent of rotational speed, torque measurement.
2. *Stress range.* Dependent on the test subject clientele:
   a) Standard ergometer 20–450 W
   b) High-performance ergometer 20–1000 W
3. *Rotational speed range*:
   a) 30–100 /min
   b) 30–130 /min
4. *Stress accuracy.* No less than ±3 W or 3% (standard 5%).
5. *Stress stages.* Smallest stress which can be manually set of 5 W.
6. *Time interval.* Smallest time interval of 1 min.
7. *Exercise tolerance testing programs.* Can be programmed freely.
8. *Seat height adjustment.* Infinitely variable.
9. *Handlebar adjustment.* Infinitely variable.

7.17 Reduced Exercise ECG Leads

As with the other ECG leads, there are currently a vast number of ergometry lead programs. The are two main requirements for the reduced ECG leads during the exercise ECG:

- Minimal susceptibility to faults due to superposition of muscle potentials and movement artefacts
- Qualitative and quantitative registration of ST segment deviations.

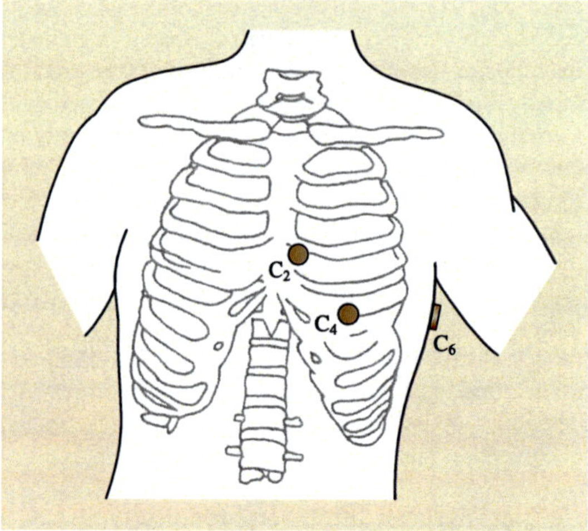

Fig. 7.15 Exercise ECG with reduced chest leads (C_2, C_4 and C_6) ◄

Six leads are generally sufficient. The measurements can be taken using multiple-use electrodes on tensioning straps and electrode belts. Appropriate use of single-use adhesive electrodes or electrode suction units is more hygienic, simpler and faster. The limb leads can be applied respectively to the right and left sides of the shoulder blade (head of the humerus) and in the lateral regions of the right and left iliac wings. These show the least superposition from muscle potentials. The chest leads are of particular interest, because ischaemic ST depressions and repolarization arrhythmias of the anterior wall appear to be most pronounced and occur most frequently in these leads. The most popular electrode sites for chest leads are C_2, C_4 and C_5 or C_6 (Fig. 7.15).

7.18 Automatic ST Measuring Programs

Elevation or depression of the ST segment in the ECG under stress allows diagnostic conclusions to be drawn regarding coronary heart conditions, so this is a major focus of measurements. Modern electrocardiographs, which usually provide an ergometry program, automatically measure and document the ST segment and amplitude. A distinction must be drawn between continuous and discontinuous ST measurement programs: in the case of continuous measurement, each QRS complex recorded is measured and analysed directly. An average beat is generated continuously for each channel from a number of beats that exhibit good correlation (usually 16). In contrast, in the case of discontinuous ST measurement, the average beat is generated from either values taken at timed intervals (e.g. per minute) or just from a start and end complex.

To obtain the average beat, the start and end point of the QRS complex are first determined. The ST amplitude and slope are measured at point $J+x$ (J point = junction point, which marks the end of the QRS complex and the transition to the full excitation phase: the ST segment). The $J+x$ interval is determined as a function of the heart rate. The isoelectric line is established between the P wave and the start of the QRS complex. The amplitude at point x is the interval from the calculated and displayed baseline. The slope at point x is given by the angle α, and is obtained by passing a regression line through the ST segment and point x.

7.19 Exercise Test

Since the diagnostic issues and the composition of the patient clientèle are assessed in very different ways, a uniform, standardized exercise tolerance test is not always possible. The guidelines compiled by the International Commission for the Standardization of Ergometry Application (ICSPE) have led to a largely standardized procedure in routine investigation and in sports medicine.

7.19.1 Stress Intensity

In ergometry, there are two different types of stress intensity in principle: maximal stress and submaximal stress. In the case of maximal stress, the exertion increases until the physical output limit of the test subject is reached (e.g. in sports medicine investigations, performance diagnosis, etc.). In the case of submaximal stress, exertion is continued until medically relevant problems occur or their correlation with the exertion can be ascertained (e.g. arrhythmias, ST elevation or depression, etc.). The level of submaximal stress is dependent on the medical problem, and may be arrived at long before maximal stress is reached.

Stress Steps

The Standardization Commission for Ergometry (ICSPE) has stipulated the following stress steps:

- 5 W for 1 min
- 10 W for 1 min
- 25 W for 2 min
- 50 W for 3 min.

Ergometry – according to standardized regulations – consists of a warm-up phase with a basic stress and at least three of these predefined stress steps as well as a recovery phase. The warm-up phase should stimulate circulation and prepare for the stress to come. It is generally of the same duration as the subsequent stress step. The wattage and the duration of the stress step are selected according to the physical capability of the test subject. A further criterion is that a so-called *steady state* is reached at the end of the stress stage. *Steady state* in this case means that at the end of the stress stage both the pulse and the blood pressure no longer deviate upwards – these values remain constant (equilibrium). (For more on the selection of the stress stage, see the section on PWC later.)

Stress Limits

The following is a rule of thumb for maximal stress with bicycle ergometry: a maximum heart rate of 220/min minus the age of the test subject. To ensure the safety of the test subject, the following so-called submaximal stress limit is recommended as a guide for routine investigations: a maximum heart rate 200/min minus the age of the test subject.

Assessment of Performance

To make sure that the ergometry procedure is suitable for diagnosis, it is very important to roughly estimate the performance of the test subject prior to the beginning of ergometry. The following rule applies to this:

The maximum nominal output for a man is 3 W/kg of body weight minus 10% for every decade over the age of 30. The maximum nominal output for a woman is 2.5 W/kg of body weight minus 8% for every decade over the age of 30.

As an example, let us consider the output settings for a 65 year old woman weighing 55 kg.
Nominal output: $(2.5\,\text{W} \times 55\,\text{kg}) - (8\% \times 3.5\,\text{decades}) = 137.5\,\text{W} - 28\% = 99\,\text{W}$; the maximum stress for this woman is therefore approximately 90 W. A basic stress and a stress interval must now be selected that will definitely enable the test subject to cope with three stress stages. In this case, the recommended approach would be to begin with a basic stress of 50 W and then carry out ergometer exercise with stress stages of 10 W and a stress duration of 1 min up to the maximum stress, which is expected to lie between 80 and 90 W. Another advantage of this method is that the output limit can be defined more precisely for test subjects that have comparatively low outputs with relatively low stress increase segments.

Pulse Working Capacity (PWC)

Derived from the rules for assessing performance and the rules for stress limits, the PWC method allows the comparison of outputs standardized to the pulse rate. The PWC value assesses the performance (i. e. the output achieved at specific pulse rates) in the submaximal range when the maximum rate has not been reached. The selected rate is added as an index (e.g. PWC_{150}). The nominal values for the PWC are not dependent on age; they are calculated based on weight and gender (for example, the nominal PWC_{150} value for men is 2.1 W/kg and for women 1.8 W/kg). These values can be found in PWC tables, which generally list PWC_{130}, PWC_{150} and PWC_{170} values.

Exercise Tolerance Test Procedure

A room temperature of 18–23 °C with a relative atmospheric humidity of 40–60% is recommended when performing an ergometry test. Aberrations should be logged. A resting ECG is recorded prior to the actual exercise tolerance test. It serves as a starting point when summarising the exercise tolerance test later. Following the application of the reduced ECG leads and the blood pressure cuff, the exercise tolerance test is initiated using a calibrated ergometer according to the exercise tolerance test method described. In modern ergometry measuring stations, the customary stress programs are stored in either the ergometer or the ECG device. In a timed ergometry investigation, one of these two devices serves as the command unit for the course of the investigation. In order to provide complete documentation, an intermediate ECG printout giving details of the heart rate, the blood pressure as well as the ST segment and the amplitude is automatically printed after each stress stage. However, it is also possible to initiate a manual ECG recording at any desired stage or to record an ECG trend over the entire duration of the test. The following actions must be performed during the exercise tolerance test:

- Constant observations of the test subject and the ECG graphs on the monitor
- The ECG strip as well as the blood pressure, the heart rate and the ST properties after each stress stage must be monitored.

Following the completion of the exercise tolerance test and a recovery phase, a resting ECG is recorded once more (and if necessary more than once), and the blood pressure is repeatedly monitored.

Types of Exercise Tolerance Testing

The general requirements for an evaluable exercise tolerance test can be summarised as follows: it should be capable of precise metering, physically measurable, and reproducible at any time. There are many types of exercise that do not satisfy these requirements, but they are mentioned below for the sake of completeness:

- *Forward bends and squats*. Insufficient for the exercise ECG because even the general requirements (*capable of metering and physically measurable*) cannot be satisfied.
- *Climbing stairs*. Unsuitable for the exercise ECG since the *reproducibility* must be questioned, there is no possibility of ECG monitoring, and the test subject cannot be subjected to maximum stress.
- *Hand crank ergometer*. Interference with the ECG recording as a result of the arm and thorax movements of the test subject and the lack of maximum stress (small muscle mass) are disadvantages of this method. However, this type of exercise tolerance test is often the only option for patients with impaired walking abilities.
- *Isometric stress (hand-grip)*. Similar disadvantages to those for the hand crank ergometer.
- *Master's step test*. The one-step or two-step test offers the following advantages: it is a relatively physiological form of exercise tolerance testing, involves a simple sequence of movements, is standardized and therefore reproducible, is simple to carry out, stresses a relatively large proportion of the total musculature, and requires a comparatively low financial outlay. Disadvantages: the timescale for maximum stress, it is virtually impossible to record the ECG and blood pressure during the exercise intolerance test, and its diagnostic informative value is very low when compared to those of bicycle and treadmill ergometry.
- *Kaltenbach and Klepzig's step-climbing test*. This test is a modification of Master's test, and the advantages and disadvantages are therefore similar.
- *Treadmill ergometer*. The treadmill is predominantly used for exercise tolerance testing of competitive athletes. The fundamental advantages of the treadmill ergometer are that it physiologically stresses the entire musculature (and this can be reproduced at any time), and it allows the stress intensity to be changed (by varying the speed at which the belt runs). Disadvantages are its relatively high investment costs, its structural requirements (in particular soundproofing), and a limited clientéle for exercise tolerance testing.
- *Bicycle ergometer*. The most popular and widely used type of exercise tolerance test is bicycle ergometry, which can be performed in a sitting position or in a recumbent position. Fundamental advantages of bicycle ergometry in the sitting position are the physiological stress and the use of the body weight; the procedure is also usually found to be pleasant by the test subject. Disadvantages are the recording of ECG and blood pressure during the exercise tolerance test and orthostatic complaints following the exercise intolerance test. The fundamental advantages of the recumbent ergometer are the unimpeded recording of the ECG and blood pressure and the quality of these recordings, the relatively high level of safety (predominantly for elderly or frail people), and the absence of orthostatic complaints following the exercise tolerance test.

Standardized Investigation Programmes

A number of individual and standardized exercise tolerance testing programmes from the early days of ergometry investigation are still employed. These are listed below, along with some examples:

- Hollman's standard test method (begins with 30 W and is increased every 3 min by 40 W up to the stress limit)
- The WHO's standard programme (begins with 25 W and is increased every 2 min by 25 W up to the stress limit)
- Knipping's vita maxima (begins with 10 W and is increased every minute by 10 W up to the stress limit)

- Kirchhoff's square wave test (stress duration of 10 min with 100 W in a sitting or recumbent position; blood pressure, heart rate and ECG are monitored every 2 min).

However, ergometry has increasingly evolved into a standardized method based on the internationally recognized regulations of the ICSPE. This has the advantage that results of investigations from different institutes and those obtained by different testing personnel can be compared and assessed according to the same standards. It is therefore the duty of all testing personnel to adhere to these regulations.

7.20 Methodological Notes

Most faults that occur during ergometry can be attributed to the patient. A large proportion of these faults are compensated for by modern equipment technology. Nevertheless, the test subject can be made aware of the fact that the necessary pedalling work can be carried out without excessive movements of the upper body. Incidents are extremely rare in the case of ergometry when performed correctly, and routine investigations are today therefore predominantly carried out by well-briefed medical assistants. The only ergometric investigations that must be supervised by the doctor over the entire duration of the exercise tolerance test are those in which – due to the general condition of the test subject – complications cannot be ruled out. A nurse or medical assistant must be present at the same time in order to assist or to provide immediate assistance if necessary.

An emergency supply of equipment and medication as well as a functioning defibrillator must be available to hand at all times.

7.21 The Diagnostic Value of Ergometry

ST segment variation during the exercise ECG is of the utmost importance for the detection of temporary myocardial ischaemias. In addition, coronary heart conditions can be established unambiguously. Coronary angiography has made an essential contribution to improving the interpretation of the exercise ECG, although there is a fundamental difference between the two: coronary angiography detects the morphological changes in the coronary arteries, whereas the exercise ECG detects the functional effects.

7.22 Indications

An indication for an exercise ECG can be given for diagnostic, therapeutic or prognostic reasons:

- Investigation of a coronary heart condition (e.g. chest pain during physical exertion)
- Investigation of the blood pressure where there is suspicion of exercise-induced hypertension
- Investigation of the stress tolerance (e.g. following a myocardial infarction, etc.)
- Assessment of cardiac arrhythmias
- Assessment of the progress and therapeutic success of medication and exercise therapy and following heart surgery
- Assessment and classification of the degree of severity of stress-induced coronary insufficiency.

7.23 Abort Criteria and Safety Measures

Exercise tolerance tests cannot be performed without risks (e.g. myocardial infarction, life-threatening arrhythmias, etc.). In order to minimize the risks, safety measures should be taken before and after the exercise tolerance test. Before the exercise tolerance test, information from the resting ECG should be analysed, including anamnesis, and the patient should be examined and informed. It is important to tell the patient to report any subjective complaints during the exercise tolerance test, and to continue to monitor the test subject following the exercise tolerance test (generally for 5–10 min); a monitoring ECG should be recorded again and the blood pressure should be measured.

Criteria for aborting ergometry are:

- Increasing retrosternal severe pain (angina pectoris)
- Horizontal or descending ST depression of >0.2 mV
- Cumulative monomorphic or polytopic ventricular extrasystoles
- Ventricular salvos
- Isolated extrasystoles that fall in the T wave of the previous cardiac action (R-on-T phenomenon)
- Atrial fibrillation or flutter
- Serious conduction defects (e.g. total AV block)
- Depolarization defects (e.g. bundle branch block)
- Systolic blood pressure >250 mmHg, diastolic blood pressure >130 mmHg, and drop in blood pressure
- Noticeable breathing difficulty (dyspnea)
- Signs of the beginning of left ventricular insufficiency
- ECG signs of a fresh myocardial infarction.

7.24 Technical Safety Aspects

Bicycle ergometers are subject to the provisions of the German Medical Devices Act (IIa) and DIN VDE 0750-238. They must be serviced regularly, as must the noninvasive blood pressure meters that may be installed on them.

7.25 Notes on Planning

A room temperature of 18–23 °C with a relative atmospheric humidity of 40–60% is recommended. As a guide, a floor space of area 16 m^2 would be suitable. There should be the option to get changed (changing cubicle ≈3 m^2), and bathroom facilities with WC/shower (approx. ≈5 m^2) should also be provided.

Further Reading

- A. Bolz, W. Urbaszek: *Technik in der Kardiologie* (Springer, Berlin Heidelberg New York Tokio 2002)
- G. Csapo: *Konventionelle und intrakardiale Elektrokardiographie*, 4th edn. (Documenta Geigy, Wehr 1980)
- E. Erdmann: *Klinische Kardiologie* (Springer, Berlin Heidelberg 2006)
- R. Klinge: *Das Elektrokardiogramm*, 8th edn. (Thieme, Stuttgart 2002)
- R. Kramme: Elektrokardiographie. In: *Handbuch der Medizintechnik*, ed. by W. Menke (ecomed, Landsberg 1991)
- K. von Olshausen: *EKG-Information*, 8th edn. (Steinkopff, Darmstadt 2005)
- J.G. Webster: *Encyclopedia of Medical Devices and Instrumentation*, Vol. 2 (Wiley, New York 1988)
- H.H. Börger: *EKG-Informationen*, 6th edn. (Steinkopff, Darmstadt 1994)
- G. Csapo: *Konventionelle und intrakardiale Elektrokardiographie* (Documenta Geigy, Wehr 1980)
- E. Erdmann: *Klinische Kardiologie* (Springer, Berlin 2006)
- K. Jung: Ergometrie in Klinik und Praxis, Wehrmed. Wehrpharm. **1**, 61–72 (1986)
- H. Hochrein, P. Bentsen, C. Langescheid, D. Nunberger: *Checkliste Kardiologie* (Thieme, Stuttgart 1988)
- H. Löllgen: *Praxis der EKG-Beurteilung* (Boehringer, Ingelheim 1991)
- H. Mellerowicz, I.W. Franz: *Standardisierung, Kalibrierung und Methodik in der Ergometrie* (perimed, Erlangen 1983)
- H.-J. Trappe, H. Löllgen: Leitlinien zur Ergometrie, Z. Kardiol. **89**, 821–837 (2000)

8. Pulmonary Function Testing

Rolf M. Schlegelmilch, Rüdiger Kramme

Pulmonary function tests (PFT) examine the functionality of the lungs. This chapter describes the working principles of major test techniques and instruments including spirometry, peak flowmetry, body plethysmography, nitrogen washout, and ergospirometry. Spirometry (Sect. 8.1), specifically the measurement of the amount (volume) and speed (flow) of air that can be inhaled or exhaled, is the most basic of the pulmonary function tests. Body plethysmography (Sect. 8.2.2) offers determination of absolute lung volume and airway resistance and is particularly sensitive in the detection of obstructive pulmonary disease requiring little patient cooperation. Nitrogen washout (Sect. 8.2.4) is a common test for measuring functional residual lung capacity. Ergospirometry evaluates the complex interaction between lung, heart, and muscle by analysing the gas exchange under exercise (Sect. 8.2.5). The chapter closes with considerations of planning and laboratory space (Sect. 8.2.8).

8.1 **Spirometry** 95
8.1.1 Flow Recording 96
8.1.2 Technology 96
8.1.3 Methodology of Spirometry 101
8.1.4 Cooperation-Free Pulmonary Function Tests 104

8.2 **Advanced Cardiopulmonary Function Testing** 105
8.2.1 Overview 105
8.2.2 Body Plethysmography 106
8.2.3 Diffusion Capacity 109
8.2.4 Nitrogen-Washout Test for Determination of FRC 111
8.2.5 Ergospirometry 111
8.2.6 Noninvasive Determination of Cardiac Output 114
8.2.7 Metabolic Activity Monitoring 115
8.2.8 Planning and Laboratory Space Considerations 116

References 116

8.1 Spirometry

Spirometers are noninvasive diagnostic instruments for screening and basic testing of pulmonary function. Offering essential diagnostic insight into the type and extent of lung function impairment, spirometry tests can be performed fast at fairly low cost. In the light of an ever-increasing prevalence of airway diseases such as asthma, bronchitis, and emphysema, pulmonary function instruments have become indispensable diagnostic tools, in clinical and office settings, in industrial and preventive medicine, as well as in epidemiology. Screening of individuals at risk, basic testing of sick patients, and treatment follow-up are key applications of spirometry.

Two essential questions of pulmonary function testing (PF testing) can be answered by spirometry:

1. What is the size of lung volume which can be inspired *or* expired?
2. What is the time it takes to exhale this volume, or what is the flow rate during exhalation?

Flow rates and resulting volumes are measured by connecting a spirometry sensor through a mouthpiece to the test subject's mouth. The most common and internationally standardized test consists of an evaluation of forced expiration after a complete inhalation, allowing the determination of forced vital capacity (FVC)

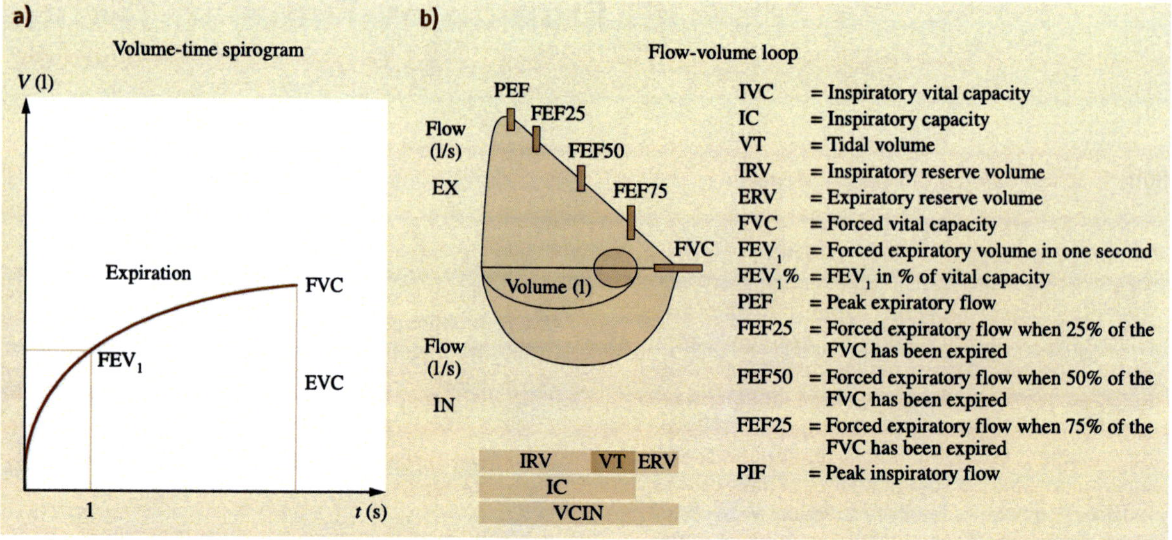

Fig. 8.1 (**a**) Forced spirogram and (**b**) flow–volume loop

and the forced expired volume during the first second (FEV_1). Recording of the test trace is taken as a forced spirogram (volume over time) or as a flow–volume loop (flow against volume). Although FVC and FEV_1 are the most common, dozens of parameters can be derived when evaluating forced expiration, all describing the shape and size of recorded traces and loops (Fig. 8.1). Besides forced spirometry, slow spirometry, i.e., the recording of slow inspiration and expiration at tidal breathing, may also be recorded, offering determination of lung-volume subdivisions such as tidal volume (VT), inspiratory and expiratory reserve volume (IRV and ERV), as well as inspiratory capacity (IC). In most cases, slow spirometry will be a part of advanced PF tests (Sect. 8.2.2).

8.1.1 Flow Recording

The basis of all PF tests consists in the recording of air flow rates against time, a procedure also called pneumotachography. Mostly, flow velocity [cm/s] is measured in a defined sensor geometry, yielding flow [l/s], also referred to as flow rate. In 1925 the Swiss physiologist Alfred Fleisch published about the clinical application of pneumotachography. While registration of spirometry traces and flow–volume loops required a recorder and manual evaluation in the past, all contemporary spirometers contain microprocessors for online data processing, displaying results on a built-in screen or transferring data to a personal computer. Measured flow is continuously accrued to lung volume by digital integration. Independent of the size and cost of any pulmonary function device, from a handheld peak flowmeter to a body plethysmograph, the flow sensor always represents the core element (Fig. 8.2).

8.1.2 Technology

While *closed systems* such as the bell or wedge spirometer were still in use two to three decades ago, only *open systems* will be found today, offering smaller footprint and size, higher precision, and more comfortable and hygienic conditions for the patient. In an open system the patient breathes from and to the ambient air through a sensor, which in case of more advanced testing can be connected to one or more valves. The spirometer detects the flow rate from the sensor and digitally integrates flow to volume

$$V = \int \dot{V} \, \mathrm{d}t \text{ (l)} ,$$

where V denotes the inspired or expired volume, and $\dot{V} = \mathrm{d}V/\mathrm{d}t$ (l/s) is the measured flow.

Figure 8.3 depicts a typical pneumotachogram (flow over time) and the resulting spirogram (flow integrated volume). The recording of breathing traces and the calculation of derived parameters are usually performed by an integrated microprocessor or a personal computer.

Measuring Range

When testing adults, the physiologically relevant flow range should reach from 25 ml/s to 10 l/s, optimally from 10 ml/s to 15 l/s, requiring a resolution of 5–10 ml/s. These conditions impose very demanding technical specifications on flow sensors. For precise detection of the first part of the forced exhalation, a linear frequency response of up to 5 Hz is required in the flow sensor characteristics.

Verification and Calibration

To verify the accuracy of a flow sensor and if needed to calibrate it, an optional verification and calibration routine should be implemented in any spirometer. Both verification and calibration are carried out by means of a manual or motorized pump of known volume of at least 2 l. When activating the calibration pump connected to the flow sensor, the spirometer will read integrated instantaneous flows. At the end of a full stroke, the spirometer should display a flow integrated volume identical to that of the pump. As digital integration can be assumed to be sufficiently precise, any deviation in volume will indicate the error in flow measurement. By moving the piston at varying speeds, flow rates between −10 l/s and +10 l/s can be simulated, covering most of the physiological flow range. Any deviation between a stroke volume resulting from low flow rates versus that produced at high flow rates will indicate a nonlinearity in the flow sensor or spirometer circuitry. Today, most spirometers offer a calibration program facilitating the calibration routine. Averaging over a given number of pump strokes, the deviation from the correct volume is displayed, and a correction factor will be considered in future measurements. Verification and calibration must be carried out without regard of body temperature pressure saturated (BTPS) correction (see later), as pump volume is measured at ambient not at lung conditions.

In recent years, disposable spirometer sensors have become common. Even if these sensors are precalibrated or calibration data are read into the spirometer, a verification routine should be available to allow for quality assurance. The same applies to the newer digital sensor systems which are advertised as calibration free. Nevertheless, accuracy and linearity should be verified at reasonable intervals.

The American Thoracic Society (ATS) and European Respiratory Society (ERS) have issued recommendations [8.1, 2] mandating testing of spirometers by independent laboratories prior to market release. By subjecting production models to a battery of tests, manufacturers can document the overall performance of their spirometers to the user. Test laboratories employ motorized, processor-controlled calibration pumps producing standardized flow–volume loops, able to simulate a wide range of clinically relevant flows. This testing procedure is part of the obligatory premarket registration procedure in the USA [Food and Drug Administration (FDA) approval]. One can assume that spirometers which are also sold in the USA fulfill these requirements. Before purchasing a spirometry device, documented compliance with the recommendations of the ERS and ATS spirometry standards [8.1, 2] should be requested from the manufacturer or supplier.

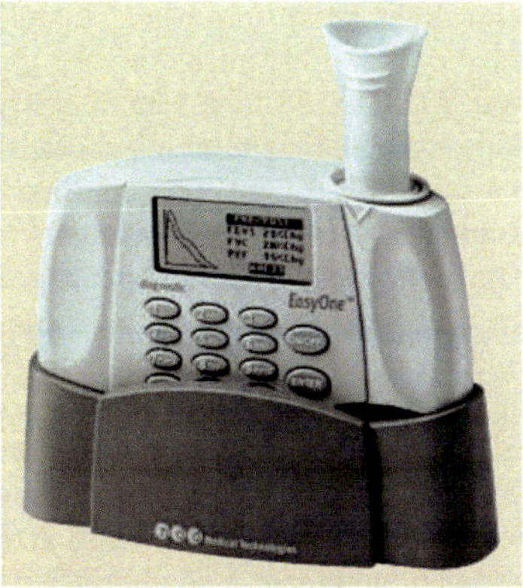

Fig. 8.2 Pneumotachograph (ndd Medizintechnik, Zurich, Switzerland)

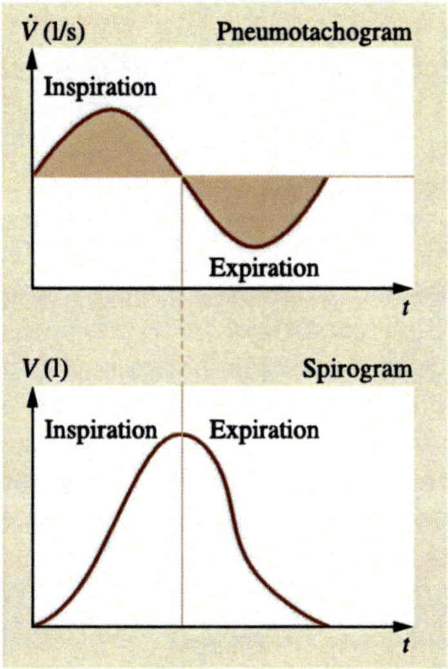

Fig. 8.3 Comparison of pneumotachogram and spirogram

Hygiene

Cross-contamination is an ever-present risk in spirometry. To effectively avoid the risk of infection in patients or individuals tested with the same spirometer, suitable preventive measures should be instituted. In case the flow sensor is a fixed installation which cannot easily be exchanged, a disposable breathing filter is indispensable. Like the sensor itself, the breathing filter must have a low resistance, minimizing any hindering effect on breathing. Usually the flow sensor can be exchanged and replaced by one that has been cleaned and sanitized. In regular clinical testing, sensors do not need to be sterile; disinfection according to manufacturer's recommendations is sufficient. Many modern spirometers incorporate disposable sensors, favorably combining adequate hygiene with ease of use. As single-use sensors should not cost much more than a good breathing filter, their design is a contradiction in terms of low cost versus good performance (in accuracy and resolution, see above). In this regard a disposable item that is not part of the sensor itself may be advantageous, as it ensures hygiene while not affecting accuracy. The ERS has published recommendations concerning the resistance of spirometry sensors and breathing filters, calibration, and hygiene of spirometers [8.2].

Conversion to Body Conditions

An *open system* requires conversion of measured flow and integrated volume from ambient to lung or BTPS conditions (37 °C, 760 mmHg, 100% relative humidity). The conversion needs to correct inspiratory and expiratory flows and volumes separately. While, during inspiration, 1 l of ambient air expands to about 1.1 l of lung air due to heating and humidification, 1 l of lung air will shrink during expiration to about 0.97 l at the sensor, both corrections depending on the sensor type and the geometry of the mouthpiece. Without BTPS correction, substantial errors in the range of 10% in flow and volume parameters may occur. Conversion formulae can be found in [8.2]. In all commercially available spirometers, conversion to BTPS conditions is implemented, automatically correcting data to BTPS conditions.

In Table 8.1, the measuring principles of currently commercially available flow sensors, their pros and cons are presented.

Differential Pressure Flowmeter

The classical method to measure respiratory flow consists in the acquisition of the differential pressure ($p_1 - p_2$) in a flow tube where a resistive element is inserted. The type of the resistor characterizes the sensor principle, and usually it will lend its name to it.

As the airstream leaving the trachea and mouth is turbulent, a short tube or a breathing filter should be inserted between the flow sensor and the mouth, acting as a flow rectifier to *laminarize* the flow pattern before entering the sensor. Only laminar flow will provide a pressure drop across the resistor which is proportional to flow rate. Depending on flow direction, the differential pressure across the resistor will be positive (expiration) or negative (inspiration). The pressure is picked up by a pressure transducer and converted into an electrical and/or digital signal. All flow-measuring devices based on differential pressure transducers are sensitive to baseline instability. Even the smallest asymmetry in the transducer leads to a drift in the flow

Table 8.1 Types of flow sensors

Type	Pros	Cons
Pneumatachograph according to Fleisch	Proven measuring principle, durable	Several sensors for different flow ranges required, sensitive to contamination, difficult to clean, frequent calibration required
Pneumatachograph according to Lilly	Proven measuring principle, durable, cleaning possible	Sensitive to contamination, calibration required
Diaphragm pneumotachometer	Proven measuring principle, durable, lightweight, simple cleaning	Calibration required
Turbine flowmeter	Small, lightweight sensor	Frequent cleaning and calibration required
Hot-wire anemometer	Offers good accuracy	Simple anemometer not accurate, calibration required
Vortex flowmeter	Simple sensor construction	Not proven in routine testing, only unidirectional, difficult to calibrate
Ultrasound flowmeter	Simple sensor construction, small, lightweight, high accuracy, calibration free	

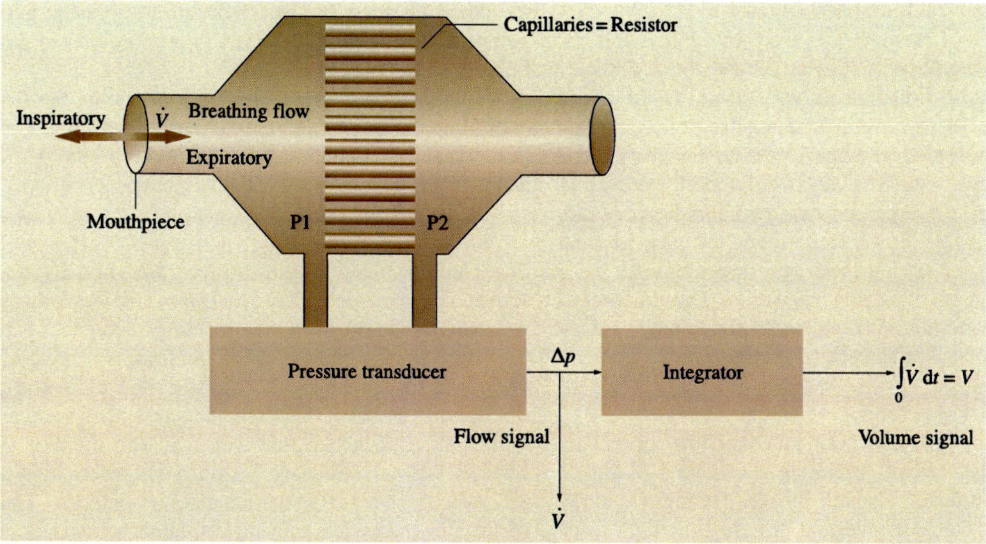

Fig. 8.4 Pneumotachograph according to Fleisch (schematic)

signal and consequently to a drift in volume. Even rather minimal flow drifts can – over time – integrate into appreciable volume offsets. To avoid this effect, a flow range around baseline needs to be defined in which no flow integration takes place (*dead zone*). As flows at the end of expirations can be rather small but are of diagnostic significance in obstructive disease, the dead zone should be kept as small as possible and may not exceed 10–15 ml/s.

The Fleisch Pneumotachometer

Until the 1980s, the pneumotachometric principle according to Fleisch was the most widely used flowmeter in PF testing, also called the Fleisch pitot tube (Fig. 8.4). The sensor consists of a bundle of capillaries, typically a roll of z-folded sheet metal acting as a fixed resistor. The bulk of flow is subdivided into small threads of flow, laminar over a certain range. The pressure difference along a capillary can be derived by applying the Hagen–Poiseuille law, which in case of laminar flow yields

$$\Delta p = c \cdot \dot{V} = \frac{8\eta l}{r^4}\dot{V} ,$$

where c is a constant dependent on the gas viscosity and sensor geometry, l is the length of the sensor tube, r is the radius of the sensor tube, η is a viscosity constant, and $\dot{V}$ is the flow rate.

In the Fleisch tube, linearity between pressure difference and flow rate is attained by use of a large number of bundled capillaries. As long as mechanical signal recorders were used, linearity in flow assessment was required to allow for direct reading of the flow rate from the trace. Today, due to fast digital online processing, also nonlinear sensor principles can be implemented for flow determination and recording. Despite the use of modern technology, sensor calibration and accuracy need to be checked regularly. Since the Fleisch tube and all other pneumotachographs employing some type of resistor are directly exposed to the exhaled air, sputum, even a small droplet, may change the resistance, leading to considerable errors in indicated flow rate. Therefore, a bend is inserted between the mouth and sensor both to retain sputum droplets and to laminarize the airstream. Alternatively, a disposable air filter can be inserted to avoid contamination and safeguard hygiene (Sect. 8.1.2).

The Lilly Pneumotachometer

Over a tube restriction, a pressure drop proportional to flow can be registered, up to a Reynolds number of ≈ 10. This relation also holds when a combination of restrictors, such as a close-mesh screen, single or in a series, characterizes the flow sensor. An essential feature of this design is the large difference between tube and mesh diameter. In practice, a layer of several sieves improves the linearity of the flow measurement but increases the breathing resistance, which needs to be considered.

Diaphragm Pneumotachometer

In this type of flowmeter a specially slotted foil diaphragm is employed, acting as the resistor. A pressure drop across the diaphragm will occur, being either proportional to flow over a certain range or needing linearization over a larger range. Depending on design, the diaphragm will exhibit a larger differential pressure at lower flows, an overproportionally smaller pressure difference at higher flows. Certain applications, such as exercise testing or intensive care measurements, may benefit from such varying resolution. The diaphragm is less prone to humidity and contamination by sputum and may be cleaned more easily.

Turbine Flowmeter

A turbine or propeller built into the flow tube characterizes this sensor, which is also called a digital volume transducer. Excited by the passing gas flow, the rotating elements interrupt or reflect the light from a light-emitting diode (LED). Photodiodes register the rotations, returning an electrical impulse frequency proportional to flow, while the total count is proportional to volume.

To minimize inertia, the turbine bearing resistance and turbine mass have to be as small as possible. Contemporary turbine flowmeters can be dismantled easily for cleaning.

Hot-Wire Anemometer

This sensor consists of a Venturi tube in which two hot wires are installed, linked in an electric bridge circuit (Wheatstone bridge). While one wire, consisting of two types of platinum, senses gas temperature, the other measures the heat flux into the passing gas. Temperature-corrected heat flux is proportional to gas flow.

A more modern version of this principle, the so-called mass flowmeter, is able to measure the amount of gas molecules passing the hot wire, featuring independence from gas temperature and mixture or viscosity.

Vortex Flowmeter

The phenomenon of fluid vortices was already described by Leonardo da Vinci as early as 1518. Vortices are usually caused by waves meeting a resistance in flow. In a vortex flowmeter the air flow is channeled towards a resistive element, the so-called bluff body, where vortices occur. The type and extent of vortices formed depend on the flow velocity. The number of vortices can be evaluated by different technologies: piezoelectric elements, thermistors, or optodes. The vortex flowmeter can only detect unidirectional flow and needs an independent sensor, such as a pressure sensor, for detection of flow direction. Therefore, this principle finds only limited acceptance in spirometry.

Ultrasound (Transit Time) Flowmeter

This method (Fig. 8.5) is based on the determination of transit times of acoustic waves traveling through a flowing medium. The velocity of acoustic waves across a flow tube increases at the same rate as the velocity of air in that tube. Three technical realizations exist:

1. Continuous phase shift analysis, where the phase shift of a continuously generated ultrasound signal is determined.
2. Impulse phase shift, where the phase shift between sinusoidal ultrasound waves is measured.
3. Time-of-flight impulse, where the transit time of the ultrasound traveling between a transmitter and receiver is determined.

While the principle of determining phase shifts has not been commercialized in medical technology, the time-of-flight method [8.3] has been successfully introduced in spirometry, with wide acceptance in clinical practice and epidemiology. The sensor technology employs two ultrasound elements which are built into the sensor housing and placed oblique to the airstream in a flow tube. Alternating ultrasound impulses are emitted in both directions, and their transit times across the tube are measured. While transit times are shorter downstream, they are longer upstream. The difference in transit time is proportional to the flow velocity, independent of any other variable such as gas tempera-

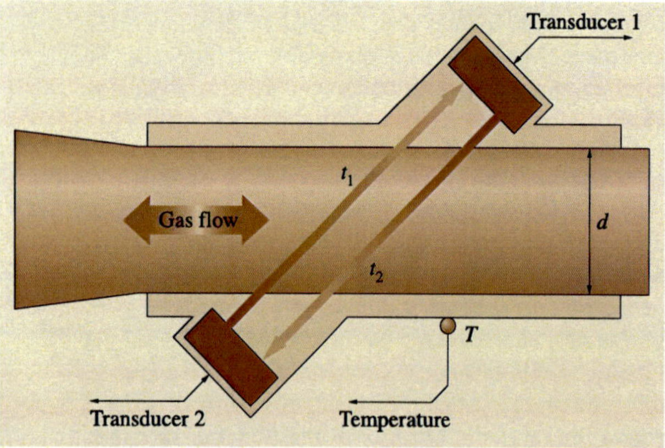

Fig. 8.5 Working principle of an ultrasound flowmeter (ndd Medizintechnik, Zurich, Switzerland)

ture, viscosity or humidity. In a commercially available ultrasound flowmeter a disposable breathing tube is employed, featuring sealed windows across from the ultrasound elements. While the windows are transparent to ultrasound, germs cannot pass, effectively avoiding cross-contamination. In addition to being calibration free, this sensor design also has the advantages of high accuracy and total hygiene. While the difference in transit time yields the flow, the sum of the transit times is directly related to the molar mass, the molecular or specific weight of the breathing gas, allowing instantaneous reading of the specific weight of the gas passing through the sensor. During expiration, the molar mass changes according to the composition of the exhaled gas, similar to a CO_2 waveform (Sect. 8.1.4).

Peak Flowmeter

For monitoring obstructive diseases such as asthma and for analyzing therapy response, peak flowmeters play an important role, in particular for home use. The classical mechanical peak flowmeter consists in a piston moving in a scaled flow tube. A forced expiration moves the piston up the vertically held tube, allowing a reading of maximum peak expiratory flow (PEF) in units of l/min or l/s. More modern peak flowmeters implement one of the above-mentioned measuring principles, most often turbines. Usually they offer electronic signal processing, display of results, memory and diary function, as well as data logging through a modem or the Internet. When applied over a longer period of time, verification and calibration of PEF meters should be warranted.

8.1.3 Methodology of Spirometry

Lung-Volume Subdivisions

Pulmonary gas transport depends on the filling capacity of the lungs with breathing gas and the speed and uniformity of gas distribution. Markers of lung filling are the lung-volume subdivisions, which are measured at slow breathing, while flow rates are of minor interest. When two or more volume subdivisions are combined, *lung capacities* result. The most important role in PF testing is played by vital capacity (VC), the volume of complete inspiration (IVC) or slow expiration volume (EVC), both directly accessible by spirometry. At the end of a deep expiration, residual lung volume (RV) remains in the lungs, which can be determined by gas dilation methods (Sect. 8.2.4) or body plethysmography only (Sect. 8.2.2). Consequently, the sum of VC and residual volume (RV) determines the volume at the end of a complete inspiration and is called the total lung capacity (TLC), an important measure of maximum lung volume.

Vital capacity consists of the subdivisions:

- Tidal volume (V_T), the volume ventilated during a regular breathing cycle
- Expiratory reserve volume (ERV), the volume that can be exhaled from breathing baseline, i. e., at the end of a regular breathing cycle
- Inspiratory reserve volume (IRV), the volume which can be inhaled above the inspiratory breath of a regular V_T cycle
- Inspiratory capacity (IC), i.e., the sum of VT and IRV.

As for TLC and RV, the functional residual capacity (FRC), being the sum of RV and ERV, can be determined by more elaborate methods only (Sect. 8.2). As determination of lung-volume subdivisions is time consuming and diagnostically less significant, subdivisions will be evaluated in combination with the more relevant FRC.

For standardization of lung volume testing refer to [8.4].

Forced Spirometry

Of the diagnostic arsenal of internal medicine, dynamic lung volumes, in particular forced vital capacity and forced expiratory volume in the first second (FEV_1), are considered essential parameters. Forced expiration not only delivers important information about an existing pulmonary obstruction in the sense of reduced airway diameter but may also indicate a loss in lung retraction, parallel to diminished lung elasticity and enhanced airway instability. Reduced elasticity mirrors the loss of functional tissue structure, resulting in reduced surface area for gas exchange and accompanied by a high demand in ventilation at reduced maximum oxygen uptake.

In contrast to the predominantly practiced forced expiration, forced inspiration can be used for differential diagnosis of extrathoracic obstruction, e.g., tracheal stenosis.

Being technically less demanding, the recording of a forced spirogram prevailed traditionally; today the flow–volume loop is registered together with the spirogram. The time-based spirogram offers observation of expiratory time, typically 3 s in a healthy subject but maybe as long as 10–20 s in an obstructed patient. As the forced spirogram represents the shape of an exponential function, analysis of its form requires a certain understanding of the underlying pulmonary

Table 8.2 The most important parameters of the forced spirogram and flow–volume loop

Parameter	Unit	Description
FVC	l	Forced vital capacity
FEV1	l	Forced expiratory volume in 1 s
FEV1/FVC	%	FEV1 in % of FVC or relative FEV1
FEVt	l	e.g., FEV0.5, FEV2, FEV3
PEF	l/s	Peak expiratory flow
FEF25-75	l/s	Mean forced expiratory flow between 25% and 75% of FVC
FEF25	l/s	Forced expiratory flow when 25% of FVC has been expired
FEF50	l/s	Forced expiratory flow when 50% of FVC has been expired
FEF75	l/s	Forced expiratory flow when 75% of FVC has been expired
MTT	s	Mean transit time
FIV1	l	Forced inspiratory volume in 1 s
FIF50	l/s	Forced inspiratory flow at 25% of FVC

mechanics. In contrast, the flow–volume loop represents a linearization of the exponential expiratory decay, as an exponential and its derivative are recorded against each other. In a healthy person the rise to peak flow is steep, while the decay is (quite) straight. The very obvious deviation from a straight line towards a concave shape means a reduction in flow, which is easy to interpret.

Table 8.2 presents the most important parameters of the forced spirogram and flow–volume loop. The list of parameters derived from forced expiration could be extended (almost) endlessly. Beside the numerical evaluation of the flow–volume loop, attention should be drawn to the analysis of its shape (Fig. 8.6).

Documentation

Online recording and evaluation of the flow–volume loop has become a gold standard in basic pulmonary function testing. When testing a patient, several trials should be registered to assess cooperation and repeatability. Loop tracings should be recorded with sufficient resolution. The aspect ratio, i.e., the flow versus volume scaling, is of importance for form analysis, e.g., in pre/post broncholytic drug medication, and should be fixed to pronounce the difference between a healthy and an obstructive loop.

Practical Considerations

The quality of most pulmonary functions tests, in particular forced expiration, depends on the cooperation of the test subject and the (incentive) instructions of the operator. After preparing the spirometer, verification or calibration, and insertion of a new mouthpiece or breathing filter, the testing procedure is explained to the subject. Before and during spirometry, breathing maneuvers are further supported by the operator, and the subject should be motivated to maximize forcing efforts and exhale completely. The subject, wearing a nose clip, should be able to breathe free and unhindered through the mouthpiece and should be encouraged to practice before recording starts. All tests should be carried out in the same body position, generally sitting. In screening and industrial testing, also a standing position is adequate. In children and seniors, use of a mouthpiece with a sealing lip is recommended. A smaller, pediatric mouthpiece should be applied in children. Unless disposable flow sensors are used, a disposable filter attached to the sensor is mandated, both as a hygienic requirement and a preventive measure to avoid intrusion of sputum into the flow sensor.

Both the ERS and ATS have issued recommendations in which at least daily volume calibration using a calibration syringe is mandated. In a calibration-free device, volume verification should be carried out at least once a month.

Slow Vital Capacity and Volume Subdivisions Instructions

The subject breathes at resting baseline. After a couple of breaths, the operator gives instructions for a slow, continuous, maximum inspiration followed by a slow, maximum expiration, followed again by a complete inspiration. The cycle should be repeated again. Parameters from repeated tests should not deviate by more than 5%.

Forced Spirometry and Flow–Volume Loops Instructions

Breathing at resting baseline, the subject is instructed to inhale completely and exhale as hard and as long as possible. This procedure should be repeated at least twice.

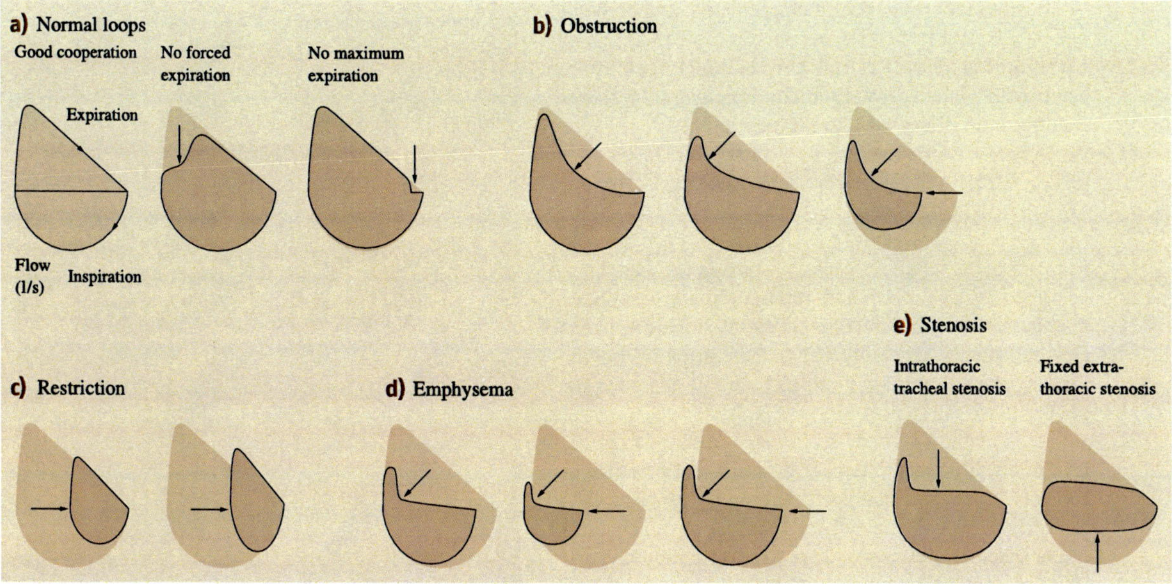

Fig. 8.6a–e Typical flow–volume loops: **(a)** normal, **(b)** obstruction, **(c)** restriction, **(d)** emphysema, and **(e)** stenosis

Breathing maneuvers need to be instructed clearly; especially the forced exhalation should be supported by incentive commands. For evaluation, maximum forced flow rates (FEFs) are picked from the recording. Contemporary software programs offer warnings to the operator in case quality standards (ERS, ATS) are not met. Either best values are taken from all forced traces or the trace with the highest sum of FVC and FEV_1 is considered for analysis.

Several errors may occur: Poor cooperation of the test subject may lead to a late or low PEF (Fig. 8.6a). In general, low FEFs may result in case no nose clip is worn or the lips did not entirely seal the mouthpiece. Deformation of loops and traces may result from coughing or swallowing. If no FEV_1 is given, exhalation was too short or interrupted. Varying loops indicate poor instruction and motivation. With good cooperation, all forced flow–volume loops should superimpose adequately.

After analyzing the results of forced spirometry, a first classification into normal, obstructive, or restrictive findings can take place (Fig. 8.6a–e). In frequently evidenced obstructive disturbances, even the severity can be assessed.

Predicted Values

Evaluation of spirometry testing is carried out by comparing measured data with predicted *norms* or reference data, derived from body weight, height, age, and gender. Predictions are calculated from equations published and recommended by scientific societies. When comparing measured with predicted values, the standard deviation, an indicator of the variation of the tested parameter in a healthy population, needs to be taken into consideration. The determination of the residual R

$$R = \frac{M - S}{\mathrm{SD}}$$

is recommended, where M is the measured value, S is the predicted value, and SD is the standard deviation, allowing comparison with a distribution in the reference population and detection of significant deviations.

The leading international societies, the ERS and ATS [8.5], have issued predicted formulae, mostly derived from Caucasian populations. These European and North American predicted values do not substantially differ from each other. When assessing pulmonary function in other ethnicities, predictions of national societies or other scientific sources should be used, or at least a percentage adaptation of Caucasian predicted values be considered.

Most processor-based test devices are equipped with different sets of prediction formulae that can be selected by the operator. The user manual should give the national background and literature references of the predicted sets.

Besides reporting measured, predicted, and percentage values, spirometry reports should contain a marker for significant deviation from certain limits, such as the standard deviation. Testing in children and adolescents requires a set of pediatric formulae.

8.1.4 Cooperation-Free Pulmonary Function Tests

For more than half a century spirometry has been a valued clinical tool and its importance has been documented in numerous publications. Forced spirometry can predict mortality better than smoking habits, existing cardiovascular morbidity, blood pressure, or gender [8.6].

As valuable as forced spirometry findings may be, it is often difficult to attain and ensure cooperation in subjects. Even well-equipped pulmonary function laboratories with well-trained staff will soon reach their limits when addressing and motivating children, geriatric or foreign-language patients. In former times, cooperation-free tests were postulated mostly for pediatric departments and epidemiological and screening studies. More recently, the cataclysmic demographic changes accompanied by a dramatic increase in the senior generation underline the urgent necessity of suitable tests. Although FEV_1 and $FEV_1/FVC\%$ are a recognized gold standard for detecting and assessing obstructive disease, more appropriate, more specific parameters might be available, correlating well with dyspnea (shortness of breath). Therefore, new, (almost) cooperation-free methods have been developed during the recent years, some of which are still pending commercial release. The next section gives an overview of proven and new developments in this area.

Ultrasound Pneumography

In Sect. 8.1.3, the ultrasound flowmeter was described and its ability mentioned to measure molar mass, the molecular or specific weight of the breathing gas, derived in real time from the sum of transit times – at no additional expenditure. While ambient air is inhaled through the sensor, the molecular weight changes during expiration as oxygen content drops and carbon dioxide increases (O_2 has much lower density than CO_2). When molar mass is registered during expiration, its waveform resembles that of CO_2. In fact, ultrasound pneumography (UPG) makes use of the scientific expertise gathered over decades of clinical research in expiratory CO_2 waveform analysis. The former approach of complicated synchronous recording of spirometry in combination with fast gas analysis, mostly by demanding mass spectrometry, can be replaced by instantaneous and genuinely synchronous measurement of flow, volume, and molar mass in UPG. Although the determination of molar mass is complicated by the dependence on temperature and humidity, recent research indicates the clinical validity of *native* molar mass derived directly from uncorrected ultrasound transit times. The subject performs regular tidal breathing into the ultrasound sensor, without the need for any special breathing maneuvers (Fig. 8.7). Molar mass waveforms are averaged and analyzed, containing information about pulmonary gas distribution. Clinical studies [8.7, 8] confirm a high correlation with FEV_1 and $FEV_1/FVC\%$ as well as the degree of obstruction.

Fig. 8.7 Ultrasound pneumography in a child (ndd Medizintechnik, Zurich, Switzerland)

Resistance Measurements

Forced Oscillation Technique. Applied in various forms since the 1970s, this testing principle combines a flowmeter, an inserted resistive element, a pressure transducer, and an oscillation generator (loudspeaker) [8.9, 10]. Low-amplitude (*forced*) pressure oscillations by the loudspeaker are superimposed on the patient's tidal breathing, resulting in a phase shift originating from the respiratory tract. Resulting flow and pressure signals are differentiated from the originating patient signals, independent from the patient's breathing pattern. The analysis yields impedance Z_{rs} with its components, resistance R and reactance X, which can be plotted in a frequency diagram. Today, this technique is most often found in a variation called *impulse oscillometry*, applied in geriatric, pediatric, and industrial medicine as well as in drug studies. This method, only requiring passive patient cooperation, i. e., tidal breathing at a mouthpiece, can complement standard pulmonary testing modalities and is suited for screening of obstructive disease [8.11]. Previous concerns about its technical complexity and limited hygiene have been cleared with the advent of modern commercial instruments. However, technical complexity and cost still exceed that of conventional spirometry equipment.

Monofrequent Oscillatory Resistance Measurement. This much simpler version of the forced modality has been proven clinically over decades [8.12]. Instead of a loudspeaker a small sinusoidal pump is used, generating small pressure swings superimposed on the patient's tidal breathing. Analysis of measured data again results in a phase diagram of resistance and reactance. In clinical practice, however, correlation with FEV_1 is just adequate, offering lower specificity and therefore only serving as an orientation or screening tool.

8.2 Advanced Cardiopulmonary Function Testing

8.2.1 Overview

In contrast to simple spirometry, advanced cardiopulmonary function testing requires a much higher level of methodology, instrumentation, and staff. This may be the reason why advanced testing can be found predominantly in internal and pulmonary departments of hospitals and specialized practices. Nevertheless, slow and forced spirometry will always remain the foundation of any type of pulmonary function instrumentation, particularly in all more complex or advanced pulmonary devices.

Among the more demanding modalities, body plethysmography takes a prime position, efficiently combining measurement of lung-volume subdivisions and airway resistance. Diffusion testing according to the *single-breath method* and determination of functional residual capacity (FRC) by N_2 washout are sometimes preferred in English-speaking countries. Historically, *closed systems* with gas dilution methods were widely applied for FRC studies but have been replaced by more hygienic devices. Besides the before mentioned, an array of more or less complex modalities can be found but without broad clinical acceptance. An overview of the most important methods is given in Table 8.3.

Finally, cardiopulmonary stress testing or ergospirometry, investigating gas exchange under physical exercise, has (in combination with ECG) become an important tool in assessing and differentiating ventilatory and cardiac disturbances. Especially in medical opinions and ratings, e.g., regarding compensation claims, this objective method allows global appraisal and staging of performance reductions in cardiopulmonary disease.

During recent years, the interest in continuous measurement of physical activity and metabolic monitoring

Table 8.3 Procedures and parameters in advanced pulmonary function testing

Procedure	Main parameters	Derived parameters
Body plethysmography	Intrathoracic gas volume (IGV)	FRC, RV, TLC
Body plethysmography	Airway resistance R_{AW}	G_{AW}, sR_{AW}, sG_{AW}
Compliance determination	Static compliance C_{stat}	P_{mi}
Compliance determination	Dynamic compliance C_{dyn}	Work of breathing
Airway occlusion test	$p_{0.1}$, p_{max}	
Single-breath diffusion test	Transfer factor T_{LCO}	K_{CO}, V_A, TLC_{SB}
Single-breath test with O_2	Membrane factor D_M	Capillary blood volume V_c
Intrabreath diffusion test	Co-diffusion D_{LCO}	Cardiac output Q_c
N_2 washout test	FRC	RV, TLC, distribution

has seen a tremendous rise. Practically all medical disciplines, from rehabilitation of obstructive patients to dementia prophylaxis, have recognized and published the invaluable importance of physical activity for improving, rehabilitating from, and particularly preventing disease. Despite a host of technical innovations facilitating its application, ergospirometry remains a demanding procedure in regards to test equipment, duration, and evaluation. This may be the reason why much more inexpensive portable devices have been developed for activity monitoring.

Medical technology, as a part of the information technology (IT) industry, undergoes the same fast-paced cycles in innovation and cost reduction that the microelectronics mass market exhibits. In the mid 1970s a computerized body box, as a body plethysmograph is called colloquially, sold for about €100 000. Today, almost 40 years later, a more operator friendly, even more accurate device can be purchased for less than €15 000. Practically all commercial instruments incorporate microelectronics, and a PC or notebook, most often implementing software under a certain operating system, featuring networking capabilities and data exchange to a hospital information system, together simplifying operation and enhancing laboratory productivity.

8.2.2 Body Plethysmography

Physical and Methodological Principles

A body plethysmograph, also called a whole-body plethysmograph or body box, consists of an airtight chamber, similar to a sealed phone booth, in which the patient is seated (Fig. 8.8). The thoracic movements created by the patient's breathing are transferred into volume and pressure swings inside the enclosure, which are measured and evaluated.

The foundations of modern body plethysmography, originally published by *DuBois* in the USA [8.13, 14] and *Ulmer* et al. in Germany [8.12], reach back to the 1950s, historically differentiating so-called constant-volume and constant-pressure modes. Applying the latter, the test person is seated in a chamber, breathing through a tube from and to the outside air, causing total thoracic volume displacements inside the enclosure. Today, virtually all commercially available body plethysmographs implement the constant-volume measuring principle, which seems to be technologically less demanding and clinically more reliable. In constant-volume mode, the subject breathes chamber air through a flowmeter and a fast-acting shutter valve. The goal of body plethysmography lies in the determination of (mean) alveolar pressure caused by compression

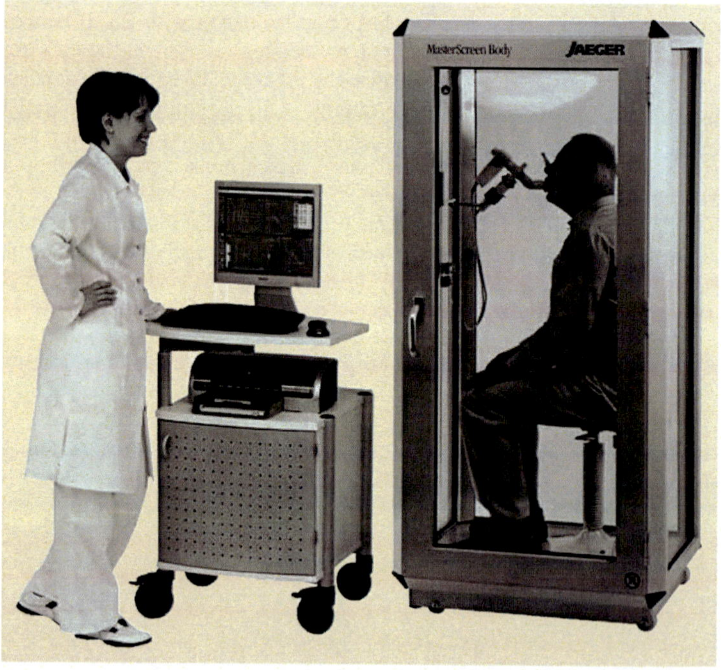

Fig. 8.8 Body plethysmograph (CareFusion 234, Höchberg, Germany)

and decompression of gas volume entrapped in the thorax. Note that this method captures the (small) compressible volume which, according to Boyle's law ($p \cdot V = \text{const}$), is proportional to alveolar pressure. Simultaneously with the compression of lung volume and the corresponding decompression of chamber volume, the alveolar pressure generates a flow and consequently a volume, which is measured at the mouth by the flowmeter. Thus, the thoracic movement causes a volume and proportional pressure change inside the lung, which is reflected by an identical volume and, again according to Boyle's law, a proportional pressure swing inside the chamber (Fig. 8.9).

In a procedure colloquially called *box calibration* the proportionality factor between chamber volume change and chamber pressure change is determined. Usually a small motorized pump is built into the enclosure, generating a constant stroke displacement, e.g., 50 ml, which is registered as a pressure swing. Calibration of the chamber is typically stable and does not require frequent checks.

Determination of Intrathoracic Gas Volume

When breathing against the shutter at the mouth, a thoracic volume change takes place, causing an alveolar pressure swing, as long as pressure equilibration between mouth and alveolar space can be assumed. When registering mouth pressure against box pressure during shutter closure, the (linear) relation between mouth pressure and box pressure can be established. This ratio of alveolar pressure to thoracic volume also enables the calculation of lung volume at the time and level of shutter closure, which is called the (intra-)thoracic gas volume (ITGV or IGV).

$$\text{ITGV} = k_\alpha \cdot \frac{p_K}{p_M} \text{ (l)} ,$$

where p_K is the chamber pressure, p_M is the mouth pressure, and k_α is a constant that depends on the chamber volume, the body volume of the subject, and the barometric pressure.

When, during tidal breathing, the shutter is triggered at the end of a normal expiration, the ITGV corresponds to the functional residual capacity (FRC).

Determination of Airway Resistance

The so-called airway resistance loop can be graphed when the box pressure is registered against flow at the mouth during tidal breathing (Fig. 8.10). According to Ohm's law,

$$U = R \cdot I \quad \text{or} \quad p_{alv} = R_{AW} \cdot \dot{V}(\text{hPa}) ,$$

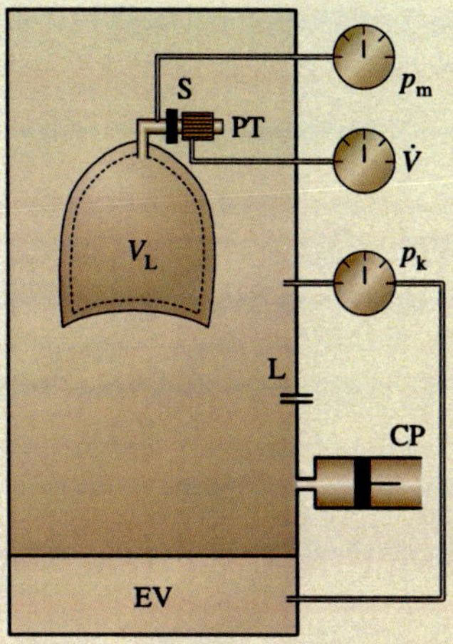

Fig. 8.9 The volume-constant plethysmographic principle

from which airway resistance (R_{AW}) can be derived as

$$R_{AW} = \frac{p_{alv}}{\dot{V}} = k_\beta \cdot \frac{p_K}{\dot{V}} \left(\frac{\text{hPa}}{1/\text{s}}\right) ,$$

where p_{alv} is the alveolar pressure, $\dot{V}$ is the flow at the mouth, p_K is the chamber pressure, and k_β is a factor comprising mainly the ratio of mouth over chamber pressure established during the shutter manoeuvre.

Technical Characteristics of Body Plethysmography

In tidal breathing, the chamber pressure changes are fairly small, only a few hPa (or cmH_2O). Disturbances caused by a patient-related temperature increase within the chamber (body heat), by breathing-related air temperature and humidity changes, and by pressure changes related to external pressure variations, to mention just a few, can be significant and may accumulate to an error greater than the sensed measuring signal. To attain adequate pressure equilibration between mouth and alveolar space, panting during the shutter maneuver should be avoided [8.15].

The construction characteristics of the body plethysmograph chamber are essential to ensure good measuring quality. Some of these features are:

- Rigidity of the enclosure
- Heat transfer characteristics of the chamber walls
- Built-in equilibration vessel
- BTPS compensation
- Calibration unit

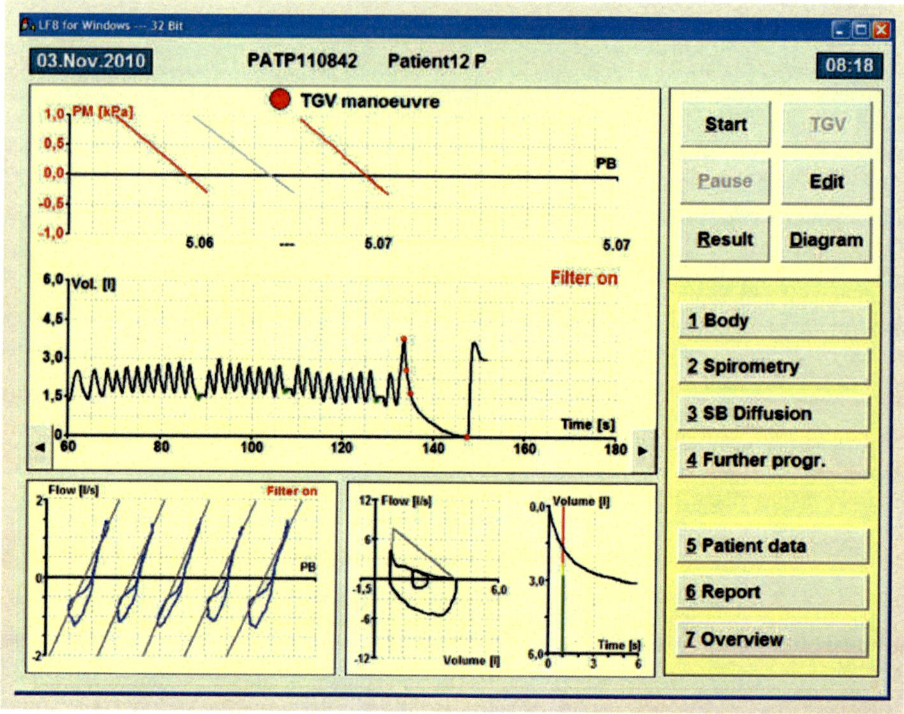

Fig. 8.10 Screenshot of a body plethysmographic test including airway resistance and shutter loops (Ganshorn Medizin Electronic, Niederlauer, Germany)

- Adjustment of a defined leak
- Type and speed of the shutter assembly.

The following remarks, although not exhaustive, should be made in this context. The enclosure should be constructed in a shape and from materials which ensure solid rigidity and fast heat transfer of the temperature rise caused by the patient. An equilibration vessel for common mode rejection of externally transmitted pressure artifacts should be built into the enclosure (Fig. 8.9). An automatic box calibration pump for dynamic simulation of thoracic volume compression and an adjustable defined leak (pneumatic low-pass filter for drift compensation) are indispensable for quality assurance. In a modern box, pressure- and flow-related BTPS compensation is carried out digitally. The resistance of a breathing filter must be considered when airway resistance is determined.

If the above-mentioned peculiarities are accounted for, body plethysmography can be implemented into clinical routine efficiently, offering fast and stable test results.

Technical service requirements for a body plethysmograph enclosure are relatively few. Except for mouthpieces and breathing filters, no other disposable material is needed. Improving hygienic requirements, modern shutter assemblies can be disassembled easily. Use of breathing filters is recommended.

A complete body plethysmographic evaluation of a patient, including repeated determination of resistance and ITGV, will take about 15 min, and in combination with spirometry determining lung-volume subdivisions, less than 30 min.

Clinical Value of Body Plethysmography

As described above, body plethysmography offers determination of airway resistance and intrathoracic gas volume in a single testing procedure. Fairly independent from patient cooperation, the test can be repeated several times, representing a truly fast and efficient modality to measure the most important parameters of pulmonary function. The full spectrum of relevant pulmonary diagnostic information can be assessed by body plethysmography when combined with slow and forced spirometry and recording of flow–volume loops, which can be carried out while the enclosure door remains open, and with testing of other parameters, such as lung compliance or diffusion transfer factor, in case an additional diffusion assembly is built in.

Airway Resistance

According to *Ulmer* et al. [8.12] and *Matthys* [8.16], airway resistance can be determined as R_{tot} or R_{eff}, offering a sensitive measure of bronchial obstruction. With high sensitivity, these parameters do not only reflect reduction in cross-section but also dynamic compression of instable airways and reduced lung elasticity (loss of retraction). Shape analysis of resistance loops enables differentiation of homogeneous obstruction (e.g., bronchitis) from inhomogeneous obstruction (e.g., emphysema), therefore documentation of resistance loops is mandated.

Intrathoracic Gas Volume/FRC

Besides FRC, intrathoracic gas volume comprises all gas compartments compressed during the shutter maneuver, also trapped air in the lungs as well as abdominal gas. This method is certainly false positive in highly obstructive patients when pressure equilibration between mouth and alveolar space cannot be attained. In contrast, the dilution and foreign gas methods will be false negative, underestimating lung volume, which tends to increase in the presence of severe obstruction. Because of its accuracy and high reproducibility, body plethysmography is considered a gold standard in lung volume assessment. Combining ERV and VC obtained during spirometry with ITGV/FRC, all volume subdivisions including RV and TLC can be determined in one sequence.

8.2.3 Diffusion Capacity

Gas exchange between alveolar space and capillary blood, i.e., the transfer of gas over the alveolar–capillary membrane, is called pulmonary diffusion. The diffusion capacity of a gas is defined by the amount of gas per unit of time exchanged over the membrane and the partial pressure difference between gas and blood phase. As the oxygen capillary partial pressure is inaccessible to noninvasive testing, the diffusion capacity of oxygen cannot be assessed. Instead, the diffusion capacity for carbon monoxide (CO) is measured. Only low concentrations of CO in the ppm range are added to the inspiratory air, as CO has high affinity to hemoglobin and patient exposure to CO should be minimized.

Single-Breath Method

For decades the single-breath method, originally developed by *Cotes* [8.17], has found broad acceptance. This modality does not assess diffusion capacity itself but the so-called transfer factor for CO (TLCO). After a deep exhalation, the patient inhales a gas mixture of air, 0.2–0.3% CO, and a low concentration of an inert gas, usually helium or methane. At full inspiration, the patient holds his breath for ≈ 10 s, allowing the gas mixture to distribute in the alveolar space. While the inert gas resides in the lung, CO will pass the alveolar–capillary membrane into the blood. After the breath-holding period, the patient exhales, and the first part of the expired air is discarded, while the center part is collected (Fig. 8.11). In more modern devices featuring fast gas analysis, the gas is analyzed continuously, and the gas concentrations of the center part are averaged. In both cases the center part is considered representative as an alveolar sample. The inert gas, not taking part in gas exchange, is diluted in the lung, characterizing the ventilatory distribution. Assuming that CO is diluted at the same rate as the inert gas, the diffusive part of the CO concentration can be estimated. The transfer factor is determined by applying an exponential decay model of alveolar CO concentration during the breath-holding period. If the initial capillary CO concentration is assumed to be zero, the transfer factor can be derived as

$$T_{LCO} = b \cdot \frac{V_A}{t_v} \cdot \ln\left(\frac{F_{ACO0}}{F_{ACO}}\right) \left(\frac{\text{mmol}}{\text{min} \cdot \text{kPa}}\right) ,$$

where

$$F_{ACO0} = F_{ICO} \cdot \frac{F_{AX}}{F_{IX}} .$$

F_I is the inspiratory gas concentration, F_A is the alveolar concentration of the expiratory gas sample, X is an index representing the inert gas, F_{ACO0} is the initial alveolar concentration of CO, V_A is the alveolar volume, t_v is the breath-holding time, and b is a constant for conversion of dimensions. Besides T_{LCO}, the alveolar volume V_A can be calculated from the dilution of the inert gas by using the same data as above, i.e.,

$$V_A = (V_{IN} - V_D)\left(\frac{F_{IX}}{F_{AX}}\right) (\text{l}) ,$$

where V_{IN} is the inspired volume prior to breath holding and V_D is the dead space volume consisting of anatomical and apparatus dead space. All volumes are reported in BTPS. Combining spirometry with V_A, RV and TLC can be derived.

Technical Characteristics of the Single-Breath Method

The technical complexity of this modality should not be underestimated, as it requires a flowmeter, analyzers for CO and the inert gas, and a valve system for switching

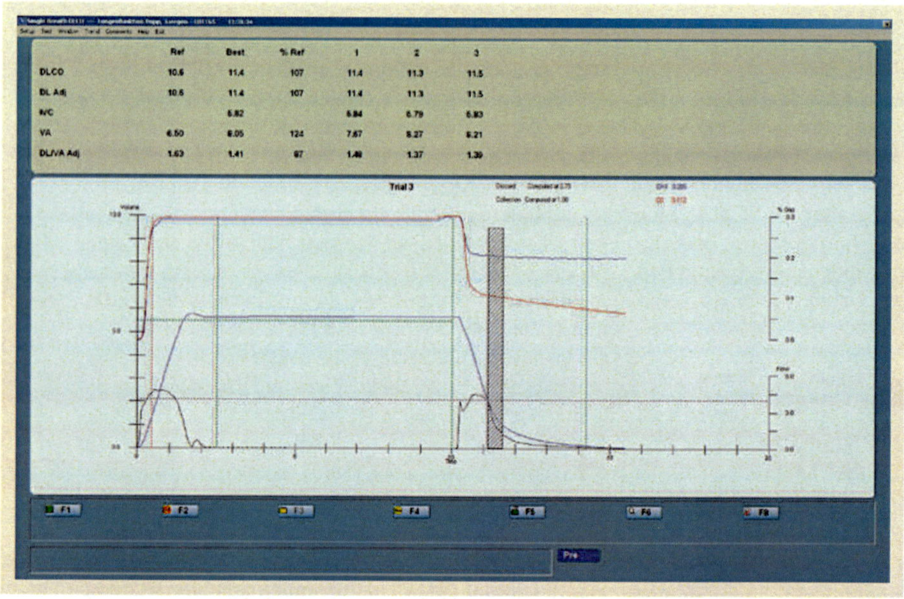

Fig. 8.11 Screenshot of a single-breath test (CareFusion 234, Höchberg, Germany)

from air to the gas mixture as well as for sampling of the alveolar gas.

Quality assurance programs could show that the quality of results depend – besides technical insufficiencies – mainly on correct operation and calibration. Avoiding frequently occurring errors, special care should be taken regarding:

- Correct and fast-acting valves
- Frequent calibration of gas analyzers with certified test gas
- Optimum patient instructions
- Performance of test according to recommendations of the European Respiratory Society [8.18].

Smokers and workers exposed to CO have higher arterial CO partial pressure, a so-called back pressure, which should be analyzed and considered in the calculation of alveolar partial pressure. If the hemoglobin content of the subject deviates from the normal range, a correction should be made when calculating the TLCO.

The test should be carried out in a sitting position.

In most cases, cleaning of the valve system is difficult, sometimes impossible, requiring the use of a breathing filter in compliance with hygiene standards. The technical service requirement for diffusion testing devices must not be underestimated, especially when complex breathing and analyzer valve systems are employed. The cost and supply logistics of testing and calibration gas should also be considered. A complete single-breath test including calibration will only need a few minutes.

Clinical Value of the Single-Breath Test

Implemented in routine clinical testing, determination of TLCO by the single-breath technique is simple and fast. When the above-mentioned errors are avoided and patients well instructed, results show acceptable accuracy and good reproducibility. The TLCO parameter does not only react to impairments of diffusion capacity in the sense of gas exchange over the alveolar–capillary membrane, but also to structural changes of lung tissue, independent from their cause, which might be either an increase or a reduction of alveolar space. A reduced TLCO cannot only be found in interstitial lung disease such as pulmonary fibrosis, sarcoidosis, alveolitis, or lung edema, but also in generalized emphysema, often characterized by loss of lung surface and increase of alveolar space. A normal or moderately reduced TLCO can also be seen in airway obstruction, while an increased TLCO may indicate obstructive hyperinflation, as attributed to asthma. Also intrapulmonary bleeding may lead to elevated TLCO levels.

TLCO is well suited for follow-up studies and therapy control and may be used as an additional parameter in the assessment of pulmonary–vascular disorders. Due to its complexity, application of TLCO in screening may be limited to specialized studies, e.g., in workers exposed to dust. In a variation of the single-breath method using a mixture of CO, inert gas, and oxygen the diffusing capacity of the alveolar–capillary membrane, the so-called membrane component, and the capillary blood volume can be assessed, an application that might be confined to more specialized laboratories.

Additional Methods in Diffusion Assessment

During recent years, in addition to the classical modality described above, the IntraBreath method was introduced, featuring a single deep inhalation but no breath-holding before exhalation, certainly advantageous in children and dyspneic patients. Using fast infrared gas analyzers for continuous reading of low-dose CO, (inert) methane, and diffusible acetylene (soluble in blood), the capillary cardiac output can be determined in combination with the TLCO. The test can also be carried out under exercise.

Due to its high patient exposure to CO, the previously used steady-state method has been abandoned. In addition to the TLCO test mentioned, a variety of other diffusion modalities exist, in particular those using the rebreathing technique. Nevertheless, their applicability in a regular laboratory and their international recognition might be limited.

8.2.4 Nitrogen-Washout Test for Determination of FRC

Widely applied for decades in the English-speaking world, this modality requires a valve and gas delivery system to allow inspiratory switching from air to pure oxygen. With each inhaled breath of oxygen, nitrogen present in the lung, not participating in the gas exchange, will be replaced. Continuous gas sampling at the mouth by either a single N_2 or combined O_2/CO_2 analyzer allows determination of expired nitrogen volume during the N_2-washout procedure. Accumulated nitrogen volume is calculated breath by breath until the N_2 concentration falls below 1%, marking the end of the washout sequence. Before the start of the washout, the N_2 concentration (including rare inert gases) is 79.2%, which allows the functional residual capacity (FRC) to be derived as

$$\mathrm{FRC} = \frac{V_{N_2}}{0.79}(l)\,.$$

For standardization of lung volume determination by N_2 washout refer to [8.4].

Technical Characteristics of the N_2-Washout Method

Historically, this method required a fast N_2 analyzer or a mass spectrometer, both expensive and demanding in operation and service. Nowadays, with the advent of fast-response combi-analyzers, the complementary gases O_2 and CO_2 can be determined instead of N_2, involving less complex and expensive technology. However, this novel approach makes synchronization of digitized gas samples necessary. As most analyzers draw a side-stream gas sample from the mouth, causing a delay between gas sample and flow, the response characteristic of the analyzer needs adjustment and correction. In a dedicated calibration routine the operator should be offered verification of the response compensation. The valve systems employed in N_2 washout are similar to those used in diffusion testing, compromising hygiene unless a breathing filter is used.

The test duration of FRC determination by N_2 washout depends on the lung distribution of the subject, taking a few minutes in a healthy and more than 20 min in a severely obstructed individual.

While classical N_2-washout devices use side-stream gas sampling, the novel ultrasound molar mass spirometry offers a less complex clinical approach with instantaneous gas analysis (Sect. 8.1.2, Fig. 8.5). The ultrasound sensor is able to determine both flow and molar mass, the specific weight of the gas, with a single transducer in the main stream of the respired gas. As molar masses of O_2 and CO_2 are distinctively different, the N_2 concentration of the breathing gas can be assessed instantaneously and synchronous to flow and volume changes [8.19].

8.2.5 Ergospirometry

Ergospirometry serves as the most prominent method to determine ventilation and gas exchange under physical exercise, in the English-speaking world often referred to as cardiopulmonary stress testing. A complete instrument (Fig. 8.12) consists of:

- A device producing a defined level of physical stress (ergometer)
- A transducer for measurement of ventilation
- Gas analyzers for O_2 and CO_2
- A computer for online processing of measured data, as well as
- A multichannel electrocardiograph (ECG).

For physical exercise, usually a bicycle or treadmill ergometer (Sect. 8.2.5) will be employed. Specialized

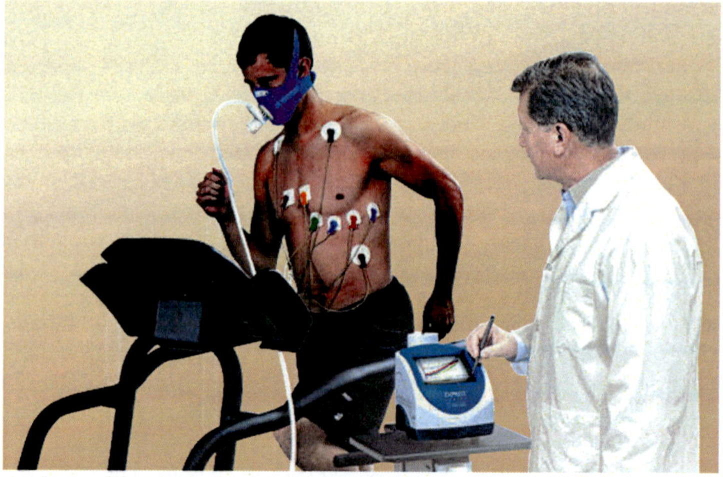

Fig. 8.12 Ergospirometry system (Medgraphics, St. Paul, USA)

Fig. 8.13 Portable ergospirometry system (CareFusion 234, Höchberg, Germany)

ergometers such as cranking or rowing ergometers will be applied in industrial and sports medicine (Fig. 8.13).

Measurement of Ventilation

In ergospirometry, flowmeters using varying measuring principles can be applied, including pneumotachographs, turbines, thermistor mass flowmeters, and ultrasound flowmeters. Calibration or verification of flow transducers before each test seems indispensable. The relative error of flow measurement should not exceed 3% of reading; the linearity in the measuring range should be within 2% of reading. For clinical purposes the range should cover at least 100 l/min, and for sports medical purposes up to 200 l/min. As the temperature in fully water-saturated expired air drops, humidity falls out and may cause erroneous reading in water-sensitive flow sensors. Supporting natural breathing through mouth and nose, a breathing mask should be used. The weight of the flowmeter should be as low as possible to allow direct docking of the flow transducer to the facemask.

Gas Analysis

Historically, mass spectrometry has been referred to as the gold standard in gas analysis, as all breathing gas concentrations, namely those of O_2, CO_2, and N_2, can be measured quickly, synchronously, and with high accuracy. As this technology requires a large investment, continuous upkeep, voluminous space, and complex handling, its clinical application has become a rarity. Today, compact to mini-sized gas analyzers are integrated into one instrument together with other electronic components. As the response and delay of gas analyzers vary widely, a compensation algorithm is implemented in the ergospirometry software to allow for precise synchronization of all signals.

Paramagnetic or fuel cells (zirconium oxide tube) are the prevalent principles in oxygen analysis, while CO_2 is determined by means of selective infrared analyzers. The absolute error of gas analyzers should not exceed 0.1% within the measuring range; linearity should lie within 1% of reading. Before each test, the calibration of gas analyzers with certified calibrations gas at two concentration levels representing inhaled and expired air is indispensable for assurance of highly precise results. Modern ergospirometry systems contain a software-controlled module for automatic calibration and verification.

Mixing Bag Method

Reference to this method is made for didactic reasons only, as practically all presently available instruments work with the breath-by-breath method. The older mixing method separates inspiratory and expiratory air of the subject by use of a so-called Y-valve. The expiratory gas is collected in a mixing bag, from which a continuous sample is drawn and analyzed. The flowmeter is mounted on the expiratory side at the entrance port of the mixing bag, allowing determination of expiratory tidal volume and minute ventilation. Oxygen uptake V_{O_2} can be calculated according to

$$\dot{V}O_2 = c \cdot V_E \left(F_I O_2 \cdot k_S - F_E O_2\right) (1/\text{min}) ,$$

where k_s is the co-called shrinking factor given by

$$k_S = \frac{(100 - F_E O_2 - F_E CO_2)}{(100 - F_I O_2)} ,$$

where V_E is the expiratory minute ventilation, and F_I the inspiratory and F_E the mean expiratory concentration of O_2 and CO_2, respectively. Additionally a factor for the conversion from BTPS to standard temperature pressure dry (STPD) conditions has to be considered.

CO_2 output can be calculated accordingly. Response of gas analyzers or delay of the gas sample are of lesser importance as the gas mixing introduces a fairly high time constant. For decades, this modality has proven its clinical reliability. Due to its inability to follow physiological responses to changing exercise levels instantaneously, the mixing bag system has all but been replaced by the breath-by-breath method.

Breath-by-Breath Method

In the breath-by-breath (B×B) mode, flow is continuously measured at the mouthpiece or a breathing mask. A continuous gas sample is drawn as proximal as possible through thin moisture-absorbing tubing and analyzed by fast-response gas analyzers. Flow and gas concentrations need to be synchronized and precisely brought into phase. Also, the response of each analyzer needs software correction. The oxygen uptake is calculated according to

$$\dot{V}O_2 = \int FO_2 \cdot \dot{V} \, dt \, (1/\text{min}) ,$$

where V_{O_2} is the oxygen uptake, F_{O_2} is the O_2 concentration or fraction, and $\dot{V}$ is the flow at the mouth. Additionally factors for conversion from BTPS to STPD must be applied by the software. The CO_2 output is computed accordingly.

As elegant as it seems, the B×B method can bring about inaccuracies and errors which need to be addressed by the operator and can only be avoided by frequent calibration and consequent verification, including precise analyzer delay adjustment. Equipment operators should demand validation and documentation of implemented algorithms from the manufacturers or suppliers of ergospirometry systems. Therefore, an *open* structured software showing the delay and response adjustment of the analyzers in real time, offering step-by-step procedures for calibration and verification, facilitates quality assurance.

Clinical Value of Ergospirometry

Modern ergospirometry is based on the work of *Hollmann* and *Wasserman* [8.20].

For clinical purposes, physical exercise is increased stepwise, e.g., using a bicycle ergometer, by increments of 25 W of 2–3 min each. This form of exercise is well suited for patients, allowing the cardiovascular system sufficient time to adapt to the exercise level, while avoiding muscular fatigue, anaerobic metabolism, and lactate production.

As a valid measure of global physical performance, oxygen uptake directly relates to the interrelated organ system of heart–lung circulation, while the maximum oxygen uptake or *vita maxima* offers an objective assessment of the maximal performance capacity of the test subject. By visualizing the dynamics of ventilation, gas exchange, and heart rate, the latter derived from ECG, reduced capacity of each involved organ may be determined and differentiated from a limited degree of motoric efficiency [8.21]. Secondary parameters such as respiratory ratio and breathing equivalent (Table 8.4) as well as exercise-related blood gases assist in detecting compensation mechanisms and evaluating deficiencies even at submaximal stress levels.

The determination of the aerobic–anaerobic metabolic transition, briefly called the anaerobic threshold, plays an important role is assessing and following the effects of training in rehabilitation and fitness centers. After addition of blood gases and blood pressure data, a conclusive report of ergospirometry test results is best given in graphical format, e.g., in the nine-field graph according to *Wasserman* [8.20].

Table 8.4 Parameters in ergospirometry

Parameter	Abbreviation/definition	Unit
Load (bicycle ergometer)	P	W
Speed (treadmill ergometer)	v	km/h
Elevation (treadmill ergometer)	s	%
Tidal volume	VT	l
Respiratory rate	R_f	/min
Minute ventilation	V_E	l/min
Oxygen uptake	VO_2	l/min
Carbon dioxide output	VCO_2	l/min
Heart rate	HR	min^{-1}
Respiratory ratio	$R_Q = VCO_2/VO_2$	–
Breathing equivalent (for O_2)	$EQO_2 = V_E/VO_2$	–
Breathing equivalent (for CO_2)	$EQCO_2 = V_E/VCO_2$	–
Oxygen pulse	$VO_2Pulse = VO_2/HR$	ml/(min kg)
Alveolar-arterial difference	$AaDO_2$	mmHg
Functional dead space	V_{Df}	%

For assessments in compensation claims and other rating cases, in particular in industrial medicine, ergospirometry is the tool of choice when an objective measure of reduced capacity is required. Also in the assessment of athletes, an ergospirometry system represents a valuable instrument which can be found in most sports medicine centers today.

A minimum of 30 min should be considered for performing a complete stress test, including patient preparation, calibration, and evaluation of data.

8.2.6 Noninvasive Determination of Cardiac Output

While ergospirometry allows only indirect assessment of cardiac function, oxygen uptake can be instrumental in determining cardiac output through the Fick equation

$$\dot{V}O_2 = Q_t \cdot avDO_2 \; (1/\text{min}) .$$

The arterio-venous O_2 content difference, which is an effort-dependent variable, requires invasive assessment by catheterization.

One of the internationally recognized noninvasive standards to determine cardiac output is the so-called CO_2-rebreathing method, an application of the Fick equation above to carbon dioxide rather than to oxygen

$$Q_t = \frac{\dot{V}CO_2}{avDCO_2} = \frac{\dot{V}CO_2}{(C_vCO_2 - C_a\,CO_2)} \; (\text{l/min}) ,$$

where $avDCO_2$ is the arterio-venous content difference, and C_vCO_2 and C_aCO_2 are the mixed-venous and arterial content of CO_2 in blood. Figure 8.14 depicts schematically the interaction of heart, lungs, and circulation.

Estimation of the terms in the latter equation requires a CO_2-rebreathing module as an expansion to a regular ergospirometry system. During rest or physical exercise the subject breathes through a valve box to which a bag is attached. The bag is filled with a mixture of CO_2 and oxygen gas, where the CO_2 partial pressure in the bag approximates the mixed-venous partial pressure of the subject at the current exercise level. Initially, the valve box allows the subject to breathe ambient air, just like during a regular exercise test, while CO_2 output is measured. Still under air breathing, a micro blood sample is

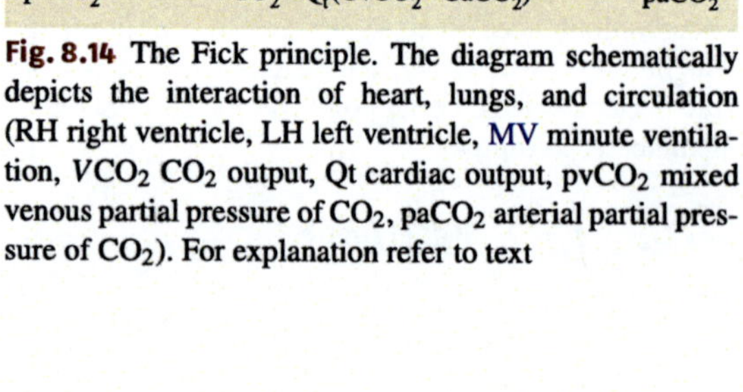

Fig. 8.14 The Fick principle. The diagram schematically depicts the interaction of heart, lungs, and circulation (RH right ventricle, LH left ventricle, MV minute ventilation, VCO_2 CO_2 output, Qt cardiac output, $pvCO_2$ mixed venous partial pressure of CO_2, $paCO_2$ arterial partial pressure of CO_2). For explanation refer to text

drawn from the subject's ear lobe and analyzed, yielding the arterial partial pressure p_aCO_2. From a digital dissociation curve, the arterial CO_2 content C_aCO_2 is derived. If micro blood gas analysis is not available, p_aCO_2 can be estimated from the reading of the end-expiratory CO_2 partial pressure $p_{et}CO_2$, an acceptable approximation in healthy subjects. After switching the subject's breathing to the bag at the end of an expiration, the CO_2/O_2 gas mixture is inhaled. While rebreathing to and from the bag continues for several breaths, CO_2 partial pressure will equilibrate between the lung and the bag as long as a constant p_vCO_2 is delivered from the right heart and no recirculation occurs. Only a few rebreathing breath cycles are needed to reach a CO_2 equilibrium, which is monitored, and the mixed-venous partial pressure determined. Applying a digital dissociation curve, the C_vCO_2 content is derived, the last component in the noninvasive determination of cardiac output.

The CO_2-rebreathing method is a very elegant procedure, bearing no risk for the subject while requiring little cooperation. Its accuracy compares well with that of invasive methods [8.22]. In contrast to most of the catheter procedures, the rebreathing method will only determine pulmonary cardiac output without right-to-left shunt. While the expansion module for an existing instrument may not be very complex, the software upgrade will determine the cost.

8.2.7 Metabolic Activity Monitoring

Although portable ergospirometry systems have seen a lot of progress in size and versatility (Fig. 8.13), their application in normal everyday life and at the workplace is still very limited. The biggest obstacle seems the restriction in testing time, as a subject cannot wear a face mask or use a mouthpiece for much more than several hours. During the last decade, improved diagnostic solutions were researched, offering assessment of physical activity over a longer period of time, up to several days, without restricting the subject in conducting a normal life. In addition the devices should be small, lightweight, and inexpensive. Even if the precision of ergospirometry is not reached with such a device, physical activity or energy expenditure in total calories spent, equivalent to oxygen uptake, should become available. Several categories of instruments based on the evaluation of heart rate (training watch), steps (pedometer), or acceleration in three axes (accelerometer) have come onto the market. Depending mostly on activity itself, these principles show good correlations with energy expenditure at moderate to high activity levels, but poor agreement in sedentary conditions in which most of our everyday life takes place. A novel multisensory device worn on the upper arm (Fig. 8.15) captures several physiological parameters, such as acceleration, and skin and ambient temperature, but also heat flux from the body and skin impedance. Employing artificial intelligence for the evaluation of raw data, energy expenditure under exercise as well as at low-level activity, rest or sleep can be assessed with reasonable to good accuracy. Data are recorded continuously for up to 3 weeks, offering a complete image of the subject's lifestyle, documenting the activity–inactivity profile minute by minute. Particularly appreciated in medical research and drug studies, the device cannot be manipulated as it turns recording on when worn and turns off when not in skin contact. Due to its small size and low weight, the wearer will forget about the moni-

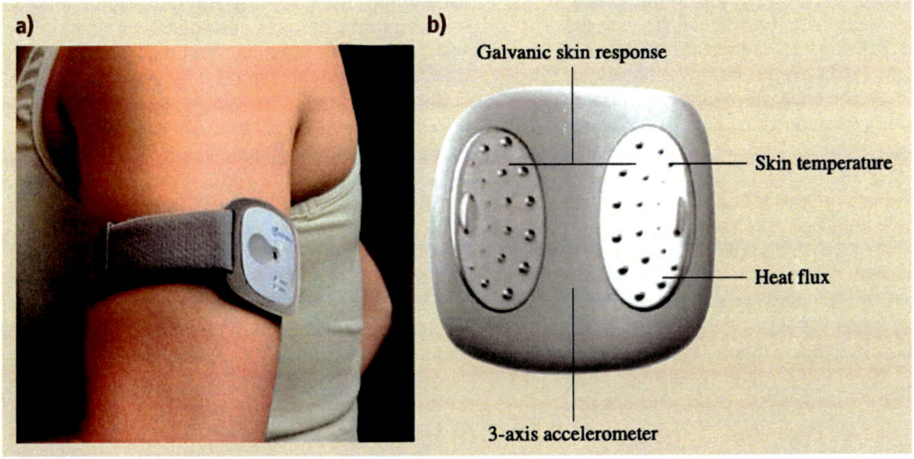

Fig. 8.15a,b Metabolic armband monitor for determination of physical activity (Bodymedia, Pittsburgh, USA)

tor on his arm and just pursue his daily routine. A broad literature base supports its application in different areas of pulmonary medicine, particularly severe chronic obstructive pulmonary disease (COPD) [8.23] and cystic fibrosis [8.24] where activity is an important outcome parameter.

8.2.8 Planning and Laboratory Space Considerations

The laboratory room in which a body plethysmograph will be installed does not need to be particularly spacious but should be in a quiet location, separated from any through traffic. Big windows adjacent to the body box may lead to pressure artifacts caused by wind gusts or sun exposure. The diffusion test can be performed in the same room as the body plethysmograph or with an add-on unit within the chamber while the box door is open. N_2-washout devices or spirometers can be placed in a laboratory room or on a trolley for mobile use.

In contrast, ergospirometry requires a spacious room with good ventilation, in particular when a treadmill or ergometer, blood gas analyzer or other laboratory equipment should be placed in the same room. A changing room and a shower should be available for the test subjects.

References

8.1 M.R. Miller, J. Hankinson, V. Brusasco, F. Burgos, R. Casaburi, A. Coates, R. Crapo, P. Enright, C.P.M. van der Grinten, P. Gustafsson, R. Jensen, D.C. Johnson, N. MacIntyre, R. McKay, D. Navajas, O.F. Pedersen, R. Pellegrino, G. Viegi, J. Wanger: Standardisation of spirometry, Eur. Respir. J. **26**, 319–338 (2005)

8.2 European Respiratory Society: Standardized lung function testing, Eur. Respir. J. **6**(Suppl 16), 3–52 (1993)

8.3 C. Buess, P. Pietsch, W. Guggenbühl, E.A. Koller: A pulsed diagonal beam ultrasonic air flowmeter, J. Appl. Physiol. **61**(3), 1195–1199 (1986)

8.4 J. Wanger, J.L. Clausen, A. Coates, O.F. Pedersen, V. Brusasco, F. Burgos, R. Casaburi, R. Crapo, P. Enright, C.P.M. van der Grinten, P. Gustafsson, J. Hankinson, R. Jensen, D. Johnson, N. MacIntyre, R. McKay, M.R. Miller, D. Navajas, R. Pellegrino, G. Viegi: Standardisation of the measurement of lung volumes, Eur. Respir. J. **26**, 511–522 (2005)

8.5 R. Pellegrino, G. Viegi, V. Brusasco, R.O. Crapo, F. Burgos, R. Casaburi, A. Coates, C.P.M. van der Grinten, P. Gustafsson, J. Hankinson, R. Jensen, D.C. Johnson, N. MacIntyre, R. McKay, M.R. Miller, D. Navajas, O.F. Pedersen, J. Wanger: Interpretative strategies for lung function tests, Eur. Respir. J. **26**, 948–968 (2005)

8.6 T.I. Petty, D.J. Pierson, N.P. Dick, L.D. Hudson, S.H. Walker: Follow-up evaluation of a prevalence study for chronic bronchitis and chronic airway obstruction, Am. Rev. Respir. Dis. **114**, 881–890 (1976)

8.7 M. Gappa, C. Buess, S.I. Fuchs: Slope indices from ultrasonic tidal molar mass-volume curves, ERS Congress, Copenhagen (2005)

8.8 R.L. Jensen, C. Buess, R.O. Crapo: Ultrasonic measurements of CO_2 during quiet breathing may diagnose degree of airway obstruction, ATS Convention (2005)

8.9 E. Oostveen, D. MacLeod, H. Lorino, R. Farré, Z. Hantos, K. Desager, F. Marchal, ERS Task Force: The forced oscillation technique in clinical practice: Methodology, recommendations and future developments, Eur. Respir. J. **22**, 1026–1041 (2003)

8.10 H.J. Smith, P. Reinhold, M.D. Goldman: Forced oscillation technique and impulse oscillometry, Eur. Respir. J. **31**, 72–105 (2005)

8.11 J. Winkler, A. Hagert-Winkler, H. Wirtz, G. Hoheisel: Die moderne Impulsoszillometrie im Spektrum lungenfunktioneller Messmethoden, Pneumology **63**, 461–469 (2009)

8.12 W.T. Ulmer, D. Nolte, J. Lecheler: *Die Lungenfunktion. Methodik und klinische Anwendung* (Thieme, Stuttgart 2003)

8.13 A.B. DuBois, S.Y. Botelho, J.H. Comroe Jr.: A new method for measuring airway resistance in man using a body plethysmograph: Values in normal subjects and in patients with respiratory disease, J. Clin. Invest. **35**(3), 327–335 (1956)

8.14 A.B. DuBois, S.Y. Botelho, G.N. Bedell, R. Marshall, J.H. Comroe Jr.: A rapid plethysmographic method for measuring thoracic gas volume: A comparison with a nitrogen washout method for measuring functional residual capacity in normal subjects, J. Clin. Invest. **35**(3), 322–326 (1956)

8.15 D.O. Rodenstein, D.C. Stanjescu: Frequency dependence of plethysmographic volume in healthy and asthmatic subjects, J. Appl. Physiol. **54**, 159–165 (1983)

8.16 H. Matthys: *Pneumologie* (Springer, Berlin, Heidelberg 1988)

8.17 J.E. Cotes, A.M. Hall: The transfer factor for the lung; normal values in adults. In: *Normal Values for Respiratory Function in Man*, ed. by P. Arcangeli (Panminerva Medica, Torino 1970) pp. 327–343

8.18 N. MacIntyre, R.O. Crapo, G. Viegi, D.C. Johnson, C.P.M. van der Grinten, V. Brusasco, F. Burgos, R. Casaburi, A. Coates, P. Enright, P. Gustafsson,

J. Hankinson, R. Jensen, R. McKay, M.R. Miller, D. Navajas, O.F. Pedersen, R. Pellegrino, J. Wanger: Standardisation of the single-breath determination of carbon monoxide uptake in the lung, Eur. Respir. J. **26**, 720–735 (2005)

8.19 S.I. Fuchs, C. Buess, S. Lum, W. Kozlowska, J. Stocks, M. Gappa: Multiple breath washout with a sidestream ultrasonic flow sensor and mass spectrometry: A comparative study, Pediatr. Pulmol. **41**, 1218–1225 (2006)

8.20 K. Wasserman, J.E. Hansen, D.Y. Sue: *Principles of Exercise Testing and Interpretation: Including Pathophysiology and Clinical Applications* (Lippincott Raven, Philadelphia 2004)

8.21 N.L. Jones: *Clinical Exercise Testing* (Elsevier, Amsterdam 1997)

8.22 N.L. Jones: Evaluation of a microprocessor-controlled exercise system, J. Appl. Physiol. **57**, 1312–1318 (1984)

8.23 H. Watz, B. Waschki, T. Meyer, H. Magnussen: Physical activity in patients with COPD, Eur. Respir. J. **33**, 262–272 (2009)

8.24 T. Troosters, D. Langer, B. Vrijsen, J. Segers, K. Wouters, W. Janssens, R. Gosselink, M. Decramer, L. Dupont: Skeletal muscle weakness, exercise tolerance and physical activity in adults with cystic fibrosis, Eur. Respir. J. **33**, 99–106 (2009)

9. Devices and Methods in Clinical Neurophysiology

Devices and

Klaus-Peter Hoffmann, Ursula Krechel

Methods in clinical neurophysiology, such as electroencephalography (EEG), electromyography/electroneurography (EMG/ENG) and the recording of evoked potentials (EP) enables the physician to evaluate the function of the central and peripheral nervous system as well as the muscular system. Thus bioelectrical potentials are recorded via measuring techniques, amplified, stored, analyzed, and evaluated. The measuring devices used for functional diagnostics are called the electroencephalograph and the electromyograph. Their respective construction and their function as well as their most widespread clinical applications are described in this chapter.

9.1 **Basics** ... 119
9.1.1 Neurophysiological Basics ... 119
9.1.2 Technical Basics ... 121
9.2 **Electroencephalograph** ... 129
9.2.1 Signals ... 129
9.2.2 Device Technology ... 130
9.2.3 Methodology ... 132
9.2.4 EEG Recording Methods ... 134
9.2.5 Evaluation and Signal Analysis ... 136
9.2.6 Special Methods ... 138
9.3 **Electromyograph** ... 138
9.3.1 Signals ... 139
9.3.2 Device Technology ... 139
9.3.3 Electromyography ... 142
9.3.4 Electroneurography ... 145
9.3.5 Evoked Potentials ... 148
9.3.6 Event Related Potentials ... 156

9.1 Basics

Excitable cells react to a stimulus with changing membrane characteristics. If a sensitive living cell is stimulated, the ion conductivity on its membrane is changed. The consisting resting membrane potential (50–100 mV) can grow to an action potential that is conducted via the nerve cell and leads to a muscle contraction. The motor neuron with all its related muscle fibers form a motor unit. The number of muscle fibers supplied by a motor unit differs to a great extent and contains 5 fibers in an eye muscle up to 1000 in the temporal muscle.

9.1.1 Neurophysiological Basics

The Nervous System

The central nervous system (Fig. 9.1) consists of the brain and the spinal canal. This is the central part of the total nervous system and displays a complex information processing system. The information perceived by the sense organs (9 bit/s) are decreased during the signal processing to the central nervous system by inhibitory and excitatory circuits on the synapses to 10^1-10^2 bit/s. After the processing and analysis they are transformed into reactions such as motion, behavior, or organic activity. The function of the brain is not only reduced to voluntary movement, but includes emotions and abilities of the mind such as remembrance and learning.

Different levels of the central nervous system participate in the control of the vegetative nervous system.

- The limbic system for the control of the emotional drives.
- The hypothalamus for the homeostatic regulation.
- The medulla oblongata for the control of the sympathetic tone.
- The spinal cord for the spinal reflex circuit.

The peripheral nervous system (Fig. 9.1) consists of all nerves and ganglia outside of the central nervous

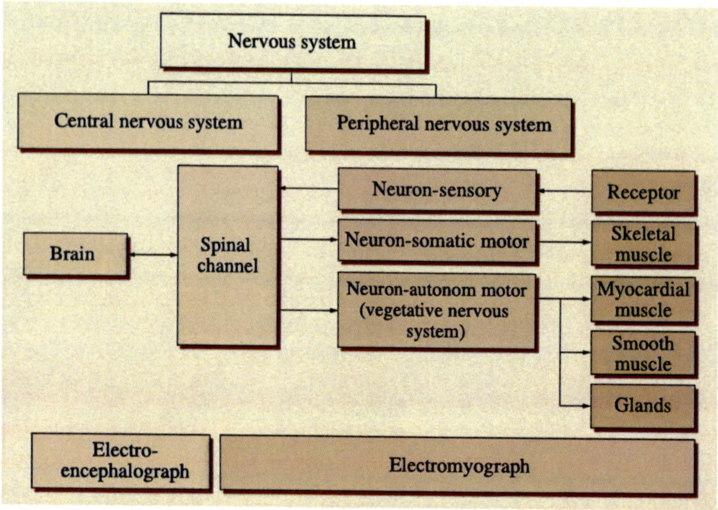

Fig. 9.1 Scheme of the central and peripheral nervous system and the application area of EEG and EMG

system. The efferent (motor) pathways innervate the skeleton muscles. The afferent (sensory) pathways conduct the stimulation from the corporal periphery to the skin receptors and the inner organs to the central nervous system.

The vegetative nervous system or the autonomous nervous system (Fig. 9.2) is a part of the peripheral nervous system. It controls the function of the organs and adapts them to their respective needs and supervises the inner milieu of the corpus. These activities function mostly unconsciously and cannot be influenced by voluntary control. Examples for controlled vital functions are heart beat, respiration, blood pressure, digestion, and metabolism.

The vegetative nervous system can be subdivided into three groups.

- Sympathetic: increasing performance (ergotropic)
- Parasympathetic: predominantly sustaining performance (trophotropic)
- Enteral: nervous system of the gastrointestinal tract, mostly independent from the central nervous system.

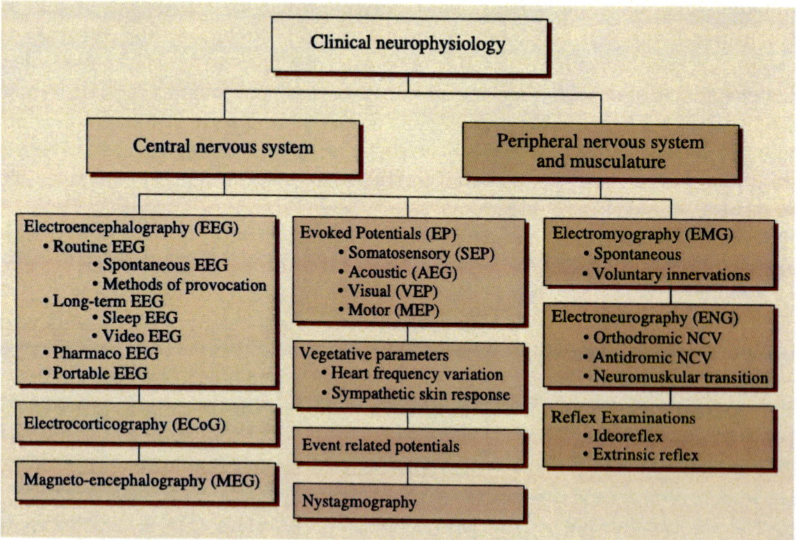

Fig. 9.2 Methods in clinical neurophysiology for the diagnostics of functional disturbances of the central and peripheral nervous system as well as the muscular system

The Muscular System

Skeletal muscles (Fig. 9.1) are able to transform chemical energy ATP directly into mechanical energy and heat. They are innervated voluntarily via motor neurons. A contraction is exclusively triggered via action potentials at the motor endplate. A significant attribute of the skeletal muscle cells are striated myofibrils with a regular succession of dark anisotropic (A-stripes) and light isotropic (I-stripes) bands.

Muscle types that are not striated are called smooth muscles (Fig. 9.2). They consist of long spindle-like cells and are loosely arranged and therefore moveable. Smooth muscles coat the inner organs such as the stomach or intestines as well as the walls of blood vessels. Smooth muscles have their own system of innervations and are involuntarily innervated by the vegetative nervous system. They carry out relatively slow movements, but have very little fatigue and are able to develop great force over quite a long time.

The heart muscle has striated and smooth musculature with its own system of innervation and is controlled involuntarily by vegetative nerves.

The Formation of Resting and Action Potentials

Different distribution of ions on the inside and outside of living cells creates a membrane potential on the cell membrane. This is continuously fed by active transport mechanisms (ion pump). The concentration of K^+ ions inside the cell is 30 times larger than outside, e.g. the inner cell in an unexcited state has a negative charge in relation to the outer fluid. The amount of this charge distortion depends on the membrane capacity.

The action potential is a voltage change on the membrane of living cells created by a cell stimulation above threshold. It performs a change in ion conductivity on the stimulated membrane and consists of three phases: brief depolarization, slow repolarization, and successive hyperpolarization. The duration of the action potential depends on temperature and the respective cell type. Motor nerves transmit the action potential along the axon, thus causing the muscle to contract.

Transmission of the Action Potential

Nerve impulse transmission differs fundamentally according to the kind of nerve fiber. Myelinated nerve fibers perform a saltatory conduction and in unmyelinated fibers they transmit impulses smoothly or continuously.

Myelinated fibers consist of axons coated by myelin, which produces an isolating effect. This myelin sheath is interrupted by nodes of Ranvier. Accordingly nerve impulse transmission jumps from node to node over a distance of 1–2 mm, thus reaching velocities up to 120 m/s.

Neuromuscular Transition

The stimulus from one nerve cell to the next or from the nerve to the muscle is transmitted via synapses. Synapses are distinguished by electrical or chemical conduction as well as by their excitatory or inhibitory effect. The chemical conduction is carried out by transmitters. Excitatory synapses, for example, release acetylcholine, adrenaline, nor-adrenaline, and serotonin, thus creating an excitatory postsynaptic potential (EPSP). Inhibitory synapses release inhibitory transmitters like, GABA thus creating an inhibitory postsynaptic potential (IPSP). A possible transmission of nerve impulses results from a summation over time and area of the excitatory and inhibitory effects of the different singular synapses.

A neuron transmits a nerve impulse to a muscle via the motor endplate. Arriving action potentials release acetylcholine, thus producing a depolarization of the endplate. Above a certain critical threshold the muscle membrane is innervated and the muscle fiber contracts. Acetylcholine is split by cholinesterase, thus producing a repolarization of the endplate, after which the initial state is reached again and is ready for a new innervation.

9.1.2 Technical Basics

The methods in clinical neurophysiology match with the spectrum of diagnostic procedures (Fig. 9.2). They complement the diagnostic findings on the basis of the anamnesis and the clinical examination, such as other diagnostic methods like laboratory diagnostics, imaging and functional diagnostics. The total view over all the necessary evaluations and findings brings the physician to the final diagnosis and thus to the therapy. Therapy control and prognostic statements reach objectivity by additional monitoring.

In clinical neurophysiology signals of electrical potential differences are evaluated. They are recorded with needle and surface electrodes. These signals are detected via measuring techniques, which are preprocessed and amplified. The respective amplifier sensitivity and bandwidth depends on the specific recording signal. These biological signals may perform amplitudes from a few microvolt up to millivolts and frequencies from direct voltage up to 30 kHz. Differential amplifiers process and record the data using various

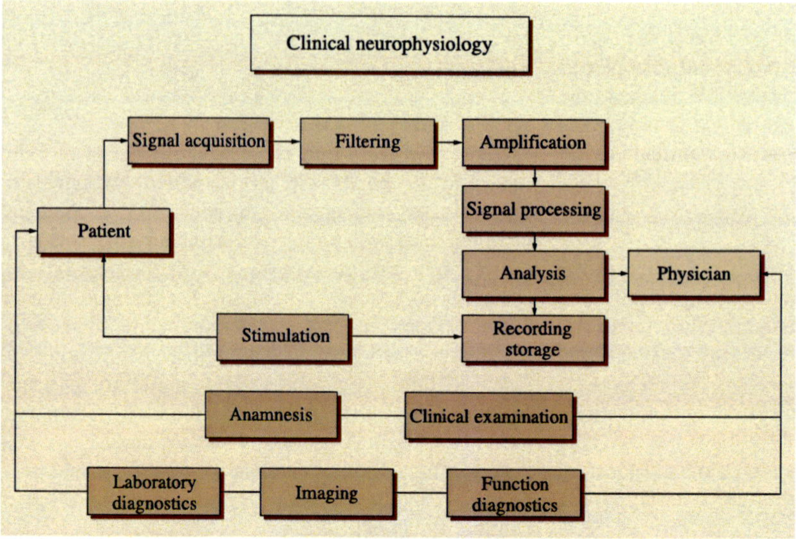

Fig. 9.3 Measuring technique for the recording of bioelectrical potentials in clinical neurophysiology in the process of diagnostic findings

different settings of sensitivities and frequency bands. The signals are evaluated visually as well as computer aided, also providing data storage facilities. The evaluation of the specific signals is supported via software adapted to the various diagnostic problems. The final diagnostic result always lies in the hands of the physician. Patient data, raw signals, analysis results, and findings are stored and printed out. In various diagnostic methods certain stimuli are applied to the patient, such as photostimulation, in the recording of evoked potentials, and in the evaluation of the nerve conduction velocity. In this case, the onset of the stimulus triggers the recording of the evoked action potentials. Different stimulators are applied for various investigations such as electrical, acoustic, optic, and magnetic stimulators (Fig. 9.3).

The purpose of signal recording, filtering, amplification, and evaluation is to record and store the data faultlessly. Reproducibility of the results is ensured by double recording e.g. in evoked potentials. The signal recording is reactionless, which means that the measurement itself does not influence the measuring values or the signal.

Electrodes

Electrodes are the interface between the biological tissue and the technical device. They make direct contact with the skin of the body via electrolytic layers (surface electrodes) or needle electrodes, which are inserted under the skin (subcutaneous) or into the muscle (intramuscular) like a cannula. Various kinds of electrode forms and materials are available depending on the application.

From the electrical point of view electrodes may be displayed as an electromotive force and a network of capacitors and resistors. Their components depend on the electrode material, the electrolytes, the geometry, the current density, and the signal frequency. In a simplified manner the equivalent circuit diagram in Fig. 9.4 can be applied. In this case, E_p means the polarization voltage, C_{tr} the capacitor part of the transition resistance (Helmholtz capacitor), R_{tr} the Ohm resistor part of the transition resistance (Faraday resistance), and R_e the electrode resistance. The Helmholtz capacitor is directly proportional and the Faraday resistance is indirectly proportional to the recording area.

If a metal electrode is put into an electrolytic solution, positively charged metal ions are released into the solution caused by the solution pressure. The osmotic pressure and the field force of the electrical field react against it. This leads to an electrical charging of

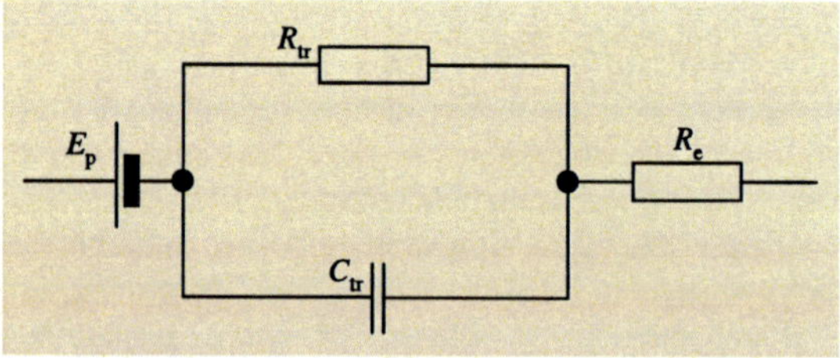

Fig. 9.4 Equivalent circuit diagram of a recording electrode

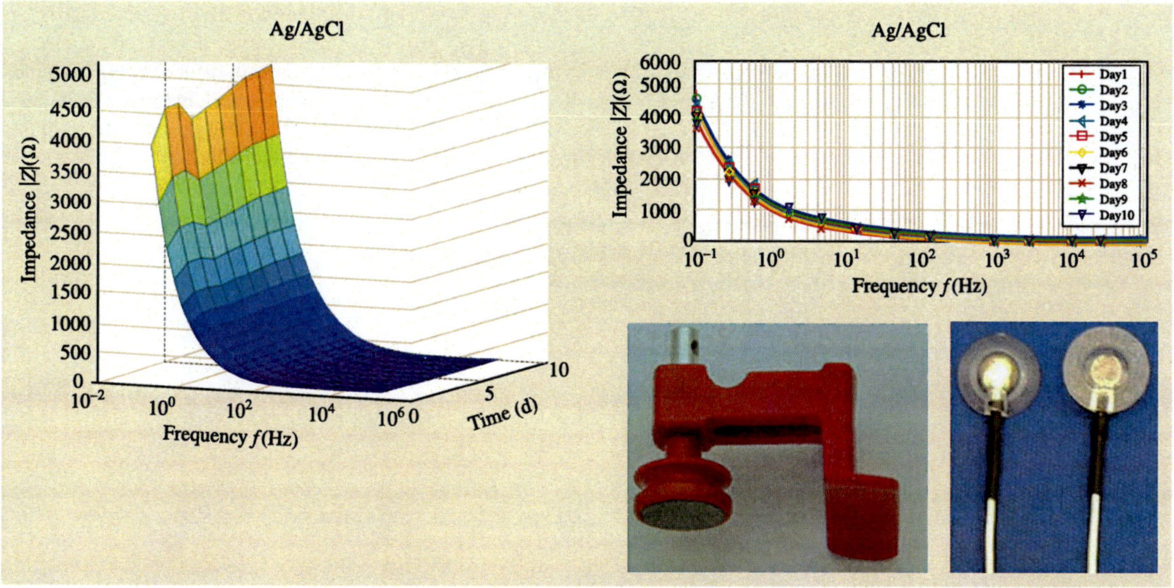

Fig. 9.5 Characteristics of 128 Ag/AgCl electrodes over a time period of 10 days. Display of mean values of the absolute value of the impedance of all electrodes as function of the signal frequency

the electrode and to a layer of opposite charge within the molecular distance of the phase limit. This so called Helmholtz double layer acts like a capacitor within a molecular plate distance, where the plate voltage is seen as an electromotive force and is called Galvani's voltage.

Pure metal electrodes are therefore called polarizing electrodes. If these metal electrodes are coated with an antisoluble salt (e.g. Ag with AgCl), where the same anion must be found in the electrolyte (e.g. in NaCl), then nonpolarized electrodes with much less and more stable Galvani's voltage are achieved. Figure 9.5 displays the characteristics of sintered Ag/AgCl electrodes over a time period of 10 d. An excellent long-term stability and a broad bandwidth down to the low frequency range of these materials are clearly shown. Therefore, Ag/AgCl electrodes are applicable for almost all signals in clinical neurophysiology.

Bipolar Recording

The bipolar recording method is very often applied in clinical neurophysiology for the measuring of bioelectrical potential differences. Two similar electrodes are placed on bioelectrical active areas. Each electrode is connected to one input of the amplifier. A differential amplifier, which amplifies the respective difference between the two electrode potentials, is used. If several electrodes are applied, such as e.g. in the EEG, rows of electrodes are formed. In this case, the output voltages equal the respective difference between the previous and the following electrode. If one assumes a potential distribution where the rising and falling trend is recorded via five electrodes, this results in a measuring set-up as shown in Fig. 9.6 with the center electrode being placed on the point of highest potential.

It is clearly displayed that the output voltages U_1 and U_2 point downwards and the voltages U_3 and U_4 point upwards. At the maximum of the potential distribution a phase shift takes place. Such maximum potentials are found e.g. on the fringe of a tumor, thus enabling a localization of the focus. In bipolar recordings the potential gradient is correctly measured.

Unipolar Recording

In unipolar recordings (Fig. 9.7) all channels are related to one common reference electrode. This reference electrode is preferably placed on a mostly inactive area, so that as few potentials as possible are recorded. Therefore, it is called neutral or indifferent. This point is more hypothetical because the influence of cerebral potentials cannot be excluded. The electrode that is placed on a bioelectrically active area is called the active, exploring or different electrode.

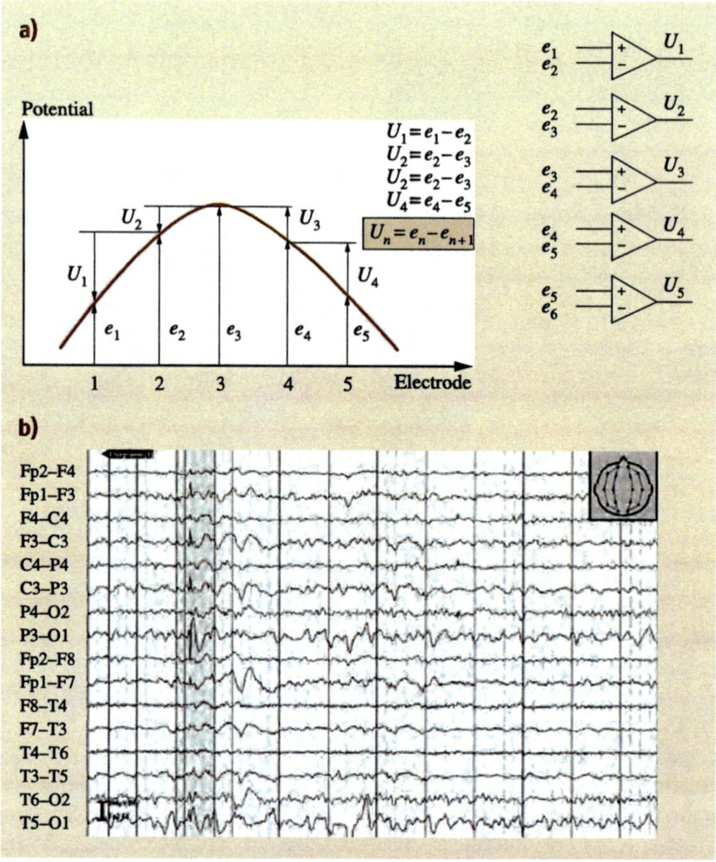

Fig. 9.6a,b Bipolar recording. **(a)** Schematic display of the electrode circuit, **(b)** Recording example. The phase shift between C3–P3 and P3–O1 is clearly demonstrated in the *marked* signal interval

Figure 9.7 shows a unipolar recording for the aforementioned potential distribution. The potential e_r was taken for the reference electrode. It is clearly displayed in the graph that no phase shift takes place in unipolar recordings. A possible focus is localized by the size of the correctly recorded potential difference.

Recording Against an Average Reference

In a recording against an average reference (Fig. 9.8) the reference point is the averaged value of all electrode potentials. This can be realized by software or by connecting all electrodes via resistors of the same value with a reference point. This potential relates exactly to the average value of all electrode potentials, because the sum of all currents at this point equals zero. In this recording a phase shift also takes place. The output voltage equals the difference between the respective electrode potential and the reference potential.

Source Recording

In source recordings the electrodes are also placed on electrically active areas. The recording is carried out against a reference that takes the immediate surrounding electrodes into consideration. In contrast to the recording against an average reference its potential does not influence the reference potential.

In source recordings (Fig. 9.9) the difference between the potential of the electrode of interest is formed with the weighted average value from the potentials of the surrounding electrodes. The weight factor w is calculated from the reciprocal value of the distance. In a square shaped electrode setting this factor is $w = 1$ for the single, $w = 0.5$ for the double and $w = 0.707$ for the diagonal. To obtain the weighted average value, the sum of the weighted electrode potentials are divided by the sum of the weight itself.

Source recordings are applied because they produce a better display of local events by eliminating signals

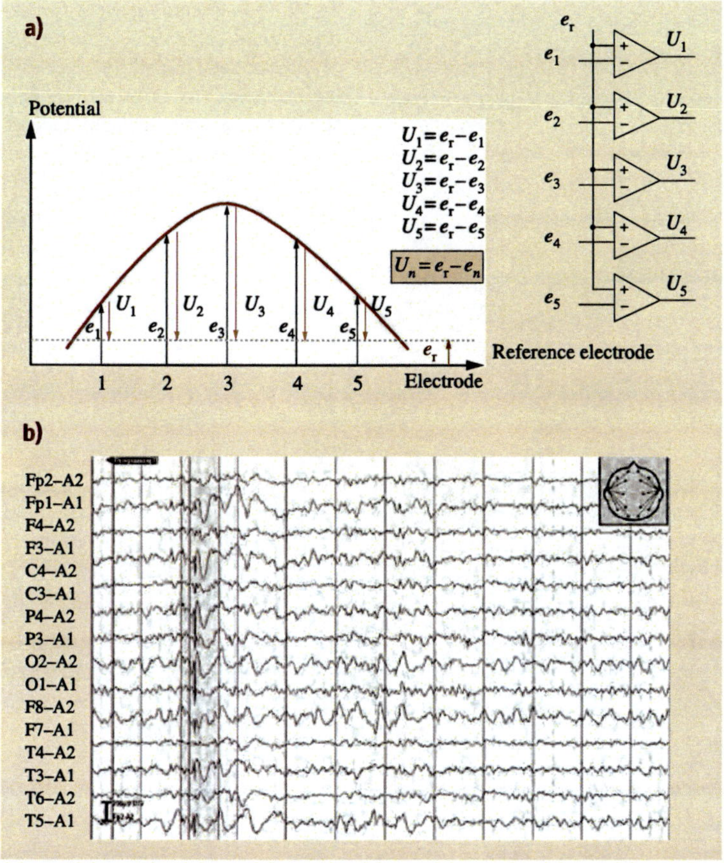

Fig. 9.7a,b Unipolar recording. **(a)** Schematic display of the electrode circuit. **(b)** Recording example. The same signal interval as in Fig. 9.6 is marked here as well. The spike shaped activity is imposed by a high amplitude in channel P3–A1

from remote areas in the actual local potential such as e.g. the influence of the EOG when opening the eyes. However, in source recordings the actual local potential itself is not recorded in a more precise manner.

Electrode Placement

For reproducible recordings and for the comparability of results in trend studies it is necessary to always place the respective electrodes in the same positions. The applied system cannot be too rigid, but must adapt to different head sizes (neonatal EEG, EEG in children and adults). Therefore, these points on the scalp must be easy to find and reliably defined. These points are the nasion (the deepest point between nose and forehead, right in between the eyes), the inion (the lower bone hunch at the middle back of the head in between the onset of the neck muscles), and both preauricular points (cavity in front of the outer auditory canal directly below the zygomatic bone and above the lower jaw joint).

The connecting lines between nasion and inion as well between the preauricular points cross at the vertex (C_Z). These connecting lines are divided into 10% and 20% sections. The electrodes are placed at the crossing points of these longitudinal and lateral rows. The designations of these positions are named according to the regions of the cerebral cortex: Fp = frontopolar, F = frontal, C = central, P = parietal, O = occipital, T = temporal, A = auricular, and also cb = cerebellar as well as pg = pharyngeal. The electrode numbers are related to hemisphere (odd numbers are on the left, even numbers on the right side) and the distance to the center line, and also z = zero. The distance between the electrodes within the various rows is equal.

The line partition in 10, 20, 20, 20, 20, and 10% sections result for the lateral longitudinal row in the electrode positions Fp_Z, F_Z, C_Z, P_Z, and O_Z, as well as for the center lateral row T3, C3, C_Z, C4, and T4. No electrodes are placed on the positions Fp_Z and O_Z no electrodes. If in the same way the lines between Fp_Z,

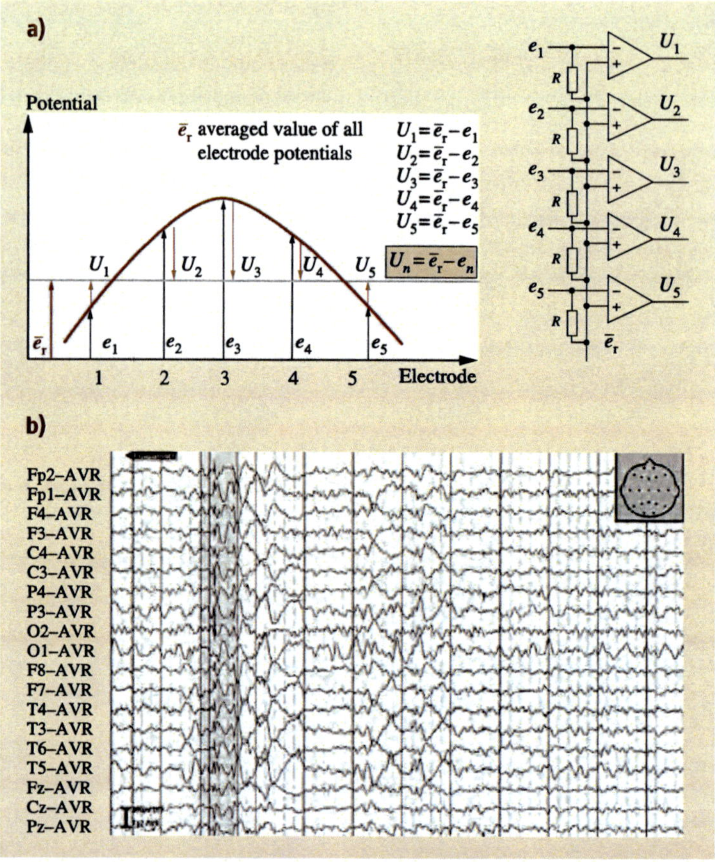

Fig. 9.8a,b Recording against an average reference. **(a)** Schematic display of the electrode circuit. **(b)** Recording example. The spike activity is shown with a very high amplitude of ca. 300 μV on channel P3–AVR

O_z, which are led via T_3 and T_4, are divided, on the left hemisphere this results in the positions Fp_1, F_7, T_3, T_5, and O_1 as well as Fp_2, F_8, T_4, T_6, and O_2 on the right hemisphere. Over the electrodes Fp_1, C_3, and O_1 as well as Fp_2, C_4, and O_2 two parasagittal longitudinal rows can be drawn, which cross the frontal (F_7, F_z, and F_8) and the lateral row in the back (T_5, P_z, and T_6). These crossing points form the electrode positions F_3, F_4, P_3, and P_4. In between these standard positions further electrode positions may be added e.g. F_1, F_2, F_5, and F_7 for the frontal lateral row. The ground electrodes are fixed on the earlobes, relating to the positions A_1 and A_2.

Amplifiers

In clinical neurophysiology differential amplifiers are used for the amplification of the recorded biosignals. These amplifiers, whose principles were developed in 1931/32 by Tönnies at the Kaiser-Wilhelm-Institute in Berlin, amplify the difference of two input signals. Practically, the amplification of common mode signals (e.g. equal phase disturbing signals, which derive from leading cables) differ from push–pull signals (e.g. a potential difference deriving from a bioelectrical generator). The relation of these two amplifications is called the common mode rejection ratio (CMRR). This CMRR should stay in the range of 80–120 dB, which means that the amplification of push–pull signals is 10 000–1 000 000 times larger than the amplification of common mode signals.

Differential amplifiers have an inverted and a non-inverted input. According to the agreement of Lyon in 1980, electrical connections are such that a negativity a different electrode is displayed with a positive polarity on the recording device. Another feature of the amplifier is the amplification factor, which is defined as the quotient between the output and the input voltage. The input impedance of the amplifier should be high (50–200 MΩ). This is necessary because the potential source should by no means be loaded by current, so the

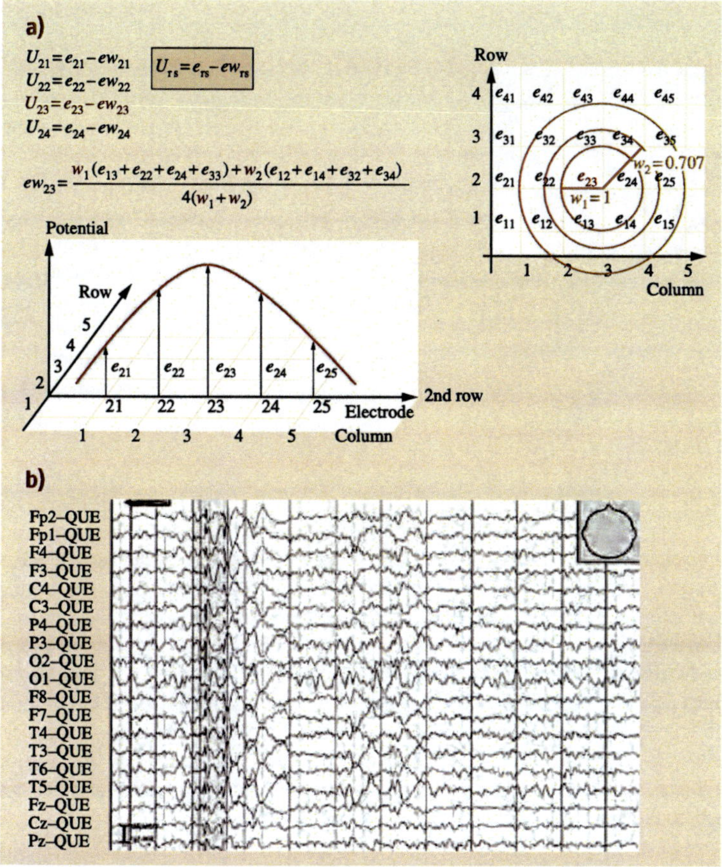

Fig. 9.9a,b Source recording. **(a)** Schematic display of the electrode circuit. **(b)** Recording example. P3–QUE shows a high amplitude here as well

measurement is carried out as reactionless. In addition, the influence of the electrode transition impedance can be minimized.

The frequency response is a measure for the dynamic transition process of electronic devices and displays the dependence of the amplification on the frequency of the signal. It can be calculated from the quotient of the complex amplitude of the output and input signal for a stationary sine wave. A high pass filter enables signals above a certain threshold frequency to pass without significant loss. The high pass filter is thus used to reduce low frequency disturbing signals, such as zero line drifting. Sometimes, a time constant value is given instead of the threshold frequency, deriving from an exponentially falling response of the high pass filter to a step function. In this case, the time constant is the time interval in which the amplitude drops to 37% of its initial value. In contrary, the low pass filter enables signals below a certain threshold frequency to pass without major loss. The low pass filter is used to reduce high frequency disturbances such as noise.

However, the phase response is based on a phase difference that usually occurs between the input and output voltage of a recording system. Another feature of the amplifier is the noise, whose name is taken from the acoustic sound image. It is actually a small disturbing signal over a wide frequency band, which can be recorded at the output when the input is short circuited. The cause of this may be found in the thermal electrode movement in resistive components.

Recording/Storing

Apart from the raw signals other values such as the patient's name, the date, the recording program, and the respective result texts are stored in a data bank. At a single glance all recorded examinations together with their measuring results are available and in trend studies all this data may be immediately recalled.

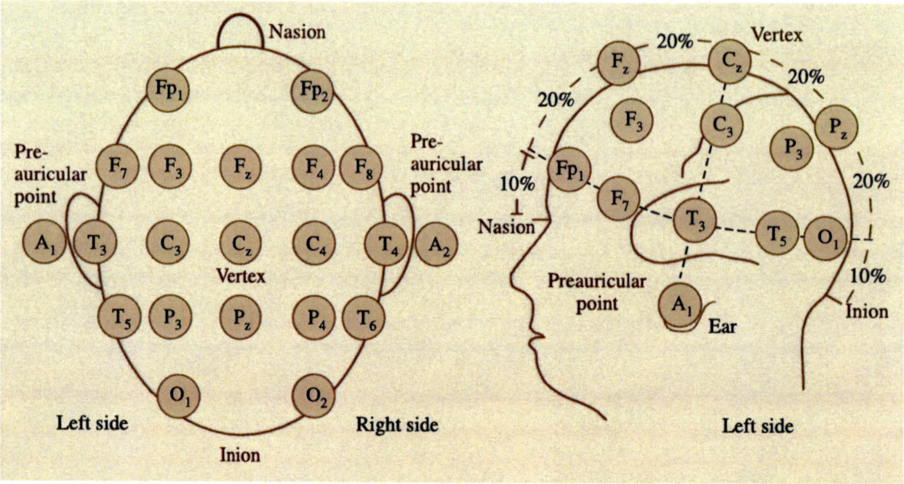

Fig. 9.10 10–20 system for electrode placement (10/20-system) Even-numbered: *right scalp*, odd-numbered: *left scalp* (F = frontal, C = central, P = parietal, O = occipital, T = temporal, A = auricular)

Network Systems

In clinical neurophysiology departments several workstations are installed for signal evaluation. In most cases, the specific recording devices are equipped for specific applications such as EEG, EMG/ENG, and evoked potentials, with the respective hardware like stimulators, examination couch, electrodes etc. There is also a difference in the workstations with respect to how the EMG examination is carried out by a specialized physician or the EEG is carried out by a medical-technical assistant in functional diagnostics. A network solution between the recording stations and establishing evaluation stations is recommended. In this case, all the data are available on each PC and for the final results the physician may recall all the information in his room.

Computer systems for neurological practice consist mostly of a recording and an evaluating station. They usually operate with a simple network function called *peer-to-peer*, where the data transfer only has to be managed between two workstations. In larger systems from four recording stations upwards and installation of a local area network (LAN) with its own server is recommended.

Signal Processing

Signal processing plays a great role in clinical neurophysiology. After preprocessing and analog/digital conversion the data are available for further evaluation. Most devices operate on a software that displays, measures, detects simple artifacts, and carries out a first analysis. The spectrum of possible analysis programs varies greatly. Fast Fourier transformation (FFT) for frequency analysis, averaging to enhance the signal–noise ratio in evoked potentials, the application of neuronal networking for pattern recognition, the application of special filters, the analysis of sleep stages, and the localization of sources of bioelectrical activity are some examples. In most cases, the machines offer direct access to the raw data, so that through using MATLAB or LabView the user's own application software can be installed and run as well.

Artifacts

Artifacts are disturbing potentials, which superimpose the measuring signal and falsify it. Their causes may be found in technical or biological processes.

Biological artifacts are generated by the patient himself. They are physiological signals that superimpose the desired potential, e.g. the EEG in evoked potentials or the ECG and EOG in the EEG. Muscle tension, movement, or sweating of the patient cause additional potentials.

Technical artifacts are generated by the machine itself or coupled in from outside. The artifacts of the measuring device are the amplifier noise, 50/60 Hz noise caused by lacking or insufficient grounding, and the application of unsuitable electrodes. Artifacts are coupled in galvanically, capacitive or inductive.

A possible artifact rejection may consist of a 50/60 Hz band filter. Averaging selects the signal responses correlated to the stimulus out of the background activity. Special algorithms and the application of neuronal networks may be used for detecting biological artifacts. Adaptive filters may serve for artifact rejection as well.

Safety Aspects

For devices used in clinical neurophysiology the general basic requirements for the application of medical technical products are validated, which means that the respective machine, the installation, and the operation must be safe. The compliance of technical parameters such as the patient's leakage current or the machine's leakage current must be checked in defined intervals.

All the above mentioned examination methods should be carried out in a separate room. Such rooms should not be located near great current consumers such as elevators, radio stations, magnetic resonance tomographs (MRT), or computer tomographs (CT), if possible. Elevators should be at a distance of 10–15 m away. In clinics MRT and CT are usually installed at some distance from other functional units. Generally speaking for neurophysiological examinations nowadays no special room shielding in the sense of a Faraday cage is necessary, because modern amplifiers have a high common mode rejection ratio. However, one should always consider electrostatic charge from shoes and synthetic fibers, which can easily be discharged through grounding. The recording seat or the examination couch as well as the patient must be connected to ground potential.

9.2 Electroencephalograph

Electroencephalography (EEG) records electrical potential differences whose source lays in cerebral processes with the help of electrodes that are normally placed on the intact scalp. EEG-machines find their application in neurophysiological functional diagnostics in neurological private practices, in neurology departments in hospitals, in neurology clinics, as well as in epilepsy centers, neurological rehabilitation clinics, and psychiatric clinics. Furthermore, they are used in the diagnostics of neurological-psychiatric sleep disorders, in drug studies, and in clinical research. They are also applied in neurosurgery and in the monitoring of intensive care units. The diagnostic value of the EEG is found especially in the area of epilepsy (classification, therapy control), the diagnostics of diffuse cerebral functional disorders (inflammatory diseases, cerebral haemorrhage, metabolism, drugs), functional disturbances because of space occupying lesions (increase of brain pressure, haemorrhage, tumors, craniocerebral injury), sleep diagnostics, brain death detection, vigilance disturbances, and the determination of the depth of narcosis. An EEG was recorded for the first time on a human in 1924 by Hans Berger in Jena.

9.2.1 Signals

The EEG is generated through surface near nerve cells of the cerebral cortex (Fig. 9.11). The cerebral cortex is about 3 mm thick and consists of six layers, among them are stellatum cells, stellate and pyramidal cells. The apicale dendrites of the latter cells are found in five of the six layers. An excitatory postsynaptic potential (EPSP) on the cell soma creates a dipole because of the depolarization with a positive pole near the surface. Inhibitory postsynaptic potentials (IPSP) on the cell soma create on opposite polarization due to hyperpolarization as well as the EPSP on the apicale dendrites. EPSP and IPSP could superimpose by successive potentials in time as well as potentials distributed in space, which occur synchronously in neighboring synapses. During a recording from an intact skin surface the potentials are deformed when traveling in between the cerebral cortex and the various layers of the skin with their different tissues. Resistors and capacitors of these layers act like a low pass filter, wherein frequency components below 1 kHz are mostly eliminated.

The EEG potentials form complex waves, which in their shape and size depend on neuronal factors in the cerebral cortex as well as on the gender, respiration, metabolism, homeostasis, oxygen and carbon dioxide concentration of the blood, blood sugar, drugs, and toxicants. They also depend on physiological factors: sleep or wake state, open or closed eyes in the wake state, on the general vigilance, and on age. The amplitudes of the EEG increase from a newly born to an infant and decrease from an adult to a person of old age, whereas the frequency of the waves increase from the infant up to old age.

Depending on the frequency alpha-waves (8–13 Hz), beta-waves (above 13 Hz), theta-waves (4–7 Hz), and delta-waves (below 4 Hz) are distinguished. The EEG is described by frequency, amplitude, occurrence, mod-

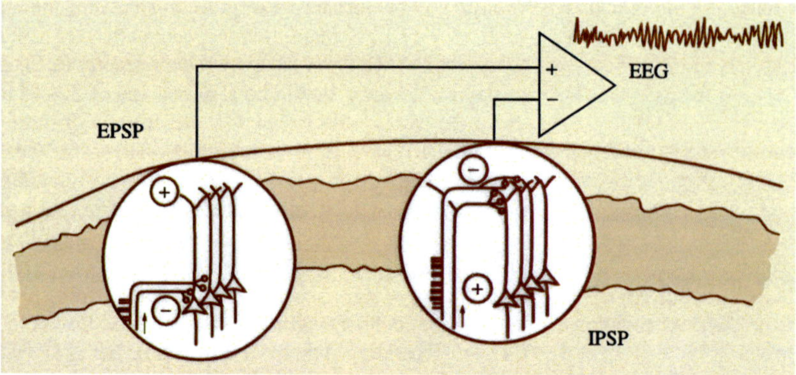

Fig. 9.11 Generation of an EEG

ulation, symmetry, and sensitivity. Special waveforms are spikes, sharp waves, spike and wave complexes (SW-complexes), and further waveforms such as μ-waves, vertex-waves or K-complexes (Table 9.1). The evaluation of the EEG is carried out under clinical aspects. Here, in his findings the physician evaluates the regularity, dominant frequency, predominant high or low amplitudes, physiological local distribution, and sensitivity, as distinguished from the basic activity continuous, discontinuous, generalized or localized EEG activity.

Figure 9.12 shows a normal EEG of a 21 year old with a blockage of alpha-waves after eye opening. In comparison, Fig. 9.13 displays an EEG with an epileptic pattern of a 31 year old male patient. The EEG-activities differ in relation to their amplitude and frequency.

9.2.2 Device Technology

In the last 15 years a considerable transition has taken place in EEG devices from paper based thermal recorders to computer EEG. This relates, on the one hand, to the hardware, consisting of a PC, A/D converter, and amplifier head box, and on the other, to computerized analysis and findings based on modern software as well as digital media for signal storage and archiving. Therefore, the general appearance of an EEG device has changed drastically (Fig. 9.14).

A standard component of an EEG device is a photostimulator. It generates short bright flashes of light with a defined luminosity and frequency. In the case of double flashes the time between the two flashes is also adjustable. The main component of a photostimulator is a stroboscope or a LED array.

Electrodes

For the recording of the EEG sintered Ag/AgCl surface electrodes are applied (Fig. 9.5). They provide an optimal recording even in the low frequency range. The contact between the electrode and the scalp is produced via an electrolyte or with sodium solution soaked felt tips. The scalp is preprocessed in order to reduce the electrode transition resistance, which should be below $10\,k\Omega$.

Table 9.1 Classification and characteristics of EEG waves and singular waves

Wave nomination	Wave frequency	Amplitude	Wave type	Occuring during
Alpha-waves	8–13 Hz	30–50 μV	Fast, physiological	Wake, eyes closed
Beta-waves	> 13 Hz	≈ 20 μV	Fast, physiological	Wake, eyes open, counting
Theta-waves	4–7 Hz	Up to 500 μV	Slow, physiological	Light sleep
Delta-waves	0.5–3.5 Hz	Many 100 μV Up to 1–5 mV	Slow, physiological	Deep sleep
Sharp waves	≥ 80 ms	Variable	Steep abnormal wave	Normal and abnormal
Spikes	≤ 80 ms	Variable	Steep abnormal	Mostly abnormal
Spike series	5–10 times with ≤ 80 ms	≥ 50 μV, often very large	Steep abnormal	For example in epileptic seizures
Spike-wave-complex	3/s	10–100 μV	Many different kinds	Always abnormal

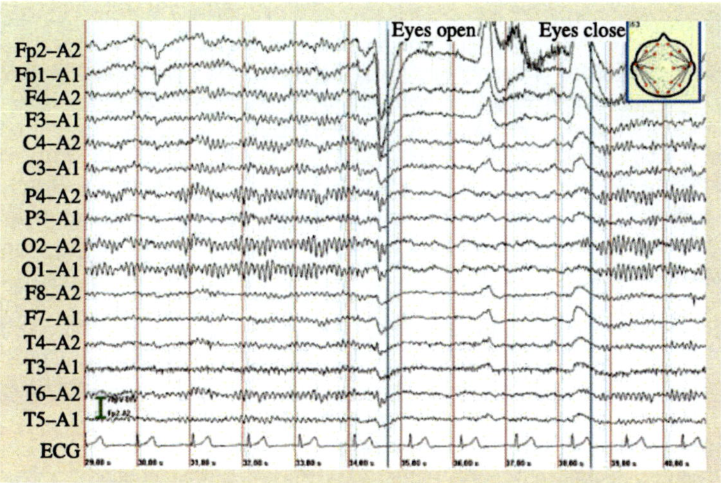

Fig. 9.12 Normal 16-channel EEG of a 21 year old male patient in unipolar recording. The connecting scheme of the electrode is displayed

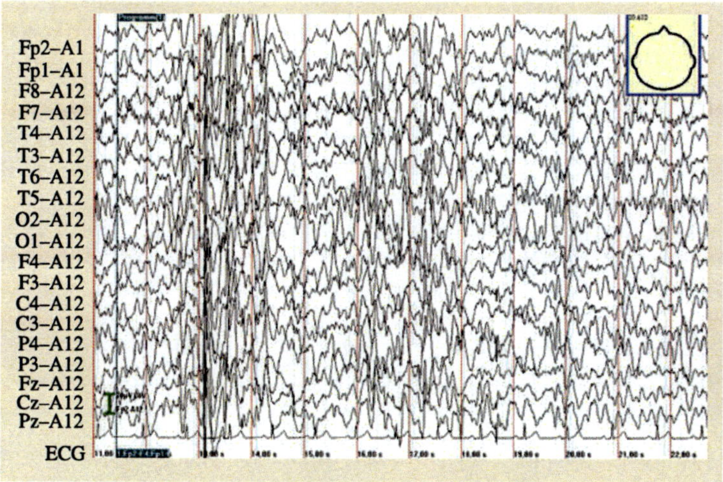

Fig. 9.13 EEG with epileptic patterns of a 31 year old male patient

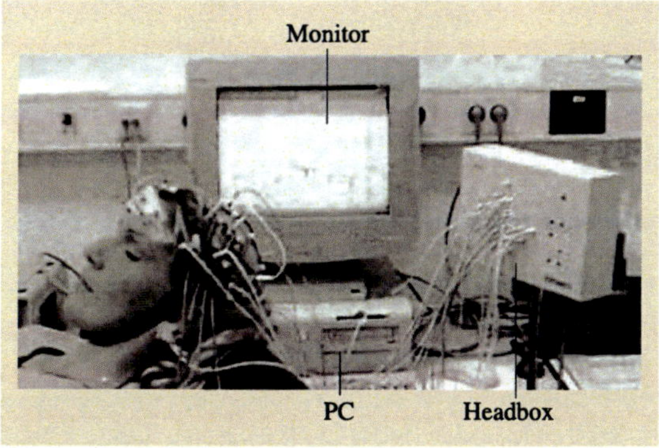

Fig. 9.14 EEG recording in a laboratory

Amplifier

A preamplifier (head box) is mounted on scaffolding near the patient's head and serves as an electrode connector and preamplifier of the EEG signals. From there they are transferred to the main device. The preamplifier has an A/D converter incorporated, therefore the digitalized signals can be faultlessly transferred via great distances without any substantial loss. The preamplifier is a differential amplifier with high demands on the signal amplification and free of distortion. Its quality criteria is a high common mode rejection ratio and a high input impedance (> 120 dB and > 100 MΩ).

For a high quality recording an active electrode, a reference electrode, and also a ground electrode must be connected to the differential amplifier. Depending on how the electrodes on are related on the respective points of the scalp and connected among each other, different EEG recordings are attained. For the interpretation of the recorded EEG the connection of these inputs per channel to the preamplifier is of vital importance; these settings are called recording programs or montages and they have to be defined and directly related to the signal. One distinguishes between: unipolar or reference recordings, bipolar recordings, and source recordings.

9.2.3 Methodology

Basic Requirements

The German Society of Clinical Neurophysiology has defined minimal demands for the recording of EEGs. An EEG device should have minimally 10 EEG amplifiers and one ECG amplifier. Electrode placement must be carried out according to the 10/20 system. Before and after a recording the device must be calibrated and the electrode transition impedance must be documented. The recording should last at least for 20 min and has to consist of reference montages as well as bipolar longitudinal and lateral row montages. The sensory sensitivity e.g. open and closed eyes (Berger effect) as well as photostimulation and hyperventilation must be included. Artifacts during the recording must be designated and possibly corrected. All important recording parameters, such as the technical settings as well as the behavior and situation of the patient, must be documented along with the recorded signals.

Recordings

During the EEG recording different montages or connections of the electrodes mounted on the scalp are applied. Principally the EEG montages displayed in Table 9.2 are available.

Usually an EEG is recorded with an amplifier sensitivity of 70 μV/cm, a paper feed and a time base of 30 mm/s, with a time constant of 0.3 s (equalling a high pass filter with a threshold frequency of 0.53 Hz) and an upper frequency limit of 70 Hz. These values should be changed only in well considered circumstances. It is recommended to avoid a mains filter of 50/60 Hz.

Table 9.2 Overview and characteristics of clinically applied EEG montages

Montages or recording programs	Pin configuration of the differential amplifier	Characteristics
Unipolar or reference recordings against a reference	Different electrode with inverting input Common reference electrode on vertex or on ear Ground electrode on contralateral earlobe	Common potential parts are highlighted, Better display of generalized activity Amplitude and phases well recognisable and comparable
Reference recordings against averaged references – common average	Different electrode with inverting input Averaged reference out of summated signal of all electrodes Ground electrode on earlobe	See above, but common potential parts are not highlighted any more Recommended for signals with many artifacts
Bipolar recordings such as lateral rows, longitudinal rows, circumferences etc.	Two neighbored electrodes on bioactive area Ground electrode on earlobe	Potential differences between the electrodes are highlighted Better display of focal activity Gradient directly comparable
Source recordings – special unipolar montages with special averaged and weighted references	Singular active electrode, here called "source", with inverting input Averaged reference from weighted summated signals of all electrodes, which are located around the active electrode (= source) Ground electrode on earlobe	Focal activities are better highlighted Amplitudes and phases remain well comparable

The concrete connection of the electrodes depends on the number of recording channels. Examples for a 16 channel recording device are given in Fig. 9.15

Methods of Provocation

In routine diagnostics photostimulation and hyperventilation are often applied as activation or provocation methods.

Hyperventilation is a provocation method that is very easy to apply. The patient is asked to activate his respiration in frequency and amplitude for 3 min, e.g. to breathe deeply and regularly at about 30 breaths per minute. During the 3 min hyperventilation the potentials are recorded mainly from the frontal, central, occipital, and middle temporal cerebral regions.

Photostimulation is carried out in a darkened room with the help of a photostimulator. The intensity of the flash lights and their frequency should be adjustable. Photostimulation is a standardized procedure: flashes are applied for at least 10 s while the patient has closed eyes. During a duration of 2 min of photostimulation mainly the frontopolar, frontal, and occipital cerebral regions are recorded.

Further routine activation methods are sleep deprivation (awakened night) or sleep recordings (after sleep deprivation, midday sleep, rare drug induced sleep).

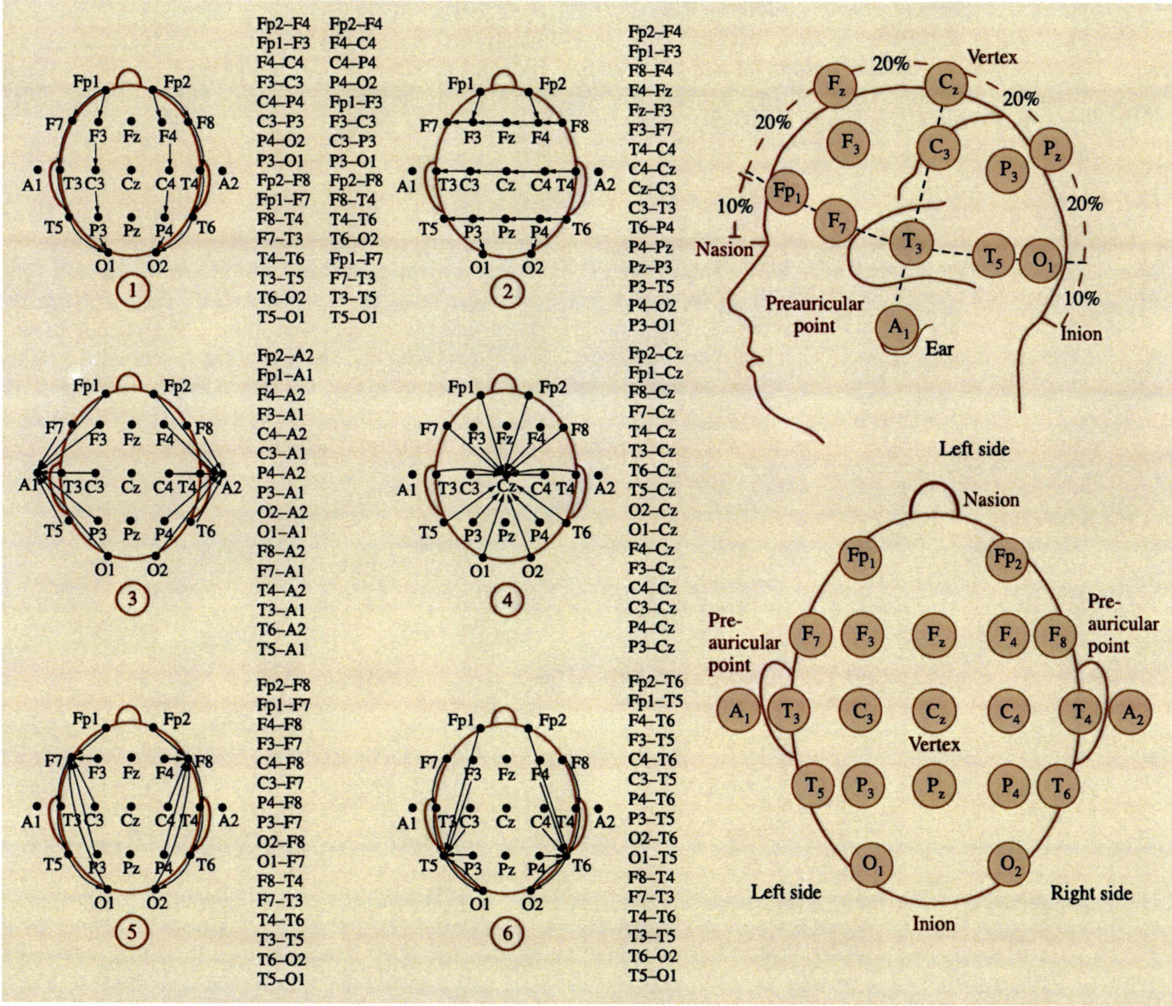

Fig. 9.15 Examples for EEG connections for a 16-channel recording device

The duration of these recordings is between about 30–60 min.

9.2.4 EEG Recording Methods

In relation to the various diagnostic applications in neurology the following methods of EEG recordings have developed been: routine EEG in neurological practice and clinics, ambulatory 24 h long-term EEG, video EEG, portable EEG, sleep EEG, and pharmaco EEG (Fig. 9.1, Table 9.3).

Routine EEG

EEG recordings should be carried out in a quiet room. The patient should position himself on an EEG couch in a half sitting, half lying position. Head and neck are to be as relaxed as possible to avoid muscle artifacts that disturb the recording. Principally, the patient's eyes are to be closed during the recording and the recording time should not fall below 20 min.

Long-term EEG

Long-term EEG recordings are carried out for 24 h in an ambulant manner with an 8-channel and 12-channel EEG amplifier. The aim is to find rarely occurring events that cannot be detected by conventional recordings. Raw EEG signals are, therefore, completely stored. The patient carries on him a small portable battery driven data recording device or the data is transmitted via telemetry to an evaluation station. All physiological artifacts such as speaking, chewing, and eating are recorded as well, thus producing a lot of disturbed signal intervals makings it more difficult to interpret the EEG.

Pharmaco EEG

A specialized area is the recording of an EEG for pharmacological purposes to evaluate the influence of drugs on the function of the central nervous system (pharmaco EEG). In particular, the reproducibility of the same or similar changes in the EEGs of different test patients is detected. Here different analysis procedures have been evaluated, for example, brain mapping. Frequency analysis calculates the respective quantified wave components in a statistic way during the recording, and the changes in the EEG are recorded and quantified over larger periods of time.

Video EEG

Parallel to the EEG routine recording a video camera records and stores the image of the patient in a time synchronized way. Also, infrared camera systems are used in order to record the video image from patients in darkened rooms. The signals of the computer EEG and the video are synchronized up to a precision of a few milliseconds. The EEG data is

Table 9.3 Application, number of channels, and recording methods in clinically standard EEG recordings

EEG method	Application	Number of channels	Recording time (average)	Evaluation
1. Routine EEG	a) Practice b) Clinic c) Epilepsy	8–12 16–24 19–32–64	ca. 10 min 20–30 min 20–40 min	Visual, computerized
2. Long-term EEG	Clinics, in addition to routine EEG, e.g. in epileptic patients	8–12	Max. 24 h	Semi-automatic computerized event recognition algorithms
3. Video EEG	Epilepsy, also more in clinics	19–24	10–60 min, event related	EEG signals directly synchronized with the video image of the patient
4. Portable EEG	Ambulant recording in intensive care, in inner medicine, and for brain death recognition	8–12	10–30 min, also 1–2 h, with display of all EEG signals	Visual, in brain death recognition additional requirements on high signal resolution
5. Sleep EEG	Sleep disorders of neurological and psychiatric origin	12–24 with polygraphy	At least 8 h at night	Semi- and automatic sleep stage recognition
6. Pharmaco EEG	Drug studies	12–24	20–30 min depending on aim of study	EEG mapping, computerized frequency analysis

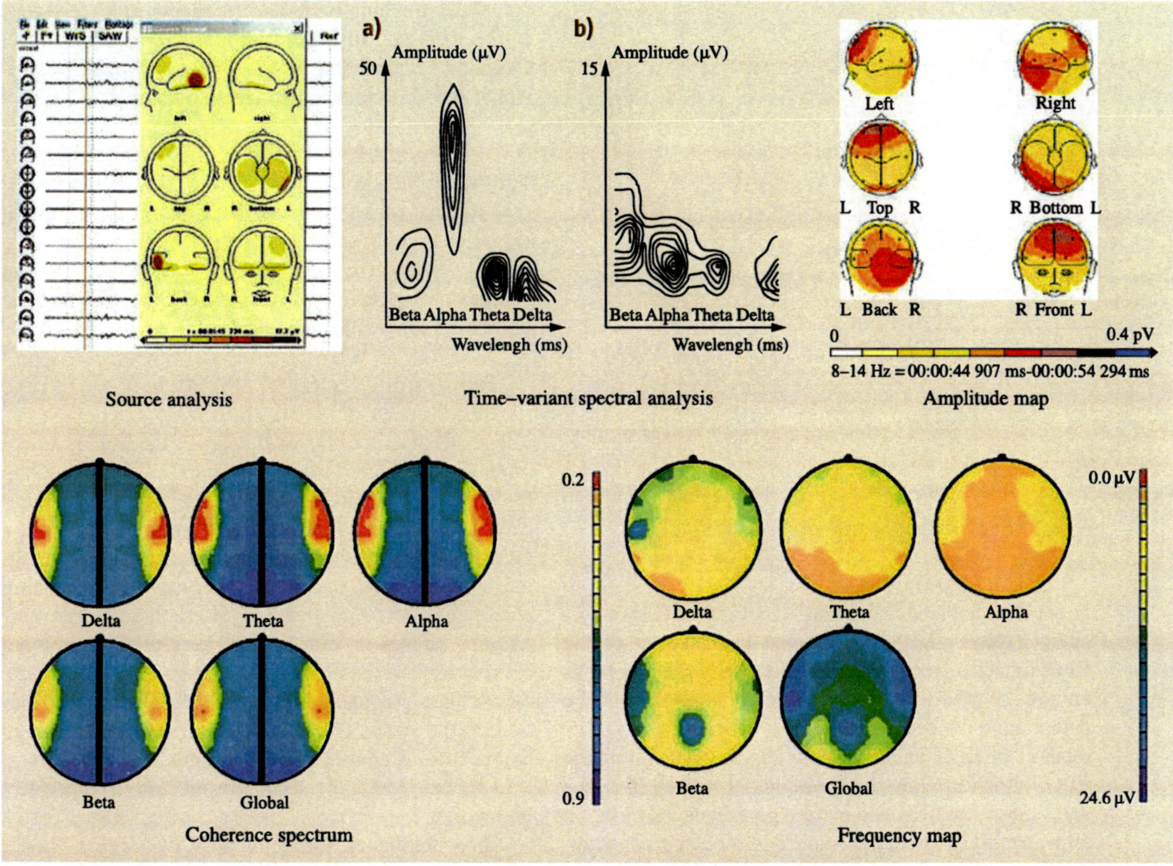

Fig. 9.16 Examples of signal analysis

displayed on the screen of a PC monitor with maximum resolution and free of flickering synchronous to the video image; even different montages may be selected.

Video EEG is considered especially important for diagnostics in epilepsy. The physician can easily evaluate from the video image when and how the onset of an epileptic seizure takes place. In relation to the respective EEG, he is able to localize the focus or the starting point of an epileptic activation. The actual recording time can be assigned as a search criteria, thus displaying the EEG and the video image at the same time.

Sleep EEG

In neurological-psychiatric sleep disorders the EEG is recorded for diagnostic purposes during the night's sleep for at least 8 h. To obtain a sleep profile with its related sleep stage determination more polygraphical channels are necessary: eye movements via the EOG, muscle tone from the chin via surface EMG, ECG, respiration and respiratory effort. Video survey of the sleeping patient is often applied.

Portable EEG

EEG recordings in the intensive care unit as well as brain death detection require a handy compact EEG unit. A notebook with a battery driven head box is used for data recording and storage. For brain death detection the German Society of Clinical neurophysiology has set special requirements for portable EEG devices: recording time should be at least 30 min, there should be double electrode distances, a 0.53–70 Hz bandwidth, a sensitivity of 2 μV/mm, and a minimum of 8 channels. The ambulant recorded EEG on the notebook may subsequently be dubbed on the central EEG archive.

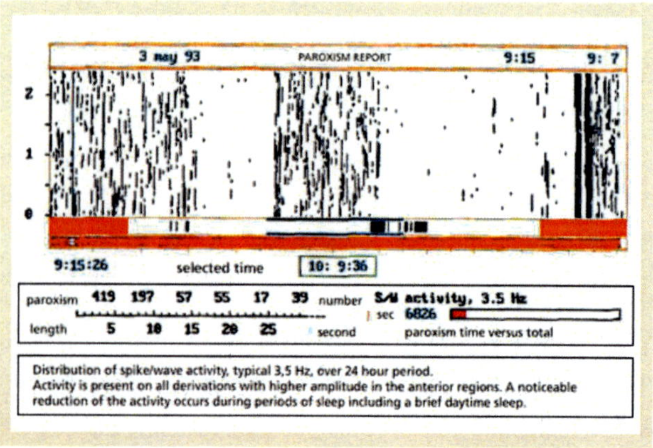

Fig. 9.17 Software for the detection of paroxysm changes in long-term EEG

9.2.5 Evaluation and Signal Analysis

EEG Signal Analysis

In addition to the visual evaluation of the EEG by a specialized physician, a computerized evaluation of the EEG signals is frequently used because of the enormous amount of raw data. Here, frequency analysis and the detection of defined characteristics are helpful. Another method of EEG analysis is the dipole analysis of the EEG, which provides a dipole model related to space and time of the EEG and evoked potentials, resulting in a localization of possible sources in bioelectrical activity. The auto correlation function of the EEG leads to the dominant frequency in the signal of an EEG epoch and its drop gives a measure of the stochastic appearance of the EEG. The cross correlation function is used to detect a common frequency within two different epochs of the EEG, e.g. between two recording channels. Time variant spectral analysis is also to be mentioned, which calculates the amplitude and frequency for each wave in the EEG and displays these in a three-dimensional frequency distribution. In Fig. 9.16 some examples for signal analysis are displayed in an overview.

Subsequent Remontage of Signal Epochs

For the evaluation of the EEG and the localization of pathological waveforms it is helpful to evaluate certain signal epochs in different montage settings. The possibility to view the same signal epoch under different *perspectives* gives the physician the chance to find the optimum montage for his findings.

Topographic Display of Amplitude and Frequency Distribution

The result of a topographic display of the EEG is a two-dimensional colored map of the cerebral activity along the surface of the scalp. Often the amplitudes, frequencies, and power spectra derived from the frequency analysis are displayed in a topographic way. The signal amplitudes of the area in between the recording electrodes are calculated via interpolation. The amplitudes gained in such a way in a two-dimensional distribution are thus quantified and given a color code, delivering a graduated picture of the potential distribution on the scalp. The display of an amplitude map in an epoch of a couple of milliseconds in a film, if e.g. a focus or a generalized activity occurs, is especially impressive.

In a frequency mapping an EEG signal epoch, whose amplitude is a function of time, is transformed via FFT into a signal, whose amplitude is a function of frequency. The result is displayed in a color code analog to the amplitude mapping, where the different colors are related to the occurring wave frequencies.

The power spectrum is calculated out of the signal transformed by the FFT. It displays the power that is

Table 9.4 Table of physiological and technical artifacts during an EEG recording

Physiological artifacts	Technical artifacts
EMG potentials from: frowning, movement of the lid, chewing, head and neck movements	Bad contact impedance of the electrodes caused by bad electrode application and slipping of electrodes
ECG-peaks or temporal pulse waves interspersing	Loose contact in the cables caused by cracks
EOG as eye movements in the frontal EEG	50/60 Hz noise from outside
Sweating – as slow fluctuations in the EEG	Large inductivities, as in neighboring elevators or radio or TV stations near-by
Respiratory movements – occipital, head movements in breathing rhythm	Electrostatically charged shoes or clothes with synthetic fibers

contained in a specific frequency or an EEG wave out of the total sum of the signal. These results are mostly displayed in the form of a histogram. The envelope of this histogram is a very obvious display of the changes of the frequency components during the different recording phases e.g. for the evaluation of drug effects or during narcosis.

Long-Term EEG Analysis

The interpretation of an ambulant long-term EEG recording is a time-consuming task because of the great amount of recorded data. An example for such a kind of analysis software is one that looks for special paroxysms by searching in the EEG raw data for the occurrence of peak values, their area, and their signal duration. The result of these calculations is a condensed display of the long-term EEG, which enables the physician to differentiate without any difficulty between paroxysmal events in the EEG from artifacts and normal EEG epochs (Fig. 9.17).

Final EEG Results

After the visual observation and evaluation of the EEG, the physician writes a final result text, with or without the help of computer-aided analysis. This result text is stored and archived together with the EEG signals.

Artifacts in the EEG

During the recording different artifacts may occur, which are differentiated as physiological and technical artifacts. They should be detected, classified, documented, and removed as soon as possible (Table 9.4). In

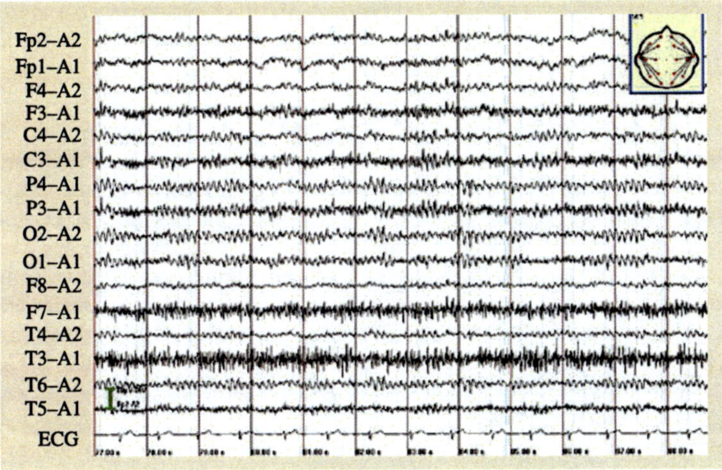

Fig. 9.18 Muscle activity in the EEG

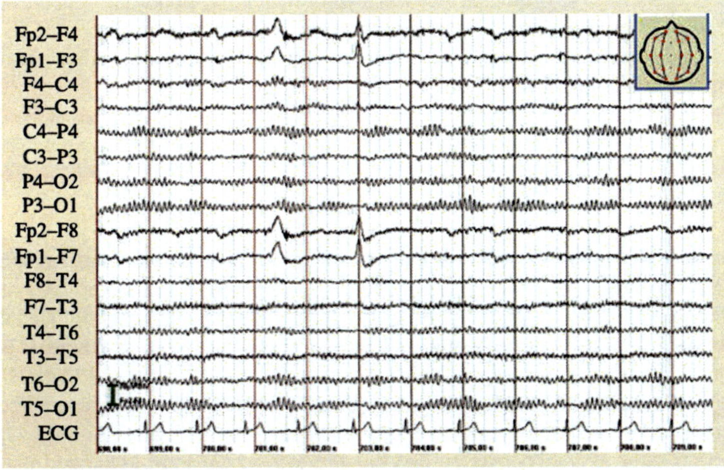

Fig. 9.19 Blink artifacts in the EEG

order to identify physiological artifacts EC, EMG and EOG are often recorded as well.

Figures 9.18 and 9.19 give two examples of physiological artifacts. Figure 9.18 displays muscle activity especially recorded from regions F7, T3, and T5. Figure 9.19 displays frontopolar eyemovements, especially recorded on Fp1–F3 and Fp1–F7, as well as Fp2–F4 and Fp2–F8.

9.2.6 Special Methods

Electrocorticography

Electrocorticography (ECoG) is used for the exact localization of lesions before a neurosurgical intervention. The electrode arrays are placed directly on the cortex. In this technique, the tissue layers between the generating location of the bioelectrical activity and the recording site are reduced considerably. The frequencies of the recorded signals have higher values.

Magneto-Encephalography

Any conductor leading electrical current is simultaneously surrounded by a magnetic field. This also applies for stimulated nerve cells. The electrical current density inside the cell is higher than outside of the cell, therefore a change in magnetic field caused by an action potential infers a stimulated cell. In contrast to electric fields, magnetic fields are not dependent on bone and tissue layers; they are only dependent on the distance between the generating area and the sensor. Therefore, this technique is able to locate the source of activity. The recorded distribution of the magnetic field along the scalp allows the location of the generated activity to be generated with sufficient accuracy, the solution of the inverse problem included.

Magneto-encephalography (MEG) in comparison to electroencephalography has the advantage of being able to determine the source of prominent activity.

Fig. 9.20 Coil system of SQUIDS (superconducting quantum interference devices) for the recording of biomagnetic fields

Therefore, it is applied in e.g. the localization of a possible focus in focal epilepsy, functional disorders caused by space occupying disorders, or the influence of clearly determinable factors on the central nervous system e.g. tinnitus. Also, examinations like the localization of specific stimulation of nerve cells (acoustical, optical) or mental activities (listen to music, thinking, speaking etc.) are carried out. In connection with other methods such as MRT, MRT, PET, SPECT further functional structures of the brain may be determined.

In any case, MEG is still a very elaborate and extensive method. The applied sensors (superconductive quantal interferometer sensors (SQUIDS) are dependent on cooling by liquid helium in order to achieve the superconductive effect. Figure 9.20 shows the coil system of a SQIDS.

9.3 Electromyograph

Electromyographs are quite universally applicable diagnostic systems. They can record, amplify, and measure the bioelectric activity of muscles and nerves under various circumstances. Their respective amplifiers perform high sensitivities, have a high input impedance, a large bandwidth, and very low noise. According to the configuration the electromyograph is equipped with an electric, acoustic, visual, and magnetic stimulator.

The principal construction of an electromyograph is displayed in Fig. 9.21 A personal computer (PC) or microcontroller (MC) acts as the central control, signal processing, and storage unit of the EMG machine. It is equipped with a hard disk as data storage, PC monitor with color display, keyboard, and laser printer. The necessary components, like amplifier, analog-to-digital converter, and the stimulation units are integrated into

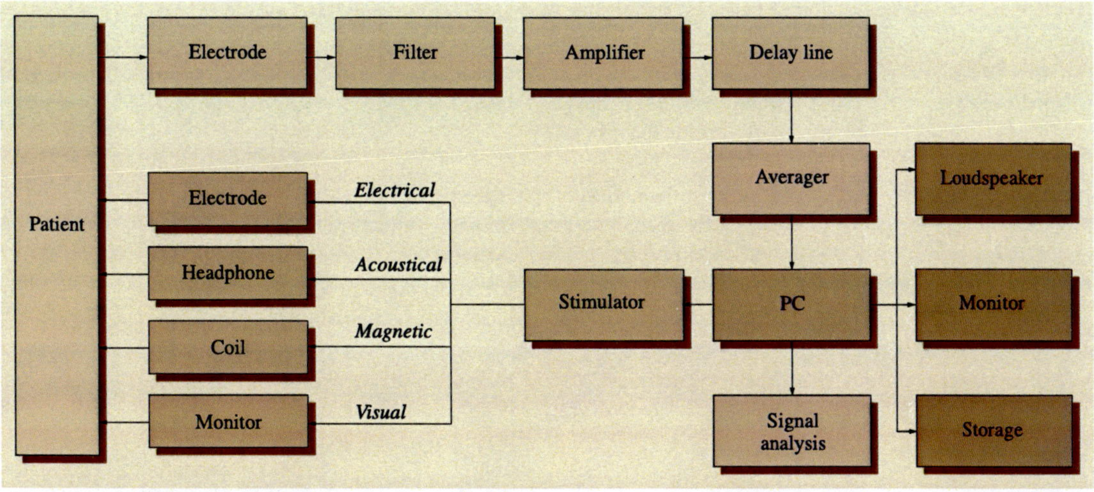

Fig. 9.21 Schematic circuit diagram of an electromyograph with four stimulators, three output devices, three hardware and software components, and the PC as the central processing unit

the PC housing. Functions like averaging, signal delay, programming of the recording programs, and evaluation of the signals are processed by software. The averaging of triggered signals serves to extract small stimulus related signals out of a surrounding signal mixture. The delay line delays signals in order to display signal components that are produced before the stimulus. A loudspeaker serves for the acoustic evaluation of the recorded signals. This is important for the immediate diagnostics, because myographic signal shapes have characteristic sound images. In addition, the loudspeaker serves as feedback for the patient, so he can judge the strength of his voluntary contraction or his relaxation.

EMG machines are applicable in:

- Neurological practices as a compact 2-channel device for EMG/EP.
- Neurological departments in hospitals as 2-channel-EMG and 4-channel EP device.
- Neurological clinics as a 4-channel-EMG and as a 4–8-channel EP device.
- Neurological rehabilitation clinics as a 2–4-channel EMG/EP device.
- Neurosurgical clinics for intra-operative EP-monitoring as a 4–8-channel EP device.
- Neurological research as an 8-channel EMG/EP device.
- Orthopedic clinics during spine operations as a 4-channel SEP/MEP device.
- Sports medicine.
- ENT as a 2-channel EMG device for larynx EMG with specialized electrodes.

To match the various requirements EMG machines are built of system components in a modular way. They especially vary in relation to the number of channels and the number and type of stimulators, software, and storage medium. Table 9.5 gives some examples.

9.3.1 Signals

Because of the great variety of an electromyograph a great number of different biosignals are recorded. Table 9.6 gives an overview of the different kinds of potentials and their measuring values that are recorded and evaluated via an EMG machine.

9.3.2 Device Technology

Electromyographs (Figs. 9.21 and 9.22) are designed in relation to the number of their channels and the kind of stimulators in a modular way. Therefore, special recording programs are available for the respective requirements. An EMG device is able to carry out a great number of neurophysiological examinations, where each examination has its own stimulation, recording, display, and evaluation characteristics. These different parameter settings are stored in the machine and may be changed or even newly designed by the user. Thus it

Table 9.5 Electrode application, stimulators, and number of channels in different EMG examinations

Examination	Recording	Stimulator	Number of channels
Needle EMG	From within the muscle with needle electrodes, only for physicians	None	1 channel
Electroneurography and reflexes	Mostly surface electrodes for stimulation and recording	Electrical stimulator for sensory and motor stimulation	2 channels
Neuromuscular transition	Surface electrodes for stimulation and recording	Electrical stimulator with pulse series stimulation	1 channel
VEP – visual evoked potentials	Surface electrodes on scalp for recording	Monitor with checkerboard reversal, stripes, bars, and colour stimulation	Practice: 1 channel Neurology: 1–3 channels Eye clinic: 1–5 channels
AEP – acoustic evoked potentials	Surface electrodes on scalp for recording	Headphones with Click, frequency specific bursts etc.	Practice: 1 channel Neurology: 1–2 channels ENT: 2–4 channels in Audiology
SEP – somatosensory evoked potentials	Surface electrodes for recording and stimulation	Electrical pulse stimulation on hands and feet	Practice: 1–2 channels Clinic: 1–4 channels also 4–8 channels
MEP – motor evoked potentials	Surface electrodes for recording on hands and feet	Magnetic stimulator with coils for cortical stimulation	1 channel

is easily possible to get to numbers of 20–40 recording programs according to the stage of extension of the particular machine. The central processing unit is often a PC, which controls the different examination programs, processes the signals, displays and evaluates them, as well as incorporating and print-out. For the signal display the international convention in neurophysiology is valid, which displays amplitudes of negative polarity on the different electrodes as positive signals.

Electrodes

The standard electrode used in electromyography is the concentric needle electrode. In a steel cannula there exists a platinum central wire coated with an isolation layer (araldite). The potential difference between the platinum wire as the active electrode and the outer steel cannula as the indifferent electrode is amplified. The electrodes are mostly 2–6 cm long and 0.3–0.6 mm in diameter. Their size and application is related to the size of the muscle to be recorded. In some cases, also bipolar needle electrodes are used. Here two central platinum wires isolated from each other are put in one cannula.

It is necessary to use an additional ground electrode, which is applied on the examined extremity near the recording site.

In neurography and in somatosensory evoked potentials (SEP) stimulation and recording electrodes are applied. In both cases, mostly surface electrodes are used, and the ground electrode is placed in between. Only in rare cases and in specialized clinics are needle electrodes for recording and stimulation applied. These needles are applied in pairs – one proximal and one distal to the examined nerve, with this technique larger and more pronounced potentials are achieved.

In evoked potentials during clinical routine examinations surface cup electrodes are attached to the scalp; they are applied in a similar way to the EEG electrodes. Only in intra-operative monitoring are small platinum needle electrodes used.

Table 9.6 Signal amplitudes and measuring values in different EMG examinations

Examination	Signal amplitudes	Analysis time	Measuring quantity
Needle EMG:			
Spontaneous activity	20–500 μV	100 ms	Potential duration
Slight innervation	200 μV–2 mV	100 ms	Amplitude
Maximum innervation	500 μV–5 mV	1 s	Discharge frequency
Electroneurography:			Latencies, motor and
Motor NCV	10 mV	20 ms	sensory nerve
Sensory NCV etc.	50–100 μV	10 ms	Conduction velocity
VEP – visual evoked potentials	50–100 μV	200 ms	Latencies and amplitudes
AEP – acoustic evoked potentials:			
– Early AEP (brainstem potentials)	0.1–0.5 μV	10 ms	Latencies
– Middle AEP	20–100 μV	100 ms	Latency differences
– Late AEP	100–500 μV	0.5–1 s	Amplitudes
SEP – somatosensory evoked potentials:			Latencies
Arm stimulation	20–100 μV	100 ms	Latency differences
Leg stimulation	10–100 μV	200 ms	Amplitudes
MEP – motor evoked potentials:			Latencies Amplitudes
Arm recording	5–10 mV	10 ms	Central conduction time

Figure 9.23 shows an overview of the different types of electrodes.

Amplifier

The amplifiers used in electromyography are differential amplifiers with a broad dynamic range and a broad frequency range. Per channel there are three input connectors for the inverted, the noninverted input, and the ground connector on the head box. An impedance measuring device for the contact impedances on the amplifier head box is recommended in the case of surface electrodes like in the recording of evoked potentials.

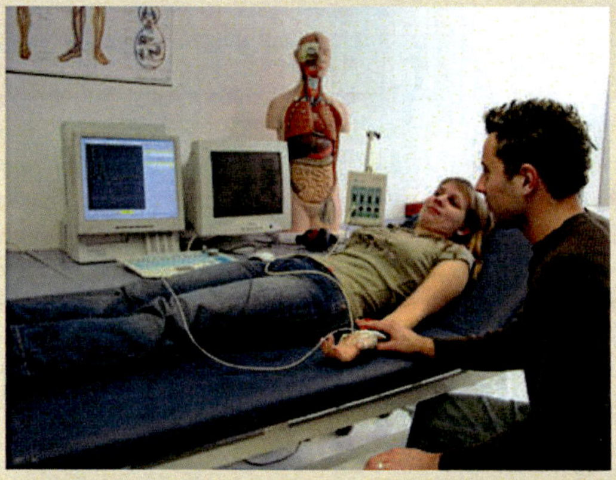

Fig. 9.22 Four-channel EMG machine with integrated stimulator. Nerve conduction velocity is measured through electrical stimulation of the motor nerves and bipolar recording of the muscle action potential with surface electrodes

Typical technical data of an EMG/EP combined machine are listed below:

- Number of amplifier channels: 2–4–8
- Preamplifier: with optically isolated input/floating input, patient connectors on the head box
- Input impedance: > 200 MΩ
- Common mode rejection ratio: > 100 dB
- Sensitivity: 1–500 μV, 1–10 mV in, e.g. 14 adjustable steps
- Noise: < 2 μV_{RMS}
- High pass filter: 1–10–30–50–100–300–500 Hz with at least 12 dB/octave
- Low pass filter: 30–100–200–500 Hz, 1.5–2.5–5–10–20 kHz with at least 12 dB/octave
- Trigger level: per stimulation signal or EMG signal with continuously adjustable amplitude or window trigger
- Averager: 1–4000 averaging steps
- Artifact rejection: selectable and adjustable
- Stimulation units: electric, visual, acoustic, magnetic

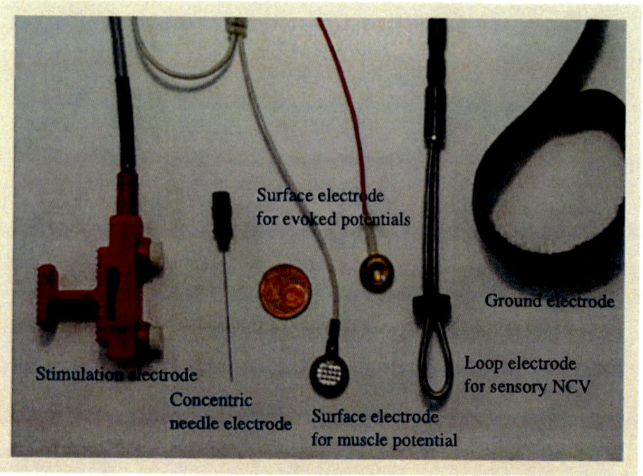

Fig. 9.23 Electrodes for electromyography

- Monitor: 17″ or 19″ monitor with a selectable waterfall display up to 32 or 64 signal traces per screen
- Time deflection: 2 ms–10 s analysis time per screen, adjustable in 10–15 steps
- Loudspeaker: two channels for listening to the raw EMG signals during the EMG needle recording with adjustable loudness control.

Stimulators

Different stimulators are mounted on the EMG machine, thus providing for various examinations such as neurography and the recording of evoked potentials. Typical values for the different stimulators are given here.

Electrical Stimulators

- Stimulation: constant current stimulator
- Intensity: 0.1–99 mA continuously adjustable
- Duration: 100–200–500 µs in three adjustable steps
- Frequency: 0.1–99 Hz adjustable
- Programs: single, pairs, pulse trains, delayed single, pairs and pulse trains, randomized stimulation.

Acoustic Stimulators

- Stimulation: click with 50–100–200–500 ms duration, tone burst, frequency specific with a plateau time of 100 ms–1 s, rise and fall slope in 1 ms–250 ms
- Intensity: click 10–130 dB spl (*sound pressure level*) for neurology
- Burst: 10–110 db spl only in audiology
- Masking: contralateral white noise of adjustable 10–110 dB
- Polarity: suction, pressure, suction/pressure alternating
- Side: left, right, both sides simultaneously.

Visual Stimulators

- TV-Monitor: 17″ with checkerboard reversal stimulation and different checkerboard sizes and steady-state-function; total field, half field and quadrant stimulation with adjustable fixation point in various positions
- LED goggles: for flash stimulation e.g. for coma patients in intensive care or in intra-operative monitoring during surgery
- LED pattern: goggles with pattern stimulation
- Ganzfeld stimulator: for electro-ophthalmological applications for the stimulation of eye movements in a determined surrounding.

Magnetic Stimulators for Motor Evoked Potentials

- Stimulation: monophasic with 100 µs, rise time 1 ms duration
- Trigger: programmable positive and negative TTL level for input and output, released by a foot panel
- Magnetic flux density:
  - 90 mm high-power coil 2.0 T
  - 70 mm double coil/butterfly coil 2.2 T
  - 120 mm curved double coil 1.6 T
  - 40 mm small coil 4.1 T
  - 70 mm medium coil 2.6 T.

9.3.3 Electromyography

The aim of an electromyographic recording is to detect differential diagnostic findings for a muscle lesion. In particular, the question has to be answered as to whether the muscle disorder or lesion is caused by a myogenic or a neurogenic process, which means whether the supplying nerve or the muscle itself is affected. Muscle action potentials are recorded via concentric needle electrodes (needle myography), thus supplying extracellular recordings from the activated muscle. Summated muscle potentials are recorded via surface electrodes on the venter but have little relevance for EMG diagnostics.

After a profound clinical examination, the physician determines, which muscle and/or nerve he will examine via EMG. EMG is an invasive examination because of the needle electrode; the physician himself has to insert the needle and has to carry out the examination. There is no valid routine procedure available for every patient, the examination is quite individual from patient to patient. An indication for an EMG recording is generally given for:

- The differentiation of neurogenic or myogenic lesions or palsy
- Muscle weakness and movement disorders with or without muscle atrophy
- Remarkable occurrence of fatigue and neuromuscular transition disorders.

A needle EMG examination usually consists of four phases:

1. Insertion activity
2. Relaxed muscle
3. Muscle with slight voluntary contraction
4. Muscle with maximum voluntary contraction.

Examples of muscle action potentials of healthy muscles are displayed in Fig. 9.25

If the physician finds hints of a pathological change in the examined muscle in a certain needle position, this finding must be verified in other parts of the same muscle. Therefore, the needle position is changed by deeper insertion or pulling out of the muscle. In any case, at least another insertion into the same muscle must be carried out. If the second insertion cannot confirm a reproducibility of the former pathological change, an insertion must again be recorded and various needle positions observed. Only pathological potentials are stored or printed out to document the comparability.

Insertion and Spontaneous Activity

In relation to the actual insertion of the needle into the muscle a transitory muscle activity may occur, the so-called insertion activity. By listening to the insertion activity via a loudspeaker the physician can determine whether the insertion activity is within the normal range of a small lesion caused by the insertion or whether a lesion can be assumed in the muscle when an especially long insertion activity is heard.

After the insertion, a healthy muscle remains in a state of *electrical silence*. However, if in the relaxed muscle potentials such as fibrillations, fasciculation, or positive sharp waves occur, this is a first hint for a pathological state of the neuromuscular apparatus. These potentials are called spontaneous activity (Fig. 9.24). They have a distinct regular discharge frequency. If they occur, they are immediately recorded and stored on the machine without any further evaluation.

Voluntary Innervation

To evaluate the innervation pattern during slight voluntary innervation (Fig. 9.25) the patient must exert the muscle very slightly, so that singular electrical discharges are seen on the display and heard via the loudspeaker. It has to be observed that the potentials are recorded close to the needle, which is the case when the discharge produces high cracking sounds and steep potentials on the screen. Ideally the needle should pick up the area of a motor unit. In the case of pathological changes, up to 20 different motor units must be recorded, compared, and evaluated (Buchthal analysis).

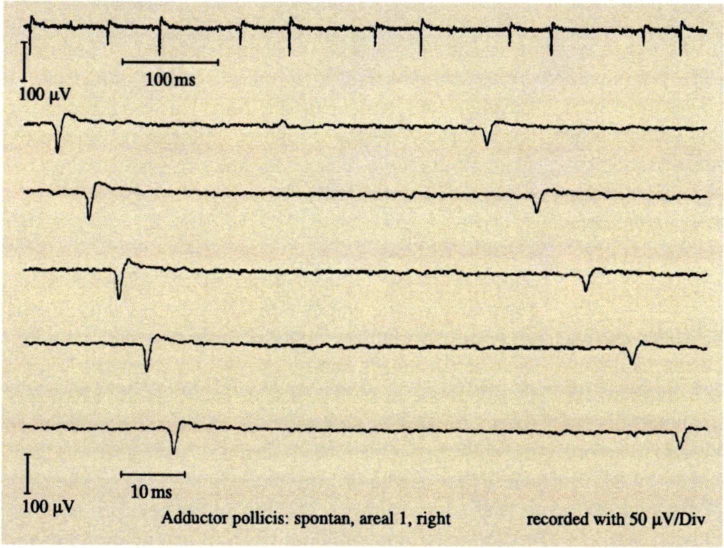

Fig. 9.24 Spontaneous activity recorded with needle electrodes from *Musculus adductor pollicis* of a 57 year old patient

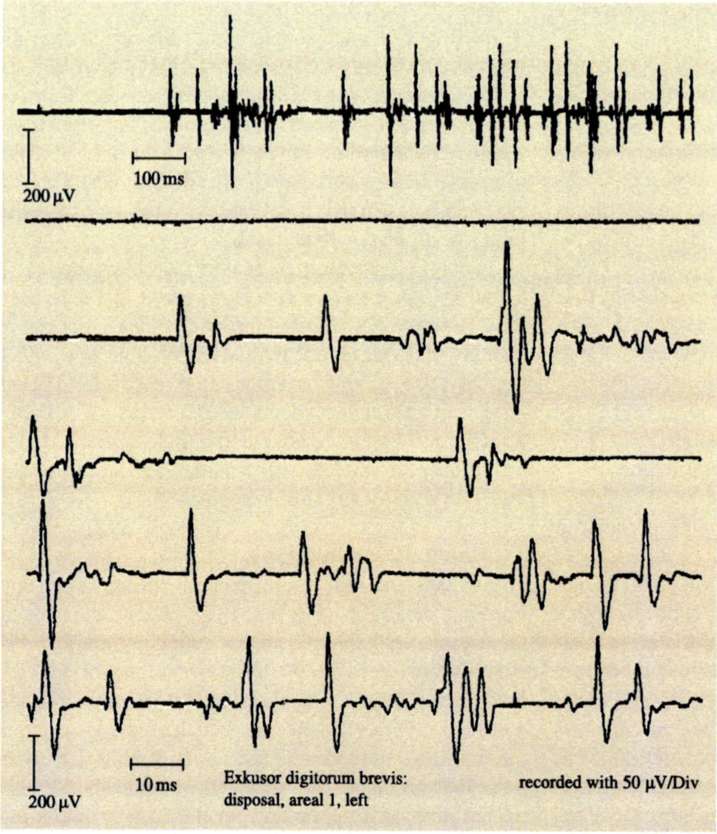

Fig. 9.25 Muscle potentials in slight voluntary contraction of *Musculus extensor digitorum brevis* in a 50 year old patient recorded with needle electrodes

Muscle potentials at maximum voluntary contraction of the muscle are recorded as an interference pattern. This is displayed on the screen as a dense potential pattern with high amplitudes. A healthy muscle shows a very dense pattern, whereas a muscle with pathological changes shows a cleared-out pattern with various different amplitudes. Neurogenic disturbances show few but high amplitude peaks and myogenic disturbances produce a dense pattern with a low mean amplitude.

Single Fiber Myography

With the help of a special single fiber electrode singular muscle fibers within a motor unit can be examined (single fiber EMG). The active platinum wire has a diameter of 0.25 mm. The recording area is not situated on the needle tip, but in the needle shaft, to make sure to record potentials from fibers that were not damaged by the insertion. Single fiber potentials are recorded during slight voluntary contraction and displayed with an analysis time of only 5 ms. Here two channels are used, on the first channel the continuous recorded potentials are observed and on the second channel the detected potentials are superimposed after signal triggering, thus enabling the physician to measure the jitter. This is a measure for the variability of the discharge behavior of the muscle fiber. With single fiber EMG one can evaluate the fiber density of a motor unit.

MACRO EMG

This is another special form of a 2-channel EMG recording: on the first channel the signal of muscle fiber with slight voluntary contraction is recorded via a single fiber electrode and on the second channel triggered by the potential on the first channel the total activity of the motor unit is summated by an averager and picked up with a special macro needle. The macro needle is a unipolar needle, in which nearly the total needle shaft acts as a large active recording area. The result is an averaged macro potential, in which potential duration and area under the averaged signal is evaluated, thus delivering a quantified value for the total amount of activity of the motor unit.

EMG Signal Analysis/Evaluation

Pathological insertion activity and spontaneous activity are not especially analyzed, because these activities are always pathological, their occurrence is simply recorded and stored.

The signals of the motor units (MUP – motor unit potential or MUAP – motor unit action potential) are evaluated in terms of signal amplitude, potential duration, occurrence of polyphasic potential, and discharge frequency and compared with normative values. These values differ in each muscle. To be able to compare these results with normative values, the EMG signals must be quantified. To record, store, and evaluate such a large amount of up to 20 MUP is a clinically very satisfying result, but also very involved and time-consuming. Therefore, a number of analysis methods have been developed. All of them agree on an overview display of the different MUPs with their evaluation values.

Online Averaging of MUPs

During the potential searching process an amplitude or window trigger level is adjusted in order to find potentials of similar size. In an automatic way the potential slope can be measured to ensure a needle position close to the motor unit, which means the optimal electrode recording position. With the help of an averager the selected potentials are averaged, automatically measured, and stored until the required number of maximum 20 MUP at different needle position is reached. Also, it is possible to quantify the discharge frequency of the examined motor unit as well as its recruitment.

Offline Pattern Recognition

Another method is the recording and storing of 1–3 s continuous MUPs. After storage, the physician marks a distinct MUP template, then this particular MUP serves as the template for an automatic pattern recognition software, searching for similar MUPs in the stored interval. Also, the measuring values are automatically calculated. This method of the subsequent analysis in an interval with MUPs also functions with pattern recognition algorithms, which are self-learning and detect reoccurring signals, averages, and measures them. This method is an offline analysis with the advantage of being able to repeat the evaluation with different parameters.

Decomposition EMG

This is a method mostly applied in the US, which takes the EMG potentials with slight voluntary innervations into an extensive mathematical algorithm, which automatically detects all repeating potential components as templates and measures them. Not reoccurring potential components remain unconsidered.

Turns/Amplitude Analysis

Muscle potentials from maximum voluntary contraction deliver an interference pattern that is quantified in a turn/amplitude graph. Here the number of reversal points are displayed over their mean amplitude. In this graph the range of normative values for the respective muscles are marked and an abnormal potential pattern is displayed as a point outside of the normative range. For a complete finding a total number of 10 EMG intervals at maximum innervations with a duration of at least 1 s in different needle positions must be recorded. This EMG evaluation is used in clinical routine examinations, because it is especially fast, simple, and concise.

Computer systems in which data banks for patient data and normative values for muscles are available and compare the results immediately with the respective normative values. This enables the physician in mostly clinical environments to examine also muscles that are not so prevalent.

9.3.4 Electroneurography

Electroneurography (ENG) is the stimulation and recording of innervations traveling along motor or sensory peripheral nerves. An important result of this examination is the nerve conduction velocity of motor and sensory nerves. Also, the examination of various reflexes, muscle fatigue testing, and the evaluation of neuromuscular transition belong to ENG.

In most cases, surface electrodes are applied for electrical stimulation and recording; needle electrodes are applied very rarely. To reduce the influence of stimulation artifacts on the recorded signal response, the ground electrode must be placed in between the stimulation and recording electrode. Machines with an automatic stimulation artifact rejection are available on the market.

The aim of a neurographic examination is the diagnosis of motor and/or sensory conduction disturbances in peripheral nerves. Often electromyographic examinations are followed by a neurographic one. Each of these various neurographic examinations work with their own stimulation programs, recording parameters, and evaluating schemes. The user may recall them from the PC, change them and/or create new ones.

Nerve Conduction Velocity

The most frequent examination of electroneurography is the determination of the nerve conduction velocity (Fig. 9.26). The motor and sensory nerve conduction velocity (NCV) is determined. For the determination of the NCV a distance as well as a time difference is needed according to $v = \Delta s/\Delta t$. Therefore, the peripheral nerves are stimulated with electrical current on two different locations and the latencies of the two stimulation sites are between the stimulation until the onset of the response signal are measured. The stimulation site and latencies more distant from the body are called distal and the stimulation site and latency closer to the body are called proximal.

The clinical indication for a measurement of the NCV are the patient's complaints about sensory disturbances or pain in the extremities, such as nerves compression syndromes. A frequent example for this is carpal tunnel syndrome (*N. medianus*) at the joint of the hand and sulcus ulnaris syndrome (*N. ulnaris*) in the bend of the elbow. The NCV is dependent on the nerve consistency, on the age, and on the skin temperature, which changes in average per 1 K to 2 m/s. The most frequent examined motor nerves on the arm are *N. medianus* and *N. ulnaris* and on the leg *N. tibialis* and *N. peroneus*.

Stimulation and recording is carried out with surface electrodes. For the determination of the motor NCV on *N. medianus* a surface stimulation electrode on the nerve near the wrist and in the bend of the arm is stimulated supramaximal. Supramaximal means that an increase in stimulation amplitude does no more increase

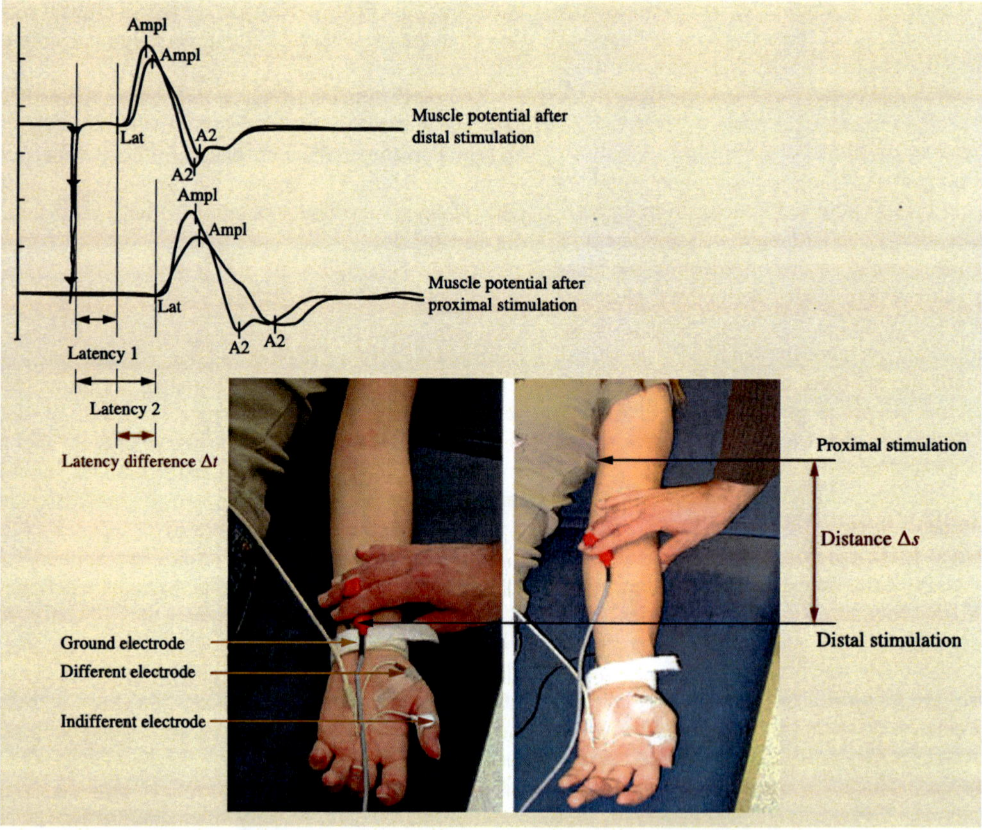

Fig. 9.26 Scheme for the determination of the motor nerve conduction velocity (NCV $= \Delta s/\Delta t$) of *N. medianus*. Stimulated in the bend of the elbow, next to the tendon of *M. biceps brachii*, and on the wrist medial to the tendon of *M. palmaris longus*. The recording is done with surface electrodes on *M. flexor pollicis brevis*

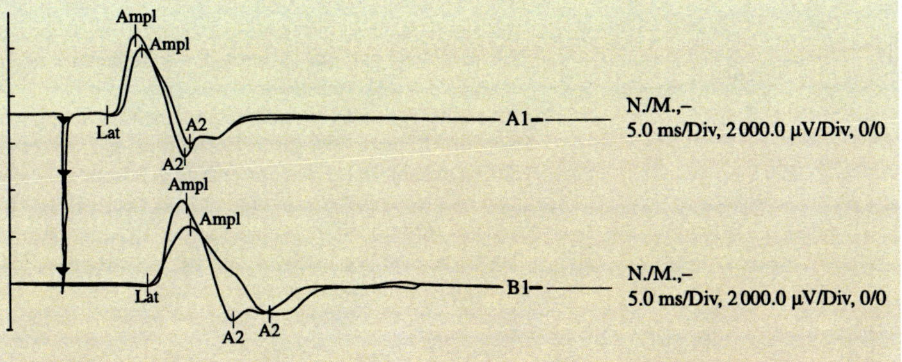

Mot. NLG

Curve	Lat. ms	Ampl. mV	NLG m/s
distal right	3.90	4.08	
proximal right	7.70	5.07	69.7

Fig. 9.27 Response potentials and determination of the motor nerve conduction velocity in the median nerve of a 27 year old patient

the response potential. The respective stimulation currents in supramaximal stimulation may reach values of 10–30 mA depending on the consistency of the top skin layer. Both response potentials are recorded on the thenar muscle with surface recording electrodes. Typical normal values for the NCV of *N. medianus* lie around 50 m/s at a skin temperature of 35 °C. Also, the absolute distal latency has a diagnostic significance; its normal value is around 4.3 ms in the median nerve.

For the determination of sensory NCV a procedure analogous to that in motor NCV is executed. The excitation continues in both directions from the stimulation site and, therefore, there is a distinction between the orthodromic (in physiological direction) and the antidromic (reverse to the physiological dispersion direction) method. In the orthodromic method fingers or toes are stimulated distally, and wrist and bend of the elbows or ankle, and bend of the knee are stimulated proximally, and recorded on wrist and bend of the elbow or ankle and bend of the knee in the direction of the physiological sensory nerve conduction. The recording is simultaneous from both recording sites, e.g. a 2-channel recording is carried out. In the antidromic method the same sites are stimulated as in motor NCV, however with considerably less stimulation intensity.

Reflex Examinations

Electroneurography also includes examinations of various reflexes: T-reflex, triggered below the patella, H-reflex, and F-wave or blink reflex. In reflex exam-

Blink-reflex

curve	R1 ms	R2 ms	R2 contra ms
N. trigeminus R	11.80	32.20	32.60

Fig. 9.28 Blink reflex (*Orbicularis oculi* reflex) of a 27 year old female proband recorded with surface electrodes

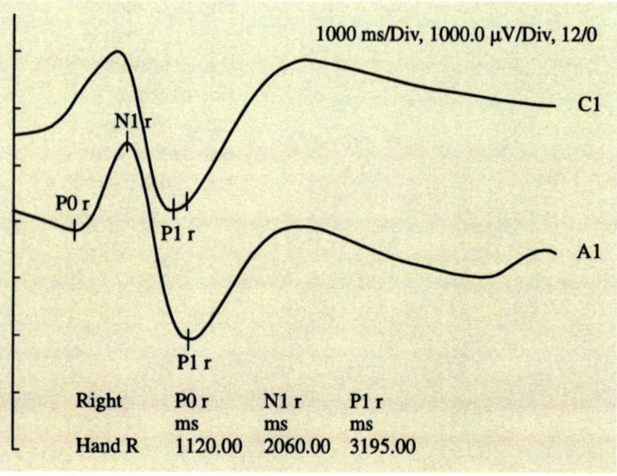

Fig. 9.29 Sympathetic skin response of a 27 year old healthy female proband recorded from the upside and underside of the hands after electrical stimulation of *N. medianus*

inations mostly latencies are evaluated, which is the time period between the stimulation and the stimulation response.

The reflex bow is a neuronal conduction of the reflectory excitation process. It travels from the stimulus detecting receptors (e.g. muscle spindles, skin receptors) along the afferent branch to the spinal or central reflex centre. With the interconnection of one or several central neurones (monosynaptic or polysynaptic reflex), the excitation travels along the efferent branches to the muscle.

The recording of the blink reflex allows a statement on the function of single core regions in the brain stem. Above one eye there is a stimulation and the recording is from the two eye simultaneously. Two stimulation responses are obtained R1 and R2 – ipsilateral and early and late response is achieved; contralateral only a late response R2.

Vegetative Parameters

An objective evaluation of disturbances of the vegetative nervous system from outside can only be carried out in an indirect way. Two examples are mentioned here: the recording of the sympathetic skin response and the measurement of heart frequency variation.

The sympathetic skin response is a potential of large amplitude, which can be recorded after a surprise stimulation e.g. between the upside and underside of the hand. This response comes from a change in the skin conduction because of the activation of the perspiratory glands and is found all over the body. The recording of the sympathetic skin response is applied in lesions of the vegetative nervous systems in the framework of the diagnosis of polyneuropathy.

Heart frequency in healthy persons varies with respiration. It increases during inhalation and decreases during exhalation. If this correlation is not found, it can lead to a respiratory arrhythmia of the heart. Heart frequency variation is recorded via the ECG, where the trigger is most practically set on the R-peak. The ECG is displayed in the time base in such a way that the variation of the following R-peaks is well displayed (Fig. 9.30).

Neuromuscular Transition

For the examination of muscle fatigue for the diagnosis of the neuromuscular transition pulse train stimulation is applied. The relation of the amplitudes and their decrement is evaluated here. In the examination of muscle fatigue and the neuromuscular transition only the amplitudes of the response potentials of supramaximum pulse stimulation are compared with another.

With double stimulation the refractory period with the absolute and relative refractory period can be evaluated. During this examination the time between the electrical double pulse is continuously decreased until the two response signals merge with each other. Then the nerve is not able to transfer the second stimulus to the muscle, because the synapses are still in their recovery phase from the first stimulus. The duration between the stimulation has to be adjustable in steps of 1 ms.

ENG – Signal Analysis/Evaluation

In the analysis of neurographic signals the response potentials to the stimulation are automatically measured. These automatic measurements are quite reliable in artifact-poor signals and increase the reproducibility of the results, because the same criteria are always applied. However, if the responses are superimposed with many artifacts or are small and hard to record due to a high degree pathology, then the physician must set the measuring cursors manually. The latency is measured as the time from the onset of the stimulation until the start of the response potential. With the help of a measuring band the nerve conduction velocity can be calculated.

9.3.5 Evoked Potentials

In the central nervous system each stimulation of receptors, afferent nerves, or brain structures in the central nervous system, the spinal channel, and in the peripheral nervous system leads to bioelectrical phenomena, which display a distinct time relation to the stimulation. These potentials are called evoked potentials (EP). They allow statements on the transformation, transition, and processing of particular stimulations and reflect the functional state of singular neuronal pathways. According to the kind of stimulation there is a distinction between acoustic (AEP), visual (VEP), somatosensory (SEP), and motor (MEP) evoked potentials. The aim of the recording of evoked potentials is to check whether the conduction of the sensory stimulation up to the cortical processing in the brain is normal, delayed, or recognizable at all. With SEP, AEP, and VEP the ascending sensory nerves pathways up to the brain stem are checked; with MEP the descending motor nerve pathways down to the peripheral musculature of arms and legs.

Device Techniques

The recording parameters of the amplifier, the analysis time, and the number of averages are characteristic for each EP examination. Table 9.7 shows an overview on the recommended parameter settings of the German Society of Clinical neurophysiology.

In addition, it is of vital importance to watch accurate electrode application thoroughly. Electrode placement here is also according to the 10–20 system. The skin of the scalp must be preprocessed so that the electrode transition impedance comes out below 2 kΩ. According to the kind of recorded evoked potential surface cup electrodes out of sintered Ag/AgCl or gold are applied.Unipolar platinum needle electrodes are only used in intra-operative application of evoked potentials. These do not need any preparation of the scalp, they are simply applied subcutaneously.

Before and during the recording the EEG raw signals are observed on the monitor in order that artifacts be recognized as soon as possible and eliminated. Especially muscle activity due to nonrelaxed patients occurs during the recording. Therefore, the patient is placed comfortably on an examination couch during the AEP, SEP, and MEP recordings, whereas he has to sit in a chair in front of the TV monitor during the VEP.

For the recording of evoked potentials averaging is indispensable. The averager is triggered simultaneously by the onset of the stimulation signal. This way,

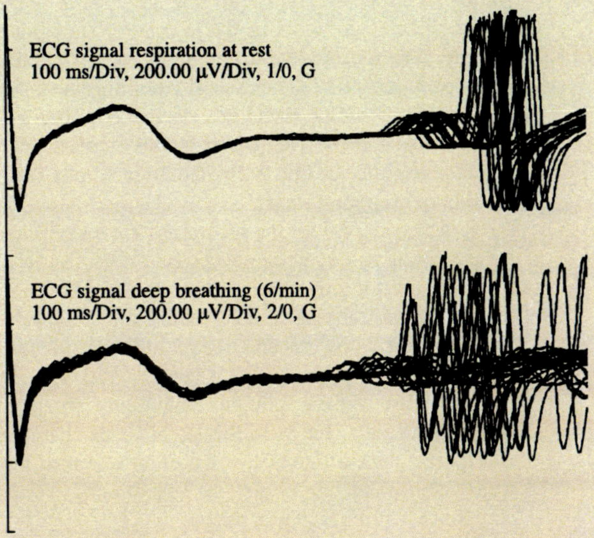

HRV respiration at rest

HR min (1/min)	HR medium (1/min)	HR max (1/min)	Variance (1/min) (max-min)	Ratio (max/min)
66.2	70.5	75.4	9.2	1.1

HRV deep breathing

HR min (1/min)	HR medium (1/min)	HR max (1/min)	Variance (1/min) (max-min)	Ratio (max/min)
62.8	74.6	87.0	24.2	1.4

Fig. 9.30 Heart frequency variation of a 32 year old male proband, displaying the change in heart frequency during respiration at rest and in deep breathing

Table 9.7 Recording parameters for the different evoked potentials

Evoked potentials	Amplifier sensitivity	High-pass filter	Low-pass filter	Analysis time	Averagering steps
VEP	5 μV/Div	0.5 Hz	100 Hz	500 ms	64–128
SEP arm	5 μV/Div	0.5–1 Hz	2–3 kHz	50 ms	128–256, rare – 1024
SEP leg	5 μV/Div	0.5–1 Hz	2–3 kHz	100 ms	128–256, rare – 1024
AEP brainstem	1–2 μV	100–150 Hz	3 kHz	10 ms	1024–2048
ERP	5–10 μV/Div	0.1–0.5 Hz	30–70 Hz	500 ms–1 s	10–32
MEP arm	10–20 Hz	5–10 kHz	1–5 mV/Div	50 ms	
MEP leg	2–5 mV/Div			100 ms	1

the signal response components that are directly related to the stimulus grow out of the stochastic background activity, thus forming the actual signal components of the evoked potential. Other signal components such as artifacts, EEG, ECG, EMG etc. do not correlate with the stimulus and are averaged out. The increase of the signal/noise ratio correlates to the square root of the number of averaging steps.

It is also helpful for the recording of evoked potentials to use an automatic artifact rejection. However, the rejection mostly only discriminates the amplitudes: if the amplitudes transcends an adjustable threshold value, this particular signal epoch is taken out from the averaging.

Normative values and the evaluation criteria for evoked potentials consist of the latencies of characteristic peaks, their difference in the comparison of both sides, inter-peak latencies, and the lack of one or several components of the response potential.

Indications for the measurement of evoked potentials are generally in:

- The diagnostics of multiple sclerosis (MS),
- Space occupying processes
- Polyneuropathies
- Intensive medicine in coma or nonoriented patients
- Intra-operative monitoring e.g. during the operation on the spinal channel
- Intoxication and/or metabolism lapses
- Posttraumatic coma after craniocerebral injury
- Brain death determination.

Acoustic Evoked Potentials

Acoustic evoked potentials (AEP) enable the examination of the auditory nerve in connection with the auditory pathways via the brainstem to the cortex. In neurology the early components within the first 10 ms after the stimulation (brainstem AEP) are of vital importance (Fig. 9.31). These early brainstem responses are generated in the acoustic nerve and in singular nucleus regions of the brainstem. It is quite an easy examination in the case of functional malfunctions due to space occupying processes in the brainstem causing obvious

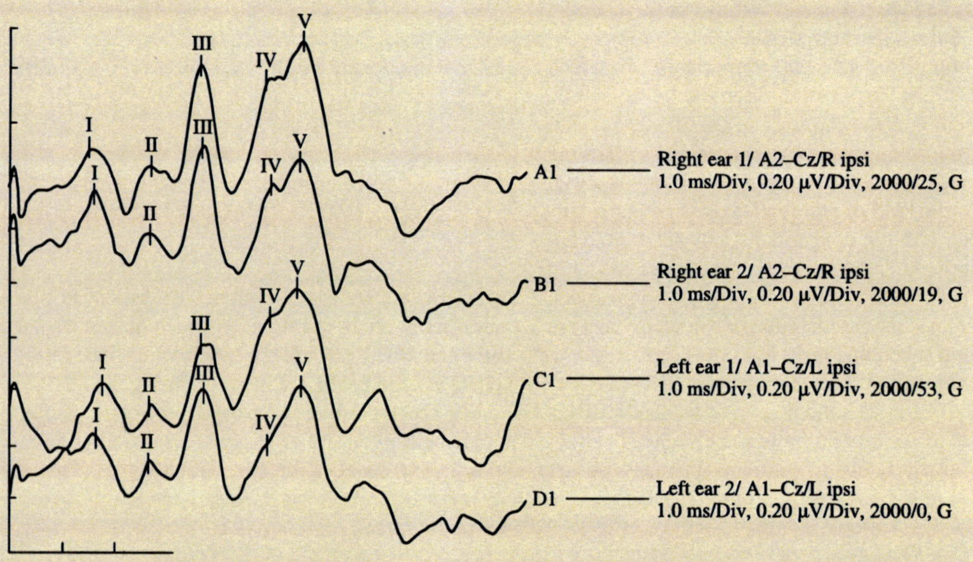

Nr	Lead	Stimulus (dB SL)	I 1.6 ms	II 2.8 ms	III 3.9 ms	IV 5.1 ms	V 5.7 ms	I–V 4.1 ms	I–III 2.3 ms	III–V 1.8 ms
A1	A2–Cz/R ipsi	70.0	1.6	2.7	3.7	5.0	5.7	4.1	2.1	2.0
B1	A2–Cz/R ipsi	70.0	1.6	2.7	3.7	5.1	5.6	4.0	2.1	1.8
C1	A1–Cz/L ipsi	70.0	1.8	2.7	3.7	5.1	5.5	3.7	1.8	1.8
D1	A1–Cz/L ipsi	70.0	1.7	2.7	3.7	5.0	5.6	3.9	2.0	1.9

Fig. 9.31 Normal brainstem AEP of a 44 year old female

changes in the BAEP, such as in tumors and hemorrhages in the brainstem.

AEP are of distinct importance also in audiology. The AEP examination is applied in cochlear and auditory nerve lesions as well as in the objective audiometry in patients who are not able to cooperate in a normal audiometry.

The acoustic stimulator consists of electromagnetic shielded headphones, which are supplied by an electronic sound generator and therefore directly applied to the ear lobe. Electromagnetic shielding is necessary to avoid electric coupling via the recording electrodes to the amplifier, because the electrodes must be applied directly under the headphones.

If the sound generator is used for the stimulation of brainstem potentials, the stimulus consists of a short rectangular acoustic signal with a pulse width of 100 μs, a so-called click.

- Click stimulation amplitude: 10–80 dB HL (hearing level), standardized 70 dB HL in neurology for objective audiometry, in ENT all sound intensities.
- Stimulation frequency: 1–30 Hz in 1 Hz steps; standard 10–15 Hz.
- Polarity: suction, pressure, alternating suction/pressure; white noise of 30 dB under the sound intensity contra-lateral to suppress of hearing the stimulus via bone conduction.

If the sound generator is used in ENT for further stimulation of middle and late AEP, then additional stimuli are selectable, such as frequency specific tone peeps and burst pulses with adjustable times for rising and falling slopes in the normal audiometry range of 125–250–500 Hz and 1–2–4–8 kHz.

The four recording electrodes are placed on:

- Two recording electrodes on the right and left mastoid (scalp bone part behind the ear lobe) ipsilateral to the stimulated ear
- Reference electrode Cz
- Ground electrode Fz.

At first the stimulation and recording are carried out on the right ear; the left ear is simultaneously masked by white noise at 30 dB below the stimulation sound intensity. Secondly the left ear is stimulated and recorded with contralateral masking of the right ear. Finally, both stimulations and recordings are repeated in order to ensure the reproducibility of the recording.

The normative values of brainstem AEP at a stimulation loudness of 70 dB HL consists of five peak latencies and one inter-peak latency: latency I: 1.5 ms; latency II: 2.6 ms; latency III: 3.6 ms; latency IV: 4.7 ms and latency V: 5.4 ms; inter-peak latency between the waves V and I: 4.0 ms. The potential peaks are evaluated in relation to their occurrence and their latency values. If one or several potential peaks are delayed or do not occur, the physician assumes a lesion in the respective location in the acoustic pathway.

In objective audiometry the amplitude relations between the IV-V complex and the wave I are also evaluated. In audiometry the recording of the brainstem potentials (brainstem evoked response audiometry (BERA)) is far more elaborate than in neurology because the loudness of the stimulus is successively decreased from 70 dB HL in 10 dB HL steps down to the hearing threshold. These recordings are time-consuming, taking up to at least half an hour.

The normative value of brainstem potentials are dependent on the stimulating loudness, on gender, on temperature, and pathological changes; however they are independent from the patient's cooperation. Therefore, their application is particularly suitable for monitoring during surgery on the inner ear and on the brainstem; they also are used in the detection of brain death.

Visual Evoked Potentials

Visual evoked potentials (VEP) (Fig. 9.32) enable the physician to examine the visual pathways. In neurology VEPs are applied to check the transition of the visual stimulation via the optic nerve to the visual pathways up to the visual cortex. Their clinical application is in the early recognition of multiple sclerosis (MS) and in central visual disturbances.

In ophthalmology their clinical application is e.g. in the objective visual acuity detection and in the objective recognition of disturbances in color vision. In electroophthalmology a great number of different visual stimuli are available, which are highly differentiated in color and gray scales.

The stimulator for routine VEP generally consists of a TV monitor, providing a checkerboard pattern over the total screen with maximum contrast, which produces each second a reversal checkerboard. The patient sits at a fixed distance from the checkerboard screen in the way that each checkerboard delivers a fixed gaze angle (standard 50 arcmin) in relation to the retina. The total visual field is 12° × 15° in relation to the gaze angle of the patient. For this recording, the patient sits in a darkened room; it is of vital importance that patients with permanent glasses/contact lenses keep them on during the recording. During the recording the patient has to

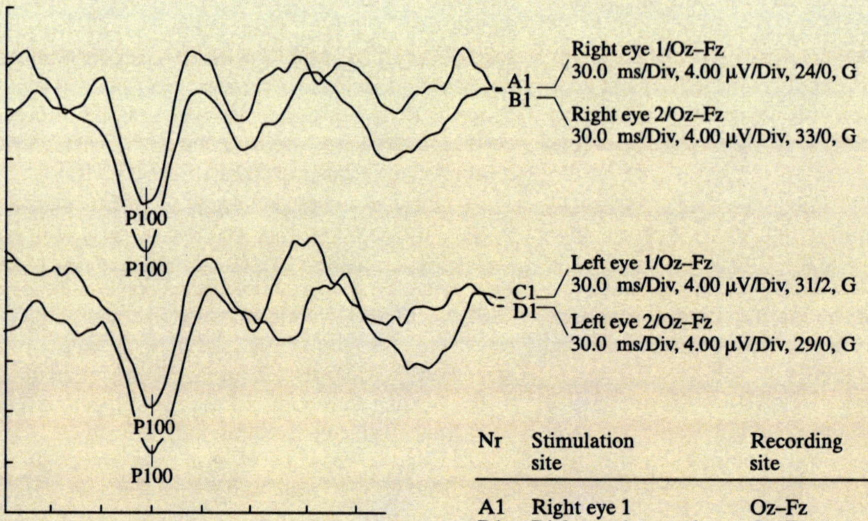

Nr	Stimulation site	Recording site	Lat P100 (ms)
A1	Right eye 1	Oz–Fz	84.6
B1	Right eye 2 control	Oz–Fz	83.4
C1	Left eye 1	Oz–Fz	85.8
D1	Left eye 2 control	Oz–Fz	85.8

Fig. 9.32 Visual evoked potential of a 57 year old female patient

fixate a small point in the checkerboard pattern (fixation point), which is located slightly above the centre of the checkerboard monitor. The patient is asked to perform great vigilance in observing the checkerboard reversal, this increases a good result of the VEP recording.

A VEP can also be recorded in coma or narcotized patients, if bright light flashes are used as stimuli. These are generated with the help of a photostimulator, which applies the flashes via LED-goggles directly to the patient's eyes.

The electrodes are placed in the following way:

- Recording electrode: 3 cm above the inion, on the centre line on the rear scalp
- Reference electrode: Fz
- Ground electrode: Fpz.

First the stimulation and recording are carried out on the right eye, the left eye is thereby covered. Then the left eye is stimulated and recorded, the right eye being covered. This is called monocular stimulation. The two response potentials are recorded once more each in order to ensure reproducibility.

The main peak in the response potential is called P100, which is the normative value of the positive peak at about 100 ms after the stimulation. Increased latencies around 10–20% are seen as pathological. The negative peak before P100 is called N75 and is often measured as well. The latencies of the responses from the left and right sides are also compared. The normative value of the VEP is dependent on the stimulus parameters and on the age. A steady-state-potential may also be recorded at a reversal frequency of 8.3 Hz.

Somatosensory Evoked Potentials

Somatosensory evoked potentials (SEP) display an objective functional check of the somatosensory system that is the pathway from the sensory nerve endings of the upper and lower extremities via the transition along the spinal channel to the brainstem, and their cortical processing up to the cortex. They have clinical value for patients who cannot or would not give valuable data on their sensory disturbances during the clinical examination. With the help of a multi-channel recording along various points on the somatosensory pathway a possible lesion can be located and diagnosed.

SEP have clinical importance in the pathology of the peripheral nervous systems, such as in polyneuropathies and compression syndrome, during surgery of the spinal cord and the spinal column, in multiple sclerosis and brainstem lesions.

The nerve endings in the extremities are repeatedly stimulated with electrical current of rectangular short pulses of 100 or 200 µs pulse width, which should be adjusted to either 3–4 mA above the motor stimulation threshold or 3–4 times the sensory threshold. This causes a considerable contraction of the related muscles but is not painful. The stimulation frequency is either 1, 2, or 3 Hz for the evaluation of the early components; for the evaluation of the late components the frequency must be reduced to 0.2–0.5 Hz.

4-channel SEP of the median nerve is recorded for each stimulation site on the right or left part of the body as follows:

- Active recording electrode: per channel,
  channel 1: Erb's point in the cavity of the clavicula,
  channel 2: lower neck on the entrance of the median nerve into the spinal column,
  channel 3: upper neck before the entrance of the vertebral canal into the brainstem,
  channel 4: scalp point C3′ (2 cm behind C3) for right stimulation, C4′ (2 cm behind C4) for left stimulation, recording contra-lateral to stimulation, because of the crossing of the pyramid pathways in the brainstem.
- Reference electrode: scalp point Fz for all channels.
- Ground electrode: at the upper forehead, in addition a large area ground electrode on the upper arm between the stimulation and recording site (Fig. 9.33).

4-channel SEP of the tibial nerve is recorded for each stimulation site on the right or left part of the body as follows:

- Active electrode: per channel,
  channel 1: inner popliteal fossa above the tibial nerve,
  channel 2: lumbal vertebra. L5 on the entrance of the tibial nerve into the spine,
  channel 3: spine on lumbal point L1,
  channel 4: median scalp point Cz′ (3 cm behind Cz) same for right and left stimulation.
- Reference electrode: scalp point Fz for all channels.
- Ground electrode: an the upper forehead, but not on a muscle, in addition a large area ground electrode on the lower leg between the stimulation and recording site (Fig. 9.35).

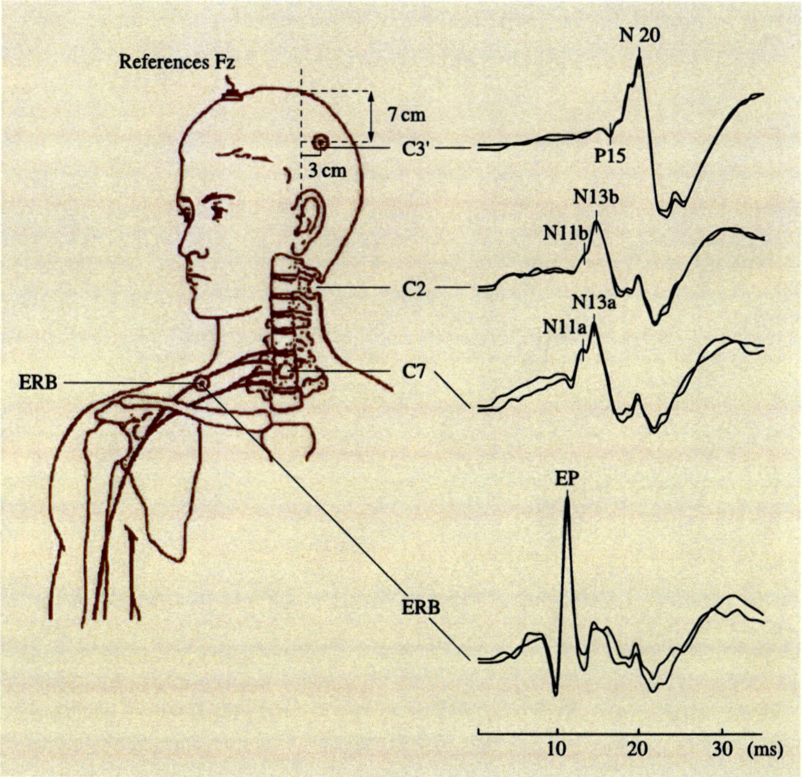

Fig. 9.33 Scheme for recording of median nerve-SEP of different recording sites with respective signals

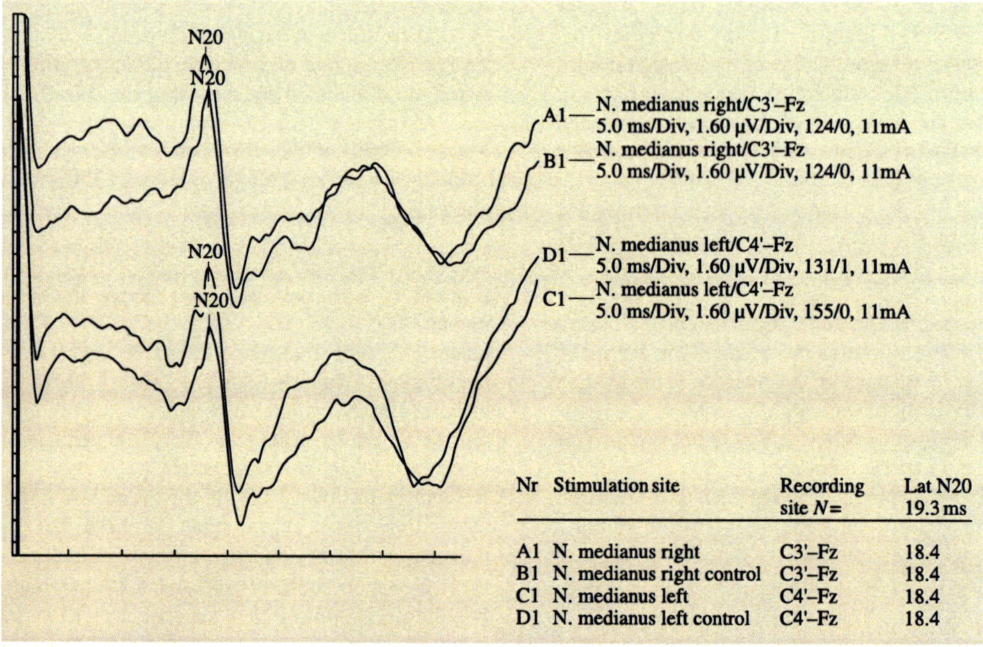

Fig. 9.34 Normal somatosensory evoked potential of the median nerve of a 50 year old patient

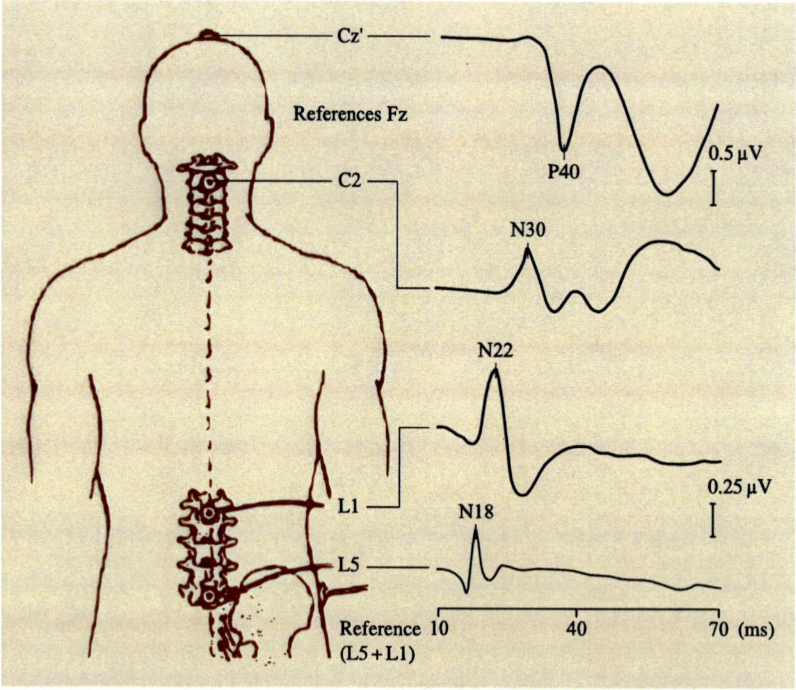

Fig. 9.35 Scheme for recording of tibial nerve-SEP of different recording sites with respective signals

SEPs are always recorded from both sides. Normative values for the cortical stimulation responses with stimulation of the median nerve are P15 and N20, which means a positive peak at 15 ms and a negative maximum at 20 ms. Normative values for early cortical stimulation responses from the tibial nerve are P30 and N33 (30 and 33 ms), for medium responses P40, N50 and P60 (40, 50 or 60 ms) and for the late responses >N75 (75 ms). Examples for responses and their measuring values for somatosensory EPs are shown after stimulation of the median nerve in Fig. 9.34 and of the tibial nerve in Fig. 9.36.

In clinics SEP is mostly recorded with four channels, so that latency differences from the singular peak latencies of the different recording sites are detected, which enables the physician to calculate the peripheral and central conduction velocity. The normative values are dependent on age, body temperature, and size of the patient. This must be considered in the conduction velocity determination.

Early cortical stimulation responses have the great advantage that they can be recorded independently from the patient's state of consciousness. They also remain mostly independent from drugs that influence the patient's consciousness, thus enabling SEP recordings in intra-operative monitoring.

Motor Evoked Potentials

Motor evoked potentials (MEP) examine the central motor pathways in an awake patient. The stimulation is carried out transcranially by a magnetic stimulator on the motor cortex in a noninvasive and pain-free way.

MEPs are often recorded in conjunction with SEP. SEP examines the ascending sensory nerve pathways together with their processing in different parts of the brain, whereas MEPs examine the descending motor pathways up to the extremities. The amplifier settings relate to the motor NCV; however longer analysis times are necessary. It is recommendable to select a waterfall display mode in order to present 10–20 MEP responses together on one screen.

The technique of a magnetic stimulator consists of a capacitor charge/discharge system with a high voltage circuit, a high power discharge switch, and a capacitor that is able to store the charged energy. The capacitor discharges via special coils; the coil construction and the geometry have great influence of the intensity, penetration depth, and the focus of the stimulation in the neuronal tissue. Therefore, coils in different shapes and forms are available for different kinds of stimulation. For the stimulation of the upper extremities a 90 mm standard coil is available. The sides of the coil are marked with the flux direction of the current. If one

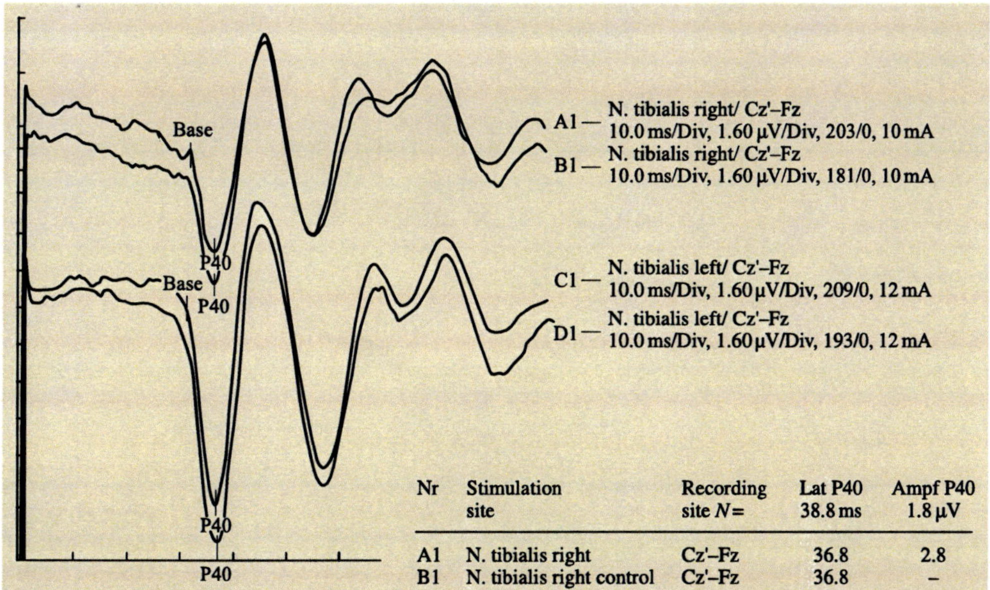

Nr	Stimulation site	Recording site N=	Lat P40 38.8 ms	Ampf P40 1.8 μV
A1	N. tibialis right	Cz'–Fz	36.8	2.8
B1	N. tibialis right control	Cz'–Fz	36.8	–
C1	N. tibialis left	Cz'–Fz	37.0	5.8
D1	N. tibialis left control	Cz'–Fz	37.0	–

Fig. 9.36 Normal somatosensory evoked potential of the tibial nerve of a 40 year old

looks at the coil from the top and the marking of the flux direction of the current is clockwise, then the left hemisphere is stimulated and the MEP response is recorded contra-lateral on the right hand. In reverse – if the flux direction is counterclockwise once the coil is turned upside down, the right hemisphere is stimulated with a contra-lateral recording on the left hand.

For the determination of the central motor conduction time (CMCT) first the motor cortex is stimulated, then the related nerve at the exit of the vertebral canal (Fig. 9.37). The MEP latencies of the cortical and cervical stimulation are measured and subtracted. The difference between them is the time, in which the stimulation pulse travels from the motor cortex to the spinal outcome of the nerve. This time is the central motor conduction time. Their normative values for the arms are between 5–9 ms, for the legs 15–19 ms. An important phenomena in cortical stimulation is facilitation. It is possible to decrease the necessary stimulation threshold to 25%, if the patient contracts the target muscle to be reached by cortical stimulation, thus carrying out a voluntary contraction. The MEP latency of the facilitated muscle then decreases to 1–3 ms.

There are some contraindications for magnetic stimulation: pace makers, metal in the scalp (e.g. aneurysma clips), and diagnosed epilepsy. Furthermore, patient and examiner must take extra care of magnetic cards and computer disks during the examination, because the data on these devices may be changed if they are in the area around 20 cm of the stimulation coil.

9.3.6 Event Related Potentials

Event related potentials (ERP) are generated in higher layers of the cortex. They occur if the patient processes certain stimuli in a conscious way. Therefore, they can only be recorded if the patient gives importance to the occurrence of a stimulus. So the patient's cooperation is of vital importance for ERP, and he is asked to differentiate between high and low tones or count the rare tones.

In ERP only the late components are evaluated. They are called P300 or potentials after cognitive stimulation. P300 relates again to the peak value of the ERP response at about 300 ms latency after stimulation onset. An indication of the recording of ERPs is given e.g. in patients with brain dysfunctions and dementia syndromes such as M. Alzheimer's, in depression and schizophrenia, as well as in brain organic psychosyndrome.

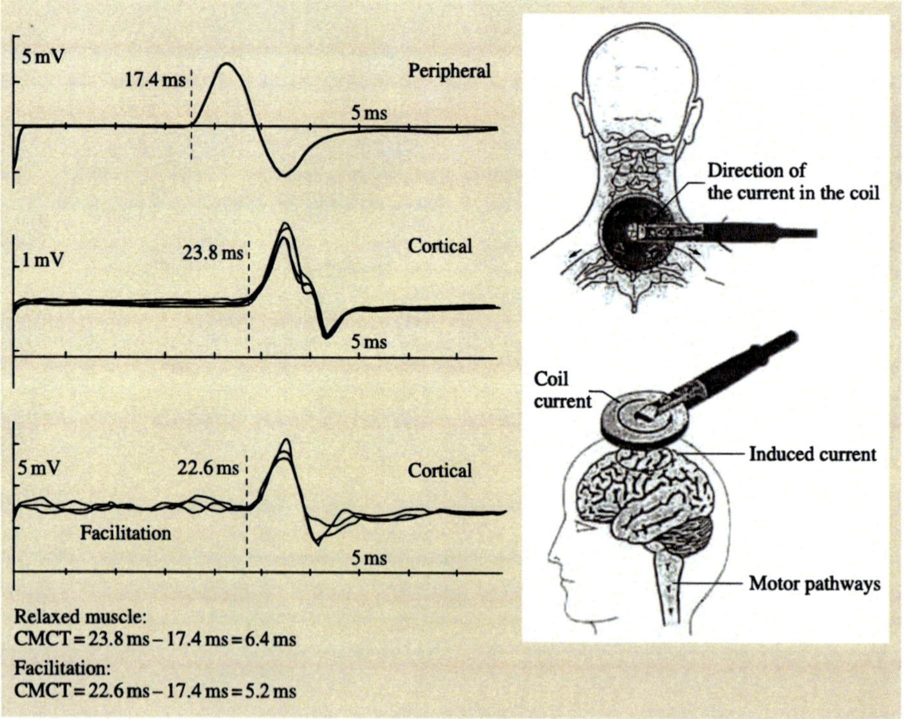

Fig. 9.37 Stimulation sites and MEP for the determination of the central motor conduction time (CMCT)

Further Reading

- C.D. Binnie: Clinical Neurophysiology: EMG, Nerve Conduction and Evoked Potentials, Vol. 1 (Elsevier, Amsterdam 2004)
- W.T. Blum, M. Kaibara, G.B. Young: Atlas of Adult Electroencephalography (Lippincott Williams Wilkins, Ambler 2001)
- J.D. Bronzio: Tissue Engineering and Artificial Organs, 3rd edn. (CRC, Boca Raton 2006)
- E. Colon, S.L. Visser: Evoked Potential Manual, 2nd edn. (Springer, New York 1990)
- R.F. Burkard, J.J. Eggermont, M. Don: Auditory Evoked Potentials (Lippincott Williams Wilkins, Philadelphia 2007)
- D. Claus, E. Kowitz-Buhler, M. Hofmann-Fox: Richtlinien und Empfehlungen der Deutschen Gesellschaft für Klinische Neurophysiologie und funktionelle Bildgebung – DGKN (Frotscher, Darmstadt 2002) pp. 39–141
- D.J. Dilorenzo, J.D. Bronzino (Eds.): Neuroengineering (Taylor Francis, New York 2008)
- J.S. Ebersole, T.A. Pedley: Curent Practice of clinical Electroencephalography, 3rd edn. (Lippincott Williams Wilkins; Philadelphia 2003)
- T.C. Handy (Ed.): Event-Related Potentials: A Methods Handbook (MIT, Cambridge 2004)
- K.W. Horch, G.S. Dhillon: Neuroprosthetics, Theory and Practice (World Scientific, Singapore 2004)
- International Federation of Clinical Neurophysiology: Recommendations for the practice of clinical neurophysiology: Guidelines of the International Federation of Clinical Neurophysiology (Elsevier, Amsterdam 1999)
- V.K. Jirsa, A.R. McIntosh (Eds.): Handbook of Brain Acitivity (Springer, Berlin, Heidelberg 2007)
- J. Kimura: Electro diagnosis in diseases of nerve and muscle, Electrodiagnosis in Diseases of Nerve and Muscle (Oxford Univ. Press, New York 2001)
- J.A. Liveson, M. Ma Dong: Laboratory Reference for Clinical Neurophysiology (Oxford Univ. Press, New York 1992)
- R. Merletti, P. Parker: Electromyography 11 (Wiley, New York 2004)
- K. Maurer, F.H. Duffy: Topographic Brain Mapping of EEG and Evoked Potentials (Springer, Berlin, Heidelberg 1989)
- T. Murofushi, K. Kaga: Vestibular Evoked Myogenic Potential, Its Basics and Clinical Applications (Springer, Berlin, Heidelberg 2009)
- E. Niedermeyer, F.H. Lopes da Silva: Electroencephalography (Lippincott Williams Wilkins, Philadelphia 2005)
- E. Stålberg: Clinical neurophysiology of disorders of muscle and neuromuscular junction, including fatigue, Vol. 2 (Elsevier, Amsterdam 2003)
- H. Preissl: Magnetoencephalography, Vol. 68 (Gulf, Amsterdam 2005)
- A.B. Ritter, S. Reisman, B.B. Michniak: Biomedical Engineering Principles (Taylor Francis, New York 2005)
- A.J. Rowan, E. Tolunsky: Primer of EEG with a Mini Atlas (Butterworth-Heinemann, London 2003)
- D.I. Rubin, J.R. Daube (Eds.): Clinical Neurophysiology (Oxford Univ. Press, Oxford 2009)
- B.E. Shapiro, D.C. Preston: Electromyography and Neuromuscular Disorders: Clinical-Electrophysiologic Correlations (Butterworth-Heinemann, London 2005)
- K. Shimoji, W.D. Willis Jr. (Eds.): Evoked Spinal Cord Potentials, An illustrated Guide to Physiology, Pharmocology, and Recording Techniques (Springer, Berlin, Heidelberg 2006)
- M.J.A.M. Van Putten: Essentials of Neurophysiology (Springer, Berlin, Heidelberg 2009)
- W. Wehrli, J. Loosli-Hermes (Eds.): Enzyklopädie elektrophysiologischer Untersuchungen (Urban Fischer, München 2003)
- S. Zschocke, E.J. Speckmann: Basic Mechanisms of the EEG Brain Dynamics Series Probability and Its Applications (Birkhäuser, Basel 1993)

10. Sleep Diagnostic Systems

Klaus-Peter Hoffmann, Robert S. Pozos

Sleep is a complex function of the brain, characterized by alternating cycles of slow waves and rapid eye movement. Sleep disorders are among the most common health problems. They have multiple causes, and different clinical disciplines collaborate to arrive at the proper diagnosis and therapy. After insomnia, sleep apnea or sleep disordered breathing is the second most common sleep disorder and is divided into two major categories: obstructive and central. This chapter discusses devices for monitoring patients at home and in sleep laboratories. Different therapeutic methods for patients with sleep apnea syndrome are discussed.

10.1 **Function and Application** 159

10.2 **Sleep Diagnostics, Sleep Laboratories, and Sleep Apneas** 160
- 10.2.1 Sleep Diagnostics 160
- 10.2.2 Sleep Laboratory 161
- 10.2.3 Sleep Apneas 162
- 10.2.4 Sudden Infant Death Syndrome 162

10.3 **Technology** 162
- 10.3.1 Devices for Monitoring of Selected Parameters for Special Applications 162
- 10.3.2 Ambulant Devices for Diagnostics 162
- 10.3.3 Devices for Diagnostics in Sleep Laboratories 163
- 10.3.4 Therapy Devices 163

10.4 **Sleep Diagnostic Procedures** 165
- 10.4.1 Ambulatory Polygraphy 165
- 10.4.2 Small Polysomnography 165
- 10.4.3 Extensive Polysomnography 165
- 10.4.4 Multiple Sleep Latency Test 165
- 10.4.5 Maintenance of Wakefulness Test (MWT) 165
- 10.4.6 Therapeutic Monitoring 165
- 10.4.7 Intensive Monitoring (Epilepsy) 166
- 10.4.8 EEG After Sleep Deprivation 166

10.5 **Signal Recording and Signal Processing** 166
- 10.5.1 Signal Recording 166
- 10.5.2 Visual Evaluation 169
- 10.5.3 Computer-Aided Evaluation 169

10.6 **Fields of Application** 171

10.7 **Methodical Instructions** 172
- 10.7.1 Handling/Application 172
- 10.7.2 Artifacts 173

10.8 **Medical Significance of Sleep Diagnostics** 173

10.9 **Therapy** 175
- 10.9.1 Behavioral Medical Psychotherapeutic Therapy 175
- 10.9.2 Chronobiological Therapy 175
- 10.9.3 Medicinal Therapy 175
- 10.9.4 Machine-Aided Therapy 175
- 10.9.5 Operational Therapy 175

10.10 **Safety Aspects** 176
- 10.10.1 Mobile Devices 176
- 10.10.2 Devices in the Sleep Laboratory 176

10.11 **Planning Advice** 176
- 10.11.1 Required Space 176
- 10.11.2 Required Time 176
- 10.11.3 Staff 176

10.1 Function and Application

With a prevalence of 15–20% of the population, sleep disorders are considered to be a common disease. Of sleep disorders, 50–60% are psychogenic, whereas for an additional 30% it is not possible to distinguish between psychic and organic and 10–20% are organically caused. The last group includes patients with sleep apnea syndrome, who on the one hand suffer from an increased vital risk, but on the other hand can be quickly and successfully treated by nasal continuous positive airway pressure (nCPAP) therapy. Overall, the international classification of sleep disorders (ICSD) published in 2001 by the American Sleep Disorders Association lists 88 different sleep disorders.

Sleep disorders have manifold causes, and their effects extend into various fields of medicine. Thus, an interdisciplinary approach is necessary for diagnostics and therapy of sleep disorders. These disciplines and faculties include: pneumology, pediatrics, neurology, psychiatry, psychology, geriatrics, cardiology, otorhinolaryngology, internal medicine, urology, nephrology, etc. In addition, due to its complicated nature, sleep medicine has developed into a field of its own, containing medical knowledge about sleep and its disorders. In Germany, specialized knowledge in this field can today be acquired in many clinical and private sleep centers.

10.2 Sleep Diagnostics, Sleep Laboratories, and Sleep Apneas

10.2.1 Sleep Diagnostics

All procedures and methods used for evaluation of disordered and normal sleep as well as for examination of an induced therapy can be summarized under the term sleep diagnostics. For practical reasons, a multistage division has been established (Fig. 10.1).

The first stage includes questions concerning antecedents, case history, and a standardized short questionnaire. The second stage consists of clinical examinations with laboratory tests and special sleep–wake anamneses. As an example, Table 10.1 shows a questionnaire for patients with suspected insomnia. The objective is the classification of the sleep disorder and, for sleep-related breathing disorders, a risk analysis. This analysis determines further diagnostic and therapeutic procedures.

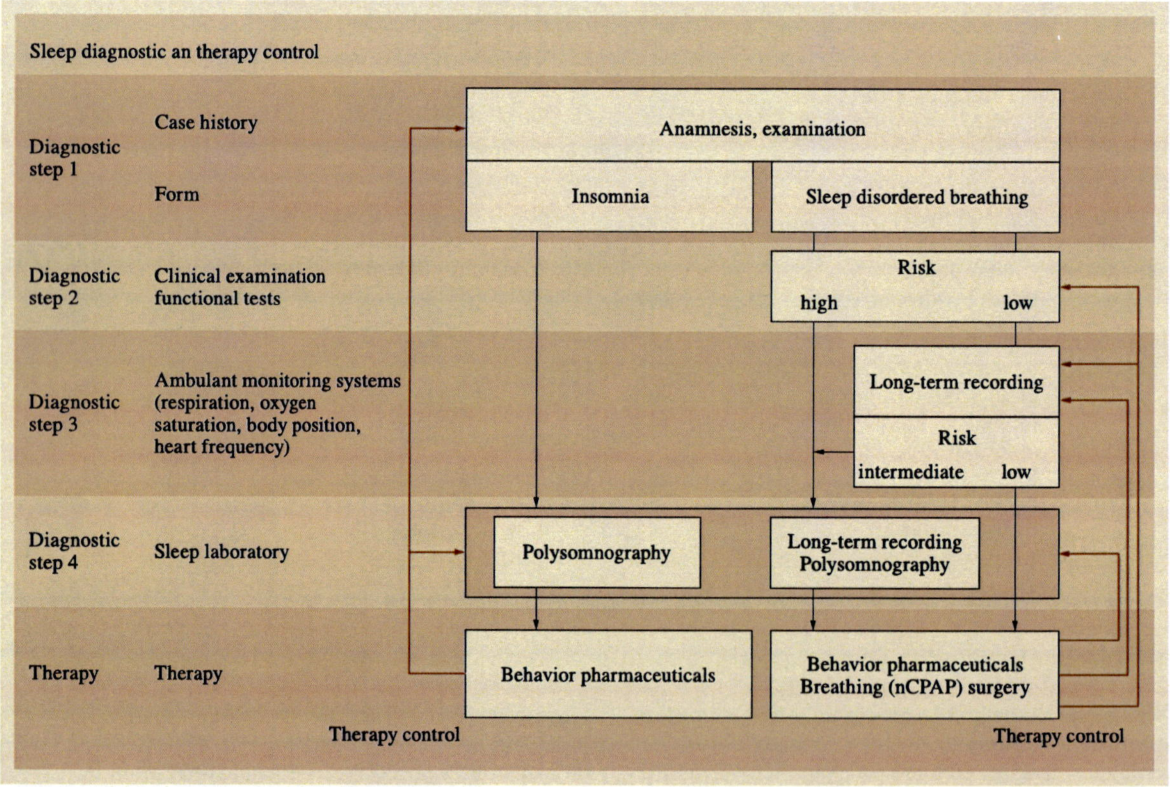

Fig. 10.1 Multistage diagnosis of sleep disorders

Table 10.1 Anamnesis, using diagnostics of insomnia patients as an example

Form of sleep disorder:	– Difficulty to fall asleep – Difficulty to stay asleep – Early awakening – Limited recreation
Pathology of sleep disorder:	– Cognitive and emotional activity in the post phase – Vegetative attendant symptoms – Special symptoms (respiratory obstructions, restless leg syndrome, nightmares, pain, anxiety)
State of health during the day:	– Vigilance – Activity and ability to cope with stress – Ability to concentrate and performance – Emotional state and general well-being
Sleep behavior:	– Time in bed and sleep duration – Sleep time – Evening habits and sleep habits
Progress and duration of the sleep disorder: Sleep prior to disease/pretreatment	– Self-medication (drugs, alcohol) – Nonmedicinal agents – Soporifics (form, time, and dose of intake)
Psychiatric and organic anamnesis and pathology: Influencing parameters	– Private and occupational life history – Living conditions and behavior affecting sleep – Physical environmental influences – General intake of pharmaceuticals – Consumption of caffeine, nicotine, alcohol, and drugs

For patients with suspected sleep-related breathing disorders, a third stage is initiated, including continuous monitoring of respiration, O_2 saturation of the blood, heart rate, and body position. For the most part, these measurements can be conducted ambulatorily.

The fourth stage includes recordings of biosignals, generally under stationary conditions in a sleep laboratory. The induced therapy is optimized and adjusted by repeated examinations to evaluate its effectiveness.

To conduct the polygraphic recordings necessary for diagnostics and therapy control, the German Sleep Society (DGSM) and the American Academy of Sleep Medicine have developed guidelines. They include respiratory mechanical measurements (respiratory gas flow, respiratory effort, nCPA pressure), cardiorespiratory measurements (blood pressure, blood gases, electrocardiogram (ECG), pulse), neurophysiological measurements (electroencephalogram (EEG), electromyogram (EMG), and electrooculogram (EOG)), and patient-related measurements (motion, position, snoring sound, video recordings). Moreover, these guidelines recommend the number of channels, the parameters to set, the application of the respective sensors, the analysis including automatic processing, the evaluation of the results, and supplemental measurements.

10.2.2 Sleep Laboratory

In a sleep laboratory, sleep is examined under laboratory conditions, recording and storing various biosignals and evaluating them with respect to manifold criteria. The cause of the disorder registered in the anamnesis is decisive for the respective diagnostic approach.

Due to the above-mentioned diversity of sleep disorders, specialized sleep laboratories have been established.

1. Psychiatric–neurological sleep laboratories are focused on psychogenic and neurological hypo- and hypersomnia, parasomnia, circadian disorders of the sleep–wake cycle, sexual disorders, and particularly epilepsy. This analysis requires additional EEG channels for seizure diagnostics including long-term EEG, EMG, body temperature, actigraphy, and registration of nocturnal erections.
2. A psychological sleep laboratory focuses on vigilance, dreams, and the causes of psychogenic sleep disorders as well as the influence of stress and noise on sleep. Neurophysiological parameters, such as skin impedance, skin and core temperature, EMG, and EEG biofeedback are additionally registered.
3. Sleep-related breathing disorders are the focus of internal–pneumological sleep laboratories. The measurement of cardiorespiratory functions including the associated change of neurophysiological parameters, such as occurrence of arousal, is typically used for diagnostics of sleep apnea syndrome. Further measured values include respiratory flow separately for nose and mouth, intrathoracic pressure, p_{CO_2}, continuous blood pressure, pH value, and long-term ECG.
4. Examinations regarding drug efficacy are conducted in a pharmacological sleep laboratory.

10.2.3 Sleep Apneas

An episode with a breath amplitude reduced by 50% for a duration of more than 10 s is denoted as hypopnea. In apnea, the amplitude is reduced by 20% for a duration of more than 10 s. A phase of apnea is accompanied by an initial decrease in heart rate and a drop in O_2 saturation after a few seconds. This phase is followed by an awakening signal of the central nervous system (CNS) and is characterized by arousal, short-term EEG activation, acceleration of the heart rate, heavy breathing, as well as an increase of blood O_2 saturation.

For obstructive apnea, the upper respiratory tracts are occluded due to the interplay of negative pressure in the pharynx and reduced activity of the pharyngeal muscles during inspiration. Here, thoracic and abdominal movements are still present and often opposed. A less frequent type of apnea is central apnea, which is due to decreased cerebral control of respiration with a decrease in respiratory flow and respiratory movements of the diaphragm. For mixed apnea, the most frequent form of apnea, the initial central apnea is followed by a phase of obstructive apnea.

10.2.4 Sudden Infant Death Syndrome

Sudden infant death syndrome (SIDS) has an exceptional position among sleep disorders, as its cause(s) cannot be explained by either clinical or extensive post mortem examinations. It is the most frequent cause of death for infants in the first year. The stage of cerebral development is a possible cause. Today, the risk groups have been successfully identified, and various prevention strategies have been developed. Long-term monitoring is still the most frequently used preventive method.

10.3 Technology

Depending on the stage of sleep disorder diagnostics, different methods and devices are applied. The choice strongly depends on the application, number of selectable parameters, sensor application, operability, evaluation strategies, software, etc.

10.3.1 Devices for Monitoring of Selected Parameters for Special Applications

To register, control, and document the vital functions of an infant in the first year with the risk of sudden infant death, special monitoring systems are provided. Monitoring may include heart rate, respiration, and movements of the infant as well as actigraphic measurements. The objective is to alarm the parents in case of loss of vital functions as well as to enable timely measures for resuscitation.

10.3.2 Ambulant Devices for Diagnostics

There are a variety of portable devices for recording and registration of polygraphic signals. The signals are recorded in the familiar domestic environment of the patient, making habituation to laboratory conditions unnecessary. After registration of biosignals, they are either transmitted telemetrically or copied to a laboratory computer by the physician the following morning, and evaluated using automated signal analysis methods. In addition, monitoring systems are used for recording, storage, and evaluation of respiration (larynx microphone), heart rate (ECG), body position, and O_2 saturation.

Enhanced diagnostic possibilities exist due to the use of an increased number of channels and the application of nCPA pressure sensors. A sleep center can service several small hospitals with a sleep laboratory and ambulatory physicians. They transmit their recorded raw data or results based on automated analysis together with the patient data and their suspected diagnosis to the center, which assists the physicians in arriving at their diagnosis. Two-way transmission involves using modems for area networks or the web. Thus, the quality of sleep disorder diagnostics can be further improved, as referrals and inpatient confinements become more efficient.

10.3.3 Devices for Diagnostics in Sleep Laboratories

Devices for multistage sleep disorder diagnostics used in a sleep laboratory include equipment for long-term recording and polysomnography. In a network, individual devices form a closed system for signal registration, transmission, analysis, evaluation, and storage. Interfaces to other analysis systems and data processing equipment are provided. Therefore, integration into the hospital information system including patient and business management is beneficial, as, e.g., all examination findings are centrally archived and are thus available to any physician in the hospital.

As an example, Fig. 10.2 shows the setup of a sleep laboratory. It consists of four recording stations, with one of them being equipped with a laptop computer to function as a mobile station. In the awake station, all data are collected. Here, the night watch is on duty, and the signals are analyzed, stored, and archived. CPAP settings, communication with the patient, and remote data transmission via modem are possible. Simultaneously with biosignal recording, a video image of the patient is recorded and stored, and can be correlated to the signals.

In the aforementioned example, an additional evaluation station is provided for the executive physician, who has access to all data. Wide, separate setting ranges for the recording parameters as well as sampling rates of 1–1000 Hz are required to enable an entire polygraphy. The equipment ranges from a 12-channel single-user station to linked 32-channel multibed stations with an evaluation unit and server.

For the application of recording devices in sleep medicine, it is also important that, during the recording, signals from previous periods can be examined at any time by the night watch or the doctor on call.

10.3.4 Therapy Devices

Sleep-related breathing disorders can be treated by nasal continuous positive airway pressure. The objective of such therapy is to prevent obstructions of the upper respiratory tracts in the inspiration phase by a positive pressure. Mainly, nasal continuous positive airway pressure (nCPAP) and nasal bilevel positive airway pressure (BiLEVEL) are applied. The use of a device with two separate pressures for inspiration and exhalation facilitates exhalation for high CPA pressures. The setting of the respective levels for inspiration and exhalation is controlled by the patient through breathing and has to be done precisely in a very short period of time.

If respiratory flow is plotted as a function of CPA pressure, the critical pressure p_{crit} can be determined (Fig. 10.3a). It is defined as the minimum pressure for which a respiratory flow can be detected, and it is a measure of the collapsibility of the pharynx during sleep. The critical pressure P_{crit} increases from snoring patients with obstructive hypopnea to patients with obstructive apnea respectively.

To increase patient acceptance of these devices, the pressure is slowly increased linearly during the phase of falling asleep, until the therapeutic pressure is reached. Newer devices, which are also denoted as automated CPAP, shift the pressure every 2 min by +2 mbar with respect to the current pressure level, and simultaneously measure the resulting change in respiratory flow. The objective of the analysis and potential adaptation of the current pressure is to reach the optimum pressure–respiratory flow window with no respiratory events (apnea, hypopnea, snoring) occurring.

Figure 10.3b shows a detail of an automated titration report with a plot of respiratory flow and pressure. Warming and humidification of the ventilation air as well as reduced noise emission by the integrated turbine are of great importance for patient acceptance of nCPAP therapy.

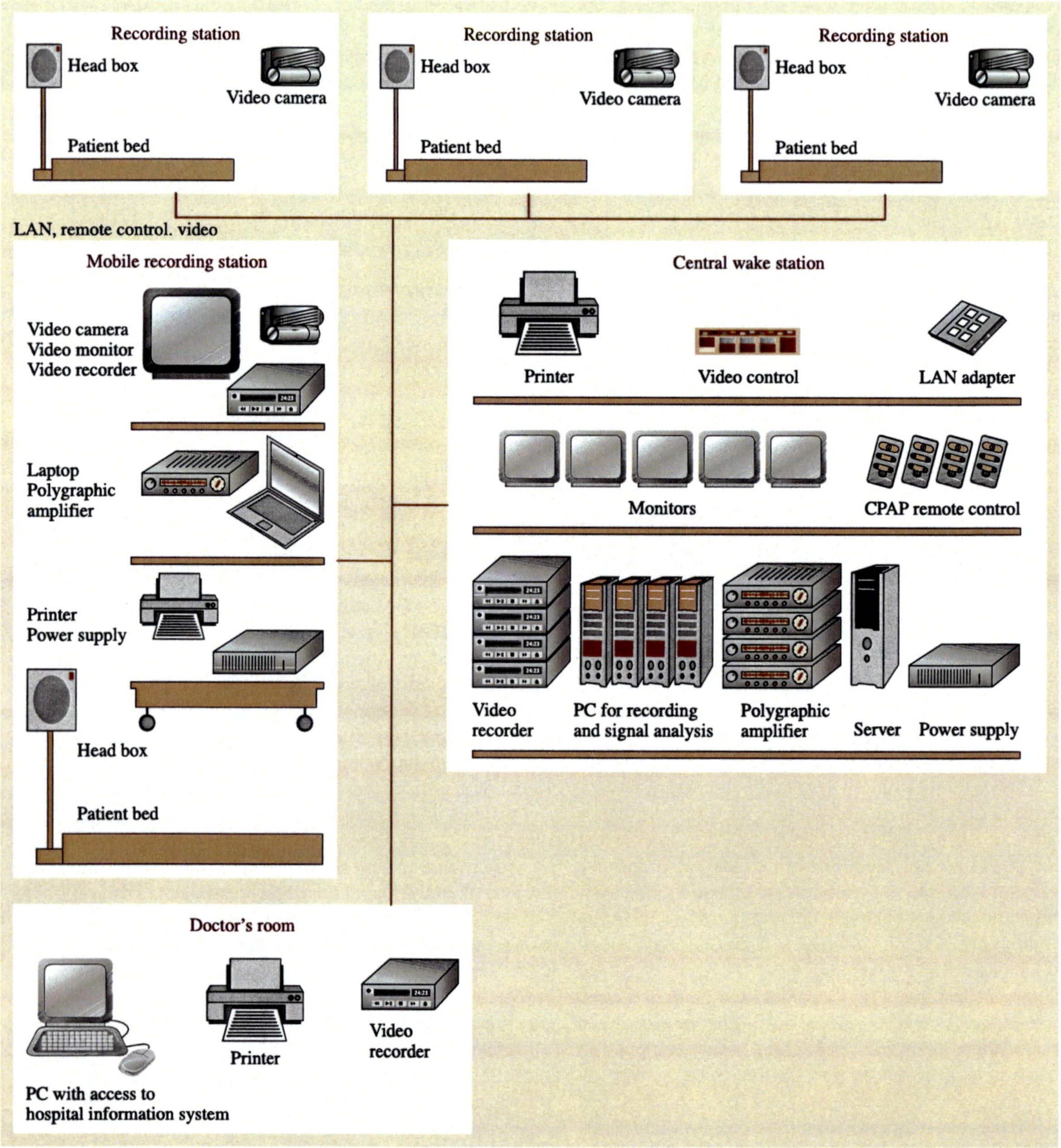

Fig. 10.2 Setup of a sleep laboratory with three stationary and one mobile recording station as well as a central wake station and an additional workstation for evaluation of the recording in the physician's room

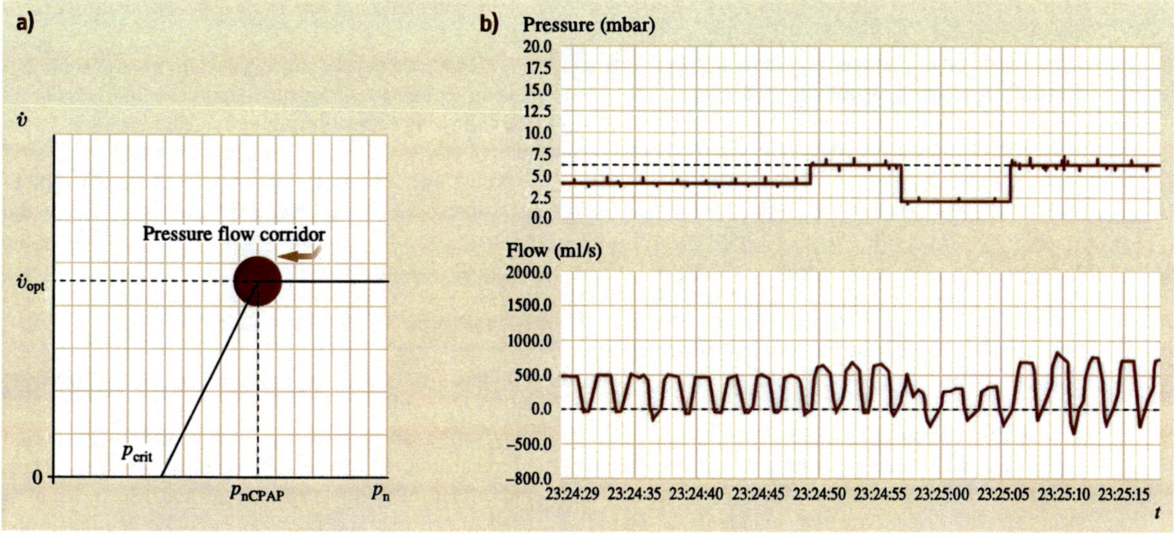

Fig. 10.3 (a) Respiratory flow–pressure curve, (b) detail from an automated titration report

10.4 Sleep Diagnostic Procedures

10.4.1 Ambulatory Polygraphy

All recordings by nonlaboratory monitoring systems are classified under the term ambulatory polygraphy. This includes devices used in the third stage of sleep diagnostics as well as all alarm systems used for domestic monitoring of suspected sudden infant death syndrome.

10.4.2 Small Polysomnography

A small polysomnography is indicated for psychogenic/psychiatric disease patterns, for differential diagnosis of epileptic seizure disorders, and for therapy of nonrespiratory diseases. At least a four-channel EEG, the EOG, EMG and ECG are required. Continuous observation of the patient via video screen and of the polygraphic recordings is required.

10.4.3 Extensive Polysomnography

Extensive polysomnography is conducted for persistent and therapy-resistant sleep disorders which at first view are not psychogenic or psychiatrically caused, for suspected sleep-related breathing disorders, and for drowsiness. Different signals like EEG, EOG, EMG, ECG, flow, effort, blood gases, snoring, blood pressure, position, temperature, movements, erection, and intrathoracic pressure can be recorded. Videometry is required.

10.4.4 Multiple Sleep Latency Test

With the multiple sleep latency test (MSLT), drowsiness can be examined. This test is conducted directly after a preceding nocturnal polygraphy, e.g., at 10 am, 12 pm, 2 pm, and 4 pm. Each recording session can take 20 min, and the patient should not resist going to sleep. The recordings include EEG, EOG, EMG, and ECG.

10.4.5 Maintenance of Wakefulness Test (MWT)

Like the MSLT, the MWT is conducted directly after a preceding nocturnal polygraphy, e.g., at 10 am, 12 pm, 2 pm, and 4 pm. The willingness of the patient to stay awake is measured. The patient is requested not to fall asleep during or between the recordings. The same parameters as for the MSLT (EEG, EOG, EMG, and ECG) are recorded.

10.4.6 Therapeutic Monitoring

The technological requirements are the same as those of extensive polysomnography, plus the possibility to record the CPA pressure. Monitoring is conducted to control the nasal CPAP or the BiLEVEL setting, which is the result of a qualified adjustment. After 3, 6 and 12 months, the setting is routinely checked.

10.4.7 Intensive Monitoring (Epilepsy)

Intensive monitoring is mainly used to detect epileptic activity. Also for this technique, a multistage procedure is applied for each patient. Routine EEG recordings (from 20 min up to 1 h), prolonged recordings (1–6 h), and long-term recordings (6–24 h) can be distinguished, which together form a complex neurodiagnostic functional unit. For simultaneous double image recording, patient behavior and EEG are displayed synchronously on a monitor.

In examinations of patients with proven epilepsy, epileptic activity are detected in 75% of patients during a 48 h recording, of which 10% occurred suddenly. Thus, ictal and interictal activity can be compared, and continuous analysis for classification can be made.

Sleep phases are particularly important for diagnosing epilepsy, as seizures increase in quantity at the following times: during early sleep between 10 pm and 12 am, 1–2 h before morning awakening, and after morning awakening. The proportion of specific potentials increases during sleep compared with the awake EEG by ≈ 20%, especially during late periods of the night and during arousals.

10.4.8 EEG After Sleep Deprivation

Previous to intensive monitoring for diagnosing epilepsy, routine EEG recordings after 24 h of sleep deprivation are conducted. During the 1–2 h period of recording, the patient is allowed to sleep. Thus, the effect of sleep deprivation and sleep itself is used to provoke and intensify epileptic activity in the EEG.

10.5 Signal Recording and Signal Processing

10.5.1 Signal Recording

The quality of signal recording is crucial for the quality of sleep disorders evaluation. Besides the choice of the sensors and electrodes necessary for each respective case, signal quality depends on proper placement and application as well as on the recording parameters set according to the signal properties, such as amplifica-

Table 10.2 Example of recording parameters for detection of polysomnographic signals according to the recommendations of the German Sleep Society

Signal	Recording site	μV/cm	f_{min} (Hz)	f_{max} (Hz)	Comments
EEG	C3:A2, C4:A1 or C3:A1; C4:A2	70	0.53	70	Instead of A1, A2 also mastoid; 10:20 system
EOG	Left, right epicanthus	70	0.53	> 30	Also horizontal and vertical recording possible
EMG	*M. mentalis* or *M. submentalis*; bipolar	20	< 1.6	> 70	
ECG	In. *Manubrium sterni*, dif. 5. intercostal area	1000			R peak as well as P and T wave readily identifiable
Respiratory flow	Mouth and both nostrils	Differentiation between hypoventilation and normal breathing			Thermistor or thermocouple
Effort	Thoracic and abdominal	Hypoventilation, normal breathing, and hyperventilation have to be identifiable			Induction, plethysmography, strain gages, piezosensors or impedance measurement
Blood gases	Saturation				Pulse oximetry, for analog output 1% per mm
EMG	*M. tibialis anterior*	20	< 1.6	> 70	Bipolar, 5 cm distance

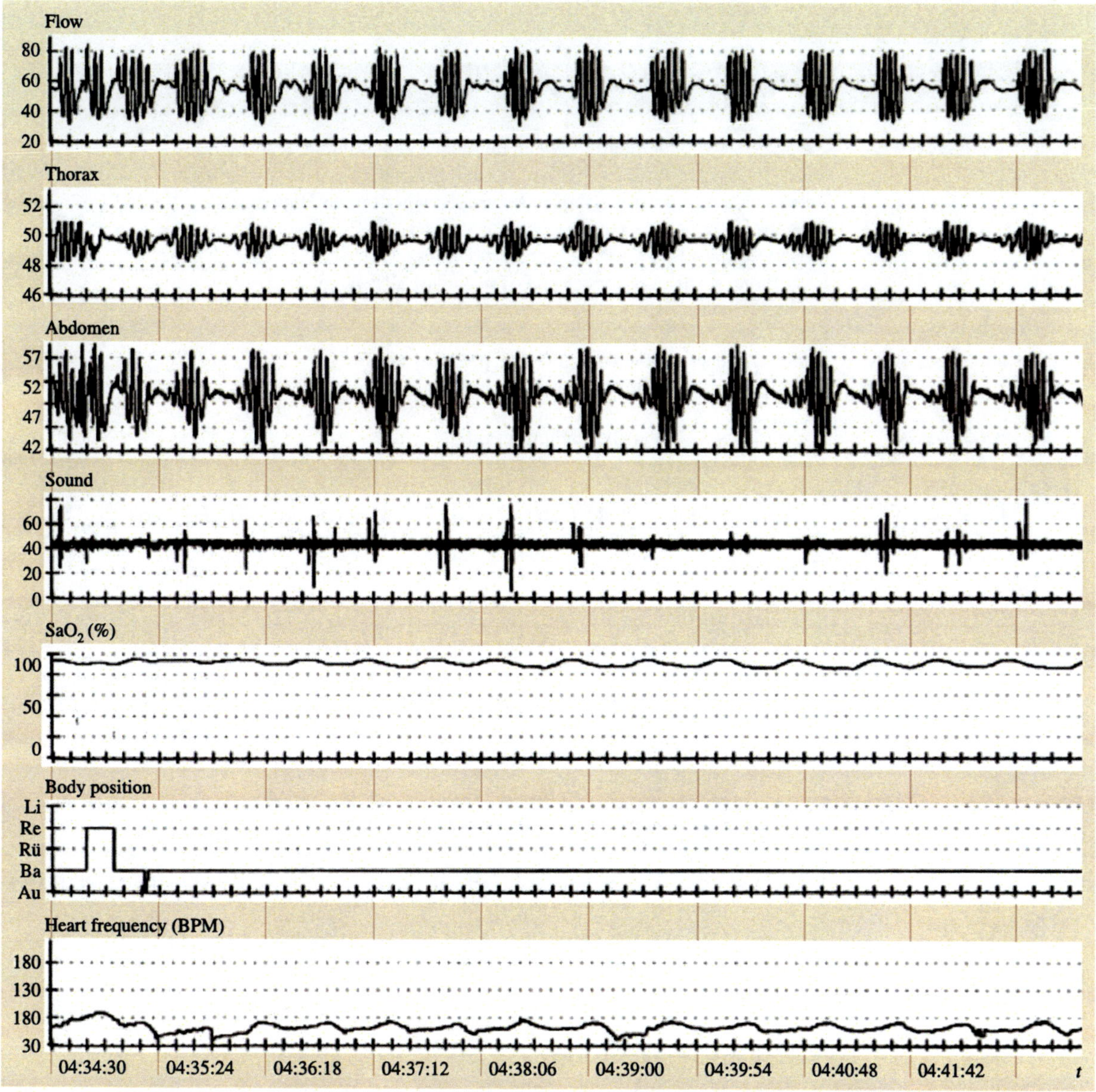

Fig. 10.4 Polysomnography (ambulatory) of a patient with obstructive sleep apnea (channels *from top to bottom*: respiratory flow, respiratory effort for thorax and abdomen, snoring microphone, oxygen saturation, body position, and heart rate). Obstructions are clearly reflected by a flattening of the respiratory flow with reduced or opposed displacements of thorax and abdomen, followed by the occurrence of snoring sounds, a reduced oxygen saturation, and an increased heart rate

tion, frequency range, and sampling rate. In Table 10.2, the device settings recommended by the German Sleep Society are summarized. The sampling rate for digitization of biosignals should be at least twice the highest frequency contained in the signal. In addition, a compromise has to be found between recording quality and the capacity of the digital storage medium.

Figure 10.4 illustrates a printout of raw data from a polygraphic recording according to the third diagnostic stage. Quality control of recording has to be done

Table 10.3 Criteria for visual evaluation of polygraphic signals in the sleep laboratory according to the recommendations of the German Sleep Society

Signal	Analysis
Respiration	Amplitude and phase shift between thoracic and abdominal respiration – Respiratory amplitude < 10% of normal respiration > 10 s ⇒ Apnea – Respiratory amplitude < 50% of normal respiration ⇒ Hypopnea – Apnea index (AI) = apneas per hour of sleep – Hypopnea index (HI) = hypopneas per hour of sleep Respiratory disturbance index (RDI) = sum of all respiratory events per hour of sleep
Blood pressure	Systolic, diastolic, and average blood pressure – Minimum and maximum values during awake state and sleep
Blood gases	A drop of > 3–4% is clinically relevant – Oxygen desaturation index (ODI) = average saturation, desaturation per hour of sleep
EEG	Visual determination of sleep phases according to the criteria by Rechtschaffen and Kales in the phases awake, 1, 2, 3, 4, rapid eye movements (REM), and movement time (MT) (Table 10.4)
ECG	Heart rate variation, cardiac arrhythmia average heart rate awake and sleeping, moment of maximum heart rate
EMG *M. tibialis*	Quantification of periodic leg movements (PMS) – Duration between 0.5 and 5 s, > double amplitude, – Periodic for four events with an interval of 4–90 s and an average of 20–40 s – Issue of arousal, PMS arousal index = number of movements with arousal per hour of sleep, from five pathological
EMG chin	Sleep phase classification, quantification with respect to phasic and transient EMG activity
EOG	Detection of REM sleep, calculation of the average REM density slow eye movements in sleep phase 1
Erection	Diagnostics of erectile impotence; temporal association to REM episodes
Interactions	Connections between sleep phases, arousals, and other EEG elements with cardiorespiratory events and movements
Body position	Apneas and hypopneas are counted for the respective body position
Temperature	Diagnostics of chronobiological disorders of the sleep–wake behavior; examination time > 32 h

online. For polysomnographic recordings according to stage 4, a greater number of recording channels are used, and depending on the case, additional biosignals are registered. Also, the video image of the sleeping patient can be shown, to enable direct correlation of patient movements and biosignals (Fig. 10.5).

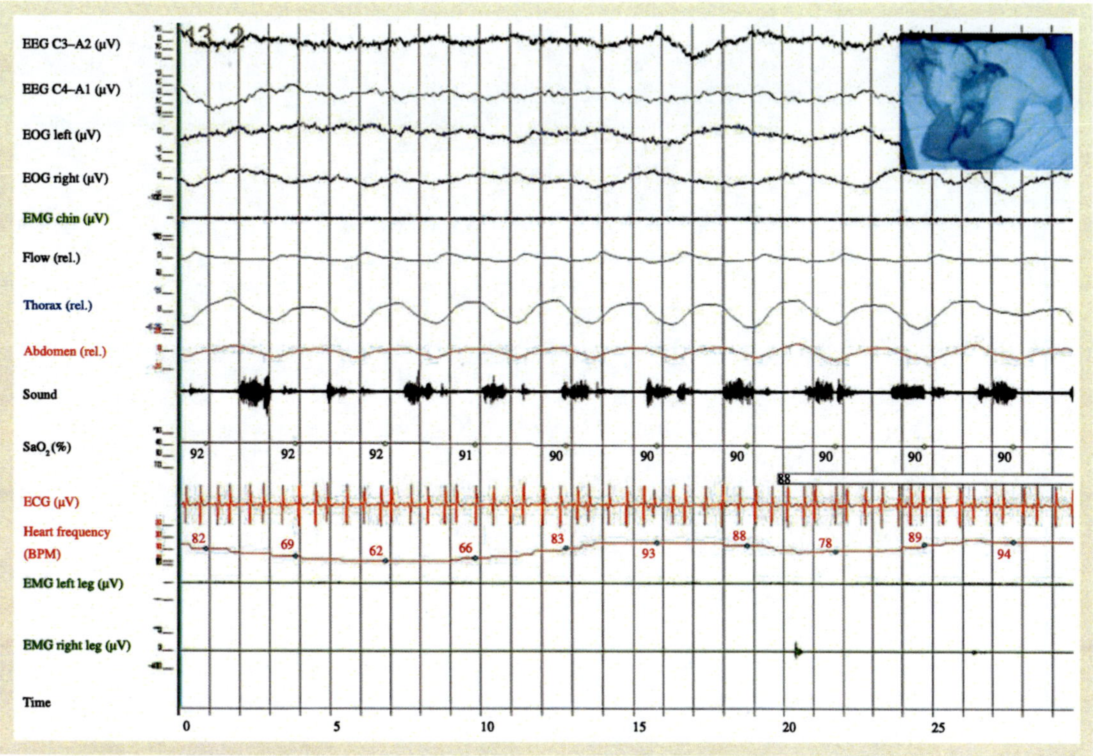

Fig. 10.5 Polysomnographic recording of a patient in the sleep laboratory with the display of the video image in the *upper window*. The following biosignals are shown: EEG C3:A1 and C4:A2, EOG, EMG from the mouth base, respiratory flow, respiratory effort, snoring sounds, oxygen saturation, ECG, heart rate, EMG from the leg and body position

10.5.2 Visual Evaluation

The first step in evaluation is a visual review of the raw data to ascertain the recording quality and the occurence of the artifacts. The whole recording is displayed using various time scales, to obtain a faster overview. Table 10.2 summarizes the criteria of a visual evaluation of polygraphic signals for sleep disorder diagnostics.

To assess the dynamics of sleep and sleep quality, the respective sleep phases and their changeovers are registered according to the the standard classification of the sleep phases by Rechtschaffen and Kales. In Table 10.4, the essential criteria of this classification are summarized.

10.5.3 Computer-Aided Evaluation

Computer-aided methods for analysis of polysomnographic signals consider both the respiratory and the bioelectric signals. Algorithms evaluate and detect apneas and hypopneas, and differentiate between obstructive and central breathing disorders (Fig. 10.6). Evaluating variation in heart rate and oxygen saturation will assist in further diagnosis. The concise display of changes of these parameters over the whole recording period is an indispensable aid to the physician.

Although the classification of sleep phases according to Rechtschaffen and Kales has become the standard and provides useful information for many applications, it is based on a quite arbitrary phase differentiation and depends on the experience of the physician. Other methods use Fourier analysis to separate the signal into its frequency components and calculate the power as a function of time. The sleep architecture can especially be registered through the slow activity in the range of delta-waves (less than 4 Hz). Thus, functional aspects of sleep regulation can be examined (Fig. 10.7).

Other methods include analysis in the time domain by detecting waves and patterns and the use of hybrid systems with analog filters. Newer methods use neural

Table 10.4 Classification of sleep phases according to the criteria of Rechtschaffen and Kales

State and % of total sleep		EEG	EOG	EMG
Awake (1%)	W	Dominating alpha activity (8–13 Hz), beta waves (> 13 Hz)	Eyeblinks, rapid eye movements	High tone, movement artifacts
Non-REM (5%)	S1	Theta activity, vertex waves	Slow, oscillating eye movements	Drop of the average muscle tone, small movement artifacts
(49%)	S2	Theta activity, K complexes, sleep spindles	No eye movements	Drop of the average muscle tone, small movement artifacts
(8%)	S3	Groups of high delta waves > 20%, < 50%	No eye movements	Drop of the average muscle tone, small movement artifacts
(13%)	S4	Groups of high delta waves > 50%	No eye movements	Drop of the average muscle tone, little movement artifacts
REM (24%)	REM	Theta activity, sawtooth wave	Conjugated rapid eye movements (saccades)	Low average tone, phasic activation
Movement time (MT)	> 50 % of the stage is interfered by movement artifacts, thus allocation to another phase is not possible			

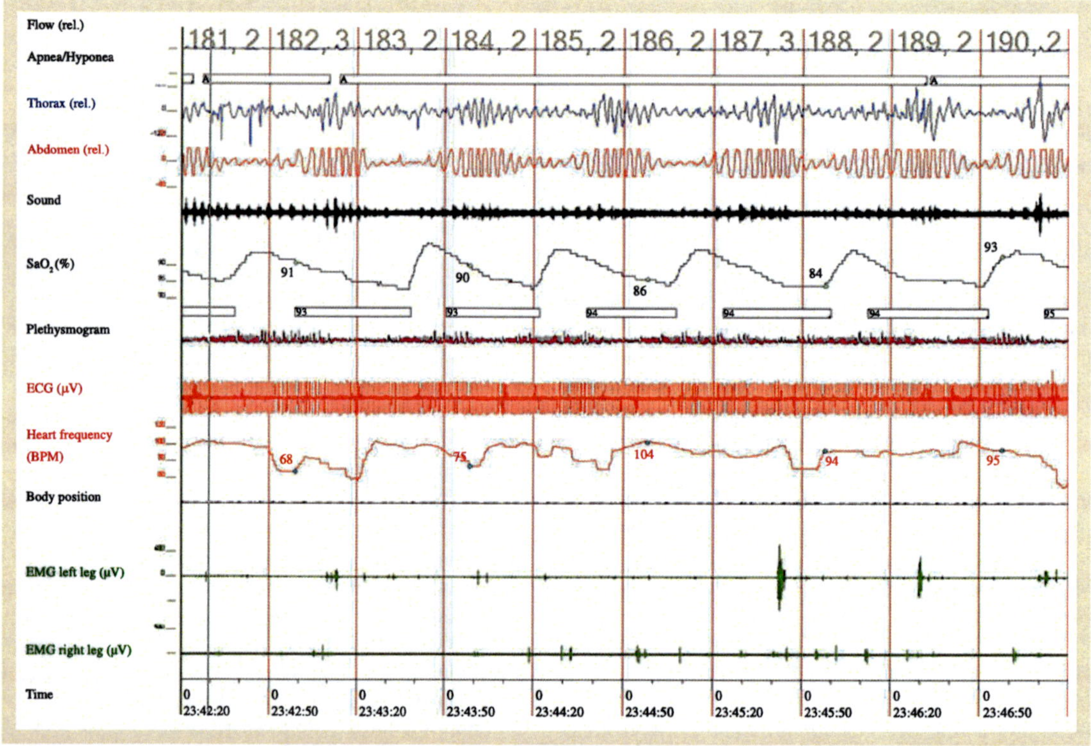

Fig. 10.6 Polysomnographic display, particularly of the respiratory parameters, including their evaluation by marking detected apneas or hypopneas and desaturation

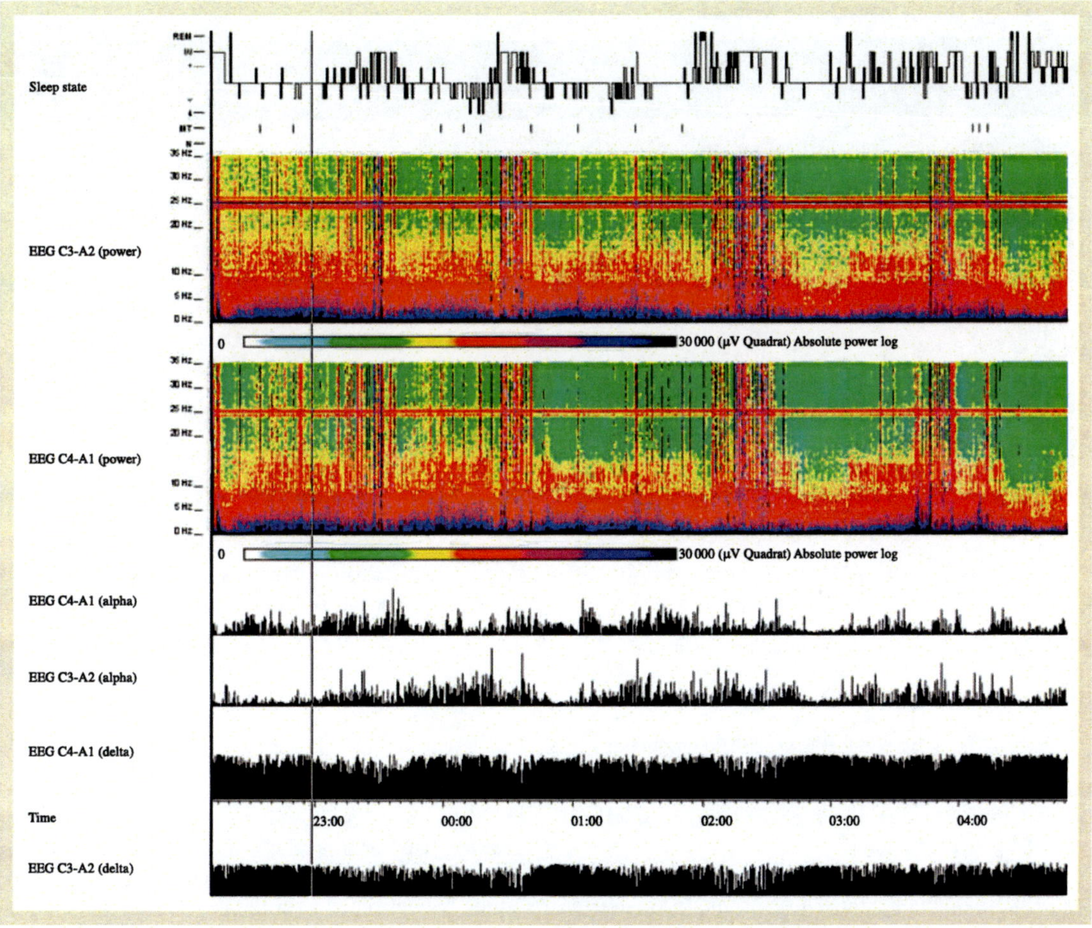

Fig. 10.7 Display of evaluation results, particularly of the neurophysiological biosignals. The classification of sleep phases according to Rechtschaffen and Kales, the frequency analysis of the EEG (two-dimensionally as a function of time), and the temporal occurrence of alpha and delta waves in the EEG are shown

networks which, after a learning stage, assess biosignals according to predetermined criteria. Currently, also algorithms based on fuzzy logic are used for classification of sleep phases.

Another parameter is the pulse transit time (PTT) which can be derived from the ECG and the pulse signal measured on the finger by photooxymetry. The pulse transit time is the time lag between the maximum of the R peak and the moment when the pulse curve reaches 50% of its amplitude. Changes in the pulse transit time correlate with both blood pressure changes and respiratory effort. Thus, an additional parameter is obtained to differentiate between obstructive and central apnea.

10.6 Fields of Application

The fields of application for diagnostics of sleep disorders are as manifold as their causes. They range from internal, neurological, and pneumological diseases to physiological reactions, e.g., jet lag due to crossing time

Table 10.5 Fields of indication in various medical subjects

Subject	Examples of functional disorders and diseases representing an indication for an examination of sleep-related breathing disorders
Andrology	Erectile dysfunctions
Endocrinology	Hypothyreosis, acromegaly, adipositas permagna
Hematology	Polycythemia
Otolaryngology	Detectable obstruction of the upper respiratory tracts for an awake patient
Cardiology	Nocturnal cardiac arrhythmia, hypertension, cardiac insufficiency of unclear genesis, dilatative cardiomyopathy
Neurology	Neuromuscular diseases, hypersomnia
Orthopedics	Kyphoscoliosis
Pediatrics	Disorders of respiratory regulation, SIDS survivors, children with suspected Pickwick syndrome, thriving disorders, big tonsils and adenoids
Pneumology	Hypoxia/hypercapnia with or without previously existing pulmonary disease, insufficiency of the right heart or global cardiac insufficiency of unclear genesis
Psychiatry	Hypersomnia, hyposomnia
Psychosomatics	Unclear *states of failure*, reduced performance
Oral surgery	Craniofacial malformations

zones during transmeridian flight or sleep disorders associated with pregnancy.

They also include all fields with respect to age, e.g., from sudden infant death syndrome to age-related sleep disorders in geriatrics. In Table 10.5, the individual subject areas and the respective functional disorders are mentioned. It is obvious that there is hardly any medical subject which is not directly or indirectly confronted with the problem of sleep disorders.

10.7 Methodical Instructions

The methodical instructions given in this chapter are related to polysomnographic recordings. For issues concerning maintenance and servicing, the instructions of the respective manufacturers of devices, accessories, and sensors have to be considered. As the type of sensors, their properties, and their application may vary, only a few, general instructions can be given.

10.7.1 Handling/Application

Which biosignals should be recorded depends on the clinical problem. At the beginning of recording, in addition to technical calibration, biological calibration has to be conducted by eye movements produced by commands, body movements, arbitrary apnea, hypopnea, and hyperventilation. This is done to control the operability of the recording technique and the fit of the sensors. For recording of bioelectric signals, only electrodes of the same material should be used, either gold or silver/silver chloride.

The German Sleep Society recommends that polysomnographic examinations be conducted on at least two consecutive nights. During the first night, the patient can acclimatize to the laboratory. Preferably, biosignals should be continuously recorded on paper with a paper speed of 10 or 15 mm/s. The German Sleep Society recommends storage of the recorded signals on optical, magnetic or digital media, so they can be analogously plotted again with a polygraph at any time.

10.7.2 Artifacts

Artifacts represent an important problem in measurement of biosignals. As a noise signal, they are superimposed onto the physiological signal, and their amplitude can be a multiple of the amplitude of the wanted signal. Biological and technical artifacts can be distinguished. The latter can be reduced by choosing the recording space, improving the sensor application (e.g., reduction of the electrode interface impedance) or changing the recording technique.

Biological artifacts can be eliminated only to some extent, e.g., by relocating the applied sensors. Those artifacts can be identified in the signal by an experienced operator. Artifacts can lead to misinterperpretation if they are analyzed by computers e.g. slow eye movements in the frontal EEG.

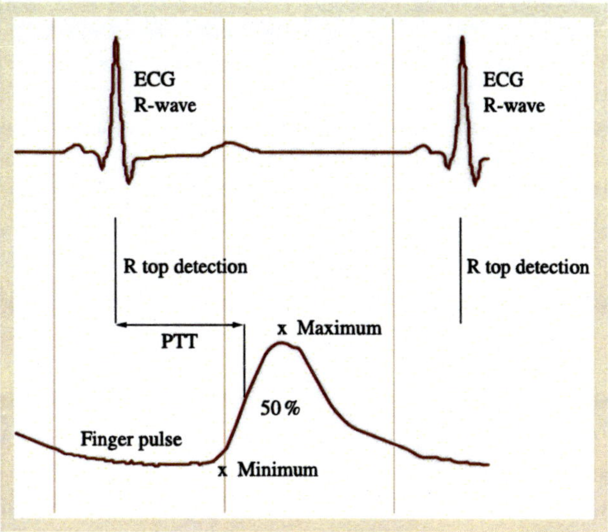

Fig. 10.8 Determination of the pulse transit time

10.8 Medical Significance of Sleep Diagnostics

The medical significance of sleep disorders emerges from their prevalence, and from the yearly indirect costs due to nondetection and nontreatment. Currently, about 20 million people in Germany suffer from sleep disorders not caused by external influences. About 3.3 million people are afflicted by tiredness during the day which may partially explain the high incidence of fatal motorway accidents which are due to dozing off behind the wheel (24%). In addition, 2.7 million people take soporifics on a regular base, although almost half of patients taking soporifics daily report that their disorders continue unchanged. Many people taking soporifics should be classified as addicted. It is absolutely essential that the disorder and the initiation of a causal therapy be medically clarified. This becomes even more urgent, as dangerous and widespread diseases such as obesity, arterial hypertension, heart insufficiency, cardia arrhythmia, sudden cardiac death, and metabolic diseases can be caused by sleep disorders.

Overall, 800 000 patients suffer from sleep apnea in Germany, of which 50% have hypertension. Three significant pathomechanisms are considered to be the major causes for hypertension:

- Blood gas changes,
- Intrathoracic pressure variations, and
- Consecutive arousal reactions.

Over time, these can lead to hypertension in the systemic and pulmonary circulation, cor pulmonale, cardiomyopathies, coronary heart diseases, cardiac arrhythmia, and stroke. Sleep research studies indicate that secondary cardiovascular and cardiopulmonary diseases do not appear or are reversible if sleep-related breathing disorders are diagnosed in time and successfully treated.

Table 10.6 summarizes the sleep disorders based on the international classification of sleep disorders published by the American Sleep Disorders Association. This international classification describes the various disorders and tries to group them with respect to their cause.

Table 10.6 International classification of sleep disorders

1. Dyssomnias	2. Parasomnias	3. Sleep disorders associated with mental, neurologic or other medical disorders	4. Proposed sleep disorders
A. Intrinsic sleep disorders	A. Arousal disorders	A. Associated with mental disorders	A. Sleep disorders of various origin
Psychophysiologic insomnia, sleep state misperception, idiopathic insomnia, narcolepsy, recurrent hypersomnia, idiopathic hypersomnia, posttraumatic hypersomnia, obstructive sleep apnea syndrome, central sleep apnea syndrome, central alveolar hypoventilation syndrome, periodic limb movement disorder, restless legs syndrome, intrinsic sleep disorder not otherwise specified (NOS)	Confusional arousals, sleepwalking, sleep terrors	Psychoses, mood disorders, anxiety disorders, panic disorders, alcoholism	Short sleeper, long sleeper, subwakefulness syndrome, fragmentary myoclonus, sleep hyperhidrosis, menstrual-associated sleep disorder, pregnancy-associated sleep disorder, terrifying hypnagogic hallucinations, sleep-related neurogenic tachypnea, sleep-related laryngospasm, sleep choking syndrome
B. Extrinsic sleep disorders	B. Sleep–wake transition disorders	B. Associated with neurologic disorders	
Inadequate sleep hygiene, environmental sleep disorder, altitude insomnia, adjustment sleep disorder, insufficient sleep syndrome, limit-setting sleep disorder, sleep-onset association disorder, food allergy insomnia, nocturnal eating (drinking) syndrome, hypnotic-dependent sleep disorder, stimulant-dependent sleep disorder, alcohol-dependent sleep disorder, toxin-induced sleep disorder, extrinsic sleep disorder NOS	Rhythmic movement disorder, sleep starts, sleep talking, nocturnal leg cramps	Cerebral degenerative disorders, dementia, Parkinsonism, fatal familial insomnia, sleep-related epilepsy, electrical status epilepticus of sleep, sleep-related headaches	
C. Circadian-rhythm sleep disorders	C. Parasomnias usually associated with REM sleep	C. Associated with other medical disorders	
Time zone change (jet lag) syndrome, shift work sleep disorder, irregular sleep–wake pattern, delayed sleep-phase syndrome, advanced sleep-phase syndrome, non-24-h sleep–wake disorder, circadian rhythm sleep disorder NOS	Nightmares, sleep paralysis, impaired sleep-related penile erections, sleep-related painful erections, REM sleep-related sinus arrest, REM sleep behavior disorder	Sleeping sickness, nocturnal cardiac ischemia, chronic obstructive pulmonary disease, sleep-related asthma, sleep-related gastroesophageal reflux, peptic ulcer disease, fibromyalgia	

Table 10.6 (continued)

1. Dyssomnias	2. Parasomnias	3. Sleep disorders associated with mental, neurologic or other medical disorders	4. Proposed sleep disorders
	D. Other parasomnias Sleep bruxism, sleep enuresis, sleep-related abnormal swallowing syndrome, nocturnal paroxysmal dystonia, primary snoring, infant sleep apnea, congenital central hypoventilation syndrome, sudden infant death syndrome, benign neonatal sleep myoclonus, other parasomnia NOS		

10.9 Therapy

Therapeutic methods for sleep disorders are as manifold as the sleep disorders themselves. A five-stage therapeutic scheme can be defined, distinguishing between behavioral medical psychotherapeutic, chronobiological, pharmacological, machine-aided, and surgical methods.

10.9.1 Behavioral Medical Psychotherapeutic Therapy

Besides recommendations regarding sleep hygiene and prevention of sleep-disturbing behavior, various relaxation exercises are used.

10.9.2 Chronobiological Therapy

This field includes measures that influence the periodic process of biological functions, such as light therapy, which can also be used to reduce the effect of an artificial time shift, such as jet lag or due to shift work.

10.9.3 Medicinal Therapy

For primary organic and chronic diseases, pharmacological measures are used. Depending on the disease pattern, stimulants, antidepressants, hypnotics, and neuroleptics are used. For medicinal treatment of mild and moderate sleep-related breathing disorders, theophylline can be used. Thus, patients can be medicinally treated, if nCPAP therapy is not available.

10.9.4 Machine-Aided Therapy

For treatment of sleep-related breathing disorders, machine-aided therapy is commonly used. Predominantly, CPAP devices are used for positive-pressure ventilation. For high expiration pressures or pressure-limited ventilation devices in the treatment of both central sleep apnea and central hypoventilation, a BiLEVEL with separate control of inspiratory and expiratory pressure can be used.

Alternative solutions include the Esmarch prosthesis to advance the mandible 2–4 mm, and the Snore Ex to pull the tongue of the patient from the back of the mouth to the front. The therapeutic effect of these methods is controversial.

10.9.5 Operational Therapy

A clear indication for oral surgery resulting in positive results is rare. Uvulopalatopharyngoplasty is used to stabilize the pharyngeal lumen. Especially in infants, tonsillectomy can be applied to eliminate obstructions.

10.10 Safety Aspects

The application of all medical devices should be based on evidence that they pose no threat to the patient, the user or third persons. This is specified in standards (e.g., regulations of the German Association for Electrical, Electronic, and Information Technologies). Basic regulations are included in the medical product law, governing the fabrication, operation, and use of medical products and their accessories. Devices have to be designed in such a way that their safety, suitability, and performance are not endangered, even when affected by malfunctions or operating errors. Besides the application-related desires, a polysomnographic system has to fulfill the medical quality assurance criteria.

10.10.1 Mobile Devices

Medical devices must not endanger the health or security of patients, users, and third persons. This is particularly important for mobile devices that the patient uses at home within the scope of therapy or ambulant diagnostics of sleep disorders. Here, operability, clarity, exclusion of potential operating errors, and long failure-free operating times have to be additionally considered. The latter is an essential precondition for use of CPAP devices in therapy of sleep-related breathing disorders, which has to be combined with a special service. Yearly maintenance should be the standard and should, besides cleaning, substitution of worn parts, and evidence of full operability, also be used to coordinate the relation between patient and machine through extensive support of and consultation with the patient.

10.10.2 Devices in the Sleep Laboratory

Devices in the sleep laboratory are used under permanent surveillance by staff. Besides device-related preconditions, training of laboratory staff is very important. The assessment procedures of the German Sleep Society for quality assurance in the sleep laboratory do not only include the control of quality and quantity of laboratory space, technical equipment, and staff, but also quality of diagnosis and therapy.

10.11 Planning Advice

10.11.1 Required Space

For the optimum design of a sleep laboratory, it should have the sleep stations and the room with the recording and processing technology (wake station) spatially separated. The sleep station area should be at least 12 m^2. Acoustic insulation, and possibilities to darken the room and control air conditioning should be provided. An infrared video system consisting of camera, microphone, infrared radiator, monitor, and recorder is necessary for simultaneous video recording of the patient together with the recorded biosignals, and is also used to keep the patient under surveillance. To communicate with the patient during recording, an intercommunication system is required.

10.11.2 Required Time

The time required for a sleep examination strongly depends on the case and can vary between 20 min (MSLT) and several days (diagnostics of epilepsy and chronobiological sleep disorders). Preparation of the patient and application of the sensors take 30–60 min on average.

10.11.3 Staff

A sleep laboratory is managed by a responsible director skilled in the diagnoses of sleep disorders. If he is not a physician, but, e.g., psychologist, a physician should be present to address medical issues. It is recommended to employ one physician (e.g., for outpatient clinic, clinical findings, interventional therapy, and on-call duty) and two assistants (e.g., for patient assignment, day test, evaluation, sensor application, and maintenance) per work station. A night watch should not observe more than three patients, among them a maximum of two CPAP patients and no more than one problem case. During the night, a physician on duty has to be available on call.

Further Reading

- American Academy of Sleep Medicine: *Casebook of Sleep Medicine. A Learning Companion to the International Classification of Sleep Disorders, Diagnostic and Coding Manual*, 2nd edn. (Westchester, IL 2008)
- American Academy of Sleep Medicine: *The International Classification Of Sleep Disorders, Revised*

Diagnostic and Coding Manual, One Westbrook Corporate Center, Suite 920, (Westchester, IL 60154-5767, USA 2001)

- J. Argod, J.L. Pepin, R.P. Smith, P. Levy: Comparison of esophageal pressure with pulse transit time as a measure of respiratory effort for scoring obstructive nonapneic respiratory events. Am. J. Respir. Crit. Care. Med. **162**, 87–93 (2000)
- I. Ayappa, R.G. Norman, J.J. Hosselet, D.M. Rapoport: Relative occurrence of flow limitation and snoring during continuous positive airway pressure titration, Chest **114**, 685–690 (1998)
- M. Berger (Ed.): *Handbuch des normalen und gestörten Schlafs* (Springer, Berlin Heidelberg 1992)
- A.A. Borbély: Two process model of sleep regulation, Human. Neurobiol. **1**, 195–204 (1982)
- D.J. Buysse, C.F. Reynolds III, D.J. Kupfer, M.J. Thorpy, E. Bixler, R. Manfredi, A. Kales, A. Vgontzas, E. Stepanski, T. Roth: Clinical diagnoses in 216 insomnia patients using the international classification of sleep disorders, Sleep **17**, 630–637 (1994)
- P. Collard, T. Pieters, G. Aubert, P. Delguste, D.O. Rodenstein: Compliance with nasal CPAP in obstructive sleep apnea patients, Sleep Med. Rev. **1**, 33–44 (1997)
- R. Ferber, R. Millman, M. Coppola, J. Fleetham, C.F. Murray, C. Iber, V. McCall, G. Nino-Murcia, M. Pressman, M. Sanders: Portable recording in the assessment of obstructive sleep apnea, Sleep **17**, 378–392 (1994)
- C. Guilleminault (Ed.): *Clinical Neurophysiology of Sleep Disorders*, Handbook of Clinical Neurophysiology Series (Elsevier, New York 2005)
- J. He, M.H. Kryger, F.J. Zorick, W. Conway, T. Roth: Mortality and apnea index in obstructive sleep apnea, Experience in 385 male patients, Chest. **94**, 9–14 (1988)
- M. Hirshkowitz, C.A. Moore: Issues in computerized polysomnography, Sleep **17**, 105–112 (1994)
- M.H. Kryger, T. Roth, W.C. Dement: *The Principles and Practice of Sleep Medicine* (W.B. Saunders, Philadelphia 1994)
- T.L. Lee-Chiong: *Sleep Medicine: Essentials and Review* (Oxford University Press 2008)
- R. Lund, P. Clarenbach: *Schlafstörungen, Klassifikation und Behandlung* (Arcis, München 1992)
- M.L. Perlis, K.L. Lichstein: *Handbook of Sleep Disorders* (Wiley 2003)
- J.H. Peter, G. Amed, S. Stephan, W. Jung, A. Ockenga, H. Himmelmann, P. von Wichert: Therapie der Schlafapnoe mit abendlich eingenommenem retardiertem Theophyllin (EUPHYLONG), Prax. Klin. Pneumol. **41**, 433–437 (1987), in German
- A. Sadeh, P.J. Hauri, D.F. Kripke, P. Lavi: The role of actigraphy in the evaluation of sleep disorders, Sleep **18**, 288–302 (1995)
- A.R. Schwartz, P.L. Smith, A.R. Gold: Clinical application of quantitative measures of pharyngeal collapsibility, ADSA News **4**, 11–12 (1997)
- R.P. Smith, J. Argod, J.L. Pépin, P.A. Lévy: Pulse transit time: An appraisal of potential clinical applications, Thorax **54**, 452–457 (1999)
- M.J. Thorpy: *Handbook of Sleep Disorders* (Marcel Dekker, New York Basel 1990)

11. Nystagmography

Klaus-Peter Hoffmann, Eduardo Fernandez

Clear vision requires both eye motion and highly accurate positioning of the fovea. Various types of oculomotor activity are distinguished (e.g., ocular reflexes, fixation, saccade, smooth pursuit). Different methods can be used to record eye movements. The most common methods for clinical application are electronystagmography, photoelectronystagmography, and videooculography. Examples for signal processing and medical significance are given.

11.1 **Application** 179

11.2 **Eye Movements** 180
- 11.2.1 Saccades 180
- 11.2.2 Smooth Pursuit Movements 181
- 11.2.3 Nystagmuses 181
- 11.2.4 Convergence Movements 181
- 11.2.5 Torsional Movements 181

11.3 **Technology and Methods** 181
- 11.3.1 Frenzel Glasses 182
- 11.3.2 Electronystagmography (ENG) and Electrooculography (EOG) 182
- 11.3.3 Photoelectronystagmography (PENG) and Infrared Oculography (IROG) 182
- 11.3.4 Magnetooculography 183
- 11.3.5 Electromagnetic Technology (Search-Coil System) 183
- 11.3.6 Videooculography (VOG) 184

11.4 **Methods** 184
- 11.4.1 Saccades 185
- 11.4.2 Smooth Pursuit Movements 185
- 11.4.3 Optokinetic Nystagmus 186
- 11.4.4 Vestibular Nystagmus 186

11.5 **Signal Recording and Signal Processing** 187

11.6 **Medical Significance** 187

11.7 **Safety Aspects** 188

11.8 **Spatial Planning** 188

References 189

Nystagmography is a method to record, analyze, and evaluate spontaneous eye movements in response to vestibular, visual, caloric, rotational or positional stimulation. Nystagmography provides an objective assessment of the oculomotor and vestibular systems. Figure 11.1 shows a simplified general schematic of the oculomotor system. Its primary task is to stabilize the images of the visual world on the retina and thus enable undisturbed vision and constant spatial perception.

On the one hand, the interesting image details have to be binocularly centered and fixed at the site of sharpest vision, the fovea ($\approx 0.8°$) of the central visual field. On the other hand, retinal image shifts occurring with movements of one's own and of the environment have to be eliminated. For this, fast and slow eye movements are available with the oculomotor system's saccades, smooth pursuit movements, vestibular and optokinetic nystagmuses, convergence movements, and fixation. The fixation phase is characterized by microsaccades.

11.1 Application

Surprisingly, even gazing at a stationary object involves smooth eye movements because the head is always in slight motion as the muscles of the body and neck attempt to maintain the posture. Only the accuracy of

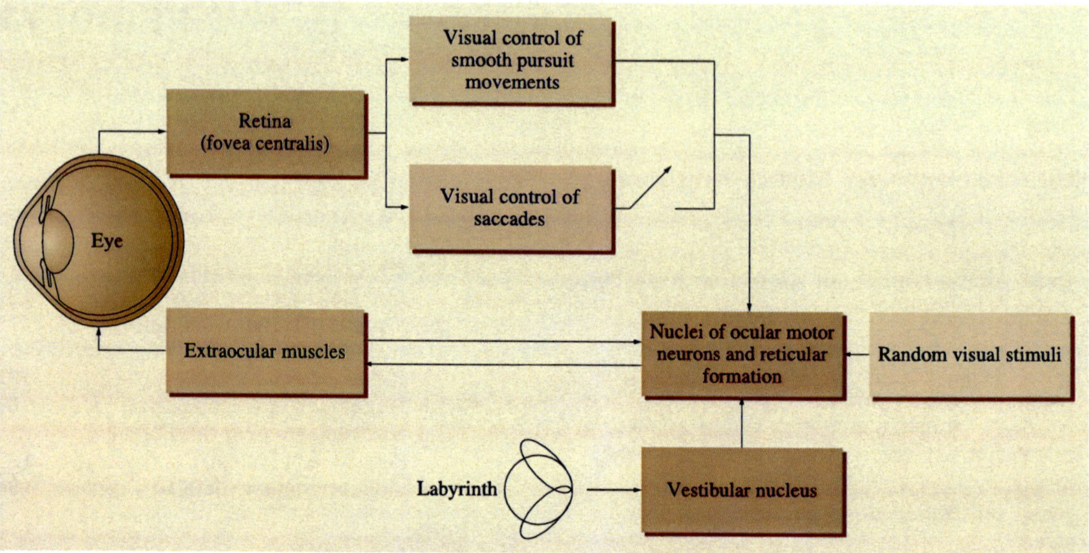

Fig. 11.1 Simplified schematic of the control of eye movements

the sensomotor connection including the possibility to compensate for interferences secures the maintenance of binocular spatial vision. This functional performance of the oculomotor activity is enabled by the central integration of the proprioceptive, vestibular, and optic afferents.

Eye movement recording has been in use for a long time. Up to the 19th century, eye movements could only be monitored by sitting closely and watching the eyes of the subject directly. Some of the first mechanical recording methods were more precise but required a level or a small mirror system attached to the sclera, thus being uncomfortable for the subject, and were rapidly abandoned. A breakthrough was achieved in 1922 when the potential difference between the cornea and the retina was used for recording of ocular nystagmus (Sect. 11.3.2). This procedure is still the most widely applied technique for eye movement recording in clinical routine, specially by otolaryngologists and neurologists to evaluate patients with dizziness, vertigo or balance dysfunction.

Thus, nystagmography is an interdisciplinary subject, which is particularly applied in neurology and otorhinolaryngology. Already since the middle of the last century, it has been established as a routine diagnostic method.

11.2 Eye Movements

A schematic of essential forms of eye movements is given in Fig. 11.2. This includes saccades, smooth pursuit movements, nystagmus, and convergence movement.

11.2.1 Saccades

Saccades are quick eye movements used to find a new fixation point. They are cerebrally controlled and are generated in the paramedian pontine reticular formation. Their maximum velocity increases with the amplitude of the movement and can be up to 700 deg/s. The duration is between 30 and 120 ms. Other parameters of a saccade include latency and overshoot defined as percentage deviation from the fixation point after a saccade. If the new fixation point is not completely achieved by a saccade, a correction saccade follows after 100–300 ms.

Saccades can be divided into internally triggered volitional saccades (e.g., memory-guided saccades, antisaccades, tracking saccades), automatic reflex saccades triggered by external stimuli (e.g., visually or acoustically induced), and spontaneous, seemingly random saccades.

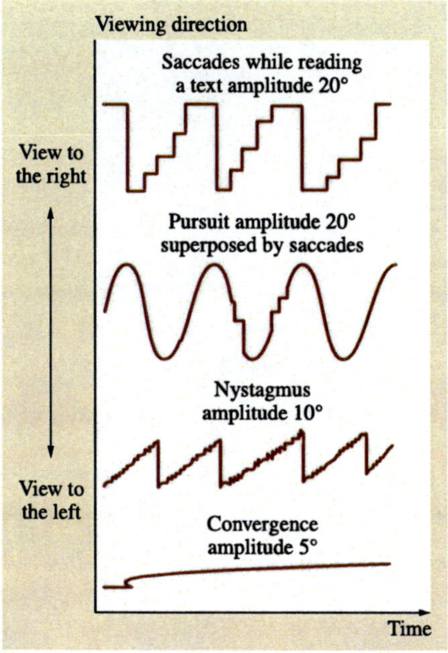

Fig. 11.2 Schematic display of eye movements

11.2.2 Smooth Pursuit Movements

Smooth pursuit movements can be described as conjugated and slow movements, by which the eye tracks a moving vision target. The outlines of the object can additionally be scanned by saccades. The average angular velocity is 30–50 deg/s. They are used to follow and stabilize a moving vision target on the retina. Care should be taken in interpretation, since these smooth movements are influenced by age and are affected by attention and patient cooperation.

11.2.3 Nystagmuses

An involuntary rhythmic ocular oscillation occurring in two phases is denoted as nystagmus. Two basic types of nystagmus can be distinguished: on the one hand the jerk nystagmus with a slow and a quick phase, with the latter determining the direction of the nystagmus, and on the other hand the pendulum nystagmus with equally quick eye movements in both directions.

The optokinetic nystagmus (OKN), also called vestibulo-ocular reflex (VOR), is a jerk nystagmus caused by large-area stimuli. An example is the so-called railway nystagmus, which can be detected when watching targets out of a moving train. The slow phase corresponds to the movement of the train (pursuit movement), the quick phase to a change of the fixation point (saccade).

A vestibular stimulation can be followed by a VOR. For instance, a vestibular nystagmus occurs after caloric or mechanical stimulation of the labyrinth. The latter can be achieved by angular acceleration. During the rotation, the nystagmus is directed in the rotational direction; after termination, it is directed in the opposite direction.

Other physiological nystagmuses include the cervical nystagmus for rotation of the cervical spine, the audiokinetic nystagmus for moving acoustic sources, the arthrokinetic nystagmus for passive arm and leg movements, and the fixation or terminal nystagmus.

Pathologic nystagmuses occur due to lesions in the optomotor system. They include rebound nystagmus, vestibular spontaneous nystagmus, head-shaking nystagmus, bearing nystagmus, muscle paretic nystagmus, periodic alternating nystagmus, etc.

11.2.4 Convergence Movements

Convergence is the simultaneous inward movement of both eyes towards each other (in the vertical axis) so that the projection of the image is in the center of the retina in both eyes. The main objective is to maintain binocular vision. Interretinal image errors possibly occurring during a fixation can be compensated by convergence movements. Those are very slow (10 deg/s) and have small amplitude (maximum 15°).

11.2.5 Torsional Movements

Rotations around the visual axis with amplitude of up to 10° and velocity of up to 200 deg/s are denoted as torsional movements. They can occur spontaneously or optokinetically due to the view of an environmental roll movement or vestibularly during body or head tilting.

11.3 Technology and Methods

The experienced clinician can conduct a differentiated examination of eye movements, especially the detection of defective eye positions, also without complex technical equipment. Quantitative registration by mea-

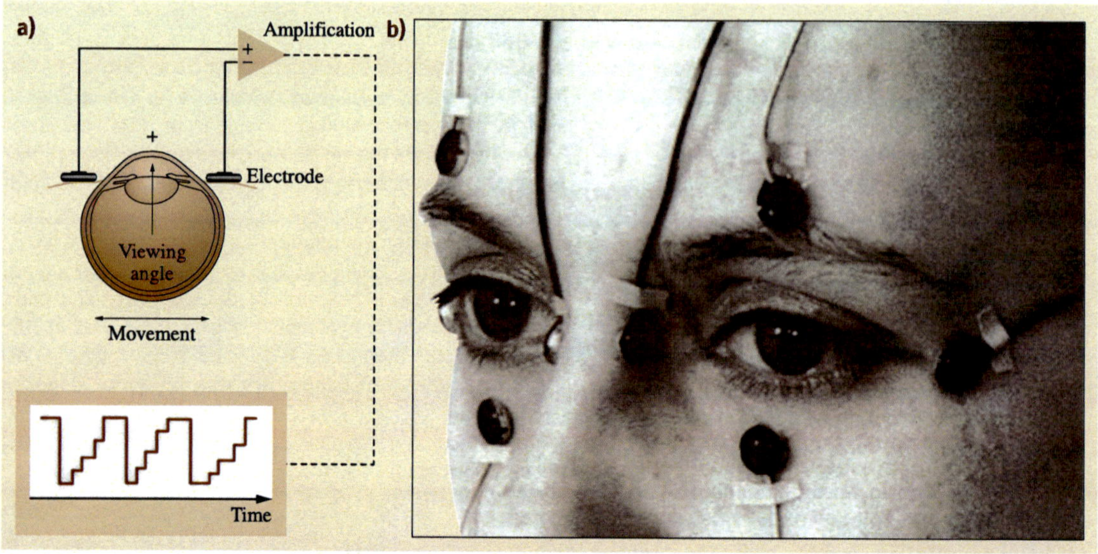

Fig. 11.3a,b Electrooculography. **(a)** Schematic of metrological registration. **(b)** Application of Ag/AgCl electrodes for bipolar recording of horizontal and vertical eye movements

suring the latency and maximum velocity, progress observation, detecting spontaneous eye oscillations, and recording of eye movements with closed lids can only be done with appropriate technology.

11.3.1 Frenzel Glasses

Frenzel glasses comprise a pair of very thick lenses which serve to completely blur the patient's observed environment. Due to the convex lenses and the illumination of the eyes, fixation is excluded, and the evaluation of eye movements with magnification becomes possible.

The advantages of Frenzel glasses over other methods to study eye movements is their low cost. This method can only be used for qualitative statements. It is widespread owing to its easy operability and sufficient evaluability.

11.3.2 Electronystagmography (ENG) and Electrooculography (EOG)

Electronystagmography is the most widespread method for registration of eye movements in clinical practice. The signal source is the corneoretinal potential which is due to concentration differences of various ions in the retinal pigment epithelium. When the eye is moved, an electric potential can be recorded close to the eye. The change of this potential is proportional to the cosine of the angle between the recording plane and the dipole axis. These potential changes can be bipolarly recorded using Ag/AgCl electrodes separately for both eyes (Fig. 11.3). A linear relation can be assumed for these values with sufficient accuracy up to 20°.

Electronystagmography can be performed with only little technical effort. Besides the electrodes and a registration component, amplifiers with a frequency range of 0.1–30 Hz for smooth pursuit movements and direct current (DC) up to 1000 Hz for saccades are required. Measurements in horizontal and vertical direction can be done for both open and closed eyes, for instance, during sleep. The resolution is about 1°. The signal is perturbed by a variety of artifacts, such as lid artifact, potential changes during light–dark adaptation of the eye, baseline drift due to changes in skin impedance, and muscle artifacts. Owing to its unlimited and simple operability, electronystagmography is still the state of the art for clinical routine diagnostics.

11.3.3 Photoelectronystagmography (PENG) and Infrared Oculography (IROG)

These methods rely on the fact that the white sclera reflects more light than the pupil and the iris. For the photoelectric method, infrared light-emitting diodes,

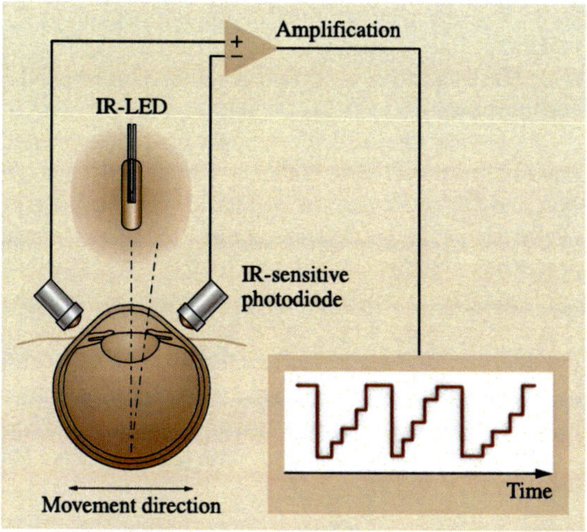

Fig. 11.4 Photoelectronystagmography: schematic of the measurement setup

which are mostly circularly arranged, are used to diffusely illuminate the eye. This light is differently reflected by the iris and the sclera. Thus, phototransistors directed towards the limbus detect potential differences during eye movement. The value of the potential difference is proportional to the movement (Fig. 11.4).

This method shows only little susceptibility to noise and is well suitable for computer-aided analysis. Also, very small eye movements down to 0.1° can be measured. Drawbacks are influences from lid movements and changes of pupil diameter. Moreover, only measurements with open eyes are possible.

11.3.4 Magnetooculography

Every current-carrying conductor is surrounded by a magnetic field. Therefore, it is possible to detect changes in the bioelectric potential by changes in the magnetic field. This is the basis of magnetooculography, which uses superconductive quantum interferometers as sensors.

Magnetooculography (MOG) is an entirely noncontact method. Currently, its resolution is still less than that of electrooculography. It demands great technical and financial effort, magnetically shielded rooms, and cooling with liquid helium to achieve superconductivity. Thus, it is only used for research purposes.

11.3.5 Electromagnetic Technology (Search-Coil System)

A contact lens with a thin wire coil is placed on the eye. The head of the patient is situated in an alternating magnetic field. Movement of the eye generates an induced potential proportional to the sine of the angle between the coil and the direction of the magnetic field. Depending on the position of the coil, both horizontal and vertical eye movements as well as torsional movements can be registered.

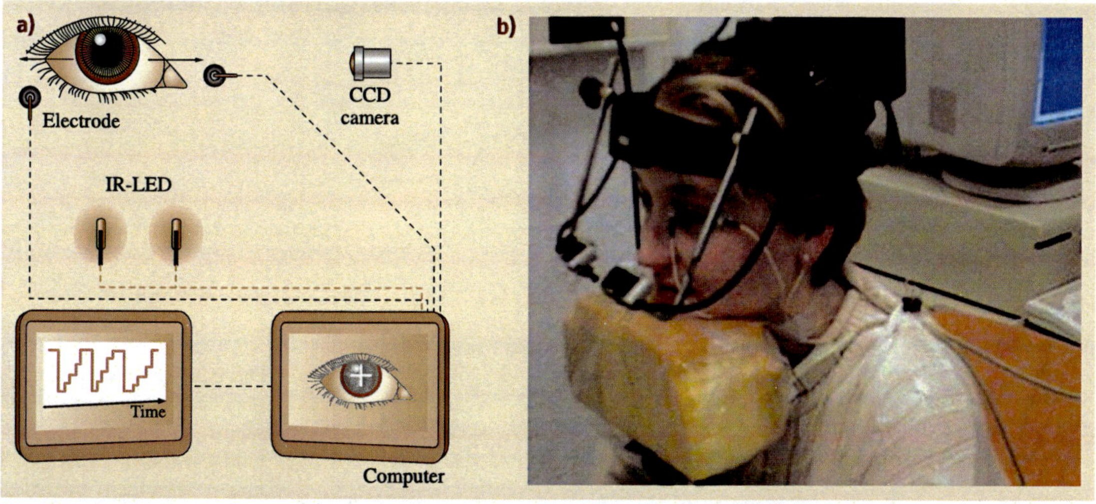

Fig. 11.5a,b Videooculography. **(a)** Schematic of the metrological registration. **(b)** Setup of a measurement station with fixation of the head. In this example, the electrooculogram was simultaneously registered by electrodes

The system provides a drift-free, low-noise signal with resolution of a few arc minutes. Owing to the application of a contact lens and the associated discomfort to the patient, electromagnetic technology has not been established in clinical practice.

11.3.6 Videooculography (VOG)

Miniaturized video cameras based on infrared-sensitive charge-coupled device (CCD) sensors enable frame rates of 250–500 Hz. This is sufficient to identify eye movements from a video image. The required illumination of the eye is achieved by at least two infrared light-emitting diodes (wavelength 850–940 nm) directed towards the eye. The pupil can be detected as the darkest point in the image, as light reflection is least there. The center of the pupil is the intersection of the maximum line segments in the horizontal and vertical directions. Its movement represents the corresponding eye movement (Fig. 11.5).

Moreover, torsional movements can be registered. For this purpose, light marks are projected onto the eye, and the movement of defined prominent patterns of the iris is analyzed. This requires the use of image processing systems.

Due to the manifold possibilities, videooculography is a modern procedure, which has recently gained in importance. With respect to temporal resolution, it still lags behind electrooculography. New clinical potentials have emerged from the registration of torsional movements. However, measurements with closed eyes are not possible.

11.4 Methods

The examination methods depend on the clinical problem, the existing oculomotor disorders, and the respective areas of origin. Qualitative statements are possible through simple clinical tests, e.g., movement of a finger by the physician, requesting the patient to fix the finger, or changes in the position

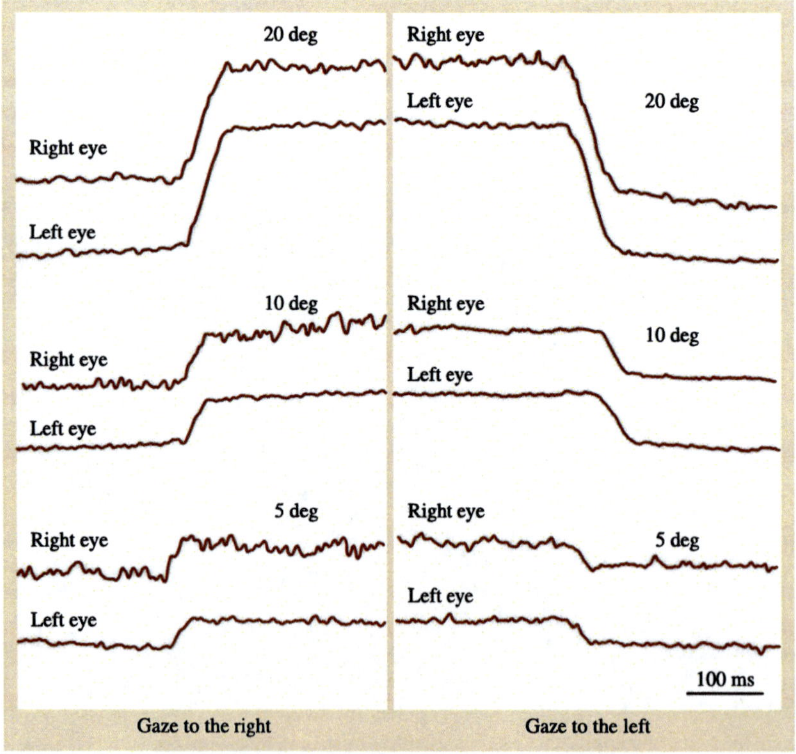

Fig. 11.6 Electrooculographically registered saccades with various amplitudes, showing that the maximum velocity increases with the amplitude of the eye movement

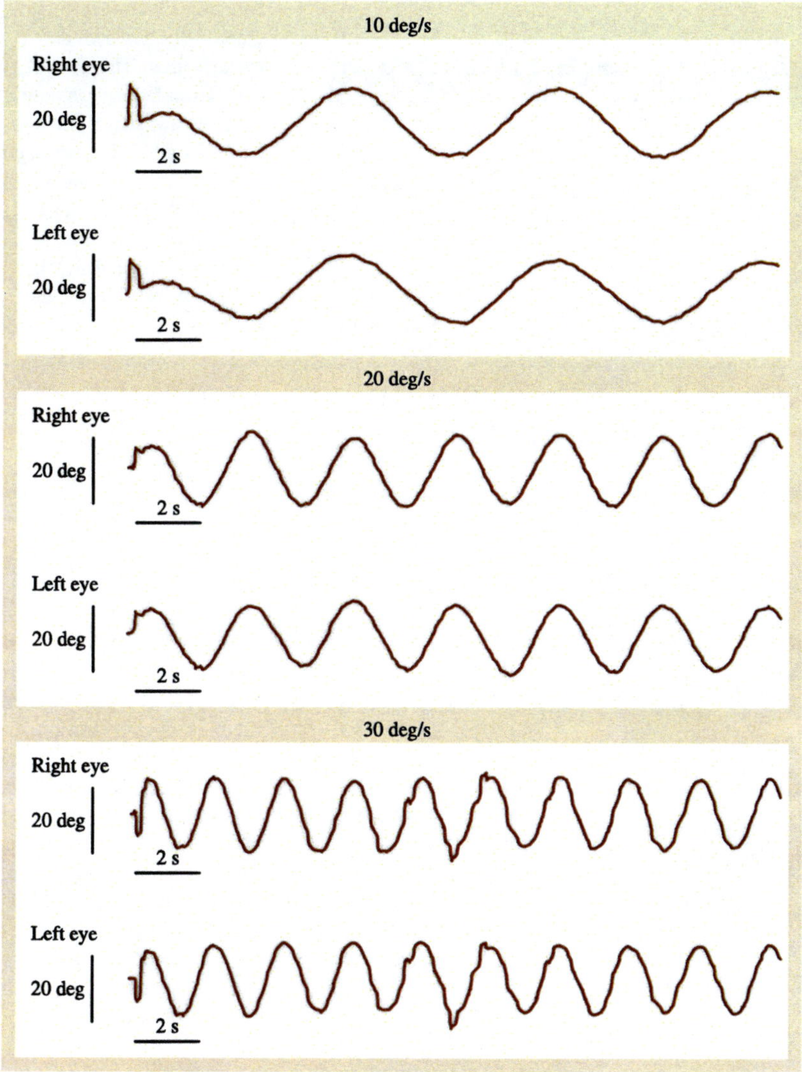

Fig. 11.7 Electrooculographically registered smooth pursuit movements of the right and left eye with various velocities of a sinusoidal moving fixation point: 10 deg/s (*top*), 20 deg/s (*middle*), and 30 deg/s (*bottom*), showing that, with increasing velocity, the smooth pursuit movements decay and the ratio of saccadic eye movements increases

of the patient. The thus provoked eye movements are observed.

In the following, some essential technical examination methods are briefly discussed.

With closed eyes, the occurrence of spontaneous nystagmuses can be examined. Eye movements are mostly registered by electrooculography.

11.4.1 Saccades

Saccades are provoked by two defined fixation points, for instance, two light-emitting diodes alternately lighting up. The patient is requested to look at the illuminated spot, respectively. As a variation, the patient can be requested to change the fixation point at an acoustic signal, or to look at a fading light spot. Figure 11.6 shows an example for registration with various amplitudes.

11.4.2 Smooth Pursuit Movements

The patient is requested to follow a moving light spot with the eyes, e.g., a spot on a monitor moving with a velocity of 10–40 deg/s (Fig. 11.7). Gain,

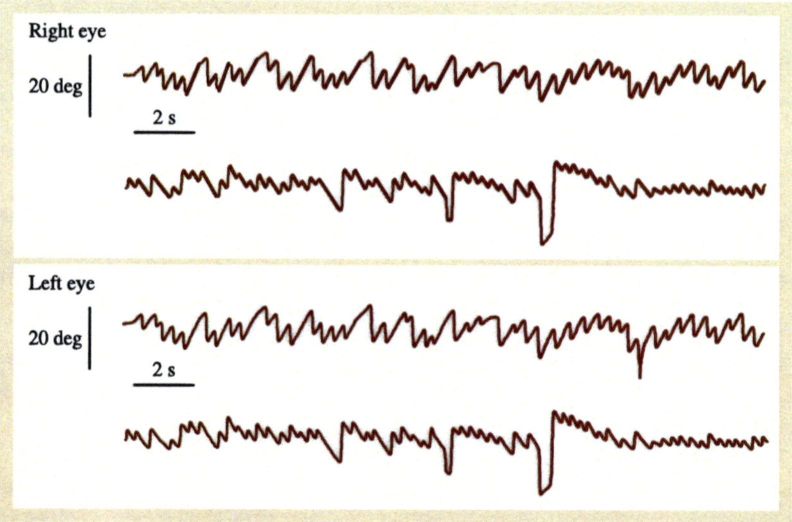

Fig. 11.8 Electrooculographically registered nystagmuses. The optokinetic stimulation was done with vertical stripe patterns (stripe width 3.5°) moving horizontally on a monitor with velocity of 20 deg/s. The movement direction changed. Detected nystagmuses are marked for the right eye

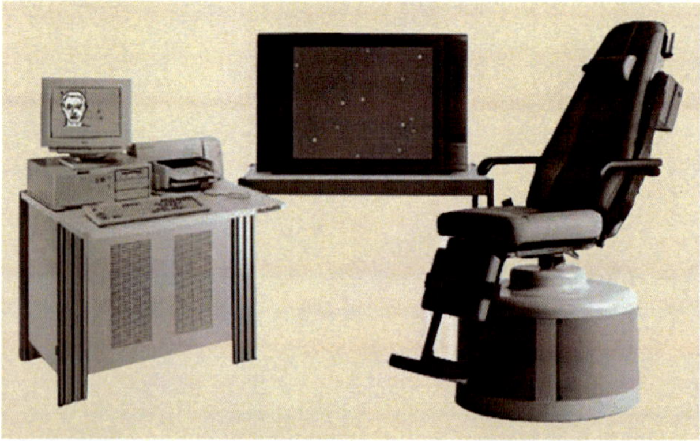

Fig. 11.9 Measurement station for the examination of the nystagmus after optokinetic stimulation with moving patterns on a television and vestibular stimulation using different movement patterns of a swivel chair (courtesy Jaeger Tönnies GmbH, Würzburg)

coherence, phase, and velocity asymmetry are determined.

11.4.3 Optokinetic Nystagmus

An optokinetic nystagmus can be provoked by a rotary drum with, e.g., attached stripes.

If horizontally moving patterns are projected on a television, nystagmuses (Fig. 11.8) can also be provoked. The velocity and size of the patterns are variable.

11.4.4 Vestibular Nystagmus

For selective stimulation of one labyrinth, the most widely used method is the caloric vestibular test. Prior to caloric testing the patient must be inspected with an otoscope to verify that there is no perforation. In the caloric vestibular test, the acoustic meatus is rinsed with warm water (44 °C) or cold water (30 °C). Thus, the two labyrinths can be examined separately. The patient's head is slightly raised. Thus, the lateral horizontal semicircular channel reaches a vertical position. Rinsing with warm water yields thermal convection currents, resulting in displacement of cupula and stereocilia, causing a nystagmus. It is measured whether or not the nystagmus is symmetrically triggered by both ears. This method is used for clinical examination of the vestibular system.

For rotational testing, the patient is seated on a rotating swivel chair with the head stabilized (Fig. 11.9). It is used to test the vestibulo-ocular reflex. Frequently applied methods include the rotating pendulum test, the rotation stop test, and the rotation test. During the rotating pendulum test, the chair is sinusoidally moved with defined amplitudes and velocities. For the rotation stop test, the patient is sufficiently adapted to a constant rotation, and then stopped within 1–2 s. The postrotary nystagmus can be detected. During a rotation test, the patient is first accelerated, then moved with constant velocity (plateau phase), and finally gradually decelerated. It is used to detect the perrotary nystagmus.

11.5 Signal Recording and Signal Processing

Signal recording is determined by the methodology applied. Most currently used systems provide computerized stimulus generation, response acquisition, and interpretation.

Signal processing depends particularly on the procedure used and the respective clinical problem. Besides comparison of the movement of the right and the left eye when looking to the right or left, respectively, and comparison with the default signal, direct calculation of parameters comes to the fore. For saccades and nystagmuses, the calculation of velocities is significant.

However, latency, overshoot, and duration of a saccade also enable statements relevant to diagnostics. For smooth pursuit movements, the gain as an expression of the ratio of response amplitude and stimulus amplitude, the determination of the decay of smooth pursuit movements, the phase relation between stimulus and eye movement, and correlation and coherence functions play an important role.

Calculation of the mentioned parameters is done both in the time domain and, after Fourier transformation, in the frequency domain. The results are displayed separately for the two eyes and for the respective line of sight, often as a function of the amplitude of the eye movement.

For videooculography, a variety of image processing techniques are applied to determine the position of the eyes. The requirements on signal processing are constantly increasing, enhancing the range of nystagmographic examination methods and improving the quality of diagnostic statements. Especially the description of the oculomotor system by mathematical models and different types of digital filtering facilitate new approaches.

11.6 Medical Significance

The oculomotor system is one of the most thoroughly studied human systems. Analysis of eye movement disorders in patients enables a degree of characterization and precise localization which up to now could only be achieved by modern imaging techniques. Besides localization, the principal information gained is the description of a functional limitation due to a lesion, a space-occupying lesion, a systemic disease or the effect of pharmaceuticals and other substances. In this context, it has to be considered that the attempt to compensate is simultaneously registered. This can be used to detect different lesions as well as to show the plasticity of brain.

Applications of nystagmography range from neurological diseases (such as multiple sclerosis and space-occupying lesions) and muscular diseases (such as myasthenia) to vestibular symptoms (such as vertigo). Table 11.1 shows some locations of lesions and their influence on eye movements.

Other applications for the analysis of eye movements are diagnostics of reading disorders, determination of sleep phases in sleep diagnostics, but also the control of events via the eyes, for instance, in paraplegics. Eye movements are also recorded in psychological examinations, for instance, during the observation of pictures (Fig. 11.10).

Table 11.1 Disorders of horizontal eye movements by various cerebral lesions

	Fixation	Saccades	Smooth pursuit movements	Optokinetic nystagmus
Cerebral cortex frontal lobe	Nonpermanent contraversive cortical gaze paretic nystagmus	Saccade paresis contraversive with gaze deviation ipsiversive	–	Contraversive (towards direction of the quick phase)
Cerebral cortex, occipital lobe	–	Gaze disorder due to hemianopsia	Ipsiversive	Contraversive
Pontomesencephalic brain stem	Ipsiversive	Slow down or paresis of saccades ipsiversive with contraversive gaze deviations	Ipsiversive	Ipsiversive
Cerebellum	Ipsiversive	Saccade dysmetry	Ipsiversive	Contraversive

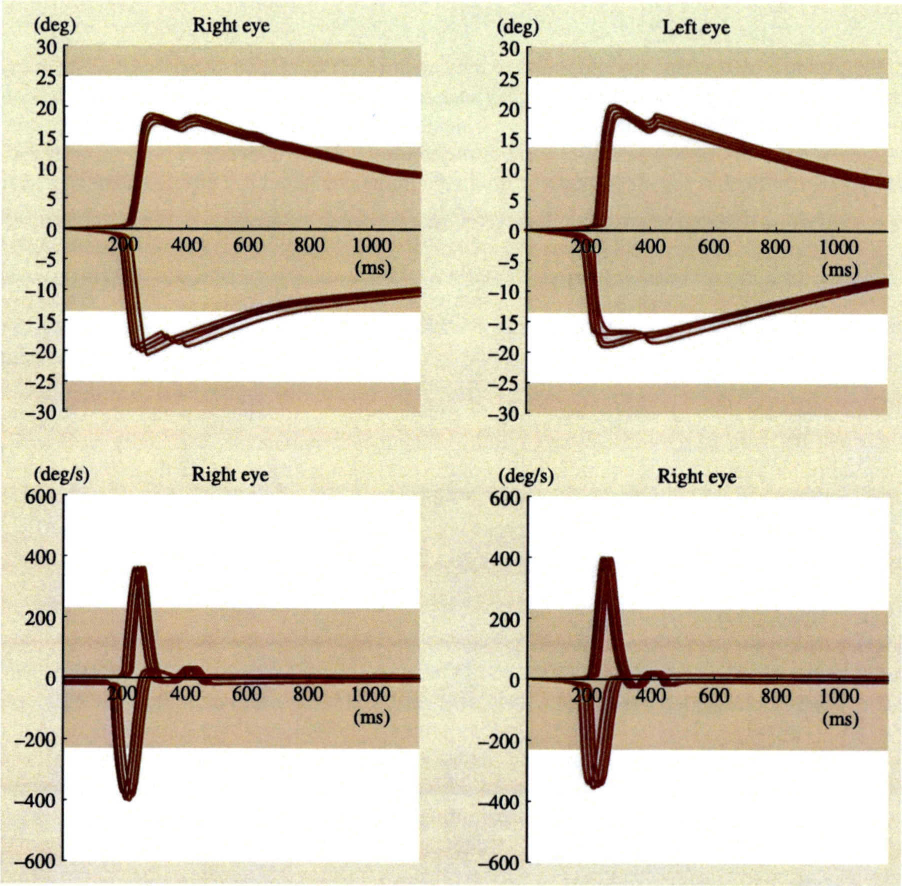

Fig. 11.10 Analysis of saccadic eye movements. The course of the eye movement velocity is displayed. The thresholds are highlighted

11.7 Safety Aspects

Nystagmography is a noninvasive diagnostic method which is applied in a medical environment. Thus, no extra safety aspects have to be considered.

11.8 Spatial Planning

The required space depends on the method applied and varies with respect to whether a swivel chair is used or not. For routine applications of electronystagmography, the required space is comparable to other diagnostic methods, such as electrocardiography or electromyography.

Further Reading

- J.H.J. Allum, F. Honegger, M. Troescher: Principles underlying real-time nystagmus analysis of horizontal and vertical eye movements recorded with electro-, infrared-, or video-oculographic techniques, J. Vestib. Res. **8**, 449–463 (1998)

- B. Brandt, W. Büchele: *Augenbewegungsstörungen* (Thieme, Leipzig 1983)
- A.M. Eckert, M. Gizzi: Video-oculography as part of the ENG battery, Br. J. Audiol. **32**, 411–416 (1998)
- J.D. Enderle: Physiological modeling. In: *Introduction to Biomedical Engineering*, ed. by J.D. Enderle, S.M. Blanchard, J.D. Bronzino (Academic, New York 1999) pp. 279–368
- P.F. Gangemi, A. Messori, S. Baldini, A. Parigi, S. Massi, G. Zaccara: Comparison of two nonlinear models for fitting saccadic eye movement data, Comput. Methods Programs Biomed. **34**, 291–297 (1991)
- S.S. Hacisalihzade, J.S. Allen, L.W. Stark: Computer analysis of eye movements, Comput. Methods Programs Biomed. **40**, 181–187 (1993)
- C.W. Hess, R. Müri, O. Meienberg: Recording of horizontal saccadic eye movements, Neuro-Ophthalmology **6**, 189–198 (1986)
- K.P. Hoffmann, A. Plewe, G. Mühlau, R. Kästner: Die sakkadischen Augenbewegungen: Methodische Aspekte und Möglichkeiten der klinischen Anwendung, EEG-EMG **22**, 172–177 (1991)
- K.P. Hoffmann, R. Both, R. Kästner, R. Weese: Zur klinischen Anwendung quantitativ ausgewerteter horizontaler Folgebewegungen des Auges, EEG-EMG **23**, 190–194 (1992)
- K.P. Hoffmann, G. Mühlau, C. Müller, R. Both, M. Jahns, W. Zahlten: Adaptive mechanisms of saccadic system in diagnostic use, Electromyogr. Clin. Neurophys. **33**, 131–135 (1993)
- M. Juhola: Detection of nystagmus eye movements using a recursive digital filter, IEEE Trans. Biomed. Eng. **35**, 389–395 (1988)
- M. Karatas: Internuclear and supranuclear disorders of eye movements: Clinical features and causes, Eur. J. Neurol. **16**, 1265–1277 (2009)
- S. Martinez-Conde, S.L. Macknik, X.G. Troncoso, T.A. Dyar: Microsaccades counteract visual fading during fixation, Neuron **49**, 297–305 (2006)
- R.M. Müri, M.T. Iba-Zizen, C. Derosier, E.A. Cabanis, C. Pierrot-Deseilligny: Location of the human posterior eye field with functional magnetic resonance imaging, J. Neurol. Neurosurg. Psychiatry **60**, 445–448 (1996)
- K. Pawlak-Osinska, H. Kazmierczak, W. Kazmierczak: Saccadic and Smooth-Pursuit Eye Movement in Neurootological Diagnostic Procedures, Int. Tinnitus **11**, 52–53 (2005)
- S. Prasad, S.L. Galetta: Eye movement abnormalities in multiple sclerosis, Neurol. Clin. **28**, 641–655 (2010)
- H. Scherer, W. Teiwes, A.H. Clarke: Measuring three dimensions of eye movement in dynamic situations by means of videooculography, Acta Otolaryngol. **111**, 182–187 (1991)

References

11.1 J.H.J. Allum, F. Honegger, M. Troescher: Principles underlying real-time nystagmus analysis of horizontal and vertical eye movements recorded with electro-, infrared-, or video-oculographic techniques, J. Vestib. Res. **8**, 449–463 (1998)

11.2 B. Brandt, W. Büchele: *Augenbewegungsstörungen* (Thieme, Leipzig 1983)

11.3 A.M. Eckert, M. Gizzi: Video-oculography as part of the ENG battery, Brit. J. Audiol. **32**, 411–416 (1998)

11.4 J.D. Enderle: Physiological modeling. In: *Introduction to Biomedical Engineering*, ed. by J.D. Enderle, S.M. Blanchard, J.D. Bronzino (Academic, New York 1999) pp. 279–368

11.5 P.F. Gangemi, A. Messori, S. Baldini, A. Parigi, S. Massi, G. Zaccara: Comparison of two nonlinear models for fitting saccadic eye movement data, Comput. Meth. Progr. Biomed. **34**, 291–297 (1991)

11.6 S.S. Hacisalihzade, J.S. Allen, L.W. Stark: Computer analysis of eye movements, Comput. Meth. Progr. Biomed. **40**, 181–187 (1993)

11.7 C.W. Hess, R. Müri, O. Meienberg: Recording of horizontal saccadic eye movements, Neuroophthalmol. **6**, 189–198 (1986)

11.8 K.P. Hoffmann, A. Plewe, G. Mühlau, R. Kästner: Die sakkadischen Augenbewegungen: Methodische Aspekte und Möglichkeiten der klinischen Anwendung, Z. EEG-EMG **22**, 172–177 (1991)

11.9 K.P. Hoffmann, R. Both, R. Kästner, R. Weese: Zur klinischen Anwendung quantitativ ausgewerteter horizontaler Folgebewegungen des Auges, Z. EEG-EMG **23**, 190–194 (1992)

11.10 K.P. Hoffmann, G. Mühlau, C. Müller, R. Both, M. Jahns, W. Zahlten: Adaptive mechanisms of saccadic system in diagnostic use, Electromyogr. Clin. Neurophys. **33**, 131–135 (1993)

11.11 M. Juhola: Detection of nystagmus eye movements using a recursive digital filter, IEEE Trans. Biomed. Eng. **35**, 389–395 (1988)

11.12 M. Karatas: Internuclear and supranuclear disorders of eye movements: Clinical features and causes, Eur. J. Neurol. **16**, 1265–1277 (2009)

11.13 S. Martinez-Conde, S.L. Macknik, X.G. Troncoso, T.A. Dyar: Microsaccades counteract visual fading during fixation, Neuron **49**, 297–305 (2006)

11.14 R.M. Müri, M.T. Iba-Zizen, C. Derosier, E.A. Cabanis, C. Pierrot-Deseilligny: Location of the human posterior eye field with functional magnetic resonance imaging, J. Neurol. Neurosurg. Psychiatry **60**, 445–448 (1996)

11.15 K. Pawlak-Osinska, H. Kazmierczak, W. Kazmierczak: title, Int. Tinnitus **11**, 52–53 (2005)

11.16 S. Prasad, S.L. Galetta: Eye movement abnormalities in multiple sclerosis, Neurol. Clin. **28**, 641–655 (2010)

11.17 H. Scherer, W. Teiwes, A.H. Clarke: Measuring three dimensions of eye movement in dynamic situations by means of videooculography, Acta Otolaryngol. **111**, 182–187 (1991)

12. Audiometry

Sebastian Hoth

The practical application of audiometric tests is determined by the needs on the one hand and the possibilities on the other hand. In clinical practice, it makes no sense to examine everything that can be tested. Only those tests are performed whose results have therapeutic consequences. This basic principle was the determinant factor for the selection of topics in the present text. The complete inventory of functional tests composed of psychoacoustic assessments and objective measures as well as the options for technical rehabilitation will be described.

12.1 **Physical, Technical and Physiological Bases of Audiometry** ... 191
12.1.1 The Physical Description of Sound .. 192
12.1.2 Anatomy and Physiology of the Auditory System ... 194
12.1.3 Hearing Disorders ... 198

12.2 **Behavioral Audiometric Assessment** ... 200
12.2.1 Stimulus and Perception ... 200
12.2.2 Temporal Aspects of Hearing ... 201
12.2.3 Pure Tone Audiometry ... 202
12.2.4 Recruitment ... 204
12.2.5 Speech Audiometry ... 206
12.2.6 Binaural Hearing ... 210

12.3 **Objective Audiometric Assessment** ... 212
12.3.1 Impedance Audiometry ... 212
12.3.2 Otoacoustic Emissions (OAE) ... 215
12.3.3 Evoked Potentials of the Auditory System ... 221

12.4 **Technical Hearing Devices** ... 228
12.4.1 Conventional Hearing Aids ... 228
12.4.2 Implantable Hearing Systems ... 236
12.4.3 Cochlear Implants ... 237

References ... 242

12.1 Physical, Technical and Physiological Bases of Audiometry

The human auditory system has at least three properties that are responsible for the comprehensiveness of modern audiometry: it is *important*, *complex*, and *vulnerable*. The outstanding *importance* of hearing among our sensory modalities requires reliable tools for its assessment. The *complexity* of the hearing system explains why we need a large number of tests for its complete exploration, and because of its *vulnerability* there will be frequent necessity to apply these tests.

Audiometric testing is expected to answer many questions:

1. Is hearing affected significantly?
2. Which parameters are suitable to describe the affection?
3. Which component of the hearing system is affected?
4. Does the hearing impairment affect the quality of life?
5. Which therapeutic options are available to overcome the impairment?

Hearing impairment can have many reasons: debris, injuries, malformations, infections, fractures, altered pressure in middle or inner ear, intoxication, aging, undersupply, dysfunction of sensory cells, atrophy of other inner ear organs, perturbed synaptic transmission, damage or degeneration of neurons, tumors, infarcts, stroke, and loss of central auditory functions for other reasons. Manifold functional deficits may go along with hearing impairment: reduced sensitivity, altered intensity processing, restricted frequency selectivity, lowered time resolution, and loss of discrimination. Not all deficits can be compensated with the medical, surgical and technical therapeutic options that are currently available. In many cases, the functional losses caused by the hearing impairment can be compensated or substituted only partially. However, in the historical context, the possibilities of technical rehabilitation have never been as abundant as at present [12.1].

12.1.1 The Physical Description of Sound

The perception of sound is one of the fundamental dimensions of our sensory systems and nearly everyone is familiar with sound phenomena and the sensations associated with them. Nevertheless, an exact definition of sound is useful and necessary. In the context of physical acoustics, sound is defined as mechanical oscillations and waves in the frequency range between 16 Hz and 20 kHz, which elicit the sensation of tones, sound or noise to the human ear. Sound can be generated and transmitted as a local deviation from the equilibrium in solid, liquid, or gaseous media. Homogeneity, isotropy and stationarity of density, pressure, and velocity are perturbed locally by the sound source. Due to the interaction between elementary particles, the local perturbation will impact the neighboring sites, while the relaxation of forces restores the equilibrium at the origin: The perturbation leaves the place of generation. In the case of a periodic event, the origin is the source of a continuous three-dimensional wave. This physical basis of forces and interactions between molecular or atomic particles, of continuity condition, adiabatic approximation, and elastic properties is the background of the mathematical description of sound phenomena by a differential equation involving time, space coordinates, and elastic constants. Its solutions describe the propagation of local perturbation in all space directions with constant velocity [12.2].

The amount of deviation from equilibrium is described by pressure, density, and particle velocity. Their gradients proceed parallel to the direction of propagation: Sound waves are *longitudinal* waves. Time and position representations of these waves are equivalent because the differential equation depends only on $(x - c \cdot t)$ but not explicitly on position x and time t. A sound wave is described by its parameters *amplitude* or intensity, *wavelength* or wave number, *frequency* or period, and *phase*. The *amplitude* describes the actual value of pressure, density or sound particle velocity. At the sites and at the time instants of maximum pressure and density (condensation), the particles are at rest and the *sound particle velocity*, i. e. the velocity of displacement from the rest position, is zero. Accordingly, a minimum of pressure and density (rarefaction) goes along with a zero crossing of particle velocity. On the other hand, the particle velocity is large if pressure and density are in equilibrium and their gradient and time derivative is maximum. The particle velocity is lower than the *sound velocity*, which in turn is in the order of magnitude of the thermal velocity. In the equations describing sound waves, the amplitude is denoted by the unspecific symbol a representing the deviation of either the pressure (in $\mu\text{Pa} = 10^{-6}\,\text{N/m}^2$), the density (in g/m^3), or the particle velocity (in m/s) from thermal equilibrium.

The definition of frequency and wavelength is based on the phase that describes the momentary state of the elementary wave or oscillation. Among other alternatives, this state can be a maximum of pressure, a minimum of particle velocity, or a positive sloping zero crossing of density. The wavelength λ indicates the spatial distance of equivalent repetitions of these unique phases. In air and under normal conditions of pressure and temperature, the wavelength of sound waves lies between 17 mm and 20 m. The reciprocal value of λ (multiplied by 2π) is the wave number k. It corresponds to the number of waves contained in the length unit. The phase of pressure or density is shifted relative to the phase of the particle velocity by a phase angle of 90°.

The frequency f corresponds to the repetition rate of a certain phase of the wave (counts per second in $\text{Hz} = 1/\text{s}$). Its reciprocal value, the period T, lies between 63 ms (at 16 Hz) and 50 μs (at 20 kHz) for audible sound. In the case of periodic oscillations containing more than one frequency (e.g. vowels), the terms frequency and wavelength are not adequate but a period can be defined as the time elapsed between two equivalent states of the waveform. It corresponds to the lowest frequency contained in the complex waveform.

The longitudinal sound wave propagates away from its origin with the *sound velocity* c, which depends from temperature, pressure, and the elastic properties of the transporting medium. In air at room temperature and normal atmospheric pressure its value is approximately 340 m/s, in liquid and solid materials the velocity is essentially higher. Frequency and wavelength are connected to each other by the relation $f = c/\lambda$, which can easily be deduced from the above mentioned tight binding between x and $c \cdot t$. If the distance λ between the wave maxima and the velocity c of their propagation are known, the frequency of their arrival at the site of observation can be determined.

The propagation of sound is associated with an energy transport that is described by the power density and given in W/m^2. The intensity needed for normal hearing persons to perceive a pure tone of 2 kHz (the hearing threshold at this frequency) is $10^{-12}\,W/m^2$. This reference intensity I_0 corresponds to the reference sound pressure $p_0 = 20\,\mu Pa$. At the upper limit of hearing the sound intensity of a 2 kHz tone amounts to $1\,W/m^2$ and the sound pressure to 20 Pa (level of discomfort). The largeness of the dynamic range – 12 orders of magnitude in the intensity scale – justifies the use of a logarithmic scale for the sound pressure level (SPL) L. Its dimension unit is the decibel (dB)

$$\frac{L}{\text{dB SPL}} = 20\,\lg\frac{p}{p_0} = 10\,\lg\frac{p^2}{p_0^2} = 10\,\lg\frac{I}{I_0}\,. \tag{12.1}$$

The level in dB indicates by how many *tenths of orders of magnitude* the considered intensity I is separated from the corresponding reference value. For the sound *pressure*, this absolute measure is only half the amount because the intensity is proportional to the square of the (effective) sound pressure

$$I = \frac{p_{\text{eff}}^2}{Z} = Zv_{\text{eff}}^2\,. \tag{12.2}$$

The constant Z appearing in this relation is the impedance of the sound wave, which on his part corresponds to the quotient of sound pressure p and particle velocity v. In air, the acoustic impedance amounts to $430\,Ns/m^3$ and in water to $1.46\times10^6\,Ns/m^3$. This demonstrates that the boundary between air and water is associated with a large discontinuity of impedance, which is of great importance for terrestrial vertebrates because it leads to the necessity for a middle ear apparatus as impedance transducer.

An acoustic signal that can be described by a sine wave in time and space contains only one frequency and is perceived as a pure tone. Since all time-dependent processes can be decomposed in contributions of different frequencies (*Ohm*'s law of acoustics), pure tones can be regarded as the *elementary particles* of acoustics. They are the basic components of sounds in the context of music (i. e. acoustic signals consisting of a few components of discrete frequencies), noises (i. e. wideband signals whose statistic properties can be stationary or time-dependent), speech (i. e. signals with rapid changes of frequency and intensity carrying a semantic meaning), and sound pulses (i. e. short signals with steep edges). The definition of the sound pressure level of complex signals containing many frequencies require the computation of an effective pressure p_{eff}, which is given by the root mean square amplitude. In the case of pure tones, p_{eff} is identical with the maximum amplitude divided by $\sqrt{2}$ (correspondingly, the same is valid for the effective value v_{eff} of the particle velocity). In the case of nonstationary signals, the effective value depends on the time constant and the complete description additionally requires the peak values.

All signals mentioned so far (tones, sounds, noises, speech, and sound pulses) are used in practical audiometry for the examination of hearing capability. For some of them, the amplitude and/or frequency is modulated in special applications. Speech signals exhibit a natural amplitude modulation whose frequency amounts to about 4 Hz according to the duration of syllables. Accordingly, some stimulation paradigms in practical audiometry involve speech simulating noise signals whose frequency spectrum corresponds to the long-term average speech spectrum and whose envelope is modulated with a frequency around 4 Hz. The extreme type of modulation encountered in sound pulses requires special corrections if a hearing level that is independent of pulse duration and repetition rate is to be defined. The level of pulses is characterized by peak values since effective values are not meaningful.

The propagation of sound is affected by objects whose impedance differs from that of the carrier medium. The impact of the obstacles depends on their dimension. If the obstacle is large in relation to the wavelength, it impedes the propagation and produces a geometrical *shadow* on its back side, whereas small objects deflect the sound wave, which is therefore present at the side opposite to the source with little or no attenuation. Consequently, the distortion of a broad band signal affects its frequency spectrum. Signal components of low frequency are present at the far side of the obstacle without attenuation, whereas the intensity of high frequencies is reduced. These effects are important in the context of spatial hearing and for the quality of perception of the own voice.

If the wavelength is small in relation to the dimensions of the obstacle (i. e. at high sound frequencies), diffraction phenomena can be neglected and the interaction between the sound field and the boundaries of the medium can be described by the laws of geometric reflection. In spaces that are limited partially or totally by walls, special oscillations can occur due to the superposition of incoming and reflected waves. These *standing waves* depend on the boundary conditions. In the case of reverberant closures (vanishing particle velocity at the sonically hard ending), only waves with nodes of the particle velocity (or rather lobes of the sound pressure) can interfere constructively with their own reflection. This condition is fulfilled only for volumes whose dimension coincides with an integer multiple of the half wavelength.

The same holds true in the case of two free closings (nodes of the sound pressure or rather lobes of the particle velocity). Some wind instruments as well as the open ear canal exhibit a free and a hard ending. Under these conditions, the wavelengths of standing waves are defined by the condition that the geometrical dimensions must coincide with an odd multiple of a quarter of the wavelength. For the ear canal with a length of approximately 2.5 cm, the lowest resonant frequency lies around 3.4 kHz. The phenomena of standing waves and resonances are important in the context of free field audiometry and they play a role in all acoustic measurements within the ear canal (impedance audiometry, recording of otoacoustic emissions, in-situ recordings) and in the provision with hearing aids (reflections within the residual ear canal volume). However, the boundary conditions (one free and one hard ending) are not fulfilled exactly since neither the walls of the ear canal nor the tympanic membrane are ideally sonically hard, and the entrance of the ear canal is not ideally sonically soft. Therefore, the real resonance frequencies do not obey the rules given here exactly.

Standing waves can arise only if the sound wave is completely reflected at the wall between two media. This presumes that the reflection coefficient

$$r = \frac{\zeta - 1}{\zeta + 1}, \tag{12.3}$$

determined by the quotient $\zeta = Z_2/Z_1$ of the impedances Z_1 and Z_2 of the media is unity. The sound wave striking the interface between two media of different impedance is partially reflected (fraction r) and partially absorbed (fraction $\alpha = 1 - r$). In the context of hearing, the most important interface is that between air (in the tympanic cavity) and water (in the inner ear). Because of the large impedance difference of these media, the reflection coefficient of the boundary is large. The middle ear has the task to overcome the impedance discontinuity. The function of the impedance transducer can be examined with impedance audiometry.

12.1.2 Anatomy and Physiology of the Auditory System

A large number of anatomic structures and very complex physiologic mechanisms are involved in processing the acoustic information within the auditory system. The actual sense organ for the perception of sound is the organ of *Corti* in the inner ear (cochlea). Up to here, the sound energy is transported from the outer ear (concha) and across the outer ear canal (meatus acusticus externus), the eardrum (tympanic membrane), and the middle ear (tympanic cavity and ossicular chain). After mechanoelectrical transduction in the inner ear, the information contained in the sound signal is transferred, processed, interpreted, and classified by the neural structures of the auditory pathway (hearing nerve, afterbrain, midbrain, thalamus, auditory cortex, and association areas). The pathways originating from both ears have projections to either sides of the brain. This provides the possibility of analyzing interaural differences of time and intensity, which play an important role in directional hearing and for the detection of signals in noisy or reverberant environments.

The peripheral parts of the auditory system are shown in Fig. 12.1. Outer ear and ear canal pick up, collect, and conduct the sound energy. They amplify the signal with a gain characteristics, which depends on direction and frequency, they protect the delicate structures of the organ against injury and they suppress the noise of flow and wind. Tympanic membrane and middle ear ossicles (malleus, incus, and stapes) feed the signal into the cochlea. They transform sound pressure and particle velocity in order to overcome the barrier between the low impedance of air and the high impedance of the inner ear liquids. The sensory cells within the inner ear are connected to ascending (afferent) and descending (efferent) fibers of the auditory nerve. The

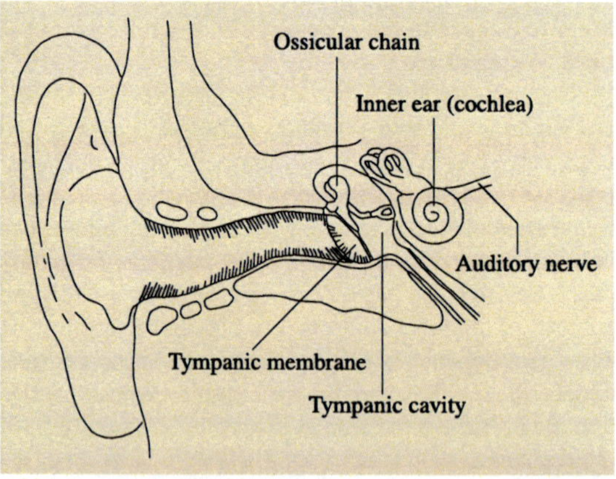

Fig. 12.1 Cross section through outer ear, ear canal, middle ear, inner ear, and hearing nerve. The tympanic cavity is connected to the nasopharynx via the *Eustachian* tube (after [12.3])

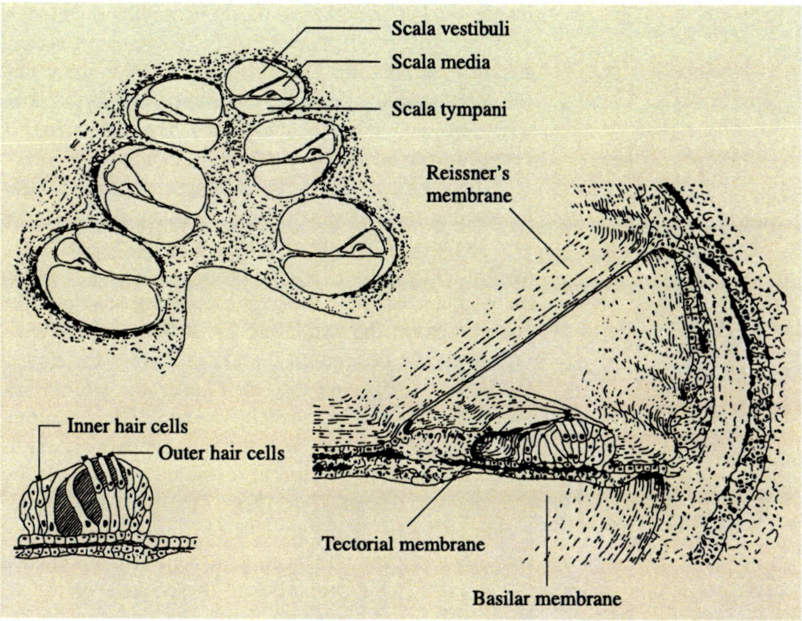

Fig. 12.2 Cross section through the inner ear and magnification of the cochlear duct with the organ of *Corti* and the hair cells (after [12.3])

movements induced by the sound wave causes them to segregate a neural transmitter, which activates the afferent fibers to produce an action potential. These elementary neural signals are the *information quanta*, which are integrated, decoded, analyzed, and exploited in the brain.

The tubular interior of the inner ear is arranged in a spiral with two and a half turns. It has a length of 32 mm in women and 37 mm in men. In longitudinal direction it is divided into three compartments by *Reissner*'s membrane and the basilar membrane (Fig. 12.2): the lower scala tympani, which is filled with perilymph, the scala media (*ductus cochlearis*), which is filled with endolymph, and the upper scala vestibuli, which is again filled with perilymph. The last mentioned scala vestibuli is separated from the tympanic cavity by the membrane of the oval window that is attached to the stapes footplate. The fluid spaces of scala tympani and scala vestibuli are connected to each other at the apex by a hole, the helicotrema. In the case of static pressure differences or subsonic vibrations, this allows the incompressible inner ear fluids to escape by retracting and bulging the round window.

The organ of *Corti* located on the basilar membrane carries approximately 3400 groups of sensory hair cells, which constitute the actual receptors for sound oscillations. They can be compared to electro-acoustic transducers (microphones). There are two different kinds of hair cells. The slim outer hair cells (OHC) are arranged in three to five rows and are connected primarily to efferent (descending) nerve fibers. The bulky inner hair cells (IHC) are located closer to the axis of the cochlea (modiolus) and are connected primarily to afferent (ascending) nerve fibers. Each hair cell carries 50–150 cilia on its upper or apical pole. The cilia of the OHC are mechanically linked to the tectorial membrane, whereas the cilia of the IHC float freely in the endolymph.

Human hair cells are highly differentiated and are not able to reproduce themselves by segmentation or regeneration. Hence, the damages and functional deficits caused by a loss of hair cells are irreversible. In recent years, some researchers succeeded in breeding hair cells out of stem cells in animal experiments, but an application in humans is not yet visible because of major methodical problems. To date and in the near future, a loss of hair cells can only be compensated by technical sound amplification with hearing aids or substituted by bypassing the sensory level of sound processing with cochlear implants.

At their lower or basal pole, the hair cells are connected via synaptic junctions to the fibers of the hearing nerve (cranial nerve VIII). This is the entrance of the central auditory system composed of neurons in serial connection to each other, which transfers the acoustic information from the inner ear to the auditory cortex.

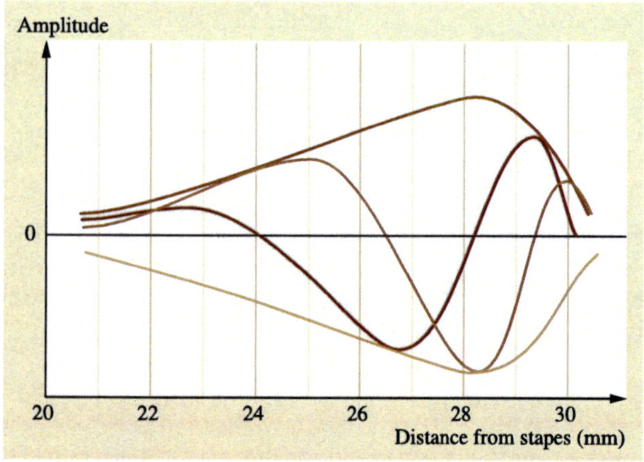

Fig. 12.3 Oscillation of the cochlear partition at a frequency of 200 Hz. Snapshots of the traveling wave for two different instants and the envelope are shown (after [12.4])

The fibers of the first afferent neuron take course firstly as a bundle in the axis of the cochlea (ganglion spirale) and later, after leaving the inner ear canal (*meatus acusticus internus*) end up in the first relay station of the brain stem, the cochlear nucleus (nucleus cochlearis). From here, the auditory pathway branches to the superior olives of both sides. A second crossing of the neural path occurs between the inferior colliculus in the midbrain and the medial geniculate body in the thalamus. This is the origin of the acoustic radiation (radiatio acustica), which ends in the temporal lobes (gyrus temporalis), the site of the auditory cortex.

The physiological processes involved in hearing are extraordinarily complex, so that only their basic principles can be described here. From the external ear canal, the sound waves reach the eardrum, a funnel-shaped membrane with a thickness of 0.1 mm and an area of 60 mm^2. The sound-induced vibrations of the eardrum are carried forward to the chain of middle ear ossicles and transferred to the membrane of the oval window. Two middle ear muscles (musculus stapedius and m. tensor tympani) keep the ossicular chain in place and exert influence on its function.

The middle ear apparatus enhances the efficiency of sound transmission from the air to the inner ear. A direct excitation of the inner ear fluids by the air borne oscillation would be associated with a loss of approximately 40 dB. The impedance transducer composed of ear drum, ossicles, and middle ear cavity increases the sound pressure by virtue of the surface ratio and reduces the particle velocity by virtue of the lever action. Theoretical considerations based on middle ear models yield variable results depending on the assumptions and boundary conditions. Experimental results show that the input impedance of the ear is reduced by a factor of 30–75, depending on frequency. The energy loss is reduced from 98% to about 40% [12.5].

The motion of the oval window membrane produces an oscillation in the perilymph, which also includes the cochlear partition composed of basilar membrane, endolymph, *Reissner's* membrane, and *Corti*'s organ and propagates along the longitudinal dimension of the cochlea. Since the width and the stiffness of the basilar membrane depend on the distance from the stapes, the velocity of propagation and hence the wavelength are a function of this distance. The oscillation resulting from this dispersion is a traveling wave whose amplitude maximum is located at a unique site defined by the sound frequency (Fig. 12.3). The tonotopic organization of the inner ear, i. e. the projection of the sound frequency to a specific location, is hence a consequence of the mechanical properties of the cochlear partition. High sound frequencies are processed at a site near the oval window at the basis of the cochlea, low frequencies at its apex.

Due to the construction scheme of *Corti*'s organ shown in Fig. 12.2, the transversal displacement of the basilar membrane is converted into a horizontal shearing of the hair cell stereocilia. This deflection causes the opening of selective potassium channels at the apex of the cilia, whose diameter is approximately 0.5 μm. The influx of potassium ions depolarizes the lateral cell membrane, which induces the activation of calcium channels. The influx of Ca ions leads to active contractions and elongations in the outer hair cell (OHC) and to the disbursement of a transmitter substance in the inner hair cell (IHC). After these processes, the electrical potential of the cell membranes returns to its equilibrium state (repolarization). The energy required for the actions of the hair cells is provided from the difference potential between endolymph and perilymph and is maintained by the *stria vascularis*. The transmitter molecules traverse the 20 nm synaptic gap within 0.5 ms by diffusion and activate special receptors in the nerve fiber. This initiates the generation of an action potential ascending the neural auditory pathway.

The excitation (depolarization) of the hair cell is followed by an inhibition (hyperpolarization) evoked by the reverse deflection of the cilia: the influx of K ions is blocked, the membrane potential is elevated, no mechanical activity of outer hair cells is induced, the discharge of the transmitter is inhibited, and the discharge rate of the hearing nerve is reduced below the spontaneous activity. Owing to the alternation of excitatory and inhibitory deflection of the cilia, the hair cell operation is similar to that of a half wave rectifier. The movement of the single cilia is coordinated by very thin (with a diameter of approximately 50 Å) filaments between them, the tip links, and the side links. This ensures that random movements like those associated with thermal noise do not induce hearing sensations. Moreover, tip links and side links are involved in the control of the ion channels (gating spring theory).

As long as the stimulus frequency does not exceed the limit of approximately 1 kHz corresponding to the refractory period of the sensory cells, these processes take place during every depolarizing phase of the sound wave. Thus, the information about the frequency is coded both in the place of generation and in the rate of action potential in one single fiber of the hearing nerve. The limits of frequency resolution are hence determined by the sharpness of the maximum of the traveling wave and the capability of the central neural system to analyze and distinguish time patterns. The spatial separation, however, is only partially explained by the theory based on the passive traveling wave. Additional sharpness is introduced by the active and locally restricted frequency selective hair cell contractions, which at the time enhance the sensibility of hearing by orders of magnitude (Fig. 12.4). These cellular processes can be measured and exploited for diagnostic purposes in the form of otoacoustic emissions.

The inner ear is embedded safely in the hardest bone of the human skeleton, the petrous portion of the temporal bone. This favors its stimulation by solid-borne sound. Hearing by bone conduction plays some role in everyday life (e.g. perception of the own voice) and it has great importance in audiology. Stimulation of the inner ear without involvement of the normal conduction path is applied for functional testing and for provision with prosthetic devices. If the footplate of a tuning fork is placed on the skull, the osseous vibration is transferred to the perilymph, basilar membrane, and sensory cells of both ears, resulting in an auditory sensation. The sound conduction via bone is nearly free from damping [12.7]. Therefore, special care has to be taken if only one ear shall be stimulated (masking of the ear contralateral to stimulation).

The hearing nerve (nervus acusticus) consists of about 30 000 fibers, most of them being afferent and connected to the inner hair cells. Starting from their synaptic endings, the single fibers proceed to the modiolus where they join the fibers originating from the other cochlear turns. The nerve takes course along the inner ear canal and enters the brain stem at the cerebello pontine angle. Along their route, the action potentials propagate with high velocity due to the electric isolation of the fibers by their myelin sheath. If this coating is damaged, the velocity of neural transmission is reduced (if all fibers are affected) or the neural synchronization is disturbed (in the case of partial affection).

The tonotopic organization of the cochlea is carried forward by the hearing nerve to the cochlear nucleus and by the following neurons up to the primary auditory cortex. The information is transferred from one to the next neuron by synaptic junctions with a delay of about 0.5 ms between pre- and post-synaptic ac-

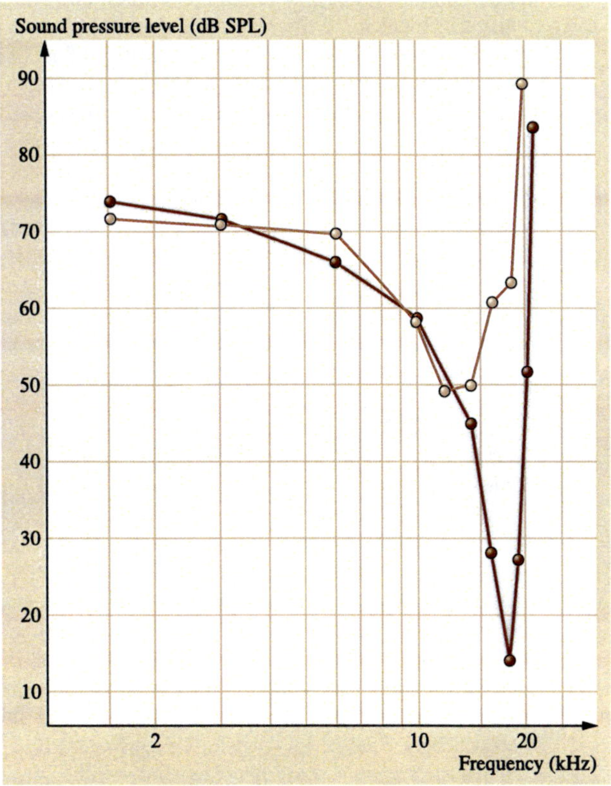

Fig. 12.4 Curves of constant basilar membrane velocity (0.04 mm/s) at one site of the basilar membrane, measured as a function of frequency with the *Mößbauer* effect of nuclear absorption of radiation. The comparison of results in a healthy ear (*full circles*) and an ear with damaged outer hair cells (*open circles*) of a guinea pig demonstrates the existence of a vulnerable frequency selective process that enhances the sensibility by about 40 dB within a restricted frequency range (after [12.6])

tion potentials. Due to the crossings to the contralateral pathways, the stimulation of one ear leads to the activation of the auditory centers in the temporal lobes of both brain hemispheres. Here, the signals are analyzed and interpreted in specialized brain regions, e.g. the speech perception field in the left hemisphere (*Wernicke*'s area) and the auditory association areas. For scientific and diagnostic purposes, the electric potentials evoked by acoustic stimuli in different levels of the auditory pathway can be observed with the electric response audiometry [12.8].

The central auditory system is subject to development and maturation during childhood and adolescence up to the age of 18 [12.9]. The number of neural connections (synapses) increases immediately following birth until the age of around 4 [12.10]. After this early development, the high synaptic density of the cortex is reduced by an elimination (pruning) of neural connections to about half of its highest value. Synaptogenesis and pruning are controlled by quality and quantity of acoustic input. It is assumed that preferably those synapses are eliminated that have not been used. In the case of auditory deprivation due to a peripheral dysfunction during the sensitive periods, the system of neural connections between cochlea and cortex will atrophy. Later restitution of the peripheral function can induce processes of neural and cortical reorganization but these are far less effective than those taking place in early childhood.

12.1.3 Hearing Disorders

Due to its complexity the auditory system can exhibit manifold functional deficits. They are usually classified according to the site of lesion. External ear and outer meatus may cause conductive losses by malformations, stenoses, ear wax (cerumen), proliferations, debris, or foreign objects. In a wider sense, conductive hearing disorders are caused by a dysfunction of outer and middle ear as opposed to sensorineural hearing losses caused by dysfunctions of the inner ear or the components of the neural system. Sensorineural hearing disorders can have (endo)cochlear and retrocochlear origin (Table 12.1).

Conductive hearing impairments can have the following reasons:

- Irregularity in the ventilation of the *Eustachian* tube
- Reduced pressure in the middle ear cavity due to inflammatory processes
- Physical defects of the tympanic membrane
- Chronic ulceration of the middle ear bone
- Middle ear infection (otitis media) with secretion in the middle ear cavity
- Proliferation of bony tissue around the stapes footplate (otosclerosis)
- Fracture or luxation of the ossicles
- Malformations.

Conductive losses can never cause complete deafness because the inner ear is excited by bone conduction even in the case of a total disruption of the middle ear apparatus. The maximum conductive hearing loss is around 50 dB. Most middle ear disorders can be corrected by conservative treatment or middle ear surgery. If this is not possible, the restoration of hearing is possible with special hearing aids delivering vibratory stimuli (e.g. bone conduction or bone anchored hearing aids).

Inner ear or (endo)cochlear hearing impairments can have the following reasons:

- Hereditary depletion of the hair cell population, in many cases dominant in the medial cochlea.
- Slowly increasing progressive hearing loss mainly at high frequencies due to hair cell damage predominant in the basal cochlea (*presbyacusis*).
- Hair cell loss mainly in the basal cochlea due to chronic noise exposition or explosion trauma.
- Deterioration of hair cells by auto immune reactions.
- Hair cell damage induced by ototoxic drugs.
- Acute degradation of hearing within minutes or hours without external reason (sudden deafness).
- Paroxysmal hypoacusis mainly at low frequencies associated with attacks of dizziness and induced by elevated endolymphatic pressure (Morbus *Menière*).
- Malformations.

Many of these inner ear diseases go along with subjective ringing of the ear (tinnitus) and in all of them

Table 12.1 Categories and nomenclature of hearing disorders

Outer ear	Middle ear	Inner ear	Hearing nerve	Neural pathway
	Transmission	Sensory	Neural	Central
		(Endo)cochlear	Retrocochlear	
Conductive losses		Sensorineural disorders		

the degree of hearing loss can be absolute (complete deafness). As long as some residual hearing is present, cochlear hearing impairment can be corrected with sound amplifying hearing aids. If the highest amplification yields no benefit, hearing can be restored only with cochlear implants.

A further attendant symptom of inner ear hearing loss is pathological loudness growth. Soft stimuli are not perceived while the loudness of strong stimuli is equal to that perceived by normal listeners. The uncomfortable level is less elevated than the perception threshold, the dynamic range is reduced. This loudness recruitment does not occur in conductive or retrocochlear damage. It is explained by the loss of outer hair cells, which act as active physiologic amplifiers for signals of low amplitude and as attenuators at high levels of stimulation (Fig. 12.5). Some patients with recruitment are extremely sensitive to noise (hyperacusis).

Loudness recruitment is encountered in practically all cases of hearing impairments caused by excessive noise exposition. The primary damage impacts primarily the outer hair cells particularly in the basal section of the cochlea and leads to a temporary threshold shift (TTS) at high frequencies, which is followed by a permanent threshold shift (PTS) in the case of sustained exposition. Not all persons are equally susceptible to noise-induced hearing loss (NIHL). Several current investigations focus on the question of whether the vulnerable inner ear can be identified before irreversible defects have developed.

Neural or retrocochlear hearing disorders are caused by a functional deficit of the hearing nerve or the auditory pathway. The acoustical information is not transmitted properly or not processed correctly. In many cases, the hearing loss is only an attendant symptom of diseases, which affect neither exclusively nor primarily the auditory system:

- Compression of the hearing nerve by vascular loops
- Tumors in the inner ear canal (e.g. acoustic neuroma or vestibular schwannoma)
- Multiple sclerosis (if portions of the auditory pathway are affected)
- Diffuse lesions of the brainstem without pronounced focuses
- Insufficient blood supply of brainstem regions, resulting from aneurysma or infarcts
- Atrophy or malformation of the hearing nerve.

Neural hearing disorders are frequently accompanied by functional deficits of the inner ear, since processes claiming space in the inner ear canal can also affect the arterial supply of the cochlea. Furthermore, many unspecific attendant symptoms (tinnitus, vertigo, affections of motor or sensory nerves) can occur. The treatment of neural disorders with hearing aids can be successful with respect to the hearing loss but it cannot be considered as adequate, especially in cases with normal inner ear function. Tumors of the hearing nerve or brainstem must be removed surgically since they can affect vital functions. Neurosurgery can lead to complete deafness. In such cases, therapy with cochlear implants will not restore hearing because the hearing nerve may have been corrupted. At best, some rudimentary hearing may be re-established with auditory brainstem (ABI) or midbrain implants (AMI).

A special class of hearing disorders, characterized by a normal function of the outer hair cells (as evidenced by normal otoacoustic emissions) and some alterations in the neural processing (as evidenced by altered auditory brainstem responses), is summarized under different notations as auditory neuropathy, perisynaptic audiopathy, auditory synaptopathy, dyssynchrony, or myelinopathy. According to this labeling, the function of the synaptic or neural transmission may be disturbed, but also a dysfunction of the inner hair cells or a loss of neural synchronization is discussed.

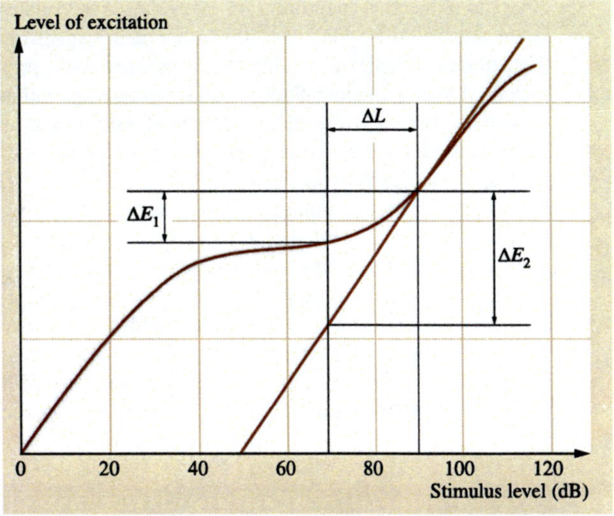

Fig. 12.5 Pathological loudness growth and reduced dynamic range (loudness recruitment) can be explained by the loss of low level amplification and high level attenuation. These are the effects of outer hair cell activity (*curved line*). If these cells are destroyed, the perception threshold is elevated and the uncomfortable level is unchanged or reduced (*straight line*). After [12.11]

Up to now, the anatomic and physiologic correlates are not fully identified. Auditory neuropathy spectrum disorders (ANSD) according to the nomenclature proposed by the American Speech and Hearing Association (ASHA) may occur unilaterally or bilaterally and they are accompanied by hearing losses ranging from moderate up to complete deafness.

Central hearing impairment is characterized by a reduced performance of the auditory system in accomplishing higher discrimination tasks. The deficit may be caused by organic reasons as e.g. cerebral hemorrhage. If an organic origin cannot be identified, an auditory processing disorder (APD) or a central auditory processing disorder (CAPD) may be suspected. Persons with (C)APD may have difficulties in localizing the source of sound and they do not recognize subtle acoustic differences of intensity, frequency, or time structure, even though the sounds themselves are loud and clear. These problems are more frequent in children and are more likely to occur when a person with APD is in a noisy environment. The diagnostics of APD or CAPD is based on the detection of abnormalities in complex signal analysis tasks without signs of organic defects of the peripheral hearing system or abnormalities in (nonverbal) mental health. Treatment with hearing aids is not adequate since the elevation of sound intensity is not beneficial for an intact sensory organ. The therapy is very individual; it includes training of concentration, attention, and auditory memory enhancement, as well as environmental modifications such as classroom acoustics.

12.2 Behavioral Audiometric Assessment

Behavioral or subjective examination of hearing is based on methods that constitute a subset of psychoacoustic testing. The subject is exposed to specifically defined acoustic signals, which elicit perceptions and reactions. From the observation and registration of responses, sensation thresholds, discrimination limens (*just noticeable differences*, jnds), masking phenomena, and speech perception can be studied and described in terms of the physical stimulus parameters. Since the quantity derived in these experiments is a subjective description of sensation, which involves the whole auditory system, the topodiagnostic power of subjective audiometry is rather poor.

12.2.1 Stimulus and Perception

The decibel scale defined above takes reference of the sound pressure of 20 μPa, which corresponds to the hearing threshold of young normal hearing persons at the frequency of 2 kHz, close to the place of largest sensibility (which is around 3 kHz due to the ear canal resonance). The sound pressure level 0 dB SPL is thus the lowest limit of sensation at one arbitrary frequency. Tones of most other frequencies are perceived only at higher levels. This is reflected in the dependence of the normal hearing threshold on the frequency, see Fig. 12.6.

The mean hearing threshold is the lowest of all isophones, which are defined as lines connecting the tones that elicit equal subjective loudness sensations. Generally, the volume level L_N in phon of a tone of arbitrary frequency is given by the sound pressure in dB SPL of the 1 kHz tone with equal subjective loudness. Two special isophones are the just audible pure tones defining the threshold and corresponding to the line labeled with 3 phon, and the discomfort level at 110 phon.

The concept of volume level is adequate for the description of absolute subjective loudness across frequencies, but it fails with respect to loudness relations associated with different levels at constant frequency. At least in the region above 40 phon, the augmentation of the volume level by 10 phon corresponds approximately to a duplication of subjective loudness. To achieve a correct description of loudness relations, a scale for the relative loudness N with the unit sone has been created on the basis of comparative scaling of pure tones [12.12]. N equals 1 sone for a tone of 40 phon and grows by a factor of 2 with each 10 phon increment. For N less than 1 sone, the simple power law is not valid and the relative loudness reaches $N = 0$ at threshold. Relative loudness plays no role in practical audiometry, where only direct scaling is applied (categorical loudness).

Besides the transformation of physical sound intensity in a subjective loudness sensation, the discrimination of intensity differences or temporal changes constitutes another field of psychoacoustic investigation. Just noticeable differences (jnd) of level can be determined with modulated stimuli or by sequential stimulation with stimuli of constant but different levels. In practical audiometry, the testing of discrimination abilities is limited to the exploration of certain hearing disorders that affect the sensibility for level differences (recruitment). In this context, also the masking of a target stimulus by a concurrent signal (e.g. narrow band noise) may be of diagnostic relevance.

12.2.2 Temporal Aspects of Hearing

The complete description of hearing requires the consideration of not only frequency and intensity but also the temporal characteristics of acoustic signals as their third dimension. For the function of hearing, as well as for its diagnostic examination, the temporal structure of acoustic stimuli is essential. Speech and many signals used in audiometry are nonstationary stimuli. Their subjective perception depends on temporal parameters such as duration: short signals appear to be softer than longer signals with the same intensity. If the duration of a stationary signal exceeds the limit of approximately 200 ms, the perceived loudness is independent of its duration; below this limit the level must be incremented by 10 dB if the duration is divided by 10 to achieve the same loudness sensation. Short pulses are thus much stronger even if they elicit the same sensation. Since the extent of hearing damage is determined by the physical intensity and not by the subjective loudness, the hazard potential of short signals is underestimated by the subjective sensation. The sound pressure of impulses may be described by the peak value (dB SPL p.e. = peak equivalent) or by the (generally much lesser) value that is relevant for subjective rating (dB HL = hearing level).

Most acoustic signals carrying information exhibit fast temporal changes of frequency and/or intensity. This holds true especially for speech, which codes a large portion of information in transients between vowels and consonants. The ability to analyze time structures is, therefore, especially important for the human ear. Physiologically, this is reflected in the existence of classes of neurons especially in the brain stem, which respond only to temporal changes. The functional integrity of these neuronal structures can be tested by psychoacoustic assessment of the time resolution capability. One of the relevant quantities is the just noticeable interval between two tone or noise pulses. It can be measured by presentation of two sequential pulses of limited duration (200 ms), one of them being interrupted by a gap of variable length. The subject has to indicate, which of the two stimuli appeared interrupted. In the case of a correct answer, the gap length is shortened, otherwise prolonged. After a defined number of reversions the mean value is determined (*gap detection threshold*). Normal hearing subjects achieve a jnd of approximately 5 ms. Similar results are obtained if the task consists in the recognition of the shorter of two or three consecutive pulses (*temporal difference limen*).

Another quantity related to temporal aspects of hearing is the processing of stationary stimuli of

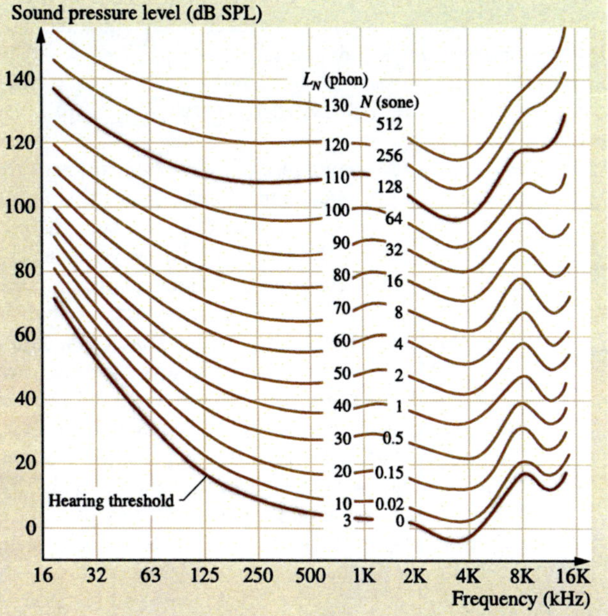

Fig. 12.6 The curves of constant loudness (isophones) of the normal hearing ear define the hearing threshold (*lowest curve*) and the volume level for pure tones. The volume level L_N in phon of a tone of arbitrary frequency is given by the sound pressure in dB SPL of the 1 kHz tone with equal subjective loudness. The loudness N is defined as 1 sone for a tone of 1 kHz with 40 dB SPL and incremented by a factor of two with each 10 dB. This power law is not valid for N less than 1 sone. After [12.13]

long duration [12.14]. It is known that the perceived intensity decreases with increasing duration. The mechanisms involved in this effect are not known in detail. Empirically, adaptation is distinguished from fatigue. *Adaptation* describes the fading at low stimulus levels. It occurs within a few minutes and normal sensibility is re-established within the same time range. *Auditory fatigue* is observed at high stimulation levels and its recovery is much slower (temporary threshold shift, TTS).

As has been mentioned above, the perception threshold for tones is elevated in the presence of noise. If the noise is switched off, the *masked threshold* returns to its normal value, the *absolute threshold of hearing* (in quiet). This masking release is not simultaneous but delayed by approximately 200 ms depending on the noise level [12.14]. In the case of inner ear hearing loss, the effect of forward masking is stronger and the ability to take advantage of the gaps in a nonstationary fluctuating noise is reduced or lost. This is one of the reasons accounting for the worse speech discrimination of persons with hearing impairment in noisy situations and reverberating environment.

12.2.3 Pure Tone Audiometry

Historically, the tuning fork was the first instrument applied to test the perception of pure tones. This crude method did not yield quantitative results, but it helped in finding out whether hearing loss was frequency-dependent and it allowed observation of its progress in time. For other purposes the stimulation of the ear with tuning forks is still integrated in practical audiometry [12.15]: if the bottom of the fork is brought in contact with the skull, the inner ear is stimulated via bone conduction. In cases of unilateral hearing loss, the lateralization of the sensation depends on the type of impairment: conductive disorders tend to emphasize the perception in the affected ear, whereas perceptive (sensorineural) disorders accentuate the perception in the better ear. In the case of mixed hearing loss with conductive and perceptive components and in cases of both-sided hearing impairment, this so-called *Weber* test is not applicable.

A second procedure based on the tuning fork – the *Rinne* test – is applied to explore the nature of impairment in one isolated ear: the comparison of the loudness elicited by the tuning fork oscillating at the entrance of the outer ear canal (air conduction) and vibrating with its foot plate brought into contact with the mastoid (bone conduction). If the air conducted stimulation is perceived stronger, a conductive disorder can be excluded. Like the *Weber* test, this test is restricted to pure forms of conductive or perceptive losses.

The hearing threshold for pure tones is defined by the lowest intensity or level of a stimulus of given frequency that leads to a perception in at least 50% of trials. In normal hearing subjects, this quantity is strongly dependent on frequency (Fig. 12.6). For clinical purposes, a curvilinear reference line is not very suitable. Therefore, the devices used for the determination of threshold (pure tone audiometers) make use of a level calibration based on the normal threshold as reference sound pressure. Thus, the normal threshold corresponding to 0 dB HL (hearing level) is a straight line in the pure tone audiogram.

Another level scale occasionally used in practical audiometry is based on the individual threshold of the ear under examination. This makes sense if the relevant quantity is given by the distance between threshold and a specific response, e.g. the acoustic reflex. Given a hearing threshold of 20 dB HL and a response threshold of 80 dB HL, the sensation level is defined by the difference 60 dB SL.

The determination of the hearing threshold, i. e. the lowest perceivable sound level, is coupled with the problem, that the limit between silence and hearing is associated with maximum uncertainty. Even if the ambient noise in the audiometric test room does not exceed the limits specified in ISO 8253-1, the determination of the hearing threshold requires a high level of understanding, cooperation, concentration, and patience of the subject. If one of these preconditions is not fulfilled, the results of subjective testing will be very questionable.

A further problem of threshold determination results from the fact that only one of the two ears of one subject shall be examined. This is accounted for by using headphones rather than loudspeakers or tuning forks for the presentation of stimuli. Nevertheless, the stimulus can be perceived in the ear opposite to the side of stimulation. This can be prevented by masking the contralateral ear with adequate signals.

In a clinical environment, the hearing threshold cannot be determined according to the strict rules applied in experimental psychoacoustics. For diagnostic purposes an accuracy of ± 5 dB is considered sufficient. Starting at -10 dB HL, the level of the continuous or pulsed test tone is incremented until the subject indicates a sound perception, e.g. by activating the response button. At this instance, the level has already surpassed the threshold by a few dB. To correct this error, the level is now reduced and again incremented. After some repetitions of this procedure, a reliable threshold can be found. All thresholds in the frequency range from 100 Hz to 10 kHz, separated by octave steps below 500 Hz and half octave steps above 1 kHz, constitute the conventional pure tone audiogram. The extended high frequency range covers frequencies up to 16 kHz in smaller steps. Traditionally, the level axis in the audiogram runs from low values at the top to higher values at the lower border, i. e. the threshold representing normal audition lies above the threshold representing an ear with hearing loss (Fig. 12.7). The vertical distance between the reference line (0 dB HL) and the value under consideration is defined as hearing loss (dB HL).

In addition to the threshold for stimulation via air conduction (AC), the complete audiogram contains another curve representing the threshold for bone conduction (BC). The transducer is placed on the mastoidal process behind the outer ear. If contact area and force are within the normal limits, a defined quantity of sound is transferred to the inner ear via BC. Thus, the function

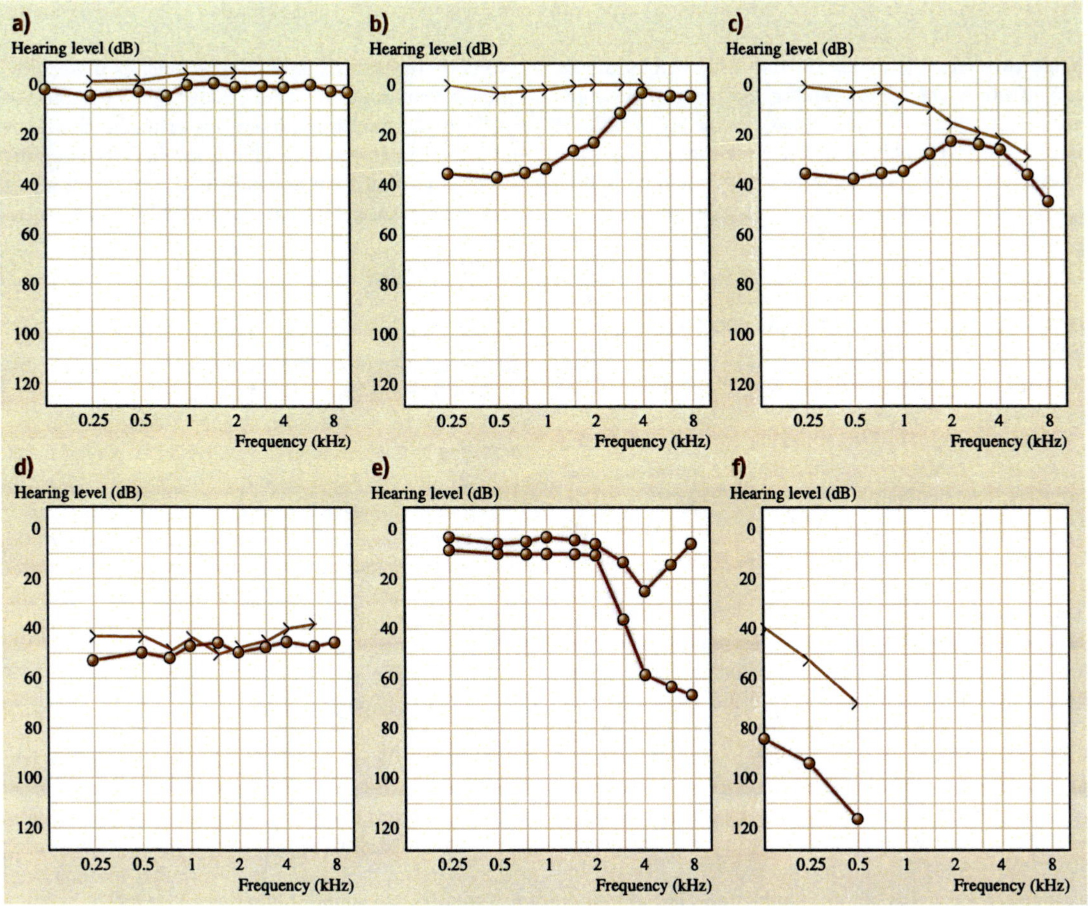

Fig. 12.7a–f Pure tone audiograms in case of normal hearing (**a**), conductive hearing loss at low frequencies (**b**), combined conductive and perceptive hearing loss (**c**), frequency independent sensorineural hearing loss (**d**), two stages of a noise induced inner ear hearing loss (**e**) and severe to complete deafness (**f**) with tactile perception of bone conducted low frequencies (*left corner audiogram*). All but one of the diagrams show the thresholds for air and bone conduction

of the inner ear can be tested without being affected by outer ear, external meatus, ear drum, and middle ear. The calibration of the transducers for air and bone conduction takes into account that very different quantities of energy are required for the excitation of the ear drum and of the skull; in both cases, 0 dB corresponds to the normal perception threshold.

The bone conduction threshold can be determined only for frequencies in the range above 250 Hz. For lower frequencies, the energy required for suprathreshold stimulation is very high and not only auditory but also tactile sensations are elicited (Fig. 12.7f). A second limitation at high frequencies above 6 kHz results from the fact that the bone stimulator produces air borne sound that can reach the inner ear via the natural sound conduction chain. In addition to these limitations, the sound intensity transferred to the inner ear by BC is much less reproducible than with air conduction because of the individual variability of force and skin flap thickness. Consequently, the accuracy of BC thresholds is typically not better than ±10 dB in a clinical environment.

The difference between the threshold for air and bone conduction is a measure for middle ear disorders. If the distance between AC and BC amounts to more than 10 dB (*air–bone gap*), the transmission of sound energy through the tympanic membrane and the ossicular chain is impacted by pathological alterations of the middle ear (Fig. 12.7b,c). From the frequency dependence of the threshold difference, information about the type and origin of the conductive loss can be deduced. If the tympanic cavity is filled with effusion following a middle ear infection, the friction and/or the mass of the middle ear apparatus is incremented and the AC threshold is elevated, preferably at high frequencies. In the case of enlarged stiffness of the ossicular chain, the low frequencies will be damped. Fractures or luxations of the ossicles, as well as otosclerosis in the advanced state or large perforations of the ear drum affect the whole frequency range. Hearing loss caused by transmission can never exceed the limit of approximately 50 dB since every air conducted sound exceeding the corresponding intensity leads to excitation of the bone and hence of the cochlea.

The stimulation of the inner ear by bone conducted sound is not limited to one side because the whole skull oscillates independently of the site of excitation. Transcranial attenuation is negligible at low frequencies and amounts to not more than 10 dB at high frequencies. However, also the application of air conducted stimuli, acoustic *cross-talk*, leads to the involuntary stimulation of the contralateral ear if the level of stimulation exceeds a certain limit. This limit is approximately 50 dB in the case of the supra-aural headphones usually applied in audiometry. If the level L_S of the stimulus lies above this limit, bone vibrations corresponding to a hearing level of $L_S - 50\,\mathrm{dB}$ are produced. Depending on the constellation of thresholds for air and bone conduction on both ears, the test tone is perceived only in the contralateral ear. To prevent this, the contralateral ear is masked by a narrow band noise via air conduction. The effective level of masking must exceed the value $L_S - 50\,\mathrm{dB}$. In the case of a conductive loss ΔL_C in the contralateral ear, the masking level L_M must be incremented accordingly

$$L_M \geq L_S - 50\,\mathrm{dB} + \Delta L_C \quad \text{with} \quad 0 \leq \Delta L_C \leq 50\,\mathrm{dB}\,. \tag{12.4}$$

Since the loss ΔL_C due to a contralateral middle ear dysfunction cannot exceed 50 dB, the maximal masking level is given by $L_M \geq L_S$. This simple formula corresponding to the worst case does not contain any unknown quantities and can, therefore, easily be applied in practice as

$$L_M \approx L_S \quad \text{if} \quad L_S \geq 50\,\mathrm{dB} \quad \text{(for air conduction testing)}\,. \tag{12.5}$$

The validity range of this rule is limited to moderate levels of stimulation (since at high levels an overstimulation of the contralateral ear may occur) and to cases with negligible middle ear damping in the ipsilateral ear (since otherwise the contralateral masker possibly affects the ear to be tested).

For the determination of bone conduction thresholds, somewhat different rules must be applied since the test tone is transferred without any losses to the contralateral ear (even at low stimulations levels L_S). Since the masker is presented by air conduction, a conductive loss ΔL_C in the contralateral ear must again be compensated

$$L_M \geq L_S + \Delta L_C \quad \text{with} \quad 0 \leq \Delta L_C \leq 50\,\mathrm{dB}\,. \tag{12.6}$$

While for the determination of the air conduction threshold, a total loss of middle ear transmission in the contralateral ear can be assumed for simplicity, this is not possible for bone conduction since a masking level L_M according to the formula $L_M \geq L_S + 50\,\mathrm{dB}$ would mask not only the contralateral but also the ipsilateral ear. Therefore, the masking level must be lower than $L_S + 50\,\mathrm{dB}$ and higher than L_S (otherwise, the test tone would be perceived in the contralateral due to the occlusion effect). Consequently, not a fixed value but only a range of admissible values can be specified for L_M,

$$L_S + 20\,\mathrm{dB} \leq L_M < L_S + 50\,\mathrm{dB} \quad \text{(for bone conduction testing)}\,. \tag{12.7}$$

The final formulas (12.5) and (12.7) are valid in most but not all cases. The only criterion characterizing the correct masking level is that the threshold should remain stable if the masking level is incremented. In any case, special care must be taken to ensure that the masking level remains below the uncomfortable level.

12.2.4 Recruitment

Tests above the threshold are performed to identify a pathological loudness growth (recruitment), which accompanies level-dependent distortions and a reduced dynamic range. One of these tests, the alternate binaural loudness balance (ABLB) described by *Fowler* in 1937 [12.16], is applicable only in cases of unilateral hearing impairment. Both ears are stimulated alternately by tone pulses of a frequency for which the interaural threshold difference amounts to at least 30 dB. Starting at threshold intensities, the level is incremented in steps of 20 dB for the impaired ear. The stimulus level at the contralateral normal ear is adjusted to achieve equal subjective loudness perception. If the slope of the lines connecting the points of equal loudness in both audiograms decreases at high levels, the recruiting ear is identified. Parallel lines indicate that no recruitment is present, or, equivalently, the dynamic range of the impaired ear is not reduced.

No normal contralateral ear is required for another recruitment test based on the short increment sensitivity index (SISI) described by *Jerger* in 1962 [12.17]. The ear under examination is stimulated by a continuous pure tone above threshold, whose amplitude is periodically incremented by 1 dB. Subjects with a recruiting ear are able to detect the increments, whereas normal subjects as well as persons with pure conductive or neural impairments are not. The increased sensibility of the impaired inner ear for amplitude modulations can be explained by the higher slope of its input/output function (Fig. 12.5). The test is performed at a frequency with a hearing loss of at least 40 dB. The continuous tone is presented with an intensity 20 dB above threshold. After conditioning of the patient, the level of this tone is incremented for 0.2 s every 5 s. If the patient detects more than 70% of the increments, a recruitment is probable (cochlear damage). Ears without recruitment (neural damage) are characterized by an index below 30%. Rates between these limits should not be interpreted, they are possibly a consequence of insufficient cooperation of the patient or of the choice of unsuitable parameters (frequency and intensity of the stimulus).

As an alternative to the SISI test, the sensibility for level increments can be examined by another procedure described by *Lüscher* and *Zwislocki* in 1949 [12.18]. In this test, the jnd is determined directly by decreasing the increments from initially 4 dB down to 0.2 dB. In accordance with the SISI test, a jnd of 1 dB is considered as the limit between recruiting and nonrecruiting ears.

Especially in the case of sloping audiograms, the perception threshold of a pure tone masked by a narrow band noise can serve as an indicator for recruitment, as described by *Langenbeck* in 1949 [12.19]. The masked threshold is determined for several pure tone frequencies covering the range from normal hearing at low frequencies to a hearing loss exceeding 45 dB at high frequencies. This permits the comparison of healthy and damaged regions in the same cochlea. The level of the narrow band noise, whose center frequency is equal to that of the target tone, is adjusted to a level between 45 and 75 dB. In the case of normal hearing, the masked threshold is found just above noise level. As the frequency region of impaired hearing is approached, the masked threshold can either join or deviate the threshold in quiet. In the first case, the discrimination capability is conserved in spite of the hearing loss, as is expected for a recruiting ear; in the latter case, the detection of a pure tone in noise is affected by the hearing loss and a neural

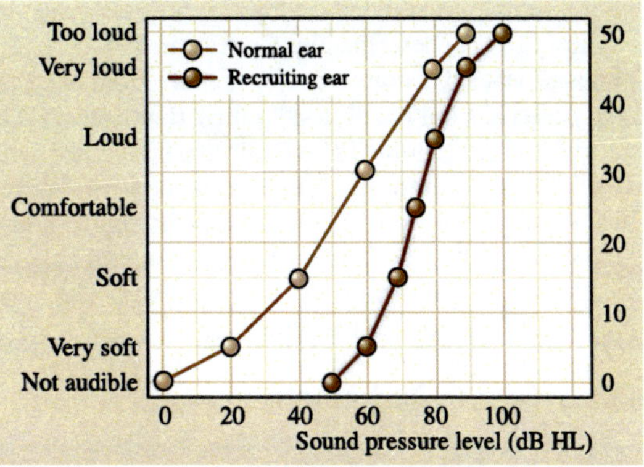

Fig. 12.8 Results of loudness scaling obtained from a normal hearing and a hearing impaired subject (schematic)

origin of impairment is probable according to the model of *busy lines* [12.20].

Some of the recruitment tests described so far are only indirect indicators of recruitment based on the sensitivity for level increments or the masked threshold. A direct assessment of the reduction of dynamic range is possible only with the ABLB test and only for single-sided hearing loss. No such limitation is imposed on the direct loudness scaling described by *Heller* in 1985 [12.21]. Narrow band pulses with center frequencies between 250 Hz and 4 kHz are presented in a randomized sequence by loudspeakers and assigned by the patient to one of the seven categories *not audible*, *very soft*, *soft*, *comfortable*, *loud*, *very loud*, and *too loud* (Fig. 12.8). Internally, these verbal categories are assigned to 0, 5, 15, 25, 35, 45, and 50 numerical units, respectively. For each frequency, a loudness-intensity function is constructed. Optionally, the data can be transformed to iso-response contours and transferred to the audiogram.

The intersection of the loudness-intensity-function with the horizontal axis is correlated to the perception threshold, the level judged as *too loud* corresponds to the uncomfortable level. In the case of a pathologic loudness growth, the slope of the function is larger than in normal ears and the distance between individual and normal values decreases with increasing stimulus level. This horizontal distance correlates with the need for amplification and it can be used as basis for a loudness-based fitting of hearing aids.

Recruitment tests and loudness scaling are used to identify the cochlear nature of hearing impairment. Neural dysfunctions can not be excluded with these tests. They are characterized by a pathologic auditory fatigue, which can be detectedby the threshold tone decay test (TTDT) described by *Carhart* in 1957 [12.22]. Tones of low intensities close to the threshold are usually perceived only within a limited time interval. In normal ears, this limit is beyond 30 s for a tone with 10 dB SL. If the intensity required for 60 s persistence lies between 15 dB and 25 dB SL, a pathological *adaptation* typical for cochlear hearing loss is probable. In the case of pathological *fatigue* typical for neural hearing loss, this level is above 30 dB SL.

12.2.5 Speech Audiometry

There is no doubt that understanding of speech is the highest level and paramount object of hearing and its preservation or restitution. The perception and discrimination of spoken speech is examined for audiometric purposes in order to assess the limitations in communication and the social handicap resulting from hearing impairment, as well as the need for amplification and the benefit of hearing aids and prostheses. Moreover, speech audiometry results may contribute to the diagnostic classification of hearing dysfunction especially in the identification of (central) auditory processing disorders (APD). For most of the other audiometric issues, speech signals turn out be too complex, but they are indispensable in the context mentioned above, since the capabilities grouped around speech perception cannot be predicted from the threshold for pure continuous tones.

In addition to the acoustic complexity of speech, speech audiometry is associated with further difficulties arising from the large number of factors contributing to the communication by spoken speech: hearing acuity, processes of neurocognition, intellectual capabilities, level of language development, short range memory, concentration efforts, readiness to cooperate, and others. It is not possible to examine all abilities that are relevant for speech understanding in all target groups with a single test. Therefore, speech audiometry is composed of several tests differing in test material, the use of noise and, of course, the specific language. One common feature of nearly all tests is the use of diagrams showing the fraction of correct items in dependence on the stimulus level or signal to noise ratio. From the discrimination function, the speech perception threshold (SRT), defined as the level of stimulus or the signal to noise ratio associated with a correct perception of 50% of the items, is derived.

The ability for perception and discrimination of speech can be embedded in the audiological context only if the acoustic properties of speech are considered. Therefore, the most important of these physical features shall be described here. Speech is produced by the human organs for *phonation* (production of voice) and *articulation* (modulation of voice). The simplest elements of speech are vowels characterized by a periodic oscillation composed of a *fundamental frequency* F_0 (corresponding to the voice pitch) and up to four *formants* F_1 to F_4 (higher frequency bands). The frequency ranges of the formants are unique for every vowel; they vary only to a low degree from one speaker to the other. The formant F_1 lies around 500 Hz, F_2 between 1 and 2 kHz, and the higher formants, which are less important for the identification of the vowel, lie between 2 and 4 kHz.

Consonants cannot be described by typical frequency spectra. As far as characteristic frequencies can be defined, they are located in the range of the formants F_3 and F_4. In contrast to vowels, consonants can be voiced and unvoiced. Voiceless consonants are produced without oscillations of the vocal chords by obstructions in the vocal tract, which contribute to the acceleration of the air stream and hence generate turbulences. In voiced consonants, these mechanisms are overlaid by vocal chord oscillations. Phonetically, consonants are further divided according to the place (e.g. labial, dental, palatal, or velar) and mode (e.g. plosive or fricative) of articulation. Many consonants are merely a frequency transition between preceding and subsequent vowel (transients). Their properties depend on the vowel with which they are fused and on their placement as initial, intermediate, or final phoneme. Transients and formants are identifiable in the spectrogram (or sonogram), which is a three-dimensional diagram showing the time dependence of frequency and intensity of a speech signal (Fig. 12.9).

Frequency and intensity of language components can be roughly translated into the audiogram to assess the impact of hearing loss on speech understanding (Fig. 12.10). In the case of high frequency hearing loss the recognition of many consonants (especially sibilants) is most affected, whereas a pronounced low frequency hearing loss affects mainly the fundamental frequency, which is not very important for speech intelligibility. If the hearing loss amounts to 40–50 dB at all frequencies, vowels and strong sibilants remain audible, while the silent consonants are not properly understood. The pathological loudness perception (recruitment) found in many cases of sensorineural hearing loss has the consequence that soft language elements are too weak, whereas the stronger components are

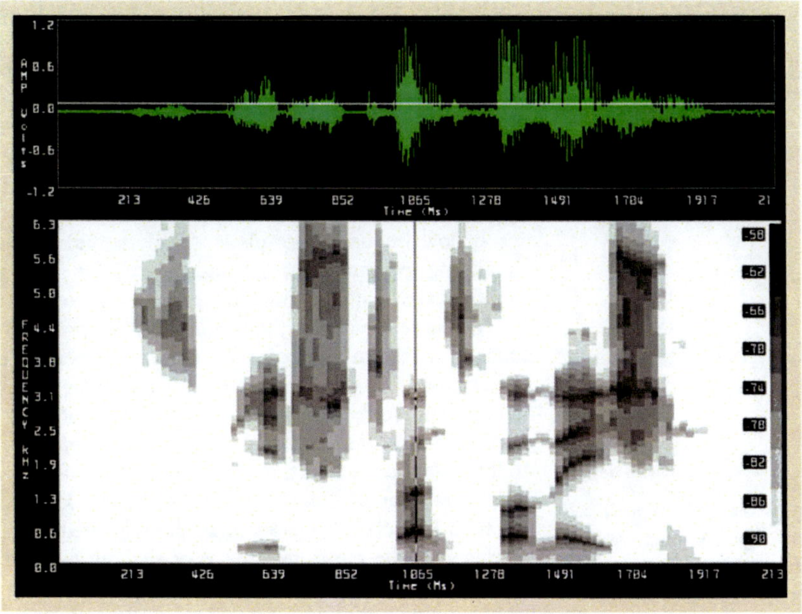

Fig. 12.9 Oscillogram (*top*) and spectrogram (*bottom*) of the utterance *speech transformation*, spoken by a male speaker

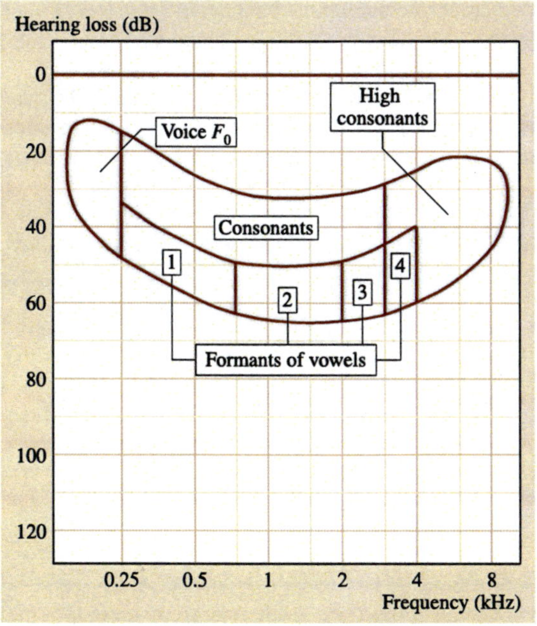

Fig. 12.10 Projection of the frequencies and levels relevant for colloquial speech into the pure-tone audiogram

perceived clearly or even uncomfortably loudly. The problems arising from the time dependence of speech signals – especially from the rapid transitions between loud vowels and soft consonants – cannot be illustrated in this diagram.

The selection of the test material used for speech audiometry is oriented on the objective in the special case under examination. Persons with rudimentary speech discrimination can only be examined if the test material contains sufficient redundancy, such as semantically meaningful sentences. The other extreme are virtually nonredundant monosyllabic words. The numerals used in the German *Freiburg speech intelligibility test* described below lie between these extremes. The responses of the subjects are part of a known and very limited word inventory. For a given material, the difficulty of the tests can be influenced by offering a closed set of response alternatives. To ensure that a subject can perform the same test several times without fear of distortion of results due to learning effects, the material must be divided into equivalent test lists (containing, for example, 10 or 20 everyday sentences, numerals, or monosyllabic words each). Whatever the objective, the statistical distribution of phonemes within the test material should be representative for the specific language.

The groups or lists of test items are presented through headphones or speakers, optionally with a simultaneously presented noise. The subject is asked to repeat the item (word or phrase) or to select it out of predefined answer alternatives, the examiner counts the number of correct answers. The result is the percent-

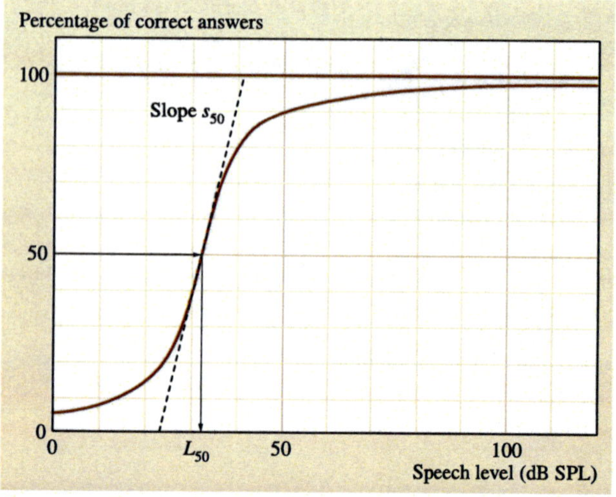

Fig. 12.11 The increase in the proportion of correctly reproduced test items with increasing stimulus level or signal-to-noise ratio is described by a sigmoid psychometric function (12.8)

age of test items reproduced correctly as a function of absolute or relative (referred to the background noise level) speech level. The relationship between stimulus level (or level difference) and the proportion of correct responses (speech intelligibility index) is called the discrimination function (performance-intensity function). It shows a step-shaped curve that can be described by two parameters: the level (or the signal-to-noise ratio) L_{50}, where the probability of a correct answer is 50% (speech perception threshold), and the slope s_{50} at this point (Fig. 12.11). For the mathematical description, the following logistic function is used:

$$p(L) = \frac{1}{1+\mathrm{e}^{-\frac{L-L_{50}}{u}}} \quad \text{with} \quad u = \frac{1}{4s_{50}} . \qquad (12.8)$$

In German-speaking countries, the test that is most used for practical audiometry is still the Freiburg speech intelligibility test according to the German standard DIN 45621, introduced by *Hahlbrock* in 1957 [12.23]. The test material consists of two-digit numbers and monosyllabic words that have been recorded by a trained professional speaker and arranged in phonetically balanced groups (10 lists of 10 numbers and 20 lists of 20 monosyllabic words). Below a certain speech level (10 dB SPL for numerals and 15 dB SPL for monosyllabic words), none of the words is correctly understood. With growing volume, the proportion of correct answers increases and finally reaches 100% at 30 dB SPL (numerals) and 50 dB SPL (monosyllables) for normal hearing listeners. For numerals, the normal values of L_{50} and s_{50} are 18.5 dB SPL and 5% per dB, respectively. Because of the high redundancy, the logistic function is very steep. For the same reason, hearing disorders affect only the location but not the shape of the function. Therefore, the impact of hearing loss can be described by the shift of L_{50}. This difference is called the *hearing loss for speech*. Since the correct perception of low frequency vowels is already sufficient for identifying the numerals, the hearing loss for speech correlates well with the low frequency hearing loss.

The curve describing the intelligibility of monosyllables is shifted to higher levels and is less steep than that corresponding to the numerals. This is due to the fact that the correct recognition of the test words requires not only the discrimination of vowels but also of the softer consonants. Hearing disorders not only affect the location but also the shape of the logistic function. Especially in cases of severe high frequency hearing loss the curve may be very shallow because of the difficulty of detecting the high consonants. Often, then, up to the discomfort level a score of 100% is not achieved, i. e. there is a *discrimination loss*. In rare cases, the speech discrimination decreases at high levels after having reached a maximum value (roll-off).

The standardized Freiburg speech intelligibility test is used in particular in the administration of hearing aids and the subsequent measure of control, and it forms the basis for the calculation of the earning capacity in the case of occupationally acquired hearing loss. It has some disadvantages, which have led to the development of alternative tests:

- In some patients with profound hearing loss, it provides no useful results since the numerals may be too easy and the monosyllables too difficult.
- The test lists are not phonetically balanced and not equivalent to each other regarding presentation level and degree of difficulty.
- The testing of speech intelligibility in noise is not standardized.
- The subject must be familiar with the German language.
- The test procedure is not automatable, because the response of the patient is evaluated by the investigator.
- The response of the subject is only rated as true or false, an evaluation of the confusions is not possible.
- The lack of an announcement stimulus leads to an unreasonably high rate of wrong answers.

One of the recent approaches to overcome these drawbacks is the *Oldenburg sentence test* developed by *Wagener* and *Kollmeier* [12.24]. It is an open set sentence test based on meaningful phrases consisting of five words (proper noun, verb, numeral, adjective and object) such as e.g. *Doris paints nine green chairs*. Each of the five words is taken from a stock of ten alternatives from which the records are assembled at random. This increases the number of usable test lists and it reduces the probability for recognition of a test item (test repeatability). At the speech perception threshold L_{50} the discrimination function of the Oldenburg sentence test has a high slope $s_{50} = 17-20\%$ per dB. Thus, the threshold can also be determined in quiet as well as in speech-simulating noise with high accuracy.

In all speech tests, the duration of the investigation can be reduced by the optimization of test materials and strategy. The time required is proportional to the number of test lists that must be tested to reach a sufficient significance. Speech tests based on redundant test material (numerals and sentences) have a steep logistic function that is shifted but not deformed in the case of hearing loss. In the ideal case of an adequate volume level, only one test list needs to be checked for a reliable determination of L_{50}. If the test material has little redundancy (monosyllables), the psychometric curve is shallower and not only shifted but also distorted intrinsically in impaired ears. For its reconstruction, at least two test lists must be checked. Regardless of the details of the curve the examiner usually aims not to carry out measurements in regions of 0 or 100% speech discrimination. Normal values and the above mentioned relations between pure tone and speech audiogram help to find reasonable starting levels in the vicinity of the inflection L_{50}. By application of adaptive procedures, the target can be achieved more accurately and faster. Here, the level of presentation is increased or reduced for each test item in accordance with the previous answers.

In addition to the speech discrimination in silence, the study of intelligibility in noise and its improvement by hearing aids is one of the subjects of speech audiometry because the combination of speech and noise constitutes one of the major challenges of many hearing impaired persons and users of hearing aids. Besides the *Oldenburg sentence test* described before, the *Göttingen sentence test* after *Wesselkamp* et al. [12.25], comprising 20 test lists of 10 sentences in combination with the voice babble noise after *Sotschek* as background noise and the *Innsbruck sentence test* after *Hochmair* et al. [12.26], composed of 30 lists of 20 sentences combined with speech simulating CCITT (Comité Consultatif International Téléphonique et Télégraphique) noise (HSM test) are suitable for this purpose. The spectrum of the international standard CCITT-noise is adjusted to the average frequency distribution of several languages. It has its highest intensity at 800 Hz. The time-modulated version of the CCITT noise is called *Fastl* noise. The modulation is not achieved by a fixed frequency but with a band-pass noise whose intensity peaks at 4 Hz corresponding to the mean syllable length. The *Göttingen sentence test*, recorded with high speech velocity by an untrained speaker is suitable for use with hearing aid users, while the HSM test, due to the slow and pronounced articulation, is primarily designed for the assessment of CI users.

The speech tests presented so far are applicable only in German speaking countries. Some of the most common tests used in other language areas are described in this paragraph. At a very elementary level of speech perception, the CUNY (*The City University of New York*) closed set nonsense syllable test is devoted to the examination of the discrimination of single phonems presented in VC (vowel–consonant) and CV (consonant–vowel) syllables [12.27]. Each test item is embedded in a carrier phrase *You will mark ..., please*, the competing noise is an equalized cafeteria noise. *Nilsson* et al. [12.28] introduced the *Hearing In Noise Test* (HINT), which consists of 25 phonemically balanced lists of 10 sentences and is suitable for the adaptive measurement of sentence SRT. Within the *European HearCom* project, the SPIN (*speech in noise*) test based on digit triplets overlaid by a rushing background noise was designed to mimic real-life circumstances. The *spondee recognition test* described by *Cramer* and *Erber* [12.29] is especially suitable for testing hearing impaired children. Test lists of ten spondaic words (consisting of two long syllables) are presented monaurally, the responses are indicated by pointing to labeled picture cards. Finally, the archetype of the *Oldenburg sentence test* shall be mentioned. In 1982, *Hagerman* published Swedish sentences suitable for clinical measurements of speech intelligibility in noise [12.30]. The test was developed for hearing aid evaluation in sound field as well as with earphones prior to hearing aid fitting. Ten lists of phonetically balanced five word sentences constitute the inventory for the construction of new lists by cutting and combining the single items. In the test situation, the speech stimulus is accompanied by modulated speech-simulating noise.

The accuracy and reliability of speech audiometry results are described by the variability σ, which depends on the fraction p of correct responses and the number n

of test items according to the formula

$$\sigma = \sqrt{\frac{p \cdot (1-p)}{n}} .$$

The 95% confidence interval is obtained by multiplication of σ with 1.96 [12.31]. If the test is performed twice (e.g. with and without hearing aid), the difference of the results is significant if the auxiliary variable Φ given by

$$\Phi = \arcsin\sqrt{\frac{x}{n+1}} + \arcsin\sqrt{\frac{x+1}{n+1}}$$

(with $x = p \cdot n =$ number of correct responses)

of the second test lies within the interval limited by

$$\Phi_{\mathrm{L}} = \Phi - 1.96\sigma_{\mathrm{diff}} \quad \text{and} \quad \Phi_{\mathrm{H}} = \Phi + 1.96\sigma_{\mathrm{diff}}$$

calculated from the first test and using

$$\sigma_{\mathrm{diff}} = \sqrt{2}\sigma\,(\Phi)$$

$$\text{with} \quad \sigma^2\,(\Phi) = \frac{1}{n+\frac{1}{2}} \quad \text{for} \quad n \geq 50$$

$$\text{and} \quad \sigma^2\,(\Phi) = \frac{1}{n+1} \quad \text{for} \quad 10 < n < 50 .$$

The application of these rules, derived under the assumption that binomial statistics is valid, shows that an unexpectedly large number of test items must be used to observe significant effects [12.31].

12.2.6 Binaural Hearing

All aspects of hearing discussed so far are the result of psychoacoustic studies on the stimulation of one ear. Without doubt, however, hearing with both ears (binaural) is to be regarded as the natural listening situation. It differs from monaural listening at first by a lower threshold of perception. The difference between the monaural and the binaural hearing threshold is 3 dB (doubling of intensity). For suprathreshold stimuli, the difference between monaural and binaural hearing is approximately 10 dB, i.e. the addition of the second ear results in a doubling of loudness. This effect is called binaural loudness summation.

Many high order capabilities of the auditory system, e.g. spatial hearing, localization of sound sources, and discrimination of speech in noise and reverberation, are result of the binaural detection and processing of acoustic signals. In addition, the outer ear plays an essential role: its acoustic transfer function acts as a filter whose output depends on direction, distance, and frequency of the sound source. Spatial characteristics of the sound field are coded into temporal features by reflection, shadowing, scattering, diffraction, interference, and resonance. The representation of these features within the central nervous system produces an impression of spaciousness that cannot be achieved by stimulation via headphones. One special aspect of spatial hearing, namely directional hearing, is based on the presence of two input signals, the crossing of the auditory pathway and the exploitation of the differences between the two signals (binaural processor). The position of the ears on both sides of the head leads to differences in level, time, and tone of the two inputs arriving from a lateral acoustic signal source. The tone differences arise from the fact that the shading effect by the head is relevant only for high frequencies. For the same reason, level differences also occur mainly at high frequencies that are not bent around the head. They contribute to the lateralization of the auditory event in so far as they amount to more than 1 dB. Furthermore, the sound waves arriving from sources that are outside the median plane reach the two ears at different times (or with different phases at the same time). The lower limit for the exploitation of these interaural time differences is less than 30 μs. Time and level differences allow for sound localization with an accuracy of 3–5°. This location, however, is not unambiguous. All sources located on the surface of the cone shown in Fig. 12.12 (cone of confusion) have the same interaural differences; they differ only in tone. Actually, the sound event sites that are most often confused in a directional hearing test in the horizontal plane are equivalent in terms of interaural time and level differences. A special case is the confusion between front and rear. The determination of the elevation, i.e. the location of sound sources that are located in the median plane (angle of the cone equal to 180°), is done without the aid of interaural differences and is, therefore, uncertain.

Interaural time differences are only useful for the location of sound sources if the duration of the signal is limited sufficiently sharp. This condition is not satisfied with continuous tones or with the occurrence of reverberation. In these cases, the sensation of both ears merges to a single auditory event. Nevertheless, the sources of continuous stimuli or of short signals accompanied by echo can be localized because the location of such auditory events is determined primarily by the first sound pressure change that reaches the listener (law of the first wave front). This mechanism is important for hearing in closed rooms, but it fails if standing waves are generated.

Binaural hearing is critical for the detection and discrimination of signals in noise. A pure tone is easier

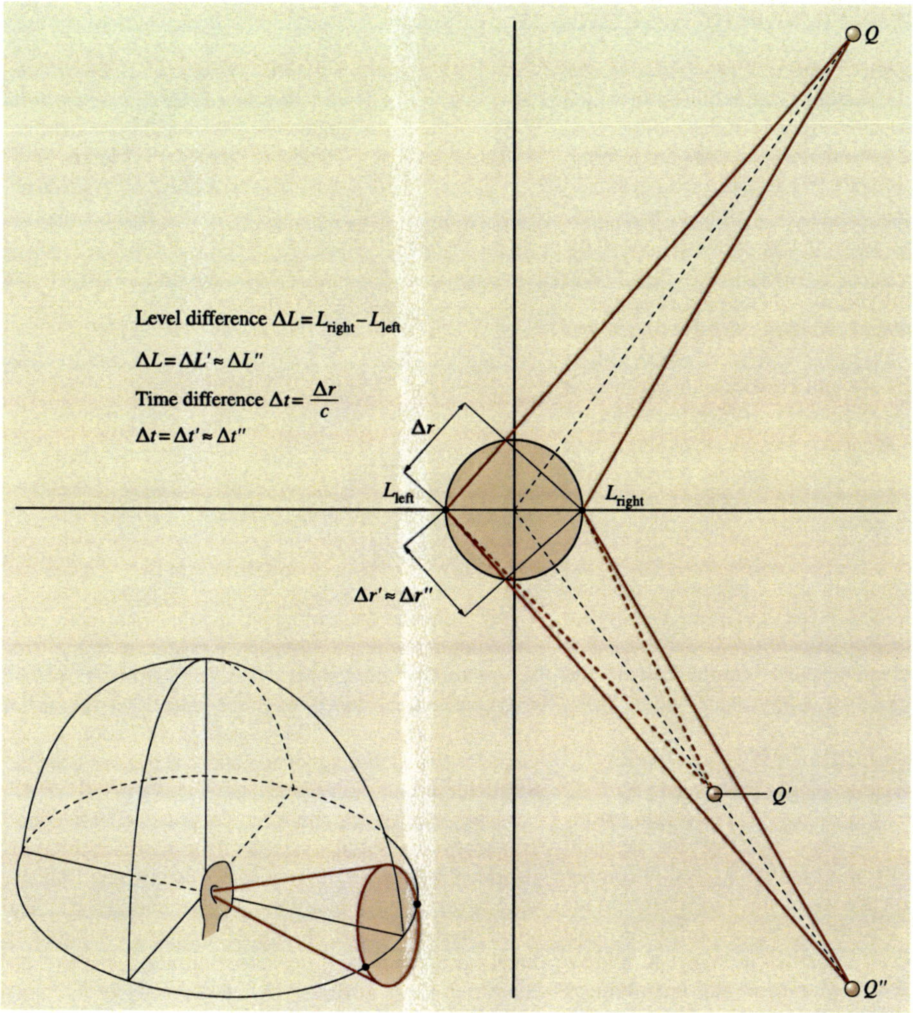

Fig. 12.12 In the case of lateral sound incidence from sources Q, Q' or Q'', the signal arrives in both ears at different times t and with different intensity L. The sound event sites, which are equivalent with regard to the interaural differences Δt and ΔL, are placed on the surface of the *cone of confusion* (the tip of which is in the middle of the head and the axis of which coincides with the line joining the ears). This approximation is valid only if the head is modeled by a sphere without ears and with diametrically opposite ear canal entrances – a very crude and unrealistic approximation

to detect in noise (i. e. at a lower signal to noise ratio) when two signal sources are separated. Experimentally, the effect of binaural hearing on the masking of a signal can be replicated simply by presenting a mixture of test tone and broadband noise to one ear via headphones. The subject adjusts the test tone level such that it is no longer perceived. If the same noise is now presented to the other ear, the noise sensation is louder but the test tone is heard again and hence the binaural hearing threshold is lower than in the monaural case (binaural masking level difference, BMLD). Similarly, the binaural threshold for the intelligibility of speech in noise can be determined and compared with the monaural threshold (binaural intelligibility level difference, BILD). Depending on the spatial arrangement of the sources of signal and noise, the threshold difference amounts to about 10 dB in normal hearing subjects. The enhancement of discrimination and intelligibility described by BMLD and BILD is closely related to the ability of to understand speech in babble noise (cocktail party effect), which requires a functioning binaural system. Mild single sided hearing loss is sufficient to affect its function seriously. The assessment of monaural and binaural thresholds in noise is, therefore, suitable for a quantitative description of the ability to follow a conversation in noise. It has been shown that BMLD and BILD are lower in hearing impaired patients and in older normal hearing subjects.

12.3 Objective Audiometric Assessment

In contrast to subjective audiometry, which makes use of psychoacoustic methods, the methods in which a physiological reaction or response of the auditory system is measured for audiometric purposes are referred to as objective audiometry. The responses depend to a much lesser degree on attention and active participation of the patient than in the subjective methods. Among the reactions to acoustic stimuli, the physical properties of the tympanic membrane (impedance audiometry), the sound emitted by the inner ear (otoacoustic emissions) and the electric processing in auditory nerve, pathways and cortex (auditory evoked potentials) are part of audiologic diagnostics. Since all of these signals are superimposed by physiologic and external interferences, the accuracy of the objective measures is limited. Therefore, some of the applications are of interest primarily in patients who are unable or unwilling to cooperate. Other applications provide specific diagnostic information that can not be obtained with subjective methods. In the context of identifying the cause of hearing impairment, objective measures are not merely a substitute, but a complement to the subjective procedures.

12.3.1 Impedance Audiometry

Impedance audiometry involves all audiologic examinations based on the measurement of the acoustic resistance opposed by the eardrum to the incoming sound wave. The dependence of this impedance on air pressure and probe tone frequency provides information about the physical properties of eardrum, middle ear, and ossicular chain. Moreover, the recording of impedance during acoustic stimulation with tonal pips allows the observation of physiological responses of the auditory system.

At the interface between two media of different acoustic impedances Z_0 and Z_1, sound waves are reflected. The measurement of the impedance is based on the relationship between the reflection coefficient r and the normalized impedance difference

$$r = \frac{(Z_1 - Z_0)^2}{(Z_1 + Z_0)^2} . \tag{12.9}$$

The impedance of the middle ear depends on the mass M (tympanic membrane, ossicles, possibly secretions), the friction R (middle ear and inner ear), the elasticity k (eardrum, middle ear tendons, tympanic air), and the sound frequency f or angular frequency ω

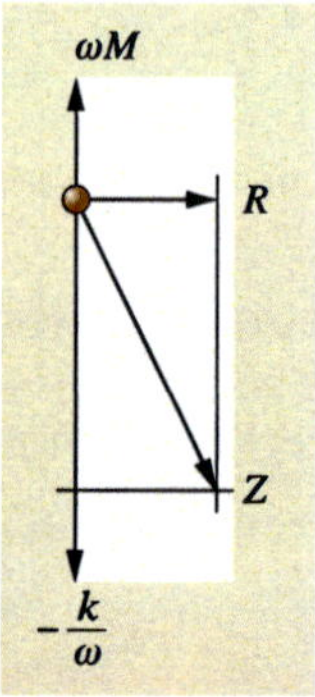

Fig. 12.13 Vector calculation of the complex impedance Z from the real components (ωM and k/ω) and the imaginary part R (R = friction, M = mass, K = elasticity, ω = frequency)

(Fig. 12.13) [12.32]

$$Z = \sqrt{R^2 + \left(\omega M - \frac{k}{\omega}\right)^2} . \tag{12.10}$$

Apart from the frequency, all variables that enter into the calculation of the impedance characterize the condition of the middle ear. This fact accounts for the diagnostic importance of impedance. At the usual low test frequencies, the largest contribution to the impedance and its diagnostic importance arises from the elasticity term.

The relationship between impedance and sound frequency plays only a minor role in practical tympanometry because the impedance is measured almost exclusively with a test frequency around 220 Hz. The main reason is that at higher frequencies, standing waves can occur ($\lambda/4$ resonance) and the result of the measurement is governed by the geometry of the outer ear canal. Measurements with higher frequencies (e.g. 600 or 1000 Hz) yield more information about the physical processes in the middle ear, especially in children, and the improvements achieved by operational interventions can be observed, but the results obtained in different patients are less comparable.

In the context of audiometry, the use of impedance (in the appropriate unit *acoustic Ohm* = $kg/(m^2s)$) is less common than the use of its reciprocal (the admittance). Since the impedance describes the resistance of the middle ear, the admittance corresponds to the willingness (compliance) of the middle ear to forward the sound to the inner ear. A large admittance is, therefore, equivalent to a low acoustic impedance and thus to a high flexibility of the eardrum. The unit of compliance is m^2s/kg or mho (the reciprocal to ohm). Since

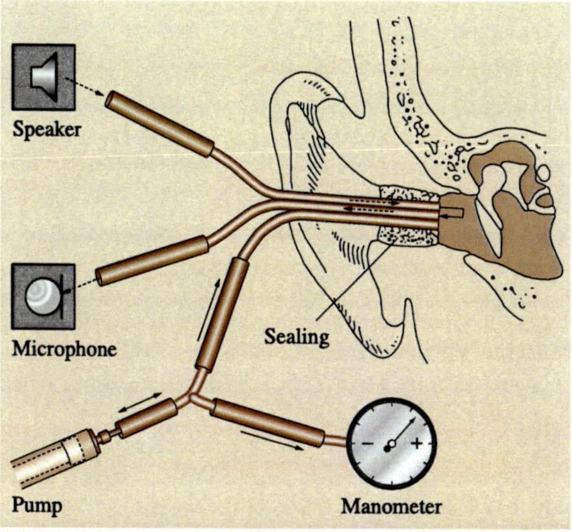

Fig. 12.14 Impedance ear canal probe with three tube lines for speaker, microphone and air pump/gauge, and an individual tip sealing the ear canal (after [12.33])

this unit is not particularly clear, the description of the compliance by an equivalent volume (whose unit is cm^3 or ml) is more usual. The ear canal and its boundary with the middle ear is understood to be a variable volume of air. If the sound is reflected, the entire system behaves like a small volume of air (with rigid walls); if the sound is absorbed, it represents a larger volume.

The measurement of the impedance is done indirectly by measuring the reflected portion of a test tone of known intensity, which is radiated into the ear canal. The key components of a device to measure the middle ear impedance are:

- A speaker for producing the test tone
- A microphone to measure the intensity of the reflected sound
- A pump and a pressure gauge for application and measuring the air pressure in the ear canal
- A device to display the results.

The probe is inserted into the ear canal and fitted with an air-tight, variable size ear tip (Fig. 12.14).

Tympanometry

The *tympanogram* reflects the compliance as a function of the pressure applied and air measured in the ear canal (Fig. 12.15). In the case of elevated pressure, the eardrum is displaced inward and it reflects a large part of the probe tone. Consequently, its elasticity decreases and thus the impedance is higher and the compliance

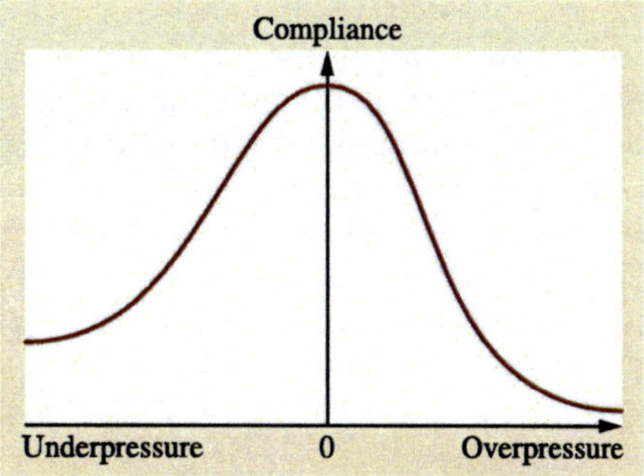

Fig. 12.15 In the case of normal middle ear function, the application of overpressure or underpressure in the ear canal results in very low values for the compliance of the tympanic membrane. At normal atmospheric pressure (no pressure difference between ear canal and middle ear), maximum compliance is recorded (schematic representation)

lower. At reduced pressure the eardrum is also more strained than at normal pressure because of the (unaltered) pressure in the middle ear, thus resulting also in a lower compliance. The vibrational properties are most favorable if the pressure in the outer canal equals the pressure in the middle ear, normally at atmospheric pressure. Therefore, the normal tympanogram exhibits a maximum of compliance at normal pressure ($p = 0$). The pressure differences applied for routine measurements are within ±30 hPa.

The most important feature of a tympanogram is the maximum of compliance, which is described quantitatively by its location, its height, and a form parameter. The height of the peak has a pronounced individual variability. If the maximum compliance is significantly higher than the normal value, an atonic cicatrice in the tympanic membrane, a fracture of the stapes, or a defect of the incus may be present. Very flat compliance maxima will be observed when the middle ear is filled with secretion (otitis media with effusion, OME) or when the tympanic membrane is reinforced by a scar. The shift of the maximum to positive or negative value indicates an overpressure or underpressure in the tympanic cavity. An excess pressure can be due to dysfunction of the *Eustachian* tube, a reduced pressure indicates oxygen-consuming inflammatory middle ear diseases – e.g. the onset of otitis media without effusion. Variations of the middle ear pressure in the range of ±10 hPa have no diagnostic significance. If a viscous middle ear effusion is present, secretions are deposited on the ear drum, which limits its mobility. The corresponding tympanogram exhibits no compliance peak, it is flat with a slight increase at negative pressure. A similar result emerges in the case of a perforated eardrum because the admittance of the large composite volume of ear canal and middle ear is independent from pressure and not influenced by the tympanic membrane. Flat tympanograms can also indicate errors of measurement (e.g., a probe blocked by debris or pointing to the wall of the ear canal).

One of the main objectives of impedance audiometry, the diagnosis of otitis media, is better achieved by multifrequency tympanometry (MFT) using additional probe frequencies higher than 226 Hz ([12.34] and references therein). Tympanograms recorded around 660 Hz and 1000 Hz exhibit several and variable peaks and notches in the pressure dependence of conductance and susceptance. Their exploration is recommended in children having a history of otitis media, or else abnormal or notched 226 Hz tympanograms. As verified by myringotomy (a tiny incision created in the eardrum in order to relieve pressure caused by the excessive build-up of fluid), the MFT method has been proved to detect middle ear abnormality at a higher rate.

Middle Ear Reflexes

The second application of impedance audiometry is the detection of the stapedius reflex, mediated by a temporal change in impedance during the acoustic stimulation of one ear (Fig. 12.16). The impedance change is caused by contractions of the stapedius muscle affixed at the head of the stapes. The contraction is activated by strong acoustic stimuli and results in a stiffening of the ossicular chain and thus in an increase in the eardrum impedance. Contraction and impedance change follow the stimulus onset with a slight delay of about 10 ms and they last for the duration of the stimulus.

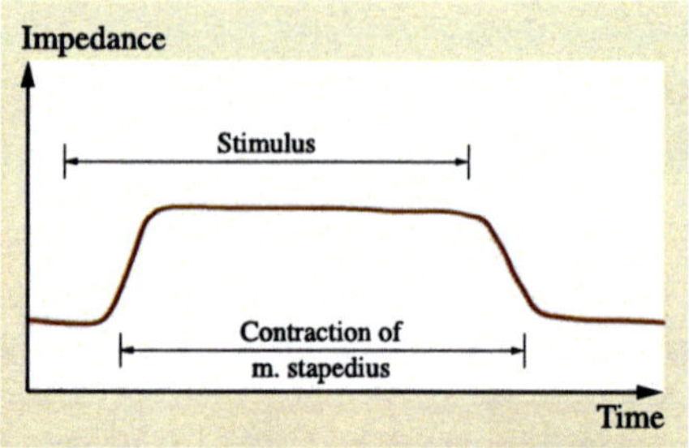

Fig. 12.16 Shortly after presentation of a sufficiently strong tonal pip, the impedance of both eardrums is temporarily increased in normal hearing as a result of the acoustic reflex

The reactive contraction of the stapedius m. evoked by strong acoustic stimuli disables the oscillation of the stapes, thereby protecting the inner ear from excessive noise intensities. In order to fully understand the reflex process and to interpret the findings, the reflex arc has to be considered (Fig. 12.17). The acoustic reflex is an acoustic-facial reflex, i. e. the triggering (afferent) branch includes the middle ear, inner ear, and the auditory nerve; the executive (efferent) limb is the facial motor nerve that innervates the middle ear muscles. Afferent and efferent limbs of the reflex circuit are connected to each other in the brainstem nuclei of auditory and facial nerves. Since the executive branch innervates the middle ear muscles of both ears, a monaural acoustic

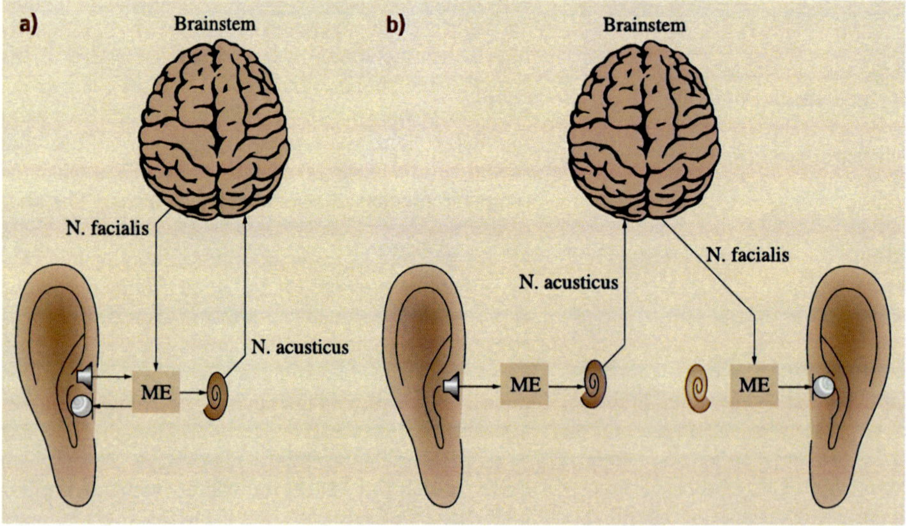

Fig. 12.17a,b Course of the ipsilateral (**a**) and contralateral (**b**) arc in the induction of the acoustic reflex (schematic representation). The acronym ME stands for middle ear

stimulation causes impedance change on both sides: the acoustic reflex can be triggered by ipsilateral and contralateral stimulation. In most practical applications, the question arises as to whether the reflex can be elicited in one ear. In others, however, the interest is not focused on the stimulus ear but on the probe ear.

For the observation of the ipsilateral acoustic reflex, the speaker integrated in the ear canal probe presents not only the continuous probe tone (e.g. 226 Hz) but also a reflex-inducing tone burst (duration e.g. 1 s) of arbitrary frequency (e.g. 500 Hz, 1 kHz, 2 kHz, or 4 kHz) and high intensity (70 to 110 dB HL) only. To measure the contralateral stapedial reflex, a headphone is placed on the stimulus side and the probe is placed in the ear canal on the probe side. In either case, the measurement is performed at the pressure corresponding to the compliance maximum. Starting at about 70 dB HL, the stimulus level is increased until the reflex threshold is exceeded. In normal hearing ears, this threshold is found typically at levels between 70 dB HL and 90 dB HL. A small percentage of normal hearing subjects exhibit no reflex.

The acoustic reflex is generally triggered only by stimuli associated with a large subjective loudness. The acoustic reflex threshold (ART) is, therefore, not primarily correlated with the level of the stimulus (dB HL), but rather with the level difference related to the individual threshold of the examined ear (dB SL). In case the of conductive dysfunctions, the reflex threshold is, therefore, increased by the amount of hearing loss. If the hearing loss amounts to more than about 30 dB, no reflex will be triggered, because that would require a stimulus level of at least 110 dB HL ($\geq$ 70 dB above threshold). In sensorineural hearing loss, the reflex threshold is elevated only when the hearing loss is more than 50 dB. Consequently, the reflex threshold is closer to the hearing threshold in many ears with sensory hearing loss. This manifests the restricted dynamic range of the hair cell-damaged ear (*Metz* recruitment [12.35]). Inner ear dysfunction impacts only the stimulus side but not the probe side.

The effect of neural hearing loss on the stapedius reflex depends on whether the lesion is located peripherally in the auditory nerve or more centrally in the brainstem, referred to the acoustic-facial reflex connection. In the first case, the reflex is absent or branded by a pathological reflex fatigue (decay); in the second case, the acoustic reflex is not impaired. Abnormal findings may be present both in the stimulus ear and in the probe ear. In many cases, the interpretation of the findings is obscured by additional cochlear pathologies.

12.3.2 Otoacoustic Emissions (OAE)

Otoacoustic emissions (OAE) are sound waves that are generated in the inner ear and radiated in the external auditory canal, mediated by the middle ear ossicles and the tympanic membrane. They were first described by *Kemp* in 1978 [12.36]. Otoacoustic emissions may be spontaneous (SOAE) and evoked (EOAE) by acoustic stimuli. The existence of OAE is attributed to the nonlinear and active cochlear processes involved in the sound preprocessing in the cochlea. These processes manifest themselves in the micromechanics of the basilar membrane and they are responsible for the high sensitivity, the wide dynamic range, and the good frequency resolution of hearing. The source of the OAEs are microscopic movements of the cochlear outer hair cells (OHC). If isolated OHCs are held alive in preparation, they can be stimulated to active contractions by chemical, electrical, and mechanical stimuli. It is supposed that OHCs at their natural site are triggered to contractions and elongations by acoustic stimuli as part of the physiological hearing process if the stimulus frequency corresponds to the targeted area of the cochlea. This enhances the displacement of the basilar membrane on the one hand and, on the other, it induces a retrograde traveling wave of small amplitude, which propagates to the oval window, through middle ear and eardrum and finally manifests itself as measurable sound pressure fluctuations in the ear canal. The evoked emissions can thus be regarded as a byproduct of a nonlinear mechanical system with active feedback. Depending on the nonlinearity of the biological transduction, quadratic, cubic, and higher effects might occur, which can be detected by the corresponding distortions of the input.

Among the different types of OAEs, only EOAEs are routinely used in audiological diagnostics. They are detectable in almost 100% of normal ears. Because of their low intensity and the inevitable presence of noise, a sensitive microphone and complex signal processing are necessary for their detection. Another difficulty arises from the necessity of an acoustic stimulus, whose intensity exceeds that of the emissions by orders of magnitude and causes a passive echo response of the eardrum and ear canal walls, which is superimposed on the physiological signal.

EOAEs are divided into post-stimulatory (delayed) transient evoked otoacoustic emissions (TEOAE) and per-stimulatory (simultaneous) emissions (Table 12.2). The latter are further differentiated as stimulus frequency otoacoustic emissions (SFOAE), whose frequency matches that of the stimulus, and distortion

Table 12.2 Classification and nomenclature of otoacoustic emissions

Otoacoustic emissions (OAE)			
Spontaneous OAE (SOAE)	Evoked OAE (EOAE)		
	Post-stimulatory OAE	Per-stimulatory OAE	
	Transitory evoked OAE (TEOAE)	Stimulus frequency OAE (SFOAE)	Distortion product OAE (DPOAE)

products or distortion product otoacoustic emissions (DPOAE), whose frequency is different from that of the stimulus. TEOAEs are firmly established in the audiological diagnosis because of their high reliability. In contrast, the detection of SFOAEs is difficult. Therefore, among the per-stimulatory OAEs only DPOAEs are used for diagnostic purposes. Only the two types of OAEs established in practical audiometry – TEOAE and DPOAE – are described in two sections of this chapter.

It is not yet clear whether the mechanisms underlying the generation of TEOAEs and DPOAEs are different and whether thus delayed and simultaneous emissions provide different information on the function of the inner ear. Most probably, the different emissions reflect only two different aspects of the same active and nonlinear cochlear amplifier. The difference of the techniques used for their detection impacts the practical application and the information outcome because at comparable stimulus levels, DPOAEs are detectable with greater sensitivity, i. e. even with a more pronounced hearing loss, than TEOAEs.

Transitory Evoked Otoacoustic Emissions (TEOAE)

For the measurement of TEOAEs (TEOAE), the acoustic signal is recorded in the ear canal immediately following a transitory stimulation. The active physiological echo is superposed by the stimulus, its passive

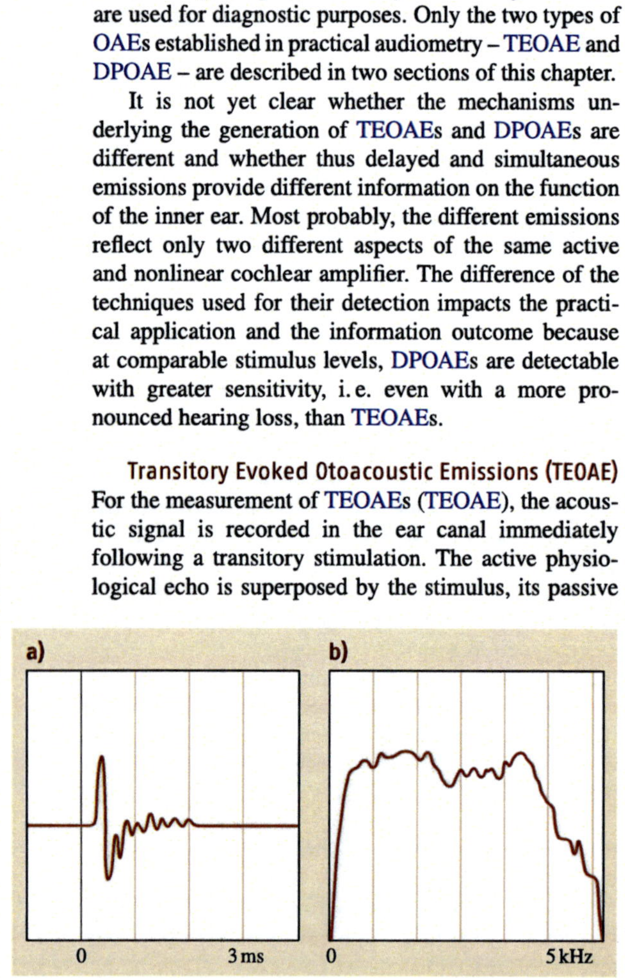

Fig. 12.18a,b Temporal oscillogram **(a)** and frequency spectrum **(b)** of the click stimulus. Ideally, the spectrum registered in the ear canal is very broad and nearly flat (after [12.37])

(mechanical) echo and ambient noise. The amplitude of these competing signals exceeds the target signal by several orders of magnitude. In order to detect the delayed emission, the signal to noise ratio must be improved by analog and digital signal processing. The main principles are effective acoustic screening and the selection and averaging of many signal segments.

In most TEOAEs applications, the acoustic stimulation is performed with a temporal sequence of clicks, i. e. single square-wave signals of a duration of about 100 μs, which are presented at regular or irregular intervals of at least 20 ms. The time course and spectrum of a click stimulus are shown in Fig. 12.18. The recording of the stimulus and its frequency spectrum in the ear canal are the base for monitoring and the eventually corrective adjustment of the probe. In the ideal case, the click stimulus exhibits a white spectrum, i. e. all frequencies are represented with the same intensity. Such a constellation is the criterion for a closed ear canal volume free from resonance and, on the other hand, it is also a prerequisite for the functional testing of the entire cochlear partition with a single stimulus.

The probe introduced into the ear canal (Fig. 12.19) contains a magnetic transducer that delivers the short-term stimulus (click or tone burst) and a sensitive electret microphone for the registration of the sound pressure signal. The analog processing of the microphone signal registered after the stimulus consists of band-pass filtering in the range of about 300 Hz to 10 kHz and a linear amplification. The amplified signal sweep is digitized with a sample rate matched to the upper frequency limit and fed to the computer. The first step of the digital signal processing is the selection of appropriate signal sections in accordance with an amplitude criterion (artifact rejection). If at least one sample within a sweep exceeds a given (but variable) amplitude limit, the whole segment is excluded from further processing. This is equivalent to an interruption of the measurement in times of less favorable signal to noise ratios. The sweeps that pass the artifact suppression are added together point by point and thus provide a time-dependent mean curve. This signal averaging emphasizes all signal components that are correlated to the stimulus, while the amplitude of

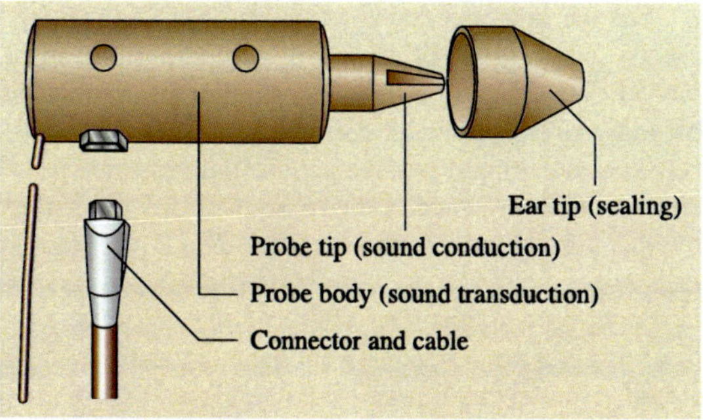

Fig. 12.19 View of the ear canal probe used for acoustic stimulation and recording of OAE (Madsen *Capella*, GN Otometrics, Münster)

stochastic components is diminished. The resulting improvement of the signal to noise ratio is based on the validity of different laws for the addition. The amplitude of the physiologic signal responding to the stimulation increases linearly, whereas the stochastic components are superposed according to their variances. If the conditions of a stable deterministic stimulus response and a stationary noise source are met, the signal to noise ratio improves by a factor of $\sqrt{N}$ after N additions or averages (Fig. 12.20).

The improvement of signal to noise ratio generated by averaging is not equivalent to a selective amplification of the physiological response, since it emphasizes all stimulus related signal components such as the acoustic stimulus itself and its time-decaying echoes. In order to separate the passive (mechanical) from the active (physiological) responses, a particular stimulus sequence is applied (the nonlinear stimulus block in Fig. 12.21).

The effect of the nonlinear stimulus sequence is based on the different behavior of physiological responses and competing signals related to the stimulus. While the growth of the OAE amplitude with increasing stimulus amplitude is nonlinear and saturates at high stimulus levels, the amplitude of the mechanical response increases linearly. Therefore, the passive mechanical responses to $k = 3$ stimuli can be canceled through a stimulus of threefold amplitude (and analogously for arbitrary values of k). For the nonlinear active signal components, this cancelation is incomplete. When using a nonlinear stimulus sequence, i. e. a sequence suitable to compensate linear signal

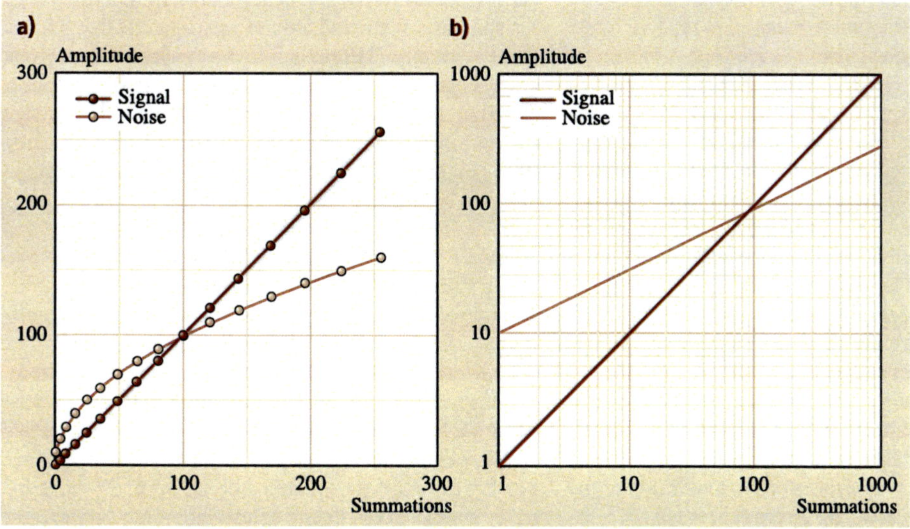

Fig. 12.20a,b By averaging or summation, invariant (stable) signal components are amplified to a greater extent than noise. While the signal amplitude grows linearly with the number of summations, the noise increases only as a root function. If the effective amplitude of the unaveraged noise is 10 times larger than the (unaveraged) signal amplitude, the amplitudes of signal and noise are equal after 100 summations **(a)**. If the same facts are considered in a double logarithmic representation **(b)**, the inversion of a negative to a positive difference of signal and noise is even more pronounced. The consideration of averaging instead of summation causes only a scale correction, but no change of the amplitude relations (after [12.37])

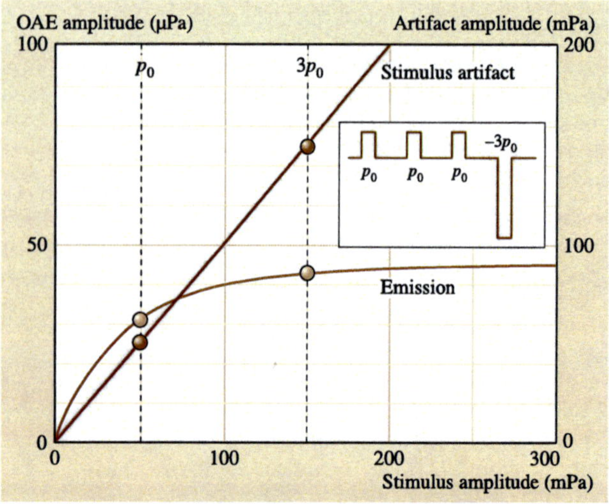

Fig. 12.21 Utilization of the nonlinearity of the TEOAEs for their separation from the stimulus artifact (which depends linearly on the stimulus level) by application of a stimulus sequence composed of several elementary events with canceling amplitude (after [12.37])

components, the resulting signal contains only the uncompensated portion. Compared with conventional stimulation (*linear mode*), the cochlear emission loses amplitude but it appears more prominent relative to the mechanical stimulus response.

The result of a TEOAE measurement consists of one or two curves that reflect the averaged time-dependent behavior of the ear canal sound pressure in a time window starting with the stimulus presentation (Fig. 12.22). The transformation of these curves in the frequency domain provides the spectrum of the TEOAE. If two equivalent (but not identical) partial averages are available, the spectra of signal and noise can be calculated separately (cross-power spectra). Furthermore, the availability of two mean curves of $A(t)$ and $B(t)$ allows us to calculate a correlation coefficient (reproducibility) and to estimate the variance of the residual noise. From the grand mean $(A+B)/2$, the effective amplitude of the compound signal (emission and noise) can be calculated. The effective amplitude of the residual noise can be estimated from the root of the variance (root mean square, RMS) of the mean deviation $(A-B)/2$. In addition to these data, the representation of the wave form and spectrum of the stimulus recorded in the ear canal, the number of averaging steps and artifacts, the position of the amplitude limit, the stimulus intensity, and the stability of the stimulus conditions are part of the documentation of a TEOAE study.

For the diagnostic evaluation of TEOAE measurements, the presence of signals of cochlear origin must be verified. This is equivalent to the differentiation of delayed emissions as compared to averaged noise (residual noise) and passive echoes (stimulus artifact). An important criterion is the reproducibility expressed by the correlation coefficient $R(A, B)$. If the reproducibility is large, it is improbable that the measured signal is pure noise. The effect of the stimulus artifact (which is not completely abolished by the nonlinear stimulus mode) can be assessed with the aid of windowing. If a high reproducibility is found only in the first few milliseconds following the stimulus, the cochlear origin of the signal is unlikely.

The reliable and unambiguous distinction between signals of cochlear origin and interferences constitutes the most essential part of TEOAE evaluation. If delayed emissions are present, the ear under examination has a near-normal hearing threshold at some frequencies. The incidence of detectable TEOAEs decreases steplike from 100 to 0% with increasing inner ear hearing loss. The position of the step depends on the stimulus intensity. For a stimulus level of $L = 80$ dB SPL, the point of 50% incidence lies at $HL_{min} = 32$ dB [12.38]. Here, HL_{min} denotes the minimum hearing loss (i. e. the lowest threshold) in the frequency range from 1 to 4 kHz. As soon as a restricted area of the organ of *Corti* functions normally, TEOAEs will be present. No regions with substantial hearing loss contribute to the response. The frequency spectrum and the reproducibility (or signal to noise ratio) calculated in individual frequency bands allows a rough assessment of the frequency range affected by hearing loss.

The relationship between hearing threshold and TEOAEs described so far is valid for the special case of pure sensory (endocochlear) hearing loss. However, of course, also conductive disorders affect detectability and amplitude of the TEOAEs. The deterioration of sound transmission through the middle ear leads to an attenuation of both stimulus and response. If the air–bone gap exceeds the limit of 20 dB at all frequencies, TEOAEs are no longer detectable. Purely retrocochlear hearing disorders do not affect the TEOAE, they are typically characterized by the constellation of poor hearing with nearly normal emissions. However, since many causes of retrocochlear hearing disorders – e.g. space-occupying lesions in the cerebellopontine angle – are associated with cochlear damages, this constellation is rather rare. Especially in the case of large or long-term existing tumors, the TEOAE measurement provides no reference to the neural genesis of hearing impairment,

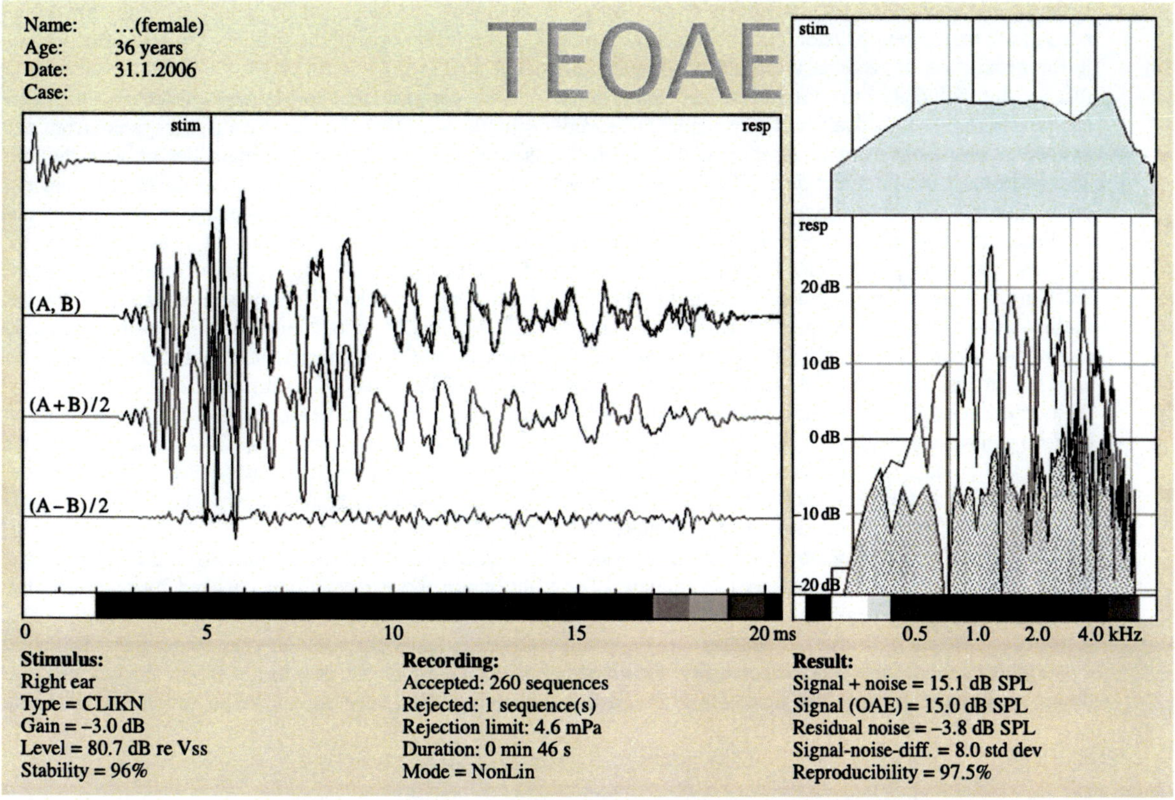

Fig. 12.22 Transient evoked otoacoustic emissions recorded in a nor mal hearing ear. The primary test results are the mean partial averages $A(t)$ and $B(t)$ (*large box*). These curves are the source for the calculation of frequency spectra, emission amplitude, reproducibility and residual noise. Moreover, the stimulus parameters and noise conditions are given

i. e. detectable emissions are limited to the frequencies of the normal threshold [12.39].

The measurement of TEOAEs does not allow for quantitative and frequency-specific determination of the hearing threshold. The spectrum of each delayed emission exhibits irregular and individual peaks and notches, facing no hearing loss in the audiogram. Only the absence of emissions in a wide frequency interval can be interpreted as an indication of a raised threshold for those frequencies. A special feature must be noted, however: if the hearing loss is limited to low frequencies, not all low frequency components are missing but only those with long latency (since the apical cochlear responses are generated later). This can be read from the OAE waveform or from the spectrogram and it can be confirmed by windowing and band-pass filtering the original response.

The spectrum of emissions evoked by broadband stimuli (click) contains all frequencies for which the cochlea has a near normal function. Functional losses exceeding the limit of about 30 dB result in the disappearance of the corresponding TEOAE components. The complete absence of physiological echoes thus means that the hearing loss exceeds this value at all frequencies. Conversely, the presence of emissions permits the conclusion that the function of at least some of the hair cells is nearly normal and thus the minimal hearing loss is less than about 30 dB. The validity of these statements is limited to the frequency range between 1 and 4 kHz, a hearing loss outside these limits cannot be detected with TEOAEs.

Distortion Product Otoacoustic Emissions (DPOAE)

The nonlinearity of cochlear signal processing is accompanied by several consequences, one of them being the inability of the inner ear to process two frequencies independently if they are sufficiently close to each other.

If the cochlea is excited simultaneously with two (primary) pure tones, their physiological processing results in the production of secondary tones with frequencies that are not included in the stimulus. These distortions can be perceived subjectively and they can be measured as acoustic emissions from the inner ear in the ear canal. This is the basis for the detection of DPOAEs (distortion product otoacoustic emissions).

From the mathematical analysis of nonlinear distortions, using a power series expansion and the calculation rules for trigonometric functions, the formulas for the frequencies present in the output of a nonlinear system processing the input frequencies f_1 and f_2 can easily be obtained. Arranged according to their frequencies, the symmetric (quadratic) distortion leads to components with $f_2 - f_1, 2f_1, f_2 + f_1$ and $2f_2$, and the anti-symmetric (cubic) distortion delivers signal components with the frequencies $2f_1 - f_2$, $2f_2 - f_1$, $3f_1$, $2f_1 + f_2$, $2f_2 + f_1$ and $3f_2$. Many of these theoretically possible combination tones were measured in the ears of experimental animals and humans. The functional testing of hearing is limited to the cubic distortion products with the frequencies $2f_1 - f_2$ and $2f_2 - f_1$ because of their good detectability, and among these, mostly the former: the tone whose frequency lies below the two tones, separated from the lower one by the difference of the primary stimulus frequencies (Fig. 12.23a).

In contrast to the delayed TEOAEs, the perstimulatory DPOAEs can be disaggregated from the stimulus not in the time domain but only in the frequency domain. Since the frequency of the cochlear emissions can be predicted accurately from the stimulus frequencies, the signal detection can be automated. The release of measurable distortion product works best with a stimulus that is composed of two suprathreshold continuous pure tones of approximately equal intensity with the frequencies f_1 and $f_2 \approx 1.2f_1$. To allow the detection of distortion effects of physiological origin, technical nonlinearities must be reduced as much as possible. For this reason, the two sine tones are presented by separate transducers. Therefore, the DPOAE probe contains two speakers and a microphone.

Prior to the DPOAE measurement, the acoustic conditions in the ear canal are checked by recording the transfer function upon excitation with broadband click stimuli. Ideally, the stimulus spectrum is flat over a wide frequency range. In the case of larger fluctuations, the probe position must be controlled and corrected. If the

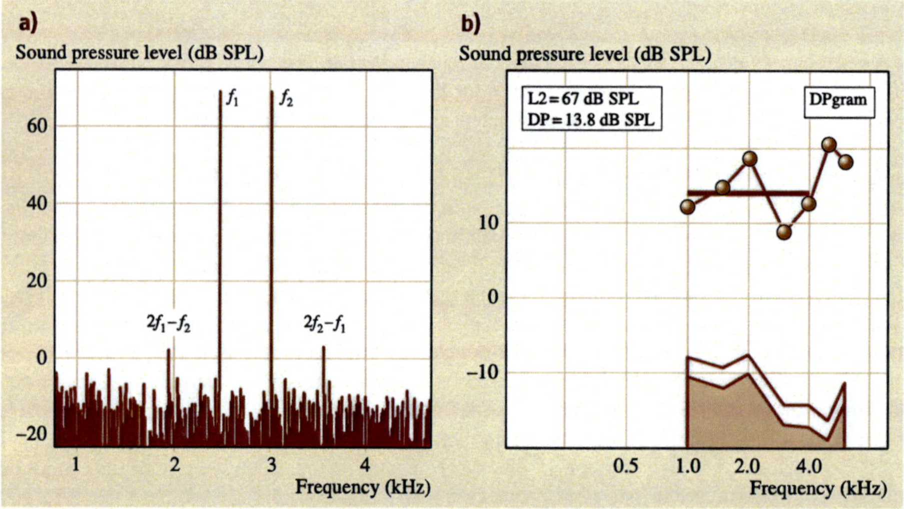

Fig. 12.23a,b Frequency spectrum measured in the ear canal when stimulated with two pure tones **(a)**. With increasing number of averages, the cubic difference tones (distortion products) grow clearly out of the noise background. **(b)** Diagram showing the amplitude of the distortion product in dB SPL as a function of the stimulus frequency f_2 in kHz (DP gram) for stimulation of a normal hearing ear with two primary tones of equal intensity ($L_1 = L_2 = 67$ dB SPL). The *shaded area* in the lower part of the diagram corresponds to the residual noise, its upper limits are given by the average noise level incremented by one and two standard deviations (after [12.40] with kind permission)

transducers are fed at different times and the responses are recorded separately, not only the ear canal response, but also the function of the two loudspeakers and their feedthrough can be checked.

During the DPOAE measurement, the microphone signal is recorded and sections of this continuous signal, which are synchronized to the phase of the stimulus tones, are summed. The phase-aligned summation results in a gain of stimulus and cochlear response relative to the unsynchronized noise. In the spectrum of the running average, the lines corresponding to the frequencies of the distortion products emerge increasingly from the background noise (Fig. 12.23a). From this spectrum, the amplitude of the distortion product at the frequency $2f_1 - f_2$ is extracted, together with the amplitude of the background noise, derived from a narrow frequency band centered around $2f_1 - f_2$. In the DP-gram, these data are displayed for several stimulus frequencies (Fig. 12.23b).

The selective stimulation with two frequencies excites only two places of the basilar membrane. According to the constant frequency ratio $f_2/f_1 \approx 1.2$, the activated regions are separated by a constant distance, independent of the stimulus frequencies. The distance is about 1.3 mm and it corresponds roughly to the equivalent rectangular bandwidth (ERB) for frequencies above 500 Hz. The distortion product is generated in the overlapping zone of the two excited areas. Because of the asymmetry of the traveling wave peak, this zone is located close to the place where the higher of the two frequencies (f_2) is processed. Accordingly, the correlation between the amplitude A_{DP} of the emission with the hearing threshold HL(f_2) is much larger than between A_{DP} and HL(f_1). Further evidence for the generation site of the distortion product can be deduced from suppression experiments: It has been shown that the suppression of the DPOAE by application of a masker with variable frequency has the greatest impact if the masker frequency is equal to f_2 [12.41]. The width of the overlap zone increases with increasing stimulation level. This has two consequences: First, the information obtained from the measurement of DPOAEs at high stimulus levels is not solely specific for the hearing threshold at the stimulus frequency, and second, the frequency ratio and/or the level difference of the primary tones have to be modified if the DPOAE are to be measured with stimulus levels close to hearing threshold (scissors paradigm) [12.42].

The relationship between DPOAEs and hearing loss parameters is controlled by rules that are very similar to those for TEOAEs:

- The incidence of DPOAEs evoked by frequencies f_1 and $f_2 = 1.2f_1$ is close to 100% in normal hearing ears.
- The amplitude of the emission decreases with increasing hearing loss.
- If the hearing loss at frequency f_2 exceeds the limit of about 50 dB, the DPOAEs are no longer detectable.
- Incidence and amplitude of DPOAEs are not influenced by purely retrocochlear hearing disorders.
- Conductive hearing loss causes predominantly an attenuation at low frequencies; in the case of larger hearing loss, the DPOAEs disappear completely.

These rules apply to all pairs of stimuli for which f_2 lies between 1 and 4 kHz. At lower frequencies, the incidence of detectable emissions is well below 100% even in normal hearing ears. At higher frequencies, the detection of DPOAEs is often impacted by technical distortions.

The current application of OAEs in practical audiometry permits a categorization of hearing impairment in the categories of mild, moderate, and severe hearing loss; an exact determination of the hearing threshold is not possible. The categorization is achieved by the combination of TEOAE and DPOAE, since in standardized test conditions, the former are detectable only if the hearing loss does not exceed 30 dB at any frequency, whereas the latter remain observable up to a threshold elevation of about 50 dB. Particularly TEOAEs, whose susceptibility for the disturbing interferences is very low, have proved to be suitable for a reliable and objective assessment of hearing, making this method optimal for newborn hearing screening and early detection of pediatric hearing loss. To date it is not yet clear whether OAEs, beyond their dichotomy, are able to yield quantitative results in terms of a frequency specific objective determination of the hearing threshold. Some current works are focused on the question of whether the threshold can be quantitatively determined using the stimulus level dependence of the amplitude (*growth function*) or other parameters of OAEs. The results of these works are encouraging, but not yet secured for practical use.

12.3.3 Evoked Potentials of the Auditory System

Auditory evoked potentials (AEP) are electrical voltages of physiological origin that are associated with auditory sensations and can be triggered by acoustic or electric stimuli and measured using electrodes. The techniques used for investigating properties of hearing with the help of the AEPs are summarized as electric response audiometry (ERA). They are objective performance tests that allow a quantitative determination of the hearing threshold. This feature is one of the columns accounting for the great importance of ERA in the audiometric test inventory [12.8]. An additional benefit arises from the fact that with the help of the ERA differential topodiagnostic statements, especially for the distinction between sensory and neural hearing loss, can be obtained.

The difficulty in measuring the AEPs is that a signal of very small amplitude (the AEPs) must be detected in the presence of strong noise (the spontaneous EEG). The amplitude of the interference can exceed the amplitude of the target signal by several orders of magnitude – especially if the EEG is superimposed by muscle activity and external electromagnetic interference. The first requirement for the measurement of AEPs is, therefore, a substantial reduction of avoidable contaminations. The following measures help to improve the signal to noise ratio:

- Relaxed and comfortable positioning of the patient
- Acoustic and electric shielding
- Linear EEG amplifier with high common mode rejection
- Filtering of the EEG signal
- Artifact suppression
- Signal averaging.

The measurement of AEPs is based on the recording of the EEG during acoustic stimulation. The patient should sit or lie relaxed during the measurement. As far as the potentials and their parameters are independent of the vigilance (as is the case with the peripheral components of short latency), the measurement under conditions of spontaneous sleep, sedation, or general anesthesia can improve the outcome. The acoustic stimuli are presented mainly via headphones, in special cases via free sound field (loudspeaker), bone conduction transducers, or insert ear phones. In most practical applications, the acoustic stimuli have a short duration or they are modulated in time. The EEG activity is recorded via surface electrodes fixed on the scalp. If an acoustically and electrically shielded room is available, patient, transducers, and EEG amplifier should be housed in it. Outside the shield are the investigator and the remaining parts of the apparatus, consisting essentially of a stimulus generator, an analog/digital converter, and a computer.

Transient Responses

Most of the electrical responses of the auditory system can be registered only as a delayed and temporarily limited (transient) response to a transitory stimulus. Stimulation and EEG analysis are synchronized with each other, whereby the intervals between two stimuli can be constant or randomized. The registration of the EEG sweep begins with the onset of stimulus or shortly before. The averaging of many signal segments yields a time-dependent curve composed of the AEPs and the reduced amplitude of the EEG noise (residual noise). The diagnostic conclusions arise from the evaluation of curves measured with different quality (e.g. frequency) and intensity of the stimuli and from the parameters derived from these curves.

For the emergence of measurable evoked potentials, a large number of nerve action potentials must be generated with a high degree of synchronization. This is possible only with stimuli that go along with rapid changes. In principle, these changes can affect every auditory differentiable characteristic of the stimulus. Most often, this characteristic is the intensity or the sound pressure of the stimulus, and the change is brought about by switching the stimulus on or off. The usual transient stimuli are the broadband click and frequency selective tone bursts. High frequency selectivity and short stimulus duration principally exclude each other. Therefore, responses that require a high degree of neuronal synchronization supply less frequency specific information. Conversely, frequency selective stimuli are not suitable for yielding precise information on the time dependence (latency) of the responses.

Transient AEPs are composed of many individual vertex positive and negative potential peaks, which are generated in different parts of the ascending auditory pathway and are classified into three groups according to their latency: early auditory evoked potentials (EAEP), also denoted as auditory brainstem responses (ABR) in the time range from 1 to 10 ms, middle latency AEP (MAEP) with latencies up to 50 ms, and late or slow AEP (SAEP) with latencies up to 500 ms. The methods used to measure these responses are denoted with BERA (brainstem electric response audiometry) for early, MLRA (middle latency response audiometry) for middle, and CERA (cortical electric response audiometry) for late AEPs. Some of these names make reference to the anatomical location of the response generators. The assignment is not unambiguous, but it may be regarded as certain that J1, the first among the EAEPs originates in the auditory nerve and the subsequent ABR components J3 and J5 in the brainstem, whereas the MAEP are generated in the thalamus (diencephalon) and the primary auditory cortex, and the SAEP in the auditory cortex.

In addition to the electrical potentials mentioned so far, later components that reflect the processes involved in auditory perception, cognition, and discrimination are known. One of them is a vertex negative wave with a latency of approximately 200 ms (N200). It can be observed as specific response to deviant stimuli that are embedded randomly in a series of standard stimuli (mismatch negativity, MMN). If the subject is asked to count the rare stimuli or to react with a motor response, a positive half-cycle wave with a latency of 300 ms (P300) can be observed (event related potential, ERP). Since the detection of this potential requires active cooperation of the subject, they are not used routinely in objective audiometry. Further, very slow changes of voltage (contingent negative variation, CNV) and still later processes (N400) reflect the expectation of a stimulus announced by a preceding warning signal and the processing of semantic information in speech signals.

Due to the different paradigms used for stimulation and signal processing, it is not possible to record all AEP components with a single measurement. In principle, the approaches for measuring early, middle, and late evoked potentials are identical, so that nearly all responses can be recorded with the same equipment by changing some parameters (stimulus type, stimulus interval, EEG gain, filter limits, time window, sampling rate, and number of averages).

The EEG signal is detected by electrodes which are usually attached to vertex and mastoid, the ground electrode being located on the forehead. After amplification, filtering, and A/D conversion, the EEG signal is subjected to an artifact control to select those sweeps that do not exceed a predetermined amplitude limit. The summation of N signal sections ($N \approx 2000$ for ABRs, $N \approx 600$ for MAEPs, and $N \approx 50$ for SAEPs) enhances the signal components related to the stimulus (such as the AEPs) linearly, while stochastic components are amplified approximately proportional to $\sqrt{N}$ in the case of stationary conditions. Thus, the signal to noise ratio improves according to the noise release gain given by $G = 10 \log N \mathrm{dB}$ (Fig. 12.20). If the interferences are not stationary, the signal quality can be improved by weighted averaging, using the reciprocal value of the variance or of the maximum amplitude of the individual sweep as weight (artifact rejection is a special case of weighted averaging making use of the weighting factors 0 and 1).

Quality and reliability of the results can be characterized by the ratio of the variances of overall response and residual noise (response to noise ratio), by the correlation coefficient calculated out of two equivalent partial averages (reproducibility) or estimated by visual comparison of the partial averages. The evaluation and diagnostic exploitation of AEPs include the identification of the response components and the determination of their coordinates (latency and amplitude). The former is used to determine the hearing threshold, the latter contributes to the differentiation of type and site of hearing disorders.

The early responses of auditory nerve and brainstem, often referred to as auditory brainstem responses (ABRs) are studied with BERA. This method includes the post-stimulus time range from 1 to about 12 ms. Due to their small amplitude (50 nV to 0.5 μV), ABRs can be measured with a sufficiently large amplitude only if the elementary neural events are highly synchronized. This can be achieved by stimulation with a short rectangular pulse (click). ABRs evoked by click stimuli are, therefore, not suitable for a frequency specific determination of the threshold. The early responses are very stable regarding the influences of vigilance and drugs and therefore easily recordable during sleep or under anesthesia. They are regularly present at birth, but their morphology and parameters (amplitude, latencies, latency differences) are different from the pattern observed in the adult. In the course of maturation of the neural auditory system, the modifications disappear during the first 18 months of life. The whole ABR is composed of several components generated in various stages of the ascending auditory pathway and referred to as waves J1 to J5 according to their temporal sequence (Fig. 12.24).

The complete response pattern is observed only at stimulus levels far above threshold. The amplitudes of all components increase with increasing stimulus intensity while their latencies decrease (i. e., stimuli of lower intensity are associated with a longer cochlear processing time). These relationships are represented in a diagram, which facilitates the identification of deviations from the normal values.

The lack of frequency specificity of ABRs evoked by broadband clicks limits a wide field of application, in spite of their otherwise favorable properties. This was motivation to intensify the efforts to find a frequency specific audiometric assessment based on early potentials. One of the approaches based on stimulation with brief tonal bursts and simultaneous masking with broadband noise passed by a filter with a notch

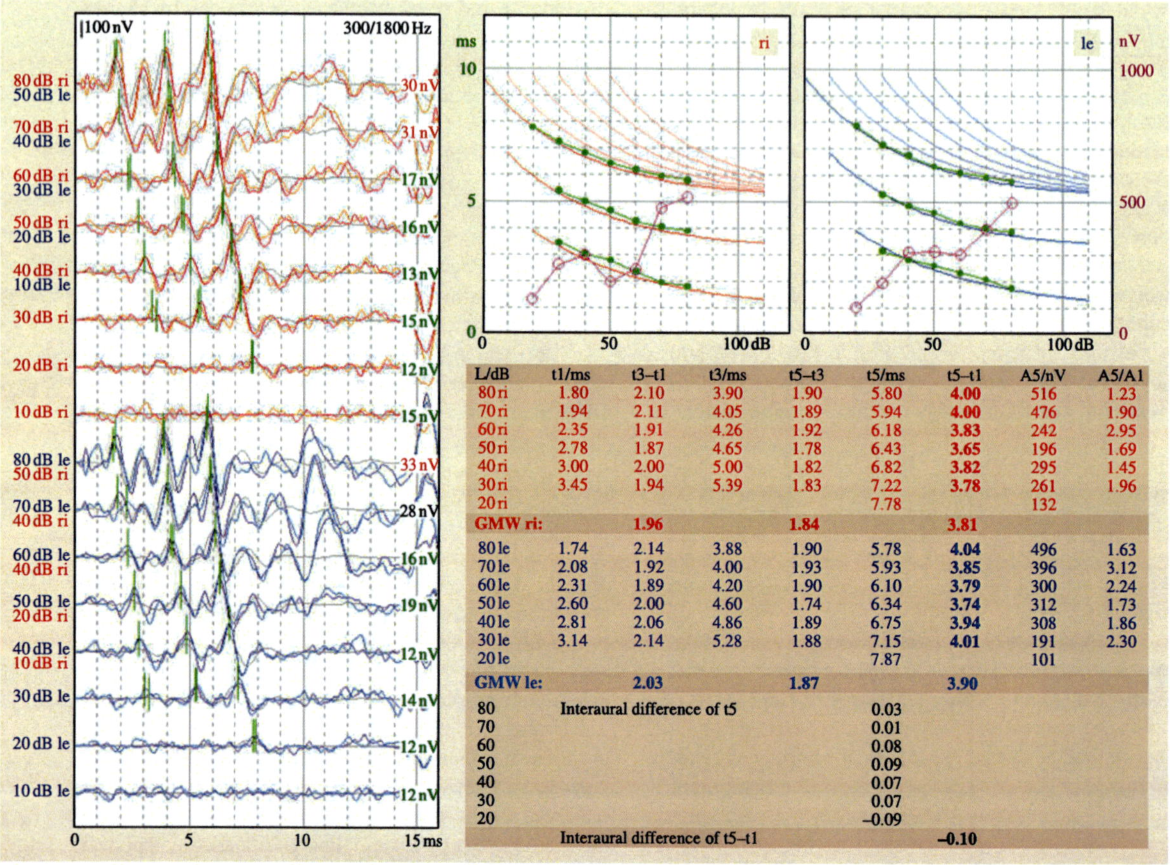

L/dB	t1/ms	t3–t1	t3/ms	t5–t3	t5/ms	t5–t1	A5/nV	A5/A1
80 ri	1.80	2.10	3.90	1.90	5.80	**4.00**	516	1.23
70 ri	1.94	2.11	4.05	1.89	5.94	**4.00**	476	1.90
60 ri	2.35	1.91	4.26	1.92	6.18	**3.83**	242	2.95
50 ri	2.78	1.87	4.65	1.78	6.43	**3.65**	196	1.69
40 ri	3.00	2.00	5.00	1.82	6.82	**3.82**	295	1.45
30 ri	3.45	1.94	5.39	1.83	7.22	**3.78**	261	1.96
20 ri					7.78		132	
GMW ri:		**1.96**		**1.84**		**3.81**		
80 le	1.74	2.14	3.88	1.90	5.78	**4.04**	496	1.63
70 le	2.08	1.92	4.00	1.93	5.93	**3.85**	396	3.12
60 le	2.31	1.89	4.20	1.90	6.10	**3.79**	300	2.24
50 le	2.60	2.00	4.60	1.74	6.34	**3.74**	312	1.73
40 le	2.81	2.06	4.86	1.89	6.75	**3.94**	308	1.86
30 le	3.14	2.14	5.28	1.88	7.15	**4.01**	191	2.30
20 le					7.87		101	
GMW le:		**2.03**		**1.87**		**3.90**		
80	Interaural difference of t5				0.03			
70					0.01			
60					0.08			
50					0.09			
40					0.07			
30					0.07			
20					–0.09			
	Interaural difference of t5–t1					**–0.10**		

Fig. 12.24 Auditory brainstem responses (ABRs) evoked by clicks and recorded in a normal hearing 15 year-old female subject. The *left panel* shows the original responses recorded at various stimulus levels in right (*ri*) and left (*le*) ear. Latencies and amplitudes derived from these curves and their level dependence are shown numerically and graphically on the *right-hand side* (GMV means *weighted average*)

around the burst frequency (notch noise ABR [12.43]). The responses obtained with this method can be evoked by frequencies between 500 Hz and 4 kHz and detected down to stimulus levels close to the hearing threshold in many cases. According to the properties of the cochlear traveling wave, the latencies of the responses are longer for low than for high stimulus frequencies. The same physiological mechanisms are the reason that the neural activity is less synchronized at low than at high frequencies, and this in turn limits the accuracy of the hearing threshold determination.

The most important applications of the ABRs are the objective determination of hearing threshold in infants and children and the topodiagnostic distinction between sensory and neural hearing disorders. The hearing threshold is extrapolated from the response threshold, which is defined as the lowest stimulus level eliciting a clear response and deduced visually from the series of curves. For the derivation of topodiagnostic statements, latencies and amplitudes are measured and plotted as functions of the stimulus level. Based on the comparison with the normal values, on the calculation of latency differences – especially the cochleo-mesencephalic latency difference (*central conduction time*) $t_5 - t_1$ – and on the evaluation of side differences, the hearing disorder can be classified and identified as conductive disorder, sensory deficit, or neural lesion. In the case of a conductive hearing loss, the latencies of all responses are prolonged and the amplitudes are reduced at all stimulus levels because of the acoustic attenuation. In cochlear hearing impairments, the alterations of the AEP depend on the frequency range affected by the disorder. In many cases, the latencies are prolonged at low stimulus levels and nearly normal at high intensities. This holds true for t_1, t_3 and t_5, the central conduction time $t_5 - t_1$ is not affected. In other cases, the prolongation of latencies is independent of the stimulus level and it may be different for t_1 and t_5, resulting in a shortened central conduction time. Hearing disorders of neural origin caused by place occupying processes or neurologic degenerations are typically labeled by a prolongation of $t_5 - t_1$, a reduction of amplitude, and significant side differences.

BERA has a high sensitivity for the detection of mass lesions in the cerebellopontine angle, in contrast to subjective tests, which are at best able to identify a cochlear involvement. In the diagnostic context, the ABRs are supplemented by the analysis of the middle ear reflex and the functional test of the vestibular system. If one or more of these studies confirm the suspicion of the existence of a retrocochlear disorder, the finding is controlled with the help of imaging techniques (computed tomography and magnetic resonance).

The middle latency auditory evoked potentials (MAEP) are only marginally used in audiological diagnostics. In terms of their properties (e.g. amplitude,

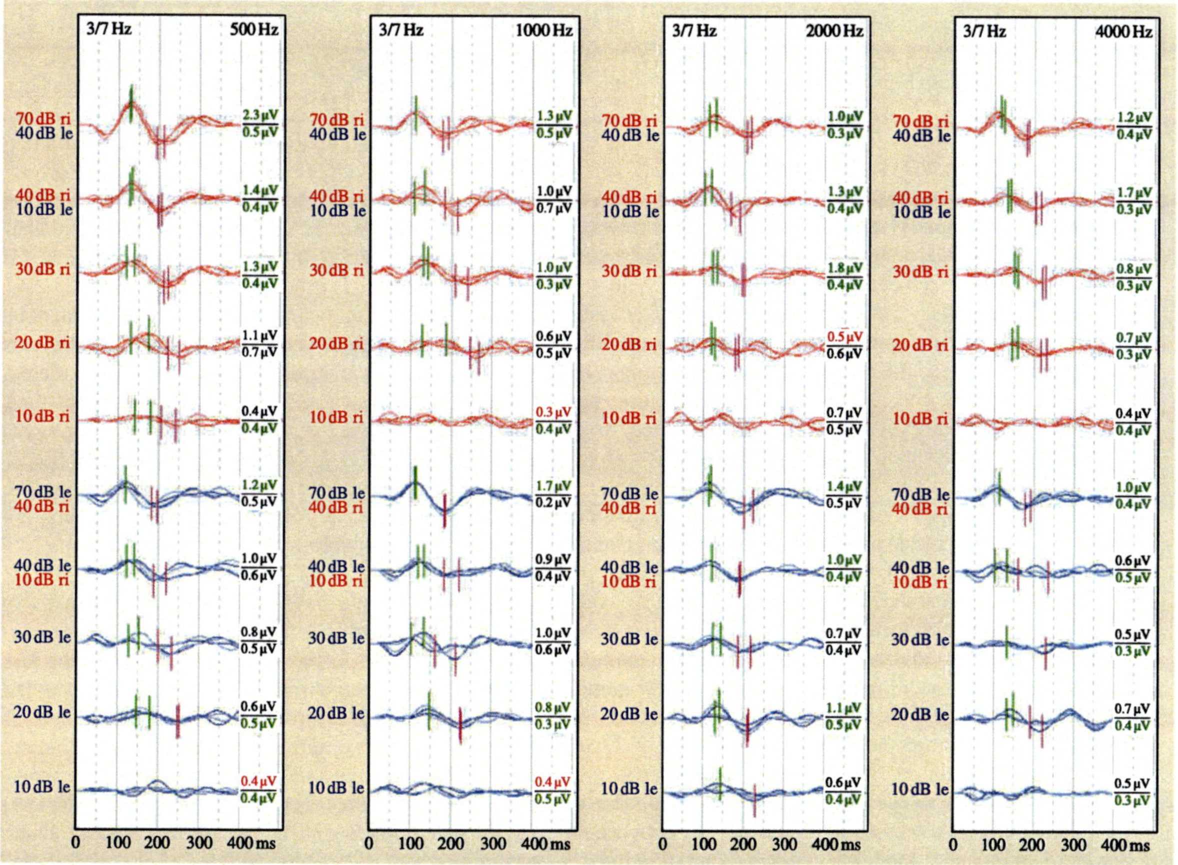

Fig. 12.25 SAEP recorded with CERA in a 51 year-old normal hearing subject at four different frequencies and various stimulus levels in both ears. Next to each curve, the level of stimulus and contralateral masking are given on the *left-hand side* and the response to noise ratio (RMS amplitude of the overall average divided by the RMS amplitude of the residual noise) on the *right-hand side*

impact of maturation, vigilance, and alertness) and their diagnostic significance (e.g. frequency specificity), they are settled between the early and the late AEP components. MAEPs possess neither the high stability typical for ABRs, nor can they compete with SAEPs in the reliability of frequency specific determination of the hearing threshold.

The late or cortical potentials measured with CERA essentially comprise the maxima and minima of voltage located at about 100, 200, and 300 ms, denoted by N1, P2, and N2 (Fig. 12.25). The amplitude of the responses increases with increasing stimulus level, their latency depends on stimulus intensity only very close to the threshold. Latency and amplitude change considerably if the subject falls asleep. Reliable results can, therefore, be expected only in awake and attentive patients. This partially explains why the application of CERA is not useful in infants and young children. Another reason is the uncompleted maturation of the cortical auditory pathways.

Auditory Steady State Responses (ASSR)

While the measurement of transient auditory evoked potentials discussed so far requires a silent break between the acoustic stimuli, the auditory steady state responses (ASSR) examine the auditory system under per-stimulatory stationary conditions. The acoustic stimulus is present without interruption while the EEG signal is recorded and analyzed. Signal processing comprises averaging and statistical analysis in the time and frequency domain with the aim to isolate a feature of the EEG signal which is closely related to the physiological stimulus processing. The adequate stimuli are continuous and stationary but not invariant in time because the neuronal activity evoked by acoustic signals without any temporal structure is not detectable in the compound EEG activity.

Among several variants of the ASSRs, the responses to amplitude modulated stimuli (amplitude modulation following response, AMFR) have recently gained particular practical importance. Acoustic stimulus and signal detection are shown schematically in Fig. 12.26. The amplitude of a continuous tone whose carrier frequency f_C lies in the range from 250 to 8000 Hz is modulated with a modulation frequency f_M between 40 and 120 Hz and a modulation depth from 80 to 100%. The amplitude modulation may be combined with a frequency modulation with the same modulation frequency but a much smaller modulation depth of typically 10%. The modulated stimulus leads to a synchronized excitation of groups of neurons in the auditory pathway, most probably located between midbrain and thalamus. Therefore, the modulation frequency is present in the EEG signal, which is recorded either laterally (e.g. vertex versus mastoid) or – especially in the simultaneous stimulation of both ears – medially (vertex to occiput). For the detection of this frequency, the electrode signal is transformed into the frequency domain after amplification and narrow band filtering and its amplitude and phase is analyzed in the vicinity of the modulation frequency. The neural response is considered to be present if the amplitude of the frequency f_M protrudes significantly from the background and/or if its phase statistics deviates significantly from random distribution. The impact of alertness, vigilance, and age on the response amplitude is stronger for high than for low modulation frequencies. It is the aim and claim of the method to allow a frequency-specific, objective and automated determination of the hearing threshold in children. For this purpose, modulation frequencies around 90 Hz have turned out to be most suitable. Since the identification of AMFRs is based on the detection of the modulation frequency of the stimulus in the EEG spectrum, several stimuli with different carrier frequencies and different modulation frequencies may be presented simultaneously, and even on both ears at the same time, as far as mutual masking is excluded [12.44].

To date, the diagnostic potential of ASSRs cannot be appraised definitely as yet. Currently, effective stimulus paradigms and statistical methods for signal detection are being tested and compared in terms of their performance. For practical applications, some systems that detect the EEG activity evoked by the amplitude-modulated stimulus and optionally seek the response threshold automatically using adaptive algorithms are commercially available.

Objective Audiometry for Screening Purposes

If audiometric testing is carried out for screening purposes, e.g. for early detection of congenital hearing loss in newborns, objective methods are preferably applied because of their independence from the individuals ability and willingness to actively cooperate. In this context, the methods are not operated at their maximum performance level (described in the preceding chapters) but in a simplified and automated procedure, which ensures high stability and sensitivity. As was shown earlier, ABRs are detectable approximately 10 dB above hearing threshold. If a hearing loss of 30 dB is considered relevant for therapeutic intervention, a stimulus level of 40 dB is adequate for detecting the relevant cases. Application of high-intensity stimuli would improve the detectability of the response, but it would also include the risk that babies with hearing loss exceeding the therapeutic limit are considered as normal (missing hits).

In the case of OAEs, the responses elicited by stimuli of high suprathreshold intensity remain detectable as long as the functional structures generating them are present and functioning. Although the response detectability decreases with lower stimulus intensity, it is still not clear whether the loss of functional elements elevates the response threshold or just reduces the response magnitude. It has been shown [12.45, 46] that TEOAEs disappear if the relevant hearing loss (which remains to be defined) exceeds the limit of 30 dB, no matter how intense the stimulus. Similarly, the loss of DPOAEs indicates a relevant hearing loss of at least 50 dB. For TEOAEs, the *relevant hearing loss* is the best value of threshold in the frequency range from 1 to 4 kHz, for DPOAEs it is the threshold at frequency f_2 (i.e. the higher among the two primary tone frequencies). In contrast to ABRs, the application of high stimulus levels does not imply the risk of missing hits since neither a low nor a high stimulus can evoke a response of hair cells that are out of order, but it improves the signal detection. For field conditions, a click level between 65 and 80 dB SPL (equals 35 to 50 dB HL) has proven suitable for the TEOAEs. DPOAEs can be recorded at $L_1 = L_2 = 65$ dB SPL with sufficient probability, but their application is less favorable for newborn hearing screening since their sensitivity limit of 50 dB hearing loss is too high if all cases of therapeutic relevance are to be identified.

If ABRs are detected, all components of the hearing system from the most peripheral site up to the auditory nuclei in the brainstem must be functioning properly. Hearing disorders of conductive, sensory, or neural nature affect the response. The elevation of the hearing threshold corresponds exactly to the elevation of the response threshold. In the case of OAEs, the presence of a response depends only on the functional state of sound conduction through the middle ear and its amplification by the outer hair cells. The relation between hearing loss and presence of a response may be somewhat obscured by the influence of conductive losses, which impede the transmission not only of the stimulus but also of the response. Because of this double impact, responses of dichotomic nature usually disappear already when the damping reaches 20 dB.

Hearing assessment within screening programs should work without intervention of qualified per-

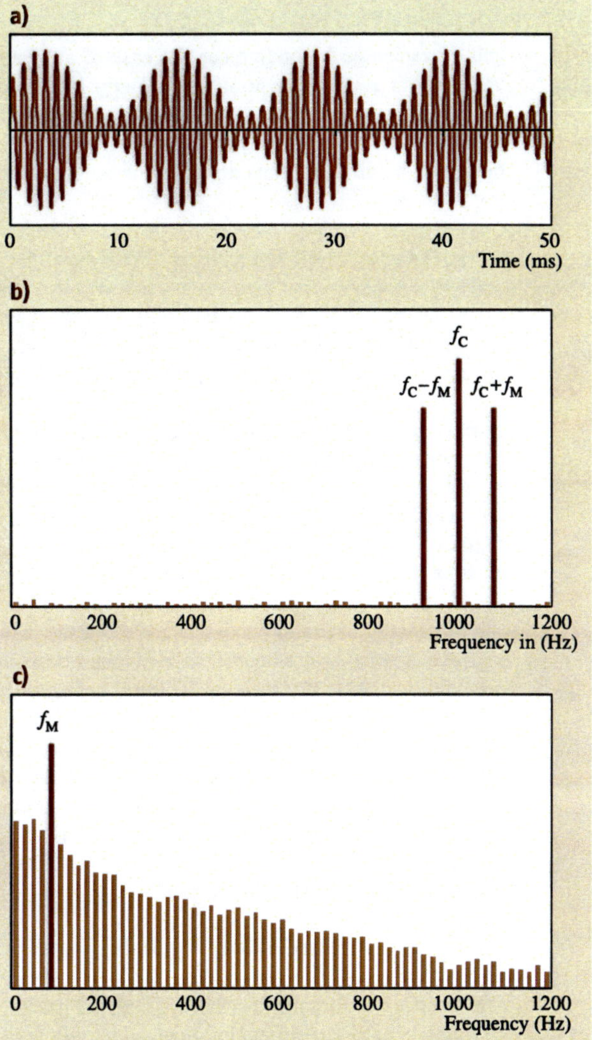

Fig. 12.26a–c Schematic illustration of stimulus properties and signal detection in the registration of AMFR. The spectrum of the amplitude-modulated stimulus tone **(a)** with carrier frequency f_C consists of three lines at the frequencies $f_C - f_M$, f_C, and $f_C + f_M$ **(b)**. The modulation frequency f_M is found in the EEG spectrum **(c)**

sonnel. This requires an automated signal detection yielding a dichotomic result (PASS or REFER). Different algorithms can be used to decide whether a response is present or not, the most simple among them being based on reproducibility or signal to noise ratio. In some commercial systems the classification is based on the averaging of signal polarity and its evaluation by statistical considerations. For each sample of the microphone or electrode signal, a digital counter is incremented by one unit in the case of a positive sample and decremented by one unit in the case of a negative value. After recording enough signal epochs, the decision whether the dominance of positive or negative values is statistically significant is made according to binomial distribution. This algorithm is very powerful in distinguishing between stochastic and nonstochastic signals and it is very stable against perturbations by artifacts. In the case of OAEs, a response is considered as present if a given probability level is exceeded by at least eight samples. In the case of ABRs, a similar procedure is applied to the averaged response after convolution with a template typical for newborn ABR.

12.4 Technical Hearing Devices

Many hearing disorders can be compensated at least partially with technical devices. These devices may be conventional amplifying hearing aids, implantable hearing systems, or cochlear implants. Since neither sound amplification nor implanted prostheses are able to restitute normal hearing, technical compensation is only considered if conservative and surgical treatment does not lead to satisfactory results. In any case, the benefit to be expected and the risk of unwanted side effects must be traded against each other. The amplification of sound through hearing aids is associated with a stronger and potentially hazardous sound exposure of the pathologic ear. Implantations bear the risk of anaesthesia and eventually loss of the residual hearing, especially in the case of cochlear implants. This makes clear that ENT physicians must be involved in the administration of hearing devices and the control of success and rehabilitation. Other disciplines engaged in the process of rehabilitation are audiologists and acousticians, furthermore logopedists and pedagogues, and experts for pediatric audiology where appropriate.

12.4.1 Conventional Hearing Aids

The decision to administer hearing aids is based on a complete audiometric exploration, which allows the identification of type and site of lesion and the amount of hearing loss. Sound amplification is the only therapeutic option in cases of irreversible cochlear damage, whereas conductive and neural impairments do not belong to the classical target group. The audiological criteria for the prescription of hearing aids are a hearing loss of at least 30 dB at frequencies between 500 Hz and 3000 Hz and a speech discrimination (monosyllables) below 80% at 65 dB SPL. If both ears are affected, the supply with hearing aids should be bilateral because less amplification is required, acoustic sources can be localized more easily, and speech discrimination in noise and reverberation is improved. Moreover, the auditory deprivation in the impaired ear can promote the emergence of tinnitus.

Hearing aids are available in different mountings, which are worn behind the ear (BTE), in the ear (ITE), or even completely in the canal (CIC). The latter helps to take benefit of the natural gain and directional filtering of the outer ear. This reduces the gain requirement and it is advantageous for directional hearing and speech intelligibility in noise. In addition, the sound signal is less distorted, as no tube is necessary to feed the sound to the external meatus. In contrast, most BTE hearing aids are combined with a tube and an ear mold in order to prevent acoustic feedback. If the hearing loss is only moderate, BTE devices can be equipped with external transducers placed in the ear canal or with microtubes, which do not require an ear mold.

The conventional sound-amplifying hearing aid is an electronic device that provides the acoustic stimulus with higher intensity. Its main technical components are a sound sensor (microphone), an amplifier with control elements, an energy source (battery) for operation of the amplifier, and an output transducer (loudspeaker). Additional components may be available, as acoustic filters, an induction coil (telecoil), an external input, and a remote control.

The basic components of a hearing aid and their operation are shown in Fig. 12.27. The sound signal is converted in a time-dependent voltage by the microphone. If the acoustic information is available additionally as an electrical signal (e.g. from telephone or TV), two conversions may be bypassed if the electrical signal is fed directly into the hearing aid either through an induction coil or an input socket. The electrical signal is amplified by a factor that may depend on frequency and intensity, optionally filtered and limited, and supplied to the output transducer (loudspeaker). The output signal reaches the eardrum either directly or through a tube. If necessary, means for the elimination of acoustic feedback can be provided.

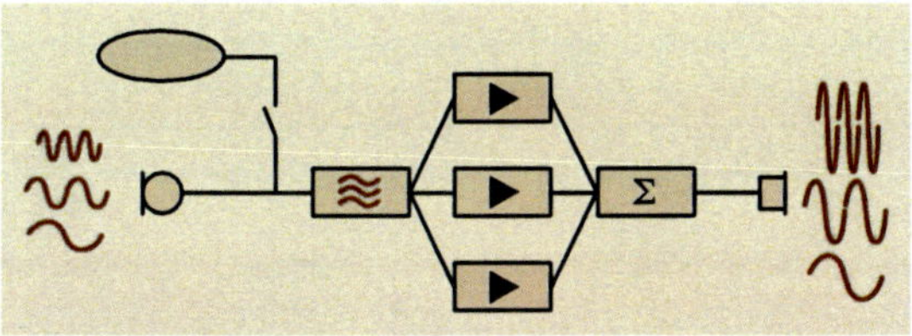

Fig. 12.27 Simplified drawing of the essential components of a hearing aid

The high requirements of a microphone suitable for hearing aids in terms of dimension, efficiency, transmission characteristics, sensitivity, energy requirements, and independence from temperature are met by an electret microphone powered by a field-effect transistor as impedance converter. Its working principle is that of an electrostatic converter whose capacitor charge is not maintained by an external power source but by a prepolarized film. In most hearing aids, omni-directional microphones (pressure transducers) are used. A directivity is given by the head and the position of the microphone entrance behind the ear, the outer ear, and the ear canal. Directional microphones with two openings (pressure gradient transducers) or arrays of multiple microphones are used optionally to improve acoustic focus and noise reduction.

The amplification of the electrical microphone signal is described by the input/output characteristic of the amplifier, i. e. by the functional relationship between input level L_i and output level L_o. In general, the gain is dependent on frequency and level of the input signal. If this is not the case, the amplifier characteristics are straight lines with slope $\Delta L_o / \Delta L_i = 1\,\text{dB/dB}$ parallel to the angle bisector for all frequencies in the double logarithmic representation (Fig. 12.28a). The gain factors corresponding to different curves are then moved parallel to each other. From the i/o-characteristics of a linear amplifier, the gain in dB can be read as the difference $\Delta L_o - \Delta L_i$ at a single point.

The output of each amplifier is limited by the operating voltage. In hearing aids, a further limit is given by uncontrollable acoustic feedback at high output levels. Feedback can be prevented by a circuit composed of diodes, which cuts the voltage spikes above a predetermined limit (peak clipping, PC). In the input/output function, the PC is characterized by a horizontal branch at high input levels (Fig. 12.28b). Unlike other techniques used in hearing aid amplifiers, the PC takes effect without any delay. Additionally to the upper limit of the output level, a lower limit is given by the noise of amplifier and converters. This noise is partially amplified, so that it is equivalent to a minimum input sound level. Signals whose level is below that limit cannot be transmitted by the hearing aid because they are at least partially covered by noise.

The limitation of the output signal protects the hearing aid user against noise intensities that exceed his discomfort level but it cannot compensate the loss of natural dynamic compression occurring in many sensorineural hearing impairments. A better approach to compensate pathological loudness growth is the reduction of the gain factor continuously at levels above approximately 50 dB. Manual operation of the volume control by the hearing aid user is neither precise nor fast enough to react efficiently to changes of sound level. Therefore, many hearing aids are equipped with an automatic gain control (AGC). It causes a reduction of gain with increasing noise level and thus a compression of the acoustic dynamic range. Unlike the limiting circuits, AGC does not react simultaneously but with a certain delay.

The attack and release time of the gain control must be greater than the period of the lowest frequency to be amplified, because otherwise the amplitude within a single sound wave would be regulated and the signal would be distorted. The lower limit is thus at about 10 ms. In the steady state of the amplifier, the sine waves of all frequencies must be transmitted as sine waves. Transient effects and decay processes do not occur in the case of slow changes of sound level (stationary signal), but they arise at rapid changes, such as those occurring in speech signals (dynamic behavior). They can cause short-term noise peaks that exceed the maximum permissible output level. To prevent this, the AGC is always combined with a PC.

Systems with very long time constants (>200 ms) react only to very slow changes of sound level and are denoted by automatic volume control (AVC). Response times below 20 ms are suitable to compress individual syllables. This improves speech intelligibility, since the forward masking of speech elements with low level (e.g. fricatives) preceded immediately by elements with high level (e.g. vowels) is prevented (syllabic compressor).

The effect of gain control on the static i/o-function of the hearing aid is shown in Fig. 12.28c,d. Depending on whether the gain is controlled by the input or the output signal, input-controlled (AGC/i) and output-controlled (AGC/o) systems are distinguished. In both cases, the dynamic characteristic deviates from linearity above the control threshold (starting point or knee point) down on the gain line. Thus, the amplifier oper-

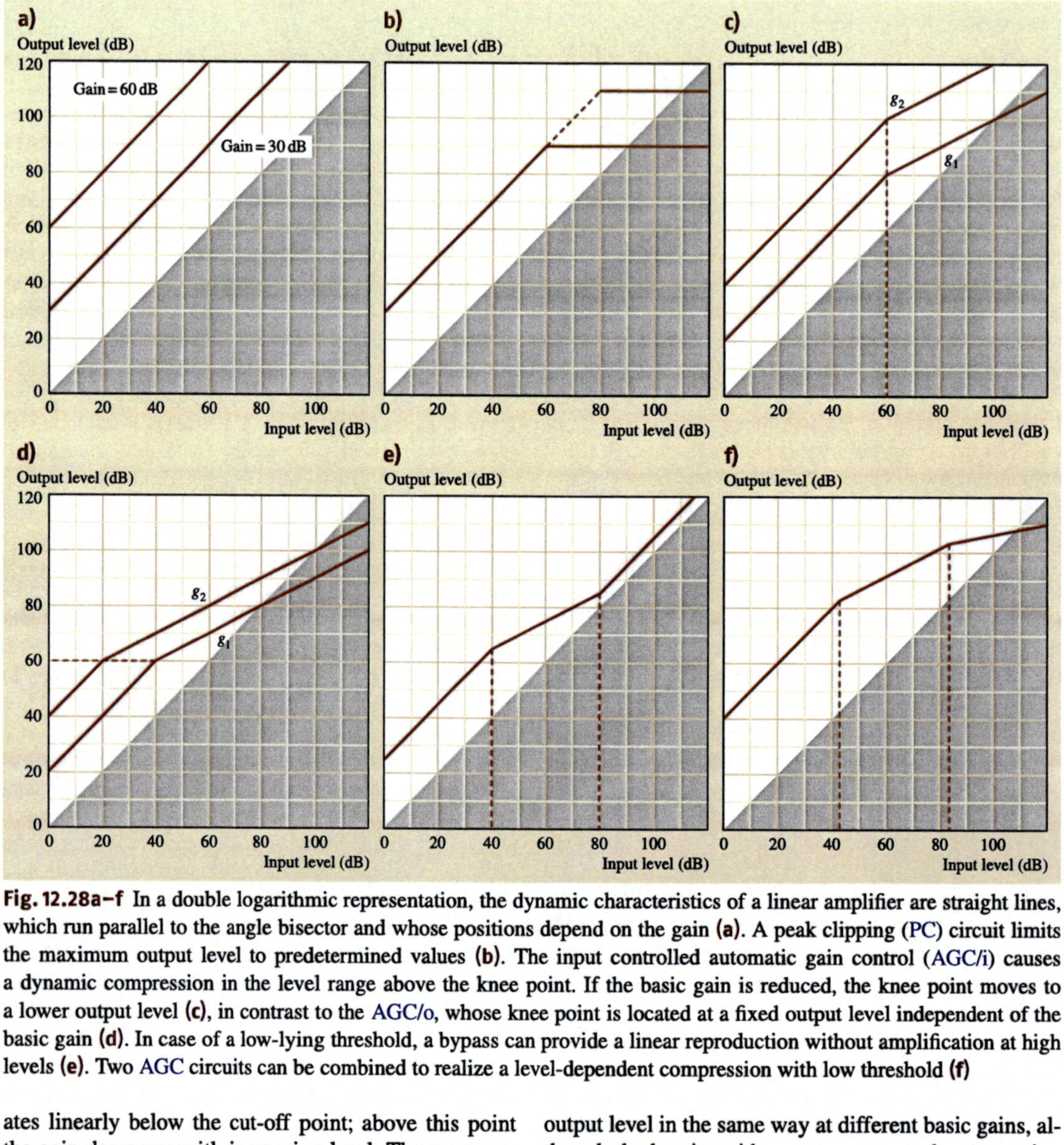

Fig. 12.28a–f In a double logarithmic representation, the dynamic characteristics of a linear amplifier are straight lines, which run parallel to the angle bisector and whose positions depend on the gain **(a)**. A peak clipping (PC) circuit limits the maximum output level to predetermined values **(b)**. The input controlled automatic gain control (AGC/i) causes a dynamic compression in the level range above the knee point. If the basic gain is reduced, the knee point moves to a lower output level **(c)**, in contrast to the AGC/o, whose knee point is located at a fixed output level independent of the basic gain **(d)**. In case of a low-lying threshold, a bypass can provide a linear reproduction without amplification at high levels **(e)**. Two AGC circuits can be combined to realize a level-dependent compression with low threshold **(f)**

ates linearly below the cut-off point; above this point the gain decreases with increasing level. The compression $c = \Delta L_i / \Delta L_o$ is defined as the slope of the static amplifier characteristics.

It is the purpose of the gain control to prevent the occurrence of high noise levels at the ear of the hearing aid user. From this point of view, only gain control according to the output signal (AGC/o) is useful. Since this is associated with significantly longer recovery times than an input-controlled AGC, the AGC/i is often advantageous. As shown in Fig. 12.28, the AGC/i controls the output level in the same way at different basic gains, although the hearing aid user may not need compression at lower levels. Thus, both AGC circuits have advantages and disadvantages. Therefore, it depends on the individual case whether the independence of the basic gain or the shortest possible response times have higher priority for the hearing-impaired patient.

Specific requirements for the dynamic behavior of the hearing aid can be fulfilled with special analog circuits, or – much more flexible – with digital signal processing. For example, to achieve a low-frequency compression, a two-channel AGC/i can be used to attenuate the predominantly low-frequency noise, including the voice of the hearing aid user. As with the conventional AGC/i, the gain increases with increasing input sound pressure level, as well as the emphasis of high frequencies. At high input levels (above 80 dB) only a slight amplification in the range of 3 kHz is effective, intended to compensate for the lost ear canal resonance. The starting point of gain control is 40 dB, the compression ratio is 2 : 1 (Fig. 12.28e). The high frequency range is amplified linearly in such two-channel devices. The improvement of speech intelligibility is based on the fact that upward masking of the high frequency signal components is avoided by the onset of compression at low levels in the low-frequency range.

An even more effective use of the residual dynamic range can be achieved with hearing aids whose compression ratio is adjustable in the entire dynamic range (wide or full dynamic range compression, WDRC or FDRC). Below the control threshold at $L_i = 45$ dB the amplifier works linearly, at higher levels up to 85 dB the compression ratio can be set individually between 1 and 3. At higher levels, a conventional AGC is effective with a fixed compression ratio (Fig. 12.28f).

In summary, the main feature of PC and AGC is the reduction of gain with increasing sound intensity. This corresponds to a nonlinear system and it is inevitably associated with signal distortions. The result of these distortions are additional tones that have a negative effect on sound quality and speech understanding.

The amplifier characteristics described so far are related primarily to the level dependence of amplifier gain. In general, however, the hearing loss and hence the need for correction, also depend on frequency. This can be effectively taken into account by a digital multichannel signal processing. For devices with dynamic compression, this is also advantageous if the gain does not depend on the frequency, because a narrow band noise of sufficient intensity would otherwise reduce the gain in the entire frequency range. In multichannel devices, the input signal is split into several components by a filter bank. The basic amplification starting point of the AGC and possibly also the compression ratio can be programmed and stored separately for each channel. The output signal is reconstructed by summation of the components processed individually.

Multichannel hearing aids were previously implemented as digitally controlled analog devices in which storage for different programs was available. In fully digital hearing aids the preamplified microphone signal is digitized and processed in a signal processor working in real-time. The systems with multichannel nonlinear amplification with a flexible setting of parameters can include optional noise reduction systems, passive or active feedback cancelation, and they may automatically adjust the listening program to the sound environment using adaptive neural networks to identify acoustic objects. The digitally processed signal is D/A converted and fed to the final amplifier.

A special feature of signal processing in some digital hearing systems is the frequency transformation. In order to make the high frequency range accessible to hearing impaired persons, either a *transposition* algorithm is applied, which shifts signals of high frequencies to lower frequency regions, or the bandwidth is *compressed* above a defined cut-off frequency (knee point), leaving the low frequency range unchanged. For mild or moderate hearing loss, a high cut-off frequency and a low compression ratio is selected, whereas severe hearing losses require a lower knee point and more compression. Clinical and practical experience shows that the recognition of high frequency consonants improves significantly by nonlinear frequency compression [12.47].

After processing, the electrical signal must be converted to an acoustic output and supplied to the ear. The electro-acoustic transducer is supposed to meet the requirements of small size, high efficiency, constant transmission factor over a wide frequency range, and a linear dynamic behavior. These requirements are partially contradictory (such as the quality of reproduction of low frequencies and small dimensions) and are, therefore, not always fulfilled at the same time. Hearing aid speakers are based almost exclusively on the electromagnetic principle, i. e. the change of magnetic induction by moving a metal diaphragm (anchor) in the field of the coil of a permanent magnet.

Practically all types of hearing aids are faced with the problem of acoustic feedback produced by the re-entry of the amplified output into the microphone entrance. If the gain factor for this loop amplification is greater than 1, the speaker produces a whistle tone whose intensity is only limited by the clipping of the amplifier. To avoid this, either the gain must be reduced or the re-entry of the output prevented. The former is rarely possible or advisable, as the required gain is determined by the degree of hearing loss and not by the feedback risk. For the latter, several solutions are available: the use of directional microphones, acoustic isolation of the output from the input using contralateral routing of signals (CROS), or acoustic isolation through an ear mold, which seals the ear canal partially or completely and thus interrupts the acoustic connection between output and input. In many digital hearing instruments, acoustic feedback is prevented by filtering or active sound compensation, thus making an ear mold unnecessary (open fit).

For the production of a custom-specific ear mold, an ear canal impression is taken and used for the production of the actual earpiece either from transparent hard plastic or soft silicon. In ITE devices, earpiece and equipment housing are identical. For BTE, the connection between speaker and ear mold is made by a tube attached to the support hook of the hearing aid. The sound is supplied to the residual ear canal volume through a hole in the ear mold.

Hole and hose lead to distortions of the sound signal. This mainly affects the high frequencies while low frequencies are transmitted almost without attenuation. The acoustic properties of the ear piece can be influenced by additional drilling with different diameters and shapes. Drilling with about 2.5 mm in diameter suspends the low frequency amplification, so that in the case of a pure high frequency hearing loss the low frequencies can still be processed by the natural ear.

The possibilities for influencing the acoustic properties of the coupling by additional drilling is even more limited, the larger the required sound amplification. Severe hearing disorders require an acoustically tight closure of the residual ear canal volume in most cases and cannot be supplied with external transducer technology, which has recently been introduced. If, however, the hearing loss is lower than 80 dB, the output transducer can be placed in the ear canal and held with a mushroom-shaped seal. A major advantage of open dome fitting with external receivers is the lower loss in high frequencies, which results in a better sound quality.

The description of the properties of individual components does not allow reliable conclusions about the acoustic behavior of the whole hearing system. To assess the overall response of the hearing aid, acoustic measurements in the coupler – a cavity that emulates the geometry and acoustics of the ear canal – or *in situ* – with a probe microphone in the ear canal – must be carried out. The result of such measurements are curves and diagrams containing the reproduction characteristics of the hearing at different frequencies and levels. Among these curves, the *reference frequency response* is the most important. It is obtained by recording the output level L_o while feeding the hearing aid with pure tones of variable frequency and fixed input level $L_i = 60\,dB$ SPL. The parameters of the hearing aid are set so as to achieve an output level lying 15 dB below the output level achieved with an input level of 90 dB SPL and maximum gain at the reference test frequency (usually 1600 Hz, but 2500 Hz in special high

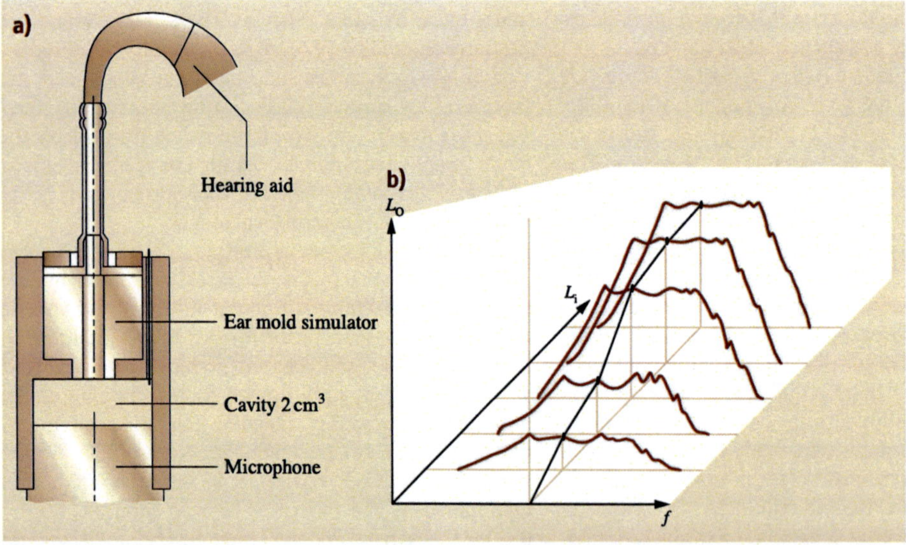

Fig. 12.29a,b Measurement of the acoustic frequency responses of a hearing aid with the 2 cm^3 coupler is shown in **(a)**. Each curve $L_o(f)$ is measured at a fixed input level L_i. The setting of the hearing aid parameters is the same for all curves. The dynamic characteristics $L_o(L_i)$ are formed out of the three-dimensional representation as cross-sectional areas of constant frequency

frequency devices). The test gain characteristics $L_o(f)$ corresponding to other input levels (50 to 100 dB SPL) are measured without changing this device setting. The complete array of frequency responses provides an overview of the frequency-dependent and level-dependent response of the hearing aid (Fig. 12.29).

In addition to the hearing aid performance curves recorded with constant parameter settings, the data sheet of a hearing aid often contains the frequency response at the largest achievable sound level. By default it is recorded at the highest amplifier position and at a constant input level of 90 dB SPL over all frequencies (OSPL90). Another common and standardized graph is the curve of maximum gain representing the gain measured as a function of frequency with an input level of 50 dB SPL in the range of linear operation of the hearing aid set to maximum amplification.

The complete graphical representation of the performance of a hearing aid is a multidimensional problem, because the output level L_o depends on the input level L_i, the sound frequency f, and the hearing aid settings (e.g. amplification gain $g = L_o - L_i$ and compression ratio $c = \Delta L_i / \Delta L_o$). From the complete array of frequency responses $L_o(f)$ shown in Fig. 12.29, the dynamic functions $L_o(L_i)$ shown in Fig. 12.28 can be constructed, each of which is valid for a fixed frequency and a particular hearing aid setting.

The array of frequency responses yields direct information on the range of input levels in which the hearing aid behaves linearly. An ideal linear signal processing by a straight frequency response $L_o(f)$, which shifts vertically when the input level L_i, is changed by ΔL_i. This is equivalent to the requirement that the dynamic functions $L_o(L_i)$ have a constant slope of 1 dB/dB at all frequencies and over the whole level range. Although (or perhaps *because*) no real hearing system fulfills this requirement, it is common to denote a hearing aid as linear if its frequency responses $L_o(f)$ measured at different input levels differ by not more than a mere vertical shift (even if this shift is different from the corresponding input level difference). In this nomenclature, a nonlinear hearing aid is characterized by frequency responses which do not run parallel to each other.

The best possible knowledge of the hearing aid properties is only one of three pillars of a rational hearing aid provision. The second column is the entity of audiometric data. As the third pillar, the individual adaptation of the hearing aid connects acoustical and audiological data, producing a particular parameter set for the perfectly fitted hearing aid. The task of this unit is to map the frequency and intensity range of normal hearing (especially the language area) to the restricted field of hearing of the ear to be supplied (Fig. 12.30).

The essential characteristics of the hearing aid are the effective acoustic gain, the compression ratio, and the maximum output level. As can be seen in Fig. 12.30, these variables can depend on frequency and level. Apart from these parameters, the temporal behavior of the hearing aid is of great importance for speech intelligibility. It plays an important role, especially if the gain

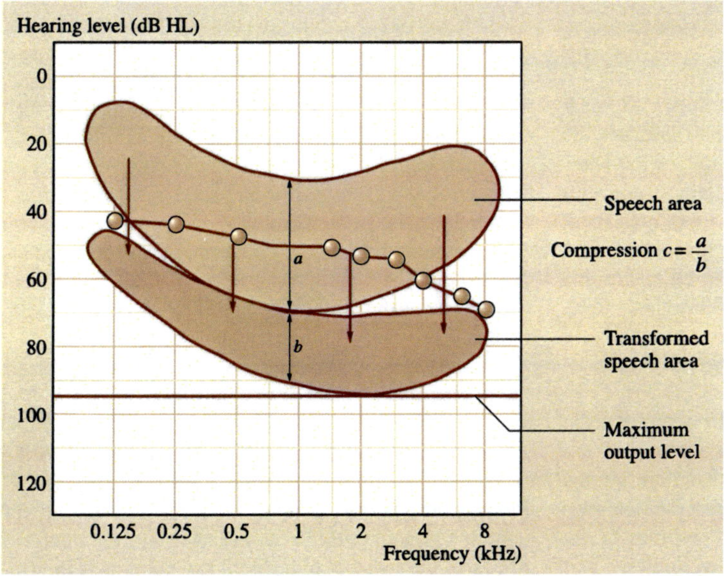

Fig. 12.30 Projection of the language area in the residual field of hearing of the hearing impaired using a hearing aid. The *interconnected circles* indicate the threshold of hearing

is controlled automatically. The audiometric data consisting of hearing threshold, discomfort level, loudness scaling, and speech audiogram determine the selection and first setting of the hearing aid.

The necessary amount of amplification and its frequency dependence can be estimated from the hearing threshold. The obvious idea that the required gain equals the degree of hearing loss (mirroring the audiogram) proves to be useless because the amplified internal noise would become audible. In general, the amplifier gain will be lower than the threshold shift. According to the simplest among the prescriptive methods, the gain g measured in the $2\,\text{cm}^3$ coupler is given by half the hearing loss HL: $g = \text{HL}/2$. A refinement of this rule is Berger's formula, according to which the necessary gain is also calculated solely from the hearing loss, divided by a frequency-dependent number N

$$g(f) = \frac{\text{HV}(f)}{N(f)} + C(f)\,. \qquad (12.11)$$

The denominator $N(f)$ lying between 1.5 and 2.0 depends on the frequency in such a way that the gain is slightly higher at medium frequencies than at low and high frequencies. This accentuation of the mid frequency is further emphasized by the additive correction $C(f)$ regardless of the hearing loss. The formula ((12.11)) proposed for BTE devices must be modified for ITE systems to take into account the lower need of amplification.

According to another formula (prescription of gain and output, POGO), the correction of hearing loss by division does not depend on frequency:

$$g(f) = \frac{\text{HV}(f)}{N} + C(f)\,. \qquad (12.12)$$

Finally, the National Acoustics Laboratories, Sydney (NAL) method is based on a formula that calculates the target gain $g(f)$ at given frequency f out of the hearing loss $\text{HL}(f)$ determined at this frequency and (with less weight) the hearing loss averaged over three fixed frequencies (0.5, 1 and 2 kHz). For a given threshold curve, the results obtained from the formulas according to Berger, POGO, and NAL can differ considerably. This demonstrates that the threshold is not a valid and reliable criterion for calculating individual amplification needs.

In the methods described so far, the target gain is derived from audiometric data and the setting of the hearing aid is adjusted on the basis of these presettings with the help of a coupler measurement (see below). This approach contains two systematic sources of error: the transducer used in pure tone audiometry and the simulation of the ear canal through a standard coupler volume. Neither meet the conditions in the real ear canal. The deviation is especially large in children whose ear is small and growing. It is the goal of the DSL method (desired speech level) after *Seewald* et al. [12.48] to limit the influence of these sources of error. The approach is to describe the hearing threshold by the sound pressure level in the ear canal, which is measured with a probe microphone system or with insert earphones. Using the individual transfer function (real ear to coupler difference, RECD) or standard correction factors, the requirements to achieve the desired speech level can be determined. This is to ensure that the broadband speech signal is audible, sounds well, and is not distorted. The success of fitting depends essentially on regular controls of the ear-coupler transfer function and the corresponding correction of the gain.

The sole focus on hearing threshold is unsuitable for a proper setting of the hearing aid in the entire area above the threshold. The correct setting requires the involvement of at least the discomfort level or preferably the entire area of hearing, which can be assessed by subjective loudness scaling. From the loudness growth function measured at different frequencies and levels, the required gain can be deduced by comparison with the normal curve (Fig. 12.31). Loudness scaling is thus the only method that considers consistently the frequency and level dependence of subjective loudness perception and provides not only the target gain but also the compression needed.

The prescribed gain of the hearing aid can be checked with the ear simulator in the test box. These measurements of output level at defined input signals provide very accurate and reproducible results, but they are not always relevant for the conditions in the target ear canal. In order to determine the real performance of the hearing aid at the eardrum, probe microphone measurements must be performed in the ear canal (in-situ measurement). In contrast to the coupler measurements, they capture the output of the hearing aid in the residual volume under the influence of sound feeding, ear mold, ear canal geometry, and middle ear impedance.

In many in-situ systems not the microphone itself but a probe tube connected with it is placed in the ear canal. The measurement is carried out both without and with the hearing aid (Fig. 12.32a,b). The probe microphone records the sound in the ear canal signal during presentation of noise or sinusoidal stimuli of defined frequency and intensity. A real-time analyzer processes the microphone signal and constructs the fre-

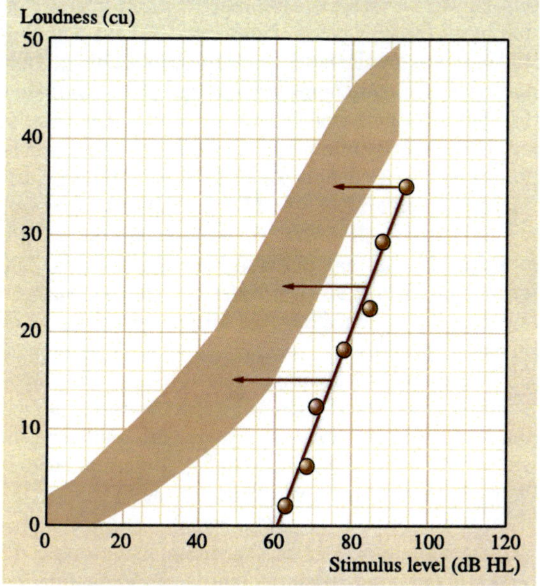

Fig. 12.31 Determination of the level-dependent need of amplification for one frequency (narrow band noise) from the comparison between the individual loudness growth function of a hearing impaired subject (*data points*) with the *shaded* normal range. The length of the *horizontal arrows* shown for the categories *soft* (15 categorical units, cu), *medium* (25 cu) and *loud* (35 cu) is equivalent to the corresponding required gain (after [12.49] with kind permission)

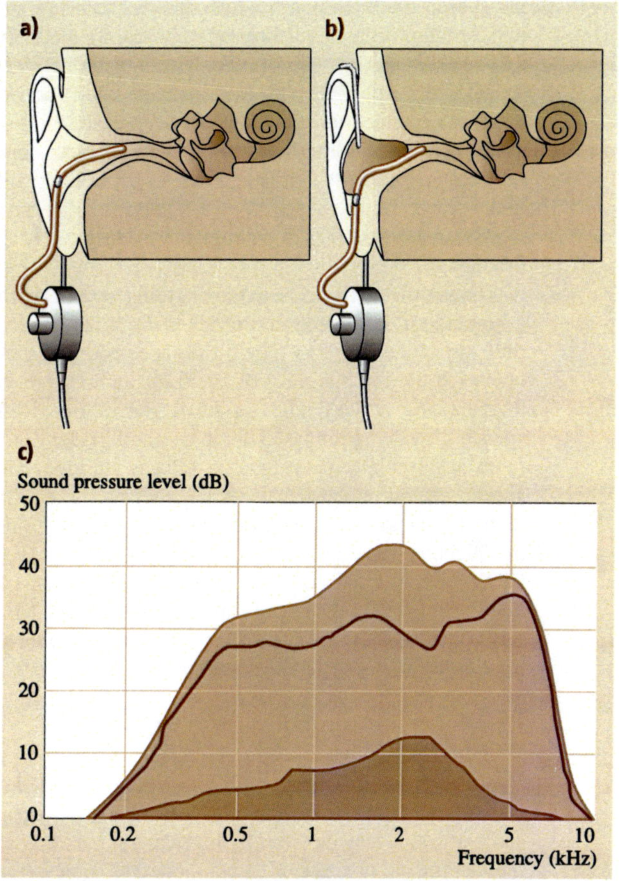

Fig. 12.32a–c In-situ measurement with probe tube in the ear canal, without **(a)** and with **(b)** hearing aid (after [12.40]). The probe measurement **(c)** yields the outer ear transfer function (*dark shaded*) with open ear canal (open ear gain) and the in-situ amplification (*light shaded*) with hearing aid. The difference between the two curves (*middle line*) represents the effective acoustic amplification (insertion gain)

quency transfer functions $L_0(f)$ as well as the dynamic characteristics $L_0(L_i)$.

At the beginning of the in-situ measurement, the external ear transfer function is measured by the probe tube placed in the open ear of the patient. Plotted as a function of frequency, several maxima in the range between 2 and 5 kHz are usually observed (dark area in Fig. 12.32c). If the input level is independent of frequency, this curve reflects the natural amplification of outer ear and ear canal (open ear gain, OEG). As a next step of in-situ fitting, the sound intensity is measured in the ear canal after insertion of the ear piece and activated hearing aid (bright area in Fig. 12.32c). The difference between the frequency responses obtained with and without the hearing aid is the effective acoustic gain (insertion gain, IG).

The in-situ measurement is an advantageous and convenient alternative to the coupler measurement. This holds true particularly in children, whose ear canals are approximated only poorly through a coupler and vary considerably. Regarding the assessment of the ear piece and its transmission properties, the in-situ measurement is even without competition. It allows the detection of the effects of additional drilling installed by the acoustician for influencing the sound quality and is hence an indispensable tool for systematic manipulations of the ear mold.

Despite these advantages, a hearing aid can not be fitted solely on the basis of in-situ measurement, since even the precise measurement of the sound field at the eardrum does not provide information about the subjective sound perception of the patient and especially on his speech intelligibility. To fine tune the hearing aid and for monitoring the outcomes, only scaling methods and speech tests are appropriate. The threshold determined with wobble tones in free soundfield with active hearing aid (aided threshold) is not very useful in this respect because it describes only the steady state conditions but not the dynamic behavior of the hearing aid in the processing of speech signals. Therefore, it is used in assessing the success of hearing aid provision only as orientation.

The profit achieved in understanding speech with a hearing aid is defined and checked in terms of free field speech audiograms. The fitting is considered successful if the intelligibility of Freiburg monosyllables improves by at least 20% at 65 dB or if the aided speech discrimination at 65 dB is as good as the optimum achievable without hearing aid. In addition, the speech discrimination at 80 dB in quiet and at 60 dB in noise should be checked (e.g. with the Göttingen or the Oldenburg sentence test). In the case of binaural rehabilitation, the gain of speech discrimination can be demonstrated with the improvement of speech intelligibility in noise, e.g. by measuring the BILD (*binaural intelligibility level difference*; see above).

12.4.2 Implantable Hearing Systems

The conventional sound amplifying hearing aids described so far are associated with various disadvantages, some of which may be offset by surgical implantation of the hearing aid or some of its components. To describe these disadvantages and to define the applications of implantable hearing aids, acoustic, audiological, medical and cosmetic aspects must be distinguished. The specific shortcomings of conventional hearing aids are the acoustic feedback, the poor transmission of high frequencies, the possible intolerance of ear molds and the stigma of the hearing aid worn as a visible prosthesis.

The common feature of all (fully or partially) implantable hearing systems is the direct mechanical transmission of the processed sound signal by vibrational excitation fed either to the ossicles or directly into the fluid of the cochlea. In comparison with acoustic stimulation, two conversions of the signal can be avoided and the final stimulus is less distorted. This improves the sound quality especially at high frequencies. In a partially implantable hearing system, the external parts comprise microphone, signal processor, and power source. A transcutaneous inductive link serves to transmit the information to the implanted unit, as well as its energy supply. The implanted amplifier module controls a magnetic vibration system attached at the incus (floating mass transducer, FMT), a coupling rod which is implanted in the surgically enlarged middle ear space and connected with the incus (*Otologics* Carina), or a hydromechanical drive that transfers the vibration directly into the inner ear fluids (direct acoustic cochlear stimulation, DACS). The FMT (MED-EL *Vibrant Soundbridge*, Fig. 12.33) consists of a small cylinder, the wall of which contains a coil driving a magnetic mass moving along the cylinder axis. The oscillation of this mass is transferred to the incus whose movement is conducted by the stirrup to the inner ear as in natural hearing.

In the case of fully implantable systems, all components including microphone, signal processor, amplifier, output transducer, and power source are implanted. In one of the current systems (*Otologics* Carina, Heidelberg, Germany), the microphone is located under the skin behind the ear. Its output signal is fed to the processing unit implanted in the mastoid, the output of which drives a piezoelectric transducer. A coupling rod whose end is fixed to the joint of malleus and incus conducts the vibration to the ossicular chain. Energy is delivered from batteries that are implanted in the

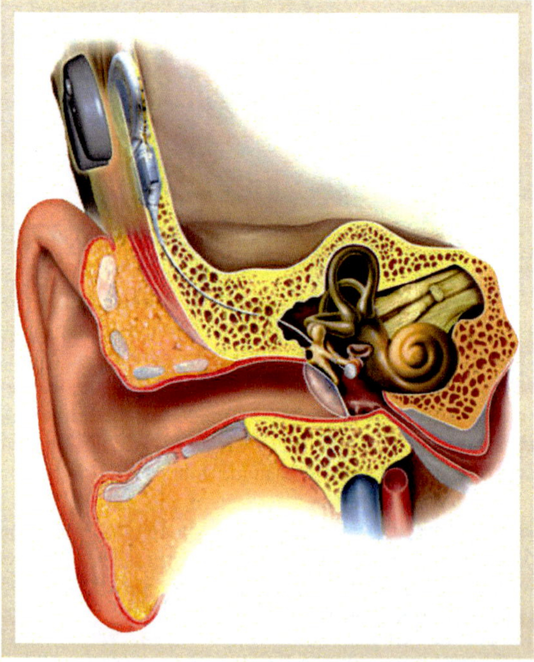

Fig. 12.33 View of a partially implantable hearing system (MED-EL Vibrant Soundbridge, Innsbruck, with kind permission)

main module and charged transcutaneously by an induction coil worn temporarily with a headband. The system is suitable for moderate to profound sensorineural hearing loss and limited by feedback problems of the hydrophone that may arise at high amplifications.

In another fully implantable hearing system (Esteem, ENVOY medical, Cologne, Germany), the sound signal is picked up by the natural eardrum, which acts as microphone diaphragm, and converted by a piezoelectric sensor attached to the hammer (malleus). The electronics module implanted behind the ear amplifies the signal and transmits it to a second piezoelectric transducer fixed to the stirrup (stapes). To avoid acoustic feedback, the junction between anvil (incus) and hammer has to be dissolved – an aspect that is currently discussed controversially among experts.

To assess the value of the implantable hearing aids, the aspects mentioned at the beginning of this section must be considered separately. The compelling advantage common to all of these systems is the superior sound quality due to the direct mechanical stimulation of the inner ear. The aspect of the occluded ear canal has lost some of its original weight since the acoustic feedback control based on digital technology has improved such that today an open dome fitting is possible with conventional hearing aids in many cases. Partially implantable systems are unfavorable from the audiological point of view because the extra-auricular placement of the microphone is disadvantageous at least in comparison to ITE devices. Cosmetic aspects and the stigmatizing effect of the conventional hearing aid should not be overestimated, as only fully implantable systems are really invisible.

The main target group of implantable hearing systems is constituted by patients with moderate to profound inner ear hearing loss. In contrast, the bone-anchored hearing aid BAHA (Cochlear Ltd. Europe) is suitable for cases of conductive hearing loss caused by outer and middle ear problems (e.g. atresia) that cannot be corrected by surgery. It consists of a bone conduction hearing aid, which is attached to a titanium screw implanted in the skull bone behind the ear. The sound is picked up by the integrated microphone and converted to a vibrational excitation of the bone so that outer ear, ear canal, and middle ear are not involved in the transmission chain. The skull vibration stimulates both inner ears with a relevant intra-cranial attenuation around 10 dB only at high frequencies.

12.4.3 Cochlear Implants

In cases of complete deafness, profound hearing impairment, or residual hearing, the function of the auditory system cannot be restored neither with conventional sound-amplifying hearing aids nor with the implantable hearing systems described above. The inner ear does not need to be aided but its function needs to be substituted. This functional substitution and hence the (re)habilitation of deaf patients is possible with the cochlear implant (CI). The CI stimulates the distal endings of the hearing nerve fibers or the spiral ganglion by using intracochlear electrodes (Fig. 12.34). The parameters of the electrical stimuli are calculated in a speech processor and transmitted transcutaneously to the implant as high frequency pulses.

The artificial electrical stimulation of the auditory nerve is fundamentally different from the physiological process, in which the action potentials are evoked by the segregation of neurotransmitters and mediated by elementary generator potentials. If the afferent radial fibers of the auditory nerve are exposed to an electric field, action potentials are generated and synchronized by rectangular pulses applied between two electrodes. The electric field distribution depends on the relative position of the electrodes to the auditory nerve and the intermediate fluids and tissues. Usually, the stimulation is monopolar, i. e. the reference electrode is located a large distance away from the stimulating electrode (e.g. in the temporal muscle).

It is the task of the CI system to transform the temporal structure of frequency and intensity of the acoustic signal in a sequence of action potentials, which is suitable for central processing in the auditory cortex. An exact imitation of the cochlear function fails because of its complexity and the incompleteness of our knowledge of details. With regard to the neural coding of frequency, it is plausible (and justified by the success) to simulate both the tonotopic organization of the cochlea and the ability of the auditory nerve to analyze the periodicity. Therefore, the stimulating electrodes are located in the inner ear where they are lined up along the cochlea and permit a selective stimulation of the region that is specific for the actual frequency. In most speech coding strategies, also the time structure of the signal is coded at least partially in the pulse sequence.

In all prostheses based on pulsatile stimulation, the stimuli are biphasic, charge-balanced current pulses that are submitted by the programmable current sources to each of the electrodes. In the case of bipolar stimula-

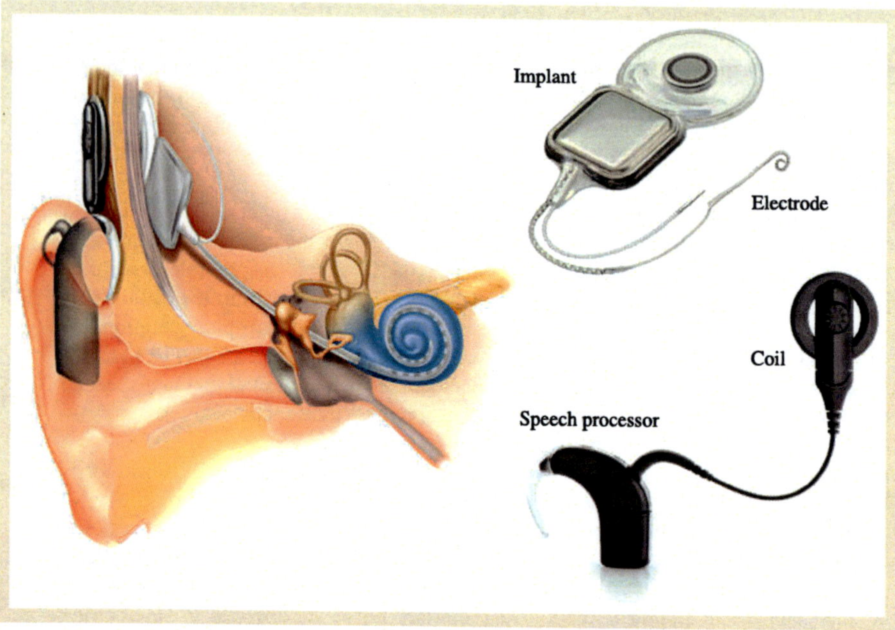

Fig. 12.34 Components of a CI system (Nucleus System N5). The actual implant consists of the encapsulated receiver and decoder circuit and the intracochlear electrode array. The external parts are the speech processor and the transmitter coil, which is magnetically held behind the patient's ear. In the *right section*, the implant CI512 and the speech processor CP810 are shown (with kind permission of Cochlear, Macquarie University, Australia)

tion, the adjacent electrode or more distant electrodes can be chosen as reference. A close bipolar mode results in a very limited local distribution of the electric field, but it will require high current to reach the distant neural structures. The lowest threshold is achieved if all nonactive electrodes are connected to a common reference point (common ground). In monopolar stimulation, the housing of the implant and/or an additional extracochlear electrode is used for as reference for the current path.

In natural hearing, the sound intensity is encoded in the discharge rate of the individual fibers and the number of active fibers. In the case of electrical stimulation, the number of action potentials grows with the current strength and duration of the electrical pulse, and the subjective loudness increases. The stimulation covers all fibers located in a region above the threshold field strength. The number of intracochlear electrodes is limited by the superposition of electric fields to about 15–20 for weak stimuli. By increasing the stimulus intensity, the spatial resolution decreases, as well as the ability to distinguish between pitches. The time resolution is limited by the duration and repetition rate of pulses. Typical values are 25 μs and 1000 pulses per second (pps) and electrode, respectively. Due to the staggered temporal sequence of electrode activation, the total pulse rate increases according to the number of electrodes, typically to 14 400 pps.

A CI system includes several components (Fig. 12.34). The actual implant is an encapsulated integrated circuit, which is located below the scalp behind the ear. Its longitudinal dimension is about 2 cm, its thickness a few millimeters. Additional parts of the implant are a receiver coil with a permanent magnet and a thin tube-shaped extension with 12–22 ring-shaped or spherical platinum electrodes located at its tip. The electrodes are inserted into the lower one and a half turns of the scala tympani. The spatial arrangement of the electrodes is adapted to the natural tonotopic organization of the inner ear. The activation of the front electrode excites apical or medial fibers of the auditory nerve and elicits the sensation of low pitch, the posterior electrodes located at the basal end of the cochlea produce high pitch sensations. In addition to this tonotopic frequency assignment (place pitch), the pitch perception is also affected by changes of the pulse rate (rate discrimination).

The external components of the CI system are the microphone, the speech processor, and the transmitter coil. Microphone, speech processor, and batteries are integrated in a housing that is worn behind the ear. The transmission coil is equipped with a permanent magnet, which fixes it opposite to the subcutaneous receiver coil.

The transformation of the acoustic signal into electrical pulses starts with the analog preprocessing of the microphone signal (amplification, high pass filtering, and possibly compression) and its conversion into a digital signal. The digital processing described below yields the parameters of the electric stimuli, which are encoded in high-frequency pulses (e.g. $f = 5\,\text{MHz}$). The RF pulses are transmitted transcutaneously and decoded by the implant, which is supplied with energy by additional power-up pulses. A control logic ensures that transmission errors are detected and unintended stimulation is prevented. The programmable current sources deliver rectangular biphasic pulses of variable duration (in the range from 20 to 500 μs) and strength (up to a maximum of 1.5 μA) to the electrode selected by a multiplexer circuit. Each electrode is assigned a frequency band whose center frequency and width increase logarithmically from apical to basal end. All current CI systems have the option to measure the voltage between two electrodes and transmit the actual value to the external components by a detuning of the RF resonant circuit (backward telemetry). In this way, the electrical impedances between electrodes and the electrically evoked action potentials of the auditory nerve can be measured (telemetry of electrically evoked compound action potentials, TECAP) [12.50].

Most of the strategies applied for signal coding are based on the pulsatile stimulation of the auditory nerve. An exception is the compressed analog stimulation strategy (CAS) in which the time course of the acoustic signal corresponds directly to the electrode current. Here, the microphone signal is split into several partial signals with a filter bank. Each band-pass filter is associated with an electrode. The individual components are digitized and transformed to the dynamic range of the associated electrode (mapping). This results in a sinusoidal time course of the electrode currents with high frequency at the basal electrodes and low frequency at the apical electrodes. The stimulation amplitude corresponds to the sound intensity in the individual frequency bands and the sound signal is transmitted without any coding loss or delay. Unlike the pulsatile strategies, several electrodes may be addressed simultaneously. However, the resulting overlap of electrical fields can affect the discrimination and lead to unpleasant side sensations.

Historically the first pulsatile coding strategies were based on the principle of feature extraction. Here, the signal was reduced drastically to very few relevant elements – the fundamental frequency F_0 (voice pitch), the first and second formants F_1 and F_2 of vowels, and the total amplitude. The frequency of the formants determined the selection of the stimulating electrode, the fundamental frequency defined the pulse rate, and the overall sound intensity was transformed to current and width of the stimulus pulses for each electrode within the patient-specific limits. Despite the obvious limitations, especially in regard to the voiceless sounds (for which F_0 is not defined), many CI users achieved a good speech discrimination up to open speech intelligibility.

In the so-called *n of m strategy*, which is less language-specific, *n* stimulation electrodes are selected from the total of *m* available channels in accordance with the ranges of largest intensity in the sound spectrum. The preamplifier with automatic gain control is followed by a programmable filter bank with band-pass filters whose center frequencies lie between 250 Hz and 10 kHz. Each band-pass is associated with an electrode in a tonotopical order. The number *n* of active electrodes and the stimulus rate (between 180 and 300 pps) depends on the overall intensity of the signal, resulting in a higher density and frequency of pulses for strong vowels than for soft consonants. The electrodes are activated sequentially from basal to apical, whereby all nonselected electrodes are skipped without pause in order to achieve a high transfer rate of information.

While the coding strategies presented so far are largely based on the principle of tonotopy, the CIS strategy (continuous interleaved sampling [12.51]) emphasizes the time structure of the sound signal and thus utilizes the ability of the auditory system to analyze periodicities (Fig. 12.35). After amplification and high pass filtering, the microphone signal is split in a device-specific number of frequency bands with a gammatone filter bank. In each channel, the envelope of the time signal is calculated by rectification and low pass filtering. From the samples of the envelope, the amplitude of the stimulus pulse is obtained by a nonlinear transformation. Due to the fast continuous staggered sampling and the sequential activation of the electrodes, the time and frequency dependence of the sound intensity is reflected in the chronology and spatial distribution of the pulses. Using a typical pulse width of 80 μs, stimulation rates of about 1 kHz per channel can be realized. Except for the very coarse grid of the frequency axis, this coding strategy is very similar to the natural model of peripheral signal processing in cochlea and auditory nerve.

Additional temporal information can be transmitted by coding the fine structure of the signal. For this purpose, the zero crossings of the signal within one frequency band are determined from the half-wave rectified signal component. The presentation of the stimulus pulses in some of the lowest frequency channels is then synchronized to the zero crossings (fine structure processing, FSP).

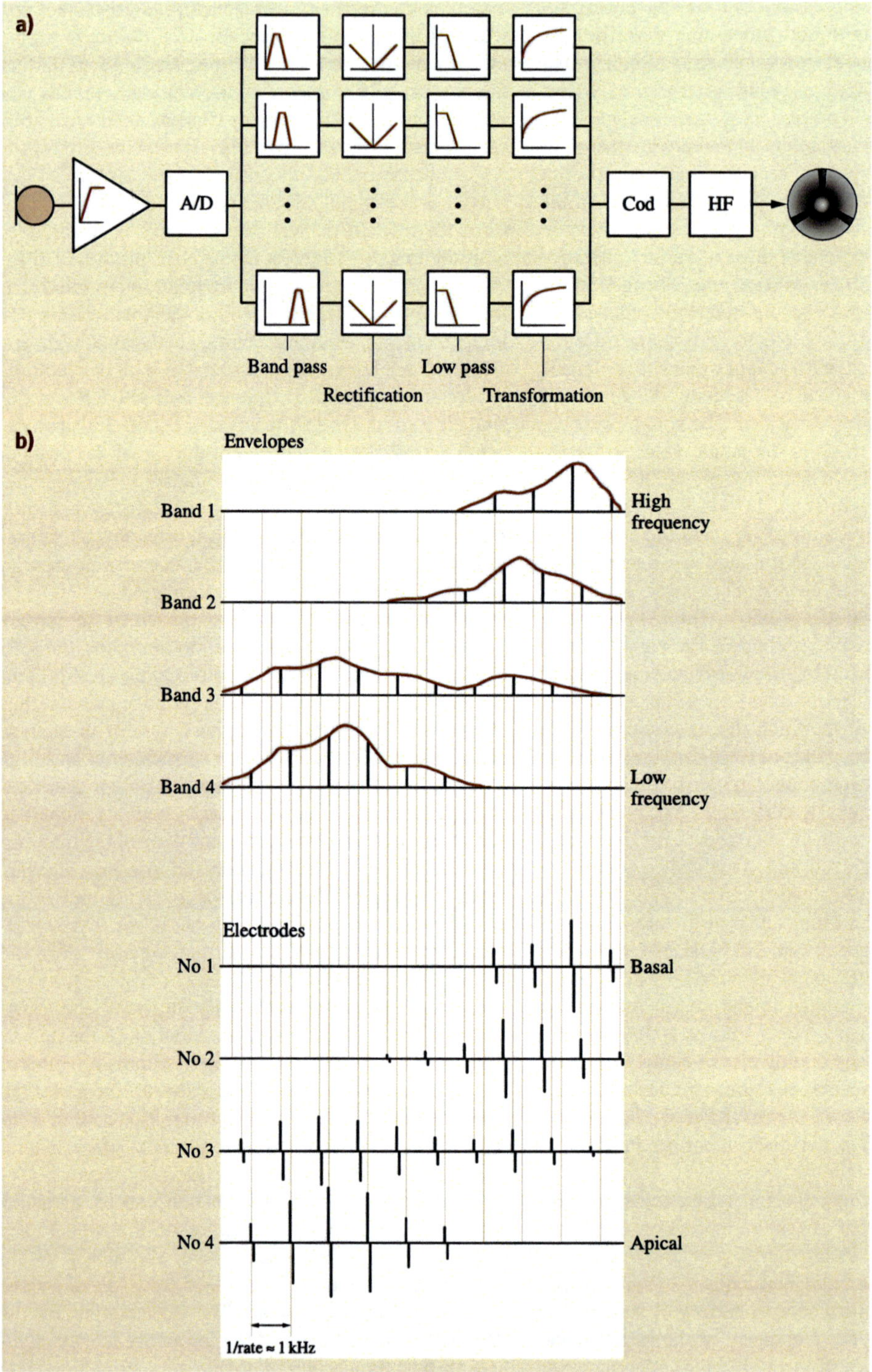

Fig. 12.35a,b Block diagram **(a)** and functional scheme **(b)** of the CIS strategy. The relationship between the amplitude of the envelope and the temporal sequence of electric pulses is shown for a sound signal developing from initially low frequencies (vowel) to high frequencies (sibilant)

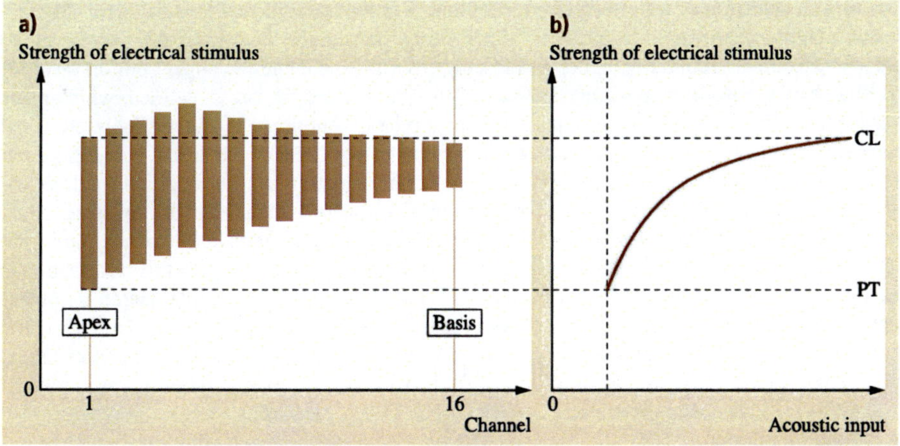

Fig. 12.36a,b Perception thresholds (PT) and comfortable stimulation levels (CL) for the 16 electrodes of a (fictional) CI system. The limits reported by the patient define the dynamic range **(a)** and the mapping rule that associates the acoustic sound pressure to the electrical stimulus intensity **(b)**

In addition to *n of m* and CIS, the ACE strategy (advanced combination encoders by Lochlear Ltd.) is applied widely in one of the current CI systems. It can be regarded as a modification of the *n of m strategy* in which the maximum number *n* of active electrodes and the channel rate are varied within the limits given by the device-specific capacity (typically 20 000 pps). Unlike the competing strategies, the *n* activated electrodes are redefined out of the total inventory *m* for each stimulation pattern (frame) so that a high-resolution spectral mapping of the signal is possible even with low values of *n*.

Generally, the speech coding in CI systems is designed to mimic the natural pitch detection and time pattern analysis by mapping the frequency spectrum to the electrode locations and by transforming the time structure of the speech signal into an appropriate pulse sequence. The realization of this goal is limited by technical shortcomings. The available systems differ in their performance, but these differences are minor in relation to the discrepancy between the natural system and its technical surrogate. In spite of the further progress to be expected – particularly in speed and performance of signal processing, but also in the number and density of the electrodes – this gap will most probably remain. The replacement of about 3400 natural receptor groups by 12–22 electrodes cannot lead to a restoration of the normal signal discrimination and speech intelligibility. This is the main reason why patients supplied with a CI need intensive technical, audiological, speech therapeutic, and educational support after surgery.

The condition to be fulfilled for the supply of a CI is a deafness or hearing impairment whose severity does not justify the expectation of rehabilitative success in terms of speech intelligibility on the basis of conventional amplifying hearing aids. The success of rehabilitation depends largely on onset and duration of deafness. For congenitally deaf or prelingually deafened patients, the late supply of a CI will generally not lead to an open speech understanding. Early intervention, however, may make a nearly normal development of auditory and language skills accessible to many deaf infants and young children.

In the border area between hearing aids and cochlear implantation, patients with residual hearing at low frequencies can be subject to a successful auditory rehabilitation by the combination of electric and acoustic stimulation (EAS). In these cases, implantation techniques which guarantee hearing conservation and short electrode arrays that do not reach the apical regions are applied. The technical equipment is a sound amplifying hearing aid combined with a CI speech processor in a single device, which delivers both acoustic output and RF pulses for the implant. It has been shown that a large number of patients take benefit from this special option in terms of their speech discrimination performance [12.52].

The supply of a cochlear implant requires extensive preoperative diagnostics. Beyond the basic audiologic tests, it includes imaging techniques such as computed tomography (CT) for the representation of bone formation in the implantation area and magnetic resonance imaging (MRI) for the detection of a fluid-filled cochlea. The functional integrity of the auditory nerve is tested in a preoperative electrical stimulation experiment. An extratympanal ball electrode is positioned in the meatus near the tympanic membrane or the tip of a transtympanic needle electrode is placed on the promontory or in the niche of the round window. During the delivery of electrical pulses of variable frequency and current strength, the patient is asked to describe his perceptions. For all stimulus frequencies that elicit auditory sensations, the stimulus currents corresponding to the subjective perception level and the discomfort level are determined. If the subjective test is not feasible, the functionality of auditory nerve and auditory pathway can be objectified in anesthesia by means of electrically evoked potentials of the auditory system.

The CI surgery [12.53] is performed under general anesthesia. The implant is placed in the mastoid behind the ear and the electrode array is inserted into the *scala tympani* through the round window or through an opening created immediately adjacent to it (cochleostomy). The opening of the inner ear is closed with tissue and the implant is secured mechanically. During the operation, the function of the implant and its effect on the auditory system are controlled by impedance telemetry, the recording of electrically evoked compound action potentials (TECAP), or the observation of the stapedius reflex during electric stimulation.

During the post-operative individual fitting of the speech processor, the acoustic input is disabled and the stimulation parameters are adjusted by a computer running a specific software. Within the preselected strategy, the main parameters of the individual signal processing program are the ranges of permissible stimulus strength for all electrodes. Along the device's internal stimulus intensity scale, the current amplitude and where necessary also the duration of the stimulus pulses is increased according to a logarithmic function valid for all electrodes. From the stimulus intensities that correspond to the subjective threshold and the upper comfort level for each electrode, a mapping law is defined, which maps the acoustic input amplitude to the available electrical dynamic range (Fig. 12.36). In addition, some shape parameters of this function can be modified to affect the perception of loudness or to improve the noise reduction. The patient-specific parameters of the speech coding are based on subjective specifications. If this information is not reliable, the adaptation is supplemented by objective methods, e.g. the measurement of the electrically induced and contralaterally registered stapedius reflex, the electrically evoked auditory brainstem responses (E-ABR) or the TECAP [12.54].

During the post-operative auditory training, the patient obtains guidance and support for the processing of the new impressions. The auditory skills are tested with a test battery graded by difficulty. The success of rehabilitation can be divided into three stages. An acoustic orientation, i. e. a perception and recognition of environmental sounds, is to be expected in all patients, an effective acoustic support of lipreading by the implant can usually be achieved also by prelingually deafened patients, and finally, an open speech understanding without eye contact to the speaker and the ability to communicate via telephone is achieved generally only by post-lingually deafened patients. Speech perception in noise and reverberation remains a challenge for many implant recipients, even for bilateral implantees. Technical approaches to solving this problem are based on digital noise reduction or delay and digital overlay of the acoustic signal recorded from multiple microphones (adaptive beam forming). In this respect, as in other areas, the technical development has not yet been completed and further progress of rehabilitation can be expected.

References

12.1 S. Hoth: Audiometrie. Die Untersuchung des Gehörs und seine technische Versorgung. In: *Medizintechnik*, ed. by R. Kramme (Springer, Berlin Heidelberg 2011) pp. 217–270, 4th edn.

12.2 S. Hoth: Medizinische Akustik und Audiologie. In: *Medizinische Physik*, Vol. 1, ed. by J. Bille, W. Schlegel (Springer, Berlin Heidelberg 1999) pp. 305–427

12.3 H.G. Boenninghaus, T. Lenarz: *Hals-Nasen-Ohrenheilkunde für Medizinstudenten*, 13th edn. (Springer, Berlin Heidelberg 2007)

12.4 G. von Békésy: *Experiments in Hearing* (McGraw-Hill, New York 1960)

12.5 H.P. Zenner: *Hören, Physiologie, Biochemie, Zell- und Neurobiologie* (Thieme, Stuttgart 1994)

12.6 P.M. Sellick, R. Patuzzi, B.M. Johnstone: Measurement of basilar membrane motion in the guinea pig using the Mössbauer technique, J. Acoust. Soc. Am. **72**, 131–141 (1982)

12.7 S. Stenfelt, R.L. Goode: Bone-conducted Sound: Physiological and clinical aspects, Otol. Neurotol. **26**, 1245–1261 (2005)

12.8 S. Hoth, T. Lenarz: *Elektrische Reaktions-Audiometrie* (Springer, Berlin Heidelberg 1994)
12.9 A. Kral: Frühe Hörerfahrung und sensible Entwicklungsphasen, HNO **57**, 9–16 (2009)
12.10 A. Sharma, P.M. Gilley, M.F. Dorman, R. Baldwin: Deprivation-induced cortical reorganization in children with cochlear implants, Int. J. Audiol. **46**, 494–499 (2007)
12.11 J.O. Pickles: *An Introduction to the Physiology of Hearing*, 2nd edn. (Academic, London 1988)
12.12 E. Zwicker, H. Fastl: *Psychoacoustics* (Springer, Berlin Heidelberg 1990)
12.13 J. Fricke, L.M. Moser, H. Scheurer, G. Schubert: *Schall und Schallschutz* (Physik Verlag, Weinheim 1983)
12.14 E. Zwicker, R. Feldtkeller: *Das Ohr als Nachrichtenempfänger*, 2nd edn. (Hirzel, Stuttgart 1967)
12.15 G. Böhme, K. Welzl-Müller: *Audiometrie*, 3rd edn. (Huber, Bern 1993)
12.16 E.P. Fowler Sen.: Measuring the sensation of loudness. A new approach to the physiology of hearing and the functional and differential diagnostic tests, Arch. Otolaryngol. **26**, 514 (1937)
12.17 J.F. Jerger: The SISI test, Int. Audiol. **1**, 246 (1962)
12.18 E. Lüscher, J. Zwislocki: A simple method for indirect monaural determination of the recruitment phenomenon (difference limen in intensity in different types of deafness), Acta Otolaryngol. (Stockh.) Suppl. **78**, 156–168 (1949)
12.19 B. Langenbeck: Das überschwellige Gehör des Schwerhörigen und seine Messung auf Grund eigener Versuche, Z. Laryngol. Rhinol. **28**, 463–480 (1949)
12.20 E. Lehnhardt, R. Laszig: *Praxis der Audiometrie*, 9th edn. (Thieme, Stuttgart 2009)
12.21 O. Heller: Hörfeldaudiometrie mit dem Verfahren der Kategorienunterteilung (KU), Psychol. Beitr. **27**, 509–519 (1985)
12.22 R. Carhart: Clinical determinations of abnormal auditory adaptation, Arch. Otolaryngol. **65**, 32–40 (1957)
12.23 K.H. Hahlbrock: *Sprachaudiometrie* (Thieme, Stuttgart 1957)
12.24 K. Wagener, B. Kollmeier: Evaluation des Oldenburger Satztests mit Kindern und Oldenburger Kinder-Satztest (Evaluation of the Oldenburg sentence test with children and the Oldenburg children's sentence test), Z. Audiol. **44**, 134–143 (2005)
12.25 M. Wesselkamp, K. Kliem, B. Kollmeier: Erstellung eines optimierten Satztestes in deutscher Sprache. In: *Moderne Verfahren der Sprachaudiometrie*, ed. by B. Kollmeier (Median, Heidelberg 1992) pp. 330–343
12.26 I. Hochmair-Desoyer, E. Schulz, L. Moser, M. Schmidt: The HSM sentence test as a tool for evaluating the speech understanding in noise of cochlear implant users, Am. J. Otol. **18**(Suppl.), 83 (1997)
12.27 S.B. Resnick, J.R. Dubno, S. Hoffnung, H. Levitt: Phoneme errors on a nonsense syllable test (A), J. Acoust. Soc. Am. **58**(Suppl. 1), 114 (1975)
12.28 M. Nilsson, S.D. Soli, J.A. Sullivan: Development of the Hearing In Noise Test for the measurement of speech reception thresholds in quiet and in noise, J. Acoust. Soc. Am. **95**(2), 1085–1099 (1994)
12.29 K.D. Cramer, N.P. Erber: A spondee recognition test for young hearing-impaired children, J. Speech Hear. Disord. **39**, 304–311 (1974)
12.30 B. Hagerman: Sentences for testing speech intelligibility in noise, Scand. Audiol. **11**, 79–87 (1982)
12.31 T. Steffens: Test-retest-difference of the Regensburg modification of the OLKI-Reimtest with speech simulating noise in children with hearing aids, Z. Audiol. **45**(3), 88–99 (2006)
12.32 J.F. Jerger, J.L. Northern (Eds.): *Clinical Impedance Audiometry*, 2nd edn. (Thieme, Stuttgart 1980)
12.33 G. Stange: *Funktionsprüfungen der Hör- und Gleichgewichtsorgane* (Thieme, Stuttgart 1993)
12.34 P.K. Harris, K.M. Hutchinson, J. Moravec: The use of tympanometry and pneumatic otoscopy for predicting middle ear disease, Am. J. Audiol. **14**(1), 3–13 (2005)
12.35 O. Metz: Threshold of reflex contraction of muscles of middle ear and recruitment of loudness, Arch. Otolaryngol. **55**, 536 (1952)
12.36 D. Kemp: Stimulated acoustic emissions from within the human auditory system, J. Acoust. Soc. Am. **64**(5), 1386–1391 (1978)
12.37 S. Hoth, T. Lenarz: *Otoakustische Emissionen*, 2nd edn. (Thieme, Stuttgart 1997)
12.38 S. Hoth, K. Neumann: *Das OAE-Handbuch* (Thieme, Stuttgart 2006)
12.39 S. Hoth, W. Heppt, M. Finckh: Verhalten der evozierten otoakustischen Emissionen bei retrocochleären Hörstörungen [Behaviour of evoked otoacoustic emissions in cases of retrocochlear hearing disorder], Otorhinolaryngol. Nova **4**, 128–134 (1994)
12.40 J. Kießling, B. Kollmeier, G. Diller: *Versorgung und Rehabilitation mit Hörgeräten*, 2nd edn. (Thieme, Stuttgart 2008)
12.41 P. Kummer, T. Janssen, W. Arnold: Suppression tuning characteristics of the 2f1-f2 distortion-product otoacoustic emission in humans, J. Acoust. Soc. Am. **98**, 197–210 (1995)
12.42 P. Boege, T. Janssen: Pure-tone threshold estimation from extrapolated distortion product otoacoustic emission I/O-functions in normal and cochlear hearing loss ears, J. Acoust. Soc. Am. **111**(4), 1810–1818 (2002)
12.43 T.W. Picton, J. Oulette, G. Hamel, A. Durieux-Smith: Brainstem evoked potentials to tone pips

in notched noise, J. Otolaryngol. **8**, 289–314 (1979)
12.44 T.W. Picton, A. Dimitrijevic, M.A. John: Multiple auditory steady-state responses, Ann. Otol. Rhinol. Laryngol. **111**, 16–21 (2002)
12.45 S. Hoth: The influence of inner ear hearing loss on transitory evoked and distortion product otoacoustic emissions, Laryngol. Rhinol. Otol. **75**, 709–718 (1996)
12.46 W. Wagner, P.K. Plinkert: The relationship between auditory threshold and evoked otoacoustic emissions, Eur. Arch. Otorhinolaryngol. **256**, 177–188 (1999)
12.47 D. Glista, S. Scollie, M. Bagatto, R. Seewald, A. Johnson: Evaluation of nonlinear frequency compression: Clinical outcomes, Int. J. Audiol. **48**(9), 632–644 (2009)
12.48 R.C. Seewald, M. Ross, M.K. Spiro: Selecting amplification characteristics for young hearing-impaired children, Ear Hear. **6**(1), 48–53 (1995)
12.49 T. Steffens: Die sinnvolle Nutzung moderner Hörgerätetechnik, HNO **44**, 213–222 (1996)
12.50 P. Abbas, C. Brown, J. Shallop, J. Firszt, M. Hughes, S. Hong, S. Staller: Summary of results using the Nucleus CI24M implant to record the electrically evoked compound action potential, Ear Hear. **20**, 45–59 (1999)
12.51 B.S. Wilson, D.T. Lawson, C.C. Finley, R.D. Wolford: Coding strategies for cochlear prostheses, Am. J. Otol. **12**(Suppl.), 56–61 (1991)
12.52 U. Baumann, S. Helbig: Hören mit kombinierter elektrischer und akustischer Stimulation [Hearing with combined electric acoustic stimulation], HNO **57**, 542–550 (2009)
12.53 A. Ernst, R.D. Battmer, I. Todt: *Cochlear Implant heute* (Springer, Berlin, Heidelberg 2009)
12.54 J. Müller-Deile: *Verfahren zur Anpassung und Evaluation von Cochlear Implant Sprachprozessoren* (Median, Heidelberg 2009)

13. Measurement Techniques in Ophthalmology

Albert J. Augustin

Modern ophthalmology uses numerous techniques to evaluate both function and pathophysiology of the organ. This chapter covers the major techniques used in ophthalmology. These include evaluation-techniques for the refractive part of the eye such as wavefront-analysis (Sect. 13.13), confocal laser microscopy (Sect. 13.5), as well as keratometry (Sect. 13.14) and the orbscan technology (Sect. 13.18). In addition, several methods for the exact determination of the intraocular pressure (Sects. 13.1–13.1.5) and methods to evaluate the optic nerve (13.2) in glaucoma are discussed. Imaging of the retina by means of OCT (Sect. 13.2) and angiography (Sects. 13.20 and 13.21) are important techniques in diseases of the vessel system, as well as age related macular degeneration. Electrophysiology (Sect. 13.8) including pattern techniques is mandatory to diagnose and to follow various diseases of the optic nerve head and the retina. Ophthalomologic ultrasound (Sect. 13.16) is needed for both imaging of pathologies as well as length measurement to determine the power of intraocular lenses before implantation into the eye.

13.1 **Measurement of Intraocular Pressure** 246
13.1.1 Goldmann's Applanation Tonometry 246
13.1.2 Schiötz's Impression Tonometry ... 246
13.1.3 Noncontact Tonometry (Air-Puff Tonometer) 246
13.1.4 Mackay–Marg Tonometry (Tonopen) 247
13.1.5 The Pascal Dynamic Contour Tonometer (DCT) 247

13.2 **Optical Coherence Tomography (OCT)** 247
13.2.1 OCT Macula 247
13.2.2 OCT of the Optic Nerve Head (ONH) 248
13.2.3 OCT with Visante 249

13.3 **Laser-Scanning Tomography with the Heidelberg Retina Tomograph (HRT)** 249

13.4 **Nerve Fiber Polarimetry with GDx** 251

13.5 **The Rostock Cornea Module (Confocal Laser Microscope)** 252

13.6 **Automatic Refractometry** 252

13.7 **Visually Evoked Potential (VEP)** 253
13.7.1 Flash VEP 253
13.7.2 Pattern VEP 254

13.8 **The Ganzfeld ERG (Ganzfeld Electroretinogram)** 254

13.9 **Pattern Electroretinography (Pattern ERG, PERG)** 257

13.10 **Multifocal ERG (mfERG)** 259

13.11 **Electrooculograms (EOG)** 260

13.12 **Adaptometry** 261

13.13 **Aberrometry (Wavefront Analysis)** 261

13.14 **Keratometry** 263

13.15 **Retinoscopy or Skiascopy** 263

13.16 **Ultrasound** 263
13.16.1 A-Scan Ultrasound, Ultrasound Biometry 264
13.16.2 Ultrasound Pachymetry A-Scan 264
13.16.3 Ultrasound B-Scan 264

13.17 **Corneal Topography** 265

13.18 **The Orbscan** 265

13.19 **Scheimpflug Examination** 267

13.20 **Fluorescence Angiography of the Retina (Sodium-Fluorescein)** 267

13.21 **Fluorescence Angiography of the Retina (Indocyanine Green)** 269

13.22 **Visual Field Measurement (Perimetry)** 269

13.23 **Exophthalmometry** 271

References 271

13.1 Measurement of Intraocular Pressure

Subject: Tonometry is used to measure the intraocular pressure.

Measurement: Different instruments can be used: Goldmanns's applanation tonometer, Schiötz's impression tonometer, noncontact-(impression) tonometer, and the Mackey–Marg tonometer. Goldmann's applanation tonometer shall be deemed to be the standard. Follow-up measurements using different methods are not advisable (the measurement results are influenced by numerous factors). In addition, the corneal thickness affects the results. Thus, thickness should be measured at all times.

13.1.1 Goldmann's Applanation Tonometry

Measurement: Goldmann's applanation tonometer is attached to a slit lamp. The clamp can vary depending on the manufacturer. The front side of the tonometer head has a standardized diameter of 3.06 mm. Fluorescent and anesthetic eye drops have to be applied before the measurement. Afterwards, the tonometer head approaches the cornea until both surface areas touch. This is done with blue light illumination. Due to a specific microsection in the measuring device two yellow semicircles are visible in the slit lamp examination. When the pressure results in a flattening of the cornea, the two inner edges of the semicircles touch. The regulation of the pressure can be adjusted with a scroller on the tonometer clamp. Assuming an average thickness of the cornea and a normal amount of lacrimal fluid by a round area (with a diameter of 3.06 mm), the following calculation results in the IOP value: subtract the adhesive power of the lacrimal fluid from the necessary power for flattening the cornea. Equate the measured pressure with the intraocular pressure (IOP).

Indication: The intraocular pressure is an important parameter for the evaluation of glaucoma and its therapy. Applanation tonometry is a simple, elegant and – apart from certain exceptions – very precise technique to determine intraocular pressure.

Evaluation: The thickness of the cornea affects the measurement. The references contain several calculation options for the correction factor. The Orssengo–Pye-formula is mainly used to do this.

13.1.2 Schiötz's Impression Tonometry

Subject: The measurement determines the impression depth by a defined (gravitation) force effect.

Measurement: The instrument equals an invert circle. The lower fraction consists of a hand holder, a base plate, and a mobile coaxial bolt. Following the application of anesthetic eye drops, the instrument (with its base plate) is attached perpendicularly to the cornea. The deflection of the bolt can be shown on the upper fraction of the tonometer by the articulation on the scale. The measurement can be distorted by moving the tonometer slightly. It is recommended to seat the patient as flat as possible.

Indication: Schiötz's impression tonometry is used mostly for measuring intraocular pressure. The method is especially used for bedfast patients.

Evaluation: The values read off the scale are transformed to the intraocular pressure with the table attached to the Schiötz tonometer. Please consider that – depending on the insert weight – different columns are valid for IOP calculation. New impression tonometers calculate the values automatically.

Relevance: By increasing the dependency of the individual variable elasticity of the sclera and cornea such as a high myopia, the danger of an incorrect measurement is much higher than with applanation tonometry. Therefore, this method is only used in those cases where applanation tonometry is not possible.

13.1.3 Noncontact Tonometry (Air-Puff Tonometer)

Subject: The measurement determines the time that is necessary to flatten the cornea at a defined force effect (air-puff).

Measurement: The measurement occurs automatically without previous anaesthesia. Generally, the patient sits in front of the instrument like in front of a slit lamp. The patient should open their eyes wide. With reduced visual acuity the eye that not has to be examined should be covered. Corneal applanation is detected via an electrooptical system. Usually the measurement is repeated three times. An average out of the values will be calculated.

Relevance: Noncontact tonometry can be easily performed by a trained ophthalmic assistant.

13.1.4 Mackay–Marg Tonometry (Tonopen)

Subject, measurement: The measurement is possible with or without topical anaesthetics. The patient should stand up or sit on a chair. The cornea is flattened with a probe in the centre of which is a small diameter plunger that senses the ocular tension. The sterilized rubber cover is replaced for each patient, which ensures sterility. The patient should open his/her eyes widely and press the probe gently perpendicularly against the cornea.

Indication: The Mackay–Marg tonometer is used by patients with an irregular corneal surface, such as corneas following chemical burns. It is the only instrument that gives exact values in such cases.

Evaluation: A typical curve shows two spikes of the generated pressure in dependence on time. The first spike reflects the wanted intraocular pressure. The second spike reflects the necessary pressure for applanation of the cornea with the plunger and the surrounding annular ring. The measurement should be repeated several times to calculate an average. With recent instruments such as the Tonopen the measurement will be averaged and interpreted automatically. In the end, a digital display shows the calculated value. Compared with Goldmann's tonometer the IOP values of the Mackay–Marg tonometer can be measured insignificantly higher.

Relevance: Even though the main principle of the Mackay–Marg tonometer has beem known for more than 50 years, its distribution is still low. The Schiötz tonometer is likely to be replaced by other tools such as the Tonopen in the future.

13.1.5 The Pascal Dynamic Contour Tonometer (DCT)

Subject: A miniaturized piezo-resistive pressure sensor senses the intraocular pressure. The voltage that is measured by the piezo-crystal is proportional to the affecting pressure.

Measurement: The dynamic contour tonometry (DCT) is a noninvasive and direct measurement of the intraocular pressure (IOP) that is not influenced by the structural characteristics of the eye. The probe has a concave surface just a little more even than the human cornea. The measuring equals the installation and progress of Goldmann's tonometry. The tonometer is installed to the slit lamp and then applied to the cornea. Due to the concave surface of the peak the probe covers the natural corneal curvature. The tangential and bend forces should be disregarded so that the measured pressure on the cornea surface is equal to the intraocular pressure. The measured values are recorded digitally. After the measurement a chart can be evaluated with the chronological pressure sequence.

Indication: The advantage over other tonometers is the dynamic measurement over a certain period of time. With the DCT it is possible to measure the diastolic and the systolic pressure IOP and their difference to the ocular pulse amplitude. This is an indirect indicator for choroidal perfusion.

Evaluation: On average, the DCT measures a value at 3.2 mmHg higher than Goldmann's measurement. This technique is independent of the thickness of the cornea.

Relevance: It may become very important for the diagnosis and therapy of glaucoma. Scientific evaluation is ongoing.

13.2 Optical Coherence Tomography (OCT)

Subject: Optical coherence tomography (OCT) allows an in vivo description of different anatomic sections of the eye. OCT has a resolution of up to $< 10\,\mu m$ longitudinal and $< 20\,\mu m$ transversal.

Measurement: During the measurement the examined object is scanned automatically with an infrared light (coherence length of 830 nm). The intensity of the reflected light is determined. The temporal hesitation gives information on the distance of the scanned object.

Relevance: Retina – During the last years the OCT examination of the macula and the optic nerve head (ONH) has been used in addition to angiography (macula) as well as in glaucoma patients. Anterior chamber – The examination of both the anterior chamber and the cornea by means of OCT is a promising alternative to ultrasound biomicroscopy.

13.2.1 OCT Macula

Indication: OCT of the macula is a standard examination in many diseases such as macular hole and macular pucker. As an upgrade to angiography OCT determines

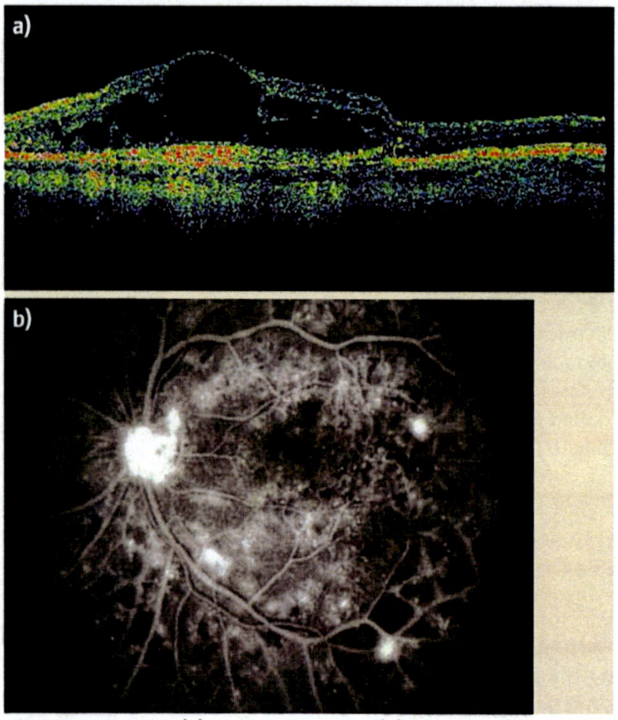

Fig. 13.1a,b OCT **(a)** and angiography **(b)** of a diabetic macular edema

the volume (quantitation of macular edema). Using the new high resolution OCT single retinal layers may be distinguished (in vivo histology) (Fig. 13.1).

Evaluation: OCT examination of the macula gives two-dimensional pictures of the retina. The intensity of the reflected light will be either color-coded (red = strong, blue = poor) or pictured in different levels of gray (light = strong, dark = poor). The locations of the strongest reflexion are nerve fibers, the choroid, and epiretinal membranes.

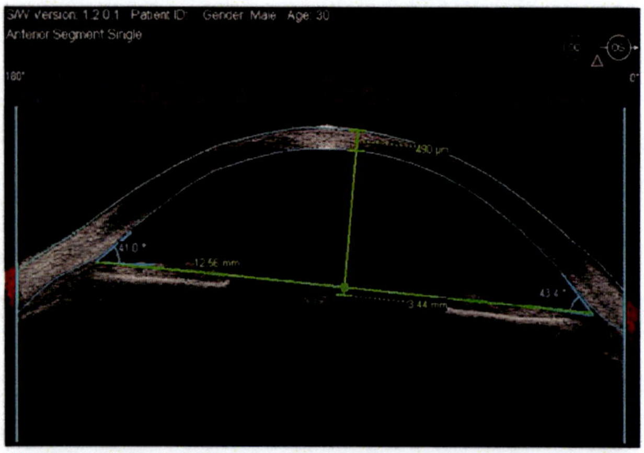

Fig. 13.2 Visante OCT showing chamber angle, anterior chamber depth, and corneal thickness

Usually, the neuro-sensory retina and the choroid have an average reflection level. Pigment epithelium damage would cause a higher intensity of the choroid underneath. The vitreous body has a low reflection level. At a strong contrast an increase in the OCT allows the imaging of vitreous body structures such vitreoretinal traction and the posterior hyloid. Via computer reconstruction a three-dimensional image of structures can be generated. The volume of a macular edema can be calculated exactly and compared to previous examinations. Due to the higher resolution of the newest generation of OCT, three-dimensional imaging of certain retina layers can be calculated. The dimension of a pigmented epithelium elevation or the loss of nerve fibres can be imaged. The quantitation will be affected by measuring the thickness or by direct volume determination.

The standard imaging of the central retinal thickness is known as a *map*. On the *map* the macula is shown as a colored round area. The normal map of the retina has a blue color with a central green or yellow circle for the foveal ridge. If the retina is thickened in the center the map becomes red to yellow (with strongly increased values even white). Besides the color-coded images the average for certain areas (central, nasal, temporal, cranial, and caudal) of the macula are calculated in microns. They will be listed next or in combination with the map on the OCT printout. A retina thickness of more than 180 µm is considered as pathological.

13.2.2 OCT of the Optic Nerve Head (ONH)

Indication: As compared to other methods (HRT) or polarimetry (GDx), OCT of the ONH is a highly sensitive and very specific examination for imaging the loss of nerve fibers in glaucoma subjects. As an advantage over GDx, OCT and HRT allow the determination of the size of the ONH and the excavation.

Evaluation: Both measurement of the ONH size and excavation are done automatically. The outline of the ONH has to be done manually only for certain abnormalities of the ONH. On the printout a vertical incision through the ONH is shown. For every single incision (vertical, horizontal, and transversal) the computer indicates an ONH diameter, excavation diameter, and the width of the ONH border. Analysing every incision the

ONH areas (*disk area*) and the excavation area (*cup area*) can be determined. The relation of the two areas illustrates the *cup disk area ratio*. Besides the cup disk area ratio a cup disk horizontal and vertical ratio is stated of the instrument. The vertical cup disk ratio should be smaller at a normal formed ONH than the horizontal cup disk ratio. The measurement of the thickness of the nerve fibers works according to the idea of the layer thickness identification.

13.2.3 OCT with Visante

Indication: The OCT of the anterior chamber is similar to ultrasound biomicroscopy. It enables a precise measurement to determine the anterior chamber depth and the chamber angle. It represents an excellent instrument for diagnosis and evaluation of the acute angle closure glaucoma. The corneal thickness and the thickness of a cornea flap can be measured exactly before and after a refractive procedure. By diagnosing a keratoconus the Visante OCT allows a diagnostic saving and a monitoring of the process. Visante OCT allows the measurement of tumor tissue on the iris front face, but it is not as good as ultrasound biomicroscopy.

Evaluation: The examiner must determine the anterior chamber depth and the chamber angle manually. Corneal thickness is shown color-coded. Light cornea areas are shown in orange to dark red. The following parameters can be read off the table: smallest and biggest corneal thickness, as well as an average of different localizations (Fig. 13.2).

13.3 Laser-Scanning Tomography with the Heidelberg Retina Tomograph (HRT)

Subject: Laser scanning tomography allows the in vivo topography of different anatomic structures of the eye (Fig. 13.3).

Measurement: With the aid of a diode laser a two-dimensional optical cross-section of 32 focal planes (z-axis) can be taken. The images are used by the computer for reconstruction of the steric structures. The user must define the standard reference level and the ONH border. These two values can essentially influence the measurement to prevent causing error (especially through the examiner); a computer model exists for pattern recognition. It automatically determines the ONH border via a combination of HRT and ophthalmoscopic images. For a higher resolution the computer uses the TruTrak software for motion artifacts. Moreover, the computer matches the position of the pictures between different measurements.

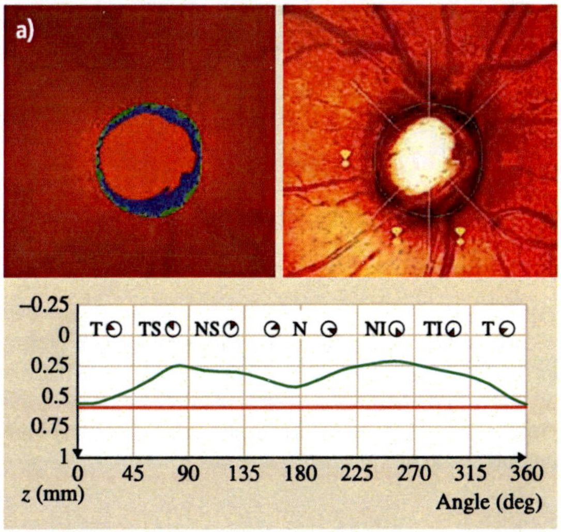

Indication: In many cases laser scanning tomography is used in the diagnosis and follow-up of glaucoma.

Evaluation: Similar to the OCT of the ONH, the following values are measured – size of the ONH, cup disk ratio, disk margin, and thickness of the retinal nerve fiber layer. In process monitoring the values can be compared it previous images. Furthermore, there is the possibility of comparing individual measurements with values of a normal population. For this purpose, a graphic image is used. It shows part of the ONH excavation and the ONH margin as the colored part of a bar. The excavation is pictured in red, the neuroretinal border in green, and the bended border in blue. Different localizations (temporal and superior, etc.) are shown.

Fig. 13.3a–c Evaluation of the optic nerve head using HRT **(a,b)** and OCT **(c)**. The patient shows significant glaucomatous damage (abnormal size of the optic nerve). In Moorfield's analysis **(b)** the nerve is judged as pathological. Burk's (RB) and Mikelberg's (FSM) dicriminant values are negative; the also indicate glaucomatous damage ◀▶

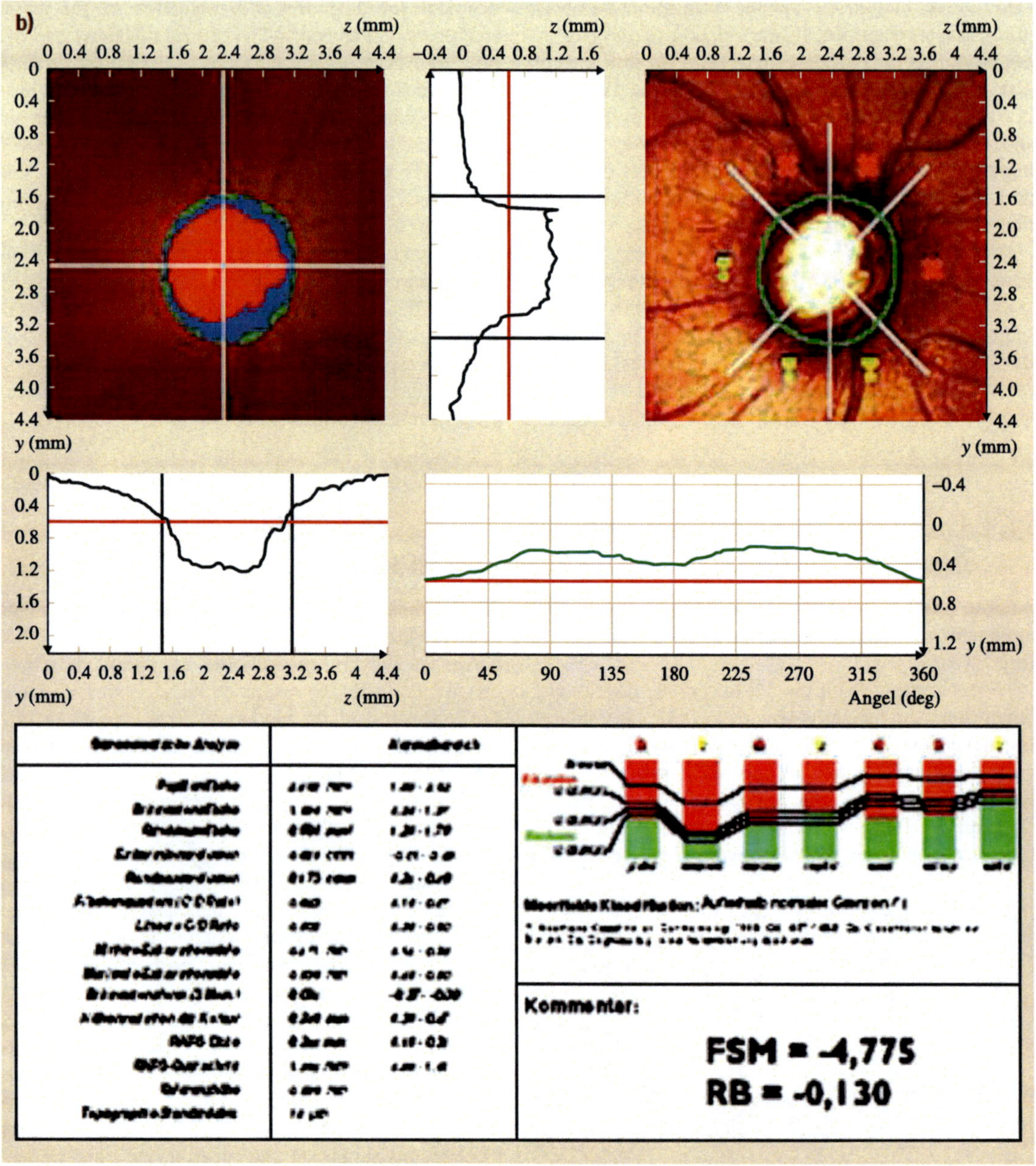

Given curves cut the bars and mark the areas of the normal (predicted) and those that differ from the collective (low 95%) bar division. Moorfield's regression analysis allows a classification out of ONH diagnostics with a sensitivity for glaucoma damage at 74–90%.

Relevance: Laser scanning tomography is a popular method for an objective diagnostic documentation of ONH and the nerve fiber layer especially in daily praxis. It has big relevance in an early diagnosis and the follow-up of glaucoma patients.

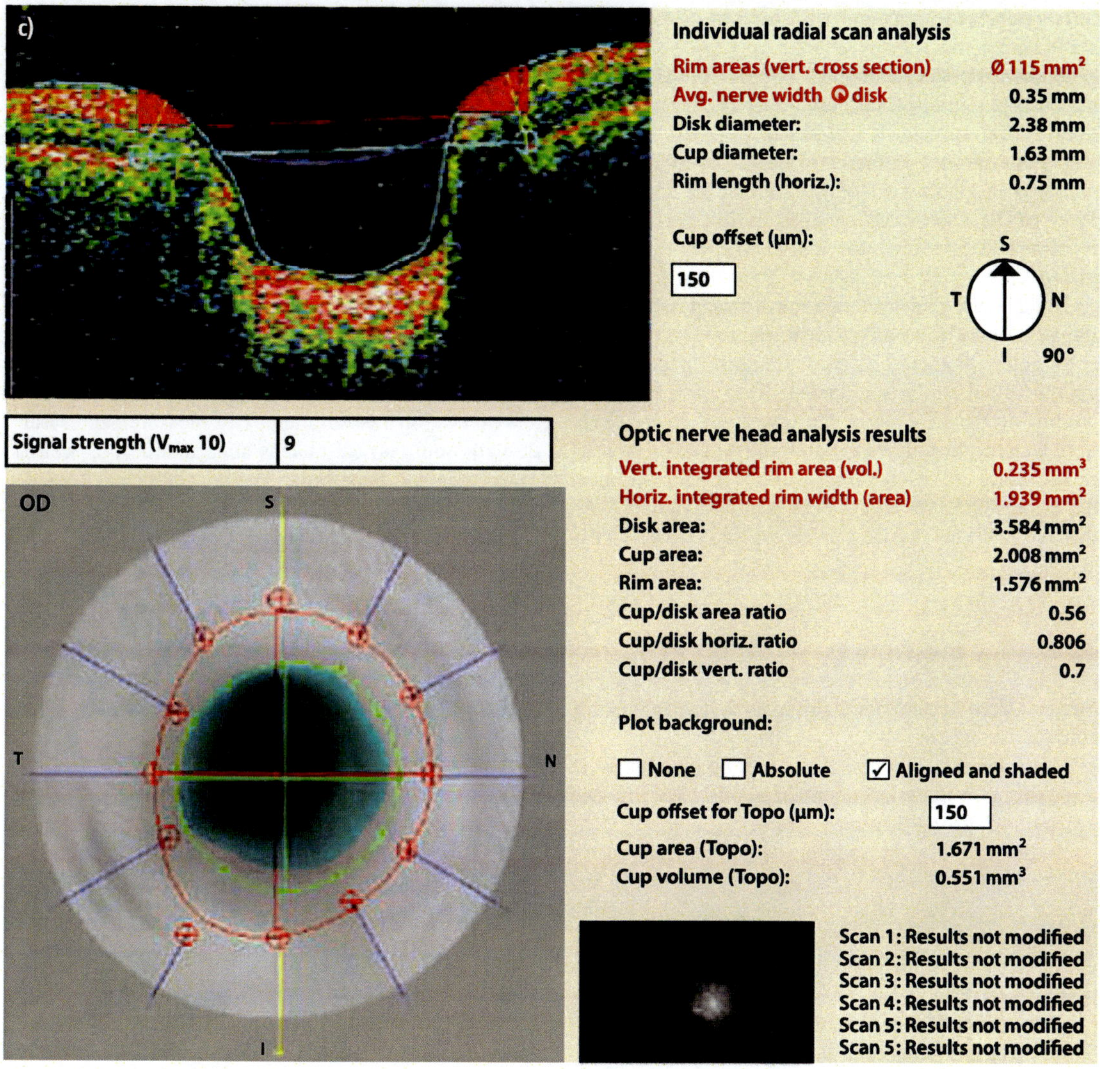

13.4 Nerve Fiber Polarimetry with GDx

Subject: During the examination the retardation of polarized light is measured and the peripapillar and macular nerve fiber layer thickness is calculated.

Measurement: A dilated pupil is recommended before the measurement but not absolutely mandatory. Within some seconds an automatic measurement on 32 768 measuring points starts. The instrument combines the characteristics of a scanning laser ophthalmoscope and a Fourier ellipsometer. The retina is scanned by a polarized infrared light (780 nm) and reflected to the pigmented epithelium (with defect to the sclera). Depending on the structure and thickness of the irradiated tissue a polarized light pervades it in different velocities. Structures with a straightened fibre orientation extend the passing time of the light with vertical polarization to the fibre direction oscillating waves. The thickness of the nerve fibre is determined by the difference between the retention periods of different polarized lights and with help of a histological comparison.

Indication: The technique is suitable for the determination of the peripapillar nerve fiber border and nerve fiber bunch damage. It has a sensitivity of 96% and a specifity of 93% in the early diagnosis of glaucoma damages. A glaucoma modification can be determined even before observable changes in the perimetry. It is especially qualified for a follow-up of glaucoma patients.

Evaluation: The values are transformed into a three-dimensional or false-color-coded picture. In the false-color-coded picture thick structures are marked from yellow to red and a thin from light to dark blue. According to the ISNTnorm the upper and lower nerve fibre bunches are colored red and yellow, and the temporal and nasal are colored blue. Moreover the instrument has an age and race database of standard values. With this the collected values can be evaluated. The analysis occurs similarly to OCT and HRT n both a table form and graphically. In the table the calculated average values can be numerated circular and after their localization (superior, inferior). If the values are pathologic they are highlighted with red. The table also shows the standard deviation of the normal collective and the difference between both eyes. This results in the probability of a glaucoma (nerve fiber indicator (NFI) = the number). The probability of a glaucoma is low at a NFI of ≤ 30, marginal from 30–50, and high from ≥ 50.

Relevance: The noninvasive diagnostic procedure is suitable for judging suspect nerve fibre damage or ONH margin damage. Compared with the HRT and OCT examinations this method does not include an ONH size and excavation measurement. This measurement is fragile with both corneal or lens abnormalities, as well as by myopia and hyperopia. The development of a software module (enhanced cornea compensation ECC) can correct the aberration to some extent.

13.5 The Rostock Cornea Module (Confocal Laser Microscope)

Subject: Used for the examination of the corneal and the conjunctival layers.

Measurement: Frontal cross-section images with a maximum impression depth of about 1500 μm and a size of 400 μm × 400 μm are taken.

Indication: The procedure allows the imaging of keratocytes, endothelial cells, and nerve fibers in the cornea. Therefore it is suitable for a good evaluation of corneal diseases and monitoring after surgery. Furthermore, it may be used to monitor filtering bleb morphology.

Evaluation: The quantitation of the endothelial cell density happens when cells in a certain area are enumerated. All corneal layers can be imaged. Thus, it helps to obtain an early diagnosis of corneal dystrophies.

Fig. 13.4 Image of the inner corneal layer using confocal optics

Relevance: The recent, noninvasive procedure has great potential in the diagnosis of corneal and conjunctival diseases. Today it is mainly used in specialized centers.

13.6 Automatic Refractometry

Subject: The quality of the optical image on the retina is measured. The lens combination with the best image results in the wanted refraction.

Measurement: The measurement works automatically and depends on the instrument; the patient may sit, stand, or lie down. For the examination it is important that the patient looks into the distance. A picture already installed in the instrument accomplishes this. Cycloplegia is recommended for children. This helps to exclude accommodation, which would falsify the measurement.

A patient with poor fixation should cover the other eye to achieve better measurements. The measurements follow different principles. The following procedures are used: Foucault's method, Scheiner's principle, and picture measurement. In Foucault's method a punctual light is changed through a lens into a parallel light bunch. The parallel light bunch hits the eye and is combined to a point to the retina. If a poor refraction of the eye exists the point is shown as a blot or bar. The reflected light strengthens the statement. Refraction abnormality further deforms the former punctual light. An optimal refraction will reflect the light to the original point. If divergent lights of the original light source are shown they can be determined and quantified with a light detector. Different conventional lenses that could adjust eye refraction can decrease the aberration of the light source. If the aberration is not detectable the refraction of the previous lenses is equal to the wanted optimal refraction.

For Scheiner's principle a lens aperture is used that produces two punctual lights. The light rays are broken and focused through the breaking media into the eye. Depending on the distance of the two aperture plates and the refraction of the eye the light rays are focused in front or behind the retina. By the use of changing the distance between the plates the refraction can be calculated by a defined globe length.

Picture measurement process is a modification of Scheiner's principle. Reflection of an incoming light ray generates a dot-shaped light source which is projected on the retinal surface. Because of the refraction of the eye the light rays are broken and can be focused with a lens again. Depending on the angle of incidence the light is projected in different widths on the detector. The determined distance is defined for an emmetropic eye. In myopia the distance increases, in hyperopia it decreases. The application of glasses enables the calculation of the refraction that leads to emmetropia.

Indication: An objective refraction should be part of every ophthalmological examination. Before and after a surgery (cataract surgery, refractive cornea surgery) this measurement is strongly recommended.

Evaluation: Depending on the instrument three automatic or manual measurements are needed for an exact refraction. An integrated software calculates a supposed refraction. A cause for falsification can be the *myopia of the machine*. This happens if a patient sees a blurred image and the immediate accommodation induces an increase of myopia. If astigmatism is only minimal or even nonexistent false values from up to 0.5 dpt may be determined. Characteristic of this is a strongly changing astigmatismus angle.

Relevance: Today, automatic refractometers are part of the standard equipment of every ophthalmological practice or hospital. They allow a fast and easy measurement of objective refraction. Generally, the measurements taken using an automatic refractometer can be done by medical assistants.

13.7 Visually Evoked Potential (VEP)

Subject: Visual evoked potentials are electric potential differences that are deduced from the scalp over the visual cortex after a light impulse. VEP are a special form of evoked potentials. The measuring of the time (latency) and height (amplitude) of the potential give information on the function of the optic tract. The visual stimulation of the retina happens either with a light flash (blitz VEP) or a pattern with contrast reversal (pattern VEP). In pattern VEP the pupil should be not dilated. A healthy person has latency for the primer cortical potential of 100 ms.

13.7.1 Flash VEP

Measurement: The stimulation occurs with single flashes from a xenon arc lamp. The lamp should be installed at a visual angle of 20° in front of the patient. Alternatively, a standard flash like the GanzfeldERG (electroretinography) or a flash with an intensity of 1.5–3 cd/m^2 (< 1.5 Hz) can be used. The flash duration should last a maximum of 5 ms.

Indication: This method is used to measure a reduction of the transduction velocity of the optic nerve, which occurs in inflammatory disorders. Moreover, other pathologies of the optic nerve or the optic tract to the visual cortex can be detected. A comparison of both sides of the nerve response impulse is recorded and measured. The flash VEP examination is helpful in children with a profound reduction of visual acuity, media opacity, or limited compliance. An impression as to the quality of the retino-cortical transmission is given.

Evaluation: Flash lights can induce a complex answer beginning at about 30 ms and ending at 300 ms. Positive and negative peaks are numerated in order of

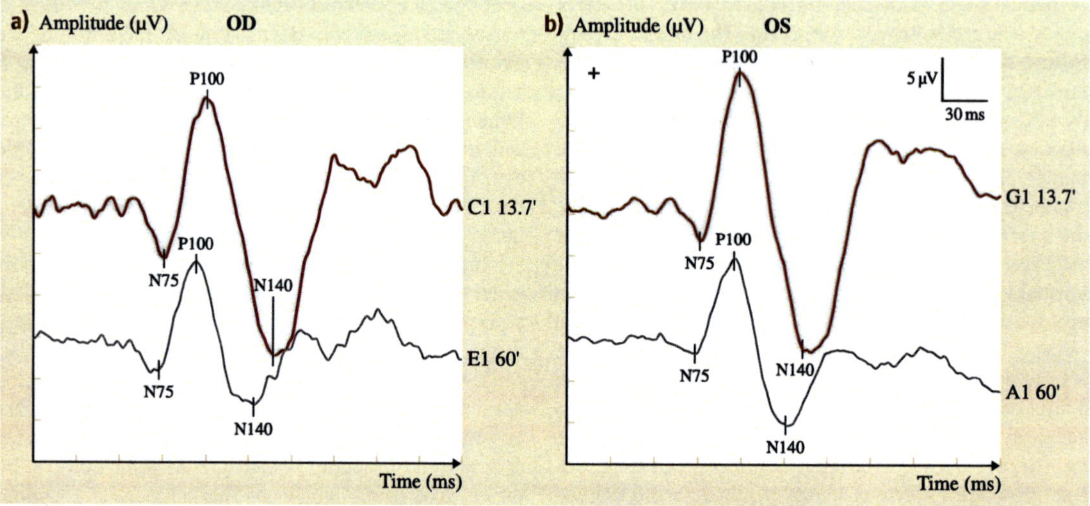

Fig. 13.5a,b Pattern VEP in a healthy subject for **(a)** the right eye (OD) and **(b)** left eye (OS)

their occurrence: N stands for a negative and P for a positive peak. The most important vertex is located in N2 and P2 with about 90–120 ms. Flash evoked VEP can be already evoked in infants in the 24th gestation week with a peak at 300 ms.

Relevance: Flash VEP is thought to be inferior to pattern VEP. The latency varies significantly interindividually. It is however, useful to follow patients over a certain period.

13.7.2 Pattern VEP

Measurement: The patient is seated in front of a display in a shaded room. For a period of 20 min a checkerboard pattern is presented on the display. A so called reversal impulse takes place so that in a rapid exchange the white array turns black and the black array turns white. One eye is covered. The other fixates a certain point on the monitor. Three electrodes are fixed on a certain area of the head to measure the impulses (small needle electrodes/acupuncture needles can be punctured underneath the scalp). After that the patient should sit comfortably and fixate the specific point on the monitor.

Evaluation: The latency can vary between different measurements and this may add up to 5%. The amplitude varies up to 25%. Therefore the latency is the more sensitive parameter in VEP. The maximum amplitudes are deduced by a stimulation of 10–20 min (visual angle). With smaller or bigger impulse patterns the amplitude of the P100 component decreases. By a large impulse pattern (> 2°) the results equal the results of the flash evoked VEP. The latency lengthens with very big and small impulse patterns (> 15°, > 2°). In general, the weaker the contrast of the stimulus, the smaller the amplitude and the larger the latency. Pattern VEP is one of the most important examinations in neuroophthalmology. However, this method is a secondary measure. It can only confirm and support a diagnosis (Fig. 13.5).

13.8 The Ganzfeld ERG (Ganzfeld Electroretinogram)

Subject: A light evoked electrical summation answer (potential modification) of neuronal and not neuronal cells can be measured.

Measurement: With an active electrode on the cornea the ERG potentials are collected and deduced against a reference electrode on the ipsilateral temple side temporal of the outer eyelid border. For the active electrode different models can be used: Burian–Allen, Henkes, Jet, and Dawson–Trick–Litzkow, and Arden. The electrodes by Burian–Allen, Henkes, and Jet are contact lenses. They are applied with a contact gel (methylcellulose fraction ≤ 5%) like therapeutic contact lens or contact lens on the cornea. When using the Burian–Allen electrode closure of the eyelid is not possible. The DTL (Dawson–Trick–Litzkow) electrode is a wire electrode. It consists of a metal coated nylon thread that is installed between the medial and lateral eyelid angle. The nylon thread should lie on the tear coating over the under eyelid border in the lower quarter of the cornea. Arden's gold membrane electrode is fixed to the lower eyelid. Contact to the conjunctiva has to be avoided. The use of an active electrode that is fixed to the skin is not recommended. As a reference electrode a silver chloride or gold cup electrode is attached temporally to the eyelid border. Before putting on the electrode the skin should be free of greasy secretion. Otherwise the skin resistance is much higher. Following the skin conductance should be improved with an electrode gel. After that the skin resistance should be $< 5\,k\Omega$ (between 10 and 100 Hz). For grounding a third electrode should be used. Usually the third electrode is applied to the forehead or to the earlobes and should have similar impedance to the reference electrode. The signals of the electrodes are collected during the measuring in an electrode box and isolated from the patient potentiated with the factor 1000–10 000. It is important that the incoming resistance at the amplifier is 100 times greater than the electrode resistance (active and reference electrode) and it is at least $10\,M\Omega$. The frequency width should be at 0.3–300 Hz and variably adjustable. For illustrating either an oscilloscope or an A/D changer is used. Normally today the last instrument is installed after the multiplier. The multiplier should have a scan frequency of at least 1000 Hz/channel to avoid loss of information. Standardized light stimuli (with a certain background lighting) are offered for measuring the potentials. The potentials are varied after the examination (e.g. photopic, scotopic ERG) in length, intensity, wavelength, frequency and the interval to obtain a differentiation between the cell specific summation results. For generating the light stimuli a complete area ball similar to the perimetry hemisphere is used. For an optimal illumination of the retina every ERG measurement should be done with dilated pupils. Before the examination a 20 min dark adaptation by scotopic and 10 min light adaptation by photopic ERG is necessary. The relation between the light stimulus (*standard flash* (SF)) and background lighting (complete area lightning) should be 3 : 34.

Indication: The most popular indications for the ERG are hereditary retina diseases and inexplicit view aggravations. With a discrepancy between the ophthalmological indication and visual acuity an electrophysiological examination is mandatory.

Evaluation: The electroretinogram consists of different components. They reflect a complex answer to the different participating neurons and mechanism. The diagram is divided into: early receptor potential, a-wave, b-wave, oscillatory potentials, c-wave, and the scotopic answer. The early receptor potential (*early receptor potential* (ERP), distal PIII-wave) consists of a cornea positive component followed by cornea negative parts. They are probably associated with a voltage transformation during the photochemical process. The positive potentials (R1) are mostly generated by the outer segment of the cones. The successive negative potentials (R2) are generated by the outer segments of the rods and the cones. Generally, a measurement is hindered because an important light intensity and a special impulse condition are needed.

The a-wave is generated by the inner segments of the photoreceptors. Depending on the examination condition (scotopic or photopic), the wave is generated either by the cones or the rods. The wave develops through hyperpolarization as an effect of phototransduction cascade (sodium ionic channel closing) in the different receptors. On the cornea a negative potential develops, which is deduced from the active electrode. In the ERG diagram only the negative branch of the negative wave of the cornea is shown. The rest is superposed through the successive b-wave. The amplitude of the a-wave is measured from the baseline right at the beginning of the a-wave amplitude until the maximum negative amplitude of the a-wave. The peak time is defined as a time from the beginning of the light stimulus until the maximum of the amplitude. A determination of the slope of the a-wave is possible by a any light intensity. With this specific information on the phototransduction process can be done.

The b-wave is developed in bipolar cells. Depending on the light stimulus and the background lighting, the results of different bipolar cells can be deduced. It is assumed that – by a scopic answer and a low stimulus (light flash period < 100 ms) – mainly the on-bipolar cells of the rods are pictured. A high stimulus adds the on-bipolar cells of the cones. The extra activated off-bipolar cells of the cones are mainly not substantial for the answer. When a measurement is done under photopic conditions the b-wave is shown as a complex answer of different (on- and off-) bipolar cells and further neurons. The b-wave follows the a-wave and is a positive amplitude. The amplitude of the b-wave is measured from the maximum negative amplitude of the a-wave until the maximum positive amplitude of the b-wave. The peak of the b-wave is defined as the time from the beginning of the light stimulus until the maximum of the amplitude of the b-wave. The ascending legs of the b-wave have fast oscillating components attached. They are called oscillating potentials. They have very high frequency (100–160 Hz) and small amplitude. The maximum of these potentials is located in the inner granular layer and the inner plexiform layer. The activity of the amacrine cells and the excitation cycle (amacrine cells – bipolar cells and amacrine cells – ganglion cells) of the inner plexiform layer are supposed to be the cause for the potentials. There is no uniform definition for the determination of the amplitude of the oscillatory potential. The amplitudes are measured depending on the certain laboratory either of a previous negative minimum to the maximum peak or from a peak of the amplitude to a later minimum. It is possible to calculate an average of the amplitude. Depending on the laboratory, the evaluation can be different.

The c-wave is a summation of the cornea positive hyperpolarization of the retinal pigment epithelium and the cornea negative hyperpolarization of the Müller cells as a cause of the decrease of the extracellular potassium concentration. The measurement should be accomplished under scotopic conditions and with high intensity stimuli (light density: 10 s). Difficulties in interpretation and the risk of eyelid artifacts finally made this a more scientific tool, which is not widely used in the clinical routine. The scotopic threshold response (STR) is a cornea negative potential that can be deduced from maximum dark adaptation conditions and from poor light intensities. They should mainly generate from amacrine cells. However, in clinics this is not of high relevance. For sufficient interpretation, every laboratory should establish its own standard values. A normal distribution of the values is assumed. Thus, a median value of 5 and 95% percentile is routinely used. Consequently, subnormal are values that are out of this 5–95%-percentile. For further evaluation, the amplitudes and the peak time of the scotopic b-wave can be imaged in a graph in dependence on the intensity. The amplitude can be described as follows:

$$\frac{V}{V_{max}} = \frac{I}{(I^n + \sigma^n)},$$

where V stands for the amplitude, I for a stimulus intensity, and σ for stimulus intensity where half maximum amplitude V_{max} is achieved; n is the value for the slope, this can be equalized with about 1. A loss of photoreceptors leads to a decrease of the maximum amplitude (Fig. 13.6).

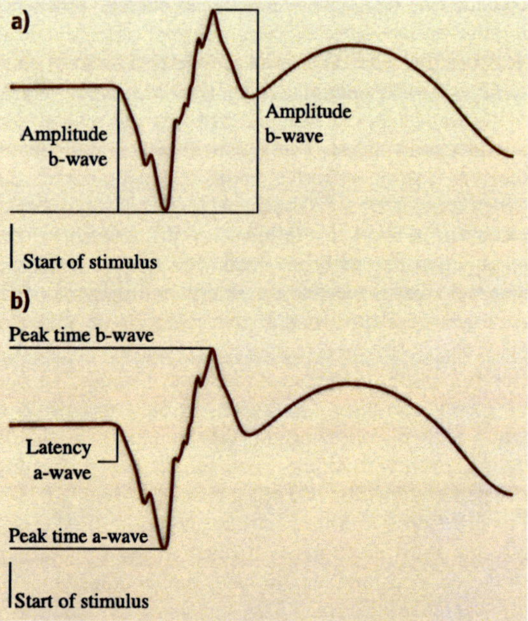

Fig. 13.6 **(a)** Evaluation of the different components of the ERG (amplitudes and peak time of the a and b-waves). **(b)** Difference between latency and peak time

Examples: For differential diagnosis typical ERG indication of retinal diseases are listed. Retinitis pigmentosa shows a decreasing amplitude and peak time changes already in an early stage of the disease even before fundus changes occur. Typically, the scotopic ERG by maximum light stimulus intensity shows a decrease of the amplitude and peak time of the a-wave. The same can be developed by an examination of the cones. With the aid of the ERG differentiation, depending on heredity pattern and characteristics, the affection and time course can be judged. In congenital Leber's amaurosis (LCA) there are only rudimentary or no answers of the receptors. Very often the ERG is pathologic and the fundus seems to be unremarkable. In cone dystrophy a rod generating ERG is normal or strongly increased (amplitudes), whereas the answer of the cone system is subnormal, or even undetectable. The cone-rod dystrophy is marked through an early progressive modification of the oscillatory potential and the rod dominated ERG. With morbus Refsum the ERG amplitude reductions are shown in both receptor types. A peak time elongation is rare. Even with a normal fundus appearance the A-betalipoproteinemia mostly shows a strongly affected answer of the rod system as compared to that of the cone system and only rarely a peak time elongation. With the ERG mucopolysaccharidoses can be classified. The mucopolysaccharidoses I-H (M. Pfaundler–Hurler) show a subnormal to zero ERG for both receptor types with a stronger affection of the rods. In the mucopolysaccharidoses I-S (Scheie) a pathological ERG answer of the cones and rods system is detectable until the third to fourth life decades. The mucopolysaccharidosis II (Hunter) shows an ERG modification with an existing retinopathy. A pigment degeneration with subnormal answer of the rods and cones system in ERG is detected in mucopolysaccharidoses III-A (Sanfilippo Type A).

Relevance: In daily routine Ganzfeld ERG is one of the most important examinations for differential diagnosis of retinal diseases.

13.9 Pattern Electroretinography (Pattern ERG, PERG)

Subject: Following constant illumination the electrical answer of the central retina is measured.

Measurement: The measuring principle is similar to Ganzfeld ERG. Only active electrodes are qualified for conduction. This is the only way to image the impulse patterns. Therefore, gold membranes or DTL electrodes should be used, and the best close-addition should be switched before. During the examination the patient should look at the impulse pattern to avoid lateral eye movements and artifacts. During the measurement the signals will be augmentedly AC-linked with a factor of 10 000. The incoming resistance for the amplifier should be at least 10 MΩ and the bandpass used should have a filter of 1–10 Hz. Because this technique has to handle relatively small answer components with a size of about 8 μV (maximum), the possibility of answer averaging (> 150) and automatic artifact suppression should be given. The latter should eliminate interfering answers of more than 100 μV. Further processing occurs automatically through a computer determining the wanted values with interpolation. The given stimulus should match a checkerboard pattern. The size of the impulse pattern is standardized. An array should add up at 0.8° and the complete area 10–16°. The contrast between the light and dark arrays should be at least 80%. The light density for the white stimulating array should be at least 80 cd/m^2 at pale background lighting. The offered impulses should change with a frequency of 1–3 Hz (2–6 reverse impulses/s) at a transient PERG and 8 Hz (16 reverse impulses/s) at a steady-state PERG. The patient should be examined at neutral pupil size. For a better fixation a binocular examination may follow. To exclude falsification of the revulsion at the worse eye, the better eye should be covered in the second examination (Fig. 13.7).

Indication: The PERG reflects the activity of the ganglion cell layer. Diseases of the third neuron such as glaucoma and ocular hypertension, diabetic retinopathy, optic neuritis, hereditary diseases of the optic nerve, drusen of the optic nerve, and compressions of the optic nerve can be diagnosed with the PERG. In combination with the pVEP diseases of the anterior visual pathway are evaluated.

Evaluation: Graphical imaging of the PERG results in a curve along the time axis. The curve includes two valleys and one peak, which are described as components. The first component (= N35) is a valley at 35 ms. This followed by a peak or positive component (= P50) at 50 ms. The second negative component (= N95) is at 95 ms. The components are investigated by using the following values: peak time and amplitude height. The amplitude of the N35 component is a deflection of the baseline until the first negative maximum (valley) of the N35-component. The amplitude of the P50-component is measured from the first maximum negative deflection of the curve (N35) to the highest peak of the curve. If the N35-component is not determined, an average out of the values at zero and the values from the beginning of the P50 component should be taken. The amplitude of the N95-component is either calculated from the peak of the P50-component or from the baseline of its own maximum negative deflection. The peak time matches the time period of the stimulus until the particular negative or positive maximum (N35, P50 and N95). In humans, the components are most likely generated by both neuronal and nonneuronal structures. The P50-component is generated by cells of the inner retina layers such as amacrine cells. The N95-components are mainly generated by ganglion cells, and the P50-component is generated mainly anterior the ganglion cells. A small portion of the ganglion cell layer may also contribute to the P50-component. The quotient N95 : P50 helps differentiate the location of the pathology. A combination of PERG values and pattern VEP is always reasonable for differential diagnosis. A pathological P100-latency in the pattern VEP helps to differentiate between dis-

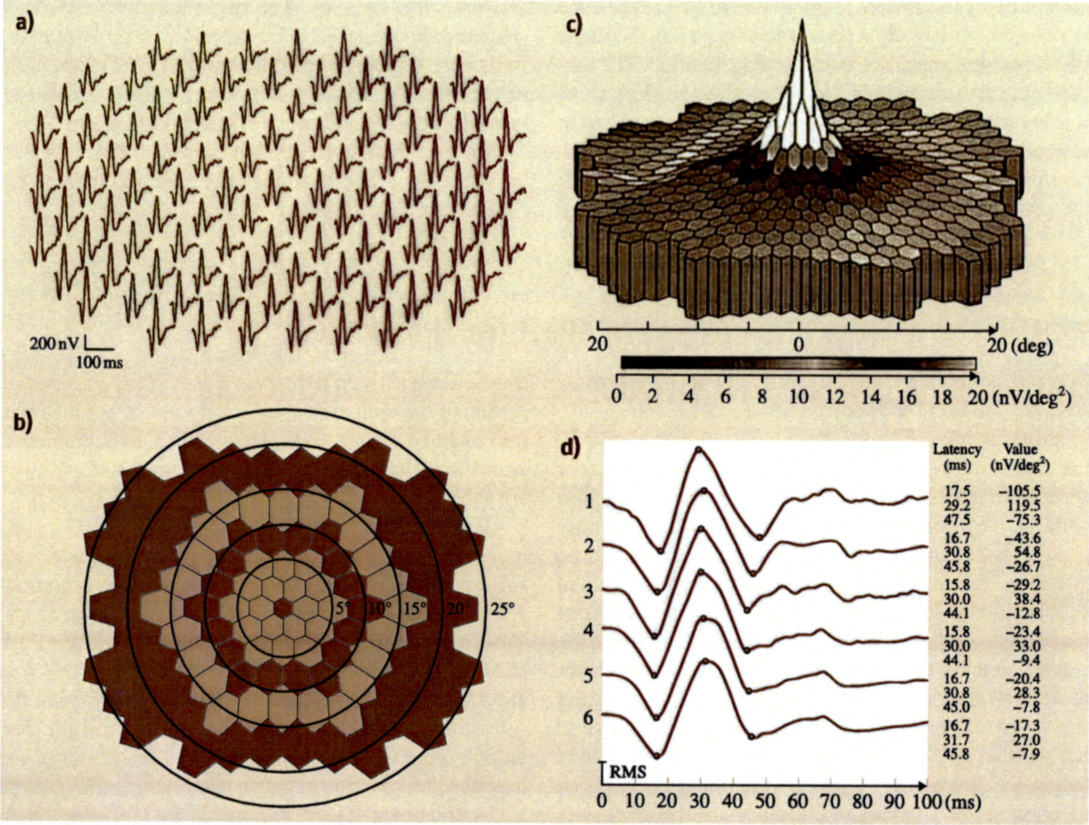

Fig. 13.7a–d Multifocal ERG. Physiological findings. **(a)** Focal ERG answers (trace arrays) calculated from 103 hexagonal stimuli in the central 25° **(b)**. **(c)** three-dimensional presentation (scalar product) of the answers shown in **(a)**. **(d)** Focal ERG answers presented in five groups around the center **(b)**. The value of the amplitude relates to the plane of the respective ring groups (nV/deg^2)

eases of the optic nerve (with and without ganglion cell alteration) or retinal diseases.

Cases study: Patients suffering from open-angle-glaucoma exhibit PERG changes (reduced amplitudes) before visual field defect can be evaluated. Particularly the N95-component is affected. A compression of the optic nerve may lead to a decrease of the N95-component and the quotient N95 : P50. In rare cases, the P50-component can be reduced. Such a subnormal PERG is combined with a bad prognosis for the visual function. In diabetics a neuronal damage shown on PERG may occur even before detection of fundus changes. Such patients show reduced amplitudes, which are more pronounced with progression of the disease. Patients suffering from diabetic retinopathy who show a progression of the P50-component alteration are judged as those with bad prognosis. PERG is qualified for a good differential diagnosis in alterations of the ganglion cell layer induced by macular diseases. Typically, the absolute values of the P50 and the N95 components are reduced in macular degeneration; however, the ratio N95 : P50 remains unchanged. In 40% of the cases diseases of the optic nerve result in alteration of the PERG. In 85% of the cases one may find an unchanged P50-component and a reduction of the N95-component. Depending on the stage of the disease, optic neuritis results in a more or less pronounced alteration of the PERG. In the acute phase (mainly in the second to the fourth week) a reduction of the amplitude of the P50-component can be found; this phenomenon is reversible. Parallel to the recovering P50-component, a reduction of the N95-component is found; this also recovers during the first 2–4 weeks. Hereditary atrophies of the optic nerve result in a decreasing amplitude of the N95-component and a decrease of the quotient N95 : P50. A subnormal amplitude of the P50-component is uncommon. Leber's disease shows an acute and subclinical stage before visual acuity is reduced and a decrease or loss of the N95-component is detected.

Relevance: PERG represents a good and efficient instrument for the evaluation of the function of retinal ganglion cells.

13.10 Multifocal ERG (mfERG)

Subject: The electroretinographical answers from different retinal areas of the central visual field (corresponding to 25° isoptere) are recorded. Both the cone system and the nerve function can be estimated.

Measurement: mfERG is similar to Ganzfeld ERG and PERG. The active electrodes can be Burian–Allen contact-len electrodes, or DTL-electrodes. It is important that both the fixation and the visual acuity are not affected by the lens. Like in Ganzfeld ERG and PERG, a reference electrode is installed at the temporal margin of the orbit of the ipsilateral site; another electrode is placed in the middle of the forehead. The signal should be intensified from 100 000 till 200 000 and filtered with a bandpass of 10–300 Hz. Afterwards it is digitalized with a scanning rate of 16/picture frequency. From the ERG summation answers at different stimuli a local ERG answer can be extracted with m-transform-algorithms. Additional calculation models help to gain more information on the inner and outer retinal layers. The stimulation is done via a computer monitor with a picture frequency of 75 Hz. The stimulation field consists of close packed hexagonal (61, 103 or 241) arrays. In the center the size of the arrays is about 4.7 times smaller as compared to the periphery. The light density of each hexagon is modulated independently of each other according to a pseudo-randomized binary m-sequence. Within this sequence each hexagon alternates independently from each other between black and white. The average light density should be 100–200 cd/m^2 for bright and < 1 cd/m^2 for dark stimulus elements. The contrast should be at least 90% and the brightness over the entire monitor field should not vary more than 15%. The stimulation lasts about 8 min divided into parts at 60 s. During this time period a light density change of each hexagon takes place at a value of 2^{14} times. The measurement should be accomplished at best possible visual acuity and with dilated pupils.

Indication: mfERG is qualified for the evaluation and differentiation of macular diseases and is, therefore, a further means for differential diagnosis of visual acuity alterations and retinal changes of unknown origin. With the aid of mfERG diseases can be detected even before morphological signs are seen. This is true for the following diseases: Morbus Stargardt, other macular dystrophies, and toxic maculapathies such as chloroquine maculopathy.

Evaluation: The graphical imaging of the mfERG can be done in many ways. The curves of each hexagon can be imaged individually in the trace array. For each hexagon a single ERG curve with two negative maximums (valleys) and a positive maximum (peak) exists. Here, the first negative maximum is described as the N1-component, the first positive maximum as the P1-component, and the second negative peak as the N2-component. The N1-amplitude is measured from the baseline to the maximum negative deflection of the N1-component. The P1-amplitude is measured from the maximum of the N1-component until the maximum of the P1-component occurs. The peak time is measured in analogy to the evaluation of the Ganzfeld ERG – from the beginning of the stimulus until the particular maximum of the curve occurs.

For facilitation of evaluation the computer creates a scalar product from the point product of the local answer and a sample. For the sample, the average focal answer of a normal population, the normalized averaged answer of the complete examination, or the area corresponding to a hexagon is used. The calculated scalar products (in rare cases the amplitude of P1) are used for one, two, or (rarely) three-dimensional pictures. The sizes of hexagons are chosen that way so that the result of each hexagon is equal. By the dividing the calculated scalar answers through the area one obtains the cone density per hexagon (nV/deg^2). In a healthy patient the three-dimensional mfERG image offers an image that is pointed in the middle and surrounded by a flat sinking area. The graphical image is a simplification of the results, which may lead to a misinterpretation. Latency shifting may lead to a decrease of the scalar product, which can again change the graphical image.

However, more difficulties occur if the recording is not done adequately. The automatic evaluation would take the measured noise as an answer and would divide this through the different array sizes and lead to a supposedly pointed mfERG image. Besides that the three- or two-dimensional evaluations of the results can also be used for the calculation of answer groups. In this case, the hexagons are combined in a ring group around the fixation center. Depending on the instrument, five to six ring groups are created. The ring groups are numbered from 1 to 6 starting from the middle. In the graph six curves are compared with each other and present a statistics with receptor density, latency, and amplitude heights per group. Besides that the mfERG can calculate a record of the time of change between light and dark. The different answers are classified in 1. kern (first order kernel) for the deduction of brightness and 2. kern (second order kernel) for the deduction of change between light and dark. As the first order kernel is considered as an answer of the outer and middle retinal layers, the second order kernel is mostly a part of the inner retinal layers. With extraction an *optic-nerve-head-component*, a pure ganglion cell component, can be conserved.

Case study: Depending on the disease, a marred function of the central retina is expressed differently. Stargardt's disease shows amplitude reduction even before the appearance of morphological changes. Peak time changes may be found in 10% of advanced stadiums. In contrast, retinitis pigmentosa shows – besides an amplitude reduction – a distinct extension of the peak time that accumulates to the periphery. Also carrier of an X-chromosomal retinitis pigmentosa can show amplitude reduction and peak time extension.

Relevance: The mfERG is mostly a clinical tool (next to fluorescence angiography and high resolution OCT).

13.11 Electrooculograms (EOG)

Subject: The electrooculogram registers a constant existing potential on the eye. The origin of the potential is a transepithelial electrical charge difference that lies over the pigment epithelium.

Measurement: Silver chloride electrodes or light polarized electrodes such as the Ganzfeld ERG reference electrodes are suitable for the measurement. The electrodes should preferably be fixed very close to the temporal and medial eye angle. The skin should be clean and treated beforehand with a conductive paste. The impedance should be less than 10 kΩ and should have a frequency between 30 and 200 Hz. For grounding an electrode should be fixed on the forehead. The deduced signal should be amplified with 5 K. An alternating current amplifier is used to avoid artifacts. The stimulation is done in analogy to Ganzfeld ERG. A stimulus light density of 50–100 cd/m^2 is recommended. To achieve a better illumination of the retina the examination should be done with dilated pupils. If dilation is not possible, it is suggested to use a higher stimulus illuminating density (400–600 cd/m^2). During the measurement smooth horizontal saccades with the duration of 2–5 s should be evoked; there should be at least 10 saccades/min. Before the examination illumination of the eye (e.g. ophthalmoscopy) should be avoided. To start the examination, a 15 min preadaptation phase should be maintained. This dark period follows a light phase. At least 5 min before the beginning of the light phase the patient should start with the saccades. At the same time the recording of the signals should start. The examination should not finish before 25 min after the beginning of the light phase. Next to this standard examination with *slow oscillation* with a slow change between dark and light, almost oscillation is established with faster changes between light and dark. Here six cycles are done throughout the light and dark phase (length: 60–80 s each). During this time the patient has to make horizontal saccades such as slow oscillation measuring.

Indication: The EOG is a noninvasive method that gives information on the function of the pigment epithelium. Therefore, the EOG can give differential diagnostic information on retinal diseases. It can be used particularly for an exclusion of pigment epithelium alterations resulting in unspecific visual acuity reduction. You may differentiate between a vitelliform Best-maculopathy and an adult vitelliform macular degeneration.

Evaluation: The deduced signal of the slow oscillation examination gives a rectangular swinging curve (per one movement of the globe). The voltage difference between the upper and lower maximum of the wave is of high interest. After dark adaptation the difference decreases until a so called dark valley occurs. For the value of the dark valley use the lowest voltage difference during the dark phase. Normally it can be measured after 11–12 min of darkness. For the determination of the remaining potential, the average from the measurement of the beginning of the light phase should be used. After the illumination, in the first 5 min there is a steep rise in the voltage difference per bulb movement (passage). At the end of the increase the light peak is reached. The relation of the light peak to the dark valley (Arden quotient) is an important estimation scale for the EOG. Besides this there is a quotient from the light peak and the rest potential that is always smaller than the Arden quotient. Latency describes the time from the beginning of the light phase until the light peak. Next to the quotient from the light peak and dark valley, the values of the continuance potential (μV/deg) can be seen. Like any other electrophysiological examination, the EOG needs its own standard values for differentiation of pathological indications. The almost-oscillation examination gives a voltage difference curve with changing potential peaks and valleys. Out of the six cycles a peak and valley average value is established. Out of it a quotient is formed. Additionally, an average latency is defined. The duration of the latency is called similar to the slow oscillation examination, the time between dark valley and light peak. It is possible for incorrect measurement to develop by head movement. Then by mistake a small quotient from light and dark values can develop. Therefore, it is important that the head of the patient always be in the *straight ahead* position.

13.12 Adaptometry

Subject: The time that is necessary to recognize a pattern is measured. The identification of the pattern depends on the light intensity. Through modification of the variables different adaptation times can be calculated.

Indication: Adaptometry is qualified for the diagnosis of cone and rod dysfunctions such as cone-rod dystrophy, retinis pigmentosa, retinis-pigmentosa associated syndrome, congenital night blindness, fundus albipunctatus, Oguchi's diseases, Stargardt's disease, and achromatopsia. In addition, inflammatory disorders such as posterior uveitis, intoxications with chloroquin, ethambutol, or iron oxide can be detected with the aid of adaptometry as a reduced retinal function.

Measurement: For the measurement an illuminated hemisphere like for the EOG and ERG examination is used. The examination starts in a state of brightest illumination with dilated pupils and with best corrected visual acuity. The pupil diameter should be documented before and after the examination. After 5 min adaptation time, the patient may recognize a focusing screen with a striped pattern; with the aid of a scanner the patient can inform the examiner about the identification of the pattern. The duration until to the signal is registered automatically. In the following steps the test light intensity is always reduced after the identification of the pattern about a constant level (about 1.8 log units). Then an approach of test light brightness until the odor detection threshold of the patient is made. If the patient recognizes the striped pattern, the wanted threshold is reached and the process can be repeated.

Evaluation: The measured values result in a sinking curve of the noticeable light intensity (relative threshold impulse energy) along the dark adaptation time. Depending on the wavelength (green or red), the curve is steep or flat. Normally, the adaptation curve shows a biphasic progress with a characteristic bend at the crossing from light to dark. The bend, called Kohlrausch's bend, develops typically after 6–7 min; it marks the end of the function of the cones. From Kohlrausch's bend a slope of sensitivity develops just through the function of the rods. The slope of 3–4 decimal power takes place typically over 30–50 min until the absolute threshold in the adaptation is reached in the end. Pathological curves are usually monophasic, because either the cone's or the rod's part is missing in the curve.

Relevance: Adaptometry demonstrates a sensible instrument for the detection and progress of retinal diseases. It is used in both clinics as well as practice. Concerning the diagnostical power, adaptometry is inferior to Ganzfeld ERG because compliance of the patient is needed and a relative long examination time is required.

13.13 Aberrometry (Wavefront Analysis)

Subject: Measurement of the alteration of retinal imaging quality by higher order refractive errors.

For the quantification of the complete image, the following terms of definition are accomplished: wavefront variance and Strehl ratio. The wavefront variance quantifies the modification of the entire optics. In the human eye it is depending on the pupil size. It is calculated directly from the single wavefront alterations (see the evaluation). From the wavefront variance (RMS) the optical quality can be calculated with the Strehl number (Strehl ratio). This is a value that reflects the image quality of the optics. It is calculated from the relation between the generated intensity maximum and the maximal possible intensity of the aberration free image.

Indication: In refractive surgery aberrometry is suggested to minimize calculation mistakes of higher order.

Principle: Aberrometry can be determined with the aid of different instruments, such as the Hartmann–Shack sensor, the Tscherning aberrometer, the *ray-tracing* aberrometer, and the *spatially resolved refractometer*. The Hartmann–Shack sensor is a widespread aberrometer that creates a sharp pictured focus point on the retina. This is reflected from the retinal surface so that a punctual light source is generated on the retina. This punctual light source is the origin of the spherical wavefront that is broken in a bunch of parallel light rays after the passage through the optics of the eye. With the aid of micro lenses the bunch is broken as a lot of small focal points that are pictured with a CCD camera. If the optic is optimal, the images of the single rays are located on the picture at a similar distance from each other. If this is not the case and the pictures differ from each other, the wavefront difference can be determined with the aid of the divergence of the reference position in x- and y-direction through deduction.

The Tscherning aberrometer uses a laser ray that is divided in many rays with an aperture mask. The part rays run through the optics in the eye and are projected on the retina. The resulting pattern on the retina can be observed with an ophthalmoscope lens. If the optic is optimal, all points are on the retina at an equal distance like before the passage in the eye. If this is not the case, out of the divergence in x- and y-directions the wavefront difference can be determined like with the Hartmann–Shack sensor.

With the ray-tracing sensor the eye is examined with a laser ray parallel to the optical axis. The light ray is broken by the cornea and the lens and at last hits the retina. In the case of optimal optics the light rays meet each other at the same location on the retina. If the striking positions differ from each other it can be determined with the *position-sensing detector*. Out of the dislocation in x- and y-directions the wavefront can be calculated analogously to the previous process.

The *spatially resolved refractometer* or Scheiner aberrometer uses two laser rays. One beam is fixed and the other is movable. The fixed laser beam processes through the center of the pupil and the second light ray is movable on different positions in the pupil. At nonoptimal optics both rays hit misplaced of each other on the retina. By adjusting the incoming angle and the incoming position, the course of the ray can be changed in such a way that the rays combine on the retina. With the angle and the pupil position the wavefront can be calculated.

Evaluation: A Strehl number of 1.0 stands for an aberration-free refraction. In reality this is almost not possible. Therefore, in practice a physiological optic is seen as aberration free if the Strehl ratio is greater than 0.8.

The results are shown in dependency of the size of the pupil. Normally, a pupil diameter of 4 mm is used. The image is mostly color-coded: red describes a positive aberration (in μm) and blue a negative aberration. The colors green and yellow represent an optimal wavefront. In addition to the complete values, the wavefront variance can be calculated.

The wavefront variance is also called RMS wavefront mistake (root mean square wavefront mistake). It is a radical from the average of the square of the RMS. For evaluation the values are set in relation to the wavelength. Afterwards, the optical quality is judged with the Marechal criterium. The Marechal criterium says that the optical image is optimal if the relation to the RMS wavefront mistake to the wave lengths is smaller than $\frac{1}{14}$. If the pupil width of an eye is just as big as the relation of the RMS and the wavelength is $\frac{1}{14}$, there is a present critical pupil size.

Relevance: Aberrometry has become an important instrument in refractive corneal surgery. In this field the demand for visual quality is very high. The aberrometry is the last step for evaluating an optical image on the retina. Navigating wavefront lasik leads to a better visual acuity as compared to a normal lasik. The numbers of aberrations of higher order can be reduced.

13.14 Keratometry

Subject: Keratometry describes the measurement of the radius of curvature of the cornea in different meridians.

Examination: With the aid of a cornea-mirror the image can be examined. Both, the power and the axis of the central corneal astigmatism are measured. With the aid of a microscope the projection of the images on the cornea can be seen. The entry angle can be changed manually. At an optimum entry angle the two cornea projections are getting in contact. If the angle is too big the images are too far apart. If the angle is too small the images are partly covering each other. The entry angle is calibrated so that the radius of the curvature can be calculated or read from the instrument.

Indication: The keratometry is important for the diagnosis of a keratokonus, for contact lens adjustment and the calculation of an intraocular lens (biometry).

Relevance: Today an automatic determination of the radius of curvature of the cornea can be achieved with modern instruments (e.g. automatic refractometer, IOL-Master, corneal topography). Therefore Javal's ophthalmoscopic keratometry or the Zeiss bomb became rare tools during the last years.

13.15 Retinoscopy or Skiascopy

Subject: The determination of the refractive state of the eye by means of a retinoscope. Retinoscopy or skiasopy is performed with the patient fixating binocularly a near object such as a letter, a word, or a picture mounted on, or held close to, the retinoscope and wearing the distance correction. No working distance lens power is subtracted or added to the finding, since the plane of regard is at the same distance as the retinoscope. The examination criteria are the observation of light-shade-phenomena in the patient's pupil. Those phenomena develop when the projection of a light beam on the retina of the patient occurs.

Measurement: The electric hand-held retinoscope projects a light-beam into the pupil of the patient. This results in a red light reflex. The examiner sits at 50 cm distance in front of the patient with a correcting glass of +2.0 dpt. This is to compensate the distance and – in the case of emmetropia – to focus the reflected light to the examination distance. By turning the skiascope around its vertical axis the examiner can observe how the light reflex drifts in the pupil. In hyperopia the light reflex moves in the same direction, in myopia it moves to the opposite. The examiner may now change the refraction by applying correction glasses. If the correct refraction has been reached the light reflex does not move when the skiascope is turned. In an emmetropic eye the retina illuminates very shortly when turning the skiascope. This flash is called the flicker point. For the determination of astigmatism, the skiascope must be used horizontally and vertically with a bar-shaped light source. Alternatively, the retinoscope can be moved to the direction of the astigmatism axis and then vertically. The difference between both corrections is calculated. This describes the cylinder value in the direction of the astigmatism axis. Single glasses may be replaced by a bar with integrated glasses at different diopter values. For fast measurements these retinoscopy bars are very convenient when examining children. Accommodation power needs to be eliminated as with other procedures (cycloplegia).

Relevance: The usage of automated refractometers (AR) to obtain the refractive power is very popular and AR have replaced the skiascope almost completely (except for children or patients without cooperation). With the help of skiascopy the refractive status of the eye can be determined very quickly.

13.16 Ultrasound

Subject: Using ultrasound technology exact one dimensional (A-Scan) or two-dimensional (B-Scan) measurements of the length of the globe, anterior chamber depth, or corneal thickness can be performed. An ultrasound signal is reflected from the boundary layers of different media. For calculating the distance in mm the following formula is used

$$\text{distance [mm]} = \text{velocity of sound [m/s]} \times \text{duration } [\mu s/2000] .$$

The velocity of sound differs depending on the respective medium. In the vitreous body it is at 1532 m/s, in silicone oil 984 m/s. The average velocity of sound in the entire eye system of a phakic eye is at 1550 m/s.

13.16.1 A-Scan Ultrasound, Ultrasound Biometry

Indication: A-Scan technology is mainly used for the biometric measurement of the axial length. This has to be performed before cataract surgery or lens implantation. Other indications are the measurement of the thickness and the evaluation of intraocular tumors (if necessary intraorbital tumors of the anterior orbita) and the imaging of the retinal, choroidal detachments, or retinoschisis.

Measurement: An 8 MHz ultrasound probe with parallel sonic rays is held over the eye surface, which is covered with gel or water. When using water a cone is placed over the cornea. The measurement results in a two-dimensional diagram. On the x-axis the distance of the ultrasound probe is shown; the y-axis represents the intensity of the reflected sound. The measurement is made between two echo peaks that develop at the boundary of two media. Therefore, for the determination of the size the difference of the echo peaks is used. An exact measurement for ocular biometry can be made by means of the so called section biometries. It is recommended that the single sections (anterior chamber depth, thickness of the lens, and distance from the lens to the retina) be determined separately. A possible error source in biometry is the age of the patient and the deformation of the globe. If the patient is younger than 14 years, the growth of the globe must be taken into account. A staphyloma in a high myopic eye can pretend a short globe. The cooperation of the patient is very important. Eye movements may result in falsification of the axis length values. Therefore, with a subject without cooperation, the investigation should be done under general anaesthesia.

For the determination of the axial length the average has to be calculated from 4–5 measurements. In 90% of the cases the axial length of the partner eye is almost equal. If not, the complete measurement procedure should be repeated.

For the calculation of the intraocular lens the axial length value, the anterior chamber depth, the keratometry results, or the refractive error and the A-constant are needed. Many formulas exist for this purpose. They can be divided into two main groups: physical or geometrical-optical formula that are based on a theoretical model of the eye and empiric (regression) formula that are calculated from statistical postsurgical data. The A-constant depends on the lens that is to be implanted. It varies depending on the design and material. The empirically determined formula are very precise in eyes with an axial length of 24 mm ± 1 mm. This matches the majority of the population. With a longer or shorter axial length it is recommended to use the theoretical formula, because in those cases the empirical formula is mostly not adequate.

Relevance: Ultrasound biometry has become almost completely replaced by laser interference biometry. Only in media opacities such as a mature cataracts is ultrasound biometry still used to calculate the intraocular lens.

13.16.2 Ultrasound Pachymetry A-Scan

Intention: The measurement of the corneal thickness.

Corneal pachymetry can be performed as follows: Measurement of the cornea thickness with ultrasound (20 MHz probe).

The determination of time that is necessary for the sound to passage the cornea enables the calculation of the corneal thickness.

13.16.3 Ultrasound B-Scan

Indication: The B-scan examination is suitable for a two-dimensional imaging and measurement of retinal or choroidal tumours, choroidal detachments, optic nerve head and vitreal abnormalities, as well as intravitreal foreign bodies.

Measurement: Similarly to the A-scan examination the contact surface of the probe is moistened with gel or water. When examining the anterior chamber a water bath should always be used. The graphical image is a cross-section through the eye or the anterior chamber. With the help of moveable calipers the diameter of the structures in the cross-section can be determined in all directions. After marking the areas of the structures examined, some instruments can determine certain values automatically.

Relevance: Because of its high resolution, the easy handling, lack of side-effects such as x-rays, the low costs, and the availability, the B-scan ultrasound is an important method for the for the evaluation and follow-up of many ocular changes and disorders.

13.17 Corneal Topography

Subject: The shape of the corneal surface is measured and imaged.

Measurement: There are different ways to measure the shape of the cornea. In the following, a short summary on the placido plate and Astra-Max technique is given.

The Placido plate imaging is based on a distortion, or if needed a superposition of light rings projected on the cornea. Pictorially, the pattern on the cornea is reminiscent of a bull's eye.

Rules: The closer the circles, the steeper the cornea, or the further the circles, the flatter the cornea. The Placido plate instruments may vary with the principle of examination. There are two types of instruments that are either working proximally to the eye (*small target type*) or examining it with a certain distance (*large target type*). Both instruments have pros and cons.

The *small target* system need less light and can examine a bigger area but is sensitive to movements. Furthermore, depending on the face anatomy, the measurement may be more difficult. The *large target* system has opposite characteristics to the *small target* system. It needs more lightning, can only examine a relatively small area of the cornea but is not as sensitive to movement artifacts. The Astra-Max three-dimensional topography uses a concentrically oriented *polar grid*. The pattern is recorded and automatically worked out from three different angles. With the three-dimensional image from geometrical calculation, the flexion of the cornea over the complete cornea *limbus to limbus* can be calculated in better quality.

Indication: Generally speaking, there are two fields of application for corneal topography. The follow-up of keratektasia and differential diagnosis as well as pre- and postsurgical assessment of the cornea following refractive surgery. Furthermore, pterygium-induced astigmatism can be imaged and the indication for surgery can be validated. Visual acuity alterations of unknown origin may also be further examined by this technique. Early or abortive forms of keratoconus can be identified and quantified.

Evaluation: The results are usually imaged color coded. If the corneal surface shows no astigmatism in the center a round or oval area of light-yellow to light-blue color is shown. By astigmatism a typical hourglass figure is shown. If the axis of the hour-glass figure is vertical it is called astigmatism rectus; if the figure is horizontal it is called astigmatism inversus. In oblique astigmatism the hour-glass figure is diagonal. If the hour-glass is distorted this may be judged as an irregular astigmatism. It can be a precursor of a keratoconus (lazy-eight) or induced by a scar. Other images that indicate a keratoconus is a pronounced eccentric ectasia that shows a red color on a green or dark-blue (later stage) area.

Relevance: Corneal topography is an essential measure before every refractive procedure. After surgery it is used to investigate the result. It is, therefore, mandatory in every refractive institution. It is also important for diagnosis and follow-up corneal ectasia in diseases such as keratoconus and for the investigation of new therapeutic approaches (cross-linking) (Fig. 13.8).

13.18 The Orbscan

Subject: The shape of both the cornea surface and the back face as well as the lens surface are measured. In addition, the following information is collected: anterior chamber depth, topography of the iris surface, pupil localization and diameter, angle κ, white to white distance, and the thickness of the cornea.

Measurement: The measurement is made with a movable chink of light that makes – like a slit lamp – a cross-section of the anterior eye segment. The images are made of the chink of light in less than a second and they are evaluated automatically. From the values the front and back face of the cornea can be calculated. This technology gives important additional information on the anterior eye segment.

Indication: The orbscan is an ideal instrument especially for refractive procedures because it enables not only simple corneal topography but also a topography of the corneal back face and the lens surface. Moreover, the orbscan can determine the cornea thickness before a laser examination.

Evaluation: The evaluation is made in analogy to the diagnosis of the corneal topography, the pachymetry, and the imaging of the anterior eye segment.

Relevance: The orbscan instrument is outstanding as compared to other procedures for corneal examinations

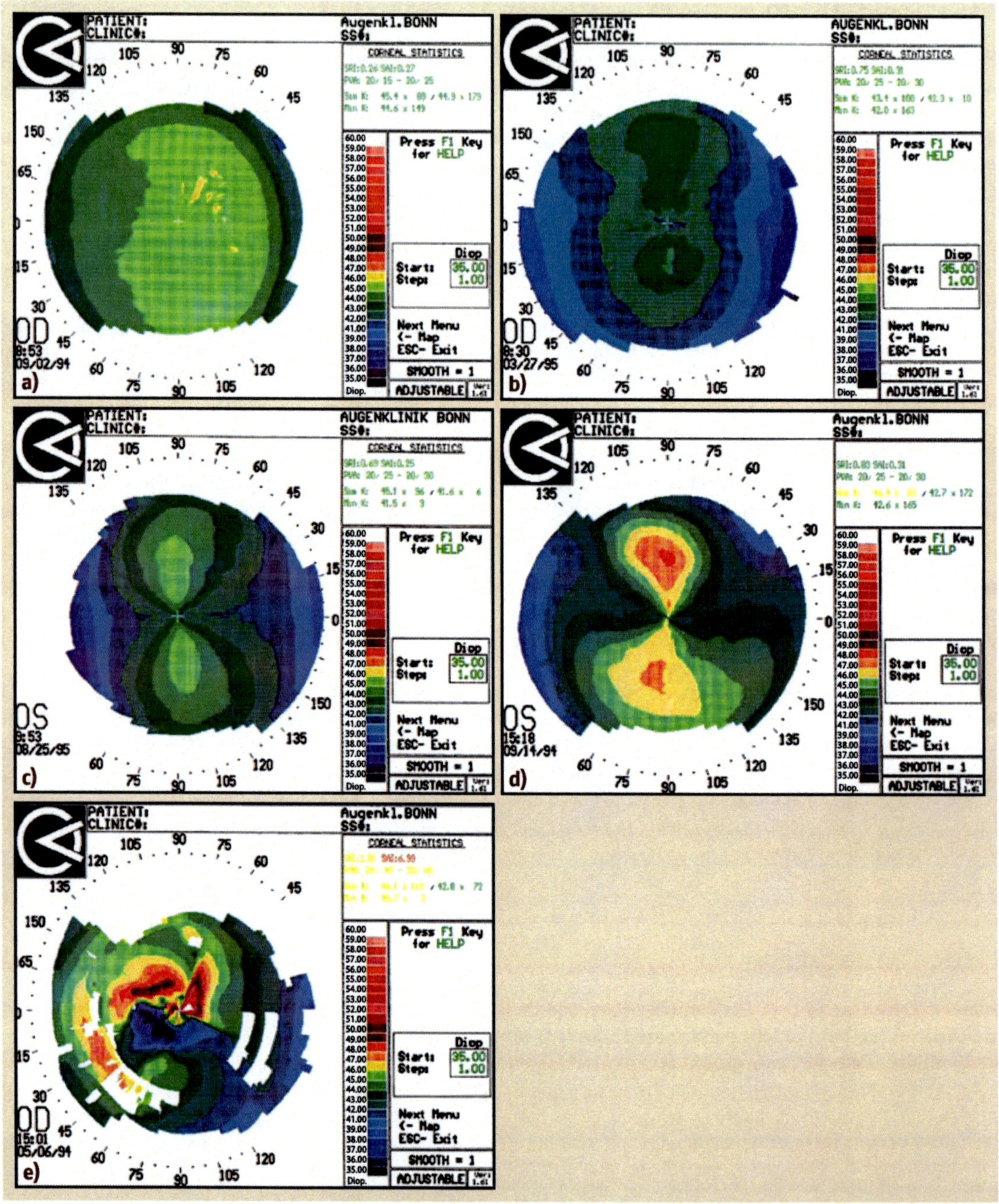

Fig. 13.8a–e Qualitative patterns in corneal topography according to *Bogan* et al. [13.1]. **(a)** Round, **(b)** oval with slight hour-glass shape, **(c)** symmetrical hour-glass shape, **(d)** asymmetrical hour-glass shape, **(e)** irregular

because it can provide more information (topography, pachymetry) in one session. In pachymetry the values vary at ≈ 23–28 μm as compared to the ultrasound examination.

13.19 Scheimpflug Examination

Subject: Imaging and measuring all visible anterior segment structures with a computer based reconstruction of 20–50 pictures.

Measurement: The measurement is made automatically with a high-resolution camera that rotates around the eye in less than a second. Afterwards, the images gained are calculated to a three-dimensional hologram.

Indication: Like the orbscan measurement the Scheimpflug camera is an ideal instrument for imaging and measuring different structures and proportions of the anterior segment.

13.20 Fluorescence Angiography of the Retina (Sodium-Fluorescein)

Subject: The time of the influx and the dispersion of the coloring in the retinal vessels and the choroid are imaged and documented with a specially designed fundus camera (Fig. 13.9).

Indication: Today fluorescein angiography is the standard imaging procedure in many retinal and choroidal diseases.

Implementation: For the examination about 500 mg sodium-fluorescein (10% solution in 5 ml or 5% solution in 10 ml) is given intravenously. Incoming blue stimulating light (excitation filter: 490 nm) lifts the electrodes of the molecule sodium-fluorescein to a higher energy level. Once the the electrodes slide back to the ground state green light (530 nm) is emitted. A yellow-green band elimination filter prevents a picture of the reflected blue light. The picture is usually documented digitally.

Sodium-fluorescein has a relatively low molecular weight (formula: $C_{20}H_{10}O_5Na_2$), the molecule diffuses through Bruch's membrane and the pores of the choroidal capillaries. Physiological diffusion barriers are the large choroidal vessels, the retinal vessels, and the retinal pigment epithelium.

The injection may lead to local irritations (by extravasations of sodium-fluorescein), illness, vomiting, syncopes, asthmatic attack (rare). In addition, anaphylaxia (very rare) has been described. Thus, an angiography room should be equipped with the usual emergency tools.

Evaluation: *Phases of the fluorescein angiography.*

Fluorescein angiography is a dynamic procedure. Thus, the temporal process of the angiography is documented and divided into different phases. It starts at about 12–25 s (arm-retina-time) after the beginning of the injection.

- Prearterial phase (filling of the choroid): The filling of the choroid occurs about 1 s before the filling of the retinal vessels. It can be observed as a homogenous background fluorescence. The intensity negatively correlates with the pigmentation level of the fundus.
- Arterial phase: This starts immediately after the filling of the choroidal capillaries and ends with the complete filling of the arteries.
- Arteriovenous phase: This is marked by a complete filling of the arterioles and capillaries and ends with the beginning of a laminar filling of the veins.
- Venous phase: This ends with a complete filling of the veins and can be divided into an early (arterioles and venules fluoresce at a similar level) and a late venous phase (obviously less fluorescence of the arterioles).
- Late phase (recirculation phase): A retinal passage (filtration) has occurred. This results in a distinct decrease of the fluorescence in healthy subjects.

Hypofluorescence

Hypofluorescence can be either the cause of a blockade of the regular fluorescence or the result of a filling defect (less diffusion).

Hyperfluorescence: Hyperfluorescence can be the result of an intensified background fluorescence following a defect of the retinal pigmented epithelium. This transmission defect develops from focal RPE atrophy or missing RPE cells with the result of *uncovering* the choroidal fluorescence. It is characterized by an early hyperfluorescence. Its intensity is increasing and then fades without modification in size or shape. In addition, hyperfluorescence can occur by extravasation of fluorescein from the choroidal vessels (defect of the outer blood–retina barrier).

Defects of the retinal vessels or abnormal newly formed vessels (such as proliferations) can lead to a fluorescein leakage. In addition, staining of the tissue can be the result of dye retention. This occurs in drusen or scars and can be seen in the late phase of the angiogram.

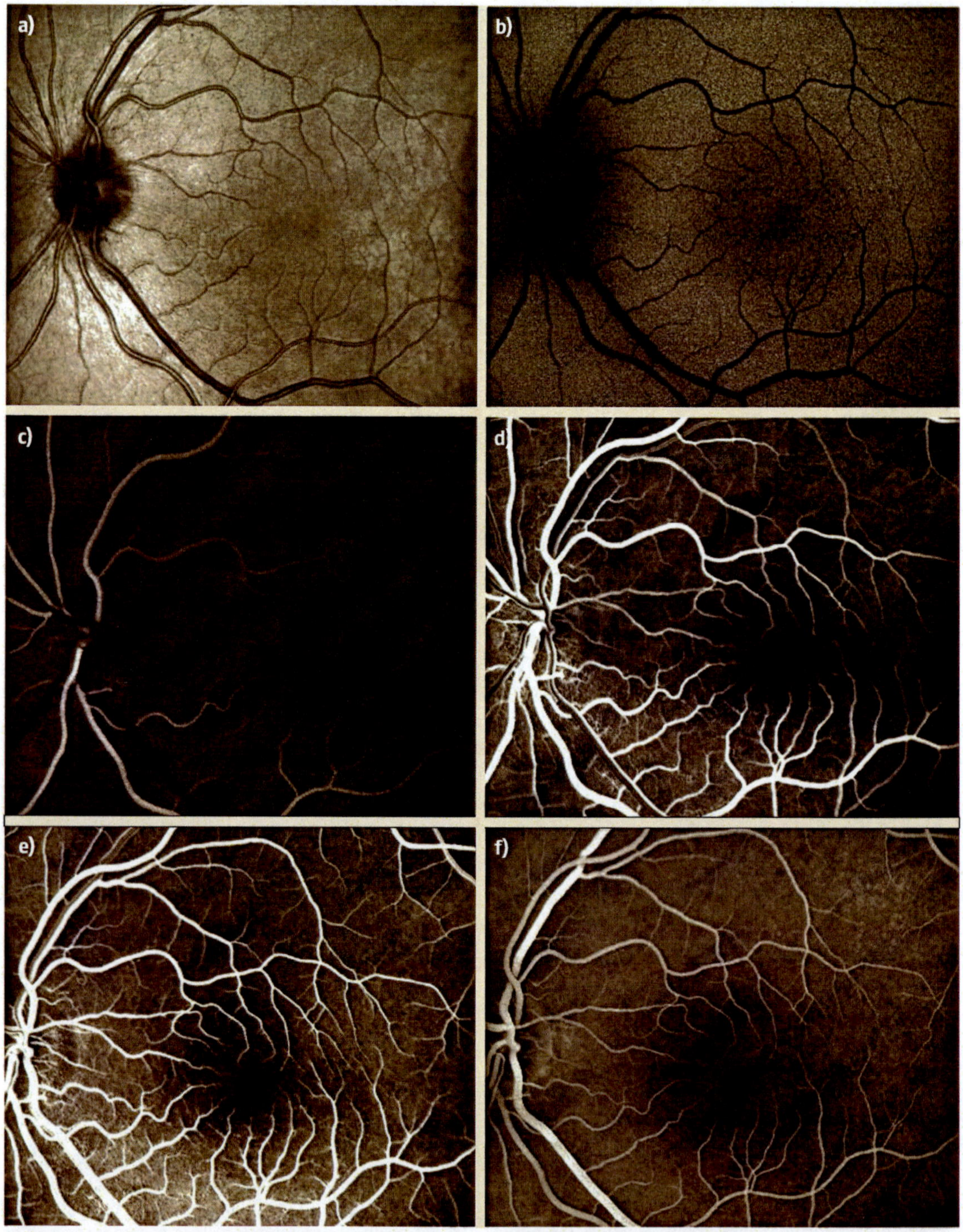

Fig. 13.9a–f Fluorescein angiography of a healthy subject. **(a)** Infrared image; **(b)** autofluorescence, **(c)** arterial phase, **(d)** early venous phase: laminar venous phase, **(e)** late venous phase; **(f)** late phase

13.21 Fluorescence Angiography of the Retina (Indocyanine Green)

Subject: see Fluorescence angiography

Indication: ICG angiography offers the advantage of a better visibility of the choroidal vessels, a less intense fluorescence blockade by the retinal pigmented epithelium and a higher affinity to the choroid vessels. Therefore, indications are diseases originating from the choroid such as choroidal neovascularizations (CNV) in age-related macular degeneration or primary choroidal diseases (such as disorder of the choroidal circulation)

Implementation: 25–50 mg of the dye are given into 2–4 ml water (for intravenous injection). After the dye has be soluted the fluid is injected quickly. Thereafter 5 ml physiological sodium-chloride solution should be injected. With this procedure a sharp coloring front can be achieved in the early phase. The fluorescence of ICG conforms about 4% of the sodium-fluorescein. Thus, very sensitive image detectors are required. ICG is a water-soluable tricarbocyanin dye. 98% of the ICG are bound to plasma proteins. It is not metabolized but elimated via the liver and biliary tract. It is relatively secure and has only few side effects (the mortality rate as a result of the application of the ICG is 1 : 333 333 as compared to 1 : 222 000 for fluorescein angiography). The most seen side effects are sickness (2.9%), vomiting, *flushing*, itching, rash, dyspnoea, and syncope (0.2%).

13.22 Visual Field Measurement (Perimetry)

General remarks: The visual field describes the area that can be noticed with the eye without moving it. It embraces the complete field of view from the center to the outer border. One may differentiate between the monocular and binocular visual field. Moreover, the visual field depends on adaptation as well as the size and the brightness of the observed object. In elderly subjects the size of the visual field decreases as a result of a normal aging process. For colors the visual field is smaller than for white light. Objects that are located at the edge of the field may, therefore, not be noticed as being colored.

Measurement: In all examinations the patient has to fixate a certain point and is not allowed to move his eyes. The examination must be performed separately on both sides (one eye must be covered). The automatic static perimetry is the method of visual field measurement that is used the most. At the static perimetry the patient sits in front of the perimeter in which randomly blinking light points are shown. The subject under examination fixates a light in the middle of the display. He/she should press a button whenever a light is shown in the surroundings. If the button is not pressed, the light intensity will be increased automatically and the point will be re-tested. This will be repeated to a certain intensity; if there is no answer a defect will be recorded. A subtype of the automatic perimetry is the so called blue-yellow perimetry. Here, the patient is presented a blue light in a yellowish illuminated hemisphere. It is especially sensitive to early visual field alterations.

The kinetic (Goldmann's) perimetry is the oldest method for visual field testing. Here, the person being tested sits in front of a hemisphere, very rarely in front of a flat test display. Light dots move from outward to inward (into the visual field). The moments at which the points are recognized are recorded. This test is done manually. For this examination light point at different (defined) brightness and size are used. Perception thresholds are measured, they are called isopteres. For the imaging of the blind spot one may start at 15° paracentrally and then move outwards to the margins of the blind spot. A detailed routine examination with Goldmann's perimeter should proof the outer edges with the following marks (spot sizes and intensities in chronology as indicated): V/4, I/4, I/3, I/2, and I/1. The blind spot is mostly proofed with the mark I/4. The determination of the outer margins is done with mark III/4 or V/4.

Finger perimetry (confrontation test, parallel try, comparison perimetry) is the most simple and fastest way to examine the visual field. The examiner and patient sit at distance of about 50 cm facing each other. Both cover one of their eyes with the hand so that the particular facing eye is fixed. Using the other hand the examiner moves the object (a pen) into the visual field. This is done out of every direction. The patient indicates when he sees the pen and the doctor can compare this with his own perception. Especially the outer borders and sideways deficiencies can be determined with this test. The test is fast and can be done at the bedside (Fig. 13.10).

Indication: Different diseases of the eye or the brain such as glaucoma or perfusion disorders (stroke) may affect the visual field. Defects of the visual field are determined by means of perimetry. The result is further compared to a standard result.

The blue-yellow perimetry is more sensitive than the automatic standard perimetry. Thus, this procedure allows the identification of early visual field defects such as early stages of glaucoma. For older patients the blue-yellow perimetry is not recommended. As a result of age-related lens opacification, the respective wavelengths are filtered. Therefore, blue-yellow perimetry is used for early diagnosis in younger patients suffering from ocular hypertension.

Evaluation: To judge if the visual field of the patient is normal or abnormal the results are compared to standard values of healthy persons. These are the visual field margins for the left eye for a white light stimulus: upper margin 50°, nasally 60°, lower margin 70°, temporally 95°. Usually the blind spot is located temporally between 10 and 20°. Depending on the examination (kinetic or static), the outer margin may vary because of the threshold of perception (higher for moved light sources as compared to static lighting). When doing a perimetry it is strongly recommended to always examine both

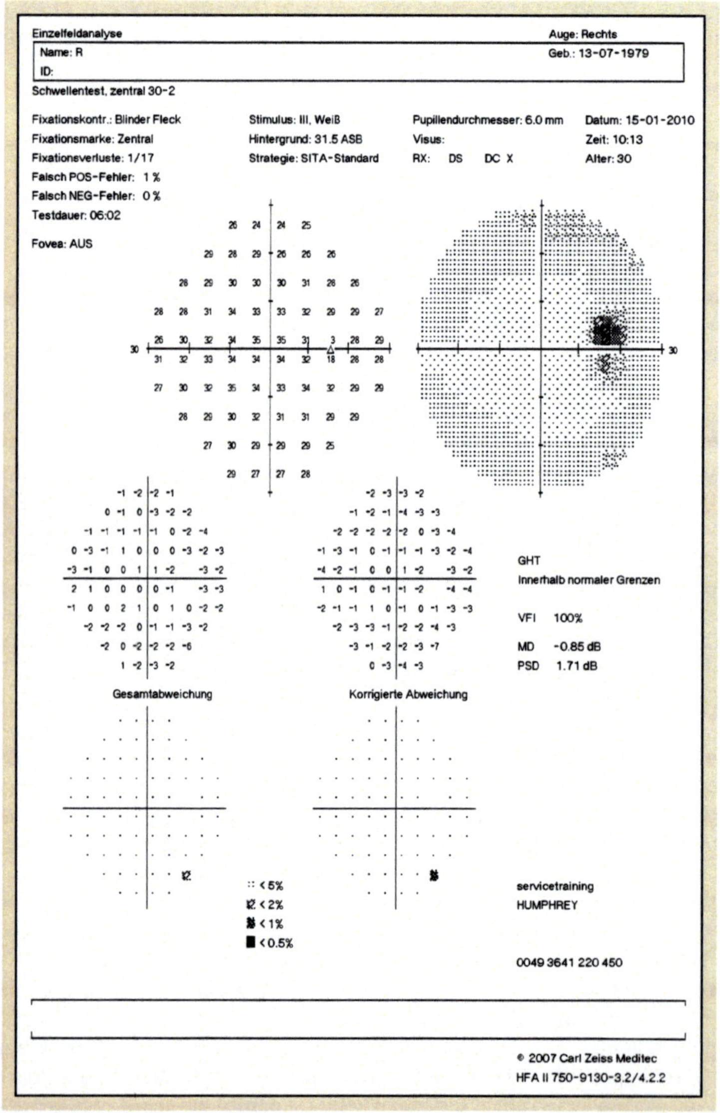

Fig. 13.10 Physiological visual field of the right eye

eyes (separately). This ensures that central (brain) damage can be differentiated from local (eye) damage. In some rare cases, the binocular visual field is measured. The results of the automatic perimetry can be imaged in different ways: look-up table, difference table, defect depth, gray scale, or shades of gray. A look-up table is a point by point printout of the threshold in decibel. The difference table is a point by point printout of the algebraic difference between the actual measured threshold and the expected or age-corrected normal values. Differences of 4 dB or less are printed normally as a symbol. Differences of 4–9 dB are marked as a small defect, of 10–19 dB as a moderate defect, and of 20 dB or greater as a deep defect. The gray scale or shade of gray is a graphical image of the measured thresholds. The complete area of the available test stimulus is divided into 9–10 intervals and each area matches a shade of gray according to a scale. Light gray tones are physiological, dark gray and black tones (exception: the blind spot) are considered pathological.

13.23 Exophthalmometry

Subject: Exophthalmometry measures the protrusion of the globe.

Measurement: Different instruments can be used – the exophthalmometer by Mutch, by Luede, and by Hertel. Hertel's exophthalmometer is the tool that is mainly used. Two movable connecting pieces touch the lateral orbital margins. The distance between the two connecting pieces is used as the base (basis line). The next step is that the examiner reads the position of the corneal apices with the deflections mirror inside the connecting pieces.

Indication: The Hertel exophthalmometer is a simple and correct method for the diagnosis of a retrobulbar mass. Exophthalmometry is part of the basic examination procedures in endocrine orbitopathy.

Evaluation: For a follow-up the same basis is used. The distance between the orbital margins and the apex of the cornea is 18–24 mm (European subjects). A difference of more than 2 mm between the two eyes or more than 24 mm is considered pathological. A possible source of error might be a face asymmetry.

Relevance: A simple method to quantify the eye position by suspicion of protrusion of the globe. This method is a standard procedure in ophthalmology.

References

13.1 S.J. Bogan, G.O. Waring 3rd, O. Ibrahim, C. Drews, L. Curtis: Classification of normal corneal topography based on computer-assisted videokeratography, Arch. Ophthalmol. **108**, 945–949 (1990)

14. Functional Force Assessment of Skeletal Muscles

Paul A. Iaizzo, William K. Durfee

Measurements of muscle function and strength are essential components of many clinical neurological or physical exams, and they are important for monitoring physiological function in various animal research studies. Evaluation of muscle strength is necessary for differential diagnosis, to determine the presence of disability, to plan potential treatments, and/or to track the effectiveness of treatments. As outcomes-based medical practice becomes more prevalent, the need for quantitative outcomes assessment of muscle strength becomes even more critical. Several assessment techniques and tools are currently available, but most require a cognitively cooperative subject. Manual muscle testing is the most widely used method to assess muscle function, however its reliability and accuracy are variable. When greater accuracy is needed, instruments that provide precise readouts of resistive forces can be employed, for example, hand dynamometers, pinch grips, and computer-controlled dynamometers. Stimulated muscle force assessment is an alternative and versatile approach for quantitative involuntary muscle torque.

14.1 **The Need for Skeletal Muscle Force Assessment** 273

14.2 **Manual Muscle Strength Testing** 275
14.2.1 Apparatus 276

14.3 **Advanced Muscle Assessment Methods** ... 278
14.3.1 Measuring Muscle Dynamics 278
14.3.2 Electromyogram 279

14.4 **Stimulated Muscle Force Assessment** 279

14.5 **Stimulated Muscle Force Assessment in Animal Models** 283

14.6 **Conclusion** 285

References 285

14.1 The Need for Skeletal Muscle Force Assessment

There is a growing need in clinical medicine to validate the quantitative outcomes of an applied therapy. In addition, the measurement of muscle function is an essential component of most neurological and/or physical exams. In general, muscle strength is often correlated to function, potential work productivity, and/or quality of life. Muscle function becomes reduced as we age, when associated with a skeletal impairment, and as a secondary consequence of many disease processes, i. e., stroke, nerve compression, or demyelination. Therefore, assessing muscle function is an important clinical skill that is routinely used by neurologists, orthopedists, general practitioners, anesthesiologists, and occupational and physical therapists. Furthermore, the evaluation of strength is used for differential diagnosis, to determine if an impairment or disability is present, to decide if a patient qualifies for treatment, and/or to track the relative effectiveness of a given treatment.

In a research setting, careful assessment of muscle function is often required to further our understanding of the normal or potentially impaired neuromuscular system. In such experiments, muscle forces can be assessed at the intact individual level (in vivo), in chronic and acute animal models (in situ), within isolated muscle strips or even within single myofibrils (in vitro), and at the molecular/biochemical level. In this chapter, only whole muscle testing (in vivo and in situ) will be discussed.

There are several components of muscle performance. The American Physical Therapy Association uses various definitions to explain the characteristics of muscle function. *Muscle performance* is the capacity of a muscle to do work, and *muscle strength* is the force exerted by a muscle or group of muscles to overcome a resistance in one maximal effort. *Instantaneous muscle power* is the mechanical power produced by the muscle relative to time (muscle force times muscle velocity), and *muscle endurance* is the ability to contract a muscle repeatedly over time [14.1]. Of these performance indicators, muscle strength is the one most commonly measured when assessing the muscle function of intact humans.

In assessing muscle strength, the conditions under which the muscle contracts must be specified so that the muscle test data can be interpreted properly. The following conditions are relevant:

- *Isometric contraction* – the muscle contracts while at a fixed length.
- *Isotonic contraction* – the muscle contracts while working against a fixed load, for example, a hanging weight.
- *Isokinetic contraction* – the muscle contracts while moving at a constant velocity; generally, isokinetic contractions are only possible with the limb strapped into a special machine that imposes the constant velocity condition.
- *Concentric contraction* – the muscle contracts against a load that is less than the force produced by the muscle so that the muscle shortens while contracting.
- *Eccentric contraction* – the muscle contracts against a load that is greater than the force produced by the muscle so that the muscle lengthens while contracting.

Isometric muscle tests are the most common, as they are the simplest to perform and reproduce and, because the test conditions are well defined, they are the most appropriate for comparing results within a population. Two considerations are important when testing muscle under isometric conditions. First, because muscle force varies with muscle length, the length of the muscle must be specified when planning and reporting a muscle test. The manual muscle test has strict and well-defined rules for the subject's posture and joint positions that must be followed if one is to make clinical decisions based on the test [14.2].

Second, isometric muscle tests of intact human muscle are conducted either with the limb held in a fixed position by the examiner or with the limb fixed to a brace or jig (Sect. 14.4). While these methods hold the limb in a fixed position, the muscle will not be strictly isometric because of tendon stretch. The mismatch between limb condition and muscle condition only causes problems when trying to infer details about muscle dynamics such as rise time or contraction speed from externally measured forces. Even if the whole muscle could be fixed at proximal and distal ends, during a twitch, the distance between z-lines in the myofibril will shorten, which means the sarcomeres are shortening due to internal muscle elasticity. This is why the length tension and dynamic properties of whole muscle deviate somewhat from those of the isolated sarcomere [14.3]. Nevertheless, length tension or ankle-angle/isometric-torque analyses can be done in vivo [14.4, 5].

Testing of intact human muscle requires that muscle output be measured external to the body and, as a result, muscle force is never measured directly. As shown in Fig. 14.1, muscles wrap around joints and attach to limbs at the proximal and distal ends. There is a kinematic relationship between the measured force and the actual muscle force that depends on the details of muscle attachment and varies with joint angle. To solve the kinematic relationship, one needs information about muscle attachment location, the geometry of the joint, and the joint angle. Such information can be readily obtained from a magnetic resonance imaging (MRI) scan, or a more generic geometry can be

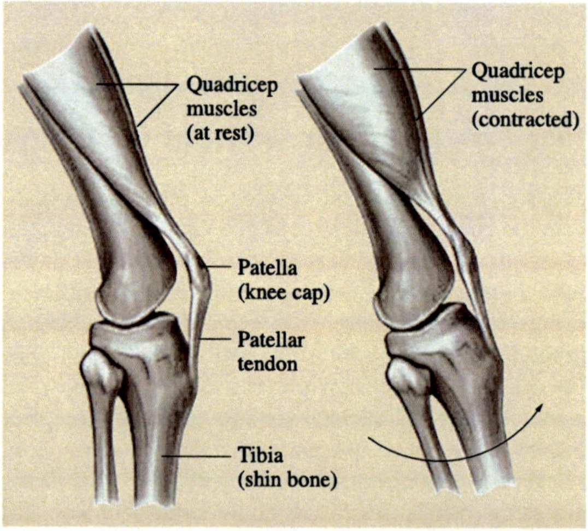

Fig. 14.1 Muscles wrap around joints. Muscle force is related to external force produced by a limb through skeletal geometry of joints and attachment points (Source: Medical Internet Solutions, http://aclsolutions.com/)

assumed through dimensions gathered from cadaver studies [14.6, 7].

While often reported as a force, external testing of muscles more correctly should be reported as a torque. Figure 14.2 illustrates how force varies with location of the resistive load along the limb, while torque does not. Reporting muscle strength as torque about a joint eliminates this difficulty. If force is reported, the distance between the joint and the resistive load point should be measured to permit conversion to torque.

External measurement of torque about a limb joint means that all of the forces acting on that joint are measured, and that the contribution of the muscle or muscle group under study cannot be easily separated out. In other words, there are confounding forces generated by synergistic muscles; for example, when testing foot plantar flexion to determine gastrocnemius strength, the soleus may also be contributing to the measured torque about the ankle. Yet another complicating factor may be undesired activations of antagonist muscles. One example is when you *flex your arm muscles*. In general, the resulting torque from the biceps and triceps are in balance, the arm does not move, and no external torque will be measured even though the muscles are contracting actively.

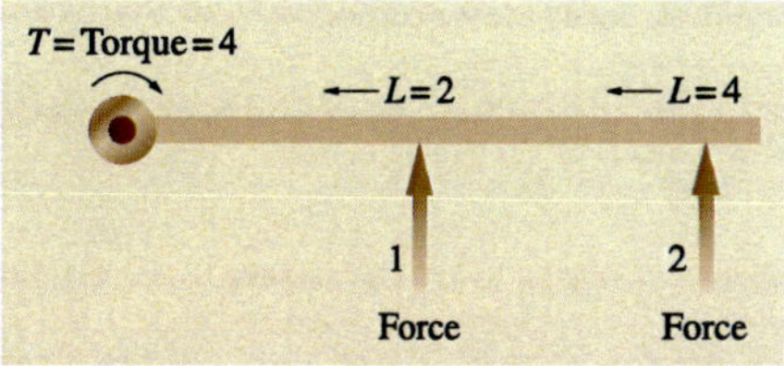

Fig. 14.2 A torque of 4 is produced about a joint. It takes an opposing force of 2 to balance the torque if the opposing force is applied at *arrow 1*. If the opposing force is applied at *arrow 2*, it only takes a force of 1 to balance the torque. Thus, the perceived torque, and therefore the scoring of muscle strength, depends on where along the limb the examiner places his or her hand

14.2 Manual Muscle Strength Testing

The simplest and most common method of assessing muscle strength is the manual muscle test (MMT). Manual muscle testing is a procedure for evaluating strength and function of an individual muscle or a muscle group in which the patient voluntarily contracts the muscle against gravity load or manual resistance [14.2, 8]. It is quick, efficient, and easy to learn; however, it requires total cooperation from the patient and learned response levels by the assessor.

The procedures for conducting the MMT have been standardized to assure, as much as possible, that results from the test will be reliable [14.2, 8, 9]. The specific muscle or muscle group must be determined, and the examiner must be aware of, and control for, common substitution patterns where the patient voluntarily or involuntarily uses a different muscle to compensate for a weak muscle being tested.

To conduct a MMT, the patient is positioned in a posture appropriate for the muscle being tested, which generally entails isolating the muscle and positioning so that the muscle works against gravity (Fig. 14.3). The body part proximal to the joint acted on by the muscle is stabilized. A screening test is performed by asking the patient to move the body part through the full available range of motion (ROM). The main test is then performed either unloaded, against a gravity load, or against manual resistance, and a grade is assigned to indicate muscle strength.

Manual grading of muscle strength is based on palpation or observation of muscle contraction, ability to move the limb through its available ROM against or without gravity, and ability to move the limb through its ROM against manual resistance by the examiner. Manual resistance is applied by the examiner using one hand with the other hand stabilizing the joint. Exact locations for applying resistive force are specified and must be followed exactly to obtain accurate MMT results [14.2]. A slow, repeatable velocity is used to take the limb

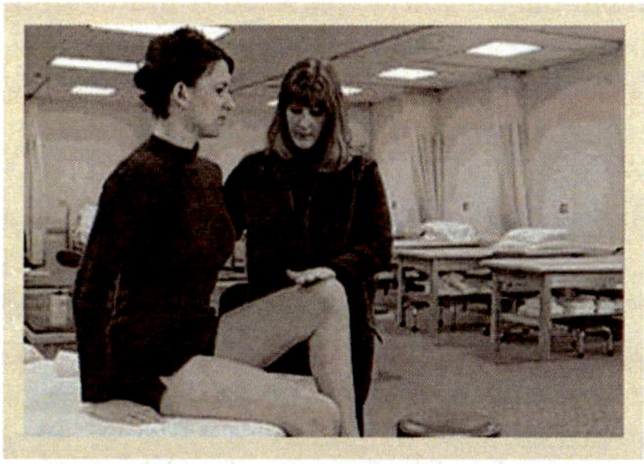

Fig. 14.3 Manual muscle test of the iliopsoas (Source: Spokane Falls Community College, Spokane)

Table 14.1 Manual muscle test scores

Score	Description
0	No palpable or observable muscle contraction
1	Palpable or observable contraction, but no motion
1+	Moves limb without gravity loading less than one half available ROM
2−	Moves without gravity loading more than one half ROM
2	Moves without gravity loading over the full ROM
2+	Moves against gravity less than one half ROM
3−	Moves against gravity greater than one half ROM
3	Moves against gravity over the full ROM
3+	Moves against gravity and moderate resistance less than one half ROM
4−	Moves against gravity and moderate resistance more than one half ROM
4	Moves against gravity and moderate resistance over the full ROM
5	Moves against gravity and maximal resistance over the full ROM

Table 14.2 Grading of isometric manual muscle test

Score	Description
3	Maintains position against gravity
3+	Maintains position against gravity and minimal resistance
4−	Maintains position against gravity and less than moderate resistance
4	Maintains position against gravity and moderate resistance
5	Maintains position against gravity and maximal resistance

through its ROM, applying a resistive force just under the force that stops motion. The instructions to patients are to use all of their strength to move the limb as far as possible against the resistance. For weaker muscles that can move the limb but not against gravity, the patient is repositioned so that the motion is done in the horizontal plane with no gravity.

MMT grades are assigned on a 0 to 5 scale with +/− modifiers: 1 = trace score, 2 = poor, 3 = fair, 4 = good, 5 = normal (Table 14.1). Grades above 1 demonstrate motion, and grades above 3 are against manual resistance. Other comparable scoring scales exist [14.7].

A common alternative to motion-based MMT is the isometric MMT in which the limb is held in a fixed position while the examiner gradually applies an increasing resistance force. The instructions to the patient are, *Don't let me move you*. The amount of force it takes to *break* the patient is used to assign a score. Scoring norms for isometric MMT are provided in Table 14.2. Because isometric MMT always involves a resistance, scores start at 3.

The assignment of scores is based on clinical judgment and the experience of the examiner. The amount of resistance applied by the examiner is also based on clinical experience and is adjusted to match the muscle being tested as well as the patient's age, gender, and body type. A major advantage of the MMT, as opposed to almost any other method, is that it can be adapted to measure the force in almost any muscle including facial muscles, tongue movements, and all of the intrinsic muscles in the hand.

While the MMT is the most widely used method to assess muscle function, its reliability and accuracy are questionable [14.10, 11]. MMT scores are least accurate for higher force levels [14.10, 12, 13]. Interrater reliability for MMT is not high, suggesting that the same examiner should perform multiple tests on one subject or across subjects [14.2]. MMT scores do correlate well with results from handheld dynamometers [14.14], implying that both are valid measures of muscle strength. However, as explained in Sect. 14.4, all tests based on voluntary activation of a muscle are prone to artifacts because of patient motivation and examiner encouragement. An additional limitation of MMT is that the scoring scale is nonlinear, making it challenging to apply longitudinally in a given patient to follow disease progression or treatment response; for example, the change in absolute muscle force between MMT scores 0 and 1 is much less than the force change between scores 3 and 4.

14.2.1 Apparatus

The appeal of the MMT is that it can be performed simply with the patient, the examiner, and a bench or table. This makes it ideal for the routine clinical environment, where specialized equipment is unavailable and time is short. It is also well suited for situations in which testing must be performed away from the clinic, for example, in nursing homes, rural areas, or remote emergency settings.

When greater accuracy of results is needed, instruments are available that provide precise readouts of the resistive force the muscle works against [14.15]. One example is a handheld dynamometer such as the one shown in Fig. 14.4. This instrument is sandwiched be-

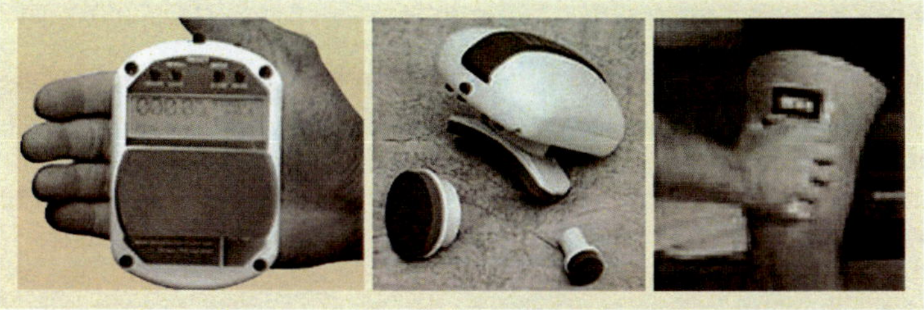

Fig. 14.4 Handheld dynamometer. Pictured is the Lafayette MMT System (Lafayette Instrument Company, Lafayette)

tween the examiner's hand and the patient's limb, and provides a readout of force. The interrater reliability for handheld dynamometers is good when used with a standard procedure [14.16–18], as is the test–retest reliability [14.14]. While the electronic dynamometer delivers a quantitative force reading, standards relating force in Newtons to the 1–5 scoring of a MMT have not been developed because the dynamometer does not take the size and build of the subject into account, an important, albeit subjective, part of assigning MMT scores.

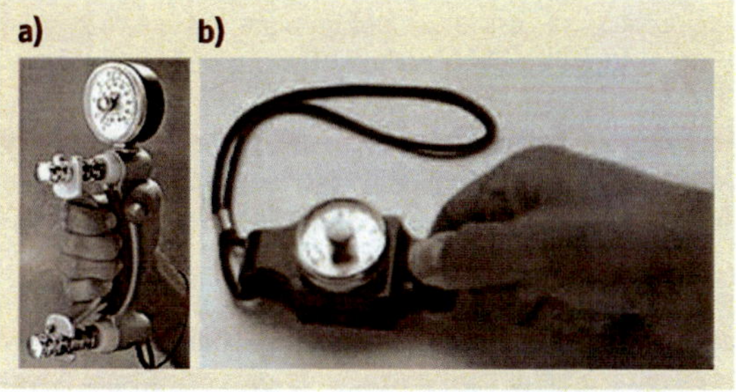

Fig. 14.5a,b The Jamar Hand Dynamometer **(a)** (NexGen Ergonomics, Canada) and the B&L Pinch Gauge **(b)** (B&L Engineering, Tustin)

Other products have been developed for specific tests of muscle strength, for example, the hand dynamometer and pinch grip devices shown in Fig. 14.5. These are easy to use and common for diagnostic tests of the hand. Despite their quantitative nature, readings between different types and brands of dynamometers can vary [14.19, 20]. Computer-controlled dynamometers offer a variety of loading conditions for muscle testing and for strengthening treatments [14.21, 22] (Fig. 14.6). Along with isometric and isotonic loading, dynamometer machines provide isokinetic conditions in which the muscle group acts against a computer-controlled resistance that moves the limb at a constant angular velocity. While these and other quantitative instruments provide accurate readings of muscle force, they are rarely found in the clinic because, for most clinical purposes, the MMT is sufficiently accurate to monitor muscle strength, is fast and easy to administer, and requires no equipment.

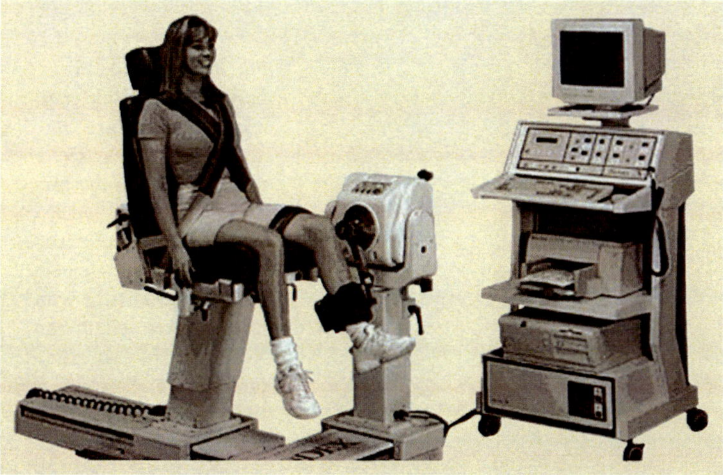

Fig. 14.6 Biodex dynamometer for computer-controlled muscle testing (Biodex Medical Systems, Shirley)

14.3 Advanced Muscle Assessment Methods

14.3.1 Measuring Muscle Dynamics

Muscle is a complex actuator whose external properties of force and motion result from the action of thousands of muscle fibers which, in turn, result from the action of millions of structural and active proteins whose interaction is triggered by biochemical events. While most muscle testing focuses on the overall strength of a muscle or muscle group, more sophisticated assessment can be useful for in-depth examination of muscle function, including its dynamic, kinematic, and fatigue properties.

The approach used to measure muscle function in more detail generally involves developing a mathematical model of muscle activity and then using experiments to identify the parameters of the model.

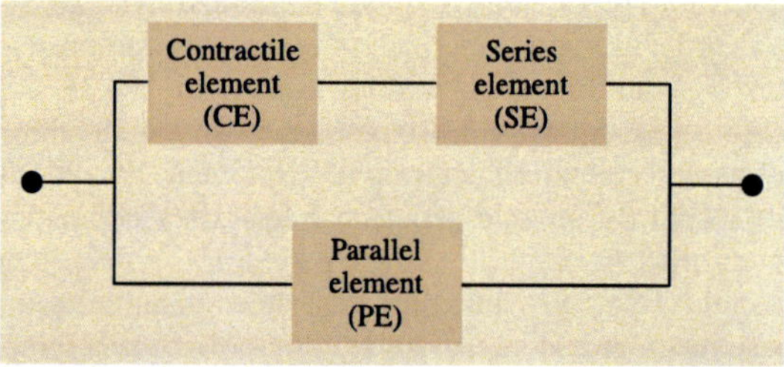

Fig. 14.7 Hill muscle model. The contractile element (CE) contains the active element with dynamics, force–velocity, and force–length properties. The series element (SE) is the inherent internal elastic element, and the parallel element (PE) represents passive connective tissue

Overviews of these methods are provided in *Zajac* and *Winters* [14.25], *Durfee* [14.23], *Zahalak* [14.26], *Crago* [14.27], and *Kearney* and *Kirsch* [14.28]. The modeler must first choose the appropriate complexity of the mathematical model. The optimum choice is a model that is sufficiently complex to reveal the behavior of interest, but not so complex that parameters cannot be identified. Generally, Hill-type input–output models [14.29, 30] are a good balance, as they capture key force–velocity, force–length, and activation dynamics at a whole-muscle level (Fig. 14.7).

Model parameters can be identified one at a time, using the approach followed by *Hill* [14.29], or all at once using modern system identification techniques [14.23, 28]. Electrical activation of the muscle is a particularly convenient means for excitation because, unlike voluntary activation, there is control over the input, an essential component for an effective system identification method. Testing can be done under isometric conditions for determining recruitment and twitch dynamic characteristics, or under arbitrary loading to find active and passive force–length and force–velocity properties.

Identification of muscle properties is most easily accomplished using isolated muscle in acute animal model studies. Here the muscle is unencumbered by joint attachments and extraneous passive tissue. Muscle tendon can be directly attached to a force sensor and placed in a computer-controlled servo mechanism to apply known length and velocity trajectories, all while being stimulated; for example, the isometric recruit-

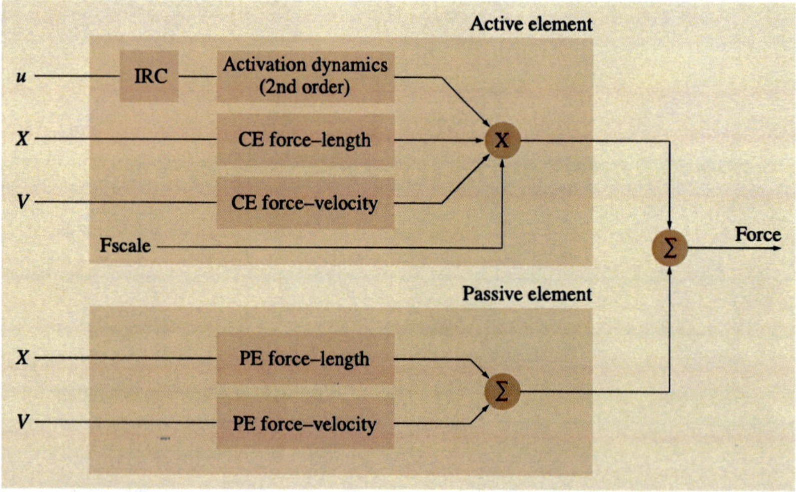

Fig. 14.8 A model that can be used for muscle property identification. The active element has recruitment and twitch dynamics which multiplicatively combine with active force–length and force–velocity properties and sum with passive force–length and force–velocity properties to produce overall muscle force (after [14.23, 24]). IRC – isometric recruitment curve; CE – contractile element; PE – parallel element

ment curve, the relationship between muscle force, and stimulus strength can be identified using either point-at-a-time or swept-amplitude methods, the latter being efficient in implementation [14.31]. Using the model shown in Fig. 14.8, active and passive force–length and force–velocity properties can be estimated using brief bouts of controlled, random length perturbations, and then verified through additional trials where both stimulation and length are varied randomly [14.23, 24] (Fig. 14.9). Simultaneous identification of active and passive muscle properties for intact human muscles is more challenging and represents an ongoing area of research [14.28].

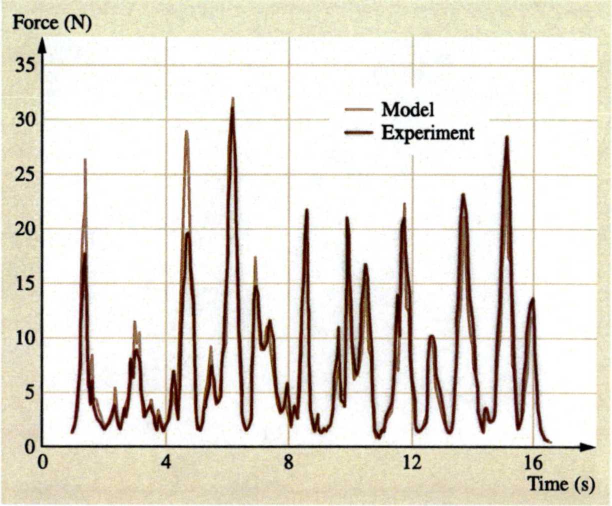

Fig. 14.9 Results from an isolated muscle experiment where muscle active and passive properties were identified, then the resulting model was verified against experiment data. Data were generated while the muscle underwent simultaneous, random, computer-controlled stimulation and length perturbations (after [14.24])

14.3.2 Electromyogram

Contracting skeletal muscle emits an electrical signal, the electromyogram (EMG). Electrical recording of the EMG using needle or surface electrodes is an important diagnostic indicator used in clinical neurology to diagnose neuromuscular disorders including peripheral neuropathies, neuromuscular junction diseases, and muscular dystrophies. EMG is also used in research studies as an estimator of muscle activity for biomechanics and motor control experiments. The reader is referred to *Merletti* and *Parker* [14.32] and *Basmajian* and *DeLuca* [14.33] for a comprehensive discussion of surface and needle EMG used in research applications, and to *Preston* and *Shapiro* [14.34], *Kimura* [14.35], and *Gnatz* [14.36] for an introduction to clinical EMG. EMG assessment is limited because it is only an indirect indicator of muscle force.

14.4 Stimulated Muscle Force Assessment

Most devices used clinically to quantify muscle force and increase objectivity rely on voluntary effort, which can be problematic. Pain, corticospinal tract lesions, systemic illness, and inconsistent motivation can significantly affect voluntarily activated muscle force. In addition, some very weak patients are unable to complete a full range of motion in voluntary force assessment tasks [14.37, 38]; for example, monitoring muscle function in patients confined to the intensive care unit is a challenge. Often such patients are on potent pain medications (e.g., morphine) and are sedated, and may have significant alterations in level of consciousness due to underlying critical illnesses [14.39, 40]. Thus, it can be extremely difficult to ask these patients to provide reproducible voluntary efforts. Even when cooperation is good, the standard measures of force assessment are qualitative. The alternative is to stimulate their muscles to contract without voluntary input.

Stimulated muscle force assessment is a versatile approach for quantitative involuntary muscle torque in humans. A muscle is activated by noninvasive nerve or motor point stimulation. A rigid apparatus is used to secure the appropriate portion of the subject's body in a predetermined position which confines movement to a specific direction, e.g., ankle dorsiflexion, thumb adduction, arm flexion, or neck flexion [14.3, 5, 41, 42] (Figs. 14.10 and 14.11). The innervating nerves or motor points of the muscle are stimulated using surface electrodes, with either a single stimulus to generate a twitch contraction or with short trains of stimuli (e.g., with 5 ms interpulse intervals) to produce tetanic contractions [14.4, 5]. Strain gauges are used to measure isometric torque and, via acquisition software, all data are displayed and online analyses are performed. Various parameters of the isometric contractions can be calculated including time between stimulus and torque onset, peak rate of torque development, time to peak

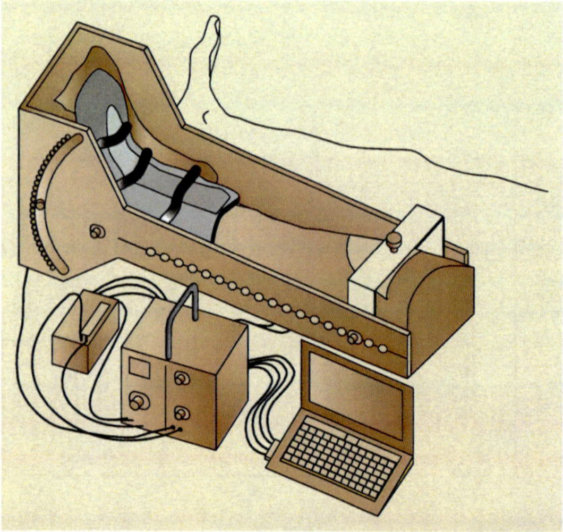

Fig. 14.10 Muscle force assessment system to determine involuntary isometric torque of the human dorsiflexor muscle. It comprises the following main components: (1) a stabilizing frame with knee supports, (2) the torque plate with mounted boot to fix the foot which can be rotated between −40° and 40°, (3) a Wheatstone bridge strain gauge system that detects the evoked torque (not shown), (4) a stimulator/amplifier unit that can supply variable stimulus pulse amplitudes and pulse durations and can amplify the voltage changes from the Wheatstone bridge, and (5) a computer with data acquisition hardware and software for recording, analyzing, and displaying all signals (after [14.40])

torque, half-relaxation time, and others (Fig. 14.12, Table 14.3).

Such information is predicted to correlate with underlying physiological conditions and/or the presence of a myopathic or neuropathic disorder. The average torque generated by healthy control subjects varies by less than 5% with repeated testing for contractions elicited from the muscle groups studied [14.5, 41, 42]. This assessment approach has utility in a number of research arenas, both clinical and nonclinical. Specifically, it has added clinical value in diagnosing neuromuscular disorders, tracking weakness due to disease progression, and quantitatively evaluating the efficacy of a therapy [14.38, 43–47]. Compared with MMT assessment methods, measuring isometric muscle torque generated by stimulation is quantitatively objective and reliable, and increases the types of patients that can be studied, including those under sedation [14.40] (Fig. 14.10). Stimulated muscle force assessment is particularly useful when studying patients with a known genetic disorder, because it can provide important information about genotype–phenotype associations [14.38].

The general configuration of the measurement system consists of the following main components:

1. A stabilizing device that holds the subject's limb in a defined position
2. A force transducer that detects the evoked torque produced by a specific muscle group
3. Hardware devices for nerve stimulation, signal amplification, and signal conditioning
4. A computer for stimulus delivery
5. Data acquisition software for recording, analyzing, and displaying signals (Figs. 14.10–14.13).

The stabilizing device system used in most of our previously published reports to study the ankle dorsiflexors is a modification of a previously described apparatus [14.4]. This device can be configured to maintain the subject's leg in a stable position while allowing access for stimulation of the common peroneal nerve lateral to the fibular head (Fig. 14.11a). The torque about the ankle joint, produced by the dorsiflexor muscles (i. e., primarily generated by the tibialis anterior with contributions from the peroneus tertius and exten-

Table 14.3 Contractile parameters calculated from stimulated muscle force assessment

Parameter	Units	Definition
Peak torque	N m	Maximum amount of torque developed
Contraction time	s	Time from onset of torque to time of peak torque
Half-relaxation time	s	Time from peak torque to time when torque decays to half of peak torque
Peak rate of development	N m/s	Maximum rate of torque development
Peak rate of decay	N m/s	Maximum rate of torque decay
Time to peak development	s	Time from onset of torque to the peak rate of development
Time to peak decay	s	Time from peak rate of development to peak rate of decay
Half maximal duration	s	Time when the generated torque is maintained at a level of half of the peak torque
Latency to onset	s	Time from the stimulus to the onset of torque development

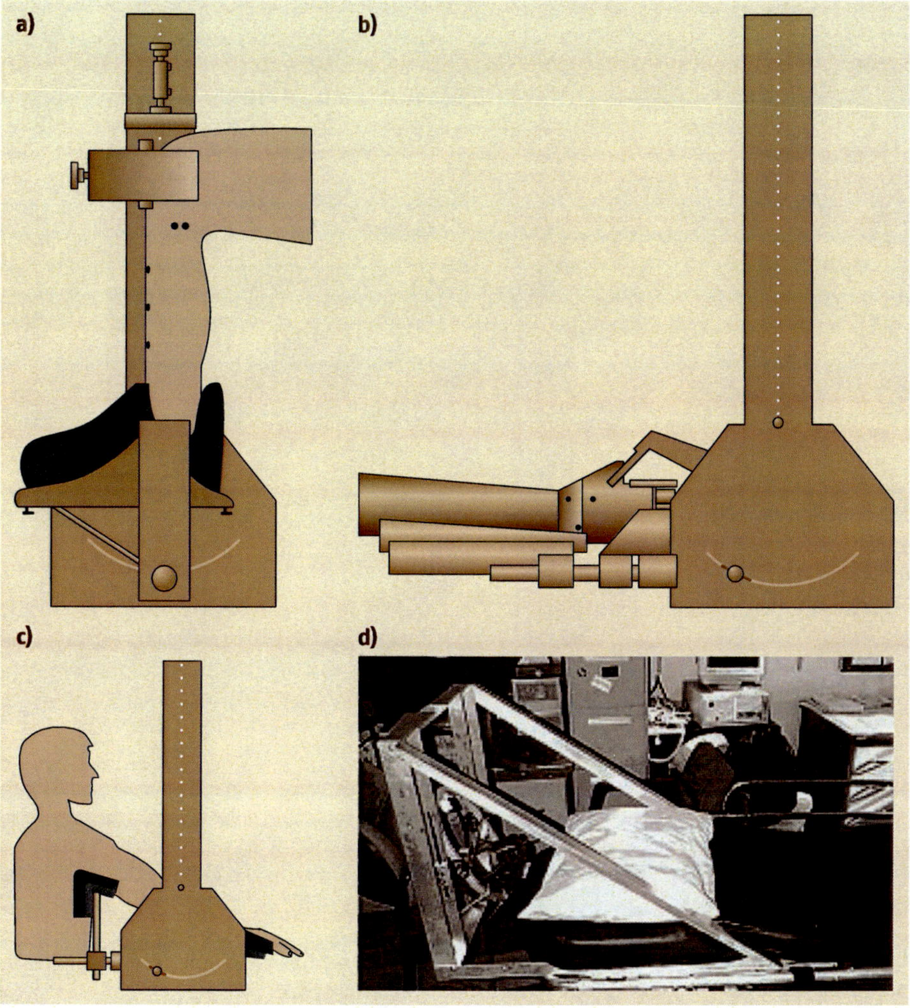

Fig. 14.11a–d Various applications of stimulated muscle force assessment: **(a)** dorsiflexor muscles in a seated individual with stimulation of the common peroneal nerve lateral to the fibular head, **(b)** adductor pollicis muscle following ulnar nerve stimulation, **(c)** activated biceps force with motor point stimulation, and **(d)** head stabilizing/force system to study sternocleidomastoid muscle function following motor point stimulation (after [14.5, 41, 42])

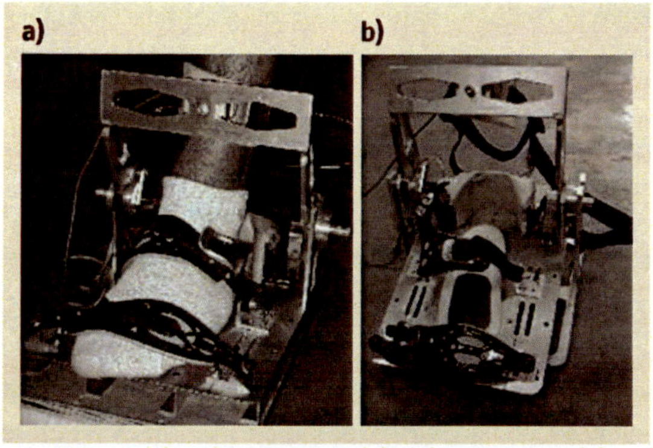

Fig. 14.12a,b Prototypes of the next-generation dorsiflexor stimulated muscle force assessment systems

sor digitorum muscles), is then measured. One or two padded, adjustable clamps are used to maintain stability of the leg (knee slightly flexed in supine position or flexed at 90° while seated). Modified inline skate boots of varying sizes are affixed to the torque plate and adapted for either the right or left foot. The foot and ankle can be rotated within a 40° range while secured in the skate boot. The device can also be used for subjects in supine position, in which case the support frame is secured to the upper leg proximal to the knee (Fig. 14.10) [14.40,46]. More recently, our laboratories have initiated redesigns of these devices to make them smaller, lighter, easier to position on a given subject, and have an easier user interface to the computer and data acquisition control system (Fig. 14.12).

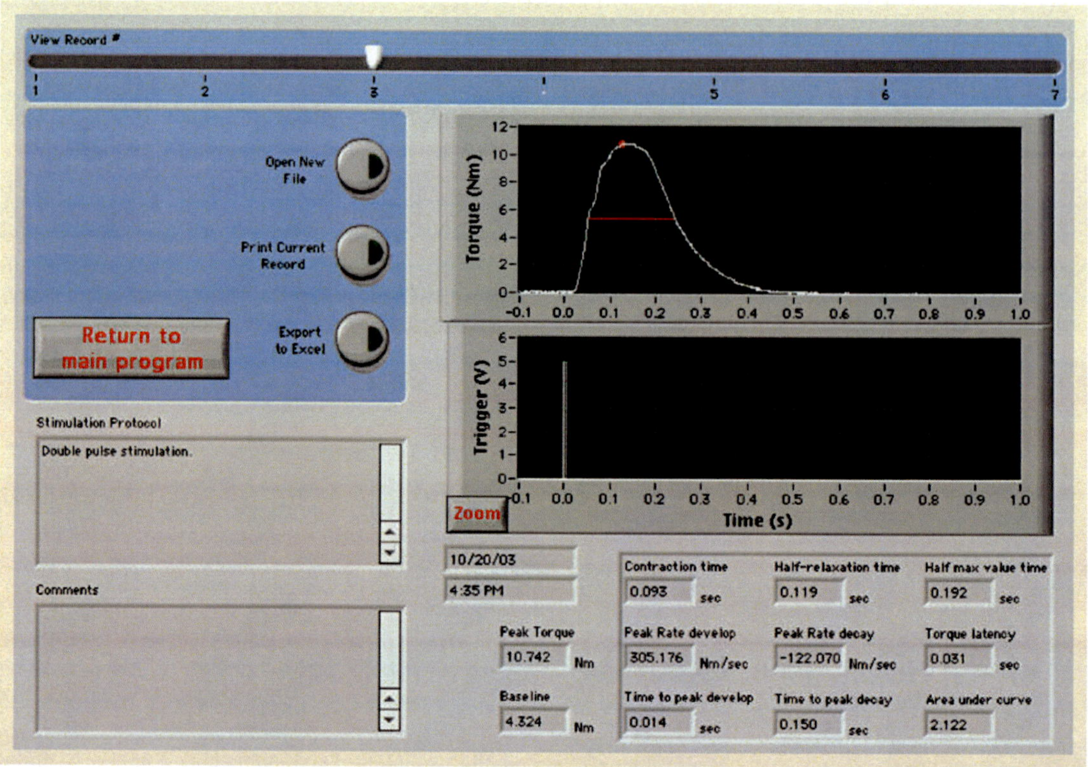

Fig. 14.13 An example of a typical data display available to the investigator during subsequent offline analyses. Graphically displayed are the muscle torque waveform and the stimulus administered (a double pulse with a 5 ms interpulse interval). Numerically displayed are various contractile parameters. Time 0 is the time of stimulation. The red line indicates the time when half of the peak torque has been generated. The display shown is the torque generated by the dorsiflexor muscles of a normal, healthy subject

A stabilizing apparatus for the arm and hand, used to measure muscle torque of the adductor pollicis, can be attached to the main stabilizing frame (Fig. 14.11b). Using straps, the forearm can be secured to the arm stabilizing unit, which can be adjusted for varying arm lengths. The digits (2–5) are placed in the hand well, and the thumb is secured to the constructed thumb bar attached to the torque plate. Shown in Fig. 14.11c is the configuration which is used to record force generated by the biceps muscle. This approach can also measure forces in the sternocleidomastoid muscle in the anterior neck [14.42], which can be used to study the effect of therapy in patients with cervical dystonia. As for the aforementioned muscles, force can be produced by peripheral nerve stimulation or by voluntary effort. In addition, motor point stimulation of the muscle itself with large surface electrodes has been employed [14.41]. The force of the isometric muscle contraction is obtained as change in torque applied to the instrument torque plate.

Stimulated force assessment has been used to study patients with a wide variety of disorders including amyotrophic lateral scoliosis, Brody's disease, chronic inflammatory demyelinating polyneuropathy, malignant hyperthermia, muscular dystrophy, myotonia, periodic paralysis, and nerve conduction blocks. Recently, stimulated force assessment was employed by our laboratory to evaluate athletes with potential overtraining syndrome. The new insights to be gained by employing this approach in a variety of healthcare and biomechanics applications will improve clinical understanding of underlying pathophysiologies, provide an accurate means to determine clinical outcomes, and contribute to the overall knowledge of muscle mechanics.

Stimulated force assessment has limitations. First, it depends on the ability to easily stimulate nerves with surface electrodes so that supramaximal activation can be achieved. Many muscles, and the nerves that supply them, are deep and cannot be activated with surface electrodes. Second, the muscles must generate torque on a limb in a measurable direction. This could be a challenge, for example, with back muscles that act on a multiple-degree-of-freedom structure. Third, it assumes no spillover of the stimulus to antagonist muscles that would contaminate the force readings. Because of these limitations, not all muscles that can be tested by voluntary force assessment methods can be tested by stimulated force assessment.

14.5 Stimulated Muscle Force Assessment in Animal Models

To emphasize the versatility of this method, a specialized version of the device was constructed and used to study ankle dorsiflexor torques in hibernating black bears in the Rocky Mountains [14.48, 49]. Shown in Fig. 14.14 are recordings made in the field; in some cases, environmental temperatures were well below freezing. Black bears (*Ursus americanus*) stay inside their winter dens for 5–7 months of the year, dur-

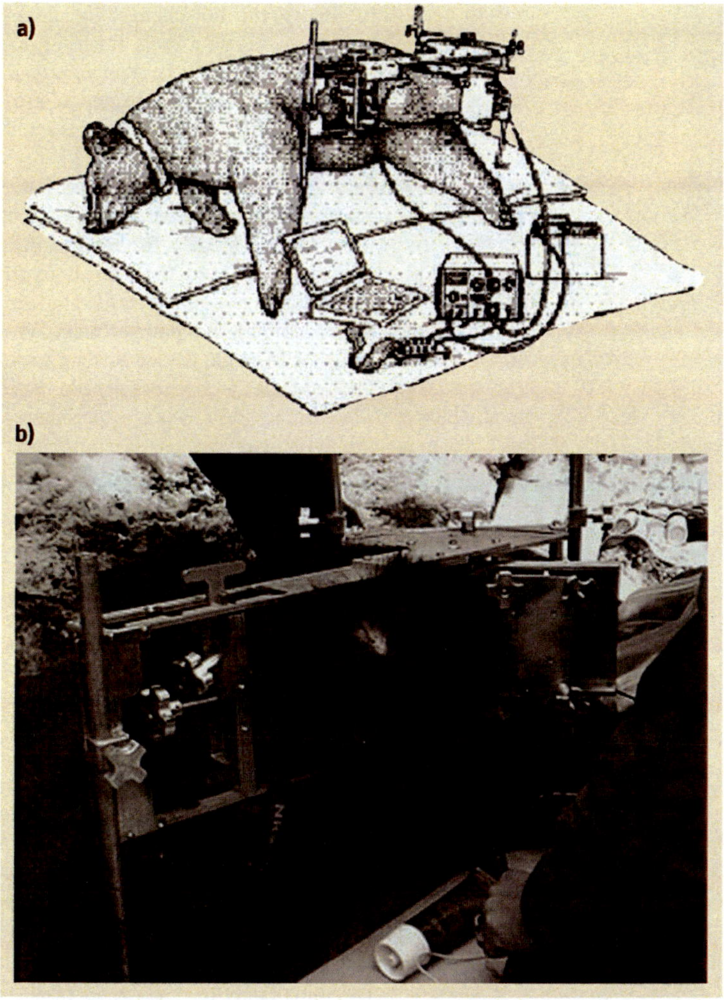

Fig. 14.14a,b The force assessment system used in the field to study dorsiflexor forces in overwintering black bears **(a)**. In this approach, we found that positioning the animals on their sides provided the most reproducible results. In panel **(b)**, one can observe that the bear fur was shaved to allow for placement of the superficial stimulating electrodes

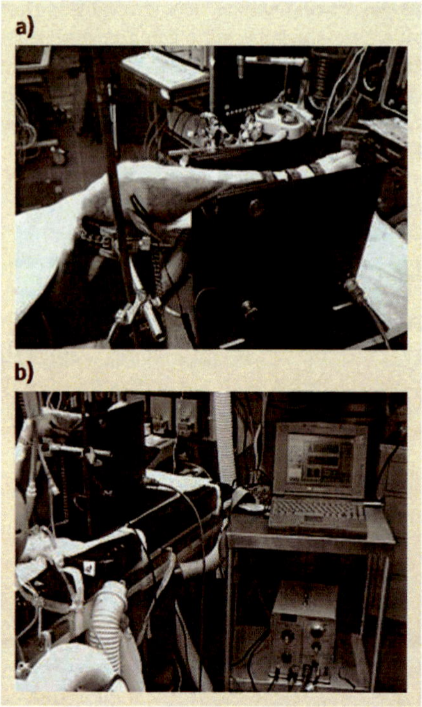

Fig. 14.15a,b Muscle force assessment in the laboratory setting. **(a)** Right hind limb of a canine secured to a force plate to record muscle torque. In this case, the nerve stimulation electrodes with clips and electrode optimize nerve/muscle activation. **(b)** Complete system of a computer, stimulation/force transducer unit, and the limb torque unit

ing which time their body temperature drops to about > 5 °C below normal and they do not eat, drink, urinate, defecate, or show any other perceptible activity [14.48]. Although inactivity in humans (i. e., as a result of confined bed rest, weightlessness, or limb immobilization) leads to atrophy of skeletal muscle, loss of muscle tone, and impaired strength, the black bear does not suffer similar deterioration. We found that overwintering black bears lose less than 23% of their strength over 130 days, unlike humans who are weakened by a 90% strength loss over the same period [14.48]. When we employed stimulation protocols to study muscle fatigue behavior as we previously used in humans to diagnose mitochondrial myopathies [14.45], we observed that black bear fatigue profiles did not alter during the hibernation period [14.49]. Additionally, muscle contractile properties, including contraction time, half-relaxation time, half-maximum value time, peak rate of development and decay, time to peak force development, and time to peak force decay, did not change, indicating that obvious alterations in whole-muscle function occurred over the winter. This study further supported our findings that black bears have a high resistance to atrophy despite being subjected to long-term anorexia and limited mobility [14.48, 49].

Our group has also designed and constructed force assessment systems that have been used to study force properties in various animal models for muscle disorders. Additionally, we have used these methods to evaluate drug and anesthetic effects, hypermetabolism (e.g., malignant hyperthermia), and compartment syndrome. Shown in Fig. 14.15 is the device that we have developed to perform muscle force assessment in large mammalian models (canine and swine) in the laboratory setting; as with the human systems, it is important to stabilize the leg.

Most methods to assess murine (mouse) muscle function are highly invasive and do not allow for long-term testing in a given animal. Our laboratory de-

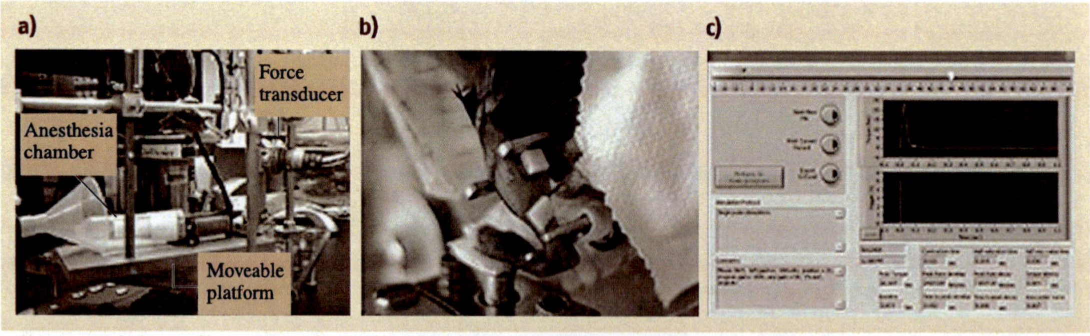

Fig. 14.16 **(a)** Device setup employed for noninvasively studying leg force properties in a mouse, **(b)** leg setup and electrode placements, and **(c)** online LabVIEW program display of measured forces (after [14.50])

scribed a novel murine assessment system developed to measure skeletal muscle contractile properties on multiple occasions over time in intact animals [14.50] (Fig. 14.16). In control mice, we measured less than 8% day-to-day variability and also detected a difference ($p < 0.003$) in peak twitch tension (P_t) due to cardiotoxin-induced muscle damage.

Duchenne muscular dystrophy (DMD) is a devastating disease caused by degeneration of striated muscle due to the absence of dystrophin. In this study, when control mice were compared with dystrophic mice (MDX), differences in normalized torque and P_t per body weight were detected ($p < 0.003$). The use of stimulated muscle force assessment in mouse models will be useful for investigators testing the efficacy of novel therapies for DMD because it can noninvasively take measurements at different time points and can sensitively detect differences in function due to pathology.

14.6 Conclusion

The assessment of a patient's muscle strength is an important vital function. Strength assessment is necessary for determining distribution of weakness, disease progression, and treatment efficacy. The particular assessment approach utilized is dictated by the clinical circumstance or the severity of illness. Several assessment techniques are currently available to the healthcare provider and researcher, each with their unique attributes. As outcomes-based medical practice becomes the norm, the need for quantitative outcomes assessment of muscle strength will grow, and stimulated muscle force assessment may become increasingly important.

References

14.1 American Physical Therapy Association: *Guide to Physical Therapy Practice* (American Physical Therapy Association, Virginia 1997)

14.2 H.M. Clarkson: *Musculoskeletal Assessment: Joint Range of Motion and Manual Muscle Strength* (Lippincott Williams Wilkins, Philadelphia 2000)

14.3 F.E. Zajac: Muscle and tendon: properties, models, scaling, and application to biomechanics and motor control, Crit. Rev. Biomed. Eng. **17**, 359–411 (1989)

14.4 J.G. Quinlan, P.A. Iaizzo, G.A. Gronert, E.H. Lambert: Ankle dorsiflexor twitch properties in malignant hyperthermia, Muscle Nerve **12**, 119–125 (1989)

14.5 T.J. Brass, M.K. Loushin, J.W. Day, P.A. Iaizzo: An improved method for muscle force assessment in neuromuscular disease, J. Med. Eng. Technol. **20**, 67–74 (1996)

14.6 D.A. Winter: *Biomechanics of Human Movement* (Wiley, New York 1979)

14.7 G.A. Yamaguchi: A survey of human musculotendon actuator parameters. In: *Multiple Muscle Systems: Biomechanics and Movement Organization*, ed. by J.M. Winters, S. Woo (Springer, New York 1990)

14.8 J.M. Winters, S. Woo (Eds.): *Multiple Muscle Systems: Biomechanics and Movement Organization* (Springer, New York 1990)

14.9 M. Neistadt, E. Crepeau: *Willard and Spackman's Occupational Therapy* (Lippincott Williams Wilkins, Philadelphia 1998)

14.10 L. Daniels, C. Worthingham: *Muscle Testing: Techniques of Manual Examination* (Saunders, Philadelphia 1995)

14.11 D.M. Iddings, L.K. Smith, W.A. Spencer: Muscle testing. 2. Reliability in clinical use, Phys. Ther. Rev. **41**, 249–256 (1961)

14.12 S.M. Schwartz, M.E. Cohen, G.J. Herbison, A. Shah: Relationship between two measures of upper extremity strength: Manual muscle test compared to hand-held myometry, Arch. Phys. Med. Rehabil. **73**, 1063–1068 (1992)

14.13 R.W. Bohannon: Manual muscle test scores and dynamometer test scores of knee extension strength, Arch. Phys. Med. Rehabil. **67**, 390–392 (1986)

14.14 C. Knepler, R.W. Bohannon: Subjectivity of forces associated with manual-muscle test grades of 3+, 4−, and 4, Percept. Mot. Skills **87**, 1123–1128 (1998)

14.15 R.W. Bohannon: Test-retest reliability of hand-held dynamometry during a single session of strength assessment, Phys. Ther. **66**, 206–209 (1986)

14.16 L. Amundsen: *Muscle Strength Testing: Instrumented and Noninstrumented Systems* (Churchill Livingstone, New York 1990)

14.17 R.W. Bohannon, A.W. Andrews: Interrater reliability of hand-held dynamometry, Phys. Ther. **67**, 931–933 (1987)

14.18 M. Horvat, R. Croce, G. Roswal: Intratester reliability of the Nicholas Manual Muscle Tester on individuals with intellectual disabilities by a tester

having minimal experience, Arch. Phys. Med. Rehabil. **75**, 808–811 (1994)
14.19 J. Dunn, M. Iversen: Interrater reliability of knee muscle forces obtained by hand-held dynamometer from elderly subjects with degenerative back pain, J. Geriatr. Phys. Ther. **26**, 23–29 (2003)
14.20 P. Click Fenter, J.W. Bellew, T. Pitts, R. Kay: A comparison of 3 hand-held dynamometers used to measure hip abduction strength, J. Strength Cond. Res. **17**, 531–535 (2003)
14.21 N. Massy-Westropp, W. Rankin, M. Ahearn, J. Krishnan, T.C. Hearn: Measuring grip strength in normal adults: Reference ranges and a comparison of electronic and hydraulic instruments, J. Hand Surg. [Am.] **29**, 514–519 (2004)
14.22 J.M. Cabri: Isokinetic strength aspects of human joints and muscles, Crit. Rev. Biomed. Eng. **19**, 231–259 (1991)
14.23 W.K. Durfee, K.I. Palmer: Estimation of force-activation, force-length, and force-velocity properties in isolated, electrically stimulated muscle, IEEE Trans. Biomed. Eng. **41**, 205–216 (1994)
14.24 W.K. Durfee, K.E. MacLean: Methods for estimating isometric recruitment curves of electrically stimulated muscle, IEEE Trans. Biomed. Eng. **36**, 654–667 (1989)
14.25 W.K. Durfee: Model identification in neural prosthesis systems. In: *Neural Prostheses: Replacing Motor Function After Disease or Disability*, ed. by R.B. Stein, P.H. Peckham, D. Popovic (Oxford Univ. Press, New York 1992)
14.26 F. Zajac, J.M. Winters: Modeling musculoskeletal movement systems. In: *Multiple Muscle Systems: Biomechanics and Movement Organizations*, ed. by J.M. Winters, S. Woo (Springer, New York 1990)
14.27 G. Zahalak: An overview of muscle modeling. In: *Neural Prostheses: Replacing Motor Function After Disease or Disability*, ed. by R.B. Stein, P.H. Peckham, D. Popovic (Oxford Univ. Press, New York 1992)
14.28 P. Crago: Creating neuromusculoskeletal models. In: *Biomechanics and Neural Control of Posture and Movement*, ed. by J.M. Winters, P. Crago (Springer, New York 2000)
14.29 R. Kearney, R. Kirsch: System identification and neuromuscular modeling. In: *Biomechanics and Neural Control of Posture and Movement*, ed. by J.M. Winters, P. Crago (Springer, New York 2000)
14.30 A.V. Hill: The heat of shortening and the dynamic constants of muscle, Proc. R. Soc. Lond. [Biol.] **126**, 136–195 (1938)
14.31 J.M. Winters: Hill-based muscle models: A systems engineering perspective. In: *Multiple Muscle Systems: Biomechanics and Movement Organization*, ed. by J.M. Winters, S. Woo (Springer, New York 1990)
14.32 R. Merletti, P. Parker: *Electromyography: Physiology, Engineering, and Noninvasive Applications* (Wiley, Hoboken 2004)
14.33 J.V. Basmajian, C.J. DeLuca: *Muscles Alive: Their Function Revealed by Electromyography* (Williams Wilkins, Baltimore 1985)
14.34 D.C. Preston, B.E. Shapiro: *Electromyography and Neuromuscular Disorders: Clinical-Electrophysiologic Correlations* (Butterworth-Heinemann, Boston 1998)
14.35 J. Kimura: *Electrodiagnosis in Diseases of Nerve and Muscle: Principles and Practice* (Oxford Univ. Press, New York 2001)
14.36 S.M. Gnatz: *EMG Basics* (Greenleaf, Austin 2001)
14.37 M.J. Fillyaw, R. Tandan, W.G. Bradley: Serial evaluation of neuromuscular function in management of chronic inflammatory demyelinating polyneuropathy, Phys. Ther. **67**, 1708–1711 (1987)
14.38 J.W. Day, C. Sakamoto, G.J. Parry, F. Lehmann-Horn, P.A. Iaizzo: Force assessment in periodic paralysis after electrical muscle stimulation, Mayo Clin. Proc. **77**, 232–240 (2002)
14.39 A.C. Jackson, J.J. Gilbert, G.B. Young, C.F. Bolton: The encephalopathy of sepsis, Can. J. Neurol. Sci. **12**, 303–307 (1985)
14.40 H.F. Ginz, F. Zorzato, P.A. Iaizzo, A. Urwyler: Effect of three anaesthetic techniques on isometric skeletal muscle strength, Br. J. Anaesth. **92**, 367–372 (2004)
14.41 J. Hong, P.A. Iaizzo: Force assessment of the stimulated arm flexors: quantification of contractile properties, J. Med. Eng. Technol. **26**, 28–35 (2002)
14.42 J. Hong, J.H. Falkenberg, P.A. Iaizzo: Stimulated muscle force assessment of the sternocleidomastoid muscle in humans, J. Med. Eng. Technol. **29**, 82–89 (2005)
14.43 J.G. Quinlan, P.A. Iaizzo, G.A. Gronert, E.H. Lambert: Twitch responses in a myopathy with impaired relaxation but no myotonia, Muscle Nerve **13**, 326–329 (1990)
14.44 J.G. Quinlan, D.J. Wedel, P.A. Iaizzo: Multiple-pulse stimulation and dantrolene in malignant hyperthermia, Muscle Nerve **13**, 904–908 (1990)
14.45 W.J. Schulte-Mattler, T. Müller, M. Deschauer, F.N. Gellerich, P.A. Iaizzo, S. Zierz: Increased metabolic muscle fatigue is caused by some but not all mitochondrial mutations, Arch. Neurol. **60**, 50–58 (2003)
14.46 H.F. Ginz, P.A. Iaizzo, T. Girard, A. Urwyler, H. Pargger: Decreased isometric skeletal muscle force in critically ill patients, Swiss Med. Wkly. **135**, 555–561 (2005)
14.47 H.F. Ginz, P.A. Iaizzo, A. Urwyler, H. Pargger: Use of noninvasive stimulation muscle force assessment in long-term critically ill patients: A future stan-

dard in the intensive care unit?, Acta Anaesthesiol. Scand. **52**, 20–27 (2008)
14.48 H.J. Harlow, T. Lohuis, T.D. Beck, P.A. Iaizzo: Muscle strength in overwintering bears, Nature **409**, 997 (2001)
14.49 T.D. Lohuis, H.J. Harlow, T.D. Beck, P.A. Iaizzo: Hibernating bears conserve muscle strength and maintain fatigue resistance, Physiol. Biochem. Zool. **80**, 257–269 (2007)
14.50 S.A. Frommer, J.M. Jacobs, D.N. Qiu, G.W. Williams, C.L. Soule, C.M. Verfaillie, P.A. Iaizzo: Assessment of murine muscle contractile properties using a novel in vivo system, Muscle Nerve (2011), in press

Part C Medical I

Part C Medical Imaging

15 Digital Radiography
Lothar Heuser, Bochum, Germany

16 Computed Tomography
Thorsten M. Buzug, Lübeck, Germany

17 Ultrasound Diagnostics
Rudolf Götz, Willich, Germany
Frank Schön, Zug, Switzerland

18 Medical Infrared Imaging
Gerald C. Holst, Winter Park, USA
Thorsten M. Buzug, Lübeck, Germany

19 Endoscopy
Martin Leonhard, Tuttlingen, Germany
Klaus-Martin Irion, Tuttlingen, Germany

20 Cone-Beam Computed Tomography and Navigation
Dirk Schulze, Freiburg, Germany
Gerhard Hoffmann, Biberach, Germany

21 Interventional Radiology – Angiography
Doris Pommi, Forchheim, Germany

22 Near-Infrared Spectroscopy (NIRS)
John McNulty, San Diego, USA
Michael Born, San Diego, USA
Robert S. Pozos, San Diego, USA

23 Magnetic Resonance Imaging
Wolfgang R. Nitz, Erlangen, Germany

24 Magnetic Particle Imaging
Jörn Borgert, Hamburg, Germany
Bernhard Gleich, Hamburg, Germany
Thorsten M. Buzug, Lübeck, Germany

25 MR-Guided Interventions and Surgery
Andreas Melzer, Dundee, UK
Erwin Immel, Dundee, UK
Rachel Toomey, Dundee, UK
Fabiola Fernandez-Gutierrez, Dundee, UK

26 Devices and Materials in MRI
Gregor Schaefers, Gelsenkirchen, Germany
Andreas Melzer, Dundee, UK

15. Digital Radiography

Lothar Heuser

Digital imaging was developed and initially used in astronomy. Computed tomography (CT) was the first digital modality used in radiology, followed by magnetic resonance imaging (MRI), digital subtraction angiography (DSA), and phosphor storage plates (computed radiography, CR).

A digital image consists of a certain number of picture element cells called pixels which are arranged like a chess board in rows and columns. Matrix sizes range from 512 × 512 up to 4000 × 4000 pixels. The grey levels are stored as binary numbers in a range of 8–16 bits (= binary digits).

In conventional film/intensifier screen radiography, the film fulfills all the roles of detector, display medium and long term storage. In digital radiography, separate units for detection or data acquisition, display, and storage are required.

There are certain advantages to the digital image, including lossless reproduction and duplication, and postprocessing, i. e., window setting, enlargement, edge enhancement, as well as subtraction and integration procedures. Advantages in workflow management are transport over networks and connectivity to radiology (RIS) and hospital information systems (HIS).

Digital subtraction angiography (DSA), phosphor storage plates (computed radiography, CR), and flat-panel detectors (planar detectors), as well as their special applications, are described in particular in this chapter.

15.1 **Historical Background** 291

15.2 **From Analog to Digital Image** 292
15.2.1 Spatial Resolution, Dynamic Range, and Dose Quantum Efficiency 292
15.2.2 Differences Between Analog and Digital Images 294

15.3 **Digital Imaging Systems in Radiology**..... 296
15.3.1 Digital Image Intensifier Radiography and Conventional DSA 296
15.3.2 Digital Storage Screens................. 297
15.3.3 Digital Flat-Panel Detectors (Planar Detectors)......................... 300

15.4 **Digital Image Processing** 305
15.4.1 Frequency Filtering....................... 306
15.4.2 Multifrequency Filtering................ 307
15.4.3 Dual-Spectrum Radiography (Energy Subtraction) and Temporal Subtraction 307

15.5 **Image Communication and Archiving**..... 307
15.5.1 Example 1 – Hospital Outpatient or Patient at a Doctor's Surgery...... 308
15.5.2 Example 2 – Hospital Inpatient...... 308

References 309

15.1 Historical Background

Ever since the discovery of x-rays, film has been used as the imaging medium in projection radiography. Digital imaging techniques were developed for astronomical photography and analysis. Computed tomography (CT), invented in 1973, was the first radiological modality which used digital imaging techniques. This was enabled by the development of more powerful, less space-consuming computers and the fall in prices of microelectronic devices. Compared with recent systems, processing speed and capacity were very low. Therefore, the first scanners, dedicated to scull and brain imaging, used a small imaging matrix of 64×64 pixels. In the first whole-body scanners, the matrix was extended to 128×128 pixels. Spatial resolution was significantly lower as compared with conventional x-ray films, but the dynamic range, i. e., contrast resolution, was more than a hundred times higher.

The introduction of this new technology caused a new problem: in conventional x-ray technology the film was both image detector and long-term storage simultaneously. Rapid-access media for long-term storage of CT data did not yet exist. Due to the low capacity of hard-disc drives, CT data had to be stored on magnetic tapes. The latter were not suitable for long-term storage and had the additional disadvantage of delayed access to stored images. Therefore, CT images had to be copied to Polaroid films and to conventional x-ray films.

Magnetic resonance tomography (MRT) was the second digital imaging modality introduced to clinical radiology, in 1976. Data acquisition was different from CT, but the algorithms for image reconstruction were similar. The image matrix was also quite small, and conventional x-ray films were used for storage.

Secondary digitization of fluoroscopic images led to the development of digital subtraction angiography (DSA), in 1981. The capacity and performance of computers had extended in the meantime, and the imaging matrix could be expanded to 512×512 pixels. Digital fluoroscopy, the variant without subtraction, used a matrix of 1000×1000 and later even 2000×2000 pixels. The dynamic range was 10 bit, corresponding to 1024 gray scale steps.

Digitalization of conventional projection radiography was achieved by the introduction of digital image plates in 1984, which used cassettes with the same formats as conventional x-ray films.

Digital image plates were developed from image intensifier screens used together with x-ray films. In intensifier screens, absorbed radiation energy is completely converted to light, whereas in image plates that use storage phosphors the absorbed energy can be stored and then released by a laser readout process (photostimulable luminescence). Matrix size was 2000×2000 in the first system, with dynamic range of 10 bit. Although the spatial resolution of photostimulable phosphor plates was significantly worse compared with conventional film–screen combinations, monitors able to display the full 2000×2000 pixel matrix did not exist at that time. Therefore, images had to be printed out on conventional films for analysis and storage. The hardware required for image processing was quite expensive, and there were no options for dose reduction. These obvious disadvantages hindered the early spread of computed radiography. After expansion of the matrix to 4000×4000 pixels, photostimulable image plates could be used for all diagnostic purposes.

The introduction of flat-panel detectors in 2000 led to the development of fully digitalized x-ray systems. Due to their better dose quantum efficiency (DQE), dose reduction could be achieved as compared with conventional film–screen combinations. Further advantages were the instant availability of images and the avoidance of transport of the cassettes to the reading station. The matrix consisted of 3000×3000 pixels.

Today, digital imaging technology is realized in the majority of radiological departments, creating a filmless workflow. Technical advances and price reductions of hardware and storage media are the reasons why digital imaging now does not suffer from any disadvantages compared with conventional film-based imaging.

15.2 From Analog to Digital Image

In an analog image, the fluoroscopic screen or the film serves as the detector. The generated signal is proportional to the radiation dose. Within the first 100 years of the discovery of x-rays, the conventional detectors consisting of film and intensifier screens were optimized. For fluoroscopy, image intensifier systems were developed. Both resulted in a marked decrease of radiation dose.

15.2.1 Spatial Resolution, Dynamic Range, and Dose Quantum Efficiency

Spatial resolution, grayscale dynamic, and dose quantum efficiency are the most important criteria for a digital image detector system. Optimized film–screen combinations with light emission adapted to the color of the film base were regarded as the most dose-effective

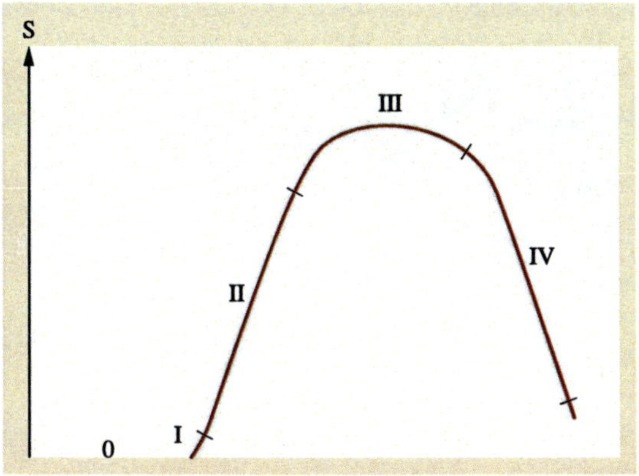

Fig. 15.1 Gradation curve of the film (0 – fog; I – toe; II – linear part; III – shoulder; IV – solarization)

detector systems. There was a reciprocal relationship between sensitivity and spatial resolution, i. e., the larger the grain size of the intensifier screen, the higher the sensitivity and the worse the spatial resolution. X-ray film contains 220 000 silver bromide grains per cm^2, which serve as picture elements. The dynamic range (latitude) of the film is limited by the maximum optical density that can be generated by exposure. The special criterion of the film is its gradation curve (Fig. 15.1). Proportionality between the radiation dose and attenuation exist only in section II of this curve. This results in about 40 gray levels. Special low-grade films can display up to 80 gray levels.

Digital images do not suffer from this limitation. They utilize a matrix of discrete numerical values to represent the image. This set of picture element cells (pixels), which are arranged like a chessboard in rows and columns, is called the matrix (Fig. 15.2). Used matrix sizes range from 512×512 (262 144) up to 4000×4000 pixels (Fig. 15.3). The gray levels are stored as binary numbers in a range of 8–16 bits (binary digits). Smaller numbers of gray scale levels (8 bit) are used for ultrasound imaging and laser film printing, and 10–14 bits are used for digital fluoroscopy and digital radiography. Since monitors can display only 64 gray levels (6 bit) and the human eye can resolve only 16 gray levels (4 bit), special window settings are

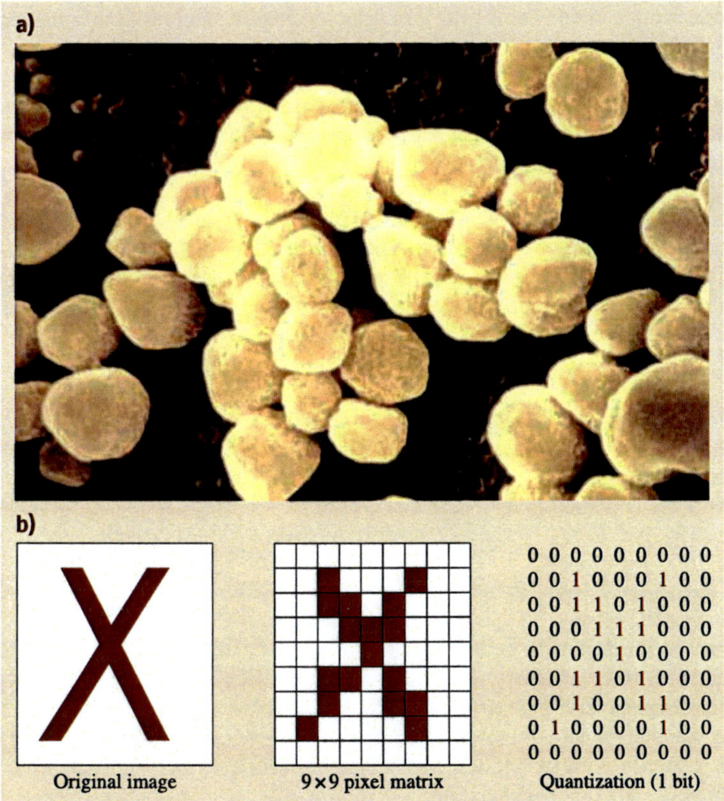

Fig. 15.2 **(a)** Scanning electron microscope image of a film emulsion, showing the silver bromide grains. **(b)** Digitization of an analog image; matrix size and image depth were chosen arbitrarily

CT	512×512
HR-CT	1024×1024
DSA	1024×1024
Digital fluoroscopy	2000×2000
Flat-panel detector	3000×3000
Digital luminescence radiography	4000×4000

Fig. 15.3 Examples of matrix sizes used by various imaging methods in radiology

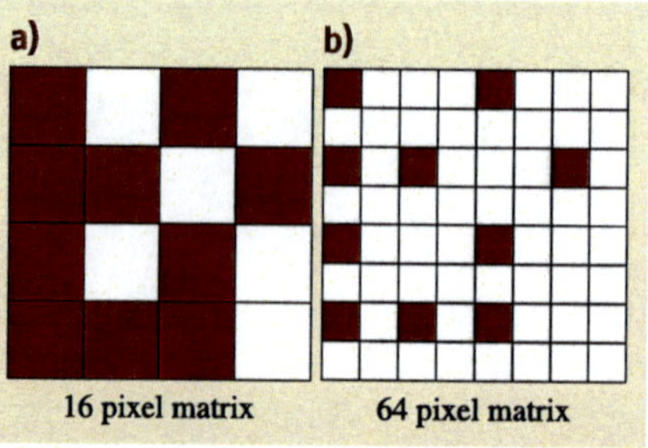

Fig. 15.4a,b Effect of image matrix on noise. **(a)** Enlarging the image matrix size to 64 pixels. **(b)** 16-pixel matrix. The number of image points without information increases from 6 to 50

required to display the whole latitude of a digital image [15.1].

15.2.2 Differences Between Analog and Digital Images

Organization

In conventional film screen technology the film serves as detector, display medium, and storage medium. These three functions are separated in digital technology. Image intensifiers, photostimulable phosphor plates, and flat-panel detectors with direct or indirect conversion are used as detectors. They all generate analog electric signals that have to be converted into digital signals by an analog-to-digital converter (ADC). The raw data of the digital image are specially processed for perception by the human eye and can then be displayed on a monitor. For short-term storage, a hard-disc drive (HDD) in the processing unit is used, while for long-term storage the data have to be transferred to a redundant array of independent discs (RAID) or saved onto a magneto-optical disc, compact disc (CD) or digital versatile disc (DVD).

Digital Image Parameters

As mentioned above, the gray scale dynamic range of a digital image is significant larger than that of film. However, the spatial resolution of film is significant better; for example, a 30×40 cm image plate contains 16 000 000 pixels in a 4000×4000 matrix. An x-ray film of the same size contains 264 000 000 grains of silver bromide that serve as picture elements. Since the overall resolution is a product of spatial resolution and latitude, all relevant details can be displayed on digital images. Expanding the matrix causes additional problems: First, computer performance and storage capacity have to be improved. Second, if all other parameters stay unchanged, the signal-to-noise ratio (SNR) decreases, as shown in Fig. 15.4. To compensate for this disadvantage, tube loading, the DQE of the detector system, or the readout process can be altered.

Many studies have compared exposure levels and image quality in conventional and digital radiography. In conclusion one can state that:

- Using equivalent exposure, diagnostic information from digital radiography is superior to conventional radiography.
- For equivalent image quality, digital radiography requires less exposure than conventional radiography.

Digital radiography has certain advantages in terms of exposure. The gray levels of the image are normalized (signal normalization), and the brightness of an image does not depend on the absorbed dose (Fig. 15.5). Therefore, there is no loss of contrast in case of over- or underexposure, unlike in conventional films, and un-

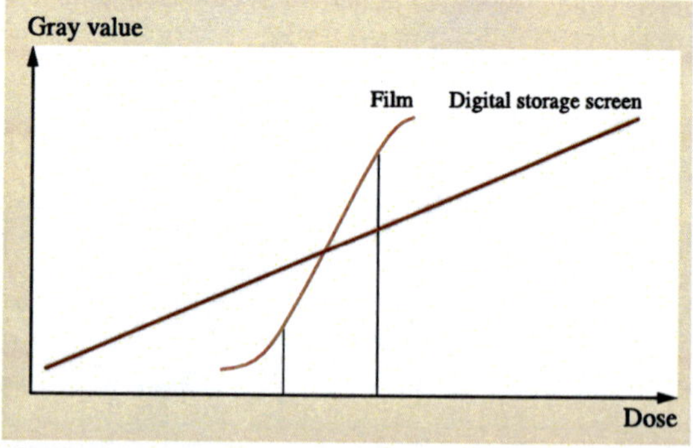

Fig. 15.5 Signal behavior with films and with digital storage screens

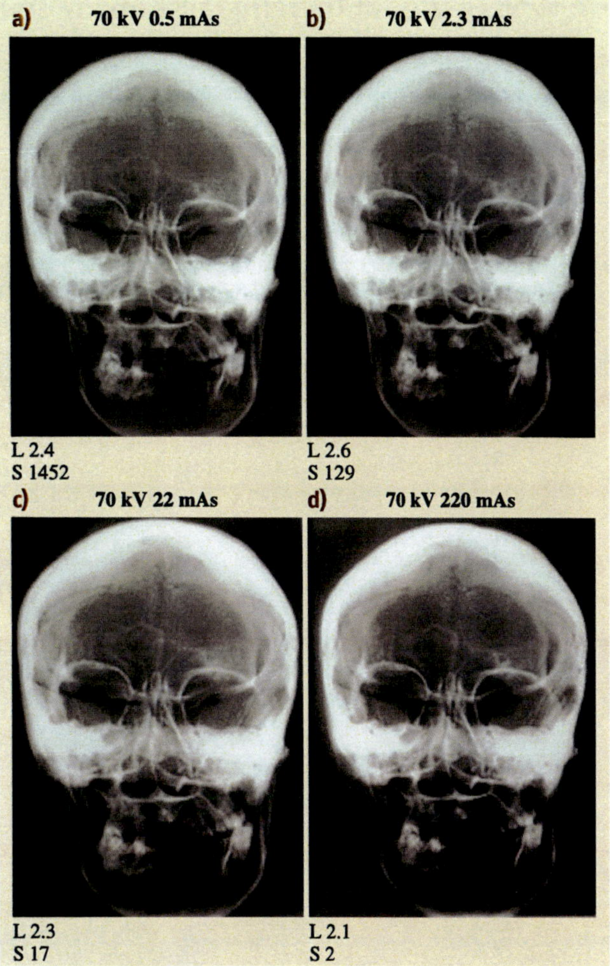

Fig. 15.6a–d Illustration of a skull specimen with four different exposure doses (**a**–**d**). Image (**d**) was obtained using a radiation dose 500 times higher than image (**a**). Image brightness and contrast remain constant under exposure doses that vary by several orders of magnitude. An excessively high dose (**d**) can no longer be recognized visually based on the brightness or contrast of the image ▶

der special conditions the exposure dose can be reduced (Fig. 15.6). However, too much reduction of dose will certainly result in increased noise and loss of detail information (Fig. 15.7). On the other hand, overexposure results in an unnecessarily elevated dose to the patient but is not seen directly in the digital image, unlike on conventional films.

Image Reproduction, Image Processing, and Image Postprocessing

The storage of a digital image as a binary-coded dataset has many advantages: Lossless reproduction and duplication are possible. Brightness and contrast can be adjusted for optimal perception by the viewing eye, which may be very useful in case of large differences of contrast or attenuation. Image processing means gray scale processing, edge enhancement, as well as subtraction and integration of frames. A main application is DSA.

In summary, digital imaging has the following fundamental differences from conventional cassette-based film–screen systems [15.2]:

- Decoupling of image acquisition, presentation, and storage, requiring further investments in workstations, storage devices, and picture archive and communication systems (PACS)
- Large dynamic range, with no effects of over- or underexposure
- Brightness and contrast can be modified independently of the absorbed dose of the digital system
- Risk of unrecorded overexposure due to signal normalization
- Image quality and detail resolution depend on the quality of the detector and display systems.

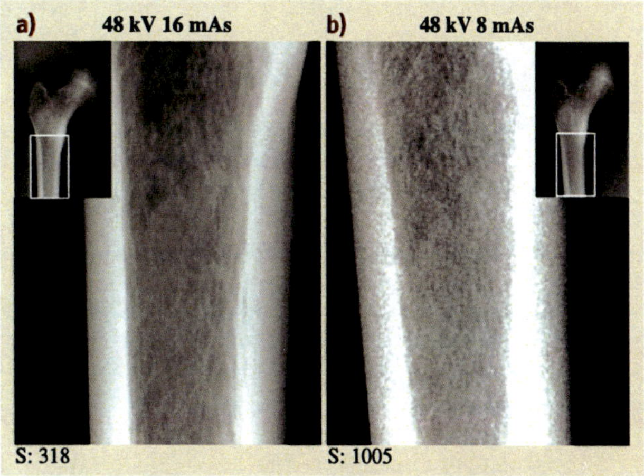

Fig. 15.7a,b Signal normalization in digital radiography. Radiographs of a femur specimen. (**a**) The dose was halved while keeping the other radiation parameters constant. The brightness remains unchanged, but there is a noticeable increase in image noise and a decrease in recognizable details ▶

15.3 Digital Imaging Systems in Radiology

Figure 15.8 shows a comparison of the different digital detector systems used in radiology.

15.3.1 Digital Image Intensifier Radiography and Conventional DSA

In digital fluoroscopy and conventional DSA, an analog image is generated on a fluoroscopic screen and is enhanced by a factor of 300 by an image intensifier. It appears rotated by 180° and demagnified on the output screen. Conversion into an analog video signal is performed either by a tube operated camera (Vidicon) or charge-coupled device (CCD) chip. So far, this forms a complete x-ray image intensifier chain. The analog signal is then converted into a digital signal by an ADC and sent to a processing unit (Fig. 15.9). For a long time, display on a monitor required reconversion to an analog signal. Today, digital monitors are available for direct display. Matrix sizes are 1000 × 1000 for a series of frames and 2000 × 2000 for single images. Compared with conventional film–screen technology there are advantages in terms of image processing, reproduction, duplication, as well as transport and archiving. Spatial resolution is limited by the resolution of the image intensifier and the matrix size.

In conventional DSA the image is also generated in a conventional intensifier chain and digitized. Processing is performed as a mask technique, with the first frames of a series prior to contrast injection serving as masks that are subtracted as single or integrated images from the frames after contrast injection. In this way, structures that are common to both the mask and post-

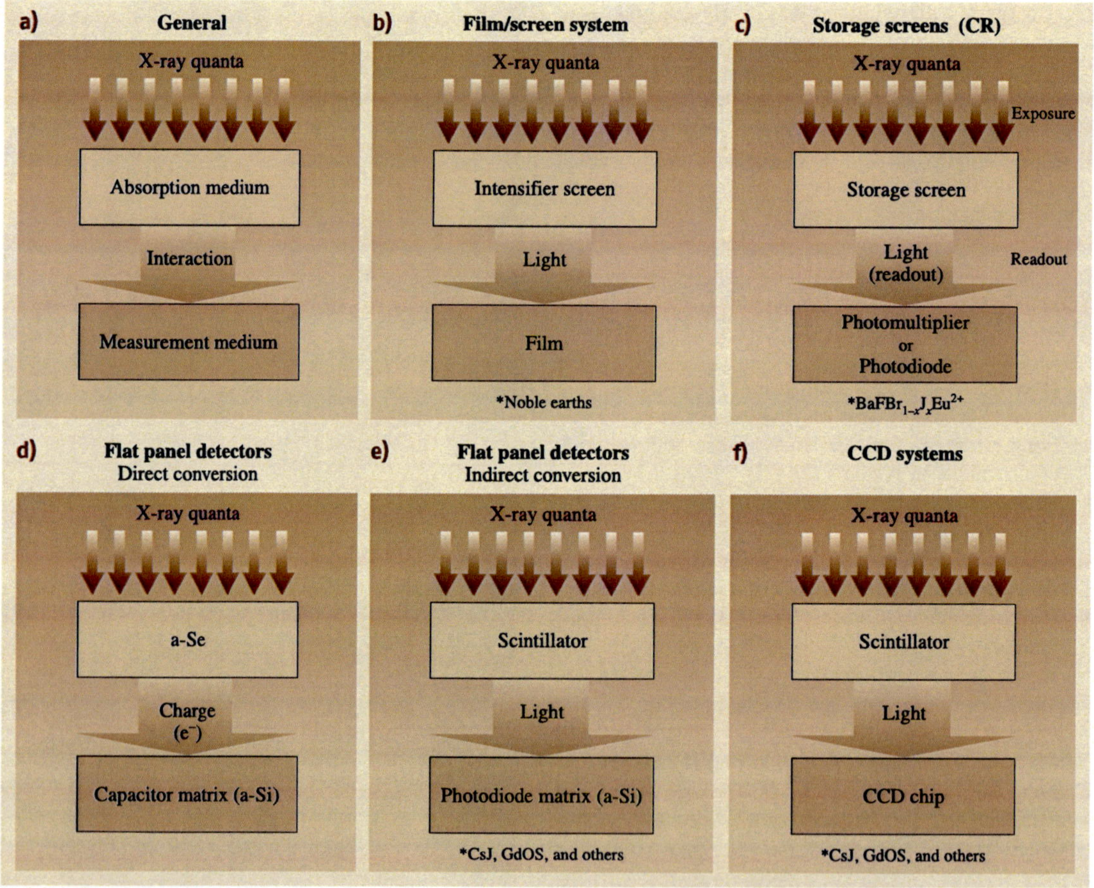

Fig. 15.8a–f Comparison of the various detector systems and their working principles (after [15.3])

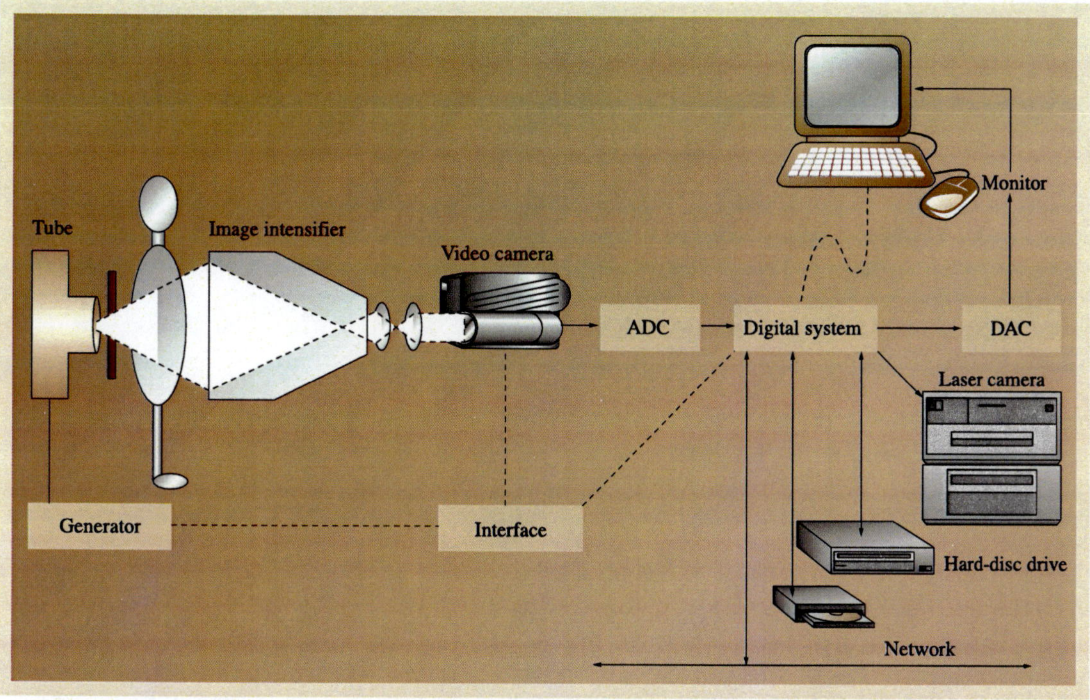

Fig. 15.9 Schematic illustration of digital image intensifier radiography. The analog image is generated in an x-ray image intensifier–television chain, and is then digitized by an analog-to-digital converter (ADC). Image processing takes place on a computer. Thereafter, transmission and storage can proceed. Conversion back into an analog signal (digital-to-analog converter, DAC) is required if the image is to be displayed on analog media (after [15.4])

contrast image are removed. The resulting (subtracted) image contains just the contrasted vessels (Fig. 15.10). The difference signal is additionally amplified, which results in enhanced contrast. Since the images are stored as binary digital datasets, the whole process is performed in real time and circulation of contrasted blood can be watched on a monitor.

This technique yields the following advantages:

- A reduced amount of contrast medium is required compared with film angiography, with a corresponding reduced risk of contrast-induced adverse effects
- Display of vessels without overlay of other structures, which is especially useful in the scull base where arteries cross dense bony structures
- Utilization of negative contrast media such as carbon dioxide for vascular imaging (Fig. 15.11), which can be used in patients with contraindications for iodine contrast media.

DSA is sensible to voluntary and involuntary movements (heart beats, bowel movements) of the patient. These cause a shift of the structures that undergo subtraction between the mask and postcontrast images. The resulting artifacts can degrade image quality to such a degree that vessel analysis becomes impossible. There are two solutions to this problem: remasking and pixel shift. In remasking, a different mask, probably closer to the contrast frames, is chosen. By using this technique, the time interval for movements is reduced, resulting in fewer artifacts. Movements after contrast injection cannot be eliminated by this method. Using the pixel-shift method, the mask and contrast images are shifted with respect to each other by pixels or subpixels until the anatomic structures match up. In most cases, this technique leads to a marked improvement in image quality.

15.3.2 Digital Storage Screens

Digital phosphor storage plates have a similar structure to intensifier screens. The storage medium (for example, BaF) is applied as a powder on a polyester base. However, in contrast to intensifier screens, in phosphor

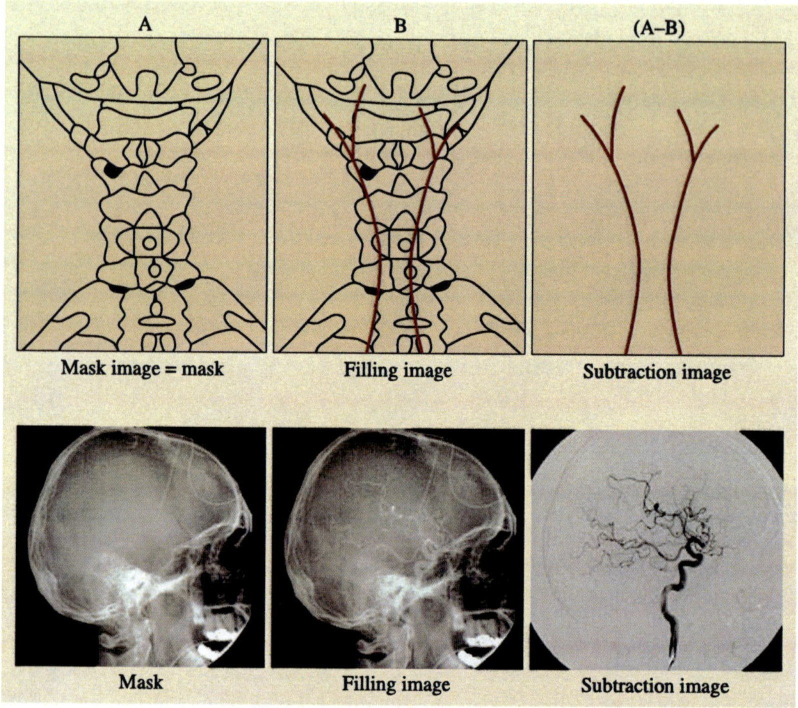

Fig. 15.10 Schematic illustration of mask technology (*top row*) and original images (A. carotis interna) in DSA

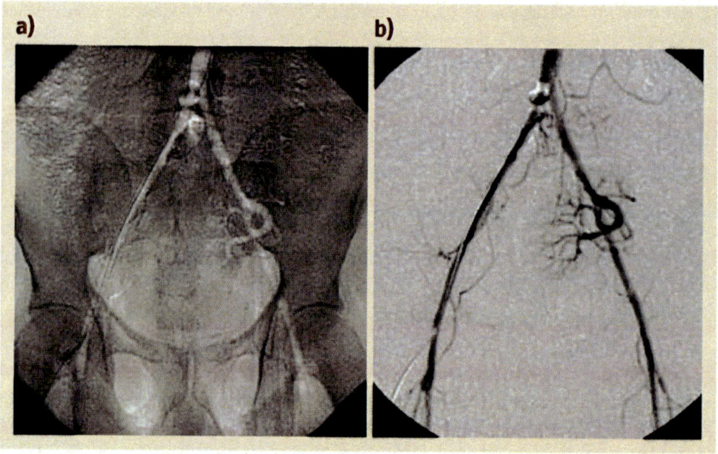

Fig. 15.11 **(a)** Digital subtraction angiography with CO_2 in a patient with kidney transplant. **(a)** Unsubtracted image after injection of 50 mL CO_2. **(b)** Subtraction image. The quality of vascular imaging is comparable to that with positive x-ray contrast medium. This method is suitable for patients with contraindications to iodine-containing contrast media

image plates only a small fraction of the released electrons are converted to visible light. The major part of the generated free electrons are stored at a high energy level. This is achieved by the phosphor being doped; i. e., it contains impurities of another element (europium) which modifies the crystal lattice structure, forming electron traps. Instead of falling back to their ground state by emitting fluorescent light, the electrons excited by exposure to x-ray quanta remain at a higher metastable energy level (Fig. 15.12). This state persists for several hours. In the readout unit, the screen is removed from the cassette and scanned point by point by a fine laser beam. The electrons now emit fluorescent light and return to their stable original state. The emitted light is measured by photomultipliers, converted into electric signals, and digitized (Fig. 15.13). After the

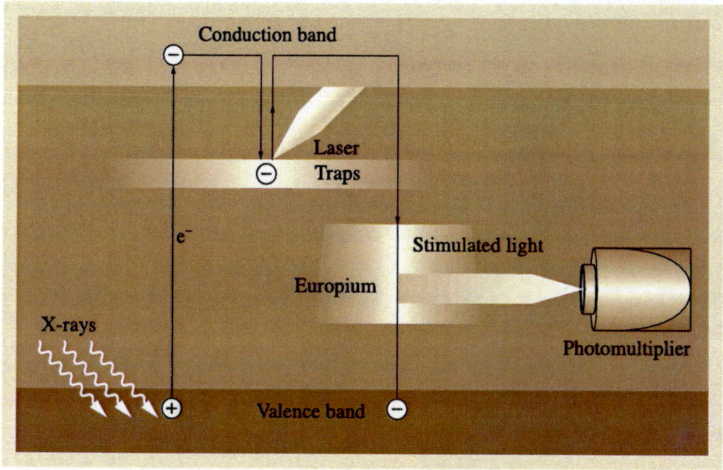

Fig. 15.12 Schematic illustration of energy changes during the exposure and readout process of a digital luminescent screen

readout process, the screen is erased with bright visible light and loaded into the cassette for the next exposure. As in conventional intensifier screens, the system suffers from the problem of the reciprocal relationship between sensitivity and image sharpness; i. e., a thicker conversion layer is more sensitive, but the increased scattering of light quanta results in lower image sharpness. This scattering by the luminescent layer needs to be minimized as far as possible, which is achieved partly by dyeing it and partly by using special binders (e.g., a synthetic resin varnish).

The computer generates a raw data image, which is processed further using special programs. Depending on the type of processing, the result is either an image similar to the one obtained using conventional film or a version with edge enhancement. Moreover, steeper or flatter gradients can be created through special look-up tables (LUT). Whereas early systems started with a 2 k matrix, matrix sizes of 4 k are common today. The system is used both in x-ray cassettes and as a built-in device in complete x-ray units (e.g., chest imaging systems for standing patients). The detector's input dose is 2.5 μGy, i. e., comparable to that of a screen–film system of sensitivity class 400 [15.2].

The advantages of this method are:

1. The use of normal cassette formats allows direct replacement of film cassettes by digital cassettes

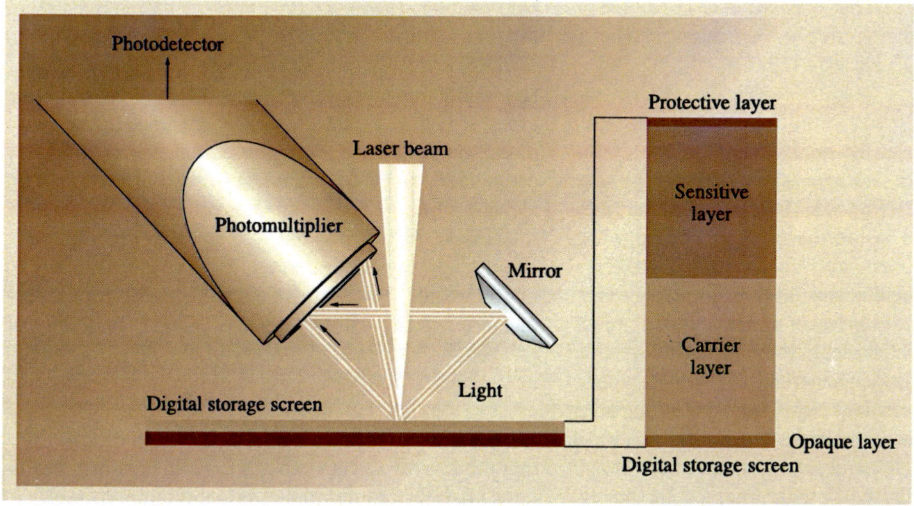

Fig. 15.13 Schematic illustration of the readout process of a digital storage screen. The laser beam's point-by-point scanning releases the trapped electrons, and the light emitted as a result is fed to the photomultiplier either directly or by means of a mirror. It is converted in the photodetector into an electric signal

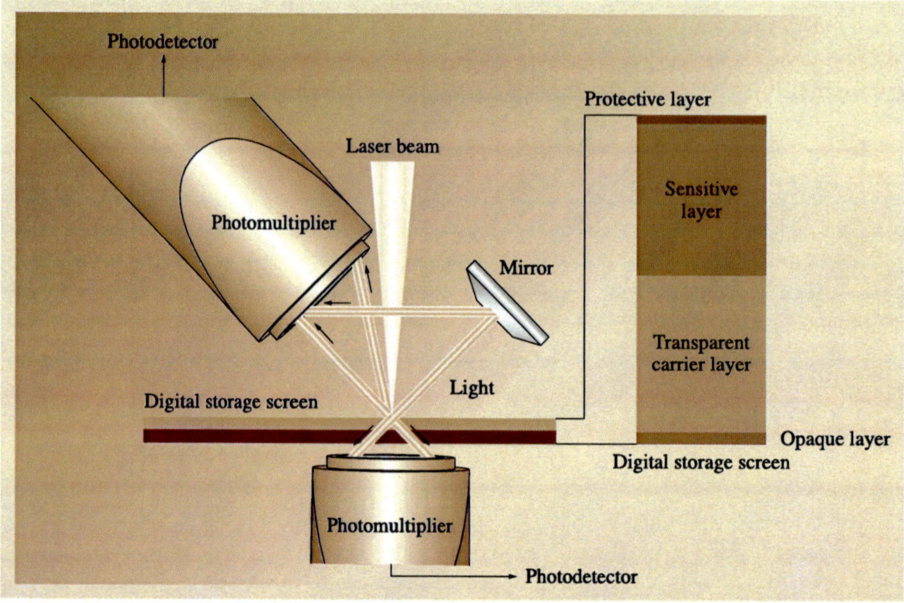

Fig. 15.14 Schematic illustration of the readout process using the dual scanning method. Unlike a simple readout process, the luminescent screens have a transparent carrier layer such that the readout process can be performed by photomultipliers on both sides

without the x-ray facility having to be upgraded or converted.

2. Signal normalization always results in correct exposure, i. e., the brightness and contrast remain unchanged even under large dosage jumps. Major dose reductions, however, are not possible with these systems with the exception of a few situations, as increased noise causes loss of image details.

A further development of this method consists in *dual scanning* (Figs. 15.14, 15.15). By using a transparent carrier material, readout is possible from both sides with an additional photomultiplier, which leads to better SNR and a 40% DQE increase [15.5, 6].

Storage Screens with Needle Image Plates (NIP)

Instead of a powder as the luminescent material, a further development of storage screen technology uses CsBr needle crystals (Figs. 15.16, 15.17). The luminescent light is guided inside the crystal needles, which function as optical fibers, thus minimizing lateral diffusion of the light. The detector layer is correspondingly thicker, thereby increasing dose efficiency while maintaining the same spatial resolution. During the readout process, there are further benefits due to deeper penetration of the laser readout beam into the luminescent layer, which ultimately reduces image noise [15.7]. Since the configuration and arrangement of the crystal are very similar to those of planar image detectors, the DQE is improved. This results in a dose decrease of up to 50% compared with powder image plates [15.7, 8].

15.3.3 Digital Flat-Panel Detectors (Planar Detectors)

Flat-panel detectors are built-in components of x-ray units. This means that the detector is connected online to the computer using either a cable or a wireless system [wireless local area network (WLAN), Bluetooth, etc.]. Whereas cassettes with digital storage screens, similar to screen–film cassettes, have to be removed from the cassette holder after exposure and transferred to the reading device, in direct radiography with flat-panel detectors the image is available within a few seconds. Two types of flat-panel detectors have been developed: with direct and indirect energy conversion. In direct energy conversion, x-ray quanta are converted directly into an electric signal, whereas in indirect energy conversion fluorescent light is generated, then converted into electric signals by photodiodes. The feature common to both methods, however, is an active matrix made of capacitors and thin-film transistors (TFTs), consisting of amorphous silicon and acting as a switch for spatial coding (Fig. 15.18).

Detectors with Indirect Energy Conversion

Incoming x-ray quanta generate visible light in a scintillator made of cesium iodide or gadolinium oxysulfide.

This light is converted into electric signals below the scintillator in a matrix of photodiodes made of amorphous silicon. One problem with this method is posed by scattering of light in the scintillator, leading to signal spreading and thus adversely affecting spatial resolution. This problem, however, can be minimized by forming the cesium iodide (CsI) crystals in the shape of needles, so that they behave like optical fibers and permit directional transmission of light. Systems with indirect conversion (optodirect systems) show the highest quantum efficiency among all digital detectors in the field of skeleton radiology (55–70 kV) (Fig. 15.20). At 70 kV, the DQE can reach 65%. The detector input dose required for posterior–anterior chest x-rays is in the range of 1–1.5 μGy, which is comparable to a screen–film system of sensitivity class 800.

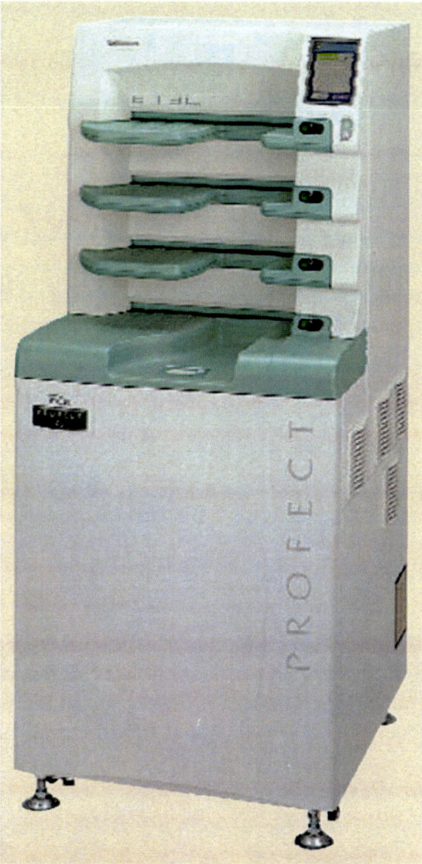

Fig. 15.15 The Fujifilm FCR PROFECT CS readout unit. The system operates with four drawers for different cassette formats

Detectors with Direct Energy Conversion

The direct method dispenses with the intermediate step of light conversion. Electric charges are generated directly through the absorption of x-rays in the 500 μm converter layer made of amorphous selenium. The schematic layout of such a detector is shown in Figs. 15.18 and 15.19.

An applied high voltage causes electron–hole pairs generated by the x-rays to separate. Under the selenium layer there is a matrix of collecting electrodes, which define the individual pixels. Below each electrode there is a capacitor for charge storage and a switching tran-

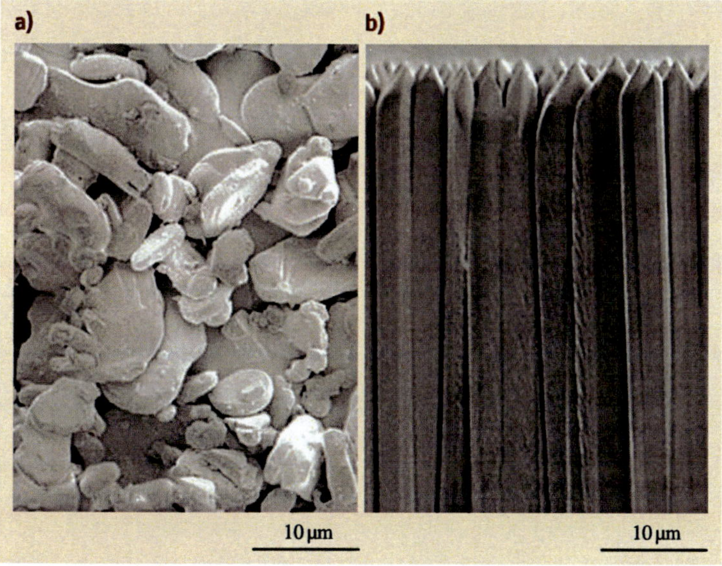

Fig. 15.16a,b Scanning electromicrographs of the conversion layers of a conventional screen with powdered luminescent material **(a)** and a needle crystal screen **(b)**. The crystals behave similarly to optical fibers and reduce light scattering. They have a better DQE

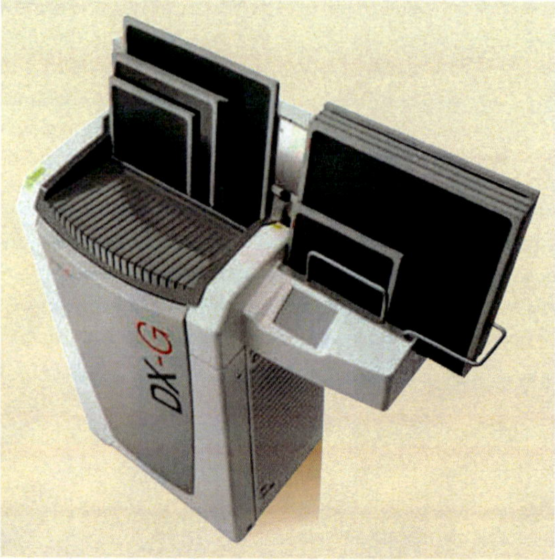

Fig. 15.17 Agfa DX-G readout unit for needle crystal screens and conventional luminescent screens

sistor for readout. The DQE of systems with direct conversion (electrodirect systems) is lower in this range, but in the range 20–30 kV it is higher. In addition, these systems have the advantage of higher signal in the case of small, high-contrast structures, therefore being especially suitable for digital mammography. However, they are used in automated chest x-ray workstations also, despite the poorer DQE in the hard beam range.

The first direct detectors used in radiology were CCD systems. Since CCD chips are only 2–4 cm^2 in size, the recording format needs to be decreased correspondingly with a lens or with fiber optics, which leads to considerable reductions in dose efficiency and image quality. Alternatively, CCD chips can be mounted in a linear array, and the image compiled sequentially from full-format rows (the slot-scan method). This slot technology decreases the scattered radiation to such an extent that a grid can be dispensed with. The resulting decrease in dose compensates as far as is possible the lower DQE encountered in CCD technology [15.9].

The advantages of direct detectors over storage plate systems consist of higher dose efficiency (at least in the case of powder-coated storage plates) and in a higher performance speed, enabling display of the recorded image within a few seconds, which facilitates monitoring of critical patients and enables a higher examination frequency (Fig. 15.21). Nowadays, there also exist mobile devices with a portable detector which transmits

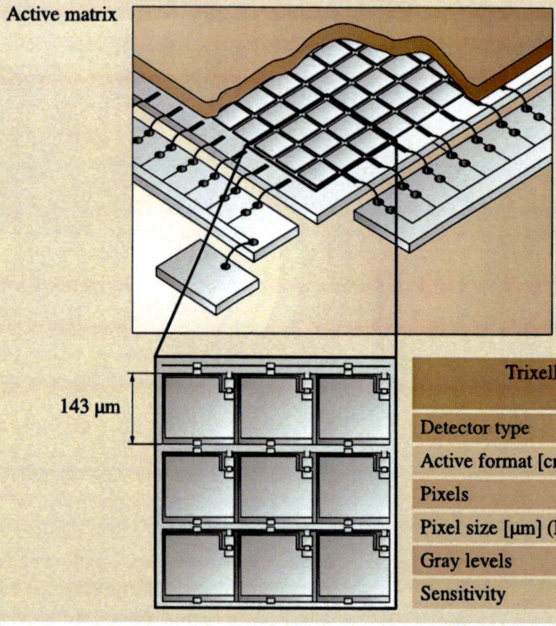

Trixell detector (Siemens/Philips) –Parameters–	
Detector type	Indirect amorphous silicon
Active format [cm]	43 x 43
Pixels	9 Mio
Pixel size [µm] (MTF)	143 (3.5 lp/mm)
Gray levels	16384 (14 bit)
Sensitivity	400–800

Fig. 15.18 Illustration of the active matrix and the characteristics of a detector with indirect energy conversion

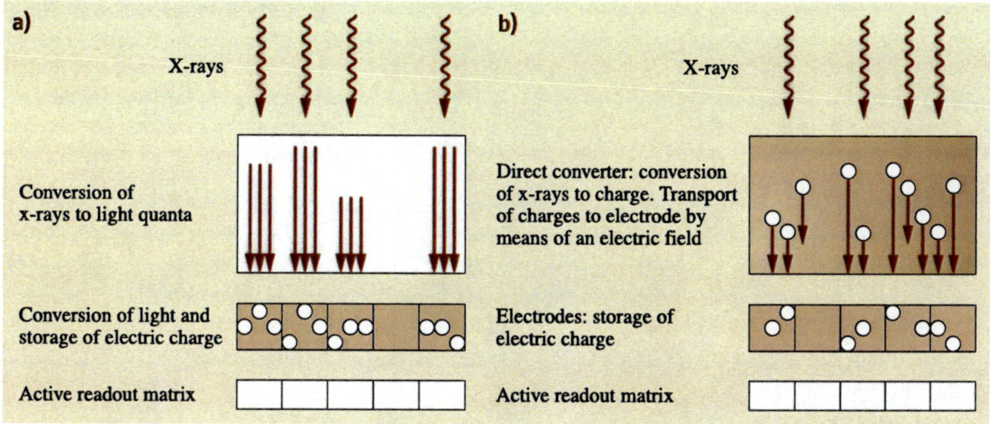

Fig. 15.19a,b Schematic illustration of the working principle of planar detectors with indirect **(a)** and direct **(b)** energy conversion (after [15.10])

image data either over a cable or using a wireless system. Thus, when x-rays are taken in the operating theater or in the intensive care unit, this information is made available directly afterwards. On the downside, this system suffers from higher procurement costs (by a factor of 3–4).

Flat-Panel Detectors for Fluoroscopy and Angiography

Flat-panel detector technology has advantages when compared with image intensifiers also, and has displaced them in newer angiographic units. Depending on the field of application (interventional radiology, neuroradiology, cardiology), detectors are available in various sizes, as single- or dual-plane systems.

As in static radiography, flat-panel detectors for angiography and fluoroscopy exist as detector types with indirect and direct x-ray conversion, as already described above. For the x-ray spectrum used in angiography and fluoroscopy, cesium iodide (CsI) offers considerable advantages as compared with selenium due to the higher atomic number of the elements involved. This means a higher DQE and results in a better signal-to-noise ratio. The technical requirements imposed on flat-panel detectors for dynamic applications are significantly greater than those in static radiography. In particular, the demands in DSA, dual-plane x-ray serial angiography, and fluoroscopy in terms of minimizing the dose pose special challenges for DQE, image readout speed, and the temporal resolution of dynamic

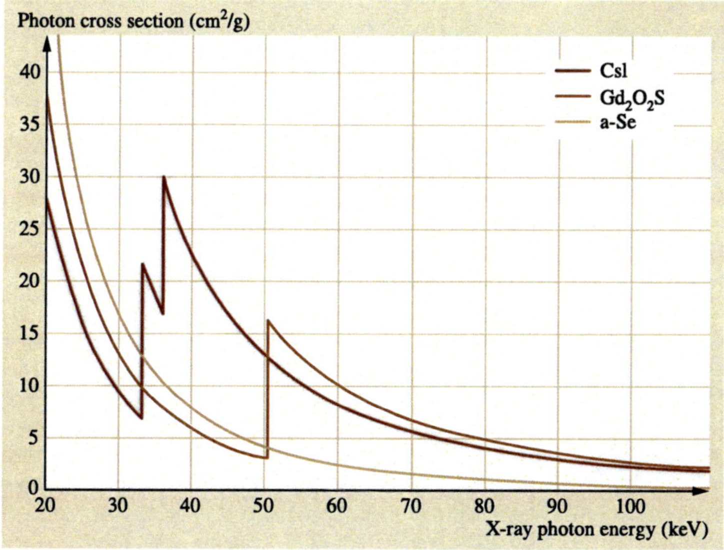

Fig. 15.20 Signal intensity as a function of quantum energy for direct energy conversion with amorphous selenium (Se) and for indirect conversion with cesium iodide (CsI) and gadolinium oxysulfide (Gd_2O_2S). It is evident that the latter detector has a better signal yield in the kV range for skeletal diagnostics, whilst the selenium detector is superior to detectors with indirect energy conversion in the mammography soft beam range

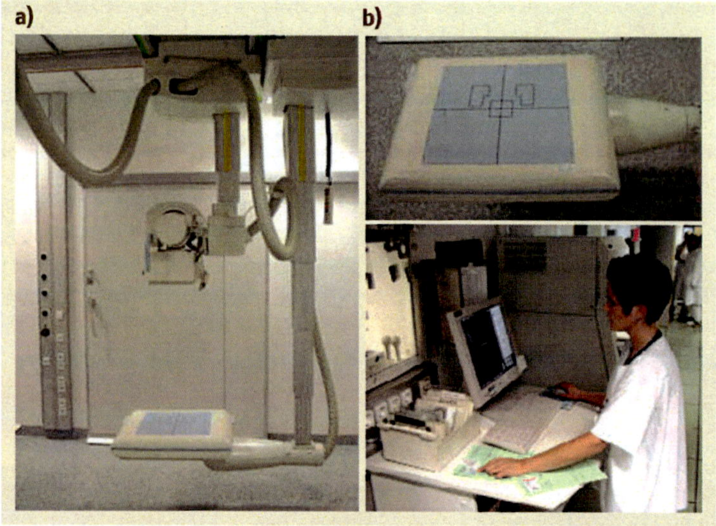

Fig. 15.21a,b Planar detector (Trixell) in the Siemens Axiom Aristos x-ray unit. The advantages of direct radiography are evident. The image appears on the monitor a few seconds after exposure

flat-panel detectors. One of the main problems is the readout process, which should be made as short as possible to avoid blurring through movement during the exposure. The duration of the readout process, however, affects the DQE. Pixel sizes lie between 150 and 200 μm. Image frequencies range from single images to 60 frames/s (in pediatric cardiology).

Compared with conventional x-ray image intensifiers, flat-panel detectors offer several technical advantages [15.10], such as:

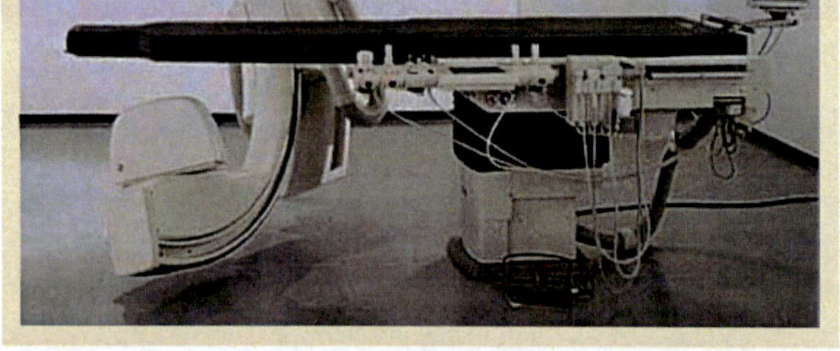

Fig. 15.22 C-arm angiography unit with planar detector (Axiom Artis, Siemens). The physical dimensions of the detector are small

- Homogeneous image quality across the entire image area (no distortion, by the Earth's magnetic field)
- Better imaging in high contrast (e.g., catheter tips, guide wires, etc.)
- Higher linear dynamic range (C-arm CT)
- Better transmission of contrast differences [higher modulation transfer function (MTF)]
- Higher DQE
- Square or rectangular geometry of the x-ray-sensitive surface [advantageous for three-dimensional (3-D) reconstructions]
- Low volume and therefore geometrical layout (Fig. 15.22) can be used in environments with strong magnetic fields (magnetic navigation).

C-Arm CT (Planar Detector CT, PD-CT)

The first reports on attempts to acquire projection data across a 180° fan beam angle go back to 1997–2001 [15.11, 12]. They were performed with C-arm systems equipped with conventional image intensifiers. The smaller dynamic range, image distortions, and susceptibility to interference from magnetic fields during rotation limit the performance of image intensifiers as detectors for CT systems and prevent their use for classic CT-type applications such as imaging of soft-tissue structures [15.13]. Therefore, image intensifier-based systems have been used only for high-contrast imaging and for 3-D rotation angiography (3-D DSA) [15.11, 12, 14, 15]. It may be assumed that these systems will gradually be replaced by PD-based C-arm CT systems (Fig. 15.23), since

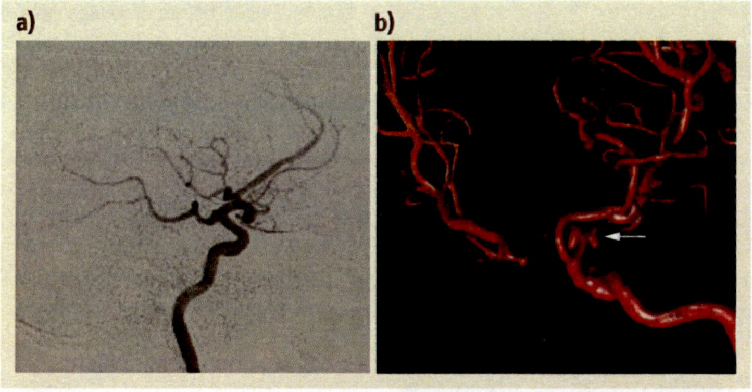

Fig. 15.23 **(a)** Rotation angiography image of the A. carotis interna. **(b)** 3-D reconstruction with the image of a small aneurysm (*arrow*)

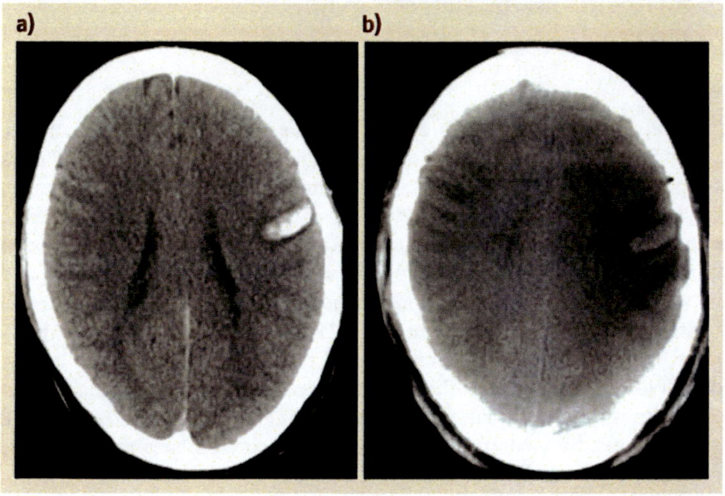

Fig. 15.24a,b Circumscribed cerebral hemorrhage with conventional CT **(a)** and with C-arm CT **(b)**. The image quality of the C-arm CT is virtually identical to that obtained using the conventional method (after [15.16])

they offer higher dose efficiency and better image quality.

Generation of 3-D DSA images through PD-CT during an intervention is already very common [15.16–18]. Compared with image intensifier-based systems, PD technology offers significantly better soft-tissue imaging, resulting in images similar to spiral CT (Fig. 15.24).

PD-CT is also referred to as cone-beam CT. It requires special image reconstruction methods, due to the pronounced divergence of the lateral rays. The reconstruction method of choice is the Feldkamp algorithm [15.19]. In principle, high image quality can only be ensured for the central layer corresponding to the central beam. Image quality decreases for regions outside the central layer, and artifacts increase with increasing distance from it. These artifacts are generally referred to as cone-beam artifacts [15.20]. To avoid them, accurate reconstruction algorithms are necessary, in general requiring high computing power and not yet being practical for everyday applications. However, their eventual use is becoming ever more realistic with ongoing advances in computer technology.

15.4 Digital Image Processing

The raw data image generated by the detector is inadequate for imaging purposes. Artifacts must be eliminated and the information processed in order to generate a high-quality image. This requires robust programs that

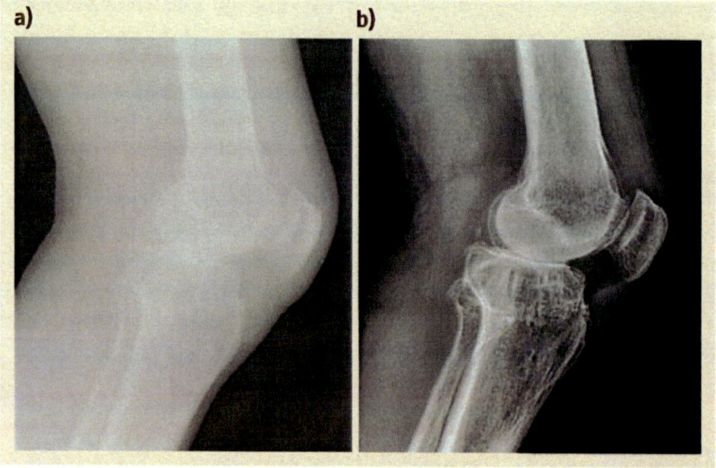

Fig. 15.25a,b Effect of image processing. **(a)** The raw image data from a digital casette (CR)/direct radiography (DR) system cannot be utilized diagnostically. **(b)** After processing with MUSICA 2 software (Agfa), imaging of bone and soft tissue contrast is optimized

provide constant brightness and contrast and that are adapted to the specific organ of interest (Fig. 15.25).

In planar detectors, there exists the problem that the crystals are not of uniform length (production tolerances) and, for this reason but also due to other errors, provide variable signal intensities. These distracting effects are eliminated by flat fielding, in which all the pixels are allocated individual correction factors. Offset effects associated with pixels are captured with dark-field images (image data acquisition without radiation) and corrected also. Differences in pixel sensitivity and variations in the signal amplification of pixels can be corrected with flood field images (image data acquisition with constant, homogeneous radiation across the entire detector area). The greatest challenge for image processing software consists in the identification of those pixels that require correction [15.21]. Nonfunctioning pixels are easy to identify (through appropriate limits being placed on the offset and gain), using the specified procedures. Partially functional pixels and those with nonlinear response are not affected by these methods. If, on the other hand, the offset and gain limits are set too narrowly, normal pixels are falsely identified as defective. Another very widely used method for eliminating bad lines (line artifacts) or speckle noise caused by pixel defects is the median filter [15.22], where the defective pixel is replaced by the median of the neighboring pixel values.

15.4.1 Frequency Filtering

Digital image processing is used, on the one hand, to compensate for the nominally lower spatial resolution of digital systems (MTF restoration), and on the other, to exploit the intrinsic advantages of digital detectors; for example, in chest x-rays, as a result of digital reduction of the dynamic range [dynamic range reduction (DRR) or dynamic range compression (DRC)], good contrast is maintained in the lungs and improved transparency and contrast are obtained in the mediastinum [15.3]. This overcomes an inherent drawback of conventional radiography, namely the inverse relationship between contrast and dynamic range. In addition, appropriate filtering creates the effect of local contrast enhancement, which facilitates recognition of low-contrast structures but allows image characteristics to deviate more strongly compared with conventional x-ray images. For this reason, such procedures for local contrast enhancement or structure emphasis (frequency modulation) should normally be used cautiously.

As a general rule, the same image processing algorithms are used for flat-panel detectors as for storage screen systems. However, there exist differences depending on the manufacturer, both in terms of features and standards and in terms of the underlying filtering. In chest x-rays in particular, it is important to adjust the image processing characteristics of different imaging systems and manufacturers uniformly where they operate alongside each other, so as not to impair the image analysis in control examinations. In systems with blurred mask filtering, a large filter kernel should be selected for dynamic compression and harmonization and only a low weighting factor (0.5) used. DRC, DRR, and comparable algorithms are mostly not as susceptible to processing errors as edge-emphasizing filters. The latter, in particular, can distort the image characteristics (excessively emphasizing vascular structures) and then mask spot or area shadows.

15.4.2 Multifrequency Filtering

Multifrequency filtering is offered nowadays by many companies, e.g., as multiscale image contrast amplification (MUSICA) by Agfa, UNIQUE by Philips, or multiobjective frequency processing (MFP) by Fuji. These are complex image processing methods that permit contrast enhancement while taking into account the original contrast and the size of the object being imaged (spatial frequency composition) [15.23–26]. By selecting the filter parameters appropriately, low-contrast structures (hard to recognize in the original image) can be enhanced and high-contrast ones (those that are easily recognizable) reduced in contrast accordingly [15.4]. In summary, through multifrequency filtering, the low image noise in direct radiography makes it possible to optimize display of even the smallest or low-contrast structures without limiting the resolution by superimposed image noise [15.27].

15.4.3 Dual-Spectrum Radiography (Energy Subtraction) and Temporal Subtraction

Two different techniques have been developed for energy subtraction: the single-exposure dual-detector and the dual-exposure single-detector techniques.

In the single-exposure dual-detector technique, there is only one exposure on two detectors positioned one behind the other, separated by a filter (Cu). This creates one exposure with an unfiltered spectrum and a corresponding percentage of low-energy quanta, and one with a hardened spectrum in which the higher-energy quanta predominate.

The dual-exposure single-detector technique needs only one detector. Two exposures are made, one at tube voltage of 120 kV and a dose corresponding to a 400 system, and one at 60 kV and a dose corresponding to a 1000 system. This technique has been employed in chest x-rays for detection and characterization of calcified loci [15.28–30]. Although this results in diagnostic advantages compared with conventional x-rays, so far this technique has failed to enter routine practice, since it does not solve the main problem in the diagnosis of focal pulmonary lesions, namely distinguishing between benign and malignant ones.

Temporal subtraction has been used to assess new lesions and changes that have occurred since the last examination, relying on direct comparison with previous x-rays [15.31]. After using software algorithms developed specially to achieve optimally artifact-free subtraction (iterative image warping), changes appear in the differential image as dark (increase) or bright (decrease) spots. Temporal subtraction has proved to be of benefit in the detection of isolated focal lesions and of low-contrast, multifocal, blotchy changes, and in the follow-up of extensive changes [15.32, 33]. Low-contrast lesions and lesions in obscured locations (e.g., infraclavicular) were diagnosed more efficiently by employing this technique.

15.5 Image Communication and Archiving

A further advantage of digital image technology is that digital datasets can be transmitted over data lines. The corresponding global standard data format is called diagnostic imaging and communication (DICOM). The transmission speed depends on the network's bandwidth. One must remember that transporting conventional films involves human resources, and that these resources are in short supply in today's hospitals. This creates difficult issues even for emergency staff. The problems multiply when moving x-ray bags between various departments or hospitals by post, by taxi, or by means of public transport. Therefore, image transmission over a data network from the imaging system to the referring physician is the only practical solution for achieving an efficient workflow. This has led to the development of local data networks in hospitals and medical practices. Regional networks that link medical institutions exist also and are being constantly expanded.

A similar situation applies to archiving of images. Conventional films require the building of heavy archival systems that take up considerable space. Bearing in mind that rapid developments in CT and MRI have increased the number of individual images by a factor of 10–100, it is easy to see that neither budgets for films nor space for the growing archives are available. The digital archive requires only a fraction of the space taken up by conventional systems. The initial high costs needed for storage media have long since dropped to such an extent that digital archiving always comes ahead in any comparison. The real advantage for staff arises from the fact that images in a digital archive are sorted alphanumerically and can be retrieved at any time. The time-consuming search for films, e.g., for a particular therapy – the critical point in bygone days – is no longer required.

Special picture archive and communication systems (PACS) have been developed for radiology. They are designed to support not only image communication and archiving, but also the workflow in radiology departments. This means that such a PACS needs to be able to do more than just storing, archiving, and distributing the data generated by the various imaging modalities. What is needed, rather, is interfacing and communicating with the radiology information system (RIS) and/or the hospital information system (HIS). Communication with the RIS and HIS takes places across an HL7 interface, while interfacing to the modalities is via special DICOM protocols. Normally, the workflow proceeds as follows.

15.5.1 Example 1 – Hospital Outpatient or Patient at a Doctor's Surgery

1. The patient's first contact is the central admission counter. Here, the master data are recorded and sent by the HIS to the RIS. In the case of an ambulatory practice, the master data are entered directly into the RIS.
2. The radiology job is generated in the RIS, and sent via the DICOM worklist server to the appropriate modality, where it appears on the worklist. Depending on the system and configuration, a prefetching order is also sent to the archive server to fetch any previous examinations of the patient.
3. The master data are sent automatically to the modality, so that the examination can be performed.
4. Once the examination is completed, the generated images are sent to the PACS and the examination data to the RIS.
5. The radiologist's workstation normally consists of three monitors: two for image viewing and one for RIS data, which enables analysis of images stored in the PACS and comparison with previous examinations or other modalities. Other add-ons include digital dictation and voice recognition, also implemented in the RIS.
6. Images and diagnoses can be stored on a CD/DVD and given to the patient. Hard copies on film or paper can also be made.
7. Images are stored initially on the PACS server, where they are kept for a period that depends on the particular requirements (e.g., 3–5 years). At the same time, they are copied into a long-term archive and saved there in the form of a compressed dataset. Image data and diagnoses can also be saved in the PACS together with the images. The corresponding originals, however, are also saved in the RIS.

15.5.2 Example 2 – Hospital Inpatient

With the master data available, notification is sent via an order entry module in the HIS. The data are sent to the RIS, where they can be inspected. Using this interface it is possible to confirm the indication or to reject the order, if the circumstances require this (indication, missing laboratory results, etc.). If the job is approved, the sequence is as in example 1 from step 2.

Once the examination has been completed and the findings approved, they are sent from the RIS to the HIS to be entered into the electronic patient file, where they are archived together with the rest of the patient's data (laboratory tests, surgery reports, outpatient and discharge letters, etc.).

Further Reading

- H.P. Busch, S. Busch, C. Decker, C. Schilz: Bildqualität und Dosis in der digitalen Projektionsradiographie, RöFo **175**, 32–37 (2003)
- H.P. Busch, K. Faulkner: Image quality and dose management in digital radiography: A new paradigm for optimization, Radiat. Prot. Dosim. **117**, 143–147 (2005)
- J.T. Dobbins, E. Samei, H.G. Chotas, R.J. Warp, A.H. Baydush, C.E. Floyd Jr., C.E. Ravin: Chest radiography: Optimization of x-ray spectrum for cesium iodide – amorphous silicon flat-panel detector, Radiology **226**, 221–230 (2003)
- O. Dragusin, R. Breisch, C. Bokou, C.J. Beissel: Does a flat panel detector reduce the patient radiation dose in interventional cardiology?, Radiat. Prot. Dosim. **139**(1–5), 266–270 (2010)
- S. Grampp, C. Czerny, C. Krestan, C. Henk, L. Heiner, H. Imhof: Flachbilddetektorsysteme in der Skelettradiologie, Radiologe **43**, 362–366 (2003)
- I.D. Honey, A. Mackenzie, D.S. Evans: Investigation of optimum energies for chest imaging using film-screen and computed radiography, Br. J. Radiol. **78**, 422–427 (2005)

- W.A. Kalender: Der Einsatz von Flachbilddetektoren für die CT-Bildgebung, Radiologe **43**, 379–387 (2003)
- Y. Kyriakou, T. Struffert, A. Dörfler, W.A. Kalender: Grundlagen der Flachdetektor-CT (FD-CT), Radiologe **49**, 811–819 (2009)
- R. Loose, M. Wucherer: Occupational exposure to radiation, Radiologe **47**, 27–40 (2007)
- E. Ludewig, W. Hirsch, B. Bosch, K. Gäbler, I. Sorge, D. Succow, A. Werrmann, D. Gosch: Untersuchungen zur Qualität von Thoraxaufnahmen bei Katzen mit einem auf einer Nadelstruktur basierenden Speicherfoliensystem – Modelluntersuchungen zur Bewertung der Bildqualität bei Neugeborenen, RöFo **182**, 122–132 (2010)
- R. Rühl, M.M. Wozniak, M. Werk, F. Laurent, G. Mager, M. Montaudon, A. Pattermann, A. Scherrer, J.-P. Tasu, M. Pech, J. Ricke: CsI-detector-based dual-exposure dual energy in chest radiography for lung nodule detection: Results of an international multicenter trial, Eur. Radiol. **18**, 1831–1839 (2008)
- R. Schaetzing: Computed radiography technology. In: *Advances in Digital Radiography*, ed. by E. Samei, N.C. Durham (Radiological Society of North America, Oak Brook 2003) pp. 7–22
- E.L. Schmidt, W. Herbig: Der Flachbilddetektor auf der Basis von amorphem Selen, Radiologe **43**, 374–378 (2003)
- E. Vano: ICRP recommendations on managing patient dose in digital radiology, Radiat. Prot. Dosim. **114**, 126–130 (2005)
- E. Vano, B. Geiger, A. Schreiner, C. Back, J. Beisse: Dynamic flat panel detector versus image intensifier in cardiac imaging: Dose and image quality, Phys. Med. Biol. **50**, 5731–5742 (2005)
- M.J. Yaffe: Digital mammography. In: *Handbook of Medical Imaging*, Vol. 1, ed. by J. Beutel, H.L. Kundel, R.L. van Metter (SPIE, Bellingham 2000) pp. 329–372

References

15.1 P.F.J. New, W.R. Scott: *Computed Tomography of the Brain and Orbit* (Williams Williams, Baltimore 1975)

15.2 M. Uffmann, C. Schaefer-Prokop, U. Neitzel: Abwägen von Dosisbedarf und Bildqualität in der digitalen Radiographie, Radiologe **48**, 249–257 (2008)

15.3 M. Prokop, C.M. Schaefer-Prokop: Digital image processing, Eur. Radiol. **7**(Suppl. 3), 73–82 (1997)

15.4 C. Schaefer-Prokop, M. Uffmann, J. Sailer, N. Kabalan, C. Herold, M. Prokop: Digitale Thoraxradiographie: Flat-panel-Detektor oder Speicherfolie?, Radiologe **43**, 351–361 (2003)

15.5 P. Monnin, Z. Holzer, R. Wolf, U. Neitzel, P. Vock, F. Gudinchet, F.R. Verdun: Influence of cassette type on the DQE of CR systems, Med. Phys. **33**, 3637–3639 (2006)

15.6 L. Riccardi, M.C. Cauzzo, R. Fabbris, E. Tonini, R. Righetto: Comparison between a built-in dual side chest imaging device and a standard single side CR, Med. Phys. **34**, 119–126 (2007)

15.7 R. Fasbender, R. Schaetzing: New computed radiography technologies in digital radiography, Radiologe **43**, 367–373 (2003)

15.8 M. Körner, S. Wirth, M. Treitl, M. Reiser, K. Pfeifer: Initial clinical results with a new needle screen storage phosphor system in chest radiograms, RöFo **177**, 1491–1496 (2005)

15.9 E. Samei, M.J. Flynn: An experimental comparison of detector performance for direct and indirect digital radiography systems, Med. Phys. **30**, 608–622 (2003)

15.10 M. Spahn, V. Heer, R. Freytag: Flachbilddetektoren in der Röntgendiagnostik, Radiologe **43**, 340–350 (2003)

15.11 M. El-Sheik, J.T. Heverhagen, H. Alfke, J.J. Froelich, J. Hornegger, T. Brunner, K.J. Klose, H.J. Wagner: Multiplanar reconstructions and three-dimensional imaging (computed rotational osteography) of complex fractures by using a C-arm system: Initial results, Radiology **221**, 843–849 (2001)

15.12 R. Fahrig, S. Fox, S. Lownie, D.W. Holdsworth: Use of a C-arm system to generate true 3-D computed rotational angiograms: Preliminary in vitro and in vivo results, Am. J. Neuroradiol. **18**, 1507–1514 (1997)

15.13 T. Bruijns, R. Bastiaens, B. Hoornart, E. von Reth, F. Busse, V. Heer, T. Ducourant, A. Cowen, A. Davies, F. Terrier: Image quality of a large-area dynamic flat detector: Comparison with a state-of-the-art II/TV system, Proc. SPIE **4682**, 332–343 (2002)

15.14 R. Baba, Y. Konno, K. Ueda, S. Ikeda: Comparison of flat-panel detector and image-intensifier detector for cone-beam CT, Comput. Med. Imaging Graph. **26**, 153–158 (2002)

15.15 U. Linsenmaier, C. Rock, E. Euler, S. Wirth, R. Brandl, D. Kotsianos: Three-dimensional CT with a modified C-arm image intensifier: Feasibility, Radiology **224**, 286–292 (2002)

15.16 T. Struffert, A. Doerfler: Flachdetektor-CT in der diagnostischen und interventionellen Neuroradiologie, Radiologe **49**, 820–829 (2009)
15.17 H.C. Becker, O. Meissner, T. Waggershauser: C-Bogen-CT-unterstützte 3-D-Navigation perkutaner Interventionen, Radiologe **49**, 852–855 (2009)
15.18 P. Huppert, G. Firlbeck, O.A. Meissner, H. Wietholtz: C-Arm-CT bei der Chemoembolisation von Lebertumoren, Radiologe **49**, 830–836 (2009)
15.19 L.A. Feldkamp, L.C. Davis, J.W. Kress: Practical cone-beam algorithm, J. Opt. Soc. Am. A **1**, 612–619 (1984)
15.20 W.A. Kalender: *Computed Tomography* (Publicis, Erlangen 2005)
15.21 J.A. Rowlands, J. Yorkston: Flat panel detectors for digital radiography. In: *Handbook of Medical Imaging*, Vol. 1, ed. by J. Beutel, H.L. Kundel, R.L. van Metter (SPIE, Bellingham 2000) pp. 223–328
15.22 E.R. Dougherty, J. Astola: *An Intruduction to Nonlinear Image Processing* (SPIE, Bellingham 1994)
15.23 S. Hoeppner, I. Maack, U. Neitzel, M. Stahl: Equalized contrast display processing for digital radiography, Proc. SPIE **4681**, 617–625 (2002)
15.24 P. Vuylsteke, E. Schoeters: Multiscale image contrast amplification (MUSICA), SPIE Med. Imaging Proc. **2167**, 551–560 (1994)
15.25 M. Stahl, T. Aach, T.M. Buzug, S. Dippel, U. Neitzel: Noise-resistant weak-structure enhancement for digital radiography, Proc. SPIE **3661**, 1406–1417 (1999)
15.26 M. Stahl, T. Aach, S. Dippel: Digital radiography enhancement by nonlinear multiscale processing, Med. Phys. **27**, 56–65 (2000)
15.27 E. Dencker, M. Stahl, S. Dippel, C.M. Schaefer-Prokop, S. Baus: Nonlinear multiscale processing in digital chest radiography: Results of a patient study with simulated chest lesions, Eur. Congr. Radiol., Book Abstracts (Vienna 2000), Abstract 1319
15.28 F. Fischbach, T. Freund, R. Röttgen, U. Engert, R. Felix, J. Ricke: Dual-energy chest radiography with a flat-panel digital detector: Revealing calcified chest abnormalities, Am. J. Radiol. **181**, 1519–1524 (2002)
15.29 F. Kelcz, F.E. Zink, W.W. Peppler, D.G. Kruger, D.L. Ergun, C.A. Mistretta: Conventional chest radiography versus dual-energy computed radiography in the detection and characterization of pulmonary nodules, Am. J. Radiol. **162**, 271–278 (1994)
15.30 J. Ricke, F. Fischbach, T. Freund, U. Teichgraber, E.L. Hanninen, R. Rottgen, U. Engert, H. Eichstadt, R. Felix: Clinical results of CsI-detector-based dual-exposure dual energy in chest radiography, Eur. Radiol. **13**, 2577–2582 (2003)
15.31 J. Ricke, F. Fischbach, U. Teichgräber, T. Freund, R. Felix: First experiences with a detector based dual energy system for thorax radiography, Eur. Radiol. **12**(Suppl. 1), 166 (2001), paper B0261
15.32 M. Tsubamato, T. Johkoh, T. Kozuka, N. Tomyama, S. Hamada, O. Honda, N. Mihara: Temporal subtraction for the detection of hazy pulmonary opacities on chest radiography, Am. J. Radiol. **179**, 467–471 (2002)
15.33 S. Kakeda, K. Nakamura, K. Kamada, H. Watanabe, H. Nakata, S. Katsuragawa, K. Doi: Improved detection of lung nodules by using a temporal subtraction technique, Radiology **224**, 145–151 (2002)

16. Computed Tomography

Thorsten M. Buzug

In this chapter, historical milestones of computed tomography (CT) (Sect. 16.2), recent technology with a focus on generation and detection of x-rays (Sect. 16.3), as well as image reconstruction (Sect. 16.4) are discussed. Furthermore, the chapter includes aspects of applications (Sect. 16.5), dose exposure in computed tomography (Sect. 16.6), and a brief overview on special CT developments (Sect. 16.7). Since this chapter gives a review, the interested reader is referred to recent literature on computed tomography including a detailed discussion of CT technology in the references section.

16.1 **Background** ... 311

16.2 **Milestones of Computed Tomography** ... 313
- 16.2.1 Analog Geometric Tomography and Tomosynthesis ... 314
- 16.2.2 Generations of CT Systems ... 314
- 16.2.3 Spiral CT ... 317

16.3 **Computed Tomography Technology** ... 318
- 16.3.1 X-Ray Generation ... 318
- 16.3.2 Detector Systems ... 321

16.4 **Image Reconstruction** ... 323
- 16.4.1 Fourier-Slice Theorem ... 323
- 16.4.2 Filtered Backprojection ... 324
- 16.4.3 Raw-Data Interpolation for Spiral CT ... 324
- 16.4.4 Artifacts ... 326

16.5 **Scan Planning and Applications** ... 329
- 16.5.1 Scan Planning ... 329
- 16.5.2 Hounsfield Units and Gray-Value Mapping ... 329
- 16.5.3 Three-Dimensional Data Representation ... 331
- 16.5.4 Clinical Applications ... 335

16.6 **Dose** ... 335

16.7 **Special System Designs** ... 338
- 16.7.1 Electron-Beam CT ... 339
- 16.7.2 Volume CT ... 340
- 16.7.3 Micro-CT ... 340
- 16.7.4 PET-CT ... 340
- 16.7.5 Dual-Source CT ... 341

References ... 341

16.1 Background

Research in computed tomography is still as exciting as at the beginning of its development during the 1960s and 1970s; however, several competing methods exist, the most important being magnetic resonance imaging (MRI). Since the invention of MRI during the 1980s, the phasing out of CT has been anticipated. Nevertheless, to date, the most widely used imaging technology in radiology departments is still CT. Although MRI and positron emission tomography (PET) have been widely installed in radiology and nuclear medicine departments, the term tomography is clearly associated with x-ray computed tomography. In the USA, computed tomography is also called computerized axial tomography (CAT).

Computed tomography has evolved into an indispensable imaging method in clinical routine. It was the first method to noninvasively acquire images of the inside of the human body that were not biased by superposition of distinct anatomical structures [16.1–4]. This is due to the projection of all the information into a two-dimensional imaging plane, as typically seen in planar x-ray fluoroscopy. Therefore, CT yields images of much higher contrast compared with conventional radiography. During the 1970s, this was an enormous step toward the advance of diagnostic possibilities in medicine.

Some hospitals have actually replaced their conventional shock rooms with a CT-based virtual shock room. In this scenario, imaging and primary care of the patient takes place using a CT scanner equipped with anesthesia devices. In a situation where fast three-dimensional imaging of a trauma patient is necessary (and it is unclear whether MRI is an adequate imaging method in terms of compatibility with the patient), computed tomography is the standard imaging modality. Additionally, due to its ease of use, clear interpretation in terms of physical attenuation values, progress in detector technology, reconstruction mathematics, and reduction of radiation exposure, computed tomography will maintain and expand its established position in radiology.

Recently, interesting technical, anthropomorphic, forensic, and archeological as well as paleontological applications of computed tomography have been developed. These applications further strengthen the method as a generic diagnostic tool for nondestructive material testing and three-dimensional visualization beyond its medical uses. Magnetic resonance imaging fails whenever the object to be examined is dehydrated. In these circumstances, computed tomography is the three-dimensional imaging method of choice.

Today, whole-body scans as well as imaging of very small vessels, for instance, vessels of the feet, belong to clinical routine (Fig. 16.1). Furthermore, the preoperatively acquired CT image stack can be used to synthetically compute projections for any given angulations. A surgeon can use this information to get an impression of the images that are taken intraoperatively by a C-arm image intensifier. Therefore, there is no need to acquire additional radiographs and the artificially generated projection images actually resem-

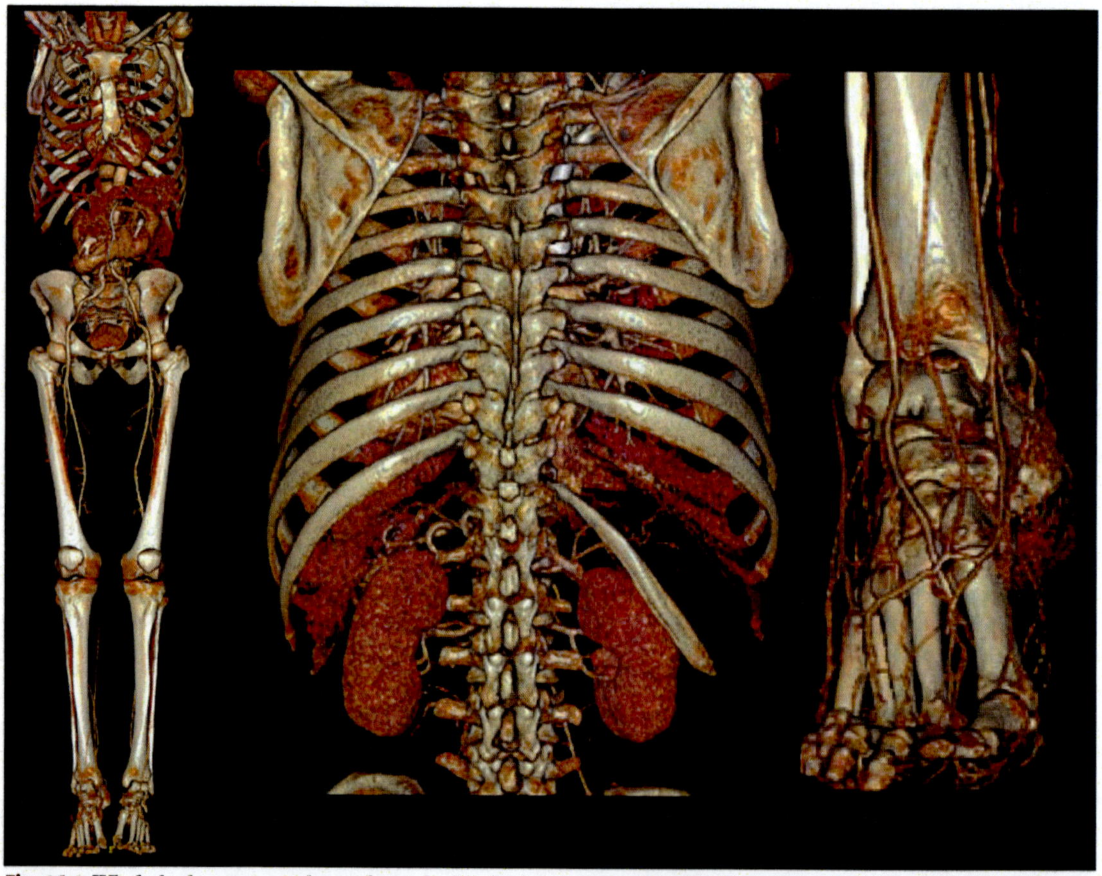

Fig. 16.1 Whole-body scans can be performed with the latest generation of CT systems, including multislice detector systems. Even very small vessels of the feet can be precisely visualized (courtesy Philips Medical Systems)

ble conventional radiographs. Additionally, the German Employer's Liability Insurance Association insists on a CT examination in severe accidents that occur at work. Therefore, CT has advanced to become the standard diagnostic imaging modality in trauma clinics. Patients with heavy trauma, fractures, and luxations benefit greatly from the clarification provided by imaging techniques such as computed tomography.

16.2 Milestones of Computed Tomography

Conventional x-ray imaging suffers from the severe drawback that it only produces two-dimensional projections of a three-dimensional object. This results in a reduction in spatial information (although an experienced radiologist might be able to compensate for this). In any case, a projection represents an averaging. The result of the averaging can be imagined if one were to overlay several radiographic sections on the light box for diagnosis. It would be difficult for even an expert to interpret the results, as averaging comes along with a considerable reduction in contrast, compared with the contrast present in one slice.

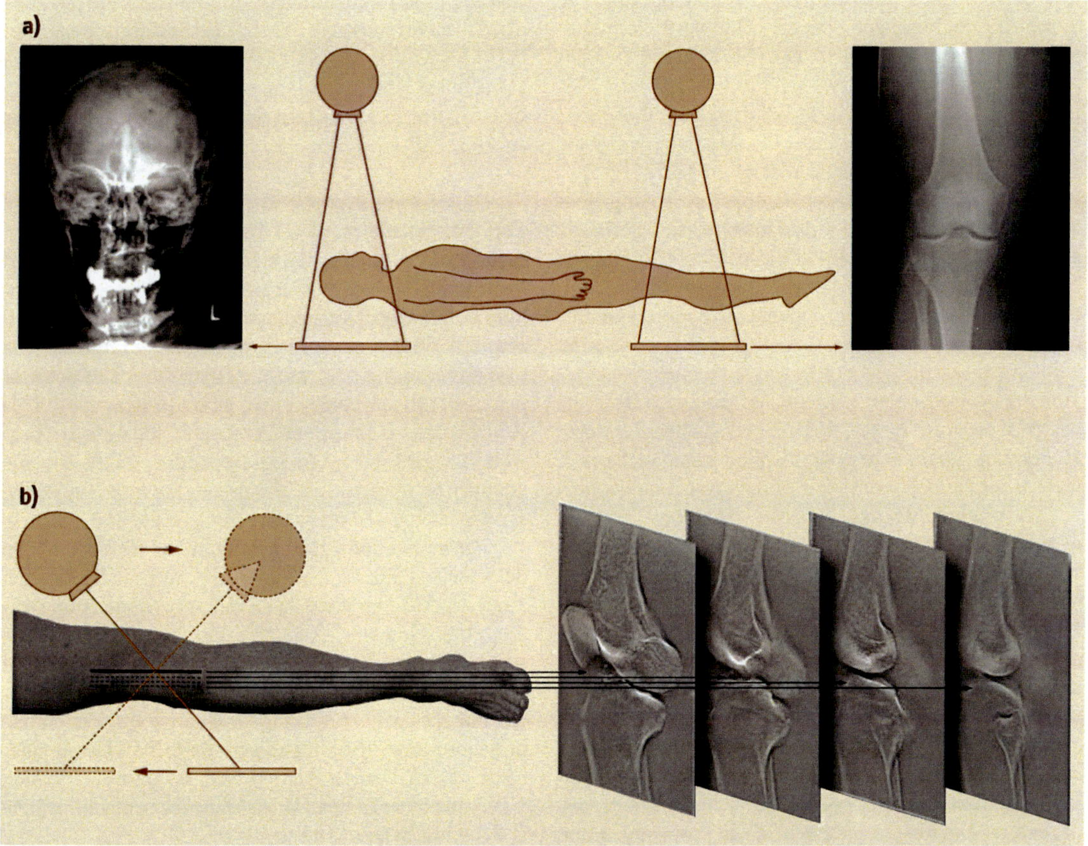

Fig. 16.2a,b Conventional planar x-ray leads to low-contrast images that do not allow clinical diagnostics involving soft tissue **(a)**. The figure shows the cranial bone (*left*), in which details of spatial structures in the brain cannot be recognized. In the image of the knee (*right*), even bone structures have low contrast. This low contrast is caused by the averaging process during the x-rays' passage through the body. The first attempts to create radiographic slices of the human body were carried out using conventional or analog geometric tomography, also referred to as tomosynthesis if the acquired x-ray images are digitally postprocessed **(b)**

Figure 16.2a shows conventional images of the cranium (left) and of the knee (right). These images show the high attenuation of x-rays within, for example, the cranial bone and most notably around the dental fillings. The small differences in attenuation that characterize soft tissue, however, are not visible at all. The morphology of the brain, in particular, is completely lost in the averaging process. In the knee, even the bone structures are imaged with poor contrast due to the superimposition.

16.2.1 Analog Geometric Tomography and Tomosynthesis

In the 1920s, the desire to undo the averaging process that characterizes conventional x-ray radiography led to the first tomographic concept. The word "tomography" itself is composed of two Greek words: *tomos* (slice) and *graphein* (draw). The use of the word "tomography" was considerably influenced by the Berlin physician *Grossmann*, whose Grossmann tomograph was able to image one single slice of the body [16.5].

The principle of the conventional or analog geometric tomography method is illustrated in Fig. 16.2b. During image acquisition, the x-ray tube is moved linearly in one direction, while the x-ray film is synchronously moved in the opposite direction. For this reason, only points in the plane of the center of rotation are imaged sharply. All points above and below this region are blurred, more so at greater distances from the center of rotation. Hence, this method can be interpreted as *blurring tomography*. It is called *tomosynthesis* if there is digital postprocessing of the projection images.

The blurred information above and below the center of rotation does not disappear, but is superimposed on the sharp image as a kind of gray veil or haze. Therefore, a substantial reduction in contrast is noticeable. However, the gain in quality compared with a simple radiograph is clearly visible in the example of the tomosynthetically acquired slice sequence of the knee in Fig. 16.2b [16.6, 7].

Due to the increased availability of electronic x-ray detectors, tomosynthesis systems are currently regaining scientific attention [16.8]. In modern systems, projection images are measured during movement and stored digitally using an image intensifier system. This allows subsequent image reconstruction that is superior to the analog blurring technique.

A related method, called *orthopantomography*, is now widely used in dental radiology. In this method, a panoramic view of rows of teeth is produced on an imaging plane that is curved to follow the jaw. However, sophisticated trajectories allow reconstruction of slices by geometric tomography as well. Nowadays, the term tomography, despite competitive modalities such as magnetic resonance imaging (MRI) or positron emission tomography (PET), is still most commonly associated with computed tomography (CT) or, more precisely, with x-ray CT.

16.2.2 Generations of CT Systems

Computed tomography avoids the superimposition of blurred planes and produces such high contrast that even soft tissue can be imaged well. The resulting leap in the quality of diagnostic imaging led to the enormous success of CT. Historically, four distinct generations of CT have emerged. Their classification relates to both the way in which x-ray tubes and detectors are constructed, and the way that they move around the patient. Figure 16.3a–d illustrates the different generations schematically.

First-Generation CT

The first generation of CT involves an x-ray tube that emits a single, needle-like x-ray beam, which is selected from the x-ray cone by means of an appropriate pinhole collimator. This geometry is referred to as *pencil beam*. A single detector is situated on the opposite side of the measuring field and the x-ray tube. The detector is moved synchronously along with the x-ray tube. This displacement is linear and is repeated for different projection angles γ (Fig. 16.3a). Depending on the specific attenuation properties of the tissue, the intensity of the x-ray is attenuated on its path through the body.

The amount of x-ray attenuation is measured by the detector and subsequently digitally recorded. For each angle γ, this step yields a simple, one-dimensional radiograph. However, from this initial radiograph, it is still not possible to determine the spatial distribution of the tissue attenuation coefficients. It is clear that, to determine the location of two consecutive objects on one projection line, the situation needs to be viewed sideways as well. This is the approach taken by CT, which views an object from all sides, since the projection angle is varied from 0° to 180°.

The first CT scanner, built by the company Electric and Musical Industries Ltd. (EMI), was based on this principle. In 1972, *Hounsfield* realized the scanner in the EMI central research laboratories [16.9]. For his invention, he jointly won the Nobel Prize for medicine together with Allen M. Cormack in 1979. The first ex-

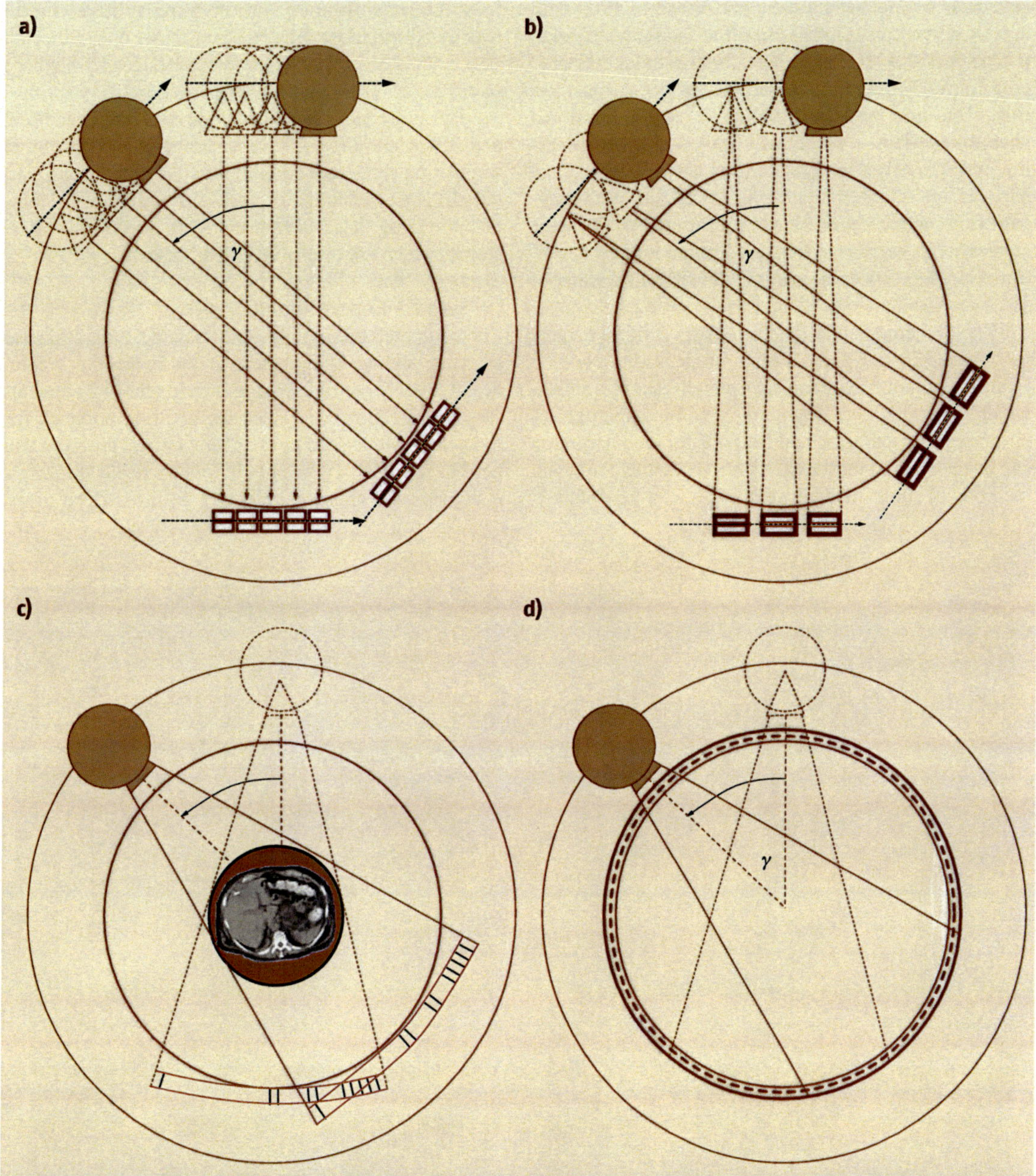

Fig. 16.3a–d First-generation CT devices are equipped with a pencil beam and a single detector. These are moved linearly, and the configuration is rotated through different projection angles **(a)**; CT scanners of the second generation have one x-ray source with fan-beam geometry, as well as a short detector array. The x-ray fan is created from a conical x-ray source by means of a slit-shaped collimator **(b)**; the third generation of CT scanners has a substantially larger angle of the x-ray fan and a longer detector array, such that the entire measuring field can be x-rayed simultaneously for one single projection angle **(c)**, and the fourth generation has a stationary closed detector ring **(d)**

periments by Hounsfield were performed in 1969 with a radioactive americium source that was linearly displaced along with the detector. Hounsfield collected data from 0° to 180° necessary for image reconstruction. The first reconstruction of a two-dimensional slice took 9 days – a clinically unacceptable processing time. However, for the first time, a two-dimensional slice image could be achieved that did not originate from averaging or blurring information, as in the conventional geometric tomography mentioned above. Figure 16.4 shows a very early EMI CT head scanner of 1972.

The first commercial scanners from EMI had a narrow focused x-ray beam and a single sodium iodide (NaI) scintillation detector. This pencil-beam principle, which is not practiced any more, is of fundamental importance as its mathematical methods for reconstruction can be understood most easily. Indeed, the mathematical methods of more modern geometries can be obtained from the pencil-beam geometry using suitable coordinate transformations. Figure 16.4a shows one of the first tomographs of the first generation produced by Siemens in 1972.

The subsequent rapid development of CT has been, and still is, driven by three essential goals: reduction of acquisition time, reduction of x-ray exposure, and, last but not least, reduction of cost. Throughout the course of optimizing these factors there are several historical stages, which are briefly described below.

Second-Generation CT

The computed tomograph of the second generation has an x-ray source with a narrow fan beam and a short detector array consisting of approximately 30 elements (Fig. 16.3b). However, since the aperture angle of the fan beam is small, the x-ray tube and detector array still need to be translated linearly before the projection angle is adjusted for another projection. In the earliest of the

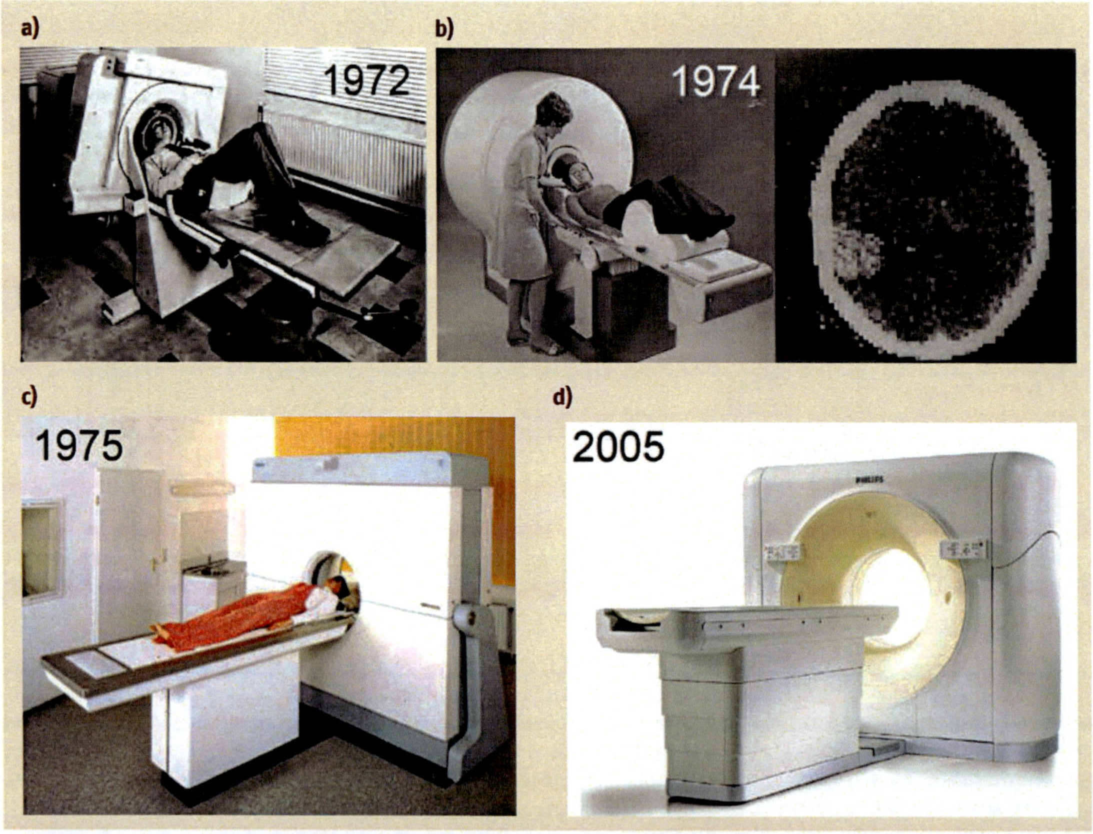

Fig. 16.4a–d Development steps of CT. **(a)** 1972: EMI head scanner (first generation); **(b)** 1974: Siemens Siretom (first generation); **(c)** 1975: Philips Tomoscan 200 (second generation); **(d)** 2005 Philips Brilliance (third generation)

second-generation scanners, the angle of the fan beam was about 10°.

Despite the need for linear displacement, the acquisition time was reduced to a few minutes per slice, as the detector array could measure several intensities simultaneously. However, the measuring field was still small. For this reason, and due to their long acquisition times, first- and second-generation scanners were mainly restricted to use in imaging the cranium. Figure 16.4b (right) gives an idea of the scale of the measuring field. The cranium can be fixed in the scanner and shows no large intrinsic movement during the acquisition time (relative to the spatial resolution that could be achieved at the time).

This is certainly not the case when imaging the area of the thorax or the abdomen, as the intrinsic movements of the heart and lung, as well as the movement of the diaphragm and soft organs of the abdomen, produce artifacts in the reconstructed images. The mathematical methods of reconstruction demand that all points of one slice are x-rayed from all angles from 0° through 180°. One object moving out of the imaging plane during rotation of the x-ray tube, due to patient movement, will inevitably result in errors in image reconstruction. Such image errors, referred to as motion artifacts, are described in detail in Sect. 16.4.4.

Third-Generation CT

The main goal of developments in the 1970s was to reduce acquisition time to less than 20 s. This was intended to give enough time to acquire an image of the abdomen with minimal error, while the patient held their breath. A major step toward achieving this goal was an extension of the second generation's fan-beam concept, i. e., the introduction of a substantially larger angle of the x-ray fan and a correspondingly longer detector array. Figure 16.3c shows a schematic illustration of this principle of third-generation scanners. Nowadays, the angle of an x-ray fan beam is typically between 40° and 60° and the detector array is usually constructed as an arc with between 400 and 1000 elements. In this way it is now possible to simultaneously x-ray the entire measuring field, which is currently wide enough to cover the torso, for each projection angle γ. As a result, the third generation of CT systems can completely abandon linear displacement of the x-ray tube.

The acquisition time for third-generation systems is reduced considerably, since continuous rotation can take place without interruption for linear displacement. The majority of CT scanners currently in use are fan-beam systems of the third generation. Figure 16.4c and d show two CT scanners of the second and third generation from 1975 and 2005, respectively.

Fourth-Generation CT

The fourth generation of CT scanners does not differ from the third generation with respect to the x-ray tube. The fan-beam source also rotates continuously around the measuring field without any linear displacement. The difference is in the closed, stationary detector ring with up to 5000 single elements. The x-ray tube rotates either outside (Fig. 16.3d) or inside the detector ring.

If the x-ray tube is outside the detector ring, it is necessary to prevent the x-rays from radiating through the detectors from behind. Therefore, the detector ring is dynamically tilted away from the path of the tube. In this way, the line of sight between the tube and the appropriate section of the detector ring only passes through the patient (and patient table) and not through the electronics behind the detectors.

Fourth-generation tomographs establish *inverse fans*, which are centered on detectors rather than on the x-ray focus. An inverse fan is also referred to as a *detector fan*. An inverse fan can be very dense, limited only by the sampling rate at which individual detectors can be read out. As a result, unlike third-generation tomographs, this scanner is not limited to the spatial resolution of a single fan beam.

16.2.3 Spiral CT

Another development, which was a great leap forward in capability from that of the third generation, led to what is identified by *Bushberg* et al. [16.10] as the sixth generation of tomography scanners. This refers to the introduction of slip-ring technology, which enables spiral or helical sampling.

As the x-ray tube must be continuously supplied with energy, the rate of circular movement was previously limited by the attachment of an electric cable, which was mounted on a spool. In this process the cable was unwound in one direction and carefully wound up in the other. This represented a huge obstacle to the reduction of acquisition time. The x-ray sampling unit had to stop and start again after a certain angle of rotation.

Although data could be collected throughout both clockwise and counterclockwise rotations, limits were placed on high velocities due to the increasing torsional moment. This problem was solved by the introduction of slip-ring technology. In this technology, the energy is provided via sliding contacts situated between the outside of the sampling unit, in what is called the *gantry*, and the rotating sampling unit. This enables the sampling unit carrying the x-ray source and, in the case of third-generation scanners, the detector array to rotate continuously. As a result of slip-ring technology, rotation frequencies of two rotations per second, i.e., *subsecond scanners*, are nowadays commonplace.

However, there are also smaller, more compact devices, in which the sampling unit is independent from an external energy supply during rotation, making use of accumulators. An example is given by the mobile Philips Tomoscan M CT scanner, where accumulators are mounted on the rotating sampling disk and a capacitive radiofrequency (RF) link is used for data transfer.

The slip-ring innovation enabled a new acquisition technique. Along with a continuous motion of the patient table through the sampling unit, it became possible to measure data in the shape of a spiral (strictly speaking, the x-ray tube trajectory is a helix). The spiral CT technique was demonstrated successfully using a prototype by *Kalender* in 1989 [16.11]. The spiral CT technique is discussed in more detail in Sect. 16.4.3.

16.3 Computed Tomography Technology

Computed tomography systems consist of a *front-end*, i.e., the scanner unit, and a *back-end*, i.e., the control console and viewing station. All components of the front-end, i. e., the x-ray tube, x-ray filter, aperture unit, collimator, detector system, high-voltage generator, cooling system, data-acquisition system (DAS), slip ring, patient table, motors, motor controllers, as well as mechanical components, are highly developed today. However, due to their key position in the system, only two main components of the front-end will be discussed here: the x-ray tube and the corresponding detector system. For a detailed discussion of the physics of x-ray generation, photon–matter interaction, x-ray detection, and photon statistics the reader is referred to [16.4].

16.3.1 X-Ray Generation

X-ray radiation is of electromagnetic nature; it is a natural part of the electromagnetic spectrum, with a range that includes radio waves, radar and microwaves, infrared, visible and ultraviolet light, to x- and γ-rays. In electron-impact x-ray sources, the radiation is generated by deceleration of fast electrons entering a solid metal anode and consists of waves with a range of wavelengths roughly between 10^{-8} and 10^{-13} m. Thus, the radiation energy depends on the electron velocity v, which in turn depends on the acceleration voltage U_a between the cathode and anode. In medical diagnostics, acceleration voltages are chosen between 90 and 140 kV, for radiation therapy they lie between 10 and 300 kV, and for material testing they can reach up to 500 kV.

Electrons are emitted from a filament, which is directly heated to approximately 2400 K to overcome the binding energy of the electrons to the metal of the filament. Filaments are usually made of thoriated tungsten with a melting point of 3683 K. Due to their thermal energy, electrons are boiled off from the filament. This process is called thermionic emission. To produce a small electron focus on the anode, the trajectories of the accelerated electrons must be controlled by electron optics. The focusing device is a cup-shaped electrode (frequently named a Wehnelt cylinder) that forms the electric field near the filaments such that the electron current is directed to a small spot on the anode.

While the acceleration voltage determines the energy interval of the x-ray spectrum, the intensity of the generated x-ray spectrum or the number of x-ray quanta is solely controlled by the anode current. The acceleration voltage and the anode current can be controlled by the user.

With the entry of accelerated electrons into the anode, sometimes also called the *anticathode*, several processes take place close to the anode surface. Generally, the electrons are diffracted and slowed down by the Coulomb fields of the atoms in the anode material. The deceleration results from the interaction with the orbital electrons and the atomic nucleus. As known from classical electrodynamics, acceleration and deceleration of charged particles creates an electric dipole, and electromagnetic waves are radiated. Usually, several photons emerge throughout the complete deceleration process of one single electron. It can happen, however, that the entire energy eU_a of an electron is transformed into a single photon. This limit defines the maximum energy of the x-ray radiation.

Due to the fact that the slowing down of electrons in the anode material is a multiprocess deceleration cascade, a continuous distribution of energies is produced, the so-called *bremsstrahlung*. Since the free electrons are unbound, their energy cannot be quantized. The continuous *bremsstrahlung* is superimposed by a characteristic line spectrum, which originates from direct interaction of fast electrons with the inner shell electrons of the anode material. If an electron on the K-shell or K-orbital is kicked out of the atom by a collision with a fast electron, i. e., the atom is ionized by the loss of an inner electron, an electron of one of the higher shells fills the vacant position on the K-shell. As the inner shells represent states having lower potential energy than the outer shells, this process is accompanied by the emission of a photon. This process creates sharp lines in the x-ray spectrum that are characteristic fingerprints for the anode material.

It should be noted at this point that the image quality of CT specifically suffers from the fact that x-ray attenuation is a complicated function of wavelength. The details of the functional behavior are given in the next section. Generally, one has to bear in mind that a low-energy x-ray, i. e., radiation with a longer wavelength, is more strongly attenuated when passing through matter than high-energy x-ray. (Low-energy quanta are generally undesired in x-ray imaging. They increase the dose to the patient, but do not contribute to imaging, because they are almost totally absorbed by the human body.) As a consequence, the center of the polychromatic x-ray is shifted to higher energies (harder radiation). This is the origin of what is called beam hardening, which produces artifacts in the reconstructed images.

Generally, a flat metal filter measuring a few millimeters is mounted onto the x-ray tube. The filtering of the useful beam reduces the number of x-ray quanta while increasing the average energy of the radiation. This prehardening of the radiation reduces beam-hardening artifacts during image reconstruction and the dose to which the patient is exposed.

Unfortunately, the quantum efficiency of the conversion from kinetic energy into x-ray radiation, for a tungsten anode (W, $Z = 74$) working with an acceleration voltage of $U_a = 140\,\text{kV}$, is roughly on the order of $\eta = 0.01$. This means that 99% of the kinetic energy is transferred locally to the lattice, heating up the anode. As a result, CT x-ray tubes have serious heating problems.

Since it is the energy deposition in the target volume that produces the heat load, the tube current and the duration of exposure, or more precisely the product of current in milliamperes and exposure time in seconds, are two important parameters of the practical scan protocol that the radiologist has to choose appropriately. The heat capacity of an x-ray tube is measured in *heat units*. Therefore, for several decades, rotating anode disks have been used to distribute the thermal load over the entire anode. The anode target material is rotated about the central axis, and therefore new, cooler anode material is constantly rotated into position at the focal spot [16.12]. In this way, the energy of the electron beam is spread out over a line, called the focal line, rather than being concentrated at one single point. In Fig. 16.5a–c a modern Philips MRC 600 x-ray tube is shown. A liquid-metal-filled spiral-groove bearing allows very high continuous power compared with conventional ball bearings. Often, a heat exchanger is placed on the rotating acquisition disk to cool the anode.

Ideally, x-rays should be created from a point source, because an increase in source size will result in a penumbra fringe in the image of any object point. The size and shape of the x-ray focus seen by the detector determines the quality of the resulting image. The effective target area, called the optical focus, depends on the orientation of the anode surface that is angulated with respect to the electron beam. The projection of the focus shape onto the detector must be minimized to obtain a sharp image. However, this surface angle increases the tube power limit, because it allows the heat to be deposited across a relatively large spot while the apparent spot size at the detector will be smaller by a factor of the sine of the anode angle [16.12]. The image quality is degraded by a large focus diameter, due to significant partial shadow areas of each object point. The mathematical expression that measures the image quality is called the *modulation transfer function* (MTF). A large angle between the incident electron beam and the anode surface normal is generally not desired, because there is a certain probability that the electrons will be elastically reflected from the surface and thus not contribute to x-ray generation. The probability of this backscattering effect increases as the atomic number of the anode material increases and as the angle between the surface normal and the anode rotation axis decreases.

Typically, the x-ray focus diameter for diagnostic tubes is found to be between 0.3 and 2 mm. The penetration depth of electrons into the anode depends on the kinetic energy and the anode material and can be found to be up to 30 μm. The radiologist may have the option of choosing between two focus sizes. This is technically realized by focusing the electron beam onto two target points with different anode orientations. In some systems the spatial resolution of the detector can also be doubled by switching between two foci during data acquisition. This concept is called a flying focus.

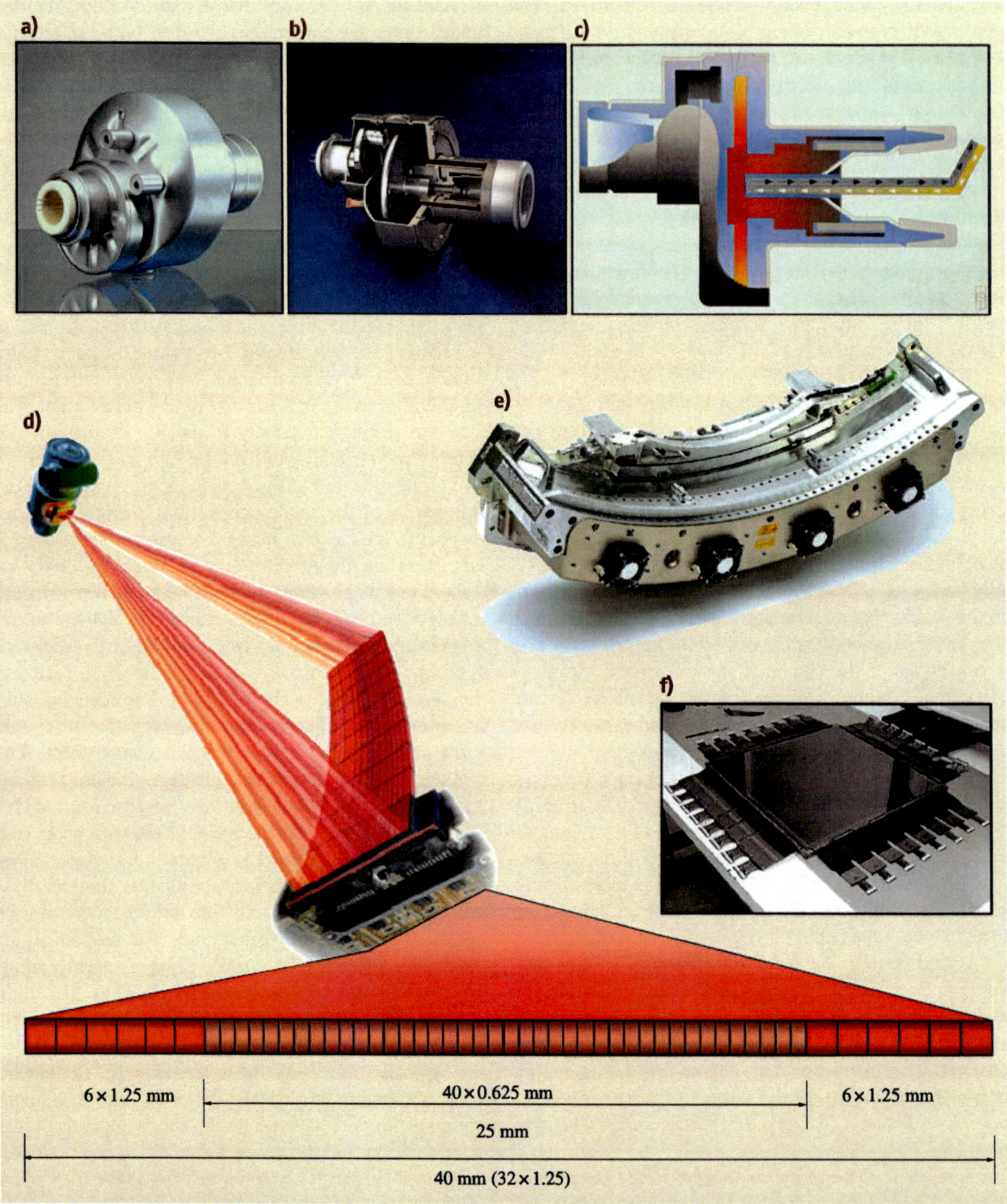

Fig. 16.5a–f Modern high-power x-ray tube (**a**,**b**) and schematic illustration of a modern x-ray tube with a rotating anode disk (**c**), schematic illustration of a multislice detector (**d**), a detector unit of a 64-row CT system (**e**), and the realization of a $41 \times 41\ cm^2$ flat-panel detector with 2048×2048 pixels (**f**)

A second effect that influences imaging quality is caused by the anode surface. X-ray beams leaving the anode tangentially to the anode surface are reduced in intensity when arriving at the detector. This intensity reduction is due to the self-absorption of photons by the anode, caused by the microscopic roughness of the anode surface. The x-ray intensity decreases gradually with a reduction in the angle between beam and anode surface. This effect, which is called the *heel* effect, grows during the lifetime of any x-ray tube, because the roughness of the anode surface increases due to erosion by electron bombardment along the focus line. In modern CT systems the cone of the utilized x-rays is becoming larger and larger. Therefore, it is important to know the radiation characteristics of the anode, because the basic assumption of the imaging principles of fluoroscopy and of CT reconstruction is that the object to be scanned is illuminated homogeneously. Slight deviations from a homogeneous beam profile can be compensated for by specially formed filters mounted on the x-ray tube and by detector calibration.

16.3.2 Detector Systems

Typically, x-ray quanta are not measured directly, but are detected via their interaction products (for example, emitted photoelectrons). The overall detection efficiency is primarily determined by the geometric efficiency (also called the *fill factor*) and the quantum efficiency (also called the *capture efficiency*). The geometric efficiency refers to the x-ray-sensitive area of the detector as a fraction of the total exposed area, and the quantum efficiency refers to the fraction of incident quanta that are absorbed and contribute to the signal. The overall detection efficiency is the product of the geometric and the quantum efficiencies.

Gas Detectors

X-ray radiation is able to ionize gases. This fact was discovered very early in the last century and led to the development of the well-known Geiger–Müller counter. In the first tomographic experiments carried out by Cormack and Hounsfield, the Geiger–Müller counter was used as a detector in pencil-beam geometry. In the early days of clinical CT, gas-based detector arrays were also manufactured for what are known as third-generation scanners in fan-beam geometry. Even today, some scanners using high-pressure xenon are in use. The photoelectric interaction, $h\nu + \mathrm{Xe} \rightarrow \mathrm{Xe}^{+} + \mathrm{e}^{-}$, describes the first part of the detection process chain. Xenon ions and electrons are attracted by high voltage to a cathode and an anode, respectively. A series of alternating cathode and anode pairs forms the detector array [16.4]. The current produced by recombination is a measure of the x-ray intensity entering the detector.

A weak quantum efficiency (a low probability for photoelectric absorption) can be compensated for by high-pressure and tall ionization chambers. Another advantage of tall chambers is improved directional selectivity of the detector element. Since the ionization probability is proportional to the travel length of x-ray quanta inside an element (one cathode–anode block), detection of x-ray quanta with an oblique entrance will be suppressed. In this way, a tall detector element has built-in collimation. However, the septa between detector elements are insensitive regions that will decrease the geometric efficiency of the detector.

Scintillation Detectors

Today, almost all modern CT systems are equipped with scintillator detectors. Such a detector essentially consists of two main components: a scintillator medium and a photon detector. In a first step, the short-wave x-ray radiation entering the detector is converted into long-wave radiation (light) inside the scintillation material. Typical scintillator materials used are cesium iodide (CsI), bismuth germanate (BGO), and cadmium tungstate ($CdWO_4$). The choice of material is critical and depends on the desired quantum efficiency for the conversion from x-ray to light and on the time constant for the conversion process, which determines the *afterglow* of the detector. For very fast fluorescence decay, i. e., a very small time constant, as required by modern subsecond scanners, ceramic materials made of rare-earth oxides such as gadolinium oxysulfide (Gd_2O_2S) are used.

On the rotating sampling unit, the series of detector elements is arranged in a circle segment with the x-ray source at its center. In Fig. 16.3c this is schematically shown for a third-generation scanner. X-rays that have been scattered may undergo deflection through a small angle and finally reach the detector. This detection is undesired because it reduces the contrast of the image. To suppress the measurement of scattered x-ray quanta, collimator lamella are attached to each element. Such an antiscatter collimator grid is directed toward the x-ray focus to filter out photons not traveling in the line of sight between the x-ray source and the detector. Without an antiscatter grid, the image quality would be significantly reduced.

However, there is an obvious disadvantage of the antiscatter grid. To block an oblique entrance of scattered x-ray photons effectively, a minimum lamella thickness of 0.1 mm is required. In practice, the detector elements have a total geometric efficiency of about 50–80% [16.13]. This decreased fill factor leads to an undesired reduction in spatial detector resolution.

Solid-State Multislice Detectors

Crystal- and ceramic-based solid-state detectors described in the previous section can be extended to multirow or multislice detector systems. The key feature of such multirow arrays is a maximum effective x-ray-sensitive area. This feature is quantified by the fill factor, which is explained below. Xenon high-pressure gas detector array systems cannot be easily extended to flat area detector systems. Therefore, all modern multislice CT systems are based on solid-state scintillation detectors.

In Fig. 16.5d, a cone-beam detector system is shown schematically. In contrast to detector systems in technical CT systems, for example, in micro-CT (where flat-panel detectors are employed), almost all clinical CT systems are equipped with cylindrical detector units. As illustrated in Fig. 16.5d, the multiarray system forms a cylindrical barrel with the x-ray source as its center. If the number of rows of such multislice detector systems is chosen to be very high, the orientation of the respective x-ray fan inside the cone beam, relative to the axial slice, becomes significant, and the requirements of image reconstruction increase. However, modern CT systems are equipped with cone-beam sampling units due to improved spatial resolution and faster image acquisition. Image reconstruction in cone-beam systems is discussed in Sect. 16.4.

Different technical configurations of multislice detector systems are available today. Usually, detector units can be partitioned and combined. This flexibility is achieved by the electronic interconnection of pairs of detector segments leading to a desired slice thickness.

Solid-State Flat-Panel Detectors

Without reconstruction mathematics that are adapted to the cone-beam situation, image artifacts arise for higher x-ray fan angulations, even in a 16-row detector system. This leads to clinically unacceptable image quality. In any case, the reconstruction algorithms used for the cone-beam geometry need to be revised for practical implementation. Therefore, a logical step forward is to leave multirow detector arrays and move in the direction of real flat-panel detector systems, which have recently become commercially available. However, they were not originally designed for use in CT systems, but rather to compete with established radiography systems using film cassettes, computed radiography cassettes, and image intensifiers.

Each sensor element of a flat-panel detector consists of a photodiode and a thin-film transistor (TFT). Both are made of amorphous silicon on a single glass substrate. The pixel matrix is coated with an x-ray-sensitive layer. Multichip modules are used as readout electronics at the edge of the detector field. The x-ray-sensitive coating is, for instance, a cesium iodide layer. The basis is the single glass substrate with a silicon matrix of 2048 × 2048 sensors, each 200 μm in size. The monolithic structuring is done with thin-film technology such that a composition of a set of medium-sized subpanels, which have undesired dead zones at the interfaces, potentially producing imaging artifacts, is not required. A final cesium iodide (CsI) coating forms the scintillator layer of the detector.

The CsI layer is applied directly onto the pixel matrix by a physical deposition process. The production technique is known from semiconductor production. Physical and chemical processing steps, i. e., the combination of photolithography and further etching phases, are applied to produce the finely structured detector elements. This leads to an x-ray-sensitive detector field with a desired high fill factor, i. e., the ratio of the *x-ray-sensitive area of the detector* to the *total area of the detector*.

X-ray quanta entering the detector are converted into visible light in the upper CsI scintillator layer. The light photons are guided to the photodiodes of the next processing layer. The photons are absorbed, thereby producing an electric charge in the photodiodes that is proportional to the intensity of the x-ray radiation.

During detector exposure, the electric charge is integrated and stored in the detection element, which acts as a capacitor. The actual readout process is initialized by the thin-film transistor, which switches the charge to the readout electronics via a data link. There, amplification and analog-to-digital conversion are performed on the same chip, resulting in fast operation with low noise. In Fig. 16.5f, a preassembly, raw detector with ribbon-cable data links [16.14] is shown. Its 4 194 304 pixels are integrated onto a 41 × 41 cm^2 active area.

A highly desired system property of the digital scintillation detector system is linear dynamics over a wide range of illumination. In this way, high- and low-dose applications have the same contrast information, i. e., excellent contrast resolution. Film systems are capable of imaging objects with high contrast within a very narrow exposure range only. If the object is over- or underexposed, the contrast of the image can easily be too low. Due to the linear response characteristic curve of the digital system, it is robust against over- and underexposure. However, within the range of low-contrast imaging, flat-panel systems do not achieve the quality of the dedicated CT detector systems described in the sections above [16.13].

In addition to a linear detector dynamics characteristic, another important advantage of the flat-panel detector is its excellent spatial resolution. To optimize the spatial resolution, CsI is evaporated onto the matrix so that direct contact between the scintillation material and the carrying photodiode matrix is established. This manufacturing step is designed in such a way that CsI grows anisotropically, forming needles on the matrix. If x-ray quanta are converted into visible light inside the CsI structure, the emerging photons are, in all likelihood, traveling along the needles, because they act as a fiber-optic cable. In this way, photons are guided directly onto the photodiode or in the opposite direction.

The photons that are traveling in the opposite direction face a mirror on the top side of the CsI layer that ensures that eventually almost all photons find their way to the photodiode. This light guidance effect of the CsI fiber structure is the reason for the high quantum efficiency of digital flat-panel detectors. The x-ray-sensitive CsI coating can be made very thick to obtain high quantum efficiency and also to suppress broad photon scattering, which would reduce the spatial resolution. The scintillation light is bundled by the CsI fibers onto a small point on the photodiode matrix. However, an isotropic CsI layer must always find a compromise between high quantum efficiency and high spatial resolution. Figure 16.12b in Sect. 16.7 shows a a prototype volume or cone-beam CT equipped with a flat-panel detector.

16.4 Image Reconstruction

The fundamental problem of computed tomography can be easily described: Reconstruct an object from its shadows or, more precisely, from its projections. An x-ray source with a fan- or cone-beam geometry penetrates the object to be examined, such as a patient in medical applications, a skull found in archeology, or a specimen in nondestructive testing (NDT). In so-called third-generation scanners, the fan-shaped x-ray beam fully covers a slice section of the object to be examined. Depending on the particular paths, the x-rays are attenuated to varying extents when passing through the object; the local absorption is measured with a detector array.

Of course, the shadow that is cast in only one direction is not an adequate basis for determination of the spatial distribution of distinct structures inside a three-dimensional object. To determine this structure, it is necessary to irradiate the object from all directions. Let $p_{\gamma_i}(\xi)$ represent the attenuation profile of the beam as a function of the x-ray detector array coordinate ξ under a particular projection angle γ_i. If the different attenuation or absorption profiles are plotted over all angles of rotation γ_i of the sampling unit, a sinusoidal arrangement of the attenuation or projection integral values is obtained. In two dimensions, these data $p_{\gamma_i}(\xi)$ represent the Radon space of the object, which is essentially the set of raw data. From a mathematical point of view, image reconstruction in computed tomography is the task of computing the spatial structure of an object from the shadows that it casts. The solution to this problem is complex and involves techniques from physics, mathematics, and computer science. The described scenario is referred to as the inverse problem in mathematics.

In computed tomography, the meaning of the mathematical term *inverse problem* is immediately apparent. The spatial distribution of attenuating objects that produce the projection shadow is not known a priori. This, actually, is the reason for acquiring the projections along the rotating detector coordinate ξ over a projection angle interval of at least 180°. It is an inversion of integral transforms. From a sequence of measured projection shadows $\{p_{\gamma_1}(\xi), p_{\gamma_2}(\xi), p_{\gamma_3}(\xi), \ldots\}$, the spatial distribution of the objects, or more precisely the spatial distribution of the attenuation coefficients $\mu(\xi, \eta)$ within a chosen section through the patient, must be estimated.

In 1961, the solution to this problem was applied for the first time to a sequence of x-ray projections by which an anatomical object had been measured from different directions. Allen MacLeod Cormack (1924–1998) and Sir Godfrey Hounsfield (1919–2004) were pioneers of medical computed tomography and in 1979 received the Nobel Prize for medicine for their epochal work during the 1960s and 1970s.

16.4.1 Fourier-Slice Theorem

Depending on the path η which the x-ray quanta take through the body – starting at the x-ray tube $\eta = 0$ and

Fig. 16.6a–g From the raw data to the reconstructed image. Sinogram of attenuation or projection values (**a**), high-pass-filtered projection values (**b**), backprojection of the filtered projection (**c**), accumulation of all backprojections of the high-pass-filtered (**d**) and the native (**e**) projection profiles; data acquisition of spiral CT (**f**), row-data interpolation for spiral CT (**g**) ▶

ending at the detector $\eta = s$ – the intensity of the rays will be attenuated differently. After taking the logarithm of the intensity measured on detector element ξ the sum of attenuation values $p_\gamma(\xi)$ is obtained along the x-ray path for the projection angle γ.

The x-ray source rotates around the object to be examined, penetrating it with x-rays. The sum of attenuation values, the so-called projection integral $p_\gamma(\xi)$, is measured and saved for each angle. Therefore, the projection integral – plotted over all angles – takes a sinusoidal course (Fig. 16.6a). This plot that collects the raw data of the measurement is called a sinogram and represents the Radon space of the object, named after the Austrian mathematician *Radon* [16.15].

The key theorem that describes the relation between the Fourier transform of the projection integral and the Fourier transform of the desired image is the Fourier-slice theorem. One is interested in the spatial distribution of attenuation values $f(x, y) = \mu(\xi, \eta)$ having measured only the projection data $p_\gamma(\xi)$.

The Fourier-slice theorem ensures that the one-dimensional Fourier transform $P_\gamma(q)$ of the measure projection integral $p_\gamma(\xi)$ can be identified with a radial line in the Cartesian Fourier space $F(u, v)$ of the object $f(x, y)$ drawn at the angle γ of the corresponding measurement, with $u = q \cos\gamma$ and $v = q \sin\gamma$. Filling the Fourier space in that way over all angles γ leads to a complete Fourier representation of the object, which can be reconstructed in principle by an inverse two-dimensional Fourier transform. For a detailed discussion the reader is referred to [16.4].

16.4.2 Filtered Backprojection

The Fourier-slice theorem is not directly implemented due to some interpolation problems that would arise. In today's CT systems, so-called filtered backprojection is widely used. Within this method, which is also called the convolution method, the projection integral data $p_\gamma(\xi)$ are digitally filtered. Figure 16.6a,b shows how this is done in the frequency domain. The linear weighting $|q|$ of the Fourier-transformed projection profile $P_\gamma(q)$ is essentially a high-pass filter. Consequently, the filtered projection profile $h_\gamma(\xi)$ shows the changes along the profile only. The result of the high-pass filtering of the measured attenuation profile $p_\gamma(\xi)$ is illustrated in Fig. 16.6b. It can be seen that the sinusoidal structure of the data is preserved. However, due to the high-pass filtering, only changes of $p_\gamma(\xi)$ are represented by $h_\gamma(\xi)$.

This filtered profile $h_\gamma(\xi)$ is projected back in the direction γ of the measurement of $p_\gamma(\xi)$. In Fig. 16.6c this is exemplarily demonstrated for a single backprojection angle γ. Obviously, a single filtered backprojection alone does not represent the image of the object. However, the errors of one direction are consistently compensated by accumulation of the backprojections of all available views in the interval $\gamma = [0, \pi)$.

The concept for reconstruction of the tomographic image $f(x, y)$ from the projection values $p_\gamma(\xi)$ is demonstrated by a successive backprojection. In Fig. 16.6d the intermediate results of the reconstruction process of a tomographic abdomen image are shown for an increasing number of projection angles $N_\text{p} = \{1, 2, 3, 10, 45, 180\}$. This sequence reveals that a sufficiently high number of projection angles is needed to reconstruct the image of the object. In Fig. 16.6e the intermediate results of the simple backprojection, i.e., the direct backprojection of the measured projection integral $p_\gamma(\xi)$, is shown. This naive strategy leads to a blurred image independent of the number of backprojection angles.

When CT images are compared, it is important to know the exact weighting function within the filtered backprojection. The linear weighting with $|q|$ as described above is the mathematically ideal case. In practice, the filter is often less progressive in the high-frequency domain. The actual type of weighting can be chosen at the console of the scanner. The operator is asked to choose a filter kernel from a filter bank including a variety of gradually changing weightings from *very sharp* to *very smooth*. The optimal choice of a filter kernel is always a compromise between a high noise level with fine spatial resolution and smooth images with low spatial resolution [16.4].

16.4.3 Raw-Data Interpolation for Spiral CT

A first step toward a true volume image is the so-called spiral CT method (Sect. 16.2.3), which was proposed by

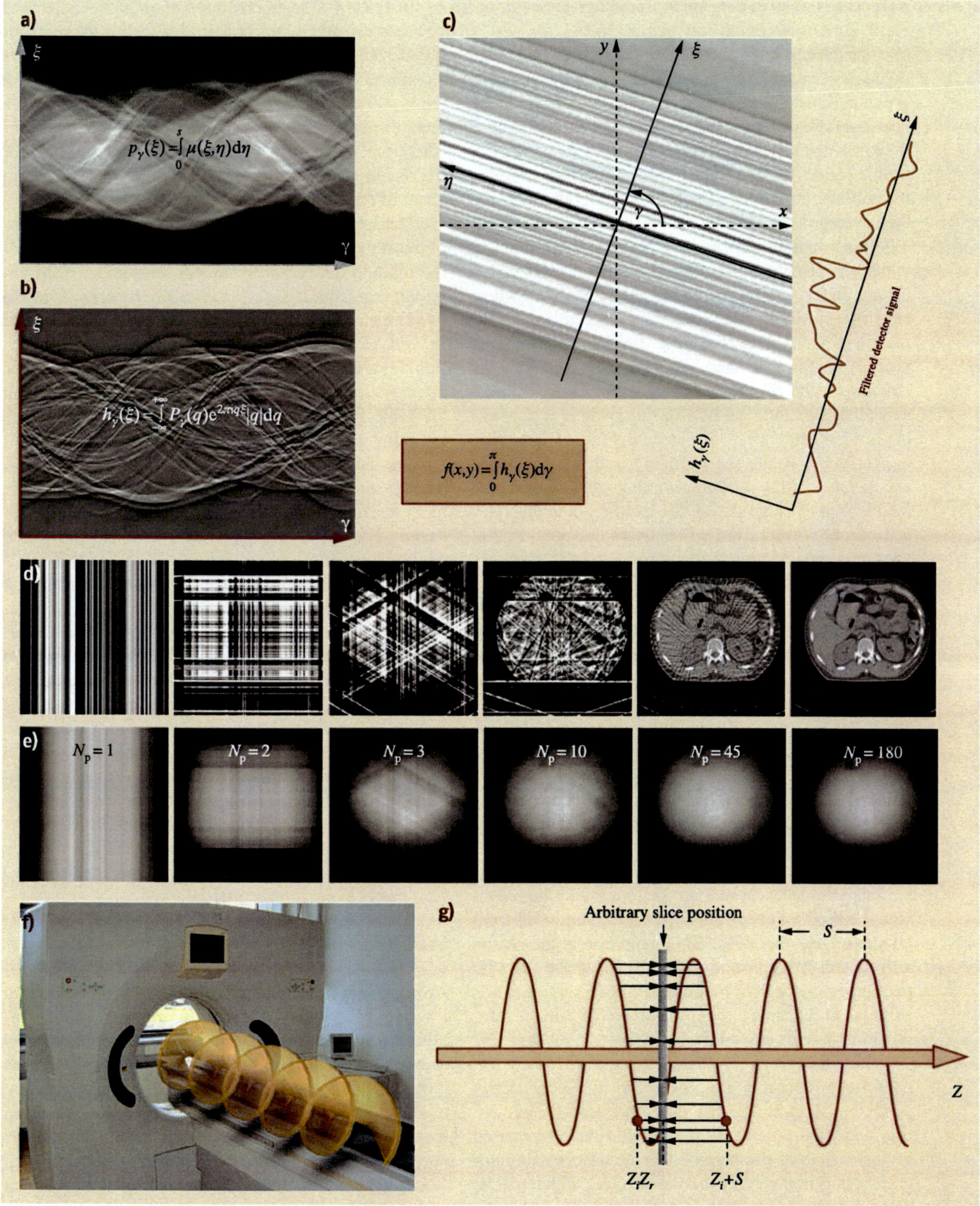
a)
ξ
$p_\gamma(\xi) = \int_0^s \mu(\xi,\eta)\,d\eta$
γ
b)
ξ
$h_\gamma(\xi) = \int_{-\infty}^{+\infty} P_\gamma(q)\,e^{2\pi i q \xi}|q|\,dq$
γ
c)
y
ξ
η
γ
x
ξ
Filtered detector signal
$h_\gamma(\xi)$
$f(x,y) = \int_0^\pi h_\gamma(\xi)\,d\gamma$
d)
e)
$N_p = 1$
$N_p = 2$
$N_p = 3$
$N_p = 10$
$N_p = 45$
$N_p = 180$
f)
g)
Arbitrary slice position
S
Z
$Z_i Z_r$
$Z_i + S$

Kalender in 1989 at the annual Radiological Society of North America (RSNA) conference [16.11].

The inadequacies of the simple slice stack produced by conventional CT are easy to understand. Due to the preset collimation, each slice has a certain width, which is also referred to as the slice thickness. Within this slice thickness the intensity is weighted with its sensitivity profile – given by the source intensity distribution inside the collimation and the detector sensitivity profile – and then averaged.

This averaging process is a problem in all cases in which the object is characterized by boundaries which are angulated with respect to the axial slice, i. e., where structures to be displayed quickly change in the direction of the table feed. In these cases, the averaging process results in a step-like slice stack so that the structure to be displayed has a staircase-like appearance. The development of slip-ring technology, which has already been briefly discussed in Sect. 16.2.3, made it possible to rotate the sampling unit, i. e., the tube and detector array system, continuously. If the table feed is kept constant during the rotation, then the x-ray source rotates around the patient on a spiral path; strictly speaking, this movement describes a helical trajectory (Fig. 16.6f).

This orbit interpretation is based on a coordinate system attached to the patient table, since the x-ray source does of course still run along a circular path. The spiral path only arises from the patient's view. With this concept, complete projection data acquisition of an object is possible, and the scan time for a volume could be considerably reduced in comparison with conventional tomography.

Indeed, it is remarkable that spiral CT technology works at all. An essential requirement for the reconstruction methods described in Sects. 16.4.1 and 16.4.2 is the completeness of the raw data. This means that an object in the measurement field can only be reconstructed if all points of the object are illuminated from all sides, i. e., over 180°. This condition is the reason why artifact-free conventional CT scans of the heart are practically impossible because the heart motion shifts parts out of the slice to be reconstructed while the sampling unit rotates around the heart.

Thus, the projection data to be used for the reconstruction process do not fulfill the consistency condition. Rather, the differences between the projection data of a complete cycle should only be caused by the change of perspective. The reconstructed tomogram is thus impaired by motion artifacts, which will be described in detail in Sect. 16.4.4.

In spiral CT scanners, the motion of the objects to be reconstructed is in fact the decisive innovation compared with conventional CT scanners. The object to be examined is no longer scanned in a single plane. The reason for this is that, due to the continuous patient table feed, the source trajectory is not a closed circular orbit. Therefore, a complete set of raw data is not available for the reconstruction process – the data are inconsistent in terms of Sect. 16.4.1.

The key idea governing the reconstruction process of the spiral CT method is based on the assumption that the missing data of one slice can be completed by interpolation (Fig. 16.6g). If this has been done, then the two-dimensional reconstruction procedures described in Sect. 16.4.2 are again available without any restriction. Figure 16.6g shows the simplest principle of a slice interpolation. The helical rise, i. e., the path along which the table is moving during one 360° rotation of the sampling unit, will be denoted with s here.

One may now select an arbitrary slice position z_r because no preferred axial position regarding the data basis exists due to the constant table feed. For the selected slice there initially is only a single projection angle γ_r for which the projection dataset $p_{\gamma_r}(\xi)$ is available. The projection data $p_\gamma(\xi)$ of all other projection angles must be provided accordingly by means of interpolation. For this purpose, the data that have not been measured under the other required projection angles in the selected slice position z_r must be interpolated on the basis of the closest neighboring angles of the helical trajectory that have actually been measured. Contrary to conventional CT, one has to consider the table speed v as a function of the rotation frequency $1/T_{rot}$ of the sampling unit. This yields the table feed per rotation or the helical rise $s = v/T_{rot}$. If one includes the width or thickness of the x-ray fan d defined by the collimator, then these parameters are usually combined to define the new scan protocol parameter pitch $p = s/d$.

16.4.4 Artifacts

Artifacts are image errors that may emerge due to a variety of reasons. Artifacts can originate from a simplification of the reconstruction method – to date usually the filtered backprojection – which assumes monochromatic radiation or continuous representation of the projection signal. Artifacts may also stem from the use of special sampling technologies and detector arrangements, or simply from defective detector elements. Corrective actions may only be taken if the causes of such artifacts are known. Such countermeasures are in fact very important, since the filtered backprojection has the disadvantage that artifacts are projected back over the entire image so that the overall diagnostic value of the image is reduced or completely destroyed. In this section a brief overview of CT artifacts is given [16.4].

Partial-Volume Artifacts

If a detail of an object consists of a sharply contrasted boundary, the limited resolution of a detector system of course becomes particularly noticeable. The boundary will usually not be located directly at the edge between one detector element and another. Therefore, the x-ray intensity at the corresponding element that has to image this boundary will be linearly averaged over the detector width $\Delta\xi$. Due to this averaging step, the object is blurred. Due to the superposition of filtered backprojections from all directions, this inconsistency leads to artifacts within the reconstructed image which are visible as streaks from the origin of the inconsistency along the backprojection path. Partial-volume artifacts are thus observed, for instance, as ghost lines that extend, particularly, from straight object boundaries. This is due to the fact that the backprojections from the other directions are not able to consistently correct an erroneously detected value, which has been projected back over the entire image. As a countermeasure, the beam may be collimated finely and the corresponding artifacts will be reduced.

Beam-Hardening Artifacts

X-rays produced by electron-impact sources, where fast electrons are entering a solid metal anode, cannot be monoenergetic or monochromatic. In Sect. 16.3.1, the different spectral x-ray components have been introduced, such as the continuous spectrum of the bremsstrahlung (from fast electrons decelerated by the Coulomb fields of the atoms in the anode material) and the characteristic emission lines. Radiation attenuation does not only depend on the path length but is also a function of the specific, wavelength-dependent interaction between x-rays and the material concerned. The reconstruction is in this case – similar to the partial-volume artifact described above – impaired by the nonlinearities which occur here. The beam-hardening artifact is caused by the nonlinear relation between the attenuation values μ and the measured values of the projection p. If an x-ray beam with broadband energy spectrum passes through an object, the spectrum changes along the path. This is due to the fact that different bands of the frequency spectrum are differently attenuated, depending on the specific attenuation coefficients $\mu = \mu(\xi, \eta, E)$ of the material being radiographed. In general, low-energy, i. e., soft, x-ray beams, are more strongly absorbed than high-energy, hard x-ray beams. This is the reason why this effect is named hardening of the x-ray spectrum and the corresponding image error is named the beam-hardening artifact. Similar to the partial-volume artifact, the beam-hardening artifact can be explained by the inconsistency of the individual projection values from different directions, which cannot complement each other correctly within the filtered backprojection method. One corrective method applied in virtually all CT scanners consists of filtering the soft radiation next to the source, i. e., before the radiation reaches the tissue. This may, for example, be done with thin aluminum or copper foils, and for a single material with known properties, it is of course possible to correct for beam hardening computationally.

Metal Artifacts

One known problem in CT is the appearance of metal artifacts in reconstructed CT images. Low-energy x-rays are attenuated more strongly than high-energy x-rays. Recall that the absorption is given by $\alpha \propto Z^4\lambda^3$. Due to the Z^4 dependence, this beam-hardening effect is prominent for metals that are introduced into the human body, such as dental fillings or hip prostheses, and leads to inconsistencies in the Radon or projection space. These inconsistencies observed in the integral attenuation values are due to the polychromatic x-ray spectrum produced by the x-ray tube. Additionally, without applying the dual-energy principle, the total attenuation of the x-ray intensity is an a priori unknown combination of the photoelectric effect and the Compton effect. This often leads to artifacts in the reconstructed images in the form of dark stripes between metal objects with light, pin-striped lines covering the surrounding tissue. Besides beam hardening, another origin of metal artifacts is a higher ratio of scattered radiation to primary radiation, causing a low signal-to-noise ratio (SNR) in the metal shadow. This effect will be discussed in the next section. However, inconsistent projection data can also be repaired with surrogate data created by interpolation or be treated as missing data in order to ignore the inconsistent data in a statistical approach. The problem with interpolated surrogate data is that they always include residual inconsistencies due to missing information during interpolation. On the other hand, within the missing data approach, a brute-force method is applied to eliminate inconsistencies. However, the problem with this strategy is that the reconstruction actually suffers from voids in the projection data. Recently, it has been shown [16.16] that both strategies may be combined. A weighted maximum-likelihood expectation-maximization (MLEM) approach can be used to reduce the influence of the residual inconsistencies from interpolation in such a way that optimal imaging quality is obtained by optimizing the compromise between residual inconsistencies and data voids.

Motion Artifacts

So far it has been assumed that the morphology in the slice to be reconstructed does not change during data acquisition. However, if one also has to take into account the temporal variation of the attenuation coefficient $\mu = \mu(\xi, \eta, E, t)$, one faces the problem of image reconstruction with a changing data basis; that is, the data measured during the rotation are inconsistent. As a countermeasure, one fundamental goal for engineers developing new CT scanner generations is the acceleration of data acquisition, particularly with respect to the time constants related to anatomical and physiological motions. The presently used scanners are multislice subsecond CTs, which, however, are not able to display perfect radiographs of beating hearts without electrocardiogram (ECG) triggering.

Sampling Artifacts

As in general in any signal processing task, Shannon's sampling theorem must not be violated in CT. This applies to both the reconstruction of an axial slice and the subsequent reconstruction of three-dimensional (3-D) data presentations by slice stacking (Sect. 16.5.3). Subsampling of a signal also results in the typical aliasing artifacts. An inherent sampling problem discussed in [16.4] arises particularly for a detector array with a rectangular sensitivity profile of the single detector elements, namely that the individual elements would have to be arranged at half the distance of their own width. As this requirement cannot be met due to obvious technical reasons, one makes use of an elegant mechanical trick. The corrective action, which is nowadays used to prevent aliasing, is based on either the quarter detector shift or the so-called *flying focus* of the x-ray tube.

Electronic Artifacts (Ring Artifacts)

There are several electronic defects that can deteriorate the image and in most cases destroy it. The most famous, or rather notorious, electronic defect is the failure of a detector channel. In third-generation CT scanners, such a detector defect will result in prominent so-called ring artifacts. As the x-ray source and the detector array are tightly joined at the sampling unit, the failure of an individual detector element, or the corresponding processing channel, becomes specifically visible. During the filtered backprojection the virtual lines connecting the corresponding detector element and the x-ray source, which are sometimes called defective beams, form the tangents of a circle. This means that all values outside the circle are seriously affected by this artifact. Inconsistencies with the measured values of the corresponding other projection directions in fact arise for each point of each line. Due to the backprojection, all image areas are again affected by the artifact.

Scatter Artifacts

For the detector element located in the unscattered, direct beam path, in principle it does not make any difference which physical mechanism of interaction actually reduces the intensity. Other detector elements located outside the direct line of sight may in fact be impaired by certain interactions. Particularly in the area of strongly attenuating anatomical objects such as the shoulder, abdomen, and pelvis, the measured values may be distorted due to scattered radiation. These scattered x-rays may become a considerable part of the overall signal. Whereas the scattered radiation is almost the same for all projection angles, it is very different for the wanted signal. In projection directions in which highly absorbing objects are located one after another, the wanted signal may become extremely weak, so that the scattered radiation dominates the signal. Within the filtered backprojection, inconsistencies then arise from this projection direction, which result in streak artifacts. With regard to the interfering radiation caused by scattering reaching the detector, third-generation CT scanners are superior to fourth-generation CT scanners. The detector array of such scanners is designed such that the row of detector elements is arranged on a circular segment with the x-ray source located at the circle center. It is therefore possible to collimate the radiation by septa so that scattered radiation – with an incident angle into a *false* detector element, which is larger than a threshold angle – is effectively shielded. The critical angle is determined by the length and the spacing of the detector segments. In fourth-generation scanners the detectors are located on a circle, the center of which is the isocenter of the measurement field. An x-ray source, to which a detector collimation might be focused on successfully, obviously cannot be located there.

16.5 Scan Planning and Applications

In the following sections, some important practical aspects of computed tomography (CT) will be discussed. Among others, these aspects concern scan planning, data processing, and representation – particularly gray value scaling. In particular, scan planning plays an important role in clinical applications of CT, since scans cannot be arbitrarily repeated due to the system-inherent radiation dose to which the patient is exposed.

Therefore, it is furthermore not only important to plan the scan properly, i. e., to prepare the image acquisition, but also to be informed about how the image data may be represented, i. e., how to postprocess the acquired data appropriately to maximize the information that is available from the images by means of modern visualization techniques.

16.5.1 Scan Planning

The first and most important step in planning a CT scan is the acquisition of an overview scan, which is called a topogram (Siemens), scanogram (Philips), scout view (General Electric) or pilot scan by the different manufacturers. To acquire this overview scan, the rotation of the sampling unit is stopped at a desired angle. In principle, any angulation is possible, but typical positions are *anterior–posterior* (a.p.), i. e., x-ray examination from the patient's front to the patient's back, and *lateral*, i. e., x-ray examination from the side.

During acquisition of the overview scan, the patient table is continuously moved through the measuring field. Figure 16.7a exemplarily shows images resulting from an a.p. and a lateral overview scan. The resulting images are quite similar to conventional x-ray images. However, both imaging techniques differ in that a parallax in the axial direction of the patient does not occur for the overview scan because of the minimal divergence of the x-ray beam due to slice collimation.

By means of the a.p. overview scan, it is possible to plan and program a particular slice plane, the slice thickness, and the number of slices or the volume. When using the lateral overview scan it is also possible to program special slice orientations by appropriately tilting the entire gantry. This is especially useful in skull radiographs and spinograms to exclude sensitive organs (e.g., eyes) from the scan. Figure 16.7a shows an a.p. overview scan of a thorax. Here, axial slices with a thickness and distance of 8 up to 10 mm are typically scanned. If the lumbar vertebrae are to be examined, the gantry must be tilted such that it is adapted to the orientation of the individual vertebral bodies. Figure 16.7a also highlights that this is the only possibility for acquiring planar images of, for instance, the intervertebral disks.

Figure 16.7b,c shows two different orientations of the skull. If one aims to acquire coronal skull images, it is necessary to place the patient appropriately on the table. In Fig. 16.7b, a lateral overview scan of the patient can be seen. The patient is lying on his abdomen with his head heavily hyperextended. This hyperextension of the head is indeed required due to the limitations in the angle of gantry tilt. Two tomographic slices are shown on the right in Fig. 16.7b. Such radiographs are frequently used either to diagnose chronic sinusitis or to evaluate bone fractures. In slice 1, artifacts due to dental fillings can clearly be identified. Slice 2 shows a section of the frontal calvarium of the skull with spongiosa. Both slices allow unobstructed paranasal sinuses to be discerned. In comparison, axial images without gantry tilt are shown in slices 3 and 4 of Fig. 16.7c.

16.5.2 Hounsfield Units and Gray-Value Mapping

In CT, the attenuation values μ from (16.1) are usually represented as gray values. In this context, an approach developed by Hounsfield has proven to be appropriate and is nowadays commonly used. Here, the attenuation values are transformed onto a dimensionless scale and are related to the attenuation value of water. The definition of these CT values reads

$$\text{CT value} = \frac{\mu - \mu_{\text{water}}}{\mu_{\text{water}}} 1000 \,. \tag{16.1}$$

In honor of Hounsfield, the unit of these values

$$[\text{CT value}] = \text{HU} \tag{16.2}$$

is called the Hounsfield unit (HU). On this scale, the CT value of -1000 HU is assigned to air and the CT value 0 HU to water. In principle, this is an open-ended scale, but in practice it ends at ≈ 3000 HU. The range of 4000 HU overall can be captured quite well by means of 12-bit gray-value images. This scaling is arbitrary, but nevertheless has practical consequences. Since the attenuation values of almost all organs – except the bones – are quite similar to those of water, the difference from the attenuation value of water that is given by (16.1) is of per mill.

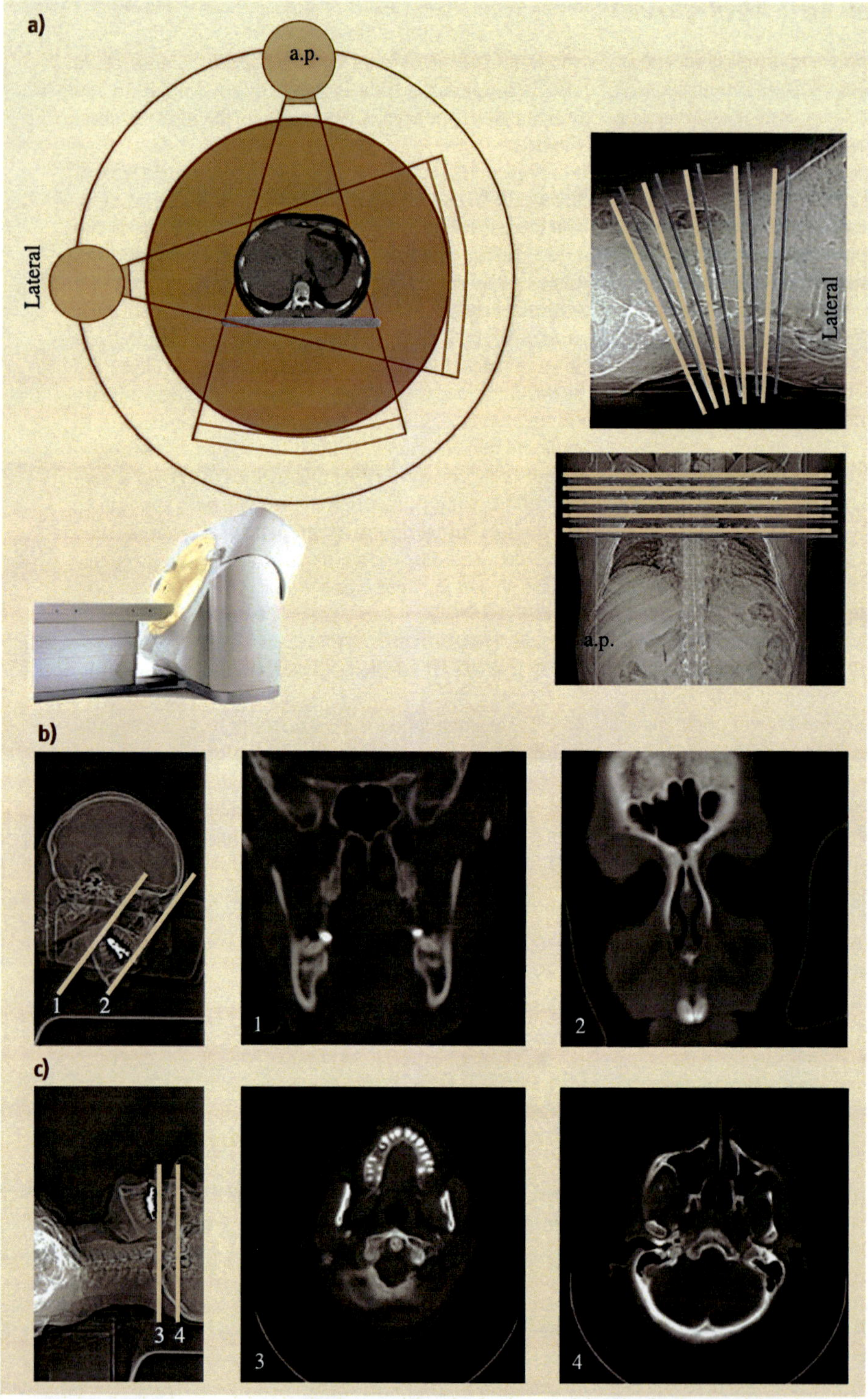

Fig. 16.7a–c Preparation of an overview scan to plan the slice plane position **(a)**. With a fixed tube detector position, the table is continuously moved through the *gantry*. This produces projection images similar to x-ray fluoroscopy. Two geometries are typically used: lateral – the patient is x-rayed from the side; anterior–posterior (a.p.) – the patient is x-rayed from the front to the back. Lateral overview scan for planning the slice position of head tomograms **(b,c)**. Due to the limited tilt of the gantry, the patient has to be bedded face-down and with his head hyperextended to the neck to be able to acquire a coronal representation of the facial cranium (slices 1 and 2). In comparison, axial images without gantry tilt (slices 3 and 4)

Radiologists are accustomed to considering CT values as absolute values which can unequivocally be assigned to the organs. Deviations of these CT values for certain organs indicate pathology. Section 16.4.4 already pointed out the dependence of the x-ray attenuation on the wavelength of the radiation and the potential artifacts arising from this. This problem, which emerges in the case of all CT scanners used for diagnostic imaging, is a consequence of the usage of polychromatic radiation spectra.

While passing through the body, the spectral distribution of the radiation changes such that unequivocal assignment of attenuation values is actually not possible. Nevertheless, the view of the radiologists is largely justified, since most organs behave like water with regard to radiation physics. Therefore, it is possible to correct the beam hardening for these objects by means of a calibration measurement performed with a water phantom. Thus, for the CT values of soft tissue, the definition of the Hounsfield value is directly linked to the tissue density ρ. In this context, Fig. 16.8b is of special interest because different organs and organ changes can be readily distinguished.

It is sensible to divide the whole Hounsfield scale into diagnostically relevant intervals as shown in Fig. 16.8. Figure 16.8a shows a histogram of the relative frequencies of the CT values of an abdominal slice. The accumulations of air, the foam plastic of the patient bed, as well as the fat and the organs can be seen. In the representation of the tomographical slice of the parenchymatous organs, the problem arises that many organs are mapped into overlapping intervals of CT values. Therefore, a sound diagnosis is actually not easy; thus, textures of organs are also important in clinical practice.

The human visual system cannot resolve the complete dynamic range from -1000 HU up to 3000 HU with 4000 gray-value steps. This is the reason why, in practice, gray-value discriminations of only 256 or 512 steps are resolved on display devices.

Recent studies have shown that the human observer is able to discriminate between 700 and 900 shades of gray for the available luminance range of current medical displays and in optimal conditions [16.17]. To be able to detect differences between organs that have rather similar visual representations in their attenuation, it is necessary to map the respective anatomically sensible Hounsfield interval appropriately to the perceptible gray-value range. For this, one uses the piecewise linear function

$$G = 255 \cdot \begin{cases} 0 \text{ for CT value} \leq \text{WL} - \frac{\text{WW}}{2} \\ \text{WW}^{-1}\left(\text{CT value} - \text{WL} + \frac{\text{WW}}{2}\right) \\ 1 \text{ for CT value} \geq \text{WL} + \frac{\text{WW}}{2} \end{cases} , \tag{16.3}$$

where WW denotes the window width and WL the window level.

Figure 16.8b,c shows the corresponding piecewise linear function for a bone window (WL = +300 HU, WW = 1500 HU) and a soft tissue window (WL = +50 HU, WW = 350 HU), as well as their effects on the representation of an abdominal tomogram. Density differences in the spinal process are visible only in the bone window, but differentiation of the soft tissue is hardly possible due to the large width of the window. In the soft tissue window, organs such as the liver and the kidney can be distinguished quite well. However, in this relatively narrow window, all CT values above +225 HU are displayed undifferentiated as white areas.

Figure 16.8a also exemplarily shows the results of applying different windows to an image slice of the thorax. Again, the relative frequency of the CT values is given. These images are especially challenging, since lung tissue, soft tissue, and bone tissue might be interesting from a diagnostic point of view. For the classification of the areas, three windows turned out to be quite useful. Thus, the lung or pleura window (WL = −200 HU, WW = 2000 HU), which allows the lung tissue with its low density to be differentiated, is added to the already mentioned soft tissue window and bone window.

16.5.3 Three-Dimensional Data Representation

In the previous sections, the reconstruction of two-dimensional images has been considered. However, today, medical applications of computed tomography are mainly related to three-dimensional imaging. In a first step, a stack of two-dimensional slices must be acquired. A conventional technique is given by a sequential procedure where the patient, lying on the patient table, is moved slightly in the axial direction of the scanner, i.e., the scanner's z-axis. The table then stops and a complete raw data set of a single slice is measured. The stack of tomogram slices is subsequently used to compute the three-dimensional representation of the depicted anatomy. This procedure is called *secondary reconstruction*.

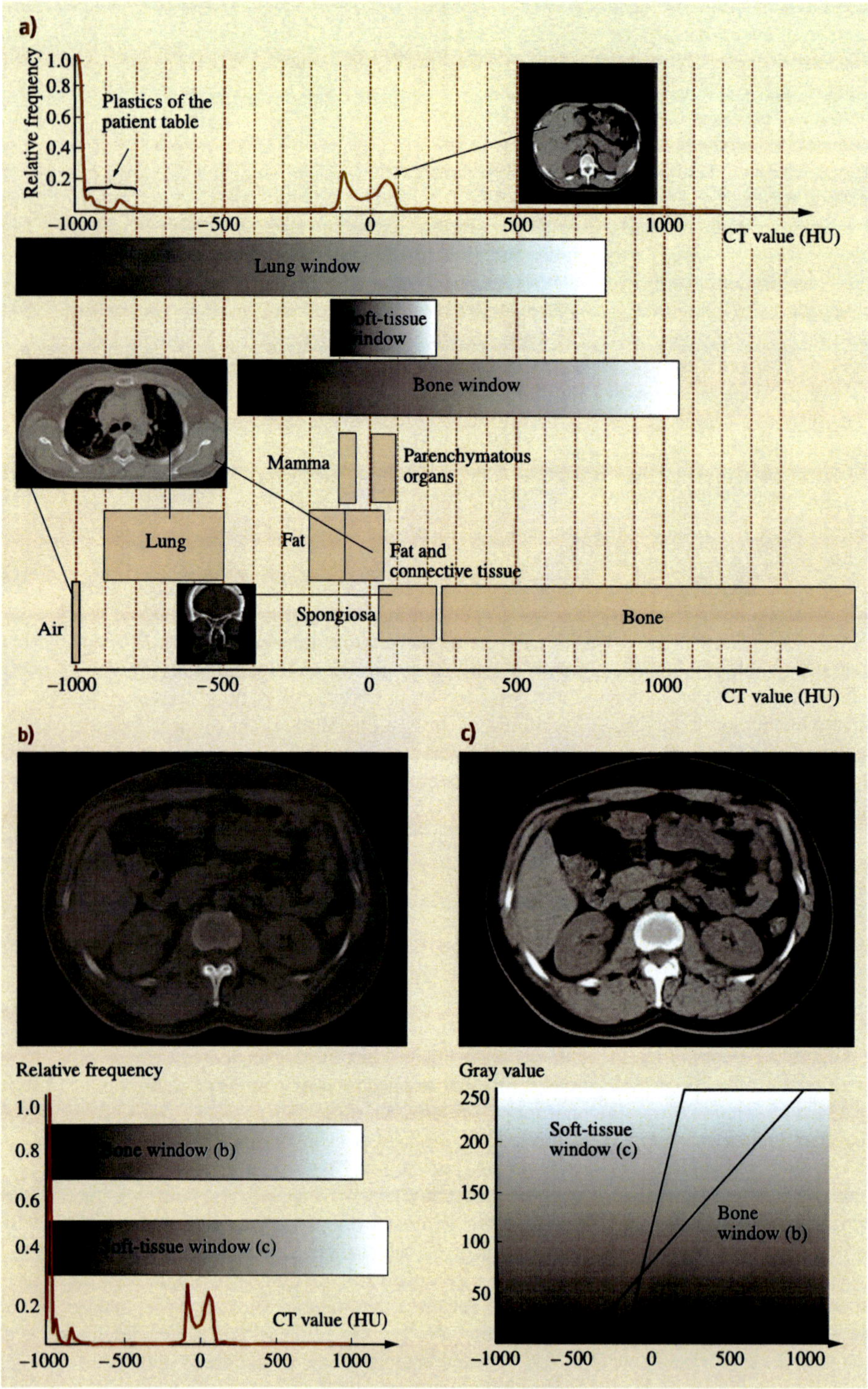

Fig. 16.8a–c Attenuation values in Hounsfield units (**a**). The data are compiled from the collections in [16.18]. The principle of windowing. The complete interval of practically sensible Hounsfield values (HU), ranging from −1000 HU up to 3000 HU (relative frequency of the values for the abdomen area – *lower left*), cannot be recognized and distinguished by the human visual system. Therefore, the different anatomically interesting Hounsfield intervals have to be mapped to appropriate grayscale intervals that can be differentiated. The *lower right* image provides the characteristic curves for two anatomically relevant windows. The *upper* images show the result of the bone window ((**b**): $WL = +300$ HU, $WW = 1500$ HU) and the soft tissue window ((**c**): $WL = +50$ HU, $WW = 350$ HU)

As for the surface visualization method, it is necessary to select a gray value threshold representing the surface. In this context one has of course a certain degree of freedom, as different objects may be visualized as long as their Hounsfield values are clearly different from one another. The gray-value isosurface will then be illuminated with a virtual light source so that the corresponding light reflections can be computed to display the result.

An alternative visualization method of 3-D CT data is multiplanar reformatting (MPR). MPR is used to show angulated sections through the three-dimensional stack of slices. Typically, the principal sections (the sagittal, coronal, and axial slices) are presented to the radiologist. In Fig. 16.9e, a coronal reformatting is shown.

Figure 16.9b,c illustrates the volume rendering technique. This method assigns a physical light reflection and scattering value to each voxel (spatial pixel). The computer is then used to illuminate this *data fog* with a virtual light source and to model the optical impression artificially. To do so, real reflection and scattering of light is simulated. If bones, organs, and contrast-media-enhanced vessels are assigned different optical properties, interesting insights into the pathological status of the patient can be gained. In a postprocessing step, certain tissue types can be suppressed. In Fig. 16.9d, for example, bone is eliminated by a spe-

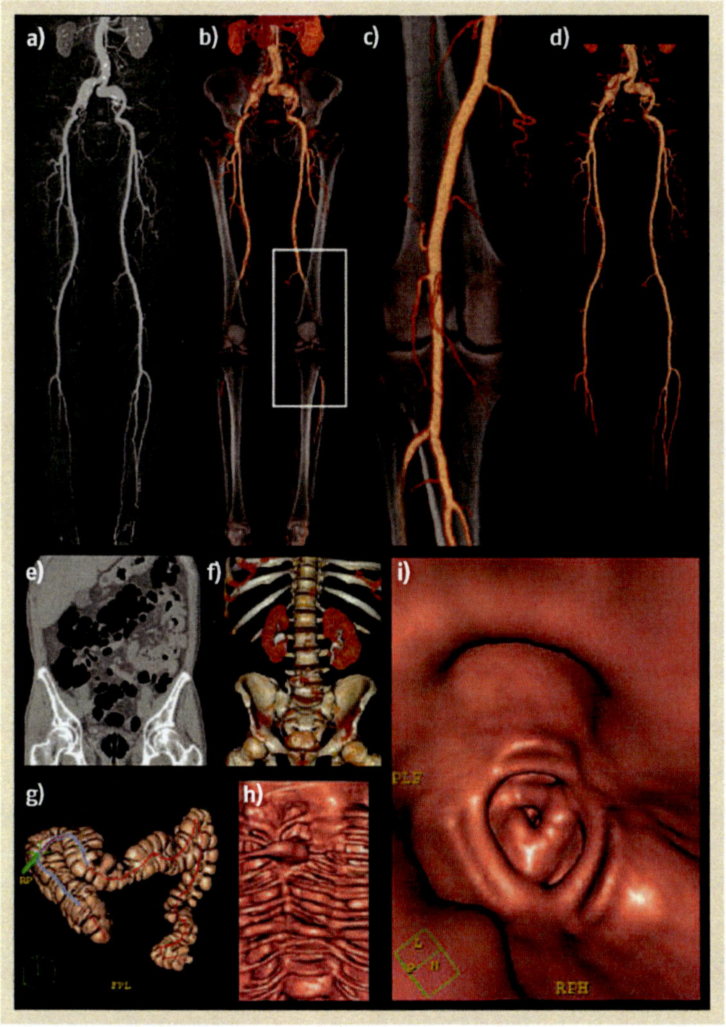

Fig. 16.9 (a–d) Angiographic image acquisition with CT (CTA). **(a)** *Maximum-intensity projection* (MIP), **(b)** volume rendering, **(c)** zoom into the knee and visualization from the back, and **(d)** virtual bone elimination with the bone removal technique. **(e–i)** Three-dimensional representation of the abdomen. **(e)** Coronal reformatting, **(f–i)** volume and surface rendering, **(g)** virtual endoscopy of the intestine together with **(i)** the representation of the virtual endoscopy trajectory, and **(h)** virtually opened intestine in cylindrical coordinates

cial bone removal technique for better visualization of the vessel tree.

An alternative to volume rendering is surface rendering, as illustrated in Fig. 16.9f–i. The individual shades of gray of the layers (images) in the data stack represent the degree of physical attenuation of the x-ray beam. In a clinical context, deviations from the normal distribution of these values may indicate pathological changes in the patient.

During visualization, it is possible to visualize certain ranges of values and to selectively suppress others. If the viewer chooses a constant gray value, i. e., a threshold, all spatial points with this value may be displayed in space as an isosurface. This surface is approximated by triangles using a technique called triangulation. Then, the mosaic of triangles is again illuminated and displayed using the virtual method described above. The larger the number of mosaic pieces

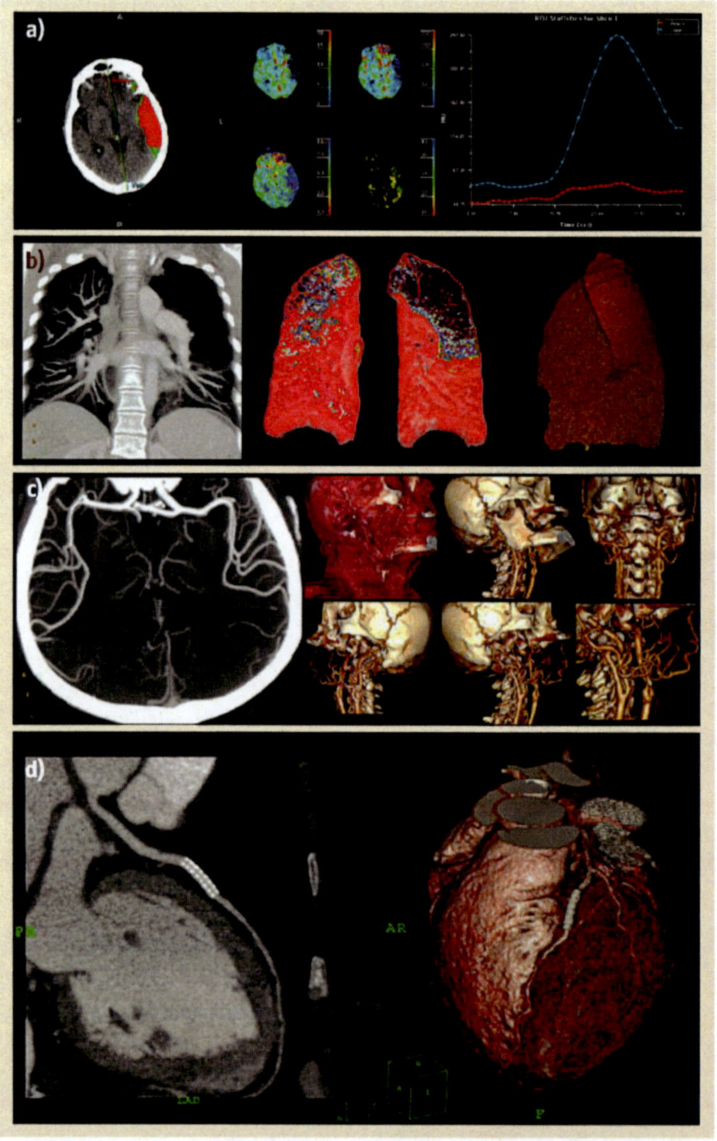

Fig. 16.10 **(a)** Perfusion measurement for stroke patients, **(b)** visualization of lung emphysema, **(c)** CT angiography of the cranial vessels, and **(d)** imaging of a coronary stent

chosen for the reconstruction of the surface, the more lifelike the result.

In addition to the three-dimensional representation of organs or bone surfaces, which may be interactively rotated on the computer screen, an interior view of hollow organs and airways of contrast-enhanced vessels can be produced. In this way, virtual flights into the body – for example, into the bronchial tubes or the intestine – can be visualized. In Fig. 16.9i, a virtual insight into the human intestine is shown. The value of such a three-dimensional diagnostic approach lies in the reduction of the large amount of data otherwise inherent to the hundreds of single slices.

16.5.4 Clinical Applications

The final section of this chapter describes the broad variety of clinical applications of modern CT imaging by means of some examples. In Fig. 16.10, four typical applications are illustrated. Important fields of application are summarized in the following.

- *Brain perfusion*: The blood flow in the brain (cerebral blood flow, CBF), blood volume in the brain (cerebral blood volume, CBV), mean transit time (MTT), and peak time of the bolus maximum (time to peak, TTP) are acquired and displayed as colored overlays on the relevant CT slices. In this way, colored maps of tissue vitality give indications of an acute or chronic infarct (Fig. 16.10a).
- *Liver perfusion*: Arterial and portal measurements of perfusion in liver studies.
- *Tumor perfusion*: Characterization of known lesions via their perfusion.
- *Lung measurement*: Diagnostics of lung emphysema (Fig. 16.10b); automatic detection of lung nodules.
- *Calcium scoring*: Quantification of coronary calcification.
- *Vessel analysis*: Visualization of vessel trees (Fig. 16.10c); analysis of stenoses and aneurysms; planning of stents.
- *Cardio CT*: Identification and quantification of stenoses; planning of stents and visualization of implanted stents (Fig. 16.10d).
- *Virtual endoscopy*: Three-dimensional CT data as a basis for anatomical interior views of hollow organs (Fig. 16.9i) and contrast-enhanced vessel trees.
- *Trauma*: Fast imaging of the entire body for diagnostics of accident injuries.
- *Dental planning*: Three-dimensional reconstruction and slices through the maxilla and mandible as a planning basis for implantation of prostheses for the oral surgeon.
- *Planning of radiotherapy*: Three-dimensional CT reconstruction as a basis for dose planning in the radiation therapy of tumors.
- *Image-guided surgery*: Three-dimensional CT reconstruction as a basis for planning and navigation of surgical interventions.
- *Interventional imaging*: Visualization of a surgical instrument tip during a biopsy.

In particular, perfusion measurements show that today CT is on its way towards becoming a functional modality. This technique allows the blood flow to be measured after contrast media injection into different organs.

16.6 Dose

The gain in diagnostically valuable information accompanying the advent of computed tomography was considered to be exceptional. This is one of the main reasons why the applied dose was not considered to be of vital importance during this developmental stage. Based on the number of devices currently installed in Germany and an average of 3500 examinations per year, an annual total number of several hundreds of millions of slices can be estimated to be acquired each year.

This approximately lies within the same range as the total number of standard projection radiographs acquired in clinical practice. Considering the different types of radiological examinations (x-ray examinations), one immediately becomes aware of the following discrepancy. Although CT examinations represent only about 4% of all radiological examinations, their share of the total dose amounts to $\approx 35\%$. In short, CT accounts for the largest portion of medically related x-ray exposure.

Considering new generations of CT scanners, such as spiral multislice CT, the applied dose indeed is not reduced. In fact, it is more likely that the dose applied during one single imaging session will increase. This is mainly due to the fact that, in modern scanners, longer sequences and thinner slices can be easily measured [16.19]. Initiated by reports about unnecessarily high x-ray exposure in pediatric CT examinations [16.20], which were actually performed on the basis of the same scanning protocols as were used in human adults, a high sensibility concerning the applied dose has emerged among both manufacturers and users. Recent developments such as automatic exposure control take the problem of dose into account. With this technology, one can easily adjust the dose according to the anatomy of the patient or likewise take into account whether the individual under examination is an adult or an infant.

Figure 16.11a schematically illustrates that one initially has to differentiate patients according to their physique and between infants and adults. If we adopt the same tube current–time product for infants as is used in scan protocols for adults (Fig. 16.11a, left), the obtained intensity of the x-ray at the detector and thus the applied dose, with respect to the achievable diagnostic quality (i. e., the SNR $= \mu/\sigma$), is unnecessarily high. Therefore, in the case of a pediatric examination, it indeed is reasonable to decrease the anode current to a level that allows a comparable diagnostic quality to be obtained as achieved in the examination of an adult (Fig. 16.11a, right). Such automatic exposure control (AEC) can be steered by means of the scanogram.

Figure 16.11b illustrates dynamic adaptation of the x-ray tube current or dose with respect to the anatomical situation observed along the axial z-axis (longitudinal dose modulation). In anatomical areas in which, as known a priori, low attenuation of the x-ray intensity is to be expected (for example, within the lung area) the tube current and thus the dose can be lowered, without decreasing the image quality. If all areas are imaged with the same tube current, this will result in either dose rates that are too high or noise levels that are too strong.

Figure 16.11c illustrates dynamic adaptation of the tube current across the projection angle γ to the integral attenuation of the radiated body area (angular dose modulation). At all sections through the body, which are oval rather than circular, the angular dose modulation is to be applied. In particular for image acquisitions of the shoulder, strong dose modulation is to be used.

Another possibility for dose reduction arises in cardiac imaging. In Fig. 16.11d, a trigger sequence of an ECG-triggered image acquisition of the heart is depicted. Since data are only acquired within the resting phase of the beating heart, the overall dose can be significantly reduced, provided that the tube current is switched off outside the data window. This effective and efficient procedure is called temporal dose modulation. Figure 16.11e shows the effect of the combination of longitudinal and angular dose modulation on the attenuation of the x-ray beams and on the tube current.

Apart from device-related measures for reduction of radiation exposure, for which the manufacturers bear responsibility, there is a set of user-related measures which also affect the applied dose. The quantities and their connection to the dose, adjustable in the scan protocol, are given in the following. For a detailed discussion on radiation exposure in CT, one is referred to the book by *Nagel* [16.19].

- *Current–time product (mAs product)*: Dose and mAs product, i. e., the product between the x-ray tube current and acquisition time, have a linear dependence. However, the standard deviation, i. e., the image noise, increases proportionally to the inverse of the square root of the mAs product.
- *Acquisition time*: At constant x-ray tube current, the dose increases linearly with acquisition time. However, the mAs product is always to be considered as a total, so that, at a constant dose, the acquisition time can be reduced along with simultaneously increasing the tube current. Thereby, the image noise is unaffected; however, motion artifacts are less likely to occur at short acquisition times (Sect. 16.4.4).
- *Tube voltage*: With increasing tube voltage U_a (Sect. 16.3.1), the efficiency of the x-ray tube as well as the penetration of the radiation are increased. The intensity of the radiation increases with U_a on average following a power of 3.5. With deeper penetration of the radiation, the image contrast obviously decreases. However, this is more than compensated for by the better quantum statistics. Therefore, the image quality generally is improved – obviously at the expense of a higher dose being applied to the patient. If the voltage is, for instance, increased from 120 to 140 kV, in order to reduce the dose whilst maintaining image quality, the mAs product must be reduced by 40%. Thus, a dose reduction of $\approx 15\%$ is obtained [16.19].
- *Thickness of the object to be imaged*: For infants and frail patients, one has to bear in mind that a smaller mAs product has to be adopted. Due to lower attenuation, the statistics of the x-ray quanta are still as good as for adults or heavier patients. Correspondingly, image quality is not affected. For heavier patients, an increase in x-ray tube voltage is prefer-

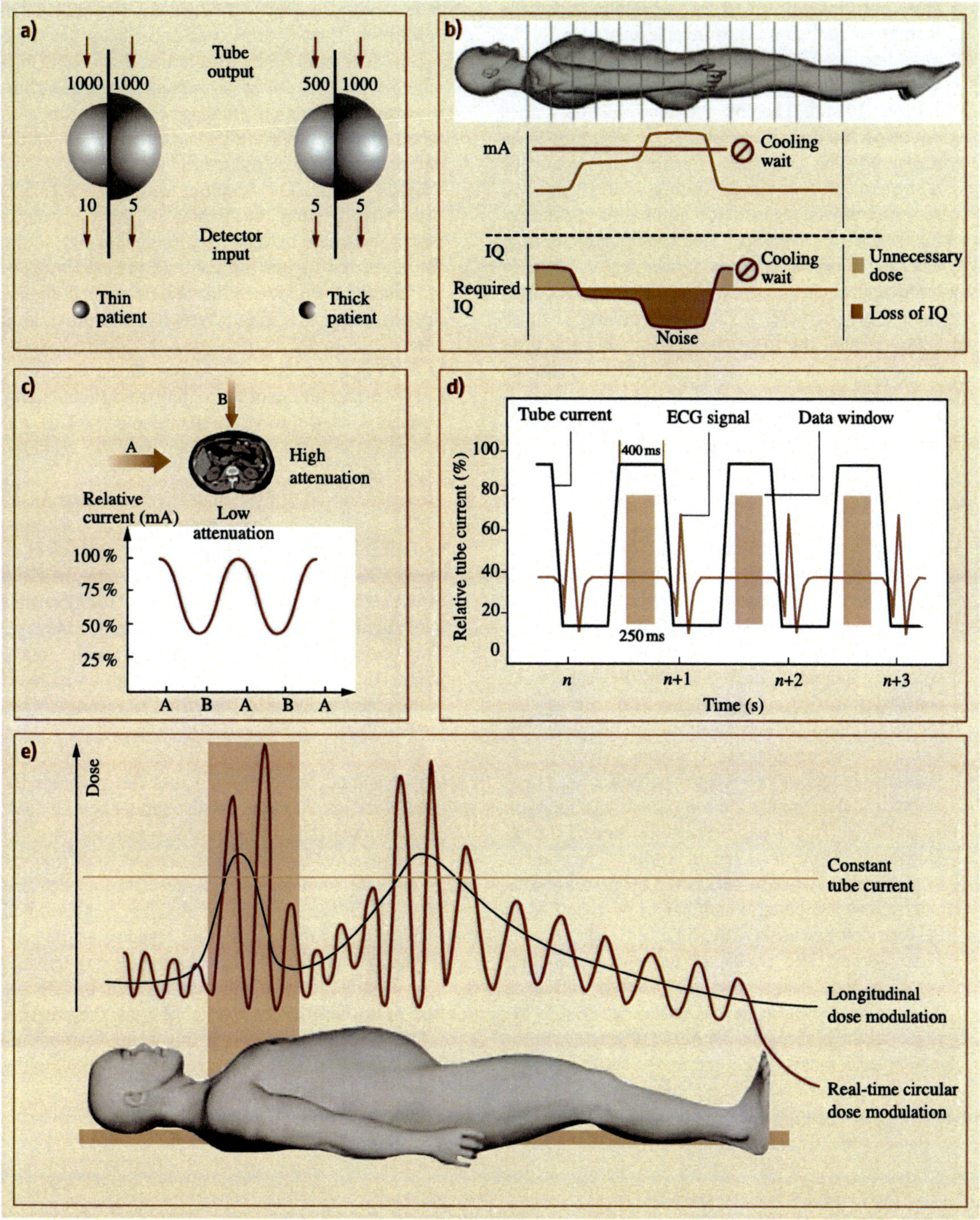

Fig. 16.11a–e Active methods for dose reduction [16.19]. **(a)** Automatic exposure control (AEC) differentiates between heavier and thinner patients. **(b)** The longitudinal dose modulation – a dynamic adaptation of the tube current or dose to the anatomical situation. **(c)** Angular dose modulation – a dynamic adaptation of the tube current or dose to the shape of the axial body section. **(d)** Temporal dose modulation. **(e)** Combination of longitudinal and angular dose modulation. (IQ – image quality)

able to an increase in the mAs product, since the increase in radiation exposure is less intense.

- *Slice thickness*: The slice thickness can typically be adjusted by means of a tube sided collimator from 1 up to 10 mm. Thereby, the slice thickness does not affect the dose if the same body section is to be measured. The advantage of a finer slice sequence is the reduction of partial-volume artifacts as well as step artifacts occurring on coronal or sagittal reformatting of the image. The disadvantage is that, at a finer collimation, fewer x-ray quanta will reach the detector such that the image noise will correspondingly increase. If the image quality is to be kept constant, the mAs product and thus the dose will have to increase inversely proportionally to the slice thickness.
- *Pitch factor*: A pitch factor of $p = 1$ means that, in the case of a rotation of the sampling unit through 360°, the patient table is moved linearly by a length equal to the adjusted slice thickness (Sect. 16.4.3). If $p < 1$, the individually measured slices have a larger overlap, such that the image quality is increased. However, this results in an increased dose. Accordingly, if $p > 1$, the dose can be reduced. In theory, an artifact-free image reconstruction should be possible up to a pitch of $p = 2$. With this, the scanning length is increased such that, for the same mAs product, fewer x-ray quanta are available for the image formation of a coevally larger body section. Therefore, the image noise increases.
- *Scanning length*: If the imaged body section is enlarged, the dose applied to the patient will increase accordingly. This is expressed by means of the effective dose or the dose–length product. The number of slices always has to be limited to the diagnostically relevant section, which has to be specified in the overview scan (Sect. 16.5.1).
- *Filter kernel*: The choice of the respective filter kernel at first glance does not directly influence the dose. However, as described in Sect. 16.4.2, the choice of the high-pass filter of the filtered backprojection does indeed affect the image resulting from the reconstruction, since the choice of the filter kernel represents a trade-off between noise and spatial resolution. If one wants to reduce the noise while maintaining high spatial resolution, this is only possible at the expense of an increase in dose. Thus, it always depends on the diagnostic question as to whether an appropriate choice of the filter kernel can be used for dose reduction.
- *Window width*: The window width (Sect. 16.5.2) used for adjusting the display of the CT images does not initially have a direct influence on the dose. However, the higher the contrast selected by constricting the window width, the stronger the noise present in the images. Conversely, it is possible to smooth the image by enlarging the window width. If there is a contrast reserve due to the diagnostic question, smoothing of the visualization can already be incorporated during the planning step. This will decrease the mAs product and thus the radiation exposure.
- *Field of view (FOV)*: When using a very small FOV, i. e., a very strong detail magnification, as a general rule, a very sharp reconstruction filter has to be used. This is due to the fact that the magnification of the section under examination is chosen because a locally more detailed image needs to be analyzed. This immediately has consequences for the image noise, which can only be reduced with an increase in the mAs product and, therefore, an increase in the applied dose.

Finally, it should be mentioned that the progress toward smaller detector elements also influences the dose. It can be shown that, for constant signal-to-noise ratio, the dose increases with the inverse fourth power of the detector element size, as described in detailed in [16.4]. If in the formula of Brooks it is assumed that $b = d = \Delta\xi$ for the sampling distance at the center of rotation and for the slice thickness, respectively, then, in consideration of the definition of the signal-to-noise ratio $\text{SNR} = \mu/\sigma$, one obtains the relation $D \propto (\text{SNR})^2/(\Delta\xi)^4$. Obviously, this limits a further reduction of the size of the detector elements, since the dose may not be arbitrarily increased.

16.7 Special System Designs

In the last few decades, special CT imaging designs have been developed for certain applications, and this chapter concludes with a brief review of five of these.

For coronal imaging, the problem of motion artifacts (Sect. 16.4.4) arises with the conventional CT imaging concepts described above. Third-generation CT systems therefore require ECG-triggered data acquisition, because with respect to the time constants of heart motion, even subsecond CT scanners without ECG-triggered data acquisition are too slow.

For this kind of clinical application, electron-beam computed tomography (EBCT) has been developed and will be described below in Sect. 16.7.1. Furthermore, a short description of volume or cone-beam CT – often abbreviated as VCT or CBCT, respectively – will be given in Sect. 16.7.2. These systems are subsequent steps beyond multislice CT (MSCT). However, if very small object details in the micrometer range must be imaged, the spatial resolution of conventional clinical CT is insufficient. For this task, special micro-CT systems have been developed (Sect. 16.7.3). Since nuclear imaging is able to visualize the metabolism of a patient, and x-ray transmission computed tomography gives complementary morphological information, a consequent idea is the combination of both modalities into so-called PET-CT scanners, which will be introduced in Sect. 16.7.4. In the last section (Sect. 16.7.5), a system extension to include two sampling units, i. e., two x-ray tubes and two corresponding multislice detector arrays in a so-called dual-source system, is described.

16.7.1 Electron-Beam CT

If there is a need for extremely short acquisition times, the concept of moving sampling systems must be left behind entirely. One approach to achieving this is the use of electron-beam computerized tomography (EBCT). This type of CT was developed for cardiac imaging.

In EBCT, there is no longer a localized x-ray tube rotating around the patient as in conventional CT technology. Instead the patient is situated, in a manner of speaking, *inside* the x-ray tube. An electron beam is focused onto wolfram target rings, which are arranged in a half circle around the patient, and generates the desired x-ray fan beam upon impact with the wolfram target. The x-ray irradiation is measured with a stationary detector ring. Such systems have mainly been sold to cardiologists by the company Imatron. The electron-beam technique is able to acquire an image slice in 50 ms. Figure 16.12a shows a diagram of an EBCT system as well as an illustration of a modern Imatron system. Further technical details can be found in [16.22].

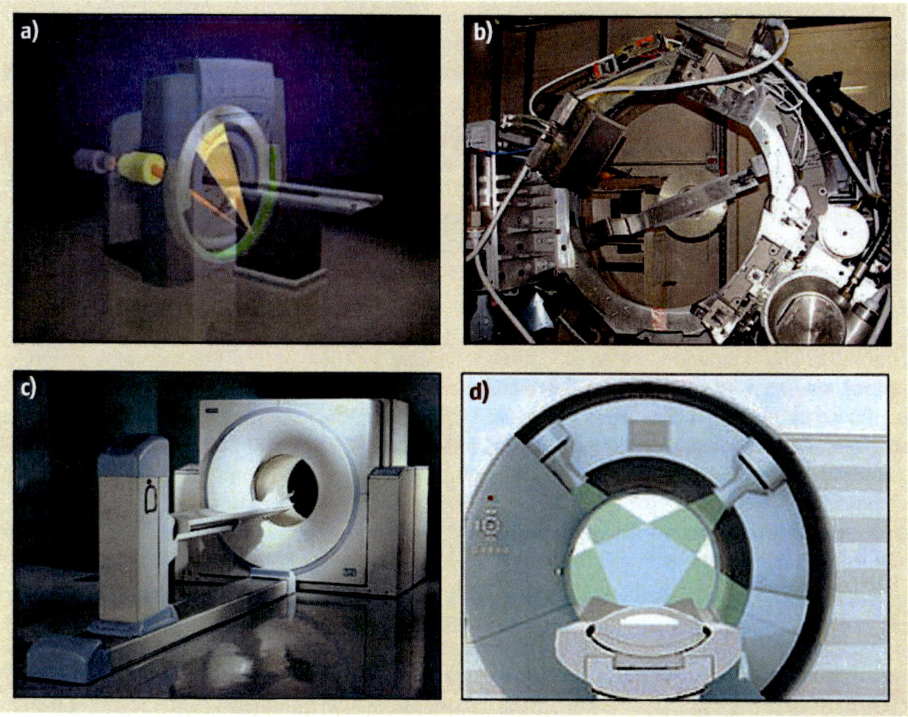

Fig. 16.12 **(a)** Electron-beam CT, **(b)** volume or cone-beam CT, **(c)** PET-CT, and **(d)** dual-source CT [16.21]

16.7.2 Volume CT

There has so far not been a common definition of the generations of development of CT. In [16.10], scanners equipped with a cone-shaped x-ray beam and a plane detector are referred to as the seventh generation. However, even within the cone-beam scanner itself, one needs to distinguish between systems that use only a small cone opening, as in the case of a multislice (multiline) detector system or indeed a symmetric x-ray cone. In particular, the necessary reconstruction methods differ extensively between these systems. To understand the motivation behind the development of cone-beam CT systems, recall that the step from the pencil-beam to the fan-beam concept came along with the advantages that the x-ray source was exploited more effectively, and that there was a reduction in acquisition time. The efficiency of the energy transformation in the generation of x-ray radiation is just about 1%.

As the heat produced inside the x-ray tube essentially defines the physical capacitance, and therefore limits the measuring time, the next straightforward step in the development of CT scanners involved the use of a cone-shaped x-ray beam, which is already produced in the x-ray tube. Both the pencil-beam and the fan-beam geometry are created by means of appropriate pinhole or slit collimators, which reshape the original x-ray source intensity profile, reducing efficiency.

Technologically, there are three important problems that had to be solved before the successful application of cone-beam geometry to CT imaging. First of all, a flat-panel detector, which did not exist at the time, had to be introduced to replace the line or multiline detector arrays. Second, the huge amount of raw data that quickly emerge on subsecond scanners in particular had to be transferred from the rotating sampling unit to the image reconstruction computer. The bandwidth required for the data transfer poses a challenge even today. Third, there is the problem of reconstruction, whose mathematics is slightly more sophisticated compared with the two-dimensional methods.

Figure 16.12b schematically shows the cone-beam geometry of an imaging system with a planar detector (flat-panel detector, Sect. 16.3.2). The planar detector geometry was first applied for technical applications in micro-CT, since the required charge-coupled device (CCD) chips were only available for this geometry. Figure 16.12b shows a prototype of a so-called volume CT (VCT) developed by General Electric Medical Systems.

16.7.3 Micro-CT

Recently, micro-CTs, which essentially comprise a miniaturized design of the cone-beam CTs mentioned in the previous section and which are typically used for nondestructive, three-dimensional microscopy, have become commercially available. The x-rayed measuring field, often as small as $2\,\mathrm{cm}^3$ in volume, is so small that medical applications might seem to be ruled out. Indeed, these scanners are more commonly used for material testing and analysis, but medical applications are on their way to taking center stage. An example in human medicine is the analysis of trabecular structures in bones. Micro-CTs are also ideal scanners for radiological examinations of small animals [16.23]. Micro-CTs are often produced as desktop devices and have a measurement chamber that is entirely shielded by lead walls against scattered x-ray beams, so that no further means of protection are necessary. The object to be examined is placed on a rotating specimen disk, which is controlled by a stepper motor.

The two most crucial components of micro-CTs are the x-ray tube and the two-dimensional detector array. In particular, it is the size of the focus and the size of the detector elements that, apart from the mechanical accuracy of the rotary motion, determine the spatial resolution. Therefore, micro-CT systems need a microfocus x-ray tube. An x-ray focus size of $b_F < 10\,\mu\mathrm{m}$ is desirable. Unfortunately, when using such a small target area for the electrons, the anode current cannot be very large. The current is typically less than $100\,\mu\mathrm{A}$. Since the current controls the intensity of the x-ray spectrum, there are certain constraints with respect to the materials being examined.

A 12-bit x-ray charge-coupled device (CCD) chip with a pixel matrix of 1024×1024 or higher may be used as a detector, which can be connected to a scintillation crystal by fiber optics. The size of the picture elements typically has an order of magnitude of around $b_D < 10\,\mu\mathrm{m}$. SkyScan specifies a resolution of about $10\,\mu\mathrm{m}$. As micro-CTs are cone-beam x-ray systems, three-dimensional reconstruction methods are required to calculate the images [16.24].

16.7.4 PET-CT

With the exception of contrast-enhanced angiography and perfusion techniques, CT, on its own, is only able to provide morphological information, i. e., information on the shape of objects. On the other hand, positron emission tomography (PET) provides information on metabolism, i. e., the biomedical function of an anatomical region [16.25]. CT, however, is based on the attenuation of x-ray radiation. Different organs, having different absorption properties, are therefore only imaged according to their shape, and the patient is considered to be passive during imaging.

In PET, the patient is injected with a radioactively marked *tracer*, which is metabolized inside the body. One very important tracer is 2-(fluorine-18)-fluoro-2-desoxy-D-glucose (^{18}F-FDG), which can be used to trace glucose metabolism. This is especially important for oncology studies, since, due to their faster metabolism, the ^{18}F-FDG uptake of tumors exceeds the uptake of nonmalignant tissue. Compared with ordinary glucose, ^{18}F-FDG differs in the presence of the tracer atom. As such, it behaves like glucose only at the beginning of the metabolic chain. Thereafter, the molecule is detected but not catabolized any further, leading to an accumulation of the tracer inside the tumor.

^{18}F-FDG is a positron emitter, so wherever the tracer accumulates, the process of positron annihilation is intensified; i. e., upon collision with an electron, the positron entirely dematerializes, becoming two gamma-ray photons that fly away in opposite directions. These gamma-ray photons are then measured by two detectors, located opposite each other, in what is called a coincidence measurement. By means of, for instance, filtered backprojection or statistical reconstruction, the location of the dematerialization can be reconstructed, and tumors are represented as *hot spots* inside the image.

An interesting approach to imaging diagnostics is the combination of both morphological and functional imaging methods. The goal of displaying function along with morphology in a single image has been realized for some time using methods of image registration. Registration is an image processing step that must overcome problems caused by the different positioning of the patient in the two different scanners and changes resulting from the time that passes between the two acquisitions.

In the case of combined techniques, the patient is successively scanned using the different imaging modalities. Using a combined PET-CT scanner, such as, for instance, the *biograph* from Siemens (Fig. 16.12c), the images are acquired almost simultaneously, with the patient effectively in the same position, so that the location of a tumor relative to the surrounding anatomy can be displayed immediately.

16.7.5 Dual-Source CT

Figure 16.12d shows a recent development in the field, the so-called dual-source CT. Two complete sampling units, both of which consist of an x-ray tube and a detector array, are installed perpendicular to each other. Best images of the beating heart are obtained when the acquisition interval is set to the diastolic phase. Since the patients are usually agitated, this resting phase of the heart is often so short that motion artifacts are inevitable. Usually, images have to be acquired across 180°, in order to obtain a complete raw dataset within the relatively short resting phase of the heart. However, when employing a sampling unit taking advantage of the dual-source technology, only a rotation of about 90° is necessary. Thereby, the sampling time is halved, such that motion artifacts are significantly reduced. If one wants to abandon the administration of beta blockers, which are used to lower the pulse, artifact-free image reconstruction by means of a single source scanner is no longer possible. Here, the strength of the halved acquisition time becomes apparent. Using these modern dual-source subsecond scanners a temporal resolution can be achieved that only differs by a factor of 2 from the acquisition speed achieved by electron-beam CT. Considering the applied dose, the ECG-supported triggering of the acquisition window is of great importance. Here, compared with the standard CT scan protocol in which the tube current is not modulated, a dose reduction of up to 50% can be achieved.

References

16.1 E. Seeram: *Computed Tomography* (Saunders, Philadelphia 2001)

16.2 J. Hsieh: *Computed Tomography* (SPIE, Bellingham 2003)

16.3 W.A. Kalender: *Computed Tomography* (Publicis, Munich 2005)

16.4 T.M. Buzug: *Computed Tomography* (Springer, Berlin, Heidelberg 2008)

16.5 G. Grossmann: Procede et dispositif pour la representation radiographique des sections des corps, French Patent 771887 (1934)

16.6 W. Härer, G. Lauritsch, T. Mertelmeier, K. Wiesent: Rekonstruktive Röntgenbildgebung, Phys. Bl. **55**, 37 (1999), (in German)

16.7 W. Härer, G. Lauritsch, T. Mertelmeier: Tomographie – Prinzip und Potential der Schichtbildverfahren. In: *Handbuch diagnostische Radiologie*, ed. by T. Schmidt (Springer, Berlin, Heidelberg 2003), (in German)

16.8 G.M. Stevens: Volumetric tomographic imaging. Ph.D. Thesis (Univ. Stanford, Stanford 2000)

16.9 G.N. Hounsfield: Computerized transverse axial scanning (tomography). I. Description of system, Br. J. Radiol. **46**, 1016 (1973)

16.10 J.T. Bushberg, J.A. Seibert, E.M. Leidholdt, J.M. Boone: *The Essential Physics of Medical Imaging* (Lippincott Williams Wilkins, Philadelphia 2002)

16.11 W.A. Kalender, W. Seissler, P. Vock: Single-breath-hold spiral volumetric CT by continuous patient translation and scanner rotation, Radiology **173**, 414 (1989)

16.12 K.M. Mudry: X-ray. In: *The Biomedical Engineering Handbook*, Vol. 1, ed. by J. Bronzino (CRC, Boca Raton 2000) pp. 61–1

16.13 W.A. Kalender: Der Einsatz von Flachbilddetektoren für die CT-Bildgebung, Radiologe **43**, 379 (2003)

16.14 G. Brunst: High resolution digital flat panel detectors for x-ray applications – Basics. In: *Medical Imaging*, ed. by W. Niederlag, H.U. Lemke (Health Academy, Dresden 2002) p. 63

16.15 J. Radon: Über die Bestimmung von Funktionen längs gewisser Mannigfaltigkeiten, Ber. Verh. Sächs. Akad. Wiss. Leipz. Math. Nat. Kl. **59**, 262 (1917)

16.16 M. Oehler, T.M. Buzug: Statistical image reconstruction for inconsistent CT projection data, Methods Inf. Med. **3**, 261 (2007)

16.17 T. Kimke, T. Tuytschaever: Increasing the number of gray shades in medical display systems – How much is enough?, J. Digit. Imaging **20**, 422 (2006)

16.18 E. Krestel (Ed.): *Imaging Systems for Medical Diagnostics* (Siemens, Berlin 1990)

16.19 H.D. Nagel (Ed.): *Strahlenexposition in der Computertomographie*, 3rd edn. (CTB, Hamburg 2002)

16.20 S. Sternberg: CT scans: A very high-dose diagnosis, USA Today **Nov. 20** (2000)

16.21 Siemens Medical Solutions: *Excellence in CT, SOMATON Definition, Computed Tomographie* (Siemens Medical Solutions, Forchheim 2005)

16.22 G. Weisser: Technische Grundlagen der EBCT. In: *MR-Angiographie und Elektronenstrahl-CT-Angiographie*, ed. by J. Gaa, K.-J. Lehmann, M. Georgi (Thieme, Stuttgart 2000) p. 145

16.23 N.M. De Clerck, D. van Dyck, A.A. Postnov: Noninvasive high-resolution μCT of the inner structure of living animals, Microsc. Anal. **1**, 13 (2003)

16.24 G. Wang, S. Zhao, P.-C. Cheng: Exact and approximate cone-beam x-ray microscopy. In: *Modern Microscopes I*, ed. by P.C. Cheng, P.P. Huang, J.L. Wu, G. Wang, H.G. Kim (World Scientific, Singapore 1998)

16.25 J. Ruhlmann, P. Oehr, H.-J. Biersack (Eds.): *PET in der Onkologie – Grundlagen und klinische Anwendung* (Springer, Berlin, Heidelberg 1998)

17. Ultrasound Diagnostics

Rudolf Götz, Frank Schön

Today, ultrasound diagnostics is an important imaging method in virtually all medical fields. The fact that it is quick, simple and in particular cost-efficient plays a major role in this. Further advantages are provided by the mobility and the broad spectrum of use of modern ultrasound diagnostic systems. Not least, these properties and also the absence of ionizing radiation make its use indispensable these days.

17.1 **Basic Physical Principles** 344
17.1.1 Principle 344
17.1.2 Generation of Sound Waves 344
17.1.3 Reflection 344
17.1.4 Spatial Mapping – Transit Time 345
17.1.5 Penetration Depth, Axial Resolution and Frequency Ranges 346
17.1.6 Influencing Factors: Pressure and Temperature 346
17.1.7 Second Harmonic 347
17.1.8 Broadband Harmonics 348

17.2 **Visualization of the Blood Flow and Vascular System** 348
17.2.1 Doppler 348
17.2.2 B-Mode 349
17.2.3 Ultrasound Contrast Medium 349

17.3 **Equipment Technology** 350
17.3.1 The Basic Design of an Ultrasound Unit 350
17.3.2 Doppler Ultrasonography 352
17.3.3 Types of Probes 356
17.3.4 Focusing 358

17.4 **Three-Dimensional Ultrasound (3-D, Real-Time 3-D)** 360
17.4.1 Acquisition Techniques 360
17.4.2 3-D Reconstruction 362
17.4.3 New and Additional Technologies 363

17.5 **Operation of an Ultrasound Unit** 367
17.5.1 General Conditions 367

According to data from the German Electrical and Electronic Manufacturers' Association (Zentralverband der Elektrotechnischen Industrie – ZVEI), the ultrasound market for new devices has a sales volume in Germany of approximately € 200 million per year. Approximately 50% of this volume is from acquisitions in clinics and hospitals, and approximately 50% is from doctors in private practices. The distribution of sales between the different specialist fields gives the following picture: approximately 30% gynaecology, approximately 30% internal medicine and approximately 20% cardiology.

The high density of devices used in everyday diagnostics is due to there being a large number of different technical and applicative methods (Table 17.1) and a number of different industrial suppliers. Ultrasound diagnostics has partly replaced or supplemented other methods, such as conventional x-ray diagnostics but

Table 17.1 Overview of the advantages and disadvantages of imaging methods

	Ultrasound	CT	MRI	X-ray	Angio	Nuclear medicine
Ionizing radiation	No	Yes	No	Yes	Yes	Yes
Real-time	Yes	No	No	No	Yes	(Yes)
Type of image	Slice	Slice	Slice	Section	Section	Section
Overall costs	Low	Very high	Very high	High	High	Very high

also computer tomography (CT) and magnetic resonace imaging (MRI).

In contrast to these methods, ultrasound is a so-called real-time method. The organs being investigated are displayed on a monitor in real time in this sectional imaging technique. They therefore correspond to a tomographic sectional image familiar from CT and MRI. Ultrasound imaging thus differs substantially from the method of looking through the body in a conventional x-ray examination.

17.1 Basic Physical Principles

17.1.1 Principle

The following explanations regarding the generation of ultrasound images follow theoretical considerations and are presented in an idealized form. They are based on physical principles which are intended to be presented to the reader in a way which is simple and easy to understand. The actual generation of ultrasound images is much more complex, however. An essential part of image generation is based on the phenomenon of scattering, which makes the process which actually takes place much more complicated, and this will only be referred to here in passing. It will not be discussed in further detail in the text below.

In ultrasound diagnostics, the transmitter and receiver are combined in the ultrasound probe. The probe is connected to the ultrasound unit via a cable, thereby enabling very free positioning of the ultrasound probe on the body and thus also virtually any desired examination plane and slice orientation.

In contrast to the x-ray method, which is a transmission method, ultrasound is a so-called reflection method. The ultrasound probe transmits a short ultrasound pulse which penetrates the body and is partially reflected at interfaces, e.g. between liver and kidney. Once the pulse has been emitted, the unit switches the ultrasound probe to receive mode. The ratio of transmission to reception time is approximately 1 : 1000.

The reflected components of the sound wave transmitted are then recorded by the ultrasound probe and fed to the unit for further processing. With the exception of the continuous wave method (Sect. 17.3.2), ultrasound units therefore basically work with what is known as the pulse method. This principle is also wide-spread in the natural world and is used, for example, by bats, dolphins and other animals for orientation.

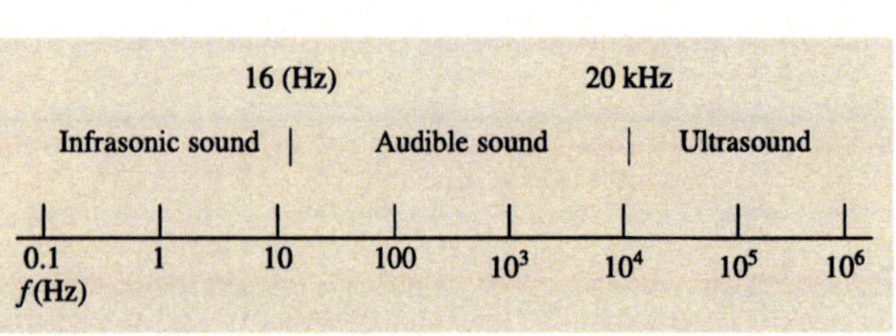

Fig. 17.1 Frequency ranges

17.1.2 Generation of Sound Waves

Special, cut piezoelectric crystals are used to generate sound waves. These synthetic crystals are manufactured industrially, for example from fractions of barium titanate, lead zirconate, lithium compounds or other ceramics. These are subject to what is known as the piezoelectric effect, which was described by the brothers Jacques and Pierre Curie using crystals of tourmaline. As soon as a voltage is applied to a crystal of this nature, it changes its form or geometry. Depending on the polarity of the voltage, dilatation or contraction of the piezoelectric crystal occurs.

If a high-frequency alternating current is therefore applied, then it oscillates at precisely that frequency and generates high-frequency sound waves. If the alternating current applied is at a correspondingly high frequency, this produces ultrasound waves which are no longer audible to humans (Fig. 17.1). In the receive mode, however, sound waves impinging on the crystal are converted into an electrical alternating current which is then processed further by the unit (Sect. 17.3).

17.1.3 Reflection

On the way through the tissue, components of the transmitted sound wave are reflected at interfaces between different organs and sections of organs. The energy of the waves which are reflected is determined in each case by the differences in the so-called wave impedance (Z) of the individual organ parts and sections scanned. There is more or less reflection at these multiple interfaces depending on the relationship between the individual wave impedances, on the angle of impact with respect to the interface and on the surface texture (scattering).

In order to contribute towards the generation of an image, the reflected wave must have sufficient energy as energy is also lost on the return path to the crystal. The ideal reflected component is approximately 1%, which corresponds to a reflection coefficient of $R = 0.01$. Higher reflection coefficients result in a greater loss of energy of the propagated wave overall, thus resulting in insufficient penetration of the ultrasonic beam into deeper tissue.

The reflection coefficient is by definition between 0 and 1, where $R = 0$ means that there is no reflection taking place and $R = 1$ corresponds to total reflection. The reflection coefficient R is calculated from the wave impedances of the two media which form the interface (Fig. 17.2). The difference between the two wave impedances is crucial here.

As can be seen from (Fig. 17.3), the wave impedance of air is vastly different from that of human tissue, with the result that there is a high reflection coefficient of nearly $R = 1$ and total reflection occurs. An interface with air is therefore an obstacle to ultrasound which cannot be overcome. In order to achieve the best possible transmission from the probe to the body without any loss of energy, ultrasound gel is used as a coupling medium. On account of its very high water content, the gel conducts sound well and should ensure air-free contact with a reflection coefficient of roughly $R = 0$.

However, it is not only air but also metal parts, bones and calcium particles which can present problems that produce virtually total reflection. This explains acoustic shadows behind bones, gall stones or endoprostheses, for example.

17.1.4 Spatial Mapping – Transit Time

The spatial mapping of the scanned tissue and reflecting interfaces is done by measuring the transit time from transmission of the pulse to receipt of the respective reflected components. Early echoes means that the waves have been reflected by interfaces close to the ultrasound probe. Late echoes means that the waves have been reflected by interfaces a long way from the ultrasound probe. This difference in time is represented on the monitor display by a corresponding distance between the pixels in the sound beam direction. However, this presupposes knowledge of the speed of sound conduction in the tissue, which is assumed to be on average 1530 m/s (Fig. 17.4). Since different tissues also have different wave impedances, this average is inevitably a generally accepted compromise and is used in standard commercial ultrasound units.

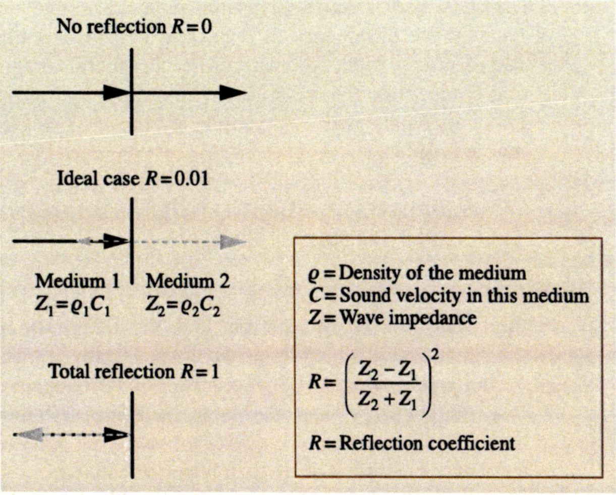

Fig. 17.2 Reflection coefficient and wave impedance

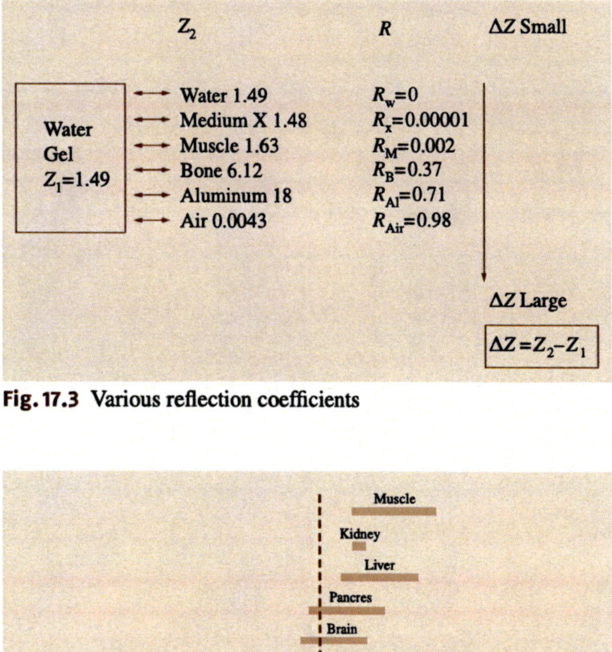

Fig. 17.3 Various reflection coefficients

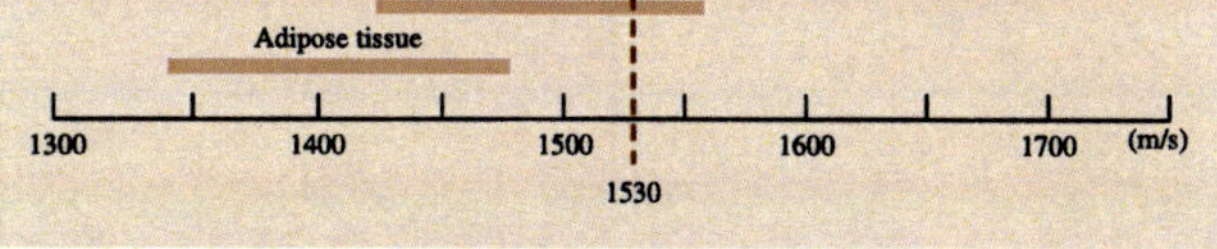

Fig. 17.4 Sound wave travel speed in tissue

Only very modern ultrasound units take into account these differences in the speed of travel of the sound waves when calculating the ultrasound images and compensate for the differences in the transit time, either manually with intervention by the user or else automatically.

17.1.5 Penetration Depth, Axial Resolution and Frequency Ranges

It is not only the number of reflecting interfaces and their reflection coefficients but also the frequency of the sound waves which influence the penetrative power of the sound waves. Higher transmission frequencies lose more energy over the same travelling distance than lower transmission frequencies and thus achieve a shorter penetration depth than lower frequencies.

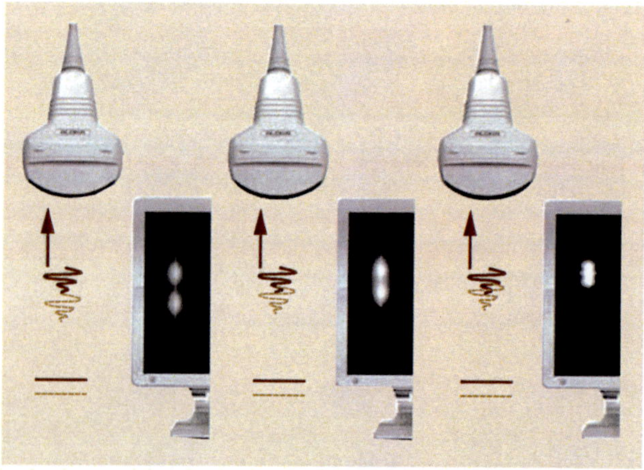

Fig. 17.5 Illustration of two reflectors at different distances

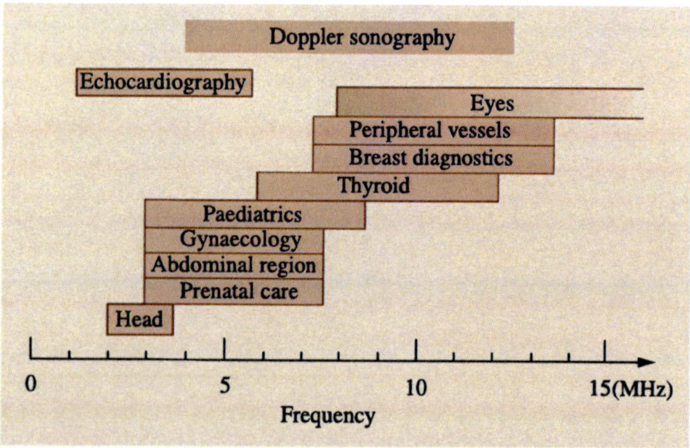

Fig. 17.6 Frequency ranges of diagnostic ultrasound

However, high frequencies have an advantage in terms of the axial resolution on account of their shorter wavelength (λ). In a theoretical best-case scenario, the minimum wavelength for separate imaging of two interfaces is a single wavelength. A frequency of 5 MHz, for example, gives a wavelength of 0.3 mm, whereas a frequency of 10 MHz corresponds to a wavelength of 0.15 mm. More individual interfaces can therefore theoretically be resolved at 10 than at 5 MHz (Fig. 17.5).

As can be seen in the figure, a pulse packet does merely not consist of a single wave with a positive and negative amplitude but also includes attack and, in particular, decay amplitudes. These lengthen the overall pulse packet by several oscillations. In modern high-end units on the market, a considerable degree of technical complexity is sought after in order to control the transient phenomena in the crystals. The aim is to generate the shortest possible pulse packets, ideally with only a single sine pulse. This can only be achieved through precise knowledge of the respective crystal characteristics of the ultrasound probe and with appropriate control using customized electrical pulse lengths and pulse forms. As a result of this *electrical behavior* of the crystals, pulse packet lengths are achieved which are nearly the size of a single wavelength. The theoretical resolution of 0.15 mm at 10 MHz can therefore approximately be achieved, as mentioned in the example above.

The axial resolution is thus proportional and the penetration depth is inversely proportional to the frequency. For practical applications of ultrasound, this provides a necessary compromise between the desire for high axial resolution and a good penetration depth.

Figure 17.6 gives an overview of the different frequencies used to produce images of different organs.

17.1.6 Influencing Factors: Pressure and Temperature

As already mentioned, the speeds at which sound waves pass through a medium are primarily dependent on the material properties of the medium. A change in the atmospheric conditions of temperature and pressure likewise influence the propagation velocity of the ultrasound pulse. Both an increase in temperature and a rise in pressure go hand in hand with faster propagation of sound waves.

In this context it is significant that the ultrasound pulse itself exerts positive and negative pressure in periodic alternation on the transmission medium. This means that the ultrasound pulse itself influences the sound-conducting properties of the medium with its pressure fluctuations. These effects on the medium and the resulting interactions continuously change the properties of the ultrasound pulse during the propagation through the tissue. This change has an effect on the signal quality and can be put to technical use.

17.1.7 Second Harmonic

During the course of time of a sinusoidal wavelength, the periodic pressure fluctuation triggered by the ultrasound pulse leads to a regularly changing sound propagation speed. The positive half-wave (pressure) moves with greater speed than the negative half-wave, with the result that the negative-going edge of the positive sound wave becomes increasingly steep as the distance from the sound source increases (Fig. 17.7).

A similar phenomenon can be observed in sea waves on the beach. The lower section of the wave, which is close to the sea bed, moves more slowly because it is braked by the ground which is getting closer all the time. The upper section of the wave is largely uninfluenced by this and thus moves more quickly. At some point, this effect becomes so great that the crest of the wave *falls* forwards, as it were, and the wave breaks.

The pulses from an ultrasound unit emitted into the body also experience this deformation on their path through the tissue. Whereas the sound wave still exhibits a virtually sinusoidal pressure curve when it is close to the crystal, as the distance from the crystal increases the shape of the wave increasingly resembles a saw-tooth. The higher the intensity of the sound emitted, the earlier and more pronounced this deformation. Every nonsinusoidal oscillation can be broken down into multiple sinusoidal oscillations of different frequencies by means of Fourier analysis (Fig. 17.8).

The saw-tooth-like shape of the changing pulse contains firstly a sinusoidal fundamental wave (a), furthermore a sinusoidal oscillation with a relatively low amplitude and half wavelength (b), and also other sine oscillations with constantly decreasing amplitudes which correspond to an integral quotient of the fundamental wavelength (c). These high-frequency components are also known as harmonic frequencies. This *harmony of sound* is familiar to us all from music, as the generation of sounds in musical instruments is subject to similar principles.

Conventional signal processing suppresses the higher-frequency signal components, so that the dominant fundamental of each reflected signal is primarily used for imaging. In the method known as *second harmonic imaging*, however, the first overtone (second harmonic) is isolated and used to obtain an image. The fundamental (first harmonic) and all other overtones are filtered out and are not used further.

Using the second harmonic frequency has a number of advantages. The main advantage is the significantly

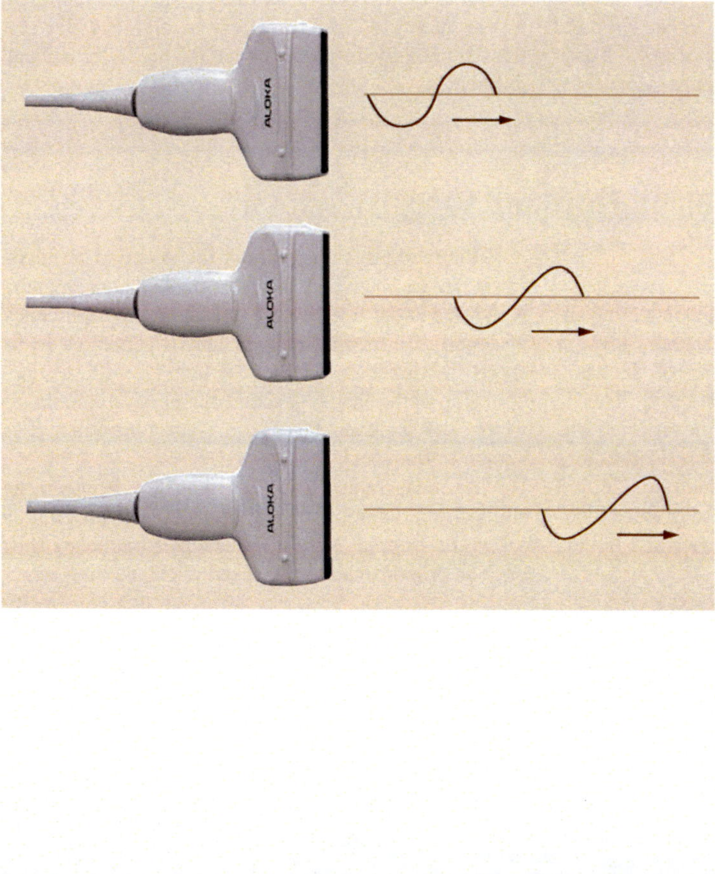

Fig. 17.7 Change in the shape of a sound wave as a result of nonlinear sound velocity

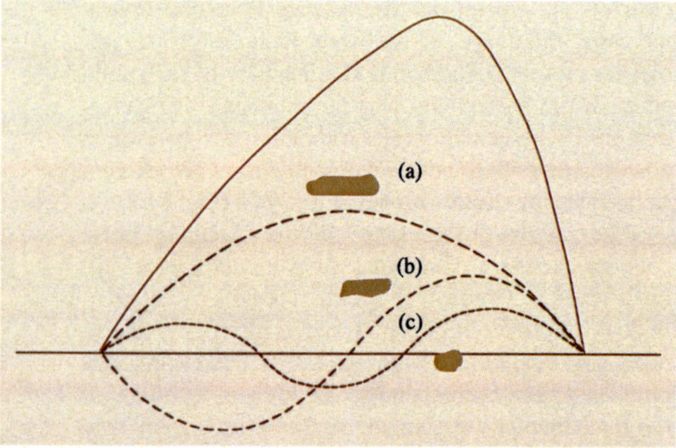

Fig. 17.8 Transformation of a nonsinusoidal oscillation into various sinusoidal oscillations

clearer differentiation between liquid and solid tissues. The reason for this is the more pronounced saw-tooth-shaped signal deformation in solid structures (tissue) and the lesser degree of deformation in liquid tissues, in which virtually no reflected harmonic waves are generated. Some companies therefore also refer to this technique as tissue harmonic imaging. Another advantage of using harmonics is that the amplitude of the interfering side lobes which occur automatically at the fundamental frequency is too low to generate overtones.

Tissue harmonic imaging is therefore a technique which reduces artifacts in the ultrasound image and makes it possible to more cleanly and clearly delimit liquid-filled cavities in particular, such as the amniotic cavity, the bladder, the cardiac cavities or cysts. These days it is also used in devices in virtually all price ranges.

17.1.8 Broadband Harmonics

One disadvantage of using the second harmonic is that the reflected second harmonic has less penetrative power as a result of the doubling in the frequency. Because these waves must cover the distance to the ultrasound probe, however, the depth of field of the ultrasound image which can be displayed is overall reduced. Use of this technology should therefore be left to the user, who differentiates depending on the organ being investigated by switching the unit on and off.

Some modern high-end devices provide another possibility for using harmonic signals, however. In addition to the harmonics already described, Fourier analysis also shows so-called half-waves (subharmonics) with half the fundamental frequency of the transmission pulse.

The simultaneous use of half-waves and second harmonics to calculate an image compensates for the disadvantage of the short depth of field because of the lower frequency of the half-wave and its better penetrative power. The reception and processing of these signals place high demands on the frequency spectrum of the ultrasound probe which can be effectively used and also on the signal processing electronics. In general, the ultrasound probe and the piezoelectric crystals used in it have optimized frequency ranges due to their material and design. The broadband capability of the ultrasound probe is indispensable for the simultaneous reception of half-waves and harmonics and is crucial to the quality of this reception. In high-end devices, this technology is found under the name broadband harmonics.

17.2 Visualization of the Blood Flow and Vascular System

17.2.1 Doppler

In addition to calculating the depth of a reflecting interface, the pulse reflection method generally used in ultrasound also provides the possibility of detecting a moving structure and measuring its speed. This is possible due to technical processing of the reflected signals, which differs from the method described before of determining the transit time and which, in addition to displaying the tissue morphology, also enables functional diagnosis of moving tissue volumes and liquids (blood, muscle, urine).

This method is named after the Austrian physicist Christian Doppler and is based on the phenomenon, which we know from everyday life, that the frequency of sound of a moving object changes according to its speed and direction. Everyone is familiar with the phenomenon of the changing sound of a passing ambulance which has its siren turned on. This phenomenon is known as frequency shifting.

The measurable change in frequency between the emitted and the reflected signal, which in ultrasound is called the Doppler shift or the Doppler frequency shift (f_D), is proportional to the speed of the moving struc-

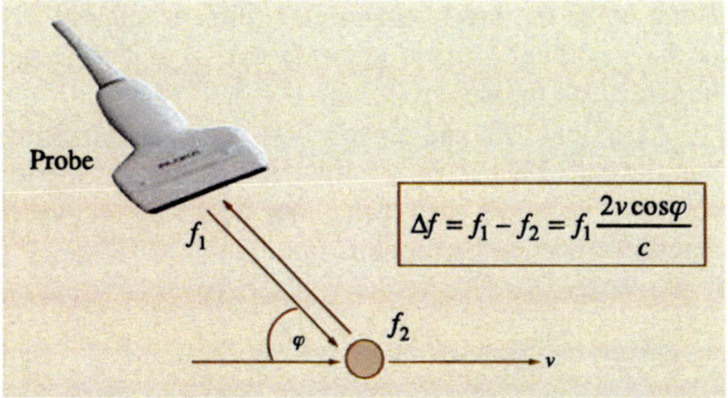

Fig. 17.9 Formula for describing the Doppler effect (f_1 transmission frequency, f_2 reception frequency, Δf measured Doppler shift, v speed of the target particle, φ angle of incidence between the sound beam direction and the direction of movement of the target particle, c sound velocity)

ture. The relationship between the Doppler shift and the speed of the target particle is illustrated in Fig. 17.9. The speed of the movement can be calculated by determining f_D in the ultrasound unit and with the aid of the known variables of sound velocity, transmission frequency and reception frequency.

A fundamental problem in the practical Doppler measurement is its reliance on angles. As can be seen in Fig. 17.9, there is a connection, dependent on the angle φ, between the moving structure and the position of the observer. Extreme cases are, firstly, direct measurement in or against the direction of movement ($\varphi = 0°$) and, secondly, measurement at right angles to the direction of movement ($\varphi = 90°$). In the first case, f_D corresponds to the actual speed, whereas in the second case the result given by the correction factor ($\varphi = 90°$, which gives cos $\varphi = 0$) does not give a measurable speed.

For practical use, it holds that an angle of between 0° and no more than 60° to the direction of movement should be used when using the Doppler technique.

17.2.2 B-Mode

In the B-mode, too, visualizations of the blood flow are also possible to a limited extent, since the reflection coefficient of blood is approximately 1000 times lower than that of tissue. On certain conditions (immediate proximity to the blood-conveying organ, slow-flow phenomenon, haematocrit, etc.), the blood flow is visualized spontaneously, which is known as spontaneous flow or spontaneous echo.

However, by subtracting two consecutively recorded image lines and subsequently superimposing them with one of the noncontrast lines, modern signal processing technology provides the possibility of imaging the movement of the reflecting structures which takes places between the two recordings. The advantage of this method (B-flow) is that it does not rely on angles. The spatial resolution of the illustration of the flow is dependent on the resolution of the B-image, however, and this can have a negative effect on the diagnosis. This method is therefore currently much less widespread than the Doppler method.

17.2.3 Ultrasound Contrast Medium

Depending on the medical indication and in order to achieve an accurate representation of the organ perfusion and the perfusion dynamics, ultrasound contrast medium is increasingly used in diagnosis. Ultrasound contrast medium is applied intravenously, passes largely unaffected through the lungs and is then pumped into the muscles and organs by the left heart in a highly diluted solution with the blood. Ultrasound contrast medium is composed of countless tiny gas bubbles with a diameter of between 2 and 4 μm and reflects ultrasound which is beamed in significantly better than the natural corpuscular constituents of the blood. There is also a second effect, which increases the visualization of the gas bubbles and therefore of the blood: the gas bubbles are considerably smaller than the spatial dimension of the sound waves.

In the compression wave of an ultrasound pulse, areas of elevated and reduced pressure periodically surround each bubble completely. In the area of elevated pressure, bubbles are compressed. Conversely, bubbles in the low-pressure area dilate. On account of these rhythmic changes in volume, the bubbles become a source of sound as they begin to oscillate. However, the acoustic pressure of the initial sound field and the change in diameter are not in a linear relationship with one another. If the pressure doubles, for example, the bubbles do not reduce in size to the same degree. As a result of this nonlinear oscillation and the slight difference in the size of the individual bubbles, they emit a broadband frequency spectrum back to the ultrasound probe. Just like the wave breaking on the beach, this broadband, nonsinusoidal signal can be understood as the sum of multiple sinusoidal single frequencies (Fig. 17.8). The second harmonic (Fig. 17.8) is usually particularly pronounced here, meaning that it can be selected in the receiving component of the ultrasound unit and used for the imaging. This method is known by the term contrast harmonic.

The so-called mechanical index (Mi) measures the mechanical effects of the sound waves on tissue and contrast medium bubbles

$$\mathrm{Mi} = \frac{p^-}{\sqrt{f}} .$$

Here, Mi is the mechanical index (dimensionless), p^- is the negative acoustic pressure (MPa) and f is the transmission frequency (MHz).

At a low Mi and thus a low acoustic pressure, bubbles still behave in a linear fashion, and at higher acoustic pressures their behaviour becomes nonlinear and they generate harmonics.

17.3 Equipment Technology

17.3.1 The Basic Design of an Ultrasound Unit

The fundamental components of an ultrasound unit are the devices shown in the diagram (Fig. 17.10).

The combination of these components constitutes a medical device which is approved in accordance with the respective applicable regulations, which satisfy EU and international agreements. Any change to these components or their parts (e.g. repair with a structural change, such as recoating of an ultrasound probe) means that this approval is automatically rescinded.

Ultrasound Probe

Specific ultrasound probe designs are available for different fields of use. They differ essentially in the design of the crystal array. The standard designs described in the section below can be found in different sizes, widths and radii of curvature, and in a wide variety of other ultrasound probe designs. The shape of the housing differs depending on the intended use of the respective *instrument*.

Base Unit with A/D Converter

The base unit of a modern ultrasound unit is essentially composed of two parts. The interface between

Table 17.2 Types of probes and their application

Organ	Application	Type of probe
Brain	Transcranial	Sector
Brain	Intraoperative/burr hole	Convex/sector
Pituitary	Transnasal	Linear, flexible
Thyroid, neck vessels	Transcutaneous	Linear
Heart, aorta	Transthoracic/transesophageal	Sector
Lung, mediastinum	Endobronchial ultrasonography (EBUS)	Convex
Pleura	Transcutaneous	Convex/sector
Breast	Transcutaneous	Linear
Spine	Intraoperative	Linear/convex
Liver, gall bladder, bile ducts, kidney	Transabdominal/endosonographic (EUS)	Convex
Pancreas, spleen, stomach, intestines	Transabdominal/endoscopic	Convex
Liver, spleen, pancreas	Laparoscopic	Linear/sector
Major abdominal vessels	Transabdominal	Convex
Uterus, ovaries, pregnancy	Transabdominal/transvaginal	Convex/RT3D
Bladder, prostate	Transabdominal/transrectal	Convex/linear
Scrotum	Transcutaneous	Linear
Hip	Transcutaneous	Linear
Peripheral vessels	Transcutaneous	Linear
Joints, muscles, skin	Transcutaneous	Linear

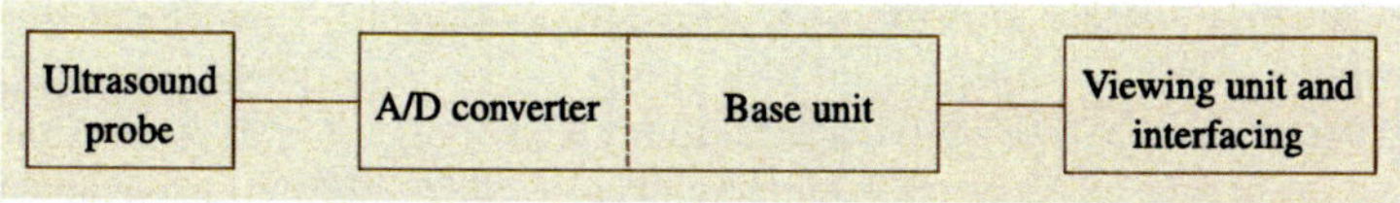

Fig. 17.10 Components of an ultrasound unit

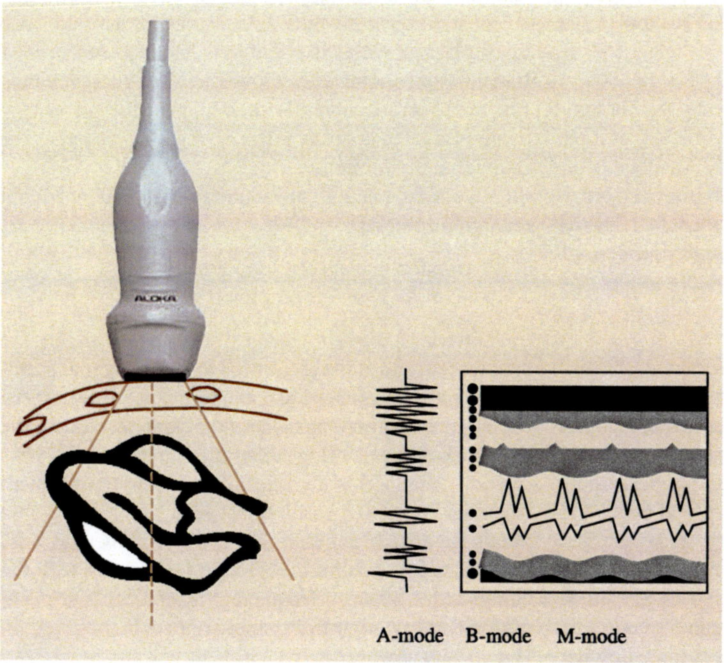

Fig. 17.11 Diagram of A-mode, B-mode, and M-mode

the analogue-functioning ultrasound probe and the digital processor is in the analogue/digital (A/D) converter. The purpose of the A/D converter is firstly to generate all of the electrical pulses and pulse sequences required for steering the sound beam and also to receive the reflected signals and convert them to digital form. It thus becomes clear that the image quality in particular is heavily dependent on the quality of the A/D converter.

The base unit also has the task of scan conversion (spatial mapping of the scan lines recorded) and further processing the digitized signals by means of image processing. Other functions include the operator-controlled computer dialogue, provision of measuring and calculation programs, as well as archive functions and generation of signals for viewing units (monitors) and interfaces.

Monitor and Interfaces

Today, digital flat-screen monitors are predominantly used. Only a few ultrasound units still work with conventional video signals. The interfaces for data transmission, printer control and archiving which are normally used today are also digital.

A-Mode

Historically, the A-mode method (amplitude mode) was the first ultrasound method used. It has been almost exclusively replaced in the medical field by the methods described below, as the diagnostic value of the A-mode is very limited. Measurements of depths and distances can only be carried out at a single point.

Along the sound propagation line, signals reflected through media boundaries are displayed as individual peaks on the depth scale depending on the transit time – the pulse reflection method (Fig. 17.11). The value of the amplitude is dependent on the ratio of the wave impedances between the media boundaries.

B-Mode

In the B-mode (brightness mode), the display of the A-mode signal is modified such that the reflected signal is no longer displayed as a peak but rather as an individual, depth-dependent (transit time) pixel. The value of the amplitude is represented by the brightness of this pixel, so that a line is generated with points of varying brightness. The simple B-mode method cannot be used diagnostically. However, it was the basis for the development of the M-mode and the 2-D B-mode.

M-Mode

The *M-mode (motion mode)* is produced by the horizontal deflection (time axis) of a B-mode line and storage and display of the resulting images. The M-mode is intended to be used for the diagnosis of moving organ parts, such as cardiac valves or cardiac muscle. It stands out in this area in particular due to its high time resolution, as just one individual line is repeatedly scanned.

2-D B-Mode

The 2-D B-mode method (two-dimensional brightness mode) is these days the most important sectional imaging technique in ultrasound diagnostics and is generally referred to as B-mode. The image is produced by quickly stringing together a number of individual B-mode lines (scanning lines) horizontally to give a flat 2-D image. Here, the image geometry is determined by the relative arrangement of the individual B-mode lines. It is also dependent on the design of the ultrasound probe. As in the traditional simple B-mode, the brightness of the individual pixels is determined by the amplitude of the reflected signals. Today, at least 256 shades of grey are required as standard.

Digitally Encoded Ultrasound

The practical applications of ultrasound require a necessary compromise between the desire for high axial resolution and a good penetration depth. Digital encoding of ultrasound pulses significantly improves this compromise, with the benefit of better penetration.

During transmission, a typical digital encoding is superimposed on the ultrasound pulse as an identification pattern. This encoding can also be detected in the ultrasound pulse after it has been reflected at the different interfaces in the patient's body, with the result that, when processing the reflected signals in the ultrasound unit, the encoded pulses can be separated from unencoded, unwanted signals (e.g. noise, artifacts). This means that only those signals which include the relevant individual code are processed further.

An example from the natural world shows that the same principle is used by bats. The individual encoding of the transmission pulses emitted by a bat allows it to identify its own pulses from a wide number of pulses from other bats. Every individual bat is therefore able, despite a large number of other pulses, to move about safely in space with its own signals.

In addition to the encoding, signal compression also takes place, making it possible to produce an encoded transmission signal with a lower level of energy, which can then form a signal with a higher level of energy again at a later point as a result of adding together the individual digital pulses.

Two fundamental advantages of digital encoding emerge from this:

1. Unwanted signals are suppressed.
2. The useful signal is amplified.

Unwanted signals such as noise and artifacts are suppressed, and at the same time the useful signal is amplified and raised above the ambient noise level. At the same frequency and thus also at the same resolution, the effects mentioned bring about an increase in the penetration depth. This significantly reduces the problem of achieving a high resolution whilst at the same time obtaining a high penetration depth. This method does not put us in a position to be able to do away with the compromise between frequency (resolution) and penetration depth. A significant improvement in the situation is possible, however.

17.3.2 Doppler Ultrasonography

As outlined in the introduction to the fundamental principles, Doppler ultrasonography is used to detect blood flow. Different methods can be used technically:

- Pulsed-wave (PW) Doppler
- High-PRF Doppler
- Continuous-wave (CW) Doppler
- Color Doppler.

With Doppler ultrasonography, a distinction is also drawn on the basis of whether a device operates solely in the Doppler mode or whether it operates in the so-called duplex mode by superimposing a B-mode. In *duplex mode*, the scanning point (*sample volume*) at which the velocity of blood flow is measured is displayed in B-mode. The anatomical mapping at the point of the velocity measurement is therefore defined precisely. A Doppler device for use in duplex mode must also include all the technology for generating a B-mode.

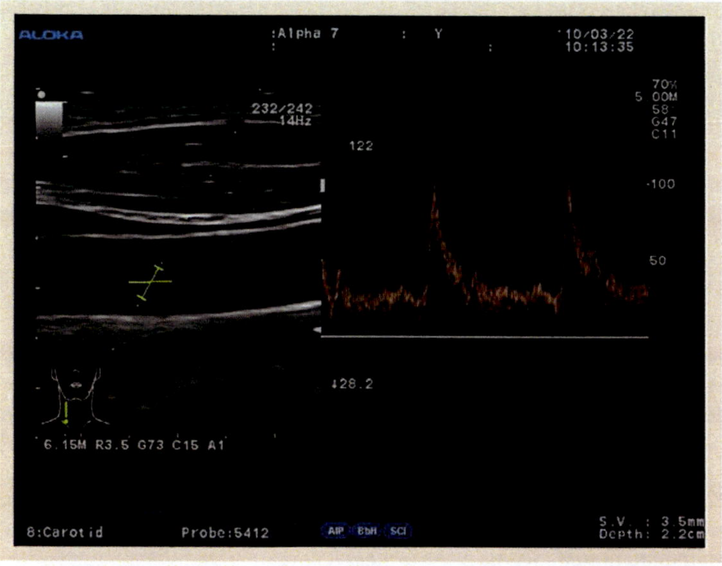

Fig. 17.12 Duplex Doppler, B-mode with sample volume (*left*), and PW spectrum (*right*)

PW Doppler

The pulsed-wave (PW) Doppler technique uses the pulse reflection method and also the A-, M- and B-mode imaging methods. The processing of the transit time information is used here to determine the sample volume position. Reflected signals from the sample volume are not used to display an image but rather to measure the velocity according to the Doppler principle.

Similar to the M-mode, the time profile of the velocity of blood flow is displayed on the monitor. The position of the sample volume can be changed and positioned accurately by sight in duplex mode (Fig. 17.12).

The maximum velocity of blood flow which can be measured using PW is limited by the repetition frequency of the pulses (PRF). By using other, additional sample volumes, this limitation – which results from what is known as aliasing – is shifted into a velocity range which is no longer diagnostically relevant.

High-PRF Doppler

The PW Doppler technique is limited by what is known as *aliasing*. Above a certain frequency and velocity of blood flow, this causes incorrect display (alias display) of the Doppler signal, which can lead to incorrect interpretation of the flow direction.

This effect is dependent on the frequency of the pulse (scanning frequency or pulse repetition frequency (PRF)), which should be as high as possible. An artifact such as this can be observed in Western films, for example, when the carriage wheels suddenly appear to rotate backwards because the scanning frequency (television frame frequency) is too low.

The high-PRF Doppler technique which should be found in all modern duplex systems was developed to avoid this effect of aliasing while maintaining the depth selectivity. It allows the aliasing cut-off frequency to be shifted upwards by increasing the PRF. The price paid for this is multiple sample volumes, which results in a slight restriction in the depth selectivity.

The use of duplex systems with automatic switching from PW to high-PRF is a useful addition to the pure PW mode.

Today, duplex devices should include frequency analysis in order to make it possible to detect pathological flows in the characteristics of the Doppler spectrum displayed. The Doppler signal is broken down and displayed in its individual frequency components by Fourier transform or similar methods (Fig. 17.8).

When the angle of incidence is known, frequency components can be equated with velocity components. Further information about the velocity is thus obtained. The most important here is the information about the degree of turbulence, which gives insights into pathological flow.

CW Doppler

The CW Doppler (continuous wave) technique is the only ultrasound method which is not a pulsed method. Two physically separate crystals, a transmission crystal and a reception crystal, are accommodated in a probe and operate simultaneously. Whereas one crystal constantly transmits, the second crystal continuously receives the reflected signals. Alternatively, it is also possible to use groups of crystals, e.g. in a phased array ultrasound probe.

Due to the continuous transmission, depth mapping is no longer possible, however, as is it is not possible to measure pulse transit times. The velocities measured are rather divided along the entire path of the measuring beam. The resulting advantage is the ease with which it is possible to measure high velocities of blood flow, such as those which are to be found in high-grade stenoses. The disadvantage of this method compared with the PW Doppler technique is the lack of depth selectivity.

Color Doppler

The industrial development of the first color Doppler in Japan (ALOKA, 1985, SSD-880) was an important step in the development of ultrasound diagnostics.

The color Doppler is in principle a PW Doppler. Here, it is not just a single sample volume which is used but several hundred. These sample volumes are combined in lines to form an extensive two-dimensional Doppler image. The Doppler image is superimposed in the correct position on the black-and-white B-mode image. These two images can be viewed together in real time in the real-time mode.

In order to be able to distinguish the Doppler information from the morphological black-and-white information (B-mode), it is displayed with color-coding (Fig. 17.13).

A generally accepted illustration of the direction of flow is given with the colors red (towards the ultrasound probe) and blue (away from the ultrasound probe). However, it must be mentioned that there is no additional standardized color coding system. In the individual color Doppler systems, too, various color coding systems (color charts) can be accessed which more or less follow subjective preferences.

The effect of aliasing mentioned above occurs as an artifact in the color Doppler technique, too. This has the effect that, in the middle of a flow which is color-coded blue, for example, a red spot appears which is completely surrounded by blue (Fig. 17.14). This can be seen when the maximum velocity which can be displayed is exceeded at precisely this point. The color then *turns* into the other color.

As it is obviously impossible for a *red island of reverse flow* to suddenly appear in the middle of a blue flow, the user can easily identify this phenomenon as aliasing.

The velocity (cut-off frequency) above which aliasing occurs is called the Nyquist limit

$$\text{Nyquist limit} = f_{\text{Ny}} = \frac{\text{PRF}}{2} .$$

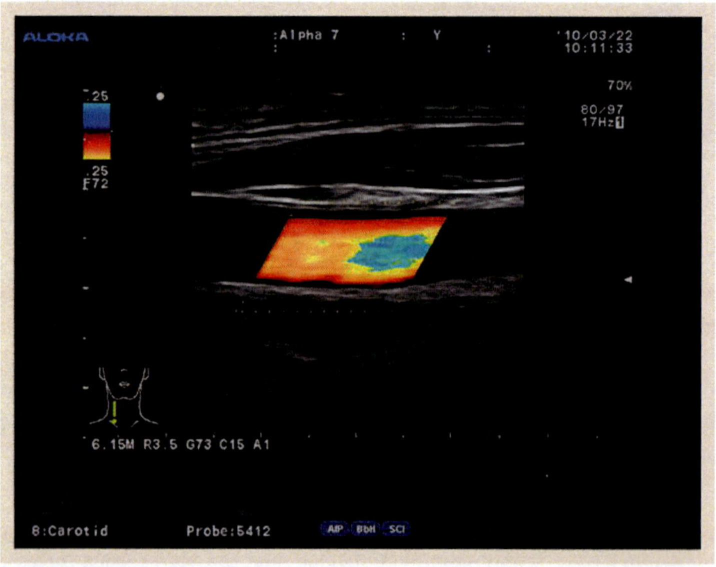

Fig. 17.13 Color Doppler with aliasing

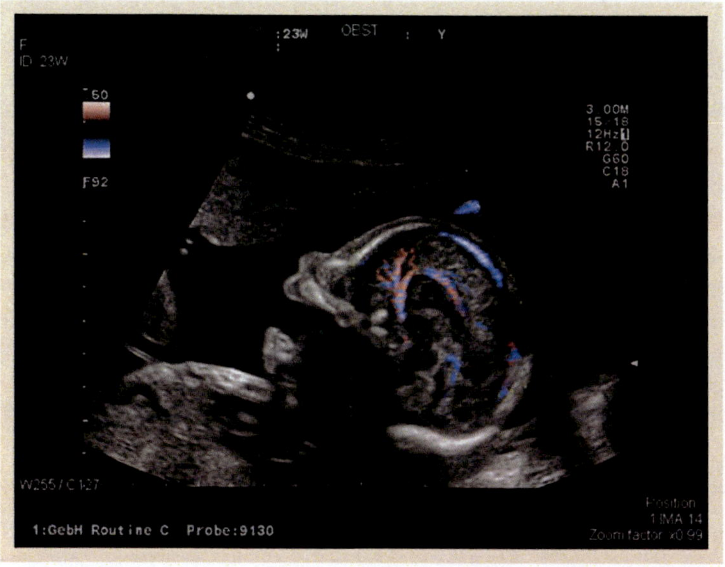

Fig. 17.14 Vascularization in the fetal brain, imaged using eFlow technology

Aliasing can even be useful here in that it shows us the point in the blood flow with the highest velocity. In order to measure the highest velocity, we position the sample volume at this alias position. Using the color Doppler technique it is then possible to detect and pinpoint the extent of pathological flows such as jets and to locate small vessels in real-time mode.

If the color Doppler technique is also added to the duplex mode so that the B-mode, color Doppler and Doppler spectrum are all displayed, this is referred to as the triplex mode.

Power Doppler

The term *power Doppler* is understood as meaning another version of the color Doppler technique. It is also known by the following synonyms:

- Amplitude Doppler
- Color angio
- Power color, etc.

The power Doppler technique also uses the Doppler shift to image blood flow. However, it does not display the velocity of the flow using color-coding, but merely shows the amplitude of a point of movement in the received signal detected by the Doppler shift present. The amplitude represents the number of reflectors (mainly erythrocytes) and is not displayed in the normal color Doppler technique. The power Doppler is generally more sensitive and can more easily detect slow-moving blood particles. It therefore fills the vascular lumen better than the normal color Doppler is able to, as the low amplitudes of the available signal information are not sufficient for further analysis with the normal color Doppler.

High-Resolution Modern Color Doppler Techniques

At this point it is worth mentioning a further development of the power Doppler technique which is known by various different names (e.g. eFLOW, dynamic flow, etc.). In some modern units, the method can be found with markedly higher spatial resolution values (<0.3 mm). It is extremely well suited for accurate, two-dimensional illustration of the vascular lumina through which blood is actually flowing. This method is of advantage particularly in the case of slow velocities or small vessels and when imaging a complex organ vascularization, but also in the case of stenoses.

It is made possible using modern pulse generators (beamformers) which shorten and optimize transient phenomena in transmission pulses. This makes extremely short pulses possible, which enable a higher spatial resolution.

Tissue Doppler Imaging

A special Doppler technique has become established for imaging the heart, and this technique does not color-code the blood flow but rather the cardiac tissue, which is to say the cardiac wall (myocardium). This is referred to as *tissue Doppler imaging* (TDI). Since the move-

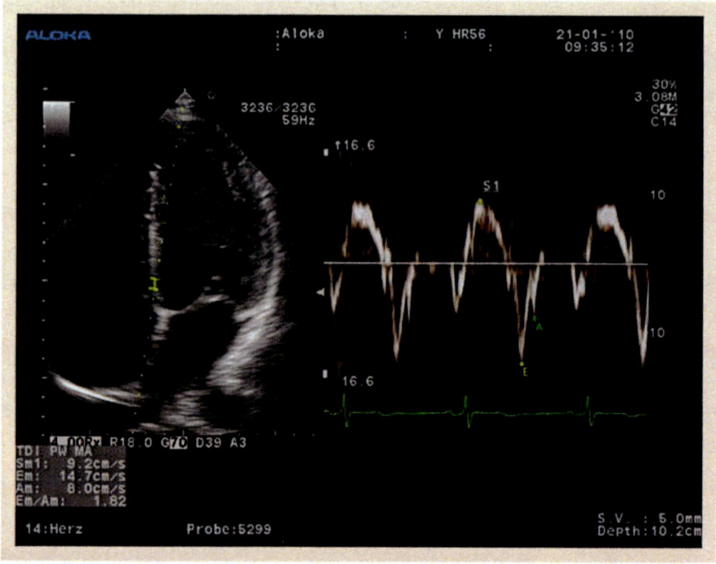

Fig. 17.15 Tissue Doppler imaging (TDI)

ment in the myocardium is considerably slower than that of the blood, low-pass filters are necessary here instead of the high-pass filters which are customary in the blood flow Doppler. This technology enables conclusions to be drawn regarding the velocity of individual segments of myocardium (Fig. 17.15) and thus allows specific diagnosis in the case of disorders concerning the movement of the wall, e.g. as a result of a myocardial infarction or diastolic dysfunction. Additional information can be found in the further reading.

17.3.3 Types of Probes

The types of probes used today are primarily divided into two groups. Mechanical probes and electronic probes, or combinations of the two, are almost exclusively used.

Mechanical Probes

Today, these are still used for special applications (e.g. radial technique, skin diagnostics). Here, the deflection

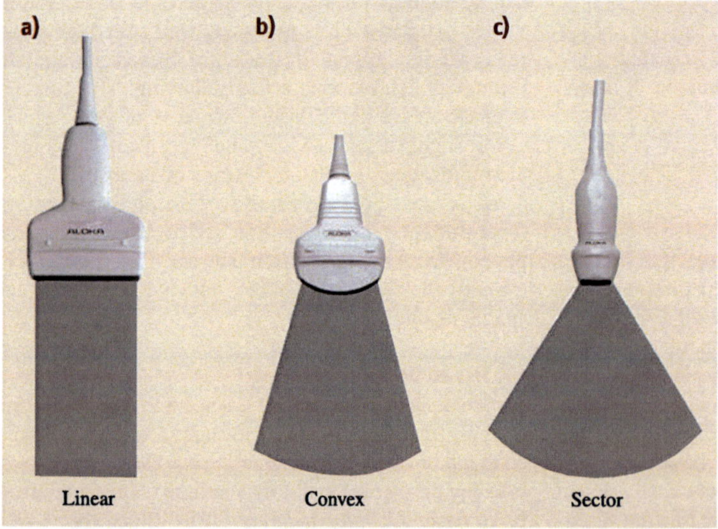

Fig. 17.16a–c Sound field geometry of various types of probes

for generating a two-dimensional image is carried out by a small electric motor in the ultrasound probe.

Electronic Probes

Electronic probes are used far more frequently, and we know these as linear, convex and sector scanners. They differ in principle due to their deflection geometry and their deflection method (Fig. 17.16).

The deflection of the ultrasonic beam is realized electronically by time-shifted activation of a large number of crystals; no mechanical wearing parts are used whatsoever. The maximum transmission frequency is usually approximately 15 MHz.

Linear Probes. The simplest form of an electronic probe is the linear scanner. It consists of a large number of crystals arranged next to one another, and depending on the size of the probe it can contain up to approximately 400 crystals (Fig. 17.17).

A scanning line is always constructed using several crystals, and in this case six crystals lying next to one another are used, for example. This group of active crystals is also called an array, which has given us the frequently used term linear array. Each crystal is activated by an electronic pulse and transmits a sound wave which merges with the others to form a wave front. As a result of various physical effects, a sound field with a focal constriction is generated in the typical sound field geometry illustrated (solid line). Another six crystals are then activated, forming the next (dashed) sound field. According to the same principle, the next (dotted) sound field is then generated in turn, and this is repeated in the direction of the arrow until a complete image can thus be constructed from the reflected pulses. In order to generate a real-time display with 20 images per second (20 Hz), all of the processes described must have taken place within approximately 1/20 s.

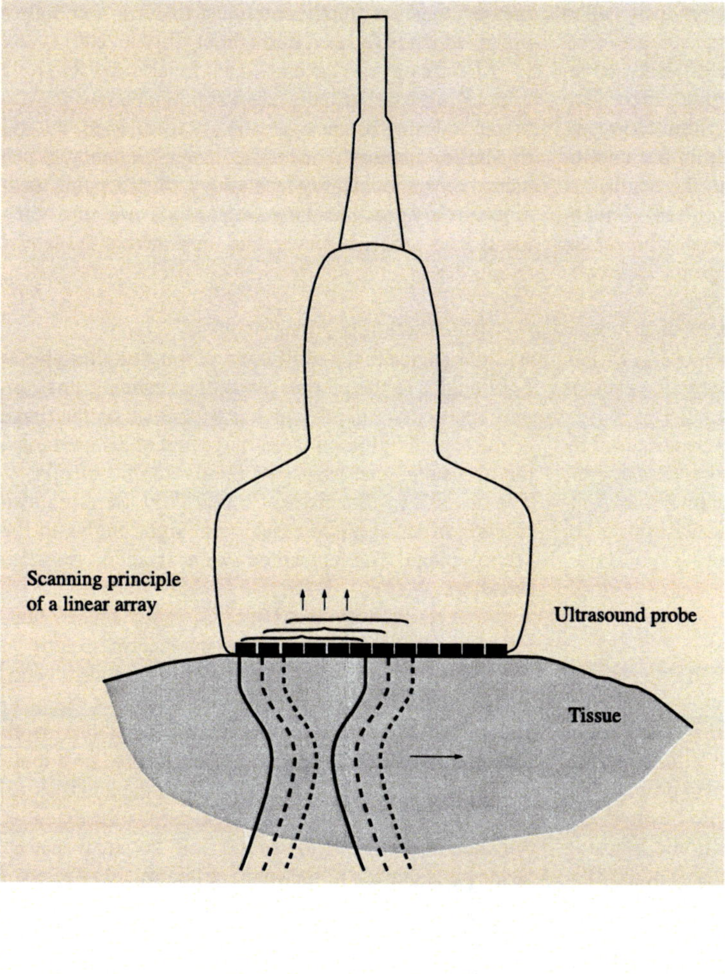

Fig. 17.17 Scanning principle of a linear probe

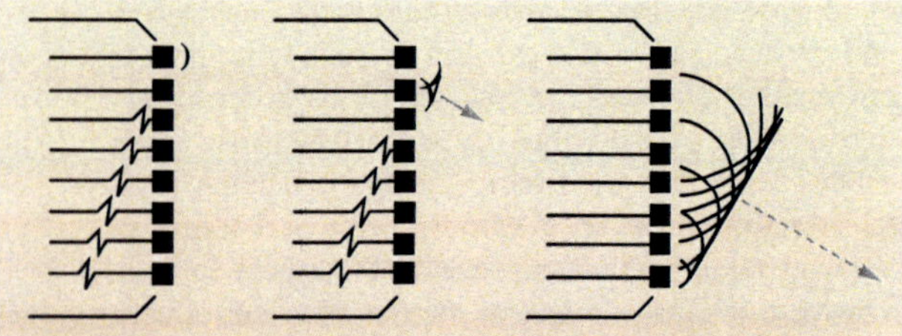

Fig. 17.18 Phased array principle

Convex Probes. In the *convex probe*, also known as a curved array, the same method is used. The only difference here is that the crystals are not arranged on a linear contact surface but on a convex contact surface. Different radii of curvature are available for different fields of use (Fig. 17.16b).

Sector Probes. In the electronic sector, also called a phased array, considerably fewer crystals are used than in the linear or convex array. The electronic activation is carried out with a time shift or phase shift, which is where the term phased array comes from. However, this is no longer done from the outside in but continuously from one side to the other, with the result that an oblique wave front is generated (Fig. 17.18). Changing the activation at an appropriate speed achieves an oscillation in the beam, which in turn produces a two-dimensional image in the form of a sector.

The advantage of sector probes can be clearly seen from the example of avoiding rib shadows (Fig. 17.19), as the small contact surface of the sector probe enables examination through the intercostal spaces.

A disadvantage of the sector technique is caused by the divergence of the ultrasonic beams at greater depths, which results in the lateral resolution becoming increasingly poor in deeper tissues. The gaps which appear are filled in by the device using interpolations.

17.3.4 Focusing

The focal constriction, mentioned above, is important in all electronic probes, and this is known as the focal area. The narrower the sound field in this focal area, the better the lateral resolution of the system, that is to say in the lateral direction, perpendicular to the sound propagation direction. The diameter of the ultrasonic beam in the focal area depends on the different electronic focusing techniques of the device. Resolution values cited by the manufacturers usually relate to the focal area. The focal position is also crucial, however. Ideally, the focal position should be precisely where the organ to be examined is located. This is where the biggest advantage of electronic probes over mechanical probes is to be found: the possibility of shifting the focal position electronically.

Dynamic Focusing

In dynamic focusing, the crystals are no longer all activated simultaneously; instead, the outer crystals are activated first, followed by the central crystals. The position of the focus is determined by the time-shifted activation between the outer and inner crystals (delay line). In addition, in electronic probes it is possible to use multiple focal zones simultaneously, which makes it possible to produce a sharp image over virtually the entire depth of field, although this comes at the expense of the image frequency. This is called dynamic focusing (Fig. 17.20). The image regions of the individual focal zones are recorded one after the other using a buffer memory and are finally combined to form a complete image which is focused throughout (Fig. 17.20b).

Modern devices have up to ten focal zones, but only four of these are usually used simultaneously for dynamic focusing because, as already mentioned, the use of a buffer memory reduces the image frequency. When using conventional array technology, this focusing technique is only possible in the longitudinal direction of the ultrasound probe, however, and only affects the lateral resolution.

Slice-Thickness Focusing

At right angles to the ultrasound probe, and thus also at right angles to the section plane, the scanning lines are only focused at a depth which is dependent on the shape of the crystals. Both at close range and at distant range, the ultrasound can propagate relatively uncontrollably. The actual two-dimensional sound field therefore also extends in an unwanted fashion at right angles to the section plane. This unwanted extent is called the slice thickness.

As a result, very small vessels, cystic hollow bodies or low-echo lesions are not intersected exactly in the middle, and every scanning beam also records echogenic structures which are behind or in front of the small structures. Other echo signals from exactly this depth therefore impinge on the ultrasound probe, and these cannot unambiguously be attributed to the small structure (Fig. 17.21a). Only in the focal area is the extent of the slice thickness of the sound field so small that no tissue parts outside of the small structure are affected.

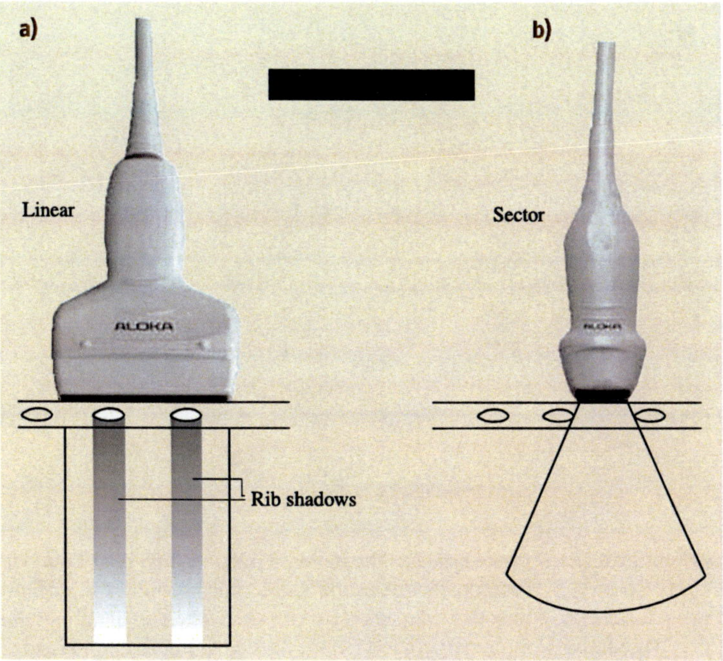

Fig. 17.19a,b Comparison of sector and linear probes

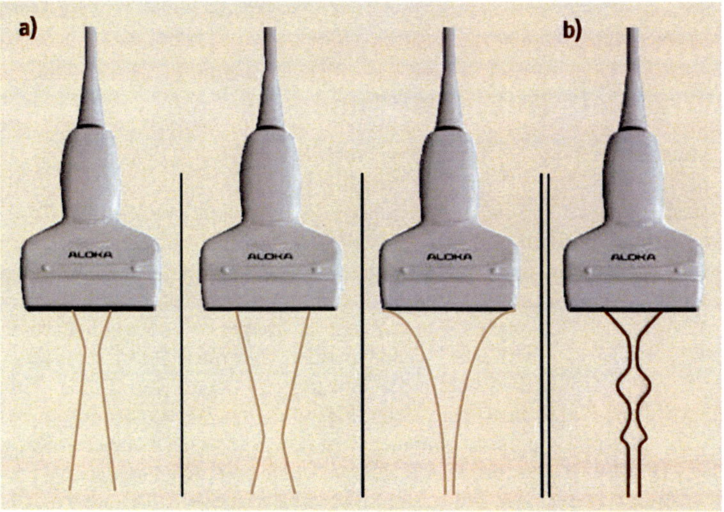

Fig. 17.20a,b Principle of dynamic focusing

The matrix array technique is a new ultrasound probe technique which also provides electronic focusing at right angles to the section plane and thus provides a reduced and uniform slice thickness from close range to about 25 cm.

Conventional ultrasound probes consist of a single row of piezoelectric crystals lying next to one another and therefore also only permit focusing in this single lateral plane. The matrix array ultrasound probe has a two-dimensional arrangement of up to approximately 1000 tiny square crystals which can be activated selectively. Electronic focusing as described by time-shifted activation of the individual elements is therefore not only possible in the lateral plane but also perpendicular to it (elevation plane). This technique largely eliminates slice thickness artifacts (Fig. 17.21b) and also enables

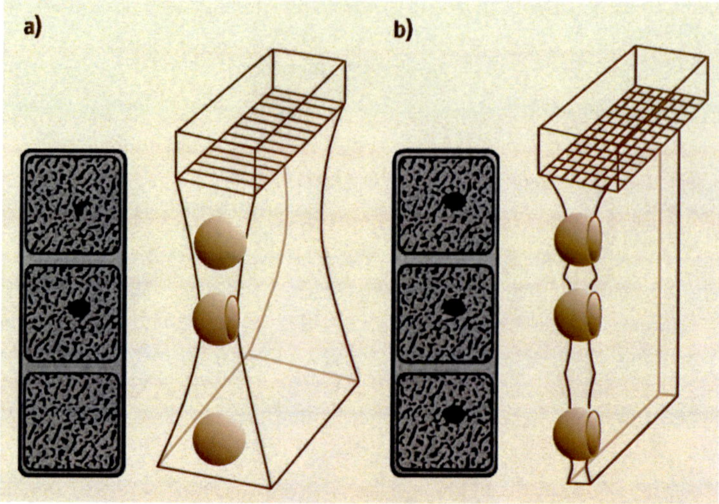

Fig. 17.21 (a) Conventional technique versus (b) matrix array

very small structures to be visualized without artifacts and with a high resolution.

Another positive effect of the matrix array technique is the considerable improvement in the penetrative power on account of the reduction in beam divergence and thus the increase in the energy density which accompanies it at the same time. It allows higher ultrasound frequencies to be used while maintaining the same penetration depth, thus enabling an improvement in the frequency-dependent axial resolution.

17.4 Three-Dimensional Ultrasound (3-D, Real-Time 3-D)

The development of three-dimensional ultrasound technology took into account the needs of many users for three-dimensional imaging of anatomical structures. The objective was among other things to simplify and improve the diagnosis of structures which can often only be assessed with difficulty in 2-D images.

17.4.1 Acquisition Techniques

Modern ultrasound diagnostics can also record and display three-dimensional blocks of data. To this end, a third dimension perpendicular to the 2-D plane is added to the two-dimensional (B-mode) ultrasound images and processed as a 3-D data set.

In everyday clinical practice, 3-D technology was only able to establish itself with the development of efficient computer processors, which enabled a true real-time 3-D display. When 3-D technology emerged in the 1980s, computing times were still in the region of hours. As a result of precisely this real-time capability of the systems, 3-D technology has recently also become known and marketed as 4-D ultrasound.

Hands-Free Technology Without Position Sensors

This simple technology uses the conventional 2-D ultrasound probe, which is moved or panned as uniformly as possible by the examiner parallel to the 2-D plane or around a fixed point in order to record an image sequence. The number of images to be recorded which will later be used to reconstruct the 3-D data set is specified before the procedure is begun (Fig. 17.22). This hands-free technology does not allow accurate three-dimensional mapping, however, as the distance actually recorded is not known and there is no guarantee of a precise and straight recording movement. This also means that no measurements can be made and volumes cannot be determined.

Hands-Free Technology with Position Sensors

This hands-free technology is a development of the technology without position sensors and is used to avoid the disadvantage of undefined recording. By attaching a position sensor to the 2-D ultrasound probe, constant

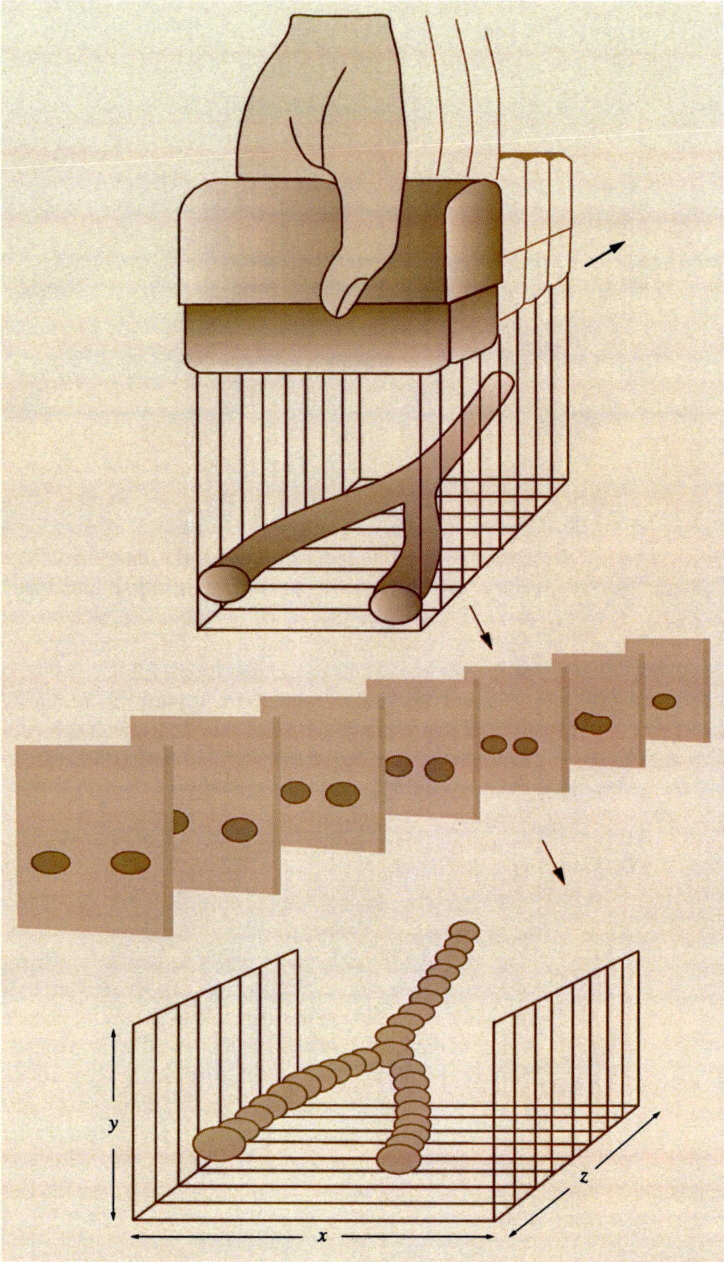

Fig. 17.22 Data block from multiple parallel sectional images using 3-D hands-free technology

determination of the position of the ultrasound probe and thus spatial mapping of the individual 2-D images becomes possible. Data sets recorded in this way allow measurements to be made and volumes to be determined.

The position detection can be achieved using different techniques and position sensors. In addition to sensors in the electromagnetic field, infrared, ultrasound and acceleration sensors (gyro sensors) are also used. These forms of sensor technology all have their own

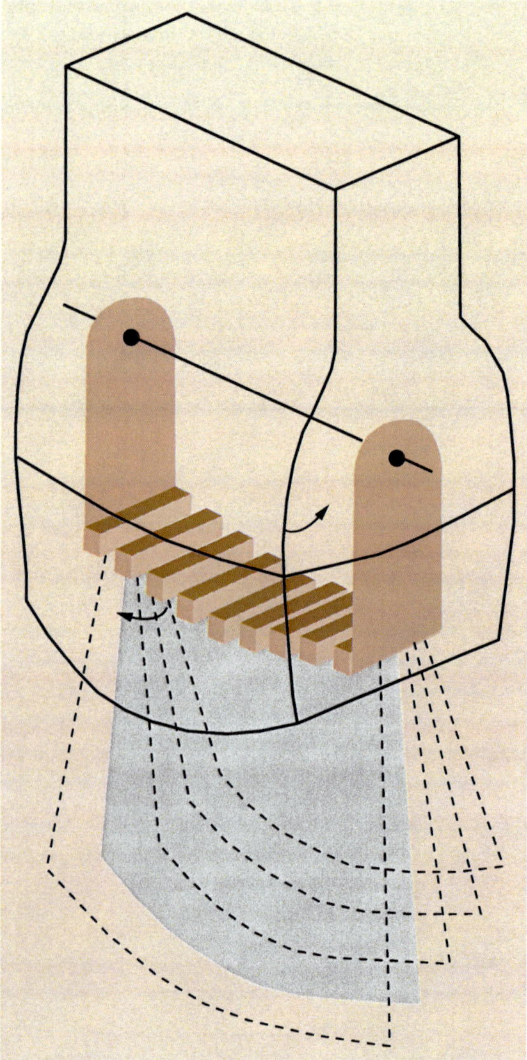

Fig. 17.23 Integrated 3-D/4-D ultrasound probe with mechanical deflection

methodological advantages and disadvantages. The disadvantages predominantly have an adverse effect on the spatial resolution.

Integrated 3-D Ultrasound Probe with Motorized Deflection

In order to overcome the problems described with hands-free technology and ensure simple handling, motor-driven 2-D crystal arrays have been combined with mechanical displacement transducers which enable reproducible recording of a 3-D data block. The array oscillates in a special liquid-filled ultrasound probe housing and scans the anatomical structures repeatedly at such a speed that a real-time representation of the volume can be calculated (Fig. 17.23).

This type of ultrasound probe is currently primarily used in gynaecology to image the foetus, as it achieves very good resolution with integrated convex arrays.

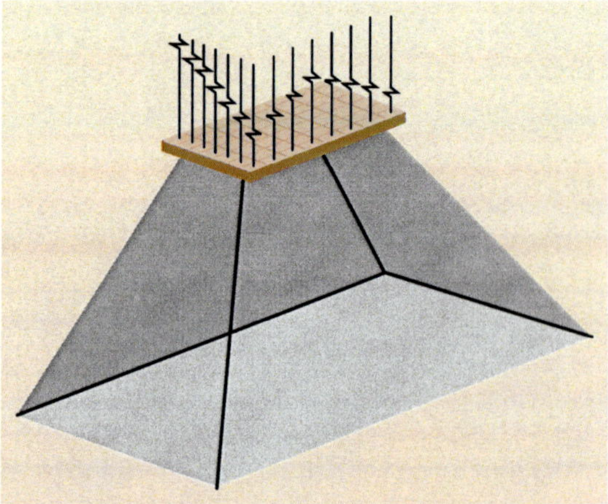

Fig. 17.24 3-D frustum of a pyramid, generated with a matrix phased array ultrasound probe

Integrated 3-D Ultrasound Probes with Electronic Deflection

This smaller and lighter construction manages without mechanically moving components. As with the matrix array ultrasound probe, the array of these probes consists of a two-dimensional, usually square arrangement of tiny piezoelectric crystals which can be activated selectively. According to the principle of the sector phased array method, the crystals are electrically activated at different points in time, so that the scanning beam forms a narrower or wider angle with respect to the array surface. By time-shifted activation of the respective crystal rows and columns, unlike with the conventional phased array ultrasound probe the direction of the beam can be determined in both horizontal axes (Fig. 17.24).

This type of ultrasound probe is currently used primarily in cardiology, as it is predestined for intercostal use due to its small contact surface.

17.4.2 3-D Reconstruction

Irrespective of the scanning method, in all 3-D methods all of the sectional images acquired are lined up during

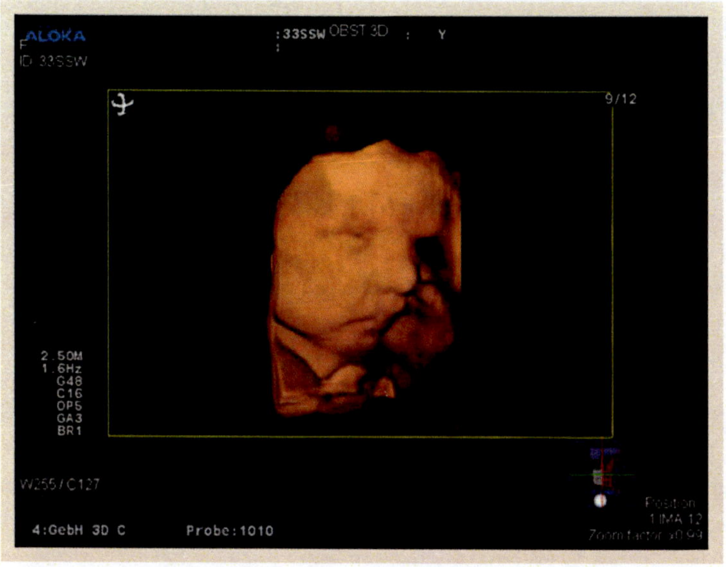

Fig. 17.25 Surface 3-D

the scanning process such that their positions and angles are correct, providing a three-dimensional data set.

This spatial data block represents the basis for all further calculations. Different display calculations are used here:

- Surface display
- Multiplanar display
- Transparent display.

Surface Display

Surface display requires a considerable difference between the wave impedances of the tissue structures (change in impedance), which generally exists between liquid and solid structures. In order to visualize the desired interfaces, it must be possible to clearly delineate the object under examination from its surroundings.

These conditions can often be found when performing diagnosis during pregnancy, as the skin of the foetus is demarcated from the amniotic fluid by a considerable change in impedance.

Specific algorithms for representing surfaces in 3-D data sets (rendering) and further calculations from the graphical image processing, such as the generation of artificial shadows, for example, allow a display which is photorealistic (Fig. 17.24).

Multiplanar Display

This technology allows the examiner to image section planes which cannot usually be displayed from an application point of view. It is possible to simultaneously display multiple sections through an organ or structures which are at any desired angles with respect to one another. The examiner thus gains the possibility of better visualizing anatomical or pathological structures spatially, measuring them and determining their volume.

Transparent Display

By using filtering, it is possible to suppress faintly reflecting structures to the advantage of structures which reflect signals strongly. This allows clear 3-D imaging of bones, for example, which is usually necessary when searching for anomalies in the foetal skeleton.

17.4.3 New and Additional Technologies

This section gives an overview of particular new technologies which have started being used in recent years in ultrasound technology:

- Extended field of view
- Trapezoidal technique
- Steered compound imaging (real-time compound imaging)
- Speckle reduction
- eTracking.

Extended Field of View

The term extended field of view (EFOV) is understood as meaning the extension or widening of the 2-D

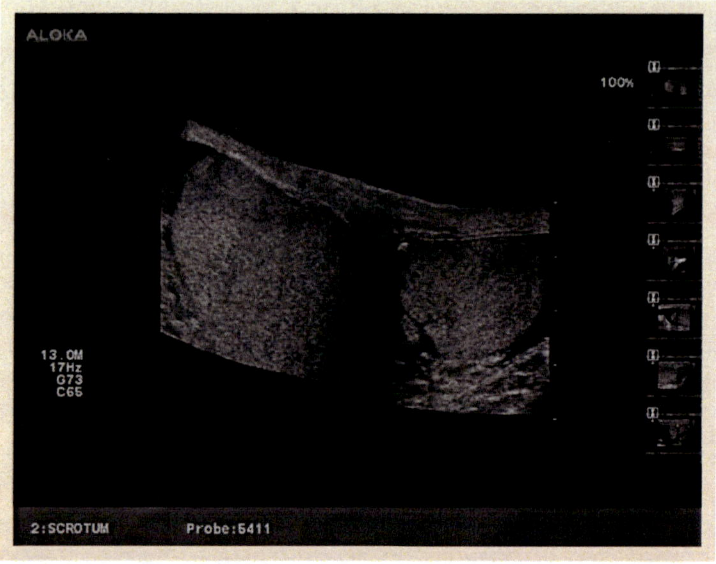

Fig. 17.26 Scrotum imaged using extended field of view

imaging plane. By shifting the ultrasound probe in its longitudinal axis, an extended two-dimensional image is generated.

This does not involve extending the information in the 3rd plane (perpendicular to the 2-D plane) as is the case with 3-D, but rather extended the 2-D plane itself. Using image processing algorithms, matching sections of the respective individual images are lined up together using pattern recognition, and these individual images therefore produce a calculated, new and coherent picture of the organ structures (Fig. 17.26).

This technology is also known by other trade names: SieScape, Freestyle, LOGIQ View, Panoramic Ultrasound, etc.

Trapezoidal

With the trapezoidal technique, the width of the sound field of a linear probe is extended. By time-shifting the activation of the outer groups of crystals (phasing or steered beam principle), a trapezoidal sound field is generated (Fig. 17.27).

This technology is known by other names, such as Virtual Convex, for example.

Steered Compound Imaging (SCI)

The conventional electrical activation of linear and convex array ultrasound probes has the effect of emitting a transmission pulse at right angles to the surface of the ultrasound probe. Anatomical structures or interfaces running parallel to the ultrasonic beam are struck by the ultrasonic beam either tangentially or at a very acute angle. Under these conditions, the portions of the ultra-

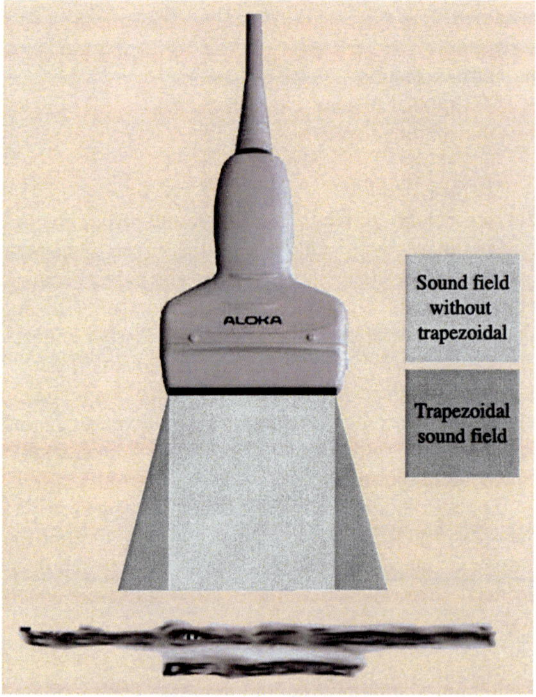

Fig. 17.27 Extension of sound field using trapezoidal technique with linear probes

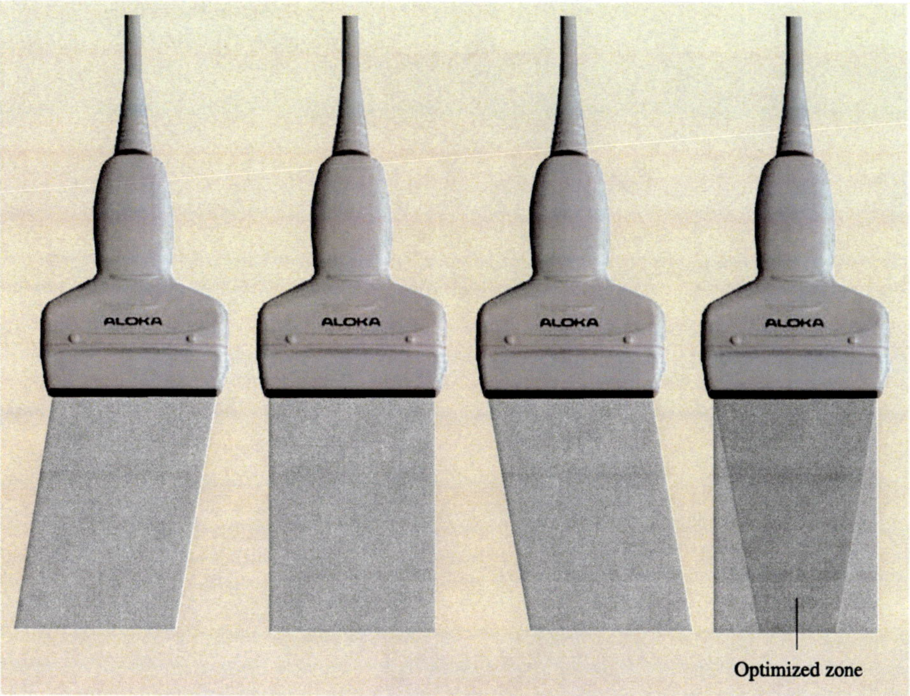

Fig. 17.28 Image optimization using steered compound imaging

sonic beam reflected in the direction of the ultrasound probe are very low. These structures therefore appear only faintly or not at all. The scanning beam also barely reaches the structures lying distal to these interfaces, because the majority of the energy of the ultrasonic beam is deflected by reflection.

In addition to the known lateral edge shadows artificial inhomogeneities also appear in the distal structures and interfaces. The consequences of these phenomena can be reduced if the structures being examined are sounded not just from one direction but from different directions using beam steering (phasing) and the resulting sound reflections are superimposed using image processing to form a compound ultrasound image. Usually, three consecutive images are generated, which have been recorded in a range between −30, 0 and +30° (Fig. 17.28). Modern devices furthermore also allow angle steering, adapted according to requirements, in various intermediate stages which makes it possible to image different depths of field. Every reflecting interface is therefore optimally sounded from different directions.

This technology is also known by other trade names, such as Realtime-Compound-Imaging, SonoCT, CrossBeam, etc.

Speckle Reduction

Because of their physical properties, ultrasound images are always accompanied by noise. The signal-to-noise ratio is in the region of approximately 1.9 and can only be visually and subjectively improved by corresponding image processing. As a result of the constantly increasing computational power of computers, further image-processing steps can be added after the actual acquisition of signal data. Special algorithms allow image characteristics to be emphasized and changed. This includes image homogenization and the accompanying subjective noise suppression, and also, for example, edge enhancement of adjacent organ structures.

These image-processing measures can in many cases aid fast and easy diagnosis, but they are not always and exclusively of use, as both false positive and also false negative image information can be generated. The trained operator should therefore always be free to decide on their use and also their degree of use.

Vessel Distension Analysis (Echo Tracking, eTracking)

The growing awareness of relationships between cardiovascular diseases and vascular function and also

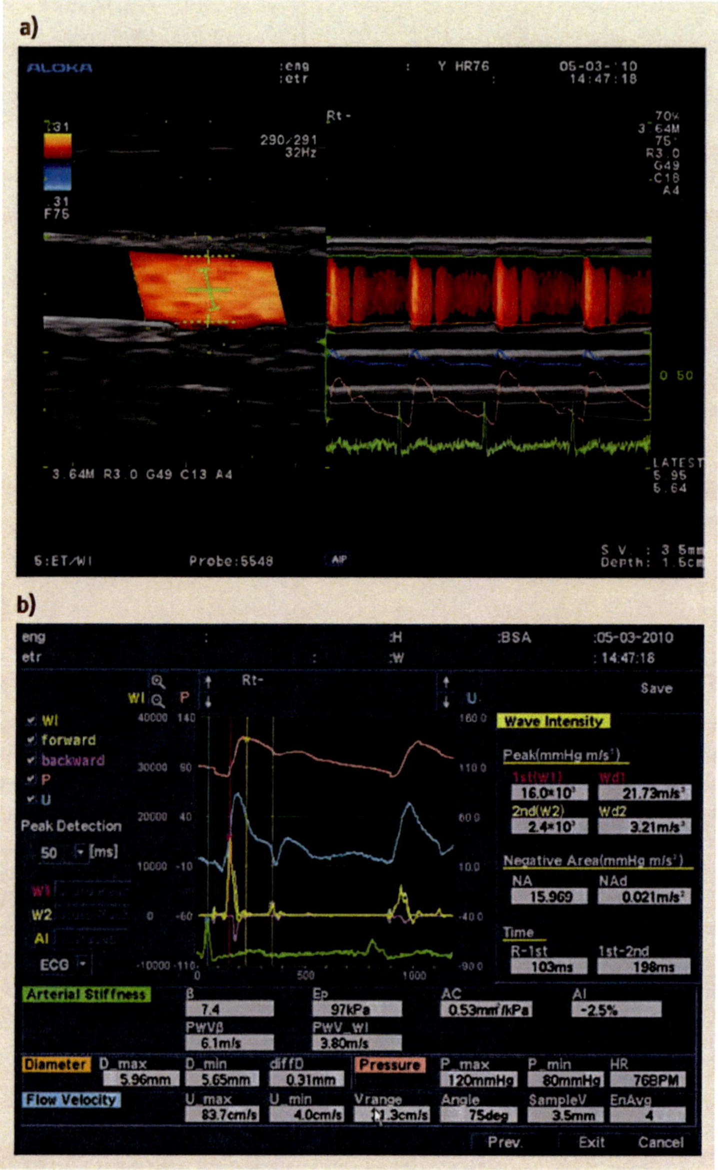

Fig. 17.29a,b Vessel distension analysis using eTracking

vascular mechanics has meant that diagnostic parameters can be raised by means of a series of different technical methods (ultrasound, tonometry, plethysmography, etc.).

Ultrasound can be used to calculate a number of relevant parameters, such as the pulse wave velocity (PWV), vascular elasticity (β-Index), augmentation index (AIx) and arterial function (FMD), from the vascular distension. The distension is determined in real time by analysing the raw data signal. Using specialist software, the movement of moving organ structures is followed by user-defined points along an ultrasonic beam (tracking). This is done with a high level of temporal (1 kHz) and spatial precision (1/16 of the wavelength).

A development of this analysis method is based on the simultaneous and colocated detection of the velocity of the blood flow using Doppler technology. This allows further parameters to be calculated, e.g. pulse wave propagation or wave intensity, which describe the coupling between the heart and the vascular system downstream (Fig. 17.29a,b).

17.5 Operation of an Ultrasound Unit

17.5.1 General Conditions

Operation of an ultrasound unit has today become standard in most medical fields. Some important general conditions should be observed and are detailed below.

Documentation

The operator of an ultrasound unit is generally required by the accounting body and by legal authorities to document findings.

All suppliers of ultrasound equipment provide relevant advice and solutions on request for the analogue and digital storage of data and images for the findings. In the simplest scenario this includes thermal printers or internal digital memory with CD/DVD/USB storage or export interfaces, such as the vendor-neutral DICOM (digital communication in medicine) standard for network-based archiving of still images or image sequences and measurement data.

The DICOM standard not only governs the archiving of medical image data as well as other modalities such as CT and MRI, but also provides further functions and procedures which support work-flow (e.g. work list, print, structured report, etc.).

Application-Specific Additions and Upgrades

Modern digital ultrasound units allow additional hardware or software to be installed to extend the scope of use of the unit. This includes, for example, units for simultaneous ECG recording, memory upgrades, specific analysis programs (stress echo, contrast medium assessment, TDI evaluation, analysis of the vessel distension and pulse wave velocity, etc.).

In addition to these device-specific upgrades, there is also always the possibility of extending the intended use of a device with additional ultrasound probes.

Maintenance and Repair, Updates

The ultrasound unit manufacturers provide recommendations for safe operation and warnings against improper use of units and probes. The user should find this information in the safety instructions provided with the equipment documents.

Important: in line with the Medical Devices Operator Ordinance (Medizinische Betreiberverordnung, MedBetreibV), the operator of an ultrasound unit has certain duties to ensure the safety of the systems used. To this end, the ultrasound unit manufacturers and retailers offer tailored maintenance and inspection agreements which conform to the law support service for their customers. These contracts often also include the option of insuring against damage to the units and probes. There are also offers for regular updates of the device software.

Hygiene

In general, the operator should inform him or herself about possible legal hygiene regulations in the context of the planned examinations. In addition, the guidelines of the professional associations include recommendations regarding hygiene which should also be observed.

For the cleaning of units and also the cleaning, disinfection and sterilization of ultrasound probes and auxiliary equipment, such as puncture adapters, the manufacturers provide information about the methods to be used and approved cleaning, disinfection and sterilization agents.

Safe Use of Ultrasound Diagnostics

According to current scientific knowledge, the diagnostic use of ultrasound which complies with standards does not have any unwanted side effects. However, it is recommended that the Doppler mode be used cautiously, particularly for the purpose of foetal examinations. Because ultrasound energy is absorbed and converted into heat at the transition between tissues in bony structures, it is not possible to rule out cell-damaging effects with certainty. It is therefore imperative that the recommendations of the relevant professional associations are observed.

Further Reading

- V.M. Shami, M. Kahaleh (Eds.): *Endoscopic Ultrasound* (Springer, Berlin Heidelberg 2010)
- F. Gress, T. Savides: *Endoscopic Ultrasonography* (Wiley, New York 2009)
- M. Hofer: *Ultrasound Teaching Manual* (Thieme, Stuttgart 1999)
- S. Delorme, J. Debus: *Sonographie* (Thieme, Stuttgart 2004)

- K.D.M. Resch: *Transendoscopic Ultrasound for Neurosurgery* (Springer, Berlin, Heidelberg 2006)
- C.F. Dietrich: *Endosonographie* (Thieme, Stuttgart 2008)
- H. Madjar: *Kursbuch Mammasonographie* (Thieme, Stuttgart 2005)
- E. Merz: *Sonographische Diagnostik in Gynäkologie und Geburtshilfe* (Thieme, Stuttgart 2002)
- M. Thelen: *Bildgebende Kardiodiagnostik* (Thieme, Stuttgart 2007)
- H. Lambertz, H. Lethen: *Transösophageale Echokardiographie* (Thieme, Stuttgart 2007)
- P. Hoskins, A. Thrush: *Diagnostic Ultrasound* (Greenwich Medical, London 2003)
- C.R. Hill, J.C. Bamber, G. Haar (Eds.): *Physical Principles of Medical Ultrasonics* (Wiley, Chichester 2004)
- K.K. Shung: *Diagnostic Ultrasound* (Taylor Francis, New York 2006)
- B. Block: *Color Atlas of Ultrasound Anatomy* (Thieme, Stuttgart 2004)

18. Medical Infrared Imaging

Gerald C. Holst, Thorsten M. Buzug

This chapter describes passive as well as active thermal methods and their applications in medicine. To keep the body temperature at a constant level, excess heat is released by evaporation, convection, conduction, and radiation. Evaporation of sweat is the most obvious mechanism. Its effectiveness depends upon the ambient temperature and relative humidity. Convective cooling is most noticeable when standing in front of a fan. Conduction occurs when physically touching an object. Conductive cooling depends on the temperature difference between the object temperature and the skin temperature. Feverish patients are sometimes immersed in cool water. Radiative cooling is a function of the skin temperature and the environmental temperature. It is experienced when placing your hand in a freezer. An infrared (thermal) imaging system is calibrated to provide surface temperature. Local variations in skin temperature is a function the above quantities as well as metabolism (cancers and exercise), inflammation, circulatory disturbances, and skin condition (scabs, moles, and hair).

18.1 **Background** 369
18.2 **Infrared System Design** 370
18.2.1 Detectors 371
18.2.2 Performance Metrics 371
18.2.3 Image Processing 372
18.3 **Infrared Physics** 372
18.3.1 Planck's Blackbody Law 372
18.3.2 Camera Formula 372
18.3.3 Emissivity and Reflectivity 373
18.4 **IR Imaging in Medical Applications** 373
18.5 **Specific Applications** 374
18.5.1 Circulatory Disturbance 374
18.5.2 Sports Medicine 374
18.5.3 Skin Cancer 375
18.6 **Limitations of IR Imaging in Medical Applications** 377
18.7 **Summary** 377
References 378

18.1 Background

Variations in skin temperature are related to local metabolic differences. These variations may be normal (e.g., during exercise) or abnormal (e.g., presence of cancer). Temperatures may be measured with either a contact or noncontact device. Contact devices include the common liquid-in-glass thermometer, resistance thermometer, and thermocouple. Noncontact devices may be either a nonimaging or imaging infrared (IR) system. A nonimaging system (such as a radiometer) simply measures the radiation. The system's calibration converts the output voltage to a temperature value. Imaging systems create a two-dimensional image of the skin. As with the radiometer, the IR imaging system measures the radiation that appears to emanate from the skin.

What is most striking about IR imaging is that it is one of the few modalities which is noninvasive and contactless. Recently, this was manifested in a case where feverish individuals potentially infected with severe acute respiratory syndrome (SARS) were identified among a huge number of persons in airport security areas by their facial temperature profile. However, since its invention it has been conjectured that medical infrared imaging is a functional imaging modality (infrared functional imaging, IRFI) that can be used to visualize pathologically increased metabolism by its temperature signature. Especially for some types of cancer, detection of tumors follows the thermographic paradigm that the strong growth of malignant tumors is accompanied by an increase of metabolism, leading to a conspicuous thermal signature. Since cancer cells have a high metabolic rate, a localized elevated skin temperature may warrant further tests for the presence of cancer.

IR imaging can be expected to yield sensible results in cases of inflammation processes, arthritis, rheumatism, circulatory disturbances, any cases of allergy with skin symptoms as well as burning and scalding, frostbite, monitoring the success of skin transplantation, etc. For qualitative measurements, IR images are also used in pain diagnostics. An overview of the general causes of fluctuations of skin temperature that can be measured with IR camera systems can be found in [18.1].

Although the medical community has known about IR imaging for years, there is now significant renewed interest. The medical IR imaging community is small compared with the overall nondestructive testing (NDT) community (SPIE has hosted *Thermosense* since 1978 [18.2]). Increased temperature is an indicator of potential failure in electrical power distribution and machinery [18.3]. Industries such as automobile, petrochemical, die casting, and paper manufacturing use IR imaging systems for process control [18.3]. Numerous manufacturers have developed IR imaging systems for NDT. These calibrated systems lend themselves to medical applications.

18.2 Infrared System Design

Due to the atmospheric spectral transmittance, electronic imaging system design is partitioned into seven generic spectral regions (Fig. 18.1) of which four are associated with thermal imaging systems. The ultraviolet (UV) region ranges in wavelength from 0.2 to 0.4 μm. The visible spectral region ranges in wavelength from 0.4 to 0.7 μm. Televisions and electronic still cameras operate in this region. The near-infrared (NIR) imaging spectral region spans ≈ 0.7–1.1 μm. Image intensifiers, starlight scopes, and night-vision goggles operate in this region. The first infrared imaging band is the short-wavelength infrared (SWIR) imaging band, which covers approximately 1.1–2.5 μm. The second infrared band is the mid-wavelength infrared (MWIR) spectral region that covers ≈ 2.5–7.0 μm. The third infrared band is the long-wavelength infrared (LWIR) spectral band. It covers the spectral region from ≈ 7 to 15 μm. The fourth infrared band is the far-infrared (FIR) or very long-wave infrared (VLWIR) region. It applies to all systems whose spectral response extends past 15 μm. The MWIR and LWIR regions are sometimes called the first and second thermal imaging bands, respectively. Note that SWIR, MWIR, and LWIR are generic terms; for example, a LWIR system's spectral response can be anywhere in the LWIR region (e.g., 8–14 μm, 7.5–10.5 μm, 8–12 μm, etc.).

For convenience, targets are labeled as either hot or cold with respect to the immediate background. For MWIR and LWIR systems, the term thermal is misleading. IR imaging systems do not sense warmth or

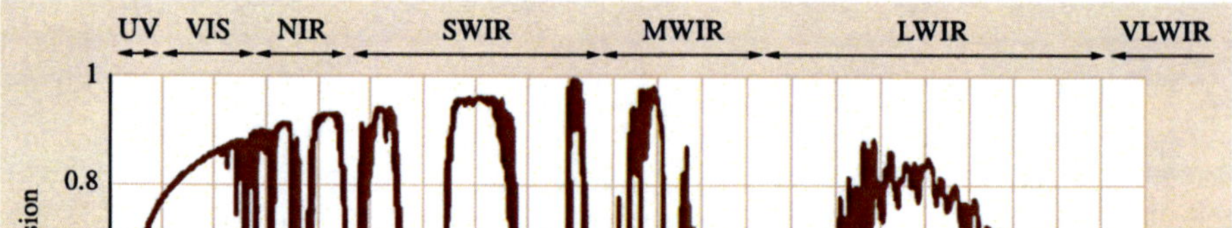

Fig. 18.1 Representative atmospheric transmittance as a function of wavelength λ over a 1 km path length. The transmittance varies with temperature, relative humidity, and airborne particulates. Note the nonlinear wavelength scale

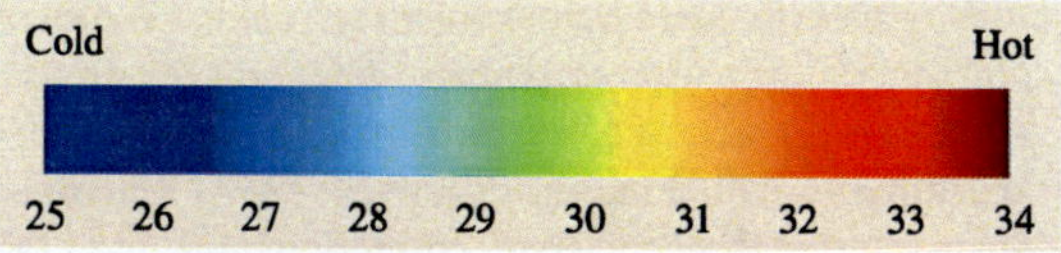

Fig. 18.2 Representative calibration (25–34 °C). The temperature scale can be adjusted to any value

cold (they are not thermometers) but rather the radiation emitted by an object. Hot refers to a target that appears warmer than its immediate background and cold means that the target appears cooler than its immediate background.

The detector output is simply a voltage. After calibration, a specific voltage level represents a temperature value. This voltage can be mapped into pseudocolors. Consistent with human feelings, cold objects are represented as blue and hot objects as red (Fig. 18.2). Whether the output is displayed on a grayscale or as pseudocolors is a matter of personal preference. However, pseudocolors generally facilitate image interpretability.

18.2.1 Detectors

The detector is the heart of an imaging system because it converts infrared radiation into a measurable electrical signal and converts target spatial information into electrical temporal information. Current terminology lists detectors generically as cooled or uncooled. Within each category, there are many different types of detectors. Early IR imaging systems contained only a few detectors, necessitating scanning mechanisms. Current systems are staring arrays that contain 320×240 detectors or 640×480 detectors. With the smaller array (320×240), data interpolation creates a 640×480 output to be consistent with most displays.

Cooled Detectors

LWIR photon detectors must be cooled below 100 K, with 77 K considered a typical temperature. These temperatures can only be reached with a mechanical cooler or with liquid nitrogen. Many MWIR detectors can operate at 200 K, and this temperature can be easily achieved with a thermoelectric cooler (TEC). TECs appear to have an infinite lifetime, whereas mechanical coolers degrade over time. Coolers add cost and bulk, and consume power.

InSb, a high-quantum-efficiency MWIR detector, has replaced the low-quantum-efficiency Pt:Si detector. Although generally labeled as HgCdTe, it is a mixture ($Hg_{1-x}Cd_xTe$). By varying the ratio, the spectral response can be tailored to the MWIR or LWIR region. The most popular is the LWIR detector with a peak response near 12 μm. A filter is used to limit the response to the LWIR region. The quantum-well infrared photodetector (QWIP) is based upon mature GaAs growth technology. The response can be tailored from 3 to 19 μm. The LWIR version typically has a spectral response from 8.3 to 10 μm. The responsivity and noise are temperature sensitive, so QWIP devices are typically cooled to less than 60 K.

Uncooled Detectors

Microbolometers with either VOx or α-Si coating can operate at room temperature and therefore are called uncooled devices. Although called uncooled, these devices may have a TEC to stabilize the detector temperature. Uncooled devices are lightweight, small in size, and easy to use. Because they use very low power, they lend themselves to handheld, battery-operated devices. Uncooled detectors generally have much lower responsivity than cooled detectors.

18.2.2 Performance Metrics

Thermal sensitivity refers to the smallest temperature differential that can be detected. It depends upon the light-gathering properties of the optical system, the detector responsivity, and the system noise. A laboratory measure of thermal sensitivity is the noise equivalent differential temperature (NEDT). It is also called the noise equivalent temperature difference (NETD) or noise equivalent temperature (NET). Generally, the NEDT for cooled detectors is less than 10 mK and may be as high as 100 mK for uncooled sensors.

Spatial resolution provides valuable information regarding the finest detail that can be discerned. It is usually specified by the instantaneous field of view (IFOV) and is also called the detector angular subtense (DAS)

$$\mathrm{IFOV} = \mathrm{DAS} = \frac{d}{l_\mathrm{f}}\,, \tag{18.1}$$

where d is the detector linear dimension and l_f is the system's effective focal length. The field of view (FOV) is simply the IFOV multiplied by the number of detectors.

The minimum resolvable temperature (MRT) is a composite metric that combines thermal sensitivity and spatial resolution [18.4]. It is used primarily to characterize military systems and, as such, has less value to the medical imaging community.

18.2.3 Image Processing

Many cameras have built-in image processing capability. Indeed, we can say the modern thermal imaging system is a microcomputer with a detector attached. Cameras come with a large variety of image processing software. For temperature-measuring devices, typical built-in functions include isotherms, color rendition, spot temperature measurements, temperature profile, and area-weighted temperature. Area-weighted functions are also called region-of-interest (ROI) functions. Most systems offer emissivity correction that allows the user to enter both the target emissivity and the surrounding temperature.

18.3 Infrared Physics

A comprehensive infrared physics equation is quite complex. For simplicity, it is assumed that the skin radiates isotropically and that the radiant exitance follows Planck's blackbody law. Reflectance of ambient radiation can significantly affect the target signature. For convenience, skin reflectance is assumed to be essentially zero.

18.3.1 Planck's Blackbody Law

The spectral radiant exitance of an ideal blackbody source, whose absolute temperature is T, can be described by Planck's blackbody radiation law

$$M_e(\lambda, T) = \frac{c_1}{\lambda^5}\left(\frac{1}{e^{(c_2/\lambda T)} - 1}\right) \text{ W/(m}^2\,\mu\text{m)} . \tag{18.2}$$

where $c_1 = 3.7418 \times 10^8 \text{ W}\,\mu\text{m}^4/\text{m}^2$ and $c_2 = 14\,388$ μm K. As shown in Fig. 18.3, as the absolute temperature increases, the spectral radiant exitance increases at shorter wavelengths. A blackbody must be heated to 600 K before it appears to glow (heated to incandescence). Equivalently, visible imaging systems can only sense the radiation from sources whose temperature is greater than 600 K. Clearly, only an IR imaging system can detect skin temperatures (nominally around 310 K).

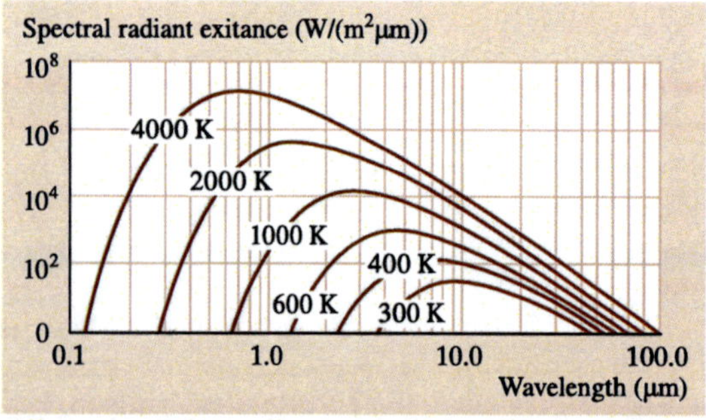

Fig. 18.3 Planck's spectral radiant exitance as a function of wavelength for different temperatures T

18.3.2 Camera Formula

When viewing an ideal target, the camera formula provides the system output voltage

$$V_{SYS} = G \int_{\lambda_1}^{\lambda_2} \frac{R_D(\lambda) A_D}{4F^2} M_e(\lambda, T)\, d\lambda , \tag{18.3}$$

where G is the system electronic gain, $R_D(\lambda)$ is the system spectral response (detector spectral response multiplied by the optical spectral transmission), A_D is the detector area, and F is the f-number (the focal length divided by the aperture diameter). The limits of integration depend upon the spectral response of the system (e.g., MWIR or LWIR)

What is of interest is the difference between a target and its immediate background

$$\Delta V_{SYS} = G \int_{\lambda_1}^{\lambda_2} \frac{R_D(\lambda) A_D}{4F^2} [M_e(\lambda, T_T) - M_e(\lambda, T_B)]\, d\lambda . \tag{18.4}$$

Although the differential voltage is proportional to the radiant flux difference, it is convenient to specify the flux difference between a target and its immediate background by an equivalent temperature difference ΔT. Using a Taylor series expansion for the flux difference and keeping the first term (valid for small temperature differentials)

$$\Delta V_{SYS} \approx \left(G \int_{\lambda_1}^{\lambda_2} \frac{R_D(\lambda) A_D}{4F^2} \frac{\partial M_e(\lambda, T_B)}{\partial T}\, d\lambda \right) \Delta T = k_{CAL} \Delta T . \tag{18.5}$$

The partial derivative is called the thermal derivative and must be evaluated at the background temperature.

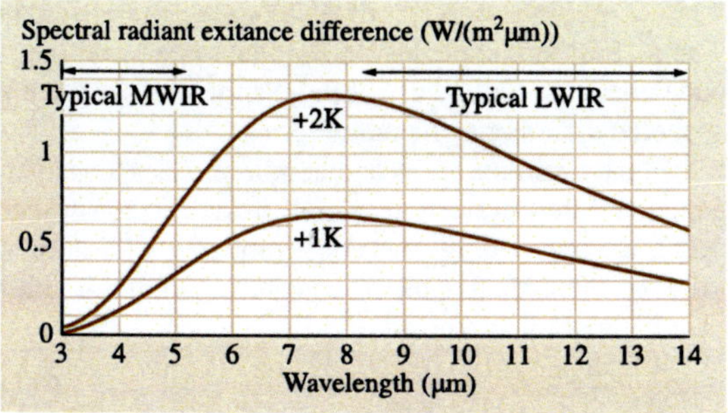

Fig. 18.4 Planck's spectral radiant exitance difference from 37 °C (310 K)

The bracketed term is a calibration term that is unique for each type of IR imaging system. After calibration, the output voltage differential is proportional to ΔT. The differential voltage is mapped into an intensity or pseudocolor. Figure 18.4 illustrates the differential radiant exitance for small skin temperature increments above 37 °C (310 K). While there is more differential signal in the LWIR, MWIR systems perform equally well; they just have different calibrations.

18.3.3 Emissivity and Reflectivity

By Kirchhoff's law, when an object is in thermal equilibrium with its environment, the absorption $\alpha(\lambda, T)$ is equal to the emissivity $\varepsilon(\lambda, T)$. This leads to the popular saying that good absorbers are good emitters. Real blackbodies do not emit all the radiation described by (18.2) but only emit a fraction of it. The ratio of actual radiant exitance to the theoretical maximum is the emissivity. The radiation that appears to emanate from a target is actually a combination of emitted and reflected radiation. Expressed as a ratio,

$$\varepsilon(\lambda, T) + \rho(\lambda, T) = 1 . \tag{18.6}$$

For an ideal blackbody, $\alpha(\lambda, T) = 1$ and $\rho(\lambda, T) = 0$. For nonideal blackbodies, the amount of radiation that appears to emanate from the target is

$$\varepsilon_T(\lambda, T_T) M_e(\lambda, T_T) + \rho_T(\lambda, T_T) M_e(\lambda, T_{AMB}) , \tag{18.7}$$

where T_{AMB} is the effective ambient temperature. The apparent background radiation is

$$\varepsilon_B(\lambda, T_B) M_e(\lambda, T_B) + \rho_B(\lambda, T_B) M_e(\lambda, T_{AMB}) . \tag{18.8}$$

For example, when imaging an expected tumor, ε_T is the skin emissivity over the tumor and ε_B is the emissivity of *normal* skin. The reflective component makes calibration difficult. In the next section it is shown that skin emissivity is sufficiently high that the reflective component can be neglected in many situations.

18.4 IR Imaging in Medical Applications

For bare skin, $\varepsilon \approx 0.98 \pm 0.01$, and this tends to be independent of skin color [18.5]. While hair does not affect skin emissivity, it acts as an insulator which prevents accurate skin temperature measurements. Neglecting reflection (i. e., assuming $\varepsilon_T(\lambda, T) = \varepsilon_B(\lambda, T) \approx 1$) introduces a small error in temperature measurement. While this error is considered minor, it is appropriate to discuss *apparent* temperature differentials rather than absolute temperature differences ΔT. While quantitative imagery is always desired, qualitative differences may suffice. It is advisable to have a baseline image (Table 18.1). This is called differential infrared thermography (DIT). The difference can be magnified by subtracting the baseline image from the desired image.

Very often IR imaging is the interpretation of subcutaneous processes from the cutaneous temperature distribution. As shown in Fig. 18.5, as the tumor depth increases, lateral heat flow provides a larger hot area than the actual tumor. Furthermore, the edges are diffuse, making precise location difficult. Core temperature (37 °C) is generally held constant for depths larger than 20 mm, and therefore the thermographic paradigm holds for near-to-skin processes only.

Table 18.1 Image pairs for qualitative imaging

Desired image	Baseline image
After or during exercise	Resting before exercise
After or during smoking	Before smoking
Potential tumor on left breast	Right breast (assumed normal)

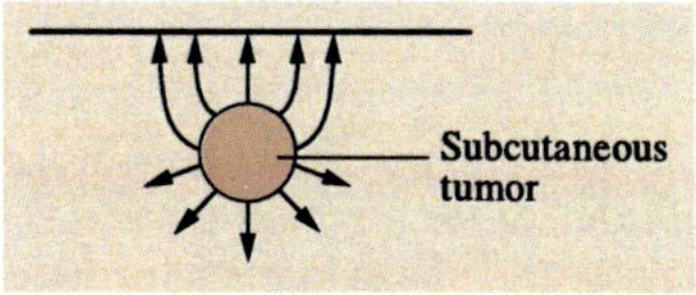

Fig. 18.5 Heat diffusion

Table 18.2 Local temperature signature of certain pathologies

Medical condition	Apparent ΔT (K)
Inflammations (e.g., rheumatism, arthritis)	1–2
Vascular diseases (e.g., thrombosis)	1–2
Breast cancer	1.5–4
Malignant melanoma	2–4

Although local temperature increases are relatively small [18.6] (Table 18.2), these values can easily be measured with current IR imaging systems. Image quality is considered very good when the signal-to-noise ratio (SNR) is greater than 10 and excellent when SNR > 40 (SNR = $\Delta T/\text{NEDT}$). For the values listed in Table 18.2, both cooled and uncooled systems will provide adequate SNR. For subtle changes ($\Delta T < 0.5$ K), a cooled system may be required.

Finally, surface scabs and certain moles act as insulators and thus prevent heat from reaching the surface. Here, the area will appear cold compared with normal skin. Mathematically, this is a negative ΔT but is commonly reported as a positive ΔT.

It is important to recognize that reflectivities in the visible and the IR are usually different. Infrared imagery will always look different than visual imagery. Features seen in thermograms may or may not be present in visual images. It is prudent to record both visual and infrared imagery.

18.5 Specific Applications

Skin variations may be measured either passively or actively (also called passive thermography and active thermography). This far, this chapter has described passive temperature methods. High metabolic rate appears as an elevated skin temperature. For surface insulators (e.g., moles), the active method is preferred. Here the mole and surrounding skin is either heated or cooled and the time history of the temperature is measured (e.g., natural thermoregulation). The *normal* skin will reach *normal* temperature before the mole does.

18.5.1 Circulatory Disturbance

As an example of a prominent circulatory disturbance, Fig. 18.6 shows the effective cooling of a hand of a person smoking a cigarette. This qualitative example shows the vascular narrowing and the subsequent reduction in blood perfusion associated with smoking. For this image sequence a simple (MWIR) line-array detector scanner (NEC TH5104 Thermo Tracer) was used. This camera employs a detector array of 255×223 pixels and a temperature resolution of NEDT = 100 mK. Obviously, smoking restricts the blood flow as a result of vasoconstriction. As a consequence of the reduced blood flow, the metabolic rate is reduced and, with lower metabolic rate, the hand cools to ambient temperature.

18.5.2 Sports Medicine

Figure 18.7 shows the temperature distribution on the femoral skin during exercise testing on a bicycle ergometer. After 2.5 min at rest, the subjects underwent a load-graded exercise test. Brake resistance (measured in watts) was increased every 3 min, thus increasing the workload up to the submaximal condition corresponding to 65% of the maximal heart rate value. The recovery phase was defined as the time necessary to restore the rest heart rate by means of recovery exercise and rest. With blood demand from the working muscles there is skin vasoconstriction and skin surface cooling. As the muscles stop working and the recovery and the restoration phases start, thermoregulatory control is invoked to limit body temperature increase. Here the skin vessels dilate to increase heat conduction to the skin.

Figure 18.7 highlights the results of *Merla* et al. [18.8], who recorded images every minute during exercise and the recovery phase. The images were

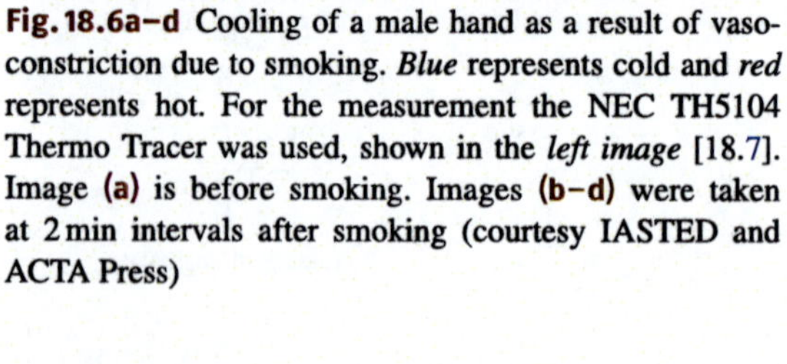

Fig. 18.6a–d Cooling of a male hand as a result of vasoconstriction due to smoking. *Blue* represents cold and *red* represents hot. For the measurement the NEC TH5104 Thermo Tracer was used, shown in the *left image* [18.7]. Image **(a)** is before smoking. Images **(b–d)** were taken at 2 min intervals after smoking (courtesy IASTED and ACTA Press)

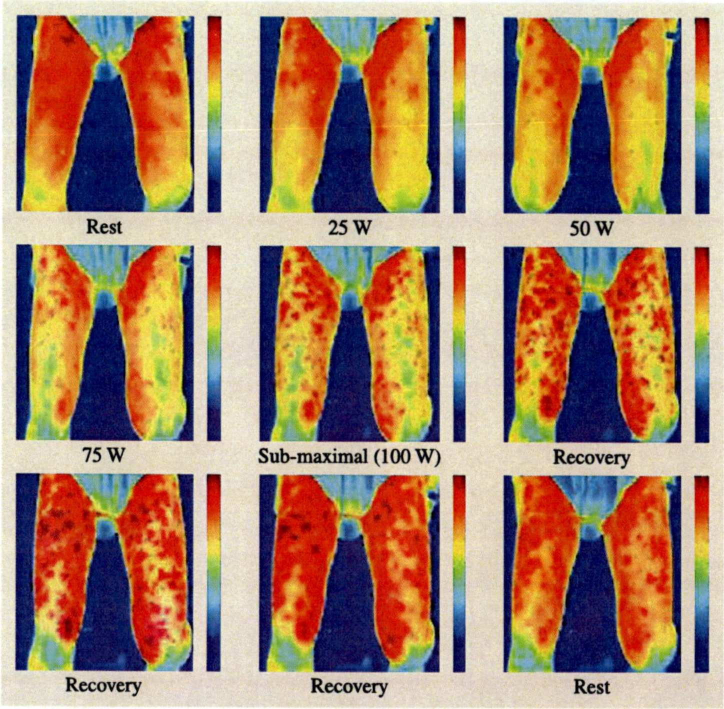

Fig. 18.7 Variation of the thigh skin temperature during exercise on a bicycle ergometer. The barely visible vertical temperature scales next to each image are identical to Fig. 18.2. Note that the skin temperature is less than the core temperature (37 °C) (after *Merla* [18.8]). The temperature scale (*blue* to *red*) is 30 to 38 K

taken with an AEG camera AIM 256 Pt:Si (MWIR). Its detector array consists of 256×256 pixels and has a temperature resolution of NEDT = 100 mK.

These sports-medical measurements were correlated with the corresponding oxygen intake. It was shown that the degree of skin cooling and the thermoregulation depends on the individual fitness level of each subject. Currently, the extent to which IR imaging is able to estimate the hemodynamics of the thigh muscles is being studied.

18.5.3 Skin Cancer

In the last 30 years the number of patients diagnosed with skin cancer has increased dramatically. However, there is no appropriate and sound way to decide noninvasively if a skin tumor is benign or malignant. Generally, the diagnosis is made with Stolz's traditional ABCD rule of dermatoscopy based on the four main criteria or lesion parameters: asymmetry, border, color, and diameter, with a semiquantitative score system [18.9, 10]. Frequently, this method is improved by computerized scanning methods based on polarized light surface microscopes [18.11]. With both methods a suspicious spot has to be observed over a period of time to obtain a reliable result, i. e., the evolution of the spot is important, which refines the ABCD method into the ABCDE method [18.10].

For the time being the only way to get an accurate diagnostic finding is an invasive histological examination. IR imaging may avoid redundant excision of tissue and potentially enables detection of malignant melanoma at an early stage. Thanks to the facts that malignant melanoma has higher consumption of glucose caused by a higher level of metabolism and augmented branching of blood vessels, so-called angiogenesis, the temperature should differ from surrounding healthy tissue. Associated with the increased demand of energy, it is conjectured that malignant melanoma shows a higher temperature (2–4 K) than its surrounding skin [18.6]. Infrared imaging has a high potential to detect the beginnings of angiogenesis, when cancer cells first try to develop their own blood supply, which is a necessary step before they can grow rapidly and metastasize. Other types of skin cancer (e.g., basalioma) will behave differently due to an encapsulation strategy of the tumor.

For screening of suspicious moles, a clinical protocol has been proposed that consists of an IR imaging sequel after provocation of the skin area by cool-

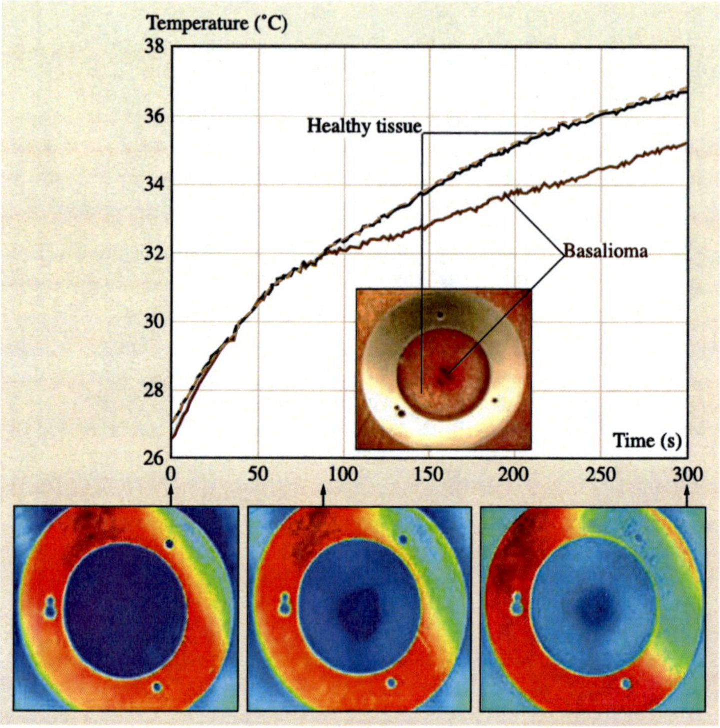

Fig. 18.8 Thermoregulation sequence of a basal-cell carcinoma (basalioma) after forced cooling [18.12]. The boreholes of the circular plate are used as fiducial landmarks for motion compensation. The temperature pattern of the marker does not disturb landmark detection (courtesy IEEE)

ing [18.12]. To be accepted into clinical routine a simple, reproducible measurement protocol of the thermographic image acquisition and patient preparation has to be set up. To get the characteristics or temperature signature, respectively, the relevant skin area has to be provoked. Generally, two directions are possible with active thermography. On one hand the skin can be warmed up or, on the other hand, the skin can be cooled down. One major risk of the warming method is that the protein denaturation process starts when the skin temperature exceeds 42 °C. For that reason the cooling-down method is used to produce a substantial temperature difference within a few minutes.

The cooling is carried out using direct contact with cooled gel packs. An area of about 10 cm by 10 cm is cooled to 20 °C. After this patient preparation step, the signature of the thermoregulation process is recorded by the LWIR FLIR SC 3000 camera. The QWIP detector array consists of 320×240 pixels. It can resolve temperature differences as low as NEDT = 30 mK.

During thermoregulation the tumor shows up as a cold spot compared with surrounding tissue (Fig. 18.8). A difficulty in the evaluation of the thermoregulation process is that a spot cannot be automatically detected on the basis of the temperature image alone, because at the starting point of thermoregulation the entire skin region of interest has a homogeneous temperature distribution (20 °C).

To overcome this problem a circular marker with boreholes used as fiducial landmarks is attached to the skin and a normal digital photograph is taken prior to the thermographic session (see the inset to the temperature regulation graph in Fig. 18.8). By assigning the corresponding marker borehole location of each IR image with the respective digital image, motion of the patient can be compensated. This is important because a small motion in an image acquisition with a macroscopic lens can cause errors in the correlation between the frames of the thermographic sequence (details of the image sequence registration process can be found in [18.7]). Without stable motion compensation, automated comparison of the temperatures of skin and spot is not possible.

The signature of the curves is exponential. However, focusing on the thermoregulation of the spot, it can be seen that there is synchronous warming to 32 °C in the first 90 s. As can be seen in Fig. 18.8, the temperature of the basalioma increases more slowly than the surrounding healthy skin when the temperature exceeds 32 °C. The basalioma is not visible in the temperature image during the first 90 s of the recording; however, it is clearly recognizable in the end. One possible explanation for the lower temperature is based on the physiological characteristics of a basalioma. A basalioma is created from cells which can produce an isolation layer, as already mentioned in the study by *Maleszka* et al. [18.13], who saw this effect for psoriatic arthritis [18.14].

18.6 Limitations of IR Imaging in Medical Applications

While the physical principles of infrared imaging are clear there are a variety of problems one has to cope with in medical applications of thermographic imaging. Typical sources of errors in diagnostic imaging are given in the following list:

- Complexity of an exact model of the medical thermoregulation process. Even the simple bioheat equation of *Pennes* [18.15]

$$\rho c \frac{\partial T}{\partial t} = \nabla(k \nabla T) + c_b w_m(T) \rho_b (T_a - T) + Q_m + P(z, t) \,, \quad (18.9)$$

  where ρ, c, and k are the density, specific heat, and thermal conductivity of tissue, respectively, c_b is the specific heat of blood, ρ_b is the density of blood, T is the local tissue temperature, T_a is a reference temperature (typically arterial blood temperature), t is time, Q_m is the metabolic heat production per volume, and $P(z, t)$ is the heat deposited per volume due to spatially distributed heating. In this general form, w_m is a function of temperature to include the specific case of temperature-dependent perfusion [18.16] which is an ill-posed problem, because of:
  - Patient-dependent variability of the thermoregulation process due to different biofeedback time constants
  - Spectral specification as well as accuracy and resolution of the infrared camera
  - The lack of a definition of a standard in active thermography concerning the methodology of thermal excitation or provocation
  - Vagueness of determination of thermal characteristics of skin cancer, i. e., variation of the emissivity coefficient of suspicious moles
  - Only the surface temperature is imaged
  - Unknown relation between thermoregulative time constants and the amount of applied energy in active thermography
  - Inhomogeneity and speed variation of energy transfer in active thermography
  - Spurious reflection of background radiation
  - Insufficient patient acclimatization before imaging (patient-induced uncertainty, e.g., time of patient's most recent warm meal, as well as ambience induced, e.g., unstable thermoregulation of the examination room, etc.).

For a discussion of some of these points see *Nowakowski* [18.17].

18.7 Summary

Undoubtedly, infrared imaging or thermography is an effective medical screening modality. However, if a tumor inside the breast is actually resulting in a specific heat signature on the breast skin, then, in most cases, the tumor is in an advanced state. Therefore, cancer diagnosis using IR imaging should be restricted to some special types of skin cancer. Currently, several studies are underway to examine the suitability of IR imaging as a screening modality for skin cancer. It is expected, for instance, that a malignant melanoma will show a faster temperature increase than its surrounding tissue. This assumption is motivated by the fact that this kind of cancer has a very active metabolism and possibly angiogenesis. However, the basalioma shows the opposite behavior because it is encapsulated from the surrounding tissue. A point which should be mentioned again is that one can only see the temperature distribution on the surface of the human body. IR imaging cannot directly reveal the inside of a body. A further discussion of technical difficulties can be found in [18.17].

References

18.1 T. Furukawa: *Biological Imaging and Sensing* (Springer, Berlin, Heidelberg 2004)

18.2 R.P. Madding, G.L. Orlove, H. Kaplan: Twenty-five years of Thermosense: A historical and technological retrospective, SPIE Proc. **5073**, 1–16 (2003)

18.3 G.C. Holst: *Common Sense Approach to Thermal Imaging* (JCD, Winter Park 2000)

18.4 G.C. Holst: *Testing and Evaluating of Infrared Imaging Systems*, 3rd edn. (JCD, Winter Park 2008)

18.5 N. Schuster, V.G. Kolobrodov: *Infrarotthermographie* (Wiley-VCH, Weinheim 2000)

18.6 O. Dössel: *Bildgebende Verfahren in der Medizin* (Springer, Berlin, Heidelberg 2000), in German

18.7 T.M. Buzug, S. Schumann, L. Pfaffmann, U. Reinhold, J. Ruhlmann: Skin-tumour classification with functional infrared imaging, Proc. 8th IASTED Conf. Signal and Image Processing (SIP 2006) (Acta, Honolulu 2006) pp. 313–322

18.8 A. Merla, L. Di Donato, G.L. Romani: Infrared functional imaging: analysis of skin temperature during exercise, Proc. 24th Annu. Int. Conf. IEEE EMBS (IEEE-EMBS, Piscataway 2002) pp. 1141–1142

18.9 W. Stolz, D. Hölzel, A. Riemann, W. Abmayr, C. Przetak, P. Bilek, M. Landthaler, O. Braun-Falco: Multivariante analysis of criteria given by dermatoscopy for the recognition of melanocytic lesions, 15th Meet. Am. Acad. Dermatol., Dallas, Book of Abstracts (1991) pp. 7–12

18.10 R.A. Fiorini, G. Dacquino, G. Laguteta: A new melanoma diagnosis active support system, Proc. 26th Annu. Int. Conf. IEEE EMBS, San Francisco (IEEE-EMBS, Piscataway 2004) pp. 3206–3209

18.11 G. Zouridakis, M. Doshi, N. Mullani: Early diagnosis of skin cancer based on segmentation and measurement of vascularization and pigmentation in Nevoscope images, Proc. 26th Annu. Int. Conf. IEEE EMBS, San Francisco (IEEE-EMBS, Piscataway 2004) pp. 1593–1596

18.12 T.M. Buzug, S. Schumann, L. Pfaffmann, U. Reinhold, J. Ruhlmann: Functional infrared imaging for skin-cancer screening, Proc. 28th Annu. Int. Conf. IEEE EMBS, New York (IEEE-EMBS, Piscataway 2006) pp. 2766–2769

18.13 R. Maleszka, M. Rozewicka, M. Parafiniuk, A. Kempinska, D. Mikulska: Trial of thermographic investigations application in patient with psotriatic arthritis, Dermatol. Klin. **5**, 11–15 (2003)

18.14 A. Zalewska, B. Wiecek, A. Sysa-Jedrzejowska, G. Gralewicz, G. Owczarek: Qualitative thermograhic analysis of psoriatic skin lesions, Proc. 26th Annu. Int. Conf. IEEE EMBS, San Francisco (IEEE-EMBS, Piscataway 2004) pp. 1192–1195

18.15 H.H. Pennes: Analysis of tissue and arterial blood temperatures in the resting human forearm, J. Appl. Physiol. **1**(2), 93–122 (1948)

18.16 T.R. Gowrishankar, D.A. Stewart, G.T. Martin, J.C. Weaver: Transport lattice models of heat transport in skin with spatially heterogeneous, temperature-dependent perfusion, BioMed. Eng. OnLine **3**, 42 (2004)

18.17 A.Z. Nowakowski: Limitations of active dynamic thermography in medical diagnostics, Proc. 26th Annu. Int. Conf. IEEE EMBS, San Francisco (IEEE-EMBS, Piscataway 2004), 2004) pp. 1179

19. Endoscopy

Martin Leonhard, Klaus-Martin Irion

Endoscopy has an established position in modern medical diagnosis and treatment. The advantages of endoscopy are important for all those involved and are decisive for the success we enjoy today: for the doctor it means improved diagnosis and treatment; for the patient it means minimally invasive approaches, reduced trauma, reduced risk of infection, quicker wound healing, and often, shorter stays in hospital; and as a result for the hospital and society this translates into reduced costs. This general statement can also be looked at in considerably more detail for individual indications, but that is not the purpose of this chapter. The age of purely diagnostic endoscopy is far behind us. Endoscopy is either replacing open surgery and microsurgical procedures or now exists in competition with them in a wide range of fields. In this respect, medical and medical technological advances are occurring continuously, and there appears to be no end to this development in sight. The field is seen as highly innovative with new approaches emerging continually, of which, however, only some will prove to be both a financial and a clinical success.

19.1 **Basics** ... 380
- 19.1.1 Terminology ... 380
- 19.1.2 Endoscopy and the Strengths of Optical Perception ... 380
- 19.1.3 Challenges ... 381
- 19.1.4 History ... 382

19.2 **Endoscopes and Endoscopic Accessories** ... 382
- 19.2.1 Light Sources and Illumination ... 382
- 19.2.2 Imaging: Endoscopes and Image Sensors ... 383
- 19.2.3 Video Chain ... 386
- 19.2.4 Monitors ... 386
- 19.2.5 Image and Video Documentation ... 386
- 19.2.6 Telemedicine in Endoscopy ... 387
- 19.2.7 Endoscopic Instruments ... 387
- 19.2.8 Preparation of Endoscopes and Instruments ... 388
- 19.2.9 Peripheral Units ... 389
- 19.2.10 Endoscopy Workstations ... 390

19.3 **Integrated Operating Theaters** ... 391

19.4 **Medical Applications** ... 393
- 19.4.1 Most Common Procedures ... 393
- 19.4.2 Special Procedures ... 393
- 19.4.3 Miniaturization ... 394
- 19.4.4 Controlled Access ... 394
- 19.4.5 Alternatives to Surgical Microscopes ... 394
- 19.4.6 Gastroenterological Endoscopy ... 394

19.5 **Tissue Differentiation** ... 394
- 19.5.1 Molecular Imaging/ Fluorescence Endoscopy ... 395
- 19.5.2 Infrared Endoscopy ... 396
- 19.5.3 Zoom Endoscopes, Endomicroscopes ... 396
- 19.5.4 Ultrasound Endoscopes ... 396
- 19.5.5 Optical Coherence Tomography ... 397
- 19.5.6 Virtual Chromoendoscopy ... 397

19.6 **Further Future Developments** ... 398
- 19.6.1 Natural Orifice Translumenal Endoscopic Surgery – NOTES ... 398
- 19.6.2 Single Port ... 399
- 19.6.3 Simulation ... 399
- 19.6.4 Endorobotics ... 400
- 19.6.5 Capsule Endoscopy ... 401
- 19.6.6 Endonavigation ... 401

References ... 402

19.1 Basics

Endoscopes can either be introduced through natural orifices (mouth, nose, ear, esophagus, trachea, urethra, rectum, vagina or even the tear duct in the eye or the lactiferous ducts of the breast) or via small, artificial incisions in order to reach the organs of the abdominal cavity or enter joint spaces.

Successful endoscopic concepts always reflect the challenge between the maximum possible reduction of trauma associated with the surgical approach, by reducing the external diameter of instruments on the one hand, and, on the other, the maximum image quality and possibility for manipulation and interventions which favor larger telescopes and large instruments or working channels. In addition, the general parameters are defined by human anatomy and technological capacity.

19.1.1 Terminology

Endoscopy concerns the visualization of concealed body cavities, as is clear from the Greek roots of the word ἐνδοσ (*endos*, Greek: inside) and σκοπεῖν (*skopein*, Greek: to look). Endoscopy is a medical technology and not a clinical speciality in its own right. Today it is daily routine in many medical fields [e.g., gastroenterology, abdominal surgery, urology, gynecology, sports medicine, ear–nose–throat surgery (ENT), pneumology] – in fact there are only a few fields in which endoscopy is not represented.

The names of different endoscopic procedures are generally composed of the term for the organ which is the object of the investigation, followed by the suffix "scopy", e.g., gastroscopy, laparoscopy, cystoscopy, hysteroscopy, arthroscopy, laryngoscopy, and bronchoscopy, to give a selection of examples from the fields listed above.

Seen from this angle, names such as *resectoscopy* and *videoscopy* are inappropriate, although we encounter them from time to time. These terms are derived from instruments or types of endoscope. In this case, the instrument concerned is the resectoscope (an instrument from urology and gynecology), or the videoendoscope, which can now be found in a wide variety of applications and describes an endoscope which, amongst other things, is equipped with a distal image sensor.

The term *keyhole surgery* is often employed in scientific journalism as a striking term for endoscopy, which restricts the possibilities of endoscopy to purportedly secretive observation. On the contrary, thanks to the modern possibilities for using images in the training of doctors and when informing the patient, endoscopy has, in fact, become very transparent and with its diverse treatment possibilities has advanced far beyond the boundaries of optical diagnosis. We regard the term endoscopy as being established; the surgical element of endoscopy is often referred to as minimally invasive surgery (MIS). We thus use these terms throughout.

19.1.2 Endoscopy and the Strengths of Optical Perception

When we are faced with a complex situation, the first thing we want to do is *to get a picture* of it. When we have correctly understood something, we say "I see." This idiomatic representation brings the optical sense and our sight into the focus of our perception. It is not merely coincidence that sight is our most important sense. The retina is one of the few places in the body where the cellular bodies of neurons can be found outside of the actual central nervous system. Based on the image processing which occurs in the retina, we can regard this from a technical point of view as a type of optical coprocessor.

For the purposes of our discussion, optical image formation is based in principle on three components [19.1]:

1. illumination with visible light,
2. an object with characteristic absorptive and reflectance properties, in special cases with fluorescence and other effects; these properties modulate the illumination light,
3. observation with photosensitive and color-sensitive sensors, either our eye or a semiconductor sensor, and processing in the retina and in the brain or technical means of image processing.

This list already identifies two central components of endoscopy for us: high-quality illumination and imaging. The details of the endoscopic image chain are discussed in Sect. 19.2.3.

Magnification and Resolution

With the naked human eye we can see objects of approximately 0.1 mm, using optical aids such as a microscope this can even be extended to view structures in the submicrometer range. Special endoscopic procedures can be used to view structures measuring just a few micrometers. In principle, it can be asserted that with modern endoscopic equipment the accuracy of detail and resolution of the image considerably surpasses the perception during open surgery and thus allows the doctor to work more precisely. This is also due to the characteristic that, at a short distance from the object, endoscopes have a magnifying effect. Furthermore, this advantage is reinforced by the continuous improvement of optical and electronic systems, most recently by the introduction of high-definition (HD) camera systems.

The resolution of an endoscopic system depends on the optical parameters. As a result, image enlargement with optical zoom is associated with an increase in the information gained, while a *digital* zoom is not.

Color Vision

Our color vision provides us with an additional dimension that other imaging procedures such as magnetic resonance imaging and ultrasound do not offer. In normal observation conditions, the human eye cannot differentiate between more than 256 grey levels (8 bits) [19.2], whereas the endoscopic image delivers 10 bits in three color channels. This is particularly important with respect to the fact that the doctor must differentiate between a multitude of red tones in tissue.

This is not to be confused with procedures such as color Doppler ultrasound, in which a further measured variable (flow speed) is superimposed over the grey value image via a false-color representation.

Significance of Surfaces

One obvious weakness of light is that it cannot penetrate tissue deeply and can only visualize surfaces. Our internal organs are composed of manifold internal surfaces, which have been made accessible by the use of endoscopes. In addition, a multitude of pathological changes occur on surfaces, making endoscopy a very effective diagnostic and treatment procedure. A carcinoma, commonly referred to as cancer, is a malignant tumor originating in epithelial tissue, which can be found on the (internal or external) surface.

It is important to understand that surface structures can be visualized considerably better and more precisely with optical procedures than with other imaging procedures thanks to the color rendering and the excellent spatial resolution. As a result, with the current technical possibilities, all procedures of virtual endoscopy via computed tomography (CT) and magnetic resonance using interpolation procedures only appear to be as accurate but are in practice considerably inferior to optical procedures.

Perception and Processing of Visual Stimuli

The biological aim of sight is to recognize objects, dangers, and surroundings, not primarily to act as an optical or spectral recording system. When it comes to recognizing what is visible in an image, the human observer is usually superior to technical procedures.

Goethe, a well-recognized German poet, focused intently on colors, sight, perception, and their complex interaction. His argument of cause and effect, i.e., *one only sees what one knows*, was expanded by *Berci* in his ground-breaking work *Endoscopy* [19.3] in 1976 into *what ones sees should become known and documented.* Even today this remains a central pillar of endoscopy. Image processing and recognition is one of the tasks for which we are still searching for a final solution.

However, current research is also focusing on the advantages of optoelectronic and biophotonic procedures over pure imaging with white light in order to achieve better, i. e., more objective, tissue differentiation [19.4].

Classification

Information which we cannot see, such as ultrasound, magnetic fields, and x-rays, must first be converted into something that we can understand. The preferred solution is an optical image, which is why these procedures are often referred to as medical imaging modalities, even though the original information cannot be *seen* at all, but is rather measured. Endoscopy is generally not defined as an imaging modality in standard nomenclature, although based on modern endoscopy this is a point which could definitely be discussed further.

19.1.3 Challenges

Sight

Endoscopy has deprived the doctor of a direct view of the site of examination or operation. Although in the meanwhile resolution has been improved to such an extent that it considerably surpasses conventional sight, stereoscopic vision remains limited.

Haptics

The digital (*digitus*, Latin: finger) feeling and *grasping* of organs with the fingers is another important sense which could be relied upon in open surgery but which is now translated by long instruments and thus barely perceivable.

Manipulation

The space available to move and manipulate is restricted by the choice of access; appropriate planning and experience are required to compensate for this. Complex procedures are sometimes performed with expensive telemanipulation systems, although this can normally not be justified from an economic point of view. In addition, the desire for miniaturization restricts the size of instruments (e.g., forceps), which can result in longer interventions.

Conclusion

The restrictions placed on the senses of sight, touch, and to a certain extent, smell as well as the restricted manipulation possibilities which a doctor is prepared to accept in endoscopy must also be clear to the engineer when designing new instruments so as not to broaden this deficit even further. The target should be to open up possibilities with endoscopy which were previously not available.

19.1.4 History

The desire to look into concealed body cavities goes back thousands of years. There is written evidence that instruments for proctoscopy were in use in ancient Egypt [19.5], and specula have been retrieved from Pompeii dating from the first century AD. Technical possibilities and medical advances were mutually dependent. A few milestones are listed here to serve as examples; more detailed information can be found in, for example, *Reuter*'s work on endoscopy [19.6]:

1806 Philipp Bozzini builds the first instrument for introducing light into body orifices, his so-called *Lichtleiter* (light conductor), in Frankfurt (Germany) [19.7].

1868 Adolf Kußmaul (Freiburg, Germany) performs the first gastroscopy on a sword swallower using an open tube and a candle.

1877 Max Nitze (Berlin, Germany) uses an optical lens system and a glowing platinum wire for illumination.

1902 Georg Kelling (Dresden, Germany) performs diagnostic laparoscopy on two female patients.

1912 The Dane Severin Nordentoft reports for the first time on knee arthroscopy. Kenji Takagi (Tokyo) begins performing arthroscopy with a cystoscope in 1919, and Otto Bircher in 1921 in Aarau (Switzerland).

1929 Heinrich-Otto Kalk (Berlin, Germany) is seen as a true pioneer of diagnostic laparoscopy through his work.

1934 John C. Ruddock (Los Angeles, CA) describes laparoscopy as a valuable diagnostic method.

1950 Publications on gynecological laparoscopy begin to appear: Raoul Palmer (Paris, France, 1950), Hans Frangenheim (Wuppertal, Germany, 1958).

1958 Basil J. Hirschowitz presents a flexible gastroscope, which still features a distal light bulb (University of Michigan, MI, USA).

1959 Harold H. Hopkins (Reading, UK) patents a rod lens system that dramatically improves image transmission in endoscopes; the invention is introduced in rigid endoscopes in 1965 by Karl Storz.

1963 Karl Storz introduces illumination with glass fibers in the endoscope as a new technology (cold light).

1985 Erich Mühe in Böblingen (Germany) performs the first laparoscopic cholecystectomy (removal of the gallbladder).

19.2 Endoscopes and Endoscopic Accessories

Modern endoscopy systems are composed of a variety of components. Basically, these core components serve to generate light and illuminate the intracorporeal treatment area optimally while producing images which are then relayed out of the body and visualized optimally. Alongside purely endoscopic visualization, numerous instruments and devices are also required to enable minimally invasive treatment. In principle, one differentiates between rigid and flexible endoscopes.

19.2.1 Light Sources and Illumination

Light transmission has been performed in both rigid and flexible endoscopes via optical fiber bundles since the 1960s. Physically, the transmission of light via a light conducting fiber is based on the principle of total internal reflection between the core and the sheath of the individual glass fiber. Halogen, halide, and xenon light sources are employed as extracorporeal light sources.

This light, focused via the proximal plug, is coupled into the endoscope. Recently, the first light-emitting diode (LED)-based light sources have also become available for endoscopic applications.

The quality of any illumination must be comparable to sunlight. One criterion is the color temperature of a light source, which is based on quantum-mechanical considerations on the radiation of ideal, black bodies. The color temperature of the sun is the same as its surface temperature of approximately 6500 K. A further measured variable is the color rendering index, which is a measure of the quality of the color reproduction.

Light sources based on xenon high-pressure short-arc lamps are now the standard light sources for endoscopy, both in the field of flexible endoscopy and in minimally invasive surgery. These systems are characterized by a very homogeneous white light spectrum and high light intensity. Due to the excellent focusing possibilities, xenon light systems can be used just as efficiently in large- and small-caliber endoscopes. The color temperature comes very close to that of daylight; the color rendering index is excellent. Systems with 180 W are now sufficient for almost all endoscopic applications. The light is guided from the light source over an orderless glass fiber bundle (light cable) to the endoscope by means of total reflection in the glass fiber.

With technical advances in the field of LEDs, the first super-bright semiconductor elements are now available, which can also be used in endoscopic illumination. The advantage of this technology lies in the considerably longer lifetime of the illuminant, the high efficiency factor (i.e., less heat generated), and the small dimensions. This means that, alongside self-contained light sources, LEDs can also be installed distally and proximally in endoscopes. The current disadvantage of this lies in the inhomogeneous white light spectrum and the poor focusability of LEDs in small cross-sections of glass fibers.

19.2.2 Imaging: Endoscopes and Image Sensors

Today, endoscopic image transmission is performed optically, fiber-optically or optoelectronically depending on the type of endoscope. The common element in all endoscopes is a distal objective, which determines the direction of view and the image field.

The image field of an endoscope or the numerical aperture (NA) of the endoscope objective is dependent on the refraction index of the ambient medium. If an examination, for example, a cystoscopy, is performed in an aqueous medium, for the same optical conditions, the image field is smaller than in applications in air or in CO_2, as is the case in laparoscopy.

Rigid Endoscopes

In traditional rigid endoscopes, images are transmitted completely optically via a lens system. In this field, the so-called rod lens system of Hopkins has managed to win widespread recognition because of its outstanding quality. Transmission systems based on rod-shaped gradient lenses have failed to gain ground. Alongside rigid endoscopes with fixed, standard directions of view at 0°, 12°, 30°, 45°, 70°, 90°, and 120°, rigid endoscopes have recently been introduced which feature adjustable directions of view [19.8]. This has the advantage of being able to adjust the direction of view during an intervention without changing the external contours of the endoscope.

Image Transmission with Glass Fiber Bundles

Image transmission over ordered optical fiber bundles occurs primarily in *flexible endoscopes*. Today we differentiate between so-called fully flexible image waveguides and semiflexible image waveguides. While semiflexible image waveguides are predominantly used in semiflexible mini-endoscopes and micro-endoscopes (Fig. 19.1) due to their small-caliber variants of approximately 1 mm (35 000 pixels) or less, traditional flexible fiber endoscopes with flexible image waveguides are increasingly being replaced by optoelectronic endoscopes.

Electronic Image Sensors

Miniature charge-coupled device (CCD) and also complementary metal–oxide–semiconductor (CMOS) image sensors are integrated in the tip of the endoscopes with an objective lens and signal processing technology, delivering electronic images with resolutions many times that of fiber endoscopes. In gastroenterologi-

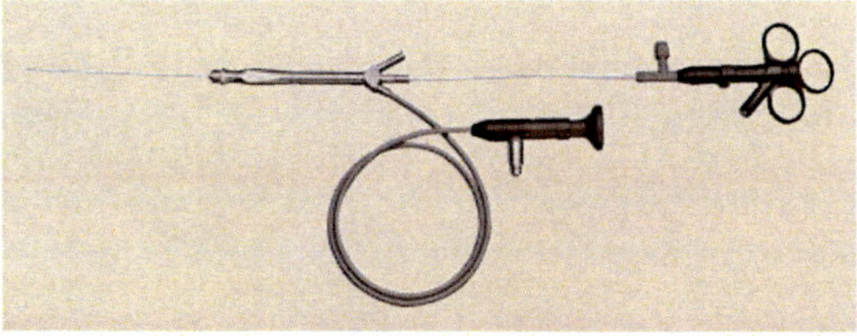

Fig. 19.1 Marchal miniature endoscope sialendoscope (Source: Karl Storz)

Part C | 19.2

cal applications, almost all traditional fiber endoscopes have already been replaced with videoendoscopes.

Video Technology

The first flexible videoendoscopes over 20 years ago featured black-and-white image sensors; the hollow organ was illuminated with sequential red–green–blue (RGB) light, and the compilation of the three color separations was used to calculate and visualize an endoscopic color image continually by means of image conversion. Modern videoendoscopes have color CCD and/or CMOS image sensors with on-chip RGB color filters or color filters for complementary colors (Fig. 19.2).

Video technology is an elementary component of endoscopy. This is not just applicable to the field of flexible videoendoscopes (Fig. 19.3), for which miniaturized image sensors are primarily installed distally, but also to rigid endoscopes, and particularly in the applications of minimally invasive surgery. The use of high-quality, proximal camera systems makes it possible to reproduce the extremely high transmission quality of Hopkins rod lens systems almost perfectly.

Endocameras: 1-Chip and 3-Chip

Proximal endocameras are equipped with one or three image sensors and usually have parfocal zoom objectives (Fig. 19.4). Color rendering is considerably better in 3-chip cameras than in 1-chip systems.

While 1-chip and 3-chip camera systems in standard phase alternating line (PAL) or National Television System Committee (NTSC) standard versions did not come close to the resolution limits of rod lens telescopes, these limits are almost being reached by the latest 3-chip HD endocameras. Alongside the HD endocamera, the remaining components of the transmission elements such as the control unit, monitors, and digital recorders have also been adapted to the high-resolution standard. There is thus no technical reason standing in the way of HD camera systems penetrating the market. The HD requirements are specified by television. There are transmission standards with 1280×720 p (progressive) and 1920×1080 i (interlaced). The progressive standards (1280×720 p/1920×1080 p) only are applied to cameras or monitors. Modern HD image sensors with CCD and CMOS technology satisfy these standards.

Stereoendoscopes

Although the image quality of today's endoscopic systems is extraordinarily high, particularly when performing endoscope-assisted MIS interventions surgeons sometimes miss the depth perception offered

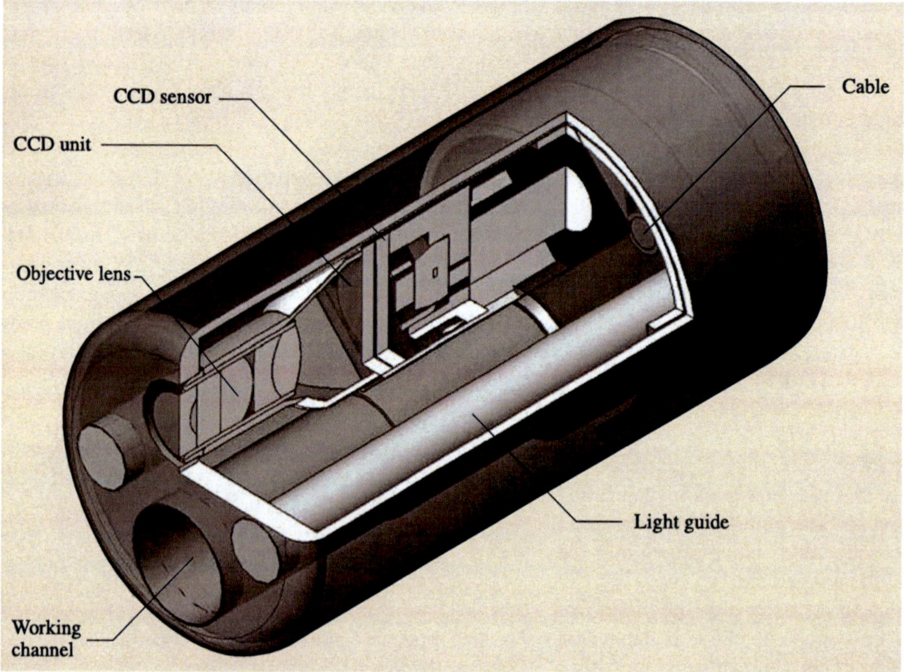

Fig. 19.2 Distal construction of a videoendoscope with miniaturized electronic image sensor (Source: Karl Storz)

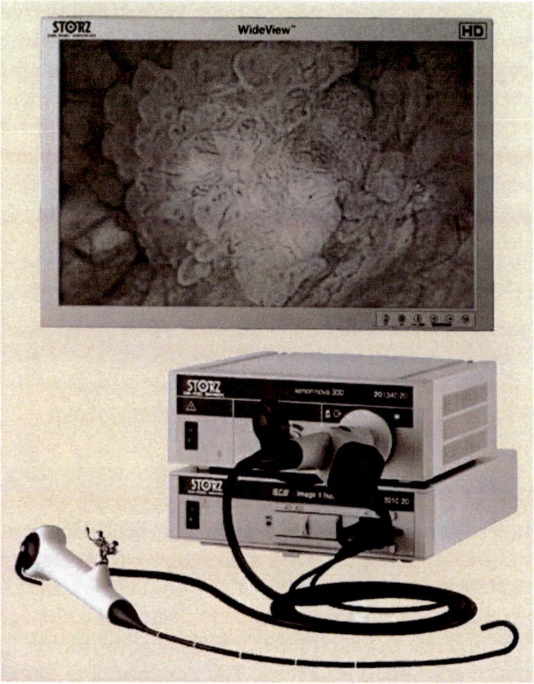

Fig. 19.3 Videocystoscope with distal electronic image sensor

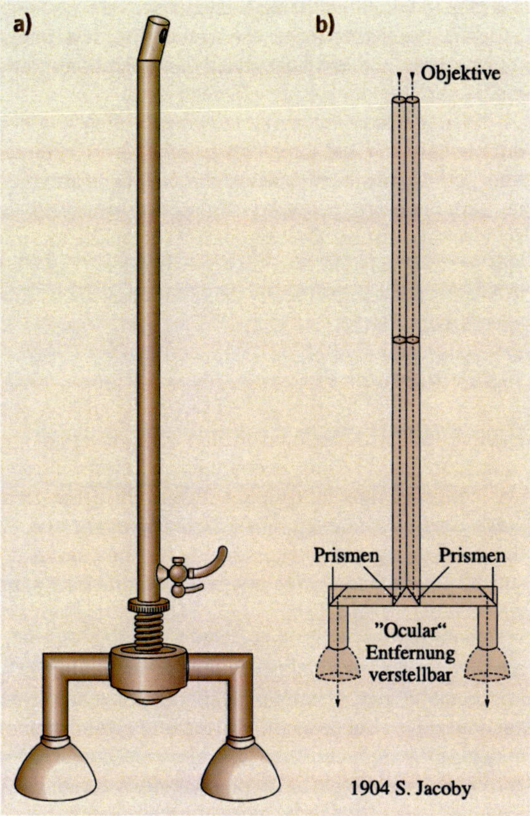

Fig. 19.5a,b Jacoby stereocystoscope (1904) (Source: Reuter, Endoscopy) (drawing adapted)

by their visual system in open operations. There was thus the desire from very early on – as with stereomicroscopy – for stereoendoscopes.

Stereoendoscopes compile endoscopic image information from two different spatial directions – similar to the human eyes. This stereo image pair is delivered to the eyes of the beholder separately via special visualization units.

The first patent for a stereoendoscope was awarded in 1905 [19.9] and was built jointly by Berlin urologist S. Jacoby (Germany), a student of Max Nitze, and the Berlin company Louis & Heinrich Loewenstein in 1904 (Fig. 19.5). It was not until the mid 1990s that a number of manufacturers introduced stereovideoendoscope systems for minimally invasive surgery onto the market. None of these systems proved successful. Studies showed that three-dimensional vision was possible

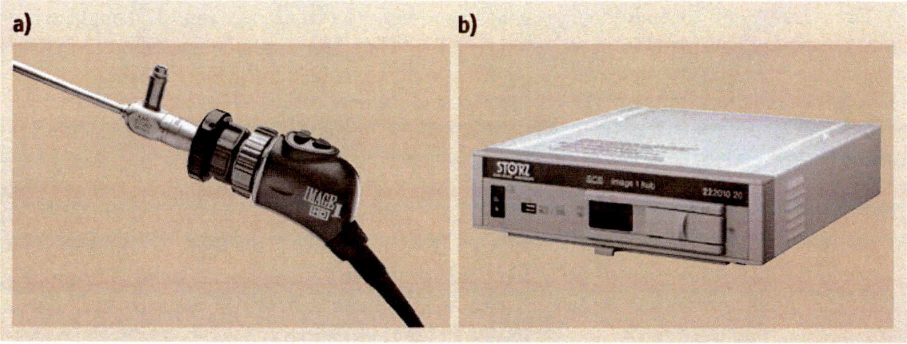

Fig. 19.4a,b High-resolution 3-chip endocamera IMAGE1 HD (**a**) endoscope and 3-chip CCD camera head (**b**) camera control unit (CCU)

over the monitor, but the associated disadvantages such as lower resolution, lower color rendering, less image transparency, and less light meant that operations could not be performed any more efficiently.

Even autostereoscopy systems were offered which did not have any real three-dimensional (3-D) information. Improvements in visualization of 3-D information in particular are necessary for stereoendoscopy to achieve acceptance. One approach is to make the image not only visible from two perspectives, but also to visualize it volumetrically using different focus levels [19.10].

19.2.3 Video Chain

Having dealt with the fundamental components of the light source, light cable, endoscope, and electronic image converter, the following will introduce the term *video chain*. As in every chain, it is important that all links are strong and of high quality, as the overall result of a good endoscopic image is determined by the weakest link in this chain.

It is often the *small* things which can bring a high-quality system to its knees. It is thus particularly important to pay attention to all coupling and connection points, for example, as regards cleanliness of optical surfaces. In the same way, electrical connections and signal cables must be checked carefully for kinks or damaged connections which can affect the result. When choosing a video signal, the highest available quality should be selected.

In the video chain, the endoscopic camera (image converter) is followed by the monitor and optionally the image and video documentation system or telemedicine possibilities.

19.2.4 Monitors

Today, thin-film transistor (TFT) monitors are used almost exclusively in the visualization of endoscopic images. Their quality has increased considerably in recent years. The change in the operating theater away from cathode-ray tube (CRT) monitors towards flat-screen monitors around 2005 occurred at a time when the image reproduction of flat-screen monitors was still considerably below the possibilities offered by CRT monitors. The aura of the new, the expectation that this deficit would be resolved, and the weight advantage, which allows easy and flexible positioning of the screen, paved the way for this premature technological change.

Important criteria when selecting a good monitor are excellent color rendering, as large as possible a viewing angle, and very low latency, i. e., no delay between the operator's actions and visualization thereof on the monitor. In particular the latter point represents an important difference in comparison with monitors for nonmedical applications. In contrast to monitors for other imaging modalities, in endoscopy excellent color resolution is decisive – particularly with respect to the color tones of human tissue. Technical specifications are good for an initial comparison, but true performance and differences between units only become visible in a direct comparison in the operating theater.

19.2.5 Image and Video Documentation

In the beginnings of endoscopy, image documentation was an almost secretive art, as only the examiner was in a position to view the examination field. However, the wish to share this view with fellow examiners was expressed from early on.

Today, the doctor is obliged to document pathological findings and the properly performed operation in image or digital video form. The commonplace, high-quality, electronic image sensors available today make excellent image documentation possible. Of course, an important factor when recording images is that there is optimal visibility and that the site is well irrigated.

Digital archiving of still images is standard today. Because, in general, an electrical connection exists between the recording unit and the image sensor and thus at least indirectly with the patient, the documentation unit must be designed as a medical device. Conventional consumer products are not permitted.

Additionally, systems tailored especially to medical applications allow the doctor to record images systematically and specifically for the intervention being performed and then to prepare these in the corresponding surgical reports. The units must be simple to handle, and in particular, images must be simple to capture [19.11]. There are a number of different systems on the market, which in some cases are tailored to very specific medical applications (Fig. 19.6).

Alongside the capture of still images, documentation of video sequences is playing an increasingly important role. With the increased requirements vis-à-vis resolution, video sequences in HD format (720 p, 16:9) in MPEG2 can achieve data rates of approximately 12 Mbit/s or memory requirements of 90 MB/min. For 1080 p, approximately 20 GB is required per hour of recording time.

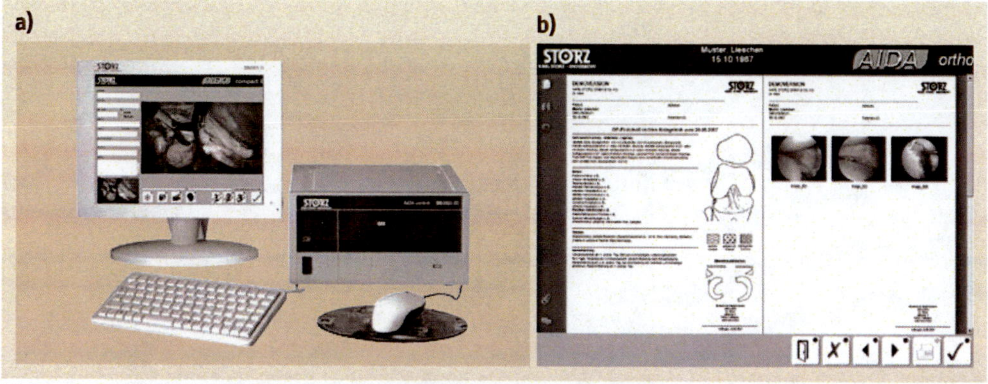

Fig. 19.6a,b Clinically approved system for digital storage of patient, image, and video data. This model is tailored to orthopedic applications: **(a)** AIDA compact ortho, **(b)** documentation mask for the surgical protocol of partial resection of a medial meniscus

The integration of patient-related (image) data in existing hospital information systems (HIS) and picture archiving and communication systems (PACS) is becoming ever more important and should be delivered by high-quality endoscopic image and video memory systems (Sect. 19.3).

19.2.6 Telemedicine in Endoscopy

Broadband data links now allow us to transmit endoscopic video images to any location. This can be in order to get a second opinion or for teaching purposes; it could be to a location within the same hospital or even another continent. The possibility to interact with the recipient is important, either simply via speech or even additionally via graphic interfaces (telestration). For telemedical applications in endoscopy, it is very important that image transmission be temporally synchronized.

19.2.7 Endoscopic Instruments

Alongside the endoscope, which takes care of optical imaging, endoscopy requires a range of further instruments for insertion and manipulation.

Access Instruments

Flexible endoscopes are generally used without further insertion aids. If required, a lubricant may be applied and/or local anesthetic administered. This applies to gastroscopes, colonoscopes, cystoscopes, nasopharyngoscopes, bronchoscopes, etc. These flexible endoscopes are inserted under visual control.

Rigid endoscopes are usually inserted via an insertion tube, which is referred to differently depending on the application. Rigid endoscopes are usually inserted under local or general anesthetic.

To examine the bladder in urology, a single *shaft* is required for a diagnostic procedure and a double shaft system for a surgical intervention, whereby the continuous irrigation provides clear visibility and removes resected tissue or liquid clouded by blood.

In laparoscopy, access to the abdominal cavity (peritoneum) is achieved using a Veress-needle. This step allows, once the peritoneum has been insufflated with gas (CO_2), the creation of a working space without visual control. A rigid endoscope is then introduced via the trocar which is equipped with a sharp trocar mandrel. Additional instruments can then be introduced into the operating region via other trocars under visual control.

Manipulation Instruments

Depending on the application, endoscopy allows the use of a wide variety of instruments which can be used to manipulate tissue, such as exploratory probes, grasping forceps, dissectors, retractors, scissors, hooks, suction and irrigation catheters, resection loops, needle holders, staplers, and clip applicators. Figure 19.7 shows examples of different grasping forceps and a complete set for laparoscopic partial nephrectomy. The range of different instruments is enormous; the leading manufacturers offer several thousand different instruments for the most diverse medical indications.

Dimensions of Minimally Invasive Instruments

The dimensions of the instruments are once again subject to two parameters: As minimally invasive as

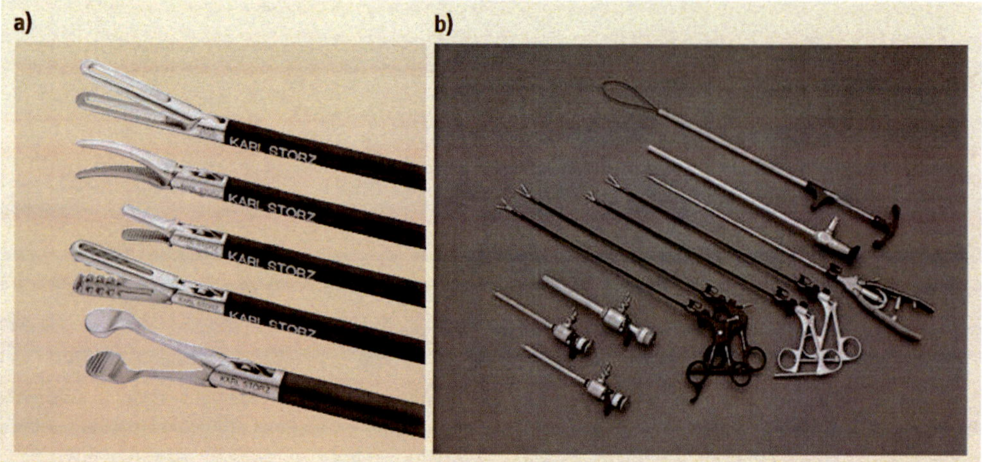

Fig. 19.7a,b Laparoscopic grasping forceps. **(a)** A selection of different jaws, and **(b)** an endoscopic set for laparoscopic partial nephrectomy

possible, and as effective as possible as far as optical image and mechanical manipulation are concerned. For this reason, purely diagnostic instruments are generally more slender and are used without or only under local anesthetic; surgical treatment endoscopes are usually larger in diameter and normally require general anesthetic. The typical diameter of minimally invasive instruments ranges from 10 to 5 mm, which can be seen as standard in laparoscopy, through to 1.9–3.5 mm for minilaparoscopy instruments. Minilaparoscopy is a rediscovered method originally used in internal medicine for liver diagnosis. The approach was first elected by the internists Georg Kelling, Heinrich-Otto Kalk, and John C. Ruddock at the beginning of the 20th century.

Instruments which are introduced into the working channels of flexible endoscopes generally have a diameter ranging from 2.2 to less than 0.5 mm.

19.2.8 Preparation of Endoscopes and Instruments

The significance of hygiene and proper instrument processing has been becoming ever more important in recent years, particularly due to new versions of pathogens which are not sufficiently deactivated by the traditional and often manually performed methods of instrument decontamination and thus potentially pose a hidden risk of human-to-human transferral of infectious microorganisms via surgical instruments. A further reason for this is the ever smaller and more complex nature of instruments, which accordingly places higher requirements as regards the creation of a hygienically sterile state. In principle, all instruments contaminated with microorganisms, particularly invasively and minimally invasively employed instruments, can cause infections. It is thus essential that they be subjected to a safe, validated reprocessing before being reused [19.12].

The reprocessing cycle usually consists of the following steps:

- Cleaning, defined by bioburden reduction by 1–4 log levels (bioburden reduction to 10^{-1}–10^{-4} of initial contamination).
- Disinfection with bacteria reduction to 10^{-4}–10^{-5}.
- Or high level disinfection (bacteria reduction to 10^{-6}), if appropriate for the procedure and surgical instrumentation.
- Sterilization with bacteria reduction to 10^{-6} (killing 99.9999% of bacteria or leaving just one out of one million alive).
- Sterilization process with a double safety factor (so-called *overkill* processes with bacteria reduction to 10^{-12}) which destroys all forms of microbial life.

Cleaning

Cleaning is primarily the mechanical depletion of contaminations, for example, by brushing the surfaces, rinsing, and/or auxiliary use of ultrasound. The effectiveness can be boosted through the use of chemical agents based on enzymes, aldehydes, and other active agents. Thorough cleaning is essential before (high level) desinfection and sterilization because inorganic and organic materials that remain on the surfaces of the instruments can interfere with the effectiveness of these processes.

Disinfection

Disinfection is based either on the use of chemical substances with the corresponding spectrum of activity or on the action principle of the thermal denaturing of organic compounds with the help of water at a temperature of 93 °C or higher, which fully coats the surface of the instrument.

In the meanwhile it has become commonplace to perform these procedures in automated washer–disinfector machines, whereby the parameters relevant for success, such as the concentration of the chemicals, their exposure times, and the temperature, are controlled and monitored accordingly. This avoids the possibility of human error. The reduced handling of contaminated instruments and aggressive chemicals thanks to such fully automated processes also helps to protect staff.

High Level Disinfection. High level disinfection (HLD) is a process that eliminates all pathogenic microorganisms, with the exceptoon of bacterial spores. HLD is normally accomplished using a chemical disinfectant solution.

Sterilization

Following properly performed cleaning and disinfection, the instrument is subjected to sterilization depending on the application at hand. The internationally preferred procedure, in health care facilities due to sterility assurance and cost efficiency, remains steam sterilization (autoclaving) with moist heat (pressurized saturated steam) at a temperature of 134–137 °C (272–279 °F) for an exposure time of 3–5 min, which is subject to different regulations in each country (in the USA, AAMI ST 79) [19.13].

For instruments declared as thermolabile (based on the materials used) there are a number of alternative, low-temperature sterilization processes available such as traditional gas sterilization with ethylene oxide or formaldehyde. Additionally, there are also newer processes available which are based on hydrogen peroxide (H_2O_2), peracetic acid or ozone, which must be assessed and selected for the individual goods to be sterilized in terms of their efficiency, suitability, and material compatibility.

Care

An additional, important aspect in the scope of the reprocessing process of a medical device is service and care, which are usually performed as part of the functional check after successful cleaning/disinfection but before sterilization, and which serve to ensure the functioning and value retention, particularly for instruments with moveable parts. In this step, the moving parts are oiled locally with a silicone-free instrument oil which is biologically harmless (biocompatible) and does not hinder the selected sterilization process (e.g., steam penetration through the care agent film to the surface beneath). The instrument is operated several times to ensure distribution of the care agent within the joint.

Structural Design

Reprocessing of medical devices is a complex process, the success of which is dependent on multiple factors including, importantly, the instrument itself and its mechanical design layout and also the selection of the materials used.

Requirements and Directives

On the topic of hygiene, the legal and regulatory parameters require qualified processes and staff who are suitably trained and qualified.

Cleaning [19.14], disinfection, and sterilization [19.15, 16] are subject to a multitude of field-specific recommendations [19.17], international standards, and national regulations and represent a continuous challenge for both users and manufacturers of medical devices.

19.2.9 Peripheral Units

Alongside the units for videoendoscopic image capture and documentation, a whole range of additional units for assisting surgery are required.

Insufflators and Suction/Irrigation Systems

All MIS interventions performed in the abdomen require the creation of a pneumoperitoneum, which allows organs and tissues to separate slightly and become visually and instrumentally accessible while the abdominal wall remains closed. Gas insufflators are used for the generation and maintenance of the pneumoperitoneum. The regular changing of instruments and suction, e.g., of liquids, result in continuous pressure and volume losses, which place great demands on modern gas insufflation units. The units should provide a maximum flow rate of 30 l/min and not exceed an intra-abdominal pressure of 15 mmHg. To avoid hypothermia in patients undergoing longer interventions, the new generation of units have an integrated module for warming and/or moistening the gas.

Alongside the insufflation of gases it is often necessary in MIS interventions to irrigate the operating site or the lens with liquid or remove blood and tissue remnants. In addition, efficient irrigation can also be used for hydrodissection, i. e., to dissolve adhesions. Modern suction and irrigation systems unite both requirements.

Energy Application Systems

A therapeutic effect is achieved by applying energy in order to cut, separate, initiate hemostasis, coagulate, disintegrate calculi, and for many other actions.

High Frequency

Amongst other things, high-frequency (HF) current (Chap. 34) is used to create thermal effects endoscopically (Fig. 19.8). This can be used to dissect soft tissue and coagulate existing hemorrhaging under endoscopic visual control. There is a general differentiation between monopolar and bipolar HF electrosurgery. Monopolar technology is used most often in endoscopy.

Laser

Lasers have been used in medicine since shortly after their technical implementation at the beginning of the 1960s. Different media allow different laser wavelengths in the visible, infrared, and ultraviolet ranges of the spectrum. Depending on the wavelength, pulse duration, and interaction duration, it is possible to achieve a truly characteristic interaction in the tissue [19.18]. Important interaction partners are water and hemoglobin, the absorption properties of which define how deeply the light penetrates (Chap. 29). Lasers are used widely in ENT [19.19] and urology [19.20], where they are employed for lithotripsy, for tumor coagulation, to separate stenoses, and for prostate enucleation.

Laser light is transmitted in endoscopy via flexible, thin glass fibers; for special applications in the far infrared, mirror systems (CO_2 lasers) are also used. Recent years have seen these systems become ever more compact and also very service-friendly thanks to the use of laser diodes.

Other Energy Delivering Systems

With photodynamic therapy (PDT) an endoscopic tumor treatment is possible which exploits the interaction of (laser) light, oxygen, and a photosensitizer that accumulates selectively in malignant tissue [19.21].

Energy can also be withdrawn endoscopically using *cryoprobes* to destroy tumor tissue by localized supercooling, or be delivered in concentrated form using *ultrasound* to separate tissue.

Systems for Further Interventional Procedures

As endoscopy is very multifaceted, special interventional systems are used for the most diverse of applications. These include

- implants such as endoluminal stents or occluders, which can be introduced endoscopically,
- motorized instruments used to ablate bone and hard tissue (shavers, drills) or to cut soft tissue (morcellators).

19.2.10 Endoscopy Workstations

Endoscopy workstations are tailored to the requirements of the respective application.

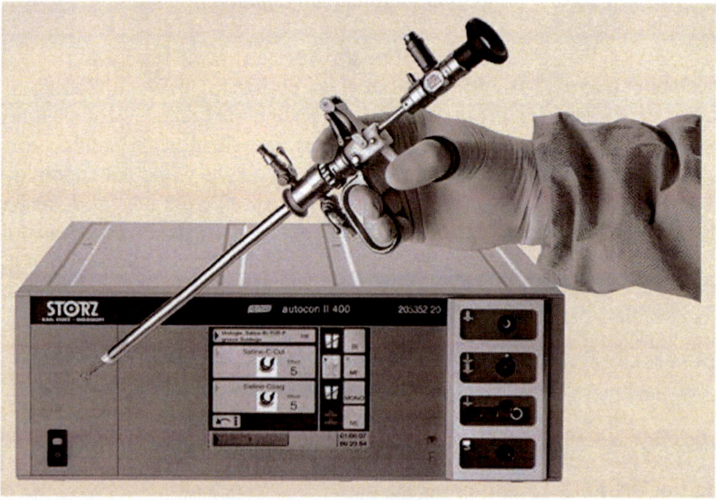

Fig. 19.8 High-frequency units (monopolar/bipolar) with an endoscopic instrument set for transurethral resection (TUR)

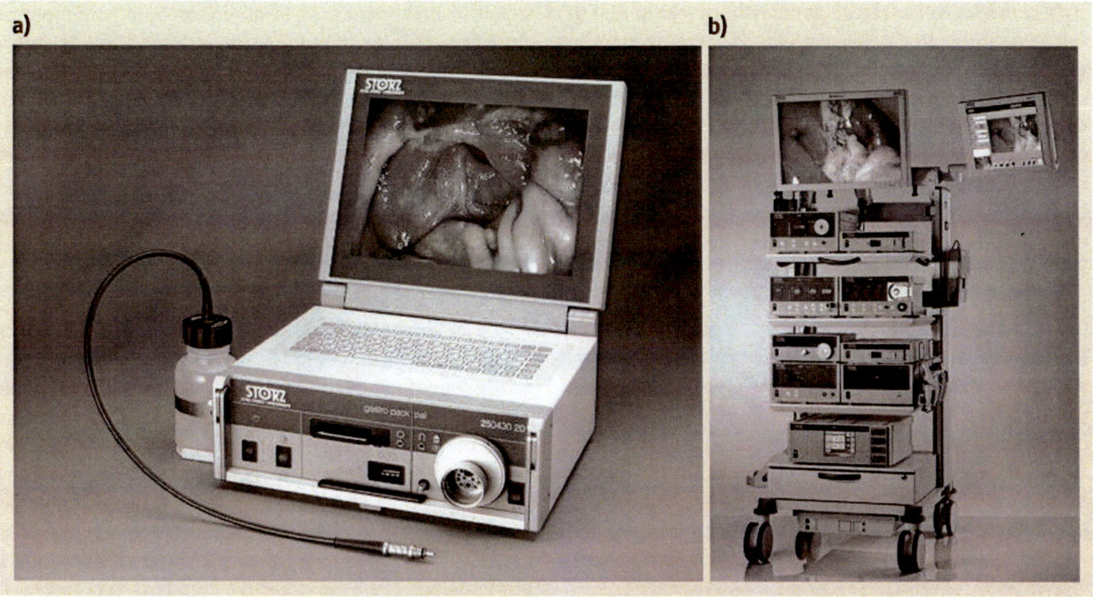

Fig. 19.9a,b Two different endoscopy workstations: **(a)** compact system for outpatient departments with camera control unit, light source, monitor, keyboard, and suction/irrigation pump (Gastropack), **(b)** equipment trolley for the operating theater additionally with HF surgery system, insufflator, and still image and video archiving system

For clinical outpatient departments, emergency medicine, and even doctors' practices specializing in endoscopy, the focus is on a compact system which is fundamentally concentrated on illumination and image rendering. Biopsies and minor interventions are possible. These systems are light and easy to transport, and have been available on the market for quite some time (Fig. 19.9a).

Day clinics, which usually perform routine operations endoscopically and under general anesthetic, and fully equipped operating theaters in hospitals require a larger range of treatment options. As a result, a modular concept is called for, exactly calibrated to the needs at hand. The components are installed on a mobile equipment trolley and can be positioned at the required location in the operating theater. Mobility over greater distances is also possible to some extent (Fig. 19.9b).

Newly designed operating theaters are now predominantly equipped with fully integrated workstations (Sect. 19.3).

19.3 Integrated Operating Theaters

As a result of the rapid development in minimally invasive surgery, there is now a wide variety of devices in the operating theater in comparison with the past. The operation and handling of these systems, which are often from different manufacturers, have thus become more and more complex and time consuming. In addition, the surgeon requires a considerable amount of time to document all the points relevant to surgery and compile a report. On top of all this, the surgeon is continuously being confronted with more complex MIS interventions, and time budgets are always being cut. These increasing requirements can only be satisfied by improving the ergonomic handling of units and instruments and comprehensive computer-assisted documentation and data provision [19.22]. The design of system workstations and new concepts for integrated operating theaters set benchmarks for the logistical harmonization of operating activities, reduce stress for the surgeon and assistants, and thus reduce operating and changeover times.

System integration allows, amongst other things, central control and operation of endoscopy units and further devices capable of being integrated such as HF units, operating tables, operating lamps, and operating light cameras, as well as documentation systems and the telemedicine unit via a central touch-sensitive monitor (touchscreen) directly from the sterile area and/or at the nurse's workstation [19.23]. There is also the possibility to preset start configurations for the system parameters and telemedicine basic settings with personal or intervention-specific presettings, so that the units set themselves automatically when the respective configuration is selected. This saves time on the one hand and contributes to quality assurance on the other. The user interfaces of the integrated devices are represented one-to-one on the system's user interface. Users thus do not have to familiarize themselves with a new concept. Alongside the realistic user interface for each individual unit there is also a freely definable overview display of the most important unit parameters. The individual, selected units can also be controlled from this display mode. In addition, the visualization of unit parameters as well as alarm and status messages from the operating system is realized in a section of the touchscreen clearly visible to the surgeon.

A further dimension of operating room (OR) integration is data management, which results from the obligatory documentation requirement. In this respect, the systems offer all necessary functions for holistic, accurate documentation of all endoscopic and open interventions. User-friendly entry systems allow the simple capture of still images, video sequences, and spoken commentaries on findings and operative interventions directly from the sterile area by simply activating the touchscreen, a footswitch or the camera head buttons.

A digital imaging and communications in medicine (DICOM) interface can be used to give the system access to DICOM worklists available on the hospital network, which allows rapid transfer of patient data.

Alongside system integration and data management, audio and video technology for communication and data transfer also form part of the holistic concept of an optimally integrated operating solution. The different video signals from endoscopic and room cameras as well as from the PACS and HIS systems can also in principle be connected and routed as required and even telecommunicated, thus allowing them to be displayed outside of the operating theater. Audio and video technologies are also controlled via the touchscreen at the nurse's workstation. The man–machine standard interfaces often already feature intelligent voice-controlled systems.

The focus on integrated OR systems (Fig. 19.10), which is increasing productivity and efficiency in many surgical procedures, has put hospital administration in

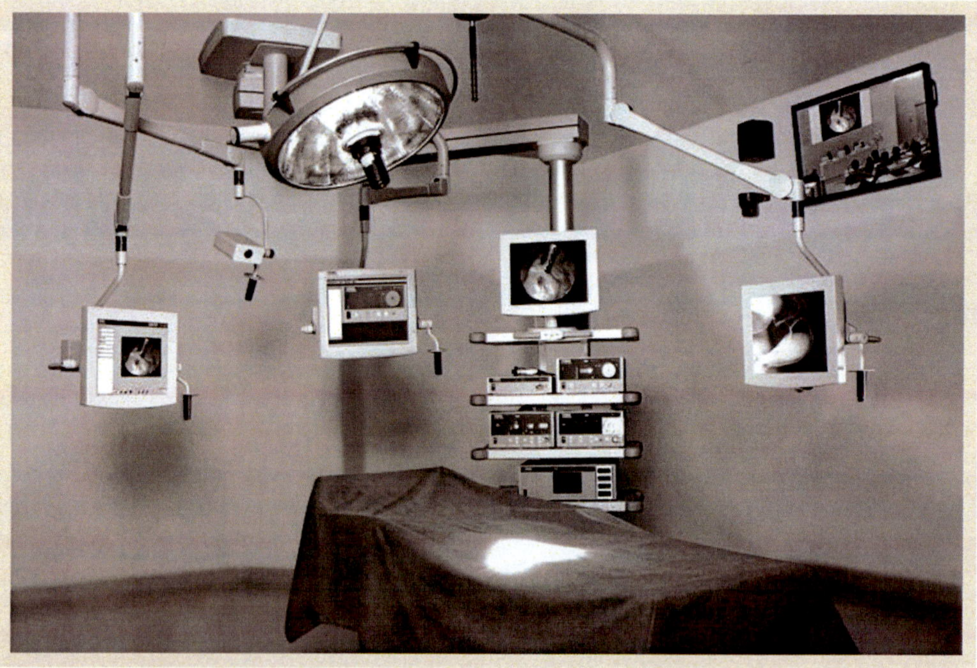

Fig. 19.10 Fully integrated operating room OR1 Karl Storz (Source: Karl Storz)

a position to see the operating theater as an important source of income. In terms of profitability, the operating theater has become a decisive factor in the success of a hospital.

These integration technologies have been recently complemented by efficient OR management systems. Efficient OR management is based on the harmonization of all processes and integrated components combined with the possibility to connect seamlessly with the hospital information system. Such a system (e.g., Orchestrion [19.24]) makes it possible to optimize the planning of operations and staff schedules, and ensures efficient flow of information between the operating theaters and the hospital information system. By optimizing the way in which the operating theater is used, the hospital achieves smoother processes and is thus in the position to offer patients a higher quality of treatment. This tool is helpful in preplanning of operations and coordinates and controls the use of available OR resources. It is thus an important element in the development of strategies for optimal utilization of limited resources. Thanks to the implementation of standardized interfaces, these tools can be easily integrated into the existing information technology (IT) structures of a hospital. The modular design of the system can be adapted to suit the individual requirements of every hospital and offers a dynamic platform to be able to integrate future developments and improvements.

19.4 Medical Applications

Endoscopy is employed in a wide variety of medical and surgical disciplines for diagnosis and treatment. It is routinely used in, amongst others, the different subdisciplines of ear–nose–throat medicine, pneumology, gastroenterology, abdominal surgery, urology, gynecology, and also sports medicine.

Users are found in all strata of the healthcare system, from specialists to ambulatory healthcare centers and day clinics as well as all types of hospitals from primary health care to university hospitals.

19.4.1 Most Common Procedures

The most common surgical endoscopic interventions are arthroscopic interventions, endoscopic sinus surgery, laparoscopic gallbladder removal, transurethral resection of bladder tumors, and benign prostate tissue (BPH) [19.25]. Hernias, bariatric surgery hysterectomy, and appendectomy are some of the most common surgical procedures, although they are only partially performed endoscopically. Additionally, in 2008, more than 16 million diagnostic endoscopic examinations were performed in the US.

19.4.2 Special Procedures

The following illustrates some special endoscopic applications to demonstrate the potential and particular possibilities, but by no means does it claim to be exhaustive.

Part C | 19.4

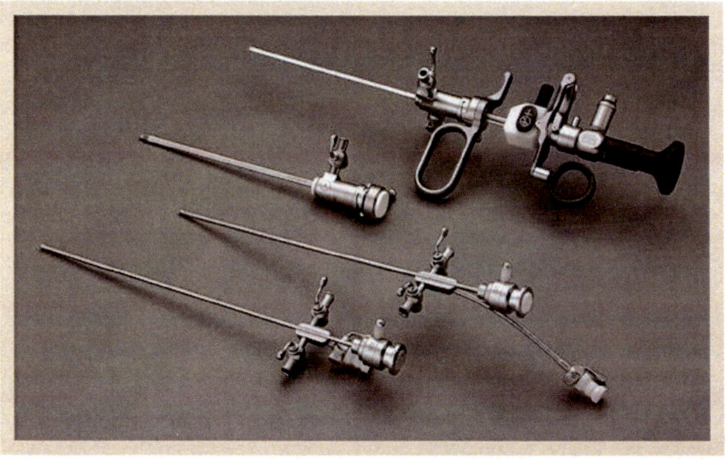

Fig. 19.11 Miniaturization in endoscopy using pediatric urology as an example. *Top*: resectoscope for tissue ablation with HF current, shaft diameter 3.7 mm; *bottom*: cystoscope shafts (Source: Karl Storz)

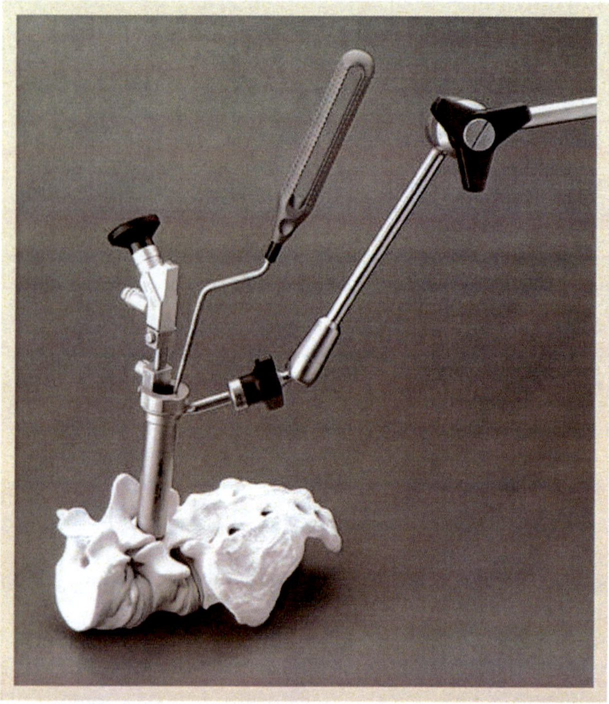

Fig. 19.12 Spinal system for treatment of disc herniations and spinal canal stenoses (Source: Karl Storz)

19.4.3 Miniaturization

Endoscopes can be miniaturized so that they are suitable for operations on children and newborns. To give an example, we cite the incision of the closed ureteral orifices in a newborn baby with hydronephrosis, whereby a small scalpel is introduced via a delicate pediatric cystoscope and used to remove an anatomical anomaly (Fig. 19.11). Today it is even possible to operate endoscopically on the fetus while it is still in the womb. Other applications for mini-endoscopy include examination of the tear duct in the eye, salivary ducts in the mouth (Fig. 19.1), lactiferous ducts in the breast or the fallopian tubes. This is where the finest, flexible fiber-optic endoscopes with diameters of 0.3 mm to approximately 1 mm come into use.

19.4.4 Controlled Access

In some medical applications it was previously not possible or usual to work under visual control. In terms of more minimally invasive and safer accesses, endoscopic assistance has become established in anesthesia and emergency medicine in recent years in attempts to find the right access in cases of difficult intubation.

Endoscopic examination of the uterus (hysteroscopy) is also on the advance in this respect, particularly for performing targeted biopsies or removal of polyps, in comparison with blind curettage. This is driven by improved reimbursement and new indications such as transcervical sterilization. Even if hysteroscopy could be used without causing pain and without anesthesia by any specialist in a practice, enhanced training and logistics is needed to support this procedure in an office environment.

19.4.5 Alternatives to Surgical Microscopes

In the fields where surgical microscopes are used, endoscopic procedures are gaining ground. Endoscopic procedures are becoming increasingly popular in laryngeal surgery, neurosurgery, and in delicate operations in spinal surgery (Fig. 19.12). The depth of field of endoscopic systems is considerably larger than with microscopes. Moreover, endoscopes can be used to look around corners, which is not possible with microscopes as a result of the large working distance.

19.4.6 Gastroenterological Endoscopy

Possibilities to examine the digestive tract are offered by a wide range of extremely differentiated endoscopy systems which can be coupled with ultrasound, zoom or fluorescence procedures. In addition, a piggyback procedure is used for examination of the bile ducts, in which a small-caliber endoscope is introduced through a large-caliber duodenoscope. Gastroscopy and colonosopy are amongst the most commonly employed diagnostic procedures. Fig. 19.13 shows a typical, flexible endoscope from gastroenterology with a distal image sensor.

19.5 Tissue Differentiation

As discussed in Sect. 19.1.2, illumination is modulated by the illuminated tissue and perceived by an image sensor. A variety of procedures now make it possible to gather information about the tissue, being considerably more detailed than the information offered by the colored white-light image.

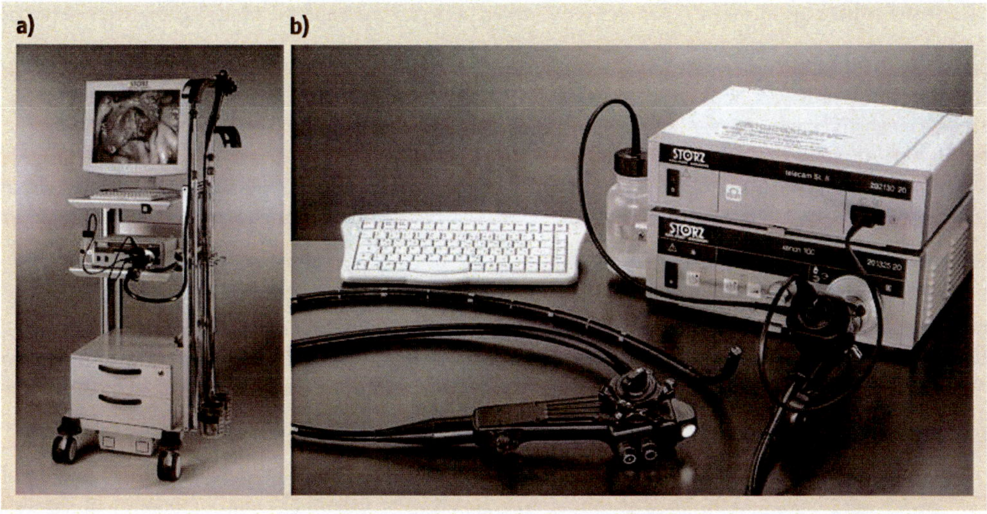

Fig. 19.13a,b Flexible videocolonoscope with distal, high-resolution solid-state image sensor: **(a)** video cart, **(b)** close up look of videocolonoscope, CCU and lightsource

19.5.1 Molecular Imaging/ Fluorescence Endoscopy

Molecular imaging is one of the great beacons of hope in the field of cancer research and early detection, as it makes pathological changes visible at the cellular level. Fluorescence-based endoscopy systems are also of great importance as optical detection units. The procedure itself uses the distinct metabolism of abnormal cells to display fluorescent substances accumulated in the tumor via a molecular mechanism. This optical form of molecular imaging, in which the suspicious area in the endoscopic image lights up in color, was initially used in bladder cancer with great success [19.26].

Carcinomata in situ with a thickness of just a few micrometers ($< 30\,\mu m$) can be visualized with this technique at a very early stage (Fig. 19.14). No other imaging procedure is in a position to do this.

Nowadays, the procedure is also used for numerous other organ systems. The target of various initiatives in molecular imaging is to have new disease-specific marker substances and tailored detection systems available in the future. This can only occur through cooperative alliances between pharmaceutical companies and device manufacturers.

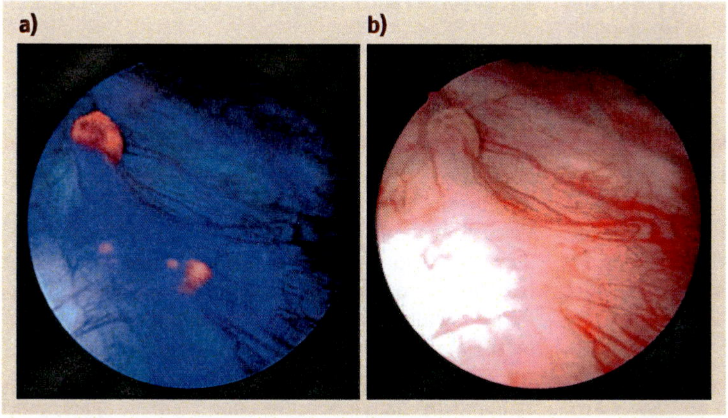

Fig. 19.14a,b Endoscopic cystoscopy of small exophytic tumor with small satellite tumors: **(a)** photodynamic diagnosis (PDD) mode, **(b)** white light (Source: Karl Storz)

19.5.2 Infrared Endoscopy

Infrared endoscopy, i.e., endoscopic representation of image information in the infrared range, has been on the wishlist of diagnostic medicine for many years. For example, it could be used to recognize inflammatory conditions better. However, this target is subject to technological limitations. Endoscopic capture and transmission of near-infrared (NIR) light from 650 nm up to approximately 1000 nm and a maximum of 2000 nm is possible with existing technologies. Image transmission beyond that cannot be achieved with standard optically transparent lenses.

Current research and development activities have thus concentrated on the NIR range. Attempts are being made, amongst other things, to determine perfusion following transplants or visualize enteroanastomotic insufficiencies at an early stage. For intraoperative examination of the anastomosis following intestinal resection, the surgeon has visual and manual control available in open surgery; in laparoscopic interventions, however, only visual, endoscopically assisted control under white light is available. For better control of intestinal perfusion, *Carus* et al. [19.27] first employed intraoperative fluorescence diagnostics for laparoscopic colorectal anastomosis. The basis for these applications is the NIR-transmitting endoscope as well as cameras and optical contrast agents, which are excited by special excitation light to exhibit NIR fluorescence. The intravenously administered marker indocyanine green (ICG) is excited to fluoresce in the infrared region at around 830 nm on exposure to red light. The diagnostic principle has been used since the 1960s to examine the retina, and as a result ICG is approved for the most diverse of applications in many countries. It is hoped that use of ICG fluorescence laparoscopy technology will significantly reduce the relatively high rate of anastomotic insufficiencies in the colon and rectum.

Technically, the ICG fluorescence system is based on an endoscopic 300 W xenon light source, into the light path of which an ICG excitation filter can be swung alternately. As a result of this deep red excitation light, the presence of ICG in blood can be detected as NIR fluorescence even in deeper tissue structures. This fluorescence is detected by an imaging laparoscope via an observation filter, separated from interfering light rays, and converted into an electronic image signal by a special camera which is sensitive to the NIR range. A type of false-color coding is used to convert the infrared (IR) fluorescence information to a visible range and display it on the monitor. Optionally, one can swap between the standard white-light mode and the IR mode.

19.5.3 Zoom Endoscopes, Endomicroscopes

Today, histological microscopic examination is the gold standard for assessing tissue with regard to its nature or malign degeneration. There has long been a desire to be able to perform this laboratory examination in situ at the time of the operation.

As early as 1980, *Hamou* [19.28] described a procedure in Paris in which he dyed the tissue to be examined with methylene blue and then viewed it with a rigid endoscope, which normally delivers a standard wide-angle image or, in contact with tissue and by altering the optical focus level, an endoscopic image with 150× optical magnification. This allowed the recognition of the size of the cell nuclei and anomalies in the cell structure. The procedure was commercialized by Karl Storz under the term contact endoscopy and used in the detection of cervical dysplasia and to examine the larynx. The procedure did not become widely used. Today, magnification endoscopy has found a new impetus in gastroenterology.

The optical magnification of extended tissue structures superimposes the information of many tissue layers on top of each other. Recent years have seen the further development of procedures from microscopy which can be used to produce a defined slice level endoscopically using the confocal principle. To do this, one requires either a distal microscanner or a proximal scanner, which is technically demanding. Alternative procedures are visible on the horizon, but clinical evidence is not yet available.

19.5.4 Ultrasound Endoscopes

Endoscopy is an unrivalled procedure for assessing the surfaces of organs. Ultrasound technology is used to assess deeper structures and processes. Endoscope-assisted ultrasound examinations allow tissue characterization and conclusions about deeper layers, which remain invisible in purely endoscopic observation. Piezoceramic ultrasound elements are integrated either as individual elements or in the form of an array into the tip of the endoscope and emit short, high-frequency ultrasound pulses into the tissue. These sonic pulses are reflected by tissue interfaces and other tissue inhomogeneities at different penetration depths, captured by the same ultrasound element again, converted into electrical signals, and transmitted to an evaluating processor unit. The information from various ultrasound echograms is compiled to create so-called ultrasound B-mode images (brightness images) and visualized for the doctor for further evaluation on the monitor. The frequencies of endosonographic systems are currently in the range 5–30 MHz. Earlier electromechanical scanner units have now been exclusively replaced by electronically controlled array transducers (linear/radial scanners). Interventional operations in particular, e.g., fine-needle biopsies, can be performed under endosonographic control.

19.5.5 Optical Coherence Tomography

Optical coherence tomography (OCT) can be seen as an optical equivalent to ultrasound imaging. The basic principle of OCT is based on white-light interferometry. In this, an arm with known optical path length is used as a reference to the measuring arm. The interference of the signals from the two arms provides a pattern, which one can use to determine the relative optical path lengths within a depth profile. In a one-dimensional raster scan, the ray is then directed transversally in one or two directions, which allows one to record a two-dimensional (2-D) or 3-D tomogram.

Today, OCT is routinely used in ophthalmology; in endoscopy, current applications are still experimental (Fig. 19.15), mostly on the larynx and in the bladder. The penetration depth is around 2 mm.

19.5.6 Virtual Chromoendoscopy

Differentiation between benign and malignant structures is an important aspect of endoscopic diagnostics. To improve conclusions with regard to tissue characterization and tissue differentiation, the application of coloring methods with, e.g., methylene blue, toluidine blue, and indigo carmine has been common for many years. Similar contrast accentuations are achieved by

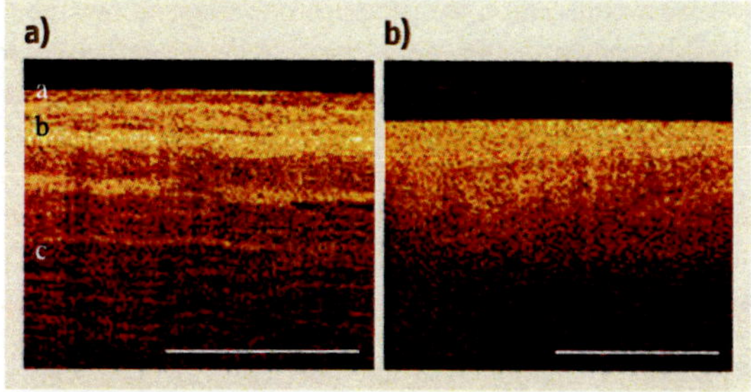

Fig. 19.15 **(a)** Endoscopic OCT images in false-color rendering. Healthy urothelium (*a* urothelium, *b* lamina propria, *c* muscularis propria); **(b)** muscle-invasive lesion, the layer structure is no longer identifiable, scale 1 mm (Source: A. Goh, S.P. Lerner, Baylor College of Medicine, Houston)

the use of acetic acid, for example, in early detection of carcinomata in the cervix, esophagus, and oral cavity.

Several years ago these technologies were joined by spectral image processing technologies, which can be used to perform special real-time color analyses of videoendoscopic images. We differentiate between purely software-based systems (virtual chromoendoscopy) (Fig. 19.16) and systems which visualize a combination of narrowband illumination with specific, electronic color assignment (narrowband imaging, NBI). The idea behind the spectral processing techniques is to improve the recognizability of special structures by limiting optical image information to a narrowband range in comparison with the whole visible spectral range, thus suppressing less relevant spectral information. For this reason, the wavelengths which are of greatest interest in endoscopic examinations are those which correspond to the range in which the blood pigment hemoglobin displays the highest absorption. These lie in the blue between approximately 410 and 440 nm and in the green between 530 and

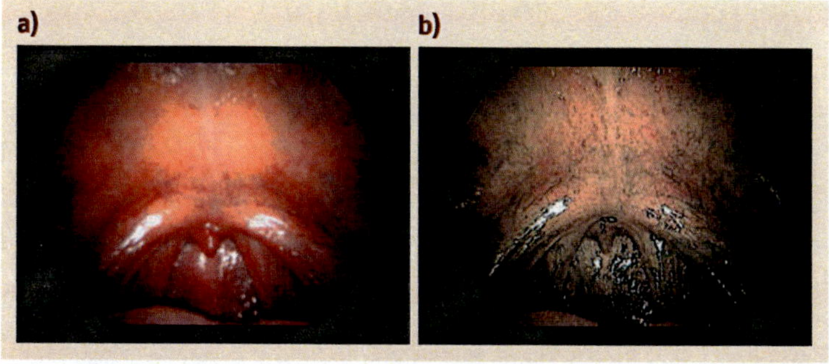

Fig. 19.16a,b Virtual chromoendoscopy: **(a)** white-light image of the palate, **(b)** defined vascular structure (Source: Karl Storz)

580 nm. Restriction to such spectral ranges allows contrast accentuation between tissue regions with higher and lower hemoglobin concentrations. This allows for better differentiation of vascular structures near to the surface. Pathological structures, e.g., in the region of the esophagus, are easier to identify.

19.6 Further Future Developments

Alongside the procedures for tissue characterization (Sect. 19.5) there are a whole host of further innovative approaches. The first clinical implementation has already occurred, but for the most part clinical acceptance cannot yet be foreseen.

19.6.1 Natural Orifice Translumenal Endoscopic Surgery – NOTES

For some years now there have been intense discussions in surgical and gastroenterological circles about further reduction of the trauma associated with the surgical approach. It is starting to look possible to cross the boundaries between internistic gastroenterology and abdominal surgery. To do this, it is accepted that a previously intact organ structure (e.g., the stomach or vagina) will need to be penetrated surgically. In contrast to trauma associated with the surgical approach through the abdominal wall, however, this is seen as acceptable by the protagonists.

Operating via natural orifices appears possible. NOTES stands for natural orifice translumenal endoscopic surgery. The aims were laid down in 2005 in a white paper [19.29] led by *Rattner* and *Kalloo*. To date, amongst other things, gallbladders and appendices have already been removed from patients transgastrically (via the stomach) and transvaginally [19.30] (via the vagina). We can assume that, by the beginning of 2010, over 500 patients have undergone NOTES procedures across the world. A large percentage have been performed on women transvaginally, as this approach has been well known for a long time and the access site can be closed easily in open surgery without any problems. In principle, the transgastric approach is preferred, but no satisfactory method of closing the stomach wall has yet been found. Some of the operations are performed as hybrids with the help of a laparoscopic port.

The very high initial expectations placed on instruments and surgical techniques have still not been met. Nevertheless, some technological innovations have been realized in this regard.

Figure 19.17 shows the ANUBISCOPE from Karl Storz as an example. It is used for transgastric appli-

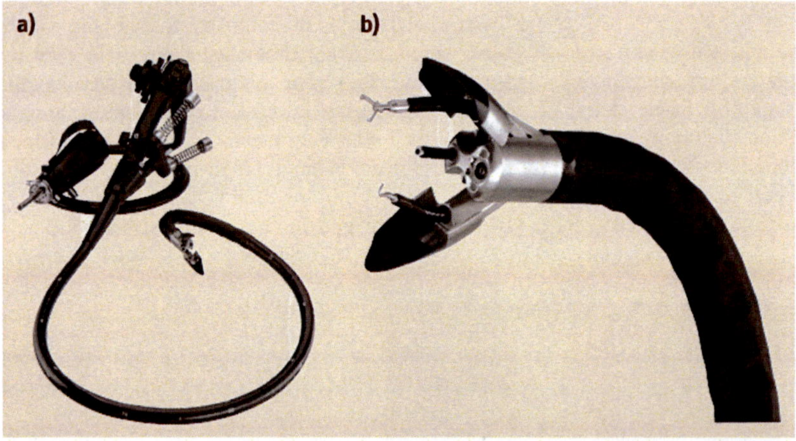

Fig. 19.17a,b Flexible, interventional platform for performing transgastric operations, ANUBISCOPE with three working channels, two of which can be angled for triangulation of flexible working instruments. The two half-shells can be moved: when closed they serve as an atraumatic introduction and dilation aid; when open they act as a retractor and secure the operating region against the nonvisible region. **(a)** Whole system, **(b)** distal close-up with coagulation hook, grasping forceps, and suction catheter (Source: Karl Storz, FDA approval pending)

cations with total reflection option (210° angle in four directions). This is a platform with a flexible shaft; the working channels are larger and intended for flexible instruments, which can be angled and used to manipulate tissue accordingly. Triangulation, which is not possible in traditional, internistic endoscopy, is made possible here by the instruments which can be angled. The special mechanism on the tip of the endoscope allows atraumatic introduction as with a blunt trocar in laparoscopy (closed state), and in the operating area, thanks to the open position of the half-shells, offers protected operating, stabilization of the instruments, and their triangulation. A third working channel can be used to introduce further instruments such as an articulating catheter for suction or as a guide for a laser fiber.

It is still too early to sum up the history of NOTES, as it is far from finished.

19.6.2 Single Port

Once the enormous challenge of NOTES (Sect. 19.6.1) became apparent to physicians and engineers working at companies active in the field, they began to search for alternative innovations. The idea of reducing the trauma associated with the surgical approach and not leaving any scars was expanded on so that work should only be performed via a single laparoscopic site or a single port. As a matter of preference, the umbilicus was suggested as the approach, because this would at least avoid creating a new scar (the umbilicus being, to a certain extent, a natural scar).

A multitude of different terms were developed, which although they might differ slightly in the details, all basically follow the same principle. ***Single-port laparoscopic surgery*** refers to an approach port which can accommodate several instruments (Fig. 19.18).

Despite its potential advantages, the procedure is associated with considerable ergonomic restrictions and limitations, as it is more difficult to perform laparoscopic interventions via this procedure and it demands extraordinary dexterity and expert knowledge on the part of the surgeon. The Dundee ENDOCONE system was realized as an integral insertion–instrument–retraction system to resolve this problem and facilitate the performance of single-port operations. The system features curved, instruments in order to enlarge the operating space between the surgeon's hands.

19.6.3 Simulation

The establishment of minimally invasive surgery (MIS) brought with it an increase in training requirements for new endoscopic techniques. Hand–eye coordination training with endoscopic instruments using simple in vitro models is not sufficient for basic MIS training. Intensive courses using in vivo large-animal models have been necessary to date in order to teach prospective MIS surgeons the requisite instrumental interactions and manoeuvres. The availability of high-performance

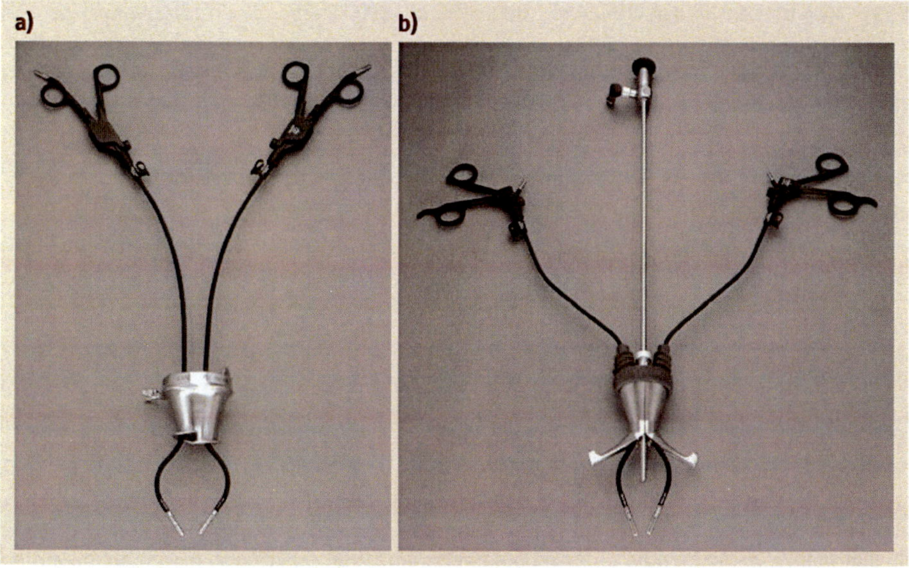

Fig. 19.18a,b Single-port approach systems and instruments display various differences, particularly when it comes to their shape. **(a)** CUSCHIERI ENDOCONE, **(b)** X-CONE consisting of two half-shells, which are locked together via a seal (Source: Karl Storz)

computers and graphic processors means that it is now possible to produce virtual simulators, which on the one hand present the operating site relatively realistically and, on the other hand, can already simulate the interactions between the instruments and the soft tissue almost as in real life. The requirements vis-à-vis real-time performance and image resolution are largely met.

There are different simulators for the different specialities, e.g., urology [19.31], gynecology, visceral surgery, coloscopy, etc. The simulator system for abdominal surgery comprises a monitor, a control unit, a simulation personal computer (PC), a patient torso with trocar modules including tactile force feedback, and endoscopic instruments (Fig. 19.19).

The available training scenarios offer the possibility to practice the individual techniques such as handling of the endoscopic camera, the endoscopic grasping forceps, and a suction/irrigation handle, etc. Combinations can also be trained. The variation of the individual scenarios happens via different difficulty levels and different directions of vision (0°, 30°, 45°) on the one hand, and different laparoscopic views on the other, which are varied within the simulation but are also dependent on the position of the trocar modules, which can be placed in almost any position on the patient module.

The simulators offer an easy-to-use user interface. All possible actions are described within the interface with text or additionally with images, so that the user can start with the system directly. The user can choose between different scenarios, tasks, and difficulty levels. In addition, there is usually also an integrated student–teacher system. Individual students then have the possibility to create user accounts in which the results of the training scenarios are saved. This presents the opportunity of monitoring training progress. Additionally, the teacher can view the results of all the created user accounts via his login.

19.6.4 Endorobotics

Robots are stationary or mobile machines which perform specified tasks according to a defined program. One possible task is the use in medical environments to move or position instruments or visual systems. Problematic, however, in the medical environment, is the autonomous movement of such a system. Today we do not speak of robots, but of handling systems, which operate under direct interactive control of the surgeon. In endoscopy, one currently differentiates between two types of *robot systems*. Firstly, there are so-called telemanipulation systems, which are used to position and activate endoscopic instruments and endoscopes by remote control. Secondly, wireless videocapsules are sometimes referred to as endorobots, as they pass through the gastrointestinal tract as an autonomous endoscopic visualization system, compiling corresponding image information for transmission to the outside (Sect. 19.6.5).

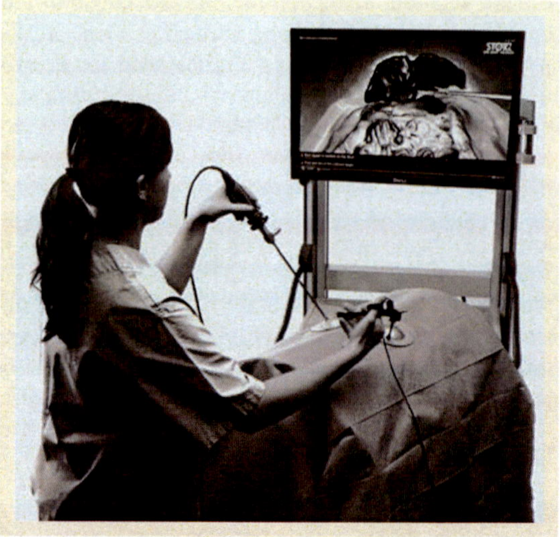

Fig. 19.19 Virtual MIS simulator for laparoscopic interventions

The following describes endoscopic telemanipulation systems [19.32–35]. The first attempts at telemanipulation systems were made by a team led by *Taylor* [19.36]. A positioning system for laparoscopes was launched onto the market in the mid 1990s by Computer Motion, Inc. [19.37]. Speech input was used to guide the endoscope in the abdomen and position it. This system was later superseded by the Da Vinci telemanipulation system from Intuitive Surgical. The Da Vinci system is composed of a control unit with stereo monitor and, on the patient side, a three-armed instrument positioning/activation unit and a positionable stereolaparoscope. During laparoscopic interventions, the surgeon sits at the control console and moves the instruments on the operating table from this position. The software-assisted 3-D visualization provides the doctor with a detailed, 3-D view of the operating field and a great depth of field. The doctor controls the instruments from the control console using control elements. In doing so, the exact movements of his wrist, hand, and fingers are acutely enacted by the remote-controlled instruments. The camera-assisted laparoscope can be moved and positioned in the body; simultaneously, the image section can be magnified. This provides the surgeon with optimal vision over the operating field. The operating and, in particular, the performance of critical manoeuvres are facilitated for the doctor, but due to the high cost of the system and the extensive time and effort needed pre- and postoperatively, the system has only enjoyed success in a single MIS application.

19.6.5 Capsule Endoscopy

The endocapsule contains an image sensor, an illumination unit based on mini LEDs, and usually a localization element which can be used to determine the position of the capsule in the gastrointestinal tract. An external control unit receives the image information and saves it in the form of a series of pictures spanning several hours. In addition to the visualization of the esophagus, this type of wireless endopill diagnostics allows inspection of the entire small intestine in sufficiently good image quality. This new procedure is easy to perform, very patient friendly, and considerably more informative than conventional radiological intestinal contrast agent examinations. Prior to capsule endoscopy, there was no satisfactory way of visualizing the small intestine, be it endoscopically or radiologically. A number of systems are currently available on the market.

However, capsule endoscopy is still not an alternative to videogastroscopy or videocolonoscopy for examining the stomach or colon. This is due to the fact that there is no way of controlling the movement and direction of the pill and that there is no possibility for direct treatment when the respective findings are detected.

Current research activities are thus concentrated on the remote control of endopills [19.38].

19.6.6 Endonavigation

The first steps in researching endonavigation were taken at the end of the 1980s in paranasal sinus surgery [19.39, 40]. The position of endoscopic paranasal sinus surgery instruments was compiled in three dimensions and then visualized in a preoperatively recorded three-dimensional CT data record. Position detection is effected via the instrument tip; the instrument itself was held in a holding arm with six degrees of freedom; the position was then determined by an angle encoder system [19.41]. Further developments at the beginning of the 1990s allowed so-called *noncontact* navigation systems. Camera systems detect the position of instruments based on reflectors mounted on the instruments (Fig. 19.20). Such systems have managed to make a place for themselves in neurosurgery, orthopedic surgery, and paranasal sinus surgery. As optical position detection of the instruments occurs via the extracorporeal portion of the instrument, use of these navigation systems is only possible in combination with rigid instruments and/or endoscopes.

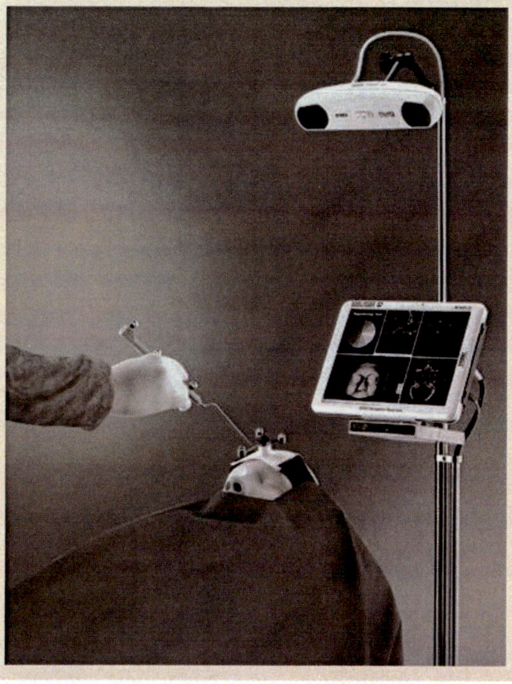

Fig. 19.20 Endonavigation system SURGICAL COCKPIT (Source: Karl Storz)

The use of innovative electromagnetic-based navigation systems has now made it possible to detect the intracorporeal position of (flexible) instruments. This requires the integration of a miniaturized coil element in the instrument. The advantage lies in the possibility of determining the intracorporeal position, i. e., the localization of flexible instruments and endoscopes [19.42]. An associated disadvantage is that ferromagnetic materials around the measurements can lead to measuring errors. In the future, navigation techniques in combination with other sensor technologies will facilitate new endoscopic visualization and documentation techniques [19.43, 44].

References

19.1 M. Leonhard, P. Cappabianca, E. de Divitiis: The endoscope, endoscopic equipment and instrumentation,. In: *Endoscopic Endonasal Skull Base Surgery*, ed. by E. de Divitiis, P. Cappabianca (Springer, New York 2003)

19.2 A. Oppelt: *Imaging Systems for Medical Diagnostics* (Siemens Publics Corp. Publishing, Erlangen 2005) pp. 258–

19.3 G. Berci: *Endoscopy* (Appleton-Century-Crofts, New York 1976)

19.4 Bundesministerium für Bildung und Forschung: Förderschwerpunkt Biophotonik, www.biophotonik.org (2007)

19.5 Grund, R. Salm: Systeme für die Endoskopie, In: *Medizintechnik*, 3rd. ed., ed. by R. Kramme (Springer, Berlin Heidelberg 2007)

19.6 M.A. Reuter, H.J. Reuter, R.M. Engel: *History of Endoscopy* (Kohlhammer, Stuttgart 1999)

19.7 P.P. Figdor: *Philipp Bozzini* (Endopress, Tuttlingen 2002)

19.8 F.H. Ebner, J.S. Marquardt, B. Hirt, G.C. Feigl, M. Tatagiba, M.U. Schuhmann: Broadening horizons of neuroendoscopy with a variable-view rigid endoscope: An anatomical study, J. Cancer Surg. **36**, 195–200 (2010)

19.9 Loewenstein L&H: Kystoskop, Patent 164966 (1905)

19.10 B.T. Schowengerdt, E.J. Seibel: True 3-D scanned voxel displays using single or multiple light sources, J. Soc. Inf. Disp. **14**(2), 135–143 (2006)

19.11 A. Endress: Entwicklung eines anwenderorientierten, workflowadaptierten und qualitätssichernden OP-Bilddokumentations- und Auswertungssystems am Beispiel der Universitäts-Frauenklinik Tübingen, Dissertation (Univ. Tübingen 2004)

19.12 H.C. Weiss: Karl Storz Fachinformation Aufbereitung und Hygiene (Karl Storz, Tuttlingen 2011)

19.13 ANSI/AAMI ST79:2006: Comprehensive guide to steam sterilization and sterility assurance (2006)

19.14 DIN EN ISO 15883: Washer – Disinfectors (2009)

19.15 DIN EN ISO 17664: Sterilization of medical devices – Information to be provided by the manufacturer for the processing of resterilizable medical devices (2004)

19.16 DIN EN ISO 17665-1: Sterilization of health care products – Moist heat – Part 1: Requirements for the development, validation and routine control of a sterilization process for medical devices (2006)

19.17 Working group instrument reprocessing: Proper maintenance of instruments („red brochure" www.a-k-i.org) (2009)

19.18 H.-P. Berlien, G.J. Müller: *Angewandte Lasermedizin* (Ecomed, Landsberg 2000)

19.19 G. Grevers, R. Baumgartner, A. Leunig (Eds.): *Laser im Kopf-Hals Bereich* (Springer, Berlin Heidelberg 2001)

19.20 A.G. Hofstetter: *Lasers in Urological Surgery* (Springer, Berlin Heidelberg 1997)

19.21 R.M. Szeimies, D. Jocham, M. Landthaler (Eds.): *Klinische Fluoreszenzdiagnostik und Photodynamische Therapie* (Blackwell, Berlin 2003)

19.22 M. Leonhard: The KARL STORZ OR1™. In: *New Aspects of High Technology in Medicine* (Monduzzi, Bologna 2000)

19.23 K. Irion, P. Novak: System workplace for endoscopic surgery, Minim. Invas. Ther. Allied Technol. **9**(3/4), 193–197 (2000)

19.24 KARL STORZ D-Cut Brochure ORCHESTRION: Bring Harmony to your OR (Karl Storz, Tuttlingen 2008)

19.25 Gesundheitsberichterstattung des Bundes: Die 50 häufigsten Operationen der vollstationären Patientinnen und Patienten in Krankenhäusern www.gbe-bund.de (2008)

19.26 M. Kriegmair, R. Baumgartner, R. Knuechel, P. Steinbach, A. Ehsan, W. Lumper, F. Hofstädter, A. Hofstetter: Fluorescence Photodetection of neoplastic urothelial lesions following intravesical instillation of 5-aminolevulinic acid, Urology **44**, 836–841 (1994)

19.27 T. Carus, H. Lienhard: Die laparoskopische Fluoreszenzangiographie mit Indocaningrün zur intraoperativen Beurteilung der Perfusion bei kolorektarlen Anastomosen, Meeting Abstracts, 126. Kongress der Deutschen Gesellschaft für Chirurgie (2009)

19.28 J. Hamou: Contact endoscopy and micro endoscopy, US Patent 4385810 (1980)

19.29 D. Rattner, A. Kalloo: ASGE/SAGES Working Group on natural orifice translumenal endoscopic surgery, Surg. Endosc. **20**, 329–333 (2006)

19.30 J. Marescaux: Surgery without scars, Arch. Surg. **142**(9), 823–827 (2007)

19.31 M. Noll: Konzeption und Evaluation eines Endoskopie-Simulators für die Urologie, Dissertation (Univ. Munich 2002)

19.32 Y. Nang: Automated Endoscope System for Optimal Positioning, US Patent 5515478 (1996)

19.33 E. Glassman, W.A. Hanson, P. Kazanzides, B.D. Mittelstadt, D. Brent, B.L. Musits, H.A. Paul, R.H. Taylor: Image-Directed Robotic System for Precise Robotic Surgery Including Redundant Consistency Checking, US Patent 5086401 (1992)

19.34 J. Funda, D.A. Larose, R.H. Taylor: Robotic System for Positioning a Surgical Instrument Relative to a Patient's Body, US Patent 5572999 (1996)

19.35 R. Taylor: System and Method for Augmentation of Surgery, US Patent 5402801 (1995)

19.36 R. Taylor, Y.-Y. Kim: Signaling Device and Method for Monitoring Positions in a Surgical Operation, US Patent 5279309 (1994/1992)

19.37 Y. Wang: Automated Endoscope System For Optimal Positioning, US Patent 5515478 (1996/1994)

19.38 EU-Projekt im 6. Rahmenprogramm: Versatile endoscopic capsule for gastrointestinal tumor recognition and therapy (VECTOR), www.vector-project.com (2006–2010)

19.39 G. Schlöndorff, D. Meyer-Ebrecht, R. Mösges, W. Krybus, L. Adams: CAS – computer assisted surgery, Arch, oto-rhino-laryngology, Suppl 2, 45 (1987)

19.40 R. Mösges, G. Schlöndorff: A new imaging method for intraoperative therapy control in skull base surgery, Neurosurg. Rev. **11**, 245–247 (1988)

19.41 KARL STORZ Firmenbroschüre: CAS – Computer Assisted Surgery, 5th ed. 1/92 (1992)

19.42 K.M. Irion, U. Hagelauer: Medical instrument for the treatment of tissue by means of a high-frequency current and medical system with medical instrument of this type, US Patent 7588569 B2 (2009)

19.43 W. Draxinger, M. Noll, H. Stepp, K. M. Irion : Method and device for creating at least one section of a virtual 3D model of a body cavity, PCT WO 2005/077272 A1 (2005)

19.44 L. Soler, S. Nicolau, A. Hostettler, J.-B. Fasquel, V. Agnus, A. Charnoz, J. Moreau, B. Dallemagne, D. Mutter, J. Marescaux: Virtual Reality and augmented reality applied to endoscopic and NOTES procedure, Proc. World Congress on Medical Physics and Biomedical Engineering (Munich 2009)

20. Cone-Beam Computed Tomography and Navigation

Dirk Schulze, Gerhard Hoffmann

Cone-beam computed tomography (CBCT) is a rather new image modality which is now especially in use in dentistry and ENT surgery. Image acquisition comprises a rotating system of x-ray tube and detector deploying 200–600 fluoroscopic image from an examination object. Image reconstruction starts subsequently where mainly filtered back projection and its modifications are used. The reconstructed dataset is displayed in specialized viewers. Moreover these DICOM (digital imaging and communications in medicine) stacks can be used for purposes in image guided surgery.

20.1 **Technical Background of Dental Digital Volume Tomography** 405
- 20.1.1 Data Acquisition 405
- 20.1.2 Data Reconstruction 406
- 20.1.3 Data Evaluation 407

20.2 **Areas of Application of Dental CBCT** 408
- 20.2.1 Preservative Dentistry 408
- 20.2.2 Prosthodontics 409
- 20.2.3 Functional Diagnostics and Therapy 409
- 20.2.4 Surgical Dentistry 410
- 20.2.5 Implantology 410
- 20.2.6 Cranio-Maxillofacial Surgery 411
- 20.2.7 Orthodontics 411
- 20.2.8 Technical Restrictions of the Procedure 412
- 20.2.9 Dose 412
- 20.2.10 Navigation 412
- 20.2.11 Prerequisites for Navigated Procedures 412
- 20.2.12 Further Processing 412
- 20.2.13 Registration 412

References 413

20.1 Technical Background of Dental Digital Volume Tomography

Since the late 1990s, devices for cone-beam computed tomography (CBCT) have been used, but already one decade before, the very first clinical prototypes had been applied for angiographic applications.

Finally, adaptations of used detectors as well as the available processing power in combination with radiographic tubes applied in dentistry resulted in a practicable design of CBCT systems.

In contrast to fan-shaped ray beams used in conventional computed tomography, CBCT utilizes conical or pyramidal ray beams. The area-shaped absorption reliefs resulting from exposure are captured by flat-panel detectors and visualized by special three-dimensional imaging algorithms (Fig. 20.1).

20.1.1 Data Acquisition

In CBCT systems, a conical or pyramidal projection geometry is used, with its shape depending on the respective aperture system. A displacement of this system in the z-axis as found in spiral computer tomography (CT) is currently not applied. Therefore, the recording system rotates around the object with a fixed z-axis. The extent of rotation is currently determined by conventional imaging algorithms – as a rule, rotation angles of $\pi + \phi$ (where ϕ is the cone or pyramidal angle of the beam) to 2π are used. For extension of the field of view (FoV), rotation may take place both horizontally and eccentrically; however, in the majority of appliances,

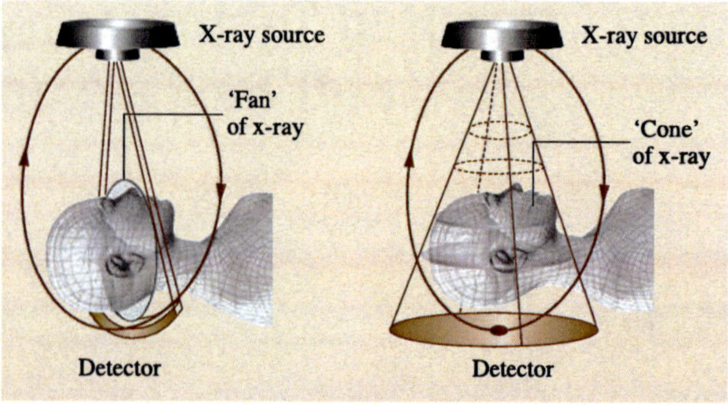

Fig. 20.1 Comparison between fan-beam and cone-beam radiation geometry

centric revolution is carried out. On the orbit described, between 200 and 600 radioscopies (fluoroscopies) of the object are acquired within 10–20 s. For instance, depending on the device and the recording mode, the effective exposure time is between 20 and 50 ms per projection. Moreover, continuous, unpulsed exposures

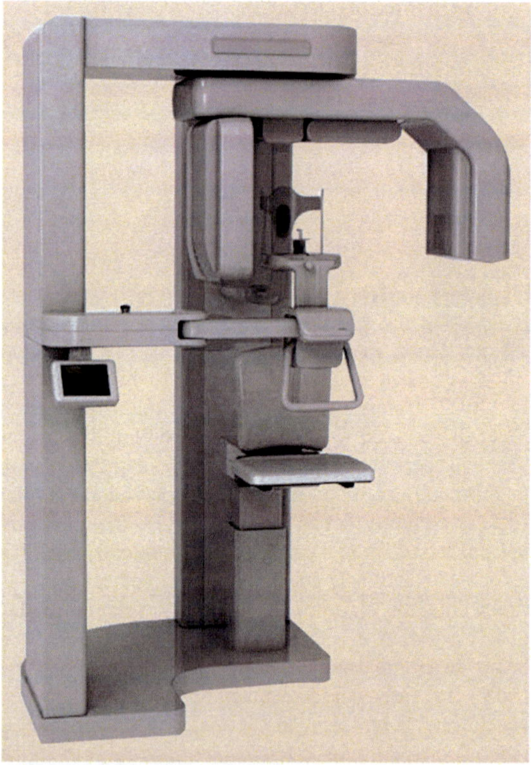

Fig. 20.2 CBCT device. Infinitely adjustable FoV: 5×5 to 24×19 cm^2 (Source orangedental)

are carried out as well. For legal reasons, the tube parameters are globally subject to substantial fluctuation ranges; in general, however, tube voltages between 80 and 120 kV and a tube current between 0.5 and 10 mA are applied [20.1].

The actual acquisition of the absorption relief is done by flat detectors or detectors coupled with image intensifiers, e.g., charge-coupled device (CCD) sensors. This creates fluoroscopies of the examination volume for each position of the orbit. As flat detectors, arrays of amorphous silicon are used that are coupled to a scintillator (e.g. Csl, GdO) via glass fibers. X-ray quanta hitting the scintillation layer create light quanta that produce a proportional quantum of electric charge in the detector layer. According to the detector configuration, they are collected line by line, analog/digital converted, and spatially encoded. Apart from this indirect detection, application of arrays of amorphous selenium also allows the use of direct quantum detection [20.2]. As far as image intensifier systems are concerned, signal amplification is connected upstream of detection, which is achieved by chain-linking of secondary electron multipliers. This procedure, which has been known for about 60 years, produces high output signals even in the event of small input signals; however, at the same time the dose decrease associated therewith directly leads to a deterioration of the signal-to-noise ratio as well [20.3]. This is why image intensifier systems are still used to a limited extent in dental CBCT.

In many cases, it is currently already possible to vary the acquired section (field of view) of the examination volume, providing FoVs between 4×4 and 24×20 cm^2 [20.1]. In terms of perspective, variable FoVs represent the most futureproof systems for the user, since acquisition may be controlled according to indication and unnecessary radiation exposure can be avoided by fading out uninteresting areas (Fig. 20.2).

20.1.2 Data Reconstruction

Fluoroscopies collected (raw data) may be reconstructed to an imaging volume via different mathematical algorithms, with the modified Feldkamp algorithm being the most frequently used. This represents a three-dimensional (3-D) modification of the filtered back-projection (FBP) method known from computed tomography (CT). In this process, the recorded information is back-projected into the acquisition zone, folded back mathematically. The overlapping of all back-projections results in a three-dimensional reflection of the absorption occurring during acquisition. The blur occurring during this process is adjusted by filtering, which at the same time helps to improve the contrast. In the event of strong absorptions, for example, due to metallic filling materials or implants, typical artifacts arise within this reconstruction algorithm, spreading out across the overall image in a linear or laminar way [20.2]. Image quality may be improved by iterative procedures.

Algebraic reconstruction (ART) represents another, more complex kind of image reconstruction. In this process, the absorption information is decomposed in matrices, and the original absorption values are reconstructed by systems of equations (Fig. 20.3). If carried out exactly, these arithmetic operations are substantially more complex than FBP and therefore are not yet regularly applied.

In general, data reconstruction in the meantime is done on specialized graphics processor units (GPU), leading to a reduction of reconstruction time to below 1 min. Reconstruction is frequently done on isolated reconstruction processors, delivering the result to a review station or a server.

Depending on the kind of detector, the result of reconstruction is a spherical (round detector) or cylindrical (rectangular detector) volume. Unlike in classic CT, in CBCT this is composed of isotropic voxels (Fig. 20.4). According to the absorption, a grayscale value can be assigned to each individual voxel. The assignment of grayscale values on a histogram is based on primary instrument calibration. There are substantial differences between CBCT systems which should be considered in subsequent data processing. Since the result of reconstruction in CBCT is usually a volume file, subsequent decomposition of the dataset in individual layers across the length of the z-axis is required.

20.1.3 Data Evaluation

Evaluation of reconstructed data is done either in viewing programs provided by the manufacturer or in software applications of third-party distributors or classic digital imaging and communication (DICOM) viewers. Usually, a view of all three spatial planes is offered in the form of secondary multiplanar reconstructions (MPR). Frequently, there is a complete three-dimensional representation in a further window, with the MPR representing the diagnostically substantial information value, since they allow exact inspection of any anatomical field marker or pathological change

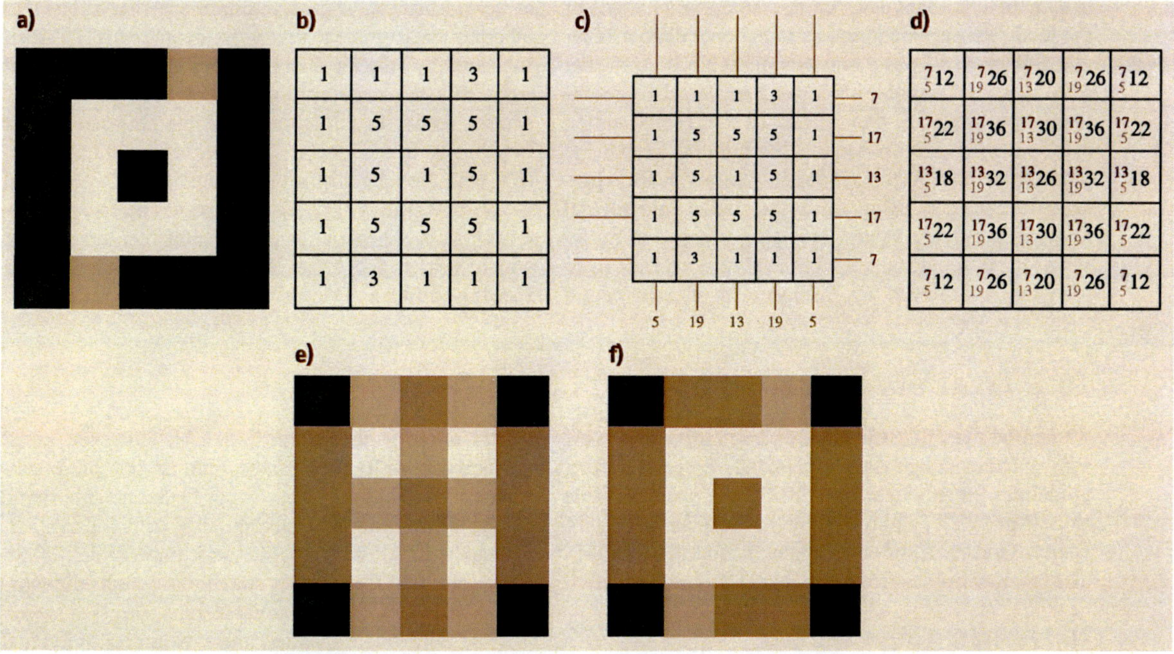

Fig. 20.3a–f Principle of filtered backprojection (**a–f**). (**a**) Examination object, (**b**) numbers representing absorption values of the object, (**c**) two projections and their results row by row and column by column, (**d**) back projected values and their sums, (**e**) image representing the sums, (**f**) image after filtering

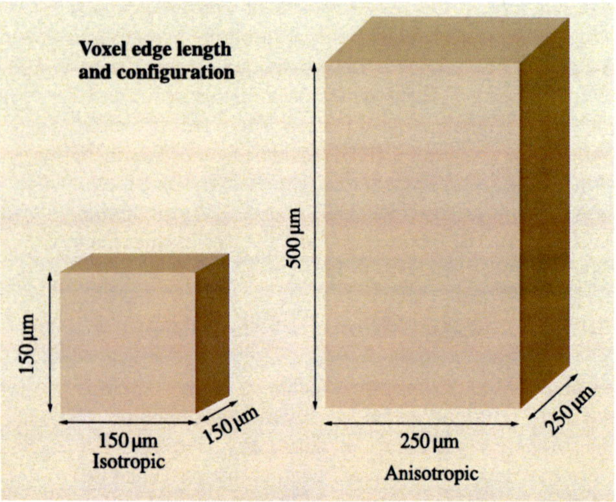

Fig. 20.4 Difference between isotropic and anisotropic voxels

in the acquired volume. Moreover, analysis of these datasets requires specialized MPR in dentistry. For example, free, non-axis-related MPR are created along the dental arch to provide a panoramic view of all the teeth. Historically, this kind of reconstruction is above all attributable to the fact that, for assessment of changes of the teeth, the periodontium, and the bone, dentists usually fall back on panoramic tomography. In a further step, it is then possible to create transversal representations positioned at right angles to the reconstruction axis, showing a cross-section through the respective jaw. Depending on the bit depth, the grayscale representation along a histogram is distributed into several hundred grayscale values. However, since this represents an inadequate number for the human eye, adequate imaging is achieved by displacement of the center of representation and a limitation of the number of grayscale values (windowing). In so doing, it must be considered that, due to the exposure parameters used, CBCT provides a poorer representation of the soft tissue than classic CT. A further spreading of absorption values of different soft tissues such as fat and muscles is only achievable by increasing the dose. Therefore, CBCT currently remains reserved for diagnostics of high-contrast structures.

The decisive advantage over projection radiography of course is the generally available three-dimensional localization by different MPR levels.

Pathological changes and geometrically exact information can only be documented and transported by screenshots or different report functions (according to software).

Three-dimensional reconstructions are frequently realized using the volume rendering technique (VRT). Within this system, the overall dataset of the volume is reconstructed to another object, which may be modified manually in its position and transparency. At the same time, modification of the grayscale values is possible by overlapping of color spectra, i.e., using a color lookup table (CLT).

Metric analyses as used for planning of implants or for measuring spatial requirements are subject to inherent measurement errors in dental CBCT, albeit substantially smaller than in classic CT. Since a voxel represents the smallest information unit in a CBCT dataset, measurement is only possible between at least two voxels. In the ideal case, a measurement may coincide with the length of a voxel edge. In the worst case, measurement may take place between the two furthermost points. There is an incremental measuring error which, in the ideal case, is already considered by the viewing software.

20.2 Areas of Application of Dental CBCT

Structured use of CBCT required a scientifically based analysis of existing data records [20.1, 4, 5]. Therefore, guidelines for application of CBCT have been available since April 2009, authorized by the German Society of Dental, Oral and Craniomandibular Sciences (GSDOM). The following indication areas are defined individually.

20.2.1 Preservative Dentistry

In the event of unfilled teeth, high-resolution CBCT examination achieves the sensitivity of film-based or digital intraoral recordings [20.6, 7]. However, metal artifacts as well as artifacts due to the hard substance of neighboring teeth occur in the neighborhood of metal restorations, making approximal caries diagnostics in clinical applications impossible [20.8]. Therefore, CBCT is hardly suitable for caries diagnostics, particularly of approximal lesions.

The following indication areas within the individual specific fields in which a CBCT examination may be carried out are currently under discussion or are already evidence-based.

Endodontics

- Apical changes in the event of existence of clinical abnormalities, if they are not detectable on two-dimensional recordings or spatially correlatable [20.9, 10].
- Root fractures, since for purely mathematical reasons their identification is safer than using two-dimensional imaging [20.11].
- Root resorption, e.g., following dental trauma [20.12] (Fig. 20.5).

Periodontics

- Visualization of the osseous periodontal situation, thanks to the excellent representation of the three-dimensional periodontal morphology [20.13–15].

20.2.2 Prosthodontics

In prosthodontics, CBCT provides additional possibilities for diagnostics and therapy planning. In the future, CBCT data in combination with digital data from intraoral scanners by integration in planning software may provide additional possibilities for therapy planning in the sense of a virtual setup. At the time of guideline creation, evidence-based data regarding this subject were available only to an insufficient extent or not at all. Currently, the following areas of indication for prosthodontics are becoming apparent where a CBCT examination may take place:

- Additional information regarding diagnostics of tooth abutment significance (root surface, furcation findings, etc.).
- Visualization of quantitative and qualitative bone material available (implant-supported dental prosthesis, removable prosthetics).
- Representation of nerve exit points (implant-supported dentures, removable prosthetics).
- Diagnostics of osseous diseases of the temporomandibular joint [20.16–20].
- Virtual planning of implant-prosthetic restorations [20.21].
- Linking of 3-D data to the construction software of computer-aided design (CAD)/computer-aided manufacture (CAM) systems (e.g., for CAD/CAM-manufactured drilling templates, long-term temporary appliances or definitive denture).

20.2.3 Functional Diagnostics and Therapy

In the area of diagnostics and therapy of craniomandibular dysfunctions, clinical and instrumental diagnostic procedures are complemented by imaging procedures. Within this process, tomographic x-ray procedures

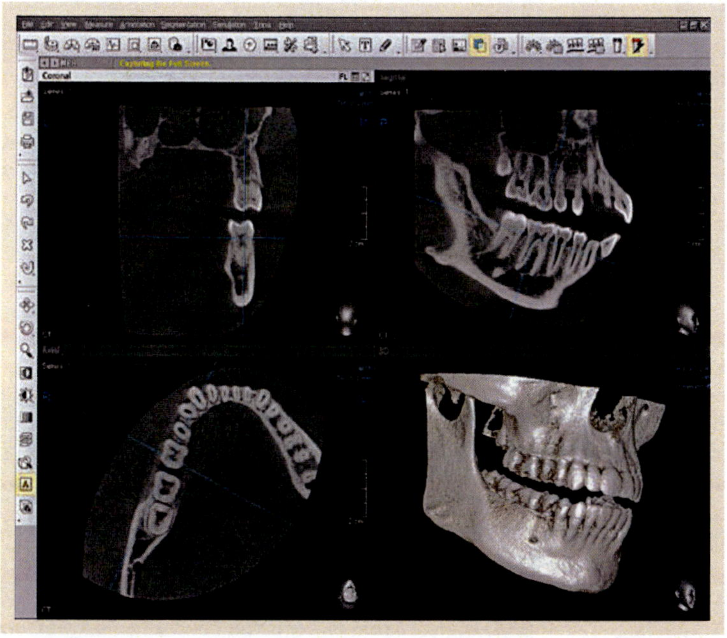

Fig. 20.5 CBCT dataset 12×8.5; root canals are clearly visible (Source orangedental)

principally are exclusively conducive to the representation of osseous changes in the region of the temporomandibular joint. Several scientific studies have revealed that use of CBCT achieves at least the same results as when using classic or computed tomographies of the temporomandibular joint [20.16, 22, 23]. With regard to quantitative analyses, CBCT shows very good consistency with actual measurements at macroscopic-anatomic preparations [20.24, 25].

For representation of cartilage structures, magnetic resonance imaging remains the procedure of choice, more so as osseous contours are also represented in three-dimensional (3-D) cross-sections. Currently notable main indications for CBCT in temporomandibular joint diagnostics are:

- Exclusion of primary temporomandibular joint diseases
- Recording of differentially therapeutically relevant findings (extent of erosive processes of the condyles, sclerosis, position of condyles, malpositions of the condyle in the mandibular fossa).

20.2.4 Surgical Dentistry

In dental surgery, CBCT predominantly serves for diagnostics of osseous (pathological) findings and spatial conditions that have been partly described already under the other special fields. At present, the possible fields of application becoming apparent within the special fields where CBCT may be used for x-ray diagnostics are:

- Root fractures (although with signs indicating that sensitivity for fresh fractures may be reduced immediately after trauma) [20.26, 27]
- Fractures of alveolar processes [20.26]
- Intraosseous pathological changes such as odontogenic tumors or larger periapical osseous lesions [20.28]
- Positional anomalies of teeth
- Preoperative tomographic diagnostics during planned surgical removal of (partly) retained wisdom teeth. Use of CBCT may be suitable here if the spatial positional relationship between the mandibular canal and the wisdom tooth cannot be interpreted with sufficient security based on already existing, conventional radiographs or is estimated to be critical [20.29, 30]. Based on its low impact in terms of therapeutic benefit, however, routine application prior to removal of wisdom teeth is not recommended [20.31].

20.2.5 Implantology

In implantology, CBCT is predominantly used in therapy planning, typically for visualization and measurement of the initial osseous situation as well as for visualization of three-dimensional implant prosthetic treatment planning (i.e., templates). This requires a metric measurement whose accuracy depends on the actually achieved spatial resolution, the contrast resolution, and the signal-to-noise ratio. Identification of the measuring points has a decisive impact on the measuring accuracy as well. All published data are based on in vitro examinations which do not show any blurring artifacts due to patient movements. For linear measuring sections typical in implantology, maximal relative errors of 3–8% have been established [20.32, 33]. For a measuring section of 10 mm this means a possible inaccuracy of approximately 0.5 mm.

Based on these recordings, using software it is possible to simulate and evaluate planned implants, superstructures, augmentations, incisions, dentures, and prosthetics. Values and findings gained in this way may be used for planning of possibly exact, prosthodontically oriented positioning of implants with the best possible utilization of the available bone material. Moreover, it is possible to detect deficits in the existing tissue range and to anticipatorily recognize the necessity of augmentations/distractions/implant-preparing measures and where appropriate plan corresponding measures. Transfer of spatial information from the planning system to surgical reality takes place via computer-based manufactured transfer and drilling templates or using direct instrument navigation. In vitro, axial deviation of up to 4° and linear deviations of up to 2.4 mm are stated [20.34]. Related in vivo data barely exist, although a preliminary analysis based on a very small number of cases states maximal linear deviations of 6 mm and 11° axially [20.35]. So far, there exist few valid data concerning clinical application of CBCT-supported implantation using drilling templates. However, first results have shown good conformity of CBCT planning with the clinical situation [20.36]. In the future, it is likely that CBCT may be used directly for intraoperative navigation procedures as well [20.37]. In principle, these processes require observation of system-immanent possible inaccuracies and keeping to safety margins, to avoid injury of sensitive neighboring structures.

Since, due to the high absorption of a titanium implant, there arise hardening artifacts in the further beam path, an evaluation of the direct peri-implantary region as well as of the region between implants in the direction of the beam path is possible only to a very restricted extent.

20.2.6 Cranio-Maxillofacial Surgery

In addition to the surgical fields of application of dental surgery and implantology already mentioned, CBCT may be applied in cranio-maxillofacial surgery for the following indications, for instance [20.38]:

- Odontogenic tumors [20.39]
- Anomalies of bone pathology and structure particularly in the event of osteitis, osteomyelitis, and osteoporosis
- Maxillary sinusitis
- Salivary stones
- (Osseous) temporomandibular diseases
- Craniofacial traumatology
- Representation of the spatial course of intraosseous structures (osseous nerve and vascular channels)
- Diagnostics and surgical planning in the event of complex malformations (Fig. 20.6).

20.2.7 Orthodontics

In general, children are subject to increased risk of suffering consequential damage due to ionizing beam exposure. This should be absolutely, and of course increasingly, considered in orthodontics, and the indication must be adapted accordingly [20.40].

Although knowledge is currently based on a relatively low level of evidence, the following orthodontic indications are currently becoming apparent for which CBCT-based diagnostics could be recommendable:

- Diagnostics of anomalies of tooth findings.
- Diagnostics of anomalies and dysplasia of dental roots.
- Differential-diagnostic assessment of tooth eruption disorders.
- Representation of the periodontal bone material for prognostic assessment of planned tooth movement.
- Diagnostics of craniofacial malformations.

Even if existing CBCT datasets in principle are suitable for calculation of two-dimensional (2-D) cephalograms, they are only convenient as a starting point for intended 3-D cephalometry, because of the inevitable loss of information [20.13, 41]. Fundamental work has already been presented [20.27, 42, 43].

Particularly with regard to the frequently juvenile patients, the choice of a suitable radiographic examination procedure requires special consideration of radiation protection. Possibilities for dose reduction must be exhausted (e.g., reduced optical resolution, re-

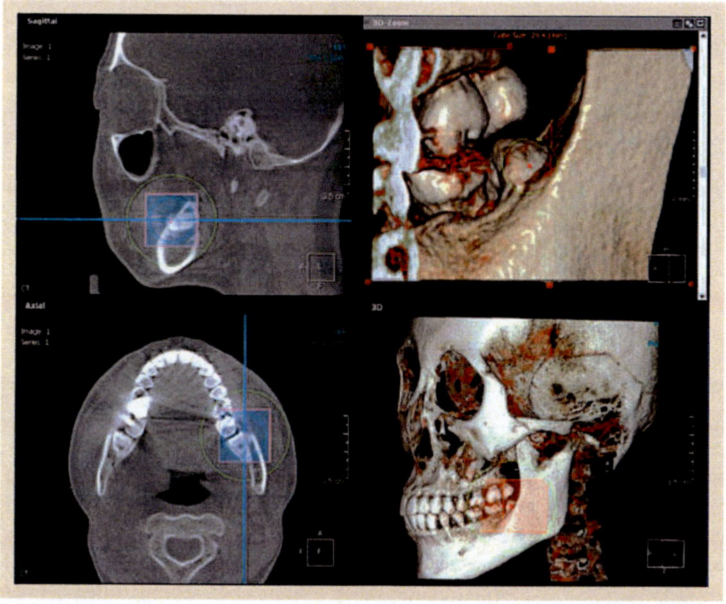

Fig. 20.6 CBCT, 20×19 dataset, overview image for surgery (Source orangedental)

duced exposure angles, reduced number of projection recordings). Moreover, examination procedures without any ionizing radiation should be considered as an alternative.

20.2.8 Technical Restrictions of the Procedure

Unlike classic CT, CBCT imaging is substantially more sensitive to patient movements. They may be initiated by random movements such as, e.g., during swallowing, or due to constant tremors.

This may have a lasting effect on the quality of the overall volume dataset, while in CT only one (in the event of sequential acquisition) or a few axial layers (in the event of spiral acquisition) are concerned. At the same time, differentiated representation of soft tissues using CBCT is impossible. However, a huge advantage of this procedure is the comparatively low occurrence of metal artifacts, enabling also diagnostic assessment of structures within a layer near metal fillings, while in conventional CT metal artifacts may render whole layers unusable.

20.2.9 Dose

Depending on the system and the imaged volume, beam exposure (effective dose) is between 30 and 150 μSv, sometimes even lower.

Even if applying *low-dose* protocols, conventional CT systems currently cannot achieve the spatial resolution of CBCT systems.

20.2.10 Navigation

Navigation combines recording of preoperative data of a patient with his intraoperative position. In such a way, it is comparable to the methodology known from the global positioning of vehicles or other objects.

Fundamentally, one distinguishes between active and passive navigation.

In the case of active navigation, real-time data to support a surgical intervention are displayed on a separate monitor or even shown in the operating field (augmented reality) [20.44]. At the same time, it is possible to bring *active* instruments into the navigation window. At an additional stage of extension, situational changes of tissue qualities can be communicated to the user via haptic (*forced feedback*) mechanisms to decide on the operative procedure [20.45].

In the event of passive navigation, preoperative data are exclusively used for planning a therapeutic intervention. As a result, computer-assisted planning, e.g., may be stored in the form of a template or a splint and released via CAD/CAM procedures. This object is then applied during surgery as a positioning guide or success control; there is no direct back-coupling as in active navigation.

20.2.11 Prerequisites for Navigated Procedures

Acquisition of a preoperative or pre-interventional dataset is absolutely required for navigation [20.46, 47]. Acquisition essentially takes place using slice-imaging procedures such as CBCT, CT, and/or MRI, with all datasets being able to be merged with each other as well. Prior to data acquisition, data registration in the navigation or planning software at a later stage should be considered, insofar as it may be required to introduce reference markers. Moreover, it is necessary to adapt the quality of the gained cross-sectional data for further applications (radiation beam length, necessity of soft-tissue information).

20.2.12 Further Processing

The cross-sectional data gained from the above-mentioned modalities is imported in planning surfaces and converted for processing. As a rule, the voxels are tetrahedronized here or abstracted in similar procedures; in general, this is called segmentation. All information that is not essential is dismissed from the dataset, and only interesting data are processed. There exist different algorithms for segmentation; for example, it is possible to filter data according to their absorption values (threshold method) [20.48]. Apart from this, there are also other methods available that are substantially more subtle (watershed algorithm, marching cubes) [20.49]. Finally, all of them lead to an anatomic model which is generated from the cross-section dataset and shows a considerably lower data volume than the original dataset.

20.2.13 Registration

With regard to registration, one distinguishes between active and passive navigation procedures. For passive procedures, the segmented datasets are usually collocated using models (plaster models in special articulators or other appliances), so that transfer of software-assisted implementations such as drillings to the model or model-supported templates becomes possible. Since the information transfer is taking place a priori here, this process is called passive navigation. The resulting information, e.g., in the form of a positioning or drilling template, is then implemented in the operative situs.

For active navigation, the prepared data are adjusted to the patient's position in a quasi-live way. This process may require integration of reference markers, which are recorded on the spot for triangulation using infrared cameras and used for collocation of the dataset. Moreover, there exist systems using existing anatomic landmarks for data matching [20.50, 51]. However, in comparison with passive navigation, corresponding implementations are carried out intraoperatively and therefore are substantially more cost intensive. However, this enables real-time analysis of preoperatively gained data and so allows minimally invasive interventions and a particularly protective procedure. Furthermore, use of intraoperative three-dimensional data acquisition is possible here, as currently available with so-called three-dimensional C-arms, to provide the preoperative dataset with a live update and investigate the preoperative planning in detail [20.52].

Consequently, navigation procedures are subject to a number of modifying influences, making accuracy in the submillimeter range currently almost impossible. Furthermore, careful selection of the preoperative imaging procedure, exact planning, and choice of suitable reference objects are indispensable for successful navigation.

References

20.1 C.H. Kau, M. Bozic, J. English, R. Lee, H. Bussa, R.K. Ellis: Cone-beam computed tomography of the maxillofacial region – An update, Int. J. Med. Robot. **5**(4), 366–380 (2009)

20.2 A.C. Miracle, S.K. Mukherji: Conebeam CT of the head and neck, Part 1: Physical principles, Am. J. Neuroradiol. **30**(6), 1088–1095 (2009)

20.3 R. Baba, Y. Konno, K. Ueda, S. Ikeda: Comparison of flat-panel detector and image-intensifier detector for cone-beam CT, Comput. Med. Imaging Graph. **26**(3), 153–158 (2002)

20.4 A.C. Miracle, S.K. Mukherji: Conebeam CT of the head and neck, Part 2: Clinical applications, Am. J. Neuroradiol. **30**(7), 1285–1292 (2009)

20.5 W. De Vos, J. Casselman, G.R.J. Swennen: Cone-beam computerized tomography (CBCT) imaging of the oral and maxillofacial region: A systematic review of the literature, Int. J. Oral Maxillofac. Surg. **38**(6), 609–625 (2009)

20.6 B.G. Akdeniz, H. Gröndahl, B. Magnusson: Accuracy of proximal caries depth measurements: Comparison between limited cone beam computed tomography, storage phosphor and film radiography, Caries Res. **40**(3), 202–207 (2006)

20.7 F. Haiter-Neto, A. Wenzel, E. Gotfredsen: Diagnostic accuracy of cone beam computed tomography scans compared with intraoral image modalities for detection of caries lesions, Dentomaxillofac. Radiol. **37**(1), 18–22 (2008)

20.8 Y. Zhang, L. Zhang, X.R. Zhu, A.K. Lee, M. Chambers, L. Dong: Reducing metal artifacts in cone-beam CT images by preprocessing projection data, Int. J. Radiat. Oncol. Biol. Phys. **67**(3), 924–932 (2007)

20.9 S. Lofthag-Hansen, S. Huumonen, K. Gröndahl, H. Gröndahl: Limited cone-beam CT and intraoral radiography for the diagnosis of periapical pathology, Oral Surg. Oral Med. Oral Pathol. Oral Radiol. Endod. **103**(1), 114–119 (2007)

20.10 A. Stavropoulos, A. Wenzel: Accuracy of cone beam dental CT, intraoral digital and conventional film radiography for the detection of periapical lesions. An ex vivo study in pig jaws, Clin. Oral Investig. **11**(1), 101–106 (2007)

20.11 C. Hannig, C. Dullin, M. Hülsmann, G. Heidrich: Three-dimensional, nondestructive visualization of vertical root fractures using flat panel volume detector computer tomography: An ex vivo in vitro case report, Int. Endod. J. **38**(12), 904–913 (2005)

20.12 N. Cohenca, J.H. Simon, A. Mathur, J.M. Malfaz: Clinical indications for digital imaging in dento-alveolar trauma. Part 2: Root resorption, Dent. Traumatol. **23**(2), 105–113 (2007)

20.13 A.G. Farman, W.C. Scarfe: Development of imaging selection criteria and procedures should precede cephalometric assessment with cone-beam computed tomography, Am. J. Orthod. Dentofacial Orthop. **130**(2), 257–265 (2006)

20.14 R. Mengel, M. Candir, K. Shiratori, L. Flores-de-Jacoby: Digital volume tomography in the diagnosis of periodontal defects: An in vitro study on native pig and human mandibles, J. Periodontol. **76**(5), 665–673 (2005)

20.15 B. Vandenberghe, R. Jacobs, J. Yang: Diagnostic validity (or acuity) of 2-D CCD versus 3-D CBCT-images for assessing periodontal breakdown, Oral Surg. Oral Med. Oral Pathol. Oral Radiol. Endod. **104**(3), 395–401 (2007)

20.16 O.B. Honey, W.C. Scarfe, M.J. Hilgers, K. Klueber, A.M. Silveira, B.S. Haskell, A.G. Farman: Accuracy of cone-beam computed tomography imaging of the temporomandibular joint: comparisons with panoramic radiology and linear tomography, Am. J. Orthod. Dentofacial Orthop. **132**(4), 429–438 (2007)

20.17 K. Fu, W. Zhang, D. Liu, H. Chen, X. Ma: Cone beam computed tomography in the diagnosis of temporomandibular joint osteoarthrosis, Zhonghua Kou Qiang Yi Xue Za Zhi **42**(7), 417–420 (2007)

20.18 A.M. Hussain, G. Packota, P.W. Major, C. Flores-Mir: Role of different imaging modalities in assessment of temporomandibular joint erosions and osteophytes: A systematic review, Dentomaxillofac. Radiol. **37**(2), 63–71 (2008)

20.19 Z. Krisjane, I. Urtane, G. Krumina, A. Bieza, K. Zepa, I. Rogovska: Condylar and mandibular morphological criteria in the 2-D and 3-D MSCT imaging for patients with class II division 1 subdivision malocclusion, Stomatologija **9**(3), 67–71 (2007)

20.20 E.L. Lewis, M.F. Dolwick, S. Abramowicz, S.L. Reeder: Contemporary imaging of the temporomandibular joint, Dent. Clin. North Am. **52**(4), 875–890 (2008), viii

20.21 S.D. Ganz: Computer-aided design/computer-aided manufacturing applications using CT and cone beam CT scanning technology, Dent. Clin. North Am. **52**(4), 777–808 (2008), vii

20.22 K. Honda, T.A. Larheim, K. Maruhashi, K. Matsumoto, K. Iwai: Osseous abnormalities of the mandibular condyle: Diagnostic reliability of cone beam computed tomography compared with helical computed tomography based on an autopsy material, Dentomaxillofac. Radiol. **35**(3), 152–157 (2006)

20.23 W.C. Scarfe, A.G. Farman, P. Sukovic: Clinical applications of cone-beam computed tomography in dental practice, J. Can. Dent. Assoc. **72**(1), 75–80 (2006)

20.24 M.L. Hilgers, W.C. Scarfe, J.P. Scheetz, A.G. Farman: Accuracy of linear temporomandibular joint measurements with cone beam computed tomography and digital cephalometric radiography, Am. J. Orthod. Dentofacial Orthop. **128**(6), 803–811 (2005)

20.25 K. Honda, Y. Arai, M. Kashima, Y. Takano, K. Sawada, K. Ejima, K. Iwai: Evaluation of the usefulness of the limited cone-beam CT (3-DX) in the assessment of the thickness of the roof of the glenoid fossa of the temporomandibular joint, Dentomaxillofac. Radiol. **33**(6), 391–395 (2004)

20.26 N. Cohenca, J.H. Simon, R. Roges, Y. Morag, J.M. Malfaz: Clinical indications for digital imaging in dento-alveolar trauma. Part 1: Traumatic injuries, Dent. Traumatol. **23**(2), 95–104 (2007)

20.27 G.R.J. Swennen, F. Schutyser, E. Barth, P. De Groeve, A. De Mey: A new method of 3-D cephalometry. Part I: The anatomic Cartesian 3-D reference system, J. Cranio-fac. Surg. **17**(2), 314–325 (2006)

20.28 J.H.S. Simon, R. Enciso, J. Malfaz, R. Roges, M. Bailey-Perry, A. Patel: Differential diagnosis of large periapical lesions using cone-beam computed tomography measurements and biopsy, J. Endod. **32**(9), 833–837 (2006)

20.29 L. Flygare, A. Ohman: Preoperative imaging procedures for lower wisdom teeth removal, Clin. Oral Investig. **12**(4), 291–302 (2008)

20.30 M. Kunkel, W. Wagner: *Leitlinie: Operative Entfernung von Weisheitszähnen* (Deutsche Gesellschaft für Zahn-, Mund- und Kieferheilkunde, Düsseldorf 2006)

20.31 H. Better, I. Abramovitz, B. Shlomi, A. Kahn, Y. Levy, A. Shaham, G. Chaushu: The presurgical workup before third molar surgery: How much is enough?, J. Oral Maxillofac. Surg. **62**(6), 689–692 (2004)

20.32 A. Suomalainen, T. Vehmas, M. Kortesniemi, S. Robinson, J. Peltola: Accuracy of linear measurements using dental cone beam and conventional multislice computed tomography, Dentomaxillofac. Radiol. **37**(1), 10–17 (2008)

20.33 S. Veyre-Goulet, T. Fortin, A. Thierry: Accuracy of linear measurement provided by cone beam computed tomography to assess bone quantity in the posterior maxilla: A human cadaver study, Clin. Implant Dent. Relat. Res. **10**(4), 226–230 (2008)

20.34 N. Van Assche, D. van Steenberghe, M.E. Guerrero, E. Hirsch, F. Schutyser, M. Quirynen, R. Jacobs: Accuracy of implant placement based on pre-surgical planning of three-dimensional cone-beam images: A pilot study, J. Clin. Periodontol. **34**(9), 816–821 (2007)

20.35 G.A.P. Di Giacomo, P.R. Cury, N.S. de Araujo, W.R. Sendyk, C.L. Sendyk: Clinical application of stereolithographic surgical guides for implant placement: Preliminary results, J. Periodontol. **76**(4), 503–507 (2005)

20.36 H. Nickenig, S. Eitner: Reliability of implant placement after virtual planning of implant positions using cone beam CT data and surgical (guide) templates, J. Craniomaxillofac. Surg. **35**(4–5), 207–211 (2007)

20.37 M. Heiland, P. Pohlenz, M. Blessmann, H. Werle, M. Fraederich, R. Schmelzle, F.A. Blake: Navigated implantation after microsurgical bone transfer using intraoperatively acquired cone-beam computed tomography data sets, Int. J. Oral Maxillofac. Surg. **37**(1), 70–75 (2008)

20.38 S. Haßfeld, M. Kunkel, H. Ulrich, W. Wagner, J.E. Zöller: Stellungnahme: Indikationen zur Schnittbilddiagnostik in der Mund-, Kiefer- und Gesichtschirurgie (CT/DVT), Der MKG-Chirurg **1**, 148–151 (2008)

20.39 J.M. Fullmer, W.C. Scarfe, G.M. Kushner, B. Alpert, A.G. Farman: Cone beam computed tomographic findings in refractory chronic suppurative osteomyelitis of the mandible, Br. J. Oral Maxillofac. Surg. **45**(5), 364–371 (2007)

20.40 K. Horner, V. Rusthon, A. Walker, K. Tsiklakis, P.N. Hirschmann, P.F. van der Stelt, A. Glenny, X.L. Velders, S. Pavitt: *European Guidelines on Radiation Protection in Dental Radiology. The Safe Use of Radiographs in Dental Practice* (European Comission, Brussels 2004)

20.41 V. Kumar, J. Ludlow, L.H. Soares Cevidanes, A. Mol: In vivo comparison of conventional and cone beam CT synthesized cephalograms, Angle Orthod. **78**(5), 873–879 (2008)

20.42 B. Grayson, C. Cutting, F.L. Bookstein, H. Kim, J.G. McCarthy: The three-dimensional cephalogram: Theory, technique, and clinical application, Am. J. Orthod. Dentofacial Orthop. **94**(4), 327–337 (1988)

20.43 G.R.J. Swennen, F. Schutyser: Three-dimensional cephalometry: Spiral multi-slice versus cone-beam computed tomography, Am. J. Orthod. Dentofacial Orthop. **130**(3), 410–416 (2006)

20.44 T.M. Peters: Image-guidance for surgical procedures, Phys. Med. Biol. **21/51**(14), R505–R540 (2006)

20.45 J. Wurm, T. Dannenmann, C. Bohr, H. Iro, K. Bumm: Increased safety in robotic paranasal sinus and skull base surgery with redundant navigation and automated registration, Int. J. Med. Robot. **1**(3), 42–48 (2005)

20.46 S. Haßfeld, J. Mühling: Computer assisted oral and maxillofacial surgery – A review and an assessment of technology, Int. J. Oral Maxillofac. Surg. **30**(1), 2–13 (2001)

20.47 R. Ewers, K. Schicho, G. Undt, F. Wanschitz, M. Truppe, R. Seemann, A. Wagner: Basic research and 12 years of clinical experience in computer-assisted navigation technology: A review, Int. J. Oral Maxillofac. Surg. **34**(1), 1–8 (2005)

20.48 T. Rodt, S.O. Bartling, J.E. Zajaczek, M.A. Vafa, T. Kapapa, O. Majdani, J.K. Krauss, M. Zumkeller, H. Matthies, H. Becker, J. Kaminsky: Evaluation of surface and volume rendering in 3-D-CT of facial fractures, Dentomaxillofac. Radiol. **35**(4), 227–231 (2006)

20.49 Q. Niu, X. Chi, M.C. Leu, J. Ochoa: Image processing, geometric modeling and data management for development of a virtual bone surgery system, Comput. Aided Surg. **13**(1), 30–40 (2008)

20.50 E.B. Strong, A. Rafii, B. Holhweg-Majert, S.C. Fuller, M.C. Metzger: Comparison of 3 optical navigation systems for computer-aided maxillofacial surgery, Arch. Otolaryngol. Head Neck Surg. **134**(10), 1080–1084 (2008)

20.51 A.M. Pham, A.A. Rafii, M.C. Metzger, A. Jamali, E.B. Strong: Computer modeling and intraoperative navigation in maxillofacial surgery, Otolaryngol. Head Neck Surg. **137**(4), 624–631 (2007)

20.52 D.M. Kahler: Image guidance: fluoroscopic navigation, Clin. Orthop. Relat. Res. **421**, 70–76 (2004)

21. Interventional Radiology – Angiography

Intervention

Doris Pommi

Interventional imaging in radiology and cardiology needs systems with enhanced imaging capabilities and cutting-edge technology. After the definition of digital subtraction angiography in Sect. 21.1, its range of application is described in Sect. 21.2. The advantages of interventional radiology procedures are outlined in Sect. 21.3 on the basis of existing angiography systems. Section 21.4 give an overview on trends in interventional radiography.

21.1 **Definition of Digital Subtraction Angiography** 417
21.1.1 Components Required for DSA........ 418

21.2 **Application Range for Angiography**........ 419

21.3 **Advantages of Interventional Radiology Procedures**.. 420

21.4 **Trends of Development**......................... 420
21.4.1 Enhanced Computer Performance... 420
21.4.2 Special Visualizations.................... 420
21.4.3 Hybrid Applications...................... 420

After radiologists had learned how to visualize the vascular system under x-ray fluoroscopy and contrast media, they had the idea of treating vascular diseases with special angiography catheters in the same way. While the catheter was first used for applying contrast media only, it now served as a tool for transporting and placing different therapeutic *devices*, implants, and chemotherapeutic substances such as balloon catheters, endoprostheses (stents), and chemoembolization material. For this procedure, image guidance and monitoring played a decisive role. So, from merely diagnostic radiology, interventional radiology evolved. Therefor the angiography suite, compared to other imaging modalities (like computertomography, magnetic resonace tomography, etc.), will be used more and more for interventional procedure and less for pure diagnostic purposes.

After the previous achievements of Sven-Ivar Seldinger (1921–1999) and Charles T. Dotter (1920–1985), it was above all Eberhard Zeitler and Andreas R. Grüntzig (1939–1985) whose developments turned interventional radiology into a minimally invasive method for treating cardiovascular diseases throughout the world.

Today in interventional radiology, a great variety of often highly complex therapeutic interventions are carried out under image guidance (x-ray, ultrasound, computer tomography, magnetic resonance imaging), including interventions of the vascular system. An important method in the field of vascular intervention is digital subtraction angiography (DSA).

21.1 Definition of Digital Subtraction Angiography

Digital subtraction angiography is a type of angiography and is used for examining and visualizing blood vessels and blood flow (Figs. 21.1, 21.2).

In DSA, two consecutive images of the body part to be examined and the vessels to be diagnosed are acquired and postprocessed digitally. First, a mask (empty) image is made. Then, contrast medium is injected and the fill image is acquired.

Visualization of the blood vessels, i.e., without surrounding tissue, is achieved by digitally subtracting the mask image from the fill image. Image parts differing in their x-ray absorption will remain – in this case the

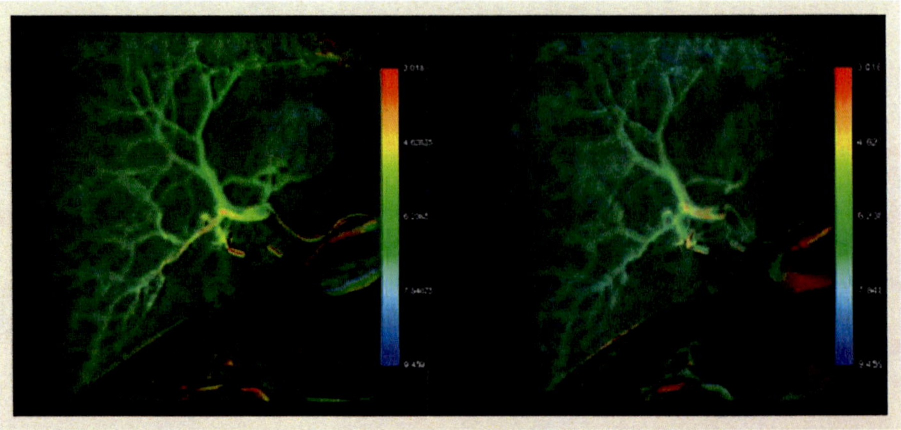

Fig. 21.1 Color-coded DSA image of blood vessels (courtesy of Siemens)

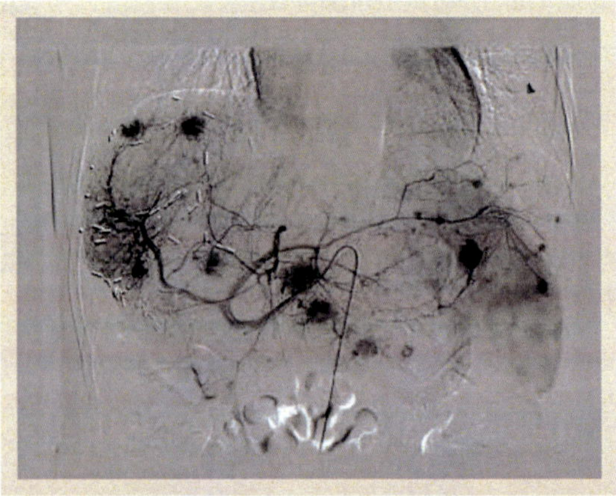

Fig. 21.2 Image of hepatic vessels (courtesy of Siemens)

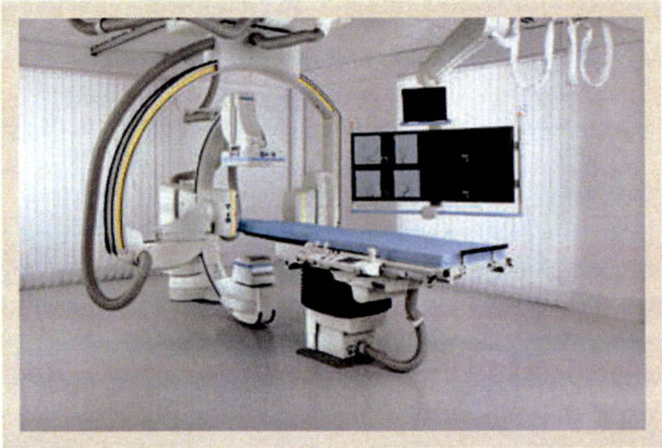

Fig. 21.3 Biplane angiography system (courtesy of Siemens)

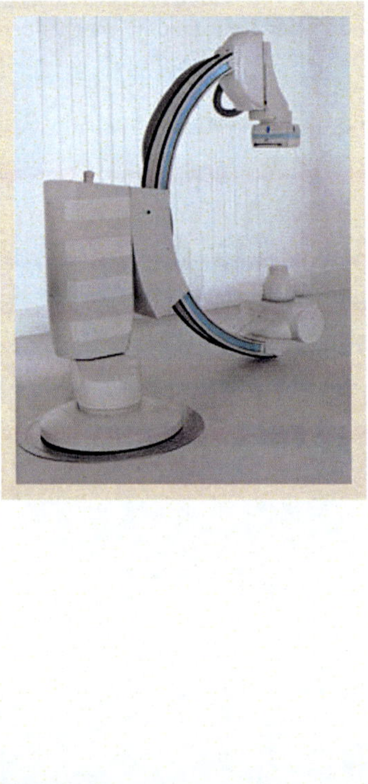

Fig. 21.4 Monoplane angiography system (courtesy of Siemens)

inner volume of the vessels filled with contrast medium (Figs. 21.1, 21.2).

21.1.1 Components Required for DSA

An angiography room for DSA includes the angiographic x-ray system, the examination table with control elements, a display suspension system for image display, and different peripheral devices.

The angiography system (Figs. 21.3–21.5) for visualizing vessels filled with contrast medium has at least one x-ray tube and a receiver, which are mostly installed on one joint C-arm and can therefore be moved synchronously within one plane. Formerly used image intensifiers are nowadays frequently replaced by flat detectors.

Biplane systems (Fig. 21.3) – systems with two independent tube/detector units in two fluoroscopy planes

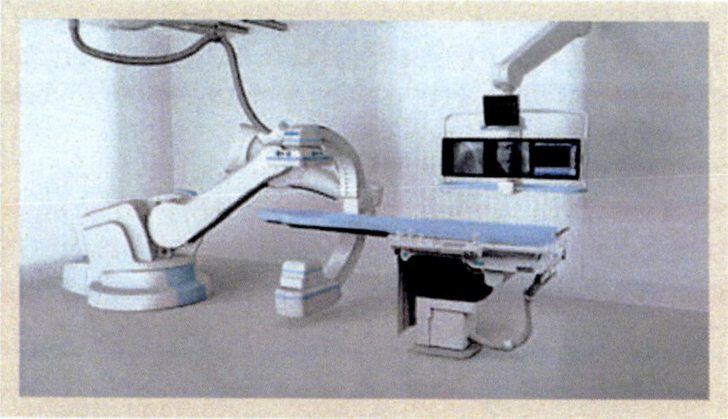

Fig. 21.5 Monoplane angiography system with robotic technology (courtesy of Siemens)

– are mainly used in neuroradiology and pediatric cardiology.

Very important for optimal and convenient patient positioning and free access for physician and staff during the sometimes very time-consuming interventions is a radiolucent, tiltable table.

The angiography system and table are operated directly in the examination room via an operating panel and foot switch; like the tube and detector, both can be covered with a sterile cover.

Monitors for image viewing [live image, reference image, three-dimensional (3-D) visualization, etc.] are mostly installed in a common holder (display suspension system) and are available in various types and different numbers.

The contrast medium is mostly delivered through a semi- or fully automatic injector, which can be synchronized with the triggering of the acquisition.

Different catheters, sheaths, coils, stents, and balloons are used for catheter work. Sterile work procedures are very important in angiography.

In the control room, a reading workplace is required for postprocessing [e.g., pixel shift, transfer to a picture archiving and communication system (PACS), 3-D postprocessing] in addition to the control computer and image processor.

21.2 Application Range for Angiography

In interventional radiology there are several focus areas regarding vascular procedures and therapy. Depending on diagnosis and thus therapeutic requirements, the main differentiators are:

- Opening of a vessel (e.g., dilatation)
- Closing of a vessel (e.g., embolization).

The most important procedures are:

- Balloon angioplasty (PTA, percutaneous transluminal angioplasty): catheter procedure under x-ray guidance for reopening constricted or occluded vessel sections by means of unfoldable balloons
- Stent implantation (stent angioplasty): catheter procedure for inserting and unfolding metal grid vessel supports (stents) under x-ray guidance for treating recanalized vessel constrictions and occlusions
- Thrombolysis: x-ray-guided catheter procedure for eliminating blood clots through local administration of thrombolytic drugs
- Chemoembolization: x-ray-guided catheter procedure for local chemotherapy and atrophying tumor-supplying arteries, particularly in the liver and uterus
- Thermal ablation: image-guided needle procedure for thermal destruction of tumor tissue by radiofrequency or laser energy.

Thanks to cutting-edge catheter and imaging technologies, angiographic development is tending toward interventional procedures, away from merely diagnostic vascular visualization. A decisive advantage here is the minimally invasive character of these therapies, which can therefore be carried out in a much more convenient way for the patient. For the different, sometimes very complex examinations, high-resolution imaging is particularly important to ensure precise navigation and

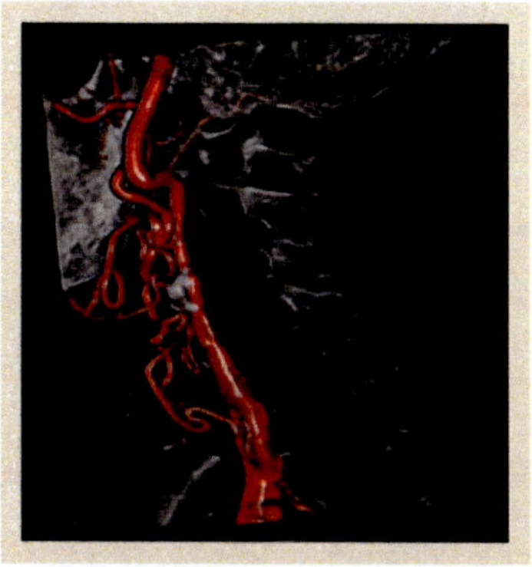

Fig. 21.6 Three-dimensional spatial image of cervical vessels (courtesy of Siemens)

placement of substances in the potentially very narrow vessels.

Apart from diseases of the cardiovascular system, other vascular systems such as the biliary system can also be treated through such interventions.

Angiography systems are frequently used today. Applications include:

- Elimination of aneurysms using stents or coils
- Stenosis treatment (angioplasty) through balloons and/or stents
- Tumor treatment through embolization or injection directly into the tumor center
- Dialysis shunts through balloon dilatation
- Interventional technology combined with surgical intervention directly in the operating room (hybrid OR).

Today, thanks to digital technology and high-performance computers, 3-D spatial images (Fig. 21.6) can be created almost in real time. Images from computed tomography (CT), magnetic resonance imaging (MRI), or functional images from nuclear medicine can immediately be overlaid with the angiography images. This helps to increase diagnostic confidence, facilitate instrument guidance, and improve follow-up.

21.3 Advantages of Interventional Radiology Procedures

A minimally invasive intervention often saves the patient an open surgery and a long length of stay associated with it. Many interventions can be performed on an outpatient basis today. This is a convenient alternative for the patient and saves time and costs during his hospital stay.

Use of catheter techniques allows for drugs, instruments, and implants to be placed directly at the site of action. Advanced imaging techniques in the pre-, intra-, and postoperative phases enable safe, fast, and seamless documentation of intervention planning, performance, and follow-up.

21.4 Trends of Development

21.4.1 Enhanced Computer Performance

Due to the use of minimally invasive procedures, it has become increasingly important to ensure optimal, fast, and detailed imaging performance. All previous diagnostic images need to be available during an intervention and, if required, it must be possible to overlay them on the monitor to provide the maximum amount of information for the intervention. For this procedure, substantial amounts of data must be transferred and displayed, requiring the use of high-performance computer systems and networks.

21.4.2 Special Visualizations

Special visualizations such as 3-D reconstructions (Fig. 21.6) or soft tissue visualization must be available for monitoring treatment success or to allow for fast reactions in case of critical incidents. This requires extension and specialization of the application range, enabling the user to implement the functions needed for his work.

21.4.3 Hybrid Applications

In radiology and surgery the tendency is toward hybrid ORs. These are operating rooms where multiple imaging modalities are used separately or simultaneously and where interventional radiologists work together with surgeons. Repositioning, which is time consuming and can jeopardize the patient, can thus be avoided, optimizing room and equipment utilization, as well. An example hybrid application is the combination of a cardiac catheter laboratory and OR.

General tendencies are for:

- Simplified system operation, e.g., through robot-assisted systems and case-specific protocols
- Comprehensive, customizable display of any required information in the angiography room/OR, for example, on large-format screens

- Reduction of radiation dose through radiographic procedures (e.g. DSA, CT angiography)
- Increased use of 3-D imaging with shorter acquisition or processing times and higher resolution (Fig. 21.6)
- Interdisciplinary use of imaging systems, e.g. in hybrid operating rooms.

Further Reading

- M. Wojtowycz: *Handbook of Interventional Radiology and Angiography*, 2nd edn. (Mosby, St. Louis 1995)
- G. Torsello, M. Steinbauer: *Ein neuer Gefäßarbeitsplatz – Der Hybrid-OP-Saal* (Schimmel Gill, Berlin 2010)
- T.J. Vogl, W. Reith, E.J. Rummeny (Eds.): *Diagnostische und interventionelle Radiologie* (Springer, Berlin, Heidelberg 2010)
- H.H. Schild (Ed.): *Angiographie* (Thieme, Stuttgart 2003)
- D.A. Gervais, T. Sabharwal (Eds.): *Interventional Radiology Procedures in Biopsy and Drainage* (Springer, Berlin, Heidelberg 2010)
- K. Valji: *Vascular and Interventional Radiology*, 2nd edn. (Saunders Elsevier, Amsterdam 2006)
- S. Baum, M.J. Pentecost (Eds.): *Abrams' Angiography: Interventional Radiology*, 2nd edn. (Lippincott Williams Wilkins, Philadelphia 2005)

22. Near-Infrared Spectroscopy (NIRS)

John McNulty, Michael Born, Robert S. Pozos

One of the greatest challenges facing medicine is to treat the patient using real-time data that accurately reflect oxygen concentration in the patient's tissue. Regulation of oxygen and its counterpart, carbon dioxide, is controlled by a large number of physiological control systems that are local, regional, and systemic. This key molecule is involved in oxidative metabolism and is an overall indicator of physiological well-being. Oxygen is attached to hemoglobin and released into cells, where it diffuses to mitochondria for utilization to produce adenosine triphosphate (ATP) that powers all cellular functions. The biomedical value of near-infrared spectroscopy (NIRS) is its ability to record oxygen levels, especially StO_2, in a noninvasive manner. In clinical cases, peripheral StO_2 values recorded from the forearm and thenar regions are frequently used as surrogates for central oxygen levels. Thus, NIRS holds the promise of being a major tool in normal and pathological functioning. Applications of this technology are far reaching and range from monitoring patients suffering from septic shock, type 2 diabetes to schizophrenia. Use of NIRS has great promise but suffers from two major types of issues: technical and physiological. There are different types of NIRS devices which detect *oxygen concentrations* using different probing strategies and different algorithms. In addition to the dissimilar NIRS technologies is the fact that the NIRS signal has a number of anatomical, e.g., fat, and physiological barriers, e.g., skin blood flow, before it reaches the muscle/organ vascular bed. Various regions of the body, e.g., thenar versus forearm, have dissimilar cardiovascular responses to physiology and pathophysiological conditions, making interpretation challenging. This chapter presents a brief overview of NIRS technology, its applicability, and promise.

22.1 **NIRS – Technical** 423
22.2 **NIRS Technology: Engineering Aspects** 425
22.2.1 Theory of NIRS Penetration 425
22.3 **Instrumentation and Equipment** 427
22.4 **New Developments: Multidepth Differential Approach** 430
22.5 **Clinical Application and Study of NIRS** 431
22.5.1 Overview 431
22.5.2 Thenar 431
22.5.3 Forearm 431
22.5.4 Brain 431
22.5.5 Lower Limb 432
22.5.6 Breast Tumors and Kidneys 432
22.6 **Does Skin Blood Flow Affect NIRS Measurements?** 433
22.6.1 Forearm Skin Blood Flow Study Versus StO_2 434
22.6.2 Thenar Skin Blood Flow Study Versus StO_2 435
22.7 **Future of NIRS** 436
References 437

22.1 NIRS – Technical

Near-infrared is part of the electromagnetic spectrum, which is described in terms of frequencies (Hz) or wavelengths (m). Some examples of such radiation include television (TV)/radio waves (10^8 Hz), microwaves (10^{10} Hz), infrared (10^{13} Hz), visible light (10^{15} Hz), and ultraviolet (10^{16} Hz). This chapter will concern itself with the near-infrared (NIR), whose frequency is $10^{14.5}$ Hz and whose wavelength range is 780–2526 nm, its application to detection of oxygen levels in tissue, and its various medical applications [22.1].

Table 22.1 Comparison of various oxygen measurements

	SaO_2	SpO_2	SvO_2	$ScvO_2$	StO_2
Measure	Arterial O_2 saturation	Arterial O_2 saturation	Mixed venous O_2 saturation	Central venous O_2 saturation	Tissue O_2 saturation
Percent hemoglobin O_2 saturation	Yes	Yes	Yes	Yes	Yes
Place of measurement	Arteries	Pulsing arteries	Pulmonary artery	Superior/inferior vena cava, right atrium	Peripheral circulation (veins, capillaries)
Method of measure	Arterial blood draw/ blood gas analyzer	Pulse oximeter	Pulmonary artery catheter	Central venous catheter	Placed on area in question: forearm, thumb, skull, etc.
Use of measure	O_2 loading in lungs	O_2 loading in lungs	Global tissue oxygenation	Surrogate for SvO_2	Tissue perfusion
What it indicates during shock	Lung/heart function	Lung/heart function	Changes in O_2 delivery and consumption	Changes in O_2 delivery and consumption	Peripheral perfusion status
Requires pulsatile flow	Yes	Yes	Yes	Yes	No

Sir William Herschel (1738–1822) is credited with the discovery of near-infrared radiation (NIR) and designed a method to filter light from a telescope to demonstrate that there was light radiation outside of the visible range [22.2]. After Herschel's discovery, NIR was not pursued by the scientific community until nearly 150 years later, when *Kaye* [22.2] described a technique called NIR spectroscopy (NIRS) for analysis of biological specimens. Although Kaye was the first to describe NIRS, Millikan developed a dual-wavelength oximeter for muscle, and in 1977, Jobsis first described an in vivo application of near-infrared spectroscopy, a technique originally designed for clinical monitoring of tissue oxygenation [22.3, 4]. Today, NIRS is flourishing in the field of medicine, because of its noninvasive nature and ease of use for detection of oxygen levels.

The emphasis of this paper is on StO_2, since it is measured using NIRS. StO_2 values (saturation of tissue oxygen) are determined by different algorithms used by competing vendors and is expressed as a percentage of the ratio of oxygenated hemoglobin to total hemoglobin in veins and capillaries. This measurement is related to, although different from, SpO_2, SaO_2, SvO_2, and $ScvO_2$. SaO_2 and SpO_2 are measurements of arterial oxygen saturation, whereas SvO_2 and $ScvO_2$ are measurements of mixed venous oxygen saturation. StO_2 is a measure of tissue perfusion. During shock, StO_2 levels indicate tissue perfusion at the site, and those values are used in certain cases to infer the condition of the patient relative to oxygen saturation.

A key point is that oxygen measurements are tied to oxyhemoglobin measurements, which are dependent on the factors that favor dissociation of oxygen. Oxygen is more easily released when pH is decreased, body temperature is increased, and when the arterial partial pressure of carbon dioxide is increased. Conversely, when pH is increased, temperature is decreased, and arterial partial pressure of carbon dioxide is decreased, oxygen is more tightly bound to hemoglobin. In addition, once oxygen reaches the mitochondria, its utilization or lack thereof is another indicator of cell integrity. The metabolic pathway for release of oxygen from hemoglobin and its diffusion to the mitochondria and subsequent oxidation must be kept in mind as one interprets StO_2 values.

To underscore the importance of the metabolic pathway of oxygen transfer by way of hemoglobin, the tissue hemoglobin index (THI) is another NIRS measurement that measures the hemoglobin signal strength and is useful for determining whether the StO_2 sensor is optimally positioned over muscle. THI levels are used to monitor the strength of the NIRS signal from tissue hemoglobin as compared with blood hemoglobin. Recently there has been interest in whether THI itself has value beyond simply verifying probe placement, and whether the correlation between THI and StO_2 could provide noninvasive data on blood hemoglobin concentration [22.5]:

THI has been studied during the clinical assessment of tissue oxygen perfusion status to convert a StO_2 *downward slope during arterial occlusion to an index of local oxygen consumption and to assess microvascular reactivity when blood flow is reestablished after arterial occlusion. Since inva-*

sive blood draws are not always feasible in patients, researchers have sought to establish a link between noninvasive continuous tissue hemoglobin measurements (THC) and blood hemoglobin concentration (Hbt).

To adequately understand the applications of NIRS, it is important to appreciate the interaction of the NIRS signal with biological tissue from an engineering and well as from an anatomical, physiological point of view. The NIRS signal must penetrate skin and fat to get to the veins and capillaries. In addition, control of the blood flow to the skin may vary from one region to another depending on the condition, e.g., exercise, hemorrhage, shock, etc., which will also influence the accuracy of NIRS.

22.2 NIRS Technology: Engineering Aspects

22.2.1 Theory of NIRS Penetration

In the NIR spectrum of radiation, light photons are capable of penetrating deep into a tissue sample, even through bone. Hemoglobin absorbs NIR light better than other molecules in a biological media.

This region of NIR also detects the oxygen-dependent metalloproteins in hemoglobin. These metalloproteins act as chromophores, responsible for the color of the molecule. Hemoglobin (Hb) is a globular protein that has an iron(III) metal cofactor that binds oxygen. Myoglobin (Mb) is another metalloprotein that plays an important role in red muscle, and the saturation level is dependent on the workload of the muscle (Fig. 22.1). NIRS cannot distinguish the difference between hemoglobin and myoglobin, but studies using model calculations on general data for human muscle have been preformed. A group at the University of Nijmegen in The Netherlands found that both deoxygenated Hb and Mb as detected by NIRS varied between 0.04 and 0.13 mol/m^3, while the variation in Mb saturation (53–86%) even exceeded that of Hb (63–84%) [22.6].

NIRS technology has advanced significantly over the last 30 years, and different instruments have been developed. Among these instruments, continuous-wave (CW) measuring instruments are the most common. These instruments are based on a modified Beer–Lambert law along with Doppler shift effect, vibrational spectroscopy, and the Frank–Condon principle [22.3]. The Beer–Lambert law provides the physical and mathematical basis for NIRS. It states that transmission of light passing through a solution of a colored compound, hemoglobin, is absorbed by the compound, resulting in a reduction in the intensity of the emerging light. Absorption and concentration of a chromophore are described by the Beer–Lambert law

$$A = \log \frac{I_0}{I} = \varepsilon c d \, ,$$

where A is the absorption of light expressed as optical density, c is the concentration of chromophore, ε is its extinction coefficient, and d is the thickness (optical path length) or depth of transmitted light in the tissue. The extinction coefficient describes the absorbance characteristics of a particular compound for a certain wavelength. As mentioned earlier, the extinction coefficients for different species of hemoglobin range from (carboxyhemoglobin) to 10.0 (methemoglobin). This enables the hemoglobin content to be measured. When NIR light illuminates a compound, some light passes through the compound. Knowing the wavelength, the extinction coefficient, and the thickness of the compound the Beer–Lambert law can be used to derive the concentration of the substance. Biological tissue poses a problem, because photons do not travel in a straight line as they do in other material. This is because living biological tissue is dynamic and constantly changing its makeup. Oftentimes, the light source and the light detector are in the same plane, and light does not com-

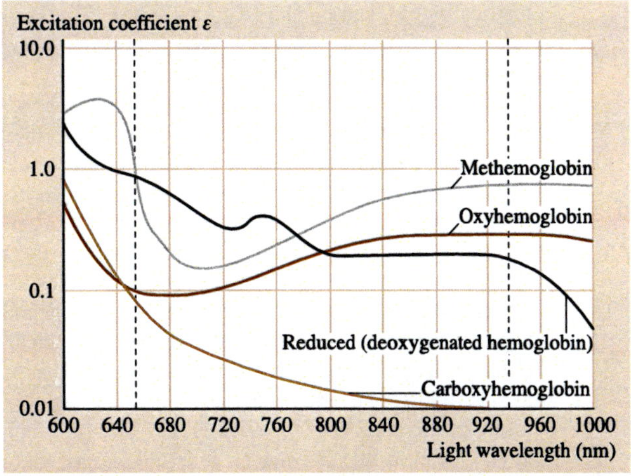

Fig. 22.1 Excitation coefficient of different hemoglobin molecules at different wavelengths

pletely traverse a tissue sample. If this is the case a correction factor is needed to compensate for the light path being greater than the probe separation. For NIRS probes, the wider the probes are separated, the deeper the light penetrates into a sample. A modification of the Beer–Lambert law is required to determine these differences in length

$$A = \sum \varepsilon c d \mathrm{DPF}\,,$$

where DPF is the differential path length and d is the interpreted distance [22.4]. DPF is a scaling factor that expresses the actual path length for a given part of the sample, and the sum is over all the absorbing chromophores (hemoglobin) that contribute to the NIRS signal. A DPF is needed to conduct quantitative spectroscopic measurements. A typical range for DPF measurements is 4–6.5, which means that the path length of photons traveling through a sample is 4–6.5 times longer than the spacing between the optodes.

A similar technology to NIRS uses the principle of laser Doppler to measure the total velocity of the local blood and the concentration of moving blood cells within the measuring volume that is being studied. This technology works by emitting a laser beam that is carried by a fiber-optic cable and penetrates the sample. The light interacts with the tissue and is scattered and partly absorbed. Light hitting moving blood cells undergoes a change in wavelength (Doppler shift), while light hitting static objects is unchanged. The frequency and magnitude of these changes are related to the amount and velocity of hemoglobin in the sample. The returning light is collected by the returning fiber, converted into an electrical signal, and displayed on a monitor. The PeriFlux System 5000 uses this laser Doppler principle and is mentioned later in this chapter in more detail.

The Frank–Condon principle describes the vibrational properties that are used for NIRS. When a molecule absorbs or emits a photon, its electronic and vibrational energy levels change. When a molecule vibrates, the probability of finding a given atom at a certain point is inversely proportional to its velocity. Atoms prefer a low kinetic energy level configuration when vibrating. This configuration is where the total energy and the potential energy of the molecule are nearly identical. When a NIR photon penetrates the sample, it is most likely to be absorbed when the nuclei of the hemoglobin molecules are stationary or moving very slowly. When the NIR photon hits the nuclei, it is not immediately excited. Thus, the excited nuclear configuration of the molecule tends to be close to the intersection of vibrational energy and potential energy because of the brief amount of time of an electrical transition. Therefore, transitions tend to take place between vibrational levels in which the nuclear configurations have minimal change, and they tend to occur when the nuclear kinetic energies are small. Hemoglobin molecules undergo this transition, and the energy of the molecules affects the returning light signal, thus providing information about the sample volume. These variations in vibrational and nuclear configuration give rise to anharmonicity, which is the deviation of a molecule's nature from a harmonic oscillator vibrating in simple harmonic motion. This causes the fundamental frequency of the molecule to be imprecise and causes combination and overtone bands. The NIR method gathers the measured spectral response caused by the fundamental vibration in the hemoglobin and correlates that to the structural information of the molecule. Depending on the quantity of NIR-absorbing molecule, absorbing type, and thickness of the sample, the harmonic vibrations occur at unique frequencies.

The molecular moieties that demonstrate the most overtones and combinations of fundamental vibrations when absorbing in the NIR region are −CH, −NH, −OH, (and −SH). The two key issues that determine the frequency and intensity of NIR absorption bands are anharmonicity and Fermi resonance [22.1]. Intramolecular interactions, such as nonsymmetrical vibrations, affect the energy curve of a molecule under NIR light. The spacing between energy levels that the molecule can attain is not identical. When the spacing of the molecule decreases, the energy level of the molecule increases. This energy increase is a quantum-mechanical model of an anharmonic oscillator. The origin of the NIR overtone bands is a result of these multilevel energy transitions that occur at multiples of the fundamental frequencies. With an anharmonicity constant χ of 0.01–0.05 the number of wave overtones can be estimated using the fundamental frequencies equation

$$\upsilon_x = \Delta \upsilon \upsilon_0 (1 - \Delta \upsilon \chi)\,,$$

where υ_x is the wavenumber of overtone x, υ_0 is the wavenumber of the fundamental vibration, and χ is the anharmonicity constant [22.1]. Fermi resonance occurs when two vibrating molecules accidently have exactly the same energy and interact with each other, producing a special type of configuration interaction between two NIR absorption bands of a polyatomic molecule. This causes the bands to split into two peaks of higher and lower frequencies that are different from the expected frequencies. In biological samples, dipole interactions and hydrogen bonding must be considered, since they can alter the vibrational energy states. This can shift the absorption bands of certain molecules, such as hemoglobin, or possibly give rise to new ones. NIR absorption bands are typically broad, overlapping, and 10–100 times weaker than their corresponding mid-infrared bands. This makes near-infrared less sensitive than mid-infrared. As a result, data processing is needed to interpret and relate the spectral information to the sample properties.

22.3 Instrumentation and Equipment

There are numerous companies that manufacture NIRS devices, but they all share six basic components:

1. A filtered light source capable of generating multiple wavelengths in the NIR region
2. Fiber-optic bundles that link the probes to a computer for data collection
3. A tissue probe containing an emitting and collection region
4. Photon detection hardware such as a photocathode/photomultiplier and signal amplifiers
5. A processing computer for multivariate algorithm processing of light data
6. A data display system, usually a computer or TV monitor [22.4].

The light source is usually a tungsten halogen lamp, since it is small and rugged, but other sources can be used. Detector types include silicon, lead sulfide (PbS), and indium gallium arsenide (InGaAs). Silicon detectors have fast signal acquisition with high signal-to-noise ratio, small size, and high sensitivity from the visible region to 1100 nm, which is well within the NIR spectrum. PbS detectors are slower but are sensitive from 1100 to 2500 nm, making them very popular in biological settings because they provide good signal-to-noise properties. The most expensive of the three detectors is the InGaAs detector. It combines the speed and size of a silicon detector with the wavelength range of the PbS detector. Since the spectrum of NIR light is polychromatic, a filter is needed to produce a single, monochromatic beam of light.

As mentioned above, a computer processing algorithm is needed to compare the delivered light with the returning light. Each NIRS device uses its own, unique algorithm to analyze the collected data. The patents for these NIRS devices describe the algorithms they used. In a general sense, an algorithm is a method to solve a problem using a sequence of instructions. Computer data processing takes a process and uses a computer program to convert, analyze or summarize the data into usable information. To interpret the overlapping portions of the NIR absorption spectra, a minimum of two NIR wavelengths are needed for the computer processor to apply its algorithm, although up to 12 NIR wavelengths can be used. A multiwavelength algorithm is needed because of the absorption peaks for the different molecules in the sample. Absorption coefficients are the basis of application of these multiwavelength algorithms, being nothing more than algebraic expressions. Each NIRS device has its own specially developed algorithm used to interpret the information collected. The basic criteria for the absorption coefficients are:

1. Each NIR spectrum should be measured independently without spectral contributions from other oxygen-dependent chromophores
2. The relative extinction coefficient should be measured for each chromophore at the same light scattering and path length conditions
3. Spectra should be acquired at optical geometries and scattering functions appropriate to the particular application used.

With the use of the principles stated above, a correct algorithm can convert the measured changes in light absorption at various NIR wavelengths into concentration or ratio data for the sample site of interest. A problem can arise if these multiwavelength algorithm principles are applied to a tissue sample where the hemoglobin content is rapidly changing. Human skin is a very complex medium and poses a great challenge to NIRS devices because the accuracy of data collected is an issue. Not only is human skin very dynamic, but the nonuniformity between person to person requires the NIRS device to be calibrated. Calibration of NIRS devices is necessary to quantify the information collected by these devices. Multivariate calibration is needed before a NIR spectrometer can be used for any quantitative analysis. That calibration process involves the following steps [22.1]:

1. Selection of a representative calibration sample set.
2. Spectra acquisition and determination of reference values.

3. Multivariate modeling to relate the *spectral variation* to the *reference values* of the analytical target property.
4. Validation of the model by cross-validation, set validation or external validation.

The two most frequently used multivariate regression methods in quantitative NIR analysis are principal component regression (PCR) and partial least-squares (PLS) regression. PCR uses the principal components to perform regression on a sample volume to be predicted. PCR is used when the variables are close to being collinear, which is problematic for quantifying a signal. Typically, only a subset of principle components in the regression are used, allowing for a normalized estimate. Oftentimes, the selection of the principle components is according to a standard protocol in NIR analysis. PLS finds a linear model between the signals by projecting the observed and predicted variables into a new space. PLS finds the directions of greatest variability by comparing both spectral and target property information. The main difference between the two methods is that the first principal component or factor in PCR represents the most relevant variations showing correlation with the target property values.

In qualitative analysis of an NIR signal, spectral variations acquire discrete values that represent the quality of the signal, which is related to the sample properties. As mentioned earlier, tissue is problematic for interpreting the returning NIRS signal. To solve this problem, multivariate classification methods are once again used, this time for grouping samples with similar characteristics such as deoxyhemoglobin and oxyhemoglobin in the microvasculature of human skin. Pattern recognition methods used in qualitative analysis are divided into two learning algorithm categories: *supervised* and *nonsupervised*. Supervised methods use a machine learning technique to deduce a function from basic data. The basic data consist of inputs and desired outputs. The output can be a function similar to a regression used to predict the probable input. The task of the supervised method of analysis is to predict the value of the function for any input after experiencing the basic data examples. Supervised methods in essence are used to build classification rules for a number of prespecified subgroups. In most cases the group structure of the training set is known, such as average hemoglobin concentration. Nonsupervised methods are different than supervised methods in that they do not require any prior knowledge about the group structure in the data, but instead produce a grouping. The ultimate goal of qualitative analysis of a NIRS signal is to develop mathematical criteria for applying parameters. This would allow similarity between samples to be expressed quantitatively.

NIRS signal quality must be determined before any data collection can occur. The noninvasive continuous nature of NIRS devices makes them very appealing, especially in a clinical setting where these devices are frequently used. Accurate data must be collected to understand what is happening in the sample volume being observed. Comparisons of NIRS signals between two different states for the sampling volume, for example, oxygen-bound hemoglobin and free hemoglobin, are required to determine whether changes are significant, commonly using a paired t-test. Paired t-tests are used for comparing means of NIRS signal changes between two states within a sampling volume. Autoregressive models are commonly used to analyze time-series data. An autoregressive model is a type of random process that tries to predict natural phenomena. It is very difficult to derive an autoregressive model for NIRS data because the signals are not constant during the dynamic processes of human skin regulation. Baseline corrections are oftentimes performed but could distort actual changes in the NIRS signal.

Recent NIRS technology is not without its problems. The major problem with NIRS is its inability to quantify concentration changes in hemoglobin using current CW-type instruments, although this has not stopped clinical use of these devices. Clinicians struggle with obtaining an accurate reading of oxygen delivery to vital organ beds. The state of organ perfusion could ideally be assessed by a tissue oxygenation device [22.7]. Quantification is the central issue surrounding use of NIRS data. NIRS technology is reliable when changes in hemoglobin are global, over the entire body, as quantification is possible and optical path length in the sample volume can be determined. Quantification of the NIRS signal becomes problematic when changes in hemoglobin are localized, such as localized skin cooling or heating. The reason for this inability to quantify localized changes is due to influencing factors that cause NIRS to become unreliable. Taking the example of localized skin cooling, some factors that influence the NIRS signal of concentration changes in hemoglobin include skin blood flow, adipose (fat) tissue, and muscle blood flow. With current CW-type instruments, deep muscle blood flow is difficult to quantify because the NIR light must pass through the skin and adipose tissue before reaching its target tissue. The skin has its own blood flow and the adipose tissue absorbs monochromatic NIR light, thus weakening its

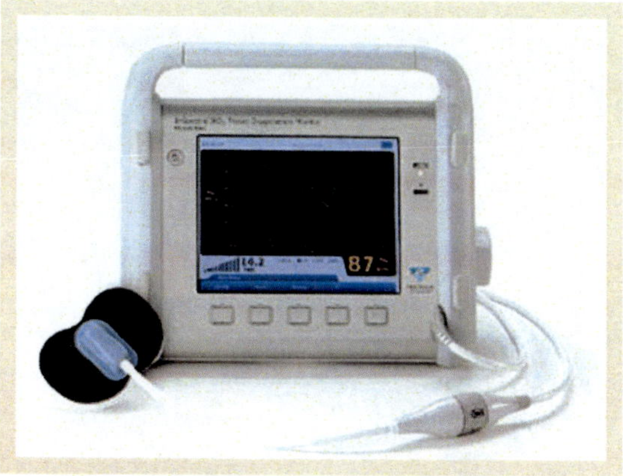

Fig. 22.2 InSpectra StO_2 tissue oxygenation monitor by Hutchinson Technologies

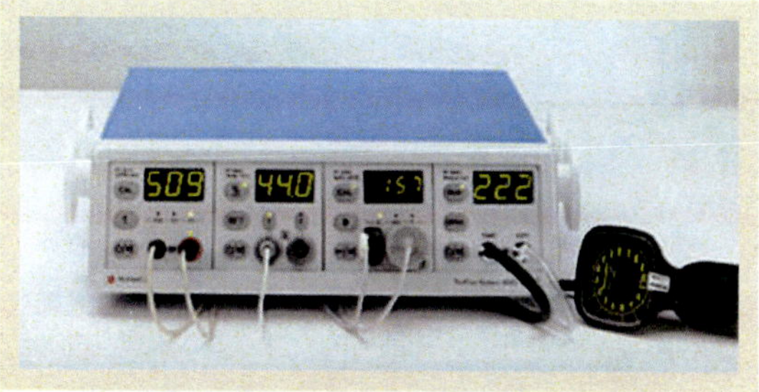

Fig. 22.3 PeriFlux system 5000 by Perimed Inc.

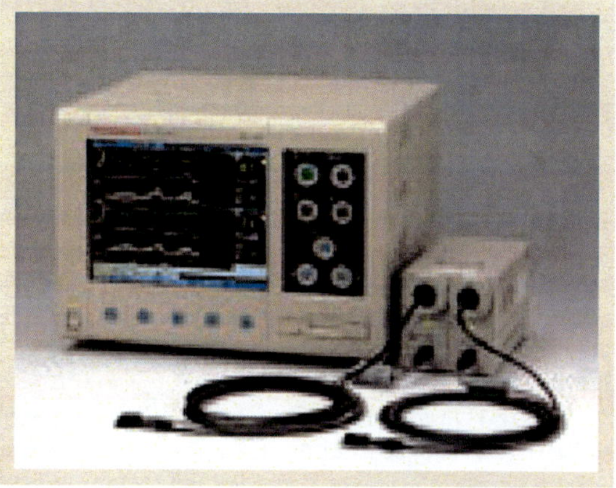

Fig. 22.4 The Niro-500 oxygenation monitor by Hamamatsu

signal. The complexity of this scenario creates a challenge for accurate measurement of localized muscle blood flow after localized skin cooling. This is just one example of the complicating factors that may influence NIRS measurements, yet the method is used in hospitals to assess patient status because studies have shown that these devices help doctors assess patient status.

Some of the modern NIRS devices include the InSpectra StO_2 Tissue Oxygenation Monitor by Hutchinson Technologies, PeriFlux System 5000 by Perimed Inc., Niro-500 Oxygenation Monitor by Hamamatsu, T.Ox Tissue Oximeter by Vioptix, and INVOS System by Somanetics, to name a few. This aspect of NIRS is rapidly changing, and the technology presented later gives an overview of the different approaches used in NIRS studies. Due to this rapid development, NIRS equipment may be completely different by the time this chapter is published. These systems may vary in a fundamental manner in that the number of wavelengths vary from 2 to 12. How effective these various systems are is difficult to ascertain.

The InSpectra StO_2 Tissue Oxygenation Monitor (Fig. 22.2) provides a noninvasive way to collect continuous, real-time information for the perfusion status of a patient by using StO_2, which is determined by the ratio of oxygenated hemoglobin to total hemoglobin in the microcirculation of the sampling volume. A strong linear correlation $r^2 > 0.93$ between StO_2 and microvascular $\%SO_2$ was observed for isolated animal hind limb, kidney, and heart [22.6]. Observation of this parameter is used to assist clinicians in early detection of inadequate tissue perfusion (hypoperfusion) of a patient.

The PeriFlux system 5000 (Fig. 22.3) is a modular system that enables simultaneous measurement of blood perfusion and transcutaneous p_{O_2/CO_2} using a laser Doppler method. One way in which the PeriFlux differs from the InSpectra device is its sampling depth. The PeriFlux system measures at a depth of 2 mm compared with 12 mm or greater depths in other systems. PeriFlux probes may also simultaneously collect temperature information from the site of interest, making this system very useful for shallow studies, e.g., skin.

The Niro-500 oxygenation monitor (Fig. 22.4) by Hamamatsu measures the tissue oxygenation index (TOI) of a sample volume. The TOI indicates the oxygen saturation level with respect to the normalized tissue hemoglobin index that reports the real-time concentration changes in oxygenated hemoglobin, deoxygenated hemoglobin, and total hemoglobin. This device allows for measurements at two sites simultaneously when monitoring patients' brain oxygenation in the intensive care unit (ICU), critical care unit (CCU), or neonatal intensive care unit (NICU).

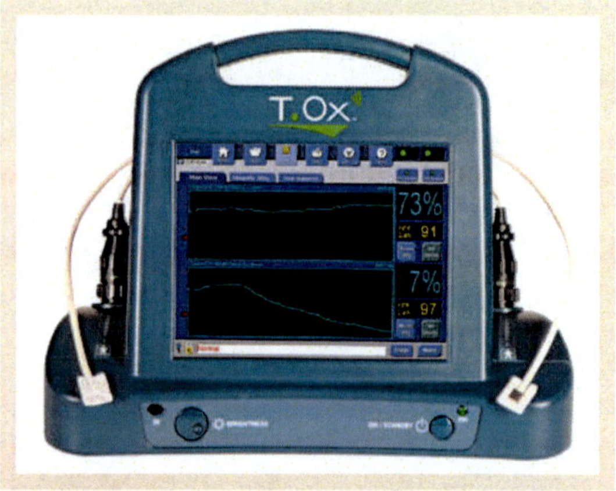

Fig. 22.5 T.Ox (tissue oximeter) by Vioptix

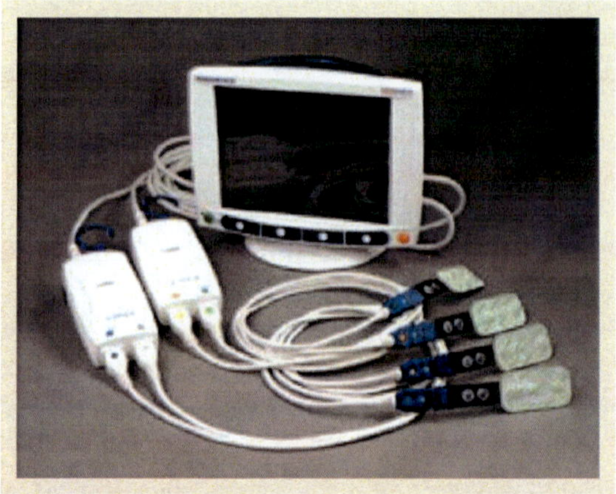

Fig. 22.6 The INVOS system by Somanetics

The T.Ox (tissue oximeter) (Fig. 22.5) is a portable unit with a liquid crystal display (LCD) touchscreen monitor with the capability to operate two fiber-optic sensors. Along with instant data acquisition of the sampling volume, this device is equipped with adjustable audible and visual alarms for StO_2 concentrations and low battery.

The INVOS system by Somanetics (Fig. 22.6) uses in vivo optical spectroscopy. Like the other devices, INVOS provides real-time information of regional oxygen levels in the sampling volume beneath the NIR-emitting probe. The newest model, the sixth generation INVOS, is the only commercially available four-channel system that reportedly monitors cerebral and peripheral blood circulation simultaneously. This is beneficial for detecting early indicators of shock or other cerebral perfusion abnormalities in patients.

22.4 New Developments: Multidepth Differential Approach

A new group of devices are being developed and evaluated to address the issue of skin blood flow influencing the NIRS signal and more broadly any situation in which the area of interest may be influenced by blood flow; for example, when monitoring StO_2 in the brain in adults, the NIRS signal may encounter blood flow from the sinuses. This new group of NIRS devices utilize a multidepth differential approach so that values from the skin can be subtracted from deeper tissues. Bezmer reports on the use of this system for measuring StO_2 values on forearm and thenar simultaneously. He utilized one multidepth probe on the thenar and another on the forearm. This device from Hutchinson Technology employs reflectance-mode probes that have one 1.5 mm optical fiber to illuminate the tissue and two optical fibers to detect the backscattered light from the tissue. Using space between the illumination fiber and the two detection fibers of 15 and 25 mm, respectively, he recorded StO_2 values at those depths. Subtracting those values from the 1.5 mm probe gives a differential value that subtracts out the skin value and theoretically gives a value for the 15 and 25 mm depths. The optical data were subsequently analyzed using the second-derivative values of the ratio of 720 to 760 nm, which is related to StO_2 [22.1]. Other devices have been developed to utilize the multidepth

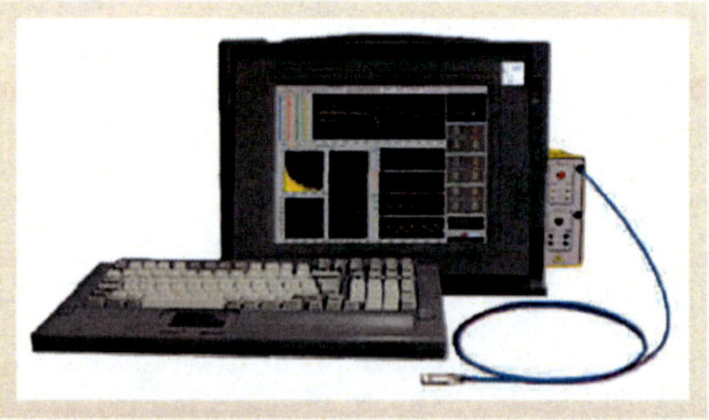

Fig. 22.7 O2C (oxygen to see) system developed by LEA Medizintechnik GmbH

differential approach [22.8]. The Lea is another example of such a system which measures blood flow and StO_2 levels simultaneously. Presently, many companies are addressing the need for multiple probes. Thus, this presentation will be outdated by the time it is published.

The O2C (oxygen to see) system developed by LEA Medizintechnik GmbH (Fig. 22.7) was designed to be a noninvasive diagnostic device for determination of oxygen supply in the microvasculature of blood-perfused tissue. It uses a glass-fiber probe that serves as a measuring point for two separate channels. One channel measures the oxygen saturation of hemoglobin of the skin while the other channel records the oxygen saturation of deeper tissues such as skeletal muscle. Probes range from 2 mm depth up to 15 mm depth and can be used with a variety of probe heads for specific applications.

22.5 Clinical Application and Study of NIRS

22.5.1 Overview

The areas of the body where NIRS is used include the thenar or thumb region [22.5], the forearm [22.9], the cranial area [22.10], the legs [22.11], the breast [22.12], and others [22.13] where oxygen transfer to tissue takes place. This technology is also being used on neonates as well as in adults [22.13]. For peripheral measurements such as in the thenar and forearm, these values are sometimes used as surrogates for central cardiovascular measurements. Since the signal is thought to reflect what is occurring at the microvascular level, a vascular occlusion test (VOT) is used [22.14]. This consists of occluding the arterial flow to the upper limb and then releasing the pressure. The VOT-StO_2 values can be divided into baseline, ischemia, reperfusion, and hyperemia [22.15]. The rate of deoxygenation can be determined from the downslope during the ischemic phase, and the rate of reoxygneation can be measured from the upslope during reperfusion [22.16]. Although the VOT procedure has promise, since it may reflect pathological conditions such as sepsis, the procedure has not been standardized due to probe position and size as well as deflation thresholds [22.14]. The following are major factors that influence StO_2 values: gender (particularly with increasing age), position of the subject, which muscle is measured, adipose tissue thickness, skin blood flow, and edema.

22.5.2 Thenar

The thenar section of the hand is a popular testing site in current research [22.7] due to its convenience for clinical studies. The hand is easily accessible while the patient is in supine position and easily manipulated when the patient is unconscious. In a recent study, StO_2 values were reported to be a good predictor of severe shock in trauma patients [22.7]. In this prospective, nonrandomized, observational, descriptive study in normal human volunteers ($n = 707$) and patients admitted to the trauma center ($n = 150$) StO_2 monitoring was performed continuously and noninvasively, with values recorded at 2 min intervals. The trauma team, blinded to StO_2 values, classified each patient into one of four groups (no shock, mild shock, moderate shock, and severe shock) using conventional clinical parameters. The mean $\pm$ standard deviation (SD) thenar StO_2 values for each group were as follows: normal, $87 \pm 6\%$ ($n = 707$); no shock, $83 \pm 10\%$ ($n = 85$); mild shock, $83 \pm 10\%$ ($n = 19$); moderate shock, $80 \pm 12\%$ ($n = 14$); and severe shock, $45 \pm 26\%$ ($n = 14$). The thenar StO_2 values clearly discriminated the normal patients and those with severe shock ($p < 0.05$) [22.7]. The results indicated that NIRS of the thenar was a quick, noninvasive, and accurate predictor of tissue dysoxia.

22.5.3 Forearm

Another common site of interest for NIRS is the forearm. Studies such as [22.17] and [22.9] used the forearm as the site for their StO_2 measurements while studying skin blood flow. However, while this site is popular in research, questions arise regarding whether this site is suitable clinically, due to possible interference from adipose tissue above the muscular tissue on the NIRS readings [22.18]. This influence will be further discussed later on, as well as whether the forearm is more sensitive to cardiovascular changes than is the thenar.

22.5.4 Brain

NIRS is popular for use in neonates at intensive care units. Oxygen exchange in the brain is essential for neonatal development, and noninvasive techniques are highly sought after when dealing with infants [22.19]. A recent study was conducted that compared NIRS technology and the intravenous ^{133}Xe clearance technique, which is the current clinical technique for measuring cerebral blood flow in newborn infants. Forty cerebral blood flow measurements using NIRS (CBF_{NIRS}) measurements were obtained during 19 ^{133}Xe measurements in 16 infants. The test–retest variation of repeated near-infrared measurements during each ^{133}Xe clearance was 17.5%. CBF_{NIRS} was closely related to CBF_{XE} ($r^2 = 0.84$, $p < 0.0001$), with a slope of 0.75 (standard error of the mean, SEM = 0.064) and an intercept of 1.58 ml/100 g/min (SEM = 0.51). The difference between the measurements obtained by the two methods ($CBF_{NIRS} - CBF_{XE}$) was negative in the high range of CBF, whereas the difference was close to zero in the low range. The conclusion was that CBF measured with near-infrared spectroscopy was in good agreement with CBF measured with the ^{133}Xe method. The near-infrared spectroscopy method has the advantage of being noninvasive, and it does not involve ionizing radiation. Because of methodological constraints, however, NIRS may underestimate CBF in the high flow range, and it may have limitations for application in clinical research [22.19]. Another question raised is the cranial placement of the NIRS probe, which may result in influence of the cranial bone on these particular StO_2 readings. The adult skull is much thicker and denser than that of an infant, which could possibly present an obstacle for use in adult patients. While it seems to be useful with babies, further studies need to be conducted before this application of NIRS is used in the adult ICU [22.20].

22.5.5 Lower Limb

NIRS probes have also been used in studies dealing with the lower extremities. A recently conducted study showed that StO_2 monitoring in patients with peripheral arterial disease was very beneficial for diagnosing those with this disease as well as for monitoring the disease once diagnosed. Thirty-five normal and 14 peripheral arterial disease (PAD) subjects from two clinical centers were evaluated in a prospective cross-sectional analysis comparing StO_2 by using the InSpectra tissue spectrometer at rest (baseline) and after treadmill exercise. Measurements were obtained at baseline and peak exercise for normal subjects and at baseline, initial claudication distance (ICD) and absolute claudication distance (ACD) in PAD subjects. Endpoint values were the mean of 15 data points. Times to 50% StO_2 recovery to baseline (T50) and complete recovery to baseline (T100) were measured. Baseline StO_2 was 65% in both groups. The peak exercise StO_2 was significantly lower and the absolute change in StO_2 and the percentage change in StO_2 were significantly greater in PAD patients ($p < 0.45$). T50 and T100 were longer in PAD patients compared with normal subjects ($p = 0.0001$ and 0.002, respectively). T50 of > 70 s yielded a sensitivity of 89% and a specificity of 85% for PAD [22.21]. The noninvasive manner showed promise for identifying patients with PAD. However, a study performed on the same group but eliminating the exercise protocol and applying a more clinical protocol could provide further answers and insight on the use of NIRS for PAD.

22.5.6 Breast Tumors and Kidneys

Other areas of interest in NIRS monitoring that show promise of clinical benefits are in monitoring of the breast and the kidneys. A recent study conducted on breast tumor tissue showed that this particular tissue relayed different and unique StO_2 and THI readings as compared with normal breast tissue. THI values in tumorous breast tissue were shown to be 2–4 times higher compared with normal breast tissue, and StO_2 levels in the tumorous tissue tended to be 5–20% lower as compared with normal breast tissue [22.12]. If validated, this would not only make breast tumors easier to diagnose, but also make the female patient much more comfortable in the clinical setting. Another study conducted on infants monitored StO_2 levels in the deltoid and kidney of the baby. The kidney was chosen for these low-birth-weight babies because patent ductus arteriosus (PDA) in low-birth-weight infants would increase StO_2 in the pulmonary vasculature and parenchyma due to increased pulmonary blood flow, while the decrease in blood flow distal to the PDA would diminish StO_2 in distal organs such as the kidney and skeletal muscle, and these relative differences would be detectable by NIRS. The results from this study showed that, in those babies treated for PDA, StO_2 values were 34% in the kidney and 35% in the deltoid (average values, $n = 13$). The average StO_2 values in babies who ended up not being treated for PDA were 55% in the kidney and 53% in the deltoid. Data analysis showed that these differences in StO_2 values were significant ($p = 0.01$) [22.13]. Results suggested that portable NIRS technology showed great promise in predicting and diagnosing for babies with patent ductus arteriosus and low birth weight.

22.6 Does Skin Blood Flow Affect NIRS Measurements?

It is obvious that the portability, ease of use, convenience, and noninvasive characteristics of NIRS StO_2 monitoring make this technology very popular when thinking of future applications in the clinical setting. However, certain questions arise that must be answered before this technology can be completely embraced. To measure tissue oxygen saturation accurately at any given site, the NIRS signal must pass through other parts of the body that might not necessarily be of interest. Some of these areas include skin, peripheral blood, adipose tissue, and bone. Perhaps the most popular question of concern, and possibly the most influential, is whether skin blood flow influences StO_2 readings.

Skin blood flow in the human body serves many functions such as thermoregulation, oxygen transport, hormone transport, CO_2 transport, and hydrogen ion transport [22.22]. Skin blood flow can also be altered in various conditions such as exercise, extreme heat, extreme cold, and even shock. During exercise, extra oxygen must be delivered to the muscle and tissue cells so that they may perform their work. To address this extra need for oxygen, skin blood flow to the relevant area is increased [22.22]. In conditions of extreme heat, skin vasculature undergoes vasodilatation to divert the flow of heated blood to the skin, where heat can be more readily given off by the body in order to cool. The opposite of this process is called vasoconstriction and is seen in extreme cold conditions for the opposite effect on the body. Both vasoconstriction and vasodilatation are controlled by the autonomic nervous system. Another state that alters skin blood flow is shock, resulting in deprivation of blood flow to the tissues. This serious medical condition can lead to serious conditions such as hypoxemia and even cardiac arrest. Shock slows and in some cases stops skin blood flow as compared with its normal rate of flow [22.23].

Limited studies have been carried out in a variety of ways, suggesting that skin blood flow has a significant influence on StO_2 readings. *Buono* et al. [22.11] reported that skin blood flow had a significant effect on NIRS measurements. Leg skin blood flow (SkBF) was increased and decreased following local heating and intradermal epinephrine injection using 30-gauge needle into the dermis and raising an intradermal wheal under direct visualization to avoid injection into the subcutaneous tissue. Epinephrine decreased muscle saturation (StO_2), and heating the leg increased StO_2. Results are shown in Fig. 22.8a,b.

The results suggest that, with a significant decrease or increase in skin blood flow to the leg, a significant decrease or increase, respectively, in StO_2 was observed as well [22.11]. The major criticism of this research was that the authors had no way to determine how deep the epinephrine or warming affected deep layers. Supplemental study would be recommended to verify the influence of skin blood flow on StO_2 measurements. In addition, *Davis* et al. reported that skin blood flow of

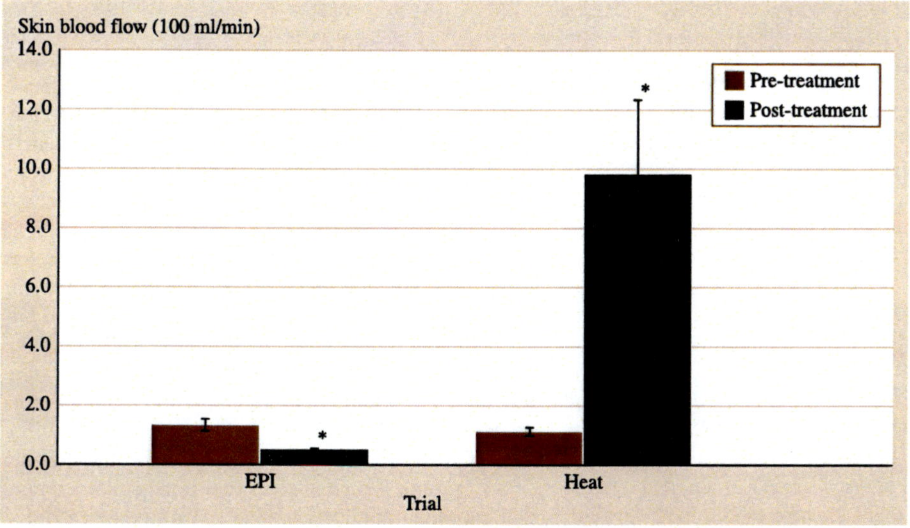

Fig. 22.8 Effect of epinephrine and heat on muscle oxygen saturation (StO_2) (mean ± SE). * Different from pretreatment measurements $p < 0.05$ (after [22.11])

Table 22.2 Effects of local cooling followed by capsaicin and occlusion on StO_2 values. * Significance at $p < 0.05$

15 mm data	StO_2		THI		Application site blood flow		Near application site blood flow	
	Average	SD	Average	SD	Average	SD	Average	SD
Baseline	76.85	11.5	7.07	3.84	15.7	2.87	14.35	4.32
After cooling	75.43*	9.91	5.64*	2.65	29.37*	17.47	28.93*	17.21
After capsaicin	90.42	4.2	6.8	2.84	78	84.6	105	44.31
After occlusion	90.57	5.62	6.99	2.83	77.43	64.06	106.14	39.36

15 mm data	Application site temp.		Near application site temp.	
	Average	SD	Average	SD
Baseline	32.2	1.36	31.86	0.98
After cooling	26.38*	1.68	25.03*	0.7
After capsaicin	28.3	1.99	26.27	0.95
After occlusion	28.32	2.11	26.05	2.51

25 mm data	Application site temp.		Near application site temp.	
	Average	SD	Average	SD
Baseline	31.88	1.25	31.66	1.03
After cooling	26.03*	1.54	25.61*	1.52
After capsaicin	26.67	2.49	25.89	1.13
After occlusion	26.96	2.39	26.51	1.58

25 mm data	StO_2		THI		Application site blood flow		Near application site blood flow	
	Average	SD	Average	SD	Average	SD	Average	SD
Baseline	82.1	4.73	13.9	4.24	14.65	6.6	13.79	4.38
After cooling	77.9*	9.87	12.48*	3.66	19.54*	17.87	17.75*	14.51
After capsaicin	85.33	6.3	12.58	3.41	39.62	48.59	55.72	37.68
After occlusion	86.31	4.85	13.4	3.76	41.96	35.83	69.87	54.29

the forearm influenced NIRS measurements. Using local heating as well as whole-body heating, they reported large increases in NIRS-derived tissue oxygenation correlated with increases in skin blood flow. They used the NIRO550 Hamamatsu Photonics unit that uses a different algorithm than the Hutchinson system used by *Buono* et al. [22.24].

22.6.1 Forearm Skin Blood Flow Study Versus StO_2

Recent studies in our laboratory suggest that skin blood flow had significant influence on NIRS StO_2 measurements taken on the forearm. Two different probes, 15 and 25 mm, were used on 11 male and female subjects who had their forearm skin blood flow altered using a cooling pad to cause vasoconstriction followed by capsaicin (0.1%) to cause vasodilatation. Controls were run in both sets of experiments, where no capsaicin was used on the forearm. Results for the 15 mm probe showed that, prior to capsaicin application, subjects averaged a 75.4% StO_2 level, whereas after capsaicin application, an average of 90.4% was recorded, corresponding to a 15% increase. Results for the 25 mm probe showed a baseline of 77.9% and, after capsaicin, an average of 85.3%, a 7.4% increase. In addition, total hemoglobin index (THI) levels were recorded, and although the 15 mm probe recorded overall lower THI levels with an average of 5.64 before capsaicin and 6.8 after capsaicin, both the 15 and 25 mm probes recorded the same trends upon cooling and capsaicin application. The 25 mm probe recorded

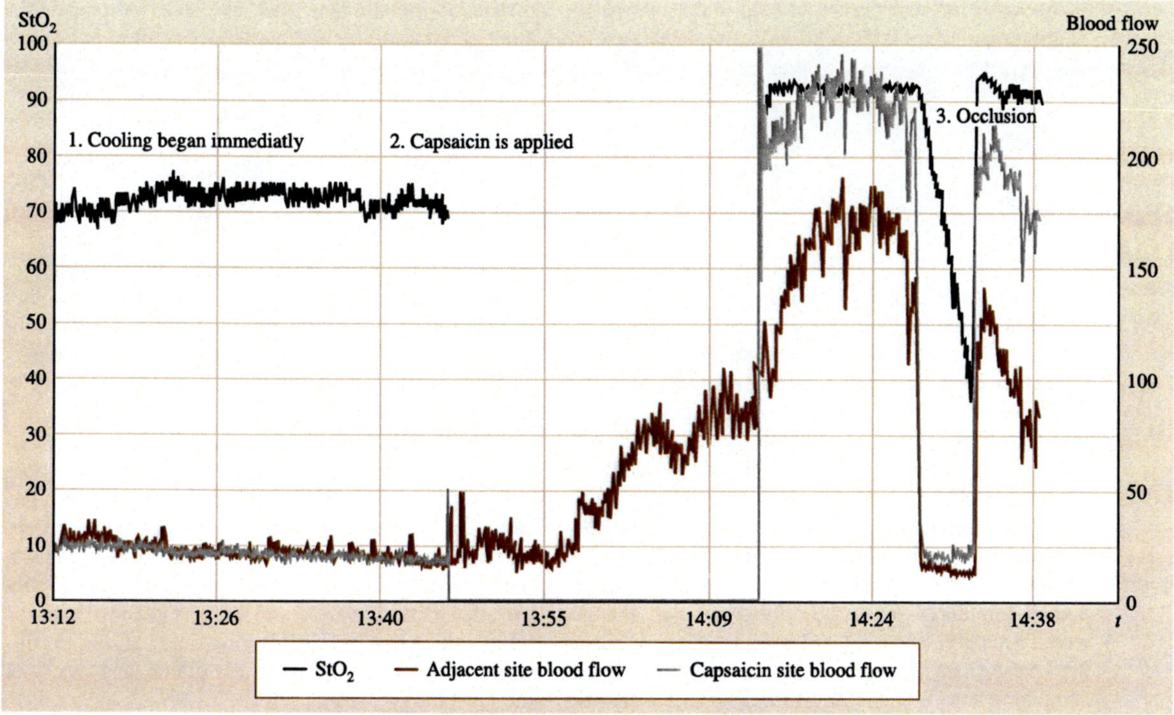

Fig. 22.9 Effects of cooling, capsaicin, and occlusion of forearm on StO_2. Initial experiment was 1. To cool forearm followed by 2. Application of capsaicin and then 3. Occlusion while monitoring StO_2 and skin blood flow

an average THI of 12.48 before capsaicin and an average of 13.4 post application. These results suggest that, since the 15 mm probe showed twice the increase in StO_2 levels compared with the 25 mm probe, since THI trends were similar between the two probes, the 15 mm probe may be more influenced by skin blood flow in the forearm. A paired T-analysis test was run, which showed that the increase in skin blood flow due to capsaicin had a statistically significant effect on the StO_2 readings. The data are summarized in Table 22.2.

A representative graph of StO_2 values and blood flow values from the study is shown in Fig. 22.9. The gap in data is due to the removal of the probes for application of capsaicin.

Analysis of the data [with 95% confidence interval (CI)] shows that, with a significant increase in skin blood flow after capsaicin in both the 15 and 25 mm probes, a significant increase in StO_2 measurements was recorded for the 15 and 25 mm probes, respectively. These results suggest an influence of skin blood flow in the forearm on NIRS StO_2 measurements.

22.6.2 Thenar Skin Blood Flow Study Versus StO_2

A similar study was performed on the thenar eminence to investigate whether local vasoconstriction/vasodilation would influence StO_2 and THI values during body-wide vasoconstriction induced by drinking a cold drink followed by local vasodilatation. Using the InSpectra StO_2 tissue oxygenation monitor (15 mm depth probe) and the PeriFlux system 5000 noninvasive near-infrared spectrometer, the study monitored volunteers who ingested the cold beverage as quickly as they could followed by application of a topical vasodilator (capsaicin) to a local area of the thenar. Baseline readings of StO_2 and THI were 84.1 ± 4.8 and 15.1 ± 1.4, respectively. After the consumption of the drink, the subjects' StO_2 readings dropped to 76.6 ± 5.01 and THI dropped to 14.8 ± 1.48, suggesting a vasoconstriction effect. The blood flow data following the drink also demonstrated vasoconstriction, as the baseline value of 143.4 PU (perfusion units) dropped to 52.4 PU. The *capsaicin effect* reversed the vasoconstriction and leveled off the values

Table 22.3 Effect of ingestion of cold drink followed by capsaicin. Skin blood flow, THI, and StO_2 were measured. * Significance at $p < 0.05$

	StO_2 (% O_2)	THI	Skin flow (PU)
Baseline	84.1±4.8	15.1±1.4	143.4±104.8
After drink	76.6±5.01*	14.4±1.23	52.4±31.6*
After cap	78.3±4.4	14.4±1.2	56.3±56.9
After cuff	80.7±5.1	14.8±1.3	81.5±99

of StO_2 and THI to 78.3±4.4 and 14.4±1.2, respectively. After capsaicin application, blood flow was occluded by automatic blood pressure cuff occlusion. The purpose was to simulate an undeniable decrease in blood flow upon pressure release and study the StO_2 and THI values. Once the cuff was released, an increase in StO_2 and THI corresponded to the increase in superficial skin blood flow [22.25]. Results of this study are shown in Table 22.2.

These results, compared with those of the previous experiment in which the forearm was locally cooled, suggest that the forearm is more responsive to capsaicin than is the thenar, since the only significant changes were found after the subject drank the drink, and even then these changes were only significant for StO_2 and skin blood flow. In Table 22.2 these significant changes are indicated with an asterisk (95% CI). These different results between the thenar and forearm could be due to the location of the probe on the body, the thenar, or possibly due to the different protocol of the experiment. However, the fact that these results agree with other similar studies that report that the forearm is more sensitive to cardiovascular changes than is the thenar may suggest that the thenar and forearm are physiologically different [22.1, 25].

Some additional questions raised about the previous studies concern the depth of the infrared probes and the correct depth for measurement of tissue oxygen saturation. Suggestions are that 15 and 25 mm probes do not penetrate deep enough to record a signal that is mostly oxygenated tissue hemoglobin. Criticisms suggest that skin thickness and tissue thickness must be taken into account [22.26]. Experimental design, different devices, NIRS device algorithms, and theory of the technology must be further reviewed and studied. Ince and *Soller* have expressed clearly opposing views on technology and by inference which technology may be superior [22.8].

22.7 Future of NIRS

Although NIRS signal derivation, accuracy, and qualitative ability are the center of controversy concerning this technology, NIRS still shows promise in monitoring oxygen balance and consumption in a sampling volume. The future of NIRS technology must address the current problems concerning transmission, reception, and quantification of the NIRS signal. Different technologies abound. Which is the best may depend on the application. In addition, the multidepth probe approach for NIRS is still in its infancy but might shed important light on the question of the influence of blood flow and oxygen concentration from interfering systems such as the skin or brain sinuses. Miniaturization of the NIRS system has been desired for quite some time. Fast signal spectroscopy offers an improvement over conventional NIRS instruments, which detect signals corresponding to relatively slow hemodynamic responses. Producing a much faster signal is beneficial for detection by a CW-type NIRS system.

In addition, understanding the complex physiological responses that occur in physiological and pathophysiological states must also be pursued. The debate as to whether the forearm or thenar is a clinically important site for monitoring patients motivates the beginning of a new set of studies that need to combine use of multidepth differential probes with different physiological experiments that mimic various clinical conditions such as hemorrhage as well as normal physiological states such as exercise. Investigators in this exciting field should keep in mind the complex physiological control of skin blood flow. *Joyner* states the problem succinctly [22.27]:

We have another example (skin blood flow) (like the coronary and skeletal muscle circulations) of a highly redundant physiological system that is critical to whole body homeostasis.

Later he adds,

The control of skin blood flow defies simple reductionist explanations ...

Muscle may also influence the NIRS signal. Although it is tempting to assume that the vasculature and tissue remain the same after some central hypoxic event, the researcher and clinician should be wary about making rapid decisions regarding an NIRS signal until they have considered the complications of the changing dynamics of skin and muscle blood flow and its local and neural control. If possible, core and peripheral temperatures should be measured to possibly explain the StO_2 values.

Presently our laboratory is investigating use of a multidepth differential system in the forearm and thenar, utilizing capsaicin to promote local vasodilation. Further studies to elucidate the mechanisms associated with vasodilation and vasoconstriction are required to assist in the development of a robust, accurate NIRS technology based on new insights into physiological control in both normal and pathophysiological states. NIRS has become an important tool in clinical management of various types of hypoxic disorders, but before it is widely used, additional clinical and basic science research is required.

References

22.1 R. Bezemer, J.M. Karemaker, E. Klijn, D. Martin, K. Mitchell, M. Heger, C. Ince: Simultaneous multidepth assessment of tissue oxygen saturation in thenar and forearm using near-infrared spectroscopy during a simple cardiovascular challenge, Crit. Care Med. **13**, S5 (2009)

22.2 S.M. Cohn, A.B. Nathens, F.A. Moore, P. Rhee, J.C. Puyana, E.E. Moore, G.J. Beilman: Tissue oxygen saturation predicts the development of organ dysfunction during traumatic shock resuscitation, J. Trauma **62**(1), 44–55 (2007)

22.3 F.E. Barton: Theory and principles of near infrared spectroscopy, Spectrosc. Eur. **14**(1), 12–18 (2002)

22.4 G.J. Beilman, J.J. Blondet: Near-infrared spectroscopy-derived tissue oxygen saturation in battlefield injuries: A case series report, World J. Emerg. Surg. **4**, 25 (2009)

22.5 D.E. Myers, L.D. Anderson, R.P. Seifert, J.P. Ortner, C.E. Cooper, G.J. Beilman, J.D. Mowlem: Noninvasive method for measuring local hemoglobin oxygen saturation tissue using wide gap second derivative near-infrared spectroscopy, J. Biomed. Opt. **10**(3), 034017 (2005)

22.6 S.M. Coyle: Brain–computer interface using a simplified functional near-infrared spectroscopy system, J. Neural Eng. **4**, 219–226 (2007)

22.7 B.A. Crookes, S.M. Cohn, S. Bloch, J. Amortegui, R. Manning, P. Li, M.S. Proctor, A. Hallal, L.H. Blackbourne, R. Benjamin, D. Soffer, F. Habib, C.I. Schulman, R. Duncan, K.G. Proctor: Can near-infrared spectroscopy identify the severity of shock trauma patients?, J. Trauma-Inj. Infect. Crit. Care **58**(4), 806–816 (2005)

22.8 B.R. Soller, K.L. Ryan, C.A. Rickards, W.H. Cooke, Y. Yang, O.O. Soyemi, B.A. Crookes, S.O. Heard, V.A. Convertino: Oxygen saturation determined from deep muscle, not thenar tissue, is an early indicator of central hypovoluminain humans, Crit. Care Med. **37**(1), 384–385 (2009)

22.9 R.S. Pozos, J.M. McNulty: Influence of skin blood flow on NIRS StO_2 readings on the forearm, Preliminary Data (2009)

22.10 A. Villringer, B. Chance: Noninvasive optical spectroscopy and imaging of human brain function, Trends Neurosci. **20**(10), 435–442 (1997)

22.11 M.J. Buono, P.W. Miller, C. Hom, R.S. Pozos, F.W. Kolkhorst: Skin blood flow affects in vivo near-infrared spectroscopy measurements in human skeletal muscle, Jpn. J. Physiol. **55**(4), 241–244 (2005)

22.12 N. Shah, A.E. Cerussi, D. Jakubowski, D. Hsiang, J. Butler, B.J. Tromberg: The role of diffuse optical spectroscopy in the clinical management of breast cancer, Dis. Markers **19**(2/3), 95–105 (2003–2004)

22.13 M.A. Underwood, J.M. Milstein, P.M. Sherman: Near-infrared spectroscopy as a screening tool for patient ductus arteriosus in extremely low birth weights infants, Neonatology **91**(2), 134–139 (2007)

22.14 N. Jones, M. Terblanche: Tissue saturation measurement-exciting prospects but standardization and reference data still needed, Crit. Care Med. **14**, 169–172 (2010)

22.15 H. Gomez, A. Torres, P. Polanco, H.K. Kim, S. Zenker, J.C. Puyana, M.R. Pinsky: Intensive Care Med. **34**, 1600–1607 (2008)

22.16 J. Creteur, T. Carollo, G. Soldati, G. Buchele, D. De Backer, J.-L. Vincent: The prognostic value of muscle StO_2 in septic patients, Intensive Care Med. **33**, 1549–1556 (2007)

22.17 K. Nagashima, T. Yoshida, H. Nose, A. Takamata, T. Morimoto: Negative pressure breathing and the control of skin blood flow during exercise in a hot environment, N. Y. Acad. Sci. **813**, 604–609 (1997)

22.18 K. Abozguia, T.T. Phan, G.N. Shivu, A.R. Maher, I. Ahmed, A. Wagenmakers, M.P. Frenneaux: Reduced in vivo skeletal muscle oxygen consumption in patients with chronic heart failure – A study using near infrared spectrophotometry (NIRS), Eur. J. Heart Fail. **10**(7), 652–657 (2008)

22.19 L. Skov, O. Pryds, G. Greisen: Estimating cerebral blood flow in newborn infants: Comparison of near infrared spectroscopy and ^{133}Xe clearance, Pediatr. Res. **30**(6), 570–573 (1991)

22.20 A. Villringer, J. Planck, C. Hock, L. Schleinkofer, U. Dirnagl: Near infrared spectroscopy (NIRS): A new tool to study hemodynamic changes during activation of brain function in human adults, Neurosci. Lett. **154**, 101–104 (1993)

22.21 A.J. Comerota, R.C. Throm, P. Kelly, M. Jaff: Tissue (muscle) oxygen saturation (StO_2): A new measure of symptomatic lower extremity arterial disease, J. Vasc. Surg. **38**(4), 724–729 (2003)

22.22 S.P. Mortensen, E.A. Dawson, C.C. Yoshiga, M.K. Dalsgaard, R. Damsgaard, N.H. Secher, J. González-Alonso: Limitations to systemic and locomotor limb muscle oxygen delivery and uptake during maximal exercise in humans, J. Physiol. **566**, 273285 (2005)

22.23 A. Kumar, J.F. Parrillo: Shock: Pathophysiology, classification and approach to mangement. In: *Critical Care Medicine: Principles of Diagnosis and Management in the Adult*, ed. by J.E. Parrillo, R.P. Dellinger (Mosby, St. Louis 2007) pp. 379–422

22.24 S.L. Davis, P.J. Fadel, J. Cui, G.D. Thomas, C.G. Crandall: Skin blood flow influences near-infrared spectroscopy-derived measurements of tissue oxygenation during heat stress, J. Appl. Physiol. **100**, 221–224 (2006)

22.25 R.S. Pozos, M.J. Born: Influence of skin blood flow on NIRS StO_2 readings on the thenar, Preliminary Data (2009)

22.26 V. Quaresima, M. Ferrari: Evaluation of the skin blood flow contribution to the noninvasive measurement of muscle oxygenation by near infrared spectroscopy, J. Physiol. Sci. **56**(3), 267–268 (2006)

22.27 M.J. Joyner: Cutaneous blood flow: Uncomfortable in our own skin?, Am. J. Physiol. Heart Circ. Physiol. **296**, H29–H30 (2008)

23. Magnetic Resonance Imaging

Wolfgang R. Nitz

Magnetic resonance imaging (MRI) represents an exciting technology not only from a technological perspective but also in view of it's clinical potential. A brief introduction will be given with respect to the historical development of *nuclear magnetic resonance* (NMR) later to be called MRI. A MRI system can be considered primarily composed of a magnet, a magnetic field gradient system, *radio frequency* (RF) coils for signal processing. The introduction of those hardware components is followed by a description of basic MRI principles and applications. Although MRI is a noninvasive technology working without ionizing radiation, it has nevertheless a few safety aspects to be considered. The chapter will close with a speculative outlook to the future of MRI and combined modalities.

23.1 **History of MRI** 439

23.2 **MRI – System Components** 441
23.2.1 The Magnet – The Magnetic Field Strength B_0 441
23.2.2 The System for the Magnetic Field Gradient 442
23.2.3 The Radiofrequency System 444
23.2.4 Measurement Control, Acquisition, and Image Reconstruction Systems 445

23.3 **MRI – Basic Principles and Applications** 446
23.3.1 Slice Selection and Spatial Encoding 446
23.3.2 The Spin-Echo Sequence 447
23.3.3 The Multi-Echo Spin-Echo Sequence 449
23.3.4 The Gradient Echo Sequence 450
23.3.5 The Sequence Family 451
23.3.6 MRI Spectroscopy 452

23.4 **MRI – Safety-Relevant Aspects** 453
23.4.1 Attraction and Torque Due to Strong Magnetic Fields 453
23.4.2 RF Interaction with the Patient's Body 454
23.4.3 Interaction with Active and Passive Implants 454
23.4.4 Interaction Based on Changes in Magnetic Field Gradients 454
23.4.5 Safety Issues in Conjunction with Loss of Superconductivity (Quenching) 455
23.4.6 Gadolinium-Containing Contrast Agents and NSF 455

23.5 **MRI – Pictures of the Future** 456
23.5.1 Magnetic Field Strength 456
23.5.2 RF Technology 456
23.5.3 Application Development 456
23.5.4 Hybrid Systems 457
23.5.5 Theranostics – Therapy Under Image Guidance 458

References 458

23.1 History of MRI

Wolfgang Pauli proposed the existence of a nuclear spin in 1924 [23.1]. His statement followed a year after the introduction of the spin of the electron by George Eugene Uhlenbeck and Samuel A. Goudsmit. In 1933, *Otto Stern* and *Walther Gerlach* were successful in demonstrating the existence of a nuclear spin in deviating a beam of hydrogen molecules in a magnetic field [23.2–5]. Supported by the experience of Gorter, *Rabi* of Columbia University in New York was successful in measuring the *nuclear magnetic moment* [23.6] in 1937. Earlier, Gorter had performed similar experiments, but with negative result. *Gorter* was the first to use the term *nuclear magnetic resonance* (NMR) in his publications [23.7].

It is general perception that the experiments launching the development of NMR were performed by *Bloch* and *Purcell* in 1946, proving the existence of the nuclear spin and the phenomenon of (nuclear) magnetic resonance [(N)MR] [23.8, 9], work for which Bloch and Purcell shared the Nobel Prize in 1952.

The spin is characterized by an angular momentum combined with a magnetic moment. The spin is a quantum-mechanical entity. As a consequence, the magnetic moment of the proton (with spin quantum number 1/2), exposed to an external magnetic field B_0, can only have two possible positions: parallel or antiparallel to the direction of the field. Those positions are characterized by a difference in energy, with the parallel position being energetically preferred.

If the energy of an electromagnetic wave corresponds exactly to the energy difference between these two possible states, parallel aligned spins can be pushed into the antiparallel position. With the return of the antiparallel spins back to the original parallel position, the energy difference is released as an electromagnetic wave, also called the NMR signal. As the wavelength λ of the stimulating electromagnetic field has to be identical to the wavelength representing the difference between the two possible states

$$\Delta E = \gamma \hbar B_0 = h\nu = h\frac{c}{\lambda}\,,$$

which is typical for a *resonance* condition, the phenomenon is also called *magnetic resonance*.

As more spins are aligned parallel to the external magnetic field, as compared with the antiparallel alignment, a longitudinal nuclear magnetization builds up. The longitudinal nuclear magnetization can be treated using the principles of classical physics. If the longitudinal nuclear magnetization is tilted away from the parallel direction of the external magnetic field, the angular momentum of the spins causes precession of the magnetization around the original alignment (the direction of the external magnetic field B_0). This precessional frequency is called the Larmor frequency and is identical to the frequency representing the energy difference between the two quantum-mechanical states. The projection of the precessing nuclear magnetization onto the transverse plane is called the transverse nuclear magnetization. If all the longitudinal nuclear magnetization is converted to transverse nuclear magnetization, the initiating radiofrequency (RF) pulse is called a 90° RF excitation pulse. Following excitation, the longitudinal nuclear magnetization within the tissue will recover within a tissue-specific timeframe. *Bloch* introduced the term T_1 within his phenomenological Bloch equations [23.10] to characterize this *recovery time*. The mechanism of recovery is called *relaxation*, and the correlated time constant is termed the T_1 relaxation time. The precessing transverse nuclear magnetization will induce a voltage in any antenna system or coil in close vicinity to the object. The signal will decay within a tissue-specific time, which Bloch termed T_2. The underlying mechanism is called T_2 relaxation, and the time constant the T_2 relaxation time. Experiments to measure tissue-specific relaxation times on living cells and animal tissues were performed as early as 1955 [23.11]. *Damadian* at the Downstate Medical Center in Brooklyn and *Hollis* at Johns Hopkins University in Baltimore evaluated the T_1 and T_2 relaxation times of normal and cancer tissue and observed that cancerous tissue demonstrated longer relaxation times as compared with normal tissue [23.12, 13]. It was *Damadian*'s belief that he had found the ultimate technology for diagnosing cancer [23.14]. Unfortunately, he gave no practical hint regarding how to perform cancer screening on a human body.

It was the American chemist *Lauterbur* at the State University of New York who, in 1971, introduced the concept of using a magnetic field gradient superimposed on a static magnetic field to spatially encode the signal induced by the precessing nuclear magnetization [23.15]. The previously mentioned resonance frequency is a function of the strength of the magnetic field. If a magnetic field gradient is superimposed on the static magnetic field, the magnetic field strength becomes a function of location, and with this also the resonance frequency becomes a function of location. Fourier transformation (FT) can be applied to analyze the signal induced by the transverse nuclear magnetization in order to identify the frequency components indicating the location, and thereby assign a brightness to the pixel on the display representing that location, proportional to the detected amplitude.

Lauterbur used two test-tubes filled with water to demonstrate spatially encoded nuclear magnetic resonance (NMR) by superimposing a magnetic field gradient on a static magnetic field, producing a signal providing a projection of the objects. Repeating the experiment with rotated magnetic field gradients and applying the reconstruction algorithm recently established for x-ray computer tomography, he was able to produce an image of the water within the test-tubes.

In April 1974, Lauterbur gave a presentation at a scientific meeting in Raleigh, North Carolina, which was also attended by Richard Ernst from the University of Zurich. Ernst recognized that the back-projection could be replaced by a combination of phase and frequency encoding of the MR signal [23.16]. This method is still the main reconstruction algorithm used in NMR. To avoid the fear associated with the word *nuclear*, NMR was more and more called magnetic resonance imaging (MRI) around 1981.

The British company EMI, which revolutionized medical imaging in 1973 with the introduction of the first x-ray computer tomography system, announced in 1976 the development of a scanner based on radio waves, and in 1978 presented the first image of a human head using MRI. This announcement gave a significant boost to the medical imaging industry, as they were not to repeat their mistake of underestimating the potential of a new imaging modality as they did with x-ray computer tomography. In the early 1980s, the first MR installations based on commercial products and aiming for routine clinical applications were seen.

23.2 MRI – System Components

23.2.1 The Magnet – The Magnetic Field Strength B_0

The phenomenon of magnetic resonance is only observed in the presence of a strong magnetic field. The strength of the static magnetic field B_0 is described in units of Tesla (T), at least within the medical community [for physicists, T means the magnetic flux density in units of $\mathrm{V\,s/m^2}$ and the magnetic field strength is measured in units of A/m or oerstedt (Oe)].

Early systems (1983) used resistive magnets with magnetic field strength in the range 0.1–0.2 T.

Within the same year it became obvious that superconductivity could be used to generate higher magnetic fields. Since then, the field strength used in clinical scanners has steadily increased. In 1985, the field strength of the majority of the systems was 0.5 T. Currently, in 2010, there is a strong tendency towards 3.0 T systems, whereas 1.5 T systems have the largest market share.

The primary reason for using a higher magnetic field strength is obvious: the ratio between signal and background noise (the signal-to-noise ratio, SNR) is, to a first approximation, proportional to the magnetic field strength [23.17].

The electromagnetic noise N, with its primary origin within the patient, is approximately linear in the magnetic field strength B_0

$$N \sim B_0 .$$

The induced MRI signal is a function of the rotating (transverse) nuclear magnetization, meaning proportional to the resonance frequency and the amplitude

$$S \sim \frac{\mathrm{d}M_{xy}}{\mathrm{d}t} \sim \nu M_{xy} ,$$
$$\nu \sim \gamma B_0 .$$

The transverse nuclear magnetization M_{xy} is generated by tilting the longitudinal nuclear magnetization M_z away from the direction of the main magnetic field. If a 90° RF excitation pulse is used, all of the available longitudinal nuclear magnetization M_z is converted to transverse nuclear magnetization M_{xy}.

The amplitude of the longitudinal nuclear magnetization is a function of the occupation probability of the different energy levels (of parallel and antiparallel alignment of the nuclear spins and their correlated magnetic moments). The amplitude of the longitudinal nuclear magnetization M_z scales with the magnetic field strength used

$$M_z \sim B_0 .$$

Considering amplitude and frequency, the induced signal is proportional to the square of the magnetic field strength used

$$S \sim B_0^2 ,$$

and the signal-to-noise ratio (SNR) is approximately linear in the magnetic field strength

$$\mathrm{SNR} = \frac{S}{N} \sim B_0 f(T_1, T_2)$$

and is of course a function of the tissue-specific relaxation times T_1 and T_2.

The increase of the SNR is only one aspect of *high-field* systems. There are a number of phenomena representing advantages and disadvantages correlated with utilization of higher magnetic field strength. T_1 relaxation times are prolonged with higher magnetic fields, and differences between tissues are often reduced, leading to a reduction of contrast in T_1-weighted images. The sensitivity to gradients in magnetic susceptibility is increased with field strength, as magnet field inhomogeneities increase proportionally to the product

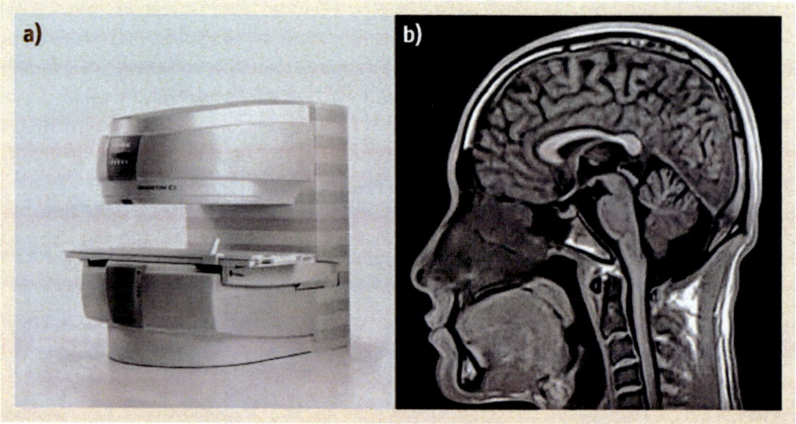

Fig. 23.1a,b A commercially available MR system using a permanent magnet with a magnetic field strength of 0.35 T (MAGNETOM C!). The T_1-weighted sagittal image of a human brain demonstrates a slice thickness of 4 mm, acquired within 5 min (courtesy Airforce General Hospital, Beijing, PR China)

of magnetic field strength and magnetic susceptibility. With the improvement in SNR, motion artifacts are no longer masked by noise and are often pronounced due to the strong signal from fat-containing moving tissue. Safety-relevant aspects need special consideration [23.18] as some potentially hazardous interactions scale with the field strength. Last but not least, the costs for an MR system scale with field strength.

Around 1991 the major vendors decided to introduce low-cost *low-field* MR systems (field strength up to 0.4 T) parallel to the *high-field* systems, as 1.0 and 1.5 T systems were called at that time. This diversity has since been maintained, and *low-field* systems are still commercially available (Fig. 23.1). A magnetic field strength of up to 0.35 T can be achieved by assembling a number of permanent magnets. Only one vendor is known to be still working with resistive magnets (stand-up MRI, FONAR Corp.). All other vendors (Siemens, General Electric, Philips, Toshiba etc.) are working with either permanent magnets or superconductive magnets. Production and maintenance of permanent systems are relatively cost effective. A disadvantage is the limitation to low field strength and the corresponding limit in SNR. Another disadvantage is the change in magnetic field strength as a function of temperature of the magnet material. Approximately 14 t of permanent magnet material is needed to achieve a field strength of 0.35 T. Resistive magnets have relatively low production costs, but are expensive in operation, have in general an extreme weight (40 t), and share the disadvantage of a temperature-dependent field strength with the permanent magnets. The production costs of superconductive magnets are relatively high, and, if helium is to be refilled regularly, they are expensive in operation, but they have the advantages of a magnetic field strength that does not change with temperature and that the generation of any reasonable magnetic field strength seems to be possible without limit and that the weight of the magnets is moderate (about 4 t for a 1.5 T system). The advantages and disadvantages of working with higher magnetic field strength will be further discussed in conjunction with safety-relevant aspects. Besides the magnetic field strength, the magnet design plays an important role that should not be underestimated. The size of the patient bore has been limited to 60 cm in the past, mainly for financial reasons, although since 2004 systems with larger patient bore openings have been available (e.g., 70 cm; Fig. 23.2). The larger opening of the patient bore is not only a comfort factor for the patient; it is also helpful in reducing the number of medical examinations refused due to claustrophobia and provides a possibility to study patients whose circumference would need more than a 60 cm bore.

23.2.2 The System for the Magnetic Field Gradient

To excite a specific slice or volume at a specific location, the MR resonance frequency, also called the Larmor frequency, has to be a function of location. For this reason, a magnetic field gradient is established in the direction of slice selection and for the duration of excitation. The slice is selected using a RF pulse whose frequency range covers the Larmor frequencies within the region to be excited. To cover all three dimensions, the magnetic field gradient coil (gradient coil) is a composition of three gradient coils arranged orthogonally to one other (x, y, z; Fig. 23.3). Sending appropriate electric currents through these coils will generate magnetic field

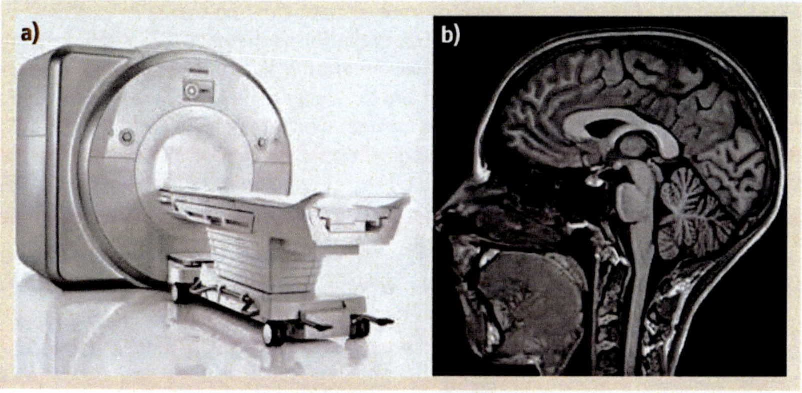

Fig. 23.2a,b Commercially available MR system using a superconductive magnet with a magnetic field strength of 3.0 T and patient bore of 70 cm (MAGNETOM SKYRA). The T_1-weighted sagittal image of a human brain demonstrates a slice thickness of 1.2 mm, acquired within 9 min

gradients that are linearly superimposed, providing the possibility of producing any arbitrary excitation plane without having to move any mechanical parts. A magnetic field gradient is also activated during the time of data acquisition. This encodes spatial information into the signal via frequency encoding, as the Larmor frequency of the emitted MR signal will be a function of location, depending on the established linear change in magnetic field strength in the direction of frequency encoding.

The electric currents I within the gradient coils will result in a Lorentz forces $\boldsymbol{F}$ in the presence of a strong magnetic field B_0

$$\boldsymbol{F} \approx \boldsymbol{I} \times \boldsymbol{B}_0 .$$

These forces cause vibration of the gradient coil and are the primary source of noise during an MR examination.

The amplitude of the magnetic field gradient is important with respect to achieving a good spatial resolution. A strong magnetic field gradient is also essential for diffusion weighted imaging (DWI). Wherever the amplitude time integral is of relevance, the stronger the magnetic field gradient, the shorter the encoding action. The speed of a gradient system (coil plus amplifier) is characterized by the time needed to achieve a specified gradient amplitude (ramp time). The speed primarily determines the time needed between excitation and data acquisition and directly influences the resulting image quality. The ratio between the maximum gradient amplitude and the ramp time is introduced as the *slew rate*.

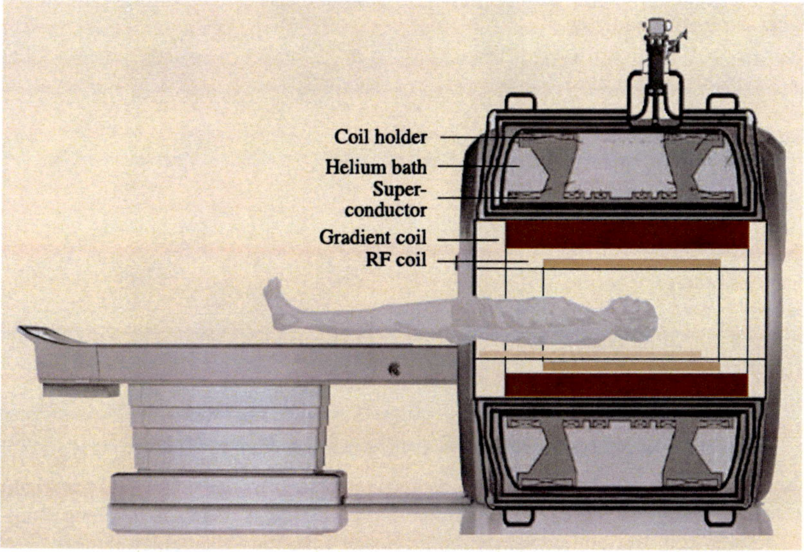

Fig. 23.3 Composition of a MR system

Air-cooled gradient coils were used in 1986 with gradient amplitudes of 3 mT/m and slew rate of 3 T/(m s). In 2003, generally available gradient amplitudes were 45 mT/m with slew rate of 200 T/(m s) using water cooling. Ramp times to establish a magnetic field gradient are between 100 and 800 μs, during which several hundred Amperes have to be maintained. To drive these currents, ≈ 2000 V are needed. The development of faster and stronger gradient systems is slightly limited by human physiology. Fast and rapid switching of magnetic field gradients induces electric potentials within the body that, with the current technology, can reach amplitudes usually used to control muscle contractions. The phenomenon is called *peripheral nerve stimulation* (PNS). As this can potentially be painful, it has to be avoided. All vendors providing strong and fast gradient systems have a *stimulation monitor* that changes imaging sequences prior to execution to avoid peripheral nerve stimulation.

Beginning with the development of MR imaging, methods have been developed and established with the aim of shortening measurement times by, e.g., undersampling of the data, with correction of the artifacts resulting from the consequent signal ambiguity using fancy algorithms. Similar approaches are currently under investigation that may allow use of faster gradient systems [23.19].

23.2.3 The Radiofrequency System

The radiofrequency system consists of a transmitter, transmit antenna, receive antenna, and receiver. Depending on the magnetic field strength used, the fundamental frequency will be 8 MHz (0.2 T) up to 128 MHz (3.0 T). The RF power amplifier has to have a peak power of several kilowatts. Since the early days of MR, a transmit antenna, also called the body coil, is located immediately behind the cover within the patient bore. The body coil is usually also able to receive the MR signal. On the other hand, placement of a receive antenna as close as possible to the patient's body (surface coil) has some significant advantages. As the patient's body emits electromagnetic noise even in the absence of excitation, and as the signal amplitude scales inversely with distance from its origin, use of a surface coil improves the SNR. Current systems utilize an array of surface coils with independent preamplifiers and/or their own receive channels for SNR optimization and for other reasons to be mentioned later. For some regions it is of advantage to use the surface coil not only as a receive coil but also as a transmit coil. In case of knee examination, an extremity coil is usually used for excitation and signal reception, exciting only the slice within the knee to be studied. The advantage is that, in transverse excitations, the adjacent knee is not excited, which otherwise would potentially lead to imaging artifacts. In case of imaging of the spine, large coil arrays, integrated into the patient table,

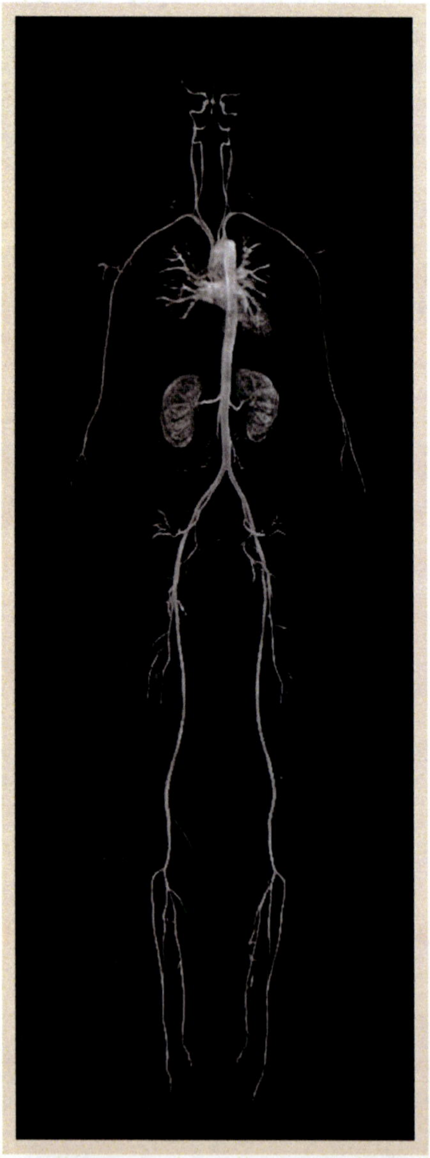

Fig. 23.4 Contrast-enhanced MR angiograph of the whole body measured within 1 min 17 s (courtesy Hong Kong Sanatorium & Hospital, Happy Valley, Hong Kong)

have been shown to be beneficial. Contrast-enhanced MR angiography studies demand an even larger surface coil coverage (Fig. 23.4) [23.20, 21]. Figure 23.5 shows a coil arrangement suitable to cover to whole vascular system from head to toe.

With the introduction of spatially distributed coil arrays, some methods evolved to utilize the spatial information provided by the coil distribution, reducing the measurement time by undersampling. As all coils receive the signal *in parallel*, the term *parallel acquisition techniques* (PAT) has been established, where the PAT factor indicates the degree of undersampling and the corresponding reduction in measurement time. The undersampling causes signal ambiguity leading to so-called *overfolding* artifacts. Prior to showing the final images to the user, a background task is to analyze the raw data or the final image of each coil to identify and remove such artifacts. The algorithm using image information for this purpose has been named *sensitivity encoding* (SENSE) [23.22]. Another algorithm applied to the signal data with the same aim of removing the above-mentioned artifacts is called *generalized autocalibrating partially parallel acquisitions* (GRAPPA) [23.23]. All these PAT techniques have one disadvantage in common: Any attempt to reduce the measurement time with the spatial resolution being held constant will result in a loss in SNR according to

$$\mathrm{SNR} \sim \frac{1}{\sqrt{\mathrm{PAT}}} \,.$$

The loss in SNR experienced when using parallel acquisition techniques can be compensated with the better SNR achievable with higher magnetic field strength. For this reason, parallel imaging techniques have shown their full potential in conjunction with high-field systems.

Besides the utilization of spatially distributed surface coils to improve the image quality or for the purpose of reducing measurement time with undersampling, it has been and still is discussed whether it would be of advantage to use spatially distributed coil arrays for transmission, so-called transmit arrays (TX arrays).

Following the term *parallel acquisition techniques* (PAT), the name *parallel transmit* (pTX) has been introduced. It seems that, with the increasing importance of high-field systems, pTX might provide some features to address some of the challenges faced when approaching higher magnetic fields. One of those challenges is the homogeneity of the excitation field, also called the B_1 field. The wavelength of the electromagnetic fields (EM) correlated with higher magnetic fields

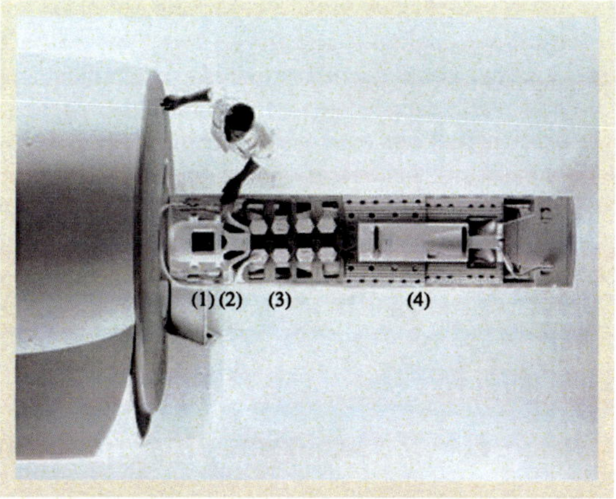

Fig. 23.5 Possible coil arrangement for a whole-body MR angiography: head coil (1), neck coil (2), body array (3), and peripheral angiography coil (4) (spine coil integrated within patient table)

B_0 is approaching the spatial dimensions of the patient's body. This leads to potential interaction of the EM field with and within the patient's body, causing B_1 inhomogeneities. The latter results in different RF excitation and/or refocusing amplitudes, leading to signal intensity changes within the image that are unrelated to anatomy. This will lead to spatial inhomogeneity of an otherwise homogeneous signal distribution within the image, unrelated to the underlying anatomy. This is of course an unwanted phenomenon. Besides other possible alternatives to compensate this artificial appearance, spatially distributed transmit coils can be used to homogenize the B_1 field.

23.2.4 Measurement Control, Acquisition, and Image Reconstruction Systems

At the beginning of the development of MR imaging, documentation of patient data, image storage, and measurement protocols were handled by a PDP-11 computer, and measurement control as well as image reconstruction were performed using vendor-specific proprietary hardware and software. In the early 1990s, most MR vendors came to the conclusion that they could benefit from the vastly evolving market of personal computers (PCs).

Most (N)MR systems today use three computer systems:

- A *host*, handling the patient data including the generated images, providing the user interface to run the measurements, and displaying the generated images.

- A *measurement control*, providing the gradient amplifier with data necessary to establish the magnetic gradient fields on time, supplying the RF power amplifier with the necessary information, and handling the synchronization with the image reconstruction system.
- An *image reconstruction system* with the primary task of performing fast Fourier transformation. The spatial information is contained within the frequency and the phase information of the received MR signal. That information is retrieved using a Fourier transformation. Secondary tasks for the image reconstruction system are inline postprocessing algorithms that run prior to or during image reconstruction, or immediately afterwards.

In theory it is also conceivable to combine all of these tasks within one computer [23.24].

23.3 MRI – Basic Principles and Applications

MRI is a well-established modality for routine clinical imaging. In conjunction with healthcare reform, there are standards and guidelines describing as and when MRI is appropriate [23.25, 26]. To ensure the quality of diagnostic studies, these standards and guidelines are supported by matching billing codes for reimbursement [23.27]. Medicare and Medicaid service centers are constantly implementing cost-savings approaches in healthcare to ensure an affordable and adequate healthcare system [23.28]. Standards and guidelines include recommendations with respect to spatial resolution, weightings, slice orientations, and coverage. The nomenclature *weighting* is used to indicate the tissue-specific MR-related parameter that dominates the image contrast. The sources for MR signals are protons (primarily the nuclei of hydrogen atoms located within relatively freely moving water molecules). If the imaging protocol is selected such that the number of protons within a voxel dominates the signal amplitude and thereby the image contrast, the weighting is called proton density (PDw). If the factor dominating the image contrast is the speed of recovery of the longitudinal nuclear magnetization following excitation, the weighting is called T_1w. If the image contrast is dominated by the effect of the tissue-specific fading of the MR signal, the weighting is called T_2w. Based on experience acquired over decades, vendors provide programs consisting of a list of protocols for different regions (e.g., head, spin, shoulder, and knee), different weightings (e.g., PDw, T_1w, and T_2w), and different orientations (axial, transverse, coronal, tilted or any arbitrary orientation). An institution running an MR system can modify these programs based on current state-of-the-art imaging techniques, past experience, and current guidelines and recommendations, and can save these modified programs to become part of their routine clinical imaging. The basis for a measurement protocol is the essential series of actions for excitation, refocusing, spatial encoding, and data acquisition, called a *sequence*.

23.3.1 Slice Selection and Spatial Encoding

The sequence starts with ramping up of the magnetic field gradient in the direction of slice selection, as shown in Fig. 23.6. As soon as the magnetic field gradient is stable, a RF pulse is applied with a frequency range matching the Larmor frequencies within the region to be excited. A user-defined field of view and matrix define the size of a single spatial volume, called a *voxel*. The sum of all signals coming out of a single voxel defines the brightness of the corresponding pixel on the screen. The only phenomenon utilized for spatial encoding is the fact that the precessional frequency of the rotating transverse nuclear magnetization is a function of the magnetic field strength at that location. If a brief period of different rotational frequencies is utilized, the phase position of the adjacent transverse nuclear magnetizations will be altered. This is called phase encoding. Such phase encoding is usually applied directly after excitation in a direction perpendicular to the direction of slice selection. At the same point in time the expected dephasing during frequency encoding is prospectively dephased in the opposite direction to get a rephasing point during data acquisition. When switching on the frequency-encoding gradient, an *echo* is formed, and the data are acquired in the presence of a magnetic field gradient that is perpendicular to the direction of slice selection and perpendicular to the direction of phase encoding. The acquired signal is digitized according to matrix size and frequency bandwidth and is stored in a raw data matrix. As each data point

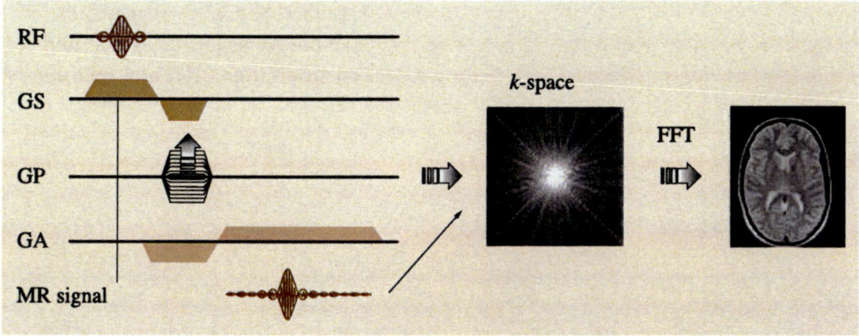

Fig. 23.6 Sequence diagram. A RF excitation pulse is applied as soon as the slice selection gradient (GS) reaches the nominal value. Following the excitation pulse, the generated but (due to the differences in resonance frequencies in the direction of slice selection during excitation) dephased transverse nuclear magnetization is rephased. During the same time period the phase-encoding gradient can be applied (GP) and the rephasing in the direction of frequency encoding can be prepared (GA) – as the transverse nuclear magnetization will dephase due to the differences in resonance frequencies as a consequence of the frequency-encoding magnetic field gradient during the readout period. Finally, the data are acquired in the presence of a frequency-encoding magnetic field gradient (GA). The data are saved into a raw data matrix, also called *k*-space. A two-dimensional fast Fourier transformation (FFT) applied to those data will lead to the final image

within the raw data matrix has an index called *k*, the raw data matrix is also called *k-space*. Depending on the matrix size in the direction of phase encoding, multiple repetitions with different phase-encoding gradients are required to obtain enough information to reconstruct an image.

23.3.2 The Spin-Echo Sequence

The spin echo was found accidentally during an experiment to measure a tissue-specific T_1 relaxation time [23.29]. Combined with the spatial encoding scheme of an imaging sequence it is also termed the spin-echo sequence. Initially introduced for imaging in 1983, it is still used today for T_1w imaging. Besides the already discussed tissue-specific parameters PD, T_1, and T_2, there is another imaging-relevant parameter, the magnetic susceptibility χ, which indicates whether the external magnetic field is increased (paramagnetic or ferromagnetic behavior) or decreased (diamagnetic behavior).

With this, the protons within a given tissue experience an effective magnetic field according to

$$B_{0(\text{eff})} = (1+\chi)B_0 \,.$$

The Larmor frequency scales with this effective magnetic field. Biological tissue is diamagnetic in general, although differences on a scale of 10^{-6} [23.30] are considered small. For slice selection, these small differences are usually negligible. For spatial encoding, the variations of the magnetic susceptibility within the spatial dimensions of a voxel will cause different resonance frequencies, resulting in faster fading of the MR signal. This faster fading is characterized by the time constant T_2^*, given by

$$\frac{1}{T_2^*} = \frac{1}{T_2} + \gamma \Delta B \,,$$

with γ being the gyromagnetic ratio of 2.675×10^8 $\text{T}^{-1}\ \text{s}^{-1}$ for the proton, and ΔB representing the magnetic field gradient caused by the difference in magnetic susceptibility. The resulting spatial distribution of different resonance frequencies is spatially fixed and constant in time. This allows the dephasing to be reversed using a so-called RF refocusing pulse. A spin-echo sequence has a 90° RF excitation pulse and also a 180° RF refocusing pulse. With this 180° RF refocusing pulse, the faster component of the transverse nuclear magnetization will be placed behind the slower component, and in the process of catching up, a spin echo is formed, which is acquired in the presence of a frequency-encoding magnetic field gradient.

Several repetitions with different phase-encoding steps are necessary to obtain enough information to reconstruct an image. The time between repetitions is called the repetition time (TR). The time between excitation and data acquisition is called the echo time (TE) (Fig. 23.7). A long repetition time (3–9 s) will suppress any influence of different T_1 relaxation times. If additionally a short echo time is selected (≤ 20 ms), dif-

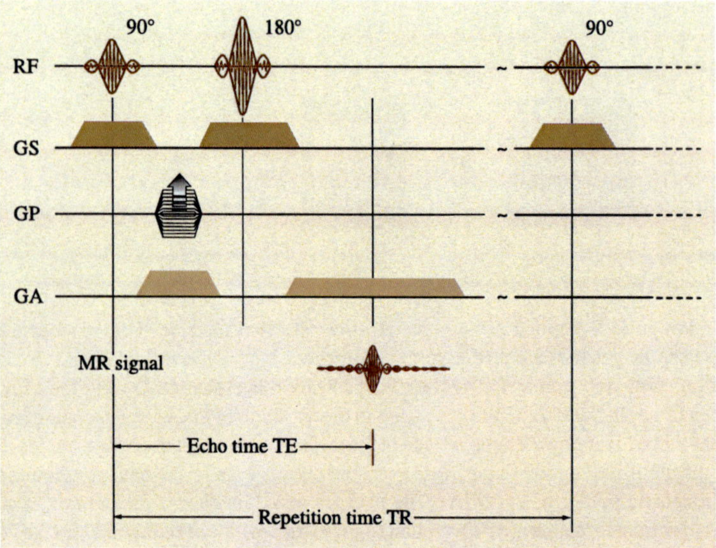

Fig. 23.7 Sequence diagram for a spin-echo sequence, illustrating the repetition time (TR) and echo time (TE)

ferences in T_2 relaxation times will be suppressed as well. Such a protocol is called proton density weighting (PDw), as primarily the number of protons per voxel will be responsible for the amplitude of the MR signal. Selecting the same repetition time but using a longer echo time will lead to a T_2-weighted (T_2w) protocol. With short echo times and relatively short repetition times (300–800 ms), T_1-weighted (T_1w) images will be acquired. The above parameters are suggested in view of the relaxation times within biological tissues (Table 23.1). The matrix size in the direction of phase encoding usually dictates the number of necessary repetitions and with this the overall measurement time. For a T_1-weighted protocol, a measurement time shorter than 5 min should be possible. As mentioned above, the excitation is a result of introducing energy into the spin system. The T_1 relaxation time is a function of how fast the spin system can release this energy to the environ-

Table 23.1 MR relaxation times for different tissues

Region		T_1 relaxation time (ms)			T_2 relaxation time (ms)
		1.5 T	1.0 T	0.2 T	
Brain	Gray matter (GM)	921	813	495	101
	White matter (WM)	787	683	390	92
	Cerebrospinal fluid (CSF)	3000	2500	1200	1500
	Edema	1090	975	627	113
	Meningioma	979	871	549	103
	Glioma	957	931	832	111
	Astrocytoma	1109	1055	864	141
Liver	Normal tissue	493	423	229	43
	Tumor tissue	905	857	692	84
Spleen	Normal tissue	782	683	400	62
Pancreas	Normal tissue	513	455	283	
	Tumor tissue	1448	1235	658	
Kidney	Normal tissue	652	589	395	58
	Tumor tissue	907	864	713	83
Muscle	Normal tissue	868	732	372	47
	Tumor tissue	1083	946	554	87

ment, the surrounding tissue. Aqueous solutions such as cerebrospinal fluid show long T_1 relaxation time and appear hypointense (dark) on T_1-weighted images. Mass lesions such as tumors often not only show displacement of normal anatomy but usually involve edema, documented by hypointense (dark) appearance on T_1-weighted images. With a few exceptions, tissues with long T_1 relaxation times usually also demonstrate long T_2 relaxation times, as a consequence of intramolecular dipole–dipole interactions within the rapidly tumbling water molecules. For this reason, pathologic tissue usually shows up as hyperintense (bright) on T_2-weighted images. The signal amplitude that later defines the brightness of the according pixel is given by

$$S \sim \frac{\mathrm{d}M_{x,y}}{\mathrm{d}t} \sim M_0 \left(1 - \mathrm{e}^{-T_\mathrm{R}/T_1}\right) \mathrm{e}^{-T_\mathrm{E}/T_2} ,$$

where M_0 represents the maximum possible longitudinal nuclear magnetization M_z that can be achieved within the given voxel. M_z is converted to transverse nuclear magnetization M_{xy} using a 90° RF excitation pulse. M_{xy} rotates with the Larmor frequency, inducing a signal in an adjacent receiver coil.

Diagnostic confidence has been increased with the introduction of T_1-shortening paramagnetic contrast agents [23.31–33]. Inflammatory lesions taking up contrast agents usually show up as very hyperintense (bright) on T_1-weighted images.

23.3.3 The Multi-Echo Spin-Echo Sequence

Images with different echo times are needed to calculate the average T_2 relaxation time within a voxel. Those images can be acquired with a single sequence containing multiple RF refocusing pulses. On viewing such images, it becomes obvious that image contrast is only slightly changing, especially for images acquired with late echoes. This observation led to the idea of acquiring another k-space line rather than producing another image. This concept will potentially significantly reduce the measurement time (Fig. 23.8).

Such a sequence has been introduced with the acronym *rapid acquisition with relaxation enhancement* (RARE) [23.34]. The image quality at the time of first introduction was rather moderate, and the method lacked attention. Seven years later, *Melki* and *Mulkern* were searching for a fast T_2-localizer and *rediscovered* the multi-echo spin-echo approach [23.35]. Likely based on progress in technology, implementation of the *localizer* revealed images with impressive quality. The surviving acronyms for the sequences that evolved out of this approach are *fast spin echo* (FSE), and *turbo spin echo* (TSE).

The advantage of this method is a significant reduction in measurement time, where the reduction is proportional to the number of echoes utilized, the so-called *echo train length* (ETL). A disadvantage should be expected due to the fact that each Fourier line has a different *weighting*. Theoretically this could lead to imaging artifacts and image misinterpretation [23.36]. In practice, the potential shortening of measurement time is not fully exploited but rather used to improve the contrast by selecting longer repetition times and increasing the spatial resolution. Both measures would prolong the measurement time, where the prolongation is now compensated by using the multi-echo approach. This leads to a significant improvement for PD- and T_2-weighted imaging and overcompensates the above-

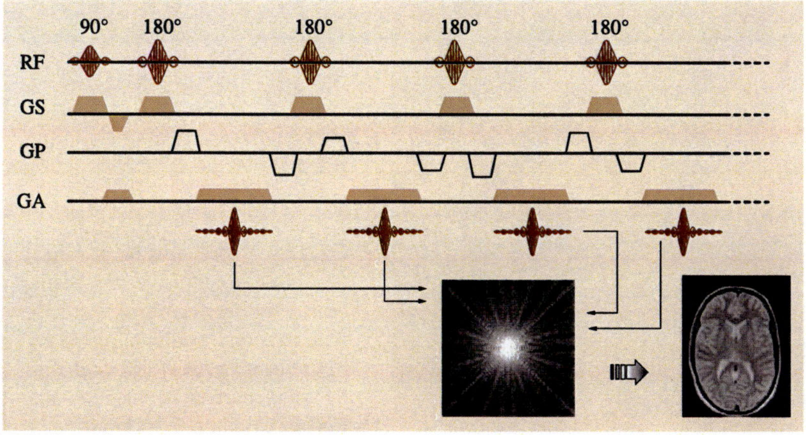

Fig. 23.8 Sequence diagram for a multi-echo spin-echo sequence [turbo spin echo (TSE), fast spin echo (FSE)]

mentioned potential disadvantages of the multi-echo approach. Today PD- and T_2-weighted imaging are solely done using FSE and TSE sequences.

23.3.4 The Gradient Echo Sequence

Omitting the 180° RF refocusing pulse leads to the generation of an *echo* by using bipolar gradient switching in conjunction with frequency encoding (Fig. 23.6). This suggested the acronym gradient-echo sequence (GRE). Different vendors use different acronyms for this generic approach, such as *fast low-angle shot* (FLASH) [23.37], *fast field echo* (FFE)-T1 or *spoiled gradient recalled acquired steady state* (SPGR). Without the use of a 180° RF refocusing pulse, the signal decay no longer follows solely the intramolecular spin–spin interaction characterized by the T_2 relaxation time, but is now also a function of local magnetic field inhomogeneities. The latter are partly introduced by the patient themselves due to differences in magnetic susceptibility of neighboring tissue. The *faster* signal decay as a consequence of all these dephasing mechanisms is characterized by the relaxation time T_2^*. It is customary in gradient-echo imaging to optimize the signal amplitude by using a low-angle excitation instead of a 90° excitation pulse. The signal response in relation to the excitation angle α in any given tissue follows

$$S \sim M_{x,y} = M_0 \frac{\left(1 - \mathrm{e}^{-T_\mathrm{R}/T_1}\right)}{1 - \cos\alpha\, \mathrm{e}^{-T_\mathrm{R}/T_1}} \mathrm{e}^{-T_\mathrm{E}/T_2^*} \sin\alpha \,.$$

GREs are used whenever short repetition times are desired. GREs are used for fast imaging and/or wherever T_2^* sensitivity is desired. As an example, GREs are applied for imaging of the beating heart (Fig. 23.9). Phase encoding is not limited to spatial encoding of the second dimension within an imaging plane, but can also be used to further *partition* a slice, in this case called a volume. This approach is called 3-D imaging. The only disadvantage is that, for every phase-encoding step in the direction of slice selection, all phase-encoding steps within the imaging plane have to be repeated. To stay within clinically practical measurement times (≤ 12 min), only gradient-echo approaches with short repetition times were suitable in the past. Gradient-echo sequences are the framework for all MR angiography methods (Fig. 23.10).

The T_2^* sensitivity is very helpful in identifying hemorrhagic lesions and is currently fully exploited

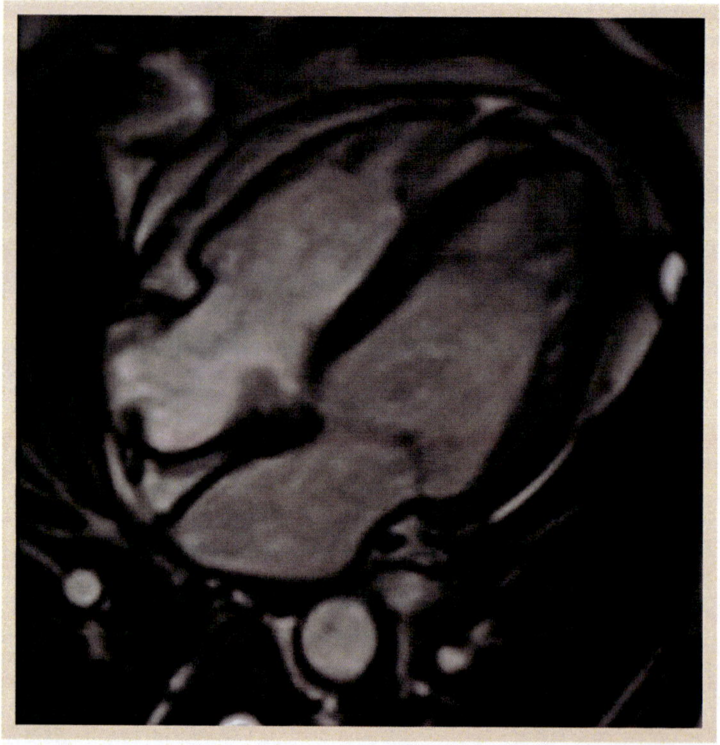

Fig. 23.9 Representation of a singe timeframe of an image of a beating heart. Measurement time was 9.28 s with temporal resolution of 43 ms (four-chamber view, myocardial infarction; courtesy of the PLA 306 Hospital, Beijing, PR China)

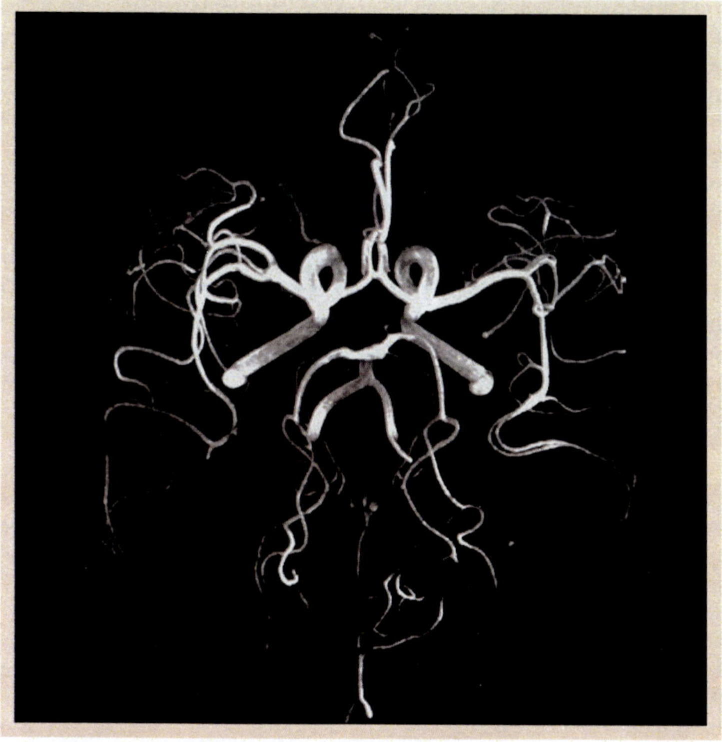

Fig. 23.10 Axial view of the intracranial vasculature generated without use of a contrast agent. The contrast is solely based on unsaturated blood flowing into the imaging volume while acquiring the data [time-of-flight MR angiography (ToF-MRA)]. Measurement time was 5.5 min

in *susceptibility-weighted imaging* (SWI) [23.38]. The change in oxygen concentration within the blood vasculature of active brain regions is correlated with alteration of the magnetic susceptibility and is used in *blood oxygenation-dependent imaging* (BOLD) to visualize active brain regions. As this allows the function of the brain to be documented, it is also called *functional MRI* (fMRI) [23.39, 40]. In conjunction with intravenous injection of paramagnetic contrast agents, the correlated changes in magnetic susceptibility as a marker for the passage of the contrast agent can be used to trace and document perfusion deficits when using T_2^*-sensitive protocols [23.41, 42]. One further feature of gradient-echo sequences is worth mentioning. This feature is important for diagnosis of benign or malignant abdominal mass lesions. Protons of hydrogen atoms within lipids have a resonance frequency that is about 3.5 ppm lower than the resonance frequency of protons contained in hydrogen atoms within freely moving water molecules. This is due to the differences in the electronic environment (oxygen versus carbon). This will cause slower rotation of the transverse nuclear magnetization within lipids as compared with water, a phenomenon called *chemical shift*. In spin-echo imaging, the 180° RF refocusing pulse places the faster component of the rotating transverse nuclear magnetization behind the slower component, and at the time the echo reaches its maximum, both transverse nuclear magnetizations are in phase again and only the image shift in the direction of spatial encoding will remain. In GRE imaging there is no RF refocusing pulse, and depending on the echo time, there will be a situation where the transverse nuclear magnetization within lipids will be in *opposed-phase* with the transverse nuclear magnetization within water, whereas at a later echo they will be *in-phase* again. In the *antiphase* case the net signal will be zero for a voxel, with fat and water showing approximately identical transverse nuclear magnetizations. The corresponding pixel will show up dark due to the signal void. As fat-containing mass lesions are usually benign, this can be used to differentiate benign adenomas from malignant metastases [23.43].

23.3.5 The Sequence Family

The subjective impression is that SE, GRE, and the multi-echo concepts of FSE or TSE are the most important imaging sequences for current routine clinical imaging. In any case, all existing imaging sequences can be characterized as belonging to either the SE or the GRE group [23.44], with different types of hybrids. Worth mentioning is the possibility of manipulating the longitudinal or transverse nuclear magnetization prior to or within an imaging sequence. A classic example is the inversion of the longitudinal nuclear magnetization prior to starting the imaging sequence. Such inversion allows the signal of a tissue with a specific T_1 relaxation time to be nulled. Tissue is only able to emit a signal if longitudinal nuclear magnetization is available at the time of excitation. The RF excitation pulse will convert the longitudinal nuclear magnetization to transverse nuclear magnetization which rotates with the Larmor frequency, inducing an MR signal in a coil adjacent to the object to be imaged. Following inversion, the inverted longitudinal nuclear magnetization will approach the parallel alignment with the main magnetic field within a time given by the T_1 relaxation time of the tissue. There is a point in time at which the longitudinal magnetization will be zero, occurring at the point of transition between antiparallel to parallel alignment. If the excitation pulse is placed at that point in time, that tissue with a specific T_1 relaxation time will not be excited. Lipids have a relatively short T_1 relaxation time. On selecting a short time period between inversion and excitation pulse, e.g., 150 ms, the signal from lipids will be suppressed. The time between inversion and excitation pulse is called the inversion time T_I. As a 150 ms inversion time was considered short at the time, the acronym short-tau inversion recovery (STIR) was established [23.45]. At the other end of the scale of relaxation times is the relative long time of freely moveable water molecules. If a long inversion time is selected (e.g., 2.5 s), the signal from fluid is suppressed, leading to the acronym *fluid-attenuated inversion recovery* (FLAIR) [23.46]. Such a protocol is helpful to identify benign cystic lesions and to identify periventricular lesions that might otherwise remain unnoticed in the presence of adjacent bright signal from cerebrospinal fluid. The recovery of longitudinal nuclear magnetization following inversion follows

$$M_z = M_0 \left(1 - 2\mathrm{e}^{-T_\mathrm{I}/T_1} + 2\mathrm{e}^{-(T_\mathrm{R}-T_\mathrm{E}/2)/T_1} - \mathrm{e}^{-T_\mathrm{R}/T_1}\right) .$$

A further important application is the preparation of the transverse nuclear magnetization in *diffusion-weighted imaging* (DWI). In the presence of diffusion, a bipolar magnetic field gradient amplitude will lead to a dephasing, resulting in a signal void [23.47]. The effect on the signal can be characterized by the so-called *b*-value

$$b = \gamma^2 G_\mathrm{DW}^2 \delta^2 \left(\Delta - \frac{\delta}{3}\right) ,$$

with G_DW being the amplitude of the diffusion-weighting magnetic field gradient, δ represents the duration of the magnetic field gradient, and Δ indicates the temporal distance between the two gradient lobes. In routine clinical imaging, *b*-values of up to $1000\,\mathrm{s/mm^2}$ are customary.

The signal is as follows

$$S \sim \mathrm{e}^{-bD} ,$$

with D being the *apparent diffusion coefficient* (ADC). As the diffusion is a tensor, *diffusion tensor imaging* (DTI) will enable the measurement of the preferred diffusional direction [23.48] with an according graphical representation [23.49]. The preferred diffusional direction of water molecules seems to be parallel to nerve sheets. The diffusional directivity of these water molecules indirectly allows the display of nerve fiber tracts.

23.3.6 MRI Spectroscopy

Different nuclei have different Larmor frequencies, and identical nuclei have different Larmor frequencies depending on their electronic environment. This phenomenon is exploited in MR spectroscopy (MRS) to measure the levels of different metabolites in body tissues. The shift in Larmor frequency as a function of the electronic environment is usually described in parts per million (ppm) with respect to a reference frequency (usually free water); for example, the resonance frequencies of hydrogen nuclei in amino acids such as *N*-acetyl-aspartate (NAA) found in neurons are shifted by 2 ppm relative to the spectral line of free water. Fourier transformation following data acquisition without a readout magnetic field gradient will lead immediately to a display of spectral lines, with the amplitudes of individual lines representing the concentration of the underlying metabolites. MRI spectroscopy enables chemical analysis of tissue without biopsy, although the number of detectable chemicals is rather limited. To achieve spatial resolution in MRI spectroscopy a dual echo can be applied, for example, a 90° RF excitation pulse and two 180° RF refocusing pulses applied in the presence of magnetic field gradients that

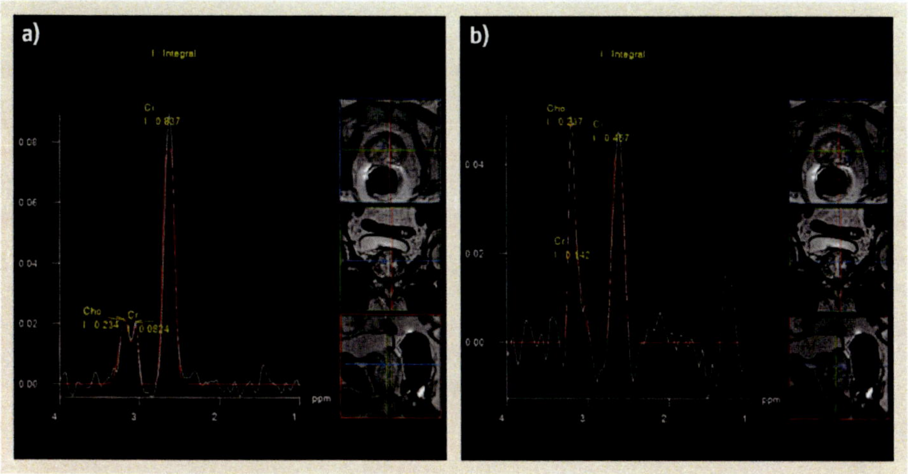

Fig. 23.11 (a) Proton MRI spectroscopy of a healthy prostate. **(b)** Proton MRI spectroscopy of a prostate carcinoma (courtesy of the University Hospital Mannheim)

are perpendicular to each other. The final signal will come out of the voxel represented by the intersection of the three orthogonal slices. The currently achievable voxel size is about 2–8 ml (for proton spectroscopy).

Besides proton spectroscopy, phosphorus spectroscopy should be mentioned. In this case, phosphorus is utilized as the signal-emitting nucleus.

Proton spectroscopy is mainly applied to the brain, whereas phosphorus spectroscopy is utilized to study muscles, as phosphorus allows detection of energy metabolism. In confirming the diagnosis of a prostate carcinoma, proton spectroscopy demonstrates a lowering of the citrate peak and an increase of the choline peak as compared with a normal prostate. Citrate is produced by prostate tissue. Prostate carcinomas consume citrate, and the intracellular content of citrate will be lowered. Choline is part of the cell membrane and in malignant prostate lesions, in conjunction will cell proliferation, the choline level is increased (Fig. 23.11).

23.4 MRI – Safety-Relevant Aspects

For the time period 1995–2005, the Food and Drug Administration (FDA) database shows 389 entries where humans have been harmed in conjunction with MRI. Ten percent of these accidents were related to ferromagnetic objects attracted by the strong magnetic field. Seventy percentage were related to burns caused by RF interactions. The most important safety issues can be summarized as follows:

- Attractive forces by the strong magnetic field
- Significant torques within the strong magnetic field
- RF interaction with the patient's body
- RF interaction with active or passive implants
- Peripheral nerve stimulation (PNS) caused by the rapid switching of the large magnetic field gradient amplitudes
- Acoustic noise due to the Lorentz forces on conductors with electric currents flowing in the presence of a strong magnetic field
- Cold gases in case of loss of superconductivity (quenching)
- Nephrogenic systemic fibrosis (NSF), in conjunction with Gd-containing contrast agents.

23.4.1 Attraction and Torque Due to Strong Magnetic Fields

Attractive forces on ferromagnetic objects are a consequence of a change in energy as a function of location

$$F = \nabla U .$$

The potential energy is the product of the magnetic moment and the applied external magnetic field

$$U = \tfrac{1}{2} \boldsymbol{M}\boldsymbol{B}_0 ; \quad \boldsymbol{M} = \frac{\chi}{\mu_0} V \boldsymbol{B}_0 .$$

The magnetic moment M of a ferromagnetic object is the product of its volume V, its magnetic susceptibility χ, and the magnetic field strength at the current location B_0. Simplified to the view of a single dimension, the attractive force on a ferromagnetic object is proportional to the magnetic field gradient at the current location and the magnetic field strength at that location

$$F_z = \frac{\chi V}{\mu_o} \boldsymbol{B}_0 \frac{\partial \boldsymbol{B}_0}{\partial z} .$$

For a pair of scissors with weight of 0.4 lb, the horizontal maximum pulling force for a 1.5 T system is approximately 20 times higher than the gravitational force, comparable to a weight of 8 lb. The critical aspect is the discontinuity of the force, which is rapidly changing as a function of location. The maximum attractive force is experienced in the vicinity of the patient bore, whereas the attractive force at the isocenter of the magnet is zero. A ferromagnetic geometrically asymmetric object will have a strong tendency to align its long axis parallel to the direction of the magnetic field. Ferromagnetic scissors will always fly into the magnet with their pointed tip entering first. The torque is proportional to the square of the magnetic field strength and is at its maximum at the isocenter of the magnet.

23.4.2 RF Interaction with the Patient's Body

The power spectrum of the electromagnetic radiation used in MRI is too weak to lead to ionization, molecular destruction, or generation of free-radical molecules. The applied energy will only lead to warming of the patient. The power W absorbed by the patient is proportional to the square of the resonance frequency of the system ω_0 (which is proportional to the magnetic field strength used) and proportional to the square of the magnetic component B_1 of the electromagnetic radiation used. The power W absorbed by the patient is proportional to the fifth power of the circumference b of the patient and inversely proportional to the internal conductivity ρ

$$W \approx \frac{\omega_0^2 B_1^2 b^5}{\rho} .$$

Acceptable levels of power deposition into a patient are documented in the international guideline on safety requirements in MR (IEC 60601-2-33). The metabolic rate of the average patient is about 90 W. This is the energy per unit time needed to maintain body temperature. IEC 60601-2-33 allows power deposition of up to 2 W/kg without medical supervision. With a body weight of 176 lb this means 160 W for the duration of the MR measurement. Doubling of the power deposition (4 W/kg) is allowed under medical supervision, and this corresponds to the metabolic rate of a marathon runner. All vendors have to ensure that their systems will refuse to start a measurement where the patient will be exposed to an energy level beyond these guidelines.

Another known potential complication is examination of patients with permanent cosmetics and tattoos. Studies have reported transient, sometimes painful skin irritation, cutaneous swelling or heating sensations, in about 1.5% of these cases. If the patient communicates painful sensations, the MR measurement can and will be aborted by the technologist running the system. There are no reports that any of this damage is of permanent nature. Swelling, reddening or blistering have so far been temporary.

23.4.3 Interaction with Active and Passive Implants

In general, all active and passive implants are of concern [23.50]. The static magnetic field strength will dislodge ferromagnetic implants, if the torque and attraction exceed the holding force. This is reported to be the case for older pacemakers, older cochlear implants, and older aneurysm clips. Orthopedic implants are in general nonferromagnetic and do not present a contraindication to MRI study. The static magnetic field is also potentially harmful to the function of active implants such as older pacemakers. The industry involved in design and production of implants is, of course, aware of the importance of MR for today's diagnostic radiology, and is anxious to introduce MR-compatible products. There are already a number of publications dealing with the MR compatibility of some pacemaker designs and neurostimulators. Especially for active implants, the concern is not limited to the static magnetic field. The coupling of the applied RF should not be underestimated. The potential coupling of the RF with, e.g., pacemaker leads, remains a risk, even if this is considered a potentially low risk.

23.4.4 Interaction Based on Changes in Magnetic Field Gradients

Peripheral Nerve Stimulation (PNS)

The law of induction indicates that a temporal change in the magnetic field strength will induce a voltage

$$\oint_{\partial S} E \, \mathrm{d}l = -\frac{\partial}{\partial t} \iint_S B \, \mathrm{d}A .$$

The human body can present conductive loops, despite its poor conduction. Depending on the orientation of the switched magnetic field gradients, the amplitudes and switching times of today's gradient systems are capable of inducing voltages and currents within the human body that mimic biochemical voltages usually applied to control muscle contraction. Vendors are obliged to provide measures to prevent painful patient experience during MR examination. Such a measure is called a *stimulation monitor*.

Noise Exposure During MR Examination

A conductor carrying a current in the presence of a magnetic field will experience a mechanical force, known as the Lorentz force after the Dutch physicist Hendrik Antoon Lorentz who described it. The current of up to 600 A flowing through a gradient coil located within a 1.5 T system will experience a force on a single wire on the order of 2.827 kN equivalent to a weight of 635 lb. That force will remain active for the duration of the magnetic field gradient being switched on and will vanish at the time the magnetic field gradient is switched off. The duration of a slice selection gradient is about 2.5 ms. The change from force to no force is 200 Hz, which generates a tone between a musical *a* (220 Hz) and a *g* (196 Hz). For a frequency bandwidth of 195 Hz/voxel, the duration of the frequency-encoding gradient will be 5.128 ms. The switching rate of the gradient would be 97.5 Hz, which is close to a *G* (98 Hz). Unfortunately the switching of magnetic field gradients is rarely sinusoidal, and different tasks require different switching frequencies, so the final noise is rarely considered harmonic. It is required that the noise level for a patient remain below 99 dB(A) with or without ear-protective devices. No vendor is allowed to introduce a scanner to the market capable of generating more than 140 dB(A).

23.4.5 Safety Issues in Conjunction with Loss of Superconductivity (Quenching)

The loss of superconductivity of the current-carrying coil of the main magnet is called a *quench*. Currently commercially available and relatively easy to handle superconducting wires are composed of niobium, NiTi, or Nb_3Sn and are placed in about 1800 l of liquid helium at a temperature of −269 °C in order to be and remain superconductive. Loss of superconductivity is rarely sporadic (usually occurring during ramping up of the main magnetic field) but may be provoked by pressing the emergency button (the quench button). As the superconductive magnet becomes resistive, the energy stored in the now decaying magnetic field will be transferred to the liquid-helium bath, resulting in vaporization of helium. After five seconds the pressure build-up in the cryostat will be high enough to activate the blowout disc, as planned in such circumstances. The evaporating coolant is directed out of the building through a *quench pipe*. Within 1 min the coil currents have reached a zero value and the magnetic field is zero. After 2 min, the vaporization of the helium is down to a negligible level. The temperature of the evaporating helium is close to −269 °C, and the surrounding environment will be cooled accordingly. There remains the danger of burns when touching covers that were affected by the evaporating helium. In case of a blockage of the quench pipes, the helium gas will be forced into the scanner room and there will be the danger of displacement of oxygen. In that case, the scanner room is to be evacuated. Measures are to be planned and exercised to prepare for such an emergency. There are no known cases where humans have been harmed in case of a quench.

23.4.6 Gadolinium-Containing Contrast Agents and NSF

In contrast to iodine-containing contrast media, gadolinium-containing MR contrast agents were long considered non-nephrotoxic with very low risk of adverse reactions or other complications. The enthusiasm even triggered suggestions to use gadolinium-containing contrast agents in conjunction with x-ray imaging [23.51, 52]. In 2006, a group of scientists at AKH in Vienna published a study of nine end-stage renal disease patients who underwent MR angiography using a gadolinium chelate (gadolinium-diethylenetriaminepentaacetic acid bismethylamide (Gd-DTPA-BMA)) over a period of approximately 2 years, of whom five patients developed nephrogenic fibrosing dermopathy (NFD), also called nephrogenic systemic fibrosis (NSF) [23.53, 54]. Nephrogenic systemic fibrosis involves fibrosis of skin, joints, eyes, and internal organs. The disease was first published in 2000 and has a mortality rate of 5%. Since 2006 it has been suspected and later confirmed that the disease is associated with exposure to gadolinium. Currently there are approximately 500 reported cases that can be linked to administration of Gd-containing contrast agents. Considering the approximate number of 21.5 million MR studies with MR contrast agents per year, the number of cases seems to be rather low. Nevertheless, the reported risk of Gd-containing contrast agents has to be acknowledged.

23.5 MRI – Pictures of the Future

For the past 20 years it has frequently been predicted that developments within the area of magnetic resonance and magnetic resonance imaging would reach maturity, and further progress would be slow and only incremental – similar to the development in x-ray computed tomography. So far we are still waiting for those times.

23.5.1 Magnetic Field Strength

About two-thirds of the market is currently working with magnetic field strength of 1.5 T. High-field systems with magnetic field strength of 3.0 T have a market share of about 20%, and the remaining devices are low-field (≤ 0.5 T) MR systems. Magnets with even higher magnetic field strength (7 T) are still limited to academic institutions, but there are already 37 installations known. It should also be mentioned that there are at least two 9.4 T installations and two 11.7 T systems, and of course this refers to magnets with a bore diameter suitable to study humans.

23.5.2 RF Technology

The signal gain when working with higher magnetic field strength (Fig. 23.12) is supplemental and supports measures to reduce measurement times in conjunction with the utilization of spatial information from distributed coil matrices. Otherwise, measurement of spatial information would involve time-consuming additional phase-encoding steps. The ability to use coil matrices for *parallel imaging* is a function of the number of coils and coil profiles. Modern commercial MR systems are currently prepared to serve coils of up to 128 independent receiver channels, and the first prototype coils are available. Direct digitization within the transmitter chain as well as the receiver chain has just been introduced into commercial systems and will enable future application developments.

23.5.3 Application Development

For development of new MR applications the following fields deserve special attention:

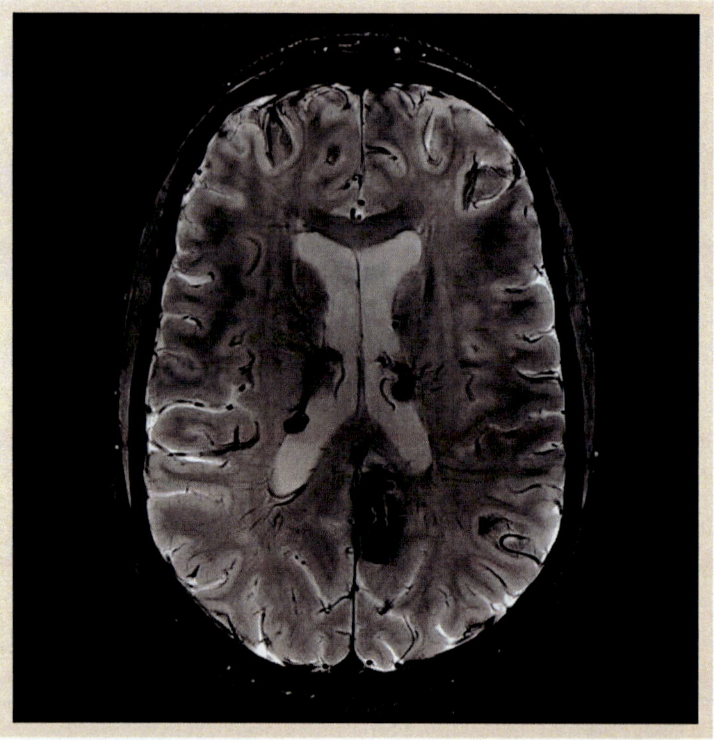

Fig. 23.12 Image of a T_2-weighted axial cut of the brain of a patient with multiple sclerosis, performed on a 7.0 T system (courtesy of the University Hospitals of New York)

- Motion correction, registration, and mapping of anatomic structures
- Mapping and visualization of different (additional) information (parametric imaging)
- Perfusion measurements without contrast agents (arterial spin labeling (ASL))
- MR angiography without contrast agents (native)
- Time-resolved contrast-enhanced MR angiography (time-resolved imaging with interleaved stochastic trajectories (TWIST) and time-resolved imaging of contrast kinetics (TRICKS))
- MR data acquisition in the presence of a continuously moving patient table (TimCT – total imaging matrix with continuous table movement, MDS – move during scan).

In general it can be stated that the current standardization and automatization is streamlining the workflow, resulting in shorter times during which patient has to remain in the scanner, while simultaneously ensuring reproducible results; e.g., for brain studies, slice orientations and coverage are recommended in specific guidelines. In modern current systems a three-dimensional localizer is usually initiated automatically after the patient has been positioned. Current modern systems are immediately able to adjust the next imaging protocol with respect to angulation and requested coverage in automatically analyzing the 3-D-localizer for anatomical structures and landmarks. Similar approaches are currently already available to automatically position required slices and volumes for studying the knee of a patient. For imaging the heart, current modern systems utilize the localizer to select the recommended short-axis or long-axis views automatically. In imaging the spine, a single mouse click can indicate to the system which intervertebral disc space is to be evaluated, and the system will automatically adjust the imaging plane to have the same angulation as the disc.

The number of selected elements of a surface coil will affect the coverage as well as the image noise. If more coil elements are selected than needed, the image noise is unnecessarily increased. If the number of coil elements selected is too low, the coverage necessary for a diagnosis may be insufficient. Current modern systems automatically select or deselect required or unnecessary coil elements.

23.5.4 Hybrid Systems

Following the success of positron emission tomography–computed tomography (PET-CT), the idea of a MR-PET system as a hybrid technology has also surfaced within the MR community. The introduction of an MR-PET system as a routine clinical modality is only a question of time. MR-PET combines complementary modalities. The superior ability of MR to display soft tissue contrast is combined with the unique feature of PET to provide biochemical information.

Potential scientific applications are:

- Simultaneous activation studies by PET and fMRI (real-time information correlation)
- PET combination with diffusion tensor imaging (DTI), three dimensional chemical shift imaging (3-D-CSI), and high-resolution structural MRI
- Dynamics of distribution of pharmaceutical products within anatomical structures
- Development and evaluation of cell therapy (stem-cell migration tracking and differentiation).

Potential clinical applications are:

- Differential diagnosis of recurrent tumors versus radiation necrosis
- Early diagnosis of Alzheimer's disease – improved prognosis due to potentially earlier medication
- Staging, restaging, and therapy monitoring in pediatric oncology
- Whole-body tumor staging and therapy monitoring
- Myocardial infarction: differentiation between hibernation and stunning.

There are a few challenges to be mastered in integrating a PET system within a MR system, starting with the development of a MR-compatible PET detector

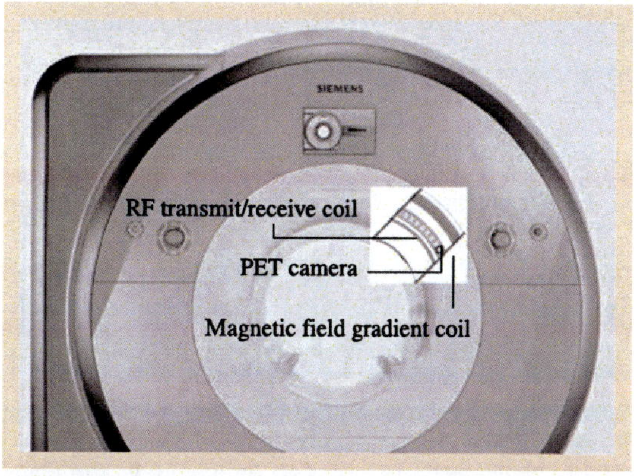

Fig. 23.13 Schematic sketch of a fully integrated MR-PET system

(Fig. 23.13). Beneficial for this development has been the early introduction of MR systems with a larger bore diameter.

23.5.5 Theranostics – Therapy Under Image Guidance

Another future possibility is the combination of an imaging modality such as MR with a therapy system such as radiation therapy. Patient position and immobilization in radiation therapy has been and still is a challenge. As of today there are no established solutions to treat cancer close to organs that permanently move due to breathing. The inherent and inevitable internal organ motion causes the tumor to be in different positions when treatment occurs, and it is impossible to determine where the dose is actually going. In the near future, continuous imaging – with MRI – will allow guidance of the focus of radiation therapy to ensure effective treatment.

As is obvious from the above, physicists and engineers as well as physicians and technologists will face an extremely exciting and interesting future in view of all these potential new and novel developments.

Further Reading

- P. Reimer, P.M. Parizel, F. Stichnoth: Clinical MR Imaging: A Practical Approach, (Springer, Berlin Heidelberg 2010)
- M.F. Reiser, W. Semmler, H. Hricak: Magnetic Resonance Tomography, (Springer, Berlin Heidelberg, 2008)
- A. Oppelt (Ed.): Imaging Systems for Medical Diagnostics, (Publicis Corporate, Erlangen 2005)
- V.M. Runge, W.R. Nitz, S.H. Schmeets: The Physics of Clinical MR Taught Through Images, (Thieme, New York Stuttgart 2009)
- C. Westbrook, C. Roth, J. Talbot: MRI in Practice, (Blackwell, Oxford 2005)

References

23.1 W. Pauli: Zur Frage der theoretischen Deutung der Satelliten einiger Spektrallinien und ihrer Beeinflussung durch magnetische Felder, Naturwissenschaften **12**(37), 741–743 (1924)

23.2 I. Estermann, O. Stern: Über die magnetische Ablenkung von Wasserstoff-Molekülen und das magnetische Moment des Protons, Z. Phys. **85**, 17 (1933)

23.3 R. Frisch, O. Stern: Über das magnetische Moment eines rotierenden Wasserstoffmoleküls, Z. Phys. **85**, 4 (1933)

23.4 W. Gerlach, O. Stern: Über die Richtungsquantelung im Magnetfeld, Ann. Phys. **74**, 673–699 (1924)

23.5 W. Gerlach, O. Stern: Das magnetische Moment des Silberatoms, Z. Phys. **9**(1), 353–355 (1922)

23.6 I.I. Rabi, J.R. Zacharias, S. Millman, P. Kusch: A new method of measuring nuclear magnetic moment, Phys. Rev. **53**, 318 (1938)

23.7 C.J. Gorter, L.J.F. Broer: Negative result of an attempt to observe nuclear magnetic resonance in solids, Physica (The Hague) **9**, 591 (1942)

23.8 F. Bloch, W.W. Hanson, M. Packard: Nuclear induction, Phys. Rev. **69**, 127 (1946)

23.9 E.M. Purcell, H.C. Torrey, R.V. Pound: Resonance absorption by nuclear magnetic moments in a solid, Phys. Rev. **69**, 37–38 (1946)

23.10 F. Bloch: Nuclear induction, Phys. Rev. **70**, 460–473 (1946)

23.11 E. Odeblad, B.N. Bhar, G. Lindström: Proton magnetic resonance of human red blood cells in heavy water exchange experiments, Arch. Biochem. Biophys. **63**, 221–225 (1956)

23.12 R.V. Damadian: Tumor detection by nuclear magnetic resonance, Science **171**, 1151–1153 (1971)

23.13 D.P. Hollis, J.S. Economou, L.C. Parks, J.C. Eggleston, L.A. Saryan, J.L. Czeisler: Nuclear magnetic resonance studies of several experimental and human malignant tumors, Cancer Res. **33**, 2156–2160 (1973)

23.14 R. Damadian: Apparatus and method for detecting cancer in tissue awarded, US Patent 3789832 (1974)

23.15 P.C. Lauterbur: Image formation by induced local interactions: Examples of employing nuclear magnetic resonance, Nature **242**, 190–191 (1973)

23.16 A. Kumar, D. Welti, R.R. Ernst: NMR Fourier zeugmatography, J. Magn. Res. **18**, 69–83 (1975)

23.17 W.A. Edelstein, G.H. Glover, C.J. Hardy, R.W. Redington: The intrinsic signal-to-noise ratio in NMR imaging, Magn. Reson. Med. **3**(4), 604–618 (1986)

23.18 C.K. Kuhl, F. Träber, H.H. Schild: Whole-body high-field-strength (3.0-T) MR imaging in clinical practice. Part I. Technical considerations and clinical applications, Radiology **246**(3), 675–696 (2008)

23.19 J. Hennig, A.M. Welz, G. Schultz, J. Korvink, Z. Liu, O. Speck, M. Zaitsev: Parallel imaging in nonbijective, curvilinear magnetic field gradients: A concept studym, MAGMA **21**(1–2), 5–14 (2008)

23.20 M.R. Prince, E.K. Yucel, J.A. Kaufman, D.C. Harrison, S.C. Geller: Dynamic gadolinium-enhanced three-dimensional abdominal MR arteriography, J. Magn. Reson. Imaging **3**(6), 877–881 (1993)

23.21 M.R. Prince, D.L. Narasimham, J.C. Stanley, T.L. Chenevert, D.M. Williams, M.V. Marx, K.J. Cho: Breath-hold gadolinium-enhanced MR angiography of the abdominal aorta and its major branches, Radiology **197**(3), 785–792 (1995)

23.22 K.P. Pruessmann, M. Weiger, M.B. Scheidegger, P. Boesiger: SENSE: Sensitivity encoding for fast MRI, Magn. Reson. Med. **42**(5), 952–962 (1999)

23.23 M.A. Griswold, P.M. Jakob, R.M. Heidemann, M. Nittka, V. Jellus, J. Wang, B. Kiefer, A. Haase: Generalized autocalibrating partially parallel acquisitions (GRAPPA), Magn. Reson. Med. **47**(6), 1202–1210 (2002)

23.24 G. Vassalo, M. Boltano, J. Linardos, J. Damadian, J.J. Cohen, R.V. Damadian: Control of MRI System, US Patent 6157194 (2000)

23.25 American College of Radiology: ACR Appropriateness Criteria, http://www.acr.org/ (last accessed June 2, 2011)

23.26 American College of Radiology: ACR Guidelines and Standards, http://www.acr.org/ (last accessed June 2, 2011)

23.27 Durable Medical Equipment (DME): Billing Codes and Reimbursement Rates, http://www.findacode.com/hcpcs (last accessed June 2, 2011)

23.28 Centers for Medicare and Medicaid Services (CMS): Healthcare Common Procedure Coding System (HCPCS), http://www.cms.gov/MedHCPCSGenInfo (last accessed June 2, 2011)

23.29 E. Hahn: How I stumbled across the Spin Echo, Third Annual Lauterbur Lecture, Proceedings of the International Society of Magnetic Resonance in Medicine (Philadelphia 1999)

23.30 P.W. Kuchel, B.E. Chapman, W.A. Bubb, P.E. Hansen, C.J. Durrant, M.P. Hertzberg: Magnetic susceptibility: Solutions, emulsions, and cells, Concepts Magn. Reson. **18A**, 56–71 (2003)

23.31 V.M. Runge, J.A. Clanton, W.A. Herzer, S.J. Gibbs, A.C. Price, C.L. Partain, A.E. James Jr.: Intravascular contrast agents suitable for magnetic resonance imaging, Radiology **153**(1), 171–176 (1984)

23.32 H.P. Niendorf, R. Felix, M. Laniado, W. Schörner, C. Claussen, H.J. Weinmann: Gadolinium-DTPA: A new contrast agent for magnetic resonance imaging, Radiat. Med. **3**(1), 7–12 (1985)

23.33 D.G. Gadian, J.A. Payne, D.J. Bryant, I.R. Young, D.H. Carr, G.M. Bydder: Gadolinium-DTPA as a contrast agent in MR imaging–theoretical projections and practical observations, J. Comput. Assist. Tomogr. **9**(2), 242–251 (1985)

23.34 J. Hennig, A. Nauerth, H. Friedburg, D. Ratzel: Ein neues Schnellbildverfahren für die Kernspintomographie, Radiologe **24**, 579–580 (1984)

23.35 P.S. Melki, R.V. Mulkern, L.P. Panych, F.A. Jolesz: Comparing the FAISE method with conventional dual-echo sequences, J. Magn. Reson. Imaging **1**, 319–326 (1991)

23.36 R.T. Constable, J.C. Gore: The loss of small objects in variable TE imaging: Implications for FSE,RARE, and EPI, Magn. Reson. Med. **28**, 9–24 (1992)

23.37 A. Haase, J. Frahm, D. Mathaei, et al.: FLASH imaging. Rapid imaging using low flip-angle pulses, J. Magn. Reson. **67**, 256–266 (1986)

23.38 E.M. Haacke, Y. Xu, Y.C. Cheng, J.R. Reichenbach: Susceptibility weighted imaging (SWI), Magn. Reson. Med. **52**(3), 612–618 (2004)

23.39 S. Ogawa, T.M. Lee, A.R. Kay, D.W. Tank: Brain magnetic resonance imaging with contrast dependent on blood oxygenation, Proc. Natl. Acad. Sci. USA **87**(24), 9868–9872 (1990)

23.40 S. Ogawa, R.S. Menon, D.W. Tank, S.G. Kim, H. Merkle, J.M. Ellermann, K. Ugurbil: Functional brain mapping by blood oxygenation level-dependent contrast magnetic resonance imaging. A comparison of signal characteristics with a biophysical model, Biophys. J. **64**(3), 803–812 (1993)

23.41 R.C. Brasch: Work in progress: methods of contrast enhancement for NMR imaging and potential applications. A subject review, Radiology **147**(3), 781–788 (1983)

23.42 L. Knutsson, F. Ståhlberg, R. Wirestam: Absolute quantification of perfusion using dynamic susceptibility contrast MRI: pitfalls and possibilities, MAGMA **23**(1), 1–21 (2010)

23.43 D.D. Stark, J. Wittenberg, M.S. Middleton, J.T. Ferrucci Jr.: Liver metastases: Detection by phase-contrast MR imaging, Radiology **158**(2), 327–332 (1986)

23.44 W.R. Nitz: Fast and ultrafast non-echo-planar MR imaging techniques, Eur. Radiol. **12**(12), 2866–2882 (2002)

23.45 G.M. Bydder, J.M. Pennock, R.E. Steiner, S. Khenia, J.A. Payne, I.R. Young: The short TI inversion recovery sequence – An approach to MR imaging of the abdomen, Magn. Reson. Imaging **3**(3), 251–254 (1985)

23.46 B. De Coene, J.V. Hajnal, P. Gatehouse, D.B. Longmore, S.J. White, A. Oatridge, J.M. Pennock, I.R. Young, G.M. Bydder: MR of the brain using fluid-attenuated inversion recovery (FLAIR) pulse sequences, Am. J. Neuroradiol. **13**(6), 1555–1564 (1992)

23.47 E.O. Stejskal, J.E. Tanner: Spin diffusion measurements: Spin echoes in the presence of a time-dependent field gradient, J. Chem. Phys. **42**(1), 288–292 (1965)

23.48 P.J. Basser, J. Mattiello, D. LeBihan: MR diffusion tensor spectroscopy and imaging, Biophys. J. **66**(1), 259–267 (1994)

23.49 S. Mori, B.J. Crain, V.P. Chacko, P.C. van Zijl: Three-dimensional tracking of axonal projections in the

brain by magnetic resonance imaging, Ann. Neurol. **45**(2), 265–269 (1999)
23.50 F.G. Shellock: Magnetic Resonance Procedures: Health Effects and Safety (2001), 450pp.
23.51 H.J. Wagner, M. Kalinowski, K.J. Klose, H. Alfke: The use of gadolinium chelates for x-ray digital subtraction angiography, Invest. Radiol. **36**(5), 257–265 (2001), ,
23.52 H.J. Wagner, M. Kalinowski, K.J. Klose, H. Alfke: Erratum for H.J. Wagner et al., Invest. Radiol. **36**(5), 257–265, Invest. Radiol. **36**(9), 553 (2001)
23.53 T. Grobner: Gadolinium – A specific trigger for the development of nephrogenic fibrosing dermopathy and nephrogenic systemic fibrosis, Nephrol. Dial. Transplant. **21**, 1104–1108 (2006)
23.54 T. Grobner: Erratum for Grobner, Nephrol. Dial. Transplant. **21**(4), 1104–1108, Nephrol. Dial. Transplant. **21**(6), 1745 (2006)

24. Magnetic Particle Imaging

Jörn Borgert, Bernhard Gleich, Thorsten M. Buzug

Magnetic particle imaging (MPI) is a quantitative imaging method that uses the nonlinear re-magnetization behavior of ferromagnetic nanoparticles to determine their local concentration. Superparamagnetic iron oxide (SPIO) particles represent such suitable nanoparticles. SPIOs are readily available as clinically approved contrast agents for liver examinations in magnetic resonance imaging (MRI), and usually administered into the bloodstream via intravenous injection. Starting from a brief overview of the history of the discovery and ongoing research on MPI in Sect. 24.2, Sect. 24.3 introduces the technical concepts of MPI. Section 24.4 will explain how to get to actual images, once data has been acquired. Section 24.5 describes alternative system designs next to traditional, symmetric geometries commonly used for medical imaging devices, and other uses of magnetic particle imaging technology, like spectroscopy. A possible combination of MPI with magnetic resonance tomography (MRT) for hybrid MPI/MRT systems is introduced in Sect. 24.5.3. Finally, Sect. 24.6 discusses potential applications for MPI and how it can provide clinical benefits, covering cardiovascular applications in Sect. 24.6.1, oncology applications in Sect. 24.6.2, cell labeling/tracking in Sect. 24.6.3, and concluding with applications that require new, modified tracer materials in Sect. 24.6.4.

24.1 **Introduction** 461
24.2 **A Brief History of Magnetic Particle Imaging** 462
24.3 **How Magnetic Particle Imaging Works** 463
24.3.1 Signal Generation and Acquisition . 463
24.3.2 Spatial Coding: Selection Field 464
24.3.3 Performance Upgrade I: Drive Field 467
24.3.4 Performance Upgrade II: Focus Field 467
24.4 **From Data to Images – Reconstruction** ... 468
24.5 **Beyond General Purpose Systems – Special Geometry** 470
24.5.1 Unlimited Access – Single Sided Magnetic Particle Imaging 470
24.5.2 Zero-Dimensional Imaging – Magnetic Particle Spectroscopy 471
24.5.3 MPI/MRI Hybrid Systems 471
24.6 **Putting MPI to Use – Applications** 472
24.6.1 Cardiovascular 472
24.6.2 Oncology, Sentinel Lymph Node Imaging and Hyperthermia 473
24.6.3 Cell Labeling and Tracking 473
24.6.4 Gastrointestinal and Lung Imaging. 474
References 474

In comparison to established methods for medical imaging, MPI is relatively new. It was invented in 2001 by *Gleich* and *Weizenecker*, who first reported on the new method in 2005 [24.1]. The method provides a unique combination of features.

24.1 Introduction

MPI measurement is inherently quantitative; its signal, therefore, is a direct measure of how much material is present at a certain location. The correlation of the image signal with material concentration is known from methods in nuclear medicine, like PET and SPECT. This similarity is the reason why contrast agents are called *tracer materials* in the MPI context.

Furthermore, MPI promises to deliver high spatial and temporal resolution. In contrast to MRI, its sampling scheme exhibits a tenfold higher voxel rate. This advantage can be exploited to deliver true real-time imaging.

MPI realizes *direct imaging* of the particles by measuring their magnetic properties. In this way, the sensitivity of MPI in detecting iron oxide can exceed that of MRI by several orders of magnitude, since only *indirect particle imaging* is realized in MRI, i. e. particles are detected by measuring their influence on the relaxation behavior of protons.

Since MPI uses various static and oscillating magnetic fields to perform its measurement, it is completely free of ionizing radiation, as used for computed tomography (CT) and x-ray based methods, and of sources of radioactivity, as used for positron emission tomography (PET) and single photon emission computed tomography (SPECT). As far as is known today, no adverse or long-term effects on patients are expected from magnetic fields like those applied for MRI or MPI.

24.2 A Brief History of Magnetic Particle Imaging

Magnetic particle imaging in its original form was invented in 2001. After a first publication as a patent in [24.2], results of static 2-D measurements of undiluted Resovist were presented in [24.1]. After extending the imaging of phantoms filled with magnetic material to dynamic two-dimensional image acquisition in [24.3], dynamic three-dimensional image acquisition in vivo with clinically approved tracer concentration was demonstrated [24.4]. Also in 2009, *Goodwill* et al. demonstrated an alternative approach, using so-called narrowband magnetic particle imaging [24.5–7], which bears potential for more sensitive imaging, while the former efforts by Gleich and Weizenecker et al. were aiming at demonstrating MPI's capabilities in real-time image acquisition.

Parallel to experimental work, important theoretical assessments have been published to determine the potential of magnetic particle imaging. Starting in 2007, *Weizenecker* et al. published a simulation study that allowed the assessment of the image quality of a virtual MPI scanner by taking into account different particle characteristics and concentrations [24.8]. In 2009, *Knopp* et al. extended this assessment to different trajectories of a field free point and their influence on the image [24.9]. To demonstrate that other system topologies can increase the performance of the method, *Weizenecker* et al. presented an approach using a field-free line instead of a field-free point in [24.10]. Additionally, *Knopp* et al. provided the proof that the realization of a field-free line in actually feasible in [24.11].

Exploiting the flexibility of magnetic particle imaging in terms of unconventional imaging system geometries, *Sattel* et al. published first results on the deployment of a single-sided MPI scanner design in [24.12].

Also in 2008, *Biederer* et al. presented the design of a magnetic particle spectrometer [24.13], and results of its use for the analysis and characterization of magnetic nanoparticles [24.14]. Details on the magnetic particle spectrometry, as well as on single-sided MPI will be presented in more detail in Sect. 24.5.

As an example of other applications of MPI, the group of *Weaver* et al. uses MPI for temperature measurements. They laid out the technical grounds and presented another approach for a magnetic particle spectrometer [24.15–19] already in 2007. They investigated an extension of MPI to improve temperature measurement by comparing the relative intensity of several individual harmonics. Independently, *Moreland* et al. presented a cantilever torque magnetometer based approach to realize a selection field gradient strength of more than 100 T/m for ultra high resolution MPI for very small samples in [24.20].

In 2008, *Bohnert* et al. started reporting on the physiological compatibility of MPI [24.21, 22].

In the important area of research on particle optimization for MPI, first efforts have been reported by *Lüdtke-Buzug* et al. [24.23, 24], *Krishnan* et al. [24.25], as well as *Markov* et al. [24.26, 27].

First results on the use of MPI for medical applications were published by *Bulte* et al. in [24.28]. They assessed the use of MPI for stem cell tracking, as detailed in Sect. 24.6.3.

24.3 How Magnetic Particle Imaging Works

In order to get MPI to work, two basic components are needed. First, one has to find a way to get the particles to emit some kind of characteristic signal. To end up at a quantitative method, this signal should contain information about the amount of magnetic material. How this is done by MPI will be explained in Sect. 24.3.1. As a second component, one needs a way to tell where the signal comes from in relation to the object that is examined. This usually is called spatial encoding, and will be explained in Sect. 24.3.2. In addition to these two basic components, Sect. 24.3.3 (drive field) and Sect. 24.3.4 (focus field) introduce means to increase the performance of MPI with respect to acquisition speed and size of the field-of-view (FOV), the volume-of-interest that can be examined.

24.3.1 Signal Generation and Acquisition

One possible magnetic material suitable for MPI is iron oxide, usually available in the form of iron oxide based nanoparticles. A basic theory for describing the magnetization curve of small mono-domain particles is the Langevin theory, which is defined under the assumption that the particles are always in thermal equilibrium. The relation between the external magnetic field and the magnetization of the particle is not linear, but exhibits nonlinear parts, as shown in Fig. 24.1. The magnetization shows a sharp increase, as the external field increases from zero, and quickly goes into *saturation*. In thermal equilibrium, hysteresis effects are not present.

In its very basic form, MPI applies a time-dependent external magnetic field to periodically change the magnetization of the magnetic material using *send coils*. It *simultaneously* detects this change in magnetization by the induced voltage in *receive coils*. Assuming for a moment that the relation between the external field and the magnetization of the particle would be linear, the induced voltage would indeed resemble the amplitude progression of the external field. In the frequency domain, the sent signal as well as the received signal would show up as a single peak at the frequency used, which is called the *fundamental frequency*. However, as the relationship is nonlinear, higher frequency components, named harmonics, are added to the spectrum of the received signal, showing up as multiples of the fundamental frequency. The presence of these harmonics is an indication of the presence of magnetic material, and the whole set of harmonics represents the *MPI signal*.

The time-dependent external field that periodically changes the magnetization of the magnetic material is called the *modulation field*. Its frequency is usually in the range of several tens to over hundreds of kilohertz, with the first results published using a frequency of 25 kHz [24.1]. These frequencies are normally not detectable to the human ear and scanner operation is thus scarcely audible. Using higher frequencies can be beneficial, as the noise in the receiver electronics is in many cases dominated by $1/f$ behavior. On the contrary, certain physiological limitations apply for the exposure of human bodies to electromagnetic waves, one of those being energy deposition. It is proportional to the square of the field amplitude and frequency, thus posing limitations to the use of higher modulation field frequencies.

To be effective, the amplitude of the modulation field should be high enough to ensure that the change in magnetization goes well into the nonlinear areas, preferably nearly into saturation. The higher the amplitude, the more pronounced the higher harmonics in the received spectrum, and thus the MPI signal, will be. Technically feasible amplitudes are in the range of several mT/μ_0 up to about $20\,mT/\mu_0$.

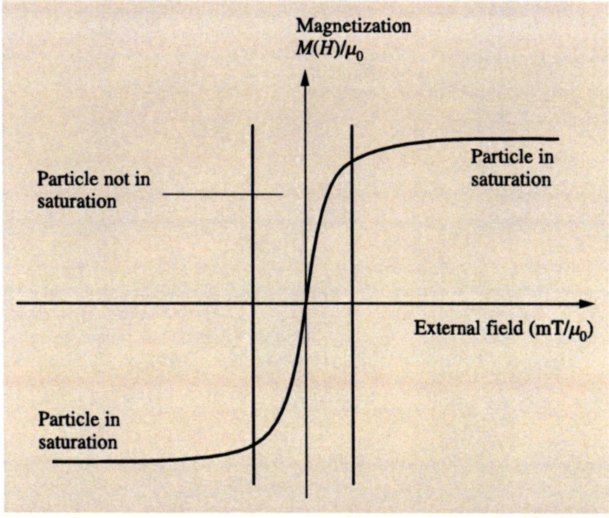

Fig. 24.1 The relation between the external magnetic field (usually measured in A/m or mT/μ_0) and the magnetization of the particle. If the external field is small, the particles are not yet in saturation, and the magnetization shows a sharp increase. For larger external fields, the particle goes into saturation, and the magnetization hardly changes with the external field

To gain information on the exact amount of magnetic material, i. e. to make a *quantitative measurement*, it is sufficient to read the amplitude of one selected harmonic from the spectrum. Given a suitable calibration measurement with a well known amount of magnetic material, the amplitude of the selected harmonic in relation to its value during the calibration measurement will be proportional to the amount of iron. It is, of course, mandatory to keep all parameters, for example the field strength of the modulation field, constant between measurements.

24.3.2 Spatial Coding: Selection Field

Using a setup as outlined above, i. e. a modulation field with sufficient amplitude that penetrates the volume-of-interest, one can easily tell whether or not magnetic material is present. However, it is not possible to determine *where exactly* the magnetic material is and how much material is present at a certain location. What has been missing up to now is a way to determine the spatial distribution of the magnetic material.

This is accomplished by introducing a local interaction field, which is constructed in such a way that it ensures signal confinement, i. e. the limitation of the origin of MPI signal to a very small region. Figure 24.2 depicts a magnetic field realized by a set of coils or permanent magnets in Maxwell configuration. In the context of MPI, this field is named *selection field*. When realized with coils, the currents in the coils simply have to run in opposite directions. When using permanent magnets, equal poles have to face each other. In contrast to the modulation field, which exhibits approximately the same field vector everywhere, the selection field exhibits field vectors that depend on the position. More so, the field contains one special location, named *field free point* (FFP), which is simply characterized by the field magnitude or field vector being zero. While veering away from this FFP, the field strength quickly rises to nonzero values. By choosing either high enough currents or powerful enough permanent magnets, the selection field can be constructed such that the particles' magnetization quickly approaches saturation as the distance to the FFP increases. In this case, the modulation field acting on the particles does not result in a sufficient change in magnetization, as depicted in Fig. 24.3. Consequently, almost no MPI signal will be measurable, and the respective spectrum will only show the fundamental frequency of the modulation field.

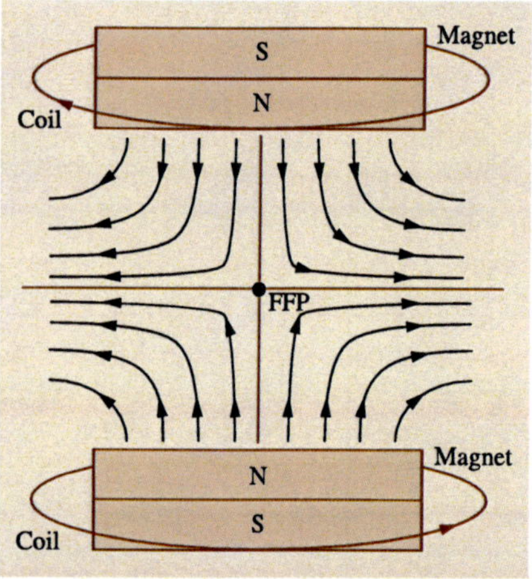

Fig. 24.2 The magnetic field generated by a Maxwell configuration of opposing magnets or coils. Mathematically, the field is zero in exactly one point, the FFP (FFP). It should be noted that for the use of coils to realize the field configuration, the current has to run in opposite directions

By moving the sample in relation to the FFP, i. e. sampling the volume-of-interest, and measuring the amount of higher harmonics at each sample point, the complete volume-of-interest can be examined. In this most simple realization of an MPI system, it is assumed that a signal only originates from particles that are within or very near the FFP and thus are not in saturation and can, therefore, react to the modulation field with a measurable change in magnetization. All other particles are silenced, as they are in saturation.

This relative movement can easily be realized in 3-D, allowing for the construction of a simple MPI scanner device. The first publication on MPI [24.1] shows two-dimensional images acquired this way, while recent work of *Goodwill* et al. [24.5–7] shows first three-dimensional images. The largest disadvantage of this method is its slowness. The relative movement between the sample and the field has to be realized mechanically. Thus the entire measurement process bears a lot of latency, either resulting in measurement times of several minutes for samples of very limited size or in very coarse spatial sampling of the object. As a result, these acquisition protocols will never be sufficient to perform in vivo measurements with satisfactory spatial and temporal resolution. This is especially true for living specimens or structures of interest that move rather fast, like for example vessels in the cardiovascular system.

Estimation of the Spatial Resolution of MPI

By sticking to the simple picture outlined above, an estimation of the spatial resolution can be given by inspection of the magnetization curve of magnetic material as depicted in Fig. 24.3. As mentioned above, un-magnetized material, either being in or near the FFP, can easily react to the external field, whereas material that is already in saturation, will not. The size of the area that enables reaction to the external field is usually given by the full width at half maximum (FWHM) of the derivative of the magnetization curve of the magnetic material. The magnetization curve is given by the Langevin theory [24.29] to be

$$M = M_0 L\left(\frac{HVM_s\mu_0}{k_B T}\right)$$
$$\text{with}\quad L(\alpha) = \coth(\alpha) - \frac{1}{\alpha} \qquad (24.1)$$

with M_0 being the saturation magnetization of the sample, H being the external field, V being the volume of the particle, M_s being the saturation magnetization of the particles' material, μ_0 being the magnetic permeability of the vacuum, k_B the Boltzmann constant, and T being the absolute temperature.

The value of M at the FWHM nicely corresponds to the value of the external field $H_{1/2}$, at which the magnetization reaches half its saturation magnetization $M_{1/2} = 1/2 \cdot M_0$. The value of the selection field has to become high enough to render the particle sufficiently saturated, i. e. outside the FWHM. The main characteristic of the selection field is its gradient strength, which is given by the change of field per unit length. This information can be used to relate the field value to a length, which can be correlated to a resolution, since two structures, or two samples of magnetic material, have to be apart at least that distance to be separable by their signal. Thus, the resolution becomes

$$R = 2\frac{H_{1/2}}{G} \qquad (24.2)$$

assuming a constant gradient of the selection field G, which is justified for small distances.

By inspection of (24.1), it becomes apparent that the resolution of MPI depends on the volume V of the particles that constitute the magnetic material. The larger the volume, the steeper the magnetization curve and the better the resolution (Fig. 24.3). In particle analysis it is common to assume spherical particles and characterize them by their diameter. Using the relationship $V = 1/6\pi d^3$, the volume V can be translated into a diameter. Figure 24.4 shows isolines of the same resolution plotted over a two-dimensional parameter space spanned by the core diameter of the particles and the selection field gradient strength. It can be seen that resolutions of the order of 1 mm, which are necessary for many medical applications, either require very high selection field gradients or very large particle core diameters. Taking a selection field gradient strength of $3\,\text{T}/(\text{m}\cdot\mu_0)$, which is feasible from the technical point of view, therefore, requires particles with a core diameter of 30 nm or more.

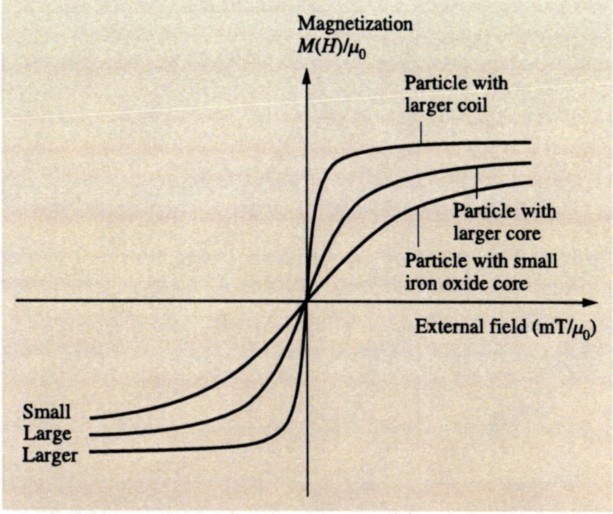

Fig. 24.3 Magnetization curves for particles with different core diameter sizes. The larger the diameter of the magnetic core, the steeper the magnetization curve

Estimation of the Detection Limit of MPI

In order to determine the detection limit of MPI, the signal that is being generated by a definite test sample of magnetic material must be compared to a noise signal. The lowest possible noise level is called *patient noise*, as it is assumed that the instrumentation itself no longer contributes to the total noise and all that is left is produced by the patient or the phantom alone (assuming a conductive object). Furthermore, from magnetic resonance imaging, it is known that a magnetization can be measured with equal sensitivity regardless of the frequency of the magnetization change [24.30]. When a magnetization oscillates faster, the induced voltage in a recording coil increases, so that the signal strength rises with the frequency. On the other hand, the thermal current fluctuations due to the conductivity of the patient are frequency-independent. For higher frequencies, these frequency-independent currents generate a higher noise voltage in the recording coil, as

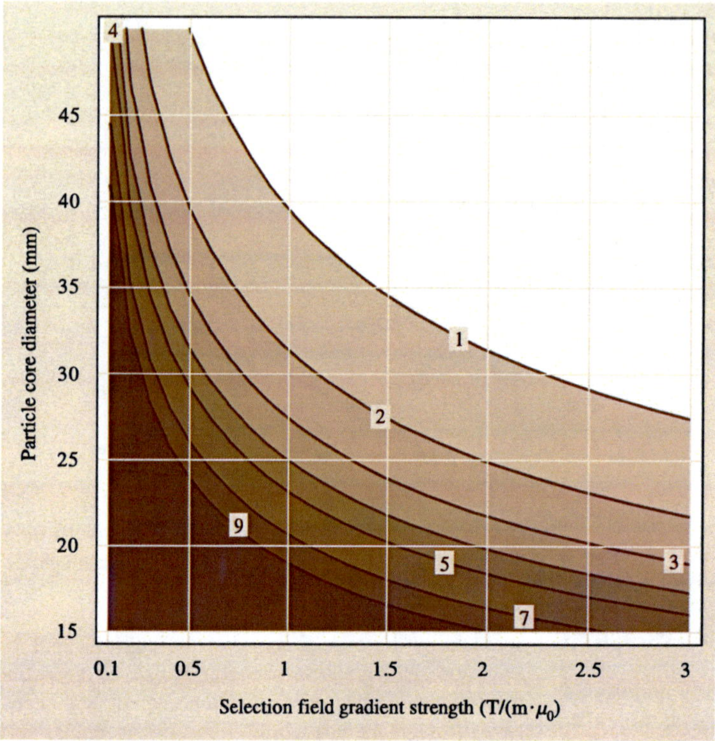

Fig. 24.4 Isolines of the same spatial resolution plotted over the selection field gradient strength in $T/(m \cdot \mu_0)$ and core diameter of the magnetic nanoparticles in nm. The *numbers in the small boxes* give the resolution in mm for the denoted isoline

faster fluctuations induce more voltage. In total, the signal to noise ratio stays (almost) constant, independent of the chosen frequency of oscillation.

To determine the signal and the noise, a square-shaped single loop coil of 10 cm edge length is being used. Placing such a coil on the chest of an adult human yields an increase of the resistance of the coil of approximately 100 mΩ at 20 MHz, equating to a noise voltage of $40\,\mathrm{pV}/\sqrt{\mathrm{Hz}}$ root mean square, calculated from [24.31]

$$U_n = \sqrt{4k_B TR\Delta f}\,. \tag{24.3}$$

To determine the signal strength, it is assumed that in a finite volume V, a given total magnetization oscillates with a given frequency, leading to

$$U_{\mathrm{ind}}(t) = s\omega MV \cos(\omega t) \tag{24.4}$$

with ω being the angular frequency, M being the magnetization, V being the volume of the object, and $s = 10^{-6}\,\mathrm{T/A}$ being the sensitivity of the test coil. The product of the magnetization M and the volume V is a dipole moment, the value of which is $92\,\mathrm{fA\,m^2}$, assuming that the test sample consists of 1 pg of iron oxide available in a finite volume. Using (24.4), this results in a signal strength of 12 pV peak and about 8 pV root mean square.

To ensure the detectability of the signal, the arbitrary choice is made that the signal must be five times higher than the noise, which would be 200 pV. This corresponds to 25 pg of iron in a given, finite volume. This derivation assumes a frequency bandwidth of 1 Hz, which corresponds to a measurement time of 1 s. Prolonging the measurement to, for example, 625 s – a little more than 10 min – leads to a frequency bandwidth of $1/625$ and, following (24.3), to a decrease in the noise by a factor of 25. This, in turn, would also lower the demand on the signal by a factor of 25 and thus lower the detection limit to 1 pg iron oxide.

To convert these amounts of material into concentrations, usually expressed by the number of iron atoms per liter of solvent/dispersion, the magnetic material contained in one voxel must be considered. As the molar mass of iron oxide (Fe_3O_4) is 231.5 g/mol, 1 g of iron oxide contains about 4.32 mmol iron oxide and thus about 13 mmol iron, and therefore, 1 pg of iron oxide contains about 13 fmol iron. Assuming that this number of iron atoms is contained in a voxel of $1\,\mathrm{mm}^3$, the concentration in this voxel is about 13 fmol(Fe)/mm^3, corresponding to a detectable concentration of about 13 nmol(Fe)/l.

Starting from 25 pg instead of 1 pg, implying a measurement time of 1 s instead of about 10 min, yields a detection limit of 324 nmol(Fe)/l for a voxel size of 1 mm^3. Resovist has an undiluted concentration of 500 mmol(Fe)/l. Applied according to prescription, 1.4 ml of undiluted Resovist are administered for one examination during an MRI scan. Assuming an adult human's blood volume to be 6 l, this results in a steady state concentration of 116 μmol(Fe)/l, about 360 times the detection limit for a voxel size of 1 mm^3 and a measurement time of 1 s.

All these estimations imply that MPI is being used as a single voxel method, i. e. the concentration of the magnetic material is measured one voxel at a time. Using more effective coding schemes, as exemplified in the next section, can also increase the detection limit.

24.3.3 Performance Upgrade I: Drive Field

The basic MPI setup described in Sect. 24.3.2 relies on the mechanical movement of the sample in relation to the FFP, while the MPI signal is generated by a modulation field. This results in a very slow image acquisition. In order to speed up the whole process, one can switch to electromagnetic fields in order to move the FFP instead of the sample. This was introduced first for two-dimensional imaging in [24.3] and finally for three-dimensional imaging in [24.4].

Taking into account the special shape of the selection field as depicted in Fig. 24.2, it becomes evident that the direction of the magnetization of a sample of magnetic material will change, once the FFP passes the sample. However, such a change also occurs when the FPP passes the material at some distance, and so the change of magnetization depends on the relative position and movement of the object with respect to the FFP. As this change in magnetization is picked up by a recording coil in the very same manner as a magnetization change induced by the modulation field, i. e. by the change in magnetization leading to the induction of a voltage in the recording coils, the explicit use of the modulation field can be omitted. This new, FFP-moving field is called a *drive field* and it replaces the modulation field. As a matter of fact, a modulation field of sufficient field strength already moves the FFP considerably and thus can be considered to act as a drive field. Already in the first publication about MPI, the use of a drive field was proposed as a means to improve the performance of the imaging method. Later realizations of MPI, for example single-sided MPI as presented in [24.32], directly use a drive field for moving the FFP and changing the magnetization of the magnetic material, never even considering the slow and intricate mechanical movement of the sample.

By comparing the gradient field strength of the selection field at the FFP and the field strength limits of the drive fields as reported in [24.1], one can conclude that the spatial range that is being covered by this electromagnetic FFP movement cannot exceed more than some tens of millimeters

$$2\frac{20\,\mathrm{mT}/\mu_0}{3\,\mathrm{T}/(\mathrm{m}\cdot\mu_0)} = 13.3\,\mathrm{mm}$$

$$\text{and}\quad 2\frac{20\,\mathrm{mT}/\mu_0}{1.5\,\mathrm{T}/(\mathrm{m}\cdot\mu_0)} = 26.7\,\mathrm{mm}\,. \qquad (24.5)$$

This kind of limitation for the volume-of-interest is clearly not acceptable for medical applications in general diagnostics, where it is expected to have systems cover a complete crosssection of a human body, like for example, magnetic resonance imaging (MRI) or computed tomography (CT). Due to the very nature of the Maxwell equations it is not possible to realize a gradient field with an FFP that has the same field gradient in all three spatial directions. In accordance with Fig. 24.2, a gradient field imposed by a Maxwell configuration of coils or by permanent magnets will have a high gradient in the lateral direction, e.g. $3\,\mathrm{T}/(\mathrm{m}\cdot\mu_0)$, to go with the example given above, and henceforward a gradient of $1.5\,\mathrm{T}/(\mathrm{m}\cdot\mu_0)$ in the axial directions. As a result, the volume-of-interest in the high gradient direction is only half of that in the other two directions, leading to noncubic voxels and MPI being intrinsically nonisotropic. In other words, based on the Maxwell configuration, the spatial resolution, which depends on the gradient strength of the selection field, is twice as high in the high gradient direction compared to the two low gradient directions.

24.3.4 Performance Upgrade II: Focus Field

In order to overcome the limitation to small volumes-of-interest, one could simply increase the field strength of the drive field. While this is technically challenging to begin with, applying fields of several hundred mT at frequencies of 25 kHz or even more might lead to energy deposition values (specific absorption rate – SAR) exceeding regulatory limits, and to peripheral nerve stimulation (PNS). For the time being it is sufficient to realize that an increase in amplitude has to be compensated by a decrease in frequency. So, in addition to the drive field, which is limited to about $20\,\text{mT}/\mu_0$, another set of orthogonal, homogeneous fields, called *focus fields*, is added. For a focus field strength in the area of $300\,\text{mT}/\mu_0$, the coverage of the field of view would be 20 cm for the high gradient direction of a selection field of $3\,\text{T}/(\text{m}\cdot\mu_0)$ and 40 cm for the low gradient direction. The frequency of the focus field, however, is low compared to the drive field, i. e. in the area of a few hertz, such that the field will not be used to move the FFP over the whole field of view for imaging, which thus would result in a rather poor performance. In fact, the movement of the FFP will be produced by a combination of the focus and the drive field. This can be done in *multistation mode*, where the focus field is used to move the whole volume that is covered by the drive field, named *cuboid*, to a certain position within the field-of-view and keep it there, while the drive field does its job. This results is covering the volume-of-interest by cuboids, which constitute small 3-D images in their own right, which are then combined into a complete 3-D data set. Another option is to combine a simultaneous variation of both, the focus and the drive field, to produce a continuous movement of the FFP, called *continuous mode*. In contrast to the multistation mode, the resulting image is rather one complete image that covers the volume-of-interest than a combination of small cuboids.

In addition to just covering the complete volume-of-interest, the focus field can also be used to realize a different, but very effective imaging mode. If, for example, the area of interest consists of only a subvolume of the complete field-of-view, and, moreover, this subvolume is not rectangular, but rather a part of an irregular 3-D shape, then the focus field can be used to image only those cuboids containing the respective subvolume. As a result, much less space needs to be scanned, leading to an effective image acquisition.

24.4 From Data to Images – Reconstruction

The role of reconstruction is to turn abstract data that are acquired during a scan into something meaningful to the user. Most often, the result is an ordered set of gray values, triplets or even multiplets, that describe color and other features of the object or phantom. These values are mapped to a rectangular grid and constitute a volume in the case of a 3-D scan or an image in the case of a 2-D scan.

When MPI is performed in its simplest form as described in Sect. 24.3, reconstruction is a simple task. The location of each single voxel is recorded together with information about the concentration, which can easily be transferred into a volume or image, depending whether the trajectory of the FFP is two-dimensional or three-dimensional in space.

If MPI is realized with performance and effectiveness in mind, i. e. by deploying a drive field to move the FFP over the object or phantom, then a dedicated reconstruction algorithm becomes a necessity. The main reason is that while moving the FFP over a distribution of magnetic material, the receiving coils do not only receive signal from the material that is directly located at the FFP, but also from material that is located within a certain distance to the FFP.

To understand this, a close examination of the two-dimensional case is sufficient. Recapitulating the geometry of the selection field from Fig. 24.2, one can easily observe that the field direction of the selection field changes, depending on whether the observed position is above or below a horizontal line through the FFP. Furthermore, limiting the observation to the z-component of the field suggests that this z-component switches from a negative value (aligned against the axis) to a positive value (aligned along the axis). So, changing position from above to below is in effect the same as having an FFP that moves from below to above. The result of the FFP passing by is that the z-component of the magnetization aligns itself along the external field, given that this field is strong enough, and the magnetization itself changes, as depicted in Fig. 24.5. This change in magnetization is, of course, detected by the z-recording coil and constitutes an MPI signal. It is obvious that such a signal can result from every piece of magnetic material that is located on a line perpendicular to the path of the FFP, albeit the signal will be lower the further away the material is from the path. Effectively, the signal recorded while the FFP is moving originates from all magnetic material on that line.

To incorporate this into the reconstruction, one must rely on the fact that MPI imaging can be described by an approximate linear imaging model. This implies that the impact of an amount of magnetic material on the resulting MPI signal is proportional to its concentration, e.g. twice the amount leads to twice the signal. This is justified, as long as the fields generated by the magnetization change in the sample can be considered small in

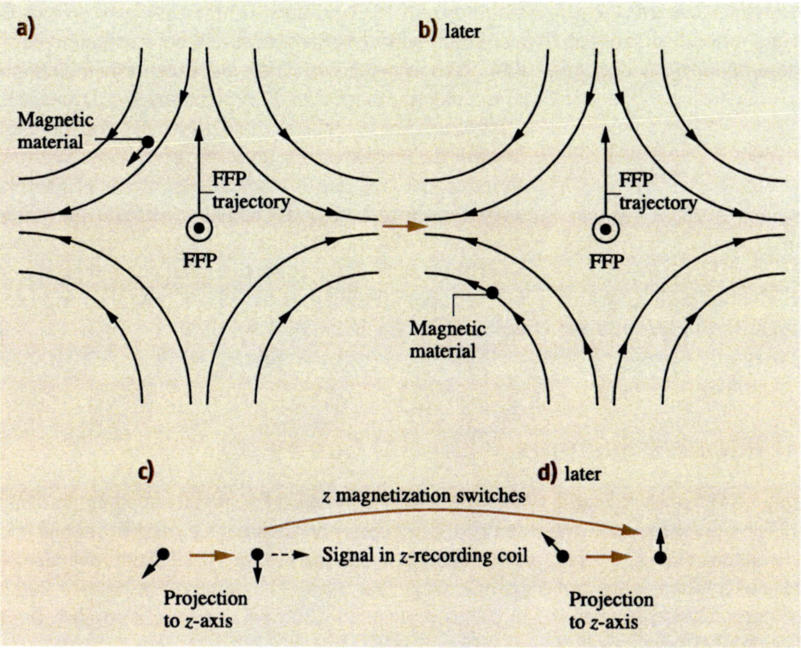

Fig. 24.5 Signal generation by magnetic material that is not directly on the path of the FFP. As the FFP passes the magnetic material, its direction of magnetization changes (a to b). Hence, limiting the observation to the z-component of the magnetization, one can conclude that the magnetization switches (c to d), even for particles that are not directly on by the trajectory of the FFP

comparison to the external fields. In that sense, knowing the impact of a single deposit of magnetic material on the signal and taking into account that this impact is proportional to the concentration, the total signal is a superposition of all signals from small deposits of magnetic material located along the line, weighted with the respective local concentration.

Taking this one step further, the knowledge about the signal caused by small deposits of magnetic material everywhere within the volume-of-interest, or at least at points on a suitable grid covering the volume, can be used to calculate the concentration values on this grid for an unknown object. The reason is that the signal caused by the object must be the same as that of the small deposits weighted with the local concentration of the unknown object. This information is collected during a calibration step using a small *delta-like* probe that is positioned at suitable locations within the whole volume-of-interest. The data that is collected during one individual measurement n of N total measurements results in a time series of K samples over time, recorded during the motion of the FFP along a certain trajectory. Recording at all N positions leads to $K \times N$ values that constitute a matrix $\mathbf{G}_{kn}$ with dimensions $K \times N$, which is usually named the *system function* or *system matrix*.

An object can be represented by the unknown concentrations C_n at all N positions, and the measurement of that unknown object, which is a time series U_k of K values, can be expressed as an expansion in terms of the entries of the system function G_{kn} as

$$U_k^{\text{meas}} = \sum_n G_{kn} \cdot C_n \,. \tag{24.6}$$

If C_n can be determined in such a way that the right-hand side of (24.3) resembles the measured values, then the unknown distribution of concentrations C_n has been reconstructed. Technically, this is done by minimizing the following terms with respect to C_n in a least squares sense:

$$\left\| U_k^{\text{meas}} - \sum_n G_{kn} \cdot C_n \right\|^2 = \min \tag{24.7}$$

which is equivalent to solving the set of linear equations $\mathbf{A} \cdot x = b$ by inversion of $\mathbf{A}$ to $x = \mathbf{A}^{-1} \cdot b$.

Depending on the number of reconstructed voxels, the mathematical problem described by (24.6) and (24.7) can be over-determined, as G_{kn} bears more information than necessary due to the large amount of measurements gathered during the acquisition of the system function. Usually, G_{kn} is not a square matrix, thus inversion is not straightforward. Being over-determined, this kind of problem has more than one solution, and it is not ensured that a numerical procedure converges towards the one solution that resembles the distribution of C_n, as G_{kn} contains systematical errors and noise. To ensure convergence towards the desired solution, i. e. to stabilize the numerical problem, Tikhonov regularization can be applied by adding an additional term [24.33]

$$\left\| U_k^{\text{meas}} - \sum_n G_{kn} \cdot C_n \right\|^2 \pm \lambda^2 \left\| \sum_n \boldsymbol{\Gamma}_{kn} \cdot C_n \right\|^2 = \min . \qquad (24.8)$$

The usual choice for $\boldsymbol{\Gamma}$ is the identity matrix, which leads to a solution that is minimal in the sense of the sum norm of C_n, and thus, in this case corresponds to a solution with the smallest possible overall concentration. Tikhonov regularization and especially the choice of the identity matrix for $\boldsymbol{\Gamma}$ is sometimes interpreted in a *Bayesian* sense, arguing that it ensures convergence towards the most probable solution, given that the errors contained in G_{kn} and the actual measurement process obey certain statistics. To actually perform the minimization of (24.8), and thus the reconstruction, several algorithms such as the conjugate gradient and algebraic reconstruction techniques have been proposed and analyzed, for example in [24.4, 9].

24.5 Beyond General Purpose Systems – Special Geometry

Strictly speaking, MPI relies only on the existence and controlled movement of an FFP or line, generated by a suitable selection field. How the other fields, i. e. drive and focus fields, are generated and applied, does not depend on the selection field geometry. Due to this flexibility, it is possible to build and deploy MPI systems that differ from the traditional geometry of medical imaging devices, which is a kind of tube, ring, or bore that surrounds the patient. One such realization, a single-sided MPI scanner, is described in Sect. 24.5.1.

For special purposes, e.g. tracer material research, it can be efficient to realize systems that only measure the signal from a bulk probe, without any spatial resolution. Section 24.5.2 introduces such a system for the concept of magnetic particle spectroscopy, using special devices that do not bear any means for spatial encoding, but only record the spectrum of a tracer, i. e. its MPI signal. Finally, Sect. 24.5.3 explains that MPI systems can also be built in ways that allow for magnetic resonance imaging, thus representing intrinsic or hybrid combinations of MPI and MRI.

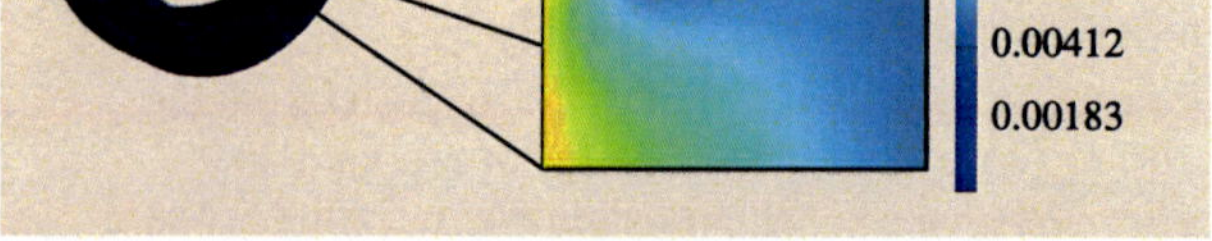

Fig. 24.6 A concept for a single-sided MPI scanner [24.12]. The setup generates two FFPs, one in front of the device, which will reside within the patient, and one within the device, which cannot be used for imaging

24.5.1 Unlimited Access – Single Sided Magnetic Particle Imaging

If one of the coils that generate the selection field is considerably smaller than the other, it fits concentrically into the other one. As a consequence, an FFP is generated on both sides of the assembly on the coil axis, as depicted in Fig. 24.6, and currents in the two transmit coils run in opposite directions. If all other components of the system can be assembled on one side of the plane defined by the selection field coils, a single-sided MPI scanner can be realized, which allows for improved patient access. No interfering structure elements are present on one side of the scanner, and the volume-of-interest is defined by the region around one of the FFPs. As the other field free point is located within the device itself, i. e. at the other side of the coil assembly, it must be assured that no magnetic material resides within the respective vicinity of its location.

Sattel et al. demonstrated first experimental results of such a single-sided MPI device using various phantoms in [24.12]. The performance with respect to resolution and contrast is lower than that of a symmetric geometry. However, the results of this new concept are promising, as the key benefit of this design lies in the flexibility. There is no limitation to patient size, because the coil assembly can be applied similarly to an ultrasound transducer. However, due to limitations in penetration depth of several centimeters for practical scanner dimensions and selection field gradient strength, only tissues and structures near the patient surface are within reach for high resolution imaging.

24.5.2 Zero-Dimensional Imaging – Magnetic Particle Spectroscopy

As introduced in Sect. 24.2, *Biederer* et al. and *Weaver* et al. in [24.14, 16, 17], reported on a magnetic particle spectrometer (MPS) that can be used to measure the nonlinear magnetization of particles and can thus be of use for tracer research. As an example, characteristics of the particles such as the size of the iron core and magnetization behavior can be calculated from spectrometer measurements. Results like these enable more realistic simulations of the imaging performance in magnetic particle imaging and can lead to an improvement in the process chain of the development of new magnetic nanoparticles [24.25].

Essentially, an MPS consists of a magnetic particle imaging device without facilities for FFP movement (zero-dimensional movement). In a shielded chamber, the probe to be measured is subjected to an oscillating magnetic field. A measurement of the decay of the harmonics in the magnetization response can be correlated to the nonlinearity of the magnetization curve given by (24.1). This, to stay within the above example, can be used to estimate the particle core size [24.23].

24.5.3 MPI/MRI Hybrid Systems

By design, MPI measures the spatial distribution of a magnetic material. In order to interpret this information in a medical/diagnostic context, it is required to provide an image of the underlying anatomy for reference. A simple solution would be the subsequent acquisition of the anatomical information with a suitable other modality, e.g. CT or MRI, and a retrospective fusion of that information with the MPI data. However, it seems theoretically feasible to realize an intrinsic combination of MPI with MRI, constituting a hybrid MPI/MRI system, that would provide both the distribution of the magnetic material and the anatomical reference information.

Both MPI and MRI use strong magnetic fields for spatial encoding and transmission and both are equipped to receive and record weak signals. The field geometries and frequencies are not identical, but with moderate technical effort, components for hybrid use of MPI and MRI can be realized. One example would be an MPI selection field generated by electromagnets. Reversing the current in one of the coils turns the selection field into a homogeneous field suitable for polarizing protons, with a possible field strength of up to 0.5 T for resistive coils and 1 T or more for superconducting coils.

However, the realization of hybrid MPI/MRI systems bears a lot of challenges, e.g. sufficient homogeneity for MRI (B_0), the lack of a transmit system for the RF-field (B_1), and a receive system for frequencies as high as several 10 MHz. One potential solution, or rather a work-around, would be the deployment of field cycling or prepolarized MRI systems [24.34].

In such a system, a strong, but not necessarily highly homogeneous field is applied for several hundred milliseconds in order to polarize the protons. After this polarizing pulse, the field is reduced to a much lower value to apply the RF pulses. In MRI, the field homogeneity needs to be roughly $10\,\mu T/\mu_0$, which would be 10 ppm for 1 T, but a moderate 1% for 1 mT, which seems feasible for realization by a selection field coil pair. For such a low field, the drive field coils can be used to apply the RF pulse. The Larmor frequency of protons at 0.6 mT is 25 kHz, which is the drive field frequency of current experimental MPI demonstrators [24.1]. The drive field amplitudes possible in an MPI scanner do exceed the necessary amplitudes for MRI by far. To ensure that the MRI signals are above a certain frequency, e.g. 1 MHz, the drive field strength is ramped up to several tens of mT/μ_0. Higher frequencies result in a higher signal to noise ratio (SNR).

Such a prepolarized MRI approach may be sufficient to create an anatomical reference for many applications of MPI. For some applications, it may even be superior to conventional MRI. For patients with implants, for example, prepolarized MRI would allow near implant imaging as long as the implant is not ferromagnetic [24.35]. The MPI/MRI combination would be able to operate at even lower frequencies than the 2 MHz used in [24.35]. Therefore, it would be possible to image inside metallic stents for the assessment of re-stenosis. Furthermore, field cycling allows for contrasts that are not accessible to conventional MRI. *Conolly* et al. demonstrated that the protein content of tissue can be imaged using field dependent T_1 relaxation times [24.36]. Other favorable applications exist in the area of tissue conductivity and multinuclei imaging.

24.6 Putting MPI to Use – Applications

As MPI is a relatively new imaging technique, the proof of clinical benefit for virtually all medical applications is still pending. In the following sections, some of those applications in medical diagnostics will be exemplified. In addition to applications in diagnostics, MPI can also offer advantages in image-guided treatment and in the field of hyperthermia.

24.6.1 Cardiovascular

As commercially available tracer materials like Resovist are formulated for intravenous injection, MPI is well suited for applications that can use tracer materials flowing in the blood stream for a certain amount of time. One important application falling into this category is the diagnosis and assessment of cardiovascular disease (CVD). One of the most important aspects of CVD is coronary artery disease (CAD), which is responsible for conditions that are known as myocardial infarction, with unstable or stable angina, commonly referred to as a *heart attack*. All of these conditions are caused by either a partial closure (stenosis) or a complete obstruction (occlusion) of vessels that supply the heart muscle with blood. If the blood supply is not restored in time, the result is necrosis of heart muscle tissue.

In acute cases, a direct referral to the catheter laboratory (cathlab) for catheterization and angiography, with the option of immediate treatment by angioplasty and stenting, is still the most common course of action. However, in subacute cases, a clear trend to replace this invasive step with noninvasive methods can be noticed, using, for example, contrast enhanced computed tomography or CT-angiography (CTA) [24.37]. Consequently, only those patients showing a treatable stenosis or obstruction in the CT exam will be subjected to an intervention.

In certain cases, these diagnostic steps are preceded by electrocardiography (ECG) to investigate heart function and determine the levels of certain enzymes released during heart muscle cell death. These can include creatine phosphokinease (CPK) and, more recently, troponin [24.38]. In some countries, the diagnostic process is combined with a cardiac stress test using single photon emission computed tomography (SPECT [24.39]) and sometimes positron emission tomography (PET [24.40, 41]) to determine myocardial perfusion. Areas of decreased perfusion can indicate the presence of a stenosis or occlusion as its cause, increasing the diagnostic evidence and indicating the need for intervention.

This whole set of information, from in vitro tests and imaging, can be used to develop a comprehensive picture of the state of the heart of the patient. It might be presumed that MPI is able to provide a large part of this information, thus simplifying the use of different modalities and the need for a complex workflow to the use of one modality.

As shown for a preclinical setting in [24.4], tracer materials in clinically approved concentrations can provide functional information while flowing through various parts of the heart and the cardiovascular system of a mouse. In a next step, it has to be proven that these results can be translated to the clinical case. During a potential examination, the tracer material would be followed while entering the right atrium via the vena cava. Subsequently, it would be possible to assess the wall motion of the right ventricle and the dynamics of the ejection into the pulmonary vessel system. In a similar manner, left ventricle wall motion and ejection dynamics would complete the collection of functional information. Additionally, the coronary blood supply would be imaged, while the tracer material passes through the coronary arteries, providing information similar to an angiography performed in the cathlab or in CT. In contrast to the traditional two-dimensional cathlab, the information delivered by MPI would be three-dimensional, preventing foreshortening and overlapping. Additionally, MPI works without using any harmful ionizing radiation, which would be beneficial in those cases that require long examination times. Immediately after imaging the coronary blood supply, the myocardial vitality can be assessed by measuring myocardial perfusion. Similarly to cardiac stress tests using SPECT or PET, this information will be correlated to potential stenosis or occlusions in the coronaries.

As a result, most of the information used today to form a comprehensive overview of the state of the cardiovascular system might be acquired with MPI in a single session while following the flow of a tracer material through the cardiovascular system. Consequently, MPI has the potential to reduce required hospital resources, to simplify workflow, and to add to the benefit of the patient, who has to undergo fewer and shorter examinations.

It still has to be analyzed whether MPI can also be used to determine the atherosclerotic plaque burden. If suitable amounts of iron oxide would accumulate in the plaque, it could be quantified by MPI. Widespread research effort is being put into the understanding of such accumulation processes already [24.42]. The coatings of the tracer materials determine the physiological properties and the pharmacokinetics of the tracer. It may be possible that improved tracer materials with custom-made coatings to support the accumulation process will lead to vast improvements in the detectability of atherosclerotic plaque by MPI.

24.6.2 Oncology, Sentinel Lymph Node Imaging and Hyperthermia

New studies suggest the importance of measuring microvascularization for tumor staging [24.43]. Particularly during therapy, monitoring of the blood supply can indicate the success of the therapy by showing a decrease in the blood supply. After intravenous injection, MPI might be used to image this local blood supply with high spatial accuracy by measuring the amount of blood per tissue. Additionally, it is investigated whether iron oxide based tracer materials can migrate into tumor tissue or lymph nodes after systemic injection [24.44]. Given that sufficient amounts of iron oxide can be accumulated in these tissues, MPI could be used to perform quantitative measurements.

One specific application is the quantitative measurement of magnetic materials accumulated in sentinel lymph nodes. These are lymph nodes near a tumor, which collect the interstitial fluid of the tumor. If a tumor forms metastases, it is likely that malign cells migrate via the lymph nodes and show up first in the sentinel lymph nodes. During surgery, usually all lymph nodes in the vicinity of the tumor are resected. However, if only the sentinel lymph nodes are resected and those present themselves as negative, i. e. no trace of metastases can be found, it is an option to refrain from further lymph node resections. As a consequence, the extent of the surgery is greatly reduced, and the trauma for the patient would be limited, leading to increased quality of life. Today, this procedure is most common in the treatment of mamma carcinoma. To locate the sentinel lymph nodes, tracer materials containing blue dye and radioactive colloids are injected into the mamma carcinoma and its circumference. The distribution of the fluid is imaged by a gamma camera prior to surgery in order to identify the sentinel lymph nodes. During surgery, a Geiger counter and visual identification of the blue dye are used to find the marked lymph nodes.

It might be assumed that an MPI scanner can image the distribution of magnetic material used instead of the radioactive colloids in the current procedure. Instead of using a whole body scanner, however, it would be more appealing to have smaller, more versatile devices, like the single-sided MPI devices presented in Sect. 24.5.1. Such small and mobile devices may also be suitable for other applications during surgery.

The use of magnetic materials and oscillating electromagnetic fields for hyperthermia is currently under investigation as an experimental therapy for certain tumor types, especially those that are inaccessible to surgery [24.45]. To perform the therapy, the magnetic material is injected into the tumor and expected to stay within the tumor or near the tumor during heating, thus minimizing the damage to surrounding tissue. In hyperthermia, MPI can provide an additional benefit by locally confining the heat generation. If a selection field is present, the heat generation will be focused on the area that is covered by the field free point. This can be helpful in those cases where the magnetic material has leaked out of the tumor or if important tissue is very near the area that should be heated. In those cases, MPI might also monitor the distribution of the magnetic material and the power dissipation to optimize the treatment and outcome.

24.6.3 Cell Labeling and Tracking

To date, cell labeling is a very exciting and broad field of research. Although many different cell types can be labeled, only two examples, red blood cells and stem cells, will be treated in detail.

Red Blood Cell Labeling

Blood is composed of almost 50% red blood cells (RBCs). Similar to labeling them with radioactive tracer materials for SPECT imaging, RBCs can also be labeled with nanoparticles made from iron oxide. *Magnani* et al. [24.26, 46] described techniques of cell loading for murine and human RBCs. These cells, as with the radioactive case, might be used as a blood pool agent with ultra long blood retention time to highlight vessel structures and blood filled compartments. This might support certain interesting applications that rely on the determination of the local blood volume, e.g. hemangiomas [24.47] and monitoring and detection of bleeding, for example in the intestine [24.48, 49].

Due to the high abundance of RBCs, it seems possible to realize a sufficient tracer concentration of 1 mmol/l in the blood, enabling fast MPI applications like those described for cardiovascular applications. In oncology, monitoring the tumor blood volume by repeated measurements might lead to interesting opportunities for therapy monitoring during radiation- and chemotherapy.

In cardiovascular applications, blood pool agents might allow for therapy response monitoring with MPI, e.g. after angioplasty as a treatment of a stenoses it could be used to monitor the restoration of the blood supply. In neurovascular applications, loaded RBCs might be used for monitoring stroke patients for bleeding in intensive care and during therapy. Bleeding can be a life-threatening condition, especially during lyses therapy. MPI could be used to detect small amounts of labeled blood, which flows into regions that have previously been free from blood, thus indicating unwanted bleeding.

Stem Cell Labeling

Among cell labeling applications, stem cell labeling is the most challenging, as it is most desirable to localize only a few cells, or ultimately single cells. Given that stem cells can be loaded with up to 10 pg of iron and taking into account a detection limit of 1 pg for MPI, as previously stated, a single stem cell should be detectable with MPI. First explorative studies to use currently available, experimental MPI systems to detect loaded stem cells have been published in [24.28]. To date, the successful detection of very small quantities is limited by the sensitivity of current experimental devices. *Goodwill* et al. report on efforts to maximize sensitivity [24.5–7]. Additionally, available tracer materials only perform at about 1 to 3% of the theoretical optimum [24.1]. An improved tracer material would also improve sensitivity and promote various applications, including the area of stem cell tracking.

24.6.4 Gastrointestinal and Lung Imaging

In contrast to the applications in the previous sections, which use tracer materials administered into the blood stream, some applications, that would make use of other pharmaceutical forms of magnetic material, seem possible and interesting. In the area of gastrointestinal imaging, it might be an option to provide the magnetic material as ready to swallow. This would facilitate the imaging of the colon, an area where patient discomfort (need for bowel cleansing) hinders widespread adoption of colonoscopy for cancer detection, and diagnosis of small bowel diseases, like Crohn's disease or ileus and bowel obstructions. In the latter area, MPI would compete with contrast enhanced x-ray or CT, but would allow for extended acquisition times and potentially better diagnosis, due to its lack of harmful radiation and insensitivity to motion.

Another area for the application of new pharmaceutical forms of magnetic material would be the assessment of lung ventilation. Combining ventilation and perfusion exams to provide comprehensive information about the status of the lung might provide new and improved diagnostic information similar to the correlation of coronary to myocardial perfusion for cardiovascular diagnosis, as presented in Sect. 24.6.1. It can be estimated that the inhalation of the iron content of 1 ml of Resovist, which is equivalent to an iron amount of 28 mg, as an aerosol would be sufficient to perform imaging with acquisition times in the area of a few seconds. This would be considerably faster than today's method of choice, SPECT imaging, which takes several minutes to produce an image.

References

24.1 B. Gleich, J. Weizenecker: Tomographic imaging using the nonlinear response of magnetic particles, Nature **435**, 1214–1217 (2005)

24.2 B. Gleich: Verfahren zur Ermittlung der räumlichen Verteilung magnetischer Partikel, German Patent DE-10151778-A1 (2001)

24.3 B. Gleich, J. Weizenecker, J. Borgert: Experimental results on fast 2-D-encoded magnetic particle imaging, Phys. Med. Biol. **53**, N81–N84 (2008)

24.4 J. Weizenecker, B. Gleich, H. Dahnke, J. Rahmer, J. Borgert: Three-dimensional real-time in vivo magnetic particle imaging, Phys. Med. Biol. **54**, L1–L10 (2009)

24.5 P. Goodwill, G.C. Lee, G. Scott, P. Stang, S. Conolly: Direct imaging of ferumoxides using magnetic particle imaging: Sensitivity and instrument construction, Proc. Intl. Soc. Mag. Reson. Med. (2009)

24.6 P. Goodwill, G. Scott, P. Stang, G.C. Lee, D. Morris, S. Conolly: Direct imaging of SPIOs in mice using magnetic particle imaging: Instrument construction and 3-D imaging, Proc. Intl. Soc. Mag. Reson. Med. (2009) p. 596

24.7 P. Goodwill, C. Scott, P. Stang, S. Conolly: Narrowband magnetic particle imaging, IEEE Trans. Med. Imaging **28**(8), 1231–1237 (2009)

24.8 J. Weizenecker, J. Borgert, B. Gleich: A simulation study on the resolution and sensitivity of magnetic particle imaging, Phys. Med. Biol. **52**, 6363–6374 (2007)

24.9 T. Knopp, S. Biederer, T.F. Sattel, J. Weizenecker, B. Gleich, J. Borgert, T.M. Buzug: Trajectory analysis for magnetic particle imaging, Phys. Med. Biol. **54**(2), 385–397 (2009)

24.10 J. Weizenecker, B. Gleich, J. Borgert: Magnetic particle imaging using a field free line, J. Phys. D: Appl. Phys. **41**, 105009–105012 (2008)

24.11 T. Knopp, T.F. Sattel, S. Biederer, T.M. Buzug: Field-free line formation in a magnetic field, J. Phys. A: Math. Theor. **43**, 012002 (2010)

24.12 T.F. Sattel, S. Biederer, T. Knopp, K. Lüdtke-Buzug, B. Gleich, J. Weizenecker, J. Borgert, T.M. Buzug: *Single-Sided Coil Configuration for Magnetic Particle Imaging*, Springer IFMBE, Vol. 25/VII (Springer, Munich 2009)

24.13 S. Biederer, T.F. Sattel, T. Knopp, K. Lüdtke-Buzug, B. Gleich, J. Weizenecker, J. Borgert, T.M. Buzug: A spectrometer for magnetic particle imaging, Proc. 4th Eur. Congr. Med. Biomed. Eng., IFMBE Proc., Vol. 25 (Springer, Berlin Heidelberg 2008) pp. 2313–2316

24.14 S. Biederer, T. Knopp, T.F. Sattel, K. Lüdtke-Buzug, B. Gleich, J. Weizenecker, J. Borgert, T.M. Buzug: Magnetization response spectroscopy of superparamagnetic nanoparticles for magnetic particle imaging, J. Phys. D: Appl. Phys. **42**(20), 205007 (2009)

24.15 J.B. Weaver, A.M. Rauwerdink, C.R. Sullivan, I. Baker: Signal dependence on frequency in magnetic particle imaging, Med. Phys. **34**(6), 2361 (2007)

24.16 J.B. Weaver, A.M. Rauwerdink: An alternative spatial encoding method for magnetic nanoparticle imaging, Med. Phys. **35**(6), 2642 (2008)

24.17 J.B. Weaver, E.W. Hansen, A.M. Rauwerdink: Estimating temperature from the magnetic nanoparticle magnetization, Med. Phys. **35**(6), 2907–2908 (2008)

24.18 J.B. Weaver, A.M. Rauwerdink, C.R. Sullivan, I. Baker: Frequency distribution of the nanoparticle magnetization in the presence of a static as well as a harmonic magnetic field, Med. Phys. **35**(5), 1988–1995 (2008)

24.19 A.M. Rauwerdink, E.W. Hansen, J.B. Weaver: Nanoparticle temperature estimation in combined AC and DC magnetic fields, Phys. Med. Biol. **54**, L51–L55 (2009)

24.20 J. Moreland, J. Eckstein, Y. Lin, S.-H. Liou, S. Ruggiero: Magnetic particle imaging with a cantilever torque magnetometer, 2007 APS March Meeting, http://meetings.aps.org/Meeting/MAR07/Event/62054, 2007

24.21 J. Bohnert, B. Gleich, J. Weizenecker, J. Borgert, O. Dössel: Evaluation of induced current densities and SAR in the human body by strong magnetic fields around 100 kHz, IFMBE Proc. (Springer, Berlin Heidelberg 2008) pp. 2532–2535

24.22 J. Bohnert, B. Gleich, J. Weizenecker, J. Borgert, O. Dössel: Optimizing coil currents for reduced SAR in magnetic particle imaging, IFMBE Proc. (2009) pp. 249–252

24.23 K. Lüdtke-Buzug, S. Biederer, T.F. Sattel, T. Knopp, T.M. Buzug: *Particle-Size Distribution of Dextran- and Carboxydextan-Coated Superparamagnetic Nanoparticles for Magnetic Particle Imaging, World Congress on Medical Physics and Biomedical Engineering*, Springer IFMBE, Vol. 25/VIII (Springer, Munich 2009)

24.24 K. Lüdtke-Buzug, S. Biederer, T.F. Sattel, T. Knopp, T.M. Buzug: Preparation and characterization of dextran-covered Fe_3O_4 nanoparticles for magnetic particle imaging, Proc. 4th Eur. Congr. Med. Biomed. Eng., IFMBE Proc., Vol. 22 (Springer, Berlin Heidelberg 2008) pp. 2343–2346

24.25 M.R. Ferguson, K. Minard, K.M. Krishnan: Optimization of nanoparticle core size for magnetic particle imaging, J. Magn. Magn. Mater. **10**, 1548–1551 (2009)

24.26 D.E. Markov, N.P.M. Haex, J. van Zanten, H. Grull, H.M.B. Boeve: Magnetic particle imaging: Quantitative assessment of tracer performance, 7th Int. Conf. on the Scientific and Clinical Applications of Magnetic Carriers (Vancouver 2008)

24.27 N.P.M. Haex, J. van Zanten, H. Grull, D.E. Markov, H.M.B. Boeve: Magnetic nanoparticles as imaging agents for magnetic particle imaging, 7th Int. Conf. on the Scientific and Clinical Applications of Magnetic Carriers (Vancouver 2009)

24.28 J.W.M. Bulte, B. Gleich, J. Weizenecker, S. Bernard, P. Walczak, D.E. Markov, H.C.J. Aerts, J. Borgert, H. Boeve: *Developing Cellular MPI: Initial Experience* (ISMRM, Toronto 2008) pp. 201–204

24.29 S. Chikazumi, S.H. Charap: *Physics of Magnetism* (John Wiley & Sons, New York 1964)

24.30 Z.-P. Liang, P.C. Lauterbur: *Principles of Magnetic Resonance Imaging: A Signal Processing Perspective* (Wiley-IEEE Press, New York 1999)

24.31 J.B. Johnson: Thermal agitation of electricity in conductors, Phys. Rev. **32**, 97–109 (1928)

24.32 T.F. Sattel, T. Knopp, S. Biederer, B. Gleich, J. Weizenecker, J. Borgert, T.M. Buzug: Single-sided device for magnetic particle imaging, J. Phys. D: Appl. Phys. **42**(2), 1–5 (2009)

24.33 W. Press, S. Teukolsky, W. Saul, B. Flannery: *Numerical Recepies in C: The Art of Scientific Computing* (Cambridge Univ. Press, Cambridge, New York 1992)

24.34 A. Macovski, S. Conolly: Novel approaches to low-cost MRI, Magn. Reson. Med. **30**(2), 221–230 (1993)

24.35 R.D. Venook, N.I. Matter, M. Ramachandran, S.E. Ungersma, G.E. Gold, N.J. Giori, A. Macovski, G.C. Scott, S.M. Conolly: Prepolarized magnetic resonance imaging around metal orthopedic implants, Magn. Reson. Med. **56**(1), 177–186 (2006)

24.36 S.E. Ungersma, N.I. Matter, J.W. Hardy, R.D. Venook, A. Macovski, S.M. Conolly, G.C. Scott: Magnetic resonance imaging with T1 dispersion contrast, Magn. Reson. Med. **55**(6), 1362–1371 (2006)

24.37 R. Schmitt, S. Froehner, J. Brunn, M. Wagner, H. Brunner, O. Cherevatyy, F. Gietzen, G. Christopoulos, S. Kerber, F. Fellner: Congenital anomalies of the coronary arteries: imaging with contrast-enhanced, multidetector computed tomography, Eur. Radiol. **15**(6), 1110–1121 (2005)

24.38 M.B. Nienhuis, J.P. Ottervanger, H.J.G. Bilo, B.D. Dikkeschei, F. Zijlstra: Prognostic value of troponin after elective percutaneous coronary intervention: A meta-analysis, Catheter Cardiovasc. Interv. **71**(3), 318–324 (2008)

24.39 D.E. Kuhl, R.G. Edwards: Image separation radioisotope scanning, Radiology **80**, 653–662 (1963)

24.40 D.A. Chesler, J.R.B. Hoop, G.L. Brownell: Transverse section imaging of myocardium with 13 NH_4, J. Nucl. Med. **14**, 623–627 (1973)

24.41 M.M. Ter-Pogossian, M.E. Phelps, E.J. Hoffman, N.A. Muullani: A positron emission transaxial tomography for nuclear medicine imaging (PETT), Radiology **114**, 89–98 (1975)

24.42 Z.A. Fayad, V. Fuster: Clinical imaging of the high-risk or vulnerable atherosclerotic plaque, Circ. Res. **89**, 305–316 (2001)

24.43 V. Saxena, I. Gonzales-Gomez, W.E. Laug: A noninvasive multimodal technique to monitor brain tumor vascularization, Phys. Med. Biol. **52**, 5295–5308 (2007)

24.44 A. Luciani, E. Itti, A. Rahmouni, M. Meignan, O. Clement: Lymph node imaging: basic principles, Eur. J. Radiol. **58**, 338–344 (2006)

24.45 A. Jordan, R. Scholz, P. Wust, H. Fähling, J. Krause, W. Wlodarczyk, B. Sander, T. Vogl, R. Felix: Effects of magnetic fluid hyperthermia (MFH) on C3H mammary carcinoma in vivo, Int. J. Hyperth. **13**(6), 587–605 (1997)

24.46 M. Magnani, L. Rossi, M. D'ascenzo, I. Panzani, L. Bigi, A. Zanella: Erythrocyte engineering for drug delivery and targeting, Biotechnol. Appl. Biochem. **8**(1), 1–6 (1998)

24.47 J.G. Zheng, Z.M. Yao, C.Y. Shu, Y. Zhang, X. Zhang: Role of SPECT/CT in diagnosis of hepatic hemangiomas, World J. Gastroenterol. **11**(34), 5336–5341 (2005)

24.48 S.I. Zink, S.K. Ohki, B. Stein, D.A. Zambuto, R.J. Rosenberg, J.J. Choi, D.S. Tubbs: Noninvasive evaluation of active lower gastrointestinal bleeding: comparison between contrast-enhanced MDCT and 99mTc-labeled RBC scintigraphy, AJR Am. J. Roentgenol. **181**(4), 1107–1114 (2008)

24.49 D.M. Howarth: The role of nuclear medicine in the detection of acute gastrointestinal bleeding, Semin. Nucl. Med. **36**(2), 133–146 (2006)

25. MR-Guided Interventions and Surgery

Andreas Melzer, Erwin Immel, Rachel Toomey, Fabiola Fernandez-Gutierrez

This chapter covers the use of magnetic resonance imaging (MRI) for guidance of interventional and operative procedures, including the basics on MRI systems design, integration, workflow principles, device visualization, and instrument positioning and tracking. MRI provides soft tissue contrast superior to x-ray technology and ultrasound. The contrast can be weighted towards water or fat, and images acquired in arbitrary, multiplanar orientations and three-dimensional (3-D) volumes. MRI's flow and temperature sensitivity alongside its lack of ionizing radiation and nephrotoxic iodinated contrast agents renders it a suitable imaging technique for vascular and percutaneous interventions. Intraoperative MR imaging allows detection of hidden structures of the tissue volume in the surgical field. Although MRI has been applied since the beginning of the 1990s during operation and intervention, the lack of approved MRI-compatible tools and the technical hurdles in integrating MRI systems into clinical applications have hampered its wider distribution. These problems can be overcome through appropriate technical solutions and the use of suitable nonmagnetic and nonconductive materials.

25.1 **MRI Basics** ... 478

25.2 **MRI Image Guidance for Interventions and Surgery in Comparison with CT and Ultrasound** ... 479

25.3 **MR Systems Design and Setup for Interventions and Surgery** ... 479

25.4 **Instruments for Interventional and Intraoperative MRI** ... 482

25.5 **MR-Applicable Endoscopic Instrument Systems** ... 482

25.6 **Instrument Representation and Tracking in MRI** ... 483

25.7 **MR-Guided Robotics and Navigation** ... 485
- 25.7.1 Technology of MR-Compatible Robotic Instrument Guiding Systems ... 486
- 25.7.2 Technique for Robotic-Assisted MR-Guided Interventions ... 487
- 25.7.3 Evaluation of Targeting Precision . 488
- 25.7.4 MRI-Compatible Robot MrBot for Prostate Biopsy ... 489
- 25.7.5 Technical Issues of MR Robotics ... 489

25.8 **Hybrid Multimodal Imaging for MR-Guided Diagnosis and Therapy** ... 491

25.9 **Therapeutic MR-Guided Imaging** ... 493

25.10 **MR-Guided Delivery of Implants** ... 494
- 25.10.1 MR-Guided Delivery of Stents and Stent Grafts ... 495
- 25.10.2 MR-Guided Heart Valve Prosthesis 496
- 25.10.3 MR-Guided Delivery of Vena Cava Filter ... 496
- 25.10.4 MR-Guided Delivery of Closure Devices for Cardiac Septal Defects . 497

25.11 **Conclusions** ... 498

References ... 498

Image-guided interventional procedures can be divided into *percutaneous interventions* performed by rigid cannulae or probes, i. e., for core biopsy, drainage, and local drug injection, and *vascular interventions* performed by flexible guide wires and catheters inserted through access ports into the arterial or venous system. Ultrasound, x-ray fluoroscopy, and computed tomography (CT) are the main imaging modalities for percutaneous interventions. Vascular interventional radiology is usually performed in the catheterization laboratory (cathlab). Apart from the occasional use of ultrasound for initial vascular access, conventional x-ray angiography and fluoroscopy systems are applied for guiding catheters and guide wires, and monitoring delivery of implants such as stents, filters or valves. The obtained pictures are a projection of all the objects in the x-ray beam path and provide only limited spatial orientation. Soft tissue and vascular structures become indirectly visible only after administration of allergenic and nephrotoxic iodine contrast agents or carbon dioxide gas. The increasing cooperation between interventional cardiologists and cardiac surgeons has led to the integration of angiographic equipment into operating rooms. Intraoperative imaging generates novel hybrid procedures of interventional, catheter-based methods combined with conventional or minimally invasive techniques. As the depiction of bony structures and bleeding is the domain of computer tomography (CT), it is successfully used in trauma surgery [25.1]. MRI provides soft tissue contrast superior to x-ray technology and ultrasound (US). The contrast can be weighted towards water or fat, and images acquired in arbitrary, multiplanar orientations and 3-D volumes. MRI's high flow and temperature sensitivity alongside the lack of ionizing radiation and nephrotoxic iodinated contrast agents renders MRI a suitable imaging technique for vascular and percutaneous interventions and surgical techniques. Although MRI has been applied since the beginning of the 1990s during operation and intervention, the lack of approved MRI-compatible tools and the technical hurdles in integrating MRI systems into clinical applications have hampered its wider distribution. These problems can be overcome through appropriate technical solutions. In the following sections a selection of interventional MR techniques and examples of instruments, devices, and MR systems are described and discussed.

25.1 MRI Basics

Magnetic resonance imaging is based on a strong static magnetic field (0.2–7 T), alternating magnetic fields (gradients), and a high-frequency (HF) system with transmitting and receiving coils (antennae). The image acquisition uses resonance effects with the magnetic moment of hydrogen nuclei (proton spins) bound in water, lipid, and protein molecules [25.2]. The spins can have two energy states (parallel and antiparallel). Due to the interaction of the magnetic moment of the nuclei with the static magnetic field B_0, the spins precess to the B_0 orientation at a certain frequency. This precession frequency of protons is described by the Larmor equation (25.1) and is about 42 MHz in 1 T, about 64 MHz in 1.5 T, and about 128 MHz in 3 T.

$$\omega_0 = \gamma \cdot B_0 , \qquad (25.1)$$

where ω_0 is the Larmor frequency and γ is the gyromagnetic ratio.

The state of the parallel and antiparallel protons in a magnetic field B_0 is excited by the absorption of electromagnetic pulse energy from the transmitting coil (high-frequency/radio-frequency (HF/RF) system), resulting in alignment of the spins at a certain angle (90°). This excitation can only be achieved if, before the excitation, the total magnetization is in the B_0 direction and the frequency of excitation is equal to the Larmor frequency (25.1).

The flip angle (FA) describes the deflection of the spins caused by a RF pulse in a simplified mechanistic model. Following the RF pulse, the protons return back to the original state (relaxation) and send a signal at the same frequency, which is detected by the receiving coil. The gradient coils change the static magnetic field to generate spatial information due to the resulting differences in the resonance frequencies. Specific sequences of RF pulses and switching of the gradients in the millisecond range in combination with certain times of echo (TE) and times of repetition (TR) of the pulses yield the optimum signal strength for reception. Spin-echo (SE) and fast gradient echo (GRE) sequences can be distinguished. The image contrast can be weighted towards a water (T_2) or lipid (T_1) signal or with the options of suppressing either the water or the fat signal. As change of temperature changes the MR signals, this can be monitored during interventional procedures. There are numerous methods of MR thermometry applied, based on the temperature-sensitive processes of MR imaging, e.g., proton resonance frequency (PRF), the diffusion coefficient (D), T_1 and T_2 relaxation times, magnetization transfer, proton density, and experimental temperature-sensitive contrast agents [25.3]. After image processing, the MR signals are defined for each volume element, and the signal intensity is calculated. The MR signal is a mathematically complex quantity. A distinction is made between amplitude (magnitude of the signal = gray level) and the phase of the complex signal. Temperature deviations are calculated using the phase change for each voxel at different times. Temperature changes can be represented in milliseconds, for example, during thermal ablation by percutaneous laser probes (LITT) [25.4] or MRI-guided focused ultrasound (MRgFUS) [25.5].

25.2 MRI Image Guidance for Interventions and Surgery in Comparison with CT and Ultrasound

MRI provides reproducible cross-sectional images in variable orientation with variable soft tissue contrast. The high flow sensitivity, as well as the representation of static fluids and the visualization of pathological tissue alteration, answer a variety of diagnostic questions during interventional or surgical procedures. Although CT guidance is similar to the approach of intervention under MR guidance, comparison with ultrasound guidance is also considered. Ultrasound (US) and MRI provide for multiplanar display, whereas CT only enables axial and para-axial views. In contrast to MRI, ultrasound, however, requires direct acoustic coupling with the tissue via liquids or gels. In addition, a needle during puncture is barely shown at full length. MR visualization of instruments in comparison with CT and ultrasound is of a different kind, since the signal originates from excited protons in the tissue itself, and not from attenuation, scattering or reflection of sound or ionizing radiation. An important requirement during interventional procedures is the need to map thermal destruction of tissue, which is only provided by MRI but not by CT or US. Disadvantages of MR guidance are the lack of space for access to the patient inside the magnet, the relatively low contrast and low spatial resolution of fast sequences, the high sensitivity to motion and material artifacts, and the difficult visualisation of instruments (Table 25.1).

Table 25.1 Comparison of imaging modalities for interventional guidance: x-ray fluoroscopy (FL), ultrasound (US), computed tomography (CT) und magnetic resonance imaging (MRI)

Characteristics	FL	US	CT	MRI
Ionizing radiation	–	++	––	+
Soft tissue contrast	––	o	+ KM	+++
Temporal resolution	+	(+)	++	(+)
Spatial resolution	–	o (+++)[1]	+	+++[2]
Instruments visualization	+	–	++	o
Usability, intraoperative	o	++	+	–
Interactive control, tracking automated image plane localization	–	–	–	++
Acquisition time	+	++	++	++
Artefacts	++	–	+	––
Real time imaging	(++) Radiation!	++	+ Radiation!	+
Reproducibility	–	––	+	+
Flow sensitivity	–– KM	++	–– KM	++
Temperature mapping	––	o	––	++
3-D data set	––[3]	+[4]	++	+++

[1] Depending on frequency: high resolution at high frequency at low penetration depth.
[2] High field strength and longer acquisition time increase spatial resolution.
[3] CT analogue 3-D reconstruction of x-ray fluoroscopy can be achieved through rotation during imaging and reconstruction.
[4] 3-D ultrasound requires spatially defined movements with reconstruction or 2-D arrays.

25.3 MR Systems Design and Setup for Interventions and Surgery

The geometry and field strength of MR systems are important to determine access to the patient and relevant for the achievable image quality. The implementation of operational processes and interventional MRI is limited due to the strong electromagnetic fields and physical limitations of access to the patient. The following technical solutions are currently being promoted.

1. MRI installed in the operating room.
2. MRI movable on a rail system connecting two operating rooms.

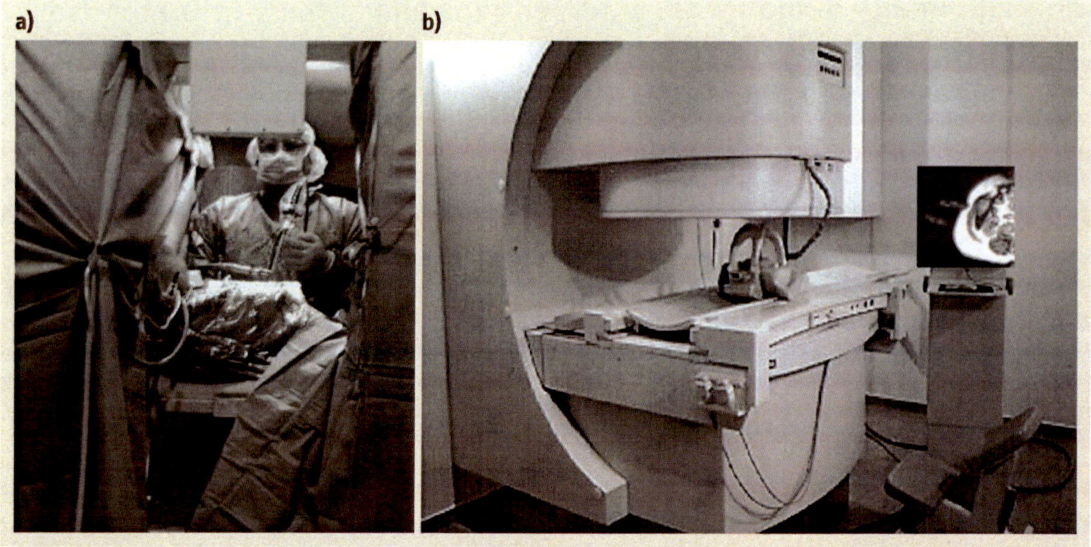

Fig. 25.1a,b Open MRI. **(a)** 0.5 T General Electric (GE) Signa SP MRI and **(b)** 0.2 T Siemens Magnetom OPEN provided the first options for intraoperative and interventional MR imaging

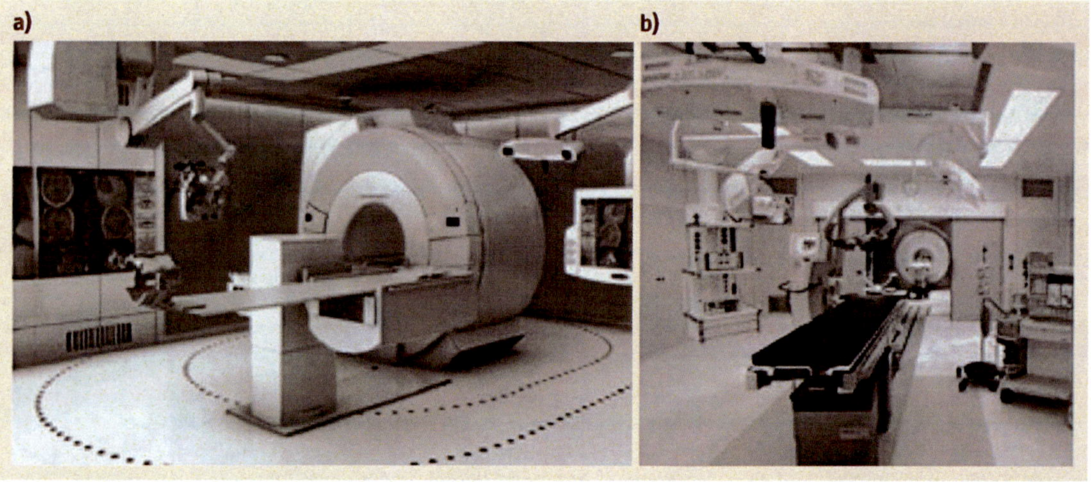

Fig. 25.2a,b State-of-the-art intraoperative MRI. **(a)** BrainSuite, Brainlab, Munich, Germany; **(b)** MR surgical Suite General Electric and Maquet, Millwaukee

3. MRI adjacent to one or more operating rooms, installed and accessible for surgical control, interventional or diagnostic techniques. The patient table is moved via a floor-mounted rail system or via wheeled cradles.

MRI systems were first applied for operations and interventions in the mid 1990s, for example, 0.5 T vertical open-configuration MR (Fig. 25.1) [25.6]. Due to the high costs, technical problems, and the limited diagnostic capabilities, this system has been discontinued. Horizontal open MRI equipment with field of 0.2–0.6 T was evaluated for interventional percutaneous techniques and for development of tools and systems for experimental procedures.

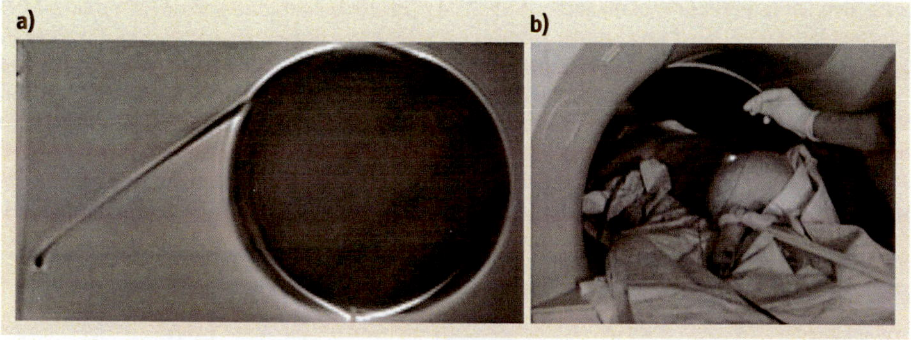

Fig. 25.3 **(a)** nitinol guide wire in MRI. The MRI image shows only a minor artifact, but the conductivity can lead to heating, and electric arcing in the worse case **(b)** (after [25.7])

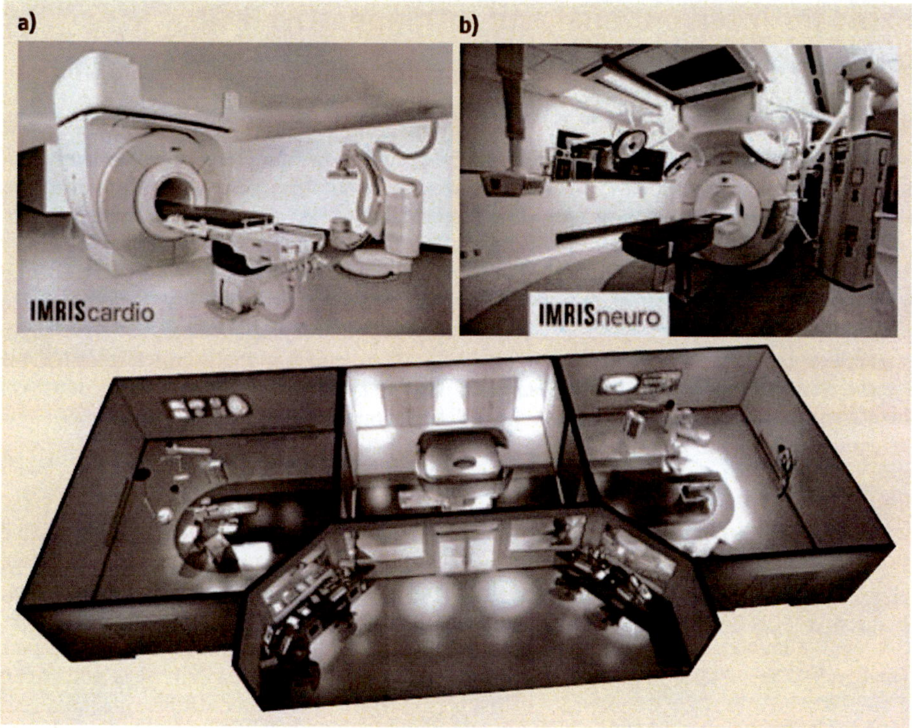

Fig. 25.4a,b Ceiling-mounted MR can be translocated between the surgical room **(a)** and cathlab **(b)** and can be used for diagnostic procedures while parked in the center

For applications in neurosurgery, the development of pivoting and linear movable operating tables successfully enabled transfer of the patient between operating room (OR) and MR [25.8]. Despite their relatively low field strength (0.2–0.6 T), open MRIs have acceptable image quality, however they are inferior in terms of spatial temporal resolution, signal-to-noise ratio, and contrast compared with closed-bore high-field MRI at 1.5 and 3 T. Therefore, operations and interventional procedures are increasingly carried out in high-field MRI systems with patient transfer cradles (Fig. 25.2).

The resonance frequency of 64 MHz at 1.5 T and 128 MHz at 3 T can produce inductive heating effects in electrically conductive instruments or implants [25.9]. In experimental studies with guide wires, temperatures above 60 °C [25.10, 11] were measured, and the formation of an electric arc from a guide wire to the skin could be documented in an animal study (Fig. 25.3) [25.12].

Standard closed-bore high-field MR systems provide bore diameter of 60 cm and length of 165–180 cm, which compromises the positioning of instruments, the conduct of the procedure on the patient within the bore, and administration of general anesthesia. Direct manual access and instrument guidance is almost impossible. To solve this problem, 1.5 T MR scanners with 70 cm bores such as the Siemens Magnetom Espree in 2005 and GE Optima MR450w in 2009, which have 125 and 145 cm bore lengths respectively, were introduced. At the same time, ceiling-mounted, moveable MR(I) system/scanner was commercialized to allow intraoperative MR imaging without moving the patient and use of conventional neurosurgical instrument systems, while the MR scanner is parked in the adjacent room (Fig. 25.4). In 2009 the combination of mobile MRI system between a catheter laboratory and the operating room was presented. In this case, the MR scanner is transferred from a central room into either the operating room or the cathlab. The angiography system must be powered off and rebooted for subsequent use. For all free-moving equipment and systems in the MR space, a consistent MR-safe structure made from predominantly nonmagnetic materials is required. Electrically conductive components in contact with the patient must be avoided [25.7].

25.4 Instruments for Interventional and Intraoperative MRI

Tools and techniques for MR-guided biopsies were published as early as 1986 [25.13]. MRI-controlled tumor ablation was described in 1996 [25.14], and the first clinical experience with MR-guided stent implantation [25.15] in 2001. The further spread of interventional MRI has been hindered by the lack of other suitable CE-marked or Food and Drug Administration (FDA)-approved instruments. The most important material for conventional interventional probes and instruments is surgical steel. Its ferromagnetic properties, however, prohibit its application in a magnetic field. Even conventional metal-based implants and instruments (for example, stents, needles or guide wires) lead to significant imaging artifacts due to their electromagnetic properties [25.16]. The following metals and alloys are suitable for MR because of their material properties such as biocompatibility, mechanical properties, ease of processing, and magnetic characteristics: tantalum, titanium and titanium alloys, nitinol (nickel-titanium shape-memory metal), nonmagnetic steel-like alloys with chromium/cobalt/molybdenum, and to a limited extent precious metal alloys (gold and platinum). Puncture needles made from the mentioned alloys are commercially available from various manufacturers.

Cannulae with diameter of 0.8–3 mm are represented at 3–5 times size magnification in the MR image (Chap. 26). From personal experience, an artifact size of 3–5 mm is required for reliable detection of the cannula in the MRI image. Therefore, metals that produce a susceptibility artifact have been preferred for cannulae and puncture needles with diameter up to about 1.5 mm. The strong electromagnetic fields used in MR imaging can induce heat in electrically conductive instruments [25.16]. The critical length of the instruments is determined by the wavelength of the RF pulse. A cannula can form a dipole antenna for the electric field of the RF pulse transmitting coil. The induced alternating current (AC) reaches a maximum when the length (L) of the instrument is a multiple of half the wavelength ($L = k\lambda/2$). According to (25.2), critical tool lengths are calculated to be about 1.88 m in 0.2 T, about 0.23 m in 1.5 T, and about 0.11 m in 3 T, for example.

$$L = \frac{\lambda_{\text{tissue}}}{2} = \frac{c}{2f\sqrt{\varepsilon}} \tag{25.2}$$

Because of these findings, preferably diamagnetic materials such as ceramics and plastics should be used to avoid inductive heating. MR visibility can be improved using metal rings or paramagnetic coatings. Ceramics and plastics, although MR compatible, were previously not suitable for thin-wall tubes with small diameters in view of the required stiffness and hardness. Adequate sharpening combined with sufficient elasticity is difficult to achieve. First experience with the use of glass-fiber-reinforced plastics has been positive [25.17], and corresponding puncture needles and guide wires are commercially available. MR visibility of polymer wires and cannula can be enhanced by dotting with paramagnetic nanoparticles.

25.5 MR-Applicable Endoscopic Instrument Systems

Intraoperative application of MRI in endoscopic surgery has advantages due to the reduced operative trauma. Limitations of endoscopic surgical procedures are, among others, the indirect visualization of the operative

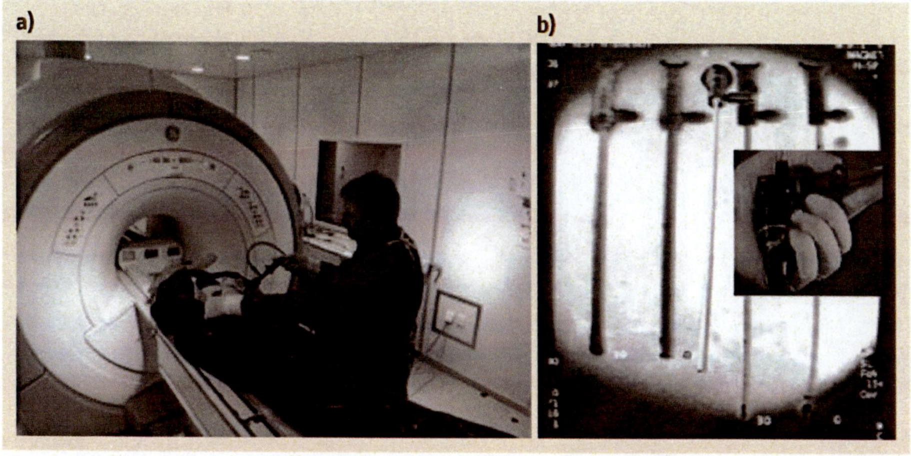

Fig. 25.5 **(a)** Experimental MRI-guided laparoscopic surgery in a human cadaver in 1.5 T MR; **(b)** rigid MR compatible rod lens endoscopes (Olympus Winter & Ibe, Hamburg)

field and the absence of tactile and haptic feedback. The endoscope enables a high-quality display of tissue surfaces in the surgical field, but the extent of a tumor and its vascular supply is beyond optical intraoperative visualization. Intraoperative MR imaging allows detection of hidden structures of the tissue volume in the surgical field. Diagnosis of areas behind or under the surfaces of the respective organs and anatomical compartments enables assessment of tumor spread and important adjacent structures such as nerves and vasculature. MR visualization of the operative tissue volume allows rigorous planning of access, sparing of surrounding tissue, and control of the surgical resection and thermal or cryogenic tissue ablation (freezing technique). Different optics for endoscopic procedures have been developed based on nonmagnetic nonferrous metals, titanium, and plastics (Fig. 25.5), and evaluated [25.18]. Metal-free optical cable of up to 4 m in length has been designed to enable removal of the light source from the MR room to eliminate disturbing influences from the light generator.

The same approach has been applied for the camera control unit to ensure the largest possible distance to the magnet outside the RF cabin for interference-free imaging. Battery-powered camera equipment showed the least interference with MRI. For observation of the endoscopic image, liquid-crystal display (LCD) or thin-film technology (TFT) with radiofrequency (RF) shielding, without significant disturbance of MR imaging, can be used in proximity to a magnetic field. Brightness and contrast are, however, reduced by the use of fine metal mesh screen for RF shielding. Modern screens and projection systems based on light-emitting diodes (LEDs) and organic LEDs are relatively brighter and offer improved usability. The first animal experiments assessing MR-guided endoscopic surgery were performed in 1998, and in the same year MR-guided neuroendoscopy was carried out on human cadavers [25.19, 20]. The open MRI working group at the Charité, Berlin, is currently developing MR-guided endoscopic surgery and MR-compatible instruments [25.21].

25.6 Instrument Representation and Tracking in MRI

Since invasive procedures entail a significantly greater risk than diagnostic methods, the appropriate technical conditions for visualization of critical structures and instruments have to be established. To obtain optimized images during the intervention, continuous adaptation of the picture settings and minimization of the acquisition time are required. Modern MR systems allow such adjustments to be made while a sequence is running. However, currently no correlation between sectional position of the MR image and the instrument is provided by MR manufacturers. Third-party products were established in the late 1990s to provide correlation of instrument position and slice orientation using optical navigation systems. These localization techniques are based on two principles: optical tracking of LEDs or light-reflecting spheres, and tracking of resonant antenna coils.

Optical tracking is based on the principle of monitoring either infrared LEDs or light-reflecting spheres (Fig. 25.6).

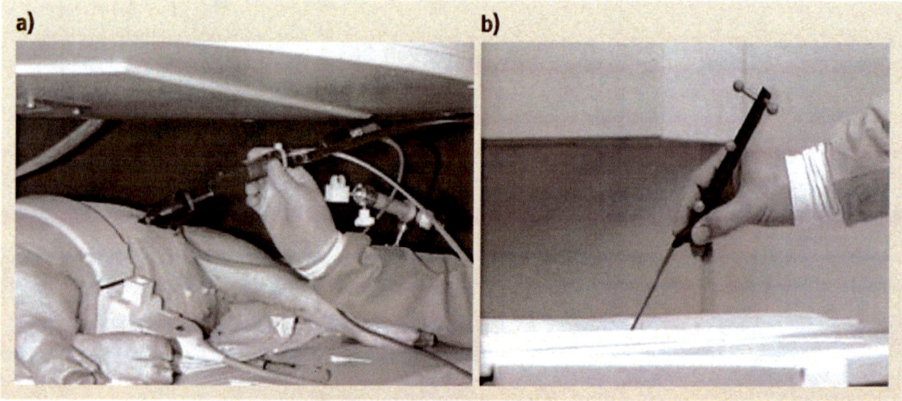

Fig. 25.6a,b Optoelectronic tracking of instruments in horizontal open MRI: **(a)** active LEDs and **(b)** light-reflecting spheres (Northern Digital, Waterloo, Canada; Siemens, Erlangen, Germany)

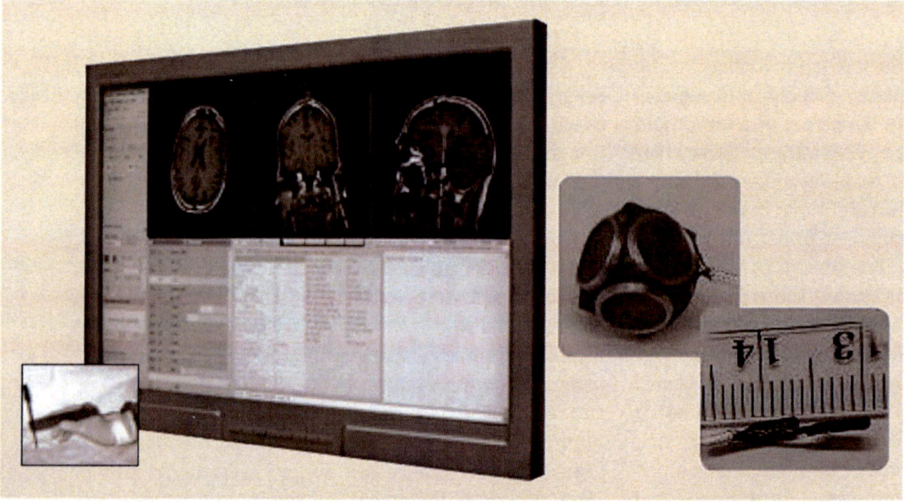

Fig. 25.7 Electromagnetic tracking in MRI enabled by a 3-D magnetic field sensor made from three orthogonal pick-up coils for a handheld guided tool and a microsensor with about 1.5 mm diameter for catheters (Robin Medical, Baltimore)

The position of the patient, the MRI, and the interventional device have to be monitored and registered into a world coordinate system. The coordinates of the device detected by the optical navigation system are provided to the MRI scanner so that the slice orientation and position of the image acquisition are set accordingly. Thereby, the MR follows the instrument position. Unfortunately, approaches realized for both the 0.5 T GE Signa open and 0.2 T Siemens horizontal open magnet have been discontinued.

Electromagnetic coils, another option for tracking of devices in MR, have been established at the same time and are commercialized by Robin Medical of Baltimore, MD (Fig. 25.7). These coils are detected based on the signal acquired by the MRI.

These two techniques are only suitable for rigid instruments such a surgical tools and cannulae and cannot be applied for guidance of flexible guide wires and catheters. Elastic bending of cannulae during insertion is also not detected. The difficulty with the mentioned tracking tools is the restriction to only one set of spatially defined layers per sequence. Unexpected patient movement or respiratory movement results in instrument displacement from the given layer. In addition, anatomy with different resilient tissues in the insertion path often leads to elastic bending of the needle. The beveled end-cut of cannulae creates a moment at the tip of the cannulae while penetrating collagenous fiber-containing layers, triggering further deterioration. When the instrument can no longer be clearly identified on the monitor, the sequence must be restarted with modified orientation. During this maneuver, lasting several minutes, there is no visual control of the instrument. The instrument tip may be in a painful

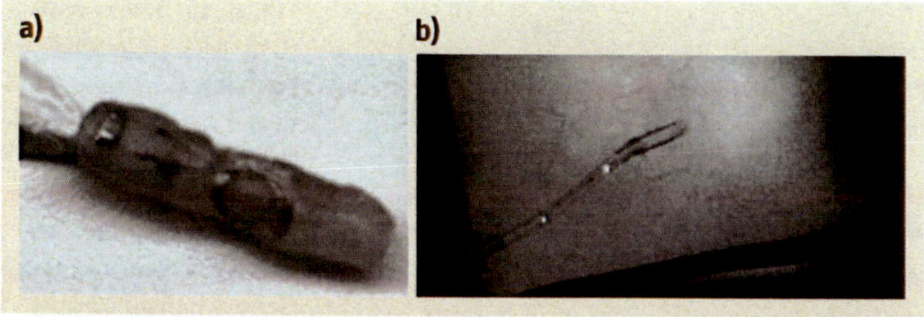

Fig. 25.8 (a) Resonant markers proximal and distal to a balloon catheter for stent implantation. The markers are tuned to the resonance (Larmor) frequency of ca. 64 MHz of a 1.5 T MRI and lead to signal enhancement (b)

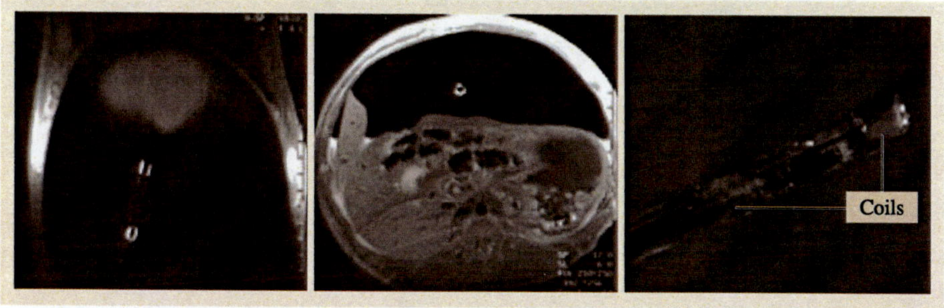

Fig. 25.9 The resonant marker requires a medium that provides an MR signal, such as water or oil. For instruments or devices such as endoscopes positioned in air, a small container with the microcoil embedded can be used for tracking (Siemens Magnetom OPEN 0.2 T at CWRU, Cleveland, OH, USA)

or critical position for the patient and must be withdrawn [25.22].

Therefore, intracorporeal tip-tracking techniques are required to adjust the MR slice position according to the instrument's position.

The visibility of instruments and implants made from nonconductive polymer composites can be improved using resonant circuits tuned to the Larmor frequency of the MR system (25.3) [25.22]

$$\omega_0 = \frac{1}{\sqrt{LC}}, \qquad (25.3)$$

where L is the inductance and C is the capacitance.

The resonant circuit consists of a coil and capacitor and undergoes inductive coupling (like an antenna) with the electromagnetic energy of the RF pulses and the energy emitted by the proton relaxation. On MR images of gradient-echo sequences (GRE) with low flip angles (FA $< 60°$), a local increase in signal intensity in and around the resonant circuit is shown (Fig. 25.8). Localization and imaging of catheters and implants are improved with this feature. This improved visualization may provide significant advantages – for example, a resonant circuit may allow detection of thrombosis or stenosis in a stent, and resonant markers may allow tracking of devices (Fig. 25.9) and automatic slice selection.

25.7 MR-Guided Robotics and Navigation

Although three-dimensional MR imaging provides a defined spatial orientation, compared with other imaging techniques (US and CT) it is difficult to determine the position of an image slice relative to the position of the instrument and locate this position within the scanner at the patient's skin surface. The problem of localization of instruments by MRI can be reduced with navigation and robotics systems. These systems capture the spatial position of the instrument and pass this data to the computer of the MR system. Based on these data, correlation between the layer orientation and the axis of the instrument becomes possible. Optoelectronic or electromagnetic tracking navigation systems are helpful; however, positioning of the instrument is performed manually. There is no feedback from the position of the MR image to the hand–arm system of the operator, or vice versa. This shortcoming can be overcome by the use of mechatronic systems.

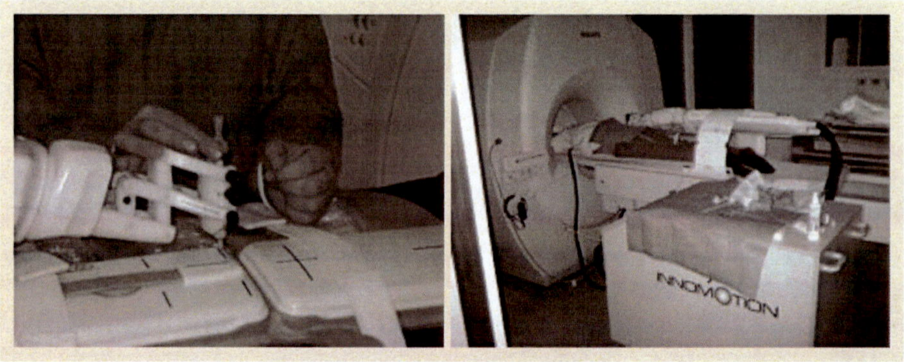

Fig. 25.10 Innomotion (Innomedic, Herxheim and KIT, Karlsruhe, Germany) MR-compatible robotic system (1.5 T GE HDX, Dundee (*left*), 1.5 T Philips Intera at Praxis Dr. Gert Lorenz, Gelsenkirchen, Germany)

Robotic systems have sensors to determine the position of the movable elements to enable a controller to drive the robotic arm to certain points in space. The sensorial position data can be transmitted by means of a coordinate transformation to the MR imaging system for corresponding imaging orientation. It is also possible to send the coordinates of the MR image position and orientation to the robotic system to move the arm to the corresponding point in space. The robotic system bridges the gap between positioning an instrument according to the image data and hands-free navigation [25.23, 24]. The deviation of the instrument during introduction into the body is not influenced, as this is determined by the structural and mechanical properties of the pierced tissue together with the material, diameter, and beveled end of the needle tip. The robotic system can, however, indicate this deviation after insertion and provide a visual indicator to correct the insertion angle (Fig. 25.10).

Conventional robotic and manipulator systems for surgical applications are usually based on the kinematics of the human hand–arm system [25.25]. These designs are not suitable due to the limited space inside the magnet and typically used ferromagnetic metals, electric sensors, and electric motors. In the last 10 years, special robotic systems for x-rays and ultrasound techniques, and in some cases also for MR imaging, have been developed [25.26]. The technologically sophisticated requirements to realize MR-compatible robotic systems with respect to nonmagnetic, nonconducting materials, sensors, and drive systems make it difficult to develop a medical device and are associated with considerable financial cost. The published systems have not been developed as medical devices, except for one [25.27], and therefore do not have the appropriate approval from the FDA or CE mark. The Innomotion system has had CE marking since 2005 and is enabled by pneumatic cylinder drives and metal-free sensors to position a nonmetallic kinematic arm with accuracy of ± 0.1 mm and $\pm 0.1°$. Use of precision-engineered plastics, composites, and ceramics eliminates interaction with the MR system [25.28]. The system is suitable for various indications, for example, MR-guided biopsies [25.27, 29], MR-guided treatment of pain, and MR-guided tumor ablation. The system can also be applied in computed tomography, for example, for CT-based positioning of osteosynthesis in pelvic fracture treatment and infiltration of joints [25.30]. At the National Institutes of Health (Bethesda, Virginia) two Innomotion systems are currently being developed for MR-guided robotic-assisted heart surgery. First experiences have been described for implantation of heart valve prostheses by minimally invasive approach through the apex of the heart in both ex vivo and in vivo porcine animal models [25.31].

25.7.1 Technology of MR-Compatible Robotic Instrument Guiding Systems

Innomotion (Innomedic and Karlsruhe Institute of Technology (KIT), Karlsruhe, Germany), the first MR-compatible assistance system, has been developed to provide precise and reproducible instrument positioning inside the magnet. MRI compatibility is achieved by testing all components and the complete system at different field strengths for MRI units including the 1.5 T Siemens Magnetom Symphony/Espree, Philips 1.0 T Gyroscan, 1.5 T Intera, and GE 1.5 T HDx. Targeting precision under MRI guidance has been determined in ex vivo organ models embedded in agarose gel. In vivo targeting precision has also been evaluated during MRI-guided percutaneous interventions in a porcine animal model.

The pneumatic robotic assistance system (Fig. 25.11) consists of a MR-safe robot arm which can be maneuvered in six degrees of freedom. The robot

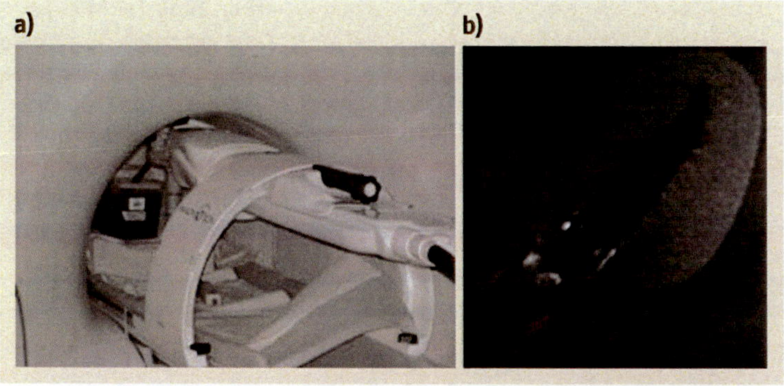

Fig. 25.11 (a) MRI-guided robotic-assisted transapical implantation of an MR-enhancing self-expanding heart valve prosthesis in an ex vivo porcine heart **(b)** MR image of the heart valve (Philips 1.5 T, Intera FFE, TR = 140 ms, TE = 4.25 ms, FA = 30°)

arm is attached to a 260° arch that is mounted onto the patient table of the scanner and that can be passively prepositioned on either side of the arch at 0, 30, and 60° according to the region of interest (e.g., spine, liver, kidney, breast). The arch is movable along the x-axis (longitudinal) of the scanner and can be firmly attached to the patient table of the MR system. Due to its exchangeable fittings, the robot can be easily attached to other MRI platforms or bore sizes. Active position detection is achieved via fiber-optically coupled limit switches, along with rotational and linear incremental sensors. The kinematics of the device has been carefully optimized for use in closed-bore MRI scanners and the CT gantry. Piezoelectric drives were tested, but due to the RF noise during MRI scanning and the risk of inductive heating of the electric power lines, they were discarded and pneumatic cylinders with slow motion control have been developed instead to drive all six degrees of freedom (DOFs).

At the front end of the arm, a module for application and insertion of coaxial probes (e.g., cannulae for biopsies, RF, cryo- or laser probes, endoscopes, etc.) provides two degrees of freedom along the x- and z-axes and is attached to a robotic arm with four degrees of freedom. This design ensures stable positioning of the instrument within a tool center point while maintaining an *invariant point of insertion* at the skin entry point. In conjunction with the two axes for movement about the tool center point (±30°), the instrument trajectory can be changed to other targets without moving the robot arm or repositioning the arm on the arch. The application module for clinical use provides manual translation and rotation of the cannula. A pneumatic drive has been developed to insert the cannula in incremental steps of 1–20 mm. Such automated cannula insertion is considered in the same product class as the interventional probe 2B or 3 (for applications near the cardiovascular or central nervous system). A graphical user interface provides trajectory planning directly on the MRI images (Fig. 25.12). MR digital imaging and communications in medicine (DICOM) images are sent via the picture archiving and communication system (PACS) to the robot personal computer (PC) and used for intervention planning and registration of the robot. A set of transverse and coronal 20 mm-thick slices are acquired, covering the contrast-agent-filled markers. Through graphical analysis of the marker alongside the coordinates of the slices listed in the DICOM header of the exported images, registration of the robot is performed.

Fig. 25.12 MR images are sent as DICOM files via PACS and can be used on the graphical user interface (GUI) of the robot for planning and registration (Prostate Biopsy, Zangos et al., University Frankfurt)

25.7.2 Technique for Robotic-Assisted MR-Guided Interventions

The patient is placed in a predetermined position suitable for the intervention (supine, prone or lateral). The system is prepositioned and firmly attached to the table with clamps. Based on the pre-interventional im-

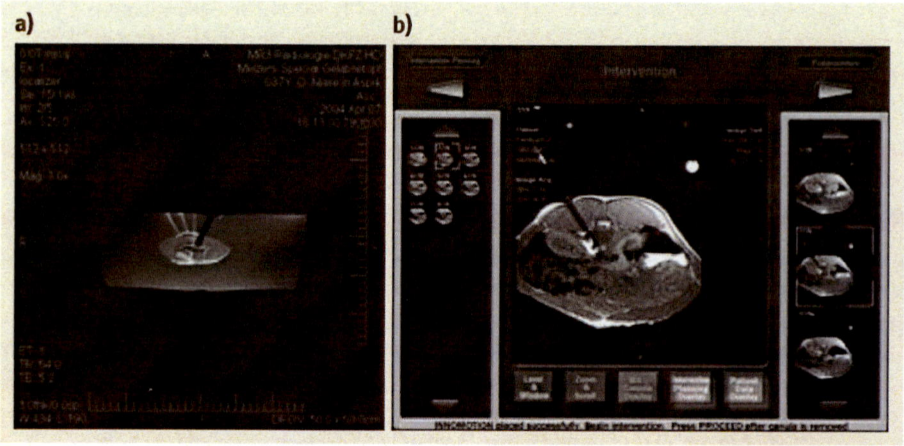

Fig. 25.13 **(a)** MRI target precision of a 20 gauge cannula insertion into a porcine kidney embedded in agarose. **(b)** Overlay images for graphical evaluation of target precision in an in vivo porcine model (Michael Bock et al. DKFZ, Heidelberg Germany)

ages and the anatomical region of interest, the table is moved using the projection of laser beams from the MRI gantry. In the first version the robot is referenced to the coordinate system of the MR scanner using the same laser line. The arm moves back and forth and returns so that the light detectors at the upper part of the application module are aligned with the laser (within ±0.5 mm). The laser light is switched off, and the table is automatically moved into the MRI bore until the position of the laser line matches with the zero position of the z-axis of the MRI scanner. The latest release just requires approximate manual positioning of the markers in the laser light. For planning of the intervention, MR imaging is performed by using fast gradient echo sequences in transverse, sagittal or coronal orientation. Suitable slices are selected and sent via the network in DICOM format to the computer of the robotic assistance system. The insertion site and a target point are selected using the graphical user interface, and the corresponding coordinates are sent to the control unit. The drives are activated, and the application module is moved with the tool center point to the insertion site on the skin. The cannula can then be inserted through a guiding sleeve or along an open angle.

25.7.3 Evaluation of Targeting Precision

Mechanical targeting precision has been determined with a FARO arm under dry-lab conditions. The MRI procedures were performed using a 1.5 T Siemens Magnetom Symphony, a Philips 1.0 T Gyroscan, and 1.5 T Intera. The test was done on ex vivo organ models consisting of fresh porcine kidney embedded in agarose and gelatine (Fig. 25.13).

Targeting precision was also evaluated during MRI-guided percutaneous interventions in a porcine animal model under general anesthesia (isoflurane). The animals (four 3 month-old domestic pigs, 30–40 kg) were placed in prone position on the patient table, and a surface coil was fixed around the planned insertion site lateral to the spine. Using T_1- and T_2-weighted planning images, the appropriate region of interest was defined using the graphical user interface of the Innomotion control computer. The robot arm then moved and oriented the needle holder to the insertion point automatically. Manual insertion of 20 and 22 gauge MR-compatible titanium grade 4 cannulae (MRI Devices-Daum, Schwerin, Germany) was carried out under real-time MRI guidance. The accuracy of the insertion point and the insertion angle were determined by overlaying the pre-interventional images with the new MRI image.

The intervention was completed within the magnet from the rear opening, where an MR-compatible in-room monitor was placed. During the insertion of the needle, real-time MR images were acquired to control the route of the cannula. To visualize the advancement of the cannula through the tissue, fast gradient echo sequences ($\mathrm{TR} = 4.4$ ms, $\mathrm{TE} = 2.2$ ms, $\mathrm{FA} = 70°$, $\mathrm{TA} = 0.7$ s) were used. At the desired region of interest (nerve root, plexus coeliacus), spin echo images were acquired for verification of the cannula position. A test injection of contrast agent solution (gadolinium diethylenetriamine penta-acetic acid 1 : 100 ml of 0.8% saline solution) was carried out under real-time MRI control ($\mathrm{TR} = 1.8$ ms, $\mathrm{TE} = 4.3$ ms, $\mathrm{TA} = 0.5$–0.8 s, $\mathrm{FA} = 20°$) to visualize the drug distribution. Final therapeutic injection of 10–25 ml with

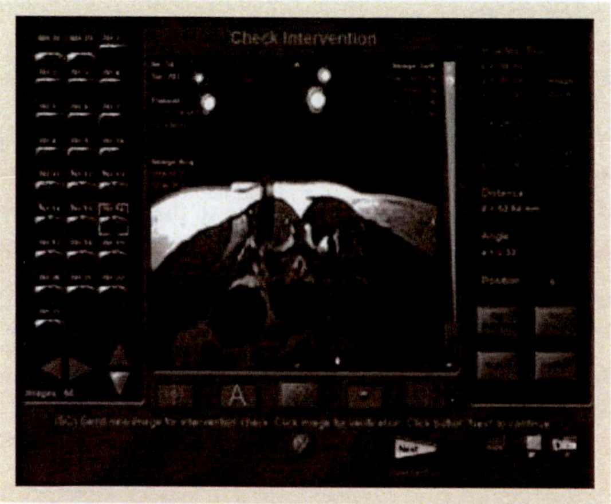

Fig. 25.14 MR guided facet joint pain treatment. Targeting precision can be evaluated by image overlay of planning line and needle artefact (1.5 T Philips Intera at Praxis Dr. Gert Lorenz, Gelsenkirchen, Germany)

contrast doped mepivacainhydrochloride (Scandicain 1%, Astra-Zeneca, Germany) was performed.

All procedures were completed successfully, including injections at the sympathetic chain, sciatic nerve, and coeliac plexus. The direct MRI control and new sequence techniques allowed correction of the insertion path in case of misdirection due to anatomical structures. The insertion site and the insertion angle were evaluated by manual measurements on overlays of the planning Innomotion image and the subsequent MR control image (Fig. 25.14).

The position and orientation of all cannula insertions were appropriately visualized on axial MRI images. The precision of the insertion site in the axial plane was ±1 mm (minimum 0.5 mm, maximum 3 mm). The angular deviation in the transverse plane of the cannulae was ±1° with a minimum of 0.5° and maximum of 3°.

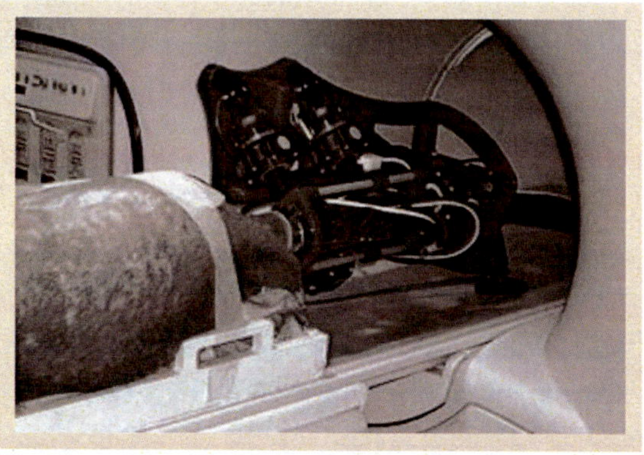

Fig. 25.15 Johns Hopkins MR-compatible robot in an animal procedure for prostate interventions (after [25.32])

For MRI-guided cannula interventions, as the cannula is currently advanced manually, access is difficult if the insertion is done inside the magnet. Therefore, direct control of the insertion under real-time MRI is recommended to allow correction of the insertion in case of misdirection of the cannula and for precise positioning of the tip of the cannula in the volume of interest. To facilitate the procedure, tip-tracking techniques have been evaluated to provide automated correlation of the robot position with the image slice orientation [25.24].

25.7.4 MRI-Compatible Robot MrBot for Prostate Biopsy

MrBot [25.33] has been developed at Johns Hopkins for image-guided access to the prostate gland. The robot is designed for transperineal cannula insertion for closed-bore tunnel-shaped scanners. The system accommodates various end-effectors for percutaneous interventions such as biopsy, serum injection, or brachytherapy with low-dose radiation seed implantation. For MRI safety the robot is constructed exclusively of nonmagnetic, dielectric materials such as plastics, ceramics, and elastomers and is free of electrical conducting wires. The system is driven by novel pneumatic step motors (PneuStep) [25.32]. This unique motor provides easily controllable, precise, safe pneumatic actuation comparable to electric step motors. Fiber-optic encoding is used for position detection, so that all electric components are located outside the MR room (Fig. 25.15).

The precision of repeated motion has been evaluated in a 1.5 T MRI scanner, showing mean errors of 0.076 mm. The robot has been tested as MR safe in all MR systems available at the time of development [25.32]. Clinical use and regulatory approval remain to be conducted. The robot has been evaluated for cannula insertion accuracy with in vitro and ex vivo experiments and in vivo animal experimental studies [25.34, 35].

25.7.5 Technical Issues of MR Robotics

The following main technical issues require careful consideration

- MR imaging safety and procedures compatibility (Chap. 26),

- coregistration of robot and image position,
- compensation of patient movement and respiration,
- options for haptic (force) feedback,
- mode of operation and level of robotic control.

MR Imaging Safety and Procedures Compatibility

MRI safety can only be achieved by using nonmagnetic and nonconductive materials. For use in computed tomography, radiolucency of the robot arm can be accomplished by the same materials so that the robot end-effector holding the instrument in the scan plane does not cause undesired artifacts. The robot system must also provide safe interfacing with the imaging system and allow rapid displacement for immediate access to the patient in emergency situations. When the robot system is actuated, it should not interfere with the imaging system and should allow passive displacement at all times.

Coregistration of Robot and Image Position

Intervention planning is based on images in certain planes and orientations. The coordinate system of the robot must therefore be registered to the coordinate system of the imaging system. If the robot is permanently attached to the patient table of the imaging device, this registration can be done through an imaging procedure or through microcoil-based tracking. Microcoil tracking has the benefits of being very fast and accurate and providing continuous registration, and immediate reregistration in case the device is moved from one imaging unit to another or shifted between two rooms.

Compensation of Patient Movement and Respiration

A limiting factor of robotic assistance during image-guided interventional procedures is inadvertent patient movement and organ displacement due to respiration motion. Compensation of these effects would first require real-time tracking of target movement due to respiration motion and accurate transformation to motion of the robot. Making such a robot clinically viable, robust, and safe would result in very high cost.

Industrial robotic systems react faster than a human and are able to compensate respiratory motion, such as the CyberKnife (Accuracy, Sunnyvale) and linear accelerator held by an industrial robot arm [25.36]. This massive, metal robot driven by powerful electric motors is not suitable for MRI-guided procedures. Therefore, other safety means need to be realized that avoid risk of instrument displacement caused by patient motion. The safest approach is to provide for mechanical release of the instrument from the holder.

Options for Haptic (Force) Feedback

Robotic systems with master–slave functionality that include active needle-driving mechanisms and force feedback may need to be provided to the operator. The importance of this feedback is a subject of ongoing debate, but there are some clinical applications where it seems desirable, e.g., if the cannula touches bone. The technical hurdle is, however, to measure the force at the tip of the instrument accurately and to transmit this to a master control unit. Existing force feedback devices are too bulky and not approved for the clinical environment. In addition, friction between the cannula and tissue during insertion is high, which compromises the accuracy of force feedback measurements [25.37]. Therefore, assessment of the need for force feedback remains at the clinical research level.

Mode of Operation and Level of Robotic Control

The adequate user interface for an interventional robot has not yet been determined. Joystick control should be designed for master–slave systems to mimic the classic handling position of a probe during conventional interventions. Scaling of motion and applied force would be helpful only if fine manipulation of tissues is required, which is not the case in interventions. During procedures such as biopsy with a straight-line trajectory through predictable soft tissue, some degree of autonomy seems appropriate if sufficient robustness can be achieved. During percutaneous radiotherapy procedures, radioactive seeds or probes are implanted into the patient, which could be performed by the robot. The main element of autonomous control is tracking of position and coregistration of MR slice orientation and robot front-end, and vice versa. The robot should also move automatically to the skin insertion point.

Undoubtedly, MR robotics carries the potential to facilitate MR-guided interventions and provide the required access to the patient while lying in the scanner bore. Successful clinical application of MR robotics, however, requires intuitive systems operation and minimal user training. The robot must be quick and easy to setup and not significantly increase the length of procedures, and preferably shorten it. The robot must also be cost effective.

At the University of Dundee, Innomotion in combination with a MR-guided focused ultrasound system (MRgFUS, Fig. 25.16) is being applied and further developed (Sect. 25.8).

Fig. 25.16 MR-compatible robotic system in combination with a MR-guided focused ultrasound system (ExAblate 2001, InSightec, Tirat Carmel, Israel)

25.8 Hybrid Multimodal Imaging for MR-Guided Diagnosis and Therapy

Despite the excellent diagnostic imaging capabilities of MRI, which does not expose patients and physicians to ionizing radiation, it is difficult to use during intervention and surgery due to spatial constraints and the strong electromagnetic fields. Ultrasound and optical imaging are harmless compared with ionizing radiation-based imaging and easier to apply compared with MRI. Ultrasound and the emerging field of biophotonic imaging are usually not MRI compatible and not designed to be used in conjunction with MRI. As a consequence, surgeons and interventionists continue to use traditional imaging and operating techniques, which are based on visual inspection of the pathological anatomy structure either by traditional open or endoscopic surgery. Radiologists still have to expose themselves and their patients to long radiation times to perform complex cardiac interventional procedures. The frequent inadequacy of fluoroscopy as a navigation guide and failure to visualize the pathological target results in long procedure times and the use of nephrotoxic and allergenic iodine contrast agents. In Germany, 1.5% of the cumulative risk of cancer has been attributed to diagnostic x-rays [25.38].

MR provides superior soft tissue contrast in different weightings and functional imaging for visualization of pathological lesions. Positron emission tomography (PET), however, visualizes biochemistry, e.g., lactate concentration or detection of specific molecular targets. Combining MR with PET/CT information and biophotonics for local imaging in combination with ultrasound for instrument guidance could therefore improve image-guided diagnosis and treatment precision. With this functional and operational integration, both time to diagnosis and time to operation or intervention could be significantly reduced, and some of the technical hurdles of MRI safety and access limitations could be avoided. Various efforts have been undertaken to make instruments and devices available for procedures performed by surgeons and interventionists, but only a few sites worldwide have so far introduced MRI/PET/CT to ORs despite the very promising outlook.

Because of the needs for intraoperative imaging at different stages of a procedure, the concepts of intra-, inter-, and perioperative MR imaging have developed [25.39]. The objective is to apply imaging immediately after or during an operation for economic and safety reasons and to improve the necessary techniques and tools. For this purpose a teaching and research operation system for integration of imaging procedures was established in 2001 at the Fachhochschule Gelsenkirchen, Germany [25.40]. The complex requirements for the integration of imaging procedures should include computer-based numerical simulation [25.41]. A first implementation of this ap-

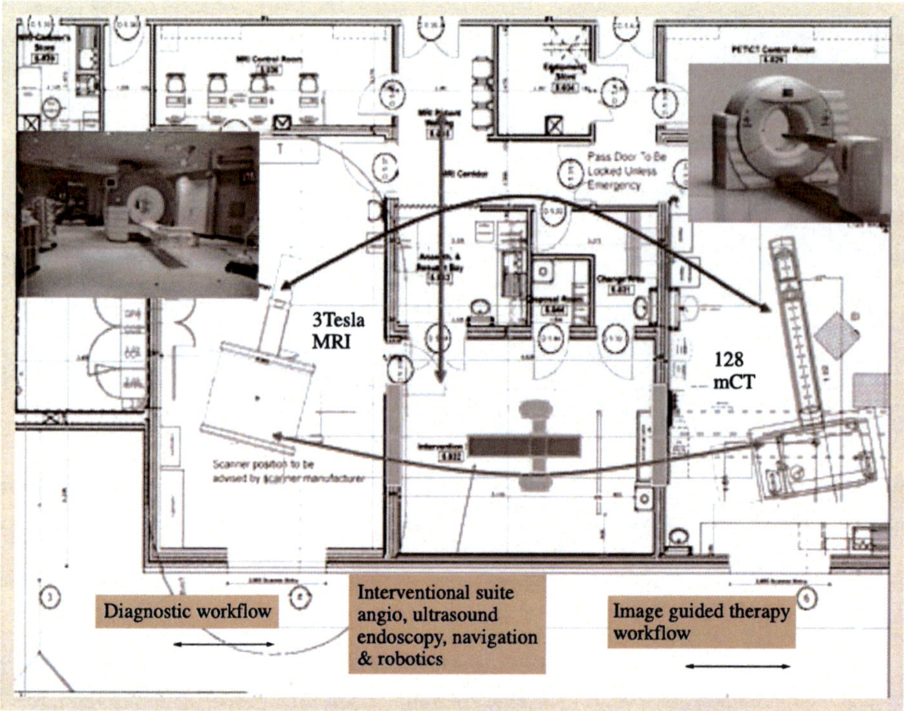

Fig. 25.17 Integrated multimodality image-guided diagnosis and therapy setup (MITOS) at the Clinical Research Center of University Dundee & NHS, Tayside, Scotland

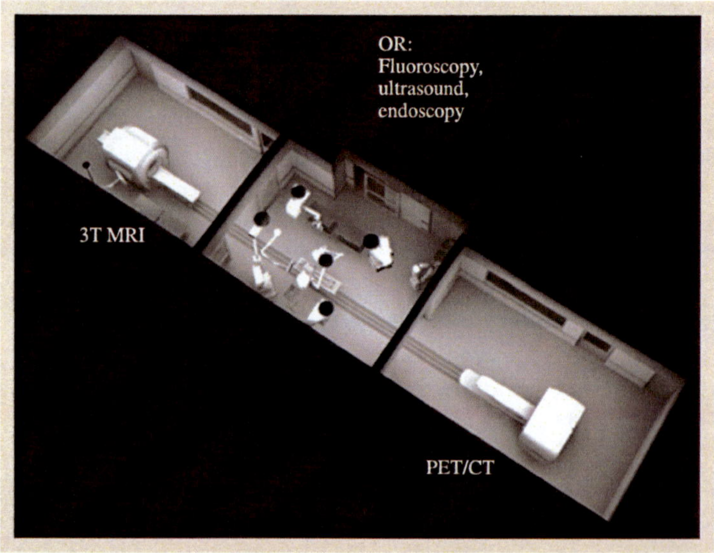

Fig. 25.18 Advanced multimodality image-guided OR (AMIGO) 1st GE concept by *Jolesz* and *Hynynen* (after [25.5]) (BWH, Harvard Med School, Boston)

proach in the clinical setting is currently underway at the University of Dundee, Scotland, by combining 3 T MRI and PET/CT with a multifunctional intervention/surgical suite (MITOS) (Fig. 25.17). The focus is on development of multimodal image-guided diagnostic and therapeutic procedures for early detection and treatment of cardiovascular and oncological diseases [25.38]. The work is closely linked to clinical requirements, supported by ISO-certified processes, and continuously assessed in view of product capability and marketing.

Jolesz and *Hynynen* [25.5] have developed a similar concept named AMIGO with GE for integration of MRI and PET/CT with surgery (to be installed at BWH, Harvard Medical School, Boston, MA) (Fig. 25.18), and Philips has announced the first XMRO suite in Japan, combining MR, Angiolab, and CT.

Installation of multimodality interventional imaging systems is difficult in the current clinical setting, and thus the potential benefit to patient care and the healthcare system is not being realized. One (and perhaps the major) reason for this is that the implementation and safety of intraoperative imaging require an interdisciplinary (*interclinical* and *inter-technical-science*) and cross-sectorial approach (academia and industry), which is difficult to establish. It is therefore the intention of the Integration of Interventional Imaging in the Future Operating System (IIIOS) consortium (www.IIIOS.eu) and the National Center for Image Guided Therapy (NCIGT) (www.NCIGT.org) to undertake research and training to develop dedicated technology, processes, and skills for safe, efficient, effective, and economic implementation of magnetic resonance (MR) with ultrasound and biophotonics in combination with imaging from computed tomography (CT) and positron emission tomography (PET).

25.9 Therapeutic MR-Guided Imaging

MRI-guided focused ultrasound (MRgFUS) is a method that combines thermal ablation with MRI. The ExAblate (InSightec, Tirat Carmel, Israel) delivers high-intensity focused ultrasound energy precisely directed from an extracorporeal (phased-array) transducer towards the target tissue through the intact skin, resulting in irreversible coagulative necrosis and apoptosis when temperature above 56 °C is maintained for at least 1 s (Fig. 25.19).

Real-time MR thermometry allows continuous temperature monitoring and precise case-dependent customization of the therapeutic energy at the focus. Due to the sharp drop in energy density outside the treatment focus, the resulting lesion is highly discrete, sparing tissues adjacent to the focus, and enabling an outstandingly high safety profile compared with other ablative therapies. The main advantages of this combination are the good delineation of the tumor for target definition, the exact three-dimensional treatment planning, and continuous temperature measurement for control of local warming. MRgFUS, which allows access to noninvasive thermal ablation, is likely to represent an alternative to interventional ablation, surgical resection, and radiation therapy [25.42]. The method is approved in Europe, Japan, and the USA for treatment of uterine fibroids and is being evaluated in clinical trials for treatment of breast, liver, prostate, and brain cancers as well as for relief of pain associated with bone metastases. Its use for bone metastases has already received CE approval [25.43].

A promising development is MR-guided, ultrasound-induced targeted drug delivery (UITDD) [25.44] in

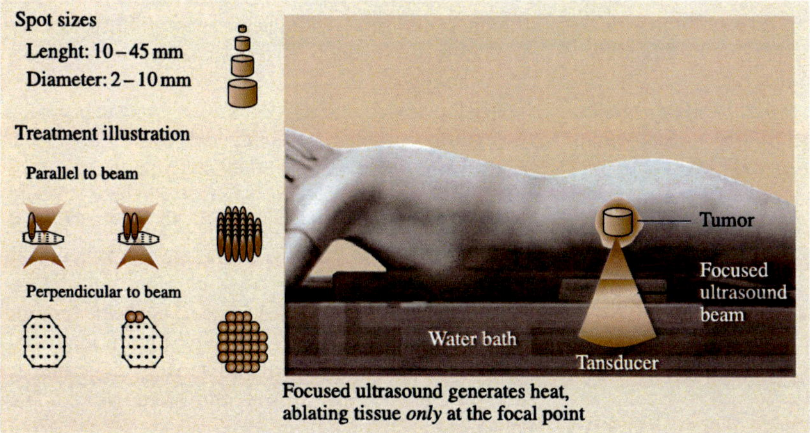

Fig. 25.19 Principle of noninvasive and no-access MR-guided focused ultrasound (MRgFUS) ablation of tumors (ExAblate 2001, InSightec, Tirat Carmel, Israel)

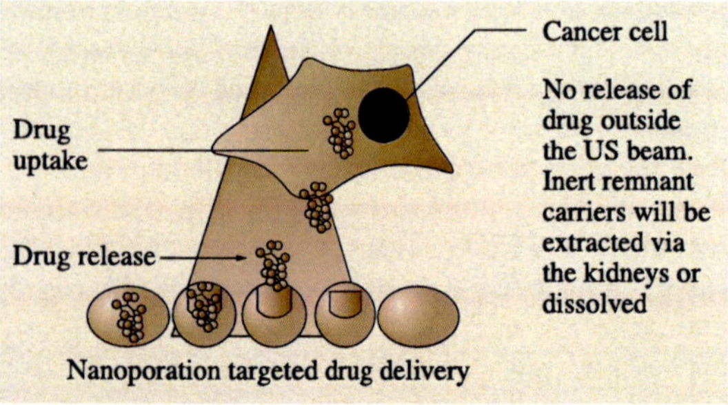

Fig. 25.20 Concept of MR-guided focused ultrasound-mediated targeted drug delivery of chemotherapy agents (www.nanoporation.eu)

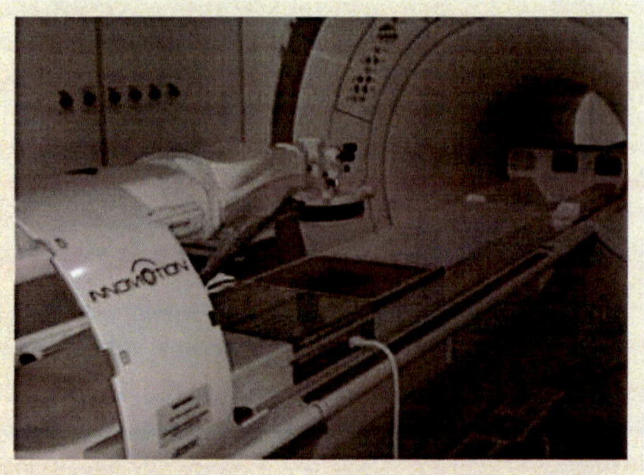

Fig. 25.21 Innomotion robot test setting on the ExAblate 2000 MRgFUS system (InSightec, Tirat Carmel, Israel), positioning the conformable bone system mobile FUS transducer ExAblate 2001 (IMSaT, Dundee, UK)

which focused ultrasound induces a temporary change in vascular and cell membrane permeability. This principle is described as sonoporation [25.45] and can lead to release of drugs from drug carriers; for example, liposomes [25.46] or polymers [25.47] are used (Fig. 25.20). The aim of current research in this area is combination of MRgFUS tumor ablation with subsequent drug release. This can result in a higher concentration of active drug in the treatment zone while maintaining a low systemic concentration and fewer side-effects.

From a clinical perspective, the combination of PET and MRI is desirable to allow better correlation of PET signals to the morphology and anatomy of the patient. A PET scan may give direct evidence of remaining active tumor tissue following ultrasound-guided ablation. Remnant tumor tissue can either be ablated or, if this is not possible due to adjacent important anatomical structures, MR-guided focused ultrasound-mediated drug delivery can be applied to provide high focal concentration of potent chemotherapy agents. The Innomotion robotic arm is being investigated as a positioning aid for the MRgFUS system (Fig. 25.21), which may be useful in ultrasound-mediated drug delivery and in tumor ablation.

25.10 MR-Guided Delivery of Implants

Implant delivery under image guidance is mainly used for endovascular implants such as stents, filter, valves, or occluders. Tumor markers are implanted via image guidance in breast and lung tumors to provide the surgeon with guidance during subsequent surgical excision of the marked lesion. Non-magnetic stainless-steel medical alloys (e.g. 316L) cause major MR image artifacts due to the material (susceptibility artifact) and/or its elctromagnetic characteristics (RF artifacts) which are undesirable in instruments for MR-guided interventional procedures. These artifacts are caused by distortion of the magnetic field and interference with the radiofrequency (RF) waves of the MR imaging process. Complete signal loss occurs in close proximity to or inside such implants (Chap. 26). The unconventional electromagnetic characteristics of nitinol (50 : 50 nickel and titanium) render it suitable for endovascular implants for MR imaging and guidance. Although nickel is a ferromagnetic metal, when combined as an intermetallic compound with a cubic-centered crystal structure of titanium and nickel atoms, it loses its magnetic moment. nitinol shows similar magnetic susceptibility to other medical-grade titanium alloys [25.48].

Although nitinol was considered the perfect choice for MRI, research into its MRI safety in recent years has revealed the problem of conductive heating of metallic implants, which occurs at about 10 cm length in 3 T MRI imaging and at about 20 cm length in 1.5 T MR systems. Most implants are below the critical length, but care has to be taken if stents are implanted with overlap and for very long stents designed for abdominal aortic aneurysm (AAA) and superficial femoral artery (15–20 cm length). Heating of more than 4 °C has been measured in phantom models according to American Society for Testing and Materials (ASTM) standards [25.16].

A secondary risk for the patient is the problem of missing MR signals in the lumen of the implants leading to lack of diagnosis. The signal void is caused by the closed-cell design of the implant structure, which functions like a Faraday cage. The MR signal cannot be received from the substrate in the lumen, thus making MRI detection of instant thrombosis and restenosis or other intra-stent complications impossible. By integrating a MRI antenna (electric resonator tuned to the resonance frequency of the MRI, i. e., 64 MHz for 1.5 T) into the stent and other vascular implants, i. e., heart valve prosthesis, improved imaging is possible, as has been proven in various animal trials.

25.10.1 MR-Guided Delivery of Stents and Stent Grafts

A variety of reports on MR imaging behavior of stents have been published in recent years. MRI in the presence of stents has been examined in terms of contrast enhancement [25.51, 52], influence of different materials and structures [25.53], and stent grafts [25.49, 50] under particular MRI settings. The effects of 0.2, 1.0, and 1.5 T sequences and imaging parameters, the orientation of the stent axis in the static magnetic field, and the orientation of the recorded slice have been systematically assessed [25.22, 54].

Preclinical evaluation of an artifact-free copper alloy based balloon expandable stent, invented by *Bücker* and *Rübben*, has been carried out. The concept of their stent is a design that avoids the RF shielding effect [25.55]. Although the stent has not shown any image interference, its complete lack of visualization does not allow guidance during delivery and makes later diagnostic MR difficult. In addition, the copper alloy is not biocompatible. MRI-guided implantation of coronary stents has been demonstrated in a porcine animal model [25.56].

The first clinical report of MR-guided stent delivery was published in 2000 [25.57]. The group used nitinol self-expanding stents, implanted under MR guidance in a 1.5 T system. MRI-guided implantation of AAA grafts (for AAA repair) has been performed in an animal trial on pigs with success [25.58, 59].

All self-expanding nitinol stents and stents for grafts show an artificial diameter increase in MRI imaging, and shielding of the stent lumen. To overcome the RF artifacts, stents have been enabled to function as electric resonators to interact with the MR imaging process [25.60]. The balloon expandable version of the resonant stents has been evaluated in a chronic animal trial, and the MR enhancing function was still present up to 7 months after implantation [25.61]. Current research and development (R&D) effort includes the design of a nitinol stent comprising a resonant circuit as a discrete component, comparable to an x-ray marker (Fig. 25.22). This stent will provide MR-guided delivery with control of the implantation process. As described by (25.3), the resonance frequency is a function of the inductivity (L) and the capacity (C). The inductivity is dependent on the surface and the length of the windings of the coil in and around the stent. The smaller the surface encapsulated by the coil, the lower the resonance frequency. The stent's resonance frequency is tuned at the expanded stage to its optimum, i. e., 64 MHz for 1.5 T. For implantation, the stent is either crimped onto a balloon or constrained within a delivery system. At this stage, the resonance frequency is below the MR Larmor frequency (25.1). During expansion, the inductor achieves its preset shape and at the same time the resonance frequency tunes to the preset value, i. e., 64 MHz for 1.5 T. As a result, the stent starts enhancing the MR signal (Fig. 25.23). This technique carries the greatest potential for adequate visualization of stents in MRI.

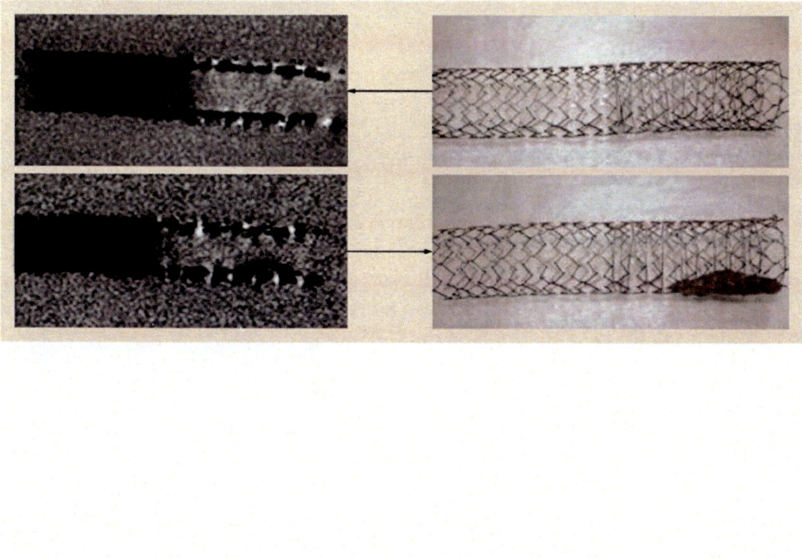

Fig. 25.22 MRI imaging of self-expanding nitinol stent. To overcome RF artifacts, the stent has been equipped with an electric resonator (fine wire coil on the right side) tuned to 64 MHz to couple inductively with the MR imaging process of a 1.5 T scanner (after [25.49, 50]) (www.vueklar.com)

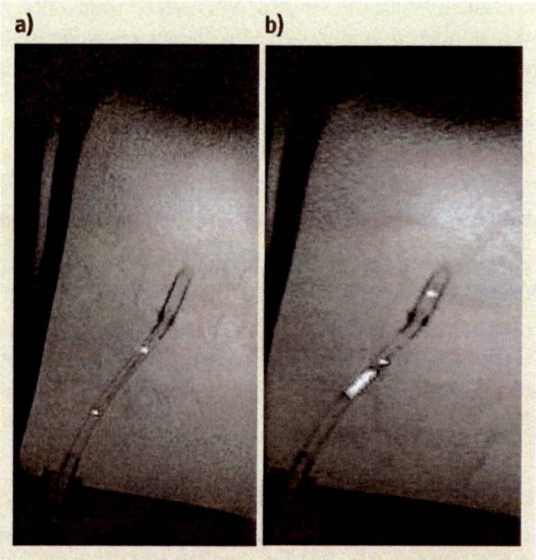

Fig. 25.23 **(a)** The crimped stent is invisible, but after balloon expansion the resonance frequency tunes to the preset value, i.e., 64 MHz for 1.5 T. **(b)** As a result the stent starts enhancing the MR signal (Siemens 1.5 T, University Frankfurt) ◀

Artifacts can be minimized by using nitinol with suitable magnetic susceptibility (Chap. 26) and a design of the holding structural element which avoids electrical conductive loops. Due to the mechanical requirement the design cannot be made loop-free and nonconductive. A resonant circuit tuned to the Larmor frequency of the MR tomography can overcome the RF artifacts and thus improve visualization of the resonant prosthetic heart valve, similar to the stents described above. We have developed a first set of prototypes using a nitinol stent and freshly excised porcine aortic valve (Fig. 25.24) [25.62]. The valves have been successfully implanted under MRI guidance (1.5 T) in an acute porcine animal model [25.31].

25.10.2 MR-Guided Heart Valve Prosthesis

Artifact-free visualization of heart valve prosthesis is one of the main conditions for safe and reliable examination with MRI and a prerequisite for MRI-guided valve repair.

25.10.3 MR-Guided Delivery of Vena Cava Filter

MRI artifacts of a vena cava filter (VCF) are dependent on both the susceptibility and the RF characteristics of the design. Closed loops of the metal struts are

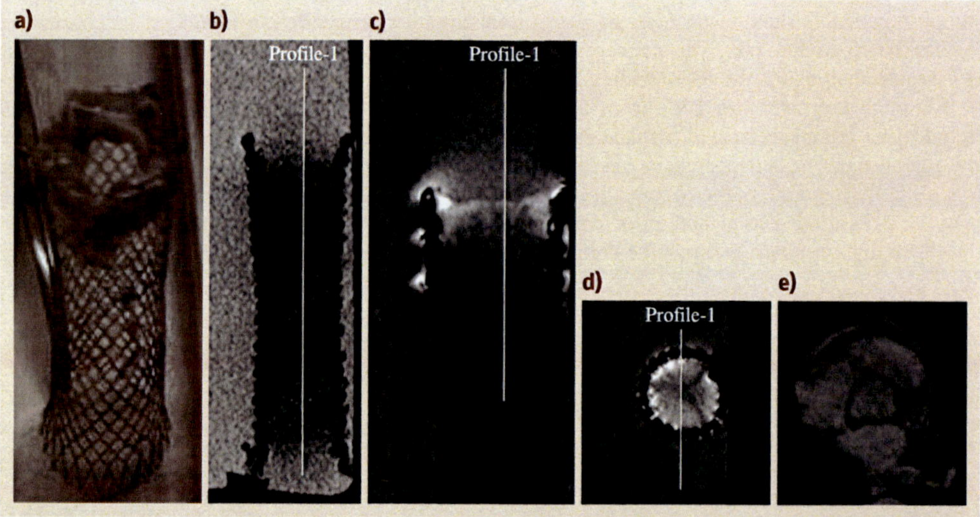

Fig. 25.24 **(a)** MR-visible self-expanding heart valve prototype using a nitinol $22\times60\,\mathrm{mm}^2$ stent and a freshly excised porcine aortic valve (after [25.27]). The nonresonant section is shielding the lumen **(b)**. The section with the bright signal contains the resonant circuit **(c, d)** and shows the valve leaflets in detail. Panel **(d)** shows the enhanced MR image of a resonant heart valve compared to the non resonant MR image of a native human valve. **(e)** The MRI-signal intensity can be measured along the profile lines shown in **(b–d)**

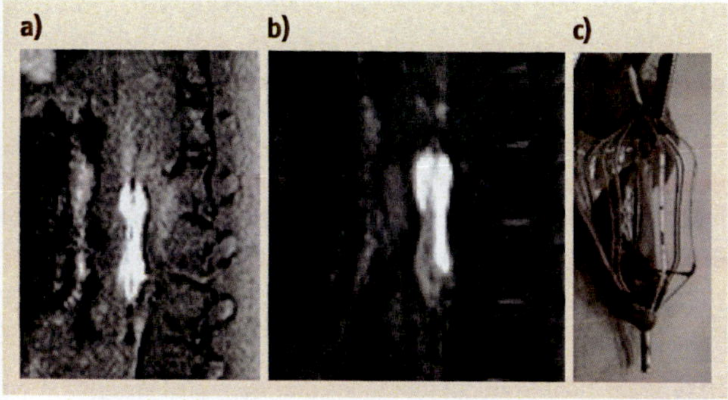

Fig. 25.25 (a) Basket-type nitinol vena cava filter demonstrates a sectional MR signal void because of a blood clot. (b) Integration of a resonant circuit provides improved imaging and depiction of clotted blood in the filter. Panel (c) shows a photo of the blood clot visualized in MRI (b) (1.5 T at RWTH, Aachen, after [25.65])

subjected to eddy-current induction and local magnetization. The TrapEase and OptEase VCFs are cut from nitinol tube in a structure of interconnected closed cells, leading to RF shielding [25.63, 64]. Two types of active MRI VCFs with integrated resonant circuits have been produced: wire based and laser cut from nitinol tube (Fig. 25.25). The active VCFs have been successfully tested in an acute porcine animal model for evaluation of MRI-guided implantation, blood clot imaging, and retrieval of the filter [25.65].

25.10.4 MR-Guided Delivery of Closure Devices for Cardiac Septal Defects

MR imaging of cardioseptal defects is an option to avoid ionizing radiation, invasive catheterization, and transesophageal ultrasound specifically for pediatric patients, although the adult patient who has received a patent foramen ovale (PFO) closure also benefits from MRI during follow-up [25.66]. MRI-guided delivery of closure devices has been successfully evaluated in an animal model [25.67] using a custom-made passive nitinol occluder.

Closure devices have been examined by MRI, e.g., the Amplatzer from AGA medical [25.22] and Starflex from NMT. The Starflex is made of stainless-steel wire, causing major artifacts, while the nitinol mesh of the Amplatzer also contributes to RF shielding with low artifacts, although the proximal threaded tube and the distal tube that are used for clamping both ends of the mesh lead to large artifacts (Fig. 25.26). Novel intratunnel devices for PFO closure such as SeptRx (NDC, Fremont, CA, USA) and Coherex feature smaller artifacts, mainly because less material is present and no stainless-steel components are applied as structural components. PFO compression via two rings connected through the PFO tunnel should provide safe sealing, and the integration of resonant circuits is equally helpful to overcome shielding artifacts and facilitates cardiac MR imaging for PFO leakage detection. During delivery the two rings are constrained, which detunes the resonance frequency; during expansion the frequency reaches 64 MHz, leading to signal enhancement.

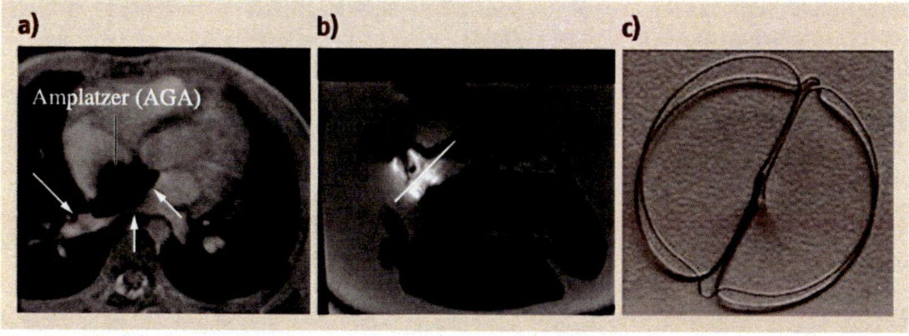

Fig. 25.26 (a) The cardioseptal occluder of AGA Medical (Amplatzer) is made of nitinol mesh connected at both ends by stainless-steel tubes which generate significant MR artifact. (b) Novel nitinol dual-ring occluder functions as a resonant circuit to overcome shielding artifacts and to facilitate cardiac MR imaging and assessment of PFO occlusion (c). Profile line was positioned on the occluder in the MR image to measure the MR signal intensity

25.11 Conclusions

Safe and efficient MR-guided interventions can be realized if important criteria of MRI safety and imaging compatibility are applied to the development of appropriate instrument systems. Today's potential indications for interventional MRI are in the field of the classic radiologic procedures, i. e., organ puncture for biopsy, interstitial tumor treatment, and vascular procedures. Intraoperative MRI has been proven beneficial in neurosurgery, leading to improved tumor resection, and cardiovascular surgery under MRI guidance is currently been explored. The advantages of MR application lie in the variable soft tissue contrast with flow and temperature sensitivity, the arbitrary slice orientation, and the lack of ionizing radiation and nephrotoxic contrast media. The majority of interventional and intraoperative MRI hardware and software components, devices, and systems are technically suitable and clinically feasible, providing an interesting key technology for the medical industry. Although thousands of MR-guided interventions and surgical procedures have been performed within the last 15 years, the ongoing lack of approved medical devices for MRI-guided interventions, particular in the vascular field, hinders widespread introduction. As soon as clinical feasibility has been proven and positive results of the evaluation of clinical procedures become public knowledge, patients' attending physicians will certainly ask for the MRI option, mainly because of the lack of radiation exposure. This can create a situation in which the patient drives the medical market (a so-called patient-driven market). The same was observed for laparoscopic cholecystectomy. Here, with the introduction of video endoscopy 20 years ago and especially with the publication of the first 100 laparoscopic cholecystectomies by *Périssat* [25.68], rapid spread occurred. Today, hardly any patient would choose surgery by conventional open approach. This possibility runs the risk of too rapid an introduction of new MRI-guided treatments and technologies. However, on the other hand, conservative structural hurdles may be overcome faster. It is therefore necessary to answer questions about the safety, efficiency, effectiveness, and economy of MR-guided interventions within multidisciplinary teams and prove these aspects in valid clinical studies.

References

25.1 A.L. Jacob, P. Regazzoni, D. Bilecen, M. Rasmus, R.W. Hügli, P. Messmer: Medical technology integration: CT, angiography, imaging-capable OR-table, navigation and robotics in a multifunctional sterile suite, Minim. Invasive Ther. Allied Technol. **16**, 205–211 (2007)

25.2 P.C. Lauterbur: Image formation by induced local interactions: Examples of employing nuclear magnetic resonance, Nature **242**, 190–191 (1973)

25.3 V. Rieke, P.K. Butts: MR thermometry, J. Magn. Reson. Imaging **27**, 376–390 (2008)

25.4 T.J. Vogl, M.G. Mack, R. Straub: Percutaneous MRI-guided laser-induced thermotherapy, Minim. Invasive Ther. Allied Technol. **6**, 53–64 (1997)

25.5 F.A. Jolesz, K. Hynynen: Magnetic resonance image-guided focused ultrasound surgery, Cancer J. **8**, 100–112 (2002)

25.6 J. Kettenbach, W.S. Grundfest: Cost-effectiveness of image-guided surgery, Acad. Radiol. **5**(Suppl. 2), 428–431 (1998)

25.7 J. Kettenbach, D.F. Kacher, A.R. Kanan, B. Rostenberg, J. Fairhurst, A. Stadler, K. Kienreich, F.A. Jolesz: Intraoperative and interventional MRI: Recommendations for a safe environment, Minim. Invasive Ther. Allied Technol. **15**, 53–64 (2006)

25.8 J.S. Lewin: Interventional MR imaging: Concepts, systems, and applications in neuroradiology, Am. J. Neuroradiol. **20**, 735–748 (1999)

25.9 W.R. Nitz, A. Oppelt, W. Renz, C. Manke, M. Lenhart, L. Johann: On the heating of linear conductive structures as guide wires and catheters in interventional MRI, J. Magn. Reson. Imaging **13**, 105–114 (2001)

25.10 W.R. Nitz, G. Brinker, G. Diehl, G. Frese: Specific absorption rate as a poor indicator of magnetic resonance-related implant heating, Invest. Radiol. **40**, 773–776 (2005)

25.11 R. Luechinger, V.A. Zeijlemaker, E.M. Pedersen, P. Mortensen, E. Falk, F. Duru, R. Candinas, P. Boesiger: In vivo heating of pacemaker leads during magnetic resonance imaging, Eur. Heart J. **26**, 376–383 (2005)

25.12 A. Bücker: Safety of MRI-guided vascular interventions, Minim. Invasive Ther. Allied Technol. **15**, 65–70 (2006)

25.13 D.G. Thomas, C.H. Davis, S. Ingram, J.S. Olney, G.M. Bydder, I.R. Young: Stereotaxic biopsy of the brain under MR imaging control, Am. J. Neuroradiol. **7**, 161–163 (1986)

25.14 J.S. Lewin, J.L. Duerk, V.R. Jain, C.A. Petersilge, C.P. Chao, J.R. Haaga: Needle localization in MR-

guided biopsy and aspiration: Effects of field strength, sequence design, and magnetic field orientation, Am. J. Roentgenol. **166**, 1337–1345 (1996)
25.15 C. Manke, W.R. Nitz, B. Djavidani, M. Strotzer, M. Lenhart, M. Völk, S. Feuerbach, J. Link: MR imaging-guided stent placement in iliac arterial stenoses: A feasibility study, Radiology **219**, 527–534 (2001)
25.16 G. Schäfers, A. Melzer: Testing methods for MR safety and compatibility of medical devices, Minim. Invasive Ther. Allied Technol. **15**, 71–75 (2006)
25.17 S. Kos, R. Hügli, E. Hofmann, H.H. Quick, H. Kuehl, S. Aker, G.M. Kaiser, P.J. Borm, A.L. Jacob, D. Bilecen: MR-compatible polyetheretherketone-based guide wire assisting MR-guided stenting of iliac and supraaortic arteries in swine: Feasibility study, Minim. Invasive Ther. **18**, 1–8 (2009)
25.18 A. Melzer, A.A. Schmidt, K. Kipfmüller, G. Grönemeyer, R. Seibel: Technology and principles of tomographic image-guided interventions and surgery, Surg. Endosc. **11**, 946–956 (1997)
25.19 A. Melzer, M. Wendt, R. Seibel, B. Kipfmüller, J. Lewin: Experimental intra- and interoperative MRI during laparo-endoscopic surgery, Radiol. Suppl. **217**, 1051 (1999)
25.20 A. Melzer, M. Scholz, R. Seibel: Feasibility of MR guided neuroendoscopy. In: *Interventional MRI*, ed. by R.B. Lufkin, D.H.W. Grönemeyer (Mosby, New York 1999) pp. 154–160
25.21 S.S. Chopra, G. Wiltberger, U. Teichgräber: Evaluation of laparoscopic liver resection with two different Nd:YAG lasers for future use in a high-field open MRI, Photomed. Laser Surg. **27**, 281–286 (2009)
25.22 M. Burl, G.A. Coutts, I.R. Young: Tuned fiducial markers to identify body locations with minimal perturbation of tissue magnetization, Magn. Reson. Med. **36**, 491–493 (1996)
25.23 J. Kettenbach, G. Kronreif, A. Melzer, G. Fichtinger, D. Stoinavoici, K. Cleary: Ultrasound, CT, MRI guided robotic interventions. In: *Image Processing in Radiology: Current Applications*, ed. by E. Neri, D. Caramella, C. Carlo Bartolozzi (Springer, Heidelberg, New York 2008) pp. 393–409
25.24 H. Zimmermann, S. Müller, B. Gutmann, A. Melzer, W. Nitz, W. Semler, M. Bock: Targeted HASTE (TASTE) imaging with automated device tracking for MR-guided needle interventions in closed-bore MR-systems, J. Magn. Reson. Imaging **53**, 481–488 (2006)
25.25 A. Melzer, M.O. Schurr, W. Kunert, G. Bueß, U. Voges, J.U. Meyer: Intelligent surgical instrument system ISIS. Concept and preliminary experimental application of components and prototypes, Endosc. Surg. Allied Technol. **1**, 165–170 (1993)
25.26 K. Cleary, A. Melzer, V. Watson, G. Kronreif, D. Stoianovici: Interventional robotic systems: Applications and technology state-of-the art, Minim. Invasive Ther. Allied Technol. **15**, 101–113 (2006)
25.27 S. Zangos, A. Melzer, K. Eichler, C. Sadighi, A. Thalhammer, B. Bodelle, R. Wolf, T. Gruber-Rouh, D. Proschek, R. Hammerstingl, C. Müller, M.G. Mack, T.J. Vogl: MR-compatible assistance system for biopsies in a highfield system: Initial results in patients with suspicious prostate lesions, Radiology **259**, 903–910 (2011)
25.28 A. Melzer, B. Gutmann, T. Remmele, R. Wolf, A. Lukoscheck, M. Bock, H. Bardenheuer, H. Fischer: INNOMOTION for percutaneous image-guided interventions, IEEE Eng. Med. Biol. Mag. **27**, 66–73 (2008)
25.29 M. Moche, D. Zajonz, T. Kahn, H. Busse: MRI-guided procedures in various regions of the body using a robotic assistance system in a closed-bore scanner: Preliminary clinical experience and limitations, J. Magn. Reson. Imaging **31**, 964–974 (2010)
25.30 M. Wiewiorski, V. Valderrabano, M. Kretzschmar, H. Rasch, T. Markus, S. Dziergwa, S. Kos, D. Bilecen, A.L. Jacob: CT-guided robotically-assisted infiltration of foot and ankle joints, Minim. Invasive Ther. Allied Technol. **18**(5), 291–296 (2009)
25.31 A. Melzer, K. Horvath, M. Guttman, M. Li, E. Immel, R. Lederman, J. Spilner, C. Lorenz, E. McVeigh: MR-guided transapical delivery of a self expanding heart valve prosthesis equipped with a resonant circuit: Initial experiences in porcine model, ISMRM Proc., Vol. 219 (2006)
25.32 D. Stoianovici: Multi-imager compatible actuation principles in surgical robotics, Int. J. Med. Robot. Comput. Assist. Surg. **1**, 86–100 (2005)
25.33 M. Muntener, A. Patriciu, D. Petrisor, D. Mazilu, H. Bagga, L. Kavoussi, K. Cleary, D. Stoianovici: Magnetic resonance imaging compatible robotic system for fully automated brachytherapy seed placement, Urology **68**, 1313–1317 (2006)
25.34 M. Muntener, A. Patriciu, D. Petrisor, M. Schar, D. Ursu, D. Song, D. Stoianovici: Transperineal prostate intervention: Robot for fully automated MR imaging–system description and proof of principle in a canine model, Radiology **247**(2), 543–549 (2008)
25.35 S. Shah, A. Kapoor, J. Ding, P. Guion, D. Petrisor, J. Karanian, W. Pritchard, D. Stoianovici, B. Wood, K. Cleary: Robotically assisted needle driver: Evaluation of safety release, force profiles, and needle spin in a swine abdominal model, Int. J. Comput. Assist. Radiol. Surg. **3**(1/2), 173–179 (2008)
25.36 J.R. Adler Jr., M.J. Murphy, S.D. Chang, S.L. Hancock: Image-guided robotic radiosurgery, Neurosurgery **44**, 1299–1307 (1999)
25.37 H. Breitwieser, S.M. Boscan, H. Becker, U. Voges, H. Junker, A. Melzer: Feasibility of manual control device for robotic assisted needle insertion in CT and MRI, Radiol. Suppl. **225**, 155–192 (2002)

25.38 A. Berrington de González, S. Darby: Risk of cancer from diagnostic x-rays: Estimates for the UK and 14 other countries, Lancet **363**, 345–351 (2004)

25.39 A. Melzer: Technology in the operating theatre of the future – Integration of imaging methods in the image guided surgery, Zentralbl. Chir. **133**, 197–200 (2008)

25.40 A. Melzer: Lehr- und Forschungs-OP zur Integration bildgebender Verfahren, Health Technol. **1**, 16–17 (2002)

25.41 A. Melzer: Prinzipien Magnet-Resonanz-geführter Intervention, Operation, Navigation und Robotik, Bundesgesundheitsblatt **53**(8), 768–775 (2009)

25.42 G.W. Bradley: MR-guided focused ultrasound: A potentially disruptive technology, J. Am. Coll. Radiol. **6**, 510–513 (2009)

25.43 D. Gianfelice, C. Gupta, W. Kucharczyk, P. Bret, D. Havill, M. Clemons: Palliative treatment of painful bone metastases with MR imaging-guided focused ultrasound, Radiology **249**, 352–362 (2008)

25.44 E.C. Pua, P. Zhong: Ultrasound-mediated drug delivery, IEEE Eng. Med. Biol. Mag. **28**, 64–75 (2009)

25.45 P. Prentice, A. Cuschieri, K. Dholakia, M. Prausnitz, P. Campbell: Membrane disruption by optically controlled microbubble cavitation, Nat. Phys. **1**, 107–111 (2005)

25.46 A. Schroeder, R. Honen, K. Turjeman, A. Gabizon, J. Kost, Y. Barenholz: Ultrasound triggered release of cisplatin from liposomes in murine tumors, J. Control. Release **137**, 63–68 (2009)

25.47 J. Kost, K. Leongt, R. Langer: Ultrasound-enhanced polymer degradation and release of incorporated substances, Proc. Natl. Acad. Sci. USA **86**, 7663–7666 (1989)

25.48 A. Melzer, G. Schäfers, S. Konak, S. Michitsch, T. Bertsch: Nitinol in magnetic resonance imaging, Minim. Invasive Ther. Allied Technol. **8**, 95–102 (2004)

25.49 P.R. Hilfiker, H.H. Quick, J.F. Debatin: Plain and covered stent-grafts: In vitro evaluation of characteristics at three dimensional MR angiography, Radiology **211**, 693–697 (1999)

25.50 P.R. Hilfiker, H.H. Quick, M. Schmidt, J.F. Debatin: In vitro image characteristics of an abdominal aortic stent graft: CTA versus 3-D MRA, MAGMA **8**, 27–32 (1999)

25.51 A.H. Matsumoto, G.P. Teitelbaum, M.J. Carvlin, K.H. Barth, M.A. Savin, E.P. Strecker: Gadolinium enhanced MR imaging of vascular stents, J. Comput. Assist. Tomogr. **14**, 357–361 (1990)

25.52 J. Link, J.C. Steffens, J. Brossmann, J. Graessner, S. Hackethal, M. Heller: Iliofemoral arterial occlusive disease: Contrast-enhanced MR angiography for preinterventional evaluation and follow-up after stent placement, Radiology **212**, 371–377 (1999)

25.53 D.J. Taylor, G. Brown: Magnetic resonance imaging of vascular stents, Proc. Int. Soc. Magn. Reson. Med., Vol. 7 (1999) p. 1892

25.54 T. Klemm, S. Duda, J. Machman: MR imaging in the presence of vascular stents: A systematic assessment of artifacts for various stent orientations, sequence types and field strengths, J. Magn. Reson. Imaging **12**, 606–615 (2000)

25.55 A. Bücker, A. Rübben: Metallic endoprosthesis compatible with magnetic resonance, US Patent 0249440 (2004)

25.56 E. Spüntrup, A. Rübben, T. Schäffter, W.J. Manning, R.W. Günther, A. Buecker: Magnetic resonance-guided coronary artery stent placement in a swine model, Circulation **105**, 874–879 (2002)

25.57 C. Manke, W.R. Nitz, M. Lenhart, M. Volk, A. Geissler, B. Djavidani, M. Strotzer, P. Kasprzak, S. Feuerbach, J. Link: Stent angioplasty of pelvic artery stenosis with MRI control: Initial clinical results, RöFo Fortschr. Geb. Röntgenstrahlen Bildgeb. Verfahr. **172**, 92 (2000)

25.58 H. Eggebrecht, M. Zenge, M.E. Ladd, R. Erbel, H.H. Quick: In vitro evaluation of current thoracic aortic stent-grafts for real-time MR-guided placement, J. Endovasc. Ther. **13**, 62–71 (2006)

25.59 H. Eggebrecht, H. Kühl, G.M. Kaiser, S. Aker, M.O. Zenge, F. Stock, F. Breuckmann, F. Grabellus, M.E. Ladd, R.H. Mehta, R. Erbel, H.H. Quick: Feasibility of real-time magnetic resonance-guided stent-graft placement in a swine model of descending aortic dissection, Eur. Heart J. **27**, 613–620 (2006)

25.60 T. Bertsch, G. Schäfers, M.M. Busch, A. Melzer: Inductively coupled active MRI stents, CARS 2001 Proc., Int. Congr. Ser., Vol. 1230 (2001) pp. 2000–2005

25.61 D. Kivelitz, S. Wagner, J. Schnorr, M. Busch, A. Melzer: A vascular stent as an active component for locally enhanced magnetic resonance imaging: Initial in vivo imaging results after catheter-guided placement in rabbits, Invest. Radiol. **38**, 147–152 (2003)

25.62 E. Immel, F. Gilbert, A. Melzer: MRI resonant prosthetic heart valves, Minim. Invasive Ther. Allied Technol. **18**, 123–126 (2009)

25.63 E. Immel, A. Melzer: Improvement of the MR imaging behaviour of vascular implants, Minim. Invasive Ther. Allied Technol. **15**, 85–92 (2006)

25.64 G. Schäfers, W. Zylka, M. Busch, A. Melzer: Wireless application of an intravascular resonator within the vena cava for MRI, CARS 2001 Proc., Int. Congr. Ser., Vol. 1230 (2001) pp. 1094–1099

25.65 N.A. Krämer, E. Immel, A. Melzer, C. Ocklenburg, R.W. Guenther, G.A. Krombach, E. Spüntrup: Evaluation of an active MR vena cava filter in an acute porcine animal model, Radiology **258**, 446–454 (2010)

25.66 S.E. Petersen, T. Voigtlander, K.F. Kreitner, P. Kalden, T. Wittlinger, J. Scharhag, G. Horstick, D. Becker, G. Hommel, M. Thelen, J. Meyer: Quantification of shunt volumes in congenital heart diseases using a breath-hold MR phase contrast technique–comparison with oximetry, Int. J. Cardiovasc. Imaging **18**, 53–60 (2002)

25.67 A. Bücker, E. Spüntrup, R. Grabitz, F. Freudenthal, E.G. Muehler, T. Schaeffter, J.J. van Vaals, R.W. Günther: Magnetic resonance-guided placement of atrial septal closure device in animal model of patent foramen ovale, Circulation **106**, 511–515 (2002)

25.68 J. Périssat, D. Collet, R. Belliard, C. Dost, G. Bikandou: Cholecystectomy by laparoscopy. Operative technic. Results of the first 100 cases, J. Chir. (Paris) **127**(6/7), 347–355 (1990)

26. Devices and Materials in MRI

Gregor Schaefers, Andreas Melzer

Magnetic resonance imaging (MRI) provides superior visualization of morphology, biochemistry, and physiology through differently weighted soft tissue contrasts (T_1, T_2, fat or water suppression, etc.). In addition, its high flow sensitivity depicts functional aspects such as blood flow, perfusion, and diffusion of tissue and organs. The physical nature of MRI provides three-dimensional, spatially defined image volumes which allow image guidance and automatic slice orientation as well as device localization (tip tracking). Therefore, magnetic resonance (MR) safety and compatibility are important issues for all items, including implants, surgical tools, and electronic or mechatronic equipment, to be used within an MR environment. MR testing of medical devices is required for device approval by regulatory agencies, e.g., the Food and Drug Administration (FDA) and the European Union (EU) Notified Bodies.

The intention of this chapter is to provide established scientific data and information, but also to cover the developing character of the field. It is useful for the growing reader community in the field of MR safety/compatibility to understand the historical path of developed methods and the complexity of this subject, especially MR labeling information derived from testing performed from approx. 1995 to 2005 and even before, which is already obsolete but has to be understood by the clinical MR operator to avoid risk to patients. The change in 2005 of the labeling "MR compatible" to the new ASTM labeling of medical devices into the categories of "MR safe", "MR conditional", and "MR unsafe" and the related revision of MR testing reflect the ongoing research in this field. The chapter provides a comprehensive overview of MR safety and compatibility interactions, issues on existing MR testing methods, and published standards [26.1–4], as well as updating the discussion on physical interactions that will be relevant in the future.

26.1 **MR Safety** 504

26.2 **Interactions in the MR Environment** 504
- 26.2.1 Magnetically Induced Displacement Force (Static, Dynamic) 505
- 26.2.2 Magnetically Induced Torque (Static, Dynamic) 505
- 26.2.3 Radiofrequency-Induced Heating .. 505
- 26.2.4 Gradient-Induced Interaction 506
- 26.2.5 Safe Operation of Devices Within the MR Environment (Dependent on Individual Demands) 506
- 26.2.6 Image Quality 507
- 26.2.7 MR Image Artifacts Due to the Medical Device 507

26.3 **Examples of MR Artifacts Caused by Medical Devices** 510
- 26.3.1 Stent Artifacts Summary 511

26.4 **Evaluation of MRI Artifacts of Implants** ... 512

26.5 **MR Safety Labeling** 514
- 26.5.1 Terms, Definitions, Icons, and Marking 515

26.6 **Interpretation of MR Labeling** 517

26.7 **Discussion** 518

References 519

Use of items and devices in MR applications such as clinical diagnosis and interventional procedures is increasing with the increasing number of MR scans carried out each year. Products have to comply with MR safety and compatibility requirements. The intention of the basic testing methods issued as ASTM standards [26.5–8] is testing of passive devices or implants for MR safety and MR image artifacts. New test methods are currently under development for active implantable medical devices and will be published in 2011 as an ISO/IEC technical specification [26.9]. After improvement of these new methods within the MR technical community, these test methods will be probably also related back to passive devices to test certain interactions such as radiofrequency (RF)- and gradient-induced voltages and heating as well as gradient-induced vibration as sources of unintended stimulation of devices. Device operation safety is mostly impacted by the electromagnetic and pulsed as well as static magnetic field described under electromagnetic interference (EMI) to a device or electromagnetic compatibility (EMC) within the upcoming ISO/IEC technical specification [26.10].

26.1 MR Safety

Although the general opinion is that there are no permanent negative health effects related to MR scanning, MR manufacturers have to take the necessary precautions to achieve and maintain this situation. Apart from the general safety aspects related to electricity, temperatures, mechanical precautions, etc., also the specific safety aspects of MR scanning must be controlled and have to fulfill the requirements formulated in safety standards. For MR, these specific safety requirements are stated in IEC 60601-2-33. Recently the third edition of this standard has been published [26.11]. It includes the specific requirements and limits formulated for electromagnetic fields applied during MR scanning of patients. The third edition now also introduces safety limits for these parameters specifically for the MR worker (the operator in the hospital, the MR engineer at the MR manufacturer, MR researcher, MR tester, etc.). The limits for patients are all based on temporary effects, i. e., dizziness, light flashes or metal taste related to the static magnetic field, heating related mainly to the RF fields, and peripheral nerve stimulation (PNS) generated by gradient application. Since these limits are set conservatively, they also apply for the MR worker as set for the patient. The fact that MR workers may be exposed more frequently does not change the limit values for short-term effects, since it is generally believed that no long-term effects are to be expected. Negative long-term effects related to exposure to electromagnetic fields generated by MR scanners have not yet been proven. Initial studies have been started recently to try to demonstrate that such effects will indeed not be observed.

It is important to realize that the third edition of IEC 60601-2-33 does not include safety requirements specifically addressing interventional applications of MR scanners. Consequently, no requirements are formulated in this standard related to the type testing needed to demonstrate the safety of accessories applied during interventional scanning or additional safety requirements formulated for MR scanners used for interventional scanning. Also, safety requirements for other equipment used in combination with MR scanning, such as medical implants, surgical tools, and electronic or mechatronic equipment, are not specified in this standard, although warnings are included. However, a number of other standards do partly address these situations, and more work is underway in the international MR community to improve this situation. IEC 60601-2-33, however, requires that:

> *When the implant device is labelled as MR safe or MR conditional, the instructions for use shall explain that further information is described in the accompanying documents of the implant manufacturer.*

26.2 Interactions in the MR Environment

The various parameters for the most important MR interactions are given in Sects. 26.2.1–26.2.7 (the list being limited due to current developments). For further details, other literature is available, where the physics of the MR interactions is explicitly described [26.12], therefore an overview level is the intention of this section.

In MRI, the expression "magnetic field" is used for the main static magnetic field. Because of the fact that the working volume of this homogeneous static magnetic field is air, the relative permeability μ_r is considered to be 1. Electromagnetic theory shows that the magnetic field strength H (magnetizing field) in units of

A/m is given by

$$B_0 = \mu_r \mu_0 H \,, \tag{26.1}$$

where the magnetic flux density B (the magnetic field) in units of T equals H if we ignore μ_0. The static field is denoted by B_0, the electromagnetic alternating field component by B_1, whereas $\mathrm{d}B_{x,y,z}/\mathrm{d}x$, y, z denotes the magnetic field gradient.

26.2.1 Magnetically Induced Displacement Force (Static, Dynamic)

Note, for diamagnetic, paramagnetic, and ferromagnetic materials in devices below their magnetic saturation when exposed to a magnetic field, the maximum deflection (angle) occurs at the point where the gradient product $|\boldsymbol{B}| \cdot |\nabla \boldsymbol{B}|$ is maximal. For ferromagnetic materials in devices that are magnetically saturated, the maximum deflection is measured where $|\nabla \boldsymbol{B}|$ is maximal [26.13]. Thus, especially for devices or parts within the magnet bore entrance, the maximum magnetically induced displacement force is dependent on the extent of the static magnetic field and is influenced by the active shielding of the magnet and the field decay as well as the magnetic saturation of the test object material and the magnetic properties of the material itself.

In summary, for ferro- and paramagnetic materials (all metals) and diamagnetic materials, the magnetically induced displacement force depends on:

- The static magnetic field B_0 (saturation)
- The static field gradient of the fringe field
- The magnetic saturation of the device material.

26.2.2 Magnetically Induced Torque (Static, Dynamic)

The magnitude of the torque induced by the static magnetic field depends on the device dimensions and geometry, and on the magnetic saturation of the material if the device is not completely saturated. The maximum static torque experienced by the device is at the isocenter of the magnet, where the static magnetic field is homogeneous and maximum.

In summary the torque depends on:

- The static magnetic field B_0 (below saturation)
- The device dimensions and geometry
- The magnetic saturation of the material.

Dynamic forces and torques are possible for electrically conductive devices and structures which are translated in the magnetic field. In addition to the static, magnetically induced force, a dynamic magnetically induced force based on Lenz's law can be detected for electrically conductive device parts of large size that are translated through the spatial gradient of the static field. This effect is potentially important for moving components. The interaction is dependent upon the speed of the movement, the magnitude of the spatial gradient of the magnetic field, and the effective area of induction, as well as on the conductivity of the device material.

Dynamic torque, which can occur in the static magnetic field, becomes detectable if electrically conductive parts are rotated within the static magnetic field. Device parts with high conductivity and a significant area of induction for eddy currents will especially exhibit a counter-torque that influences the handling of the device.

This interaction depends on:

- The speed of movement
- The magnitude of the spatial gradient of the magnetic field
- The effective area of induction
- The conductivity of the device material.

26.2.3 Radiofrequency-Induced Heating

Radiofrequency (RF)-induced heating and induced voltages are a complex issue, dependent on various parameters. RF pulses are in the MHz range and are the main source of energy amongst the interactions responsible for heating [26.14]. Not only device properties such as electrical conductivity, dimensions, shape, etc. have to be considered, but also the geometric arrangement relative to the specific MR environment of a given MR scanner. This includes also the geometric arrangement relative to the specific MR coil and the parameter settings of the MR sequence. It is necessary to know exactly, for each MR system, which coil and pulse sequence produces which RF output (spatially localized power deposition) and how the specific absorption rate (SAR) (see IEC 60601-2-33 for limits [26.11]) is calculated, measured, adjusted, and displayed by the system software [26.14]. In the area of RF fields, computer simulation of electromagnetic fields, SAR, and temperature distribution is currently being developed to assist in heating testing, at least for implants or instruments [26.15, 16]. Loop-closing contacts or the insertion of an instrument can be dangerous for the patient due to induction from and coupling with the RF field. Electrically conductive structures as well as the patient's body can suffer from induced voltages and thus currents through the patient tissue. These currents, as well as sparking due to high voltage, may cause skin burns or even fire [26.17] and harm the patient or the MR interventionist.

This interaction depends on:

- The electric conductivity and permittivity of the device materials (and the impedance of connected device parts if electronic)
- The geometric dimensions of the device
- The relative configuration of the device or its components
- The conductivity and permittivity of surrounding tissue
- The power of the RF pulses (SAR)
- The geometric arrangement relative to the RF transmit coils (or to the tangential electric field)
- The position and architecture of the patient's body relative to the RF transmit coil
- The specific MR coil used (i.e., its electromagnetic field characteristics)
- The center frequency of the specific MR system.

26.2.4 Gradient-Induced Interaction

Switched gradient magnetic fields of currently, for instance, 40 mT/m gradient strength and 200 (mT/m)/ms gradient slew rate (see IEC 60601-2-33 for limits [26.11]) are used in MRI to provide spatial encoding of MR signals. Voltages induced by these switched gradient fields interfere with medical devices and may also, via the device, indirectly impact on the patient (through stimulation). If specific coupling criteria are met for interaction with the switched gradient magnetic fields, heating may also take place in electrically conductive structures.

Voltages induced due to this switched gradient field environment can interfere with medical devices in terms of their operation or by stimulating pulses in the patient.

In addition, dynamic forces and torques can be generated by switched gradient magnetic fields, resulting in vibration of the device or manipulator arms due to eddy currents in electrically conductive device parts, thus causing interaction also with the static magnetic field. The magnitude of such vibration (amplitude and frequency) is a function of the geometric position and arrangement of the device relative to the gradient coils (x, y, z) and depends on the gradient parameters (gradient pulse frequency, gradient strength, and slew rate) as well as on the size and weight of the device itself. The source of this vibration is inside the device materials acting on the outside and considering each components source vibration.

Interactions induced by switched gradients include:

- Gradient-induced voltages (stimulation, activation)
- Gradient-induced heating
- Gradient-induced vibration.

This interaction depends on:

- The gradient amplitude (x, y, z) expressed in units of mT/m
- The effective stimulation time of the gradient pulses
- The combination of these two parameters resulting in the gradient slew rate expressed in (mT/m)/ms
- The gradient pulse shape
- The device position within the gradient coil
- The device orientation within the gradient coil
- The effective area of induction
- The conductivity of the device materials
- The static magnetic field B (providing the counterpart field).

26.2.5 Safe Operation of Devices Within the MR Environment (Dependent on Individual Demands)

Some examples of how an MR system can affect safe device function are inhibition of electrical circuits or mechanical components such as springs or levers of devices. For cases where specific parts, actuators or sensors cannot be designed to be suitable for high-field application, MR testing has to determine specific conditions under which a device can operate safely in the MR environment.

The maximum peak acoustic noise level of the MR system during scanning is not allowed to exceed 140 dB relative to 20 μPa in the accessible area of the MR equipment [26.11]. Thus, acoustic noise could also be an issue for certain sensors. Finally, it is important to examine the interaction among different devices or systems before MR safety labeling. The aforementioned examples show the importance of a specific and comprehensive functional testing protocol.

For evaluation of safe operation, the following interactions have to be considered:

- Static magnetic field
- Switched gradient magnetic field
- RF electromagnetic field

including parameters such as:

- Device configuration
- Device orientation
- Possible or intended device operational status (on, off, standby or other function modes).

26.2.6 Image Quality

Image quality issues, such as signal-to-noise ratio (SNR) and B_0 homogeneity, are important for safe diagnostic operation of MR systems.

The effect on image quality is a major consideration. RF emission by actuators or other active parts of a device can significantly decrease the image quality of the MR system (e.g., by decreasing the SNR) and can therefore affect the clinical diagnosis. Interference can be visible in the MR image, appearing as, for example (but not limited to), noisy images due to spikes, dots or stripes and spurious proton signals [26.19]. Other degradations, disturbances or inaccuracies of the MR image are likely (decrease in SNR or B_0 field homogeneity, etc.).

26.2.7 MR Image Artifacts Due to the Medical Device

In most cases, MR image artifacts do not primarily affect patient safety but rather have high clinical relevance for adequate diagnosis, e.g., stent patency. If exact positions are needed, e.g., for needle insertion or implantation of x-ray markers or radioactive seeds, safety becomes critical for the whole diagnostic procedure. Susceptibility [26.20] and RF artifacts depend on various parameters such as the magnitude and orientation of B_0, the alignment of the medical device relative to B_0, and the MR pulse sequence parameter settings (Fig. 26.1).

MRI susceptibility artifacts are caused by field inhomogeneity around an object (Fig. 26.1c,d) and show a particular pattern due to the magnetic field lines [26.18, 21]. Comprehensive artifact testing is necessary to provide artifact information for all relevant configurations. With respect to MR imaging compatibility, artifacts influencing image quality can hide information or result in incorrect depiction of the position of a device, e.g., a needle, in the MR image [26.19, 22]. Thus, artifacts can lead to diagnostic misinterpretation due to missing information resulting from susceptibility artifacts due to the device materials. Furthermore, RF artifacts [26.23] are likely, due to coupling of electrically conductive structures with the electromagnetic field, e.g., for an instrument introduced into the patient. Such coupling can result in signal shielding (Faraday cage effect), distortion or even intended amplification of the signal (mainly by designed structures, e.g., coils), which can be used for active device visualization such as tip tracking. For external devices placed close to the patient, image distortion in different dimensions can occur. The image distortion can extend into the patient tissue even if there is no physical contact of the device with the patient.

Application of low-susceptibility materials such as polymers and ceramics minimizes such artifacts. However, plastic parts can contain hydrogen protons that disturb the image due to MR signals generated inside the imaging volume. On the other hand, signals from plastic parts outside the imaging volume can degrade image quality if folded into the field of view. Therefore, MR artifact testing should be implemented at an early stage of the development process of devices to check the materials used within or close to the imaging volume of the MR system. Polymers can also contain dotting with

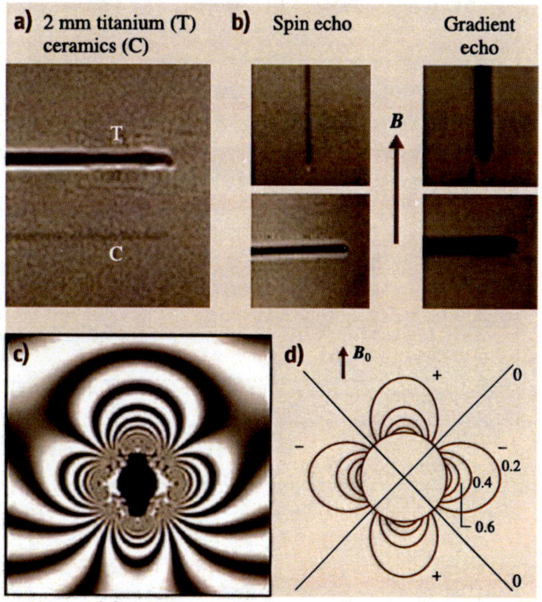

Fig. 26.1 **(a)** Titanium (T) and ceramic (C) cannulae in interventional MRI. **(b)** Artifact size in relation to orientation and sequence type (according to User Manual Magnetom OPEN and A. Oppelt, Siemens, Erlangen) **(c)** Field mapping of an artifact, cross-section of a 1 mm titanium probe. **(d)** Schematic view of the magnetic field perturbation; spins near the object to the left and right are at lower resonance frequency, whereas those above and below the object are at higher resonance frequency (modified according to [26.18])

Sample	EP	400 °C	500 °C	600 °C	700 °C	800 °C	1000 °C
Oxide thickness	0.011 µm	0.0338 µm	0.0244 µm	0.081 µm	0.550 µm	5.775 µm	21.9 µm
COR SE	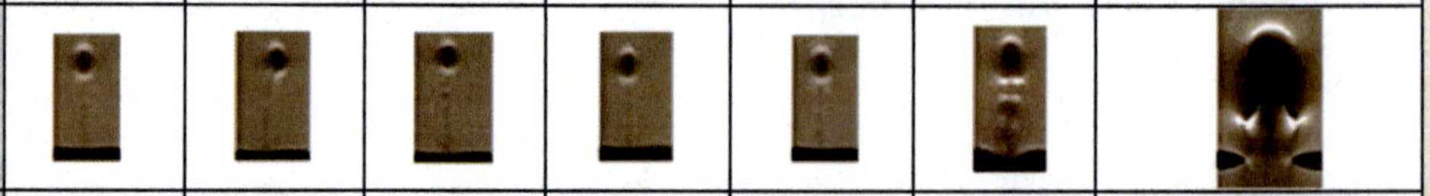						
Thickness (mm)	4.9	4.9	5.0	4.9	4.8	10.1	33.2 (187 % in length)
COR GE	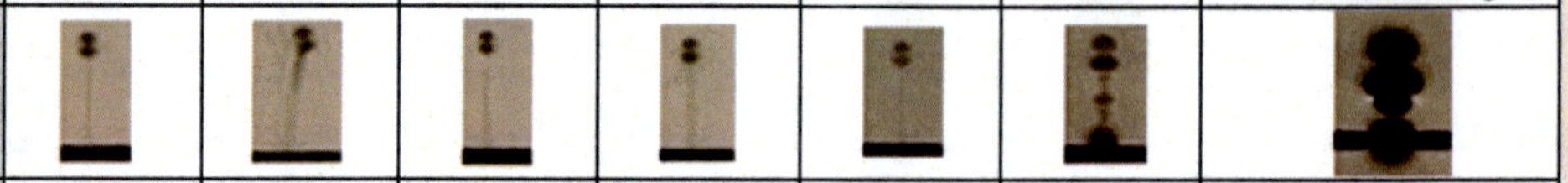						
Thickness (mm)	Tip 4.4	4.4	4.5	4.4	4.4	Up to 7.8	Up to 25.4 (187 % in length)
TRA SE	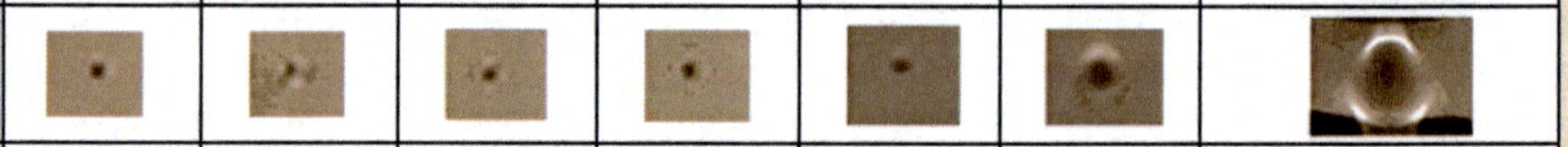						
Diameter (mm)	<0.2	<0.2	<0.2	<0.2	<0.2	6.0	27.1
TRA GE	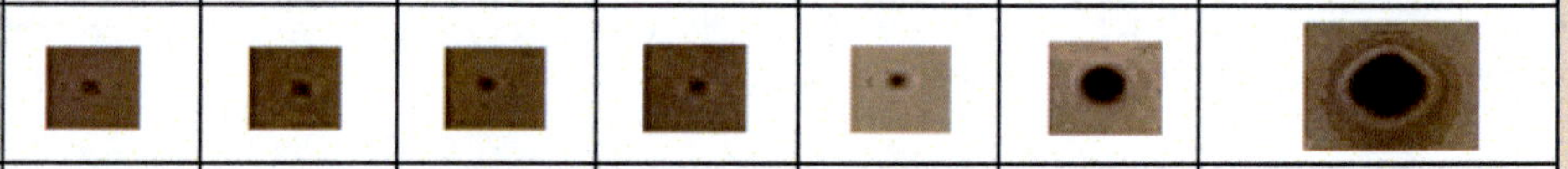						
Diameter (mm)	<0.2	<0.2	<0.2	<0.2	<0.2	4.9	18.7

Fig. 26.2a Artifact size of samples electropolished (EP) to 1000 °C at 1 T parallel to the main B_0 field

Sample	EP	400 °C	500 °C	600 °C	700 °C	800 °C	1000 °C
Oxide thickness	0.011 µm	0.0338 µm	0.0244 µm	0.081 µm	0.550 µm	5.775 µm	21.9 µm
COR SE	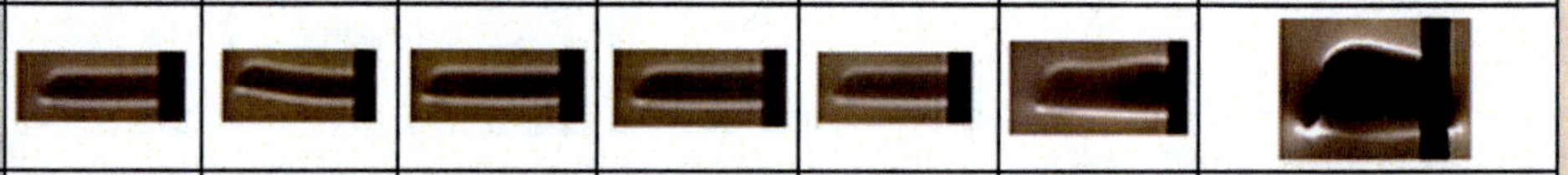						
Thickness (mm)	6.7	6.8	6.7	6.8	7.4	Up to 12.7	Up to 39.7 (163 % in length)
COR GE	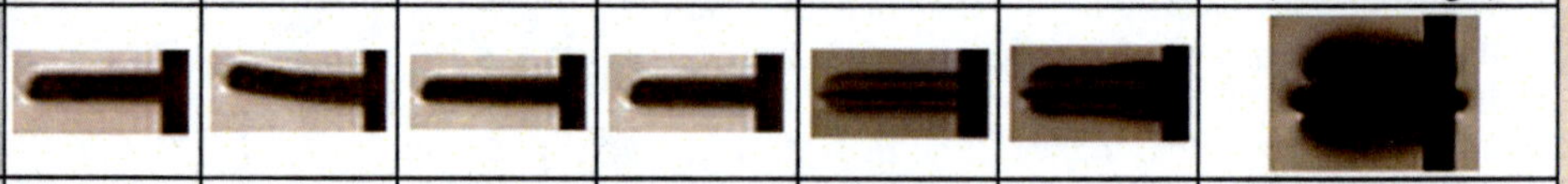						
Thickness (mm)	7.8	7.8	7.9	7.8	8.8	Up to 13.7	42.4 (165 % in length)
TRA SE	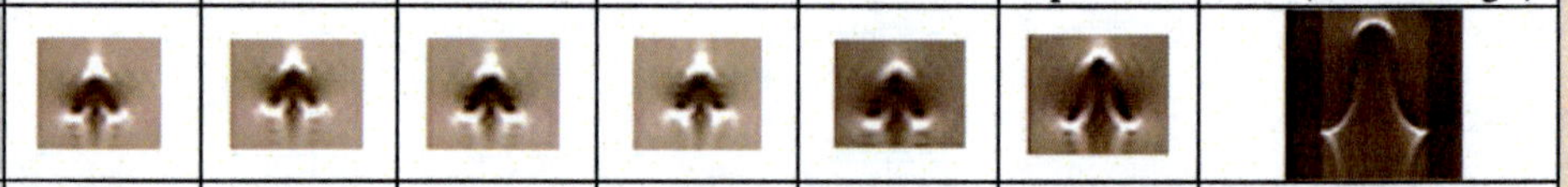						
Diameter (mm)	7.7	7.9	7.8	7.9	8.4	12.5	38.5
TRA GE	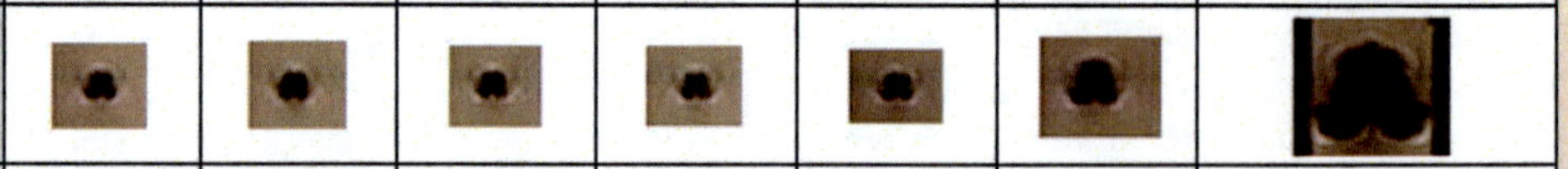						
Diameter (mm)	6.6	6.8	6.8	6.7	6.9	9.7	31.1

Fig. 26.2b Artifact size of samples electropolished (EP) to 1000 °C at 1 T, perpendicular to the main B_0 field

Sample	EP	400 °C	500 °C	600 °C	700 °C	800 °C	1000 °C
Oxide thickness	0.011 μm	0.0338 μm	0.0244 μm	0.081 μm	0.550 μm	5.775 μm	21.9 μm
COR SE	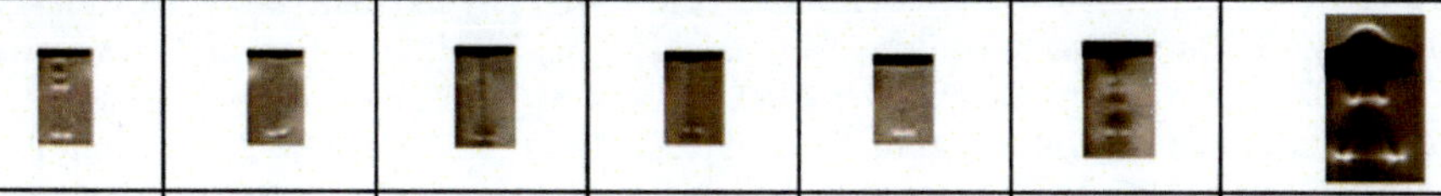						
Thickness (mm)	<0.2	<0.2	<0.2	<0.2	<0.2	Up to 7.7	Up to 23 (160 % in length)
COR GE	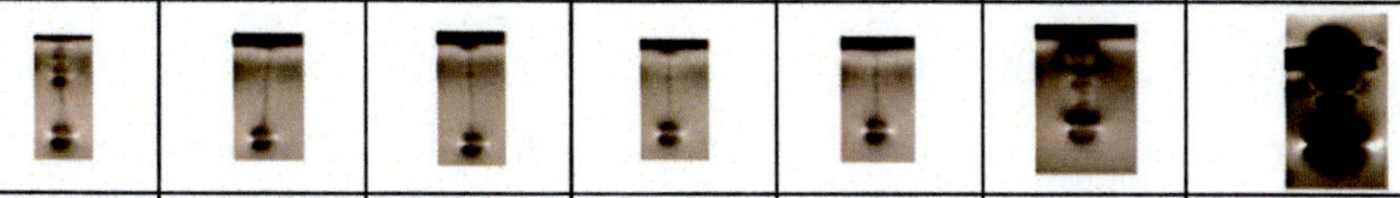						
Thickness (mm)	Tip: 7.3	Tip: 7.3	Tip: 7.3	Tip: 7.3	Tip: 7.3	Up to 12.7	Up to 30 (190 % in length)
TRA SE							
Diameter (mm)	1	1.2	1	1	1	8	15.5
TRA GE							
Diameter (mm)	1	1.2	1	1	1	8.3	19.4

Fig. 26.2c Artifact size of samples electropolished (EP) to 1000 °C at 1.5 T, parallel to the B_0 field

Sample	EP	400 °C	500 °C	600 °C	700 °C	800 °C	1000 °C
Oxide thickness	0.011 μm	0.0338 μm	0.0244 μm	0.081 μm	0.550 μm	5.775 μm	21.9 μm
COR SE							
Thickness (mm)	7.9	8.1	8.5	8.6	8.6	13.3	36
COR GE							
Thickness (mm)	12.4	12.5	12.5	12.5	12.5	18.7	47.7
TRA SE							
Diameter (mm)	8.2	8.5	8.7	9	9	12.6	33.6
TRA GE							
Diameter (mm)	8.3	8.4	8.4	8.6	8.7	12.7	35.6

Fig. 26.2d Artifact size of samples electropolished (EP) to 1000 °C at 1.5 T perpendicular to B_0 field

Part C | 26.2

ferromagnetic debris from manufacturing, causing significant artifacts. The presence of an oxide layer can play a role for the artifact size, but only if the oxide has certain paramagnetic characteristics different from the core metal or alloy. On Nitinol samples, increasing thickness of oxide layers have been determined in MRI. Figure 26.2a–d shows 1 mm × 40 mm Nitinol wire samples which were placed in 0.9% NaCl doped with 2 mmol/l gadolinium MR contrast agent, (Multihance, Schering AG, Berlin) parallel to the main field of a scanner with 1.0 and 1.5 T, and perpendicular to it at 0.3 T. Coronal (TE) and transversal (TR) orientated spin echo (SE: TR/TE = 300/30 ms) and gradient echo (GE: TR/TE = 100/10 ms, FA = 40°) images were obtained with a Hitachi (0.3 T)/Philips Gyroscan NT (1.0 T/1.5 T) using a quadrature head coil with 27 cm inner diameter.

26.3 Examples of MR Artifacts Caused by Medical Devices

Instruments as well as implants can cause significant image artifacts. The problem of imaging artifacts will be illustrated using the example of stents. These are tube-like, expandable devices which are placed via catheters, mainly in narrowed or occluded arteries to reestablish or maintain blood flow. The metallic material and the stent shape can lead to severe changes of the image information, leading to artifacts [26.10, 12, 24, 25].

Conventional imaging sequences are often compromised and signals complete vanish inside the stent lumen and in close proximity. A variety of reports on experimental [26.26–28] and clinical studies [26.29, 30] of postinterventional MR imaging of stents have been published in recent years, plus safety aspects and the influence of different materials and structures for MRI in the presence of stents [26.31] and stent grafts [26.32, 33] for particular MRI settings. Evaluation of imaging parameters and the effects of different field strengths, sequence types, parameters, and stent orientations have been reported. We have utilized field strengths of 0.2, 1.0, and 1.5 T and a variety of different material processing techniques with a particular focus on nitinol and implementation of resonant stent structures. MR image quality of the material (tissue) outside and inside eight different common vascular stents with nominal diameter of 8 mm and length of up to 80 mm after expansion was examined. Experiments were performed with unchanged sequence parameters at static magnetic field strengths B_0 of 0.2, 1.0, and 1.5 T. Effects of the type of sequence and imaging parameters, the orientation of the stent axis in the static magnetic field, and the orientation of the recorded slice were systematically assessed [26.34]. Suitable strategies for recording diagnostic MR images during interventional procedures and in follow-up examinations were derived from the experiments. The results should help to choose the optimal imaging modality and, in the case of MRI, suitable sequence types and imaging parameters in examinations of patients during and after stent implantation. As a result of these experiments, specific strategies have been developed to reduce susceptibility artifacts by selecting other materials such as niobium and tantalum for expanding balloon stents and to improve the behavior of nitinol for self-expanding stents. To overcome RF artifacts, stents that work as electric resonators to interact with the MR imaging process have been developed. This RF antenna effect of stents provides enhanced signal intensity of the stent lumen. By inductive coupling, improved imaging of the stented area has become possible [26.35, 36].

MR systems have field strengths that vary from 0.2 up to 3 T. Basically the same sequence types are used, such as spin echo and gradient echo [26.37]. The sequence choice has an important influence on the size of the artifacts in MRI. Another important factor is the field strength of the MR system. The field strength has a significant influence on the generation of artifacts from devices and implants such as stents. The higher the field strength, the greater the influence on the MR image. Changes of the susceptibility also become more relevant. Thus, devices that show only minimal artifacts in low-field (0.2–0.5 T) MRI demonstrate significantly increased artifact size for higher fields. Susceptibility artifacts are well understood and have been examined extensively [26.34, 38, 39].

Flow artifacts [26.40–42] depend on the field strength, and the basic material characterization and the electromagnetic behavior are more relevant to the MR imaging of stents. Although 1 and 1.5 T differ by 50%, the artifact size does not increase by 50%. As 1.5 T is the current state of the art in coronary MR angiography as well as cardiac imaging, stent samples have been examined in 1.5 as well as 1 T. RF artifacts are more relevant in this context, because the resonant frequencies are different, leading to different imaging behavior [26.23, 43]. All stents that consist of interconnected segments are subjected to eddy current induction as well as parasitic resonant frequencies, both of which lead to shielding of the lumen (Fig. 26.3).

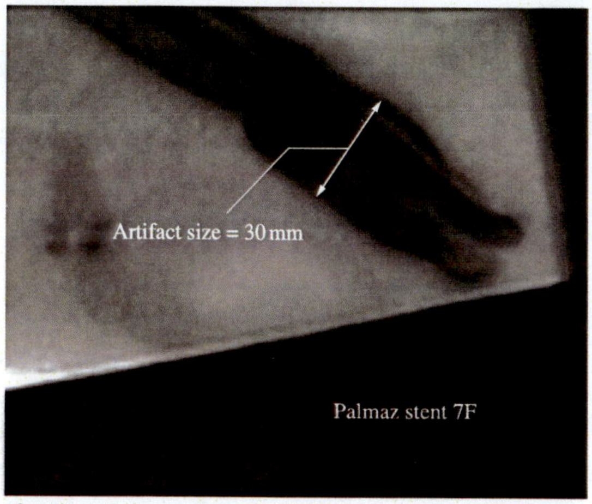

Fig. 26.3 Palmaz 3 mm stainless-steel stent (316L) in MRI

As described above, susceptibility artifacts and RF artifacts prevent lumen visualization.

The MRI Larmor resonance frequency is defined as

$$\omega_L = \gamma_H * B_0 ,$$

where $\gamma_H = 42.58\,\text{MHz/T}$, i.e., 63.87 MHz at 1.5 T (Fig. 26.5).

In addition to the pulse sequence (gradient or spin echo) and the *susceptibility* of the material, the geometric shape of an implant, e.g., a stent, has a significant influence on the distortions of the static magnetic B_0 field. As a result, a nonuniform pixel intensity is generated in areas which should have a uniform pixel intensity.

Copper has magnetic susceptibility close to those of tissue and water. Thus, susceptibility artifacts can be neglected when using copper wire rings. To prove this concept, an experiment has been performed using four copper rings: a noninsulated and insulated ring closed by soldering, an open ring, and a ring tuned using a capacitor to the resonant frequency (42.58 MHz) of a 1.0 T magnet (Figs. 26.6, 26.7).

26.3.1 Stent Artifacts Summary

The following simplified description for stent artifacts can be defined:

- The material causes a local inhomogeneity of the stent lumen, whereby the spatial encoding of the MR signal is altered.
- Eddy currents cause magnetization of the stent. These local magnetic fields lead to additional local inhomogeneity of the magnetic field, causing spatial miscoding of the spin frequency similar to the effect of susceptibility.
- RF pulses from the MRI device do not enter the stent lumen because of the Faraday cage shielding effect and the induction of eddy currents.

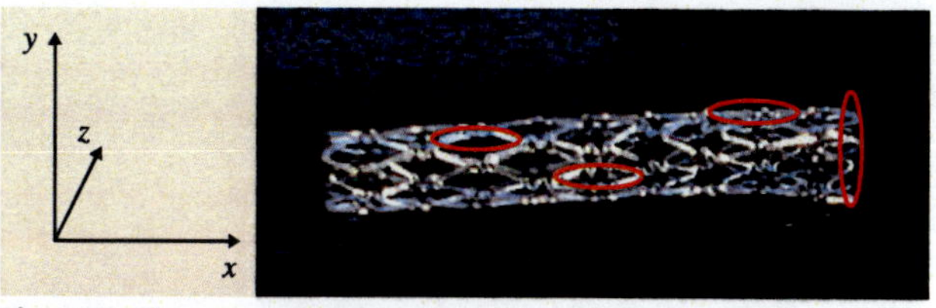

Fig. 26.4 Stent sample with schematic pathways for eddy currents depending on the direction of gradient fields and RF fields. Eddy currents are also induced in the circumference of the stent, because all segments are interconnected

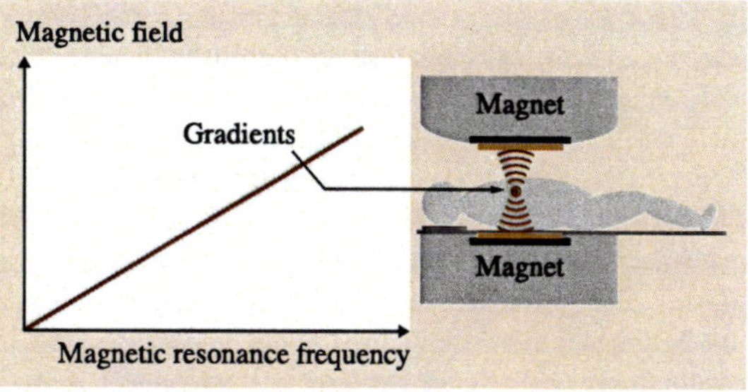

Fig. 26.5 In MRI, biological tissue is exposed to strong static, permanent magnetic field and switched gradient magnetic fields (schematic view with MRI cut open)

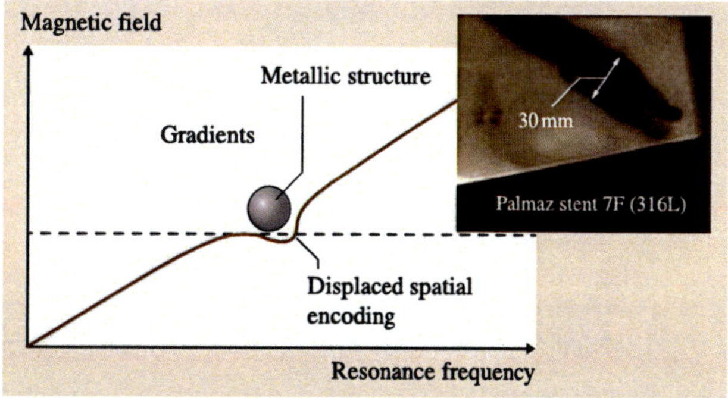

Fig. 26.6 Schematic representation of the principle of a susceptibility artifact. Alteration of the magnetic field homogeneity by a metallic rod causes a change of the local resonance frequency and thus a mismatch of the spatial encoding, leading to the artifact on the right

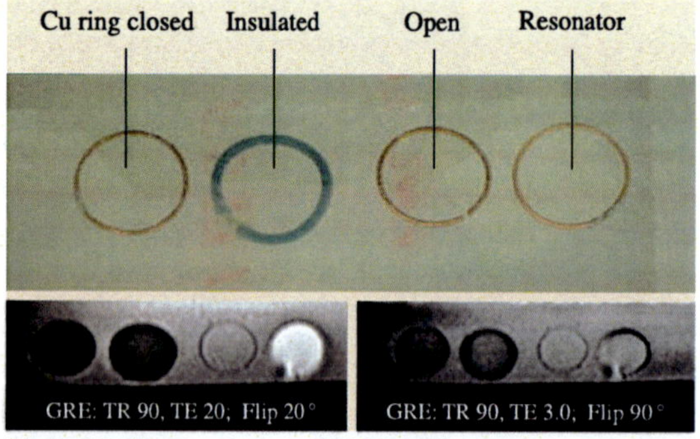

Fig. 26.7 Copper rings in MRI. The first and second closed Cu rings on the *left* demonstrate the phenomenon of local magnetization with induction of eddy currents. The field inhomogeneity shields the inner section of the ring. Whereas the open ring is neutral and does not alter the MRI signal, the resonant ring (*right*) shows significant signal enhancement depending on the flip angle of the gradient echo (GRE). This proves that the resonant structure can enhance the MR signal intensity within the lumen of a stent. Most stents can be characterized as a set of interconnected rings

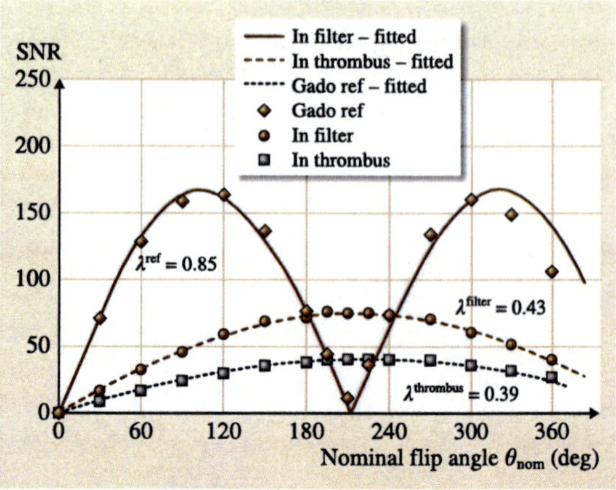

Fig. 26.8 Dependence of the signal-to-noise ratio on the flip angle for a vena cava filter (*open circles*), a thrombus clot (*open squares*), and a tube filled with 1 mol/l gadolinium-doped water (*solid circles*) (after *Bartels* et al. [26.23])

- If excitation RF pulses enter the stent, escape of the signals emitted due to proton relaxation inside the stent is hindered.

To evaluate the capability of a resonant circuit to overcome stent artifacts, we selected the stent with the smallest visible MR artifacts. The stent was equipped with a wire coil tightly wound around the stent, and the wire was soldered to a surface-mount device (SMD) capacitor and thus tuned close to the resonance Larmor frequency: 42.58 MHz at 1 T and 63.87 MHz at 1.5 T. In addition to the resonant circuit, the technique described by *Bartels* [26.23] has been used, i.e., using the flip angle to increase the pulse energy within the stent to overcome the shielding (Figs. 26.8, 26.9).

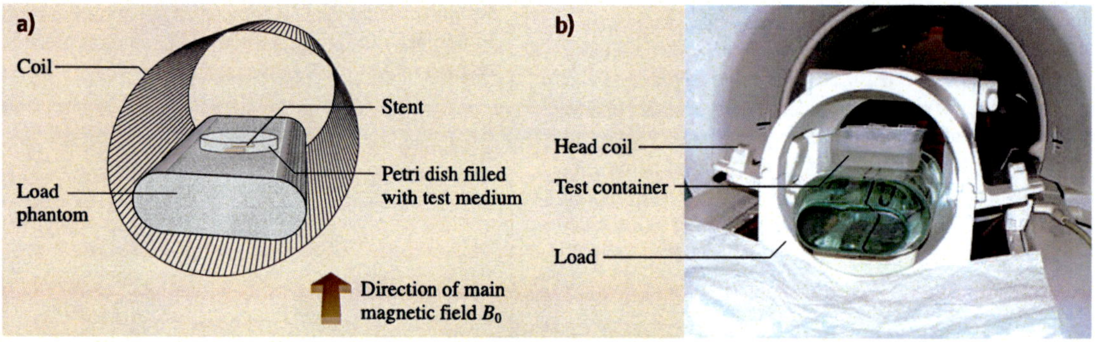

Fig. 26.9a,b Setup schematic and photograph of MRI testing of stent samples. Measurements took place in a Petri dish as a test container in 1 T and 1.5 T Philips and Siemens MR using the standard head coil. The test medium was water and 2 mmol/l Magnevist (Schering, Berlin), equivalent to 1 mmol/l Gd-diethylenetriamine penta-acetic acid (DTPA)

26.4 Evaluation of MRI Artifacts of Implants

Another example is the vena cava filter. These devices are placed via the femoral vein into the infrarenal vena cava to filter blood clots stemming from pelvic veins or deep vein thrombosis in or-

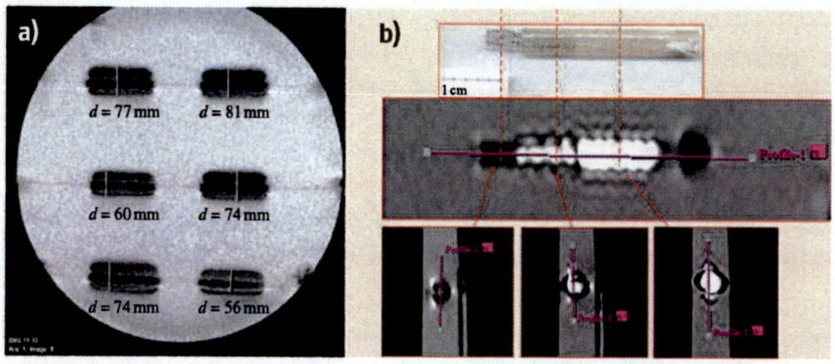

Fig. 26.10 **(a)** Various 3 mm × 20 mm stent samples in MRI (Siemens Magnetom Vision, fl2d, TE = 11 ms, TR = 300 ms, FA = 40 °C, slice 3 mm, matrix = 256 × 256, FoV = 128 mm). **(b)** Nitinol 3 mm × 20 mm stent partially covered by a resonant circuit tuned to 64 MHz in Philips 1.5 T Intera, FFE 3d, TE = 6 ms, TR = 200 ms, FA = 10 °C, # of Acq: 2, Slice = 2 mm, matrix: 256 × 256, FoV: 90 mm

der to prevent pulmonary embolism. Depending on the field strength, pulse sequence, and material, artifacts of different size can occur (Figs. 26.10a,b; 26.11a,b).

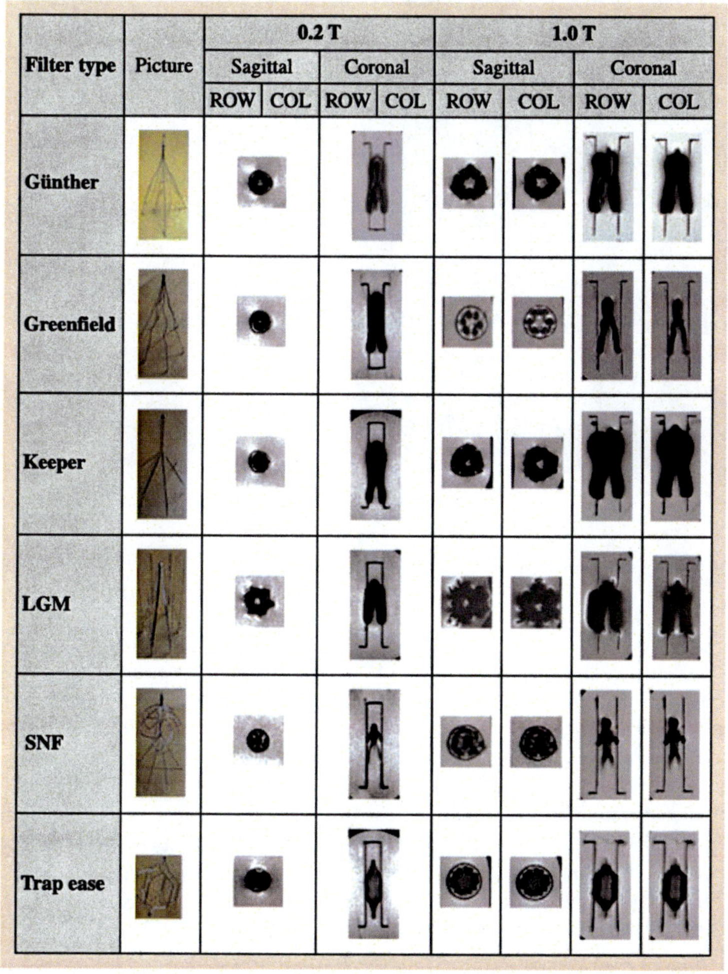

Fig. 26.11a Vena cava filter in MRI, 0.2 and 1 T, gradient echo sequence

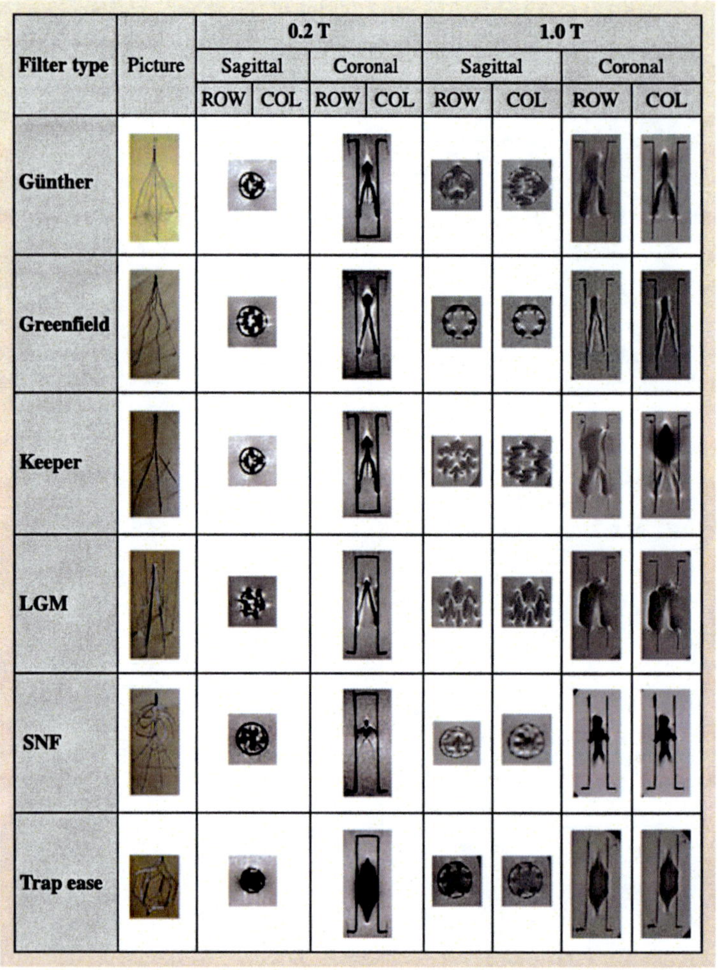

Fig. 26.11b Vena cava filter in MRI 0.2 and 1 T spin echo sequence

26.5 MR Safety Labeling

In medical practice, the question arises of whether a patient with a specific implant can be safely MR-scanned or whether a specific device can be used close to or on the patient during MR scanning. This information is often not available, and the clinical staff have to make the decision. It is very important to base such clinical decisions for MR safety or conditional scanning on original information from the instructions for use from the MR implant manufacturer. Some information can be found on the Internet, for example, at www.mrisafety.com; however, often the reader is pointed to advices of e.g. or similar *consult the manufacturer of the particular device for the latest safety information* (e.g. Conditional 5, whereas only *MR Conditional* is the official term according to ASTM F2305). Information from the Internet only may not be complete, can be outdated or may not cover the responsibility of the device manufacturer or the MRI operator. MR safety labeling provided by the device manufacturer is mandatory to ensure patient safety.

www.magresource.com for example is listing original MR labeling information from the device manufacturer. Information based on old MR standards as well as new MR standards is clearly marked. The database is however restricted to registered users.

Currently, ASTM F2503 [26.9], which defines marking practice for items brought into the MR environ-

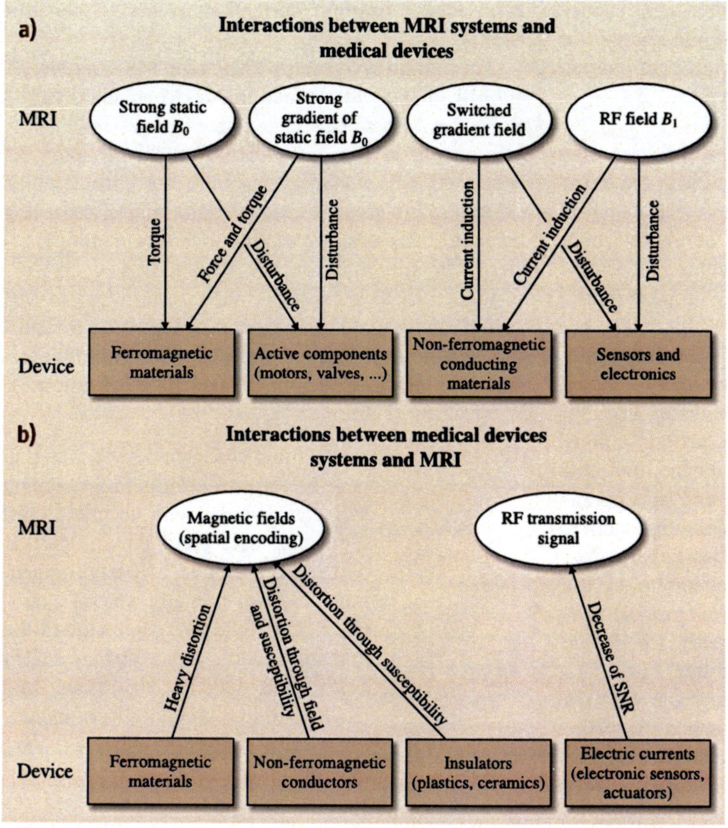

Fig. 26.12 **(a)** Interactions of MRI with devices and **(b)** interaction of a device with MRI (after *Gutmann*)

ment, requires MR safety labeling using the following categories and icons: MR Safe, MR Conditional, and MR Unsafe. Besides implants and interventional instruments commonly used with diagnostic imaging modalities, other devices, e.g., surgical setup, frames or robots, are being used with MRI as well, and enter the MR environment. Considering complex systems, basic and specific tests are needed for evaluation of the MR interaction between device components and the MR environment. Thus, medical device manufacturing requires not only standard technical and clinical knowledge but also MR safety-specific knowhow for device use in combination with a specific MR environment (Fig. 26.12)

In the early 1990s, standards were established for safety testing of passive devices brought into the MR scanner during scanning.

ASTM F2052 [26.5], ASTM 2213 [26.6], and ASTM F2182 [26.7] are regularly updated and therefore are still of value, although the limits proposed in these standards are still based on practical experiences rather than scientific data. Also, ASTM F2119 [26.8] has been updated very recently.

26.5.1 Terms, Definitions, Icons, and Marking

ASTM F2503 [26.9], which describes terms, definitions, and icons for marking items (including medical devices) for use in MR environments, superseded the *historical* definitions of MR safe and MR compatible [26.44], which are listed here for clarity but no longer in use:

MR safe: The device, when used in the MR environment, has been demonstrated to present no additional risk to the patient or other individuals, but may affect the quality of the diagnostic information.

MR compatible: The device, when used in the MR environment, is MR safe and has been demonstrated to neither significantly affect the quality of the diagnostic information nor have its operations affected by the MR device.

The MR conditions under which the device was tested should be specified in conjunction with the terms MR safe and MR compatible, since a device which is safe or compatible under one set of conditions may not be found to be so under more extreme MR conditions. The definitions were extracted, with permission, from [26.44].

In addition to several MR safety issues that apply to items and components such as magnetically induced displacement forces and torques, RF-induced heating, induced voltages, acoustic noise, interactions among devices, as well as a considered safe device, the term MR compatible covers image quality and safe operation of the device according to the definition above. MR image artifacts are not considered a direct safety issue but, on the other hand, safe operation of the device is a direct safety issue. MR compatible was often misinterpreted and used incorrectly for device labeling. Thus, the ASTM subcommittee F04.15.11 for MR standards decided to withdraw the term in 2005 based on the marking practice defined in ASTM F2503 [26.9]. MR compatible was not redefined and is no longer used in current standards or for MR labeling.

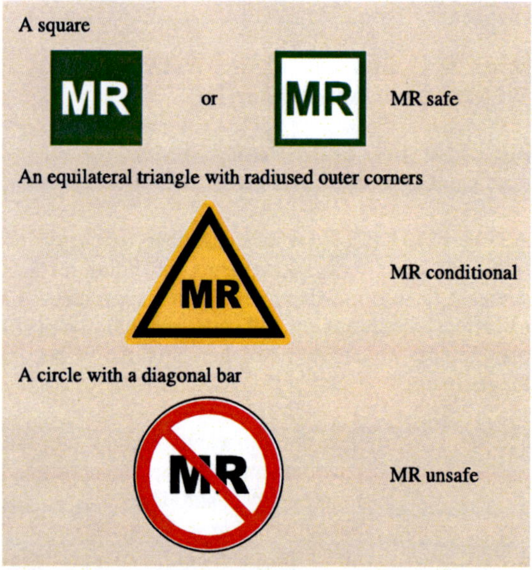

Fig. 26.13 Current icons for MR marking of devices and items used within the MR environment (extracted with permission from ASTM F2503-08 [26.9])

The current *active* terms (MR safe, MR conditional, and MR unsafe) are defined in ASTM standard F2503 and cover MR safety-related interactions in the MR environment as discussed earlier in this chapter. As mentioned, MR image artifacts are not considered as a direct safety issue. However, information regarding image artifacts and degradation of image quality is requested in order to inform the MR operator.

Figure 26.13 shows the three current icons of MR safety, which should be used for direct marking on items if applicable and/or together with more detailed information in their instructions for use. The definitions of the terms are as follows:

MR conditional: an item that has been demonstrated to pose no known hazards in a specified MR environment with specified conditions of use. Field conditions that define the specified MR environment include field strength, spatial gradient, $\mathrm{d}B/\mathrm{d}t$ (time rate of change of the magnetic field), radiofrequency (RF) fields, and specific absorption rate (SAR). Additional conditions, including specific configurations of the item, may be required.

MR safe: an item that poses no known hazards in all MR environments. NOTE 1 – MR safe items include nonconducting, nonmagnetic items such as a plastic Petri dish. An item may be determined to be MR safe by providing a scientifically based rationale rather than test data.

MR unsafe: an item that is known to pose hazards in all MR environments. NOTE 2 – MR unsafe items include magnetic items such as ferromagnetic scissors. The definitions were extracted with permission from [26.9].

All devices including interventional tools that cannot be marked MR safe or MR conditional are marked as MR unsafe. MR unsafe devices may not enter the controlled access area, which can be larger than, but is often identical to, the RF shielded room or the MR room. Considering the definitions of the terms above, most items (and devices) will be subject to the MR conditional label, due to the fact that more or less specific conditions – result from MR testing that determine under which conditions the device's use is safe inside the MR environment. Basically, these devices contain metal components that can exhibit a certain level of force and torque, but also have the possibility of induced voltages and induced heating because of the materials' conductivity. The latter issues can be encountered for other conductive materials, e.g., carbon fiber. Other safety issues may exist that affect the safe operation of the device (e.g., electrical drives and circuits) or the MR system (degradation of image quality by, e.g., RF emission). Devices that consist of metal or other electrically conductive materials cannot be marked MR safe unless testing has shown no known hazards in all MR environments. This rigorous requirement is mainly satisfied by devices that consist solely of nonmetallic and nonconductive materials such as plastic, wood, and glass. The historical definition of MR safe could be used only in conjunction with the MR conditions under which the device was tested. Therefore, a device considered MR safe under at least one condition according to this historical term was (and still is today) in reality an MR conditional device, and should be labeled as MR conditional instead of MR safe using the F2503 terminology. Replacement of old device MR labeling is considered important by responsible agencies to ensure clear understanding of safety by the MR operator.

The content of MR labeling is based on the interactions mentioned above as include in ASTM F2503 [26.9] and in the FDA guidance on *Safety and Compatibility of Passive Implants* [26.45]. The following provides an extended example of such labeling.

Through nonclinical testing it has been demonstrated that the DEVICE is MR Conditional:

Force and Torque. *By testing of magnetically induced displacement force and torque the DEVICE showed a magnetically induced force of* 3% *(equal to* 2° *deflection) of the limit and a magnetically induced torque of* <1% *of the limit; static magnetic field of* 3 T *with a spatial gradient magnetic field of* 8.4 T/m *and a spatial gradient field product of* 18.0 T^2/m. *According to these test results, entering the MR magnet can be considered safe directly after implantation without safety discussion only for static magnetic fields of* 3 T *and less with a spatial gradient magnetic field of* <100 T/m *and a spatial gradient magnetic field product* <100 T^2/m *(values extrapolated and cut off; extrapolation exceeds 100 by far). Nonclinical testing has not been performed to rule out the possibility of implant migration at static gradient magnetic fields stronger than mentioned above.*

RF Heating. *In nonclinical testing with a* 3 T *MR SYSTEM NAME and MODEL and implant in worst-case phantom position, the DEVICE produced a maximum temperature rise of* 4.5 °C *in a static human torso-shaped phantom with a background temperature increase of* 3.4 °C *at a whole body averaged specific absorption rate (WBA-SAR) software display of* 3.8 W/kg (3.4 W/kg *in a phantom calorimetric test) for* 20 min *of MR scanning with transmit/receive body coil. The* local body *SAR shall be limited to* 5.0 W/kg *for using the MR body coil (theoretical estimated WBA-SAR* < 1.5 W/kg*).* Note that the WBA-SAR is inappropriate to scale exact local temperature increases. *Local SAR can deviate and result in much higher values than the WBA-SAR software display. Measurement inaccuracies and additional safety margins should be taken into account. Before each individual MR procedure, it might be necessary to discuss the situation with regard to patient benefit by consulting medical experts and MR physicists.*

Gradient Interactions. *(to be considered for induced voltages/heating/vibrations (standardization work in progress))*

MR Image Artifacts *can affect the surroundings of the implant with a distorted DEVICE* length *of* +20% *in a standard spin echo sequence and* +50% *in a standard gradient echo sequence. The image artifact of the DEVICE* diameter *can be* +300% *in a standard spin echo sequence and* +540% *in a standard gradient echo sequence.*

26.6 Interpretation of MR Labeling

Old and obsolete MR labeling information exists for several medical devices from the past. Therefore, the MR user should be aware of risks and pitfalls when MR scanning a patient with an implanted or externally attached medical device. It should also be verified whether there is a contraindication to MR scanning of patients with devices expressed in the instructions for use from the MR system manufacturer. Important information that is needed for interpretation is as follows:

- Exact and secure data about the medical device
- Written original MR labeling information from the instructions for use (from the device manufacturer or cooperating reliable MR safety online database) according to the example above
- Compatibility datasheet of the MR system used
- Exact and secure data about the RF coils of the MR system.

The MR labeling should have a section about magnetic forces and torques. Important parameters to be controlled based on the compatibility datasheet of the MR system are the static magnetic field with its spatial gradient magnetic field in T/m and the spatial gradient magnetic field product in T^2/m. Note that static magnetic fields of less (1 T) than that displayed in the labeling (e.g., 3 T) exist, but with higher spatial gradients, which is not covered by the static magnetic field only. Also it might be possible that the MR labeling only displays the maximum technically accessible spatial gradient magnetic field used for MR testing. Usually, the MR manufacturer presents the global maximum spatial gradient magnetic field. Comparing these values, two options for the decision that entering the MR magnet is safe are technically meaningful (but very different from liability point of view):

1. The MR labeling reports also the extrapolated spatial gradient magnetic field (e.g., < 100 T/m in the example above) which is directly comparable with the compatibility datasheet of the used MR system, or
2. MR labeling only reports the maximum technically accessible spatial gradient magnetic field used for MR testing, but in addition the corresponding deflection angle, which allows the experienced MR user to carry out the extrapolation for comparison with the compatibility datasheet. Note: A lot of MR labeling information are providing 700 or 720 G/cm. This value is low compared to the spatial gradient magnetic field inside the magnet bore of most modern MR systems. Furthermore, this value is technically meaningless if not a deflection angle is provided to make the user aware of how far the item might be away from the limit of 45° deflection and which safety margin is still remaining under the given value of 720 G/cm obviously used for testing.

The RF heating section of the MR labeling should provide sufficient data [MR system, field strength, temperature increases, scan duration, local and average SAR (software displayed, calorimetric, background), location, configuration, orientation, MR coil, pulse sequence, etc.] from MR testing of the sensitivity of heating of a device under so-called worst-case considerations in a specific test phantom.

This information provides an experimental example. Such data for safe MR scanning are difficult to transfer to a specific patient, especially if high sensitivity for RF heating is shown in the MR labeling. The specific absorption rate (SAR) and related B1/RF mode selections on the MR system are today the only parameters to limit the RF power and tissue heating of the patient. Note that the WBA-SAR is inappropriate to scale to exact local temperature increases at implanted devices. Only under the specific assumptions stated in the MR labeling is MR scanning intended to be safe. Therefore the device manufacturer should have considered the whole product matrix to be covered by the worst-case(s). Additionally; a transfer of test results from the phantom test to human is necessary to apply, using for example computer simulation.

Before each individual MR procedure, it might be necessary to discuss the situation with regard to patient benefit by consulting medical experts and MR physicists.

Future test methods currently under development will increase the reliability of data, resulting in less uncertainty for the patient population.

Data considered for the interaction with switched magnetic gradients should include both, parameters about software and hardware (device location and orientation).

Information from MR labeling about MR image artifacts should show dimensional data to inform the MR user about the extent of artifacts created by the device when using standard spin and gradient echo sequences. Optimization to decrease artifacts is feasible by adjusting the individual MR pulse sequence.

26.7 Discussion

Overall patient safety during an MR scan is obviously very important. Formulation of safety-related requirements in international standards enables this safety to be obtained and maintained. The risks related to MR scanning are therefore minimal. Nevertheless, careful attention by the operator is required before each scan, because interaction between the MR environment and materials or devices brought into the MR scanner with the patient may introduce additional risks to the patient. When these devices are accessories, labeling on the devices should provide enough information regarding allowable MR scan parameters, but the label cannot cover all potential circumstances.

Interviewing the patient prior to each scan remains very important, whereby it must be realized that sometimes the patient himself does not know the relevant information about the materials or devices which he brings into the scanner, because they are implanted. Therefore, patients should be subject to further clinical examination or imaging by other modalities if the situation remains unclear.

Currently, many studies are ongoing and results have sometimes been published on the MR safety of specific devices, although such results must always be interpreted with care. Often, these studies apply a limited set of MR parameters, defining in fact the conditions for safe scanning. Also, the number of patients studied is usually very limited, therefore the conclusions cannot be extrapolated to form general safety statements. Careful testing of MR safety for devices by the manufacturer or testing laboratory working according to ISO 17025, GLP or other quality standard therefore remains essential.

It is good to realize that, for reasons mentioned in this chapter, MR manufacturers in general still do not approve scanning of patients with implants, and formulate warnings in their instructions for use regarding the significant risks associated with scanning of patients with active or passive implants containing conductive materials.

When the implanted device is labeled as MR safe or MR conditional, the operator is informed via the instructions for use of the implant regarding the safety and possible conditions to be taken into account during scanning. The instructions for use from the MR manufacturer should explain according to IEC 60601-2-33 that further information is described in the documents from the implant manufacturer.

References

26.1 ISO/IEC TS 10974: International Technical Specification for, Requirements for the safety and compatibility of MRI for patients with an active implantable medical device; to be published 2010/2011

26.2 J. Nyenhuis, S.-M. Park, R. Kamondetdacha, A. Amjad, F.G. Shellock, A.R. Rezai: MRI and Implanted Medical Devices: Basic Interactions with an Emphasis on Heating, IEEE Tran. Device Mater. Reliab. **5**(3), 467–480 (2005)

26.3 G. Schaefers, A. Melzer: Testing methods for MR safety and compatibility of medical devices, Minim. Invasive Ther. **15**(2), 71–75 (2006)

26.4 G. Schaefers: Testing MR safety – An overview of the methods and current standards and compatibility, IEEE Eng. Med. Biol. Mag. **27**(3), 23–27 (2008)

26.5 ASTM F2052-06e1: Standard Test Method for Measurement of Magnetically Induced Displacement Force on Medical Devices in the Magnetic Resonance Environment (2006)

26.6 ASTM F2213-06: Standard Test Method for Measurement of Magnetically Induced Torque on Passive Implants in the Magnetic Resonance Environment (2006)

26.7 ASTM F2182-09: Standard Test Method for Measurement of Radio Frequency Induced Heating On or Near Passive Implants During Magnetic Resonance Imaging (2002) www.astm.org

26.8 ASTM F2119-01: Standard Test Method for Evaluation of MR Image Artifacts from Passive Implants (2001) www.astm.org

26.9 ASTM F2503-08: Standard Practice for Marking Medical Devices and Other Items for Safety in the MR Environment (2008)

26.10 F.G. Shellock, E. Kanal: *Magnetic Resonance Bioeffects, Safety, and Patient Management* (Raven, New York 1994)

26.11 IEC 60601-2-33: Medical electrical equipment – Part 2-33: Particular requirements for the safety of magnetic resonance equipment for medical diagnosis; 3rd edition, (2010)

26.12 K.M. Ludeke, P. Roschmann, R. Tischler: Susceptibility artefacts in NMR imaging, J. Magn. Reson. Imaging **3**, 329–343 (1985)

26.13 S.M. Boyd, W.S. Boivin, J.N. Coletta, K. Kempa, L.N. Kerr: Mechanical Forces on Implanted Medical Devices in MRI Units, Lab. Inf. Bull. **4208**, 1–10 (2000)

26.14 K.B. Baker, J.A. Tkach, J.A. Nyenhuis, M. Phillips, F.G. Shellock, J. Gonzalez-Martinez, A.R. Rezai: Evaluation of specific absorption rate as a dosimeter of MRI-related implant heating, J. Magn. Reson. Imaging **20**, 315–320 (2004)

26.15 S. Schwarzer, U. Jorczyk, A. Melzer, W. Zylka: Numerical Investigation of Electromagnetic Properties of MR Active Implants, Biomed. Tech. **49**(1), 184–185 (2004)

26.16 J.P. Thiele, M.-A. Golombeck, O. Dössel: Thermal heating of human tissue induced by electromagnetic fields of magnetic resonance imaging, Biomed. Tech. **47**(1), 743–746 (2002)

26.17 H. Kugel, C. Bremer, M. Puschel, H. Lenzen, B. Tombach, H. Van Aken, W. Heindel, Hazardous situation in the MR bore: induction in ECG leads causes fire, Eur. Radiol. **13**, 690–694 (2003)

26.18 K. Butts, J.M. Pauly, B.L. Daniel, S. Kee, A.M. Norbash: Management of Biopsy Needle Artifacts: Techniques for RF-Refocused MRI, J. Magn. Reson. Imaging **9**, 586–595 (1999)

26.19 A. Oppelt (Ed.): *Imaging Systems for Medical Diagnostics*, 2nd edn. (Publicis Corporate Publishing, Erlangen 2005) p. 592

26.20 J. Schenck: The role of magnetic susceptibility in magnetic resonance imaging: MRI magnetic compatibility of the first and second kinds, Med. Phys. **23**(6), 815–850 (1996)

26.21 B. Müller-Bierl, H.J. Graf, G. Steidle, F. Schick: Compensation of magnetic field distortions from paramagnetic instruments by added diamagnetic material: Measurements and numerical simulations, Med. Phys. **32**(1), 76–84 (2005)

26.22 A. Oppelt, I. Delakis: Safety aspects in interventional MRI, Z. Med. Phys. **12**(1), 5–15 (2002)

26.23 L.W. Bartels, C.J. Bakker, M.A. Viergever: Improved lumen visualization in metallic vascular implants by reducing RF artifacts, Magn. Reson. Med. **47**(1), 171–180 (2002)

26.24 C.J.G. Bakker, R. Bhagwandien, M.A. Meorland, M. Fuderer: Susceptibility artifacts in 2DFT spin-echo and gradient-echo imaging: The cylinder model revisited, J. Magn. Res. Imaging **11**, 539–548 (1993)

26.25 A. Melzer: MRI safety of medical devices and procedures, Minim. Invasive Ther. Allied Technol. **15**(2), 51–52 (2006)

26.26 G.P. Teitelbaum, H.V. Ortega, S. Vinitsky, H. Stern, J.S. Tsuruda, D.G. Mitchell, M.D. Rifkin, W.G. Bradley Jr.: Low-artifact intravascular devices: MR imaging evaluation, Radiology **168**, 713–719 (1988)

26.27 G.P. Teitelbaum, W.G. Bradley Jr, B.D. Klein: MR imaging artifacts, ferromagnetism, and magnetic torque of intravascular filters, stems and coils, Radiology **166**, 657–664 (1988)

26.28 A.H. Cragg, S.C. De Jong, W.H. Barnhart, S.K. Landas, T.P. Smith: Nitinol intravascular stents: results of preclinical evaluation, Radiology **189**, 775–778 (1993)

26.29 A.H. Matsumoto, G.P. Teitelbaum, M.J. Carvlin, K.H. Barth, M.A. Savin, E.P. Strecker: Gadolinium enhanced MR imaging of vascular stents, J. Comput. Assist. Tomogr. **14**, 357–361 (1990)

26.30 J. Link, J.C. Steffens, J. Brossmann, J. Graessner, S. Hacketal, M. Heller: Iliofemoral arterial occlusive disease: contrast-enhanced MR angiography for preinterventional evaluation and follow-up after stent placement, Radiology **212**, 371–377 (1999)

26.31 D.J. Taylor, G. Brown: Magnetic resonance imaging of vascular stents, Proc. Int. Soc. Magn. Reson. Med. **7**, 1892 (1999)

26.32 P.R. Hilfiker, H.H. Quick, J.F. Debatin: Plain and covered stent-grafts: in vitro evaluation of characteristics at three dimensional MR angiography, Radiology **211**, 693–697 (1999)

26.33 P.R. Hilfiker, H.H. Quick, M. Schmidt, J.F. Debatin: In vitro image characteristics of an abdominal aortic stent graft: CTA versus 3D MRA, MAGMA **8**, 27–32 (1999)

26.34 T. Klemm, S. Duda, J. Machann, K. Seekamp-Rahn, L. Schnieder, C.D. Claussen, F. Schick: MR imaging in the presence of vascular stents: a systematic assessment of artifacts for various stent orientations, sequence types, and field strengths, J. Magn. Reson. Imaging **12**, 606–615 (2000)

26.35 E.M. Haacke, R.W. Brown, M.R. Thompson, R. Venkatesan: *Magnetic Resonance Imaging, Physical Principles and Sequence Design* (Wiley-Liss, New York 1999)

26.36 E. Immel, C. K. Naber, H. H. Quick, F. Breuckmann, D. Daumgart, S. Aker, G. M. Kaiser, K. Thonak, A. Melzer: Resonant balloon expandable stent designed for the improved visualization and implantation in MRI, Am. J. Cardiol. **100**(8A), 14L (2007)

26.37 E. Immel, A. Melzer, C. Thonak: Self expanding stents designed for the improved visualization and implantation in MRI, Am. J. Cardiol. **100**(8A), 85L (2007)

26.38 K.M. Ludeke, P. Roschmann, R. Tischler: Susceptibility artefacts in NMR imaging, J. Magn. Reson. Imaging **3**, 329–343 (1985)

26.39 C.J. Bakker, R. Bhagwandien, M.A. Moerland, M. Fuderer: Susceptibility artifacts in 2DFT spin-echo and gradient-echo imaging: the cylinder model revisited, J. Magn. Reson. Imaging **11**, 539 (1993)

26.40 A.J. Evans, R.A. Blinder, R.J. Herfkens, C.E. Spritzer, D.O. Kuethe, E.K. Fram, L.W. Hedlund: Effects of turbulence on signal intensity in gradient echo images, Invest. Radiol. **23**, 512–518 (1988)

26.41 S. Fabregues, K. Baijens, R. Rieu, P. Bergeron: Hemodynamics of endovascular prostheses, J. Biomech. **31**, 45–54 (1998)

26.42 H.S. Muller, H. Schwarzenberg, F. Wesner, R. Drost, C.C. Gluer, M. Heller: Visualization of flow patterns from stents and stent-grafts in an in vitro flow-model, Invest. Radiol. **33**, 762–770 (1998)

26.43 C.R. Camacho, D.B. Plewes, R.M. Henkelman: Nonsusceptibility artifacts due to metallic objects in MR imaging, J. Magn. Reson. Imaging **5**, 75–88 (1995)

26.44 Historical standard, ASTM F2052-02: Standard Test Method for Measurement of Magnetically Induced Displacement Force on Medical Devices in the Magnetic Resonance Environment (ASTM International, West Conshohocken 2002) A copy of the complete standard may be purchased from service, www.astm.org

26.45 Guidance for Industry and FDA Staff; Establishing Safety and Compatibility of Passive Implants in the Magnetic Resonance (MR) Environment; Document issued on: August 21, 2008; US Department of Health and Human Services, Food and Drug Administration, Center for Devices and Radiological Health; http://www.fda.gov/cdrh/osel/guidance/1685.pdf

Part D Therapeu

Part D Therapeutic Devices

27 **Long-Term Ventilators for Intensive Therapy**
Thomas Peyn, Lübeck, Germany

28 **Defibrillators and ICD Systems**
Rüdiger Kramme, Titisee, Germany

29 **Laser Systems**
Frank Zgoda, Berlin, Germany

30 **Inhalational Anesthesia Devices**
Erich Siegel, Lübeck, Germany

31 **Extracorporeal Blood Purification Systems**
Jörg Vienken, Bad Homburg, Germany

32 **Heart-Lung Machines**
Ulrich Böckler, München, Germany
Andreas Hahn, München, Germany

33 **Application of Shock Waves and Pressure Pulses in Medicine**
Friedrich Ueberle, Hamburg, Germany

34 **High-Frequency Surgery**
Bernhard Hug, Umkirch, Germany
Reiner Haag, Fridingen, Germany

35 **Medical Radiation Therapy**
Peter H. Cossmann, Wettingen, Switzerland

36 **Mechanical Circulatory Support Systems**
Roland Hetzer, Berlin, Germany
Ewald Hennig, Berlin, Germany

37 **Neural Interfaces for Implanted Stimulators**
Xiao Liu, London, UK
Andreas Demosthenous, London, UK
Nick Donaldson, London, UK

38 **Cardiac Pacemaker Systems**
Armin Bolz, Karlsruhe, Germany

39 **Introduction to Neuroprosthetics**
Klaus-Peter Hoffmann, St. Ingbert, Germany
Silvestro Micera, Zürich, Switzerland

40 **Implantable Microsystems**
Prashant Tathireddy, Salt Lake City, USA
Florian Solzbacher, Salt Lake City, USA
Robert Hitchcock, Salt Lake City, USA
Klaus-Peter Hoffmann, St. Ingbert, Germany

41 **Visual Prostheses**
Eduardo Fernandez, Elche, Spain
Klaus-Peter Hoffmann, St. Ingbert, Germany

42 **Rehabilitation and Therapeutic Robotics**
Loredana Zollo, Rome, Italy
Dino Accoto, Rome, Italy
Silvia Sterzi, Rome, Italy
Eugenio Guglielmelli, Rome, Italy

43 **Cardiac Devices and Testing**
Michael Bateman, Minneapolis, USA
Stephen A. Howard, Minneapolis, USA
Christopher Rolfes, Minneapolis, USA
Timothy G. Laske, Mounds View, USA
Paul A. Iaizzo, Minneapolis, USA

44 **Functional Electrical Stimulation in Rehabilitation and Neurorehabilitation**
Kei Masani, Toronto, Canada
Milos R. Popovic, Toronto, Canada

45 **Treatment Planning and Patient Treatment**
Golam Abu Zakaria, Gummersbach, Germany
Wilhelm Schütte, Gummersbach, Germany
Stephan Garbe, Bonn, Germany

27. Long-Term Ventilators for Intensive Therapy

Thomas Peyn

Long-term ventilators are life-sustaining devices. Depending on patient conditions, intensive care ventilators must support or replace spontaneous breathing. In this context, optimizing mechanical ventilation and weaning the patient off the ventilator are typical challenges that caregivers face on the intensive care unit every day. To meet all clinical requirements, current ventilators provide many different ventilation and weaning modes, some of which are explained in this chapter.

27.1 **Tasks of the Ventilator** 525

27.2 **Function and Components of a Long-Term Ventilator** 526
- 27.2.1 Power Supply 527
- 27.2.2 Gas Mixer 527
- 27.2.3 Pressure Generator 527
- 27.2.4 Breathing System 528
- 27.2.5 Gas Humidifier 528
- 27.2.6 Expiratory Valve 528
- 27.2.7 Operating and Display Unit 528
- 27.2.8 Alarm System 529
- 27.2.9 Patient Monitoring 529

27.3 **Technical Implementation** 529
- 27.3.1 Continuous Flow Systems 529
- 27.3.2 Demand Flow Systems 530
- 27.3.3 Combined Flow Systems 530

27.4 **Controlling the Ventilator** 530
- 27.4.1 Start of Inspiration 530
- 27.4.2 Switchover Behavior 531

27.5 **Ventilation Procedures** 531
- 27.5.1 Controlled Ventilation (Total Respiratory Substitution) 531
- 27.5.2 Supported Spontaneous Breathing 534
- 27.5.3 Spontaneous Breathing (Without Respiratory Support) 536
- 27.5.4 Mixed Ventilation 536

27.6 **Ventilation Extras and Special Functions**. 540
- 27.6.1 Inverse Ratio Ventilation 540
- 27.6.2 Sigh 540
- 27.6.3 AutoFlow 540
- 27.6.4 Automatic Tube Compensation 542
- 27.6.5 Noninvasive Ventilation 542
- 27.6.6 Apnea Ventilation 543

27.7 **Patient Monitoring and Alarm Limits** 543

27.8 **Weaning Strategy and SmartCare/PS** 543

27.1 Tasks of the Ventilator

In order to supply the body with air containing oxygen, the volume of the lungs increases when breathing in. When breathing in, the pressure in the lungs drops, and in this way atmospheric air is sucked into the lungs. The reverse applies when breathing out: thoracic and respiratory muscles return to their idle state, and the lung volume reduces. The resulting intrapulmonary increase in pressure results in the expiratory gas, which contains carbon dioxide, being expired. In an idle state, this procedure is repeated about 10–20 times per minute, in the case of an adult. For bodyweight of 75 kg, the volume moved with each breath varies between about 350 and 850 ml. Taking into account additional physical activities, which can result in a significant increase in respiration, it is more than possible that our respiratory musculature shifts a total volume of 10 000–15 000 l of air over the course of a day. These figures underline the fundamental importance of our respiratory system for the entire organism. If the interplay of the organs and muscles involved in respiration is interrupted, ventilators can completely take over breathing or, if the patient is simply breathing too flatly, a share of the breathing

Table 27.1 Reasons for using ventilators

Reason for ventilation	Examples of possible causes
Functional limitations of the respiratory musculature	Due to a muscular injury or paralysis
Pathophysiological changes to the respiratory system and/or pulmonary tissue	Due to increased airway resistances or to a reduced lung compliance
Disruptions to respiratory mechanism	In case of thorax injuries
Disruptions to gas exchange/diffusion	Through the accumulation of pulmonary liquid or changes to the alveolar membrane
Disorders of the respiratory drive	Due to neurological disruption, cranial illness or injury

work. The many reasons that may require the use of a ventilator can, in principle, be allocated to five different issues (Table 27.1).

As well as the examples named here, use of a ventilator can also be required due to undesired respiratory depression caused by strong pain relievers or use of muscle relaxants. The examples listed make clear that patients can be dependent on breathing gas being supplied to them via special gas dosage equipment for many different reasons. A differentiation must be made here between a patient who is still breathing spontaneously and only needs to be offered an increased concentration of oxygen from a ventilation therapy device and a patient who is no longer able to perform the necessary breathing work and must be ventilated by a ventilator. Regardless of the breathing disruption, a long-term ventilator performs a life-saving function in all cases. Against this background, the ventilator must fulfill three different tasks:

1. Oxygenation of the patient: Provide and supply the patient with a mixture of oxygen and air.
2. Partial or total assumption of respiratory work: Generate and dose defined gas flow and respiratory pressure.
3. Monitoring of the device and patient: Generate alarms and visualize changes.

Ventilation therapy is primarily aimed at supporting or replacing the patient's breathing, thus guaranteeing efficient gas exchange. For this reason, the user of a ventilator must ensure that the selected device configuration

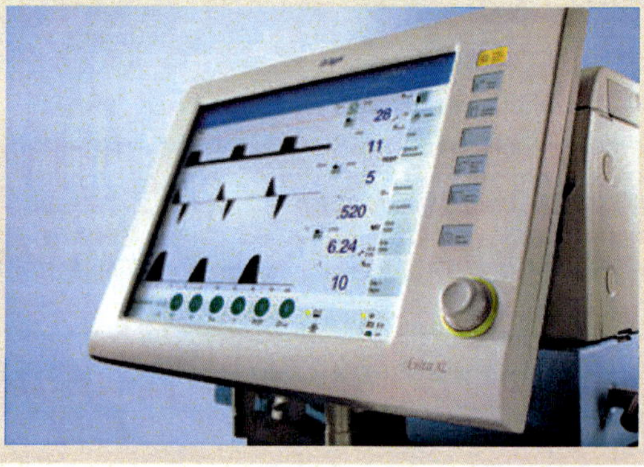

Fig. 27.1 Intensive care ventilator (Evita XL, Dräger Medical)

supplies the patient with adequate oxygen and that the ventilation sufficiently reduces the amount of carbon dioxide in the body. The required efficiency control is performed using arterial blood-gas analysis (Chap. 41). With a long-term respiratory device (Fig. 27.1) the therapist is given a tool with the help of which he can simultaneously train and strengthen the respiratory musculature of the patient, so that the patient gradually becomes used to breathing independently. This weaning process can last hours, days or even weeks. One reason for this wide range is the different duration of ventilation, during which atrophy of the respiratory musculature can occur. Furthermore, the duration of the weaning phase is influenced by the clinical picture of the patient, the recovery process, and the overall clinical situation.

27.2 Function and Components of a Long-Term Ventilator

If the patient's own respiratory activity is insufficient, some form of external substitution or complementary intervention is required to ensure that the breathing gas is transported to the lungs. Artificial respiration can be achieved by the delivery of defined gas flows and by the creation of positive pressures. There is a significant difference from spontaneous breathing: When the patient breathes spontaneously, the intrapulmonary pressure stays roughly at the baseline of atmospheric pressure. During artificial respiration, in contrast, the breathing gas is supplied to the lungs with a positive pressure. A hose or breathing circuit forms the interface between the ventilator and the patient. It is generally designed as a dual-hose circuit (consisting of an inspiratory and expiratory hose) in the case of clinical intensive care ventilators. In principle, the patient can be connected with the ventilation system either via a face mask or via a tube inserted into the trachea. The user will decide which method to use based on the respective circumstances such as the expected ventilation period and take the appropriate medical measures to prepare the patient.

Different system components are required for the technical implementation of a long-term ventilator. Their systematic arrangement is shown in Fig. 27.2.

27.2.1 Power Supply

Ventilators require electric power, oxygen, and compressed air to operate properly. The devices are usually supplied via an external power source as well as via the hospital's central gas supply (with a supply pressure of approximately 3–6 bar or 39–87 psi). In functional areas without a central gas supply or during transportation of patients within the hospital, it is necessary to ensure the functioning of the device by other means. Potential solutions include the use of separate compressors, compressed gas cylinder packs, and accumulators. Increasingly, ventilators not dependent on compressed air, which provide ventilation by filtering and using ambient air, are also being used. Only an oxygen source and a power supply are needed to operate such types of ventilators.

27.2.2 Gas Mixer

The gas mixer allows the user to vary the oxygen concentration of the inspiratory gas between 21% and 100% by volume. While ventilators often used to be supplied by external mechanical gas mixers, technological developments have resulted in an electronically controlled gas mixer integrated in the ventilator becoming standard. Gas mixers usually have a dual function in that they are also responsible for ensuring that the breathing gas to be supplied is prepared and delivered in the required quantity and at the correct rate. It is often the threshold ranges which pose the greatest challenges to these metering systems. If, for example, a volume of 20 ml with an oxygen concentration of 30% by volume is required for ventilation, 17.7 ml of gas must be delivered via the compressed air valve and 2.3 ml via the oxygen valve.

27.2.3 Pressure Generator

The pressure or flow generator is responsible for delivering the mixed gas prepared by the gas mixer according to the ventilation parameters set in the operating unit. At the simplest level, the gas flow or pressure can be created manually using a manual ventilation bag. Technically comparable methods are to be found in older

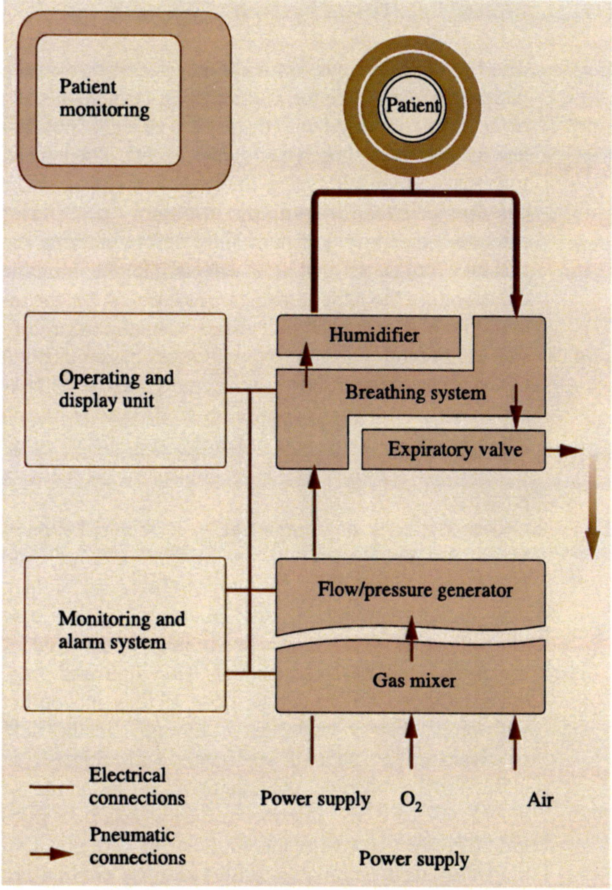

Fig. 27.2 Components of an intensive care ventilator

ventilators and anesthesia workstations. These devices generate the inspiratory gas flow using mechanically moving parts, for example, bellows or piston pumps. Modern intensive ventilators, on the other hand, feature combined gas mixture and metering devices which provide the required gas quantities directly from the gas supply systems. From a technical point of view, these components can be designed as a flow or a pressure generator. A flow generator is a controlled valve whose output provides a defined gas flow; the output pressure is not specified. A pressure generator, on the other hand, behaves similarly to a compressor, whose output provides a defined pressure with an unspecified gas flow. Pressure generators are often used to drive ventilators which are not dependent on compressed air and that use suctioned ambient air for patient ventilation.

27.2.4 Breathing System

The breathing system forms the interface between the patient and the ventilator. Clinical long-term ventilators are usually connected with the patient via an inspiratory and an expiratory hose (dual-hose circuit). The expiratory valve is closed during the inspiratory phase. The gas flow delivered through the inspiratory port passes through a breathing gas humidifier before entering the patient's lungs, so that it is adapted to the climatic conditions in the patient's lungs (see later). After the inspiratory phase, the patient exhales when the expiratory valve is opened. Thus, the expiratory gas passes through the ventilator again, but is not reused for the following inspiration as with anesthesia workstations. Based on this characteristic, the breathing systems of long-term ventilators are also referred to as *non-rebreathing circuits*.

For the sake of completeness, it should be mentioned that ventilators with *one-tube circuits* are used in individual clinical function areas. These circuits are also non-rebreathing circuits; however, in this case the expiratory valve is not positioned on the side of the device but near the patient. The improved handling provided by the single hose is thus in conflict with an expiratory valve that is arranged in the field of vision of the patient. Ventilators with single-hose circuits have become the standard in the areas of emergency care and intermediate care based on individual product requirements. In intensive care units, on the other hand, ventilators with two-hose circuits are more common due to the required expiratory volume monitoring.

27.2.5 Gas Humidifier

Breathing gas humidifiers are used to warm and humidify the inspiratory gas. Since the dry and relatively cool supply gas would dry out the patient's airways, it is absolutely vital to humidify and warm the breathing gas to avoid the risk of causing irreversible damage to the ciliated epithelium. Active or passive breathing gas humidifiers are available for this purpose. Active breathing gas humidifiers are located in the inspiratory limb and use electrical energy to heat a water bath. When the cold, dry gas passes over the water surface it absorbs water molecules and is thus warmed and humidified. Passive breathing gas humidifiers, termed heat and moisture exchangers (HMEs), in contrast, are placed close to the patient, between the tube and the Y-piece. These products are designed to buffer a significant fraction of the moisture and heat expired by the patient. This retained moisture is then used to condition the inspired gas passing through the HME during the next inspiration. Using a HME together with an active breathing gas humidifier in a single breathing circuit is not permitted as it would significantly impair the resistance of the HME. HMEs are also unofficially referred to as *filters*, which is a misnomer as they have nothing in common with each other, unless they also fulfill a hygienic function (for example, HMEs with an integrated microbial filter).

27.2.6 Expiratory Valve

Besides the above-mentioned switchover function, the expiratory valve is also responsible for two additional important tasks. If the valve is not opened completely during expiration, a positive end-expiratory pressure (PEEP) is created in the lungs. This PEEP pressure is therapeutically important because it increases the gas exchange surface of the lungs. An adequate PEEP can also prevent the collapse of individual alveolar areas. The expiratory valve can also fulfill another function providing that the ventilator also controls the valve during the inspiratory phase. In this case, the expiratory valve can compensate for undesired pressure rises in the breathing system caused, for example, by the patient coughing.

27.2.7 Operating and Display Unit

The operating and display unit forms the interface between the ventilator and the user. These units are often touchscreens designed to display pressure and flow curves as well as multiple menus for setting different ventilation modes, adjusting alarm limits or measured value overviews, etc. The parameter settings entered in the operating unit control the other device components and therefore significantly influence the ventilation pattern applied to the patient.

27.2.8 Alarm System

The monitoring and alarm system ensures that the ventilation parameters set in the operating and display unit are actually applied. It also issues audible and visual alarms to alert staff to critical changes in the patient's condition or technical malfunctions. The following is measured: the inspiratory oxygen concentration (controlled by the gas mixer), and the ventilation pressure and ventilation volume (to monitor the pressure/flow generator). In addition, the use of an active breathing gas humidifier makes inspiratory breathing gas temperature measurement with an alarm system obligatory (to control the breathing gas temperature). Please note that this chapter does not provide a detailed explanation of the sensor technology or the different measurement principles as this information can be found in Chap. 47.

27.2.9 Patient Monitoring

Patient monitoring is used to monitor the patient's vital functions. This includes, for example, electrocardiogram (ECG), blood pressure (noninvasive and/or invasive), oxygen saturation, and often also the measurement of the carbon dioxide concentration in the breathing gas. Although patient monitors do sometimes display ventilation data, these devices are to be seen as an independent display unit with an alarm facility. This makes patient monitoring a significant component of the intensive care workstation, but it is not part of the ventilator. For more information on this topic please refer to Chap. 48.

27.3 Technical Implementation

Ventilators differ according to their gas flow behavior. Devices which provide ventilation by generating a continuous gas flow (continuous flow systems) are used as well as demand flow devices which only provide a gas flow during the inspiratory phase.

27.3.1 Continuous Flow Systems

A continuous flow of oxygen and compressed air is supplied from the pressurized gas sources via metering valves. This gas flow is guided via an inspiratory hose to the patient and from there via an expiratory hose to an expiratory valve. Thus, a patient who is breathing spontaneously can always cover his air requirements from the gas flow. If the patient needs to be ventilated, the expiratory valve is closed for a defined period. This leads to a pressure rise in the breathing system and the gas flow is supplied to the patient's lungs. The inspiratory ventilation pressure is limited by an adjustable pressure control valve. With

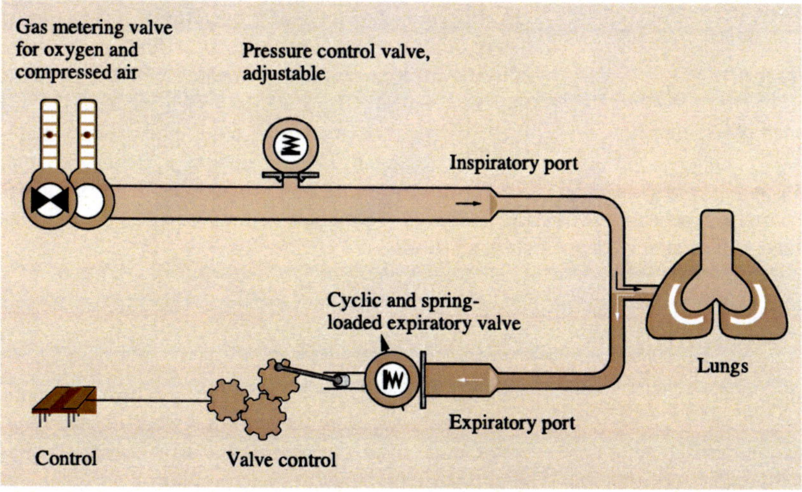

Fig. 27.3 Continuous flow system

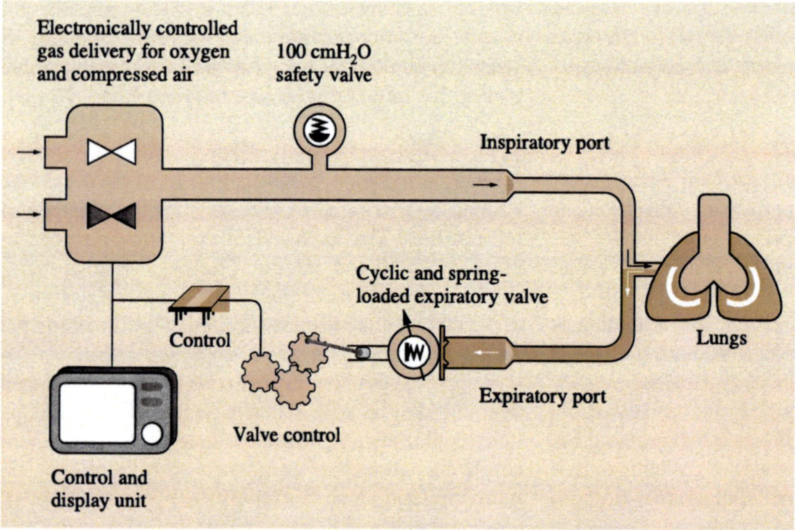

Fig. 27.4 Demand flow system

the switchover to expiration, the expiratory valve is opened and the pressure in the breathing system returns to the ambient or the PEEP pressure. The breathing gas exits the lungs through the existing decrease of pressure, is mixed with the constant flow, and flows through the expiratory valve. In practice, continuous flow systems are preferably used in neonatal respiration and in respiration therapy. Respiration therapy or also continuous positive airway pressure (CPAP) devices differ from ventilators in that their expiratory valve can only generate a constant, positive pressure. The use of a respiration therapy device is thus reserved for spontaneously breathing patients, who are, for example, dependent on an increased oxygen concentration (Fig. 27.3).

27.3.2 Demand Flow Systems

In contrast to the above-mentioned continuous flow systems, demand flow systems only supply a gas flow during the inspiratory phase. In the case of controlled ventilation, the gas flow is generally synchronized in accordance with the user-defined time pattern (time-cycled ventilation). This can be defined with controllers, for example, for the inspiratory and expiratory time according to the operating philosophy of the device. Demand flow systems respond to spontaneous breathing activities by the implementation of a trigger criterion (Sects. 27.4, 27.5). Demand flow systems reduce the gas consumption considerably and are therefore also used in inhalators which promote the breathing depth of the patient and/or apply aerosols synchronized with inspiration (Fig. 27.4).

27.3.3 Combined Flow Systems

Some ventilators use combined flow systems for technical reasons or due to special operational requirements. These deliver a low constant or base flow on the order of some liters per minute, compared with the patient's gas requirement. These systems then increase the gas flow to meet the respective requirement when inspiration is triggered. Combined flow systems are used, for example, in ventilators which need to cover the entire spectrum of patients ranging from premature babies through to adults.

27.4 Controlling the Ventilator

27.4.1 Start of Inspiration

The start of an inspiratory phase is mandatory in controlled ventilation modes after the expiratory time has elapsed. In ventilation modes during which the ventilator merely supports the patient's work of breathing, on the other hand, the ventilator must recognize the patient's breathing activities and provide an adequate gas flow. The trigger of the ventilator acts as a detector for the spontaneous breathing efforts. This is a device function with a user-adjustable threshold (known as the trigger threshold). Although the patient's spontaneous breathing activity is reflected in pressure and flow changes in the breathing system and both signals could in principle be used as a trigger criterion, flow trigger systems are mainly used in current ventilators because the tendency for artifacts is low. In this case, the ventilator responds to the flow signal generated by the patient and the trigger threshold can be adjusted in units of l/min.

27.4.2 Switchover Behavior

The ventilator's switchover behavior from inspiration to expiration or vice versa is influenced significantly by the long-term ventilator control. In this regard, a distinction can be made between four different types of control:

- Time-cycled (occasionally also referred to as time-time-cycled) mode. The duration of the inspiratory and expiratory phase and thus also the switchover time is defined by user-adjustable time parameters. Controlled (mandatory) ventilation modes are often implemented with this control type, as it enables patient ventilation with a defined ventilation ratio (inspiratory to expiratory time).
- Pressure-cycled mode. In pressure-cycled modes, the switchover from inspiration to expiration takes place when an adjustable ventilation pressure has been reached. Inspiration takes place after a time phase has elapsed (pressure-time-cycled mode) or by a patient activity and on reaching a trigger criterion. Nowadays, pressure-cycled mode is mainly used with impermissibly high pressure rises in connection with an audible alarm as a safety element.
- Volume-cycled mode. The switchover from inspiration to expiration occurs in this case after the delivery of a ventilation volume (tidal volume) to be defined by the user. As with the pressure-cycled mode, the next inspiration takes place after a time phase has elapsed (volume-time-cycled mode) or by a patient activity. The volume-cycled mode is sporadically used in supporting ventilation modes. Moreover, in US and American-influenced markets it is used in mandatory ventilation modes, too.
- Flow-cycled mode. This control is primarily used for supporting ventilation modes and initiates the expiratory phase when the gas flow decreases to a defined value (e.g., 25% of inspiratory peak flow). In flow-cycled modes, the start of inspiration is mostly initiated by a patient activity and on reaching a trigger criterion.

27.5 Ventilation Procedures

Basically, ventilators can only follow two different behavior patterns despite the virtually endless number of ventilation modes. These are controlled and supported ventilation methods. The main difference between these techniques is as follows: During controlled (mandatory) ventilation, the behavior of the ventilator is defined by the user-configured ventilation parameters. In supported ventilation modes, on the other hand, the ventilator follows the patient's breathing activities and only provides support by increasing the pressure in the breathing system. Variations of these two ventilation methods are pure spontaneous breathing without any support pressure and mixed ventilation modes with alternating mandatory and supporting phases. Figure 27.5 provides an overview of the different respiration/mechanical ventilation methods and illustrates the medical necessity of having an individual ventilation therapy.

27.5.1 Controlled Ventilation (Total Respiratory Substitution)

The ventilation modes belonging to this category are characterized by the fact that the patients do not breathe spontaneously (SPN) and that the ventilator performs the total work of breathing. The different modes can be divided into volume-controlled (VC) modes and pressure-controlled (PC) modes. Volume-controlled modes are characterized by the fact that the patient is ventilated by a user-defined volume. In volume-controlled modes the airway pressure is a resulting parameter and provides information on lung expansibility (compliance) and airway resistances (resistance). In contrast to this, the user specifies an inspiratory ventilation pressure in pressure-controlled modes. Conversely, in these modes it is not the ventilation pressure but

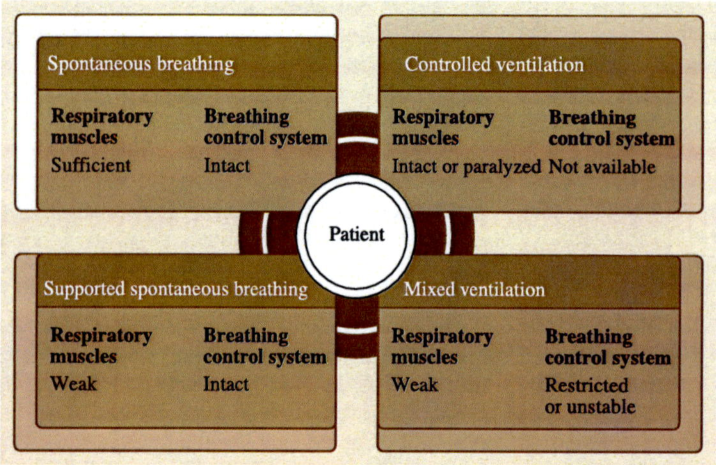

Fig. 27.5 Forms of respiration and mechanical ventilation

the tidal volume which provides information on the condition of the patient's lungs. In some cases ventilation modes are manufacturer specific but naming also varies in international linguistic usage. To enhance usability, particular operating philosophies or a prefix to the ventilation mode have been implemented by the manufacturers.

Volume-Controlled Ventilation

This mode is suitable, for example, for patients with healthy lungs but without spontaneous breathing. It is important to note that in this mode the device does not respond to any spontaneous breathing activities by the patient.

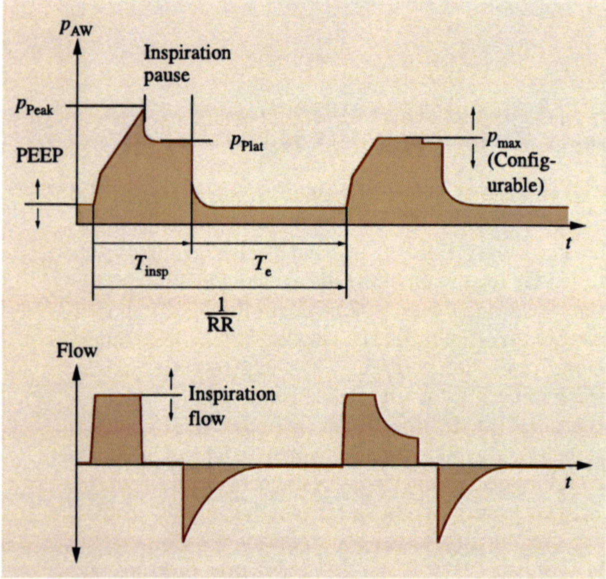

Fig. 27.6 Volume control, continuous mandatory ventilation

The chronological sequence of inspiration and expiration (Fig. 27.6) is as follows: At the beginning of inspiration the ventilator closes the expiratory valve and delivers the (constant) inspiratory flow according to the *flow setting*. This leads to an increase in pressure in the breathing system and in the lungs. If the inspiratory flow setting is so high that the set tidal volume (V_T) is reached before the inspiratory time T_{insp} elapses, the inspiratory valve closes and the delivery of breathing gas stops. Owing to pressure compensation between the breathing system and the patient's lungs, the ventilation pressure decreases from the peak pressure P_{Peak} to the plateau pressure value P_{Plat}. The expiratory valve remains closed until the inspiratory time T_{insp} has elapsed. This inspiratory pause (insp. pause) can be recognized as a plateau pressure phase P_{Plat} in the pressure curve. The expiratory valve opens when the inspiratory time T_{insp} has elapsed. The pressure in the breathing system therefore falls to the PEEP pressure level generated by the expiratory valve. The expiration gas leaves the lungs and is transported via the expiratory branch of the ventilator into the atmosphere. The next inspiration follows after the expiratory phase, whose length is determined by the respiratory rate setting (RR or f), has elapsed. To avoid pressure peaks, the maximum inspiratory working pressure of the ventilator can be limited via the P_{max} controller. If provided by the device, pressure limitation is available in all volume-controlled ventilation modes. The term pressure-limited ventilation (PLV) used in this context does not therefore provide any information on the ventilation mode that is currently being applied. The abbreviation PAW used in the graphic stands for *airway pressure*.

The user should keep in mind that the sudden resumption of spontaneous breathing in this ventilation mode can lead to serious dyssynchronies between the patient and the ventilator. In such an event, the patient would literally fight with his spontaneous breathing against the mandatory ventilation pattern. The activation of the trigger function (VC-AC = volume control-assist control or $IPPVA_{Assist}$) can compensate for this deficiency to an extent. However, in this operating mode, the device will respond to every inhalation effort by the patient by applying the set ventilation stroke. The increase of the respiratory rate and the minute volume then lead inevitably to a decrease of the arterial carbon dioxide concentration. The breathing drive is thus reduced and may lead to respiratory standstill (apnea). Volume-controlled ventilation (VC-CMV) or intermittent positive pressure ventilation (IPPV) is becoming less important, as the VC-SIMV ventilation mode (mixed ventilation) provides an identical ventilation and, moreover, also features a weaning mode (Sect. 27.5.4).

Independent Lung Ventilation. Independent lung ventilation (ILV) is the independent ventilation of each lung using a special double-lumen tube whose ends are each connected to a separate ventilator. To chronologically synchronize the actions of both ventilators, Evita ventilators, for example, can be electrically connected with each other using a special connection cable. One of the ventilators assumes the *master* function, determining the breathing rhythm. Depending on the selected presetting, the second ventilator, also referred to as the *slave*, performs ventilation in parallel, delayed or inversely to the actions of the master. ILV is only available as volume-controlled ventilation.

Pressure-Controlled Ventilation

During classic pressure-controlled ventilation, user-defined pressure alteration intervals are used for patient ventilation. Pressure-controlled ventilation dominates in neonatology intensive care units and is extremely important for treating adults with pulmonary damage, but is not only reserved for this group. There are major differences between the conventional pressure-controlled ventilation (PCV) and pressure control-biphasic positive airway pressure (PC-BIPAP)/pressure control-continuous mandatory ventilation (PC-CMV) as regards the spontaneous breathing characteristics. While the patient's activities in PCV can create undesired pressure rises and lead to corresponding alarms, the PC-BIPAP and PC-CMV ventilation modes give patients the complete freedom to inhale or exhale at any time. For more information on this topic please refer to Sect. 27.5.4.

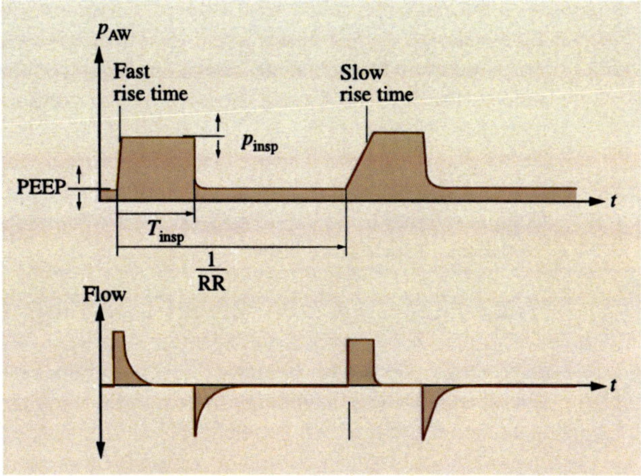

Fig. 27.7 Pressure control, biphasic positive airway pressure (PC-BIPAP)

The chronological sequence of inspiration and expiration (Fig. 27.7) is as follows: At the beginning of inspiration the ventilator closes the expiratory valve and delivers an initial flow, which causes a rise in pressure in the breathing system. Although the user-defined inspiratory pressure P_{insp} is built up very rapidly in the breathing system, the time required until the prevalent pressure in the breathing system can be built up in the lungs is much longer because of the resistance of the tube and the expansion of the lungs. The continuously decreasing pressure difference between the pressure in the breathing system and the pressure in the lungs in this phase is reflected in a decelerating gas flow delivery. The ventilator stops the gas flow delivery as soon as a total pressure balance has been achieved. The expiratory valve opens when the inspiratory time T_{insp} has elapsed. The pressure in the breathing system therefore falls to the PEEP pressure level generated by the expiratory valve. The expiration gas leaves the lungs and is transported via the expiratory branch of the ventilator to the atmosphere. The next inspiration follows after the expiratory phase, whose length is determined by the respiratory rate setting (RR or f), has elapsed. The time in which the ventilator changes the pressure from PEEP pressure to the inspiratory pressure level can be adjusted by the *ramp* controller.

In pressure-controlled ventilation it is also possible to activate the trigger function and to synchronize the rise of pressure to spontaneous breathing efforts of the patient. The PC-CMV mode then changes to pressure control-assist control (PC-AC). The result of this is that the ventilator responds to every inspiration effort by the patient with a mandatory inspiratory phase. The PC-AC mode is referred to as $BIPAP_{Assist}$ on some ventilators but behaves as described above.

27.5.2 Supported Spontaneous Breathing

In principle, the modes belonging to this group have several similarities, including, for example, that the spontaneously breathing patient triggers the start of inspiration and that the ventilator responds by producing a positive pressure in the breathing system. For this reason, supported spontaneous breathing is often known as *pressure-supported spontaneous breathing*. Internationally, the use of the abbreviations for pressure support (PS) and assisted spontaneous breathing (ASB) differs, but generally users are familiar with both names. Switchover from inspiration to expiration takes place when mode-specific criteria such as flow or volume target values have been met. Supported spontaneous breathing is suitable for patients with weakened respiratory muscles and an intact respiratory drive.

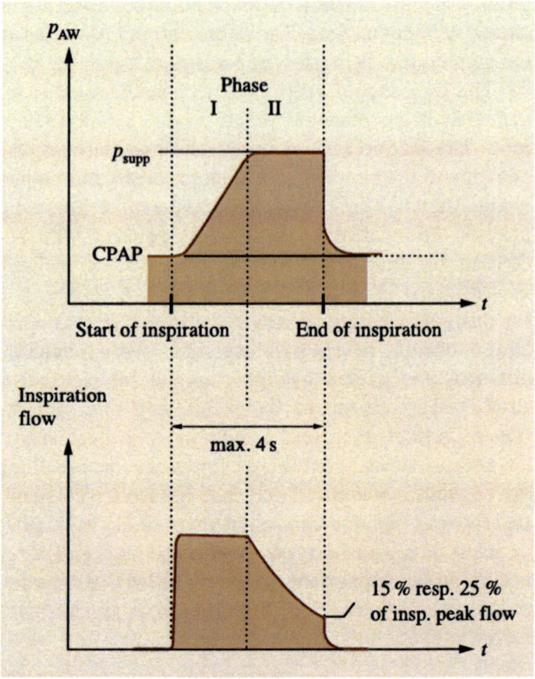

Fig. 27.8 Pressure support (PS)

Conventional Pressure Support

The chronological sequence of inspiration and expiration (Fig. 27.8) is as follows: As soon as the ventilator is triggered by an inhalation effort, the ventilator delivers a gas flow and generates the inspiratory pressure defined by the adjuster Δp_{Supp}. The speed at which the ventilator achieves the set pressure support level can be influenced by the *ramp* controller. A rapid rise time results in a high initial flow and reduces the patient's work of breathing. The switchover to expiration takes place when the level of pressure support has been reached *and* after the inspiratory gas flow has decreased to a defined inspiratory flow (e.g., 25% of inspiratory peak flow). Although in this method the patient is always provided with the same pressure support irrespective of the intensity of his breathing efforts, the spontaneously breathed volumes can vary depending on patient activity.

Proportional Pressure Support

The proportional pressure support (PPS) and proportional assist ventilation (PAV) modes are further developments of the pressure support mode described in the last section. While in conventional pressure support the ventilator provides the support pressure itself when the patient no longer actively breathes, in PPS (PAV) mode the ventilator supports the patient in proportion to the work of breathing performed by the patient himself. This also results in the markedly high variability of tidal volumes which follow the natural course of spontaneous breathing. The main idea and objective of proportional pressure support is that the ventilator should only take over that part of the work of breathing which the patient is not capable of performing himself. Since the work of breathing to be performed is determined by the airway resistance and the elastic recoil force of the lungs, PPS can be individually adjusted to the patient's respective disease by means of two different adjustment parameters ($Vol._{Assist}$ and $Flow_{Assist}$). This differentiation is necessary because a patient with increased airway resistance has to cope with a fundamentally different problem than a patient with a pathologically changed elasticity. In the first case the gas flow is impaired, whereas in the second case the volume of spontaneous breathing is insufficient. A different pressure curve profile is therefore allocated to the controllers. When the resistive work of breathing increases, support is provided by means of a flow proportional pressure characteristic (FlowAssist). If, on the other hand, the disease is characterized by a marked elastic work of breathing, the pressure curve is proportional to the tidal volume (Vol.Assist). For better understanding, the different pressure curve profiles are shown separately in the lower part of Fig. 27.9. The pressure support which is actually effective for the patient results from the overlap of the two curve functions. Support that has been calculated too generously can result in the patient receiving a greater volume than intended (*runaway*) because of the reinforcing characteristics of the system. For this reason, the pressure and tidal volume alarm limits are extremely important here.

Variable Pressure Support

By observing one's own breathing activities it can be noticed that breathing volumes are not constant but vary within a certain range. Variable pressure support is available in the spontaneous breathing mode SPN-CPAP and tries to mimic the physiologic course of respiration. For this purpose in this mode the level of pressure support changes from breath to breath (Fig. 27.10). Variations in pressure result from the settings Δp_{supp} and Press. var. (pressure variation from 0% to 100%). Preliminary findings indicate that this mode might have beneficial effects on oxygenation.

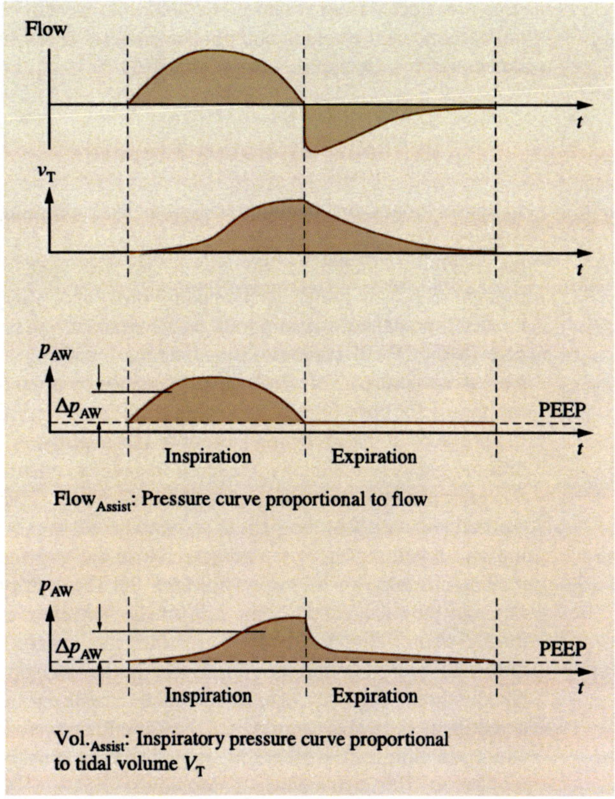

Fig. 27.9 Spontaneous proportional pressure support (SPN-PPS)

Volume Support

This method varies the level of pressure support as well. However, the main objective and the circumstances differ from the modes described above. Activating volume support ensures that the patient receives the

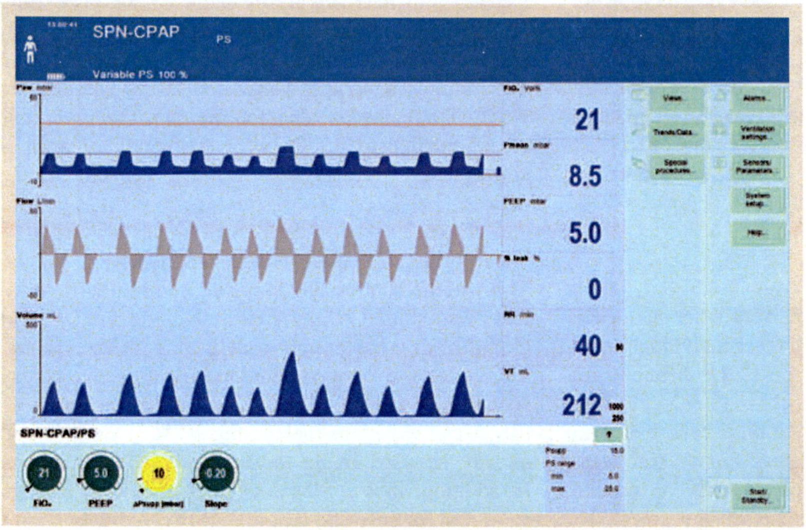

Fig. 27.10 Variable pressure support

user-preselected tidal volume. The tidal volume will be delivered even if the lung compliance changes or spontaneous efforts decrease.

27.5.3 Spontaneous Breathing (Without Respiratory Support)

In these modes the patient receives neither pressure support nor mandatory ventilation strokes. In fact, the modes allocated to this group are not ventilation modes but rather spontaneous breathing (SB) modes. Spontaneous-continuous positive airway pressure (SPN-CPAP) facilitates spontaneous breathing at a certain pressure level which the user can set by adjusting the PEEP controller. If the patient inhales, causing the pressure in the breathing system to fall slightly, additional gas is immediately delivered in order to ensure the pressure level remains constant at all times. This method requires that the patient's spontaneous breathing and breathing drive are sufficiently stable and the patient himself can be responsible for the respiratory rate and the tidal volume. The task of the ventilator is limited to providing the required breathing gas, creating the PEEP pressure, and monitoring the patient's spontaneous breathing activities. Spontaneous breathing on a ventilator is always useful when the patient's physical condition requires the breathing activities to be closely monitored. This also means it is possible:

- to prevent the collapse of individual alveolar areas
- to increase the gas exchange surface of the lungs by a positive airway pressure and
- to provide the patient with an increased oxygen concentration.

To increase efficiency, CPAP can be combined with a support method. This turns a purely spontaneous breathing mode into a supported spontaneous breathing mode. Figure 27.11 shows spontaneous breathing in combination with pressure support.

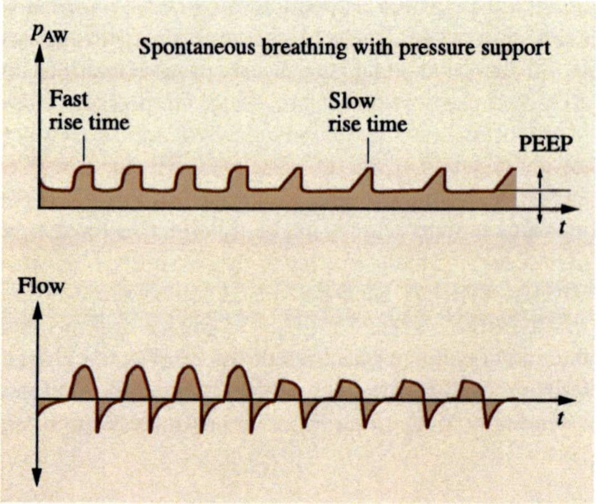

Fig. 27.11 Spontaneous continuous positive airway pressure/pressure support (SPN-CPAP/PS)

27.5.4 Mixed Ventilation

As already described, different ventilation modes allow adaptation of ventilators to meet individual patient requirements. In a figurative sense, the different ventilation modes also represent the patient's transition from controlled ventilation to spontaneous breathing. Reducing respiratory or ventilatory support (the weaning process) should be started as early as possible to allow the patient to breathe on his own. Not only is this in the patient's own interest, it is also important from an economic point of view. The ventilation modes (VC-SIMV, PC-BIPAP, etc.) belonging to mixed ventilation combine elements from controlled ventilation and supported spontaneous breathing. The mandatory phases take place in either pressure- or volume-controlled form, as already known from controlled ventilation, and ensure a certain minimum ventilation. The user will determine the weaning strategy based on the type of controlled ventilation (volume or pressure controlled) and gradually reduce the work of breathing performed by the ventilator. In practice, this reduction is achieved by reducing the mandatory frequency or alternatively by reducing the inspiratory pressure. The ventilation modes belonging to this category are especially suitable for patients whose breathing drive is not stable and whose respiratory muscles are weak. These modes also require that the user pays close attention to the fact that the ventilator settings correspond to the physical working capacity of the patient.

Volume-Controlled Mixed Ventilation

The volume control-synchronized intermittent mandatory ventilation, with or without pressure support (VC-SIMV) mode ensures a defined, constant, mandatory, volume-controlled ventilation irrespective of any spontaneous breathing activities by the patient. VC-SIMV can therefore replace the VC-CMV mode and already be used in the controlled ventilation phase. In addition, in VC-SIMV mode the ventilator is responsible for ensuring that mandatory ventilation is as

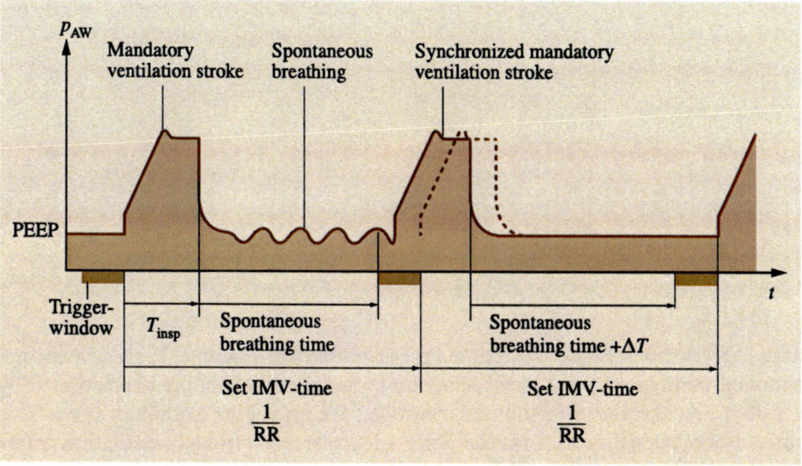

Fig. 27.12 Volume control synchronized intermittent mandatory ventilation (VC-SIMV)

comfortable as possible for the patient. For this reason in VC-SIMV mode the ventilator is free to vary the time between the mandatory ventilation strokes, while maintaining the number of ventilation strokes per minute. As with VC-CMV, the mandatory minute volume (MV) results from the respiratory rate and the set tidal volume. The patient is supplied with additional breathing gas at the start of spontaneous breathing so that the total minute volume consists of mandatory and spontaneous minute volume (Fig. 27.12).

The chronological sequence of mandatory and supported phases is as follows: In VC-SIMV mode, a trigger window is present before the mandatory ventilation stroke. If the patient inhales during this phase, the ventilator will already start the ventilation stroke, which was about due anyway. The time difference between the regular default cycle time (IMV time) and the actual cycle time is added to the next spontaneous breathing phase so as not to exceed the user-defined number of ventilation strokes per minute. The ventilator supports spontaneous breathing with pressure support from the end of a mandatory phase until the trigger window is activated again. This produces a series of alternating *mandatory phases* in which the ventilator delivers a set ventilation pattern and *supported phases* in which the patient is allowed maximum freedom. Although, in principle, there is nothing wrong with this method, unfortunately overlaps frequently occur between mandatory and spontaneous activities in clinical practice. This results in unpleasant dyssynchronies and an increase in the number of alarms for the medical staff because the ventilator does not respond to the patient's requirements in the mandatory phase. This type of situation is therefore equally undesirable for the patient and medical staff and should thus be avoided by all means, also in view of the weaning duration. In this context, ventilation modes [e.g., pressure control-synchronized intermittent mandatory ventilation plus (PC-SIMV+) or PC-BIPAP] or additional functions (e.g., AutoFlow), which also tolerate spontaneous breathing during the mandatory phase, can provide a remedy.

If the patient is not breathing spontaneously, the volume control-mandatory minute volume (VC-MMV) ventilation mode (Fig. 27.13) behaves in the same way as a volume-controlled ventilation mode and, under these conditions, is apparently not different from the previously described VC-SIMV mode. The only crucial difference between the two modes is that the mandatory part of ventilation in VC-MMV can vary depending on the spontaneous breathing activities. If the patient is breathing a sufficiently large volume independently, the controlled ventilation stroke is not applied. VC-MMV thus provides the patient with *unlimited* freedom to switch from pure controlled ventilation to total spontaneous breathing. The mode has proven itself, for example, during postoperative ventilation, but is not a substitute for individual breathing training. As during other volume-controlled ventilation modes, unpleasant overlapping of spontaneous and mandatory phases can also occur during VC-MMV. The implementation of an additional function (e.g., AutoFlow), if available, may improve tolerance of these interferences, like in VC-SIMV.

The chronological sequence of mandatory and supported phases is as follows: The mandatory minute

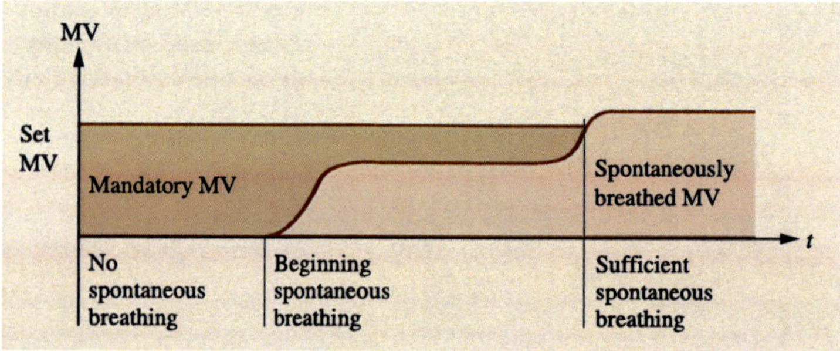

Fig. 27.13 Volume control mandatory minute volume (VC-MMV)

volume is determined by setting the respiratory rate and tidal volume controller. Controlled ventilation takes place according to the known pattern. At the start of spontaneous breathing, the ventilator continuously calculates and forecasts which spontaneous volume the patient will reach in a minute. If the calculation is positive (the portion breathed spontaneously by the patient is greater than the mandatory portion which the device would deliver), the ventilator automatically stops mandatory ventilation. If the calculation is negative, the ventilator delivers a corresponding number of synchronized mandatory strokes in order to reach the user-defined minute volume. In case of apnea, ventilation takes place according to the predefined adjustment parameters. Pressure support can be used to support spontaneous breathing. Furthermore, adequate adjustment of the spontaneous breathing rate alarm limit RR_{spon} is especially important in VC-MMV to detect abnormally fast respiration (tachypnea).

Pressure-Controlled Mixed Ventilation

PC-BIPAP is equally suitable for use in patients with pulmonary function disorders and those with healthy lungs. The PC-BIPAP ventilation mode can be described as a time-cycled and pressure-controlled form of ventilation that allows the patient to inhale or exhale at any time. PC-BIPAP is therefore also often compared to the time-controlled switchover between two different SPN-CPAP levels. The mandatory and spontaneous activities do not therefore alternate in chronological succession as in VC-SIMV, but take place simultaneously. This device behavior, also referred to as a *free breathing system*, significantly improves patient comfort and facilities the smooth transition from controlled ventilation through to complete spontaneous breathing. The technical basis of the mode is the dynamic interplay of the inspiratory gas metering device and an active expiratory valve. In the case of a pressure loss, the gas delivery unit supplies the additionally required gas flow. A sudden pressure rise (for example, caused by the patient coughing), on the other hand, is compensated by limited controlled opening of the expiratory valve. This special form of pressure-controlled ventilation offers different benefits to both patient and user.

- The ventilator responds to the patient's requirements at all times; eliminating the struggle against closed valves; ensuring patient ventilation is less stressful.
- The harmonized interaction between the device and the patient enables shallower sedation; the reduction of medication might reduce costs and increases the patient's breathing drive.
- An awake and responsive patient can be activated and motivated, both positively influencing the healing process.
- Communication with the patient allows pain medication or ventilator settings to be optimized.
- The patient's spontaneous breathing allows the reduction of mandatory ventilation and improves the pulmonary gas distribution and the gas exchange.
- Spontaneous breathing decreases intrathoracic pressure forces, relieves the heart, and has a positive effect on the venous return flow to the heart.
- The patient contributing to the work of breathing as early as possible can reduce the ventilation period and thus the risk of infection.
- The *free breathing system* reduces the number of alarms triggered by dyssynchronization, making ventilation less stressful not only for the patient but also for medical staff.

These benefits are also available to the user in volume-controlled ventilation when using the AutoFlow function (Sect. 27.6).

The chronological sequence of mandatory and supported phases in PC-BIPAP is as follows: As with VC-SIMV, a trigger window is present before the

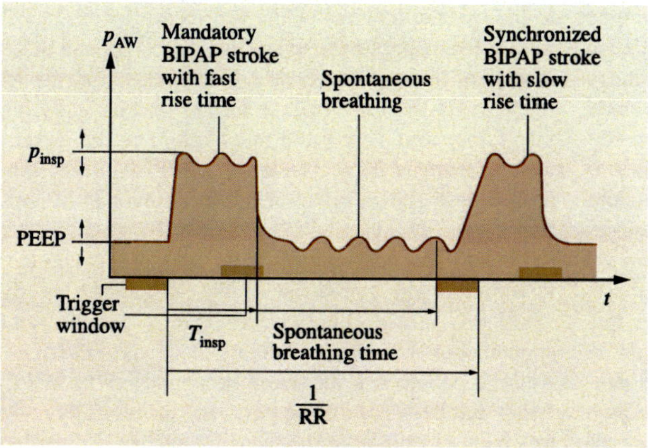

Fig. 27.14 Pressure control-biphasic positive airway pressure combined with spontaneous breathing (PC-BIPAP)

mandatory ventilation stroke. If the patient inhales during this phase, the ventilator will already deliver the pressure change, which was about due anyway. The time difference between the regular default cycle time and the actual cycle time is added to the next spontaneous breathing phase. This offset ensures that the user-defined respiratory rate remains constant. The ventilator supports spontaneous breathing with pressure support from the end of a mandatory phase until the trigger window is activated again. The pressure rise time defined by the user with the *ramp* setting influences both the supported and mandatory pressure rise. Thus, the patient always receives the breathing gas with a uniform flow characteristic, whose decelerating course is generally perceived to be significantly more pleasant than a constant flow. As a further feature, PC-BIPAP is provided with an additional trigger window which is used to synchronize the pressure change from inspiratory pressure to the PEEP level with spontaneous breathing. This may mean that the length of the inspiratory period can also vary to a certain extent. Like for the ventilation frequency, the intervals are also corrected and compensated accordingly here by the inspiratory period adjustment in the next mandatory ventilation phase (Fig. 27.14).

Pressure Control-Synchronized Intermittent Mandatory Ventilation

This ventilation mode behaves like the volume-controlled pressure control-synchronized intermittent mandatory ventilation (VC-SIMV) variant as regards the chronological switchover between mandatory and spontaneous ventilation phases. The only difference from volume-controlled VC-SIMV is that the mandatory ventilation strokes are pressure controlled in this case. The *free breathing system* described under PC-BIPAP is not obligatory here or has only limited availability depending on the manufacturer.

Pressure Control-Airway Pressure Release Ventilation (PC-APRV)

Although the main characteristics of PC-APRV (Fig. 27.15) are very similar to the PC-BIPAP ventilation mode, this mode is based on a completely different ventilation philosophy. As an interesting contrast to all the other modes in which the mandatory inspiration produces a pulmonary pressure rise, the volume change in PC-APRV occurs by a brief pressure relief. This *release phase* enables an extremely short and incomplete expiration. To better understand this situation, it may be useful to imagine a patient who breathes spontaneously at an increased pressure level (SPN-CPAP). If spontaneous breathing subsequently deteriorates, resulting in an increase of arterial carbon dioxide, it will be necessary to increase ventilation. Until recently the

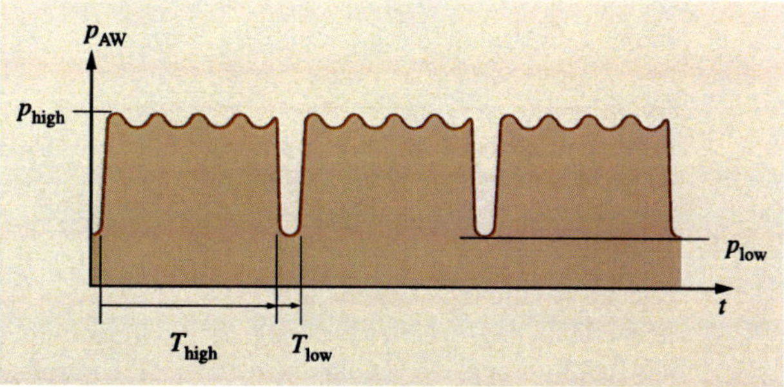

Fig. 27.15 Pressure control airway pressure release ventilation (PC-APRV)

pressure in the breathing system, and thus the volume transported to the lungs, was increased. PC-APRV, on the other hand, utilizes the fact that the patient's lungs are already filled by the positive pressure and causes a brief expiration by reducing the pressure in the breathing system. The pressure relief phase in PC-APRV is extremely brief, typically 0.5 s. Since this time does not allow for a complete expiration in adult ventilation, only the lung compartments which can be quickly emptied and filled are ventilated in the release phase. The regions to which the gas only reaches slowly due to increased airway resistance are hardly influenced by the brief pressure change. However, often it is precisely these lung regions which can only be ventilated with difficulty using conventional ventilation and/or are susceptible to collapse. By stabilizing the pressure conditions in these critical parts of the lungs, PC-APRV can usually provide for gentler ventilation and improved gas exchange. The PC-APRV adjustment parameters differ slightly from the usual designations and parameters due to the specific features of the ventilation mode. In PC-APRV, for example, the expiratory time T_{low} can be set instead of the respiratory rate. This makes it easier for users to set the brief release phases. Accordingly, the inspiratory time is referred to as T_{high} and the inspiratory pressure as p_{high}. The term "PEEP" would also be absurd here and is therefore replaced by the designation p_{low}.

PC-APRV does not provide pressure support as it would be in conflict with the philosophy of the ventilation mode. Otherwise PC-APRV has the same spontaneous breathing characteristics as PC-BIPAP, i.e., the patient can inhale or exhale at any time. PC-APRV has been used in the past very successfully in the treatment of patients with severe gas exchange disorders and is gaining in importance, at least in this application. PC-APRV can, in principle, be used for patients with healthy lungs, but is only seldom observed due to the described deviations from physiological respiratory patterns.

27.6 Ventilation Extras and Special Functions

27.6.1 Inverse Ratio Ventilation

Normally, the inspiratory period is shorter than the expiratory phase. However, the inspiratory time may need to be increased at the expense of the expiratory phase when treating diseases with serious oxygenation disorders. The term inverse ratio ventilation (IRV) merely states that ventilation takes place with an inverse inspiration-to-expiration ratio, in other words, the inspiratory time is longer than the expiratory time. The designation IRV alone does not provide any information on whether the ventilation is volume or pressure controlled.

27.6.2 Sigh

By observing one's own spontaneous breathing you will have noticed that breathing is not always even. Occasionally we inhale a larger quantity of air than usual, i.e., we sigh. Originally, the sigh function was introduced in ventilation with the aim of reproducing the typical variations in volume that occur during spontaneous breathing, thereby breaking up the monotonous and unphysiological ventilation pattern of controlled ventilation. The sigh function, which is also referred to as intermittent PEEP, is accomplished by temporarily increasing the PEEP pressure to a higher value. In the Evita ventilator the PEEP pressure is increased, for example, every 3 min for two breathing cycles. The term *sigh* results from the deep drawn inspiration whose discharge creates a noticeable flow murmur similar to that of a spontaneous sigh. The use of a sigh could not become established as originally conceived. However, today the sigh function has become important as a therapeutic tool to reopen collapsed alveoli (recruitment).

27.6.3 AutoFlow

The AutoFlow function is an optional addition which is available in all volume-controlled ventilation modes on Dräger ventilators (with the exception of ILV). The ventilation mode is set using the ventilation parameters which are valid for the respective mode. When AutoFlow is activated, the ventilator replaces the constant flow typical of volume-controlled ventilation modes with the decelerating flow pattern known from pressure-controlled ventilation modes. This combines the benefits of pressure-controlled ventilation with the strengths of volume-controlled ventilation. The modified volume-controlled ventilation can be characterized as follows:

- AutoFlow ensures that the user-defined tidal volume is applied at the lowest possible ventilation pressure.
- AutoFlow automatically adjusts the ventilation pressure to pulmonary compliance changes, eliminating the need to manually correct the inspiratory pressure in pressure-controlled ventilation.
- AutoFlow tolerates spontaneous breathing activities by the patient at any time, thus adding the benefits of a *free breathing system* listed under PC-BIPAP to volume-controlled ventilation.
- AutoFlow eliminates the peak pressure usually occurring during volume-controlled ventilation, which might cause partial overinflation of individual lung areas.
- AutoFlow takes over the responsibility for setting the "Flow" and p_{max} controllers. The reduction of parameters makes it easier to set volume-controlled ventilation.

The sequence of inspiration and expiration and control characteristic is as follows: The chronological sequence of inspiration and expiration remains unaffected by the AutoFlow function and corresponds to the behavior of the respective ventilation mode. After activation of AutoFlow, the ventilator applies the next mandatory ventilation stroke with a minimum constant flow. The ventilation pressure measured at the end of this inspiration is used as the inspiratory pressure in the following inspiratory phase. The flow is then delivered with a decelerating flow profile. After the switchover to expiration, the ventilator compares the applied tidal volume with the user-defined set value. The inspiratory pressure of the following mandatory phase is slightly increased or decreased to offset the differences. The adjustment mechanism is limited to a fluctuation margin of ±3 mbar from breath to breath. The pressure increase is stopped as soon as the inspiratory pressure approaches the upper airway pressure alarm limit of up to 5 mbar. If this pressure is not sufficient to apply the set tidal volume completely, the device generates the alarm *volume inconstant*. Although spontaneous breathing activities can lead to fluctuations in the expiratory volume, AutoFlow ensures that a constant tidal volume is applied on average over time. Using AutoFlow is always possible and useful provided that there are no special pulmonary restrictions and the patient is receiving volume-controlled ventilation anyway. Irrespective of the current ventilation situation, the upper airway pressure alarm limit is very important because of the above-mentioned dual function. Figure 27.16 shows a screenshot of EvitaXL. From left to right, you can recognize a conventional, volume-controlled ventilation stroke, the stroke applied with a minimum constant flow after activation of AutoFlow, as well as the first of the subsequent AutoFlow ventilation strokes.

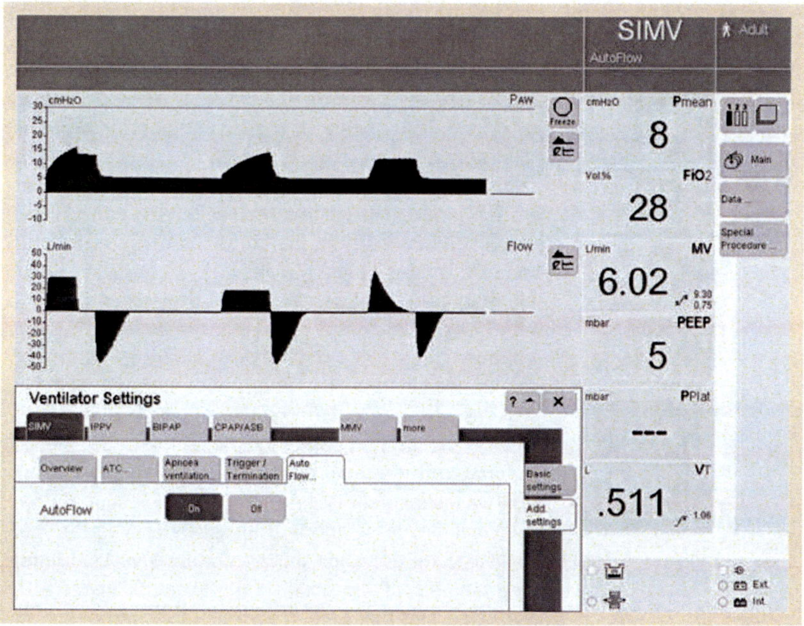

Fig. 27.16 AutoFlow

Volume guarantee (VG) is an adjunct in pressure-controlled ventilation modes and is comparable to the above-mentioned AutoFlow function. Depending on the manufacturer, such specific control algorithms are available in separate ventilation modes only (e.g., PRVC = pressure regulated volume controlled).

27.6.4 Automatic Tube Compensation

Intubated patients have to perform considerably more work to breathe due to the artificial narrowing of their airways. This additional work of breathing (WOB) results from the pressure difference arising along the tube, which also has to be counteracted by the respiratory muscles. The tube type (endotracheal tube or tracheal cannula), the tube diameter, and the gas flow currently delivered through the tube are major determining factors. To better understand the importance and the effect of the gas flow, simply try to breathe for several minutes through a 0.2″ straw while holding your nose closed. You will notice that considerably greater breathing efforts are required to create larger gas flows. Consequently, this means that the patient should be supported in relation to the flow. Based on this idea, automatic tube compensation (ATC) increases the pressure in the breathing system during inspiration by the pressure drop induced by the flow between the tube connector and tube tip. During expiration, compensation takes place by a brief reduction of the PEEP level. This gives the patient the impression that he is not actually intubated. However, the dynamic pressure changes in the breathing system lead to significant changes in the pressure–time curve. These curve progressions, which are initially unusual for the user, merely reflect the variations in pressure before the tube and do not thus correspond with the pulmonary pressure curve. More detailed information on the pressure at the tube tip can be provided by devices which calculate this pressure and display it as a separate pressure curve. Tube compensation is available in all ventilation modes and can be adjusted to the current ventilation situation by entering the tube type, the tube diameter, and the desired compensation as a percentage. Since ATC greatly simplifies the patient's work of breathing, the user should significantly reduce the pressure support, or even switch it off temporarily, before activating the ATC function. Tube compensation (TC) is, in principle, suitable for all intubated patients, but can be configured to be limited to the inspiratory phase, for example, for patients with obstructive respiratory diseases (Fig. 27.17).

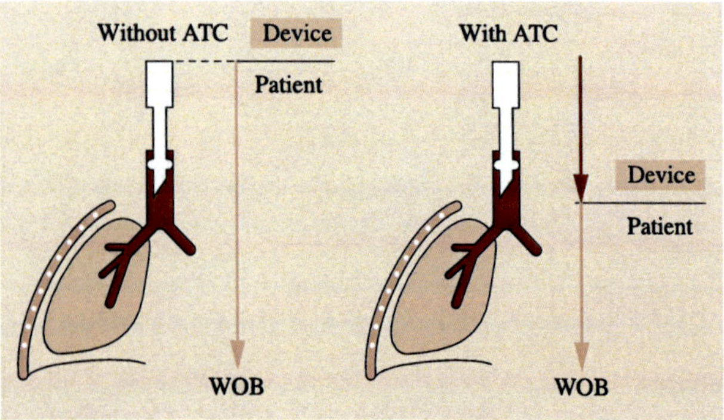

Fig. 27.17 Automatic tube compensation (ATC)

27.6.5 Noninvasive Ventilation

Noninvasive respiratory therapy is gaining in importance as a means of avoiding intubation or as a supporting respiration therapy after extubation, not to mention the tremendous drive to cut costs in the healthcare sector. However, mask ventilation presents the user, patient, and ventilator with special challenges. These include, for example, the availability and acceptance of suitable face masks and the communication and cooperation between the user and the patient. There are also special technical challenges concerning the performance of the ventilator, since leaks and disconnections occur frequently during mask ventilation. Noninvasive ventilation (NIV) is an optional addition and should be available as an adjunct to all standard ventilation modes. Due to the wide range of manufacturer-specific features, only a few basic aspects of mask ventilation are considered here. The quality of mask ventilation is primarily determined by whether the ventilator is prone to self-trigger due to leaks at the breathing mask. Depending on the ventilator, the trigger sensitivity either is adjusted automatically or has to be manually adapted by the user. In this context, it must be mentioned that not all ventilators provide volume-controlled ventilation in NIV. However, even if this feature is available, applied volumes might vary significantly based on the respective device performance. In any case, high-performance leak compensation is essential in such applications. Leak compensation continuously compares the set tidal volume with the expiratory measured volume and increases the inspiratory gas delivery in the event of leaks. Last but not least, monitoring and alarm limits must meet special demands during noninvasive ventilation. For example, if the position of the mask needs to be corrected and this involves temporarily removing and refitting the mask, it would usually result in a near-immediate disconnect alarm under normal conditions. Such an alarm can be avoided by an adjustable alarm delay time $T_{\text{Disconnect}}$, which usually is only available in the NIV operating mode.

27.6.6 Apnea Ventilation

During the weaning process, the ventilator and patient share the responsibility for the total minute volume. If apnea occurs in this phase because the patient cannot cope, the remaining mandatory volume applied may not be able to ensure sufficient ventilation anymore. For this reason, to avoid hypoventilation, the apnea alarm can be combined with apnea ventilation in modes which allow or support spontaneous breathing. In this case, the ventilator ensures the defined controlled ventilation of the patient, until the user can attend to the current patient situation. In adult ventilation, apnea ventilation is generally implemented as volume-controlled ventilation, whereas in neonatology pressure-controlled ventilation is used.

27.7 Patient Monitoring and Alarm Limits

Although the ventilator automatically monitors many device functions, the user has to set alarm limits manually for other different measured values. The importance of patient monitoring has already been shown using the example of apnea ventilation. Despite this, many users often view alarms as an unpleasant side-effect of ventilation. A more positive picture emerges when you compare the adequate adjustment of the alarm limits to a *protective fence* surrounding the patient. As long as patient activities remain within the limits of the *fenced-off* area, there is no cause for worry or reason for an alarm. However, as soon as the patient tries to leave the *protective zone* which has been created for him, the device will report this *breakout attempt* by generating a corresponding alarm. Inevitably, this raises the issue of what importance should be given to the different alarm limits and how they should be set. Basically, the following can be stated:

- There are no unimportant alarm limits. It must be noted that the patient's situation may change at any time. The alarm limits which do not currently appear relevant should therefore also be adjusted according to the patient's needs.
  - Alarm limits should always be set based on the current respective measured values with the aim of allowing as much freedom as possible while giving priority to safety.
  - All alarm limits must always be checked and adjusted to meet the patient's current needs whenever a ventilation mode is set or changed. These steps can avoid predictable alarms and unnecessary stress.
  - There are no useless alarms. Each alarm is generated as a result of a change in the patient's condition, a misjudgment by the user or a technical fault requiring the user's immediate attention.

The task of the monitoring and alarm system changes as the work of breathing is transferred to the patient in the weaning process. During controlled ventilation the monitoring and alarm system is mainly used to monitor the functioning of the device and its effect on the patient. However, it also assumes responsibility for monitoring respiratory sufficiency in the weaning phase. Obviously, the physiological fluctuations of the respiratory rate and the tidal volume require that the alarm limits, based on the current measured value, are set much more tightly in controlled ventilation than during mixed ventilation or (supported) spontaneous breathing.

27.8 Weaning Strategy and SmartCare/PS

The total time during which a patient is ventilated and the length of the weaning process can vary greatly. While some postoperative patients may only still need to be ventilated for a few hours and can be weaned almost immediately, this process may prove to be extremely problematic and protracted for other patients. Particularly in the latter case, the user is required to assess the patient's situation and strength intensively to achieve the correct balance. While the patient should not be overexerted, the weaning phase should not be allowed to continue for an unnecessary length of time. For this reason, assessing the patient and adjusting the ventilation parameters accordingly has the highest priority, requiring immediate attention. In clinical practice, this demand cannot always be met or only partially met by the medical personnel. Furthermore, assessing the patient and making the corresponding ventilator adjustments used to be based just on the personal experience of the responsible physician. The wide range of different ventilation modes available also meant that a consistent weaning strategy was not necessarily recognizable in all cases. Despite that, there are signs that potential for optimization has already been identified, and it can be expected that the situation will develop positively. Nowadays, users can apply ventilation modes which, for example, allow spontaneous breathing from the start of ventilation. Since it is possible to ventilate and wean the patient in one mode, ventilation can be adapted more effectively to the patient's needs and it is no longer necessary to switch back and forth between different modes. Furthermore, protocols or guidelines have been developed within the framework of quality control procedures describing and regulating the gradual weaning process. Many users are therefore instructed to conduct weaning adhering to a defined procedure. If any deviations are necessary, they are documented and integrated into the further development of these protocols as part of a continuous improvement process. Despite that, this does not seem to have helped clinical personnel to a significant degree, so that the time intervals between the individual weaning steps are still often longer than actually necessary.

A highly promising approach to weaning therapy is provided by the optional SmartCare/PS function available on some Dräger ventilators. SmartCare/PS can be used for weaning intubated or tracheotomized patients and is only accessible in the SPN-CPAP/PS ventilation mode. This knowledge-based system automates the weaning process on the basis of an implemented therapeutic strategy. SmartCare/PS compares the current measured values for the respiratory rate, the tidal volume, and the expiratory carbon dioxide with the different threshold values defined in the weaning protocol. In this way, SmartCare/PS can independently assess the ventilation situation and can even recognize *normal* ventilation (normoventilation). The measured values are interpreted every 2 or 5 min, and the pressure support is increased or reduced as a result. If the support pressure reaches a minimum value which is dependent on the setup of the breathing system, the system carries out a spontaneous breathing trial with the patient. During this phase, the system observes and assesses the patient's activities as well as potential instabilities. An instability exists when the patient is no longer normoventilated, and SmartCare/PS responds by increasing the pressure support. If the observation phase is completed successfully, SmartCare/PS will suggest that the patient should be disconnected from the ventilator. If extubation can or should only occur at a later time, SmartCare/PS will continue to respond to changes in the three input values and adjust the pressure support accordingly. At this stage, SmartCare/PS can still revise or uphold its recommendation that the patient should be disconnected from the ventilator. This decision is based on the number and relative duration of instabilities registered by the device. The duration of the previous stabile normoventilation is used to assess the time of an instability.

This brief description of SmartCare/PS clearly shows that the system is based on medical knowledge and not mathematical algorithms. The classifications performed at intervals of several minutes and automatic changes to the support pressure release intensive care medical staff from routine tasks, thus enabling the weaning phase to be reduced. Besides these reasons, other factors such as increasing cost-cutting pressure, numerous attempts at standardization, and efforts to improve the quality of patient care indicate that the symbiosis of clinical knowledge and innovative technology is soon set to capture other day-to-day clinic operations.

Further Reading

- D.W. Chang: *Clinical Application of Mechanical Ventilation* (Delmar, Albany 2001)
- R. Kuhlen, J. Guttmann, R. Rossaint: *New Forms of Assisted Spontaneous Breathing* (Urban Fischer, Munich 2001)
- D. Oakes, S. Shortall: *Ventilator Management* (Health Educator, Orono 2009)
- S.P. Pilbeam, J.M. Cairo: *Mechanical Ventilation* (Mosby, St. Louis 2006)
- M. Dojat, L. Brochard: Knowledge-based systems for automatic ventilatory management, Resp. Care Clinics 7(3), 379–396 (2001)
- A. Hasan: *Understanding Mechanical Ventilation* (Springer, Berlin Heidelberg 2010)
- A.M. Esquinas (Ed.): *Noninvasive Mechanical Ventilation* (Springer, Berlin Heidelberg 2010)
- A.S. Slutsky, L. Brochard (Eds.): *Mechanical Ventilation*, Update Intens. Care Emerg. Med. 40 (Springer, Berlin Heidelberg 2004)

28. Defibrillators and ICD Systems

Defibrillators

Rüdiger Kramme

Defibrillation is the definitive treatment for the life-threatening cardiac arrhythmias, ventricular fibrillation, and pulseless ventricular tachycardia. Defibrillation consists of delivering a therapeutic dose of electrical energy to the affected heart with a device called a defibrillator. Defibrillators can be external, transvenous, or implanted, depending on the type of device used or needed. Some external units, known as automated external defibrillators (AEDs), automate the diagnosis of treatable rhythms, meaning that lay responders or bystanders are able to use them successfully with little, or in some cases no, training at all except safety precautions also covered in this chapter (Sects. 28.1–28.5). An implantable cardioverter-defibrillator (ICD), described in Sect. 28.6, is a small, battery-powered electrical impulse generator which is implanted in patients who are at risk of sudden cardiac death due to ventricular fibrillation and ventricular tachycardia. The device is programmed to detect cardiac arrhythmia and correct it by delivering a jolt of electricity.

28.1 **Defibrillator Technology** 546
28.1.1 Physical Principles 546
28.1.2 System Properties 548

28.2 **Therapeutic Intervention** 549
28.2.1 Defibrillation/Cardioversion 549
28.2.2 External and Internal Defibrillation 550

28.3 **Methodological Notes** 550
28.3.1 Electrodes and Contact Agents 550

28.4 **Complications** 551

28.5 **Technical Safety Aspects** 551
28.5.1 Use 551
28.5.2 Device 551

28.6 **Implantable Cardioverter-Defibrillators** .. 551
28.6.1 ICD Development 552
28.6.2 System Technology 553
28.6.3 Algorithms 554
28.6.4 Electrodes 554
28.6.5 Complications 555
28.6.6 Function Checking 555

References 556

Variable impulse generation or conduction can cause arrhythmias, with the result that the coordination of the myocardial fibers is impaired, suspended or even plunged into chaos (fibrillation). Ventricular fibrillation is uncoordinated myocardial fibrillation with no ejection from the ventricles, being characterized on an electrocardiogram (ECG) by irregular and disorganized depolarizations with high frequency. Pulseless ventricular tachycardia is characterized by a regular and rapid sequence of broad QRS complexes, and as with ventricular fibrillation, there is no ejection. Both types of rhythm are life-threatening conditions of the cardiovascular system and are thus indications for defibrillation. Defibrillation is understood as meaning a brief, phasic pulse of energy which is intended to bring about simultaneous depolarization of all the myocardial fibers, which means that, after approximately 5 s of administering an electrical pulse, no ventricular fibrillation or ventricular tachycardia can be detected any longer in the ECG. The objective of this measure is to terminate tachycardic ventricular and supraventricular arrhythmias so that, following a refractory period (no excitation is possible in this phase) which generally lasts between 200 and 500 ms, the sinus nodes once more assume the pacemaker function as the primary center of excitation.

Defibrillators are electrotherapeutic high-voltage devices which are used within the course of resuscitation and to terminate tachycardic ventricular and supraventricular arrhythmias (Fig. 28.1).

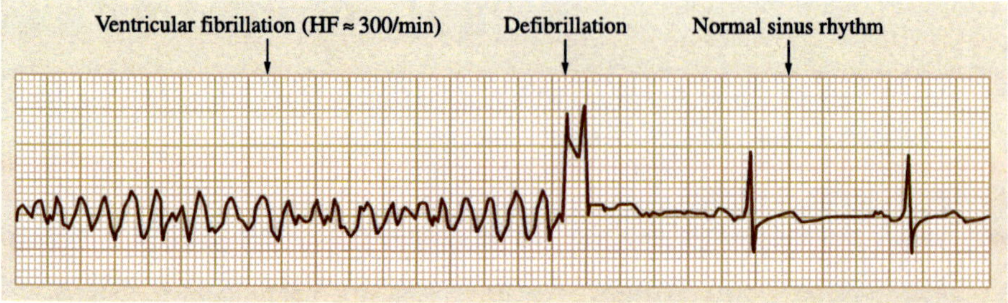

Fig. 28.1 ECG documentation of a defibrillation

28.1 Defibrillator Technology

28.1.1 Physical Principles

The portable defibrillator is a direct-current (DC) voltage system which is usually not dependent on mains electricity (Fig. 28.2) and which is essentially composed of the following system components:

- The energy supply via mains connection or rechargeable batteries
- A capacitor as an energy store (capacity = n pulses at 360 J)
- A charging circuit for the capacitor (duration of charging, i. e., when the maximum energy is reached, is on average 10^{-8} s, considerably less in some manufacturers' models), and a discharge circuit which delivers the current pulse at different, preselectable energy levels (e.g., 2–360 J).

The DC pulse stimulation ranges between 3 and 8 ms at current of 10–27 A (internally) and 22–60 A (externally).

Automatic safety discharge should occur when no shock is triggered (after ≈ 10 s), when the defibrillator shock is triggered, and when a new energy level is preselected, as well as in the event of technical malfunction.

The energy (E) which can be stored in the capacitor can be determined from the capacity (C) and the available voltage (U) as

$$E = \left(\frac{1}{2}\right) CU^2 , \quad (\text{VA} = \text{W}) .$$

Waveform of the Energy Shock

The terms waveform and curve technology are understood as referring to the time-based sequence of energy output. The shape of the wave dictates firstly how much energy is supplied to the patient and secondly over what period this energy is administered. The optimum amount of energy for the defibrillator pulse is the amount of energy which causes least myocardial damage. A distinction is made between mono-, bi-, and triphasic defibrillation shock configurations (Fig. 28.3). Whereas, in the case of the monophasic waveform, the current only flows in one direction and the polarity does not change, with the biphasic waveform the current is delivered in one direction, interrupted, and continued in the opposite direction. Here, the polarity changes with every phase. The main forms of the biphasic discharge characteristic are the truncated exponential curve and the square pulse. Today, particularly in the case of implantable and automatic external defibrillators, biphasic shock forms are preferably used, which differ as a result of varied adaptation to the thoracic impedance of the patient (e.g., different pulse output, peak-to-peak voltage, and pulse duration) and are more effective on the first shock administered than monophasic shocks. Biphasic defibrillations are generally more effective and gentler on the heart than monophasic ones and have the advantage that the shock-induced dysfunction of the myocardial cells is less and that defibrillation success can be achieved with lower energy and voltage. The device-dependent amount of energy is 150–200 J for the first defibrillation and 200–360 J for all others, whereas it is always 360 J in the first monophasic defibrillation. In addition, biphasic pulse forms allow devices to be further miniaturized.

Whereas the optimum energy flow in monophasic defibrillation is somewhere in the range from 30 to 40 A s, with biphasic shock it is believed to be in the range from 15 to 20 A s. Defibrillators whose administered shock corresponded to the real transthoracic flow would be preferable. Some of the biphasic defibrillators which are available on the market use very different

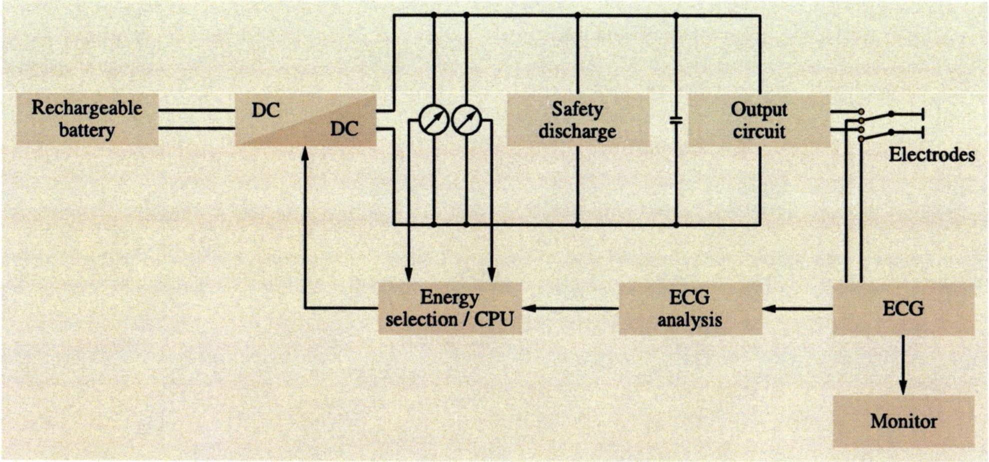

Fig. 28.2 Block diagram of a semiautomated DC defibrillator (after [28.1])

manufacturer-specific pulse forms. There is currently no general consensus about the effectiveness and potential negative effects of the various pulse forms on offer. The operating modes are divided into asynchronous and synchronous operation: whereas the heart's own pulses are taken into account in the synchronous operating mode (so-called sync pulse or QRS triggering), this is not done in the asynchronous operating mode. The latter operating mode should be reserved for *strictly emergency defibrillations*. Only the synchronized operating mode is suitable for cardioversion.

Thoracic Impedance

Thoracic impedance is the resistance in the body which opposes the energy pulse or flow from the defibrillator. It ranges between 15 and 150 Ω; usually it is 70–80 Ω. The impedance should therefore be taken into consideration when the necessary energy is administered, as the patient's thoracic impedance is crucial to the amount of energy required. Because the impedance varies to a large degree in humans, dynamic adaptation of the waveform of the energy pulse is an important feature. In modern devices, the thoracic impedance is automatically measured and taken into account before defibrillation, meaning that energy can be delivered more accurately. Larger electrodes incidentally reduce the impedance.

Defibrillators can be divided into manual defibrillators, semiautomated and automated external defibrillators (AED), fully automated defibrillators, and defibrillator implants (Fig. 28.4). The criterion for

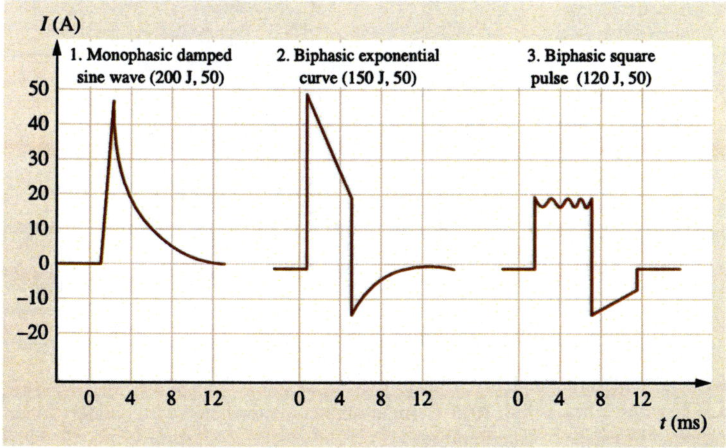

Fig. 28.3 Monophasic and biphasic pulse forms

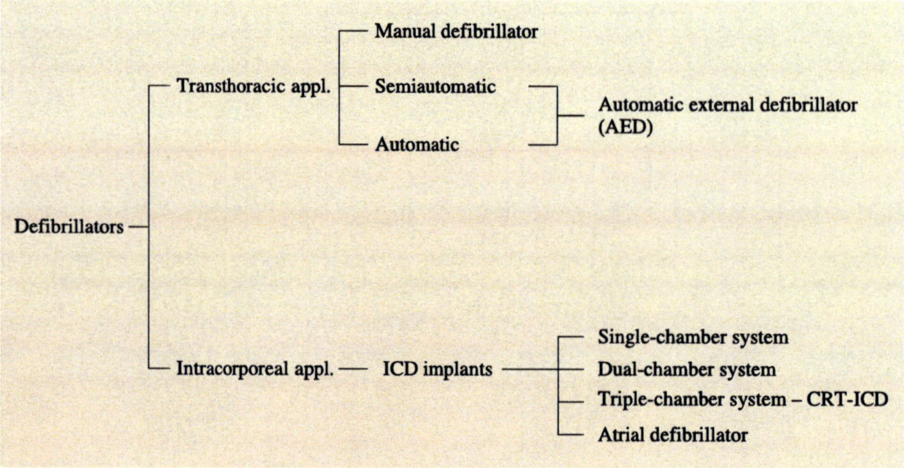

Fig. 28.4 Overview of types of defibrillator devices

distinguishing between semi- and fully automated defibrillators is that, with the semiautomatic defibrillator, the user is shown a defibrillation recommendation and the administration of the pulse is triggered by the user, whereas this is done automatically with a fully automated defibrillator. The electrodes used are so-called paddles, adhesive electrodes or internal paddles.

28.1.2 System Properties

In addition to the conventional manual defibrillators, there are also semiautomated and automated systems (AED), which perform an automatic ECG analysis via attached electrodes and have an integrated display. Defibrillation without a monitor or display and without rhythm analysis is known as *blind defibrillation*. Modern defibrillators have a *quick look* monitoring device in the paddles. When ventricular fibrillation is detected, in a semiautomated device (Fig. 28.5) the user is given a recommendation to defibrillate.

The expandable functionality, which is usually achieved using pluggable modules, makes the real system properties possible in semiautomated defibrillators and modules: integrated external pacer (transthoracic pacemaker), noninvasive blood pressure, S_pO_2, capnography, arrhythmia detection, and recording component for documentation. Within the scope of quality assurance for a defibrillation which has been performed, complete recording by the device is of particular interest, including ECG recording and 12-channel ECG reports as well as recording of all other relevant data for later evaluation and documentation. The possibility of remote transmission of data, for example, from the ambulance to the hospital's emergency department, will also play a relatively large role in the future.

Form Analysis of Ventricular Fibrillation

To determine the optimum time to deliver the defibrillation pulse when fibrillation characteristics are definitely present, computer-aided form analysis of ECG signals is a promising and innovative new development which is currently still being researched. It involves evaluating the frequency and amplitude characteristics of the surface ECG using so-called fibrillation scoring algorithms. This algorithm is based on linear (e.g., Fourier analysis, wavelet theory) and nonlinear methods (such

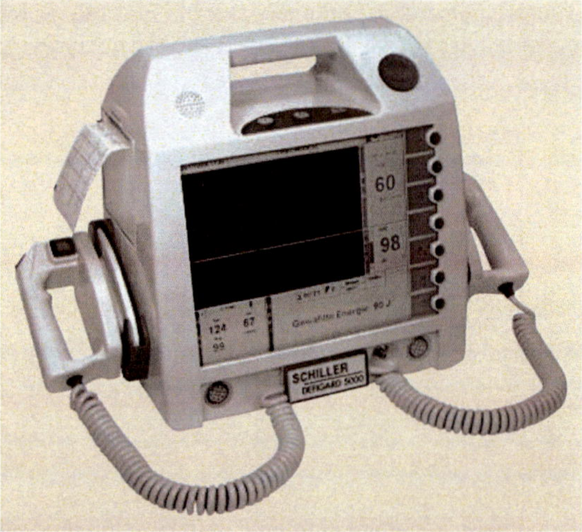

Fig. 28.5 Semiautomated defibrillator with paddles (courtesy Schiller)

as an $N(\alpha)$ histogram, which can be used to investigate the degree of randomness in a signal). Further advantages of this technology may be that ineffective energy pulses are avoided and damage to the myocardium is thereby limited, and that information is provided about the quality of the resuscitation performed.

Automated External Defibrillator

AED systems – predominantly using two-phase curve technology – only deliver a defibrillation pulse if there is a life-threatening cardiac arrhythmia present which is recognized as such by the device. The user (or also semiskilled lay helper in the case of *public access defibrillation*) is also guided by means of a control menu. Modern systems additionally offer a voice-controlled user interface. The American Heart Association (AHA) takes the view that up to a third of all deaths resulting from sudden cardiac death could be prevented by widespread AED use. The use of AED systems is classified as an integral part of basic resuscitation measures.

28.2 Therapeutic Intervention

Synchronized defibrillation (Fig. 28.6) is referred to as electrical cardioversion, i. e., the pulse of energy is triggered by the R wave in the ECG.

28.2.1 Defibrillation/Cardioversion

This synchronization is carried out to prevent the pulse being delivered in the vulnerable phase (T wave) and to prevent the risk of ventricular fibrillation being triggered. In contrast, asynchronous cardioversion, which takes place independently of the R wave, is what is known as defibrillation (Fig. 28.7). From a hemodynamic point of view, blood is no longer or is hardly pumped into the circulatory system during ventricular flutter and fibrillation, as the individual muscle fibers do not contract in a coordinated fashion. In the case of atrial flutter and fibrillation, approximately 20–30% less blood is ejected since the filling capacity of the ventricles is not optimized as a result of atrial contraction being stopped. Indications for electrical cardioversion are:

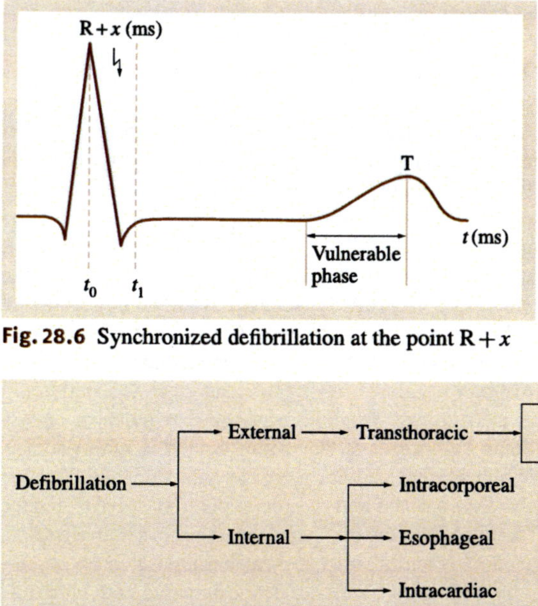

Fig. 28.6 Synchronized defibrillation at the point R + x

- Atrial flutter and fibrillation
- AV junctional reentrant tachycardia
- WPW (Wolff–Parkinson–White-syndrome) and ventricular tachycardia.

With ventricular fibrillation, defibrillation is the only therapy which can be used that has any prospect of success. Electrical cardioversion is contraindicated in digitalis overdose, hypokalemia, and atrial fibrillation (if there is deficient anticoagulation for at least 2–3 weeks and arrhythmia for longer than 48 h).

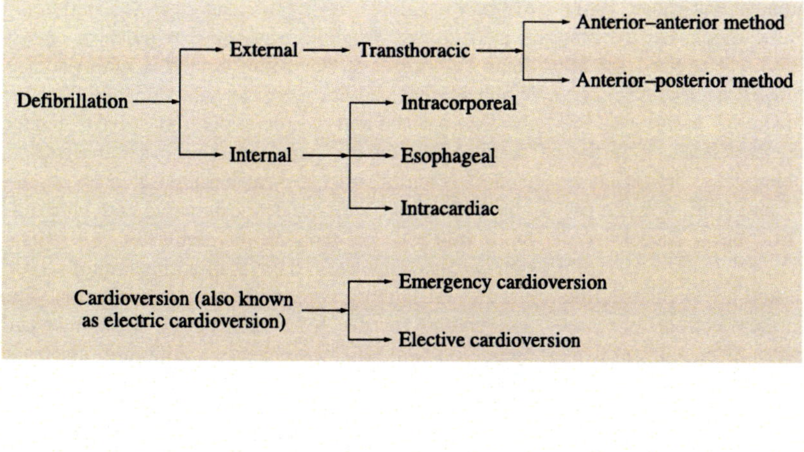

Fig. 28.7 Method overview for defibrillation/cardioversion

28.2.2 External and Internal Defibrillation

External, transthoracic defibrillation is performed by means of electrodes and pulses of energy from $\approx 2\,J$ up to 360 J. As a guide value which is dependent on the body weight, a pulse strength of 3 J/kg is recommended for adults and 2 J/kg for children. In terms of electrode application, a distinction is made between the anterior–anterior method and the anterior–posterior method: whereas in the anterior–anterior method, which is mainly used in emergency situations, both electrodes are placed on the thorax (at the base and apex of the heart), in the anterior–posterior method a large-area electrode is placed below the back. The anterior–posterior method is preferred for electrotherapy in the case of arrhythmias, since the largest proportion of the administered pulse of energy flows directly through the myocardium. For internal, that is to say intracorporeal, defibrillation, sterile internal paddles are used which in the case of an open thorax cover the exposed heart (in this case the pericardium), i. e., the surfaces of the electrodes must be in contact with the myocardium over its entire area. For internal defibrillation the energy pulse is a maximum of 50 J.

Special Applications

Esophagus. An esophageal pulse electrode (cylindrical electrode at the end of a catheter) is placed at the level of the left atrium. The counterelectrode is attached to the thorax. An additional electrode to the cylindrical electrode enables bipolar stimulation and recording of an esophageal ECG signal which is used to synchronize the energy pulse.

Intracardiac. This application is likewise carried out using a catheter. One electrode lies at the apex of the right ventricle, while the second electrode is placed in the superior vena cava.

28.3 Methodological Notes

In principle, all commercially available defibrillators are operated in the same way and are suitable for both internal and external defibrillation. As a rule, a visual and/or acoustic signal is generated when the defibrillator is operational, i. e., when the capacitor is charged. The electrodes, which are provided with gel, are placed firmly on the thorax and pressed on, and the preselected energy dose is triggered. This is usually done directly via a trigger on the handles. Owing to the ERC Guidelines of 2005, the practice of performing defibrillation three times within the space of a minute using mono- and biphasic defibrillators is obsolete. This application model has been replaced by delivery of a single shock at full energy (so-called one-shock strategy); i. e., with a monophasic shock, 360 J is recommended for delivery of the shock. For the first biphasic shock, at least 150–200 J is advised for all discharge characteristics. Following each shock, cardiopulmonary resuscitation should be performed for 2 min before administering the next shock. If further shocks are necessary, with monophasic defibrillators the energy level is kept at 360 J and with biphasic defibrillators the energy level is in contrast successively increased.

The energy pulse delivered and its amplitude, the form and polarity of the shock, the electrode size and position, and also the homogeneity of the current density in the cardiac muscle (myocardium) are all crucial for efficient defibrillation or cardioversion.

28.3.1 Electrodes and Contact Agents

Correct application of the adhesive electrodes is necessary to achieve faultless ECG recording and ultimately to prevent equipment-related interpretation errors. Despite being more expensive, self-adhesive defibrillation electrodes (so-called pads) are safe and effective and should therefore be preferred over normal plate electrodes (so-called paddles). Another advantage of adhesive electrodes from the viewpoint of the operator is that it is possible to defibrillate from a *safe distance*, without having to lean over the patient. It is also quicker to administer the first pulse of energy using self-adhesive electrodes than it is with paddles. Whereas with paddles a contact gel is needed between the skin surface and the metal plate in order to reduce the skin impedance – and thus improve the electrical contact – and also to prevent burns, this is not necessary with adhesive electrodes. Compared with electrode paste and gels, gel pads are favorable for safe operation because they avoid the risk of arcing and of a short circuit and thus prevent ineffective defibrillation. The Association for the Advancement of Medical Instrumentation (AAMI) recommends a minimum electrode area of 150 cm^2, since larger electrodes have a lower impedance or even lead to a decrease in the transmyocardial flow. The diameter of customary electrodes is in the region of 8–12 cm for adults and children with body weight > 10 kg and 4.5 cm for children with body weight < 10 kg.

28.4 Complications

Cardioversion and defibrillation can result in the following serious and minor complications:

- Induced ventricular fibrillation, e.g., as a result of incorrect triggering, which can ultimately lead to asystole (cardiac arrest) (currents > 10 mA flowing through the heart can cause fibrillation in the ventricles).
- Postdefibrillation arrhythmias such as ventricular and supraventricular extrasystoles and ventricular flutter.
- Arterial embolisms.
- Burns and irritation of the skin, for example, due to an insufficient amount of electrode contact paste being used on the electrode surface.

28.5 Technical Safety Aspects

28.5.1 Use

- Avoid direct contact with the electrodes (life-threatening), conductive contact with the patients or other people (safe distance).
- There should be no moisture on the patient's skin (electrical bridge), and the patient should also be positioned such that he is electrically isolated.
- Only perform cardioversion if the ECG is free from artifacts and if reliable ECG monitoring is possible. When too much electrode contact paste is used on the paddles there is the possibility of an electrical bridge forming (risk of short circuit).
- All additional devices which are connected to the patient must be defibrillation proof; otherwise, they must be disconnected from the patient during cardioversion/defibrillation.
- Caution should be exercised with patients with energized implants: the functioning of the implant may be restricted or suspended, and the implant itself may be damaged or may even become unusable.

28.5.2 Device

- Defibrillators belong to class IIb of the German Medical Devices Act (MPG, Medizinprodukte Gesetz).
- Defibrillators must only be used in an explosion-proof atmosphere.
- Devices which are not defibrillator proof must be disconnected from the patient, otherwise
- Equipment should be labeled according to DIN-IEC 601 as defibrillator proof.
- Maximum energy 360 J.
- Trigger buttons only on both paddles (connected in series).
- Protective circuits, which ensure a reduced power setting when the defibrillator is switched off and ensure energy recovery no later than 1 min after defibrillator charging.
- Because of their unforeseeable and frequently changing use, defibrillators should always be connected to mains electricity at their device base locations so that they are operational and ready for use on an ad hoc basis.

28.6 Implantable Cardioverter-Defibrillators

Implantable cardioverter-defibrillators (ICD) are primarily used for reliable detection and treatment of life-threatening cardiac arrhythmias (ventricular tachycardia and fibrillation) which originate in the ventricles and which cannot be treated with medication (Fig. 28.8). The American and European Societies of Cardiology (American College of Cardiology (ACC)/AHA and ESC) also recommend ICD implantation in patients who have survived sudden cardiac death which did not occur as a result of myocardial infarction and also for primary prophylaxis in patients with restricted cardiac pumping function (EF < 30%). When necessary,

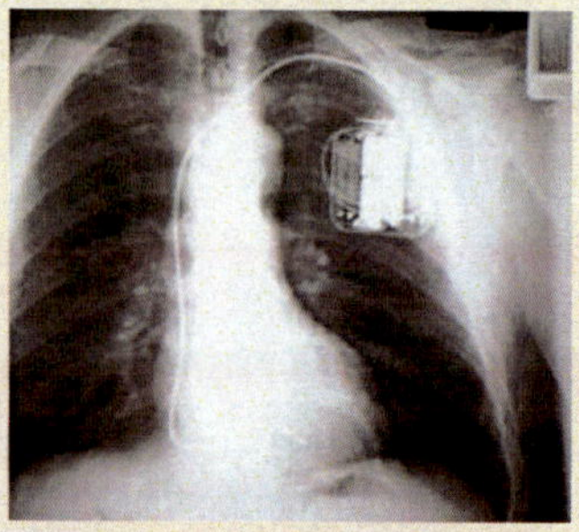

Fig. 28.8 Chest x-ray with implanted ICD system (courtesy Medtronic)

ICD systems can also administer pacemaker pulses in bradycardic phases. ICD systems can also provide antitachycardic stimulation, cardioversion, and defibrillation.

ICD implantations have increased significantly in numbers. The following reasons are vital to this development:

- A simplified implantation technique reduces the rates of mortality resulting from surgery.
- Very low morbidity.
- High level of patient acceptance.
- ICD therapy has considerable advantages over medicinal therapy for prevention of sudden cardiac death – within the scope of secondary prophylaxis, for example, following hemodynamic ventricular tachycardia.
- For idiopathic ventricular fibrillation there is currently no alternative to ICD therapy.
- A further indication is primary prophylaxis for high-risk patients with no symptoms.
- The costs of ICD therapy have decreased considerably as a result of a reduction in product prices alongside increasing product life.

28.6.1 ICD Development

The following chronological events are of particular significance in ICD development. The first transthoracic defibrillation with direct current was carried out by Lown in 1962. The first model of an implantable automated defibrillator (*automated implantable cardioverter-defibrillator*, AICD) was presented by Mirowski as early as 1969 and was implanted in a human for the first time in 1980. This system was used to detect ventricular flutter and fibrillation using the criteria of heart rate and/or absence of isoelectric ECG portions. Detection of arrhythmias was done using myocardial screw-type electrodes. A programmable ICD system was used in clinical practice in 1988 and a multiprogrammable ICD system in 1989. The first dual-chamber ICD came onto the market in 1997, while an atrial defibrillator was implanted as early as 1996. ICD electrodes placed intra-atrially are a safe and effective method for terminating atrial fibrillation. The advantages of this are firstly the low energy output and secondly that the patient does not require short anesthesia.

A combined atrioventricular defibrillator was first implanted in 1997. More sophisticated atrioventricular defibrillators have the advantage that, besides providing fully automated detection of atrial and ventricular signals, they also deliver electrical pulses to both ventricles in order to terminate arrhythmias which occur. So-called implantable CRT defibrillators (CRT-ICD for short) for cardiac resynchronization therapy (CRT) have been available since 2001. These devices deliver a shock to interrupt and therefore stop tachycardia. The right and left ventricles are also stimulated simultaneously so that the cardiac beats are coordinated (resynchronized, in the true meaning of the word) (Fig. 28.9).

a)

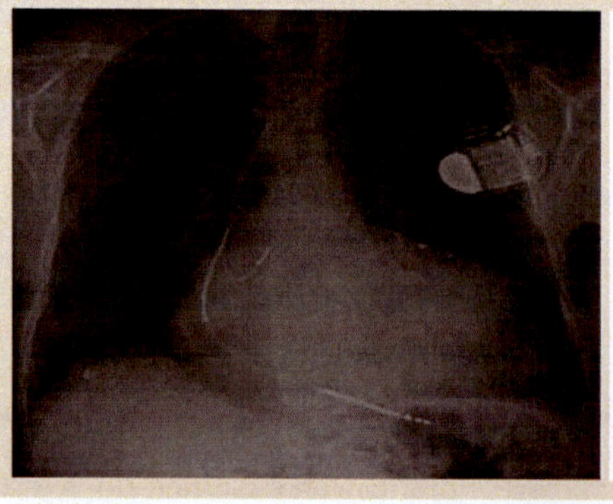

b)

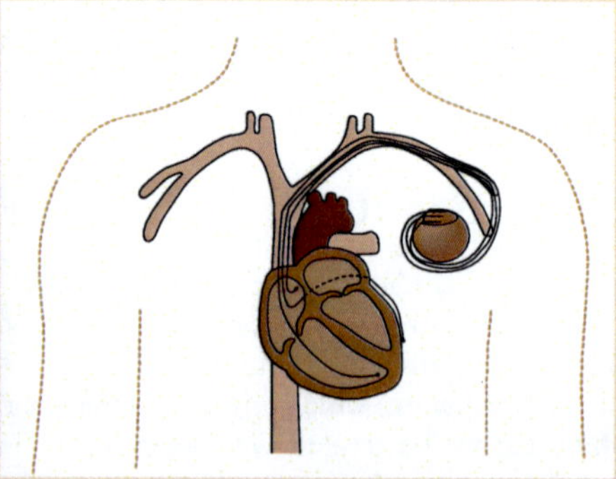

Fig. 28.9a,b Implantable biventricular or CRT-ICD system. **(a)** Position checking in the x-ray image. **(b)** Schematic electrode positioning (courtesy of Biotronik)

28.6.2 System Technology

Miniaturized construction, more effective defibrillation as a result of low-energy pulses, and improved and updated software enable subpectoral implantation with transvenous electrode insertion under local anesthetic to be standard practice today (Fig. 28.10).

Conventional products have volume $< 40\,cm^3$ and weight $< 80\,g$, with average device lifespan of 4–6 years. Today, ICD technology provides fine-tuned, integrated systems:

- Single-chamber ICD (ventricular rate adaptive/rate modulation (VR) system) with the therapeutic possibility of antitachycardic stimulation, cardioversion, and defibrillation. The defibrillator electrode is placed in the right ventricle.
- Dual-chamber ICD (dual rate adaptive/rate modulation (DR) system) with an additional atrial electrode or *single-lead ventricular pacing with atrial tracking-ICD (VDD-ICD)*.
- Triple-chamber ICD (biventricular defibrillator system, CRT-ICD) with the additional possibility of cardiac resynchronization therapy (CRT). The electrode fixed in the right ventricle is supplemented by a further electrode (a so-called coronary sinus electrode) which selectively stimulates the left ventricle (Fig. 28.9a,b).

Just as there is the international pacemaker code (North American Society of Pacing and Electrophysiology/British Pacing and Electrophysiology Group generic Pacemaker Code (NBG), Chap. 38), there is also a code for ICDs (Table 28.1).

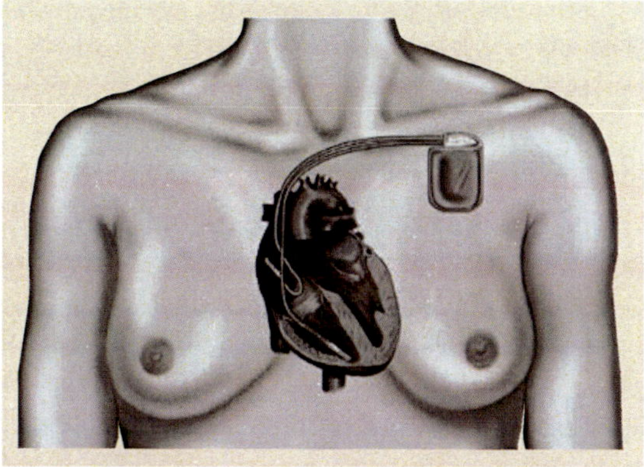

Fig. 28.10 Implantable ICD system (schematic) (courtesy Medtronic)

An ICD system consists of a generating unit (or pulse generator) and an electrode. The generating unit includes the housing, the battery (lithium iodide), capacitors, microprocessor, amplifier, magnetic switch, audio transducer for signals, antenna for telemetry, and a connection head for electrodes. The electrode (or probe) consists of an isolated electrode body with a multilumen construction. The conductors for detection, stimulation, and defibrillation are arranged inside the electrode, isolated from one another.

Table 28.1 The NASPE/BPEG defibrillator (NBD) code (after [28.2])

I Shock chamber	II Antitachycardia stimulation chamber	III Tachycardia detection	IV Antibradycardia stimulation chamber
0 = none	0 = none	E = intracardiac EG	0 = none
A = atrium	A = atrium	H = hemodynamics (H includes E)	A = atrium
V = ventricle	V = ventricle		V = ventricle
D = dual (A + V)	D = dual (A + V)		D = dual (A + V)
Short-form code			
ICD-S	ICD with shock capability only		
ICD-B	ICD with bradycardia pacemaker as well as shock		
ICD-T	ICD with tachycardia and bradycardia pacemaker as well as shock		

Additional note:
Site of detection and stimulation:
Single-chamber system = atrium *or* ventricle, dual-chamber system = atrium *and* ventricle.
Reaction to detection:
1. Triggered: synchronous stimulation at available and nonadjustable frequency when there is no cardiac activity.
2. Inhibited: when cardiac activity is present, delivery of a pulse is suppressed.
3. Frequency adaptation: stimulation frequency is adapted to the physical strain at the time.
The coding is listed on the ICD housing and in the ICD documentation.

Depending on the ICD model, the therapy possibilities are:

1. Pacing (pacemaker function in the case of bradycardic cardiac arrhythmias)
2. Antitachycardic pacing (ATP, the intention is to terminate ventricular tachycardias with overstimulation – a so-called burst with every n-th stimulus)
3. Cardioversion (when ATP is unsuccessful, a shock of energy of 15–20 J is automatically administered, and up to 35 J for further shocks)
4. Defibrillation (in the case of ventricular fibrillation, a shock of energy is initiated immediately with an upper limit, depending on the device, of 35–40 J).

Integrated ventricular demand rate-responsive pacing (VVIR) or dual chamber rate-responsive pacing (DDDR) pacemaker circuits (Chap. 38) are standard in modern ICD technology. The operational lifespan of dual-chamber systems is somewhat shorter than that of single-chamber systems. New shock forms, improved electrode systems, and optimized shock delivery are leading to a considerable reduction in the defibrillation threshold (DFT), which today is generally below 15 J. There is also the fact that short and constant charge times prevent the DFT from increasing. Biphasic or sequential defibrillator pulses are currently preferred in clinical use. Studies have confirmed that the pulse of energy required increases with the duration of ventricular fibrillation and with the increase in conduction fronts, and conversely that the sooner the ventricular fibrillation is terminated after it arises, the lower the energy. Comprehensive diagnostic program memories, which in the context of patient aftercare enable validation of the therapeutic intervention, are part of today's standard equipment. Some ICD devices provide the option of telemetry transmission of recorded intracardiac electrograms via an antenna in the head of the ICD implant. The periodic or event-triggered reports are transmitted via mobile telephony to an Internet-based database by means of a mobile patient unit. The attending doctor can then access the transmitted data via the Internet.

28.6.3 Algorithms

Internal ECG signal detection algorithms analyze the amplitude, frequency, and slew rate of the ECG. Errors in the interpretation of tachycardic arrhythmias can lead both to false ICD discharges and to erroneous failure to detect ventricular arrhythmias (so-called undersensing). Detecting ventricular arrhythmias and differentiating them from supraventricular arrhythmias is based on different algorithms for detection and differentiation. The cardiac rhythm frequencies which emerge are classified and assigned to up to six detection zones: a bradycardia zone, a sinus rhythm zone, a fibrillation zone, and one to three tachycardia zones.

ICD algorithms distinguish between supraventricular and ventricular tachycardias on the basis of criteria such as frequency stability, QRS width criterion (if, for example, six out of eight QRS complexes are wider than in the sinus rhythm), sudden-onset rapid heartbeat, the relationship between atrial and ventricular signals, the site of acceleration and AV association, or distinguishes between different variants in the AV relationships by means of detection of a pattern over a period between two RR intervals. Detection criteria can be programmed in the sensing system responsible for detection and redetection. Furthermore, it is also possible to include atrial events in the detection algorithms. The mode switch algorithm induces an automatic change of operating mode from the DDDR to the dual chamber demand rate-responsive pacing (DDIR) mode to ensure that, after high atrial frequencies have been detected, the triggered mode for ventricular stimulation is actuated in aid of inhibited operation. To increase the sensitivity and specificity of the detection of ventricular tachycardias in ICD therapy, in addition to the electrical criteria it is also a good idea to take into consideration information about the hemodynamic situation, such as the contractility of the left ventricle.

28.6.4 Electrodes

If ICD implantation was still a complex surgical procedure at the beginning of the 1990s, then the development of a transvenous electrode was a considerable simplification. The combination of transvenous electrodes with subcutaneous defibrillation electrodes is a vital step forwards.

The electrically active generating unit housing (*active can*) and/or proximal section of the defibrillation probe is used as the anode, while the section of the electrode in the right ventricle functions as the cathode. A distinction is drawn between two configurations: the single-coil probe from the right ventricle to the ICD housing, and the dual-coil probe with a defibrillation surface between the right ventricle, the superior vena cava, and the ICD housing (triad configuration), which is the most favorable in terms of energy. If the section of the electrode in the right ventricle functions as an anode due to programming, this reversal in polarity likewise results in a reduced energy requirement during defibrillation.

This measure is particularly suitable for patients with a very high defibrillation threshold. The signal detection function and antibradycardic stimulation are performed via the tip of the probe, which is constructed in true bipolar fashion (between the electrode tip and a proximally positioned ring) or in integrated bipolar fashion (between the electrode tip and the distal shock coil).

Modern ICD electrode bodies have a multilumen construction, and a few are coaxial. Silicone is predominantly used (or less often polyurethane) as the insulation for the external layer of the probe and offers a high level of biocompatibility and flexibility but is very sensitive to mechanical stress. Other distinguishing features of transvenous probes are the number of defibrillation probes (one or two) and the number of coils (single or dual). Active or passive fixing of ICD electrodes can be done using fixed screws or extensible screws and using anchors.

Atrial and epimyocardial electrodes and coronary sinus electrodes (CS or superior vena cava (SVC) electrodes) are used as additional pacing/sensing electrodes. The CS electrode has at its distal end an acceleration sensor, which is used to record and differentiate acceleration values in at least two different directions.

A special shock probe is the array electrode (also known as a finger electrode or SQ array), which is implanted subcutaneously. This electrode has up to three shock coils and is used additionally when there is an insufficient defibrillation threshold. The individual coils of the finger electrode are connected in parallel with the right-ventricular shock coil via a Y-connector.

The single-lead VDD-ICD is an electrode system for a dual-chamber ICD. The atrial signals are recorded at this electrode, which is fixed in the apex of the right ventricle, by means of two electrode rings. This electrode thus replaces the otherwise additional atrial electrode. Bipolar and tripolar ECG leads in ICD systems enable reliable ischemia detection, which has considerable advantages over the conventional surface ECG owing to the early detection of ischemias and also their severity and duration.

28.6.5 Complications

The spectrum of ICD complications ranges from harmless complications, such as skin irritations and minor burns on the areas where the electrodes have been applied, to serious complications, such as the emergence of extrasystoles following defibrillation, ventricular tachycardia, ventricular fibrillation (in the case of incorrect triggering) or the occurrence of asystole following electric cardioversion in the case of sick sinus syndrome. ICD-specific complications can be caused on the one hand by the generating unit:

- Pressure-related necroses
- Risk of infection
- Technical errors and malfunctions as a result, for example, of strong external magnetic fields or battery depletion
- Perforation of the generating unit

and on the other hand by the electrodes:

- Electrode dislocation can cause stimulation of muscles
- Inadequate administration of shocks as a result of detection of muscle potential or external sources of interference
- Fractures in the electrodes, defects in the insulation or a rise in the sensitivity threshold can cause electrode malfunction.

One malfunction in the detection function (sensing) is undersensing or oversensing. In the case of undersensing, life-threatening arrhythmias are not detected or are detected too late, whereas with oversensing the shocks administered are inadequate.

28.6.6 Function Checking

Whereas in the early days of ICD device technology it was still necessary for technical reasons to test the functioning of an ICD patient's implant every 2 months, this check is now carried out independently by the device itself. Function checking of the system (ICD query, analysis of stored ECGs, battery status, detection test, etc.) every 3–6 months is sufficient. If there is an indication of incorrect detection (*so-called* over- or undersensing), a chest x-ray is generally performed to identify insulation defects in the probe. Because of the improvement in battery technology, the shelf-life of single-chamber systems should be ≈ 9 years, dual-chamber systems should last ≈ 7 years, and triple-chamber systems ≈ 5 years. The lifespan of the device will therefore be considerably longer than that of previous ICD systems. Replacement of the generating unit is also only necessary at a later point in time. Stored data, such as treatment episodes, electrograms, battery state, and electrode resistance, can be queried via telemetry. For the purpose of function checking, various device parameters are measured intraoperatively during pacemaker and ICD implantations and, if necessary, adjusted (Table 28.2).

Table 28.2 Limit values for intraoperative measurements during implantation of pacemakers and ICD systems (after 28.3)

	Ventricle	Atrium
Signal amplitude	> 5 mV, optimally > 8 mV	> 2 mV, optimally > 3 mV
Impedance (electrode-dependent)	300–1200 Ω	300–1200 Ω
Slew rate	> 0.5 V/s	> 0.3 V/s

Further Reading

- American Heart Association (AHA): Guidelines for cardiopulmonary resuscitation and emergency cardiovascular care, Circulation **112**(Suppl. I), IV-1–IV-5 (2005)
- A.D. Bernstein, J.-C. Daubert, R.D. Fletcher, D.L. Hayes, B. Lüderitz, D.W. Reynolds, M.H. Schoenfeld, R. Sutton: The revised NASPE/BPEG generic code for antibradycardia, adaptive-rate, and multisite pacing, PACE **25**, 260–264 (2000)
- CoSTR: International consensus on cardiopulmonary resuscitation (CPR) and emergency cardiovascular care (ECC) science with treatment recommendations, Circulation **112**(Suppl. I), b2–b5 (2005)
- ISO: *ISO 11318:2002, Cardiac Defibrillators* (ISO, Geneva 2002)
- ISO: *ISO 14708-6:2010, Implants for Surgery – Active Implantable Medical Devices* (ISO, Geneva 2010)
- ERC European Resuscitation Council: Guidelines for resuscitation, Resuscitation **67**(Suppl. 1), S1–S190 (2005)
- A.W.C. Chow, A.E. Buxton (Eds.): *Implantable Cardiac Pacemakers and Defibrillators* (Blackwell, Oxford 2006)
- R.X. Stroobandt, S.S. Barold, A.F. Sinnaeve: *Implantable Cardioverter* (Wiley-Blackwell, New York 2009)
- T. Kenny: *The Nuts and Bolts of Cardiac Pacing* (Wiley-Blackwell, New York 2005)
- W. Haverkamp, G. Breithardt: *Moderne Herzrhythmustherapie* (Thieme, Stuttgart 2003)
- The International Liaison Committee on Resuscitation (ILCOR): Guidelines for resuscitation, Resuscitation **67**(2/3), 157–342 (2005)
- B. Lüderitz: *Herzrhythmusstörungen. Diagnostik und Therapie*, 6th edn. (Springer, Berlin, Heidelberg 2010)
- D. Wietholt, L.J. Ulbricht, H. Gülker (Eds.): *Implantierbare Kardioverter-Defibrillatoren* (Thieme, Stuttgart 1997)

References

28.1 A. Bolz, W. Urbaszek: *Technik in der Kardiologie* (Springer, Berlin, Heidelberg 2002)

28.2 A.D. Bernstein: *Pacemaker, Defibrillator and Lead Codes* (Heart Rhythm Society, Washington 1993)

28.3 H. Roskamm, F.-J. Neumann, D. Kalusche, H.-P. Bestehorn (Eds.): *Herzkrankheiten*, 5th edn. (Springer, Berlin, Heidelberg 2004)

29. Laser Systems

Frank Zgoda

This chapter gives a short introduction to the laser and its history providing an insight into the physics and technology of laser systems. For the use of lasers in medicine, not only knowledge of the application methods and biophysical effects of the laser radiation on tissue is essential but also detailed information about the specific characteristics of the various laser systems. The fields of use are manifold and can be classified according to parts of the body and special treatment modalities. The chapter concludes with a section on safety aspects and future prospects with relating to technological developments in endoscopy and smart systems.

29.1 **History of the Laser** ... 558

29.2 **Physics and Technology** ... 558
- 29.2.1 Laser Medium ... 559
- 29.2.2 Laser Radiation ... 560
- 29.2.3 Laser Systems ... 562
- 29.2.4 Beam Delivery Systems ... 563

29.3 **Application Methods** ... 564
- 29.3.1 Noncontact Mode ... 564
- 29.3.2 Contact Mode ... 564
- 29.3.3 Gas or Liquid Flushing During Laser Cutting ... 564
- 29.3.4 Special Applicators ... 564

29.4 **Biophysical Effects on Tissue** ... 564
- 29.4.1 Absorption and Penetration Depth 564
- 29.4.2 Photochemical Effects ... 566
- 29.4.3 Photothermal Effects ... 566
- 29.4.4 Photodecomposition Effects ... 567

29.5 **Laser Types in Medicine** ... 567
- 29.5.1 The Ruby Laser ... 569
- 29.5.2 The Neodymium:YAG Laser ... 569
- 29.5.3 The Frequency-Doubled Nd:YAG Laser (KTP Laser) ... 569
- 29.5.4 The Erbium:YAG Laser ... 569
- 29.5.5 The Holmium:YAG Laser ... 569
- 29.5.6 The Alexandrite Laser ... 569
- 29.5.7 The Diode Laser ... 570
- 29.5.8 The CO_2 Laser ... 570
- 29.5.9 Argon-Ion and Krypton-Ion Lasers ... 570
- 29.5.10 The Excimer Laser ... 570
- 29.5.11 The HeNe Laser ... 571
- 29.5.12 The Dye Laser ... 571
- 29.5.13 The Free-Electron Laser ... 571

29.6 **Fields of Use** ... 571
- 29.6.1 The Eye ... 571
- 29.6.2 The Body Surface ... 571
- 29.6.3 The Vascular System ... 572
- 29.6.4 Open Surgery ... 572
- 29.6.5 Endoscopy ... 573
- 29.6.6 Photodynamic Therapy ... 573
- 29.6.7 Lithotripsy ... 573
- 29.6.8 Laser-Induced Thermotherapy ... 573

29.7 **Safety Aspects** ... 574
- 29.7.1 Permissible Exposure and Laser Classification ... 574
- 29.7.2 Safety Requirements for Laser Devices ... 576
- 29.7.3 Protective Measures and Application Safety ... 576

29.8 **Future Prospects** ... 577

References ... 577

The laser has been linked to medical applications from the very start. Light has been used in healing processes for some time, for example, for treating diseases of the skin, supporting the production of vitamin D by UV light and for treating eye diseases (Meyer-Schwickerath 1948). The defining feature of laser radiation is that it is an easy to control energy source that can be used not only to heat specific tissues and to destroy autologous stones but also to selectively excite photochemical processes [29.1,2].

29.1 History of the Laser

The word LASER is an acronym for light amplification by stimulated emission of radiation. This process was postulated by Einstein as early as 1917 [29.3] and forms the basis for the production of monochromatic, coherent and collimated light with a higher power density and a greater spectral purity than can be attained with any other light source [29.4,5].

Working according to the same principle, in 1954 Townes built the first maser, an amplifier for microwaves. The principle of the laser for optical radiation was published for the first time four years later by Schawlow and Townes (USA) and Basov and Prokhorov (USSR). Maiman made an important breakthrough in 1961 with the first working laser emitting visible light (ruby laser), and Javan followed in the same year with the first infrared gas laser (HeNe).

The medical use of lasers began when Campbell (1961) introduced the use of the ruby laser into ophthalmology, who was soon followed by Goldman in dermatology. Although the laser quickly became established in ophthalmology (initially the argon-ion laser, L'Esperance 1968), its use in other areas of medicine was more gradual. Polanyi first pioneered the use of the carbon dioxide laser (CO_2 laser) in 1965, to be followed by Kaplan in 1967. Since then the CO_2 laser has become established in all surgical disciplines.

The argon-ion laser and the Nd:YAG laser became interesting for a wider range of medical applications when it became possible to couple the lasers to glass fibers. From this point on, further developments progressed very rapidly. Hematoporphyrin derivatives (HpD) were discovered in 1960 but they were not used for photodynamic therapy until 1972 by Diamond. Q-switching and mode-locking were introduced into pulsed systems in ophthalmology in 1977 by Fankhauser and Aron-Rosa, and it was in this field that Trokel used the excimer laser for the first time in 1983.

Today, physicians have a wide variety of lasers to choose from for a multitude of uses. The advantages offered by the laser have led to it being used as a standard procedure in many medical fields, which are constantly being added to.

29.2 Physics and Technology

A laser is a device for the conversion of electrical energy into *ordered* light energy. Conventional light sources such as light bulbs or fluorescent lamps have as much in common with lasers as a spark gap has with a radio transmitter. Lasers are ideal light emitters in the terahertz region and are analogous to radio transmitters in the megahertz region. Einstein postulated that induced emission is rudimentarily apparent in every luminescent effect of matter; however, it can only be brought to dominance in selected materials [29.6].

Different medical lasers are employed according to the tissue to be treated and the specific therapy required much in the same way as special instruments are used in medicine for different operations (Fig. 29.1). Lasers differ from one another by their emission wavelength, beginning in the UV region at a wavelength of 0.2 μm and extending to 10 μm in the infrared. As most lasers are based on a narrow spectral transition of electrons, in general they cannot be tuned in the same way as a radio transmitter. The only exceptions to this are dye lasers, vibronic solid-state lasers and the free-electron laser. Some gas lasers have several fixed wavelengths according to the type of selected gaseous medium.

Laser systems vary in their technical and physical configuration, more specifically in their constructional design and the temporal characteristics of the emission. On its own, the laser is just a light source. However, what is required is a complete system made up of a beam transmission system and an end device to direct the radiation to the tissue to be treated, together with a means of controlling the effect. According to the wavelength, a quartz glass fiber, a fiber bundle, a hollow waveguide, or an articulated arm is used for transmission of the radiation. The end device can be, for example, a surgical microscope with a micromanipulator, an endoscope, or a slit lamp (Fig. 29.2).

Monitoring of the effect is dependent on the laser-induced processes and is, in many cases, based on the visual control of an experienced therapist. Imaging systems, such as ultrasound or magnetic resonance tomography, are used for highly differentiated procedures where precise dosage control is required e.g. interstitial thermotherapy of tumors or metastases.

29.2.1 Laser Medium

In order to be able to understand the need for the wide variety of appliances available, some principles of physics that form the basis for the different *light amplifiers by induced emission* are explained here.

Spontaneous emission of light results when energy is fed into a gaseous, liquid, or solid body so that *luminous electrons* are raised from the ground state energy level n_1 to an excited state n_2. After a characteristic time interval, the electron spontaneously reverts to the ground state, thereby emitting a photon. The distribution of the electrons between the ground state and the excited state (ratio of the occupation numbers) is dependent on the temperature (Fig. 29.3) i. e. the higher the temperature, the more electrons find themselves in an excited state.

In a thermal light source, such as an electric light bulb, energy is continually fed by heat in order to transfer the electrons to higher levels. The life span of the excited states is short; in order for the electron to be able to return to the ground state this cannot be completely filled. The excitation of an electron can also be effected by irradiation of light. In this case, the higher level is populated by absorbing the corresponding energy difference ΔE.

In addition to the direct transition from a higher level directly to the ground state, a return to the ground state can also occur stepwise through several intermediate levels. As the lifetimes of the levels are different, i. e. the levels empty at different rates, the electrons can collect at higher levels.

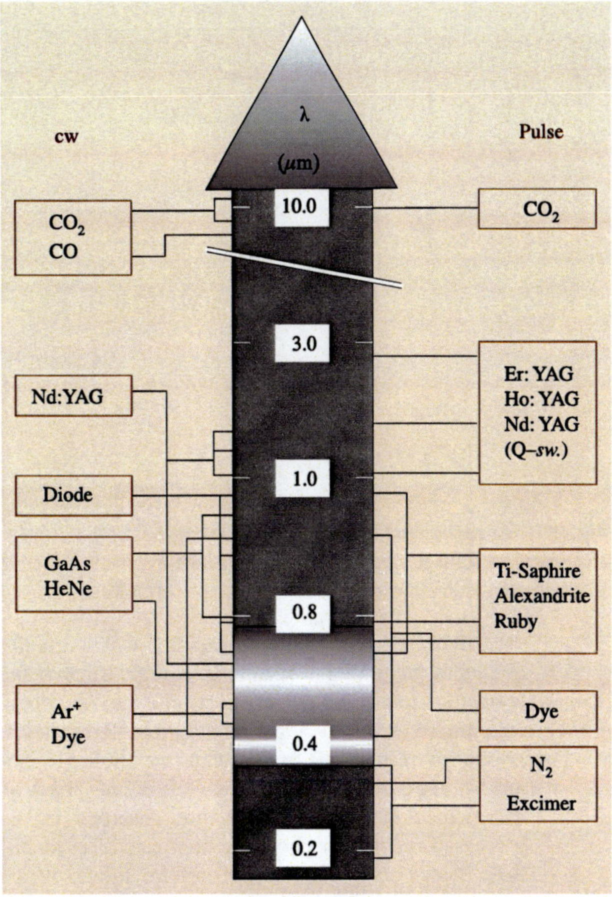

Fig. 29.1 Medical laser systems and their wavelengths

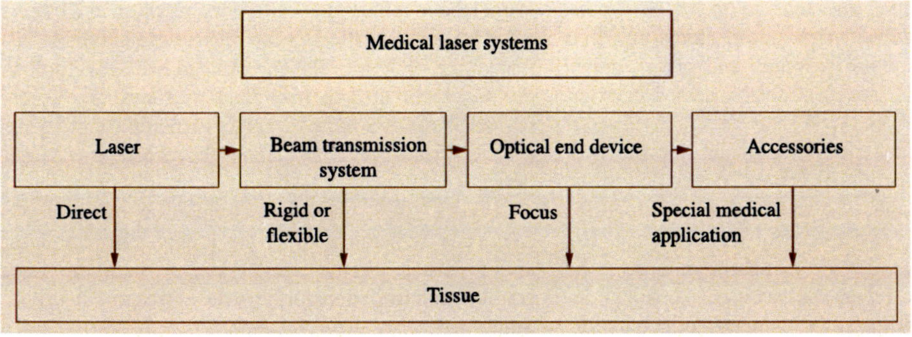

Fig. 29.2 Components of a medical laser system

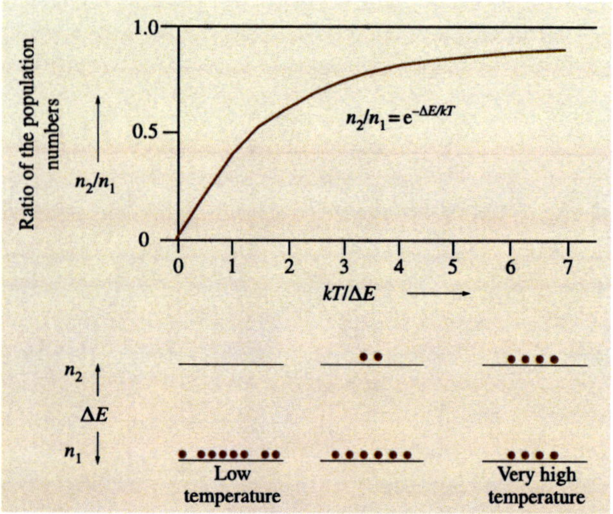

Fig. 29.3 According to Planck's law the ratio n_2/n_1 of the occupation numbers shifts toward 1 during thermal heating. It is impossible to reach population inversion

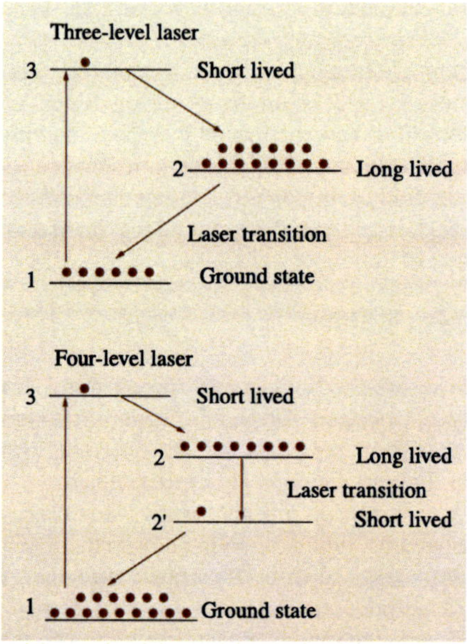

Fig. 29.4 Energy levels in laser systems

Induced emission is a process by which irradiation of a photon of a matching energy triggers the premature transition of an electron from an occupied higher level to a partially empty lower level. A special feature of this is that the emitted photon has the same wavelength, direction and phase as the incident photon, which is not absorbed in the process. The incident photon is duplicated i. e. amplified. Most lasers make use of several levels, which vary both in lifetime and in occupation number. Figure 29.4 shows schemes for a three-level or a four-level laser.

With regard to the laser process, the number of atoms with occupied upper laser levels must always be larger than the number of atoms in the ground state. All substances in which this so-called occupation inversion is possible can be used as a laser medium e.g. free atoms, molecules and ions in gases or vapors, dye molecules dissolved in solution, atoms and ions incorporated into a solid (crystals and glass), doped semiconductors, and free electrons harnessed in external magnetic fields.

The relevant excitation mechanism for the production of an occupation inversion is specific for the particular type of laser. The most important processes are optical pumping and electrical gas discharge. Excitation of a semiconductor is achieved directly by an electric current (charge injection). Chemical reactions can also be used as a means of excitation. In the case of continuous-wave lasers, the occupation inversion must be constantly maintained i. e. pumping energy must be fed into the system continually.

In the laser medium excited atoms already emit spontaneous photons in all directions. These can produce more photons in the laser medium by induced emission, which in turn leads to more induced photons being set free etc. This process continues like an avalanche until the photons leave the laser medium. If the laser medium is allowed to expand in one direction, the random photon avalanches that are produced in this direction have a significantly higher amplification than in any other direction because they run through the laser medium longer.

Using two mirrors placed at each end, the photon avalanche is reflected back into the medium and effects a further amplification. If the mirrors are optimally adjusted, the photon avalanche can run back and forth in this *resonator* up to several hundred times. One of the resonator mirrors must be semitransparent in order for a fraction of the photons to leave the resonator as a laser beam.

Figure 29.5 shows a comparison of a high-frequency transmitter with the components amplifier and feedback circuit with a laser, where the laser medium and the resonator have the corresponding functions.

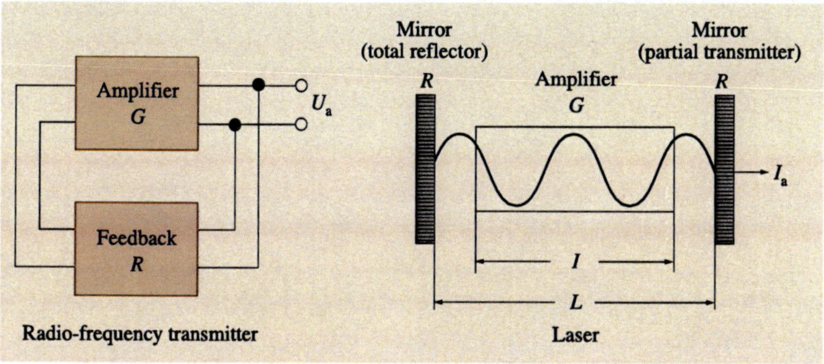

Fig. 29.5 Comparison of the functional principles of a high-frequency transmitter and a laser using the laser medium as an amplifier

29.2.2 Laser Radiation

In contrast to the radiation of a thermal light source, laser radiation has three important characteristics (Fig. 29.6).

1. Coherence: The laser radiation has a defined spatial and temporal order (set phase relation).
2. Collimation: The laser radiation is well collimated (low divergence).
3. Monochromaticity: The laser emits a high intensity, narrow (monochromatic) spectral band.

These three characteristics determine the good focusing ability necessary to attain high energy densities enabling precision work with a beam of very small cross sectional area.

In addition to the basic physical characteristics of laser light, the optical characteristics of biological tissue i. e. absorption, fluorescence, and scattering also play a crucial role in medical laser applications and in the use of laser radiation for diagnostics. These characteristics can be evaluated both photometrically and spectroscopically.

For therapy it is the collimation of the radiation and the absorption characteristics of the irradiated tissue that are the properties that are used the most. The interaction of the radiation with different tissue types is influenced by two main parameters: the exposure time on the tissue and the effective power density (Fig. 29.7). The power density can be varied over a wide range by selecting the appropriate focal length. A smaller focus is achieved with a shorter focal length rather than a longer one; however in this case, the irradiation area is more affected by small deviations in the optimal distance. The exposure time can be regulated using either continuous-wave mode (CW), switched mode (with variable duty cycle), or short pulses.

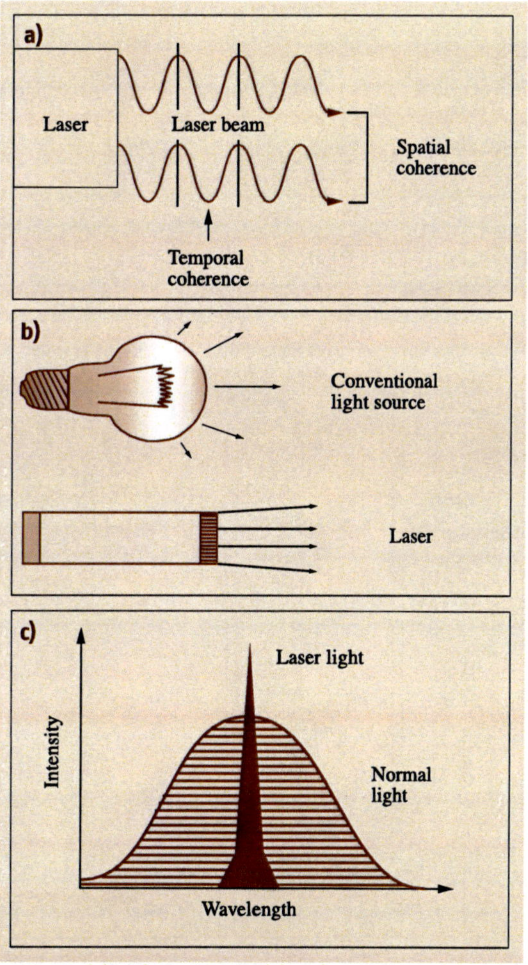

Fig. 29.6a–c Properties of laser light: **(a)** coherence, **(b)** collimation, **(c)** monochromaticity

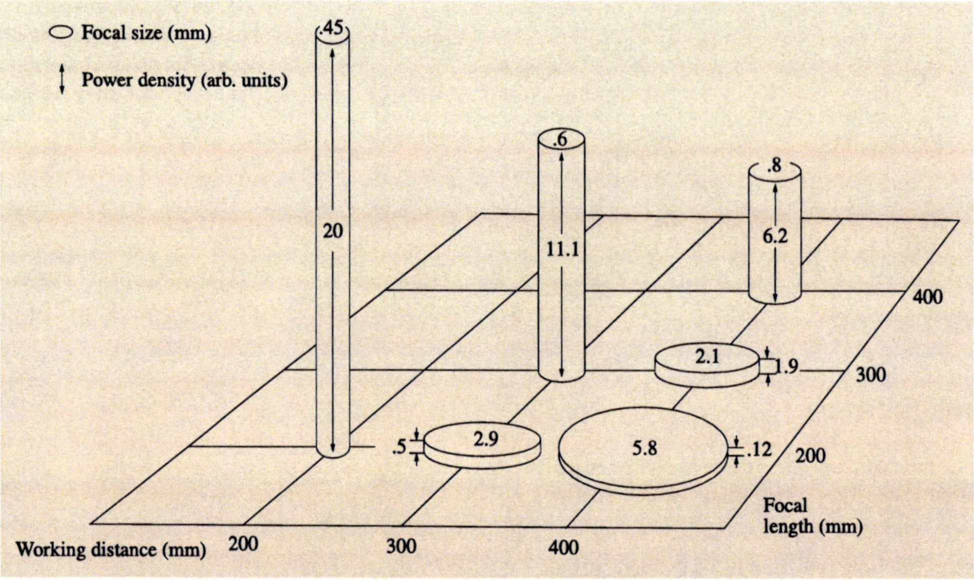

Fig. 29.7 Relationship of focal size and power density for different focal lengths and working distances

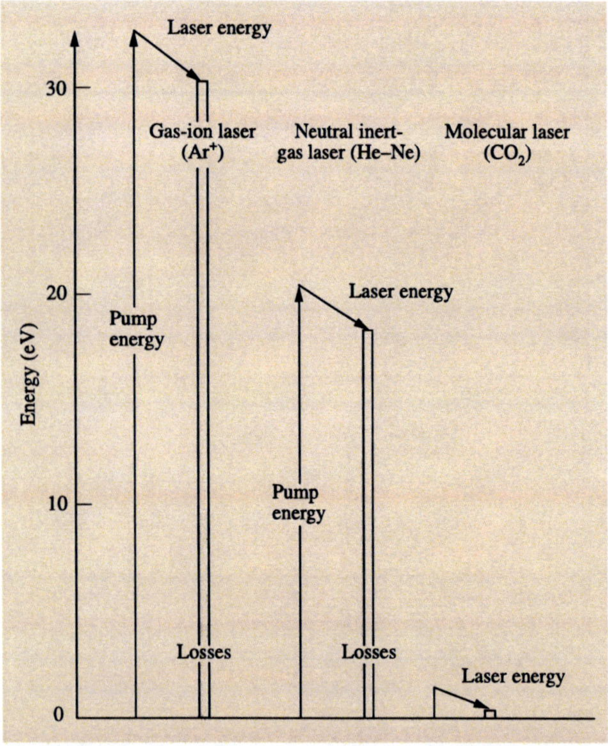

Fig. 29.8 Some examples of the ratio between pumping energy and laser energy; the losses in heat must be removed by cooling

This allows the ratio of therapeutic effects to side effects to be controlled according to the tissue interactions required.

29.2.3 Laser Systems

Laser systems are not only made up of a laser medium, a resonator and a pump source. Only a small amount of the energy used is converted into laser radiation, leaving a major part to be dissipated as heat. Figure 29.8 shows examples of the ratio of pump energy to laser energy produced for lasers in common use. It can be seen that the efficiency of an argon-ion laser is less than 0.1%, whereas that of a Nd:YAG laser is about 1–3%, and for a CO_2 laser it is 5–20%. Diode lasers have the highest efficiency with 30–40%, which is on a par with light-emitting diodes (LED). For comparison, the light yield of a common light bulb is only 3%.

The most important components making up medical laser systems are shown in Fig. 29.9. The schematic representation shows the laser resonator with the pumping source, the electrical power supply and cooling system together with a detector for the power measurement, a safety shutter for the release of the laser radiation, and the in-coupling point of a pilot or aiming beam for those lasers operating in the invisible UV or infrared range.

29.2.4 Beam Delivery Systems

The distance between the laser device output and the patient must be bridged in order to deliver the laser radiation to the tissue (Fig. 29.2). Flexible quartz glass fibers, specially doped if necessary, can be used for the transmission of visible laser radiation and the adjacent spectral range of approx. 0.3–2 μm. Beyond this spectral range articulated mirror arms are used to cover the range 0.19–0.3 μm (excimer lasers) and 3–10 μm (erbium and CO_2 laser, Fig. 29.10). CO_2 laser radiation can be transmitted over short distances by waveguides but the beam quality is not comparable with that achieved using articulated arms. The transmission of short pulsed energy-rich laser radiation such as that required for laser lithotripsy and laser angioplasty, makes special demands on the optical fibers.

The current development of photonic bandgap fibers has met with considerable interest for devices with wavelengths and power densities that cannot be transmitted with normal fibers or, if at all, with only poor quality. These special fibers have a hollow core surrounded with a periodically microstructured cladding. The two-dimensional periodic structure creates a photonic bandgap in the fiber material, preventing the radiation of a particular wavelength from propagation in the material. As a result the radiation is confined to the hollow core and guided with very low attenuation and is practically dispersion free.

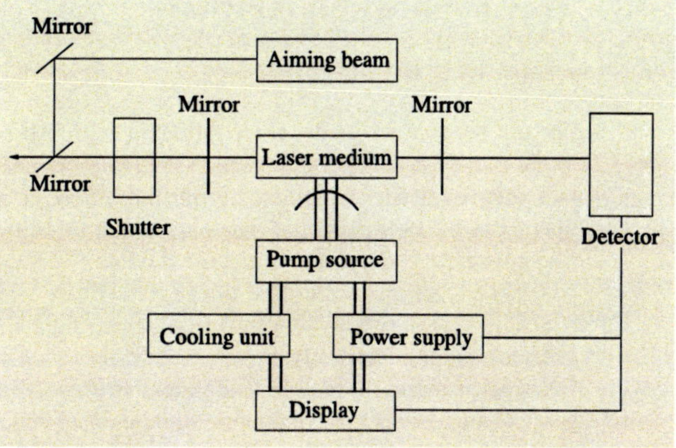

Fig. 29.9 Technical components making up a laser device. The laser beam is apparent by a visible aiming beam if it emits at a wavelength outside the visible region. The detector measures the power or energy at the laser head

Laser		λ(μm)	Operating mode	Fiber
Excimer	ArF	0.193	Pulsed	
	KrF	0.248	Pulsed	
	XeCl	0.308	Pulsed	Quartz glass
	XeF	0.351	Pulsed	
Dye		0.488–0.900	cw: Pulsed	
Argon		0.488/0.514	cw	
Nd:YAG/SHG		0.532	cw: Pulsed	
HeNe		0.541/0.543	cw	
		0.604/0.633	cw	
Krypton		0.647/0.676	cw	
Ruby		0.694	Pulsed	
Alexandrite		0.710–0.820	Pulsed	
Ti-Saphire		0.700–1.000	cw: Pulsed	
Diode, Ga(Al)As		0.750–0.950	cw: Pulsed	
Nd:YAG		1.061/1.210	cw: Pulsed	Fluoride glass
Nd:YLF		1.047/1.313	cw	Chalcogenide glass
Ho:YSSG		2.088	Pulsed	
Ho:YAG		2.0075	Pulsed	
Er:YSGG		2.795	Pulsed	
Er:YAG		2.94	Pulsed	
CO		5.0–6.0	cw	Polycrystalline silver halides
CO_2		9.6/10.6	cw: Pulsed	Hollow waveguides

Fig. 29.10 Laser wavelengths and the availability of optical fibers for laser systems

29.3 Application Methods

29.3.1 Noncontact Mode

For the noncontact mode, laser radiation is delivered to the tissue via a beam guidance system without contact with the tissue. Articulated arms and quartz glass fibers can be used for this, whereby handling of the fibers is easier because they are more flexible. With the possible exception of the bare fiber (glass fibers with a cleanly broken or polished end) from which the radiation is emitted divergently, optical end devices such as focusing handpieces, micromanipulators (for use with surgical microscopes or ophthalmic slit lamps), or endoscope coupling devices (for rigid endoscopes) are fitted to the end of beam delivery systems.

29.3.2 Contact Mode

The tissue to be cut is brought into direct contact with the end of the fiber. The tissue is vaporized with the hot fiber end and a homogenous carbonization zone remains as a sharply demarcated cut edge. The blackening of the fiber end by adherent carbonization is important for effective cutting. A quick change to noncontact mode e.g. coagulation, is possible by *burning off* the tip. The small diameter of commercially available fibers allows endoscopical use in the contact mode.

29.3.3 Gas or Liquid Flushing During Laser Cutting

In order to protect the optics of end devices at the patient end or the distal end of the fiber, the applicator or the fiber tip can be flushed with inert gas. Both Nd:YAG and diode lasers can also be used with liquid flushing. In this case, the laser beam is guided away by total reflection in the stream of water making it possible to effect coagulation of the tissue *around the corner* as it were. Flushing enlarges the depth of the cut in soft tissue because vaporization and ablation products are removed from the cutting channel, allowing unhindered penetration of the laser radiation. Photoablation, in the contact mode, can increase the photohydraulic effect and result in the formation of ablation channels that allow the fiber to pass through without any resistance.

29.3.4 Special Applicators

In addition to standard applicators, other delivery systems have been developed for special purposes. The scattering applicator ensures homogenous and isotropic irradiation and is used for the interstitial coagulation of tumors and metastases. For intracardial and urological coagulation and ablation of tissue special catheters that allow laser radiation to irradiate sideways are available. A multifiber catheter composed of a number of thin quartz fibers is used for laser angioplasty with the excimer laser (308 nm). This construction includes a flushing channel and facilitates the transmission of the laser radiation over a large cross section whilst maintaining a high level of flexibility. The light guides for photodynamic therapy are fitted with micro optics at the distal end, allowing homogenous irradiation of a sharply defined area.

29.4 Biophysical Effects on Tissue

For therapeutical purposes, the effect of the radiation on the different types of tissue is mainly determined by two parameters: the exposure duration of the radiation and the effective power density, taking account of the tissue's specific absorption and scattering [29.1, 2].

29.4.1 Absorption and Penetration Depth

Tissue specific chromophores such as hemoglobin and melanin, which both absorb in the visible region of the spectrum, limit the penetration depth of the laser radiation in the tissue. In contrast, radiation in the near infrared region is distributed homogeneously in the tissue, leading to penetration depths of more than 5 mm, depending on the wavelength.

Light energy is mainly absorbed in water in the mid to far infrared (3.0–10.6 μm). As the coupling is 10–100 times more effective compared to laser absorption in the visible range, the penetration depth and the distribution volume is very small. The fact that tissue is vaporized in this wavelength range, even at low power densities, means that tissue can be cut and/or removed without causing significant thermal damage to the surrounding area.

Figure 29.11 shows the spectral profile of the absorption coefficient for water and soft tissue. In the

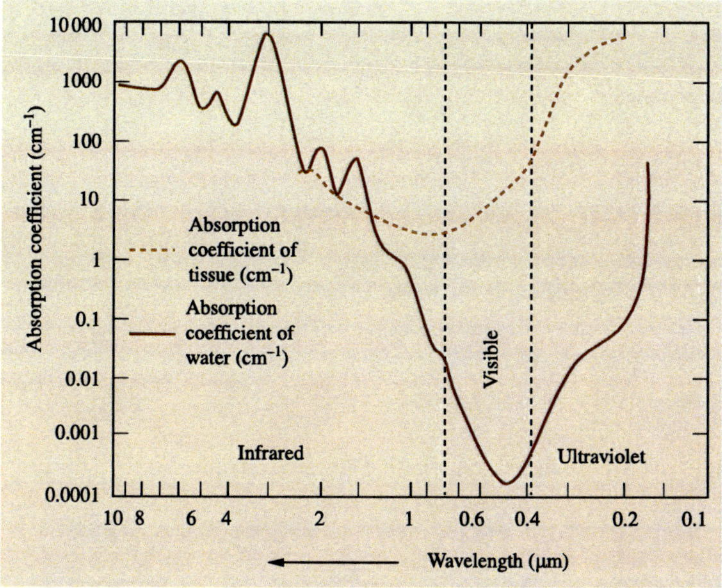

Fig. 29.11 Absorption coefficient as function of wavelength for water and tissue

visible region, soft tissue exhibits a clearly higher absorption due to autologous chromophores such as hemoglobin, melanin and coenzymes etc., whereas in the infrared region the spectral profile of water exhibits a good approximation.

It should be taken into consideration that during irradiation of body tissue with progressive exposure, photothermal effects have a considerable impact on optical characteristics such as absorption and scattering [29.1].

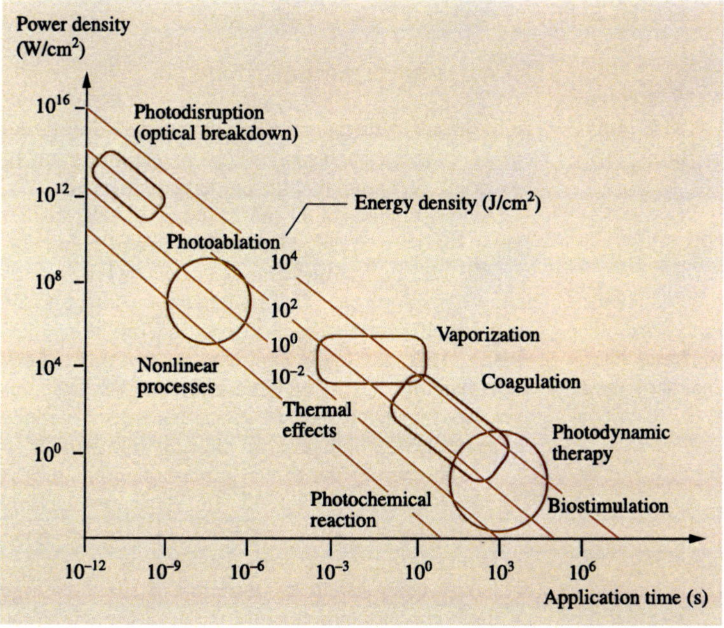

Fig. 29.12 Energy density regions as a function of the application time for different laser radiation effects on tissue

Table 29.1 Laser–tissue interaction

Photochemical effects	
Photoinduction	Biostimulation
Photoactivation of drugs (POD)	
Photochemotherapy	Photodynamic therapy (PDT)
	Black light therapy (PUVA)
Photothermal effects	
Photohyperthermia	37–43 °C: Damage of normal tissue is reversible
	43–65 °C: Edema forms in cells, tissue welding, protein precipitation
Photothermolysis	Thermal dynamic effects
	Microscopical slight overheating
Photocoagulation	65–100 °C: Coagulation, necrosis
Photocarbonization	100–300 °C: Drying out, vaporization of water, carbonization
Photovaporization	> 300 °C: Pyrolysis, vaporization of tissue
Photodecomposition effects	
Photoablation	Rapid thermal explosion
Photodisruption	Optical breakdown, mechanical shockwaves
Photofragmentation	Lithotripsy

For medical treatment methods, depending on the exposure duration and power density, the effects can be divided into three main classes (Table 29.1, Fig. 29.12) [29.1,2].

- Photochemical effects (10–1000 s; 10^{-3}–1 W/cm^2),
- Photothermal effects (1 ms–100 s; 1–10^6 W/cm^2),
- Photodecomposition effects (10 ps–100 ns; 10^8–10^{12} W/cm^2).

29.4.2 Photochemical Effects

The main effects are photoinduction or photoactivation, also known as biostimulation, and photodynamic therapy (PDT) including photosensitization. Laser energy is used to bring about photochemical reactions by absorption either in the body's own, or in extraneous, chromophores.

There are four basic types of photochemical reaction mechanisms

- Photo-induced isomerization e.g. bilirubin decomposition
- Photo-induced charge generation e.g. in visual processes
- Photo-induced synthesis e.g. photosynthesis in plants
- Photo-induced dissociation e.g. in photodynamic therapy (PDT).

29.4.3 Photothermal Effects

Here the incident radiation is converted into heat and causes coagulation, vaporization, or carbonization of the tissue, depending on the temperature reached.

At temperatures of 43–65 °C the membrane disintegrates, followed by enzymatic induction and edema formation. Thermodynamic inflammatory response occurs in this transition zone. The irradiated tissue reacts to the treatment after several weeks by sealing off the smallest vessels. This reaction is exploited for the treatment of naevi flammei. Tissue welding also occurs in this temperature range but it is based on another vaguely defined tissue reaction, which is as yet not fully understood.

Coagulation of the tissue (protein precipitation) is effected at a temperature of 65–100 °C; hemostasis and tumor denaturation are examples of processes using coagulation. Here, the tissue is damaged but is retained as a tissue composite to be decomposed at a later stage by the body and subsequently replaced by new tissue growth.

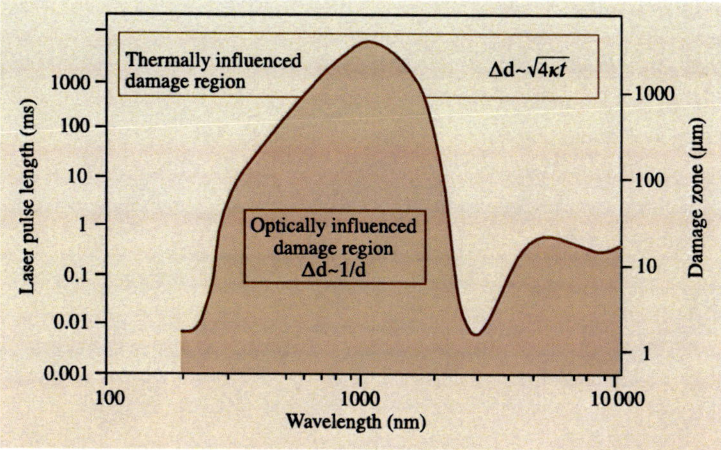

Fig. 29.13 Depending on wavelength and laser pulse length, it is possible to recognize two regions separated by the critical time. The thermally altered border zones of the regions are determined either directly by laser radiation or by thermal conduction (α: absorption coefficient in cm^{-1}; κ: temperature conduction $1.2\times10^{-7}\,m^2/s$)

Carbonization of the tissue in the temperature range 100–300 °C is the result of vaporization of water and desiccation of the tissue. The vaporization of tissues occurs at temperatures over and above 300 °C (Fig. 29.1). Bone material *melts* at temperatures above 700 °C.

29.4.4 Photodecomposition Effects

Nonthermal effects such as photoablation and photodisruption can be summarized as the photodecomposition of material. This third class of interaction is based on nonlinear optical effects because of the higher order components of the absorption coefficient becoming effective at the high peak intensities of energy-rich laser pulses. Pulsed lasers can produce peak powers from megawatts to terawatts (10^6–10^{12} W) in the nanosecond to femtosecond range. This principle is currently used in ophthalmology for the destruction of turbid posterior lens capsules and uses the same method as the destruction of calculi in laser lithotripsy.

If tissue is exposed to pulsed UV-laser radiation, it is absorbed by a very thin layer on the surface. The tissue layer absorbs so much energy that it is separated with force from the lower surface layers. This effect, known as photoablation, uses light induction to separate material fragments from a surface without causing a secondary heating effect in the surrounding tissue. However, a specific repetition rate of the laser pulses should not be exceeded in order to avoid subjecting the surrounding tissue layers to thermal damage. Photoablation is used in the laser recanalization of arteriosclerotic vessels and in refractive corneal surgery.

Depending on the optical penetration depth of the radiation, the energy from each laser pulse must have a tissue specific intensity in order to exceed the threshold for photoablation and to ablate tissue. As the damage threshold of the delivery system limits the transmitted energy density, the energy per pulse can be raised accordingly by increasing the pulse length. On the other hand, a greater pulse length means a longer continuous exposure time, allowing heat to be conducted from the heated volume to the neighboring tissue, ending with the explosive ejection of tissue particles.

Figure 29.13 can be used to calculate at which wavelength and laser pulse duration the damage zone Δd is either mainly determined by the optical penetration depth or by heat conduction [29.1]. This knowledge can be used to optimize the parameters of a photoablation laser and, as a result, limit the degree of thermal damage to an absolute minimum.

29.5 Laser Types in Medicine

The specific characteristics of different laser systems are used according to the tissue interaction required (Fig. 29.14, Table 29.2). Laser systems are usually classified by the aggregate state of the laser medium into solid-state lasers (e.g. ruby, Nd:YAG, Er:YAG, Ho:YAG, alexandrite lasers), diode lasers, gas lasers (e.g. CO_2, argon- and krypton-ion, excimer and HeNe lasers), and liquid lasers (dye lasers). The free-electron laser (FEL) is not classifiable according to this pattern.

Table 29.2 Medical application fields for different lasers

Laser	Mode	Medical application field
Ar+, (Kr+)	CW	Dermatology, ophthalmology, ENT
KTP	Pulsed	Plastic surgery, dermatology
FDL	Pulsed	Dermatology (585 nm), urology (504 nm)
Nd:YAG	CW	Surgery, urology, gynecology, neurosurgery, gastroenterology, pulmology
Nd:YAG	Pulsed	Ophthalmology, lithotripsy
CO_2	CW	Surgery, dermatology, ENT, gynecology, neurosurgery, plastic surgery
Er:YAG	Pulsed	Dermatology, plastic surgery, dentistry
Diode	CW	Surgery, dermatology (810, 940, 980 nm); PDT (630 nm); dermatology (800 nm)
Excimer	Pulsed	Angiology, ophthalmology
Ruby	Pulsed	Dermatology
Alexandrite	Pulsed	Dermatology

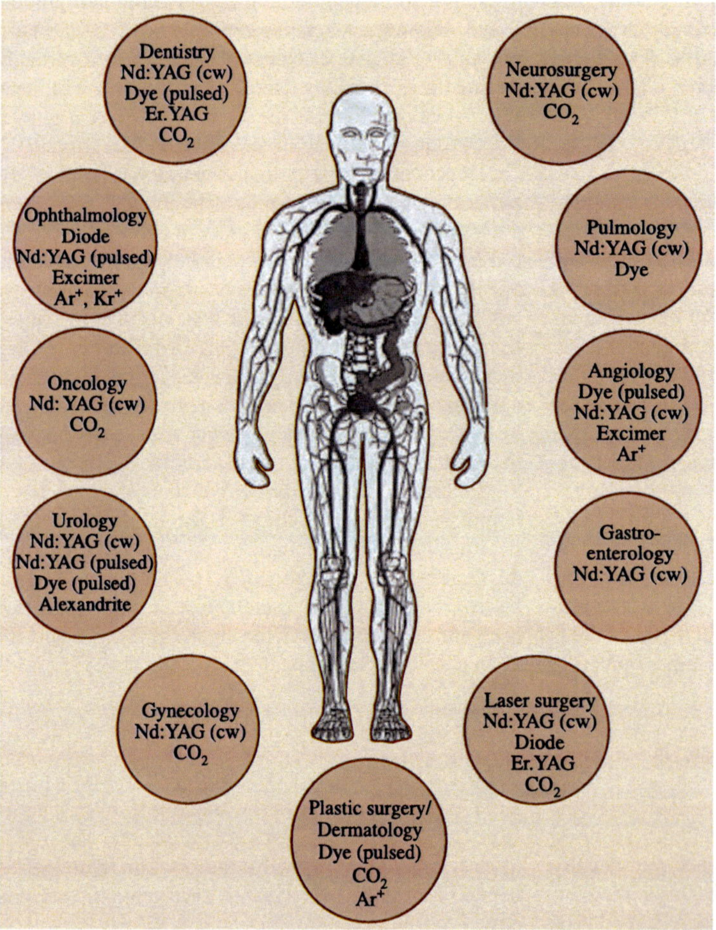

Fig. 29.14 Medical application fields of the laser

29.5.1 The Ruby Laser

The ruby laser, operating at a wavelength of 694 nm, was the first laser to be built. In the early stages of their development ruby lasers had a number of technical weaknesses, such as low efficiency or local intensity spikes, which led to destruction of the delivery fibers. However, this has now changed and modern medical ruby lasers are compact and reliable systems. Designed as Q-switched lasers (Q-switched ruby laser, QSRL) these lasers deliver high-energy pulses with durations of 20–40 ns. The QSRL is currently used both for the treatment of superficial pigmented skin lesions and for the removal of tattoos. Ruby lasers with a longer pulse length (0.3–5 ms) have recently become available and these devices are the preferred laser for epilation treatment (hair removal).

29.5.2 The Neodymium:YAG Laser

The continuous-wave Nd:YAG laser is currently the most important representative of solid-state lasers. At a wavelength of 1064 nm (less frequently at 1320 nm) it is a typical volume coagulator. It is possible to effect a deep unspecific coagulation, vaporization, or to cut tissue, according to the power density used. Transmission using glass fibers allows ubiquitous application of the radiation. For example, it can be used via a flexible or rigid endoscope for the coagulation of hemorrhages, minor malformations or tumors, and at higher power densities for the recanalization of tumor stenoses. Using a focusing handpiece and correspondingly higher power densities, resection of parenchymatous organs such as liver, spleen, pancreas and kidney is possible with simultaneous hemostasis. The wavelength 1320 nm is used particularly for lung parenchymal tumor resection. The Nd:YAG laser is also used very frequently in the contact mode (Sect. 29.3.2).

Pulsed Nd:YAG laser systems (Q-switched Nd:YAG) in the nanosecond and picosecond region are used to effect photodisruption and have established themselves in ophthalmology for the treatment of posterior capsular opacification and for peripheral iridotomy in cases of acute angle-closure glaucoma. The laser can also be employed in lithotripsy.

29.5.3 The Frequency-Doubled Nd:YAG Laser (KTP Laser)

By positioning certain crystals in the optical path of a laser, the nonlinear optical effects can be utilized to generate the harmonics of the basic wavelength. The frequency doubling of Nd:YAG laser radiation is usually effected with the help of a potassium titanyl phosphate crystal (KTP, $KTiOPO_4$). The resulting wavelength, 532 nm, is absorbed well by hemoglobin and its clinical use corresponds to that of the argon-ion laser. Devices with a high output power (> 15 W) have the advantage of allowing the use of shorter pulse durations, thereby reducing discomfort during the treatment. A partial frequency doubling of short-pulsed Nd:YAG laser radiation is utilized in lithotripsy laser systems. In this case, the higher absorption of the shorter wavelength facilitates ignition of the plasma necessary to produce the laser-induced shockwaves (FREDDY principle).

29.5.4 The Erbium:YAG Laser

The Er:YAG laser is a flashlamp pumped solid-state laser emitting infrared light at 2940 nm, which corresponds to the local absorption maximum of water. Articulated arms are usually used as laser beam delivery system, and less frequently zirconium fluoride or sapphire fibers. If short pulse durations (approximately 1 ms) are used, it is possible to ablate extremely thin skin layers (depending on the energy density, up to 10 μm) almost athermally. This property makes the Er:YAG laser highly suitable for use in dermatology and plastic surgery. The laser is also used as a dental drill, usually in combination with a water spray to cool the enamel.

29.5.5 The Holmium:YAG Laser

The Ho:YAG laser emits infrared light at a wavelength of 2100 nm. These lasers are usually flashlamp pumped pulsed systems, although more recently, diode pumped continuous-wave low-power Ho:YAG lasers have also been developed. Ho:YAG lasers are used for the treatment of hard substances such as cartilage and bone, and because the radiation can be transmitted via glass fibers, they can also be used endoscopically (arthroscopy, lithotripsy).

29.5.6 The Alexandrite Laser

The alexandrite laser emits light at 755 nm and operates by virtue of its Q-switching properties at pulse durations in the nanosecond region. Both the clinical fields of use and application techniques are similar to those of the Q-switched ruby laser. A long pulsed version of the alexandrite laser, with pulse lengths in the millisecond region, has been developed for use in laser epilation.

29.5.7 The Diode Laser

Diode lasers in the near-infrared region have been employed in ophthalmology since the end of the 1980s. Further development of the spectral emission range (630–980 nm) and the output power (medical lasers up to 100 W), has led to a wider use of the diode laser in medicine. Its advantages over other laser systems are not only its compact design, but also its mobility, water-free cooling system, easy installation and low-key maintenance.

Diode lasers with output powers of up to 1 W are used diagnostically in a variety of ways, ranging from optical imaging in laser Doppler sensors to determine blood flow, to use in fluorescence diagnostics.

Applications such as laser-induced coagulation using the Nd:YAG laser can be carried out equally well by the diode laser operating at an emission wavelength of 980 nm. The lower penetration depth at 980 nm can be compensated for in most tissues by making relevant changes to the application parameters, i. e. higher output power and longer exposure times. The depth of the coagulation is identical for both 1064 and 980 nm because the regions beyond the optical penetration depth are only influenced by wavelength independent heat conduction.

Urology is currently the main indication field for new high-power diode lasers although they are gradually also becoming established in other disciplines, such as ENT. Diode lasers that operate at 630 nm are used as the light source for photodynamic therapy.

29.5.8 The CO_2 Laser

The continuous-wave CO_2 laser (10 600 nm), by virtue of its high absorption in water, is a very precise cutting instrument with a very low penetration depth. It has found use in microsurgical procedures and areal removal because it does not cause significant thermal damage to the surrounding area. However, its hemostatic properties are very limited and it can only be used to stop minor bleeding from small vessels such as capillaries.

As it is not possible to transmit the radiation by fibers, the laser beam is transmitted using the articulated mirror arm, which is currently this laser's biggest disadvantage. Hollow waveguides are sporadically used, which allow flexible beam guidance; however their beam quality and focusability are poor. Further progress is expected along the lines of photonic fibers, but as yet, no products are commercially available (Sect. 29.2.4).

Pulsed CO_2 lasers emit a very quick series of short impulses with a high energy density instead of a continuous beam. As less heat is transferred to the surrounding tissue, cutting or vaporization can be performed with considerably less thermal side effects to the surrounding tissue (the so-called *super-pulse* mode). Pulsed or continuous-wave CO_2 lasers fitted with special scanner systems are used in dermatology and plastic surgery for the superficial removal of thin layers of skin [29.1, 7].

29.5.9 Argon-Ion and Krypton-Ion Lasers

Argon-ion lasers are continuous-wave lasers that can principally emit at several wavelengths in the range of 250–530 nm, whereby the systems usually used in medicine emit blue/green mixed light (488 and 514 nm, *all-lines* mode) or green light (514 nm alone). The argon laser's high selectivity for the body's own chromophores (hemoglobin, melanin) makes it suitable for use in ophthalmology and dermatology. The radiation is guided via an optical fiber with a focusing handpiece or a slit lamp fitted to the distal end. Special scanners are available to facilitate the automated and uniform treatment of larger areas. The krypton-ion laser can also emit at several wavelengths within the range 350–800 nm, whereby the most intensive emission lines are 531, 568 and 676 nm, and the main area of use is in dermatology and photodynamic therapy (PDT).

Gas ion lasers have the disadvantage of not only being sensitive to vibration, but also of being bulky and relatively expensive to maintain. Moreover, the laser tubes have an average lifetime of between 1000 and 10 000 operating hours, which is relatively short for a laser. As a result, in many applications gas ion lasers have now mostly been replaced by frequency multiplied solid-state lasers or diode lasers. However, there are situations where their excellent beam quality is required and for these special cases they remain indispensable.

29.5.10 The Excimer Laser

Excimer lasers are gas lasers with a laser medium that is a mixture of a noble gas (argon, krypton, xenon), a halogen (chlorine or fluorine) and a buffer gas (helium or neon). According to the gas mixture the exclusively pulsed laser radiation lies in the ultraviolet range between 157 and 351 nm. The most common appliances are those with the mixtures ArF (193 nm), KrF

(249 nm), or XeCl (308 nm). Using these wavelengths with short pulse lengths (10 ns) and very high peak power densities (10^8 W/cm^2), the threshold for the process of photoablation can be exceeded. These lasers are not only used in corneal surgery for the operative correction of defective vision but also in laser angioplasty.

29.5.11 The HeNe Laser

The red helium-neon laser (632 nm) is used therapeutically for low-level laser therapy (LLLT, soft laser, biostimulation) but its main use in the past was as a pilot laser for invisible radiation of surgical laser appliances such as CO_2 and Nd:YAG laser in noncontact mode. However, more recently this function has been taken over by the smaller and much less expensive diode laser.

29.5.12 The Dye Laser

The functional principle of the dye laser is based on the excitation of an organic dye solution to fluorescence by energy-rich flashes of a flashlamp (for pulsed systems; FDL, flashlamp pumped dye laser) or by a pumping laser (continuous-wave systems). In contrast to other lasers, dye lasers emit over a relatively broad spectrum. However, by using a wavelength selective filter, only a selected wavelength, fixed or variable, according to the laser device, is amplified and transmitted through a fiber optic system. Rhodamine 6G is a typical representative of the dyes used and, with the appropriate tuning, emits wavelengths in the visible range from 570–630 nm. These specialized lasers are used clinically in dermatology [29.7] and lithotripsy.

29.5.13 The Free-Electron Laser

The laser medium of the free-electron laser (FEL) is a high-energy electron beam that reaches relativistic velocities (i. e. near the speed of light). Compared to the materials that make up conventional laser media in which the electrons are bound to atoms or molecules, these free electrons oscillate in a spatially periodic magnetic field. This very complex technology allows the tunable production of radiation ranging in wavelength from microwaves, through the visible and ultra violet spectral range to x-rays, whereby the emission duration can extend from quasi continuous to femtosecond. However, with the systems that have as yet been technically realized, it has only been possible to use a fraction of the possibilities this technology promises. A medical FEL system emitting in the mid-infrared range has already been employed as a surgical tool in ophthalmology, otorhinolaryngology and neurosurgery. Other research is investigating the use of FEL for the spectroscopy of biological molecules, as a tool to kill pathogenic microorganisms and as a radiation source for optical coherence tomography (OCT) [29.8].

29.6 Fields of Use

29.6.1 The Eye

Due to the optical transparency of the cornea, lens and vitreous body, the laser was predestined for use in ophthalmology. Laser radiation over a broad wavelength range can pass through the transparent media to reach the back of the eye. Photocoagulation of the detached retina was the first application to be used on the eye and is still one of the most frequently used laser procedures in ophthalmology. There are various other standard procedures such as laser trabeculoplasty (LTP) and laser iridotomy (LI), both of which are used to effect a lowering of intraocular pressure for the treatment of glaucoma.

Short pulsed excimer lasers that utilize photoablation are widely used for refractive corneal surgery. One example of photodisruption is the opening of the after-cataract membrane using the Q-switched Nd:YAG laser (laser capsulotomy), which utilizes of the optical breakdown effect.

29.6.2 The Body Surface

The body surface was one of the first application fields to use the medical laser. The indications can be divided into the two main tasks

- removal or coagulation of skin and skin appendages, and
- therapy of intracutaneous vessel lesions and malformations.

Coagulation with the Nd:YAG laser or removal with the CO_2 laser are currently the methods of choice for skin tumors such as basaliomas, spinaliomas and metastases of malignant melanomas. Pigment anomalies are treated with the ruby or alexandrite laser although the argon laser can also be used. The argon, Nd:YAG, and CO_2 laser are used to remove virally induced tumors such as condyloma, molluscum lesions and verrucae. Due to its low penetration depth, treatment with the CO_2 laser leads to a vaporization of these structures, whereas the argon and Nd:YAG lasers, with their high penetration depth, coagulate the tissue and this is subsequently rejected by the body.

Use of the CO_2 laser leads to immediate vaporization of the skin, and is therefore the laser of choice for the treatment of epithelial dysplasia (leukoplakia, morbus Bowen), for the therapy of chronic ulcers and for cleaning wounds. The ruby, alexandrite, pulsed Nd:YAG and CO_2 lasers are used for the removal of tattoos.

29.6.3 The Vascular System

Argon-ion, dye and Nd:YAG lasers have proved to be very successful for the therapy of spider nevus, nevus flammeus and cutaneous planotuberous hemangiomas. The argon-ion laser is particularly successful because of its low penetration depth and high absorption in hemoglobin. A thermal dynamic reaction is induced in the laser treated vessels in the form of an angiitis, which results in an occlusion of the teleangiectasia. The occluded vessels are then left to the natural resorption processes of the body.

The dye laser is used in the treatment of port-wine stains and particularly good results are achieved in the treatment of homogenous port-wine stains in children.

Hemangiomas belong to the most common pediatric vascular tumors. The majority of hemangiomas heal spontaneously before the age of eight, which is why a *wait-and-see* approach is advisable. If however, an unfavorable localization leads to considerable functional disturbance or disfigurement of the patient, then there is a case for an earlier therapy indication, even if the patient is very young i. e. a baby or a small child. The Nd:YAG laser is used for the treatment because of its larger penetration depth. Superficial burns to the skin are avoided by using special ice cubes to cool the skin during the laser procedure [29.1].

A further application of the Nd:YAG laser is the percutaneous interstitial irradiation of voluminous cavernous angiomas using a bare fiber. The lesion to be treated is punctured and the fiber positioned in the tissue. The laser output of 4–5 W and an exposure time of 1 to 5 s culminate in an intraluminal thrombosis and damage to the vessel wall is followed by sclerotization [29.9]. If a higher laser output (8–10 W) is used, the end of the fiber should be flushed during the laser operation to prevent damage to the fiber tip.

29.6.4 Open Surgery

Even with the best preoperative planning, surgical interventions are never free from possible intraoperative or postoperative complications. Bleeding can be a cause for concern both during an operation and postoperatively, as can also bile leaks or biliocutaneous fistula after liver resection or complications after operations on the pancreas, spleen, kidneys, mammary tissue, and the brain in particular. A loss of 100 ml of blood in pediatric surgery can already lead to a life-threatening condition, which then urgently requires blood transfusion.

Under these circumstances, the advantages of lasers in surgical disciplines can be summed up in five points.

1. Inherent hemostasis
2. Precision work
3. A decrease in the number of instruments used in the operation area
4. Noncontact tissue removal (i. e. aseptic application)
5. Minimal traumatization of the surrounding tissue due to zero-force application.

Neurosurgeons use the Nd:YAG laser and CO_2 laser in the cranial cavity to resect tumors, for angioma sclerotization in stereotactical operations and for plexus coagulation.

In surgery of the thorax, the laser is used for parenchyma resection, for treatment of fistulas and for decortications. The Nd:YAG laser is an established tool in open surgery for the resection of parenchymatous organs in the abdominal cavity. Small veins, with a diameter of 3–5 mm, and arteries up to 1.5 mm are sealed when they are cut by the laser beam, but larger vessels must be ligated beforehand.

The laser has also established itself in gynecological abdominal surgery. The CO_2 laser is used in a number of operative refertilization methods such as intrapelvic adhesiotomy or tube implantation, as well as for the operative removal of myomas. CO_2 lasers and Nd:YAG lasers are also used for mammary amputation and subcutaneous mastectomies.

A further indication for the CO_2 laser is for the treatment of mechanically irritating malformations or deformities. Some examples of these are interdigital neuromas, ganglion cysts, heel neuromas, bony heel spurs, bone and hip joint operations for hemophiliacs, and the tarsal tunnel syndrome. The use of the CO_2 laser makes it possible to completely remove the structure by vaporization and thereby help prevent it from growing again.

29.6.5 Endoscopy

The introduction of endoscopic procedures has marked a major change in surgery. Not only does the risk of infection decrease when a body cavity is not opened, but also the danger of postoperative adhesions is considerably less and the operation stress for the patient is much reduced.

Laser light can be transmitted via a quartz glass fiber to the application site through the working channel of an endoscope under visual control. This technical development has led to further miniaturization and increased flexibility of the minimal invasive operation method. The use of this technique makes it possible to access and open malignant stenoses of the esophagus and the alimentary canal.

The Nd:YAG laser can be used in contact with a bare fiber for the treatment of congenital scarred nonmalignant changes. CO_2 laser radiation can currently only be delivered to the site of treatment using an articulated arm, which restricts application to the use of rigid endoscopes with endoscope couplers.

29.6.6 Photodynamic Therapy

Photodynamic therapy (PDT) is an increasingly important treatment modality. After intravenous injection of a photosensitizer such as a hematoporphyrin derivative (HpD) and subsequent intratumoral enrichment for 24–48 h, the tumor is irradiated either superficially or interstitially using a dye laser or diode laser (630 nm). The photosensitizer absorbs the radiation, decomposes and releases oxygen radicals that have a local destructive effect on the tumor.

Photodynamic therapy is currently used for the treatment of superficial lesions such as dysplasia and early stages of carcinomas in dermatology, gastroenterology, urology, gynecology and neurosurgery. Of late PDT has also been used in ophthalmology for the treatment of neovascular processes. The development of new dyes that exhibit a better selectivity for tumor tissue and stronger absorption in the red and infrared range of the spectrum to penetrate deeper into the tissue is currently the subject of intensive research.

29.6.7 Lithotripsy

The destruction of concrements has boomed since the development of extracorporeal shockwave lithotripsy (ESWL), which prior to this was carried out by mechanical methods. In contrast to the first generation ESWL devices, it is now possible to focus the shockwaves with considerably more precision. However, it is not possible to perform extracorporeal shockwave lithotripsy at every site in the body where stones occur. Examples of this are the intra and extra hepatic bile ducts, pancreatic duct, and ureters that are shadowed by bones. These are particularly expedient areas for the use of laser lithotripsy (LLT). In contrast to ESWL, the shockwaves produced in laser lithotripsy are not generated externally and focused on the concrements from outside the body but are produced directly in front of or in the concrements themselves. This means that the shockwaves do not have to pass through all the surrounding structures, the light energy being converted to mechanical energy in-situ.

This has been made possible by the development of Q-switched Nd:YAG lasers, flashlamp pumped dye lasers, and alexandrite lasers. The use of these very strong laser impulses has opened up the area of nonlinear interaction, photodisruption and ablation for medical purposes. Concrement fragment size can be varied by changes to the wavelength, pulse duration, pulse energy, and interval length. LLT is not yet used on a routine basis but its development is a dynamic one. The main indications for its use are for bile duct concrements, in combination with lysis or extraction, and concrements of the oral cavity and the pancreas.

29.6.8 Laser-Induced Thermotherapy

Laser-induced thermotherapy (LITT) has been used for some time to treat pathological tissue changes in different regions of the body in the contact mode. It includes the intraluminal coagulation of vascular changes as well as the palliative coagulation and hyperthermia of metastases in the liver, lung, and brain, which have both become very important over the last few years. For the subcutaneous LITT of smaller congenital malformations (congenital vascular disorder, CVD), a bare fiber is used to puncture and access the area to be treated. In order to effect larger coagulation volumes of up to 3 cm in diameter, special scattering applicators have been developed that deliver laser radiation from a Nd:YAG or diode laser over a relative long exposition time of up

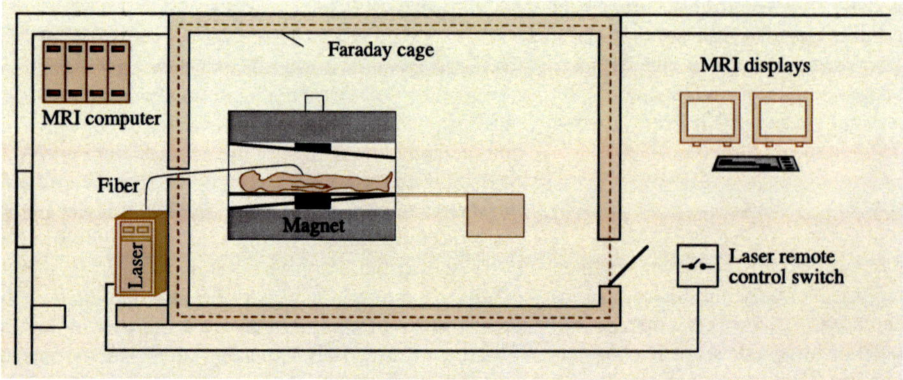

Fig. 29.15 Devices for MRT-controlled laser-induced thermotherapy (LITT)

to 20 min. Water-cooled sheaths on the applicators allow laser outputs of up to 30 W without carbonization at the interface between applicator and tissue. Minimal invasive procedures such as LITT of isolated liver metastases, specific brain tumors, or benign prostate hyperplasias require real-time monitoring, such as the imaging modalities ultrasound or MRT, for precise control of the therapy (Fig. 29.15) [29.9].

29.7 Safety Aspects

A laser system that generates radiation energy powerful enough to remove biological tissue is also a potential source of danger to both the patient and the operator, making it in effect no different to a sharp scalpel or an RF surgical device. The long-distance effect of the laser radiation presents an added danger to the eyes for which the protective measure of simply keeping a distance from the divergent radiation exit (fiber end, focusing optics) is usually overestimated. However, by implementing simple safety measures it is possible to achieve an even higher level of safety than for many other devices. This requires knowledge of the basic physical interactions and strict adherence to the respective safety specifications [29.10–14].

29.7.1 Permissible Exposure and Laser Classification

The danger to both the eyes (Fig. 29.16) and to the skin (Fig. 29.17) has led to the specification of maximum permissible exposure (MPE) levels [29.12]. Some examples for medical lasers are shown in Table 29.3. All safety measures should be implemented in such a way that should an unintentional irradiation occur, the MPE values cannot be exceeded.

The damage that laser radiation can incur is dependent on the output power, exposure time and wavelength. As a result, laser systems are divided into classes according to these parameters, thus giving an indication of the danger potential for each individual device. A simplified version of the laser classes according to IEC 60825-1 is shown in Fig. 29.18. The divisions are such that the class 1 comprises the so-called *intrinsically safe* systems i. e. looking directly into the laser beam is either not dangerous or technically impossible.

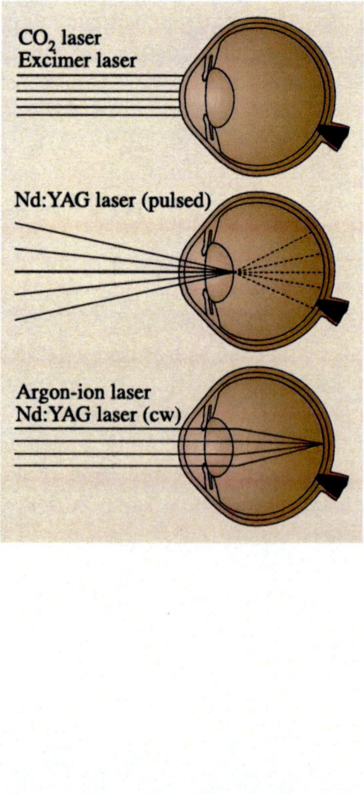

Fig. 29.16 Possible hazards to the eyes from different laser types

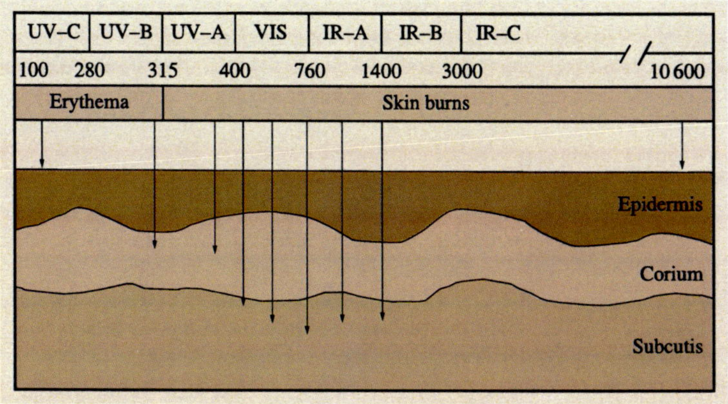

Fig. 29.17 Penetration depth of laser radiation in the skin and possible skin damage dependent on wavelength (nm)

Table 29.3 Frequently used lasers and their maximum permitted exposures (MPE)

Laser type	Wavelength (nm)	Maximum permitted exposure of the eyes (MPE) according to IEC 60825-1:2007		Maximum permitted exposure of the skin (MPE) according to IEC 60825-1:2007	
		Power density (W/cm^2)	Time (s)	Power density (W/cm^2)	Time (s)
Argon-ion	488, 514	0.0018	1	1.1	1
Helium-Neon	633	0.0018	1	1.1	1
Diode (GaAs)	970	0.0062	1	3.8	1
Nd:YAG	1064	0.0090	1	5.5	1
CO_2	10 600	0.56	1	0.56	1

Class 2 lasers are lasers emitting in the visible spectral range (400–700 nm). For this class it is assumed that the reflex response (including the blink reflex) limits the exposure duration to about 0.25 s. However, new research has shown that these figures are unreliable. For lasers of classes 1 M and 2 M, a glance in the beam with visual aids (telescopes, magnifying glasses, surgical microscopes, etc.) can be dangerous although there is no danger for the naked eye under normal predictable circumstances. The radiation from the lasers of class 3R

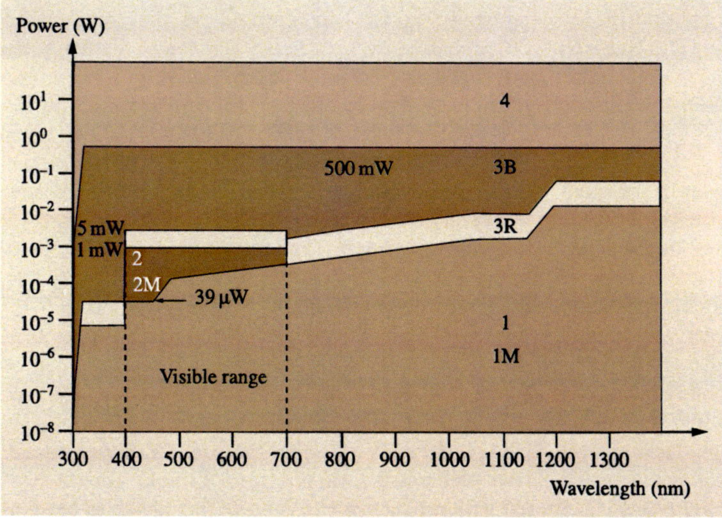

Fig. 29.18 Laser classification overview (for an emission time of 10^3 s)

can be dangerous for the eye, although less so than for class 3B. For lasers of class 3B, looking at a diffuse reflection of the laser beam (e.g. on skin) is permitted but a glance in the direct beam or the specular reflex is dangerous. For class 4 lasers, even diffuse reflections are dangerous to the eye, and the direct beam can cause burns to the skin. Needless to say, the lasers used in surgery and ophthalmology are all attributed to class 4.

29.7.2 Safety Requirements for Laser Devices

Safety requirements for medical laser systems of classes 3B and 4 are specified in the international standards IEC 60601-2-22 *Medical electrical equipment – Part 2-22: Particular requirements for the safety of therapeutic and diagnostic laser equipment* and IEC 60825-1 *Safety of laser products – Part 1: Equipment classification and requirements*. The essential requirement *protection against radiation* of the European Medical Devices Directive 93/42/EEC (MDD, Chap. 4) is implemented from these standards by formulation of a series of constructive requirements. According to these requirements, all medical laser systems must incorporate the features

- A key switch (or equivalent) to protect against unauthorized use.
- An emergency switch for the immediate interruption of emission.
- A stand-by/ready circuit to protect against accidental triggering off of the system.
- An interlock connection for remote interlocking.
- An optical and/or acoustic laser-ready and emission indicator.
- An aiming device to show the radiation's incidence point on the target before triggering.

Furthermore, the actually delivered power or pulse energy should only deviate from the previously set value within definite limits. These settings must be displayed in SI-units (W, J). The failure of an explicit list of components (e.g. shutter, trigger switch, attenuator, monitoring circuits, timer) should not pose any hazard. All protective measures must be effective within tolerance times of approximately 130 ms if tissue damage is to be avoided due to excessive laser output.

The optical end device of the beam guidance system can be a further source of danger. A badly constructed focusing handpiece can heat up to temperatures that can burn the operator's hands or a micromanipulator with a defective or the wrong integrated safety filter can reflect dangerous levels of radiation directly into the eyes of the operator. As a result, every medical laser accessory must go through an EC conformity assessment procedure for medical devices.

It is important for the manufacturers of medical laser devices to know that these systems are regarded as medical devices of Class IIb according to the classification criteria in Appendix IX of EC Directive 93/42/EEC. Only devices with a successfully completed conformity assessment procedure can be put on the market in the European Union (Chap. 4).

29.7.3 Protective Measures and Application Safety

Regardless of more specific national regulations, some safety measures must be taken if the laser radiation exceeds the maximum permissible exposure (MPE).

- Safety glasses or goggles must be worn.
- Dangerous specular reflections on medical instruments are to be avoided by using roughened surfaces.
- Protective measures should be taken to prevent fire and explosions [29.15].
- Protective filters should always be used when using optical observation instruments.
- Disposable and covering materials should be flame resistant.

The requirements on laser protective filters and laser protective glasses are covered by the standard EN 207 *Personal eye-protection equipment – Filters and eye-protectors against laser radiation (laser eye-protectors)*. The labeling of standardized laser protective glasses is shown in Fig. 29.19. It should be noted that there is no one pair of glasses that offers protection against all wavelengths. Every pair of protective glasses protects exclusively against the wavelength (or range of wavelengths) and the mode of operation (continuous wave, pulsed, or mode coupled) as specified on the glasses themselves!

As a result of pyrolysis of the tissue during thermal laser applications, patients and operators can be confronted with substances that have both toxic and infectious potential [29.1]. According to the latest technology, the only effective protective measure against laser plume is the use of a special smoke evacuator fitted with a handpiece to remove the smoke as close as

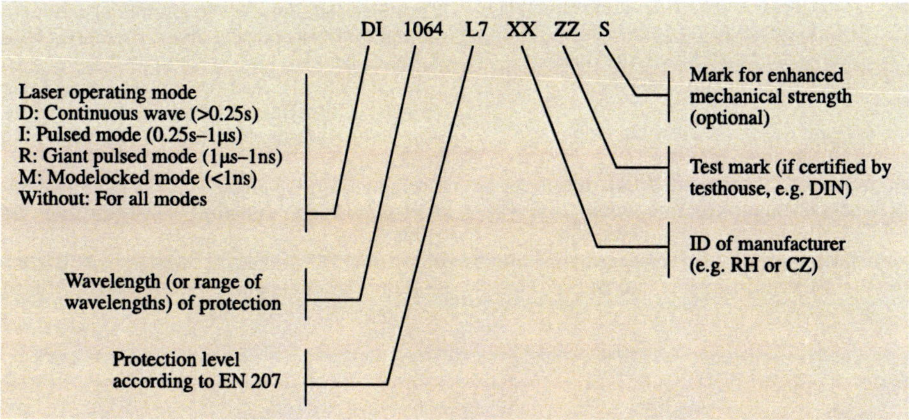

Fig. 29.19 Labeling of laser protective glasses according to EN 207

possible to where it was produced. These devices are equipped with a two-stage filter system: a membrane filter to catch particles and droplets, and a downstream active carbon filter to reduce unpleasant smells.

29.8 Future Prospects

Against the background of laser applications in medicine, future prospects for further developments are becoming apparent. Technologically there is an increasing trend towards the use of diode lasers, which will allow the realization of more compact and simpler systems. This development will be accompanied by a reduction in costs that will lead to an even wider acceptance of lasers in medicine. The development of suitable components for endoscopic laser surgery will make the realization of newer and enhanced applications possible. In this context, the availability of flexible optical transmission systems for all spectral ranges is a central development aim.

Further developments are to be expected in the field of in-vivo or in-situ diagnostics in which the same applicator is used for both modalities. This combination of diagnosis and therapy to localize the tissue that is to undergo therapy or even for therapy control itself, will characterize the next generation of *minimal invasive* medical laser systems. The development of these *smart systems* will occur gradually, whereby the new developments in endoscopic methods make a significant contribution.

In addition to the therapeutic applications, there are a large number of diagnostic applications that have not been dealt with here. These will become increasingly more important as will also laser surgery at a cellular level. The last few years have not only seen an increase in indications but also the spread of medical laser systems. It can now be safely said that the laser no longer competes with the scalpel!

References

29.1 H.P. Berlien, G. Müller: *Applied Laser Medicine* (Springer, Heidelberg Berlin 2003)

29.2 V. Knappe, F. Frank, E. Rohde: Principles of lasers and biophotonic effects, Photomed. Laser Surg. **22**, 411–417 (2004)

29.3 D. Kleppner: Rereading Einstein on radiation, Phys. Today **58**, 30–33 (2005)

29.4 D. Meschede: *Optics, Light and Lasers*, 2nd edn. (Wiley-VCH, Weinheim 2007)

29.5 M. Young: *Optics and lasers*, 5th edn. (Springer, New York 2000)

29.6 F. Träger (Ed.): *Springer Handbook of Lasers and Optics* (Springer, Heidelberg Berlin 2007)

29.7 M.P. Goldman, R.E. Fitzpatrick: *Cutaneous Laser Surgery* (Mosby, St. Louis 1999)

29.8 G.S. Edwards, R.H. Austin, F.E. Carroll, M.L. Copeland, M.E. Couprie, W.E. Gabella, R.F. Haglund, B.A. Hooper, M.S. Hutson, E.D. Jansen, K.M. Joos,

D.P. Kiehart, I. Lindau, J. Miao, H.S. Pratisto, J.H. Shen, Y. Tokutake, A.F.G. van der Meer, A. Xie: Free-electron-laser-based biophysical and biomedical instrumentation, Rev. Sci. Instrum. **74**, 3207–3245 (2003)

29.9 G. Müller, A. Roggan: *Laser-Induced Interstitial Thermotherapy* (SPIE, Bellingham 1995)

29.10 D. Sliney, M. Wolbarsht: *Safety with Lasers and other Optical Sources* (Plenum, New York 1980)

29.11 B.A. Rockwell, R.J. Thomas, A. Vogel: Ultrashort laser pulse retinal damage mechanisms and their impact on thresholds, Med. Laser Appl. **25**, 84–92 (2010)

29.12 K. Schulmeister: Present and alternative dosimetry concept for laser exposure limits, Med. Laser Appl. **25**, 111–117 (2010)

29.13 K. Schulmeister, M. Jean: The risk of retinal injury from Class 2 and Class 3R lasers, including medical laser aiming beams, Med. Laser Appl. **25**, 99–110 (2010)

29.14 A. Schirmacher: Eye protection for short and ultrashort pulsed laser systems, Med. Laser Appl. **25**, 93–98 (2010)

29.15 W. Wöllmer, G. Schade, G. Kessler: Endotracheal tube fires still happen – A short overview, Med. Laser Appl. **25**, 118–119 (2010)

30. Inhalational Anesthesia Devices

Inhalational

Erich Siegel

Anesthesia devices are used in operating rooms in hospitals by medical staff to ensure that operative and diagnostic procedures can be performed on a patient without pain in an unconscious and relaxed state. This chapter provides an overview on the concept of these devices. It describes the intended medical use and based on this the necessary technical components. The principles of these main components are explained in detail: the mechanical or electronic dosing of the gases O_2 and N_2O or air, the dosing of anesthetic agent using the vaporizer principle, the ventilator with a rebreathing system, and the ventilation modes used in anesthesia. To ensure safe anesthesia, both the device and the patient have to be monitored. The measuring principles for the necessary device monitoring parameters are also discussed, such as the concentrations of oxygen, nitrous oxide, anesthetic agent, and carbon dioxide. In addition, the exhaled volume and the pressure in the breathing system are described.

30.1 **Anesthesia Devices in General Anesthesia** 579

30.2 **Functional Principle, Medical Aspects** 580

30.3 **Functional Principle of the Main Components** 582
- 30.3.1 Drug Dosing 582
- 30.3.2 Ventilator and Breathing System .. 585
- 30.3.3 Monitoring 590
- 30.3.4 Sensor Principles for Device Monitoring 592

30.4 **Safe Operation Prerequisites** 594
- 30.4.1 Connection Prerequisites 594
- 30.4.2 Training and Continuing Medical Education 594
- 30.4.3 Cleaning 594
- 30.4.4 Maintenance and Pre-use Check .. 595

30.1 Anesthesia Devices in General Anesthesia

Anesthesia devices are designed to assist trained medical anesthesia staff, so that operative and diagnostic procedures can be performed without pain, the consciousness of the patient can be suppressed, and the oxygen supply can be guaranteed.

To meet these aims, the following drugs are delivered by the device:

- Oxygen (approximately 30 vol. %) to ensure adequate oxygenation of the patient during the intervention
- Nitrous oxide (approximately 70 vol. %) or intravenous administration of an opioid, for example, remifentanil, to prevent the patient from feeling any pain
- An anesthetic agent (isoflurane, sevoflurane or desflurane) administered via the lung, or propofol administered intravenously, to suppress consciousness.

Anesthesia is called inhalation anesthesia if the analgesic (nitrous oxide) and the agent used to achieve unconsciousness (for example, isoflurane) are delivered to the body via the lung.

Anesthesia is called total intravenous anesthesia (TIVA) if the analgesic (for example, remifentanil) as well as the drug for unconsciousness (propofol) are administered intravenously.

Anesthesia is called balanced anesthesia if one drug is administered intravenously (for example, remifen-

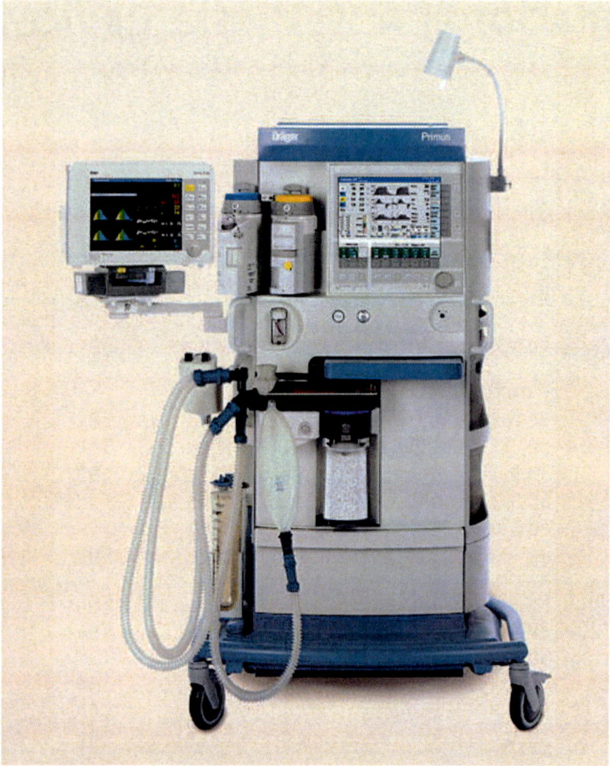

Fig. 30.1 Example of an anesthesia device (Primus, Dräger, Lübeck)

tanil) and the other drug is delivered via the lung (for example, isoflurane).

Since patients are usually relaxated using muscle relaxants during surgery, their breathing musculature is paralyzed, which means that every anesthetic device must be equipped with a device for automatic, controlled ventilation. The possibility of manual ventilation is also important to allow the physician to intervene during induction and termination of anesthesia and also to provide assisted ventilation during these phases if required.

To ensure that anesthesia is safe and transparent, both the device and the patient are monitored. For this purpose, the parameters of the delivered anesthetic gas, i. e., the inspiratory oxygen concentration and the nitrous oxide and anesthetic agent concentrations, are measured. In addition, the exhaled CO_2 and the exhaled volume are determined and the pressure in the system is controlled. These device monitoring parameters are the parameters to be measured in accordance with current standards.

To test cardiovascular function, the electrocardiography (ECG) and noninvasive or invasive blood pressure are measured. Blood oxygenation is monitored by the oxygen saturation. In addition, the effect of anesthesia must be monitored, i. e., the level of patient unconsciousness, the degree of painlessness, and the level of relaxation.

Therefore, an anesthesia device (Fig. 30.1) consists of the following components, according to its intended medical purpose:

- Drug dosing unit
- Ventilator with breathing system
- Monitoring unit consisting of 3 subunits:
  - One which monitors drug dosing and the ventilator called device monitoring
  - One which monitors the patient called patient monitoring
  - One which monitors the *depth of anesthesia* called anesthesia effect monitoring.

30.2 Functional Principle, Medical Aspects

Figure 30.2 shows the three main components (drug dosing unit, ventilator, and device monitoring) of an inhalation anesthesia machine. During inspiration, the ventilator forces the gas from the breathing bellows through the CO_2 absorber and the inspiration valve into the patient's lung. During this time, the drugs (oxygen and nitrous oxide at concentrations of approximately 30 vol. % O_2 and 70 vol. % N_2O as well as, e.g., 2 vol. % isoflurane) flow continuously from drug dosing to the reservoir bag, which also acts as the manual breathing bag at the same time. During expiration, the path to drug dosing is opened by the controlled valve. The gas from the reservoir bag flows together with the exhaled gas from the patient to the breathing bellows of the ventilator. All gas that exceeds the desired pressure at the end of expiration escapes through the anesthetic gas scavenging valve. During inspiration, the path to the reservoir bag is closed by the control valve, and the ventilator forces the gas back to the patient.

The amount of fresh gas flow is based on the following considerations.

- How much oxygen and nitrous oxide is absorbed by the patient?
- How much leakage is there from the device?
- Which rebreathing rate should be used?

Taking a 70 kg patient as an example, uptake of 250 ml/min oxygen and about 100 ml/min N_2O is normal in steady state. With a leakage rate of about 50 ml/min, a minimum flow of 400 ml/min is needed in equilibrium. In the induction phase, more than 1300 ml/min is needed. In addition to the minimum level of fresh gas, the physician also has to adjust the desired gas rebreathing rate. The patient must also be provided with an adequate respiratory minute volume during ventilation and also during spontaneous breathing; for a patient with body weight of 70 kg, this is approximately 6 l/min. It is therefore necessary for the patient to receive this amount of gas per minute. However, that does not mean that the fresh gas which is provided by drug dosing is always required to supply this amount of fresh gas to the system. If, as shown in Fig. 30.2, part of the exhaled gas is reused for inspiration, a fraction of the respiratory minute volume is sufficient as the fresh gas flow.

The type of breathing system which reuses a part of the expiratory gas is called a rebreathing system. Often, a fresh gas flow of 3 l/min with approximately 1 l oxygen and 2 l nitrous oxide is used. Depending on the degree to which the system is sealed and if monitoring is used, the fresh gas flow can be reduced further. If the fresh gas flow is reduced to approximately 1 l/min, this is known as a low-flow system. The type of fresh gas flow used does not have any effect on the anesthesia. The factors which are decisive for the magnitude of the fresh gas flow include the experience of the anesthetist, the degree to which the system is sealed, and whether monitoring is used. In a rebreathing system there are two sources of water: first the patient, who exhales warm (37 °C) water with 100% humidity in the system, and second the soda lime in the CO_2 absorber, which produces water and heat through a chemical reaction with CO_2. Therefore, no active humidification is necessary in anesthesia care.

If a low fresh gas flow is used, adjustments induced by drug dosing are slow acting. In other words, it takes a long time until these adjustments take effect in the patient. If, on the other hand, a high flow is used, the system is flushed quickly and adjustments to the agent concentrations are passed on quickly to the patient.

A primary objective of anesthesia is to ensure that the oxygen (O_2) supply to the patient is maintained.

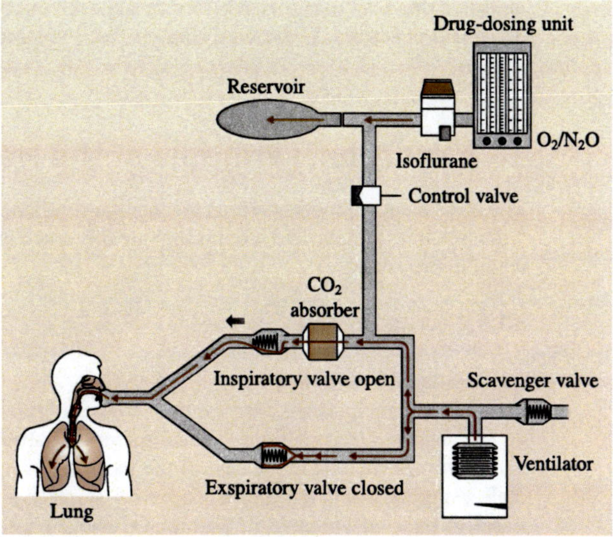

Fig. 30.2 Schematic layout of gas flow of an anesthesia device

This is determined by the inspiratory O_2 concentration and the mean pressure caused by ventilation. In this case, the O_2 supply must be set in such a way that the O_2 partial pressure measured by the blood gas analysis reaches approximately 100 mmHg in patients up to the age of 60 years with normal hemoglobin content. For adults, this corresponds to an oxygen saturation of arterial blood of more than 95%.

The anesthetic agent concentration determines the depth of anesthesia (depth of hypnosis). The lower the end-expiratory concentration, the shallower the anesthesia. The higher the value, the deeper the anesthesia. Determination of the depth of anesthesia itself is difficult and requires an experienced physician using clinical signs such as skin color, heart rate, and blood pressure. Continuous monitoring of parameters derived from the electroencephalography (EEG), such as the bispectral index (BIS) index, allows the depth of hypnosis to be determined objectively.

The ventilator parameter settings are determined both by the patient's oxygen demand and by the fact that the CO_2 needs to be removed. This is achieved as follows. Proportional to the patient's weight, in a first step, approximately 10 ml/kg body weight is set as the breath or tidal volume at a frequency of approximately 10 /min. Thus, a patient weighing 70 kg receives approximately 7 l/min as the respiration volume per minute at a tidal volume of 700 ml and a frequency of 10 breaths per minute. In a second step, the current end-tidal CO_2 value is monitored and compared with the target value. A CO_2 partial pressure of approximately 35 mmHg is desired. In the third step, the tidal volume is adjusted to meet the current needs of the patient using the measured end-expiratory CO_2 value.

30.3 Functional Principle of the Main Components

The three main components (drug dosing, the ventilator with breathing system, and monitoring) are discussed in detail in this section.

30.3.1 Drug Dosing

Dosing of O_2 and N_2O

Today, two principles are used for delivering the gases O_2, air, and N_2O, namely mechanical metering valves and electronic mixers. In the case of the metering principle using metering valves, the respective single flow is set in units of l/min (e.g., 1 l/min O_2 and 2 l/min N_2O); however, in the case of mixers, the gas selection O_2/N_2O or O_2/air, the O_2 concentration in %, and the total flow in units of l/min are set (e.g., 33% O_2 and 3 l/min total flow for O_2 and N_2O, to receive the same 1 l/min O_2 and 2 l/min N_2O).

Part D | 30.3

Mechanical Dosing

With this principle (Fig. 30.3), high-precision mechanical needle valves are used for dosing. The pressure in the gas supply is approximately 5 bar and is reduced to approximately 3 bar by pressure reducers, which are arranged in the gas flow direction upstream from the high-precision needle valves. This prevents pressure variations in the gas supply from disrupting the set flow.

The gas flow is set by turning the high-precision needle valve, which in turn changes the area of a ring-shaped opening. In some machines, the flow is measured using a measuring tube (tapered glass tube) in which a specially shaped float moves up and down according to the magnitude of the flow. The measuring tube must be positioned vertically to allow for precise measurements. This is the only way of ensuring that the float can move freely. The flow can also be measured electronically, so that the flow set using the high-precision needle valve can be displayed digitally and also be represented as a bar graph on a screen.

The metering valve buttons are provided with different profiles so that the oxygen valve can also be distinguished by touch (or *blind*) from the other metering valves. They are also protected against accidental readjustment.

Electronic Dosing

In the case of electronic gas dosing (Fig. 30.4), a switchable two-gas mixer is used. Depending on operator selection, it supplies O_2 and N_2O (selection N_2O) or O_2 and air (selection air) in accordance with a preset O_2 concentration and a preset total fresh gas flow. Therefore, the desired O_2 concentration and the magnitude of the total fresh gas flow are set directly and do not have to be calculated from single flows like in mechanical mixers.

In the case of electronic dosing, the gases are dosed successively in time, each one individually, and subsequently mixed in a storage tank. If the mixed gas tank is partially emptied through the withdrawal of gas, the following process takes place. If, for example, 33% O_2 should be delivered and 300 ml is missing from the mixed gas tank, the control unit opens the O_2 gas inlet valve in the first step and 100 ml O_2 is fed into the

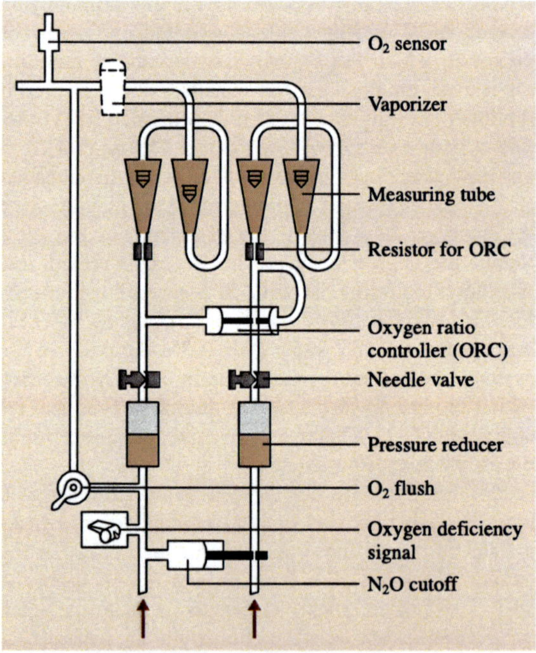

Fig. 30.3 Mechanical gas dosing principle

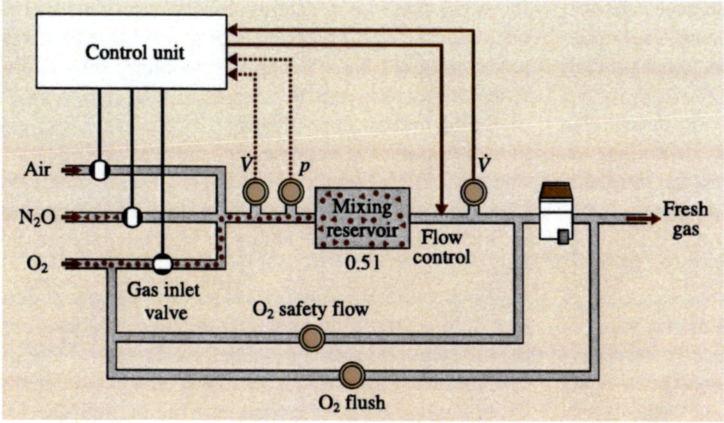

Fig. 30.4 Electronic gas dosing principle

mixed gas tank via flow metering using flow and time measurement. As soon as 100 ml has been delivered, the O_2 valve is closed. All gas inlet valves are temporarily closed. In a third step, the N_2O inlet is then opened and 200 ml N_2O is fed into the tank via the flow and time measurement. This process creates a 33% O_2 mixture in the tank. The N_2O valve is closed again. This type of filling cycle lasts approximately 1 s. If gas is extracted, the pressure falls in the gas tank and the process starts over again.

Elements for Monitoring O_2, N_2O Dosing

Oxygen Deficiency Signal. The oxygen deficiency signal issues an alarm in the event of an oxygen supply failure. This is regardless of whether, for example, it is caused by an empty oxygen cylinder in the machine, a central gas supply disruption or the accidental removal of the oxygen coupling (Fig. 30.3).

This alarm is issued when the minimum supply pressure in the 5 bar oxygen supply line of the machine falls below approximately 2 bar. The alarm lasts for at least 7 s and cannot be switched off.

Nitrous Oxide Cutoff. An oxygen supply failure conceals the danger that pure nitrous oxide might suddenly enter the breathing system. To prevent this, the nitrous oxide in the nitrous oxide supply line is shut off by the nitrous oxide cutoff in the event of a pressure failure in the oxygen supply line (5 bar) (Fig. 30.3). The device's nitrous oxide supply is only reactivated when the oxygen supply has been restored.

Proportional Valve Oxygen Ratio Controller. The oxygen ratio controller (ORC) is a safety feature which monitors the oxygen and nitrous oxide flows to the patient. It reduces nitrous oxide in the low pressure range and ensures that at least approximately 20–25 vol. % O_2 is present in the fresh gas line.

The ORC works by comparing the N_2O flow with the O_2 flow in order to control the administration of nitrous oxide to ensure that the ratio of nitrous oxide to oxygen never exceeds approximately 3 : 1.

The ORC is called a *proportional valve* because this safety feature is designed to monitor the ratio of O_2 to N_2O. In mechanical gas metering, ORC is implemented pneumatically, and, in electronic mixers, in control software.

Inspiratory Oxygen Measurement. The oxygen deficiency signal, nitrous oxide cutoff, and ORC monitoring elements can only be used to prevent errors when the correct gas is flowing through the provided line. These elements are useless if O_2 is inadvertently mixed with N_2O in the line. This type of error can only be detected through direct measurement of the gas type. Inspiratory oxygen measurement is therefore mandatory in the standards for anesthetic machines because it can detect errors associated with insufficient or incorrect dosing of O_2.

Delivery of Anesthetic Agents Isoflurane, Sevoflurane, and Desflurane

Operating Principle. Today, isoflurane, sevoflurane, and desflurane are used for inhalation or balanced anesthesia as anesthetic agents to achieve unconsciousness. They are also called volatile anesthetic agents, because they are exhalable and evaporate quickly. An anesthetic agent vaporizer converts the anesthetic agent isoflurane or sevoflurane from a liquid to a vapor state and adds them to the fresh gas at a preset concentration. The anesthetic agents isoflurane and sevoflurane have relatively high vapor pressure at room temperature. Isoflurane, for example, has a vapor pressure of approximately 300 mbar at 20 °C.

The saturation concentration (vapor pressure/air pressure) for isoflurane is 30 vol. % at 20 °C. The vapor pressure depends on temperature, so the saturation concentration increases at higher temperatures. In the case of isoflurane, for example, the concentration of saturated vapor increases from 30 vol. % at 20 °C to approximately 60 vol. % at 35 °C.

At 20 °C, the concentration of saturated vapor is much higher (20 fold) than therapeutically necessary. For isoflurane and sevoflurane, an anesthetic agent concentration in the range 1–3 vol. % is used. Consequently, the anesthetic agents cannot be inhaled directly and need to be diluted accordingly.

The vaporizer is primarily designed to reduce the high saturation concentration of, e.g., 30 vol. % to the concentration required during anesthesia, e.g., 2 vol. %. For this purpose, the fresh gas flow is divided into two partial flows (Fig. 30.5): a vaporizer flow with gas containing anesthetic agent and a flow which bypasses the vaporizer without anesthetic agent. At the end, both flows are joined again.

The vapor pressure of the anesthetic agent changes with temperature. As a result, if the anesthetic agent vaporizer were only based on this principle, changes in temperature would also affect the delivered anesthetic agent concentration. If a vaporizer based on this principle was set to, e.g., 3 vol. % at 20 °C and the temperature was then increased to 35 °C under equilibrium conditions, the delivered concentration of isoflurane would rise to 6 vol. %. However, anesthetic agent vaporizers should ensure that the delivered anesthetic agent concentration in the respective temperature range should be temperature independent. Therefore, the physical effect of the temperature dependence of the vapor pressure or saturation concentration needs to be compensated. The principle of temperature compensation is based on reducing the vaporizer flow as the temperature increases to compensate for the increase of the saturation concentration. In this way, less gas passes through the vaporizer chamber as the temperature increases. Technically, this problem is solved by using the different thermal expansions of two metals to widen the bypass opening as the temperature increases.

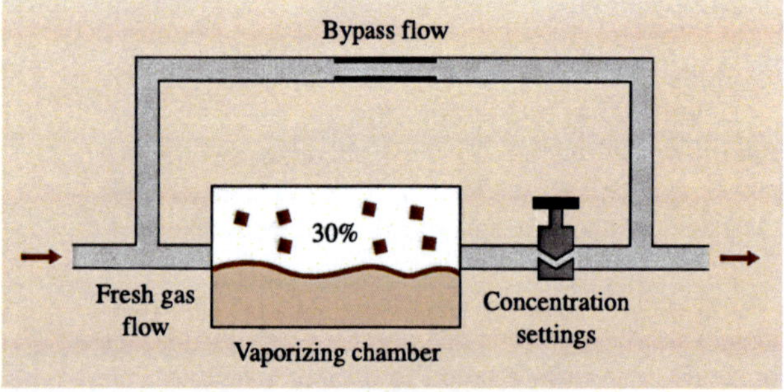

Fig. 30.5 Anesthetic agent vaporizing principle

Each anesthetic agent needs its own separate vaporizer due to the different saturation concentrations and the different effect on the patient.

Anesthetic agent vaporizers must only be used in the fresh gas line because the gas passing through the vaporizer repeatedly would result in uncontrollably high concentrations in the rebreathing system. The vaporizer must also always be used below the boiling point (48 °C for isoflurane). Concentrations cannot be controlled above the boiling point.

Since the boiling point of desflurane is 22 °C, this anesthetic agent requires the use of a different principle from that mentioned above. Desflurane is heated in a heating chamber to 40 °C and the desflurane is added as a vapor to the fresh gas. It is also necessary to know the fresh gas flow to determine the quantity to be added. This is determined via a control loop.

Safety Features

Coded Filling Device. If an incorrect anesthetic agent is administered, under- or overdosage is possible depending on the anesthetic agent and the vaporizer. To avoid this, vaporizers are provided with a coded filling device. These devices ensure that the anesthetic agent is only delivered to the intended anesthetic agent vaporizer. The geometric coding of the anesthetic bottle, filling hose, and vapor filling opening is designed to ensure this.

Measuring the Anesthetic Agent Concentration. To detect vaporizer malfunctions, and thus any possible associated under- or overdosage, the anesthetic agent concentration is measured during inspiration. It is important to ensure that the monitor is always set to the used agent.

Dosing the Intravenous Anesthetic Agent. Since the intravenous hypnosis agent propofol and the intravenous analgesic remifentanil are metabolized quickly in the body, within minutes, continuous administration of these drugs is required. For this reason, during anesthesia these drugs need to be administered by syringe pumps, which need to be suitably attached to the anesthesia machine.

A bolus (a certain amount of a drug given in a specific period of time) is administered to raise blood-level concentrations to the desired therapeutic level. The decreasing effect of the drug caused by drug clearance is compensated by a constant infusion rate. The blood concentration cannot be measured directly in the patient. This means that the physician has to depend on monitoring the effect on the patient.

30.3.2 Ventilator and Breathing System

The second component of an anesthesia device is the ventilator. A distinction between its parts must be made according to their tasks: the input unit, the control unit, the breathing system, the bellows unit with drive, the fresh gas supply, and ventilator monitoring (Fig. 30.6). These components facilitate patient ventilation. In anesthesia, controlled ventilation is the most frequently used form of ventilation.

Subcomponents of a Ventilator

Input Unit (Human Interface). The input unit of the ventilator, also called the human interface, is the interface between the user and the device. It allows the medical personnel via the control unit to set the values which are appropriate for the respective patient, according to the operating philosophy of the device; for example, the frequency and inspiration-to-expiration ratio can be used to set the inspiration and expiration time, as widely used in anesthesia. In volume-controlled ventilation, the controlled ventilation stroke is determined by the specification of the volume to be administered and the inspiratory pause time or also by the volume and the inspiratory flow.

Today, the parameters are usually entered indirectly. In other words, each adjustment is only carried out by the control unit and applied to the patient if confirmed again by the user. This is a means of detecting erroneous settings before they are applied.

Control Unit. The control unit converts the set values preset by the input unit into machine data. It controls the valves and defines the times at which the inspiration valve, the expiration valve, and the fresh gas decoupling valve are opened or closed. It also triggers the drive unit to administer the tidal volume during inspiration to the patient. These activities depend on the respective operating mode.

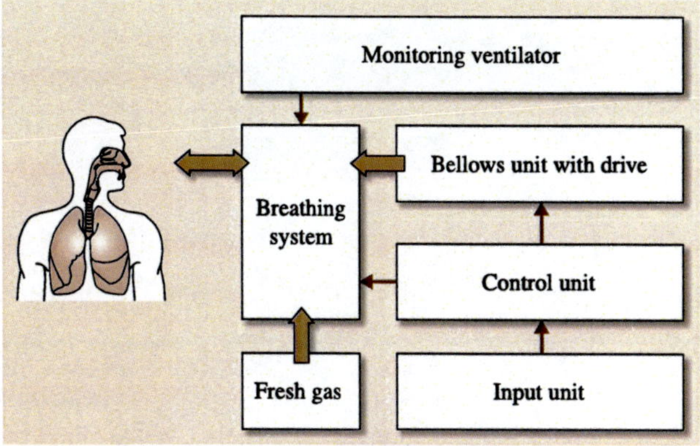

Fig. 30.6 Subcomponents of a ventilator

Today, anesthesia ventilators are operated as time-time-cycled devices in controlled ventilation modes. In such time-time-cycled devices, the switchovers from both inspiration to expiration and from expiration to inspiration take place at defined times, independently of the condition of the patient's lungs.

Breathing System. In clinical practice there are two different types of breathing systems: the rebreathing system and the non-rebreathing system.

Non-rebreathing systems are used in intensive care ventilators. With this system, the fresh gas is supplied directly to the patient's lungs. The exhaled gas is released, without diversion, into the ambient air.

Rebreathing systems are used in anesthesia because of the costs of the gases in anesthesia devices (Fig. 30.7). By recirculation of the expiratory breathing gas into the inspiration gas flow, the anesthetic agents contained in the gas are used to best effect. This allows for a reduction of the delivered doses of inhalation narcotics. In this way, the fresh gas volume per minute can be reduced far below the minute volume.

Since the exhaled gas contains CO_2, this system requires the integration of a CO_2 absorber to ensure that the patient is supplied with CO_2-free gas. The absorber is the characteristic feature of a rebreathing system. The soda lime (primarily $Ca(OH)_2$) contained in the absorber bonds the CO_2 in the exhaled air. At the same time, the chemical reaction produces heat and moisture, which contributes to gas heating and humidification.

The inspiratory concentration of the breathing gas delivered to the patient depends significantly on the set fresh gas concentration and the concentration of the expiratory gas.

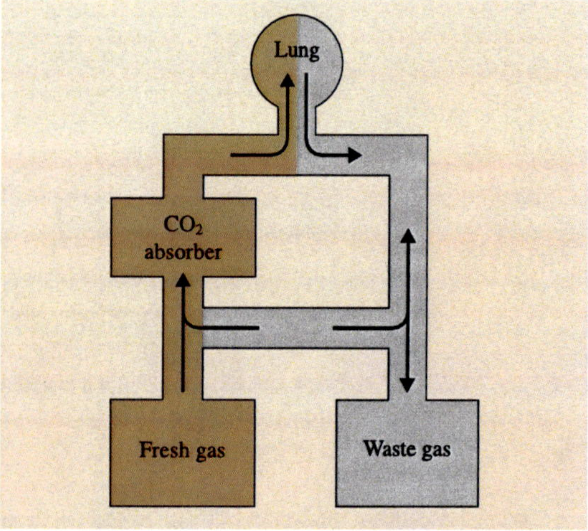

Fig. 30.7 Principle of a rebreathing system

Today, rebreathing systems have become standard in anesthesia because they consume less gas, therefore having lower operating costs and being less environmentally harmful, and provide for improved gas climatisation.

The systems are classified as follows depending on the magnitude of the fresh gas flow:

- High flow, approximately 3–6 l/min
- Low flow, approximately 1 l/min
- Minimum flow, approximately 0.5 l/min.

However, the lower the fresh gas flow setting, the higher the degree to which the system must be sealed and the greater the demands placed on device monitoring.

Bellows Unit with Drive. The bellows unit is the main component of a ventilator. It is used to supply the patient with the required tidal volume. The bellows can be pneumatically emptied using compressed air, for example, in the case of Ventilog, AV1, and Julian devices, or also electronically using a motor, for example, in the case of Cicero, Fabius or Primus devices.

In a pneumatically driven ventilator (also called the *bellows-in-bottle principle*), the bellows is fastened in a pressure-tight chamber. The application of compressed air in the chamber causes the bellows to be emptied in the inspiratory phase, and the gas contained in the bellows is delivered to the patient. In the expiratory phase, the pressure in the chamber is reduced, so that the pressure returns to ambient pressure again and the bellows returns to its initial position.

The bellows may be moved back and forth by means of an electric motor with connected gears, which results in the contained gas being emptied. This effect can also be implemented with a piston cylinder unit, which can be compared to a large syringe. The piston is moved back and forth via a spindle by a step motor. Since the cylinder has a defined cross-sectional area and the drive via the spindle is very precise, this provides for very accurate information on the currently delivered volume being administered. As drive gas is not needed when an electric drive is used, costs are reduced accordingly.

Fresh Gas Supply. As shown in Fig. 30.2, the O_2/N_2O mixture from the dosing unit is added to the anesthetic agent vaporizer. After passing through the vaporizer, the gas mixture consisting of O_2, N_2O, and the anesthetic agent is also called fresh gas. This gas mixture is delivered to the inspiration side of the breathing system. The fresh gas can be supplied in two different ways:

- Continuous (conventional system)
- Discontinuous (fresh gas decoupled system).

Continuous Fresh Gas Supply into the Inspiration Line. This conventional system (Fig. 30.8), implemented, for example, in Sulla 808 with Ventilog 2 or Narkomed devices, features the characteristics that the manual ventilation bag is not included in the gas flow during controlled ventilation and that gas flows continuously into the breathing system independently of inspiration and expiration.

During inspiration, fresh gas is delivered to the patient in addition to the volume from the bellows. Therefore, the volume delivered to the patient consists of the bellows volume plus the fresh gas volume. Therefore, the system has the following characteristics:

- The minute volume is fresh gas flow dependent. This means that, if the fresh gas flow is reduced from, e.g., 5 l/min to 2 l/min, without adjusting the ventilation, the minute volume (MV) is reduced by 1.5 l at an inspiration/expiration ratio of 1 : 1, i.e., MV change $= 1/2(5\,\text{l/min} - 2\,\text{l/min}) = 1.5\,\text{l/min}$.
- If a large flow is present, the pressure curve in the plateau phase of volume-controlled ventilation shows a continuous pressure rise.
  The pressure curve exhibits a second pressure peak as a result.
- If a very small fresh gas flow is present, with descending bellows, a negative pressure phase is produced at the start of expiration because the

conventional system is not equipped with a gas reservoir. Therefore, to avoid this effect, a sufficiently high fresh gas flow must be maintained.

Today, some devices on the world market feature a steady state in which a fresh gas compensation eliminates this fresh gas flow dependency. By measuring the volume, for example, during inspiration, consisting of a fresh gas and a ventilator part, a control algorithm can be used to reduce the part from the bellows unit in such a way that the total of the fresh gas and bellows parts corresponds to the set volume.

Discontinuous Fresh Gas Supply into the Inspiration Line. This system (Fig. 30.9) is also called a fresh gas decoupled system and is implemented, for example, in Cicero and similarly in Fabius and Primus devices. A characteristic feature of this system is that the fresh gas flow can only flow into the inspiration line of the breathing system during the expiratory phase of the patient. During the inspiratory phase, the gas flows into a reservoir bag (manual ventilation bag), which allows the bellows to be filled during the expiratory phase.

Due to this discontinuous fresh gas supply, the system has the following characteristics.

- The minute volume is independent of the fresh gas flow, i. e., if the flow is adjusted from, e.g., 5 l/min to 2 l/min, the minute volume remains unchanged. This characteristic makes fresh gas decoupling essential for low-flow anesthesia applications.
- Due to the decoupling between drug dosing and the patient, the pressure curve is not affected by the fresh gas flow; in particular, no negative phase is present for low flow.
- The reservoir bag is included in the system during controlled ventilation and is filled and emptied with the rhythm of ventilation. This feature also allows recognition of fresh gas decoupling in volume-controlled ventilation.

Monitoring the Ventilator. Monitoring of device parameters is essential to make malfunctions in individual components (such as the ventilator including the breathing system) identifiable and transparent and to provide the user with the option of intervening. Three sensors – for pressure, volume, and CO_2 – are responsible for monitoring the correct functioning of the ventilator and the breathing system. Usually, these sensors are integrated into general device monitoring and also perform additional patient monitoring tasks.

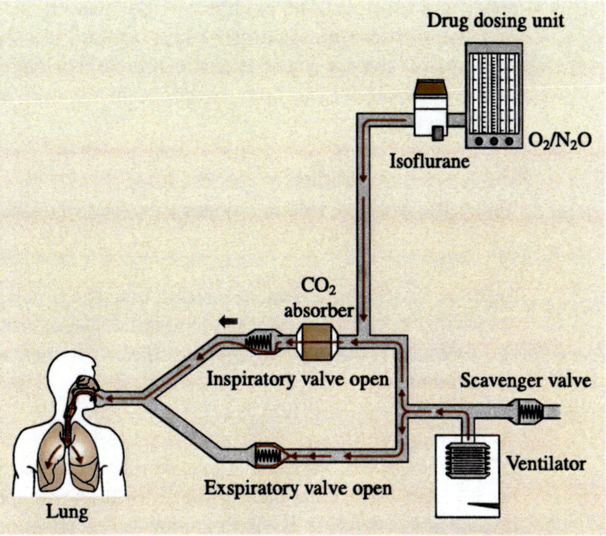

Fig. 30.8 Conventional breathing system with continuous fresh gas supply

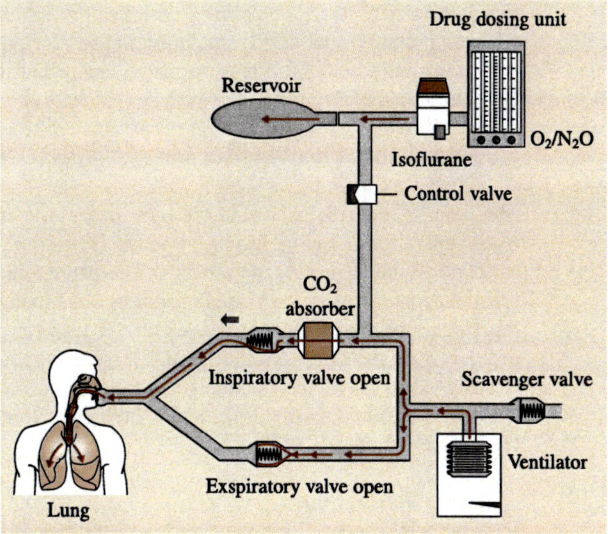

Fig. 30.9 Fresh gas decoupled system through discontinuous gas supply

Volume measurement can be used to check the set tidal volume. The correct ventilation frequency can be determined from the volume, pressure or CO_2 measurement. The pressure measurement warns against too large pressure in the breathing system (stenosis) and is designed to protect against the risk of barotrauma. The inspiratory CO_2 measurement detects an exhausted soda lime absorber. The expiratory CO_2 measurement and the pressure measurement warn against disconnections (as the set lower pressure limit is no longer periodically reached in both directions in the case of disconnection).

Ventilation Modes in Anesthesia

Today, four main ventilation modes are used in anesthesia.

Manual Ventilation. Manual ventilation enables the anesthetist to initiate anesthesia, to ventilate the patient, and to wake up the patient from anesthesia for spontaneous breathing.

Controlled Ventilation. This is the main form of ventilation in anesthesia, because a muscle relaxant is often used during anesthesia, having the side-effect that the breathing musculature is also paralyzed. A distinction is made between two forms: volume-controlled ventilation and pressure-controlled ventilation. Both are used for different applications. Volume-controlled ventilation is primarily used for patients with healthy lungs, whereas pressure-controlled ventilation is used for patients with pulmonary problems, neonates, and pediatric patients.

Spontaneous Breathing Support (Pressure Support)

The use of muscle relaxants is not necessary in many operations, and a laryngeal mask is used instead of a tube to secure the airways. The patient can breathe spontaneously with stable spontaneous breathing frequency, however, the musculature is weakened depending on the drugs administered. Pressure support is designed to assist the breathing musculature which has been weakened by the drugs and to help the patient take deeper breathes.

Mixed Ventilation

In cases where operations are performed without use of muscle relaxants, but the doses of hypnosis and analgesic drugs are so high that spontaneous breathing frequency is not stable, there may be periods without spontaneous breathing. During such periods, minimum ventilation must be maintained to ensure that the patient is provided with sufficient oxygen. The patient's breathing is basically spontaneous. The ventilator, which has a low safety frequency setting, delivers controlled ventilation cycles synchronously with the patient's inhalation efforts.

To assist patients in overcoming the difficulties caused by the tube and the side-effects of the drugs, spontaneous breathing is supported by pressure support.

During the anesthesia termination phase with muscle relaxants, where the concentration of the drugs is gradually reduced and spontaneous breathing returns, mixed ventilation can be used to help support emerging spontaneous breathing and ensure minimum ventilation. This allows the anesthetist to use his hands for other tasks. Usually, one hand is busy providing manual ventilation.

Volume-Controlled Ventilation

Volume-controlled ventilation, also called intermittent positive pressure ventilation (IPPV), is a time-cycled, volume-controlled ventilation mode. In this controlled ventilation mode, the ventilator administers the preset tidal volume (e.g., 10 ml/kg body weight) with a fixed defined constant inspiration flow with the specified frequency (e.g., for adults, 10 /min).

The pressure in the breathing system and in the lung results from the set parameters and the resistance and compliance of the patient's lungs (Fig. 30.10).

High pressure peaks can occur depending on the settings. Therefore, a maximum pressure limit P_{max} must be set which ensures that the flow is automatically reduced when the pressure limit has been reached. Additionally, pressure monitoring is also essential. An advantage of the IPPV mode is that the patient is always provided with a defined minute volume even with changes in lung properties.

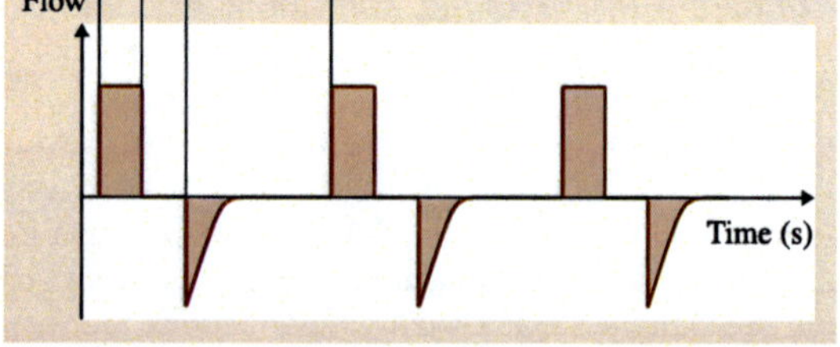

Fig. 30.10 Pressure and flow curve for volume-controlled ventilation

This mode is primarily used for patients with healthy lungs and is very popular in anesthesia. It is also used when volume constancy and thus maintenance of arterial CO_2 partial pressure within strict limits is a primary objective of ventilation, for example, in traumatic brain injury. Since cerebral perfusion is dependent on CO_2, the volume must be kept constant to set the optimal intracerebral pressure.

Pressure-Controlled Ventilation

Pressure-controlled ventilation (PCV) is a time-cycled, pressure-controlled ventilation mode. In this controlled ventilation mode, the ventilator delivers a ventilation stroke at a constant preset pressure level, e.g., 15 mbar, with a decelerating inspiratory flow pattern during the entire inspiration time at a specified frequency (Fig. 30.11). The administered volume cannot be preset and results from the applied pressure and lung compliance. Depending on the pressure setting, the delivered volume may be too large or too small. Therefore, expiratory volume monitoring is essential. The main advantage of this mode is that the pressure values which have been set for the patient are never exceeded. This mode is primarily used for neonates, infants, and children, where the thorax is not stabile and also the endotracheal tube is not blocked, resulting in a leak. It is also helpful for patients with inhomogeneous lungs, where it is necessary to prevent overinflation of the normal areas. This mode is also suitable for ventilating patients with bronchopleural fistulas. PCV is also increasingly used for ventilation with laryngeal masks.

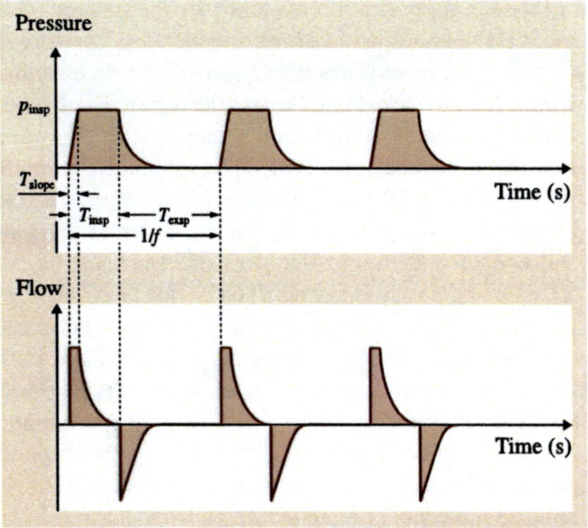

Fig. 30.11 Pressure and flow curve for pressure-controlled ventilation

Spontaneous Breathing Support (Pressure Support)

Pressure support is a pressure-supported ventilation mode which can be used in cases where spontaneous breathing is present but insufficient (Fig. 30.12). It sup-

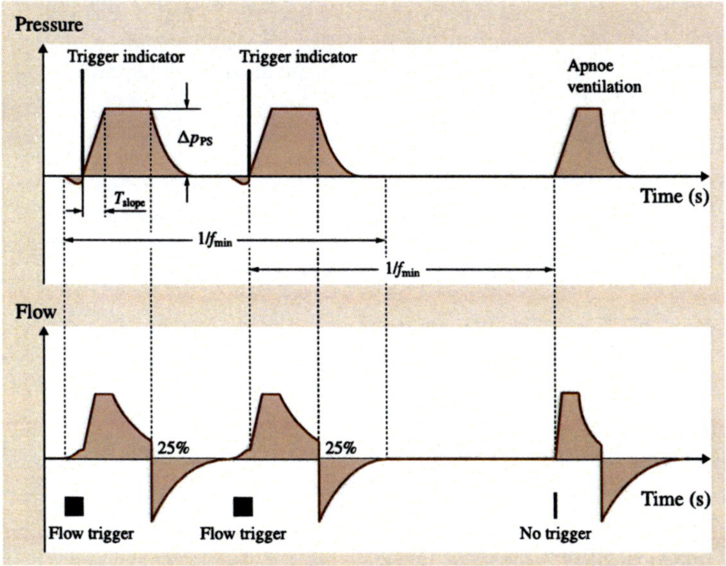

Fig. 30.12 Pressure and flow curve for supported spontaneous breathing (pressure support, PS)

ports each individual breath taken by the patient. This mode is also called *assisted spontaneous breathing* (ASB) or the *inspiratory help system* (IHS). In a similar way to how the anesthetist senses the restarting of spontaneous breathing in the patient at the breathing bag and supports it manually, the device can support insufficient spontaneous breathing. The device takes over part of the breathing work; however, the patient controls the start and end of inspiration. The start of inspiration is detected using a trigger system. The device then provides a flow and thus creates a pressure rise which increases up to the preset support pressure and also remains constant. Expiration starts as soon as the inspiration flow falls below a specific value (e.g., 25% of the maximum flow). By reducing the support pressure, the support provided by the device is reduced and the patient is induced to take over more breathing work.

Mixed Ventilation (SIMV/PS)

The basic idea of synchronized intermittent mandatory ventilation (SIMV) (Fig. 30.13) is that the patient basically breathes spontaneously and the ventilator intermittently delivers controlled ventilation strokes (volume controlled or pressure controlled) with a very low safety frequency synchronously with the patient's inhalation efforts. This prevents mechanical ventilation strokes from taking place during spontaneous expiration. No spontaneous breathing is possible during a controlled stroke.

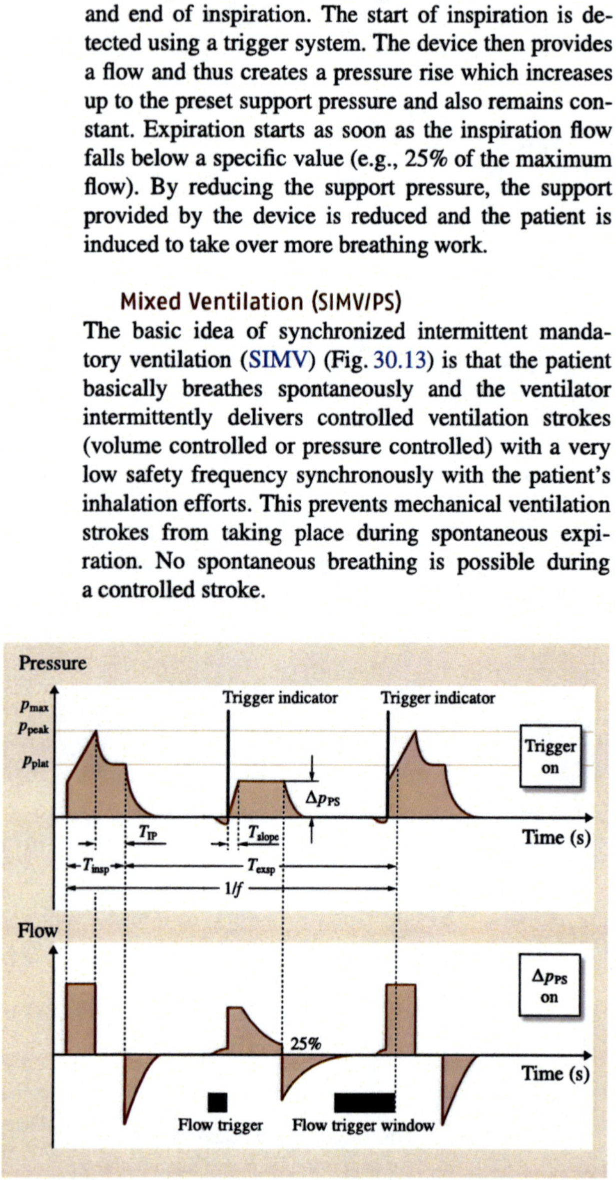

Fig. 30.13 Pressure and flow curve for mixed ventilation SIMV (VC) with PS

SIMV is a mixed form between pure spontaneous breathing and controlled ventilation. The ventilation strokes of the device are synchronized with the patient's breathing, for which the trigger function of the machine is used. The ventilation stroke is triggered when the patient tries to inhale again at the end of the spontaneous breathing phase and thus triggers the trigger pulse. The consequence of these is a minimum ventilation which results from the product of tidal volume (VT) and the frequency. The patient can breathe freely and spontaneously between these controlled strokes. In these spontaneous breathing phases, the patient can be additionally supported with pressure support by a suitable pressure setting, thus relieving him of part of his breathing work.

30.3.3 Monitoring

The third significant component of an anesthesia device is monitoring. It has two main purposes: firstly, to constantly monitor the correct administration of drugs and the correct functioning of the ventilator, and secondly, to constantly monitor the effects of the anesthesia on the physiological functions of the patient. The corresponding monitoring values are used to determine whether the device is functioning properly and whether the patient is being anesthetized and ventilated as desired.

In the event that the limits of a previously defined range are exceeded, the corresponding audible and visual alarm of the monitored parameter is triggered. To meet the two most important tasks, namely the measurement and monitoring of a specific parameter, different system components in the monitor are required.

Measuring Unit

Using a sensor, this unit converts the physical values, such as pressure, flow, oxygen concentration, temperature or absorption, into a measurable electrical signal; for example, a sensor is used to create an electrical current from the breathing gas flow.

Display Unit

This unit shows the measured result after it has been converted into the corresponding units, such as volume in ml. The display can be shown as a digital numerical sequence or as a curve.

Monitoring Unit

This unit checks whether the actual values are within the target range. To achieve this, lower and upper limit values are defined for the monitoring unit by the user.

The monitoring unit compares the currently measured value with the lower and upper limit value. If limit values are exceeded, an alarm is issued to allow the user to intervene.

The output values from the different sensors can be collated in such a way that the screen can be arranged differently with curves, data, and alarm displays.

Tasks of Different Sensors

Generally, anesthesia devices are equipped with nine sensors, including five device monitoring sensors for monitoring drug dosing and the ventilator:

- Oxygen
- Pressure
- Volume
- Carbon dioxide (CO_2)
- Anesthetic agent

and four patient monitoring sensors for monitoring patient health during uncomplicated operations:

- Electrocardiogram (ECG)
- Noninvasive blood pressure (NIBP)
- Oxygen saturation (SpO_2), and
- Body temperature.

Display of Dosed Gases

The gas dosing unit and the anesthetic agent vaporizer deliver the drugs into the fresh gas flow of the breathing system. Due to the rebreathing system, the inspiratory concentrations of oxygen, nitrous oxide, and anesthetic agent differ from the concentrations set on the dosing devices. This difference increases when the fresh gas flow is reduced. To determine the concentrations being delivered to the patient, the inspiratory concentrations of oxygen, nitrous oxide, and anesthetic agent (isoflurane, sevoflurane or desflurane) need to be measured and displayed.

Monitoring the Device Components

Five different sensors with upper and lower alarm limits ensure that the device components are constantly monitored. The purpose of this type of monitoring is to allow qualified and trained medical personnel to intervene immediately in the event that the device fails to function properly.

Oxygen measurement is used to monitor gas dosing, to identify, for example, whether the O_2 and N_2O lines have been mixed up. Measuring the nitrous oxide concentration is another way of checking this. Oxygen measurement is also used to detect hypoxia, if, for example, too low an O_2 concentration was set. Anesthetic agent measurement is responsible for monitoring the anesthetic agent vaporizer and ensuring that it is functioning correctly. CO_2 measurement allows detection of rebreathing if, for example, an inspiration or expiration valve is defective. Inspiratory CO_2 measurement monitors the CO_2 absorption by the soda lime and indicates the point when the soda lime has to be replaced at the latest. Expiratory CO_2 measurement monitors adequate ventilation and thus also breathing system integrity to ensure that the device and patient are not disconnected.

Pressure and volume measurement are used to monitor the proper functioning of the ventilator. Volume measurement indicates whether the ventilator is working properly, in particular whether the set volume is actually delivered. Pressure measurement ensures that disconnection is detected and also that the physician is alerted if the pressure is too high, to prevent barotrauma.

Patient Monitoring

The different sensors designed to monitor both the device and the patient allow the physician to assess the state of the patient. The inspiratory oxygen concentration and the mean pressure produced during ventilation are used to determine patient oxygenation. The difference between the inspiratory and expiratory O_2 concentrations provides an indication of the patient's O_2 uptake. Volume measurement together with the expiratory CO_2 measurement is used to set the patient's ventilation correctly. The expiratory anesthetic agent concentration provides a reference point for the depth of anesthesia. The difference between the inspiratory and expiratory anesthetic agent concentration values provides an indication of the patient's anesthetic agent uptake.

The difference between the peak pressure and plateau pressure in volume-controlled ventilation indicates the airway resistances. Patient and device compliance, on the other hand, are reflected in the plateau pressure in volume-controlled ventilation.

The oxygen saturation sensor shows the blood oxygenation level and also provides information on the heart's pump function. Monitoring the blood circulation is also necessary for patient monitoring. The heart's pump function can be observed by measuring the heart rate via evaluation of the oxygen saturation level as well as by measuring the noninvasive blood pressure. The ECG indicates whether and how the heart is electrically activated.

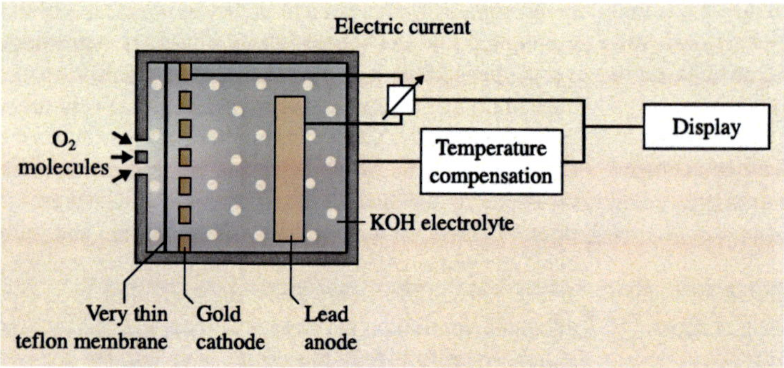

Fig. 30.14 Principle of a fuel cell for measuring oxygen concentration

For operations on the heart, and in neurosurgery and other related areas, enhanced monitoring with invasive blood pressure measurement is required in addition to the form of monitoring described before (Chap. 48).

30.3.4 Sensor Principles for Device Monitoring

Fuel Cell for Oxygen Concentration Measurement

The inspiratory oxygen concentration measurement monitors the oxygen delivery. A method often used for O_2 measurement is the fuel cell (Fig. 30.14).

The electrochemical reaction system of the fuel cell is accommodated in a housing which is sealed by an extremely thin membrane. This membrane prevents the alkaline electrolytes from escaping from the housing but allows oxygen molecules to diffuse through.

The O_2 concentration in the gas determines the O_2 concentration in the electrolyte of the fuel cell. The alkaline electrolytic solution contains a lead anode and a gold cathode. After the oxygen molecules of the gas to be measured have passed through the membrane, basically the following electrode reactions take place. On the gold cathode, oxygen releases electrons from the cathode material, forming OH^- ions and positively charging the cathode. On the anode, lead reacts with the OH^- ions to form lead oxide and water. During this process, the anode is negatively charged. If the cathode and anode are connected with each other, an electrical current proportional to the oxygen concentration is produced.

The lifetime of this system is limited by two processes: lead is transformed to lead oxide, and water as the catalyst in the cell diffuses out.

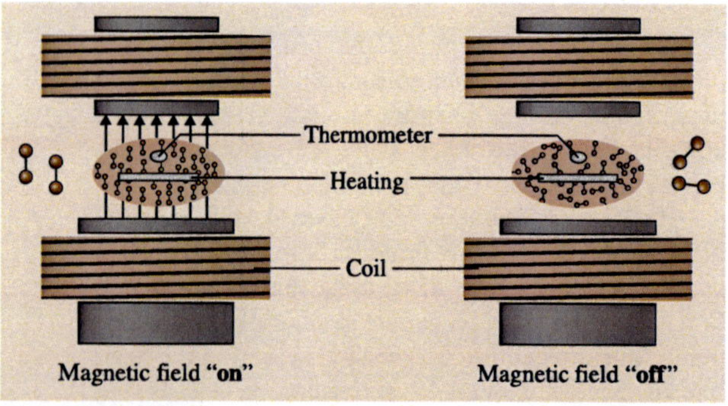

Fig. 30.15 Principle of a paramagnetic sensor cell for measuring oxygen concentration

Paramagnetic Sensor for Oxygen Concentration Measurement

The gas mixture used in anesthesia consists of oxygen, nitrogen, nitrous oxide, carbon dioxide, and the volatile anesthetic agent. Only the oxygen molecule has a paramagnetic moment, which can be influenced by a magnetic field. Therefore, oxygen can be measured selectively in this gas mixture. The measurement principle (Fig. 30.15) is based on the thermal conduction of oxygen. The thermal conduction of a gas depends on the degrees of freedom of the molecules, with more degrees of freedom contributing more to the thermal conduction. A magnetic field is used to reduce the degrees of freedom of the oxygen molecules. The oxygen molecules are aligned like tiny magnets and their freedom of movement is restricted.

Therefore, the thermal conduction depends on the external magnetic field and has a small but measurable influence on the overall thermal conduction of the gas mixture. A heating element is brought to a constant temperature in a magnet and conducts heat to the gas in the cuvette. A nearby temperature sensor measures the temperature, which depends on the thermal conduction of the gas. The magnetic field is periodically turned on and off at a specific frequency and the temperature sensor consequently measures small temperature fluctuations whose amplitude is proportional to the oxygen concentration.

This paramagnetic thermal-conductive oxygen analyzer has a short time constant, no moving parts, and long lifetime durability.

Piezoresistive Sensor for Pressure Measurement

Inspiratory and expiratory pressure measurements can be used to create pressure–time diagrams to make ventilation more transparent. This ensures that disconnections and apnea phases can be detected and also that warnings are issued if the pressure is too high.

An electrical pressure signal can be produced from a mechanical–electrical (piezoresistive) converter (Fig. 30.16). For this purpose, a pressure cell is sealed with a movable membrane. The membrane is attached to a solid-state device, whose electrical resistance depends on the elongation of the membrane. The electronic pressure measurement is determined by the actual stretching of the membrane. The stretching of the membrane causes the solid-state device to bend, leading to a variation in resistance. In this way, the current prevalent pressure is uniquely correlated with the measured resistance.

Hot-Wire Principle for Volume Measurement

The expiratory volume measurement detects the total quantity of exhaled gas. In particular, in the case of pressure-controlled ventilation, volume measurement is the only way of obtaining information on the volume delivered to the patient. An additional inspiratory measurement can be used to display the inspiratory and expiratory flow curve.

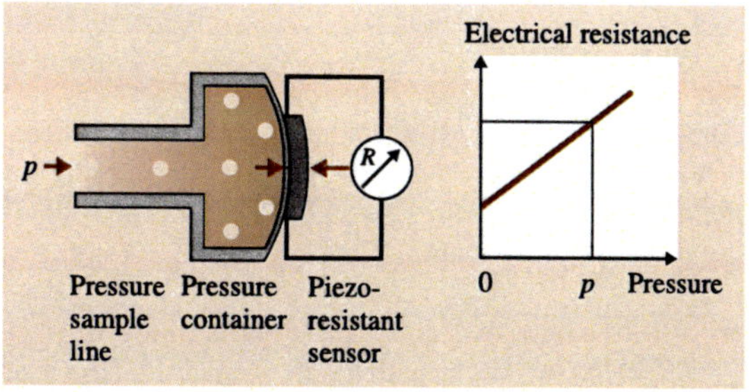

Fig. 30.16 Principle of a piezoresistive sensor for pressure measurement

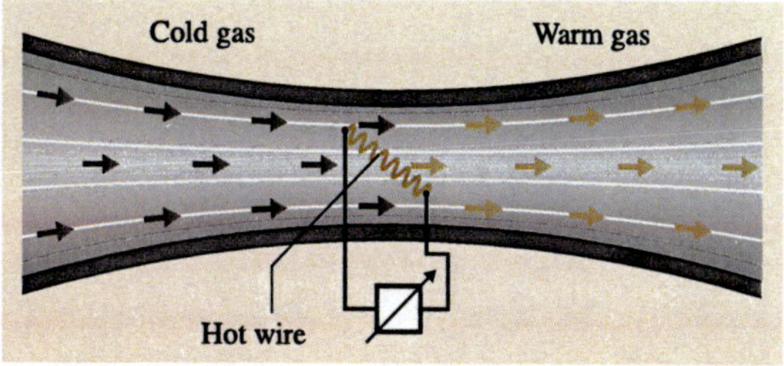

Fig. 30.17 Principle of a hot-wire anemometer for flow and volume measurement

The hot-wire anemometer method is a purely electrical method for volume measurement (Fig. 30.17). An extremely thin platinum wire is heated to a temperature of approximately 180 °C using an electrical current. When gas flows past this wire, the wire is cooled. The greater the volume per time flowing past the wire, the more the wire is cooled. If the temperature of the platinum wire is kept constant by a control circuit, the required heating current can be used as an indication for the gas flow. A high flow requires a high electrical current to keep the temperature of the heating wire constant; a lower flow requires a small electrical current to reach the same temperature. The volume is obtained by integrating current over time electronically.

Infrared Absorption Measurement for CO_2, N_2O, and Anesthetic Agent Concentration Measurement

Infrared absorption spectroscopy (Fig. 30.18) is based on the physical principle that polyatomic gases absorb infrared radiation at characteristic frequencies. The level of absorption depends on the concentration of molecules.

For measurement, the molecules are channeled into a cuvette with a defined length and illuminated with an infrared light source. The detection element (the detector) detects the remaining residual radiation after absorption. If the light source has an intensity of I_0 and the beam is weakened by the molecules to the intensity I, then the concentration can be calculated according to the Lambert–Beer law

$$\text{Concentration} = \text{Absorption constant} \times \ln(I_0/I)\,.$$

CO_2 and N_2O are determined in the 4 μm range always using *one* specific light frequency for each gas. Today, the anesthetic agents halothane, enflurane, isoflurane,

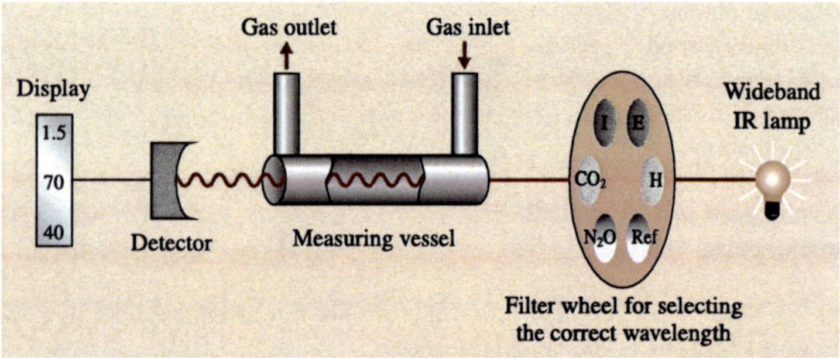

Fig. 30.18 Infrared (IR) absorption for measuring CO_2, N_2O, and anesthetic agent concentrations

sevoflurane, and desflurane are identified by measuring the absorption at three wavelengths in the range 8–9 μm for determination of the corresponding concentration.

Sensor Principles for Patient Monitoring
The principles of ECG, oxygen saturation, blood pressure, and body temperature measurement are described in Chap. 48.

30.4 Safe Operation Prerequisites

30.4.1 Connection Prerequisites

The proper functioning of the anesthesia device is only guaranteed in rooms equipped with appropriate supply connections. Therefore, it must be ensured that the anesthesia device as a life-support system is connected to the hospital's internal emergency power system. It is equally important to ensure that oxygen, compressed air, and, where used, nitrous oxide from the central supply at a suitable pressure (e.g., 5 bar) are always available. If there is no redundancy, it must be ensured that backup cylinders are readily available to maintain anesthesia in the event of disruption to the central supply system. For cardiological procedures, it is also required that the device be connected to the central ground in the operating room. It must also be considered that mobile telephones can disturb the proper functioning of these devices.

Since nitrous oxide and anesthetic agents (isoflurane, sevoflurane, desflurane) are used in the device, being intended only for patient inhalation and not for inhalation by operating room personnel, it is necessary to provide for an anesthetic gas scavenging system to remove the gas. Since no device is totally leak-tight and, for example, anesthetic gases are released into the environment when mask anesthesia is used, a recirculation system with a suitable air exchange rate is required to keep unwanted levels in the workplace to a minimum.

30.4.2 Training and Continuing Medical Education

Training provides users with the knowledge of the procedures required to operate the device as designed. The instructions for use form the basis of training because they contain the main source of information for users. However, in addition to receiving initial training and studying the instructions for use, users should also receive further instruction and training on a regular basis to ensure safety. All training and advanced training programs must be directed towards instructing users how to use the respective anesthesia workstation properly, enabling them to make best use of the functions and to apply the knowledge and skills they have acquired to ensure safe operation in routine and emergency situations.

30.4.3 Cleaning

Anesthesia devices can be a source of infection. Therefore, these devices must be prepared according to defined protocols before use on the patient. The following points are important:

- All breathing-gas-conducting parts must be disinfected for each patient as standard. This can be achieved, for example, by replacing the corresponding parts of the breathing system and ventilator for each patient. Nowadays, filters are often used on the Y-piece to meet these requirements.

- Parts for invasive use such as invasive pressure measurement lines or also endotracheal tubes must be sterilized.
- The surface of the machine must be disinfected at least once a day to reduce bacteria and to prevent the transfer of bacteria by hand from the machine to the patient.

To ensure devices are hygienic, it is essential that a fixed replacement regime is used and that this routine is verified by conducting hygiene tests at the site. If the test result is not satisfactory, the regime must be adapted accordingly.

30.4.4 Maintenance and Pre-use Check

The manufacturer integrates safety features into the device in the design, development, and production phases. The CE mark indicates that all applicable safety requirements of all pertinent EC directives which have to be observed for the medical device have been met. However, this does not guarantee the required safety during clinical use when the device is used for years, even if the device has been used as intended and operated within its design limits in accordance with the instructions for use. During operation, these types of devices are subject to wear; i. e., the current state (actual state) differs increasingly from the desired state (target state). Therefore, users and operators are required to organize and carry out tests and checks on a routine basis, e.g., one or two times a year, to ensure device safety. It can be assumed that safety is maintained between two such regular inspections. However, users must always ensure before using any medical devices that they are functioning properly and operating correctly. Pre-use checks are designed to detect errors which may have been caused by disassembly, cleaning, disinfection, sterilization, transportation, and assembly.

Further Reading

- J.A. Baum: New and alternative delivery concepts and techniques, Best Pract. Res. Clin. Anaesthesiol. **19**(3), 415–428 (2005)
- J.-U. Meyer, G. Kullik, N. Wruck, K. Kück, J. Manigel: Advanced technologies and devices for inhalational anesthetic drug dosing. In: *Handbook of Experimental Pharmacology*, Vol. 182, ed. by J. Schüttler, H. Schwilden (Springer, Berlin, Heidelberg 2008) pp. 451–470
- J.A. Baum: *Low Flow Anaesthesia*, 4th edn. (Butterworth-Heinemann, Oxford 2000)
- EN IEC 60601-2-13 (2003-05) International Standard: Particular requirements for safe and essential performance of anaesthetic systems

31. Extracorporeal Blood Purification Systems

Jörg Vienken

The increase in the number of chronic kidney disease patients has encouraged the development of hemodialysis as a routine therapy. Basic improvements in the dialysis system, from the membrane and the tubing system to the dialysis monitor, have laid the basis for this development. The understanding of underlying control variables and their settings in terms of optimal therapy has also led to reduction of side effects that used to occur and therefore improved the quality of life and reduced mortality [31.1]. Innovative processes associated with the analysis of individual physiological patient parameters use feedback systems and control realized in the dialysis machine. These are currently state of the art. Using bioimpedance procedures, precise monitoring of the water balance of patients is possible. Even patients with liver failure currently use extracorporeal blood purification procedures. In contrast to hemodialysis, one uses protein-permeable membranes with which albumin-bound toxins are filtered. Their purification of the filtrate is then achieved by adsorber columns. Clinical studies have confirmed the efficacy of this procedure.

Extracorporeal blood purification will, therefore, play a central role in renal and hepatic organ failure in the future.

31.1 **Historical Perspective** ... 598
31.1.1 Prologue ... 598
31.1.2 Blood Purification: The Search for Principles ... 598
31.1.3 Membranes, Transport Processes, and Dialyzers ... 598
31.1.4 Preparatory Work for Extracorporeal Blood Circulation for Blood Purification ... 599
31.1.5 Removal of Nephrotoxic Substances Through Dialysis as an Experimental Therapy ... 600

31.2 **Blood Purification for the Therapy of Renal Failure** ... 600
31.2.1 Principles, Control Variables, and Conditions for Hemodialysis ... 600
31.2.2 Dialysis Membranes ... 601
31.2.3 Dialyzer Construction ... 604
31.2.4 Performance Parameters ... 605
31.2.5 Low-Flux, High-Flux, and High-Performance Dialyzers ... 608

31.3 **Dialysis Machines and Additional Equipment: Use and Conditions** ... 611
31.3.1 Determination of Fluid Status with Whole-Body Bioimpedance Spectroscopy ... 613
31.3.2 Description of the Technology and Methodology ... 613
31.3.3 Validation of Clinical Data ... 614

31.4 **Blood Purification in Liver Replacement Therapy** ... 614
31.4.1 Detoxification ... 615
31.4.2 Synthetic Liver Function ... 616

References ... 617

The pioneers in the field of blood purification systems could scarcely dream that by 2010 there would be 1.8 million patients with kidney disease who are undergoing standardized blood purification with dialysis procedures. This should be considered in the light of the fact that, at the beginning of research into extracorporeal blood circulation, seemingly insurmountable technical difficulties existed. Routine treatment with an artificial kidney was even inconceivable. The famous clinician Franz Volhard held this to be even useless and dangerous in the 1920s.

The understanding of physiological processes in the body as well as the high state of technology now available for extracorporeal blood purification processes has led to an annual increase in dialysis patient numbers of 7 to 8% at this time, a figure growing faster than the world population, which is increasing at 1.2% per year. Linked to this are the increasing survival rates with dialysis patients since the beginning of the new millennium, which can be attributed to improvements in dialysis procedures, related therapies, and also with better prevention of kidney diseases [31.2].

Blood purification processes can be regarded as one of the most successful treatment procedures in the area of organ replacement therapies. One can assume in the future that in the Western industrial countries, on average about 1000 patients with kidney disease per one million inhabitants will have to undergo hemodialysis (HD). This places high demands on the corresponding medical technology, but also with the payers who must ensure that all patients can receive this vital treatment.

31.1 Historical Perspective

31.1.1 Prologue

William Harvey (1578–1657) presented his idea of blood circulation for the first time to a skeptical audience in 1628. His ideas were so unusual, and so counter to prevailing wisdom, that his followers were ridiculed soon after by his contemporaries with the nickname *circulatores*. Circulatores were peddlers, stray dogs, and vagrants and the epitome of speculation and lack of seriousness. It was impossible to believe at the time what Harvey, and before him Giordano Bruno, asserted, that the blood flows through the organism in a circuit. Leonardo da Vinci had already reported that the heart pumped out blood at high pressure, but without coming up with the idea that the consequence of this observation must be blood circulation. This idea was advanced by William Harvey for the first time.

Objective scholars of the time shied away from the obvious consequences of blood circulation theory for physiology and pathology. One countered Harvey's thesis with the statement that "... if the blood circulates at the speed asserted by Harvey, absorbable and harmful matter would be distributed in a chaotic mess!" and tried in a last attempt to rescue the respected teacher Galenus of Pergamon (129–199) with his idea of a blood circulatory time of 12 to 15 h [31.3].

31.1.2 Blood Purification: The Search for Principles

How can bodily poisons be removed from the blood? With blood circulation such as advanced by William Harvey, this should have been simple. It meant that access to the blood for detoxification could be obtained anywhere in the body. Anyway, the now well-known physical principle of diffusion was not known at that time. Already in the early eighteenth century, René Dutrochet (1776–1847) had undertaken the first examinations of the transport of water through *biological dividing walls*, independent of the earlier and similar observations of Jean Antoine Nollet (1700–1770) and George Frédérique Parrot (1767–1852). The real progress, however, took place with the work of Thomas *Graham* (1805–1869) in Glasgow published as his Bakerian lecture on *osmotic forces*. In this publication, Thomas Graham coined the terms *dialysis* and *dialyzer* for the first time [31.4]. Graham's experiments were performed in 1854 with the aid of membranes made of bladders, which he later replaced with membranes made of parchment. Such membranes, however, were often fragile and uneven in terms of thickness and size. They were often contaminated with bacteria, and therefore unsuitable for precise scientific investigations.

In Zurich, and in parallel to Graham, Adolph Fick from the German city of Kassel in Hesse worked on the mathematical description of transport processes through membranes. These investigations led to the fundamental *Fick's laws*. They were published in 1855 and described selective material transport through a semipermeable membrane as a consequence of concentration gradients [31.5]. Fick's laws remain today the principle for the description of movement of molecules through membranes and therefore still form the basis for the clearance properties of dialysis membranes for blood purification.

31.1.3 Membranes, Transport Processes, and Dialyzers

It became important to pose the question about the optimum *membrane* material for transport investigations fairly early. The solution was devised by an agricultural chemist named *Schumacher* of Bonn, Germany. He introduced a new type of membrane material called collodium. Collodium is a syrupy fluid that forms after evaporation of a solvent – acetone, ether, or alcohol. It subsequently constructs a membrane film upon drying [31.6]. From a chemical point of view, collodium, a cellulose nitrate, is therefore an ester of nitric acid. Due to its chemical configuration and its contents of nitrate groups, it explodes at a temperature of 186 °C and is also therefore known as *gun cotton*. *Schumacher* was a good chemist. He succeeded in casting extremely thin membrane skins of collodium with which he could study the processes that were significant for the transport of molecules through membranes. His work in 1860 carried the title *On Membrane Diffusion* [31.6]. That was the first time this term appeared in the scientific literature, and it prevailed as the driving force for many developments from biochemistry and medicine to reverse osmosis. Later, in 1921, *Eggerth* of Hoagland Laboratories in New York was able to make membranes of collodium with different permeabilities by using different concentrations of alcohol solutions as a solvent [31.7]. Many of the details given by Schumacher and Eggerth about which materials could be separated still form the basis for the technology of modern membrane production, both for the production of commonly used low-flux dialysis and for high-flux membranes.

Cellulose nitrate in the form of collodium and, after 1945, the regenerated cellulose from cotton became the dominant membrane material for HD up to the end of the twentieth century. The production of the famous Cuprophan membrane, which led the dialysis market for 50 years, however, was finally abandoned in 2006.

At the end of the nineteenth century, Emil Abderhalden (1877–1950) of Switzerland attempted to separate substances from the blood of pregnant women using dialysis. It appears that an American with German ancestors, Johan Jacob Abel (1857–1938) of Baltimore, had heard of these attempts and began to develop a dialysis machine that he described in a report on Monday, August 11, 1913 in the London Times as the *artificial kidney*. Abel is called the father of dialysis due to the term he introduced, even though he had not intended to use this treatment on patients with kidney disease or kidney failure. Rather, at the time he only applied animal trials on dogs to produce an *artificial urine* in order to isolate substances such as salicylates, phenolsulfonphthalein, and aromatic amino acids by dialysis and to then purify these substances [31.8]. In any case, Johan Jacob *Abel* should be respected for the technical development of the first dialysis machine, which he called the *vividiffusion apparatus* [31.8]. This machine, built in 1913, was already very similar to today's capillary dialysis systems. At that time, *artificial kidneys* were made of a hollow glass cylinder in which the inflowing blood flowed over a header consisting of 8 or even 32 glass distribution tubes through approximately 20-cm-long membrane tubes made of collodium. These membrane tubes had a cross-section of about 2 cm. The end of the dialyzer consisted of a second distribution header as with the blood input area with which the blood was returned to the dog's blood circulation. The perfect design of the glass distribution headers must be attributed to the skillful biochemist Benjamin B. Turner (1871–1945), who was also the coauthor of the work along with Abel.

Abel is credited with having already carried out examinations about possible extractable substances from the colloidin (collodium) membranes he produced. It was shown that, under the circumstances, colloidin extracts could influence the blood and therefore render the experiment more difficult. In none of his experiments could differences in comparison to control experiments with saline solution be shown [31.8]. Studies of extractable substances from polymers for medical technology are now routinely carried out and are even the subject of ISO standards [31.9].

31.1.4 Preparatory Work for Extracorporeal Blood Circulation for Blood Purification

At the end of the nineteenth century the first studies were carried out on organ perfusion in order to determine isolated organ functions in vitro. A leader in this area was Franz Hofmeister's laboratory at the Physiological Institute of the University of Strasbourg. Strasbourg was also the way-station for the scientific careers of two scientists who later played a great role in the development of extracorporeal blood circulation. Johan Jacob Abel received his doctorate at Strasbourg in 1888. The second, the German Georg Haas (1886–1971), who was to work later in Giessen, was a close collaborator of Hofmeister.

One can rightly argue that the foundations for the theory and practice of extracorporeal circulation were laid in Strasbourg, from which Georg Haas profited later in carrying out the first extracorporeal blood purification in Giessen/Germany in the 1920s on patients with kidney disease.

31.1.5 Removal of Nephrotoxic Substances Through Dialysis as an Experimental Therapy

Johan Jacob Abel conducted his research exclusively on animals, whereas *Haas* conducted the first procedure on a human being [31.10]. The physician Haas began his experiments on dialysis in Strasbourg at the beginning of the First World War in 1914. No doubt, the main idea of HD for the select diffusion of crystalline blood substances on artificial exchange surfaces was born in *Haas'* laboratory at Hofmeister's institute [31.10].

In contrast to Abel, Haas began with the goal of removing uremic toxins for the treatment of renal failure. He described his first observations [31.11]:

> *Now, as to the therapeutic success of this case, we clearly had the impression of a detoxification process. Immediately following blood purification, the patient's mood was visibly better. Yes, I would say, superior. He was the talk of the ward during the night of the purification, and in the following nights he slept quite well without sleeping pills. Nausea and headaches, which were intractable in the days before blood dialysis, which had given an immediate indication for a procedure, had completely disappeared. Above all, he had regained his appetite, which had been nearly absent before. The improvement in his status lasted about six days, until headaches, sleeplessness and vomiting reappeared.*

Haas abandoned his experiments on the use of a dialyzer before the Second World War for a number of reasons. A major reason for this was the great demand for him as a director of the university hospital and the nursing school of St. Josef's Hospital in Giessen, and also the lack of recognition and acceptance by his colleagues, especially by Volhard, who was probably the most famous clinician of the time.

Only Willem *Kolff* in Kampen, the Netherlands, and later in the USA was able to conclude the work of Haas in the 1940s on his rotating drum kidney, from which kidney replacement therapy could be developed from an experimental to a standard therapy today [31.12]. Further technical developments in the area of membrane technology, e.g., through the use of capillary membranes instead of flat or tubular membranes, and also in the area of dialysis machines, e.g., through volume-control, analyzed and controlled through feedback systems incorporated into the extracorporeal blood circuit, additionally contributed to the success of extracorporeal treatment systems.

31.2 Blood Purification for the Therapy of Renal Failure

31.2.1 Principles, Control Variables, and Conditions for Hemodialysis

If the kidneys fail to perform their functions, then the body must be detoxified through extracorporeal blood circulation containing blood pumps and filters. Following intense investigations, the kidney patient must undergo a 4-h blood purification three times a week. In this procedure, blood is taken from a shunt between the veins and arteries of the patient's forearm. The blood then passes through blood tubes into the dialyzer (artificial kidney). The control of blood flow, its temperature, its blood pressure, blood volume, etc. are taken over by a monitor in the dialysis machine during the treatment process (Fig. 31.1).

The actual blood purification or blood washing then takes place in the dialyzer. This filter mimics the filtering unit of the natural kidney, the glomerulus. A typical dialyzer contains about 10 000 capillary membranes which are arranged in a bundle. These hollow fiber membranes typically have an inner diameter of 200 μm. The wall of the capillaries consists of membranes with a thickness of about 40 μm through which the material exchange takes place (Fig. 31.2). The interstices between the capillary membranes are perfused with an isotonic rinsing solution (dialysis fluid or dialysate) in countercurrent flow to the direction of blood flow. This creates a concentration gradient for material from the blood because the dialysate only contains some electrolytes to maintain tonicity. Various blood toxins (uremic toxins) are filtered out of the blood into the dialysate path using these concentration gradients. They are thus removed and later disposed of. This blood purification procedure is called *hemodialysis* (Fig. 31.3a). The process of hemo*dia*filtration, which is used more and more for blood purification, additionally uses a process controlled by convective forces in addition to the diffusively controlled transport of uremic toxins. Here, high ultrafiltration volumes are used in order to provoke a *solvent drag*, whereby uremic toxins with higher

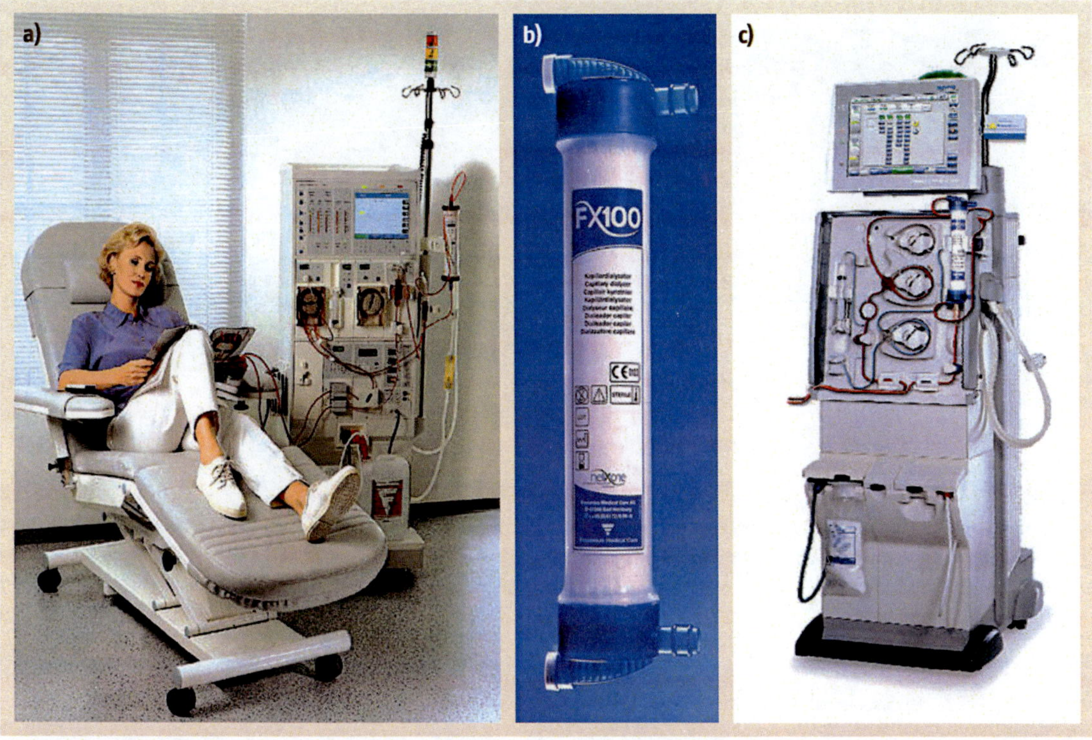

Fig. 31.1a–c Blood purification by hemodialysis is achieved in patients with kidney disease when blood is passed and filtered extracorporeally through a dialyzer. Treatment parameters, such as blood flow, dialysate flow, blood pressure, blood volume, pH, and the performance parameters, can be assessed and appropriately monitored by the dialysis machine. Patients (**a**) undergo this procedure three times per week for 4 or more hours nearly as a rule. Dialyzers (**b**) are the key functional elements of blood purification systems. Dialysis machines (**c**) (5008 machine from Fresenius Medical Care) are able to control a patient-specific therapy with the assistance of feedback systems. Even highly efficient blood purification systems, such as hemodiafiltration, can be operated cost-effectively, as the machine offers online availability of the necessary substitution solution directly from the dialysis fluid as a standard

molecular weights are filtered out in the secondary circuit (Fig. 31.3b). Obviously, high-volume ultrafiltration of the patient requires compensation of the lost volume by substitution with an infusion solution prior to or after the dialyzer (pre- or postdilution). In the past, such a substitution solution was very costly due to the requirements for microbiological and chemical purity. Modern dialyzers enable the introduction of highly pure substitution fluids online directly from the circulation of the dialysis fluid (Fig. 31.3c). Thus, *online hemodiafiltration* (OL-HDF) has developed into an accepted and affordable blood purification procedure.

Treatment of renal failure should be understood as a therapy under the premise of a systems approach. It simultaneously uses a dialysis machine, a tubing system, a dialyzer, and the circuit of dialysis fluids. All components of this system act under synergistic conditions and have to be appropriately controlled in order to perform a blood purification therapy.

31.2.2 Dialysis Membranes

In nephrology, water is seen as a uremic toxin and must be removed by HD because patients with kidney disease are limited in their ability to excrete water. This process, known as ultrafiltration, specifies the chemical composition of the dialysis membrane. A membrane must be at least in part hydrophilic in order to allow water permeation. Consequently, a series of hydrophobic polymers such as polypropylene, polyethylene, etc. are excluded as membrane materials for HD. In many cases, it helps to blend originally hydrophobic polymers with hy-

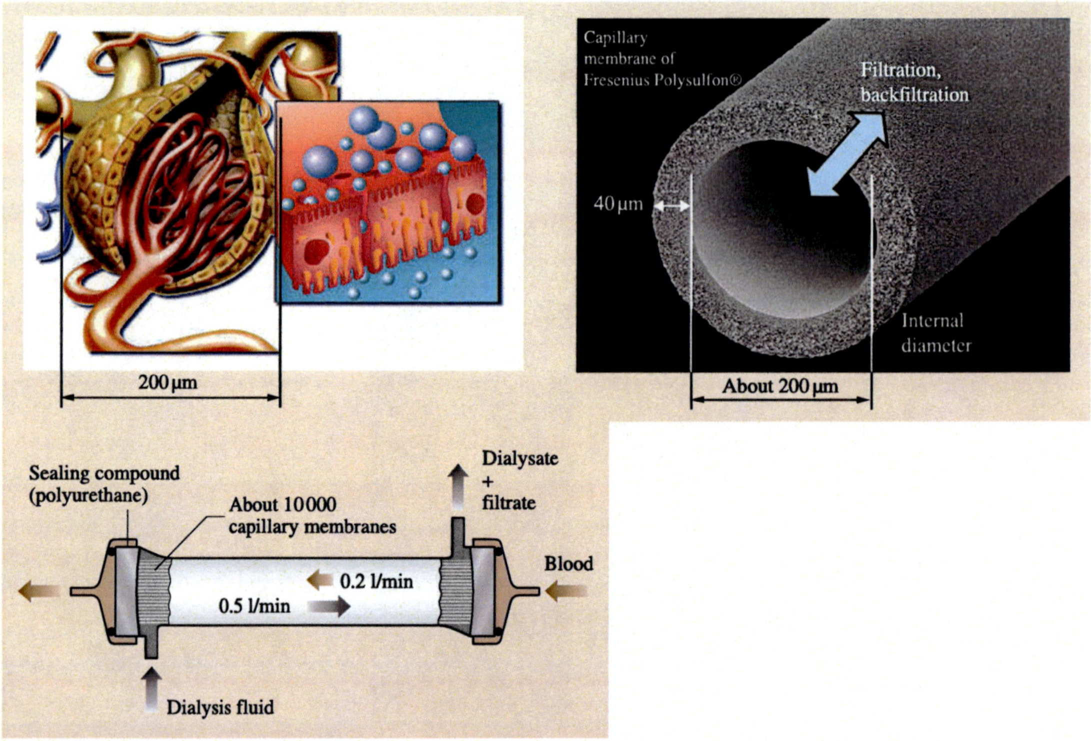

Fig. 31.2 The dialysis membrane assumes the function of cleaning the blood from the filtration unit of the kidneys, the glomerulus. The capillary membranes assume the membrane functions in the walls of the capillaries. It is interesting to observe that the dimensions of the inner diameter of the glomerulus and the inner diameter of the capillary membranes are similar. A dialyzer contains more than 10000 capillary membranes through which the patient's blood is passed. The dialysis fluid is passed in a countercurrent flow to the patient's blood

drophilic polymers. As an example, polysulfone (PSu) and polyamide (PA) can be rendered partially hydrophilic by the addition of polyvinylpyrrolidone (PVP) or polyethylene glycol (PEG), or, as in the case of polyacrylonitrile (PAN), by the addition of methallyl sulfonate.

The basic polymer and the additives of a dialysis membrane dictate its chemical and physical behavior. This includes resistance and stability against chemical steriliants and sterilization processes, its adsorptive characteristics for bacterial contaminants, and its removal characteristics for both uremic toxins and inflammatory mediators. In addition, the blood compatibility and biocompatibility of a membrane is influenced by its polymeric additives, e.g., in relation to blood coagulation, the activation of red and white blood cells, and the generation of vasoactive mediators such as bradykinin.

An ideal polymer membrane for dialysis should show physical stability and an excellent diffusive permeability and, if necessary, allow for high ultrafiltration rates. High ultrafiltration rates enable the application of convective forces, which provoke solvent drag for large molecules in hemodiafiltration. It should have performance and biocompatibility characteristics that promote effective treatment conditions and avoid acute and chronic clinical sequelae.

Recent evidence indicates that an ultraclean dialysis fluid that is free of microbial contamination not only helps with acute inflammatory reactions but can also reduce or eliminate long-term complications of dialysis [31.13]. Therefore, it is particularly important to ensure no contamination in the dialysis fluid with chemical, bacterial, or other contaminants. The adsorption capability of a dialysis membrane for organic contaminants such as endotoxins is, therefore, of paramount importance. Membranes made of polysulfone show efficient adsorption capability for endotoxins. They offer additional protection for patients and are also approved for the filtration of dialysis fluid for online HDF [31.14].

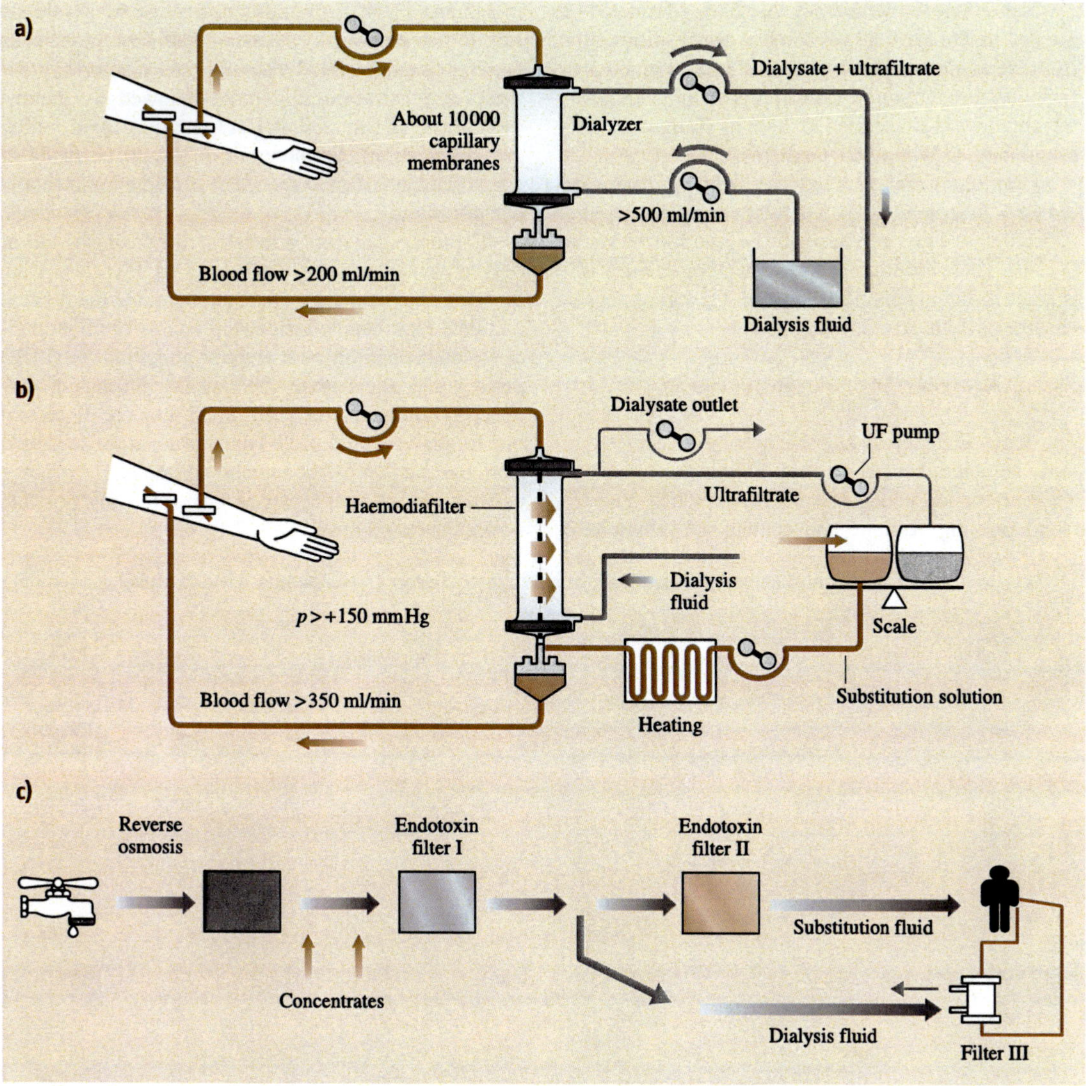

Fig. 31.3a–c Schemes of blood purification procedures. **(a)** hemodialysis (HD); **(b)** hemodiafiltration (HDF), by which the high ultrafiltration rate is compensated by a substitute solution. **(c)** In the online HDF procedure, the substitution solution is directly obtained from the dialysis machine from the dialysate circulation. Endotoxin filters in the circuit guarantee the necessary purity of the substitution solution

Synthetic hollow fiber membranes have a diameter between 185 and 230 μm, with a wall thickness between 20 and 40 μm (Fig. 31.2). Typical polymers for the dialysis membrane are polysulfone (PSu, Fresenius Medical Care, and others), polyacrylonitrile (PAN, AN69, AN69ST, Gambro Hospal), polymethylmethacrylate (PMMA, Toray), and polyamide (PA, Gambro).

Membranes for dialyzers are now consistently offered as hollow fiber membranes made from synthetic polymers. Flat and tubular membranes indeed played a role only in the beginning of dialysis therapy, but the treatment of larger numbers of patients with the necessary voluminous dialysates would be inefficient, impracticable, and costly.

The cellulose membranes that have dominated the market in the past 50 years have nearly disappeared from the market due to their alleged lack of blood compatibility. Production of the Cuprophan and Hemophan membrane was abandoned in 2006 by its manufacturer, Membrana, in Wuppertal, Germany.

The range of dialyzers and filters with capillary membranes is far-reaching and difficult to grasp. Six hundred fifty different dialyzer types could be found on the market in 2010. These are equipped with membranes made of at least ten different polymers. Table 31.1 gives an overview of some of the most common filters.

31.2.3 Dialyzer Construction

The ideal dialyzer should be able to effectively remove soluble substances from the blood throughout the treatment period with consistent performance, have low blood priming volume, and contain a biocompatible membrane [31.15]. In addition, it must be absolutely safe to use and ideally be steam-sterilized in order to avoid problems caused by possible sterilization byproducts (e.g., extracted oligomers followed by gamma irradiation, or residues of ETO (ethylene oxide) after gas sterilization). Furthermore, it should be made of a material that can be disposed of in an environmentally friendly way.

Capillary dialyzers consist of a housing made of polycarbonate or polypropylene (e.g., FX class from Fresenius Medical Care, Bad Homburg, Germany) with a hollow fiber bundle inside (Fig. 31.2). The fiber bundle is embedded on both ends in polyurethane (PUR) potting and attached in the housing in such a way that two compartments are formed, one for blood and one for dialysis fluid. Both compartments are separated from one another by the membrane. Material exchange between the compartments is therefore only possible across this membrane.

Table 31.1 Overview of some of the most common filters. Data from chosen dialyzers (manufacturer listed in alphabetical order)

	Performance, clearance data					Technical data				
Dialyzer manufacturer	KUF (ml/mmHg)	Urea (ml/min)	Creatinine (ml/min)	Phosphate (ml/min)	Vit B_{12} (ml/min)	Sterilization	Membrane material	Wall thickness (μm)	Lumen (μm)	Application
ASAHI PAN 650 SF	30	194	189	174	146	ETO	Polyacrylonitrile	35	250	HD/HDF/HF
BBraun Diacap HI PS15	50	190	178	176	127	γ-rays	Polysulfone – PVP – blend	40	200	HD/HDF
Fresenius Medical Care F8 HPS	18	252	224	193	118	Inline steam	Polysulfone low flux	40	200	HD
Fresenius Medical Care FX60	46	193	182	177	135	Inline steam	Polysulfone helixone	35	185	HD/HDF/HF
Gambro Polyflux-17S	71	191	179	176	136	Steam	Polyamide/PES/PVP	50	215	HD/HDF
Hospal Nephral-ST 400	50	189	176	156	111	γ-rays wet	PAN/AN69ST	42	210	HD/HDF/HF
Nikkiso FLX-15GW	50	190	179	175	143	γ-rays wet	Polyether sulfone	30	210	HD/HDF
Nipro 150FH	67	198	194	193	155	γ-rays wet	Cell.-tri-acetate	15	185	HD/HDF
Toray BK-1.6U	31	187	169	153	108	γ-rays wet	PMMA	30	200	HD/HDF

* Information from the manufacturer's data sheets
HD – haemodialysis; HDF – haemodiafiltration; HF – haemofiltration; PES – polyethersulfone; PAN – Polyacrylonitrile; PVP – Polyvinylpyrrolidone; PMMA – Polymethyl methacrylate

A sophisticated means of microtome cutting technique gives extremely smooth surfaces at the surface of the PUR potting at the ends of the bundles. The activation of blood coagulation pathways and of blood cells can thereby be significantly reduced. The inlet and outlet for the dialysis fluid are found in most dialyzers at the respective ends of the housing.

Housing Material

The housing material for earlier dialyzers was generally made of polymers such as polycarbonate, whereas for the new generations of dialyzer housings are made from polypropylene (FX-class from Fresenius Medical Care, Bad Homburg). Such polymers can be disposed of in an environmentally friendly manner. The housing and the fiber bundles must be structured in such a way that the volumes of the blood compartments are as small as possible in order to prevent the blood of the dialysis patient from being damaged.

Dialyzers and their housings are regarded as disposable and are not to be reused for the most part in Europe or the USA. In a recently published report, the EU commission stated that reuse of medical disposables has no proven advantage in cost, ecological impact, and performance over single-use devices [31.16].

Fiber Bundles

The performance characteristics of a dialyzer are primarily based on the size and type of the fiber bundle inside. The best performance, and hence the best exploitation of diffusive and convective transport from the blood to the dialysis compartment, can only be guaranteed if the flow of the dialysis fluid can occur around each fiber, even in the center of the membrane bundle. The latter is achieved with some newer housing options by a special design of the cap area and with a Moiré structured capillary membrane (FX-class, Fresenius Medical Care). The loosening of the membrane bundle is carried out through undulation of the capillary membrane in the sense of a Moiré structure or through irregular insertion of textile fibers or so-called spacer yarns into the bundle [31.17].

Potting Material and Sealing Compounds

The potting material affixes the capillary bundle and also makes it possible to separate the blood and dialysis fluid compartments. It consists in most cases of PUR. Epoxy resins can also be used. The amount of necessary sealing compound has been continuously reduced over the years. A reason for this was the observation that PUR has a high binding capacity and low degassing kinetics for ethylene oxide (ETO) under ETO sterilization. In the case of contact of the potting compound with blood, such as in HD, ETO can be extracted from the dialyzer by human blood and be released into the circulating blood in the body. Blood proteins, such as albumin, have a high binding affinity to ETO. Once bound to albumin, the so-formed albumin–ETO complex provokes and stimulates the development of IgE antibodies by plasma cells. Serious allergic reactions among hypersensitive patients or among patients who have been primed by a previous contact with ETO can develop as a consequence [31.18, 19]. For this reason, the majority of dialyzer manufacturers have stopped the sterilization of their filters with ETO for safety reasons.

31.2.4 Performance Parameters

The treating physician checks the efficiency of the dialysis treatment for the removal of water, urea, creatinine, and phosphate in everyday clinical practice. In many countries, the Kt/V value is used as an indicator or measure for adequate dialysis. The *urea reduction rate* (URR) is also still used as a parameter in some hospitals.

The abbreviation Kt/V describes the clinically targeted dialysis dose, where K stands for urea clearance, t for the effective dialysis time, and V for the urine distribution volume of the body of the individual patient. The last corresponds at a first approximation to the water content of the body and adds to about 60% of total body weight.

Kt/V must be determined at regular intervals. It is recommended that it be determined at least once a month. In Germany and the USA, this measure serves as the basis for determining dialysis reimbursement by health insurers. Depending upon the dialysis unit and patient groups, the measurements are also made more frequently (once a week) or less often (every 6 weeks to every 3 months) [31.20, 21]. As an example, Table 31.2 provides conditions for the assessment of dialyzer performance. In addition, the development of technically better systems for online determination of urea clearance has led to carrying out these measurements automatically during dialysis treatment (e.g., OnLine Clearance Monitor, Fresenius Medical Care), and therefore the dialysis dose of Kt/V can be continuously monitored [31.22].

In stark contrast, predialysis measurements of urea levels with an estimation of dialysis efficiency besides Kt/V or URR are considered to be a bit imprecise and are not efficient over time [31.23].

Table 31.2 Conditions for the assessment of dialyzer performance

Clearance test parameter	Standard	Differences with some manufacturers
Type of test	In vitro	
Test medium	Dialyzer fluid, enriched with urea, creatinine, phosphate, and vitamin B_{12}	
Concentration of test substances	Urea: 100 mg/dl Creatinine: 10 mg/dl Phosphate: 5 mg/dl Vitamin B_{12}: 5 mg/dl	
Test medium temperature	$37 \pm 1.5\,°C$ (USA) $37 \pm 1.0\,°C$ (Europe)	
Test medium pH (important for the determination of phosphate)	7.4 ± 0.1 *AAMI Standard, Vol. 3, Dialysis RD16, 1996	
Blood flow (Q_B)	200 ± 4 ml/min	300 ml/min
Flow dialysis fluid (Q_D)	500 ± 10 ml/min	
Filtrate flow (Q_F)	0 ml/min (*EN 1283)	7, 10, and 60 ml/min

*AAMI: Association for the Advancement of Medical Instrumentation (USA)

With a planned increase in dialysis efficiency, first the state of vascular access should be reviewed in order to maximize performance. In the next step, one should – if organizationally feasible – consider extending the dialysis time or increasing blood flow. If there are opportunities to improve urea clearance (K), as a rule it is recommended to use a dialyzer with increased performance, e.g., through the use of a larger membrane surface area.

The assessment of dialysis performance is still carried out on the basis of urea clearance, whereby urea serves as a low-molecular-weight molecule to quantify dialysis dosage. This does not imply that urea is considered to be a uremic toxin. Toxins with a molecular weight of 300–500 are known as low-molecular-weight entities. For the last few years, however, so-called midsized molecules, known as middle molecules, have come back as a central point of interest for nephrologists. In the 1970s to the 1980s, middle molecules were considered to be uremic toxins with a molecular weight of up to 5000. Today in nephrology, one includes in these molecular classes molecules up to a molecular weight of even 20 000. The small protein β_2-microglobulin is a representative and model molecule for this class. It has a molecular weight of 11 818 and should not be removed by HD from patient blood by diffusion. Only high ultrafiltration, based on convective transport with high-flux membranes, allows for the efficient removal of this molecule. As a consequence, high-flux dialysis and hemodiafiltration were applied in order to remove these midsized molecules (HDF, Fig. 31.3b).

The European Uremic Toxin Working Group (EUTox) and its director, *Vanholder*, from Ghent/Belgium, recently published a list of molecules classified as uremic toxins [31.24]. Of the total 92 identified molecules and compounds, 68 belong to the class of low-molecular-weight substances, and another 23 are protein-bound. Twelve toxins are shown to have middle molecular weight, and the remaining 12 substances possess molecular weights of about 15 000. Toxic molecules with a molecular weight above albumin (66 000) should be filtered only to a limited extent during HD according to current doctrine.

UF Coefficient

The ultrafiltration coefficient (UF_{coeff}) of a dialyzer, often denoted as *flux*, determines the membrane's permeability for water. It is not only used as a measure of hydraulic permeability, but also serves as an indicator of the ability of a dialyzer to remove the medium- or high-molecular weight substances. The ultrafiltration coefficient of a membrane (UF_{coeff}) is defined as the product of hydraulic permeability (L_p) and the membrane surface (A) in the dialyzer

$$\underset{\text{Ultrafiltration coefficient}}{UF_{coeff}} = \underset{\text{Hydraulic permeability}}{L_p} \cdot \underset{\text{Surface size}}{A} \, . \quad (31.1)$$

Multiplying the ultrafiltration coefficients by the transmembrane pressure (TMP), one obtains the resulting ultrafiltration rate (Q_F, also called filtrate flow) during dialysis:

$$\underset{\text{Ultrafiltration rate}}{Q_F} = \underset{\text{Ultrafiltration coefficient}}{UF_{coeff}} \cdot \underset{\text{Transmembrane pressure}}{TMP} \, . \quad (31.2)$$

The ultrafiltration rate per surface unit of the membrane is somewhat proportional to the fourth power of the average pore radius [31.25]. The pore size of dialysis membranes, which normally is in the area of 1.5 nm for low-flux membranes and of 3 nm for high-flux membranes, has no significant effect on the membrane's ultrafiltration coefficients.

It follows from (31.1) that the ultrafiltration coefficient is a direct function of the size of the membrane surface. Therefore, one can simply compare the ultrafiltration coefficient of membranes with varying surfaces by dividing the values by the size of the *effective membrane surfaces*. This implies the surface of the membrane that lies within the two parts of the potting material. Because the size of the membrane surface is already an integral part of the definition of the required ultrafiltration coefficient of a membrane, the attending physician – with knowledge of the ultrafiltration coefficient – requires no additional information about the individual membrane's surface area. This does not apply, however, for the membrane's capability to remove dissolved substances. Figure 31.4 depicts the dependence of the ultrafiltration rate of the applied TMP for low-flux (Fig. 31.4a) and high-flux (Fig. 31.4b) membranes.

It is easy to see that for a given ultrafiltration, e.g., for 3 l/h, the necessary TMP with low-flux membranes is significantly higher than with high-flux membranes. Lower TMPs have an additional advantage as they preserve the solid components of blood such as erythrocytes, leukocytes, and platelets. With high-flux membranes, lower TMPs are applied for high ultrafiltration rates and, thus, enable an easy realization of convective therapies for the removal of midsized molecules.

Standard Values

Table 31.2 captures the conditions under which the manufacturer of dialysis membranes measures the respective ultrafiltration coefficients. The relationship between TMP and ultrafiltration rate is linear for relatively low TMP values for all membranes. However, the deterioration of membrane function during clinical use limits the achievable filtrate flow through the formation of a secondary membrane layer consisting

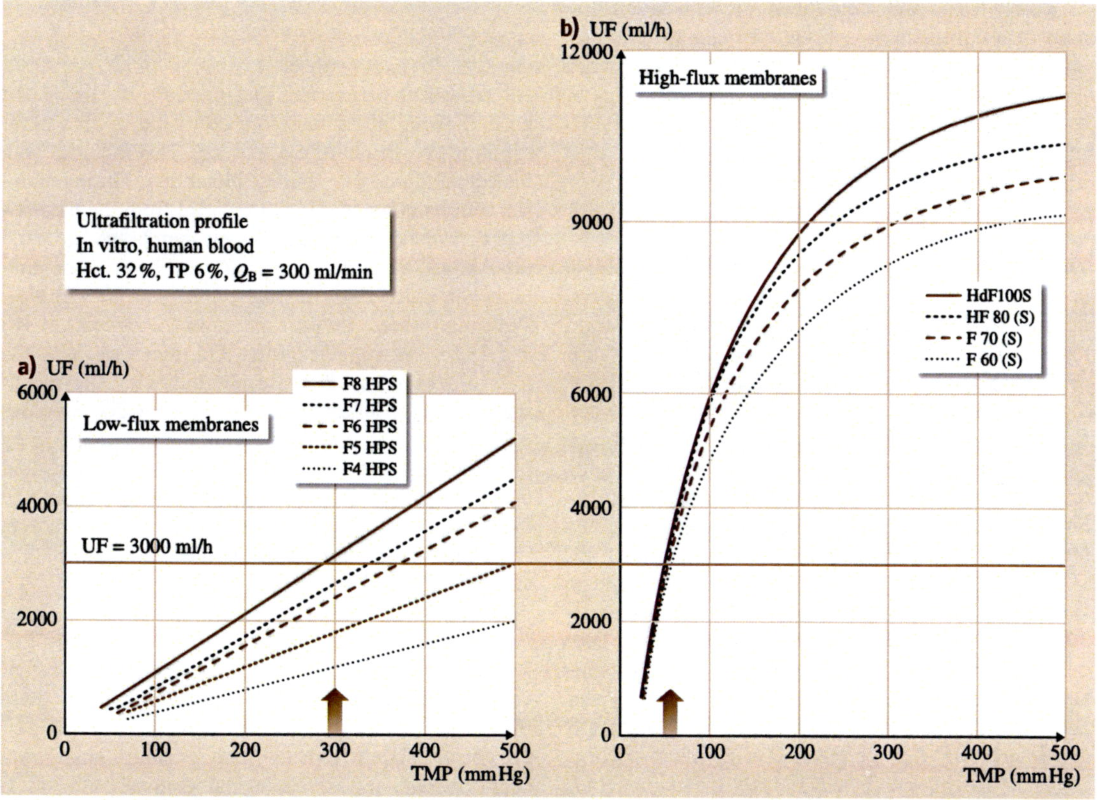

Fig. 31.4a,b Ultrafiltration profiles of low-flux (**a**) and high-flux (**b**) dialysis membranes. Ultrafiltration is a function of transmembrane pressure (TMP) whereby the dependencies in low flux are linear; for high-flux membranes, they are exponential. TMPs to achieve an ultrafiltration rate of 3 l/h are significantly less if high-flux membranes are used. F4–F8 HPS denote to low-flux dialysers from Fresenius Medical Care (**a**) whereas F60–HDF100S refer to high-flux dialysers from this manufacturer

of blood protein and blood cells. Normally, one measures the ultrafiltration rate (Q_F) with several different TMP values. The ultrafiltration coefficient (UF_{coeff}) is then defined as the increase of the linear portion of the curve resulting from Q_F to TMP. Alternatively, some manufacturers calculate the ultrafiltration coefficients according to (31.2) from the ultrafiltration rate that is measured with a fixed TMP value.

Values for ultrafiltration coefficients can differ by up to ±20% from information provided by the manufacturer. If one would like to use the ultrafiltration coefficient derived from standard conditions in order to compare different dialyzers, the blood flow rate used (Q_B) must be identical with all comparable measurements. Information is often lacking. Most manufacturers provide the values for ultrafiltration coefficients measured at blood flows of $Q_B = 200$ ml/min, analogously to the corresponding rules for the measurement of clearance.

Lately, UF coefficients measured with blood flow rates of 300 ml/min have been used and are provided in product leaflets, as nephrologists prefer higher blood flows due to the higher performance that can thus be achieved.

31.2.5 Low-Flux, High-Flux, and High-Performance Dialyzers

The classification of low-flux, high-flux, and high-performance dialyzers using their ultrafiltration coefficients varies from country to country. In Europe, low-flux dialyzers are those that have an ultrafiltration coefficient under 10 ml/h mmHg [31.26]. In the USA, low-flux dialyzers are defined with ultrafiltration coefficients of less than 8 ml/h mmHg. High-flux dialyzers vary in their respective values in a large range, beginning from a minimum of 8–10 to over 50 ml/h mmHg. Some organizations and countries, however, set higher low limits as minimums: for example, the HEMO study group (American National Institute of Health, Hemodialysis (HEMO) Study Group) sets > 14 ml/h mmHg, in Germany the regulations have been > 20 ml/h mmHg [31.27] since 2000, and some other authors give > 40 ml/h mmHg for high-flux dialyzers. Some facilities and the Japanese authorities frequently and additionally rate the removal of β_2 microglobulin. Here, the sieving coefficient for this molecule should be greater than 0.6. The general classification is made, however, on the basis of ultrafiltration coefficients.

The ultrafiltration coefficients of high-flux dialyzers may also exceed 50 ml/h mmHg with some FX dialyzers from Fresenius Medical Care, the APS-SYNTRA, or Polyflux series from the companies Asahi, Baxter, or Gambro when applied in hemofiltration (HF) or hemodiafiltration (HDF). These dialyzers are especially appropriate for convective blood purification procedures. Today one describes high-flux dialyzers with very high ultrafiltration coefficients as a rule as *hemofilters* or *hemodiafilters* as soon as they are used for convective therapy using a substitution fluid.

During the passage of blood through the dialyzer, water is continuously removed through filtration, which is associated by a continuous hemoconcentration along the length of the dialyzer. As a result, plasma oncotic pressure locally increases, so that a reduction of the in vivo ultrafiltration coefficient of the membrane follows. In addition, the hydraulic permeability, or water permeability, of a membrane (ρ) depends on the number and, above all, on the size of the pores as [31.28]

$$\underset{\text{Membrane porosity}}{\rho} = \underset{\text{Number of pores}}{N} \cdot \pi \cdot \underset{\text{Pore diameter}}{r_p^2} . \quad (31.3)$$

All treatment parameters that promote or inhibit the pore-blocking effect therefore affect the in vivo ultrafiltration coefficient of the dialyzer. This includes both the actual filtrate flow and the blood flow. During dialysis, cellular components and proteins from the patient's blood are deposited on the dialysis membrane and build up a so-called secondary membrane layer. This phenomenon is described in biotechnology as *filter fouling*. With high filtrate flows, such as in the context of HF and HDF, hemoconcentration and, therefore, also the pore blockade can increase significantly. The resulting increase in plasma oncotic pressure then opposes the (much larger) mechanically generated pressure by blood pumps, which is needed for the removal of fluid during treatment. This phenomenon can be countered by increasing blood flow velocity. As a consequence, shear forces at the membrane surface increase and protein deposits are reduced.

Ultrafiltration Coefficient and Backfiltration

The phenomenon of backfiltration emerged with the advent of high-flux dialyzers. During clinical use, these filters may face a potential chemical or microbiological contamination from the dialysis fluid. Apart from heavy metals, mostly biological contaminants originating from cell wall components of gram-negative bacteria (endotoxins) may enter patients' blood through backfiltration. Since this observation has been described in 1991 [31.29], a *higher UF coefficient* and *backfiltration* have been inseparable terms.

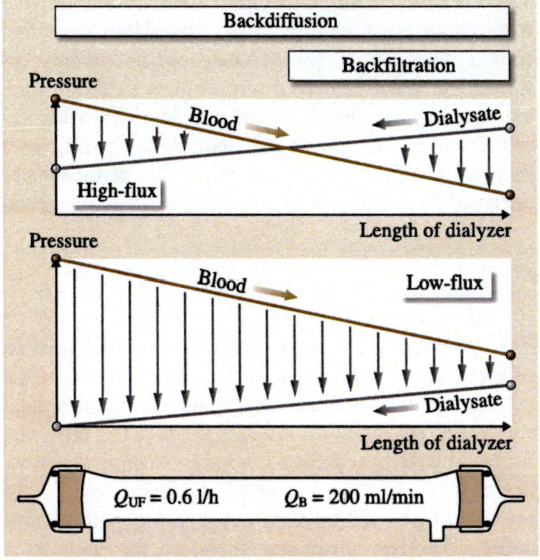

Fig. 31.5 Pressure profiles for blood and dialysis fluid as a function of the length of a dialyzer. If the pressure of the dialyzer fluid exceeds that of blood, backfiltration can take place. Backdiffusion due to available concentration gradients between the dialysate and the blood compartment have also been observed even against an ultrafiltration flux over the entire length of the dialyzer

The term backfiltration describes the transition of fluid from the dialysate to the blood side of the dialyzer. At the blood inlet of the dialyzer, the TMP in the blood compartment exceeds the pressure in the dialysate compartment. As a consequence, ultrafiltration occurs from the blood into the dialysate. With increasing distance from blood inlet, however, the pressure in the blood compartment continuously sinks along the long axis of the dialyzer to the point where the pressure in the dialysate compartment is higher than that in the blood compartment [31.29]. At this point, a reverse fluid filtration into the patient's bloodstream can result (Fig. 31.5).

Originally the volume of backfiltered fluid was calculated using the pressure relationships of blood and dialysate in the respective inlet and outlet positions. Today we know that these calculations often overestimate the volume of backfiltration, as the running pressure profiles of blood and dialysate are not similar along the long axis of the dialyzer. They don't fall linearly, as was earlier assumed. This nonlinear form of profile course is especially characteristic of high-flux membranes [31.30]. It depends on the change in blood viscosity and on small variations in fiber diameter along the long axis of the dialyzer and is also a consequence of the continuously increasing oncotic pressure due to ultrafiltration of water. The danger of contamination of blood through backfiltration with high-flux membranes is countered by the high adsorption capabilities of polymers with aromatic structures containing benzene rings such as polysulfone and, to a lesser extent, polyamide (Fig. 31.6, [31.14]).

Clearance, Dialysance, and K_0A

The clearance of a dialyzer is defined as its volumetric removal rate for a substance dissolved in blood. It corresponds to the amount of this substance that the dialyzer removes from the blood over a unit of time, divided by the respective concentration in the streaming blood. The clearance is calculated as

$$K = Q_{Bi}[(C_{Bi} - C_{Bo})/C_{Bi}] + Q_F(C_{Bo}/C_{Bi}) \quad [\text{ml/min}] \tag{31.4}$$

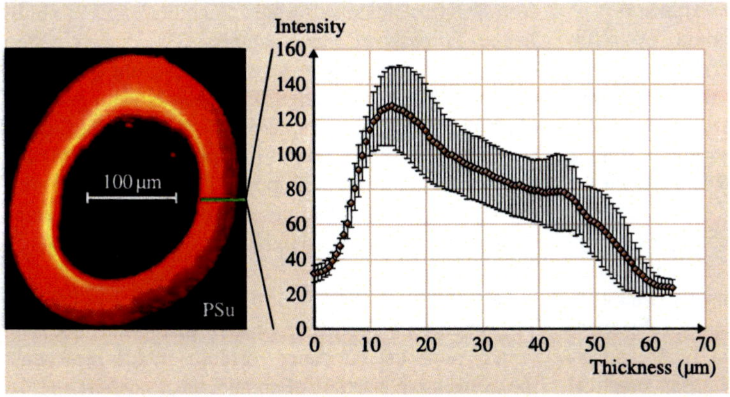

Fig. 31.6 Polysulfone membranes are able to adsorb endotoxins that are transported from the dialysate side into the membrane. This was shown with an experiment using fluorescent-marked endotoxins. These endotoxins were finally adsorbed within the membrane (after [31.14])

or

$$K = Q_{DFi}[(C_{DFi} - C_{DFo})/C_{Bi}]$$
Clearance diffuse portion
$$+ Q_F(C_{DFo}/C_{Bi}) \quad [ml/min] .$$
Convective portion (31.5)

If the adsorption of blood components at the membrane equals zero, the ultrafiltration rate (Q_F) is calculated from (31.4):

$$Q_F = Q_{Bi} - Q_{Bo} ,$$

Ultrafiltration rate = Blood flow at blood inlet − Blood flow at blood outlet (31.6)

where K is the urea clearance, Q the flow, and C the respective solute concentration. The subset indices B, DF, and F stand for blood, dialysis fluid, and filtrate. Fluids that stream into the dialyzer contain the additional index i, such as the out-streaming index o. The first two parts of the equations represent the clearance percentage caused by diffusion of a substance, while the last part represents the convective components of the mass exchange. If no ultrafiltration takes place, the last factors are removed. Under standard conditions with dialyzers, mostly the diffusive clearance values are given. This implies that the respective measurements are carried out without additional filtration.

Relevance of Diffusive Clearance Values

For low-flux dialyzers, diffusive clearance values give a reliable measure of clearance for diffusible substances. This does not apply, however, for high-flux dialyzers. For these types of filters, the pressure profiles of blood and dialysis fluid cross in the dialyzer. Even with a Q_F set at 0 ml/min, there is always a filtration of fluid from the blood into the dialysate (*internal filtration*) as well as a reverse filtration from the dialysis fluid into the blood compartment (*internal backfiltration*).

Measurements for the determination of diffusive clearance, i. e., clearance determination with $Q_F = 0$ ml/min, are always subject to the influence of internal filtration or backfiltration. Finally, internal filtration is affected by the concentration profile of soluble substances in the membrane itself as well as along the length of the capillary membrane.

Because the diffusion rate of substances is inversely proportional to the square root of the molecular weight, following the analyses of Einstein, measurements of only the diffusive clearance of substances such as vitamin B_{12} (molecular weight 1355) and inulin (molecular weight 5200) are of limited practical value. Various experiments have shown, however, that even under conditions of high dialysis flow rates and thinner membranes, these substances are mainly removed by diffusion. This also applies to drugs such as Vancomycin, which is often used in in vitro experiments in the place of vitamin B_{12}. Still, with increasing molecular size, the proportion of substances removed by convection plays an increasingly important role.

Effects of Flow Rates on Dialyzer Clearance – the K_0A Concept

The clearance of soluble substances is a direct function of blood flow, dialysate flow, and filtrate flow. For the determination of dialyzer performance in in vitro measurements, the flows in a dialyzer to be applied are given as fixed by the Standards Authorities. In clinical reality, however, these values depend greatly on the actual composition of patient blood. Different flow conditions will affect a dialyzer's clearance. Therefore, the factor K_0A has been introduced in order to get normalized values under various clinical conditions.

K_0A stands for the so-called mass transfer area coefficient of a dialyzer and serves to describe the clearance characteristics of a dialyzer under different blood and dialysate flow conditions. A standard formula applies for the conditions of $Q_F = 0$ ml [31.31]:

$$K_0A = [Q_B/(1 - Q_B/Q_D)]$$
Mass transfer area coefficient = Factor considers the blood and dialysate flow
$$\cdot \ln[(1 - K/Q_D)/(1 - K/Q_B)] .$$
Factor considers the flow – adapted in vitro urea clearance (31.7)

Alternatively, one can also read the K_0A value from a nomogram (clearance versus K_0A), which is created from in vitro clearance values. Such diagrams are published by dialyzer manufacturers. The determination of K_0A values not only takes into account blood and dialysate flows but also offers other advantages over clearance measurements. With the help of K_0A measurements, a comparison of dialyzers is easily possible that accounts for different membrane surfaces: one simply divides the K_0A values by the individual surface dimensions (A) of the dialyzers.

Membrane Sieving Coefficients

The sieving coefficient (SC) of a membrane is determined from the clearance formula, which represents the convective portion of membrane transport (31.4).

Because the removal of midsized molecules occurs primarily by convective ultrafiltration (solvent drag), one defines the sieving coefficient as the relationship between the concentration of a substance in the ultrafiltrate (C_F) to the average value from the sum of concentrations in blood inlets (C_{Bi}) and outlets (C_{Bo}) in the dialyzer, divided by two:

$$\underset{\text{Sieving coefficient}}{SC} = \underset{\text{Concentration relationship feed to filtrate}}{2C_F/C_{Bi}+C_{Bo}} . \quad (31.8)$$

If the SC of a substance is *one*, it refers to an unlimited and unhindered passage of a substance through the membrane. With an SC value of *zero*, the membrane is impermeable to this substance. The area left to the SC curve represents the area for which molecules in so-defined molecular-weight ranges can pass through. To the right of the SC curve are molecules in a molecular-weight range that cannot pass through the filter (Fig. 31.7).

The SC is dimensionless and is mostly only given if the removal of midsized molecules is considered, or if high-flux membranes for use in convective treatment modes, e.g., HF or HDF, are applied.

Unfortunately, measurements of SCs have the disadvantage that they are difficult to recognize through the elimination of molecules by membrane *adsorption*. The SC describes only the portion of the adsorbed molecules that is also found in the filtrate during therapy.

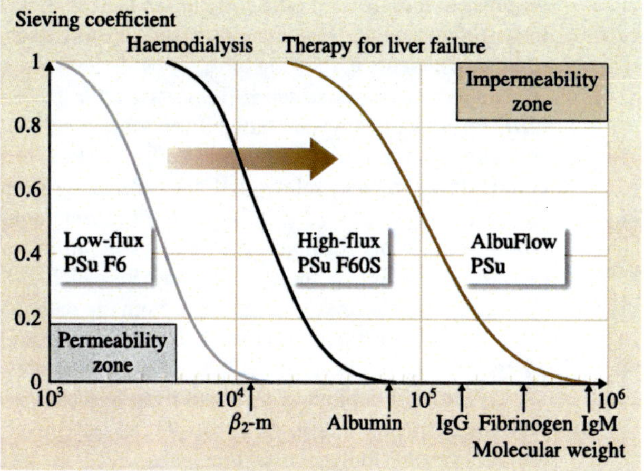

Fig. 31.7 Sieving coefficient curves for low-flux and high-flux membranes, as well as membranes for liver replacement therapy, all made of polysulfone (PSu). Molecules whose molecular weight is located to the *left of the curve* can pass through the membrane, while molecules with molecular weight to the right cannot penetrate the membrane

31.3 Dialysis Machines and Additional Equipment: Use and Conditions

In contrast to the complex physiological regulatory tasks of the natural kidneys, extracorporeal kidney replacement therapies, such as HD, HF and HDF are limited today to the correction of solute concentrations in a patient's blood induced by uremia, primarily correction of changes of metabolites, electrolytes and buffers, and the removal of excess body fluid. Despite the fact that knowledge of uremic retention substances has multiplied in recent years [31.32], the current therapy approach still applies a given exclusion range (cutoff point of the membrane). This is why unselective dialyzers or filters are still state of the art.

With regard to membranes' transport phenomena, renal replacement therapy is still almost exclusively concentrated on diffusive and convective solute transfer. This is also reflected in the design of dialysis machines and their therapeutic management. The use of adsorbents, which in 1980 was projected to be widely used for direct purification of whole blood, has not yet taken place despite different and repeated approaches, at least in the framework of chronic kidney replacement therapy. Hemoperfusion with adsorber cartridges has only proven itself as an appropriate adsorbent therapeutic procedure in serious poisoning cases.

The technical possibilities for chronic hemodialysis have expanded since its inception in 1961 to a considerable degree, but not all approaches since this time have proven successful. For now, all trends in this area attempt to pursue a maximum removal of uremic retention substances combined with the goal of a reduced patient exposure to inflammatory stimuli. In order to protect patients from the influence of inflammatory stimuli, blood- as well as biocompatible disposables (dialyzers, blood tubing) are successfully used and manufacturing procedures, e.g., use of materials with low leachables and adequate sterilization procedures, have been optimized. Further, the use of chemically and microbiologically highly pure dialysis fluids (dialysates) has become a focus of dialysis therapy. Over the past 15 years, the medical device industry has made significant progress in this area in both substance as well as cost of these products and, thus, enabled a successful therapy of kidney patients in the long term.

Modern dialyzers, which are appropriate for the OL-HDF form of therapy, combine all the aforementioned characteristics in terms of treatment effectiveness, biocompatibility, patient safety, and cost efficiency in an exemplary manner. The term *hemodiafiltration* (combining the terms *hemodialysis* and *hemofiltration*) therefore describes a form of therapy in which uremic retention substances are removed from the patient's blood through diffusive (dialysis) as well as through a large volume of convective (filtration) methods.

The parallel deployment of both these solute transport mechanisms makes it possible to effectively remove smaller-molecular-weight substances as well as the maximum removal of higher-molecular-weight substances. These most effective forms of treatment according to current standards to treat renal failure were first practiced by *Leber* et al. and described in [31.33].

The use of HF requires the substitution of large filter volumes (15–40 l, in special cases even more) in the form of sterile infusion solutions. The treatment costs resulting from this have prevented the significant spread of this procedure in the years between 1980 and 2000. A way out, as described above, is the so-called *online* production of the requested substitution solutions using cold filtration of the dialysate through the dialyzer itself. It was first described in the literature in 1978 by *Henderson* et al. [31.34]. Approval ambiguities and obstacles, however, have led to a very slow and limited deployment of this promising technique. Commercial manufacturers have offered equipment for this procedure since around 1984, but up to the late 1990s one could not establish a significant equipment base for the general use of HDF.

At least in Europe, however, new authorization rules have led to a continuously increasing inventory of such equipment. Since 2005, Fresenius has presented in its 5008 dialysis machine (Fig. 31.1c) a medical device that offers OL-HDF for the first time not just as an option, but as standard equipment in each unit. The decision to render OL-HDF as a modern standard procedure was based on the overall positive past experiences with this class of equipment and the results of the safety analyses related to these [31.35, 36].

In contrast to the otherwise preferred safe solution realized by a diversity of safety systems, e.g., by means of the combination of an operating and a protection system, the OL-HDF system uses redundant suitable filters with the appropriate return stop rate for viruses, bacteria, endotoxins, and exotoxins. Such membrane filters (e.g., DIASAFE, Fresenius Medical Care) with effective surfaces $> 2\,m^2$ allow an operating clinical life of up to 3 months. Depending upon the technical execution of the individual device types, two or three filters are used that can be operated alternatively in multiple uses as well as in disposables. In the cost-saving version, only multiply-used filters are deployed, so that their integrity must be established through a strictly prescribed test before the beginning of treatment. Modern equipment conducts such tests within the context of an automated procedure prior to the start of treatment.

The now widespread availability of such systems for OL-HDF have led to an increasing market presence and thus to increasing treatment figures derived on the basis of tentative but mounting evidence about the clinical significance of this technique [31.37]. Retrospectively gathered clinical data indicate, even with cautious interpretation, that the use of effective HDF treatment over longer time periods demonstrates a reduction of patient mortality by roughly 30% [31.37]. Subject to a methodologically comprehensive study, these relationships are reason enough to regard HDF as the method of choice for extracorporeal treatment and clinical implementation for chronic renal failure. This particularly applies because technical developments in recent years have meant that appropriate substitution solutions that meet quality standards can be delivered directly from the dialysate (OL-HDF). This makes it possible to overcome the traditional cost barriers.

A central question in extracorporeal dialysis treatment concerns the need to set the necessary dialysis dose for individual patients whose individual treatment plan must be successfully implemented in individual procedures. This is judged primarily on the basis of urea generation in treatment-free intervals and the subsequent urea removal during treatment based on the adequacy of the treatment. The possibilities and limits of this approach, described as urea modeling, is multilayered, and its discussion is beyond the scope of this chapter. Treatment planning has traditionally been carried out for the individual patient. It allows for the monitoring of the actually administered dialysis dose for individual treatment only on a sporadic basis, since this monitoring was associated with significant labor and laboratory costs.

Extensive analysis of the data in national and international databases, e.g., United States Renal Data System (USRDS), EDTA/ERA, etc., confirm both the relationship between dialysis dose and mortality as well as the fact that this ideal was not achieved in a significant number of individual treatments. It is explained by the notion that the proposed dialysis dose was not achieved for a variety of reasons. The now widespread introduction of technical systems to measure dialysis dose in real time through the dialysis machine itself allows for an almost complete review of treatment success, at least in this respect.

The technique used for this purpose mainly measures the conductivity differences in the dialysate before and after use of the dialyzer [31.35]. This allows for effective in vivo determination of urea clearance of the dialyzer and then performs the calculation of the relevant Kt/V dose parameter. The procedure is extremely cost-effective and can be done automatically by the device. The now widespread use of this technology in the devices made by many manufacturers can be considered as one of the main contributions to the quality of dialysis treatment in recent years.

Dialysis patients suffer complications in 20–30% of treatments; the complications include headaches, nausea, vomiting, and loss of consciousness due to a drop in blood pressure. The causes of these events are certainly multifactorial and may vary from patient to patient. However, the process of dialysis itself presents a massive intervention in the various homeostases of the organism, which determines the aforementioned symptoms. Since the mid-1980s, medical research and the industry have endeavored to adjust the intelligent rules systems of dialysis treatment to the individual physiological reactions of patients, so that the incidence of these complications can be reduced to a minimum. In connection with the hemodialysis procedure itself, technical solutions concentrate on approaches that impact the intradialytic cardiovascular stability of patients [31.38].

The relative change in blood-volume-controlled ultrafiltration and the regulation of intradialytic thermal energy of the body are clinically accepted monitoring parameters today. The use of these techniques allows for the reduction of the estimated incidence of dialysis-induced blood pressure drops in patient populations who are predisposed to this condition by about 30–50% [31.39, 40]. At present, studies have focused on easily measurable, but difficult-to-interpret, conductivity differences pre- and postdialysis to determine and, finally, to regulate the electrolyte balance of patients under dialysis. In contrast to the established procedure of automatic ultrafiltration regulation and the control of the thermal energy balance, there are still no reliable clinical data yet available. Without doubt, however, an assessment of this will be possible in the near future, so that the clinician will be given these three effective ways to make dialysis treatment largely free of complications.

31.3.1 Determination of Fluid Status with Whole-Body Bioimpedance Spectroscopy

Limited renal function strongly influences the water balance of patients with renal failure. Fluid is withdrawn as part of the HD procedure through the ultrafiltration of fluid in order to normalize hydration. Many patients, however, suffer from chronic overhydration, as all current clinical methods are very insensitive to the objective determination of hydration status [31.41].

Pathological changes of the heart, e.g., left ventricular hypertrophy, are triggered by this overabundance of water and increase the mortality of affected patients. It is possible to prevent these dangerous changes in the heart through better hydration management or even to reverse them [31.42]. A clinically deployable diagnostic method to determine hydration status is therefore a prerequisite for this. Extensive analysis of different measurement methods have shown that in particular *whole-body bioimpedance spectroscopy* is an appropriate method [31.41, 43]. The results of these studies were the starting point of the body composition monitor (BCM, Fresenius Medical Care) for objective determination of hydration status.

31.3.2 Description of the Technology and Methodology

Using whole-body bioimpedance spectroscopy, the impedance response of patients is determined over a frequency range of a few kilohertz to the megahertz range (Fig. 31.8). The frequency range of beta dispersion is thereby sampled by a frequency sweep over 50 frequencies. In the lower kilohertz range, the measurement of current flows is only through the extracellular space; at high frequencies, however, both the extra- as well as the intracellular spaces are measured. By modeling as per *Cole* [31.44], the resistance of the extracellular space (R_{ECW}) and the resistance of total body water (R_{TBW}) are calculated.

In order to present a clinically applicable method, these resistances will need to be converted into easy-to-interpret and clinically relevant variables. This is achieved in two steps by the combination of the fluid model [31.45] with a body composition model [31.46] and makes it possible to differentiate overhydration from muscle mass. The clinically usable output vari-

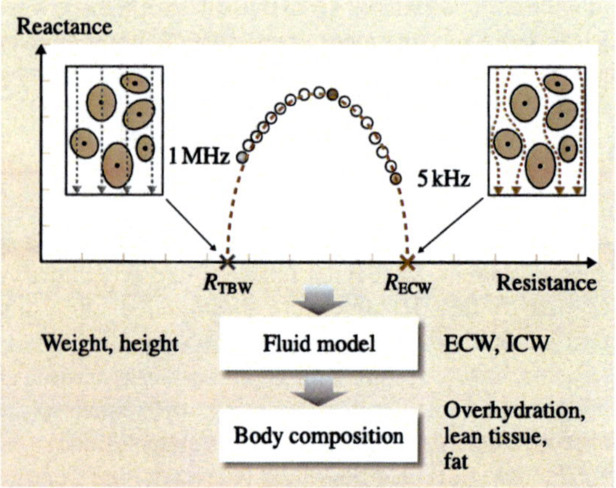

Fig. 31.8 Calculation steps from measured resistances and reactances over the fluid volume for body composition. Extracellular resistance, or R_{ECW}, and the total resistance R_{TBW} are determined on the basis of Cole–Cole modeling using 50 measured resistances and reactances at 50 frequencies. This is possible as at lower frequencies (e.g., 5 kHz) the current flows only around the cells, while at very high frequencies (about 1000 kHz) they can cross the cell membrane. The two resistances thus calculated (R_{ECW}, R_{TBW}) are converted in the fluid model to extracellular (ECW) and intracellular water (ICW). In the next step, the determination of body composition is made (overhydration, lean tissue mass, fat mass) using the body composition model

ables are overhydration, *lean tissue mass*, and adipose tissue mass.

Using BCM, the standard hydration of a patient can be shown as a hydration target. This compares to the hydration status of a uremic patient with a patient with healthy kidney function and the same body composition. The attending physician decides whether to bring the patient to target with a positive or negative offset to standard hydration. It is possible to achieve the intended target after only weeks or months, because only a very cautious withdrawal of fluid can be tolerated by the patient. However, it is crucial that this long goal be kept

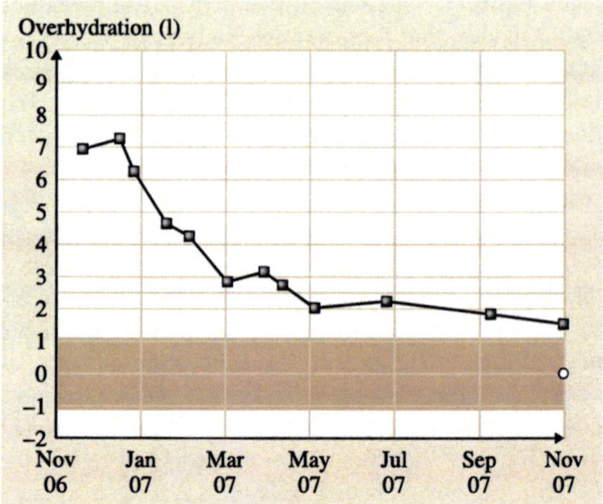

Fig. 31.9 Slow reduction of fluid status in a 73-year-old, heavily overhydrated patient over a period of 12 months

in mind even with a changing body composition of the patient (e.g., increased fat mass).

31.3.3 Validation of Clinical Data

The validation of the measured and calculated values is very important, as the BCM allows, for the first time, a quantitative calculation of the hydration status. Validation is carried out in several stages in comparison to then-available gold-standard reference methods.

This includes, for example, the indicator dilution method, total body potassium measurement, and dual x-ray absorptiometry. A complete overview of the validation studies conducted so far can be found in [31.46]. An analysis of 500 European hemodialysis patients further demonstrated that at least 25% of these patients were clearly overhydrated [31.47]. An overhydration of 2.5 l is already associated with a significantly increased mortality [31.48]. It is possible to reduce this overhydration through the design of long-term hydration management programs. Figure 31.9 shows the hydration process in a hemodialysis patient for whom a 7 l overhydration was reduced to 1.5 l within a year.

31.4 Blood Purification in Liver Replacement Therapy

In an issue of the American magazine *Newsweek*, hepatitis C virus (HCV) was termed an *insidious killer virus*, which has already infected several hundred million people globally [31.49]. HCV is responsible for 20% of acute and 70% of chronic liver inflammation. The associated hospital costs are immense. They amount in the USA alone to more than $ 5.5 billion [31.50]. Liver transplantation, which is the only possible therapy for many patients, is, however, not an alternative in the short term due to the insufficient supply of available donor organs, and the waiting list for liver transplants is steadily increasing.

Detoxification procedures on the basis of extracorporeal blood circulation have therefore become the basis for liver support systems. Initial work on membrane processes started in 1992 [31.51]. A basis for recent developments is the dual function of the liver, considered both as a detoxification organ and as an organ for synthesizing clotting factors, proteins, and hormones. Extracorporeal circulation for liver replacement therapy should ideally support or replace both functions.

31.4.1 Detoxification

In contrast to most uremic toxins, which are removed by dialysis because they are water soluble by nature, liver toxins are mostly protein (albumin)-bound and accordingly have hydrophobic (lipophilic) properties. Because of the relatively low-molecular-weight cutoff for typical hydrophilic dialysis membranes (membrane cutoffs are below the molecular weight of albumin), even modern dialysis membranes are not suitable for detoxification in liver replacement therapy. New therapeutic procedures such as molecular adsorbent recirculating system (MARS) (Gambro, Fig. 31.10) and the Prometheus System (Fresenius Medical Care, Fig. 31.11) use membranes that are able to remove albumin-bound liver toxins from the bloodstream and discard them in a secondary circuit with the help of adsorbers [31.52, 53]. Following the postulated mechanisms of the respective

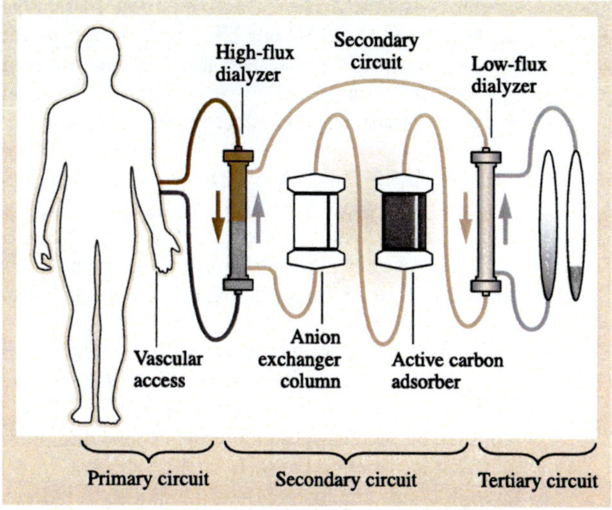

Fig. 31.10 Schematic principle of MARS. The blood of a patient with liver disease is transmitted over the *MARS flow filter*, whereby, according to the inventor, the albumin-bound toxins are led through the membrane into the secondary circuit. They are then bound to the *free* albumin molecules available in the secondary circuit. Two adsorbers in the secondary circulation, an active charcoal and an anion exchanger, clean this albumin solution online, so that free albumin molecules can continue to be used for further absorption of toxins

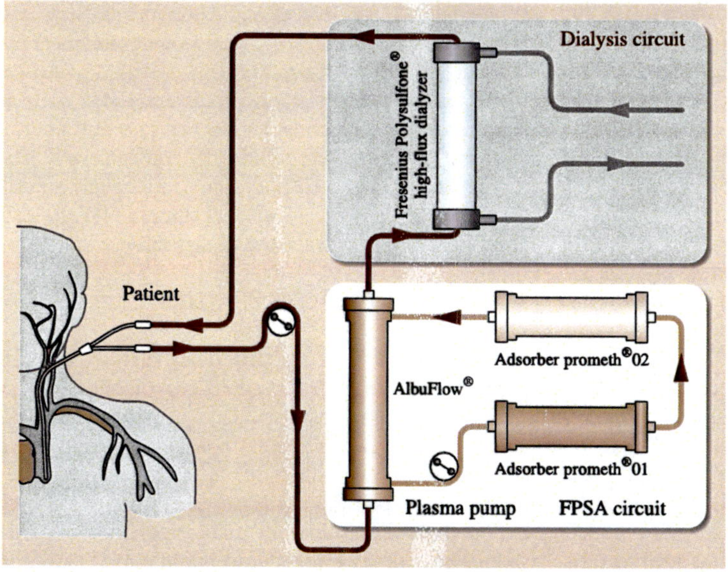

Fig. 31.11 In extracorporeal circulation of the Prometheus system, also known in the literature by the abbreviation FPSA – fractionated plasma separation and adsorption (Fresenius Medical Care, Bad Homburg, Germany) – albumin-bound liver toxins are depleted by a highly permeable plasma filter in the primary circuit. Albumin is then purified in two adsorber columns of the secondary circuits (prometh 01 and 02). It is then returned to the patient's bloodstream. Subsequent serial dialysis continues to provide the removal of small-molecule liver toxins, such as ammonia (NH_3)

inventors of MARS (see below), high-flux membranes are used here [31.52, 54] that allow the passage of albumin, while in the Prometheus system protein-permeable membranes are used [31.53, 55]. Clinical trials have already demonstrated the effectiveness of these procedures [31.52–57].

The MARS system (Fig. 31.10) has already been available since the 1990s [31.52, 54, 56, 57]. More than 10 000 patients with liver disease have been treated by the MARS system, and therefore, the largest database in the area of liver replacement therapy is available in this realm.

MARS (originally produced by the Teraklin company, Rostock, now offered by GAMBRO AB, Hechingen) is an extracorporeal blood purification system with a primary, a secondary, and a tertiary blood circuit. As in HD, the patient's blood is circulated for 4 to 6 h through a high-flux filter (MARSflux Filter). A buffered 5–10% commercial albumin solution (150 ml/min) is circulated in a closed secondary circuit, which cleans with two adsorber columns made of active carbon and an anion exchanger. Furthermore, the secondary circuit is supplemented by a tertiary circuit that contains an additional low-flux dialysis filter. This filter permits the diffusive removal of small molecular toxins.

With MARS it is possible [31.52, 56, 58] to separate albumin-bound toxins from the albumin in the blood and to transfer them through a not yet completely understood mechanism into the secondary circuit. Here, it is picked up by the then-circulating free albumin and is removed through the adsorber column. Thus, the secondary circuit always presents free albumin to take up toxins.

The Prometheus system [31.53, 55, 56] for extracorporeal liver replacement therapy (Fig. 31.11) contains, in contrast to MARS, an albumin-permeable filter whose SC curve has been shifted to higher molecular weights in comparison to dialysis, as shown in Fig. 31.7.

By shifting the SC curve to higher molecular weights, one obtains membranes with a cutoff of $> 200\,000$. Using this, the patient's own plasma and toxins bound to albumin can be filtered. Blood cells and large proteins, e.g., IgM, are retained, as their SC is *0*.

Using the two adsorbers *prometh 01 and 02*, which are set in series in the secondary circuit, it is possible to clean the patient's own and toxin-laden albumin. Blood can then be led back into the patient. The adsorber particles consist of neutral resins (prometh 01) or are anion exchangers (prometh 02) with a large internal surface area (about 120 000 m^2 per cartridge with a particle diameter of about 400 μm). Lipophilic as well as electrically charged toxins, such as bilirubin, can thereby be removed.

Anticoagulation is carried out in liver replacement therapy similarly to HD, with the administration of heparin. Recent studies in the field of HD have shown that anticoagulation with sodium citrate may better address the specific pathological conditions for liver patients [31.59–61]. For regional coagulation, which applies exclusively to extracorporeal blood circulation, two pumps replace the necessary heparin pump. The first pump advances a citrate solution (4% trisodium citrate) from the albumin filter into circulation. Citrate forms a chelate complex with ionized calcium and magnesium. Blood clotting can be reduced by minimizing Ca^{2+}/Mg^{2+} volumes, at best to a concentration under 0.2 mmol/l, because these ions are cofactors in the coagulation cascade. In contrast to work in the older literature, which showed no dependence of activated clotting time (ACT) on Ca^{2+} levels in blood [31.62], observations by ***Falkenhagen*** at the Donau University in Krems, Austria [31.63] have shown that a lengthening of clotting time has been observed if one is able to reduce the Ca^{2+} level to below 0.2 mmol/l. The initial coagulation status after blood has passed through the extracorporeal circuit is achieved again if the patient is infused with a diluted solution of calcium ions through the second pump. It is important to know that the sodium citrate needed to reduce the calcium depends upon the hematocrit of the treated patient [31.58]. A significant and positive side effect of anticoagulation with sodium citrate solutions and the related reduction of calcium and magnesium concentrations is the lower complement activation [31.58].

In order to prevent hypercalcemia as well as hypocalcemia, it is necessary to measure and control calcium concentrations within defined intervals. The Prometheus system's CiCa device continuously adjusts the necessary physiological concentration of the calcium solution through the use of algorithms.

31.4.2 Synthetic Liver Function

Biological cells have characteristics that are so well designed that no engineer could copy them easily. It is feasible to treat liver patients by the supplementation of necessary proteins, hormones, and clotting factors, but the costs, also due to the related purity requirements, would be immense. Therefore, the use of liver cells in bioreactors for extracorporeal circulation should be seen as a possible solution. Liver cells can be sensitive and be synthetically continuously active if the conditions of the cell culture in the bioreactor are at an optimum. In the past this has not been possible as the survival of isolated hepatocytes in vitro has been quite short (less than 1 week). The development of three-dimensional bioreactors in which liver cells can be cultivated in their entirety, i. e., all the cells present in the liver are grown, and are not just exclusively hepatocytes, allows for treatment periods of 1 to 2 months.

Thus, a bioreactor for extracorporeal treatment of patients with liver disease is a realistic therapeutic medium. Clinical applications of bioreactors with liver cells have so far not been satisfactory [31.64], but they offer promising prospects. However, a broad application of such bioreactors has not yet met the necessary European criteria for approval for a bioreactor for human therapies, as such a device would be categorized as a *medicinal product* (pharmaceutical drug) due to its *principal mode of action*. The newly established regulations for advanced therapy medicinal products (ATMP) by the European Commission exclude the possibility of considering such a bioreactor a *medical device* [31.65]. As a consequence of this regulation, expensive clinical trials with large patient numbers would be required for approval. This creates difficulties for small and medium-sized enterprises (SMEs) for cost reasons.

Conclusion for Liver Replacement Therapy

It is possible that the two major liver functions, detoxification and synthesis, can be at least partially replaced through extracorporeal blood circuits, so that therapeutic success would be possible. Clinical trials with adsorber systems show promising results that have already led to approval for a detoxification system in liver replacement therapy.

References

31.1 J. Himmelfarb, T.A. Ikizler: Hemodialysis, N. Engl. J. Med. **363**, 1833–1845 (2010)

31.2 A. Kramer, V. Stel, C. Zoccali, J. Heaf, D. Ansell, C. Grönhagen-Riska, T. Leivestad, K. Simpson, R. Palsson, M. Postorino, K. Jager: An update on renal replacement therapy in Europe: ERA-EDTA Registry data from 1997–2006, Nephrol. Dial. Transplant. **24**, 3557–3566 (2009)

31.3 C. Goddemeier: Die Entdeckung des Blutkreislaufs, Deutsch. Ärztebl. **104**, 1168–1170 (2007)

31.4 T. Graham: The Bakerian lecture: on osmotic forces, Philos. Trans. R. Soc. London **144**, 177–228 (1854)

31.5 A. Fick: Über Diffusion, Ann. Phys. Chem. **94**, 59–86 (1855)

31.6 W. Schumacher: Über Membrandiffusion, Ann. Phys. Chem. **110**, 337–370 (1860)

31.7 A. Eggerth: The preparation and standardisation of collodion membranes, J. Biol. Chem. **18**, 203–221 (1921)

31.8 J.J. Abel, L. Rowntree, B. Turner: On the removal of diffusible substances from the circulating blood of living animals by means of dialysis, J. Pharmacol. Exp. Ther. **5**, 275–283 (1913)

31.9 International Standard Organisation, ISO 10993-16: Biological evaluation of medical devices, Part 16. Toxicokinetic study design for degradation products and leachables (1997)

31.10 G. Haas: Dialysieren des strömenden Bluts am Lebenden, Klin. Wochenschr. **2**, 1888 (1923)

31.11 G. Haas: Über Blutwaschung, Klin. Wochenschr. **7**, 1356–1362 (1928)

31.12 W. Kolff, H. Berk, M. ter Welle, J. van der Leij, E. van Dijk, J. van Nordwijk: The artificial kidney: a dialyser with a great area, Acta. Med. Scand. **117**, 121–134 (1944)

31.13 G. Lonnemann: The quality of dialysate: an integrated approach, Kidney Int. **76**, S112–S119 (2000)

31.14 M. Henrie, C. Ford, M. Andersen, E. Stroup, J. Diaz-Buxo, B. Madsen, D. Britt, H. Chih-Hu: In vitro assessment of dialysis membrane as an endotoxin transfer barrier: geometry, morphology, permeability, Artif. Organs. **32**, 701–710 (2008)

31.15 M. Kessler, B. Canaud, L. Pedrini, J. Tattersall, P. ter Wee, R. Vanholder, C. Wanner: European best practice guidelines for haemodialysis (Part 1). Biocompatibility, Nephrol. Dial. Transplant. **17(7)**, 32–44 (2002)

31.16 EU Commission: Report from the commission to the European parliament and the council: Report on the issue of the reprocessing of medical devices in the European Union, in accordance with Article 12a of Directive 93/42/EEC from August 27 (2010)

31.17 J. Vienken, C. Ronco: New developments in hemodialysers, Contr. Nephrol. **133**, 105–118 (2001)

31.18 W. Ansorge, M. Pelger, W. Dietrich, U. Baurmeister: Ethylene oxide in dialyser rinsing fluid: effect of rinsing technique, dialyser storage time and potting compound, Artif. Organs. **11**, 118–122 (1987)

31.19 H. Lemke: Mediation of hypersensitivity reactions during hemodialysis by IgE-antibodies against ethylene oxide, Artif. Organs. **11**, 104–110 (1987)

31.20 F. Gotch, H. Peter, F. Panilio, T. Folden, M. Keen: Online measurement of delivered kt/V during dialysis, J. Am. Soc. Nephrol. **6**, 600 (1995)

31.21 National Kidney Dialysis Outcome Quality Initiative (DOQI): Clinical Practice guidelines: hemodialysis adequacy and peritoneal dialysis adequacy, J. Am. Kidney Dis. **30**(2), 1–64 (1997)

31.22 U. Kuhlmann, R. Goldau, N. Samadi, T. Graf, M. Gross, G. Orlandini, H. Lange: Accuracy and safety of online clearance monitoring based on conductivity variation, Nephrol. Dial. Transplant. **16**, 1053–1058 (2001)

31.23 F. Gotch, J. Sargent, M. Keen: Whither goest Kt/V?, Kidney Int. **58**, 3–18 (2000)

31.24 R. Vanholder, Eutox Gruppe: Review on uremic toxins: classification, concentration and interindividual variability, Kidney Int. **63**, 1934–1943 (2003)

31.25 M. Lysaght: Haemodialysis membranes in transition, Contrib. Nephrol. **61**, 1–17 (1988)

31.26 C. Ronco, P. Ghezzi, N. Hoenich, P. Delfino: Computerized selection of membranes and hemodialysers, Contrib. Nephrol. **133**, 119–130 (2001)

31.27 Dialysestandard der Arbeitsgemeinschaft klinische Nephrologie e.V. und der Dialysegesellschaft niedergelassener Ärzte e.V (2000)

31.28 W. Clark, R. Hamburger, M. Lysaght: Effect of membrane composition and structure on solute removal and biocompatibility, Kidney Int. **56**, 2005–2015 (1999)

31.29 J. Leypoldt, B. Schmidt, H.J. Gurland: Measurement of backfiltration rates during hemodialysis with highly permeable membranes, Blood Purif. **9**, 74–84 (1991)

31.30 J. Leypoldt, B. Schmidt, H.J. Gurland: Net ultrafiltration may not eliminate backfiltration during hemodialysis with highly permeable membranes, Artif. Organs. **15**, 164–170 (1991)

31.31 J. Sargent, F. Gotch: Principles and biophysics of dialysis. In: *Replacement of Renal Function by Dialysis*, 4th edn., ed. by C. Jacobs, C. Kjellstrand, K. Koch, J. Winchester (Kluwer Academic, Dordrecht 1996) pp. 188–230

31.32 R. Vanholder, N. Meert, E. Schepers, G. Glorieux: Uremic toxins: do we know enough to explain uremia?, Blood Purif. **26**, 77–81 (2008)

31.33 H.W. Leber, V. Wizemann, G. Goubeaud, P. Rawer, G. Schütterle: Simultaneous hemofiltration/hemodialysis: an effective alternative to hemofiltration and conventional hemodialysis in the treatment of uremic patients, Clin. Nephrol. **9**, 115–121 (1978)

31.34 L.W. Henderson, M.L. Sanfelippo, E. Beans: "On line" preparation of sterile pyrogen-free electrolyte solution, Trans. Am. Soc. Artif. Intern. Organs. **24**, 465–467 (1978)

31.35 H. Polaschegg: Automatic, noninvasive intradialytic clearance measurement, J. Int. Artif. Organs. **16**, 185–191 (1993)

31.36 H.D. Polaschegg, T. Roy: Technical aspects of online hemodiafiltration, Contrib. Nephrol. **158**, 68–79 (2007)

31.37 B. Canaud, J. Bragg-Gresham, M. Marshall, S. Desmeules, B. Gillespie, T. Depner, P. Klassen, F. Port: Mortality risk for patients receiving hemodiafiltration versus hemodialysis: European results from the DOPPS, Kidney Int. **69**, 2087–2093 (2006)

31.38 M. Krämer: Physiological monitoring and control in hemodialysis: state of the art and outlook, Expert. Rev. Med. Devices. **3**, 617–634 (2006)

31.39 Q. Maggiore, F. Pizzarelli, A. Santoro, G. Panzetta, G. Bonforte, T. Hannedouche, M.A. Alvarez de Lara, I. Tsouras, A. Loureiro, P. Ponce, S. Sulkovà, G. van Roost, H. Brink, J.T. Kwan: Study group of thermal balance and vascular stability. The effects of control of thermal balance on vascular stability in hemodialysis patients: results of the European randomized clinical trial, J. Am. Kidney Dis. **40**, 280–290 (2002)

31.40 A. Santoro, E. Mancini, C. Basile, L. Amoroso, S. DiGiulio, M. Usberti, G. Colasanti, G. Verzetti, A. Rocco, E. Imbasciati, G. Panzetta, R. Bolzani, F. Grandi, M. Polacchini: Blood volume controlled hemodialysis in hypotension-prone patients: a randomized, multicenter controlled trial, Kidney Int. **62**, 1034–1045 (2002)

31.41 M. Kraemer, C. Rode, V. Wizemann: Detection limit of methods to assess fluid status changes in dialysis patients, Kidney Int. **69**, 1609–1620 (2006)

31.42 R.M. Fagugli, P. Pasini, F. Pasticci, G. Ciao, B. Cicconi, U. Buonchristiani: Effects of short daily hemodialysis and extended standard hemodialysis on blood pressure and cardiac hypertrophy: a comparative study, J. Nephrol. **19**, 77–83 (2006)

31.43 P. Chamney, P. Wabel, U. Moissl, M. Müller, A. Bosy-Westphal, O. Korth, N. Fuller: A whole-body model to distinguish excess fluid from the hydration of major body tissues, J. Am. Clin. Nutr. **85**, 80–89 (2007)

31.44 K.S. Cole, R.H. Cole: Dispersion and adsorption in dielectrics, J. Chem. Phys. **9**, 341–351 (1941)

31.45 U.M. Moissl, P. Wabel, P. Chamney, I. Bosaeues, N. Levin, A. Bosy-Westphal, O. Korth, M. Müller, L. Ellegard, V. Malmros, C. Kaitwatcharachai, M. Kuhlmann, F. Zhu, N. Fuller: Body fluid volume determination via body composition spectroscopy in health and disease, Physiol. Meas. **27**, 921–933 (2006)

31.46 P. Wabel, P. Chamney, U. Moissl, T. Jirka: Importance of whole-body bioimpedance spectroscopy for the management of fluid balance, Blood Purif. **27**, 75–80 (2009)

31.47 P. Wabel, U. Moissl, P. Chamney, T. Jirka, P. Machek, P. Ponce, P. Taborski, C. Tetta, N. Velascon, J. Vlasak, W. Zaluska, V. Wizemann: Towards

improved cardio-vascular management: the necessity of combining blood pressure and fluid overload, Nephrol. Dial. Transplant. **23**, 2965–2971 (2008)
31.48 V. Wizemann, P. Wabel, P. Chamney, W. Zaluska, U. Moissl, C. Rode, T. Malecka-Masalska, D. Marcelli: The mortality risk of overhydration in haemodialysis patients, Nephrol. Dial. Transplant. **24**, 1574–1579 (2009)
31.49 Newsweek Magazine: Hepatitis C, The insidious spread of a killer virus, from May 20, pp. 75–79 (2002)
31.50 J.P. Leigh, C.L. Bowlus, B.N. Leistikow, M. Schenker: Cost of hepatitis C, Arch. Intern. Med. **161**, 2231–2233 (2001)
31.51 P. Konstantin, J. Chang, V. Otto, G. Brunner: Artificial Liver, Artif. Organs. **16**, 235–242 (1992)
31.52 S. Mitzner, J. Stange, S. Klammt, P. Peszynski, R. Schmidt, G. Nölde-Schomburg: Extracorporeal detoxification using the molecular adsorbent recirculating system for critically ill patients with liver failure, J. Am. Soc. Nephrol. **12**, 75–82 (2001)
31.53 J. Vienken, H. Christmann: How can liver toxins be removed? Filtration and adsorption with the Prometheus System, Ther. Aph. Dial. **10**, 125–131 (2006)
31.54 J. Stange, S. Mitzner: A carrier-mediated transport of toxins in a hybrid membrane: safety barrier between a patient's blood and a bioartificial liver, J. Int. Artif. Organs. **19**, 677–691 (1996)
31.55 K. Rifai, T. Ernst, U. Kretschmer, C. Hafer, H. Haller, M.P. Manns, D. Fliser: The Prometheus device for extracorporeal support of combined liver and renal failure, Blood Purif. **23**, 298–302 (2005)
31.56 K. Rifai, C. Hafer, J. Rosenau, C. Athmann, H. Haller, M.P. Manns, D. Fliser: Treatment of severe refractory pruritus with fractionated plasma separation and adsorption. (Prometheus), J. Scand. Gastroenterol. **41**, 1212–1217 (2006)
31.57 J. Stange, S. Mitzner, T. Risler, C. Erley, W. Lauchart, H. Goehl, S. Klammt, P. Peszynski, J. Freytag, H. Hickstein, M. Löhr, S. Liebe, W. Scharek, U. Hopt, R. Schmidt: Molecular adsorbent recycling system (MARS): clinical results of a new membrane-based blood purification system for bioartificial liver support, Artif. Organs. **23**, 319–330 (1999)
31.58 U. Heemann, U. Treichel, J. Loock, T. Philipp, G. Gerken, M. Malago, S. Klammt, M. Loehr, S. Liebe, S. Mitzner, R. Schmidt, J. Stange: Albumin dialysis in cirrhosis with superimposed acute liver injury: a prospective controlled study, Hepatology **36**, 949–958 (2002)
31.59 J. Hartmann, K. Strobl, D. Falkenhagen: Anticoagulation in combined membrane/adsorption systems Contributions, Sec. Biol. Med. Sci. MASA **2**, 39–49 (2008)
31.60 S. Herget-Rosenthal, U. Treichel, F. Saner: Citrate anticoagulated modified fractionated plasma separation and adsorption: first clinical efficacy and safety data in liver failure, J. Am. Soc. Nephrol. **14**, 729A (2003)
31.61 J. Kozik-Jaromin, V. Nier, U. Heemann, B. Kreymann, J. Boehler: Citrate pharmacokinetics and calcium levels during highflux dialysis with regional citrate anticoagulation, Nephrol. Dial. Transplant. **24**, 2244–2251 (2009)
31.62 M. Janssen, J. Deegens, T. Kapinga, J. Beukhof, P. Huijgens, A. van Loenen, J. van der Meulen: Citrate compared to low molecular weight heparin anticoagulation in chronic hemodialysis patients, Kidney Int. **49**, 806–813 (1996)
31.63 D. Falkenhagen, Donau University Krems, Austria (personal communication)
31.64 L. Kjaergard, J. Liu, B. Als-Nielsen, C. Gluud: Artificial and bioartificial support systems for acute and acute-on-chronic liver failure, JAMA **289**, 217–222 (2003)
31.65 European Commission: Regulation (EU) No 1394/2007 of the European Parliament and of the Council of November 13, 2007 on Advanced Therapy Medicinal Products. J. Official, European Union, L324/121 from 10.12.2007 (2007)

32. Heart–Lung Machines

Ulrich Böckler, Andreas Hahn

During cardiac surgery and interventions on the large vessels, it is often necessary to disconnect the heart and lungs from the natural circulatory system and to stop them from working temporarily. When open heart surgery or coronary interventions are carried out, the heart and lung function is usually taken over by an external heart–lung machine (HLM). This procedure is also referred to as extracorporeal circulation (ECC) (Sect. 32.2) or cardiopulmonary bypass. The function and technical details of HLMs are decribed in this chapter.

32.1 **Historical Development of Extracorporeal Circulation** 621

32.2 **Extracorporeal Circulation** 623

32.3 **Structure and Function of the Heart–Lung Machine** 623
- 32.3.1 Blood Pumps and Their Function.. 624
- 32.3.2 Oxygenator and Gas Exchange Function......... 624
- 32.3.3 Tubing Systems for Extracorporeal Circulation....... 625
- 32.3.4 Blood Filters 625
- 32.3.5 Cardiotomy Reservoir.................. 625
- 32.3.6 Cannulae 626

32.4 **Components of the Heart–Lung Machine** 627
- 32.4.1 Basic Components..................... 627
- 32.4.2 Additional HLM Components........ 629

32.5 **Extracorporeal Circulation** 629
- 32.5.1 Preparing the ECC....................... 630
- 32.5.2 Carrying Out the Perfusion........... 630
- 32.5.3 Patient Monitoring and Therapy During ECC.............. 631
- 32.5.4 Complications During Extracorporeal Circulation 633

32.6 **Differentiation of Heart–Lung Machines**. 635
- 32.6.1 Modular Systems........................ 635
- 32.6.2 Nonmodular Systems.................. 636

32.7 **Aspects of Technical Safety**.................... 636

32.8 **Prospects for Further Development** 637

References 638

32.1 Historical Development of Extracorporeal Circulation

The French physiologist Julien Jean César *Le Gallois* wrote down the first realistic principles of ECC in his monograph [32.1] in the early nineteenth century. He suggested replacing the function of the heart with a continuous injection of natural or, if possible, artificially produced blood.

This concept was, however, only implemented in 1828 by *Kay* [32.2] and, several years later, by *Brown–Séquard* [32.3]. In 1849, the German scientist *Loebell* [32.4] also experimented with isolated organ perfusion. He examined the secretions from explanted kidneys that had been artificially perfused. Ernst Bidder repeated these experiments using a pressurized container that served as a volume reservoir for artificial perfusion. In 1867, *Schmidt* [32.5] from the Institute of Physiology in Leipzig refined this perfusion technique. He managed to influence perfusion flow and pressure during artificial perfusion using a container that could be elevated and lowered. However, the enrichment of blood with oxygen was yet to be achieved.

Artificial oxygenation of the blood during the perfusion of organs was not achieved until 1882 when *von Schröder* [32.6] invented a device for dispensing air directly into the bloodstream for oxygenation. However, this only partially met the basic requirement for artificial oxygen enrichment since extreme foaming made it unsafe to use this method at this point in time.

In 1885, *von Frey* and *Gruber* [32.7] developed the first forerunner of the modern HLM at the Institute of Physiology in Leipzig. It contained a film oxygenator with a rotating glass cylinder that oxygenated blood artificially. Von Frey and Gruber were the first to use a syringe pump with a corresponding valve control for perfusion in a closed system.

In 1890, *Jacobj* [32.9] from the Institute of pharmacology in Strasbourg (France) presented his invention, the hematisator, a blood pump that was capable of generating a pulsatile flow. Controlled by two valves, a rubber balloon was compressed rhythmically by a seesaw lever driven by a water motor. Jacobj was aware of the problems associated with the oxygenation processes of his time and focussed on a new technique. He used explanted isolated lungs, explaining that these were much easier to manage. This involved perfusion of two organs, and for this reason he would later name his device the double haematisator.

It was half a century later that *Gibbon* [32.10] finally achieved the revolutionary breakthrough. After 20 years of development, the surgeon from Jefferson Medical College in Philadelphia produced an HLM. He had constructed a pump oxygenator in 1934, which he used to take over the heart and lung function of cats that had had their pulmonary artery clamped for 25 min. Experiments on rabbit and monkey lungs carried out in the 1950s in the USA were less successful. In 1950, Gibbon received financial and technical support from the computer company IBM.

The rapid technical advancement and improved knowledge in the field of physiology made it possible to finally use the device on a human patient. Gibbon's device, based on a roller pump and his screen oxygenator, was used for the first time in 1952. His first patient, however, died as a result of a misdiagnosis.

On 6 May 1953, Gibbon used the HLM successfully on a human for the first time. The 18-year-old female patient was suffering from a congenital atrial septal defect. The machine was used for 45 min in a partial bypass, including 26 min in a total bypass, completely taking over the heart and lung function. That year, Gibbon performed four further operations that were, however, unsuccessful.

Despite Gibbon's major breakthrough in the history of cardiac surgery, scientists concentrated on the results published by in 1949 *Bigelow* [32.11], who stipulated that the body core temperature had to be lowered as a prerequisite for corrective surgery with circulatory arrest. Hypothermia slowed down the metabolism, which allowed the venae cavae to be clamped for a short time without causing damage to the sensitive organs as a result of oxygen deprivation.

In 1953, *Lewis* [32.12] from Minneapolis successfully managed to close an atrial septal defect using surface hypothermia. Several years later, *Sealy* [32.13] combined profound hypothermia and partial cardiopulmonary bypass to perform corrective surgery.

In early 1954, *Lillehei* [32.14] made a name for himself with his concept of cross-circulation (Fig. 32.1). Artificial blood oxygenation and the use of an HLM were not required for this procedure. Cross-circulation involved the continuous reciprocal exchange of blood between two individuals. The donor, who had an identical blood type to the patient, served as the oxygenator for the patient's blood. The arterial donor blood was pumped into the recipient's circulation using a Sigmamotor pump.

On 26 March 1954, Lillehei carried out the first surgical procedure using cross-circulation. As a result, more complex congenital heart defects, such as ventricular septum defects and Fallot's tetralogy, could be treated surgically with success for the first time.

However, as the disadvantages of cross-circulation, such as reduced pump volume, outweighed the advantages when compared to the HLM, the procedure was

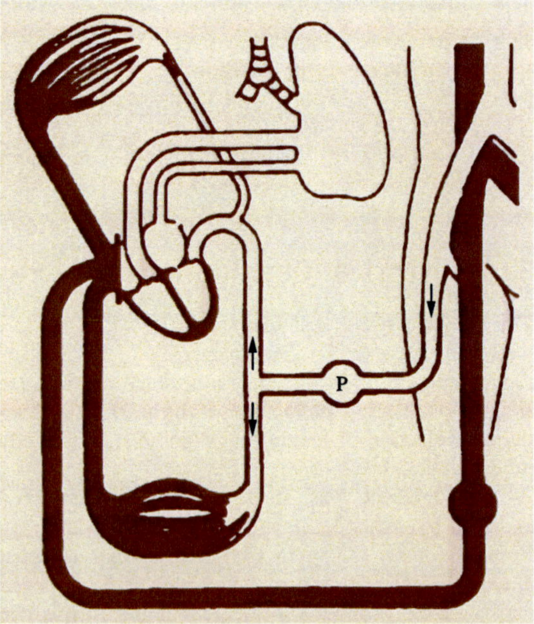

Fig. 32.1 Principle of cross-circulation according to Lillehei using a donor to oxygenate the patient's blood (after *Galetti* and *Brecher* [32.8])

not widely used. Cross-circulation also put both patient and donor at great risk and had a high mortality rate.

In 1955, *Kirklin* [32.15] carried out a further series of surgical procedures on patients with congenital heart defects at the Mayo Clinic in Rochester, MN, USA. He used the Mayo–Gibbon HLM, which had been improved in the intervening years. The biggest problems were the amount of fresh blood preserves (5–11 units) needed to prime the HLM and the risk of lethal embolisms caused by foaming blood, which meant that a controlled low blood flow through the oxygenator was a prerequisite. At first, Sigmamotor pumps were available as blood pumps. Later, Gibbon adopted the roller pumps first described by *Beck* in 1924 [32.16], which were utilized for transfusions. These roller pumps were later named after their advocate *De Bakey* [32.17].

The development of different types of oxygenators such as bubble, film, disk, spiral coil, and membrane oxygenators in the mid-1950s led to further advances in cardiac surgery. The first commercially manufactured membrane oxygenator went on sale in 1967.

Alongside *Zuhdi*'s [32.18] realization in 1961 that an HLM could be primed without blood, the use of an integrated heat exchanger to induce hypothermia by directly cooling the blood was an important development.

In 1962, the authors *Galetti* and *Brecher* [32.8] described the *ideal perfusion* as an intervention in which the heart and lung function is taken over completely by a machine with minimal detrimental effect on the body. Although advances in device technology and material science, new surgical techniques, and improved perfusion techniques have led to significant strides toward achieving this ideal, perfection has yet to be achieved.

32.2 Extracorporeal Circulation

ECC or perfusion is defined as a procedure carried out with the aid of an HLM.

In the early stages, cardiac surgery was mainly concerned with the treatment of congenital heart defects. Today, the main focus is on acquired heart defects. If heart disease cannot be treated with drug therapy or cardiological intervention, cardiac surgery becomes necessary.

The term minimally invasive cardiac surgery relates to an alternative method in which the access path to the operation area is kept as small as possible to reduce the patient's surgical trauma. This primary surgical technique usually has a direct influence on the configuration of the HLM and on the way the perfusion is carried out. OPCAB (off-pump coronary artery bypass) as well as off-pump have also become established terms and refer to cardiac surgeries carried out without the support of an HLM.

ECC has evolved from its beginnings in the area of experimental organ perfusion to a routine procedure for open heart surgery. It will always be necessary when surgical intervention is carried out on an arrested heart or when assisted perfusion is required to support circulation on a beating heart.

The HLM can never be looked at in isolation. Additional components, mainly in the form of sterile disposables such as an oxygenator, heat exchanger, or reservoir, take on other important functions during ECC and are also described below. This is essential for understanding the extremely complex overall system of the HLM and its functions.

32.3 Structure and Function of the Heart–Lung Machine

The HLM is the basis of ECC and assumes two important organ functions once the cardiopulmonary bypass has been established:

- Pump function of the heart
- Gas exchange function of the lungs

A sufficient perfusion volume that corresponds to the normal cardiac output of the patient under anesthetic and an adequate perfusion pressure (50–90 mmHg) are of particular importance during any ECC. The HLM must also ensure sufficient oxygenation, elimination of CO_2, and control of the blood temperature.

The following components make up the basic equipment of the ECC that is used during modern cardiac surgery:

- Blood pumps
- Oxygenator
- Tubing system with various tubing diameters

- Blood filters with various functions
- Cardiotomy reservoir
- Cannulae and intracardiac suction tubes

Because the HLM and the disposables must be regarded as a closed and functional unit, they must be set up in a certain way. ECC is always based on the same principle.

Venous blood is transferred from the venous vascular system (usually both venae cavae or the right atrium) to the cardiotomy reservoir using cannulae. The so-called arterial blood pump (roller pump or centrifugal pump) then actively pumps the blood from the cardiotomy reservoir through the so-called oxygenator and through another cannula back into the arterial vascular system (usually the aorta ascendens).

An additional arterial filter is located between the oxygenator and the arterial line leading back to the patient to filter-aggregated leukocytes and platelets, cell and tissue debris, denatured protein, air, and particle abrasion from the tubing system. Additional pumps are needed for cardiotomy and vent suction or as cardioplegia pumps.

During the intervention, cardiotomy suction tubes can be used to aspirate blood from the operation area in order to keep it free of blood.

Vent suction tubes aspirate directly from the heart chambers to prevent overextension of the arrested heart with excessive blood volume, which could cause extensive damage. The aspirated blood is collected, filtered, and defoamed in the cardiotomy reservoir and is usually fed back directly into the extracorporeal system.

Cardioplegia pumps are used to administer cardioplegic solutions into the aortic root or selectively into the two coronary ostia with volume, time, and pressure control.

32.3.1 Blood Pumps and Their Function

Blood pumps can generally be categorized into two groups, depending on their function:

1. Roller pumps
2. Centrifugal pumps

Essentially, both pump types should meet certain criteria when used with the HLM:

- Dosed delivery of liquids, with precise display of the actual flow rate
- Ideal ratio of size to delivered volume
- External control mechanism (e.g., for monitoring functions)
- Sufficient pressure or vacuum generation
- Minimal blood damage
- Adjustable occlusion settings (roller pumps only)
- Pulsatile flow generation
- High efficiency
- High reliability and safety
- Option for emergency operation (e.g., manual operation)

Roller Pumps

De Bakey blood pumps are based on the displacement principle and deliver blood through a tubing segment from the pump housing using rotating rollers. The roller pump consists of a rotating pump arm with two attached cylindrical rollers and a pump housing into which a semicircular silicone tubing segment is inserted and then secured using special tubing inserts.

The rotating rollers alternately compress the tubing segment and deliver the liquid contained in the tubing in accordance with the rotational speed and direction. The exact adjustment (occlusion) of the two rollers determines the delivery. When a central thumbwheel is turned, the occlusion rollers move outward symmetrically and occlude the inserted tube evenly. This reduces erythrocyte damage caused by shear stress or direct crushing.

Centrifugal Pumps

Centrifugal pumps do not use direct displacers and are therefore nonocclusive pumps. Such pumps use centrifugal forces to transport the blood instead of tubing compression. Because of the technical principle of operation, the centrifugal pump has a limited application. Although it can be used as an arterial pump, it has no application as a suction pump. Centrifugal pumps have the advantage of delivering limited amounts of air and causing less blood damage in the long run. The fact that an additional flow meter is required to determine the delivered flow is a disadvantage.

32.3.2 Oxygenator and Gas Exchange Function

The artificial lung, also called the oxygenator, takes on the lung function during ECC and is therefore responsible for the exchange of vital gases. So-called membrane oxygenators are now used as standard. They contain a semipermeable membrane in the shape of a microporous hollow fiber. This liquid-impermeable membrane separates the gas side from the bloodstream.

Due to the partial pressure gradients, O_2 and CO_2 diffuse through the microporous membrane. A correspondingly adjusted supply of air and O_2 on the gas side allows for selective and exact control of paO_2 and $paCO_2$ on the blood side.

Today, most membranes are made of polypropylene or polyethylene and permit operating times of 6–8 h.

32.3.3 Tubing Systems for Extracorporeal Circulation

The tubing system is used to connect the individual components of the extracorporeal system and forms a closed circuit with the vascular system of the patient. Depending on the location, either PVC or silicone tubing is used, which is available in different diameters and wall thicknesses. Tubing sizes range from a diameter of 1/8″ and a wall thickness of 1/16″ to 3/16″ and 1/4″ for pediatric applications (with correspondingly low flow rates). Tubing for adult applications is usually 3/8″ or 1/2″ with a wall thickness of 3/32″, facilitating flow rates of more than 10 l/min. In order to achieve the desired flow when small tubing diameters are used, active pressure or suction (within certain limits) is used in the respective line.

Tubing with the smallest possible lumen and shortest possible length is used to keep the priming volume and system foreign surface as small as possible. In contrast to the pioneering days of ECC, modern medicine uses completely preassembled disposable systems manufactured under clean room conditions (Fig. 32.2).

Although the number of HLMs, components, and applications on the market is relatively small, there is practically no standardization and a wide range of customer-specific preassembled tubing systems is available.

32.3.4 Blood Filters

Blood filters are integrated into the ECC system mainly to avoid micro-embolisms caused by autologous influences, foreign particles, and microbubbles.

Depth Filters

Depth filters, made of Dacron wool or polyurethane foam, are inserted into cardiotomy reservoirs and are used mainly for particle filtration. The pore size can vary from 80–100 μm (coarse separation of particles) to 20–40 μm (micropore range).

Mesh Filters

Mesh filters are used as arterial blood filters and work on the principle of sieves. The pore size of mesh filters ranges from 20 to 40 μm. The material is a mesh of woven polyester strands. Unlike depth filters, mesh filters have little adhesive force. Their air retention properties are excellent as air bubbles can only pass the filter medium when a certain pressure difference (bubble point pressure) is reached.

32.3.5 Cardiotomy Reservoir

The cardiotomy reservoir collects and filters the blood aspirated from the operation area and feeds it back into the ECC as required. At the same time, it provides volume storage. The transparent housing facilitates continuous level control. A detailed scale makes it easy to quantify the level and detect corresponding changes in volume.

A minimum residual volume always remains in the reservoir during the entire ECC to prevent air delivery into the extracorporeal system. The blood flow or the blood–air mixture passes through a defoaming module

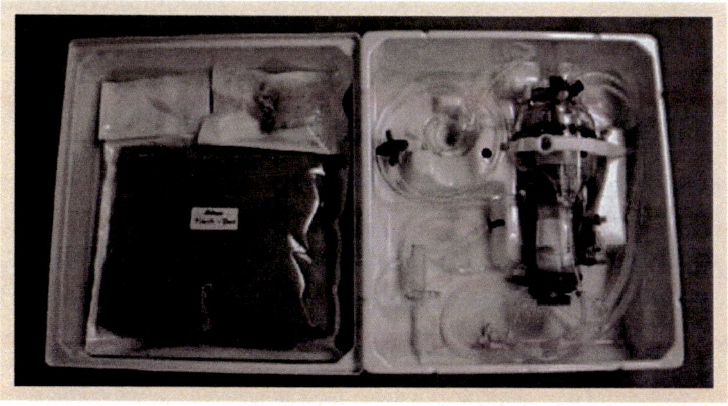

Fig. 32.2 Completely preassembled tubing system with infant oxygenator (Sorin Group S.p.A., Milan)

integrated into the reservoir and a combination of depth and mesh filters for optimal filtration.

The incoming and outgoing tubing can be plugged or screwed into the different types of connectors on the cardiotomy reservoir.

Hard-shell reservoirs that are exposed to air are used alongside pure venous reservoirs in the shape of a collapsible bag. The advantage of these soft-shell reservoirs is that the blood volume does not come in direct contact with air. In this case, the suction tubes are connected to a separate reservoir. These systems are referred to as *closed* or *semi-open*.

32.3.6 Cannulae

Cannulae are the interface between the ECC system and the vascular system of the patient. They are inserted into the relevant vessels by a surgeon, where they are secured and deaerated before their sterile connection to the tubing system of the HLM.

Arterial Cannulation

Arterial cannulae are used to return the oxygenated blood to the systemic circulation of the patient. The aorta ascendens is the most frequently used site. Depending on the type of intervention and surgical access path, an alternative site such as the femoral artery or iliac artery may be preferable or necessary.

The type and size of the selected cannula depends on both the required blood flow and the anatomic conditions.

Arterial cannulae are available in different sizes, with straight or bent tips, suture or flange collars, and optionally with wire reinforcement or length markings.

Venous Cannulation

Venous cannulae drain the patient's blood from the venous vascular system to the extracorporeal system.

The most commonly used cannula is the so-called two-stage cannula. The tip of the cannula with its side openings is placed in the vena cava inferior, and the blood from the vena cava superior and the right atrium is drained through the side openings in the second stage.

The type and size of the selected cannula is determined by conditions that are similar to those on the arterial side. However, venous cannulae are actually larger because the blood flow in the venous branch is usually caused by hydrostatic suction alone.

Two cannulae may be used on the venous side, if required, depending on the surgical access path or on the fact that the right atrium of the heart needs to be bloodless. These drain the blood from the two large venae cavae or from the vena femoralis and vena jugularis separately.

Vent Catheters

A so-called vent suction tube is also inserted to protect the temporarily arrested heart and especially the left ventricle from overextension due to blood flowing back from the Thebesian veins and bronchial circulation.

Using this vent suction tube, excessive blood is drained usually via the auricle of the left atrium or a side port of the cardioplegia cannula into the cardiotomy reservoir of the HLM.

Cardioplegia Cannulation, Myocardial Protection

Cardioplegia is defined as a reversible, artificially induced arrest of the heart. This is required for most cardiac interventions to create an operation area free of movement for the surgeon or, if the heart muscle or vessels close to the heart need to be opened, to allow access to the heart valves or to the internal structures of the heart.

The induction of a cardioplegic heart arrest usually requires a process called myocardial protection. As the name implies, this involves protection of the arrested heart muscle. The most common method of plegia and myocardial protection is the administration of cardioplegic solutions. The administration of hyperkalemic solutions interrupts the conduction pathway and hence the mechanical activity of the heart.

Blood, electrolytes, colloids, and crystalloids are some of the base solutions available. Myocardial arrest, the administration of cold cardioplegic solution, and to a certain extent additional external cooling lead to drastically reduced oxygen consumption and thus represent the most important mechanisms for protecting the myocardial cells.

Immediately before the heart arrest is induced, the aorta ascendens is clamped between the aortic valve and the aortic cannula. The cardioplegic solution is then administered directly or indirectly into the coronaries. Depending on the nature of the surgery planned and other factors, there are two methods of administration. Antegrade administration delivers the solution via the aortic root or directly into the coronary ostia on the arterial side. Retrograde administration delivers the solution via the coronary sinus on the venous side.

A variety of different cannula types is available for both methods. If the aortic valve is sufficient, the cardioplegic solution can be administered via the aortic root using an aortic root cannula. After the cardioplegic solution is administered, the cannula serves as an aortic root vent for left ventricular relief and deaeration. If the aortic valve is insufficient, the cardioplegic solution cannot be infused via the aortic root. With the aorta clamped, the incompletely closed aortic valve would cause a retrograde flow of the solution into the left ventricle. In this case, myocardial protection has to be carried out by direct selective catheterization of the coronary ostia using coronary perfusion cannulae.

In the case of retrograde administration, the cannula is inserted via the right atrium. The cardioplegic solution is then administered via the coronary sinus. This is an effective way of reaching the poststenotic areas of the heart muscle in particular as stenoses are not prevalent in the venous coronary system. This leads to optimum distribution of the cardioplegic solution and considerably improves myocardial protection.

The infusion of cardioplegic solutions against the normal coronary vascular blood flow has proved particularly successful during coronary reoperations when inadequate native coronary flow is a factor, in the case of insufficient coronary artery development, as well as for patients who are at risk of embolization caused by arteriosclerotic material.

Intracardiac Suction Tubes

Intracardiac suction tubes are used to keep the operation area free of blood by aspirating blood from this area into the cardiotomy reservoir.

Cardiotomy suction is an important factor as it contributes greatly to blood damage. Shear stress at the tubing tips is not the only, or even most important, cause of blood damage; it is caused mainly by the constant mixing of blood and air, which leads to massive blood damage through hemolysis.

32.4 Components of the Heart–Lung Machine

Section 32.4.1 lists the most important basic components of modern HLMs and provides a brief description.

32.4.1 Basic Components

Basic components are online gas and blood gas analysis in the form of measuring devices for the measurement and display of gas flows, gas concentrations in the gas flow, and saturation or partial pressures directly in the blood flow:

- Mobile console for mounting multiple pumps; includes the power supply, emergency power supply, and electronics.
- Adjustable mast system for mounting the holders for the oxygenator, filters, cardiotomy reservoirs, external pumps, and additional devices.
- Blood pumps
  - Large roller pump: pump with a longer tubing path in the pump housing that produces higher flow rates, required for the arterial blood flow, suction, or vent pump (Fig. 32.3).
  - Small roller pump: pump with a shorter tubing path in the pump housing for lower flow rates, e.g., for the perfusion of infants and children or for the administration of cardioplegic solutions, usually a double pump (Fig. 32.4, Table 32.1).
  - Centrifugal pumps: pumps for the arterial blood flow.

Fig. 32.3 Fifth-generation roller pump (Sorin Group, Munich)

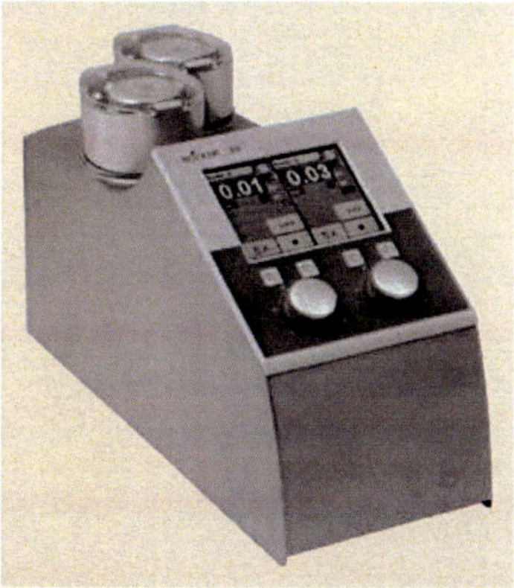

Fig. 32.4 Fifth-generation double pump (Sorin Group, Munich)

Table 32.1 Technical specifications of roller pumps (Sorin Group, Munich; model Stöckert S5)

Display	S5 roller pump	S5 double pump
RPM display range	0–250 RPM	0–250 RPM
RPM control range	3–250 RPM	3–250 RPM
RPM display accuracy	±0.5% for full scale value 250 RPM	±0.5% for full scale value 250 RPM
LPM display range	0–3.29 LPM for 1/4″ tubing	0–1.6 LPM for 1/4″ tubing
	0–6.98 LPM for 3/8″ tubing	0–2.28 LPM for 5/16″ tubing
	0–11.3 LPM for 1/2″ tubing	0–16.2 LPM for 5/8″ tubing

- Control and monitoring devices
  - Pressure monitor, including sensors, for measuring different pressures in the extracorporeal system and for controlling the flow rates accordingly.
  - Temperature monitor, including sensors, for measuring and displaying different system temperatures and, if required, patient temperatures.
  - Level monitor including sensors, for measuring and controlling the volume level in the cardiotomy reservoir and for regulating the arterial pump flow correspondingly.
  - Bubble monitor in the form of ultrasonic sensors that regulate the flow rate of the affected pump when air is detected in the system.
  - Pulsatile flow control for creating and controlling a pulsatile flow profile.

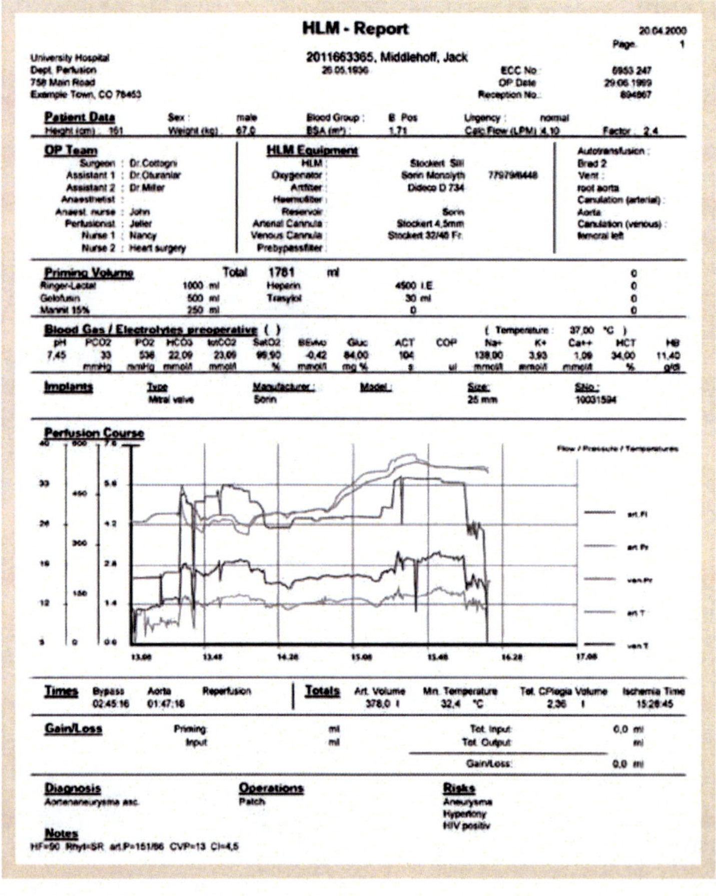
HLM - Report
20.04.2000
Page 1
University Hospital
Dept. Perfusion
758 Main Road
Example Town, CO 78453
2011663365, Middlehoff, Jack
26.05.1936
ECC No: 6953 247
OP Date: 29.06.1999
Reception No.: 694867
Patient Data Sex: male Blood Group: B Pos Urgency: normal
Height (cm): 161 Weight (kg): 67.0 BSA (m²): 1.71 Calc Flow (LPM) 4.10 Factor: 2.4
OP Team
Surgeon: Dr.Cottogni
Assistant 1: Dr.Oturanlar
Assistant 2: Dr.Miller
Anaesthetist:
Anaest. nurse: John
Perfusionist: Jeller
Nurse 1: Nancy
Nurse 2: Heart surgery
HLM Equipment
HLM: Stockert SIII
Oxygenator: Sorin Monolyth 7797948448
Artfilter: Dideco D 734
Haemofilter:
Reservoir: Sorin
Arterial Cannula: Stockert 4,5mm
Venous Cannula: Stockert 32/40 Fr.
Prebypassfilter:
Autotransfusion:
Brad 2
Vent:
root aorta
Canulation (arterial):
Aorta
Canulation (venous):
femoral left
Priming Volume Total 1761 ml
Ringer-Lactat 1000 ml Heparin 4500 I.E. 0
Gelofusin 500 ml Trasylol 30 ml 0
Mannit 15% 250 ml 0 0
Blood Gas / Electrolytes preoperative ( ) (Temperature: 37,00 °C)

pH	PCO2	PO2	HCO3	totCO2	SatO2	BEvivo	Gluc	ACT	COP	Na+	K+	Ca++	HCT	HB
7,45	33	536	22,09	23,09	99,90	-0,42	84,00	104		138,00	3,93	1,09	34,00	11,40
	mmHg	mmHg	mmol/l	mmol/l	%	mmol/l	mg %	s	ul	mmol/l	mmol/l	mmol/l	%	g/dl

Implants Type: Mitral valve Manufacturer: Sorin Model: Size: 25 mm SNo.: 10031594
Perfusion Course

Times Bypass 02:45:16 Aorta 01:47:16 Reperfusion
Totals Art. Volume 378,0 l Min. Temperature 32,4 °C Tot. CPlegia Volume 2,36 l Ischemia Time 15:28:45
Gain/Loss Priming ml Input ml Tot. Input 0,0 ml Tot. Output ml Gain/Loss: 0,0 ml
Diagnosis: Aortenaneurysma asc.
Operations: Patch
Risks: Aneurysma, Hypertony, HIV positiv
Notes: HF=90 Rhyt=SR art.P=151/66 CVP=13 CI=4,5

Fig. 32.5 Extract from a perfusion log with graphics and various patient and intraoperative data

- Cardioplegia control, including pressure and bubble sensor, for controlling cardioplegia delivery.
- Timer for measuring important times and intervals during perfusion such as total perfusion time, aortic clamping time, and reperfusion time.

- Display and control panel, providing access to all display and control elements described above as well as to additional system information and alarm management.
- Electronic or mechanical gas blender, for controlling and displaying the ventilation gases (air, O_2, and CO_2) that are delivered to the gas side of the oxygenator; vacuum controller, for the precise control and display of a permanent vacuum on the venous reservoir.
- Anesthetic gas vaporizer, allowing and displaying the administration of a precisely dosed amount of anesthetic gas to the oxygenator.
- Electronic documentation system that displays and stores all relevant data during perfusion. All data are stored centrally and displayed instantly as a perfusion report (Fig. 32.5), which allows for data evaluation at a later stage for statistical and scientific purposes.

Figure 32.6 shows a modern HLM.

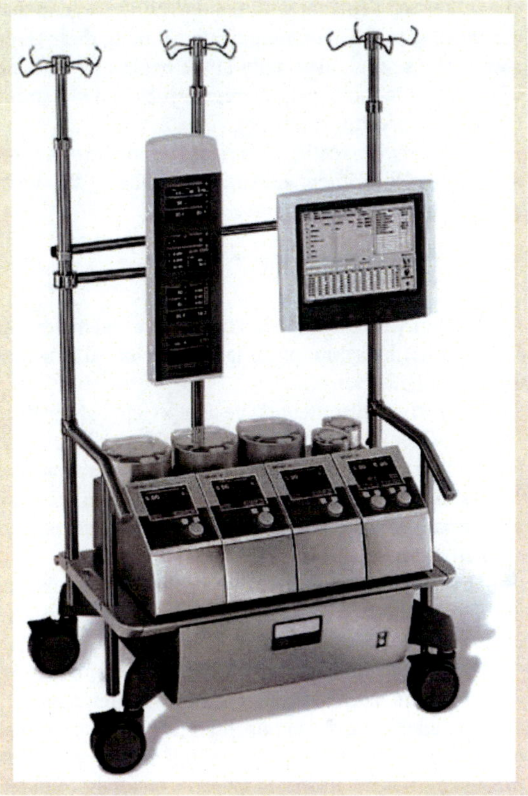

Fig. 32.6 Industrial heart–lung machine with panel PC and data management system (Sorin Group, Munich)

32.4.2 Additional HLM Components

During ECC, the heat exchanger of the oxygenator indirectly regulates the patient's body temperature by heating or cooling the blood flow. Patients can also be placed on a water mattress that increases or decreases body temperature.

Heater–Cooler Device

For this purpose, the heater–cooler device delivers water with a temperature ranging from 2 to 42 °C. Two-chamber pumps ensure that the delivered water does not cause a buildup of overpressure in the circuits.

Modern heater–cooler devices feature several circuits with different temperatures. Two circuits supply the oxygenator and the heating/cooling mattress, while a third is connected to the heat exchanger in the cardioplegia system. The heat exchanger is supplied from two independent tanks with different temperatures, which allows instant switching from cold to warm or vice versa.

Ideally, a remote control is integrated into the control panel of the HLM, which facilitates operation of the device and allows it to be placed outside the operating theater, leading to considerably lower noise levels in the OT.

Heater–cooler devices have always presented a hygiene challenge because the system parts that transport water inevitably become contaminated by germs. Regular water changes and the use of appropriate chemical cleaning agents are the two most effective cleaning and maintenance strategies.

32.5 Extracorporeal Circulation

Perfusion in ECC is an extremely complex process and plays a central role during cardiac surgery. The operation and monitoring of the HLM is the exclusive responsibility of the perfusionist, who discusses every action and decision with the operating surgeon and the anesthetist. Clear communication among these three parties is essential, especially at the beginning and end of ECC and in the event of irregularities and situations that require coordinated action.

The following section describes the course of a standard perfusion without making reference to the wide range of modified procedures.

32.5.1 Preparing the ECC

ECC begins with the selection of the required materials, followed by the setup, priming, and technical function test of the HLM.

The perfusionist mounts the components of the sterile disposable set in the relevant holders, inserts the tubing into the pump heads, and places temperature probes and bubble, flow, and pressure sensors in the appropriate positions. All connections are checked and secured as necessary.

The dry, sterile system is then primed with a mixture of different solutions and drugs, completely vented, and checked for leaks.

Using a checklist, the relevant settings, connections, functions, and alarms are checked and documented.

The arterial and venous lines are connected using the appropriate cannulae, which the surgeon has previously placed in the large vessels of the patient, connecting the blood circuit of the patient to the extracorporeal system.

The coagulation system is deactivated by the systemic administration of approximately 300–350 IU/kg of heparin before the cannulae are placed and the perfusion is started.

32.5.2 Carrying Out the Perfusion

The initiation of ECC is a major intervention that affects the entire organism and all vital functions of the patient. The following steps take place simultaneously at the start of perfusion.

- The perfusionist opens the venous line in consultation with the surgeon and the anesthetist and, by virtue of the hydrostatic differential, drains the blood from the venous vascular system of the patient into the venous reservoir of the HLM.
- The arterial pump is started and the blood flows from the HLM into the patient's arterial vascular system. The flow is increased until inflow and outflow are balanced. A previously calculated perfusion flow serves as a theoretical point of reference and corresponds to an artificial heart time volume.
- Depending on the blood flow, the oxygenator is supplied with a mixture of air, O_2, and anesthetic gas or CO_2 as required. This mixture is determined and controlled by mechanical or electronic gas blenders.
- The heater–cooler device supplies the heat exchanger in the oxygenator with warm or cold water and regulates the temperature.

The perfusionist continuously monitors and regulates the parameters and keeps them within the desired range.

- Perfusion flow.
- The HLM generates the blood flow that is required to perfuse the organs and the microcirculation sufficiently. This blood flow is first calculated by multiplying the body surface area with a so-called flow index (similar to the natural heart index). This theoretical index ranges from 2.4 and 3.0 l/m^2 body surface/min and depends on several factors, e.g., temperature, hematocrit value, body fat percentage, and depth of anesthesia. The perfusion flow must then be adjusted on the basis of parameters such as oxygen saturation, partial O_2 pressure, and other situational factors.
- Temperatures:
  - Blood temperature in the arterial and venous lines of the extracorporeal tubing system
  - Set and actualize temperatures of the heater–cooler device
  - Temperature of the cardioplegic solution
- Pressures in the extracorporeal tubing system.
- During perfusion, pressure measurements before and after the oxygenator and, if required, before the arterial cannula in the cardioplegia line or before the hemofilter result in an automatic flow adjustment of the assigned arterial pump (visual and acoustic alarm).
- Pressure monitors have two modes of operation:
  - Start–stop mode, i. e., the assigned pump is stopped when the set pressure value for the pump stop is reached and is restarted automatically only after the pressure has decreased;
  - Control mode, i. e., the pump slows down when the set pressure value for the pump control is reached and is only stopped when the stop pressure is reached.
- Perfusion times:
  - Total perfusion time
  - Aortic clamping time

- Reperfusion time
- Circulatory arrest times, head perfusion times, or hemofiltration times (where applicable)
- Level monitoring in the cardiotomy reservoir.
- Level monitoring protects against complete depletion of the reservoir, as this would cause delivery of air into the arterial branch of the extracorporeal system and lead to a severe air embolism in the patient's aorta. The level sensor is positioned at the desired minimum level on the cardiotomy reservoir. If the liquid level in the reservoir reaches the level sensor, the level monitor automatically decreases the flow of the assigned arterial pump and triggers a visual and acoustic alarm.
- Level monitors have two modes of operation:
  1. Start–stop mode, i. e., the pump is stopped when the stop level is reached and is restarted automatically only after a release level has been reached.
  2. Control mode, i. e., the pump slows down when the control level is reached and is only stopped when the stop level is reached.
- Bubble detector. Bubble monitoring also protects against air delivery. The ultrasonic sensor detects air bubbles that exceed a certain size and automatically stops the assigned pump, triggering a visual and acoustic alarm at the same time. The system can also be set to generate acoustic and visual warnings if microbubbles below a certain size are detected. Similar to level monitoring, bubble monitoring protects the patient from severe air embolisms in the arterial vascular system.
- Cardiotomy suction strength. Optimum visibility in the operation area is essential during cardiac surgery. The surgeon or the assistant can use one or more cardiotomy suction tubes to keep the operation area free of blood. The perfusionist, who is in constant communication with the rest of the team, adjusts the suction strength to meet the needs of the surgeon.

32.5.3 Patient Monitoring and Therapy During ECC

Because the cardiovascular system of the patient and the ECC system form a unit during perfusion, the perfusionist must fully focus on the patient. This means that he or she is responsible for controlling and regulating the gas exchange and the resulting partial O_2 and CO_2 pressure, the acid–base balance, renal function, coagulation, perfusion pressure (MAP, CVP), temperature, and a number of other blood gas values and electrolytes.

The perfusionist uses these parameters to get an overall view, interprets them in the relevant context, and reacts in an appropriate manner.

The section below describes a few essential aspects of the complex perfusion procedure in more detail.

Hemodynamics

The arterial blood pressure is the most important parameter of hemodynamic monitoring. During nonpulsatile perfusion, only the mean arterial pressure (MAP) can be measured, while pulsatile perfusion techniques also allow for measuring the systolic and diastolic pressure. Every clinic has its own standard regarding the ideal perfusion pressure. As a rule, the average perfusion pressure target is 60–90 mmHg for adults and 30–50 mmHg for pediatric applications.

Compared to perfusion flow, which is vital to adequate organ perfusion and microcirculation, perfusion pressure plays a relatively minor role.

The MAP is controlled by the two parameters pump minute volume (PMV) and total systemic vascular resistance (SVR). A reduction in PMV inevitably leads to a decrease in MAP, while the MAP increases subject to SVR when the PMV rises. The SVR can also be regulated with special drugs that dilate (vasodilatators) or constrict the vessels (vasopressors) as well as with anesthetics.

O_2 Supply

The O_2 required for the organs needs to be delivered during ECC in a way that corresponds to the physiological gas exchange in the lungs. The supply of oxygen to the organism is determined by the arterial O_2 content (aO_2) and the flow (Q)

$$O_2\ \text{supply} = Q \times aO_2\,. \tag{32.1}$$

O_2 consumption is calculated by multiplying the flow (Q) and the arteriovenous O_2 concentration difference ($avDO_2$)

$$O_2\ \text{consumption} = Q \times avDO_2\,. \tag{32.2}$$

The oxygen supply to organs depends on the position of the sigmoidal O_2 binding curve or on the P_{50} value. The P_{50} value is the so-called half-saturation oxygen tension where 50% of the hemoglobin (Hb) is saturated with O_2.

The more limited the O_2 transport capacity during ECC, the greater the importance of the P_{50} value for the O_2 supply to organs and tissue. If there is a left shift in the O_2 binding curve (e.g., caused by hypothermia), then the P_{50} value decreases due to an increased Hb$-O_2$ affinity. As a result, the O_2 delivery to organs and tissue deteriorates. In contrast, a right shift in the O_2 binding curve (e.g., due to a temperature increase or acidosis) leads to decreased Hb$-O_2$ affinity and an increased P_{50} value, which facilitates O_2 delivery to tissue.

A simple and safe method to ensure adequate O_2 delivery during ECC is the continuous monitoring of the arterial saturation (SaO_2) and mixed venous saturation (SvO_2) on the HLM. Arterial O_2 saturation is an excellent parameter for judging the oxygenation performance, while the mixed venous O_2 saturation can indicate an imbalance between O_2 supply and O_2 demand.

Blood Gas Analysis and Management of the Acid-base Status. Two different methods can be applied to regulate the acid–base status during hypothermic ECC. Opinions differ as to which of the two methods represents a more physiological approach to the regulation of the acid–base balance.

The pH-stat measuring method is based on the fact that a pCO_2 value of 40 mmHg and a pH value of 7.40 are regarded as normal values, with the actual temperature not taken into account. The blood gas analysis carried out at 37 °C is adapted to the actual blood temperature. With the pH-stat method, the CO_2 partial pressure must be kept constant by feeding CO_2 into the oxygenator to replace the additional CO_2 dissolved in the blood. This corresponds to a state of respiratory acidosis and hypercapnia.

In contrast, the a-stat measuring method measures the pH value at 37 °C, but does not adjust this value in relation to the actual blood temperature of the patient. Under hypothermic conditions, this would mean a decreased pCO_2 value and an increased pH value and, as a consequence, hypocapnia or respiratory alkalosis.

It has not yet been established which of the two methods is superior for regulating the acid–base status in a hypothermic circuit. However, the a-stat method has a number of advantages over the pH-stat method:

- With the a-stat method, the cerebral blood flow correlates to the cerebral O_2 consumption during hypothermia, which means that the cerebral autoregulation is not affected.
- A physiological pH value is maintained in the intracellular space.

As the pH-stat method also offers advantages during deep hypothermia due to improved cerebral blood circulation and cooling, both methods may be used (pH-stat during the hypothermic phase and a-stat during the warmup phase).

Blood Coagulation. Systemic heparinization of the patient needs to be performed before starting ECC. The initial dose of heparin administered for the purpose of anticoagulation ranges from 200 to 400 IU/kg body weight. Anticoagulation efficiency during cardiopulmonary bypass is primarily determined from the activated clotting time (ACT).

To avoid coagulation complications at the start of ECC, part of the heparin dose is added during the priming of the HLM – this varies between 2500 IU and 10 000 IU for adults, depending on clinic and coagulation management. The initial dose is administered via central venous access or is injected into the right atrium or the aortic root.

Before the ECC can be started, the ACT value must be checked a few minutes after administration of the heparin. ECC should never be started if the ACT is < 400 s. If this is the case, the ACT has to be controlled again and, if required, additional heparin must be administered. If the AT III value is too low, AT III is substituted to ensure the anticoagulatory effect of the heparin.

During cardiopulmonary bypass, ACT values should stay over 400 s and heparin is substituted when the value falls below 400 s. After completion of the ECC, protamine is used as a heparin antagonist. The purpose of the antagonization is to restore a normal ACT (< 130 s). Unlike heparin, protamine is not administered as a bolus but as a short infusion given over the course of approximately 5–10 min. If protamine is administered too quickly, it can cause peripheral vasodilation in combination with hypotension and cardio depression. Once the administration of protamine has been started, cardiotomy suction is stopped to avoid complications from embolisms.

As a general note, the method briefly described above applies primarily to ECC systems without coating. Coagulation is a highly complex topic that cannot be adequately covered within the limitations of this article.

Hypothermia

Artificial lowering of the body temperature (induced hypothermia) during ECC slows down the metabolism of the organism, facilitating a reduction in tissue O_2 con-

Table 32.2 Hypothermia grade depending on body core or rectal temperature

Hypothermia grade	Temperature (°C)
Light hypothermia	37–32
Moderate hypothermia	32–28
Deep hypothermia	28–18
Profound hypothermia	18–4

sumption for the duration of hypothermia. Hypothermia is achieved by cooling flowing blood with a heat exchanger that is integrated into the oxygenator. Even a moderate drop in body core temperature, e.g., to 30 °C, decreases O_2 consumption by approximately 50% because hypothermia has a protective effect on organs and leads to an improved ischemic tolerance in the various tissues.

Most modern surgeries are carried out under normothermic or light hypothermic conditions. Deep systemic cooling of the patient is called for in case of aortic arch surgery or a small number of complex surgical interventions on infants and may be combined with temporary circulatory arrest or low-flow perfusion when required.

Hypothermia is clinically graded according to the target temperature measured rectally or at the body core (Table 32.2).

Urine Production

During ECC, the renal function must be continuously monitored by measuring urine output. Urine output of 0.5–1.0 ml/kg body weight/h is considered adequate during ECC. Sufficient perfusion flows and pressures are required to ensure adequate diuresis. If the urine output decreases during ECC, the perfusion pressure should be increased with a higher perfusion volume or by administration of drugs. If renal function does not improve, diuresis can be treated with loop diuretics or substituted temporarily by a hemofilter.

32.5.4 Complications During Extracorporeal Circulation

Continuous advancements in the area of device technology, the introduction of quality assurance measures, the improvement and enhancement of monitoring functions, and, last but not least, the standardization of training ensure that the number of incidents during perfusion is kept to a minimum. However, the possibility of incidents can never be ruled out completely.

All incidents can be categorized as being caused by technical problems or human operation. Incidents with technical causes result mainly from defective devices or components, material fatigue, or manufactured products of inferior quality. Incidents caused by human operation are almost always the result of misinterpretation of situations and values or inappropriate reactions.

It is essential to protect and preserve the life and health of the patient under all circumstances and using all available means. Preventive measures for early detection of such incidents, appropriate training with regard to countermeasures, and clear communication among all parties can help prevent irreversible damage to the patient even in the case of a very serious incident, such as total failure of the arterial blood pump, rupture of tubing, or failure of the power or gas supply.

Incorrect Placement of the Arterial Cannula

Proper arterial cannulation of the relevant vessel (ascending aorta, femoral artery, etc.) and the correct choice of cannula (size, type) are vital for adequate body perfusion. Arterial perfusion cannulae that are too small and material defects are the most common causes of cannula-related complications. Complications that could have serious consequences for the patient at a later stage can also occur during cannulation and surgery. These include:

- Aortic dissection
- Displacement of the arterial cannula
- Dislocation of the arterial cannula

Incorrect Placement of Venous Cannula(e)

Because venous blood is usually drained from the patient into an HLM using the hydrostatic differential, the reflux of the blood depends on the central venous pressure and the resistance of the venous cannula(e) as well as on the tubing size of the venous line. Manufacturers' flow charts list the flow capacity of the individual cannulae and make it relatively easy to determine the correct size of cannula to meet the respective requirements.

A cannula of insufficient size may severely obstruct the venous reflux during a total cardiopulmonary bypass, which may cause blood stasis and dilation of the right heart. The drop in blood level in the oxygenator or reservoir then forces the perfusionist to reduce or adjust the pump minute volume depending on the situation. If such an event occurs, the cause of the error should always be addressed with measures such as recannulation or additional cannulation so that adequate perfusion can continue.

Additional factors that can obstruct venous drainage are as follows:

- Dislocation of the venous cannula(e)
- Displacement of the venous cannula(e)
- Large amounts of air in the venous line (air block)
- Relocation of the venous line
- Luxation of the heart
- Hypovolaemia
- Insufficient hydrostatic differential

Arterial Air Embolisms

Systemic arterial air embolisms are among the most serious complications that can occur during cardiopulmonary bypass and can have many different causes. The most common causes of arterial air embolisms are insufficient reservoir levels, air embolization via cardioplegia cannulae, air application caused by incorrect flow direction in the vent suction tube, oxygenator malfunction, disconnection of the tubing system, material fatigue leading to rupture of the arterial pump tubing segment, and aggressive warming of the patient.

A large number of complications described in the literature could be added to this list. Depending on the amount and intensity of air delivery, an arterial air embolism can have catastrophic consequences for the patient, including irreversible organ damage. Arterial air embolisms can lead to hypoxic brain damage. For this reason, measures must be taken immediately after an embolism is detected in order to minimize subsequent damage.

If an air embolism occurs despite the applied preventive measures (level sensor, bubble sensor, arterial filter), immediate corrective action is required that must go well beyond removing the air from the systemic circuit:

- Immediate interruption of the cardiopulmonary bypass and positioning of patient in maximum Trendelenburg position (head-down position)
- Decannulation of the aortic cannula and deaeration via the incision point using a suction tube
- Removal of air from the arterial line and aortic cannula
- Hypothermic, retrograde perfusion via the superior vena cava until air can no longer be detected above the aortic root
- Intermittent compression of the two carotid arteries
- Hypertension using vasoconstrictors
- Drug therapy with steroids and barbiturates
- Antegrade, hypothermic perfusion (at $\approx 20\,°C$)
- Hyperbaric oxygen therapy as a last resort

Oxygenator Failure

Inadequate oxygenator performance during cardiopulmonary bypass can have major consequences and lead to hypoxic organ damage. Various monitoring options (arterial O_2 saturation, arterial blood gases, and visual evaluation) allow for relatively fast interpretation of the oxygenator's gas exchange performance, enabling the user to take instant action if complications should occur.

The perfusionist can visually evaluate the returned arterial blood from the start of extracorporeal perfusion. If a significantly reduced arterial O_2 saturation and unacceptable arterial pO_2 values are measured at the beginning of the perfusion despite sufficient ventilation of the delivered gases, it must be assumed that the oxygenator failed because of an oxygenator module material defect or that the surface required for the gas exchange has been reduced by protein or thrombus deposits.

Depending on the remaining oxygenation capacity and the stage of the surgery, the oxygenator may have to be replaced. This requires a temporary interruption of perfusion (with the corresponding risk to the patient).

Incidences Caused by Technical Problems

In the past, an interruption to the power supply in the operating theater during ECC was a serious problem because both arterial pump and overall system were affected. Modern HLMs have a built-in independent emergency power supply. Using the emergency power supply (uninterruptible power supply, UPS), the overall system can be battery-operated for a period of approximately 20–130 min (depending on the load).

If the power supply or power pack fails, the UPS switches automatically to battery operation via an internal electrical switch. When the power supply returns, it switches back automatically.

An additional UPS control module monitors the battery charge status during power supply failure and displays the remaining battery life, which depends on the load. If a total roller pump failure occurs, the rotation and, hence, the required flow can be maintained with a hand crank. In modular systems, the defective pump can be replaced by a functioning pump, if required. In the case of centrifugal pumps, the pump head can be removed and inserted into a mechanical drive that is also operated using a hand crank.

If the central gas supply fails, the oxygenator is supplied with gas from transportable gas cylinders.

32.6 Differentiation of Heart–Lung Machines

Functionality, maximum reliability, safety and ergonomic aspects, optimum user-friendliness, and intuitive operation are factors for the design and development of HLMs.

The most important design feature that distinguishes HLMs is the modular or semimodular configuration.

In addition, the integration of blood gas analysis devices, flow sensors, electrical remote-controlled tubing clamps, and other systems is gaining importance.

32.6.1 Modular Systems

The modular concept of HLM systems (Fig. 32.7) offers maximum user flexibility. The overall system can be custom-tailored, resulting in HLM configurations that meet individual clinical requirements. Individual components, modules, and accessories can be added to the system at any time according to the requirements of the user or replaced when defective. Modifications or system extensions that include additional monitoring, control, and measuring devices facilitate the use of the HLM for a variety of perfusion tasks.

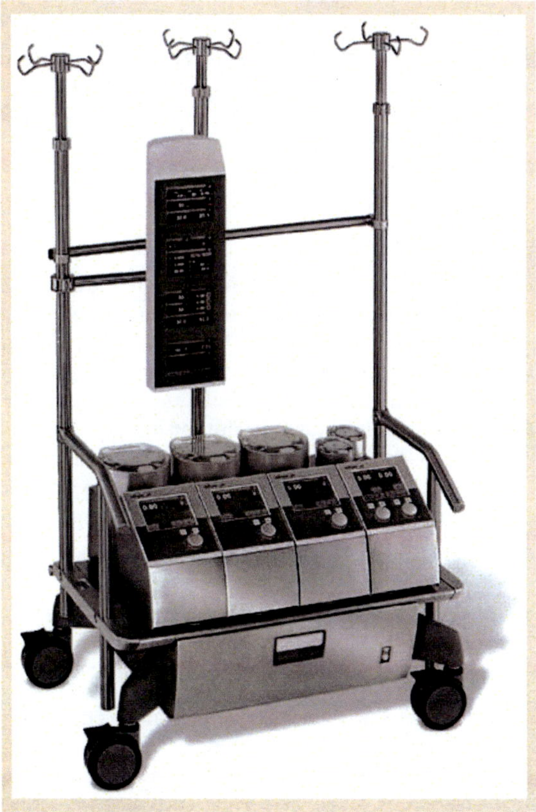

Fig. 32.7 Modular HLM

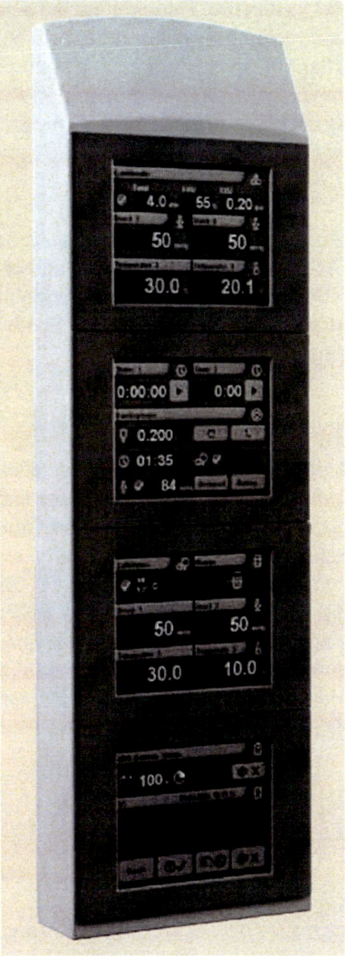

Fig. 32.8 Control panel for connecting monitoring, control, and measuring devices and the central display and control module (CDM). The control modules for the monitoring, control, and measuring devices can be arranged on the control panel as required

Modular systems consist of a basic mobile console that supports 3–6 roller pumps, depending on the configuration. The modular roller pumps form the basis of the perfusion system. They are placed on the pump table of the console and connected to the power supply and electronic control via plug connectors. The control panel (Fig. 32.8) accommodates all modular monitoring, control, and measuring devices. As different types of surgery require various levels of monitoring, control panels accommodate a number of control modules.

The modular design also includes pumps that can be mounted in any position on the mast system, preferably separate from the control module. These pumps are also available for the semimodular systems described below. The main advantage of the freely positioned pumps are shorter tubing paths that lead to a reduction in the foreign surface and priming volume of the ECC system.

32.6.2 Nonmodular Systems

In contrast to modular designs, nonmodular systems are a configuration in which the number, type, and arrangement of the roller pumps as well as the scope of monitoring and control functions are predetermined by the concept and design (Fig. 32.9).

The system includes all important monitoring and control functions, such as bubble monitoring, level monitoring, pressure monitoring, cardioplegia monitoring, temperature monitoring, and timers. In such a system, the roller pumps are built into the housing and the standard monitoring functions are integrated fully into the HLM control panel without any option of reconfiguration. In comparison to modular systems, the options for extending the system with additional roller pumps and monitoring, control, and measuring devices are limited. Another marked difference from modular systems is that the main components of the HLM can only be replaced by authorized service technicians when technical defects occur.

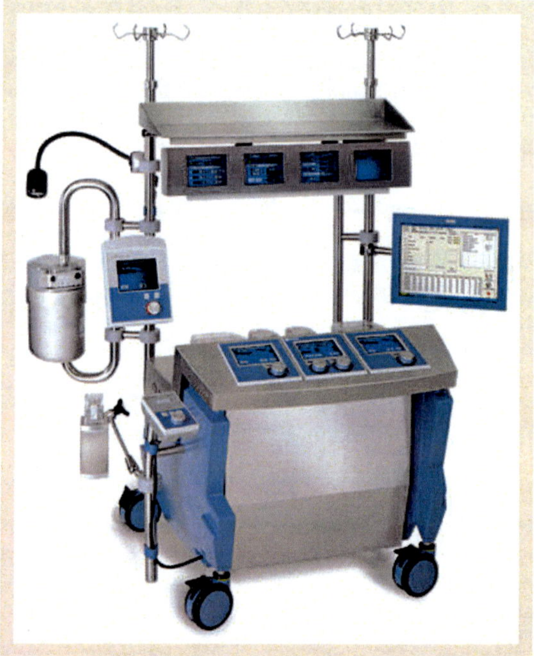

Fig. 32.9 Compact HLM with semimodular configuration

32.7 Aspects of Technical Safety

HLMs should be designed, constructed, and manufactured with an emphasis on optimum safety. The user must always be in command of the system technology, especially when adverse events occur. For this reason, the main focus of any safety assessment must be on the detection of risks that are inevitable when technical devices are used or when man and machine interact. The most important safety requirement for an HLM is the maintenance of the patient's blood circulation. Perfusionists contribute greatly to the safety of such procedures.

The objective is to find solutions that prevent the occurrence of one or more adverse events. The minimum safety requirement is compliance with the laws and regulations (basic requirements) and the relevant standards.

Risk analysis is an important preventive measure that, based on EN ISO 14971, has developed into risk management. Risk management goes beyond the scope of pure risk analysis and requires decisions on the acceptability of risks and on the measures to reduce risks as well as a review of the measures taken.

If an adverse event occurs despite the measures that were taken, a targeted reaction is required. This means that a safe status must be determined and defined in such a way that any danger to the health and safety of patients and users is kept to an absolute minimum. Because the evaluation of probability of risks depends on their exact allocation, these must be classified:

- *Unintended adverse event*, such as a stopped pump – caused by a technical defect
- *Intended adverse event*, such as a stopped pump – caused by the detection of an air bubble in the tubing system

A life-sustaining system is not safe if the function is interrupted and an alarm is triggered. A temporary pump stop can be acceptable under certain conditions to avoid a risk caused by an adverse event such as air entering the tubing system.

The following standards are mandatory for the design of HLMs: Directive 93/42/EEC or MPG (Medical Device Act), EN 601-1, EN 601-1-1, EN 601-1-2, EN 601-1-4, EN 601-1-6, EN 62304.

All requirements must be determined at the start of the development process and must be implemented in system, device, and component specifications. A risk analysis must be generated for each function and for the overall system. Error control measures need to be stated and verified by prediction and testing.

All development tests as well as the entire product documentation have to be stored in a master file.

HLMs must always be constructed to be first-fault tolerant, i. e., a fault such as component failure caused by short-circuiting or interruption should never bring the device into an uncontrollable operational state. In any event, the error must be detected and indicated by an alarm and the function must be stopped.

The perfusionist then decides which measures to take so that the perfusion can continue. For example, if a pump stops due to motor failure, the affected tubing segment can be inserted into a backup pump or, in an emergency situation, the defective pump can be operated using a hand crank. If the level sensor on the cardiotomy reservoir or on the oxygenator fails, the perfusionist can monitor the level manually and stop or reduce the flow rate of the relevant assigned pump.

It goes without saying that HLMs must meet all applicable requirements for electrical safety, electromagnetic compatibility, explosion safety, and fire safety as well as the environmental requirements in the operating theater (e.g., temperature range). All devices must undergo stress testing under transport and storage conditions. In addition to these characteristics, users also evaluate cleaning and disinfection properties and look for adequate ergonomic design.

The operating instructions are an integral part of the device and are also inspected during the approval process. They contain information about safety and are therefore an important factor for the safe operation of the device. For this reason, they not only contain normal operating instructions but also information on error prevention as well as warnings regarding operational states that are not permitted or use in combination with devices that have not been tested or approved.

HLMs are Class IIb medical devices in accordance with MPG (Medical Device Act) and are subject to technical safety checks. The official transfer of ownership to the operator also includes an inspection that must be documented. Furthermore, the operator must be given formal instruction on the functions of the device as part of the transfer of ownership.

32.8 Prospects for Further Development

Cardiac surgery has developed greatly in recent years due to the increasing use of minimally invasive and video-assisted endoscopic and robot-assisted technology. The development of new, less invasive surgical techniques aims to reduce surgical trauma by reducing the size of surgical access and avoiding the use of an HLM. The implantation of vessel prosthetics without the use of an HLM is already established as a standard procedure in aortic surgery. In contrast, the trend toward implanting heart valves without ECC is still at an early stage of development.

Contrary to initial expectations, surgical interventions on a beating heart, especially in the area of bypass surgery, are still only being carried out at a small number of heart centers because of the complex and time-consuming nature of the surgery involved.

On the other hand, the use of extracorporeal systems outside heart surgery centers is increasing rapidly. Terms like ECMO or ECLS refer to minimalist systems consisting only of a pump and oxygenator that carry out the main functions of the HLM, namely, the pump function of the heart and the gas exchange function of the lungs.

These systems can be used for several days or even weeks on account of the small number of components and the resulting minimum foreign surface. Along with the extended period of operation, the compact design makes the system practically autonomous and therefore independent of the operating-theater infrastructure. As a result, the use of these systems is on the increase in other departments, such as cardiological institutions. One of the main application areas of these mobile mini ECC systems will most likely be patient transport from clinic to clinic by road or air with partial or complete cardiopulmonary support that will present completely new possibilities and challenges.

Running parallel to the development of equipment technology is the advancement in the area of disposables such as tubing, oxygenators, filters, reservoirs, and cannulae. The focus here is on increased efficiency and also on materials that have direct contact with and cause damage to blood cells. The aim is to use different coatings to create surfaces on the individual components and complete tubing sets that are almost biocompatible. A number of companies are offering various products that are, in some cases, a definite improvement, but to date no company has succeeded in producing a surface that is completely biocompatible with blood.

Further Reading

- H.S. Baraka, M.A. Baroody, S.T. Haroun: Effect of α-stat versus pH-stat strategy on oxyhemoglobin dissociation and whole-body consumption during hypothermic cardiopulmonary bypass, Anaest. Analg. **74**, 32 (1992)
- G.D. Buckberg: Antegrade/retrograde blood cardioplegia to ensure cardioplegic distribution: operative techniques and objectives, J. Card. Surg. **4**, 216 (1989)
- K.G. Davidson: Cannulation for cardiopulmonary bypass, Cardiopulm. Bypass **5**, 115–169 (1986)
- T.G. Hattersley: Activated coagulation time of whole blood, JAMA **196**, 436 (1966)
- M. Kurusz, V.R. Conti, J.F. Arens, B. Brown, S.C. Faulkner, J.V. Manning: Perfusion accident survey, Proc. Am. Acad. Cardiovasc. Perfus. **7**, 57–65 (1986)
- J.B. Magner: Complications of aortic cannulation for open heart surgery, Thorax **26**, 172–273 (1971)
- W.A. McAlpine, M.W. Selman, T. Kawakami: Routine use of aortic cannulation in open heart operations, Am. J. Surg. **114**, 831–834 (1967)
- N.L. Mills, J.L. Ochsner: Massive air embolism during cardiopulmonary bypass, J. Thorac. Cardiovasc. Surg. **80**, 798–817 (1980)
- C.T. Mora: *Cardiopulmonary Bypass* (Springer, Berlin Heidelberg 1995)
- K.M. Taylor: *Cardiopulmonary Bypass* (Chapman Hall, London 1988)

References

32.1 J.J.C. LeGallois: *Expériences sur le principe de la vie, notamment sur celui des mouvemens du coeur, et sur le siège de ce principe; suivies du rapport fait à la première classe de l'Institut sur celles relatives aux mouvemens du coeur* (D'Hautel, Paris 1812)

32.2 J.P. Kay: Physiological experiments and observations on the cessation of the contractility of the heart and muscles in the asphyxia of warm-blooded animals, Edinb. Med. Surg. J. **29**, 37–66 (1828)

32.3 E. Brown-Séquard: Recherches expérimentales sur les propriétés physiologiques et les usages du sang rouge et du sang noir et leurs principaux éléments gazeux, l'oxygène et l'acide carbonique, J. Phys. l'Homme Anim. **1**, 95–735 (1858)

32.4 C.E. Loebell: De conditionibus quibus secretiones in glandulis perficiuntur, Dissertatio Inauguralis Marburgenis, University Marburg (1849)

32.5 A. Schmidt: Die Atmung innerhalb des Blutes. Zweite Abhandlung, Ber. Verh. königl. sächs. Ges. Wiss. Leipz., Mathe.-Phys. Cl. **19**, 99–130 (1867)

32.6 W. von Schröder: Über die Bildungsstätte des Harnstoffs, Arch. Exp. Pathol. Pharmakol. **15**, 364–402 (1882)

32.7 M. von Frey, M. Gruber: Untersuchung über den Stoffwechsel isolierter Organe. 1. Ein Respirationsapparat für isolierte Organe, Arch. Physiol. **6**, 519–532 (1885)

32.8 P.M. Galetti, G.A. Brecher: *Heart-Lung Bypass* (Grune Stratton, New York 1962)

32.9 C. Jacobj: Apparat zur Durchblutung isolierter überlebender Organe, Arch. Exp. Pathol. Pharmakol. **26**, 388–391 (1890)

32.10 J.H. Gibbon Jr.: Application of a mechanical heart and lung apparatus to cardiac surgery, Minn. Med. **36**, 171–180 (1954)

32.11 W.G. Bigelow, W.K. Lindsay, W.F. Greenwood: Hypothermia – its possible role in cardiac surgery: An investigation of factors governing survival in dogs at low body temperatures, Ann. Thorac. Surg. **132**, 849 (1950)

32.12 F.J. Lewis: Closure of atrial septal defects with the aid of hypothermia. Experimental accomplishments and the report of one successful case, Surgery **33**, 52 (1953)

32.13 W.C. Sealy, I.W. Brown, W.G. Young: A report on the use of both extracorporal circulation and hypothermia for open heart surgery, Ann. Thorac. Surg. **147**, 603 (1958)

32.14 W.C. Lillehei: Controlled cross circulation for direct vision intracardiac surgery correction of ventricular septal defects, atrio ventricularis communis, and tetralogy of Fallot, Postgrad. Med. **17**, 388–396 (1955)

32.15 J.W. Kirklin, J.W. Du Shane, R.T. Patrick: Intracardiac surgery with the aid of a mechanical pump oxygenator system (Gibbon type): Report of eight cases, Proc. Staff Meet. Mayo Clinic **30**, 201–207 (1955)

32.16 A. Beck: Über Bluttransfusion, München Med. Wochenschau **72**, 1232 (1925)

32.17 M.E. De Bakey: Simple continuous-flow blood transfusion instrument, New Orleans Med. Surg. J. **87**, 386–389 (1934)

32.18 N. Zuhdi, J. Carey, J. Cutter, L. Rader, A. Greer: Intentional hemodilution, AMA Arch. Surg. **87**, 554–559 (1963)

33. Application of Shock Waves and Pressure Pulses in Medicine

Friedrich Ueberle

Since 1980 extracorporeal shock waves and pressure pulses are used to disintegrate kidney stones. Since then, ESWL – Extracorporeal Shock Wave Lithotripsy – is the most important treatment for kidney stones. In the past years, more medical applications for these special pulsed acoustic waves were developed, e.g. for orthopedic pain therapy (ESWT – Extracorporeal Shock Wave Treatment). Shock waves and pressure pulses have high peak pressures (up to 150 MPa) and very short time duration, typically 2–5 s. They are mostly generated by one of three principles: The electrohydraulic (spark gap) source, the electromagnetic (EMSE) source or the piezoelectric source. Some thousand pulses are released at a slow rate. They propagate through water into the body of the patient by an acoustic coupling means. In order to achieve the intended medical effects, they are focused on the treatment site. A number of physical parameters can be attributed to the effects of the pulses both on stones and tissue, the most important being the effective acoustic energy and the energy flux density. Stone disintegration is governed by direct wave effects, squeezing and cavitation. The waves also act on tissue, which may cause healing effects, but also some side effects. Targeting and treatment control is usually done by imaging systems using x-rays or ultrasound imaging. The result of the treatment can be described by efficiency quotients, which comprise stone-free rate as well as side effects.

33.1 **Introduction – Historical Development** ... 642

33.2 **Definitions of Physical Terms: Acoustics – Sound Waves – Pressure Pulses – Shock Waves** ... 643
- 33.2.1 Wave Forms Used in Acoustic Therapies ... 643
- 33.2.2 Sound Waves – Mechanical Waves in Media ... 644
- 33.2.3 Definition of Shock Wave and Pressure Pulse ... 645
- 33.2.4 Mechanical Effects of Sound Waves at Interfaces ... 645
- 33.2.5 Mechanical Effects of Sound Waves Caused by Nonlinear Properties of the Media ... 647
- 33.2.6 Thermal Effects Caused by Pressure Pulses ... 649

33.3 **The Acoustic Field of a Lithotripter – Basics of Measurement Technology** ... 649
- 33.3.1 Sensors and Measurement Technology ... 649
- 33.3.2 Measurement Parameters of Lithotripter Sound Fields ... 651
- 33.3.3 Energy Parameters of the Sound Field ... 653
- 33.3.4 Significance of the Pressure Pulse Parameters in Lithotripsy ... 655
- 33.3.5 Which Physical Effects Play a Significant Role in Stone Comminution? ... 657
- 33.3.6 Effect of the Pressure Pulses on Tissue ... 658

33.4 **Generation of Pressure Pulses for Extracorporeal Lithotripsy (ESWL) and Extracorporeal Shock Wave Therapy (ESWT)** ... 660
- 33.4.1 Point Sources and Extended Pressure Pulse Sources ... 660
- 33.4.2 Generation of Spherical Pressure Pulses and Shock Waves ... 660
- 33.4.3 Generation of Planar Sound Waves ... 662

33.5 **Extracorporeal Lithotripsy (ESWL) and Extracorporeal Shock Wave Therapy (ESWT) in Practice** ... 664
- 33.5.1 System Components of a Lithotripter ... 664

33.5.2 Approaches to Online Disintegration Tracking During Treatment 667

33.6 **The Patient** 667
33.6.1 Patient Bedding for ESWL and ESWT Treatment 667
33.6.2 Stone- and Patient-Related Influences 667

33.7 **Assessment of the Clinical Efficiency of Lithotripters** 668
33.7.1 The Controversy About the Right Focal Size 669
33.7.2 Side Effects and Safety 669

33.8 **Associations and Societies for Lithotripsy and Pressure Pulse Therapy** 670

References 671

33.1 Introduction – Historical Development

On 26 February 1980, for the first time, a patient with kidney stones was treated successfully with an extracorporeal shock wave lithotripter. Research and development for this moment had taken about 10 years from the first ideas to successful clinical application.

The first proposals to treat concretions with sound reach back to the 1940s. A British working group tried to disintegrate gallstones in vitro using continuous wave (CW) ultrasound at 400 kHz lasting between 5 and 60 s. Their success rate was about 80% [33.1].

Unfortunately, animal experiments in vivo resulted in heavy tissue damage, which was lethal for the animals. Consequently, further efforts were cancelled in 1956 [33.2–4].

The discovery that shock waves were able to penetrate soft tissue without harming it was made by chance in 1966 when a Dornier technician was touching a metal plate under fire, which was serving as a target in high-speed gas cannon experiments [33.5]. He felt a kind of electric shock without suffering any visible damage. This phenomenon of transmitting shock waves through tissue was a subject of research by Dr. Hoff at Dornier Systems, Friedrichshafen, Germany, until 1971. During a scientific collaboration with the Dornier team, Prof. Häusler (Saarbrücken, Germany) raised the idea of disintegrating kidney stones using sound pulses generated by multiple sound sources, which would be transmitted through the body. Subsequent research activities soon led Prof. Häusler away from the originally unfocused high-speed projectiles generating shock waves when they hit a steel target. Instead, spark gaps (almost like spark plugs in a car) were used to generate shock waves by underwater discharge [33.6]. In 1972 urologist Prof. Schmiedt at the Klinikum Grosshadern of the University of Munich, with his team Chaussy (Munich) and Eisenberger (now Stuttgart), joined the working group. Research funding was secured from the German government (BMFT), which supported the subsequent animal experiments and later on the development of an efficient x-ray targeting.

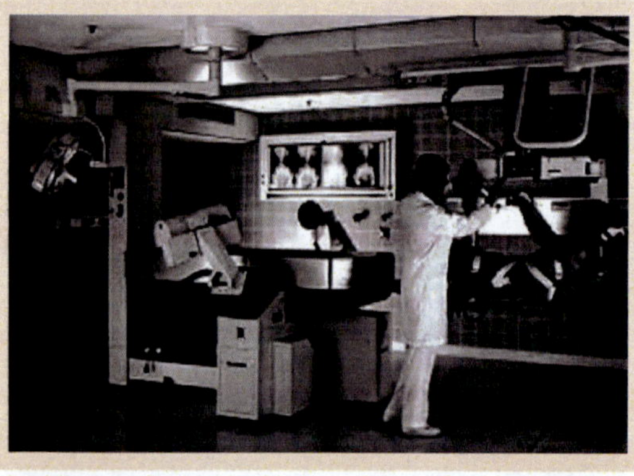

Fig. 33.1 The first series lithotripter Dornier HM3 was introduced in 1983. The patients were treated by the spark-gap source in a large bathtub filled with degassed water. Targeting was done using two crossed x-ray tubes

After 3 years of clinical studies involving about 200 patients, the first series kidney stone lithotripter Dornier HM 3 was installed in Stuttgart in 1983. Soon the method was extended beyond kidney stones to the treatment of stones in the entire urinary tract. In 1985 experiments to disintegrate gall stones were started. This minimally invasive method is still in use today under certain circumstances, but it was made obsolete by the forthcoming minimally invasive surgical cholecystectomy in most cases.

Since its first introduction of noninvasive shock wave treatment, the medical applications of this energy have been continuously extended (Table 33.1). At present extracorporeal shock wave therapy (ESWT) for the treatment of soft tissue pain (e.g., tennis elbow pain, heel spur pain) is in clinical evaluation and approval worldwide. The sound sources used in ESWL

Table 33.1 Applications of shock waves and pressure pulses in medicine and biology. (Some applications are only in the experimental stage and may not be approved for clinical use in certain countries)

Application	Clinically applied since	Literature
Ureter stone lithotripsy	1980/1983	[33.10]
Gall stone lithotripsy	1985	[33.11]
Gall duct stone lithotripsy	1985	[33.12, 13]
Pancreatic duct stones lithotripsy		[33.14]
Salivary stone lithotripsy	1989	[33.15]
Pseudarthrosis treatment	1989	[33.16]
Shockwave effect on tumors		[33.17]
Tendinosis calcarea treatment	1993	[33.18]
Pain therapy of tendons (tennis elbow, heel spur pain, shoulder pain)	1992	[33.19, 20]
Therapy of motion infringement	1994	[33.21]
Shockwave induced transfection of cells		[33.7, 8]
Peyronie's disease treatment	1997	[33.22]
Coronary vessels treatment	2002	[33.23]

and ESWT are similar – no wonder, because the first clinical tests to treat soft tissue pain were undertaken using standard or slightly modified ESWL machines. Presently the use of shock waves for the application of biotechnological agents is under research. The first results are promising; as it turned out those shock-wave-induced cavitations are able to open up molecular pathways into cells for a short time. During this time, molecules (e.g., genes, cytotoxics) are able to penetrate the cells [33.7–9].

Due to the close relationship between the ESW disciplines, the following technical description may be mostly restricted to the vocabulary and the knowledge learned from ESWL.

33.2 Definitions of Physical Terms: Acoustics – Sound Waves – Pressure Pulses – Shock Waves

33.2.1 Wave Forms Used in Acoustic Therapies

The simplest wave form is a sinusoidal wave comprising positive and negative pressures (overpressure and rarefaction pressure phases).

Each sinusoidal wave has one precisely defined frequency – in contrast to a shock wave. The wave is called an ultrasonic wave if its frequency is above approximately 16 000 oscillations per second. If only one or a limited number of sinusoidal cycles are generated, which is common in ultrasound imaging systems, the result is called a sound pulse.

A pressure pulse typically consists of only one positive pressure wave, which decays in an exponential slope to ambient pressure within one to 5 μs. Immedi-

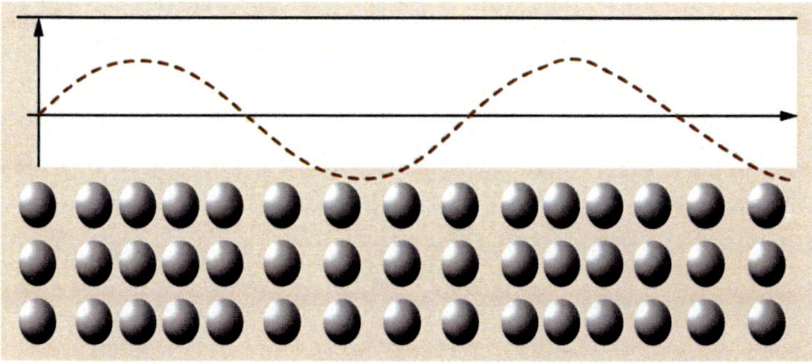

Fig. 33.2 Acoustic waves propagate in fluid and gaseous media as longitudinal compression waves

ately a rarefaction period lasting a few microseconds follows. When the rise from ambient pressure to the positive pressure maximum takes place in a very short period of time (a few nanoseconds), the resulting signal is called a shock wave. After the pressure pulse or shock wave has passed, the medium has returned to ambient pressure conditions.

33.2.2 Sound Waves – Mechanical Waves in Media

A sound wave is a mechanical wave that propagates in a medium like gas, fluid, or a solid body. During the passing of the wave, the molecular distances in the medium undergo a slight local change. In particular, in fluids and gases the waves propagate as longitudinal compression waves (Fig. 33.2).

In solids more wave forms like transversal waves, shear waves, surface waves, etc. can propagate.

Acoustic Properties of Media

Media differ by their mechanical properties like elasticity and compressibility. These properties determine the velocity of sound c, and the acoustic impedance $Z = \rho c$, which is calculated as a product of the medium density ρ and the sound velocity c. Table 33.2 shows

Table 33.2 Acoustic properties of some important media (after [33.24, 25])

Material	Density (kg/m^3)	Sound velocity (m/s)	Impedance ($N s/m^3$)
Air	1.293	331	429
Water	998	1483	1.48×10^6
Fat tissue	920	1410–1479	1.33×10^6
Muscle tissue	1060	1540–1603	1.67×10^6
Bone tissue	1380–1810	2700–4100	$4.3–6.6 \times 10^6$

Table 33.3 Mechanical and acoustic parameters of some stone materials (more data, in particular for kidney stones, are published in [33.27])

Material	Pressure strength (MPa)	Tensile strength (MPa)	Vickers hardness (kg/mm^2)	Speed of sound (m/s)	Density (g/cm^3)	Impedance (10^6 Rayl)	Disintegration energy (mJ/cm^3)	Comment
Model stones								
Plaster balls with 10% micro spheres	11	1.4	18.8	2096–3195	1.1	2.4–3.7	1599–1764	[33.26] [33.27] [33.28]
Calcite stone							5800	[33.29]
Graphite							22 500	[33.29]
Glimmer							67 000	[33.29]
Glass							1500	[33.29]
Glass marbles							3669	[33.30]
Breeze blocks							620–670	[33.30]
Gall stones	2.2–3.2	0.4–1.0		1700–2100	1.1–1.5	1.9–3.1	8050 (4336–17 850)	[33.28,30]
Kidney stones				2600–3800	1.8–3.0	4.8–7.8		[33.30]
Magnesiumammonium-phosphat	8	0.6	22–40	2050				[33.10,31]
Tricalciumphosphat	4.8	0.6						[33.10]
Harnsäure		1.8	31.2–41.6	3318–3471	1.48–1.54	5.4–6.2		[33.10,27]
Calcium oxalate		1.1	98–125	3000				[33.10,31]
Mixed data of CaO_x, Ca apatite, Mg ammon, P	2.0–17.6	0.1–3.4		1875–3390				[33.32]

some data of biological materials, and Table 33.3 summarizes data of different stones.

33.2.3 Definition of Shock Wave and Pressure Pulse

A shock wave is a mechanical wave in a fluid (or a gas) that rises from ambient pressure to peak positive pressure in a few nanoseconds (10^{-9} s). The cause for this almost instant rise is the nonlinear properties of fluids, which lead to a local increase of the velocity of sound in the positive pressure zone. During the propagation, trailing positive pressure parts of the wave approach on the leading front. The front steepens. This effect is saturated when the steepening is balanced by the energy loss by dissipation. Only in this saturated case is it physically valid to use the term shock wave. Besides the direct generation of a shock wave by a sound source, which moves at a higher speed compared to the velocity of sound in an ambient medium, shock waves are also generated by the steepening of a strong positive pressure pulse. This may happen, for example, in the converging sound field of a focused lithotripter source. Nowadays pressure amplitudes of 10 MPa to over 100 MPa (Mega-Pascal, 1 MPa = 10 bar, which equals ten times the atmospheric pressure) are used in pressure pulse therapies. As a shock wave – like all kinds of sound waves – propagates with a velocity, which is determined by the medium and the shock wave pressure, one can calculate a shock front thickness, which is the spatial distance between the point of ambient pressure and the point of peak positive pressure. In tissue, the shock front thickness is about 1.6 to 6 μm (micrometers, 10^{-6} m). Therefore, significant forces may act on a cell membrane that is only a few molecule layers thick when a shock wave is passing. The pressure gradient, i. e., the change in pressure between two places, is maximized when shock waves occur instead of other kinds of acoustic wave forms, like sinusoidal high-power ultrasound.

It should be pointed out that in lay language the term shock wave frequently is used for all kinds of pressure signals, even if their rise times are much longer than a few nanoseconds. In this case the term pressure pulse, according to the IEC Standard 61846 [33.34], is more appropriate. The measurement problems of the fast pressure rise times will be discussed below.

33.2.4 Mechanical Effects of Sound Waves at Interfaces

When a sound wave passes an interface between media with different acoustic impedances, the propagation of the sound may be significantly changed. In most cases, knowledge of the magnitudes of the impedance of the different media (which is given by the product of density and sound velocity) is sufficient to characterize their properties.

Transmission and Reflection of Sound Waves

A fraction of sound energy is reflected into medium 1 at the interface when the media impedances are different, for example when the sound wave is passing from fat into muscle tissue. The remaining sound energy is transmitted into medium 2. The examples in Table 33.4 demonstrate the fraction of transmitted energy, depending on the properties of the tissues (e.g., muscle → bone).

Another very important property of the reflected sound wave can be learned from the reflection formula: If the impedance of medium 2 is less than that of medium 1, then the algebraic sign of the reflected pres-

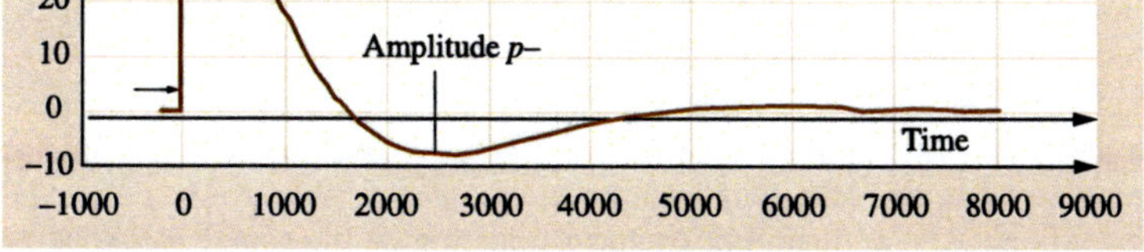

Fig. 33.3 A typical shock wave is characterized by a positive pressure pulse with an extremely short rise time t_r, followed by an exponential decay to ambient pressure. A trailing rarefaction phase lasting a few microseconds follows. Typical shock wave simulation (after [33.33])

Table 33.4 Examples of the amount of reflected and transmitted sound energy at the interface of two media calculated using the reflection formula for pressure $r = (Z_2 - Z_1)/(Z_2 + Z_1)$ with magnitudes of impedance Z_1 and Z_2

Interface	Reflected pressure (%)	Reflected sound energy (%)	Transmitted sound energy (%)
Water – fat tissue	−5	0.25	99.75
Fat tissue – muscle	11	1.2	98.8
Muscle – fat tissue	−11	−1.2	98.8
Muscle – bone	44–60	19–36	64–81
Muscle – model kidney stone	22	5	95
Model kidney stone – muscle	−22	5	95
Muscle – air	−99	98	2

sure becomes negative. The physical meaning of this is that a positive pressure is reflected as negative pressure back into medium 1. Almost the complete pressure amplitude is inverted during reflection at plane interfaces with a large difference in impedance. The pressure gradient at the interface takes twice the value of the gradient of the traveling wave in a homogeneous medium. No energy is transmitted to medium 2. These effects occur in particular at all interfaces from tissue to air, e.g., at the lung. As almost the complete energy is reflected at the interface, the tissue is no longer able to tolerate the resulting forces, and it tears. Thus, when the high-pressure path of a pressure pulse device is directed toward the lung, severe harm can occur. Similar problems can occur at the surface of other organs, e.g., gas in the intestine or at the skin/air interface on the opposite side of the body. If air bubbles are trapped at the impinging site of the sound waves, small bleedings (petechiae) frequently occur.

The effect of pressure inversion also occurs at another interface, which is important in lithotripsy: at the rear side of the stone, because the adjacent medium (e.g., kidney parenchyma) has lower impedance than the stone. Thus part of the leaving shock wave is inverted and reflected back into the stone. It overlays with the trailing rarefaction part of the shock wave, resulting in very strong traction forces at the rear side of the stone. These forces cause a spallation of the stone, occurring somewhere at the rear surface. This effect is named after its discoverer, Hopkins. In ESWL it can be observed in particular when stones are impacted, e.g., in the urethra or the shoulder (Fig. 33.5) [33.35, 36].

The Hopkins effect can be observed at impacted stones. The concrement is damaged not at the sound entry side, but at the adjacent side. The reflected pulse overlays with the rarefaction portion of the impinging pulse, leading to high strain forces inside the stone. This effect explains findings on impacted concretions that are disintegrated from the back toward the front.

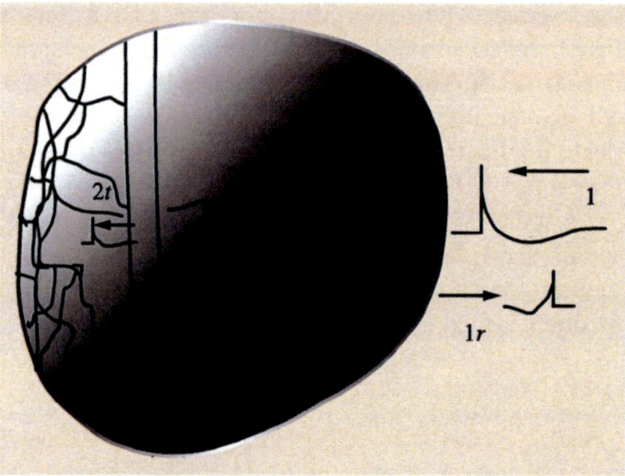

Fig. 33.4 If a sound wave impinges at the interface between two media, a part of the acoustic energy is reflected. If the acoustic impedance of the second medium is less than the impedance of the first medium, then the polarity of the wave is inverted. Positive pressures become negative and vice versa. At impacted stones the Hopkins effect can be observed. The concrement is damaged not at the sound entry side but at the adjacent side. The reflected pulse overlaps with the rarefaction portion of the impinging pulse, leading to high strain forces inside the stone. This effect explains findings on impacted concretions that are disintegrated from the back toward the front

Refraction of Sound Waves at Boundaries

A wave that is impinging on an interface at an angle other than normal to the surface continues its path at a different angle. According to Snell's law, this angle depends on the speed of sound in both media. This refraction effect is used to design acoustic lenses for the focusing of sound.

Refraction can also occur at interfaces between tissues, and at the boundary between the coupling fluid of the sound source and the patient skin. Refraction is the main reason for a smearing of the sound focus inside the body of the patient. Furthermore, refraction may be the cause of a potential aberration between the focus display of the localization device (x-ray or ultrasound imager) and the actual focus position of the pressure pulse source [33.37–40].

Diffraction of Sound Waves

Comparable to the diffraction of optical waves, diffraction of sound waves occurs at edges. Therefore, a certain portion of the sound not only propagates straight ahead but is bent around the edge.

Therefore, an edge-bending wave occurs at the fringe of any sound transducer as well as at interfering mechanical structures, e.g., an inline scanner device. This portion of the wave, which is unavoidable in sound sources of finite extensions, results in an inverted (rarefaction) pressure, which follows the primary transmitted positive pressure pulse. The maximum of this rarefaction portion occurs on the axis close to the focus, even if the sound source only transmits positive pressure pulses.

Absorption and Attenuation

Absorption and attenuation of sound waves lead to a loss of acoustic energy in every kind of tissue. The attenuation is usually frequency dependent ($0.5-0.7\,\mathrm{dB\,cm^{-1}\,MHz^{-1}}$, which means a bisection of the pressure amplitude of 1 MHz after a 12 cm path length, or bisection of a 2 MHz wave after 6 cm). Thus, not only does the amplitude of a sound pulse change, but also the pulse itself changes its shape, resulting in less steep rise and decay [33.40]. Positive portions (P+) of the pressure pulses are reduced by 15–30% at penetration depths of 100 to 120 mm (typical for kidney stones) [33.42]. In contrast, the rarefaction amplitudes P− are only reduced by 6% [33.43]. This can be explained from the different frequency contents of the (steepened) positive portion and the nonsteepened rarefaction portion (e.g., 5–8 μs duration, Fig. 33.9). This rarefaction portion has a quasisinusoidal shape and contains frequencies in the range of 120 to 200 kHz, which are minor attenuated. As a contrast, the positive steepened signal portions have a saw tooth – like shape-containing frequencies up to the high megahertz range, which are particularly attenuated in tissue.

33.2.5 Mechanical Effects of Sound Waves Caused by Nonlinear Properties of the Media

So far, we have discussed the propagation of sound in a linear medium that does not change its properties no matter how high the pressure amplitudes are. Essentially all waves in a medium propagate nonlinearly. Therefore, we need to discuss some of the nonlinear effects in ESWL and ESWT.

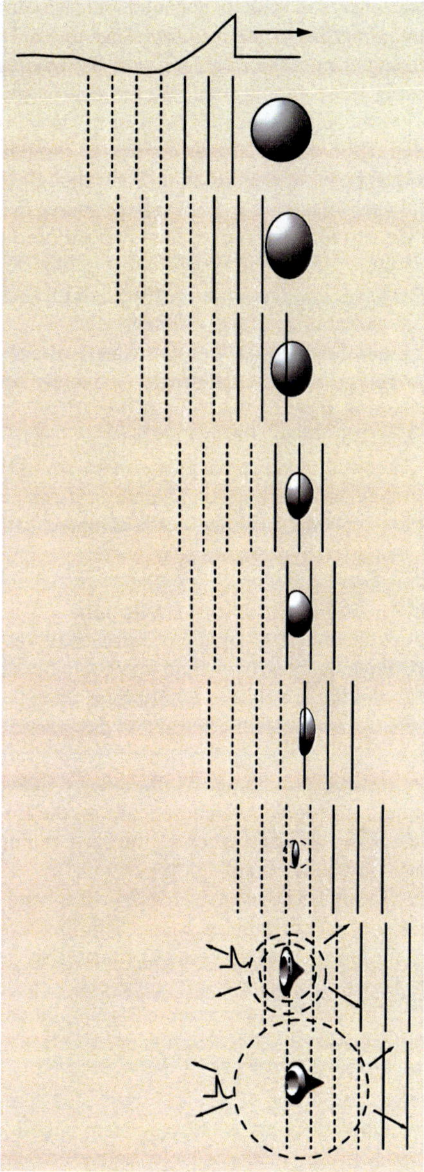

Fig. 33.5 If a cavitation bubble is hit by a shock wave, then it collapses. During the collapse a water jet is generated in the direction of the shock wave propagation or toward a nearby surface (after [33.41])

Steepening of the Positive Pressure Portions of Sound Waves

High local positive pressures result in a strong local compression and thus a local increase in medium density. Simultaneously the velocity of sound increases due to the elastic properties of the medium. Thus, the trailing positive signal portions catch up with the leading portions during wave propagation. At the end of this process an ideally steep (shocked) front of the wave is built up, even if the wave starts as a sinusoid (assuming the starting positive amplitude is high enough to feed the nonlinear effects). The time and the source distance until the signal is fully steepened depend on the starting pressure of the sound wave, focusing (convergence) of the wave, and properties of the medium such as attenuation and nonlinearity coefficient.

In media with low attenuation (like water), already at rather low energy settings, steepening frequently occurs somewhere in front of the focus of the ESWL or ESWT devices. But the energies at higher frequencies, which are generated by steepening, are more strongly affected by attenuation in tissues. Therefore, it can be assumed that steepening in tissue needs a longer path length or higher starting pressures as in water.

Generation and Excitation of Cavitations

Cavitations are characterized by the occurrence of vacuum or gas-filled bubbles in a fluid medium. Stable cavitation bubbles are in a state of equilibrium when the vapor pressure inside the bubble is equal to the pressure of the fluid.

If the rarefaction pressure of the sound wave exceeds a certain threshold, which depends on the frequency of the wave and properties of the surrounding medium, then chemical bonding forces can be exceeded even in homogeneous media [33.24]. The bonds break up and cavitation bubbles occur. The theoretical cavitation threshold of tissues was estimated to be -16 MPa [33.33, 44]. though cavitation events can be observed at significantly lower rarefaction amplitudes during lithotripsy. Most likely cavitation occurs close to stones and inside vessels [33.45].

The rarefaction portion of a sound wave is also responsible for another cavitation effect – micro bubbles, which may exist as remainders of larger collapsed bubbles start to grow during the rarefaction pressure portion of the sound waves. They may reach a stable size, up to three orders of magnitude larger than the original size. These bubbles may remain stable for several hundred microseconds.

After this stable phase, oscillations of the bubble occur until it is reduced to its final size. Every time during oscillation when the bubbles reach their minimum volume, they emit a shock wave to the surrounding environment, called a collapse wave. Afterwards, they start growing again and the cycle is repeated. The number of oscillations as well as the maximum size of the bubble depends largely on the properties of the surrounding medium. Occasionally these bubbles may be observed for more than 1 s [33.37]. In large vessels, bubble clouds could be identified by ultrasound images [33.46]. Other research groups have used an external sensor to measure cavitation signals extracorporeally during lithotripsy [33.47].

When a shock wave hits a cavitation bubble, the excess external pressure leads to a shrinking of the bubble, while the bubble absorbs a part of the sound energy. When the exciting forces and energies are large enough, this leads to a forced bubble collapse. During the collapse, a part of the energy stored inside the bubble is released as a new acoustic wave to the surrounding medium. The maximum collapse energy is released when the original bubble radius is around 500 μm (these values are valid for water and typical lithotripter shock waves). The bubble collapse takes place about 2–3 μs after the shock wave has passed by. The collapse pressure amplitude is about 1/10 of the original shock wave pressure and its duration is typically 30 ns. Thus, the acoustic energy of the collapsing bubble is 1000 times less than that of the exciting shock wave. Larger bubbles do not collapse that hard because they are slowed down by the rarefaction phase of the shock waves.

As a shock wave usually hits a bubble only from one side, the bubble collapse tends to be asymmetric, in particular close to interfaces. In this case, a water jet in the direction of the nearest surface is created (Fig. 33.5). The water jet can reach velocities as high as 400 to more than 800 m/s. Thus, the jet has enough power to perforate aluminum foils and plastic membranes.

During a bubble collapse, local effects may harm the tissue, which leads to small bleedings (petechiae) inside the tissue and at the skin. At cell level, an increase in the permeability of cell membranes is attributed to the effects of cavitation [33.9, 48].

Around the year 1990 researchers tried to find out if a bubble collapse would be able to create free radicals. To date, free radicals have only been identified in extracellular space. The question of whether free radicals caused by collapsing cavitation bubbles might also appear inside cells is still controversial [33.49, 50].

Trailing after the positive pressure pulses, the signal of every pressure pulse source also includes rarefaction portions. Due to these rarefaction portions, bubble seeds several micrometers in size grow to large bubbles of several hundred micrometers. Due to these bubble seeds, new active cavitation bubbles are created. However, the exact thresholds for this process are not defined. Depending on the water purity and gas content, values between −0.5 and −20 MPa can be found. For tissue a theoretical cavitation threshold of −12 to −16 MPa was estimated [33.51]. Furthermore, the duration of the rarefaction signals additionally plays an important role. The shorter the signal duration, the higher rarefaction pressures are tolerated by the tissue. In animal experiments with lithotripter pulses, the first damage occurred at about −1.5 MPa in mouse intestine [33.52]. Notably, lung tissue is very susceptible to damage [33.53].

Bubbles can be observed by high-speed photography at quite a large distance in front of the focus on the axis of a lithotripter pressure pulse source in water [33.54]. Recent measurements during stone treatments also demonstrated the occurrence of cavitation around the stone [33.55].

The amount of cavitation occurring in different tissues apart from the kidney is largely unknown. In water, bubbles may reach several millimeters in size. In tissue cavitation bubbles behave much differently. Most likely cavitation occurs in vessels and capillaries with a diameter of about 100 μm. These vessels contract immediately after the first shock waves [33.56]. It can be expected that the maximum bubble size is limited by the diameter of the vessels [33.53, 56, 57]. The collapse of such small bubbles is significantly less powerful. Even a minor static excess pressure of 0.1 MPa is able to suppress cavitation completely [33.58].

It is of vital importance to restrict the maximum pulse repetition rate in order to avoid undesired side effects due to cavitation and its interaction with the trailing shock wave [33.59]. Besides reduced tissue damage, the number of shocks needed to disintegrate a given stone mass is also reduced when the pulse repetition rate is limited to 60/min or even less [33.60].

33.2.6 Thermal Effects Caused by Pressure Pulses

The pulse duration of a clinical pressure pulse or a shock wave is very short; as a rule, the time duration of the complete wave is only 3 to 5 μs. Nevertheless, a peak power of more than one megawatt per pulse can be reached. The average sound energy per pulse in the focal region is 10 to 150 mJ. Depending on the pulse intensity, the energy setting of the device, the pulse repetition rate, the stone material, and the heat conduction properties of the surrounding medium, the stone may warm up during the course of a treatment. In vitro, a maximum temperature elevation of 1.5 °C was measured [33.61].

33.3 The Acoustic Field of a Lithotripter – Basics of Measurement Technology

The following description of the acoustic field of lithotripters will be restricted to those parameters that are most relevant in practice. They are defined in the international measurement standard IEC 61846 Pressure Pulse Lithotripters [33.34, 62]. The correlation of these parameters to stone comminution is shown in Table 33.5. The most relevant parameter for stone comminution is the acoustic energy that hits the stone, i. e., the energy distributed in the cross-sectional area of the stone [33.63, 64]. The biological importance of the pressure pulse parameters is less understood and will be discussed in a later chapter.

To evaluate the pressure pulse parameters, measurements must be made in degassed water according to the International Standard IEC 61846 (because no internationally acknowledged standard tissue phantom is defined). Sound waves are less attenuated in water as compared to tissue and the nonlinearity parameters of water are different from those of tissues. Therefore, the measurement results cannot be directly applied to the situation in an individual patient. If a wave propagates in tissue, then the sound signals are influenced by the tissue layering, the attenuation of sound, and the nonlinearity parameters. As a rule, the greater the attenuation of the amplitudes, the greater the reduction of the amplitudes of higher frequencies in the pressure pulse signals. Thus the rise time of the pulses increases, resulting in an individual change in focal geometry [33.39, 40].

33.3.1 Sensors and Measurement Technology

New sensors for shock waves and pressure pulses continue to be developed. The most important parameters for these sensors are as follows:

- Small effective diameter to allow for measurement of parameter changes in small-scale highly focused fields, where the pressure amplitudes may vary by > 50% within a range of 2–5 mm;

- Broadband from < 100 kHz to > 10 MHz with low sensitivity variation (< 3 dB) [33.76, 77];
- Resistant to at least several thousand pulses;
- Known linearity over a broad pressure range (less than −10 MPa to more than 50 MPa);
- Constant calibration and easy recalibration during use;
- Appropriate price per pulse;

Table 33.5 Hydrophones used for lithotripter measurements. A comparison of the wave forms of the most frequently used hydrophones is given in Fig. 33.6

Sensor type	Advantages (+)/disadvantages (−)	In use since	Literature/ comments
Piezoceramic sensor	+ Simple handling + Long service life − No good signal reproduction	1970	Used as hydrophone for quality control (long term stability of lithotripters)
PVDF membrane	+ Good signal reproduction − Cavitation damage − Short service life − Very high prize − Large sensor head dimensions	Ca. 1982	[33.65] [33.66] Focus hydrophone according to IEC 61846; Reference hydrophone according to FDA
Contact-free PVDF membrane	+ Long service life − Relatively large effective aperture	1992	[33.67]
PVDF needle (IMOTEC)	+ Long service life + Reasonable price per pulse − P-reproduction varies − Cavitation damage	1987	[33.68, 69] Mostly used hydrophone in the past years; field hydrophone according to IEC 61846
Fiber optic hydrophone using the reflected light at the fiber tip, which is modulated by a transient change of water density by the pressure pulse	+ Good signal reproduction + Long service life + Reasonable price per pulse + Good P-reproduction + Easy calibration − Fragile fiber needs frequent maintenance − Handling requires much training	1988	[33.70] Focus Hydrophone and field hydrophone according to IEC 61846
Fiber optic hydrophone based on interferometer principles	+ Good signal reproduction + Good P-reproduction + Can be used in vivo − Handling only by trained personnel − Fragile fiber needs frequent maintenance − Service life?	1998	[33.71, 72]
Light spot hydrophone (LSHD), using the reflected light at a glass–water interface, which is modulated by a transient change of water density by the pressure pulse	+ Good pressure signal reproduction + Good reproduction of rarefaction pressure + Robust and durable, glass block can be easyly exchanged, usually only slight adjustments of the glass block position required after damage, without the need to leave the measurement position + Calibration can be checked at any time − Large sensor head dimensions	2004	[33.73]
Steel ball sensor	− No good signal reproduction + Long service life + Reasonable price per pulse	1991	[33.74] Quality control hydrophone and measurement of cavitation signals
Capacitive steel block sensor	− Signal fidelity low + Very durable + Low price per measurement	1990	[33.75] For quality control

- Hydrophones must satisfy the requirements of IEC 61846 for a focus hydrophone or field hydrophone depending on the intended use;
- High adhesion to water to avoid cavitation at the hydrophone surface.

With a hydrophone made of plastic material like PVDF the last point in particular is very difficult to achieve. Rarefaction pressures of more than approximately −5 to −10 MPa often cannot be measured with sufficient reliability with such hydrophones due to the signal distortions caused by cavitation.

An additional major source of problems is the low cavitation resistance of the PVDF membrane hydrophone electrodes. They are very sensitive to pits caused by cavitation jets, which results in electrical interruption of the sputtered leads and significantly decreased sensitivity.

PVDF needle hydrophones are made of a PVDF-coated steel needle. Therefore they are less sensitive to damage [33.68]. In recent years, this type of hydrophone has frequently been used as a standard shock wave hydrophone because of its comparably long service life and reasonable price [33.26, 30, 69], but it has significant disadvantages. The metal needle is mechanically excited by shock waves and thus generates additional signals interfering with the trailing parts of the pressure pulse, in particular with the rarefaction part. Therefore, the reproduction of rarefaction portions and most of the signal energies are not reliable with this type of hydrophone (Fig. 33.6).

Hydrophones made of glass fibers are particularly advantageous because the adhesion of water to glass is higher than the cohesion of water itself [33.70]. Thus, as cavitation at the probe surface is less likely, rarefaction pressures can be measured with high reliability. If any cavitation events occur at an optical hydrophone due to gas bubbles in the fluid, they can be identified by a rapid leap in the hydrophone output voltage, as the reflectivity of the laser light will change instantaneously in the presence of a gas bubble. As an alternative to the fragile glass fiber, the active surface of the light spot hydrophone (LSHD) is composed of a thick glass block, which also is able to tolerate very large rarefaction pressure amplitudes without causing cavitation [33.73].

In conclusion, because of all these problems and the extremely high price of fragile PVDF membrane hydrophones, the state of the art in measurement technology are optical hydrophones such as the fiber-optic hydrophone and the LSHD.

33.3.2 Measurement Parameters of Lithotripter Sound Fields

International Standard IEC 61846 describes the measurement parameters in detail. Further requirements, in particular for ESWT devices, are specified in a consensus paper of the Technical Working Group of the German Society for Shock Wave Lithotripsy (DGSL) [33.78]. The DIGEST (German and International Shock Wave Therapy Society) also accepted the parameters and gave further advice on their application.

The correlation of the parameters with the comminution of stones and with biological effects shall be discussed in the following sections.

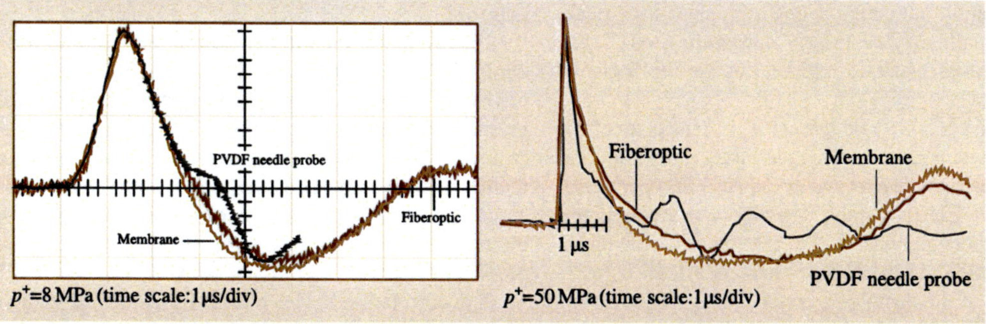

Fig. 33.6 Comparison of the focal wave form of a pressure pulse source at different energy settings, measured using three different hydrophone types. The amplitudes were calibrated to read the same $p+$ value. While the rise time and the $p-$ amplitude are reproduced comparably, the pressure-time curves and thus the energy flux density values measured by the PVDF needle probe are significantly different. In contrast, the PVDF membrane hydrophone and the fiber-optic hydrophone display almost the same curves

Pressure – Time Parameters of Single Waves: Pressure, Intensity, Energy Flux Density, Pulse Duration, Rise Time

The *amplitudes* of the *positive* and the *rarefaction pressure* are read from oscilloscope recordings of single pressure pulses (Figs. 33.2 and 33.6). The significance of the peak positive pressure values for the efficacy of lithotripters were often overestimated in the past. The actual pressure value seems not to have a high correlation to stone comminution as long as the pressure exceeds a certain (stone-material-dependent) threshold.

The *rise time* is determined as the difference in time between the 10 and 90% points of the positive pressure amplitude. In order to determine a typical shock wave's rise time of $< 5-10$ ns, it is necessary to use oscilloscopes and sensors with a frequency bandwidth of 500 MHz and sampling rates of 1 GSample/s. Additionally, the effective sensor area has to be adjusted exactly parallel to the shock wave front. Otherwise, parts of the sensor receive the wave with a temporal delay and thus the rise time of the wave appears to be longer. On the other hand, rise time has not been identified up till now as a very important parameter in the usual ESWL and ESWT regimes (< 5 ns to > 250 ns); thus the measurement efforts can be slightly relaxed. Measuring with a reduced bandwidth (e.g., 20 MHz) will reduce the noise, thereby enhancing the measurement quality.

The *pulse width* is defined as the signal duration between those points in time when a level of 50% of the maximum positive pressure amplitude is reached. When the wave form is complicated, this pulse width definition may lead to misinterpretations, e.g., if the wave has a long duration but contains one high short-time peak, its pulse width may be largely underestimated. These kinds of waves occur in regions outside the focus of a lithotripter.

The *pulse intensity integral* (PII) is a measure of the average power per pulse. A synonymous, more common term in lithotripsy is *energy flux density* (ED) per pulse. The unit of ED is J/mm^2. Usually a manufacturer's technical specifications of the ED state the maximum value measured in the focus. There is a high correlation of ED to the depth of a crater that is eroded out of model stones after a number of shocks. On the other hand, in vitro experiments with perfused pig kidneys also showed the relevance of ED+ (ED of the positive portion of the pressure pulse) in determining the risk of tissue damage [33.79].

Spatial Parameters of Signals in the Pressure Pulse Field

When the location of the measurement is only several millimeters away from the focal maximum, the shape of the pulses changes significantly [33.80]. The rise time of the signal increases quickly at positions other than the focal maximum [33.81]. The pressure-time curves before and after the focus depend on the kind of pressure pulse source. In the case of spark-gap sources, a shock wave is found before the focus. In the case of electromagnetic shock wave emitter (EMSE) and piezo sources that emit less steep signals, the shock front may only occur several millimeters before the focal region, depending on the energy setting. It should be noted that the steepness of the front is of minor significance for the disintegration of stones [33.82].

Axial and Lateral Pressure Distribution, Full Width Half Maximum or (−6 dB) Width

Axial and lateral distributions of the positive pressure in the focal region usually are plotted in a graphic representation (Fig. 33.7). From these curves single values are extracted as a compact description of the focus according to the standard IEC 61846. According to the standard, the *full width half maximum* (FWHM) values of the axial and lateral pressure distributions are stated. The FWHM values give the spatial distance between those points where the pressure amplitude has 50% of the focal peak value. The term −6 dB width is used frequently instead of the term FWHM.

The *focal volume* is calculated from these axial and lateral −6 dB values under the simplifying assumption that the focal region is shaped like an elliptical cigar that extends along the direction of the wave propagation path.

However, the medical significance of these (standardized) values is not clear. All those values are based on measurements relative to the peak positive focal pressure, which depends on many technical design parameters. Besides the energy setting, the peak pressure mainly depends on the aperture angle of the source (the angle of a cone between source aperture and focus) – the larger the aperture angle, the higher the focal peak pressure and the smaller the focal FWHM values.

The correlation between pressure-based values and stone comminution effect is low. In contrast, it is shown below that the effective energy is very important for stone comminution. Also, in most publications, biological effects are described in terms of intensity and ED values [33.79, 83].

33.3.3 Energy Parameters of the Sound Field

The ED is calculated by integrating the squared pressure pulse signal; therefore, ED is also called the PII. Lateral from the axis, the ED values decrease at about the same rate or faster than the peak pressure values because ED depends on the square of the pressure.

The focal energy is calculated by integrating the ED values in the −6 dB focal area. So the term *focal energy* specifies the energy that is evaluated in a circle around the focal spot (point of maximum pressure) with a diameter of the FWHM [(1) in Figs. 33.7 and 33.8].

Effective energies are calculated by integrating over a certain effective area, which is defined from certain effects. An important effective area is a circle with a radius of 6 mm around the focus, which is a typical stone radius in ESWL. It is important to note that a comparison of the energies of different pressure pulse sources is only possible based on the same effective areas, as a comparison based on the focal energies within the FWHM would be misleading when the focal dimensions are different.

The *total energy in the focal area* is approximated by $E_{5\,\mathrm{MPa}}$. This energy is calculated by integrating the EDs over the focal area, which is delimited by the radius where the pressure just exceeds 5 MPa.

The total energy of larger sources (EMSE, piezo) is about the same as the emitted primary acoustic energy. In contrast, in sources that use reflecting mirrors for focusing purposes, the total energy additionally depends on the portion of the (spherical) wave that is reflected by the mirror. Thus in reflector systems the total energy sometimes is less than half of the primary acoustic energy. The nonfocused portion also hits the patient but does not contribute significantly to stone comminution. It is possible that the unfocused wave portions contribute to the dynamics of cavitation bubbles.

In general, the larger the aperture angle of the system, the larger is its focal pressure as well as the focal ED at a given level of primary acoustic energy. In contrast, the lateral focal dimensions decrease with increasing aperture angle. Thus, a pressure pulse source

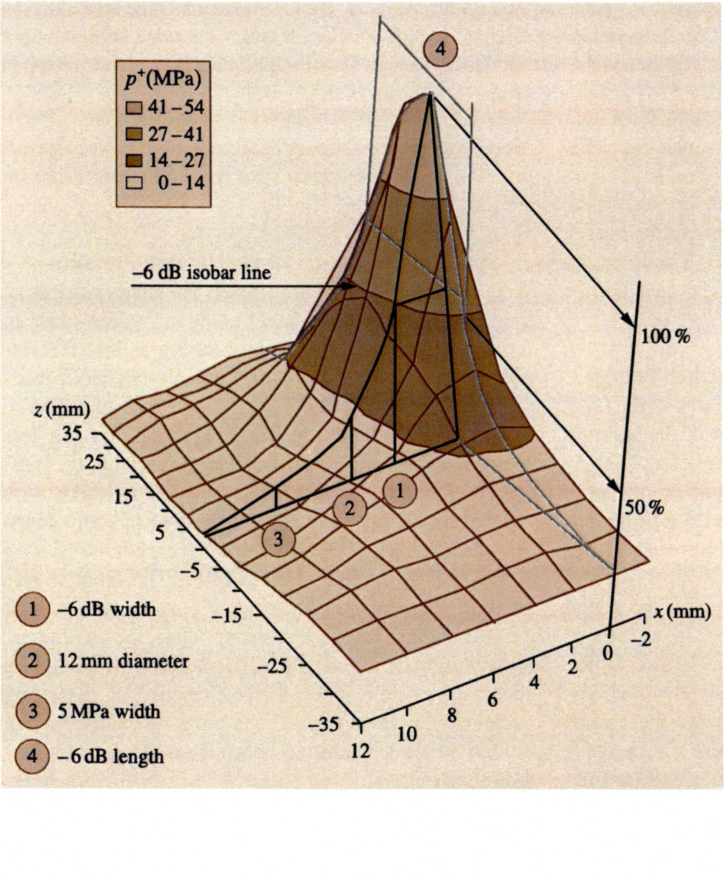

Fig. 33.7 Many spatial pressure field parameters are derived from measurements of the axial and lateral $p+$ distributions, according to the standard. In contrast, for the assessment of the disintegration efficacy, the effective energy $E_{12\,\mathrm{mm}}$ in the circle delimited by a 6 mm radius and the effective energy $E_{5\,\mathrm{MPa}}$ in the circle delimited by the 5 MPa pressure value are much more important

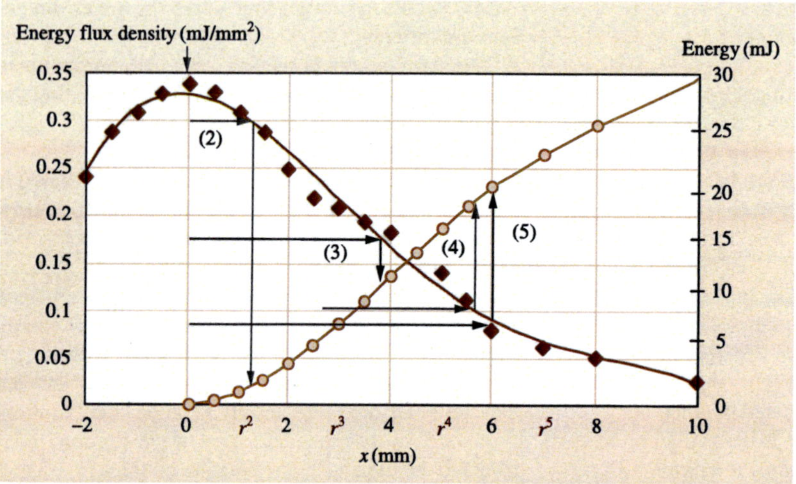

Fig. 33.8 Energies (*circles*) are calculated by spatial integration of the energy flux density values (*diamonds*). In circular symmetric pressure pulse sources, the energy can be calculated by integration over a certain radius. Biologically effective areas are delimited by the radius where the ED threshold (0.1 mJ/mm^2) for the generation of stress fibers or the detachment of endothelium (0.3 mJ/mm^2) occurs. Radii important for the stone comminution are 6 mm, and the 5 MPa limit. (1) Focal energy flux density (ED) (2) Radius r_2 delimiting the area where ED is larger than 0.3 mJ/mm^2; (3) Radius r_3 delimited by the half-maximum focal pressure value FWHM. Integrating inside this area calculates the focal energy; (4) Radius r_4 that encloses ED values larger than 0.1 mJ/mm^2; (5) Radius r_5 (here: 6 mm) that delimits the area of a typical model stone or stone concretions. Integration yields the energy $E_{12\,\mathrm{mm}}$, which is relevant for stone comminution. Integration up to the boundary of the measured area yields the total focal energy. Usually this value is well approximated by an integration in the area delimited by the 5 MPa pressure values (not shown in the figure)

with a large aperture angle may have a high focal ED even though the total energy is low.

A significant pressure pulse effect on biological tissues and on stones can only occur if the energy thresholds of the material are exceeded. These energy thresholds are difficult to evaluate; some values will be given below.

Primary Acoustic Energy and Electrical Energy

The primary acoustic energy is concentrated in the focal area by focusing. Depending on the type of the pressure pulse source, it is not possible to focus the complete primary acoustic energy as in systems with a point source focused by a mirror. An (additional) portion of the energy can be shaded by inline scanners or other structures in the sound field [33.30].

The electrical energy necessary to drive the sound source is usually stored in a capacitor. It is calculated from the capacitance multiplied by the square of the charging voltage. Only a fraction of this electrical energy is transformed into acoustic energy; the rest is dissipated as heat in the source. Different source types have differing electroacoustic efficiencies and use different capacitor voltages. Therefore, neither voltage nor primary energy is appropriate for comparing different sources.

Spatial Distribution of Rarefaction Pressure

In the early years of lithotripsy, the measurement of the rarefaction pressure amplitudes in regions close to the focus was unsatisfactory. Already at low rarefaction pressures of several megapascals cavitation may occur [33.24], in particular at the surface of a PVDF hydrophone. In addition, some hydrophones are less sensitive to rarefaction pressures. Furthermore, the trailing rarefaction signal parts sometimes interfere with reflections of the wavefront at the hydrophone. Since the introduction of the fiber-optic hydrophone as well as the LSHD, the reproduction of rarefaction pressures has been significantly more reliable, as the adhesion at the water–glass interface is strong. Thus, the occurrence of cavitation on a glass surface is unlikely. If cavitation occurs nevertheless due to the presence of cavitation nuclei on the glass surface, the onset of cavitation can be immediately identified by a signal jump in the hydrophone output.

The primary acoustic signal generated by an EMSE or a spark-gap source does not contain significant rarefaction pressures. Nevertheless, rarefaction occurs in the focal region along the axis. It is caused by the diffraction wave at the rim of the sound source or reflector. This rarefaction wave appears to be a mirror image of the positive pressure wave, which is emitted by the outer ring of the aperture.

In the design phase of a pressure pulse source the suppression of rarefaction signal portions is one important point. As the rarefaction signal is not affected by nonlinear steepening, its amplitude in the focus remains lower than the steepened positive pressure amplitude (ca. 1/10 to 1/5). The usual duration of the rarefaction signal is about the same as the duration of the primary pressure pulse (2–5 μs).

Due to nonlinear effects, the positive pressure focus is axially shifted away from the geometric focus of the source by some small amount. In contrast, the focus of nonsteepened signals is shifted slightly closer to the source. Therefore, the maximum of the rarefaction pressure may occur several millimeters before the positive pressure maximum.

The piezo source generates a trailing additional rarefaction pressure signal that merges with the edge diffraction wave in the focus. This additional rarefaction pressure can be minimized by certain electrical and mechanical design features.

33.3.4 Significance of the Pressure Pulse Parameters in Lithotripsy

In the early years of urethral stone lithotripsy, most of the sound field parameters shown in Table 33.6 could not be measured with sufficient accuracy and usually were not published. Either the peak positive pressure or the capacitor voltage of the spark-gap source was used to characterize the treatment parameters. Of course, as soon as different generator types (electrohydraulic, EMSE, piezoelectric) appeared, this led to confusion and misinterpretation. The characterization of clinical study parameters based on capacitor voltages does not provide sufficient information to compare different treatment regimes and devices. One of the major advantages of the international measurement standard IEC 61846 is that it improves the comparability of data and study results.

Table 33.6 shows the correlations between different standardized physical parameters and the comminution of stones. It can be concluded that the effective energy impinging on the stone is the most important parameter for the eroded volume (i.e., $E_{12\,\text{mm}}$ and $E_{5\,\text{MPa}}$), whereas the pulse intensity distribution determines the depth and shape of the eroded craters. The 5 MPa value, which was used to determine the boundary of the eroded area, is a fair estimate of the disintegration threshold of stone materials (ca. 2 to 10 MPa [33.84, 85]).

All energy values include a dose effect. The number of shocks S_z per stone volume V increases proportionally as the effective acoustic energy per shock E'_{eff} decreases, given by $E'_{\text{eff}}/V = \text{const}_{\text{Material}}/S_z$ [33.28]. The material constant ($\text{const}_{\text{Material}}$) depends on the stone composition, and its value for model stones is ca. 2 mJ/ml [33.63].

Stone comminution only occurs when a certain threshold energy per shock E_0 is exceeded, which is material dependent. Thus the effective energy that contributes to comminution is $E'_{\text{eff}} = E_{\text{eff}} - E_0$. The endpoint of comminution is reached when all concretions are crushed to less than 2 to 3 mm because these fragments can pass the urethra spontaneously without causing many problems.

Other parameters like peak pressure, rise time, and pulse width seem to play minor roles in lithotripsy – at least as long as they are in the typical ranges shown in Table 33.6.

The most important parameters for other treatment modalities have not yet been identified. In pain therapy (ESWT), thresholds of the ED seem to play a major role (see below). Therefore, in the future it may become important to introduce energies based on different areas whose boundaries are determined by ED (PII) threshold values [33.81, 86, 87]. These threshold values need to be evaluated by biological and other model experiments.

According to the standards, the −6 dB focal dimensions are calculated from those positions where the positive pressure has decreased to 50% of the focal maximum value. Thus, all those focal dimensions strongly depend on the focal peak pressure. This has a number of implications. The stronger the focusing of a sound source, the higher is its peak pressure and consequently the smaller its focal diameter becomes. However, as the disintegration, and thus the effective focal diameter, does not depend on peak pressure but on a threshold value (e.g., 5 MPa), this implies that the statement of a −6 dB focal extent is not a valid measure for the efficacy of a lithotripter or an ESWT device. Nevertheless, there is an ongoing discussion about which device is better: one with a small or one with a large focus (see below).

Symbol	Parameter	Unit	Typical value range	Significance	Comment
Focal parameters				According to standard IEC 61846 measured at the location where $p+$ is maximum in the sound field (acoustic focus) [33.34]	
$p+$	Peak positive pressure	Megapascal, MPa	7 to > 80 MPa (ESWT) 20 to > 100 MPa (ESWL)	Low correlation to stone disintegration* (0.704**) (0.54***)	
$p-$	Peak rarefaction pressure	Megapascal, MPa	−3 to −15 MPa	Correlation to disintegration unknown* (0.394**)	Significant for the creation of transient cavitation
T_r	Rise time (measured from 10 to 90% of $p+$)	Nanoseconds, ns	< 5–500 ns	Low correlation to stone disintegration* (0.006**)	Varies significantly in the sound field lateral of the focus
T_w	Pulse width (measured from 50 to 50% of $p+$)	Nanoseconds, ns	< 150–> 500 ns	Important to determine the integration limits of PII and energy parameters	
P_{II+}, ED+	Pulse intensity integral or energy flux density of the positive signal portion	Millijoules per millimeter squared, mJ/mm^2	< 0.02–0.8 mJ/mm (ESWT), < 0.1– > 1.2 mJ/mm (ESWL)	Correlates to the depth of a crater in stone material as well as to effects on cells*	Lateral distribution of P_{II+} correlates to lateral distribution of $p+$ [33.81]
P_{II}, ED	Pulse intensity integral or energy flux density of the complete wave (until the absolute value is less than 10% of the peak amplitude)	Millijoules per millimeter squared, mJ/mm^2	< 0.03–1.0 mJ/mm^2 (ESWT), < 0.1– > 1.2 mJ/mm^2 (ESWL)	Correlates very good to the depth of a crater in stone material (0.99***) as well as to effects on cells*. Correlates fairly to the eroded volume. (0.801**) (0.75***)	Lateral distribution of P_{II} correlates to lateral distribution of $p+$ [33.81]
Field parameters					
w_x, w_y, w_z	Focal extensions measured from pressure distributions in X, Y (lateral) and Z (axial) directions between the locations of 50% of $p+$	Millimeters, mm	Lateral: < 2–> 8 mm Axial: < 15–> 100 mm	Low correlation to stone disintegration*	Often misinterpreted as significant for Disintegration power of a source
V_f	Focal volume, usually approximated by $V_f = \pi/6 w_x w_y w_z$	Cubic millimeters, mm^3	< 100– > 1 000 mm^3	Correlation to stone disintegration unknown	Often misinterpreted as significant for disintegration power and sensitivity towards misplacement of a target (see below)
$E+$, E	Focal energy derived by integrating the P_{II} distribution to the lateral focal extensions w_x, w_y. The time integration limits either include the positive signal portion for $E+$ or the complete signal for E	Millijoules, mJ	$E+$: 0.5–180 mJ E: < 1.5–230 mJ	Low correlation to stone disintegration* (0.442**) (0.62****)	Parameter must not be used to compare different pressure pulse sources, if their focal width w_x, w_y is different
$E_{eff}+$, E_{eff}	Effective energy derived by integrating over a predetermined area, e.g. a circle with 12 mm diameter: $E_{12\,mm}$ or a circle including pressures > 5 MPa: $E_{5\,MPa}$	Millijoules, mJ	$E+$: 0.5–65 mJ E: < 1.5–80 mJ	Highest correlation of all energy parameters to disintegrated stone volume ($T_{total\,0}$: $r = 98$***) ($E_{5\,MPa}$: $r = 0.97$****)	Most important parameters to determine stone disintegration, see text

Table 33.6 Parameters describing the sound field of a pressure pulse source [33.34, 78]. The correlation of the parameters with stone disintegration and biological effects marked * were taken from [33.26, 82] and [33.89]. The ** numbers in brackets are correlation coefficients, composed of experiments with six different pressure pulse sources [33.90]. The correlation coefficients marked *** were taken from [33.26], those marked **** are from [33.82]. The parameters are shown in Figs. 33.2 and 33.6–33.10 ◀

Instead of using the −6 dB focal width for the characterization of a lithotripter, a more relevant parameter will be proposed: the *disintegration diameter*. This value is determined by the diameter of the area that is bordered by an increase, by a factor of 2, in the number of shocks necessary for the disintegration of a typical concretion, as compared to the shock number in the focus (Fig. 33.9). Model stone (12 mm diameter) measurements with different lithotripters have demonstrated that the disintegration diameter does not depend on the standardized (−6 dB) focal dimensions. The disintegration diameter seems to depend only weakly on the pressure pulse source design (spark gap or EMSE); in particular there is only weak dependence on the aperture angle of the source (angles between 54° and 74° were tested). The disintegration diameter turned out to be 18 to 22 mm [33.88]. The experimental setup and measurement results are shown in Figs. 33.9 and 33.10.

33.3.5 Which Physical Effects Play a Significant Role in Stone Comminution?

A number of effects that contribute to stone comminution were published recently:

- Direct action of the pressure pulse forces on a stone leads to cracks due to stress and to the growth of cracks [33.91].
- Strongly focused waves lead to a cone-shaped erosion of the stone [33.92].
- Spallation effects are caused by the reflection of the pressure pulse at the rear side of the stone, including polarity reversal of the reflected wave (Hopkins effect) [33.91]. The reflected pulse overlaps with the rarefaction portion of the impinging pulse, leading to high strain forces inside the stone. This effect explains findings on impacted concretions that are disintegrated from the back toward the front.
- Cavitation effects are responsible for the expansion of cracks. They also induce extremely high amplitude collapse pressure pulses of a few nanoseconds' duration directly on the stone surface, and, due to water jets with speeds of 200 to 800 m/s, they are able to drill pits into the stone surface [33.71, 91]. Homologous craters can be identified in microscopic images at the stone front facing the wave.

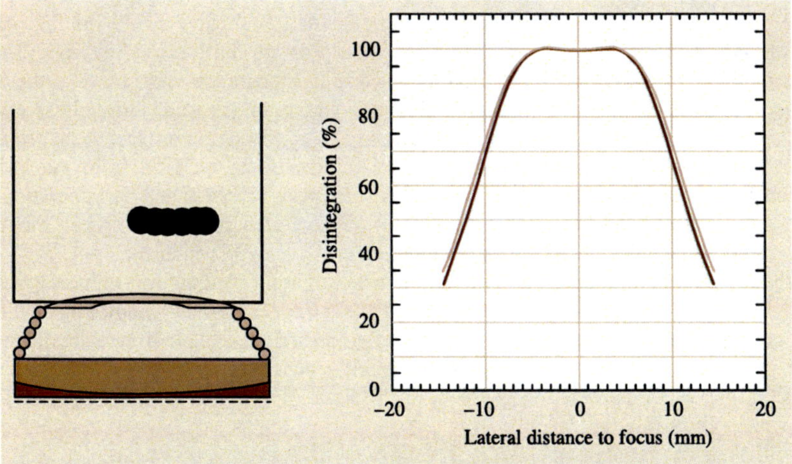

Fig. 33.9 The number of pulses for the comminution of a 12 mm stone to concretions < 2 mm in size is doubled when the stone is displaced by ±11 mm beside the focus. The experiment was carried out with different pressure pulse sources (spark gap, aperture angle 54°, and EMSE 62°). Nevertheless, the measured differences in disintegration are minor. The *disintegration diameter* $f'x$, which is defined by a decrease in the disintegration efficacy of 50%, which means doubling of the number of pulses is 22 mm in both sources

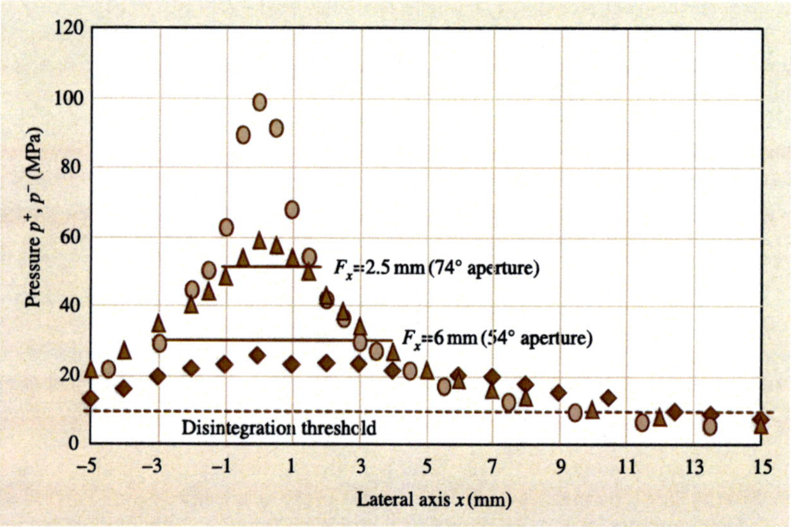

Fig. 33.10 The focal pressure distributions of stronger and weaker focusing sources mainly show large differences in the area of ±2 mm around the focal maximum. Thus the standardized (−6 dB) focal width F_x of the strongly focused source (*circles*) is only 2.5 mm, the standardized (−6 dB) focal width F_x of the next weaker focused source (*triangles*) is 6 mm, and that of the weakest focused source is 18 mm (*diamonds*). Nevertheless, the effective disintegration diameter of all three sources is almost identical (18 to 22 mm), as the disintegration threshold of ca. 10 MPa is reached at almost the same diameter with all three sources

- Clusters of cavitation bubbles collect around the stone. During collapse, they act as a water hammer [33.93].
- Due to different velocities of wave propagation inside and outside the stone, mechanical strains are created by the passing waves, which lead to binary fraction of the stones (fraction into two pieces) due to growing micro cracks [33.84].

The contribution of each effect to the final stone comminution is not yet clear. It can be estimated that the importance of every effect varies depending on the surrounding medium around the stone (fluid, in part or completely impacted, e.g., in the shoulder tendon or in the urethra). If a reasonably low static positive pressure restricts cavitation, then stone comminution is reduced by ca. 20%, i. e., the number of shocks necessary to crush the stone into pieces smaller than 2 mm increases by 20%. If the stone is surrounded by a medium of higher viscosity, e.g., castor oil, then the number of shocks also increases and the final concrements remain larger. This occurs due to the suppression of cavitation in the oil [33.94]. At higher surrounding static pressure or impacted stones, the additional static forces can lead to slower growth of cracks [33.84].

Different stone materials need typical energies for the cracking of a unit volume into pieces smaller than 2 mm [33.30, 95]. The typical energy value of model stones made of plaster, which are frequently used for lithotripter test purposes, is about 2 mJ/ml [33.63]. At the same time, the pressure must exceed at least 2 to 10 MPa [33.84, 85]. Disintegration takes place at those parts of the stone where the pressure threshold is exceeded. The amount of transgression seems to be of minor importance, which can be seen from the low correlation between peak pressure and disintegration. The amount of disintegrated stone volume per shock depends on the effective energy per shock.

However, when treating patients the influences of the tissue as well as the geometrical conditions of the sound path play a fundamental role in overall treatment success. These influences will be discussed in a later section of this paper.

33.3.6 Effect of the Pressure Pulses on Tissue

It is of great interest to distinguish between unintended side effects (e.g., bleeding during lithotripsy or irritation of nerves after ESWT), and desired effects. While the

Table 33.7 Energy flux density thresholds for biological effects of pressure pulses on cells

ED, PII (energy flux density)	Finding	Note	
0.3 mJ/mm^2	Complete detachment of endothelial cells, occurring at the entrance side of the pressure pulse Most likely caused by positive pressure	Thrombosis, occlusion of a vessel ED based on PVDF needle hydrophone measurements	[33.83]
0.22 mJ/mm^2	Rounded mitochondria in attached cells	ED based on PVDF needle hydrophone measurements	[33.83]
0.1 mJ/mm^2	Stress fiber formation in adhering endothelium	Reaction to shear forces, likely a reaction to cavitation forces. ED based on PVDF needle hydrophone measurements	[33.83]
0.045 mJ/mm^2	Bleeding of intestine	Mouse model, ED values were estimated from the published signals	[33.52]
0.007 mJ/mm^2	Skin damage	Mouse model, ED values were estimated from the published signals	[33.52]

mechanisms of stone comminution in ESWL are largely known, the healing effects in ESWT are not well understood in a biophysical sense. The shock wave might generate micro defects, which lead to rebuilding processes in the diseased tissues. The focusing of shock waves on soft tissue leads to a release of substance P and the triggering of cellular cascades [33.96].

The following mechanisms are identified:

- A dose effect, which increases with an increasing number of pressure pulses [33.52];
- An energy-dependent effect, which increases with increasing energy per shock;
- The effect of cavitation, which is mainly attributed to a shock wave–bubble interaction [33.97, 98].

The biological pressure limits that can be found in the literature were mostly determined for side effects of shock waves during lithotripsy. Other side effects of lithotripsy have rather biological causes, e.g., by obstruction of the urethral pathways. Altogether severe side effects that require treatment like kidney or subcapsular hematoma occur only in a small number (0.1–0.7%) of treatments [33.99, 100]; a transient hematuria is a more or less frequent effect that is accepted as a common side effect. Cavitation has been identified as the main cause of damage.

The ED thresholds for the generation of stress fibers and stronger effects on cells were evaluated using a human umbilical cord model [33.83, 101]. The effects and associated ED values are shown in Table 33.7. It should be noted that the ED values were determined using PVDF needle type hydrophones and integrating over the complete pressure signal. The large potential errors of this kind of sensor were already mentioned in the measuring technology section above. From the lateral ED distributions of a pressure pulse source it is possible to estimate the biologically effective zones using the threshold values. These zones can be characterized by the diameter of areas comparable to the disintegration diameter given above. Presently the values serve as a coarse guide to the adjustment of devices. In order to assign these values to real treatment situations, they still need to be verified with a better pressure pulse measurement technology and more realistic tissue models. Recent experiments with perfused pig kidneys showed a strong correlation between damage and the ED+ values, while no good correlation to other shock wave parameters were found [33.79].

The limits of unwanted bone damage are only indirectly known. Microscopic and macroscopic fractures were found in most cases at extremely high energy settings, which are rarely used in lithotripsy [33.7, 102–104].

No significant long-term side effects of ESWT or damage that could be shown by diagnostic means have been reported in the orthopedic literature to date, although some damage has been demonstrated in animal tests. Still, there are a number of directions that it is vital to verify if ESWT really is the appropriate treatment for the diagnosed disease [33.105].

The pressure limit for skin damage was determined at 0.6 to 1.6 MPa and for intestinal bleeding at 1.6 to 4 MPa [33.52]. Thermal effects were excluded; thus the most likely damage mechanism is cavitation. The pressure pulses affected blood and other abdominal tissue much less. The ED values may be estimated from published graphical measurement data. The values are between ED = 0.007 and 0.045 mJ/mm^2.

33.4 Generation of Pressure Pulses for Extracorporeal Lithotripsy (ESWL) and Extracorporeal Shock Wave Therapy (ESWT)

In practice, all ESW sources are composed of a pulsed electrical energy source, an electroacoustical converter, and means for the focusing of the pulsed acoustic waves. Focusing is necessary to target the pressure pulse application area. With pointlike sources (e.g., spark gaps, laser discharges) and cylindrical sources, a focusing mirror is the first choice. Lenses are usually applied to focus plane sources (e.g., EMSE), while spherical curved sources (EMSE, piezo) are self-focusing.

At the end of each section, specific differences between the source concepts are discussed in terms of the generation of the pressure pulses and their effect on the wave fronts. However, it must be pointed out that it is not yet possible to make statements on the medical and biological relevance of these parameters.

33.4.1 Point Sources and Extended Pressure Pulse Sources

Depending on the generation principle, one must differentiate between point sources and extended sources. Point sources generate spherically expanding acoustic waves. Extended sources generate planar waves, cylindrically propagating waves, or spherically converging waves, depending on the source shape.

All those propagation modes of the primary sound waves require different means for focusing. Although, due to the nonlinear propagation, from the pressure signal measured in the focus of any source it is not possible to determine the kind of source and the focusing means. As will be discussed below, specific differences rather occur in the sound field outside the focus.

33.4.2 Generation of Spherical Pressure Pulses and Shock Waves

Spark-Gap Source (Electrohydraulic Source)

The first kidney lithotripter devices were equipped with spark-gap sources. This classical principle is still in use today. Comparable to the spark plugs of a car motor, a high voltage (usually 12 to 30 kV) applied to two electrodes a few millimeters apart generates an underwater spark, thereby generating an expanding plasma bubble, which leads to a compression of the surrounding water. After the electrical energy pulse has faded, a vapor bubble remains, which expands further with easing velocity and finally collapses after several hundred microseconds.

The effective primary compression wave is only generated during the existence of the plasma bubble. As this bubble expands at a speed that is greater than the velocity of sound in the surrounding medium, the wave front of this compression wave is extremely steep from the beginning and thus represents a shock wave by physical definition (Fig. 33.11). Its rise time is about 1 to 3 ns.

When the secondary collapse of the vapor bubble occurs, another, much weaker, shock wave develops.

The high voltage is stored in a capacitor of 40 nF to some 100 nF. Its complete energy content is transferred to the plasma bubble at the instant of the spark generation. The acoustic part of this energy is only a fraction of less than 1%.

Ellipsoidal Reflector for the Focusing of Spherical Waves. A rotational ellipsoid has the properties to reconcentrate in the second focus every acoustic event that happens in its first focus. If a target, e.g., a stone, needs to be placed in the second focus (the so-called *therapy focus*), then the ellipsoid must be cut somewhere at a suitable location (Fig. 33.12). The remaining mirror encloses a certain spatial angle of the spherical wave emerging at the first focus. Thus, this part of the spherical wave is focused, while the remaining part penetrates the tissue unfocused. The unfocused wave reaches the second focus earlier than the focused wave. Its maximum amplitude is less than 1 to 2 MPa, which is significantly less than the amplitude of the focused wave.

Nonlinear steepening results in an increasing axial shift of the pressure maximum of the focused wave by ca. 2 to 5 mm beyond the geometric focus F_2.

Special Properties of the Spark-Gap Source. The small preliminary unfocused pressure pulse reaches the focus ca. 40 μs before the main focused shock wave. Almost 1 ms after the focused signal a third, rather weak, pressure pulse may occur, which is generated by the collapse of the primary vapor bubble. It should be noted that this vapor bubble is generated due to the spark-gap principle and should not be confused with the occurrence of cavitation bubbles somewhere else in the field.

Pressure pulses generated by spark gaps usually have the properties of shock waves, which means that the pressure rise at the wave front is extremely rapid. Thus during the propagation in tissue the shock wave is particularly affected by the attenuation of its higher fre-

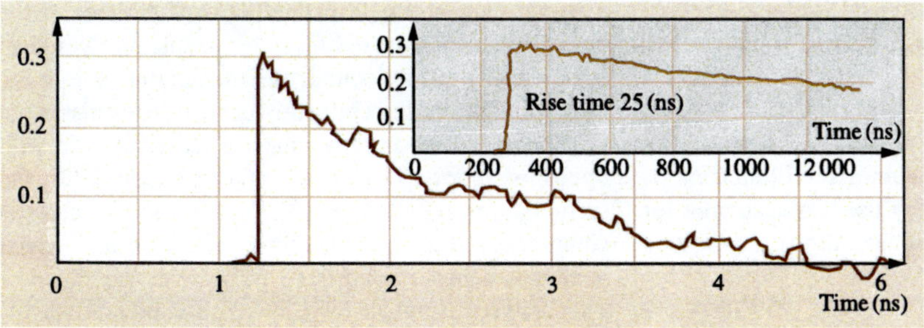

Fig. 33.11 Primary pressure pulse of an electrohydraulic spark gap source. *Small image*: enlarged view of the pressure pulse front with a steep rise typical for shock waves (rise time value delimited by the properties of the hydrophone used)

quencies. It can be assumed that part of the wave energy is completely attenuated before it reaches the focus. Due to the steepening process the high-frequency parts are constantly renewed, which again leads to further energy loss.

Recent Developments in Electrohydraulic Pressure Pulse Sources. The treatment concept of the first lithotripter generation included the use of a new electrode for every patient. This has the undisputed advantage that the electrode wear is well known over the course of treatment. Due to the increasing burn-up of the electrode tips, there are some disadvantages:

- An increasing electrode gap is the cause of an increasing ignition delay. During this delay part of the electrical charge drains off the capacitor. Thus, this part of the primary electrical energy is no longer transformed into acoustic energy.
- Due to the irregular erosion of the electrode surface, which is a side effect of the burn-up, the actual path of the spark, as well as the instant of ignition, fluctuates from shock to shock. This results in a fluctuation of the primary spherical pressure pulse amplitude, as well as a variation of the exact location of the spark. Therefore, the focal pressure varies largely from shock to shock – sometimes up to ±30%.
- It is difficult to determine the true focal dimensions. According to the IEC standard, the measurement has to be repeated several times to get statistically relevant data. However, due to the statistical evaluations, the focal widths appear to be larger, while the peak pressures and EDs seem to be smaller.

Newer developments try to reduce the longer ignition delay due to increasing electrode gaps by increasing the conductivity of the surrounding fluid. For this purpose sometimes an electrolyte is used [33.106]. Due to these measures, the pressure variation from shock to shock is significantly reduced, treatment efficacy increases, and the measured values of the focal dimensions become notably smaller.

Other manufacturers have developed adjustable electrodes. As the electrodes are placed in the center of explosions, the demands for safety and positional accuracy of the tips with respect to the ellipsoid focal spot play a very important role.

A very different approach was published by a Czech research group. They used a metal cylinder covered by a thin porous ceramic layer. A high number of ignition paths penetrate the pores of the ceramic layer in the direction of the surrounding highly conductive water. This setup represents an electrohydraulic cylindrical wave source, which may be focused using a rotational symmetric paraboloid mirror [33.107].

Explosives and Laser-Generated Shock Waves. As early as 1986, a Japanese lithotripter was built that used tiny explosive pills in the first focus of a rotational symmetric elliptic mirror. The device was used for the treatment of kidney stones, and reports about its application in ESWT therapy were published.

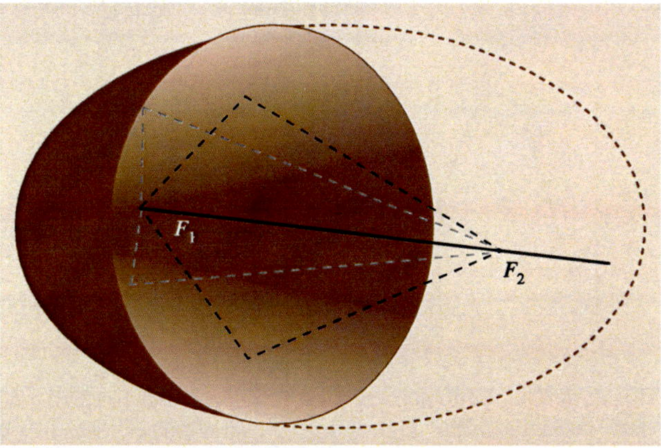

Fig. 33.12 The primary spherically expanding shock wave of the spark-gap source in the focus F_1 is refocused in the focus F_2 by a rotationally symmetric ellipsoidal reflector

The laser-generated thermal and dielectric breakdown of a fluid was also under research for the generation of shock waves [33.108]. Although this method is able to generate highly precise and reproducible pressure pulses, the high cost of applicable lasers makes this approach less feasible for commercial lithotripters. Therefore, the method is mostly used in academia for the generation and excitation of cavitations.

33.4.3 Generation of Planar Sound Waves

Electromagnetic Shock Wave Emitter

The principle of the electromagnetic shock wave emitter (EMSE) was published first in 1962 [33.109]. The source uses a flat coil that is energized by an electric current impulse of several kilo-Amperes. This results in a strong magnetic field extending over the area of the coil. The coil is covered by a thin insulator. On top of it is placed a highly conductive membrane. Due to the magnetic field of the coil, eddy current flows inside the membrane, which in turn creates a magnetic field with opposite polarity. Due to the two magnetic fields, the membrane is repelled (Fig. 33.14). Thus, the water is compressed in the volume above the membrane, which generates a positive pressure pulse with the shape of a squared semisinusoid (Fig. 33.13). The pressure distribution above the membrane is approximately constant. The duration of the pulse is $\approx 2-5\,\mu s$.

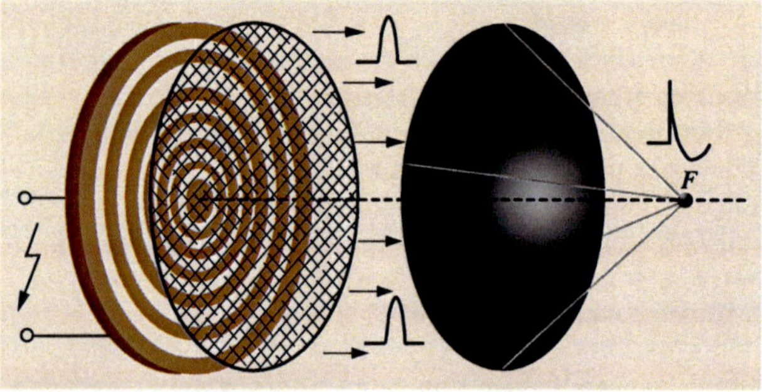

Fig. 33.13 The shape of the primary pressure pulse of the electromagnetic pressure pulse source (EMSE) is the square of the driving current pulse [33.109]. The wave contains only weak high-frequency components. Therefore, it is only slightly attenuated in tissue. A shock front occurs by steepening during propagation to the focus

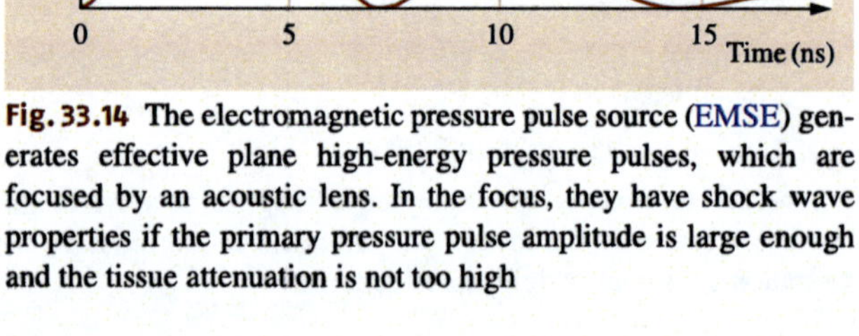

Fig. 33.14 The electromagnetic pressure pulse source (EMSE) generates effective plane high-energy pressure pulses, which are focused by an acoustic lens. In the focus, they have shock wave properties if the primary pressure pulse amplitude is large enough and the tissue attenuation is not too high

A capacitor again is used for the storage of the driving electrical energy. Its capacity is several hundred nanofarad to one microfarad, which is significantly more than the capacitor of a typical spark-gap source. Therefore, the charging voltage can be less; typical values for an EMSE are ca. 8 to 20 kV. Nevertheless, the electroacoustic efficacies of EMSE and spark gap (and piezo) are not comparable due to the different principles. Thus, a comparison of charging voltages or even electrical energies does not give useful information on the efficacy of different sources.

Acoustic Lenses for the Focusing of Sound Waves

Acoustic lenses are used for the focusing of planar pressure pulse sources like the EMSE or planar piezoelectric transducers. Depending on the velocity of sound of the lens material, its shape differs. If the velocity of sound is higher than that of the surrounding water, the lens shape becomes concave (e.g., perspex lenses). If the velocity of sound is less than that of the water, the lens shape becomes convex (e.g., silicon lenses).

Special Properties of the Electromagnetic Shock Wave Emitter

The primary acoustic signal of the EMSE has the shape of a squared sine half-wave at first. Its rise time is several hundred nanoseconds. Thus, it only contains a small amount of higher-frequency energy and is therefore only slightly attenuated in tissue. The higher spectral components occur during the propagation to the focus caused by the steepening process, when the primary acoustic pressure is high enough (ca. > 1 MPa) and the propagation path is long enough.

Steepening by a Shock Tube

In order to generate a steep signal in front of the lens, a long tube can be inserted between the EMSE and the lens. During propagation inside the tube, the rise time of the wave gets shorter, and a shock wave develops. Disadvantages of the shock tube are its large length of 20 to > 50 cm, its weight, and the fact that some of the waves that travel along the wall are coupled inside the wall and thus are lost. Therefore, shock tubes are not used in newer shock source designs.

Steepening by Focusing

On the way to the focus, the pressure pulse steepens due to the collimation of pressure. Thus as a rule with appropriately designed systems (distance lens – focus, primary pressure) shock waves develop in water in the focal area. In contrast, it is hard to prove that the same steepening occurs in tissue in vivo. As the tissue properties (e.g., attenuation, nonlinearity, and layering) are different in every patient, an exact prediction of the steepness in the focal area is not possible. Nevertheless, experiments in vivo demonstrate that the waves may be fully steepened.

Pressure pulses with a focal pressure of less than 25 MPa are usually unable to generate fully steepened signals at the usual geometries (focal distance 5 to 10 cm), even in plain water. At typical ESWT apertures (80–90° angle) shock waves can be observed in lens-focused EMSE systems at higher energy settings.

In an ESWL system with a self-focusing spherical calotte (EMSE of 12 cm aperture), shock waves were measured to develop even at relatively low focal pressures (≈ 20–30 MPa) due to a long propagation path and relatively high primary pressures [33.92].

Cylindrical EMSE

In this special EMSE, the coil is wound in a cylindrical shape and is covered by a cylindrical membrane. The pressure pulse is emitted radially symmetrical. It has about the same shape as the pressure pulse of the flat membrane EMSE. The waves of the cylindrical EMSE are focused by a section of a rotationally symmetric parabolic reflector [33.110].

Piezoelectric Pressure Pulse Source

In a piezoelectric pressure pulse source the transformation of electrical energy into acoustic energy is done by a number of polarized ceramic platelets, made of lead-zirconate-titanate, that elongate or contract due to a charge transferred by an electrical impulse (up to ca. 5 kV).

Usually several dozen to several thousand ceramic platelets are used that are arranged adjacent to each other on a plane or on a spherical carrier (the first sources were cuts of a sphere with a diameter of 50 cm) [33.111, 112]. The platelets are excited simultaneously and emit a plane wave front, which converges in case the carrier is curved.

The time signature of the pressure pulse depends largely on the coating material of the front and rear side (backing) of the transducer. It also depends on the electrical excitation signal. If the backing is well matched to the properties of the piezo material, then the shortest pressure pulse is shaped as a sinusoidal wave composed of a single positive pressure and a trailing negative portion [33.111, 112]. The time duration of the signal thus depends on the length of the piezoelectric elements (e.g., 5 mm [33.113]). Usually it is chosen between 1 and 2 μs per half wave.

Pulse shapes that comprise a number of positive and negative portions can also be generated by adequate construction of the transducer. By oscillating between positive and rarefaction pressures, strong cavitation occurs in the focal area, which results in gross tissue damage. Possible applications for these kinds of signals could be the disintegration of tumors [33.114–116] and the transfection of cells in biotechnological and medical applications [33.8, 9].

Self-Focusing Assemblies – Spherical Sound Sources

Piezoelectric elements are favorably arranged on a spherical surface. Thus, all pressure pulses emitted from each single element arrive at the center of the sphere at the same instant, which results in the formation of a focal maximum. The smaller the aperture angle of the radiating surface, the closer the maximum pressure is shifted toward the source. This shift is compensated by an outward shift due to nonlinear steepening. When the sound waves are fully steepened, then the geometrical focus and the positive maximum often appear at the same position.

Self-focusing arrangements can also be found in EMSE sources using a spherically curved spiral coil and a corresponding membrane [33.80, 92].

New Approaches to Increase the Efficiency of Piezoelectric Pressure Pulse Sources

The primary peak pressure of a single-layer piezoelectric generator is < 1 MPa and thus lower than the primary pressure of an EMSE source. Therefore, the sound intensity at the piezo transducer surface also is less, so a larger area is necessary to achieve the necessary focal energies.

New approaches [33.117, 118] are based on layered transducer structures that are composed of two layers of piezo elements with time-delayed excitation. Basically, the front layer is excited in phase when the sound waves of the rear layer penetrate it. This approximately doubles the surface pressure, so the energy is quadrupled [33.117]. Therefore, the diameter of the transducer can be reduced to ca. 30 cm (for ESWL), which makes new system concepts of the lithotripter feasible. A layered arrangement may be obtained by bonding the front and rear sides of a metal sphere with piezo elements. For ESWT purposes handheld transducers of < 10 cm diameter are available using double-layer technology.

Special Properties of the Piezoelectric Pressure Pulse Source

Like the EMSE, a piezoelectric transducer generates primarily not shock waves but triangular rising signals. As the spectrum of these piezo pulses only contains minor high-frequency components, about the same propagation and attenuation conditions apply for both piezo and EMSE sources.

Additionally, the typical pressure pulse signal of a piezo transducer includes a rarefaction portion that has about the same amplitude as the positive portion. This generated rarefaction part adds up to the rarefaction generated by the edge diffraction wave in the focus. However, as the rarefaction part does not steepen during propagation, the focal amplitude of the rarefaction portion of the focus signal has less amplitude than the positive pulse, which may have developed a shock front.

An optimal signal shaping of the piezo pulse by electrical and mechanical means is mandatory. If the transducers are allowed to oscillate and to emit multiple positive and rarefaction pressure periods, then this leads to tissue damage by cavitation [33.112, 114]. Such transducer configurations are used for the targeted destruction of tissue, but not for lithotripsy. For lithotripsy, successful efforts were undertaken to generate both monopolar positive pressure pulses and to minimize the edge diffraction waves [33.119, 120].

Due to the many possibilities for pulse shaping, piezoelectric transducers offer a wide field for the evaluation of different bioeffects. A bioeffect study of pressure pulses starting with a rarefaction phase demonstrated that these inverse pulses caused significantly larger damage in agar models and in rabbit liver [33.121]. In contrast, other authors reported clearly less hemolysis when the rarefaction pressure was generated using a reflector with pulse-inverting properties as compared to a standard spark-gap arrangement [33.55, 122]. Stone comminution in the rarefaction case was significantly less at comparable reflector geometries [33.123].

33.5 Extracorporeal Lithotripsy (ESWL) and Extracorporeal Shock Wave Therapy (ESWT) in Practice

33.5.1 System Components of a Lithotripter

A typical lithotripter system as in Figs. 33.1 and 33.15 is composed of the following components:

- Pressure pulse source with high-voltage pulse generator and coupling means (water circuit and bellows)
- Source positioning means
- Localization means including video monitor
- Enclosure with power supplies, control computer, electrical high-voltage generator, etc.
- Patient stretcher with supplies for leg rests, arm rests, etc.
- ECG monitor for triggered pressure pulse release

Coupling Means

Between the pressure pulse source and the patient a water path is usually inserted that is enclosed in a bellows at the patient's side. The water pressure can be adjusted to support the coupling. Between the bellows and the skin an ultrasound coupling gel is used. It is important to avoid the trapping of air bubbles in the gel layer in order to prevent disturbance of the pressure pulse propagation and possible skin petechiae due to cavitation effects.

Localization

During lithotripsy of urinary stones, gallstones, and shoulder calcification, as well as for bone treatment (pseudarthrosis), the target can be effectively visualized using ultrasound or x-ray systems.

During ESWT of tennis elbow pain, heel spur pain, etc., the treatment spot cannot be identified by these means and the doctor needs the cooperation of the patient. Usually he asks the patient to identify the exact localization of the maximum pain spot and directs the pressure pulses at this spot during treatment. In most cases, ultrasound images are able to show soft tissue structures like tendons and thus are appropriate for treatment control.

Some systems offer dual imaging. This means the simultaneous use of x-ray and ultrasound imaging for the precise localization of the target and for a real-time ultrasound control of the treatment. Thus a continuous treatment control with minimal radiation is possible. However, for anatomical reasons, ultrasound localization is not possible with all stones and treatment situations (e.g., deep ureter stones, or if the stones are hidden behind rib shadows). Therefore, most lithotripters are equipped with an x-ray device, or lithotripters at least have provisions to add an external x-ray C-arm.

The requirements for both the precision and accuracy of localization are very high. As discussed above, the effective spot in the focal plane has a maximum diameter of 18–22 mm. The axial extent of the effective focus often has comparable dimensions. Thus the cross hair of the localization display needs to point to the focus with the highest precision with deviations of preferably less than 1 mm. To assure the quality of the localization device adjustment, a focus phantom is used that can be attached to the therapy head. It shows the exact therapy focus position in the localization display and should be used frequently to verify the correct adjustment of therapy head focus and localization systems.

X-ray Localization

In general a localization device must be able to display the volume around the focus from all spatial directions. In older lithotripters this task was solved with two x-ray systems whose central axes crossed in the therapy focus. Newer lithotripters have a movable x-ray arm that swivels isocentrically around the therapy focus. With the x-ray tube in an upright position perpendicular to the patient stretcher (AP position), the target is positioned horizontally to the position of the focal marker. Afterward, the x-ray tube is swiveled isocentrically around the therapy focus into an oblique position (in the cranial–caudal, CC, plane at ca. 30°). In this plane the vertical position of the target is adjusted to the focal marker (Fig. 33.16) by elevating the patient stretcher or the therapy head. Instead of the CC plane, sometimes a lateral position is used.

During treatment, frequent localization controls are necessary even at the expense of some additional x-ray dose (which is quite low in modern systems), particularly if it is suspected that the patient has moved.

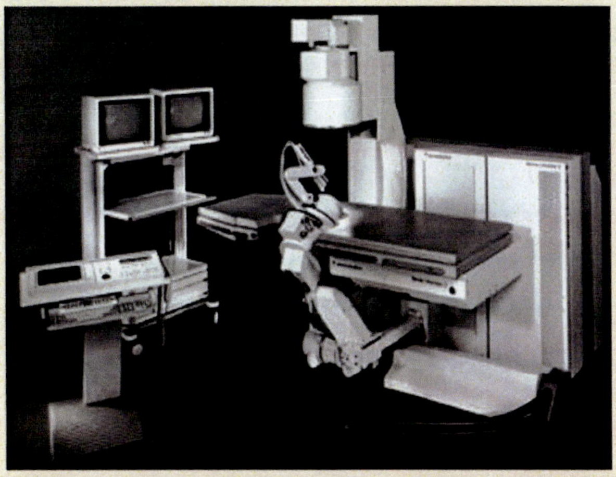

Fig. 33.15 A modern lithotripter, which can be used as a urological work station at the same time. The figure shows all system components such as the pressure pulse source including source positioning mechanics and attached ultrasound outline system, the x-ray system attached to a rotating C-arm, including a monitor cart, the patient stretcher, which can be moved in three orthogonal axes, and the control console (Dornier Lithotripter S)

Variations of the X-ray Localization

The first lithotripters (Fig. 33.1) used two independent x-ray systems for localization, which were adjusted in a fixed position with respect to the shock wave focus. Both x-rays crossed the patient from an oblique angle, which made the orientation difficult. To account for these problems, the patient could be moved in planes parallel to the x-ray imaging planes.

Newer lithotripters sometimes use the so-called inline x-ray targeting method. In this case one of the x-ray projections is taken through a radiation-transparent opening in the center of the shock source. Under certain conditions, the relatively small image sector may be disadvantageous. To avoid excess radiation due to the penetration of the water path, an air-filled balloon can be inflated inside the shock source when an x-ray is taken.

If no direct access of the x-ray projections through (or along) the pressure pulse source is possible due to its large size, offline x-ray targeting can be used. In this case the x-ray arm is arranged at a position that is shifted along the long patient stretcher axis by some distance. For targeting, the patient is moved to the x-ray center position by a defined shift of the stretcher. For treatment, he is shifted back to the focus position. It is understood that the requirement for mechanical

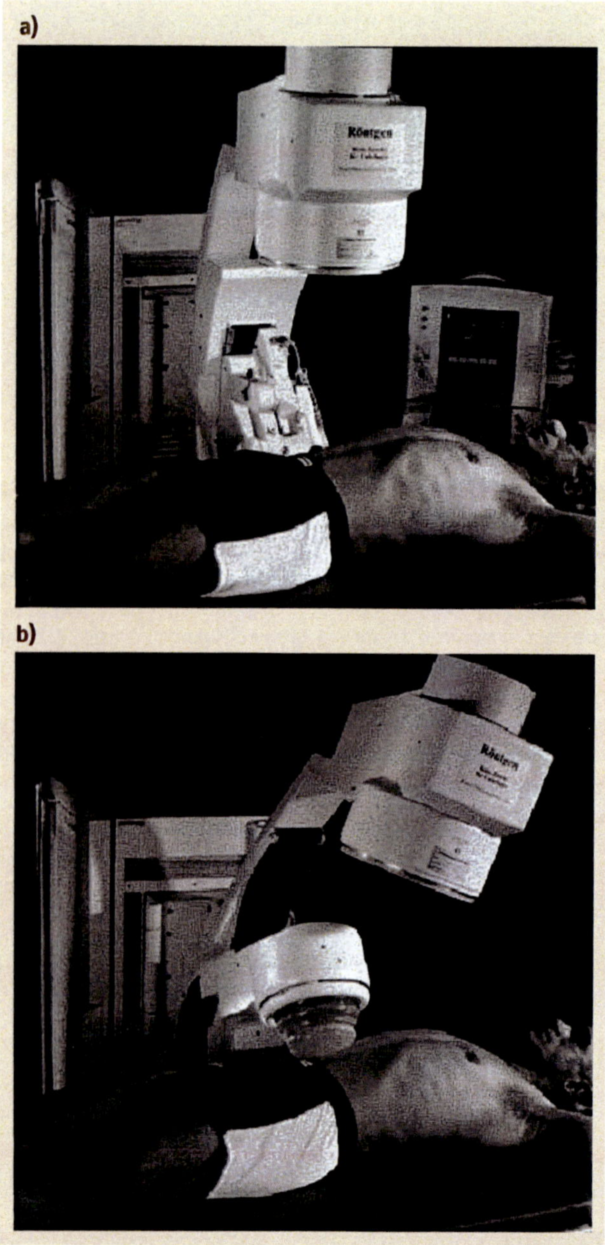

Fig. 33.16a,b During x-ray targeting at first the AP direction is used (**a**). The target is adjusted to the therapy focus by moving the patient in the horizontal x-y-plane (**b**). Afterward, the C-arm is tilted 30° (CC) and the target is adjusted to the therapy focus in the z-direction by elevating the patient stretcher or the therapy source ◂

precision of such arrangements is extremely high, particularly as the stone position and the progress of stone comminution cannot be tracked online by x-ray without moving the patient.

In recent devices localization systems have been built that do not require a mechanical link to the therapy head. Instead, an electronic navigation system using ultrasonic or camera-based sensors measures the mutual positions of the x-ray device and therapy head. The attending physician is informed of the position of the therapy focus via numerical displays or a video overlay in the x-ray image [33.124].

Another system is significantly less expensive but just as accurate. It uses laser spots projected from the therapy head onto small target marks at the x-ray device. With the help of these laser spots, it is possible to use almost every x-ray C-arm device available at the hospital for targeting purposes.

Ultrasound Targeting

Depending on the arrangement of the ultrasound transducer, inline and outline systems are distinguished.

The inline ultrasonic transducer is built into the center of the pressure pulse source along its axis. Usually it can be moved axially and rotated around its axis. The advantage of the inline scanner is the use of the same propagation path and entrance window as the path of the therapy waves. The disadvantage of the inline arrangement is the shading of parts of the therapy wave and the occurrence of diffraction waves at the transducer housing. The resulting loss in therapy energy needs to be compensated by a higher primary pressure pulse energy.

Moreover, the ultrasound image may be impaired by the water path between transducer and patient inside the coupling bellows, causing multiple reflections and artifacts. To minimize these problems, the inline transducer is usually moved toward the patient for localization and retracted during the pressure pulse release.

In the ultrasound outline system the transducer is fastened at a mechanical arm, which can be rotated isocentrically around the therapy focus. Thus it is possible to choose the optimum windows for coupling the therapy source and the imaging transducer to the patient. As the localization transducer axis does not comply with the therapy source axis, the attending physician needs to have a good anatomical orientation. The outline transducer has the advantage that it allows one to visualize the anatomy in standard imaging planes, which is important for the correct assessment of the state of the disease and the tracking of treatment success. Additional advantages are the easy exchange of the transducers for optimum adaptation to the treatment situation and the artifact-free ultrasound image. Misalignments of a few millimeters between the localization display and the therapy focus, caused by the different sound pathways, can sometimes occur [33.38]. Such declinations can usually be compensated by the real-time monitoring of the treatment progress, as the exact site of the stone hits can be estimated from the ultrasound images by trained persons.

Sensor-based navigation systems have also been used for the adjustment of the ultrasound imaging system. The position of the therapy focus is displayed as an overlay in the ultrasound image. In principle, the ultrasound transducer could be used freehand, but for continuous treatment monitoring a mechanical holding arm is necessary.

33.5.2 Approaches to Online Disintegration Tracking During Treatment

Early on, attempts were made to track the disintegration progress online. As piezotransducers are capable of receiving ultrasonic signals, the echoes of test signals or high-energy pulse echoes from the focal region can be received by piezo lithotripters. A distinction between cavitation and stone signals is feasible by evaluating the polarity of the echoes [33.37]. By overlaying the signals of the lithotripsy pulse echo and standard ultrasonic images the focal echoes could be amplified in order to implement a hit-or-miss monitoring [33.125].

Recent clinical trials demonstrated a highly significant indication of a hit or miss by evaluating the time duration of a Doppler shift when test signals were sent shortly after the pressure pulses. It was also feasible to track the progress of stone comminution by changes in the Doppler signals [33.126].

Nevertheless, none of these methods is (yet) used in actual clinical lithotripsy. In particular, solutions that interrupt the excitation of high-energy pressure pulses in the absence of stone echoes from the focus lead to irregular triggering of the shocks (comparable to the effects when respiration triggering was used), which is very irritating for both the patient and the physician [33.79]. Therefore, to date, a regular, continuous triggering of the pressure pulses is preferred and seems tolerable even if a certain number of pulses are missing the stone (up to ca. 30%).

A continuous and well-targeted shock wave application could be achieved by using an online target-tracking method. Mechanical tracking is hard to realize due to the large moved mass of the therapy head. A possible option would be the use of electronic focusing, e.g., by a phased array composed of a number of synchronized small single sources (16–128). Such designs were published as acoustic mirrors. They were used in clinical trials, but the high technological expenditures, and thus high price, prevented further steps toward widespread introduction [33.127].

33.6 The Patient

33.6.1 Patient Bedding for ESWL and ESWT Treatment

The treatment of stones usually is done with the patient in a lying position, which facilitates analgesia, supervision of blood pressure and ECG, and, in some cases, the use of sedation or anesthesia.

ESWT treatment may be done in a sitting or lying position, depending on the treatment site. A lying position may be advantageous for avoiding blood circulation problems, in particular if treatment is done under local anesthesia.

The extremities can be supported by appropriate holders. In this way the limbs can be positioned in a comfortable treatment position. These tools help to avoid decoupling of the target and patient movements due to reflexes caused by fear or pain.

33.6.2 Stone- and Patient-Related Influences

In the section on stone disintegration, the mechanisms were discussed from a physical view. The starting point was idealized conditions, which only occur when measurements are made in a water bath. In order to account for the influences of the patient, the Technical Working Group of the Deutsche Gesellschaft für Stoßwellenlithotripsie (DGSL) started the design of a patient phantom [33.78]. Probably due to the large number of parameters to be taken into account, the development is yet to be completed.

During actual stone treatment a number of real-world influences play a significant role in treatment outcome, the necessary pulse energy, and the overall energy dose. These influences are as follows:

1. Attenuation due to absorption in tissue;
2. Dissipation of energy and losses due to nonlinear distortions of the pressure pulses;
3. Reflections at tissue boundaries (skin, fat, muscle, kidney capsule, urethra, etc.);
4. Shadowing by bone structures (ribs, spine, pelvic bone) [33.30];
5. Shadowing by components installed inside the pressure pulse source propagation path, e.g., inline ultrasound transducers [33.30, 110];
6. Shadowing by gas-filled structures (e.g., gas in intestines or bubble layers in the coupling gel);
7. Shadowing and absorption by sludge, viz., already comminuted stone mass [33.63];
8. Aberration at boundaries by the diffraction of sound (which also leads to a declination of localization);
9. Influence of the medium surrounding the stone (tissue, urine, biles, etc.) on the excitation and enhancement of cavitation [33.30, 57, 71, 91];
10. Impacted stones, e.g., in the urethra [33.128, 129], or shoulder calcifications [33.35], which require markedly higher energy doses for complete disintegration as compared to free-floating stones in the kidney. In environments of higher viscosity (e.g., PVA) 8 to 10 times more pressure pulses are needed for the complete destruction of a stone as compared to water [33.30];
11. Gas bubbles in the pressure wave path (e.g., in the coupling gel between bellows and patient, or in greasy layers at the skin), which significantly increase the attenuation in the propagation path;
12. Different stone compositions, which can multiply the necessary pulse count for comminution [33.27];
13. Mispositioning due to localization errors, which may lead to misalignments of stone and focus in the range of centimeters;
14. Motions of the organ and the stone due to breathing, which typically lead to an excursion of around 30 mm; thus the stone leaves the focal region in ca. 30% of cases [33.130];
15. Uncontrolled patient movements, which may lead to gross mispositioning; thus frequent localization control is necessary (despite the moderately increased x-ray dose).

These influences can create a strong cumulative influence on the total number of pressure pulses per treatment: The number of pulses of one treatment usually is 5 to 10 times higher (1000 to > 5000 pulses per treatment at a retreatment rate of 10 to > 40%) than the number of pressure pulses needed for a comparable model stone in an idealized in vitro model using degassed water (e.g., to disintegrate 0.9 ml model stone material into pieces < 2 mm, 50 to 500 pressure pulses are needed [33.131]).

33.7 Assessment of the Clinical Efficiency of Lithotripters

For the assessment of the clinical efficiency of lithotripters the disintegration efficacy is not the only important parameter. Patient-related factors also need to be taken into account. These factors include the need for sedation, analgesia, or anesthesia, as well as auxiliary posttreatment measures like, e.g., endourological interventions to remove urinary tract obstruction or multiple lithotripsies (retreatments, re-ESWL). All these factors have been combined in the definition of the clinical efficiency quotient (EQ) [33.132, 133]

$$EQ_A = \frac{\%\,\text{stone-free patients}}{100\% + \%\,\text{retreatments} + \%\text{auxiliary measures post ESWL}} .$$

Modern treatment strategies also include auxiliary measuers before lithotripsy, like catheterization or pushing a ureter stone up the urethra back into the calyx. These measures can be included in the equation as an additional addend in the denominator of the above equation, which is then termed EQ_B.

Some first-generation lithotripters had EQs from 25%, whereas modern lithotripters reach up to 67% [33.134]. Thus both the increased stone-free rate (usually controlled after 3 months) as well as the continuing reduction of auxiliary measures and the lower pain level of the newer sources play a major role. But different authors often publish significantly different EQs for the same lithotripter type (e.g., HM3: 0.25 to 0.67). It is interesting to note that just this lithotripter, which was introduced as the first device 25 years ago, still is deemed the most efficient lithotripter by some doctors [33.135]). According to a recent study, both the applied treatment strategy (energy, pressure pulse count, localization, anesthesia) and the training and experience of the physician play a crucial role in treatment efficacy [33.136]. Another reason for large differences in the EQs may be different patient populations used in the studies (e.g., exclusion or inclusion of ureter stones, restriction of stone size, etc.).

Due to the progress of measurement technology, physical data of the pressure pulse sources can be measured with satisfactory quality today. Thus a pressure pulse dosimetry should be introduced, taking into account the number of pulses per treatment times the effective energy per pulse $E_{12\,\mathrm{mm}}$ [33.64].

With respect to the ESWT, the sum of effective energies per pulse or the sum of EDs is under discussion [33.86].

33.7.1 The Controversy About the Right Focal Size

Statements on the focal size traditionally (and according to the standard) refer to the axial and lateral (−6 dB) focal extent. Unfortunately, these values have no viable correlation to the disintegration efficacy, as shown above. Even recent publications that compare the devices of different manufacturers with smaller and larger focal extents do not reveal the effective energies $E_{12\,\mathrm{mm}}$ or $E_{5\,\mathrm{MPa}}$ which have significant correlation to the stone comminution efficacy. This is due to the lack of obligation to state such parameters. In a consensus report [33.78] driven by the DGSL, technical experts agreed on an adequate data set in Germany, but the actual implementation has only taken place for ESWT devices on the DIGEST Web site. There the technical data of ESWT devices are published, which were consistendly measured using fiber-optic hydrophones or other acknowledged focus-type hydrophones according to IEC 61846.

For lithotripters such a database does not yet exist. Comparisons of devices using model stones may give a certain orientation [33.137], but they are not accepted by all manufacturers. An international standard of these kinds of tests is lacking.

A comparison of clinical and technical data of devices with a wide focus and lower peak pressure and devices with a smaller focal width but significantly higher peak pressure on the basis of the EQ values is difficult due to incomplete data sets. Still, two examples shall be given.

The clinical EQ_As are 0.65 and respectively 0.62 in the case of a wide-focus, low-amplitude device (18 mm lateral −6 dB width at max. 33.8 MPa) compared to a small-focus, high-amplitude type (approximately 6 mm lateral −6 dB width at ca. 90–100 MPa), and thus would be comparable [33.84]. But in another publication concern was raised that the patient populations might have had too large differences to justify comparison of the data [33.79].

Another comparison of the EQ_As between a low-pressure, wide-focus spark-gap lithotripter (33 MPa at 16 mm −6 dB width) and a high-pressure, small-focus EMSE (90 MPa at 5.4 mm −6 dB width) using data from different publications reveals that the published EQ_A values are (up to) 0.67 (low-pressure source) versus 0.71 (high-pressure source) [33.64, 138], but the patient populations are also not clearly reported.

In conclusion, from the available data a clear advantage of one of the philosophies – wide-focus, low-pressure or small-focus, high-pressure – cannot yet be demonstrated.

So far the side effects of different lithotripters are rarely taken into account. This may be in part because the fraction of severe side effects seems to be too low to make statistically significant statements (at least for the devices that are actually on the market). The supporters of the wide-focus lithotripters point to very limited side effects and, in the case of the first example above, the complete absence of analgesia or anesthesia despite high efficacy at a low overall pressure pulse number. The reason might be the exceptionally low rarefaction amplitude of less than 5 MPa, which avoids tissue-damaging cavitation almost completely [33.84].

The supporters of the high-pressure, smaller-focus philosophy point to the positive experience in several million successful treatments, where subcapsular hematoma as the most frequent severe complication occurred in only 0.1–0.7% of cases, and other organs were affected even less [33.139].

Due to the unsolved controversy, recently some manufacturers have started to offer lithotripters with adaptable focal properties. In earlier devices this was achieved by exchangeable reflectors. In recent lithotripters, the pulse duration of the EMSE pulse can be modified; or it is possible to switch the delay times between the two layers in one piezo lithotripter, resulting in different focal properties. Recommendations by the manufacturers suggest using the larger focus for kidney stones and the smaller one for targeting ureter stones.

33.7.2 Side Effects and Safety

Direct side effects of the ESWL include pain, urinary tract infections, obstruction of urinary pathways, hematuria, cardiac rhythm disturbances, kidney hematoma, subcapsular hematoma, and harm to surrounding organs. Only the last three side effects are estimated to be severe. Delayed complications include loss of kidney function, hypertension (controversial), steinstrasse, residual fragments, and recurrent stones. Most of these complications are very rare, some are rated differently by different authors, and almost all can be minimized by properly following the treatment guidelines of the urological societies [33.140].

ESWT side effects are reported to be low at low-energy treatments like tennis elbow: transient skin reddening, pain, small hematoma, and seldom migraine [33.141]. In higher-energy regimes also large area hematoma were observed in single cases.

Sound energy is almost completely reflected at gaseous tissues. In particular, at the lung severe damage can occur. Therefore, it is mandatory during therapy planning to exclude lung tissue from the pressure pulse path, even at large distances from the focus. The safety distance of the focus from lung tissue depends on the design parameters of the pressure pulse source.

The effects of pressure pulses on nerves remain undetermined. It was demonstrated that nerves could be activated by cavitation [33.48]. But there are no actual results on nerve damage available. As a safety measure, the inadvertent insonification of nerves in the vicinity of the focus or at the propagation axis should be avoided. Signs of numbness or other neurological reactions should be watched carefully. In the meantime, patents can be found that claim the insonification of nerves with high-intensity sound pulses.

Increased hematoma were reported when blood-thinning medication was taken. These medications should therefore be discontinued in time before the pressure pulse treatment. The Deutsche Gesellschaft für Stosswellenlithotripsie specifies discontinuing blood-thinning medication at least 10 d before treatment [33.142].

Pregnant women should not be exposed to x-rays. Additionally, the effect of shock waves and pressure pulses on the development of the fetus is not well understood (including indirect effects caused by pain during treatment).

In patients with heart pacers, it must be ascertained that the pacers are not influenced by the strong electromagnetic pulses of the therapy source as well as acoustic pulses during treatment; otherwise, treatment should be cancelled.

While lithotripters of the first generation generally used ECG triggering, in newer generation lithotripters the pulse repetition frequency can be chosen from 60 (seldom 30) up to a maximum repeat rate of 120 to (rarely) 240 pulses per minute. Due to the excitation of cavitation, the disintegration effect of the pulses is reduced at increasing pulse repetition rates, whereas the danger of kidney hematoma increases [33.59, 64, 143]. Recent animal experiments also point to reduced kidney trauma at slower pulse repeat rates.

In general the ECG must be watched in patients with a predisposition for cardiac rhythm disturbance. If necessary, the release of the pressure pulses should be triggered by ECG. The same holds if heart tissue is somewhere near the axis of pressure pulse wave propagation.

International Standard IEC 601-2-36 describes the general requirements for the safety of lithotripters. Medical guidelines are given by local urological or orthopedic societies and are constantly updated to incorporate the recent state of knowledge.

33.8 Associations and Societies for Lithotripsy and Pressure Pulse Therapy

Some societies were founded in Germany that deal with all aspects of pressure pulse lithotripsy (ESWL) and pressure pulse therapy (ESWT). The Deutsche Gesellschaft für Stosswellenlithotripsie (DGSL) within the scope of the Deutsche Gesellschaft für Urologie DGU is particularly competent for lithotripsy and related applications of therapeutic energy applications in urology. The DGSL develops consensus papers and guidelines for treatment and quality assurance. The society watches actual technical and medical developments of extracorporeal and intracorporeal lithotripsy and related topics.

The Deutsche und Internationale Gesellschaft für Stosswellentherapie (DIGEST) is dedicated to the needs of pain therapy and the treatment of pseudarthrosis by pressure pulses [33.144]. The International Society for Musculoskeletal Shockwave Therapy (ISMST) also supports the development of ESWT applications.

References

33.1 H. Lamport, H.F. Newman, R.D. Eichhorn: Fragmentation of biliary calculi by ultrasound, Fed. Proc, Balt. (9) (1950), pp. 73–74

33.2 M.L. Berlinicke, F. Schennetten: Über Beeinflussung von Gallensteinen durch Ultraschall in vitro, Klin. Wochenschr. **21/22**, 390 (1951), in German

33.3 W.P. Mulvaney: Attempted disintegration of calculi by ultrasonic vibrations, J. Urol. **70**(5), 704–707 (1953)

33.4 C.E. Coats: The application of ultrasonic energy to urinary and biliary calculi, J. Urol. **75**(5), 865–876 (1956)

33.5 M. Hepp: *Überblick über die Entwicklung der Stoßwellenlithotripsie* (Dornier Medizintechnik, Friedrichshafen 1984)

33.6 F. Rieber: Shock Wave Generator, Patent 2559227 (1951), US Patent

33.7 M. Delius, P.H. Hofschneider, U. Lauer, K. Messmer: Extracorporeal shock waves for gene therapy? Lancet **345**, 1377 (1995)

33.8 K. Tschoep, G. Hartmann, R. Jox, S. Thompson, A. Eigler, A. Krug, S. Erhardt, G. Adams, S. Endres, M. Delius: Shock waves: a novel method for cytoplasmic delivery of antisense oligonucleotides, J. Mol. Med. **79**, 306–313 (2001)

33.9 F. Ueberle: Cell transfection by pulsed sound wave effects, Proc. Microtec (VDI, Berlin 2000)

33.10 C. Chaussy: *Berührungsfreie Nierensteinzertrümmerung durch extrakorporal erzeugte, fokussierte Stoßwellen* (Karger, Basel 1980)

33.11 T. Sauerbruch, M. Delius, G. Paumgartner, J. Holl, O. Wess, W. Weber, W. Hepp, W. Brendel: Fragmentation of gallstones by extracorporeal shockwaves, N. Engl. J. Med. **314**, 818–822 (1986)

33.12 T. Sauerbruch, M. Stern: Fragmentation of bile duct stones by extracorporeal shock waves, Gastroenterology **96**, 146–152 (1989)

33.13 M. Delhaye, A. Vandermeeren, M. Baize, M. Cremer: Extracorporeal shock wave lithotripsy of pancratic calculi, Gastroenterology **102**, 610–620 (1992)

33.14 T. Sauerbruch, J. Holl, M. Sackmann, G. Paumgartner: Disintegration of a pancreatic duct stone with extracorporeal shock waves in a patient with cronic pancreatitis, Endoscopy **19**, 207–208 (1987)

33.15 H. Iro, N. Nitsche, H.T. Schneider, C. Ell: Extracorporeal shock wave lithotripsy of salivary gland stones, Lancet **II**, 115 (1989)

33.16 V. Valchanou, P. Michailow: High energy shockwaves in the treatment of delayed and nonunion fractures, Int. Orthopedics (SICOT) **15**, 181–184 (1991)

33.17 M. Delius, G. Adams: Shock wave permeabilization with ribosome inactivationg proteins: a new approach to tumor therapy, Cancer Res. **59**, 5227–5232 (1999)

33.18 M. Loew: Die Wirkung extrakorporal erzeugter hochenergetischer Stoßwellen auf den klinischen, röntgenologischen und histologischen Verlauf der Tendinosis calcarea der Schulter – eine prospektive Studie. In: *Die Stoßwelle, Forschung und Klinik*, ed. by C. Chaussy, F. Eisenberger, D. Jocham, D. Wilbert (Attempto, Tübingen 1993) pp. 153–156

33.19 G.P. Dahmen, L. Meiss, V.C. Nam, B. Skruodies: Extrakorporale Stoßwellentherapie (ESWT) im knochennahen Weichteilbereich der Schulter, Extrem. Orthop. **11**, 25 (1992)

33.20 J.D. Rompe, C. Hopf, K. Küllmer, J. Heine, R. Bürger: Analgesic effect of extracorporeal shock wave therapy on chronic tennis elbow, J. Bone Joint Surg. B **78**, 233–237 (1996)

33.21 H. Lohse-Busch, M. Kraemer, U. Reime: The use of extracorporeally induced shock waves in the treatment of muscular malfunctions of different etiologies. In: *Extracorporeal Shock Waves in Orthopedics*, ed. by W. Siebert, M. Buch, J. Chinnery, A. Faust (Springer, Berlin Heidelberg 1997), Chap. 14

33.22 M.S. Michel, T. Ptaschnyk, A. Musial, P. Braun, S.T. Lenz, P. Alken, K.U. Köhrmann: Objective and subjective changes in patients with Peyronie's disease after management with shockwave therapy, J. Endourol. **17**(1), 41–44 (2003)

33.23 T. Nishida, H. Shimokawa, K. Oi, H. Tatewaki, T. Uwatoku, K. Abe, Y. Matsumoto, N. Kajihara, M. Eto, T. Matsuda, H. Yasui, A. Takeshita, K. Sunagawa: Extracorporeal cardiac shock wave therapy markedly ameliorates ischemia-induced myocardial dysfunction in pigs in vivo, Circulation **110**, 2977 (2004)

33.24 H. Kuttruff: *Physik und Technik des Ultraschalls* (Hirzel, Stuttgart 1988)

33.25 S.K. Stranne, F.H. Cocks, R. Gettliffe: Mechanical property studies of human gallstones, J. Biomed. Mater. Res. **24**, 1049–1057 (1990)

33.26 B. Granz, G. Kohler: What makes a shock wave efficient in lithotripsy?, J. Stone Dis. **4**(2), 123–125 (1992)

33.27 D. Heimbach, R. Munver, P. Zhong, D. Jacobs, A. Hesse, S.C. Müller, G.M. Preminger: Acoustic and mechanical properties of artificial stones in comparison to natural stones, J. Urol. **164**, 537–544 (2000)

33.28 H. Koch, M. Grunewald: Disintegration mechanisms of weak acoustic shock waves, Ultrason. Int. Conf. Proc. (1989) pp. 1136–1141

33.29 M. Hepp: Work bond index values, personal communication (1989)

33.30 M. Delius, F. Ueberle, S. Gambihler: The destruction of gall stones and model plaster stones by extracorporeal shock waves, Ultrasound Med. Biol. **20**(3), 251–258 (1994)

33.31 V.R. Singh, R. Agarwal: Mechanical and ultrasonic parameters of kidney stones, J. Lithotr. Stone Dis. **2**(2), 117–123 (1990)
33.32 C.J. Chuong, P. Zhong, G.M. Preminger: A comparison of stone damage caused by different modes of shock wave delivery, J. Urol. **148**, 200 (1992)
33.33 C. Church: *On Nucleation Theory*, DKE 821.3 Document 55/99 (VDE, Frankfurt 1999)
33.34 IEC: Norm 61846 Pressure Pulse Lithotripters (International Electrotechnical Commission, Geneva 1998)
33.35 M. Loew: Die Wirkung extrakorporal erzeugter hochenergetischer Stoßwellen auf den klinischen, röntgenologischen und histologischen Verlauf der Tendinosis calcarea der Schulter, Habilitationsschrift (Heidelberg 1994)
33.36 P. Zhong, G. Preminger: Mechanisms of differing stone fragility in extracorporeal shock wave lithotripsy, J. Endourol. **8**, 163–168 (1994)
33.37 F. Ueberle: Ein Konzept zur Ortung und Erkennung von Zielen für Schallpulse hoher Energie. Dissertation (Karlsruhe 1988)
33.38 W. Folberth, D. Hassler: Die Wertigkeit von Inline- und Outline Ultraschall-Lokalisation in der extrakorporalen Stoßwellen-Lithotripsie, Z. Urol. Poster **I**, 46 (1990)
33.39 H. Vergunst, O.T. Terpstra, F.H. Schroder, E. Matura: In vivo assessment of shock-wave pressures, Gastroenterology **99**, 1467–1474 (1990)
33.40 H. Vergunst, T. Onno, O.T. Terpstra, F.H. Schroder, E. Matura: Assessment of shock wave pressure profiles in vitro: Clinical implications, J. Lithotr. Stone Dis. **1**(4), 289–298 (1989)
33.41 A. Philipp, M. Delius, C. Scheffcyk, A. Vogel, W. Lauterborn: Interaction of lithotripter-generated shock waves with air bubbles, J. Acoust. Soc. Am. **93**, 2496–2509 (1993)
33.42 R.O. Cleveland, M.R. Bailey, L.A. Crum, M.A. Stonehill, J.C. Williams Jr., J.A. McAteer: Effect of overpressure on dissolution and cavitation of bubbles stabilized on a metal surface, 135th ASA Conf. Proc. (1998) pp. 2499–2500
33.43 A.J. Coleman, T. Kodama, M.J. Choi, T. Adams, J.E. Saunders: The cavitation threshold of human tissue exposed to 0.2 MHz pulsed ultrasound, Ultrasound Med. Biol. **21**, 405–417 (1995)
33.44 J. Herbertz: Spontaneous cavitation in liquids free of Nuclei, Fortschr. Akust. DAGA 88 (DPG, Bad Honnef 1988) pp. 439–442
33.45 F. Fedele, A.J. Coleman, T.G. Leighton, P.R. White, A.M. Hurrell: A new sensor for detecting and characterising acoustic cavitation in vivo during ESWL, Proc. Inst. Acoust. **26**(2), 422–432 (2004)
33.46 M. Delius, S. Gambihler: Sonographic imaging of extracorporeal shockwave effects in the liver and gallbladder of dogs, Digestion **52**(1), 55–60 (1992)
33.47 A.J. Coleman, M.J. Choi, J.E. Saunders: Detection of acoustic emission from cavitation in tissue during clinical extracorporeal lithotripsy, Ultrasound Med. Biol. **22**(8), 1079–1087 (1996)
33.48 G. Schelling, M. Delius, M. Gschender, P. Grafe, S. Gambichler: Extracorporeal shock waves stimulate frog sciatic nerves indirectly via a cavitation-mediated mechanism, Biophys. J. **6**, 133–140 (1994)
33.49 D. Suhr, F. Brümmer, D.F. Hülser: Cavitation-generated free radicals during shock wave exposure: Investigations with cell-free solutions and suspended cells, Ultrasound Med. Biol. **17**, 761–768 (1991)
33.50 S. Gambihler, M. Delius: Influence of dissolved and free gases on iodine release and cell killing by shock waves in vitro, Ultrasound Med. Biol. **18**(6/7), 617–621 (1992)
33.51 J. Herbertz: Physikalische Grenzwerte für die sichere medizinische Anwendung des Ultraschalls am Menschen, DAGA Fortschr. Akust. (DPG, Bad Honnef 1993)
33.52 D.L. Miller, R.M. Thomas: Thresholds for hemorrhages in mouse skin and intestine induced by lithotripter shock waves, Ultrasound Med. Biol. **21**(2), 249–257 (1995)
33.53 M. Delius, G. Enders, G. Heine, J. Stark, K. Remberger, W. Brendel: Biological effects of shock waves: Lung hemorrhage by shock waves in dogs – Pressure dependence, Ultrasound Med. Biol. **13**, 61–67 (1987)
33.54 M. Keller, R.E. Riedlinger: Vergleich der Kavitation bei verschiedenen Stoßwellengeneratoren, Biomed. Tech. **35**, 233–234 (1990)
33.55 B. Jordan, M.R. Bailey, R.O. Cleveland, L.A. Crum: Detection of lithotripsy induced cavitation in blood, 135th ASA Conf. Proc. (1998) p. 2809
33.56 M. Delius, R. Denk, C. Berding, H.-G. Liebich, M. Jordan, W. Brendel: Biological effects of shock waves: Cavitation by shock waves in piglet liver, Ultrasound Med. Biol. **16**, 467–472 (1990)
33.57 P. Zhong, Y. Zhou, S. Zhu: Dynamics of bubble oscillation in constrained media and mechanisms of vessel rupture in SWL, Ultrasound Med. Biol. **27**(1), 119–134 (2001)
33.58 M. Delius: Minimal static excess pressure minimizes the effect of extracorporeal shock waves on cells and reduces it on gallstones, Ultrasound Med. Biol. **23**, 611–617 (1997)
33.59 H. Wiksell, A.-C. Kinn: Implications of cavitation phenomena for shot intervals in extracorporeal shock wave lithotripsy, Br. J. Urol. **75**, 720–723 (1995)
33.60 K.T. Pace, D. Ghiculete, M. Harju, R.J. Honey: Shock wave lithotripsy at 60 or 120 shocks per minute: A randomized, double-blind trial, J. Urol. **174**(2), 595–599 (2005)
33.61 S. Lovasz, I. Palfi, Z. Romics: Temperature elevation of stones during extracorporeal shock wave lithotripsy (ESWL): A hypothesis for possible cause

of complications, Abstracts, Tagung der Deutschen Urologischen Gesellschaft DGU (1999)
33.62 FDA: *Draft of Suggested Information for Reporting Extracorporeal Shock Wave Lithotripsy Device Shock Wave Measurements* (FDA, Rockville 1991)
33.63 H. Lobentanzer: The concept of acoustic energy in lithotripsy, Dornier User Lett. **7**, 22–26 (1991)
33.64 B. Forssmann, F. Ueberle, C. Bohris: Towards a new EMSE generation, J. Endourol. **16**(Suppl 2), 18–21 (2002)
33.65 A.J. Coleman, J.E. Saunders: A comparison of PVDF hydrophone measurements in the acoustic field of a shock wave source. In: *Extra- und intrakorporale Lithotripsie bei Harn-, Gallen-, Pankreas- und Speichelsteinen*, ed. by C. Ell, M. Marberger, P. Berlien (Thieme, Stuttgart 1990) pp. 15–22
33.66 M.E. Schafer: Cost effective shock wave hydrophones, J. Stone Dis. **5**, 73–76 (1993)
33.67 W. Schätzle: Spezielle Fokusdruck-Sensoren für die Lithotripsie und deren Kalibration, Fortschr. Akust. DAGA (1992) pp. 309–312, in German
33.68 M. Müller, M. Platte: Einsatz einer breitbandigen Piezodrucksonde auf PVDF-Basis zur Untersuchung konvergierender Stoßwellen in Wasser, Acustica **58**, 215 (1985)
33.69 M. Müller: Dornier-Lithotripter im Vergleich: Vermessung der Stoßwellenfelder und Fragmentationswirkungen, Biomed. Tech. **35**, 250–262 (1990)
33.70 J. Staudenraus, W. Eisenmenger: Fibre-optic probe hydrophone for ultrasonic and shock wave measurements in water, Ultrasonics **31**, 267–273 (1993)
33.71 C. Koch, W. Molkenstruck, R. Reibold: Shock-wave measurement using a calibrated interferometric fiber-tip sensor, Ultrasound Med. Biol. **23**(8), 1259–1266 (1997)
33.72 A.J. Coleman, E. Draguioti, R. Tiptaf, N. Shotri, J.E. Saunders: Acoustic performance and clinical use of a fiberoptic hydrophone, Ultrasound Med. Biol. **24**, 143–151 (1998)
33.73 B. Granz, R. Nanke, J. Fehre, T. Pfister, R. Engelbrecht: Light spot hydrophone, innovation in lithotripsy, Med. Solut. **June**, 86–87 (2004)
33.74 S.D. Pye, N.J. Parr, E.G. Munro, T. Anderson, W.N. McDicken: Robust electromagnetic probe for the monitoring of lithotripter output, Ultrasound Med. Biol. **17**(9), 931–939 (1991)
33.75 L. Filipczynski, J. Etienne: Capacitance hydrophone for pressure determination in lithotripsy, Ultrasound Med. Biol. **16**, 157–165 (1990)
33.76 P.A. Lewin, M.E. Schafer: Shock wave sensors: I. Requirements and Design, J. Lithotr. Stone Dis. **3**(1), 3–17 (1991)
33.77 G.R. Harris: Lithotripsy pulse measurement errors due to non-ideal hydrophone and amplifier frequency responses, IEEE Trans. Ultrason. Ferroelectr. Freq. Control **39**(2), 256–261 (1992)
33.78 O. Wess, F. Ueberle, R.N. Dührssen, D. Hilcken, W. Krauss, T. Reuner, R. Schultheiss, I. Staudenraus, M. Rattner, W. Haaks, B. Granz: Working group technical developments – consensus report. In: *High Energy Shock Waves in Medicine*, ed. by C. Chaussy, F. Eisenberger, D. Jocham, D. Wilbert (Thieme, Stuttgart 1997) pp. 59–71
33.79 J. Rassweiler, T. Bergsdorf, S. Ginter, B. Granz, A. Häcker, A. Lutz, O. Wess, D. Wilbert: Progress in lithotripter technology. In: *Therapeutic Energy Applications in Urology*, ed. by C. Chaussy, G. Haupt, D. Jocham, K.U. Köhrmann, D. Wilbert (Thieme, Stuttgart 2005)
33.80 J. Staudenraus: Erzeugung und Ausbreitung freifeldfokussierter Hochenergiedruckpulse in Wasser, Dissertation (University of Stuttgart, Stuttgart 1991), in German
33.81 F. Ueberle: Acoustic parameters of pressure pulse sources used in lithotripsy and pain therapy. In: *High Energy Shock Waves in Medicine*, ed. by C. Chaussy, F. Eisenberger, D. Jocham, D. Wilbert (Thieme, Stuttgart 1997) pp. 76–85
33.82 T. Dreyer, R.E. Riedlinger, E. Steiger: Experiments on the relation of shock wave parameters to stone disintegration, J. Acoust. Soc. Am. **103**(5), 3072 (1998)
33.83 P. Steinbach, K. Wörle, M. Seidl, R. Seitz, F. Hofstädter: Effekte hochenergetischer Ultraschallstoßwellen auf Tumorzellen in vitro und humane Endothelzellen in situ. In: *Die Stoßwelle*, ed. by C. Chaussy, F. Eisenberger, D. Jocham, D. Wilbert (Attempto, Tübingen 1995) pp. 104–109, in German
33.84 W. Eisenmenger: The mechanisms of stone fragmentation in ESWL, Ultrasound Med. Biol. **27**(5), 683–693 (2001)
33.85 K. Sass, K. Steffen, E. Matura, W. Folberth, H. Dreyer, J. Seifert: Experiences with lithotripters: Measurement of standardized fragmentation, J. Stone Dis. **4**(2), 129–140 (1992)
33.86 M. Maier, F. Ueberle, G. Rupprecht: Physikalische Parameter extrakorporaler Stoßwellen, Biomed. Tech. **43**, 269–274 (1998)
33.87 F. Ueberle: Shock wave technology. In: *Extracorporeal Shock Waves in Orthopedics*, ed. by W. Siebert, M. Buch (Springer, Berlin Heidelberg 1997) pp. 59– – 87
33.88 F. Ueberle: Pressure pulses in medicine. In: *Shock Focussing Effects in Medical Science and Sonoluminescence*, ed. by R.C. Srivastava, D. Leutloff, K. Takayama, H. Grönig (Springer, Heidelberg 2003)
33.89 M. Seidl, P. Steinbach, K. Wörle, F. Hofstädter: Induction of stress fibres and intercellular gaps in human vascular endothelium by shock-waves, Ultrasonics **32**(5), 397 (1994)
33.90 S.F. Mishrik, N.P. Cohen, A.C. Baker, M.I. Wills, H.N. Whitfield, R.C.L. Feneley: Choosing a powerful lithotripter, Br. J. Urol. **71**, 653–660 (1993)
33.91 N. Vakil, S.M. Gracewski, E.C. Everbach: Relationship of model stone properties to fragmentation

mechanisms during lithotripsy, J. Lithotr. Stone Dis. **3**(4), 304–310 (1991)
33.92 W. Eisenmenger: Die Physik der akustischen Nierensteinzertrümmerung und neue klinische Resultate, Vortrag anlässlich der Verleihung der Hermann von Helmholtzmedaille, DAGA Tagung Aachen (DPG, Bad Honnef 2003)
33.93 V.K. Kedrinskii: On a mechanism of target disintegration at shock wave focusing in ESWL, 135th ASA Conf. Proc. (1998) pp. 2803–2804
33.94 S.L. Zhu, P. Zhong: The role of stress waves and cavitation in stone comminution in shock wave lithotripsy, Ultrasound Med. Biol. **28**, 661–671 (2002)
33.95 D. Holtum: Eigenschaften und Desintegration von menschlichen Gallensteinen unter Stoßwelleneinwirkung, Dissertation (Stuttgart 1993)
33.96 M. Maier, B. Averbeck, S. Milz, H.J. Refior, C. Schmitz: Substance P and prostaglandin E2 release after shock wave application to the rabbit femur, Clin. Orthop. **0**, 237–245 (2003)
33.97 M. Delius, F. Ueberle, W. Eisenmenger: Extracorporeal shock waves act by shock wave – Gas bubble interaction, Ultrasound Med. Biol. **24**(7), 1055–1059 (1998)
33.98 P. Zhong, I. Cioanta, S. Zhu, F.H. Cocks, G.M. Preminger: Effects of tissue constraint on shock wave-induced bubble oscillation in vivo, Proc. 135th ASA Conf. Seattle (1998) p. 2495
33.99 D. Jocham: Report at the meeting of the German Society for Shockwave Lithotripsy, 1998
33.100 G.W. Drach, S. Dretler, W. Fair, B. Finlayson, J. Gillenwater, D. Griffith, J. Lingeman, D. Newman: Report of the United States cooperative study of extracorporeal shock wave lithotripsy, J. Urol. **135**, 1127–1133 (1986)
33.101 P. Steinbach, F. Hofstädter, H. Nicolai, W. Rössler, W. Wieland: Determination of energy-dependent extent of vascular damage caused by high-energy shock waves in an umbilical cord model, Urol. Res. **21**, 279–282 (1993)
33.102 D.M. Kaulesar-Sukul, E.J. Johannes, E.G. Pierik, G.J. van Eijck, M.J. Kristelijn: The effect of high energy shock waves focused on cortical bone, J. Surg. Res. **54**, 46–51 (1993)
33.103 G. Haupt, A. Haupt, A. Ekkernkamp, B. Gerety, M. Chvapil: Influence of shock waves on fracture healing, Urology **39**(6), 529–532 (1992)
33.104 M. Delius, K. Draenert, Y. Al Diek, Y. Draenert: Biological effects of shock waves: In vivo effect of high energy pulses on rabbit bone, Ultrasound Med. Biol. **21**(9), 1219–1225 (1995)
33.105 Deutsche Gesellschaft für Stoßwellenlithotripsie (DGSL): Konsensus-Workshop 4 der DGSL. In: *Die Stoßwelle* (Attempto, Tübingen 1995)
33.106 D. Cathignol, J.-L. Mestas, F. Gomez, P. Lenz: Influence of water conductivity on the efficiency and the reproducibility of electrohydraulic shock wave generation, Ultrasound Med. Biol. **17**, 819–828 (1991)
33.107 P. Sunka, V. Babicky, M. Clupek, J. Benes, P. Pouckova: Localized damage of tissues induced by focused shock waves, IEEE Trans. Plasma Sci. **32**(4), 1609–1613 (2004)
33.108 E. Steiger: Extracorporal laser induced shock wave lithotripsy (ESWL). In: *Laser* (MZV-EBM, Kaufering 1987) pp. 201–206
33.109 W. Eisenmenger: Elektromagnetische Erzeugung von ebenen Druckstößen in Flüssigkeiten, Akust. Beihefte, Acustica **1**, 185–202 (1962)
33.110 O. Wess, E.H. Marlinghaus, J. Katona: Lars, eine großaperturige Leistungsschallquelle für medizinische Anwendungen, Fortschr. Akust. DAGA (DPG, Bad Honef 1989)
33.111 R. Riedlinger, F. Ueberle, H. Wurster, W. Krauß, P. Vallon, G. Konrad, B. Kopper, H.P. Stoll, R. Goebbels, T. Gebhardt, M. Ziegler: Die Zertrümmerung von Nierensteinen durch piezoelektrisch erzeugte Hochenergie-Schallpulse, Urologe A **25**, 188–192 (1986)
33.112 F. Ueberle: Piezoelektrisch erzeugte Hochenergiepulse und ihre Eignung zur Eignung zur Lithotripsie. In: *Die extracorporale und laserinduzierte Stoßwellenlithotripsie bei Harn- und Gallensteinen*, ed. by M. Ziegler (Springer, Berlin Heidelberg 1991)
33.113 T. Dreyer, R.E. Riedlinger: Modeling of piezoceramic composite transducer structures generating strong sound pulses in therapy, IEEE Ultrason. Symp. Proc. (2001) pp. 1027–1030
33.114 T. Feigl, J. Waldfahrer, B.A. Völklein, H.T. Schneider, J. Zenk, R.E. Riedlinger, H. Iro, C.H. Ell: Destruction of normal and malignant human cells by high-energy pulsed ultrasound, Proc. World Congr. Ultrason. (1995) pp. 1087–1090
33.115 H.T. Schneider, T. Feigl, M. Löhr, R. Riedlinger, E.G. Hahn, C. Ell: In vitro effects of high energy pulsed ultrasound on human tumor cells, Eur. J. Gastroenterol. Hepatol. **6**, 257–262 (1994)
33.116 K. Joechle: Kavitationsdosimetrie in hochenergetischen Ultraschallfeldern, Dissertation (Univ. Heidelberg, Heidelberg 1996)
33.117 T. Dreyer, W. Krauss, E. Bauer, R.E. Riedlinger: Investigations of compact self focusing transducers using stacked piezoelectric elements for strong sound pulses in therapy, IEEE Ultrason. Symp. Proc. (2000) pp. 1239–1242
33.118 J.Y. Chapelon, D. Cathignol, C. Cain, E. Ebbini, J.U. Kluiwstra, O. Sapozhikov, G. Fleury, R. Berriet, L. Chupin, J.L. Guey: New piezoelectric transducers for therapeutic ultrasound, Ultrasound Med. Biol. **26**(1), 153–159 (2000)
33.119 R.E. Riedlinger, S. Weiß, F. Ueberle: Nichtlinearitäten des transienten Schallfeldes eines piezoelektrischen Hochenergie-Pulssenders, Fortschr. Akust. DAGA (1987) pp. 489–493, in German

33.120 D. Cathignol, J.Y. Chapelon, J.L. Mestas, A. Birer, P.A. Lewin: Minimization of the negative pressure in piezoelectric shock wave, Ultrason. Int. Conf. Proc. (1989) p. 1142
33.121 D. Cathignol, J. Tavakkoli, A. Arefiev: Influence of the pressure time waveform on the transient cavitation effect, 135th ASA Conf. Proc. (1998) p. 2799
33.122 A.P. Evan, L.R. Willis, B.A. Connors, J.A. McAteer, J.E. Lingeman, R.O. Cleveland, M. Bailey, L.A. Crum: Separation of cavitation and renal injury induced by shock wave lithotripsy (SWL) from SWL-induced impairment of renal hemodynamics, 135th ASA Conf. Proc. (1998) p. 2487
33.123 M.R. Crum, M. Bailey, P. Kaczkowski, I. Makin, P. Mourad, K. Beach, S. Carter, U. Schmiedl, W. Chandler, R. Martin, S. Vaezy, G. Keilman, R. Cleveland, R. Roy: Therapeutic ultrasound: A promising future in clinical medicine, 135th ASA Conf. Proc. (Seattle 1998) p. 719
33.124 K. U. Köhrmann, M.S. Michel, P.M. Braun, A. Weber, P. Alken: New interactive navigation system for integration of fluoroscopic and ultrasound imaging, J. Endourol. **13** (Suppl 1), A32 FP3-5, 3-6 (1999)
33.125 M.A. Kuwahara, K. Kambe, K. Taguchi, T. Saito, S. Shirai, S. Orikasa: Initial experience using a new extracorporeal lithotripter with an anti-misshot control device, J. Lithotr. Stone Dis. **3**, 141–146 (1991)
33.126 C. Bohris, T. Bayer, C. Lechner: Hit/miss monitoring of ESWL by spectral Doppler ultrasound, Ultrasound Med. Biol. **29**, 705–712 (2003)
33.127 M. Fink: Time-reversed acoustics, Sci. Am. **281**, 91–97 (1999)
33.128 N.J. Parr, S.D. Pye, A.W.S. Ritchie, D.A. Tolley: Mechanisms responsible for diminished fragmentation of uretreal calculi, J. Urol. **148**, 1079–1083 (1992)
33.129 R. Nitsche, A. Amelsberg, T. Berg, U.R. Fölsch: Extracorporeal shock wave lithotripsy of gallstones in different biles and water in vitro, Digestion **55**, 175–178 (1994)
33.130 C.A. Carlson, P.G. Boysen, M.J. Banner, J.S. Gravenstein: Stone movement during ESWL. In: *ESWL for Renal Stone Disease*, ed. by J.S. Gravenstein, K. Peter (Butterworths, Boston 1986) pp. 77–85
33.131 K.U. Köhrmann, F.U. Kahmann, A. Weber, J. Rassweiler, P. Alken: Vergleich verschiedener Lithotripter anhand der Desintegrativen Effektivität (DE) und Desintegrativen Bandbreite (DB) am In-vitro-Steinmodell, Akt. Urol. **24**, 320–325 (1993), in German
33.132 G.G. Tailly: Consecutive experience with four Dornier lithotripters: HM4, MPL 9000, Compact and U/50, J. Endourol. **13**(5), 48–51 (1999)
33.133 J. Rassweiler, T.O. Henkel, K.U. Köhrmann, D. Potempa, K.P. Jünemann, P. Alken: Lithotripter technology: Present and future, J. Endourol. **6**, 1–13 (1992)
33.134 K. U. Köhrmann: Comparison of Lithotripters, Business Briefing, Eur. Pharmacother. (2005), pp. 90–93, www.touchbriefings.com/pdf/1134/kohrmann.pdf (last accessed 08 May 2011)
33.135 R. Gerber, U.E. Studer, H. Danuser: Is newer always better? A comparative study of 3 lithotriptor generations, J. Urol. **173**(6), 2013–2016 (2005)
33.136 N.F. Logarakis, M.A. Jewett, J. Luymes, R.J. Honey: Variation in clinical outcome following shockwave lithotripsy, J. Urol. **163**, 721–725 (2000)
33.137 J.M. Teichman, A.J. Portis, P.P. Cecconi, W.L. Bub, R.C. Endicott, B. Denes, M.S. Pearle, R.V. Clayman: In vitro shock wave lithotripsy comparison, J. Urol. **164**, 1259–1264 (2000)
33.138 http://www.urologyklina.com/subspecialisaties.aspx?name=niersteenverbrijzelingNL (last accessed 08 May 2011)
33.139 N. Hirata, Y. Kushida, T. Ohguri, S. Wakasugi, T. Kojima, R. Fujita: Hepatic subcapsular hematoma after extracorporeal shock wave lithotripsy (ESWL) for pancreatic stones, J. Gastroenterol. **34**(6), 713–716 (1999)
33.140 P.N. Maheshwari, M.G. Andankar, A.L. Saple, A.T. Oswal: Extracorporeal shock wave lithotripsy: Complications and their prevention, Bombay Hospital J. **44**(2) (2002), www.bhj.org/journal/2002_4402_apr/endo_181.htm (last accessed 08 May 2011)
33.141 M. Haake, I.R. Böddeker, T. Decker, M. Buch, M. Vogel, G. Labek, M. Maier, M. Loew, O.M. Maier-Boerries, J. Fischer, A. Betthuser, H.C. Rehack, W. Kanovsky, I. Müller, L. Gerdesmeyer, J.D. Rompe: Side-effects of extracorporeal shock wave therapy (ESWT) in the treatment of tennis elbow, Arch. Orthop. Trauma Surg. **122**(4), 222–228 (2002)
33.142 DGSL 1999: Discussion at the general meeting, Deutsche Gesellschaft für Stoßwellenlithotripsie (1999)
33.143 M. Delius, M. Jordan, H. Eizenhoefer, E. Marlinghaus, G. Heine, H.G. Liebich, W. Brendel: Biological effects of shock waves: Kidney haemorrhage by shock waves in dogs – Administration rate dependence, Ultrasound Med. Biol. **14**, 689–694 (1988)
33.144 DIGEST: http://digest-ev.de (2010) (last accessed 08 May 2011)

34. High-Frequency Surgery

Bernhard Hug, Reiner Haag

High-frequency surgery (HF surgery) has been the dominant form of electrosurgery for many years. HF surgery can be defined as the application of electrical energy in surgery for effecting a thermally induced change or destruction of tissue cells with the aim of hemostasis (stopping bleeding), cutting tissue, or sealing it. In HF surgery, high-frequency alternating current (preferably 0.3–4 MHz) is delivered by special applicators (or active electrodes) to the tissue to be treated, where a thermal tissue interaction takes place due to the electrical resistance of the tissue.

34.1 **Development of High-Frequency Surgery** 678

34.2 **Physical and Technical Principles** 681
34.2.1 Bioelectrical and Biothermal Effects on Tissue 681
34.2.2 Coagulation 683
34.2.3 Electrotomy (Cutting) 683

34.3 **Technology and Techniques** 684
34.3.1 HF Generator Technology 684
34.3.2 HF Application Technique 685
34.3.3 Leakage Currents 686

34.4 **Types of Current and Their Application** ... 687
34.4.1 Cutting Currents 687
34.4.2 Coagulation Currents 691

34.5 **Methodical Instructions for Application and Safety** 697
34.5.1 Patient and User Safety 697
34.5.2 Neutral Electrode 698
34.5.3 HF Instruments and Cables 699
34.5.4 HF Surgical Unit 699
34.5.5 Incidences During Application of HF Surgery 699

34.6 **Outlook** 700

References 701

In contrast, in the case of the electrosurgical operating method galvanocautery – which is insignificant today – a direct current is or was used to heat a cautery (Greek *kauter*) directly as a surgical instrument in order to transfer heat from it to the tissue. Today, this method is only applied for a small number of indications, i. e., whenever a flow of current through the body tissue is to be avoided (e.g., ophthalmology) [34.1,2].

Today, HF surgery has become an indispensable tool for all surgical disciplines, whether for in-patient care or private medical practices.

A considerable advantage of cutting tissue using HF-assisted surgery, as compared with conventional cutting techniques using a scalpel or scissors, is that hemostasis takes place at the same time the cut is made due to the respective vessels becoming sealed. Other advantages lie in the prevention of cross-contamination, mechanical tissue protection,and the possibility of endoscopic application. When applying the method for a targeted hemostasis of bleeding vessels, selective hemostasis – as opposed to the alternative vessel ligature – can be achieved quickly and simply by boiling off (coagulating) spatially limited tissue without having to use any foreign substances.

Due to the history of its development, a large number of synonymous terms are used for HF surgery. Although some of the terms are based on totally different techniques historically, in most cases the same application method is meant today. With no claim as to its completeness, a list of some of the internationally common terms is given as follows: HF surgery, RF surgery, radiosurgery, electrosurgery, cautery, electrocautery, diathermy, endothermy, transthermy, electrotomy or – frequently in the USA – also referred to as *Bovie* after an American pioneer in HF surgery.

34.1 Development of High-Frequency Surgery

The development of this method originates from the advantageous of the therapeutic application of heat. In the papyri written by the Egyptians in the second millennium BC (Papyrus Ebers, Papyrus Edwin Smith), information can already be found on the selective, therapeutic use of heat. The treatment of battle wounds using heated stones, opening festering inflammations using small sticks of wood heated in hot oil, or cauterizing wounds using the *fire drill* (glowing piece of charcoal) were recommended there. Even Hippocrates, who is often referred to as the *father of medicine* (≈ 400 BC), used the *ferrum candens* (cautery) in just the same way as Arabian or Roman physicians to reduce bleeding during amputations with glowing cauteries or knives. Even into the nineteenth century the ferrum candens, heated in different ways, was the method of choice in surgery.

New techniques have always been readily adopted in medicine. Thus, the predecessor of the electric battery developed by Count Volta around 1800 (voltaic pile) had only been known for a few years when the era of electrosurgery was heralded – any ideas about electric street lighting were still a long way off. In this connection, the recommendations of the Englishman Humphrey Davy (1807) for using electricity to decompose organic parts played a less decisive role than the suggestions of the Munich physicist Karl August Steinheil, according to which the Viennese dental surgeon Moritz Heider deadened a dental nerve using an electrically heated platinum wire. Throughout Europe physicians especially were now working on the new method; the Finnish physician Gustav Crusell then coined the term galvanocautery. Other pioneers worthy of note during this period are the Frenchmen Alphonse Amussat, Charles Sedillot, Auguste Nelaton, and Leroy d'Etoiles and the Englishmen John Marshall, Thomas Harding, George Waite, and Robert Ellis. Although he cannot be regarded as the originator of galvanocautery, Breslau physician Albrecht Theodor Middeldorpf (1824–1883) took much credit for the new method by standardizing surgical methods and described them in great detail in his book *Galvanocautery: A Contribution Towards Operative Medicine* in 1854. The basic instruments used in galvanocautery originate from him, i.e., the galvanocautery and the galvanocautery cutting loop – ligatura candens. One of the most famous surgeons of that time, Theodor Billroth (1826–1894), who even recognized Middeldorpf as the inventor of galvanocautery, wrote in 1878 in his *Handbuch der allgemeinen und speziellen Chirurgie* (Manual of General and Specialized Surgery) [34.3]:

> *Middeldorpf not only invented galvanocautery; in his book he also completely exhausts it, so it seems to me, by inventing the most practical battery and the most practical instruments, and he additionally formulated the indications for the usability of this operating method with the utmost precision.*

One of the few from the first group of German surgeons whose work on this topic deserves special mention after Middeldorpf was Victor von Bruns. Technical improvements of the ligatura candens trace back largely to him, specifically the contractible, loop-shaped platinum wire (snare) that was heated using direct currents of 10–20 A at voltages of 3–6 V. It was now possible to operate on places that were difficult to access and allow the glowing heat to take effect after having previously applied the cold wire without any hindrance – ideal conditions for applying the method endoscopically, and it was ultimately in the endoscopic area that galvanocautery was able to establish itself. In general surgery, hot cautery prevailed, not electrically as in the form of galvanocautery, but in the form of the thermocautery invented by Claude A. Paquelin. Differently shaped, hollow pieces of platinum were heated until they glowed and were then kept glowing by blowing in a mixture of petrol and air that burned on the glowing platinum surface. That thermocautery was the preferred method wherever it was possible to work with open access for laparotomies (open surgery) may be due to the fact that galvanocautery was complicated and expensive. In addition, handling the batteries existing at that time (mostly of a zinc–platinum type) was expensive and required intensive maintenance.

The requirements for implementing electric current not only as a form of energy for operating a modern ferrum candens, but also for applying it directly for use in the tissue, were similarly developed during the mid-nineteenth century. From the electrophysiological experiments conducted by Duchenne de Boulogne, whose results were published in 1855, the fact was established that low-frequency, alternating currents lead to muscle contractions and nerve irritation and are thus useless for surgical purposes. At the end of the nineteenthth century, though, Nikola Tesla and Jacques-Arsene d'Arsonval showed that alternating currents in a frequency range of 2 kHz to 2 MHz lead to heating the tissue without causing any stimulation of the muscles or nerves. In 1899 Walther Nernst demonstrated the relevance of these results and proved the inherent working principle. He formulated the Nernst law of electrical nerve stimulus threshold named after him, which established the correlation between the threshold value of the electric current intensity necessary for stimulating a nerve and the frequency of the alternating current producing the stimulus.

These fundamental findings formed the basis for Clive Rivière being able to report on successes in the spark treatment of tumors and tuberculous skin diseases in Paris in 1900. The era of HF surgery had begun. First, a device designed by d'Arsonval was used; its damped oscillations with high-voltage peaks enabled tissue to be destroyed by sparks arcing to the tissue. Simon Pozzi referred to this method as fulguration (Lat.: *fulgur*, lightning), a term used up to the present day to describe the virtually noncontact form of power transmission to a patient. In addition to Pozzi's work, de Keating-Hart's work was trendsetting. He used a fulguration device (Fig. 34.1) – as also recommended by Professor Czerny – for the irradiation of malignant tumors in combination with resection in Marseille and Paris (Broca Hospital).

Electrodesiccation (according to William L. Clark) has also been known since 1907. For this method, needle electrodes are used; they are positioned on or inserted into the tissue, and under the influence of an HF current the tissue cells subsequently reach the boiling point and dehydrate. In 1909, Eugène L. Doyen reported a two-pole method for the *electrocoagulation* of malignant tumors. The term *voltaisation bipolaire* was then introduced. The applications for this method were limited to begin with, due to the generators available (Oudin resonators and spark-gap generators with a spark frequency < 3 kHz). For, as Nernst had already demonstrated, the lower frequencies are accompanied by strong faradic tissue stimulation. Even though spark-gap generators with frequencies of up to ≈ 1 MHz had been available since 1907, and their application could also be extended to neurosurgical indications, it was not possible to cut tissue (electrotomy) to a satisfactory extent using this given technology. It was later demonstrated that unmodulated or slightly modulated HF currents are required to successfully achieve the desired

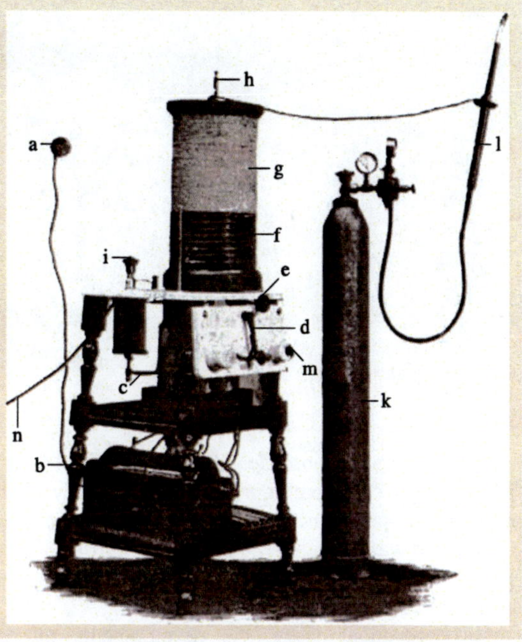

Fig. 34.1 Fulguration device as invented by de Keating-Hart for *cancer fulguration*. For ≈ 10 min strong *fulguration spark bundles* were applied to the tumor tissue. Excessive heating was prevented by a flow of carbon dioxide conducted coaxially via the electrode (after [34.4]). Short explanation: a mains supply, b transformer, c Leyden jar, d variable resistor, e regulating screw for solenoid, f solenoid, g resonator, h connector for applicator, l adjusting screw for spark gap, k CO_2 gas cylinder, l applicator, m circuit breaker, n grounding cable

effect, which could not be generated from spark-gap generators. Such current generation was not possible until tube generators came into use, in the form used by George A. Wyeth (endotherm) in the mid-1920s. However, the hemostatic effect of these generators was not as strong as that offered by spark-gap generators. For a long time, two separate generators were therefore used for electrotomy and hemostasis (cutting with tube generators, coagulating with spark gap). The patent application of American inventor *William T. Bovie* in 1931 [34.5] marked another milestone in the development of HF surgery. Bovie proposed to offer surgeons the option of different kinds of HF surgical currents using the same generator with the same applicator. A technical level had now been reached where faradic (neuromuscular) stimulation was largely avoided and

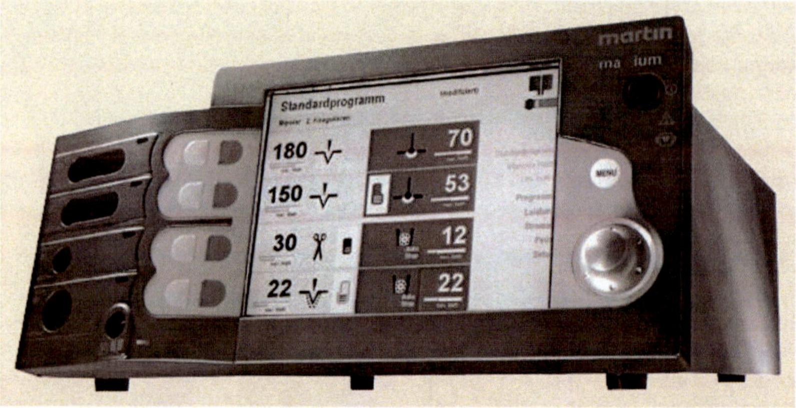

Fig. 34.2 Advanced HF surgical unit with a menu-guided user interface. The treatment parameters are clearly arranged on a screen next to the respective patient output

smooth cutting was possible. Whereas in the USA both Bovie and his collaborator, neurosurgeon Harvey Cushing, deserve special mention for this new application, in Germany it was *von Seemen* (1898–1972), from Alsace, who wrote a guidance paper [34.6] on the topic during his time in Munich.

In subsequent years, the developments of HF surgery were concentrated more on the use of higher frequencies and thus on treatment using radiated electric or magnetic fields. In this context, parts of the body to be moderately heated were exposed to the electrical influence of a capacitor field (*Schliephake* [34.7]) or the magnetic influence of a coil field (*Esau* [34.8]). The shortwave devices (27.12 MHz) and decimeter devices (433.92 MHz) available in the 1930s and 1940s were completed by microwave devices (2.45 GHz) after the magnetron in radar technology became available after World War II. Although some of the shortwave devices were also suitable for surgical applications, it was HF hyperthermia – i. e., the artificially produced, temporary, and localized *fever* – that was the main focus of attention. It was probably due to the high price of tubes that these devices did not gain in popularity until after 1945. For surgical applications, most devices still contained a spark-gap generator for coagulation and a tube generator for cutting tissue. It was not until 1955 that the first fully electronic, universal device built on a tube principle was offered for HF surgery. The development of modern transistor technology in the 1970s offered a technological evolution of the technical design of the device. With this technology, devices could be built considerably smaller and more compact. However, as regards new potential applications, the method did not offer any new developments. Then, however, in the mid-1970s, the development of the so-called argon beamer marked an important milestone. Based on a *plasma scalpel* principle, as suggested by Americans *Shaw* [34.9], *Goucher* [34.10], and *Incropera* [34.11], for vaporizing tissue, *Morrison* [34.12] developed the *argon beamer* which has been used in basically the same way up to the present day. Whereas with the plasma scalpel a high-energy plasma beam – comparable to a cutting torch – was directed toward tissue, the argon beamer merely uses an ionized argon beam as a conducting medium in order to deliver the HF energy to the tissue without contact. And while the plasma scalpel was not of any real medical significance, the argon beamer firmly established itself in various medical applications.

Microprocessor-controlled HF surgical units have been in use since the early 1990s. As a result, feedback control technology has opened up completely new possibilities. A large number of different types of therapeutic current could be developed and adapted to specific indications and applications. Due to the great variety of application options, each with its process-specific treatment parameters, using these new devices soon became quite complex, however. Future developments took this into account, and today units with a display screen and a menu-guided user interface (Fig. 34.2) are offered to facilitate use.

34.2 Physical and Technical Principles

34.2.1 Bioelectrical and Biothermal Effects on Tissue

When electric current flows through biological tissue, three different effects occur, depending on the type of current, current intensity, and frequency.

Electrolytic Effect

With direct current and low-frequency alternating currents, the electrolytic effect dominates, i. e., ion migration takes place in the tissue. Positively charged ions travel to the negative pole (cathode) and negatively charged ions to the positive pole (anode). This effect is used in medicine in ionophoresis for transporting certain drugs into the body. In HF surgery this effect is undesirable as the tissue's cytochemical content can become damaged.

Faradic Effect

When alternating currents flow through biological tissue with a frequency of up to 20 kHz, the faradic effect occurs. Currents within this frequency spectrum stimulate nerves and muscle cells, which can lead to muscle contraction. This stimulus effect peaks at frequencies between 10 and 100 Hz (Fig. 34.3). The faradic effect is currently used effectively in stimulation-current diagnostic procedures and the stimulation-current therapy (e.g., for muscle paralysis). In HF surgery this effect is undesirable as muscle contractions are painful, and possibly even dangerous for the patient, and are a hindrance for the surgeon.

Thermal Effect

With HF alternating currents, both the electrolytic and faradic effects are largely prevented in biological tissue, and thus the thermal effect dominates. The frequency of the alternating current is then at least 300 kHz, hence the term HF surgery. This desired thermal effect is mainly used for two different applications, namely, for cutting and, much more frequently, coagulating. The amount of heat created in the tissue mainly depends on the specific resistance of the target tissue, the current density, and

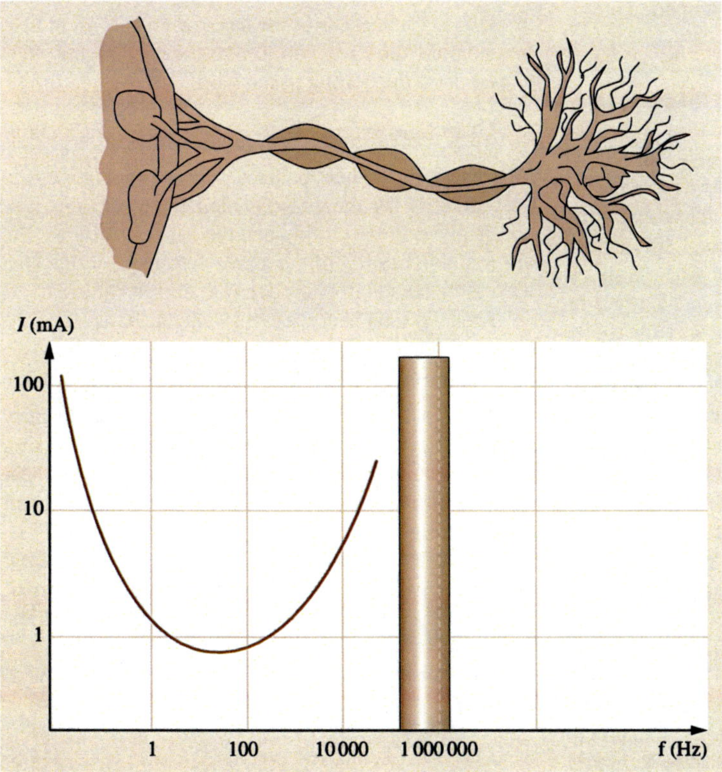

Fig. 34.3 The Nernst law of electrical nerve stimulus threshold, which states the frequency-dependent threshold value of the currency intensity necessary for stimulating a nerve

the duration of exposure. The thermal effect is achieved through conversion of electrical energy into thermal energy.

Heat can be produced in every conductive or current-carrying matter. The conversion of electrical energy into thermal energy takes place effectively without loss of energy.

According to Joule's law, the following correlations result

$$Q = Pt = UIt = I^2Rt = \frac{U^2t}{R} \quad (\mathrm{J = W\,s})\,, \qquad (34.1)$$

where Q is the heat, P the power, U the voltage, I the current, R the resistance, and t the time.

The HF power P_{coag} required for coagulation can be calculated approximately by using both the heat quantity Q_{coag} and the coagulation time t_{coag}

$$P_{\mathrm{coag}} = \frac{Q_{\mathrm{coag}}}{t_{\mathrm{coag}}} \quad (\mathrm{W})\,. \qquad (34.2)$$

The heat quantity Q_{coag} depends on the quantity m_{coag} of the tissue to be coagulated, on its specific heat capacity c_{coag}, and on the temperature difference Δt_{coag} within the coagulum (of $\approx 37\,°\mathrm{C}$ to $60{-}100\,°\mathrm{C}$) between the start and end of the coagulation period

$$Q_{\mathrm{coag}} = m_{\mathrm{coag}} c \Delta t_{\mathrm{coag}} \quad (\mathrm{W\,s})\,. \qquad (34.3)$$

In Q_{coag} only the heat quantity required for coagulation is represented. Depending on the coagulation technique used, an additional quantity of heat Q_{env} also must be taken into account to represent the unintentional dissipation of heat in the surrounding current-carrying tissues. Though Q_{env} is negligibly small for bipolar coagulation applications in comparison with Q_{coag}, in monopolar coagulation techniques Q_{env} can at times be very large relative to Q_{coag} if unfavorable circumstances obtain. These unfavorable circumstances include, e.g., monopolar coagulation techniques where a large part of the HF current flows past the tissue to be coagulated, for example through the irrigation fluid used in transurethral resection (TUR) procedures. The heat quantity Q_{env} always poses a risk of undesired secondary effects, and it is absolutely necessary to take this into account during use. This also explains why a higher power value is required in monopolar applications as opposed to bipolar applications.

Apart from the heat quantities Q_{coag} and Q_{env}, the heat quantity Q_{AE} – which is the heat created within the active electrode during coagulation – also must be taken into consideration for some coagulation procedures. The temperature of the active electrode should not rise during coagulation because this creates a layer of coagulum that can stick to the active electrode; if the active electrode were to directly contact the coagulum, the electrode would unavoidably be heated.

Thus the entire heat quantity created is calculated

$$Q_{\mathrm{tot}} = Q_{\mathrm{coag}} + Q_{\mathrm{env}} + Q_{\mathrm{AE}} \quad (\mathrm{W\,s})\,, \qquad (34.4)$$

where Q_{coag} is the heat quantity required for coagulation, Q_{AE} the heat quantity active electrode, Q_{env} the unintended surrounding heat quantity, Q_{tot} the total heat quantity, t_{coag} the coagulation time, P_{coag} the power for coagulation, Δt_{coag} the temperature difference, m_{coag} the quantity of tissue to be coagulated, and c_{coag} the specific thermal capacity.

The HF power P_{S} required for cutting can similarly be calculated approximately by using both the heat quantity Q_{S} and the cutting duration t_{S} as

$$P_{\mathrm{S}} = \frac{Q_{\mathrm{S}}}{t_{\mathrm{S}}} \quad (\mathrm{W})\,. \qquad (34.5)$$

When cutting, a tissue volume proportional to the length, the average depth, and the width of the cut is heated so strongly that its water content vaporizes. The heat quantity required for vaporizing the water content in tissue (Q_{S}) consists of the heat quantity Q_{100} to heat tissue fluid from 37 to 100 °C plus the heat quantity Q_{D} to evaporate the boiling tissue fluid.

$$Q_{\mathrm{S}} = Q_{100} + Q_{\mathrm{D}} \quad (\mathrm{W\,s})\,. \qquad (34.6)$$

In analogy to the heat balance during coagulation, the heat quantity for the unavoidable heating of the tissue not involved in the cutting and the heat quantity for the unavoidable heating of the active cutting electrode also have to be taken into consideration during cutting.

Thus the total heat created is calculated as

$$Q_{\mathrm{tot}} = Q_{\mathrm{S}} + Q_{\mathrm{U}} + Q_{\mathrm{AE}} \quad (\mathrm{W})\,, \qquad (34.7)$$

where P_{S} is the power for cutting, Q_{S} the heat quantity required for cutting, Q_{D} the heat quantity for vaporizing the water content of the tissue volume, Q_{100} the heat quantity from 37–100 °C, V_{S} the cut-tissue volume, and t_{S} the cutting time.

The current density J plays a key role in HF surgery; only if the current density is sufficiently high (normal: $1{-}6\,\mathrm{A/cm^2}$) can the desired cutting or coagulation effect be achieved.

The current density decreases quadratically with distance r

$$J \propto \frac{I}{r^2} \quad (\mathrm{A/cm^2})\,. \qquad (34.8)$$

Table 34.1 Specific resistance ρ of biological tissue (0.3–1 MHz), modified as described by *Reidenbach* [34.13]

Tissue	Specific resistance ρ (Ω cm)
Blood	160–300
Muscle, kidney	160–260
Spleen	270–300
Heart	200–230
Liver	200–380
Brain	670–700
Lung	160–1000
Fat	1600–3300

When tissue properties are homogeneous, the temperature-increase capabilities are reduced by the distance r at the fourth power

$$\Delta t \propto \frac{l}{r^4} \quad (\mathrm{K})\,. \tag{34.9}$$

Biological tissue itself mainly behaves like an ohmic resistor (Table 34.1). Whereas the specific resistance in muscle tissue and well-vascularized tissue is relatively low, tissues with little fluid content such as bones, cartilage, and fat tissue have a high specific resistance, which in turn translates to a relatively low current flow. In muscle tissue, the resistance is $\approx 150–300\,\Omega$, whereas in fat tissue it lies between 500 and $1000\,\Omega$.

Regardless of the method used to heat the tissue (HF current, laser, IR coagulator, ultrasound, ferrum candens, etc.), the thermal effects can be classified qualitatively as shown in Table 34.2.

34.2.2 Coagulation

The term coagulation (Lat. *coagulare*) generally describes the precipitation, flocculation, or coagulation of a material, i. e., the transition of colloidal materials (colloid: from Greek *kolla*, glue, and *eidos*, form, appearance) from a solution state to a gel state. This process could also be described simply as a tissue's boiling point. During this process, the tissue is heated relatively slowly by HF electric current such that intracellular fluid and extracellular fluid vaporize and cause the tissue to shrink. The resulting *tissue glue* and the tissue shrinkage generated lead to the contraction of the perforated blood vessel and subsequent hemostasis. In many cases, such as with small vessels, this fast and effective coagulation can replace expensive fibrin glues or extensive ligatures. In daily use of HF surgery, the term coagulation is used not only for the tissue reaction described above but also more widely for a certain operating mode of HF surgical units. Depending on the quality and quantity of the HF current and technique used, a distinction is made between contact coagulation, forced coagulation, desiccation, fulguration (also referred to as spray coagulation), argon-assisted coagulation, and vessel sealing (Sect. 34.4.2).

34.2.3 Electrotomy (Cutting)

Even though the biophysical process occurring during a cutting procedure has not yet been fully explored in detail to date, it can be plausibly explained by a phenomenological description and through theoretical considerations. In doing so, two totally different initial situations must be taken into account. On the one hand, the cutting electrode is already in contact with the tissue at the beginning of the cutting procedure, while on the other hand the cutting electrode gradually approaches the tissue in an activated condition. Using a HF alternating current, it is possible to produce relatively high, but very concentrated, current densities in tissue using knife-shaped or needle-shaped electrodes with small electrode surface areas. As a result, the tissue

Table 34.2 Thermal tissue damage depending on temperature

Temperature	Tissue reaction
Up to ca. 40 °C	No significant cell changes
From ca. 40 °C	Reversible cell damage (depending on the duration of exposure)
From ca. 49 °C	Irreversible cell damage
From ca. 60–65 °C	Coagulation: Collagens are transformed into glucose, the tissue containing collagen shrinks and produces a hemostasis of the bleeding vessels
From ca. 90–100 °C	Dehydration/desiccation (drying out): Transition of intracellular and extracellular fluids to the vaporous state. Glucose can produce a sticking effect due to dehydration; the coagulum shrinks
From ca. 200 °C	Carbonization: The tissue carbonizes as in a 4th degree burn; unpleasant smell of the burnt tissue; the healing process can be impaired
Several hundred °C	Vaporization (evaporation of the tissue): Emission of plume and gases

heats up to over 100 °C in a flash, resulting in a vapor pressure increase and the rupture of cell membranes. The resulting isolative vapor between the electrode and the tissue subsequently prevents the unhindered ohmic current flow into the tissue; a cutting voltage builds up between the electrode and the tissue, creating sparks between them. The energy transfer subsequently takes place through these sparks. As extremely high energy densities occur in their very small base points having a size of just a few microns ($r = 10–20\,\mu m$), the targeted tissue cells vaporize. It is irrelevant whether this process takes place openly under atmospheric conditions or in an electrically nonconducting liquid. This model concept is also supported by the work being done by the study group headed by *Thiel* and *Fastenmeier* [34.14].

If, on the other hand, one approaches tissue with an electrode that is already active, the voltage present ignites sparks at a certain proximity to the tissue. These sparks, which are indispensable for a cutting process, create a bridge over the *last* air gap before subsequently producing the steam layer that maintains the process as described above. However, in the first scenario, the reduced current flow into the tissue occurring at the beginning of the desired cutting application leads to a delay in starting the cut. This was overcome in the past using a pulse-type power increase when activating the current (to facilitate starting the cut); today, power control technology is used to compensate the phenomenon.

Cutting tissue with an *HF knife* involves no mechanical effort. The cutting electrode virtually glides through the tissue, similar to a hot wire gliding through butter or wax. As a result of the high temperature generated at the edge of the cut, the danger of contamination is reduced. In addition, hemostasis is produced simultaneously with the cut. The precisely controllable cut can be utilized by the surgeon both for open surgery and endoscopically.

34.3 Technology and Techniques

34.3.1 HF Generator Technology

The main task of an HF generator is to convert the electric current of the commercial power supply into HF current. The reliability of the electrical (galvanic) insulation between the mains circuit and the patient circuit has absolute priority to ensure safety. The current generation of microprocessor-controlled HF generators was developed on the basis of a careful analysis of the tissue effects created, taking into account also the various special requirements of the different surgical disciplines. Thus, HF surgical devices are now available with which consistent cutting and coagulation effects can be achieved. This provides, for example, a gentle coagulation effect without any carbonization, or reproducible cutting properties largely unaffected by external conditions such as the cutting speed or electrode size. The output power levels of automatically controlled HF surgical units are also considerably lower than those of conventional devices. Before 1975 devices were still being offered with an output power of up to 1000 W, even though no more than 150 W was typically required at the time (as also today), even for applications with a relatively high power requirement (applications in liquid environments) [34.15]. This former trend toward higher output power was reversed by appropriate standardization (IEC 60601-2-2) [34.16], whereby the maximum HF output power for HF surgical devices was limited to 400 W.

A series of monitoring features implemented in today's devices, such as neutral electrode monitoring, overdose protection circuits, visual and audible activation indicators, tissue impedance monitoring and HF leakage current compensation, translate into an increase in safety for the patient and surgeon. For all the manufacturers of HF surgical devices, international standards stipulate that controlling devices with an HF output power of over 50 W must be equipped with a monitoring system that detects a break in the neutral electrode contact and is able to prevent an energy output if a fault occurs. However, it is recommended to use devices that not only detect a cable break but additionally monitor the correct application and pad positioning of the neutral electrode on the patient.

The patient output circuit (applied part) must not have any direct connection to ground when using HF surgical devices. In other words, a *floating* (insulated) applied part must be used. HF devices are therefore not constructed as devices with an applied part of type B protection rating against electric shock but are defined as type CF (defibrillation-proof). In a completely floating output – devices with an applied part of type CF (cardiac floating) – neither of the two patient HF leads has an HF ground contact by design, whereas in the case of devices with a type BF (body floating) protection rating, the neutral electrode connection is designed to have ground contact via a capacitor. For application on the open heart and for the combination with intracardial catheters, HF surgical devices with an applied part of protection type CF, and thus defibrillation-proof output, are available.

Since capacitive leakage currents – i. e., HF currents diverted to ground due to partial capacitances, e.g., generated by the housing, leads, and patient – can occur in an uncontrolled fashion and cause patient burns, standardized limits must not be exceeded. Since the leakage currents become higher with an increase in frequency, the application of HF surgery virtually has an upper limit. Low frequencies, on the other hand, lead to nerve and muscle stimuli, as Nernst has demonstrated. Considering this, IEC 60601-2-2 only recommends working with frequencies over 300 kHz, and, due to the potential for leakage current, it recommends not using frequencies above 5 MHz.

However, the Faraday effect can also occur in the above-mentioned higher working frequencies. Their effects are derived from the rectifying process of sparks – which are absolutely vital in electrotomy, for example. In order to avoid faradic irritation, a so-called *antifaradization capacitor* of low capacity is positioned in the patient circuit. At certain areas, however, the occurrence of muscle twitching cannot be completely eliminated (e.g., stimulation of the obturator nerve during resection in the urinary bladder).

34.3.2 HF Application Technique

The various types of application can basically be divided into monopolar, monoterminal, and bipolar methods. The monopolar application technique is used most frequently.

Monopolar Technique

In the monopolar technique, an active and a so-called neutral electrode must be connected to the HF surgical device. The physical effects required for electrosurgery (cutting, coagulation) are produced at the active electrode. The neutral electrode, which compared to the active electrode covers a far greater surface area, is normally placed on the thigh or upper arm of the patient. The contact area between the neutral electrode and the skin of the patient is large, in order to ensure that the current density (current per unit area) remains relatively low. In contrast, the active electrode only has a small contact area, which is why it produces a very high current density. The selective heating of tissue using the high temperature generated by the active electrode would not be possible without such a high current density and the target tissue resistance.

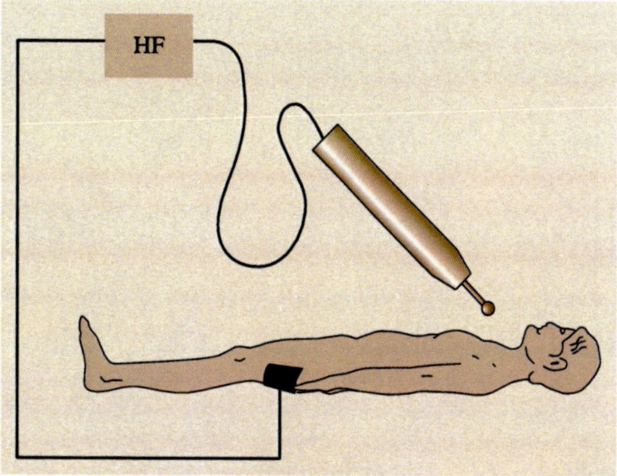

Fig. 34.4 Monopolar circuit with large-surface neutral electrode attached to the thigh of the patient

The neutral electrode is also referred to as a plate electrode, passive return electrode, dispersive electrode, indifferent electrode, or, incorrectly, also as a grounding electrode. When the surgeon touches the tissue with the monopolar active electrode of an activated HF surgical device, HF electric current flows to the patient and generates the desired physical effect (Fig. 34.4) directly at the point of contact between the tissue and the relatively small-surfaced active electrode.

Monoterminal Technique

With this technique, the circuit is closed via the patient's body capacitive contact to ground (Fig. 34.5). No neutral electrode is used in this case. Basically, this is a special form of the monopolar mode. This technique is only safe for small working currents and is therefore only suitable for minor surgical interventions, e.g., in dentistry and in dermatology. For this application technique, only units with a maximum HF output power of 50 W should be used. Since in this working mode the HF current is intended to flow back to the HF unit via ground, an increase in electromagnetic interference with other electrical units can occur.

Units with an output power greater than 50 W must not be used monoterminally, as incorrect operation could cause severe patient burns.

Bipolar Technique

The bipolar technique is best described as the integration of both electrodes (active and neutral) in a single instrument, such as bipolar forceps with branches or

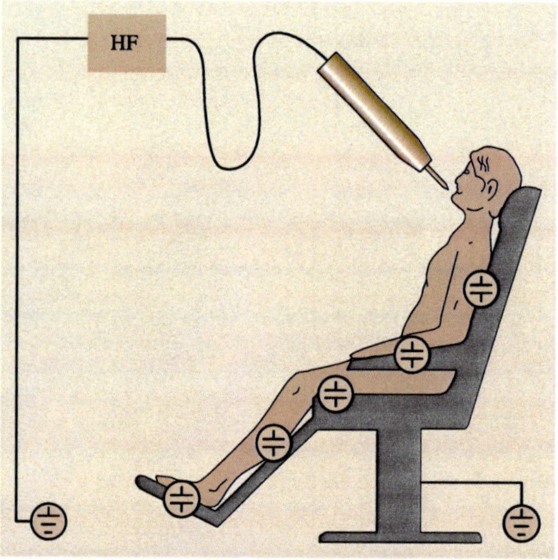

Fig. 34.5 Monoterminal circuit. The grounding of the HF unit on the HF side allows the reverse flow of current via a grounded treatment chair to which the patient is capacitively coupled

tines that are insulated from each other (Fig. 34.6). In this case, the current flows into the tissue via one electrode and back to the HF surgical unit via the other. This means that the current only flows in the narrowly defined tissue area between the two electrode tips. The bipolar technique offers an increase in safety particularly in the case of fine dissection in neurology. A separate neutral electrode is not necessary for operation.

In comparison with the monopolar technique, this offers the following advantages:

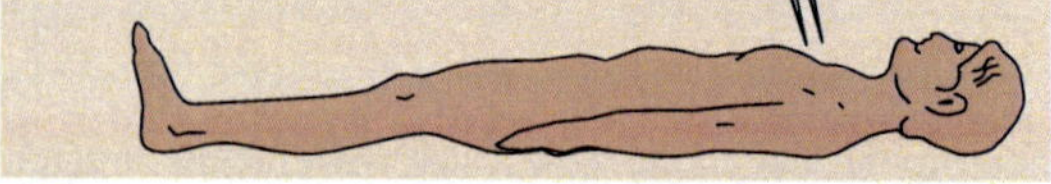

Fig. 34.6 Bipolar circuit. The current flow remains limited to the tissue grasped between the two tips of the forceps

- The current only flows through the tissue held between the two electrodes where the thermal effect is intended.
- The danger of patient burns by touching conductive objects during the operation is negligible.
- There are no stray currents.
- It has reduced influence on cardiac pacemakers and units that are connected to the patient in addition to the HF unit (e.g., for monitoring).

Whereas the bipolar technique mainly used to be applied in neurosurgery, it is successfully used today in ENT (ear, nose, throat), gynecology, and particularly for minimally invasive surgery (MIS). The bipolar technique is primarily used for coagulation with bipolar forceps and for sealing vessels and tissue. Despite several attempts to apply it in other areas, bipolar cutting continues to be limited to a few *special cases* thus far. The basic problem in using the bipolar mode for the cutting function is that the spark formation required for targeted cutting relies on the constant presence of an active electrode and a nonactive electrode. With the bipolar technique, both electrodes are more or less the same size. This means that the active/nonactive function can change during the cutting process and the whole process will come to a halt. An exception to this is bipolar cutting in a conductive saline solution for transurethral resections (Sect. 34.4.1). However, in this case one would correctly have to describe this as a semibipolar technique, as the two electrodes are of a significantly different size and it is always clear which of the two electrodes is to assume the function of the active electrode. Basically, the neutral electrode – as we know from the monopolar technique – is highly reduced in size here and held at very close proximity to the active electrode.

34.3.3 Leakage Currents

When applying HF surgery in flexible endoscopy and MIS, one should keep in mind that a capacitive current can flow due to the high frequency. This is the case whenever a capacitive reactance is present (capacitor effect).

The capacitive reactance follows the relationship

$$X_C = \frac{1}{2}\pi fC \quad (\Omega)\,, \qquad (34.10)$$

where C is the capacity of a capacitor, f the frequency on the capacitor, and X_C the capacitive reactance. In practice this effect is best known as capacitive coupling.

Capacitive coupling refers to a physical effect that always occurs when HF current flows from one electrically conductive medium to another while separated by an insulator. This can take place because an insulator for HF current does not have the same insulating properties as an insulator for low-frequency current (e.g., 50 Hz mains frequency). However, since our knowledge of the properties of insulative materials is based on those commonly used in the low-frequency range, the user expects these materials to have a similar insulating behavior when HF current is applied. The user is therefore not aware of the capacitively coupled current. This capacitive coupling depends on a series of factors:

- The distance between the two electrical conductors: the smaller the distance, the higher the capacitive current
- The cross sections of the electrical conductors: the larger the cross sections, the higher the capacitive current
- The level of the electric voltage: the higher the electric voltage, the higher the capacitive current
- The frequency level: the higher the frequency, the higher the capacitive current
- The form of the electric voltage: voltage characteristics with few harmonics cause a lower capacitive current than voltages with a high harmonic content.

For applications in open surgery, the capacitive coupling effect plays a small role. In the case of laparoscopic surgery and in flexible endoscopy, however, capacitive coupling is a more frequent phenomenon due to the long instruments used in these procedures. Since leakage currents are unavoidable, undesirable, unnoticeable, and unpredictable, it is particularly important for users to bear in mind that capacitive coupling can occur when using such instruments. In practice, this means that a current can also flow through a plastic insulating layer such as is present, e.g., in trocars or MIS instruments and that in the event of concentrated contact with the tissue, a current density can be produced that is high enough to cause a burn.

34.4 Types of Current and Their Application

The effect that an HF current can cause in the tissue is determined mainly by its application time, voltage level, and the degree of amplitude modulation. The frequency (0.3–5 MHz) does not play any decisive role in this case. However, the crest factor is an important parameter here. It describes the relationship between the peak value and the effective value of an alternating electric current and thus indicates how strongly a current is modulated in its amplitude. Thus, a sinusoidal alternating voltage with an effective value (U_{eff}) of 230 V (mains voltage), for example, has a peak value (U_{p}) of ≈ 325 V; the crest factor (C_{F}) here is 1.41 ($\sqrt{2}$). For the same output power, the output voltage must be higher for a current with a high crest factor. As to its influence on the tissue effect, the following holds true: the higher the crest factor, the more pronounced the coagulation effect; the lower the crest factor, the greater the cutting effect. Due to the sinusoidal forms of current used in HF surgery, the value of the crest factor cannot fall below 1.41

$$\text{Crest factor } C_{\text{F}} = \frac{U_{\text{p}}}{U_{\text{eff}}}\,, \qquad (34.11)$$

with C_{F} (sinus) = 1.41, C_{F} (symmetrical rectangle) = 1.

34.4.1 Cutting Currents

It is only possible to cut tissue using HF alternating current if the electric voltage between the active electrode and the tissue is sufficiently high to generate electric sparks (Fig. 34.7). When an active electrode penetrates the tissue, electric sparks are automatically produced wherever the distance between the active electrode and the tissue is small enough. However, sparks do not occur until a minimum voltage of ≈ 200 V is exceeded. If this voltage increases, the spark intensity increases

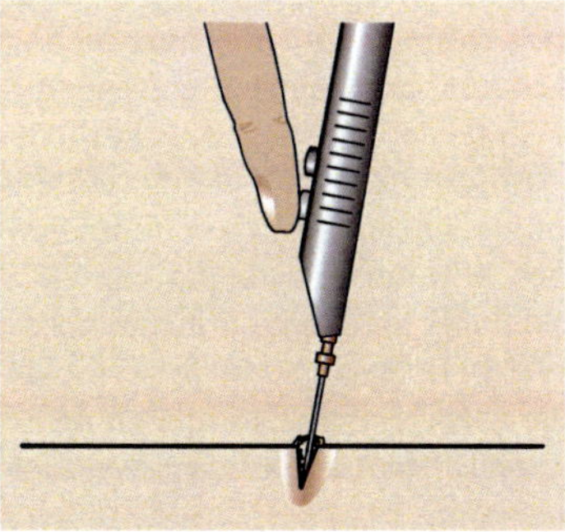

Fig. 34.7 Dissection: HF surgical cutting (electrotomy) is based on sparking between the cutting electrode and the tissue

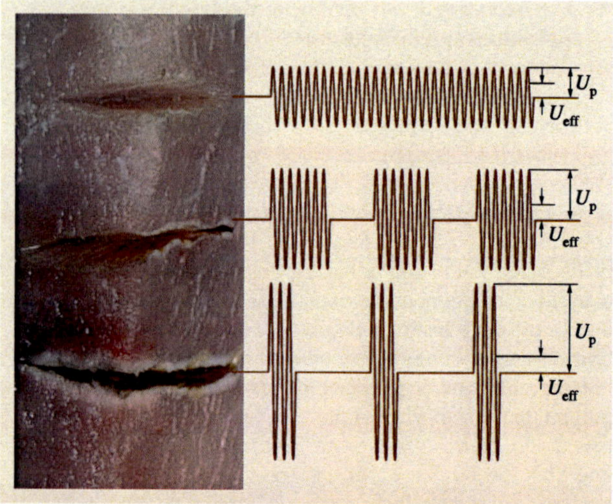

Fig. 34.8 HF surgical cuts with different crest factors. An unmodulated current produces a pure cut that only shows a very low thermal effect at the cutting edge (*top*). A modulated current causes a blend cut, whereby its thermal effect produces a narrow coagulation halo at the cutting edge (*center*). Strongly modulated current produces a cut with high eschar formation (super blend cut). The powerful thermal effect involved creates a wide coagulation margin along with slight carbonization on the cutting edge (*bottom*)

proportionally. Depending on the appearance of the cutting surfaces, a distinction is made between a *pure cut* and a *blend cut*. For a pure cut, where the cutting surfaces do not show any discoloration as a result of the heat effect, an HF current with a crest factor of ≈ 1.4 is used. For a blend cut, where the cutting surfaces show a brownish discoloration, an HF current with a crest factor of between 2 and 4 is used, depending on the required degree of coagulation. The more pronounced the eschar formation on the cutting surfaces, the greater the hemostatic effect accompanying the cut. A current with a crest factor higher than 4 no longer allows for satisfactory cutting. With the introduction in 1920 of tube generators, unmodulated currents did not initially provide sufficient hemostasis during cutting. Therefore, a spark-gap-based fulguration current was superimposed on (or mixed into) the cutting current. This is described as a mixed or blend current – a term that is still commonly used to describe unmodulated current today. A term frequently used for a smooth cut is *pure cut*, for an average eschar formation *blend cut*, and for a greater eschar formation *super blend cut* (Fig. 34.8).

Various parameters impacting the quality of the cut are minimized today using feedback control technology. The following factors play a considerable role here:

- Size and shape of the cutting electrode: for HF generators, it makes a great difference whether a large-blade electrode or a microneedle is used for cutting.
- Type of cut and cutting speed.
- The cutting quality depends on whether cutting is superficial or deep, incision speed is fast or slow.
- Tissue properties: if the cutting electrode penetrates tissue with a low electrical resistance (muscles, vessels), the output voltage sometimes breaks down; for tissue with a high electrical resistance (fat), this effect is less pronounced.

Depending on the circumstances, different control algorithms are used based on the output power, the output voltage, and the spark intensity. The aim here is to offer the user reproducible cutting properties that are independent of external conditions to the greatest possible extent.

Controlled Cutting Currents

For voltage control and regulation, sinusoidal voltages ranging between 200 and 650 V are used. In this case, the output voltage level also has an influence on the spark intensity and, consequently, on the degree of coagulation achieved during the cutting process (apart from the degree of modulation, the spark intensity thus also influences the degree of coagulation). No cut is possible below 200 V, as no spark ignition can occur between an active electrode and tissue at this voltage. On the other hand, voltages of over 650 V lead to an undesirable carbonization of the tissue and excessive wear of the active electrode. In homogeneous tissue, the voltage control compensates for variable cutting depths, cutting speeds, and electrode sizes. However, since mostly heterogeneous tissues (or different types of tissue) are cut during surgical interventions, *constant voltage regulation* alone is not sufficient. Low-impedance tissue (e.g., muscle) requires a much lower cutting voltage for a certain cutting effect than high-impedance tissue (e.g., fat). A purely voltage-controlled cutting process can therefore come to a halt when cutting through fat tissue. In order to prevent this, the control algorithm in modern HF generators uses another control criterion – e.g., the output power. The output voltage is *allowed* to fluctuate but the output power is kept constant. If an increase occurs in the tissue impedance (cut transition from muscle tissue into fatty tissue), the output voltage is increased in order to keep the amount of power delivered to the tissue constant. This guarantees a consistent cutting effect even in heterogeneous tissue with highly variable tissue impedances.

The spark intensity has already been established as a main factor in the cutting process. As the existence of sparks inevitably has a retroactive effect on the generator itself, this can be utilized for control technology. On the one hand, sparks cause harmonics in the current, while on the other hand, sparks have a rectifying effect. This was already mentioned above in connection with faradic stimulation that can occur in electrotomy. During cutting, electrical variables are generated that are directly related to the spark intensity and thus may impact the cutting process. If one of these variables is regulated, the spark intensity can be kept constant. In this way, the cutting quality can similarly be reproduced and is largely independent of the cutting speed and type of tissue. However, spark regulation alone is not sufficient as an overall control algorithm to compensate for all the aspects involved. If the contact area of the active electrode is reduced – e.g., at the end of a cut when the electrode is lifted from the tissue – both a *constant power* algorithm and a *constant spark* algorithm try to maintain the process right to the end, even though the effective electrode area becomes steadily smaller. As a result, the tissue response would be far too strong. However, this phenomenon can then be controlled again by the *constant voltage* algorithm. An all-inclusive control algorithm, in contrast, links the three individual criteria to form one complex in which the best control algorithm is used as the situation demands in terms of impedance, voltage, and current. In this case, the required output power level is as low as possible, but as high as necessary.

Microneedle, lancet, and blade electrodes are used as cutting electrodes in open surgery; in transurethral, hysteroscopic, and cystoscopic resections, rigid loop electrodes are used, whereas hook electrodes are the best option in laparoscopic surgery.

Fractionated Cutting Currents

HF surgical interventions in gastroenterology present a particular challenge. The focus here is on endoscopic polypectomy (removal of an adenoma or a tumor on the mucous membrane, mostly in the intestines), mucosectomy (surface removal of gastric or intestinal mucosa), and papillotomy (widening of the biliary and pancreatic duct orifice in the duodenum, also referred to as sphincterotomy). In these indications, collateral thermal damage should be as superficial as possible, but as large as necessary for sufficient hemostasis. The wall thicknesses of the target organs fluctuate considerably according to their position and elastic properties (stomach 2–8 mm, colon 0.5–4 mm), which make a resection very difficult due to the constant risk of perforation. Another factor to be considered is that not only is an open breakthrough to be classified as a perforation, but also a coagulation penetrating the wall. From a physical point of view the initial conditions are very unfavorable for electrotomy, particularly in polypectomy. Typically, prior to HF activation, the polyp is completely trapped within the polypectomy snare loop and held under tight mechanical tension. This results in a tissue contact situation that is very unfavorable to the electric cutting process as this process cannot take the form of a smooth cut due to the constant mechanical pressure that the cutting electrode exerts on the tissue. Once the cutting process has begun, the cut will continue in a jerky, hesitant manner. In order to provide an adequate clinical effect in this situation, a very HF power must first be applied in order to *condition* the tissue prior to cutting. However, this leads to the accompanying coagulation effect that is a great deal stronger than the required cutting power needed for the cutting process – which is much lower (by several orders of magnitude) than the initial cutting power needed for starting the cut. As a result, a highly inhomogeneously coagulated resection surface is produced which is too pronounced at the outer edges (risk of perforation) and too weak in the middle (risk of bleeding).

This problem is solved by temporarily discontinuing the cutting process right after it has been started. The short initial cutting phase is followed directly by a pure coagulation phase without any cutting effect. Thereafter, cutting pulses of varying intensity, automatically adjusted to the requirements of the procedure, are successively delivered to the tissue precoagulated between the pulses. The result is a *fractionated*, step-by-step cutting process by which a homogeneously coagulated resection surface is achieved. Depending on the type of polyp (sessile or pedunculated), the degree of coagulation can be controlled by selecting the most appropriate type of current.

In the case of endoscopically assisted manipulation of a cutting electrode during papillotomy, the surgeon frequently must correct the direction of the cutting electrode as he must assess the tissue condition regularly while cutting in order to determine whether the cutting length obtained is sufficient. In this case it is very helpful if the cutting procedure is not continuous but pulsed. This means that the pulse-type cutting procedure used progresses *millimeter by millimeter*, giving the physician a chance to decide during the cutting pauses whether to continue or discontinue the procedure.

Argon-Assisted Cutting

If argon gas is used in combination with the normal cutting mode of the HF surgical unit, this method is referred to as argon-assisted cutting. For this purpose a special applicator is used. An active cutting electrode with argon gas flowing coaxially around it during the cutting procedure protrudes from the applicator shaft. Argon-assisted cutting is performed with the normal cutting voltages provided by the HF surgical unit, i. e., the argon gas is not ionized and is therefore not electrically conductive. The cutting site is surrounded by an inert argon-gas atmosphere, which adopts the function here of a protective gas atmosphere. Due to the argon gas, less oxygen reaches the cutting surface, leading to:

- A reduction in plume formation – the surgeon thus has a clearer view of the operating site
- Less carbonization – this means a faster healing process.

Argon-assisted cutting is particularly suitable for breast and liver surgery.

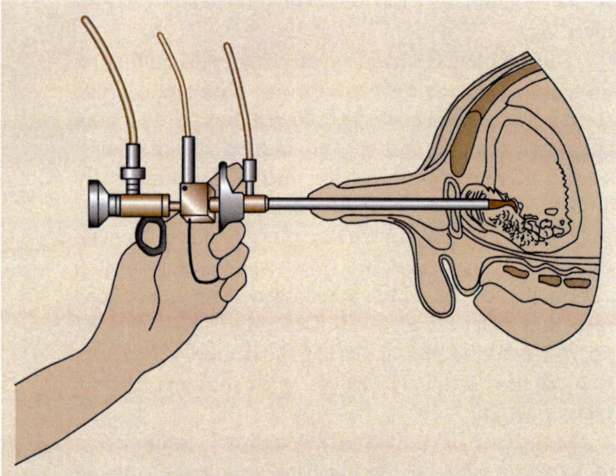

Fig. 34.9 TUR application. Via a rigid resectoscope, bladder or prostate tissue is resected transurethrally with a loop electrode, using the HF surgical monopolar mode with an electrically insulating sugar solution, or the bipolar mode with an electrically conductive NaCl solution

Bipolar Cutting

Generally the same physical laws governing the monopolar cutting process apply to bipolar cutting as well. Therefore, the current qualities and the possibilities created by control technology as shown for monopolar cutting currents can basically also be applied in the bipolar mode of application. The restrictions mentioned in connection with bipolar cutting in Sect. 34.3.2 are primarily due to the limitations of the application instruments available. Although great efforts were taken from the mid to late 1990s to improve the performance of bipolar cutting instruments and enhance the range of available products, none of the instruments have thus far been able to establish themselves as a real standard. The attempts at integrating additional electrodes in an instrument (in tripolar or multipolar arrangements) have also not led to the hoped-for breakthroughs.

Bipolar Transurethral Resection (Bipol-TUR)

The transurethral resection (TUR) of the hypertrophic prostate (TUR-P), resections in the bladder (TUR-B) in urology, or the transcervical resection of the endometrium in gynecology are standard procedures that are traditionally carried out by HF surgery using the monopolar technique in electrically nonconductive irrigation fluid (Fig. 34.9). A sugar solution serves as an irrigation medium (sorbitol + mannitol) that, however, due to irrigation pressure, can be washed into the circulation as a result of intraoperative bleeding. This absorption of hypotonic irrigation fluid can subsequently provoke a change in the electrolyte composition (hyponatremia) that – better known as TUR syndrome – leads to a brain or pulmonary edema and must be treated in intensive care. Since this risk increases with the length of the surgery, the duration of the surgery is limited.

When applying the bipolar TUR, as opposed to monopolar resection, a physiological saline solution can be used as an irrigation medium. With this medium the TUR syndrome is avoided. In bipolar TUR instruments, the two electrodes are combined within the resectoscope; either the shaft of the resectoscope functions as the nonactive electrode or it is arranged as the opposite electrode in close proximity to the active loop electrode. In both cases, the arrangement is quasibipolar, as described in Sect. 34.3.2. Since saline is electrically conductive, however, resection poses a problem because this fluid creates an electrical shunt, causing the electric current to be discharged from the cutting electrode and bypass the target tissue – which means that the required cutting voltage cannot be achieved. In order to overcome this situation, a very high power must be produced at the loop electrode, at least for a short time. The high power level causes the irrigation fluid to evaporate abruptly along the entire cutting electrode, thus forming a plasma layer along the loop electrode. It is obvious that the energy required for forming the plasma will be dependent on the surface size of the loop electrode. The plasma layer produced in this way can then be maintained at a considerably lower power level. The loop is then ready for resection and any tissue in contact with it can be cut immediately – there is no problem *starting the cut* and no need to avoid the pressure exerted by the loop (as described above).

The method offers the following advantages:

- Reduced risk of the TUR syndrome
- Statistically significant but low risk of an obturator nerve stimulation. Due to the bipolar arrangement, the current does not flow through the body to the neutral electrode attached to the thigh but only in close proximity to the resection site
- Less blood loss
- Lower risk for pacemaker patients due to lower interference
- No thermal damage to the tissue in deeper layers
- Slightly lower cost of irrigation fluid
- No neutral electrode needed.

34.4.2 Coagulation Currents

The aim of coagulation is to denature tissue using HF current, or to constrict vessels to an extent where *bleeding stops*. The coagulation effect mainly depends on the level and form of the output voltage, the current density in the tissue, the tissue resistance, the form and size of the active electrode, and the application time. To coagulate biological tissue, a temperature of $\approx 70\,°C$ (158 °F) is required.

If the coagulation temperature of $\approx 70\,°C$ is exceeded, the glucose within the coagulate dehydrates and the tissue can stick to the active electrode. At even higher temperatures, a carbonization of the tissue can result (Table 34.2).

The demands made on an HF surgical unit are very complex for coagulation; it is therefore important to be able to utilize different coagulation modes in order to meet all surgical needs. The following modes should be available: contact coagulation, forced coagulation, desiccation, spray coagulation, argon-assisted coagulation, bipolar coagulation and bipolar vessel sealing. The user can then select the coagulation mode best suited to the respective surgical intervention and optimize the intensity setting accordingly.

Contact Coagulation

A distinction is made here between two basically different forms of application:

- Direct application of the HF current to the tissue via an HF surgical instrument (handle) incorporating special contact coagulation electrodes such as ball or plate electrodes.
- Indirect application of the HF current to the tissue via an additional surgical instrument to which the current is transmitted from the HF surgical handle; a cutting electrode (needle or blade electrode) is usually used for this purpose.

The direct application mode is used to a far lesser extent than the indirect application mode. Among general practitioners, where very little cutting takes place, it plays a more significant role than in daily routine application in the OR. There, selective coagulation goes hand in hand with cutting and takes place in constant alternation, *cutting/coagulation*. The source of spontaneous bleeding can be quickly grasped and compressed, and thus mechanical hemostasis can be achieved using surgical dissecting forceps. The needed *boiling off* of the targeted tissue can then be attained by simply applying HF current to these forceps (therefore, indirect application). Since the surgeon holding the forceps in his hand is frequently in contact with the patient with his other hand, he can virtually become *part of the patient* from an electrical point of view since he is only insulated by very thin surgical gloves. The higher the coagulation voltage used for the application, the greater the risk of the glove failing to insulate and thus the risk for the surgeon. As a result, he would notice and feel the coagulating effect of the HF current at the instrument contact point on his hand. In order to minimize this risk, only HF voltages of $\approx 600–700\,V$ should be applied when using the indirect form of application. To support the user here, specially optimized currents are offerred today for use with clamp forceps (so-called clamp currents). In order to reduce the risk even further, electrically insulated instruments are suggested that protect the grasping area (Fig. 34.10).

The diversity of differing types of tissue (for instance, muscle, fat) and their states (bleeding, dry, intimate, loose electrode contact, etc.) to which HF current is applied is reflected in a very wide impedance

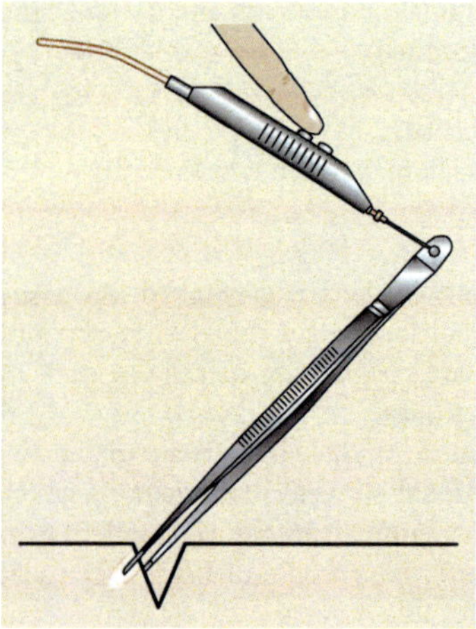

Fig. 34.10 Indirect contact coagulation via insulated surgical forceps. Alternatively, the HF surgical electrode can also be used in direct contact with the tissue for contact coagulation

spectrum (≈ 50–1000 Ω). A broadband adjustment of the HF generator, i. e., constant power output across a wide tissue impedance range, is therefore required here. Since the maximum output voltage is limited for indirectly applied contact coagulation, only coagulation currents with relatively low crest factors are permitted for use here. These currents have a partially cutting characteristic that many users intentionally use for cutting in situations where the cut is to be accompanied by a distinct hemostatic effect, or if the risk of contamination is relatively high (e.g., in visceral surgery).

So-called care coagulation represents a special form. In *care* coagulation (also referred to as *soft* coagulation) the peak voltage between the active electrode and the tissue remains restricted to values < 200 V. At these low voltages, no electric spark can be produced between the active electrode and the tissue. Any risk of unintentional cutting effect or undesirable carbonization is thus eliminated. Another advantage of care coagulation is the lower degree of eschar on the active electrode, which therefore needs to be cleaned less often during a surgical intervention. Care coagulation is suitable for all surgical interventions where the active electrode is brought directly into contact with the tissue to be coagulated (e.g., in neurosurgery, gynecology, ENT and particularly in minimally invasive surgery (MIS)).

As the level of the current depends on the coagulation progress, this can be used as a criterion for switching off the power supply. During the coagulation process the tissue impedance decreases considerably during the initial phase, and thus the current increases. Once the intracellular and extracellular fluids have vaporized, however, the impedance increases rapidly due to dehydration and the current is abruptly reduced to a minimum as a result. A dynamic sensor system registers this change and switches the generator off automatically as soon as a certain limit value is reached. One speaks of contact coagulation with an auto-stop function in this context. For the surgeon this means that optimal coagulation is still possible even if he cannot observe the tip of the electrode in the tissue. Using the auto-stop coagulation mode, the user can adjust the switching-off threshold and thus preselect different coagulation degrees. The overall speed of the coagulation process varies with the output voltage used. A higher output voltage dehydrates the tissue more rapidly, whereas a lower output voltage means that less power is applied to the tissue and more time is required. This also implies that, due to heat conduction, the coagulation can continue to spread until the process is completed. Depending on the power setting on the HF surgical unit, different coagulation depths can therefore be achieved with electrodes of the same size.

Desiccation

The term desiccation (Lat.: *desiccare*, to dry out) is used mainly to define a very special form of application rather than a specific form of current. In this case, a needle-shaped electrode is pricked into the tissue and left in place for as long as the the current is applied. At the given current intensity, the tissue then boils off and dries out in the area immediately surrounding the needle. As desiccation is a process that takes place deep inside the tissue, and since the thermal tissue reaction cannot be controlled visually, this form of application is advantageously combined with the auto-stop function already described in Sect. 34.4.2.

In the treatment of tumors, desiccation is used to reach tissue volumes as large as possible. Therefore, a suitable current to be used is the care coagulation current or a special desiccation current that is offered today for HF-induced interstitial tumor therapy (HFITT). In HFITT, special monopolar, bipolar, or multipolar applicators are brought into the center of the tumor or metastasis under imaging control (MRI, x-ray, ultrasound). HF current is applied over a longer period of time (typically 20 min) and the spreading of the thermal lesion is observed using imaging. Once the coagulation edges have extended into the healthy tissue, the power supply is discontinued. In order to be able to generate the largest possible coagulation volume, the needle electrodes are partially cooled to prevent the tissue close to the needle from desiccating too early. If this tissue were to dry out too soon, any further energy transfer would be prevented due to the high impedance of the dried-out tissue and the tumor could not be completely ablated. HFITT is mostly used today in tumor therapy for liver tumors and liver metastases.

Forced Coagulation

In this coagulation mode, high, pulsed, strongly modulated output voltages of up to 3 kV are used (crest factor 5–7). This coagulation current is referred to as *forced* because, thanks to its high voltage, it even overcomes high impedances and can therefore generate coagulation even in *unfavorable* situations. The term *forced* also emphasizes the exclusive coagulation character of the current. Due to the relatively high voltages, sparks can occur between the active electrode and the tissue during forced coagulation, and thus extensive coagulation can be achieved within a short period of time. This mode fulfills the requirements for a standard coagulation and enables the surgeon to work fast and effectively – but with the compromise that tissue carbonization might also possibly occur. If a forced coagulation is to be applied indirectly via a surgical instrument, the instrument must be electrically insulated. This current is also applied in a slightly modified form for transurethral and gynecological interventions using an electrically nonconductive irrigation liquid (Fig. 34.9). Here it is particularly important that the current does not tend to cut, as the coagulations are carried out using very thin wire loop electrodes with wire thicknesses of 0.3–0.4 mm. If a cutting coagulation current were used, the loop electrode would sink into the tissue as soon as slight pressure was exerted; this cannot be tolerated for a resection in the bladder, where the tissue walls are thin.

Spray Coagulation or Fulguration

Due to the type of apparatus available in Pozzi's day, fulguration, which is largely referred to today as spray coagulation (and in very few cases also as sideration), represented the first possibility for applying HF current to living tissue (Fig. 34.11).

For spray coagulation very high pulsed and extremely strongly modulated output voltages of several thousand volts (up to 8 kV) are used (crest factor up to 20). If the user approaches the tissue with a small-area electrode (needle electrode) under spray voltage, the air between the tip of the electrode and the tissue is ionized at a distance of 3–4 mm from the tissue. Via the ionized air in the electric field, a spark discharges to the tissue, followed by further spark discharges in a process of *spraying* energy to the tissue surface and coagulating a relatively large tissue area – hence the term spray coagulation. If a ball electrode is used instead of a needle

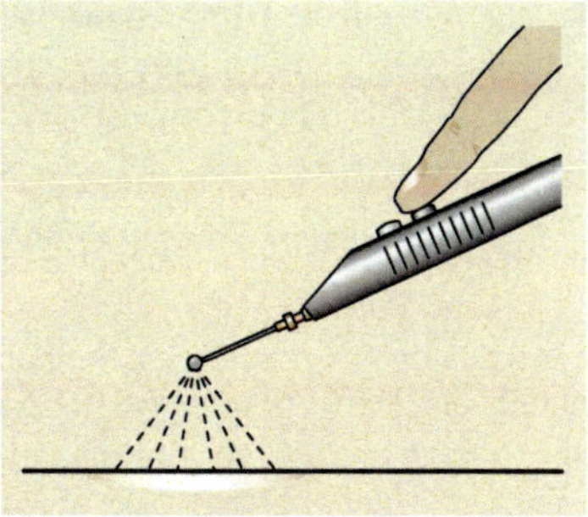

Fig. 34.11 Spray coagulation – also referred to as fulguration

electrode, a weaker electric field strength is produced at the blunt *electrode tip* at the same distance from the tissue. This only increases again as the distance to the tissue gets smaller; so when using a ball electrode, the ionization of the air with the accompanying spark discharge only takes place at a closer distance

$$\text{Electric field strength } E = \frac{\text{Voltage } U \text{ (kV)}}{\text{Distance } l \text{ (mm)}}\,.$$

Using spray coagulation, tissue can therefore be coagulated without touching it (noncontact coagulation). Whenever surface coagulation is necessary or diffuse bleeding must be stopped, the application of spray coagulation is advantageous, e.g., for coagulating parenchymal tissue (liver, kidney, spleen, lung). And if direct contact with the tissue is to be avoided for heat reasons, or if energy application is not possible by contact coagulation due to high tissue impedance – e.g., for hemostasis on the osseous sternum in thoracic surgery – spray coagulation is frequently the only possible application form for hemostasis.

For endoscopic applications, spray coagulation currents of lower voltage (such as endo spray) are used (typically limited to 4 kV) as a result of the dielectric strength requirements applying to the instruments.

Argon-Assisted Coagulation (Argon Beam)

There are clinical situations that cannot be sufficiently dealt with using conventional monopolar and bipolar coagulation methods. Problematic, for instance, is the open or endoscopic coagulation of large tumoral or hemorrhagic areas, which can only be addressed via technically challenging techniques and with a considerable expenditure of time when using conventional methods. Any plume formation obstructing the view, carbonization of the tissue, or coagulated tissue sticking to the electrode should also be excluded, particularly where endoscopic application is concerned. Since the mid-1980s, the argon beamer HF method has been available for hemostasis in open surgery and since the mid-1990s in flexible endoscopy as well. Today auxiliary argon-gas modules can be added to advanced HF surgical units to have the argon-beam type of current readily available whenever it is needed.

The argon beam is a monopolar coagulation method in which HF alternating current is transmitted to the tissue to be coagulated through ionized, and thus electrically conductive, argon gas (argon plasma). Like spray coagulation, this method requires a high voltage (e.g., > 2000 V) and a relatively short distance between the electrode and the tissue. Basically, the whole process is similar to that of spray coagulation but differs where the electric field strength required for ionizing the argon gas is considerably less than for air (argon 5 kV/cm, air 25 kV/cm). In comparison with spray coagulation, the applicator can therefore be held at a considerably greater distance from the tissue (Fig. 34.12). In addition, the characteristics of the thermally generated tissue eschar are distinctly different from the effect that can be achieved using spray coagulation. The argon beam produces a focused, controlled current distribution in the beam, which in turn leads to a smooth, homogeneous coagulation of the tissue surface. In addition to this, the argon-gas beam blows tissue fluids away from the site to be coagulated. This ensures that the energy reaches the tissue directly, not only a film of blood possibly covering the tissue.

Due to its physical and electrical composition, the plasma beam will automatically deflect from coagulated (high-resistance) areas toward bleeding or insufficiently coagulated (low-resistance) tissue areas of the application site. Thus a precise and uniform coagulation effect is achieved with superficial thermal damage limited in both depth and width; the typical depth of penetration is limited to 2–3 mm.

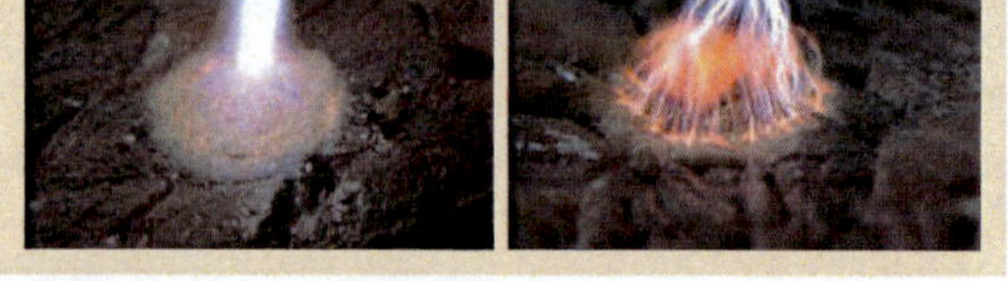

Fig. 34.12a,b Comparison of **(a)** argon-assisted and **(b)** spray coagulation

Compared with conventional monopolar and bipolar methods, argon-assisted coagulation offers significant advantages:

- Fast, large-surface coagulation of superficial hemorrhages.
- Automatically flow of current to tissue that has not yet been (sufficiently) coagulated, thus producing an eschar of even depth without any gaps.
- Possibility of lateral coagulation.
- Reliable hemostasis with minimal traumatization of the organ.
- Contact-free (*noncontact*) application, therefore no electrode charring.
- Maximum penetration depth of 3 mm, thus lower risk of perforation during coagulation in thin-walled organs.
- Faster wound healing due to minimal necrosis and less carbonization.
- No vaporization.
- The argon flow blows any blood produced away to the side; thus, the source of the bleeding is reached directly, which in turn improves the coagulation effect.
- Reduced smoke plume generation (better view), less unpleasant odor.
- Shorter procedural times.
- Reduced complication rate.

The endoscopic application of the argon beam has proved its worth for many indications, among them the hemostasis of tumor bleeding or bleeding after dilatation or bougienage, the treatment of anastomosis and scar stenosis, as well as partially obstructive tumors or tumors close to walls. Tumor tissue and granulation tissue growing through a mesh stent after stent insertion can be devitalized by endoscopic argon-beam application.

Necrosis can be obtained with the argon beam in tumor tissue located in areas subject to a risk of perforation (duodenum, colon). For bronchoscopy, thin, flexible probes are available that have been specially adapted to the working channel of the bronchoscope. In the tracheobronchial system, the argon beam can be used to stop bleeding and treat intrinsic obstructive tumors. Due to the numerous advantages that this coagulation method offers, the argon beam has established itself relatively quickly in the operating room. The areas of application are mainly limited at present by the special applicators required for the discipline/indication involved. It is to be expected that new developments of combination instruments will enable additional indications to be added using this method.

Bipolar Coagulation

As in bipolar cutting, the physical laws governing tissue interaction are the same for bipolar coagulation currents as for monopolar coagulation currents. Due to the integration of both poles in one instrument, however, only currents with a limited output voltage (max. 500–800 V) can be used for reasons of dielectric strength. Strongly amplitude-modulated currents with high crest factors and high-voltage pulses – as required for noncontact coagulation – are therefore ruled out for practical reasons. As a result, bipolar coagulation is nowadays performed exclusively as contact coagulation and desiccation. Indirect application of current via surgical instruments is not an option in this case, as the tissue must come into direct contact with the bipolar instrument. In the majority of cases, bipolar forceps are used, which are offered in a large variety of shapes and sizes specially adapted to the requirements of the surgical disciplines concerned.

Auto Start and Auto Stop Functions

As in the case of monopolar contact coagulation, there is also the possibility in bipolar coagulation of terminating the coagulation process automatically by using the automatic-stop (auto-stop) function. This function can be beneficial for coagulating hypertrophic nasal concha and myomas, for example. A bipolar needle arrangement is pricked into the tissue, and the target tissue is then ablated by the application of HF energy (desiccation). In addition to the auto-stop function, automatic activation without using the foot switch (auto start) is possible for bipolar coagulation. A tissue contact sensor registers when the tissue to be coagulated is pinched between the two electrodes and automatically activates the generator. To ensure that the surgeon has sufficient time for positioning and preparation before the generator is activated, individually programmable delay times can be selected as needed. The auto-start function is only recommended and permitted for open surgery. Jointly using the auto-start and auto-stop functions offers a high degree of operating convenience as the generator is automatically activated with the preselected delay time, automatically supplies the correct power, and automatically switches off as soon as the optimum coagulation result is obtained. In addition, undesirable side effects such as tissue carbonization and electrode charring are avoided.

Nonstick Technology

Similarly to searing in a frying pan, tissue tends to stick to a metal contact surface when heated for coagulation via electrosurgical forceps. In an attempt to detach the forceps from the tissue, it is possible for tissue to be *torn off* and the source of the bleeding broken open again. The chemical–physical process that is ultimately responsible for the sticking is largely unknown. What can be observed, however, is that the effect is all the more pronounced the higher the temperature of the forceps tips rises, and it is particularly pronounced when sparks occur between the tissue and the electrodes of the forceps.

One possibility for limiting the electrode temperature is to dissipate the thermal energy produced at the electrode contact point as effectively as possible. This can be achieved, for example, by selecting a material for the electrodes with a high specific-heat storage capacity and a good thermal conductivity. While the stainless steel normally used for forceps has a relatively high heat-transfer capacity (0.5 kJ/(kg K), whereas it is only 0.235 for silver and 0.383 for copper), thus offering a relationship between the amount of heat applied and the temperature increase obtained that is very favorable for coagulation electrodes, the thermal conductivity of the material is rather poor. In fact, it is only 15 W/(m K) and as such 25 times lower than that for copper. As a result, the process of removing the heat from the contact point and conducting it into the body of the forceps is too slow – heat builds up. The best conductor of heat is silver with 418 W/(m K). For this reason, a special silver alloy is used for the distal electrode area (tissue contact area) of nonstick forceps – one that is mechanically robust and combines the excellent thermal and electrical properties of silver with biocompatibility. This allows the heat energy to be quickly transferred, via the distal tines of the forceps, from the tissue contact point to the stainless steel composite where the thermal energy is readily absorbed due to the heat capacity of the material.

With the method developed by *Reidenbach* [34.17], which has become known as electrohydrothermosation (EHT), fluid (e.g., NaCl) is run through the tines of the forceps to the electrode contact points on the tissue surface. This fluid surrounding the electrodes ensures that, due to evaporative cooling, the temperature does not exceed 100 °C. This temperature is sufficiently high for the required tissue coagulation but does not cause complete drying out of the contact surfaces or even any tissue carbonization. Although the EHT method is very effective, it is very expensive, as special forceps and additional irrigation fluid are needed. In addition, some indications do not allow an irrigation fluid to be applied.

Nonstick forceps therefore are an attractive alternative. They are particularly advantageous for use in neurosurgery, and also wherever tissue adhesion to the forceps is not tolerated.

Bipolar Vessel Sealing

Although the bipolar application form had already been introduced at the beginning of the twentieth century [34.18] and the coagulation mechanisms effective for occluding or sealing blood vessels had been sufficiently investigated in the early 1960s [34.19], *only* smaller vessels with a diameter of $\approx 2-3$ mm could be sealed reliably with the bipolar method until the end of the twentieth century. The development achieved by an American study group led by *Buysse* [34.20] and *Ryan* [34.21] must therefore doubtlessly be regarded as a milestone in bipolar technology. At the end of the 1990s they introduced a system that enabled them to seal considerably larger vessels by using a specially controlled coagulation process applied via bipolar clamp forceps (Fig. 34.13).

While being heated, the tissue undergoes different stages. In the first phase, the connective tissue is coagulated, thereby maintaining the fiber structure; this is followed by amorphous coagulation – i.e., structures disintegrate – and finally its complete destruction in the form of carbonization with loss of substance. When sealing vessels, it is important to fuse the body's own collagen and elastin (i.e., the basic fibrous connective tissue structure), thereby maintaining the morphological tissue structure. Mechanical pressure is necessary for connecting these fibers during heating in order to ensure that the vessel walls are in close proximity with each other. Insufficient pressure merely leads to thrombosis inside the vessel, which does not ensure a reliable vessel occlusion for larger vessels. Insufficient heating produces a weak seal or only fusion of the adventitial layers (the lax connective tissue surrounding the blood vessel). If heated too strongly, the physical properties of the connective-tissue fibers are destroyed. The amorphous coagulate would tend to break and detach from the remaining tissue; the fiber structure would be destroyed and thus the mechanical stability and elasticity lost. Ideal sealing can only take place if the physical properties of the tissue are retained. When sparks are applied to the tissue, however, temperatures of more than 1000 K are reached, destroying the integrity of the tissue.

Consequently, the bipolar vessel-sealing current must not be allowed to reach voltages that could cause sparking ($< 180-200$ V). To ensure that sufficient energy can be delivered into the tissue for adequate coagulation at such low HF voltages, the energy transfer process must be supported by control technology. To this end, one utilizes the fact that the tissue impedance drops again after the first desiccation phase directly after discontinuing the HF power supply, which in turn enables repeated energy transfer at permissibly low HF voltages. In other words, the tissue is coagulated by applying the energy in pulsed mode. The process is automatically stopped only after the intended sealing level has been reached.

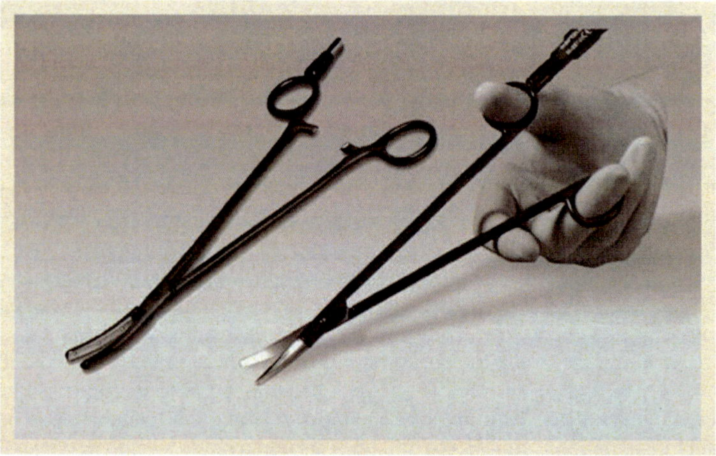

Fig. 34.13 Reusable bipolar instruments. Bipolar clamp forceps for vessel sealing in open surgery. Prior to cutting the tissue it is sealed intensively using the forceps (*left*). Bipolar scissors enable the application of coagulating, bipolar HF current via the two scissor blades during mechanical cutting (*right*). In contrast to vessel sealing, the hemostatic effect achieved in the process is only sufficient for smaller vessels

Reliably sealed tissue typically has a translucent, parchmentlike appearance, with the thermally damaged area limited to the margins.

The main advantages of the method are as follows:

- No foreign material – such as sutures and clips – is left inside the patient
- Cost reduction by saving suture material
- Shorter operating times (gentler treatment of patients and lower costs)
- Bipolar procedure with its typical advantages (no aberrant currents, no heating of distant tissue structures, absence of a neutral electrode, electromagnetic compatibility).

34.5 Methodical Instructions for Application and Safety

Modern HF surgical units equipped with automatic monitoring functions combine operational ease of use and functional diversity with the highest possible safety and protection for the patients. As with all HF surgical units, their safety also depends on proper handling. There are some basic rules that need to be observed to ensure maximum safety. They are just as important as the instructions for use of the HF unit and should therefore be part of the standard procedure.

The following summary has been drawn up from practical experience. It focuses on four main areas:

1. Patient,
2. Neutral electrode
3. Instruments and cables
4. HF surgical unit.

34.5.1 Patient and User Safety

General Rules for Application

- The patient must be positioned in a dry and electrically insulated area. He should not come into contact with electrically conductive objects that are grounded or have a high capacity to ground, e.g., the operating table or infusion stands.
- During longer operations, the urine must be drained away through a catheter.
- If possible, any wet operating table covering, drapes, or towels should be replaced by dry ones.
- When positioning the patient, no skin-to-skin contact should occur at any points such as between the fingers and thighs.
- The cables of the active electrode handle and the neutral electrode should be positioned so that they do not touch either the patient or other leads.
- Do not coil the HF cable; use a plastic clamp and not a metal clamp to attach it to the surgical drape.
- A sufficient distance of at least 15 cm should be maintained between the active HF electrode and an ECG electrode.
- Proper precautions should be taken for patients with implantable devices such as pacemakers in order to avoid interference with or damage to the implanted device. We recommend consulting a cardiologist.
- Pay attention to potential metal parts in or on the patient's body. Body-piercing jewelry should be removed wherever possible.
- Take particular care with anesthetics, degreasing agents, and disinfectants! They must be completely evaporated and removed from the HF target area. An electric spark could ignite the alcohol they contain.
- Prior to HF application in the gastrointestinal tract, make sure that no combustible, endogenous gases are present. A quick insufflation/suction combination is suggested to minimize any possible concentration.
- If irrigation is insufficient during a TUR procedure, H_2O molecules can dissociate to H_2 and O_2 and accumulate in the upper part of the urinary bladder. A resection carried out in this gas mixture could cause an explosion.
- The active electrode handle should not be placed down on the patient; it should always be positioned in the instrument holster intended for this purpose.
- For surgical interventions where the HF current flows through organs or vessels with a relatively small cross section, undesirable thermal damage can be safely avoided by using the bipolar technique.
- Do not make an electrical connection to the instrument until just prior to the desired HF application. This is particularly important when using the foot switch for activation.
- Always set the acoustic signal loud enough so that all users may hear the tone indicating the activation of the HF generator.

Rules for Application when Using the Argon Beamer

If argon gas is used on the patient, the following applies:

- Argon-assisted coagulation is a monopolar form of HF current application. Consequently, the safety regulations for monopolar HF surgery must be observed.
- Do not blow any argon gas into open blood vessels.
- Do not press the tip of the applicator against tissue as this could cause a gas emphysema.
- High HF voltages are required for gas ionization; make sure that the electrical insulation of leads and applicators is fully intact and undamaged.
- Argon is heavier than air (1 : 1.38); it can accumulate close to the ground and displace oxygen.
- Argon gas cylinders are under high pressure (up to 200 bar). Handle these pressurized tanks with care!

Rules for Application in Endoscopy

Apart from the general application rules given above, the following points must be additionally observed in endoscopy:

- The HF unit must not be activated if the electrode is not in the field of view.
- The argon gas flowing into the respective body cavity increases the pressure already present there due to air or CO_2 insufflation. Only use insufflators with regulated pressure releases.
- Guide wires used in combination with an HF application must be electrically insulated (e.g., Terumo wires).
- Prior to HF application in the gastrointestinal tract, make sure that no combustible, endogenous gases are present.
- Do not introduce any combustible, oxygenated breathing gas into the tracheobronchial system prior to or during argon-beam application (if necessary, keep oxygen concentration levels below 30% and ensure intermittent activation of the HF current).
- When using a video endoscope, prevent chip destruction by making sure that the argon plasma is not directed at the camera chip.

34.5.2 Neutral Electrode

In the monopolar application mode, a thermal effect is required exclusively at the active electrode. However, no thermal reaction should take place underneath the neutral electrode (NE).

Table 34.3 Contact quality monitors (CQM) of different suppliers

Acronym	Supplier	Type
PCS	Martin	Patient control system
REM	Valleylab	Remote electrode monitoring
ARM	Conmed	Aspen return monitor
NESSY	Erbe	Neutral electrode safety system
PDM	Aesculap	Permanent dynamic monitoring
PACC	Integra	Patient contact control system

To prevent patient burns, potential heating in the NE application area must be kept below 6 °C (42.8 °F) (ANSI AAMI HF 18) [34.22]. This requires correct attachment of the NE and making absolutely sure that it does not detach during the course of an operation. Advanced HF surgical units offer NE monitoring technology that keeps proper NE attachment under constant control and disarms monopolar HF energy delivery if a problem is detected (use of contact quality monitors). However, this requires the use of a split neutral electrode. The acronyms used by different manufacturers for such monitoring technologies vary depending on their monitoring philosophies (Table 34.3).

The original assumption that the return flow of current from deep tissue layers was equally distributed across the surface of the neutral electrode was not correct. In fact, the current distribution shows a distinct current concentration at the edges of the NE (edge effect). This effect is caused by the layered structure of the human skin, with the dermis offering good conductivity lying over the poorly conducting fatty tissue layer. If rectangular neutral electrodes are used, the edge closest to the target site will offer the least resistance to the HF current, leading to a *hot spot*. An electrode as round as possible, with a sufficiently long peripheral edge directed toward the operating field, is therefore ideal.

Since an even distribution of the current in both halves of the NE is not possible due to the irregular flow of current on the surface of the tissue, monitoring systems based on an even current distribution have not proved successful in practice. In contrast, monitors that continually evaluate the dynamic impedances between the two halves of the NE offer the highest level of patient safety. Advanced monitors are capable of recognizing whether a single-part NE or a split NE is connected. Additionally, given sufficient safety awareness, users can choose to have nonsplit electrodes locked and thus barred from use.

Rules for Neutral Electrode Application

- Ensure that the insulation of the NE is undamaged.
- Connect the NE cable to the surgical unit and the NE with the utmost care.
- Occasionally check the NE alarm of the HF surgical unit for proper functioning.
- Attach the NE with the long edge directed toward the operating field.
- Position the NE on one of the suitable extremities (upper arm, thigh) as close to the operating field as possible.
- Ensure good tissue contact; this may require shaving any hair preventing proper contact.
- Keep fluids away from the NE area, as these can adversely affect both the adhesion and the electrical properties of the NE.
- Do not reuse disposable (single-use) NEs. For safety and hygiene reasons, disposable NEs should never be reused.
- Observe the expiration date of the NE.
- The NE must not be trimmed or reduced in size.
- In any circumstance, additional contact gel should never be applied to the NE.
- Bony or uneven surfaces, implant sites, places with thick layers of fat (such as the abdomen or buttocks), and scarred tissue are unsuitable for NE application.
- Use a contact quality monitor that requires the exclusive use of split NEs.
- Do not subject NE cables lying on the floor to any mechanical stress; in particular, avoid rolling the equipment trolley or even the C-arm over them.

34.5.3 HF Instruments and Cables

- Only use original accessories recommended by HF surgical unit's manufacturer. In case of doubt, request a certificate of compatibility from the manufacturer or dealer.
- Prior to each application, perform a visual check to ensure that the insulation of plugs, cables, and instruments is not damaged in any way. A short functional check of the instruments is recommended as well.
- During the operation, it is best to cleanse the electrodes with a moist cloth (swab). Do not scratch as this could damage the surface and insulation of the electrode.
- It is advisable always to keep the electrodes clean immediately following each use. Should the electrodes have any bits of old, dried tissue on them, soak them in a solution before cleansing.
- The manufacturer's instructions for reprocessing (cleaning, sterilization) must be duly observed.
- Disposable instruments must always be discarded immediately after the operation (multiple use is not permitted!).
- Keep reusable instruments in a safe place. Observance of the manufacturer's instructions for handling and care increases both the safety of the patient and the service life of the instruments.
- Accessories with visible defects must be checked or returned to the manufacturer for inspection or repair.

34.5.4 HF Surgical Unit

- The use of any HF surgical unit requires the user to be familiar with its correct handling.
- Only use HF units equipped with a contact quality monitor.
- The HF power setting should always be selected as low as possible and as high as necessary. Automatically controlled HF surgical units are the best option.
- The cutting or coagulation function should only be activated for as long as it is really needed.
- Always conduct a functional test before starting the operation. When the NE has been applied, briefly activate the HF current by pressing the active electrode handle button.
- Pay attention to error messages and alarm signals.
- Repeat training courses on the principles of HF surgery and their safe use each year (make use of training courses offered by the manufacturer).
- To avoid video disturbances in older systems, the HF cable should run as far away from the camera cable as possible.

34.5.5 Incidences During Application of HF Surgery

As part of the enforcement of the Medical Devices Act, the German Federal Institute for Drugs and Medical Devices (BfArM, Berlin) has evaluated the incidents reported in the field of HF surgery over a number of years. It concluded that the few complications reported were mainly caused by user errors and were not due to a malfunction of the medical devices. The BfArM has set guidelines from these user errors that have been taken into consideration in the above-mentioned rules.

In the few incidents reported, the damage mainly involved the NE. In this connection the BfArM made the following statement:

... A distinct accumulation of incidents can be observed for single-part neutral electrodes, for which there is no safety monitoring. The very few "burns" observed in conjunction with split electrodes are due to causes other than high current densities ... chemical burns due to disinfectants that have not evaporated.

Where injuries in the coccyx and dorsal areas were concerned, the BfArM made the following determination [34.23]:

... The users sometimes spoke of burns, but the investigations carried out by the manufacturer, or an assessment of the photos sent in, revealed that they were actually acid burns or inflammations. During the operation, the area above the sacrum is particularly subjected to a pressure load that leads to a slight perfusion of the respective skin areas.

It is therefore absolutely vital to ensure that HF units are only put into operation and used by qualified personnel who have received special training and are fully familiar with the device.

34.6 Outlook

The automatically controlled HF surgical units currently available are of a very high standard, both in terms of technical safety and usability. The growing number of technical possibilities will contribute to further improvement of these standards as well. In addition, application-specific improvements will mean that the handling of the HF surgical units will become even easier. Instrument recognition and the resulting automatic parameterization of the devices with suitable settings will play a role here. Only logical and meaningful changes to the automatic presettings will then be possible within predetermined limits (expert systems). Consequently, there will be no standardized interfaces for HF instruments but specific solutions provided by each manufacturer.

The integration of HF units into OR systems via communication networks will also contribute toward enabling the control of all the devices required for an operation using a common user interface. Touch-screen monitors and voice control are relevant keywords here.

As HF currents are being optimized increasingly toward specific applications, HF generators will be showing a trend toward the specialist user instead of the general user. Solutions with special devices and systems will be offered for the different specialist fields, e.g., gastroenterology, urology, neurosurgery – optimized in terms of costs as well as handling.

Argon-assisted applications will gain even greater importance in daily OR use. The range of applications has by no means been exhausted and is currently restricted only by the available applicators. Combination instruments – such as polypectomy snares, sclerosing needles, or biopsy forceps offering optional argon-assisted coagulation – will enable the method to handle new indications.

Greater importance will be attached to the continuing development of surgical instruments, particularly with regard to specialized instruments and instruments for specific indications. It will be a tremendous challenge to increase the quality of the instruments and simplify their handling in daily use. Despite the enormous cost containment pressure that affects the public health sector on a global scale, there will be a trend toward the use of disposables. Hepatitis, AIDS, rabies virus, Creutzfeldt–Jakob disease (CJD), and bovine spongiform encephalopathy (BSE) are just a few of the dreaded, contagious diseases now running rampant around the globe – and new diseases continue to extend the list year after year. As a result, instrument processing becomes a growing challenge for hospitals.

Vessel sealing will play an important part particularly in laparoscopic surgery. Surgical sutures and ligatures are particularly difficult to use in a laparoscopic environment, and foreign material is unavoidably left in the patient. Reliable vessel sealing will radically change this field and help establish standardized surgical techniques. The large number of different methods used for tissue dissection will be greatly reduced as a result. Ultrasonic dissection and water-jet dissection leave much to be desired as regards their performance, and these procedures are also expensive. They will be replaced by bipolar, HF-induced tissue sealing. To ensure the greatest possible safety for patients, it will also be important for users, surgeons, and surgical staff to keep themselves well informed about the current and future developments taking place in the industry.

References

34.1 R. Haag: Grundlagen der HF-Chirurgie I, Krankenhaustechnik **3**, 36–37 (1993)

34.2 R. Haag: Grundlagen der HF-Chirurgie II, Krankenhaustechnik **4**, 36–38 (1993)

34.3 H. Sudermann: Albrecht Theodor Middeldorpf (1824–1868) und die Elektrochirurgie, Würzburger medizinhist. Mitt. **19**, 59–110 (Königshausen Neumann, Würzburg 2000)

34.4 B. Wiesner: Über Fulguration nach de Keating Hart, Münch. Med. Wochenschr. **11**, 569–570 (1908)

34.5 W.T. Bovie: Electrosurgical Apparatus, Patent 1813902 (1931), Liebel Flarsheim Co, Cincinnati

34.6 H. von Seemen: *Allgemeine und spezielle Elektrochirurgie* (Springer, Berlin 1932)

34.7 E. Schliephake: Über die Tiefenwirkung und elektive Gewebswirkung kurzer elektrischer Wellen, Strahlentherapie **38**, 655–664 (1930)

34.8 A. Esau: Einrichtung zur Behandlung des menschlichen Körpers im elektromagnetischen Wechselfeld eines Kurzwellengenerators, Deutsche Reichspatentschrift Nr. 560139, Jena (1932)

34.9 R. Shaw: Plasma arc scalpel, US Patent 3434476, San Francisco (1969)

34.10 R. Goucher: Surgical instrument employing electrically neutral, D.C. induced cold plasma, US Patent 3838242, Hogle-Kearns International, Salt Lake City (1974)

34.11 F. Incropera: Plasma arc scalpel, US Patent 3991764, Purdue Research Foundation, West Lafayett (1976)

34.12 C.F. Morrison Jr., F.W. Harris, M.D. Patzer: Electrosurgical Method and Apparatus for Establishing an Electrical Discharge in an Inert Gas Flow, Patent 4060088 (1977)

34.13 H.-D. Reidenbach: *Hochfrequenz- und Lasertechnik in der Medizin* (Springer, Berlin Heidelberg 1983)

34.14 C. Thiel, K. Fastenmeier: Der Lichtbogen beim Schneiden in der HF-Chirurgie – Ein elektrophysikalisches Modell, Biomed. Tech. **41**(1), 494–495 (1996)

34.15 E. Roos: *Elektrochirurgie im modernen Krankenhaus I–V* (Gebr. Martin, Tuttlingen 1972)

34.16 IEC 60601-2-2: *Medical Electrical Equipment – Part 2-2: Particular Requirements For The Safety Of High-Frequency Surgical Equipment* (IEC, Geneva 2006)

34.17 H.-D. Reidenbach: Entwicklungstendenzen in der HF-Chirurgie, Biomed. Tech. **37**, 134–140 (1992)

34.18 Norisan Apparatebaugesellschaft m.b.H.: Deutsche Reichspatentschrift Nr. 479179, Nürnberg (1929)

34.19 B. Sigel, M. Dunn: The mechanism of blood vessel closure by high frequency electrocoagulation, Surg. Gynecol. Obstet. **121**, 823–831 (1965)

34.20 S. Buysse: Vascular tissue sealing pressure control, US Patent 5776130, Valleylab Inc., Boulder (1998)

34.21 T. Ryan: Energy delivery system for vessel sealing, US Patent 5827271, Valleylab Inc., Boulder (1998)

34.22 ANSI: ANSI/AAMI HF18, Electrosurgical Devices (ISO, Geneva 2001)

34.23 D. Schröder: Vorkommnisse bei der Anwendung der HF-Chirurgie, Medizintechnik **123**, 145–147 (2003)

35. Medical Radiation Therapy

Peter H. Cossmann

Historically speaking, medical radiation therapy is a relatively old medical branch as it originated two centuries ago with the discovery of x-radiation by W. C. Röntgen in 1895. Shortly after this historical event the first treatment attempts were carried out by some courageous physicians. Since then radiation therapy evolution has undergone several major leaps due to essential developments in mechanical engineering, electronics, and computer science.

An introductory part gives an overview of these developmental steps through history. This is followed by some basic information on the physical and technological principles of radiation physics. Here radiation types, radiation sources, and the interaction of radiation with biological tissue are covered and finally rounded out with a more detailed look at dosimetry.

Furthermore, the three therapy fields for radiation treatments are discussed: percutaneous therapy, brachytherapy, and radionuclide therapy.

The main part covers equipment technology for the generation of radiation. In radiation therapy, ionizing radiation is used, which either results from radioactive decay processes or is generated by the acceleration of charged particles. The technologies described are x-ray equipment, cobalt unit, linear accelerator, brachytherapy sources/treatment units, and radioactive implants. A main focus is on the technology of linear accelerators, which serve as the daily workhorses nowadays.

Finally, special techniques and newer developments in teletherapy are discussed. Besides stereotactic and tomotherapy approaches, technological developments and refinements in the field of linear accelerator technology are described in more detail.

35.1 **X-Radiation** ... 704

35.2 **Historical Development of Radiation Therapy** ... 704

35.3 **Physical and Technical Principles of Radiation Physics** ... 705
- 35.3.1 Types of Radiation ... 705
- 35.3.2 Radiation Sources ... 706
- 35.3.3 Interactions with Matter ... 706
- 35.3.4 Dosimetry ... 707

35.4 **Forms of Therapy** ... 708
- 35.4.1 Percutaneous Therapy ... 709
- 35.4.2 Brachytherapy ... 709
- 35.4.3 Radionuclide Therapy ... 709

35.5 **Equipment Technology for the Generation of Radiation** ... 710
- 35.5.1 X-ray Equipment ... 710
- 35.5.2 Cobalt Unit ... 711
- 35.5.3 Linear Accelerators ... 711
- 35.5.4 Miniature Accelerators ... 712
- 35.5.5 Brachytherapy Sources and Brachytherapy Equipment ... 713
- 35.5.6 Radioactive Implants ... 714

35.6 **Special Techniques and Newer Developments in Teletherapy** ... 716
- 35.6.1 Intensity-Modulated Radiotherapy ... 716
- 35.6.2 Breathing-Adapted Radiotherapy (BART) ... 716
- 35.6.3 Image-Guided Radiotherapy ... 717
- 35.6.4 Dynamic Adaptive Radiotherapy ... 718
- 35.6.5 Stereotactic Radiosurgery/Radiotherapy (SRS/SRT) ... 718
- 35.6.6 Tomotherapy ... 719
- 35.6.7 Intensity-Modulated Arc Therapy ... 719
- 35.6.8 Current Developments in Linear Accelerator Technology ... 720

References ... 721

35.1 X-Radiation

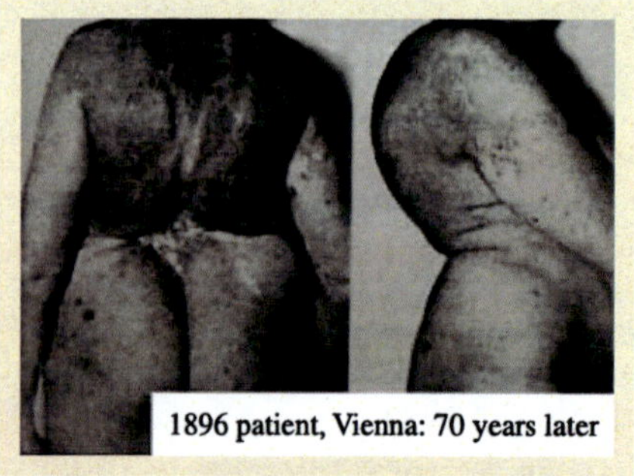

Fig. 35.1 First radiation treatment to remove a nevus pigmentosus et papillomatosus (after [35.1])

Fig. 35.2 First treatment of carcinomas of the lip and cheek using X-radiation (after [35.2])

On 8 November 1895 at Würzburg University's Institute of Physics, W. C. Röntgen discovered a new type of radiation. In his first disclosure, Röntgen spoke of so-called x-rays because he didn't know what type of radiation he was dealing with. With this he laid the foundations for medical radiation therapy (RT). Therapy with ionizing radiation began in 1896 in Vienna, only a few months after Röntgen's discovery. Initial reports regarding the successful therapeutic application of X-radiation can be traced back to Freund, Sjögren, and Stenbeck. Based on newspaper reports of erythema and alopecia following exposure to x-rays, Professor Freund from Vienna had in 1896 treated a young girl suffering from nevus pigmentosus et papillomatosus with radiation. In 1966, the patient was presented at the German Radiology Congress. The result, which is illustrated in Fig. 35.1, clearly showed that the radiation had undeniably had the desired effect, which is to say that the hair on that patch had definitively disappeared. However, presumably also as a result of a lack of knowledge about adequate dosage, there was a significant side effect – considerable curvature of the spine in the radiation area. This illustrated in a striking manner where the main problem lies with this form of treatment: achieving a desired effect while avoiding undesirable side effects.

In 1899 in Stockholm, T. Sjögren and B. Stenbeck reported the first successful treatment of carcinomas of the lip and cheek using X-radiation. Figure 35.2 shows examples of the treatments of carcinomas of the lip and cheek using X-radiation.

35.2 Historical Development of Radiation Therapy

Initially, gas ion tubes consisting of a partially evacuated glass bulb with a cold cathode and anode were used as therapy systems. The first x-ray tube was a Hittorf tube as shown in Fig. 35.3.

The high voltage necessary for generating x-rays was produced using induction coils, which were, in principle, a transformer with a primary and secondary coil on a ferrite core. The cathode is in the form of a concave mirror, so that the cathode rays that emerge from the cathode perpendicular to the cathode surface meet at a point. At this focal point is the anticathode, a piece of sheet metal composed of platinum or tungsten that is inclined by about 45° and conductively connected to the anode. X-radiation is generated as a result of the metal ions and electrons impinging on the electrodes (cathode and anode), and there is simultaneous sputtering of the electrode material. The gas molecules present in the Hittorf tube are absorbed by the sputtered metal particles, and the vacuum in the tube increases. Ever higher voltages must be applied in order to ensure the flow of current. Because of the higher voltage, the x-rays become increasingly hard and finally the tube ceases to work. To make them usable again, gas had to be let into the tube. Soft x-rays are rays with a relatively

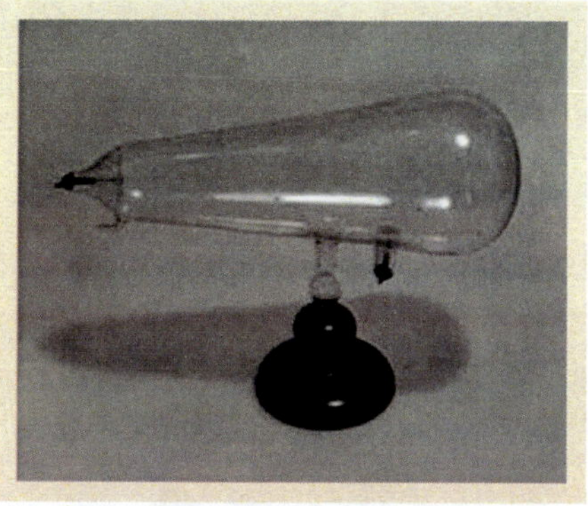

Fig. 35.3 Hittorf tube (cold cathode tube) (courtesy Electrosuisse Museum of Technology and Engineering, Fehraltdorf)

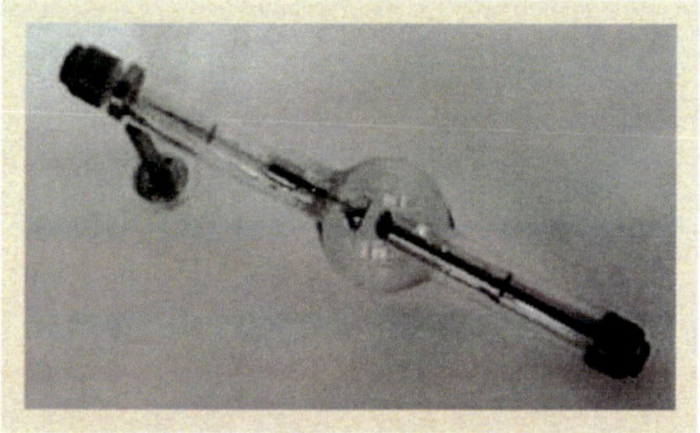

Fig. 35.4 Coolidge tube (hot cathode tube) (courtesy Siemens, Erlangen)

long wavelength, and thus with low penetrating power, which are generated by relatively low voltage (approx. 30 000–50 000 V). Hard x-rays have a short wavelength with a high penetrating power and with low absorbability (100 000–300 000 V). Modern x-ray tubes such as the so-called Coolidge tube are high-vacuum hot cathode tubes, which are pure electron tubes [35.3]. The hot cathode is a filament made from tungsten, which has a very high melting point. The tube has no special anticathode; rather, the anode serves as the anticathode facing the cathode. It is composed of copper, a good thermal conductor, and, because of the buildup of heat due to the cathode rays impinging on it, it is provided with a cooling attachment, e.g., a water cooling system. At the point where the cathode rays impinge upon it, the anode has a tungsten disc to protect it. A Coolidge tube of this kind is illustrated in Fig. 35.4.

Hot cathode tubes have various advantages over cold cathode tubes. Firstly, soft or hard X-radiation can be generated using variable voltages, and the intensity of this radiation can be regulated by adjusting the filament current. Their properties only change in the long term. Modern tubes are operated exclusively using alternating current (AC) power and a transformer. For this purpose, the voltage of 220 V is raised to 30 000–300 000 V.

The discovery of radium radiation by Curie in 1898 led by chance to a therapeutic application of radioactive substances, because in 1901 Becquerel acquired dermatitis from a radium source, and this dermatitis was no different to radiation dermatitis, which had been observed in connection with x-rays. Therapy for the treatment of dermatological conditions was consequently begun at the St. Louis Hospital in Paris using a radium source provided by Curie. Initially it was mainly the beta rays of radium that were used for therapeutic purposes. Subsequently, from 1906 onward, the radiation sources were encapsulated in order to utilize only gamma radiation. Radium was of great significance, particularly in gynecological RT, but it has since been completely superseded by methods of afterloading with synthetic radioactive emitters, partly for reasons of radiation protection. Telecurietherapy using synthetic radioactive substances began after the discovery of synthetic radioactivity by Joliot-Curie in 1934 and, among other things, once it was possible to produce ^{60}Co and ^{137}Cs in nuclear reactors. The first use of a telecobalt machine for RT followed in 1952. It continued to be used until the middle of the twentieth century, until the first clinical applications of a circular electron accelerator (betatron) were able to be realized. Use of the linear accelerator (linac), first developed in 1930, began in the early 1970s. Since then linacs have almost completely superseded the betatrons, and today they are the most important radiation source in the field of teletherapy.

35.3 Physical and Technical Principles of Radiation Physics

35.3.1 Types of Radiation

Radiation that can generate ions when passing through matter as a result of impact interactions is described as ionizing radiation. A distinction is drawn between directly and indirectly ionizing radiation. Directly ionizing radiation consists of charged particles such as electrons, protons, α-particles, and heavy ions; their kinetic energy is sufficient to achieve impact ionization of an atom. In the case of indirectly ionizing radiation, charged particles are produced from uncharged particles such as photons or neutrons, which then transfer their energy once more to the surrounding atoms through impact ionization [35.4–6].

Photons – in the form of x-rays and γ-radiation – and also electrons are predominantly used in RT. In addition, protons – as well as isolated heavy ion radiation – are also increasingly being used, with the generation/application of heavy ions in particular having high technological requirements, which currently are not yet practicable except in a research center environment. In recent times, moreover, boron neutron capture therapy has again brought neutrons for therapeutic use into the discussion, this being a form of therapy that is limited to a few fields of application and that is furthermore technologically complex and therefore not yet very widespread.

Since RT works mainly with photons for deep therapy and electrons for superficial and semideep therapy, the following sections are limited to these types of radiation.

35.3.2 Radiation Sources

The generation of photons in the form of X- and γ-radiation is achieved either by means of deceleration/energy dissipation of accelerated particles in a target (Fig. 35.5) or by radioactive decay. The technical basis for this is x-ray tubes in the low to middle energy range (kV) and accelerators in the high energy range (MV). The latter are also used for the generation of high-energy electrons.

Radioactive substances serve as further radiation sources. The three possible types of decay are illustrated

α-decay	β-decay	γ-radiation
${}^{A}_{Z}X \to {}^{A-4}_{Z-2}Y + {}^{4}_{2}He$	${}^{A}_{Z}X \to {}^{A}_{Z+1}Y + \beta^- + \bar{\nu}$	${}^{A}_{Z}X^+ \to {}^{A}_{Z}X + \gamma$
	${}^{A}_{Z}X \to {}^{A}_{Z-1}Y + \beta^+ + \nu$	

Fig. 35.6 Types of radioactive decay

in Fig. 35.6. Following α- or β-decay, the daughter nucleus is often in an excited state. The majority of these excited states have very short half-lives and, by emitting therapeutically usable γ-radiation, pass into a state of lower excitation or into the ground state.

35.3.3 Interactions with Matter

Photons are attenuated by matter on the one hand by absorption (photoelectric effect or pair production) and on the other by scattering (excitation or Compton effect). Absorption is dependent on the linear attenuation coefficient, a matter constant, and also on the layer thickness. In the case of the photoelectric effect, the photon transfers all its energy to an electron, and this photoelectron leaves the atomic shell. With photon energies above 1022 MeV, the photon can be converted into an electron–positron pair. In the case of elastic scattering of photons, the electrons in the atomic shell are excited and the energy is then emitted in another direction in the same quantity. The Compton effect involves inelastic scattering, i. e., the photon releases an electron from the shell with some of its energy and then continues to move in a different direction. Within an energy range of approx. 20 keV–20 MeV, which is to say practically in the entire energy spectrum of RT, the Compton effect is dominant for all human tissue. The interaction of photons with tissue takes place in two stages. In

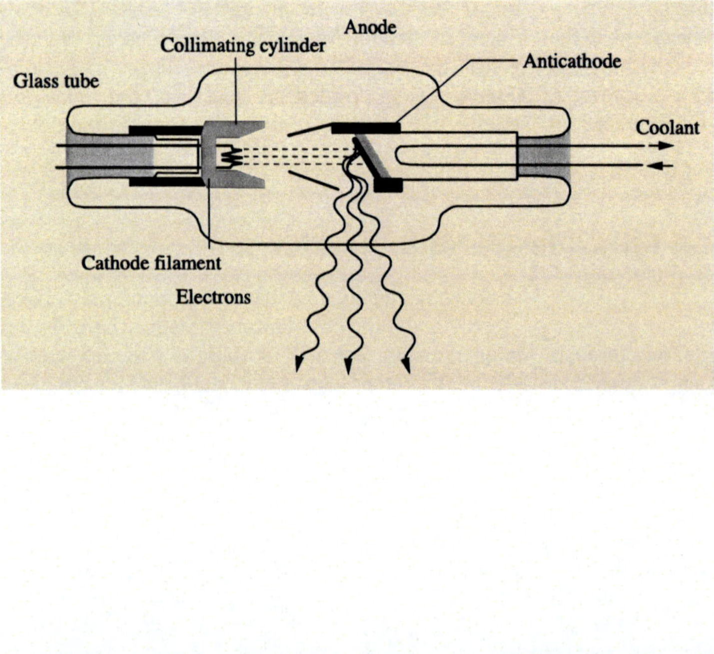

Fig. 35.5 Schematic view of an x-ray tube

the first stage, charged particles are released (electrons), and in the second stage these particles either transfer their energy to the matter or release electrons. Since the effect of photons is indirect in this way, it is also referred to as indirectly ionizing radiation. The higher the energy of a photon, the deeper it is able to penetrate into the tissue before it releases an electron. In the case of photons generated by linear accelerators and having energies of 4–25 MV (ultrahard x-radiation), a higher dose is generated in the deep tissue than at the skin surface; this is known as the so-called buildup effect, which does not occur with x-rays in an energy range of 200–400 kV. The higher the energy, the less the photons are decelerated; they are scattered in a forward direction as well as other directions. In this way, a higher dosage is achieved using ultrahard x-rays for target areas that are situated deeper. At the same time, this leads to better protection of the skin and, as a result of the lower proportion of lateral radiation, to a more sharply defined lateral limit of the dosage.

The interaction of electrons with matter takes place exclusively by means of scattering processes. Elastic scattering leads to the particles changing direction. In the case of inelastic scattering, the particles change their direction and lose energy due to the excitation of the atomic shell, the ionization of atoms, or the generation of x-ray Bremsstrahlung. Electrons are electrically charged particles that transfer their energy directly to tissue, and this therefore involves directly ionizing radiation. This results in a relatively high exposure of the skin, a large area in which the maximum dose is deposited, and a rapid decrease in dose after the maximum dose, which means that the region behind the target area situated at the surface is better protected than with photons.

35.3.4 Dosimetry

Dosimetry, in the modern sense of the word, was unknown in the early stages of RT. The following three effects were used for the semiquantitative determination of the level of radiation: the redness of the patient's skin as a *biological dosimeter*, the blackness of a photographic plate, or the luminescence of various substances (usually barium cyanide). In order to estimate the dose, it was important to know the hardness of the radiation, as all three methods give different results with different mixtures of rays.

In the field of RT, *dose* is understood as meaning the energy that is absorbed by tissue when it is treated with radiation, since only the energy remaining in the tissue can have a radiation effect. We talk of the so-called absorbed dose, measured in *gray*, with 1 gray corresponding to an energy of 1 Joule per kilogram of tissue.

In order to make the dosages used by different hospitals and clinics in international cooperation or used in publications more comparable, there are standards governing how they should be referred to. In the case of the reference point concept used for this purpose, reference is commonly made to ICRU Report 50 [35.7]. In everyday clinical practice, the *ICRU reference point*, or the *dose according to ICRU*, is therefore often spoken of.

Any material in which a signal is generated as a result of exposure to radiation can be used to measure the dose of ionizing radiation. In principle, any physical, chemical, or even biological effect that is caused by ionizing radiation can therefore be utilized.

To this day, ionization dosimetry is of the greatest significance among methods for radiation detection and measurement. It is based on the primary interaction process of radiation and matter, ionization. Ionization chambers in two designs are primarily used: cylinder chambers and flat chambers. Cylinder chambers, which are also known as compact chambers, consist of a cylindrical chamber wall of a conductive material and an axial, usually metal, electrode.

The main advantages of this dosimetry approach are that it is possible to use tried and tested electrical and electronic measurement technology and that the system possesses a very high degree of sensitivity, measuring accuracy, reproducibility, and long-term stability, is not damaged by radiation, and can be used in the entire range of applications of ionizing radiation while being less energy dependent [35.4]. The chambers are also convenient, user-friendly, and very robust. The ionization of solids, including semiconductors, is also routinely used in dosimetry. Mobile electrons and also mobile positively charged *holes* are generated in the semiconductor by means of radiation, i. e., it is made into an n-type or p-type semiconductor. The sensitivity in semiconductors is around 20 000 times greater than in ionization chambers with a measuring volume of equal size. The advantage of semiconductor detectors is their high spatial resolution, but in the longer term the sensitivity decreases as a result of irreversible radiation damage in the material. Recently, diamond detectors have also been used. They belong to the so-called conductivity detectors and, after obligatory preirradiation, exhibit detector currents that can exceed those of ionization detectors by several powers of ten – and with especially high spatial resolutions, too. Scintillation detectors are based on the physical phenomenon of excitation and are predominantly used for the detection of radiation. The shell electrons that are raised to a higher orbit fall back to the ground state, emitting light; the light pulses emitted are usually amplified by the use of photomultipliers. Thermoluminescent dosimetry uses special ion crystals, known as thermoluminescent dosimeters (TLD), as detectors. These lithium fluoride or calcium fluoride crystals are selectively doped with foreign atoms, such as magnesium, manganese, or titanium, and are characterized by the fact that they do not immediately return to their ground state after excitation. A long-term luminous effect therefore remains following irradiation, which is termed luminescence. Some of the absorbed energy is stored in the crystals and can be retrieved in the form of the emission of light by controlled heating. TLDs exist in various forms and are suitable both for in vivo and also for phantom measurements. Simultaneous irradiation of several of these miniaturized dosimeters also enables the dose distributions to be measured. Spatial resolutions in the submillimeter range and with measuring accuracies of up to 1% can be achieved, although measuring accuracies of this level can only be achieved with a very large amount of experimental outlay.

Films are also very highly suitable as dosimeters, as was recognized at an early stage. Following irradiation with subsequent development, the films are densitometrically assessed with special equipment (densitometers, film scanners). The reproducibility of the process conditions, from the manufacture of the film through development to assessment, play a fundamental role in the accuracy of the measurement, and this means that films are of only limited suitability for routine use. Films are the dosimeters with the highest spatial resolution, as practically every individual silver particle acts as a dosimeter. Ultimately, however, the resolution is limited by the diameter of the light beam of a densitometer or by the CCD detector density of the film scanner used.

Ferrous sulfate dosimetry, also known as Fricke dosimetry, belongs to the group of chemical methods and was at the same time the first approach of what was known as gel dosimetry. The measuring process is based on the fact that divalent iron in an air-saturated aqueous sulfuric acid solution is irreversibly oxidized by ionizing radiation to form trivalent iron. The amount-of-mass concentration of the trivalent iron ions is determined by means of extinction measurement. The method exhibits proportionality of amount-of-mass concentration and absorbed dose in a wide dose range, it is independent of the dose rate, and absolute determinations of the dose can be achieved with an accuracy of 1%. The method is thus also used by calibration service providers for the calibration of dosimeters. Gel dosimetry has in recent years been gaining increasingly in significance, as new gels that are less difficult to process have in the meantime become available. This group includes the BANG (bis acrylamide nitrogen gelatin) gels, which have been available for a short period of time. This involves a monomer that is incorporated into a gelatin matrix. When subjected to radiation, polymerization is induced, resulting in a change in the spin–spin relaxation time T_2 as a function of the respective locally absorbed dose. Magnetic resonance imaging (MRI) was initially exclusively used as an evaluation method; now there are also optical systems based on computer tomography (CT) that make use of the change in optical density that occurs simultaneously.

For the sake of completeness, calorimetry should also be discussed as another approach. It is based on the principle that, in the event of complete conversion of the absorbed energy into heat, it is possible for this energy to be directly and absolutely measured. In addition to graphite calorimeters, water calorimeters are increasingly being used in national standard laboratories, as the absorbed dose in water has since become the basis of clinical dosimetry. Owing to the high experimental outlay, calorimetry cannot be used in routine clinical applications.

In recent years in particular, various biological dose-measuring methods have also been further developed and used in clinical settings. In addition to the purely qualitative method already mentioned of assessing the redness of skin, which has of course been superseded, these methods include quantitative ring chromosome counting and fluorescence analysis of chemically labeled DNA strands.

35.4 Forms of Therapy

In general, modern RT can be divided into three forms: percutaneous RT, brachytherapy, and radionuclide therapy. In percutaneous therapy, also called teletherapy, the radiation source is situated outside the body of the patient to be treated with radiation, and the rays penetrate through the skin and into the body. Examples are irradiation using a teletherapy unit or an accelerator. In brachytherapy, one or more small radiation sources are inserted into each region of the body that is to be treated with radiation. Irradiation thus takes place directly at the location of the tumor, that is, just a short distance from it (from the Greek *brachy*, short). Brachytherapy can be *intracavitary*, e.g., in the uterus, *intraluminal*, e.g., in the vagina, the esophagus or the bronchus, or *interstitial*, using needles or thin catheters implanted in the tumor or in the tumor bed for that purpose. In radionuclide therapy, radioactive substances are administered, which preferably accumulate in a certain organ or body part [35.6, 8, 9].

35.4.1 Percutaneous Therapy

In percutaneous therapy, a distinction is drawn, depending on the depth of the region to be radiation treated, between superficial therapy, semideep therapy, and deep therapy. Superficial therapy includes skin radiation therapy on the one hand and intraoperative radiation therapy on the other. These are performed using therapy systems specifically designed for the purpose, which are usually in an energy range of between 10 and 50 kV. Fields of application for semideep therapy are mainly the treatment of benign conditions such as inflammatory processes of the ligaments and joints. Photons in the range of 150 to 300 kV from so-called orthovoltage units are used for this, and sometimes low-energy electrons of linacs are also used. Deep therapy using high-voltage equipment is today performed mostly using linacs, which have virtually replaced the cobalt unit. The energy range for this so-called high-voltage therapy is between 1 and 25 MeV. Typical doses applied in teletherapy are approx. 2 Gy per radiation session; varying numbers of sessions ultimately lead to total doses of up to approx. 75 Gy.

35.4.2 Brachytherapy

In the field of brachytherapy, distinctions are made both on the basis of the dose rate used and the region being treated. If the dose rate is taken as the value of measurement, then a distinction is drawn between low-dose rate (LDR), medium-dose rate (MDR), high-dose rate (HDR), and pulsed-dose rate (PDR). According to the definition based on ICRU, the dose rates are 0.4 to 2.0 Gy/h for LDR, 2.0 to 12.0 Gy/h for MDR, and greater than 12.0 Gy/h for HDR, the usual dose rates with modern HDR brachytherapy systems being between 100 and 300 Gy/h. The PDR is a relatively new variant, which for reasons of radiation protection appears to be establishing itself increasingly as a replacement for LDR approaches.

An alternative classification arises from the nature of the application: here, a distinction is drawn between contact, intracavitary, endoluminal, and interstitial therapy. In the case of contact therapy, therapies are administered close to the surface using specially shaped radioactive applicators or radioactive preparations inserted into boluses. In intracavitary therapy, cavities in the human body that are present naturally or are artificially created are used to insert radioactive substances. In endoluminal therapy, also referred to as intravascular therapy, radioactive sources are introduced into the vessel lumen either temporarily (HDR) or permanently in the form of stents (LDR). Interstitial therapy involves temporary or permanent insertion of radioactive emitters directly into the tumor tissue, these emitters being applied in catheters or hollow needles positioned in advance.

35.4.3 Radionuclide Therapy

Familiar radionuclide therapies include radioiodine therapy for the treatment of thyroid conditions. It has been used worldwide since the 1950s and belongs not to the specialist field of RT/radiooncology but to the discipline of nuclear medicine. Based on statistical studies of over 100,000 patients treated, it has been possible to guarantee that radioiodine treatment will not cause an increase in thyroid cancer or other malignant diseases in the body. The exposure to radiation that the body undergoes in the process corresponds to that of a thorough x-ray examination. Radioiodine therapy is carried out in the case of thyroid disorders (e.g., hyperthyreosis associated with Graves' disease, uni- or multifocal autonomy, latent hyperthyreosis) to reduce the size of large thyroid glands (*goitre*) and also for the follow-up treatment of thyroid carcinomas. For reasons of radiation protection, the therapy must be carried out on an in-patient basis, as the patients emit radioactivity in their excretions (urine, stools, perspiration, and saliva) and also in their respiratory air. All sewage from the therapeutic ward must be collected and must only be released into the public sewerage system once the radioactivity has decayed. The radioactive iodine is administered to the patient orally in an amount individually determined in advance by means of a radioiodine test. The ^{131}I isotope used for radioiodine therapy is a beta minus emitter with subsequent gamma radiation. It emits 90% beta radiation and 10% gamma radiation. The beta radiation is therapeutically effective and has an average range into the tissue of just 0.5 mm. As a result, the exposure of the neighboring organs to the radiation is minimal. The nuclide has a half-life of around 8 d and a gamma energy of 360 keV. The amount of radioiodine required for RT is calculated, depending on the desired focal dose, from the maximum thyroidal iodine uptake, the effective half-life and the thyroid volume to be treated with radiation therapy. As a target dose, 150–200 Gy is applied for Graves' disease, 200–400 Gy for thyroid autonomy, and 150–180 Gy for reducing the size of a goitre. In cases of hyperthyreoses, permanent elimination of the metabolic disorder can be expected in roughly 80% of cases.

The radiation protection measurements and the posttherapeutic calculations of the radiation dose are performed with the aid of the gamma radiation that is emitted from the body. Gonadal exposure in radioiodine therapy of benign thyroid conditions is 0.02–0.05 Gy. The gonadal dose that leads to a doubling of the spontaneous mutation rate is 0.2–2.0 Gy. The iodine beta radiation has the following radiobiological effects in the cell system of the thyroid gland: restriction of function, loss of regenerative ability, and cell death. In addition to influencing function, the degeneration of individual thyroid cells also results in a decrease in the size of the goitre, usually by approx. 40%.

35.5 Equipment Technology for the Generation of Radiation

In RT, ionizing radiation is used that either results from radioactive decay processes or is generated by the acceleration of charged particles. In addition to x-ray tubes, various other technologies are used for the acceleration of charged particles, such as linear and circular accelerators – apart from the betatron, which is no longer in use, circular accelerators include the microtron, the cyclotron, and the synchrotron. Since linacs and afterloading units are in particular used in clinical application for the generation of radiation, the chapters that follow will include in-depth discussion of these technologies.

35.5.1 X-ray Equipment

Today, a distinction is made between soft x-ray and hard x-ray therapy, the latter also referred to as orthovoltage therapy. The boundary between the two forms of therapy is at a tube voltage of 100 kV. The therapy distance – also called the focus-skin distance – is between 0 and 5 cm in the case of short-distance radiation, and otherwise between 10 and 30 cm, or in exceptional cases up to 50 cm. In addition to the classics from Siemens (Dermopan, Stabilipan 2) and Philips (formerly C. H. F. Müller, RT50, RT100, RT250), which are over 50 years old and are still in use in some places today, microprocessor-controlled systems with integrated patient databases are now used and cover a tube voltage range of 10 to 300 kV. A modern x-ray therapy unit is depicted in Fig. 35.7.

Therapeutic x-ray units consist of a cooled tube, which is integrated in a tube head fixed to a stand, a generator, a control panel, a patient couch, and a set of applicators. Depending on the spectrum of use, systems with overhead installation or combined wall and floor installation are used as stands. As the currents of between 20 and 30 mA that are used are significantly lower than those of diagnostic x-ray tubes, there is less heat generated, and stationary, i. e., structurally simple, anodes can therefore be used. As the X-radiation generated has a broad, continuous photon energy distribution, basic filtering is performed, which eliminates the low-energy components and thus hardens the radiation. In addition to the traditional applicators – also called tubes – some modern devices of this type are also equipped, for contact therapy, with a collimator having adjustable diaphragms for field definition. In addition, individually manufactured lead cutouts are introduced into the tube

Fig. 35.7a,b X-ray therapy unit Gulmay D3225 (**a**) with control console (**b**) (courtesy Gulmay Medical, Camberley)

for the shaping of irregular fields. The use of a multileaf collimator is also a relatively new approach that is used instead of the aperture collimator. This is a device that enables irregular field shaping by means of manual positioning of the individual leaves, which move independently of one another.

35.5.2 Cobalt Unit

Cobalt units belong to the group of equipment known as teletherapy units, i. e., they are used for teletherapy. In order for sufficiently high dose rates to be achieved at a therapy distance of usually 80 cm, the sources must have a very high activity. So that the highly active source only has to be replaced very rarely, radionuclides are desirable that have a half-life that is as long as possible. Radionuclides used are therefore ^{60}Co and, only in rare cases, ^{137}Cs. In comparison with ^{137}Cs, ^{60}Co has a higher photon energy of approx. 1 MeV and therefore has therapeutic advantages, even though its much shorter half-life of 5.25 years generally means that it is necessary to replace the source after just 5–10 years. A modern cobalt unit is shown in Fig. 35.8. The tube head of the cobalt unit is connected to a stand by means of a support arm that can be rotated around a horizontal axis. This design is also known as a gantry and enables arc RT around a virtual point lying fixed in space, the so-called isocenter. The tube head itself consists of the source, a source valve or a rotary shutter, the lead shielding, diaphragms for field definition, a light-beam localizer, and a distance-measuring device. In addition to the support arm, the diaphragm can also be rotated. The treatment couch, which is fixedly connected to the system, can be moved linearly in all three spatial directions and can be rotated both isocentrically and eccentrically around its own axis.

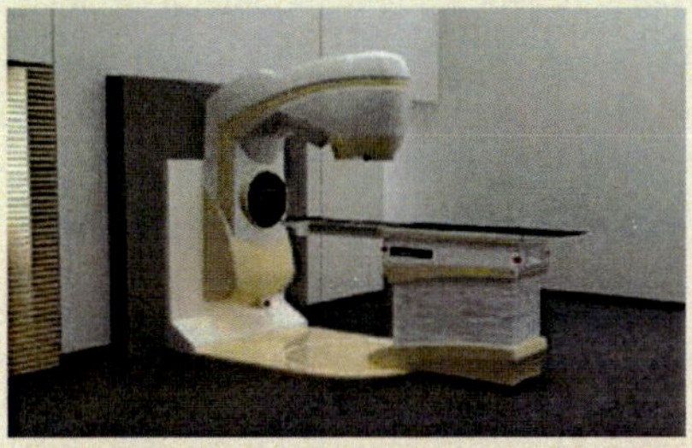

Fig. 35.8 UJP Terabalt 80 cobalt unit (UJP PRAHA, Prague)

35.5.3 Linear Accelerators

These days, modern linacs are available for the generation of therapeutic electron radiation with a maximum energy of 2–25 MeV and ultrahard photon radiation with a maximum energy of 4–25 MeV. There are several electron energy levels that can be set and – depending on the manufacturer – up to three photon energy levels. In addition, there is also a sort of special form, the *photon-only machines*, which have low maximum photon energies of between 4 and 6 MeV. They are mostly used as a replacement for cobalt units that have been taken out of service, which is to say that they can be installed in existing therapy rooms without any great additional measures being necessary in terms of radiation protection. Systems of this type manage without moving parts in the accelerator segment and also have accelerator tubes that are a maximum of 50 cm in length. This makes a significantly more compact and more simple design possible, without beam deflection or beam analysis, which lowers the costs accordingly.

Linacs are radiofrequency accelerators that accelerate electrons in straight accelerator tubes by means of radiofrequency waves of 3 GHz with a wavelength of 10 cm. The systems essentially consist of five subassemblies: the energy supply unit, the modulator, the accelerator unit, the tube head, and the control unit. Figure 35.9b shows a schematic representation of the structure of a modern high-energy linac. The modulator contains the source for the radiofrequency generation, as well as the control electronics and electrical power supply. The radiofrequency generators originally come from radar technology; they generate microwaves in magnetrons or klystrons and amplify them. Magnetrons are preferably used with low to medium electron energies and are less expensive than klystrons, but they also

a) b)

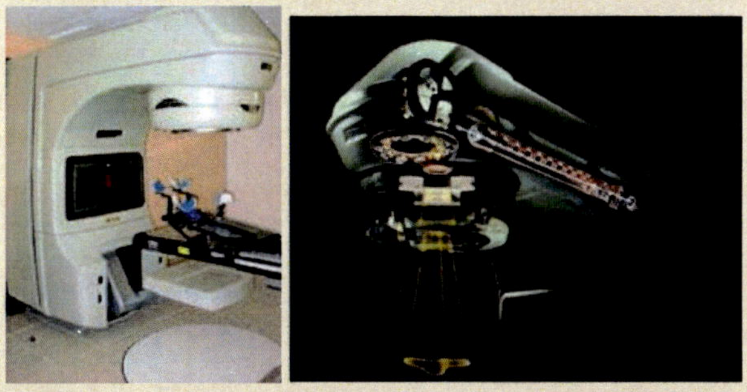

Fig. 35.9a,b Varian Clinac 21 EX linear accelerator **(a)** with beam guidance **(b)** (courtesy Varian Medical Systems, Palo Alto)

have a significantly shorter operational life span than klystrons. With klystrons, which are used with relatively high electron energies, a distinction is made between single-chamber and two-chamber systems.

The accelerator unit is composed of an electron gun and an accelerator tube, as well as cooling units and vacuum pumps. The electron gun is either a directly heated hot cathode made from tungsten coils or an indirectly heated tungsten matrix. Unheated cathode grids, known as cold cathodes, are also used, which are coated with barium sulfate, for example. These coatings reduce the work function of the electrons such that they can be extracted simply by applying a high voltage. The extraction voltages are between 15 and 50 kV. Depending on the manufacturer, traveling waves or standing waves are used as the acceleration medium. In the traveling wave principle, the radiofrequency wave travels through the tube, taking the electrons with it; at the end of the tube the electromagnetic traveling wave is then broken up. In contrast to this, when operating with the standing wave, the electromagnetic longitudinal wave is reflected at the end of the tube such that a standing wave develops. Differences in design are mainly in the length of the accelerator tube, which is longer for traveling waves than for standing waves. On the other hand, in a traveling wave system there is a lower dependence of the acceleration power on the frequency and fewer high demands on the energy level of the electrons fired.

The tube head has the task of deflecting the pencil beam, which points in the direction of the accelerator tube and has a diameter of just a few millimeters, and of adapting it geometrically to the requirements. The necessary steps for this are:

1. Concentration, focusing, and deflection
2. Primary collimation
3. Homogenization
4. Beam monitoring (position and symmetry)
5. Dose monitoring and control of the dose rate
6. Field shaping

The components used for this are the 270° deflecting magnet, the beam-steering device, the primary collimator, the carousel with photon compensating filters and electron scattering foils, a double ionization chamber, collimator diaphragms, and a multileaf collimator. In the case of photons, the field is shaped by means of collimator diaphragms as well as a multileaf collimator (MLC), and in the case of electrons, so-called tubes are used for this purpose that collimate in a cascading fashion and at the lowest level provide the option for an individually shaped insert to be accommodated for the generation of irregular field shapes. Diaphragm systems with variable, and if appropriate asymmetrical, apertures, which can also be operated dynamically, are used to delimit the radiation field. The limitation of such systems to rectangular-shaped fields has led to the integration of blocks that, despite having to be manufactured individually, can be used to generate virtually any desired irregular field shapes. More modern accelerator systems today also provide MLCs that – depending on the width of the leaves – make field configurations of virtually any shape possible. Depending on the manufacturer, these MLCs can also be operated dynamically. It is possible to modulate the field. This is done either by means of wedge filters or compensators, these being produced either in terms of software or hardware.

External wedge filters and also physical compensators are produced from heavy metal. The former are standard components of a linac, whereas the latter are manufactured individually for the patient. In modern linacs, they can also both be generated by dynamic operation of the collimator diaphragms and the multileaf collimator. In order to do this, the diaphragms and leaves move during the irradiation process, thereby modulating the intensity distribution within the field as desired.

35.5.4 Miniature Accelerators

A compact miniaturized version of the x-ray source first became a reality not that long ago, which makes possible a mobile system that can be transported from one operating theater to another. The source (Fig. 35.10b), which weighs a mere 2 kg, is a compact linac with an energy of 50 kV, which can be operated at voltages of 110–230 V and if necessary even with a battery of 9.6 V. The acceleration path is approx. 40 cm and dose rates of 2.8 Gy/min can be achieved at a distance of 1 m.

The structure of a system such as this is illustrated in Fig. 35.10a, b; the device is integrated in a stereotactic robot and thus permits intraoperative high-precision radiation. As a result of miniaturizing the source and the energy used, the radiation is also limited to a small volume. The investment in shelters and personal protection measures that are necessary in the case of conventional equipment (linacs, radioactive emitters) is therefore not necessary. As a result of the spatial restriction and isotropic distribution of the radiation, only the affected tissue around the tumor bed is irradiated and healthy tissue is spared. Heavy exposure of healthy tissue to radiation is, in contrast, a critical side effect of treatment

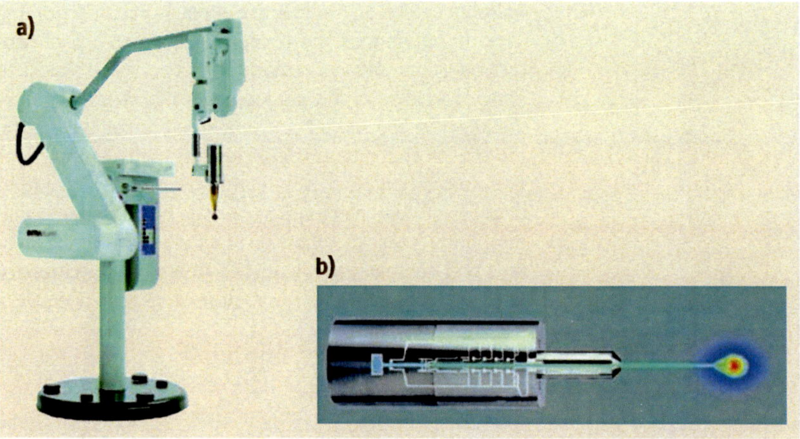

Fig. 35.10a,b Zeiss INTRABEAM miniature accelerator (**a**) with accelerator section (**b**) (courtesy Zeiss, Oberkochen)

with conventional methods of RT, as there are limits in radiation protection for medical reasons.

35.5.5 Brachytherapy Sources and Brachytherapy Equipment

In brachytherapy, closed radioactive sources are exclusively used, in which the radioactive substance is encased in a solid inactive shell or is embedded in it such that it is impossible for it to escape in normal use. In the field of LDR, radioactive wires are predominantly used that contain the isotope ^{192}Ir. The long treatment period of the forms of application initially used, combined with the high exposure to radiation of the staff, has led to remote-controlled afterloading processes being developed. To this end, various remote-controlled afterloading systems have been designed that work with ^{192}Ir, ^{137}Cs, or ^{60}Co as radioactive emitters. High dose rates, and as a result very short therapy times, can be achieved with the gamma emitter ^{192}Ir. The half-life of iridium is 74 d, and the mean gamma energy is 0.35 MeV. Iridium has therefore become prevalent, with the exception of a few applications. In the course of optimizing costs, however, multiple-source systems are now being offered that can be equipped with a ^{60}Co source of the same size but with approximately only a quarter of the activity. It is usually only necessary to exchange the emitter once in the first 10 years of operation, compared with 35 to 40 times in the case of conventional iridium emitters.

Figure 35.11 shows a modern multichannel afterloading device. Two different drive systems are used that drive the practically punctiform source by remote control from a vault into the applicator: steel cores in the form of flexible shafts or solid wires. The former have a particularly high degree of flexibility and a relatively large diameter, whereas the latter are considerably less flexible, despite being very thin. The Ir source is a combined beta–gamma emitter whose gamma radiation component is used for the therapy. The encapsulated emitters are encased in one or two high-grade steel sleeves that on the one hand screen out the undesirable beta radiation fractions and on the other hand reduce the wear on the source experienced during transport through the guide tube and during movement in the applicators. An indexer system, such as the one illustrated in Fig. 35.11b, allows – depending on the manufacturer – up to 40 channels to be actuated sequentially, and thus allows spatially complex dose distributions to be generated in combination with corresponding applicator configurations. Prior to the actual radiation procedure, the radiation session is simulated

Fig. 35.11a,b Varian GammaMedPlus HDR afterloader (**a**) with indexer (**b**) (courtesy Varian Medical Systems, Palo Alto)

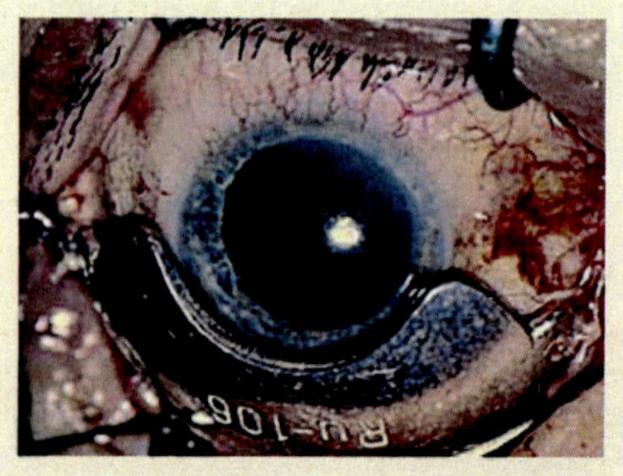

Fig. 35.12 ^{106}Ru applicator with a cutout for the cornea

with a nonradiative metal wire (dummy) in order to check the route in advance for any possible resistance resulting from extreme bends or connection problems in the connections to the indexer or applicator. In addition to the Ir-HDR devices, which have an initial source activity of 10 Ci, there are also PDR systems that have a very similar design.

These systems are designed for regular repeat RT and have an initial source activity of 1 Ci. Isolated systems exist for specific applications that also include the so-called Curietron. The Curietron is a remote-controlled afterloading unit for low to medium dose rates for the insertion of ^{137}Cs sources for intracavitary gynecological brachytherapy. Special systems are used for endovascular RT with afterloaders: the application of gamma rays is currently performed by means of ^{192}Ir, while different sources are used for beta rays. The clinically relevant application of beta rays is to date carried out using a ^{90}Y wire or using a ^{90}Sr/^{90}Y active line. The systems also differ with respect to their centering in the vessel lumen: some use a special centering balloon, while others are self-centering (radioactive balloon catheters, radioactive stents). At this point it must be mentioned, however, that even so-called centered systems always remain eccentric with respect to what is known as the plaque, such that fluctuations in the dose in the vessel are unavoidable. In the future, the use of endovascular afterloading with gamma rays will probably include application in vessels with a relatively large diameter such as venous bypass vessels and peripheral vessels, whereas brachytherapy using beta rays will become the established treatment for coronary arteries, including in the case of in-stent restenosis.

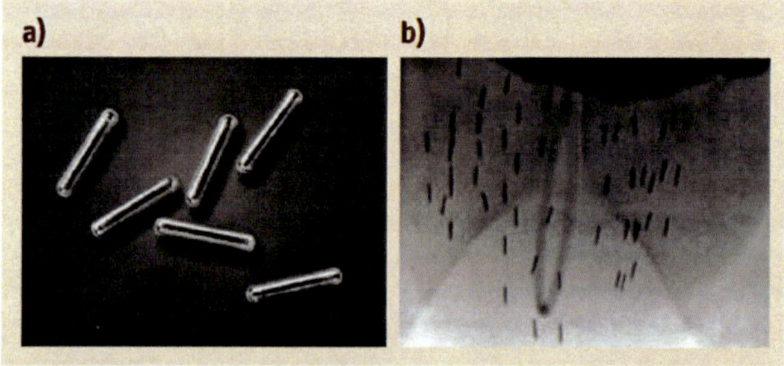

Fig. 35.13a,b Iodine seed therapy of carcinoma of the prostate ((**a**) seeds, (**b**) fluoroscopic image, taken after application) (courtesy Eckert & Ziegler Bebig, Berlin)

35.5.6 Radioactive Implants

The use of radioactive implants is actually a subform of brachytherapy. Because this form of therapy has quite specific applications in the field of cancerous and vascular diseases, however, these will be discussed separately below.

Eye Applicators

In contact therapy for tumors of the eye, small dome-shaped plates are used that contain the beta emitter ruthenium (^{106}Ru) as the radionuclide. The nominal surface dose rate is 120 mGy/min and the activity varies between 10 and 50 MBq. These applicators (Fig. 35.12) allow radiation treatment of up to 100 Gy to a depth of up to 2 mm, without damaging adjacent sensitive structures.

Seeds

Permanent brachytherapy or seed implantation is a curative and at the same time particularly gentle form of RT for locally limited carcinoma of the prostate. Because it is highly effective, it is increasingly being used as an alternative to radical prostatectomy, which is the removal of the entire prostate gland. In this method, the radiation sources are introduced directly into the prostate in the form of tiny radioactive implants (seeds) with the aid of special cannulae.

Seeds of this kind are shown in Fig. 35.13a, and the situation following application is shown in Fig. 35.13b. Iodine (^{125}I) or palladium (^{103}Pd) is used as the radioactive source, which is applied to a support matrix and embedded in laser-welded titanium capsules. Significantly higher radiation doses can in this way be administered directly in the prostate, while sparing surrounding organs, than in the case of external radiation treatment. The implants remain in the prostate and emit the radiation directly into the tumor tissue, with the radiation slowly decreasing over several months (in the case of ^{125}I, over about 12 months, and with ^{103}Pd over about 3 months). The prostate carcinoma is thus gradually destroyed, without the healthy tissue suffering any appreciable damage. Precise determination of the position during the application by means of ultrasound is essential when implanting seeds. Between 40 and 200 seeds with an activity of 20–80 MBq are typically implanted in a tumor. Therapy planning with the aid of ultrasound ensures that a dose of 144 Gy is achieved at the prostate surface. Two different forms of application are used: in addition to the application of individual isolated seeds, in the so-called rapid-strand technique the seeds are applied in a chain. As a result, a relatively low degree of mobility of the sources can in particular be expected.

Stents

Coronary heart disease is one of the most common causes of death in countries in the West. Healthy coronary arteries normally have a smooth, elastic wall. In patients with coronary heart disease, deposits of cholesterol, a natural component in blood, begin to collect on the vessel wall and reduce the blood flow, and a stenosis forms. Besides the option of operative therapy, known as a bypass operation, first and foremost nonoperative methods of dilating the constricted coronary vessels have become important for the treatment of vasoconstrictions. Balloon dilatation or percutaneous transluminal coronary angioplasty (PTCA) was introduced as early as 1979 as a nonoperative treatment method for coronary heart disease. Very small lesions form in the vessel wall as a result of dilating the coronary artery using a balloon. These lesions heal again without any problems. However, the cell growth induced by the healing process reduces the size of the cross section of the vessel once again (scar formation). In medicine, we then talk of a proliferation, i. e., extra tissue growth. As a result, in around a third of cases of successfully dilated coronary arteries, a restenosis, that is, renewed vascular occlusion, appears after a short period of time. The main causes of restenoses are the contraction of the vessel wall following removal of the balloon catheter and the increased cell growth as a result of healing processes in the damaged vessel wall.

Since stents only prevent the contraction of the vessel walls following balloon dilatation, it has been necessary to look for new ways to prevent extra tissue growth. By using radioactivity, which is introduced by ion implantation into the surface of the stent, the intention is to prevent the proliferation of the tissue. Studies have shown that cell growth following balloon dilatation can be significantly reduced by treating the vessel wall with radiation. However, a certain minimum activity (> 100 kBq) is necessary in order for the stent to be medically effective, i. e., for it to prevent cell growth. It has also been possible to demonstrate that a reduction in the rate of restenosis can also be achieved by temporary introduction of a strong source into the coronary vessel following stent implantation by means of HDR afterloading technology. Compared with radioactive stents, however, this method has the disadvantage that a very strong radioactive source (up to 400 MBq) must be introduced into the patient's body in order to deliver a correspondingly high dose of radiation to the vessel wall in a short space of time. The operating team and the patient are therefore subject to a higher level of radiation exposure than when using a radioactive stent whose activity is significantly lower (< 1 MBq).

Figure 35.14 illustrates the effect of a radioactive coating. Following balloon dilatation with subsequent fitting of an inactive stent, the blood vessel slowly becomes constricted again (restenosis), whereas irradiation by means of a radioactive stent damages the cells of the vessel wall and as a result prevents the growth of

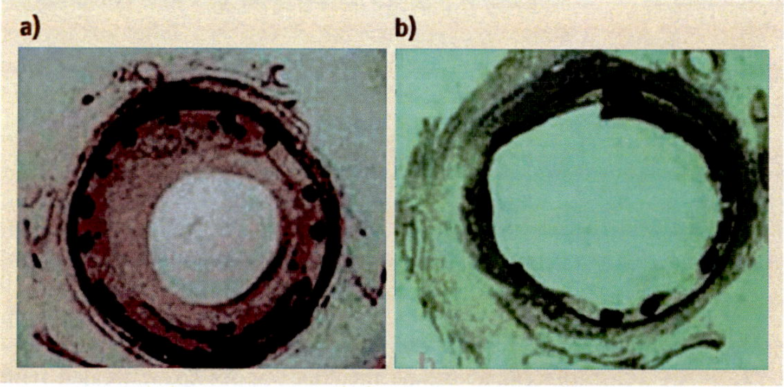

Fig. 35.14a,b The influence of a stent on cell growth; (**a**) without and (**b**) with radioactive coating (courtesy ETH, Zurich)

the outer wall and renewed vascular occlusion. The first radioactive stent was produced in 1992 at the Karlsruhe Research Center (Forschungszentrum Karlsruhe, FZK). This stent was produced from steel whose alloying elements had been activated.

The most important nuclides today in terms of stent coating are palladium (^{103}Pd) and phosphorus (^{32}P). The radioactive isotope of phosphorus was first introduced into the base material of the stent by means of ion implantation. This achieved a homogeneous distribution through the entire stent and good adhesion to the base material. The effectiveness of this product was demonstrated in clinical studies. In order to supply a stent with palladium, a gold layer is first applied to the stent by means of electroplating. This acts as an adhesive layer for the palladium, which is likewise deposited on the stent by means of electroplating. The palladium is then covered with a final gold layer to prevent leaching. In this way, the undesirable low-energy X-radiation of the palladium is also absorbed.

35.6 Special Techniques and Newer Developments in Teletherapy

As a result of, among other things, the continuous progress in computerization, RT is achieving rapid development in the implementation of novel therapeutic technologies. Various approaches of this kind, based for the most part on the technology of linacs, are described below.

35.6.1 Intensity-Modulated Radiotherapy

Intensity-modulated radiotherapy (IMRT) involves modulation of the fields, which in principle can be realized technologically in two different ways: either by means of a physical compensator or with the aid of an MLC. Nowadays, field modulation is performed almost exclusively using MLCs; Fig. 35.15 shows a modern MLC with increased resolution, that is, with a reduced leaf width of 5 mm instead of the leaves 1 cm in width that are generally customary. Depending on the manufacturer of the accelerator, there are two different application techniques: the step-and-shoot (SS) technique and the dynamic MLC technique (dMLC).

Fig. 35.15 Multileaf collimator (courtesy Varian Medical Systems, Palo Alto)

In the SS technique, different subfields – segments, as they are known – are superposed sequentially and field modulation is thus achieved, i. e., a defined MLC position is first assumed and some of the dose is then applied, followed by repetitions with other MLC layouts. In the dMLC technique, the individual leaves move during the entire period of radiation treatment. Because this technique operates using windowlike apertures that move smoothly across the field, it is also known as the sliding-window technique. A paradigm shift in the dosimetric therapy planning occurs as a result of IMRT. In previous advance planning, the planner interactively changes the field configuration until a satisfactory dose distribution has been achieved. The IMRT, in contrast, allows inverse therapy planning. Certain restrictions are defined in the planning system in the form of target doses (for the tumor) and maximum doses (for critical organs) as well as associated weighting factors. The planning computer then calculates from this the optimum field configuration in an iterative backward process.

Early on, IMRT was treated as a new silver bullet. However, experience has shown that, although the dose distribution can be significantly better adapted to the target volume with the use of this technique, the dose distribution is in the process spread over a relatively large region.

35.6.2 Breathing-Adapted Radiotherapy (BART)

Particularly in the thorax and upper abdominal region, the susceptibility of the target volume to respiratory displacement means that the regions to be treated with radiation must be selected such that they are larger than the actual volume of the tumor in order not to jeopardize the control of the tumor. This sometimes results in very severe side effects. Newer approaches have now led to the possibility of controlling the irradiation via the breathing, i. e., only activating the therapy beam when the target volume is situated in the irradiation field. This therapeutic technique is also called gating. Various systems are used to monitor breathing that are either based on a spirometer, and therefore measure the respiratory volume, or else record the thoracic excursion using external measuring technology. Figure 35.16 shows a system that records the movements of the chest wall by means of an external marker box. Two breath coaching techniques are used to ensure that the patient's breathing is stable and constant. The patient is first of all assisted in maintaining a regular breathing cycle by means of digitized audio commands (*breathe in/breathe out*) over a loudspeaker. There are also various systems (prism glasses/mirror glasses, LCD glasses/LCD monitor) to help the patient stabilize the breathing depth and thoracic excursion using visual information (biofeedback). This is particularly important when tumors are being irradiated that move simultaneously with the thoracic wall.

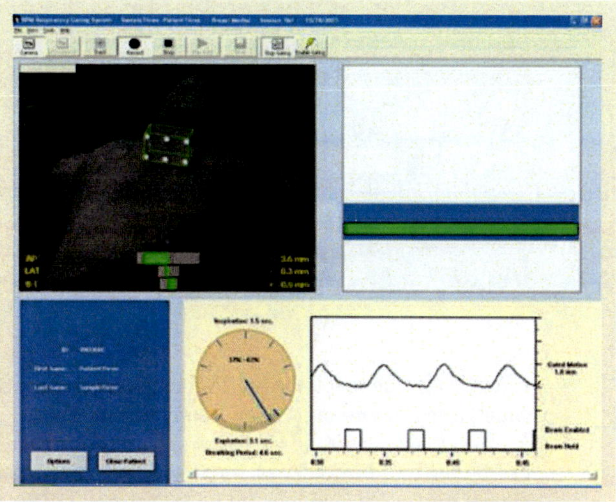

Fig. 35.16 Varian RPM Gating breathing control system (courtesy Varian Medical Systems, Palo Alto)

35.6.3 Image-Guided Radiotherapy

Linacs have recently been developed that are equipped with dedicated image-guided radiotherapy (IGRT) systems. The fundamental concept of systems such as this is to provide imaging technology on the radiation unit that primarily serves to verify the position of the patient. In addition to a radiological mode, which allows digital x-ray images to be taken, there is a fluoroscopy mode for live fluoroscopic images and also a computed tomography mode, which allows a CT scan to be performed with a conical beam arrangement and is therefore called cone-beam CT (CBCT).

Figure 35.17 shows a system of this kind. It is composed of two different imaging systems. One uses MV therapeutic radiation and has been known for some time as portal imaging. The second system is based on traditional kV imaging and consists of an x-ray tube as well as a high-resolution detector. Systems of this type, which are all based on the most modern semiconductor technology (amorphous silicon), now have a resolution of up to 4 million pixels with detector surfaces of up to $40 \times 40\,\mathrm{cm}^2$.

In the case of the kV imaging systems, a distinction is drawn between gantry-based systems (Fig. 35.17) and room-based solutions. A room-based solution is illustrated in Fig. 35.18. Whereas the gantry-based systems rotate correspondingly as the gantry rotates and thus do not permit imaging that is simultaneous with the application of the dose, room-based systems allow ver-

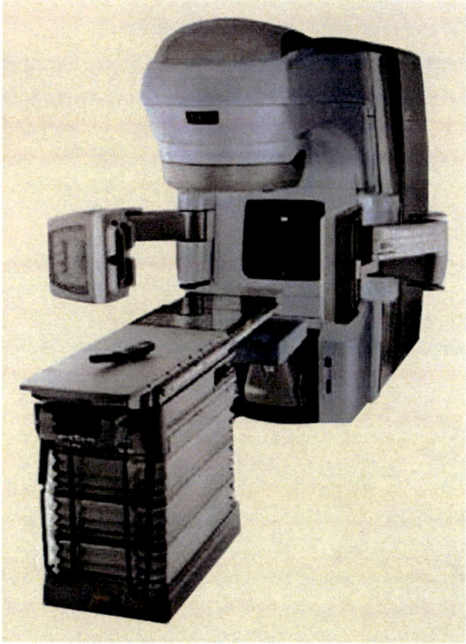

Fig. 35.17 Varian Clinac iX linear accelerator with gantry-based imaging system onboard imager (courtesy Varian Medical Systems, Palo Alto)

Fig. 35.18 Brainlab ExacTrac room-based imaging system (courtesy Brainlab, Feldkirchen)

ification that, although limited to fixed solid angles, is simultaneous by means of fluoroscopy.

Either MV-based or kV-based monitoring images can be produced in order to verify the position of the patient on the basis of internal bone structures. These images can then be compared online with the existing reference images using digital superposition tools, known as matching, and this enables immediate repositioning based on a 2-D/2-D match, with the images having to be taken orthogonal to one another. The CBCT mode enables matching on the basis of three-dimensional data records that also show soft tissue structures. Such approaches are of particular advantage in the pelvis, as in this region the target volume that is to be irradiated is not in a fixed position in relation to bony structures. The fluoroscopy mode provides a live fluoroscopic image and thus allows moving structures in the thoracic region, such as a lung tumor, to be analyzed in terms of their movements and allows any corrections to be made.

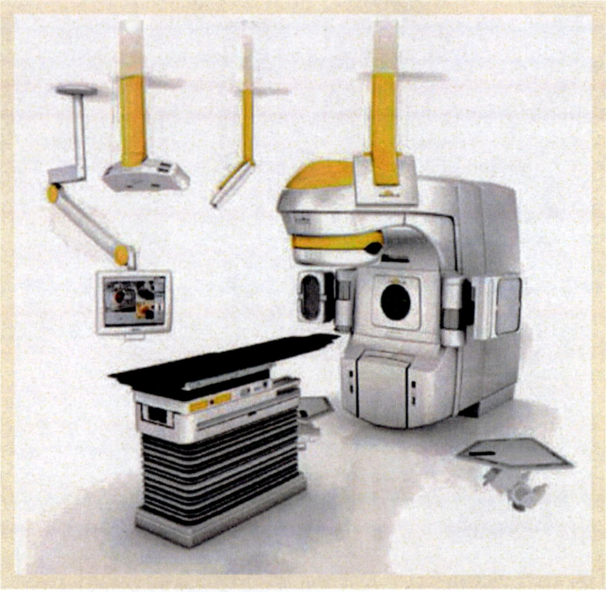

Fig. 35.19 Varian Novalis TX radiosurgery unit (courtesy Varian Medical Systems, Palo Alto)

35.6.4 Dynamic Adaptive Radiotherapy

Dynamic adaptive radiotherapy (DART) is a new type of approach whose aim is to adapt the therapy continuously to the changing parameters. The previous approaches were based on the assumption that the anatomical relationships only change slightly during the course of therapy. However, IGRT has in particular demonstrated that this need not be the case. The concept of dynamically adapted radiotherapy is based on the fact that a new plan is drawn up with the help of regularly acquired CBCT data, this only being possible as an offline process with the technology currently available. This means that the therapy plan is adapted accordingly on the basis of the new data, such as the shrinkage of the target volume, for example. Among other things, this allows healthy tissue to be increasingly conserved. New avenues are now being pursued in which this approach is implemented online. This is particularly recommended when changes in the patient's anatomy or physiology have profound consequences in the case of highly conformal forms of therapy. Online adaptation presupposes fast acquisition of the CT data, followed by very extensive automatic contouring and accelerated calculation of the plan. Implementation places very high demands on the clinical workflow and, because of technical limitations that still exist and other limitations, is still in its infancy.

35.6.5 Stereotactic Radiosurgery/Radiotherapy (SRS/SRT)

High-precision radiation treatment in the cranial region places very high demands on the accuracy of the radiation field and also patient positioning. In order to achieve accuracies in the region of 1 mm or less, a special piece of therapy equipment, the so-called gamma knife, which is equipped with a stereotactic ring for fixing the cranium, was developed as early as the 1960s. Ultimately it is a cobalt unit, but with the special feature that it contains 201 sources all of which are collimated at one point. There are now more modern systems for intra- and extracranial stereotaxy that are either based on a conventional linac or consist of a compact linac

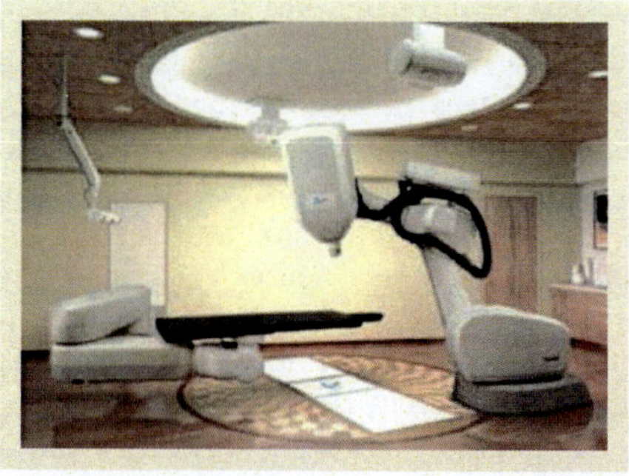

Fig. 35.20 Accuray CyberKnife robotic radiosurgery system (courtesy Accuray, Sunnyvale)

fixed to a robotic arm. Figure 35.19 shows a system of this type that is based on a gantry-mounted linac, while Fig. 35.20 illustrates a Cyberknife. The latter is a completely new development, in which the linac is mounted on a robotic arm – in the same way as is used in the automotive industry. The two approaches have a similar level of precision, which is in the submillimeter range.

Unlike the gamma knife, these systems are equipped with additional imaging hardware that enables the position of the patient to be monitored. The imaging systems used are based on digital x-rays, except that they involve two permanently installed kV units, arranged at a defined angle with respect to one another, each consisting of an x-ray tube and a digital imaging panel. Depending on the technological concept, the imaging technology also includes cone-beam functionality.

35.6.6 Tomotherapy

In this generation of RT units, a linac is integrated in a gantry that is similar to a CT gantry. This approach is based on IMRT and IGRT. Similarly to an x-ray tube in a CT unit, the linac circles around the patient and is also used for medical imaging in this case, with the result that the position of the tumor can be checked and the subsequent treatment can be performed without any complex recalculation and adaptation processes. The therapy itself is carried out on a cross-sectional basis, that is, in small slices – hence also the name tomography. Figure 35.21 shows the two systems currently in clinical use. Besides the Hi-ART system from the American manufacturer Tomotherapy, which was developed originally, there is now also a version by Mitsubishi/Brainlab that enables noncoplanar therapy techniques by inclination of the gantry and table. The latter version also contains a kV imaging system that is used in place of MV imaging (as in the Tomotherapy system). This more recent approach also allows online verification by means of fluoroscopy during the therapy.

Due to the highly conformal distribution of the radiation dose achieved with this technology, a further reduction in the side effects is expected, which means destruction of the tumor while providing even better conservation of the healthy tissue surrounding the tumor and of the organs at risk. However, many questions on this front still remain unanswered, as tomotherapy is still in the early stages of its development.

35.6.7 Intensity-Modulated Arc Therapy

Intensity-modulated arc therapy (IMAT) combines traditional arc therapy with intensity modulation. This means not only field aperture correction – also known as conformal arc therapy – but also true field modulation.

This combined technology places extraordinarily high demands on the linac control system since the dynamic leaf movements of the MLC have to be coordinated, i. e., synchronized, with the gantry rotation, and the control of the dose rate also has to be adjusted to it. Clinical introduction has since been successfully realized in the case of one manufacturer. Figure 35.22

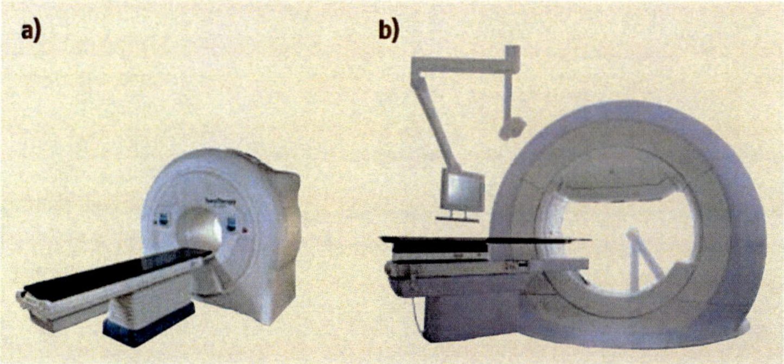

Fig. 35.21 (**a**) Tomotherapy Hi-Art tomotherapy system and (**b**) Mitsubishi MHI TM 2000/Brainlab Vero (courtesy: a) Tomotherapy, Madison; b) Mitsubishi Heavy Industries, Tokyo and Brainlab, Feldkirchen)

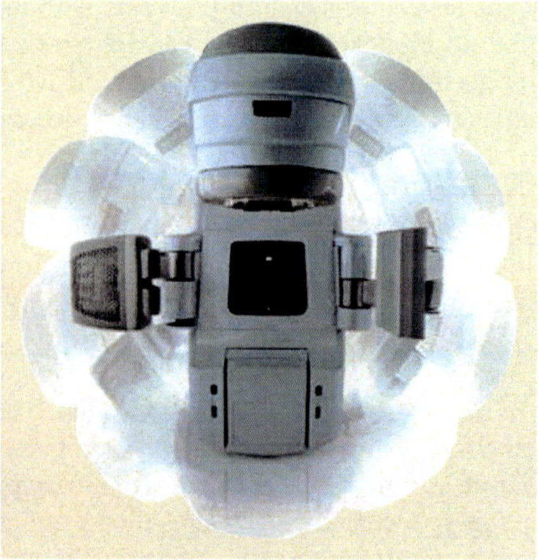

Fig. 35.22 Varian RapidARC intensity-modulated arc therapy (courtesy Varian Medical Systems, Palo Alto)

shows a schematic representation of the course of therapy of this kind. As a result of this combined application form that allows even better optimized, highly conforming dose distribution, just as with the tomotherapy already discussed, better chances of recovery are expected with fewer side effects. In addition, one chief advantage is the faster application of the dose. On the one hand, this leads to a higher patient throughput, and on the other hand the therapy dose can be applied immediately and quickly following image verification, which means that movements by the patient or changes in the internal body topography (e.g., movement of gas in the intestine) during the radiation treatment are less likely.

35.6.8 Current Developments in Linear Accelerator Technology

Forms of therapy with intensity modulation of the treatment areas have now virtually become standard, and until now, in the case of linacs, the pencil beam with a diameter of only a few millimeters has been expanded and modified by means of compensating filters such that a symmetrical beam geometry is obtained up to a field size of 40 cm × 40 cm. As a result, this beam profile that was complex to produce was then remodulated again by intensity modulation by means of an MLC in order to obtain the desired intensity distribution. A new approach is now to dispense with the compensating filters entirely, especially since compensating filters significantly lower the dose rates and it would therefore be possible to significantly shorten the duration of the therapy using systems without compensating filters. A new linac of this type is illustrated in Fig. 35.23. This system allows irradiation of intensity-modulated fields in an assembly that does not have compensating filters.

A further step in the direction of a comprehensively dynamic dose application that also makes it possible to compensate for the movements of target areas and patients makes it necessary to be able to dynamically adjust not only the MLC and the gantry rotation but also other relevant linac parameters, such as the position and rotation of the collimator, during the radiation process. This is now possible for the first time with the new system shown in Fig. 35.23.

Considering the comprehensive dynamically operated parameters, it is all the more important to have appropriate verification imaging available during the radiation process. Where there is a combination of respiratory triggering and IMAT, because of the complexity it makes sense to provide monitoring by means of fluoroscopy during the radiation process. An approach of this kind can be implemented for the first time using the linac depicted in Fig. 35.23.

This new development in the field of linac technology offers various functions, although the potential is still to undergo clinical testing. In particular, the ques-

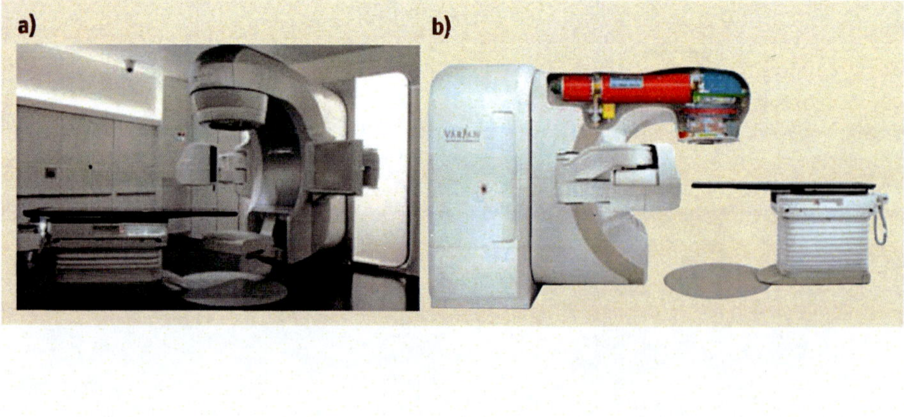

Fig. 35.23 **(a)** Linear accelerator Varian TrueBeam, **(b)** with beam generation and beam guiding components (courtesy Varian Medical Systems, Palo Alto)

tion remains unanswered regarding the extent to which the increase in the dose rates by up to four times compared with the dose rates of the systems previously used has radiobiological consequences. Relevant clinical studies are necessary to clarify these and other aspects.

References

35.1 L. Freund: Ein mit Röntgenstrahlen behandelter Fall von Naevus pigmentosus pilferus Demonstrations in der Gesellschaft der Ärzte in Wien, 15.1.1897, Wiener Klin. Wschr. **147** (1937)

35.2 F. Lennmalm (Ed.): Förhandlingar vid Svenska Lakare-Sallskapets Sammankomster ar 1899. Stockholm, Sweden: Isaac Marcus, **205**, 9 (1900), (In Swedish)

35.3 E. Konecny, V. Roelke, B. Weiss (Eds.): *Medizintechnik im 20. Jahrhundert* (VDE, Berlin 2003)

35.4 H. Krieger (Ed.): *Strahlenphysik, Dosimetrie und Strahlenschutz, Part I, Strahlungsquellen, Detektoren und klinische Dosimetrie* (Teubner, Stuttgart 2001)

35.5 E. Podgorsak (Ed.): *Radiation Physics Handbook for Medical Physicists* (Springer, Berlin Heidelberg 2005)

35.6 J. Richter, M. Flentje (Eds.): *Strahlenphysik für die Radioonkologie* (Thieme, Stuttgart 1998)

35.7 ICRU, International Commission on Radiation Units, Measurements: Prescribing, recording, and reporting photon beam therapy, ICRU Report 50 (ICRU, Bethesda 1993)

35.8 W. Schlegel, J. Bille (Eds.): *Medizinische Physik, Part II, Medizinische Strahlenphysik* (Springer, Berlin Heidelberg 2002)

35.9 E. Wintermantel, S.W. Ha (Eds.): *Medizintechnik – Life Science Engineering*, 5th edn. (Springer, Berlin Heidelberg 2009)

36. Mechanical Circulatory Support Systems

Roland Hetzer, Ewald Hennig

After many years of intensive interdisciplinary research and development, mechanical pump systems to support or replace the failing heart (Ventricular Assist Devices/VAD and the Total Artificial Heart/TAH) are firmly established in the treatment of end-stage heart failure.

Initially, the systems followed the pulsatile principle, i. e. pumps with valves imitated the function of the native heart. The later developed continuous-flow rotary blood pumps have the advantages of being considerably smaller and noiseless and working without valves. They have become the current standard in mechanical circulatory support (MCS).

The current state of technology and medical expertise allows the application of MCS systems with acceptable risks and quality of life. With small, mobile, implantable electromechanical blood pumps patients can return home and take part in *normal* life.

Today, MCS is considered a definitive, permanent therapy when contraindications rule out heart transplantation and in view of the growing shortage of donor organs. Because of their unlimited availability the systems will play a more important role in future medicine, whereby further improvements to the systems are required.

36.1 Introduction – History 724

36.2 Indications for Application of MCSS 725
36.2.1 Bridge to Transplant 726
36.2.2 Bridge to Recovery 726
36.2.3 Bridge to Bridge = Bridge to Decision 727
36.2.4 Permanent Support or Replacement of the Cardiac Pump Function (Destination Therapy) 727

36.3 Classification of MCSS 727
36.3.1 Physical Pumping Principles and Concepts 727
36.3.2 Pulsatile Systems 727
36.3.3 Nonpulsatile Systems – Rotary Blood Pumps 727

36.4 Today's Systems 728
36.4.1 Pulsatile Systems 728
36.4.2 Nonpulsatile Systems (Axial and Centrifugal Rotary Blood Pumps) .. 732
36.4.3 MCSS for Pediatric Patients 736
36.4.4 Summary 737

36.5 Complications 739
36.5.1 Device Malfunction 739
36.5.2 Hemolysis 740
36.5.3 Thromboembolism 740
36.5.4 Bleeding 740
36.5.5 Infection 740
36.5.6 Problems at Anastomoses 741
36.5.7 Right Ventricular Insufficiency 741
36.5.8 Anatomic Obstruction of Flow 741
36.5.9 Arrhythmias 741

36.6 Technical Follow-Up and Care 742

36.7 Psychosomatic Syndromes and Quality of Life During Treatment with MCS 742
36.7.1 Psychosomatic Syndromes During MCS 742
36.7.2 Profound Psychological Processes. 743
36.7.3 Coping Skills 743
36.7.4 Quality of Life 743
36.7.5 Therapeutic Consequences 743
36.7.6 Summary 743

36.8 Overview and Outlook 744

References 747

Today, after years of intensive interdisciplinary research and development and initial studies on application in humans, mechanical pump systems for supporting or replacing the failing heart (ventricular assist devices (VAD) and the total artificial heart (TAH)) are firmly established in the treatment of end-stage heart failure.

36.1 Introduction – History

In 1951, Dennis performed the first open heart operation on a human being using a self-built heart–lung machine, followed in 1953 by Gibbon, who used extracorporeal circulation to correct an atrial septal defect (ASD) in open surgery. The machine was built by IBM engineers. This led to the development of pumps and gas exchangers (oxygenators) that were able to take over the function of the heart and lungs for a limited period. The application of machines such as the heart–lung machine (HLM) provided the basis for today's open heart surgery.

In 1957, pioneers Kolff and Akutzu implanted the first total artificial heart (TAH) in a dog. The animal survived for 90 min. The first ventricular assist device (VAD) was implanted in 1963 by Liotta and Crawford, followed in 1966 by the first successful clinical application of a VAD to support the heart until it recovered (bridge to recovery, BTR) by Liotta and DeBakey.

After the first heart transplant (HTX) was performed successfully in 1967 by *Barnard* in Cape Town, the artificial heart no longer seemed so important [36.1]. However, in 1969, *Cooley* implanted an artificial heart designed by *Liotta* for human application to bridge a patient to transplantation (BTT), thereby opening up the new field of VADs and TAHs [36.2]. The first clinical application of a pneumatically driven TAH as a permanent implant was accomplished in 1982 by *Jarvik* and *DeVries* with a device they had developed further at the Institute for Artificial Organs in Salt Lake City (Kolff) [36.3].

However, media coverage of clinical courses that were fraught with complications had a negative impact on public opinion; it seemed that the TAH was not going to yield positive results in the near future.

In Germany, an artificial heart program led by *Bücherl* in cooperation with AEG, Siemens, MBB, and Hoechst was set up in Berlin in 1971 [36.4]. A range of different assist devices and heart replacement systems were developed, one of which is still in clinical use today (BerlinHeart EXCOR). The first clinical application of this extracorporeal pneumatic blood pump (Bücherl VAD) as a BTR in a patient after heart surgery took place in Berlin in 1985.

The artificial heart designed by Bücherl and Hennig (Berlin TAH) was implanted for the first time as a BTT in 1986. The newly established *Deutsches Herzzentrum Berlin* (DHZB, German Heart Institute Berlin) continued the program with the implantation of TAHs in 1987, and the application of cardiac support systems (ventricular assist devices) in 1988 [36.5].

With the successful, temporary application of VADs as a BTT these systems met with growing acceptance and a new era opened up in which it was possible to gain a wealth of experience in terms of indications, control of complications, technical advancements, and long-term care. This experience ultimately helped to make permanent application – the original goal – a clinical reality.

Following on from these first experiences with adults, the requirement for a system for children became apparent. Miniaturized devices that are also suited to the needs of small children and infants had been under development since the early 1990s [36.6].

Initially, these systems followed the pulsatile principle, i. e., pumps with valves imitated the function of the native heart; this dictated the size, weight, noise generation, and high energy consumption of these devices. The later developed continuous-flow rotary blood pumps, which are considerably smaller, work without valves, and are noiseless, are without these drawbacks. The first pump of this type, the MicroMed system developed by *DeBakey*, was implanted in a patient at the DHZB by Hetzer on 13 November 1998 [36.7]. Since it was shown that such a different blood flow is well-tolerated by the body, the rotary blood pump has developed to become the current standard in mechanical circulatory support.

Nowadays, left ventricular assist devices (LVADs) can provide patients with an acceptable quality of life over several years. With small, mobile, externally driven or implantable electromechanical LVADs, patients can be allowed to return home and take part in *normal* life. Some have been able to return to work [36.8]. Thus, today, it is legitimate to consider mechanical circulatory support (MCS) as a definitive, permanent therapy when contraindications rule out heart transplantation. In view of the growing shortage of donor organs, more and more often MCS systems are the alternative to heart transplantation. Because of their unlimited availability they will play a more important role in future medicine, whereby further improvements to the systems are required.

Another important aspect stems from the observation that long-term support of the heart in cases of myocarditis, after cardiac surgery, and also in cases of chronic cardiomyopathy resulted in complete recovery, allowing the pump to be explanted (bridge to recovery, BTR) [36.9, 10]. This most fascinating route to recovery came about without detailed knowledge of the mechanisms of myocardial recovery or its predictability, and should be the subject of intensive research.

Chronic heart failure resulting from coronary artery disease is the fastest growing category of heart disease. The challenge facing future heart surgeons is to help these patients with a combination of organ-preserving surgery and, where appropriate, temporary or permanent mechanical cardiac support. So, in the future, MCS is likely to become a routine therapy, particularly in older patients, comparable to the current situation with implantable pacemakers and defibrillators.

Of course, the increased use of VADs and TAHs raises a number of ethical, legislative, and regulatory questions; economic aspects must also be discussed. However, these cannot be covered in the following short composition. Here, proven MCS systems are described, with an emphasis on the systems available today (2010).

Based on the experience of the DHZB, the institution with the widest range of systems worldwide and the highest number of patients treated with medium- and long-term MCS systems, indications and specific concomitant therapies are presented [36.11].

It is not possible to give a detailed description of the numerous systems here; links are provided to the manufacturers' websites, which can also be used to follow the steady progress in the refinement of implants and the development of new concepts and to obtain the current application statistics.

36.2 Indications for Application of MCSS

In 2009 a total of 120 000 heart operations were performed in Germany. Over 65 000 coronary artery bypass surgeries and roughly 18 000 heart valve operations were carried out in 84 cardiac surgery centers. Around 95 000 of these heart operations were performed using the heart–lung machine.

In some cases, the patient cannot be taken off the heart–lung machine and an MCSS is applied as an emergency measure. Normally, patients are temporarily connected to the system, i. e., for a number of days or weeks, until their own heart is able to function efficiently.

A heart transplant (HTX) is the final, life-saving option available to patients with severe heart disease. Given the current indication criteria, about 1000 donor hearts are needed every year in Germany. In 2009, however, fewer than 350 hearts were transplanted. The discrepancy between the number of patients affected and the number of donor organs available is glaringly obvious. Patients who are on the waiting list for an HTX might go into life-threatening heart failure at any time. Application of an MCSS offers the last possible therapy option.

A range of different devices have been implanted at the *Deutsches Herzzentrum Berlin* under the leadership of Roland Hetzer to bridge the waiting time to possible recovery of the heart (BTR) or, if a suitable donor organ is available, to bridge the waiting time to a potential subsequent transplantation (BTT). Since 2000, these devices have also been implanted for permanent support, known as *destination therapy* (DT), in patients who cannot or do not wish to receive a heart transplant.

Table 36.1 shows the range of devices used successively at the DHZB since 1988, with the corresponding number of implantations performed and the maximum observation period. Thirteen different systems have been implanted in a total of 1355 patients (as of 30 September 2010) with maximum therapy durations of more than 5 years. Over the resulting accumulated period of more than 590 patient years, the DHZB has gained the widest range of experience worldwide at a single institution.

Not included in this list is the number of patients who have received a *short-term device* as an emergency measure to prevent acute cardiogenic shock, in order to gain time before making a final decision (bridge to decision, BTD). These short-term devices include intra-aortic balloon pumps, catheter turbines, and certain types of extracorporeal circulatory support (ECMO, etc.).

Because the existing NYHA (New York Heart Association) classification of heart failure does not discriminate between different levels of cardiogenic shock, the US Assist Device Registry INTERMACS created a new system for the differentiated categorization of patients.

Table 36.1 Long-term circulatory support devices used at the DHZB

System	Date of first implant	No. of patients	Max. duration days / years
EXCOR	01 June 1988	703	1837 / 5.0
EXCOR pediatric	01 February 1990	110	432 / 1.2
Novacor*	01 November 1993	113	1965 / 5.4
Heartmate I*	01 May 1994	23	390 / 1.1
DeBakey*	13 November 1998	41	518 / 1.4
LionHeart*	01 September 2000	6	1258 / 3.4
INCOR	16 June 2002	185	1852 / 5.1
DuraHeart	27 May 2004	16	1700 / 4.7
CardioWest	30 November 2004	49	775 ongoing / 2.2
CorAide*	26 November 2005	1	221 / 0.6
Heartmate II	09 May 2006	112	1105 / 3.0
Jarvik 2000	06 September 2007	10	745 / 2.0
VentrAssist*	04 March 2008	3	339 / 0.9
HeartWare	26 August 2009	76	400 ongoing / 1.1
HeartWare BVAD	25 September 2009	15	369 ongoing / 1.0
Total experience	30 September 2010	1463	> 590 pat. years

* No longer used at the DHZB

The INTERMACS levels are defined as follows:

1. Critical cardiogenic shock
2. Progressive decline
3. Stable but inotrope-dependent
4. Recurrent advanced heart failure
5. Exertion intolerant
6. Exertion limited
7. Recompensated heart failure.

Levels 1–4 correspond to NYHA Class IV. Currently, we do not consider INTERMACS levels 5–7 to be indications for MCS. Based on earlier experiences, different system configurations are indicated to achieve stabilization of the circulatory system.

Deep cardiogenic shock (level 1) always requires a biventricular assist device (BVAD) or a TAH. In primary left ventricular failure (levels 2, 3, 4) implantation of a left ventricular assist device (LVAD) may be sufficient. In cases of imminent right ventricular failure the additional support of the right ventricle is indicated, at least temporarily, more recently with the application of an implantable BVAD.

Today, the risk of right ventricular failure after LVAD implantation can be very reliably predicted by echocardiographic criteria. Potential contraindications include the presence of septic conditions, brain damage, and advanced malignancies, and the acknowledged impossibility of rehabilitation.

36.2.1 Bridge to Transplant

Patients listed for a heart transplant are at risk of developing unmanageable heart failure as the waiting time progresses, and they then require an MCSS to survive long enough to be able to receive the transplant. In the same way, patients who have received an MCS system as a result of acute heart failure may be listed for a transplant if they satisfy the criteria for an HTX.

Impaired organ function caused by heart failure prior to MCSS can often be relieved by stabilizing the patient's circulatory system. Congestion-induced pulmonary dysfunction and liver or kidney failure caused by the lack of perfusion are then no longer contraindications for subsequent HTX. The overall condition of the patient and, therefore, also the initial situation for a now elective heart transplantation is greatly improved.

36.2.2 Bridge to Recovery

In certain types of heart disease that lead to heart failure, such as some forms of myocarditis or in a small number of patients with dilative cardiomyopathy, recovery of the myocardial function can be achieved with temporary cardiac support using MCS over a period of days, weeks, or months. Afterwards the assist device can be explanted.

Even in cases of heart failure after cardiac surgery, the critical situation can be managed initially by effectively supporting the left ventricle, which is most often affected, with an LVAD.

36.2.3 Bridge to Bridge = Bridge to Decision

When cardiogenic shock leads to impaired organ function, particularly of the brain, and it is uncertain whether recovery is possible, an MCS system designed for short-term application should be used, which is generally less invasive and easier to connect. Stabilization of the circulatory system wins extra time for further diagnosis and potential improvement of the situation. If the organism recovers, then the switch can be made to a longer term or even permanent system.

36.2.4 Permanent Support or Replacement of the Cardiac Pump Function (Destination Therapy)

The goal was to develop systems for permanent application, to either support or replace the heart. After extensive experience with so-called temporary MCS systems with application times of several years, a number of systems available today allow destination therapy. Patients who cannot receive a heart transplant due to known contraindications or who themselves refuse such a transplant can be offered life-prolonging support with extracorporeal or implantable VADs or with an older TAH model, with a reasonable outlook. With complete aftercare and regular monitoring of the device, including the replacement of wear and tear components, device application times of several years are called for and have, in the meantime, been realized.

36.3 Classification of MCSS

36.3.1 Physical Pumping Principles and Concepts

Blood pumps can function according to different hydromechanical principles, as pulsatile displacement pumps or as rotary pumps with continuous axial or radial flow. The pumps themselves can be located outside the body (extracorporeal) or inside the body (intracorporeal). All systems are dependent on external sources of power. In the case of implanted pumps, power is delivered via percutaneous leads. It is possible to supply power through the undamaged skin by induction with an external and an implanted coil.

TAHs are implanted to replace the explanted heart; the driving system can remain outside the body or it can be placed inside the thorax, forming a unit with the pump.

Depending on the objective of the therapy, systems can be designed for short-term (1–30 days), medium-term (up to 6 months), or long-term (5 years or more) application. In the case of long-term systems, the aim is to accommodate all the required system components such as the pumping chamber, activator (motor), power transmission and power storage, the control and monitoring unit and, where necessary, the volume-equalizing chamber inside the body. This aim has not yet been realized with the systems available today.

36.3.2 Pulsatile Systems

Like piston pumps, blood pumps can be designed as displacement pumps with elastic blood chambers, activated by pneumatic, hydraulic, or mechanical compression devices, which are usually powered by electric motors. A disadvantage of this concept is that volume compensation must take place for every volume of blood moved, either through a connection to the free atmosphere (vented system) or to a volume-equalizing chamber implanted in the body. These pumps can generate a flow of blood that is similar to the natural blood flow, with periods of relaxation (filling, diastole) and ejection (emptying, systole).

To be able to direct the bloodstream, pulsatile pumps require valves that are similar to heart valves. Stresses placed on the numerous components are problematic with regard to the individual system's long-term fatigue resistance.

36.3.3 Nonpulsatile Systems – Rotary Blood Pumps

The first rotary blood pump used in the clinical setting was designed around 1982 by Wampler at Nimbus Corporation. This catheter-mounted, intra-aortic axial flow pump (hemopump) was about the size of an eraser on an ordinary pencil. It was inserted through a small incision in the femoral or external iliac artery, advanced to the aorta, and positioned across the aortic valve. An Archimedes screw rotated 17 000 to 25 000 times per minute, drawing blood from the left ventricle and ejecting it into the descending aorta. This system now belongs to history [36.12].

Rotary blood pumps create a continuous flow and do not require valves. They are smaller than displacement pumps, making them easier to implant, do not require volume compensation, and normally only have one moving component, the rotor, with low mechanical load. As a result, the long-term operational safety of these pumps is significantly higher than that of pulsatile pumps. Contact-free positioning of the rotor by magnetic bearings or a combination of hydrodynamic and magnetic bearings means that there is no abrasion, so that, in theory, the life of the system is unlimited.

A distinction is made between axial and radial pumps, depending on the direction of the flow of blood at the rotor. The different hydrodynamic designs determine speed range, positioning possibilities, and the possible outlet orientation (180° for axial, 90° for radial pumps), as well as the characteristics of the pump performance curve.

The electric motor for operating the pump is integrated. The rotor is usually used as a direct impeller, which transmits the required force to the blood, allowing a flow of 10 l/min under normal arterial afterload.

The rotating magnetic field generated in the stator, i. e., the coil assembly located in the pump housing can – depending on the design – produce speeds of over 10 000 rpm.

Initial concerns over the possible adverse effects of continuous, rather than normal, pulsatile perfusion on organ function have since been dispelled by clinical experience with thousands of implantations of this type of system and application times of up to several years.

36.4 Today's Systems

Table 36.2 gives an overview of the MCSS approved for clinical application in Europe and the United States in 2010.

A number of clinically proven systems not included in this list have been withdrawn from the market by the manufacturers either for economic or for technical reasons (Arrow LionHeart, Arrow CorAide, Ventracor VentrAssist).

The technical construction, flow dynamics, and the composition of the materials used for these variously activated electropneumatic, electromechanical, eletromagnetic, and electrohydraulic systems are so complex that a detailed description is beyond the scope of this chapter [36.13].

Once again, detailed information can be obtained from the manufacturers' websites. Links to these websites are included in the system descriptions below. Only the functional principles are summarized here.

36.4.1 Pulsatile Systems

Extracorporeal Pulsatile MCSS for Short-Term Application

These relatively inexpensive and easily connectable systems are used in cases of acute failure of one or both ventricles, when the patient goes into cardiogenic shock, and when there is a chance of recovery if the heart is supported. If the patient's condition fails to improve, these systems can be replaced by longer term, implantable devices. Up to now, extracorporeal systems are the therapy of choice for biventricular heart failure.

Abiomed BVS 5000 VAD. BVS 5000 manufactured by the company Abiomed (www.abiomed.com) is a pneumatically driven pulsatile pump used in intensive care to support the left, the right, or both ventricles. This device has been on the market since 1993 and has been used in over 6000 cases. It is primarily used for BTR therapy.

More recently, there has been a sharp drop in the number of applications of this 17-year old system, as its efficiency is limited by inflow problems caused by the required length of the tube lines and the fact that it has to compete with newer systems with better hemodynamic characteristics.

Abiomed AB 5000 VAD. AB 5000 is a smaller, pneumatically driven membrane pump. It can be used to provide left, right, or biventricular support. The recently developed mobile driving unit facilitates patient transport and also allows a certain degree of mobility. With a chamber volume of 80 ml, it can pump 6 l/min The period of application is limited to 30 days.

Extracorporeal Pulsatile MCSS for Long-Term Application

The BerlinHeart EXCOR VAD system. EXCOR is a pneumatic extracorporeal system produced by BerlinHeart GmbH (www.berlinheart.de). With its wide range of

Table 36.2 MCSS approved in Europe and the United States in 2010 (AbioCor only for clinical studies in the USA)

Extracorporeal	
Short-term application	
Pulsatile	**Nonpulsatile**
Abiomed AB 5000	CardiacAssist, TandemHeart PTVA
Abiomed BVS 5000	Levitronix CentriMag
	Abiomed Impella
Long-term application	
Pulsatile	**Nonpulsatile**
Thoratec PVAD	
Berlin Heart EXCOR	
Berlin Heart EXCOR Pediatric	
Medos VAD	
Intracorporeal	
Long-term application	
Pulsatile	**Nonpulsatile**
Novacor LVAS	HeartAssist 5 (formerly MicroMed DeBakey VAD)
Thoratec P-IVAD	Berlin Heart INCOR
HeartMate XVE	Thoratec HeartMate II LVAS
CardioWest temp. Total artificial heart (TAH-t)	DuraHeart
AbioCor Implantable Replacement Heart	Jarvik 2000
	HeartWare HVAD

pumps offering stroke volumes of between 10 and 80 ml, apical, atrial, and arterial cannulas in numerous shapes and sizes, and stationary as well as small, mobile electropneumatic driving units, this system is unique in that it can be adapted to the needs of each individual patient, from newborns and infants right through to adults.

The blood pump consists of a transparent polyurethane housing, the interior of which is divided into a blood chamber and an air chamber by a highly flexible, triple-layer membrane. The membrane on the blood side is seamlessly integrated into the housing to prevent the adhesion of blood clots. All blood contact surfaces are coated with heparin (CARMEDA coating, Carmeda AB, Upplands Väsby, Sweden).

The polished titanium connectors fixate the Sorin tilting disc valve; the air chamber is equipped with a Delrin connector for the PVC air driving tube. Polyurethane tricuspid valves are used for the EXCOR pediatric pumps as their dynamic characteristics are better suited to small stroke volumes.

The pump can be connected to one or both ventricles (Fig. 36.1). Silicone cannulas that connect at the apex of the left ventricle or at the left or right atrium send the blood from the heart to the pump from where it is directed back to the ascending aorta or the pulmonary artery through arterial silicone cannulas.

The IKUS stationary driving unit is used for large blood pumps (50–80 ml stroke volume) and for the EXCOR pediatric (10–30 ml stroke volume) system for children. In biventricular operating mode the pumps can be driven separately, with selectable synchronous pul-

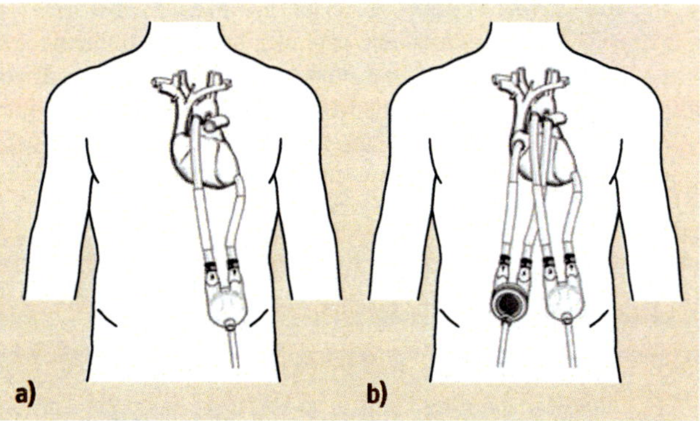

Fig. 36.1a,b Berlin Heart EXCOR as LVAD and BVAD with cannulation. **(a)** Apical – aortic; **(b)** atrial – pulmonary artery

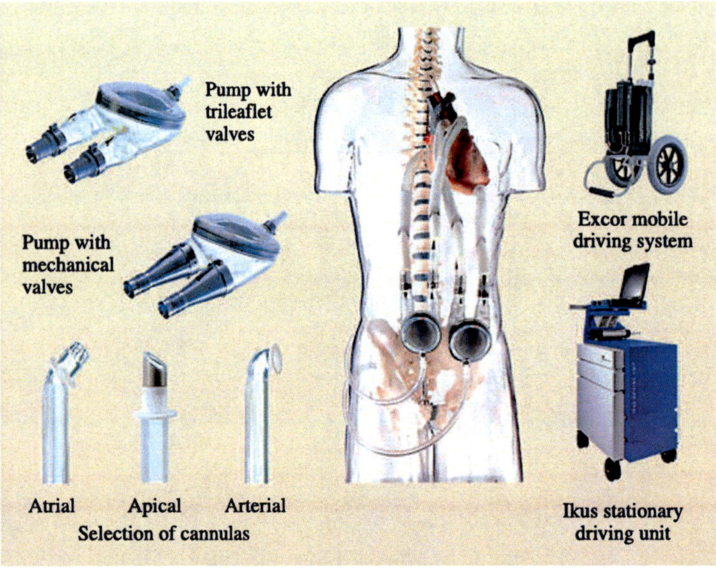

Fig. 36.2 EXCOR system overview

sation or counterpulsation, and the frequencies can be controlled independently.

For blood pumps of 50 ml and more the small driving unit EXCOR Mobile can be used, allowing the patient full freedom of movement without having to rely on an external power supply.

Figure 36.2 gives an overview of the BerlinHeart EXCOR system components.

Thoratec PVAD. The pump of the paracorporeal ventricular assist device (PVAD/Thoratec Corp., www.thoratec.com) consists of a rigid plastic housing containing a flexible, polyurethane blood sac, which is filled and emptied of blood by a compressed air pulse. Two Delrin tilting disc heart valves direct the blood flow. A stroke volume of 65 ml achieves a flow rate of 7 l/min. Polyurethane cannulas that can be connected to the left ventricular apex or to the left atrium direct the blood to the pump from where it is forced back into the ascending aorta through an arterial prosthesis. To provide right ventricle support, the right atrium and the pulmonary artery are cannulated.

Instead of a large stationary driving system, mobile patients can be fitted with a miniaturized unit with which they can be discharged.

Medos VAD. Like the EXCOR device, the Medos system (www.medos-ag.com) is a VAD for pulsatile cardiac support in children and adults. The pumps are available in a range of different sizes and, as is the case for all other systems, are approved for single use only. This system is also driven by electropneumatic units with battery or mains supply.

Implantable Pulsatile MCSS for Long-Term Application

Implantable long-term MCSS are designed for BTR and BTT therapy but based on experiences gained to date in exceptional cases they may also be selected for DT therapy.

Thoratec P-IVAD. The original Thoratec PVAD was modified to produce an implantable model. The extracorporeal PVAD pumps described above were encased in titanium, so now they can be implanted inside the body. As a result, the length of the blood-transporting cannulas is reduced, and the number of transcutaneous injuries to the surface of the body is halved because only the thinner driving line has to be introduced for each pump. The risk of infection at these sites is lowered. Like the earlier PVAD, P-IVAD can be used as an LVAD, RVAD, or BVAD (Fig. 36.3).

Thoratec HeartMate XVE. The HeartMate XVE is an electromotorically driven LVAD displacement pump. Inside a titanium case, an elastic polyurethane (PU) membrane is moved by an electric motor and a cam-and-follower gearing in the direction of its longitudinal axis by a circular pusher plate, emptying the rigid blood chamber. Filling of the chamber takes place passively due to the pressure of the inflowing blood. Two bioprosthetic heart valves ensure the unidirectional flow of blood. The pump is operated by an external controller, which supplies power to the electric motor of the implanted pump through a percutaneous driveline. The volume of blood displaced by each pulse must be compensated. This is achieved through a tube connected from the inside of the pump of the driving chamber to the outside along the electric driveline (vent) [36.14].

The pump can be implanted into a pre-peritoneal pocket in the left upper abdominal wall or it can be implanted intra-abdominally (Fig. 36.4). In 2003, because of the positive results of long-term clinical trials, this system became the first LVAD to be approved by the FDA as DT in the United States.

WorldHeart Novacor LVAS. WorldHeart's Novacor LVAS is an implantable, electromagnetically driven LVAD [36.15] (www.worldheart.com).

A polyurethane blood sac with a seamless interior is symmetrically compressed by two pusher plates to propel the blood into the aortic vascular graft, and then released again to fill passively via the apical inflow cannula.

The pump is implanted through the peritoneum into the left upper abdominal wall and, like the HeartMate XVE, is connected to the external components by a percutaneous lead with an integrated venting tube (Fig. 36.5).

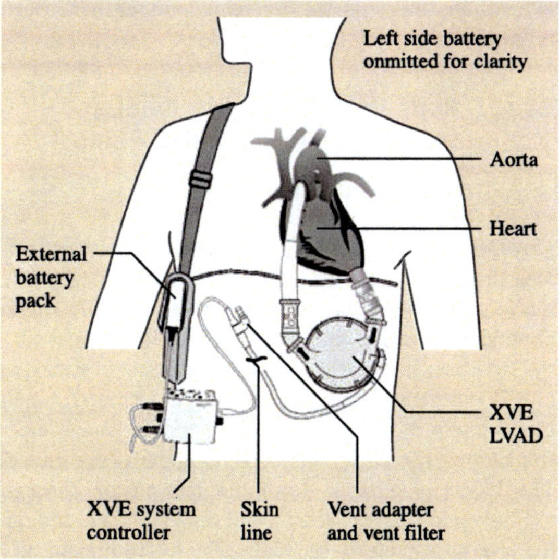

Fig. 36.4 Thoratec HeartMate XVE with external components

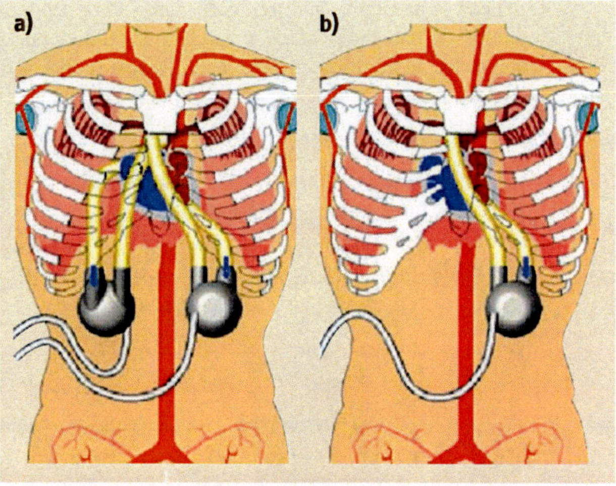

Fig. 36.3a,b P-IVAD. Position of the implantable pneumatic blood pump for biventricular (**a**) or univentricular (**b**) support

This system can automatically adjust its pumping rate to complement the residual function of the native heart, thus adapting to the respective circulatory requirement.

Implantable Pulsatile TAH for Temporary or Permanent Application

TAHs, or artificial ventricles for orthotopic cardiac replacement, occupy a special position among the implantable systems in clinical use today. For implantation of these devices the native heart is removed, as for

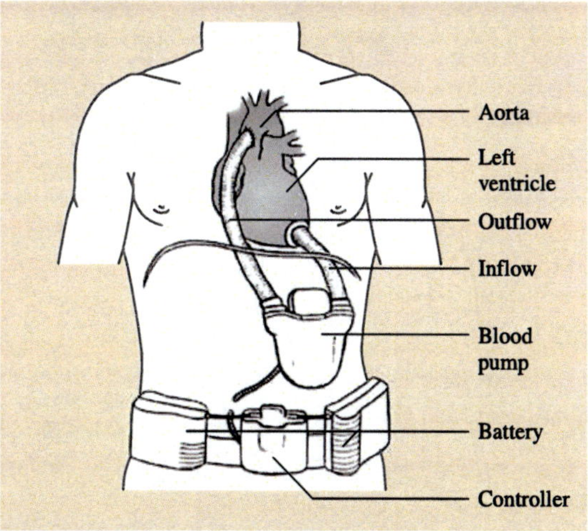

Fig. 36.5 WorldHeart Novacor LVAS with external components

a heart transplant, and replaced with two separate pumps or one dual-chamber mechanical blood pump connected to the atria and the great vessels (aorta, pulmonary artery).

SynCardia Systems, Inc. CardioWest Temporary Total Artificial Heart (TAH-t). This pneumatically driven TAH is a further development of the JARVIK 7 TAH first applied in the clinical setting in 1982. Worldwide this is the only FDA-approved and CE-certified artificial heart for temporary implantation as a bridge to transplantation (BTT) today (www.syncardia.com).

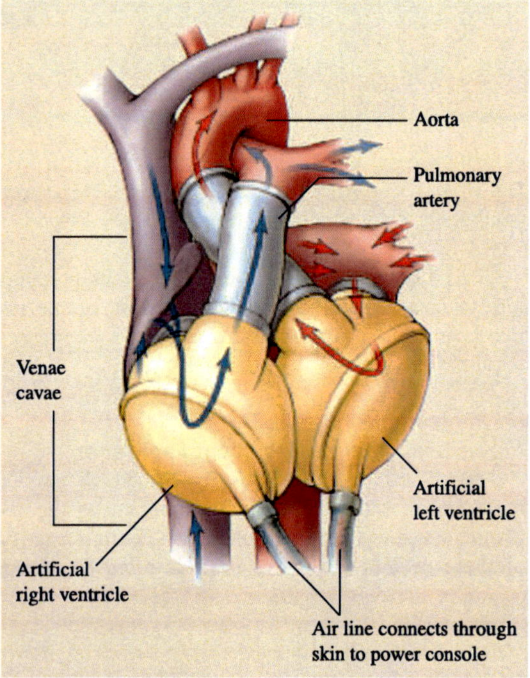

Fig. 36.6 SynCardia CardioWest TAH. Schematic implant position

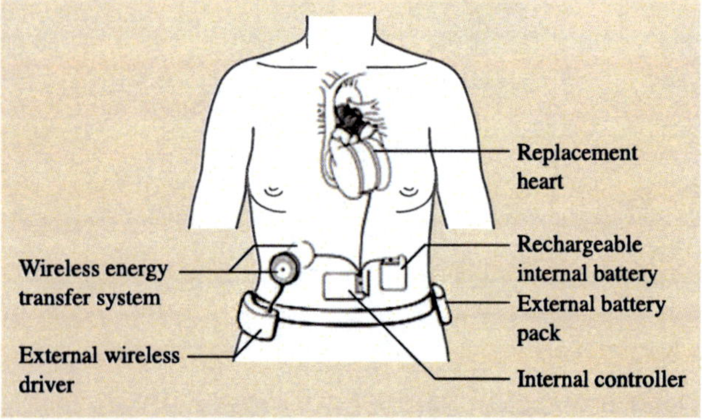

Fig. 36.7 Abiomed AbioCor. Totally implantable TAH with TET

The displacement diaphragm pumps, which are pneumatically activated by an extracorporeal power console, are similar in structure to the EXCOR design and generate a flow rate of up to 9.5 l/min against normal circulatory afterloads (Fig. 36.6). The electropneumatic drive system is newly developed. The small mobile unit replaces the previous mobile driving system EXCOR TAH-t. Unfortunately, the pumps are so bulky that they only fit inside the thorax of relatively large patients. A smaller version is currently being developed and tested in animal studies.

Abiomed, AbioCor Implantable Replacement Heart. The electro-hydraulically driven Abiocor TAH is the first fully implantable TAH [36.16] (Fig. 36.7).

All components such as pumping chambers, drivers, the volume compensator, control unit, battery pack, and transcutaneous energy transmission systems (TETs) are implanted inside the body. This reduces the risk of infection since the skin does not have to be penetrated for the leads and venting tubes to be connected. However, the safety requirements for the many highly technical and sensitive components of this life-sustaining system are very stringent. After a preliminary clinical study was conducted in 2001, with questionable results, the first commercial application was attempted in 2009; the patient died one month later. Since then, no further implantations have been reported.

36.4.2 Nonpulsatile Systems (Axial and Centrifugal Rotary Blood Pumps)

Extracorporeal Systems for Short-Term Application

CardiacAssist, Inc. TandemHeart pVAD. The TandemHeart pVAD (CardiacAssist, Inc.; Pittsburgh, PA, www.cardiacassist.com) is a percutaneous LVAD with continuous flow for temporary ventricle support.

It is used to treat patients in cardiogenic shock as a BTD therapy. The components consist of a small blood pump driven via magnetic coupling by an electric motor. The rotor is hydrodynamically suspended. The blood is pumped from the left atrium through a transseptal venous cannula and directed back through the arterial catheter (Fig. 36.8). The hydrodynamically suspended rotor spins at a rate of 3000–7500 rpm and can deliver a continuous flow of up to 4.0 l/min.

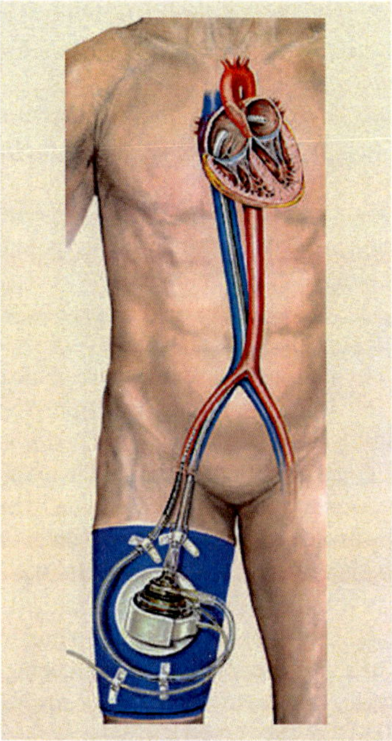

Fig. 36.8 CardiacAssist PTVA (TandemHeart). Cannulation of the extracorporeal pump

Levitronix CentriMag. The Levitronix CentriMag (Levitronix LLC; Waltham, Mass., www.levitronix.net) is a VAD with a magnetically suspended impeller that, like the TandemHeart, is driven by magnetic coupling. It can be connected directly to the cannulation of the heart–lung machine, and with the appropriate connection technology can be applied as an LVAD, RVAD, BVAD, or in a set for extracorporeal membrane oxygenation. At 5500 rpm it can achieve a flow rate of up to 9.0 l/min, depending on the type of cannulation used. The system is also used as a temporary RVAD, in addition to an implanted LVAD, in cases of acute right ventricular failure or for left–right or biventricular support as a BTB (BTD).

Abiomed Impella. Impella catheter pumps support the heart with a flow rate of up to 5.5 l/min. The pumps can be connected directly in a minimally invasive procedure by insertion via a blood vessel or after thoracotomy, e.g., during cardiac surgery. With the Impella mobile console patients can be moved around the hospital or transported to another institution.

The LP2.5 systems, which were designed primarily for application by cardiologists, carry an electrically driven microaxial pump with a pigtail extension at the tip of a catheter and are inserted into the heart through the femoral artery via the aortic valve. They take the blood from there and pump it into the aorta. With a flow rate of 2.5 l/min they stabilize the heart function for up to 2 days. The larger version LP5.0 delivers a flow rate of up to 5.5 l/min for up to 5 days.

The modification LD (left direct) is implanted by cardiac surgery. It supports the left ventricle with a maximum flow rate of 5 l/min. The blood is pumped directly into the aorta through a vascular prosthesis (left ventricle and pump in sequence). The pump weighs only 17 g.

The RD (right direct) works in a similar manner. To support the right ventricle the blood is conveyed from the right atrium into the pulmonary artery (right ventricle and pump in parallel). Both systems can be applied for up to 10 days.

Implantable Long-Term MCSS with Axial Flow

The axial pumps clinically available today are produced by various manufacturers and have similar system configurations. The pump is implanted into the thorax, while the electronic control unit, the power supply for stationary and mobile operation, the battery charger, and the monitor for setting the operating parameters (rotational speed, alarm limits) and surveillance of the function remain outside the body. A percutaneous lead connects the pump to the mobile external components; a Dacron velour cover fixes the lead in the tissue and protects the wound from infection. The pump itself is a small electric motor with hermetically encapsulated coils; the rotor (impeller) is located in a tight blood-conducting cylinder.

In systems with a mechanically suspended impeller, inflow and outflow stators inside the cylinder support the bearings. The rotor contains one or more small permanent magnets, so that it can be set in rotation by the rotating electromagnetic field of the coil. As it is the only moving component of the system, these constructions can provide high functional reliability and long-term durability.

There are, however, significant differences in the individual systems with respect to the following components:

- Anatomic design and the conduits to the heart and great vessels
- Finish of the titanium blood contact surfaces
- Fluid dynamic shape of the rotor
- Rotor suspension
- Design of the percutaneous driveline
- Performance diagrams (rotary speed – flow – differential pressure)
- Measurement and control technology

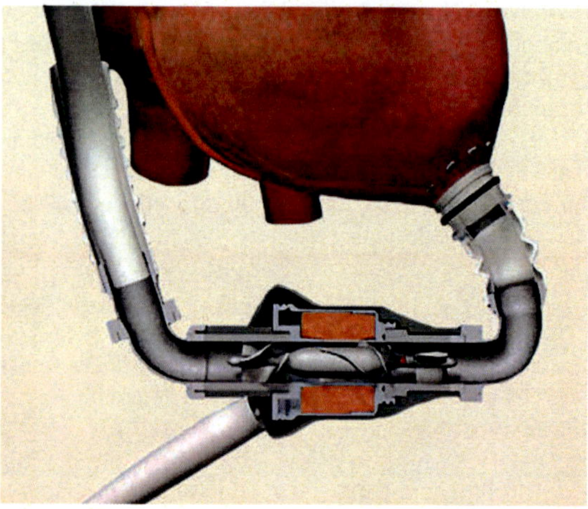

Fig. 36.9 Thoratec HeartMate II LVAS. Schematic of the interposition of the pump between left apex and aortic vascular graft

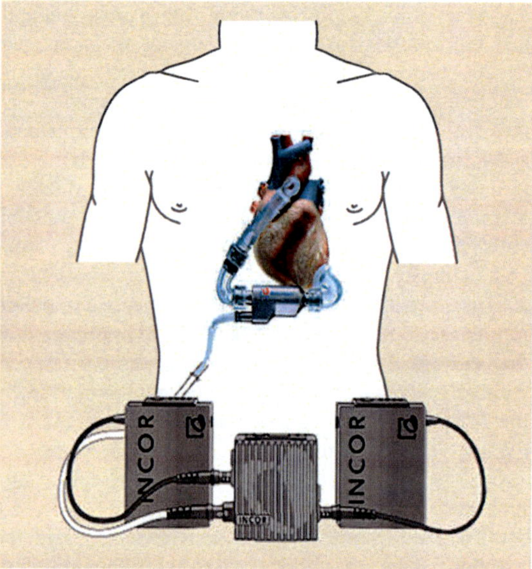

Fig. 36.10 Schematic of an implanted Berlin Heart INCOR system with external components

- Documentation of the operational parameters and their evaluation.

MicroMed Technology HeartAssist 5, Formerly DeBakey VAD. The HeartAssist 5, formerly MicroMed (www.micromedtech.com) DeBakey VAD, is a small, electrically driven implantable turbine, which in patients with terminal heart failure can pump up to 10 l/min at a rate of 12 000 rpm from the left ventricle into the ascending aorta. The rotor is held on both sides by ceramic bearings.

As the only rotary LVAD, blood flow is measured here directly using a Transsonic ultrasound flow probe that ensures reliable system monitoring.

An elaborate documentation and evaluation system allows early detection of imminent malfunctions and provides a subsequent detailed analysis.

Thoratec HeartMate II LVAS. The HeartMate II is similarly constructed, but it uses a different cannulation technology for the left ventricular apex. In contrast to the rigid, curved inflow cannula of the MicroMed system, the HM II apex connector is connected to the pump inflow by a flexible vascular graft. This has proven to be beneficial for pump positioning and for the ingrowth of the connector located in the myocardium (Fig. 36.9).

As this system is a little bigger than the HeartAssist 5, during implantation a pocket must be prepared in the abdominal wall to accommodate the pump housing. The rotor is held by two jewel bearings. With speeds between 8000 and 15 000 rpm the system can achieve flow rates of up to 10 l/min

Berlin Heart INCOR. A special feature of the BerlinHeart INCOR LVAD (Fig. 36.10) is the contact-free suspension of the turbine. A combination of permanent magnets in the stators and in the rotor, and electromagnets and position sensors in the stators make it possible to actively control the position of the rotor. This means that the bearings are free of wear and tear.

Position sensors allow the calculation of many of the operating parameters including the pulsating differential pressures (in front of and behind the rotor) according to the residual cardiac activity, the pulsating momentary flow and the filling state of the heart. As a result, valuable information on the circulatory status of the patient can be gained, which can contribute to therapy management.

Possible malfunctions such as flow obstructions or the formation of thrombotic deposits in the pump can be detected at an early stage and appropriate countermeasures can be initiated.

All blood contact surfaces of the pump are coated with heparin (the CARMEDA method).

Jarvik Heart, Inc., Jarvik 2000 Flow Maker. This LVAD (www.jarvikheart.com) differs significantly from the previously described systems in terms of positioning, technical structure, electronic concept, and performance. The pump was deliberately designed only to support an insufficient but still functional left ventricle. By reducing the performance requirements it was possible to greatly simplify the design.

The body of the pump simultaneously serves as the apical connector; the pump is implanted intracardially. As with the DeBakey pump, the rotor is supported by mechanical bearings. The device is entirely without automatic monitoring or an electronically complex regulation system. The pump is implanted directly into the left ventricular apex through a left lateral thoracotomy and is secured with a sewing cuff. Blood is pumped through a vascular graft into the descending aorta (Fig. 36.11).

A unique feature of this system is the percutaneous connector for the electric driveline power cable. A titanium pedestal, similar to those that have been in use for many years in cochlear implants for infection-free power transmission, is mounted on the skull behind the left ear and connected to the power cable bundle, which runs subcutaneously to the pump. The external components can then be hooked up with the corresponding connector. Patients can shower, and the skin exit site will remain free of infection even with minimal care [36.17].

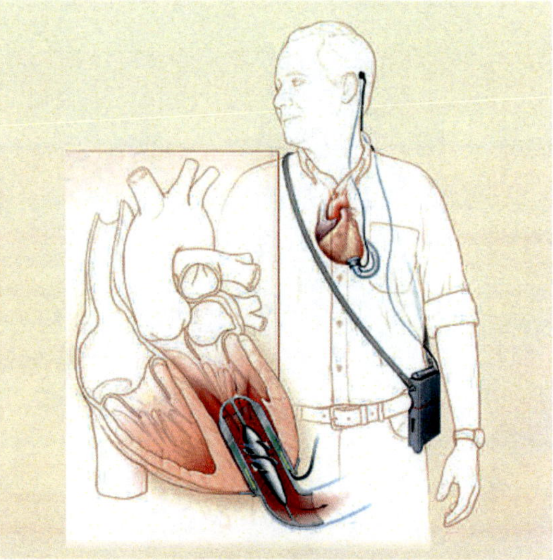

Fig. 36.11 Jarvik Heart 2000 Flow Maker. Schematic of the intracardial implantation; auricular pedestal for the electric driveline

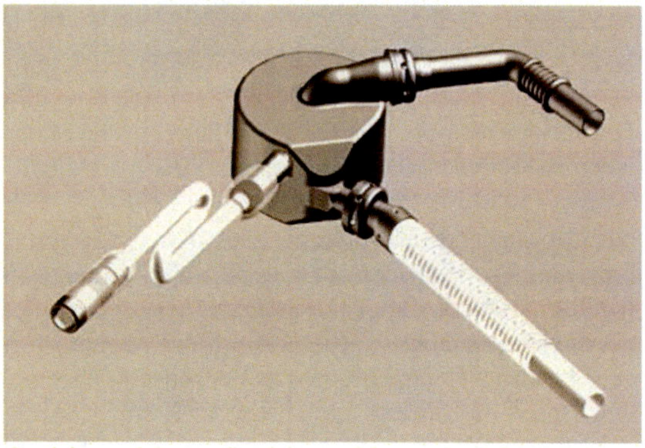

Fig. 36.12 Terumo Heart DuraHeart LVAS blood pump with rigid apical cannula

Implantable Long-Term MCSS with Radial Flow

TerumoHeart, Inc., DuraHeart Left Ventricular Assist System (LVAS). The DuraHeart (www.terumo-europe.com) is an implantable centrifugal pump made from titanium. In three hermetically separated chambers there are an electronic motor, a rotor magnetically coupled to the motor with an integrated permanent magnet, and a regulated electromagnetic system for contact-free magnetic suspension. Should the magnetic suspension fail, the device can fall back on hydrodynamic emergency bearing features, although these only function for a limited period.

Like all rotary pumps, the DuraHeart system propels the blood from a cannula implanted in the left apex through a vascular graft into the aorta. The complex magnetic suspension technology accounts for the large size of the pump (Fig. 36.12). For this reason it must be implanted in the abdominal wall.

HeartWare International, Inc., HVAD System. The HeartWare (www.heartware.com) HVAD is the smallest left ventricular assist device with radial flow currently available that has the capacity to take over the total pumping capacity of the left ventricle under normal physiological pressure. The pump has three main components:

- A front housing with inflow cannula
- A rear housing with a center post
- A wide-bladed impeller.

Each half of the housing has a separately connected stator coil assembly, making the motor redundant. Large permanent motor magnets can be embedded in the wide blades of the impeller. The combination of passive magnetic and hydrodynamic suspension without complex bearing electronics increases functional reliability,

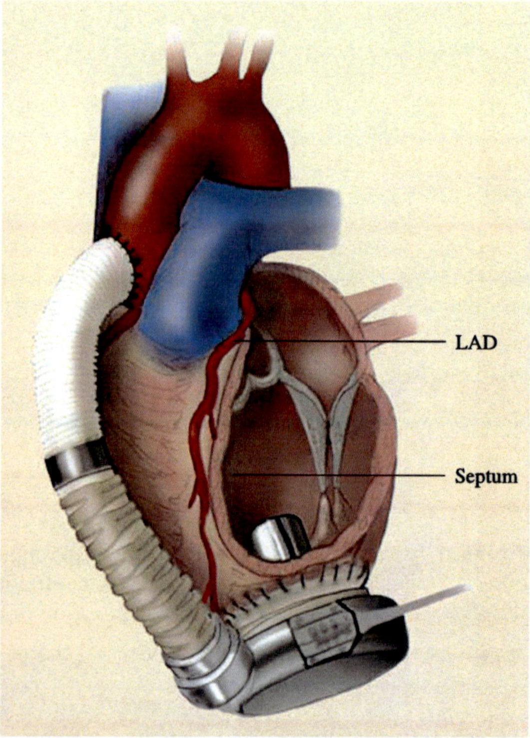

Fig. 36.13 HeartWare HVAD system. Direct apical cannulation and aortic outflow graft to the ascending aorta

but also demands the highest level of manufacturing precision. The blood contact materials are titanium, a titanium nitrite coating of the rotor and center post, and zirconium ceramics.

Flow is calculated on the basis of power consumption, speed of rotation, blood viscosity, and with the help of empirical data of a performance chart. The variations in power consumption and the shape of the curve of the calculated flow displayed on a mobile monitor allow an evaluation of the operating conditions and the filling, or activity, of the supported heart.

During implantation a sewing ring is fixed to the apex. The pump with the inflow connectors is introduced here and thus lies directly at the heart, finding space in the pericardium (Fig. 36.13). The outflow graft can be connected to the ascending or descending aorta.

36.4.3 MCSS for Pediatric Patients

The first clinically available MCSS were designed in terms of performance and anatomic adaptation to fit the requirements of adults. Soon, however, there was a need to develop systems for small patients that could be adapted to body weights of 1–30 kg with flow rates of 1–4 l/min. Connections to the circulatory system had to be miniaturized and new types of valves were required for pulsatile pumps.

Berlin Heart EXCOR, an extracorporeal pneumatic system, was applied as the first pediatric system for BTT with a 50 ml pump in 1990 (DHZB). In 1992, the first implantation of a pediatric system, EXCOR Pediatric with a 10 ml pump took place at the DHZB. Today, a range of pumps (10, 25, 30 and 50 ml) with corresponding miniaturized cannulas are available. This system is also approved for clinical application in the USA. The company MEDOS now also offers a CE-certified pediatric assist system.

In 2004, in the USA, the National Institutes of Health contracted five research groups to develop cardiopulmonary assist devices for pediatric patients with body weights of 2–25 kg (contract HHSN268200448192C (NO1-HV-48192), *Pediatric Circulatory Support*).

The projects were: *PediaFlow VAD*; *PediPump*, an axial turbine for intravasal or extravasal placement; *Pediatric Jarvik 2000 Flowmaker*, a miniaturized version of the Jarvik 2000; and Thoratec's modified pneumatic *pVADs*. A compact heart–lung support system *pCAS* is also included in the program.

A very detailed description of a prescriptive design approach was published recently [36.18]. This explains the multi-disciplinary program of different departments

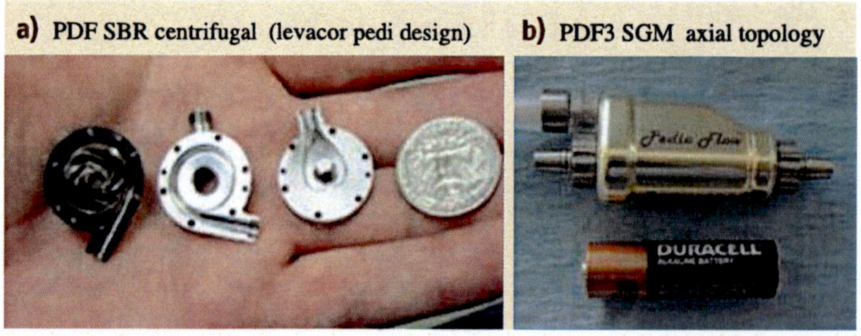

Fig. 36.14a,b PediaFlow. **(a)** First centrifugal design; **(b)** new axial design

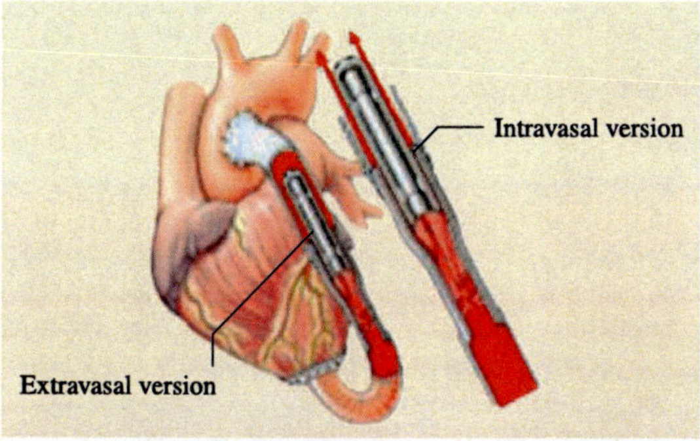

Fig. 36.15 Axial PediPump (extravasal and intravasal version), Cleveland Clinic, Foster–Miller Technologies

at the universities of Pittsburgh in collaboration with the industry to develop the PediaFlow VAD that resulted after 5 years in three different pump topologies with axial or radial flowpath. Judged on the basis of a multicomponent analysis including criteria for anatomic fit, performance, biocompatibility, reliability, and manufacturability the mixed-flow magnetically levitated pump, the PF3, has been chosen as the best-suited design (Fig. 36.14).

The PediPump design is shown in Fig. 36.15.

Despite concentrated efforts, in 2010 none of these five NIH systems had yet reached the status of clinical application.

36.4.4 Summary

Whereas initially, in the 1980s, the market was led by pulsatile pneumatic extracorporeal systems, followed by first-generation implantable electromechanical pulsatile VADs, today it is the second and third-generation nonpulsatile implantable systems for mechanical circulatory support that dominate the market (Table 36.3).

The new technology's initial *teething problems* were soon overcome, and new concepts and further miniaturizations with improved reliability and biocompatibility prevailed. Before long the implantation of rotary pumps was favored. The clinical success of these devices inspired manufacturers to continue developing new concepts. Figure 36.16 gives a scaled overview of the various systems available in 2010.

The number of applications of pulsatile displacement systems quickly fell in favor of continuous-flow systems. DHZB application statistics illustrate this trend. Since 1998, the proportion of pulsatile extracorporeal and intracorporeal systems applied has dropped almost linearly from 100 to 3%, while the total number of applications has doubled. Until recently, pulsatile extracorporeal systems and the temporary TAH were only unrivalled in the case of indications for BVAD. The fact that some rotary pumps with continuous flow can be implanted as BVADs has since been demonstrated by the first applications at the DHZB [36.19] and at the Medical University of Hannover [36.20]. A laboratory study funded by the Friede Springer Foundation made it possible to carry out the first clinical applications of these HVADs as BVADs.

Because the terms *nonpulsatile* or *continuous flow* are often misinterpreted, a brief explanation is given here. The flow of the rotary pumps at a constant rpm (revolutions per minute) is only nonpulsatile when the heart does not affect any changes in the inflow pres-

Table 36.3 Clinical implantable VADs 2010

Implantable ventricular assist devices – clinical 2010			
Manufacturer	**First clinical**	**Volume (cm^3)**	**Weight (g)**
First generation – volume displacement			
WorldHeart Novacor	1984	700	1800
Thoratec HeartMate I XVE	1991	600	1200
Second generation – rotary, mechanical bearings			
MicroMed DeBakey	1998	38	115
Jarvik 2000	2000	25	90
Thoratec HeartMate II	2003	63	180
Third generation – rotary, wearless, no bearings			
BerlinHeart INCOR	2002	60	200
Terumo DuraHeart Active MagLev	2004	150	420
HeartWare HVAD Axial hydrodynamic support plus passive MagLev	2006	45	145

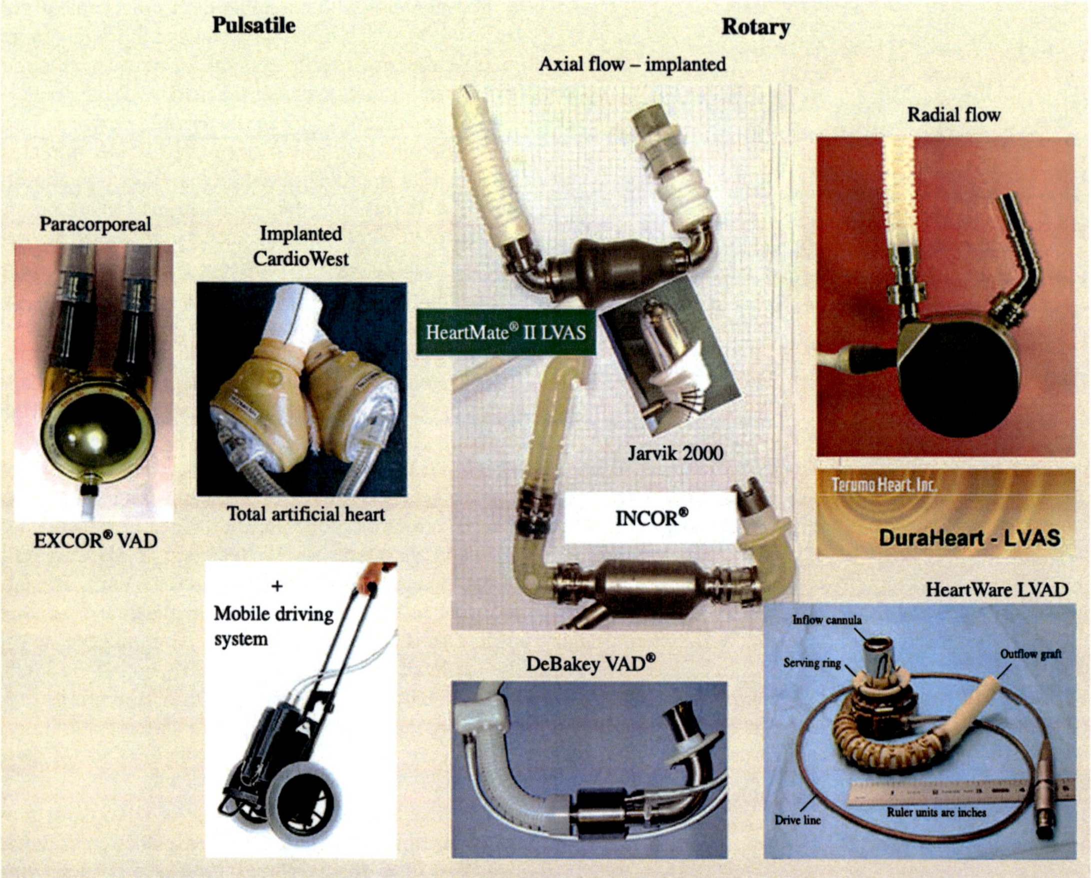

Fig. 36.16 Scaled overview of today's clinical pulsatile and continuous flow VADs

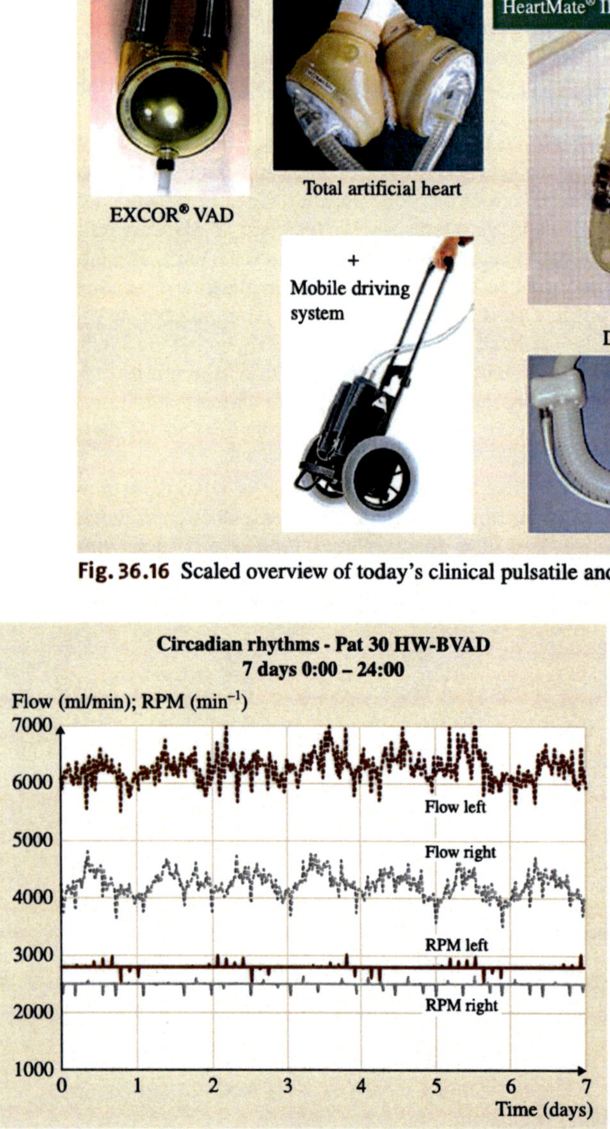

sure to the pump, i. e., in the absence of contractions or when the heart is asystolic or fibrillating. Any change in differential pressure in the pump causes a more or less pulsatile flow at a constant rpm, corresponding to the pressure–flow characteristics of the pump design. In that case, the flow is normally continuous. The diastolic standstill typical to displacement pumps does not occur. These flow pulsations give rise to corresponding pressure amplitudes; the term *pulseless perfusion*, therefore, can only be used to a limited extent, particularly when the physiological effects of nonpulsatile VAD perfusion on organ function are being discussed. Ultimately, we still do not have any reliable long-term observations relating to this question.

Fig. 36.17 Circadian rhythm of the left and right pump flow in a patient with HeartWare BVAD; flow variations over 7 days with constant rotational speed ◄

Another example shows how the continuous flow in a biventricular VAD adjusts approximately to physiological conditions even without external regulation of the pump parameters (Fig. 36.17). Even at constant rpm, signs of some kind of autoregulation corresponding to the recognition of the body's perfusion requirements can be clearly recognized in the circadian biorhythm of the cardiac output of a patient with a BVAD giving *pulseless* support. This is an example of the interaction between the biotechnical hybrid system consisting of an unregulated blood pump and the naturally regulated circulation [36.21].

It seems certain that the future of mechanical circulatory support belongs to rotary pumps. With ever smaller devices implantation will become less and less invasive. New fluid dynamic designs and optimized blood contact surfaces will ensure even greater biocompatibility, and the concomitant drug therapies still necessary today, for instance for controlling coagulation, might in the future be reduced or dispensed with entirely.

36.5 Complications

When describing the complications that follow the application of MCSS it must be taken into account that even today this therapy is only selected for patients in the final stage of an imminently life-threatening illness. The evaluation and classification of early postoperative complications is limited because existing concomitant circumstances in themselves represent complications or involve a high risk of complications.

Often, cause and effect cannot be clearly distinguished. For this reason, BTB or BTD (bridge to bridge; bridge to decision) can be implemented initially to stabilize and evaluate cardiogenic shock patients using less invasive (and less expensive) mechanical circulatory support before other more invasive therapies are applied. Based on our experience at the DHZB and supplemented by published data and data received through personal communication, the most commonly occurring problems in the various post-implantation periods are presented below.

36.5.1 Device Malfunction

Device malfunction can affect all components of the system:

- The blood pump, as the central component of the MCS system
- Connections to the patient's vascular system
- The pneumatic leads or electric cables that supply power to the pump
- The control units to which the pumps are connected
- In pneumatic systems, the driving unit where the pneumatic pressure pulse is generated
- The integrated system monitor/alarm system
- The power supply.

Blood pumps, as the key component of any life-sustaining VAD, are subject to special requirements with regard to fatigue strength and reliability [36.22]. The membranes or blood sacs of pneumatically driven systems are subjected to high mechanical stress by the vast number of alternating load cycles that bend and stretch the material (ca. 40–50 million cycles per year). To improve safety, the membranes are designed as multi-layered membranes. As a result, the risk of rupture with blood escaping into the air chamber or the risk of an air embolism in the blood chamber is extremely low [36.23].

Mechanical valves are another potential source of trouble. Cracks can appear in the valve discs or in the retaining struts of the valve ring; jamming of the valve in the valve ring has also been observed. However, these defects have only occurred occasionally.

A more frequent observation is the formation of blood clots, resulting from either insufficient or unadjusted anticoagulation therapy, for instance in the event of infection or inadequate perfusion of the pump (e.g., hypovolemia). In such cases the pump can be exchanged.

Rotary pumps have only one moving part – the impeller. Depending on the type of suspension, this is either wear-free (magnetic suspension, hydrodynamic suspension) or very low-wear, so that operation times of more than 10 years can be expected.

Extracorporeal systems require comparatively long, percutaneous blood-conveying cannulas to connect the blood pump to the vascular system of the patient. In addition to the potential complications of infection at the skin exit site (Sect. 36.5.5), the patient's movements may, depending on the position of the blood pumps, cause the cannula to bend. If this happens repeatedly at the same site on the cannula, cracks may form, thus increasing the risk of bleeding and air embolisms.

In implantable systems the power electric driveline to the pump is the component subjected to the most mechanical stress. In many patients every movement, even the heartbeat itself, causes a slight movement of the pump, so that the power cable is permanently subjected to bending stress. If tensile stress occurs at the same time, whether accidentally, e.g., the bag containing the controller and the heavy batteries falls, or consistently as a result of a gain in weight, and therefore also circumference of the patient, the conducting wires break, causing the pump to malfunction or fail. Cable defects that necessitate the exchange of the pump have been repeatedly observed in various models.

The control and driving units normally have a redundant design, or at least an operating mode for emergencies. Patients are also equipped with replacement components and receive instruction on exchanging parts.

Even with systems that are durable, easy to operate, equipped with sophisticated alarm systems and, in some cases, emergency operating features or redundancies as well as good usability, patient operating errors can still lead to fatal accidents. Increasingly, VADs are being used to treat older patients who – sometimes for the first time during therapy – experience cognitive or manual function limitations that cannot be compensated, even with extensive and repeated instruction on operating the system.

36.5.2 Hemolysis

The use of mechanical pumps in the circulatory system is always associated with the risk of damage to the corpuscular components of the blood. Contact between the blood and foreign surfaces and the stress placed on the blood cells by the pressure-dependent and flow-dependent shear stresses generated by the pumping all have a damaging effect on the blood. This trauma can be minimized through the choice of materials and surface finishing, and by optimizing fluid dynamics in terms of the extent and exposure time of the mechanical load.

Depending on the particular system, our patients' serum lactate dehydrogenase (LDH) values lie in the 180 U/l (normal) to 400 U/l range. Any sudden increase in this regularly measured parameter indicates VAD-induced hemolysis. This may be caused by an obstruction in the pump inflow or outflow, resulting in a local increase in speed and a corresponding rise in shear stress. Possible aspiration of the pump with a partial or complete occlusion of the inflow causes turbulence around the rotor and a sharp increase in flow-dynamic stress on the blood, resulting in cell trauma.

It should be noted that hemolysis is not a problem generally associated with today's pumps, in contrast to the systems of the pioneering days. This is also proven by the fact that it is irrelevant whether a patient is fitted with one or two well-functioning pumps. However, patients with MCS systems can exhibit hemolysis due to other effects such as infection or medication.

36.5.3 Thromboembolism

Even with carefully monitored anticoagulation therapy that takes into account the varying requirements of the individual systems, blood contact with the biomaterials used (titanium, silicone, polyurethane, etc.) can result in the formation of blood clots. Unfavorable flow conditions in areas with stagnating blood flow and, therefore, poor washout contribute to this complication.

36.5.4 Bleeding

After implantation of an MCSS there is always a risk of secondary bleeding in the postoperative phase, requiring renewed surgical intervention. This bleeding tendency is exacerbated by the profound shock, which in itself induces coagulation disorders. In the later course there is a risk of bleeding due to over-anticoagulation, trauma, and infection.

36.5.5 Infection

Implantations of VADs and TAHs are associated with a high incidence of infection. An early infection rate of 5–20% is reported; in long-term applications this rate increases to up to 90% depending on the system. Circulatory failure and concomitant diseases promote infection.

Today's systems all require external access through the surface of the body to the vascular system (extracorporeal pumps: large-bore blood-conveying cannulas) or to the implants (leads). At the skin exit site, the cannulas or leads are coated with a porous synthetic material to give the injured tissue a chance to grow into it, thereby forming an infection barrier along the wound tunnel. However, generally speaking, the wound area around these access sites is always slightly inflamed and can become infected, even with meticulous care of the wound, thus providing a gateway for bacteria to enter the body. Infections can penetrate right through to the implant where they can no longer be controlled with medication.

Infections that spread as far as the inflow cannula at the tip of the left ventricle are particularly serious. This can cause the cannula to migrate away from the chamber resulting in the formation of a mycotic aneurysm with the risk of rupture and recurrence, even after surgical intervention.

Years ago, inductive *transcutaneous energy transfer systems* (TETs) were developed to supply power and to transmit data and control signals through the undamaged surface of the body. An implanted coil, electromagnetically coupled with an external supply coil conveys the signals through the intact skin to a similarly implanted control and power storage unit, which activates the pump. The long-term capacity of the required implanted power storage unit still presents a problem today; after a limited number of loading and unloading cycles, a further operation is needed to exchange the unit.

36.5.6 Problems at Anastomoses

Impaired wound healing at the anastomosis of the apical cannula is common, but not often described. This problem was examined in over 200 autopsies in DHZB patients treated with 11 different long-term devices.

Even in the absence of infection, granulation tissue can quite often form around the cannula in the ventricle, spreading to such an extent that the cannula becomes obstructed or the tissue embolizes into the pump. This complication often occurs with pumps that have a rigid connection to the inflow cannula. When the pump is connected to the cannula with a flexible conduit there is obviously less opportunity for infection or granulation.

A microporous surface structure allows the cells of the myocardium to form a fixed neointima. In all cases, after only a short time, smooth metallic surfaces initially cause the formation of thrombotic deposits, which later develop into granulation tissue (Fig. 36.18). At the aortic anastomosis, in individual cases, aneurysms have sometimes been seen.

36.5.7 Right Ventricular Insufficiency

More recent publications report right ventricular insufficiency in 25–35% of LVAD patients immediately after implantation. The rate at the DZHB is somewhat lower at 15% [36.11].

If the problem of right ventricular insufficiency cannot be reversed by a sophisticated drug regimen, application of a temporary extracorporeal right ventricular support remains the only option. If there is a disturbance of gas exchange in the lung, the RVAD can be combined with an oxygenator.

36.5.8 Anatomic Obstruction of Flow

Cramped anatomic conditions in the thorax, e.g., due to small patient size or the more bulky blood pumps, can result in compression of the venous, blood-conveying vessels or a dislocation of the pump, resulting in apical cannula inflow obstruction. Therefore, pump function can already be impaired during surgery, and may even come to a complete standstill. Poor positioning can be identified by intraoperative transesophageal echocardiography (TEE) and surgical correction can then be performed.

36.5.9 Arrhythmias

Arrhythmias are common in patients with indications for an assist implantation. Many of our patients had already received an implantable cardioverter/defibrillator (ICD) prior to surgery.

In the event of ventricular arrhythmia, tachycardia or fibrillation, the pumping capacity of the left pump may also fall due to the then reduced right ventricular function and the pulmonary hypertension often present in these patients. This can lead to suction of the apical cannula and even pump failure. This situation may even require the implantation of an ICD. Patients with

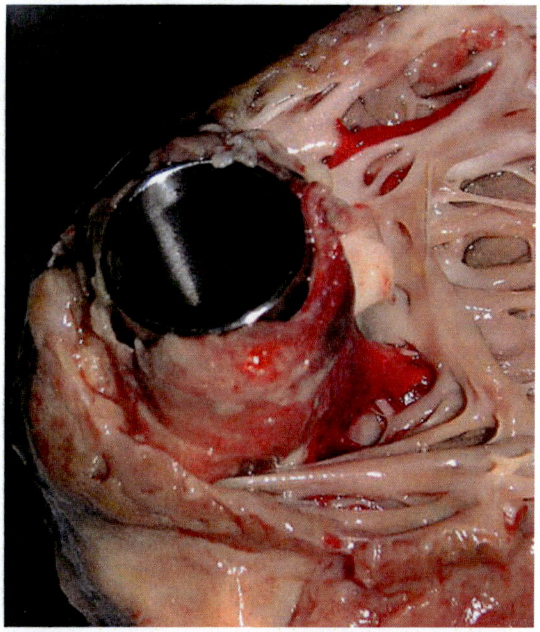

Fig. 36.18 Apical cannula of the DuraHeart in the left ventricle with thrombotic deposits and granulation tissue

BVADs and good flow conditions are not compromised by serious heart rhythm disturbances of this kind; some of our patients with lasting cardiac fibrillation have survived for longer periods without symptoms.

36.6 Technical Follow-Up and Care

Patients who have been allowed to return home and who only visit the clinic as outpatients – in some cases at intervals of several months – require technical aftercare and monitoring. First and foremost, a 24-h/7-day-a-week hotline is needed to deal with acute problems.

The first attempts at establishing remote monitoring of the system function have been made. Similar to telemedicine centers, a dialog may be set up whereby patient data can be accessed and analyzed via cordless telephone networks and the Internet. Remote monitoring requires internal logging of the system parameter data; this enables the diagnosis and, if necessary, the temporal reconstruction of events when alarm signals are given.

The Thoratec HeartMate II records certain conditions, each with a corresponding set of parameters containing all real-time values. However, its memory storage is limited to 120 data sets.

With the DeBakey system two memory areas are constantly updated with high-resolution data. If an error occurs, the respective data are recorded in such a way that the beginning of an event is stored in the first area. The second area is overwritten with any newly occurring events so that in the clinic the first and last occurrences can be examined in detail.

With the BerlinHeart INCOR system an unlimited number of events can be recorded as long as the patient is connected to the home equipment monitor. A comparatively small amount of event-triggered data, only 400 parameter sets, can be stored in the control unit itself.

The HeartWare system stores data sets in 15 min intervals, covering a maximum of 30 days.

In the Terumo DuraHeart control unit detailed short-term storage and a long-term time protocol are combined over a 30-day period. In the event of a malfunction, the changes of specific signals are recorded.

Experience shows that optimal monitoring depends on detailed data storage that allows an analysis of past events.

36.7 Psychosomatic Syndromes and Quality of Life During Treatment with MCS

36.7.1 Psychosomatic Syndromes During MCS

The implantation of an MCSS means an intensive psychological stressor for the patients and induces emotional distress and various concerns. Patients are confronted with the threat of sudden death, very often experience a high degree of anxiety and depression before and after implantation of the assist device, and suffer in a psychological sense from a kind of self-infarction. The psychiatric–psychological reactions after surgery are very complex in nature due to the simultaneous impact of organic and emotional factors.

In the initial postoperative period organic factors are of specific importance:

- Psychotropic side effects of the extracorporeal circulation, cerebromicroembolic events, metabolic disorders, pre-existing neuropsychological dysfunctions, and cerebrotoxic medication effects
- Besides these, initially acute psychological shock reactions resembling a *possum reflex* or states of pronounced psychovegetative activation and an excess of diffuse bodily sensations are observed.

In the following days or weeks processes of adaptation to the assist device develop and organic psychotropic factors gradually lose their significance in favor of emotional reactions and defence or coping processes. Patients tend to restore their psychic balance and to adapt to the new reality.

In psychological care multifaceted treatment modules are required to support patients adequately. It is necessary to attend to (often hidden) mental confusion, delirium, and short psychotic disorders, and to treat these states with basic supportive interventions and/or neuroleptic medication or tranquilizers. Over time there is an increase in affective disorders such as depression, anxiety, or mixed disorders, as well as adjustment disorders. *Neurocognitive dysfunctions* are very often underdiagnosed: they comprise impairment of the ability to maintain attention and concentration, deficits in learning and short-term memory, and mental flexibility, and are often only to be detected by neurocognitive testing.

The most prominent reactions are depressive mood, hopelessness, feelings of inadequacy, anxiety, and sleep disorders with often rapid cycling from day to day. These affective and bodily sensations are often induced by assist-device-specific side effects such as noise, vibrations, altered cardiac rhythms, etc. Furthermore patients suffer from isolation from their family or partners and are confronted with the demands of the clinical setting. Deprivation effects emerge but on the other hand patients also feel themselves to be safe and secure in the department and realize that the threat to their lives has been overcome and that they can now look to the future.

In accordance with these conflicting psychological feelings patients are ambivalent and their reactions range from latent suicidal ideas or impulses to confidence, intensive future planning, or even euphoric states.

36.7.2 Profound Psychological Processes

All these psychosomatic reactions can be understood as transient regressive developmental processes.

The assist device implantation represents a breach of the body's boundaries. The internal world of the organs and abdominal cavities (anchored in the *body image*) is forced open, the body is invaded from the outer world, the motor of life has lost its vitality and has been violated. A new *replacement heart* is implanted, and this high-tech apparatus cannot be controlled by myself, it is controlled from outside the body and is under the care of initially unfamiliar people. In the intimate heart cavity – is my heart still there? What importance does it have now? Patients are forced to form a union between an internal and partly external life center, a union between living and dead matter.

This induces the regressive processes mentioned above and reactualizes very early developmental periods where rhythms, internal bodily sensations, and basic perceptual and affective patterns create the basis for the more complex and more mature self structuring processes.

36.7.3 Coping Skills

During treatment with the assist device *depressive withdrawal* is the predominant coping strategy during the initial period, as well as *denial and wishful thinking*. After some months on MCS patients rely upon *active problem-solving* i. e., they tend to reorganize their daily lives and overcome difficulties. They are engaged in fighting against their illness, encourage themselves and look for success and self-confirmation. They intensively seek *social support* from their family and friends and emphasize as the most important single strategy: *placing trust in the doctors*. In contrast, nearly half the patients hesitate to follow the doctors' advice to some degree.

36.7.4 Quality of Life

Quality of life, mostly measured with the SF-36 Questionnaire, is reduced during life with MCS and the mean scores in most of the studies are about 25% below the scores of chronically ill patients; there are no relevant differences comparing the physical and the psychological domains.

The patients have generally positive overall perceptions of the MCS but express several specifics pertaining to infections, sleep disturbances, pain related to the driveline, device noise, and fears of MCSS malfunction. An important concern relates to stroke and the correlated long-term impairment. These specific concerns are responsible for the reduced quality of life.

36.7.5 Therapeutic Consequences

It seems very important to offer patients continuous psychological treatment:

- To treat the transient psychosomatic or psychiatric disorders according to the current guidelines
- To analyze the sources for regressive behavior, feelings of guilt and suicidal ideas
- To support the patients in developing individual plans for active, self-contained behavior and for the caregivers to sustain support.

36.7.6 Summary

There is a high degree of psychosomatic or psychological disorders during the initial weeks after MCS implantation. Once the first emotional threat is overcome, and due to the reduction of external organic factors, we observe a rapid remission of psychological syndromes and an increase in active problem-solving behavior instead of regressive withdrawal. Nevertheless, neurocognitive impairment remains an important issue for longer time periods.

More detailed information on the psychiatric-psychosomatic disorders and coping mechanisms associated with mechanical circulatory support is available in [36.24].

36.8 Overview and Outlook

Artificial cardiac assist devices for longer-term application have been in clinical use since the 1980s and were initially employed in life-threatening situations as a bridge to subsequent heart transplantation, i. e., for a limited, foreseeable period. However, then came the development of systems that could function for several years and, for some patients who were not able to receive a transplant, the permanent use of these pumps became inevitable.

Technical improvements and higher numbers of temporary artificial heart implantations as a BTT do not inherently contribute to solving the donorship problem, nor do they contribute to improving the care of patients who today can only be helped to achieve better life expectancy with a heart transplant. In Germany, the number of organ donors is falling steadily. In 2009 over 1000 patients were listed for a heart transplant, but the number of donor organs was only 350. The increased use of bridging systems (BTT) merely leads to higher numbers of patients on the waiting list and lower chances of receiving a transplant.

Nevertheless, the concept of bridging to subsequent transplantation can be viewed as a historic achievement, because for the first time it became possible to use artificial cardiac assist devices in a meaningful and acceptable way. The only justifiable solution for the future is the provision of technically safe, long-term implantable systems that offer significant improvements in life expectancy without greatly limiting quality of life and that can, therefore, be seen as an alternative to transplantation [36.25].

Syncardia's current development of a scaled down version of the well-established temporary artificial heart CardioWest is only an interim solution that will enable its implantation in smaller patients. Due to the limited life of the membranes and the external pneumatic driving unit it will remain a BTT system.

Given the recognized need for a long-term implantable biventricular mechanical cardiac support or heart replacement, research groups around the world have for years been trying to refine TAH systems for clinical use – with a range of different concepts and often at enormous financial expense covered by state funding and industrial support – but so far without further clinical success.

ABIOCOR, the first fully implantable pulsatile electrohydraulic artificial heart system developed by the American company ABIOMED, Inc., was approved by the FDA for limited commercial clinical application, but considerable anatomic and also technical problems and the lack of clinical success have raised doubts about its continued development.

In Europe, two new projects on pulsatile implantable artificial hearts have recently been introduced; both are based on the initial developments of the 1980s. Just a short time ago, the French heart surgeon Alain Carpentier announced that, together with researchers from the aerospace company EADS, he had created a fully implantable artificial heart (Fig. 36.19). It is reported to have functioned well in animal trials with calves. The French artificial heart weighs 1 kg and is made of biolyzed tissue, titanium, and plastic. A special blood contact surface derived from porcine cartilage is designed to prevent the formation of blood clots. Details have not yet been released, but the device is certainly not small, as could be seen from a published video. According to the company CARMAT, which was established to produce the system, the de-

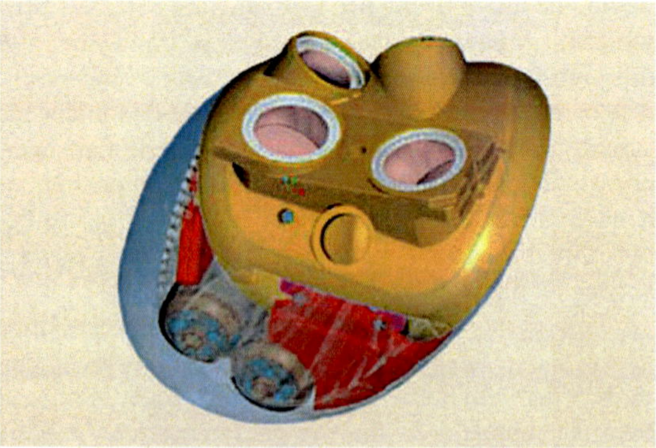

Fig. 36.19 CARMAT artificial heart, designed by A. Carpentier

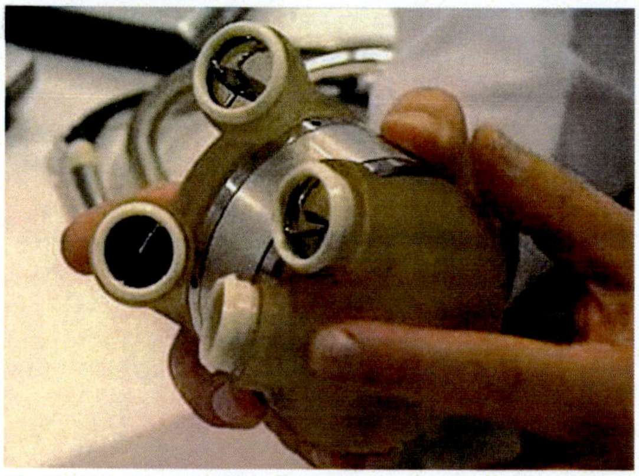

Fig. 36.20 The ReinHeart, under development by RWTH Aachen University and R. Körfer

vice should be ready for clinical application in 4–6 years.

This year, together with the RWTH University in Aachen, Reiner Körfer presented a concept for a permanent artificial heart, ReinHeart. This is the continuation of the MiniACcor development project, which has been under way there for several years. ReinHeart is larger than the normal heart. It is expected to fit into the thorax of 80% of adults and should weigh no more than 800 g. Comparable to the native heart, the device consists of two chambers and four valves, as well as an electromagnetic linear drive (Fig. 36.20). These elements and further components are designed to be implanted into the body. It will take several more years of development before this system can be applied in the clinical setting.

It is now time to consider the realization of total heart replacement systems based on two rotary blood pumps of existing permanent LVAD systems and thus using the clinical experience gained from thousands of patients with these devices [36.26]. The long-term suitability of different second or third generation rotary blood pumps with contact-free magnetic or magnetic–hydrodynamic suspension of the rotor and accordingly high fatigue resistance and reliability has been demonstrated in clinical applications over periods of up to several years with acceptable complication rates.

With the support of the *Friede Springer Foundation*, investigations into the suitability of clinically proven LVADs for biventricular cardiac support or heart replacement were initiated in 2009 at the DHZB with the TAHROT study. Now, in the first stage of this study, rotary pumps are being implanted as BVADs in patients with terminal biventricular heart failure at the DHZB (Figs. 36.21, 36.22).

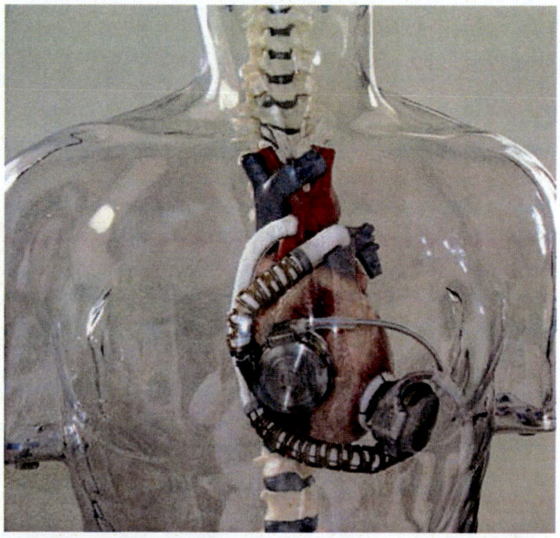

Fig. 36.21 Anatomical positioning of two HeartWare HVAD blood pumps

The working area of the pump, which functions as an RVAD and was originally conceived as an LVAD, is transformed by banding the pulmonary outflow graft; this increased resistance moves the flow–pressure area (afterload of the RVAD) into that of the natural systemic circulation. The pump itself does not need to be redesigned (Fig. 36.23).

The development of even smaller rotary LVADs and possibly also BVADs shows promising progress. Three modified miniaturized systems from the company HeartWare are already being tested in animal experiments (Fig. 36.24). These new axial pumps are said to have a similar capacity but less than 50% of the weight of an HVAD, with a correspondingly small size. Implantation will be greatly simplified; a combination for total heart replacement seems very promising.

Thoratec, the world leader in the market for VADs, has new miniaturized axial (HM X) and centrifugal (HM III) rotary blood pumps in its development program (Fig. 36.25). The aim is to achieve long-term application with only minimal or no anticoagulation therapy.

To improve patient quality of life, a transcutaneous energy transmission system (TET) was developed to supply fully implantable blood pumps (Fig. 36.26). The TET allows two-way data and energy transfer through the intact skin, thereby ruling out the risk of infection at a skin exit site wound.

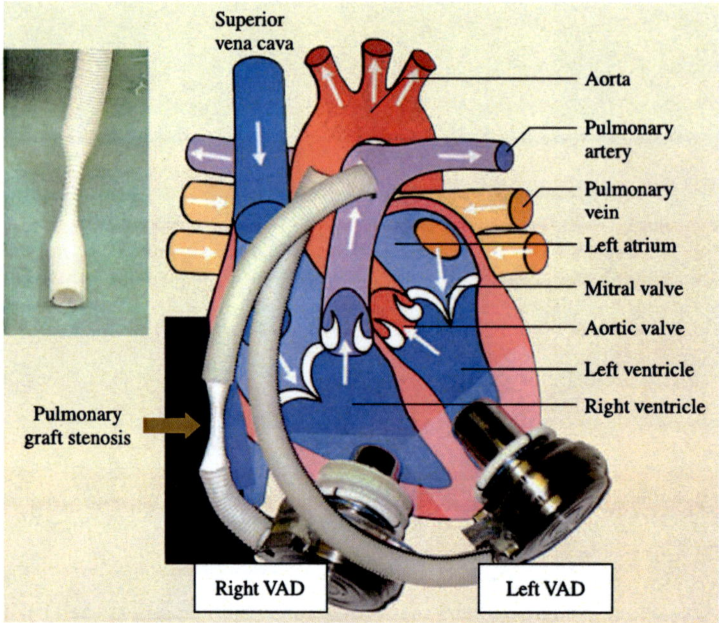

Fig. 36.22 Schematic of the left and right ventricular cannulation and the aortic and banded pulmonary artery conduits

On 10 December 2009, *Euromacs* (EUROMACS VAD registry; European Register for Mechanical Circulatory Support e.V.) was founded in Berlin to allow comprehensive pooling and documentation of all data on applications of MCSS in Europe. The evaluation of experience from a larger patient pool is expected to result in a more rapid transfer of findings, both to hospitals and to system manufacturers. This collection of data will provide a valuable research tool.

The current state of technology and medical expertise allows the application of mechanical circulatory

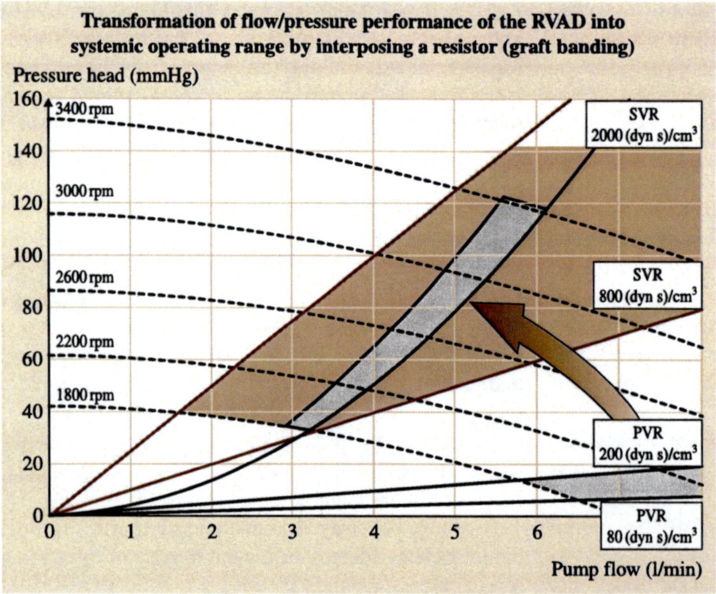

Fig. 36.23 Performance diagram of the HeartWare HVAD: transformation of the working area of the RVAD into the field of systemic resistance by banding the pulmonary artery graft

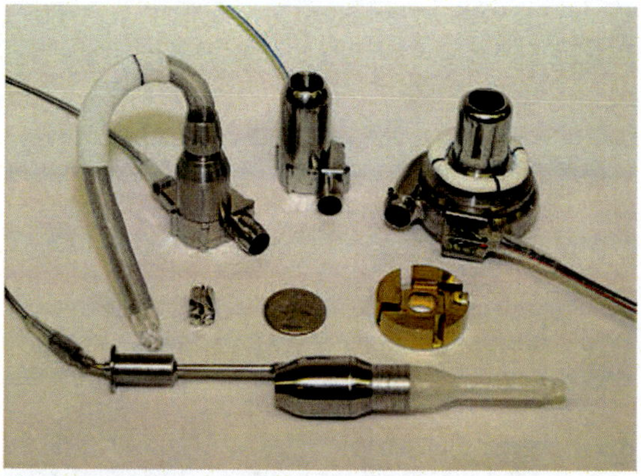

Fig. 36.24 New projects of HeartWare International. From *left to right* (*top*): MVAD for right lateral thoracotomy; MVAD axial flow pump; today's HVAD. *Bottom*: intracardial trans AV-MVAD

Fig. 36.25 New projects of Thoratec Corporation: miniaturized axial pump HeartMate X in comparison with the HeartMate II

support systems for a number of indications, extending right through to the permanent replacement of the heart function with acceptable risks and quality of life.

The provision of patient aftercare by specially trained technical and medical staff is of growing importance, on the one hand so that complications can be recognized and averted as they develop, and on the other hand, to give patients a sense of security in their day-to-day living with the device. Telemedical monitoring of external VAD patients will play a helpful role here. Given that MCSS can now be applied as an alternative to heart transplantation, the numbers of patients requiring outpatient care from the implant institutions are set to increase rapidly. Additional capacities will need to be created to satisfy the requirements for aftercare.

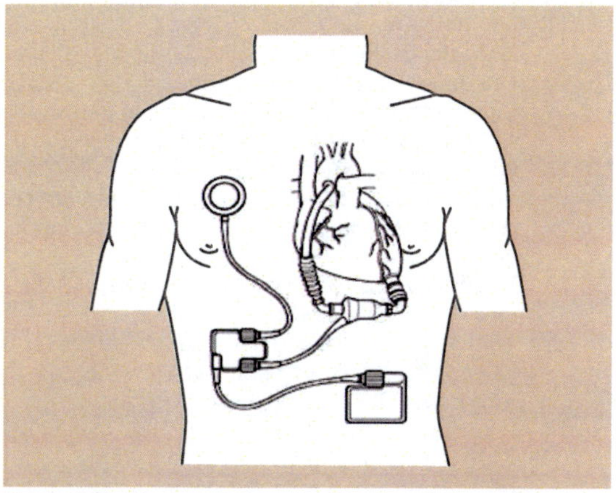

Fig. 36.26 New projects of Thoratec Corporation: FILVAS, a fully implantable LVAD with transcutaneous energy transmission (TET)

References

36.1 C.N. Barnard: The operation. A human cardiac transplant: An interim report of a successful operation performed at Groote Schuur Hospital, Cape Town, S. Afr. Med. J. **41**, 1271–1274 (1967)

36.2 D.A. Cooley, D. Liotta, G.L. Hallman, R.D. Bloodwell, R.D. Leachman, J.D. Milam: Orthotopic cardiac prosthesis for two-staged cardiac replacement, Am. J. Cardiol. **24**, 723–730 (1969)

36.3 L.D. Joyce, W.C. DeVries, W.L. Hastings, D.B. Olsen, R.K. Jarvik, W.J. Kolff: Response of the human body to the first permanent implant of the Jarvik-7 total artificial heart, Trans. Am. Soc. Artif. Intern. Organ. **29**, 81–87 (1983)

36.4 E.S. Bücherl, E. Hennig: 20 Jahre Kunstherzforschung in Berlin. In: *Chirurgie im Wandel der Zeit 1945–1983*, ed. by H.W. Schreiber, G. Carsten (Springer, Berlin Heidelberg 1983) pp. 86–108, in German

36.5 R. Hetzer, S. Bücherl: Herztransplantation und Kunstherzentwicklung in Berlin. In: *Koronare Herzerkrankung und dilatative Kardiomyopathie: Diagnostik und Therapie; Ergebnisse eines Symposiums am Deutschen Herzzentrum Berlin*, ed. by

E. Fleck (MMV Medizin, München 1988) pp. 143–144, in German

36.6 R. Hetzer, E. Hennig, M. Loebe (Eds.): *Mechanical Circulatory Support in Children* (Steinkopff, Darmstadt 1997) pp. 1–61

36.7 M.E. DeBakey: A miniature implantable axial flow ventricular assist device, Ann. Thorac. Surg. **68**, 637–640 (1999)

36.8 R. Körfer, L. Arusoglu: Herzunterstützungssysteme und Kunstherzimplantation. In: *Herzchirurgie*, ed. by G. Ziemer, A. Haverich (Springer, Berlin Heidelberg 2010) pp. 847–872, 3rd edn., in German

36.9 J. Müller, R. Hetzer: Left ventricular recovery during left ventricular assist device support. In: *Cardiac Assist Devices*, ed. by D.J. Goldstein, M.C. Oz (Futura, Armonk 2000) pp. 121–135

36.10 J. Müller, G. Wallukat, Y.G. Weng, M. Dandel, S. Spiegelsberger, S. Semrau, et al.: Weaning from mechanical cardiac support in patients with idiopathic dilated cardiomyopathy, Circulation **96**(2), 542–549 (1997)

36.11 E.V. Potapov, A. Loforte, Y. Weng, M. Jurmann, M. Pasic, T. Drews, M. Loebe, E. Hennig, T. Krabatsch, A. Koster, H.B. Lehmkuhl, R. Hetzer: Experience with over 1000 implanted ventricular assist devices, J. Card. Surg. **23**(3), 185–194 (2008)

36.12 R.K. Wampler, J.C. Moise, O.H. Frazier, D.B. Olsen: In vivo evaluation of a peripheral vascular access axial flow blood pump, Am. Soc. Artif. Intern. Organ. Transplant. **34**, 450–454 (1988)

36.13 J. Courtney, B.S. Gemmato, D. Matthew, B.S. Forrester, J. Timothy, B.S. Myers, et al.: Thirty-five years of mechanical circulatory support at the Texas Heart Institute an updated overview, Tex. Heart Inst. J. **32**, 168–177 (2005)

36.14 O.H. Frazier, E.A. Rose, M.C. Oz, W. Dembitsky, P. McCarthy, B. Radovancevic: Multicenter clinical evaluation of the HeartMate vented electric left ventricular assist system in patients awaiting heart transplantation, J. Thorac. Cardiovasc. Surg. **122**, 1186–1195 (2001)

36.15 P.M. Portner, P.G.M. Jansen, P.M. Oyer, D.R. Wheeldon, N. Ramasamy: Improved outcomes with an implantable left ventricular assist system: A multicenter study, Ann. Thorac. Surg. **71**, 205–209 (2001)

36.16 R.D. Dowling, L.A. Gray Jr, S.W. Etoch, H. Laks, D. Marelli, L. Samuels, J. Entwistle, G. Couper, G.J. Vlahakes, O.H. Frazier: The AbioCor implantable replacement heart, Ann. Thorac. Surg. **75**(6 Supp.), 93–99 (2003)

36.17 M.P. Siegenthaler, O.H. Frazier, F. Beyersdorf, J. Martin, H. Laks, J. Elefteriades, A. Khaghani, U. Kjellman, B. Koul, J. Pepper, R. Jarvik, S. Westaby: Mechanical reliability of the Jarvik 2000 heart, Ann. Thorac. Surg. **81**(5), 1752–1758 (2006), (discussion pp. 1758–1759)

36.18 J.F. Antaki, M.R. Ricci, J.E. Verkaik, S.T. Snyder, T.M. Maul, J. Kim, D.B. Paden, M.V. Kameneva, B.E. Paden, P.D. Wearden, H.S. Borovetz: PediaFlow Maglev ventricular assist device: A prescriptive design approach, Cardiovasc. Eng. **1**(1), 104–121 (2010)

36.19 R. Hetzer, T. Krabatsch, A. Stepanenko, E. Hennig, E.V. Potapov: Long-term biventricular support with the Heartware implantable continuous flow pump, J. Heart Lung Transplant. **29**, 822–824 (2010)

36.20 M. Strueber, A.L. Meyer, D. Malehsa, A. Haverich: Successful use of the HeartWare HVAD rotary blood pump for biventricular support, Thorac. Cardiovasc. Surg. **140**, 936–937 (2010)

36.21 C.S. Hayward, S. Hu, A.M. Keogh, P.S. Macdonald, E. Kotlyar, D. Robson, M. Harkess, P. Jansz, P. Spratt: Interaction between activity and LVAD function in outpatients. Heart Failure and Transplant Unit, St. Vincent's Hospital, Sydney, Australia, Int. Soc. Rotary Blood Pumps, Annu. Meet. (Berlin 2010), priv. commun.

36.22 E.J. Birks, P.D. Tansley, M.H. Yacoub, C.T. Bowles, M. Hipkin, J. Hardy, et al.: Incidence and clinical management of lift-threatening left ventricular assist device failure, J. Heart Lung Transplant. **23**, 964–969 (2004)

36.23 J. Martin, O. Friesewinkel, C. Benk, S. Sorg, S. Schulz, F. Beyersdorf: Improved durability of the HeartMate XVE left ventricular assist device provides safe mechanical support up to 1 year but is associated with high risk of device failure in the second year, J. Heart Lung Transplant. **25**, 384–390 (2006)

36.24 W. Albert, A. Bittner, R. Hetzer: Psychiatric-psychosomatic disorders and coping mechanisms during mechanical circulatory support with the biventricular Berlin Heart assist device. In: *Quality of Life and Psychosomatics*, ed. by W. Albert, A. Bittner, R. Hetzer (Steinkopff, Darmstadt 2000) pp. 29–43

36.25 M.S. Slaughter: Long-term continuous flow left ventricular assist device support and end-organ function: Prospects for destination therapy, J. Card. Surg. **25**, 490–494 (2010)

36.26 O.H. Frazier, E. Tuzun, W.E. Cohn, J.L. Conger, K.A. Kadipasaoglu: Total heart replacement using dual intracorporeal continuous-flow pumps in a chronic bovine model: A feasibility study, Am. Soc. Artif. Intern. Organ. J. **52**(2), 145 (2006)

37. Neural Interfaces for Implanted Stimulators

Xiao Liu, Andreas Demosthenous, Nick Donaldson

For a paraplegic, direct control of certain body parts is destroyed by the injury, leading to the loss of sensation and muscle control. However, most of the nerves and muscles below the injury site may still be functional and can be artificially stimulated by an appropriate neural stimulator to restore the lost functions or to facilitate rehabilitation. Thanks to CMOS technologies, the vast majority of stimulation electronics can be integrated onto a small silicon chip, able to drive many electrodes and generate complex stimulation patterns. However, it is of paramount importance that any advance in circuit functionality should not compromise electrical safety. This chapter looks into several safety issues, especially on the parts near the stimulation interface, i. e., stimulating electrodes, implantable cables, and stimulator output stages. Various methods and solutions for building reliable and robust stimulation interfaces are reviewed and compared in terms of system complexity, physical size, level of safety, and circuit performance.

37.1 **Stimulating Electrodes** 751
37.1.1 Electrode–Electrolyte Interface and Circuit Model 751
37.1.2 Safety Issues 753

37.2 **Implantable Cable Management** 753
37.2.1 Current Limitations.................... 753
37.2.2 Solutions.................................. 755

37.3 **Design of Stimulator Output Stage** 756
37.3.1 Stimulation Pulses...................... 757
37.3.2 Current Generator Circuits 757
37.3.3 Safe Output Stage Circuits........... 759

37.4 **Conclusions** 763

References 764

There is a large population worldwide suffering from neural disabilities and dysfunctions due to injuries and diseases. For example, *spinal cord injury* (SCI), which disrupts the normal pathways between the brain and the muscles or sensory cells, is one of the most disastrous medical conditions. The worldwide annual estimate of SCI is 2.5–57.8 new cases per million population [37.1]. SCI is not only physically and psychologically devastating for individual patients, but also puts a large economic burden on the society's health care system. Thus there is constant need for new technologies for treatment and to facilitate rehabilitation.

The simple fact that electrical changes, described either as ionic currents or neural voltage signals, are the cause of neural communications in the body makes it possible to use electrical stimulation to mimic the normal neural activities of muscles. By placing stimulating electrodes within excitable neural tissue and passing electrical stimuli through these electrodes, it is possible to generate excitation or inhibition of neural targets. If the electrical stimulation targets nerves that innervate peripheral muscles, in order to replace or restore lost functions in neurologically impaired patients, it is referred as *functional electrical stimulation* (commonly abbreviated as FES).

In FES, the excitation of nerves and muscles is caused by artificial electrical stimuli rather than the voluntary neural driving signals. In the last few decades, FES has been approved as a useful method to restore neural functions in many applications. These include the restoration of hand grasp [37.2], standing and

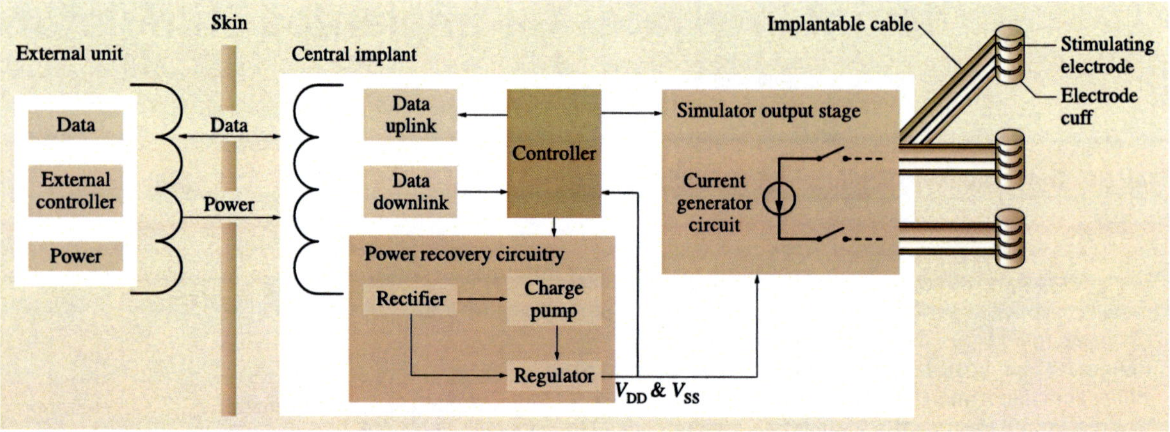

Fig. 37.1 A simplified block diagram of an implantable stimulation system

walking [37.3], bladder and bowel control [37.4], and breathing [37.5].

A device that supplies the desired electrical stimuli for an FES application is known as a *neural stimulator*. During the last century, the design of neural stimulators has progressed enormously. These are categorized as surface stimulators (external stimulation unit with stimulating electrodes placed on the surface of the skin), semi-implantable stimulators (external stimulation unit with implanted stimulating electrodes), and fully implantable stimulators (implanted stimulation unit with implanted stimulating electrodes), depending on the location of the stimulating electrodes and the stimulation unit. Compared with surface stimulators, implantable stimulators offer several unique advantages:

1. Compared to surface stimulation, an implantable stimulator greatly simplifies the prestimulation and poststimulation procedure by avoiding the repeated placement and removal of stimulating electrodes. Thus implanted stimulators are more convenient for patients.
2. The implantation allows near-target placement of stimulating electrodes, providing better selectivity.
3. A full implantation avoids transcutaneous cables, which can introduce infection.

Implantable stimulators are currently widely used in clinical trials and have resulted in several successfully commercialized products for chronic applications. By 2004 there were more than 1000 different types of stimulation systems implanted into more than 200 000 patients [37.6].

The development of implantable stimulators has greatly benefitted from the advances in integrated circuit and system-on-a-chip (SOC) technologies, which offer a high-density integration of electrical components with small power consumption. The size of implantable stimulators has shrunk significantly over the years, making the devices easier to implant.

Figure 37.1 shows a simplified block diagram of an implantable stimulation system for FES applications. The complete stimulation system is composed of an external unit and an implanted device, outside and inside the human body, respectively. Figure 37.1 also shows the core modules in the external and implanted devices.

Due to the high power requirement (> 1 mW) of FES applications, the limited capacity of a small internal battery may not be sufficient for stimulations on a regular basis (daily). Thus, implanted devices are usually inductively powered at radio frequency during stimulation, or the internal battery is recharged during idle time (nonstimulation period, e.g., at night when the patient is sleeping) by an inductive link. The inductive power link is essentially a transformer made of two magnetically coupled coils. One of the coils is embedded in the implantable device while the external coil aligns with the internal coil for efficient power transmission. The received alternating voltage across the tuned coil is further rectified to DC voltage by a rectifier in the power recovery unit. Then the rectifier's output is further regulated to one or more stable DC voltage levels or pumped to an even higher voltage level before regulation if a supply voltage higher than the rectifier's output is required. The recovered DC supplies provide the power required for the operation of other circuit modules in the implanted device, such as data uplink/downlink, digital control unit, and stimulator output stage.

The data downlink in the implanted device allows the implanted device to be remotely controlled by an external unit. The controller interprets the stimulation commands from the received data and guides the stimulator output stage to generate appropriate stimulation pulses. The stimulation commands usually accommodate a wide range of different stimulation profiles (e.g., the amplitude and duration of stimulus current) in order to meet the different requirements of each individual patient with specific SCI conditions.

Data downlink is only a one-way communication, conveying information from the external unit to the implanted device. It is also desirable to feed internal sensed information back to the external unit via a data uplink. Typical data transmitted to the external unit include humidity information inside the implant package, electrode impedance, and status of the power supply. This information helps to study the efficacy of stimulation and monitor the status of the implant. The sensed information is analyzed by an external device that may modify the stimulation commands on the data downlink if necessary. Thus the duplex communication mode allows closed-loop control of an implanted device.

Another important module in implanted devices is the stimulator output stage, which is functionally the closest unit to the biological interface (the stimulating electrodes and the tissue). It is responsible for generation of electrical stimulation pulses according to the commands received from the control unit and delivers the stimulus to the tissue impedance.

In this chapter, we intend to describe some of the common principles in the design of integrated neural stimulators, emphasizing the safe stimulation interface between neural tissue and implantable electronics. The topic covers stimulating electrodes, implantable cable management, and stimulator output stage.

37.1 Stimulating Electrodes

37.1.1 Electrode–Electrolyte Interface and Circuit Model

An electrode is an interface between a biological environment and a stimulator. During stimulation, electrical currents in metallic conductors, such as the implantable electronic circuits, are carried by free electrons while electrical currents in biological tissue and body fluids are carried by the motion of ions. The electrode, which itself is a metallic conductor, carries current by free electrons. At the electrode–tissue interface there is a transduction of charge carriers in order to continue the relay of current delivery.

To perform neural stimulation, at least two electrodes (known as a *dipole*), a positive anode and a negative cathode, are required. However, three or more electrodes are commonly used in order to confine the electrical field in a more localized volume of tissue for better selectivity [37.7]. Figure 37.2a shows the electrical circuit model for an electrode–electrolyte

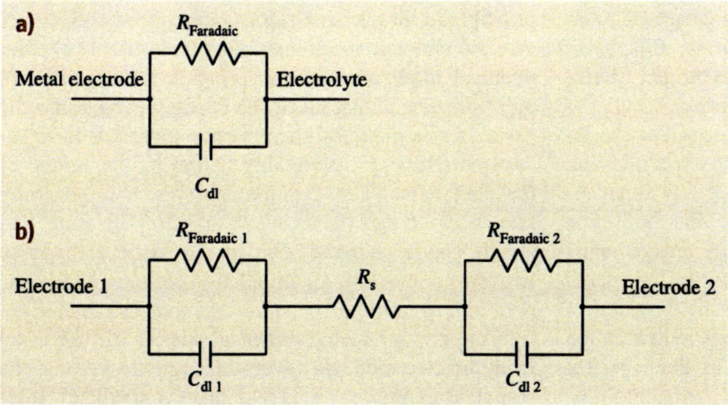

Fig. 37.2a,b Electrical circuit modes for **(a)** an electrode–electrolyte interface **(b)** the interelectrode impedance for a dipole configuration

interface. The half-cell potential [37.8] for the electrode has been omitted from the model because the same electrode material is used to implement the dipole electrode configuration in which the electrodes' half-cell potentials are the same and have been cancelled out. In the model, C_{dl} represents the double-layer capacitor at the interface. Depending on the type of metal conductor used and the roughness of electrode surface, the C_{dl} value is on the order of $\mu F/cm^2$ [37.9]. $R_{faradaic}$ represents the Faradaic resistance, which is in parallel with C_{dl}. Because of the double-layer capacitor effect, the impedance of the electrode is frequency dependent. At the same time, C_{dl} and $R_{faradaic}$ themselves are both nonlinear time-varying components [37.10] that make the accurate modeling of electrode impedance difficult. Figure 37.2b shows the electrical circuit model for the interelectrode impedance in a dipole. R_s represents the electrolyte or tissue resistance between the two electrodes, known as *access resistance*.

The equivalent model shown in Fig. 37.2 is important from the safe stimulation point of view, because it shows that the stimulation reaction at the electrode–electrolyte interface is a combination of several electrochemical processes occurring at different pathways, each of which results in different charge delivery mechanisms. There are two primary mechanisms involved for passing current from one medium to the other: capacitive charging (also known as non-Faradaic) or the Faradaic process (also known as oxidation–reduction reactions, or electrochemical reactions).

Capacitive charging involves the redistribution of charged chemical species in an electrolyte. It charges and discharges the so-called electrode double-layer capacitance. Ions in the tissue are attracted or repelled by charge on the electrode to produce pulses of stimulus current (ionic). No charge actually crosses the electrode–electrolyte interface, but rather charge builds up on either side, such as on the two plates of a capacitor. For a perfect capacitive electrode, $R_{faradaic}$ in Fig. 37.2 is infinite, indicating there is no net charge across the interface via the Faradaic process.

Since there is no net charge transfer across the interface in a capacitive electrode, both electrode and electrolyte maintain electroneutrality at any time. As long as the charge density is below the limit that a capacitive electrode can tolerate, only charge redistribution occurs and capacitive charging is considered safe in the neural environment. When the direction of the applied current is reversed, the charge redistribution is also reversed and the charge stored by the capacitor in the first phase is recovered.

In a Faradaic process, the charge is injected from the electrode to the electrolyte and vice versa. There is a reduction process that acquires additional electrons occurring at the negative electrode and an oxidation process that removes excess electrons at the positive electrode. One may generally think of capacitive charging as representing charge storage and the Faradaic process as representing charge dissipation [37.11] or leakage.

Unlike the capacitive electrode, the products formed in the electrolyte in Faradaic charge injection cannot be recovered upon reversing the direction of current if the products diffuse away from the electrode–electrolyte interface. Hence, Faradaic reactions are divided into reversible and irreversible reactions. The degree of reversibility depends on the relative rates of kinetics (electron transfer at the interface) and mass transport [37.11]. If the kinetic process is fast relative to the rate of mass transport, the Faradaic reaction is reversible. With fast kinetics, the electrochemical products formed at the interface do not diffuse away from the surface and are still stored near the electrode interface. If the direction of stimulus current is reversed, those products can be reversed to their original forms. However, in a Faradaic reaction with slow kinetics, the electrochemical products are able to diffuse away from the electrode in the lengthy time frame and there is no effective storage of charge at the electrode interface. If the direction of stimulus current is reversed, those products cannot be reversed to their original forms. Thus, such a Faradaic reaction is irreversible. Irreversible Faradaic reactions result in a net change in the electrochemistry of the environment that leads to electrode dissolution, gassing, variation in the pH of the electrolyte, etc. It potentially creates harmful chemical species that are deleterious to the surrounding tissue as well as damaging the electrode. Thus, an important principle in long-term electrical stimulation is that irreversible Faradaic reactions should be avoided. Another practical implication of this principle from a circuit's point of view is that all of the charge injected into the tissue in one phase of a stimulation pulse has to be recovered afterward, preferably before the beginning of the next pulse. This requirement, which is generally known as charge balancing, is a safety measure that avoids charge accumulation in the tissue and has to be incorporated into all stimulation circuits [37.9]. The reverse charge should be applied as soon as possible before the newly created chemical species diffuse away from the electrode. However, this is not to say that the recuperation phase should take place immediately after the first stimulation phase. A small interphase gap between the two phases is often useful in maximizing the stimulation effect [37.12].

37.1.2 Safety Issues

The use of stimulating electrodes in a neural environment should ensure that desired physiological effects are obtained and maximized while negative effects remain below a dangerous level. This section will not discuss the general safety requirements that apply to all the implanted electrodes, such as biocompatibility, but rather electrode performance under stimulation.

During stimulation, a large amount of charge is passed through the stimulating electrodes. A good stimulating electrode should be able to carry enough stimulus charge for neural excitation without exceeding the safety limit of the electrode. The maximum net charge (Q_{max}) that an electrode can tolerate without causing any irreversible electrochemical reactions at the electrode–tissue interface and the electrode itself is given by $Q_{max} = qA$, where q and A are the geometric charge density and the surface area of the electrode, respectively. For example, platinum (Pt) electrodes have a charge density limit of $0.05 \sim 0.15\,mC/cm^2$ [37.13], while the limit for activated iridium oxide is up to $3\,mC/cm^2$ [37.14]. For any clinical application, the maximum charge density for the electrode must be higher than its practical usage, and it is suggested that as large a safety margin between the practical usage and the limit as possible be reserved for safe stimulation.

Besides the charge density, the charge per phase is also important in terms of safe stimulation. The stimulus charge per phase is calculated by the integration of the stimulus current over the stimulation phase (pulse width) and is simply equal to the product of current amplitude and pulse width if the stimulus current is constant. Statistical results showed a significant correlation between first-order tissue damage and charge per phase [37.15]. McCreery found that the thresholds for cochlear neuron excitation and damage were 1 and 3 nC/phase, respectively. Stimulating at charges greater than 3 nC/phase resulted in damage in myelinated axons. Since the iridium electrodes used in McCreery's stimulation experiment were pretreated to have an electrode charge density much larger than the charge per phase during stimulation, the neural damage was unlikely due to the electrode dissolution. The severity of the tissue damage increased with higher charge per phase [37.15].

Thus it is always better for a stimulator to supply a charge per phase that is just above the excitation threshold. This not only saves power consumption but is also considered safer by providing a wide safety margin to the limit of neural damage.

However, meeting these two requirements is not enough to guarantee safety. Electrode dissolution or gassing at the interface may still occur, mainly due to charge imbalance. The charge imbalance here indicates stimulation pulses in which the charge in the stimulation phase is different from the discharge phase. The previous discussion on charge density and charge per phase assume stimulation pulses with balanced charge. There is a maximum amount of charge that an electrode can tolerate. The accumulated net charge will cause an increase in the electrode voltage. The region of safe electrode voltage is known as the *water window* beyond which gassing and electrode dissolution will occur. Taking a platinum electrode, for example, the lower and upper bounds of the water window are about +0.8 and −0.6 V, respectively, when using an Ag/AgCl reference electrode [37.13], or +1.2 and −0.8 V, respectively, when using a calomel reference electrode [37.16]. Once the practical electrode voltage exceeds the safety limit (the breakdown voltage) defined by the water window, unintended biochemical reactions will occur at the interface. The byproducts formed could result in deleterious reactions with nearby tissue.

37.2 Implantable Cable Management

37.2.1 Current Limitations

For most implanted stimulation systems, the small stimulating electrodes are placed at the stimulation site, near the targeted muscles or nerves, while the stimulator itself is placed at a more spacious or convenient location. The stimulator is connected to the electrodes via implantable cables (Fig. 37.1). These cables are composed of several wires that carry stimulus current to or from the electrodes. Assuming a dipole electrode configuration, each stimulation channel will require two dedicated wires to provide a stimulation path. Thus an

N-channel stimulator will have $2 \times N$ wires between the stimulator and the electrodes. If each cable contains m wires, it will subsequently require $2 \times N/m$ cables. The cable count will be even higher if a tripole (two anodic electrodes and one cathodic electrode) configuration is used. Due to the surgical difficulty of implanting many cables, the number of stimulation channels is often limited by the maximum number of cables that can be implanted. In this section, two detailed examples are given to illustrate how this limitation affects the design of implant systems.

The first example that shows the importance of reduced cable count is the usage of a grommet in the Finetech–Brindley bladder control implant [37.17], which is one of the most successful implants in humans with both CE and FDA (US Food and Drug Administration) approvals [37.18]. The CE marking is a mandatory conformity mark placed on many products in the European Union (EU). A CE-marked medical device meets all the regulations and requirements in the EU for medical devices. The Finetech–Brindley implant is able to stimulate sacral anterior roots of patients with SCIs to empty their bladder or urinate on demand. In the implantation procedure of the Finetech–Brindley implant, silicone rubbers were used for adhesion and sealing. However, the adhesion of silicone rubbers to the surface of implant parts could be degraded by the presence of a thin fibrous membrane between them, which compromised the hermeticity of the sealing [37.19]. This can be disastrous if the implantable cables run through a subarachnoid space because any hole in the seal will allow cerebrospinal fluid leakage. The solution in the Finetech–Brindley bladder implant was to apply a small grommet encompassing all the implantable cables to the subarachnoid space. The gap between the threaded cables and the inner wall of the grommet was filled with silicone rubber. By having a grommet to encompass all the cables, the growth of fibrous tissue is kept on the outside of the grommet and the inside is not affected. Additionally, the suitable stress tension provided by the grommet helps to achieve a water-tight seal and solves the potential leakage problem. A high cable count not only requires a larger grommet for implantation, but also increases the complexity of cable insertion and the difficulty of filling all the gaps during implantation.

Another example is the sacro-lumbar anterior root stimulator implant (SLARSI) [37.20], a stimulator designed by the Implanted Devices Group, University College London, to stimulate certain root levels for leg movements and bladder control. Ten completely independent channels are available in the stimulator (see detailed root-channel mapping in [37.20]). In the SLARSI, the stimulating electrodes are implanted in the spinal canal at the cauda equina, while the stimulator is placed in the abdomen. Besides the same possible leakage problem as in the Finetech–Brindley implant, the functions that can be provided by the SLARSI are also limited by the interconnecting cables. In SLARSI, the stimulating electrode pairs are configured in a pseudotripolar manner with the two anodes in every stimulation channel shorted together. Thus the number of wires required by this pseudotripole configuration is the same as a dipole configuration. Therefore, 10 channels will need 20 wires in total, requiring 5 four-wire *Cooper* cables [37.21]. More functions would be obtained for a SLARSI system with more channels, but at present this is impossible because the available space in the inner body for cable implantation is very limited and surgeons are reluctant to implant more than 5 cables entering the spinal canal through the dura.

Besides space constraints, complex cable management approaches are usually rejected by implant surgeons due to the high risk of mechanical failure and infection [37.22]. Usually, several wires are packed into an implantable cable. Each wire is used to deliver a stimulus current, a power-supply voltage, a data stream, or a clock, depending on the application. If the insulation fails on only one wire inside the cable and the stimulator is off at the time, it will still be safe because there is no applied potential (voltage difference) between the exposed part of the broken wire and the internal body. Hence, there will be no pathway for the current to flow from the exposed part of the broken wire in such a condition. However, if the stimulator is on, current could flow between the exposed part of the wire and the electrodes. When two or more wires are exposed in a failure, current will also flow between the exposed parts because of the voltage difference between them, causing electrolysis. These would be severe safety hazards for the tissue near the exposed area.

It is desirable for future implants to be able to stimulate at many sites [37.23]. For example, by applying FES on many muscles at the same time, the paralyzed patient will not be limited to restoring simple leg movement but may be able to transfer or walk again. Much effort has been devoted to increasing the functionalities of stimulators without complicating the implantation procedure while maintaining a high level of reliability. It is desirable to have a better stimulation system that is able to manage implantable cables in a more efficient way, reducing the cable count.

37.2.2 Solutions

One approach to reducing the number of wires is to have a separate device at each site, the *BIONs* method [37.26] shown in Fig. 37.3a [37.24]. BION is an abbreviation for bionic neurons. The term BION refers to a family of wireless, injectable microelectronic modules developed by researchers led by Loeb [37.26] at the University of Southern California. The BION microstimulator is available from Advanced Bionics Corp., Valencia, CA. Each BION contains all the units in a central implant, and since each is placed at the stimulation site, there is no wiring. However, BIONs are not a panacea, not only because of their high unit cost but also because there are sites in the body where BIONs would be too large or too nonspecific (current is not confined to the target nerve). Some BIONs contain rechargeable cells that significantly increase their total size.

Another approach to reducing the wire count is to connect some of the anodes on the electrode cuffs to a common wire, but this leads to cross-talk between the channels [37.27]. In that case, an N channel stimulator will require $N+1$ wires for linking the electrodes with the stimulator, provided all the anodes are connected together.

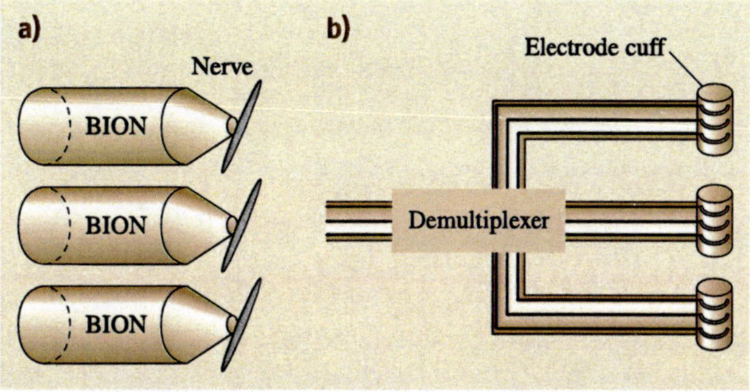

Fig. 37.3a,b Different cable management solutions: **(a)** BIONs approach, **(b)** demultiplexer approach (after [37.24] with permission from IEEE 2007)

The third approach is to introduce a demultiplexer [37.28, 29] into the wire routing, as shown in Fig. 37.3b. The demultiplexer controls the destination of the stimulus current and is able to drive multiple electrode cuffs. This approach reduces the number

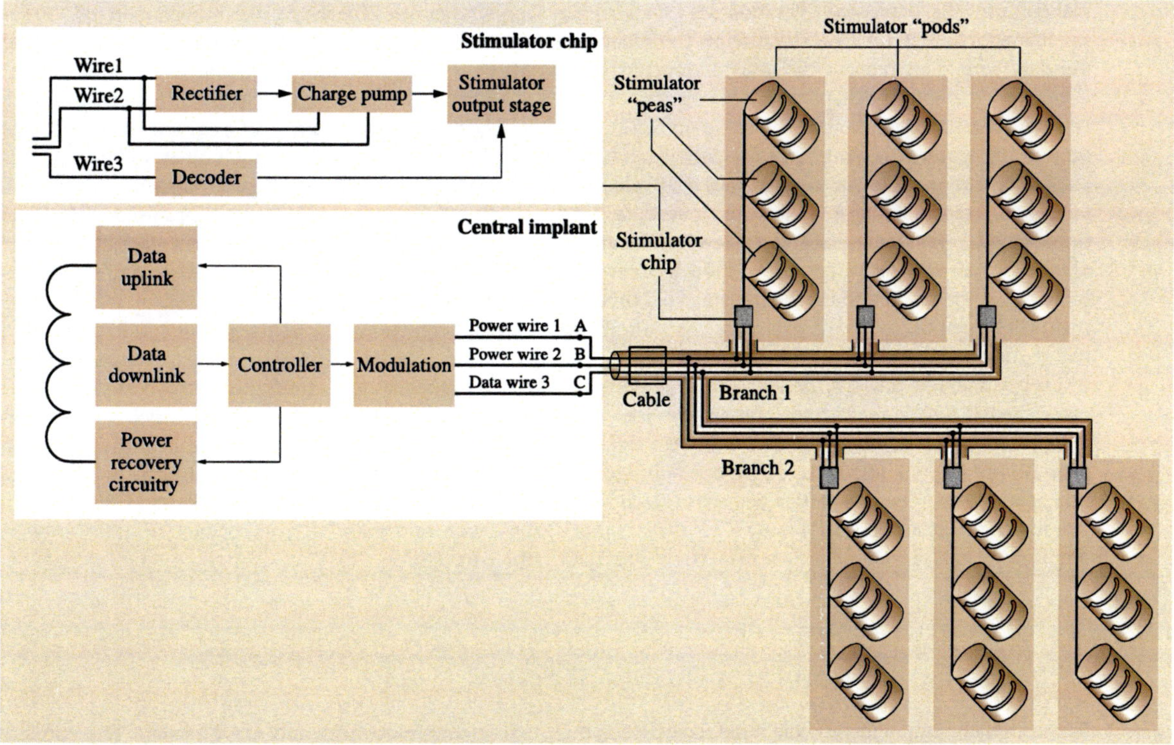

Fig. 37.4 Schematic illustration of branch–pod–pea structure (after [37.25] with permission from IEEE 2007)

Synchronization word	Address			Cathodic current	Anodic current	Cathodic duration	Inter-phase delay	Anodic duration	Error check
	Branch	Pod	Pea						

Fig. 37.5 An example of a data packet sent in a data wire

of wires from the central implant, having short local wires between the demultiplexer and the electrode cuffs. However, stimulating many cuffs simultaneously will still require many wires between the central implant and the demultiplexers. Furthermore, in order to make the demultiplexer work, extra power wires and a data wire that contains the information for the cuff selection are required.

The fourth approach moves the stimulator output stage to the stimulation site. Thus, stimulus currents are no longer sent distantly to stimulation sites but generated locally at the stimulation site. When this is done, each stimulation channel only has to be allocated a few wires that are responsible for power and data (stimulation commands). To distinguish different stimulation channels, each channel will be assigned a unique address. This requires an extra address wire that identifies each individual channel. Thus different channels can be activated sequentially by appropriate stimulation commands from the data wire. The address can also be modulated together with the stimulation commands in the data wire. This reduces the total wire count, with a small price of longer data packets that contain both the selected channels and the stimulation commands. Although the required driving wire for each stimulation channel is increased from two (inward and returning wires) to three (two power wires and one data wire in serial bits), the same three wires can be shared by all stimulation channels. Figure 37.4 [37.25] shows the branch–pod–pea structure for the transmission of stimulation commands and power from the central implant to the stimulation sites. One stimulator chip is attached to each stimulator pod that contains several electrode cuffs (peas). All the stimulator chips share the same three wires as the input. In the data wire, the same bit stream is sent to all stimulator chips, but only the bits that specifically point to a certain pea are utilized (each pea in the pod is addressed). Figure 37.5 shows an example of a data packet sent in a data wire. The data packet is processed only if the synchronization word in the packet matches the predefined sequence in the stimulator ASICs.

The address section in a multibit data packet locates the corresponding stimulator ASIC and the channel (i. e., the pea) to be stimulated. The rest of the data packet defines the amplitude and duration of the cathodic and anodic currents generated by the ASIC. The resolution of the current amplitude and phase duration decides the length of the data packet. A fine-controlled stimulus strength/intensity requires more bits in a data packet. Additionally, a long data packet usually features an error-check section, such as parity check, to check for errors from either transmission or data decoding. A data packet that fails the error check is discarded and not utilized by the stimulator ASIC. In this scheme, the stimulus current is generated onsite instead of in the central implant. Thus, it avoids sending different stimulus currents directly to the peas and using many implant cables. Therefore, the branch–pod–pea structure of the implant system greatly reduces the number of wires routed to the stimulation sites and is able to stimulate more stimulator peas. The feasibility of such a system requires the design of a stimulator output stage small enough to fit at a stimulation site, which is usually spatially tight. The integrated circuit method is an existing solution to minimize the size of many electrical components by providing high-density integration.

In order to ensure fail-safe operation, the current between interconnecting wires exposed in a failure has to be charge balanced. If the average voltage of each wire is identical, there may be current from one exposed wire to another at a certain time. But shortly there will be current in the reverse direction, which will neutralize the net charge. Since the charge is always balanced over a long time period, it will not cause any harm to nearby tissue. Thus, instead of dangerous DC, AC signals with the same DC components are transferred in the wires [37.7].

37.3 Design of Stimulator Output Stage

The stimulator output stage is the interface between the digital control unit and biological environment (i. e., stimulating electrodes and nerve tissue). The stimulator output stage delivers stimuli (stimulation pulses) to the neural interface at a suitable frequency and is composed of a stimulus current generator and an output stage circuit (see below).

37.3.1 Stimulation Pulses

Stimulus waveforms are generally either monophasic or biphasic in shape. A monophasic stimulus (Fig. 37.6a) consists of repeating unidirectional pulses. Biphasic waveforms can be realized in various forms, as shown in Fig. 37.6b–d. Each of them contains a cathodic phase where stimulus current flows through neural tissue in one direction and an anodic phase in which the current flows in the reverse direction.

Monophasic pulses are more efficient in stimulation, requiring a smaller stimulus current to recruit muscle contractions than biphasic pulses. Additionally, the hardware used to generate biphasic pulses is more complicated than that for monophasic pulses. However, biphasic pulses are used almost exclusively in current implantable stimulators because they prevent or minimize damage to stimulating electrodes and the underlying tissue. It is important that the charge injected into the electrode–tissue interface in one phase should be equal to the charge that comes out of the interface in the following phase. Thus there is no net charge accumulation at the interface. In some literature, the net charge accumulation is measured by an equivalent DC current. The product of the DC current and the stimulation time gives the net charge for a certain period. The symmetrical cathodic and anodic phases in Fig. 37.6b are easy to generate, and the net charge is zero at the end of a stimulation pulse. However, a large anodic current that has the same amplitude as the cathodic current may elicit another stimulation response. This complicates the stimulation effect. Thus the anodic current is usually limited to a certain percentage of the cathodic current (Fig. 37.6c). In Fig. 37.6b, c the anodic phase is active and square in shape. It is also possible to have a passive anodic phase for charge balance. The passive phase is achieved by shorting the stimulating electrodes via a discharging switch in series with either a blocking capacitor or a capacitive electrode whose electrode–electrolyte interface serves as a good capacitor. In the passive phase, the anodic current is at its maximum at the start of the anodic phase and decreases exponentially with time. Passive discharge is able to achieve a high degree of charge balance if the passive phase is long enough. However, compared with active discharging, passive discharging requires a longer recuperation time, which limits the maximum stimulation frequency.

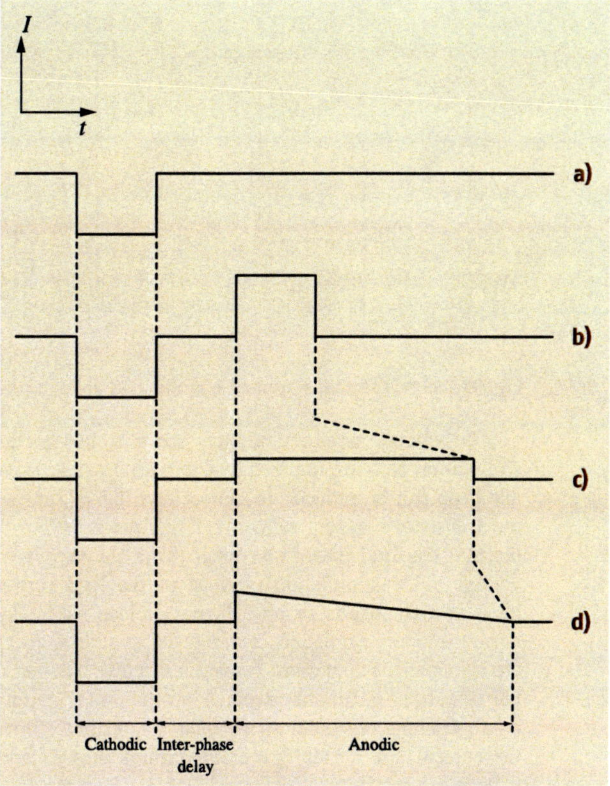

Fig. 37.6a–d Different stimulus shapes. Monophasic (**a**); biphasic (**b**)–(**d**)

However, for FES applications, which usually lie in the range of several Hertz to 50 Hz, passive discharging is able to achieve a charge balance as good as active discharging.

37.3.2 Current Generator Circuits

The current generator circuit is a key block in the stimulator output stage. It converts the digital control signals into an analog stimulus current or voltage and passes it to the output stage circuit, which is in direct contact with stimulating electrodes and nerve tissue. The current generator circuit could be either current mode [37.30] or voltage mode [37.31], realized by corresponding circuits as shown in Fig. 37.7a, b. In voltage mode the stimulation load is connected to a bipolar voltage source that generates both positive and negative voltages across the load, while in current mode a biphasic current from the source is supplied to the load.

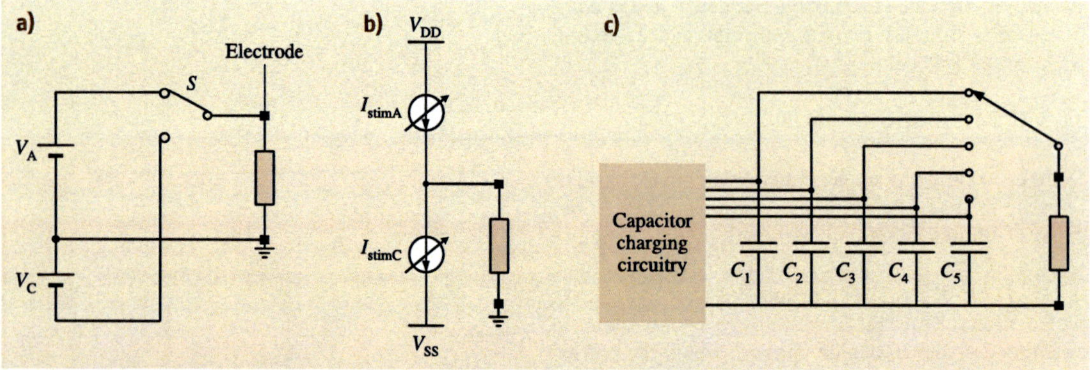

Fig. 37.7a–c Different stimulation modes: **(a)** voltage mode, **(b)** current mode, **(c)** charge mode

For voltage-mode stimulation, the switches on the stimulation path are realized by transistors whose voltage drop can be reduced to almost zero by increasing the transistor's aspect ratio (W/L). Thus it is possible that the total power consumed from the supply of voltage-mode stimulation is close to the load power itself, which results in an efficiency close to 1. For current-mode stimulation, besides the switches, there are also one or two current generator circuits. The number of current generators depends on the power-supply configurations. In order to provide a constant stimulus current, the current generator circuit should have a large output impedance, much larger than the load impedance. This is usually done by cascoding several transistors in a stack. The current generator circuit realized by a cascode topology requires an additional voltage drop across the current generator circuit and dissipates substantial power in the output transistors. This is a waste of energy and reduces the voltage headroom available for the stimulation load. In current mode, the voltage supply rails and the stimulus current are usually kept constant during stimulation. Thus the power consumption from the supply (stimulus current × supply voltage) is also constant. This wastes a significant amount of power at the beginning of stimulation. Since power efficiency is of importance in most implant applications, voltage mode seems to be the right option. However, the magnitude of the current delivered to the tissue is dependent on the interelectrode impedance (Ohm's law). Thus in voltage-mode stimulation, the stimulus current is affected by the interelectrode impedance variation. Hence it is difficult to control the exact amount of charge to the electrode and tissue because the charge is a product of the stimulus current and the stimulation time. Theoretically, voltage mode may find applications in which the interelectrode impedance is fairly stable. But in practice, due to tissue growth over electrodes [37.32], the interelectrode impedance is difficult to predict. To address the safety associated with charge delivery, voltage-mode stimulators usually feature extra complexity to measure the charge flowing through the load, such as an integrator circuit. However, in current mode, the current generator circuit presents a much larger output impedance than the electrode–tissue impedance; the influence of interelectrode impedance variation to the stimulus current is small. Therefore, the amount of charge delivered per stimulus pulse is easily controlled. Hence, the safety level of current mode is higher than that of voltage mode.

Recently, some researchers have proposed another stimulation mode, known as charge mode. It combines some characteristics from both voltage-mode and current-mode stimulation. In charge mode, the load absorbs energy either from the electrode capacitance of previously charged electrodes [37.33] or from a bank of capacitors [37.34, 35], as shown in Fig. 37.7c. The capacitance provides neither constant current nor constant voltage, but rather transfers the charge to the load. It applies a voltage waveform that is an approximation of the waveform that would appear at the load if a current mode was applied. Since the charge is delivered to the electrode directly from intermediate voltage supplies (the capacitors), it avoids the substantial unnecessary power consumption in the current source circuit as well as providing large voltage headroom. However, the impedance of the energy-storage capacitors may be comparable to the electrode impedance, which limits the amount of charge that can be transferred. As a solution, it may be required to precharge the storage capacitors to a higher voltage level before sharing the charge with the stimulation load. Charge-mode stimula-

Table 37.1 Comparison of different stimulation modes

Performance		Voltage mode	Current mode	Charge mode
Voltage compliance		Low	High	Low
Power efficiency		High	Low	High
Safety on charge delivery		Low	High	Medium
Stimulation-induced neural response		Better	Good	Better
System complexity	Size	Small	Small	Large
	Resolution	Low	High	Low

tion saves power because some of the charge supplied to the stimulation load is recycled from some other capacitance instead of being taken directly from the power supply. Therefore, it achieves a power efficiency higher than current-mode but lower than voltage-mode stimulation due to extra losses during precharging and sharing [37.36]. Similar to voltage mode, the exact amplitude of stimulus current in charge mode is not controlled except for limiting the maximum amplitude. In charge mode, the charge supplied to the load is from intermediate sources, so the charge to the load does not exceed the charge on the capacitor source, which is limited by the capacitance and the voltage across the capacitance. Thus the charge to the load is somehow controlled. It provides a better safety standard than voltage mode, but not as good as current mode. The disadvantage of charge mode is that it requires large capacitors (which act as voltage sources) that tend to be bulky and occupy lots of physical space. Kelly et al. used five 1 μF capacitors for a 15-electrode stimulation system [37.34], which would increase for more electrodes. It is also difficult to achieve fine resolution in both voltage-mode and charge-mode compared to current-mode stimulation since the former two are an approximate method of producing a current pulse, and increasing the resolution requires more voltage sources or capacitors.

Unlike current mode in which the stimulus current is constant and lasts for a certain period of time, voltage mode and charge mode are comprised of both high- and low-frequency components at the output spectrum. Since the true neural stimuli also contain high- and low-frequency components, voltage mode and charge mode are believed to be physiologically more efficacious in stimulation [37.35].

The ultimate goal of FES stimulation is to achieve more natural control of movement of paralyzed muscles. This certainly requires a multichannel stimulator that is able to stimulate many nerves. The size constraint excludes charge-mode stimulation because of the bulky storage capacitors. A more natural stimulation effect also requires a wide range of different stimulation strength that is best given in the current-mode configuration. Additionally, current-mode stimulation also gives the best safety standard in terms of charge delivery. Table 37.1 summarizes the performance of three different stimulation modes in terms of voltage compliance, power efficiency, safety on charge delivery, stimulation-induced neural response, and system complexity. The voltage compliance is the minimum voltage drop across the stimulus source. In current mode, the voltage compliance is the minimum voltage across the current generator circuit at which the current generator still provides a high output resistance [37.37].

Due to the aforementioned merits offered by current-mode current generators, current generator circuits are extensively used for implantable neural stimulators [37.30, 38–54]. The full-scale output current varies from about 100 μA to 25 mA and the resolution from 3 to 8 bits, depending on the application. Desirable features for a current generator circuit for use in implantable stimulators are (1) small output voltage compliance, (2) high output impedance, (3) good linearity, (4) low power consumption, and (5) small silicon area. Table 37.2 summarizes the performance of current generator circuits in various topologies.

37.3.3 Safe Output Stage Circuits

A current generator is only responsible for generating stimulus currents while it is the output stage circuit that guides the stimulus current to the neural tissue in a safe manner. A safe stimulation pulse received at the tissue end should be charge balanced. However, perfect charge balance is difficult to obtain. Various factors, including semiconductor failures, software faults in the implant microcontroller, parametric shift after implantation, crosstalk between adjacent stimulation channels, and cable failures, can all lead to charge imbalance [37.55]. The outcome of charge imbalance is devastating because any excess charge accumulation over time leads to damage at both the stimulating electrodes and neural tissue. The electrode itself could dissolve into the solution if overdriven. The metallic cations could form toxic products with the anions in the solution. Usually, a high level of charge imbalance also involves gassing at the electrode–tissue interface, which disturbs the normal pH balance in the solution. A shifted pH level can have a significant effect on cell function, such as destruction of tissue [37.56].

Charge imbalance can be corrected by actively monitoring the change at the neural interface or by the use of capacitive coupling.

Monitoring

Monitoring can be applied to at the stimulator-electrode–tissue interface (Fig. 37.8a). Approaches include using a window comparator to monitor the inter-electrode voltage [37.57], an on-chip sinusoidal wave-

Table 37.2 Summary of various current generator circuits for implantable stimulators

Current generator	Full-scale current	Power supply (V)	Resolution (bit)	Tracking error [a]	Linearity [b]
Binary-weighted transistors with single bias [37.46]	600 μA	±7	4	7.24%	No data
Binary-weighted transistors with single bias [37.41]	600 μA	±7	4	< 5%	Only waveform illustration
Binary-weighted transistors with single bias [37.50]	6.3 mA	No data	6	source = sink [c]	DNL and INL < 0.5LSB
Binary-weighted transistors with single bias [37.43]	126 μA	±3	6	No data	Deviate from the ideal current by 2 LSB
Binary-weighted transistors with single bias [37.40]	140 μA	1.8 or 3.3	5	< 2%	DNL: 0.38LSB; INL: 0.55LSB
Binary-weighted transistors with single bias [37.47]	2 mA	5	5	No data	Only waveform illustration
Binary-weighted transistors with single bias [37.48]	3 mA	5	5	No data	DNL: −0.54LSB; INL: 1.66LSB;
Binary-weighted transistors with single bias [37.49]	1.5 mA	5	7	No data	No data
Binary-weighted transistors with single bias [37.53]	4 mA	5	5	source = sink	DNL: 0.10LSB; INL: 0.37LSB
Binary-weighted transistors with digital driving [37.54]	1 mA	5	4	active sourcing + passive sinking	DNL: 0.032LSB; INL: 0.065LSB
An array of unity current generator [37.52]	5 mA	3	8	active sourcing + passive sinking; 5% error between cathodes (sources)	1%
Identical transistors with binary-weighted bias [37.44]	400 μA	±5	8	5.74%	DNL: 2.11LSB; INL: −3.11LSB
Active feedback [37.30]	600 μA	±6.5	6	1.2%	DNL: 0.15LSB; INL: −0.16LSB
Voltage follower [37.51]	1.3 mA	No data	No data	source = sink	No data
Voltage-controlled resistor with analogue DAC [37.38]	210 μA	5	5	< 1%	Degraded linearity in saturation region

[a] Tracking error is the mismatch between sourcing (cathodic) and sinking (anodic) currents in one current generator circuit or between sourcing currents in different current generators.

[b] Two parameters are commonly used to describe the linearity. The differential nonlinearity (DNL) defines the difference between the ideal step amplitude and the real step amplitude and the integral nonlinearity (INL) defines the maximum difference between the ideal output and the real output of the current source. Those nonlinearities are usually expressed in LSB units (where one LSB corresponds to the amplitude of an ideal step).

[c] The same DAC current is guided to the stimulation load by a switch array as either cathodic or anodic current.

Table 37.2 (continued)

Current generator	Output impedance (MΩ)	Voltage compliance (V)	Silicon Area (mm^2)[d]	Process	Year
Binary-weighted transistors with single bias [37.46]	No data	No data	No data	1.2 μm CMOS	2005
Binary-weighted transistors with single bias [37.41]	> 2	1	No data	1.2 μm CMOS	2000
Binary-weighted transistors with single bias [37.50]	No data	No data	No data	1.2 μm CMOS	1999
Binary-weighted transistors with single bias [37.43]	No data	> 1	0.14	2 μm CMOS	1997
Binary-weighted transistors with single bias [37.40]	No data	0.29	No data	0.18 μm CMOS	2007
Binary-weighted transistors with single bias [37.47]	No data	No data	No data	2 μm CMOS	2001
Binary-weighted transistors with single bias [37.48]	No data	No data	0.01	1.2 μm CMOS	1995
Binary-weighted transistors with single bias [37.49]	No data	1	No data	2 μm CMOS	1995
Binary-weighted transistors with single bias [37.53]	No data	1	0.026	0.8 μm BiCMOS	1997
Binary-weighted transistors with digital driving [37.54]	40	0.41	0.09	1 μm CMOS	2008
An array of unity current generator [37.52]	2.5	0.85	No data	0.8 μm CMOS	2004
Identical transistors with binary-weighted bias [37.44]	> 20	2.48	$\frac{0.177}{n}+0.0265$[e]	1.2 μm CMOS	2003
Active feedback [37.30]	1.3	0.5	No data	1.5 μm CMOS	2005
Voltage follower [37.51]	No data	No data	No data	No data	2000
Voltage-controlled resistor with analogue DAC [37.38]	> 10	< 0.75	0.05	1.5 μm CMOS	2005

[d] The silicon area refers to the silicon occupation for realizing one current generator circuit. In the case of actively biphasic stimulations, the current generator circuit includes both current source and current sink parts. If there are certain circuits shared globally by different current sources/sinks, such as a voltage bias generator, the silicon area for the global circuit is divided and proportionally allocated to each current generator circuit. The area data is not listed in the table if the literature data includes more circuits than the current generator circuit.

[e] n in the formula is the number of channels among which the multi-bias generator is shared.

form to measure the interelectrode impedance [37.58, 59], and an integrator to monitor the stimulus charge delivered over stimulation periods [37.60]. The measured result is compared with a predefined reference value and the output is used as an input to control the next stimulation phase. Thus, in the next phase, the output stage circuit may be subsequently switched off to prevent neural damage or supply additional current in the desired direction to neutralize the net charge. The advantage of this approach is volume saving since the monitoring circuit can be integrated with the output stage circuit. However, the monitoring circuit increases the stimulator complexity, which itself increases the probability of semiconductor failure.

Capacitively Coupled Shorting

Capacitively coupled shorting places a blocking capacitor (known as a coupling capacitor in some literature [37.23, 61]) in the stimulation path. One end of the capacitor is connected in series with a stimulating electrode and the other end of the capacitor goes to the stimulation circuitry, which supplies the current, as shown in Fig. 37.8b. During capacitively coupled stimulation, the difference between the cathodic charge and the anodic charge (i.e., the net charge) is accumulated on the blocking capacitor. By periodically discharging the capacitively coupled load, the charge imbalance is corrected.

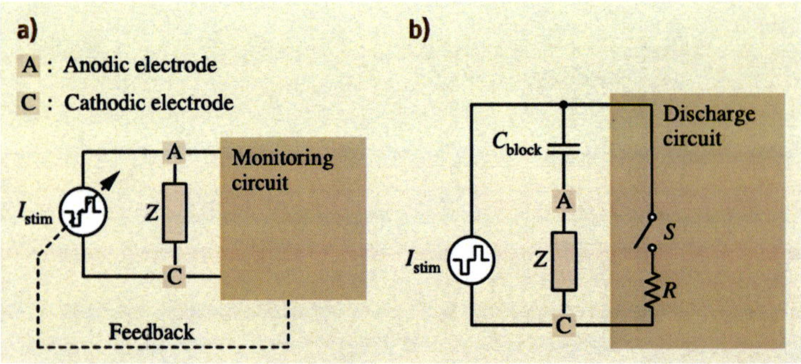

Fig. 37.8a,b Different safety features against charge imbalance in stimulator output stages: **(a)** monitoring and active control, **(b)** capacitively coupled shorting

The presence of the blocking capacitor provides several unique safety functions [37.62] to ensure safe stimulation. (i) Due to the *pass AC, block DC* characteristic of capacitors, they prevent direct current passing through the electrodes and tissue due to the absence of a complete DC stimulation path. (ii) The blocking capacitor helps to limit the maximum net charge and charge per phase during stimulation. For a given power supply V_{DD}, the worst-case charge density that a blocking capacitor C would allow is $V_{DD} \cdot C/A$, where A is the electrode area. If this is less than the maximal allowable charge density for the electrode material, the electrode cannot be overcharged by the stimulator.

Compared with monitoring techniques, the capacitive coupling approach with a blocking capacitor is relatively simple because of its unique safety features, which provide solutions to ensure charge balance and prevent excess stimulus charge in the event of failure.

However, for high-intensity stimulation, blocking capacitors are large in volume. For example, to recover partial leg movements, stimulus currents of about 1 mA intensity and 1 ms pulse width may be required. The aim is to minimize the voltage *wasted* across the blocking capacitor so that most of the power supply voltage can be made available to the load. To calculate the required capacitance, the following elementary equation may be used:

$$C = I_{stim}\frac{\Delta t}{\Delta V}, \quad (37.1)$$

where I_{stim} is the stimulus current amplitude, Δt is the stimulus current pulse width, and ΔV is the change in voltage across the blocking capacitor during stimulation. For this numerical example, to limit the capacitor voltage drop, to, say, 0.5 V, a 2 μF capacitor is required. Clearly, it is impractical to implement such a large capacitor on silicon due to the size and cost constraints, thus the use of off-chip surface mount capacitors. The blocking capacitor value may be reduced at the expense of a larger voltage drop across it, but this will result in a higher supply voltage. Due to the large physical size of blocking capacitors, stimulator designers would like to avoid their usage for applications with many channels.

In [37.63], a high-frequency current-switching (HFCS) stimulation scheme was proposed to mini-

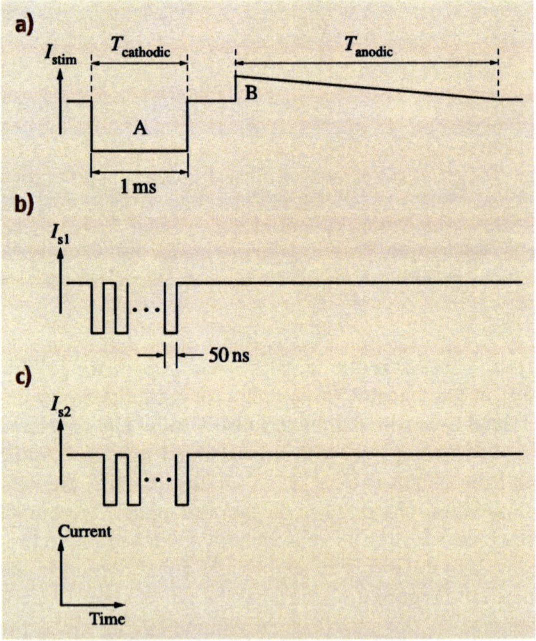

Fig. 37.9a–c Waveform illustration of HFCS technique. The long cathodic phase **(a)** is divided into two complimentary high-frequency pulse trains **(b)** and **(c)** while the discharging phase remains a long and passive period (after [37.54] with permission from IEEE 2008)

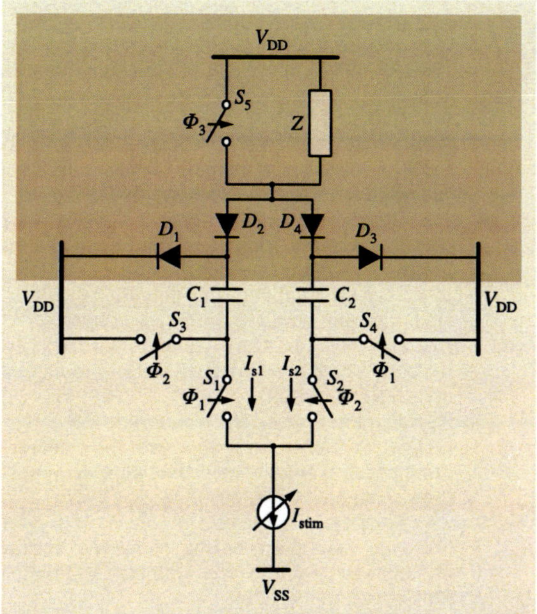

Fig. 37.10 Circuit diagram for HFCS output stage using parallel-blocking-capacitor configuration

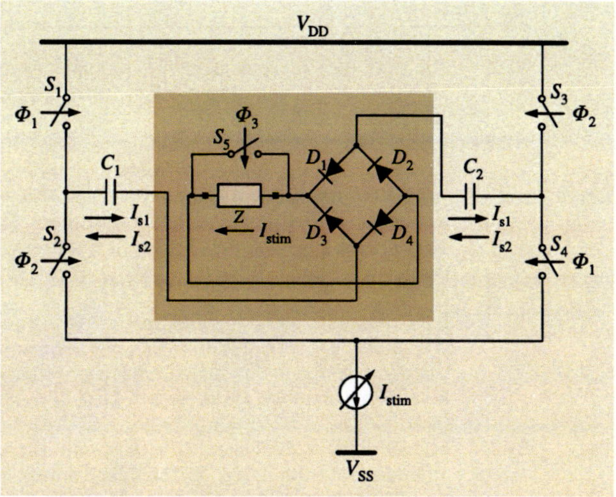

Fig. 37.11 Circuit diagram for HFCS output stage using series-blocking-capacitor configuration

mize the blocking capacitor from the microfarad range to tens of picofarad. It is based on the idea that a long continuous stimulation phase can be divided into two complementary high-frequency trains, as shown in Fig. 37.9 [37.54]. Thus, instead of a large off-chip capacitor that can accommodate a stimulation pulse in milliseconds, only two small capacitors that only need to accommodate pulses tens of nanoseconds in duration are required. According to (37.1), the output stage based on HFCS only requires subnanofarad capacitors. Thus, the complete stimulator output stage can be fully integrated on a single silicon chip. Figures 37.10 and 37.11 show two different circuit topologies to implement the HFCS output stages, namely, the parallel-blocking-capacitor configuration [37.54] and the series-blocking-capacitor configurations [37.64]. A complete multichannel stimulator output stage utilizing an HFCS technique is reported in [37.54] with a silicon occupation of only 6 mm^2.

The charge balance of an HFCS stimulator relies on passive discharging using the capacitance of platinum electrodes. The measured direct current in the HFCS stimulator output stage under various FES stimulation profiles were all smaller than 12 nA, well below the safety limit for implantable stimulation [37.54].

37.4 Conclusions

One of the most important requirements for implantable stimulator design is its reliability at neural interfaces in both long-term and short-term usage. A safe stimulation system should maintain the original properties of implanted devices, stimulating electrodes and neural tissue. This is necessary to avoid any deleterious outcomes that might harm the stimulation environment. This chapter has provided an overview of the safe usage of stimulating electrodes, implantable cables, and stimulator output stages. Various methods and solutions have been discussed in terms of system complexity, physical size, level of safety, and circuit performance. As technology evolves, new solutions in circuit and system domains are expected to emerge. However, before any new solution can be applied in human beings, it must be subjected to thorough performance and reliability evaluations and in vivo and in vitro trials.

References

37.1 A.A. Burt: (iii) The epidemiology, natural history and prognosis of spinal cord injury, Curr. Orthop. **18**(1), 26–32 (2004)

37.2 P.H. Peckham, M.W. Keith, K.L. Kilgore, J.H. Grill, K.S. Wuolle, G.B. Thrope, P. Gorman, J. Hobby, M.J. Mulcahey, S. Carroll, V.R. Hentz, A. Wiegner: Efficacy of an implanted neuroprosthesis for restoring hand grasp in tetraplegia: A multicenter study, Arch. Phys. Med. Rehabil. **82**(10), 1380–1388 (2001)

37.3 D.N. Rushton, N.D. Donaldson, F.M. Barr, V.J. Harper, T.A. Perkins, P.N. Taylor, A.M. Tromans: Lumbar root stimulation for restoring leg function: Results in paraplegia, Artif. Organ. **21**(3), 180–182 (1997)

37.4 G.S. Brindley, C.E. Polkey, D.N. Rushton: Sacral anterior root stimulators for bladder control in paraplegia, Paraplegia **20**(6), 365–381 (1982)

37.5 T. Taira, T. Hori: Diaphragm pacing with spinal cord stimulator: Current state and future directions, Acta Neurochir. Suppl. **97**(1), 289–292 (2007)

37.6 N.J.M. Rijkhoff: Neuroprostheses to treat neurogenic bladder dysfunction: Current status and future perspectives, Child. Nerv. Syst. **20**(2), 75–86 (2004)

37.7 A. Demosthenous, I.F. Triantis, X. Liu: Circuits for implantable neural recording and stimulation. In: *VLSI Circuits for Biomedical Applications*, ed. by K. Iniewski (Artech House, Boston 2008) pp. 207–240

37.8 J.G. Webster: *Medical Instrumentation – Application and Design*, 3rd edn. (Wiley, New York 1998)

37.9 M. Ghovanloo: Integrated circuits for neural interfacing: Neural stimulation. In: *VLSI Circuits for Biomedical Applications*, ed. by K. Iniewski (Artech House, Boston 2008) pp. 191–205

37.10 A.M. Dymond: Characteristics of the metal-tissue interface of stimulation electrodes, IEEE Trans. Biomed. Eng. **23**(4), 274–280 (1976)

37.11 D.R. Merrill, M. Bikson, J.G.R. Jefferys: Electrical stimulation of excitable tissue: Design of efficacious and safe protocols, J. Neurosci. Methods **141**(2), 171–198 (2005)

37.12 P. Prado-Guitierrez, L.M. Fewster, J.M. Heasman, C.M. Mckay, R.K. Shepherd: Effect of interphase gap and pulse duration on electrically evoked potentials is correlated with auditory nerve survival, Hear. Res. **215**(1/2), 47–55 (2006)

37.13 P.R. Troyk, S.F. Cogan: Sensory neural prostheses. In: *Neural Engineering*, ed. by B. He (Kluwer/Plenum, Dordrecht 2005) pp. 1–48

37.14 X. Beebe, T.L. Rose: Charge injection limits of activated iridium oxide electrodes with 0.2 ms pulses in bicarbonate buffered saline (neurological stimulation application), IEEE Trans. Biomed. Eng. **35**(6), 494–495 (1988)

37.15 D.B. McCreery, T.G.H. Yuen, W.F. Agnew, L.A. Bullara: Stimulus parameters affecting tissue injury during microstimulation in the cochlear nucleus of the cat, Hear. Res. **77**(1/2), 105–115 (1994)

37.16 N.D. Donaldson, P.E.K. Donaldson: When are actively balanced biphasic ("Lilly"), stimulating pulses necessary in a neurological prosthesis? I Historical background; Pt resting potential; Q studies, Med. Biol. Eng. Comput. **24**(1), 41–49 (1986)

37.17 Finetech Medical Ltd., Welwyn Garden City, UK (2010) http://www.finetech-medical.co.uk (last accessed 13 April 2011)

37.18 P.E.V.V. Kerrebroeck, E.L. Koldewijn, F.M.J. Debruyne: Worldwide experience with the Finetech-Brindley sacral anterior root stimulator, Neurourol. Urodyn. **12**(5), 497–503 (1993)

37.19 G.S. Brindley: A grommet for preventing cerebrospinal fluid from leaking along the outside of implanted tubes and cables, J. Physiol. **320**(Suppl.), 1P–13P (1981)

37.20 A. Vanhoestenberghe: Implanted devices: Improved methods for nerve root stimulation. Dissertation, University College London (UCL) (2007)

37.21 P. Donaldson: The Cooper cable: An implantable multiconductor cable for neurological prostheses, Med. Biol. Eng. Comput. **21**(3), 371–374 (1983)

37.22 G.S. Brindley: The first 500 sacral anterior root stimulators: Implant failures and their repair, Paraplegia **33**(1), 5–9 (1995)

37.23 P.R. Troyk, N.D. Donaldson: Implantable FES stimulation systems: What is needed?, Neuromodulation **4**(4), 196–204 (2001)

37.24 X. Liu, A. Demosthenous, M. Rahal, N. Donaldson: Recent advances in the design of implantable stimulator output stages, Proc. 18th Eur. Conf. Circuit Theory Des. (2007) pp. 204–207

37.25 X. Liu, A. Demosthenous, N. Donaldson: A safe transmission strategy for power and data recovery in biomedical implanted devices, 2007 Proc. IEEE Int. Symp. Circuits Syst. (IEEE, New York 2007) pp. 2367–2370

37.26 G.E. Loeb, F.J.R. Richmond, J. Singh, R.A. Peck, W. Tan, Q. Zou, N. Sachs: RF-powered BIONs for stimulation and sensing, Proc. 26th Annu. Int. Conf. Eng. Med. Biol. Soc., Vol. 2 (2004) pp. 4182–4185

37.27 N.D. Donaldson, D.N. Rushton, T.A. Perkins, D.E. Wood, J. Norton, A.J. Krabbendam: Recruitment by motor nerve root stimulators: Significance for implant design, Med. Eng. Phys. **25**(7), 527–537 (2003)

37.28 M.S. Humayun, E. Juan, J.D. Weiland, G. Dagnelie, S. Katona, R. Greenberg, S. Suzuki: Pattern elec-

trical stimulation of the human retina, Vis. Res. **39**(15), 2569–2576 (1999)
37.29 D.B. McCreery: Cochlear nucleus auditory prostheses, Hear. Res. **242**(1–2), 64–73 (2008)
37.30 M. Sivaprakasam, W. Liu, M.S. Humayun, J.D. Weiland: A variable range bi-phasic current stimulus driver circuitry for an implantable retinal prosthetic device, IEEE J. Solid-State Circuits **40**(3), 763–771 (2005)
37.31 M. Schwarz, M. Maschmann: Area saving stimulator cells for multielectrode arrays featuring adaptive waveform generation and monitoring, Proc. 26th Annu. Int. Conf. IEEE Eng. Med. Biol. Soc., Vol. 2 (2004) pp. 4314–4317
37.32 W. Grill, T. Mortimer: Electrical properties of implant encapsulation tissue, Ann. Biomed. Eng. **22**(1), 23–33 (1994)
37.33 S.K. Kelly: A system for efficient neural stimulation with energy recovery. Ph.D. Thesis (Massachusetts Institute of Technology, Cambridge 2003)
37.34 S.K. Kelly, J. Wyatt: A power-efficient voltage-based neural tissue stimulator with energy recovery, 2004 IEEE ISSCC Tech. Dig. Pap. (IEEE, New York 2004) pp. 228–524
37.35 M. Ghovanloo: Switched-capacitor based implantable low-power wireless microstimulating systems, 2006 Proc. Int. Symp. Circuits Syst. (2006) pp. 2197–2200
37.36 J. Simpson, M. Ghovanloo: An experimental study of voltage, current, charge controlled stimulation front-end circuitry, Proc. 2007 IEEE Int. Symp. Circuits Syst (IEEE, New York 2007) pp. 325–328
37.37 K.R. Laker, W.M.C. Sansen: *Design of Analog Integrated Circuits and Systems* (McGraw-Hill, New York 1994)
37.38 M. Ghovanloo, K. Najafi: A compact large Voltage-compliance high output-impedance programmable current source for implantable microstimulators, IEEE Trans. Biomed. Eng. **52**(1), 97–105 (2005)
37.39 J.D. Weiland, D.J. Anderson: Chronic neural stimulation with thin-film, iridium oxide electrodes, IEEE Trans. Biomed. Eng. **47**(7), 911–918 (2000)
37.40 J. Coulombe, M. Sawan, J.-F. Gervais: A highly flexible system for microstimulation of the visual cortex: Design and implementation, IEEE Trans. Biomed. Circuits Syst. **1**(4), 258–269 (2007)
37.41 W. Liu, K. Vichienchom, M. Clements, S.C. DeMarco, C. Hughes, E. McGucken, M.S. Humayun, E. De Juan, J.D. Weiland, R. Greenberg: A neuro-stimulus chip with telemetry unit for retinal prosthetic device, IEEE J. Solid-State Circuits **35**(10), 1487–1497 (2000)
37.42 B. Ziaie, M.D. Nardin, A.R. Coghlan, K. Najafi: A single-channel implantable microstimulator for functional neuromuscular stimulation, IEEE Trans. Biomed. Eng. **44**(10), 909–920 (1997)
37.43 K.E. Jones, R.A. Normann: An advanced demultiplexing system for physiological stimulation, IEEE Trans. Biomed. Eng. **44**(12), 1210–1220 (1997)
37.44 S.C. DeMarco, W. Liu, P.R. Singh, G. Lazzi, M.S. Humayun, J.D. Weiland: An arbitrary waveform stimulus circuit for visual prostheses using a low-area multibias DAC, IEEE J. Solid-State Circuits **38**(10), 1679–1690 (2003)
37.45 G. Gudnason, E. Bruun, M. Haugland: An implantable mixed analog/digital neural stimulator circuit, 1999 Proc. IEEE Int. Symp. Circuits Syst., Vol. 5 (1999) pp. 375–378
37.46 M. Sivaprakasam, W. Liu, G. Wang, J.D. Weiland, M.S. Humayun: Architecture tradeoffs in high-density microstimulators for retinal prosthesis, IEEE Trans. Circuits Syst. I, **52**(12), 2629–2641 (2005)
37.47 G.J. Suaning, N.H. Lovell: CMOS neurostimulation ASIC with 100 channels, scaleable output, and bidirectional radio-frequency telemetry, IEEE Trans. Biomed. Eng. **48**(2), 248–260 (2001)
37.48 R.S. Amand, Y. Savaria, M. Sawan: Design optimization of a current source for microstimulator applications, Proc. 37th Midwest Symp. Circuits Syst., Vol. 1 (1994) pp. 129–132
37.49 C.M. Zierhofer, I.J. Hochmair-Desoyer, E.S. Hochmair: Electronic design of a cochlear implant for multichannel high-rate pulsatile stimulation strategies, IEEE Trans. Rehabil. Eng. **3**(1), 112–116 (1995)
37.50 K. Arabi, M.A. Sawan: Electronic design of a multichannel programmable implant for neuromuscular electrical stimulation, IEEE Trans. Rehabil. Eng. **7**(2), 204–214 (1999)
37.51 S. Boyer, M. Sawan, M. Abdel-Gawad, S. Robin, M.M. Elhilali: Implantable selective stimulator to improve bladder voiding: Design and chronic experiments in dogs, IEEE Trans. Rehabil. Eng. **8**(4), 464–470 (2000)
37.52 J.D. Techer, S. Bernard, Y. Bertrand, G. Cathebras, D. Guiraud: New implantable stimulator for the FES of paralyzed muscles, Proc. 30th Eur. Solid-State Circuits Conf. (2004) pp. 455–458
37.53 S. Bourret, M. Sawan, R. Plamondon: Programmable high-amplitude balanced stimulus current-source for implantable microstimulators, Proc. 19th Annu. Int. Conf. IEEE Eng. Med. Biol. Soc., Vol. 5 (1997) pp. 1938–1941
37.54 X. Liu, A. Demosthenous, N. Donaldson: An integrated implantable stimulator that is fail-safe without off-chip blocking capacitors, IEEE Trans. Biomed. Circuits Syst. **2**(3), 231–244 (2008)
37.55 X. Liu: Novel methods for fail-safe integrated neural stimulators. Dissertation, University College London (2009)
37.56 C. Huang, P. Carter, R. Shepherd: Stimulus induced pH changes in cochlear implants: An in vitro and in vivo study, Ann. Biomed. Eng. **29**(9), 791–802 (2001)

37.57 M. Ortmanns, A. Rocke, M. Gehrke, H.J. Tiedtke: A 232-channel epiretinal stimulator ASIC, IEEE J. Solid-State Circuits **42**(12), 2946–2959 (2007)

37.58 C. Donfack, M. Sawan, Y. Savaria: Implantable measurement technique dedicated to the monitoring of electrode-nerve contact in bladder stimulator, Med. Biol. Eng. Comput. **38**(4), 465–468 (2000)

37.59 M. Ghovanloo, K. Najafi: A wireless implantable multichannel microstimulating system-on-a-chip with modular architecture, IEEE Trans. Neural Syst. Rehabil. Eng. **15**(3), 449–457 (2007)

37.60 X. Fang, J. Wills, J. Granacki, J. LaCoss, A. Arakelian, J. Weiland: Novel charge-metering stimulus amplifier for biomimetic implantable prosthesis, 2007 Proc. IEEE Int. Symp. Circuits Syst. (IEEE, New York 2007) pp. 569–572

37.61 C.Q. Huang, R.K. Shepherd, P.M. Center, P.M. Seligman, B. Tabor: Electrical stimulation of the auditory nerve: Direct current measurement in vivo, IEEE Trans. Biomed. Eng. **46**(4), 461–469 (1999)

37.62 X. Liu, A. Demosthenous, N. Donaldson: Five valuable functions of blocking capacitors in stimulators, Proc. 13th Annu. Int. FES Soc. Conf. (IFESS, Sidney 2008) pp. 322–324

37.63 X. Liu, A. Demosthenous, N. Donaldson: A stimulator output stage with capacitor reduction and failure-checking techniques, 2006 Proc. IEEE Int. Symp. Circuits Syst. (IEEE, New York 2006) pp. 641–644

37.64 X. Liu, A. Demosthenous: A fail-safe ASIC for implantable neural stimulation, Proc. 33rd Eur. Solid-State Circuits Conf. (IEEE, New York 2007) pp. 460–463

38. Cardiac Pacemaker Systems

Armin Bolz

Pacemaker technology is one of the largest milestones in Medical Technology. Over the past five decades millions of patients benefited from the advantages of this masterpieces of engineering.

Today all types of *bradycardia* and most *tachycardias* can be treated successfully. Pacemakers improve the quality of life significantly. They help to avoid syncopes and dizziness. Pacemakers also support the heart and improve the *cardiac* output. State of the art pacemakers are no longer simple pulse generators but cardiac rhythm management systems.

38.1 **Structure of a Pacemaker System** 768
38.1.1 The Programming Device............ 768
38.1.2 The Pacemaker 769
38.1.3 The Pacemaker Electrode............ 769

38.2 **The Functionality of a Cardiac Pacemaker** 771
38.2.1 The Stimulation Function 772

38.3 **Stimulation Modes**............................... 774
38.3.1 International (NBG) Pacemaker Code.. 775
38.3.2 VVI Pacemaker 775
38.3.3 Two-Chamber Pacemaker........... 776
38.3.4 Frequency-Adaptive Pacemakers . 780
38.3.5 Antitachyarrhythmia Pacemakers. 781
38.3.6 Biatrial Stimulation Systems........ 781
38.3.7 Biventricular Stimulation Systems 782
38.3.8 Selection Criteria 782

References 782

Pathological changes in the excitation formation or conduction within the heart are divided into bradycardiac (decelerating) and tachycardiac (accelerating) clinical characteristics. If there are bradycardiac changes, the cardiac output is usually too low, causing the patient to suffer from dizziness, a low maximum stress level, and even problems with consciousness. In these cases, cardiac pacemaker therapy can allow the sinus rhythm to be restored and the symptoms can either be relieved or eliminated altogether.

The idea of electrically stimulating the heart dates back to the beginning of the last century with the work of Burns and Aldini (a nephew of Galvani). But it was 1927 by the time Hyman built the first functioning external pacemaker, a small electric clockwork-driven generator. In 1948, Shockley, Bardeen, and Brattain invented the transistor and made it possible to drastically reduce the size of electric switching units, which also advanced the development of pacemaker design. The first implantable pacemaker was implanted in 1958 by a Swedish doctor named Elmquist. This device was made of 20 discrete components and weighed about 180 g. Pacemakers today weigh about 60 g and have the functionality of a small computer.

Pacemaker therapy is based on the delivery of current pulses, which lead to the artificial depolarization of some cardiac cells. They go across the conduction system of the heart, as well as across the *gap junctions* (intercellular ionic bonding channels, which serve to transmit the excitement directly from cell to cell), triggering a complete contraction of the heart. Based on this triggering effect of artificial pulses, this is referred to as the *all-or-nothing law*. In principle, the current pulses can be applied in four different ways:

Transcutaneous stimulation: In rare cases, the heart is stimulated via externally applied electrodes. In principle, this is possible, but this way is largely ineffective due to the relatively long distance to the heart. This requires high current intensity, which, in turn, triggers unwanted contractions in the skeletal muscles. External stimulation only makes sense in emergencies, e.g., when using an external defibrillator.

Esophagus stimulation: Esophagus stimulation is a minimally invasive procedure. Here, a catheter is advanced via the mouth or nose into the esophagus so that the electrodes come to rest near the atrium of the heart. This approach is mainly used for diagnostic purposes since the atrial position allows better differentiation of atrial and ventricular action. Stimulation within the esophagus does cause pain, however, which kept this method from gaining widespread practice.

Transient intracardial stimulation: In many cases, bradycardia is only temporary, requiring only temporary therapy (in general terms, this is referred to as transient stimulation). This is used for the following indications:

- Asystoles or external ventricular bradycardia with acute sinoatrial (SA) and atrioventricular (AV) blocks
- Acute conduction defect in the case of a recent anterior myocardial infarction
- Complicated pacemaker exchanges
- As a prophylactic during cardiological operations
- Acute poisoning, especially with pharmaceuticals, such as digitalis or antiarrhythmics

In these cases, a catheter is pushed into the right ventricle via a venous access. Alternatively, so-called heartwires (flexible, insulated wires with a needle at both ends) are fixed in the heart through the chest wall. This is especially suitable for open-heart surgery, since there is already direct access. Pulses are generated via an external stimulator.

Intracardial stimulation using implants: Especially common, however, is permanent stimulation for the rest of the patient's life, for example, when the patient suffers from the following conditions:

- Considerable defects in the auricular stimulus conduction system, such as SA or AV blocks of the second or third degree or fascular blocks
- Bradycardias and arrhythmias after cardiac infarction
- A sick sinus node (sick sinus syndrome)
- Carotis sinus syndrome (this clinical condition is characterized by an oversensitivity of the pressoreceptors in the carotis sinus, where, for example, heart activity is limited when one turns one's head)

Technically, all methods are very similar. Clinically, however, the use of implants is especially important, so in what follows, implant technology is discussed exclusively.

38.1 Structure of a Pacemaker System

A pacemaker system consists of an external programming device, the actual cardiac pacemaker, which contains a battery and electronic elements, as well as a pacemaker electrode (Fig. 38.1). The electrode is advanced almost exclusively via a venous access (e.g., the vena cava sup.) underneath the collarbone into the right half of the heart and anchored there. Afterward, the pacemaker is connected and implanted subcutaneously in a pocket of skin.

38.1.1 The Programming Device

When the pacemaker is programmed, the system should be optimally adapted to the needs of the patient, and the

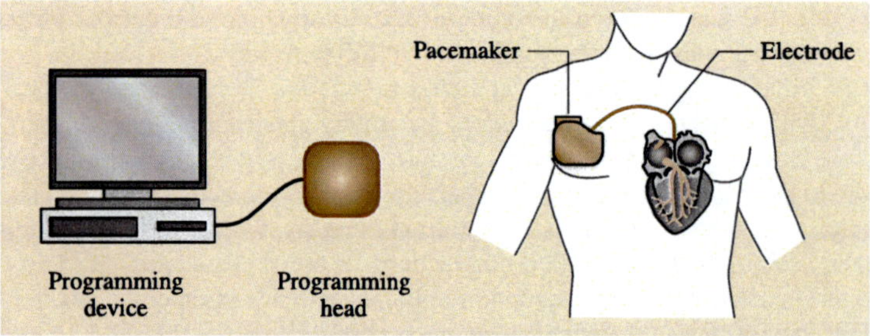

Fig. 38.1 Structure of a pacemaker system

lifetime of the battery should be maximized. With the long running times of today, it can't be assumed that the conditions on the day of implantation will remain constant over the entire lifetime of the pacemaker. It is therefore necessary to be able to adapt the pacemaker to changing conditions.

External devices that are usually based on a PC platform and drive manufacturer-specific short-range telemetric equipment are used for programming. This data transmission, usually based on inductive near-field coupling in a range below 100 kHz, allows the implant to be queried/reprogrammed. To do this, the programming head is first placed on the skin above the pacemaker housing. Afterward, the current pacemaker parameters, as well as diagnostic values (markers, event histograms, ECG, etc.), are transmitted out and evaluated. Finally, the updated parameters are reloaded into the implant. The most significant programmable parameters include the following:

- Stimulation frequency (lower and upper limits)
- Pulse amplitude and duration
- Input sensitivities
- Refractory periods
- Hysteresis (difference between the intervention and stimulation frequency)
- Av delay and hysteresis
- Pacemaker mode

The number of programmable parameters is meanwhile in the triple digits, and optimum pacemaker programming requires special knowledge, requiring supplemental education, even for cardiologists. For this reason, automatic functions are being increasingly implemented that allow the implant to adapt itself. In addition, Web-based expert systems are being considered, or are already in the test phase, that are supposed to ensure the constant availability of the most current knowledge. For more detailed information, one must refer to the respective pacemaker manuals or the corresponding special literature [38.1–4].

38.1.2 The Pacemaker

The pacemaker itself consists of a battery, the electronics, a hermetically enclosed titanium housing, and an epoxy resin connector for accommodating the electrode (Fig. 38.2). Nowadays, lithium-iodide batteries (open-circuit voltage about 2.8 V, capacitance about 1 Ah, inner resistance a few 100 and 50 kΩ) are usually used as a power source. With this type of battery, the anode is

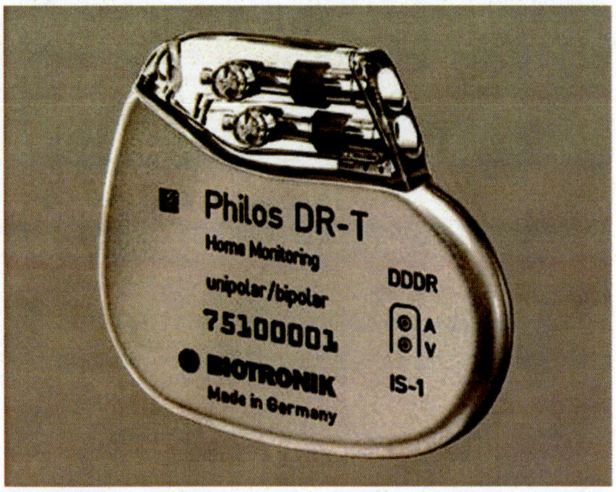

Fig. 38.2 Cross-section of a DDD pacemaker (source: Biotronik)

made of lithium and the cathode of iodine. In addition to its extraordinarily low self-discharge (< 1%/year), the lithium-iodide battery offers high stability of the inner resistance beyond the discharge time. Only at the end of the service life does the open-circuit voltage rapidly drop. Lithium-iodide batteries thus offer the greatest possible reliability with their long lifetime, small dimensions, and light weight. Today, they have running times ranging from 5 to 10 years with this technology [38.3].

The trend toward primary cells, or rechargeable batteries, which are at the beginning of their technical development and are occasionally discussed even today, offer no major advantages. The supposed prolonged service life must be weighed against the drastic increase in aging-related failures after about 10 years, which makes an implant change recommendable after this time, anyway.

The electronics are usually designed as a hybrid, multichip module. The once-used thick film technique is being increasingly replaced by multilayer ceramics due to the increasing complexity of the circuits. Recently, brand new PCB technologies (e.g., Dycostrate) have been attracting interest, which, thanks to their flexibility, allow electric connection technology to be considerably simplified, thus improving the device quality.

38.1.3 The Pacemaker Electrode

The pacemaker electrode or probe establishes the connection between the pacemaker and the heart. It consists of a connector, i. e., the plug for connection to the

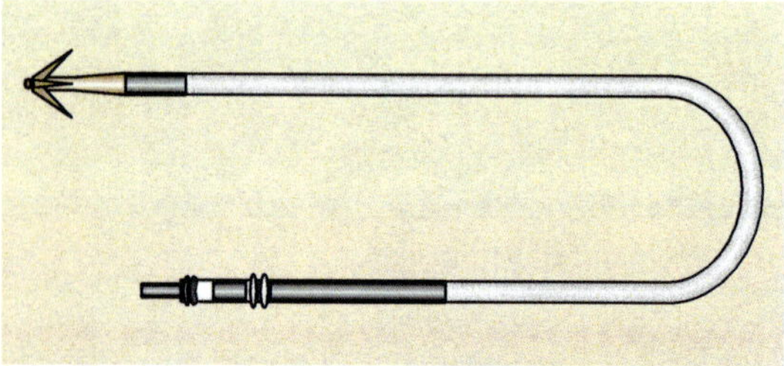

Fig. 38.3 Schematic structure of a pacemaker electrode

pacemaker, the electrode conductor, and the electrode tip (Fig. 38.3). The connector configuration has been standardized according to the IS-1 standard [38.4], so that all modern electrodes can be connected to any pacemaker.

One of the main technical challenges is the high bending stress of about 40 million load changes per year. For this reason, electrode conductors nowadays are made of four individual coiled wires, which provide the highest possible reliability with high flexibility. With a coiled wire, the bending radius is reduced, which reduces the alternating load stress on the feed line material (usually MP35N stainless steel). For this reason, electrode breakage is no longer a major problem.

The newest developments in the area of multifocal stimulation (Sects. 38.3.6 and 38.3.7) sometimes require very thin electrodes. In these cases, so-called DFT wires are used, which are special wires with a sheath made of MP35N and a silver core. These guarantee low electrical resistances while also having high mechanical strength.

Special demands are placed on the insulation material. It must be biocompatible and also be able to withstand the mechanical and chemical loads in the body. Right now, insulation made of silicone and polyurethane are being used [38.3, 4]. Polyurethane has a much lower friction coefficient, which is why it is especially preferred when implanting several electrodes via the same vessel. Clinical experience has shown, however, that when there is minor contamination, which can hardly be controlled in production, this can lead to hydrolysis, and thus to possible aging/brittleness, which ultimately leads to a life-threatening short-circuit. For this reason, several tens of thousands of electrodes from different manufacturers had to be removed/replaced a few years ago, which is why silicone is the preferred insulation material today.

Furthermore, a distinction is made between uni- and bipolar pacemaker electrodes (Fig. 38.4). Unipolar means that the electrode tip acts as the cathode and the pacemaker housing (or another counterelectrode with a large surface area) acts as the anode. Due to the considerably greater surface area of the pacemaker compared to the electrode tip, the housing is also called an *indifferent pole*. Bipolar probes also work with a cathodic tip. The anode, however, is also placed in the distal electrode area (about 2.5 cm away from the tip). Bipolar probes are somewhat thicker and more rigid than unipolar electrodes since they have two feed line coils. It is hardly possible to repair electrode breakage.

Important advantages of bipolar probe technology are the insensitivity to electric potentials whose origins lie outside the area where the electrode tip was implanted (potentials of the skeletal muscle, P-wave for ventricular probes, R-wave for atrial probes) and the lower risk of muscle contractions, even if higher energy is delivered [38.1, 3]. In addition, interference due to electromagnetic couplings in the feed lines cancel each

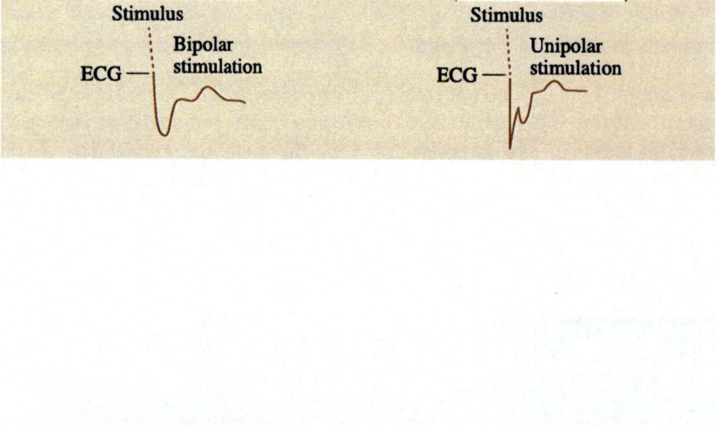

Fig. 38.4 Electric field distribution and ECG for uni- and bipolar stimulation

other out. A widespread compromise is the implantation of bipolar probes in the atrium of the heart (a smaller potential requires a higher sensitivity and the suppression of interference signals) and unipolar probes in the ventricle area.

The electrode head has direct contact with the endocardium. Its size, geometry, and surface structure are major determining factors for the stimulation and detection properties. A porous or fractal surface has proven advantageous [38.5]. The probe head is anchored using various fixation systems (Fig. 38.5). One distinguishes between electrodes with passive fixation (barblike anchor systems out of the insulation material of the electrode) and those with active fixation (system of screws on the electrode head). The former are preferred for electrode anchoring in the strongly divided surface of the right ventricle (trabeculae structure), and the latter is used for anchoring in the smoother right atrium. A hybrid type is the so-called J-electrodes, which ensure stable contact with the wall thanks to the J-shaped hook at the beginning of the electrode feed line. They are mainly used in the atrium. Probes without a fixation mechanism are used for transient pacemaker stimulation in emergency situations and can be easily removed again.

All described electrodes are fed transvenously to the right heart. They stimulate the inside of the heart and are therefore referred to as endocardial electrodes. Epicardial electrodes, which are applied to the outside of the heart, are no longer used for permanent pacemaker therapy. At most, they play a minor role in pediatric applications.

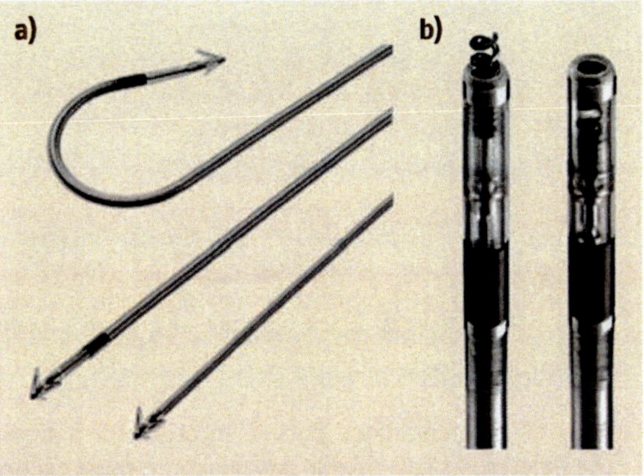

Fig. 38.5a,b Passive (a) and active (b) anchorable probe heads of cardiac pacemaker electrodes. The top passive electrode is a so-called J electrode. Below this follow a bipolar and a unipolar standard electrode. The screw electrode, which must be actively anchored, is shown once with an extended tip (ready for screwing into the myocardium) and, next to this, with a retracted tip (to avoid complications when inserting the electrode in the heart)

Pacemaker electrodes are hollow on the inside. This allows a reinforcement wire (mandrin) to be inserted during implantation, which makes it easier to push in the electrode. The mandrin is removed again before the electrode is connected to the pacemaker.

38.2 The Functionality of a Cardiac Pacemaker

A cardiac pacemaker not only blindly stimulates at a defined rate but attempts to synchronize itself to the still-existing heart rhythm and only to intervene when needed. Its structure can thus be divided into three basic modules: (1) the input stage for detecting intracardial electrical signals, and thus physiological events, (2) the output stage for stimulating the heart, and (3) a control unit, which takes the physiological demands, such as the heart rate, synchronicity, etc., into account (Fig. 38.6). The complexity of the control unit is determined to a great degree by the type of illness and, therefore, the stimulation mode to be used. A special section is dedicated to this aspect (Sect. 38.3).

The Sensing Function

The input sensitivity of a pacemaker system describes its capability to detect electric signals from the heart in the millivolt range and to interpret them as a contraction of the atrium/ventricle (perception, detection or sensing function). The fundamental structure of the input stage required for this is equivalent to that of an ECG amplifier [38.2], whereby the boundary conditions of an implant (size, supply voltage, energy requirement) sometimes demand special solution approaches.

To differentiate between physiological heart signals and interference signals, input filters are used in pacemakers. The filter characteristics have now been standardized (e.g., DIN 7505, Part 9) in order to establish binding safety standards. The greatest input amplification of a pacemaker after this stage lies between 30 and 70 Hz [38.6].

Furthermore, in clinical practice, an indirect measure for the frequency content of the electrical signal, the maximum slew rate, which is easy to determine by

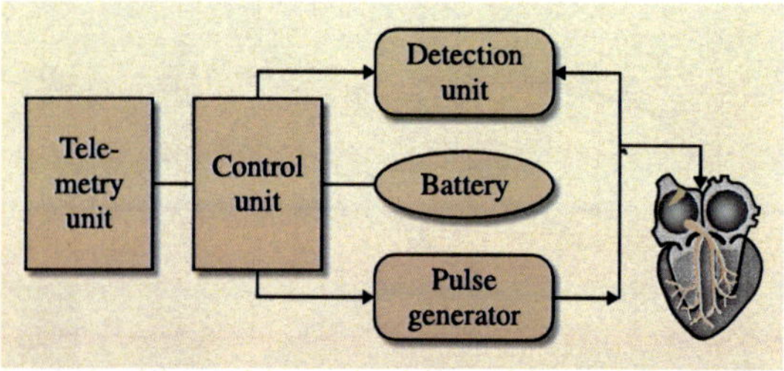

Fig. 38.6 Fundamental block diagram of a cardiac pacemaker

means of differentiation, is used. In order for a signal to be detected as such by the pacemaker, it must exceed a certain slew rate. The limit values of the slew rate lie at 0.5 V/s for the atrium area and at 1 V/s in the ventricle.

Another parameter for defining the perception function is the signal's amplitude. The so-called perception threshold or sensitivity of a pacemaker is usually programmable. In the atrium, one strives for an amplitude of 2.5 mV. In the ventricle, values up to 10 mV are to be registered. Often, real signals lie considerably below this, however [38.1].

Another special feature is so-called *blanking*. The detection channel is connected to the same electrode as the output stage. The amplifier would be overdriven by the stimulation impulse and be put out of operation over a longer period of time. For this reason, shortly before delivering a pulse, the inputs are decoupled by a transistor, which is generally referred to as *blanking*. A few milliseconds after the stimulation pulse, the blanking switches are closed again.

Logical blanking follows physical blanking in the sense of a switch. This means that no detection events within a certain window after the pulse are evaluated by the following circuit. There are several reasons for this. For one thing, vibration processes, which are unavoidable when an amplifier is switched on again, for example, due to small voltage differences, are not misinterpreted as physiological events. For another thing, refractory periods can be introduced this way, similar to cellular behavior, where an input amplifier is switched to become insensitive. The advantages of such a refractory period are explained in more detail in Sect. 38.3 with the introduction of various pacemaker modes.

38.2.1 The Stimulation Function

The behavior of a stimulation system is determined by the actual impulse generator, the electrodes required for coupling, and the tissue through which the excitation current flows. A very detailed description of this system can be found in [38.5].

The Pulse Generator

Nearly all pulse generators are based on the principle of capacitor discharge, i. e., a so-called reservoir capacitor C_{res} is charged to a certain voltage via the switch S_1 during the passive phase and discharges itself into the tissue to be excited in the active phase via switch S_2 and an additional coupling capacitor C_c (Fig. 38.7). If the required stimulation voltage lies above the battery voltage U_{bat}, a charge-pumping circuit downstream multiplies the voltage by the factor n.

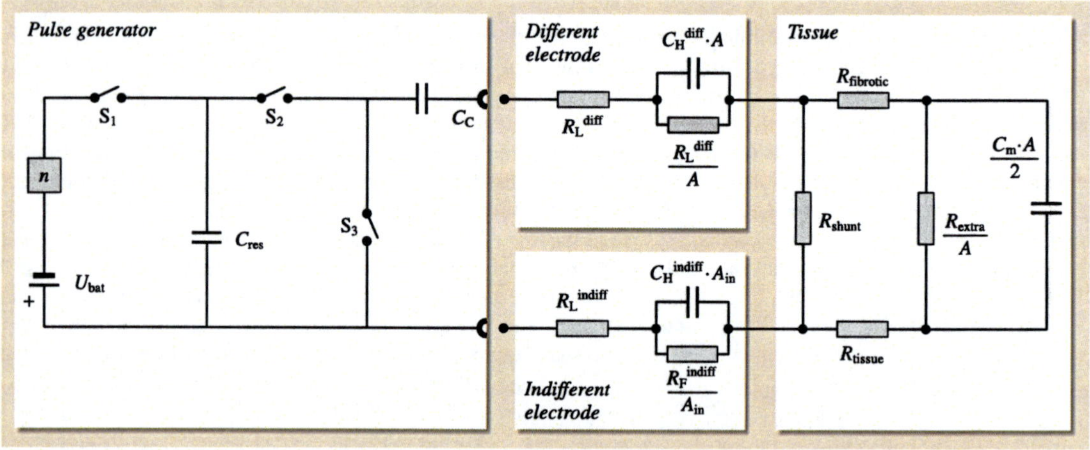

Fig. 38.7 Equivalent circuit of a stimulation system, consisting of a pulse generator, electrodes, and tissue

This setup prevents the battery from being directly coupled with the myocardium in the event of a single component defect (e.g., defective switch), which would lead to electrolysis and thus to damage to the surrounding tissue. This principle increases the safety of the stimulators and is referred to as *single failure safety*. The output voltage and current are thus in no way constant with respect to time, but drop nearly exponentially, depending on the impedance of the external components.

Since capacitive components (C_c as well as the limiting phase capacitances of the electrodes) are charged by the stimulation pulse, effective discharging between the individual stimuli must be ensured. This is done using a so-called autoshorting switch S_3, which is closed during the passive phase.

The Electrodes

From an electrical engineering point of view, pacemaker electrodes are solid-state electrolyte interfaces. A detailed description of such systems can be found in [38.2]. They therefore have capacitive properties due to the Helmholz layer and possibly oxidized top layers. C_H describes the specific Helmholz capacitance in what follows, which only delivers the capacitance that is interesting for the circuit after multiplying with the geometrical electrode surface A. This way, surface modifications, which lead to a roughening and thus to a larger electrochemical contact surface, can be easily considered in the model. R_F correspondingly describes the Faraday resistance and R_L stands for the lead resistance.

The Tissue

The excitation of at least one heart muscle cell (myozyte) is required for electrically stimulating the heart, i. e., a depolarization of its cell membrane. In its resting state, the transmembrane potential is about −90 mV. If this value is increased by a depolarization voltage U_{dep} of about 30 mV, automatic depolarization, and thus the contraction, is initiated.

For stimulation using extracellular electrodes, a displacement current is therefore necessary beyond the cell membrane. The first approximation of the electrical equivalent circuit for the target organ consequently consists of a specific cell membrane capacitance C_m and a parallel ohmic resistance R_{extra}, which takes the extracellular space into account. There are three other approximately ohmic contributions in addition to this. $R_{fibrotic}$ represents the resistance of the thin layer of fibrotic tissue, which develops after implantation between the electrode and the place of stimulation. R_{tissue} describes the resistance of the tissue between the place of stimulation and the counterelectrode. R_{shunt} stands for a leakage current, which is not coupled into the target organ via the blood or other tissues, but flows directly into the counterelectrode.

All named elements are at least known in terms of their order of magnitude, so that quantitative estimations are also possible (Table 38.1). Using the Laplace transformation and simplifications

$$C_1 = \left(\frac{1}{C_{res}} + \frac{1}{C_c} + \frac{1}{C_H^{diff} A} + \frac{1}{C_H^{indiff} A_{in}} \right) , \quad (38.1)$$

Table 38.1 Summary of the used parameters and their number values (after [38.7–10])

Parameter		Range	Average	Source
Reservoir capacitance C_{res}		5–20 μF	10 μF	10
Coupling capacitance C_c		5–20 μF	10 μF	10
Spec. Helmholtz capacity C_H	Polished surface	0.1–0.4 μF/mm²	0.2 μF	11
	Fractal surface	10–500 μF/mm²	40 μF	12
Lead resistance R_l		10–100 Ω	50 Ω	10
Geom. electrode surface A		1–20 mm²	10 mm²	10
Surface of the indifferent electrode A_{in}	Unipolar	5–15 cm²	10 cm²	10
	Bipolar	20–100 mm²	50 mm²	10
Pulse duration T		0.1–2 ms	0.5 ms	10
Membrane capacitance C_m			0.01 μF/mm²	13
Tissue resistance R		30–70 kΩ	40 kΩ	14
Shunt resistance R_{shunt}		0.1–2 kΩ	600 Ω	10
Depolarization voltage U_{dep}			30 mV	15

$$C_2 = \frac{C_m A}{2}\,, \tag{38.2}$$

$$R = R_{\text{fibrotic}} + R_{\text{tissue}}\,, \tag{38.3}$$

$$R_1 = R_F^{\text{diff}} + R_F^{\text{indiff}}\,, \tag{38.4}$$

$$R_2 = R_{\text{shunt}}\,, \tag{38.5}$$

$$a = (R_1 R_2 + R_1 R + R_2 R)\, C_1 C_2\,, \tag{38.6}$$

$$b = (R_2 + R)\, C_2 + (R_1 + R_2)\, C_1\,, \tag{38.7}$$

and the two poles p_1 and p_2

$$p_1 = -\frac{b - \sqrt{b^2 - 4a}}{2a}\,, \tag{38.8}$$

$$p_1 = -\frac{b + \sqrt{b^2 - 4a}}{2a}\,, \tag{38.9}$$

yield the minimum pulse voltage, referred to as the stimulation threshold, required for a successful stimulation

$$U_{\text{thr}} = \frac{2U_{\text{dep}}}{C_1 R_2} \frac{\sqrt{b^2 - 4a}}{\exp(p_1 T) - \exp(p_2 T)}\,. \tag{38.10}$$

Similarly, one also gets a relationship for the charge (important for calculating the service life), which must be taken from the battery for every stimulus [38.5]. The specific Helmholz capacitance has special meaning for the development of new systems. With increasing C_H, both the stimulation threshold and the charge threshold decrease. Clinically, this means that electrodes with fractal surfaces have the lowest energy consumption, and thus guarantee the longest implant service life. In order to maximize the ohmic parts of the tissue at the same time, the electrode head should be as small as possible. This design – a geometrically small head with a high specific interface capacitance – has become a standard under the term *highly ohmic electrodes* in the last 5 years.

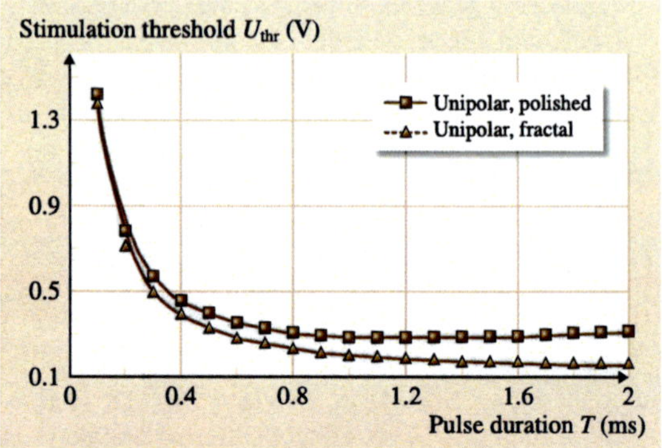

Fig. 38.8 The stimulation threshold as a function of pulse duration. An effective stimulation can be achieved for amplitude values lying above the curve

Figure 38.8 shows the dependence of the stimulation threshold on the pulse duration. From the output voltage perspective, it is recommended to choose a large pulse duration in order to avoid switching on a charging pump, which involves high conversion losses. At the same time, however, it has become evident that longer pulse durations increase the charge consumption nearly linearly. From a clinical aspect, then, it makes sense to select the shortest pulse duration that is still possible without using a charging pump. To be able to specify a simple rule of thumb for optimizing the pulse parameters, the two terms chronaxy and rheobase have been empirically introduced, which can be determined from the stimulation threshold curve. The rheobase is a theoretical value and describes the minimum voltage that triggers an electric cardiac response for an infinite pulse duration. The chronaxy is the pulse duration at the stimulation threshold for a voltage twice the strength of the rheobase. Based on experience, the energetic optimum lies near the chronaxy [38.1].

38.3 Stimulation Modes

From a physiological standpoint, the two atria/two ventricles are always excited together. For this reason, a maximum of two stimulation sites – one in the atrial or one in the ventricular plane – and, correspondingly, a maximum of only two detection sites are required. Based on this, the generally recognized NBG code was developed for defining stimulation modes.

The multifocal modes, which were recently introduced for special heart disorders (interatrial conductive disturbance, cardiac insufficiency), allow biatrial/biventricular stimulation, but are not yet included in the NBG code.

38.3.1 International (NBG) Pacemaker Code

NBG code (NASPE/BPEG Generic Pacemaker Code), which has been applied since 1988, describes the global function of a pacemaker with the specification of a maximum of five letters, of which the first three are always used, and the fourth and fifth optionally [38.3].

The first letter refers to the stimulation site:

V: Ventricle: Stimulation in the ventricle only
A: Atrium: Stimulation in the atrium only
D: Dual: Stimulation in both the atrium and the ventricle
S: Single: Single-chamber stimulation in the atrium or ventricle
0: No stimulation

The second letter stands for the place of perception:

V: Ventricle: Detection in the ventricle only
A: Atrium: Detection in the atrium only
D: Dual: Detection in both the atrium and the ventricle
S: Single: Single-chamber perception in the atrium or ventricle
0: No perception

The third letter defines the operating mode, i. e., the pacemaker function, which is triggered by a perceived signal:

I: Inhibited: The pacemaker stimulation is suppressed
T: Triggered: A perceived signal leads to the pacemaker delivering the pulse
D: Dual: Inhibition and triggering
0: No inhibition and no triggering

The fourth letter describes the programmability, telemetry, and frequency adaptation:

0: Not programmable
P: Programmable: Up to two functions are programmable
M: Multi programmable: More than two functions are programmable
C: Communication: Data telemetry possible
R: Rate modulation: Adaptation of the pacemaker frequency to a load-induced signal

The fifth letter refers to the antitachyarrhythmia function:

0: No antitachyarrhythmia function
P: Pacing: Antitachyarrhythmia stimulation
S: Shock
D: Dual: Pacing and shock

A variety of possible pacemaker modes result from combining different detection/stimulation channels together with different operating modes. A detailed explanation can be found in [38.2]. In what follows, the two most common modes, the VVI and the DDD modes, will be described in more detail.

38.3.2 VVI Pacemaker

The VVI pacemaker is a single-chamber pacemaker, where one single electrode is anchored in the right ventricle. It detects the ventricular activity in addition to the fixed-rate ventricular stimulation of the V00 and inhibits its output when an intrinsic ventricular event occurs. Thus, a VVI system only stimulates when required, i. e., when the natural excitement fails, which is why it belongs to the group of so-called *demand pacemakers*. A schematic diagram is shown in Fig. 38.9.

The flow diagram in Fig. 38.10 shows the principal mode of operation. The basic interval and the refractory period of the ventricular detection channel are started at the same time. If an intrinsic event is detected within the refractory period, it is interpreted as interference and is discarded. If it is detected outside of the refractory period, the basic interval and refractory period counters are reset. If no R-wave is detected, the VVI pacemaker behaves like a V00 and stimulates after the basic interval is over.

In the functional diagram in Fig. 38.11, the ventricular stimuli are marked with "V". As compared to the V00 principle, R-events are added now, which indicate

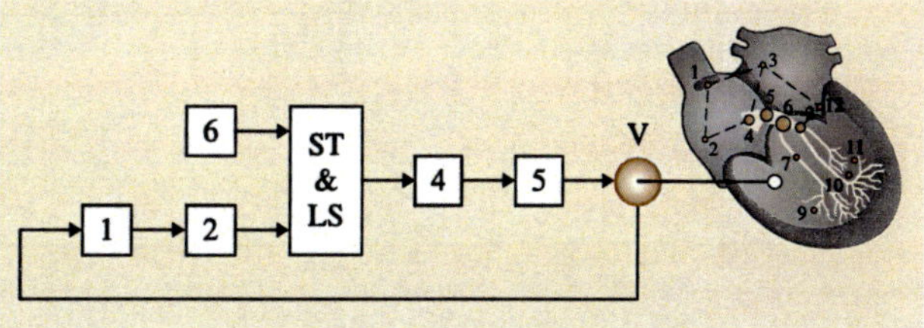

Fig. 38.9 Schematic diagram of the VVI pacemaker. The ventricular electrode (V) leads to a control and logistics circuit via an input amplifier **(1)** and a monoflop for checking the ventricular refractory period **(2)**, which also evaluates the output of the basic interval counter **(6)**. The stimulation channel is equivalent to that in the V00 pacemaker

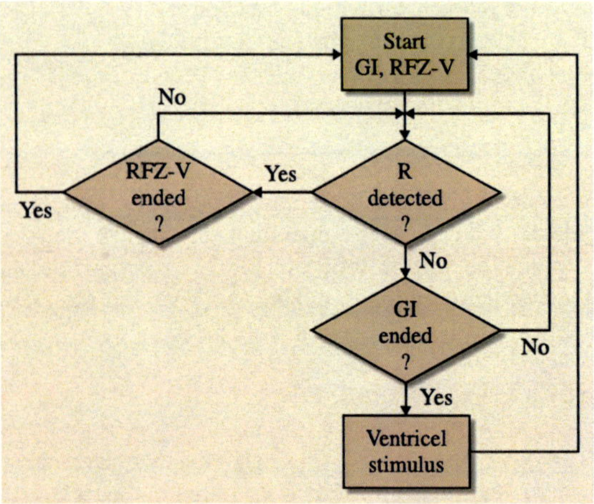

Fig. 38.10 Flow diagram of a VVI pacemaker

the detection of an R-wave. A third marking, plotted below the counter display, indicates the end of the artificial refractory period (RFZ-V) of the ventricular detection channel.

In the ECG, the first two stimulated events (V) are shown, which were triggered by the end of the basic interval. A natural event (R) follows this, which was detected outside of the refractory period and resets the counter. The pacemaker doesn't deliver a stimulation pulse, i. e., it is in inhibiting mode. Afterwards, there is an escape interval, since no intrinsic event is detected. This is followed by another normal basic interval. The ventricular extrasystoles (VES) that follow the stimulated event (V) are detected as an intrinsic cardiac excitement since they are outside of the refractory period, which causes the counter to be reset. Afterward, another escape and a basic interval follow with the associated ventricular stimuli.

Clinical Assessment of the VVI Pacemaker

The VVI pacemaker has an energy-saving mode. It only becomes active when there is no intrinsic cardiac activity (demand pacemaker). It also does not lead to parasystoles and avoids stimulation in the vulnerable phase. Like the V00 pacemaker, however, it is not able to adapt the heart rate to physical exertion. The main problem, however, is the loss of atrioventricular synchrony, which can trigger pacemaker syndrome.

VVI Pacemaker Application

Bradycardias can be generally treated with a VVI pacemaker. Extensive studies have proven in the last few years, however, that the mortality of patients with a VVI pacemaker are considerably higher than that of synchronously stimulated patients. Therefore, the use of VVI pacemakers only makes sense for patients with atrial flutter or fibrillation in connection with an AV block, as well as atrial paralysis (atrium which is not able to contract).

38.3.3 Two-Chamber Pacemaker

In the case of the two-chamber cardiac pacemakers, electrodes are applied to both the atrium and the ventricle to ensure synchronous stimulation. Two principal function options result from this – atrium- and ventricle-based actions. In the case of the atrium-controlled sequence, the basic interval counter and the newly added AV counter (which simulates the natural PR interval) always start at the same time. The ventricular-based sequence, on the other hand, uses a series connection of the two counters, whereby the AV counter must first run down before the timer can be started, and vice versa. The different approach can be especially well explained using the example of a DDD pacemaker.

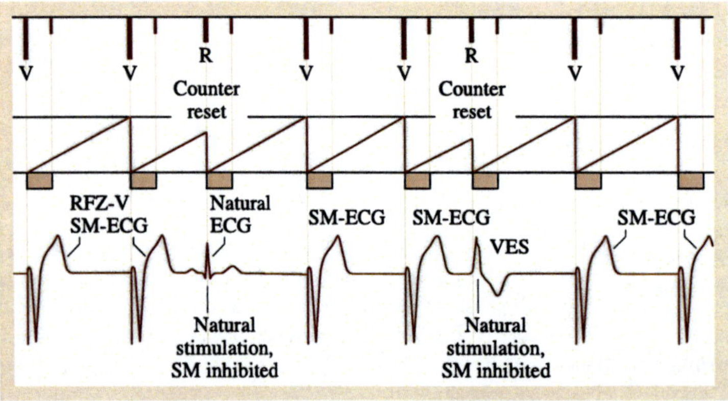

Fig. 38.11 Functional diagram of a VVI pacemaker

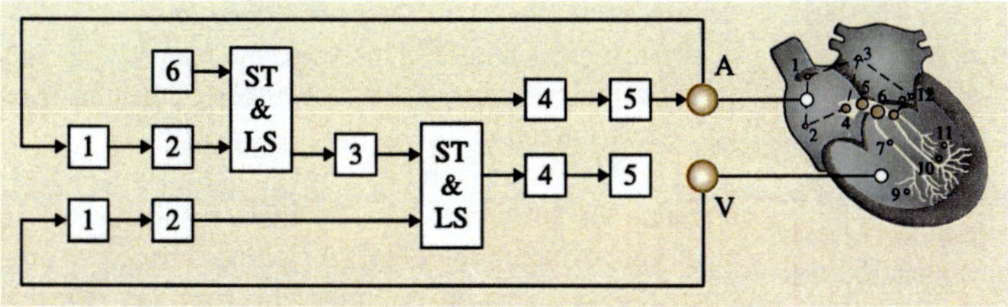

Fig. 38.12 Atrium-controlled DDD pacemaker. Both the atrium and the ventricle have a detection and stimulation channel. The basic interval counter **6** is connected to the atrium control unit, which influences the ventricular control unit via the AV counter

The Atrium-Controlled DDD Pacemaker

The disadvantages of the DVI pacemaker, especially the danger of stimulation of an intrinsic atrial event in the vulnerable phase, are remedied with the DDD pacemaker (Fig. 38.12), which, from the rhythmology point of view, is a kind of *all-rounder* among pacemakers. It combines the AAI, VAT, and VVI functional modes.

Of all the pacemakers, the DDD pacemaker comes closest to simulating the physiological function of the heart. DDD means that stimulation and detection are both possible in the ventricle and the atrium. The pacemaker also masters the trigger and inhibition mode.

The AV interval is started after every atrial excitation (natural or after the basic interval has been artificially induced). If no ventricular heart activity is detected, the ventricle is stimulated at the end of the AV interval. Ventricular extrasystoles (VES) reset the basic interval counter without starting an AV interval. This

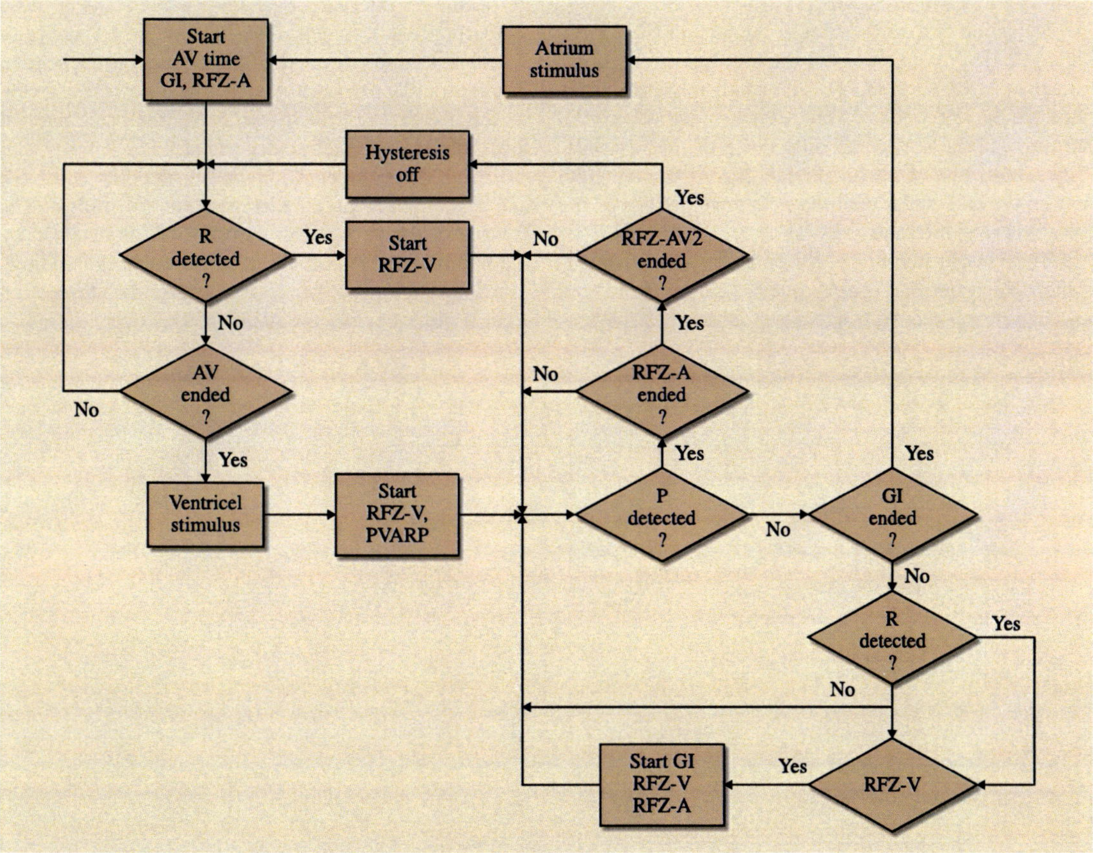

Fig. 38.13 Flow diagram of an atrium-controlled DDD pacemaker

automatically leads to a compensation pause. The basic interval counter is reset as a consequence when one of the following three events happens

- An atrial stimulus is delivered,
- A natural atrial excitation has been detected,
- A natural spontaneous ventricular excitation occurs outside of the AV interval (i. e., a ventricular extrasystole).

In addition, after each event, the respective refractory periods are started. As a special feature, an atrial refractory period should be mentioned here, which begins after ventricular events, in addition to those already described. The task of this so-called postventricular atrial refractory period (PVARP) is to detect ventricular stimuli or to suppress their retrograde feedback to the atrium. Without these protective measures, this could easily lead to one's own output signal being self-detected, and thus to pacemaker-induced tachycardia. Details are shown in Fig. 38.13.

Detection and stimulation in the atrium (P, A) and ventricle (R, V) are shown in the first line of the functional diagram (Fig. 38.14). The small upward dashes indicate the end points of the refractory periods of the atrium, and the downward ones those of the ventricle. The second line of the functional diagram contains the basic interval counter with the refractory periods of the ventricles and the atria.

The bottom line is the AV counter, which has so-called AV hysteresis. The AV interval can be prolonged by means of a second trigger level as soon as the preceding atrial event has been stimulated and was not of a physiological nature. This way, the transit-time differences are compensated, which ensures constant good AV synchronization.

The third line in the functional diagram (Fig. 38.14) contains the surface ECG. The ECG first indicates an atrial pacemaker stimulation (A) with a subsequent artificial P wave. After the AV interval is over, the ventricle is stimulated and the PVARP is begun. After the timer has expired, the above-described cycle repeats itself.

Now a detected natural atrial excitation (P) follows. The basic interval counter is reset and an AV interval is started. Since this is a natural P wave, no AV hysteresis is used. During the AV interval, no ventricular excitation is detected, and therefore, after it is over, a ventricular pacemaker stimulus is delivered (V) and the PVARP timer is started.

Then the basic interval timer runs out, after which a pacemaker pulse is generated in the atrium (A). Even before the AV interval is over, the R wave of the intrinsic ventricular event is detected. The intended ventricular stimulus is inhibited and the PVARP timer is not started. The ventricular extrasystole that now follows is detected by the pacemaker, which then resets the basic interval timer and starts the refractory period.

After the compensation pause, another atrial (P) and a ventricular pacemaker stimulus (V) follow. The P wave of the subsequent natural ECG is detected by the pacemaker (P) and the atrial stimulus is inhibited. The detected R wave prevents a ventricular stimulation since it occurs before the AV interval is over.

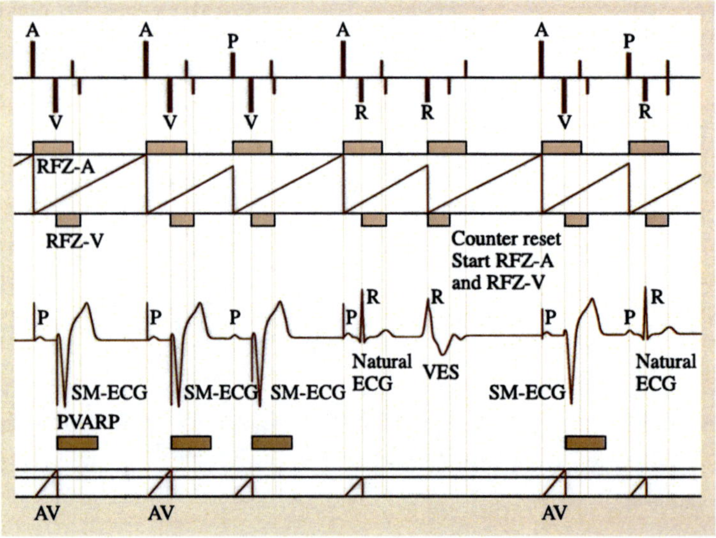

Fig. 38.14 Functional diagram of an atrium-controlled DDD pacemaker

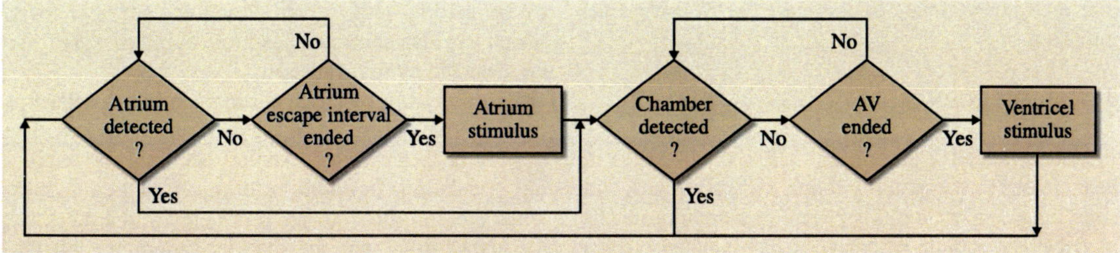

Fig. 38.15 Flow diagram of a ventricle-controlled DDD pacemaker

DDD Pacemaker (Ventricle-Controlled)

The decisive advantage of the DDD system – the complete synchronicity of both chambers – can also be achieved with time control on the ventricular level, of course. In the case of ventricular-based timing, the two counters, AV and the basic interval (here also referred to as the atrial escape interval), are in series and are never active at the same time (Fig. 38.15).

First, measurements are made in the atrium. If an intrinsic atrial event occurs, the AV counter is started and an attempt is made to detect signals in the chamber. If no atrial excitation occurs, an artificial atrial stimulation is delivered at the end of an atrial escape interval, and then measurement continues in the chamber. If a natural excitation occurs in the chamber, measurements are taken in the atrium and the associated timer is started. If this is not the case, another ventricular stimulus is delivered after an AV interval, the counter is reset, and measurements are taken from the atrium. For reasons of simplicity, refractory periods were omitted here.

Clinical Assessment of the DDD Pacemaker. This pacemaker comes closest to simulating the physiological function of the heart. With this pacemaker, no pacemaker syndrome is possible. It has an energy-saving mode of operation. Only in the case of chronotropic insufficiency is the system unable to adapt to the physical stress.

Detecting the atrial electric action is dangerous when the atria have tachycardiac arrhythmia. To prevent ventricular tachycardia from occurring here, the 1:1 atrium/ventricle conduction frequency is limited. If the atrial frequency exceeds this limit, referred to as the maximum synchronization frequency, then there is only partial conduction to the ventricle. For example, the AV delay increases from beat to beat until a beat is no longer conducted and finally the cycle starts over (Wenckebach mode). Another possibility is to just conduct every second atrial action to the ventricle. A retriggerable atrial refractory period was introduced under the term *dual demand*, which leads to a variable prolongment when tachyarrhythmias occur, and is therefore also considered to be a protective function. Most of these methods, however, have the disadvantage that they abruptly change the ventricle frequency when an atrial tachyarrhythmia occurs, which patients find unpleasant.

For this reason, more complex methods for reacting to tachycardiac atrial arrhythmia were introduced recently under the term *mode switching*. The precondition for this is an algorithm for automatically detecting atrial tachyarrhythmias. The state of the art nowadays is the so-called "x out of y" algorithm, where a programmable minimum number x of PP intervals out of a total number y of consecutive intervals must lie under a certain limit. In this way, even irregular atrial tachycardias are reliably detected. When such an arrhythmia sets in, the pacemaker switches over to a ventricular stimulation independent of the atrial action (e.g., VVI). After the end of the atrial tachyarrhythmia, there is another stimulation in DDD mode.

Application of the DDD Pacemaker. DDD pacemakers are suitable for:

- Patients with normal sinus node function and AV block,
- Rare atrial arrhythmia,
- Congenital long QT syndrome and torsade des pointes (special type of ventricular tachycardia).

A special indication for using two-chamber stimulation is hypertrophic obstructive cardiomyopathy. Due to the formation of muscle contractures in the outflow tract of the left ventricle, blood is prevented from flowing out. By choosing a very short AV delay and sometimes also a stimulation in the outflow tract, the muscular excitement can be modified such that the outflow tract obstruction is considerably reduced [38.1]. This pacemaker should not be used for supraventricular tachyarrhythmias. In this case, a VVI system is indicated.

38.3.4 Frequency-Adaptive Pacemakers

Frequency-adaptive pacemakers allow the heart rate to be increased in stressful situations, even in patients with sinus node dysfunction. They are therefore indicated for physically fit patients who have an insufficient heart rate increase under stress (chronotropic insufficiency). The frequency adaptation is achieved via sensors, which determine the stress-related parameters and relate these to the stimulation frequency (so-called *interference coupling*). Up to now, a multitude of different parameters have been tested [38.3], e.g.:

- Physical activity (vibration or acceleration sensors)
- Respiratory minute volume
- Blood temperature
- Blood oxygen content

Other parameters were tested at the same time, but only the physical activity measurement found practical application. Thanks to miniaturized acceleration sensors, it is technically easy to execute and responds immediately. The disadvantage is the nonphysiological sensor information. Furthermore, passive movements (e.g., driving over cobblestones) is misinterpreted. The state-of-the-art technology is therefore made of so-called *dual-sensor systems*, where the movement information is linked with a respiratory sensor signal based on the transthoracic impedance. This way, the reaction times are adapted and nonphysiological effects can be avoided by comparing both signals. Emotional stress or fever, which can also cause the heart rate to increase, are also not detected by this, however.

The currently most promising solution approach is the development of regulating pacemaker systems. The healthy body has several control elements for stabilizing circulation. In addition to an affected sinus node when there is chronotropic insufficiency, the AV node, the myocardium, and the vascular system also act regulatively. Regulatory systems are based on the basic assumption that the control variables of intact control elements can be used to determine the physiologically correct heart rate. For example, an increased inotropism indicates that a greater cardiac output is required, which is used to increase the stimulation frequency. Figure 38.16 makes these relationships clear. The following are currently being tested as input variables:

- AV delay (dromotropism)
- QT interval of the intracardial ECG signal
- Contractility of the myocardium ($\mathrm{d}p/\mathrm{d}t$)
- Motion dynamics of the myocardium (acceleration, impedance)

The dromotropic pacemaker is impressive with its elegance, but it can only be used with limitations, since some people with chonotropic insufficiency also have A-V dysfunction. The QT interval for detecting myocardial contractility has already proven itself clinically but under certain boundary conditions can exhibit positive feedback since other parameters also influence the chamber action potential. The two latter approaches seem to be especially promising now and are already be-

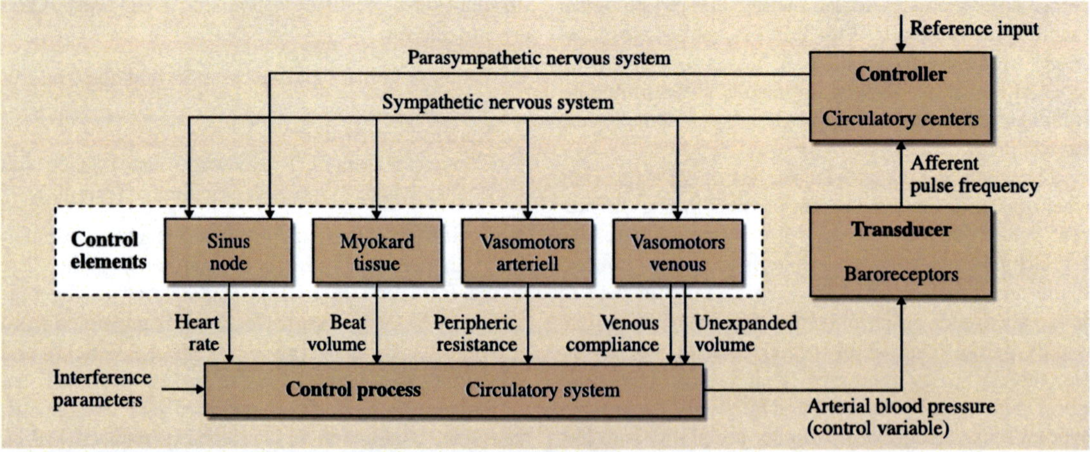

Fig. 38.16 Simplified control diagram of the baroreflex (after [38.7])

ing clinically tested in prototype implants. Their further clinical acceptance will depend on the degree to which they are able to guarantee long-term stability. Due to the sensors growing into the tissue, myocardial restructuring, or just simply aging, these systems are subject to drift, which requires recalibration.

38.3.5 Antitachyarrhythmia Pacemakers

One can attempt to end tachycardia both via a certain sequence of stimulation pulses as well as via electric shock. Since atrial tachycardia is being increasingly successfully treated with catheter-supported ablation methods, pacemaker treatment only plays a subordinate role here. The main indication for electric rhythmization treatment are dangerous tachycardiac ventricular arrhythmias. The corresponding systems are usually offered in connection with a defibrillation option (so-called ICDs).

38.3.6 Biatrial Stimulation Systems

It recently became clear that a pathological prolongment in the interatrial transition period can lead to atrial reentry tachycardia. For this reason, special pacemaker systems have been developed that allow several consecutive stimulations at short intervals in both atria. These so-called *biatrial stimulation systems* have meanwhile proven themselves in patients with prolonged P waves. Technically, they are like a DDD system, whereby, however, the second electrode comes to rest in the left atrium. The challenge here is the development of

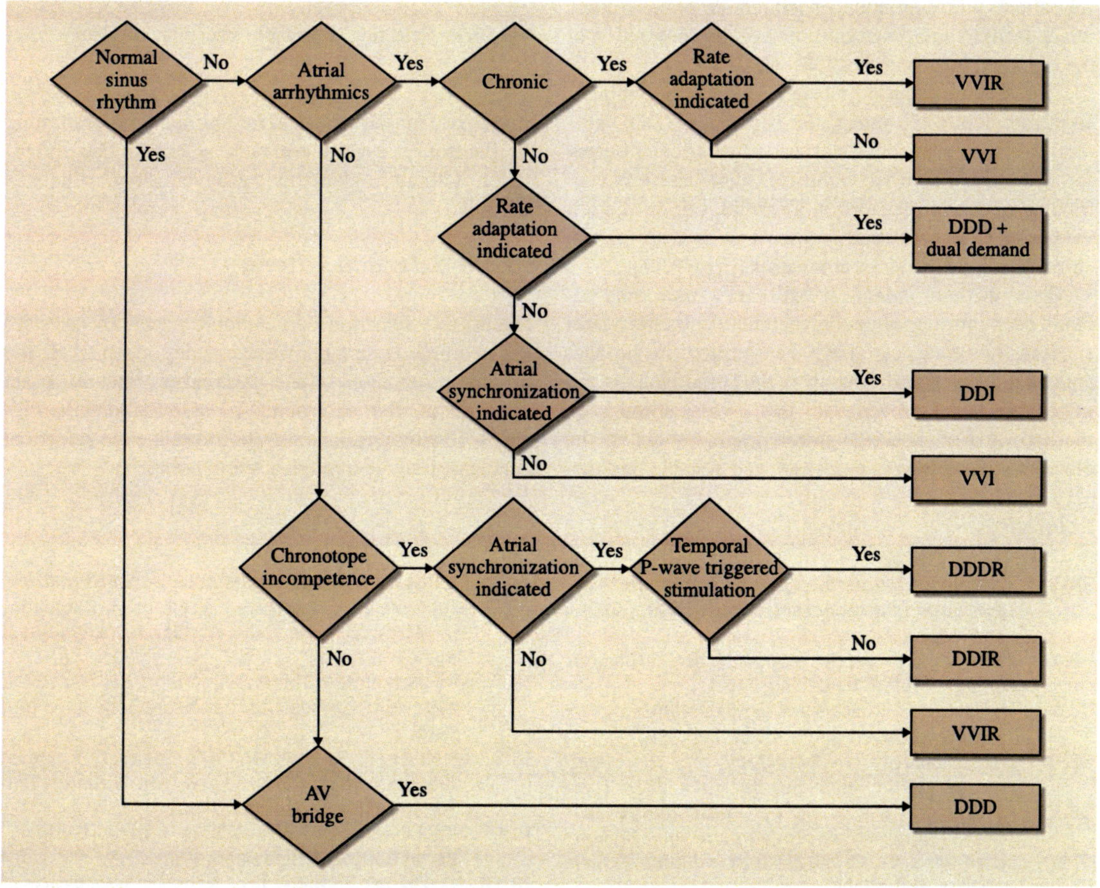

Fig. 38.17 Semischematic diagram of biventricular stimulation. Stimulation electrodes are located in the right atrium, right ventricle, and in a cardiac vein (V. cordis), which runs laterally over the left ventricle (LA left atrium, RA right atrium, LV left ventricle, RV right ventricle, V. cordis)

suitable electrode systems. The state of the art for stimulating the left atrium is access via the sinus coronarius. This way, the implant remains in the low-pressure system. This puts high demands on the flexibility of the electrodes and the skills of the implant surgeon, however.

38.3.7 Biventricular Stimulation Systems

The objective of biventricular stimulation is the treatment of serious myogenic cardiac insufficiency, independent of the cause that led to damage of the heart muscle. Although optimizing the heart rate and atrioventricular synchronization have already had favorable effects on patients with cardiac insufficiency these methods are often unsuccessful in achieving sufficient hemodynamic improvements. This is especially the case for patients with a pronounced left bundle-branch block, which leads to a desynchronization of the contraction of the right and left ventricles [38.11].

The technical basis of biventricular stimulation is an atrium-controlled ventricular pacemaker (DDD system), which has an additional second ventricular output for detection in the left ventricle. Here, the left ventricular electrode is placed in a cardiac vein in the left ventricle. The access path is the same as with biatrial stimulation, via the sinus coronarius (Fig. 38.17).

This method, which is still very new, has already been introduced in clinical practice. Its long-term success, however, can only be verified in broader studies [38.12]. Improvements in the hemodynamic situation are to be proven with this special stimulation method. In further development, improvement in the electrode technology is necessary, and suitable parameters for patient selection and for optimally programming the pacemaker system are to be selected. The combination with an implantable cardioverter/defibrillator is possible.

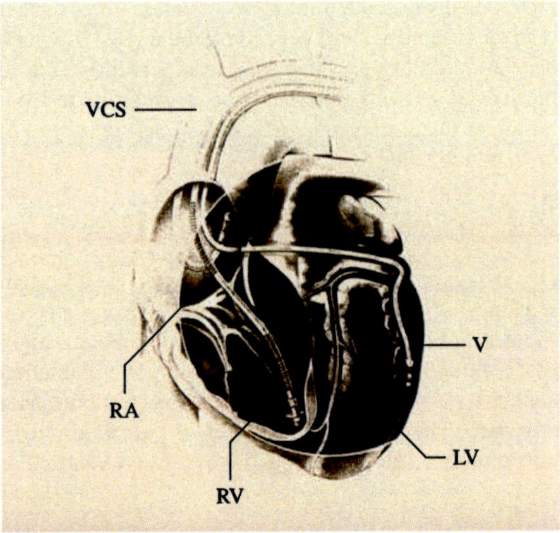

Fig. 38.18 Selecting a suitable pacemaker system (after 38.4)

38.3.8 Selection Criteria

The correct selection of a suitable pacemaker system still requires expert knowledge today, despite all the automatic functions of the pacemaker, since an exact diagnosis is often determined by essential details. Figure 38.18 attempts, however, to formalize the selection, to at least provide a rough starting point.

References

38.1 E. Alt: *Schrittmachertherapie und Defibrillatortherapie. Band 1 Schrittmachertherapie* (Spitta, Balingen 1995)

38.2 A. Bolz, W. Urbaszek: *Technik in der Kardiologie* (Springer, Berlin, Heidelberg 2002)

38.3 W. Fischer, P. Ritter: *Praxis der Herzschrittmachertherapie* (Springer, Berlin, Heidelberg 1997)

38.4 M. Schaldach: *Electrotherapy of the Heart* (Springer, Berlin Heidelberg New York 1992)

38.5 A. Bolz: *Die Bedeutung der Phasengrenze zwischen alloplastischen Festkörpern und biologischen Geweben für die Elektrostimulation* (Schiele & Schön, Berlin 1994)

38.6 M. Hubmann, R. Hardt, B. Priessnitz, R. Thull, E. Lang: Beeinflussbarkeit von Herzschrittmachern durch Warensicherungssysteme. In: *Schrittmachertherapie und Hämodynamik*, ed. by M. Hubmann, R. Hardt, E. Lang (MMV Verlag, München 1993) pp. 143–152

38.7 R.F. Schmidt, G. Thews: *Physiologie des Menschen*, 20th edn. (Springer, Berlin Heidelberg New York 1980)

38.8 F.M. Snell, S. Shulman, R.P. Spender, C. Moos: *Biophysikalische Grundlagen von Struktur und Funktion* (Hirzel, Stuttgart 1968)

38.9 K.J. Vetter: *Elektrochemische Kinetik* (Springer, Berlin Göttingen Heidelberg 1961)

38.10 E. Zheng, S. Shao, J.G. Wenster: Impedance of sceletal muscle from 1 Hz to 1 MHz, IEEE Transact. Biomed. Eng. **31**, 477–481 (1984)

38.11 A. Auricchio, C. Stellbrink, M. Block, S. Sack, J. Vogt, P. Bakker, H. Klein, A. Kramer, J. Ding, R. Salo, B. Tockman, T. Pochet, J. Spinelli: Effect of pacing chamber and atrioventricular delay on acute systolic function of paced patients with congestive heart failure, Circulation **99**, 2993–3001 (1999)

38.12 P. Ritter: Cardiac stimulation in heart failure: are we going too fast, are we going to far?, Europace **2**, 1–3 (2000)

39. Introduction to Neuroprosthetics

Klaus-Peter Hoffmann, Silvestro Micera

Neuroprosthetics is a comparatively young, dynamically developing subject with double-digit sales growth rates. Due to the preconditions on implantability, biocompatibility, and miniaturization, it is strongly linked to the development of microsystems technology, nanotechnology, information technology, biotechnology, and the application of new materials. The fields of application of neuroprostheses are diseases associated with impairments of myogenic or neurogenic functions. These can lead to the loss of the whole function. Neuroprostheses use electric stimuli to stimulate neural structures, muscles or receptors, in order to support, augment or partly restore the respective disordered or lost function. Functional disorders include paralysis after stroke, reduced hearing, tremor as an example of movement disorders, or the loss of an extremity. Often, the use of a neural prosthesis can improve the quality of life of the person concerned. The objective is to help the patient to participate in everyday life. Thus, cosmetic, ethical, and social aspects always have to be considered.

39.1 **Neuroprostheses** 785

39.2 **Application of Neural Prostheses** 787
39.2.1 Cardiac Pacemaker 787
39.2.2 Cochlear Implant 787
39.2.3 Deep Brain Stimulation 787
39.2.4 Bladder Stimulation 788
39.2.5 Human–Computer Interfaces 788
39.2.6 Extremity Prostheses 788

39.3 **Specific Technological Features** 788
39.3.1 Biocompatibility 788
39.3.2 Encapsulation 789
39.3.3 Energy Supply 789
39.3.4 Electronics 789
39.3.5 Signal Processing 790
39.3.6 Signal Transmission 790

39.4 **Biological–Technical Interface** 790
39.4.1 Electrodes 790
39.4.2 Interfaces for a Bionic Hand Prosthesis 792
39.4.3 Thin-Film Longitudinal Intrafascicular Electrode (tf-LIFE) 793

39.5 **Future Developments** 798

References 799

39.1 Neuroprostheses

Neural prostheses are applied with the objective of compensating an existing motor or sensory neural dysfunction. They electrically stimulate myogenic areas and neural structures in the peripheral, spinal, and central nervous system to supplement, replace, or restore neurological function [39.1]. However, neural prostheses can also be used to acquire and analyze bioelectrical activity from the mentioned neural structures to control technical systems (Fig. 39.1).

Two information transfers can be distinguished. Inward information transfer can be accomplished by activating any of the intact senses by appropriate stimuli (cochlear or retinal implant) or by direct electrical stimulation of neural tissue (brainstem implant). Outward information transfer can be done after recording bioelectrical signals associated with neural or muscular activation. These signals can be used for electrical stimulation of muscles (cardiac pacemaker) or neural tissue (bladder stimulator) or to control the movement of technical devices (hand prostheses) [39.2]. In some cases information transfer is not needed, for example in basic electrical stimulation of muscular or neural tissue (neuromodulation).

Figure 39.2 shows an overview of common neural prostheses. Cardiac pacemakers, cochlear implants,

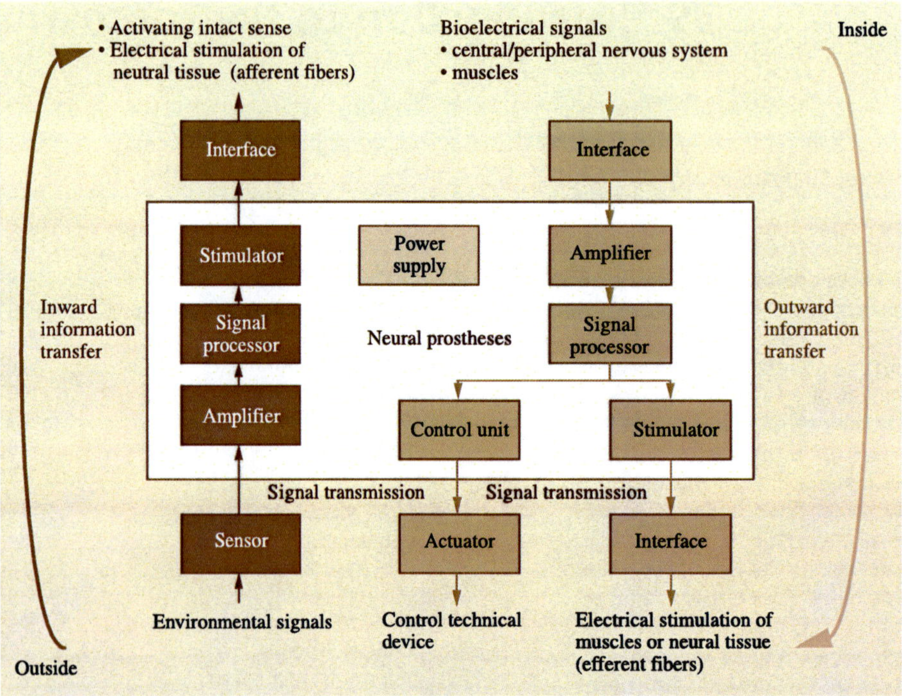

Fig. 39.1 Schematic of a neural prosthesis

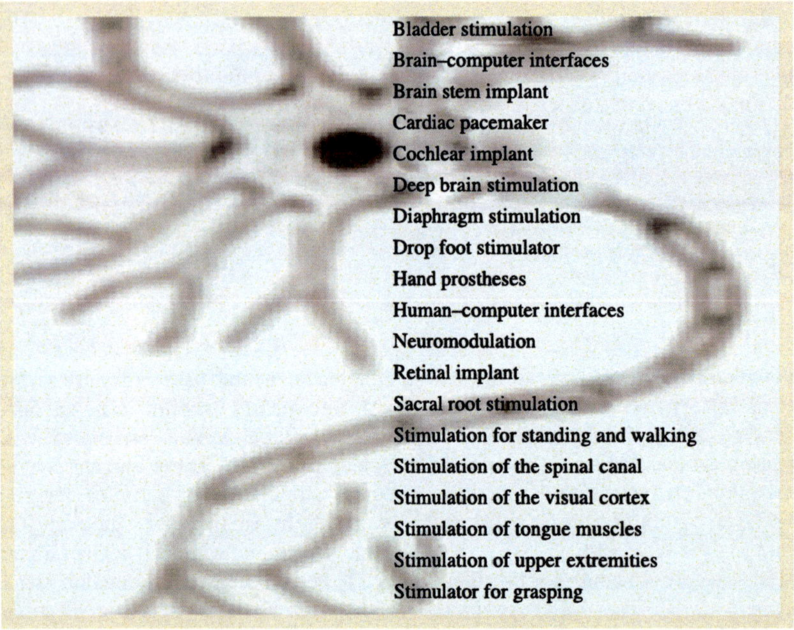

Fig. 39.2 Examples of the application of neural prostheses

and implants for deep brain stimulation have been established for years in clinical practice. Stroke patients, paraplegics, and drop-foot patients can often be helped by functional electrical stimulation. Implantable electrical stimulators are increasingly applied for the therapy of chronic pain or incontinence using neuromodulation. Other applications, for instance the retinal implant, are in clinical trials. The research field of neuroprostheses is strongly characterized by preclinical experimental work.

Cardiac pacemakers, cochlear implants, and deep brain stimulation represent about 95% of the currently implanted neural prostheses [39.1]. The state of the art is characterized by progressing miniaturization and the application of new biocompatible materials and manufacturing technologies.

39.2 Application of Neural Prostheses

39.2.1 Cardiac Pacemaker

So far, the most frequently used neural prosthesis is the cardiac pacemaker. Electrical stimulation of the heart is applied to compensate for pathological changes of the generation and transmission of stimuli, including cardiac arrhythmia. With appropriate stimulation, the respective symptoms and thus the effects on the patient can be positively influenced.

Classical cardiac pacemakers have been applied for many years. They are designed as a single-chamber system for stimulation in the right atrium or the right ventricle, or as a dual-chamber system. The latter enables the correction of disorders in atrioventricular transmission. As frequency-adaptive systems, they adapt their stimulation frequency to the respective physical strain of the patient.

The therapeutic concept also includes regular aftercare of implant patients. In the future, this can be supplemented by the use of telemedical systems for remote monitoring. Event-related data can be transmitted to a medical center by mobile radio and internet technologies. If necessary, this can be supplemented by automated alerting of the physician. Moreover, future developments for cardiac pacemakers include stimulation of the left chambers. Also, vagally controlled cardiac pacemakers are undergoing preclinical testing.

39.2.2 Cochlear Implant

In many cases, cochlear implants enable the restoration of auditory impressions and speech comprehension. Thus, deaf children can often be introduced to hearing and speaking. For the application of a cochlear implant, the auditory nerve has to be intact and healthy. A cochlear implant substitutes lost functions of the inner ear. It is placed in the cranial bone behind the ear. The stimulating electrode is positioned in the cochlea. The number of separate channels including the corresponding electrodes determines the information bandwidth that can be achieved. An external speech signal processor with microphone and coil receives the acoustic signals and transmits them to the implanted receiver. The required energy is telemetrically transmitted together with the converted acoustic signals.

Although the speech processor has to be set individually for each patient, the comprehension of acoustic signals and thus hearing and understanding has to be relearned. For many patients, lip-reading is significantly facilitated after the operation, and their ability to follow conversations is improved. It is recommended to supply the patient with cochlear implants on both sides, to improve spatial hearing.

If both auditory nerves have lost their function, the application of a cochlear implant is not reasonable. Here, acoustic sensations can be achieved by stimulating the higher tracts, e.g., in the nucleus cochlearis (in the brain stem), to enable rudimentary hearing. Stimulation of the auditory center is also conceivable.

39.2.3 Deep Brain Stimulation

Severe movement disorders can be treated by electrical stimulation deep inside the brain. Especially for patients with Parkinson's disease, this yields a therapeutic approach. Charge-compensated short rectangular electrical stimuli, for instance in the nucleus subthalamicus, inhibit the pathologically overactive regions. Similar to the cardiac pacemaker, the neurostimulator is implanted between the skin and the pectoral muscle.

Previous applications of deep brain stimulation have often been associated with an improved psychic situation. First results of stimulation of the nucleus accumbens in obsessive-compulsive disorder patients indicated improvements with respect to speech, mood, and motor behavior.

A promising approach is the electrical suppression of epileptic seizures. Several studies indicate that electrical high-frequency stimulation can inhibit a spacious synchronization of neurons. However, this method is only applied in patients whose seizure disorders cannot be treated by pharmaceuticals or surgery.

39.2.4 Bladder Stimulation

Besides impaired mobility, paraplegic patients can suffer from reduced control of internal body functions, for instance control of urinary flow. Accompanying surgical means can restore the storage function of the urinary bladder of paraplegics. Electrical stimulation of the sacral roots enables controlled bladder voiding without catheter. Besides the positive social aspects, the bladder stimulator also yields a reduced risk of infection.

39.2.5 Human–Computer Interfaces

Patients with locked-in syndrome, amyotrophic lateral sclerosis or paraplegia often have difficulties in communicating with their environment. With human–computer interfaces, technical systems can be controlled by recorded biosignals of the peripheral or central nervous system. Thus, lost communication channels can be restored, and the patients become able to interact with the environment. Often, recording, amplification, analysis, and the generation of the control or communication signals is computer-aided, including integrated microprocessors (embedded systems).

One branch is the brain–computer interface controlled by cerebral activity changes. Here, systems recording the signals with surface electrodes (EEG-controlled), plate electrodes placed on the cortex (ECoG-controlled) or implanted multielectrodes (electrode arrays) can be distinguished. The objective of all mentioned systems is the control of special output devices (for instance the cursor on a computer monitor) or the control of technical systems or actuators (e.g., technical grapplers or prostheses).

39.2.6 Extremity Prostheses

Hand prostheses belong to the group of prostheses of the upper extremities. The realization of the various grasping movements of the hand, e.g., cylindrical grasp, pincer grasp, lateral grasp, spherical grasp, is a technical challenge. Myoelectrically controlled hand prostheses are the state of the art. They use muscle action potentials from the remaining muscles of the forearm. The envelope of the recorded activity is proportional to the number of activated muscle fibers and thus to the muscle force. Therefore, it can be used as a control signal for prostheses. The precondition for a coordinated movement is the ability of the prosthesis wearer to selectively and independently tense various muscle groups.

The introduction of microcontroller systems integrated in the prostheses enables controlled movements with regulated grip velocity and grip force proportional to the muscle signal. The proportional control makes high demands on the quality of the recorded surface EMG signal.

39.3 Specific Technological Features

39.3.1 Biocompatibility

One of the main preconditions for the application of neural prostheses is their biocompatibility. This means the execution of a desired function in a specified application with an appropriate host reaction. Often, the provocation of biological functions and reactions is desired and the aim of the use of neural prostheses. This desired compatibility between a technical system and the biological tissue in the sense of an adaptation of the structure and surface of the implant to the receiving tissue is denoted as biocompatibility. Surface compatibility includes the adaptation of the chemical, physical, biological, and morphological surface properties of the implant. The adaptation of the implant structure to the mechanical behavior of the receiving tissue yields structure compatibility.

The verification of the biocompatibility for the complete system is done in three stages:

- In vitro tests with isolated cells, cell cultures, and organ structures with the objective to collect possible changes of the implant surface, to observe a possible dissolution of material components, e.g., ion release from metals, to determine corrosion and so on. Short-term as well as long-term exposure is observed.
- In vivo tests with animal models (mice, rats, dogs, rabbits) as long-term tests with the objective to observe the occurrence of toxic effects, the change of the tissue morphology, and the carcinogenic poten-

tial of the material. Histological examinations are used for the evaluation.
- Clinical studies are performed to monitor allergenic and inflammatory reactions.

Implantation site, implantation time, applied materials and their function determine the application of a neural prosthesis. In any case, the production process can influence the material biocompatibility. Plasma activation is one example of changing the biocompatibility of surfaces.

39.3.2 Encapsulation

Neural prostheses have to be encapsulated for two reasons: first to protect the implant against influences of body fluids, and second to protect the organism in the sense of biocompatibility. Moreover, electronic circuits have to be shielded from corrosion and possible occurrence of short circuits. Another function is the mechanical protection of the implant against forces produced by the body. It is essential that the encapsulation must not influence the function of the implant. Also, the sterilizability of the encapsulated implant has to be ensured.

For cardiac pacemakers, titanium housings have been applied for many years. Here, the metal housing can often be simultaneously used as counter electrode. Other materials include ceramics, (e.g., for cochlear implant) or glass (e.g., BIONsTH). For flexible microimplants, silicone, parylene, polyimide or other polymers are used as encapsulation materials.

39.3.3 Energy Supply

Lithium iodide batteries are the classical form of energy supply for implantable neural prostheses. For a sufficiently small design, they provide energy for a duration of 5–10 years. These batteries are for instance applied to cardiac pacemakers and deep brain stimulators. For other applications, e.g., retinal implant or cochlear implant, alternative energy sources have to be found. Often, the required energy is transmitted telemetrically. Future concepts are based on energy generation from the immediate environment of the implant. This can for instance be achieved using biochemical, piezoelectric, thermoelectric, electromagnetic, capacitive, or thermomechanical generators. These concepts require the design of an adaptation of the electrical properties of transducer, storage, and consumer. The specific parameters depend on the transduction mechanism from high voltage capability for piezoelectric to ultralow voltage for thermoelectrical transducers.

39.3.4 Electronics

The application of neural prostheses makes high demands on the respective electronics. They have to ensure the function of the implant over a very long period. Thus, they have to be self-verifying and have to transmit the respective operating state as well as potential errors to the outside. The design of the electronics has to be space-saving. An interaction with other diagnostic or therapeutic methods, for instance MRI, HF surgery or defibrillator, should have only little influence on the function. In recent years, the fast-paced development in microelectronics has led to smaller and smaller volumes. In order to minimize galvanic processes at the electrode, an additional circuit component is necessary for decoupling of DC components. Examples are monolithic circuits and low-voltage applications. It has to be mentioned that especially in the field of neuroprosthetics, low-voltage and low-power integrated circuits (IC) are of great benefit for recording of nerve and muscle signals. For stimulators, ICs with high-voltage technology or discrete components, which can be used as an end stage to provide the necessary current, have to be generated. In order to avoid corrosion of electrodes, very low input bias currents are necessary. Moreover, the general demands on a medical product and the electromagnetic biocompatibility have to be met.

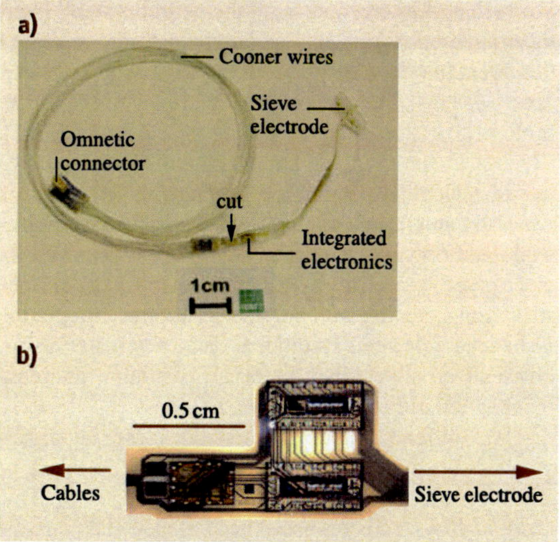

Fig. 39.3a,b Sieve electrode with multiplexer. **(a)** Electrode structure; **(b)** embedded multiplexer (after [39.3])

Microfabrication processes enable the design of novel kinds of electrodes including active electronics. The preprocessing of recorded signals is possible very close to the recording electrode. This can minimize the influence of superimposed artifacts and increase signal quality. Integrated electronic circuits can reduce the number of connection lines to the electrode. An example of such an electronic integration is the sieve electrode with a multiplexer [39.3,4] (Fig. 39.3).

39.3.5 Signal Processing

On the one hand, neural prostheses stimulate myogenic or neurogenic tissue; on the other hand, they record bioelectric signals. Here the biosignal or the stimulus has to be adjusted to the technical system or the biological tissue, respectively. For instance, the cochlear implant receives external acoustic signals, analyzes them and, after an appropriate interpretation, converts them into a stimulation signal. This requires an immediate signal processing at the implantation site. Integrated digital real-time signal processors (DSP) allow the implementation of sophisticated analysis routines and algorithms. This includes filtering, detection of certain signal features (e.g., spikes), and their classification based on Fourier transformation, wavelet analysis, neural networks etc. First, analog signals are converted into digital series and thus become computer-usable. These data are further processed in microprocessors, by the use of suitable algorithms.

In the example of the cochlear implant, the acoustic signal is decomposed into individual frequency bands, and after signal adaptation, each frequency band is assigned to one stimulation electrode. Thus, tonotopic stimulation of the cochlea is enabled.

39.3.6 Signal Transmission

The transfer of data from and to the implant is often realized by inductive interfaces. This enables a concurrent energy supply of the implant. The field of an external transmitting coil is coupled to the receiving coil of the implant. For data transmission, the carrier wave of the transmitter is modulated. Taking the cochlear implant as an example, a power of 30 mW can be transmitted, with a simultaneous data rate of 400 kBit/s. The communication of the implant to the outside can generally be achieved by the same means.

39.4 Biological–Technical Interface

39.4.1 Electrodes

For technical interactions with the neural system, interfaces are needed. Implantable microelectrodes represent the direct interface between biological tissue and technical systems. They transform the ion current of the biological system into a current of electrons [39.5]. Electrodes are used for both stimulation and recording of biological potentials. Although their surface is generally small, low contact impedance between electrode and tissue is required. This can be achieved by electrolytic treatment of the metal electrodes, surface structuring, coating with polymer structures, integration of nanoparticles etc. Traditional electrode materials include silver, silver/silver chloride, platinum, platinum black, gold, iridium, and iridium oxide.

Two different kinds of electrodes can be distinguished: flexible and stiff electrodes. One example of stiff electrodes is the Utah electrode array [39.6]. This array of 10×10 silicon-based needle-shaped electrodes is used for measuring of the extracellular potentials of neurons and single-neuron action potentials. The length of the electrodes is currently limited to 1.5 mm. By using μ-wire electrical discharge machining from a pure tungsten block the length can be increased to 5 mm [39.7].

The advantage of flexible electrodes is that they are able to move with the tissue in which they are implanted. Compared with silicon electrodes the Young's modulus is much lower. The materials used most for flexible electrodes are polyimide and polysiloxane (silicone). Polyimide is light (1.42 g/cm^3) and can be fabricated as very thin films (10–15 μm). The water absorption is smaller than 0.5%. Using photolithography arbitrary geometric structures can be realized.

The design depends on the application (Fig. 39.4). The cuff electrode is a circumneural electrode. It is placed around the peripheral nerve like a tube. Thereby the electrodes are positioned inside the cuff to get in close contact with the nerve.

The concept of the shaft electrode is to insert it into the neural tissue. It has a needle shape with multiple electrode sites. This results in a closer contact between the electrode site and the nerve fibers.

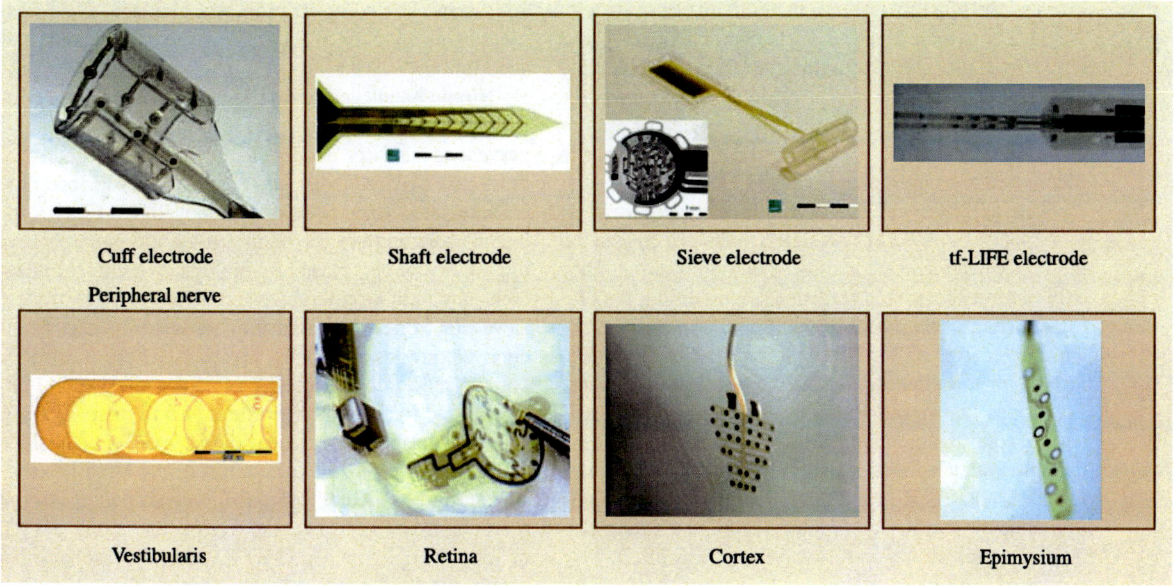

Fig. 39.4 Examples of flexible implantable microelectrodes made of polyimide

The sieve electrode is placed between two cut ends of a nerve trunk. As a guidance and fixation for the nerve, silicone tubes are placed at both sides of the sieve. The nerve fibers regenerate through the holes of the sieve electrode. The sieve area is circular in shape with a diameter of 1.8 mm. The 571 holes of the sieve are 40 μm in diameter and hexagonally arranged. Ring-shaped electrodes encircle 27 of the holes [39.3]. The advantage of the sieve electrode is the possibility to selectively contact sensory and motor nerve fibers. Thus it is possible to record signals from afferent fibers as well as to stimulate efferent fibers.

A kind of intrafascicular electrode is the thin-film longitudinal intrafascicular electrode (tf-LIFE). It com-

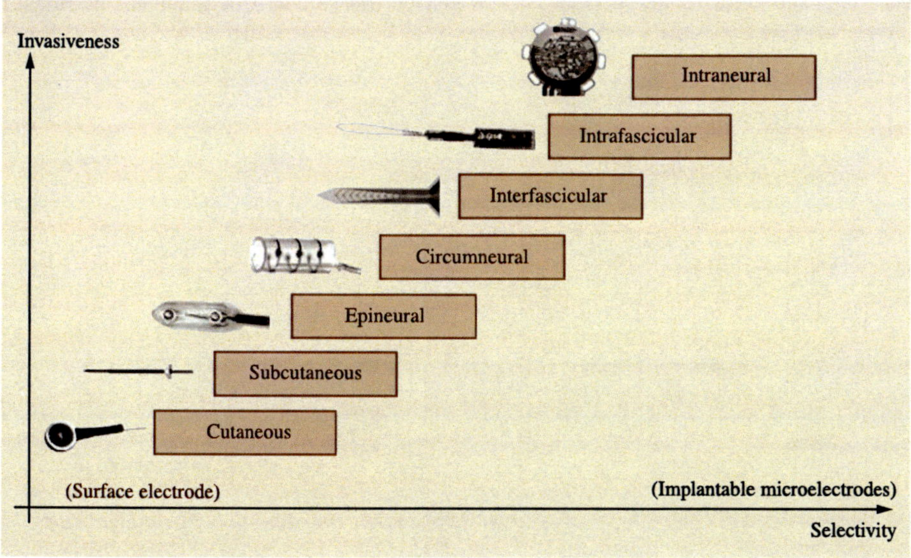

Fig. 39.5 Selectivity versus invasiveness of electrodes

bines a loop of a thin polyimide multisite electrode with Kevlar fibers including a thin needle. This needle can be used as guidance to implant the thin-film electrode longitudinally into the nerve. Additional electrodes are placed directly on vestibularis, retina, cortex or epimysium. So they can be used to stimulate electrically or to record neural or muscular signals [39.8].

The selectivity of the electrode increases with its invasiveness [39.9]. Surface electrodes are placed on the intact body surface and are thus noninvasive. They exhibit a very restricted selectivity. In addition, the biological tissue, which is situated between the electrode and the source of bioelectrical activity or the structure to be stimulated, changes the form of the signal. Electrodes introduced intimately in the biological tissue are considerable more selective. However, they have the disadvantage of invasiveness with all its side effects. Also, the design and the implantation method can influence the selectivity. Anyway, there is a strong correlation of selectivity and invasiveness (Fig. 39.5)

39.4.2 Interfaces for a Bionic Hand Prosthesis

The natural hand is a very complex system. With a weight of only 400 g, a volume of 50 cm^3, and the possibility of 22 degrees of freedom, humans can generate a grip force of more than 500 N. Force, velocity, and precision can be controlled according to the type of grasp. Approximately 17 000 different sensors are used for this regulation. So the loss of an extremity has always been a dramatic event for the person concerned. For compensation of a lost hand, both cosmetic and functional aspects have to be taken into account.

Bionic hand prostheses use neuronal signals to control hand movement and give information to peripheral nerves as sensory feedback. Therefore bidirectional interfaces and sensors are necessary. Proprioceptive sensors detect hand position and movement. Exteroceptive sensors acquire the nature of the environment and the quality of the grasp. As biological–technical interface, implantable microelectrodes like cuff electrodes, sieve electrodes, and thin-film longitudinal intrafascicular electrodes (tf-LIFEs) are used. These electrodes allow a bidirectional interface for recording neural activities and for stimulating afferent nerve fibers [39.10]. The concept of bidirectional bionic hand prosthesis is shown in Fig. 39.6. Motor control of the hand prosthesis can be established by placing electrodes inside the motor cortex [39.6, 11], the peripheral nerve [39.12] or directly epimysially [39.13]. Sensory feedback requires electrodes placed on afferent fibers or skin areas.

In this system, implanted microelectrodes record the innervation rate transmitted by the motor cortex to the peripheral nerve. The signals are amplified, A/D converted, and telemetrically transmitted to the actual hand prosthesis. Here, a control signal corresponding to the innervation rate is used to generate the desired movement. The hand prosthesis is equipped with sensors to record, e.g., temperature, holding pressure, and surface morphology of the held object. These measurement values can be transmitted to a stimulator electrically

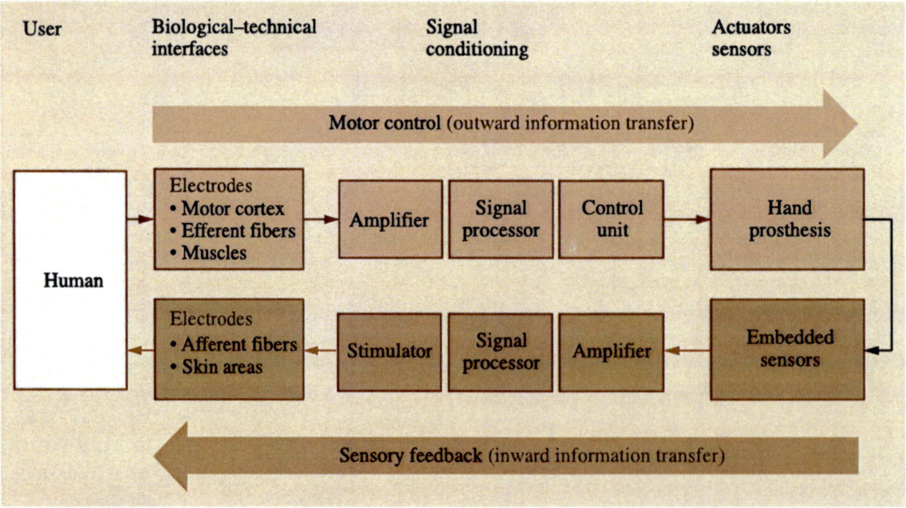

Fig. 39.6 Schematic of a bidirectional bionic hand prosthesis

stimulating the afferent fibers via the implanted electrodes. In the sensory cortical projection areas, the corresponding sensation is generated. Thus, it can become possible to provide a feedback on the grasp and hold function of the hand prosthesis, including the properties of objects, to the patient.

39.4.3 Thin-Film Longitudinal Intrafascicular Electrode (tf-LIFE)

Design

The microelectrode system is based on a double-sided longitudinal flexible multielectrode, the so-called tf-LIFE (thin-film longitudinal intrafascicular electrode), which was basically designed by *Yoshida* [39.14]. This electrode combines a loop of a thin-film electrode with a filament loop including a thin needle. This needle can be used as guidance to implant the thin-film electrode longitudinally into the nerve. It is intended as a selective electrode system for recording and stimulation applications in the nerve fascicle. To this end, the system is penetrated through a nerve with the aid of a sharp tungsten needle and an additional supporting wire such that it can be fixed around the application area. The needle and the supporting wire are directly removed after the implantation procedure. Only the thin-film electrode is placed into the nerve. Depending on the implantation of the electrode a high selectivity can be achieved [39.9].

The active part of the electrode is made of flexible polyimide thin films. The polyimide acts as substrate and as insulator on which platinum tracks are sputtered and the active sites are realized. The mechanical properties of the polyimide give great flexibility to the active part of the electrode, decreasing relative drifts between the tissue and the electrode associated with fibrous encapsulations. The basic requirement for all components that directly interact with the biological tissue is biocompatibility of the used materials. Especially for electrodes, biocompatibility has to be investigated under stimulation as well as recording conditions. Electrode materials with stable electrochemical properties like electrochemical impedance and corrosion stability are needed [39.8].

The tf-LIFE microstructure consists of a polyimide substrate with an overall thickness of 10 μm and an

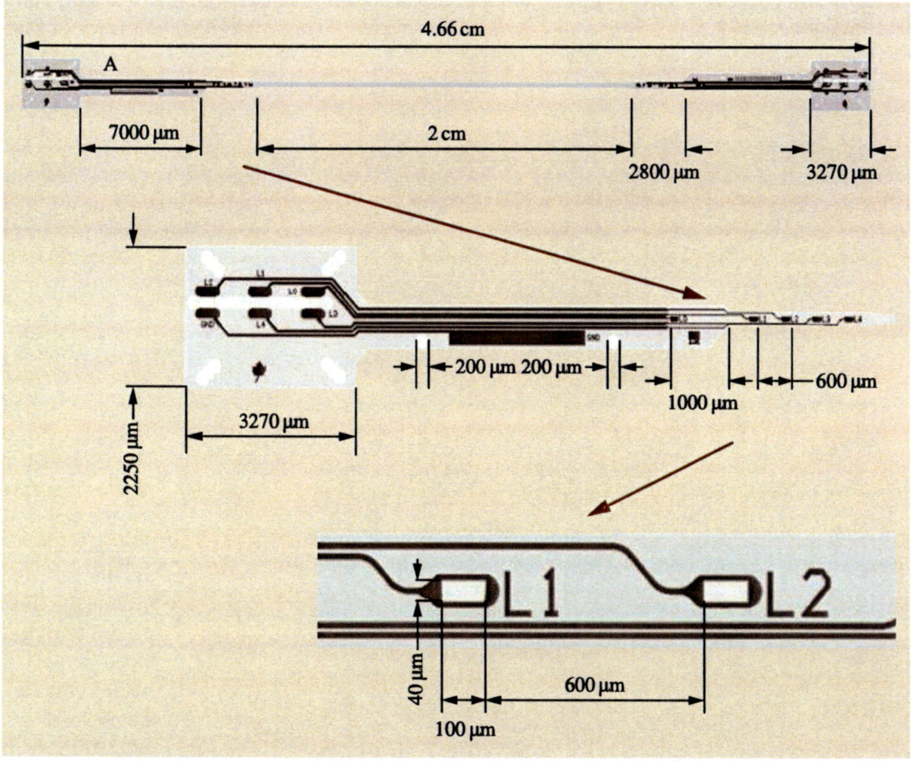

Fig. 39.7 Electrode assignment and dimensions of the tf-LIFE. GND are the ground electrodes, L0 and R0 are the counter electrodes, and L1–L4 and R1–R4 are the working electrodes for each side of the loop, respectively

overall length of over 5 cm (Fig. 39.7). On both ends of the structure there are bonding pads to contact the flexible part with an adaptor (made of Al_2O_3 ceramic via a screen printing process) for applying additional wires. These bonding areas are attached opposite to each other on each side of the adaptor (mirrored), which results in a looped structure of the thin-film electrode. The working electrodes (four electrodes on each side) and the proximal electrodes (counter and ground electrodes on each side) are on the outer part of the loop. The ground electrode has a width of 200 μm and a length of 2500 μm, and the counter and working electrodes have a rectangular shape of 100 μm × 40 μm with a semicircle of 40 μm diameter attached on each side. The connection lines and electrodes are made of a 300-nm-thick platinum layer.

The additional guiding wire is also made of polyimide with a width of 200 μm, an overall thickness of 20 μm, and a total length of 4 cm (2 cm as a loop). This guiding wire is pulled around the assembled flexible electrode loop on the adaptor and then glued on a tungsten needle with medical-grade instant adhesive. Several investigations have shown the reliability of the polyimide guiding wire regarding its mechanical strength as well as the sufficient bonding strength of the glued wire/tungsten connection.

Micromanufacturing

Microelectrode processing consists of different process and development steps and needs well-defined design parameters [39.8].

Four main steps can be distinguished:

1. Design phase
2. Processing of thin-film microelectrodes
3. Assembling of the flexible microelectrodes to wires and/or plugs
4. Encapsulation of the assembled system regarding the needs of the application.

Designing of microelectrodes starts with CAD development of photolithographic masks, which defines the implant shape and the electrode design. Important points are the connection lines and electrode sides, depending on application, location, and implantation method. For effective system integration to combine the microscopic with the macroscopic world the adaptors should be carefully designed.

Based on a micromachining process polyimide (PI 2611, Du Pont) is used as carrier layer for connection lines and pads as well as the electrode material. After applying a 5-μm-thick substrate layer of polyimide on a silicon wafer, a single metallization layer of plat-

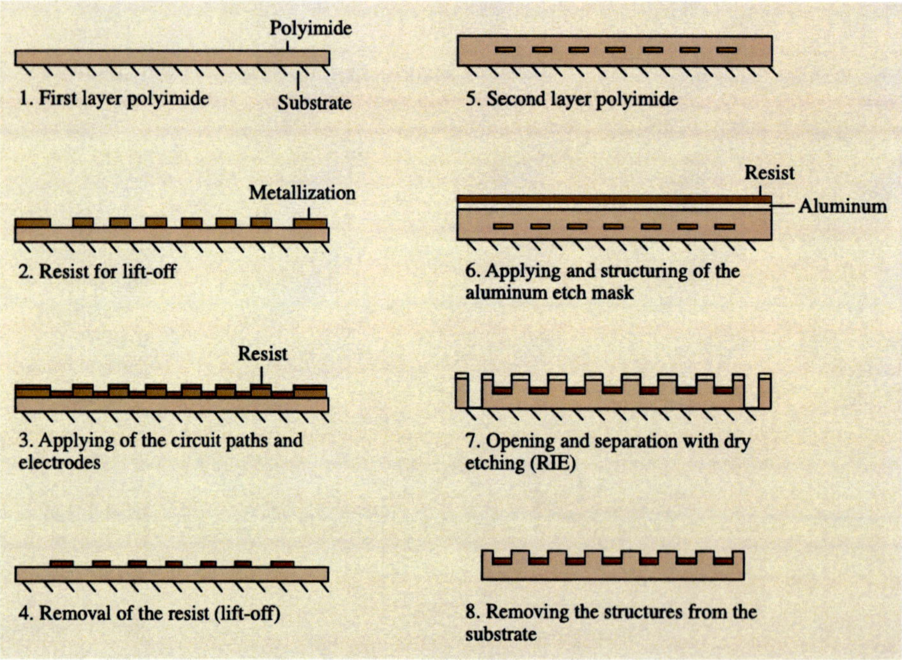

Fig. 39.8 General process sequence for the micromachining development of flexible thin-film microelectrodes based on polyimide substrate material (after [39.8])

inum is sputtered. After a lift-off process the polyimide top layer of 5 μm thickness is applied and then coated with aluminum as etching mask for the subsequent dry etching process to open the electrode area and the connection pads. After the removal of the etching mask, the wafer is cleaned and the single structures (thin-film electrodes) are removed from it. The process flow is shown in Fig. 39.8.

After removing the single structure from the silicon carrier wafer the thin-film electrode is bonded on a ceramic-based adaptor, which is made via screen printing. On the other side of the bonded adaptor 12 cooner wires (AS631) are soldered on the connection pads of the ceramics (six wires on each side). The wire bundle is covered with a medical-grade silicone tube with a diameter of 2 mm and an overall length of 50 cm. On the other side of the wires an Omnetics plug (Type NPD-12-WD-12.0-CSMT) is assembled which is also connected and encapsulated to the silicone tube.

The micro flex interconnection technique (MFI) is used for bonding of connection lines on flexible foils to a substrate (Al_2O_3, FR4 or glass) to build up a strong and reliable connection between a flexible foil and the adaptor [39.4]. By this, easy soldering or welding of wires and plugs on the adaptor system is possible. For more complex approaches electronics have to be placed directly at the electrode. Integrated circuits (ICs) should be used to reduce the size of the system. A special bonding technique was developed to connect electronics to a flexible thin-film structure. Gold bonds can be placed through a hole of the thin film on the contact pad of the IC. On the side of the gold ball, the thin-film electrode needs a ring contact around the hole. By this method a mechanical as well as an electrical contact between electrode and IC can be achieved. Additional components like capacitors or resistors can be placed in a similar way or can be glued by conductive epoxy or soldered to pads of the flexible substrate.

For protection of the electrode and to ensure biocompatibility of the electrode system, coating of the structure with Parylene C and silicone is necessary. Parylene C is a biocompatible polymer that is deposited directly on the substrate from the vapor phase in microscale gaps. So it can be cured to a wide range of geometrics [39.15]. The sterilization method is gas sterilization with ethylene oxide.

Characterization and Optimization

Electrode characterization is used to determine the behavior in a physiological environment and to optimize mechanical and electrical parameters. It is an important

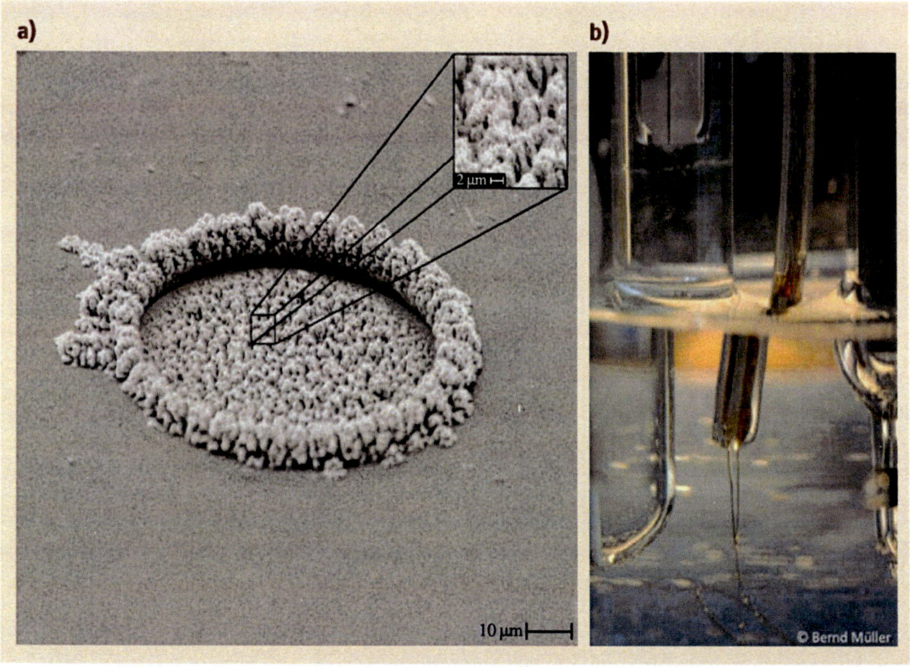

Fig. 39.9a,b Microstructured platinum. **(a)** Scanning electron microscopy picture of a microrough platinum electrode surface. **(b)** tf-LIFE during the coating process. The same setup is used for impedance spectroscopy

step in the fabrication process of implantable microelectrodes. The methods used are valuable tools for quality management to evaluate the different process steps and parameters.

To increase the effective area of a microelectrode the electrode contacts can be coated with microrough platinum (Fig. 39.9). This leads to lower impedance as well as a higher charge injection capacity [39.16].

For electrical stimulation with microelectrodes, charge injection capacity has to be considered. This means the maximum amount of charge per area that can be reversibly injected with the respective electrode. If this value is exceeded, irreversible Faradaic reactions occur with corrosion of the electrode, shift of the local pH value, air bubbles at the electrode surface, and finally damage of the surroundings. For an electrode made of platinum a maximum charge of approximately 64 $\mu C/cm^2$, and after coating with microrough platinum of approximately 524 $\mu C/cm^2$, can be injected [39.17].

For recording of biosignals the impedance of the electrode in comparison with the amplifier input impedance is important for the signal quality. After the electroplating process to coat platinum electrodes with microrough platinum, the electrodes' impedance is significantly reduced by a factor of approximately 100, depending on the signal frequency (Fig. 39.10).

Electrical impedance spectroscopy is a useful method to determine the specific impedance by in vitro investigation in 0.9% NaCl solution. This test specifies the quality of the electrode and its functionality. Using a three-electrode setup a current in a frequency range of 10–100 000 Hz is applied between the working and the counter electrode, mostly a platinum electrode. The resulting electrical potential is measured between the working electrode and the reference electrode, usually a Ag/AgCl electrode. The relation between input signal (current) and output signal (voltage) is used to calculate impedance and phase. Figure 39.10 shows a typical impedance investigation of a tf-LIFE structure with electroplated platinum electrodes.

An additional electrochemical characterization method is cyclic voltammetry. A triangular voltage is applied and the current through the electrode is recorded. If the measured current is plotted versus the potential, the reversible potential limits can be identified. The charge storage capacity of the electrode can be estimated by calculating the area inside the resulting hysteresis loop. Furthermore, pulse tests, corrosion tests, and leakage current tests are very helpful to estimate the long-term behavior of the electrodes.

Typical methods for optical characterization are light microscopy, and Fourier transform infrared (FTIR) spectroscopy. With laser profilometry the surface of the

Part D | 39.4

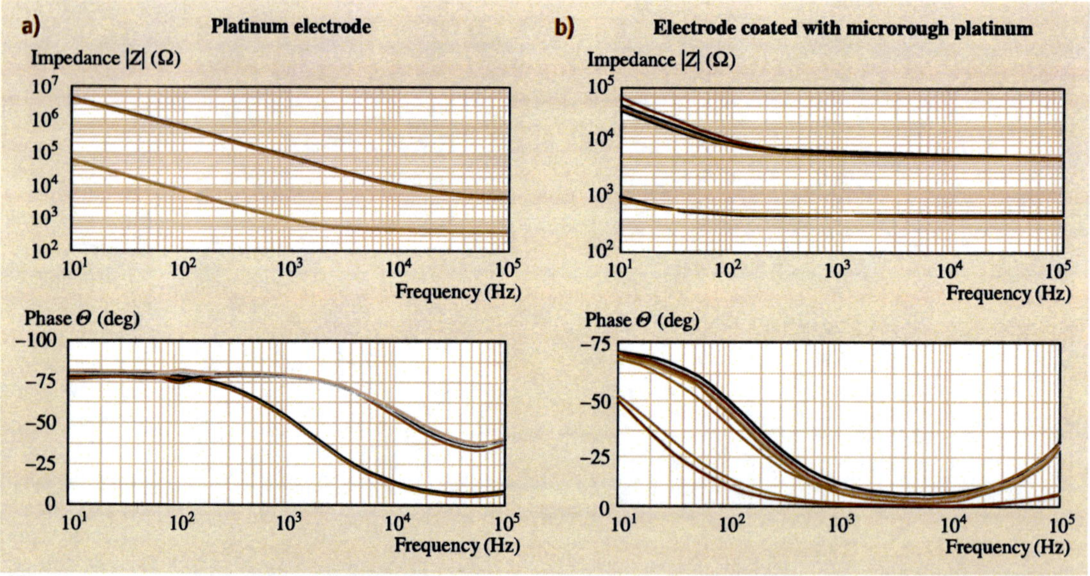

Fig. 39.10a,b Typical impedance investigation of a tf-LIFE structure (fourth generation) with a pure sputtered platinum surface (**a**) and after an electroplating process with a microrough platinum surface (**b**)

Table 39.1 Tests to be performed according to ISO 10993

Raw material tests (e.g., electrode material, Kevlar fiber)	
Cytotoxicity	ISO 10993-5
Acute toxicity test	ISO 10993-11
Intracutaneous reactivity	ISO 10993-10
Final product investigation (electrode implant with adaptor system, wires, and equipment)	
Cytotoxicity	ISO 10993-5
Genotoxicity	ISO 10993-3
Implantation	ISO 10993-6
Ethylene oxide residuals	ISO 10993-7
Irritation test	ISO 10993-10
Sensitization test	ISO 10993-10
Systemic (acute) toxicity	ISO 10993-11
Bioburden test	ISO 11737

electrode is scanned with a laser beam to obtain its profile. Scanning electron microscopy (SEM) is used to analyze the microrough structures of the electrode.

Mechanical characterizations are mainly focused on the structural parts of the electrodes like the stability of the substrate.

Biological Characterization and Tests in Animals

Before an electrode structure can be implanted in humans or animals, biological characterization including the assessment of biocompatibility has to be done. According to the ISO 10993 European standard, the following tests have to performed to ensure that the electrode system is suitable for implantation (Table 39.1).

The tf-LIFE was implanted in the sciatic nerve of rats and rabbits. They can be inserted longitudinally or transversally into the peripheral nerve tissue (Fig. 39.11b). Thus the electrodes are placed inside and parallel to the nerve fibers or perpendicularly to them. Histological studies after 3 months of chronic application in rat peripheral nerve have shown that the tf-LIFE caused only minimal damage to the nerve [39.18]. The electrodes can selectively collect nerve signals and provide a good interface to analyze sensory information from peripheral nerve fascicles [39.19].

Different kinds of algorithms are approached for processing of signals recorded with tf-LIFEs. Especially, the wavelet is a very efficient tool. Various sensory stimuli were applied to the hind leg of the animal and the elicited signals were recorded. After wavelet denoising and spike sorting, vector machines were successfully trained to use the spike waveforms of the signals to achieve decoding information regarding various sensory stimuli [39.20].

Using tf-LIFEs, it was shown that specific information is provided on the possibility of extracting and appropriately interpreting the neural code for motor commands and of delivering sensory feedback by stimulating afferent fibers [39.19]. The result showed that both the extraction of motor information and the restoration of sensory function are possible.

Four tf-LIFEs (Fig. 39.11) were implanted in the median and ulnar nerves of an amputee. Thirty-two

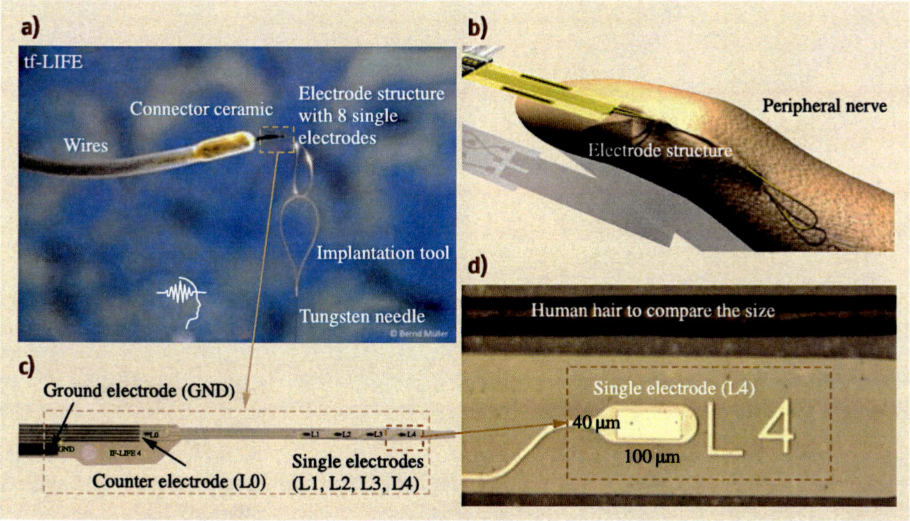

Fig. 39.11a–d Thin-film longitudinal intrafascicular electrode (tf-LIFE)

channel signals were recorded with Grass amplifiers in a frequency range of 100–10 000 Hz. To stimulate the afferent fibers the Grass stimulator delivered trains with square pulses. Rectangular cathodal pulses of duration 10–300 μs and current intensity 10–100 μA were employed [39.12]. Because the maximal current density for platinum electrodes is 1.5 mA/cm^2 the stimulation current for each working electrode of the tf-LIFE structure was limited to approximately 80 μA. The optimal active sites for sensory feedback were characterized starting with stimuli of 10 μA amplitude and 10 μs duration. Electrical characterization of the electrodes has been performed showing impedance values in the range of 5500–7500 Ω for active and 400–500 Ω for ground electrodes at 1000 Hz [39.21].

The bionic hand is a stand-alone version of the CyberHand prototype with five fingers actuated by six motors. Proprioceptive and exteroceptive sensors are embedded [39.22].

Activity classes were identified using denoising and spike-sorting algorithms. Up to 85% of individual movements were correctly classified. Discrete tactile sensation was elicited from different stimulation sites in the median as well as the ulnar nerve. The sensation magnitude was modulated by pulse frequency. The results showed that tf-LIFEs implanted in peripheral nerves can be used for tactile feedback and to control independent types of handgrip of a bionic hand prosthesis. Tf-LIFEs can be implanted and used in humans for several weeks with a high success rate, picking up signals with a good signal-to-noise ratio. Multiple electrodes in different nerves with numerous contacts guarantee a reliable flow of signals [39.12, 21].

39.5 Future Developments

Future developments within the field of neuroprosthetics are characterized by proceeding miniaturization, the application of new materials and technologies, as well as the integration of cognitive technical systems. Also, misunderstood basic issues related to the coupling of neurons with technical materials have to be clarified. Bioactive substances released after implantation can enhance biocompatibility. Thus, research in the field of neuroprosthetics is still characterized by experimental and preclinical work.

Current fields of application include:

- Central paralysis, e.g., paralysis due to stroke
- Paralysis due to lesions of the spinal canal
- Hypoventilation of central origin, sleep apnea
- Deafness, blindness

Part D | 39.5

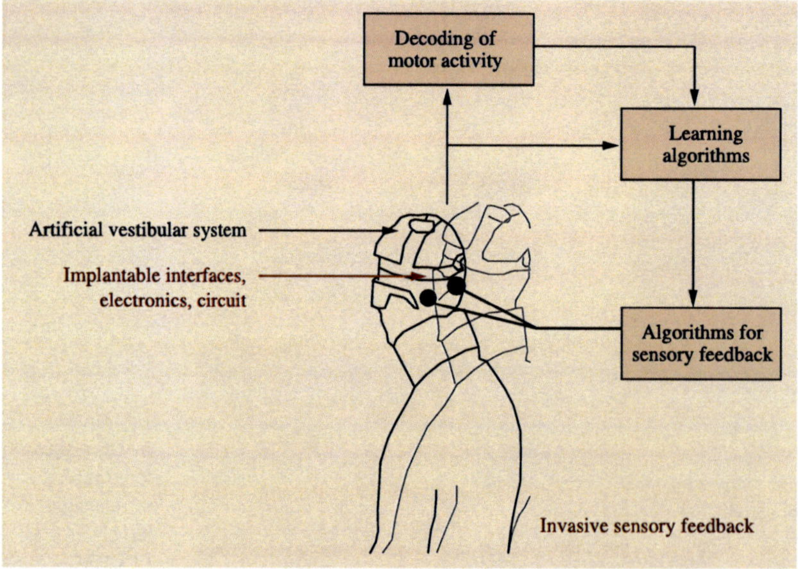

Fig. 39.12 Closed-loop neural prostheses for vestibular disorders: CLONS Project (after [39.23])

- Multiple sclerosis, seizure disorders, depression, Parkinson's disease
- Pain, cluster headache
- Weakness of the bladder, incontinence, impotence
- Vocal cord paralysis.

Autonomous active implants working as closed-loop systems could be developed for different therapeutic applications or rehabilitation using cognitive technical systems. The association of an implanted smart stimulator to stimulate the vagus nerve with an interacting and smart monitoring system can be used for control of the cardiovascular system [39.24].

Another new research area is the 3-D vestibular neural prosthesis [39.23]. An innovative closed-loop sensory neural prosthesis is able to restore vestibular information by stimulating the semicircular canals electrically. Inertial sensors embedded in a device attached to the head and donned by the user provide the information for stimulation. Innovative learning algorithms process concurrent sensory and motor data (Fig. 39.12).

Future neural prostheses will be smaller and more intelligent. New biocompatible materials will establish additional chances. Microactuated electrodes on the basis of shape memory will allow movable interfaces. In this way, the optimal position of electrical contacts can be searched inside the nerve and a loss of connection with neural cells can be counteracted [39.25]. Bioactive implants with embedded neurochemical stimulation will allow a more inartificial way to stimulate neural tissue. These new stimulators could be based on microfluidic or solid-state drug delivery systems. Owing to new materials, undesired side effects will occur more rarely or with lower intensity. The combination of technical systems with biological components will increase the long-term stability of biological connections to the nervous system.

Further Reading

- M. Akay (Ed.): *Handbook of Neural Engineering* (Wiley IEEE, New York 2007)
- J. Black (Ed.): *Biological Performance of Materials – Fundamentals of Biocompatibility* (CRC, Boca Raton 2006)
- J.D. Bronzino (Ed.): *The Biomedical Engineering Handbook*, 3rd edn. (CRC, Boca Raton 2006)
- B. Bhushan (Ed.): *Handbook of Nanotechnology*, 2nd edn. (Springer, Berlin Heidelberg 2007)
- P. Dayan, L.F. Abbott: *Theoretical Neuroscience: Computational and Mathematical Modeling of Neural Systems* (MIT Press, Massachusetts 2001)
- D.J. DiLorenzo, J.D. Bronzino: *Neuroengineering* (CRC, Boca Raton 2008)
- R.S. Greco, F.B. Prinz, R.L. Smith: *Nanoscale Technology in Biological Systems* (CRC, Boca Raton 2005)
- B. He: *Neural Engineering* (Kluwer Academic, Boston 2005)
- K.P. Hoffmann, J. Dehm: *VDE-Studie zum Anwendungsfeld Neuroprothetik* (VDE, Frankfurt 2005)
- K.W. Horch, G.S. Dillon: *Neuroprosthetics – Theory and Practice* (World Scientific, Singapore 2004)
- V.K. Jirsa, A.R. McIntosh (Eds.): *Handbook of Brain Connectivity* (Springer, Berlin Heidelberg 2007)
- N.H. Malsch: *Biomedical Nanotechnology* (CRC, Boca Raton 2005)
- Society for Neuroscience: *Brain Facts – A Primer on the Brain and Nervous System*, 4th edn. (Society for Neuroscience, Washington 2002)

References

39.1 K.P. Hoffmann, J. Dehm: *VDE-Studie zum Anwendungsfeld Neuroprothetik* (VDE, Frankfurt 2005)

39.2 G. Adelman, B.H. Smith: *Encyclopedia of Neuroscience*, 3rd edn. (Elsevier, Amsterdam 2004)

39.3 A. Ramachandran: Design, in vitro and in vivo assessment of a multi-channel sieve electrode with integrated multiplexer, J. Neural Eng. **3**, 114–124 (2006)

39.4 T. Stieglitz, M. Schüttler, K.P. Koch: Implantable biomedical microsystems for neural prostheses, IEEE EMBS Magazine **2005**, 58–65 (2005)

39.5 S. Micera: Experiments on the development and use of a new generation of intra-neural electrodes to control robotic devices, Proc. 28th IEEE EMBS Annu. Int. Conf. (IEEE, 2006) pp. 2940–2943

39.6 L.R. Hochberg: Neural ensemble control of prosthetic devices by a human with tetraplegia, Nature **442**, 164–171 (2006)

39.7 P. Tathireddy, S. Krummenacker, S. Kammer, K.P. Hoffmann, F. Solzbacher: Towards high aspect ratio tungsten micro electrode array for neural recording and stimulation applications, Proc. 13th Annu. IFESS Conf. (IFESS, 2008) pp. 235–237

39.8 K.P. Hoffmann, K.P. Koch, T. Dörge, S. Micera: New technologies in manufacturing different implantable microelectrodes as an interface to the

peripheral nervous system, Proc. IEEE EMBS Int. Conf. Biomed. Robotics (BioRob) (IEEE, 2006), 414–419
39.9 X. Navarro, T. Krüger, N. Lago, S. Micera, T. Stieglitz, P. Dario: A critical review of interfaces with the peripheral nervous system for the control of neuroprostheses and hybrid bionic systems, J. Peripher. Nerve Soc. **10**, 229–258 (2005)
39.10 K.P. Hoffmann, W. Poppendieck: New aspects in hand prostheses, mst news **4**, 11–14 (2008)
39.11 M.A. Nicolelis: Brain–machine interfaces to restore motor function and probe neural circuits, Nat. Rev. Neurosci. **2003**, 417–422 (2003)
39.12 P.M. Rossini: Double nerve intraneural interface implant on a human amputee for robotic hand control, Clin. Neurophysiol. **2010**, 777–783 (2010)
39.13 R. Ruff: Acquisition of myoelectric signals to control a hand prosthesis with implantable epimysial electrodes, Proc. 32nd IEEE EMBS Annu. Conf. (IEEE, 2010) pp. 5070–5073
39.14 K. Yoshida, K. Hennigs, S. Kammer: Acute performance of the thin-film longitudinal intrafascicular electrode, Proc. IEEE EMBS Int. Conf. Biomed. Robotics (BioRob) (IEEE, 2006) pp. 296–300
39.15 A. Ramachandran, M. Junk, K.P. Koch, K.P. Hoffmann: A study of Parylene C polymer deposition inside microscale gaps, IEEE Trans. Adv. Packag. **30**, 712–724 (2007)
39.16 W. Poppendieck, T. Dörge, K.P. Hoffmann: Optimization of microporous platinum coatings for neural microelectrodes, Proc. 13th Conf. IFESS (IFESS, 2008) pp. 319–321
39.17 W. Poppendieck, K.P. Koch, S. Steltenkamp, K.P. Hoffmann: A measurement setup to determine the charge injection capacity of neural microelectrodes, IFMBE Proc., Vol. 25 (IFMBE, 2009) pp. 162–165
39.18 N. Lago, K. Yoshida, K.P. Koch, X. Navarro: Assessment of biocompatibility of chronically implanted polyimide and platinum intrafascicular electrodes, IEEE Trans. Biomed. Eng., Vol. 54 (IEEE, 2007) pp. 281–290
39.19 S. Micera: On the use of longitudinal intrafascicular peripheral interfaces for the control of cybernetic hand prostheses in amputees, IEEE Trans. Neural Syst. Rehabil. Eng. **16**, 453–472 (2008)
39.20 L. Citi: On the use of wavelet denoising and spike sorting techniques to process electroneurographic signals recorded using intraneural electrodes, J. Neurosci. Methods **172**, 294–302 (2008)
39.21 A. Benvenuto: Intrafascicular thin-film multichannel electrodes for sensory feedback: Evidences on a human amputee, Proc. IEEE EMBS 32nd Annu. Int. Conf. (IEEE, 2010) pp. 1800–1803
39.22 M. Carrozza, B. Massa, S. Micera, R. Lazzarini, M. Zecca, P. Dario: Design of a cybernetic hand for perception and action, Biol. Cybern. **96**, 629–644 (2006)
39.23 S. Micera: A closed-loop neural prosthesis for vestibular disorders, Proc. DGBMT (2010)
39.24 S. Steltenkamp, K. Becher, T. Dörge, R. Ruff, K.P. Hoffmann: Electrode structures for acquisition and neural stimulation controlling the cardiovascular system, Proc. IEEE EMBC (IEEE, 2009) pp. 5478–5481
39.25 S. Bossi, S. Kammer, T. Dörge, A. Menciassi, K.P. Hoffmann, S. Micera: An implantable microactuated intrafascicular electrode for peripheral nerves, IEEE Trans. Biomed. Eng. **56**, 2701–2706 (2009)

40. Implantable Microsystems

Prashant Tathireddy, Florian Solzbacher, Robert Hitchcock, Klaus-Peter Hoffmann

A microsystem is a collection of electromechanical and electronic elements that have been reduced in size using advanced lithographic and machining techniques. Many types of medical devices rely on microsystem architecture to improve the health and well-being of millions of people. These systems are found in devices which are used for short and long periods of time, in diagnostic and therapeutic applications, and in devices which have limited contact with the patient and devices that are permanently implantable. Many types of implantable devices rely on microelectromechanical systems (MEMS) to achieve their functionality. Integrated circuit technology has extended beyond the fabrication of electronic elements to millimeter-scale mechanical structures with subelements in the nanometer to micrometer range. These MEMS devices often possess a high level of integration and provide numerous functions in one monolithic package. In addition, these devices allow the creation of sensors and actuators with enhanced stability and low power consumption. However the MEMS device on its own is not able to function in the body without the requisite power source, physiologic interface, and packaging that allow it to perform its intended function over the intended lifetime. This chapter describes different implantable microsystems and there use in clinical application.

40.1 **Market, Applications, and Common Requirements** 801

40.2 **Sensors** 805
40.2.1 Pressure Sensors 805
40.2.2 Voltage Sensors 805
40.2.3 Biochemical Measurement Sensors . 810

40.3 **In vitro and in vivo Testing** 814

References 816

40.1 Market, Applications, and Common Requirements

Microsystems integrate two broad classes of structures, electromechanical and electronic, into a single assembly with the appropriate size, biocompatibility, and safety to be used in implantable medical devices and provide diagnostic or therapeutic assistance to a patient. MEMS devices such as pressure sensors and accelerometers are now relatively inexpensive components that medical device designers can specify into designs to enable direct physiologic sensing or to provide on-device diagnostics. An example of direct measurement is seen with pressure transducers that are permanently implanted in an aortic aneurysm sac to measure intrasac pressure and detect leakage after aortic aneurysm repair. In other applications, implantable drug pumps, for example, pressure sensors can be used to detect outlet occlusion. Accelerometers that are integrated into pacemakers can sense the activity level of the patient and adjust the rate of the pacemaker accordingly. This capability has catalyzed an entire new generation of cardiac rhythm management devices. As new sensor technology matures and component costs are reduced, these miniature electromechanical structures will also catalyze next-generation medical devices and provide future revenue growth for device companies.

Implantable microsystems for medical applications rely on the integration of MEMS components with electronic components in a package or packages to perform their intended function. Active electronic components, MEMS components, and passive components are used in combination with application-specific integrated circuits (ASICS) to provide complete system-level performance. Predefined electronic components such as batteries, capacitors, voltage regulators, transistors, memory, microprocessors, transceivers, and oscillators are commonly specified off-the-shelf by microsystem designers. In order to reduce implant size, many designers opt for miniaturization through component size reduction or integration rather than custom ASIC design. Component size reduction is often considered a less expensive and more flexible design approach than integration, and thus we see the continued miniaturization of many passive electronic devices. These components are routinely available in ultrasmall package sizes such as the submillimeter 0201 (0603 metric) rectangular, two-lead package.

The length of time the implant is resident in the body is a key requirement in the design and testing of implantable microsystems. All devices must be developed with thorough consideration of application, outcome, and duration. ISO 10993-1:2009 describes the general principles regarding the biological evaluation of medical devices and categorizes the devices based on the nature and duration of their contact with body tissues. The ISO standard describes the implant duration as limited (less than 24 h), prolonged (24 h to 30 days), and permanent (greater than 30 days). Understanding each application is critical not only to fulfill the regulatory requirements but also to design a safe and effective device. Implantable microsystems can fall into any of these three categories. Consider a MEMS-based sensor that is attached to the fetal scalp during birth to continuously monitor infant pH. This device would have a duration of less than 24 h, whereas an intracranial pressure sensor would have a prolonged duration of between 1 and 30 days. A pacemaker is considered a permanent implant with a duration lasting much longer than 30 days. Duration is therefore a key requirement when developing any type of implantable medical device. Although on the surface there appears to be little difference in regulatory scrutiny between an implant whose duration is 1 month and an implant whose duration is many years, this is not an accurate assessment. Regulatory compliance is often dictated by data on safety and effectiveness, scrutinizing the design, manufacturing, and testing processes. Testing must be conducted not only to verify performance but also to validate safety and effectiveness over the life of the implant. For implantable microsystems this often involves lengthy and sophisticated testing.

Implantable microsystems cover a broad range of medical devices that provide very different functions over different time points and come into contact with various tissue types. These devices range in terms of integration from low-component-count systems such as neuroelectrode arrays to high-component-count systems including pacemakers, deep brain stimulators, and implantable cardioverter-defibrillators (ICD). These devices can provide physical sensing as seen with the Cardiomems pressure sensor [40.1] or they can provide continuous core temperature as can be monitored with the ingestible Cortemp body temperature device [40.2]. Other devices provide chemical sensing as demonstrated by the Integra P02 intracranial probe [40.3]. In addition to physical and chemical sensing, other devices function by stimulating various tissues. Pacemakers electrically stimulate cardiac tissue to promote normal heart rhythm. Implantable cardioverter-defibrillators continuously monitor electrical impulses in the heart and respond to ventricular tachycardia or ventricular fibrillation by delivering an electrical shock to restore normal heart rhythm. Another family of implantable microsystems that provide stimulation is cochlear implants which stimulate the inside of the cochlea and trigger the auditory cortical functions in deaf persons. Other implantable tissue stimulation systems include bone growth stimulators, phrenic nerve stimulators, and deep brain stimulators.

Because of the variety of applications for implantable microsystems, these systems also come into contact with a number of different tissue types. Subcutaneous glucose sensors come into contact with the skin, neuroelectrodes interface with neural tissue, ingestible pH and temperature sensors are in contact with the gastrointestinal mucosa, and endolumenal pressure sensors reside in proximity to the vascular system.

Despite the variety of applications, durations, and tissue types seen by these implantable microsystems, they have many common characteristics. The devices must all function in a complex and oftentimes nonpermissive environment. Virtually all implantable devices are exposed to acute inflammation, which often extends to chronically persistent inflammation. This can lead to the formation of a fibrotic capsule of scar tissue that compromises the intended function of the device. Proteins are often deposited on the surface of the device, and this accumulation of proteins can also cause unwanted drift in sensors or build an incompatible interface between the device and host tissue. Many devices are surgically implanted and have to be compatible with the surgical procedure, incision site, and postsurgical healing process. In addition, permanently implanted devices must function for extended periods of time.

Premature failure of these devices implies costly and sometimes risky revision surgery.

No matter the intended use, duration or type of tissue in contact, the devices must be designed with patient safety at the forefront. All implantable devices must be sterilized prior to placement, and the mode of sterilization brings up another challenge for the designer. Electronic systems used in medical devices may not withstand radiation sterilization nor the high temperatures associated with autoclaving. A frequent choice is ethylene oxide or hydrogen peroxide plasma sterilization, and these must be considered during the design and development process. Devices must also be designed to integrate with the host tissue and minimize inflammation. Judicious selection of materials, surface finishes, and geometry can minimize the host inflammatory response and reduce the possibility that the implant will harm the patient. An additional consideration for any permanent implant is magnetic resonance imaging (MRI) compatibility. MRI and other imaging modalities provide an important diagnostic function in current medical practice, and many new-generation implants are designed to be compatible with diagnostic imaging systems.

Implantable microsystems must be integrated into a type of package that is tolerated by the host, protects the electronics from the ionic milieu of the physiologic environment, and minimizes leakage currents from the device to the host. These packages can be as simple as a conformal coat such as is provided with neuroelectrode arrays or as complex as a pacemaker pulse generator that encases the electronics within a hermetically sealed titanium enclosure with ceramic feedthroughs for the pacing leads. In addition to host compatibility, implantable medical electronics must be compatible with electromagnetic (EM) environments. This includes the EM emission from the device and the ability of the device to withstand EM environments such as electrostatic discharge (ESD), radiated emissions, electrical fast transients (EFTs), and surges.

Several types of fully implantable transducers based on MEMS technology have been developed for biosensor applications. These systems utilize telemetry to communicate information from within the body to an externally located receiver. Commercial systems are available from several manufacturers for both research and clinical applications. Wireless systems, in which both nodes (i.e., the implanted sensor and the receiver) have internal power, provide high data bandwidth and ranges of 0.2–10 m. These systems require batteries which must be charged or replaced and necessitate larger packages. Smaller implantable systems based on active telemetry have been developed where the primary node powers the secondary node via electromagnetic coupling [40.4]. These systems do not require batteries and therefore have a longer lifetime and smaller size. Other systems based on passive telemetry systems use coupled inductors where the secondary node passively loads the primary node. The primary node oscillation changes as a function of pressure and alters the frequency spectrum of the transmitted signal [40.5]. Passive telemetry systems for biomedical pressure measurement have been approved by the Food and Drug Administration (FDA) for endovascular monitoring [40.1].

Medical microsystems oftentimes are required to communicate information from the body to external monitors or data collection systems. This communication can typically be handled in one of three modes: onboard storage, wired or wireless communication. Onboard storage consists of semiconductor memory storage that is part of the microsystem assembly. The memory records data or patient events, the medical device is retrieved, and the data from the onboard electronics are downloaded to a local host computer. Onboard storage is used in nonimplantable microsystems such as cardiac monitors, activity monitors, and gait/load monitoring systems. Implantable microsystems typically use wired or wireless communication. Wired systems are typically seen with catheter-type implants such as intracranial pressure and pO_2 monitors. Wireless communication is becoming increasingly used as the components become more available and the cost of integration decreases. In 1999 the US Federal Communications Commission (FCC) allocated the 402–405 MHz band for the Medical Implant Communication Service (MICS). Although it is a relatively new standard, it is being designed into many new generations of cardiac devices such as pacemakers and ICDs as well as implantable drug delivery systems, hearing aids, and neurostimulators. Key characteristics of MICS transceivers are 402–405 MHz, 25 μW maximum effective isotropically radiated power (EIRP), specific base station/implant negotiation, and frequency agility. MICS transceivers can achieve up to 250 kbit/s at a range of approximately 6 feet [40.6].

According to market researchers at the Freedonia Group, the current US market for all implantable medical devices is US $ 33 B, and this is expected to increase by 8.3% annually through 2014. Within the overall implantable market are implantable microsystems, which comprise approximately one-half of this market. The largest market segment of implantable microsystems corresponds to the cardiac rhythm management devices: pacemakers and ICDs. Globaldata estimates that the global market for pacemakers could be valued at US $ 4.2 billion in 2008. The market is expected to reach US $ 6.1 billion in 2015 with an annual growth rate of 5.4%. According to Reuters, the global ICD market was approximately US $ 6 B in 2008 with 7% growth. Bone growth stimulators is a smaller market, and implantable stimulators account for only a fraction of the total bone growth stimulator market. In 2006 this was a US $ 400 M market, although this number includes implantable, semi-implantable, and external stimulators. According to reports by Neurotech, a market research firm, the cochlear implant market will grow from US $ 725 M in 2008 to US $ 1.59 B in 2012. The same firm reports that the deep brain stimulator market will grow from US $ 461 M in 2008 to US $ 1.36 B in 2012 and the cochlear implant market will grow from US $ 289 M in 2008 to US $ 914 M in 2012. The worldwide infusion pump market is expected to grow from US $ 3.3 B in 2010 to US $ 4.5 B in 2015 (6.2% compound annual growth rate, CAGR) according to the *Infusion Pumps Worldwide Market April 2009* report published by Marketstrat, Inc. Although this number includes all infusion pumps, the implantable infusion pump market is expected to achieve a compound annual growth rate of around 15% and is recognized as a potentially high-volume, high-growth segment.

Various microsystems technologies have been touted as game changers in the medical device area, and the last three decades have seen a tremendous number of devices conceived, tested, and even used to motivate new companies. Only a small percentage of these inventions are truly driving market acceptance and market share today. One of the largest disappointments has been in implantable glucose monitoring systems. Whether permanently implantable or implanted for only a few days, few of the myriad proposed systems are accurate enough to provide reliable measurements over the range needed for those with significant swings in blood glucose [40.7].

MEMS-based microsystems were heralded in the 1970s as an enabling technology for implantable systems. Many technology forecasters looked into the not too distant future and predicted that multifunction sensors, microrobotic surgery, and neural interface devices would usher in a new frontier of medicine. Although the technology was not adopted at the pace that the forecasters predicted, microsystems have slowly proven to drive both improved patient outcomes and increased company revenues. Adoption of the technology has been slower than forecasted due in part to the conservative and regulatory nature of medical device development. In addition, physiologic environments are a hard place for MEMS devices to function due to an immune response that is robust to virtually all implanted materials.

Market forces that will drive new microsystems into the implantable device market include emerging technologies, new indications for use, increased regulatory scrutiny, and customer demand. Emerging technologies require time to prove reliability and cost-effectiveness. This has been shown historically, and because of safety and regulatory requirements, this technology latency will continue to be a characteristic of technology-driven medical device innovation. New indications for use will continue to be an attractive growth strategy for medical device companies. This approach can open new markets through an established device platform. For example, cochlear implants will continue to be prescribed for the profoundly deaf but will also be used in people who have severe hearing impairment and no longer respond to hearing aids. Deep brain stimulators which were developed and used for Parkinson's disease will see future applications in treating psychiatric disorders, brain injury, epilepsy, and stroke. The regulatory pathway for new indications is oftentimes significantly less arduous than for a brand new technology; however, regulatory agencies will continue to increase the levels of testing and quality for all medical devices. This requires additional company resources to respond and additional time as regulatory pathways are lengthened.

An aging and educated patient population will create demand for new technologies on both the diagnostic and therapeutic sides of healthcare delivery. As our population ages, there will be increased demand, particularly in the cardiac, neural, orthopedic, and diabetes markets, for implantable microsystems that can lead to earlier disease diagnosis and improved patient outcomes. Patient education through readily accessible information is creating a consumer-like demand for medical technology. Websites, blogs, advertisements, and word of mouth provide information that often drives patients towards specific device technologies, manufacturers, and healthcare providers.

There are large unmet needs in the medical device sector, and implantable microsystems can address many of these needs. Patients want better outcomes in virtually all areas of medicine, and patient outcomes will drive new technology into the marketplace. Implantable microsystems technology can improve patient outcome and reduce healthcare costs. These changes, however, will take place slowly and deliberately. Proving patient safety, improved outcomes, and reduced cost requires extensive bench and clinical testing for validation. These data are not only required by regulatory agencies but also by insurance companies in order to provide cost reimbursement, a critical necessity for virtually all medical devices. Combined with body area networks and telemedicine, implantable microsystems devices with features such as onboard sensing, drug delivery, tissue stimulation, microcontrollers, and wireless telemetry will be significant drivers in improving healthcare.

40.2 Sensors

Key drivers for implantable sensors can be found in personalized medicine needs for monitoring of biological markers in patients. These include neurological, metabolic, cardiovascular, and pulmonary disorders and pathologies. Measurement of, e.g., pH, pCO_2, and glucose can allow home treatment and continuous monitoring and reduce laboratory costs. The number of diabetes patients alone will increase to over 360 million patients worldwide by 2030 according to the World Health Organization. Implantable sensors will in most cases have to allow wireless transmission of data and/or power, requiring the integration of electronics, telemetry, and power modules into a small implantable device. They all have in common that they need to comprise biocompatible materials and be hermetically sealed so as to protect the device from the biological environment and vice versa without losing function of the device.

40.2.1 Pressure Sensors

Starting with their integration into microcatheters, implantable pressure sensors were amongst the first microsensor devices to become implantable. Driving applications include cardiovascular applications such as cardiac pressure, pulmonary pressure, intracranial pressure, abdominal pressure, and intraocular pressure in acute (e.g., pre- or postsurgery monitoring) or chronic use. Continuous implantable pressure sensors can be divided into wired and wireless sensors. An example of a wireless sensor could be a wireless continuous glucose monitoring system shown in Figs. 40.1 and 40.2.

Piezoresistive and capacitive effects constitute the two predominant sensing mechanisms for this type of sensor. The majority of devices in this realm are presently in research.

Pressure-sensing contact lenses are currently able to detect changes in the radius of curvature of enucleated porcine eyes of approximately 3 μm/mmHg of pressure using a temperature-compensated strain gage embedded in the circumference of a silicone contact lens. The contact lens makes use of wireless power and communication. Its weaknesses are still:

- Movement of the eye with respect to the pressure sensor causes noise in measurement
- It works better with contact lenses custom-modeled to the eye.

A current clinical intraocular pressure (IOP) sensing technique is tonometry, which is not continuous but rather an in situ sensing method. A lot of researchers have developed small-size, accurate IOP sensors, some of which are also implantable and suitable for pressure monitoring.

Although there are piezoresistive sensors and strain gage sensors, the most popular sensing mechanism is still capacitor based. Capacitive sensors have the advantages of high accuracy, good stability, and low power consumption.

Some other problems encountered with current IOP sensing designs are:

- The device implant process is not entirely reversible. Original tissues may be removed forever to offer more space for implantation
- Current IOP sensor designs are less practical because of the spatial constraints in the eye and surgical complexity.

40.2.2 Voltage Sensors

Measurement of electrical signals (primarily voltages/potentials) inside the human body is typically associated with electrophysiological measurements, which originated in external recordings through electroencephalogram (EEG) and electromyogram (EMG) recordings. One of the key driving forces for these measurements inside the body was the emergence of neuroprosthetics in the late 1980s in which electrical signals from the central or peripheral nervous system or selected muscles are used to allow patients with neurological disorders, injuries or amputations to regain motor function through biological or prosthetic limbs. Conventional EEG and EMG signals are still far from providing sufficient information about,

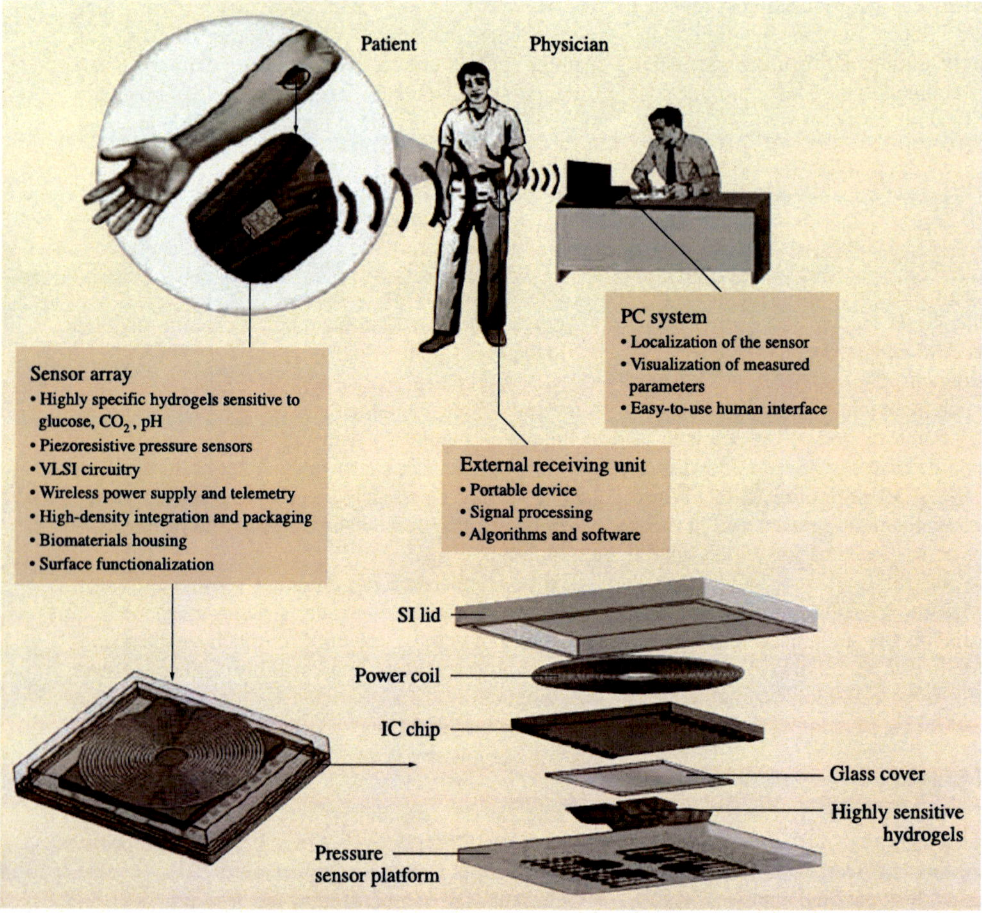

Fig. 40.1 Schematic and components for wireless implantable sensor for personalized medicine and diagnostic applications being developed by the researchers at the University of Utah (SI – Silicon, VLSI – very large scale integrated circuits)

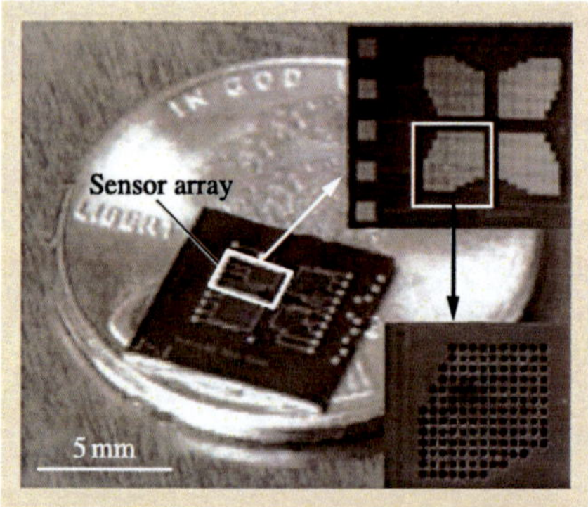

Fig. 40.2 Microchip sensor array with four independent pressure sensors (1.5 × 1.5 mm each)

e.g., motor intent, nor do they allow high-resolution sensory feedback. Bridging of broken nerve connections is not possible without tapping into the neural tissue directly. This has led to a multitude of research efforts, originating in the USA and Europe, to develop acutely and chronically implantable devices that can reliably record signals from locations as diverse as areas of the brain, peripheral nerves, and muscles. These devices exhibit much higher temporal and spatial resolution of neural/electrophysiological signals. The emergence of these devices and technologies has also provided a novel and advanced tool for basic neuroscience research, including the development of new pharmaceutics targeting the nervous system.

Today, the successful providers of this technology are offering a system-level solution that includes implantable electrodes and electrode arrays, insertion and manipulation tools (pneumatic inserters, microdrives, etc.), external electronics for control, data recording, signal processing and storage, software, and surgical and user training protocols, all of which contribute to

Fig. 40.3a,b Examples of penetrating (courtesy of Microsystems laboratory, University of Utah and Microprobes for Life Science, Inc.) and surface electrodes: implantable signal processor chips (courtesy of Reid Harrison, Intan technologies LLC.) and coils (**a**) for use in integrated implantable neural interfaces (**b**) (courtesy of Microsystems laboratory, University of Utah)

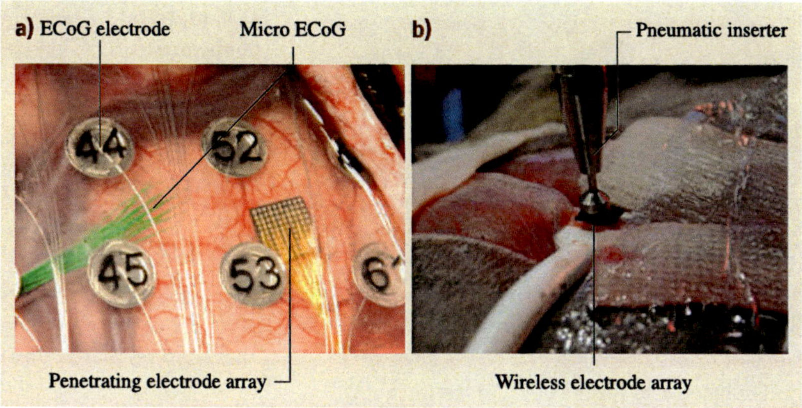

Fig. 40.4 (**a**) Wired 100-channel silicon-based penetrating electrode array on the cortex of a human epileptic patient next to a conventional 8×8 electrocortical grid array (ECoG) and micro ECoG. The smaller silicon array provides for higher temporal and spatial resolution of signals. (**b**) Pneumatic insertion of an active wireless electrode array into a sciatic nerve (animal model) (courtesy of University of Utah)

the overall performance and long-term stability of the recorded signals (Figs. 40.3, 40.4).

The continued improvement in our understanding of nervous system function, and its dysfunction due to disease and trauma, has motivated an expectation that this understanding should allow us to develop new therapeutic solutions to a variety of nervous system disorders. The emerging discipline of neuroprosthetics is one new direction that has proven to be effective in mitigating some nervous system disorders. Cochlear prosthesis [40.8], deep brain stimulation for disorders of the motor system [40.9, 10], and vagal nerve stimulation for epilepsy [40.11] and chronic depression [40.12] are specific examples of how electrical stimulation of the nervous system can allow severely disabled individuals to return to relatively normal activities of daily living. As we learn more about the normal and pathological function of the nervous system, we can expect new therapies to be developed based on neuroprosthetic techniques. Neuroprosthetic interventions in the central nervous system (CNS) are expected to restore to a limited, but functional, degree lost sensory functions (sight, hearing, and vestibular-mediated balance), and to correct dysfunctional motor functions such as Parkinson's disease, essential tremor, dyskinesia, and cognitive behaviors such as depression, schizophrenia, and bipolar disorder, and to acquire signals related to desired volitional control of the skeletal musculature. Neuroprosthetic devices applied in the peripheral nervous system (PNS) may provide new therapeutic interventions such as bladder and bowel control, the production of graceful stance and gait in subjects with stroke and spinal cord injury, volitional control of prosthetic devices (arms, hands, and legs), as well as pacing functions of the heart and diaphragm. Although only a few of these applications have begun to reach the clinical stage [40.13], many are currently being investigated in animal models, and some have passed proof-of-concept testing of the proposed therapy.

Even the simplest skeletal movement requires the coordinated activation of thousands of sensory and motor neurons in the CNS and PNS. Sophisticated movements associated with grasp and targeted movements are likely associated with complex spatiotemporal firing patterns of hundreds of thousands of neurons. In spite of this architectural complexity, four electrodes implanted

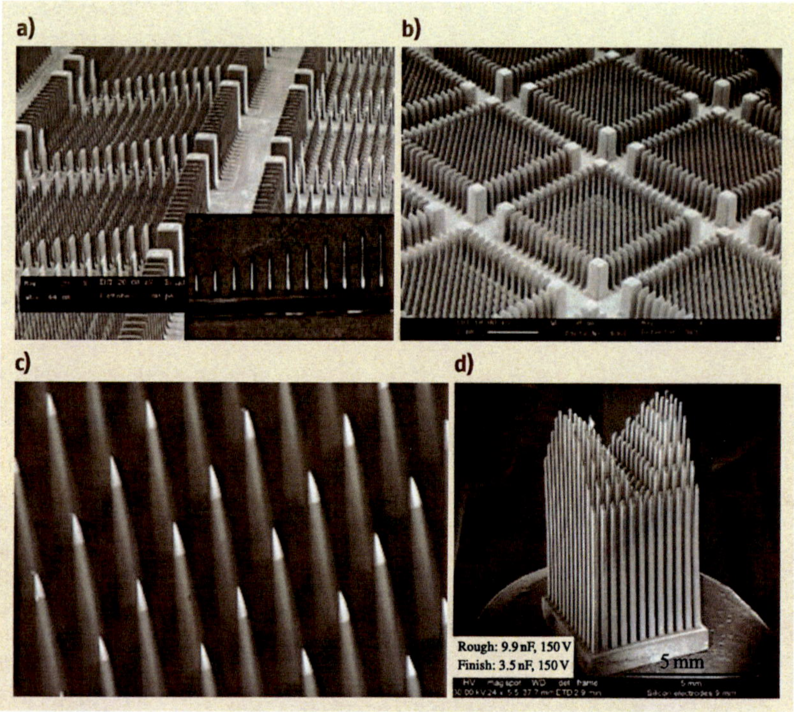

Fig. 40.5a–d Utah slant and flat arrays in wafer-scale production: arrays on a wafer after etching process (**a**,**b**), tips of the electrodes coated with iridium oxide metal and bottom of the electrodes electrically isolated by glass grid which mechanically holds them together (**c**), and a high-aspect-ratio 12 × 12 array with electrodes up to 9 mm long with 400 μm pitch (**d**) ((**a**)–(**c**) courtesy of Microsystems laboratory, University of Utah; (**d**) courtesy of Eberhard Bamberg, Viteris LLC)

in the cochlea of profoundly deaf subjects can restore functional hearing, even though normal hearing results from selective stimulation of subsets of the 30 000 auditory nerve fibers. Researchers at Case Western Reserve University have demonstrated functionally useful, walker-assisted stance in spinal-cord-injured subjects with as few as eight epimysial electrodes positioned over appropriate extensor muscles of the legs [40.14]. Continued progress in the development of such neuroprosthetic interventions will be related to improved ways to communicate directly with neurons of the central nervous system and with the afferent and efferent nerve fibers of the PNS. The concept and value of selective neuronal communication has been appreciated for decades, but only two major research programs have pursued the development of complex and integrated electrode array architectures that can begin to provide such highly selective neural communication: at the University of Utah and the University of Michigan [40.15,16]. Both of these programs have understood that highly selective neural communication can only be achieved with electrode arrays that penetrate into the nervous tissues, and both these programs have produced arrays containing on the order of 100 electrodes (Fig. 40.5).

State of the Art

Over the past two decades, the fields of neuroscience and neuroprosthetics have gained tremendous momentum through the development of novel architectures for neural interfaces. Some of these devices have great advantages over conventional neural electrodes: surface areas and dimensions are on the same order of magnitude as the biological structures they interface with (columnar structure in the cortex, neuron cell size, etc.). This enables greatly improved recording and stimulation selectivity compared with conventional neural electrodes. Key drivers in this field are the Michigan electrode array and the Utah electrode array (UEA). The commercialized version of the UEA, fabricated by Cyberkinetics Neurotechnology, Inc., was granted the first FDA investigational device exemption (IDE) for human trials [40.13, 17]. While there is little debate on the potential of this new generation of microelectrode arrays to restore lost sensory or motor function, this potential has yet to be realized in a clinical application.

Key Requirements and Challenges

Whilst detailed requirements depend on the specific application and location of implantation, a few common requirements can be listed.

A number of core issues have so far prevented a breakthrough of these devices in clinical application:

1. The long-term biocompatiblity and stability of these systems remains an unsolved problem. The materials used in novel microelectrode arrays and interfaces generally tend to be fairly bio-inert and provoke only a minimal response. Long-term data are, however, characterized by a vast distribution ranging from only days to years, rather than reliably lasting as chronic implants.
2. The signals recorded from the electrodes need to be amplified and potentially processed. Conventionally, the electronic systems needed for this purpose are bulky and power consuming, which is why processing is carried out outside the human body. The necessary skin-penetrating connectors and cables not only pose a constant risk for infection but also provide a number of mechanical and physiological failure modes due to the large number of mechanical and electrical interfaces, wires, and connectors that are exposed to the tissue. Tethering forces can displace the electrodes, changing the recording characteristics and provoking further immune system responses.

The delicate mechanical/electrical interconnects and wires can fail because of mechanical stress and exposure to a wet ionic environment. Cosmetic aspects of the experimental system used previously as proof-of-concept [40.13] would be in most cases unacceptable for real-world clinical use.

Technologies and Solutions

Implantable devices for this category can be separated into categories of penetrating and nonpenetrating (surface) electrodes, and single and array electrodes. Nonpenetrating electrodes are mostly polymer-based

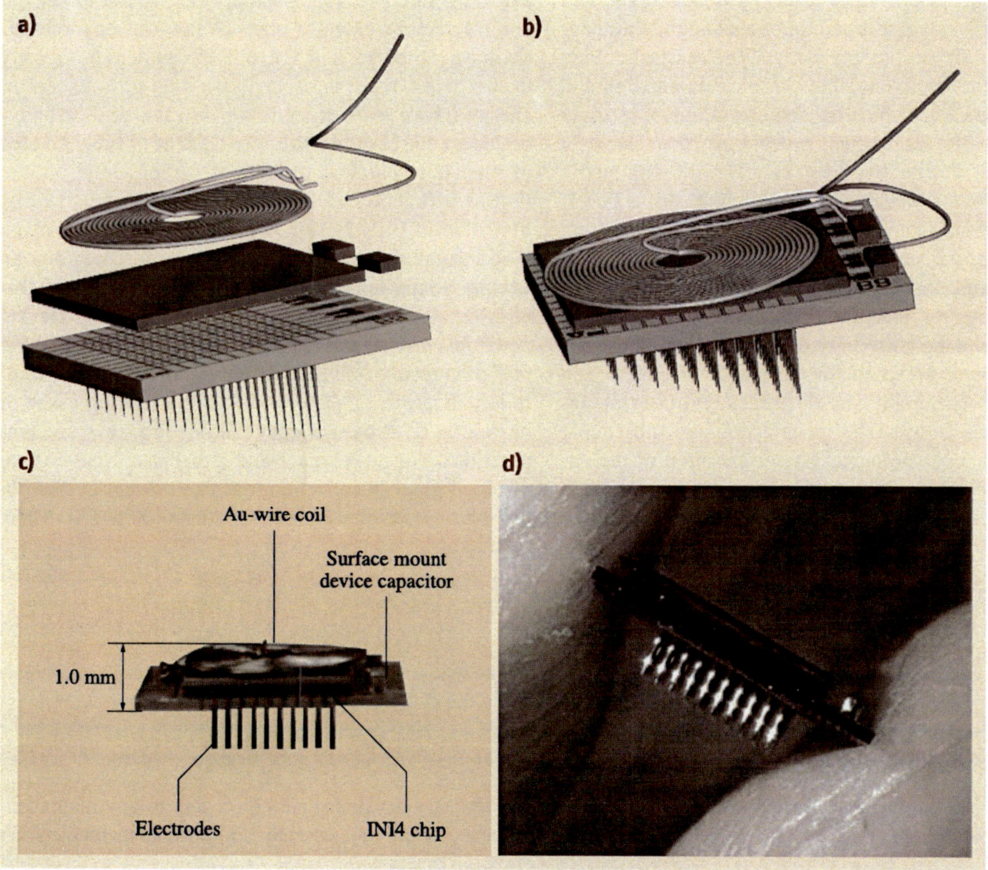

Fig. 40.6 Drawings showing different components of the Utah wireless neural interface device **(a,b)** and picture of a completely integrated wireless cortical neural recording system **(c,d)** (SMD – surface mount device) (courtesy of Microsystems laboratory, University of Utah)

(polyimide, parylene C, liquid-crystal polymer) flexible devices with Au, Pt or recently IrO_x thin-film electrodes, which come in planar or cuff form. Most of these devices are passive electrodes that are connected via a microribbon cable, bond wires or other solutions to a skin-penetrating connector. Some effort has been made to couple these devices with active electronics for amplification, data processing, and wireless telemetry (Figs. 40.6, 40.7). Alternatively, pill-shaped active devices with wireless data transmission have been developed and demonstrated. Penetrating electrodes come in three basic configurations: as single-wire electrode (Pt or IrO_x wires), tetrode (four wires bundled to one four-channel device), and arrays. For the generation of larger arrays, electrode wires are tethered together, e.g., by a ceramic platform, and manually assembled. Alternatively, silicon-based devices have been developed by various groups.

Chronically implantable neural interfaces can only be realized if the devices have a high channel count, can be made independent of wired connections, and provide long-term stable and functional neural interconnections inside the human body. Wireless technology will be necessary in enabling chronic implantation and clinical use. Since the first demonstration of a wireless system for acquisition, processing, and telemetry of biomedical data by *Song* in 1997 [40.18], about a dozen groups worldwide (Table 40.1) have developed individual components or systems for recording and wireless transmission of neural data from the central or peripheral nervous system [40.19–40]. With the exception of recent efforts at the University of Michigan and the University of Utah, none of these devices was ever designed for chronic implantation and use of high-channel-count electrodes and their potential later clinical use. The specification and subsequent technical and physiological characterization of the devices presented in literature are far from complete. Often only specific parts (electrode, amplifiers, telemetry units, etc.) were demonstrated. Table 40.1 gives an overview of the core radiofrequency (RF) wireless neural recording or stimulation activities presented in literature worldwide [40.19–21,41–46].

40.2.3 Biochemical Measurement Sensors

The measurement of biochemical species in vivo is important in a number of application areas. Monitoring of metabolic and cardiovascular markers in vivo has been driven largely by diabetes and other metabolic disorders and the need to measure, e.g., glucose levels continuously for therapeutic and research purposes. Recently, other relevant markers, including pO_2, pCO_2, pH, and others, have been subject to intensive research and development activities. Two core technologies are being pursued today: electrochemical (amperometric) sensors and hydrogel-based sensors.

The incidence of the metabolic syndrome that links obesity, heart disease, and diabetes is increasing at an alarming rate [40.47]. Popular approaches to weight control such as exercise plans or low-carbohydrate diets have failed to halt the obesity epidemic. Obesity occurs when there is a persistent long-term excess in energy intake over expenditure, and fats and carbohydrates are the body's two major sources of fuel. An area of intense research is the rate of carbohydrate utilization as a function of age, diet, exercise intensity, and fitness level [40.47–49]. A recent study that involved blood sampling showed remarkable differences among healthy young males in the dynamics of the blood glucose response after glucose ingestion [40.49]. Hence, new tools are needed for tracking metabolism that give real-time results and that are either noninvasive or minimally invasive. Diabetes mellitus is a leading cause of heart attacks, strokes, kidney failure, blindness, and amputations. For both type 1 and 2 diabetes, tight control of blood glucose concentrations can prevent and/or delay the onset of these complications [40.50–53]. Many

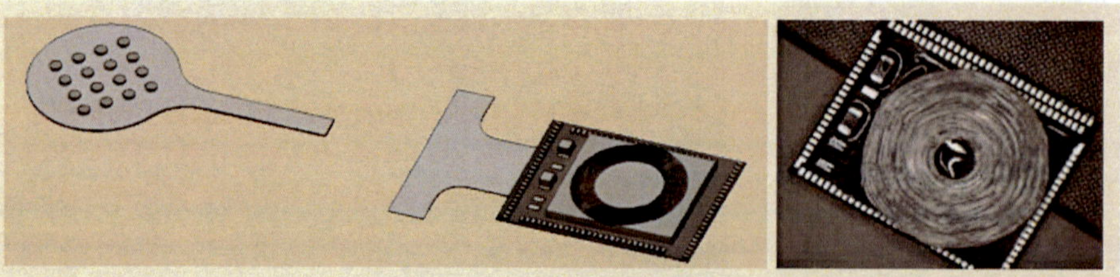

Fig. 40.7 Wireless interface module (WIM) to serve as a data telemetry and inductive power link to any multichannel neural recording device, e.g., ECoG; mean action potentials are overlaid on a surface plot of inflammatory marker isolectin B_4

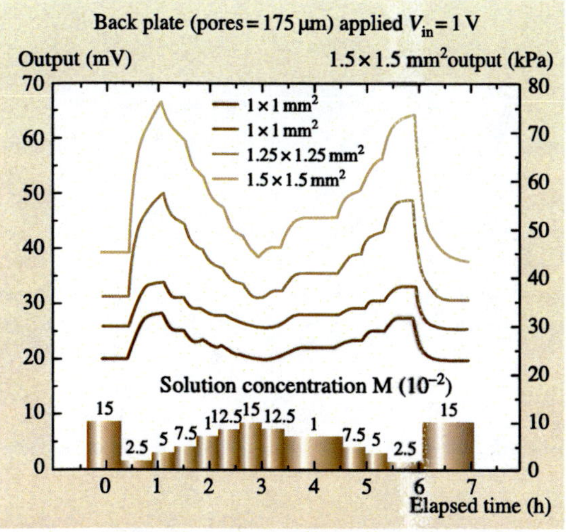

Fig. 40.8 Response of osmolality sensor developed at the University of Utah to changes in ionic strength (= 1/2 osmolality); 0.025 to 0.15 M; pH = 7.4 (gel thickness ≈ 400 μm); response time ≈ 10 min

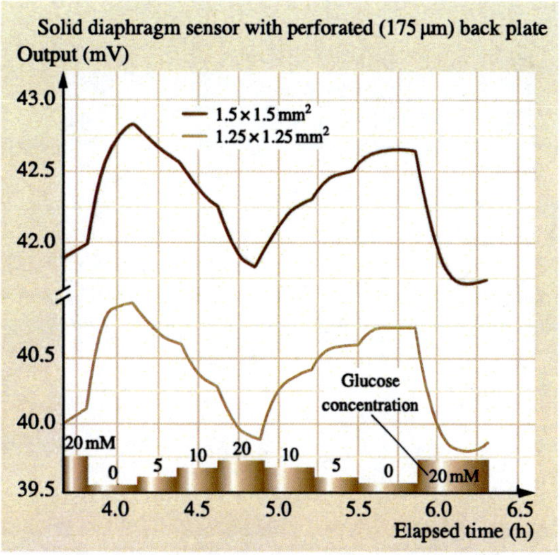

Fig. 40.9 Response of hydrogel based glucose sensor to changes in glucose concentration (0–20 mM); pH = 7.4 (hydrogel thickness ≈ 400 μm) in bovine serum

aspects of the disease can only be understood in the context of an intact organism. Progress in research on the disease is therefore highly dependent upon animal models.

Electrochemical Sensors

For human use, the FDA has approved four continuous short-term glucose sensors [40.54, 55]. All four are subcutaneous amperometric sensors such as the continuous glucose monitoring system (CGMS) from MiniMed [40.56–58]. All four and the only commercial mouse glucose sensor (Bluebox Sensors, Inc.) use an enzymatic reaction based on glucose oxidase (GOx) that continuously consumes glucose within the surrounding interstitial fluid (ISF). The electrons exchanged in this reaction are transferred to an anode and read as a current. Even though this type of sensor has been intensely researched for over 20 years, it is still not suitable for chronic implantation (> 14 days), and hence insurance companies are reluctant to pay for it. According to pioneers in electrochemical glucose sensors, all four FDA-approved sensors still exhibit instability over the approved period (3–7 days), and the calibration is good for only 12 h [40.55]. Many of the electrochemical sensors exhibit a run-in period in which the sensitivity drops by 10–30% immediately after implantation [40.55,59,60], followed by a period of stability of 1–7 days. There are many possible explanations for the instability of electrochemical sensors. The consumption of glucose may cause problems if the supply of glucose is reduced near the implant area due to its consumption by inflammatory cells [40.61,62]. All four electrochemical sensors employ an enzyme (GOx) that exhibits activity loss after 14 days in vivo [40.55]. All four are transport sensors, hence sensor calibration varies with the glucose diffusion coefficient [40.63]. All four employ nanoporous membranes to block endogenous redox species (e.g., ascorbate) from oxidizing at the anode. Some of the electrochemical sensors also employ nanoporous glucose-restrictive membranes [40.64] to avoid the oxygen-deficit problem, which arises because the molar concentration of glucose in ISF greatly exceeds that of oxygen [40.57, 65]. If either of these nanoporous membranes becomes blocked with protein fragments, glucose diffusivity will drop, changing the sensor calibration constant [40.63].

Hydrogel-Based Sensors

A smart hydrogel is a crosslinked polymer network that reversibly swells or changes electrical or optical properties in response to changes in the environment, such as pH, temperature or concentration of an analyte [40.66]. Several research groups worldwide have become interested in this topic [40.67–79]. The science is now reasonably well understood but has not yet been translated into commercial products. For the

Integrated neural interface IC	UofU INI5	UofU INI3/INI4	UofU INI1, INI2	UCLA	UCLA	Michigan (old)	Michigan (new)
Measured performance							
Design	Custom	Custom	Custom	Custom/ off the shelf	Off the shelf	Custom	Custom
Function	Recording or stimulate	Recording	Recording	Recording	Recording	Recording	Recording
System integration	Fully integrated	Fully integrated	Fully integrated	Discrete	Discrete	Hybrid	Hybrid
Implantation	Complete system	Complete system	Complete system	Electrodes only	Electrodes only	Electrodes only	Complete system (headstage)
Channels							
Number of channels/electrodes	100 signal, 1 ground	100 signal, 1 ground	88 signal, 12 ground	6	1	7	8
Simultaneously transmitted channels	100 (spike), 1 (streaming)	100 (spike), 1 (streaming)	88 (spike), 1 (streaming)	6	1	7	8
Switching time between streaming channels	< 1 ms	< 1 ms	< 1 ms				
Telemetry frequency link							
Communication protocol	FSK	FSK	FSK	MCFSK	Analog FM	TDMA/ Analog FM	TDMA/ Analog FM
Bandwidth				600 Hz	10 kHz	10 kHz	
FSK data transmission frequency	902–928 MHz	902–928 MHz	433 MHz	916 MHz	3200 MHz	94–98 MHz	
FSK data rate	345.6 kbps or higher	345.6 kbps	330 kbps				
Received signal power at distance of 13 cm	−79 dBm at 10 cm	−79 dBm at 10 cm	−86 dBm at 13 cm				
Transmission range	tbd.	0.6 m (5 m with booster)	0.3 m	9 m	< 1 m	∼ m	∼ m
Power							
Concept/supply	Wireless/ inductive	Wireless/ inductive	Wireless/ inductive	Wired 3 V	Wireless/ inductive	Wired ±1.5 V	Wireless/ inductive
Command in	Wireless	Wireless	Wireless				
Power/command signal frequency	2.765 MHz	2.765 MHz	2.64 MHz				
Minimum required receive coil voltage amplitude	5.7 V (peak)	5.7 V (peak)	5.7 V (peak)				
Total chip power dissipation	< 10 mW	8 mW	13.5 mW	66 mW	13.8 mW	2.05 mW	1.9 mW

Integrated neural interface IC	UofU INI5	UofU INI3/INI4	UofU INI1, INI2	UCLA	UCLA	Michigan (old)	Michigan (new)
Signal processing							
Maximum command input data rate (ASK)	20 kbps or higher	20 kbps	6.5 kbps				
Neural signal amplifier gain	60 dB (0.2–7 kHz)	60 dB (0.2–7 kHz)	60 dB (1.1–5 kHz)				38 dB (0.1–10 kHz)
ADC resolution (LSB = 2 μV electrode referred)	10 bits	10 bits	9 bits				
ADC sampling rate	15.7 kSamples/s	15.7 kSamples/s	15.0 kSamples/s				
ADC INL/DNL error (codes 50-511)	±0.8 LSB/ ±0.6 LSB	±0.8 LSB/ ±0.6 LSB	±0.8 LSB/ ±0.6 LSB				
Spike detector threshold resolution (LSB = 4.8 μV electrode referred)	7 bits	7 bits	7 bits				
Dimensions							
System dimension	6×6×1 mm^3	8×9×0.8 mm^3	8×9×0.8 mm^3	25×25×10 mm^3	50×50×10 mm^3		
Total IC chip area	25.4 mm^2	25.4 mm^2	27.3 mm^2 (0.5 μm 2P3M CMOS)			2.2×2.2 mm^2	5.2×3.1 mm^2
Weight (w/battery)				12.8 g	< 1 g	1.1 g	

Table 40.1 State of the art of systems developed for RF wireless neural recording or stimulation in the central or peripheral nervous systems [40.19–29, 31–46, 80] (UofU – University of Utah, UCLA – University of California, Los Angeles, FSK – frequency shift key, MCFSK – Manchester-coded FSK, TDMA – time division multiple access, ADC – analog to digital converter, ASK – amplitude shift key, LSB – least significant bit, INL – integral nonlinearity, DNL – differential nonlinearity) ◀

Table 40.2 Summary of leading candidate implantable glucose sensors

	Implantable sensor	
	Electrochemical	Hydrogel array
Requires nanoporous membrane for diffusion restriction?	Yes	No
Calibration constant varies with changes in glucose diffusivity?	Yes	No
Contains enzymes that may deactivate after 14 days?	Yes	No
Continuously consumes the analyte (glucose)?	Yes	No
Easy to implant using catheter?	Yes	No
Exhibits run-in period with drop in sensitivity?	Yes	Probably no
Response time	< 1 min	7–20 min for 400 μm-thick gels. < 5 min possible
Simultaneous measurement of glucose, pH, pCO_2?	No	Yes

measurement of electrical properties, the gel is placed on an (interdigitated) electrode structure and exposed to changes in pH or concentration of analyte. Significant progress has also been made with fluorescent glucose-sensitive hydrogels, which are boronic acid based. Binding the sugar to a receptor increases the fluorescence of the die embedded in the hydrogel. For the measurement of swelling, a smart hydrogel is confined between a porous membrane and the diaphragm of a piezoresistive pressure transducer. An increase in the analyte concentration, sensed through the pores of the membrane, is detected by a change in the swell pressure of the hydrogel. Thinner hydrogels lead to faster sensor response. Microfabrication of highly sensitive piezoresistive pressure sensors is a mature technology. The selectivity of the sensor arises from the binding specificity of the hydrogel, not from the porous membrane. Hence the pore size can be used to optimize the lifetime of the sensor in vivo.

A hydrogel is also an ideal biocompatible material for detecting changes in fluid osmolality, because the equilibrium swelling degree is determined as a trade-off between osmotic forces that favor swelling and the polymer chain-stretching entropic penalty that opposes swelling [40.73, 81]. Thus, almost all hydrogels swell, all other factors being held constant, when the osmolality of the surrounding environment decreases. A glucose-responsive hydrogel also swells or deswells in response to changes in glucose concentration. The signal must therefore be corrected for fluctuations in environmental osmolality. In sensor array developed at UofU (Fig. 40.2), this correction is performed by comparing the signal of the glucose-responsive hydrogel to that of a reference hydrogel that is unaffected by changes in glucose concentration and pH. For glucose sensing (Table 40.2) a polyampholytic hydrogel containing phenylboronic acid which is reversibly crosslinked by the pair of diols present in glucose results in hydrogel shrinkage due to increase in the entropic chain-stretching penalty [40.78]. The polyampholytic gel does not use enzymes, does not require oxygen, and is relatively insensitive to pH and ionic strength because it is polyampholytic. Endogenous diols such as in fructose and glycerol cannot crosslink the gel. Thus, glucose sensors that employ this gel are unaffected by physiological levels of fructose [40.78, 82].

40.3 In vitro and in vivo Testing

Microsystems can be designed, manufactured, and rigorously evaluated in laboratory environments yet fail prematurely after implantation due to environmental conditions such as moisture, ionic contamination, protein deposition, and blood clotting, all of which are compounded by the host inflammatory response. A complex set of physiologic responses to implanted materials leads to both short- and long-term changes to the local tissue environment that can have a profound and oftentimes deleterious effect on the implant. In virtually all implants a cascading series of events occurs in vivo, starting with the adsorption of blood proteins which coat the implant material surface with serum proteins leading to the formation of a protein matrix overlying the implanted material. This matrix subsequently becomes the site of attachment for migrating immune-responsive cells that attempt to integrate, eliminate or isolate the foreign implant. This temporally diverse inflammatory response leads to local tissue changes that include chemistry, cell type, and mechanical and electrical properties [40.83]. The dynamic physiologic response is one of the unique challenges of implanted microsystem oxygen sensors.

In vitro testing of implantable microsystems generally involves biocompatibility tests, environmental tests or performance tests. Because of the cost and complexity of in vivo tests, it is much more practical for implant developers to use in vitro methods to gain an understanding of device function in vitro before engaging in animal or human evaluation or for providing a chemical milieu that mimics a small portion of the physiologic environment to test basic performance parameters of a device. Biocompatibility tests are a regulatory requirement and typically follow ISO 10993. This international standard includes guidelines for duration of implant and type of tissue contacted, and outlines in vitro tests such as cytotoxicity, genotoxicity, hemolysis, and pyrogenicity (ISO 10993-1:2009). In vitro testing that attempts to mimic the physiologic environment has been performed with electrochemical sensors and neuroprosthetic electrodes where serum proteins and reactive scar-forming cells are placed in culture media and allowed to interact with the sensing interface [40.84, 85]. Other in vitro tests include long-term environmental conditioning in isotonic saline solutions.

In vivo testing of implanted microsystems can generally be considered from three perspectives: host compatibility, device compatibility, and device function. Host compatibility is determined through a series of tests that indicate that the device is tolerated or safe and will not cause adverse biologic effects on the host. Device compatibility tests evaluate device performance to determine the effects of protein deposition, cell adhesion, tissue encapsulation, and other unique tissue environmental and biologic factors. Device function tests consider the functional requirements of the implanted system and how it meets those requirements after short-term (< 30 days) or long-term (> 30 days) implantation. Host compatibility test design is guided by ISO 10993 but is always considered in light of the characteristics and end-use application of the device under consideration. In vivo biocompatibility testing considers requirements such as sensitization, irritation, intradermal reactivity, systemic toxicity, chronic toxicity, carcinogenicity, hemocompatibility, reproductive toxicity, biodegradation, and immune responses. The device must function in the host tissue environment and therein lies one of the greatest challenges of implanted microsystems. The Achilles' heel of most implanted microsystems is the interface between the implanted device and the host tissue. The lack of long-term reliability at the biotic–abiotic interface leads to the lack of reliability in oxygen sensors [40.86], recording and stimulating electrodes [40.87], glucose sensors [40.88], and pH sensors [40.89]. In vivo evaluation not only looks at the critical performance interface between the physiologic environment and device but also environmental conditions unique to the implanted environment. This unique environment is often hostile to, or at the very least marginally permissive of, implantable devices. Biological processes can lead to mechanical changes in the surrounding environment, material fatigue due to constant motion, bioerosion due to fluid flow or tissue friction, corrosion due to the various ions in solution, encapsulation due to immune response, local chemical changes due to inflammation or other changes directly caused by the unique physiochemical environment of the body. In vivo tests are expensive and time consuming, and present unique challenges. Implanted devices are typically evaluated for performance during the implant period. During the implant period, devices can also be evaluated for host integration using imaging techniques such as x-ray, computerized axial tomography (CAT) scan, MRI, and ultrasound. These evaluations are often limited by the compatibility of the imaging modality with the medical device. Histological evaluation of tissue adjacent to implanted materials has been the most routine measure of assessing host compatibility. The implanted device is removed along with the surrounding tissue, and the tissue, the interface between the device and the tissue, and the device are analyzed for cell type, cell density, vascularity, and fibrous encapsulation. More in-depth histology can be used to look for specific biomarkers. In vivo animal testing is required for prehuman validation of new designs or design changes.

One of the many challenges is the long-term care that is required to maintain the host in a healthy state over extended periods of time. Unlike many types of physical and environmental testing, in vivo testing cannot currently be accelerated. Although in vivo testing cannot currently be accelerated, this type of testing is often performed on the bench. As in the electronics industry, accelerated aging is conducted according to the Arrhenius formula. These types of tests validate both performance and shelf-life parameters; however, the maximum temperatures must be carefully considered because of the use of thermoplastic polymers in many medical devices.

References

40.1 T. Ohki, D. Stern, M. Allen, J. Ydav: Wireless pressure sensing of aneurysms, Endovasc. Today **3**, 47–52 (2004)

40.2 M.F. Bergeron, K.S. McLeod, J.F. Coyle: Core body temperature during competition in the heat: national boys' 14s junior tennis championships, Br. J. Sports Med. **41**, 779–783 (2007)

40.3 R.M. Stuart, C. Madden, A. Lee, S.A. Mayer: Intracranial Monitoring. In: *Bedside Procedures for the Intensivist*, ed. by H.L. Frankel, B.P. de Boisblanc (Springer, New York 2010) pp. 307–322

40.4 A. DeHennis, K. Wise: A wireless microsystem for the remote sensing of pressure, temperature, and relative humidity, J. Microelectromech. Syst. **14**, 12–22 (2005)

40.5 M. Fonseca, M.G. Allen, J. Kroh, J. White: Flexible wireless passive pressure sensors for Biomed. applications (2006)

40.6 H. Savci, A. Sula, Z. Wang, N. S. Dogan, E. Arvas: MICS transceivers: regulatory standards and applications [medical implant communications service], 179–182 (2005)

40.7 T. Vriesendorp, J.H. Devries, F. Holleman, M. Dzoljic, J.B.L. Hoekstra: The use of two continuous glucose sensors during and after surgery, Diabetes Technol. Therap. **7**, 315–322 (2005)

40.8 G.M. Clark: Cochlear implants in the third millennium, Otol. Neurotol. **20**, 4 (1999)

40.9 P. Boon, K. Vonck, V. De Herdt, A. Van Dycke, M. Goethals, L. Goossens, M. Van Zandijcke, T. De Smedt, I. Devaele, R. Achten, W. Wadman, F. Dewaele, J. Caemaert, D. Van Roost: Deep brain stimulation in patients with refractory temporal lobe epilepsy, Epilepsia **48**, 1551–1560 (2007)

40.10 L. Slowinski, J.D. Putzke, R.J. Uitti, J.A. Lucas, M.F. Turk, B.A. Kall, R.E. Wharen: Unilateral deep brain stimulation of the subthalamic nucleus for Parkinson disease, J. Neurosurg.: Pediatr. **106**, 626–632 (2007)

40.11 J. McHugh, H.W. Sing, J. Phillips, K. Murphy, C.P. Doherty, N. Delanty: Outcome measurement after vagal nerve stimulation therapy: proposal of a new classification, Epilepsia **48**, 375–378 (2007)

40.12 D.A. Groves, V.J. Brown: Vagal nerve stimulation: a review of its applications and potential mechanisms that mediate its clinical effects, Neurosci. Biobehav. Rev. **29**, 493–500 (2005)

40.13 R. Hochberg, M.D. Serruya, G.M. Friehs, J.A. Mukand, M. Saleh, A.H. Caplan, A. Branner, D. Chen, R.D. Penn, J.P. Donoghue: Neuronal ensemble control of prosthetic devices by a human with tetraplegia, Nature **442**, 164–171 (2006)

40.14 J.A. Davis Jr., R.R. Triolo, J. Uhlir, C. Bieri, L. Rohde, D. Lissy, S. Kukke: Preliminary performance of a surgically implanted neuroprosthesis for standing and transfers – where do we stand?, J. Rehabil. Res. Dev. **38**, 609–617 (2001)

40.15 A. Hoogerwerf, K.D. Wise: A three-dimensional microelectrode array for chronic neural recording, IEEE Trans. Biomed. Eng. **41**, 1136–1146 (1994)

40.16 K.D. Wise: Silicon microsystems for neuroscience and neural prostheses, IEEE Eng. Med. Biol. Mag. **24**, 22–29 (2005)

40.17 J.P. Donoghue, J.N. Sanes, N.G. Hatsopoulos, G. Gaál: Neural discharge and local field potential oscillations in primate motor cortex during voluntary movements, J. Neurophysiol. **79**, 159–173 (1998)

40.18 H. Song, D.R. Allee, K.T. Speed: Single chip system for bio-data acquisition, digitization and telemetry, 1848–1851 (2002)

40.19 P. Mohseni, K. Najafi: A battery-powered 8-channel wireless FM IC for biopotential recording applications, Solid-State Circuis Conference ISSCC, Vol. 1 (IEEE, 2005) pp. 560–617

40.20 S. Farshchi, P.H. Nuyujukian, A. Pesterev, I. Mody, J.W. Judy: A TinyOS-enabled MICA2-BasedWireless neural interface, IEEE Trans. Biomed. Eng. **53**, 1416–1424 (2006)

40.21 P. Irazoqui-Pastor, I. Moudy, J.W. Judy: Transcutaneous RF-powered neural recording device, 2105–2106

40.22 A. Bragin, J. Engel Jr., C.L. Wilson, I. Fried, G.W. Mathern: Hippocampal and entorhinal cortex high frequency oscillations (100–500 Hz) in human epileptic brain and in kainic acid treated rats with chronic seizures, Epilepsia **40**, 127–137 (1999)

40.23 M. Chen, C. Majidi, D.M. Doolin, S. Glaser, N. Sitar: Design and construction of a wildfire instrumentation system using networked sensors (poster) (Network Embedded Systems Technology (NEST) Retreat, Oakland 2003)

40.24 Crossbow Technology, Inc.: Available http://www.xbow.com/

40.25 G. DeMichele, P. Troyk: Integrated multichannel wireless biotelemetry system, 3372–3375 (2004)

40.26 J. Hill, M.D. Culler: A wireless platform for deeply embedded networks, IEEE Micro **22**, 12–24 (2002)

40.27 J. Hill: System architecture for wireless sensor networks (Citeseer 2003)

40.28 M. Modarreszadeh, R. Schmidt: Wireless, 32-channel, EEG and epilepsy monitoring system, (2002) pp. 1157–1160

40.29 P. Mohseni, K. Najafi: A wireless FM multi-channel microsystem for biomedical neural recording applications (2003) pp. 217–222

40.30 P. Mohseni, K. Najafi, S.J. Eliades, W. Xiaoqin: Wireless multichannel biopotential recording using an integrated FM telemetry circuit, Neural Syst. Rehabil. Eng., IEEE Trans. **13**, 263–271 (2005)

40.31 D. Niculescu, B. Nath: DV based positioning in ad hoc networks, Telecommun. Syst. **22**, 267–280 (2003)

40.32 A. Nieder: Miniature stereo radio transmitter for simultaneous recording of multiple single-neuron signals from behaving owls, J. Neurosci. Methods **101**, 157–164 (2000)

40.33 I. Obeid, M.A.L. Nicolelis, P.D. Wolf: A multichannel telemetry system for single unit neural recordings, J. Neurosci. Methods **133**, 33–38 (2004)

40.34 A. Oppenheim, R. Schafer: *Discrete-time signal processing* (Prentice Hall Signal Processing, 2009) p. 1120

40.35 J. Parramon, P. Doguet, D. Marin, M. Verleyssen, R. Muñoz, L. Leija, E. Valderrama: ASIC-based batteryless implantable telemetry microsystem for recording purposes, Proc. 19th Int. Conf. IEEE/EMBS (IEEE/EMBS, 1997) pp. 2225–2228

40.36 J. Polastre: *Design and implementation of wireless sensor networks for habitat monitoring, Master Thesis* (University of California at Berkeley, Berkeley 2003)

40.37 R. Sarpeshkar, W. Wattanapanitch, S.K. Arfin, B.I. Rapoport, S. Mandal, M.W. Baker, M.S. Fee, S. Musallam, R.A. Andersen: Low-power circuits for brain–machine interfaces, IEEE Trans. Biomed. Circuits Syst. **2**, 173–183 (2008)

40.38 S. Takeuchi, I. Shimoyama: A radio-telemetry system with a shape memory alloy microelectrode for neural recording of freely moving insects, IEEE Trans. Bio-Medical. Eng. **51**, 133 (2004)

40.39 Vista Integrated Systems: Available http://www.visys.biz

40.40 R. Wise, D.J. Anderson, J.F. Hetke, D.R. Kipke, K. Najafi: Wireless implantable microsystems: high-density electronic interfaces to the nervous system, Proc. IEEE **92**, 76–97 (2005)

40.41 R. Harrison, R.J. Kier, S. Kim, L. Rieth, D.J. Warren, N.M. Ledbetter, G.A. Clark, F. Solzbacher, C.A. Chestek, V. Gilja, P. Nuyujukian, S.I. Ryuz, K.V. Shenoy: A low-power integrated circuit for a wireless 100-electrode neural recording system, IEEE J. Solid-State Circuits **42**, 123–133 (2006)

40.42 R. Harrison, P.T. Watkins, R.J. Kier, R.O. Lovejoy, D.J. Black, B. Greger, F. Solzbacher: A wireless neural interface for chronic recording, Proc. 2008 IEEE Biomed. Circuits Syst. Conf. (BioCAS 2008), Baltimore (2008)

40.43 F. Solzbacher, R. Harrison, R.A. Normann, L. Rieth, S. Chakravarty, J.M. Hsu, M. Klein, H. Oppermann: A next generation chronically implantable wireless neural interface, in Neural, Interfaces Workshop, Bethesda (2005)

40.44 S. Kim, R. Bhandari, M. Klein, S. Negi, L. Rieth, P. Tathireddy, M. Toepper, H. Oppermann, F. Solzbacher: Integrated wireless neural interface based on the Utah electrode array, Biomed. Microdev. **11**, 453–466 (2009)

40.45 L.W. Rieth, A. Sharma, S. Kim, P. Tathireddy, R. Kier, R. Harrison, R. Normann, G. Clark, F. Solzbacher: Recent progress on fully integrated wireless neural interfaces based on the Utah electrode array, Program 370.6., 2009 Neuroscience Meeting Planner. Chicago (Society for Neuroscience 2009)

40.46 Y. Yao, M.N. Gulari, J.A. Wiler, K.D. Wise: A microassembled low-profile three-dimensional microelectrode array for neural prosthesis applications, J. Microelectromech. Syst. **16**, 977–988 (2007)

40.47 D. Nath, M.-T. Heemels, L. Anson: Obesity and diabetes, Nature **444**, 839 (2006)

40.48 E. Rosen, B.M. Spiegelman: Adipocytes as regulators of energy balance and glucose homeostasis, Nature **444**, 847–853 (2006)

40.49 M. Kussmann, S. Rezzi, H. Daniel: Profiling techniques in nutrition and health research, Curr. Opin. Biotechnol. **19**, 83–99 (2008)

40.50 Glucose tolerance and mortality: comparison of WHO and American Diabetes Association diagnostic criteria. The DECODE study group. European Diabetes Epidemiology Group. Diabetes Epidemiology: Collaborative analysis Of Diagnostic criteria in Europe, Lancet **354**, 617–621 (1999)

40.51 The effect of intensive treatment of diabetes on the development and progression of long-term complications in insulin-dependent diabetes mellitus. The Diabetes Control and Complications Trial Research Group, N. Engl. J. Med. **329**, 977–986 (1993)

40.52 Intensive blood-glucose control with sulphonylureas or insulin compared with conventional treatment and risk of complications in patients with type 2 diabetes (UKPDS 33). UK Prospective Diabetes Study (UKPDS) Group, Lancet **352**, 837–853 (1998)

40.53 Effect of intensive blood-glucose control with metformin on complications in overweight patients with type 2 diabetes (UKPDS 34). UK Prospective Diabetes Study (UKPDS) Group, Lancet **352**, 854–865 (1998)

40.54 K.-C. Liao, T. Hogen-Esch, F.J. Richmond, L. Marcur, W. Clifton, G.E. Loeb: Percutaneous fiber-optic sensor for chronic glucose monitoring in vivo, Biosens. Bioelectron. **23**, 1458–1465 (2008)

40.55 G.S.Z. Wilson: Y, Introduction to the Glucose Sensing Problem. In: *In Vivo Glucose Sensing*, ed. by N.J. Hoboken (John Wiley 2010)

40.56 K. Jungheim, K.-J. Wientjes, L. Heinemann, V. Lodwig, T. Koschinsky, A.J. Schoonen: Subcutaneous continuous glucose monitoring, Diabetes Care **24**, 1696 (2001)

40.57 H. Koschwanez, W. Reichert: In vitro, in vivo and post explantation testing of glucose-detecting biosensors: Current methods and recommendations, Biomaterials **28**, 3687–3703 (2007)

40.58 G. Wilson, Y. Hu: Enzyme-based biosensors for in vivo measurements, Chem. Rev. **100**, 2693–2704 (2000)

40.59 N. Wisniewski, F. Moussy, W.M. Reichert: Characterization of implantable biosensor membrane biofouling, Fresenius' J. Anal. Chem. **366**, 611–621 (2000)

40.60 R. Gifford, J.J. Kehoe, S.L. Barnes, B.A. Kornilayev, M.A. Alterman, G.S. Wilson: Protein interactions with subcutaneously implanted biosensors, Biomaterials **27**, 2587–2598 (2006)

40.61 U. Klueh, M. Kaur, Y. Qiao, D.L. Kreutzer: Critical role of tissue mast cells in controlling long-term glucose sensor function in vivo, Biomaterials **31**, 4540–4551 (2010)

40.62 X. Mou, M.R. Lennartz, D.J. Loegering, J.A. Stenken: Long-term calibration considerations during subcutaneous microdialysis sampling in mobile rats, Biomaterials **31**, 4530–4539 (2010)

40.63 J.S. Schultz: Biosensors. In: *Encyclopedia of Biomaterials and Biomedical Engineering*, Vol. 1, 2nd edn., ed. by G. E. B. Wnek (Taylor & Francis, New York 2008)

40.64 W.K. Ward, J.L. Louse, J. Birck, E.M. Anderson, L.B. Jansen: A wire-based dual-analyte sensor for glucose and lactate: in vitro and in vivo evaluation, Diabetes Technol. Therap. **6**, 389–401 (2004)

40.65 D. Gough, J. Armour: Development of the implantable glucose sensor: What are the prospects and why is it taking so long?, Diabetes **44**, 1005 (1995)

40.66 S. Gehrke: Synthesis, equilibrium swelling, kinetics, permeability and applications of environmentally responsive gels, Advances Polymer Sci. **110**, 81–144 (1993)

40.67 G. Gerlach, M. Guenther, J. Sorber, G. Suchaneck, K.-F. Arndt, A. Richter: Chemical and pH sensors based on the swelling behavior of hydrogels, Sens. Actuators B: Chemical **111**, 555–561 (2005)

40.68 G. Gerlach, M. Guenther, G. Suchaneck, J. Sorber, K.-F. Arndt, A. Richter: Application of sensitive hydrogels in chemical and pH sensors, Macromolecular Symposia **210**, 403–410 (2005)

40.69 M. Guenther, G.U. Gerlach, D. Kuckling, K. Kretschmer, C. Corten, J. Weber, J. Sorber, G. Suchaneck, K.-F. Arndt: Chemical sensors based on temperature-responsive hydrogels. In: *Proc. SPIE*, Vol. 6167 (SPIE, 2006)

40.70 S. Herber, W. Olthuis, P. Bergveld: A swelling hydrogel-based PCO_2 sensor, Sens. Actuators B: Chemical **91**, 378–382 (2003)

40.71 S. Herber, J. Eijkel, W. Olthuis, P. Berfeld, A. van den Berg: Study of chemically induced pressure generation of hydrogels under isochoric conditions using a microfabricated device, J. Chem. Phys. **121**, 2746 (2004)

40.72 S. Herber, J. Bomer, W. Olthuis, P. Bergveld, A. van den Berg: A miniaturized carbon dioxide gas sensor based on sensing of pH-sensitive hydrogel swelling with a pressure sensor, Biomed. Microdev. **7**, 197–204 (2005)

40.73 G. Lin, S. Chang, C.H. Kuo, J. Magda, F. Solzbacher: Free swelling and confined smart hydrogels for applications in chemomechanical sensors for physiological monitoring, Sens. Actuators B: Chemical **136**, 186–195 (2009)

40.74 M.P. Orthner, G. Lin, M. Avula, S. Buetefisch, J. Magda, L.W. Rieth, F. Solzbacher: Hydrogel based sensor arrays (2 × 2) with perforated piezoresistive diaphragms for metabolic monitoring (in vitro), Sens. Actuators B: Chemical **145**, 807–816 (2010)

40.75 V. Schulz, M. Guenther, G. Gerlach, J.J. Magda, P. Tathireddy, L. Rieth, F. Solzbacher: In vitro investigations of a pH- and ionic-strength-responsive polyelectrolytic hydrogel using a piezoresistive microsensor. In: *Electroactive Polymer Actuators and Devices (EAPAD) 2009*, Vol. 7287 (SPIE, 2009) pp. 728712-1–728712-10

40.76 T. Trinh: Hydrogel-based piezoresistive pH sensors: design, simulation and output characteristics, Sens. Actuators B: Chemical **117**, 17–26 (2006)

40.77 R. ter Steege, S. Herber, W. Olthuis, P. Bergveld, A. Van den Berg, J.J. Kolkman: Assessment of a new prototype hydrogel CO_2 sensor; comparison with air tonometry, J. Clin. Monit. Comput. **21**, 83–90 (2007)

40.78 G. Lin, S. Chang, H. Hao, P. Thahireddy, M. Orthner, J. Magda, F. Solzbacher: Osmotic Swelling Pressure Response of Smart Hydrogels Suitable for Chronically-Implantable Glucose Sensors, Sens. Actuators B: Chemical **144**, 332 (2010)

40.79 V. Schulz, G. Gerlach, M. Guenther, J.J. Magda, F. Solzbacher: Piezoresistive pH microsensors based on stimuli-sensitive polyelectrolyte hydrogels, Technisches Messen **77**, 179–186 (2010)

40.80 Y. Yao, M.N. Gulari, J.F. Hetke, K.D. Wise: A low-profile three-dimensional neural stimulating array with on-chip current generation, (2005) pp. 1994–1997

40.81 R.A. Siegel, Y. Gu, A. Baldi, E.E. Nuxoll, B. Ziaie: Hard and soft micro- and nanofabrication: An integrated approach to hydrogel-based biosensing and drug delivery, J. Control. Release **141**, 303–313 (2010)

40.82 P. Tathireddy, M. Vaula, L. Genyao, C.H. Seung, M. Guenther, V. Schulz, G. Gerlach, J.J. Magda, F. Solzbacher: Smart Hydrogel Based Microsensing Platform for Continuous Glucose Monitoring, 32nd Annu. Int. IEEE EMBS Conf. (IEEE, 2010)

40.83 J. Anderson: Biological responses to materials, Annu. Rev. Mater. Res. **31**, 81–110 (2001)

40.84 S. Polikov, E.C. Su, M.A. Ball, J.S. Hong, W.M. Reichert: Control protocol for robust in vitro glial scar formation around microwires: Essential roles of bFGF and serum in gliosis, J. Neurosci. Methods **181**, 170–177 (2009)

40.85 W. Kerner, M. Kiwit, B. Linke, F.S. Keck, H. Zier, E.F. Pfeiffer: The function of a hydrogen peroxide-detecting electroenzymatic glucose electrode is markedly impaired in human sub-cutaneous tissue and plasma, Biosens. Bioelectron. **8**, 473–482 (1993)

40.86 M.C. Frost, M.E. Meyerhoff: Implantable chemical sensors for real-time clinical monitoring: progress and challenges, Curr. Opin. Chem. Biol. **6**, 633–641 (2002)

40.87 R. Biran, D.C. Martin, P.A. Tresco: Neuronal cell loss accompanies the brain tissue response to chronically implanted silicon microelectrode arrays, Exp. Neurol. **195**, 115–126 (2005)

40.88 M. Gerritsen, J.A. Jansen, A. Kros, R.J.M. Nolte, J.A. Lutterman: Performance of subcutaneously implanted glucose sensors: a review, J. Invest. Surg. **11**, 163–174 (1998)

40.89 C. Mahutte: On-line arterial blood gas analysis with optodes: current status, Clin. Biochem. **31**, 119–130 (1998)

41. Visual Prostheses

Eduardo Fernandez, Klaus-Peter Hoffmann

Visual impairment is one of the ten most prevalent causes of disability that afflicts millions of people worldwide. This chapter provides an overview of some of the progress that has been achieved in the field of visual prosthesis. While the full restoration of vision seems to be impossible, it is expected that an implanted microelectronic device can create truly meaningful visual percepts that can be translated into functional gains such as the recognition, localization and grasping of objects or skillful navigation in familiar an unfamiliar environments resulting in a substantial improvement in the standard of living of blind and visually impaired persons. However we should be aware that the scientific and engineering problems are much more complex than originally believed and that there are still many unresolved issues than could cause a delay in its development.

41.1 **The Case for Artificial Vision** 822

41.2 **Visual Pathways: From Real Vision to Visual Neuroprostheses** 822

41.3 **Current Approaches to Visual Prostheses** . 824
41.3.1 Retinal Stimulation 824
41.3.2 Optic Nerve Stimulation 826
41.3.3 LGN Stimulation 826
41.3.4 Intracortical Stimulation 826

41.4 **Engineering Visual Neuroprostheses** 827

41.5 **Safe and Effective Stimulation of Visual Pathways Through Multiple Microelectrodes** 829

41.6 **Selection of Suitable Subjects for a Visual Prosthesis** 830

41.7 **Challenges and Future Perspectives** 831

References 832

Living with acquired blindness not only lowers the quality of life of these individuals, but also strains society's limited resources for assistance, care, and rehabilitation. For decades, the possibility of restoring sight to blind individuals has been a subject of intense scientific research. Drug development and genetic engineering have had only marginal success in this context, but new hope has been generated by recent advances in microfabrication technologies, neurosciences, biomaterials, neuromorphic engineering, and information technologies leading to the development of highly sophisticated neural prosthetic devices designed to interact with the nervous system. Such assistive devices have already allowed thousands of deaf patients to hear sounds and acquire language abilities, and the same hope exists in the field of visual neurorehabilitation.

Loss of vision affects millions of people worldwide and poses extraordinary challenges to individuals in our society that relies heavily on sight. Although in recent years the techniques of molecular genetics have led to a rapid identification of a great number of genes involved in visual diseases [41.1], once damaged the nervous system is capable of little functional regeneration, and currently there is no effective treatment for many patients who are visually handicapped as a result of degeneration or damage to:

1. the retina,
2. the optic nerve, and/or
3. the brain.

While pharmacological interventions provide therapeutic solutions to many physiological problems, a pharmacological approach to the mechanisms of blindness has not been discovered. Furthermore, there remains the pervasive question as to how these molecular approaches will actually restore functional vision after it is completely lost in a given individual. Therefore, there are compelling reasons to pursue the development of sophisticated microelectronic prosthesis as viable rehabilitative and therapeutic options to substitute, and ultimately, restore sight.

41.1 The Case for Artificial Vision

The possibility to restore visual perceptions in blind individuals has a long history in biomedical engineering. A German neurosurgeon, Forester, was the first to expose the human occipital pole under local anesthesia and to electrically stimulate it. In 1929 he noted that electrical stimulation induced the perception of points of light, called phosphenes and usually described as *stars in the sky*, *clouds*, and *pinwheels*, which were dependent on where the stimulation probe was placed. These findings, together with the earlier studies of *Penfield* and co-workers during the course of neurosurgical interventions for the treatment of epilepsy [41.3] settled the physiological basis for present efforts to develop a visual prosthesis for the blind.

Subsequent experiments by different research groups [41.4, 5] showed that stimulation of multiple electrodes simultaneously allowed blind volunteers to recognize simple patterns, including letters and Braille characters (Fig. 41.1). The results of these studies supported the premise that patterned electrical stimulation of the visual cortex can evoke patterned percepts. However, these early efforts did not culminate in the restoration of a useful visual sense. Problems with this early work were associated mainly with the large surface electrodes that were used to evoke phosphenes. Thus, relatively high electrical currents were required to evoke phosphenes, and when multiple electrodes were stimulated, these large currents could interact in a nonlinear fashion, evoking phosphenes with unpredictable spatial properties. Furthermore, they also found that a visual prosthesis based on relatively large surface electrodes implanted subdurally would have limited usefulness because of occasional elicitation of pain due to meningeal or scalp stimulation, and the risk of inducing epileptic seizures. This body of work suggests that relatively large surface electrodes may not be a tenable approach to stimulation of the visual cortex, and led a number of investigators to develop a visual prosthesis that could be implanted directly into the retina, the brain, or other parts of the visual pathways. One of the pioneers in this field was *Tassicker*, an Australian researcher who in 1956 was the first to patent a method of implanting a light-sensitive selenium cell in a blind person's retina to transiently restore the patient's ability to perceive the sensation of light [41.6, 7].

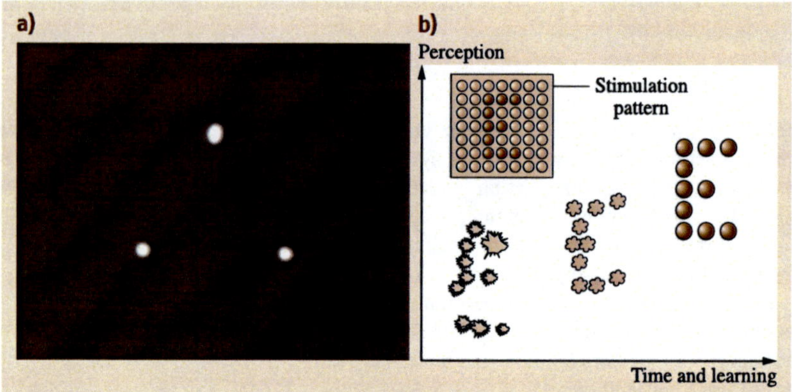

Fig. 41.1a,b Examples of patterned phosphenes. **(a)** Possible perception generated by simultaneously stimulating three electrodes arranged as a triangle. **(b)** The neural plasticity of the visual system can contribute to ever-improving correlation between the physical world and evoked phosphenes. Immediately after implantation the evoked phosphenes are likely to induce a poor perception of an object (the letter E in this example). However, appropriate learning and rehabilitation strategies will contribute to provide concordant perceptions (after [41.2])

41.2 Visual Pathways: From Real Vision to Visual Neuroprostheses

The concept of artificially producing a visual sense in blind individuals is founded on our present understanding of the structure of the mammalian visual system, its processing elements, and the relationship between elec-

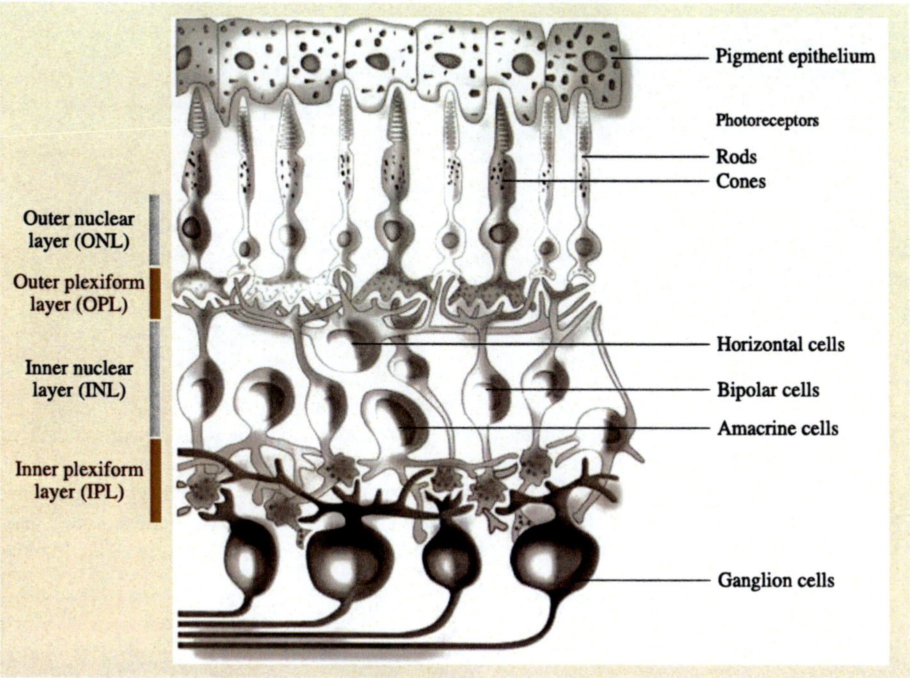

Fig. 41.2 Crosssection of the retina. Light must travel through the thickness of the retina before striking and activating the rods and cones. Subsequently, the absorbtion of photons by the visual pigment of the photoreceptors is translated into an electrical message that can stimulate all the succeeding neurons of the retina. The retinal message concerning the photonic input and some preliminary organization of the visual image are transmitted to the brain from the spiking discharge pattern of the ganglion cells (image courtesy of Dr. Markus Bongard)

trical or mechanical stimulation of any part of the visual pathways and the resulting visual sensations [41.8].

Our entire experience of the external visual world derives from the concerted activity of a restricted number of retinal ganglion cells (Fig. 41.2), which have to send their action potentials through the *bottleneck* optic nerve to the brain (Fig. 41.3). The retina is essentially a piece of brain tissue that gets direct stimulation from the outside world's lights and images. Visual input to the retina consists of a stream of photons, which can be unequivocally quantified in space and time. The retina performs spatial, temporal, and chromatic processing on visual information and converts it into a compact *digital* format composed of neural impulses [41.9]. In addition, starting at the retinal level, vision is an active process and eye movements are essential for information processing [41.10]. Thus our entire experience of the external visual world derived from the concerted activity of a restricted number of retinal ganglion cells, which have to send their information, via the optic nerve, to higher visual centers. This representation has to be unequivocal and fast, in order to ensure object recognition for any single stimulus presentation within a few hundreds of milliseconds [41.11]. Therefore, the question of how the information about the external world is compressed in the retina, and how this compressed representation is encoded in spike trains is an important challenge for the success of any visual prosthesis.

The next processing center is the lateral geniculate nucleus (LGN), which begins to integrate information from both eyes into a binocular representation of visual space. Once retinal signals arrive within the LGN, they are in a position to be influenced by a wide variety of other inputs, as well as the complex feed forward and feedback circuits. These inputs include not only visual information, but also nonvisual information that comes from cortical areas, superior colliculus, pretectum, parabigeminal nucleus, reticular nucleus, and other brain stem pathways [41.12]. The number of synapses provided by these nonretinal inputs greatly outnumbers the number of parallel pathways coming from the retina, but their function is not fully understood.

The LGN neurons mainly project to the primary visual cortex (also known as the striate cortex or V1), which was the first cortical area clearly identified in the human brain by *Gennari* in 1782 [41.13]. Area V1 is located at the back of the head, in the occipital lobe, and from here the information is distributed to a number of higher cortical centers for further processing [41.14]. The main point, as stated by *Hubel* and *Wiesel*, is that the primary visual cortex is in no sense the end of the visual path [41.15]. It is just one stage, probably an

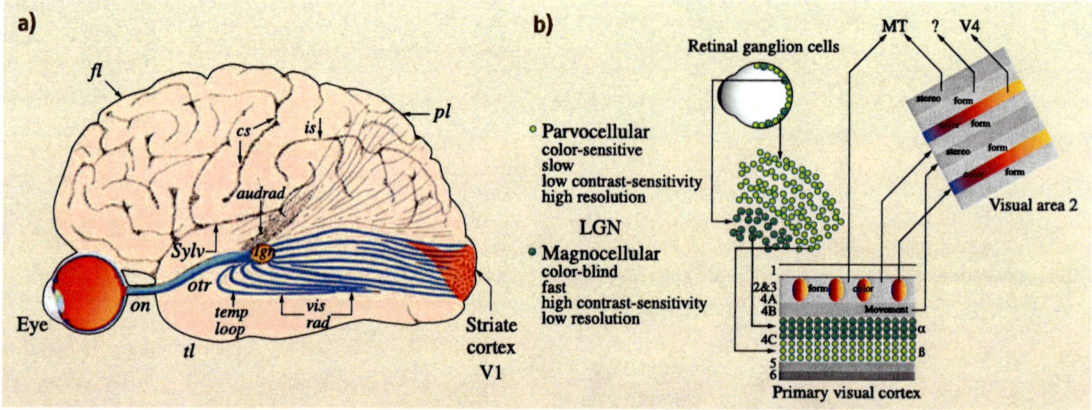

Fig. 41.3a,b Human visual pathways. **(a)** The optic nerve transmits the information from the retinal ganglion cells to the lateral geniculate nucleus (LGN), which relays information to the primary visual cortex (image courtesy of Webvision, http://webvision.umh.es). **(b)** Functional segregation of the visual system, presenting the projection of large retinal ganglion cells to the magnocellular layers of the LGN and further to layer 4 Ca of the primary visual cortex. Small ganglion cells project to the parvocellular layer and hence to the layers 4 A and 4 Cb of V1 (image courtesy of Dr. Markus Bongard)

early one in terms of the degree of abstraction of visual information processing.

As blindness results from an interruption in the normal flow of signals along the visual pathways, a visual prosthesis has to excite the neurons of the pathway at some point after the damage site [41.2]. The only requirement is that the device should have contact with still functioning neural elements.

41.3 Current Approaches to Visual Prostheses

Figure 41.4 shows the main approaches for the design of a visual neuroprosthesis. Since retinal diseases frequently reduce visual acuity and result in noncurable blindness, several groups worldwide are working on the development of different prosthesis designed to interact with the remaining healthy retina [41.6, 16–19], optic nerve [41.20, 21], and LGN [41.22, 23]. However, the output neurons of the eye, the ganglion cells, often degenerate in many retinal blindness cases [41.24, 25], and therefore a retinal, optic nerve, or LGN prosthesis would not always be helpful. This extensive degeneration usually spares the neurons in the higher visual regions of the brain, which suggests the enormous potential of a cortical prosthesis designed to stimulate cortical neurons [41.2,26]. Due to the complexity of the interconnectivity and the receptive field characteristics, visual areas beyond V1 have not been proposed as sites for vision prostheses.

In general, all the approaches share a common set of components: a bioinspired retina-like encoder able to transform the visual scene into patterns of electrical stimulation, a second stage that processes the information and transmits it through a radio-frequency link to the implanted device, and an electrode array implanted at some level in the visual pathways and located near the target neurons.

41.3.1 Retinal Stimulation

Retinal prostheses are being developed to apply electrical stimulation to the retina in order to restore vision in patients with age-related macular degeneration and retinitis pigmentosa, and more than five groups are already pursuing human clinical trials of diverse retinal prosthesis [41.27, 28]. There are significant advantages to the retinal approach as it is less invasive and the implants are lower in the visual pathways, which means that they are closer to the photoreceptors and have more opportunities for natural processing of the visual information. However, there are also some problems related with the encapsulation, fixation of the implants, power dissipation, and the production of sufficient

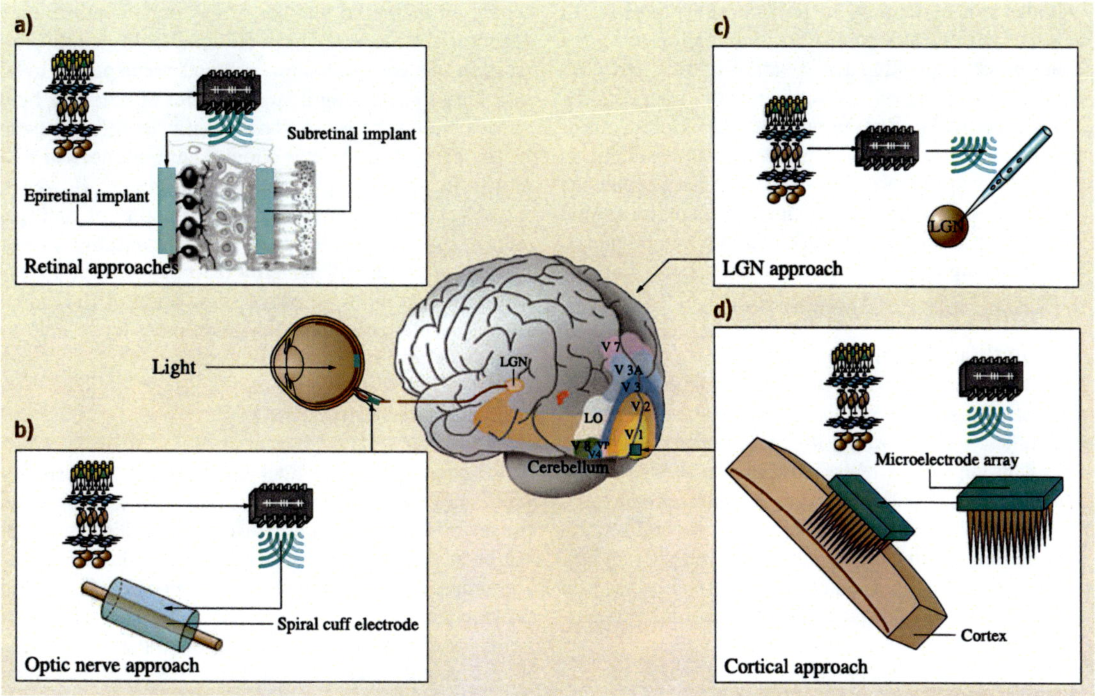

Fig. 41.4a–d Summary diagram of different approaches to restore vision. The *inset figures* illustrate the approaches in detail. **(a)** Schematic diagram of a retina crosssection showing two methods of stimulating ganglion cells. The epiretinal device is attached to the inner surface of the retina and stimulates the inner retinal layer. In the subretinal approach, photodiodes are implanted underneath the retina and used to generate currents that stimulate directly the retina. **(b)** Stimulation of the optic nerve by implanting a cuff electrode around the nerve. **(c)** LGN microstimulation. **(d)** Cortical approach

currents [41.29, 30]. Furthermore, for a retinal visual prosthesis to work, there must be some functional retinal ganglion cells, and therefore these systems are not capable of treating conditions such as glaucoma and diabetic retinopathy [41.28, 31].

Currently, there are mainly two kinds of retinal implants under development: subretinal and epiretinal. Furthermore, some groups are studying the efficacy of the suprachoroidal-transretinal stimulation to generate focal excitation in retinal ganglion cells [41.32, 33]. In addition, a novel photostimulation approach has been proposed, which involves re-engineering retinal ganglion cells or bipolar cells to become light sensitive by incorporation of artificial opsins [41.34].

Subretinal Implants

As some of the neural retina, with the exception of the photoreceptors layer, is preserved for both age-related macular degeneration and retinitis pigmentosa, a reasonable approach to the restoration of vision under these pathologies is the replication of the function of the photoreceptors [41.6, 35, 36]. One of the main advantages of this location is that the remaining retinal cells in the outer plexiform layer (vertically running bipolar cells and horizontally oriented horizontal cells) may be utilized to stimulate the amacrine and ganglion cells directly.

In a general sense all research groups focusing on the subretinal approach have proposed similar methods. Thousands of light-sensitive microphotodiodes equipped with microelectrodes are assembled on a very thin plate and placed in the subretinal space. The subretinal device or microphotodiode array (MPDA) is implanted between the pigment epithelial layer and the outer layer of the retina, which contains the photoreceptors. Then light falling on the retina generates graded currents that stimulate the dendrites of the bipolar cells in the outer plexiform layer to induce a visual sensation.

The schema is tantalizingly simple, requiring no external power or control signal. Furthermore the photodiodes can be tailored to provide either positive or negative current in response to illumination, with the intention of mimicking the function of ON and OFF bipolar cells. However, the success of this approach depends upon three assumptions. First, the bipolar cells of the dysfunctional retinal should persist and function in a somewhat physiologically normal manner. Second, the photodiodes must generate enough current under normal illumination levels. Third, the electrodes can be placed in close enough proximity to the bipolar cells. Encouraging progress has been reported in each of these areas.

Epiretinal Implants

As an alternative approach to stimulate the retina, other research groups have proposed placing the stimulating electrodes in close proximity to the retinal ganglion cells. This approach attempts to stimulate the remaining retinal neurons of patients who are blind from end-stage photoreceptor diseases. In this context, several reports indicate that significant portions of the retinal ganglion cells survive even in end-stage of pathologies such as age-related macular degeneration and retinitis pigmentosa [41.37, 38].

The epiretinal approach implies the use of implanted multielectrode arrays (MEAs) and transcutaneous telemetry to transfer data and power to the implanted MEAs. To date, a number of experiments performed in sighted and blind human subjects demonstrate the potential for epiretinal electrical stimulation to provide patterned visual perception [41.39–42]. There are, however, issues relevant to the epiretinal approach that should be solved. Thus, the epiretinal devices need to be firmly affixed in place in order to efficiently stimulate the retina and provide a consistent visual perception. Another issues of concern are power dissipation, cross-talk between electrodes, and the viability of the tissues under the implant. Nevertheless, epiretinal implants appear to be a potentially viable approach for restoring functional vision in some blind individuals.

41.3.2 Optic Nerve Stimulation

The next visual prosthetic approach is to stimulate the optic nerve fibers. The optic nerve, composed of retinal ganglion cell axons, can be reached surgically and may be a suitable location for implanting a simulation electrode providing that the retina cannot safely support a prosthesis, for example in long-standing retinal detachments.

The rationale for using a nerve cuff electrode is completely distinct from all the retinal approaches where an increased number of electrode contacts may theoretically increase the resolution [41.43–45]. In contrast, in this approach a few contacts are expected to be capable of generating a large number of potential field shapes within the nerve that can activate different parts of the nerve and thereby allow a large number of distinctly located phosphenes to be generated. However, the selective and simultaneous stimulation of multiple subsets of axons requires cuffs with a prohibitively large number of contacts. Furthermore, the present cuff electrodes do not allow selective stimulation of individual fibers, so that this approach is unlikely to be able to recreate a high-resolution visual scene.

41.3.3 LGN Stimulation

The lateral geniculate nucleus (LGN) is the primary processing center for visual information, meaning that the neural signals encoding visual information have not yet been extensively processed. The LGN receives information directly from the retinal ganglion cells and projects mainly to the primary visual cortex. The LGN holds promise as a target of electrical stimulation for artificial visual percepts since it has been demonstrated that LGN microstimulation produces predictable visual percepts [41.22, 23]. Therefore, a visual prosthesis based on LGN stimulation may be helpful to restore sight to those who have become blind because of trauma of the eye or diseases such as glaucoma, macular degeneration, and retinitis pigmentosa.

A possible disadvantage of this approach is the limited number of electrodes that can be inserted in the LGN and, hence, the limited visual resolution. Moreover the neurons in the LGN are too close together to be stimulated individually, which would be important when trying to reproduce natural vision. Furthermore, most of the projections to the LGN are from the primary visual cortex making it difficult to create complex visual percepts.

41.3.4 Intracortical Stimulation

If the primary visual cortex can be stimulated with visual information in a format somewhat similar to the way it was stimulated before the onset of blindness, a blind individual may be able to use this stimulation to extract information about the physical world around him/her [41.2, 46]. Although a cortical prosthesis based on relatively large subdural electrodes has a limited usefulness, a promising approach, which can activate populations of neurons with greater spatial specificity and lower levels of stimulation than is possible with larger electrodes on the surface of the brain, is the use of intracortical microelectrodes. In this context, *Schmidt* et al. [41.26] described the implantation of 38 floating microelectrodes within the right visual cortex of a 42 year old woman who had been blind over 22 years. 34 of the microelectrodes were able to elicit phosphenes for a period of 4 months, and most of the microelectrodes had stimulation thresholds below 25 mA. Unfortunately, these microelectrodes were not well suited for a long-term application. Thus, due to the breakage of lead wires early in the experiment, only limited tests could be done to evaluate pattern recognition. Nevertheless, taken as a whole, the previous results suggest that passing electrical currents through an array of electrodes inserted into an appropriate location in the visual pathway, is able to produce the perception of phosphenes, and that these phosphenes may be appropriate for restoring some limited but useful sense of vision to the profoundly blind. This cortical approach requires intracortical penetrating electrodes with exposed tips located in layer 4 and with sizes of the same order of magnitude as the neurons that are intended to be stimulated.

This approach is the only treatment available for blindness caused by glaucoma, optic atrophy, or diseases of the central visual pathways, such as brain injuries or stroke.

The main problem of a cortical implant is the lack of preliminary processing by the brain, particularly in the retina where much of the information reduction takes place [41.31]. Furthermore, it is unclear what kind of percepts will be evoked by simultaneous stimulation of a large number of microelectrodes.

41.4 Engineering Visual Neuroprostheses

Regardless of the differences in the approaches, all vision prostheses (with the exception of subretinal prostheses and the photostimulation approach) share a common set of components [41.8], which are illustrated in Fig. 41.5. One or two cameras provide image acquisition, which is then processed by a bioinspired retina-like encoder in order to transform the visual world in front of a blind individual into electrical signals. This first stage performs a multichannel spatio-temporal filtering, to extract and enhance the most relevant features of the scene and also re-encodes this information into a neuromorphic stream of electrode addresses [41.47]. The second stage serializes the information and transmits it through a radiofrequency link to the implanted device. This telemetric transmission provides a wireless transfer of power and data to the internal system (Fig. 41.6). The implanted electronics package must decode the signals, identify the target electrodes, and generate the stimulation signals applied using multiple microelectrodes.

These transformations must take into account many issues that are normally performed by the retina. For example, sampling across the retina is not uniform and, therefore, retinotopic gradients and magnification factors must be introduced to match image representation. On the other hand, it is clear that several streams of information are processed in parallel from any retinal point by several dozens of interneuronal subtypes before contrast, brightness, orientation movement, and color are finally coded as modulation of ganglion cell action potential series [41.48]. While chromatic information is not of utmost priority, a differential characterization will nevertheless be required when designing *achromatic* processing modules for basic representation of image components. The goal of developing such a bioinspired retinal encoder is not simply to record a high resolution image, but to transmit visual information in a meaningful way to the appropriate site(s) in the retina, optic nerve, LGN, or brain.

Although it is not the purpose of this chapter to present a detailed study of the problem of coding/decoding of retinal images by ensembles of retinal ganglion cells, increasing evidence suggests that the retina and the brain utilize distributed codes that can only be analyzed by simultaneous recording of the activity of multiple neurons [41.49–52]. Far from a simple transducer of light into electrical neural impulses, the retina performs a locally-computed spatio-temporal contrast enhancement function and a very efficient compression of visual information. These tasks are essential to provide a high adaptation capability to very different lighting conditions, high noise immunity, and to efficiently communicate the visual information by means of a limited number of optic nerve fibers. Thus, our entire experience of the external visual world derives from the concerted activity of a limited number of ganglion cells in the retinal output layer that have to represent all the features of objects in the visual world, namely

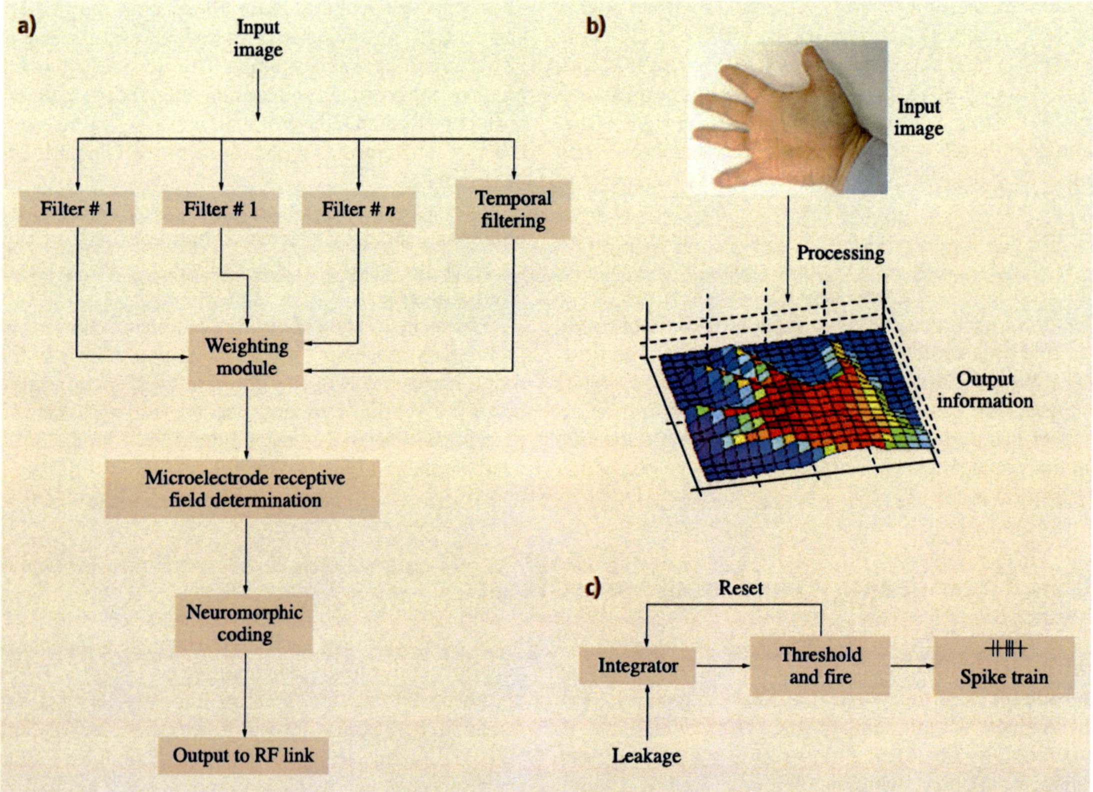

Fig. 41.5a–c Schematic organization of a bioinspired retinal model. **(a)** The input images are captured by a photosensor array and processed by a combination of several spatial and temporal filters that enhance specific features of the captured information. **(b)** Performance of the encoder/stimulator in transforming an image of a hand into an electrode stimulus pattern. **(c)** Leaky integrate-and-fire spiking neuron model showing the response to a stimulus for a given component of the activity matrix

their color, intensity, shape, movement, and the change of these features in time. This representation must be unequivocal and fast in order to ensure object recognition for any single stimulus presentation within a few hundreds of milliseconds.

The development of a bioinspired visual encoder, therefore, poses an information processing challenge without parallel in neuroscience, computer science, or communication technologies. Consequently, several researchers are developing complex retina encoders for visual implants [41.53, 54].

Figure 41.5 summarizes the basic architecture of a typical bioinspired retinal model. It is based on information about the photoreceptor distribution within the human/primate retina and on electrophysiological recordings from populations of retinal ganglion cells. Through a set of parametrized filters and functions, that simulate the complex operations of the neural retina, a portable model is obtained that can be easily translated into a hardware description for automatic synthesis using the appropriate tools. The input images are captured by a photosensor array (preferably a logarithmic response camera that has a similar response to the human visual system and can reduce saturation in high contrast visual scenes) and are processed by a set of separate spatial and temporal filters that enhance specific features of the visual information captured. The model can take into account the irregular distribution of photoreceptors within the human retina: a high density of pixels and smaller receptive field sizes in central areas and lower density and bigger receptive fields in peripheral areas. A gain factor can be specified for every individual channel as well as a global gain. The next stage reduces the information to the resolution of the electrode

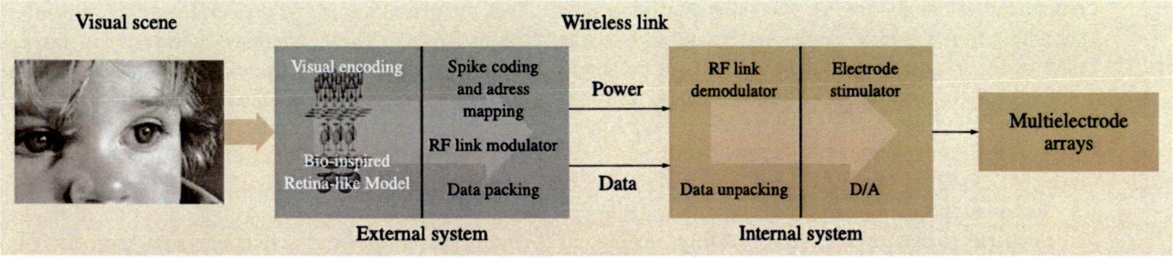

Fig. 41.6 Functional blocks of a visual neuroprosthesis. This includes a bioinspired visual information block (artificial retina) to extract and enhance the most relevant features of the visual scene. A coding block that translates the retina-like output into pulses modulated with different cadences and temporal lags and whose projection onto the electrode matrix must be fully configurable. An RF block providing a wireless transfer of power of data and the implanted electronics package that must decode the signals, identify the target electrodes, and conform the voltage shape (D/A converter) of the final waveforms to be applied to the electrodes located near the target neurons (after [41.46])

matrix, with the option of defining specific receptive field shapes and sizes. Finally, a mapping and neuromorphic coding (into output charge-balanced pulses that can be sent to each electrode) is carried out and feeds the radiofrequency link that goes to the microelectrode array. Thus, the continuously varying input video stream is converted into neuromorphic pulse-coded signals through a circuit that emulates the function of retinal neurons [41.46,55].

The implementation of the model in digital hardware provides a flexible design, achieving a high performance with response times several orders of magnitude lower than those of biological systems. The present version of the whole system is able to work properly up to 40 MHz [41.29,46,47]. This means that 40 000 electrodes may be stimulated with an inter-spike temporal resolution equal to or lower than 1 ms. The use of reconfigurable circuitry (FPGA) permits one to adjust or even change the spiking model easily [41.47,54]. Furthermore, this technology can be also used for psychophysical experimentation in order to determine the best image processing strategy, and to obtain insights into the optimal number of electrodes, gray levels, and grid size [41.56,57].

Finally the output of the bioinspired retina must be transmitted to a remote stimulating device that electrically connects with the implanted microelectrode array (Fig. 41.6). Ideally, this should be done telemetrically (i. e. without the need for attached wires) to reduce the risk of infections and an opto-coupling stage should be incorporated to protect the patient against electrical risks. Furthermore, for a durable system material must be biocompatible and electrical-charge displacements must be in margins of capacitive work, never in an irreversible-faradaic working zone [41.58].

Several factors should be taken into account, especially safety, size, and power consumptions. Thus, this stimulating device should be able to operate in real-time for a high number of electrodes, have low power requirements and be flexible enough to generate different waveforms and adapt to different stimuli and image processing conditions. Basically, it should be able to solve the problem of transmitting data and power to multiple microelectrodes, while simultaneously having the possibility of remote control and sensing of microelectrode status. This implies a compromise between the power level requirements of the stimulation circuits and the signal bandwidth requirements [41.47,59].

41.5 Safe and Effective Stimulation of Visual Pathways Through Multiple Microelectrodes

Any visual neuroprosthetic system must be implanted into the retina, optic nerve, LGN, or visual cortex and remain fully functional for periods that will eventually extend to many decades. Therefore, these devices must be highly biocompatible and be able to resist the attack of biological fluids, proteases, macrophages or any substances of the metabolism. Furthermore, it is necessary to take into account the possible damage of neural tissues by permanent charge injection and the most effective means of stimulating the visual pathways. On the other hand, although there are no major safety concerns stemming from the pilot studies carried out to date, the risks of the surgical procedures involved are not negligible. All these considerations place unique constraints on the architecture, material, and surgical techniques used in the implementation of visual prosthesis interfaces [41.29,42,46,60,61].

Important problems reported with all available microelectrodes to date are long-term viability, biocompatibility, and biostability. Recent studies of the response of neural tissue to stab wounds have shown that there are acute and chronic inflammatory reactions that affect both the neural tissue and the surface of the microelectrodes [41.62]. These reactions often result in damage to neurons and microelectrodes and lead to the proliferation of a glial scar around the implanted probes, which prevents recording or stimulation of neurons [41.63,64]. The reasons for the inflammatory response lie in molecular and cellular reactions at foreign surfaces [41.65]. These responses can be controlled, and one of the big challenges in this field is to create new, more biocompatible surfaces. The solution may involve coating the electrodes with bioactive molecules that are slowly released into the surrounding neural tissue [41.60].

41.6 Selection of Suitable Subjects for a Visual Prosthesis

The selection of a specific person for a visual implant is not straightforward. There are no strict standardized criteria for accepting or rejecting a candidate or for the best rehabilitation procedure for every type of blindness. Generally, a choice may be made between different approaches and/or rehabilitation procedures depending on availability, efficacy or rejection of invasive methods [41.28, 66, 67], but a presurgical protocol and improved methods for predicting success with a visual neuroprosthesis need to be developed [41.46,56,68].

In general, it is considered crucial, at least in these preliminary stages, that the subject have no residual vision and has not had any significant benefit from a conventional visual aid. However, these seemingly straightforward criteria have not always worked well in practice, and there are different definitions of residual vision and significant benefit from a visual aid. Needs and wishes of individual subjects are also significant variables for implant candidacy. This issue is further complicated because it is not possible to predict success with a visual implant in a specific person. Clearly, our knowledge regarding visual system anatomy and function may allow for crude bioinspired models and strategies of stimulation. However, what has not been discussed is how the type, onset, duration, and temporal profile of an individual visual loss may have repercussions on the success of the device. To decide what time is the most suitable in that matter is a difficult double-faced ethical decision [41.66].

Recognizing these limitations, it is generally acknowledged that previous visual experience is necessary for the patient to interpret and recognize the visual patterns that are generated. In this context, it has been presumed that if blindness occurs after the age of ten years of age (i. e. with respect to the critical period), visual pathways should develop normally and thus remain excitable [41.28, 66].

Improved networking among research groups would clearly help in developing a common body of standardized test and standardized selection criteria [41.56]. Among them, possible criteria to specify the characteristics of adults who are potential users of a visual neuroprosthesis [41.68] are listed here.

- Visual function criteria and pre and postimplantation testing. Indications in favour of a cortical visual implant may include profound and bilateral visual loss, but more objective and quantitative criteria should be developed. These studies need to incorporate a more quantifiable means of estimating risk and benefit for a given candidate.
- Electrophysiological criteria. Registration and analysis of electroretinogram (ERG) as well as visual evoked potentials (VEP) should be a basic component in candidate selection. Improved methods for predicting success with a visual implant need to be developed. These methods may include test using data from transcranial magnetic stimulation (TMS) of visual pathways and new imaging techniques such as functional magnetic resonance imaging (fMRI).
- Medical, anatomical, and surgical criteria. The usual candidate should be a healthy adult. The medical history should include the onset and evolution of the blindness, physical examination, and laboratory tests to include or exclude candidates and to assist the implant team in planning a total program, including postimplant training.

- Evaluation of the risks and limitations associated with the surgery and every specific approach. These are usually small for persons in good general health but increase with age and other pathological conditions.
- Quantitative evaluation of the subjects' needs and performance using standardized psychophysical and behavioral methods.
- Psychological testing to discard psychiatric disorders and other possible mental status that may affect the success of the visual prosthesis.

There are other important issues that should be taken into account and case studies of surgical sight restoration following long-term visual deprivation [41.69, 70] provide a relevant insight. For example, following ocular surgical procedures aimed at regaining some degree of functional vision, patients blinded for many years experience profound difficulty in various visual tasks, particularly those requiring the identification and recognition of objects. Interestingly, if patients are allowed to explore the same object through touch, they can recognize it immediately and register their newly acquired visual percepts with their existing senses. These findings demonstrate that even when vision reaches the brain physiologically, visual perception nonetheless remains impaired. Thus, the simple restoration of a lost sensory input may not itself suffice in reconstituting the sense it normally provides, and active visual rehabilitation may be necessary to maximize the adaptation and get the most from these devices. Moreover, some form of spatial remapping between the bioinspired encoder and the stimulation pattern delivered to the array of electrodes implanted in the visual cortex is highly recommendable [41.46, 50, 53, 59]. As a result, the right learning, remapping, and rehabilitation strategies may potentially help to modulate the plasticity of the brain and contribute to ever-improving performance and more concordant perceptions.

41.7 Challenges and Future Perspectives

Visual prosthesis is a rapidly emerging field that requires extraordinary diverse, lengthy, and intimate collaborations among basic scientists, engineers, and clinicians. Significant challenges to the development of a successful visual prosthesis include limitations to the number of electrodes, biocompatibility, encapsulation, electrode degradation, power interference, signal and imaging processing, interference with residual vision, functional evaluation, and training [41.31, 46, 68]. Despite all these issues several research groups are developing sophisticated microelectronic devices designed to stimulate viable neuronal tissue in the hopes of generating functional vision artificially.

To date, at least 23 different devices are under development and 5 groups are pursuing human clinical trials either with acute or chronic implantation of neurostimulating devices, and increasing numbers are expected within the next years [41.27, 28]. However, even the most advanced of these experimental devices has only been able to provide evidence that such an approach may indeed be feasible. What remains to be conclusively demonstrated is whether or not the visual percepts produced by implanted microelectronic devices can create meaningful visual perceptions that can be translated into functional gains such as the recognition, localization and grasping of objects, or skilful navigation in an unfamiliar environment, thus, ultimately, having the potential to improve the quality of life of a patient.

The greater impediments to future progress in implementing a visual neuroprosthesis approach are not only the technical, engineering, and surgical issues that remain to be solved, but also the development and implementation of strategies designed to interface with the visually deprived brain specifically tailored for an individual patient's own needs. This would particularly involve an improved patient selection and a custom-tailoring of visual prosthetic devices. A key issue in this context that has often been utterly underrated is the role of neural plasticity. Thus, these strategies should take into consideration not only standardized methods and employ current clinical and technological expertise, but also consider newly emerging developmental and neurophysiological evidence. For example, there is considerable evidence that adaptive and compensatory changes occur within the brain following the loss of sight [41.46, 71–73]. These studies suggest that in some patients the occipital parts of the brain that sighted subjects utilize to process visual information are transformed and utilized to process tactile and auditory stimuli. This plastic change in the brain probably allows blind subjects to extract greater information from touch and hearing, thus improving living standard and enhancing the integration of the blind in the social and working environment of a sighted society. The modulation and understanding of these neuroplastic processes is crucial for the success of any visual neuroprosthesis and can, therefore, provide the neuroscientific foundation for improved rehabilitation and teaching strategies for the blind.

It is hoped that the advances in medicine, ophthalmology, and genetics will be able to devise new ways of preventing diseases of the retina and visual pathways or in transplanting neurons that have been lost. However, genetic science and treatment will not help in injuries due to accidents and probably will not eliminate all the visual impairments that are due to aging. Therefore, progress in vision prosthesis technologies is regarded as a necessity for the future. We hope that the progress in medical technologies, material science, and bioengineering, together with the increase of intelligence in these visual neuroprosthetic devices will foster the improvements in the quality of life of people that are affected by visual impairments, and result in the development of new custom-tailored neuroelectronic systems for restoring functional sight in the blind.

References

41.1 L.R. Pacione, M.J. Szego, S. Ikeda, P.M. Nishina, R.R. McInnes: Progress toward understanding the genetic and biochemical mechanisms of inherited photoreceptor degenerations, Annu. Rev. Neurosci. **26**, 657–700 (2003)

41.2 R.A. Normann, E.M. Maynard, K.S. Gullory, D.J. Warren: Cortical implants for the blind, IEEE Spectrum **33**(5), 54–59 (1996)

41.3 W. Penfield, H. Jaspers: *Epilepsy and the Functional Anatomy of the Human Brain* (Churchill, London 1974)

41.4 G.S. Brindley, W.S. Lewin: The sensations produced by electrical stimulation of the visual cortex, J. Physiol. **196**, 479–493 (1968)

41.5 W.H. Dobelle, M.G. Mladejovsky, J.P. Girvin: Artificial vision for the blind: Electrical stimulation of visual cortex offers hope for a functional prosthesis, Science **183**, 440–444 (1974)

41.6 E. Zrenner: Will retinal implants restore vision?, Science **295**(5557), 1022–1025 (2002)

41.7 G.E. Tassicker: Retinal Stimulator, Patent 2760483 (1956)

41.8 E.M. Maynard: Visual prostheses, Annu. Rev. Biomed. Eng. **3**, 145–168 (2001)

41.9 H. Kolb, E. Fernandez, R. Nelson: Webvision – The Neural Organization of Retina and Visual System (2005), http://www.umh.es/frame.asp?url=/busquedas/resultados.asp?qu=webvision, last accessed 21.04.2011

41.10 M. Greschner, M. Bongard, P. Rujan, J. Ammermüller: Retinal ganglion cell synchronization by fixational eye movements improves feature estimation, Nat. Neurosci. **5**(4), 341–347 (2002)

41.11 S.M. Smirnakis, M.J. Berry, D.K. Warland, W. Bialek, M. Meister: Adaptation of retinal processing to image contrast and spatial scale, Nature **386**, 69–73 (1997)

41.12 S. Sherman, R. Guillery: *Exploring the Thalamus* (Academic, San Diego 2001)

41.13 C.G. Gross: From Imhotep to Hubel and Wiesel: The story of visual cortex. In: *Extrastriate Cortex in Primates*, ed. by K.S. Rockland, J.H. Kass, A. Peters (Plenum, New York 1997) pp. 1–58

41.14 K. Grill-Spector, R. Malach: The human visual cortex, Annu. Rev. Neurosci. **27**, 649–677 (2004)

41.15 D.H. Hubel, T.N. Wiesel: Brain mechanism of vision. In: *The Perceptual World*, ed. by I. Rock (W.H. Freeman, New York 1990) pp. 3–24

41.16 P. Walter: Successful long-term implantation of electrically inactive epiretinal microelectrode arrays in rabbits, Retina **19**(6), 546–552 (1999)

41.17 R. Eckmiller: Learning retina implants with epiretinal contacts, Ophthalmic Res. **29**(5), 281–289 (1997)

41.18 M.S. Humayun, J.D. Weiland, G.Y. Fujii, R. Greenberg, R. Williamson, J. Little, B. Mech, V. Cimmarusti, G. Van Boemel, G. Dagnelie, E. de Juan Jr.: Visual perception in a blind subject with a chronic microelectronic retinal prosthesis, Vis. Res. **43**, 2573–2581 (2003)

41.19 J.F. Rizzo III, J. Wyatt, J. Loewenstein, S. Kelly, D. Shire: Perceptual efficacy of electrical stimulation of human retina with a microelectrode array during short-term surgical trials, Investig. Ophthalmol. Vis. Sci. **44**(12), 5362–5369 (2003)

41.20 J. Delbeke, D. Pins, G. Michaux, M.-C. Wanet-Defalque, S. Parrini, C. Veraart: Electrical stimulation of anterior visual pathways in retinitis pigmentosa, Investig. Ophthalmol. Vis. Sci. **42**(1), 291–297 (2001)

41.21 C. Veraart: Pattern recognition with the optic nerve visual prosthesis, Artif. Organs **27**(11), 996–1004 (2003)

41.22 J.S. Pezaris, R.C. Reid: Demonstration of artificial visual percepts generated through thalamic microstimulation, Proc. Natl. Acad. Sci. USA **104**(18), 7670–7675 (2007)

41.23 J.S. Pezaris, E.N. Eskandar: Getting signals into the brain: Visual prosthetics through thalamic microstimulation, Neurosurg. Focus **27**(1), E6 (2009)

41.24 B.W. Jones, C.B. Wyatt, J.M. Frederick, W. Baehr, C.K. Chen, E.M. Levine, A.H. Milam, M.M. Lavall, R.E. Marc: Retinal remodeling triggered by photoreceptor degenerations, J. Comp. Neurol. **464**(1), 1–16 (2003)

41.25 R.E. Marc, B.W. Jones: Retinal remodeling in inherited photoreceptor degenerations, Mol. Neurobiol. **28**(2), 139–147 (2003)

41.26 E.M. Schmidt, M.J. Bak, F.T. Hambrecht, C.V. Kufta, D.K. O'Rourke, P. Vallabhanath: Feasibility of a visual prosthesis for the blind based on intracortical microstimulation of the visual cortex, Brain **119**(Pt 2), 507–522 (1996)

41.27 I. Wickelgren: Biomedical engineering. A vision for the blind, Science **312**(5777), 1124–1126 (2006)

41.28 G. Dagnelie: Visual prosthetics 2006: assessment and expectations, Expert Rev. Med. Devices **3**(3), 315–325 (2006)

41.29 R.A. Normann, B.A. Greger, P. House, S.F. Romero, F. Pelayo, E. Fernandez: Toward the development of a cortically based visual neuroprosthesis, J. Neural Eng. **6**(3), 035001 (2009)

41.30 R.A. Normann: Technology insight: future neuroprosthetic therapies for disorders of the nervous system, Nat. Clin. Pract. Neurol. **3**(8), 444–452 (2007)

41.31 J. Dowling: Current and future prospects for optoelectronic retinal prostheses, Eye **23**(10), 1999–2005 (2009)

41.32 K. Nishida, M. Kamei, M. Kondo, H. Sakaguchi, M. Suzuki, T. Fujikado, Y. Tano: Efficacy of suprachoroidal-transretinal stimulation in a rabbit model of retinal degeneration, Investig. Ophthalmol. Vis. Sci. **51**(4), 2263–2268 (2010)

41.33 K. Nakauchi, T. Fukikado, H. Kanda, S. Kusaka, M. Ozawa, H. Sakaguchi, Y. Ikuno, M. Kamel, Y. Tano: Threshold suprachoroidal-transretinal stimulation current resulting in retinal damage in rabbits, J. Neural Eng. **4**(1), S50–S57 (2007)

41.34 P. Degenaar, N. Grossman, M.A. Memon, J. Burrone, M. Dawson, E. Drakakis, M. Neil, K. Nikolic: Optobionic vision – a new genetically enhanced light on retinal prosthesis, J. Neural Eng. **6**(3), 035007 (2009)

41.35 J.P. Rizzo III, J. Wyatt, M. Humayun, E. de Juan, W. Liu, A. Chow, R. Eckmiller, E. Zrenner, T. Yagi, G. Abrams: Retinal prosthesis: an encouraging first decade with major challenges ahead, Ophthalmology **108**(1), 13–14 (2001)

41.36 A.Y. Chow, N. Peachey: The subretinal microphotodiode array retinal prosthesis II, Ophthalmic Res. **31**(3), 246 (1999)

41.37 J.L. Stone: Morphometric analysis of macular photoreceptors and ganglion cells in retinas with retinitis pigmentosa, Arch. Ophthalmol. **110**(11), 1634–1639 (1992)

41.38 A. Santos, M.S. Humayun, E. de Juan Jr, R.J. Greenburg, M.J. Marsh, I.B. Klock, A.H. Milam: Preservation of the inner retina in retinitis pigmentosa. A morphometric analysis, Arch. Ophthalmol. **115**(4), 511–515 (1997)

41.39 M.S. Humayun, J.D. Dorn, A.K. Ahuja, A. Caspi, E. Filley, G. Dagnelie, J. Salzmann, A. Santos, J. Duncan, L. daCruz, S. Mohand-Said, D. Eliott, M.J. McMahon, R.J. Greenberg: Preliminary 6 month results from the Argus II epiretinal prosthesis feasibility study, Conf. Proc. IEEE Eng. Med. Biol. Soc. 2009 (2009) pp. 4566–4568

41.40 A. Horsager, R.J. Greenberg, I. Fine: Spatiotemporal interactions in retinal prosthesis subjects, Investig. Ophthalmol. Vis. Sci. **51**(2), 1223–1233 (2009)

41.41 S.H. Greenwald, A. Horsager, M.S. Humayun, R.J. Greenberg, M.J. CaMahon, I. Fine: Brightness as a function of current amplitude in human retinal electrical stimulation, Investig. Ophthalmol. Vis. Sci. **50**(11), 5017–5025 (2009)

41.42 G.J. Chader, J. Weiland, M.S. Humayun: Artificial vision: needs, functioning, and testing of a retinal electronic prosthesis, Prog. Brain Res. **175**, 317–332 (2009)

41.43 I. Obeid, C. Veraart, J. Delbeke: Estimation of phosphene spatial variability for visual prosthesis applications, Artif. Organs **34**(5), 358–365 (2010)

41.44 M.E. Brelen, V. Vince, B. Gérard, C. Veraart, J. Delbeke: Measurement of evoked potentials following electrical stimulation of the human optic nerve, Investig. Ophthalmol. Vis. Sci. **51**, 5351–5355 (2010)

41.45 C. Veraart, F. Duret, M. Brelén, M. Oozeer, J. Delbeke: Vision rehabilitation in the case of blindness, Expert Rev. Med. Devices **1**(1), 139–153 (2004)

41.46 E. Fernandez, F. Pelayo, M. Bongard, C. Marin, A. Alfaro, L. Merabet: Development of a cortical visual neuroprosthesis for the blind: the relevance of neuroplasticity, J. Neural Eng. **2**(4), R1–R12 (2005)

41.47 F.J. Pelayo, S. Romero, C.A. Morillas, A. Martinez, E. Ros, E. Fernández: Traslating image sequences into spike patterns for cortical neuro-stimulation, Neurocomputing **58–60**, 885–892 (2004)

41.48 H. Kolb, R. Nelson, P. Ahnelt, N. Cuenca: Cellular organization of the vertebrate retina, Prog. Brain Res. **131**, 3–26 (2001)

41.49 E. Fernandez, J.-M. Ferrandez, J. Ammermüller, R.A. Normann: Population coding in spike trains of simultaneously recorded retinal ganglion cells, Brain Res. **887**(1), 222–229 (2000)

41.50 R.A. Normann, D.J. Warren, J. Ammermüller, E. Fernandez, S. Guillory: High-resolution spatiotemporal mapping of visual pathways using multi-electrode arrays, Vis. Res. **41**(10–11), 1261–1275 (2001)

41.51 D.J. Warren, E. Fernandez, R.A. Normann: High-resolution two-dimensional spatial mapping of cat

striate cortex using a 100-microelectrode array, Neuroscience **105**(1), 19–31 (2001)
41.52 M.J. Schnitzer, M. Meister: Multineuronal firing patterns in the signal from eye to brain, Neuron **37**(3), 499–511 (2003)
41.53 R. Eckmiller, D. Neumann, O. Baruth: Tunable retina encoders for retina implants: why and how, J. Neural Eng. **2**(1), S91–S104 (2005)
41.54 C.A. Morillas, S.F. Romero, A. Martinez, F.J. Pelayo, E. Ros, E. Fernández: A design framework to model retinas, Biosystems **87**(2-3), 156–163 (2007)
41.55 C.A. Morillas, S. Romero, A. Martinez, F. Pelayo, L. Reyneri, M. Bongard, E. Fernández: A neuroengineering suite of computational tools for visual prostheses, Neurocomputing **70**, 2817–2827 (2007)
41.56 G. Dagnelie: Psychophysical evaluation for visual prosthesis, Annu. Rev. Biomed. Eng. **10**, 339–368 (2008)
41.57 G. Dagnelie, P. Keane, V. Narla, L. Yang, J. Weiland, M. Humayun: Real and virtual mobility performance in simulated prosthetic vision, J. Neural Eng. **4**(1), S92–S101 (2007)
41.58 S.F. Cogan, P.R. Troyk, J. Ehrlich, C.M. Gasbarro, T.D. Plante: The influence of electrolyte composition on the in vitro charge-injection limits of activated iridium oxide (AIROF) stimulation electrodes, J. Neural Eng. **4**(2), 79–86 (2007)
41.59 S.F. Romero, C.A. Morillas, F.J. Pelayo, E. Fernández: Computer-controlled neurostimulation for a visual implant, Biodevices **1**, 84–91 (2008)
41.60 C. Marin, E. Fernandez: Biocompatibility of intracortical microelectrodes: current status and future prospects, Front. Neuroengineering **3**, 8 (2007)
41.61 S.F. Cogan, J. Ehrlich, T.D. Plante, A. Smirnov, D.B. Shire, M. Gingerich, J.F. Rizzo: Sputtered iridium oxide films for neural stimulation electrodes, J. Biomed. Mater. Res. B Appl. Biomater. **89B**(2), 353–361 (2009)
41.62 A.B. Majji, M.S. Humayun, J.D. Weiland, S. Suzuki, S.A. D'Anna, E. de Juan Jr: Long-term histological and electrophysiological results of an inactive epiretinal electrode array implantation in dogs, Investig. Ophthalmol. Vis. Sci. **40**(9), 2073–2081 (1999)
41.63 J.P. Seymour, D.R. Kipke: Neural probe design for reduced tissue encapsulation in CNS, Biomaterials **28**(25), 3594–3607 (2007)
41.64 G.C. McConnell, H.D. Rees, A.I. Levey, C.-A. Gutekunst, R.E. Gross, R.V. Bellamkonda: Implanted neural electrodes cause chronic, local inflammation that is correlated with local neurodegeneration, J. Neural Eng. **6**(5), 56003 (2009)
41.65 P. Heiduschka, S. Thanos: Implantable bioelectronic interfaces for lost nerve functions, Prog. Neurobiol. **55**, 433–461 (1998)
41.66 C. Veraart, F. Duret, M. Brelén, M. Oozeer, J. Delbeke: Vision Rehabilitation in the Case of Blindness, Expert Rev. Med. Devices **1**(1), 139–153 (2004)
41.67 J. Dowling: Artificial human vision, Expert Rev. Med. Devices **2**(1), 73–85 (2005)
41.68 L.B. Merabet, J.F. Rizzo III, A. Pascual-Leone, E. Fernandez: 'Who is the ideal candidate?': Decisions and issues relating to visual neuroprosthesis development, patient testing and neuroplasticity, J. Neural Eng. **4**(1), S130–S135 (2007)
41.69 R.L. Gregory: Seeing after blindness, Nat. Neurosci. **6**(9), 909–910 (2003)
41.70 I. Fine, A.R. Wade, A.A. Brewer, M.G. May, D.F. Goodman, G.M. Boynton, B.A. Wandell, D.I.A. MacLeod: Long-term deprivation affects visual perception and cortex, Nat. Neurosci. **6**(9), 915–916 (2003)
41.71 D. Bavelier, H.J. Neville: Cross-modal plasticity: where and how?, Nat. Rev. Neurosci. **3**(6), 443–452 (2002)
41.72 L.B. Merabet, J.F. Rizzo, A. Amedi, D.C. Somers, A. Pascual-Leone: What blindness can tell us about seeing again: merging neuroplasticity and neuroprostheses, Nat. Rev. Neurosci. **6**(1), 71–77 (2005)
41.73 A. Bernabeu, A. Alfaro, M. García, E. Fernández: Proton magnetic resonance spectroscopy ((1)H-MRS) reveals the presence of elevated myo-inositol in the occipital cortex of blind subjects, Neuroimage **47**, 1172–1176 (2009)

42. Rehabilitation and Therapeutic Robotics

Loredana Zollo, Dino Accoto, Silvia Sterzi, Eugenio Guglielmelli

The successful introduction of robotic technologies in the rehabilitation arena critically depends on the possibility to design machines able to operate in symbiosis with patients, i. e., adapting the level of assistance to their residual abilities. Safety, easiness of use, and flexibility are key factors for these systems, which typically operate in continuous physical contact with the human body.

Rehabilitation robotics aims at enabling new and effective therapeutic approaches. Highly repetitive, intensive, structured motor exercises carried out while the patient is actively interacting with an assistive robot can have a direct, positive impact on the restoration of motor functions. Active involvement of the patient in planning and execution of motor exercises is expected to empower cortical reorganization, e.g., after stroke. Rehabilitation robotic systems can also provide quantitative, accurate measurements about the patient's motor performance.

42.1 **Background** 835

42.2 **Human-Centered Approach to Rehabilitation Robot Design** 837
- 42.2.1 CBM-motus Mechanical Design 839
- 42.2.2 CBM-motus Control System Design 840
- 42.2.3 Experimental Validation............. 841

42.3 **Robot-Based Measure of Patient's Performance** 844
- 42.3.1 Kinematic and Dynamic Performance Indicators.............. 845
- 42.3.2 Case Study................................ 846

42.4 **Conclusions and Further Readings**.......... 850

References 851

This chapter has a twofold aim: (1) to introduce human-centered design criteria for rehabilitation robotic systems by analyzing a specific case study; (2) to present an example of robot-based assessment of patients' sensorimotor abilities.

42.1 Background

Rehabilitation robotics aims at developing novel solutions for assisted therapy and objective functional assessment of patients with reduced motor and/or cognitive abilities [42.1,2]. These solutions augment existing therapeutic systems to improve patients achievable functional recovery. They are intended as therapeutic tools for temporary use (i. e. to be employed only for the duration of the therapy at home or at the clinic center) and are designed to maximize the objective clinical effectiveness of the therapy and the efficiency of the entire clinical process.

Recent neuroscientific achievements (mainly regarding the mechanisms of neurogenesis [42.3,4] stimulated by voluntary movements during motor training, and cerebral plasticity underlying the motor learning and the functional recovery after cerebral injury [42.5, 6]) point out the potential of robotic technologies to create a real discontinuity in the clinical procedures of the (neuro)rehabilitative treatment. Clinical evidence in physical medicine and rehabilitation clearly demonstrates that there is an important and increasing demand for innovative therapeutic solutions to address a wide variety of pathological conditions; for instance, patients experiencing severe stroke events have a high probability, in most cases higher than 50%, of retaining severe disabilities for the rest of their lives. The positive correlation of the prevalence of many neuromotor diseases with age can give an idea of the social relevance of this research area. Novel, cost-effective solutions able to significantly improve the outcome of the rehabilitation process or to assist the patient in coping with residual abilities after the rehabilitation process are needed urgently in the several industrialized countries facing the challenges posed by their *ageing society*. Such societies predict that soon 20–35% of their population will be over the age of 65 and in need of rehabilitation support.

But how can robots be actually useful to support the functional recovery process? What is the rationale for their introduction in the rehabilitation arena?

To answer these basic questions, at least four main fundamental issues have to be considered:

1. *The technological evolution of machines for physical exercise and cognitive training*. Modern fitness and rehabilitation gyms are more and more being populated with machines equipped with a variety of proprioceptive and exteroceptive sensors for recording multimodal data on the performance of the user as well as on her/his psychophysiological conditions. In some cases, these machines also have one or more actuators and simple feedback control systems, as in the case of automatic treadmills or continuous passive motion systems. At the same time, the concept of *brain training* by means of interactive exercise/games is gaining rapid popularity. No doubt this trend is paving the way for a massive introduction of real robotics technology also in this area of rehabilitation medicine, just like what has already happened in diagnostics and surgery.
2. *Evidence-based rehabilitation*. Modern medicine is based on objective evaluation and quantitative comparative analysis of the impact of different therapeutic approaches. Robotics technology provides accurate, precise, and very sensitive tools for assessing and modeling human behavior, well beyond the capability of a human observer. This is of paramount importance for enabling appropriate initial diagnosis and early adoption of corrective clinical strategies, and for identifying verifiable milestones as well as prognostic indicators of the recovery process.
3. *Rehabilitation therapy is a time-consuming and intensive activity for health-care operators*. Typically, one or more therapists should operate at the same time to administer rehabilitation therapy to a single patient. In some cases the role of the therapist is physically very demanding, involving physical activities that can lead to professional disorders such as low back pain and strain. This is particularly true in the case of severely disabled patients who need direct physical assistance even to perform very simple exercises. Proper introduction of robotics and automation technologies in this scenario can produce a dramatic reorganization of the working procedures within the rehabilitation process. Machines can take over most of the unpleasant and physically demanding tasks, leaving to the health-care operators the possibility to mainly concentrate on the quality of the therapy that is administered. The existence of these robots creates relief and opportunity. On one hand, the problem of the shortage of specialized personnel is mitigated by the possibility that a single operator can effectively supervise multiple patients, locally or even remotely (i. e., in telerehabilitation). On the other hand, patients' access to rehabilitation is improved by the opportunity to increase the duration and the frequency of their therapy experience, with limitations depending on clinical considerations and not other organizational or economic limitations imposed by the health-care system.
4. *Empowerment of the patient*. Recent findings on neural reorganization phenomena (neuroplasticity) related to functional recovery clearly demonstrate that patients suffering from neuromotor diseases can greatly benefit from activity-dependent rehabilitation therapies, which typically require the execution of goal-directed, repetitive exercises dominated by temporal and spatial constraints [42.7, 8]. During the exercise, the patient is supposed to play an active role, so that the whole sensorimotor coordination system is solicited and trained, including high-level brain functionality, such as imagination, planning, and anticipation of motor actions. In this perspective, rehabilitation robots can represent an ideal solution to implement training environments where this technology is able to continuously provide the patient with the minimal level of physical and/or cognitive support needed to repetitively initiate, execute, and complete a given exercise, while accurately complying with predefined spatial and temporal constraints.

On the basis of this analysis, the main functional requirements that rehabilitation robots should meet to properly address the targeted application domain can be derived. In detail:

- Rehabilitation robots are among the very few robotic systems that are typically devised to operate in continuous physical interaction with the human body. The main challenge is to harmonize their behavior in constrained motion with the patient's residual abilities, which are unpredictable in nature and dynamically varying, even within the same single therapeutic session.

- In many cases, the robot should implement assist-as-needed strategies, so that the whole set of sensorimotor control functionalities underlying the execution of the motor task is adequately stimulated and trained. Typical working conditions often require the robot to challenge the patient by using haptic interaction (e.g., spatially varying force fields) for opposing or favoring motion in predefined directions. In other situations, e.g., during evaluation sessions, the robot could be required to become fully *transparent* to the patient, so as to assess human motion parameters without introducing any perturbations on the physiological system.

These requirements, and especially the key issue of enabling an active role of the patient to reach a real physical and/or cognitive symbiosis with the machine, pose major technical challenges for designing dependable, flexible, and effective robotic platforms. Typical main technical requirements include high back-driveability, easy adaptation of the mechanical structure to different anthropometric parameters, adaptive schemes for physical human–robot interaction control, and friendly human–machine interfaces for involving and motivating the patient, and for allowing customization of the robot performance, especially in terms of the level of assistance and feedback provided to the patient during the exercise.

This chapter will try to provide some direct insights into robot-aided rehabilitation by focusing on two main issues strictly related to each other to achieve the challenging objective of creating a real symbiosis between the patient and the machine by increasing the level of machine adaptability to the patient's actual status.

In Sect. 42.2, the problem of *human-centered rehabilitation robot design* will be tackled. We propose adopting a truly biomechatronic approach, by harmoniously merging mechanics, electronics, and adaptive interaction control techniques to properly address technical and functional requirements coming from the required operative condition of a tight human–machine physical interaction. The proposed design approach is outlined by presenting a specific case study, the CBM-motus, which is a novel planar rehabilitation robot recently developed at the Laboratory of Biomedical Robotics and Biomicrosystems of the Università Campus Bio-Medico, Rome, Italy.

Then, in Sect. 42.3, the possibility of exploiting the typical capabilities of a robotic system to carry out *objective, rigorous measurements* of the patient's sensorimotor abilities is analyzed. This is a crucial issue, because early detection of the positive or negative impact of a given therapeutic protocol when applied to a patient will allow the physician to change/modify/customize the therapy as appropriate to achieve the best possible clinical outcome. The potentiality of current robotic machines to provide a quantitative evaluation of patient's performance is explored by means of the definition and pilot application of a core set of kinematic and dynamic indicators. Also, the prospective use of such information to modify robot behavior in accordance with specific patient characteristics is envisaged for future developments.

42.2 Human-Centered Approach to Rehabilitation Robot Design

The design of a robotic machine for robot-aided neurorehabilitation develops in a highly collaborative research scenario where roboticists, neuroscientists, and physicians contribute to define system specifications. The main reason is that the human subject (i. e., the patient exposed to the rehabilitation therapy) plays a key role in the design in view of the tight and continuous physical human–robot coupling [42.9–13]. The robot helps the subject to carry out part of the task that she/he is no longer able to perform autonomously, with a level of assistance that can be adapted to her/his residual abilities [42.14, 15]. Requirements such as accuracy, repeatability, preprogrammed movements, and task specificity (typical of industrial or service robotics scenarios [42.16, 17]) yield to priorities imposed by the close physical contact with the user, such as safety, reliability, robustness, adaptability, and back-driveability (i. e., low mechanical impedance).

All these features depend on the robot mechanical and control design [42.18–20] and can be addressed by searching for the optimal combination of mechanics and control in order to answer requirements coming from the specific application context.

More specifically, the design of a rehabilitation machine is based on a typical human-centered, biomechatronic design approach, which is a top-down process

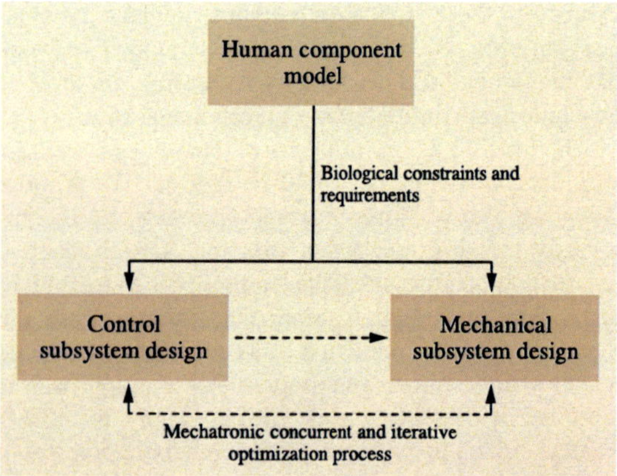

Fig. 42.1 The general scheme of the top-down biomechatronic design approach

consisting of two main steps, as depicted in Fig. 42.1 and briefly discussed in the rest of this section.

The first step includes an accurate characterization and modeling of the biological system, i. e., the human body and the human sensorimotor control system, by using the best knowledge available from different disciplines, such as biomechanics, neuroscience, and behavioral and motor sciences, and the full set of analytical, computational, and experimental methods and tools made available in bioengineering. Developing a model of the biological system that directly interacts with the machine offers the possibility to:

- Model and study the coupled dynamics between the biological system and the machine
- Evaluate robot performance in the interaction with the biological system
- Model and evaluate pathological abnormalities or alterations in the behaviors of the biological system
- Assess and optimize the behavior of the machine, by introducing and simulating modifications to the mechanical and control design hypotheses and evaluating the effects of these changes on the interaction with the biological system.

By exploiting the human (patient) component model, one can identify the initial set of constraints and requirements deriving from the biological domain that will represent the starting point of the subsequent design phases.

Obviously, developing the model of a biological system that could be, at the same time, *detailed and simple* enough to comply with the resource constraints of a machine design process, implies a series of simplifications with a level of approximation that depends on the specific application. For instance, when the goal is to investigate issues related to the interaction between the robotic machine and the patient from the mechanical and control point of view, it is typically conceivable to consider only the basic mechanisms (originated in the central nervous system) of generation of muscular forces and joint torques needed to move the limb, thus dramatically reducing the complexity of the description of the biological phenomena underlying human voluntary motor actions.

The second step is the mechatronic design of the machine, in which the functional and technical specifications are detailed and all the components of the system (interfaces, mechanical structure, actuators, sensors, and control unit) are concurrently defined and designed via an iterative process.

Although rehabilitation robotic machines are a wonderful paradigm of biomechatronic systems, quite surprisingly the harmonious merging of mechanics and control is often not achieved, mainly because highly nonlinear phenomena able to strongly degrade robot performance, such as friction and stiction, are typically not accounted for in their design.

Several examples of robotic machines for rehabilitation can be found in the literature [42.21–33]. From the perspective of robot design approaches, they can be classified into two main groups.

The first group consists of systems derived from adapting or reconfiguring industrial robots for use in rehabilitation [42.22–24]. This approach has the consequent critical drawback that low impedance, comparable to the human arm, cannot be practically be obtained, because machines are intrinsically position-controlled. Despite the use of active force feedback to enhance robot responsiveness, back-driveability required to move smoothly and rapidly in compliance with patients' actions [42.34] is not achieved. High inertia, anisotropy of dynamic properties, and low acceleration capabilities are often the main factors responsible for that [42.35, 36].

The second group consists of robots specifically conceived for tight human–machine interaction and includes two main classes of systems [42.23]:

1. Class I machines. They resort to mechanical solutions able to intrinsically enhance system back-driveability, with the main purpose of making the user perceive a very low mass, as for haptic interfaces [42.37]. They have low mechanical inertia and

friction, fine-tuning of viscoelastic properties, and high cost [42.31, 33, 38, 39].

2. Class II machines. They have a simple mechanical structure, no back-driveability, non-negligible inertia, possible presence of friction, and low cost. Although having some limitations, class II machines are very interesting for their applicability to remote rehabilitation (i. e., telerehabilitation) [42.40], justified by the low cost and the simplicity of the functioning mode.

For both class I and class II systems, control design typically resorts to traditional approaches, e.g., stiff proportional–integral–derivative (PID) voltage control (i. e., a proportional–integral–derivative action), compliance control (i. e., a proportional–derivative action plus gravity compensation) [42.41, 42], or sometimes impedance control (i. e., inverse dynamics interaction control). They can ensure the two important properties of system robustness and ease of implementation, especially in the case of PID and compliance control (which do not require an estimation of robot dynamics); however, friction and stiction are generally disregarded even though they can notably affect system performance in the interaction with the patient and degrade back-driveability.

In this section, the CBM-motus rehabilitation robot design is presented as a case study of biomechatronic system design: mechanics and control are concurrently designed and integrated to address the specific requirements coming from the rehabilitation application domain; it combines the advantages of class II machines (coming from simplicity in the mechanical structure and low cost) with the potentialities of inverse dynamics interaction control to compensate for mechanical drawbacks, such as high friction and reduced capability of viscoelastic regulation. The resulting machine is described in detail in the following.

42.2.1 CBM-motus Mechanical Design

The design criteria of the CBM-motus were inspired by the need of achieving:

- Low and isotropic apparent inertia when back-driven; the goal was pursued by designing a robot having inertia ellipses with small radius and unitary eccentricity, which are independent of the robot configuration in the workspace [42.43, 44].
- A large workspace to allow the administration of several rehabilitative treatments (target: 0.5 m × 0.5 m); workspace dimensions were determined in collaboration with therapists in order to allow a typical session of robot-aided rehabilitation therapy to be carried out.
- Interaction forces up to 50 N; the value was extracted from the analysis of the maximum force applicable by the patient at the robot end effector.

In addition, the robot was conceived for applications of telerehabilitation [42.45], thus aiming to be highly dependable, low cost, and portable in order for it to be moved to and mounted at the patient's site with no or little need for specialized skills.

The CBM-motus has a Cartesian kinematic structure consisting of two modules, each corresponding to an actuated axis. As shown in Fig. 42.2, each module includes six pulleys with the same radius ($R = 0.025$ m) and two timing belts (9.4 mm wide, reinforced with a glass fiber cord). Two couples of pulleys (on the left in Fig. 42.2) are mounted on the same shaft, and one pulley per module is directly driven by the motor, with no reduction gearing being used. Two belts for each module are mounted in such a way that the points along segments AB and CD move vertically with the same speed. A ground stainless steel bar is fixed to a couple of such points, e.g., P and P′ in Fig. 42.2. The second module is connected to the robot frame with a rotation of ±90° with respect to the first module. The end effector of the robot, i. e., the handle grasped by the patient, is connected to both bars, at their minimum-distance points,

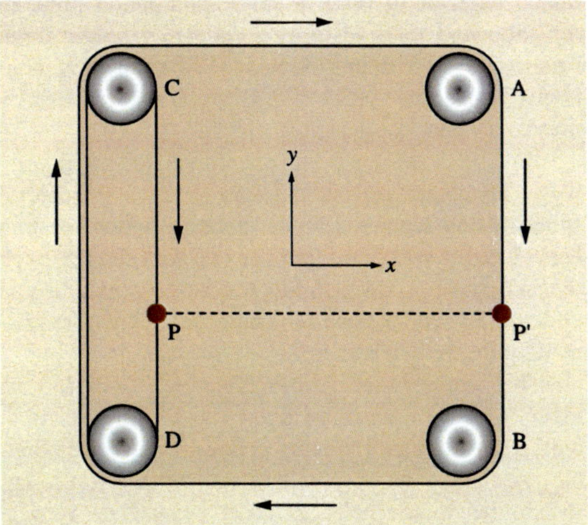

Fig. 42.2 A single kinematic module comprising six pulleys and two belts. P and P′ move vertically with the same speed

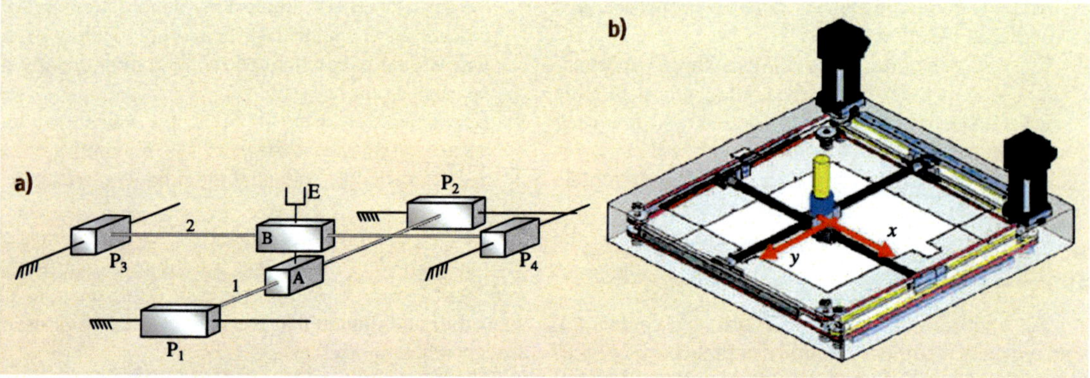

Fig. 42.3a,b Overview of the kinematic scheme **(a)** and of two assembled kinematic modules **(b)**. The rigid bars slide in a double prismatic joint and are connected to the driving belts

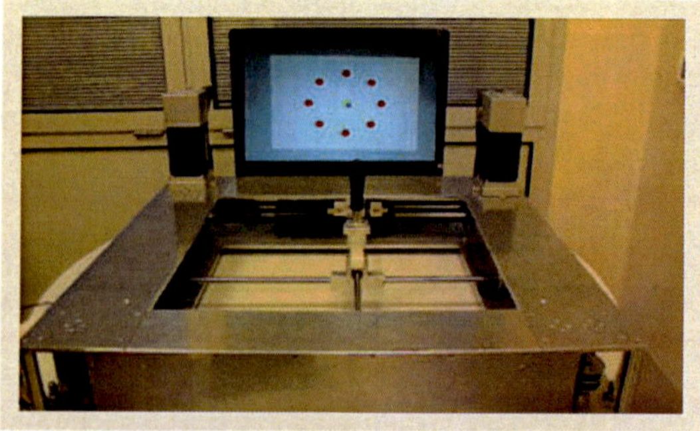

Fig. 42.4 Overview of the complete robot

through a couple of orthogonal prismatic joints, rigidly linked together to make a single compound joint, in the following referred to as a *double prismatic joint*. Friction at each prismatic joint (still remaining non-negligible because of the sliding motion) is reduced by means of linear ball bushings.

Both ends of each bar are connected to driving belts. The outer prismatic joints (P1, ..., P4) correspond to the segments of the belts to which the two bars (1 and 2) are connected. The two bars slide through the compound prismatic joint (A + B), to which the end effector (E) is connected. The bars and the double prismatic joint are depicted in Fig. 42.3.

This patented kinematic architecture [42.46] has the main feature of ensuring good rigidity of the robot with relatively small moving masses, because the double prismatic joint ensures that only tensile forces are transmitted to the belts. In order to balance vertical loads and axial forces caused by friction at prismatic joints, both ends of the bars are supported by a ball-bearing-supported wheel running in a rail. The stiffness of the system and the low moving mass ensure a high resonance frequency, which helps avoid spurious mechanical stimuli, which may perturb haptic rendering. Anyhow, the resonance frequency can be tuned by regulating screw-based belt tensioners. The two modules are actuated by DC servomotors (Aerotech BM 250) with a rated torque of 2 N m and a peak torque of 5 N m. The radius of the pulleys being 0.025 m, the maximum force which the robot is able to exert is 80 N for each axis. For safety reasons, it was limited via software to 50 N. The overall dimensions of the robot frame are $0.83 \times 0.82 \times 0.11\,\mathrm{m}^3$. The total mass (frame and motors included) is less than 30 kg. An overview of the system is shown in Fig. 42.4.

CBM-motus inertia is independent on the robot configuration and is isotropic. Thus, the mass perceived at the end effector has the same value everywhere in the workspace and is 2.59 kg.

As expected, friction is not negligible. The static friction is around 4–5% of the maximum force that the robot can generate (50 N); it requires 0.25 A along the x-axis and 0.30 A along the y-axis to be overcome, that is, about 60% of average current values in normal operating conditions (the mean value of the current is 0.42 A along the x-axis and 0.47 A along the y-axis). On the other hand, viscous friction reaches 2.1 and 2.7% of the maximum force along the x-axis and the y-axis, respectively, in normal operating conditions of 0.3 m/s.

42.2.2 CBM-motus Control System Design

The control system of a rehabilitation robotic machine has the twofold objective of (1) assisting the patient when she/he is able to move autonomously, by making the patient perceive a very low impedance (close to zero), and (2) helping or forcing the patient's limb toward the target in accordance with her/his residual motion capabilities when she/he is not able to move or complete the motor task. To pursue these objectives the control system is required to ensure a high level of adaptability to the different motor capabilities of the patients, different for each patient, and to devote maximum priority to safety in the interaction, also at the expense of accuracy in the execution of the motor task.

Viscoelastic regulation and adaptable impedance are keywords of human–robot interaction during a rehabilitation session. To achieve them, two main issues need to be addressed: the first one is CBM-motus friction compensation, friction being an undesirable phenomenon that terribly degrades back-driveability; the second one is robot impedance adaptation to specific patient requirements.

An interaction control able to cope with these two issues is proposed in the following. Key points of the control system (named *current-based impedance* control) are:

- Compensating for friction and tuning robot compliance by means of an interaction control law based on inverse dynamics.
- Closing the control loop on electric currents, in lieu of traditionally used force sensors, in addition to position feedback. Current monitoring plays a fundamental role in fine-tuning of robot impedance during interaction by providing an indirect force feedback in the control loop. It does not require force sensors mounted at the robot end effector and solves problems related to the increase of the apparent inertia perceived by a human user (due to force sensor mass), wiring issues, and the costs of the sensors.

The choice of impedance control in lieu of compliance control (as typically done in the literature) is motivated by the need to compensate for static and dynamic friction, which are nonnegligible from the identification procedure of the CBM-motus dynamic parameters. Hence, the core of the control law is the robot dynamics compensation, which allows us to significantly reduce the effects of robot friction and inertia during interaction and properly achieve impedance regulation.

Consider the robot dynamic model as

$$B(q)\ddot{q}+c(q,\dot{q})+F_{\mathrm{v}}\dot{q}+f_{\mathrm{s}}(q,\dot{q})+g(q) = \tau - J^{\mathrm{T}}(q)h\,,$$

where $B(q)$ is the inertia matrix, $c(q,\dot{q})$ is the vector of centrifugal and Coriolis torques, $F_{\mathrm{v}}\dot{q}$ is the viscous friction, $f_{\mathrm{s}}(q,\dot{q})$ is the static friction, $g(q)$ is the vector of gravitational torques, and $J^{\mathrm{T}}(q)h$ is the torque contribution due to interaction force h exerted by the patient on the robot, $J(q)$ being the robot Jacobian matrix.

The control law can be written as

$$\tau = B(q)y + c(q,\dot{q}) + F_{\mathrm{v}}\dot{q} + f_{\mathrm{s}}(q,\dot{q}) + g(q) + J^{\mathrm{T}}(q)h(I)\,, \qquad (42.1)$$

where $h(I)$ is the interaction force extracted from sensed current I. The addition of $J^{\mathrm{T}}(q)h(I)$ allows decoupling and linearizing the robot dynamics, thus yielding

$$\ddot{q} = y\,.$$

Choosing for acceleration y the following expression

$$y = J^{-1}(q)M_{\mathrm{d}}^{-1}\big(M_{\mathrm{d}}\ddot{x}_{\mathrm{d}} + K_{\mathrm{D}}\dot{\tilde{x}}_{\mathrm{d}} - M_{\mathrm{d}}\dot{J}(q,\dot{q})\dot{q} - h(I)\big)\,, \qquad (42.2)$$

where $\tilde{x} = x_{\mathrm{d}} - x$ and $\dot{\tilde{x}} = \dot{x}_{\mathrm{d}} - \dot{x}$ are position and velocity errors, respectively, between the planned trajectory x_{d} and the actual trajectory x, yields

$$M_{\mathrm{d}}\ddot{\tilde{x}} + K_{\mathrm{D}}\dot{\tilde{x}} + K_{\mathrm{P}}\tilde{x} = h(I)\,.$$

The robot behaves as a mass–damper–spring mechanical system through mass matrix M_{d}, stiffness matrix K_{P}, and damping matrix K_{D}, thus allowing modeling of patient–robot interaction as the parallel of two mechanical impedances (one for the human, the other for the robot). Robot impedance can be varied by means of control gains M_{d}, K_{P}, and K_{D} and, at the equilibrium, the robot elastic term directly balances the force exerted by the patient.

42.2.3 Experimental Validation

The idea behind current-based impedance control is to measure interaction force by means of current monitoring instead of direct force monitoring. This entails the need to identify a relation between electric current and interaction force in the whole robot workspace, by distinguishing current values during free motion (i. e., without interaction) from current values in constrained motion (i. e., during interaction).

To this purpose, the robot was programmed to perform 500 point-to-point movements along different directions of the workspace, in conditions of free motion and constrained motion. Synchronized acquisitions

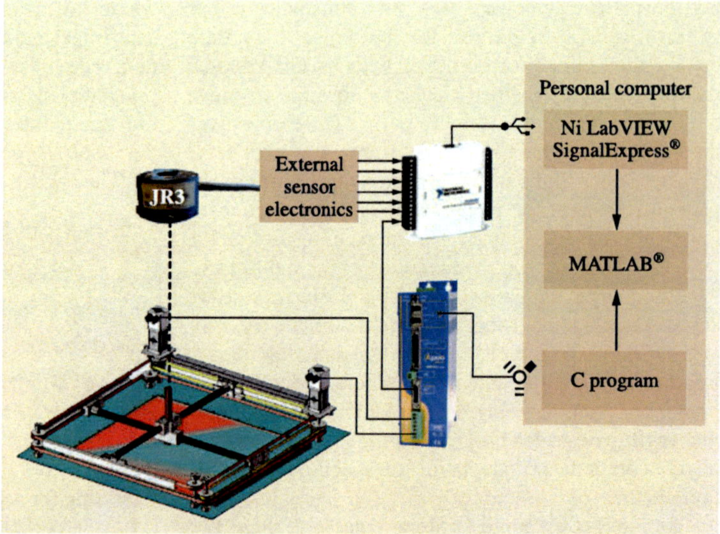

Fig. 42.5 Experimental setup for force and current recordings

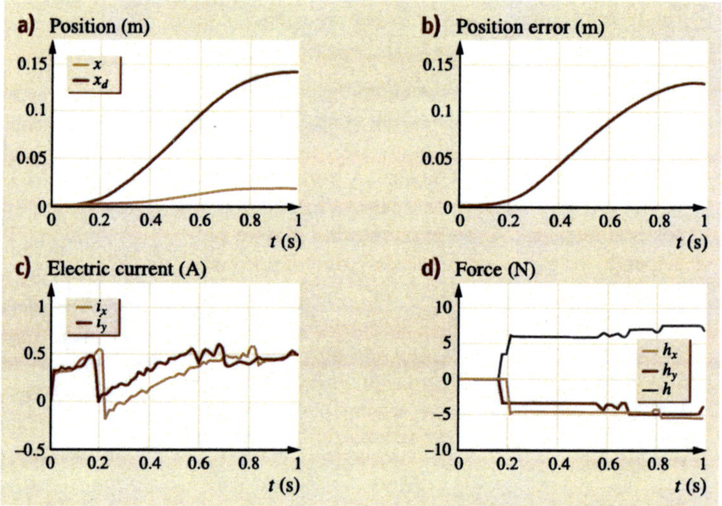

Fig. 42.6a–d Experimental results for the current-based impedance control during a constrained task with 1 kg applied at the end effector: desired and actual end-effector trajectories **(a)**, position error norm **(b)**, recorded electric currents **(c)**, and force values extracted from currents **(d)**. Control gains are $K_P = \text{diag}(40, 40)$ and $K_D = \text{diag}(4, 4)$

of force and currents were carried out (Fig. 42.5). In the case of constrained motion, the robot was required to perform a point-to-point motion with a known mass applied to the end effector by means of a pulley–string system. Four different experimental conditions were analyzed, corresponding to four different values of applied mass, (0.2, 0.4, 1.0, and 2.0 kg), in addition to the case of zero mass.

Because of the anisotropic friction distribution in the robot workspace, four separate linear regression analyses were required for directions toward positive and negative x-axes and y-axes (i. e., x^+, y^+, x^-, y^-). The analytical expressions for the four linear regression functions are

$$\begin{aligned} F_{x^+} &= -9.84 I_{x^+} + 3.73 \,, \\ F_{x^-} &= -10.42 I_{x^-} - 4.73 \,, \\ F_{y^+} &= -10.94 I_{y^+} + 4.58 \,, \\ F_{y^-} &= -9.97 I_{y^-} - 5.29 \,, \end{aligned} \tag{42.3}$$

where forces are expressed in newtons and currents in amperes. It is worth noting that they are valid when motion occurs along axes. Solving (42.3) with respect to current in the absence of interaction force provides

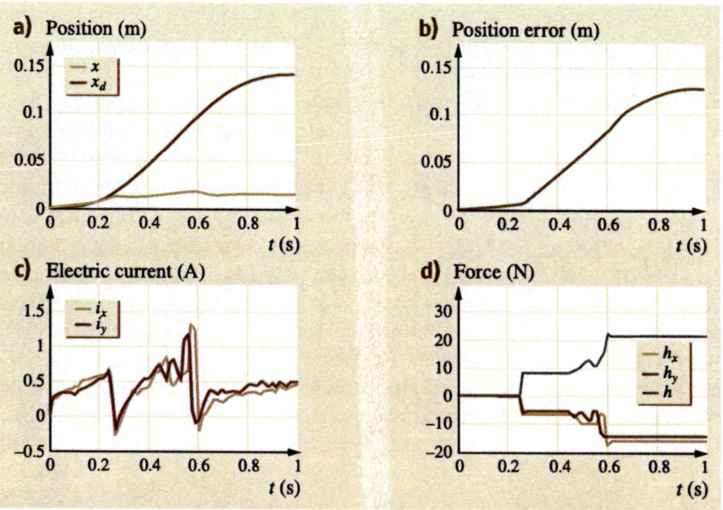

Fig. 42.7a–d Experimental results for the current-based impedance control during a constrained task with 1 kg applied at the end effector: desired and actual end-effector trajectories **(a)**, position error norm **(b)**, recorded electric currents **(c)**, and force values extracted from currents **(d)**. Control gains are $K_\mathrm{P} = \mathrm{diag}(100, 100)$ and $K_\mathrm{D} = \mathrm{diag}(10, 10)$

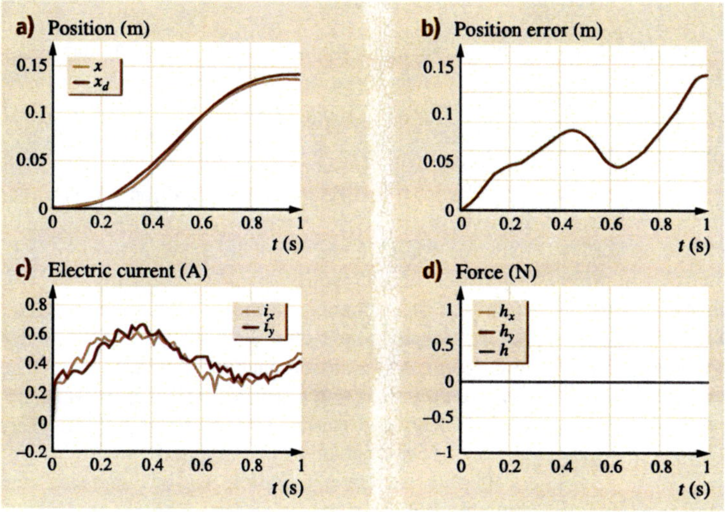

Fig. 42.8a–d Experimental results for motion in the free space: desired and actual end-effector trajectories **(a)**, position error norm **(b)**, recorded electric currents **(c)**, and force values extracted from currents **(d)**. Control gains are $K_\mathrm{P} = \mathrm{diag}(100, 100)$ and $K_\mathrm{D} = \mathrm{diag}(10, 10)$

mean values of the current in the free space for each direction. On the other hand, when one axis is commanded to move while the other one is commanded to maintain its position, the force–current relation for the axis that is not moved changes as

$$F_{x\pm} = -10.13 I_{x\pm} \, ,$$
$$F_{y\pm} = -10.46 I_{y\pm} \, .$$

For the moved axis, relations (42.3) are still valid.

Force–current relations in (42.3) were used to implement and test current-based impedance control in (42.1) and (42.2). Experimental tests of free motion and interaction with an external environment were performed with the control law. They were representative of control capability to adapt to external forces. In the case of free motion, the robot was commanded to track a minimum-jerk trajectory, i. e., a quintic polynomial function, from (0; 0) to (0.0; 0.10) m; for constrained tasks, the same trajectory was planned while a weight was applied at the end effector. Different mass values were tested by varying control gains and measuring the corresponding end-effector position and motor currents.

Figures 42.6 and 42.7 report desired and actual end-effector trajectories, position error norm, and recorded electric currents when a weight of 1 kg is applied and two different sets of control gains are chosen (i. e., $K_P = \mathrm{diag}(40, 40)$, $K_D = \mathrm{diag}(4, 4)$, and $K_P = \mathrm{diag}(100, 100)$, $K_D = \mathrm{diag}(10, 10)$). Control gains were empirically chosen in order to set two different levels of robot compliance within system stability limits. Also, force values extracted from the electric currents are shown in Figs. 42.6d and 42.7d. It can be observed in Fig. 42.7 that the computed force value is constant as long as the robot end effector holds the position imposed by the applied weight. A threshold check is made on the electric current, so that when during interaction it drops below 0.57 A for the x-axis and below 0.62 A for the y-axis (as an effect of the torque control command), interaction force is maintained constant with the position. When the current is beyond the threshold, the force values vary according to relations (42.3). The threshold value corresponds to the maximum current values for each axis in free space motion.

It is worth noting that, thanks to current feedback providing information about the applied force, the same set of control gains, $K_P = \mathrm{diag}(100, 100)$ and $K_D = \mathrm{diag}(10, 10)$, can be successfully used in free space as well as in constrained space (Figs. 42.7 and 42.8); this set can ensure a high level of adaptability to the external constraint ($h = 21.41$ N) and, contemporarily, a good level of accuracy in free motion (0.0093 m is the maximum value) for the application addressed. This achievement can be especially fruitful in managing situations of shared control between the patient and the machine without requiring one to change control gains.

42.3 Robot-Based Measure of Patient's Performance

In order to enhance the active role of the patient, by enabling even severely impaired patients to actively intervene in the decision regarding the execution of a motor task, it is desirable that the robot plans its motion on the basis of the patient's motion intentions and assists the patient to carry out that part of the task that she/he is no longer able to perform autonomously, with a level of assistance that can be adapted to her/his residual abilities. In this perspective, the robot sensory system (possibly in addition to other sources of information) plays a fundamental role. Data from the patient (such as kinematic, dynamic, and physiological data) could be processed for continuous analysis of her/his motion intentions and physiological state. This information could be used to:

1. Update robot control online during the execution of a motor exercise in order to guide, help, or force the patient's limb toward the target in accordance with her/his residual motion capabilities
2. Apply corrective actions in case of incorrect motion
3. Provide therapists with objective, accurate measurements of the subject's body functions, thus enabling therapists to track the subject's progress in therapy, evaluate the efficacy of various interventions, and customize the machine for each particular user.

In spite of the number of works on design and development of robotic devices, and a few clinical studies, differing in design and methods, on the effect of robotic devices on stroke rehabilitation in a clinical setting [42.7, 30, 47, 48], body function and structure assessment was typically carried out by means of traditional clinical impairment scales [42.49], which can suffer from being subjective, operator-dependent, and qualitative. In recent years, studies on movement smoothness [42.50] and decomposition of complex movements in submovements [42.27] used kinematic data recorded by the robot to analyze motion characteristics of unimpaired patients. Recently, the first examples of evaluation metrics were provided [42.51, 52] for quantifying motor recovery of stroke patients undergoing robot-aided rehabilitation. However, they were both limited to a characterization of patient kinematics during unperturbed point-to-point motion, where no active force regulation was required.

In this section a complete set of kinematic and dynamic indices is provided for an objective measure of the effect of robot-aided therapy [42.53]. They are used to characterize human performance during unperturbed motion [42.51, 52] as well as in resistive motion, when active force regulation is required to successfully achieve the motor task by resorting to adaptation mechanisms to external force fields, as studied in [42.54]. A robot-aided motor therapy of the upper limb was administered to 15 chronic poststroke patients. Two robotic machines were used to administer therapy: InMotion2 (also known as MIT-Manus [42.38]) and InMotion3 (i. e., MIT wrist robot [42.39]). Patients were evaluated by means of the performance indices extracted from kinematic and dynamic data measured by the MIT-Manus planar machine.

42.3.1 Kinematic and Dynamic Performance Indicators

Kinematic indices aim at characterizing the following temporal and spatial features of subject motion:

- *Duration*. This is measured by means of the *execution time*, defined as the time for performing a point-to-point movement, elapsed from movement onset (i. e., time instant where velocity exceeded a threshold of 10% of peak velocity [42.55, 56]) and movement termination (i. e., time instant where velocity went below a threshold of 10% of peak velocity). Movement duration is expected to reduce with recovery.
- *Accuracy*. This is assessed by means of the *area index*, defined as the area between the desired straight line and the curve actually performed by the patient in the XY plane during a point-to-point movement. Accuracy increases when the area tends to zero.
- *Direction*. This is measured through the *aiming angle*, defined as the angular difference between the target direction and the direction of travel from the starting point to peak speed point [42.55, 56]. The angular displacement is expected to reduce with the therapy.
- *Path length*. A normalized measure of path length is used, named the *length ratio*, and is formally defined as the length ratio between the actual patient curve and the desired straight line. An improvement of patient motion capabilities makes the actual path tend to the straight line and consequently makes the length ratio tend to 1. The length ratio provides a measure of the ability of the patient to reach the target, by means of a threshold value of 0.5; index values under the threshold indicate patient inabilities to reach the final point.
- *Velocity*. This is analyzed by means of
  - *Mean velocity error*. This is defined as the mean value of the distance vector between the ideal velocity profile for point-to-point motion, based on the minimum-jerk hypothesis [42.57], and patient velocity. A reduction of mean velocity error with training is expected.
  - *Peak speed*. This is the peak value of patient velocity. In the unperturbed motion the patient is stimulated to improve motion smoothness; thus, globally, a reduction of peak speed is expected.
- *Smoothness*. The following two measurements of smoothness are summarized from [42.50]:
  - *Jerk index*. This is calculated by dividing the mean jerk magnitude by the trajectory length and is expressed on a logarithmic scale; using a logarithmic scale has the advantage of reducing the range of variation of the index and facilitating its use in clinical practice. Note that increases in the jerk index correspond to decreases in smoothness. A decrease of this index following robot training is expected.
  - *Speed index*. The speed index is defined as the normalized mean speed (i. e., the mean value of the speed divided by the peak speed). As subjects recover, it is expected that the normalized mean speed will be significantly higher because of a reduction of the series of peaks with deep valleys in between in the subject's speed profile.

Force applied and work expended during motion are important features of recovery that clinical studies on robot therapy have often disregarded. They are especially meaningful in resistive tasks. In the following, a set of dynamic performance indicators based on force measurements are proposed.

Force exerted during motion. A basic distinction is provided between total force and useful force exerted by the patient during motion because they can exhibit different evolutions with training. Although an increase of force can be measured, the direction of force application could be wrong. Hence, with *useful force* the amount of total force directed toward the target is intended. This distinction led to the definition of four different force indices:

1. *Mean force*. This is the mean value of the force exerted by the patient during motion. Because of the rigidity of the limb on admission, patient mean force will be quite high at the beginning. In free motion it is expected to reduce with training; otherwise it should increase in resistive motion.
2. *Peak force*. This is the peak value of the force exerted by the patient during motion. It will behave similarly to mean force, but with larger variations.
3. *Useful force*. This index measures the amount of mean force directed toward the target. The useful force is calculated by weighting the mean force value with the aiming angle index normalized with respect to its maximum (i.e., 90° in our case). It ranges between zero, when the aiming angle is more than 90°, and the mean force value, when the aiming angle is 0°. An increase with recovery is expected.

4. *Useful peak force*. This measures the amount of peak force directed toward the target, and is calculated similarly to useful mean force. It varies between zero and the peak force value.

Work expended during motion. By analogy with force, a basic distinction is provided between total work and useful work expended by the patient during motion. Work is measured in addition to force because it provides a measure of the energy expended in the execution of the task by combining motion and force information in a unique parameter:

1. *Total work*. This index is calculated as the line integral of the force along the path described by the patient. In free motion it is expected to reduce with training. In resistive motion total work should increase as a consequence of force increase.
2. *Useful work*. This index measures the amount of total work directed toward the target and is extracted from the total work value by means of the aiming angle. It will vary between zero and the value for total work. An increase with recovery is foreseen, in free motion as well as in resistive motion.

42.3.2 Case Study

Experimental Protocol

Fifteen community-dwelling persons with chronic stroke (ten men, five women) met inclusion criteria and volunteered to participate in this study. Six weeks of shoulder-elbow therapy was administered to all subjects with the InMotion2 robotic machine and 6 weeks of wrist therapy was administered to all subjects with the InMotion3 machine, for a total of 12 weeks of robot-aided therapy (Fig. 42.9). The subjects received approximately 1 h of robotic therapy three times a week.

Training consisted of three games of 320 assisted-as-needed point-to-point movements from the center to eight outbound targets distributed along a circle at a distance of 0.14 m. Patients were required to move with a self-paced speed in a maximum time slot of 3 s. The assistance was tuned on the basis of the patient's performance [42.14]. In wrist training, the first two assisted games trained wrist flexion/extension, abduction/adduction, and a combination of these movements. The last assisted game exercised exclusively pronation and supination [42.39].

Evaluation was carried out on admission to and discharge from the robotic treatment for a period of 12 weeks. It consisted in performing unperturbed and resistive motion exercises with the InMotion2 robotic machine. The robot was completely passive while position and force sensors recorded the subject's kinematic and force data. Each subject was asked to perform five blocks of unassisted 16 point-to-point movements in free space (in the following briefly named as unperturbed motion) and, subsequently, one block of 16 point-to-point movements in the presence of a force field generated by the robot (in the following briefly referred to as resistive motion). The force field can be modeled as a viscoelastic force field applied against patient motion. It is generated by means of a proportional-and-derivative control (with constant gains) pushing the robot end effector to the central position of the workspace.

Results of Unperturbed Motion

Figure 42.10 reports motion trajectories of a representative patient during a point-to-point evaluation task in pretreatment and in posttreatment phases. One can see the improvement toward more and more linear trajectories over the 80 trials and, mainly, the improved capability of extending the arm toward the more distal targets (at the top of the plot).

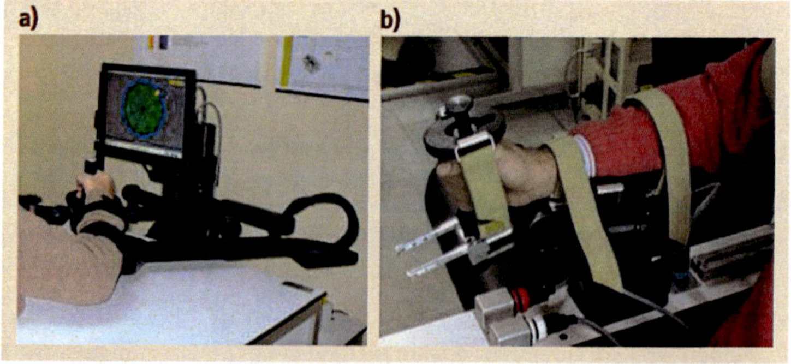

Fig. 42.9a,b InMotion2 (**a**) and InMotion3 (**b**) robotic machines

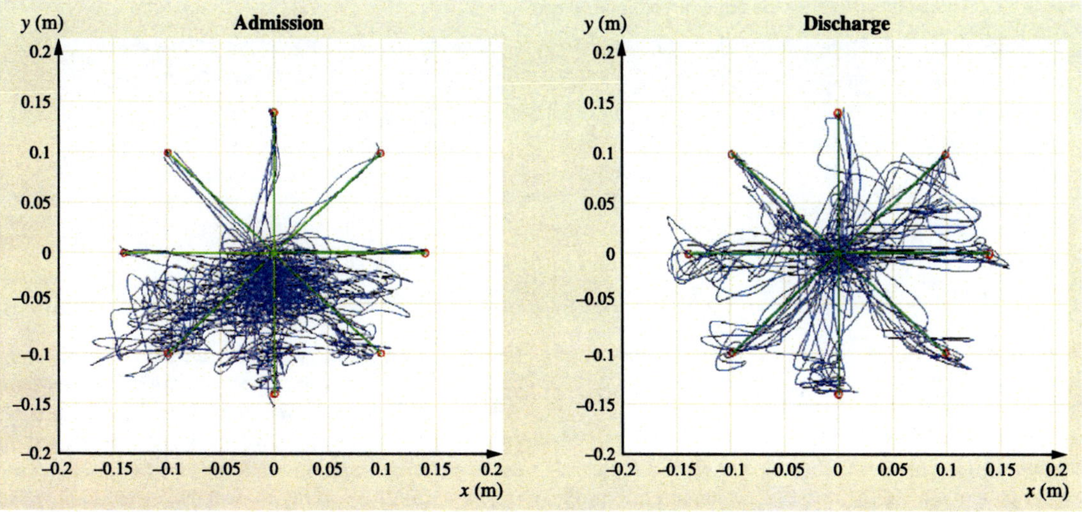

Fig. 42.10 Plot of 80 trials of unperturbed motion of a representative subject. The reference straight line is in *green* and the actual patient motion is in *blue*

Table 42.1 Kinematic indices for the unperturbed motion task

Kinematic index	Admission (mean ± SD)	Discharge (mean ± SD)	t_{14}[a]	P	r[b]
Movement duration					
Execution time (s)	3.960 ± 1.463	3.051 ± 1.20	2.307	0.018	0.703
Motion accuracy					
Area (m^2)	0.089 ± 0.0532	0.0588 ± 0.0307	4.443	0.0003	0.720
Motion direction					
Aiming angle (deg)	27.372 ± 18.641	16.253 ± 11.684	4.667	0.0002	0.740
Path length					
Length ratio	0.643 ± 0.1428	0.743 ± 0.0971	−3.21	0.0031	0.850
Motion velocity					
Mean velocity error (m/s)	0.0783 ± 0.030	0.0608 ± 0.0162	3.378	0.002	0.751
Peak speed (m/s)	0.170 ± 0.0748	0.147 ± 0.0334	1.398	0.092	0.412
Motion smoothness					
Jerk index ($1/s^3$)	2.126 ± 0.215	2.009 ± 0.157	2.968	0.005	0.645
Speed index	0.397 ± 0.687	0.458 ± 0.0659	−5.690	2.79×10^{-5}	0.934

SD standard deviation

[a] A negative t value indicates that discharge scores were higher than the admission scores.

[b] The strength, or magnitude, of our findings was determined by calculating the effect size r [42.58]:
for a small treatment effect $r = 0.1–0.23$, for a medium treatment effect $r = 0.24–0.36$, for a large effect $r \geq 0.37$

Tables 42.1 and 42.2 report the results of robot-based evaluation during unperturbed motion. Patients globally improved kinematic performance and all kinematic indices varied in the expected direction with statistically significant changes, except for *peak speed* ($t_{14} = 1.398$, $p = 0.092$, $r = 0.412$). As regards dynamic performance, it can be observed that, with training, mean force ($t_{14} = 0.768$, $p = 0.227$, $r = 0.174$), peak force ($t_{14} = 1.431$, $p = 0.0871$, $r = 0.404$), and total work ($t_{14} = 0.613$, $p = 0.275$, $r = 0.182$) did not significantly decrease; on the other hand, the amount of total force (namely, the useful work $t_{14} = -2.065$, $p = 0.0290$, $r = 0.410$) and total work (namely, the useful work, $t_{14} = -2.04$, $p = 0.030$, $r = 0.407$) directed

Table 42.2 Dynamic indices for the unperturbed motion task

Dynamic index	Admission (mean ± SD)	Discharge (mean ± SD)	t_{14}	P	r
Force exerted during motion					
Mean force (N)	1.952±0.688	1.848±0.545	0.768	0.227	0.174
Peak force (N)	4.408±1.993	3.781±1.092	1.431	0.0871	0.404
Useful force (N)	1.317±0.531	1.513±0.454	−2.065	0.0290	0.410
Useful peak force (N)	3.046±0.998	3.125±0.807	−0.425	0.338	0.091
Work expended during motion					
Total work (J)	0.217±0.156	0.195±0.062884	0.613	0.275	0.182
Useful work (J)	0.141±0.0672	0.165±0.0525	−2.04	0.030	0.407

toward the target significantly increased as a consequence of the improvement of motion direction.

Results of Resistive Motion

Figure 42.11 reports motion trajectories of a representative patient during the resistive evaluation task in the pretreatment and in the posttreatment phases. After the robot therapy, the patient exhibited improved capability of opposing the external force field and extending the arm toward the more distal targets.

Kinematic and dynamic robot-based evaluations for resistive motion are reported in Tables 42.3 and 42.4. As for unperturbed motion, patients globally improved kinematic performance in a statistically significant manner. However, in accordance with [42.55, 56], indices related to motion velocity and smoothness were not reported, being meaningless for perturbed movements. On the other hand, the dynamic analysis showed that in resistive motion force and work did exhibit a different trend with therapy with respect to unperturbed motion. All force indices increased after the robotic treatment, even if not significantly, whereas they decreased in the unperturbed motion. Only changes of *useful force* ($t_{14} = -1.599$, $p = 0.066$, $r = 0.394$) and *useful peak force* ($t_{14} = -1.728$, $p = 0.0529$, $r = 0.345$) tended to approximate the threshold value $p = 0.05$ of statistical significance. Furthermore, work expended during motion significantly increased with respect to that on admission.

Discussion

The results of this study reinforced earlier findings that short-term, goal-directed robotic therapy could significantly improve functions and structures of the exercised limb segments in persons with chronic stroke [42.30, 47, 48], and provided a measure of their improved mo-

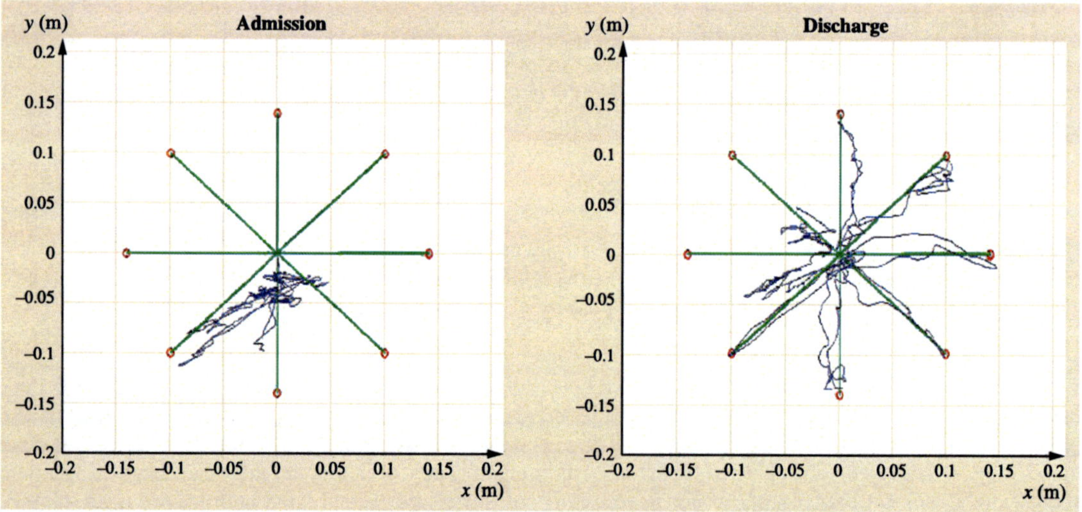

Fig. 42.11 Plot of 16 trials of resistive motion of a representative subject. The reference straight line is in *green* and the actual patient motion is in *blue*

Table 42.3 Kinematic indices for the resistive motion task

Kinematic index	Admission (mean ± SD)	Discharge (mean ± SD)	t_{14}	P	r
Movement duration					
Execution time (s)	4.849±2.193	3.914±1.758	2.450	0.0139	0.487
Motion accuracy					
Area (m^2)	0.126±0.0220	0.117±0.130	2.334	0.0175	0.510
Motion direction					
Aiming angle (deg)	32.380±27.388	20.923±22.032	2.679	0.0089	0.477
Path length					
Length ratio	0.543±0.196	0.618±0.144	−2.589	0.0107	0.503

Table 42.4 Dynamic indices for the resistive motion task

Dynamic index	Admission (mean ± SD)	Discharge (mean ± SD)	t_{14}	P	r
Force exerted during motion					
Mean force (N)	9.967±4.123	10.945±3.674	−0.897	0.193	0.260
Peak force (N)	19.306±7.333	20.922±6.488	−0.987	0.170	0.241
Useful force (N)	7.848±4.229	9.424±4.054	−1.599	0.066	0.394
Useful peak force (N)	15.502±7.805	18.060±7.519	−1.728	0.0529	0.345
Work expended during motion					
Total work (J)	0.965±0.6804	1.171±0.7396	−1.957	0.0353	0.300
Useful work (J)	0.784±0.6094	1.044±0.7415	−2.834	0.0066	0.398

tion capabilities. Kinematic changes were measured by means of eight indices describing temporal and spatial features of subject motion. Dynamic changes were measured through six indices describing the ability to modulate forces and energy and properly direct them.

The analysis of motion kinematics (Tables 42.1 and 42.3) showed that all the performance indicators changed in the expected direction. This can be regarded as a result of the relearning process which improved subject capabilities of moving and coordinating all the degrees of freedom of the upper limb, with the main effect on extension degrees of freedom (Figs. 42.10 and 42.11). Moreover, the combination of different indices, for example, smoothness, with the other spatial indices (aiming angle, area, path length) also assessed the progress of the patient's spatial and temporal efficiency, thus providing new insights into the association of motion ideation and execution (i. e., ideomotor apraxia) and envisaging the possibility of extracting prognostic indices of pathological disorders related to altered motion perception [42.59].

The analysis of motion dynamics (Tables 42.2 and 42.4) proved that whereas useful force and work (both including the dependence on a kinematic variable, i. e., the aiming angle) exhibit significant changes, mean force and total work applied during motion did not undergo a striking variation from admission to discharge, especially in unperturbed motion tasks. The main reason is supposed to be found in the type of robotic training consisting of assisted kinematic-goal-directed exercises, which stimulate the improvement of motion kinematics more than force control. When the patient was required to reach only a kinematic target, she/he applied compensatory strategies which avoided her/his controlling other independent elements [42.60–62], such as force.

Improvements of motion dynamics (i. e., force and work), mainly obtained in resistive motion, can be regarded as an indirect effect. It is worth noting that a significant change of force was not obtained, but an improvement of energy (i.e., the combination of force and position) was measured. When the patient was asked to oppose a force field, she/he was stimulated to control a major number of independent elements (i. e., motion and forces). As a consequence, owing to the contemporary improvement of position and force control, *total work* and *useful work* underwent a significant change.

The use of performance indicators can therefore provide fruitful indications of the appropriateness of the therapy and new insights into the corrective actions to apply. Furthermore, the online extraction of performance indicators from robot data is envisaged for future developments. This would enable, on one hand, patient quantitative assessment as performed in this study and, on the other hand, online change of robot impedance during therapy in order to address specific patient requirements. The immediate expected results would be an increase of patient–machine interactiveness, a really customized machine able to fit the therapy for each specific patient, and a disruptive increase of patient motivation.

42.4 Conclusions and Further Readings

Design of rehabilitation robotic machines, exposed to tight and continuous physical interaction with humans, continues being a very challenging and attractive area of robotics research. Moreover, the use of robotic systems in rehabilitation for therapy administration and sensory-motor assessment is an interesting possibility to discover and fruitfully exploit.

This chapter provided an overview of current evidence of the effect of robot-aided rehabilitation on patient recovery and pointed out the main functional and technical requirements coming from the application of robotic machines to motor therapy. Two important issues were faced in detail in this chapter, the *rehabilitation robot design approach* and the *objective measure of patient performance*, both focused on the central role of the patient and her/his active involvement in the therapy. The case study of the CBM-motus robotic system was reported as a paradigm of a design approach fully merging mechanics and control. On the other hand, an example of an objective measure of patient performance for motor and body function assessment was presented. It relies on the extraction of a core set of quantitative indicators from sensory information directly recorded by the robotic machine and their use in the kinematic and dynamic analysis of patient motor recovery.

An important challenge in the design of a robotic system for neurorehabilitation is to carefully adapt the behavior of the machine to patients' specific motor capabilities, intentions, and performance. To this purpose, it is desirable that an additional module (which was not dealt with in this chapter), i. e., the patient–machine interface, is carefully conceived. It should be able to provide the patient with an attractive graphical interface for promoting her/his active involvement in the therapy, acquire and process information on the current status of the patient, from the motion analysis to recordings of physiological signals and the cerebral activity, and provide inputs to the machine regarding the actual patient motion, her/his motion intention, and her/his responsiveness to the therapy. The ability of the machine to modulate its own viscoelastic characteristics as well as to generate force fields associated with a given workspace during the rehabilitative exercise is a fundamental feature for controlling the interaction between the patient and the machine during therapy. Future rehabilitation robotic systems are evolving toward multimodal robotic platforms that, harmoniously integrating mechanics, control, and a multimodal human–machine interface, allow one to:

- Record signals from different sources and reconstruct patient global status (i. e., physiological, neurological in addition to kinematic and dynamic).
- Extract performance indices, to be used for updating the control system (and correspondingly modify the therapy), as well as for evaluating the effects of the therapy.
- Increase the level of customization and adaptability to each specific patient of current rehabilitation machines. The multimodal analysis will allow:
  a) The online update of robot control during the execution of an assisted task in order to guide, help, or force the patient limb toward the target in accordance with her/his residual motion capabilities.
  b) Application of corrective actions in the case of incorrect motion.
  c) Change of the motor exercise and the graphical interface in order to increase patient active involvement and maximize her/his motivation.
- Investigate basic research issues related to the recovery process, such as:
  a) Neuroplasticity phenomena, defined as the recruitment of areas usually not involved in the task, and changes in the intensity, temporal sequence, and functional hierarchy of neuronal excitability.
  b) Efficiency of the sensorimotor feedback circuits, e.g. through the acquisition of indices measurable via hand sensory stimulation [42.63] and movement during functional MRI (fMRI) recording, as well as recording EEG or MEG signals during isometric contraction in both a relaxed condition and sensory stimulation [42.64].

Finally, it is worth mentioning that the use of robotic systems in conjunction with instrumentation for data acquisition and measurement of neurophysiological signals (EMG, EEG, MEG, or fMRI) is opening another important area for rehabilitation robotics research, i. e., the development of fMRI-compatible systems [42.65, 66] or, more generally, of brain-imaging-compatible robotic systems (i. e., brain-imaging-compatible systems that can be used in conjunction not only with fMRI but also with MEG, transcranial magnetic stimulation (TMS), repetitive TMS, near-IR spectroscopy, and other brain imaging tools). Whereas using EMG and EEG does not involve the issue of disturbance and interference with the functioning mode of the electrical actuated devices, the joint use of tools such as MEG and fMRI with robotic or mechatronic devices necessarily requires one to handle problems of electromagnetic compatibility and interaction. For several years there has been intensive research on MRI-compatible systems in surgical robotics [42.67–69] to exploit the great potential of MRI for guiding insertion of tools and monitoring and controlling therapy. Only recently the issue is being addressed in rehabilitation robotics, thus giving rise to preliminary studies on fMRI-compatible rehabilitation robots [42.70–72]. This is an important research pathway because brain-imaging-compatible machines can ultimately lead to *closing the loop* between the rehabilitation process and motor recovery: physicians and bioengineers could have a direct picture of the specific effects of the application of a rehabilitation protocol in terms of brain-induced modifications/adaptations, thus opening novel avenues for the identification of patient-specific, highly effective therapies based on these new platforms integrating robotic and brain imaging technologies.

References

42.1 S.C. Cramer: Brain repair after stroke, N. Engl. J. Med. **362**, 1827–1829 (2010)

42.2 E. Guglielmelli, M.J. Johnson, T. Shibata: Guest editorial, special issue on rehabilitation robotics, IEEE Trans. Robot. **25**, 477–480 (2009)

42.3 G. Kempermann, H. Van Praag, F.H. Gage: Activity-dependent regulation of neuronal plasticity and self repair, Prog. Brain Res. **127**, 35–48 (2000)

42.4 T.A. Jones, C.J. Chu, L.A. Grande, A.D. Gregory: Motor skills training enhances lesion-induced structural plasticity in the motor cortex of adult rats, J. Neurosci. **19**, 10153–10163 (1999)

42.5 R.J. Nudo, K.M. Friel: Cortical plasticity after stroke: Implications for rehabilitation, Rev. Neurol. **155**, 713–717 (1999)

42.6 F. Gomez-Pinilla, Z. Ying, R.R. Roy, R. Molteni, V.R. Edgerton: Voluntary exercise induces a BDNF-mediated mechanism that promotes neuroplasticity, J. Neurophysiol. **88**, 2187–2195 (2002)

42.7 J. Mehrholz, T. Platz, J. Kugler, M. Pohl: Electromechanical and robot-assisted arm training for improving arm function and activities of daily living after stroke, The Cochrane Libr. **4**, 1–44 (2009)

42.8 A.C. Lo, P.D. Guarino, L.G. Richards, J.K. Haselkorn, G.F. Wittenberg, D.G. Federman, R.J. Ringer, T.H. Wagner, H.I. Krebs, B.T. Volpe, C.T. Bever Jr., D.M. Bravata, P.W. Duncan, B.H. Corn, A.D. Maffucci, S.E. Nadeau, S.S. Conroy, J.M. Powell, G.D. Huang, P. Peduzzi: Robot-assisted therapy for long-term upper-limb impairment after stroke, N. Engl. J. Med. **362**, 1772–1783 (2010)

42.9 D. Reinkensmeyer, N. Hogan, H.I. Krebs, S.L. Lehman, P.S. Lum: Rehabilitators, robots and guides: New tools for neurological rehabilitation. In: *Biomechanics and Neural Control of Posture and Movement*, ed. by J. Winters, P.E. Crago (Springer, Berlin Heidelberg 2000) pp. 516–534

42.10 N. Hogan, H.I. Krebs, B. Rohrer, J.J. Palazzolo, L. Dipietro, S.E. Fasoli, J. Stein, R. Hughes, W.R. Frontera, D. Lynch, B.T. Volpe: Motions or muscles? Some behavioral factors underlying robotic assistance of motor recovery, J. Rehabil. Res. Dev. **43**, 605–618 (2006)

42.11 E. Gallotta, G. Magrone, A. Romanelli, M. Milazzo, L. Zollo, D. Formica, E. Guglielmelli, S. Sterzi: Neurorehabilitation of the upper limb using robotic systems, J. Rehabil. Med. **47**, 276 (2008), Suppl.

42.12 D. Reinkensmeyer, J.L. Emken, S.C. Cramer: Robotics, motor learning, and neurologic recovery, Annu. Rev. Biomed. Eng. **6**, 497–525 (2004)

42.13 B. Siciliano, O. Khatib (Eds.): *Springer Handbook of Robotics* (Springer, Berlin Heidelberg 2008), Part G, Chaps. 56–64

42.14 H.I. Krebs, J.J. Palazzolo, L. Dipietro, M. Ferraro, J. Krol, K. Rannekleiv, B.T. Volpe, N. Hogan: Rehabilitation robotics: Performance-based progressive robot-assisted therapy, Auton. Robot. **15**, 7–20 (2003)

42.15 L. Zollo, M. Passalacqua, D. Formica, E. Guglielmelli: Performance Analysis of Adaptive Interaction Control Laws for Rehabilitation Robotics, 2nd Bienn. IEEE/RAS-EMBS Int. Conf. Biomed. Robot. Biomechatron. (2008), Scottsdale, AZ, USA, 2008

42.16 B. Siciliano, L. Villani: *Robot Force Control* (Kluwer, Boston 1999)

42.17 D.M. Gorinevsky, A.M. Formalsky, A.Y. Schneider: *Force Control of Robotics Systems* (CRC Press, Boca Raton 1997)

42.18 L. Zollo, B. Siciliano, C. Laschi, G. Teti, P. Dario: An experimental study on compliance control for a redundant personal robot arm, Robot. Auton. Syst. **44**, 101–129 (2003)

42.19 D. Formica, L. Zollo, E. Guglielmelli: Torque-dependent compliance control in the joint space of an operational robotic machine for motor therapy, ASME J. Dyn. Syst. Meas. Control **128**, 152–158 (2006)

42.20 L. Zollo, B. Siciliano, A. De Luca, E. Guglielmelli, P. Dario: Compliance control for an anthropomorphic robot with elastic joints: Theory and experiments, ASME J. Dyn. Syst. Meas. Control **127**, 321–328 (2005)

42.21 D.J. Reinkensmeyer, L.E. Kahn, M. Averbuch, A. McKenna-Cole, B.D. Schmit, W.Z. Rymer: Understanding and treating arm movement impairment after chronic brain injury: Progress with the ARM guide, J. Rehabil. Res. Dev. **37**, 653–662 (2000)

42.22 L.E. Kahn, M. Averbuch, W.Z. Rymer, D.J. Reinkensmeyer: Comparison of robot-assisted reaching to free reaching in promoting recovery from chronic stroke. In: *Integration of Assistive Technology in the Information Age*, ed. by M. Mokhtari (IOS Press, Amsterdam 2001) pp. 39–44

42.23 S. Micera, M.C. Carrozza, E. Guglielmelli, G. Cappiello, F. Zaccone, C. Freschi, R. Colombo, A. Mazzone, C. Delconte, F. Pisano, G. Minuco, P. Dario: A simple robotic system for neurorehabilitation, Auton. Robots **19**, 271–284 (2005)

42.24 C.G. Burgar, P.S. Lum, P.C. Shor, M. Van der Loos: Development of robots for rehabilitation therapy: The palo alto VA/stanford experience, J. Rehabil. Res. Dev. **37**, 663–673 (2000)

42.25 P. Lum, D.J. Reinkensmeyer, R. Mahoney, W.Z. Rymer, C. Burgar: Robotic devices for movement therapy after stroke: Current status and challenges to clinical acceptance, Top. Stroke Rehabil. **8**, 40–53 (2002)

42.26 P.S. Lum, E. Taub, D. Schwandt, M. Postman, P. Hardin, G. Uswatte: Automated constraint-induced therapy extension (AutoCITE) for movement deficits after stroke, J. Rehabil. Res. Dev. **41**, 249–258 (2004)

42.27 H.I. Krebs, N. Hogan, M.L. Aisen, B.T. Volpe: Robot-aided neurorehabilitation, IEEE Trans. Rehabil. Eng. **6**, 75–87 (1998)

42.28 B.T. Volpe, H.I. Krebs, N. Hogan, L. Edelstein, C. Diels, M. Aisen: A novel approach to stroke rehabilitation: Robot-aided sensorimotor stimulation, Neurology **54**, 1938–1944 (2000)

42.29 K. Kiguchi, K. Iwami, M. Yasuda, K. Watanabe: An exoskeletal robot for human shoulder joint motion assist, IEEE/ASME Trans. Mechatron. **8**, 125–135 (2003)

42.30 G.B. Prange, M.J.A. Jannink, C.G.M. Groothuis-Oudshoorn, H.J. Hermens, M.J. IJzerman: Systematic review of the effect of robot-aided therapy on recovery of the hemiparetic arm after stroke, J. Rehabil. Res. Dev. **43**, 171–184 (2006)

42.31 J. Furusho, K. Koyanagi, Y. Imada, Y. Fujii, K. Nakanishi, K. Domen, K. Miyakoshi, U. Ryu, S. Takenaka, A. Inoue: A 3-D rehabilitation system for upper limbs developed in a 5-year NEDO project and its clinical testing, Proc. 2005 Int. Conf. Rehabil. Robot. (2005), Chicago, Illinois, pp. 53–56, 2005

42.32 M. Mihelj, T. Nef, R. Riener: A novel paradigm for patient-cooperative control of upper-limb rehabilitation robots, Adv. Robot. **21**, 843–867 (2007)

42.33 S.J. Ball, I.E. Brown, S.H. Scott: A planar 3DOF robotic exoskeleton for rehabilitation and assessment, 29th Annu. Int. Conf. IEEE Eng. Med. Biol. Soc. (2007), Lyon, France, pp. 4024–4027, 2007

42.34 D.A. Lawrence: Impedance control stability properties in common implementations, Proc. 1988 Int. Conf. Robot. Autom. (1988) pp. 1185–1191, Philadelphia, PA, USA, 1988

42.35 C.S. Cai, B. Roth: On the spatial motion of a rigid body with point contact, Proc. 1987 Int. Conf. Robot. Autom. (1987) pp. 686–695

42.36 A. Sharon, N. Hogan, D.E. Hardt: The macro/micro manipulator: An improved architecture for robot control, Robot. Comput.-Integr. Manuf. **10**, 209–222 (1993)

42.37 T.H. Massie, J.K. Salisbury: The PHANTOM Haptic Interface: A Device for Probing Virtual Objects, Proc. ASME Winter Annu. Meet. Symp. Haptic Interfaces Virtual Environ. Teleoper. Syst. (1994), Chicago, IL, 1994

42.38 N. Hogan, H.I. Krebs, A. Sharon, J. Charnnarong: Interactive robotic therapist, Patent 5466213 (1995), Massachusetts Inst. Technol., Cambridge

42.39 H.I. Krebs, B.T. Volpe, D. Williams, J. Celestino, S.K. Charles, D. Lynch, N. Hogan: Robot-aided neurorehabilitation: A robot for wrist rehabilitation, IEEE Trans. Neural Syst. Rehabil. Eng. **15**, 327–335 (2007)

42.40 C.R. Carignan, H.I. Krebs: Telerehabilitation robotics: Bright lights, big future?, J. Rehabil. Res. Dev. **43**, 695–710 (2006)

42.41 H. Kazerooni, P.K. Houpt, T.B. Shecridan: Robust compliant motion for manipulators. Part 1. The fundamental concepts of compliant motion. Part 2. Design methods, IEEE J. Robot. Autom. **2**, 83105 (1986)

42.42 N. Hogan: Impedance control: An approach to manipulation, Part I–III, ASME J. Dyn. Syst. Meas. Control **107**, 1–24 (1985)

42.43 H. Asada: Dynamic analysis and design of robot manipulators using inertia ellipsoids, IEEE Int. Conf. Robot., Atlanta 1984 pp. 94–102, Atlanta, 1984

42.44 O. Khatib: Inertial properties in robotic manipulation: An object-level framework, Int. J. Robot. Res. **14**, 19–36 (1995)

42.45 D. Accoto, L. Zollo, E. Guglielmelli: Design of a planar robotic machine for tele-rehabilitation of elderly patient, Gerontechnology **7**, 65 (2008), (doi:10.4017/gt.2008.07.02.002.00)

42.46 D. Accoto, F. Torchiani, E. Guglielmelli, L. Zollo, E. Cecchini, M. Orsini: Dispositivo per la terapia motoria robot-mediata dell'arto superiore, Patent RM2008A000242 (2008), filed on May 6, 2008

42.47 B.R. Brewer, S.K. McDowell, L.C. Worthen-Chaudhari: Poststroke upper extremity rehabilitation: A review of robotic systems and clinical results, Top. Stroke Rehabil. **14**, 22–44 (2007)

42.48 C.D. Takahashi, L. Der-Yeghiaian, V. Le, R.R. Motiwala, S.C. Cramer: Robot-based hand motor therapy after stroke, Brain **131**, 425–437 (2008)

42.49 M.F. Levin, J.A. Kleim, S.L. Wolf: What do motor "recovery" and "compensation" mean in patients following stroke?, Neurorehabil. Neural Repair **23**, 313–319 (2009)

42.50 B. Rohrer, S. Fasoli, H.I. Krebs, R. Hughes, B. Volpe, W.R. Frontera, J. Stein, N. Hogan: Movement smoothness changes during stroke recovery, J. Neurosci. **22**, 8297–8304 (2002)

42.51 C. Bosecker, L. Dipietro, B. Volpe, H.I. Krebs: Kinematic robot-based evaluation scales and clinical counterparts to measure upper limb motor performance in patients with chronic stroke, Neurorehabil. Neural Repair **24**, 62–69 (2010), (DOI:10.1177/1545968309343214)

42.52 R. Colombo, F. Pisano, S. Micera, A. Mazzone, C. Delconte, M.C. Carrozza, P. Dario, G. Minuco: Assessing mechanisms of recovery during robot-aided neurorehabilitation of the upper limb, Neurorehabil. Neural Repair **22**, 50–63 (2008)

42.53 L. Zollo, E. Galotta, E. Guglielmelli, S. Sterzi: Robotic technologies and rehabilitation: New tools for upper-limb therapy and assessment in chronic stroke, Eur. J. Phys. Rehabil. Med. **47**(2), 223–236 (2011)

42.54 F. Gandolfo, F.A. Mussa-Ivaldi, E. Bizzi: Motor learning by field approximation, Proc. Natl. Acad. Sci. USA **93**, 3843–3846 (1996)

42.55 M.A. Smith, J. Brandt, R. Shadmehr: Motor disorder in Huntington's disease begins as a dysfunction in error feedback control, Nature **403**, 544–549 (2000)

42.56 M.A. Smith, R. Shadmehr: Intact ability to learn internal models of arm dynamics in Huntington's disease but not cerebellar degeneration, J. Neurophysiol. **93**, 2809–2821 (2005)

42.57 T. Flash, N. Hogan: The coordination of arm movements: An experimentally confirmed mathematical model, J. Neurosci. **5**, 1688–1703 (1985)

42.58 J. Cohen: *Statistical Power Analysis for the Behavioral Sciences*, 2nd edn. (Lawrence Erlbaum Associates, Berlin Heidelberg 1988)

42.59 D.G. Kamper, A.N. McKenna-Cole, L.E. Kahn, D.J. Reinkensmeyer: Alterations in reaching after stroke and their relation to movement direction and impairment severity, Arch. Phys. Med. Rehabil. **83**, 702–707 (2002)

42.60 M.C. Cirstea, M.F. Levin: Compensatory strategies for reaching in stroke, Brain **123**, 940–953 (2000)

42.61 M.L. Latash, J.G. Anson: What are "normal" movements in atypical populations?, Behav. Brain Sci. **19**, 55–106 (1996)

42.62 B. Steenbergen, E. Van Thiel, W. Hulstijn, R.G.J. Meulenbroek: The coordination of reaching and grasping in spastic hemiparesis, Hum. Mov. Sci. **19**, 75–105 (2000)

42.63 C.D. Takahashi, L. Der-Yeghiaian, V.H. Le, R.R. Motiwala, S.C. Cramer: Robot-based hand motor therapy after stroke, Brain **131**, 425–437 (2008)

42.64 M. Tombini, F. Zappasodi, L. Zollo, G. Pellegrino, G. Cavallo, F. Tecchio, E. Guglielmelli, P.M. Rossini: Brain activity preceding a 2D manual catching task, Neuroimage **47**(4), 1735–1746 (2009)

42.65 T.W. Boonstra, H.E. Clairbois, A. Daffertshofer, J. Verbunt, B.W. van Dijk, P.J. Beek: "MEG-compatible force sensor, J. Neurosci. Methods **144**, 193–196 (2005)

42.66 R. Gassert, R. Moser, E. Burdet, H. Bleuler: MRI/fMRIcompatible robotic system with force feedback for interaction with human motion, IEEE/ASME Trans. Mechatron. **11**, 216–224 (2006)

42.67 K. Chinzei, R. Gassert, E. Burdet: Workshop on MRI/fMRI compatible robot technology – A critical tool for neuro science and image guided intervention, IEEE Int. Conf. Robot. Autom. (2006), Orlando

42.68 A. Krieger, R.C. Susil, C. Menard, J.A. Coleman, G. Fichtinger, E. Atalar, L.L. Whitcomb: Design of a novel MRI compatible manipulator for image guided prostate intervention, IEEE Trans. Biomed. Eng. **52**, 306–313 (2005)

42.69 A. Hribar, M. Munih: Development and testing of fMRI-compatible haptic interface, Robotica **28**, 259–265 (2010)

42.70 N. Yu, C. Hollnagel, A. Blickenstorfer, S.S. Kollias, R. Riener: Comparison of MRI-compatible mechatronic systems with hydrodynamic and pneumatic actuation, IEEE/ASME Trans. Mechatron. **13**, 268–277 (2008)

42.71 S.E. Mendelowitz: Design of an MRI compatible robot for wrist rehabilitation. Ph.D. Thesis (MIT, Cambridge 2005)

42.72 D. Mintzopoulos, A. Khanicheh, A.A. Konstas, L.G. Astrakas, A.B. Singhal, M.A. Moskowitz, B.R. Rosen, A.A. Tzika: Functional MRI of rehabilitation in chronic stroke patients using novel MR-compatible hand robots, Open Neuroimaging J. **2**, 94–101 (2008)

43. Cardiac Devices and Testing

Michael Bateman, Stephen A. Howard, Christopher Rolfes, Timothy G. Laske, Paul A. Iaizzo

From the earliest reported cardiac surgeries in the late 19th century, exponential improvements in surgical techniques and operating times have opened the door for the development of novel cardiac devices. The design and development of cardiac devices in the 21st century is driven by the innovation and resources of an industry forecast to be worth $ 266 billion by 2012. This market growth is demonstrated by the many patent applications submitted each year to create novel devices or embellish existing products.

This chapter seeks to provide a brief description of the history of cardiac surgery, the evolution of cardiac devices, and a glimpse into the future of the industry as a background to the design process. The fundamental principles involved with the design and development of cardiac devices are discussed, from an explanation of the many stages involved with device design and the biological factors and material properties that engineers must consider when prototyping and testing devices to the steps taken by a medical device company to successfully deliver a product to market. Additionally, the chapter includes examples of market available cardiac devices and illustrates the novel devices and therapies currently in development.

43.1 **Background** 856

43.2 **Selected Landmark Events in Cardiac Devices and Surgery** 856

43.3 **Market Released Cardiac Devices** 857

43.4 **Device Development** 857
- 43.4.1 The Six Phases of Device Development 858
- 43.4.2 Intellectual Property 859
- 43.4.3 Device Prototyping 860
- 43.4.4 Device Testing 861
- 43.4.5 The Visible Heart 864
- 43.4.6 Animal Testing 864
- 43.4.7 Clinical Testing and Regulatory Approval 866

43.5 **The Anatomy of a Device** 867
- 43.5.1 Functionality 867
- 43.5.2 Biocompatibility 868
- 43.5.3 Durability 870
- 43.5.4 Design for Manufacture 871

43.6 **Emerging Cardiac Device Technology** 871
- 43.6.1 Atrial Fibrillation 871
- 43.6.2 Heart Failure 872
- 43.6.3 Left Ventricular Assist Devices and Artificial Hearts 872
- 43.6.4 Heart Transplant 872
- 43.6.5 Valves 872
- 43.6.6 Vascular Disease 873
- 43.6.7 Hybrid Operating Rooms 873
- 43.6.8 Novel Devices 873

43.7 **Conclusions** 874

References 874

Driven by the fields of surgery and cardiology, the rapidly expanding cardiac device industry continues to produce innovative and revolutionary devices and therapies. Although many devices are conceived from a distinct need within the operating room or cardiac catheterization laboratory, their design and development often lead to novel procedures and techniques that continue to improve treatment of cardiac pathologies. There is little doubt that these innovative improvements will extend and enhance the overall quality of the life for patients worldwide.

43.1 Background

The roots of cardiac device design and development are found in fairly recent history (in the past 60 years), and are associated with the expanding fields of interventional cardiology and cardiac surgery. Innovative surgeons and interventionists in the 1950s and 1960s pioneered and revolutionized treatments for cardiac trauma and/or defects. These early procedures may seem tentative and cautious by today's standards, however the operations performed by these individuals were revolutionary. These brazen clinicians pushed the limit beyond what was considered acceptable and thus often were not well respected by their peers.

The earliest reported heart surgeries were conducted out of necessity to repair potentially fatal cardiac wounds. From *Rehn*'s repair of a stab wound in a Frankfurt hospital in 1896 (considered to be the first successful human heart surgery) [43.1] to the treatment of field wounds during the two world wars, the increasing success of cardiac repairs finally opened the door to a whole new field of cardiac surgery.

Throughout the 20th and 21st centuries there have been rapid advancements in heart surgery, thanks to concurrent progress in cardiac medical technology. The use of hypothermia, heart-lung bypass machines, and cardioplegia solutions has provided surgeons with a motionless and bloodless operating field. These advances greatly increased the possible operating time from minutes to hours. Whereas previously heart surgeries were limited to simple procedures, such as the repair of minor congenital defects, these advances then enabled the development of novel surgical techniques and technologies. More recently, this has been accompanied by the necessary testing and approval requirements. The modern day process of product development, from the initial steps of brainstorming and idea generation to bench top and clinical testing, continues to be refined. Throughout this process, the innovation and resources of an industry forecast to be worth $ 266 billion by 2012 [43.2] continue to drive the development of novel technologies in the field of cardiac devices.

43.2 Selected Landmark Events in Cardiac Devices and Surgery

Cessation of the heart for 4 min or more is known to cause permanent brain damage, thus requiring the earliest cardiac surgeries to be performed on a beating and blood-filled heart. The first step to increase this surgical window was conceived by the Canadian surgeon *Bigelow* when, in a 1950 paper, he described the use of hypothermia on dogs [43.3]. This idea was first clinically used in 1952 at the University of Minnesota, when *Lewis*, Varco, and Lillehei used hypothermia to successfully repair an atrial septal defect in a five-year-old girl [43.4]. Though hypothermia extended the window during which blood flow could be stopped, physicians considered that cardiac surgery still had to be completed in less than six minutes, thus not allowing enough time for surgeons to attempt more complicated operations.

John Gibbon originally came up with the idea for a heart-lung machine in 1930. When charged with monitoring a patient slowly dying from a massive pulmonary embolus, he realized that her condition could be improved if he could provide some assistance to help oxygenate and pump the blood [43.5]. He worked for over 20 years on his machine before attempting the first use in humans, which unfortunately failed. Subsequently, due to limited success, Gibbon halted work on his bypass machine after only a few attempts. Yet, from this groundwork, similar concepts were attempted by others around the world. One notable approach was the biological heart-lung procedure employed by *Lillehei* and colleagues in 1954–1955, in which they used a second person, typically a relative of the patient, as the life support system [43.6]. This procedure, known as cross-circulation, though ground-breaking was short lived primarily due to criticisms that it carried a risk of 200% mortality.

It was during these procedures that the consequence of surgically induced heart block also was identified. The solution to this problem was to electrically pace the heart using an external stimulator. It was Johnson at the University of Minnesota who suggested this approach to his clinical colleagues, including *Gott* and Lillehei [43.7]. Once this was found to be an effective solution for keeping such patients alive, the need for a battery powered system was also identified; this became a more immediate need when a patient being supported by an in-house power source died during a power outage in the operating room at the University of Minnesota. At that time, Earl

Bakken, cofounder of Medtronic, Inc., was repairing equipment in the hospital operating rooms, and was recruited to build the first wearable, battery powered pacemaker that was then used on a patient in 1958. During this period, further developments and refinements of mechanical means of circulation occurred, including the Dodrill-GMR, created by *Dodrill* at Harper Hospital in Michigan [43.8], a heart-lung machine developed by *Kirklin* at the Mayo Clinic [43.9], and the bubble oxygenator developed by *DeWall* and Lillehei at the University of Minnesota [43.10]; these developments offered greater advantages compared to the cross-circulation procedure. Importantly, these mechanical heart-lung machines, along with the development of the membrane oxygenator and the pacemaker in the 1960s, allowed cardiac surgeons hours in which to operate in a motionless and bloodless heart and provide pacing therapy as needed.

As the allowable surgical time for cardiac surgery increased, so did the complexity of the procedures that could be attempted. Subsequently, innovations flourished as the rapidly changing field of cardiac medicine stimulated the formation of hundreds of medical device companies. Some companies were founded to commercialize newly invented devices, such as St. Jude Medical, a business formed around development of the bileaflet mechanical heart valve. Other companies, like Boston Scientific, were started to create new markets for less invasive medicine [43.11], designing catheters that facilitated minimally invasive procedures. In modern medicine, these companies have developed a multitude of devices that are used to correct a wide variety of cardiac pathologies, and many additional innovations are submitted to the United States Patent and Trademark Office every year. For example, a search of their website [43.12] produced the following numbers of published patent applications based on specific key words:

- CARDIAC (18 920 as of July 2004; 54 155 as of May 2009; 72 690 as of April 2011),
- CARDIOLOGY (1480 as of July 2004, 4557 as of May 2009, 5988 as of April 2011),
- CARDIAC REPAIR (32 as of July 2004, 149 as of May 2009, 244 as of April 2011).

It should be noted that this list does not include many international patents, but demonstrates the exponential acceleration of device innovations in regards to the assessment of cardiac health and treatment of cardiac pathologies [43.13]. The number of patent applications submitted each year is indicative of the conception of ideas that ultimately create novel devices or embellish existing products.

43.3 Market Released Cardiac Devices

Table 43.1 displays a list of some of the most common cardiac devices available on the market today. This list exemplifies the vast number of products designed and produced for the treatment of cardiac pathologies and illustrates the diverse nature of these devices.

43.4 Device Development

The invention of a cardiac device is the same as many other well known and widely used products; both must come from an idea and/or recognized unmet clinical need. In many cases, a device can have a beginning as humble as a rough sketch on the back of a napkin, thus beginning the iterative process of device development, which may even allow such an idea to evolve into a marketable product.

Today, it is generally considered that most new medical devices are not created by large organizations, but by small startup companies. These companies are funded by either a venture capital firm or an angel investor (investment from a single person), and are typically focused on developing one primary product. Larger organizations, on the other hand, will generally focus on improving devices currently available rather than developing new devices, although they do partake in their own device development from start to finish. Academic institutions also foster device design using grants or federal funding, yet very few are able to develop a device beyond the concept (or an early prototype) stage due to the lack of resources and general capabilities of the institution. However, the intellectual property (IP) of an institution can be licensed to an external company to further develop the concept and perform animal and clinical testing [43.14]. Generally,

Table 43.1 Examples of market released cardiac devices used to treat cardiac pathologies

- Pacemakers
  - Atrial pacing systems
  - Ventricular pacing systems
  - Biventricular pacing systems
  - 3 chamber system (e.g., right atrial and each ventricle separately)
  - Activity sensing for rate control
- Implantable cardioverter defibrillators (ICD), with pacing capabilities
- Pacemaker and ICD leads
  - Active fixation
  - Passive fixation
- Ventricular assist devices
- Prosthetic valves
  - Mechanical valves
  - Tissue valves
    - * Surgically implanted
    - * Transcatheter delivered
- Prosthetic coronary arteries
  - Bypass graft anastimosis devices
  - Engineered grafts
- Valve annuloplasty rings
- Septal defect closure devices
- Stents
  - Coronary artery stents
  - Endovascular and extravascular stent grafts
- Angioplasty balloons
- Valvuloplasty balloons
- Ablation catheters
  - Cryoablation catheters
  - Radio-frequency ablation catheters
- Heart rate monitors
  - External
  - Implantable
- Drug delivery pumps and catheters
- Vascular aspiration devices
- Artificial hearts
- Positioners and stabilizers

academic institutions are not focused on device development, and thus are not fully equipped to address all of the issues involved with creating, testing, and/or marketing a cardiac device.

43.4.1 The Six Phases of Device Development

Device development can begin anywhere, but there is a general sequence for how a device is developed. There are a number of different systems to categorize the steps taken for development and testing of any cardiac device and one in particular involves breaking the process up into six phases of development. These phases are shown in Fig. 43.1.

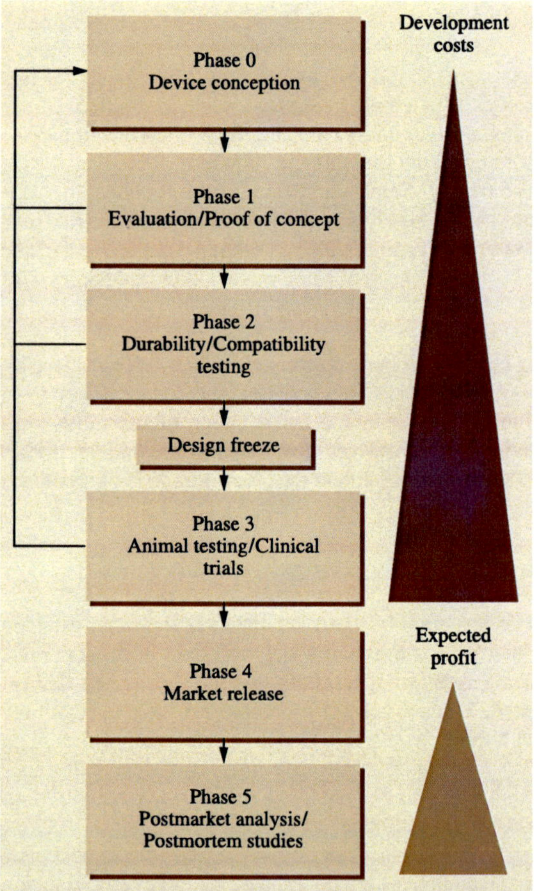

Fig. 43.1 Flow chart of device development process. The six phases of device design are shown, as well as the expected cost and profit with marketing a medical device

Phase 0 is thought of as the planning phase. During this phase much of the groundwork is completed – creating a product platform, assessing market opportunity (determining if the product will be worth pursuing financially), and determining product constraints associated with IP.

Phase 1 is considered the concept development phase. This is when the development process as well as bench top testing of the device begins. For example, feasibility studies are performed to determine whether it is technically possible to manufacture the product and/or if there is a potential market of adequate size for it to be profitable. It should be noted that in the field of cardiac devices there remains a need for pediatric-sized devices, but often it is considered not profitable to pursue these opportunities.

Phase 2 occurs after the device meets the previous criteria from phase 1. This is where further bench top tests are performed for both accelerated failures and required function of the devices. In vitro work or acute animal studies can also be initiated to assess potential biocompatibility. The end of phase 2 generally involves a design freeze, where nothing can be changed on the device without going back to phase 1 or 0.

Phase 3, which occurs after the design freeze, assures that testing will provide an accurate assessment of device function in a living organism. Testing often includes chronic implantation in appropriate animal models, which is then followed by regulatory approvals before the onset of clinical trials in humans. Subsequent to a successful clinical trial, final approval for market release of the product is sought.

Phase 4 is initiated when the product is market released; usually this is the first phase of the overall process during which the company expects to see return on investment. Yet, profitability assumes that all clinical trials were successful, that the regulatory body of the country has approved the device for market release, that the company can begin to commercialize and market the device, and that they can receive payment (reimbursement) for sales.

Phase 5 is considered to be the postmarket assessment. Even though a product is fully marketed and approved by a regulatory agency, the company is still required to perform follow-up studies on their products to ensure that they are not causing any unforeseen issues with patients over time. This also gives the company more information for future improvement of their device. Types of studies that can be performed in this phase include postmortem studies, where the company may receive the explanted device and then analyze its postimplant status or condition.

Importantly, throughout all of these phases, a number of different elements need to be checked and rechecked to achieve successful device production. For example, during *voice of customer* (VOC) research, feedback on device design is obtained from consumers of the product, including professionals who implant the device, patients, and their caregivers; this information will help ensure that the concept or device fulfills the needs of all those considered as customers. Business feasibility and marketability will also drive the device design, i. e., to be as cost efficient as possible. The initial stages of research and design are generally the cheapest portion of designing a device. When animal studies need to be undertaken, especially chronic implant studies, the costs dramatically increase. Subsequent human clinical trials can cost many millions of dollars.

43.4.2 Intellectual Property

At the outset of any device development project, it is critical to review published literature and issued patents for similar products or ideas that may be currently protected. Thus, a major part of the product design process is to maintain IP related to the technologies being developed. A continually changing design can be determined to be novel, non-obvious, and useful and may therefore warrant a new patent application. It is considered that there are a number of ways that developers can keep their IP safe from others infringing on their idea and using it as their own. For example, a *trademark* is a name or symbol that is associated with a product or brand, which means the company and only the company has full rights to use that name or symbol. This prevents others from using the name and associated marketing as a means of promoting their products or services. Similarly, a *copyright* may be given for any written and graphical material to an individual or group of individuals to protect their work and to reduce the risk of it subsequently being plagiarized [43.15]. Another suggested way to protect an idea or product is to keep it a trade secret, simply by not divulging any information regarding how a product is made, or how it operates or performs (e.g., the device runs on proprietary software or novel circuitry). One downfall to this type of IP protection is that if anyone successfully duplicates the product or procedure (even without knowing the trade secret), there are no legal consequences [43.15].

The most noteworthy way of protecting one's IP in the medical device industry is to file and obtain a *patent* with broad claims. A patent is a legal document that explicitly explains how a device works or how a procedure is done, with enough information that anyone within the field would be able to duplicate the device or process. Most medical devices are physical entities that can generally be taken apart. Thus, a competing company could theoretically, but not ethically, obtain a marketed device, disassemble the device into its core components, and rebuild it (reverse-engineering). By doing this they may improve upon the design, create a better product, and take away market revenue from the inventors. Hence, as early as possible in the device development process, a patent should be drawn up to protect the device design. Most patents provide legal protection for 20 years (e.g., US patents), thus guaranteeing exclusive marketing rights for that period. However, upon expiration, the device can be copied without legal recourse [43.15]. With cardiac medical devices this is often of little concern, since the product life of the original design is often less than 5 years, and thus is typically updated with new patent protection. Nevertheless, with all new product development, it is important to avoid infringement of all protected patents (and subsequent infringement lawsuits). Thus, for any product developer it is vital to know and understand how to read patents. Typically, patents are classified according to their device type and use and will first provide the filing numbers, inventor(s), and the date filed. Next a description of the device or process is presented, generally involving sketches and other images of IP. The most pertinent information is contained within the claims section, which specifically lays out what part of the IP is novel, and hence what is officially patented and protected by the patent. The claims section is generally the portion that legal teams will address when reviewing a patent case, as it gives the specifications of the patent and what is fully covered by the patent.

In many cases, finding and utilizing patent information can benefit medical device designers. According to the European Patent Office, there are a number of reasons and ways to use patents to your benefit, i. e., "find out what currently exists and build on it ... keep track of who's doing what ... [and] avoid infringing on other people's patent rights" [43.16]. To find patents relative to the cardiac devices you hope to develop, there are many online databases that can search for patents with specific information. A few examples of such databases include the European publication server [43.17], free patents online [43.18], and Google [43.19].

43.4.3 Device Prototyping

In the early stages of creating a medical device, it is best to fully investigate all potential options of the device design. Brainstorming and ideation are key aspects of product design. Using the philosophy that any idea is a good idea, anything from very quick drawings to off the wall ideas can be used to generate novel concepts. It is generally accepted that promoting idea generation is important because, even if a particular idea may not be possible or useful, it may spark ideas from someone else, and provide another angle or approach to solving the problem at hand. Typically once ideas are abundant, a team can begin to create a list of best concepts for the device.

Additionally, a very important part of phase 0 in the development process is the creation of quick mock ups of prototype devices. The goal is to generate many tangible ideas to help fully visualize and understand the product or concept. In other words, a sketch can give one the ability to depict the idea from one particular view point, but it requires a 3-D physical rendering (e.g., a moveable and/or moldable prototype to portray the workings of the device and the human factor interactions) to fully appreciate the idea and understand the nature in which it will interact with its environment. Additional purposes of phase 0 are to develop a proof of concept and to confirm that the ideas put forth can be constructed and eventually manufactured to perform as intended.

After a proof of concept has been attained and market potential assessed, the design team should select a handful of devices that will be fully prototyped and shown to customers for feedback. This is considered to be part of phase 1 during which feedback is obtained by collecting the *voice of customer* from a large sample size of potential users with varied backgrounds and clinical experiences; this can be obtained through direct questioning, observing, and/or discussing the product with the customer. Customers may consist of many groups of people, including those who use the device, benefit from the device, pay for the device (reimbursement, insurance companies), and those who may potentially profit from the device. These groups of individuals will offer assistance in the design process and likely improve the overall design of the product, however, it can be difficult to fully meet the expectations of all primary customers. For instance, a user may want the device to be easy and intuitive to use, in contrast to payers who may want the device to be inexpensive and beneficial to the patient in reducing costs (e.g., a reduction in the number of required future procedures). Those who expect to profit from the product (e.g., a company's chief financial officer) may demand that the costs of product development be minimized. Hence, all of these concerns must be proactively considered and addressed by the design team to create a product that satisfies the majority of their potential customers.

In addition to brainstorming, it follows that one needs to track all potential device defects and failures. Note that early on in such a process, one needs to consider the risks and benefits of employing any new procedure that utilizes a newly developed device (Fig. 43.2).

In order to complete the risk analysis on an entire device, one typically employs *failure modes and effects analysis*. These activities are generally defined as procedures in product development and operations management for identifying potential failure modes within a system, i. e., classification by the severity and likelihood of failures. Any successful analysis activity helps the research and development team identify potential failure modes based on past experience with similar products or processes, enabling the team to design those failures out of the system with a minimum of effort and resource expenditure; in turn, this helps to reduce both development times and costs. Figure 43.3 depicts an example data sheet of such an activity. Complete failure modes and effects analysis on a new cardiac device, such as a prosthetic valve, typically creates a 150-page spreadsheet of potential problems, including everything from leaflet material breakdown to misalignment of the prosthesis within the heart. Each of these failure modes must be assessed as to how likely it will happen and how severe the impact on the patient could potentially be. If one particular failure mode may feasibly occur 1 out of every 100 000 patients and, as a result, have minimal impact on the patient's health, typically this failure mode will not create a drastic design change. However, if within the same occurrence rate, a given patient may experience serious health problems from the complications, this would be a failure mode that would need to be addressed.

43.4.4 Device Testing

When a device design moves from phase 0 into phases 1 and 2 of the design process, basic testing of a cardiac device begins. During this process, it is important to consider the standards set forth by various international governing organizations with respect to testing and manufacturing medical devices. The most notable organization is the International Organization for Standardization (ISO); others include the medical device directive (MDD) and the International Electrotechnical Commission (IEC). Each of these regulatory standards are readily available to the design team to ensure that the medical product being developed will "remove or minimize as far as is possible the risk of injury, in connection with their physical features" [43.20]. Many regulatory bodies require that a company abides by these sets of standards for creating a medical device; statements such as ISO certified mean that the company, for instance, is in ISO compliance with the policies and regulations in regards to all phases of testing, manufacturing, and developing their medical devices.

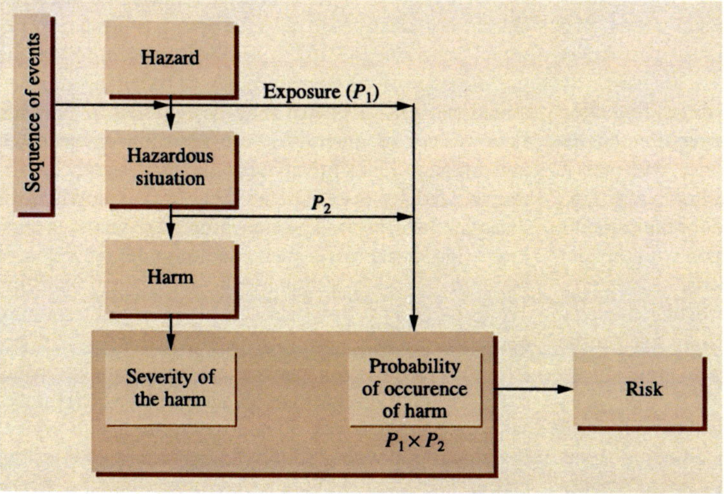

Fig. 43.2 Sample flow chart to perform risk estimations. P_1 is the probability of a hazardous situation occurring. P_2 is the probability of a hazardous situation leading to harm

FMEA number:	Part name: New rigid bileaflet heart valve substitute design	Prepared By:
Design/Mfg. responsibility:	Part number: 123456	FMEA Orig. Date:
Team members:		FMEA Rev. Date:

Part name & number/ function	Failure mode	Potential effects of failure (hazard and/or harm)	SEV[1]	Potential cause(s)/ mechanism(s) of failure	Prob[2]	Design verification	RPN[3]	Recommended action(s)	Responsibility & target completion date	Actions taken	SEV[1]	Prob[2]	RPN[3]
Leaflet (occluder) P/N 111111 Stops backward flow, opens in response to fluid flow	Rough surface	Loss of biocompatibility resulting in haemolysis, platelet aggregation, thrombus	4	Wrong specification for surface roughness	1	Compare surface to clinically acceptable samples; surface profilometry, SEM, blood flow loop testing	4						
				Inadequate production controls	2	Process validation	8						
	Contaminated surface	Loss of biocompatibility resulting in platelet aggregation, thrombus	4	Inadequate cleaning process	2	Characterize surface chemistry, validate cleaning process	8						
	Shape of leaflets create poor flow in open position	Turbulent flow resulting in haemolysis platelet aggregation, thrombus	4	Poor design	2	CFD, Flow visualization	8						
	Fracture	Embolism of small chips resulting in stroke	4	Sharp edges with high stress concentrators	2	Specify appropriate radius of all edges	10						

Notes [1]Severity rank from Table C.4; [2]Probability rating from Table C.2 or C.3; [3]Risk priority number = Severity rank X probability rating

Fig. 43.3 Example of a *failure modes and effects analysis* (FMEA) data matrix used to study a new rigid bileaflet heart valve design

Apart from biological and mechanical properties, there are other aspects of the device development process that must be taken into consideration. Obviously financial feasibility is a critical consideration, but there are other additional factors as well. For example, if a specific patient requires a pacemaker and s/he is likely to need an implantable defibrillator/pacing system in 5 years, there is no need for the battery in the pacemaker to last 20 years; in fact, many pacemakers are built with a battery lifespan between 6–8 years, depending upon the output settings of the system. This requirement allows physicians to change out the device with newer devices with better, more advanced technologies (e.g., new algorithms that may be customized for the patient's needs) [43.21]. This, in turn, is associated with the recommended longevity of the product. Most importantly, if the device is labeled as having a particular lifespan, then assessments on the product will only need to be tested for that period of time as noted by ISO regulations (ISO5840:2005(E)).

When testing the properties of a medical device, it is important that one isolates each of the components that require testing. More specifically, cardiac devices can be very complex in nature and have multiple associated parts fabricated from different composite materials. Even something as simple as a bare metal stent has multiple components. Such products may be required to hold an exact shape at physiological temperatures, the surfaces should be biocompatible, and the materials should not fracture or fail after being subjected to external pressures [43.15]. It is also important to test the performance of the developed device for all the intended situations, e.g., a device may be required to function in a variety of anatomies and maintain an accepted performance level for all manufactured sizes. This is emphasized by the ISO standards for development of a cardiac prosthesis. The standardized regulations state that a rigid heart valve, such as a tilting disc valve, must undergo durability testing over no less than 400 million cycles on at least three of each of the largest, medium, and smallest size valves (ISO 5840:2005(E)). To ensure that the device achieves this goal, tests are often designed around the critical conditions to which the device may be exposed. Shown in Fig. 43.4 is a typical flow chart one would employ to perform such a testing process.

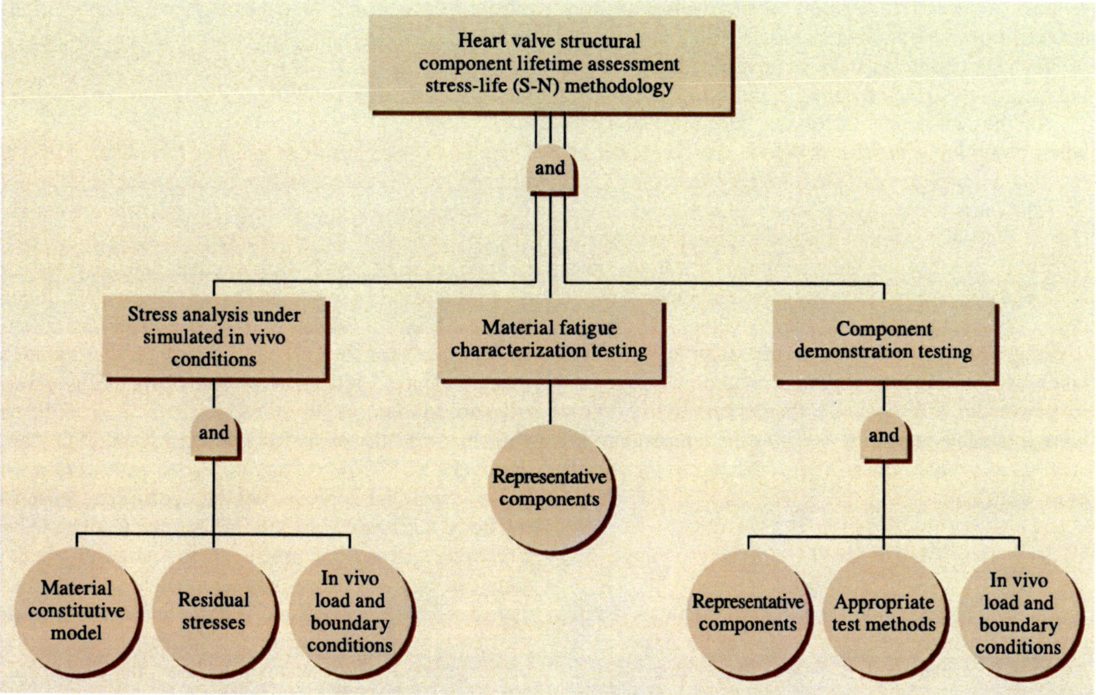

Fig. 43.4 Typical flow chart of a preclinical structural evaluation process that a valve research and development team might utilize to ensure proper durability of design. Note that typically these criterion can be evaluated in series

In many cases, the properties of cardiac devices can be tested without the use of native biological tissues. For instance, silicone, nylon, sponge, or foam can be used to create a mock testing environment for the device [43.15]. For example, if one wishes to investigate whether or not a pacing lead tip (active fixation helix) will properly engage with the endocardial surface, the properties of lead fixation into a silicone simulated tissue may be assessed, i. e., active fixation into a mock biological substrate followed by assessment of relative damage to the simulated myocardium (silicone). This is not only a more repeatable procedure due to a decrease in substrate variability, but it is also generally cheaper than obtaining live tissue or animals to perform initial basic testing.

Performing device testing with animal models or living biological tissues is not only expensive, but it also requires much effort and expertise. As such, to address questions during the early design stages of the device, alternate methods can be devised to ascertain device performance. Sometimes the testing approaches can be as simple as employing fresh meat obtained from a local grocer or meat packing plant; this can provide useful insight during initial device testing. For example, one could test an ablation catheter using fresh meat and then readily assess visual changes in the meat's appearance to determine the relative region of ablation [43.15]. Additionally, it is often possible to purchase animal organs such as hearts from meat packing plants and directly test a device in a more relevant cardiac anatomy.

Biocompatibility is a primary issue when developing a device which will be implanted within a human body. More specifically, in the case of cardiac devices, most of the current therapies and monitoring tools need to be inserted into the body and thus be exposed to the biological environment. In some cases, this is for only a short period (ablation catheters), while some devices will be implanted for a few years (invasive heart rate monitors), and yet others for the rest of the patient's life (mechanical heart valves). Nevertheless, in each of these scenarios the device is required to display an appropriate level of biocompatibility. In general, the potential for device rejection can be assessed with in vitro immunological responses to the device; a strong immunological response would be indicative of a possible problem with biocompatibility and possible device rejection. However, today most cardiac devices are constructed from well studied materials with known levels of biocompatibility; as such, adverse reaction testing is predominantly assessed during chronic animal studies.

Another important factor that one must consider when designing a cardiac device is the combined effect that temperature and pressure may have on a given device. There are many polymers or compounds that change elasticity properties when exposed to temperatures at or near normal body temperature. On the other hand, this can be advantageous depending upon how the device is designed. The informed engineer may choose to design the device, a catheter for example, to be stiff when entering into the vein/artery, thus allowing for easier placement. Subsequently, the device might become more malleable inside the body, as the temperature increases, so as not to cause internal damage to the vessels or myocardium.

43.4.5 The Visible Heart

Before embarking upon the expensive animal testing protocols that are required to prove the efficacy of a potential cardiac device, there is exceptional value in testing the device in a reanimated beating heart model. For example, the Visible Heart apparatus was developed by our laboratory at the University of Minnesota to reanimate large mammalian hearts that replicate the working physiological conditions of the heart in vivo. Reanimated hearts are perfused and then actively pump a clear crystalloid perfusate rather than blood, thus allowing the opportunity to uniquely and visually document the tissue device interactions of a large variety of cardiac devices [43.22]. Additionally, this approach has allowed researchers to employ an isolated, living heart as a model to visualize what occurs inside the heart during a device deployment procedure and how the device interacts with the organ throughout all the phases of the cardiac cycle.

In addition to the reanimation of swine hearts for early prototype testing using visible heart methodologies, our laboratory has been privileged to study the deployment of cardiac devices within reanimated human hearts. This reanimation of donor hearts, deemed not viable for transplant, enables the visualization of the specific device interaction with the varied endocardial anatomy of human hearts, both relatively healthy and those eliciting heart failure. With an increasing interest in the use of transcatheter delivered therapies, the ability to visualize the deployment and function of these devices using comparative imaging modalities has proved exceptionally beneficial to the engineers and physicians involved with evaluating novel concepts and designs.

43.4.6 Animal Testing

Once all in vitro testing techniques have been properly utilized and the device design has been locked, the development process moves into phase 3, whereupon the device must then be proven safe in appropriate preclinical testing; usually this requires extensive testing on animal models (acute and/or chronic) before it can be moved into a human clinical trial. In order to ensure proper care and use of the animals during such experimentation, a detailed prestudy protocol is developed to address proper animal welfare in accordance with one's Institutional Review Board and ISO standards (ISO 10993-2). Typically, such protocols must be reviewed and approved by the appropriate governing body, such as an Institutional Animal Care and Use Committee. In general, preclinical in vivo testing is designed to critically assess the performance of the cardiac device when implanted in a beating heart. These analyses include assessments of the device's effect on the hemodynamic performance of the heart, the biocompatibility and mechanical durability of the device, and the efficacy and ease of the implantation procedure. It is essential that any preclinical in vivo testing be designed to replicate the environmental factors that the device will experience when implanted in humans.

In general, animal testing essentially falls into two distinct categories – acute and chronic. Acute animal testing usually focuses on the immediate function of the device. Since the animal is sacrificed shortly after device implant, it is typical to perform multiple methods of invasive monitoring. This may include measuring right-sided and left-sided pressures, intracardiac electrical monitoring, the use of ultrasonic crystals to create pressure–volume loops, or blood flow and metabolic measurements. This instrumentation can help to determine if the device is working properly and assess its influence on physiological parameters, such as the effect of an artificial valve on ventricular pressures.

The protocol for a typical chronic animal experiment is designed to test the long-term biocompatibility, function, and durability of a cardiac device. In such studies, cardiac devices are implanted in the same manner as the intended human procedure, and the animals are monitored postoperatively until a predetermined endpoint. If the animal does not reach the experimental endpoint, a postmortem investigation will be conducted to determine whether the implanted device directly caused the mortality. Commonly in surviving animals, upon reaching the planned experimental endpoint, the device function is often retested both before and after the animal is sacrificed. Typically, the primary parameters assessed include the location and function of the device (i. e., did the implanted device move?), the immune response of the host and the overall health of the animal (i. e., was the systemic health of the animal compromised by the device?).

When developing a research protocol, it is important to use the appropriate species for the specific testing. Great strides have been made into the investigation of the pathophysiological mechanisms of cardiac disease with the use of transgenic mice and mice with gene deletions. However, while mice and rats can work well for specific tissue and metabolic work, most cardiac device testing of devices to be implanted in humans takes place on large animal models due to obvious technical limitations. It should be noted that within most of these large animals, the natural occurrence of cardiac disease is rare and often associated with comorbidities that preclude breeding efforts. Nonetheless, it is possible to surgically or otherwise induce various disease states (e.g., heart failure) or anatomical abnormality (e.g., an incompetent valve) within animal models. For example, to thoroughly test a device that treats heart failure induced by coronary disease, it is possible to induce a myocardial infarction in the animal prior to treatment by ligation or occlusion (e.g., the partial deployment of a covered stent) of a coronary artery [43.23].

For years, the use of canines for such experimentation has provided an excellent model for animal work since dogs typically have good temperament and can be trained to sit still during postoperative examinations. Yet, it should be recognized that canine hearts have an unusually large amount of collateral coronary circulation, which is only similar to humans who present with a slowly progressing end-stage chronic heart failure. Furthermore, this collateral circulation in the canine model can also lead to inconsistent or minimal ischemic regions that cannot be reproducibly induced [43.23].

Our laboratory and many others consider that swine are excellent models for acute cardiac device testing. For instance, swine have very similar cardiac anatomy to humans with respect to the conduction system, coronary arteries, and great vessels. Additionally, pig valves are very similar to human valves and are commonly used as tissue valve material. Yet, swine are not often used for long-term chronic studies due to frequent postoperative complications such as arrhythmias and sudden death. Additionally, it should be noted that mongrel swine grow at a fast rate, making chronic implantation difficult and often inconsistent from animal to animal [43.23]. In contrast, the use of mini pigs for chronic studies is possible, yet availability and cost can be prohibitive.

Sheep are commonly employed for chronic valve implantation studies since there is little growth in the adult sheep. These animals have been used as the historic model for such work and have valve orifice size and function very similar to a human, in addition to relatively large atria allowing for straightforward surgical approaches to the atrioventricular valves. The ovine model is also characterized by a higher immune response than other animals. This means that, in many cases, a thrombolytic event will be observed in sheep more prominently than in swine or dogs, and therefore may provide investigators with a better idea of what the failure will look like in humans [43.23]. There is vast literature available on the use of the ovine model for chronic valve implantation studies.

Today, primates are rarely used in cardiac device research due to prohibitively high cost and animal care. Thus, little cardiac work has been done and the lack of historical data provides little motivation to begin work with this model [43.23].

In summary, a well written preclinical testing protocol defines that in order to predict the safety and performance of clinical use, a sufficient number of animals of the same species, and preferably the same gender and age, should have experimental and control devices implanted. This number may be best determined based upon the risk analyses of the device and the statistical significance of the experimental design. The duration of the experiment is typically specified in accordance with the parameter under investigation, and each animal must undergo a macroscopic and microscopic postmortem examination.

Common goals of a preclinical study are to report:

- Any detectable pathological consequences around the device implant or in the major organs of the body.
- Any macroscopically or microscopically detectable structural alterations in the device itself.
- Histological assessment of any thromboembolic material, inflammatory reactions, or degenerative process.

Once the preclinical animal testing is completed using *good laboratory practices*, a third party observer (outside auditor) of the laboratory tests will produce a report to summarize all data collected and a recommendation of the clinical safety and performance of the device.

43.4.7 Clinical Testing and Regulatory Approval

Once a device has been proven safe and efficacious during rigorous animal testing, developers will embark upon clinical trials of the device before it can be properly market released. It should be noted that current regulatory processes in the United States and the European Union differ significantly in the requirements and guidelines laid out for clinical testing and market approval. However, both regulatory committees share the same fundamental principles and apply frameworks designed to ensure the safe and effective release of medical devices into the market. This section will briefly highlight current features of both approval processes.

From a legal standpoint, in the United States human clinical testing of an unapproved cardiac device cannot be initiated without pre-approval from the Food and Drug Administration (FDA) in the form of an *investigational device exemption* (IDE). The IDE is designed to provide the FDA with all the relevant data on device design and preclinical testing, as well as the intended study protocol. A device company must also apply for an IDE if they wish to expand the indication of an existing device. These clinical investigations may begin at an approved site 30 days after the FDA receives the IDE application, assuming that in-house institutional review board approval has already been obtained and the FDA has not notified the sponsor that the investigation may not begin [43.14].

To date, FDA approval is contingent on many factors. One of the most important factors determining the ease of starting a clinical trial is the device classification (Class I, II, or III), which is briefly defined as:

1. *Class I devices* pose the lowest risk to the patient and include noninvasive devices such as surgical bandages and tongue depressors. These devices are placed under the general rules applied to all medical devices and nothing more. The controls include prohibition of adulteration and misbranding, requirements on establishing registration and device listing, adverse event reporting, and *good manufacturing practices* [43.24].
2. *Class II devices*, such as cardiac catheters, are deemed to pose a high enough risk that regulation through the general controls alone is not sufficient. The majority of Class II devices require a premarket notification in the form of a 510(k) to provide data demonstrating that the described device is *of substantial equivalence* to an existing product with regard to its safety and effectiveness. Although a 510(k) can be substantiated through preclinical testing, approximately 10% of applications include clinical data [43.14].
3. *Class III devices*, such as implanted cardiac devices, are used to support and sustain human life, and also present a high risk of injury or fatality if the device fails. Almost all Class III devices require *premarket approval* by the FDA before being legally marketed, thus requiring clinical data demonstrating that the device is safe and effective in the target population [43.14]. Recently, the development of the *humanitarian device exemption* has created a pathway for accelerated market release of Class III devices; this exemption is intended for devices that address diseases or conditions that affect fewer than 4000 patients/year in the United States. Nevertheless, approval of a humanitarian device exemption requires the sponsor to prove that the device is safe and effective, and that all possible associated risks are outweighed by foreseen benefits. Typically, such approval requires smaller clinical trials in fewer institutions, allowing smaller companies to develop Class III devices beyond preclinical investigations.

Commonly, full scale clinical trials of Class III devices are preceded by a series of pilot trials typically limited to 100 patients spread over a few institutions, in order to establish the initial safety of the device and help in the design of the pivotal phase of clinical trials. The pivotal phase will be a large multicenter trial designed to generate the appropriate data required to define the patient populations in which the device functions safely and effectively. Many devices for which developers are seeking market approval are similar to current market released products and consequently can be measured against the standards set by these products. However, when entirely novel devices are introduced into clinical trials, the FDA requires randomized control studies to be completed. Yet, depending on the primary and secondary endpoints defined by the FDA, pivotal trials can require the enrollment of up to 1000 or more patients at 30–50 sites over a period of 1–2 years, with postoperative evaluations up to a year after implant. As such, these trials represent the largest commercial risk to a device developer, due to the recruitment and coordination of multiple test centers, core laboratories (for impartial data analysis), and a data safety monitoring board (to determine if continuation of the study is appropriate scientifically and ethically) [43.14].

Within the European Union, there are three major differences in the regulatory process which include 1) use of *notified bodies*; 2) general criteria for product approval, and 3) local site regulations. Notified bodies are independent commercial organizations established to regulate the process of device market approval in the European Union. The responsibilities of these organizations include:

1. Device classification
2. Continuing assessment of the trials to ensure conformity and quality
3. Approval of design dossiers for high risk devices
4. Postmarket regulatory control [43.24].

To receive market approval for a high risk device in the European Union, the manufacturer must prove that the device is safe and functions to the guidelines laid out by the manufacturer. The difference in the criteria required to gain market approval for a high risk device in the United States and the European Union is best demonstrated by the following example.

The GuardWire temporary occlusion and aspiration system (Medtronic, Inc., Minneapolis, MN), a device designed to recover liberated plaque during angioplasty or stent placement, underwent clinical trials in the United States and the European Union before market release. The device was classified as a Class II medical device by the FDA and, therefore, required 510(k) clearance. To satisfy this clearance, an 800-patient multicenter randomized trial comparing the device against a control was performed to prove a statistically significant reduction in major adverse cardiac events [43.25]. A trial of such magnitude can cost up to $ 12 million and may take up to 2 years [43.14]. To gain market approval within the European Union, the manufacturer had to demonstrate the device's safety and effectiveness. This was achieved by running a 22-patient single site study requiring significantly less investment than the clinical trial requested by the FDA [43.26].

To summarize, gaining market approval in either the United States or the European Union can be a lengthy, expensive, and time consuming endeavor. Before embarking on the development of a cardiac device, it is critical that the designer develop an understanding of these pathways and the intricacies of each regulatory body, in order to complete the process in the most timely and cost effective manner possible. To date, the differences between the United States and the European Union regulatory bodies help explain why a large portion of pilot clinical trials and early device testing occurs outside of the United States. These differences also help to explain why a typical cardiac device is introduced into general clinical practice in the United States 1–3 years after market release in the European Union [43.14].

43.5 The Anatomy of a Device

Recently, much has changed in the field of cardiac devices with the improvement of established therapies, the development of novel devices, and the increase of associated regulatory issues. In order for any device to be considered acceptable for clinical use, it must display a number of essential characteristics that illustrate the anatomy of a cardiac device. This section will outline the key criteria that need to be fulfilled when developing a cardiac device and, in the interest of brevity, will focus on implantable devices.

43.5.1 Functionality

Although it may be obvious, it is vital that any product performs to the standards set by the manufacturer's claims. Many medical devices are born from the desire to satisfy a previously unmet clinical need. So to be successful, a new product will have to complete this functionality above and beyond the currently available solutions and techniques. The functionality of a cardiac device also encompasses its ability to perform the desired task without compromising any other biological process. For example, a right-sided apically placed pacemaker lead should not compromise the function of the tricuspid valve. These specifications are often outlined by regulatory committees such as the International Organization for Standardization as seen in Table 43.2, which describes the operational specifications for cardiac valve prostheses.

For implantable cardiac devices, the question of functionality typically includes the way in which the device is delivered. Surgically implanted devices require a suitable method of fixation to the heart, such as the sewing ring of a prosthetic heart valve. How-

Table 43.2 Specifications for a prosthetic cardiac valve

Define that the device has reproducible function • Allowing forward flow with acceptably small mean pressure difference • Preventing retrograde flow with acceptably small regurgitation • Remaining fixed once placed • Having an acceptable noise level • Maintaining its functionality for a reasonable lifetime, consistent with its generic class • Maintaining its functionality and sterility for a reasonable shelf life prior to implantation
Define that the device is biocompatible • Resisting embolization • Resisting hemolysis • Resisting thrombus formation
Define that the device is compatible with in vivo diagnostic techniques
Define that the device is deliverable and implantable in the target population
ISO 5840:2005(E)

ever, these matters can be greatly complicated when one chooses to deliver such devices through minimally invasive techniques, e.g., either through the wall of the heart or through the vasculature without placing the patient on cardiopulmonary bypass. Hence, not only must the delivery system be able to place the device in its intended position, but additionally it must not induce any adverse affects on the patient. Examples of such devices include stents, septal occluders, valves, and pacing/defibrillation leads. These devices can all be delivered in the cardiac catheterization lab rather than a surgical suite, and thus greatly broaden the patient populations that can receive such therapies.

Another, more recent example involves the development of transcatheter delivered heart valves to treat patients with highly stenotic or regurgitant aortic valves. Importantly, once implanted correctly, any prosthetic aortic valve must allow blood to exit the left ventricle during ventricular contraction (systole), and close to inhibit the return blood to the heart during ventricular relaxation (diastole). Still today, the majority of prosthetic aortic valve replacements are implanted via open heart surgery. Further, such valves can be placed into two main categories that describe the materials that act as the valve leaflets – mechanical or tissue. Both designs have advantages and disadvantages, and there has been much debate over which performs better in various patient populations. However, the high surgical risk associated with patients with severe aortic valve disease due to multiple comorbidities, often considered as too sick to undergo an open heart surgical repair, has recently driven the need for transcatheter delivered, minimally invasive aortic devices [43.27, 28]. This transcatheter approach to aortic valve repair offers clear advantages because the prostheses can be implanted without open heart surgery or cardioplegia to stop the heart. To date, due to fairly complex delivery techniques, the valve leaflets have been primarily constructed from pericardial tissues; due to the nature of this approach, which requires seating either a balloon expandable or self expanding Nitinol frame in the aortic annulus on top of the native valve leaflets, questions have surfaced regarding how the implanted valve will function. It is currently considered that the options available to a cardiac surgeon while working on the heart during an open heart procedure – the ability to remove the native leaflets, size the aortic annulus directly and suture in the appropriate prosthesis – greatly decrease the likelihood of improper implantation. Consequently, there is an ongoing debate as to whether the performance of transcatheter delivered aortic valves can be considered comparable to continually improving surgically implanted valve procedures, and whether the advantages of minimally invasive implantation outweigh possible limitations in the functionality of the prosthesis [43.29, 30].

To date, companies developing these new transcatheter delivered technologies have been very specific in describing the patient populations for which the valves are designed and the conditions under which the valves should be implanted. By doing so, they have ensured that the devices are used only in situations that correspond to the conditions under which the device was designed to function.

43.5.2 Biocompatibility

Biocompatibility related to medical devices is generally defined as the ability of a material to perform with an appropriate host response in a specific application [43.31].

In 1987, Williams described how the biocompatibility of a material can be qualitatively evaluated to assess its performance when implanted. The succinctness of this term is often misleading when applying the principles of biocompatibility to cardiac devices due to the number of different materials that are commonly used during manufacture. For example, tilting disc valve prostheses such as the St. Jude Medical Masters series (St. Jude Medical, St. Paul, MN) are manufactured from a variety of materials including pyrolytic carbon for the discs and Dacron for the sewing cuff (Fig. 43.5). All of these materials elicit different responses from the host body, and need to be individually assessed and quantified before the device is implanted. Examples of such tissue responses can be seen in Table 43.3.

The mechanism of how the host responds to the different components of a cardiac device must be extensively evaluated to ensure that the correct materials are selected for the final device design. This evaluation can be complicated by the fact that it is hard to replicate the human immune response in vitro. Consequently, all cardiac devices also undergo rigorous animal testing, yet it is important to note that this too can sometimes furnish misleading results as to bioreactivity. An example of this can be seen in the design and development of the Braunwald–Cutter heart valve ball and cage prostheses (Fig. 43.6), whereby cloth-covered cage struts were designed to encourage endothelialization and hence decrease any chance of thrombolytic events [43.33]. Extensive testing in the mitral position of pigs, sheep, and calves showed promising results, thus the device was subsequently approved for human clinical trials. However, the device did not elicit the same host response when implanted in humans, in-

Table 43.3 Potential patient–device interactions causing clinical complications (after [43.32])

Adverse local tissue interactions • Inflammation • Toxicity • Carcinogenic response • Calcification • Embolization or lymphatic spread of material fragments
Induced device migrations: encapsulation or a foreign body response
Inappropriate or altered healing responses
Associated infections
Thrombosis • Thrombotic occlusion • Thromboembolism

Fig. 43.5 St. Jude Medical bileaflet valve

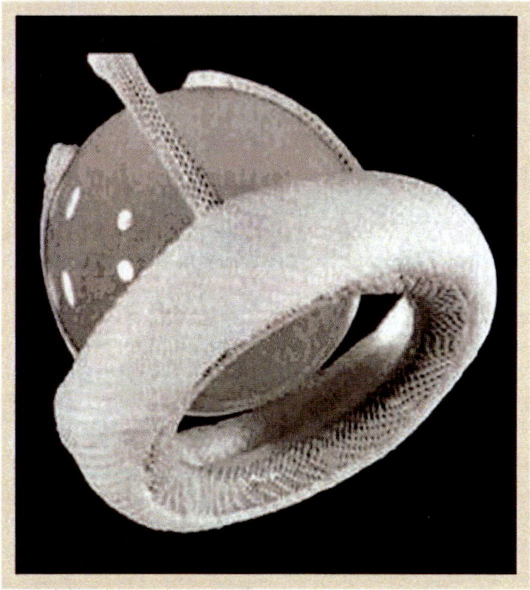

Fig. 43.6 Braunwald–Cutter valve

stead resulting in aggravated wear on the cloth cage struts and more critical debris embolization. It should also be emphasized that host responses are not limited to immune reactions; for example, implanted tissue heart valves remain susceptible to accelerated prostheses leaflet calcification. More specifically, this type of calcification is initiated by reactions between the extracellular fluid and the leaflet membranes, creating calcium phosphate mineral deposits [43.34]. As a result, much research is currently underway to enhance leaflet materials to include calcification inhibitors, thereby limiting mineral deposition on the implanted materials.

In addition to the possibility of inducing mineral deposition or endothelialization, material selection has a critical impact on the overall biocompatibility of a particular device, due to the wide variety of mechanisms by which the host body can affect the implanted device. For example, initially there were several reports concerning insulation fracture of the Medtronic 6792 pacing lead (Medtronic, Inc.) [43.35, 36]. Initially, these leads were manufactured using Pellethane 2363-80A [43.37], which has since been proven to be particularly susceptible to calcification, environmental stress cracking, and/or chain scission when implanted in humans [43.38]. Newer versions of the polyurethane coating, such as Pellethane 2363-55D, have displayed much improved biocompatibility when used to coat pacing leads [43.39]. Although material selection itself may not be the sole reason for a particular mechanism failure, how a given material is applied in the device design may also play a critical role.

A device's material can also result in unexpected interactions with the host, as evidenced by the earliest versions of the Starr–Edwards caged-ball mechanical heart valve. The valve design used a silicone ball that was found to absorb lipids from the blood and swell [43.40]. In addition to poor valve function, this resulted in the silicone balls becoming brittle and increased the possibility of ball fracture and consequent embolization of small fragments into the arteries downstream of the valve position.

Many modern implanted devices incorporate specific coatings to assist in reducing biological responses. A prime example of this is the drug-eluting stent, designed to minimize cell growth and scar formation, thus decreasing the likelihood of restenosis. To date, in treating coronary artery disease, the use of these drug-eluting stents is considered to have shown great improvement over angioplasty alone [43.41]. In a second example, steroid-eluting pacing leads have helped to manage the acute inflammation associated with lead implantation and, therefore, reduce the energy required to pace the heart [43.13]. In summary, it can be assured that the combination of pharmaceutics with implantable devices will continue to expand as novel biologically engineered coatings are identified and consequently continue to improve biocompatibility.

43.5.3 Durability

As well as exhibiting the appropriate level of biocompatibility, any device implanted into a human heart must also display sufficient durability and longevity. It should be noted that this poses a unique problem in the field of cardiac devices due to the extremely large number of cycles that such prostheses are expected to endure; of course, there are potentially life-threatening implications due to device failures as well. Furthermore, the heart is itself a dynamic organ which can elicit threefold to fourfold increased cardiac outputs (internal pressures during exercises), hence the mechanical stresses on an implanted device can increase fourfold to sixfold.

The case studies of pacing leads offer important insights into the sheer magnitude of this problem. A lead will move approximately 100 000 times a day, or 37 000 000 times annually, in multiple locations along its length, usually in more than one degree of freedom [43.13]. This continuous movement exerts forces on the lead body that, in turn, can accentuate the mechanical breakdown of the device. The underlying mechanisms can include material creep or erosion due to contact with other components. Coronary placed stents represent another important example; standard dynamic mechanical stresses due to bending, axial loading, and torsion can be induced during contraction of the heart. Additionally, stents are influenced by the dynamic changes in the coronary arteries during exercise such as circumferential forces due to changes in the lumen diameter.

The durability of any cardiac device is further challenged when the device directly interacts with the turbulent and pulsatile flow of blood, as is the case for prosthetic cardiac valves. More specifically, the flow of blood through mechanical prostheses, particularly tilting disc valves, can lead to a particular type of erosion called cavitation, whereby gaseous bubbles form and violently collapse on the surface of the valve discs due to the sudden temporary decrease in pressure during closure. This phenomenon can lead to the fracture of the valve disc and was originally observed in the 1980s in association with the failure of the Edwards-Duromedics valve (Edwards CVS, Irvine, CA) [43.42]. Subsequent tilting disc valves, such as the St. Jude bileaflet valve (St. Jude Medical), have been designed specifically to reduce this particular material erosion by reducing the area of the tilting disc.

As with other cardiac devices, prosthetic valves are expected to endure a large number of cycles in their lifetime. For example, in order to minimize the possibility of failure when embarking upon a human clinical trial, the ISO has set up a series of strict guidelines for the performance of prosthetic heart valves. The current ISO guidelines require rigid devices, such as tilting disc mechanical valves, to remain functional for 400 million cycles (approximately 10 years in vivo) and flexible devices, such as tissue valves, to remain functional for 200 million cycles (approximately 5 years in vivo). This testing must be done under the following minimum performance conditions: beat rate is 70 cycles/min, simulated cardiac output is 5.0 l/min, mean aortic pressure is 100 mmHg, and systolic duration is 35% (ISO 5840: 2005(E)). No valve can be approved for clinical trial before demonstrating this level of durability over a suitable sample size. Even today, due to the relative infancy of the field of prosthetic valves, there is little long-term clinical data regarding durability. Yet, a recent study by *Mykén* et al. regarding the durability of surgically implanted tissue valves such as the St. Jude Epic stented tissue valve (St. Jude Medical) has shown promising data 20 years post implantation, hence demonstrating the durability of these prostheses [43.43].

43.5.4 Design for Manufacture

During the cardiac device design process, it should be specifically noted that the design team must not neglect the fact that the device will eventually need to be manufactured in an efficient and cost effective manner. Such methodology is known as *design for manufacture* (DFM) and emphasizes the fact that an economically successful design should also ensure a high quality product while minimizing manufacturing cost. By utilizing manufacturing cost estimates, this process proactively helps to guide and prioritize cost reduction efforts involved with device design, and thus should have significant effects on product lead times, development costs, and ultimate product quality. As such, design for manufacture methods require input from a multidisciplinary team in addition to the core design team, thus drawing upon the expertise of manufacturing engineers, cost accountants, and production personnel [43.44]. Yet, when applying this principle to cardiac devices, it is critical to understand that the quality of care impacted by the device must not be compromised by the need for a more cost effective manufacturing process, but rather production costs can be controlled by using existing technologies and established manufacturing techniques.

43.6 Emerging Cardiac Device Technology

Rapid innovations in medical technology continue to build and advance with many new and emerging technologies that are being developed and improved constantly. For example, while specific diseases such as atrial fibrillation and heart failure have been known to exist for a long time, new therapies and devices are continuously being tested and developed on a daily basis.

43.6.1 Atrial Fibrillation

Atrial fibrillation is the most common type of cardiac arrhythmia in humans worldwide. While drug therapies and ablation procedures (radio- and cryoablation) are highly effective in treating this arrhythmia, there are associated side effects. The clotting risk within the patient's left atrial appendage is usually treated with the anticoagulant warfarin, which is effective in preventing strokes [43.45], yet another suggested approach for preventing such clots is to seal off the appendage. While surgical options and specialized tools for sealing the appendage can be used, another suggestion recently promoted by Atritech Inc. (Plymouth, MN) is to implant a tiny filter in the appendage to prevent blood clots from ever leaving the heart.

Additionally, other types of implanted devices can provide ongoing treatment for atrial fibrillation. For example, pacing implants have continuously improved algorithms that can now help with treatment and prevention. Alternatively, for patients who have a difficult time managing drug treatment either because of noncompliance or harmful side effects, implantable drug pumps may be able to deliver treatment directly into the pericardial space. Our laboratory has employed animal models to show that this type of targeted drug delivery can dramatically reduce side effects of the drug by minimizing the amount of the drug in the bloodstream while providing well-controlled doses directly to the heart (i. e., drugs can often be administered in greater concentrations than those recommended to be delivered intravenously).

43.6.2 Heart Failure

In general, congestive heart failure and dilated cardiomyopathy are associated with increased morbidity and mortality [43.46]. First line treatments typically involve therapeutic methods, including lifestyle changes (cessation of smoking, increasing exercise regimens, etc.) and medications (ACE inhibitors, beta blockers, etc.). When therapeutic methods are not enough, surgical repair or implanted devices may be required with the specific course of action being determined by the cause of heart failure (e.g., the use of an assist device; see following section).

43.6.3 Left Ventricular Assist Devices and Artificial Hearts

The concept of a completely artificial heart has been discussed since the early days of cardiac surgery. In its most basic form and function, the heart is simply two pumps that work in series. However, over the last 50 years, there have been many attempts to create an artificial heart with only limited success. In the United States, certain left ventricular assist devices are approved as a *bridge to transplant* therapy as well as for compassionate use. Currently, clinical trials are underway in the United States for the EXCOR pediatric ventricular assist device (Berlin Heart, Berlin, Germany) which can be used for medium- to long-term pediatric support. While left ventricular assist devices are typically implanted in patients with the intent to *bridge to transplant*, some patients, particularly pediatric patients, have been observed to recover intrinsic cardiac function to the extent that they can be completely weaned off the device [43.47, 48].

43.6.4 Heart Transplant

When heart disease in a given patient has advanced beyond the point where pharmaceuticals or surgery can help, a complete heart transplant may be indicated. While the procedure has changed little in the last 30 years, one area that has received focus for improvement is myocardial preservation. While organs like the kidney and liver can be safely transplanted up to 24 h after explantation from the donor, the heart is only viable for up to 6 h. Hence, new devices are being tested to increase this time, like the organ care system (Transmedics, Andover, MA), which keeps the heart beating at normothermic temperatures. This system has been reported to greatly improve heart health at 12 h of storage [43.49]. It is considered that devices and approaches like this will allow for improved transport of donor hearts and consequently increase the number of hearts available to potential recipients. Other methods to preserve heart function are also being explored. For example, our laboratory is investigating targeted delivery (pretreatment) with cardioprotective anesthetics, hibernation induction trigger, and omega-3 fatty acids using perfusion pumps, which we have observed to help improve transplanted heart function.

43.6.5 Valves

Damaged or defective functioning of valves is a well known cause of heart failure. In such cases, valve repair or replacement is often recognized as the most effective treatment. While replacement valves have been around since the 1950s, new valves and delivery systems are constantly being developed, such as the transcatheter Melody valve (Medtronic, Inc.; Fig. 43.7). While the Melody valve is designed to work when surgery is not an option (e.g., when the patient has already had numerous open heart procedures to correct congenital defects), today a surgically implanted valve is the gold standard among replacement valves. Hence, research will continue to make surgically implanted valve replacements better, with the ultimate

Fig. 43.7 Melody valve (Medtronic, Inc.)

goal of an artificial valve that will not require lifelong anticoagulation or replacement in 10–20 years, problems typically encountered with current mechanical and tissue valves. Such a valve would drastically reduce the prospect of reoperation and improve patient quality of life by not requiring concurrent anticoagulant therapy.

In addition to valve replacement, there is continued research into the development of annuloplasty devices designed to correct regurgitation by reducing the annular size of the atrioventricular valves. For example, two companies, MiCardia Corporation (Irvine, CA) and MitralSolutions, Inc. (Fort Lauderdale, FL) are developing surgically implanted annuloplasty rings that can be adjusted postoperatively to optimize their therapeutic effects. Conversely, Cardiac Dimensions, Inc. (Kirkland, WA), Edwards LifeSciences (Irvine, CA), and Viacor, Inc. (Wilmington, MA) are developing transcatheter annuloplasty devices deployed within the coronary sinus which lie adjacent to the mitral annulus to minimize dilations [43.50].

43.6.6 Vascular Disease

Today, the primary interventional treatments for atherosclerosis in coronary arteries include angioplasty and stenting. Early studies with drug-eluting stents showed them to be more effective at preventing restenosis [43.51], yet a more recent study indicated that late thrombosis rates were higher than with bare metal stents [43.52]. Hence work continues relative to developing the ideal stenting solution, including optimization of drug coatings. Alternatively, recent focus in the device developmental phase has been on biodegradable stents and coatings. For example, these stents, like the Igaki–Tamai stent (Igaki Medical Planning Co, Ltd, Kyoto, Japan) provided the short-term support required to prevent early restentosis, however did not provide adequate long-term support; yet, there was a considered lack of or limited long-term risk after stent degradation. Alternatively, biodegradable coatings, such as those used in the BioMatrix stent (Biosensors International Group, Singapore), provided short-term protection from restenosis, with the degradation of the coating leaving a bare metal stent, which may be further validated to display lower rates of late thrombosis.

For patients who are not good candidates for stenting, or if restenosis has occurred, *coronary artery bypass grafting* (CABG) is most often recommended. Since the first CABG procedure 50 years ago, it has become one of the most frequently performed cardiac surgeries. Importantly, new tools and technologies allow cardiac surgeons to perform the procedure in a less invasive manner. Equipment like the daVinci robotic surgical system (Intuitive Surgical, Inc., Sunnyvale, CA) has allowed surgeons to use smaller incisions and novel techniques to perform a wide array of cardiac procedures. Similarly, the development of heart stabilizing devices has allowed surgeons to perform cardiac surgery on the beating heart without cardiopulmonary bypass. It is envisioned that the continuing advancements of surgical techniques and tools and the development of both anastomosis devices and prosthetic coronary arteries will continue to improve patient outcomes and reduce overall hospital costs for CABG procedures.

43.6.7 Hybrid Operating Rooms

With many procedures blurring the line between cardiac surgery and cardiology, the development of hybrid operating rooms has led to increased equipment needs for both fields. This new type of operating room is built with the fluoroscopic imaging capabilities of a catheterization lab. This provides advantages as simple as a post-CABG angiogram, but also allows for multi-tiered treatments, such as stenting and valve replacement, all in the same room. This will be the ideal setting for procedures like transcatheter valve replacement, which is a surgical-like procedure requiring fluoroscopic guidance. As these centers of excellence develop, with a team of surgeons working side-by-side with cardiologists, one must consider that new devices and technologies may require design features that meet the requirements of each specialty.

43.6.8 Novel Devices

Around the world, thousands of innovators are busily developing the next generation of novel cardiac devices. For example, one potential treatment for heart failure, under clinical trials in the United States, is the Acorn CorCap Cardiac Support Device (Acorn Medical, St. Paul, MN). This device is implanted around the ventricles and provides support to the weakened heart walls and reduces ventricular wall stress. The use of this device is also expected to improve the heart's ability to pump blood, provide relief from heart failure symptoms, and potentially allow the heart muscle to heal. If this proves to be effective, it may be used in conjunction with other treatments.

43.7 Conclusions

It was the general aim of this chapter to lay out the fundamental principles involved with the design and development of cardiac devices, from an explanation of what design features engineers should integrate into their devices to the steps taken by a medical device company to successfully deliver a product to market. Additionally, the chapter provided a brief description of the history of cardiac surgery, the evolution of cardiac devices, and a glimpse into the future of the industry as a background to the design process.

References

43.1 L. Rehn: Über penetrirende Herzwunden und Herznaht, Arch. Klin. Chir. **55**, 315–329 (1897)

43.2 Espicom Business Intelligence: *World Medical Market Forecasts to 2012* (December 2007)

43.3 W.G. Bigelow: General hypothermia for experimental intracardiac surgery: The use of electrophrenic respirations, an artificial pacemaker for cardiac standstill, and radio-frequency rewarming in general hypothermia, Ann. Surg. **132**, 531–539 (1950)

43.4 F.J. Lewis, M. Taufic: Closure of atrial septal defects with the aid of hypothermia experimental accomplishments and the report of one successful case, Surgery **33**, 52–59 (1953)

43.5 P. Pastuszko: John H. Gibbon Jr., The inventor of the first successful heart-lung machine, J. Card. Surg. **19**, 65–73 (2004)

43.6 C.W. Lillehei, M. Cohen, H.E. Warden, R.L. Varco: The direct-vision intracardiac correction of congenital anomalies by controlled cross circulation results in thirty-two patients with ventricular septal defects, tetralogy of Fallot, and atrioventricularis communis defects, Surgery **38**, 11–29 (1955)

43.7 V.L. Gott: Critical role of physiologist John A. Johnson in the origins of Minnesota's billion dollar pacemaker industry, Ann. Thorac. Surg. **83**, 349–353 (2007)

43.8 F.D. Dodrill: Temporary mechanical substitute for the left ventricle in man, J. Am. Med. Assoc. **150**, 642–644 (1952)

43.9 J.W. Kirklin: The middle 1950s and C. Walton Lillehei, J. Thorac. Cardiovasc. Surg. **98**, 822–824 (1989)

43.10 R.A. Dewall: Origin of the helical reservoir bubble oxygenator heart-lung machine, Perfusion **18**, 163–169 (2003)

43.11 Boston Scientific, Natick (2010), http://www.bostonscientific.com/templatedata/imports/HTML/AboutUs/History/history.html (accessed March 5, 2010)

43.12 US Patent and Trademark Office, Alexandria, VA (2010), http://www.uspto.gov/ (accessed March 17, 2010)

43.13 P.A. Iaizzo: Emerging cardiac devices and technologies. In: *Handbook of Cardiac Anatomy, Physiology, and Devices*, ed. by P.A. Iaizzo (Springer, New York 2009) pp. 631–643

43.14 A.V. Kaplan, D.S. Baim, J.J. Smith, D.A. Feigal, M. Simons, D. Jefferys, T.J. Fogarty, R.E. Kuntz, M.B. Leon: Medical device development: From prototype to regulatory approval, Circulation **109**, 3068–3072 (2004)

43.15 S. Zenios, J. Makower, P. Yock (Eds.): *Biodesign* (Cambridge Univ. Press, New York 2010)

43.16 European Patent Office, Munich, Germany (2010), http://www.epo.org/patents/patent-information/about.html (accessed March 21, 2010)

43.17 European Patent Office, Munich, Germany (2010), https://data.epo.org/publication-server/?lg=en (accessed March 21, 2010)

43.18 FreePatentsOnline, Miami Beach, FL (2010), http://www.freepatentsonline.com/ (accessed March 21, 2010)

43.19 Google (2010), http://www.google.com/patents (accessed March 19, 2010)

43.20 D. Sawyer: *An Introduction to Human Factors in Medical Devices* (FDA, Silver Spring 1996), http://www.fda.gov (accessed March 21, 2010)

43.21 T.G. Laske, A. Legreid Dopp, P.A. Iaizzo: Pacing and defibrillation. In: *Handbook of Cardiac Anatomy, Physiology, and Devices*, ed. by P.A. Iaizzo (Springer, New York 2009) pp. 443–473

43.22 E. Chinchoy, C.L. Soule, A.J. Houlton, W.J. Gallagher, M.A. Hjelle, T.G. Laske, J. Morissette, P.A. Iaizzo: Isolated four-chamber working swine heart model, Ann. Thorac. Surg. **70**, 1607–1614 (2000)

43.23 R.W. Bianco, R.P. Gallegos, A.L. Rivard, J. Voight, A.P. Dalmasso: Animal models for cardiac research. In: *Handbook of Cardiac Anatomy, Physiology, and Devices*, ed. by P.A. Iaizzo (Springer, New York 2009) pp. 393–410

43.24 J.Y. Chai: Medical device regulation in the United States and the European Union: A comparative study, Food Drug Law J. **55**, 57–80 (2000)

43.25 D.S. Baim, D. Wahr, B. George, M.B. Leon, J. Greenberg, D.E. Cutlip, U. Kaya, J.J. Popma, K.K.L. Ho, R.E. Kuntz: Randomized trial of a distal embolic protection device during percutaneous intervention of saphenous vein aorto-coronary bypass grafts, Circulation **105**, 1285–1290 (2002)

43.26 J.G. Webb, R.G. Carere, R. Virmani, D. Baim, P.S. Teirstein, P. Whitlow, C. McQueen, F.D. Kolodgie, E. Buller, A. Dodek, J. Mancini, S. Oesterle: Retrieval and analysis of particulate debris after saphenous vein graft intervention, J. Am. Coll. Cardiol. **34**, 468–475 (1999)

43.27 A. Cribier, H. Eltchaninoff, C. Tron, F. Bauer, C. Agatiello, L. Sebagh, A. Bash, D. Nusimovici, P.Y. Litzler, J.-P. Bessou, M.B. Leon: Early experience with percutaneous transcatheter implantation of heart valve prosthesis for the treatment of end-stage inoperable patients with calcific aortic stenosis, J. Am. Coll. Cardiol. **43**, 698–703 (2004)

43.28 E. Grube, J.C. Laborde, U. Gerckens, T. Felderhoff, B. Sauren, L. Buellesfeld, R. Mueller, M. Menichelli, T. Schmidt, B. Zickmann, S. Iversen, G.W. Stone: Percutaneous implantation of the CoreValve self-expanding valve prosthesis in high-risk patients with aortic valve disease: The Siegburg first-in-man study, Circulation **114**, 1616–1624 (2006)

43.29 J. Ye, A. Cheung, S.V. Lichtenstein, S. Pasupati, R.G. Carere, C.R. Thompson, A. Sinhal, J.G. Webb: Six-month outcome of transapical transcatheter aortic valve implantation in the initial seven patients, Eur. J. Cardiothorac. Surg. **31**, 16–21 (2007)

43.30 E. Grube, G. Schuler, L. Buellesfeld, U. Gerckens, A. Linke, P. Wenaweser, B. Sauren, F.-W. Mohr, T. Walther, B. Zickmann, S. Iversen, T. Felderhoff, R. Cartier, R. Bonan: Percutaneous aortic valve replacement for severe aortic stenosis in high-risk patients using the second-and current third-generation self-expanding CoreValve prosthesis device success and 30-day clinical outcome, J. Am. Coll. Cardiol. **50**, 69–76 (2007)

43.31 D.F. Williams (Ed.): *Definitions in Biomaterials* (Elsevier, New York 1987)

43.32 B.D. Ratner, A.S. Hoffman (Eds.): *Biomaterials Science*, 2nd edn. (Elsevier, London 2004)

43.33 F.J. Schoen, S.H. Goodenough, M.I. Ionescu, N.S. Braunwald: Implications of late morphology of Braunwald–Cutter mitral heart valve prostheses, J. Thorac. Cardiovasc. Surg. **88**, 208–216 (1984)

43.34 F.J. Schoen, R.J. Levy: Calcification of tissue heart valve substitutes: Progress toward understanding and prevention, Ann. Thorac. Surg. **79**, 1072–1080 (2005)

43.35 D.L. Hayes, D.R. Holmes Jr., J. Merideth, M.J. Osborn, R.E. Vlietstra, S.A. Neubauer: Bipolar tined polyurethane ventricular lead: A four-year experience, Pacing Clin. Electrophysiol. **8**, 192–196 (1985)

43.36 K.B. Stokes, T. Church: Ten-year experience with implanted polyurethane lead insulation, Pacing Clin. Electrophysiol. **9**, 1160–1165 (1986)

43.37 J.S. Hanson: Sixteen failures in a single model of bipolar polyurethane-insulated ventricular pacing lead: A 44-month experience, Pacing Clin. Electrophysiol. **7**, 389–394 (1984)

43.38 R.E. Phillips, M.C. Smith, R.J. Thoma: Biomedical applications of polyurethanes: Implications of failure mechanisms, J. Biomater. Appl. **3**, 207–227 (1988)

43.39 G.H. Crossley: Cardiac pacing leads, Cardiol. Clin. **18**, 95–112 (2000)

43.40 H.P. Chin, E.C. Harrison, D.H. Blankenhorn, J. Moacanin: Lipids in silicone rubber valve prostheses after human implantation, Circulation **43**, 51–56 (1971)

43.41 A. Prasad, C.S. Rihal, R.J. Lennon, H.J. Wiste, M. Singh, D.R. Holmes Jr.: Trends in outcomes after percutaneous coronary intervention for chronic total occlusions: A 25-year experience from the Mayo Clinic, J. Am. Coll. Cardiol. **49**, 1611–1618 (2007)

43.42 P. Johansen: Mechanical heart valve cavitation, Expert Rev. Med. Device **1**, 95–104 (2004)

43.43 P.S. Mykén, O. Bech-Hansen: A 20-year experience of 1712 patients with the Biocor porcine bioprosthesis, J. Thorac. Cardiovasc. Surg. **137**, 76–81 (2009)

43.44 K.T. Ulrich, S.D. Eppinger (Eds.): *Product Design and Development* (McGraw-Hill, Boston 1995)

43.45 R.L. McNamara, L.J. Tamariz, J.B. Segal, E.B. Bass: Management of atrial fibrillation: Review of the evidence for the role of pharmacologic therapy, electrical cardioversion, and echocardiography, Ann. Intern. Med. **139**, 1018–1033 (2003)

43.46 L.H. Curtis, M.A. Greiner, B.G. Hammill, J.M. Kramer, D.J. Whellan, K.A. Schulman, A.F. Hernandez: Early and long-term outcomes of heart failure in elderly persons, 2001–2005, Arch. Intern. Med. **168**, 2481–2488 (2008)

43.47 D. Levi, D. Marelli, M. Plunkett, J. Alejos, J. Bresson, J. Tran, C. Eisenring, A. Sadeghi, A. Galindo, D. Fazio, A. Gupta, C. Burch, B. George, H. Laks: Use of assist devices and ECMO to bridge pediatric patients with cardiomyopathy to transplantation, J. Heart Lung Transplant. **21**, 760–770 (2002)

43.48 F.A. Arabía, P.H. Tsau, R.G. Smith, P.E. Nolan, V. Paramesh, R.K. Bose, D.S. Woolley, G.K. Sethi, B.E. Rhenman, J.G. Copeland: Pediatric bridge to heart transplantation: Application of the Berlin Heart, Medos and Thoratec ventricular assist devices, J. Heart Lung Transplant. **25**, 16–21 (2006)

43.49 W.H. Hassanein, L. Zellos, T.A. Tyrrell, N.A. Healley, M.D. Crittenden, V. Birjiniuk, S.F. Khuri: Continuous perfusion of donor hearts in the beating state extends preservation time and improves recovery of function, J. Thorac. Cardiovasc. Surg. **116**, 821–830 (1998)

43.50 N. Piazza, R. Bonan: Transcatheter mitral valve repair for functional mitral regurgitation: Coronary sinus approach, J. Interv. Cardiol. **20**, 495–508 (2007)

43.51 M.C. Morice, P.W. Serruys, J.E. Sousa, J. Fajadet, E. Ban Hayashi, M. Perin, A. Colombo, G. Schuler, P. Barragan, G. Guagliumi, F. Molnàr, R. Falotico, RAVEL Study Group: A randomized comparison of a sirolimus-eluting stent with a standard stent for coronary revascularization, N. Engl. J. Med. **346**, 1773–1780 (2002)

43.52 W.H. Maisel: Unanswered questions – Drug-eluting stents and the risk of late thrombosis, N. Engl. J. Med. **356**, 981–984 (2007)

44. Functional Electrical Stimulation in Rehabilitation and Neurorehabilitation

Kei Masani, Milos R. Popovic

Individuals with spinal cord injury and stroke may be unable to voluntarily move different body parts and perform functional movements. However, as long as the nerves innervating the muscles and the joints are intact, electrical stimulation can be used to generate joint movements by contracting muscles that actuate the joint. The electrical stimulation used for this purpose is called neuromuscular electrical stimulation (NMES). An organized and patterned NMES that aims to generate coordinated limb or body movements instead of isolated muscle contractions is called functional electrical stimulation (FES). The FES technology is not only used as a prosthetic device, as a permanently used assistive technology, but has also recently been used as a therapeutic tool. This chapter summarizes the basics of electrical stimulation, discusses various prosthetic applications including walking, standing, reaching/grasping, bladder voiding, respiration and hearing, and describes therapeutic applications. The dramatic progress made taking advantages of recent technologies such as miniaturization of electronic components, new sensory systems, new delivery methods for electrical stimulation is introduced.

44.1 **The Basis of Electrical Stimulation** 878
44.1.1 History – The Use of Electrical Stimulation in Medicine 878
44.1.2 Physiology 878
44.1.3 Technology 879
44.1.4 Muscle Fatigue 882

44.2 **Neuroprosthetic Use of FES** 883
44.2.1 Neuroprosthesis for Drop Foot 883
44.2.2 Neuroprosthesis for Walking 883
44.2.3 Neuroprostheses for Standing 885
44.2.4 Neuroprostheses for Reaching and Grasping 885
44.2.5 Neuroprostheses for Bladder Function 887
44.2.6 Neuroprostheses for Respiration .. 887
44.2.7 Neuroprostheses for Hearing 888
44.2.8 Therapeutic FES Besides FES Therapy 888

44.3 **FES Therapy** 889

44.4 **Other Uses of Electrical Stimulation** 890
44.4.1 Pressure Ulcers 890
44.4.2 Diabetes 890

44.5 **Concluding Remarks** 890

References 890

Individuals with spinal cord injury (SCI) and stroke have injuries that prevent the central nervous system from generating the desired motor command and/or from transmitting the desired motor command to the parts of the peripheral nervous system that innervate muscles. As a result, these individuals are frequently unable to voluntarily move different body parts and perform functions such as sitting, standing, reaching, grasping, and bladder voiding. However, as long as the nerves innervating the muscles, the muscles themselves, and the joints and soft tissues supporting the muscle-joint structures are intact, electrical stimulation can be used to generate joint movements by contracting muscles that actuate the joint. The electrical stimulation used for this purpose is called neuromuscular electrical stimulation (NMES). An organized and patterned NMES that aims to generate coordinated limb or body movements instead of isolated muscle contractions is called functional electrical stimulation (FES). One possible application of FES technology is to artificially generate body movements such as grasping and walking. In such a context, the FES technology is used as a prosthetic device; in the literature, this use of FES technology is referred to as neuroprosthesis.

Up to now, two prevalent modes of applying FES technology have emerged. In one application, FES is used to develop neuroprostheses that are used to permanently substitute the impaired function [44.1–4], i. e., a consumer that needs the neuroprosthesis would use the device each time he/she wants to generate a desired function. Good examples of such devices are neuroprostheses for hearing (i. e., cochlear implants) [44.5–7], grasping [44.8–10], bladder voiding [44.11–13], and walking [44.14–16]. In the second application, the FES technology is used as a therapy to retrain motor functions such as the FES cycle ergometer [44.17–22], which is used to improve and/or maintain muscle volume and the cardiovascular system, grasping [44.23–29], reaching [44.23,25,26,28,29], and walking [44.30–32]. In other words, FES therapy is used as a short-term intervention to help the central nervous system of the consumer to relearn how to execute these functions voluntarily instead of making the consumer dependent on neuroprostheses for the rest of his/her life.

These are two very different applications of the FES technology, which are not mutually exclusive. On contrary, one should first use the FES technology as a short-term therapy to restore as much voluntary function as possible, and only if an adequate return of function is not accomplished with FES therapy should clinicians resort to permanent FES systems (i. e., neuroprostheses) that the consumer would use as a permanent assistive device.

44.1 The Basis of Electrical Stimulation

44.1.1 History – The Use of Electrical Stimulation in Medicine

In 46 AD, Scribonius Largus described what may be considered the first neuroprosthesis: the torpedo ray, which was capable of generating an electric potential of 25–30 V [44.34]. For centuries, torpedoes were prescribed for all sorts of ailments, including headaches, hemorrhoids, and even mental illness.

Following the invention of the electrostatic generator in the late 17th century, electrical discharges were found to excite animal muscles. Stronger discharges, and hence stronger biological responses, were made possible when the first capacitor was invented in 1745. Physicians began treating a wide range of diseases by applying electrical discharges to their patients. Benjamin Franklin pioneered some of these techniques.

In 1791, Luigi Galvani published his discovery that dissected frog legs could be stimulated by touching a bimetallic rod to nerve and muscle. Michael Faraday built the first electric generator in 1831. It introduced the possibility of applying a series of high-frequency electrical pulses to nerves, which is the basis for all modern electrical stimulation. G.B. Duchenne utilized Faradism extensively in the latter part of the 19th century to treat various neurological disorders. Duchenne developed electrodes for localizing currents, and he produced a set of maps of the body indicating locations called motor points, where electrodes can be positioned to excite specific muscles.

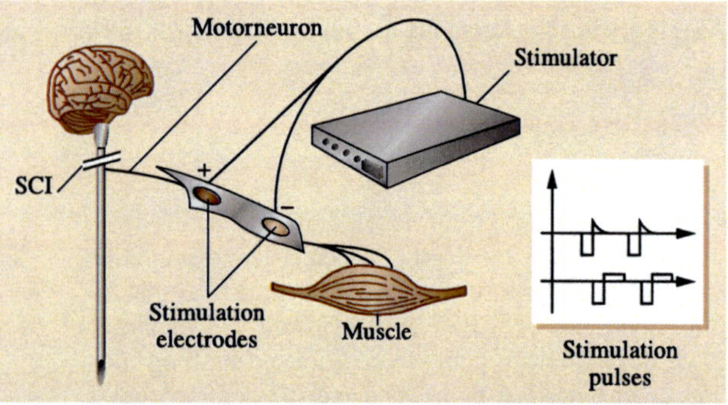

Fig. 44.1 A concept of FES using surface electrodes. The FES system causes a muscle contraction by electrically stimulating the motor axons that are connected to the muscles. The electrical stimulation generates action potentials in the motor neurons, which propagate along the motor neurons toward the muscle. When the action potentials reach the muscle, they cause the muscle to contract (after [44.33])

44.1.2 Physiology

In nerve cells, information is coded and transmitted as a series of electrical impulses called action potentials, which represent a brief change in the cell electric potential of approximately 80–90 mV. Nerve signals are frequency modulated; that is, the number of action potentials that occur in a unit of time is proportional to the intensity of the transmitted signal. Typical action potential frequency is between 4 and 12 Hz. An electrical stimulation can artificially elicit this action potential by changing the electric potential of a nerve cell or a nerve axon by inducing electrical charge into the cell (Fig. 44.1).

The stimulated nerve bundle includes motor nerves (efferent nerves, descending nerves from the central nervous system to muscles) and sensory nerves (afferent nerves, ascending nerves from sensory organs to the central nervous system). In some applications, FES can be used to directly stimulate muscles, where peripheral nerves have been severed or damaged (i. e., denervated muscles) [44.35]. However, the majority of FES systems used today stimulate the nerves or the points where the junction between the nerve and the muscle occurs. The main reason for this is the fact that direct muscle stimulation requires considerably more energy to generate contractions (at least three orders of magnitude more [44.36]), which makes these systems more challenging to implement at home and in clinical settings. An electrical charge can stimulate both motor and sensory nerves. In some applications, the nerves are stimulated to generate localized muscle activity, i. e., the stimulation is aimed at generating direct muscle contraction. In some other applications, stimulation is used to activate simple or more complex reflexes. In other words, the afferent nerves that evoke a reflex are stimulated, which is typically expressed as a coordinated contraction of one or more muscles in response to the sensory nerve stimulation.

When a nerve is stimulated, i. e., when sufficient electrical charge is provided to a nerve cell, a localized depolarization of the cell wall occurs, resulting in an action potential that propagates toward both ends of the axon. Typically one *wave* of action potentials will propagate along the axon towards the muscle (orthodromic propagation) and concurrently the other *wave* of action potentials will propagate towards the cell body in the central nervous system (antidromic propagation). Note that while the direction of propagation in the case of the antidromic stimulation and the sensory nerve stimulation is the same, i. e., towards the central nervous system, their end effects are very different. The antidromic stimulus has been considered an irrelevant side effect of FES, although in recent years a hypothesis has been presented that suggests a potential role of the antidromic stimulation in neurorehabilitation [44.37]. Typically, FES is concerned with orthodromic stimulation and uses it to generate coordinated muscle contractions.

In the case when sensory nerves are stimulated, the reflex arcs are triggered by the stimulation on sensory nerve axons at specific peripheral sites. One example of such a reflex is the flexor withdrawal reflex. The flexor withdrawal reflex occurs naturally when a sudden, painful sensation is applied to the sole of the foot. It results in flexion of the hip, knee, and ankle of the affected leg, and extension of the contralateral leg in order to get the foot away from the painful stimulus as quickly as possible. The sensory nerve stimulation can be used to generate desired motor tasks, such as evoking flexor withdrawal reflex to facilitate walking in individuals following stroke, or they can be used to alter reflexes or the function of the central nervous system. In the later case, electrical stimulation is commonly described by the term *neuromodulation*.

44.1.3 Technology

Nerves can be stimulated using current or voltage regulated pulses. Although both types of stimulation pulses are effective, current regulated pulses have a few advantages over voltage regulated stimulation pulses. One advantage is that changes in the tissue resistance (which occur all the time, in particular with surface stimulation systems) do not affect the amount of charge that is delivered to the nerves. In other words, the current regulated pulses always deliver the same amount of charge to the targeted tissue. This in turn reduces the need for persistent stimulation amplitude adjustments, which is a common problem with voltage regulated stimulators. Another major advantage is that in current regulated stimulation pulses one can accurately control how much charge is delivered to the tissue and how much of the delivered charge is extracted back from the tissue. This is an important feature because accumulation of charge may cause galvanic processes that can damage the tissue. Current regulated stimulation pulses, if applied so that they are balanced and biphasic, will ensure that the delivered charge is extracted out of the targeted tissue at the end of every single stimulation pulse.

The stimulation pulses can be either monophasic or biphasic. When the pulses are monophasic, they essentially deliver the charge to the tissue, without an attempt to remove the charge from the tissue. When the pulses are biphasic, the charge is delivered to the tissue but is also removed from it. If the biphasic pulses are balanced, the amount of charge delivered to the tissue is removed from it during the same pulse. If the biphasic pulse is unbalanced, residual charge is left in a tissue. The balanced biphasic pulses can be symmetric or asymmetric. Present day sophisticated FES systems apply asymmetric balanced biphasic pulses to ensure that muscle contractions occur only in the mus-

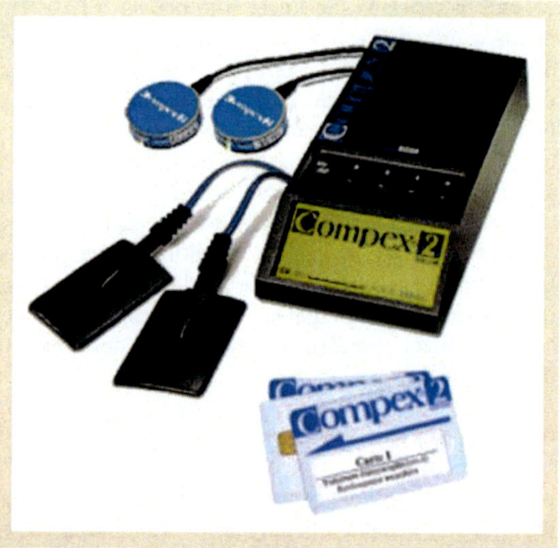

Fig. 44.2 Portable FES system with surface electrodes. This FES system is programable using the memory chip cards and can make flexible stimulation patterns with four channels (after [44.38])

cles over which the anode electrode is placed. If the pulses are symmetrical balanced biphasic, both muscles under cathode and anode may be contracted.

Nerves can be stimulated using either surface (transcutaneous) or subcutaneous (percutaneous or implanted electrodes). Surface electrodes are placed on the skin surface above the nerve or muscle that needs to be *activated* (Fig. 44.2). They are noninvasive, easy to apply, and generally inexpensive. Due to the electrode–

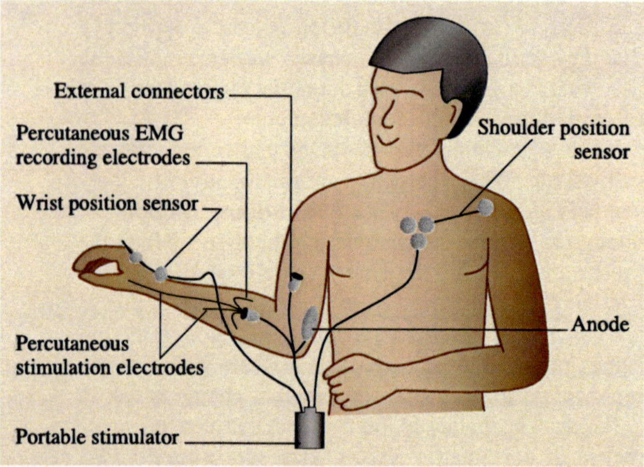

Fig. 44.3 Scheme of a percutaneous hand neuroprosthesis (after [44.39])

skin contact impedance, skin and tissue impedance, and current dispersion during stimulation, much higher intensity pulses are required to stimulate nerves using surface stimulation electrodes as compared to the subcutaneous electrodes. A major limitation of the transcutaneous electrical stimulation is that some nerves, for example, those innervating the hip flexors, are too profound to be stimulated using surface electrodes.

Subcutaneous electrodes can be divided into percutaneous and implanted electrodes. The percutaneous electrodes consist of thin wires inserted through the skin and into muscular tissue close to the targeted nerve. These electrodes remain in place for a short period of time and are only considered for short-term FES interventions (Fig. 44.3). One of the drawbacks of using percutaneous electrodes is they are prone to cause infection and special care must be taken to prevent such events.

The other class of subcutaneous electrodes is implanted electrodes. These are permanently implanted in the consumer's body and remain in the body for the remainder of the consumer's life (Fig. 44.4). Compared to surface stimulation electrodes, implanted and percutaneous electrodes potentially have higher stimulation selectivity with much less electrical charge applied, both of which are desired characteristics of FES systems. To achieve higher selectivity while applying lower stimulation amplitudes, it is recommended that both cathode and anode be in the vicinity of the nerve that is stimulated. The drawbacks of implanted electrodes are they require an invasive surgical procedure to install, and, as is the case with all surgical intervention, there exists a possibility of infection following implantation. The implanted electrodes come in different shapes and sizes. The most notable implanted electrodes are:

1. Intramuscular electrodes [44.40]
2. Nerve cuff electrodes [44.41, 42]
3. Epimysial electrodes [44.1, 2]
4. A miniature electrode-stimulator system called BION, which is implanted using a hypodermic needle [44.8, 43] (Fig. 44.5).

FES systems come in many different shapes and sizes and serve many different purposes. The common components in all FES systems are:

1. A stimulator that consists of a power source, some type of microcontroller, and a stimulator output stage that produces electrical pulses
2. A user interface
3. Electrodes.

Most modern FES systems use batteries, disposable or rechargeable, as the power source. Implantable FES systems are as big as a pacemaker and surface FES systems are typically not bigger than a hand-held voltmeter. Various user interfaces have been applied to control FES systems. The simplest are push buttons [44.38,44], joysticks [44.45], force sensitive resistors, potentiometers [44.46], accelerometers, inclinometers [44.30], and proximity sensors [44.1]. The next level in complexity are the FES systems that are controlled using electromyography (EMG) signals. EMG-based user interfaces can be proportional (i. e., as the EMG signal intensity increases the FES stimulus increases proportionally) [44.33] or EMG triggered (i. e., as the EMG signal exceeds a particular threshold the FES stimulus is delivered) [44.33]. Some sophisticated EMG triggered user interfaces use complex EMG patterns to activate the FES systems [44.38], but these are seldom used in clinical practice. The main challenge with using the EMG controlled FES systems is the fact that the electrical stimulation induces voltage noise in the body, which is often significantly larger in magnitude than the EMG signal itself [44.33]. Hence, recording an EMG signal from a muscle at the same time that the same or neighboring muscle is stimulated is not a trivial task. Finally, voice control (our unpublished results), brain machine interfaces [44.45,47], sophisticated gate phase detection sensors [44.48], and reaching motion detection sensors [44.49] have been used to control FES systems.

As discussed earlier in this article, FES can be used for neuroprosthetic and therapeutic purposes. If FES is used as a neuroprosthesis, the purpose of this device is to generate a body function that the consumer is unable to perform him/herself, such as walking, biking, bladder voiding, grasping, etc. In this application the FES system needs to be worn/used each and every time the consumer needs to perform the desired function. In essence, the consumer uses the FES device as a permanent orthotic system. Some of these neuroprosthetic systems are also used for cardiovascular conditioning and muscle strengthening. Although the ultimate goal of this type of application is therapeutic, this is not FES therapy (FES therapy is discussed later in this chapter). Good examples of these FES systems are neuroprostheses for rowing and biking. Each time the consumer would like to row or bike he/she needs to use the neuroprosthetic system, without which he/she would not be able to perform this task at all.

When FES technology is used to deliver FES therapy, the purpose of that intervention is to restore a voluntary function. In other words, FES is used only temporarily as a short-term intervention with the objective to help the neuromuscular system relearn to execute a function that has been impaired due to neurological injury or disorder. In this application, the ultimate goal of the FES intervention is to recover voluntary functions of the consumer as much as possible, so that the consumer does not need to use the FES system for the rest of his/her life. In this application, the central nervous system essentially relearns how to control the impaired muscles and how to contract them in a temporarily appropriate manner to generate desired body function. To date, FES therapies have been used to restore walking functions [44.30,31,50–52] and voluntary reaching/grasping [44.23,25,26,28,29,32,53,54].

Implanted FES systems are primarily used as permanent neuroprostheses, and some attempts have been

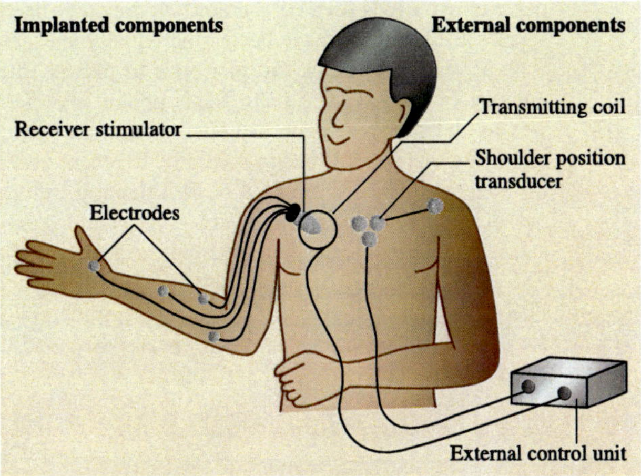

Fig. 44.4 Scheme of an implant electrode (after [44.39])

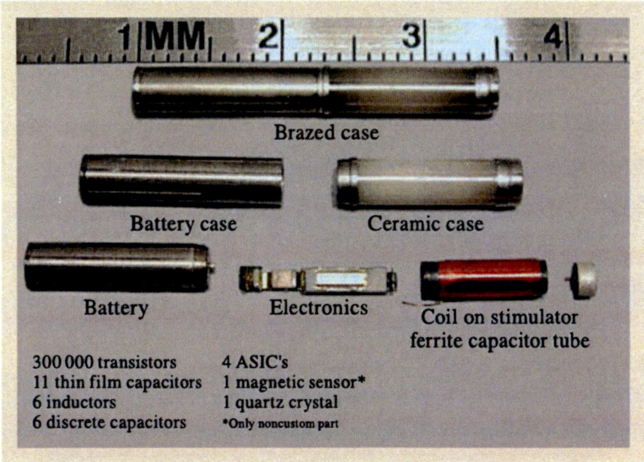

Fig. 44.5 BION implant electrodes (after [44.43])

made to use BION for FES therapy. On the other hand, surface FES systems have been used equally successfully as neuroprostheses and platforms to deliver FES therapy. In the past, the main focus of the FES field was on developing neuroprosthetic systems, in particular those that patients had to use daily. In recent years, the advances made in the field of FES therapy and the use of neuroprostheses for muscle strengthening and cardiovascular exercises have shifted the focus of the FES field, at least partially, towards the use of surface FES systems. As a result, a number of commercially available surface FES systems have been developed in last decade.

44.1.4 Muscle Fatigue

While FES has succeeded in assisting individuals with neuromuscular disorders, there is one major limitation of FES: stimulated muscles tend to fatigue rapidly, thus limiting the benefits of FES [44.55]. The problem of muscle fatigue is further compounded by the fact that paralyzed muscles show greater fatigability than healthy ones [44.56–61]. Two basic mechanisms of muscle fatigue are known: central and peripheral [44.62]. Central fatigue results from the failure of motor neuron excitation, whereas peripheral fatigue results from failure of transmission of neural signals or of the muscle to respond to neural excitation [44.63, 64]. Two major sources of peripheral fatigue are:

1. Electromechanical coupling related to propagation of action potentials within the muscle membrane and conduction of the sarcoplasmic reticulum to Ca^{2+} ions
2. Phosphometabolite concentration affecting actin-myosin cross-bridge creation and Ca^{2+}-troponin kinetics [44.65].

In the case of individuals with neuromuscular disorders, especially with SCI and stroke, the central component of fatigue is minor or absent. In this population, fatigue results mostly from within the motor unit, such as depletion of substances, accumulation of catabolites, or problems in excitation–contraction coupling [44.55]. Rapid fatigue in electrically stimulated muscle can be attributed to two major factors: mode of activation, and reverse recruitment order.

Mode of Activation

In electrically induced contractions, the same nerve fibers are usually stimulated all the time, and all the motor units are fired synchronously. This is contrary to what occurs in normal muscle movement, where the work during contraction is shared between different motor units of the same muscle [44.66]. Thus, in stimulated muscles, a synchronized and massive fiber contraction replaces the normal physiological mechanism of motor unit recruitment and firing rate regulation [44.67].

Reverse Recruitment

Normal muscles are usually recruited in an *orderly pattern*, known as the *size principle* of motor neurons, i. e., small alpha-motor neurons (small diameter axons) innervating slow motor units are recruited at a lower excitation level than larger alpha-motor neurons (large diameter axons) innervating fast units [44.68, 69]. Thus, in a normal skeletal muscle containing a mixture of muscle fiber types, low force tasks requiring fatigue resistance, such as force generation to maintain posture, are accomplished by recruiting the small motor units. The rapidly fatiguing fast motor units are recruited only at higher and more dynamic force levels. FES is characterized by a *reverse recruitment*, whereby large axons that innervate the more easily fatigable fibers are recruited at low stimulus magnitudes, and the smaller axons follow with increased stimulation levels [44.2, 39]. Due to these two factors, the large motor units are fired synchronously and at stimulation frequencies above the usual physiological values in normal consumers, which results in a rapid reduction of the muscle force, i. e., rapid fatigue [44.6, 49, 55, 64, 70–73].

Prevention of Fatigue in FES

To date, different methods have been proposed to mediate the muscle fatigue induced by FES such as modulation of stimulation parameters [44.74–76], variable frequency trains [44.70, 77–82], random stimulation trains [44.73], and sequential stimulation [44.83, 84]. Research done in the field of *stimulation parameters modulation* suggests that using lower frequency stimulations with longer pulse durations is one possible approach to reduce the FES induced muscle fatigue [44.75]. Also, the variation of the stimulation parameters over the time course of stimulation, in both an open-loop [44.85–89] and a closed-loop manner [44.90, 91], has also been found to be helpful. Some researchers have proposed the use of *variable frequency trains*, i. e., the interpulse intervals are not kept constant, as a means of remedying the muscle fatigue problem, but up to now, both positive and negative outcomes have been reported. Some time ago, it was suggested to use *random stimulation trains* (randomize pulse duration or pulse frequency or pulse amplitude) to reduce muscle fatigue. However, a recent systematic trial has demonstrated that this method is not effective [44.51,73]. The last method that was investigated as a means to address FES induced muscle fatigue is *sequential stimulation*. In this particular method, different muscle compartments are stimulated in asynchronous manner, i. e., each muscle compartment is activated at another stimulation interval and these compartments are contracted with the frequency rate much lower than if continuous FES were applied. This stimulation method mimics, in a way, the natural muscle contraction patterns observed in healthy individuals. Our recent experiments with sequential stimulation (yet unpublished results) suggest that this approach has great potential.

44.2 Neuroprosthetic Use of FES

44.2.1 Neuroprosthesis for Drop Foot

Liberson and colleagues developed a simple neuroprosthesis to correct drop foot [44.92]. The drop foot is a common symptom in hemiplegia, characterized by a lack of dorsiflexion during the swing phase of gait, resulting in short, shuffling strides. Liberson's device, which has the distinction of being the first neuroprosthesis to receive a patent, consisted of a power supply worn on a belt, two surface electrodes positioned for stimulation of the common peroneal nerve, and a heel switch. The stimulation was activated whenever the heel lost contact with the ground, and was deactivated when the heel regained contact with the ground. Stimulation of the common peroneal nerve causes contraction of the muscles responsible for dorsiflexion (i. e., tibialis anterior and extensor hallucis longus, among others. Following Liberson's invention and Vodovnik's revisions, a number of drop foot stimulators were developed. The first commercially available implanted drop foot stimulator was developed by Rancho Los Amigos Medical Center and Medtronic Inc. [44.93]. The surgically implanted components were a radio-frequency receiver, a pulse train generator, and one bipolar electrode implanted adjacent to the peroneal nerve. An external unit worn on the belt delivered power via the RF coil and received input commands from a wireless foot switch. Despite some problems with electrode migration and infections, the device was considered a success.

The Odstock Dropped Foot Stimulator [44.31], the WalkAide [44.30] and the NESS L300 (formally known as NESS Drop Foot System) [44.94] are new generations of drop foot stimulators that use surface FES technology; they are commercially available and have been FDA (USA Food and Drug Administration) approved. The ActiGait [44.15] and the STIMuSTEP [44.95] are implantable drop foot stimulators that are also commercially available and have the CE mark in Europe (Fig. 44.6).

Most modern drop foot stimulators continue to use a heel switch for active input. *Burridge* et al. tried using the foot switch on the unaffected leg, but found that it was not preferable as the patient was unable to reliably achieve heel contact on the affected leg [44.50]. Other alternatives to the heel switch include a heel/toe switch [44.3], an array of four single-axis accelerometers positioned on the shank [44.96], a tilt sensor positioned on the shank [44.97], and the Gait Phase Detection System [44.48] (a derivative of this system is presently used to detect gait phases in the above knee powered orthosis Power Knee from Ossur hf – personal communication). Drop foot stimulators are one of most successful neuroprosthesis to date. Overall, consumer perception of drop foot stimulators is they are superior to the ankle-foot orthosis [44.98].

44.2.2 Neuroprosthesis for Walking

As early as 1960, *Kantrowitz* demonstrated paraplegic standing by applying continuous electrical stimulation

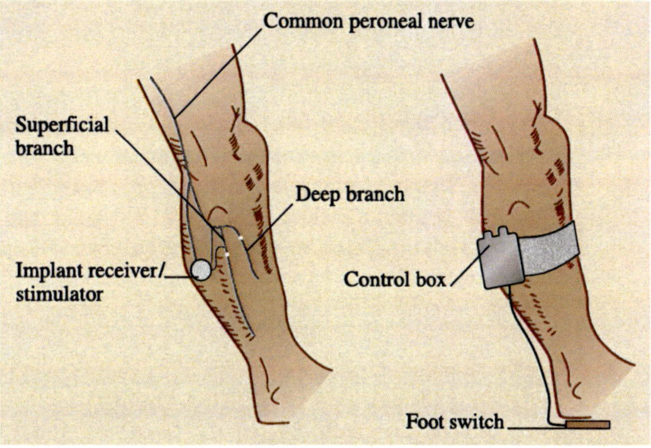

Fig. 44.6 Scheme of STIMuSTEP implant drop foot stimulator. The FES system allows stimulation of the common peroneal nerve (after [44.39])

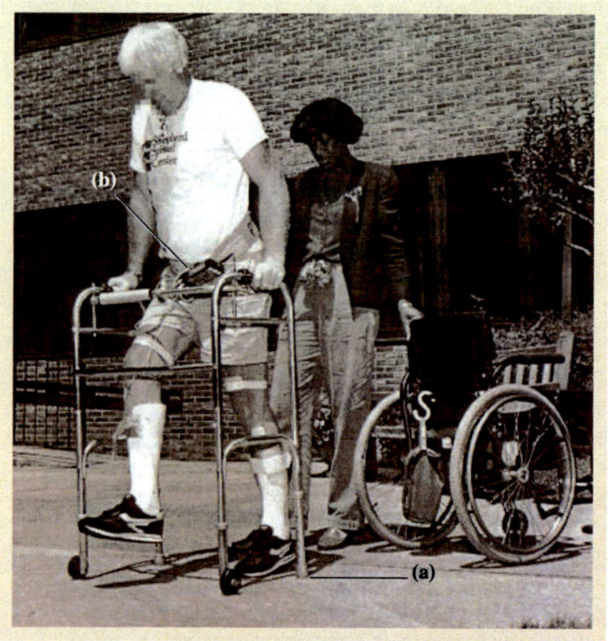

Fig. 44.7 The Parastep electrical stimulation system (after [44.101])

to the quadriceps and gluteus maximus muscles of a patient with complete spinal cord injury, using surface FES technology [44.99]. This earliest neuroprosthesis for paraplegic gait provided continuous stimulation to the quadriceps to produce a mode of gait similar to long leg-brace walking, by inducing stiffened legs. Later systems used alternating bilateral quad/glut stimulation (during the stance phase) out of phase with peroneal nerve stimulation to induce the flexor withdrawal reflex (during swing phase) [44.3]. Later, *Kralj* and colleagues described a technique for paraplegic gait using surface stimulation, which remains the most popular method in use today [44.100]. Electrodes are placed over the quadriceps muscles and peroneal nerves bilaterally. The user controls the neuroprosthesis with two push buttons attached to the left and right handles of a walking frame, or on canes or crutches. When the neuroprosthesis is turned on, both quadriceps muscles are stimulated to provide a standing posture. The left button initiates a swing phase in the left leg by briefly stopping stimulation of the left quadriceps and stimulating the peroneal nerve. This stimulation is applied suddenly, so as to trigger the flexor withdrawal reflex, resulting in simultaneous hip and knee flexion, as well as dorsiflexion. After a fixed period of time, peroneal nerve stimulation is stopped and quadriceps stimulation is initiated while the reflex is still active to complete the stride. Similarly, the right button initiates a swing phase in the right leg. Many current FES systems for walking have employed this technique as the basic concept.

As microprocessor technology developed, neuroprostheses became more portable and flexible. Examples of this type of neuroprosthesis are Parastep [44.44, 102], HAS [44.9], RGO [44.103], Praxis [44.104], and the Compex Motion neuroprosthesis for walking [44.32, 38]. The Parastep system is one of most popular products and uses Kralj's technique [44.44, 102] (Fig. 44.7), which was the first neuroprosthesis for walking to receive approval from FDA. The HAS and the RGO walking neuroprostheses are devices that also apply active and passive braces, respectively. The braces were introduced to provide additional stability during standing and walking. The Praxis system is an implantable system, which was developed to provide better stimulation selectivity and a more natural walking pattern. This system was also designed to provide a bladder-voiding function as well. Compex Motion neuroprosthesis for walking is an 8 to 16 channel surface FES system used to restore walking in stroke and SCI individuals. The system uses push button control strategy, similar to that one used in the Parastep system and a gate phase detection sensor [44.48] to trigger the FES sequences.

A major limitation of neuroprostheses for walking that are based on surface stimulation is that the gait is slow, awkward, and unnatural looking, e.g., poor stance phase posture due to hip flexion induced by rectus femoris activity when the quadriceps are stimulated. Further, the flexor withdraw reflex is subject to rapid habituation and is highly variable. Perhaps a major reason for this is that the hip flexors cannot be stimulated directly. Therefore, hip flexion during walking must come from voluntary effort, which is often absent in paraplegia, or from the flexor withdrawal reflex. Implanted systems have the advantage of being able to stimulate the hip flexors, and therefore, to provide better muscle selectivity and more natural gait patterns. Hybrid systems with exoskeleton devices have also been proposed to solve this problem [44.105]. These technologies have been found to be successful and promising, but at the present time these FES systems are mostly used for exercise purposes and seldom as an alternative to wheelchair mobility. The main reason for this is the high metabolic cost of FES assisted walking as compared to wheelchair mobility [44.39].

44.2.3 Neuroprostheses for Standing

While neuroprostheses for walking assists in keeping a standing posture in individuals with SCI, several FES systems and studies especially focused on maintaining the standing posture because standing itself is advantageous for those patients. The Case Western Reserve University/Department of Veterans Affairs (CWRUVA) standing neuroprosthesis uses an implanted stimulator with 8–16 channels to bilaterally activate muscles of the trunk, hips, and thighs making the body stiff in an upright posture. With external orthoses, such as walkers, which are used to stabilize the body, CWRUVA can offer more than 10 min of uninterrupted upright stance [44.4, 106, 107]. However, this system does not provide balance control, and the consumer has to rely on a walker or a similar device to maintain balance. In addition, this device uses an open-loop control scheme, which tends to result in quicker fatigue. The second generation of this system is presently under development [44.108].

Despite numerous attempts to develop a neuroprosthesis for unsupported, arm-free standing, a device that is actually capable of performing this task does not yet exist. To realize this technology, a closed-loop multiple joint control system is required. So far, most of the FES systems for unsupported standing that have been applied to individuals with neurological disorders were focused on stimulating ankle joints. *Hunt* et al. [44.109], *Holderbaum* et al. [44.71], and *Gollee* et al. [44.110] developed and evaluated a series of closed-loop nested feedback control systems that used FES to actuate the ankle torque of a paraplegic subject with a complete lesion at T7/8. During the experiments, the subject was standing in an apparatus that acted as a full body cast, allowing only the ankle joints to move in the anterior–posterior direction. *Abbas* and *Chizeck* [44.111] compared the performance of a closed-loop feedback system with an open-loop stimulation strategy for two paraplegic subjects (T7 and T9) during standing; both systems regulated the movement of the hip joints in the coronal plane. *Vette* et al. [44.112, 113] applied a liner feedback controller mimicking a physiological controller and succeeded in better stabilizing a patient's standing balance. *Kim* et al. [44.114, 115] recently performed a series of dynamic kinematics and stability analyses suggesting that FES assisted unsupported stance is feasible, as long as 6 degrees of freedom in the leg joints (out of 12) are controlled with the FES system. This work still needs to be properly validated in experiments with individuals with complete SCI.

44.2.4 Neuroprostheses for Reaching and Grasping

The neuroprostheses for grasping and reaching are FES systems designed to restore or improve function in stroke and tetraplegic individuals. Some notable grasping and/or reaching neuroprostheses are the Freehand system [44.116], the NESS H200 (formally known as NESS Handmaster) [44.117], the Bionic Glove [44.46], the ETHZ-ParaCare neuroprosthesis for grasping [44.33, 38], the systems developed by Rebersek and Vodovnik [44.118], the Belgrade Grasping-Reaching System [44.49], and Compex Motion neuroprosthesis for reaching and grasping [44.38]. The Freehand system is an implantable system; all the other devices use surface FES technology.

The key element for achieving the synergistic activity of muscles that results in reaching and/or grasping is the appropriate sequencing of the stimulation pulses. The available neuroprostheses for grasping can restore the two most frequently used grasping styles: the palmar and the lateral grasp. The palmar grasp is used to hold bigger and heavier objects such as cans and bottles, and the lateral grasp is used to hold smaller and thinner objects such as keys, paper, and compact discs. The lateral grasp is generated by first flexing the fingers to provide opposition, followed by the thumb flexion. The palmar grasp is generated by first forming opposition between the thumb and the palm, followed by simultaneous flexion of both the thumb and the fingers.

The neuroprosthesis developed by *Rebersek* and *Vodovnik* was one of the first neuroprosthesis for grasping [44.118]. This neuroprosthesis has three stimulation channels, which are used to generate grasping by stimulating the finger flexors and extensors and the thumb flexors. Although this device was developed almost three decades ago, it is one of the rare systems that allows a subject to control the stimulator via different sensory interfaces such as an EMG sensor, a sliding resistor, or a pressure sensor. This option is important because it allows the neuroprosthesis to be tailored to the subject. The main disadvantages of this neuroprosthesis are that donning and doffing times are long and selectivity of stimulation is quite limited. *Mereltti* et al. modified this system and used it for stroke subjects [44.119]. They applied two channels to augment the elbow and fingers/wrist extension. They concluded that the neuroprosthesis contributed greatly to recovery of hand and elbow movements in five stroke subjects, but in the remaining three the improvement was significant only at the elbow joint.

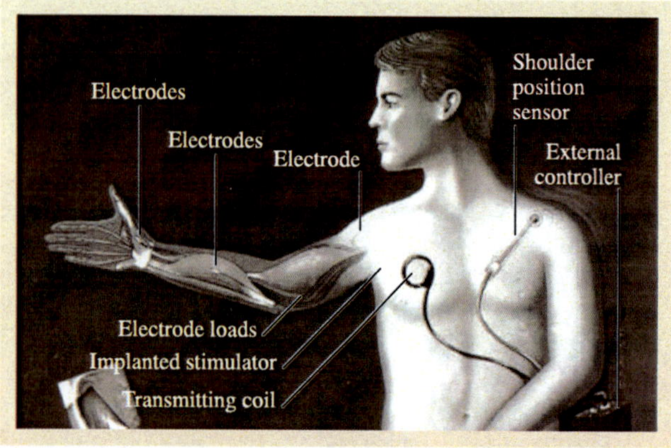

Fig. 44.8 The Freehand system (after [44.101])

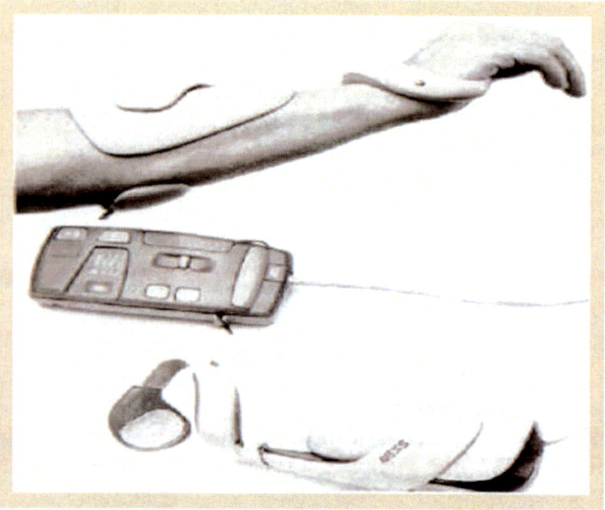
Fig. 44.9 The NESS Handmaster system (after [44.10])

The Freehand system [44.116], consists of eight implanted epimysial stimulation electrodes that stimulate flexion and extension of the fingers and the thumb in order to provide lateral and palmar grasp (Fig. 44.8). Commands are given by an external position sensor placed on the shoulder of the subject's opposite arm. An additional external switch allows the user to choose between palmar and lateral grasp. This sequence is then sent, via a radio frequency coil, to the implanted unit, which generates the stimulation sequences for each channel. The electrode leads are tunneled subcutaneously to the implanted stimulator located in the pectoral region. Surgical procedures to enhance both voluntary and stimulated hand functions are often performed in conjunction with the stimulator implantation.

The Freehand system is the first neuroprosthesis for grasping approved by the FDA. The main advantage of the Freehand system is that it is implanted, and the time needed to don and doff the system is shorter compared to most of the surface FES systems.

The NESS H200 consists of an orthosis that has built-in flexibility to enhance and control freedom of movement within the forearm and hand, while supporting the wrist joint at a functional angle of extension [44.117] (Fig. 44.9). The NESS H200 multiplexes a single channel of stimulation through a selected combination of surface electrodes on the inner surface of the orthosis; this effectively transforms the device into a three-channel neuroprosthesis. One stimulation channel is used to stimulate the extensor pollicis brevis. The second channel stimulates the flexor digitorum superficialis. The third channel stimulates the flexor pollicis longus and thenar muscle group. The setup position of the electrodes depends on the device user's specific needs. The NESS H200 is controlled with an array of push buttons allowing the subject to select the operating mode and to trigger programmed movement sequences. Using the buttons, the subject can also control stimulation intensity and thumb posture, thereby adjusting the grasp to the size and the shape of the target object. One of the major advantages of the NESS H200 is that it is easy to don and doff. It is exceptionally well designed and one of the best neuroprostheses for grasping presently available on the market.

The Bionic Glove is a neuroprosthesis designed to enhance the tenodesis grasp (by extending their wrist, users can cause passive finger flexion due to the limited length of the finger flexors) in subjects who have good voluntary control over wrist flexion and extension [44.46]. The Bionic Glove stimulates finger flexors and extensors, significantly enhancing the strength of the tenodesis grasp. Four self-adhesive surface stimulation electrodes provide stimulation, and the stimulator and a wrist position sensor are located on the forearm of the glove. An easy-to-use interface, with three push buttons on the stimulator, is used to set the stimulation parameters, and the optional audio feedback facilitates faster learning.

The Belgrade Grasping-Reaching System proposed by *Popovic* et al. is a neuroprosthesis for grasping that also provides a reaching function [44.49]. It consists of four stimulation channels, three of which are used to generate the grasping function. The fourth channel is used to stimulate the triceps brachii muscle augmenting elbow extension. Reaching is achieved by measuring the subject's shoulder velocity with a goniometer and by generating a synergistic elbow motion by stimulating the triceps brachii muscle. This neuroprosthesis, similar to the system proposed by Rebersek and Vodovnik, requires more time to don than the Handmaster.

The Compex Motion neuroprosthesis is a very flexible device designed to improve reaching, grasping, and walking functions in individuals with SCI and stroke [44.33, 38]. This multichannel surface stimulation system is programmable and can be interfaced with any sensor system. As a four-channel neuroprosthesis for grasping, it provides both palmar and lateral grasps. It can be controlled with proportional EMG, discrete EMG, push buttons, sliding resistor and brain machine interface control strategies [44.120]. By combining two Compex Motion stimulators, an eight-channel neuroprosthesis for reaching and grasping was developed [44.29]. One of the main disadvantages of this system is that it requires about 8 min to put on or take off.

It is already well established that the neuroprostheses for reaching and grasping have been successful in restoring arm and hand functions in individuals with stroke and SCI. It is the authors' opinion that this technology has much potential and in the near future, one may expect the emergence of multiple commercially available surface FES systems besides the NESS H200, as well as a revival of the Freehand system.

44.2.5 Neuroprostheses for Bladder Function

Neuroprostheses have been very successful in treating lower urinary tract dysfunctions commonly associated with SCI, such as urge incontinence and urinary retention. The first attempts to electrically stimulate the bladder were made in the 1950s, when researchers sought ways to induce bladder emptying. At that time, a bladder wall stimulator was developed and implanted in three individuals [44.121], and animal studies of pelvic nerve stimulation were carried out [44.122]. Later, it was found that electrical stimulation of the sacral anterior roots produces excellent bladder voiding, and this led to the development of the Finetech–Brindley stimulator, which is the most widely used neuroprosthesis for bladder management today [44.16].

Attempts to manage incontinence using electrical stimulation began in the 1960s [44.123]. It was found that urethral resistance could be increased by stimulating the muscles of the pelvic floor, vagina, and rectum using external electrodes [44.124]. Eventually, fully implanted systems were developed to suppress the detrusor muscle, thus preventing reflex incontinence and increasing bladder volume [44.11]. Most SCIs result in reflex incontinence. Typically, detrusor-sphincter dyssynergia develops, in which the detrusor and urethral sphincter contract simultaneously rather than reciprocally. The detrusor also becomes hyperreflexic, and the bladder becomes overactive. The standard treatments are anticholinergic medication, which blocks the neuromuscular junctions, and sensory rhizotomy (surgical transection of the posterior sacral roots). Neuroprostheses for bladder management serve as a practical alternative to these treatments. They can also augment sensory rhizotomy.

The Finetech–Brindley stimulator has been implanted in a large number of consumers, usually individuals having had a rhizotomy [44.12]. The electrodes are positioned on the second, third, and fourth sacral roots, bilaterally and extradurally. If a rhizotomy has not been performed, the electrodes must be implanted inside the dura to prevent crossover stimulation of the sensor neurons, which will trigger the detrusor reflex. A portable external controller transmits power to the implant via radio frequency coil, and the user initiates bladder voiding by pushing buttons on the external unit. Micturition is usually achieved with residual volumes of less than 50 mL, contributing to a dramatic reduction in urinary tract infections [44.125]. The Finetech–Brindley stimulator has proven to be extremely robust [44.126].

The Medtronic Interstim stimulator is a sacral root implant for incontinence, using neuromodulation to correct inappropriate reflex behavior [44.127]. It consists of fine wire electrodes inserted into the sacral foramina. When active, these electrodes inhibit the detrusor, but the mechanism of this inhibition is not yet properly understood. Thorough testing must be done using a temporary implant before permanent implantation can be recommended. The clinical success rate of this device is about 50%. Bladder emptying has to be achieved either voluntarily or by means of intermittent catheterization.

44.2.6 Neuroprostheses for Respiration

Phrenic nerve pacing is a clinically accepted technique to provide artificial ventilatory support in patients with trauma and with respiratory failure secondary to cervical SCI [44.39]. The stimulating electrodes are implanted on each phrenic nerve. Bilateral phrenic nerve stimulation results in the descent of each diaphragm and a decrease in intrathoracic pressure resulting in inspiration. Cessation of stimulation results in diaphragm relaxation, an increase in intrathoracic pressure, and exhalation.

Phrenic nerve pacing is an effective means of providing ventilatory support with an advantage over mechanical ventilation [44.128, 129], suggesting that phrenic pacing provides important health and lifestyle benefits compared to mechanical ventilation. Presently, new approaches for delivering phrenic nerve pacing are being developed and tested in a clinical setting [44.130].

44.2.7 Neuroprostheses for Hearing

Cochlear implants are neuroprostheses for the hearing impaired who have severe (70–90 dB) or profound (> 90 dB) hearing loss. A long wire electrode is implanted directly into the cochlear duct, and electrical stimulation is applied to the residual spiral ganglion cells of the cochlear nerve. Note that this neuroprosthesis primarily stimulates the auditory sensory system, and as such, does not strictly belong to the FES class of neuroprostheses. The cochlear implants were first developed in France in 1957. Since then, cochlear implants have been refined and miniaturized, and now they have received widespread acceptance, more so than any other class of neuroprostheses. More than 75 000 patients have received cochlear implants worldwide. Originally, few hearing impaired people were eligible for cochlear implantation, but as the technology has improved, the selection criteria have expanded greatly to include a wide range of hearing impairments [44.7].

Due to the success and popularity of cochlear implants, there are many different brands on the market. Most brands, however, are essentially similar. Differences between the currently available cochlear implants mainly involve the number of electrode channels (12–22), speech coding strategies, and the mode of electrode stimulation [44.5]. A recent study carried out at the University of Toronto concluded that the Clarion CI (Advanced Bionics, USA) and the Nucleus 22 (Cochlear Corp., Australia) cochlear implants were totally comparable in function and performance, although their modes of operation are not identical [44.6]. Both devices succeeded in reducing tinnitus, thereby increasing word and sentence recognition, but there was no significant reduction in vestibular function. Among the implantees, 76% reported that they were satisfied with their implants, and 96% reported an overall positive impact on quality of life. Some other brands of cochlear implant are the COMBO 40+ system (MED-EL, Austria), Digisonic (MXM, France), and the SOUNDTEC direct system (SOUNDTEC, USA), most of which are FDA approved.

Cochlear implants generally consist of:

1. An external earpiece
2. A speech processor
3. An internally implanted unit.

The earpiece, usually very small and lightweight, is worn comfortably behind the ear, much like a hearing aid. It contains an ear-level or in-ear microphone and a radio frequency coil to transmit signals to the implanted components. The speech processor can be in the form of a small box worn somewhere on the body or, in some models, it is contained in the external unit worn behind the ear. The internally implanted unit consists of a receiver coil located underneath soft tissue in a cavity drilled in the temporal squama and a 20–24 mm insulated wire ending in a multichannel electrode array, which is inserted into the cochlear duct. Sound waves are received by the external microphone and converted into electrical signals that are input into the external speech processor. There, the signals are digitally encoded and transmitted to the internal unit via radio frequency coil. The internal unit decodes the radio signals back into elementary electrical signals to stimulate each channel of the electrode array. Therefore, the multichannel device provides a complex sound analysis similar to the physiological analysis of sound in healthy individuals.

44.2.8 Therapeutic FES Besides FES Therapy

Patients with motor disabilities, such as SCI, tend to experience considerable deterioration of their physical condition and fitness. This often precipitates development of various secondary complications that ultimately result in a shorter life span. A good example of a secondary complication that results from loss of muscle tone in individuals with SCI is pressure ulcers. One possible application of FES is to help maintain or improve muscle tone in these individuals and by doing so prevent occurrence of secondary complications such as pressure ulcers or cardiovascular deconditioning. In these applications, neuroprostheses are used as a means of providing regular exercise and cardiovascular system conditioning. This, in turn, results in enhanced muscle strength, retarded muscle atrophy, and reduced spasticity. Good examples of neuroprostheses that are used in these applications are arm and/or leg crank ergometry assisted with FES [44.17–22] and FES systems for walking [44.131–133]. To date, it has been shown that these systems have positive effects on cardiovascular fitness and muscle tone. However, consumers must use these devices continuously to maintain their muscle strength and cardiovascular fitness. As soon as the consumer stops using such a device, his/her muscle strength and cardiovascular fitness tend to deteriorate again.

NEMS systems have also been used to maintain and enhance muscle strength and muscle performance [44.24, 29, 90, 117, 134]. This modality of electrical stimulation-based interventions have been used not only with individuals with neuromuscular disorders, but also with consumers who have diabetes [44.135, 136], knee arthroplasty [44.137, 138], amyotrophic lateral sclerosis [44.134], and able-bodied athletes recovering from injuries [44.72] or trying to enhance their performance [44.139, 140].

Some of the FES systems discussed above, such as FES systems for walking and FES systems for leg ergometry, have been suggested as having positive effects on bone health. For example, individuals with SCI tend to experience massive bone loss in their legs, pelvis, and parts of the spine that are immobile due to SCI. It has been reported that individuals with SCI may lose up to 1/3 of the bone structure when compared to age-matched healthy individuals. Several studies reported bone mineral density increase in individuals with SCI following FES cycle ergometry training [44.141, 142] and FES walking [44.143]. It has been also suggested that larger effects have been observed in females as compared to males [44.144]. Although these preliminary findings are encouraging, a proper evaluation of the FES interventions with respect to bone mineral density improvement need to be carried out.

44.3 FES Therapy

Since the 1970s, some FES researchers and practitioners have observed that following an FES intervention some patients using the neuroprostheses for grasping or walking experienced recovery of voluntary function. Most of these reports were anecdotal in nature. One of the first papers where this phenomenon was specifically discussed was by *Merletti* et al. in 1975 [44.119]. It took almost two decades before this phenomenon started being examined seriously.

First, it was examined in FES systems for drop foot, where scientists explored the ability of the system to restore voluntary walking function in individuals with stroke and incomplete SCI [44.30, 31, 50–52]. These studies were then followed by studies that examined use of neuroprosthesis for grasping and later neuroprostheses for reaching and grasping for restoring voluntary arm and hand functions in individuals with stroke and SCI [44.23, 25, 26, 28, 29, 32, 53, 54]. Finally, the neuroprosthesis for walking was used to investigate restoration of voluntary walking function in individuals with incomplete SCI [44.32]. Among these studies, few were registered randomized control trials that clearly demonstrated the utility of the FES interventions as a means of restoring voluntary upper and lower limb functions. It is worth mentioning that the study published by *Thrasher* et al. [44.29] showed that both reaching and grasping functions can be restored, even in severe stroke individuals who were unable to reach and grasp prior to joining the study; since they had upper extremity Chedoke–McMaster stages of motor recovery scores 1 and 2 they were not expected to improve these functions at all. The improvements published in that study have not yet been matched by any other therapeutic interventions, including rehabilitation robotics therapies and constraint induced movement therapy. The use of neuroprostheses as a means of providing short-term therapeutic intervention for improving and restoring voluntary function has been termed FES therapy (FET).

Although at present, the exact mechanisms to explain the observed carryover effect are not known, there are ongoing, or recently published, studies that may provide an insight into the mechanisms behind FET. First, we know that FET is able to completely restore voluntary control over paretic muscles, regardless of whether they were spastic or flaccid before the FET was administered to the patient [44.145]. This effect has been observed even in individuals with chronic stroke [44.145] and SCI [44.47]. It is important to stress that if FET is administered to individuals with stroke, the strength of the muscles in previously impaired limbs following FET does not reach the same levels as in able-bodied individuals or in less impaired limbs [44.145]. Second, we know that a temporal order of muscle activations in complex tasks such as grasping [44.145] and walking are restored to a large extent. Third, the tone and spasticity in the impaired limbs drops considerably, sometimes to less than 50% of the original tone and spasticity observed prior to FET [44.145]. In addition to these initial findings, it has also been suggested that through forced repetitive movements, FET may activate sensory nerves and through sensory nerve stimulation it may promote the neuroplasticity in the central nervous system [44.27, 29, 32, 37, 146]. It has been also suggested that the antidromic firing of stimulated motor nerves may enhance the cortico-muscular connectivity via Hebb's law [44.37, 146]. In any event, the carryover effect is probably multifactorial and needs to be fully examined. However, what is certain is that the FET is an effective method of restoring voluntary upper limb function in individuals following stroke and SCI, and is able to remedy the drop foot problem in hemiparetic stroke and SCI individuals by improving voluntary control of the muscles of the impaired leg. It is our impression that the FET is a very promising intervention that is only now being seriously examined and has the potential to revolutionize the way we rehabilitate individuals with diverse neuromuscular disorders.

44.4 Other Uses of Electrical Stimulation

44.4.1 Pressure Ulcers

It is a well known fact that persons with SCI are at high risk of developing pressure ulcers. Up to 95% of individuals with SCI will experience a pressure ulcer at least once in their lifetime and 31–79% may experience pressure ulcers multiple times. NMES has been shown to facilitate healing of stages II, III, and IV pressure ulcers [44.147, 148]. FES also shows positive effects in preventing pressure ulcers due to the fact it induces postural movement and increases blood flow [44.149, 150].

44.4.2 Diabetes

Physical activity, especially aerobic endurance exercise, can reduce the incidence and prevalence of diabetes. Thus, physical exercise induced by FES has the potential to reduce the risk of diabetes and potentially provide a treatment for diabetes. It has been reported that muscle training using NMES on thighs reduced blood glucose, suggesting a positive effect on patients with diabetes [44.135]. It has been also shown that the FES cycle ergometry training reduces blood glucose levels and improves glucose utilization [44.136].

44.5 Concluding Remarks

This chapter summarizes the basics of electrical stimulation, explains the difference between NMES and FES, discusses various neuroprosthetic systems, and talks about recent developments in the field of FES therapy (FET). Taking advantages of recent technology such as miniaturization of electronic components, new sensory systems, new delivery methods for electrical stimulation, and improved diagnostics techniques, this field has shown dramatic progress in recent years. Many neuroprosthetic systems are already commercialized and many more are in the process of being developed and/or commercialized. This is a booming field that is slowly, but surely, creating new technologies and interventions; many academic and commercial entities, who are championing this technology, have already emerged. We feel strongly that this field has enormous potential. We anticipate that this field, as well as rehabilitation robotics, prosthetic robotics, and brain–machine interfaces are slowly converging towards creating a combined force that will change the way we treat individuals with disabilities.

References

44.1 P.H. Peckham, K.L. Kilgore, M.W. Keith, A.M. Bryden, N. Bhadra, F.W. Montague: An advanced neuroprosthesis for restoration of hand and upper arm control using an implantable controller, J. Hand Surg. **27**, 265–276 (2002)

44.2 P.H. Peckham, J.S. Knutson: Functional electrical stimulation for neuromuscular applications, Annu. Rev. Biomed. Eng. **7**, 327–360 (2005)

44.3 P. Strojnik, A. Kralj, I. Ursic: Programmed six-channel electrical stimulator for complex stimu-

lation of leg muscles during walking, IEEE Trans. Biomed. Eng. **26**, 112–116 (1979)
44.4 R.J. Triolo, C. Bieri, J. Uhlir, R. Kobetic, A. Scheiner, E.B. Marsolais: Implanted functional neuromuscular stimulation systems for individuals with cervical spinal cord injuries: Clinical case reports, Arch. Phys. Med. Rehabil. **77**, 1119–1128 (1996)
44.5 W. Gstoettner, O. Adunka, J. Hamzavi, W.D. Baumgartner: Rehabilitation of hearing-impaired patients with cochlear implants: A review, Wien. Klin. Wochenschr. **112**, 464–472 (2000)
44.6 K.M. Higgins, J.M. Chen, J.M. Nedzelski, D.B. Shipp, L.D. McIlmoyl: A matched-pair comparison of two cochlear implant systems, J. Otolaryngol. **31**, 97–105 (2002)
44.7 T. Lenarz: Cochlear implants, selection criteria and shifting borders, Acta-Rhino-Laryngol. Belg. **52**, 183–199 (1998)
44.8 D. Popovic, L.L. Baker, G.E. Loeb: Recruitment and comfort of BION implanted electrical stimulation: Implications for FES applications, IEEE Trans. Neural Syst. Rehabil. Eng. **15**, 577–586 (2007)
44.9 D. Popovic, R. Tomovic, L. Schwirtlich: Hybrid assistive system – The motor neuroprosthesis, IEEE Trans. Biomed. Eng. **36**, 729–737 (1989)
44.10 M.R. Popovic, D.B. Popovic, T. Keller: Neuroprostheses for grasping, Neurol. Res. **24**, 443–452 (2002)
44.11 N.J. Rijkhoff, H. Wijkstra, P.E. van Kerrebroeck, F.M. Debruyne: Urinary bladder control by electrical stimulation: Review of electrical stimulation techniques in spinal cord injury, Neurourol. Urodyn. **16**, 39–53 (1997)
44.12 B. Schurch, B. Rodic, D. Jeanmonod: Posterior sacral rhizotomy and intradural anterior sacral root stimulation for treatment of the spastic bladder in spinal cord injured patients, J. Urol. **157**, 610–614 (1997)
44.13 G.S. Brindley, C.E. Polkey, D.N. Rushton, L. Cardozo: Sacral anterior root stimulators for bladder control in paraplegia: The first 50 cases, J. Neurol. Neurosurg. Psychiatry **49**, 1104–1114 (1986)
44.14 S.N. Bailey, E.C. Hardin, R. Kobetic, L.M. Boggs, G. Pinault, R.J. Triolo: Neurotherapeutic and neuroprosthetic effects of implanted functional electrical stimulation for ambulation after incomplete spinal cord injury, J. Rehabil. Res. Dev. **47**, 7–16 (2010)
44.15 J.H. Burridge, M. Haugland, B. Larsen, N. Svaneborg, H.K. Iversen, P.B. Christensen, R.M. Pickering, T. Sinkjaer: Patients' perceptions of the benefits and problems of using the ActiGait implanted drop-foot stimulator, J. Rehabil. Med. **40**, 873–875 (2008)
44.16 E. Hardin, R. Kobetic, L. Murray, M. Corado-Ahmed, G. Pinault, J. Sakai, S.N. Bailey, C. Ho, R.J Triolo: Walking after incomplete spinal cord injury using an implanted FES system: A case report, J. Rehabil. Res. Dev. **44**, 333–346 (2007)
44.17 J.C. Baldi, R.D. Jackson, R. Moraille, W.J. Mysiw: Muscle atrophy is prevented in patients with acute spinal cord injury using functional electrical stimulation, Spinal Cord **36**, 463–469 (1998)
44.18 T. Cameron, J.G. Broton, B. Needham-Shropshire, K.J. Klose: An upper body exercise system incorporating resistive exercise and neuromuscular electrical stimulation (NMS), J. Spinal Cord Med. **21**, 1–6 (1998)
44.19 S. Hooker, S. Figoni, R. Glaser, M. Rodgers, B. Ezenwa, P. Faghri: Physiologic responses to prolonged electrically stimulated leg-cycle exercise in the spinal cord injured, Arch. Phys. Med. Rehabil. **71**, 863–869 (1990)
44.20 S. Hooker, A. Scremin, D. Mutton, C. Kunkel, G. Cagle: Peak and submaximal physiologic responses following electrical stimulation leg cycle ergometer training, J. Rehabil. Res. Dev. **32**, 361–366 (1995)
44.21 P. Jacobs, M. Nash, K. Klose, R. Guest, B. Needham-Shropshire, B. Green: Evaluation of a training program for persons with SCI paraplegia using the Parastep 1 ambulation system: Part 2. Effects on physiological responses to peak arm ergometry, Arch. Phys. Med. Rehabil. **78**, 794–798 (1997)
44.22 S. Pollack, K. Axen, N. Spielholz, N. Levin, F. Haas, K. Ragnarsson: Aerobic training effects of electrically induced lower extremity exercises in spinal cord injured people, Arch. Phys. Med. Rehabil. **70**, 214–219 (1989)
44.23 J. Kowalczewski, V. Gritsenko, N. Ashworth, P. Ellaway, A. Prochazka: Upper-extremity functional electric stimulation-assisted exercises on a workstation in the subacute phase of stroke recovery, Arch. Phys. Med. Rehabil. **88**, 833–839 (2007)
44.24 S. Mangold, T. Keller, A. Curt, V. Dietz: Transcutaneous functional electrical stimulation for grasping in subjects with cervical spinal cord injury, Spinal Cord **43**, 1–13 (2005)
44.25 D. Popovic, M.R. Popovic, T. Sinkjaer, A. Stefanovic, L. Schwirtlich: Therapy of paretic arm in hemiplegic subjects augmented with a neural prosthesis: A cross-over study, Can. J. Physiol. Pharmacol. **82**, 749–756 (2004)
44.26 M.B. Popovic, D.B. Popovic, T. Sinkjaer, A. Stefanovic, L. Schwirtlich: Restitution of reaching and grasping promoted by functional electrical therapy, Artif. Organs **26**, 271–275 (2002)
44.27 M.R. Popovic, T.A. Thrasher, M.E. Adams, V. Takes, V. Zivanovic, M.I. Tonack: Functional electrical therapy: Retraining grasping in spinal cord injury, Spinal Cord **44**, 143–151 (2006)
44.28 M.R. Popovic, T.A. Thrasher, P. Zivanovic, M. Takaki, P. Hajek: Neuroprosthesis for retraining reaching and grasping functions in severe hemiplegic patients, Neuromodulation **8**, 58–72 (2005)
44.29 T.A. Thrasher, V. Zivanovic, W. McIlroy, M.R. Popovic: Rehabilitation of reaching and grasping function in severe hemiplegic patients using func-

tional electrical stimulation therapy, Neurorehabil. Neural Repair **22**, 706–714 (2008)
44.30 R.B. Stein, D.G. Everaert, A.K. Thompson, S.L. Chong, M. Whittaker, J. Robertson, G. Kuether: Long-term therapeutic and orthotic effects of a foot drop stimulator on walking performance in progressive and nonprogressive neurological disorders, Neurorehabil. Neural Repair **24**, 152–167 (2010)
44.31 P.N. Taylor, J.H. Burridge, A.L. Dunkerley, D.E. Wood, J.A. Norton, C. Singleton, I.D. Swain: Clinical use of the Odstock dropped foot stimulator: Its effect on the speed and effort of walking, Arch. Phys. Med. Rehabil. **80**, 1577–1583 (1999)
44.32 T.A. Thrasher, H.M. Flett, M.R. Popovic: Gait training regimen for incomplete spinal cord injury using functional electrical stimulation, Spinal Cord **44**, 357–361 (2006)
44.33 M.R. Popovic, T. Keller, I.P. Pappas, V. Dietz, M. Morari: Surface-stimulation technology for grasping and walking neuroprosthesis, IEEE Eng. Med. Biol. Mag. **20**, 82–93 (2001)
44.34 D.S. Schechter: Origins of Electrotherapy, N. Y. State J. Med. **71**, 997–1008 (1971)
44.35 M. Reichel, T. Breyer, W. Mayr, F. Rattay: Simulation of the three-dimensional electrical field in the course of functional electrical stimulation, Artif. Organs **26**, 252–255 (2002)
44.36 H. Kern, C. Hofer, M. Mödlin, C. Forstner, D. Raschka-Högler, W. Mayr, H. Stöhr: Denervated muscles in humans: Limitations and problems of currently used functional electrical stimulation training protocols, Artif. Organs **26**, 216–218 (2002)
44.37 D. Rushton: Functional electrical stimulation and rehabilitation – An hypothesis, Med. Eng. Phys. **25**, 75–78 (2003)
44.38 M.R. Popovic, T. Keller: Modular transcutaneous functional electrical stimulation system, Med. Eng. Phys. **27**, 81–92 (2005)
44.39 L.R. Sheffler, J. Chae: Neuromuscular electrical stimulation in neurorehabilitation, Muscle Nerve **35**, 562–590 (2007)
44.40 N. Hoshimiya, Y. Handa: A master-slave type multichannel functional electrical stimulation (FES) system for the control of the paralyzed upper extremities, Automedica **11**, 209–220 (1989)
44.41 W.M. Grill, J.T. Mortimer: Stability of the input-output properties of chronically implanted multiple contact nerve cuff stimulating electrodes, IEEE Trans. Rehabil. Eng. **6**, 364–373 (1998)
44.42 D.J. Tyler, D.M. Durand: A slowly penetrating interfascicular nerve electrode for selective activation of peripheral nerves, IEEE Trans. Rehabil. Eng. **5**, 51–61 (1997)
44.43 J.H. Schulman: The feasible FES system: Battery powered BION stimulator, Proc. IEEE **96**, 1226–1239 (2008)
44.44 D. Graupe, K.H. Kohn: Functional neuromuscular stimulator for short-distance ambulation by certain thoracic-level spinal-cord-injured paraplegics, Surg. Neurol. **50**, 202–207 (1998)
44.45 P.H. Peckham, M.W. Keith, K.L. Kilgore, J.H. Grill, K.S. Wuolle, G.B. Thrope, P. Gorman, J. Hobby, M.J. Mulcahey, S. Carroll, V.R. Hentz, A. Wiegner, I.N.R. Group: Efficacy of an implanted neuroprosthesis for restoring hand grasp in tetraplegia: A multicenter study, Arch. Phys. Med. Rehabil. **82**, 1380–1388 (2001)
44.46 A. Prochazka, M. Gauthier, M. Wieler, Z. Kenwell: The bionic glove: An electrical stimulator garment that provides controlled grasp and hand opening in quadriplegia, Arch. Phys. Med. Rehabil. **78**, 608–614 (1997)
44.47 R. Miller, M.R. Popovic, T.A. Thrasher, M. Verrier: Functional electrical stimulation therapy improves grasping in chronic cervical spinal cord injury: Two case studies, J. Autom. Control **18**, 53–61 (2008)
44.48 I.P. Pappas, M.R. Popovic, T. Keller, V. Dietz, M. Morari: A reliable gait phase detection system, IEEE Trans. Neural Syst. Rehabil. Eng. **9**, 113–125 (2001)
44.49 M.B. Popovic: Control of neural prostheses for grasping and reaching, Med. Eng. Phys. **25**, 41–50 (2003)
44.50 J.H. Burridge, P.N. Taylor, S.A. Hagan, D.E. Wood, I.D. Swain: The effects of common peroneal stimulation on the effort and speed of walking: A randomized controlled trial with chronic hemiplegic patients, Clin. Rehabil. **11**, 201–210 (1997)
44.51 M.H. Granat, D.J. Maxwell, A.C. Ferguson, K.R. Lees, J.C. Barbenel: Peroneal stimulator; evaluation for the correction of spastic drop foot in hemiplegia, Arch. Phys. Med. Rehabil. **77**, 19–24 (1996)
44.52 M. Wieler, R. Stein, M. Ladouceur, M. Whittaker, A. Smith, S. Naaman, H. Barbeau, J. Bugaresti, E. Aimone: Multicenter evaluation of electrical stimulation systems for walking, Arch. Phys. Med. Rehabil. **80**, 495–500 (1999)
44.53 G. Alon, K. Sunnerhagen, A. Geurts, A. Ohry: A home-based, self-administered stimulation program to improve selected hand functions of chronic stroke, NeuroRehabilitation **18**, 215–225 (2003)
44.54 J. Chae, R. Hart: Intramuscular hand neuroprosthesis for chronic stroke survivors, Neurorehabil. Neural Repair **17**, 109–117 (2003)
44.55 J. Mizrahi: Fatigue in muscles activated by functional electrical stimulation, Crit. Rev. Phys. Rehabil. Med. **9**, 93–129 (1997)
44.56 H.L. Gerrits, A. de Haan, M.T. Hopman, L.H. van Der Woude, D.A. Jones, A.J. Sargeant: Contractile properties of the quadriceps muscle in individuals with spinal cord injury, Muscle Nerve **22**, 1249–1256 (1999)

44.57 H.L. Gerrits, M.T.E. Hopman, C. Offringa, B.G.M. Engelen, A.J. Sargeant, D.A. Jones, A. Haan: Variability in fibre properties in paralysed human quadriceps muscles and effects of training, Pflugers Arch. **445**, 734–740 (2003)

44.58 A.J. Lenman, F.M. Tulley, G. Vrbova, M.R. Dimitrijevic, J.A. Towle: Muscle fatigue in some neurological disorders, Muscle Nerve **12**, 938–942 (1989)

44.59 R.K. Shields: Fatigability, relaxation properties, and electromyographic responses of the human paralyzed soleus muscle, J. Neurophysiol. **73**, 2195–2206 (1995)

44.60 C.K. Thomas: Fatigue in human thenar muscles paralysed by spinal cord injury, J. Electromygr. Kinesiol. **7**, 15–26 (1997)

44.61 C.K. Thomas: Contractile properties of human thenar muscles paralyzed by spinal cord injury, Muscle Nerve **20**, 788–799 (1997)

44.62 E. Asmussen: Muscle fatigue, Med. Sci. Sports **11**, 313–321 (1979)

44.63 B. Bigland-Ritchie, D.A. Jones, G.P. Hosking, R.H. Edwards: Central and peripheral fatigue in sustained maximum voluntary contractions of human quadriceps muscle, Clin. Sci. Mol. Med. **54**, 609–614 (1978)

44.64 B. Bigland-Ritchie, C.G. Kukulka, O.C. Lippold, J.J. Woods: The absence of neuromuscular transmission failure in sustained maximal voluntary contractions, J. Physiol. **330**, 265–278 (1982)

44.65 R.H. Fitts: Cellular mechanisms of muscle fatigue, Physiol. Rev. **74**, 49–94 (1994)

44.66 A.R. Kralj, T. Bajd: *Functional Electrical Stimulation* (CRC, Boca Raton 1989) p. 198

44.67 C. De Luca: Myoelectrical manifestations of localized muscular fatigue in humans, Crit. Rev. Biomed. Eng. **11**, 251–279 (1984)

44.68 E. Henneman, G. Somjen, D.O. Carpenter: Excitability and inhibitability of motoneurons of different sizes, J. Neurophysiol. **28**, 599–620 (1965)

44.69 E. Henneman, G. Somjen, D.O. Carpenter: Functional significance of cell size in spinal motoneurons, J. Neurophysiol. **28**, 560–580 (1965)

44.70 S.A. Binder-Macleod, T. Kesar: Catchlike property of skeletal muscle: Recent findings and clinical implications, Muscle Nerve **31**, 681–693 (2005)

44.71 W. Holderbaum, K.J. Hunt, H. Gollee: H-infinity robust control design for unsupported paraplegic standing: Experimental evaluation, Control Eng. Pract. **10**, 1211–1222 (2002)

44.72 D.A. Lake: Neuromuscular electrical stimulation. An overview and its application in the treatment of sports injuries, Sports Med. **13**, 320–336 (1992)

44.73 A. Thrasher, G. Graham, M. Popovic: Reducing muscle fatigue due to functional electrical stimulation using random modulation of stimulation parameters, Artif. Organs **29**, 453–458 (2005)

44.74 C.M. Gregory, W. Dixon, C.S. Bickel: Impact of varying pulse frequency and duration on muscle torque production and fatigue, Muscle Nerve **35**, 504–509 (2007)

44.75 T. Kesar, S.A. Binder-Macleod: Effect of frequency and pulse duration on human muscle fatigue during repetitive electrical stimulation, Exp. Physiol. **91**, 967–976 (2006)

44.76 D. Russ, K. Vandenborne, G. Walter, M. Elliott, S.A. Binder-Macleod: Effects of muscle activation on fatigue and metabolism in human skeletal muscle, J. Appl. Physiol. **92**, 1978–1986 (2002)

44.77 S.A. Binder-Macleod, I. Barker: Use of a catchlike property of human skeletal muscle to reduce fatigue, Muscle Nerve **14**, 850–857 (1991)

44.78 S.A. Binder-Macleod, W.B. Scott: Comparison of fatigue produced by various electrical stimulation trains, Acta Physiol. Scand. **172**, 195–203 (2001)

44.79 B.M. Doucet, L. Griffin: Variable stimulation patterns for poststroke hemiplegia, Muscle Nerve **39**, 54–62 (2009)

44.80 N. Mourselas, M.H. Granat: Evaluation of patterned stimulation for use in surface functional electrical stimulation systems, Med. Eng. Phys. **20**, 319–324 (1998)

44.81 W.B. Scott, S.A. Binder-Macleod: Changing stimulation patterns improves performance during electrically elicited contractions, Muscle Nerve **28**, 174–180 (2003)

44.82 C.K. Thomas, L. Griffin, S. Godfrey, E. Ribot-Ciscar, J. Butler: Fatigue of paralyzed and control thenar muscles induced by variable or constant frequency stimulation, J. Neurophysiol. **89**, 2055–2064 (2003)

44.83 M. Pournezam, B. Andrews, R. Baxendale, G. Phillips, J. Paul: Reduction of muscle fatigue in man by cyclical stimulation, J. Biomed. Eng. **10**, 196–200 (1988)

44.84 J.S. Petrofsky: Sequential motor unit stimulation through peripheral motor nerves in the cat, Med. Biol. Eng. Comp. **17**, 87–93 (1979)

44.85 L.-W. Chou, T.M. Kesar, S.A. Binder-Macleod: Using customized rate-coding and recruitment strategies to maintain forces during repetitive activation of human muscles, Phys. Ther. **88**, 363–375 (2008)

44.86 L.-W. Chou, S.C. Lee, T.E. Johnston, S.A. Binder-Macleod: The effectiveness of progressively increasing stimulation frequency and intensity to maintain paralyzed muscle force during repetitive activation in persons with spinal cord injury, Arch. Phys. Med. Rehabil. **89**, 856–864 (2008)

44.87 L. Griffin, B.G. Jun, C. Covington, B.M. Doucet: Force output during fatigue with progressively increasing stimulation frequency, J. Electromygr. Kinesiol. **18**, 426–433 (2008)

44.88 M.B. Kebaetse, S.C. Lee, T.E. Johnston, S.A. Binder-Macleod: Strategies that improve paralyzed human quadriceps femoris muscle performance during repetitive, nonisometric contractions, Arch. Phys. Med. Rehabil. **86**, 2157–2164 (2005)

44.89 T. Kesar, L.-W. Chou, S.A. Binder-Macleod: Effects of stimulation frequency versus pulse duration modulation on muscle fatigue, J. Electromygr. Kinesiol. **18**, 662–671 (2008)

44.90 B. Lau, L. Guevremont, V.K. Mushahwar: Strategies for generating prolonged functional standing using intramuscular stimulation or intraspinal microstimulation, IEEE Trans. Neural Syst. Rehabil. Eng. **15**, 273–285 (2007)

44.91 R.K. Shields, S. Dudley-Javoroski, A.E. Littmann: Postfatigue potentiation of the paralyzed soleus muscle: Evidence for adaptation with long-term electrical stimulation training, J. Appl. Physiol. **101**, 556–565 (2006)

44.92 W.T. Liberson, H.J. Holmquest, D. Scot, M. Dow: Functional electrotherapy: Stimulation of the peroneal nerve synchronized with the swing phase of the gait of hemiplegic patients, Arch. Phys. Med. Rehabil. **42**, 101–105 (1961)

44.93 R.L. Waters, D. McNeal, J. Perry: Experimental correction of footdrop by electrical stimulation of the peroneal nerve, J. Bone Jt. Surg. Am. **57**, 1047–1054 (1975)

44.94 J.M. Hausdorff, H. Ring: Effects of a new radio frequency-controlled neuroprosthesis on gait symmetry and rhythmicity in patients with chronic hemiparesis, Am. J. Phys. Med. Rehabil. **87**, 4–13 (2008)

44.95 L. Kenney, G. Bultstra, R. Buschman, P. Taylor, G. Mann, H. Hermens, J. Holsheimer, A. Nene, M. Tenniglo, H. van der Aa, J. Hobby: An implantable two channel drop foot stimulator: Initial clinical results, Artif. Organs **26**, 267–270 (2002)

44.96 A. Willemsen, F. Bloemhof, H. Boom: Automatic stance-swing phase detection form accelerometer data for peroneal nerve stimulation, IEEE Trans. Biomed. Eng. **37**, 1201–1208 (1990)

44.97 R. Dai, R.B. Stein, B.J. Andrews, K.B. James, M. Wieler: Application of tilt sensors in functional electrical stimulation, IEEE Trans. Rehabil. Eng. **4**, 63–72 (1996)

44.98 R. van Swigchem, J. Vloothuis, J. den Boer, V. Weerdesteyn, A.C.H. Geurts: Is transcutaneous peroneal stimulation beneficial to patients with chronic stroke using an ankle-foot orthosis? A within-subjects study of patients' satisfaction, walking speed and physical activity level, J. Rehabil. Med. **42**, 117–121 (2010)

44.99 A. Kantrowitz: *Electronic Physiologic Aids* (Brooklyn, New York 1960)

44.100 A. Kralj, T. Bajd, R. Turk: Enhancement of gait restoration in spinal injured patients by functional electrical stimulation, Clin. Orthop. Relat. Res. **223**, 34–43 (1988)

44.101 M.R. Popovic, A. Curt, T. Keller, V. Dietz: Functional electrical stimulation for grasping and walking: Indications and limitations, Spinal Cord **39**, 403–412 (2001)

44.102 D. Graupe, R. Davis, H. Kordylewski, K. Kohn: Ambulation by traumatic T4-12 paraplegics using functional neuromuscular stimulation, Crit. Rev. Neurosurg. **8**, 221–231 (1998)

44.103 M. Solomonow, R. Baratta, S. Hirokawa, N. Rightor, W. Walker, P. Beaudette, H. Shoji, R. D'Ambrosia: The RGO generation II: Muscle stimulation powered orthosis as a practical walking system for thoracic paraplegics, Orthopedics **12**, 1309–1315 (1989)

44.104 T.E. Johnston, R.R. Betz, B.T. Smith, B.J. Benda, M.J. Mulcahey, R. Davis, T.P. Houdayer, M.A. Pontari, A. Barriskill, G.H. Creasey: Implantable FES system for upright mobility and bladder and bowel function for individuals with spinal cord injury, Spinal Cord **43**, 713–723 (2005)

44.105 R. Kobetic, C.S. To, J.R. Schnellenberger, M.L. Audu, T.C. Bulea, R. Gaudio, G. Pinault, S. Tashman, R.J. Triolo: Development of hybrid orthosis for standing, walking, and stair climbing after spinal cord injury, J. Rehabil. Res. Dev. **46**, 447–462 (2009)

44.106 R. Kobetic, R.J. Triolo, J.P. Uhlir, C. Bieri, M. Wibowo, G. Polando, E.B. Marsolais, J.A. Davis, K.A. Ferguson: Implanted functional electrical stimulation system for mobility in paraplegia: A follow-up case report, IEEE Trans. Rehabil. Eng. **7**, 390–398 (1999)

44.107 J.P. Uhlir, R.J. Triolo, R. Kobetic: The use of selective electrical stimulation of the quadriceps to improve standing function in paraplegia, IEEE Trans. Rehabil. Eng. **8**, 514–522 (2000)

44.108 S.J. Gartman, M.L. Audu, R.F. Kirsch, R.J. Triolo: Selection of optimal muscle set for 16-channel standing neuroprosthesis, J. Rehabil. Res. Dev. **45**, 1007–1017 (2008)

44.109 K.J. Hunt, H. Gollee, R.P. Jaime: Control of paraplegic ankle joint stiffness using FES while standing, Med. Eng. Phys. **23**, 541–555 (2001)

44.110 H. Gollee, K.J. Hunt, D.E. Wood: New results in feedback control of unsupported standing in paraplegia, IEEE Trans. Neural Syst. Rehabil. Eng. **12**, 73–80 (2004)

44.111 J.J. Abbas, H.J. Chizeck: Feedback control of coronal plane hip angle in paraplegic subjects using functional neuromuscular stimulation, IEEE Trans. Biomed. Eng. **38**, 687–698 (1991)

44.112 A. Vette, K. Masani, J. Kim, M.R. Popovic: Closed-loop control of functional electrical stimulation-assisted arm-free standing in individuals with spinal cord injury: A feasibility study, Neuromodulation **12**, 22–32 (2009)

44.113 A.H. Vette, K. Masani, M.R. Popovic: Implementation of a physiologically identified PD feedback controller for regulating the active ankle torque during quiet stance, IEEE Trans. Neural Syst. Rehabil. Eng. **15**, 235–243 (2007)

44.114 J.-Y. Kim, J.K. Mills, A.H. Vette, M.R. Popovic: Optimal combination of minimum degrees of freedom to be actuated in the lower limbs to facilitate arm-free paraplegic standing, J. Biomech. Eng. **129**, 838–847 (2007)

44.115 J.-Y. Kim, M.R. Popovic, J.K. Mills: Dynamic modeling and torque estimation of FES-assisted arm-free standing for paraplegics, IEEE Trans. Neural Syst. Rehabil. Eng. **14**, 46–54 (2006)

44.116 B. Smith, Z. Tang, M.W. Johnson, S. Pourmehdi, M.M. Gazdik, J.R. Buckett, P.H. Peckham: An externally powered, multichannel, implantable stimulator-telemeter for control of paralyzed muscle, IEEE Trans. Biomed. Eng. **45**, 463–475 (1998)

44.117 H.T. Hendricks, M.J. IJzerman, J.R. de Kroon, F.A. in 't Groen, G. Zilvold: Functional electrical stimulation by means of the 'Ness Handmaster Orthosis' in chronic stroke patients: An exploratory study, Clin. Rehabil. **15**, 217–220 (2001)

44.118 S. Rebersek, L. Vodovnik: Proportionally controlled functional electrical stimulation of hand, Arch. Phys. Med. Rehabil. **54**, 378–382 (1973)

44.119 R. Merletti, R. Acimovic, S. Grobelnik, G. Cvilak: Electrophysiological orthosis for the upper extremity in hemiplegia: Feasibility study, Arch. Phys. Med. Rehabil. **56**, 507–513 (1975)

44.120 C. Márquez-Chin, M.R. Popovic, T. Cameron, A. Lozano, R. Chen: Control of a neuroprosthesis for grasping using off-line classification of electrocorticographic signals: Case study, Spinal Cord **47**, 802–808 (2009)

44.121 W.H. Boyce, J.E. Lathem, L.D. Hunt: Research related to the development of an artificial electrical stimulator for the paralyzed human bladder: A review, J. Urol. **91**, 41–51 (1964)

44.122 E. Bors, E. Comarr: *Neurological Urology: Physiology of Micturition, Its Neurological Disorders and Sequelae* (Karger, Basel 1971)

44.123 K.P. Caldwell: The electrical control of sphincter incompetence, Lancet **2**, 174–175 (1963)

44.124 S. Alexander: Research and clinical experience in the treatment of the neurogenic bladder by electronic implant and prostheses, Paraplegia **6**, 183–193 (1968)

44.125 G.H. Creasey: Restoration of bladder, bowel, and sexual function, Top Spinal Cord Inj. Rehabil. **5**, 21–32 (1999)

44.126 J.M. Vastenholt, G.J. Snoek, H.P.J. Buschman, H.E. van der Aa, E.R.J. Alleman, M.J. IJzerman: A 7-year follow-up of sacral anterior root stimulation for bladder control in patients with a spinal cord injury: Quality of life and users' experiences, Spinal Cord **41**, 397–402 (2003)

44.127 S.W. Siegel: Management of voiding dysfunction with an implantable neuroprosthesis, Urol. Clin. N. Am. **19**, 163–170 (1992)

44.128 W.W. Glenn, J.F. Hogan, J.S. Loke, T.E. Ciesielski, M.L. Phelps, R. Rowedder: Ventilatory support by pacing of the conditioned diaphragm in quadriplegia, N. Engl. J. Med. **310**, 1150–1155 (1984)

44.129 W.W. Glenn, J.F. Hogan, M.L. Phelps: Ventilatory support of the quadriplegic patient with respiratory paralysis by diaphragm pacing, Surg. Clin. N. Am. **60**, 1055–1078 (1980)

44.130 A.F. DiMarco, R.P. Onders, K.E. Kowalski, M.E. Miller, S. Ferek, J.T. Mortimer: Phrenic nerve pacing in a tetraplegic patient via intramuscular diaphragm electrodes, Am. J. Respir. Crit. Care Med. **166**, 1604–1606 (2002)

44.131 D. de Carvalho, M. de Cassia Zanchetta, J. Sereni, A. Cliquet: Metabolic and cardiorespiratory responses of tetraplegic subjects during treadmill walking using neuromuscular electrical stimulation and partial body weight support, Spinal Cord **43**, 400–405 (2005)

44.132 D. de Carvalho, A. Cliquet: Energy expenditure during rest and treadmill gait training in quadriplegic subjects, Spinal Cord **43**, 658–663 (2005)

44.133 D. de Carvalho, C. Martins, S. Cardoso, A. Cliquet: Improvement of metabolic and cardiorespiratory responses through treadmill gait training with neuromuscular electrical stimulation in quadriplegic subjects, Artif. Organs **30**, 56–63 (2006)

44.134 I. Handa, N. Matsushita, K. Ihashi, R. Yagi, R. Mochizuki, H. Mochizuki, Y. Abe, Y. Shiga, N. Hoshimiya, Y. Itoyama: A clinical trial of therapeutic electrical stimulation for amyotrophic lateral sclerosis, Tohoku J. Exp. Med. **175**, 123–134 (1995)

44.135 T. Hamada, H. Sasaki, T. Hayashi, T. Moritani, K. Nakao: Enhancement of whole body glucose uptake during and after human skeletal muscle low-frequency electrical stimulation, J. Appl. Physiol. **94**, 2107–2112 (2003)

44.136 J. Jeon, C. Weiss, R. Steadward, E. Ryan, R. Burnham, G. Bell, P. Chilibeck, G. Wheeler: Improved glucose tolerance and insulin sensitivity after electrical stimulation-assisted cycling in people with spinal cord injury, Spinal Cord **40**, 110–117 (2002)

44.137 M. Lewek, J. Stevens, L. Snyder-Mackler: The use of electrical stimulation to increase quadriceps femoris muscle force in an elderly patient following a total knee arthroplasty, Phys. Ther. **81**, 1565–1571 (2001)

44.138 J.E. Stevens, R.L. Mizner, L. Snyder-Mackler: Neuromuscular electrical stimulation for quadriceps muscle strengthening after bilateral total knee arthroplasty: A case series, J. Orthop. Sports Phys. Ther. **34**, 21–29 (2004)

44.139 N. Babault, G. Cometti, M. Bernardin, M. Pousson, J.-C. Chatard: Effects of electromyostimulation training on muscle strength and power of elite rugby players, J. Strength Cond. Res. **21**, 431–437 (2007)

44.140 F. Brocherie, N. Babault, G. Cometti, N. Maffiuletti, J.-C. Chatard: Electrostimulation training effects on the physical performance of ice hockey players, Med. Sci. Sports Exerc. **37**, 455–460 (2005)

44.141 M. Belanger, R. Stein, G. Wheeler, T. Gordon, B. Leduc: Electrical stimulation: Can it increase muscle strength and reverse osteopenia in spinal cord injured individuals?, Arch. Phys. Med. Rehabil. **81**, 1090–1098 (2000)

44.142 T. Mohr, J. Podenphant, F. Biering-Sorensen, H. Galbo, G. Thamsborg, M. Kjaer: Increased bone mineral density after prolonged electrically induced cycle training of paralyzed limbs in spinal cord injured man, Calcif. Tissue Int. **61**, 22–25 (1997)

44.143 B. Needham-Shropshire, J. Broton, K. Klose, N. Lebwohl, R. Guest, P. Jacobs: Evaluation of a training program for persons with SCI paraplegia using the Parastep 1 ambulation system: Part 3. Lack of effect on bone mineral density, Arch. Phys. Med. Rehabil. **78**, 799–803 (1997)

44.144 S. Ott: Osteoporosis in women with spinal cord injuries, Phys. Med. Rehabil. Clin. N. Am. **12**, 111–131 (2001)

44.145 N. Kawashima, M.R. Popovic, V. Zivanovic: Effect of intensive functional electrical stimulation therapy on the upper limb motor recovery after stroke: Single case study of a chronic stroke patient, Proc. 13th Annu. Conf. Int. FES Soc. (2008) pp. 252–254

44.146 A.K. Thompson, R. Stein: Short-term effects of functional electrical stimulation on motor-evoked potentials in ankle flexor and extensor muscles, Exp. Brain Res. **159**, 491–500 (2004)

44.147 J. Feedar, L. Kloth, G. Gentzkow: Chronic dermal ulcer healing enhanced with monophasic pulsed electrical stimulation, Phys. Ther. **71**, 639–649 (1991)

44.148 J. Griffin, R. Tooms, R. Mendius, J. Clifft, R. Vander Zwaag, F. el-Zeky: Efficacy of high voltage pulsed current for healing of pressure ulcers in patients with spinal cord injury, Phys. Ther. **71**, 433–442 (1991)

44.149 S.P. Levine, R. Kett, M. Gross, B. Wilson, P. Cederna, J. Juni: Blood flow in the gluteus maximus of seated individuals during electrical muscle stimulation, Arch. Phys. Med. Rehabil. **71**, 682–686 (1990)

44.150 S.P. Levine, R.L. Kett, P.S. Cederna, S.V. Brooks: Electric muscle stimulation for pressure sore prevention: Tissue shape variation, Arch. Phys. Med. Rehabil. **71**, 210–215 (1990)

Part E Monitoring

46 Recording and Processing of Biosignals
Klaus-Peter Hoffmann, St. Ingbert, Germany
Florian Solzbacher, Salt Lake City, USA

47 Monitoring Systems
Ullrich Hieronymi, Lübeck, Germany
Rüdiger Kramme, Titisee, Germany

48 Cardiovascular Monitoring
Ullrich Hieronymi, Lübeck, Germany
Rüdiger Kramme, Titisee, Germany

49 Respiratory Monitoring and Pulse Oximetry
Ullrich Hieronymi, Lübeck, Germany
Rüdiger Kramme, Titisee, Germany
Harald Kronberg, Michelstadt, Germany

50 Temperature Monitoring
Rüdiger Kramme, Titisee, Germany
Ullrich Hieronymi, Lübeck, Germany

51 Cerebral Monitoring
Barbara Schultz, Hannover, Germany
Arthur Schultz, Hannover, Germany
Harald Kronberg, Michelstadt, Germany

52 Brain Computer Interface
Günter Edlinger, Graz, Austria
Cristiano Rizzo, Mogliano Veneto (TV), Italy
Christoph Guger, Schiedlberg, Austria

53 Fetal Monitoring
Birgit Wacker, Böblingen, Germany
Martin Maier, Böblingen, Germany

54 Neonatal Monitoring
Roland Hentschel, Freiburg, Germany

55 Intraoperative Neuromonitoring
Werner Kneist, Mainz, Germany
Daniel W. Kauff, Mainz, Germany

56 Ionic Neural Sensing
Iasonas F. Triantis, London, UK
Anna Radomska-Botelho Moniz, London, UK
Kostis Michelakis, London, UK
Sanjiv Sharma, London, UK
Jakub Trzebinski, London, UK
Belinda Garner, London, UK
Amir Eftekhar, London, UK

45. Treatment Planning and Patient Treatment

Golam Abu Zakaria, Wilhelm Schütte, Stephan Garbe

Treatment planning and patient treatment are essential parts in the medical technology of external radiotherapy. For patients undergoing such a radio-therapeutic treatment, the prescribed and absorbed dose to the target volume and the organs at risk need to be determined and notified. Furthermore, in radio-therapeutic treatments, registrants and licenses shall ensure, within the ranges achievable by good clinical practice and optimized functioning of equipment, that the prescribed absorbed dose with the respective beam quality is delivered to the planning target volume as well as doses to other tissues and organs is minimized. Therefore in this chapter modern and standard equipment, treatment procedures, basics of treatment planning and quality control in radiation therapy will be presented.

45.1 **Principles of Radiotherapy and Treatment Planning** 898
45.1.1 Principles of Radiotherapy 898
45.1.2 Principles of Treatment Planning and Treatment 899

45.2 **Imaging in Treatment Planning** 901
45.2.1 Computed Tomography 901
45.2.2 Magnetic Resonance Imaging 902
45.2.3 Positron Emission Tomography 902

45.3 **Basic Techniques in External Beam Therapy** 902
45.3.1 Simple Treatment Planning 902
45.3.2 Three-Dimensional Conformal Treatment Planning (3-D CRT) 903
45.3.3 Inverse Planning 903
45.3.4 Intensity Modulated Radiation Therapy 903
45.3.5 Dynamic Delivery Techniques 904
45.3.6 Tomotherapy 904
45.3.7 Image Guided Radiotherapy (IGRT) 906
45.3.8 Respiratory Guided Radiotherapy 906
45.3.9 Intraoperative Electron Radiation Therapy 906
45.3.10 Stereotactic Irradiation 907
45.3.11 Total Body Irradiation with Photons 907
45.3.12 Total Skin Electron Beam Therapy 908

45.4 **Target Volumes and Organ at Risk** 909

45.5 **Modern Treatment Planning Systems** 910
45.5.1 Basic Components of a Treatment Planning System 911
45.5.2 Physics Module 911
45.5.3 Import Module 912
45.5.4 Target Definition or Contouring Module 912
45.5.5 Beam Modeling Module 913
45.5.6 Dose Calculation Module 913
45.5.7 Evaluation and Optimization Module 913
45.5.8 Plan Analysis Module 914
45.5.9 Export Module 914

45.6 **Simulation of the Patient and the First Treatment** 915
45.6.1 Conventional Simulation 916
45.6.2 Virtual Simulation 916

45.7 **Quality Control in Radiation Therapy** 917
45.7.1 Personnel 917
45.7.2 Planning system 917
45.7.3 Medical Linear Accelerator and Simulator 917

References 919

45.1 Principles of Radiotherapy and Treatment Planning

Radiation therapy is a medical field that deals with the use of ionizing radiation to humans and on animals by treating benign and malignant diseases. According to the disease and its stage, radiotherapy has either a curative or palliative intention.

45.1.1 Principles of Radiotherapy

Curative radiotherapy is carried out with the aim of healing; a palliative radiotherapy treats symptoms such as pain, occlusions, and bleeding and prevents the progression of disease.

Today high-energy, ionizing rays are commonly applied in modern radiation therapy like gamma radiation, x-ray radiation and electron radiation. In large therapy centers with appropriate radiotherapy equipment, proton therapy and heavy ions (e.g., carbon ions) are increasingly being used.

With radiotherapy, benign as well as malignant diseases can be treated. Before a radiotherapy treatment is performed, treatment planning occurs. Sometimes, one distinguishes, on the one hand, simple planning with programs without graphic representation of the patient and target volume that calculates the monitor prefix in water merely for a given field size and tumor depth from, on the other hand, extensive radiotherapy planning using 3-D planning systems. While the simple programs are programmed by the medical physicist in radiotherapy, it is within the 3-D radiotherapy planning system, it is within the 3-D radiotherapy planning system that commercial software is used that calculates the dose distribution in the CT slice and offers numerous possibilities for the graphic representation of the result.

In many cases, radiotherapy is performed on tumors in combination with chemotherapy (known as radiochemotherapy, RCTX). Chemotherapy could lead to radiosensitization of the cancer cells with subsequent strengthening of the radiation effect to the tumor. Hence a combination of radiotherapy and chemotherapy for the treatment of malignant tumors is more effective than radio- or chemotherapy alone. Combined RCTX can appear as curative, adjuvant or palliative radiotherapy. One distinguishes between the following kinds of combined RCTX:

- Simultaneous RCTX: both treatments occurring in parallel
- Sequential RCTX: chemotherapy before or after radiotherapy
- Definite RCTX: treatment of tumor exclusively by RCTX
- Adjuvant RCTX: aftercare of tumor or tumor bed after its complete or partial removal
- Neoadjuvant RCTX: reduction of tumor by RCTX with subsequent surgical removal of tumor

Other treatments, for example hyperthermia, can cause a strengthening of the radiotherapy effect. Tumors with low blood supply or low in oxygen are mostly resistant to radiation damage. Conversely, one can observe that such tissue is especially sensitive compared with therapeutic hyperthermia. Also, special substances can raise the resistance of normal tissue. An example is Amifostine, which is still used today, though only experimentally [45.1, 2].

Another area of application of radiotherapy treating benign diseases is so-called stimulative radiotherapy. It is applied on chronic-inflammatory and degenerative diseases like heel spur, tennis elbow, shoulder pains and arthrosis of different joints. Nevertheless, the main area of application of radiotherapy is the treatment of malignant tumors. Here the different radio sensitivities of tumor tissue and surrounding normal tissue are utilized. In contrast to cancer cells, normal tissue cells recover after radiotherapy relatively fast. While the fraction doses of benign diseases are about 0.5 Gy and the total dose varies from 3 to 8 Gy, the fraction doses required for the treatment of malignant tumors are about 2 Gy and the total dose varies from 30 to 80 Gy according to the kind of tumor [45.3, 4].

Radiotherapy is divided into two kinds of usage: teletherapy, where the radiation penetrates the patient's body from outside by a distant radiation source, and brachytherapy, where the radiation source is placed in cavities of the patient's body, in the tissue or directly on the external body shell.

The energy used in brachytherapy is about 1 MeV maximum, while in teletherapy using conventional medical linear accelerators the energy of the electron beams and x-ray bremsstrahlung goes up to 25 MeV. In Fig. 45.1 four depth dose curves for different kinds of radiation and different energies in water are shown.

The mechanism of interaction of radiation with cells can be explained by energy transfer to the penetrated tissue using scatter processes. The direct penetration of biomolecules does not have a larger impact on cell growth than the ionization of water molecules. The ionization of water delivers free high-toxic radicals. For

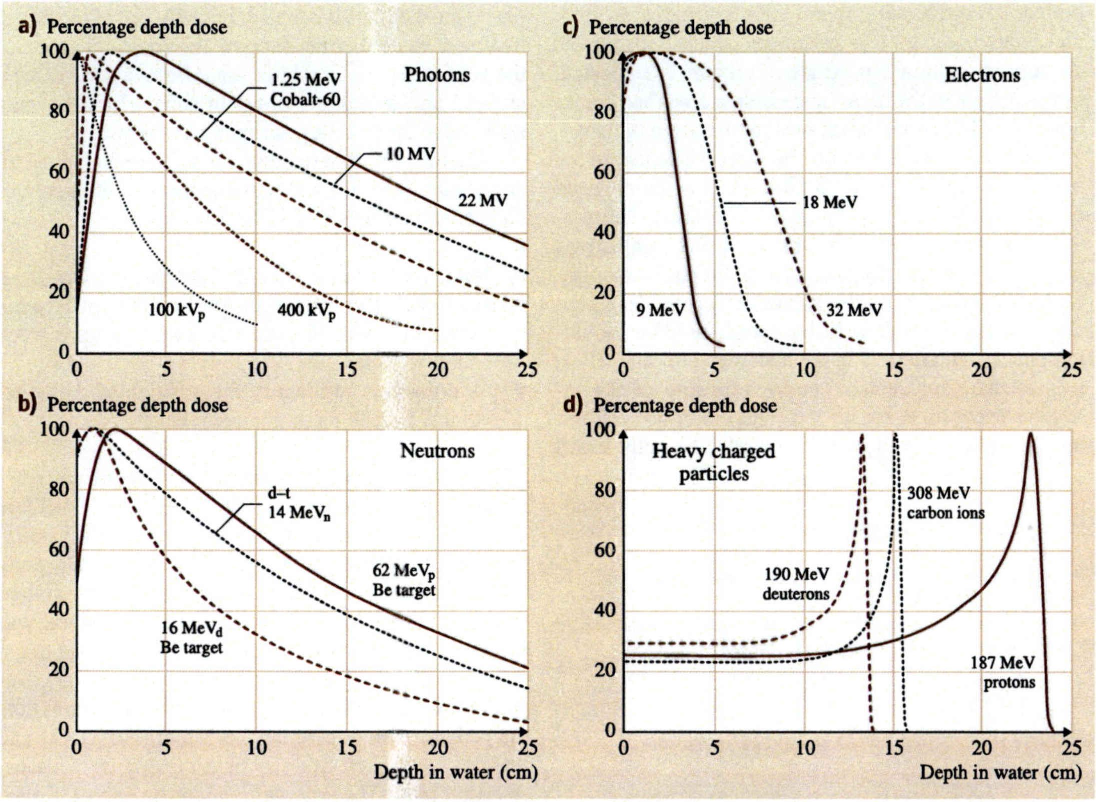

Fig. 45.1a–d Depth dose curve in water for radiation beams of various types and energies. **(a)** Photons, **(b)** neutrons, **(c)** electrons, **(d)** heavy charged particles (after [45.5])

an anticancer (tumor) effect, damage occurs to the heritable information of the DNS in the cells, especially damage to the double helix. Damage exceeding the repair capabilities of the tumor cells prevents cell proliferation and leads directly to apoptosis.

45.1.2 Principles of Treatment Planning and Treatment

In this section, the principles of treatment planning of malignant tumors with teletherapy will be described.

For the application of a radiotherapy treatment a justifiable indication must be given according to radioprotection regulation. Therefore, the tumor and extension will be localized by image-guided modalities like ultrasound, x-ray fluoroscopy, computed tomography (CT), magnetic resonance tomography and positron emission tomography (PET). The diagnosis will be substantiated by laboratory and histological investigation.

Commonly, the patient's case is presented at the tumor conference by the respective medical stuff. Meanwhile there are tumor centers in many medical clinics like breast cancer, cervical cancer, intestinal cancer and prostate gland cancer centers. In these centers, tumor conferences are held that include all physicians in their respective disciplines who determine the guidelines for the corresponding therapy. Depending on the kind of tumor, extension and staging, surgery, chemotherapy, radiotherapy or a combination of the three treatment modalities are applied. If radiotherapy or a combined RCTX is indicated, the following procedure will be carried out.

Depending on the region of treatment, immobilization support aid of the patient will be needed. This will be utilized for an optimized fixed position of the patient during radiotherapy. This is necessary especially for irradiations near organs at risk in order to protect them, e.g., patients with tumors in the head and neck region. A precise positioning is extremely important in order to protect the surrounding spinal cord. If irradiation from all angles is required, a treatment couch with almost no dose absorption, i. e., made of carbon, is indicated. For treatment of the breast and chest, a so-called combi board could be applied. It allows for variation in the angle between chest and couch and can be adjusted to the patient's size. The patient's arms need to be elevated behind the head and can repose on an armrest. This prevents the arms from being irradiated. For irradiations in the head and neck region, individualized thermoplastic head or head–shoulder masks are available. These masks are then fastened to head and shoulder mounting plates. With tumors in the abdomen and the pelvis, i. e., prostate and rectum cancer, exposure of the intestines needs to be minimized. This could be achieved by a prone position on a belly board. The belly board

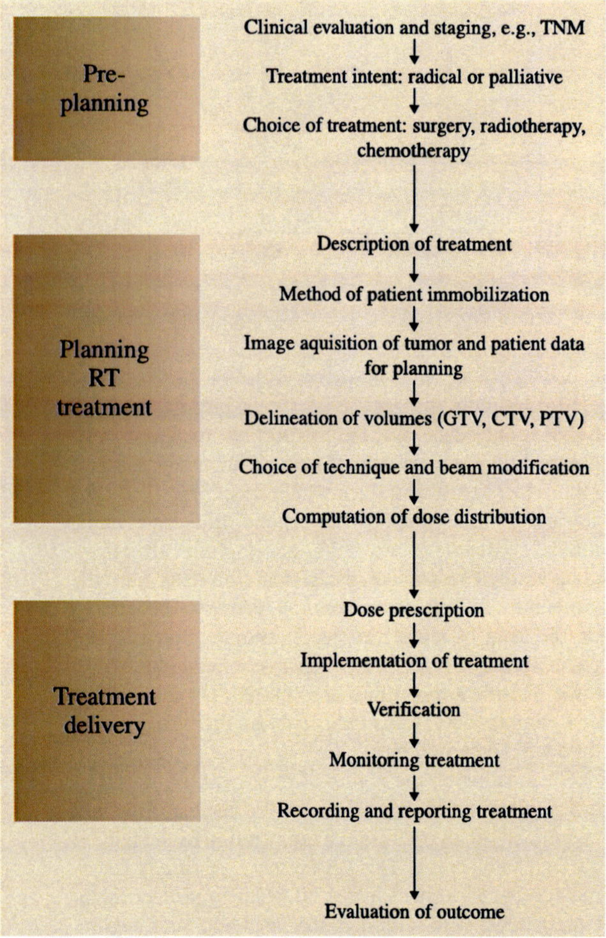

Fig. 45.2 Planning process in radiotherapy (after [45.6])

places most of the abdomen, by gravity, out of the lateral treatment beam. Further support aids for stabilization of the patient are the combifix for knee and feet, as well as head and wedge cushions. In conclusion, one may summarize the patient positioning as follows.

The well-directed treatment of a tumor, the so-called target volume, requires maximum precision in the tumor and patient setup. This will be achieved by:

- Immobilization (no motion during treatment)
- Reproducibility of the patient setup (typically the patient receives 30 dose fractions 5 times a week over 6 weeks)
- Skin markers to identify the position of the target volume in relation to treatment beams.

3-D treatment planning systems calculate the dose distribution mainly on a kV-CT slice. The kV-CT is generated in a common spiral computer scanner utilizing support aid dedicated to the patient. The CT study is then exported to the 3-D treatment planning system mainly via DICOM. On the CT study the physician will contour the gross tumor volume (GTV), clinical target volume (CTV) and planning target volume (PTV). Depending on the institution, many other structures like the patient outline (external), organ at risk (OAR) and margins are defined either by physicians or technical assistants (TAs). The external contour on each CT slice determines the region as a discrete matrix for which the calculation can be performed. If the target cannot be determined in the kV-CT study, an MRI or PET/CT study will be imported to delineate the target (volume). In those studies, target volumes will be transferred via case matching by bone structures, landmarks, external markers, etc.. MRI is used in cases of brain tumors. PET/CT is often used in cases of lung tumors to obtain information about the extension of the tumor and its vital parts. The treatment planning cannot be performed directly on an MRI or PET scan. A CT scan is necessary due to the need for a correct representation of the electron density in the grayscale in the image.

After the last structures have been contoured, treatment planning can start. Therefore, in the case of forward planning (a major part of treatment planning system (TPS) until now), the treatment beams will be built conformal to the target volume. The direction of the beams depends strictly on the OAR close to the target volume. By optimizing the weighting factors of each beam, the dose distribution can be modified with the use of compensation beams, wedges, MLC, lead blocks, mechanical compensators, etc..

According to ICRU 50 [45.7], the target volume should receive between 95 and 107% of the prescribed dose. High-energy photon irradiation ($E > 1$ MeV) does not have a dose maximum at the surface but at a certain depth depending on the energy (Fig. 45.1a). For 6 MeV photons the maximum is at about 16 mm, for 18 MeV photons at about 32 mm in water. As results of this the surface dose is significantly lower than recommended by [45.7]. For tumors close to the surface, the dose can be enhanced by using a tissue equivalent bolus with a sufficient thickness.

Before the application of the treatment plan, the treatment planning needs to be verified. The verification of all treatment beams is performed at a therapy simulator. During the verification process, all treatment parameters are transferred to the patient as there are two different kinds of therapy simulator used. One is used exclusively for verification and localization and is equipped with a kV x-ray generator for imaging and a rotating gantry. The other is a modified CT scanner with positioning room lasers (virtual simulation). During simulation all reference points and field shapes are usually marked on the skin. All beams are verified by making x-ray images that is sent via DICOM to the treatment planning system and the linear accelerator in order to compare the patient's setup with the original treatment plan and to find any deviations in the setup that must be corrected.

After simulation, the patient is transferred for treatment. The patient will be placed on the treatment couch and the skin markers adjusted to the positioning room lasers. To verify the correct patient setup, an x-ray-sensitive film is exposed, an electronic portal image is made or an MV-CT cone beam is performed. The images made are then compared to the x-ray images of the simulation and the digitally reconstructed radiographs (DRR) of the treatment plan. After correction of misalignments, the first fraction will be delivered to the patient. At least those setup verifications are repeated weekly. If the last fraction is delivered, the patient will be released to follow-up care. In Fig. 45.2 a schematic overview of a radiotherapy treatment process including treatment planning is shown.

45.2 Imaging in Treatment Planning

The definition and delineation of the target volume has been based primarily on the experience of radiotherapists, whose knowledge about anatomy, lymphatic drainage, and frequency and placement of recurrences and side effects determines the individual effectiveness and balance between volume definition and dosage for each patient.

The basic principle in treatment planning is the chosen radiation type and the intensity in its direction to the target volume. The target volume definition and the dose calculation are based on a specific 3-D representation of the properties of human tissue. The radiotherapist defines the target volume and tolerance doses for all delineated contours. The contouring of the organs and target volumes is essentially performed directly on the CT images. For more diagnostic information one can turn to the MRI and PET studies, although the dose calculation can only be performed in CT images. By putting the image studies on top of each other the target volume can be delineated on the PET or MRI study and then transferred to the CT study. MRI and CT are the standard imaging techniques for the representation of anatomical relations in the human body.

Meanwhile special MRI imaging techniques like functional MRI, diffusion, perfusion and special PET imaging methods are focusing more on radiotherapy. Those imaging techniques contain information about the biological structure and extension of tumors and the circumventing tissue.

In the following sections different diagnostic imaging techniques and their application in radiotherapy planning will be described in more detail.

45.2.1 Computed Tomography

CT plays an important role in imaging, as only CT images represent the electron densities of tissue in the grayscale (Hounsfield units) necessary for dose calculation. The relationship between electron densities and Hounsfield units requires periodic calibration. This calibration is usually performed using a CT scan of a phantom consisting of material with well-known densities. All other imaging methods only have diagnostic information and can not be used for dose calculations.

In the present status of CT scanner technology, the scanner is equipped using a detector array of up to 256 lines. CT has been used routinely for radiotherapy treatment planning since the early 1980s. Because of the widespread availability of modern third- or fourth-generation CT scanners, they are commonly used for contouring target volumes and OARs. CT studies display nearly perfect inhomogeneities, especially in humans. Geometrical distortions do not have a major influence on the application of CT studies, unlike artifacts caused by metal implant or prosthesis, which could reduce the CT's applicability. To overcome those problems, adaptive filtering of the CT raw data could improve CT images.

The definition of the target volume could be complicated by the low soft tissue contrast and inter- and intrafractional organ motion, although internal or external markers have been introduced. This problem of the uncertainty in location is overcome by introducing safety margins according to ICRU 50 [45.7].

Organ motion could be minimized by introducing respiratory triggering and abdominal pressing.

Using contrast agents, morphological differentiation between infiltrated and normal tissue could be improved. While measuring the temporal contrast agent enrichment, the capacity for perfusion and diffusion of tissue could be determined.

45.2.2 Magnetic Resonance Imaging

The principle of nuclear magnetic resonance is described in the pertinent literature. Recently, one can observe differences in magnetic properties between normal and cancerous tissue. Magnetic resonance imaging (MRI) has assumed a position of growing importance in medical imaging. Today, MRI and angiography, along with functional imaging, hold great promise for revealing characteristics of the human body. Improved soft tissue contrast has made MRI indispensable for visualizing and contouring several tumor entities. Today 1.5 Tesla MRI is a clinical standard. Spatial resolution is improved by increasing the magnetic field, and the scan time is decreased. This enables the visualization of (neuro)functional processes. Due to inhomogeneities of the magnetic field, distortions of several millimeters could occur. These distortions need to be determined and corrected before introducing MRI into radiotherapy treatment planning. This could be accounted for using phantom measurements.

Several studies have investigated the differences in target definitions due to contoured volumes by CT and MRI. It was found that in more than 40% of the investigated cases, the target volume could be shrunk down. These investigations were performed in cases of prostate and brain cancer.

45.2.3 Positron Emission Tomography

The basic principles of PET are described in the pertinent literature. PET imaging with 18F-FDG is very useful in radiation oncology for assessing the spread of the primary cancer, detecting metastases and monitoring the patient during and after treatment.

Particularly useful for this purpose are scanners that combine PET and CT in the same unit so the PET images can be overlaid on x-ray CT images. PET imaging is rapidly becoming recognized as an essential imaging tool for radiation oncology.

A good example of the application of PET in radiation oncology is lung non-small-cell carcinoma. For those tumors, PET distinguishes between atelectasis and tumor, and the involved mediastinal lymph nodes can be detected with a higher accuracy. As a consequence one can spare the treatment of uninvolved lymph nodes and escalate the dose to the primary tumor.

45.3 Basic Techniques in External Beam Therapy

Depending on the region to be irradiated, the tumor type, and the different radiotherapy technologies being used, the most important technologies are as follows [45.8–10]:

- Basic treatment planning
- Conformal treatment planning
- Intensity modulated radiotherapy (IMRT)
- Tomotherapy
- Image guided radiotherapy (IGRT)
- Respiratory guided radiotherapy (RGRT)
- Intraoperative radiotherapy
- Stereotactic irradiation
- Total body irradiation (TBI) with photons
- Total skin irradiation with electrons

45.3.1 Simple Treatment Planning

Basic one-dimensional treatment planning is used in cases of treatment, where only one simple beam or two opposed beams are used. Typical applications are stimulation irradiation (single beam) or bilateral whole brain irradiation in the case of brain metastasis (opposed beams).

For such one-dimensional treatment, the calculation becomes very simple. Assuming a quadratic field, the applied monitor units (proportional to the dose after calibration) could be calculated as

$$\mathrm{MON}(\mathrm{FS}, d) = [\mathrm{Ref.MON}/\mathrm{FSR}(\mathrm{FS})\mathrm{TMR}(\mathrm{FS}, d)] \times [\mathrm{FIA}/(\mathrm{FIA} + d_{\max})]^2 ,$$

where Ref.MON = Ref.Mon. for a 10×10 cm beam, $d_{\max}$ is the depth at dose maximum, FIA the focus isocenter distance, FS the field size, d the depth, FSR the field size ratio and TMR the tissue maximum ratio.

For calculations of applied monitor units of rectangular fields, the equation must be reformatted in equivalent-quadratic fields whereby the side length must be recalculated and a and b are the side length of the rectangular field.

The side of an equivalent square field is $2ab/(a+b)$ [45.11, 12].

45.3.2 Three-Dimensional Conformal Treatment Planning (3-D CRT)

Conformal irradiation is routinely used as a standard method in external beam therapy. Treatment planning is commonly performed in three dimensions. In order to achieve conformal treatment for the target volume, a multileaf collimator (MLC) will be shaped as the target volume appears in the beam eye view for any beam direction and sometimes a margin will be applied (forward planning).

Due to the narrow containment of the target volume, exposure to surrounding tissue can be drastically minimized. The rate of the resulting side effects at the OAR adjacent to the target volume is reduced. An escalation of the dose to the target volume is accompanied by a narrowing of the margin to the target volume and a lower dose to the surrounding OAR, resulting in a lower rate of side effects.

Image-guided contouring using CT, MRI and PET of the target volume plays a major role in treatment. Only this combination of imaging guarantees a precise delineation of a tumor in relation to surrounding tissue and OAR.

The optimization and quality control of a treatment plan is based on several criteria; one is the ICRU 50 conformality of the dose to the target volume $D_{\mathrm{PTV}}(95\% < D_{\mathrm{PTV}} < 107\%)$. Furthermore, the dose constraints for $D_{\max}$, $D_{\min}$, D_{mean} and D_{50} and the accepted tolerance doses to the OAR are adjusted utilizing a dose volume histogram (DVH) to preserve the OAR by minimizing the resulting side effects caused by overdosage. Accepted values for tolerance doses are taken from the literature. In addition to dose constraints like $D_{\max}$, there are side effects depending on the partial volume dose. The tolerance then depends on the volume receiving a certain dose and the organ's function; either the organs act in a serial-like manner, such as the spinal cord, or in a parallel manner, such as the liver, lungs or kidney. Unfortunately, most tumors have a parallel functioning structure so that an perceptible damage is only caused after a certain threshold dose.

Serially acting organs have a different threshold dose; if their function is significantly affected, the organs suffer irreversible damage. In Fig. 45.3 conformal treatment of bronchial cancer is shown.

Most treatment plans are coplanar plans; all beams are in the same plane. Noncoplanar planning is mainly used in stereotacic treatment (Sect. 45.3.10).

45.3.3 Inverse Planning

Radiation therapy inverse planning involves the development of a treatment plan whereby the planning computer is given a set of objectives and beam parameters to adjust in iterative fashion to arrive at a satisfactory dose distribution. For example, the treatment planning computer could iteratively adjust the weights of the beams to provide more uniform dose to the target.

45.3.4 Intensity Modulated Radiation Therapy

Aside from 3-D CRT treatment, IMRT has become more and more acceptable as routine radiation therapy [45.13, 14]. In IMRT the adjustment of the dose distribution to the target volume and limitation at the same time with the dose to normal tissue is performed not only by modulating the beam shapes, but also by modulating the dose dividing the beams into small beamlets (e.g., 2×2 mm). The weight of each beamlet can then be adjusted. This timely intensity modulation could be realized in two different kinds of delivery using the movable multileaf blades either in a *sliding window* technique or in a *step and shoot* technique. The MLC conversion refers to the process of transferring the ideal fluence map to deliverable MLC beamlets [45.15, 16].

This IMRT technique allows one to treat complex target volume scenarios where the PTV is very close to an OAR, i. e., head and neck case, and spare the spinal cord. In order to meet all treatment objectives (constraints) set by the physician, the inverse planning calculation becomes time consuming. The delivery of the plan (monitor units used) takes longer and the qual-

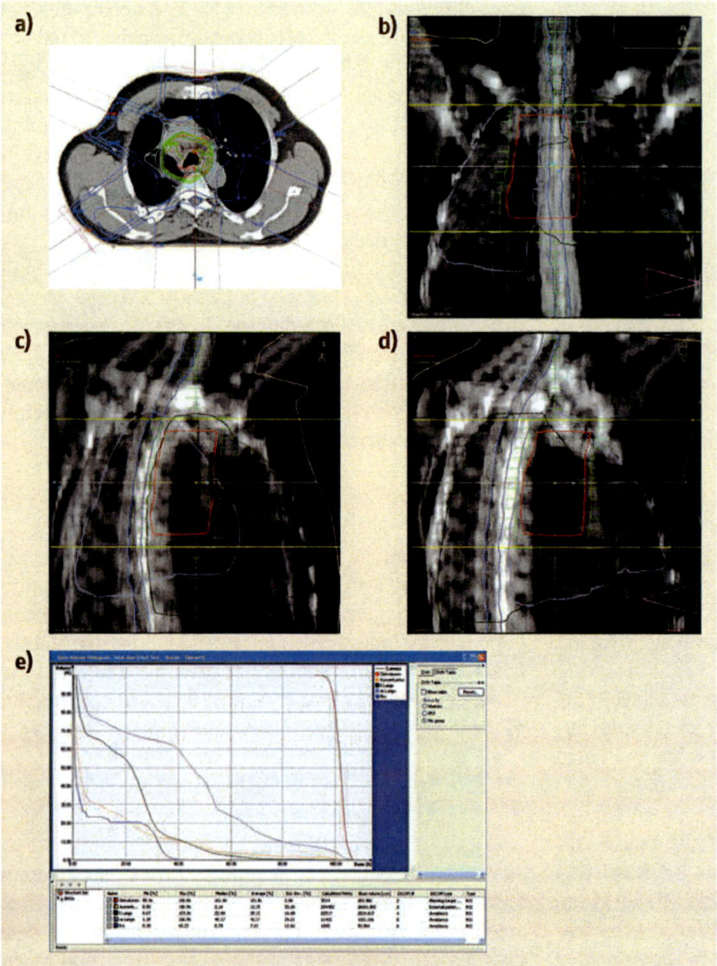

Fig. 45.3 (a) Dose distribution of three-isocentrical-beams technique (b)–(d). Pictures showing eye beam views from different beam directions (e). DVH showing relative dose for PTV and organs at risk (i. e., spinal cord and lungs) (after [45.17])

ity assurance is more tedious than in 3-D CRT. As a consequence before the first treatment of the patient, the treatment plan must be verified. Figure 45.4 shows a medical linear accelerator with an MLC.

45.3.5 Dynamic Delivery Techniques

The *sliding window approach* to treatment delivery involves moving the MLC leaves while the x-ray beam is on. This approach is more complex than step and shoot for the following reasons: (a) the window is often very narrow, resulting in narrow beam geometry, and (b) the MLC quality assurance protocol for dynamic delivery is more stringent, requiring leaf speed characterization. With the sliding window method very complex distributions can be achieved.

Allowing the gantry to arc while the leaves are moving is a technique known as intensity modulated arc therapy (IMAT) or dynamic arc therapy (RapidArc and VMAT). This approach may require a few arc passes to completely deliver the treatment because a given gantry angle can have only one MLC shape per arc. This method allows for very complex dose distributions but is challenging to plan and verify.

45.3.6 Tomotherapy

Serial Tomotherapy

Serial tomotherapy, literally meaning *slice therapy*, has been in clinical use since 1994. The commercial system, Peacock, combines a narrow MLC attachment to the linear accelerator (MIMiC collimator)

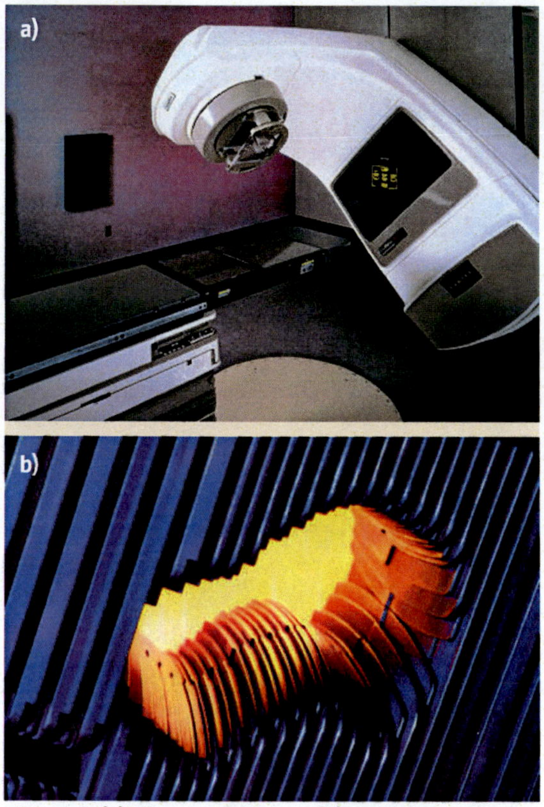

Fig. 45.4 (a) Linear accelerator, (b) multileaf collimator (after [45.18], Cinac 2100C, HD-MLC2)

with precise couch increments to deliver a slice-by-slice treatment. The MLC operates in binary mode, with the leaves either fully open or fully closed. Thousands of patients have been treated with this system, most of whom are patients with brain or head and neck cancers. The Peacock system has paved the way for other planning and delivery systems that are now available because it provided much of the early clinical data demonstrating the benefits of IMRT.

Helical Tomotherapy

In the same manner that CT scanners have moved from axial to helical delivery, Tomotherapy Inc. has adopted a helical delivery system [45.20]. The treatment unit looks much like a CT scanner, with a compact 6 MV linear accelerator replacing the x-ray tube (Fig. 45.5). With this design, many of the components in a standard linear accelerator are not needed. For example, (a) the compact size eliminates the long cantilevered gantry and the bending magnet; (b) there are no electron energies and the target is fixed, so there is no carousel; and (c) since the beam is modulated by the MLC, there is no need for a flattening filter. The absence of a flattening filter provides a higher beam output compared with a conventional linear accelerator. A binary MLC is used, similar to the MIMiC collimator previously mentioned. A primary advantage of the tomotherapy system is that it fully integrates real-time patient imaging with the delivery of radiation treatments. The clinical success of conformal (and conformal avoidance) radiotherapy relies upon the ability to accurately localize the beam on a day-to-day basis. Onboard axial image information is obtained using a megavoltage beam and a bank of CT detectors. So-called megavoltage CT (MVCT) provides surprisingly good image quality and is not subject to artifact generation in high-density objects (e.g., dental fillings) [45.21]. Also, an electron density conversion is not required because the imaging beam and treatment beam are one and the same. The imaging capabilities of this device offer a distinct advantage for comparison with virtual simulation. In helical tomotherapy, the exit fluence at the detectors is used to reconstruct the dose delivered during the actual treatment. This information is fed back to the TPS to monitor the dose delivered over the patient course of therapy. This method of providing information from the actual treatment back to the planning computer is called adaptive planning. Changes to the treatment to achieve the intended constraints can be made as treatment progresses.

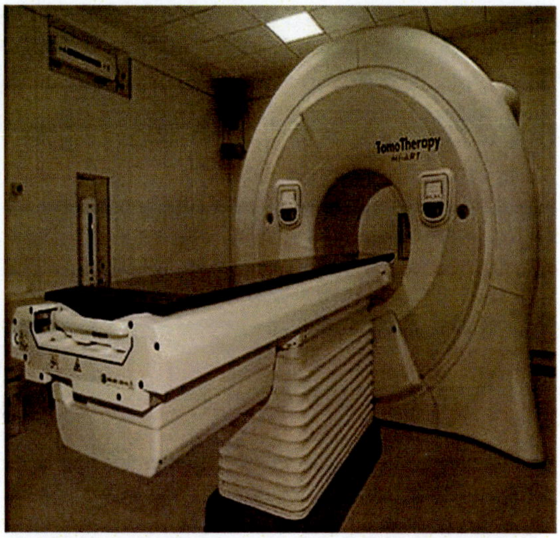

Fig. 45.5 Tomotherapy HI-Art II system (after [45.19])

45.3.7 Image Guided Radiotherapy (IGRT)

IGRT is the process of frequent two- and three-dimensional imaging, during a course of radiation treatment, used to direct radiation therapy utilizing the imaging coordinates of the actual radiation treatment plan. The patient is localized in the treatment room in the same position as planned from the reference imaging data set. An example of 3-D IGRT would include localization of a cone-beam computed tomography (CBCT) dataset with the planning CT dataset from planning (Fig. 45.6). Similarly 2-D IGRT would include matching planar kV radiograph fluoroscopy or MV images with DRRs from the planning CT scan.

This process is distinct from the use of imaging to delineate targets and organs in the planning process of radiation therapy. However, there is clearly a connection between the imaging processes as IGRT relies directly on the imaging modalities from planning as the reference coordinates for localizing the patient. The variety of image-gathering hardware used in planning includes CT, MRI and PET, among others. Through advancements in imaging technology, combined with a further understanding of human biology at the molecular level, the impact of IGRT on radiotherapy treatment continues to evolve.

45.3.8 Respiratory Guided Radiotherapy

The use of four-dimensional CT (4-D CT) in lung cancer radiotherapy reduces geographical miss and reduces normal tissue toxicity by individualizing the margin from the clinical target volume. The main steps in the target definition for 4-D CT are as follows. First, a composite of the gross tumor in all phases of the respiration cycle is created, then a margin for microscopic disease is added to create the clinical internal target volume; finally, a margin for setup error is added to create the planned target volume (4-D PTV). This 4-D PTV is designed to ensure satisfactory irradiation of the tumor in all positions throughout the respiration cycle. With the introduction of 4-DCT, there has been a great deal of interest regarding the possibility of using the 4-D CT to plan and deliver respiratory guided radiotherapy (RGRT). The patient's respiration cycle, for example, is monitored continuously by an external surrogate, an infrared marker box placed on the sternum. The movement of the marker box is picked up by a camera and a respiratory trace is seen in the control room. This trace enables the selection of a respiratory phase or *gate* for treatment delivery and the treatment beam is switched on only during this interval. RGRT has been shown to reduce the size of the PTV when compared to the standard 4-D PTV [45.23].

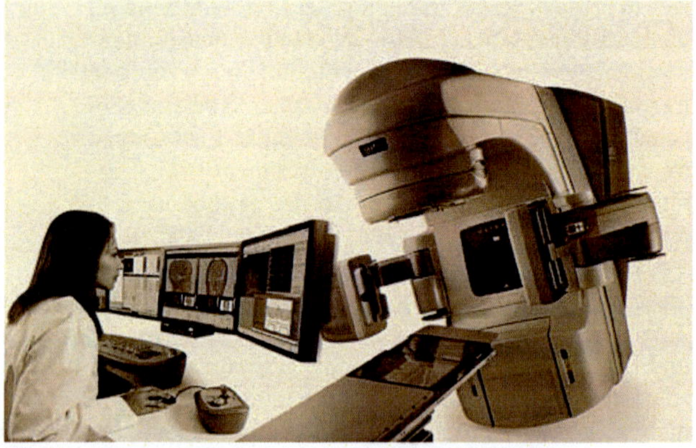

Fig. 45.6 Linear accelerator with IGRT (after [45.22])

45.3.9 Intraoperative Electron Radiation Therapy

Intraoperative electron radiation therapy (IOERT) is the application of electron radiation directly to the residual tumor or tumor bed during cancer surgery. IOERT has also been called precision radiotherapy, as the physician has direct visualization of the tumor, can exclude normal tissue from the field and can also protect critical structures within the field by narrowing the irradiated volume. IOERT is a method that provides dose escalation as a precision component of radiotherapy. IOERT can also be combined with radiation sensitizers to better treat hypoxic cells and can be given at the time of surgery when microscopic residual tumor cells are most vulnerable to destruction. IOERT has been used in combination with curative external beam therapy as it results in lower integral doses and shorter treatment times in boost dose concept (20–30% of the dose is dedicated to the IOERT). It has been proved that local tumor control and the 10 year survival rate could be improved for rectum cancer, soft tissue sarcoma, stomach cancer, pancreatic cancer, kidney cell cancer and gynecological tumors without an increase in side effects.

In principle IOERT is applied as a primary radiation therapy or directly after a presurgery external radiation therapy. If the tumors are inoperable, then IOERT is used as a palliative treatment, especially for pancreatic tumors (Fig. 45.7). The prescribed dosage varies between 10 and 30 Gy. The biological effectiveness of this

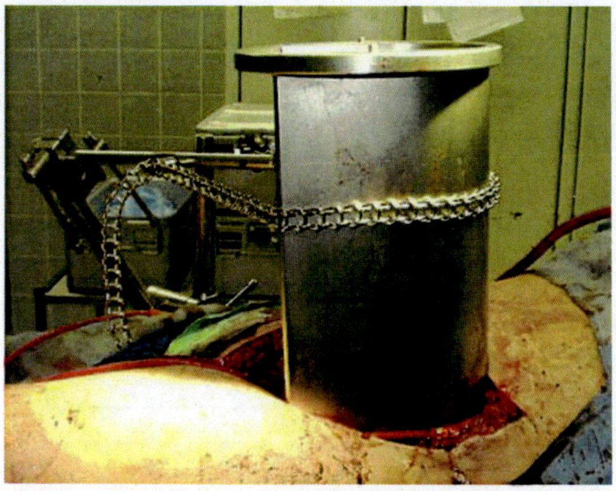

Fig. 45.7 IOERT using an electron tube during surgery directly mounted on medical linear accelerator (after [45.24])

single treatment is increased by two or three times the conventional fractionation of 2 Gy per day. A significant dose escalation in the target volume can be achieved by combining IOERT and conventional external beam therapy. In parallel, the late side effect rate in normal tissue will be reduced and local tumor control will be improved.

45.3.10 Stereotactic Irradiation

Stereotactic irradiation is a specialized type of external beam therapy. It uses focused small radiation beams targeting a well-defined tumor using extremely detailed imaging scans. In using this technique one can deposit a very high dose to the tumor and spear the normal tissue. Radiation oncologists perform stereotactic treatments, often with the help of a neurosurgeon for tumors in the brain or spine. There are two types of stereotactic radiation. Stereotactic radiosurgery (SRS) is when doctors use a single or several stereotactic radiation treatments of the brain or spine. Stereotactic body radiation therapy (SBRT) refers to one or several stereotactic radiation treatments with the body, such as the lungs.

Some physicians say that an advantage of stereotactic treatments is that they deliver the right amount of dose (12–18 Gy per fraction or single dose) to the target volume in a shorter amount of time than conventional radiation therapy, which can often take 6 to 11 weeks. In addition, treatments are given with extreme accuracy, which should limit the effect of the radiation on healthy tissues. One problem with stereotactic treatments is that they are only suitable for certain small tumors.

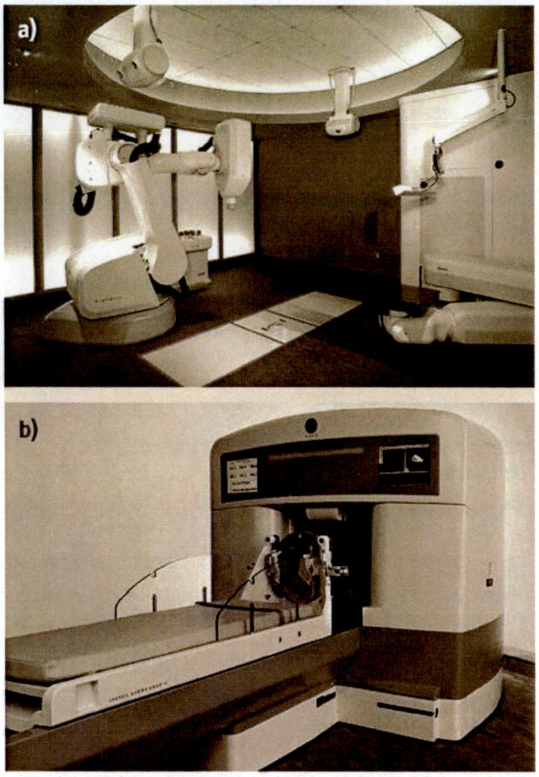

Fig. 45.8 **(a)** CyberKnife (after [45.25]), **(b)** Gamma Knife (after [45.26])

Stereotactic treatments can be confusing because many hospitals call the treatments by the name of the manufacturer rather than calling it SRS or SBRT. Brand names for these treatments include Axesse, remote-controlled radiosurgery system Cyberknife (Fig. 45.8a), Co^{60}-source Gamma Knife (Fig. 45.8b), Novalis, Primatom, and others. While the gamma knife is used on brain tumors, a medical linear accelerator with stereotactic properties can perform extracranial stereotactic treatments, and the Cyberknife could be applied for extracranial stereotactic treatments.

45.3.11 Total Body Irradiation with Photons

Total body irradiation (TBI) is primarily used as part of the preparative regimen for hematopoietic stem cell (or bone marrow) transplantation. As the name implies, TBI involves irradiation of the entire body, though in modern practice the lungs are often partially shielded to lower the risk of radiation-inducing lung injury. TBI

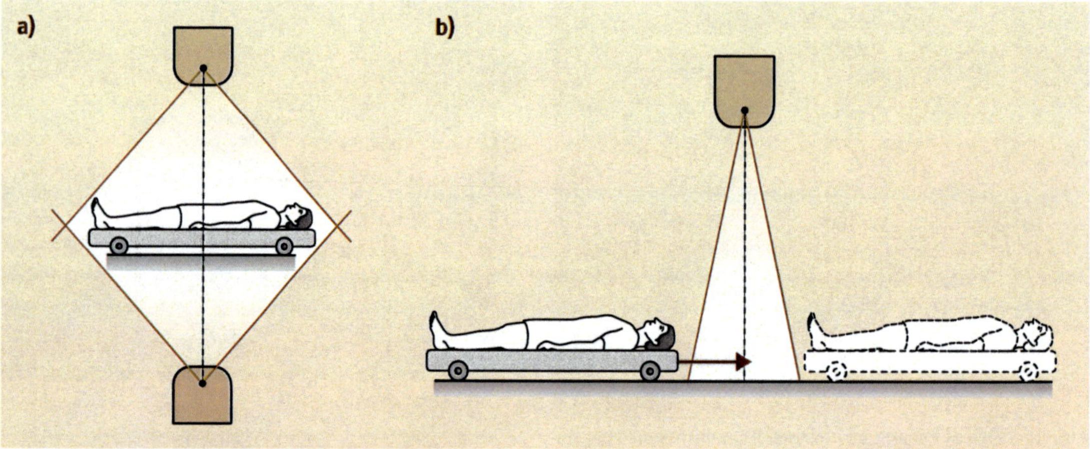

Fig. 45.9a,b Two arrangements for total body irradiation using a medical linear accelerator, parallel-opposing-beams technique (**a**) and translational technique (**b**) (after [45.8])

in a context of bone marrow transplantation serves to destroy or suppress the recipient's immune system, preventing immunologic rejection of transplanted donor bone marrow or blood stem cells. Additionally, high doses of TBI can eradicate residual cancer cells in the transplant recipient, increasing the likelihood that the transplant will be successful.

Doses of TBI used in bone marrow transplantation typically range from 10 to 12 Gy. For reference, a dose of 4.5 Gy is fatal in 50% of exposed individuals without aggressive medical care. At these doses, TBI both destroys the patient's bone marrow (allowing donor marrow to engraft) and kills residual cancer cells. Nonmyeloablative bone marrow transplantation uses lower doses of TBI, typically about 2 Gy, which do not destroy the host bone marrow but do suppress the host immune system sufficiently to promote donor engraftment [45.27].

In modern practice, TBI is typically fractionated. That is, the radiation is delivered in multiple small doses rather than one large dose. Early research in bone marrow transplantation demonstrated that this process of splitting TBI into multiple smaller doses resulted in lower toxicity and better outcomes than a single, large dose.

Indications for TBI are different kinds of leukemia in pediatrics and for adults:

- Acute lymphatic leukemia (ALL)
- Acute myeloic leukaemia (AML)
- Chronical myeloic leukemia (CML)
- Myelodysplastic syndrome (MDS)

Several techniques for delivery of TBI have been developed. One of the most common techniques is the translational technique (Fig. 45.9b), which utilizes a conventional medical linear accelerator. Therefore, large beams are needed. An extension of the focus-skin distance to about 2 m enables the width body to be captured by the beam. For the longitudinal direction the patient is moved onto a motorized couch through the beam. As an alternative, the focus-skin distance can be extended to ca. 3.5 m, so that the beam covers the entire body. Then either the patient is placed in two positions, a prone and supine setup, or two accelerators are used (Fig. 45.9a).

45.3.12 Total Skin Electron Beam Therapy

Total skin electron beam therapy (TSEBT) is a radiation treatment method to irradiate the skin in case of melanoma using fast electrons. Typical electron energies are between 6 and 12 MeV, and the total dose is in a range of 30–40 Gy.

Typical treatment techniques are as follows:

- Several field arrangement: the patient stands hands up facing the medical linear accelerator. After the first irradiation the patient must turn by a certain angle in order to irradiate the other part of the skin (Fig. 45.10).
- Continuous rotation: the patient stands hands up facing the medical linear accelerator on a continuous rotating platform and performs a full 360° rotation.

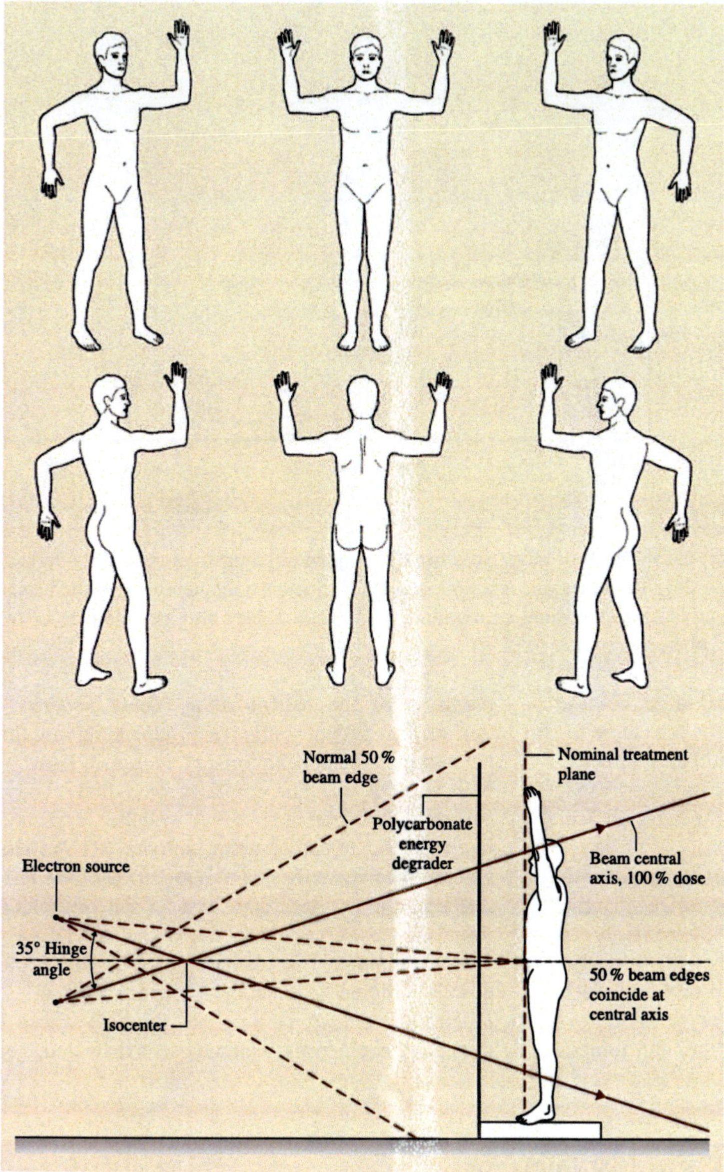

Fig. 45.10 Total skin irradiation with electrons (after [45.28, p. 83], courtesy of Dr. P. Rudd)

45.4 Target Volumes and Organ at Risk

In the ICRU 50 and the ICRU 62 as supplement, the target volume definitions are described in detail (Fig. 45.11). A target volume defines the region to be treated or to be irradiated. The following target volumes are differentiated.

- *Tumor volume*: The tumor volume is described as a macroscopic palpable region or as a tumor visible in the imaging. According to ICRU 50, the tumor volume corresponds to the gross tumor volume (GTV).

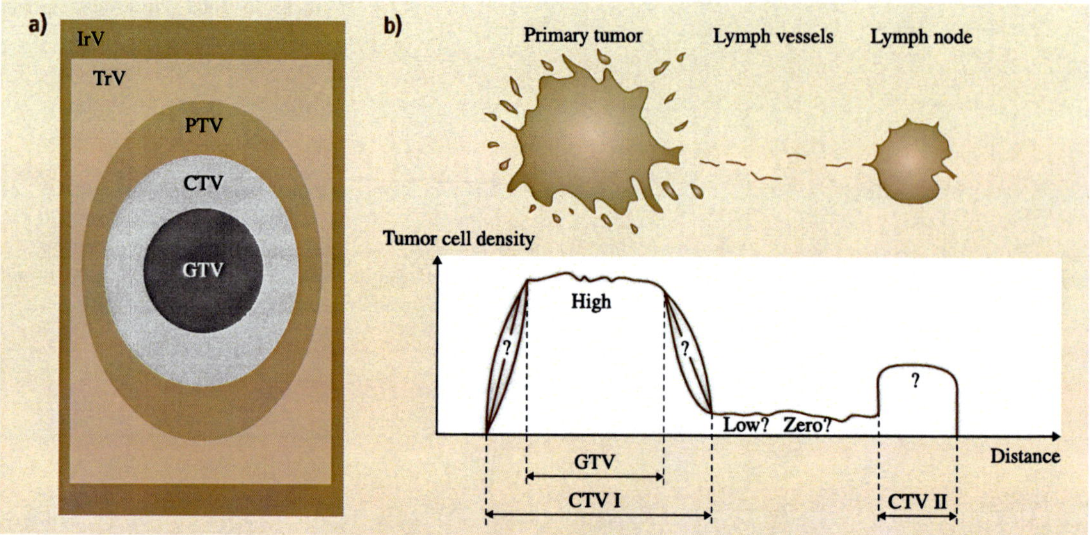

Fig. 45.11 **(a)** GTV – gross tumor volume; CTV – clinical target volume; PTV – planning target volume; TrV – treated volume; IrV – irradiated volume (after [45.28]). **(b)** Schematic example of a GTV (demonstrable tumor = striated area) and also for CTVs to care for suspected, subclinical extensions surrounding GTV (CTV 1, here also including the GTV) and the regional lymph nodes (CTV II, = open area) (after [45.29])

- *Clinical target volume*: The clinical target volume contains, in addition to the clinically detectable tumor, the area where tumor cells could exist or could not be verified. The area is called the tumor propagation area. ICRU 50 defines the volume as the clinical target volume (CTV).
- *Planning target volume*: The planning target volume contains the CTV including a safety margin. The safety margin describes the uncertainty of the determination of the CTV. The safety margin contains uncertainties due to organ motion caused by breathing, an interfractional change of the organ fillings, i. e., bladder or intestine, and limited precision of the patient setup. These uncertainties of the patient setup vary depending on the positioning. Support equipment such as a thermoplastic mask, vacuum matrices or a belly board is used. The planning target volume is named in the ICRU planning target volume (PTV). The volume receive neither more than 107% of the prescribed dose nor less than 95% of the prescribed dose.
- *Treated volume*: This defines the area where the dose is sufficient to reach the therapeutical goal.
- *Irradiated volume*: This defines the area where a direct, not exclusively scatter, dose is delivered.

45.5 Modern Treatment Planning Systems

Modern conventional 3-D treatment planning systems for IMRT and 3-D CRT of patients receiving external beam therapy consist of different modules. The fundamental modules of these treatment planning systems are, on the one hand, the beam data entry for photons and electrons and, on the other hand, the different calculation algorithms for photons and electrons. All treatment planning systems require the entry of data that describe the radiation beam. The amount of data required depends on the computational algorithm, varying from almost none to thousands of measurements that completely map the radiation beam for a variety of field sizes. Systems that require the entry of almost no measured beam data rely upon a detailed description of the design of the accelerator head, collimator and accessories. The dependence of a system on a large amount of measurements is thought by some to be a disadvantage because substantial effort is required to accumulate

Fig. 45.12 Two 3-D treatment planning systems; (*front*) Helax-TMS, (*back*) Oncentra-MasterPlan (after [45.33])

the large amount of data. The accuracy of any treatment planning system is ultimately determined by comparison with measured data. Complete validation of a treatment planning system requires a comparison between calculations and measurements over a wide variety of treatment conditions, necessitating a large amount of measured data. For photon beam calculation, pencil beam algorithms such as precalculation and collapsed cone algorithms for final calculations are recommended for forward planning. For electron calculations the Monte Carlo method is widely used. An explanation of the differences and precision of those algorithms can be found in the literature [45.9, 30–32]. For inverse planning, where iterative algorithms find an optimal solution for the objective function, the AAA superposition algorithm is commonly used [45.10]. In Fig. 45.12 two different 3-D treatment planning systems are shown.

45.5.1 Basic Components of a Treatment Planning System

A treatment planning system is a combination of hardware and software components that allow the user to produce and display the calculated dose distribution for the physician who prescribes a patient's radiation treatment. The basic components of treatment planning system are as follows [45.3, 34]:

- Software
  - Utility software for entering, printing and accessing an external beam treatment unit and measured absorbed dose data (physics module);
  - Utility software for transferring patient CT, MRI, PET and other imaging modalities and converting the CT data into electron densities. Software for creating and organizing patient data (import module);
  - Contouring software for entering the external contour of the patient, internal contours, target volumes and landmarks pertinent to the treatment (contouring module);
  - Dose calculation initializing software for establishing a calculation grid and method of calculation (beam modelling module);
  - Dose calculation software (dose calculation module);
  - Isodose display software including normalization, DVH and beam weighting (analyzing and optimizing module);
  - Printer and plotter for (scaled) documentation on paper (DIN 6827-1) or as a PDF file [45.35];
  - DICOM export of the dose plan to record and verify system;
  - Archiving system to guarantee for 30 years;
  - Backup.
- Hardware
  - Central processing unit, sufficient (virtual) memory to accommodate and operate the application software and host the database;
  - High-resolution graphics capability including high-resolution monitors;
  - (External) mass storage (hard disk) capacity sufficient to easily retain all current patients in treatment for backup, archiving, removable storage, CD, DVD, tape and network drives;
  - Digitizer and scanner to enter patient contours and phantom manual;
  - Uninterruptible power supplies (UPS) in order to prevent data loss due to electrical shut down;
  - Ethernet card (connection) for import and export of DICOM data from external modalities like CT, simulator or PACS system, and remote access. In order to overcome the large amount of data transferred, the minimum recommended transfer rate today is 100 MB/s.

45.5.2 Physics Module

All treatment planning calculation algorithms require input beam data of some form. For conventional treatment planning systems, beam data need to be measured for each beam quality available at the side. The accuracy and quality of the input data are dependent on the measured or calculated data that are produced by the user. For practical reasons, the data are generally determined over a limited range of conditions, e.g., limited depths and field sizes. Whenever calculations extend beyond the range of measured data, the output results should be scrutinized since the algorithms can perform inaccurate extrapolations. In addition, the measured data have their own inherent uncertainties or inconsistencies that depend on the care taken by the individual person generating the data, the types and sizes of the detectors that are used, as well as on the stability of the linear accelerator producing the radiation beam (e.g., variations in flatness, PPD and symmetry with gantry angle or with time). Generally both relative data and absolute data need to be determined – relative data in the form of dose ratios and absolute data in terms of the linear accelerator output calibration. The latter is a requirement of the treatment planning system if it is used for monitor unit or time calculations. Before use, the data need to be approved and locked from unauthorized use.

45.5.3 Import Module

There are at least three major sources of input data. The first includes measured beam data from a 3-D water phantom system. The second includes image data, usually from a CT scanner, MRI, ultrasound or PET. The third source includes data from keyboard and mouse and involves details of the plan input such as field size, gantry angle, collimator rotation, beam energy, etc.. The latter is standard practice for any treatment planning system. The first two usually require network connections or magnetic media such as discs or tape and must have file formats that are compatible with the treatment planning system. As part of the commissioning process, it is important to check for hardware compatibility, especially when the data sources are from different manufacturers. For image data DICOM version 3 or DICOM RT (for radiation therapy) formats must be used. While file compatibility should be defined in the specifications, compatibility of file formats can only be assessed by going through a file transfer process. It is important that the data from the water phantom system and the patient image sources be checked for accuracy and that they have been properly transferred to the radiation therapy planning system. This can be simply done by performing analysis of the input data for well-known configurations, i. e., that the geometries are correct with no magnification errors and no spatial coordinate errors.

The conversion to electron density and scattering power is often performed with a user-defined lookup table. The data within the table are then used with a linear interpolation to determine the relative electron density for any CT number. Usually such tables are generated using a water equivalent circular phantom with a number of different materials inserted into the phantom of known electron densities representative of normal tissues within the patient, especially lung- and bone-equivalent materials. The import module should also be able to read the DICOM RT format from previous irradiations of the patients.

45.5.4 Target Definition or Contouring Module

The contouring module is used for the delineation of the target volume, OAR and external contours. In most treatment planning systems, the dose calculation is limited to external contours drawn. The delineation of the target volume is the responsibility of the attending physician. The organ at risk and other contours are drawn by a medical physicist depending on the regulations of each institution. Furthermore, the reference point (isocenter) should be marked. If margins, help contours or boluses (including its density) are needed, they must be contoured in the module. The definition of margins differ from institution to institution depending on known setup errors, respiration organ fillings and uncertainties in the imaging of the tumor. The following organs are typically at risk depending on the body region:

- Head and neck: parotids, brain stem and spinal cord;
- Chest: heart, lungs and spinal cord;
- Abdomen: intestine and kidney.

The risk of side effects caused by radiation therapy was investigated by *Emami* et al. [45.36] and the tolerances found were categorized for different OARs for the appearance of side effects of 5 and 50% in 5 years (called TD5/5 resp. TD50/5).

Not only are the structures of the target volume and OAR the treatment aim and region of risk but also, after successful dose calculation, dose information about $D_{\max}$, $D_{\min}$ and D_{mean} could be derived by accessing the so-called dose volume histogram (DVH). It is recommended to delineate all OARs and critical structures within and outside the treatment area.

In the target definition module, the reference point is defined as being related to retraceable anatomic structures and defines a displacement vector to the isocenter of the medical linear accelerator. This isocenter is defined by a stationary laser coordinate system and coincide with the laser systems at the simulator, CT-scanner and medical linear accelerator. The alignment of the patient for the isocenter or the reference point is marked or tattooed on the patient. These markers allow a precise realignment of the patient relative to the isocenter at a known displacement vector.

45.5.5 Beam Modeling Module

In the beam modeling module (in forward planning), the beam configuration is selected for a target-volume-conforming arrangement by sparing the OAR. The beam quality and beam energy depend mainly on the tumor depth. For a surface near tumor one would select either electrons or photons with a tissue-equivalent bolus. For tumors deeper in the body ($d > 1$ cm) photons are the selection of choice. The beam size is shaped either by MLC or lead blocks to be conformal to the target volume. As beam modifier, it acts as either a fixed or dynamic wedge filter or a combination of them. A typical beam arrangement is a combination of several beams of different directions and different energies, and perhaps different beam qualities as well. Many TPSs are capable of calculating treatment plans for brachytherapy and have the ability to overlay or add plans from brachytherapy and external beam therapy.

The determination of the calculation grid defines the precision and resolution of the calculation, but a coarse calculation grid resulting in short calculations allows a quick overview of the dose distribution according to the prescribed dose and the dose to the OAR. If the overview is satisfactory, then the calculation grid becomes finer and the resolution could be dropped by a fraction of a millimeter. It is handled different among the TPS to which contour a voxel belongs in case of an overlapping between the planning target volume (PTV) and the OAR. The voxel is taken into account among the TPSs is different; either the voxel belongs to the volume set with higher priority or it belongs to both volumes. In the case of a DVH comparison, the qualities of different plans made in different TPSs must be taken into account.

Treatment planning using electrons can be directly added to photon planning in terms of the biological effectiveness of the beam, but on the other hand electrons mostly come with an electron applicator of a certain size, and a shaped beam needs to be constructed using lead inlays for applicators. The other difference is the shape of the percentage depth dose (PDD) of electrons in tissue. The electron beams do have a flatter $D_{\max}$ than photon beams but a higher D_{surface} and a very shape decline of a certain distance, where 97% of the dose vanishes and just bremsstrahlung remains.

In inverse planning the beam modeling module is completely different from that in forward planning. Here the constraints to the PTV and the OAR need to be defined. The dose prescription must become more precise in terms of the permission for $D_{\max}$, $D_{\min}$, D_{mean}, D_{50} etc.. One could restrict the number of gantry angles and the minimum opening size of the MLC. Problems occur when the constraints cannot be fulfilled simultaneously. Then the constraints must be reset and relaxed.

45.5.6 Dose Calculation Module

In the dose calculation module, the actual calculation of the dose distribution takes place. The calculation algorithm and its precision are chosen, i. e., pencil beam, collapse cone or Monte Carlo method, which is becoming more and more fashionable, while the CPU is improved [45.37]. The patient contour can be extended, (i. e., if the beam enters the end of the CT-scan but does not leave the patient contour before. Due to the calculation time, the inhomogeneity correction can be set off, so that with large density differences like between bone and lung tissue the calculation becomes incorrect by several percentage points.

45.5.7 Evaluation and Optimization Module

In the evaluation and optimization module, the final point of normalization is defined. The isodose setup is placed in relation to that point. According to ICRU 50, the normalization level is set to 100% (deviations need to be denoted). The prescribed dose is then set to 100% (deviations need to be denoted). In order to calculate the monitor units for a treatment fraction, the fraction dose must be configured. In the case of an integrated boost, the fraction dose for target volumes needs to be determined, although certain thresholds should neither be exceeded (2.5 Gy) nor undershot (1.7 Gy). One target volume is defined as the leading target volume, which determines whether the monitor unit is irradiated. The DVH reveals $D_{\max}$, $D_{\min}$, D_{mean} and other important dose values like D_{10}, D_{20} and D_{50}. Figure 45.13 shows the influence of different beam arrangements on the dose distribution.

If ICRU 50 for the PTV in forward planning is not fulfilled, the beam weights, gantry angle or wedge must be varied; otherwise beams for saturation or vertex beams must be added. In the case of inverse planning, constraints for PTV and OAR must be relaxed or the modulation factor [MLC openings (time or size)] changed. If this approach does not have the desired success, then one must consider loosening the ICRU 50 demand, and then a change in the constraints or, in very

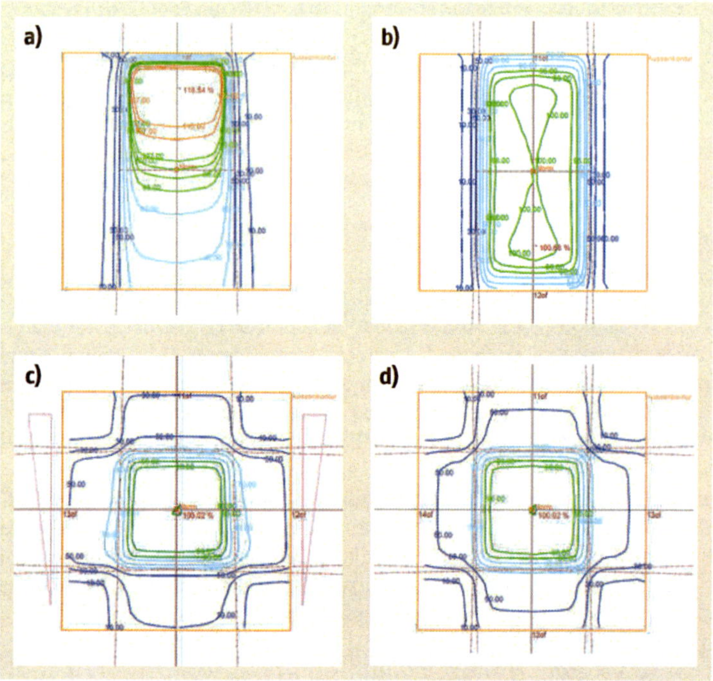

Fig. 45.13 **(a)** Anterior beam, **(b)** two opposing beams, **(c)** three-beam plan (one beam without a wedge and two beams with wedges), **(d)** four-beam box (after [45.17])

difficult cases, the contouring must be reconsidered. This is a task for the medical physicists and physicians. There will always be a tradeoff between increasing the risk of side effects and the loss of control over the (local) tumor. The general condition of the patient and the treatment goal are important boundary conditions.

If the reconsideration leads to a successful treatment plan and the ICRU 50 [45.7] could be fulfilled, the plan must be approved by medical physicist and the documentation (i. e., according to DIN 6827-1 [45.35]) must start with a printout of the plan report, requested isodose distributions (i. e., slice showing D_{max}, isocenter, reference point, normalization point, etc.), DVH and DRR.

45.5.8 Plan Analysis Module

In the analysis module, an evaluation of the treatment is performed. Here treatment plans can be added, subtracted, or overlaid. Of particular interest will be the target volume and the ICRU conformality as well as the dose at the OAR and the fulfilled constraints. If the patients had been preirradiated, then the addition of the previous irradiation is needed to see the total dose. By adding plans from different TPSs the calculation grid plays an important role because if the resolution between two plans is not identical, this might cause problems, especially with preirradiated patients, even though areas of over- and underdosage in the OAR or target volume can be visualized. Furthermore, the presentation of the isodoses, i. e., color wash, lines, the dose levels must be shown etc.. In Fig. 45.14 the radiation therapy of a head and neck case with DVH is shown.

45.5.9 Export Module

The export module becomes increasingly important when different TPSs are involved in treatment planning, where treatment plan data need to be archived or transferred to portable discs, i. e., dose verification in the case of IMRT. The export consists of either a DICOM format or some standardized data format, i. e., export to a record and verification system like MOSAIQ or Varis. These formats contain the planning CT, the structure set (PTV, external, OAR, etc.) as well as DRR or the plan report as a PDF file. Typical addresses are:

- Archive systems,
- PACS system,
- Simulator (conventional or virtual),
- Record and verification systems of the medical linear accelereator,

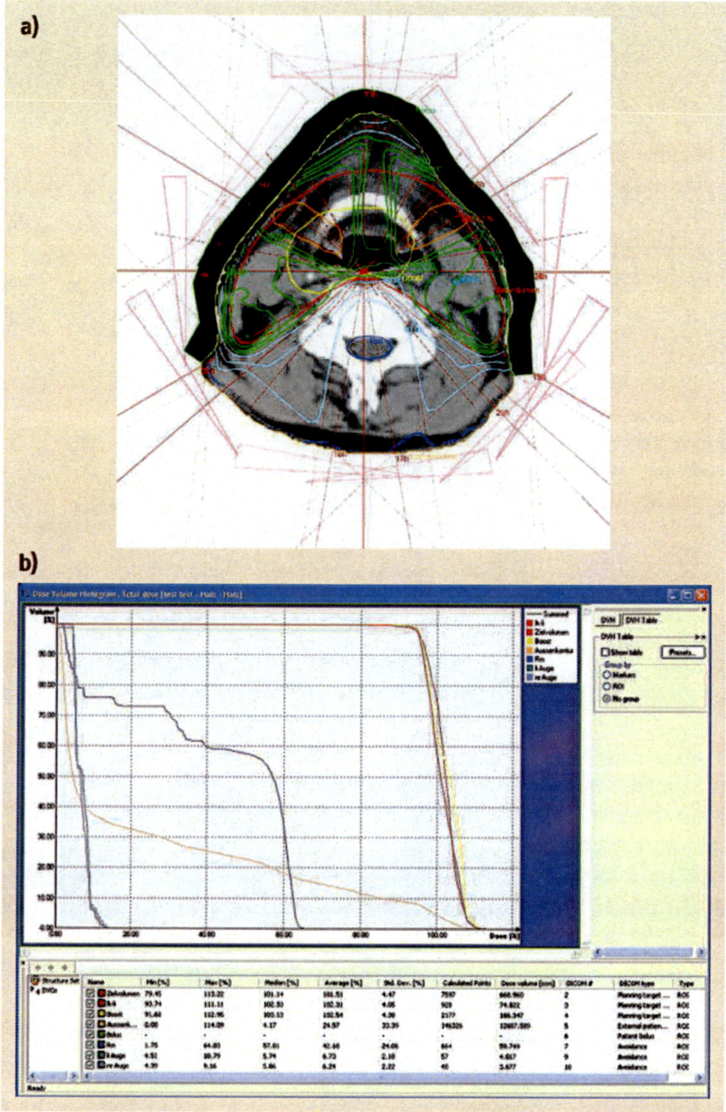

Fig. 45.14 (a) Dose distribution shown on a CT slice of a patient with a pharyngeal tumor. The technique is based on a complex combination of 19 fields (7 main fields and 12 subfields in this special case) in order to achieve an optimal dose in the target volume taking into consideration all organs at risk (spinal cord, eyes, lens, etc.). The tolerance must be considerably high. (b) Dose volume histogram showing minimum and maximum dose % inside the target volume and in the organ at risk. The ICRU requirements are not maintained in this type of case due to the positions of the organs at risk. In such cases higher or lower doses can be accepted depending on the tumor localization inside the target volume. It is up to the doctor to accept a plan or not (after [45.17])

- Portable discs for dose verification or other health institutions,
- Other treatment planning systems (for comparison, addition or futher processing).

45.6 Simulation of the Patient and the First Treatment

During conventional simulation the treatment plans are directly transferred to the patient either with an x-ray equipped machine or a linac (using a MVCT) where the plans can either be corrected or approved. This data transfer is an approval for the correctness of the patient beam data transferred by the record and verification system [i.e., MOSAIQ (IMPAC, Elekta), VARiS (Varian), LANTIS (Siemens)]. The created im-

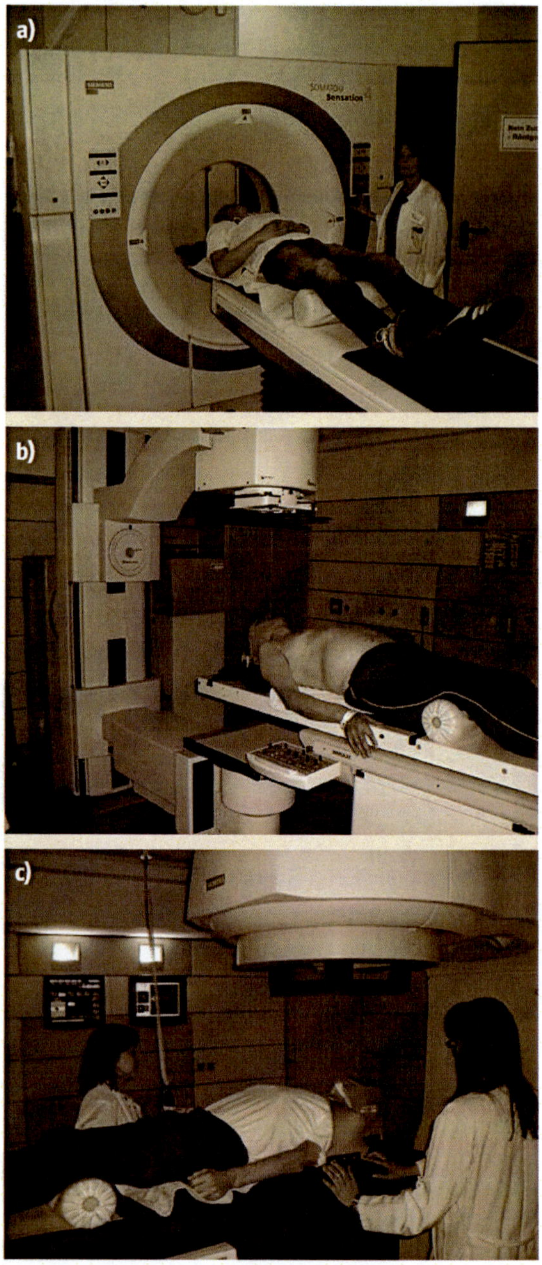

Fig. 45.15a–c From image acquisition to dose delivery, (**a**) computer tomography, (**b**) simulation, (**c**) linear accelerator (after [45.38]) ▶

ages for each beam is then compared to the DRRs from the treatment plan. The detector is either an image intensifier (conventional simulation) or a solid-state detector (in linear-accelerator-based plan verification). The modern method of performing a simulation using a CT scanner (virtual simulation) is explained in Sect. 45.6.2.

45.6.1 Conventional Simulation

After localization of the isocenter the beam shapes are transferred to the patient's body. The transfer of the beam data is supported by a room-based laser system that has the same isocenter as that of the medical linear accelerator and the CT scanner.

The marks on the patient's body using the laser system enable a precise setup for the first treatment at the medical linear accelerator because the laser system there will be a clone of the simulation laser system. The achieved precision of the patient's setup can be less than 3 mm. The precision of the setup will be determined during external radiation therapy at least every week using an EPID or a solid-state detector. The verification images are overlaid with the verification of the first treatment day or the simulation and corrected for. Some medical linear accelerators like Tomotherapy can perform linac-based CT in order to assure the correct position of the PTV and OAR (Sect. 45.3.7).

45.6.2 Virtual Simulation

Virtual simulation is the process by which a detailed 3-D model of a patient is built from a sequence of closely spaced transverse CT (or MRI) images. Since the grayscale of the CT images is directly related to tissue density, the computer model can be used to calculate the image that would be produced on a piece of film by a conventional diagnostic x-ray machine or a DRR. The advantage of the DRR is that it can be calculated almost instantly for any angular projection through the body. Use of a mechanical x-ray machine and films requires several minutes of setup and film development for each projection. When planning high-energy radiation treatments for cancer patients, it is a distinct advantage for the physicians and physicists to be able to rapidly evaluate many potential projections in order to find those that minimize the exposure of healthy organs to the radiation beams. Linear accelerators equipped with MLCs can be then programmed to selectively irradiate the tumor from many different projection angles.

45.7 Quality Control in Radiation Therapy

External radiation therapy comprises a chain of different modules starting from diagnostics, setup of the patient, treatment unit and dose and geometry verification methods. Several aspects in the quality assurance in radiation therapy are important such as, on the one hand, the education of the people carrying out the external radiation therapy and, on the other hand, the machine used for radiation therapy.

45.7.1 Personnel

Depending on the country, the education of the staff performing the external radiation therapy differs in terms of duration and content of qualifications, but the dangerous properties of ionization radiation require a special qualification of the staff by law. In Germany the law for using and handling ionization radiation concerns the medical physicist. The radiotherapist, medical physicist and the MTRA are requested to attend special courses on radiation protection and other related subjects to achieve technical qualification. This qualification has to be renewed every five years. To become a radiotherapist is regulated by legislation comprising a credit system. To become medical physicist radiation protection legislation regulates the minimum requirements.

Guidelines set up by DEGRO and ESTRO for the treatment of different malignant diseases enable the radiotherapist to remain current in modern treatment methods.

45.7.2 Planning system

For calculations treatment planning systems use beam data taken at each medical linear accelerator.

These beam data are gathered during commissioning of the linac, and then transferred, converted and implemented into the treatment planning system. This procedure is followed directly by an intense check of the planning system whereby the calculations are compared to measurements in a water tank for each medical linear accelerator.

Firstly, standard plans in the water tank or solid-state phantom are calculated and verified. These phantoms must have installations for hosting ionization chambers for absolute dosimetry or films for dose distributions. Nowadays dose distributions are verified by electronic devices like detector arrays (diodes or ionization chamber).

After the initial check of the beam data, periodical tests of these treatment plans and typical dose distributions need to be performed. This is required by legislation. The treatment planning system itself must undergo periodic checks of the intrinsic beam data. This is requested in order to identify uncertainties or changes in the beam data in an early state, especially after software updates. The frequency of those periodic checks is either monthly or after changes to hardware or software [45.39].

The requested properties in the treatment planning system to be regularly checked are as follows:

- Calculation of different treatment plans on different phantoms (PMMA, water, Alderson)
- Check of coincidence of the report and the treatment plan by checkpoints
- Transfer of treatment plans to external machines (simulator, record and verify system) and verify the data consistency with original treatment plan:
  - Collimator/gantry position
  - Shape of MLC
  - Dose values and monitor units
  - Hounsfield conversion to grayscale

45.7.3 Medical Linear Accelerator and Simulator

The treatment plans calculated by the treatment planning system during commissioning are defined as the gold standard. Therefore, changes by time of the relevant properties of the treatment plan are not permitted. As the planning site undergoes periodic check, the medical linear accelerator must undergo regular checks on its properties. Those checks can be performed daily, before patient treatment, weekly, monthly, quarterly, semiannually and annually.

The properties that need to be surveyed are as follows:

- Safety interlocks due to operating failures and malfunctions;
- Dosimetry (dose-monitor calibration, dose profiles, PDD, ...) compared with the gold standard;
- Transfer of treatment plans to the linac and check on whether the values coincide with the original plan.

Figure 45.16 shows a water tank to gather the beam data and examples of absolute dosimetry as well as an electrometer and an exposed film for checking the isocenter

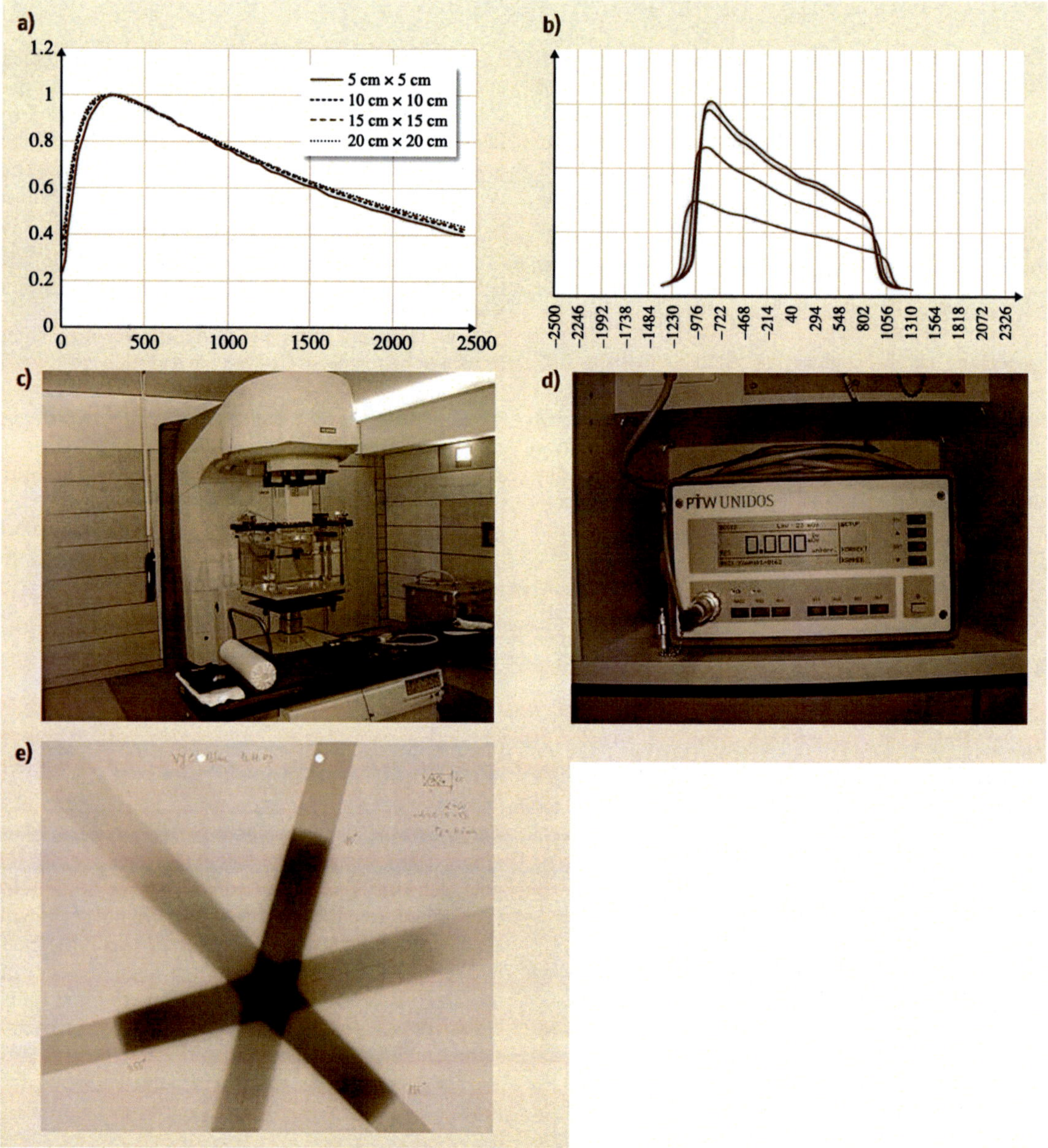

Fig. 45.16 **(a,b)** Relative dose distribution, **(c)** water phantom, **(d)** absolute dosimeter, **(e)** star shot at different gantry angles on a film after irradiation (after [45.38])

using five beams in different gantry positions of a medical linear accelerator [45.40, 41].

A conventional simulator requires periodic checks of the geometry and the isocenter like at the linac because the setup must be the same.

The data in Fig. 45.16a,b needed for the beam data include the relative dose distribution (depending on the TPS vendor) and the absolute dosimetry under reference conditions, i. e. according to DIN 6800-2 [45.42].

References

45.1 P. Wust, B. Hildebrandt, G. Sreenivasa, B. Rau, J. Gellermann, H. Riess, R. Felix, P.M. Schlag: Hyperthermia in combined treatment of cancer, Lancet Oncol. **3**(8), 487–497 (2002)

45.2 C.M. Spencer, K.L. Goa: Amifostine – A review of its pharmacodynamic and pharmacokinetic properties and therapeutic potential as a radioprotector and cytotoxic chemoprotector, Drugs **50**, 1001–1031 (1995)

45.3 M. Bamberg, M. Molls, H. Sack: *Radioonkologie*, Vol.1 (Zuckerschwerdt, München 2003), (In German)

45.4 R. Sauer: *Strahlentherapie und Onkologie* (Urban Fischer, München 2003), (in German)

45.5 Slides to Radiation Oncology Physics handbook, Produced for the Division of Human Health, last updated on Thursday, October 18 (2007)

45.6 J. Dobbs, A. Barrett, D. Ash: *Practical Radiotherapy Planning* (Arnold, London 1999)

45.7 ICRU Report 50: Prescribing, Recording and Reporting Photon Beam Therapy (International Commission of Radiation Units and Measurements – ICRU, Bethesda 1993)

45.8 E.B. Podgorsak (Ed.): *Radiation Oncology Physics: A Handbook for Teachers and Students* (IAEA, Vienna 2005)

45.9 P. Mayles, A. Nahum, J.C. Rosenwald: *Handbook of Radiotherapy Physics* (Taylor & Francis, New York 2010)

45.10 J. Van Dyk: *Modern Technology of Radiation Oncology*, Vols. 1, 2 (Medical Physics, Madison 2005)

45.11 BIR Supplement 25: Central axis depth dose for use in radiotherapy (British Inst. Radiology, London 1996)

45.12 W. Angerstein (Ed.): *Grundlagen der Strahlenphysik und radiologischen Technik in der Medizin* (Hoffmann, Berlin 2005), (in German)

45.13 S. Webb: *The Physics of Three-Dimensional Radiation Therapy* (IOP, London 1993)

45.14 S. Webb: *The Physics of Conformal Radiotherapie* (IOP, London 1997)

45.15 L.D. Potter, S.X. Chang, T.J. Cullip, R.A.C. Siochi: A quality and efficiency analysis of the IMFAST segmentation algorithm, Med. Phys. **29**, 275–283 (2002)

45.16 W. Que: Comparison of algorithms for multileaf collimator field segmentation, Med. Phys. **26**(11), 2390–2396 (1999)

45.17 Oncentra MasterPlan, Treatment-planning system, Kreiskrankenhaus Gummersbach, Department of Medical Radiation Physics (2010)

45.18 www.varian.com

45.19 Abteilung Radiologie Uniklinikum Bonn

45.20 R. Jeraj, T. R. Mackie, J. Balog, G. Olivera, D. Pearson, J. Kapatoes, K. Ruchala, P. Reckwerdt: Radiation characteristics of helical tomotherapy, Med. Phys. **31**, 396–404 (2004)

45.21 S. Yartsev, T. Kron, J. Van Dyk: Tomotherapy as a tool in image-guided radiation therapy (IGRT): theoretical and technological aspects, Biomed. Imag. Interv. J. **3**(1), e16 (2007)

45.22 Clinac iX, Varian

45.23 R. Muirhead: The potential clinical benefit of respiratory gated radiotherapy (RGRT) in non-small cell lung cancer (NSCLC), Radiother. Oncol. **95**, 172–177 (2010)

45.24 IORT.ppt, M. Treiber, S. Oertel, R. Krempien, Department of Clinical Radiology, University of Heidelberg

45.25 accuray.com, CyberKnife

45.26 elekta.com, Gamma Knife 4C

45.27 U. Quast, H. Sack: *Whole Body Radiotherapy*, DGMP-Rep. 18 (German Soc. for Medical Physics (DGPM), Berlin 2004)

45.28 A. Barrett, J. Dobbs, S. Morris, T. Roques: *Practical Radiotherapy Planning* (Hodder Arnold, London 2009)

45.29 ICRU Report 62: Prescribing, Recording and Reporting Photon Beam Therapy (International Commission of Radiation Units and Measurements – ICRU, Bethesda 1999)

45.30 F.M. Khan: *The Physics of Radiation Therapy* (Lippincott Williams Wilkins, Philadelphia 2003)

45.31 W.R. Hendee, G.S. Ibbott, E.G. Hendee: *Radiation Therapy Physics*, 3rd edn. (Wiley, New York 2005)

45.32 M. Buchgeister, U. Gneveckow, N. Hodapp, J. Salk, O. Sauer: Grundsätze zur Bestrahlungsplanung mit Computern, Neufassung 2003, DGMP Rep. 118 (German Soc. for Medical Physics (DGPM), Berlin 2003) (in German)

45.33 Department of Medical Radiation Physics, Kreiskrankenhaus Gummersbach (2008)

45.34 W. Schlegel, J. Bille: *Medizinische Physik* (Springer, Berlin Heidelberg 2002), (in German)

45.35 DIN 6827-1: Protokollierung bei der medizinischen Anwendung ionisierender Strahlung – Teil 1: Therapie mit Elektronenbeschleunigern sowie Röntgen- und Gammabestrahlungseinrichtungen (DIN, Berlin 2000)

45.36 B. Emami, J. Lyman, A. Brown, L. Coia, M. Goitein, J.E. Munzenrieder, B. Shank, L.J. Solin, M. Wesson: Tolerance of normal tissue to therapeutic irradiation, Int. J. Radiat. Oncol. Biol. Phys. **21**, 109–122 (1991)

45.37 AAPM Task Group: Issues associated with clinical implementation of Monte Carlo-based photon and electron external beam treatment planning, Report No. 105, Med Phys. **34**(12), 4818–4853 (2007)

45.38 Department of Radiology and Clinic of Radio therapy, Kreiskrankenhaus Gummersbach (2010)
45.39 DIN 6873-5: Bestrahlungsplanungssysteme, Konstanzprüfung von Qualitätsmerkmalen (DIN, Berlin 1993) (in German)
45.40 DIN 6847-5: Medizinische Elektronenbeschleuniger-Anlagen – Teil 5 Konstanzprüfung von Kennmerkmalen (DIN, Berlin 2005) (in German)
45.41 DIN 6874-5: Therapiesimulatoren – Teil 5 Konstanzprüfung von Kennmerkmalen (DIN, Berlin 2001) (in German)
45.42 DIN 6800-2: Dosismessverfahren nach der Sondenmethode für Photonen- und Elektronenstrahlung – Teil 2: Dosimetrie hochenergetischer Photonen- und Elektronenstrahlung mit Ionisationskammern (DIN, Berlin 2008) (in German)

46. Recording and Processing of Biosignals

Klaus-Peter Hoffmann, Florian Solzbacher

Measuring is the experimental determination of a measured value by quantitative comparison of the measurand with a comparison value. The measured value obtained by this procedure is given as a product of a numeric value and a dimensional unit. It can be recorded continuously as a temporal variation of a physical value or discontinuously at particular moments. The deviation of the measured value from the measurand is the measurement error. It depends on the measurement procedure, the measurement device, and environmental effects. Systematic and random errors are distinguished.

The measurement can be done directly or indirectly, depending on the chosen measurement procedure. For instance, the concentration of substances is determined by the measureable value extinction, ion activity, dielectric permittivity, or magnetic susceptibility. Here, the measurand is transformed into one or several intermediate values in order to further process the signal. The result of a measurement is the measured value itself or a combination of several measured values.

46.1 **Measuring in Medicine** 923
46.1.1 Biosignals 924
46.1.2 Biological Measuring Chain 924
46.1.3 Artifacts 933

46.2 **Registration of Biological Signals** 934
46.2.1 Bioacoustic Signals 934
46.2.2 Biochemical Signals 934
46.2.3 Bioelectric and Biomagnetic Signals 935
46.2.4 Biomechanical Signals 937
46.2.5 Biooptical Signals 940
46.2.6 Biothermal Signals 940

46.3 **Measurement and Signal Analysis from a Metrological Point of View** 940
46.3.1 Biostatistical Methods 941
46.3.2 Biosignal Analysis 941

46.4 **Test Planning and Clinical Studies** 944

This chapter covers the acquisition of different biosignals and the specifics of the measurement process in medical use.

46.1 Measuring in Medicine

The aim of measuring in medicine is the objective description of the state of a patient who might possibly not be able to cooperate. It should help the physician to define the respective therapy and to evaluate the therapy process. Moreover, the prognosis of the course of disease can often be estimated. The long-term monitoring of physiological parameters is combined with an alarm function if preset limiting values are exceeded or undershot. Further developments include closed-loop systems which directly intervene in the patient's state after analysis of the measured values, for instance by raising or reducing the infusion rate in infusion therapy. But also the examination of patients who are not able to cooperate often becomes only feasible by the determination of objective parameters (Fig. 46.1).

Often, various procedures are combined in medicine, e.g. the combination of functional, metabolic and morphologic methods for functional brain imaging in brain diagnosis. Here, the combined analysis of EEG, CT, MRI, PET and MEG can yield new information, for instance in presurgical diagnosis of epileptic seizure diseases and epileptic focus localization.

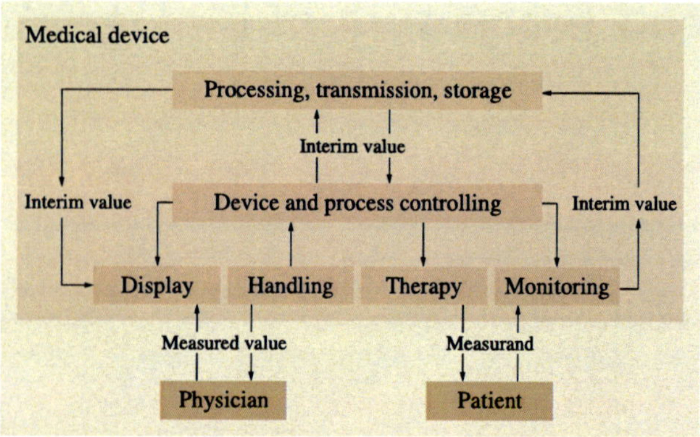

Fig. 46.1 Measuring in medicine

For measuring in medicine, the following five objectives can be distinguished:

1. Metrological acquisition, conversion, processing and telemetric transmission of biological signals.
2. Measuring the reaction or the behavior of the biological object to an external stimulus.
3. Measurements during the application of extra- or intracorporal assist systems to support organ functions or as organ compensation, as well as manipulators for therapeutic means.
4. Application of substances, irradiation or waves and measurement of reflection, absorption, scattering, distribution or fluorescence to display structures and functions in the organism.
5. Extraction of body fluids, substances and tissues, as well as tests and analysis in clinical and chemical laboratories.

In contrast to technical measurements, interindividual and intraindividual deviations occur for biological measurements, owing to biological variability. This means, that the obtained measured values vary from patient to patient. But also for one and the same patient, deviating measured values (e.g. blood pressure) occur in the course of the day.

The extent of inconvenience for the patient and the measurement procedure directly influences the reliability of the measured values. Moreover, biological sources of interference (biological artifacts with physiological origin) superimposing the measurand have to be considered. The measurement duration and the reproducibility of an examination are limited for most methods. In addition, the wide variability of the examined persons has to be taken into account, ranging from fetus, infants and trained athletes to aged people.

Opposed to the objective methods are the subjective methods, requiring cooperation of the patient. Those include audiometry, vibration tests and temperature sensation.

46.1.1 Biosignals

In the biological measuring chain, the biosignal is the actual measurand that should be metrologically determined for diagnostic purposes. Biosignals can be defined as phenomena to describe functional states and their variations in a living organism. They provide information about metabolic, morphological and functional changes, describe physiological and pathophysiological states as well as process dynamics. To analyze them, the generation locus and thus the spatial and temporal correlation is significant. Biosignals are acquired from living organisms, organs and organ parts down to single cells.

Biosignals can be distinguished with respect to their properties. On the one hand, there are structural parameters, for instance length, area, amount, volume, elasticity or viscosity. On the other hand, functional parameters can be found, among them temperature, pressure, flow, electric potentials and acoustic sounds. Biosignals are described by their frequency, amplitude, shape and the time of occurrence. They can be displayed in the time domain and the frequency domain, but also as two- or three-dimensional pictures. Their occurrence can be stochastic, stationary, periodic or discrete. Figure 46.2 shows some examples for biosignals.

With respect to the physical properties, biosignals can be divided into:

1. Bioacoustic signals (heart sound, lung sounds, speech)
2. Biochemical signals (substance compositions, concentrations)
3. Bioelectric and biomagnetic signals (electric potentials, ion currents)
4. Biomechanical signals (size, shape, movements, acceleration, flow)
5. Biooptical signals (color, luminescence)
6. Biothermal signals (body temperature).

46.1.2 Biological Measuring Chain

The biological measuring chain is used for the metrological acquisition of biosignals. It includes the registration (sensor, transducer), the processing (amplification, filtering, linearizing, transmission), the analysis (bio-

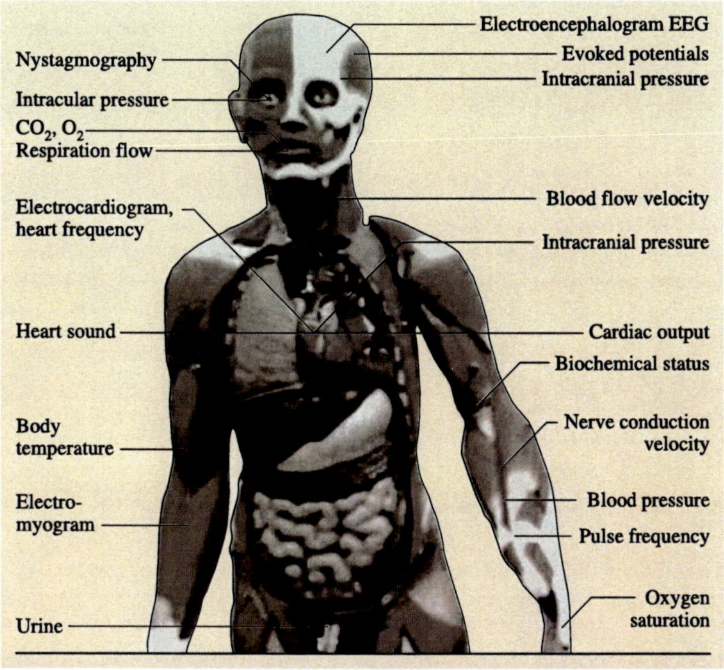

Fig. 46.2 Examples of biosignals that can be obtained by measurements

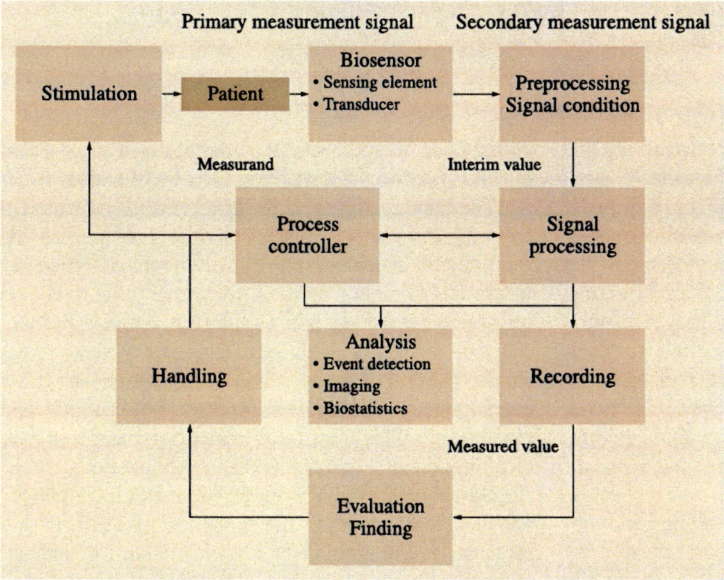

Fig. 46.3 Structure of the biological measuring chain

statistics, biosignal analysis, image processing) and the display of the biological signals. The objective is to provide assistance to the physician for diagnostics. Figure 46.3 shows the structure of a biological measuring chain in principle.

Sensors

A biosensor is a probe to register biological events and morphological structures. Often, it is directly connected to a transducer, or it transduces the primary measurement signal into a secondary signal itself (e.g.

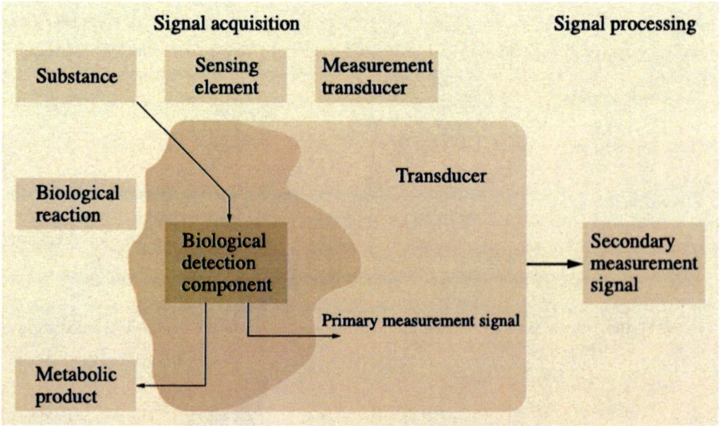

Fig. 46.4 Schematic of a biosensor

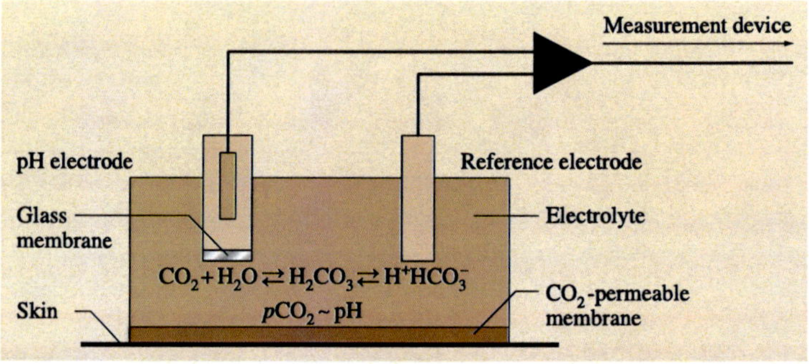

Fig. 46.5 Design and principle of a p_{CO_2} electrode

electrode). It can be distinguished between sensing and transfer elements. In e.g. a piezoresistive pressure transducer the membrane would be a transfer element, transferring pressure into mechanical stress. The piezoresistor would be the sensing element that turns the mechanical stress into an electrical signal. The secondary measurement signal is mostly an electric signal, which can be easily processed electronically. Examples for signals which can be registered are blood pressure, body temperature, heart rate, bioelectric potentials (EEG, ECG, EMG, EOG), metabolic parameters (glucose, pH-value) and the size and locus of space-occupying lesions.

Biosensors in the narrower sense include a biological detection component, which acts as a probe directly connected to the transducer. Its function is to specifically and sensitively react with the substance to be analyzed. In doing so, heat, electrons, protons, light, ions, gases, fluorescence and mass deviations can be generated. The biological component as the actual signal generator can consist of enzymes, microorganisms, organelles, cell compounds or antibodies. It is immobilized at the transducer. Figure 46.4 shows the schematic of a biosensor.

The main requirement for a biosensor is a feedback-free registration of the signals. It has to provide reproducible measurement results. The transmission behavior has to remain constant for a long time, to enable long-term recordings and monitoring. Narrow production tolerances, a high biocompatibility, low stress to the patient, low mass and low volume are other requirements. The application should be simple and manageable. This is especially related to possibilities of cleaning, disinfection and possibly sterilization.

Development objectives for sensors are reliability enhancement including the reduction of artifacts, the increase of the amount of channels (multi-transducer technology) and complexity enhancement (closed-loop systems). On the one hand, this becomes possible by miniaturization including the reduction of mass, volume and power input, by new biomaterials and by the improvement of incorporation possibilities. On the other hand, new non-contact measurement procedures have to be developed to enable a completely feedback-free registration.

Transducers

Chemoelectric Transducers. Chemoelectric transducers are used for the measurement of individual chemical components in the blood, in body tissues, in the exhaled air or on the skin. Distinctions are drawn between:

- Potentiometric sensors, based on the measurement of cell potential
- Amperometric sensors, based on cell current
- Conductometric sensors, based on admittance.

Potentiometric transducers generally consist of a metal electrode surrounded by a selectively permeable membrane, which is dipped into the electrolytes to be examined. The ions or molecules reaching the electrode change the electrochemical potential difference between this measuring electrode and a reference electrode dipped into a defined reference solution. The corresponding potential change is proportional to the logarithm of the ion concentration. As an example, Fig. 46.5 demonstrates the principle of a p_{CO_2} electrode.

Ion-selective field effect transistors have a gate which is surrounded by a membrane. Diffusion of ions or molecules through the membrane causes a change in the drain current, which can be registered as measurement signal.

Electric and Magnetic Transducers. Electric transducers are recording electrodes, which transduce an electric signal (ion current) into an electric signal (electron current). They are selected based on the properties of the corresponding biosignal, its locus of generation, the dimensions of the bioelectric generator and environmental influences. The electrodes are distinguished into two groups: microelectrodes (glass capillary electrodes, metal microelectrodes) and macroelectrodes (surface electrodes, subcutaneous needle electrodes and depth electrodes).

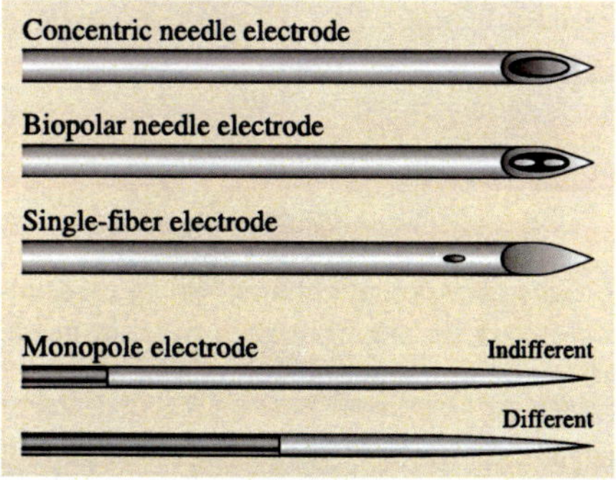

Fig. 46.6 Needle electrodes for electromyographic examinations. *Top to bottom*: concentric needle electrode, single-fiber electrode, monopole needle electrode (*top* indifferent, *bottom* different)

In Figure 46.6, schematics of various types of needle electrodes are shown. For pure metal electrodes or polarizable electrodes, positively charged metal ions move into the electrolyte solution, thus creating the Helmholtz double layer at a molecular distance. This is the reason for the high electrode impedances especially in the low-frequency range. Thus, the field of application for such electrodes is the recording of high-frequency signals, such as evoked potentials.

For unpolarizable electrodes, the anion in the electrolyte solution is part of the metal electrode (e.g. Ag/AgCl electrode, electrolyte NaCl, anion Cl^-). This leads to a reduction and stabilization of the Galvani potential and thus to lower interface impedances in the whole frequency range. Figure 46.7 shows the equivalent circuit diagram of an electrode including the polarization voltage.

Biomagnetic examinations can be done with a superconductive quantum interferometer. Such a device contains one or two Josephson elements, which enable a current flow without a potential drop. They are built up as gradiometers, to inhibit the homogeneous sensor-remote magnetic fields, especially the terrestrial magnetic field (Fig. 46.8).

Mechanoelectric Transducers. To measure length changes, strains, pressure changes in tissue, body fluids and organs as well as for the measurement of sounds, microvibrations and blood flow, mechanoelectric transducers are used. Resistive, inductive and capacitive transducers have a membrane receiving the force change. This causes the membrane to deflect, which can be detected by a deformation of strain gauges or semiconductor resistances, by a shift of small iron or ferrite cores in a coil, or by a change in the distance of the two plates of a capacitor.

For a strain gauge, the resistance can be calculated from the geometry.

$$R = \rho \frac{l}{A} , \tag{46.1}$$

with R the Ohmic resistance, ρ the specific resistance, l the length, and A the cross-section of the conduc-

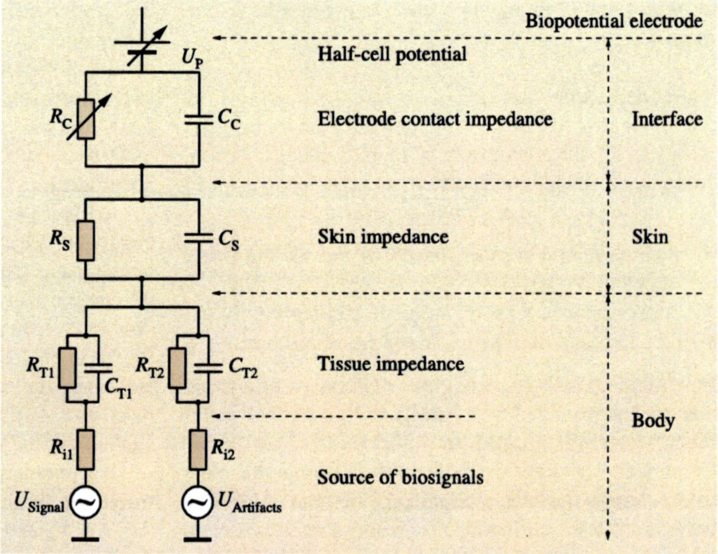

Fig. 46.7 Equivalent circuit diagram for the recording of biopotentials with surface electrodes (U voltage source for signal and artifacts, R_i internal resistance, $R_T C_T$ tissue impedance, $R_S C_S$ skin impedance, $R_C C_C$ electrode contact impedance, U_P polarization voltage)

tor. Alternatively, using piezoresistive elements as strain gauge in a Wheatstone bridge configuration, changes in resistivity can be observed that are up to one hundred times larger than the geometric effect yielding a more sensitive strain gauge and subsequently more sensitive pressure or force sensor.

If a force is applied to a capacitor yielding a change in the distance between its two plates, its capacity also changes

$$C_X = \varepsilon_0 \varepsilon_\gamma \frac{A}{x} , \qquad (46.2)$$

with C the capacity, A the area, x the plate distance, and ε the dielectric permittivity. The measured value is determined using suitable measuring bridges.

Piezoelectric transducers consist of piezoelectric contacts. Mechanical stress in the direction of a polar electric axis causes the generation of electric charges due to a shift of the atoms, at very small deformations. The induced surface charge is the product of the piezoelectric constant and the external force

$$\Delta q = k \Delta F , \qquad (46.3)$$

where Δq is the induced surface charge, ΔF the force change, and k the piezoelectric constant.

Hence, the potential change can be calculated to be

$$\Delta U = \frac{\Delta q}{\Delta C} = \frac{k \Delta F}{\varepsilon_0 \varepsilon_\gamma A} x , \qquad (46.4)$$

with ΔU the potential change, ΔC the capacity change, A the area, x the plate distance, and ε the dielectric permittivity.

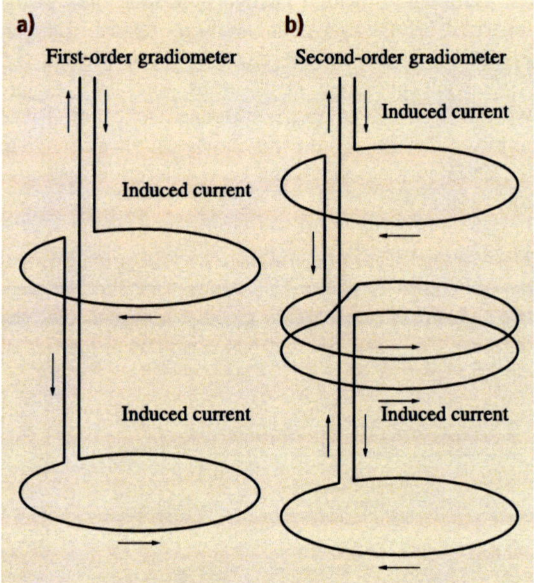

Fig. 46.8a,b Schematic design of a first- **(a)** and second-order **(b)** gradiometer for the registration of biomagnetic fields in nonshielded environments

The change of the surface charge can be measured. Figure 46.9 shows the schematic setup of a piezoceramics.

Hall sensors are used to measure magnetic fields and thus to register all parameters generating or influencing magnetic fields. The Hall voltage results from the

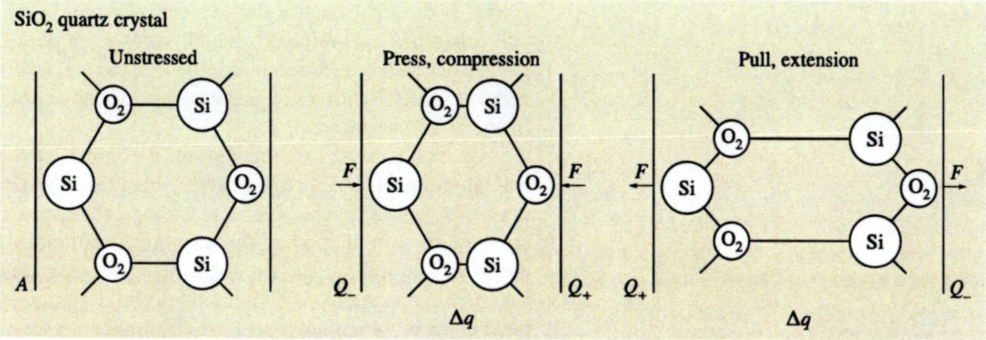

Fig. 46.9 Basic principle of the piezoelectric effect using a SiO_2 quartz crystal as an example. Owing to the positive Si ions and negative O_2 ions, a surface potential is generated in the deformed state (*A* surface, *F* force, Q_+ and Q_- electric charges)

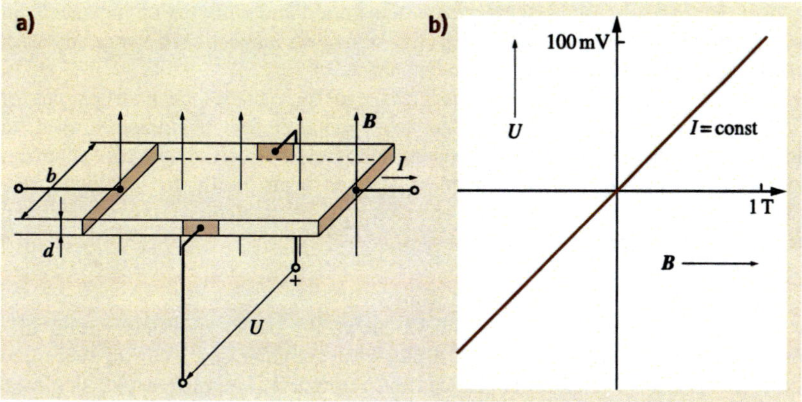

Fig. 46.10a,b Schematic setup of a Hall sensor (**a**) and its corresponding characteristic curve (**b**) (*U* Hall voltage, *B* magnetic flux density, *d* depth of plate, *I* current across plate length, *b* breadth)

Lorentz force and is proportional to a control current longitudinally to a small plate subjected to a magnetic field and its magnetic flux density. It decreases with increasing electron concentration, and increases with increasing electron velocity and electron mobility. Figure 46.10 shows the setup and the characteristic curve of a Hall sensor.

Photoelectric Transducers. Photoelectric transducers include photoresistors, photodiodes, photoelements and phototransistors. Depending on the incident light intensity, the current flowing through the device is changing. Thus, light absorptions and light reflections of body tissue and biological substances can be detected, to measure blood circulation and oxygen saturation.

Thermoelectric Transducers. Using a thermocouple consisting of 2 different metal wires, a thermoelectric voltage (Seebeck effect) can be measured between the contacts for a temperature difference

$$U_T = \alpha_T \cdot (T_2 - T_1) \,, \qquad (46.5)$$

with U_T the thermoelectric voltage, T the temperature, and α_T the Seebeck coefficient.

The thermoelectric voltage depends on the temperature and the selected metals (Fig. 46.11).

Thermistors (NTC resistors or high-temperature conductors) are sintered elements based on oxide composites. Their Ohmic resistance depends on the temperature (NTC, negative temperature coefficient). By connecting a resistor in parallel, it is possible to linearize the exponential run of the characteristic curve. Thermistors are used for the routine measurement of the respiration flow and the body temperature (Fig. 46.27).

Infrared-sensitive semiconductor materials, such as HgCdTe, InSb and PtSi, enable the conversion of the infrared radiation emitted by a patient into an electrically analyzable signal. Thus, temperature distributions can

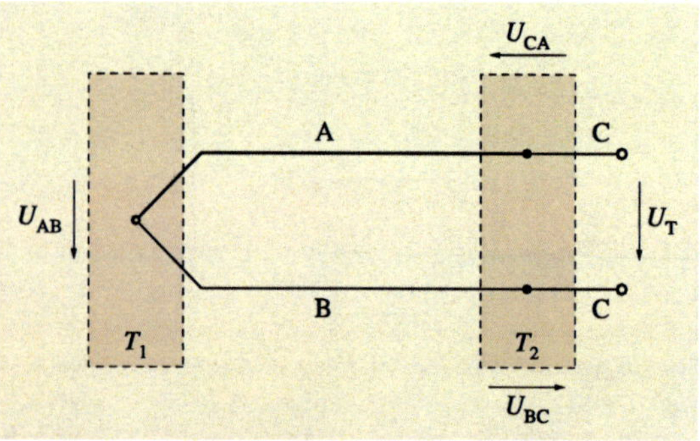

Fig. 46.11 Picture of a thermocouple consisting of metals A, B, and C. The thermoelectric voltage is proportional to the sum of the three individual voltages at the metal interfaces

be measured. The materials have to be cooled, which can be done using liquid nitrogen, a Stirling cooler or a Peltier element.

Other Transducers. For the indirect measurement of biosignals, also methods are applied which do not immediately register the biosignal itself, but measure the influence of irradiance and waves as well as the distribution of radioactive substances etc. These include particularly imaging techniques using radioactive isotopes (PET, SPECT), magnetic fields (MRI) and x-rays (CT, angiography). If isotopes are used, a scintillation camera acts as the transducer. It consists of a NaJ monocrystal transducing gamma radiation into a light signal. If the light quanta are incident on a photocathode, electrons are released, which can be amplified using a secondary electron multiplier. For x-rays, luminescent screens with an adjacent photocathode and an image amplifier are used.

For biochemical examinations in a clinical chemical laboratory, a variety of transducers and methods are used. For instance, chemical and biological reactions are applied with antibodies and enzymes, cell cultures are grown in culture media, substances are separated into their components, and interactions with light, magnetic fields or flames are registered. Transducers have to be able to count particles, cells and blood cells, and to register the temporal and spatial fractionation of substances. This also includes changes in the chemical composition of material mixtures, changes in color or temperature, absorption and emission of radiance, and the registration of nuclear magnetic resonance and electron spin resonance.

Transducers can for instance be microphones of a photoacoustic spectrometer, photodiodes used to count flowing blood cells which temporarily intercept or attenuate a laser beam, coils for magnetic resonance spectroscopy or semiconductors for infrared spectroscopy.

Amplifiers

Figure 46.12 shows the basic equivalent circuit diagram of a biosignal amplifier consisting of two input impedance transformers and a differential amplifier. The amplifier is characterized by a high input imped-

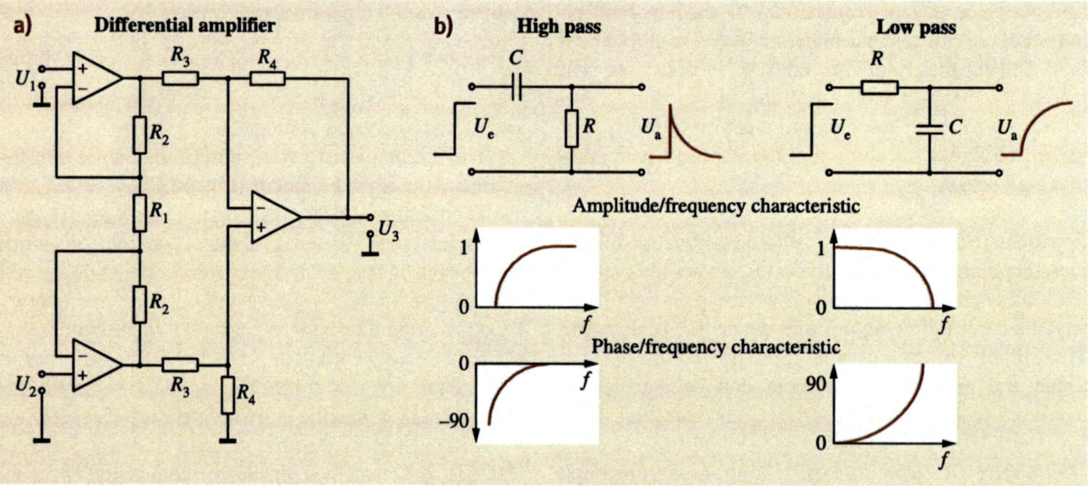

Fig. 46.12a,b Amplification of biosignals. **(a)** Setup of a differential amplifier; **(b)** control of the frequency response (amplitude and phase) with high-pass or low-pass filter

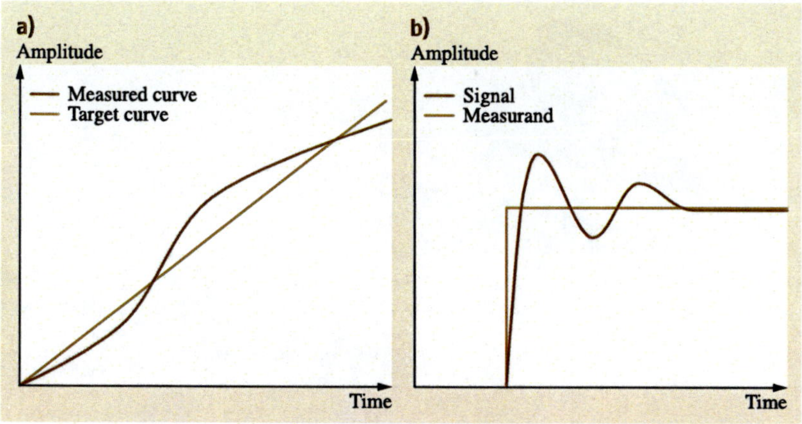

Fig. 46.13a,b Dynamic properties of a sensor. **(a)** Input and output relation for nonlinear transfer function; **(b)** sensor with time delay (dead time) and oscillating behavior

ance and a low output impedance. It is set up as differential amplifier with a high common mode rejection, so that the output voltage (U_3) is a multiple of the difference between the two input voltages (U_1 and U_2). Other properties of the amplifier are sensitivity or amplification, linearity between the input and the output signal, frequency response, phase response and noise. Depending on the properties of the biosignal, the amplification and the frequency response have to be selected. The latter is controlled with a high-pass or a low-pass filter.

Signal Processing

The ideal transmission behavior of a measuring chain is linear, i. e. a change in the measurand is directly and unambiguously related to the corresponding measured value. In reality, the relation is not linear, owing to the selected measurement process, the nonlinearity of the transducers, the influence of noise causing measurement errors, and the indirect registration of the measurand (Fig. 46.13).

Influence quantities are changes due to temperature deviations, pressure deviations and transducer deterioration. An example for a nonlinear transmission behavior of a measurement system is the invasive measurement of the arterial blood pressure with an external transducer. The system consists of the arterial catheter, the reusable valve, the pressure-sensitive transducer and the rinsing system. This coupling path is filled with physiological saline. Owing to the compliance of the tubes and the transducer membrane, to the flow resistance and the volume, it is an oscillatory system. To minimize the influence on the amplitude and the phase of the wanted signal, the catheters and the connecting tubes should be as rigid, short and large-volume as possible, air bubbles should be avoided, the transducer membrane should be rigid and the coupling path should be fixated.

The objective of signal processing is the correction of the nonlinear transmission behavior of the measuring chain. By a generally multistage signal processing, the signal is transformed into a mainly electric output signal. Digitalization enables a variety of complex algorithms and control modes. Besides measurand conversion, signal transduction, filtering and amplification, these methods particularly include linearization. This is achieved to a large extent by stabilizing the zero point and the sensitivity as well as avoiding the drift of signals. By tracing the characteristic curve, it is possible to control the transmission behavior of the measuring chain. Further approaches include models of the measuring chain and the identification of coefficients as well as the determination of the parameters for adaptive control.

In order to do so, the signal has to be A/D-converted, i. e. the continuous signal is temporarily sampled and its amplitude is quantified. The Shannon sampling theorem states that the sampling rate has to be at least twice the highest frequency contained in the signal (Nyquist frequency). In order to optimally convert the curve shape, the sampling rate should be even higher (5 or 10 times the maximum signal frequency).

A violation of the sampling theorem causes a falsification of the signal in the time domain and the occurrence of reflections in the frequency domain (*aliasing*), with the latter being displayed as apparent amplitudes in the frequency spectrum.

Storage and Registry

Direct Writing. Direct writing enables direct and immediate signal output on a graph paper. This can be

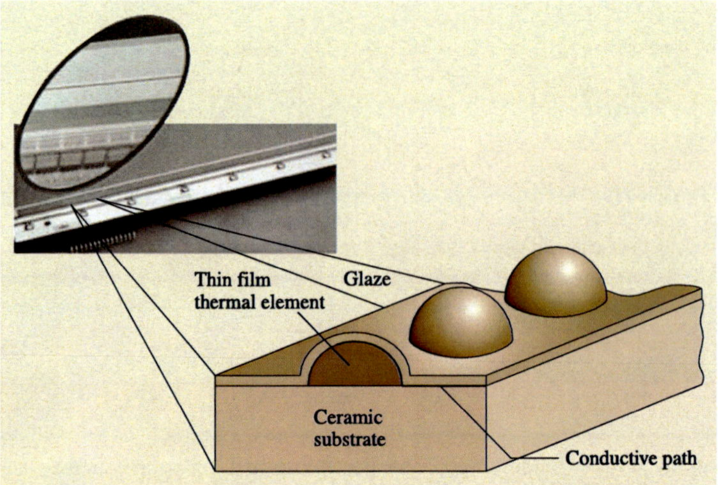

Fig. 46.14 Setup of a thermal comb

achieved with a write pointer or a writing lever, whose deflection can be controlled by a measuring instrument, often a rotary magnet in a coil field. The write pointer can be heated and thus writes a graph on thermosensitive paper. But it can also be an ink jet nozzle with a piezoelectric element activated by voltage pulses or a heating element generating a steam bubble by current pulses. In both cases, the resulting pressure causes ink to be sprayed on the graph paper.

In pigment printing, carbon paper and graph paper move in opposite direction over a writing edge. The write pointer is pressed against this edge. Thus, its movements create an orthogonal output.

Other techniques are laser printers and thermal comb writers (Fig. 46.14). The latter include thermal elements on a substrate, which can be controlled digitally. They write a graph on thermosensitive paper. Besides curves, also alphanumeric and graphic outputs are possible.

Indirect Writing. Indirect writing techniques include photo writers and UV writers. Here, a moving light spot is projected onto special photo paper by a mirror. The result is only available after developing the photo paper.

Monitor. Measurement results can also be shown on a monitor or on digital or analog displays. As the data is easily readable, especially the latter are of great importance for medical diagnostics and for the monitoring of intensively treated patients.

Storage. Biological signals can be stored on conventional data storage media, such as hard disk, DVD, CD-ROM, MOD, ZIP disks and magnetic tape. Here, it must be guaranteed that the storage media keep their data for a period of 10 years in good condition. Different database management systems help to operate the databases, provide storage, access, security, backup etc. For the exchange, management and integration of electronic healthcare information standards like the Health Level Seven (HL7) should be used.

Transmission (Telemetry)

For medical objectives, signals are increasingly transmitted telemetrically. This may concern signals generated directly in the body of the patient, which provide information about the patient's state or the functional state of an implanted device. For evaluation, both signals have to be displayed outside the body (e.g. for cardiac pacemaker patients the intracardial ECG and the charge of the pacemaker battery). This is done with magnetic transmission coils. In addition, it is possible to program a pacemaker via this path. Figure 46.15 shows a programmable cardiac pacemaker as an example for telemetry. Here, the read-out data can also be transmitted to a diagnostic station via internet.

Another application of telemetry is the examination of patients moving freely in the meantime. Here, the examination results have to be transmitted to an evaluation station or a diagnostic center. Such applications are particularly important for long-term monitoring of EEG, ECG, blood pressure and labor pains as well as for sports medicine and occupational medicine. For telemetric transmission, a transmitter and appropriate modulation of the signal are required. For amplitude modulation, the low-frequency signal is converted

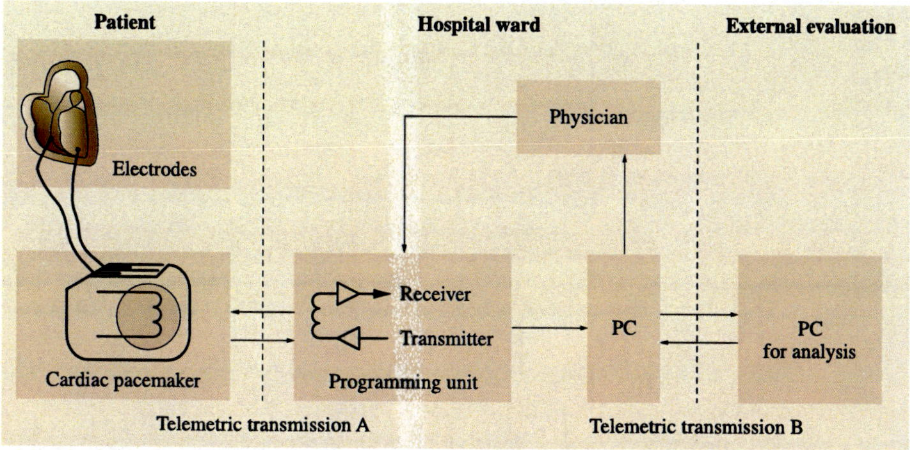

Fig. 46.15 Telemetric transmission paths for a programmable cardiac pacemaker (A via transmitting and receiving coils, B via Internet to a diagnostic computer)

into a high-frequency signal with constant frequency, but varying amplitude. For frequency modulation, the amplitude change of the biosignal is coded by the frequency change, for pulse-width modulation by the duration of square-wave pulses, and for pulse-position modulation by the temporal position of two impulses to each other. For evaluation and display, demodulation is required on the receiving side.

Transmission via telephone and cell phone is also possible. Digital transmission of biosignals has opened up new possibilities of signal transmission, also via internet. Parallel and serial data transmission are the two main possibilities. Standardized interfaces have been created requiring data transmission protocols.

Via telemetric transmission, it is possible to directly influence the patient. One example is the application of surgery robots in sparsely populated regions of the earth. Here, telemedicine offers new opportunities.

Application of Stimuli, Substances,

For medical objectives, it is often important to observe the reaction of the patient towards stimuli with respect to diagnostics. These stimuli can be mechanical, for instance to provoke reflexes, electrical or magnetic to determine the nerve conduction velocity, acoustic, optical or electrical for the stimulation to record evoked potentials and to test sensory performance. By using substances, which may also be radioactively labeled, their participation in metabolism can be examined, and thus their transport velocity, their accumulation site or their expulsion dynamics. The application can be systemic or localized.

The reflection of ultrasound waves enables the display of moving interfaces to view anatomical structures, but also work movements, for instance of heart valves, and to measure the blood flow velocity. By absorption measurement of x-rays, anatomical structures can be measured. Physical stress is used for cardiac stress tests, and for physiological examinations to measure e.g. the sympathetic skin response or the heart rate.

Process Control

The temporal correlation within the measuring chain is controlled by the process control. This includes signal processing with the setting of the sampling rate for the A/D converter, the synchronization of the averager and the stimulator, and the choice of the corresponding time-frame. Process control also includes the setting of parameters for amplification, signal analysis, storage and registry.

46.1.3 Artifacts

The registered biosignals are often superimposed by interferences, so-called artifacts. Those are registered together with the wanted signal and impede or preclude its evaluation. With respect to their point of origin, artifacts can be divided into biological and technical artifacts. They are caused by metrology, the applied method and by the patient. In the following, electroencephalography is used as an example to briefly explain artifacts:

1. *Physiological artifacts* are biological signals superimposed to the EEG, but with non-cerebral origin. Examples include various forms of electromyographical signals (EMGs) such as eye movements (EOG), tense neck (muscle artifacts), ECG, as well as pulse waves if an electrode is placed directly on

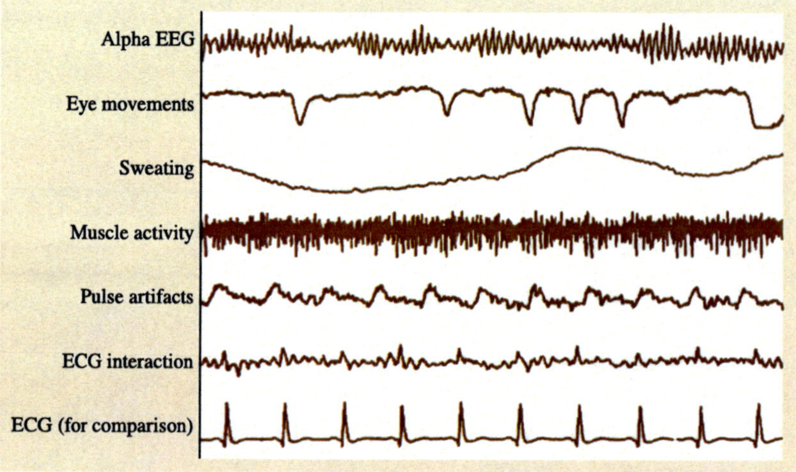

Fig. 46.16 Biological artifacts in EEG. *Top to bottom*: eye movements, sweating, muscle contractions, pulse, and ECG

a pulsating vessel, heavy sweating and spontaneous movements of the patient. Figure 46.16 shows examples of biological artifacts.

2. *Technical artifacts* are errors in the recording and registration technique. They can also be found at contact sites (electrode–skin, cable–electrode, cable–connector, connector–socket), if different electrode types and materials are used simultaneously, or for cable defects, cable movements and in the absence of grounding.
3. *Externally caused technical artifacts* are galvanically coupled interferences via direct, and high-resistance circuits, capacitively coupled currents due to alternating electrostatic fields, alternating magnetic fields and high-frequency alternating electromagnetic fields.

46.2 Registration of Biological Signals

Many biological processes are related to acoustic phenomena (Table 46.1).

46.2.1 Bioacoustic Signals

This includes sounds of the upper respiratory tracts (snoring, speech), lung sounds and heart sound. The latter evolves from the opening and closing of the heart valves and is registered phonocardiographically. For the bloodless blood pressure measurement, oscillations of the vessel walls are registered as signal. They occur due the squeezing and reach a maximum at a cuff pressure between the systolic and the diastolic value. Bioacoustic signals can be registered with a microphone and a stethoscope.

Table 46.1 Bioacoustic signals

Signal	Specification	Frequency (Hz)
Heart sound	Adults	15–1000
	Fetus	15–150
Lung		0.2–10

46.2.2 Biochemical Signals

Biochemical signals can be determined both in vivo, i. e. inside the body of a patient, an in vitro in a laboratory. They are the result of metabolic processes and can be described by identifying their structure, composition, concentration of their constituents and partial pressure. They can be registered directly or by reaction with enzymes, antibodies or cells.

The methods for examination of tissue or body fluids range from electrophoresis, gas chromatography, flame absorption spectroscopy, mass spectroscopy and infrared spectroscopy to scanning electron microscopy. The applied chemical transducers are used to determine the structure, the composition and the respective concentrations. This also includes cell counting and blood cell counting.

One example for the detection of biochemical signals is glucose identification. The reaction

$$\text{Glucose} + O_2 \xrightarrow{\text{GOD}} \text{Gluconolactone} + H_2O$$
$$\xrightarrow{H_2O_2} \text{Gluconic acid} + H_2O_2 \quad (46.6)$$

catalyzed by glucose oxidase can be amperometrically detected by the O_2 consumption or the hydrogen peroxide formation. The schematic of an amperometric glucose electrode is shown in Fig. 46.17.

Infrared spectrometers measure the intensity attenuation of infrared radiation after passing a measuring cuvette and compare it with a reference

$$I_a = I_0 e^{-kcl} \quad (46.7)$$

with I_a the output intensity, I_0 the input intensity, c the concentration, l the layer thickness, and k a constant of proportionality.

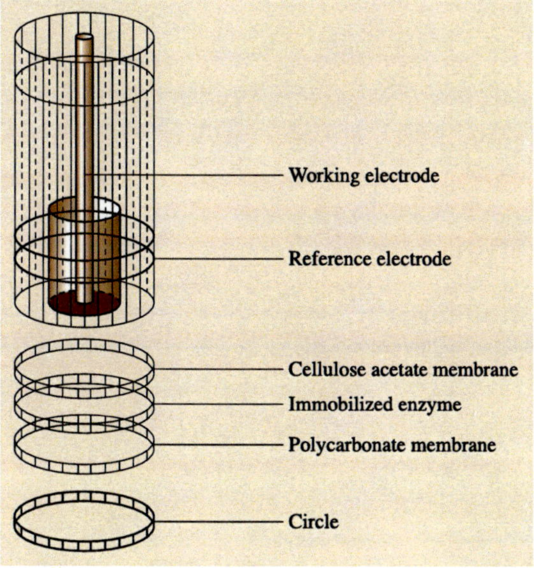

Fig. 46.17 Schematic setup of an amperometric glucose electrode

Owing to the chopper, the detector is alternatingly irradiated with both intensities. Before reaching the detector, the polychromatic light is filtered to the measurement wavelength. The schematic setup is depicted in Fig. 46.18.

46.2.3 Bioelectric and Biomagnetic Signals

Bioelectric signals (Table 46.2) are potential differences originating from processes at the membranes of nerves and muscle cells, leading to rest and action potentials. Excitations are routed as action potential and, in the case of a motor unit, cause muscle contractions. Depending on the source, EEG, EMG, EOG, ERG, ECG etc. can be distinguished.

Bioelectric potentials can be recorded with electrodes. Registering the potential difference between two electrodes on electrically active sites is denoted as bipolar recording. If one of the electrodes is placed on a site which is considered to be rather electrically inactive, for instance at the earlobe for EEG recordings, the recording is called unipolar.

One example is the ECG with the bipolar recording from limb leads (Fig. 46.19). For the latter, the reference is the neutral point of the interconnection of several electrodes via equal resistances, in order to create an

Table 46.2 Bioelectric signals

Signal	Frequency (Hz)	Amplitude (mV)
ECG (heart)	0.2–200	0.1–10
EEG (brain)	0.5–100	2–1000 μV
EMG (muscle)	10–10 000	0.05–1
EGG (stomach)	0.02–0.2	0.2–1
EUG (uterus)	0–200	0.1–8
ERG (retina)	0.2–200	0.005–10
EOG (eye)	0–100	0.01–5
FAEP (brain stem)	100–3000	0.5–10 μV
SEP (somatosensory system)	2–3000	0.5–10 μV
VEP (visual system)	1–300	1–20 μV

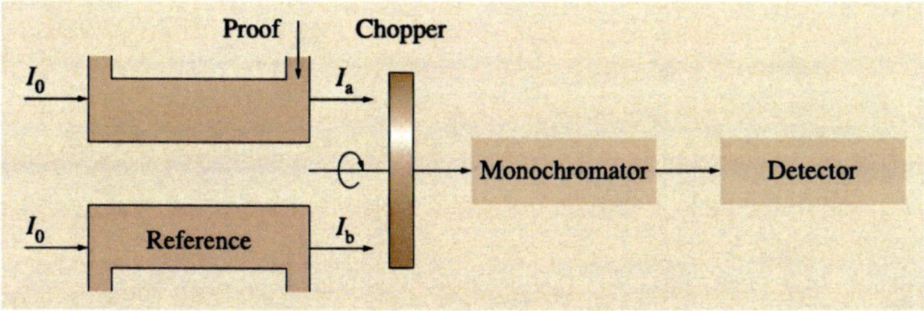

Fig. 46.18 Setup of an infrared spectrometer (I_0 input intensity, I_t output intensity, I_b reference intensity)

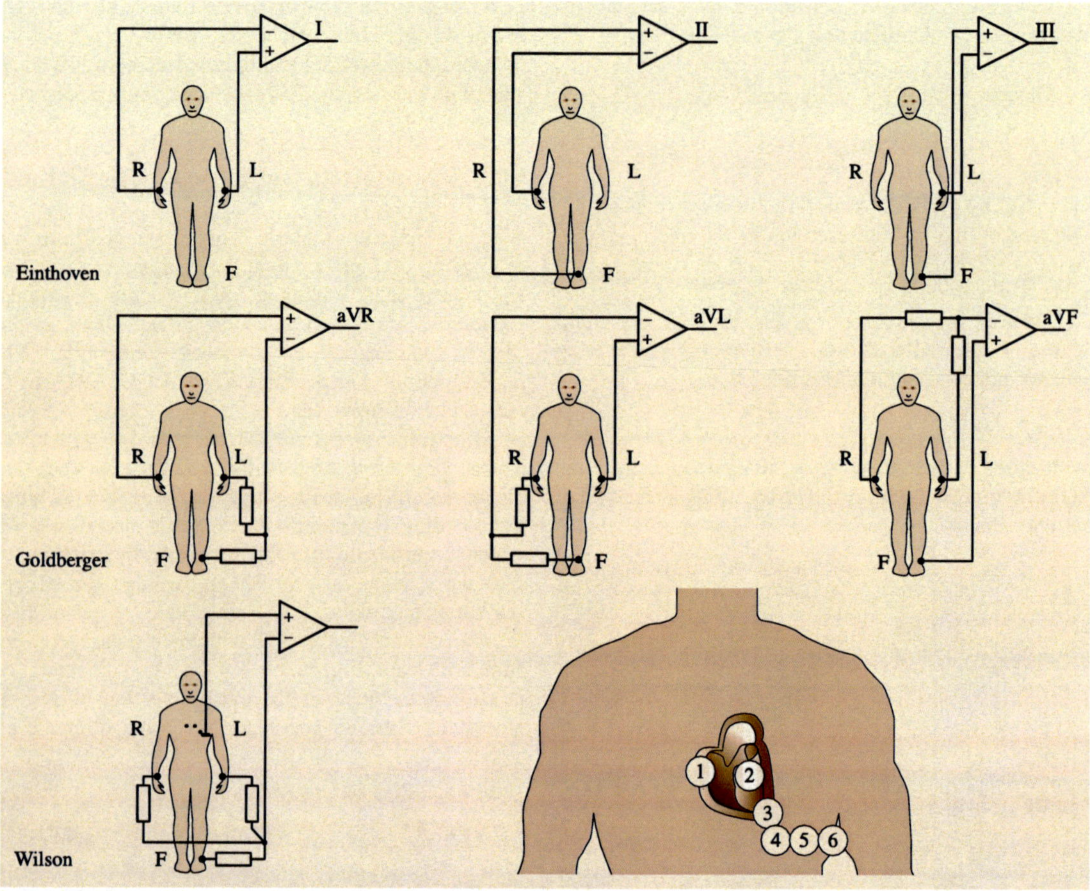

Fig. 46.19 Possibilities of ECG recording with surface electrodes (standard recordings)

Table 46.3 Biomechanical signals

Signal	Spezification	Amplitude	Conversion
Pulse (rate)		720–200 min^{-1}	
Breathing (rate)		5–60 min^{-1}	
Blood pressure (arterial)	Systole	8–33 kPa	60–250 mmHg
	Diastole	5–20 kPa	40–150 mmHg
Blood pressure (venous)		0–4 kPa	0–30 mmHg
Intraocular pressure		0–7 kPa	0–50 mmHg
Blood flow		0.05–5 l/min	
Blood flow velocity		0.05–40 cm/s	
Respiratory flow velocity		20–120 cm/s	
Cardiac output		3–8 l/min	
Respiratory volume		200–2000 ml/gasp	
Muscle work		10–500 W	
Blood volume	Adults	7000 ml	
Amount of urine	Adults	1500 ml/d	
Nerve conduction velocity	Median nerve	50–60 m/s	

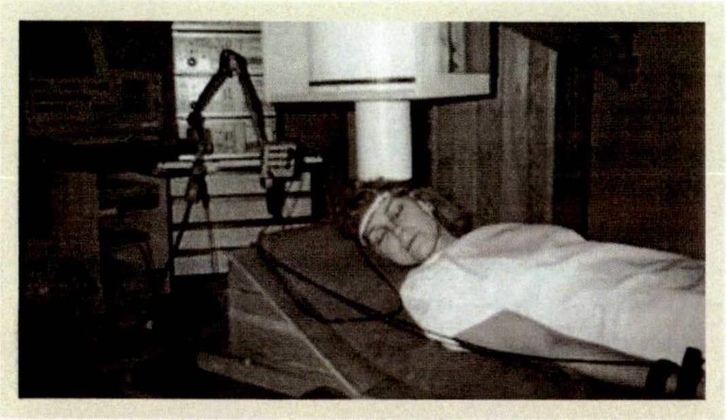

Fig. 46.20 Simultaneous registration of a somatosensorily evoked potential and the somatosensorily evoked magnetic field after stimulation of the median nerve of the right hand

average reference with the average potential of those electrodes. Recordings from the limbs and from the chest wall are perpendicular to each other. Thus, the spatial potential distribution can be calculated.

As electric phenomena are always accompanied by a magnetic field, they can also be detected by registering the magnetic field. With superconductive quantum interference devices (SQUID), these magnetic fields can be detected. The sources can be exactly located, which makes this method advantageous. As an example, Fig. 46.20 shows the simultaneous registration of a somatosensorily evoked potential and its corresponding magnetic field after electrical stimulation of the median nerve. For the magnetic registration, five second-order gradiometers were used. Thus, measurements in unshielded environment are possible.

46.2.4 Biomechanical Signals

Biomechanical signals represent the mechanical functions of the biological system. Some examples are mentioned in Table 46.3.

Movement

The measurement of movements is important for diagnostics, but also for the optimization of motion sequences. Thus, the motion sequence during walking is registered for both the fitting of lower extremity prostheses and post-fracture physical therapy. The extremities can be marked with optical markers, and the motion sequence can be recorded with a video camera. Subsequently, image processing techniques can be used to determine the position of the light spots to each other and to calculate the movement.

For tracking examinations of the upper extremities, a goniometer is fixed, e.g., at the elbow joint to measure the angle between the upper arm and the forearm. Quantitative measurements of tremor, for instance in Parkinson's patients, and of spontaneous periodic leg movements during sleep, is also diagnostically important. Eye movements can be measured electrooculographically, photoelectrically and videooculographically. From the registered movements, the velocity and acceleration of the motion can be calculated.

Pressure

Various forms of pressure measured for clinical purposes are blood pressure, cerebral pressure, intraocular pressure, pressure during labor pains etc. Blood pressure can e.g. be measured invasively by a catheter with an attached pressure transducer consisting of a resilient membrane connected to strain gauges. In a different setup, the transducer is directly positioned inside the blood vessel. Figure 46.21 shows two pressure transducers.

Blood pressure can also be measured indirectly from the cuff pressure squeezing an artery. At the point when the blood begins to flow again, measureable sound occurs, which disappears for unobstructed blood flow. Thus, systolic and diastolic blood pressure can be measured.

Flow

The detection of respiratory flow is important for the monitoring of risk patients. It can be detected by placing a thermistor below the nose measuring the temperature difference between the exhaled and the inhaled air. Respiration flow can be quantitatively measured with a pneumotachograph. With this method, it is also possible to calculate the respiratory voume by integrating the flow (Fig. 46.22)

$$\dot{V} = \frac{\pi r^4 \Delta p}{8 l \eta}\,, \tag{46.8}$$

with $\dot{V}$ flow volume, r radius, l length, η viscosity, and Δp the pressure difference, where

$$V = \int_0^t \dot{V}\,\mathrm{d}t\,, \tag{46.9}$$

where V is the volume.

Blood flow can be determined by measuring the magnetic-inductive flow and with the ultrasound Doppler method (Fig. 46.23). The measurement signal is the frequency shift between transmitted and received signal due to the movement of blood cells

$$\Delta f = f_1 - f_2 = f_1 \frac{2v \cos\varphi}{c}\,, \tag{46.10}$$

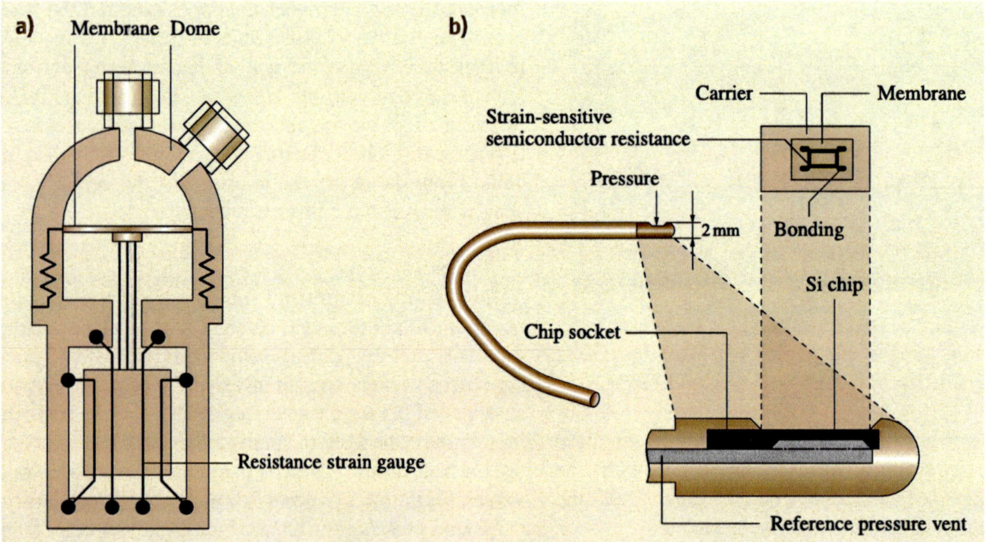

Fig. 46.21a,b Sensors and transducers for invasive measurement of blood pressure. **(a)** External transducer, **(b)** internal transducer

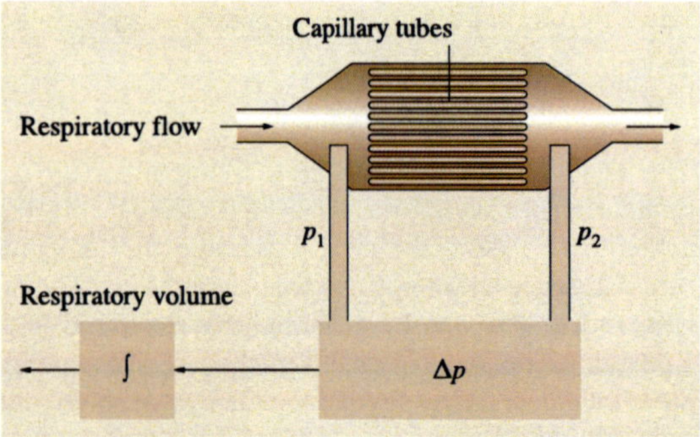

Fig. 46.22 Functional principle of a pneumotachograph. The respiratory volume can be calculated by integrating the respiratory flow over time

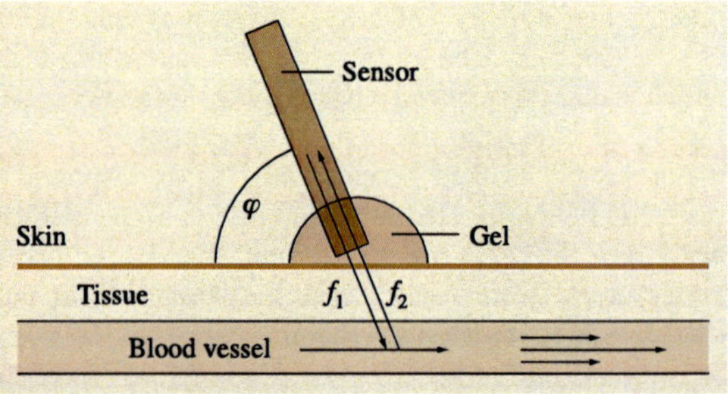

Fig. 46.23 Measurement of the blood flow velocity using the Doppler effect

where Δf is the frequency shift, f_1 the transmitted ultrasound frequency, ϕ the angle of incidence, v the particle velocity, and c the speed of sound in tissue.

Measurement of volume in form of volume flow is very important. For instance, the amount of urine within a certain time is measured as well as the forced expiratory volume in 1 s (FIV1) and the cardiac output. To determine the cardiac output, indicator dilution methods are applied. Commonly used indicators include dyes, radioactive substances and cooled NaCl/dextrose solution. Impulse-type injection of the indicator is preferred, to obtain a homogeneous mixture with the flowing blood. Downstream, an indicator dilution curve is registered reflecting the velocity distribution in the considered segment of the flow path and corresponding to the transfer function (Fig. 46.24).

The cardiac output can be determined using the relation

$$\mathrm{CO} = \frac{m_0}{\int_{t_0}^{\infty} c(t)\,\mathrm{d}t}\,, \tag{46.11}$$

where CO is the cardiac output, m_0 the amount of injected substance, and c the indicator concentration.

Velocity

Velocities have already been discussed for the registration of movements and the measurement of volume flow. Measurement of nerve conduction velocity ex-

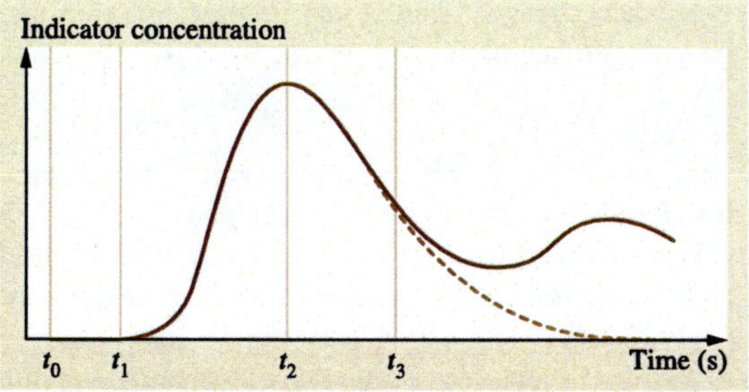

Fig. 46.24 Indicator dilution curve to detect the cardiac output (t_0 time of injection, t_1 beginning of concentration increase, t_2 maximum concentration, t_3 beginning of recirculation)

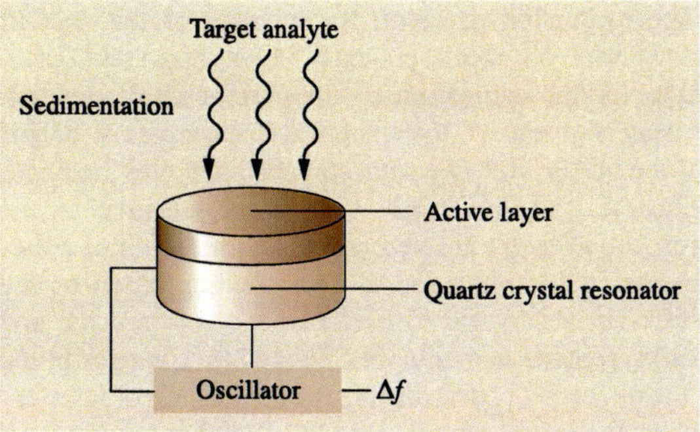

Fig. 46.25 Functional principle of a quartz microbalance

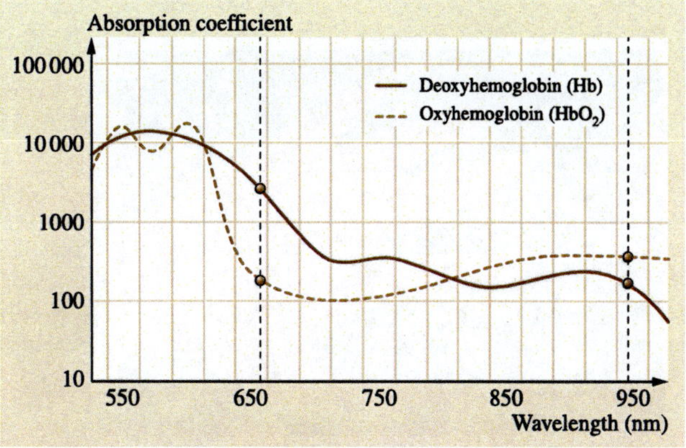

Fig. 46.26 Absorption spectrum of deoxyhemoglobin and oxyhemoglobin. The various absorptions are used to measure the oxygen saturation with a pulse oximeter

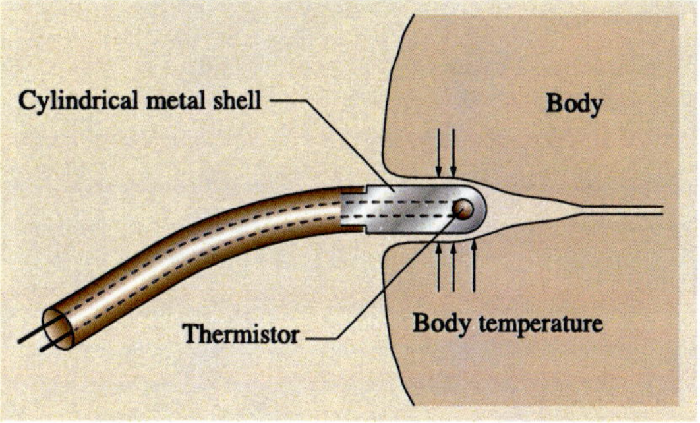

Fig. 46.27 Measurement of the body temperature using a thermistor

hibits a few special features. It is determined electroneurographically after stimulation of a nerve at two sites and registration of the corresponding motor response for motor nerves. From the difference of the latency periods and the distance between the stimulation sites, the nerve conduction velocity can be calculated as

$$\mathrm{NCV} = \frac{\Delta s}{\Delta t}\,, \tag{46.12}$$

with NCV the nerve conduction velocity, Δs the distance between the stimulation electrodes, and Δt the difference of the latency periods of the motor response potentials.

For sensory nerves, orthodromic and antidromic nerve conduction velocity are distinguished.

Size, Shape, Volume, Mass

Size and body weight are of great importance to evaluate the development of a human. This concerns also individual organs. Tape-measures and scales are used for the measurement. But also for standardization of laboratory values, e.g. for the determination of creatinine clearance, weight and size have to be known. They can be used to determine the body surface area, corresponding to the size of the glomerular membrane.

The localization, size and dimension of space-occupying lesions as well as changes in these parameters are also important. Those may be determined using imaging procedures, also including the evaluation and measurement of fractures and the size and location of organs.

Small masses with a detection limit of 10^{-12} g can be registered with a quartz microbalance (Fig. 46.25). The measurement is based on the resonance frequency shift of an oscillating crystal due to deposition of substances on the crystal surface

$$\Delta f = \frac{2.3\times 10^6 f_0^2 \Delta m}{A}\,, \tag{46.13}$$

with Δf the frequency shift, f_0 the resonance frequency of the quartz without deposit, A the area with deposit, and Δm the mass change.

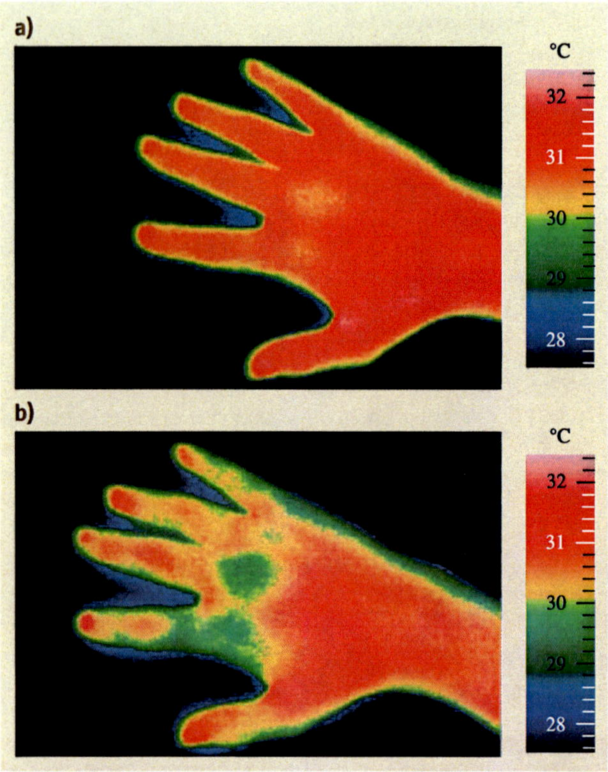

Fig. 46.28a,b Thermographic measurement of the surface temperature of the back of the hand of a 20 year old female test person before (**a**) and 2 min after smoking a cigarette (**b**). The temperature reduction due to a reduced blood circulation can clearly be seen

The surface can also be (bio-)chemically active, enabling the adsorption of substances to increase the mass.

The electrical properties of biological tissue can be described by the dielectric permittivity and the specific resistance. They vary depending on the blood circulation. Using impedance plethysmography, volume changes and pulsations are registered by the associated impedance change. Thus, it can be used to determine the heart stroke volume.

Force

Force can be measured using mechanical dynamometers indicating an elastic deformation due to the contraction of a muscle or a muscle group. The simplest method is a force measurement from the elongation of a spring. Other applied devices include hydraulic dynamometers indicating the pressure created by a piston in a liquid, and quartz dynamometers with a piezoelectric crystal.

46.2.5 Biooptical Signals

The evaluation of color, for instance of the skin of newborns or shock patients, is an important criterion for the estimation of the current vital situation. Using fluorescent dyes, optical measurement of pH value, CO_2 und O_2 with so-called optodes becomes feasible. For this method, reagents sensitive to the respective target are necessary. Another field of application is the evaluation of O_2 saturation based on the different absorption characteristics of oxygenated and deoxygenated hemoglobin (Fig. 46.26). Changes in the fluorescence of tissue during the application of laser enables the intraoperative discrimination between healthy tissue and tumor tissue.

46.2.6 Biothermal Signals

The most important biothermal signal is the body temperature measured with a thermometer, thermistor or an infrared thermometer (Fig. 46.27). Using thermography, the temperature distribution on a skin area can be determined. Changes occur due to superficial space-occupying lesions as a consequence of increased blood circulation or circulatory disorders. Figure 46.28 demonstrates the influence of nicotine on blood circulation and thus a simulated circulatory disorder.

46.3 Measurement and Signal Analysis from a Metrological Point of View

The statistical evaluation of measurement data includes on the one hand the description of results which can be directly extracted from samples, on the other hand an analytical comparison of the results of the concrete sample of the population. Here, estimation procedures provide approximations of the characteristic values of the population. For instance, the t test can be used to test if two average values belong to the same population. Other test methods include the χ^2 test, the F test and the U test. Further information can be gained from plotting the measured values as frequency distributions, diagrams or time series.

46.3.1 Biostatistical Methods

Univariate Statistical methods

Table 46.4 summarizes the formulas used to calculate the average value, the standard deviation, the test statistics for the t test, the correlation coefficient and the coefficients for a linear regression function. In the t test, the null hypothesis, claiming that both average values belong to the same population with a predefined probability value, is rejected if the test statistic is greater than the quantile of the t distribution.

Multivariate Statistical Methods

In contrast to the univariate methods, multivariate methods simultaneously include several measured values or parameters in the analysis. Important methods are cluster analysis, discriminant analysis and factor analysis.

Cluster Analysis. Cluster analysis aggregates the sample cases in several clusters. This is done based on the rule that the cases within only cluster should be as homogeneous as possible, while each cluster should be significantly different from the others. Similarity is the criterion for the aggregation of cases to clusters. Different methods have been developed for this procedure. The cluster membership provides information to the physician with respect to the allocation of a patient to a patient group.

Discriminant Analysis. Discriminant analysis is done to predict the assignment of new observations to a group. The simplest case is the assignment to a positive or negative test result, i. e. pathological or healthy. The dependent variables represent the membership to two or more groups.

Discriminant analysis consists of two steps:

1. The estimation of a discriminant function
2. The classification of the cases with respect to this function.

Often, the discriminant function is a linear combination of criteria essential for the separation

$$D = b_0 + \sum_{i=1}^{n} b_i x_i \,, \qquad (46.14)$$

where D is the discrimination value, b the coefficients, and x the measured values.

The discriminant function is derived in a multistage process, and is optimized each time based on the indispensability and the separation measure. Figure 46.29 shows the result of a reclassification.

Table 46.4 Parameters of univariate statistics

Parameter	Formula
Average value	$\bar{x} = \frac{1}{n}\sum_{i=1}^{n} x_i$
Standard deviation	$s = \sqrt{\frac{1}{n-1}\sum_{i=1}^{1}(x_i - x)^2}$
Test statistic for student test (t test)	$t = \frac{\bar{x}_1-\bar{x}_2}{S_G}\sqrt{\frac{n_1 n_2}{n_1+n_2}}$ $S_G = \sqrt{\frac{(n_1-1)s_1^2+(n_2-1)s_2^2}{n_1+n_2-2}}$
Degree of freedom	$f = n_1 + n_2 - 2$
Regression line	$y = a + bx$
Slope of the regression line calculated from the minimum square deviation of the measured values y_i from the line	$b_{yx} = \frac{\sum_{i=1}^{n}(x_i-\bar{x})(y_i-\bar{y})}{\sum_{i=1}^{n}(x_i-\bar{x})^2}$ $a_{yx} = \bar{y} - b_{yx}\bar{x}$
Linear correlation coefficient	$r_{xy} = \frac{\sum_{i=1}^{n}(x_i-\bar{x})(y_i-\bar{y})}{\sqrt{\sum_{i=1}^{n}(x_i-\bar{x})^2 \sum_{i=1}^{n}(y_i-\bar{y})^2}}$

46.3.2 Biosignal Analysis

The fast development in computer technology has opened up many possibilities for measurement analysis in time domain, frequency domain and complex variable domain. Often, results of the methods are combined and plotted together. The connection to hospital information systems and the direct access to patient data is a very important backing. This concerns also the already mentioned biostatistical methods.

Time Domain. Biosignal analysis in the time domain improves the processability of the measured biosignals and extracts features and parameters from the signal.

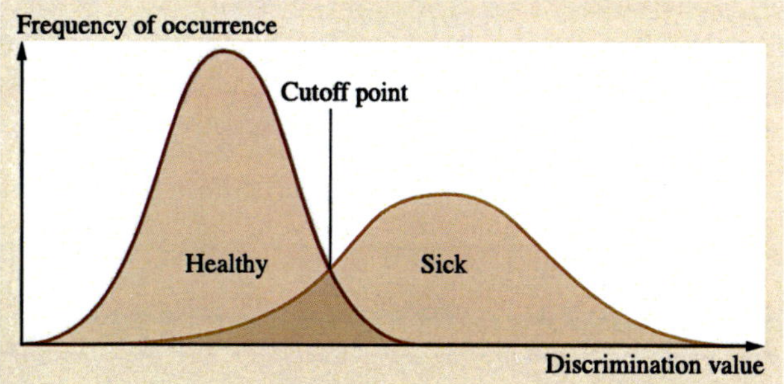

Fig. 46.29 Reclassification of cases based on the calculation of the values of the discriminant function. The frequency distribution is depicted. The healthy and the patient groups can be easily separated

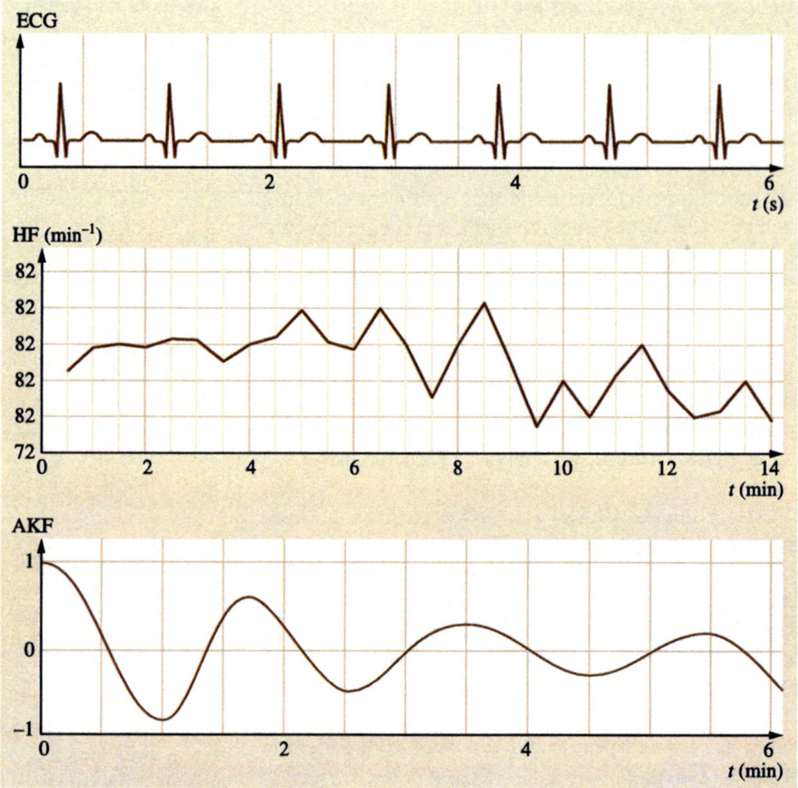

Fig. 46.30 Calculation of the autocorrelation function from the ECG. The varying distance between the R peaks due to breathing is reflected in the periodic component of the function (HF heart rate, AKF autocorrelation function)

The main fields of application are time variate signals describing the functional state or its change in functional diagnostics. Examples are heart-rate change and the occurrence of epileptic activity in the EEG.

The first stage of signal analysis is the detection and elimination of biological and technical artifacts and high-frequency noise. The latter can be easily minimized using moving averaging or smoothing algorithms such as

$$y_n = \frac{y_{n-2} + 2y_{n-1} + 3y_n + 2y_{n+1} + y_{n+2}}{9}, \quad (46.15)$$

where y is the amplitude.

The influence of biological artifacts can be minimized by additionally recording the respective biological signal. In consideration of the amplitude ratio and possible changes in shape, the artifact (e.g. EOG or pulse in EEG) can be subtracted from the wanted signal.

Averaging is often applied to improve the signal-to-noise ratio. It can be used if the wanted signal is temporally correlated with an event, e.g. a stimulus. Using continuous averaging, the signal-to-noise ratio increases proportionally to the square root of the number of averaging steps. Averaging is indispensable for the recording of evoked potentials superimposed by the EEG.

Other methods include the detection of patterns and events and their measurement with respect to temporal occurrence, duration, amplitude, frequency, shape and individual components. Examples are the measurement of waves and peaks in the ECG, the determination of the heart-rate from the distance of the R peaks, and the latency of individual waves in an evoked potential. Often, the detection of features and parameters is connected with a classification.

To determine the similarity of two signals, the cross-correlation function can be calculated. If the signal is correlated to itself, the autocorrelation function is obtained. It can be used to gain information about a periodic component in the signal, and about the stochastic component using the exponential decay. Figure 46.30

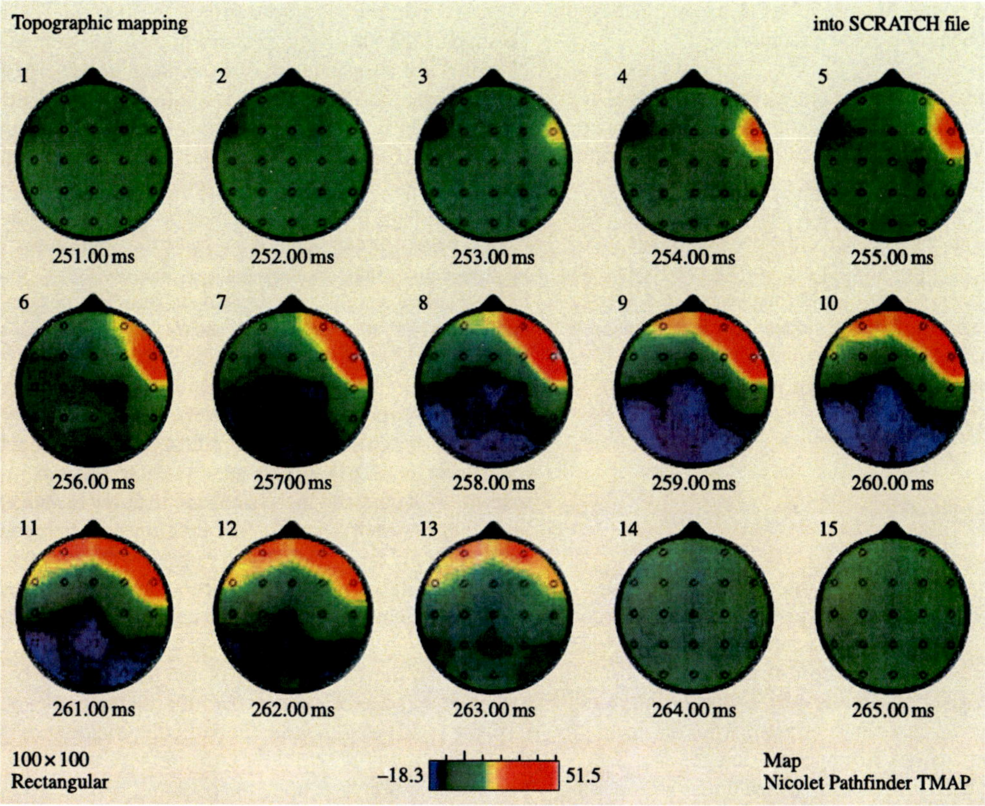

Fig. 46.31 Mapping of the cortical bioelectric potentials during a movement of the field of vision to the right. The dynamics of the potential distribution is displayed in steps of 1 ms for the period of 5–9 ms after saccade initiation

shows an example of the periodic variation of the R peak distance due to breathing, which can be determined from the rest ECG using autocorrelation.

In order to plot two-dimensional potential distributions on a body region, the potential values between the electrodes are interpolated. Subsequently, points with equal potential are connected (isopotential lines) or color-coded. The thus obtained maps can be used to get a general idea of the activity distribution and the localization of events. As an example, Fig. 46.31 shows the map of the cortical potential distribution during a saccadic eye movement. Mapping is also applied in the frequency domain.

Mathematical models can be used to perform calculations to locate bioelectric sources, and to solve the inverse problem. This means that the measured magnetic field distribution is compared to a field distribution calculated based on an assumed dipole, thus optimizing the dipole location.

Frequency Domain. Signal analysis in the frequency domain is done to gain information about the frequencies contained in the signal. The biosignal, whose amplitude is a function of time, is transformed into a signal, whose amplitude is a function of frequency. Often, Fourier transformation or fast Fourier transformation (FFT) is used. Amplitude spectra and power spectra can be calculated.

The latter provide information about the power of a certain frequency in the signal. If the frequency spectrum of an interference is known and outside the spectrum of the wanted signal, it can be subtracted from the signal. After back transformation into the time domain, an almost noise-free signal can be obtained. This procedure is called optimal filtering. But also high-pass filters and low-pass filters can be achieved using this method.

Other transformations include the z transformation, Walsh transformation and wavelet transformation. The latter does not consider the signal as a combination of sine and cosine functions, but as a combination of square functions or wavelet functions.

Image Processing. Images are two-dimensional representations which can be obtained using imaging techniques (CT, MRI, PET, thermography) or mapping of biopotentials. During their generation and processing, they are represented as matrices of gray scale values. One of the primary objectives of image analysis is to detect and locate space-occupying lesions and other processes which can be displayed, and to measure geometric structures. Images are created based on various physical methods.

In image processing, the stages:

- Image editing
- Image improvement
- Detection
- Segmentation

can be distinguished.

Linear methods of image improvement are related to contrast, focus and noise suppression. Various methods of convolution, filtering, rotation and transformation are used. But also the subtraction of images, for instance in digital subtraction angiography, is part of image processing. Here, a mask is subtracted from an image, to increase the contrast of important image components, which can hardly be distinguished from surrounding structures, and thus to highlight the desired structure. By determining the time-dependent behavior and the concentration of contrast agents in a part of the vascular system, dynamic information can be gained.

Knowledge-Based Systems

To support the physician in decision making, methods of artificial intelligence are often applied. Thus, expert systems based on knowledge are established, and the examiner can use these to find a solution by creating links. Often, rules are executed to simulate the special knowledge and the approach of experts. The explanation component pointing out the solution path is vital. Neural networks can be used for pattern recognition after a learning stage. A drawback of this method is the fact that they do not provide an explanation component. Thus, the interpretation of the results is sometimes problematic.

Biological Models and Simulation. The modeling of biological systems and the identification of the model parameters is useful to gain new insights (system research), to understand the system and to apply technical knowledge (system control). The developed models can be used for PC simulation. By experiments and the measurement of target values, the model can be compared to the biological system, validated and improved on this basis.

But also the measurement of the transmission behavior of biological systems and the identification of its order yields new models. Of particular importance is the description of control loops, for instance the regulation of blood pressure, body temperature, heart activity, the pupils, eye movements and reflexes. Knowledge of blood sugar regulation and its simulation is one of the preconditions for the development and the application of implantable insulin pumps.

46.4 Test Planning and Clinical Studies

In order to conduct clinical studies, the required measurements have to be planned systematically. Here, data can be collected retrospectively, prospectively or experimentally. It is necessary to estimate the amount of samples and to ensure that samples are collected stochastically. This includes the consideration of the independence of the elements of a sample, each having the same chance to be included in the sample.

To avoid influence by the patient on the examination result (placebo effect), blind tests are run to ensure that the patient has no information about the influencing factors. In medication studies, this means that he does not know if he receives a placebo or the active drug. If this is also concealed to the examiner, the experiment is denoted as double blind method.

Observation units, which are largely homogeneous with respect to interference, can be combined to blocks. Thus, connected or paired samples can be obtained. Pharmacokinetics and the biological and

Table 46.5 Contingency table for evaluation of diagnostic methods

	Patient diseased K	**Patient healthy** $\bar{K}$	**Sum**
Test positive T	True positive TK	False positive $T\bar{K}$	$TK + T\bar{K}$
Test negative $\bar{T}$	False negative $\bar{T}K$	True negative $\overline{TK}$	$\bar{T}K + \overline{TK}$
Sum	$TK + \bar{T}K$	$T\bar{K} + \overline{TK}$	

Table 46.6 Parameters for evaluation of a diagnostic method

Parameter	Description	Formula
Sensitivity	Correctly recognized patients	$p\left(\frac{T}{K}\right)=\frac{TK}{TK+\bar{T}K}$
Specificity	Correctly recognized healthy subjects	$p\left(\frac{\bar{T}}{\bar{K}}\right)=\frac{\overline{TK}}{\overline{TK}+T\bar{K}}$
Efficiency	Correct test result	$p\left(\frac{T}{K},\frac{\bar{T}}{\bar{K}}\right)=\frac{TK+\overline{TK}}{TK+\bar{T}K+\overline{TK}+T\bar{K}}$
Relevance	Probability of disease for positive test	$p\left(\frac{K}{T}\right)=\frac{TK}{\overline{TK}+T\bar{K}}$
Negative predictive value	Probability of absence of disease for negative test	$p\left(\frac{\bar{K}}{\bar{T}}\right)=\frac{\overline{TK}}{\overline{TK}+\bar{T}K}$

physical half-life period have to be considered in particular. The latter is especially important for the application of radioactive markers participating in metabolism.

For drug testing, 4 stages can be distinguished:

- *Stage I*: Clarification of pharmacokinetic issues and dose for healthy probands.
- *Stage II*: Testing of the basic efficacy against various diseases.
- *Stage III*: Exact comparison of the efficacy among competing therapies and registration of adverse effects from a large patient collective.
- *Stage IV*: Testing of rare adverse effects and exact discrimination of the field of application with approved drugs.

For long-term examinations, the drop-out of several probands and patients, the so-called *drop-out rate*, has to be taken into consideration. This can lead to systematic distortions. The quality of a diagnostic method can be derived from the contingency table (Table 46.5). For this, sensitivity, specificity and efficiency of the method can be determined (Table 46.6). Of particular importance is the location of the cut-off point determining the ratio of sensitivity and specificity (Fig. 46.29).

Further Reading

- M. Akay: *Biomedical Signal Processing* (Academic, New York 1994)
- M. Akay: *Detection and Estimation Methods for Biomedical signals* (Academic, London New York 1996)
- J.D. Bronzino (Ed.): *The Biomedical Engineering Handbook* (CRC, Boca Raton 2000)
- J.J. Carr, J.M. Brown (Eds.): *Introduction to Biomedical Equipment Technology* (Prentice Hall, Upper Saddle River 1998)
- R.R.H. Coombs, D.W. Robinson (Eds.): *Nanotechnology in Medicine and the Biosciences* (Gordon Breach, Amsterdam 1996)
- J. Eichmeier: *Medizinische Elektronik* (Springer, Berlin Heidelberg 1997)
- J. Enderle, S. Blanchard, J. Bronzino (Eds.): *Introduction to Biomedical Engineering* (Academic, New York 2000)
- D.L. Hudson, M.E. Cohen: *Neural Networks and Artifical Intelligence for Biomedical Engineering* (IEEE, New York 2000)
- S. Mackay: *Biomedical Telemetry* (IEEE, New York 1993)
- W.W. Orrison, D. Lewine, J.A. Sanders, M.E. Hartshorne: *Functional Brain Imaging* (Mosby, St. Louis 1995)
- T.M. Peters, J. Williams: *The Fourier Transform in Biomedical Engineering* (Birkhäuser, Basel 1998)
- R.A. Robb: *Three-Dimension Biomedical Imaging* (Wiley, New York 1998)
- T. Scheper: *Bioanalytik* (Vieweg, Braunschweig 1991)
- H. Schubert: *Sensorik in der medizinischen Diagnostik* (TÜV Rheinland, Köln 1989), in German
- T. Togawa, T. Tamura, P.A. Öberg: *Biomedical Transducers and Instruments* (CRC, Boca Raton 1997)
- J.G. Webster (Ed.): *Medical Instrumentation Application and Design* (Wiley, New York 1998)

47. Monitoring Systems

Ullrich Hieronymi, Rüdiger Kramme

The term patient monitoring describes the measurement of vital signs and organ functions of the patient by a dedicated device with automatic detection and alarming of abnormal or life-threatening conditions (e.g., violation of the selected parameter limits and arrhythmia events). The physiological parameters are acquired continuously, at automatic intervals or else in individual measurements, analyzed, displayed as waveforms and/or numerical values, stored in trends, and, if required, printed out or, for example, forwarded to patient data management systems. After an introduction to the fields of use for patient monitoring systems (Sect. 47.1), the types of monitors and displays are described (Sects. 47.2, 47.3), followed by sections on handling aspects such as controls, user guidance, and operating philosophy (Sect. 47.4). Section 47.5 covers the important features of handling alarms and events, followed by a discussion of trend display and automated calculations (Sects. 47.6, 47.7). The chapter concludes with an overview on monitor networks (Sect. 47.8).

47.1 **Fields of Use for Patient Monitoring Systems** 948

47.2 **Types of Monitors** 949
- 47.2.1 Stand-Alone Monitors 949
- 47.2.2 Transport Monitors 949
- 47.2.3 Telemetry Devices 949
- 47.2.4 Preconfigured or Compact Monitors 949
- 47.2.5 Modular Monitors 950
- 47.2.6 Information Monitors with Integrated Transport 950
- 47.2.7 Integrated Acute Medical System 951

47.3 **Monitor Screen Content** 951
- 47.3.1 Display 951
- 47.3.2 Channels 951
- 47.3.3 Waveform Representation 951
- 47.3.4 Screen Configuration 951

47.4 **Handling** 952
- 47.4.1 Controls 952
- 47.4.2 User Guidance and Operating Philosophy 952

47.5 **Alarms and Events** 952

47.6 **Trend Display** 952

47.7 **Automatic Calculations** 953

47.8 **Advanced System Properties** 953
- 47.8.1 Data Integration at the Workstation 953
- 47.8.2 Monitor Network 953

47.9 **Central Monitoring and Documentation** 953

Traditionally, the monitoring function which is primarily used at the bedside or in the operating theater is carried out over a relatively short period of time (short-term monitoring, e.g., during an operation) or a relatively long period of time (long-term monitoring, e.g., in intensive care units), and is intended to reliably draw the medical staff's attention to physiologically unstable conditions as early as possible. This requires reliable and reproducible detection and display of measurements and includes both self-monitoring of the unit and monitoring of the effectiveness of life-supporting systems to which the patient is connected.

Table 47.1 Classification of patient monitoring systems

Features			
Type of monitoring	Basic monitoring with non-invasive parameters	Basic and advanced monitoring with non-invasive and invasive parameters	Advanced and highly specialized monitoring with non-invasive, invasive, and special parameters
Parameters measured	1–5 with trend	3–7 with trend and calculations	>8 with trend, calculations, and communication with other systems
Graphs displayed simultaneously	1–4, (color) LCD	3–8, if required also 12 lead ECG, color LCD/TFT	6–32, color LCD/TFT
Communication	Stand-alone or networkable	Networkable (hard-wired or wireless), devices connected to the workstation	Networked, optionally with web browser and advanced communication and various connections
Field of use	Induction and postanesthetic rooms, recovery rooms, A&E units, outpatient clinics, step-down units, transport	Induction and postanesthetic rooms, OT, recovery rooms, A&E units, intensive care units, intermediate care, transport	General and special intensive care units, special OT units, research areas

The conventional electronic measuring and monitoring unit or monitor has become a multifunctional, software-controlled system with extensive data collection and storage capabilities which is used for diagnosis and, in combination with other clinical observations and assessments, both for therapeutic decision-making and for monitoring the effects of the therapy. In standards the general term used for such systems is programmable electrical medical system (PEMS).

For users the ease of operation of both monitor and its accessories is top priority, including their versatile use and the reliability of all measurements and alarms. There should be as few false alarms as possible, and above all, a life-threatening situation should never remain undetected.

Modern monitors are expected to be scalable, which is to say that they can be adapted to the respective requirements at any time by simply adding or removing extra modules or software options. By means of updates and upgrades, they must be forward-looking, keeping pace with new developments, and must be capable of being extended, for example, with new measuring techniques.

Modern monitors feature a highly intuitive operating philosophy and communicate with each other and with central monitoring stations via network connections. Communication with other units and with networks in the hospital is one of their advanced system properties. Table 47.1 lists some basic classifications of patient monitors and systems.

47.1 Fields of Use for Patient Monitoring Systems

Patient monitoring systems are mainly used in general and specialized operating theaters (OT) (especialliy cardiac, vascular, and neurosurgical):

- Induction and postanesthetic rooms, recovery rooms, and observation wards
- General and special intensive care units (ICU), including neonatal (NICU) and coronary (CCU) care units as well as stroke units, accident and emergency (A&E) units, procedure rooms, and outpatient clinics, emergency services and patient transport.

Furthermore, monitors are regularly used in wards and units which take patients from intensive care units at a very early stage (step-down units).

47.2 Types of Monitors

47.2.1 Stand-Alone Monitors

A stand-alone monitor does not communicate with other monitors or with a central station and is often used only for selective measurement or short-term monitoring of a few patient parameters.

47.2.2 Transport Monitors

This type of monitor is portable and a device as light as possible with an integrated display, which can be fastened to the bed or to the transport bed and that is used in battery operation to monitor the patient during transport (Fig. 47.1c).

47.2.3 Telemetry Devices

These are portable transmitters (often without a display) for wireless monitoring of patients in a defined coverage area of an antenna system installed in the clinical environment. Unlike the transport monitor, such devices are carried by the ambulant patient himself and are used for monitoring the electrocardiogram (ECG) and, if necessary, the oxygen saturation and (non-invasively) blood pressure at a central station. Modern telemetry devices no longer use the usual ultrahigh-frequency (UHF) and very high-frequency (VHF) bands which require proprietary antenna and receiver systems, but instead use a Wi-Fi radio network called WLAN (wireless local area network) using certified generally available components complying with industry standards.

47.2.4 Preconfigured or Compact Monitors

Such devices have an integrated display and a fixed and limited number of measurement parameters. Semimodular (compact) monitors can furthermore be expanded by a module for additional measurements. A recorder can be integrated. Integration in networks is usually possible nowadays. When equipped with accumulators, there is no strict distinction from use as a transport monitor. When integrated in a WLAN, the monitor can work in the same radio network as a WLAN telemetry unit. If

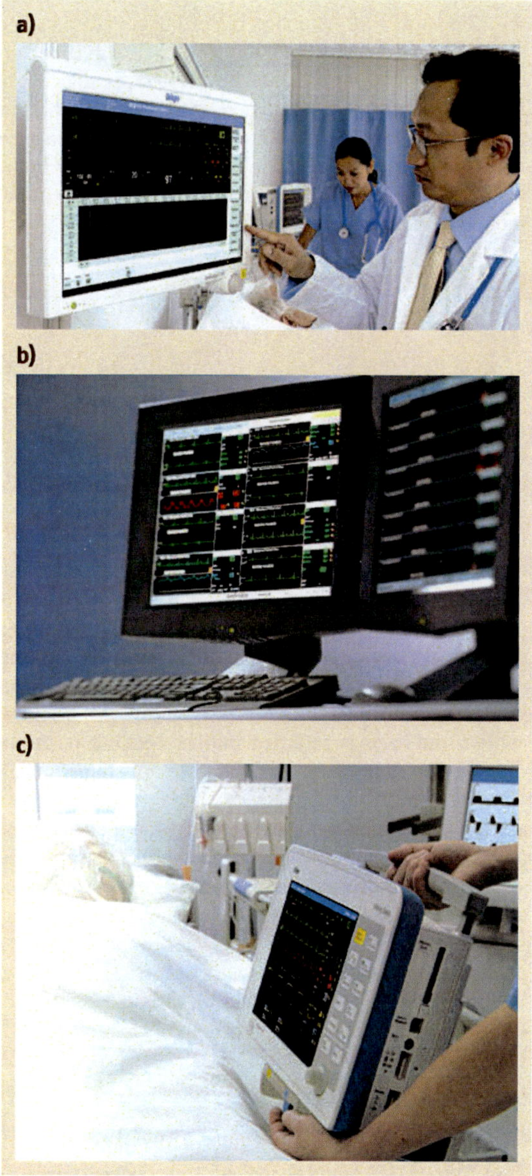

Fig. 47.1a–c Components in a modern monitoring system. **(a)** Widescreen information monitor (*foreground*) with integrated transport monitor (in the background on the docking station). **(b)** Central station for 16 patients. **(c)** Monitor is securely locked on a bed mount for transport (© Dräger Medical, Lübeck, Germany) ▶

the monitor is further miniaturized, it will without doubt no longer be possible to draw such strict boundaries between a telemetry device and monitor in the future.

47.2.5 Modular Monitors

The performance of these monitors is freely scalable by exchanging or adding modules (plug-in units or external extensions) for detection of further parameters and also by means of software options (e.g., for advanced arrhythmia monitoring) and can be adapted in each case to the specific requirements. These monitors, which have an integrated or separate screen, can often only be used in a stationary position at the patient's bedside or in the operating theater (OT) and as a rule are combined with monitor systems in a network.

47.2.6 Information Monitors with Integrated Transport

A modular monitor which is suitable for transport throughout the hospital, combined with a logically associated medical-grade panel personal computer (PC) (Fig. 47.1a). The monitor is docked at the bedside on a so-called docking station which provides secure mechanical mounting and contains all the connections to the power supply, network, and, if required, to external devices and other display units. Simply by turning a lever, the mechanical connection and all electrical connections are disconnected and the monitor is used as a transport monitor, without any interruption in monitoring (pick-and-go principle). On return from transport, the monitor is simply put on the docking station, the lever is returned into its locking position, thereby re-establishing all of the connections which were previously disconnected. There is no need for any cables and/or modules between the bed monitor and transport monitor to be reconnected, and absolutely seamless monitoring of the patient is ensured, without any restrictions on the parameters. The dedicated bedside panel PC is firstly used for advanced monitoring, to present the abundance of monitor information clearly, intelligibly, and in a large size in various screen layouts. Secondly, it acts as an information technology (IT) portal for doctors and nurses and provides access to all electronically available information in the hospital about their patients [images, e.g., x-ray, magnetic resonance tomography (MRT) and ultrasound, test results,

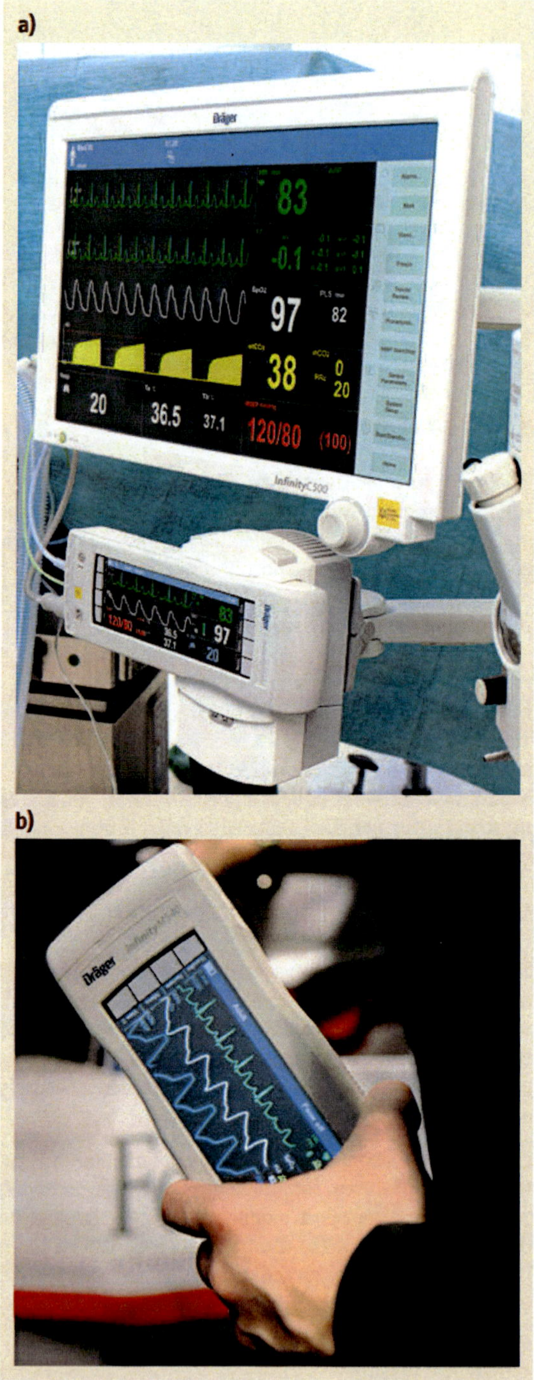

Fig. 47.2a,b Monitor components of the Infinity Acute Care System. **(a)** Medical cockpit with docked transport monitor, **(b)** hand-held transport monitor M540 (© Dräger Medical, Lübeck, Germany) ▶

medical histories, and pre-existing conditions] and to databases (red list, care guidelines). Thereby essential information for decision-making on further therapy is available where it is needed – at the bedside, the *point of care* – without a time-consuming search. Thirdly, the PC can simultaneously be used for paperless documentation by running a patient data management system.

47.2.7 Integrated Acute Medical System

This system represents a further development of the information monitor mentioned above and of the pick-and-go principle (Fig. 47.2a). A convenient and light but very powerful monitor with a touchscreen (Fig. 47.2b) records all the vital signs and stays with the patient at all times (including during transport). At the bedside it can be inserted into the automatically locking docking station with one hand and thereby connected to the information monitor which is used for comprehensive operation. The information monitor acts at the same time as an IT portal for all of the possible web-based applications or Citrix applications in the hospital. The standardized hardware provides a user interface which has basically the same design, not only for monitoring but also, for example, for respiratory therapy and anesthetic equipment. The aim is to always provide the user with the same operating philosophy in an integrated system, and as a result of this standardization to minimize both the cost of training and the sources of operator errors throughout all the wards and departments, from accident and emergency (A&E) units to the OT, and from the recovery room to intensive care units. The system is scalable according to the various requirements in the different wards and departments, facilitates intelligent data transfer, and ensures effective transport without any loss of time.

47.3 Monitor Screen Content

47.3.1 Display

Apart from very simple blood pressure and temperature measuring units, which only display numerical values, virtually all monitors have a screen for displaying waveforms of the continuously monitored parameters (usually ECG, pulse curve, respiration, and invasive blood pressure) with their associated labels, measured values, and alarm limits, as well as for displaying trends and other information.

Color flat-screen monitors based on liquid-crystal display (LCD) and thin-film transistor (TFT) technology with screen sizes between 6 inches (15 cm) and 17 inches (43 cm) and above, with resolutions of 640×480 to 1024×768 pixels, have generally become widespread. The trend is towards a wide screen of 20 inches (51 cm) and 1680×1050 pixels. Monochrome LCD or electroluminescent (EL) screens today may only be found in small stand-alone monitors and transport monitors. Good screens are flicker free, high contrast, and antiglare and have a sufficiently large viewing angle (>160°), such that they can easily be read, even if there is excessive ambient light and when viewed from a awkward perspective.

47.3.2 Channels

In addition to the size and quality of the screen, the number of possible monitor channels also describes the performance and capability of a monitor. This is the maximum of waveforms that can be selected for monitoring and their associated parameters which are displayed in parameter windows or boxes. Transport monitors and compact monitors usually have three or four channels. Modular monitors traditionally have five, six or eight channels. Modern monitors with 17 inch or 20 inch screens offer as many as 32 channels.

47.3.3 Waveform Representation

Either the waveforms constantly move from the right to the left of the screen (source mode or paper mode) or they are drawn and redrawn anew from left to right on the screen, with an erase bar deleting the waveform of the previous sweep (erase bar mode). The sweep speed can be selected, at least in modular monitors. Conventional speeds are 25 or 50 mm/s for ECG and blood pressure waveforms and 12.5 or 6.25 mm/s for respiratory ones. Many monitors have a freeze function, which enables the image on the screen to be frozen to be viewed more accurately. Waveforms showing physical measurements (ECG and blood pressure) have a scale.

47.3.4 Screen Configuration

The arrangement of the waveforms and parameters on the screen (positioning, layout) as a rule can be set by the user. At least one such configuration can be stored and is used as a standard when the monitor is switched on. In modular monitors, 5 or 10 layouts can usually be stored.

47.4 Handling

47.4.1 Controls

The controls on the monitor can be fixed buttons, a rotary knob, and touchscreen, which can be found single or in every possible combination. Some monitors additionally allow for wired or infrared remote control. Monitors with an integrated web browser may have a keyboard and mouse.

47.4.2 User Guidance and Operating Philosophy

Simple, logical, and highly intuitive usability is required to ensure good acceptance by users and to avoid faulty operation. It is user-friendly to have direct access to important function keys (all alarms off, standby, main screen, and preferably highlighted in color the most important key: alarm audio pause) and a logical, easily understandable menu design with no more than two or three levels. Some monitors provide auxiliary functions and built-in simulation programs for learning how to operate them.

47.5 Alarms and Events

An important function of the monitor is to provide audible and visual alarms with at least three different levels. The highest priority is given to alarms for life-threatening situations (such as cardiac arrest or ventricular fibrillation), to which attention must urgently and clearly be drawn acoustically and visually in the form of a red alarm. Serious warnings, e.g., when a limit value is violated, are displayed in yellow. Technical and/or advisory alarms appear in white. Audible alarms are intended to attract attention. The alarm levels must sound distinctly different and be capable of being differentiated by staff from the alarms of other medical equipment. Visual alarms are used to quickly recognize: (1) which monitor is sounding an alarm (alarm light on the monitor), and (2) what is causing the alarm (parameter value flashing colored).

An alarm can be printed automatically on a (laser) printer or a special alarm strip recorder and stored in the monitor and/or in a central station as an *event*. Storing alarms allows subsequent evaluation and, if necessary, selective documentation. This can save printing out all events.

A monitoring system can be configured in a way that the alarm from one patient bed is also displayed on all other monitors and the medical staff can immediately identify the nature and location of the alarm. In the simplest case, this is a brief indication of the bed and the parameters causing the alarm in a message bar displayed on the other monitors (simple bed-to-bed communication). Optionally, even the display content of the monitor which is emitting the alarm can appear automatically on the other monitors (full bed-to-bed communication). Either way, an alarm will sound from and be displayed on a connected central station. Particularly in wards which have been built in a manner which gives staff a very poor view of the whole ward, it may be necessary to transmit important alarms to an existing nurse call system and to display them on information displays in corridors which are visible from a long distance.

47.6 Trend Display

All values measured are stored by the monitor in the trend memory. Mean minute values (or median values) from the last 24 h or more of continuously measured parameters are usually stored (96 h is quite normal). The trend is displayed as a table or as a graph, and the user can select a period from 24 h up to a few minutes (zoom).

47.7 Automatic Calculations

Physiological calculations of variables which are derived from the continuously and discontinuously measured parameters are customary for hemodynamics, respiratory values, and oxygen supply. These results are then also available as a trend. Many monitors also permit dose calculations for various medications and infusions.

47.8 Advanced System Properties

47.8.1 Data Integration at the Workstation

By integrating data from external devices (e.g., specialized monitors from other manufacturers) and/or ventilators or anesthesia systems, it is possible to display additional waveforms and parameters together with the data from the monitor, and a common database (trend memory) is created. The data are transmitted, for example, via a medical information bus (MIB) or serially.

47.8.2 Monitor Network

Networking (local area network, LAN) is the step from an individual monitor to a monitoring system. Using industry standards for hard-wired LAN or wireless LAN (WLAN), the monitors communicate with each other from bed to bed and with central monitoring stations and also, for example, with commonly shared printers.

Data (e.g., demographic data from the hospital information system or test results) can be imported and exported (e.g., to a patient data management system) from the monitoring system via the network and special gateway computers. Using the medical grade computer of the information monitors mentioned before, it is possible, for example, that all patient-related information that is electronically available throughout the hospital can be displayed at the bedside and integrated into the monitoring process using a web browser or a dedicated application. Thus, less time is wasted searching for information, and no data are lost for documentation.

Screen contents and trend information from monitors can be accessed web-based via the hospital intranet and a gateway server so that authorized doctors can get this information from any PC in the hospital network using a standard web browser. Consultations can therefore take place and diagnoses be made anywhere in the hospital. It is also possible to trigger pager systems, in order to also provide doctors or nursing staff, for example, with alarm information outside a ward and independently of hard-wired monitors and central stations.

Since vitally important alarms must not be delayed in a monitoring network and gaps in transmitted waveforms are not acceptable, a strict physical separation between the hospital network and the monitor network has always been required in the past. Modern technology for ensuring the priority of time-critical data now also allows a monitor network to be operated as part of the network installation of a hospital as a result of logical separation of the network segments by means of virtual LANs (VLAN), with corresponding system stability and the guarantee of a minimum bandwidth for the prioritized application as quality of service (QoS). Only as a result of this, it is now possible for monitors and, for example, computers used on ward rounds to separately communicate via WLAN using the same access points without interfering with each other.

47.9 Central Monitoring and Documentation

Via network, monitoring central stations provide all alarms visually and acoustically at central workplaces and also provide an overview of the state of the connected patients using one or more screens. Limit values on the bed monitor and, if applicable, telemetry units can be changed from the central station. Central stations also have to be scalable and must be capable of being adapted, in terms of their performance, to requirements by means of hardware and/or software options, e.g., in terms of the number of patients to be monitored and displayed (usually between 4 and 32) with or without telemetry, the number of waveforms per patient, long-term storage for waveforms (e.g., 1–3 days full disclosure of 4–16 curves) and events (e.g., 1000 per patient as event disclosure), and other functions, e.g., view of beds from other wards or of waveforms stored in other central stations, as well as exporting the full disclosure waveforms and events to the intranet to be viewed on doctors' PCs when needed for diagnosis. Another important function is that of documentation, i.e., the option of printing out trends, events, and reports in adjustable formats and time periods. A laser printer with standard paper format is the standard today. Network printers can also be used without central stations directly from the monitors.

Further Reading

- R. Kaplow, S. Hardin: *Critical Care Nursing* (Jones Bartlett Learning, Sudbury 2007)
- E. Gaura, L. Girod, J. Brusey, M. Allen, G. Challen (Eds.): *Wireless Sensor Networks* (Springer, Berlin Heidelberg 2010)
- J. Dyro (Ed.): *Clinical Engineering Handbook* (Academic, New York 2004)

48. Cardiovascular Monitoring

Ullrich Hieronymi, Rüdiger Kramme

The generic term *cardiovascular monitoring* covers the monitoring of heart and circulatory functions. Cardiovascular monitoring is vital and of crucial importance in the case of cardiological diseases, unstable circulatory states, and shock as well as for pre, intra, and postoperative monitoring of the cardiovascular system. Firstly, it must be possible to commence interventions quickly in the event of any impairment, and in addition the measurements are used to assess the condition of the patient, reach a diagnosis, decide on the therapy, and monitor the therapy. It covers cardiac function (Sect. 48.1) in the form of an electrical phenomenon (ECG) and its mechanical effect, i. e. the pressure build-up and the volume delivery, which is dependent on heart rate, contractility, preload, and afterload. In particular, when the patient is in a state of shock, it is important to maintain or restore the main function of the circulatory system (Sect. 48.2), which is to supply sufficient oxygen to the body. In addition to artificial respiration and volume management, doctors have a wide variety of means available to them for this purpose, which act differently on the cardiovascular system and are only used selectively with meaningful measurements and can be checked for their effectiveness.

48.1 **Monitoring the Cardiac Function** 955
48.1.1 Electrocardiogram (ECG) and Heart Rate (HR) 955
48.1.2 Arrhythmia Monitoring 957

48.2 **Monitoring the Circulatory Function (Hemodynamic Monitoring)** 957
48.2.1 Pulse 958
48.2.2 Discontinuous Noninvasive Blood Pressure (NIBP) 959
48.2.3 Continuous Noninvasive Arterial Blood Pressure Measurement (CNAP) 961
48.2.4 Invasive Pressure Measurement in the High-Pressure and the Low-Pressure Systems 962
48.2.5 The Ideal Pressure Measurement System – Requirements 963
48.2.6 Determining the Cardiac Output (CO) 966
48.2.7 Calculation of Hemodynamic Variables 969

References 969

48.1 Monitoring the Cardiac Function

48.1.1 Electrocardiogram (ECG) and Heart Rate (HR)

In emergency and intensive care medicine, in surgical fields, and in the emergency services, continuous ECG monitoring is part of basic monitoring. The ECG provides information about the heart rate and rhythm, the excitation, conduction, and repolarization and disturbances in these functions, but does not provide any direct information about the pumping capacity of the heart (i. e. the mechanical cardiac function).

The ECG is a record of sum potentials of the electrical cardiac activity and can be measured very easily directly from the surface of the body using electrodes (Fig. 48.1). It is customary to use 3, 5, 6, or (e.g. for monitoring in cardiology) 10 electrodes. Only one ECG lead can be displayed when three lead wires are used. 7, 8, and 12 ECG leads can be obtained from 5, 6, and 10 electrodes, from which generally up to three leads can

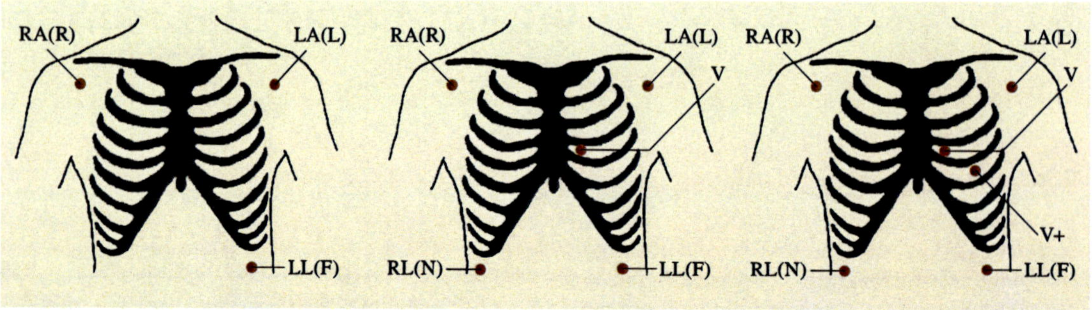

Fig. 48.1 Standard electrode positions for ECG monitoring with three, five, and six electrodes. Color code and labels according to IEC2 (AHA/US): RA: white, LA: black, LL: red, RL: green, V: brown, V+: gray/brown; in parenthesis according to IEC1 (Europe): R: red, L: yellow, F: green, N: black, V: white, V+: gray/white

be selected on conventional 5 to 8 channel monitors for constant display on the screen channels.

Three lead wires are sufficient to determine the heart rate (HR) and are often used for basic monitoring, e.g. in the normal operating theater (OT). Three particularly small neonatal electrodes are found exclusively in neonatology. Simultaneous monitoring of multiple leads is recommended for better artifact suppression, for reliable recording of cardiac arrhythmias, and for pacemaker detection. Special ECG interval measurements can also be performed. The automatic analysis and monitoring of the ST segments in the ECG is of particular importance in the case of myocardial ischaemia and infarction. In these cases, it is important to *view* the heart as far as possible from all sides, i.e. to utilize up to 12 leads so that local ischaemias are not overlooked. For this purpose, methods have also been developed that make highly reliable ST monitoring possible from eight true and four computed leads (TruST algorithm) using just six electrodes (reduced lead set) which are placed entirely as normal. Using 10 lead wires, modern monitors can display, and if necessary automatically evaluate, a 12-channel resting ECG of diagnostic quality. It is vital to keep in mind here that correct application of the electrodes is the basis for achieving a diagnostic quality for the ECG. The placing of electrodes on the thorax, which is performed in the monitoring process, cannot completely replace the true limb leads, which are necessary for diagnosis. It can simulate apparent pathological deviations of the cardiac axis, which are not present at all when electrodes are placed correctly on the limbs.

The heart rate (typical measuring range of 15–300 beats/min) is determined as a moving average over a specific time (e.g. 10 s), or a specified number of QRS complexes, and displayed numerically. As an alternative to the ECG, the heart rate can be obtained (automatically switched over by the monitor when there are interferences in the ECG, e.g. during cauterization in the OT) in the form of the pulse rate from the arterial blood pressure or the pulse curve from the SpO_2 signal. The basis of heart rate measurement is safe detection of the QRS complexes in the ECG and assessment of the intervals between them (RR intervals) using digital filters and detection techniques. This way, it is possible for impulse noise or movement artifacts to be eliminated and not be included when determining the heart rate. Pulses from pacemakers are already detected prior to the actual QRS detection on the basis of their special characteristics (very short and very steep) and can be added to the ECG as a marker. An alarm can be activated when adjustable limit values are violate, and these events can be stored as a brief ECG episode in an event memory in the monitor or in a central station.

Numerous factors can interfere with ECG monitoring: Patient-related interferences are primarily due to ECG electrodes and lead wires falling off, insufficient electrode contact with the skin (poor preparation of the skin, dried out electrode gel), sweaty or greasy skin, and also muscle tremors and movement artifacts in the patients. In the case of highly elevated movement artifacts, it is also necessary to consider the possibility of there being too low a contact force at the electrode connections. Technical problems that interfere with the ECG can, in particular, be caused by radio frequency (e.g. RF surgical equipment and, in particular, also the low-frequency modulated emissions from RF surgical equipment), by coupled AC voltage or by impulses from transcutaneous electrical nerve stimulation (TENS). In operating theaters, the use of special lead wires and/or filters to reduce interference is recommended or prescribed by the manufacturer.

48.1.2 Arrhythmia Monitoring

Continuous arrhythmia monitoring is used to detect cardiac arrhythmias and their precursors and symptoms in real time. A detected arrhythmia can trigger an alarm, be stored as an ECG strip in the event memory, and/or be printed out automatically. Life-threatening conditions such as asystole, ventricular fibrillation, or ventricular tachycardia must be reliably detected and an unambiguous acoustic and optical alarm (red alarm) must be triggered. These three forms of arrhythmia are basic or lethal arrhythmias, which every modern monitor should fundamentally detect and respond to by generating an alarm. The asystole alarm is triggered if no QRS complex is identified within a certain period of time (generally 4 s). If the reason for no QRS complex being identified is that electrodes have fallen off, a corresponding technical note is generated and the asystole alarm is not triggered. In neonatology, bradycardia (excessively low heart rate) is a lethal alarm and should fundamentally be recognized by every monitor in neonatal operating mode as life-threatening.

The monitored ECG leads form the basis for analysis of arrhythmias. Cardiac arrhythmias are frequently linked to heart conditions and can also occur sporadically in patients without heart problems. The nature and number of events are of particular clinical interest. Monitor systems often provide the option of continuously storing and analyzing the ECGs of multiple patients over several days.

Arrhythmia monitoring begins with a learning period in order to ascertain the normal QRS complex that is currently typical for the patient, and this normal QRS complex is learnt as an individual normal beat and is stored as a template or reference complex. All incoming QRS complexes are subsequently compared with the template (template matching). This involves an online decision as to whether it is a normal beat, a ventricular beat, or an artifact. Various features and detection criteria of the ECG cycle are used for this (feature extraction), e.g. polarity, width and shape (morphology) of the QRS complex, slope and area of the ST segment, as well as the time interval (RR interval) of the complexes. The distinction between normal and ventricular is no longer made according to a set of rules of fixed measurement criteria (rule-based system); the monitors instead imitate human approaches to template matching, e.g. using probability criteria or fuzzy logic. Frequency analysis (detection of flutter and fibrillation) and artifact analysis (movement, muscle tremors, etc.) is performed in parallel. After the classification of the individual QRS complexes, the distinction is made, using rhythm logic, as to whether the event concerned is an isolated ventricular extrasystole (VES) or whether it concerns linked events (pairs, runs, ventricular tachycardia, bigeminy, etc.). If a pacemaker pulse has been detected, the only important thing is to judge whether or not there is a physiological response from the heart. Because pacemaker pulses can give rise to transient electrical phenomena in the heart, which can be very similar to a natural heart action, this judgment is by no means trivial. Pacemaker patients should, therefore, always be given additional monitoring of the circulatory function, e.g. the pulse rate from SpO_2 monitoring or continuously measured arterial blood pressure.

The analytical quality of the monitors is determined, among other methods, by testing using reference ECG recordings from international societies (American Heart Association, AHA) or recognized institutions.

48.2 Monitoring the Circulatory Function (Hemodynamic Monitoring)

Hemodynamics is the study of the flow of blood in the circulatory system. This flow is driven by the pressure generated by the heart. Since the pressure in the vascular system is highly dependent on activity and the position of the body (hydrostatic pressure), blood pressure measurements are always taken at rest and are based on the height of the heart (right atrium).

The vascular system is functionally divided into a low-pressure system (small, pulmonary circuit) and a high-pressure system (large, systemic circuit), which are connected to one another by the central driving element, the heart. The heart generates pressure in its contraction phase (systole), by means of which the stroke volume (SV) is expelled from the ventricle and into the arterial vascular system. Every stroke volume conveyed generates a pulse wave. The peak pressure during the expulsion of the stroke volume from the ventricle is the systolic blood pressure (highest point of the pressure curve). The pressure at the end of the relaxation phase (diastole) is referred to as diastolic blood pressure (lowest point of the pressure curve). The difference between the systolic and diastolic pressure is the blood pressure amplitude. The pressure that maintains the blood flow in the vascular system, that is to

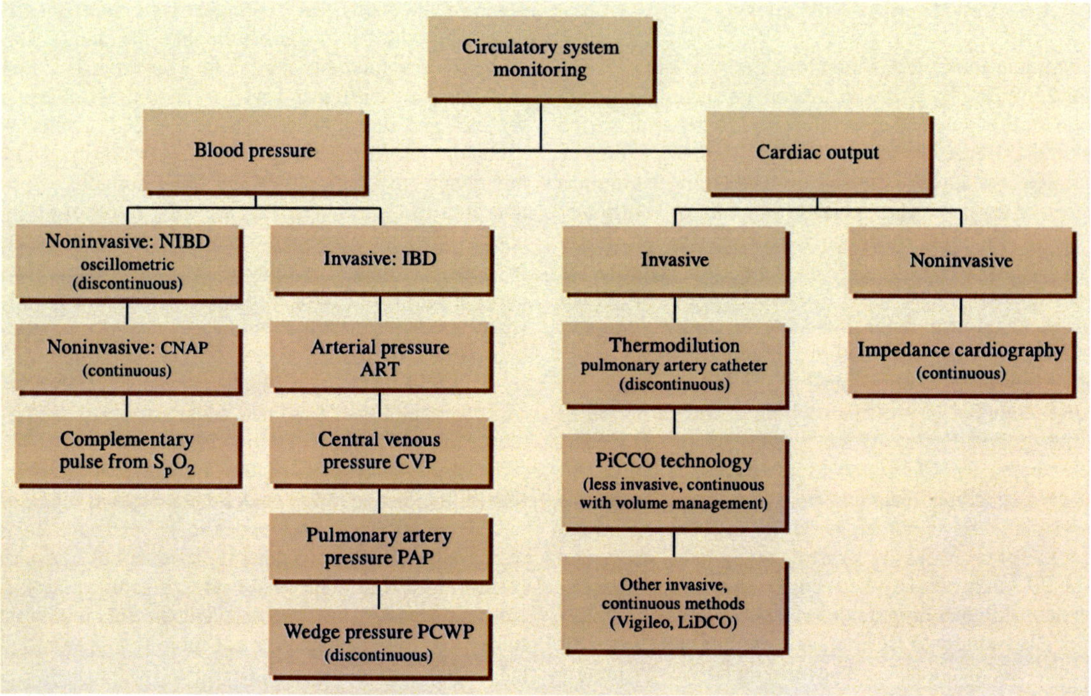

Fig. 48.2 Overview of circulation monitoring

say the driving force of perfusion, is the mean pressure (in the systemic circuit the mean arterial pressure APm (also MAP), and in the pulmonary circuit the mean pulmonary artery pressure PAPm).

The stroke volume is dependent on the preload, the contractility of the myocardium and the afterload. The preload is the stretching of the myocardium, which is brought about by passive filling of the ventricle at the end of diastole and is best described by the end-diastolic volume. The afterload is the force exerted by the cardiac muscles to overcome the resistance in the outflow tract of the left ventricle and of the peripheral circuit. The mean arterial pressure and the vascular resistance are used as measures of the afterload.

In addition to the pressure-volume work of the heart, factors that also determine the blood pressure behavior are the elasticity of the vascular system, the circulating blood volume and, in particular, the peripheral vascular resistance, which is influenced by the wall tension of the vessels (vessel tone) controlled by the sympathetic nervous system.

The delivery volume of blood per minute is known as the cardiac output (CO) and is the product of the stroke volume and heart rate.

Control mechanisms in the body regulate circulation with the aim of adjusting the cardiac output to the circulation that is required to supply oxygen to the organism and eliminate CO_2, keeping the blood pressure largely constant and adjusting the circulation in the individual organs and tissues to the functional state in each case. Because doctors have means available to them for specifically treating the individual determinants of the cardiac output (chronotropic medications for heart rate, inotropic medications for contractility, vasodilators and vasoconstrictors for vessel tone, and diuretics or fluid administration for preload), it is therefore important to maintain this information, as reliably as possible, as the basis for decision-making and therapy monitoring in the monitoring process.

Monitoring of the circulatory function involves noninvasive and invasive measurements of vital signs using indirect and direct methods, which are used in high-pressure or low-pressure systems (Fig. 48.2).

48.2.1 Pulse

Pulse monitoring is performed either invasively from the arterial pressure curve or noninvasively from the plethysmogram of pulse oximetry (Chap. 49). Pulse monitoring has particular importance as a safety measure when monitoring pacemaker patients (see above).

48.2.2 Discontinuous Noninvasive Blood Pressure (NIBP)

Noninvasive blood pressure measurements are part of the minimum requirements for pre, peri, and postoperative patient monitoring, but are also used in the field of intensive care medicine. Patient monitors have a measuring module or include the measuring parameter NIBP (synonymous abbreviation NBP) in the basic configuration. Monitors and stand-alone units work discontinuously (after being started manually or automatically at selectable intervals) according to the oscillometric measuring principle.

The auscultatory method is today performed almost exclusively *by hand* and is only an indirect part of the monitoring process. However, due to its wide distribution and historical significance it should be briefly explained here.

Sphygmomanometer

In 1896, the Italian doctor Riva Rocci developed a sphygmomanometer that was easy to handle (a device for measuring blood pressure; an inflatable cuff with an attached mercury manometer) for noninvasive measurement of the blood pressure in the upper arm. The cuff pressure was increased considerably beyond the point at which a pulse was no longer detectable at the wrist (palpatory method). When the first pulse became perceptible as the air was slowly released from the cuff, the manometer pressure read off was taken as the systolic blood pressure.

It was only after a further discovery that it also became possible to measure the diastolic blood pressure noninvasively. In 1905, the Russian doctor Korotkoff discovered characteristic turbulence noises that could be auscultated (listened to using a stethoscope) when releasing the pressure from a compressed artery. The description of the blood pressure values measured in this way as RR (after Riva Rocci), which is still customary today is historically false, because no diastolic value could be measured according to Riva Rocci, nor was a stethoscope used.

Functional Principle of the Auscultatory Method

The cuff must be placed on the exposed upper arm at the level of the heart such that the middle of the rubber bladder is positioned over the brachial artery. After palpating the brachial artery, the cuff is inflated to ≈ 30 mmHg above the pressure at which the pulse can no longer be detected. The stethoscope is placed against the brachial artery and the cuff pressure is slowly released. The pulsation that then begins causes knocking noises (Korotkoff sounds phase 1). At this point, the systolic blood pressure is read off from the manometer. In the other phases, the sounds change (see list below) until a sound can no longer be heard. This is the fifth phase and is approximately equal to the diastolic pressure.

Korotkoff sounds [48.1]:

Phase 1: A clear knocking sound begins, the intensity of which increases during the course of the phase.
Phase 2: The knocking sound from phase 1 is accompanied by an additional murmuring sound.
Phase 3: A knocking sound can be heard on its own again, which is now especially loud and clear.
Phase 4: Sudden change: the sounds appear muffled and quieter and they now lack the typical knocking quality.
Phase 5: The noises disappear completely.

Oscillometric Method

In 1909, the French doctor Pachon discovered the oscillometric measurement method. He recognized that, when the cuff pressure is released once the systolic pressure has been reached, the vessel walls begin to oscillate and maintain this behavior until the vessel is no longer occluded. These oscillations are transmitted to the air in the cuff and are read on the manometer. Today, these oscillations are measured electronically using pressure sensors.

The Functional Principle of the Oscillometric Method

In the oscillometric method, the cuff is inflated at the beginning of the measurement to ≈ 25–30 mmHg above the point at which virtually no oscillations are measured any longer (very low oscillations can come from the part of the artery lying proximal to the cuff). The pressure in the cuff is automatically released at 2–3 mmHg/s [48.2]. When the cuff pressure drops below the value of the systolic blood pressure, the artery begins to open and the oscillations become stronger. They reach their peak when the cuff pressure corresponds to the mean arterial pressure. This is the actual measured value of the oscillometric method. As the cuff pressure drops further, the oscillation amplitudes decrease once more and finally remain constant

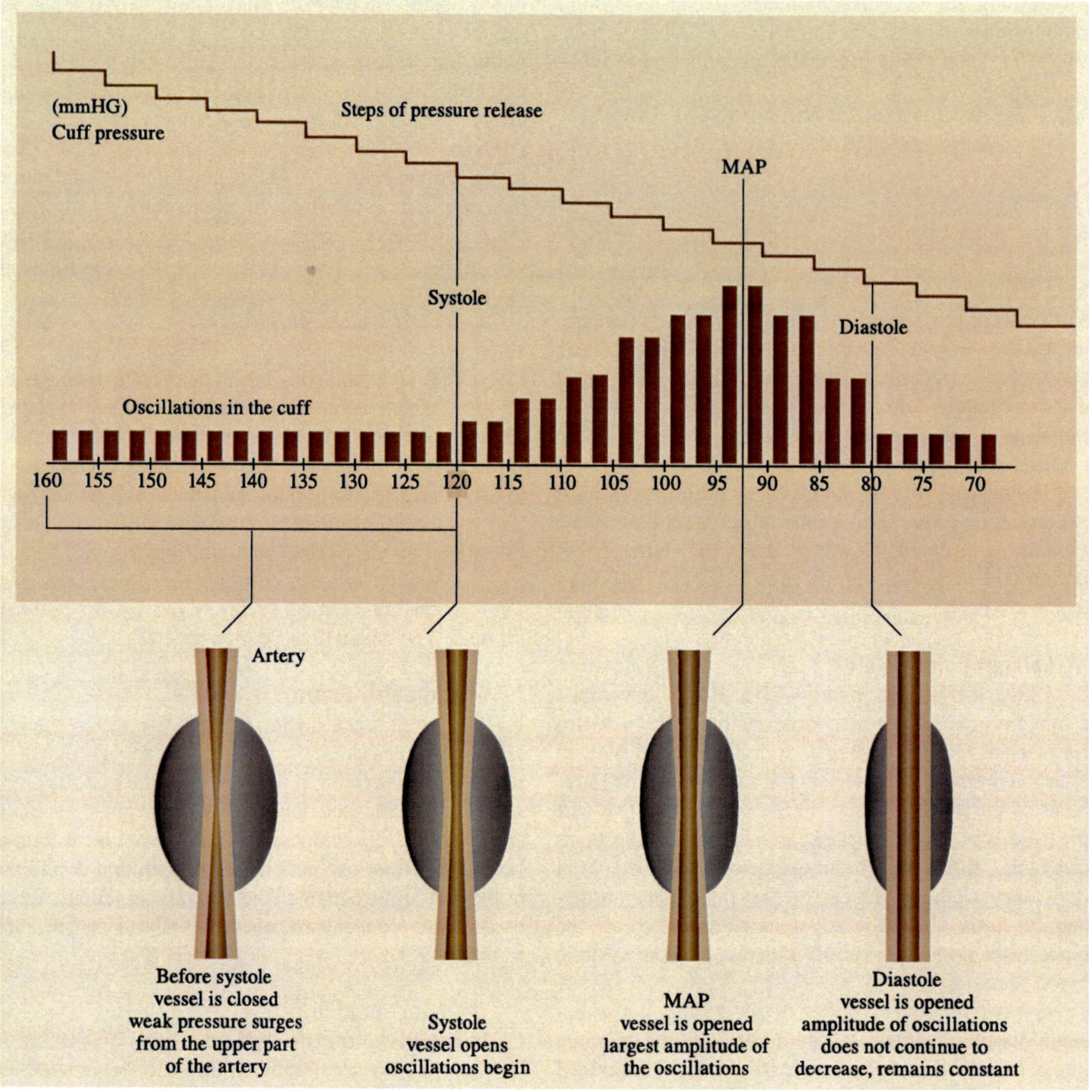

Fig. 48.3 The principle of the oscillometric method

when the cuff pressure has reached the diastolic value (Fig. 48.3). The values for the diastole and systole are determined from the progression of envelope curves around the oscillations detected and are thus calculated values.

Single-Tube and Double-Tube Systems

Cuffs with single-tube or double-tube systems are normal. In order to avoid the release of pressure influencing the measurement results, a pressure tube may be composed of two lumina. One leads directly from the cuff to the pressure sensor, and the other is only designed for inflation and pressure release. Single-tube systems are more flexible on account of their lower total cross section and can, therefore, be better handled in practice.

Plausibility Checking

Artifact detection is the most important requirement for reliable oscillometric measurement. Mechanical influences on the cuff or blood pressure tube during measurement must not be recorded by the device as relevant oscillations, as incorrect blood pressure values could otherwise be displayed. Modern equipment eliminates such artifacts by releasing the pressure in stages. Only when at least two successive pulses of the same amplitude are detected at each stage is the pressure released to the next level. If artifacts occur, then the device remains at a pressure level until two identical pulses are identified. As a result, even the smallest artifacts can be detected, e.g. in neonates. Although methods that in contrast operate with a continuous and linear release of pressure provide a result more quickly, they often give incorrect values when there is interference.

When there is insufficient plausibility, a reliable device will tell the user that it has not been possible to take an exact reading. In the case of shock patients with extremely low pressure values in peripheral vessels, the systolic and diastolic value can only be ascertained with a large degree of uncertainty as a result of the flat progression of the oscillation curves. The oscillation peak is still detected, however, and the mean arterial pressure can, therefore, be measured and displayed. The systole and diastole would be displayed as not measurable (e.g. as a dash).

Methodological Notes

Cuff Size. One of the most common sources of faults when measuring blood pressure is the incorrect choice of cuff size. The cuff width should be $\approx 40\%$ of the circumference of the limb, and the rubber bladder should surround $\approx 80\%$ of the circumference [48.3]. If the correct width is not used, this causes considerable measuring errors. If the cuff is too narrow the measurements are too high, and if the cuff is too wide the measurements are too low. It is therefore absolutely essential, particularly in obese patients, children and neonates, to use the correct cuff size in each case.

Taking into Account Hydrostatic Pressure. Blood pressure measurement is always based at the level of the heart. Measurement on the upper arm is, therefore, not a problem either in the case of seated patients or in patients in the dorsal position. Measurements can be taken from the thigh or lower leg if the upper arm is not available as a measuring site, and if the patient is lying flat. If the measuring site is not at the level of the heart, due to the hydrostatic pressure difference a lower measurement is given for the blood pressure above the level of the heart, and a higher measurement is given below the level of the heart. Measuring errors of approximately 7.5 mmHg occur for every 10 cm difference in height.

Other sources of errors in noninvasive blood pressure measurement are user errors, such as too loose application of the cuff, touching the cuff during measurement, movement of the tube (also, for example, as a result of directly touching the respiratory tubes). Technical errors are usually leaks in the tube or cuff system. Movement artifacts occur because of patient movements and must be mainly taken into consideration as a source of error in the case of children or in postoperative monitoring.

Limits to NIBP Measurement. During the measuring procedure, the tissue lying below the cuff on the limb being measured is compressed. There is a risk of nerval pressure lesions forming, particularly in cachectic patients, premature neonates, and patients with neuromuscular diseases, because the muscle tissue and adipose tissue protecting the nerves are reduced. With these patients, noninvasive pressure measurements should, therefore, be performed with the largest possible time intervals between measurements. In all other cases it is recommended that the time intervals between noninvasive blood pressure measurements be selected to be no less that 5–10 min, in order to avoid pressure damage in the region of the cuff and also in the supply area of the artery [48.4].

48.2.3 Continuous Noninvasive Arterial Blood Pressure Measurement (CNAP)

Blood pressure is not always constant but can change within a matter of seconds. In particular, during anaesthesia and its induction, variations in blood pressure can arise and require immediate medical attention. However, such episodes can be detected only unsatisfactorily or not at all using the normal NIBP measurement [48.5].

The noninvasive technique of the relaxed arterial wall (*vascular unloading technique* or *volume clamp* method), which has become known as the Penaz method [48.6], uses an optical sensor in a small cuff around the finger to measure the volume pulses that occur with each heart beat. The pressure in the cuff is regulated by means of feedback such that the optical measuring path always remains constant. When a pulse occurs, the cuff pressure is, therefore, increased accordingly, and when the pulse subsides the cuff pressure is reduced. The cuff pressure, therefore, reflects the pressure occurring in the enclosed finger artery with a high degree of accuracy. Following various techni-

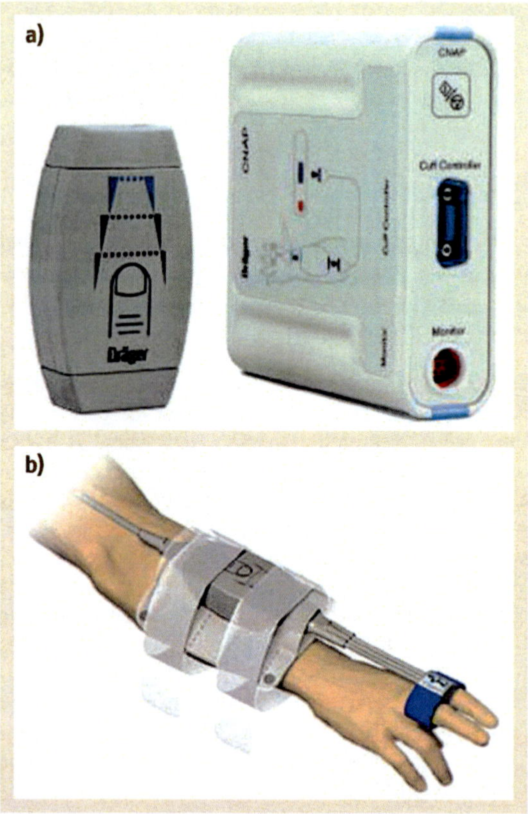

Fig. 48.4a,b CNAP Pod **(a)** with controller, **(b)** applied sensor cuff and controller fixed to the lower arm (© Dräger Medical AG, Lübeck, Germany)

cal developments in this field (Finapres, Amsterdam), CNAP then emerged (continuous noninvasive arterial pressure from CNSystems, Graz), and it is the first system of this type to be integrated as a module into a patient monitoring system (CNAP Pod from Dräger Medical, Lübeck; Fig. 48.4). CNAP uses pairs of sensor cuffs, which are placed on adjacent fingers. Care must be taken to ensure the correct size is used. Only one cuff is ever used for measurement at any one time, and after no longer than half an hour the continuous measurement is automatically switched. The venous stasis that naturally occurs during measurement on the finger very quickly decreases once more after the switch. The great advantage of this method is the calibration of the continuous measurement with the normal NIBP measurement, so that correct values are displayed even when the fingers are not level with the heart. This technique, therefore, bridges the gap between discontinuous NIBP measurement, which involves the risk of overlooking rapid drops in blood pressure, and continuous invasive measurement, the use of which is not indicated for every preferred event and requires an arterial catheter that may only be inserted by a doctor. CNAP satisfies the demand for correct, continuously measured blood pressure values including a pressure curve, instantaneously shows any occurring changes in pressure, can be applied by any instructed member of nursing staff, and involves virtually no risk of lesions for the patient. CNAP is, therefore, suitable for increasing patient safety – above all in the case of interventions that require sedation [48.7].

48.2.4 Invasive Pressure Measurement in the High-Pressure and the Low-Pressure Systems

Invasive blood pressure measurement is a direct method for determining the pressure conditions in the high-pressure and low-pressure systems. The fundamental advantage over conventional noninvasive blood pressure measurement is the continuous availability of the measurement signal (pressure curve) and the pressure values. This provides the possibility of triggering an alarm if predefined limit values violated and of further signal processing of the measurement data obtained.

The High-Pressure System

A connection is set up by means of an intra-arterial catheter between the intravasal blood column and a liquid-coupled pressure transducer, which converts the mechanical variable of pressure into an electrical signal. The arterial pressure is dependent on the stroke volume of the left ventricle and on the vascular resistance.

Links in the pressure-measuring chain (Fig. 48.5) are essentially:

- Cannula or arterial catheter
- Liquid-filled tube system with a three-way stopcock
- Pressure transducer (reusable: with pressure dome, with or without terminal membrane, and with separate or integrated flushing system; disposable transducer always including flushing system) with monitor cable
- Monitor.

Pressure Sensors and Pressure Transfer

Reusable pressure sensors (Fig. 48.6) consist of a mechanoelectrical pressure transducer and a removable, liquid-filled cavity (pressure dome). The pressure dome (single use per patient) is directly coupled to the measuring surface (transducer membrane) mechanically via an integrated thin plastic membrane. The arterial compression wave is transmitted to the pressure dome via the liquid medium (sodium chloride solution) in the pressure-measuring line and the catheter and causes a corresponding deflection in the pressure transducer membrane. The mechanical membrane deflection is converted, in the case of reusable and disposable pressure transducers, by means of pressure-sensitive chips whose electrical properties change in proportion with the pressure. An applied voltage causes an electrical signal linearly proportional to the pressure, which is conveyed to the monitor and is displayed there on the screen following amplification.

The greater the distance between the tip of the catheter and the heart, the higher the pressure amplitudes that are recorded. A time delay in the recording is determined by the transit time. The systolic, diastolic, and mean values are calculated electronically from the respective pressure curves and displayed.

Connectors

The pressure dome usually has two Luer-lock connectors, each of which has a three-way stopcock fitted to it. They are used to lead in the pressure, remove air from the system during filling, and flush the pressure-measuring line. The pressure transducers are generally fit into mounting plates that are used for fastening, for example, to an IV stand at the correct height relative to the heart. The inlet tubes and/or the slots on the mounting plates are often color-coded for the various invasive pressure measurements as:

Red: arterial pressure measurement
Blue: venous pressure measurement
Yellow: pulmonary artery pressure measurement
White: other pressure measurement (e.g. intracranial pressure measurement).

48.2.5 The Ideal Pressure Measurement System – Requirements

An idealized invasive pressure measurement system has the following characteristics, which comply with modern clinical requirements:

- Simple and fast setup of the system
- Easy to fill, with simple and secure removal of air as well as visual inspection for air bubbles

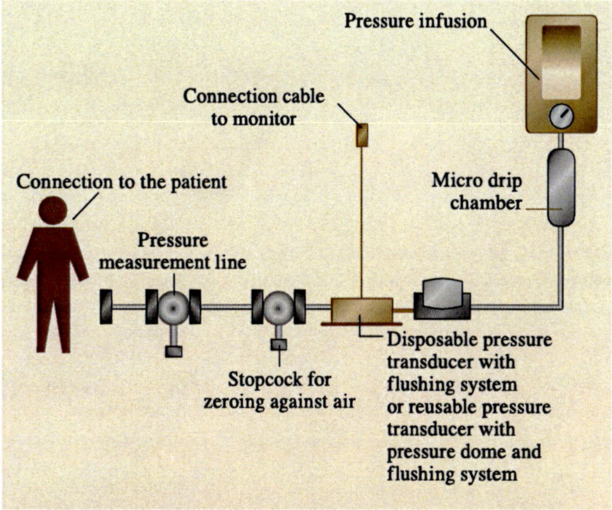

Fig. 48.5 Measuring chain for invasive pressure measurement

- Integrated system for constant flushing with *one-handed operation*, reliable flushing rate, even in the case of rapid flush
- Closed overpressure safety valve
- High reliability of all the individual components in the pressure-measuring chain
- Good resonance frequency
- High long-term stability
- Possibilities for negative calibration
- Mounting on the arm and on a bracket
- Low input resistance (impedance).

Metrological Requirements

Invasive blood pressure measurement is technically a measurement of the differential pressure between the blood vessel and the ambient pressure (atmospheric pressure). The pressure transducer must, therefore, be adjusted before each measurement (zero compensation, with the transducer adjusted to the level of the heart in order to avoid measuring errors as a result of hydrostatic pressures from the tube system). Pressure transducers are temperature-dependent, which is to say that the zero point of its actual value can vary when the temperature changes (zero drift).

The weakest link in the measuring chain is the mechanical pressure transfer. The pressure can only be determined exactly if all of the components in the system meet specific requirements, such as high static and dynamic accuracy of the measuring system (transmission behavior, natural frequency, damping, etc.). The transmission properties are dependent on the length and

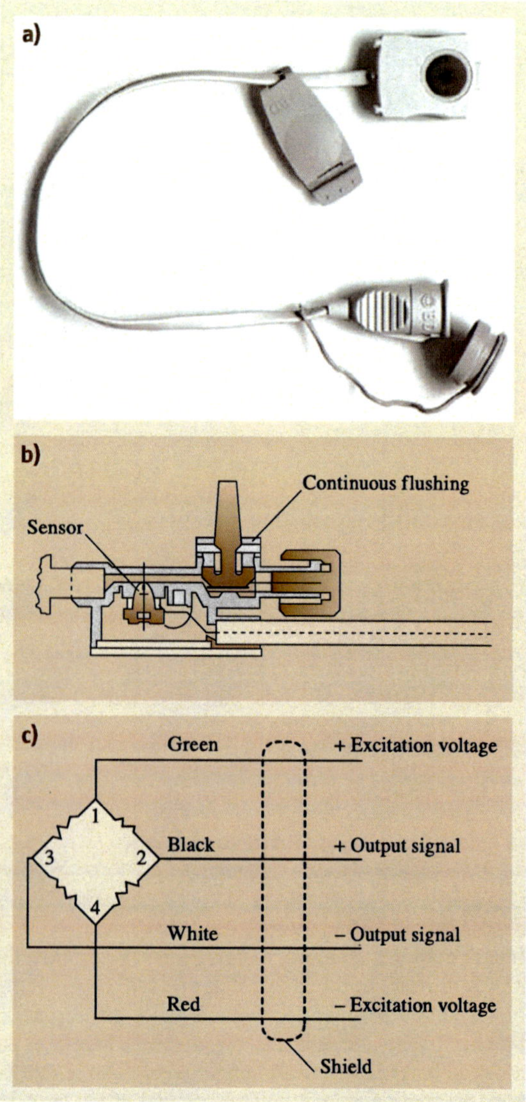

Fig. 48.6 (a) Reusable pressure transducer, (b) diagram of a disposable pressure transducer, (c) Wheatestone bridge (courtesy of Becton Dickinson, Heidelberg)

elasticity (compliance) and also on the internal diameter of the pressure tube. It is imperative to use tube connections that are as short and pressure-resistant as possible, without any additional connectors.

The functionality of all of the components should be checked prior to each measurement (bubble-free measuring system, secure connection of the individual system components, no leaks). When using reusable pressure transducers, calibration is recommended to check the functionality before each measurement.

The pressure system contains a section that is capable of oscillating (hydraulic system: from the tip of the cannula to the membrane of the pressure transducer) and can be induced to natural oscillations that alter the pressure signal. The metrological aim is to represent the pressure curve unaltered at frequencies of up to 12 Hz. Below 20 Hz – here the pressure signal has frequency components – the natural and resonance frequency of the measuring system becomes particularly noticeable, which is to say that undesirable interferences and deformations of the actual pressure signal occur. As a result of air bubbles in the measuring system, clots at the tip of the catheter or in the tube system, a disconnected or loose pressure dome of the pressure transducer, movement artifacts caused by the patient, or a change in the catheter position, the pressure curve is damped or changed and the measurement information impaired.

Pressure Measurements in the Low-Pressure System

The aim of pressure measurements in the low-pressure system is to obtain information about the right ventricular function, the pulmonary circuit, and the filling of the vascular system on the basis of the measurement results obtained.

Central Venous Pressure (CVP)

Measurement of the central venous pressure is by means of a liquid manometer or a pressure sensor, using a central venous catheter (CVC), which is placed in the superior vena cava at the entrance to the right atrium of the heart. Unlike water column manometry, pressure transducers have the advantage that the measurement information is available continuously and certain signal characteristics of the CVP curve are additionally available. Because of the small pressure values, it is particularly important that the pressure measurement system is situated level with the heart (correct zero point positioning) in order to avoid falsification of the measurement reading by hydrostatic pressure differences. The progression of the CVP curve shows the atrial contraction (a-wave), the beginning of the ventricular contraction (c-wave), and the relaxation phase (v-wave). The CVP is only shown as a mean value (CVP waveform and other pressures, Fig. 48.7) and is fundamentally influenced by the capacity of the vascular system, the cardiac output, the blood volume, the compliance of the myocardium, and also the afterload of the right ventricle.

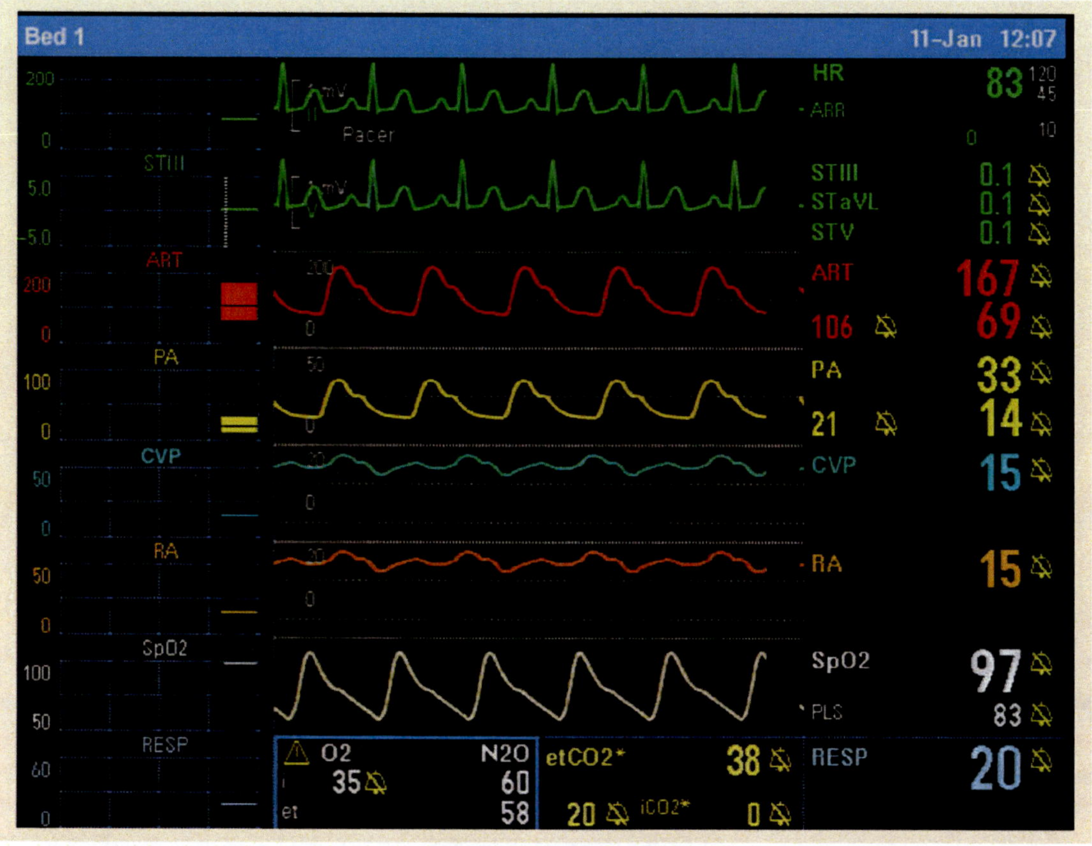

Fig. 48.7 Different pressure waveforms and mini-trends in the monitor image

Pulmonary Artery Pressure (PAP) and Pulmonary Capillary Wedge Pressure (PCWP)

In order to monitor and measure the hemodynamics of the right ventricle, a balloon catheter is pushed through the venous system into the right atrium, the right ventricle, and from there through the pulmonary valve into the arteria pulmonalis. The path of the catheter can be followed on the basis of the different typical pressure curves (Fig. 48.8).

The correct catheter position has been reached once the so-called wedge position (Fig. 48.9) has been taken, which is to say once the inflated balloon of the catheter blocks off the pulmonary artery branch. If the tip of the catheter rests against the wall of the pulmonary artery (in a so-called pseudo-wedge position), this causes damping of the pressure curves when the balloon is filled and there is a continuous rise in pressure. Although the catheter is in the right ventricle, in the wedge position the pressure in the left atrium can be inferred via the distal lumen: the PCWP value corresponds in the first approximation to the left atrial pressure (LAP) and thus to the end-diastolic filling pressure in the left ventricle, since the left atrium, pulmonary capillaries, and pulmonary artery under normal conditions form a common pressure connection during diastole.

Balloon Catheters

Using different balloon catheters (so-called flow-directed catheters, pulmonary artery catheters, or Swan–Ganz catheters), which differ in terms of their length and thickness, number of lumina, position of the lumen exit sites, and other characteristics, the CVP, PAP, and core body temperature can be measured simultaneously, and the PCWP and CO can be measured intermittently.

In addition, specialized balloon catheters provide additional possibilities such as intracardial ECG measurement (e.g. His bundle ECG), supraventricu-

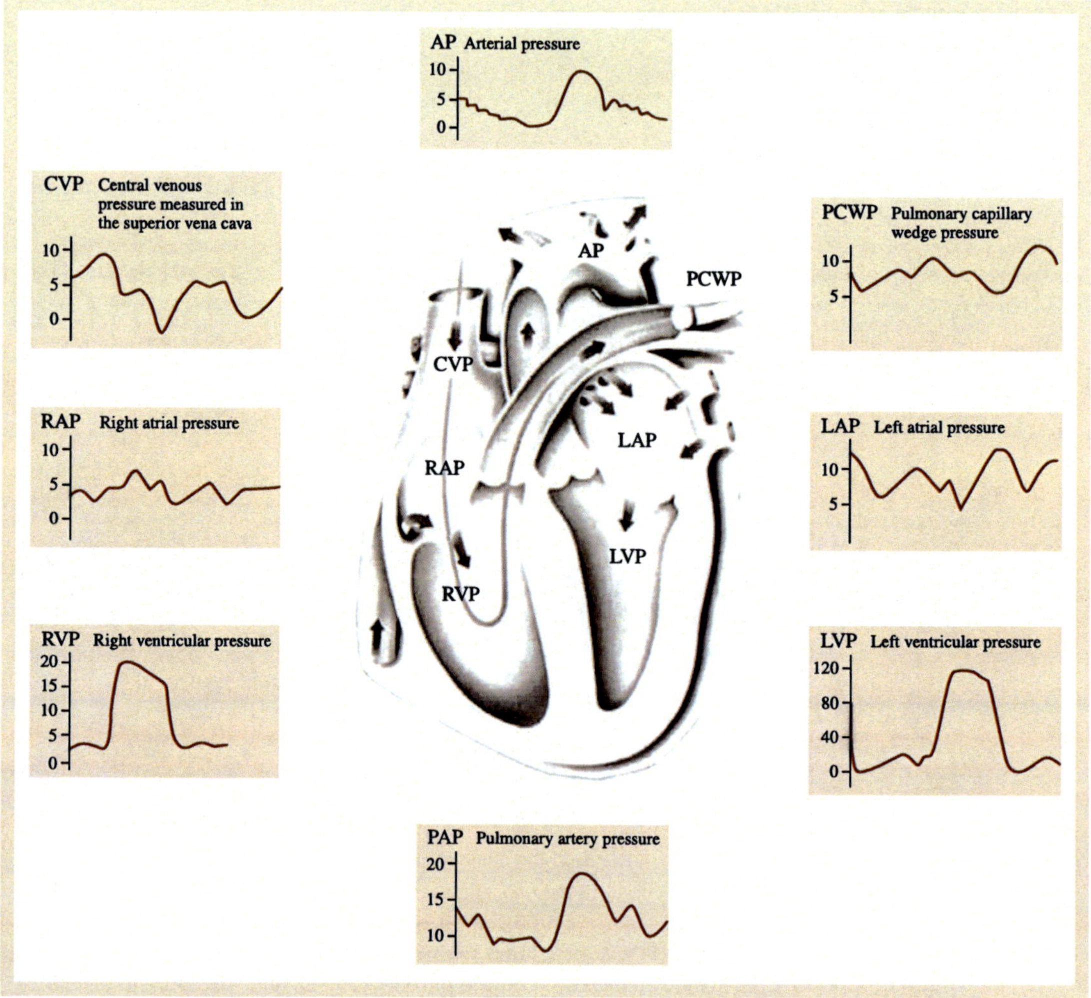

Fig. 48.8 Measuring sites and typical pressure curves in various heart and vessel sections

lar and ventricular stimulation, measurement of the mixed venous oxygen saturation S_vO_2 as a result of integration of fibre optics, introduction of a transluminal stimulation probe, or additional infusion lumina.

Balloon catheters that are newly inserted or have already been in place intravasally for hours or days are not free of risks and can cause complications, which partly depend on the catheter access site such as:

- Supraventricular and ventricular arrhythmias
- Ventricular tachycardias or ventricular fibrillation (rarely)
- Venous thromboses (particularly with a low CO)
- Sepsis (the risk rises as the duration of catheterization increases)
- Pulmonary infarction (as a result of catheter occlusion of a peripheral pulmonary artery)
- Pulmonary artery rupture (by balloon inflation or the catheter tip).

48.2.6 Determining the Cardiac Output (CO)

The cardiac output, the volume of blood conveyed per minute, is given in l/min. The classical way of determining the CO is by a method according to Fick (Fick's principle), whose calculation is based quite simply on the law of conservation of mass: CO is the quotient

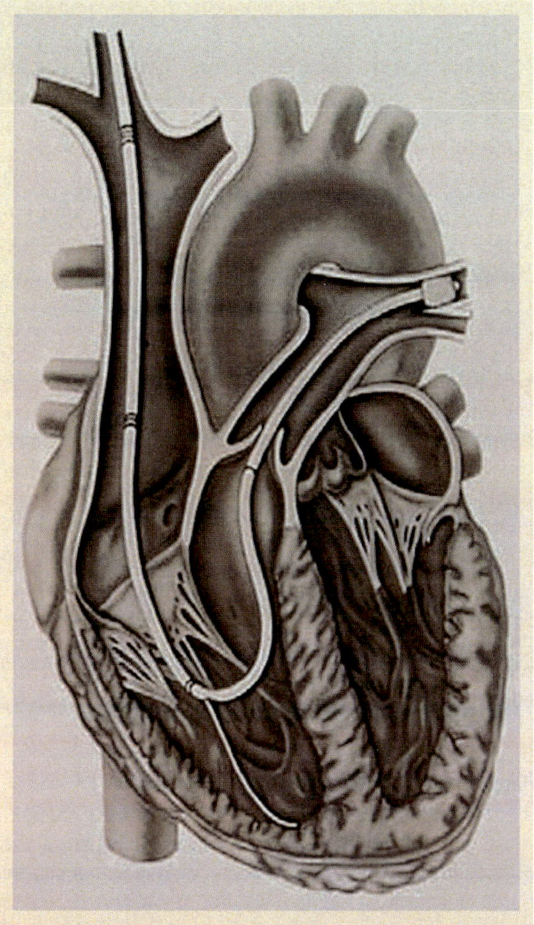

Fig. 48.9 Pulmonary artery catheter in the wedge position

from oxygen consumption (VO_2) in the body and difference in oxygen content ($avDO_2$) between arterial blood flowing to the body and mixed venous blood returning from the body: $CO = VO_2/avDO_2$. However, under routine conditions oxygen consumption cannot be measured with sufficient accuracy in the clinical environment.

Dilution Methods

The 1920s saw the development of the fundamental work of Stewart and Hamilton to determine the CO by means of dye dilution. With the introduction of the thermistor catheter by Swan and Ganz in the 1970s, of all the known indicator dilution methods (dye dilution, cold, ions, radioisotopes), thermodilution by means of a pulmonary artery catheter had established itself as

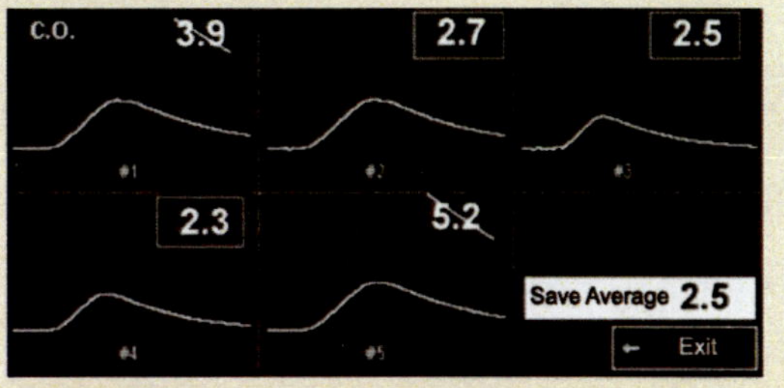

Fig. 48.10 Monitor window for the CO measurement. Erroneous measurements are deleted from a series of measurements and a mean value is formed

very practicable in the clinical environment. A defined amount of physiological sodium chloride solution (at a temperature of 0–25 °C; the lower the temperature, the more accurate the measurement) is injected into the right atrium via the proximal port of the multilumen pulmonary artery catheter. Because the injected fluid is mixed with the warm flowing blood (37 °C) and is, therefore, diluted, the change in temperature in the blood stream can be measured by the thermistor, which is situated close to the tip of the catheter. The shape and the area of the dilution curve change with the cardiac output. With a known temperature of the injected fluid and blood as well as a known volume of injected fluid, the measuring system determines the cardiac output from the area of the thermodilution curve (Fig. 48.10). The disadvantages of this method are the discontinuity and the need for the pulmonary artery catheter, the indication of which is viewed particularly critically since the study by *Connors* [48.8], if not before. The disadvantage of discontinuity was overcome by emitting heat pulses to the blood using a special pulmonary artery catheter and by evaluating their dilution curve. Since heat pulses can be applied at very short intervals (30–60 s), this virtually provides continuous measurement.

PiCCO Technology

Thermodilution can, in principle, also be performed transpulmonarily, i. e. the cold bolus passes through the lungs and a thermistor in the arterial system records the dilution curve [48.9]. The cold bolus is injected into the right atrium in the same way as in normal thermodilution except with a normal central venous catheter that is customary in intensive care patients, but the temperature profile is measured (preferably) in the arteria femoralis. The advantage of this method is that it is less invasive (omission of the pulmonary artery

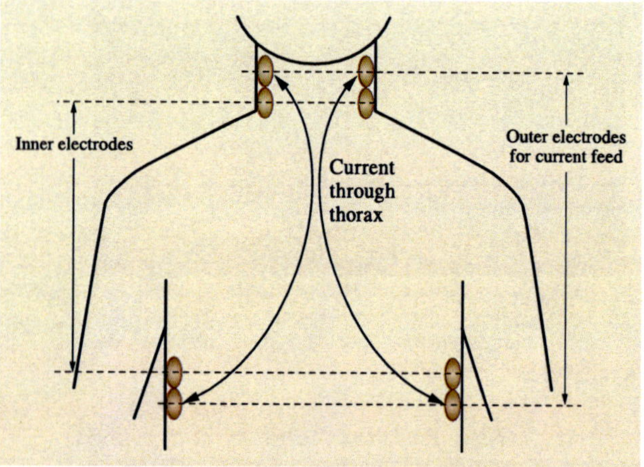

Fig. 48.11 Two double electrodes are applied to each side of the throat and thorax. The impedance is determined from the voltage drop between the inner measuring electrodes

catheter and, therefore, of the dangers mentioned above) and that respiratory influences on the measurement are reduced. PiCCO technology (introduced into the market in 1997 by Pulsion Medical Systems, Munich; PCCO, pulse contour cardiac output) is a combination of transpulmonary thermodilution (in the form of CO measurement) and pulse contour analysis (in the form of continuous determination of the changes in the CO). The CO is inferred beat-for-beat from the invasively measured arterial pressure curve. Calibration must be performed in advance by means of thermodilution, and this should be repeated at least every 8 h. The great value of the PiCCO method lies not only in the less invasive and continuous CO measurement but in the extraordinarily valuable parameters for assessment of the circulation and the volume situation of the patient, which can be read from the monitor without any further effort at all. These parameters are, in particular, the preload indicators global end-diastolic volume (GEDV) and intrathoracic blood volume (ITBV), which are significantly more sensitive and more specific than the filling pressures CVP and wedge pressure, the contractility measures $\mathrm{d}P/\mathrm{d}t_{\max}$ (maximum pressure increase) and GEF (global ejection fraction), the systemic vascular resistance SVR as a measure of afterload, the parameter EVLW as a measure of the extravascular lung water, and (in the case of patients without arrhythmia requiring artificial respiration) the stroke volume variation SVV. The last parameters are used for optimum fluid management, in particular in shock patients where it is important to stabilize the circulation and thus the oxygen consumption and to avoid and quickly recognize pulmonary oedemas.

Other invasive procedures use the pulse contour method or pulse wave analysis in isolation (Vigileo, from Edwards Lifesciences, Irvine) or in combination with lithium ion dilution (LiDCO, from LiDCO, Cambridge). Furthermore, continuous measurement of the mixed venous saturation SvO2 by means of fibre optics in a pulmonary artery catheter is used to estimate changes in the CO.

Impedance Cardiography

It has long been known that the blood volume expelled with each heart beat (stroke volume) leads to measurable variations in the thoracic impedance. There have also been attempts to determine the stroke volume and CO from the variations in the thoracic electrical bioimpedance (TEB), which are detected using four electrode arrangements [48.10]. A weak high-frequency constant current (e.g. 2.5 mA, 70 kHz) is passed through the thorax by means of the external ring electrodes (or special double electrodes, Fig. 48.11) arranged on the neck and thorax. The current seeks the path of least resistance, which is essentially the blood-conducting aorta. The voltage drop is measured currentlessly by means of the inner measuring electrodes, such that the thoracic impedance can be substantially ascertained independently of the contact resistance.

The basic impedance Z_0 is about 35 Ω. Fluctuations that originate from the breathing and from the pulsating blood flow in the aorta are superimposed on it. The respiratory signals are not of interest here and are filtered out. The higher-frequency impedance fluctuations $\Delta Z(t)$ (amplitude approximately 0.15 Ω) are represented by the impedance cardiogram (ICG), from which the stroke volume and the CO can be determined by means of certain model assumptions and refined methods [48.11].

In the 1990s, the impedance cardiograph BioZ.com from the Californian company CardioDynamics, Bothel, USA, was developed. Its measuring accuracy and reproducibility have been proven in clinical studies [48.12]. To date this is the only impedance cardiograph to have been incorporated in a patient monitoring system as a special module (Solar from GE Medical Systems).

Impedance cardiography has the advantages of:

- Being completely noninvasive
- No risk to the patient
- Continuous beat-to-beat measurement
- Being easy to apply.

The method is not indicated in patients with septic shock, in patients with aortic valve defects with regurgitation, in patients <1.20 m in height, in patients <30 kg or >155 kg in weight, when an aortic balloon pump is used, or during operations on the open thorax.

Despite all of the advantages of ICG, its nonavailability – for example in the case of septic shock when accurate observation of the hemodynamic and volumetric parameters is of real importance – is probably particularly responsible for that fact that it still has not established itself in the field of patient monitoring to this day. PiCCO-technology is clearly the more advantageous.

48.2.7 Calculation of Hemodynamic Variables

The vascular resistances can be calculated from the measurements of the pressure and flow. The total peripheral resistance (TPR), also called the systemic vascular resistance (SVR), is the resistance of the systemic circuit and mathematically it is simply the quotient of the propulsive pressure difference (mean arterial pressure APm minus central venous pressure CVP) and the flow (CO)

$$\mathrm{SVR} = (\mathrm{APm} - \mathrm{CVP})/\mathrm{CO}\,.$$

The pressure difference in the small circuit is the mean pulmonary pressure minus the wedge pressure PCWP (as a measure of the left atrial pressure). The pulmonary vascular resistance is, therefore,

$$\mathrm{PVR} = (\mathrm{PAPm} - \mathrm{PCWP})/\mathrm{CO}\,.$$

The international unit of measurement has remained dyn s/cm^5 in this case (resulting from the old unit of pressure dyn/cm^2 divided by the volume flow cm^3/s).

Monitors generally provide calculation programs that can also be used to calculate other circulatory parameters (e.g. cardiac work) and also – if blood gases and respiratory gases are additionally measured – to calculate the oxygen supply (DO_2) and oxygen consumption (VO_2).

Further Reading

- C.L. Lake, R.L. Hines, C.D. Blitt: *Clinical Monitoring: Practical Applications for Anesthesia and Critical Care* (Saunders, Philadelphia 2001)
- W.C. Shoemaker, G.C. Velmahos, D. Demetriades: *Procedures and Monitoring for the Critically Ill* (Saunders, Philadelphia 2002)

References

48.1 S. Eckert: 100 Jahre Blutdruckmessung nach Riva-Rocci und Korotkoff: Rückblick und Ausblick, J. Hyperton. **10**(3), 7–13 (2006), (in German)

48.2 S. Mieke: Non-invasive blood pressure measurement, Bulletin OIML **125**(12), 18–28 (1991)

48.3 G.T. Pickering, J.E. Hall, L.J. Appel, B.E. Falkner, J. Graves, M.N. Hill, D.W. Jones, T. Kurtz, S.G. Sheps, E.J. Rocella: Recommendations for blood pressure measurement in humans and experimental animals: Part 1: Blood pressure measurement in humans: A statement for professionals from the subcommittee of professional and public education of the American Heart Association Council on High Blood Pressure Research, Hypertension **45**, 142–161 (2005)

48.4 W.F. List: Kardiovaskuläres Monitoring. In: *Intensivmedizinische Praxis*, ed. by W.F. List, P.M. Oswald (Springer, Berlin Heidelberg 1989) pp. 41–62, (in German)

48.5 R. Dueck, L.C. Jameson: Reliability of hypotension detection with noninvasive radial artery beat-to-beat versus upper arm cuff bp monitoring, Society for Technology in Anesthesia, Annual Meeting (Proc. Society for Technology in Anesthesia, San Diego 2006)

48.6 J. Penaz: Photoelectric measurement of blood pressure, volume and flow in finger, Dig. 10th Int. Conf. Med. Biolog. Eng. (1973) p. 104

48.7 S. Siebig, F. Rockmann, K. Sabel, I. Zuber-Jerger, C. Dierkes, T. Brünnler, C.E. Wrede: Continuous non-invasive arterial pressure technique improves patient monitoring during interventional endoscopy, Int. J. Med. Sci. **6**, 37–42 (2009)

48.8 A.F. Connors, T. Speroff, N.V. Dawson, C. Thomas, F.E. Harrell, D. Wagner, N. Desbiens, L. Goldman, A.W. Wu, R.M. Califf, W.J. Fulkerson, H. Vidaillet, S. Broste, P. Bellamy, J. Lynn, W.A. Knaus: The effectiveness of right heart catheterization in the initial care of critically ill patients, J. Am. Med. Assoc. **276**, 889–897 (1996)

48.9 G. Della Rocca, M.G. Costa, L. Pompei, C. Coccia, P. Pietropaoli: Continuous and intermittent cardiac output measurement: pulmonary artery catheter versus aortic transpulmonary technique, Br. J. Anaesth. **88**(3), 350–356 (2002)

48.10 J. Nyboer: Electrical impedance plethysmography. A physical and physiologic approach to peripheral vascular study, Circulation **2**, 811–821 (1950)
48.11 M.J. Osypka, D.P. Bernstein: Electrophysiological principles and theory of Stroke volume determination by thoracic electrical bioimpedance, AACN Clinical Issues: Adv. Pract. Acute Crit. Care **10**(3), 385–399 (1999)
48.12 G.L. Yung, C.C. Fletcher, P.F. Fedullo: Noninvasive cardiac index using bioimpedance in comparison to direct Fick and thermodilution methods in patients with pulmonary hypertension, Chest **116**(4, Suppl. 2), 281 (1999)

49. Respiratory Monitoring and Pulse Oximetry

Ullrich Hieronymi, Rüdiger Kramme, Harald Kronberg

Respiratory monitoring includes the measurement, evaluation, and monitoring of parameters of the respiratory system, which are grouped firstly into parameters of respiratory mechanics (Sect. 49.1) and of the mechanics of machine-assisted artificial respiration (e.g., respiration rate, tidal volume, and airway pressure as part of the settings at the ventilator), and secondly into parameters of gas exchange (Sect. 49.2). Special consideration is given to pulse oximetry (Sect. 49.2.2) – a basic non-invasive monitoring procedure of great importance – and its recent further development to pulse CO-oximetry, to transcutaneous blood gas monitoring (Sect. 49.2.3) which has its focus in neonatal monitoring, and to capnography (Sect. 49.2.5) which is a standard monitoring to be performed during anaesthesia in the induction and the OT.

49.1 **Respiratory Mechanics** 971
49.1.1 Monitoring of the Respiration Rate. 971
49.1.2 Pressure and Flow 972

49.2 **Gas Exchange** 973
49.2.1 Gas Monitoring: Preliminary Remarks 973
49.2.2 Pulse Oximetry 976
49.2.3 Transcutaneous Blood Gas Measurement 979
49.2.4 Measuring the Respiratory Gases.... 981
49.2.5 Capnography 983

References 985

In the case of machine-assisted artificial respiration in the OT or in intensive care units, the monitoring is focussed on ventilation and oxygenation (Chaps. 27 and 30). Monitoring of the artificial respiration parameters and of the inspiratory oxygen concentration usually is part of the ventilator. Cardiovascular parameters, blood gases, and also, if necessary, anaesthetic gases (when not measured by the anaesthesia device itself) are generally monitored by means of a separate patient monitor.

49.1 Respiratory Mechanics

In monitoring, the respiratory mechanics is primarily the description of the mechanical processes of the thorax that are necessary for ventilation, in the form of pressure/volume relationships. In the case of patients who are given artificial respiration, the airway pressure and tidal volume are of particular interest, and the variables and behavior of these parameters are dependent, in particular, on the compliance (elasticity) of the lungs and on airway resistance. The aspects of pulmonary function testing (Chap. 8) are not central to monitoring but are closely related.

49.1.1 Monitoring of the Respiration Rate

The respiration rate (RR) is a vital parameter that plays a large role in neonatal monitoring, in particular in the case of spontaneously breathing patients (Chap. 54). It is preferably determined from:

- Impedance pneumography (indirect method)
- Capnography (indirect method).

The so-called *acoustic respiration rate* is measured using an entirely new method and sensor (Masimo 2010);

it is not yet possible to estimate the significance of this here.

The Principle of Impedance Pneumography

An increase and decrease in the thoracic volume is linked to inspiration and expiration. As a result of the increase in volume during the inspiration phase, the impedance of the thorax increases, whereas a decrease in the volume means a reduction in the thoracic impedance. In order to detect this, a very low, high-frequency constant current (e.g., 10 μA at approximately 40 kHz) is conducted through two thoracic electrodes. ECG electrodes are used for this, generally those of lead II or optionally, if appropriate, those of lead I. When the current is constant, a change in the impedance causes a synchronous change in the voltage drop. These fluctuations are displayed on the monitor in the form of an impedance pneumogram (or respiration waveform) and this is used to ascertain the respiration rate. The signal is highly vulnerable to movement artifacts, as every movement of the body influences the thoracic impedance.

Determining the Rate

A running average of the breaths counted per minute is displayed. With some monitors, the respiration waveform must only exceed a single adjustable trigger threshold, and this is already classed as a valid breath. Even small artifacts can, therefore, easily lead to a miscount. The combination of an upper and a lower threshold, the interval between which can be adjusted, is better and safer than simply counting the positive passages of the impedance curve through the single trigger threshold. The upper and lower thresholds span a minimum amplitude which must be exceeded by the impedance signal in order to intersect both. Thus, the depth of the respiratory action and not merely its single trigger passage decides whether it is classed as a valid breath and is included into the determination of the respiration rate.

Apnoea Monitoring

Counting the respiratory actions also makes it possible to monitor any pause in breathing or apnoea that occurs. If no breath is detected within the time frame defined by the adjustable monitoring limit *apnoea time*, then an apnoea alarm is triggered.

It is important to take into account that every heart beat causes a measurable change in impedance in the thorax, which may be visible in the impedance pneumogram. If the trigger threshold or the minimum amplitude is set such too low such in cases, then it is not the respiration rate that is counted but the heart rate (cardifact). Cases of apnoea can then be overlooked. Particularly in the neonatal monitoring mode, monitors are capable of determining whether the respiration rate and heart rate closely resemble one another or are identical and of triggering an alert in the event of such cardifacts. (For the significance of respiratory monitoring please also refer to Chap. 54.)

49.1.2 Pressure and Flow

Tidal volume, airway pressure, and the temporal profile thereof (slopes and respiration rate) are specified in the case of patients given artificial respiration by adjusting settings on the respirator or anaesthetic equipment. For those cases where the equipment displays do not provide sufficient diagnostic information, the monitoring includes modules for supervision of the respiratory mechanics. A special flow sensor measures the inspiratory and expiratory gas flows and volumes in the mainstream and also the airway pressure. Both the waveforms of flow, pressure and volume, and also the flow-volume and pressure-volume loops (F-V and P-V loops) can be displayed on the monitor. Parameters that can be displayed and provided in trends, for example, inspiratory and expiratory tidal volumes, minute volumes, maximum flows and airway resistance, maximum and average airway pressure, PEEP (positive end-expiratory pressure), respiration rate and inspiration/expiration ratio (I/E), as well as the lung compliance.

If these measurements are combined with the capnography, the CO_2 production, the alveolar ventilation as well as anatomical and physiological dead space ratios can also be ascertained. All these parameters are used for diagnosis and therapy monitoring in the case of patients given artificial respiration.

Flow Sensor with a Fixed Orifice

When measurements are to be taken close to the patient directly in the mainstream between the endotracheal tube and the Y-piece of the ventilation hoses, it is important to have a robust sensor with small dead space and low flow resistance that can be easily applied and provides reliable measurements without having to undergo calibration procedures, despite the constantly moist gases. The pneumotachograph usually used for measuring flow in functional diagnostics (Chap. 8) is not suitable for this purpose as it would quickly become blocked.

A special sensor has proved successful for use in long-term artificial respiration, which measures the gas flows in both directions by measuring the differential pressure at a flow resistance point in the form of a fixed orifice. This system is affected less quickly by the negative effects of condensate and can be constructed with such a small amount of dead space that it can even be used for neonates. The nonlinear relationship between the pressure difference and flow is compensated by using a microprocessor.

Different Measurements from the Module and Ventilator

Ventilators generally measure the pressure and volume in the equipment remote from the patient, whereas the respiratory mechanics module takes measurements close to the patient. In addition to the compliance of the ventilation hoses, the measuring temperature also plays a particularly important role for the volumes displayed. Measurements taken directly from the patient are effectively taken at body temperature ($\approx 37\,°C$). These gas conditions are described as BTPS (body temperature and pressure, saturated). The condition ATPS (ambient temperature and pressure, saturated) is present in the equipment for expiration, and strictly speaking ATPD (ambient temperature and pressure, dry) for inspiration, because dry gases are present, which are only humidified on the way to the patient. There is, therefore, a considerable difference in temperature between the two measurement points, which leads to noticeable differences in volume. The gas laws give the following equation for an ambient temperature of 22 °C

$$V(\mathrm{BTPS}) = V(\mathrm{ATPS}) \times \frac{273.15+37}{273.15+22} = 1.05 \times V(\mathrm{ATPS})\,.$$

Volume measurement at body temperature is thus approximately 5% higher than at an ambient temperature of 22 °C.

The condition STPD (standard temperature and pressure, dry) should in addition be mentioned, which was determined as being at standard temperature 0 °C, standard pressure 760 mmHg and dry (water vapor partial pressure = 0 mmHg).

49.2 Gas Exchange

Measurement methods and measurement parameters for respiratory gases, blood gases, and anaesthetic gases can be summarized as so-called gas monitoring under the generic term gas exchange.

49.2.1 Gas Monitoring: Preliminary Remarks

There are very differing conditions under which the respiratory gases and blood gases of a patient must be monitored: spontaneous respiration or artificial respiration with or without oxygen enrichment. In the case of anaesthetic ventilation, nitrous oxide N_2O or volatile anaesthetics are added.

In the pulmonary alveoli, the respiratory gases are in contact with the blood over a short diffusion path through the alveolar and capillary walls. Oxygen (O_2) first dissolves in the blood plasma and then diffuses further into the red blood cells (erythrocytes), which contain the red blood pigment hemoglobin (Hb) as an oxygen-transporting agent. Oxygen is not directly bound to Hb, but is only attached to the iron atom of each of the four haem groups. A molecule of Hb can, therefore, transport up to four oxygen molecules and easily release them again when it has reached surrounding tissue structures, which have a lower oxygen partial pressure via the bloodstream. Hb is, therefore, not oxidized, but rather oxygenated and in turn deoxygenated.

Carbon dioxide (CO_2) is constantly produced in the body as a product of metabolism and is likewise partly dissolved in the blood plasma, but is for the most part transported back to the alveoli as a bicarbonate chemically bound by attachment to an NH_2 group of the hemoglobin.

Whereas blood gas analysis (BGA) only ever measures the gases oxygen and carbon dioxide, the concentration of nitrous oxide and volatile anaesthetics in the respiratory gases must, if necessary, also be monitored. The respiratory gases are often measured as a concentration in percent by volume (vol.%), whereas the blood gases are measured as partial pressures in mmHg (= torr) or kPa. Both measurements can be converted from one to the other. In principle it would be physiologically meaningful to present partial pressures under BTPS conditions only. But – historically introduced by the gas-mixing technology – especially the inspired gases are given as volume concentration of a dry gas mixture under standard conditions STPD.

In a mixture of ideal gases with the concentrations c_i (measured in vol.%), the partial pressure p_i of one com-

Table 49.1 Composition of the air

Constituent	Abbreviation	Fraction (vol.%)
Oxygen	O_2	about 21
Nitrogen	N_2	about 78
Argon	Ar	about 1
Carbon dioxide	CO_2	about 0.03

ponent is the fraction of the total pressure (here always the atmospheric pressure p_{baro}) which is proportional to its concentration:

$$p_i = c_i \times p_{baro} .$$

The total pressure is accordingly the sum of the partial pressures of each of the individual gas components

$$p_1 + p_2 + \ldots + p_n$$
$$= c_1 \times p_{baro} + c_2 \times p_{baro} + \ldots + c_n \times p_{baro} = p_{baro} .$$

Normal respiratory air with the familiar composition (Table 49.1) is heated to 37 °C during inhalation and is mixed with a newly added gas as a result of humidification in the airways: water vapor. At full water vapor saturation, the water vapor partial pressure p_{H_2O} is entirely dependent on the temperature and is 47 mmHg at 37 °C. This is 6.2 vol. % (= (47/760) × 100 vol. %) at standard atmospheric pressure $p_{baro} = 760$ mmHg. In its otherwise unchanged composition, the respiratory air now takes up only 93.8% of the volume, and in this wet gas oxygen, therefore, only has a concentration of 19.7% (= 0.938 × 21%) or a partial pressure of 19.7% × 760 mmHg = 150 mmHg.

The remaining expiration gas in the alveoli mixes with this fresh gas, such that a noticeably lower alveolar O_2 partial pressure pAO_2 results, with a standard value of ≈ 100 mmHg (= 13.33 kPa) (conversion: 1 kPa = 1 kN/m^2; 100 kPa = 750 mmHg; 1 mmHg = 0.133 kPa). The difference between the partial pressures of a gas is the driving force behind diffusion. Venous blood arriving in the lungs has the comparatively low venous oxygen partial pressure pvO_2, which in the capillaries, in ideal ventilation, becomes equal to the alveolar partial pressure as a result of gas exchange. Because not all blood is involved in gas exchange, in practice there is an alveolar-arterial oxygen difference $AaDO_2$ of 10–20 mmHg in healthy individuals.

Venous blood transports CO_2 to the lungs and the inhaled air contains virtually no CO_2. The gas exchange leads to a normal alveolar CO_2 partial pressure $pACO_2$ of ≈ 40 mmHg (= 5.3 kPa).

Spontaneous respiration primarily aims to stabilize the CO_2 partial pressure and pH of the blood, whereas large fluctuations in the O_2 partial pressure are tolerated. It also worthwhile to measure the carbon dioxide in the respiratory gas as a partial pressure and not as a volume concentration, because the latter is dependent on air pressure.

As mentioned above, the oxygen is essentially transported by the erythrocytes. Hemoglobin binds approximately 70 times more O_2 than the blood plasma is capable of dissolving. The amount of oxygen in the blood is called the oxygen content and is measured in ml O_2 per dl blood. Together with the cardiac output, it decisively determines the transport of oxygen in the blood stream. How much hemoglobin is loaded with oxygen is dependent on the O_2 partial pressure in the blood and is given as the oxygen saturation, the percentage ratio of oxygenated hemoglobin to the total amount of hemoglobin. One gram of hemoglobin (Hb) can transport ≈ 1.34 ml of oxygen; the solubility factor for 1 ml of oxygen in 1 dl of blood is about 0.003 per mmHg. The arterial and venous oxygen content can accordingly be given by

$$caO_2 = 1.34 \times Hb \times \frac{SaO_2}{100} + paO_2 \times 0.003$$

and, correspondingly,

$$cvO_2 = 1.34 \times Hb \times \frac{SvO_2}{100} + pvO_2 \times 0.003 ,$$

where SaO_2 is the arterial and SvO_2 the venous oxygen saturation as a percentage, Hb is the hemoglobin content of the blood in g/dl, and paO_2 and pvO_2 are, respectively, the arterial and venous oxygen partial pressure in mmHg. The arteriovenous oxygen difference $avDO_2$, unlike the $AaDO_2$, is, therefore, not a partial pressure difference but a content difference that provides evidence about the oxygen consumption of the organism. Table 49.2 shows a summary of terms and units of measurement for oxygen transport.

In addition to the functional hemoglobin Hb (deoxyhemoglobin), which is available for oxygen uptake, and O_2Hb (oxyhemoglobin), which is already saturated with oxygen, there is also a series of dysfunctional Hb derivatives; the most frequent representatives of these, COHb (carboxy-Hb, Hb poisoned with carbon monoxide) and MetHb (methemoglobin), are mostly present in the blood in fractions of less than 5 and 1%, respectively. Whereas the amount of oxygen physically dissolved in the blood is simply proportional to the oxygen partial pressure (second summand in the content calculation above), the significantly more complex relationship between the pO_2 and the degree of hemoglobin oxygenation, i. e. the oxygen saturation, is represented by

Table 49.2 Terms and units of measurement in connection with oxygen transport

Measurement parameter	Abbreviation	Unit
Hemoglobin content	Hb	g/dl
Oxygen partial pressure		mm Hg or kPa
• Arterial	paO_2	
• Alveolar	pAO_2	
• Venous	pvO_2	
• Transcutaneously measured	$ptcO_2$	
Alveolar-arterial oxygen difference	$AaDO_2 = paO_2 - pAO_2$	mm Hg or kPa
Oxygen saturation		%
• Arterial	SaO_2	
• Mixed venous	SvO_2	
• Central venous	$ScvO_2$	
• Determined by pulse oximetry	SpO_2	
Oxygen content		ml/dl
• Arterial	caO_2	
• Mixed venous	cvO_2	
Arteriovenous oxygen difference	$avDO_2 = caO_2 - cvO_2$	ml/dl
Oxygen availability	$DO_2 = CO \times caO_2 \times 10$	ml/min (CO = cardiac output in l/min)
Oxygen consumption	$VO_2 = CO \times avDO_2 \times 10$	ml/min

the oxygen–hemoglobin dissociation curve (synonyms: oxygen dissociation curve, oxyhemoglobin dissociation curve) of the hemoglobin (Fig. 49.1).

The position and shape of the oxygen–hemoglobin dissociation curve are dependent on various factors in the blood, such as:

- pH and pCO_2
- Temperature
- Hemoglobin type (foetal or adult)
- Proportion of dysfunctional Hb derivatives (COHb, MetHb)
- Enzyme content (2,3-DPG).

Some deviations from the normal value lead to an improved O_2 supply. With a fever or an increased pCO_2, for example, the dissociation curve is displaced to the right, such that oxygen is released to the cells more easily. Mountain dwellers have an increased 2,3-DPG content, which also results in the dissociation curve being displaced to the right. Carbon monoxide poisoning with the formation of high concentrations of COHb is, in contrast, accompanied by displacement of the dissociation curve to the left, such that the remaining functional hemoglobin, which is already depleted, also releases the oxygen with greater difficulty.

Dangers resulting from decreased supply of oxygen can be detected in good time in the arterial blood by monitoring both the arterial oxygen partial pressure paO_2 and the oxygen saturation SaO_2. Whereas arterial blood samples could previously only be examined for this purpose in the laboratory analyzer at relatively long time intervals, noninvasive continuous monitoring procedures are now available for both measurement parameters: pulse oximetry (Sect. 49.2.2) for determining the SpO_2 as a measure of the arterial O_2 saturation, and

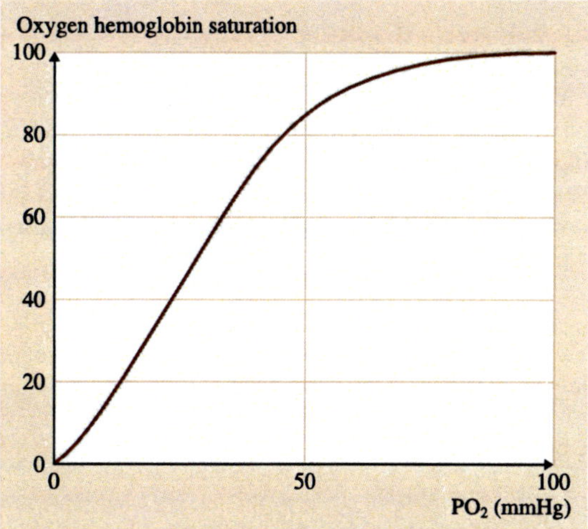

Fig. 49.1 Oxygen–hemoglobin dissociation curve of the hemoglobin under normal conditions (temperature 37 °C, pH 7.4)

transcutaneous blood gas measurement (Sect. 49.2.3) of ptc O_2 (and ptc CO_2) as correlates of the arterial gas partial pressures paO_2 (and $paCO_2$).

Important terms relating to physiological states of oxygen deficiency are:

Hypoxia	low oxygen partial pressure (paO_2)
Hypoxygenation	low oxygen saturation of the hemoglobin (SaO_2)
Hypoxaemia	low oxygen content in the blood (caO_2)
Anaemia	low hemoglobin content in the blood (Hb).

It can be seen from the oxygen-hemoglobin dissociation curve in Fig. 49.1 that, in the case of oxygen deficiency, even small changes in the paO_2 signify large changes in the SaO_2, and O_2 saturation is, therefore, a variable that reacts very sensitively in the case of oxygen deficiency, which means that pulse oximetry is particularly suitable for monitoring oxygen deficiency. On the other hand, even with a normal paO_2 of approximately 90–95 mmHg virtually all of the hemoglobin is oxygenated. Monitoring of an upper limit by means of pulse oximetry is no longer conclusive, particularly if the patient is being given oxygen, and transcutaneous blood gas measurement is considered preferable. Transcutaneous blood gas monitoring is, therefore, most commonly used to avoid hyperoxia in premature infants; relatively long-term hyperoxia in their case leads to irreversible blindness.

49.2.2 Pulse Oximetry

Pulse oximetry is a noninvasive method for continuously measuring the oxygen saturation in the arterial blood. The parameter is not called SaO_2 but SpO_2, to signify that the oxygen saturation is determined by means of pulse oximetry and to point towards the principal difference.

Functional and Fractional Saturation

The value SpO_2 from pulse oximetry is always what is known as the *functional* saturation: the oxyhemoglobin as a percentage of the *total functional* hemoglobin, which can transport oxygen:

$$SpO_2 = \frac{O_2Hb}{O_2Hb + Hb} \times 100\% .$$

Fractional saturation is determined using laboratory CO-oximeters, however: the hemoglobin as a percentage of the *total measured* hemoglobin, including the dysfunctional hemoglobin derivatives, e.g. carboxyhemoglobin or methemoglobin:

$$SaO_2 = \frac{O_2Hb}{(O_2Hb + Hb + COHb + MetHb)} \times 100\% .$$

When significant amounts of MetHb are present (in the case of methemoglobinaemia), this fundamental difference between SpO_2 from pulse oximetry and SaO_2 from the laboratory CO-oximeter naturally leads to differing results, the cause of which is not known to every user. The situation is worse when COHb is present (e.g. in the case of smoke poisoning): pulse oximetry is not capable of distinguishing between O_2Hb and COHb and in this case provides erroneously high SpO_2 values.

Conventional Pulse Oximetry

Conventional pulse oximetry was developed in 1974 by Aoyagi and has been commercially available since 1981. Pulse oximetry monitoring has since become an indispensable monitoring standard.

Using an easily applied sensor (finger sensor for adults and children or, for newborn infants, an adhesive sensor on the hand or foot, for example), the oxygen saturation SpO_2 and the pulse rate are determined and a plethysmographic curve is displayed on the monitor. Pulse oximetry uses the different absorption properties of oxyhemoglobin (O_2Hb) and deoxyhemoglobin (Hb) at different wavelengths of light (Fig. 49.2). Two LEDs transmit red and infrared light through the measurement point. A detector registers the intensities I_{RD} and

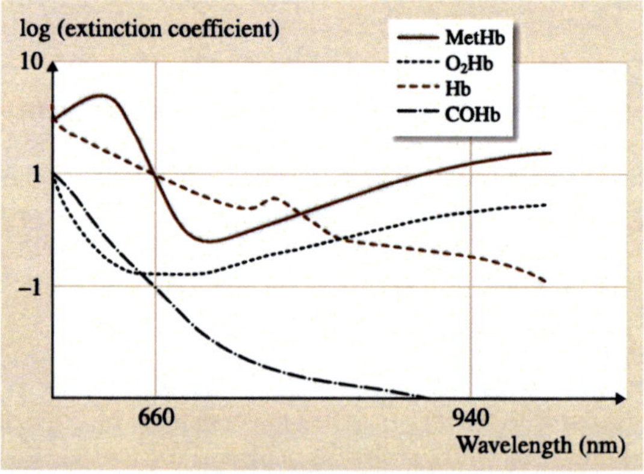

Fig. 49.2 Absorption spectra of the functional Hb forms Hb (deoxyhemoglobin) and O_2Hb (oxyhemoglobin) and also of the dysfunctional Hb derivatives COHb and MetHb, which are not capable of binding oxygen

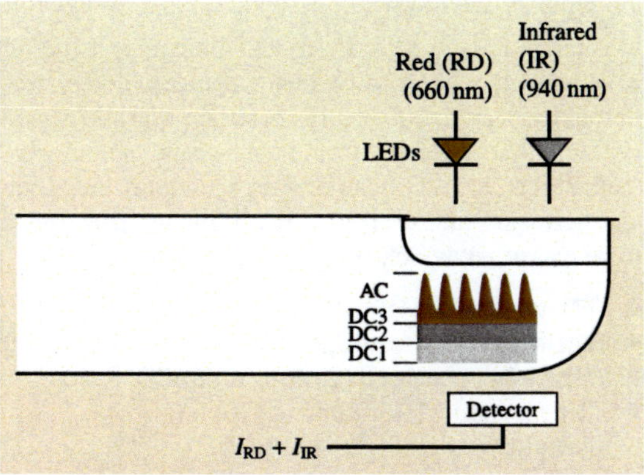

Fig. 49.3 Conventional pulse oximetry. Two LEDs transmit red (RD) and infrared (IR) light through the measurement point, and this light undergoes pulsatile absorption (AC) through arterial blood and constant absorption through tissue (DC1), venous blood (DC2) and nonpulsatile arterial blood (DC3). The detector measures the intensities I_{RD} and I_{IR} of the transmitted light as a function of time (transmission photometry)

I_{IR} of the penetrating red (RD) and infrared (IR) light (Fig. 49.3).

Conventional pulse oximetry is based on the assumption that only arterial blood has a pulsatile AC component of the light absorption, whereas tissue, venous blood and nonpulsatile arterial blood always constitute a constant DC component of the light absorption at the measurement point (Fig. 49.3). The intensities I_{RD} and I_{IR} registered by the detector thus include an AC and a DC component in each case. These are then compared with one another, and from the quotient

$$r = \frac{I_{RD}\,\mathrm{AC}/I_{RD}\,\mathrm{DC}}{I_{IR}\,\mathrm{AC}/I_{IR}\,\mathrm{DC}}$$

of these two relationships the percentage value of the oxygen saturation SpO_2 is empirically determined. This quotient from two quotients has the advantage that no calibration procedure is necessary for the pulse oximetry measurement and the process is, therefore, particularly user-friendly.

Movement in the Case of Conventional Pulse Oximetry

If the patient moves, then the venous blood at the measurement point also moves. This inevitably results in considerable interference components in the signal being investigated, because pulsatile elements no longer originate from arterial blood alone. The AC components are distorted and an imprecise saturation value emerges. Even with normal patient movements or low perfusion, the interference components in the detected signals can be far higher than the desired signals. In these cases, the quotients are primarily determined by the interference components and r has a value close to 1. Many systems then show a saturation value of approximately 85%.

As a countermeasure, in conventional pulse oximetry systems, for example, the time for determining the average is increased or the display is *frozen* if movement is detected. This does not provide clinicians with continuous information, however, and can in the worst case scenario lead to misinterpretations.

Masimo SET Pulse Oximetry

In 1989 *Diab* and *Kiani* [49.1–3], founders of Masimo Corporation, developed signal extraction pulse oximetry, which made it possible for the first time to differentiate the arterial signal from nonarterial interferences. This pioneering technology has been used in clinical practice since 1998.

Masimo signal extraction technology (SET) rejects the conventional assumption that pulsatile signal components originate from arterial blood alone. By means of adaptive noise suppression, the venous component can be separated. The principles of this are explained in more detail in [49.2].

Important features and advantages of Masimo SET pulse oximetry are:

- The saturation algorithm is independent on the pulse rate algorithm (whereas, in conventional pulse oximetry systems, an accurate pulse must be detected in order to calculate the exact arterial oxygen saturation).
- Using Masimo SET, arterial oxygen saturation and the pulse rate can even be monitored when movement has already begun before the pulse oximeter has been switched on (whereas conventional technology relies on *clean* data at the beginning of monitoring).
- As an aid for optimum sensor placement, an additional parameter is supplied: the perfusion index (PI). Displaying the signal IQ serves as signal quality control.

Perfusion Index

The infrared signal, which is influenced less by the oxygen saturation, is used to calculate the PI. The index is determined from the pulsatile (I_{IR} AC) and the nonpulsatile IR signal (I_{IR} DC) and is given as a percentage

$$\mathrm{PI} = \frac{I_{IR}\,\mathrm{AC}}{I_{IR}\,\mathrm{DC}} \times 100\% \,.$$

The value of the PI relates to the respective measurement point on the patient. According to the physiological conditions, it differs from one measurement point to the next and from patient to patient. The PI is suitable for assessing the suitability of the measurement point. A large PI means optimum measurement for pulse oximetry. The PI display covers values from 0.02 to 20%. A PI above 1% is desirable.

SpO_2 Signal Quality Control – Signal IQ

Masimo pulse oximetry provides a visual display for the quality of the plethysmographic signal. It is referred to as signal IQ (signal identification and quality indicator). Using signal IQ it is possible to identify the occurrence of a pulse as well as the associated signal quality of the measurement. When the patient moves, the plethysmographic pulse curve is frequently distorted by artifacts. The signal IQ, which is represented as a vertical line (pulse bar), reliably shows the time of the arterial pulsation, and its height indicates the quality of the signal measured. When a sharp reduction in the signal quality can have a negative influence on the accuracy of the SpO_2 measurement, a warning message (signal IQ low) is shown on the display.

Sensitivity Modes

Masimo pulse oximetry enables the measurement sensitivity to be set at three levels.

Normal: recommended mode for typical monitoring, such as in intensive care units, for example.

Adaptive probe off detection (APOD): recommended mode when, because of staffing levels, it is not possible to detect immediately if a sensor has become detached. This mode offers better protection against erroneous measurements being displayed, even though the sensor has become detached from the measurement point (e.g. as a result of the patient moving).

Maximum: this mode is used to obtain and display data even when the signal is very weak due to impaired perfusion (can be used, for example, during treatment or examination, i. e. when someone is with the patient). If the sensor becomes detached from the patient, this mode makes virtually no provision for measurements that are displayed erroneously.

Pulse CO-Oximetry

Masimo Rainbow SET Pulse CO-Oximetry, a further development of the two-wavelength pulse oximetry, was first introduced in 2005 and is based on Masimo Signal Extraction Technology. It is a continuous, noninvasive method that, in addition to oxygen saturation, pulse rate and perfusion index, enables the following parameters to be measured, which are of great interest clinically:

- The total hemoglobin (SpHb)
- The oxygen content (SpOC)
- The carboxyhemoglobin content (SpCO)
- The methemoglobin content (SpMet)
- The pleth variability index (PVI).

Masimo pulse CO-oximetry uses multiwavelength sensors that transmit more than seven different wavelengths of visible light and infrared light from LEDs through the measurement point and correspondingly detect the respective absorption.

Arterial Oxygen Content SpOC

With typical paO_2 values of around 100 mmHg, the oxygen content dissolved in the plasma is only approximately 0.3 ml/dl and can in practice be assumed to be a constant. For normal concentrations of carboxyhemoglobin and methemoglobin, the functional saturation determined using a pulse oximeter is about 2% higher than the fractional saturation: $SpO_2 = 1.02 \times SaO_2$.

For the arterial O_2 content, this approximation gives a simplified calculation for SpOC from SpHb and SpO_2:

$$\mathrm{SpOC} = 1.31 \times \left(\mathrm{SpHb} \times \frac{\mathrm{SpO_2}}{100}\right) + 0.3 \,.$$

The measurements from the pulse CO-oximetry can be compared with laboratory measurements from blood samples. Interpreting comparisons of this type requires a great amount of care, as the values ascertained from the blood gas sample may under some circumstances differ from the SpO_2, SpCO and SpMet measurements. For SpO_2, differences from the arterial blood gas sample occur when the effect of the influencing variables on the oxygen-hemoglobin dissociation curve have not been corrected. In the case of SpCO and SpMet, in addition to temperature and pH abnormal values for the oxygen saturation ($SaO_2 < 90\%$) and/or for the concentration of methemoglobin (MetHb $> 2\%$) in the blood gas sample can also influence the results.

Pleth Variability Index (PVI)

The heart and lungs interact physiologically in different ways. Intrathoracic pressure changes as a result of respiration have a direct influence on the filling of the heart with blood and on its pumping capacity. These pressure changes are visible as a cyclical change in the arterial blood pressure. A reliable, continuously and noninvasively measurable indicator of such cyclical changes would be of great clinical value.

Masimo has developed the pleth variability index, which shows the cyclical changes in the pulse curve as a result of changes in the physiological conditions. The perfusion index (PI), which has already been mentioned, is used for this purpose. The pleth variability index (PVI) is a measure of the dynamic changes in the PI during the respiratory cycle and is calculated as a percentage value from the PI changes during the respiratory cycle or multiple complete respiratory cycles:

$$\mathrm{PVI} = \frac{\mathrm{PI}_{\max} - \mathrm{PI}_{\min}}{\mathrm{PI}_{\max}} \times 100\%\,.$$

The lower the PVI, the lower the PI variability in the course of a respiratory cycle. The PVI can provide potentially useful information about the balance between the intrathoracic airway pressure and the intravascular fluid volume. Its trend can be helpful in estimating the hydration status. An increase can indicate hypovolaemia [49.1].

Overall, the parameters from pulse CO-oximetry introduced in recent years have significantly broadened the possibilities for noninvasive monitoring of patients and are attracting increasing interest among their users.

Mixed Venous Oxygen Saturation SvO_2 and Central Venous Oxygen Saturation $ScvO_2$

In pulse oximetry, the measurement point is transilluminated. This is not possible in large vessels, e.g. in the pulmonary artery, where the venous blood from all the organs arrives thoroughly mixed together. A catheter with fiber optics placed here can detect the mixed venous oxygen saturation SvO_2 by means of reflection photometry, however. If a pulmonary artery catheter is to be avoided, the central venous saturation $ScvO_2$ can in the first approximation be measured using a central venous catheter, situated a little upstream of the right atrium, and fiber optics. The two values may differ, but react to the same extent to changes in the haemodynamics [49.4]. Both parameters may, therefore, be used in monitoring as immediate indicators for haemodynamic changes.

By taking account of the arterial oxygen saturation SaO_2 and the cardiac output, important information can be obtained about the oxygen utilization of the organism.

49.2.3 Transcutaneous Blood Gas Measurement

Miniaturized electrochemical sensors of the Clark type, known in principle from blood gas analyzers, are placed directly on the skin. A measuring membrane encloses a very small volume of electrolyte, into which the oxygen and/or carbon dioxide can diffuse from the cutaneous capillary bed. Since only arterial values are of interest, the capillary bed is heated to 42–45 °C by means of a heating element in the sensor and thus arterialized: the high temperature reduces the vascular tone, the arterioles become slack, and arterial blood penetrates into the capillary bed.

The transcutaneous oxygen partial pressure $ptcO_2$ is measured according to the polarographic method (Fig. 49.4). Microelectrodes made from platinum or gold are situated in the electrolyte and are biased towards a silver anode with a fixed plateau potential (typically 0.8 V). A small $ptcO_2$-proportional measurement current,

$$I \propto ptcO_2\,,$$

then flows, which is generated as a result of the reduction of O_2 molecules at the microelectrodes

$$O_2 + 2H_2O \rightarrow 4OH^- - 4e^-\,.$$

At the same time, silver at the anode is converted to silver chloride

$$4Ag + 4Cl^- \rightarrow 4AgCl^- + 4e^-\,.$$

The response time (T90, based on partial pressure peaks of 10–90%) is approximately 8 s and is, therefore, not

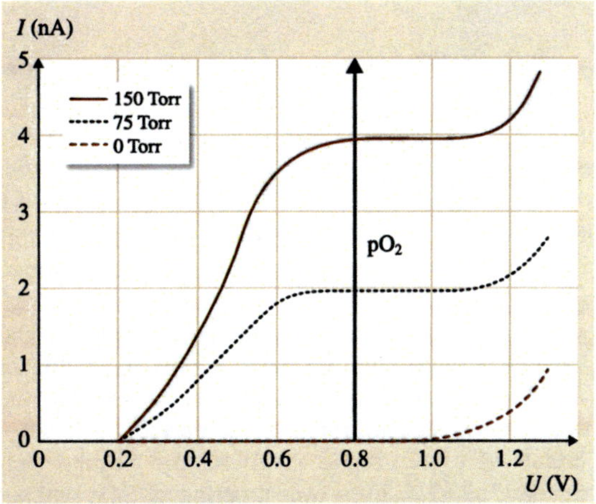

Fig. 49.4 Polarogram of an electrochemical O_2 converter

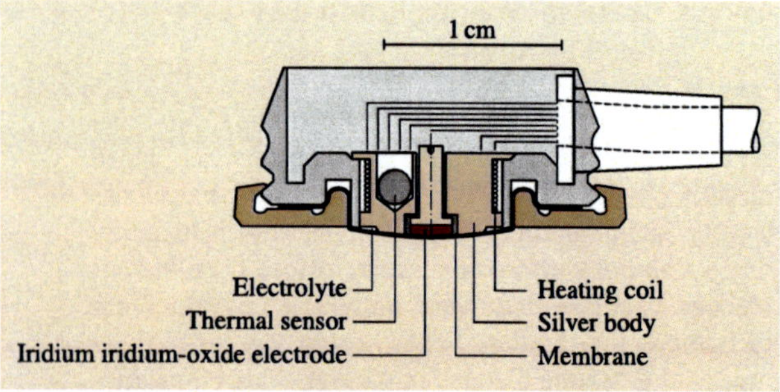

Fig. 49.5 Sectional drawing through a transcapnode

slower-acting than pulse oximetry, which uses similar averaging times to stabilize measurements.

The transcutaneous carbon dioxide partial pressure $ptcCO_2$ is measured according to the potentiometric method using a pH electrode. By way of illustration, Fig. 49.5 shows a diagram of a transcapnode with an iridium/iridium oxide pH electrode.

This type of electrode is considerably more stable mechanically and has a significantly lower and more noise-free output impedance than the previously customary glass electrodes. The CO_2, which diffuses into the measuring cell, shifts the pH value by forming carbonic acid:

$$CO_2 + H_2O \leftrightarrow H_2CO_3 \leftrightarrow H^+ + HCO_3^- \leftrightarrow 2H^+ + CO_3^{2-} .$$

When the partial reactions are at equilibrium, there is a logarithmic relationship between the CO_2 partial pressure and the pH, such that the potential present at the pH electrode is likewise logarithmically dependent on the $ptcCO_2$,

$$U \sim \log(ptcCO_2) .$$

The response time (T90) for the $ptcCO_2$ measurement has been considerably shortened over the years and is now typically 40 s.

Transcutaneous oxygen monitoring requires a minimum heating temperature of 43 °C; for CO_2 monitoring, a temperature of 42 °C is sufficient. A temperature of 43.5 °C provides a good compromise between the desired high correlation with arterial measurements and the risk of local burns to the skin. A temperature of 45 °C must not be exceeded and at this temperature the sensor must be repositioned approximately every hour; infants have been found to withstand a heating temperature of 42 °C even over a period of 24 h without burning.

Measuring Accuracy and the Influence of Errors in Transcutaneous Blood Gas Measurement

Transcutaneous blood gas sensors are intended to react as quickly as possible to changes in partial pressure and, therefore, contain only a very thin electrolyte layer. Heating the sensors leads to the drying out and sweating of the patient, and to thinning of the electrolyte, resulting in unstable measurements. Such sensors must, therefore, be calibrated at least once daily to air ($pO_2 = 0.21 \times p_{baro}$) and if necessary to at least one defined CO_2 mixture (5 or 10% CO_2). From experience, the electrolyte and measuring membrane must also be replaced at least once a week.

In a similar way to ECG electrodes, the sensors are attached to the skin using adhesive rings. A drop of aqueous contact agent is applied to the application site. This provides better diffusion and prevents inclusion of any of the surrounding atmosphere which, despite the low volume, would impede fast adjustment to real transcutaneous measurements.

In principle, transcutaneous and arterial blood gas values do not correspond with each other, but a good measuring system guarantees a high correlation ($r > 0.8$) of the $ptcO_2$ and $ptcCO_2$ values with the paO_2 and $paCO_2$ values, respectively. Heating of the capillary blood leads to significant temperature-dependent increases both in the O_2 and CO_2 partial pressure (capillary arterial reference values: $pcapO_2 + 6\%/°C$, $pcapCO_2 + 4.8\%/°C$). This is because an increase in temperature shifts both the oxygen–hemoglobin dissociation curve of the hemoglobin and the chemical equilibrium between dissolved carbon dioxide and bound CO_2 (bicarbonate or bound to hemoglobin).

During diffusion of the blood gases through the epidermis, due to the increased cellular metabolism oxygen is consumed and carbon dioxide is also released. In newborns, the compensatory oxygen effect, therefore, gives $ptcO_2$ values that differ slightly from the paO_2 (and which in older patients are always below the paO_2), whereas the $ptcCO_2$ values are elevated by about 50%. The $ptcCO_2$ values are, therefore, mostly given as corrected values, because the individual effect on the skin is relatively small here compared with the systematic temperature effect. Some manufacturers also allow in vivo correction, which is used to adapt the transcutaneous measurements to the results of blood gas analysis. In the case of stable peripheral perfusion ratios, there is then no longer any difference between the displayed transcutaneous and central arterial measurements.

In the case of low perfusion – e.g. as a result of the onset of peripheral hypoperfusion – transcutaneous O_2

measurement gives misleadingly low values. It can then be dangerous to ventilate the newborn with an increased O_2 concentration in order to compensate. However, the transcutaneous measurement method itself provides an excellent key to explaining this situation, which unfortunately is repeatedly forgotten: a sudden drop in the electrical heat output indicates hypoperfusion. This is because the sensor is heated to a predetermined stable temperature, which is always above the temperature of the blood that is flowing in (e.g. 43.5 °C). In the case of strong perfusion, the sensor is, therefore, cooled more substantially by the blood than in the case of low perfusion. In order to maintain the temperature, a higher heat output is, therefore, necessary in the case of strong perfusion. The trend in the heat output, therefore, runs parallel to the trend in perfusion. If the $p\text{tcO}_2$ trend and the trend in the heat output therefore drop simultaneously, this indicates that the $p\text{tcO}_2$ drop can be attributed to a decrease in perfusion due to the method of measurement.

Transcutaneous blood gas measurement is also possible in special cases in adults, when it is important to detect changes. It has not proved to be suitable for monitoring trends because the skin perfusion probably plays an even greater role. Known applications are monitoring in cases of hyperbaric oxygen therapy and measurements in angiology, particularly in the case of peripheral occlusive disease and when monitoring the effectiveness of measures to eliminate angiostenoses in the limbs.

49.2.4 Measuring the Respiratory Gases

In addition to oxygen and carbon dioxide, respiratory gases here should also be understood to mean nitrous oxide and volatile anaesthetics (halothane, enflurane, isoflurane, desflurane, and sevoflurane). They are either measured in the respiratory flow of the intubated patient (mainstream method) or in the sidestream method, in which the mixture of respiratory gases is drawn off from the patient's ventilation hose by means of a sampling tube at a gas flow rate of between 50 ml/min and 250 ml/min and, if necessary, is fed to the measuring cell in the monitor via a water trap and drying stage (Nafion tube).

The mainstream method is becoming increasingly widely used for isolated monitoring of carbon dioxide (Sect. 49.2.5), in which the sensor is seated directly on a special airway adapter. The advantage of the mainstream method is its faster reaction to changes in the composition of the respiratory gas, whereas the advantage of the sidestream method is its more robust construction and the lower amount of additional dead space in the ventilation hose.

Measuring Cells

Oxygen is measured either using inexpensive but slow-acting fuel cells (e.g. in the inspiratory respiratory gas in the case of artificial respiration) or in synchrony with breathing using paramagnetic measuring cells. As an example, Fig. 49.6 shows a paramagnetic measuring cell with a dumbbell-shaped design in a shock-proof construction. Two thin-walled glass balls filled with nitrogen are suspended in the strongly inhomogeneous magnetic field of a permanent magnet. Oxygen in the atmosphere surrounding the magnet gap displaces the glass balls. The resultant torsional moment is compensated with a regulating current through coils on the glass balls and the dumbbell is held in a fixed position. The oscillatory system is, therefore, on the one hand strongly damped, and on the other hand the regulating current is proportional to the O_2 partial pressure in the respiratory gas,

$$O_2 \text{ concentration: } \%O_2 = \frac{pO_2}{p_{\text{baro}}} .$$

Unlike oxygen, carbon dioxide, nitrous oxide (N_2O), and anaesthetic gases have polar molecules, which can be excited to mechanical oscillations by infrared radiation in the wavelength range of between 2.5 and 15 μm and which in the process absorb radiation of characteristic wavelengths. Figure 49.7 shows the absorption bands of the respiratory gases of interest, including water vapor.

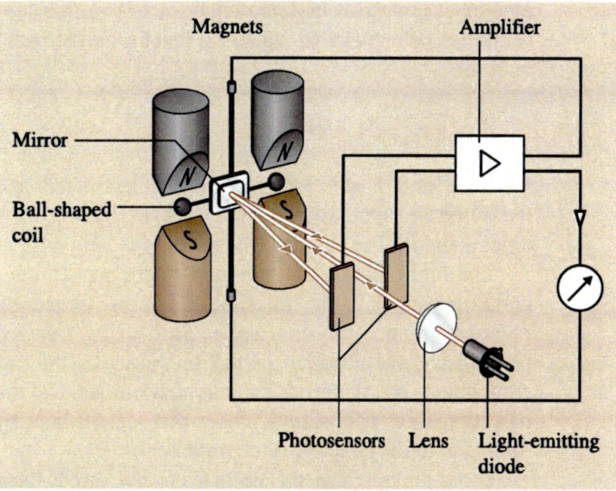

Fig. 49.6 Paramagnetic O_2 measuring cell

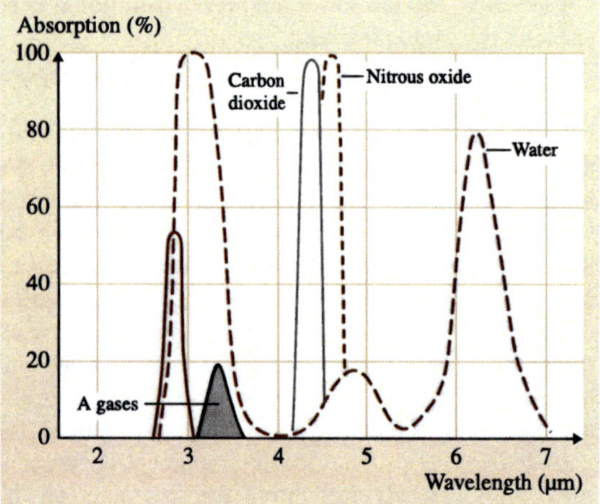

Fig. 49.7 Absorption spectra of the respiratory gases (simplified, without details of the fine structure)

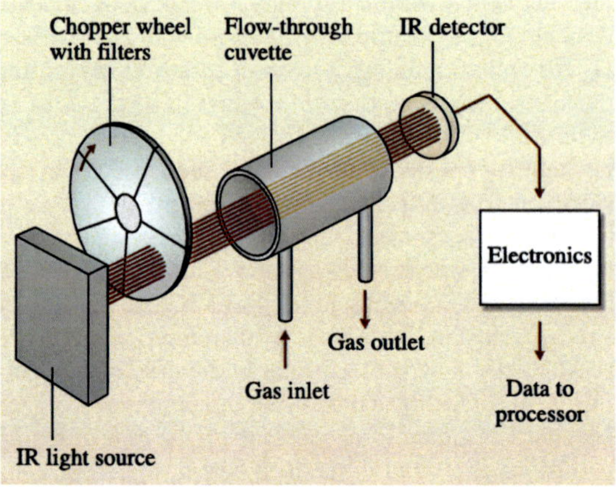

Fig. 49.8 Single-beam infrared measuring cell with chopper wheel

These very specific absorption bands are utilized for measurement. A radiation source emits an entire spectrum of infrared light. In nondispersive infrared measurement or spectroscopy (NDIR), originally a detector was used that was sensitive only to specific wavelengths absorbed by the gas being investigated. Today, by means of filtering usually only those wavelengths are selected at which there is the greatest absorption of the gas to be measured. In the dispersive infrared measuring technique, a measuring wavelength that is suitable for the gas being sought is generated by dispersion of the light (e.g. by refraction in a prism).

According to the basic law of photometry after Lambert–Beer, the absorption A is proportional to the amount c (frequently inaccurately referred to as the concentration) of molecules capable of reaction in the beam path

$$A = \log\left(\frac{I_0}{I}\right) = \varepsilon l c\,.$$

Here, I_0 and I are, respectively, the irradiated and measured beam intensity, ε is the molar extinction coefficient, and l is the length of the reacting beam path.

The higher the partial pressure of the gas components measured in the mixture, the more frequently photons impinge on such molecules. Using infrared photometry, the absolute partial pressures of the gas components are, therefore, initially determined, and the percentage gas concentrations are then determined by reference to the effective total pressure.

If the gas mixture sucked in were not dried, there would be a broad water vapor band (2.7–3.6 µm) directly above the anaesthetic gas bands (A-gases), which would then disappear. The absorption bands of the anaesthetic gases also overlap such that, when using relatively simple analyzers, it must be entered which gas is used. Measuring devices with automatic identification therefore use other spectral ranges with wavelengths that are above those shown.

The absorption bands of CO_2 (4.26 µm) and N_2O (4.55 µm) lie in a gap in the water vapor spectrum and can be reliably separated using appropriate filters in the filter plate of the measuring apparatus. The filter plate contains a filter in an additional position, which transmits radiation in a gap in the total absorption spectrum through the flow-through cell. This is used as a reference measurement to stabilize the measurements and, therefore, prevents a drift in the measurements when there is gradual contamination of the measuring cell or a change in the radiation source.

Figure 49.8 shows the structure of an infrared measuring cell with a chopper wheel, which is used to alternately move filters for the measurement wavelength and reference filters into the beam path. Today, particularly in monitoring, solid-state sensors without moving parts are used.

Measuring Accuracy

Relatively large anaesthetic gas measurement errors occur in sidestream measuring systems when sampling tubes made from the wrong material are used. So, for example, as a result of surface absorption of halothane, a polyethylene sampling tube can completely distort the erratic concentration changes that are in synchrony with breathing and can give false inspiratory and expiratory measurements. It is, therefore, imperative to ensure that only those sampling tubes that are recommended or endorsed by the manufacturer of the measuring system are used.

Similar effects arise in relation to measurement of the other gases, too, if a sampling tube is repeatedly disinfected for reasons of thrift. The inner surfaces are very rough and there is no longer any laminar flow in the sampling tube, with the result that concentration levels are distorted on the way to the analyzer.

In routine clinical practice, questions are repeatedly raised about the measuring accuracy of the end-expiratory CO_2 measurements ($etCO_2$). Deviations in the respiratory gas measuring results from the blood gas values are the reason for this: the $petCO_2$ is generally lower than the $paCO_2$ by 2–13 mmHg. Differences in the measurements of equipment from different manufacturers are also unsettling. There are technical and physiological explanations for this.

There is cross-sensitivity between the CO_2 measurement and the other gases, partly due to spectral overlapping (with N_2O) and partly because of an indirect influence of O_2 on the CO_2 absorption band (collision broadening effect). If a capnograph does not also measure the other gases simultaneously, for correction purposes it must be specified whether the gas mixture contains O_2 and/or N_2O in relatively large concentrations. Failure to observe this can lead to errors of about 2 mmHg.

Specific NDIR measuring cells are very sensitive to the CO_2 fine structure spectrum and do not require any correction. There are also infrared analyzers whose CO_2 measurements are automatically corrected with the anaesthetic gas concentrations. If the anaesthetic gas is not detected automatically, indication of an incorrect anaesthetic gas then leads to relatively big errors, not only in the anaesthetic gas but also in the pCO_2. Example: If isoflurane is stated but halothane is used, the pCO_2 is up to 7 mmHg too low.

These days there is a requirement to state the CO_2 partial pressure as a lung value in accordance with the BTPS standard, since only CO_2 lung values can be compared with blood gas values. Because the infrared measuring cells measure in dry gas, however, the CO_2 lung value must be reduced again by around 2.5 mmHg

$$pCO_2\,(\text{lungs}) \approx pCO_2\,(\text{IR cell}) \times \frac{p_{\text{baro}} - 47}{p_{\text{baro}}}\,.$$

(In actual fact, this correction is just a little too severe, as the gas in the IR cell is not completely dry.)

The influence of dead space ventilation as a result of a ventilation–perfusion disproportion is frequently underestimated, and feeds directly into the discrepancy between $petCO_2$ and $paCO_2$. Alveolar shunts also contribute to the fact that end-expiratory values can be lower than arterial values. Where there are sudden changes in the $petCO_2$, blood gas analysis must therefore be used to investigate whether the $paCO_2$ has also changed to the same degree or whether a change in the ventilation-perfusion ratio is (also) responsible for this.

It is not always known that in major operations, the patient's blood is noticeably hypothermic, but blood gas analysis is always carried out at 37 °C. The blood gas analysis values are thus elevated by approximately 4.8% per °C compared with the lung values! Therefore, if the patient were to be cooled down to 34 °C, for example, the uncorrected blood gas analysis value would be about 4.3 mmHg above the actual partial pressure in the patient's blood.

49.2.5 Capnography

Capnography is to be understood as the measurement and graphic representation of the CO_2 in the respiratory gas. Spontaneous respiration is primarily aimed at stabilizing the arterial CO_2 partial pressure to values around about 40 mmHg (Sect. 49.2) and represents the fundamental respiratory drive. Relatively large deviations in the $paCO_2$ have considerable physiological consequences, such as vasodilatation in the case of hypercapnia, with the possible consequence of an increase in the intracranial pressure. Hypocapnia leads, in contrast, to vasoconstriction with the risk of organ hypoperfusion, which primarily puts the brain at risk.

This means that, as a noninvasive and continuous monitoring process for ventilation, capnography is of high importance for the safety of the patient while under general anaesthetic and in intensive care medicine. The capnogram plots the CO_2 partial pressure over time. The end-expiratory CO_2 ($petCO_2$) is the maximum at the end of the alveolar plateau, which correlates with the $paCO_2$.

Due to different recording speeds, CO_2 curves can be displayed in high resolution (12.5 mm/s) or as a trend (25 mm/min) (Fig. 49.9). In monitoring, 6.25 mm/s is customary and trends are as a rule displayed as a course of values (1 value per minute).

In the mainstream method, the measuring sensor is situated close to the patient on an airway adapter between the tube and ventilation hose. It is preferred in the case of intubated patients because of the very accurate results and lack of temporal latency.

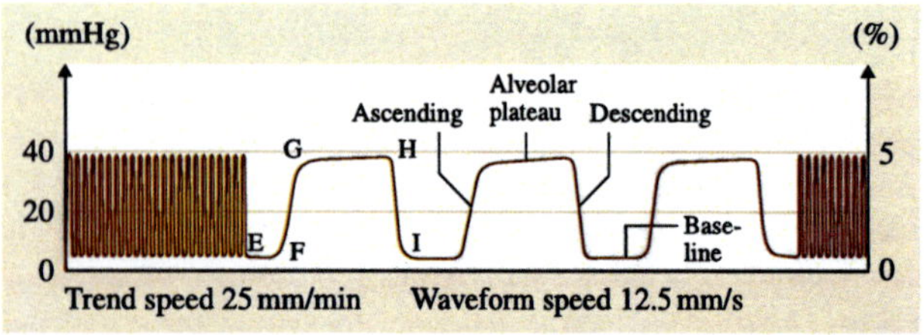

Fig. 49.9 Normal capnogram. F Beginning of exhalation (expiratory phase). Gas is exhaled from the anatomical dead space. F-G Steep rise in the CO_2 concentration, predominantly from the lower airways. G-H Alveolar plateau, consisting of the CO_2 concentration and CO_2 partial pressure of the alveolar gas. H End-expiratory or end-tidal PCO_2 ($PetCO_2$). H-I Downward curve; the inspiratory phase exhibits a rapid decrease in the CO_2 concentration as a result of the fresh gas taken in

In the sidestream method, the respiratory gas is sampled by means of a fine tube line at a sampling rate of between 50 ml/min and 250 ml/min. The measurement chamber is situated in the patient monitor. This leads to a delay in the display of the capnogram.

However, the sidestream method also allows monitoring of nonintubated patients wearing a respiratory mask, which is constructed from fine tubes and draws off respiratory gas from the nasal openings.

Using microstream technology, a sidestream method that also operates with a high level of accuracy has established itself. The method uses a laser light source at the precise wavelength, which is absorbed most by CO_2. As a result, there is no interference due to other gases with adjacent absorption bands. The special design of the measuring tubes has made the method very reliable in its handling to counter ingress of moisture or secretions into the measuring chamber. Due to its low

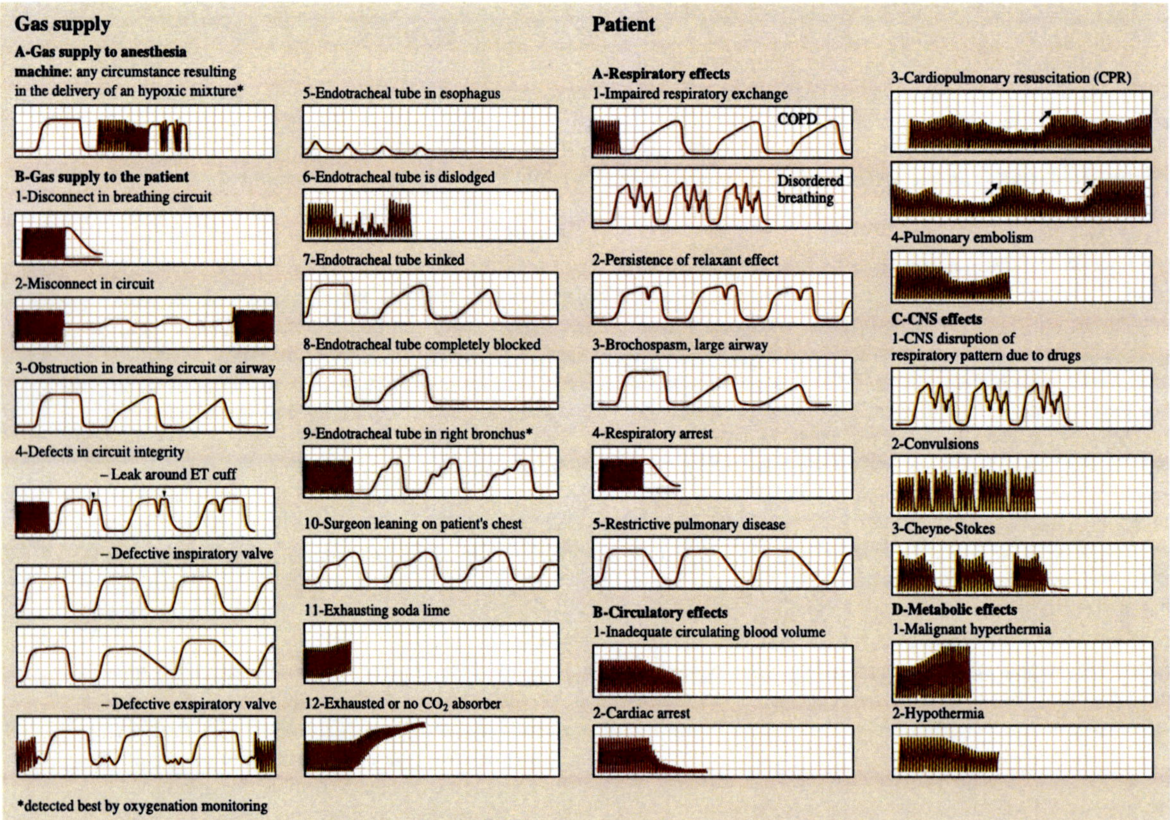

Fig. 49.10 The capnograms show influences of the anaesthesia and of the patient, which are reflected in the respective capnograms (courtesy of Hewlett Packard and Agilent Technologies)

sampling rate (50 ml/min) it can also be used for newborn infants.

Indications for capnography are:

- Monitoring of the ventilation during machine-assisted artificial respiration in anaesthesia and intensive care treatment
- Monitoring during cardiopulmonary resuscitation.

Benefits of capnography are:

- Early detection of an intraoperative air embolism, malignant hyperthermia, disconnection of the tube system and stenosis
- Detection of an obstruction, dead space ventilation, rebreathing, malfunction in the absorber or valve system, tube dislocation during narcosis, and haemodynamic changes (e.g. drop in blood pressure, cardiac arrest)
- Indirect information about the muscle relaxation.

The limitations and sources of errors in capnography are essentially:

- Faults in the respiratory physiology: ventilation–perfusion disproportion, dead space ventilation, shunt augmentation
- Faults of a technical nature: calibration errors, condensate or secretions in the sampling tube, stenosis or bending of the sampling tube
- Methodological limitations: high respiration rate, mixed gas analysis, interference of foreign gases (e.g. nitrous oxide, water vapor)
- Errors on the part of the user: interpretation errors.

Figure 49.10 shows how influences on anaesthesia and on the patient are reflected in the respective capnograms.

Further Reading

- P. Baumbach: Verständnis transkutaner pO_2- und pCO_2-Messungen (Radiometer A/S, Kopenhagen 1988)
- M.B. Jaffe: Infrared measurement of carbon dioxide in the human breath: "Breathe-through" devices from Tyndall to the present day, Anesth. Analg. **107**, 890–904 (2008)
- J.T.B. Moyle: *Pulse Oximetry*, 2nd edn. (BMJ Books, London 2002)
- M. Stücker, U. Memmel, P. Altmeyer: Transkutane Sauerstoffpartialdruck- und Kohlendioxidpartialdruckmessung – Verfahrenstechnik und Anwendungsgebiete, Phlebol. **29**, 81–91 (2000)

References

49.1 Masimo Corporation: Technical Bulletin 3, Pleth Variability Index (PVI) 7577-4583A-0607 (Masimo, Irvine 2007), www.masimo.com

49.2 Masimo Corporation: Technical Bulletin: Signal Extraction Technology, 7005-1035R-0808 (Masimo, Irvine 2008), www.masimo.com

49.3 Masimo Corporation: Acoustic Respiration Rate, 7961-6132A-1209 (Masimo, Irvine 2010), www.masimo.com

49.4 F. Bloos, K. Reinhart: Sauerstoffangebot und Sauerstoffverbrauch. In: *Monitoring in Anästhesie und Intensivmedizin*, ed. by A. Hoeft, H. Metzler, T. Pasch (Springer, Berlin Heidelberg 2008) p. 185

50. Temperature Monitoring

Temperature

Rüdiger Kramme, Ullrich Hieronymi

Temperature measurements are indispensable as a form of standard monitoring in intensive care medicine, emergency medicine, and anesthesia. This chapter covers various aspects of hyperthermia and hypothermia and discusses the measuring sites for temperature measurement as well as temperature sensors and probes.

50.1 **Hyperthermia and Hypothermia**............ 987

50.2 **Measuring Sites for Temperature Measurement**....................................... 987

50.3 **Temperature Sensors and Probes** 987

50.4 **Methodological Notes** 989

Body temperature is a vital sign and is closely related to metabolism. Monitoring of the body temperature is of particular importance in intensive care medicine, because fever elevation of the core temperature above 38 °C is an early sign of sepsis, one of the most common causes of death in intensive care units. The types of fever are important for monitoring of intensive care patients are:

- Continued fever (fluctuations in temperature >1 °C)
- Remittent fever (fluctuations in temperature >2 °C)
- Intermittent fever (fluctuations in temperature >2 °C over the course of a day, with possible drops below normal temperature)
- Undulant fever (constantly changing body temperature which does not follow a trend).

50.1 Hyperthermia and Hypothermia

If the heat input exceeds the heat loss of the body (e.g. in the case of heat stroke, alcohol or drug withdrawal, hyperthyreosis, malignant neuroleptic syndrome or malignant hyperthermia), this state is described as hyperthermia. As a result of an increase in heat loss, a core temperature below 35.5 °C is described as hypothermia (e.g. following surgical interventions, in the case of extracorporeal circulation and haemodialysis procedures and at the onset of sepsis).

50.2 Measuring Sites for Temperature Measurement

The measuring site is crucial for the assessment of body temperature (Table 50.1), because there may be a temperature difference of up to 5 °C between the different areas of the body. There can also be a temperature gradient between the core body temperature and the temperature at the periphery (e.g. shock).

50.3 Temperature Sensors and Probes

Various temperature sensors and probes are available for determining the temperature as a thermic measurement parameter (typical measurement range 24–42 °C, optimum measurement range 10–45 °C, in order also

Table 50.1 Overview of temperature sensors/probes and measuring sites for determining body temperature (Figs. 50.1–50.4)

Localization	Comments
1. Surface body temperature sensors	
Axilla	Roughly 1 °C lower than in the rectum; less suitable for temperature measurements in intensive care units (ICU) and in the operating theater (OT)
Skin	Applied to various parts of the body (e.g., forehead and soles of the feet, among others); temperature values are dependent on the perfusion quality of the skin (Fig. 50.1)
2. Temperature sensors and probes for body orifices	
Sublingual	Less suitable for temperature measurements in intensive care units and in the OT
Rectum	The simplest approximation to the core temperature but not representative, because the temperature is dependent on the perfusion of the intestinal mucosa and on spatial temperature gradients, additionally feces can be taken into account as a temperature isolator (Fig. 50.2)
Nasopharynx	Application: mucosa of the nasopharyngeal space; good conditions for temperature measurement due to the spatial proximity to the internal carotid artery
External auditory canal	Application close to the tympanic membrane; measurement result comparable to esophageal temperature (Fig. 50.4)
3. Intracorporeal temperature probes	
Esophagus	Application in the lower third by means of an esophageal stethoscope with an integrated temperature probe; temperature conditions comparable to those of the brain (corresponds approximately to the core temperature)
Pulmonary artery	Online measurement of the temperature by means of a Swan–Ganz catheter; infusion of cold solutions may interfere with the measurement
Bladder	By means of a Foley catheter; temperature conditions correspond approximately to the core temperature (Fig. 50.3)

to be able to measure extremes of body temperature, such as hypothermia or hyperthermia), which present virtually no problems technically (Table 50.1).

Thermistors (also known as negative temperature coefficient (NTC) resistors) are thermosensitive semiconductor elements whose electrical resistance decreases as the temperature increases. If the resistance changes – even very small temperature changes cause changes on account of the high responsiveness of the sensor – then the voltage in the bridge circuit changes in proportion with the temperature change.

This voltage is amplified and displayed via a calibrated (°C) electrical temperature measuring device. Thermistors are the most used temperature sensors in practice.

Thermoelements are composed of two metals which are connected to one another. One connection point is maintained at a constant temperature, while the second forms the actual temperature sensor. When heated, a difference in temperature arises between the metal connections, and this temperature difference is proportional to the electrical voltage.

Thermal measuring tapes are impregnated with cholesterol crystals whose molecular structure changes as a function of the temperature. Other sensors and probes for determining temperature are:

- Thermal radiation sensors
- Needle thermometers
- Deep-body temperature sensors.

In clinical practice it is widely customary to monitor the temperature difference between the core temperature and a peripheral temperature (surface temperature). Temperature difference is an essential criterion for early detection of the beginning of states of shock, since the temperature at the body peripheries drops more quickly than the core temperature.

Fig. 50.1 Single-use probe for skin surface measurements

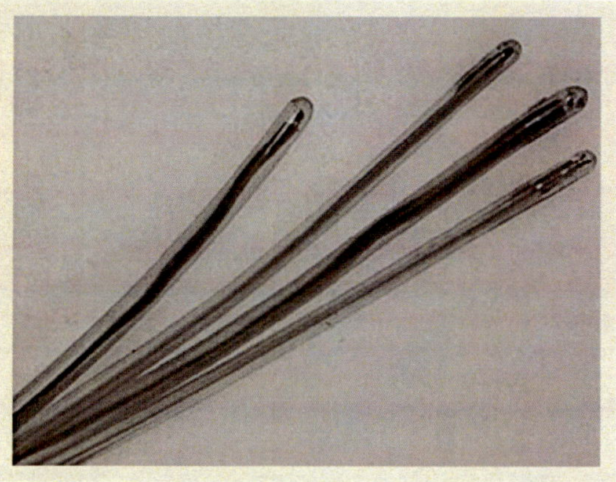

Fig. 50.2 Single-use probes for rectal or esophageal temperature measurements

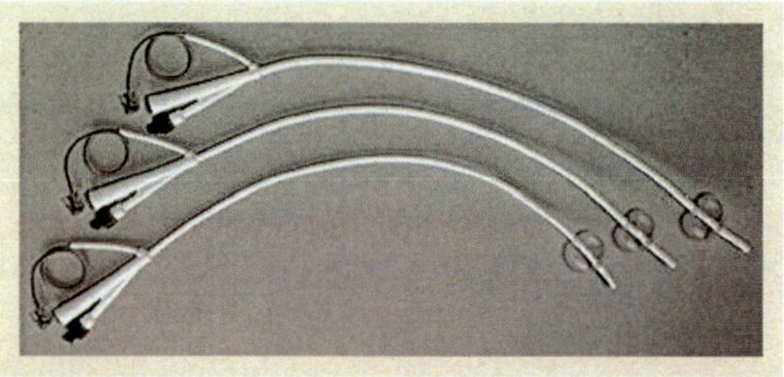

Fig. 50.3 Foley catheter for temperature measurements in the bladder

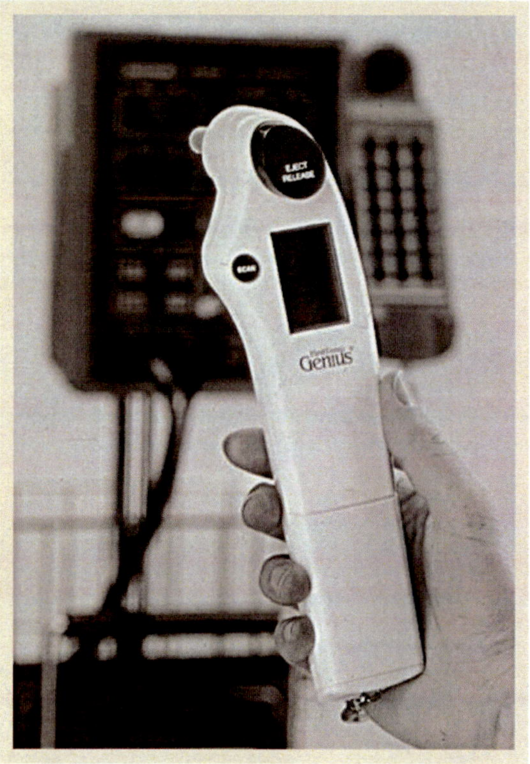

Fig. 50.4 Digital thermometer for measurement in the external auditory canal (courtesy of Johnson & Johnson, USA)

50.4 Methodological Notes

Reusable temperature sensors and probes must not be autoclaved or subjected to hot-air sterilization. Gas (ethylene oxide) or plasma sterilization is possible, however.

Temperature sensors and probes must bear the CE marking. In accordance with the German Medical Devices Operator Ordinance, they are subject to the requirement for metrological checks every two years (for infrared radiation thermometers, the period is one year).

Further Reading

- R.S. Irwin, J.M. Rippe: *Irwin and Rippe's Intensive Care Medicine* (Lippencott Williams Wilkins, New York 2007)
- P.E. Marik: *Handbook of Evidence-Based Critical Care*, 2nd edn. (Springer, Berlin Heidelberg 2010)
- J. Briegel, M. Haller, E. Kilger: Fieber als Leitsymptom in der Intensivmedizin, Intensivmed. Dialog **1**, 1–4 (1994)
- L. Negri, W. Weber: Körpertemperatur. In: *Monitoring in Anästhesie und operativer Intensivmedizin*, ed. by L. Negri, G. Schelling, U. Jänikke (Abbott, Wiesbaden 1989), Sects. 7.1, 7.8
- J.F. Zander: Fiebersenkende Maßnahmen in der Intensivtherapie, Intensivmed. Dialog **1**, 5 (1994)

51. Cerebral Monitoring

Barbara Schultz, Arthur Schultz, Harald Kronberg

In this chapter, two kinds of cerebral monitoring are presented: electroencephalography (EEG) monitoring and intracranial pressure (ICP) monitoring. EEG monitoring is used in the operating room to assess the depth of hypnosis during anesthesia and to detect effects of hypoxia and of induced hypothermia. In intensive care unit patients, EEG monitoring can be used for the control of sedation, for therapy control, e.g. in status epilepticus, for the assessment of the patient's current clinical status and trends thereof, and as an exploratory diagnostic tool with respect to epileptiform activity and focal brain disorders. Hypnotic drug effects are accompanied in a dose related manner by a slowing of the EEG. These EEG changes can be classified automatically. Especially through the automatic interpretation, EEG monitoring can be carried out as a routine method for patient monitoring with little effort. The aim of detecting and treating elevated ICP is to avoid secondary damage to the brain. The most common indication for ICP monitoring is trauma to the head. Methods for ICP measurement with higher invasiveness include intraventricular, intraparenchymal, or subdural catheter localisations. A less invasive method is epidural cerebral pressure measurement.

51.1 **EEG Monitoring** 991
51.1.1 Kinds of EEG Activity 991
51.1.2 Spectral Analysis 992
51.1.3 Other Parameters 992
51.1.4 EEG Stages of Anesthesia and Sedation 992
51.1.5 Course of EEG Stages During Anesthesia 992
51.1.6 Technical Requirements 994
51.1.7 Benefit of EEG Monitoring in the Operating Room 994
51.1.8 Benefit of EEG Monitoring in the Intensive Care Unit 995
51.2 **Intracranial Pressure** 995
51.2.1 The Physiology of Intracranial Pressure 996
51.2.2 Intracranial Pressure Measurement to Monitor Treatment 997
References 1000

51.1 EEG Monitoring

The electroencephalogram (EEG), a record of the electrical activity of the brain, can be used for patient monitoring in the operating room and in the intensive care unit. Indications for the use of EEG monitoring in the operating room are the assessment of hypnotic drug effects, the early detection of the effects of hypoxia, and the documentation of the cerebral effects of induced hypothermia. In the intensive care unit, EEG monitoring can be used for the control of sedation, for therapy control, e.g., in status epilepticus, for the assessment of a patient's current clinical status and trends thereof, and as an exploratory diagnostic tool with respect to epileptiform activity and focal brain disorders.

51.1.1 Kinds of EEG Activity

Attenuation of brain function is mostly accompanied by a slowing of the EEG. The frequency composition of the EEG can therefore be used to support the assessment of the severity of alterations or impairments of brain function. Furthermore, EEG potentials with a specific shape, e.g., epileptiform potentials, are of interest in EEG monitoring.

In principle, generalized and focal EEG alterations have to be distinguished. Generalized EEG alterations have a similar representation over different areas of the head. Typical examples are hypnotic drug effects, and alterations of the EEG caused by hypothermia, hypoglycemia or global hypoxia. Focal EEG changes are restricted to a circumscribed area of the head; they can be caused, e.g., by tumors, bleeding or localized hypoxia.

Interpretation of conventional multichannel EEG recordings takes considerable practical effort. Nevertheless, generalized EEG changes can be assessed on the basis of a single EEG channel with high reliability. This is an important precondition for the use of the EEG as a monitoring method.

51.1.2 Spectral Analysis

The conventional visual assessment of the EEG requires expert knowledge and experience. Therefore, it is a considerable simplification for the user in the operating room and in the intensive care unit if the EEG is analyzed automatically by means of computer-based methods and if support for interpretation is given.

Spectral analysis is a method for signal evaluation. The signal is subdivided into segments, which are analyzed with regard to their frequency components.

The result is a power spectrum, which can be displayed graphically and can be used as a basis for further evaluations. Single spectral parameters can be extracted, e.g., the 50% and the 90% (or 95%) quantile of the spectrum, which are also called the median and the spectral edge frequency. Other commonly used spectral parameters are the total power and the power in the different frequency bands. The following frequency bands are distinguished: delta (δ: 0.5–3.5 Hz), theta (θ: 3.5–7.5 Hz), alpha (α: 7.5–12.5 Hz), and beta (β: > 12.5 Hz). Such monoparameters are often able to show general trends within the EEG, but the complex information contained in an EEG cannot be described comprehensively or adequately by monoparameters.

51.1.3 Other Parameters

For the purpose of EEG monitoring in the operating room and in the intensive care unit, also other mathematical and statistical parameters are used.

An example is the bispectrum, which includes information from spectral analysis as well as information about phase coupling. Bispectral analysis is a component of the functions that are used by monitors which were developed by Aspect Medical Systems (Natick, MA, USA) [51.1, 2].

Entropy is a measure of the order or disorder, thus randomness or predictability, of a system. Spectral entropy is the basis of the algorithm which is implemented in monitoring units of GE Healthcare (Helsinki, Finland) [51.3–5].

Autoregressive parameters can be used to describe the course of a raw EEG. In a regression model, a given value on a time axis is calculated as the weighted sum of one or more past values and a random term. Autoregressive and other parameters are used by Narcotrend (MT MonitorTechnik, Bad Bramstedt) [51.6, 7].

A review of different methods for EEG analysis and of different monitors is given by *Jameson* and *Sloan* [51.8].

51.1.4 EEG Stages of Anesthesia and Sedation

Hypnotic drug effects are accompanied in a dose-related manner by a progressive slowing of the EEG. Criteria to classify these EEG changes visually into the stages A (awake) to F (very deep anesthesia/sedation) (Fig. 51.1) were proposed by *Kugler* [51.9].

A scale from A to F which was modified for the purpose of EEG monitoring served as the basis for the development of an automatic EEG classification. It was realized by means of a multivariate approach using special pattern recognition algorithms and is a component of the EEG monitor Narcotrend, which was developed for use in the operating room and in the intensive care unit. For the assessment of the hypnotic component of anesthesia and sedation with this device, one EEG channel is used as a standard. For this, three electrodes have to be attached to the patient's forehead. Depending on the kind of surgery performed or if injuries to the head exist, other electrode types (e.g., needle electrodes) or electrode positions can be used. Two-channel recordings are also possible.

The Narcotrend analyzes the EEG in real time; the assessments are updated every second. The central part of the main screen of the Narcotrend is depicted in Fig. 51.2.

51.1.5 Course of EEG Stages During Anesthesia

As an example for the use of automatic EEG classification, Fig. 51.3 shows a course of EEG stages recorded during anesthesia with the hypnotic propofol and the opioid remifentanil. Due to their short half-lives, the dosing of both drugs can be adapted easily and quickly. Both were given in an intravenous and continuous manner using perfusors. After the induction from stage A to stage E/F, anesthesia was maintained in the deep

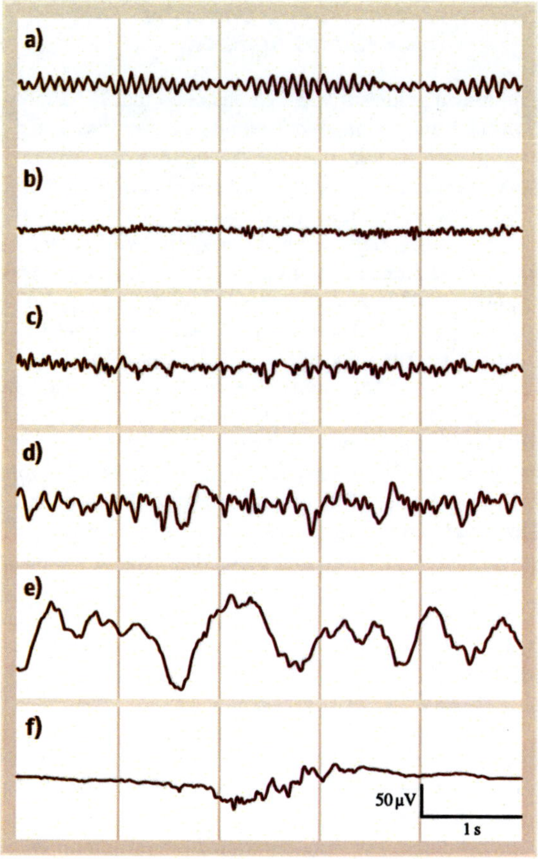

Fig. 51.1 Typical examples for EEG stages A–F

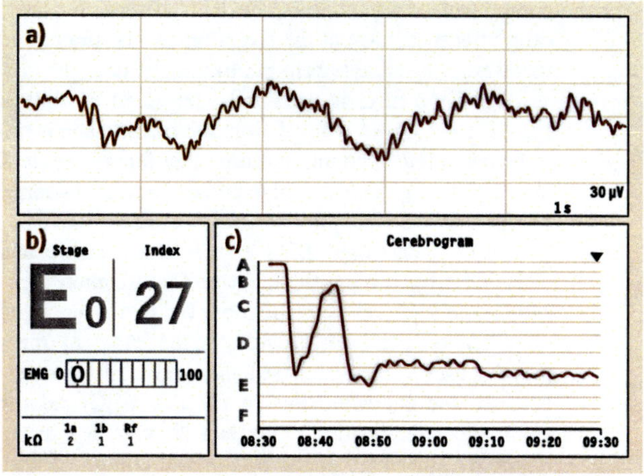

Fig. 51.2a–c A part of the main screen of the Narcotrend. **(a)** Raw EEG, **(b)** current stage (E_0) and index (27) value, EMG index (0), and electrode impedances. **(c)** Course of the EEG stages (cerebrogram). After the hypnotic (propofol) for induction of anesthesia was given to the patient, there was a quick transition from the awake state (stage A) to a deep hypnotic stage (stage E/D, ≈ 8.35 h), the muscle relaxant was given, and the patient was intubated. Due to the short half-time of the hypnotic, within a few minutes stage B was reached again (≈ 8.40 h). A propofol bolus (≈ 8.43 h) caused a deepening to level D/E, which was maintained during the steady state. The opioid used for induction and maintenance of anesthesia was remifentanil

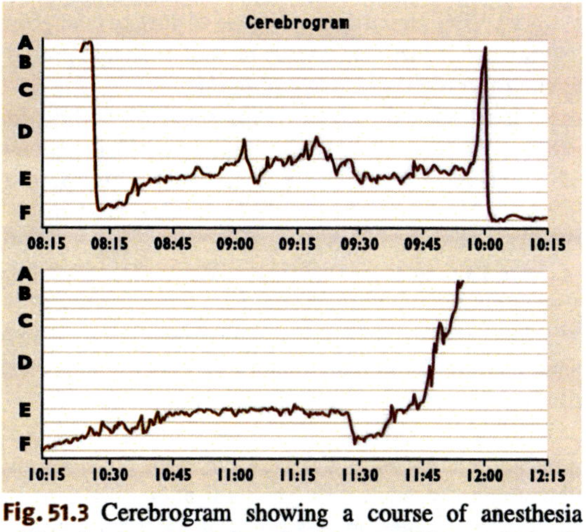

Fig. 51.3 Cerebrogram showing a course of anesthesia with a short unexpected awakening caused by a technically determined interruption of the drug supply (≈ 10.00 h)

stages D/E. At approximately 10:00, a sudden shift to B stages was noticed. This was caused by a short interruption of the drug supply due to technical reasons. When anesthesia is maintained in such light stages the possibility of acoustic perceptions is increased. Subsequently, anesthesia was maintained in stage E. At about 11:30, emergence from anesthesia began, indicated by continuous lightening of the EEG stages from stage E to the awakening in stage A/B. Postoperatively the patient did not remember any intraoperative events.

The dosage of the hypnotic for maintenance of anesthesia was titrated individually according to the EEG in this patient, therefore it was possible to avoid longer periods of over- and underdosage. With the help of the EEG, the short, unwanted lightening of anesthesia could be quickly recognized and immediately corrected.

Parameters for patient monitoring during anesthesia customarily used in daily clinical practice are, among others, blood pressure and heart rate. Especially during intravenous anesthesia, cardiovascular parameters are not a reliable basis for the assessment of the hypnotic component of anesthesia. The EEG detects hypnotic drug effects directly at the target organ, the brain, and therefore is a valuable help in titrating the dosage.

When the EEG is used and interpreted in clinical practice, the whole clinical situation of the patient has to be taken into account, because other factors, such as hypoxia, hypoglycemia, and hypothermia, can influence the EEG in a manner similar to anesthetics/sedatives.

Visually classified EEG epochs are a stable and clear basis for the development of classification algorithms and support easy validation. Using a visual EEG classification as a basis provides methodological transparency for the user, because a clear relationship exists between the characteristics of the raw EEG and the automatic classification. This can be regarded as an important advantage of the Narcotrend compared with other concepts for automatic EEG classification which do not comprise a defined relationship between the raw EEG and mathematical/statistical parameters, e.g., indexes.

51.1.6 Technical Requirements

In practical use, the following technical details of the monitors have proved to be important: The quality of the EEG signal and, as a consequence, the validity of the subsequent automatic analyses depend on a constantly good electrode–skin impedance. Therefore, it is necessary to carry out impedance checks frequently.

The EEG signal can easily be disturbed, therefore automatic artifact detection is an indispensable prerequisite for further processing of the signal.

Visual analysis has been the standard method for EEG assessment for decades, and it is still indispensable in clinical practice (e.g., assessment with regard to artifacts, identification of special potentials, e.g., epileptiform potentials). Therefore, it is necessary to show the raw EEG in a very clear way with different scaling options.

Age-related changes of the EEG have to be considered when the EEG is automatically classified and when it is displayed.

Different operating fields, patient positions, and kinds of injuries require variable electrode positions and the possibility to use different kinds of electrodes.

For an exploratory examination for the purpose of comparing EEG signals from both hemispheres, at least two channels should be available.

For documentation purposes it should be possible to make printouts of raw EEGs and of courses of spectral parameters as well as of classifications.

51.1.7 Benefit of EEG Monitoring in the Operating Room

In clinical studies it was demonstrated that EEG monitoring leads to improved quality of anesthesia, e.g., with regard to individualized dosing of drugs, avoiding awareness situations, and to fast postoperative recovery. This shall be illustrated by some examples.

The high interindividual variability with regard to dosing requirements of drugs with hypnotic action was demonstrated in a multicenter study comprising 4,630 patients [51.10]. Additionally, it was shown that EEG monitoring supports age-adjusted dosing [51.11]. In a subgroup in this study, the EEG was recorded in a blinded manner, i. e., the classification results were not displayed to the attending anesthetist. During the steady state of anesthesia, only 68.8% of the patients of this subgroup had an EEG corresponding to the medium to deep Narcotrend stages D/E. The remaining courses of anesthesia were either very light, very deep or unstable. Postoperatively, 2 out of 603 patients with blinded EEG registrations reported intraoperative awareness [51.12]. On the contrary, there was no intra- or postoperative indication of awareness in the group with maintenance of anesthesia in stages D/E.

When data from 32 study sites were analyzed it was shown that EEG monitoring caused changes of the propofol doses from −28.4 to +86.2% in the different centers [51.13].

In studies with EEG-controlled propofol/remifentanil anesthesia it was demonstrated that EEG monitoring supports gender-adapted dosing. Women had higher propofol requirements but, nevertheless, awake more quickly from anesthesia [51.13, 14].

EEG monitoring supports dosing by means of target-controlled infusion (TCI). TCI pumps incorporate dosing algorithms which are based on pharmacokinetic models. Deviations between calculated drug target levels and measured drug levels were described [51.15]. By means of the EEG, the propofol dosage can be adjusted to individual requirements, taking into account the opioid dosage as well [51.16].

In another study, neurosurgical patients undergoing brain tumor operations were monitored intraoperatively with the Narcotrend. These patients, with Narcotrend stages D/E during the steady state of anesthesia, received only half of the propofol dose which was given to patients without EEG monitoring. Most of the patients with EEG monitoring could already be extubated in the operating room. This facilitated neurological assessment immediately after the operation [51.17]. The patient care requirements in the intensive care unit were reduced.

By means of EEG, very short recovery times were achieved in patients after thoracic surgery. This resulted in more effective use of the resources in the operating room [51.18].

In a study of *Rundshagen* [51.19], patients with EEG monitoring during anesthesia showed faster postoperative recovery of their psychomotor functions than patients without EEG monitoring. According to the authors' interpretation, EEG monitoring improves the adequate control of the intraoperative depth of anesthesia, causing a reduction of the signs of drug accumulation and improved psychomotor function in the early postoperative period.

In studies of *Weber* [51.20, 21] it was demonstrated that anesthetics can be dosed individually by means of EEG.

In children undergoing cochlear implant surgery the EEG helped to maintain reliably very light EEG stages during a short period in which the implant was tested. The modification of brain function, and consequently of acoustic processing, by drugs with hypnotic action should be as small as possible during this period to enable the use of the measurement results as a basis for postoperative individual fitting of the speech processor [51.22].

51.1.8 Benefit of EEG Monitoring in the Intensive Care Unit

Including the EEG as a method for patient monitoring enables, not only in the operating room but also in the intensive care unit, more differentiated assessment of the patient in situations with limited possibilities for clinical examination and assessment. This applies to patients having received muscle relaxants and to persons in deep stages of sedation or coma with weak remaining reactions to external stimuli.

A 2 year comparison with regard to the length of stay of ventilated patients in an anesthesiological and surgical intensive care unit demonstrated that, in a year with EEG examinations carried out at regular intervals, the length of stay was 24.4% shorter than in a year without EEG examinations [51.23].

Experience shows that especially geriatric patients, who often suffer from preexisting illnesses and have limited compensation potential, benefit from EEG monitoring during sedation. The risk of secondary infections due to unnecessarily long ventilation times is reduced.

Control of sedation is an important indication for EEG monitoring in patients in the intensive care unit. Besides this, there are numerous other applications supporting diagnosis and therapy, e.g., during the therapy of increased intracranial pressure, in the diagnosis and therapy of epileptic disorders, in the assessment of the current state and the course of the disease of comatose patients, and in the exploratory examination of focal brain disorders [51.12, 24–27].

In summary, it can be concluded that, due to its modern technical implementation and especially through automatic interpretation, EEG monitoring can be carried out as a routine method for patient monitoring with little effort. It is a valuable addition to the monitoring methods used to date. The use of EEG monitoring results in considerable improvement of the quality of anesthesia and intensive care.

51.2 Intracranial Pressure

Space-occupying processes in the brain can lead to dangerous increases in intracranial pressure (ICP) if the compensation capacity of the homeostatic regulatory mechanisms concerned is exhausted. Elevated intracranial pressure causes cerebral hypoxia as a result of inadequate blood perfusion, because the drop in pressure between the systemic arterial pressure (SAP) and the ICP (CPP; cerebral perfusion pressure) is reduced. Ultimately, these underlying space-occupying processes carry the risk of life-threatening brain stem herniation with respiratory paralysis.

Space-occupying processes can proceed slowly (tumor), develop over the course of hours or days (cerebral edema) or can also appear suddenly (massive hemorrhage). There are also wave-like pathological ICP increases which are probably connected to fluctuations in the cerebral blood volume.

The most common indication for ICP monitoring is trauma to the head. This is because the usual emergency treatment of the patient with artificial respiration, sedation, and relaxation conceals the clinical symptoms of ICP increases. There is no prophylactic therapy for elevated intracranial pressure without long-term side-effects. Treatment of these patients must therefore be carried out under constant monitoring of the intracranial pressure, usually for a period of 3–6 days.

Various methods have been introduced for ICP measurement, and these can essentially be divided into two groups of different invasiveness. Invasive methods, which have an increased risk of infection when longer-term monitoring is necessary, are direct catheter measurement of the intraventricular pressure (gold standard), the intraparenchymal pressure or the subdural pressure; a less invasive method, whose risk of infection can be well controlled but which suffers from frequent measurement artifacts, is epidural cerebral pressure measurement. Lumbar pressure measurement with the possibility of drawing off fluids, a method which sometimes used to be preferred, is now no longer recommended owing to the risk of brain stem herniation. The fact that the significance of and indication for cerebral pressure measurement are barely controversial any longer is evident from the large number of publications in the last 35 years. Although measurement of cerebral pressure has been practiced for around 100 years, with the introduction of the lumbar puncture by Quincke in clinical research, ICP monitoring was finally first established in the clinical routine of cerebral monitoring with the classic study by *Lundberg* [51.28] and the definition of intracranial pressure waves which he introduced. The article by *Allen* [51.29] is recommended for an overview of the physiology and measurement technique, a comprehensive account including examples of measurements can be found in *Gaab* and *Heissler* [51.30], while *Steinbereithner* [51.31] gives a brief overview of treatment for head injuries with an evaluation of intracranial pressure measurement. *Betsch* [51.32] deals specifically in his dissertation with the technical problems of epidural ICP measurement, which is preferred in routine clinical practice in Germany. The cited works include comprehensive information on further reading, and no quotations are therefore given below.

51.2.1 The Physiology of Intracranial Pressure

The central nervous system (CNS) is suspended in the cerebrospinal fluid (CSF) and is enclosed by the dura mater. After closure of the fontanelles, the cranium forms a solid shell around the brain, while the lumbar sac, which protects the spinal cord, has a limited degree of elasticity. The adult brain contains approximately 1500 ml of parenchyma and 120 ml each of blood and CSF. The interstitial fluid makes up 10–15% of the weight of the brain. CSF is constantly formed in the ventricles as a blood filtrate (≈ 500 ml/day) and flows into the subarachnoid space, a layer between the CNS and dura mater, which is a few millimeters thick. The CFS is finally resorbed by the venous circulatory system by means of pressure-sensitive valve-like villi (Pacchioni's granulations). The result of production and resorption is a normal intracranial pressure of 5–15 mmHg. Since the production of the CSF is for the most part independent of the ICP, the venous pressure essentially determines the average intracranial pressure.

Space-occupying processes in one of the brain's components, e.g., in the case of edema, require a reciprocal change in volume with at least one of the other components if the intracranial pressure is to remain constant. Up to an ICP of around 15 mmHg, homeostatic regulatory mechanisms in these cases prevent hypoperfusion of the brain. CSF can first be forced into the extensible lumbar sac, and if further space is needed more CSF is also resorbed into the venous system. Finally, the venous system can also be compressed and the volume of blood in the brain thus reduced. Once these compensation steps are exhausted, with an ICP which increases quickly there is the threat of herniation of the cerebral tissue in the tentorial notch. Figure 51.4 shows a schematic representation of this pressure–volume relationship.

If the homeostatic regulation is exhausted (ICP > 15 mmHg) or defective, then the cerebral blood perfusion and thus the supply of oxygen to the brain follow the pressure difference between arterial inflow and venous drainage. This pressure difference largely corresponds to the cerebral perfusion pressure CPP = SAP –

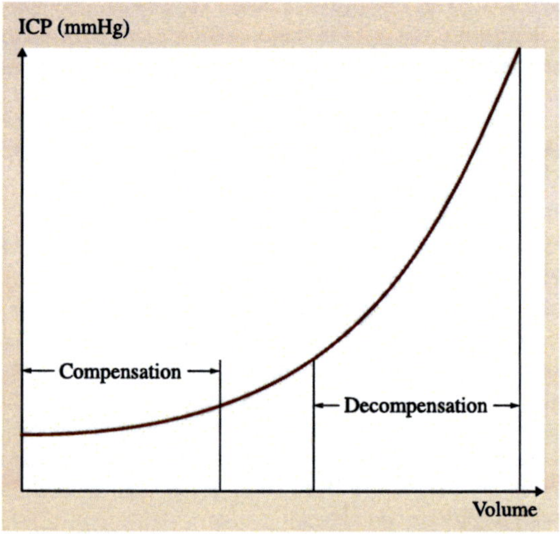

Fig. 51.4 Intracranial pressure–volume characteristic (schematic)

Table 51.1 Intracranial pressure ranges in adults

Status	Pressure (mmHg)
Normal	5–15
Slightly elevated	15–20
Requiring treatment	20–40
Very serious elevation	> 40

Table 51.2 Perfusion pressure (CPP) ranges

Status	Pressure (mmHg)
Normal	50–150, typical 80
Hypoperfusion	< 50

ICP as the average difference between the systemic arterial pressure and the intracranial pressure.

Therapeutic measures to lower the intracranial pressure must not lower the CPP to below 50 mmHg, and the CPP must therefore always be monitored in addition to the ICP. The normal intracranial pressure is not static but is affected by dynamic components which stem from arterial blood pressure waves and from fluctuations in the respiratory pressure which are conveyed by the venous system. The ICP drops during spontaneous inspiration and increases again during expiration (Fig. 51.5).

In the case of normal intracranial pressure, the pressure curve of individual ICP pulses still exhibits a marked fine structure which changes and gradually disappears as the ICP increases (Fig. 51.6). A review of the clinical relevance of this phenomenon is still the subject of research, however.

Pathological values for the intracranial pressure frequently appear as waves. Plateau or A-waves are ICP increases to a level of 50–100 mmHg for a period of 5–20 min. A-waves are obviously linked to phases of

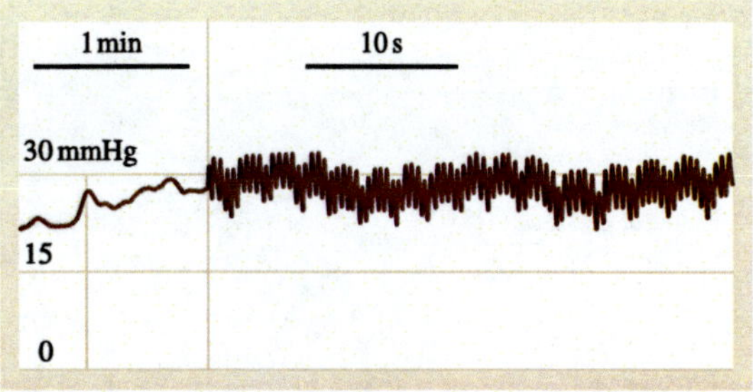

Fig. 51.5 Dynamic components of the intracranial pressure of an infant with hydrocephalus

cerebral vasodilation and therefore to hyperemia. B-waves are sudden ICP leaps from normal values to values in the range 30–60 mmHg with a frequency of approximately one leap per minute. The waves are frequently observed in head trauma patients and are caused by problems with CSF circulation and likewise by fluctuations in the cerebral blood volume. C-waves are pressure waves with a low amplitude of up to 20 mmHg and frequencies of about 6 per minute. They correlate with Traube–Hering–Mayer waves in the arterial blood pressure.

Intracranial space occupation easily leads to circulation problems or even tissue herniation in the region of the septal foramina, particularly in the case of faster processes. Pressure gradients then appear between the craniospinal compartments, which can be an indication for ICP monitoring at numerous locations, and the ICP should likewise be monitored in the region of the lesion.

51.2.2 Intracranial Pressure Measurement to Monitor Treatment

The aim of treating elevated intracranial pressure is to avoid secondary damage to the CNS. It is vital to

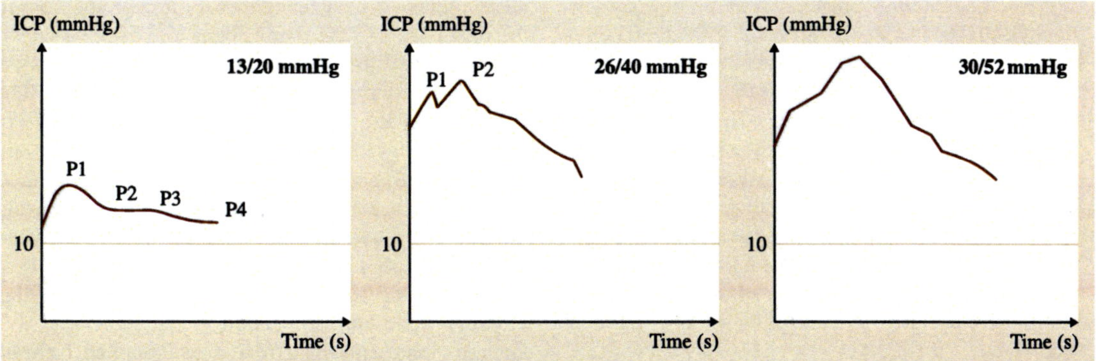

Fig. 51.6 Fine structure of the ICP pulse curve, dependent on pressure

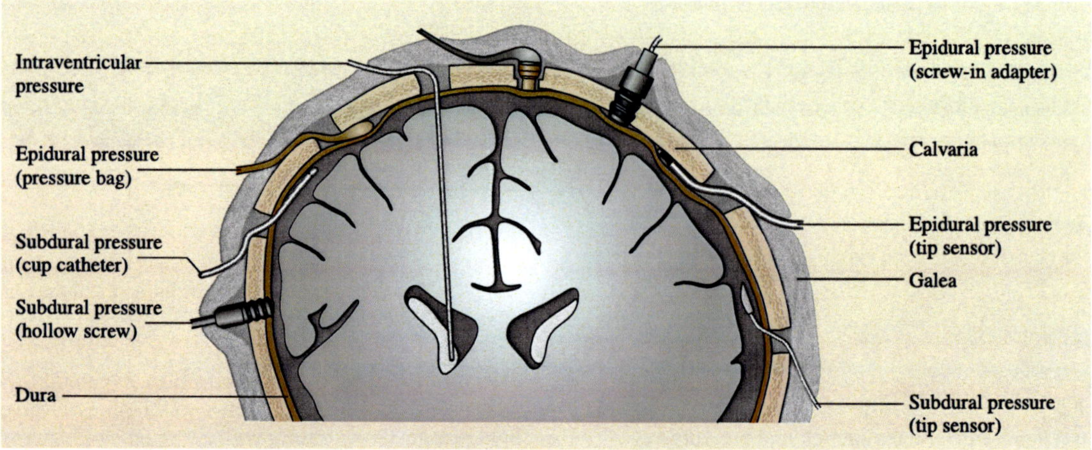

Fig. 51.7 Technical options for clinical measurement of intracranial pressure

take into account the position of the head and neck, as this can have a significant influence on the cerebral venous blood volume and thus on the ICP. A slightly elevated head position reduces the cerebral venous pressure and ICP. Hyperventilation leads, by means of the hypocapnia which accompanies it, to vasoconstriction with a reduction in the intracranial blood volume. In the event of longer hyperventilation, however, the reduced blood perfusion carries the risk of hypoxia, which can result in cerebral edema, increasing the ICP. Dehydrating and diuretic medications for osmotherapy are only suitable for acute lowering of an elevated ICP, but are not suitable for treatment of cerebral edema due to undesirable side-effects. Treatment with barbiturates with a decrease in the cerebral metabolism reduces cellular oxygen consumption and provides protection from long-term consequences of hypoxia. High doses of barbiturates over the longer term reduce the arterial blood pressure, however, with the risk of a perfusion pressure CPP which is too low. There is therefore no long-term prophylactic treatment for elevated intracranial pressure. The various forms of treatment should always be carried out while monitoring intracranial pressure. In the case of head trauma, the first 48 h following injury are particularly critical, and in most cases the ICP need not be monitored for longer than 1 week. Various studies have shown that the introduction of intracranial pressure monitoring has lowered mortality rates by around 40%, without increasing morbidity. The modern state-of-the-art methods for intracranial pressure measurement are still invasive but differ in terms of their risk of infection (Fig. 51.7):

1. Measurement of the intraventricular, intraparenchymal or subdural pressure via a catheter
2. Measurement of the epidural pressure using miniature pressure transducers or by means of pneumatic sensors.

Unusually high requirements must be placed on the zero-point stability of the sensors (drift ≤ 2 mmHg/day) since the ICP itself is a low-pressure signal. However, many sensors are less stable following application, particularly because of humidity drift, and therefore require regular zero-point adjustment. If the fine structure of the pulse waves is of interest, the measurement system must have a bandwidth of around 20 Hz (40 Hz in infants), but a bandwidth of around 3 Hz is sufficient in order to record the ICP pulse amplitude with adequate dynamics.

All methods require trepanation for the application of sensors. In order to perform intraventricular ICP measurement, a catheter is passed through the hole in the meninges and the parenchyma and into the lateral ventricles, and the pressure is usually recorded externally as in the case of blood pressure measurement. The external pressure sensor must be at the precise level of the lateral ventricles in order to avoid hydrostatic errors (1 mmHg = 13 mm fluid column). This method is the gold standard and has the advantage that acute increases in intracranial pressure can often be treated by draining off CSF. Problems include the fact that, in the case of advanced space occupation and thus a narrow ventricle, application of the catheter is difficult to impossible and that a well-applied catheter

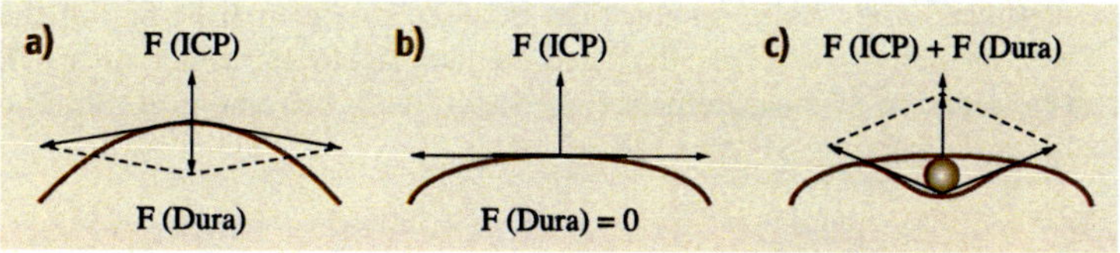

Fig. 51.8a–c The principle of applanation tonometry. **(a)** Free dural balloon, which envelops a positive pressure volume. Membrane in equilibrium between the force pushing out F(ICP) and the resulting counteracting force from the dural tension F(Dura). **(b)** Rigid sensor-measurement membrane applanates the dura, and the dural tension forces offset each other [F(Dura) = 0], so that only the force F(ICP) which corresponds to the ICP remains effective for the measurement. **(c)** Dural inhomogeneities, bone fragments or blood clots introduce a dural tension component F(Dura) which overlaps with the intracranial pressure force F(ICP) with the effect of increasing the measurement value

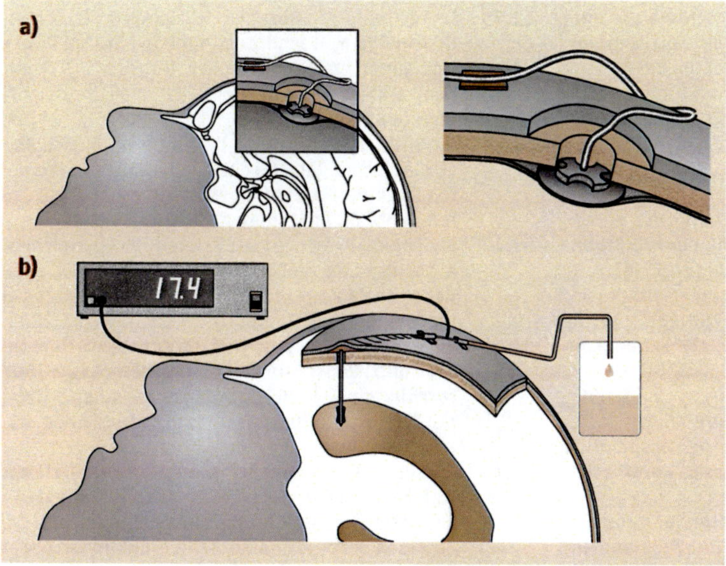

Fig. 51.9 **(a)** Epidural application of a pneumatic single-use sensor, **(b)** intraventricular pressure measurement

can become blocked; above all, the risk of infection in the immunocompromised intracerebral space increases considerably after about 2 days. Infections creep in via the oscillating fluid column in the catheter (pulse wave and respiratory wave), in which bacteria migrate against the irrigating flow and against the force of gravity at approximately 1.5 m/24 h. Epidural measurement of intracranial pressure (2) on the closed dura mater has therefore become largely prevalent in routine clinical practice as opposed to methods (1). If infections nevertheless occur (rarely), they are usually easy to control. Although this procedure also requires a burr hole in the calvaria, it is therefore considered to be a comparatively low-invasive method. The problem with this method is that epidural application of sensors requires increased surgical attention in order to avoid dural artifacts. This is often neglected in routine practice, resulting in erroneously elevated epidural ICP values. Pressure measurement on the closed dura is based on the principle of applanation tonometry, which also makes ophthalmotonometry possible. The pressure sensor itself flattens the dura at the application site, so that the dural tension can no longer have any influence on the result of the measurement.

Figure 51.8 provides a schematic representation of the principle, including the effects of measuring errors as a result of bone fragments or blood clots on the dura, which have the effect of increasing the measurement value in a William Tell effect by superimposing components of the dural tension with the pure pressure signal. Prerequisites for correct epidural ICP measurement are therefore: to apply the burr hole on the side of the lesion; to detach the dura with a blunt hook in the region of ≈ 10 mm from the internal lamina of the skull so that it can rest against the measurement membrane without any tension on the edges; and careful hemostasis in the area of the application site (bone wax, etc.) and rinsing of the dura. It is therefore an advantage if the operating surgeon can view the application site directly (as opposed to application of the sensor between the dura and the internal lamina of the skull). By way of example, Fig. 51.9a shows the technique of epidural intracranial pressure measurement using a pneumatic system. The sensors are single-use sensors which pneumatically transmit the pressure in the sensor head via a tube which is similar to a catheter. There are three types of sensors for this intracranial pressure measurement system: for application concentrically in the burr hole, for insertion between the internal lamina of the skull and the detached dura, and for intraventricular ICP measurement (Fig. 51.9b), with the option of draining CSF via a second lumen.

References

51.1 J. Bruhn: BIS-Monitor. In: *Überwachung der Narkosetiefe*, 2nd edn., ed. by W. Wilhelm, J. Bruhn, S. Kreuer (Ärzte-Verlag, Köln 2006) pp. 89–107

51.2 J.W. Johansen: Update on bispectral index monitoring, Best Pract. Res. Clin. Anaesthesiol. **20**, 81–99 (2006)

51.3 B. Bein: Entropy, Best Pract. Res. Clin. Anaesthesiol. **20**, 101–109 (2006)

51.4 J. Bruhn: Entropie. In: *Überwachung der Narkosetiefe*, 2nd edn., ed. by W. Wilhelm, J. Bruhn, S. Kreuer (Ärzte-Verlag, Köln 2006) pp. 53–63

51.5 J. Bruhn: Datex-Ohmeda Entropie-Monitor. In: *Überwachung der Narkosetiefe*, 2nd edn., ed. by W. Wilhelm, J. Bruhn, S. Kreuer (Ärzte-Verlag, Köln 2006) pp. 133–145

51.6 S. Kreuer, A. Biedler, R. Larsen, S. Schoth, W. Wilhelm: Narcotrend monitoring allows faster emergence and a reduction of drug consumption in propofol-remifentanil anesthesia, Anesthesiol. **99**, 34–41 (2003)

51.7 B. Schultz, S. Kreuer, W. Wilhelm, U. Grouven, A. Schultz: The Narcotrend monitor. Development and interpretation algorithms, Anaesthesist **52**, 1143–1148 (2003)

51.8 L.C. Jameson, T.B. Sloan: Using EEG to monitor anaesthesia drug effects during surgery, J. Clin. Monit. Comput. **20**, 445–472 (2006)

51.9 J. Kugler: *Elektroenzephalographie in Klinik und Praxis* (Thieme, Stuttgart 1981)

51.10 W. Wilhelm, S. Kreuer, R. Larsen, Narcotrend-Study Group: Narcotrend EEG monitoring during total intravenous anaesthesia in 4.630 patients, Anaesthesist **51**, 980–988 (2002)

51.11 A. Schultz, U. Grouven, I. Zander, F.A. Beger, M. Siedenberg, B. Schultz: Age-related effects in the EEG during propofol anaesthesia, Acta Anaesthesiol. Scand. **48**, 27–34 (2004)

51.12 B. Schultz, N.A. Büttner, G. Schönberg, C. Bezler, A. Schultz: EEG-based control of anaesthesia: analysis with regard to an EEG-adapted titration of propofol, Klin. Neurophysiol. **37**, 1–5 (2006)

51.13 M. Willig, B. Schultz, T. Kneif, A. Schultz: The impact of EEG monitoring on the dosing practice during intravenous anaesthesia – a multicentre analysis, Klin. Neurophysiol. **41**, 28–32 (2010)

51.14 K. Haensch, A. Schultz, T. Krauß, U. Grouven, B. Schultz: Women need more propofol than men during EEG-monitored total intravenous anaesthesia, Biomed. Technol. **54**, 76–82 (2009)

51.15 J. Fechner, S. Albrecht, H. Ihmsen, R. Knoll, H. Schwilden, J. Schüttler: Prädiktivität und Präzision einer "target-controlled infusion" (TCI) mit dem System "Disoprifusor TCI", Anaesthesist **47**, 663–668 (1998)

51.16 N. Büttner, B. Schultz, U. Grouven, A. Schultz: EEG-adjusted target-controlled infusion. Propofol target concentration with different doses of remifentanil, Anaesthesist **59**, 126–134 (2010)

51.17 B. Schultz, A. Schleppers, T. Kneif, D. Scheinichen, A. Schultz: Effects of EEG monitoring, age, and gender on the dosage of propofol during neurosurgical procedures, Klin. Neurophysiol. **39**, 189–193 (2008)

51.18 K. Schulze, G. Kraus, R.M. Nordhausen Bartlog, A. Schultz: DRGs: Fast-Track-Anästhesie in der Thoraxchirurgie – total intravenöse Anästhesie (TIVA) mit EEG-Monitoring, Anasthesiol. Intensivmed. **45**, 746–747 (2004)

51.19 I. Rundshagen, I. Dressler, N. Thattamanil, T. Hardt, T. Fritzsche, C.D. Spies, W.J. Kox: Hat die EEG-gestützte Narkoseführung einen Einfluss auf die Erholung der psychomotorischen Funktionen in der frühen postoperativen Phase?, Abstracts DAC 140 (2004)

51.20 F. Weber, H. Hollnberger, M. Gruber, B. Frank, K. Taeger: The correlation of the Narcotrend Index with endtidal sevoflurane concentrations and hemodynamic parameters in children, Paediatr. Anaesth. **15**, 727–732 (2005)

51.21 F. Weber, F. Pohl, H. Hollnberger, K. Taeger: Impact of the Narcotrend Index on propofol consumption and emergence times during total intravenous anaesthesia with propofol and remifentanil: A clinical utility study, Eur. J. Anesthesiol. **22**, 741–747 (2005)

51.22 B. Schultz, F.A. Beger, B.P. Weber, O. Niclaus, E. Lüllwitz, U. Grouven, A. Schultz: Influence of EEG monitoring on intraoperative stapedius threshold values in cochlear implantation in children, Paediatr. Anaesth. **13**, 790–796 (2003)

51.23 A. Schultz, T. Kneif, U. Grouven, B. Schultz: EEG monitoring for the control of sedation: shortening of treatment in the ICU, Klin. Neurophysiol, **38**, 198–202 (2007)

51.24 L.J. Hirsch: Continuous EEG monitoring in the intensive care unit: An overview, J. Clin. Neurophysiol. **21**, 332–340 (2004)

51.25 M. Rijsdijk, F.S.S. Lijten, A.J.C. Slooter: Continuous EEG monitoring in the intensive care unit, Neth. J. Crit. Care **12**, 157–162 (2008)

51.26 B. Schultz: EEG-Monitoring auf der Intensivstation. In: *Überwachung der Narkosetiefe*, 2nd edn., ed. by W. Wilhelm, J. Bruhn, S. Kreuer (Ärzte-Verlag, Köln 2006) pp. 300–317
51.27 S. Zschocke: *Klinische Elektroenzephalographie*, 2nd edn. (Springer, Berlin Heidelberg 2002)
51.28 N. Lundberg: Continuous recording and control of ventricular fluid pressure in neurosurgical practice, Acta Psychiatr. **36**(149), 1–193 (1960)
51.29 R. Allen: Intracranial pressure: a review of clinical problems, measurement techniques and monitoring methods, J. Med. Eng. Technol. **10**, 299–320 (1986)
51.30 M.R. Gaab, H.E. Heissler: ICP monitoring, Crit. Rev. Biomed. Eng. **11**(3), 189–250 (1984)
51.31 K. Steinbereithner, W. Mauritz, P. Sporn: Therapie des Schädel-Hirn-Traumas – aktueller Stand, Notfallmedizin **11**, 1002–1013 (1985)
51.32 H.M. Betsch: *Experimentelle und klinische Untersuchungen über Fehler und Artefakte der epiduralen Hirndruckmessung*, Dissertation (Universität Heidelberg, Heidelberg 1993)

52. Brain Computer Interface

Günter Edlinger, Cristiano Rizzo, Christoph Guger

This chapter introduces the current state of the art of brain–computer interface (BCI) technology based on noninvasive surface electroencephalogram (EEG) recordings. The basic idea of a BCI is to enable a new communication channel that bypasses the standard neural pathways and output channels in order to control an external device. The ultimate goal of BCI technology is to enable lost body or communication functions in handicapped persons. Persons suffering from, e.g., amyotrophic lateral sclerosis (ALS), stroke or spinal cord injuries might lose the ability to fully control (peripheral) muscle activity. Depending on the disease either the neural pathway might be affected or the muscle itself. In a first attempt one can substitute the neural pathways or the affected muscles with still functional pathways or muscles. This approach might be very beneficial to the subjects, though the approach might also have limitations. Subjects can use, e.g., eye movements for communication or control. In the BCI approach body functions are restored by detecting the proper neural or muscle activity above the level of injury. These signals can serve as input to the BCI, which properly encodes the patterns and converts the activity into control commands. After a certain time of training the BCI can predict the user's intentions and the user can operate, e.g., the closing/opening of a robotic hand or control a wheelchair.

52.1 **Introduction to BCI** 1003

52.2 **Measuring Brain Activity** 1004
52.2.1 Invasive BCIs............ 1004
52.2.2 Partially Invasive BCIs 1004
52.2.3 Noninvasive BCIs............ 1004
52.2.4 Other Functional Imaging BCIs 1005

52.3 **BCI System Structure** 1005
52.3.1 Noninvasive (EEG) BCI Systems 1006
52.3.2 SMR-based BCIs 1007
52.3.3 P300-based BCI............ 1010

52.4 **Conclusions**............ 1014

References 1015

52.1 Introduction to BCI

Since the early 1990s the brain–computer interface (BCI) research field started growing constantly driven by relatively high performance and low cost computer power as well as electroencephalogram (EEG) instrumentation capable of being used in real-time and closed-loop data processing. However, a first systematic discussion of possible brain–computer communications [52.1] based on EEGs can be found in *Vidal* [52.2] in the early 1970s; *Farwell* and *Donchin* described in another pioneering work the usage of evoked potentials for communication [52.3]. Figure 52.1 displays the principle of a BCI system.

Since then, the performance and usability of BCI systems have advanced dramatically over the last

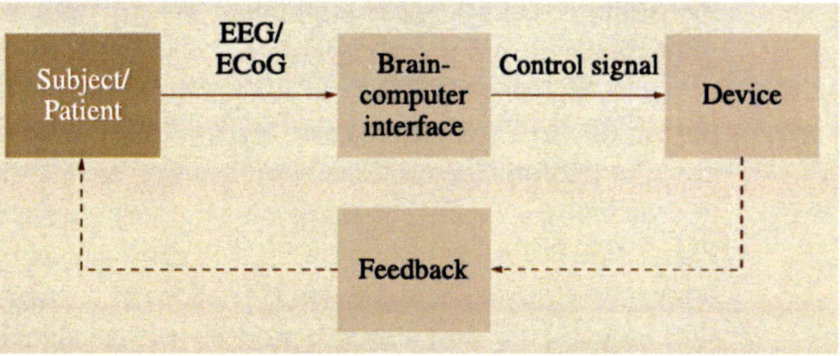

Fig. 52.1 Principle of a BCI system. EEG/ECoG/neural spike signals are measured from a subject. The BCI converts these signals into control signals utilized to construct various applications, e.g., a spelling program for BCI communication. A feedback loop provides information about, e.g., the selected character or word in the speller application so the user can change or correct the output

several years. Only about 10 years ago, one of the pioneering laboratories in BCI research in Europe published the first BCI that could provide communication for disabled users in their homes. However, the system was only validated with two users, required months of training, and was still slow and inaccurate [52.4]. Thereafter research laboratories in the USA and Europe were among the first to describe BCIs that could provide real benefit to handicapped people without extensive training [52.5–8]. Still the BCI users needed a training of several weeks to operate the system with acceptable accuracy [52.6,9]. Recently the training time of BCI systems dropped down to only minutes and some BCI systems do not even need any training [52.10, 11]. However, BCIs require the user to engage in some conscious, intentional activity to convey information. Work has shown that immersive feedback, which may include virtual reality, can reduce training time and improve accuracy [52.12, 13].

The confluence of ICT techniques (brain/neuronal–computer interfaces, affective computing, virtual reality, ambient intelligence) and neuropsychology allows their integration into an advanced platform which will improve the quality of life of people not only by providing means for communication, but also by performing advanced and user-orientated analysis of deficits and providing individual training scenarios. Still, the principal focus of many BCI research teams is to provide user/patient groups with very special needs with a new skill for communication and control. Recently there has been remarkable attention to facilitate BCI control to new user groups, such as healthy subjects as well as to less disabled people. Furthermore, BCIs have been recently evaluated to serve as an advanced neurofeedback tool in stroke rehabilitation, ADHD (attention-deficit/hyperactivity disorder) or other diseases [52.1, 14]. However, the focus and the ambitious goal of BCI research has been since the very beginning to support and help locked-in patients to facilitate communication with their environment again, thus enabling social interaction. Therefore the most popular BCI application is for spelling purposes [52.4,7, 11, 15–18].

52.2 Measuring Brain Activity

The basic principle of any BCI is to detect specific voluntarily modulated changes in brain activity in order to control a device. Different sensors and imaging modalities are available to detect in principle such changes. Measurements of the electrical activity of the brain – though the scale changes from the microscopic level to the macroscopic level – can be performed in different ways.

52.2.1 Invasive BCIs

Recording electrodes are implanted directly into the gray matter in order to better detect signals from the neurons, like single-cell action potentials or local field potentials (LFP). In this case electrodes commonly come in the form of a microelectrode array, tetrodes or microwires with different shape and size, as shown in Fig. 52.2a. Problems may arise for this approach as the body may react to the implanted electrodes in the brain, thus reducing the possibility to properly collect signals. This approach requires neurosurgery procedures for the implantation of the electrodes and thus the practical application is limited. However, the spatial resolution of invasive measurements is rather high and is in the order of several micrometers. Examples of the electrodes used are shown in Fig. 52.2a.

52.2.2 Partially Invasive BCIs

Recording electrodes are located inside the skull but do not penetrate the gray matter, they are simply applied on the surface of the cortex. This recording technique is referred to as electrocorticography (ECoG) and gives a spatial resolution of several millimeters. This resolution is better by far than the resolution of any noninvasive technique, but is still poor if compared to invasive techniques. Still, ECoG requires neurosurgery procedures for the positioning of the electrodes. BCI research has taken place in this way by simply taking advantage of patients who had such kinds of recording electrodes because of other disease (i. e., epilepsy). Examples of the electrodes used are shown in Fig. 52.2b.

52.2.3 Noninvasive BCIs

Recording electrodes are located on the scalp and the collected signal is the EEG of the subject. This technique provides lower spatial resolution signals, in the range of several centimeters, as the skull as well as other compartments (like cerebrospinal fluid) or tissue act as a sort of damping filter for lots of signals that could be collected with invasive and partially invasive techniques. On the contrary, the lack of invasiveness

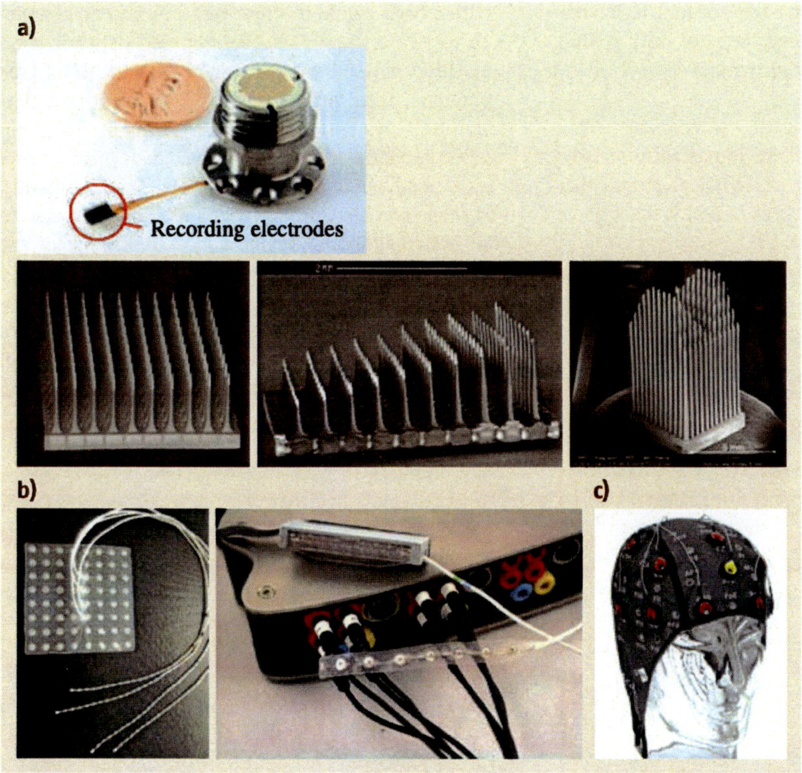

Fig. 52.2a–c BCI recording electrodes. **(a)** Microelectrode arrays (Blackrock, Salt Lake City) in three different sizes/shapes and the recording electrodes with their wire and connector. **(b)** EEG amplifier (g.USBamp – g.tec, Graz) connected to an eight-channel ECoG electrode grid and a 64-channel ECoG electrode grid. **(c)** EEG recording cap (g.tec, Graz)

has made this technique easily accessible and widely used. EEG signals are recorded typically from electrode positions mounted according to the so-called international (extended) 10/20 system. An example of an EEG cap with prefixed electrodes at the standard positions as defined by the international 10/20 system is shown in Fig. 52.2c.

52.2.4 Other Functional Imaging BCIs

Various approaches, such as PET (positron emission tomography), fMRI (functional magnetic resonance imaging), functional NIRS (near-infrared spectroscopy), and MEG (magnetoencephalography), have been evaluated as possible brain activity data sources for BCI. However, PET, fMRI, and MEG are technically demanding and bulky devices that are typically available at very special laboratories and hospitals only. Additionally, PET, fMRI, and NIRS depend on metabolic changes in the cerebral blood flow. These changes have intrinsically longer time constants compared to the electrical changes in the brain. Therefore a high data transfer rate – which is one important criterion for usability in BCI – might not be accomplished using these methods.

EEG, ECoG, and single-cell recordings are therefore still the only methods that are relatively easily accessible and have the desired temporal resolution for setting up BCI systems. Most of the BCIs used in laboratories around the world rely on the EEG and ECoG, which measures electrical brain activity from the intact human scalp or from the surface of the brain.

52.3 BCI System Structure

A subject interacts with the BCI system in a closed-loop setup. It is important to note that for a successful BCI operation the BCI system must be adaptive, i. e., it adapts itself to the subjects' specific brain activity patterns. On the other hand, the subjects learn over time to operate the BCI system. Hence an appropriate and real-time feedback is important for optimal BCI control performance and for learning the new skill. A BCI consists in principle of four major parts:

1. *Signal acquisition and signal conditioning*. This component of the BCI system is basically an amplifier including an analog to digital converter that aims to measure EEG/ECoG or neural spike signals with a high signal to noise ratio. From one channel up to 512 channels can be recorded depending on the kind of BCI experiment to be performed (see previous paragraph). Figure 52.3b displays a typical BCI system setup for 32 channels for scalp EEG recordings with an EEG cap (g.USBamp, g.tec medical engineering GmbH, Graz, Austria). Signals are also properly conditioned in order to be better analyzed by the successive block of the process.
2. *Feature extraction and classification*. This component of the BCI system is generally a signal-feature to control-signal converter which extracts distinct and stable features from the recorded signals in real time and translates them into corresponding control signals. Depending on the BCI technology, features in both the frequency domain and/or time domain can be extracted. Spatiotemporal preprocessing algorithms like common spatial patterns can be used to improve the signal to noise ratio and to maximize the difference between the signal patterns. Using supervised learning strategies nonlinear classifiers (e.g., neural network based) or linear classifiers (e.g., linear discriminant analyzer) are trained with the individual user's EEG features during a learning phase. User-specific features evaluation and optimization steps are usually investigated off-line based on training data. Thereafter most reactive parameters can be identified and used for the real-time application of the BCI. Figure 52.3a displays the different components of the BCI data processing part.
3. *Controlled device and application*. After successful training an external device like an orthosis is operated by the BCI which acts on the converted control signals.
4. *Feedback loop*. Results of the application are sent from the external device back to the user in the feedback loop. This information can be simply the position of a cursor on a screen but can also be a tactile or electrical signal fed back to the user. The results of the BCI as well as the feedback can be sent remotely via a network UDP connection to remotely located caregivers or BCI specialists.

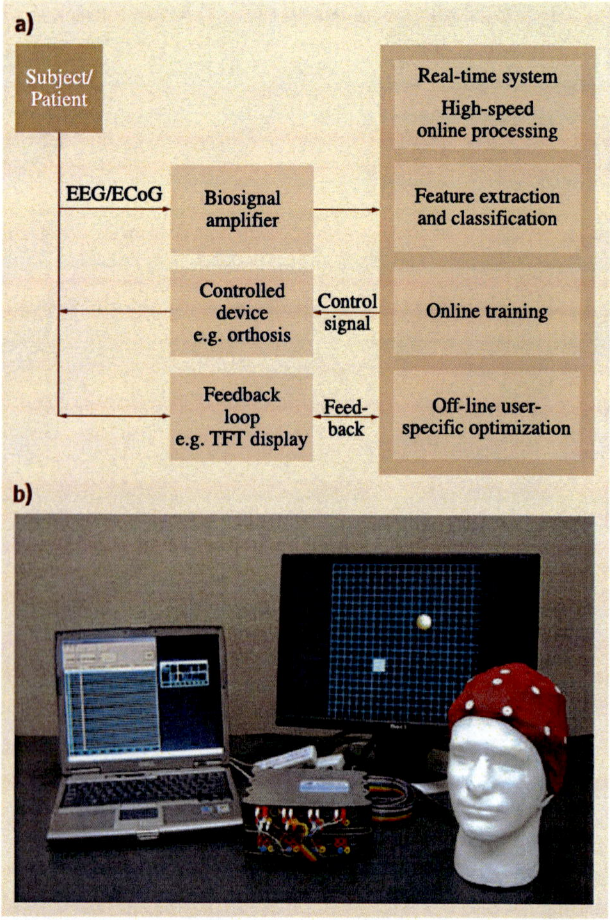

Fig. 52.3 **(a)** Components of a BCI system. Here the biosignal amplifier system g.USBamp (g.tec medical engineering, Graz) is connected to the BCI processing part running under MATLAB/Simulink on a laptop. In this case the high-speed online processing of the Simulink toolbox was used for the signal processing and paradigm presentation in real time. **(b)** BCI setup using a 32-channel g.USBamp system for data acquisition and the BCI2000 platform for real-time processing on a laptop (after [52.19]). The TFT screen displays the actual application as a 2-D cursor control task

52.3.1 Noninvasive (EEG) BCI Systems

Since this chapter aims at providing an overview of the widely used BCI technologies that are easy to use and could possibly become home monitoring device systems, EEG-based BCIs are discussed more thoroughly.

BCI systems have been successfully realized based on different EEG phenomena whereby two major groups of BCI can be distinguished.

Endogenous BCIs

In this type of BCI subjects learn and train to perform specific mental tasks to change willingly brain activity. This type of BCI includes slow cortical potentials (SCP) and sensorimotor rhythmic (SMR) or event-related desynchronization/synchronization (ERD/ERS) based BCIs:

1. *SCP based BCIs*. Very early approaches of BCIs included the use of SCP [52.4]. This approach required months of training. Today the SCP approach is not widely used anymore for BCI control.
2. *SMR- and ERD/ERS-based BCIs*. BCI systems based on the oscillatory brain electrical activity use motor imagery strategies that generate ERD and ERS in the alpha and beta frequency ranges of the EEG. In ECoG also gamma band activities have been used to construct BCI control [52.20]. More specifically, changes in sensorimotor rhythms associated with imagined hand or feet movements are mostly used to realize this type of BCI. However, even less specific movement imagery developed via training can be used [52.1, 10, 21]. Applications of this so-called SMR BCI are found for cursor control on computer screens, for navigation of wheelchairs or controlling virtual environments.

Exogenous BCIs

In this type of BCI the presentation of external stimuli evokes a specific change in the brain activity. Typical evoked potentials that are found in the ongoing EEG, depending on focused or selective attention to an external stimulus, are the P300 response and steady-state visually evoked potentials (SSVEP).

1. *P300-based BCIs*. The P300 BCI approach requires the user to focus on a visual or tactile stimulus, whereby the brainwaves differ for stimuli that the user attends versus ignores. Such a system uses the effect that an unlike event induces a P300 component in the EEG, i. e., a positive deflection in the EEG signal occurring around 300 ms after the event [52.3, 22]. In spelling applications typically several letters are displayed on a computer screen in a row–column format. All the letters are flashed transiently. The user selects and attends the letter she/he wants to select by simply counting the number of times it is flashed. Then the BCI system can determine which of several visual targets the user attends. In a similar way, recently P300 BCIs have been realized on tactile stimulation. Several tactors are mounted to different parts of the body and transiently switched on. The BCI system can determine which of several tactile targets contains the desired information. Applications here are supposed to aid in situations where tactile stimulations are more suitable than, e.g., visual cues [52.23].
2. *SSVEP-based BCIs*. SSVEP approaches use the fact that flickering light sources with flickering frequencies in the range of 5–20 Hz induce brain oscillations of the same flickering frequency. Similarly to the P300 BCI, here the brainwaves differ again for stimuli that the user attends versus ignores. Applications so far comprise, e.g., robot control or mobile phone control [52.24, 25].

52.3.2 SMR-based BCIs

Event-related synchronization and desynchronization of oscillatory components in the EEG are well-known phenomena of the electrical brain activity [52.26–29]. If, for example, the EEG is measured over the visual cortex in a relaxed person with eyes closed, then typically a very prominent rhythmic activity around 10 Hz with amplitudes of up to 100 μV can be observed. This activity is a so-called alpha rhythmic activity. If the person suddenly opens their eyes, then the amplitude of the alpha wave drops down. Compared to the relaxed state an event-related desynchronization, i. e., an amplitude decrease with respect to the relaxed phase, can be observed. Similarly, desynchronization and synchronization phenomena, i. e., a decrease and increase in amplitudes, can be observed over the corresponding motor representation areas for tongue, hand, or foot movement execution and similarly also for the movement imagery. Figure 52.4a displays a typical EEG amplitude decrease by evaluation of the power spectra for a right hand motor imagery in comparison to a resting condition measured from electrode position C3.

An imagination of the left hand is typically accompanied by an amplitude decrease over the contra lateral (right) hemisphere around electrode position C4. An imagination of the right hand is accompanied by a corresponding amplitude decrease over the left hemisphere (near electrode position C3). Figure 52.4b and c give an example of a typical ERD pattern measured during motor imagery of a left and right hand motor movement. Two distinct areas of maximal ERD over the left and right hand representation areas can

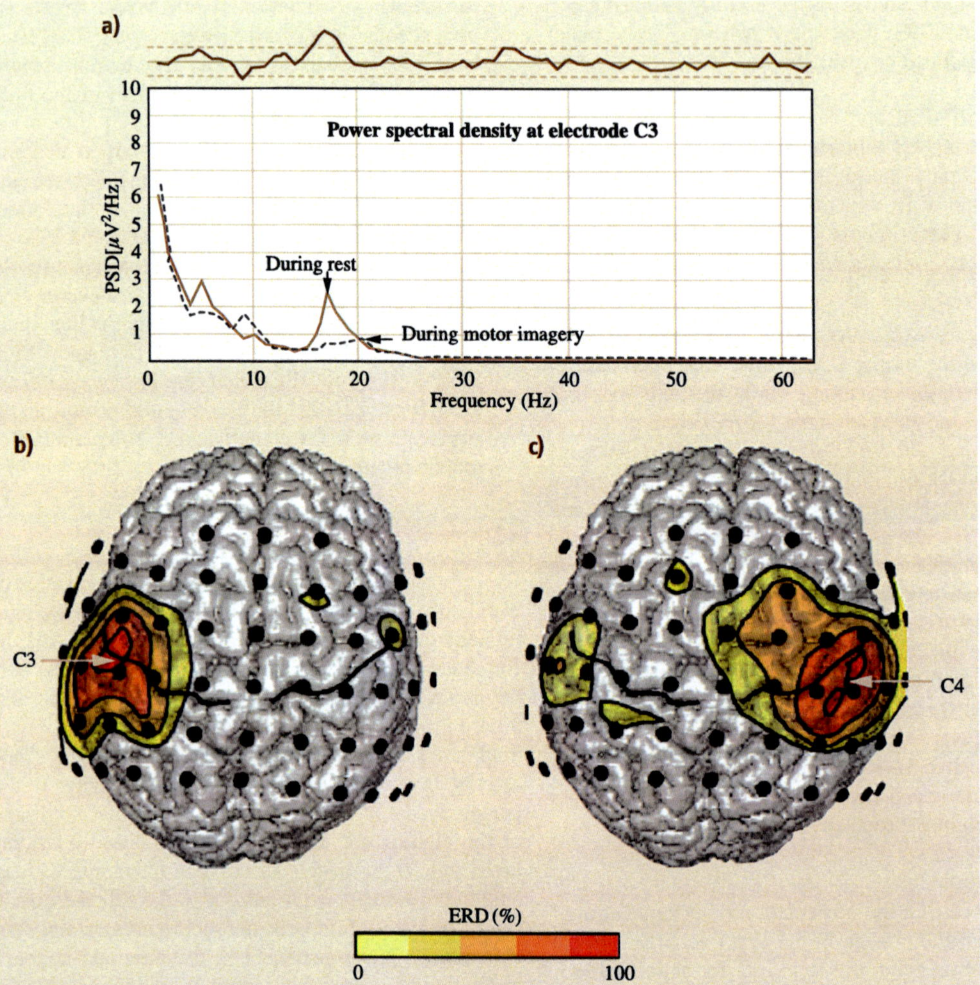

Fig. 52.4a–c Spectra and overlay display of event-related desynchronization patterns. **(a)** Display of the power spectral density during rest condition (*solid line*) and during motor imagery of the right hand (*dashed line*) recorded over electrode position C3. The *top line* indicates significant differences (95% level indicated by the *dashed brownish line*) between the two spectra around 17–18 Hz as well as at the first harmonic band around 34–36 Hz. During motor imagery the amplitudes in the specific frequency bands drop down. **(b)** and **(c)** Event-related desynchronization patterns computed from a dense 64-electrode array (positions are indicated by *small black disks*) and individual MR-based reconstructed cortical surfaces. The nose of the subject points to the top of the page. *Dark red color-coded areas* indicate largest desynchronization. *Light yellow areas* indicate smallest values of desynchronization. *Black solid lines* indicate the *Sulcus centralis*. **(b)** represents results for right hand motor imagery; **(c)** represents results for left hand motor imagery (after [52.35])

be seen. A foot movement imagery would induce an amplitude decrease in EEG measured over the vertex position. Hence these distinct and specific spatiotemporal changes in the rhythmic brain activity can be used as input to a motor imagery based BCI system. In practice amplitude changes in different frequency bands (alpha and beta bands for EEG) are utilized to train the BCI system. However, the changes in amplitudes related to motor imagery do not contain additional information about the type of the imagined movement, the velocity or the direction of the movement. Such information could be found only in more invasive recordings in ECoG [52.20, 30–33] or by observing changes in firing rates of action potential in *motor cortex in monkeys* [52.34].

It is well described in the literature that BCIs based on brain oscillatory activities might need a long training time. Hence people are trained in several sessions in order to operate a BCI with a high accuracy. Therefore it is of interest in general to investigate in how many people a BCI could work within only a short period of training.

In a study done in 99 untrained naïve subjects recruited at a local public fair in Graz, Austria, *Guger* et al. [52.36] described a simple ERD/ERS-based BCI experiment.

Subjects imagined either moving their right hand or left hand to cues provided on a monitor. Figure 52.5a,b display the corresponding experimental paradigm for the training and application phase. Figure 52.5c and d dis-

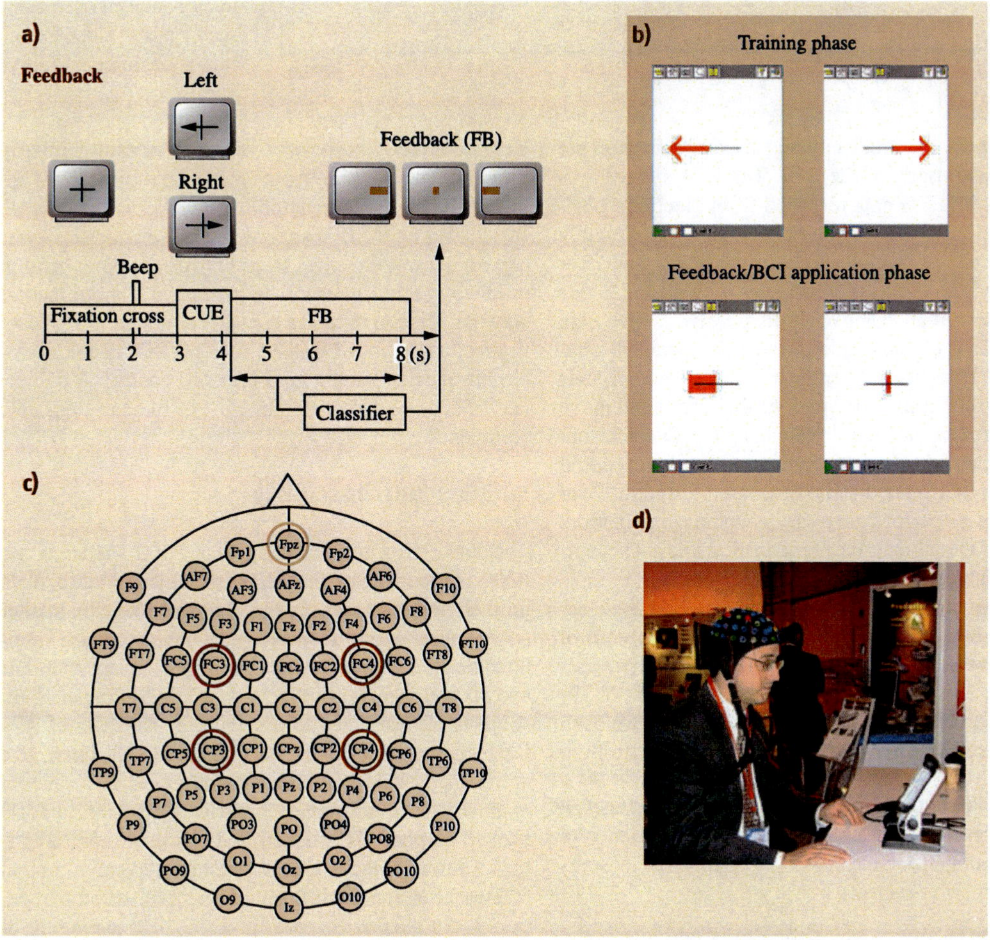

Fig. 52.5a–d Typical BCI application session. **(a)** and **(b)** yield the experimental paradigm as a function of the time. At the beginning of each trial a fixation cross is shown on the screen. At second 3 a cue, i. e., an arrow pointing to the left or right side, is displayed and the subjects perform a corresponding motor imagery of the left or right hand. The accuracy of the BCI is fed back as a bar extending to the left or right side depending on the classification accuracy. **(c)** displays the electrode positions used for the motor imagery task (after [52.36]). Two bipolar EEG channels (*brownish circles*) measured over the left and right hand representation area are sent to the BCI system. The *light brown circle* at Fpz indicates the position of the ground electrode. **(d)** shows a subject at the local fair operating the BCI

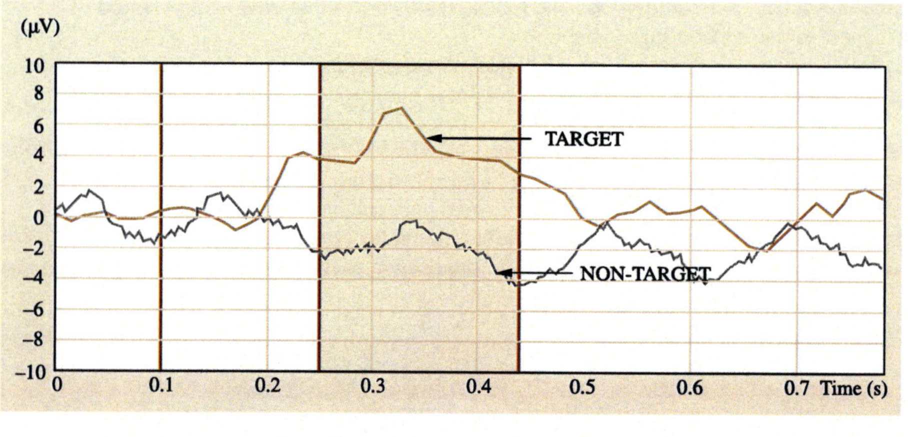

Fig. 52.6 P300 evoked brain signal response for target versus nontarget stimuli. A clear positive deflection in the EEG averaged 15 times for the target character *(brown line)* can be observed around 300 ms after stimulus onset (indicated by the *vertical brown line*). In all nontarget characters such a positive deflection cannot be observed *(gray line)*

play the bipolar electrode montage used and an example of a practical usage of the BCI. The BCI system was trained on the EEG data recorded from electrode positions C3 and C4. The training time was approximately 6 min and band-power features in the alpha band between 8 and 12 Hz and beta band between 20 and 24 Hz or parameters of an adaptive autoregressive model were estimated. For classification a linear discriminant analyzer (LDA) was used. After the training the participants had to control a computer cursor on the screen and move it to the left or right side. In this simple 1-D center out cursor task about 6% of the participating subjects could operate the BCI immediately. In about a further 20% of the subjects the performance was acceptable, but in more than half of the subjects the performance was not acceptable (see Table 52.1 for details of the accuracy).

Changes in EEG amplitudes of mu rhythmic activity and beta activity have been utilized to control cursor movement in 2-D and 3-D cursor task scenarios [52.37, 38], controlling orthotic devices [52.6, 39] or wheelchairs [52.40]. Such BCI approaches have been used rather in a *process control* way, i. e., controlling

Table 52.1 Percentage of sessions which were classified with a certain accuracy for motor imagery classified with the RLS algorithm or band power (BP) estimation. N specifies the number of subjects

Classification accuracy in %	RLS + BP percentage of sessions ($N = 99$)
90–100	6.2
80–89	13.0
70–79	32.1
60–69	42.0
50–59	6.7

a cursor directly in the x, y, or z directions or similarly moving a robotic arm. Another maybe more natural approach is the implementation of *goal-oriented control*. In a goal-oriented BCI approach it is then not necessary, e.g., to control the exact trajectory of a robotic hand in order to reach a glass of water, but in a more natural way humans can initiate a control command like "I want to grasp the glass of water". A P300 BCI is one example introducing such a goal-oriented control. A subject can select one target command out of many possible commands.

52.3.3 P300-based BCI

The underlying phenomenon of a P300 speller is the P300 component of the EEG, which is induced if an unlikely event occurs. In the classical Donchin speller setup subjects are presented 6×6 characters and numbers on a computer screen arranged in six columns and six rows, respectively. In an oddball-like paradigm subjects are instructed to concentrate on only one TARGET character out of all 36 characters. All characters are flashed on and off for a certain time in a random manner. When the target character flashes on, a P300 evoked response is induced and the maximum amplitude found in the EEG is reached typically around 300 ms after the flash onset (Fig. 52.6). Several repetitions are needed to perform EEG data averaging to increase the signal to noise ratio and accuracy of the system.

However, the layout of the character matrix, the color of the flashing and nonflashing characters, and the number of targets influence the accuracy of the speller [52.18, 22, 41, 42]. The P300 signal response is more pronounced, e.g., in the single-character (SC)

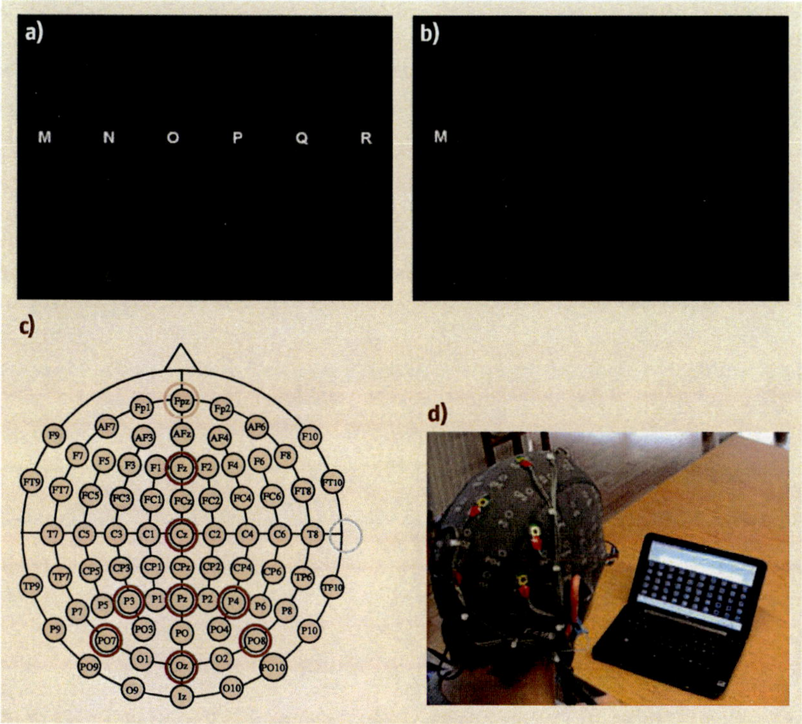

Fig. 52.7a–d P300 speller application. **(a)** and **(b)** display the screen layout of the 36-character speller. Either all characters of one row or column are highlighted at the same time in the row/column speller or only one single character is highlighted for a certain time in the single-character speller. **(c)** displays the electrode layout (after [52.22]). A total of eight electrode positions distributed mostly over occipital and parietal regions are used. *Brownish circles* indicate the electrode positions Fz, Cz, P3, Pz, P4, PO7, Oz, and PO8. The *brownish ring* indicates the ground electrode mounted on the forehead at Fpz and the *gray ring* indicates the reference electrode attached to the right ear lobe. **(d)** displays the home system setup intendiX with the portable wireless EEG device g.MOBIlab+, an active electrode system g.GAMMAsys, and the standalone P300 speller application intendiX (courtesy of g.tec medical engineering GmbH, Austria)

speller than in the row/column (RC) speller and is therefore easier to detect [52.22, 43]. Figure 52.7 shows an example of a P300 speller setup.

Spelling with a P300 BCI

An interesting question in general is how many people could learn the new skill of operating a P300 BCI. As discussed above ERD/ERS-based BCIs need typically a longer time of training and some subjects cannot use such a BCI. However, it is described in the literature that a P300-based BCI needs less training. A recent study explored the P300 BCI across 100 subjects. In addition, 35 subjects received questionnaires to explore demographic issues [52.43].

In the experiment a P300 spelling device based on a 6×6 matrix of different characters was displayed on a computer screen. Subjects had to perform predefined evaluation steps of a RC speller or of a SC speller. This yields of course different communication rates; with a 6×6 matrix, the row/column approach increases

Table 52.2 Percentage of sessions which were classified with certain accuracy. *N* specifies the number of subjects participating

Classification accuracy in %	Row–column speller: Percentage of sessions ($N = 81$)	Single-character speller: Percentage of sessions ($N = 38$)
100	72.8	55.3
80–100	88.9	76.3
60–79	6.2	10.6
40–59	3.7	7.9
20–39	0.0	2.6
0–19	1.2	2.6
Average accuracy of all subjects	91.0	82.0
Mean of subjects who participated in RC and SC ($N = 19$)	85.3	77.9

speed by a factor of 6. For training, EEG data are acquired from the subject while the subject focuses on the appearance of specific letters in the copy spelling mode. In this mode, an arbitrary word like LUCAS is presented on the monitor. First, the subject counts whenever the L flashes. Each row, column, or character flashes for, e.g., 100 ms per flash. Then the subject counts the U until it flashes 15 times, and so on. Then, EEG data elicited by each flashing event are extracted within a specific interval length of 800 ms and divided into sub-segments. The EEG data of each segment are averaged and sent to a stepwise LDA. The LDA is trained to separate the target characters, i. e., the characters the subject was concentrating on (15 flashes × 5 characters), from all other events ($15 \times 36 - 15 \times 5$). It is very interesting for this approach that the LDA is trained only on five characters representing five classes and not on all 36 classes. This is in contrast to the motor imagery approach where each class must also be used as a training class. The P300 approach allows minimization of the time necessary for EEG recording for the setup of the LDA. However, the accuracy of

Fig. 52.8 CAVE setup. *Left*: CAVE system at University College London. *Right*: Smart home VR realization from Chris Groenegress and Mel Slater from Universitat Politècnica de Catalunya, Barcelona

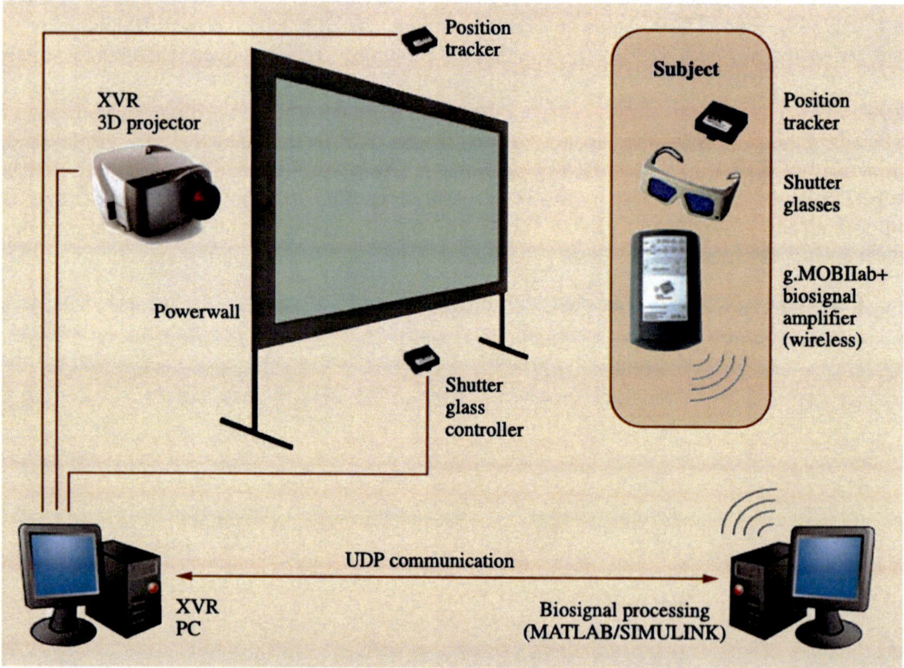

Fig. 52.9 Components of a VR setup

the spelling system also increases with the number of training characters.

In this experiment a total of 72.8% ($N = 81$) of the subjects were able to spell with 100% accuracy in the RC paradigm and 55.3% ($N = 38$) spelled with 100% accuracy in the SC paradigm. Less than 3% of the subjects did not spell any character correctly. People who slept less than 8 h performed significantly better than other subjects. Age, sex, education, working duration, and cigarette and coffee consumption were not statistically related to differences in accuracy. The disturbance of the flashing characters was rated as 1.5 on a scale from 1 to 5 (1 – not disturbing, 5 – highly disturbing). See Table 52.2 for the details of the evaluation.

Virtual Reality Smart Home Control with the P300 BCI

BCI systems were also combined with virtual reality (VR) systems both for SMR BCI and P300-based BCIs [52.13, 44]. VR systems use either head-mounted displays (HMDs) or highly immersive back-projection systems (CAVE-like systems) as shown in Fig. 52.8. Such a CAVE has three back-projectors for the walls and one projector on top of the CAVE for the floor. The system projects two images which are separated by shutter glasses to achieve 3-D effects.

There are several issues that must be solved to use a BCI system in such an environment: the biosignal amplifiers must be able to work in such a noisy environment, the recordings should ideally be done with

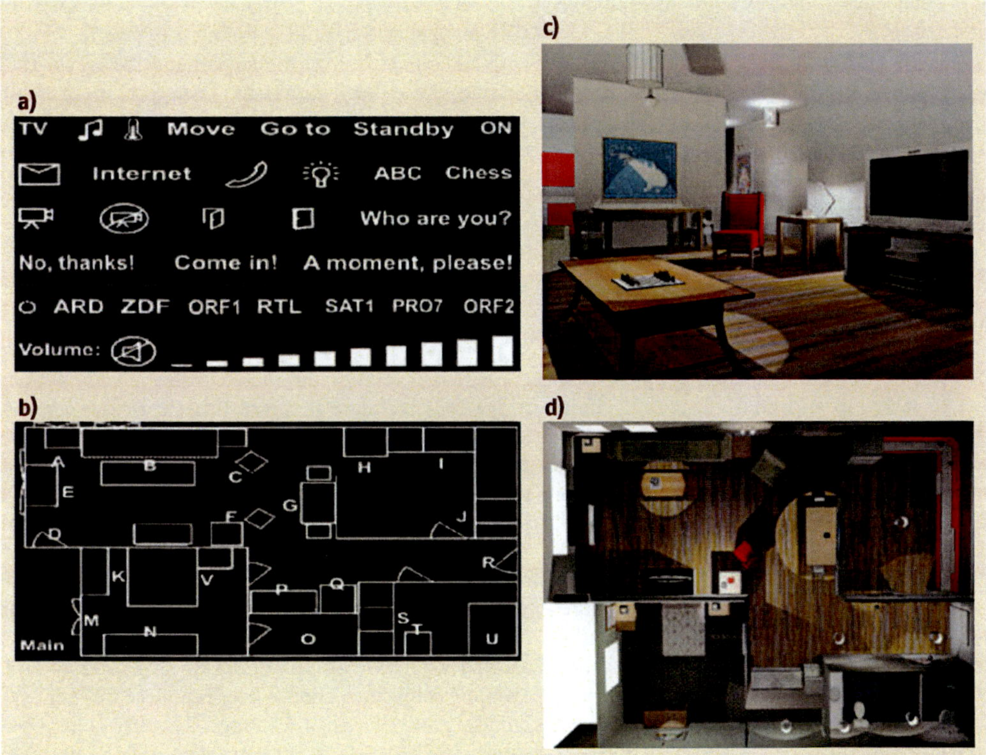

Fig. 52.10a–d Smart home interface masks and VR representation. **(a)** displays the main interface mask consisting of 41 different icons arranged in a rectangular grid. Similarly to the P300 speller matrix the individual icons are flashed on and off in a random manner. **(b)** displays a 3-D view of the living room including some of the devices that can be controlled via the BCI like the TV set, room light or telephone. **(c)** and **(d)** display a bird's eye view of the apartment. **(c)** represents the control mask to be *beamed* to 21 different locations in the apartment. Here 21 characters represent all user-selectable positions. In the *top left corner* the living room can be found, the *top right corner* represents the kitchen, the *bottom left corner* represents the sleeping room, and on the *bottom right corner* the bathroom is located as well as the entrance door of the apartment

data sent via a radio to avoid collisions and irritations within the environment, the BCI system must be coupled with the VR system for real-time experiments, and a special BCI communication interface must be developed to have enough degrees of freedom available to control the VR system.

Figure 52.9 illustrates the components in detail. A 3-D projector is located next to a projection wall for back projections. The subject stands in front the projection wall to avoid shadows and is equipped with a position tracker to capture movements, shutter glasses for 3-D effects, and a biosignal amplifier including electrodes for EEG recordings. The XVR (eXtreme VR, VRmedia, Pisa) PC controls the projector, the position tracker controller, and the shutter glass controller. The biosignal amplifier transmits the EEG data to the BCI system that is connected to the XVR PC to exchange control commands.

In order to show that such a combination is possible, a virtual version of a smart home was implemented in XVR. The smart home consists of different rooms whereby each room is equipped with several different devices that can be controlled: TV, MP3 player, telephone, lights, doors, etc. Therefore, all the different commands were summarized in seven control masks: a light mask, a music mask, a phone mask, a temperature mask, a TV mask, a move mask, and a go to mask. Figure 52.10 shows the TV mask and as an example the corresponding XVR image of the living room. The subject can, e.g., switch on the TV by selecting the TV symbol. Then, the TV station and the volume can be regulated. The bottom row of Fig. 52.10 shows the go to mask with an underlying plan of the smart home enabling the subject to move to a spot in the room where he wants to go. After the decision of the BCI system, the VR program moves to a bird's eye view of the apartment and zooms to the spot that was selected by the user. This is a goal-oriented BCI control approach, in contrast to a navigation task, where each small navigational step is controlled individually [52.44].

The control masks contained 13–50 commands. In such an application, precise timing between the appearance of the symbol on the screen and the signal processing unit is very important. Therefore, the flashing sequence was implemented under Simulink where the real-time BCI processing was also running. A UDP interface was also programmed that sent the data to the XVR system to control the smart home, as shown in Fig. 52.9.

52.4 Conclusions

BCI enabled control and communication is a new skill a subject has to learn. In an initial adaptation phase the BCI system is trained to the specific subject's brain activity. In addition the subjects have to get used and adapt to the BCI system as well. The time needed for a subject to adapt to the system is shorter by far in exogenous BCIs like P300 and SSVEP approaches. Such BCI systems yield higher accuracies in a higher number of subjects and therefore give for control purposes more reliable results. However, subjects have to look at flashing or flickering light sources or pay attention to tactile stimulations. Hence external stimulations might interfere with daily life situations and may distract subjects from other ongoing activities.

ERD/ERS-based BCIs rely on intrinsic changes of alpha and beta power activities. In principle no external stimuli or cues are necessary to operate the BCI. However, it has been shown that this type of BCI needs adaptation and training from the subjects' side over many hours [52.45]. The accuracy and speed of an ERD/ERS-based BCI is below that of a P300 BCI.

However, one of the most consistent observations in the BCI literature is the fact that a certain percentage of the population cannot operate a specific type of BCI for various reasons. Inter-subject as well as intra-subject variability often leads to a so-called BCI illiteracy [52.5, 13, 25, 46]. Across the different BCI approaches (ERD/ERS, P300, SSVEP) around 20–25% of subjects are unable to control one type of BCI in a satisfactory way. Therefore, the usage of *hybrid* BCIs has been introduced using the output of an SMR BCI as well as P300- or SSVEP-based BCIs [52.44, 47], enabling subjects to choose between these different approaches for optimal BCI control. Recently is has been shown that progress in BCI research terms of usability and robustness made it possible to move the technology to the patients' homes. *Nijboer* et al. [52.48] showed that severely handicapped people can use a P300 speller. Signals were stable over months as well as the ability to operate a 6×6 spelling application with stable accuracy. However, in order to operate a BCI system on a daily basis, well-trained staff members supported the participants. One of the participants quit the study as result of the experimental setup and discomfort of the procedure.

Therefore attempts have been made to enable a more easy usage of BCI systems at home. The system intendiX (www.intendix.com) is designed to be installed and operated by caregivers or the patient's family at home. It consists of active EEG electrodes to avoid abrasion of the skin, a portable biosignal amplifier, and a laptop or netbook running the software under Windows. Electrodes are prefixed into the cap to allow a fast and easy montage. The intendiX software allows viewing of the raw EEG to inspect data quality, but indicates automatically to the inexperienced user if the data quality on a specific channel is good or bad. If the system is started up for the first time, 5–10 training characters are entered and the user has to copy spell the characters. Then the software switches automatically into the spelling mode and the user can spell as many characters as wanted. The number of flashes for each classification can be selected by the user or the user can also use a statistical approach that detects automatically if the user is working with the BCI system. The latter one has the advantage that no characters are selected if the user is not looking at the matrix or does not want to use the speller.

References

52.1 J.R. Wolpaw, N. Birbaumer, D.J. McFarland, G. Pfurtscheller, T.M. Vaughan: Brain-computer interfaces for communication and control, Clin. Neurophysiol. **113**(6), 767–791 (2002)

52.2 J.J. Vidal: Toward direct brain-computer communication, Annu. Rev. Biophys. Bioeng. **2**, 157–180 (1973)

52.3 L.A. Farwell, E. Donchin: Talking off the top of your head: Toward a mental prosthesis utilizing event-related brain potentials, Electroencephalogr. Clin. Neurophysiol. **70**, 510–523 (1988)

52.4 N. Birbaumer, N. Ghanayim, T. Hinterberger, I. Iversen, B. Kotchoubey, A. Kübler, J. Perelmouter, E. Taub, H. Flor: A spelling device for the paralysed, Nature **398**, 297–298 (1999)

52.5 C. Guger, G. Edlinger, W. Harkam, I. Niedermayer, G. Pfurtscheller: How many people are able to operate an EEG-based brain-computer interface (BCI)?, IEEE Trans. Neural Syst. Rehabil. Eng. **11**, 145–147 (2003)

52.6 G. Pfurtscheller, G.R. Müller, J. Pfurtscheller, H.J. Gerner, R. Rupp: 'Thought'-control of functional electrical stimulation to restore hand grasp in a patient with tetraplegia, Neurosci. Lett. **351**, 33–36 (2003)

52.7 G. Townsend, B.K. LaPallo, C.B. Boulay, D.J. Krusienski, G.E. Frye, C.K. Hauser, N.E. Schwartz, T.M. Vaughan, J.R. Wolpaw, E.W. Sellers: A novel P300-based brain-computer interface stimulus presentation paradigm: Moving beyond rows and columns, Clin. Neurophysiol. **121**, 1109–1120 (2010)

52.8 T.M. Vaughan, D.J. McFarland, G. Schalk, W.A. Sarnacki, D.J. Krusienski, E.W. Sellers, J.R. Wolpaw: The Wadsworth BCI research and development program: At home with BCI, IEEE Trans. Neural Syst. Rehabil. Eng. **14**(2), 229–233 (2006)

52.9 C. Neuper, G.R. Müller, A. Kübler, N. Birbaumer, G. Pfurtscheller: Clinical application of an EEG-based brain–computer interface: A case study in a patient with severe motor impairment, Clin. Neurophysiol. **114**, 399–409 (2003)

52.10 B. Blankertz, F. Losch, M. Krauledat, G. Dornhege, G. Curio, K.R. Muller: The Berlin brain–computer interface: Accurate performance from first-session in BCI-naïve subjects, IEEE Trans. Biomed. Eng. **55**(10), 2452–2462 (2008)

52.11 Z. Haihong, G. Cuntai, W. Chuanchu: Asynchronous P300-based brain-computer interfaces: A computational approach with statistical models, IEEE Trans. Biomed. Eng. **55**(6), 1754–1763 (2008)

52.12 C. Neuper, R. Scherer, S. Wriessnegger, G. Pfurtscheller: Motor imagery and action observation: Modulation of sensorimotor brain rhythms during mental control of a brain–computer interface, Clin. Neurophysiol. **120**, 239–247 (2009)

52.13 R. Leeb, F. Lee, C. Keinrath, R. Scherer, H. Bischof, G. Pfurtscheller: Brain–computer communication: Motivation, aim, and impact of exploring a virtual apartment, IEEE Trans. Neural Syst. Rehabil. Eng. **15**(4), 473–482 (2007)

52.14 A. Nijholt, D. Tan: Brain–computer interfacing for intelligent systems, IEEE Intell. Syst. **23**(3), 72–79 (2008)

52.15 E. Donchin, K.M. Spencer, R. Wijesinghe: The mental prosthesis: Assessing the speed of a P300-based brain–computer interface, IEEE Trans. Rehabil. Eng. **8**, 174–179 (2000)

52.16 D.J. Krusienski, E.W. Sellers, F. Cabestaing, S. Bayoudh, D.J. McFarland, T.M. Vaughan, J.R. Wolpaw: A comparison of classification techniques for the P300 Speller, J. Neural Eng. **3**, 299–305 (2006)

52.17 D.J. Krusienski, E.W. Sellers, D.J. McFarland, T.M. Vaughan, J.R. Wolpaw: Toward enhanced P300 speller performance, J. Neurosci. Methods **167**, 15–21 (2008)

52.18 H. Zhang, C. Guan, C. Wang: Asynchronous P300-based brain–computer interfaces: A computational

approach with statistical models, IEEE Trans. Biomed. Eng. **55**, 1754–1763 (2008)
52.19 G. Schalk, D.J. McFarland, T. Hinterberger, N. Birbaumer, J.R. Wolpaw: BCI2000: A general-purpose brain–computer interface (BCI) system, IEEE Trans. Biomed. Eng. **51**(6), 1034–1043 (2004)
52.20 G. Schalk, J. Kubanek, K.J. Miller, N.R. Anderson, E.C. Leuthardt, J.G. Ojemann, D. Limbrick, D. Moran, L.A. Gerhardt, J.R. Wolpaw: Decoding two-dimensional movement trajectories using electrocorticographic signals in humans, J. Neural Eng. **4**(3), 264–275 (2007)
52.21 G. Pfurtscheller, C. Neuper, G.R. Muller, B. Obermaier, G. Krausz, A. Schlogl, R. Scherer, B. Graimann, C. Keinrath, D. Skliris, M. Wortz, G. Supp, C. Schrank: Graz-BCI: State of the art and clinical applications, IEEE Trans. Neural Syst. Rehabil. Eng. **11**(2), 177–180 (2003)
52.22 E.W. Sellers, D.J. Krusienski, D.J. McFarland, T.M. Vaughan, J.R. Wolpaw: A P300 event-related potential brain–computer interface (BCI): The effects of matrix size and inter stimulus interval on performance, Biol. Psychol. **73**, 242–252 (2006)
52.23 F. Cincotti, L. Kauhanen, F. Aloise, T. Palomaki, N. Caporusso, P. Jylanki, F. Babiloni, G. Vanacker, M. Nuttin, M.G. Marciani, R.M. Del, D. Mattia: Preliminary experimentation on vibrotactile feedback in the context of mu-rhythm based BCI, Conf. Proc. IEEE Eng. Med. Biol. Soc. 2007 (IEEE, 2007) pp. 4739–4742
52.24 O. Friman, I. Volosyak, A. Graser: Multiple channel detection of steady-state visual evoked potentials for brain–computer interfaces, IEEE Trans. Biomed. Eng. **54**(4), 742–750 (2007)
52.25 G.R. Muller-Putz, G. Pfurtscheller: Control of an electrical prosthesis with an SSVEP-based BCI, IEEE Trans. Biomed. Eng. **55**, 361–364 (2008)
52.26 G. Pfurtscheller, A. Aranibar: Evaluation of event-related desynchronization (ERD) preceding and following voluntary self-paced movement, Electroencephalogr. Clin. Neurophysiol. **46**, 138–146 (1979)
52.27 G. Pfurtscheller, C. Neuper, C. Andrew, G. Edlinger: Foot and hand area mu rhythms, Int. J. Psychophysiol. **26**, 121–135 (1997)
52.28 G. Pfurtscheller, F.H. Lopes da Silva: Event-related EEG/MEG synchronization and desynchronization: Basic principles, Clin. Neurophysiol. **110**, 1842–1857 (1999)
52.29 G. Pfurtscheller: Spatiotemporal ERD/ERS patterns during voluntary movement and motor imagery, Suppl. Clin. Neurophysiol. **53**, 196–198 (2000)
52.30 E.C. Leuthardt, G. Schalk, J.R. Wolpaw, J.G. Ojemann, D.W. Moran: A brain–computer interface using electrocorticographic signals in humans, J. Neural Eng. **1**(2), 63–71 (2004)
52.31 E.C. Leuthardt, K.J. Miller, G. Schalk, R.P.N. Rao, J.G. Ojemann: Electrocorticography-based brain–computer interface – The Seattle experience, IEEE Trans. Neural Syst. Rehabil. Eng. **14**(2), 194–198 (2006)
52.32 L.R. Hochberg, M.D. Serruya, G.M. Friehs, J.A. Mukand, M. Saleh, A.H. Caplan, A. Branner, D. Chen, R.D. Penn, J.P. Donoghue: Neuronal ensemble control of prosthetic devices by a human with tetraplegia, Nature **442**, 164–171 (2006)
52.33 E.A. Felton, J.A. Wilson, J.C. Williams, P.C. Garell: Electrocorticographically controlled brain–computer interfaces using motor and sensory imagery in patients with temporary subdural electrode implants. Report of four cases, J. Physiol. **106**, 495–500 (2007)
52.34 J.M. Carmena, M.A. Lebedev, R.E. Crist, J.E. O'Doherty, D.M. Santucci, D.F. Dimitrov, P.G. Patil, C.S. Henriquez, M.A. Nicolelis: Learning to control a brain–machine interface for reaching and grasping by primates, PLoS Biology **1**(2), 192–208 (2003)
52.35 G. Edlinger: *Towards the Realization of a High-Resolution EEG*, Dissertation (University of Technology Graz, Graz 1998)
52.36 C. Guger, G. Edlinger, W. Harkam, I. Niedermayer, G. Pfurtscheller: How many people are able to operate an EEG-based brain–computer interface (BCI)?, IEEE Trans. Neural Syst. Rehabil. Eng. **11**(2), 145–147 (2003)
52.37 D.J. McFarland, W.A. Sarnacki, J.R. Wolpaw: Electroencephalographic (EEG) control of three-dimensional movement, J. Neural Eng. **7**(3), 036007 (2010)
52.38 J.R. Wolpaw, D.J. McFarland: Control of a two-dimensional movement signal by a noninvasive brain–computer interface in humans, Proc. Natl. Acad. Sci. USA **101**, 17849–17854 (2004)
52.39 G. Pfurtscheller, C. Guger, G. Müller, G. Krausz, C. Neuper: Brain oscillations control hand orthosis in a tetraplegic, Neurosci. Lett. **292**(3), 211–214 (2000)
52.40 J.R. Millan, F. Renkens, J. Mourino, W. Gerstner: Noninvasive brain-actuated control of a mobile robot by human EEG, IEEE Trans. Biomed. Eng. **51**(6), 1026–1033 (2004)
52.41 T. Komatsu, N. Hata, Y. Nakajima, K. Kansaku: A non-training EEG-based BMI system for environmental control, Neurosci. Res. **61**(1), S251 (2008)
52.42 H. Serby, E. Yom-Tov, G.F. Inbar: An improved P300-based brain–computer interface, IEEE Trans. Neural Syst. Rehabil. Eng. **13**, 89–98 (2005)
52.43 C. Guger, S. Daban, E. Sellers, C. Holzner, G. Krausz, R. Carabalona, F. Gramatica, G. Edlinger: How many people are able to control a P300-based brain–computer interface (BCI)?, Neurosci. Lett. **462**(1), 94–98 (2009)
52.44 G. Edlinger, C. Holzner, C. Guger, C. Groenegress, M. Slater: Brain-computer interfaces for goal orientated control of a virtual smart home envi-

ronment, 4th Int. IEEE/EMBS Conf. on Neural Eng. NER 09 (IEEE/EBMS, 2009) pp. 463–465
52.45 G. Pfurtscheller, R. Leeb, C. Keinrath, D. Friedman, C. Neuper, C. Guger, M. Slater: Walking from thought, Brain Res. **1071**, 145–152 (2006)
52.46 B. Allison, T. Luth, D. Valbuena, A. Teymourian, I. Volosyak, A. Graser: BCI Demographics: How many (and what kinds of) people can use an SSVEP BCI?, IEEE Trans. Neural Syst. Rehabil. Eng. **18**(2), 107–116 (2010)
52.47 G. Pfurtscheller, B.Z. Allison, C. Brunner, G. Bauernfeind, T. Solis-Escalante, R. Scherer, T.O. Zander, G. Mueller-Putz, C. Neuper, N. Birbaumer: The hybrid BCI, Front Neurosci. **4**, 42 (2010)
52.48 F. Nijboer, E.W. Sellers, J. Mellinger, M.A. Jordan, T. Matuz, A. Furdea, S. Halder, U. Mochty, D.J. Krusienski, T.M. Vaughan, J.R. Wolpaw, N. Birbaumer, A. Kübler: A P300-based brain–computer interface for people with amyotrophic lateral sclerosis, Clin. Neurophysiol. **119**, 1909–1916 (2008)

53. Fetal Monitoring

Birgit Wacker, Martin Maier

Cardiotocography (CTG) is the process of monitoring the fetal heart rate and uterine activity, and is used both prior to (antepartum, prenatal) and during (intrapartum) childbirth in the delivery room. The equipment is operated by midwifes and nurses, who also analyze the graphs along with the attending physicians. This chapter covers cardiotocography (Sect. 53.1) and obstetric monitoring systems (Sect. 53.2).

53.1 **Cardiotocography (CTG)**.......................... 1019
53.1.1 Areas of Application 1019
53.1.2 Measuring Procedures................ 1019
53.1.3 Direct ECG................................. 1020
53.1.4 Toco Procedure 1020
53.1.5 Intrauterine Pressure Measurement 1021
53.1.6 Additional Functions and Options 1021
53.1.7 CTG Selection Criteria 1022

53.2 **Obstetric Monitoring Systems**................. 1022
53.2.1 Areas of Application 1022
53.2.2 System Setup 1022
53.2.3 Collection and Processing of Data from the Connected CTGs............. 1023
53.2.4 Assessment of the CTG Data and Raising the Alarm 1024
53.2.5 Presentation of the CTG Data 1026
53.2.6 Patient Data Management........... 1026
53.2.7 Archiving and Retrieving Data via Long-Term Storage Media 1028
53.2.8 Computer Communication Within Obstetrics........................ 1029
53.2.9 Computer Communication with Other Hospitals.................. 1029
53.2.10 Communication with the Hospital Administration Computer 1029
53.2.11 Exporting Patient Data to Other Hospital Systems............ 1029
53.2.12 Connection to Lab Computer........ 1030
53.2.13 Access to Internal Hospital Computer Systems 1030

53.1 Cardiotocography (CTG)

53.1.1 Areas of Application

Cardiotocographs are primarily used to monitor the fetal heart rate and uterine activity via external and internal measurements taken by what are referred to as transducers (Fig. 53.1). Some devices also measure the maternal heart rate, as well as the blood pressure and oxygen saturation of the parturient. Table 53.1 compares the various measuring procedures.

53.1.2 Measuring Procedures

Ultrasound

This procedure uses quartz crystals fitted in the transducer to transmit high-frequency acoustic signals. These sound waves are reflected by body tissue at various intensities and then received and amplified by the transducer. The shift in frequency among the reflected acoustic signals (Doppler effect) is used to detect movement. By optimizing the movement pattern of the fetal heart and using additional signal processing procedures (autocorrelation), it is possible to gain an extremely accurate reading of the fetal heartbeat. Extra procedures such as precision signal track and hold (according to the time the ultrasound waves spend in the tissue) can be used to filter out interference signals from movements unrelated to the fetal heart.

Fetal Movement Profile

The ultrasound procedure described above can also be used to determine the child's movement within the womb. The signals that are filtered out when determining the fetal heart rate can be analyzed separately to provide a valuable insight into the fetal movement pat-

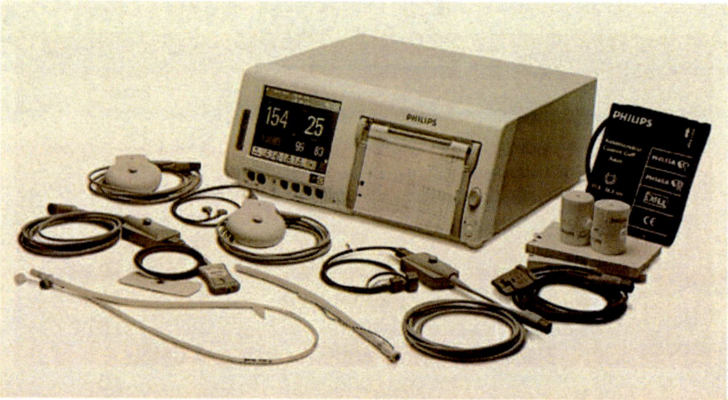

Fig. 53.1 Sample CTG with transducer and accessories. *Left to right*: Patient module with adapter cable for the scalp electrode and leg electrode, an ultrasound transducer, a toco transducer with connected cable for the maternal ECG, a patient module with adapter cable for the IUP catheter, a sensor to measure the mother's oxygen saturation and pulse. *Foreground*: IUP catheter and a scalp electrode. *Background (right)*: CTG paper with a preprinted scale, belt to attach the transducer to the mother's abdomen, and a blood pressure cuff

Table 53.1 External and internal cardiotocograph measuring procedures

Patient	Parameter	External measuring procedure			Internal measuring procedure		
		Transducer type	Area of application	Unit	Transducer type	Area of application	Unit
Fetus	Heart rate	Ultrasound sensor	Antepartum, intrapartum	rpm	Scalp electrode	Intrapartum, high-risk pregnancy	rpm
	Fetal movement profile	Ultrasound sensor	Antepartum, intrapartum	%			
Mother	Heart rate/pulse	Skin electrode, also indirect via blood pressure cuff or light sensor	Antepartum, intrapartum	rpm			
	Uterine activity	Pressure sensor (toco transducer)	Antepartum, intrapartum	Relative pressure	Catheter with pressure sensor	Intrapartum, high-risk pregnancy	mmHg or kPa
	Blood pressure	Blood pressure cuff	Under anesthesia	mmHg			
	Oxygen saturation	Light sensor	Under anesthesia	%			

tern, namely information on movement of the body and limbs of the fetus. As well as purely marking out periods of activity, the equipment can also generate a percentage value to provide information on fetal vitality without the need to calculate the movement pattern manually.

53.1.3 Direct ECG

This procedure uses a scalp electrode, which is fixed to the child's head and measures the fetal ECG to determine the heart rate (Fig. 53.2). This is what is known as an internal measurement, which can only take place during childbirth, and only then once the amniotic membranes have broken and the cervix is dilated by at least 2–3 cm. It is primarily used when the more typically used ultrasound transducers experience a loss of signal and cannot give a reliable reading of the heart rate trace. This can become necessary, particularly during the later expulsion stage (e.g., if the patient is moving a lot). A direct fetal ECG measurement also provides a qualitative analysis of the fetal ECG, the ST segment of which ST waveform analysis (STAN) can be of particular significance.

53.1.4 Toco Procedure

This procedure is used to record uterine activity by means of pressure-sensitive sensors that detect the hardening (tensioning) of the uterine muscle. The value measured is only relative and depends on many factors, such as the position of the transducer. For this reason, the display must first be calibrated with a base value once the transducer is in place. As this procedure cannot produce an absolute value, another procedure must be used where this is required. This procedure is described in more detail in the next section.

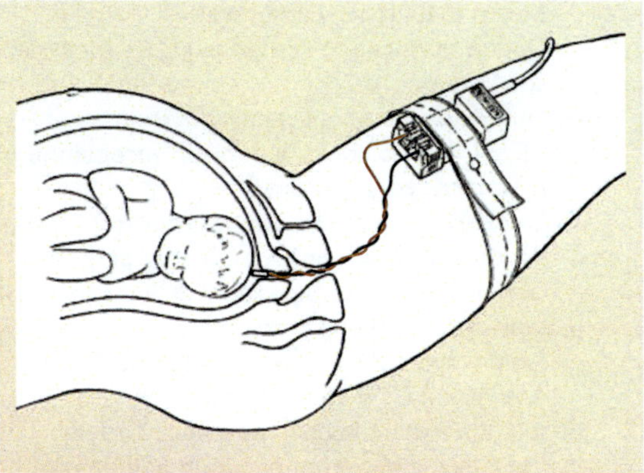

Fig. 53.2 Direct ECG

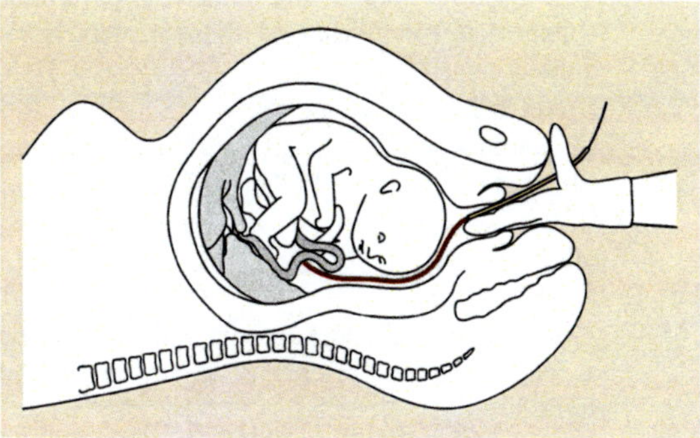

Fig. 53.3 Intrauterine pressure measurement

Fig. 53.4 Cordless transducer system

53.1.5 Intrauterine Pressure Measurement

This procedure involves inserting a catheter into the uterus via the vagina. The catheter is filled with a neutral solution and connected to a pressure sensor. This method produces an absolute pressure value, but is quite complex and its usage is in decline. Today's increasingly widespread transducers have a pressure sensor at the end of the air column or directly on the tip of a flexible probe, which is also inserted into the uterus via the vagina (Fig. 53.3). The laborious and training-intensive catheter procedure is no longer required.

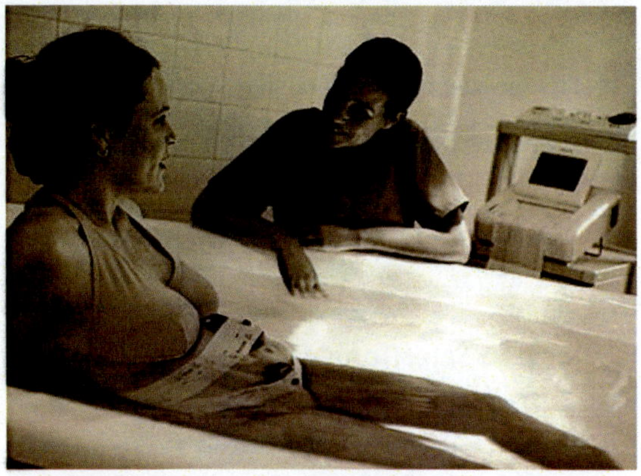

Fig. 53.5 Underwater monitoring using a cordless transducer system

53.1.6 Additional Functions and Options

Equipment manufacturers offer a range of other functions and options for the above procedures to facilitate day-to-day work and enhance quality. These extras include:

- Automatic cross-channel heart rate monitoring. The CTG constantly checks to make sure the maternal heart rate or that of the other fetus (in the case of multiple pregnancies) is not accidentally being monitored on the fetal channel (known as cross-channel verification or coincidence detection). New developments have seen the function for measuring the mother's pulse integrated into the toco transducer to simplify use.
- Battery-operated, watertight transducers also enable use in the bathtub to ensure that continuous monitoring is maintained, be it while having a relaxing bath or during an underwater birth.
- Option to connect to a telemetry system or a cordless transducer (Fig. 53.4) for monitoring outpatients and for use in underwater monitoring (Fig. 53.5).
- Integrated maternal measurements (fetal/maternal monitor).
- Options to connect to obstetric monitoring systems for a central overview of several beds with an intelli-

gent alarm system and automatic long-term storage of monitoring logs and patient records.
- Storage of graphs with remote transmission option. This can be used for home deliveries or for consultations at smaller affiliated hospitals.
- Integrated battery within a CTG with a printer for supporting scenarios involving transport; this is intended for use in the increasing range of applications for home visits, and for bridging power failures.

53.1.7 CTG Selection Criteria

CTGs should meet the following requirements.

1. Antepartum requirements:
   - One to three ultrasound channels (two/three channels if monitoring of twins/triplets is required)
   - One toco channel
   - Fetal movement profile
   - Recorder with adjustable paper speed (1, 2, 3 cm/min)
   - Digital display for heart rates and toco
   - Volume adjustment for heart sounds
   - Low level of energy for ultrasound signal
   - Autocorrelation of ultrasound signal
   - Cross-channel signal comparison (of all fetal and maternal heart and pulse rates available)
   - Method to reduce artifacts (e.g. precision signal track and hold)
   - Means to fix to the cart or wall
   - Option to extend to system interface for system connection
   - Option to connect a telemetry system
   - Coded connectors or automatic recognition of the connected transducers
   - Latex-free transducers and accessories
   - Watertight transducers for cleaning purposes
   - Option to integrate measurement of the mother's heart rate and blood pressure
   - Simple operation (e.g. via touchscreen)
   - Automated CTG assessment.
2. Intrapartum requirements: The features listed in this section either differ from or are in addition to the features listed in 1.
   - One to three ultrasound channels (two/three channels if monitoring of twins/triplets is required)
   - Direct ECG measurement
   - Channel for intrauterine pressure measurement
   - Channel for maternal ECG (MECG)
   - Option to integrate the mother's vital signs (blood pressure, SpO2, MECG)
3. Optional functions:
   - Alarm systems for tachycardia and bradycardia as well as for the presence of any maternal vital signs
   - System interface
   - Additional input options (barcode scanner, mouse, keyboard, etc.).

53.2 Obstetric Monitoring Systems

53.2.1 Areas of Application

Obstetric monitoring systems are used in the antepartum, intrapartum and postpartum sections of hospitals dealing with 500 plus births per year. Depending on individual requirements, one or more of the listed functions are required:

- Central CTG monitoring with a bed overview
- Automatic alarm system
- Patient data management functions, e.g. creating and managing electronic patient records
- Long-term storage media for data archiving and retrieval
- Computer communication within the obstetrics department
- Computer communication with other hospital systems and access to their data.

53.2.2 System Setup

This section is not intended to give a detailed description of an individual system; the commercial system referenced here is used solely to illustrate certain system features. It is essential for the user to have a workstation, which these days is usually a personal computer (PC) connected to a networked system. This PC will serve different functions and may require different configurations depending on where it is used and by whom; for example, a midwife needs to be able to input patient test results in situ, whereas the physician may like to access all data from his office. Here

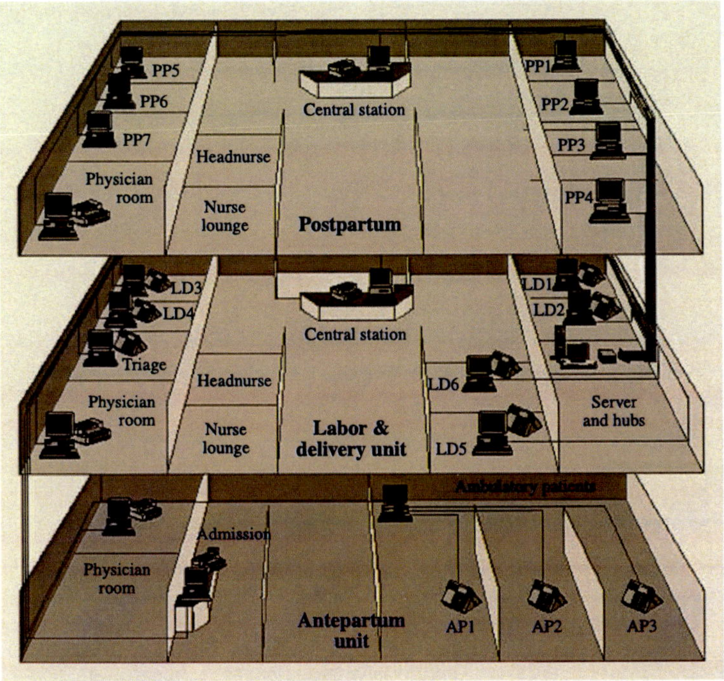

Fig. 53.6 Working area for obstetric monitoring

are some examples of where the PC may be used (Fig. 53.6):

- Central station
- Delivery room
- Ambulance
- Doctor's room
- Nurses' lounge
- Midwife's room
- Postpartum area
- Newborn nursery.

A PC will serve one or more tasks depending on which of the above areas it is used in as well as individual requirements and configurations. These tasks include:

- Collection and processing of data from the connected CTGs
- Assessment of the CTG data and raising the alarm
- Presentation of the CTG data
- Patient data management, enabling electronic patient records to be created, managed and output, and supporting the workflow within the care unit
- Archiving and retrieval of all data via long-term storage media
- Computer communication within obstetrics
- Computer communication with other hospital systems and access to their data.

No matter what the function of the obstetric monitoring system, the protection of patient data is the top priority. This is assured by:

- Restricting access rights within a role model
- Logging all access and transactions, especially changes to patient data and configuration
- Encrypting patient identification and patient data in the database
- Encrypting communication via VPN
- Preventing physical access to central system components (database server, web/terminal service server, network components such as switches and routers)

The following sections outline these individual functions in more detail.

53.2.3 Collection and Processing of Data from the Connected CTGs

One of the most important aspects of obstetric monitoring systems is the central monitoring of the CTG

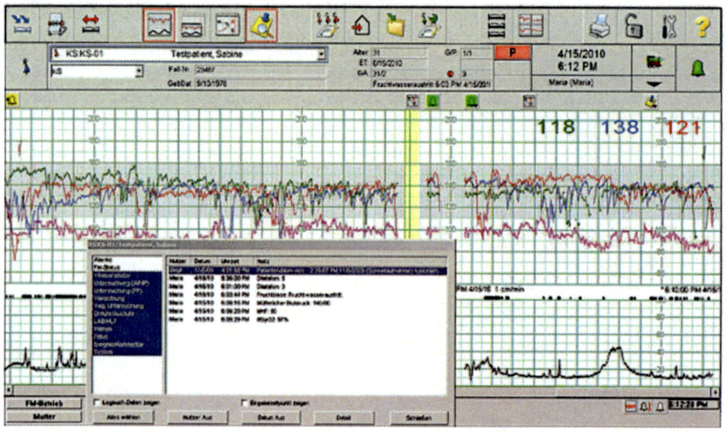

Fig. 53.7 Example of notes input on the CTG

data of pregnant women, which is why the CTGs are connected to the system. Virtually all CTGs available today feature interfaces to enable data to be transferred to a monitoring system.

There are two connection methods:

1. Digital connection with
   - Serial data transfer logs (standardized RS232, RS422, RS485 interfaces with proprietary data formats)
   - Local area network (LAN) (standardized network logs with proprietary data formats)
2. Analog connection with
   - Signal transfer over several lines corresponding to the parameters supported
   - Proprietary signal levels

The digital method is superior to the analog method in the following ways.

1. More comprehensive data
2. Greater protection against transmission interference
3. Adaptable to changes in CTG functionality
4. Reduced equipment, installation and maintenance costs

Nowadays, the analog connection is only used with older CTG equipment. When selecting the CTG connection system, the parameters to be transferred are of paramount importance.

- The following must be transferred as a minimum:
  - Fetal heart rate
  - Second fetal heart rate (for monitoring twins)
  - Maternal heart rate
  - TOCO
  - Automatic time synchronization of the CTG with the system
  - CTG changes, e.g. change in transducer
  - Recorder status
- The following are desirable:
  - Third fetal heart rate (for monitoring triplets)
  - Additional maternal parameters (e.g. SpO_2, noninvasive blood pressure, heart rate)
  - Fetal oxygen saturation
  - Notes input on the CTG (Fig. 53.7) (e.g. vaginal examinations)
  - Fetal movement profile
  - Result of the cross-channel signal comparison to avoid mix-ups (mother–fetus, fetus–fetus)
  - ST segment analysis of maternal ECG (STAN)
  - Measurement of the forces present during ventouse extraction
  - Automatic transmission of cervical dilation value and fetal station
  - CTG errors/interference

53.2.4 Assessment of the CTG Data and Raising the Alarm

One key function of obstetric monitoring systems is the assessment of CTG data so that the user can be made aware, when necessary, of any problems the patient is experiencing. The simplest method is an alarm system that is triggered whenever a threshold is breached; in other words, the system alerts the user if the fetal heart rate rises above or falls below a certain threshold for a given period. One weakness of this method is that the user is alerted more frequently than is required, for example due to false alarms or signal loss. However, there are better methods available these days; *intelli-*

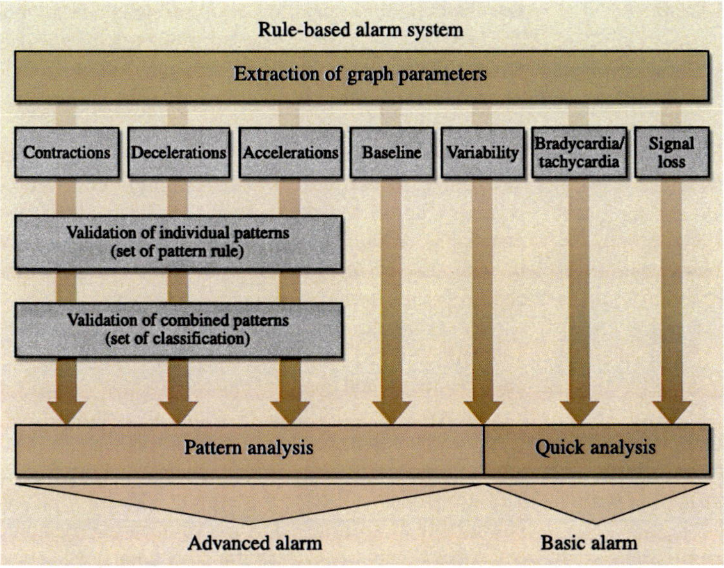

Fig. 53.8 Schematic representation of an alarm system using a rule-based algorithm as an example

gent alarms are able to monitor the profile of the heart rate signal, examine it for certain suspect, dangerous or even pathological patterns, and then alert the user. They can even take into account dependencies between different parameters, such as the profile of the fetal heart rate in relation to uterine activity. Figure 53.8 illustrates the two alarm system principles, and Fig. 53.9 shows an example of an intelligent alarm.

The following list contains the minimum requirements for the different methods.

1. Threshold breach alarm system:
   - Tachycardia alarm system (adjustable threshold values and time delay)
   - Bradycardia alarm system (adjustable threshold values and time delay)
   - Signal lost for a given time period (to be defined)
2. Intelligent alarm system:
   - Detection of accelerations and decelerations and their types (e.g. early deceleration, late deceleration, etc.)
   - Detection of contractions
   - Determination of baseline
   - Detection of variability

The operating concept and type of alarm are also key components of the alarm system. In principle, an alarm can be visual or acoustic in nature. A visual alarm system usually involves part of the screen flashing conspicuously, e.g. an alarm bell; an acoustic alarm system outputs an audible signal, which should have an adjustable volume. Many systems also offer the option of connecting the alarm to the central call system. Any alarm being issued should (unless configured otherwise) always alert the user, regardless of which workstation he or she is currently logged onto or which screen he is currently viewing. The type of alarm system should depend on the user's responsibility and be configurable, e.g. visual alarm only or acoustic alarm only.

Alarms should always be confirmed. The confirmation should be automatically documented and there should potentially be an option to add a comment.

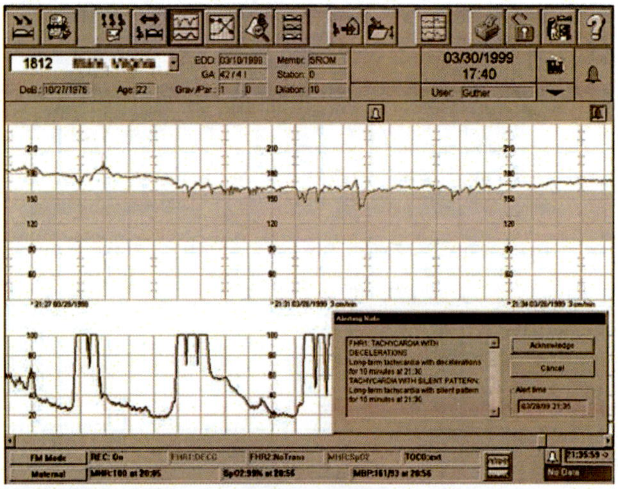

Fig. 53.9 Example of an intelligent alarm

53.2.5 Presentation of the CTG Data

As stated previously, the central monitoring of pregnant women is one of the most important tasks performed by the system. As such, the system should offer a range of ways to display the CTG graphs, such as an overview using various configurations (overview of 2, 4, 6, 9, 12 or 16 beds), a single-bed display, and a compressed single-bed display (Fig. 53.10). It should be possible to configure each PC separately to ensure maximum flexibility. The single-bed display at the very least should be able to offer a quick review of the graphs (quick scroll).

The following must be taken into account when selecting the system:

- The timeline-to-heart rate ratio must be maintained (1:1 mapping of the CTG recording) and adjusted to the respective screen resolution. This applies both to the on-screen display and print-outs.
- The overview screens should be configurable such that they offer different display formats (e.g., an overview of 2, 4, 6, 9, 12 or 16 beds) while maintaining the aspect ratio (CTG recording).
- The normal and compressed single-bed displays should, at the very least, have a scrolling function. It should be noted that it is possible to scroll back to the start (back to the first CTG graph after admission), and not just to the CTG graphs for the last 12 or 24 h.
- The normal single-bed display at the very least should provide an option to add notes directly.

53.2.6 Patient Data Management

The requirements of modern patient data management are extremely far-reaching, beginning with the patient's first visit and then spanning the entire pregnancy, labor and labor outcome, ending for the mother in the postpartum follow-up area and for the newborn in the nursery. Access to previous pregnancies and births is of course also available (Figs. 53.11, 53.12).

Patient data management in obstetrics is characterized by the fact that we are dealing with multiple *patients* (Fig. 53.13): the mother, the fetus and then the neonate(s). Administration for the fetus is typically tied in with that of the mother, with the neonate(s) then acquiring their own patient records/identity.

The tasks required in patient data management are generally divided into three categories: organization, documentation and quality assurance.

All patient data that is input or automatically recorded is stored, providing a seamless way to manage patient records. Patient management is improved as a result because all past and current data on a patient is available at all times. The entire department should of course be able to access the data both quickly and easily. Seamless management of patient records also enables statistical assessments to be performed and reports to be generated for administration, research and quality assurance purposes. It should be possible to adapt the data input to meet the requirements of the patient in question without any problems, and also to generate a flowsheet-based patient record. This record should generally contain the following information:

- Data collection/input
  - Predefined/configurable input screens that cover everything required from the necessary patient records
  - Labor progress chart (Fig. 53.14)
  - Plausibility checks during input

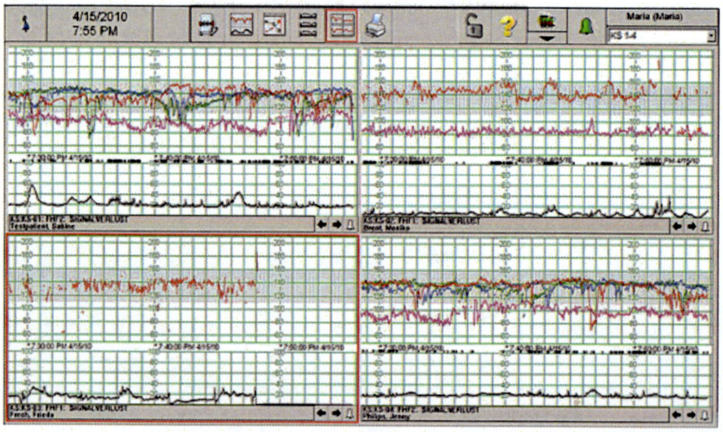

Fig. 53.10 Example of an overview display

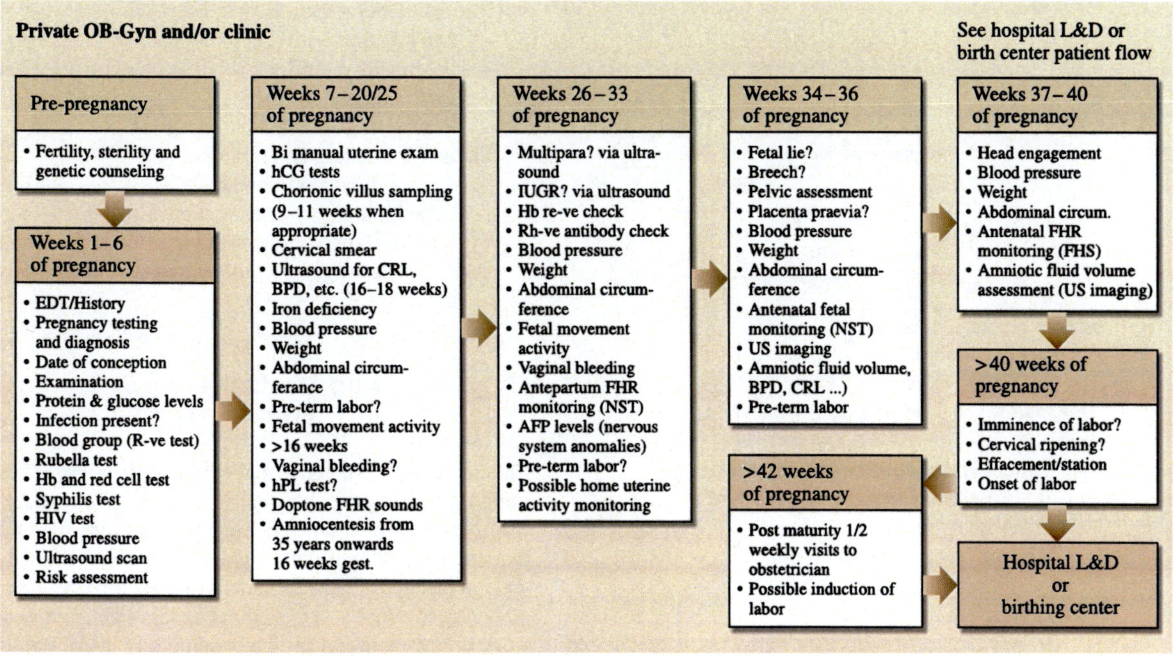

Fig. 53.11 Continuum of care – pregnancy

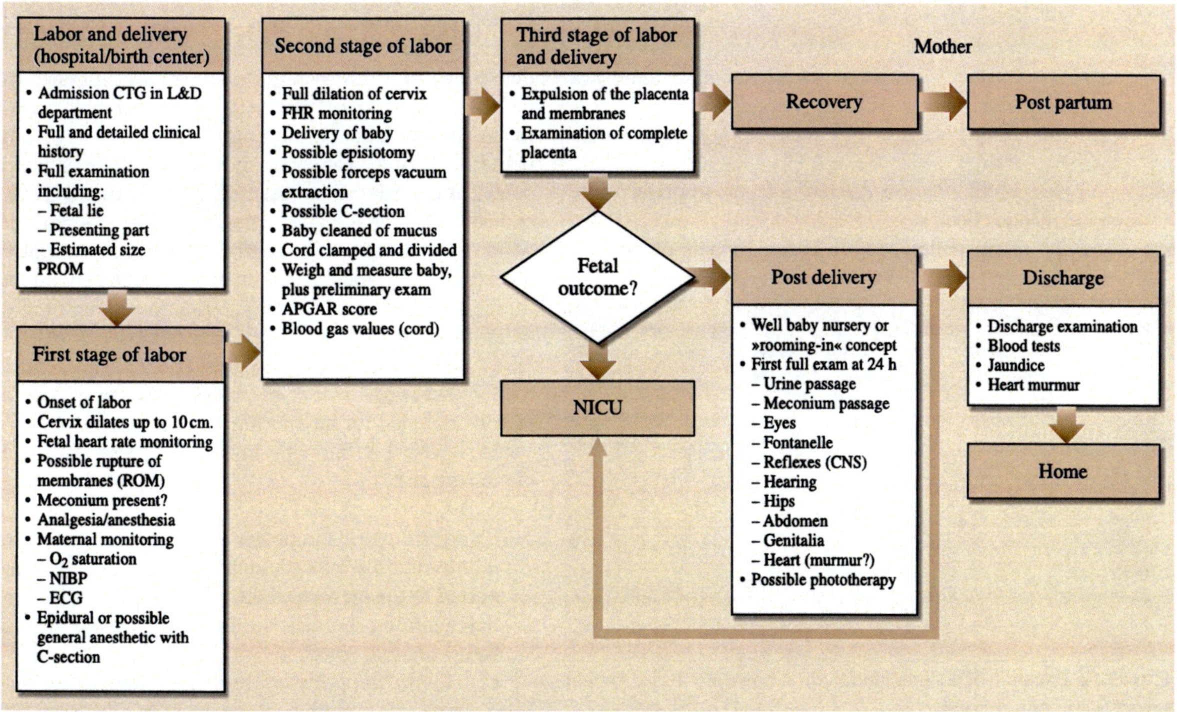

Fig. 53.12 Continuum of care – birth process, postpartum care and newborn nursery

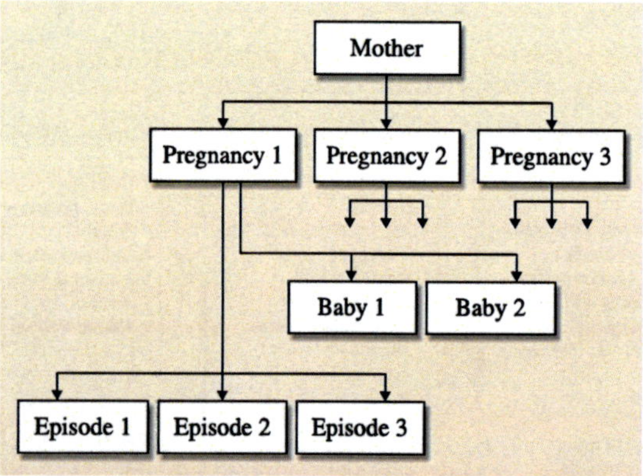

Fig. 53.13 The three patients (mother, fetus, neonate)

- Patient administration
  - Admission, transfer and discharge
  - Activity input
  - Consent to cesarean section
  - Handover of child to parents.
- Production of report
  - Generation of doctor's letter with predefined/ configurable screens and text
  - Transport service request
  - Transfer report
  - Birth report
  - Mother's discharge note to hospital administration
  - Neonate's discharge report to hospital administration
  - Mother/child's discharge report to private practice doctor
  - Registration of birth at civil register office.

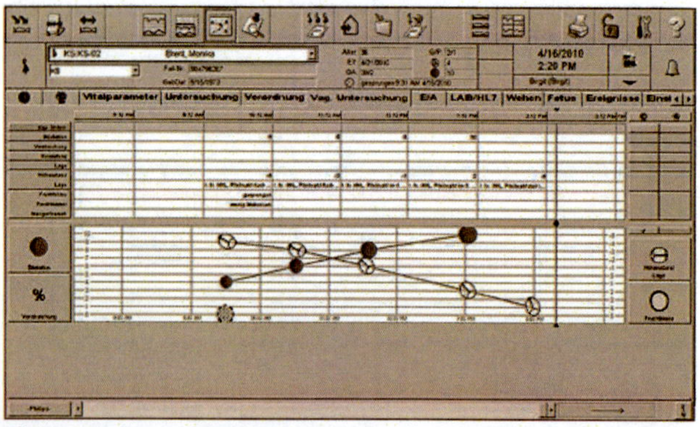

Fig. 53.14 Example of documentation for a flowsheet-based patient report (labor progress chart)

- Routine documentation
  - Tag creation
  - Pregnancy record
  - Birth register
  - Print-out of full patient records.
- Statistics, trending and reports
  - Ad hoc database queries
  - Weekly, monthly or annual birth statistics
  - Special statistics: e.g. cesarean sections for each doctor, average length of patient stay, anomalous incidents
  - Special QA statistics

53.2.7 Archiving and Retrieving Data via Long-Term Storage Media

Electronic data archiving and retrieval have grown in importance at many hospitals in the last few years, mainly due to the increased number of lawsuits filed by parents of injured children, but also because of the medical interest in the subsequent analysis of births. Modern archiving is usually achieved via access-restricted network drives managed by the hospital's IT department, for example through storage on NAS devices (network attached storage) or a SAN (storage area network). The IT department manages the access rights to ensure that no unauthorized users can access the archive or even delete it entirely. Long-term storage such as this enables patient records to be archived in full (for both inpatients and outpatients), and minimizes the risk of not being able to trace patient records or CTG records. Hard copies of records can be lost, and thermal printed CTGs can become illegible after a number of years – both of which have already led to court cases being lost because the exonerating documentation could not be submitted. Straightforward access to the information stored should be ensured for review and research purposes or to call up a closed case. A good security concept is essential for the archiving system, but is less vital for a system merely intended for monitoring CTG graphs.

However, the following must be satisfied under all circumstances:

1. Safeguarding of the power supply by means of an uninterruptible power supply (UPS) that is connected to the server and the network components.
2. Safeguarding the data on the server by one of the following two methods:
   a) *High-availability concept.* A second server operates in parallel to the first. In the event of

a defect, the fault-free server takes over all tasks until the other is functional again.

b) *RAID technology.* This is a group of hard drives that save the data redundantly and independently. Should one of the hard drives fail, it can be exchanged during operation and the system automatically ensures that the repaired drive is updated to reflect the current status.

3. Creation of a security concept to protect the long-term archive. Regular backups ensure that the archive can be restored in the event of a disaster. These backups can also be stored at different locations for extra protection, against fire for instance.

53.2.8 Computer Communication Within Obstetrics

This section covers communication between obstetrics departments in different hospitals within a town, communication with smaller outpatient clinics that work in conjunction with a hospital, the midwife recording CTGs on site, and the doctor who needs to assess patient data either at home, on the move or from his practice (Fig. 53.15).

Obstetric monitoring systems should not only enable patients to be transferred from bed to bed within the same hospital, but also to other hospitals to ensure direct access to patient records. This particularly occurs if the woman was unsatisfied with her previous delivery or the level of urgency means that the delivery location can no longer be accessed (blockage). Web access to patient records plays a key role when communicating with the doctor at home or in his practice; the hospital must have the necessary infrastructure to ensure that the communication is secure. If the doctor is out and about and his interpretation of the CTG is required as a matter of urgency, the relevant sections of the CTG can be sent to mobile telecommunication devices such as an iPhone for his consultation (Fig. 53.16).

53.2.9 Computer Communication with Other Hospitals

The obstetric monitoring system is just one of many systems within a hospital, and it is extremely important for it to be integrated within the overall network, with the result that standalone solutions are receding into the background. Normally, an HL-7-based interface is used to connect to other computer systems, with integration occurring via direct data transfer or via access at the workplace to other systems (Fig. 53.17).

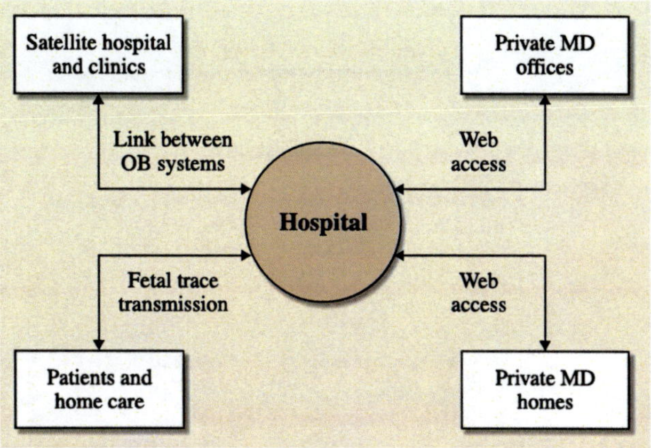

Fig. 53.15 Obstetrics outside the hospital

53.2.10 Communication with the Hospital Administration Computer

The bidirectional connection in this area enables a straightforward patient admissions procedure through the automated importation of personal data, and notifies the hospital administration computer of discharges and transfers.

53.2.11 Exporting Patient Data to Other Hospital Systems

Because data are exported to a data archive, this means that they are also available for other applications, such as for hospital-wide patient records.

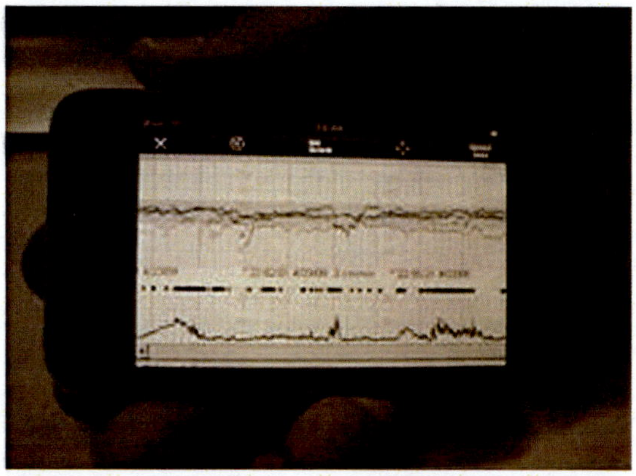

Fig. 53.16 CTG can be sent to mobile telecommunication devices such as an iPhone for consultation

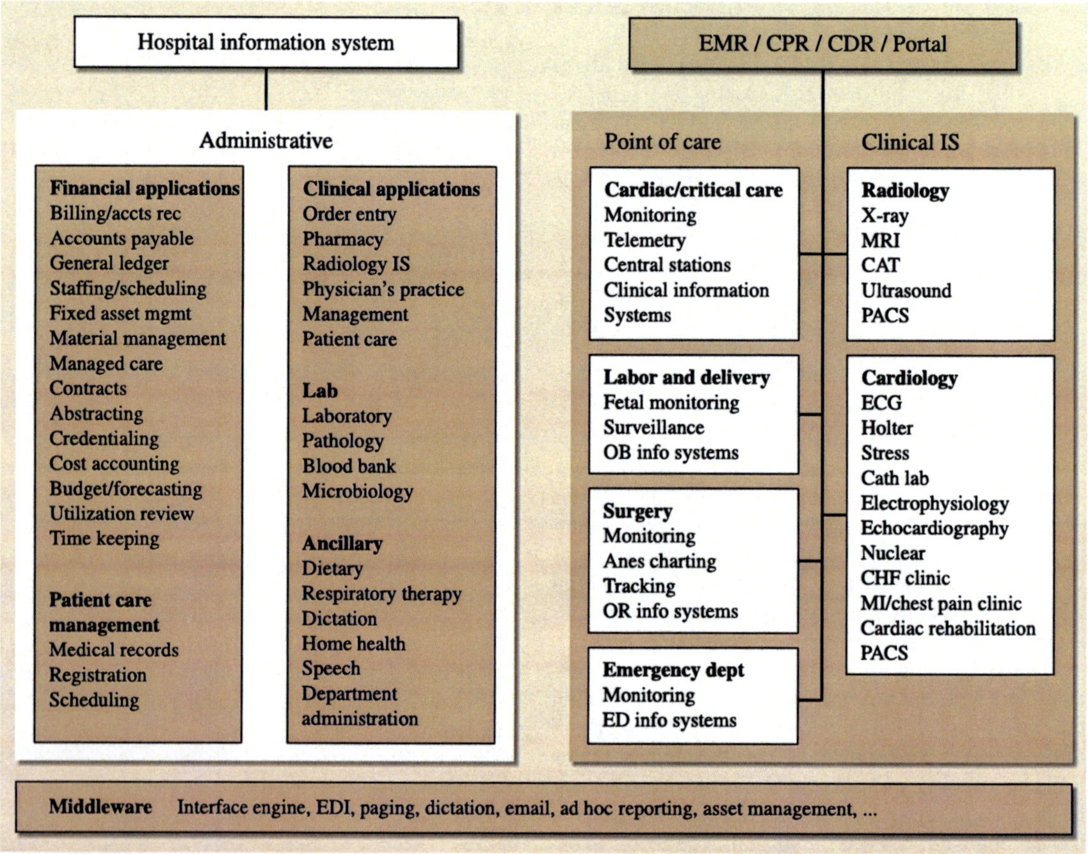

Fig. 53.17 Obstetrics and the hospital

53.2.12 Connection to Lab Computer

Lab results can be documented automatically in the patient's records by importing the data from a lab computer.

53.2.13 Access to Internal Hospital Computer Systems

Various methods have been developed in this regard:

- Direct access via a client application installed on the obstetrics workstation
- Direct access via Internet Explorer using a web-based application
- Indirect access via web-based hospital portals

A visual integration of obstetrics and other applications is possible, as long as the applications are able to exchange their context (user context and/or patient context). If a user switches from one patient to another within the program, the other system will detect this and also switch to the relevant patient.

Further Reading

- M. L. Cabaniss, M.G. Ross: *Fetal Monitoring Interpretation* (Lippincott Williams, Philadelphia 2009)
- M. Murray, G. Huelsmann, P. Romo: *Essentials of Fetal Monitoring*, 3rd edn. (Springer, New York 2007)
- D. Gibb, S. Arulkumaran: *Fetal Monitoring in Practice* (Churchill Livingstone, Kidlington 2008)

54. Neonatal Monitoring

Roland Hentschel

Because they are naturally helpless compared to adults, sick newborns are reliant on careful observation and, due to their characteristically pathophysiological qualities, the monitoring performed is sometimes quite different from that done for adult patients. In order to understand this, it is necessary to consider some of the characteristic peculiarities of neonatology.

The most important diagnoses of neonatal patients are premature birth, disturbance of adaptation (Sects. 54.1, 54.2) in the respiratory and the circulatory system, sepsis, and congenital malformations, in particular of the cardiovascular system (Sect. 54.3). All newborns have instability in the respiratory and cardiovascular systems in common. The more immature the premature infant, the more likely it will be that it needs artificial respiration or respiratory assistance due to having underdeveloped lungs (respiratory distress) or marked central respiratory regulation disorders. Respiratory pauses (central apnoeas), obstruction of the upper airways (obstructive apnoeas), or the typical respiratory pattern of the premature infant (periodic breathing) result in a drop in the O_2 saturation, with subsequent bradycardia (Sects. 54.4–54.8).

54.1 **Electrocardiogram** 1032

54.2 **Impedance Pneumography** 1033

54.3 **Combined Cardiorespiratory Analysis** 1034

54.4 **Pulse Oximetry** 1035

54.5 **Transcutaneous Measurement of the Partial Pressure** 1037

54.6 **Measurement of the $PtcCO_2$ (Transcapnode)** 1038

54.7 **Measurement of the $PtcO_2$ (Transoxode)** 1038

54.8 **Monitoring the Oxygenation – Which Method?** 1039

54.9 **Setting Alert Limits and Limit Values** 1040

References 1041

Unlike with adults, undersaturation of the blood sets in extremely quickly because the oxygen stores (as measured against the oxygen demand) are very low. On the other hand, bradycardia can occur spontaneously as a result of the activation of vagus reflexes in the pharyngeal region (accumulation of secretions). Central, obstructive or mixed apnoeas are a key problem in the monitoring of premature infants.

Newborns have two particular characteristics in terms of their respiration: they are obligate nasal breathers and, in contrast to adults, they breathe predominantly with the diaphragm. Because a newborn infant's thorax is mechanically very unstable, see-saw breathing frequently occurs: the lung volume which is gained because the diaphragm descends during inspiration is for the most part lost again as a result of inward movements of the sternum and upper thorax.

Various pathophysiological disorders, such as:

- Cerebral immaturity
- Cerebral haemorrhages
- Respiratory distress
- Infections
- Congenital metabolic disorders
- Electrolyte imbalances
- Convulsions
- Hypo- or hyperthermia
- Hypoglycaemias, or
- Anaemia

manifest themselves in cardiorespiratory disorders in premature and newborn infants. It is possible to gather from this that the following parameters are central to neonatal monitoring:

- Single-channel ECG
- Respiration
- O_2 saturation
- Transcutaneous blood gas values
- Blood pressure.

Furthermore, on account of the particular thermal instability of premature and newborn infants, temperature monitoring also plays a large role. Premature infants are therefore cared for in incubators. In contrast to adults, premature infants are at particular risk in terms of their oxygen status. As with adults, a toxic (excessively high) inspiratory oxygen concentration can trigger a sterile inflammatory response in the lungs, leading to chronic lung damage. In addition, the feared retinopathy of the retina (vascular proliferation which can lead to blindness) can also be triggered in premature infants as a result of an excessively high inspiratory oxygen concentration with subsequent hyperoxia; this risk is present up until the point at which a premature infant reaches its previously expected delivery date. Continuous monitoring of the oxygen status is therefore imperative.

In patients who are given artificial respiration, the inspiratory oxygen concentration is adjusted using the ventilator and also monitored. Also, premature infants who are capable of spontaneous breathing or those connected to a CPAP machine must always be connected to an oxygen monitoring unit. If a nasal cannula is used to administer a specific gas flow at a specific oxygen concentration, the knowledge of these variables does not allow the inspiratory oxygen concentration to be estimated, as the child can breathe additional air. It is better to set a certain oxygen concentration in the *microclimate* of the incubator and to use a suitable measuring and alert system for monitoring. As with adults, in neonatology, modular universal monitors are mostly used which record different parameters using a dedicated measurement module for each, and display these parameters on a multichromatic multichannel monitor. Alert systems for monitoring the oxygen concentration are usually integrated ex factory into ventilators and CPAP machines and also into incubators.

54.1 Electrocardiogram

The electrocardiogram (ECG) records the electrical activity of the heart in terms of its changing amplitude and the changing direction of the electrical principal axis. As with adults, the single-channel ECG is conventionally recorded by means of ECG electrodes at the typical positions on the chest. It is occasionally recommended that all of the electrodes are placed on the lateral thorax so that the lungs are covered as minimally as possible in x-ray images (which must be taken regularly), meaning that the electrodes do not have to be removed in order to perform an x-ray. For this purpose, electrodes can be attached to the upper arms and abdomen. Because the single-channel ECG is not usually used for cardiological diagnosis, standard electrode placements are unnecessary, and the red and yellow electrodes are otherwise arranged in the electrical axis of the heart. The electrodes for newborn infants are smaller than those for adults, and they are usually coated on the back with an adhesive which simultaneously acts like a contact gel in order to reduce the contact resistance. For premature infants, it is necessary to consider the skin tolerance of the material used and the x-ray transparency. In fully developed newborn infants, problems occur shortly after birth, in particular if they have a lot of vernix caseosa; it is often possible to obtain an ECG recording only after degreasing measures and by using electrodes which adhere well.

Violent (respiratory) movements frequently cause *shaking* of the ECG; this is not a critical issue when recording the heart rate, but it is for the recording of arrhythmias, which is occasionally necessary. By interconnecting the three chest electrodes, it is possible to obtain three different leads, provided that all the electrodes adhere well. The amplification selected should always result in the amplitude of the complete ECG signal occupying the entire bandwidth of the monitor channel; it is only in this way that interference with normal depolarisation and repolarisation (e.g. in the case of hypokalaemia) can be reliably detected.

However, the only diagnoses which can reliably be made from the single-channel ECG are: bradycardia, tachycardia, asystole, supraventricular and ventricular extrasystoles, ectopic rhythm, and atrial and ventricular flutter or fibrillation. Automatic evaluations such as the diagnosis of arrhythmias, measurements of the ECG portion and the like are not customary in neonatology. As already mentioned, the positioning of the electrodes is usually not critical when recording an ECG, but it is crucial when recording the thoracic impedance curve, and must therefore be selected with great care in this respect (see below). The single-channel ECG is primarily used to detect tachycardias and (more commonly) bradycardias. To this end, the alert limits for the heart rate must be set such that they correspond with the standard values for the age group (Table 54.1).

The majority of monitors these days provide memory recording of the heart rate and other parameters over a period of at least 24 h. The duration of a heart rate anomaly can then be accurately traced retrospectively over a time axis which must span a different length of time. In addition, event recording is desirable in which between five and six cardiac cycles are automatically recorded when an alert based on the heart rate is triggered, with the result that later analysis of the event is possible using a graph which is true to the original. Unfortunately, most monitors only display these events with a compressed time axis, making accurate analysis of the rhythm impossible. It should be possible to store at least 20 events.

Table 54.1 Standard values for the heart rate (average and simple standard deviation (SD)) (after [54.1])

Age	Average + 1 SD (beats/min)	
0–24 h	133	22
2nd–4th week	163	20
1st–3rd month	154	19
3rd–6th month	140	21
6th–12th month	140	19
1st–3rd year	126	20
3rd–5th year	98	18
5th–8th year	96	16
8th–12th year	79	15
12th–16th year	75	13

54.2 Impedance Pneumography

This method detects changes in the electrical impedance occurring between two electrodes during a respiratory cycle, probably as a result of changes in the vascular filling. Due to the susceptibility of premature and newborn infants to respiratory disorders, this part of the monitoring assumes great significance in neonatology. The method can show the respiration rate, respiratory pattern, amplitude and respiratory pauses via the existing ECG electrodes.

In the case of unstable respiration in premature infants, a recording speed that is as slow as possible should be selected (e.g. 6 mm/s), as this is most likely to give information about the respiratory pattern (e.g. periodic breathing). The electrodes should be positioned in the area of the greatest change in size during a respiratory cycle. This area is between two electrodes on the anterior right and left axillary lines in the region of the costal arch.

The large degree of variability in the respiratory patterns of premature and newborn infants is problematic. Since the respiratory regulation is immature at various levels, respiration is not as autonomic (*machine-like*) as in adults. The result of this is not only pronounced respiratory pauses but also variations in the amplitude and the respiratory starting position. The recording and processing of the thoracic impedance curve therefore places considerable demands on the software of the monitor.

The respiration rate (Table 54.2), the frequency of apnoeas and occasionally also the durations of the individual apnoeas are monitored using limits which can be adjusted according to the age of the patient. Sub-threshold respiratory pauses that do not reach the selected duration of a defined apnoea can be added together on dedicated monitors to give *respiratory deficits* via adjustable limit values in order to facilitate the detection of periodic breathing.

According to its definition, an apnoea exists when there is respiratory arrest for a period of at least 20 s.

Table 54.2 Standard values for the respiration rate (average and simple standard deviation (SD)) (after [54.1])

Age	Average + 1 SD (beats/min)	
Premature infant	50	10
Newborn infant	40	10
Infant	31	8
1st–4th year	24	4
5th–9th year	20	2
10th–14th year	19	3
14th–16th year	17	3

It should be possible, however, to freely select other intervals for monitoring too.

Due to the greatly oversized volume of the heart (in comparison with the thoracic volume) compared with adults, fluctuations in the size of the heart during a normal cardiac cycle, with a corresponding amplification in the impedance pneumography, occasionally appear as normal respiratory cycles. This is particularly true of the cardiac hyperactivity which is very frequently observed in the context of cardiac problems (e.g. ductus arteriosus). In this situation, the monitor can under some circumstances incorrectly interpret the heartbeat – which continues for a while even when breathing has completely stopped – as continuing respiratory activity. By activating a detection mode which compares the recorded heart rate with the incorrectly recorded *respiration rate*, this signal problem can be detected so that an alert is triggered.

In order to interpret causal relationships, it is necessary to obtain a retrospective evaluation with respect to current measures such as nutritional probing, care times or administration of medication. Only then is it possible to make any statements about the possible aetiology of a respiratory disorder. The recording of a complete respiratory cycle requires a curve profile, in opposite directions in each case (inspiration and expiration), where a defined minimum amplitude, based on a minimum or maximum value, has always been reached. The amplitude can either be input permanently or adapted to the changing depth of respiration in accordance with a (usually very complicated) algorithm.

Displacement of the normal respiration is usually not a problem when detecting a breath. However, an apnoea is occasionally recorded erroneously. Modern monitors are capable of varying or automatically adapting the size of the amplitude shown and the zero point within certain limits, such that a diagram of the respiratory pattern which is to some degree realistic can be achieved, and thus a visual recording of apnoeas is possible. After every change in the position of the electrodes, but also after the patient has been moved, the monitor setting for the impedance pneumography must be readjusted (the settings for the depth of respiration and the amplitude size or the selection of the automatic setting function). In the case of infants, for undisturbed monitoring of respiration, the electrodes are usually attached at a markedly lateral and low position (based on the entire projection field thorax–abdomen). An undisturbed recording of the single-channel ECG should also be possible in this position.

It is desirable to have the option to record the thoracic and the abdominal impedance pneumography in two channels superimposed on one another on the time axis. Ideally, the amplitude maxima will lie exactly above each other, with the leads orientated in the same direction. Each phase shift between two leads signifies a reduction in respiratory economy, and in extreme cases where there is an opposing deflection there is paradoxical respiration with entirely ineffective respiratory work (*thoracoabdominal asynchrony*).

This sometimes cyclically occurring phenomenon is found in certain chronic lung conditions of premature infants (*bronchopulmonary dysplasia*). Because it can improve under some circumstances as a result of various measures (medication, inhalations, inflation manoeuvres, CPAP), this diagnostic possibility is of great significance. Unfortunately, such a diagnosis is not yet possible using standard monitors, but there are special devices which, in addition to displaying the phenomenon, can also precisely calculate the angle of the phase shift.

54.3 Combined Cardiorespiratory Analysis

In order to recognise the aetiology of a cardiorespiratory disorder in premature infants, it is important to be able to retrospectively analyse the precise chronological sequence of respiration (respiratory pattern), heart rate and saturation curve. Only then is it possible to differentiate whether a primary O_2 undersaturation has led to apnoea and then to bradycardia, whether a primary reflex bradycardia has led to undersaturation with consequent apnoea, or whether a primary apnoea has triggered undersaturation with subsequent bradycardia (Fig. 54.1). It would also be desirable to see the graph displayed in addition to the retrospective analysis of the O_2 saturation. Unfortunately, the memory capacities of standard monitors are usually insufficient for a data set of this size.

Cardiorespiratory monitoring or analysis is important, particularly in the field of home monitoring of infants which are at risk of sudden infant death syndrome (SIDS). These are, in particular, siblings of infants who have suffered from sudden infant death syn-

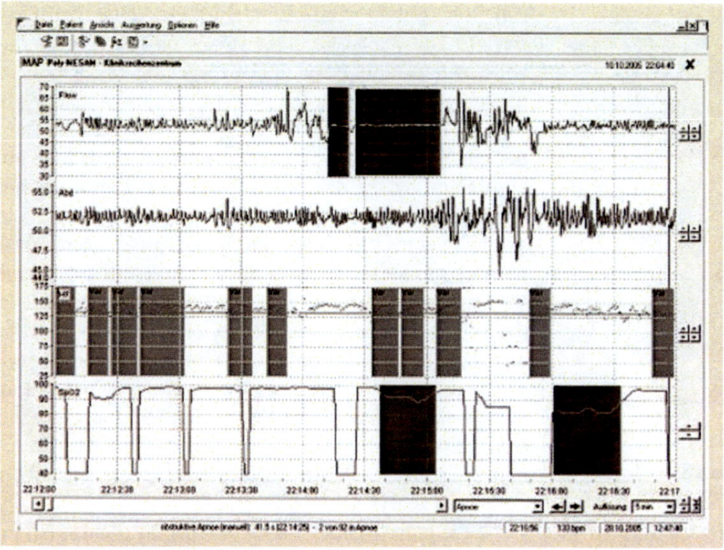

Fig. 54.1 Impedance pneumotachography. Nasal flow, abdominal movement, heart rate, oxygen saturation *(from top to bottom)*. Pronounced, prolonged, obstructive apnoea with an absence of nasal flow and with good autorespiration. Note the artefact-related decreases in saturation

drome, children with severe neurological disorders, and infants who have constant respiratory disorders.

Exclusive monitoring of the respiration is carried out, for example, using a mechanical applanation sensor (Graseby capsule). This sensor records pressure changes within a closed, air-filled system when a convexly curved balloon that is stuck to the skin in the abdominal region is slightly compressed as a result of a respiratory movement. Because it is not capable of distinguishing unsuccessful respiratory movements or a final gasp with the accompanying bradycardia from effective respiration, exclusive monitoring of the infant using this method is no longer standard these days.

Most SIDS monitors have the option of recording alert sequences and trend curves of the monitored parameters and analysing them offline using special analysis software.

According to the most recent recommendations, the monitoring of apnoeas should preferably be done using a pulse oximetry device (see below) which also records the pulse rate; additional monitoring of the respiration is then not necessary. Note: *Devices which are freely available on the market for monitoring SIDS are based on different measurement principles; they should generally be rejected on account of their high susceptibility to faults or unproven effectiveness.*

54.4 Pulse Oximetry

Pulse oximetry has developed as a continuous, noninvasive, indirect method of measurement to become the standard method of monitoring the oxygen status. It is a method which uses transmission photometry to measure the concentration of oxygenated and deoxygenated haemoglobin according to the Beer–Lambert law. These two forms of haemoglobin have different spectral properties. Two light-emitting diodes conduct light – in the range of the absorption maxima of the two substances (visible range approximately 660 nm and infrared range approximately 940 nm) – through a peripheral body part, and a photodiode is located at the exit point opposite. The percentage of saturated haemoglobin is calculated from the attenuations of the two light beams.

What is crucial here is that the measuring system records a pulse wave and thus only evaluates the pulsatile fraction of the measurement signal. The pulsatile fraction is detected via the increase in the distances covered by the light beams in the arterial filling phase in the tissue. If no pulse wave can be detected, the device does not display any reliable values; the recording of the pulse curve (plethysmogram) is therefore a fundamental part of any pulse oximetry system, and preference should be given to displaying an authentic pulse curve on a time axis (x-axis) as opposed to a simple LED dis-

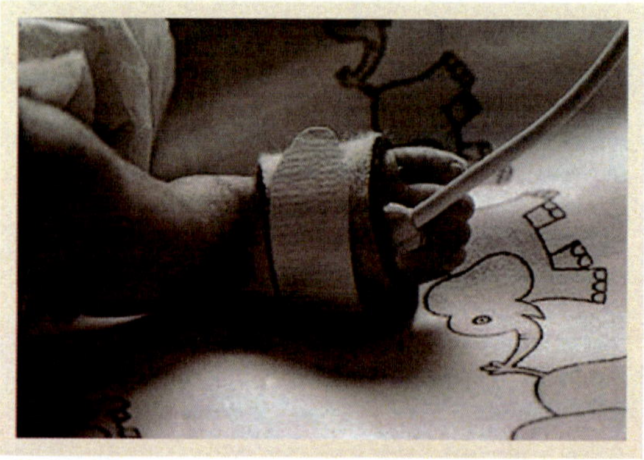

Fig. 54.2 Pulse oximetry in a premature infant

play with rising and falling *level values*. The pulse curve is generally also used to calculate a pulse rate.

The pulse curve is also suitable for use as a diagnostic tool in order to show the quality of the peripheral pulse wave, e.g. in the event of resuscitation. In neonatology, adhesive sensors or soft plastic sensors are used exclusively instead of clip sensors, and these are placed loosely around the preferred recording sites and fixed with a small piece of hook-and-loop tape (Fig. 54.2). Preferred positions are the wrist, forefoot, arm and leg.

If the patient has poor peripheral microcirculation (e.g. also in the context of catecholamine administration), application of the sensor is difficult; often no pulse wave is recorded, since the light pressure of the sensor on the skin and on the underlying tissue is sufficient to stop the perfusion in this measurement area. This can be detected by a nonphysiological plethysmography curve or by a discrepancy between the pulse rate measured and the heart rate based on the ECG recording.

Reliable measurement is also often not possible in oedematous tissue. In order to avoid pressure necroses in very small premature infants, in some circumstances the recording sites must even be changed at intervals of 3–4 h. The sensors can cause burns on the skin, namely when the surface is damaged or when sensor and device are not optimally tuned to one another.

In the event of scattered light falling on the sensors, excessively low saturation values are usually erroneously simulated. In some circumstances, however, a normal saturation value can be displayed even when there is complete loss of skin contact, although, a pulse curve is not recorded then. If the two diodes are not in exact contact with the skin, this can likewise lead to incorrect high or low saturation values. When the blood pressure falls below 30 mmHg, some devices no longer function reliably. Unfortunately, most pulse oximetry devices show measurement values on the display despite a detectable error in the recording as a result of movement artefacts or the like.

It is imperative that all pulse oximetry values which do not originate from an artefact-free recording are eliminated by the device, and at the very least are not displayed as a numerical value. All devices calculate average values from a specific – sometimes selectable – number of cardiac beats, or they calculate the averages over a specific time frame. The longer the duration of the averaging, the slower the device is to react to acute changes, but the lower the rate of (false) alerts, too. There are improvements in the software which allow movement artefacts to be better detected and excluded from the computer analysis. These devices filter the venous pulse waves which result from movement artefacts out of the analysis.

If the pulse oximeter does not display a pulse curve, or if the values are unreliable, it is first necessary to rule out the possibility of scattered light (especially from heat lamps) falling on the sensor. The next step should be to try to loosen the fixing. If this does not solve the problem either, then another recording site should be selected where less tissue is transilluminated.

Considering the sources of errors identified, pulse oximetry values must also be counter-checked regularly with arterial or hyperaemised capillary blood gases. The sites used to take blood samples and record the pulse oximetry should be identical with respect to the positional relationship to the ductus arteriosus.

The characteristic curve of this measuring technique is not linear: in the region below 75% the sensor displays unreliable values, and in the upper region the validity of the measured values is restricted by the physiologically flat profile. The devices mostly take into account general average values of haemoglobin molecules which do not serve as oxygen carriers (COHb and MetHb), but which taken together usually comprise less than 3%. This percentage can be calculated exactly by cooximetry analysis in modern blood gas analysis systems. Elevated COHb and MetHb values can lead to incorrect pulse oximetry measurements. The MetHb value can rise when nitric oxide gas is used.

Due to the asymptotic relationship between the oxygen partial pressure and the associated Hb saturation, pulse oximetry is only suitable for monitoring premature infants up to a point: hyperoxia cannot be reliably detected, whereas a decrease in oxygen can be recorded very well. Since hyperoxia must be avoided at all costs in premature infants, pulse oximetry is not a suitable method for monitoring premature infants requiring oxygen. In the case of premature infants without an additional oxygen requirement (room air), pulse oximetry is a good method for recording decreases in saturation and, to be exact, the upper limit value in this situation can be set at 100%, because hyperoxia can be virtually ruled out in room air (although it cannot be completely prevented in any case).

Pulse oximetry offers a good compromise as a simple monitoring device insofar as both the heart rate and the saturation can be recorded as central cardiorespiratory variables using a single cable and the associated sensor, meaning that there is no need to attach ECG cables. This is an advantage in particular for a very immature premature infant with very sensitive skin that does not tolerate the application of adhesive ECG electrodes. However, the recording of the pulse is not very reliable because of its susceptibility to movement artefacts.

Pulse oximetry is also of essential importance in the detection of pathophysiological perfusion disorders, in which the upper and lower halves of the body receive blood which is saturated to different degrees. In rare, complex heart defects, the lower half of the body can be better oxygenated than the upper half, but the opposite situation is more common: in persistent foetal circulation (*PFC syndrome*), unoxygenated blood is conducted – as a result of high pressure in the pulmonary circuit – from the pulmonary artery via the ductus arteriosus into the lower half of the body, thus leading to undersaturation in this region, whereas the upper half of the body is supplied with oxygen-rich blood via the section of the aorta before the junction with the ductus arteriosus.

Table 54.3 Recommended limit values for monitoring the oxygen status in neonatology

Parameter	Recommended limit values
Pulse oximetry	
Premature infants	83–93%
Sick newborn infants	92–98%
Transoxodes	
Premature infants	45–65 mmHg
Sick newborn infants	65–90 mmHg

This pathophysiological situation, which can often vary widely, is of great significance and is therefore monitored with the aid of pulse oximetry sensors on both the right hand and one of the feet. Unfortunately, most universal monitors do not provide the option to record two pulse oximetry values.

It is very difficult to standardise pulse oximetry devices; they vary from one manufacturer to the next, and from one device to the next. Even the reconstruction algorithms vary between different manufacturers depending on whether the fractional or the functional oxygen saturation is quoted. For the most common devices, however, there is now good data in the scientific literature for the correlation between oxygen saturation and oxygen partial pressure. Recommended alert limits for the monitoring of premature and newborn infants are reproduced in Table 54.3.

54.5 Transcutaneous Measurement of the Partial Pressure

This method of transcutaneous partial pressure measurement for carbon dioxide ($PtcCO_2$) and oxygen ($PtcO_2$) is peculiar to paediatrics, as the greater skin thickness in adults usually prevents accurate measurement. The local hyperthermia at the location of the measuring sensor causes arterialisation at the measurement site. The blood gases, which under these conditions are in equilibrium with the partial pressure values in the peripheral tissue, can thus be monitored noninvasively. However, the values must be regularly calibrated with (preferably arterial) blood samples. Both gases ($PtcCO_2$ and $PtcO_2$) are frequently measured together in a combination electrode.

In order to rule out burns to the skin, the monitors monitor the duration of the application of the sensor at the measurement site, request a change in the application site and recalibration and, under some circumstances, even switch off the heating of the sensor automatically when a specified time span is exceeded. The heat output needed to maintain a constant sensor temperature is usually displayed, and its profile gives valuable information about the centralisation of the circulation, provided that the ambient temperature is kept constant.

If there is a sharp rise in the heat output, the application site should be checked especially closely for signs of heat damage. With a heating temperature of 43 °C, in the case of very small premature infants, for example, the placement of the sensor should be changed at intervals of 2–3 h.

The local overheating leads to a shift in the haemoglobin dissociation curve, which results in an increase in both the $PtcO_2$ and the $PtcCO_2$. This is counteracted in the calculations by the oxygen consumption in the tissue, resulting in a realistic oxygen measurement which is close to the arterial values. However, due to the production of CO_2 in the heated tissue, the $PtcCO_2$ values are mostly too high. These effects can be compensated for in the calculations by in vivo correction. When the heat output varies to a large degree, arterial blood gas values must be taken continuously to check the displayed measurement values.

The sensors are attached to the skin using adhesive rings and a drop of contact solution, avoiding air pockets (Fig. 54.3). At intervals of approximately one week, the sensors must be covered with a new membrane. Calibration should be performed at least once a day.

Fig. 54.3 Transcutaneous sensor for measuring the $PtcCO_2$ in a premature infant

54.6 Measurement of the $PtcCO_2$ (Transcapnode)

The sensor attached to the skin generates a local skin temperature of 43 °C. CO_2 then diffuses from the skin – in which it has the same concentration as in the capillary terminal vascular bed of the blood – via a membrane with hydrophobic properties (silicone or Teflon) into an electrolyte solution. There, the CO_2 molecules become H^+ ions, which change the pH. This change is recorded by a pH electrode and compared with a reference electrode, and the voltage potential is proportional to the pCO_2. A potential source of errors is the CO_2 formed in the skin itself, which is dependent on metabolic processes. Since the partial pressure difference between the arterial and mixed venous blood is very much lower and CO_2 is better diffused than O_2 on account of their different solubility coefficients, the degree of (arterial) tissue perfusion is less crucial than in the measurement of the $PtcO_2$ (transoxode). For these reasons, the $PtcCO_2$ is frequently higher than the arterial pCO_2. A two-point calibration is performed at fixed intervals using a special calibration gas. The transcapnode is less dependent on the measurement site, blood pressure, pH, body temperature and haematocrit, and a sensor temperature of 42 °C is often sufficient. At 30–50 s, the response time is longer than that of transoxodes.

54.7 Measurement of the $PtcO_2$ (Transoxode)

The reliability of this method is dependent on the thickness of the skin, the measurement site and the peripheral perfusion at this site. It is also influenced by the sensor temperature selected and the fixing of the sensor. Moreover, it is unreliable at a pH of 7.0 and below, and also in cases of severe anaemia (haematocrit < 28%).

These conditions are rarely found, in contrast to systolic blood pressure values of below 32 mmHg, which are common in very small, sick, premature infants, and which then limit the reliability of the $PtcO_2$ measurement considerably. This can also be interpreted differently, however: under some circumstances, the discrepancy between the results of an arterial blood gas analysis with standard values and an unreliable $PtcO_2$ measurement with lowered values points to the beginning of a breakdown in the peripheral circulation. In premature infants with chronic lung disease (bronchopulmonary dysplasia), the measurement of $PtcO_2$ is often unreliable, probably on account of strong fluctuations in the skin perfusion.

For optimum correlation with the arterial pO_2, a sensor temperature of 44 °C is recommended, and in the case of small premature infants, a temperature of 43 °C is a workable compromise between the reliability of the measurement values and the risk of heat damage. Calibration is performed with the ambient air, taking into consideration the current air pressure. After every sensor repositioning, a time span of 10–15 min is required before stable steady-state conditions are reached. The response time for acute changes in the pO_2 varies between 20 and 60 s depending on the conditions.

54.8 Monitoring the Oxygenation – Which Method?

Opinions about how the oxygen status should be monitored vary greatly from one hospital or clinic to the next. There are good arguments for various standards. The fluctuations in the oxygen saturation in premature and newborn infants, which are in any case much larger than in adults, must first be kept constant within narrow limits. Hyperoxia is dangerous in premature infants in the first instance due to the potential development of retinopathy. Chronic lung damage can also arise as a result of unnecessarily high inspiratory oxygen concentrations, however. Finally, the formation of oxygen radicals is also important for other pathophysiological processes.

Hypoxia is naturally equally unwelcome; after all, all cells in the body – but in particular those in the developing brain – are highly dependent on oxygen.

The dilemma over the narrow oxygen limits is increased by the fact that the optimum target range for the oxygenation of premature and newborn infants has not yet been conclusively identified, and can only be defined by a method with great difficulty. It is known that the intrauterine foetus has a paO_2 value of only about 30 mmHg in the arterial blood, and grows without any problems. For extrauterine life, additional oxygen is required for purely pragmatic reasons as additional organ functions (e.g. respiration), metabolic activities and also the generation of heat *initiate*. For small premature infants at risk of retinopathy, according to currently prevailing scientific opinion, the arterial pO_2 should not be above 60 mmHg or below 40 mmHg. It is sensible to set the limits using the pO_2 as, due to peculiarities in the oxygen–haemoglobin dissociation curve, the upper limit cannot be monitored exactly by means of saturation (pulse oximetry) (Fig. 54.4).

Aside from the fact that complete saturation of the haemoglobin of 100% can be dangerous for premature infants anyway, the additional oxygen dissolved in the blood must be regarded as toxic. Noninvasive monitor-

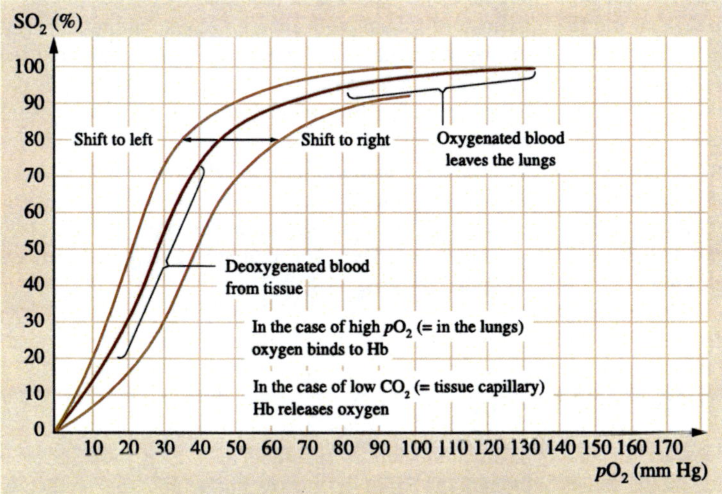

	Shift to left	Shift to right
Causes	Severe immaturity (increase in HbF) Alkalosis Hypocapnia Hypothermia Reduction in 2.3-DPG	Exchange transfusion (increase in HbF), transfusion Acidosis Hypercapnia Fever Increase in 2.3-DPG
Consequence	Increased binding of O_2 to Hb: poor O_2 release into the tissue (i.e. only in the case of lower pO_2)	Reduction of O_2 affinity: More favourable O_2 release into the tissue due to an increased pressure drop

Fig. 54.4 Oxygen–haemoglobin dissociation curve (after [54.2])

Table 54.4 Advantages and disadvantages of monitoring systems for oxygen status

	Pulse oximetry	Transoxodes
Advantages	Fast to become operationally ready No calibration necessary Short response time Virtually no local irritation Low dependency on the circulation Suitable for all ages	Measurement of the relevant variable (pO_2), Good correlation with the arterial pO_2, Few artefacts
Disadvantages	Unsuitable for hyperoxia monitoring Inaccuracy at < 75% saturation Movement artefacts common Sensitive to scattered light Frequent alerts	Warm-up time of up to 15 min Long response time, Not sensitive to brief variations in the oxygenation Reliability limited in the case of certain pathophysiological circumstances Localised burns Labour-intensive due to frequent calibration Susceptible to faults on the part of the membrane Rigid cable

ing of the oxygen status should therefore be performed by means of transoxodes in the case of small premature infants who require an elevated inspiratory oxygen concentration. If short-term under/oversaturation is suspected, the faster-reacting pulse oximeter should be used in addition. A pulse oximeter is sufficient for premature infants who do not require extra oxygen and for mature newborn infants.

For very small premature infants with poor perfusion and very sensitive skin, it may be necessary to do without transcutaneous gas sensors (and any ECG electrodes) and to monitor exclusively using pulse oximetry. In more mature patients, movement artefacts can cause false alerts in the pulse oximetry, meaning that transoxode monitoring is more suitable. On the other hand, in patients who are increasingly mobile, the rigid cables of the transoxodes and transcapnodes are a real hindrance.

If oxygen monitoring is necessary in the case of an acute emergency, pulse oximetry should in principle initially be performed preferentially because of its fast availability, regardless of the maturity of the patient. In premature infants with severe bronchopulmonary dysplasia, pulse oximetry should again preferably be used on account of its greater degree of reliability in this patient group. If there are doubts as to the relevance of measurement values with one of the two methods for an individual patient, the alternative method should be used in addition.

The advantages and disadvantages of the two alternative methods for monitoring oxygenation are compiled in Table 54.4.

54.9 Setting Alert Limits and Limit Values

When setting limit values, it is necessary to distinguish between the target range and alert values (alert limits). The target range includes the values that are aimed for, which means that the continuously recorded values should predominantly be within this range. In contrast, the alert values specify those values which, if they are exceeded or undershot, will illicit a response from the staff by triggering an alert (stimulation of the child in the event of an apnoea, adjustment of the inspiratory oxygen concentration, etc.).

If the alert range selected is too *generous* and the recorded values only show minor fluctuations, it could be that no alert is recorded over long periods, even though the child is mostly not within the medically based target range – which must be defined more narrowly – during this period. The setting of alert values and limit values is an important medical task in the case of neonatal patients, and while this task can sometimes be standardised, it must be carried out individually at other times. There are many studies in which the sensitivities and specificities of various alert limits for the diagnosis of hyperoxia and hypoxia have been tested in various pulse oximetry devices. These studies are recommended for more detailed study in this area and for issuing guidelines.

Every neonatal department should adapt the standards for its own devices to the patient, however, and specify individual limits. Standardisation of this nature for oxygen saturation and partial pressures for O_2 and CO_2 should be carried out on the basis of an arterial blood sample. This is best done using an existing vascular cannula: if a vasopuncture is performed at the time (and particularly if this is difficult), the child presents rapid changes in the blood gases even in the intubated state due to the bout of crying connected with the reaction to the pain. The resulting acutely elevated pCO_2 values and the lowered pO_2 values are then still not recorded correctly as a result of the latent periods due to the noninvasive nature of the methods used (in particular transoxodes and transcapnodes) (Table 54.3).

Further Reading

- J. Stocks, P.D. Sly, R.S. Tepper, W.J. Morgan (eds.): *Infant Respiratory Function Testing* (Wiley-Liss, New York 1996)
- H.W. Taeusch, R.A. Ballard, C.A. Gleason (eds.): *Avery's Diseases of the Newborn*, 8th edn. (Elsevier Saunders, Philadelphia 2005)

References

54.1 H. Stopfkuchen: *Pädiatrische Intensivpflege* (Wiss. Verlagsges., Stuttgart 1991)

54.2 M. Obladen, R.F. Maier: *Neugeborenenintensivmedizin* (Springer, Berlin Heidelberg 2006)

55. Intraoperative Neuromonitoring

Werner Kneist, Daniel W. Kauff

Intraoperative neuromonitoring (IONM) was described as early as 1898 when facial muscles were visually assessed for activation during surgery near the facial nerve. Nowadays, it is based on the measurement of important physiological parameters for observation of the central and peripheral nervous system. The main focus of this chapter is IONM performed by the surgeon for prevention of direct nerve damage during the surgical procedure. IONM has been developed for nerve identification and observation of neural pathways during the surgical procedure (Sect. 55.1). Aside from the prevention and risk reduction of nerve damage, IONM serves as guidance for the surgeon. Nervous tissue is not able to compensate its cell death by formation of new nerve cells. Therefore, monitoring methods should reliably observe critical pathophysiological circumstances with potential for irreversible nerve damage (Sect. 55.2). Finally, IONM leads to improved surgical outcome and avoids postoperative function disturbances, maintaining patients' quality of life. Currently, several types of neuromonitoring systems are available and IONM is used in almost all surgical centers (Sect. 55.3).

55.1 **General Principles** 1043
55.1.1 Intraoperative Nerve Mapping 1045
55.1.2 Intermittent Intraoperative Neuromonitoring 1046
55.1.3 Continuous Intraoperative Neuromonitoring 1046

55.2 **Neuromonitoring Signals** 1046
55.2.1 Signal Recording and Processing .. 1046
55.2.2 Signal Analysis 1047

55.3 **Scope of Application** 1048
55.3.1 Thyroid Surgery 1048
55.3.2 Neurosurgery 1050
55.3.3 Orthopedic Surgery 1052
55.3.4 Otolaryngology and Head and Neck Surgery 1053
55.3.5 Pelvic Surgery 1054

55.4 **Quality Management** 1056

55.5 **Guidelines and Legal Aspects** 1056

55.1 General Principles

IONM refers to intermittent or continuous observation of functional nerve integrity. Its purpose is to prevent direct (mechanical, thermal) and indirect (blood circulatory disturbances) damage of the central and peripheral nervous system (Fig. 55.1). IONM has evolved to address risks to neural structures in a wide variety of surgeries. The field continues to grow as the benefits of neuromonitoring become apparent and as the techniques and methods used are expanded (Table 55.1).

The underlying principle of IONM for prevention of direct nerve damage is to monitor the neural pathways *crossing* the operative field. The stimulus results in neural excitations which can be recorded with suitable electrodes on innervated organs or muscles (Fig. 55.2).

IONM has several objectives (Table 55.2):

- Localization of functional neural areas or neural pathways, e.g., identification of eloquent brain areas, supporting the choice of the appropriate surgical approach in neurosurgery (intraoperative nerve mapping).
- Avoidance of neurological injury and early detection of impairment to prevent irreversible damage, e.g., monitoring of the recurrent laryngeal nerve in thyroid surgery (primary prevention).
- Intraoperative verification of nerve damage for determination of subsequent surgical steps and early postoperative therapy, e.g., in pelvic surgery a uni-

Table 55.1 Intraoperative neuromonitoring methods and scopes of application

Methods	Application
Manometry	• Pelvic surgery
Intravesical pressure	
Intraurethral pressure	
Intracavernosal pressure	
Tumescence measurement	• Pelvic surgery
Electrically evoked compound action potential (ECAP)	• Neurosurgery
Electrocochleography	• Neurosurgery
Electroencephalography (EEG)	• Cerebrovascular surgery
Conventional EEG	• Cardiovascular surgery
Processed EEG	• Interventional neuroradiology
Electroneurography (ENG)	• Peripheral nerve surgery
	• Nerve root surgery
	• Plexus surgery
Electromyography (EMG)	• Thyroid surgery
Spontaneous EMG	• Spine surgery
Triggered EMG	• Pelvic surgery
	• Parotis surgery
	• Neurosurgery
	• Peripheral nerve surgery
Evoked potentials (EP)	
Auditory evoked potentials (AEP)	• Neurosurgery
Brainstem AEP (BAEP)	• Cardiovascular surgery
Mid-latency AEP (MLAEP)	• Interventional neuroradiology
Long-latency AEP (LLAEP)	
Descending neurogenic evoked potentials (DNEP)	• Spine surgery
Percutaneous stimulation (PERC)	
Epidural stimulation (EPI-DNEP)	
Spinous process stimulation (SP-DNEP)	
Motor evoked potentials (MEP)	• Spine surgery
Transcranial electric MEP (TCeMEP)	• Neurosurgery
	• Parotis surgery
Somatosensory evoked potentials (SSEP)	• Peripheral nerve surgery
Cortical SSEP	• Spine surgery
Spinal SSEP	• Cerebrovascular surgery
Dermatomal SSEP	• Interventional neuroradiology
Visual evoked potentials VEP	• Orbital surgery
	• Pituitary gland surgery
Intraoperative language monitoring	• Neurosurgery

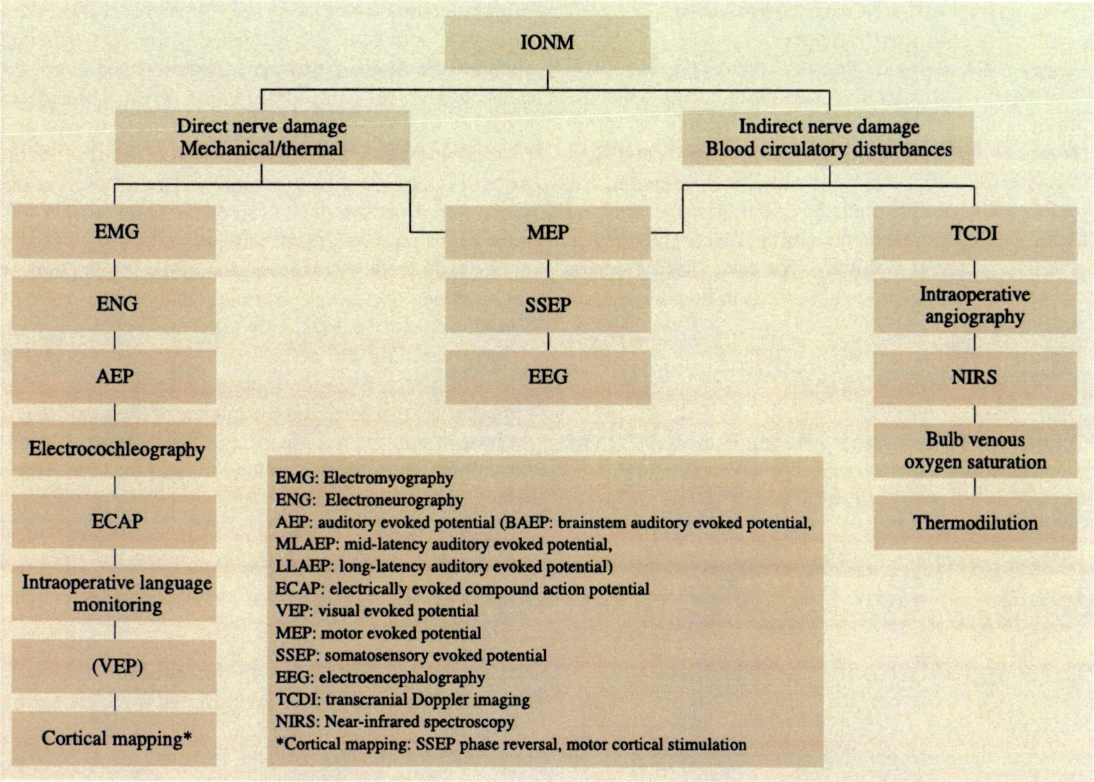

Fig. 55.1 Intraoperative neuromonitoring (IONM) for prevention of direct and indirect nerve damage

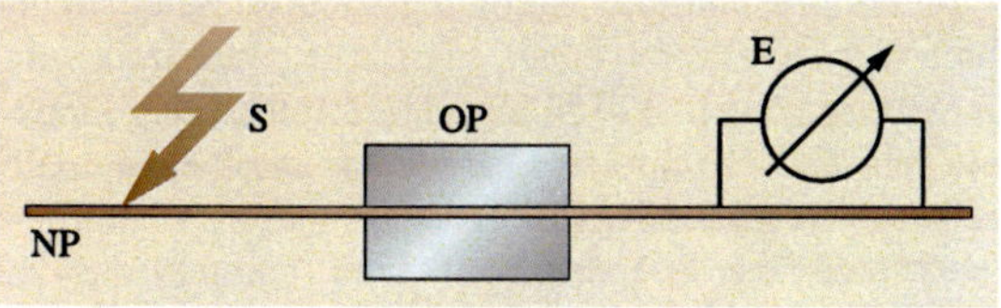

Fig. 55.2 Concept of intraoperative monitoring of neural pathways (NP). Arrangement of stimulus (S) and electrodes (E) with regard to the operative field (OP)

Table 55.2 Objectives of intraoperative neuromonitoring

⇒	Intraoperative nerve mapping provides security for the surgeon
⇒	Primary prevention of postoperative dysfunction due to controlled nerve sparing
⇒	Secondary prevention of postoperative dysfunction due to early specific therapy

lateral damaged inferior hypogastric plexus might be evaluated to determine urological therapy immediately, to prevent early and long-term urinary dysfunction or impotence (secondary prevention).

55.1.1 Intraoperative Nerve Mapping

Intraoperative nerve mapping is a technique for identification, tracking, and verification of nervous tissue. This anatomic testing enables the creation of a map providing the localization of eloquent neurological areas, for instance, speech and motor function (brain) or any organ-specific neural pathways. To combine radicality and functional integrity, the surgeon uses the acquired information for the choice of an appropriate approach for tumor excision without damaging eloquent nervous tissue. Even deep neural pathways can be identified by direct electrical stimulation (intraoperative subcortical mapping) with consecutive adaption of the neurosurgical procedure.

55.1.2 Intermittent Intraoperative Neuromonitoring

Intermittent intraoperative neuromonitoring (iIONM) provides electrophysiological nerve testing in certain phases of the operation. The control of nerve function is limited to the periods of neurostimulation. It is performed with a handheld stimulation probe, enabling identification and verification of the functional integrity of different neural pathways. For each iIONM performance, the surgical procedure needs to be interrupted.

55.1.3 Continuous Intraoperative Neuromonitoring

Continuous intraoperative neuromonitoring (cIONM) provides sustained observation of selected neural pathways during the whole critical phase of the procedure where nervous tissue is at risk for damage. The principle is to stimulate the identified nerve segments and monitor the neural pathways innervating the target organs or muscles. This presupposes great demands in terms of technique, e.g., probe design, electrodes, and methodological setup. cIONM overcomes the limitations of intermittent neuromonitoring, as it observes the *same* neural pathways and closes the gap between surgical preparation with and without monitoring. Surgery only needs to be interrupted for initial application or replacement of suitable stimulation probes in case of further areas of interest. However, there is an ongoing debate regarding the definition of cIONM concerning several aspects, e.g., continuous stimulation, repetitive stimulations in a predefined sequence of the *same* neural pathways, continuous recording, target organ or muscle fatigue, permanent application, and location of the stimulation probe.

55.2 Neuromonitoring Signals

55.2.1 Signal Recording and Processing

For IONM, the used electrodes or pressure senors are placed in the corresponding optimal areas (e.g., target muscles or organs) to record their signal response. Neurostimulation is carried out with a stimulation probe (e.g., a handheld stimulator for iIONM). If probe fixation is needed for cIONM, a suitable probe holder should meet the following requirements:

1. Prevent probe movement
2. Avoid trauma to nervous and other tissues
3. Permit access to critical regions of the operative field.

The electrodes and the probe are connected to a multichannel box which is part of the neuromonitoring device. Stimulation parameters, e.g., currents, pulse trains, pulse duration, and event thresholds, are set (Table 55.3).

Correct electrode placement is ensured by visual checking, confirmed by impedance measurement. For the interpretation of IONM signals, recordings of signal baselines are necessary. The time from the patient's arrival in theater until exposure of the at-risk nervous tissue allows the assembly of the neuromonitoring setup, initial evaluation of signals, and multiple system checks to ensure accuracy. If signals are of poor quality, there should be further optimization of the neuromonitoring setup. There are different troubleshooting protocols in use to cover anatomical (e.g., external branching of the nerve) and technical aspects [e.g., proper position of electrodes (main cause of equipment failure), the impedance measurement, signal baselines]. Furthermore, communication regarding anesthetic regimen (e.g., total intravenous anesthesia, anesthesia with inhalational agents) or bolus administration (e.g., muscle relaxant) is important, as certain anesthetics could affect the recorded IONM signals.

Monitoring is performed with comparisons of IONM signals made against the baseline data. When these signals are recorded on the hard drive of the device, it is necessary to print out a record, which has to be filed for future reference.

IONM signals suffer from a poor signal-to-noise ratio. Advanced signal processing is required to separate *noise* from signal for correct intraoperative assessment. The processing enables automatic analysis and extraction of the relevant signal from the raw signal. Thus, the recorded IONM signals could be visually observed during the operation on the monitor of the device, facilitating online verification of functional nerve integrity. The use of specific amplifiers and additional filter settings (e.g., frequency filters) enhances the quality of signals, for instance, in pelvic surgery for improvement of the reliability of IONM signals (Fig. 55.3).

Table 55.3 Scopes of application and stimulation parameters (examples)

Scopes	Method	Stimulation site	Recording site	Parameters (frequency, pulse duration, intensity)
Thyroid surgery	iIONM: triggered EMG	Vagal nerve, recurrent laryngeal nerve	Vocal muscle	4 Hz, 100 μs, 1–2 mA (vagal nerve)/ 0.8–1 mA (recurrent laryngeal nerve)
	cIONM: triggered EMG	Vagal nerve	Vocal muscle	3–4 Hz, 100–200 μs, 0.1–5 mA
Pelvic surgery	Triggered EMG	Inferior hypogastric plexus, pelvic splanchnic nerves	Internal anal sphincter	30 Hz, 200 μs, up to 150 mA
	Intravesical pressure	Inferior hypogastric plexus, pelvic splanchnic nerves	Urinary bladder	30 Hz, 200 μs, up to 15 mA
Spine surgery	SSEP	Lower limb: tibial posterior nerve, sciatic nerve, peroneal nerve Upper limb: median nerve, ulnar nerve	Somatosensory cortex (C1, C2, CZ, C3, C4), midline scalp (Cz'-Fz), contralateral scalp (C3'-Fz, C4'-Fz)	2.35–5.1 Hz, 200–300 μs, 15–40 mA
	Dermatomal SSEP	Dermatomal field stimulation	Scalp electrodes	
	MEP	Motor cortex (C3/C4), alternative C1/C2 (lower limb)	Lower limb: tibialis anterior, medial gastrocnemius, and extensor hallucis longus muscles, abductor hallucis Upper limb: abductor digiti minimi and abductor pollicis brevis, thenar	Interstimulus interval 1–5 ms/50–200 Hz, pulse duration 50 μs to 2–5 s, stimulation intensity 75–500 V/25–30 mA
	DNEP			
	PERC	Consecutive cervical lamina	Sciatic nerve (popliteal fossa)	4.7–5.1 Hz, 500–700 μs, 100–400 V
	EPI-DNEP	Spinal catheter, epidural	Sciatic nerve (popliteal fossa)	4.7–5.1 Hz, 500–700 μs, 2–40 mA
	SP-DNEP	Spinous process (of consecutive spinal levels within the proximal portion of the surgical incision)	Sciatic nerve (popliteal fossa)	4.7–5.1 Hz, 500–700 μs, 100–400 V
	EMG			
	Spontaneous EMG		Muscles innervated by corresponding nerve roots	
	Triggered EMG	Top of the pedicle screw	Appropriate muscle groups	≤ 20 mA
Skull base surgery	Triggered EMG	Nerve trunks	Corresponding muscles and muscle groups	4–4.7 Hz, 50–100 μs, up to 1 mA

55.2.2 Signal Analysis

The interpretation of IONM signals and any recommendations regarding the consequences or intervention are the responsibility of a qualified physician (e.g., surgeon) or a clinical neurophysiologist. Intraoperatively, the observer has to differentiate true neuromonitoring signal events from artifacts, for instance, such as electric stimulation, interfering electrosurgical units or contact between surgical instruments in the operative field. The recorded IONM signals are interpreted with regard to their baseline data and must be seen in relation to the specific pre- and postoperative diagnostic nerve tests. Numerous factors influence the quality of the recorded signals, such as use of stimulation probes and electrodes, localization, impedance of tissue, contact surface of the stimulation probe to the nerve, and stimulation parameters. In addition, mechanical and thermal phenomena, anesthetic agents, and other systematic variables may affect the IONM signal quality throughout the operation or even result in signal loss. Occurrence of any false neuromonitoring signals should be excluded, as this could be misleading and may give the surgeon a false sense of security during the procedure with a probably higher risk for nerve damage than without IONM. The absence of neuromonitoring signals during electric stimulation does not exclude the possibility that stimulation of tissue other than nervous

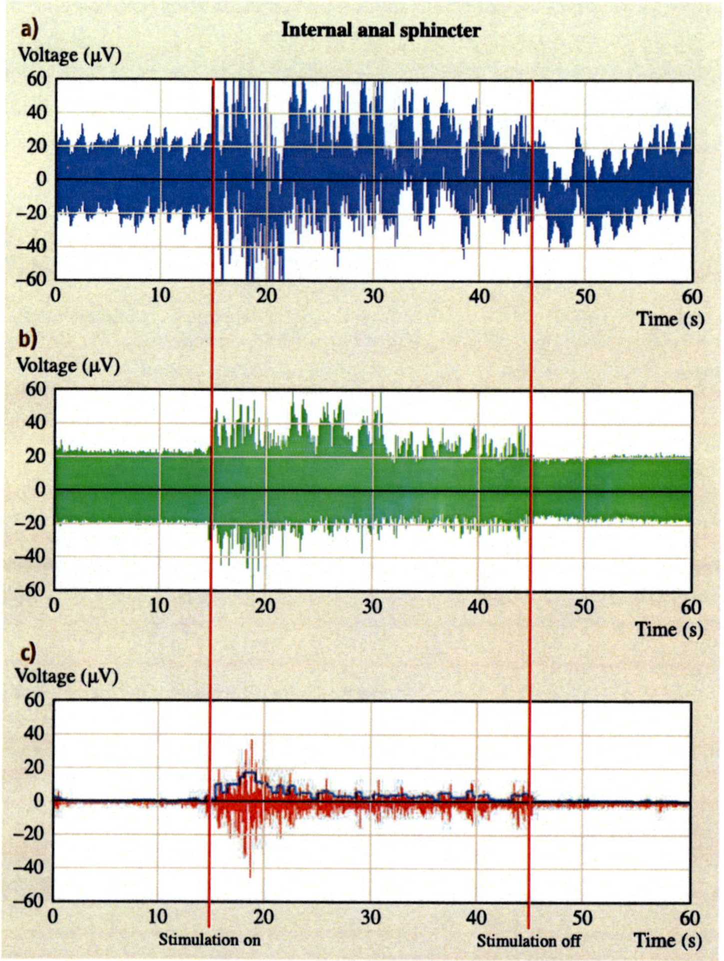

Fig. 55.3a–c Signal processing of internal anal sphincter activity for monitoring of pelvic autonomic nerves: (**a**) time-based electromyographic signal of the internal anal sphincter before and during bipolar electric stimulation as originally recorded, (**b**) original signal after high-pass filtering (cutoff frequency 5 Hz), and (**c**) signal after additional low-pass filtering (cutoff frequency 20 Hz); blue curve represents the mean absolute value of the amplitudes computed over windows with 5024 sample points

tissue was performed. The surgeon has to distinguish between signal loss by nerve damage and signal loss due to incorrect placement of the stimulation probe or newly occurred probe or electrode dislocation.

Standardization of IONM procedures is useful as it may help to eliminate these pitfalls, elucidate the mechanisms of neurological injuries, and reduce the rates of postoperative nerve-related dysfunctions.

55.3 Scope of Application

55.3.1 Thyroid Surgery

Recurrent laryngeal nerve (RLN) palsy is the most common and serious complication in thyroid and parathyroid surgery (Table 55.4). Since routine identification of RLN is an established technique, permanent palsy rates are reduced to less than 2% for first-time operations. In reoperative settings (Table 55.5) or in operations for malignant disease palsy rates are higher, ranging from 2% to 20%. Bilateral RLN palsy after total thyroidectomy is a life-threatening event. It has to be taken into account that injury of the RLN is the most common reason for legal complaints in this field.

Table 55.4 Nerve-related postoperative dysfunctions

Scope of application	Nerves/structures	Postoperative dysfunctions
Thyroid surgery	Branches of vagal nerve	
	Recurrent laryngeal nerve	Dysphonia Dyspnea Dysphagia
	Laryngeal superior nerve	Voice alterations
Skull base, posterior fossa, and brainstem surgery	Spinal cord	Hypesthesia Hemiparesis Paraplegia Quadriplegia Coma
Skull base, posterior fossa and brainstem, and head and neck surgery	Cranial nerves CN I-XII	
	Olfactory nerve CN I	Reduced or absent smell sensation
	Optic nerve CN II	Visual loss
	Eye movement nerves CN III, IV,VI	Syndrome of superior orbital fissure Eyelid ptosis Eyeball fixation Exophthalmus Pupil dilation Low vision limitation Diplopia
	Trigeminal nerve CN V	Trigeminal neuralgia Sensory disturbances in the face Jaw muscle paresis Activation of trigeminocardiac reflex → (transient) asystole
	Facial nerve CN VII	Paralysis of the mimetic musculature
	Acoustic nerve CN VIII	Hearing reduction or loss
	Glossopharyngeal nerve CN IX	Sensory disturbances of the palate Taste quality bitter will not be perceived Loss of gag reflex
	Vagal nerve CN X	Uvular deviation Aphonia Dysphagia
	Spinal accessory nerve CN XI	Shoulder drop
	Hypoglossal nerve CN XII	Tongue atrophy
Spine surgery	Brachial plexus	Paralysis of shoulder and arm muscles Pain Atrophia
	Lumbosacral plexus	Paralysis of lower limb muscles Inguinal pain Lower limb sensory disturbances
	Spinal cord	Hypesthesia Hemiparesis Paraplegia Quadriplegia Coma
	Conus medullaris	Conus syndrome
	Cauda equina	Cauda syndrome
Pelvic surgery	Pelvic autonomic nerves (superior hypogastric plexus, hypogastric nerve, inferior hypogastric plexus, splanchnic nerves, neurovascular bundles)	Sexual dysfunction Bladder dysfunction Anorectal dysfunction

Mechanisms of RLN injury are compression, stretching, constriction, ligature, clipping or transection. Nerve paralysis also occurs after thermal alteration.

For more than 10 years several studies have demonstrated the benefits of IONM in thyroid surgery. IONM has been used to localize and identify the RLN, but also

Table 55.5 Rates of temporary and permanent recurrent laryngeal nerve palsy after surgery and redo surgery for benign multinodular goiter

	Moalem 2008	Thomusch 2003
Temporary palsy	1–10% TT	1.7% sT+sT
	0.9–6% sT	2.0% sT+HT
	0–22% RDS	4.5% TT
Permanent palsy	0–1.4% sT	0.8% sT+sT
	0–1.4% TT	1.4% sT+HT
	0–13% RDS	2.3% TT

sT+sT: bilateral subtotal resection, sT+HT: subtotal resection with contralateral lobectomy, TT: total thyroidectomy, RDS: redo surgery

to predict cord function and analyze the mechanism of RLN injury.

For intermittent IONM a handheld stimulation probe (generally bipolar) is used for localization and functional testing of the RLN and vagal nerve (Figs. 55.4 and 55.5). To sense posterior cricoarytenoid muscle contraction in response to ipsilateral nerve stimulation, an EMG is recorded from endotracheal surface electrodes (tubus electrodes) or from transligamentarily inserted needle electrodes.

A high negative predictive value of more than 97% means that patients with an intact IONM signal after thyroid resection generally have normal vocal cord function. Therefore, this method is superior to macroscopic assessment of RLN alone for prediction of postoperative vocal cord function.

Conversely, studies found a lower and highly variable positive predictive value of 10–90% for IONM signals. This might result in misleading information limiting the value of IONM technology.

Pitfalls associated with IONM during thyroid surgery have been reported: equipment malfunction, improper setup, misuse of anesthetic drugs, and anatomic variations of RLN. This highlights the importance of a standardized approach.

It has to be taken into account that a significant reduction of postoperative RLN palsy due to IONM is described only for redo operations.

The IONM technique needs a learning curve. However, experts in this field have hypothesized that surgical trainees and less experienced surgeons benefit from IONM as a helpful teaching aid, whereas the experienced surgeon may benefit in difficult cases.

In all operations with IONM a pre- and postoperative laryngeal examination (flexible laryngofibroscopy to document vocal cord mobility) is mandatory. Troubleshooting requires systematic checking of IONM equipment and neuroanatomy (Sect. 55.2.1).

During the operation a systematic stepwise IONM procedure is necessary to evaluate signals from the vagal nerve and RLN and to eliminate false IONM results. Important steps are:

1. Vagal nerve stimulation before thyroid dissection
2. RLN stimulation at initial identification
3. RLN stimulation at the end of thyroid dissection, and
4. Vagal nerve stimulation after complete thyroidectomy and hemostasis.

Loss of IONM signal is an important intraoperative decision criterion for the surgeon, especially when bilateral lobectomy should be performed. An amplitude response of less than 100 μV or total loss of the primary normal biphasic stimulation waveform obtained during or after resection of the first thyroid lobe supports the indication of a two-stage operation to prevent bilateral RLN palsy (Fig. 55.6a,b).

Currently, surgeons are applying continuous IONM devices to overcome the limitations of unrecognized RLN palsy with intermittent IONM. Several vagal nerve electrodes (anchor, cuff, and adhesive tube electrodes) are being tested for real-time monitoring of RLN. Recent pilot studies have demonstrated the feasibility of cIONM for observation of RLN.

55.3.2 Neurosurgery

The potential postoperative dysfunctions after neurosurgical interventions are listed in Table 55.4. In the following, three major areas of neurosurgery are described.

Monitoring methods for removal of spinal cord and intradural tumors are comparable to those in spine deformity surgery such as SSEPs, MEPs, and EMG (see also Sect. 55.3.3).

In cranial surgery, neuromonitoring is used for translabyrinthine and middle fossa acoustic neurectomy and various procedures of the skull base and neck, e.g., resection of tumors of the lower cranial base, repair of aneurysms of the posterior circulation, and removal of brainstem tumors. Intraoperative mapping is performed to detect and preserve cranial nerves. The monitoring includes electric stimulation of nerve trunks and recording of different kinds of evoked potentials such as AEPs and visually evoked potentials for sensory cranial nerve monitoring and EMG methods for cIONM of

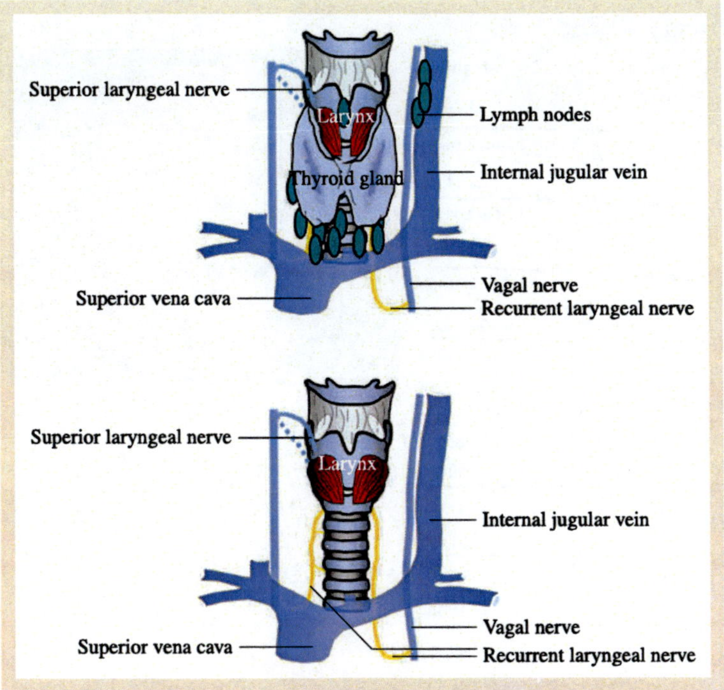

Fig. 55.4 Neuroanatomy and topography of larynx and thyroid gland

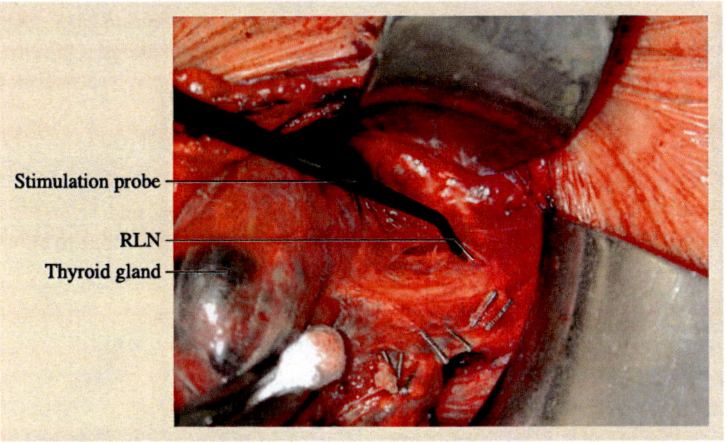

Fig. 55.5 Intraoperative neuromonitoring for identification of left recurrent laryngeal nerve (RLN) during thyroidectomy

motor cranial nerves (V, VII, IX, X, XI, XII). For signal recording, electrodes are placed in different muscles or muscle groups corresponding to the innervation pattern of the relevant cranial nerve. The orbicularis oculi and orbicularis oris muscles are suitable for facial nerve monitoring and can be observed by mechanomyogram and/or electromyogram.

Some studies have indicated increasing sensitivity and specificity with the use of transdermal intramuscular electrodes. The cochlearis nerve could be monitored by AEP, electrocochleography, and electrically evoked compound action potential. Another aspect of monitoring in skull base and brain surgery is cortical mapping. Central sulcus can be monitored by SSEP phase reversal and precentral gyrus by motor cortical stimulation.

During interventions in speech-related areas, cortical stimulation is used for speech testing throughout

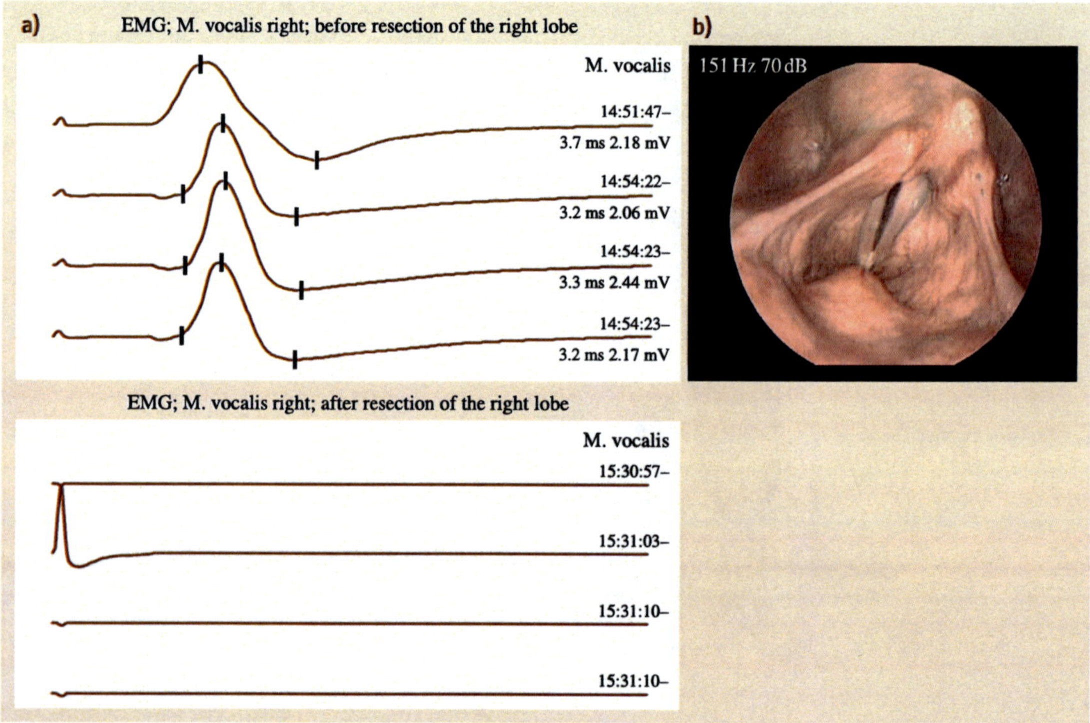

Fig. 55.6 (a) Intended thyroidectomy for benign multinodular goiter. Negative IONM result after resection of the right thyroid lobe. The surgeon decided on a two-stage operation in order not to risk bilateral vocal cord paralysis. **(b)** Postoperative laryngofibroscopy demonstrates incomplete closure of glottis. Temporary right vocal cord paralysis vanished after logopedics

the operation. If these areas are intact, the stimulation slows down speech function until speech arrest in awake patients. Visual evoked potentials are also more meaningful if raised in awake patients.

In peripheral nerve surgery, extracranial parts of motor cranial nerves could be observed, similar to intracranial parts. Important nerves are the facial nerve and vagal branches in parotis and thyroid surgery (see also Sects. 55.3.1 and 55.3.4). The monitoring is performed under electroneurography, electromyography, SSEPs, and MEPs and should ensure the control of peripheral nerve, nerve root, and plexus function. IONM aims at:

1. Identifying peripheral nerves
2. Localizing preexisting disease processes along the course of a nerve
3. Determining functional continuity across a preexisting lesion, prior to detection by other electrophysiological means
4. Determining nerve root avulsion
5. Identifying targets for nerve biopsy, and
6. Monitoring and thereby preventing damage to intact nerves during surgery.

55.3.3 Orthopedic Surgery

Perioperative neurological injury, in particular spinal cord injury, is one of the most feared complications in spinal surgery. IONM in this operative field is frequently selected for scoliosis, deformity correction, degenerative cases, instrumentation in the cervical and thoracic spine, spinal stenosis, and discectomy. There are several structures placed at risk for potential injury, including the spinal cord, nerve roots, lumbar plexus, and all relevant vascular supply to these elements. Damage to these structures could result in motor and sensory dysfunctions (Table 55.4).

Several electrophysiological modalities are currently available for monitoring various aspects of the

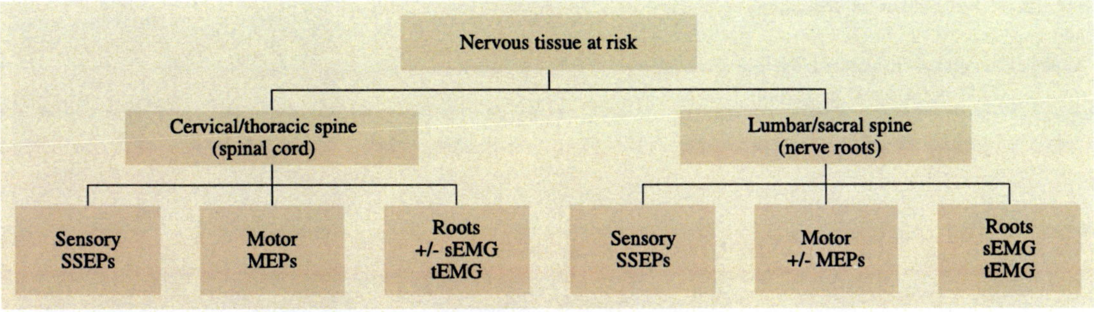

Fig. 55.7 Intraoperative neuromonitoring methods for spine surgery

central and peripheral nervous system, each offering a unique set of benefits, limitations, and sensitivity/specificity as diagnostic techniques. The monitoring methods in spine surgery are somatosensory evoked potentials (SSEPs), motor evoked potentials (MEPs), descending neurogenic evoked potentials (DNEPs), spontaneous and triggered electromyography (sEMG and tEMG) (Fig. 55.7). Multimodal IONM provides higher specificity and sensitivity for detecting nerve damage than any modality used in isolation.

SSEPs provide cIONM of the large fiber sensory tracts (dorsal column), which mediates tactile discrimination, vibration sensation, form recognition, and joint/muscle sensation. It does not involve the spinothalamic (pain and temperature) pathway. Upper limb SSEPs are obtained by stimulation of the median and ulnar nerve. Lower limb SSEPs are elicited by posterior tibial, peroneal, and sciatic nerve stimulation. The electrodes for signal recording are placed on the somatosensory cortex. Dermatomal SSEPs are used to verify nerve root function. Data are obtained by dermatomal field stimulation and recorded from the electrodes for standard SSEPs.

The MEPs monitor function of the motor tracts (corticospinal tract) which are not covered by the SSEP. They can be obtained by transcranial stimulation (electrically or magnetically) or via direct cortical stimulation. Signal recording is performed at the level of the spinal cord, nerves (e.g., spinal nerve) or muscles (e.g., tibialis anterior muscle), if stimulated transcranially or only at the level of the nerves or muscles, if cortical stimulation is carried out. MEPs have become the gold standard for IONM of the motor tracts. Recent studies have demonstrated that MEP monitoring can be performed intermittently and continuously.

DNEPs are used in all surgeries involving the thoracic spine. Different stimulation methods have been described:

1. Percutaneous stimulation via needle electrodes placed onto consecutive cervical lamina
2. Epidural stimulation with a flexible spinal catheter
3. Spinous process stimulation via needle electrodes placed in the spinous process of consecutive spinal levels within the proximal portion of the surgical incision. Data were recorded bilaterally from the sciatic nerve at the level of the popliteal fossa. Full muscle relaxant is required to prevent stimulation-induced movement and to reduce muscle artifacts.

Spontaneous EMG is primarily used to monitor nerve root function. However, it can also demonstrate subtle signs of spinal cord damage. Evoked EMG is most often used to determine if breach of the spinal pedicle wall has occurred by stimulating the pedicle screw directly or by stimulating the canal made for the screw.

Recently, it has been described that IONM could be used for identification and verification of axillary nerve function during arthroscopic shoulder stabilization.

IONM has become an important component of many types of surgery. However, a valid criticism is that scientific justification for its use does not include any blinded, placebo-controlled studies. Unfortunately, such studies can no longer be ethically performed.

55.3.4 Otolaryngology and Head and Neck Surgery

A major focus of IONM in otolaryngeal and head and neck surgery is monitoring of the facial nerve during operation for acoustic neuroma and parotidectomy. Facial nerve damage can cause cosmetic and functional morbidity diminishing patients' quality of life and resulting in medical malpractice litigation. Up to 20–70% of patients suffer from more or less pronounced postoperative facial nerve functional deficits.

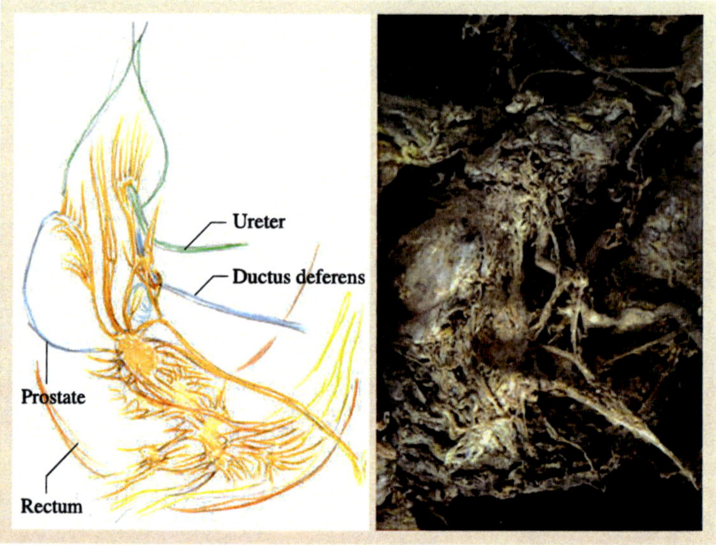

Fig. 55.8 Overview of neuroanatomy and topography of the inferior hypogastric plexus illustrated by drawing (*left*) and cadaver dissection (*right*)

In general, the IONM is performed intermittently. Nevertheless, recent studies have demonstrated the feasibility of continuous monitoring of the facial nerve.

There are several methods for monitoring. One possibility is to visually observe facial movements during the surgical procedure. The assistant gives feedback to the surgeon if facial movements are evoked, for instance, electrically by neurostimulation or mechanically during dissection. Another approach is EMG of facial muscle activity. Electrodes are placed on optimal positions for recording, typically in the frontal, zygomatic, buccal, and marginal mandibular areas. Neurostimulation is carried out with a conventional stimulation probe. Both a graphic signal observed on an oscilloscope screen and an acoustic signal are generated. In general, electrophysiologic monitoring is preferred, as it is more sensitive and specific than visual observation. It enables quantification of the recorded EMG activity.

Studies have demonstrated that functional outcome could be improved by use of IONM, although the difference was not always significant. Most patients with postoperative facial palsy demonstrate good functional recovery within 1 year. However, certain patients do not show this recovery and require nerve reconstruction. Early reconstruction demonstrates better functional results than if surgery is delayed. Neuromonitoring provides intraoperative assessment of functional nerve integrity with the possibility of reconstruction in the same operation.

55.3.5 Pelvic Surgery

Autonomic nerve injury during rectal resection or other types of pelvic surgery may significantly deteriorate functional outcome and patients' quality of life. Aside from sexual and bladder disturbances, nerve damage may also affect anal sphincter function.

The efferences of the inferior hypogastric plexus contain three up to five secondary plexus innervating the pelvic organs (Fig. 55.8). Therefore, mainly damage to the different plexus segments is responsible for the different combinations of function disturbances. Despite potentially nerve-sparing operations, varying degrees of urogenital dysfunction are found in up to 32–70%, and anorectal dysfunction occurs in up to 47% of rectal cancer patients. The costs of diagnostics and treatment for short- and long-term postoperative urogenital dysfunction and incontinence are immense.

IONM of pelvic autonomic nerves could quantify and reduce the risk of such detrimental injuries. Intermittent bipolar electric stimulation of the pelvic autonomic nerves with observation of bladder motor response or penile tumescence are thus by far the most frequently used techniques for nerve identification and functional testing (Table 55.1).

IONM with assessment of penile tumescence predicts postoperative erectile dysfunction with high specificity. The low sensitivity rate of approximately 50% may be explained by preoperative impotence, influence of anesthetic procedure, and non-stimulation-related tumescence response. In particular, combined with in-

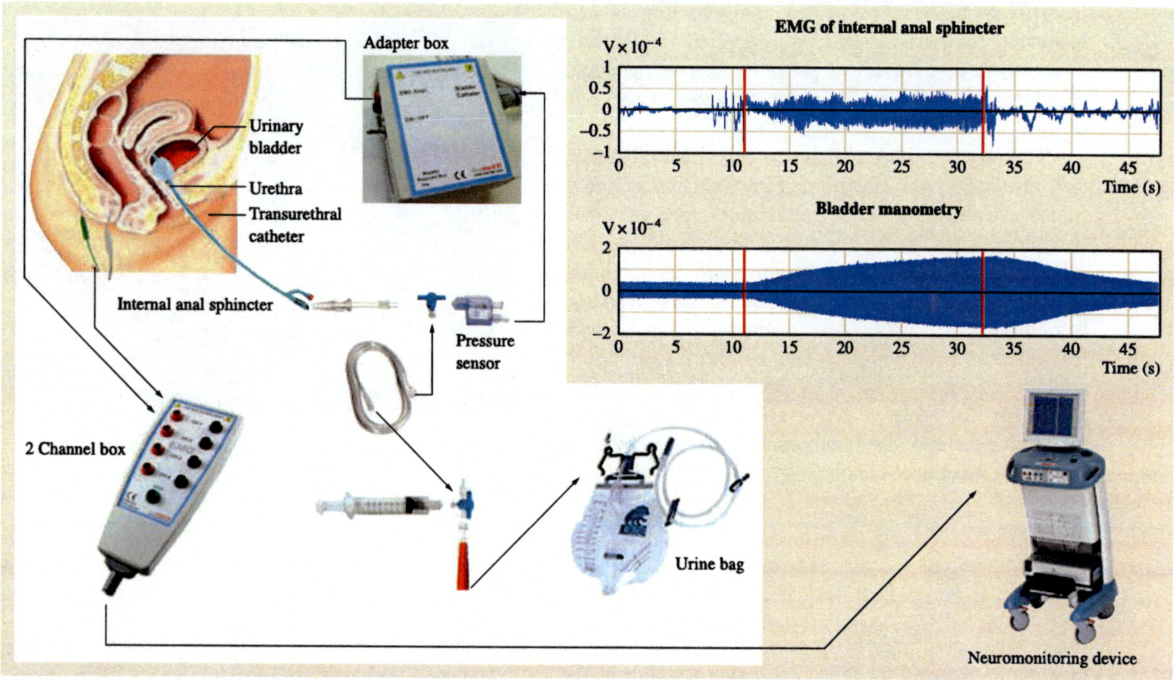

Fig. 55.9 Neuromonitoring setup for pelvic surgery

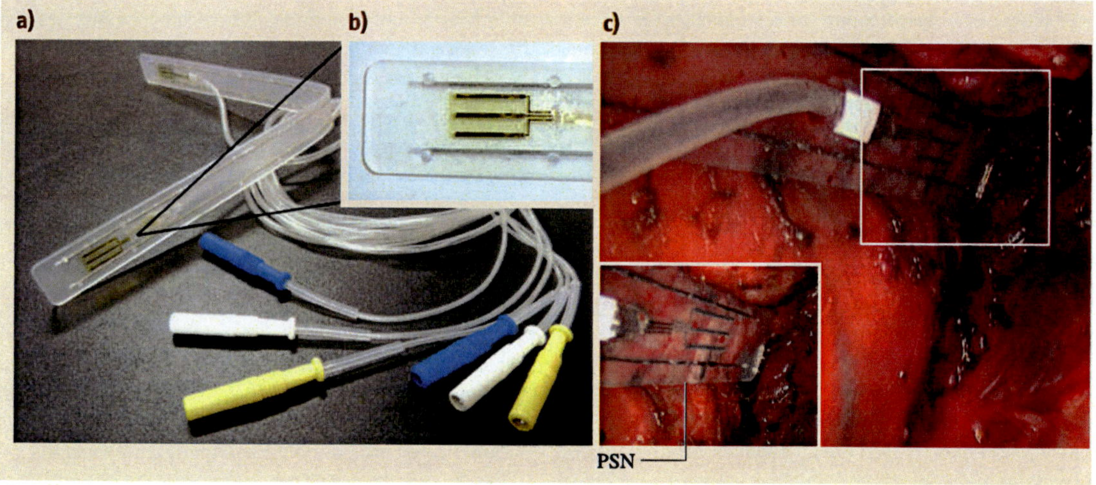

Fig. 55.10a–c Tripolar surface electrodes (polyimide and gold layers) on both ends of a V-shaped electrode holder **(a)**. Magnified image of the electrode with silicone knobs and rigid wires **(b)**. IONM in a male patient during mesorectal excision for rectal cancer [electric stimulation of pelvic splanchnic nerves (PSN)] **(c)**

tracavernous and intraurethral pressure observation, the method is nevertheless recommended for assessment of cavernous nerve preservation after prostatectomy and rectal resection in male patients.

In rectal cancer surgery, it has been demonstrated that the confirmation of incomplete pelvic autonomic nerve preservation is more accurate using IONM compared with macroscopic assessment by the surgeon (sensitivity 82% versus 46%). This finding is corroborated by gynecologists, who applied IONM with observation of intravesical pressure after radical hysterectomy and obtained similar diagnostic accuracy (88%).

Recently, it has been shown in experimental studies that intraoperative verification of internal anal sphincter function by stimulation of pelvic autonomic nerves is feasible.

Moreover, it has been demonstrated in a clinical setting that the internal anal sphincter EMG response in combination with bladder manometry can be used for intraoperative mapping and monitoring of pelvic autonomic nerves (Figs. 55.9–55.11). This offers the possibility for more selective and continuous IONM with the potential for improvement of surgical technique and functional results.

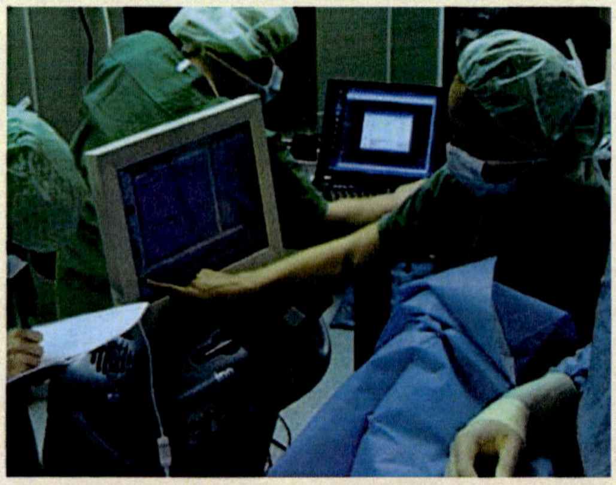

Fig. 55.11 Documentation of intraoperative neuromonitoring results

55.4 Quality Management

Nerve-sparing surgery for prevention of postoperative dysfunction caused by injury to nervous tissue in the operative field is a major goal and widely established as *standard*.

However, classification or categorization of the quality of *nerve sparing* on the basis of intraoperative macroscopic assessment, functional outcome, and postoperative diagnostic testing is problematic. For this reason, nerve sparing is not generally established as a benchmark criterion.

Based on strong evidence that IONM is sensitive and specific for detecting neurologic injury during surgical procedures, it is recommended that use of such a method should be considered in operations with high risk of injury to the nervous system, especially if there are effective intervention options. In cases of detected nerve damage, there will be an undoubted benefit from IONM.

The results of intermittent IONM at the end of a surgical procedure (based upon changes from patients' own earlier baseline IONM signals) could be used as a reference for evaluating surgical quality. Benchmarks for *controlled* nerve sparing may be drawn from the departments' own experience, from the experience of others working in the same clinical subspecialty or from professional societies and certification organizations.

The objectives of IONM are:

1. To determine what and where improvements of nerve sparing are required
2. How other surgeons/departments achieve their high rates of nerve-sparing surgical procedures, and
3. Receiving reliable information to improve the department's operative standard.

To use IONM results as a surrogate parameter for quality assessment of the surgical procedure, evidence-based protocols are needed. More prospective studies dealing with changes, cutoff values (thresholds, alarm criteria), and reference intervals of IONM signals should be performed to validate the IONM methods.

55.5 Guidelines and Legal Aspects

Professional and technical levels should be defined and could be administered by certification.

On the basis of current clinical literature and scientific evidence, IONM should be examined and evaluated by consensus conferences and position statements promoted by scientific surgical associations.

However, at this point there are only a small number of recognized professional organizations providing guidelines or certification specifically for the unique skills involved in IONM. Thus, guideline committees are encouraged to develop practice guidelines for IONM in the various surgical settings. The primary goals are to

1. delineate the objectives of IONM,
2. characterize the responsibilities and behaviors of the surgical team during IONM, and
3. describe methodological and ethical issues uniquely relevant to IONM.

Performance of IONM presupposes that patients have to be informed of the intent to use a neuromonitoring system with its potential to aid in localization and identification of nervous tissue and assessment of nerve function during the operation.

Interestingly, surgeons who employ IONM in their practice have been found to be less likely to have a history of a surgery-associated lawsuit. For expert assessment purposes, documentation of standardized IONM application is of utmost importance (for instance, EMG printouts of pre- and postprocedural IONM results).

Further Reading

- H. Ahn, M.G. Fehlings: Prevention, identification and treatment of perioperative spinal cord injury, Neurosurg Focus **25**, E15 (2008)
- W.F. Chan, C.Y. Lo: Pitfalls of intraoperative neuromonitoring for predicting postoperative recurrent laryngeal nerve function during thyroidectomy, World J Surg **30**, 806–812 (2006)
- B.A. Crum, J.A. Strommen: Peripheral nerve stimulation and monitoring during operative procedures, Muscle Nerve **35**, 159–170 (2007)
- H. Dralle, K. Lorenz: Intraoperative neuromonitoring of thyroid gland operations. Surgical standards and aspects of expert assessment, Chirurg **81**, 612–619 (2010)
- H. Dralle: Identification of the recurrent laryngeal nerve and parathyroids in thyroid surgery, Chirurg **80**, 352–363 (2009)
- H.K. Eltzschig, M. Posner, F.D. Moore: The use of readily available equipment in a simple method for intraoperative monitoring of recurrent laryngeal nerve function during thyroid surgery, Arch Surg **137**, 452–457 (2002)
- A.A. Gonzalez, D. Jeyanandarajan, C. Hansen, G. Zada, P.C. Hsieh: Intraoperative neurophysiological monitoring during spine surgery. A review, Neurosurg Focus **27**(4), E6 (2009)
- P.E. Goretzki, K. Schwarz, J. Brinkmann, D. Wirowski, B.J. Lammers: The impact of intraoperative neuromonitoring (IONM) on surgical strategy in bilateral thyroid disease: is it worth the effort?, World J Surg **34**, 1274–1284 (2010)
- L. Guo, P. Jasiukaitis, L.H. Pitts, S.W. Cheung: Optimal placement of recording electrodes for quantifying facial nerve compound action potential, Otol Neurotol **29**, 710–713 (2008)
- M.R. Isley, H.L. Edmonds, M. Stecker: Guidelines for intraoperative neuromonitoring using raw (analog to digital waveforms) and quantitative electroencephalography: a position statement by the American Society of Neurophysiological Monitoring, J Clinical Monit Comput **23**, 369–390 (2009)
- J. Jonas: Continuous vagal nerve stimulation for recurrent laryngeal nerve protection in thyroid surgery, Eur Surg Res **44**, 185–191 (2010)
- D.W. Kauff, K.P. Koch, K.H. Somerlik, A. Heimann, K.P. Hoffmann, H. Lang, W. Kneist: Online signal processing of internal anal sphincter activity under pelvic autonomic nerve stimulation: a new method to improve reliability of intraoperative neuromonitoring signals, Colorectal Dis (2010), DOI: 10.1111/j.1463-1318.2010.02510.x
- L. Klotz: Intraoperative cavernous nerve stimulation during nerve sparing radical prostatectomy: how and when?, Curr Opin Urol **10**, 239–243 (2000)
- W. Kneist, T. Junginger: Intraoperative electrostimulation objectifies the assessment of functional nerve preservation after mesorectal excision, Int J Colorectal Dis **22**, 675–682 (2007)
- W. Kneist, T. Junginger: Male urogenital function after confirmed nerve-sparing total mesorectal excision with dissection in front of Denonvilliers' fascia, World J Surg **31**, 1321–1328 (2007)
- W. Kneist, T. Junginger: Long-term urinary dysfunction after mesorectal excision: A prospective study with intraoperative electrophysiological confirmation of nerve preservation, Eur J Surg Oncol **33**, 1068–1074 (2007)
- W. Kneist, D.W. Kauff, R. Rahimi, A.D. Rink, A. Heimann, K. Somerlik, K.P. Koch, T. Doerge, H. Lang: Intraoperative pelvic nerve stimulation performed under continuous electromyography of the internal anal sphincter, Int J Colorectal Dis **25**, 1325–1331 (2010)
- W. Kneist, A. Heintz, T. Junginger: Intraoperative identification and neurophysiologic parameters to verify pelvic autonomic nerve function during to-

tal mesorectal excision for rectal cancer, J Am Coll Surg **198**, 59–66 (2004)

- F. Krauze: *Surgery of the brain and spinal cord* (Rebman, New York 1912)
- K. Lorenz, C. Sekulla, J. Schelle, B. Schmeiß, M. Brauckhoff, H. Dralle: What are normal quantitative parameters of intraoperative neuromonitoring (IONM) in thyroid surgery?, Langebecks Arch Surg **395**, 901–909 (2010)
- T.F. Lue, C.A. Gleason, G.B. Brock, P.R. Carroll, E.A. Tanagho: Intraoperative electrostimulation of the cavernous nerve: technique, results and limitations, J Urol **154**, 1426–1428 (1995)
- D.D. Lydiatt: Medical malpractice and facial nerve paralysis, Arch Otolaryngol Head Neck Surg **129**, 50–53 (2003)
- M.E. Mackey, Y. Wu, R. Hu, J. DeMaro, M. Jacquin, G.K. Kanellopoulos, C.Y. Hsu, N.T. Kouchoukos: Cell death suggestive of apoptosis after spinal cord ischemia in rabbits, Stroke **28**, 2012–2017 (1997)
- J. Maurer, H. Pelster, R.G. Amedee, W.J. Mann: Intraoperative monitoring of motor cranial nerves in skull base surgery, Skull Base Surg **5**(3), 169–175 (1995)
- E.T. Mishler, P.G. Smith: Technical aspects of intraoperative monitoring of lower cranial nerve function, Skull Base Surg **5**(4), 245–250 (1995)
- J. Moalem, I. Suh, Q.Y. Duh: Treatment and prevention of recurrence of multinodular goiter: an evidence-based review of the literature, World J Surg **32**, 1301–1312 (2008)
- W.M. Moore, T.C. Naslund, L.H. Hollier: Neurologic outcome following transient spinal cord ischemia during thoracoabdominal aortic aneurysm repair. In: *Vascular Surgery 2000*, Research strategies for the new millenium, Vol. 1, ed. by J.R. Cohen (RG Landes, Austin 1991)
- A.R. Möller: *Intraoperative Neurophysiologic Monitoring* (Luxembourg, Harwood Academic 1995)
- W.K. Ng, K. Mukhida, J.T. Rutka: Image guidance and neuromonitoring in neurosurgery, Childs Nerv Syst **26**, 491–502 (2010)
- M.R. Nuwer, E.G. Dawson, L.G. Carlson, L.E. Kanim, J.E. Sherman: Somatosensory evoked potential spinal cord monitoring reduces neurologic deficits after scoliosis surgery. Results of a large multicenter survey, Electroencephalogr Clin Neurophysiol **96**, 6–11 (1995)
- R.L. Prass: Iatrogenic facial nerve injury: the role of facial nerve monitoring, Otolaryngol Clin North Am **29**(2), 265–275 (1996)
- R. Prass, H. Luders: Constant-current versus constant-voltage stimulation, J Neurosurg **62**(4), 622–623 (1985)
- J.M. Ryzenman, M.L. Pensak, J.M. Tew Jr: Facial paralysis and surgical rehabilitation: a quality of life analysis in a cohort of 1,595 patients after acoustic neuroma surgery, Otol Neurotol **26**, 516–521 (2005)
- M. Samii, C. Matthies: Management of 1000 vestibular schwannomas (acoustic neuroma): the facial nerve – preservation and restitution of function, Neurosurgery **40**, 684–695 (1997)
- H. Silverstein, S. Rosenberg: Intraoperative facial nerve monitoring, Otol Clin North Am **24**, 709–725 (1991)
- T.B. Sloan: Anesthetic effects on electrophysiologic recordings, J Clin Neurophysiol **15**, 217–226 (1998)
- M. Stöhr, W. Wagner, K. Pfadenhauer, K. Scheglmann: *Neuromonitoring* (Steinkopff, Darmstadt 1999)
- E.D. Thuet, J.C. Winscher, A.M. Padberg, K.H. Bridwell, L.G. Lenke, M.B. Dobbs, M. Schootman, S.J. Luhmann: Validity and reliability of intraoperative monitoring in pediatric spinal deformity surgery, Spine **35**, 1880–1886 (2010)
- O. Thomusch, C. Sekulla, H. Dralle: Is primary total thyroidectomy justified in benign multinodular goiter? Results of a prospective quality assurance study of 45 hospitals offering different levels of care, Chirurg **74**, 437–443 (2003)
- C. Ulmer, K.P. Kock, A. Seimer, V. Molnar, U. Meyding-Lamade, K.P. Thon, W. Lamde: Real-time monitoring of the recurrent laryngeal nerve: an observational clinical trial, Surgery **143**, 359–365 (2008)
- A.S. Youssef, A.E. Downes: Intraoperative neurophysiological monitoring in vestibular schwannoma surgery: advances and clinical implications, Neurosurg **27**(4), E9 (2009)

56. Ionic Neural Sensing

Iasonas F. Triantis, Anna Radomska-Botelho Moniz, Kostis Michelakis, Sanjiv Sharma, Jakub Trzebinski, Belinda Garner, Amir Eftekhar

Neural disorders or malfunction can result to medical conditions such as epilepsy, Alzheimer's, chronic pain, spinal cord injury and many others, affecting the quality of life for millions of patients worldwide. Neural damage is often irreparable and when such conditions occur it is often very difficult to assess exactly the way in which neural operation is affected. Such disorders are characterised by interrelated chemical and electrical changes in the brain. At present, it is possible to monitor neural activity through the detection of electrical activity. The instrumentation required for this detection ranges from large clinical recording equipment to implantable devices. Nerve electrodes can be penetrating (needle) or non-penetrating (cuff) and depending on their shape and position they often act as spatial filters to the electrical signals they pick up due to the ionic currents involved in neural conduction. Electrical neural recordings are often severely distorted by interference from other bio-signals and vital information is lost. Moreover, trade-offs between noise and electrode dimensions impact on multi-electrode schemes. Neural amplifiers require large input dynamic range to accommodate for interference, whilst maintaining linearity, high gain and low power consumption. Although these neural recording platforms have contributed valuable insights into neuro-genic disorders and neuro-pathological studies, they mostly remain within the boundaries of research environments. Even in such environments, neural monitoring technology, as of yet, has not addressed the ionic aspect of the neural signal, thus missing out on vital neurological information. This is almost entirely due to the conventional apparatus for ion sensing in chemical laboratories being cumbersome and inappropriate for bio-signal recording. This chapter covers some of the main nerve electrode configurations used for neural monitoring, followed by an assessment of the possible merits of monitoring the ionic aspect of neural activity. Some of the latest techniques for developing and testing ionic sensors for such an application are overviewed.

56.1 **Central and Peripheral Nervous System Monitoring** 1060
56.1.1 Monitoring the Brain Function 1060
56.1.2 Peripheral Nerve Monitoring Methods 1061
56.1.3 Neural Amplifier Specifications 1063

56.2 **Chemistry of Neural Activity** 1064
56.2.1 Ionic Aspect of Brain Activity 1064
56.2.2 Ionic Aspect of Peripheral Nerve Signals 1065

56.3 **Chemical Neural Sensing Technology and Challenges** 1065
56.3.1 Requirements for a Novel Method 1065
56.3.2 State of the Art in Chemical Neural Recording Platforms 1065
56.3.3 Ionic Sensing – State of the Art and Challenges 1066

56.4 **Conclusion** 1069

References 1070

Devices for monitoring neural activity have a range of applications including diagnostics, neurophysiology and neuropathology studies, drug monitoring, and rehabilitation – through the use of the obtained signal as feedback to neurostimulation. Today, there is a massive surge in neural interfacing research, mainly due to recent advances in microelectronic and electrode technology. Various platforms are being developed for neural recording, both for the central and the peripheral nervous systems (CNS and PNS, respectively). Research in the area requires a combination of expertise on microelectronics, microfabrication, biology, medicine, and often chemistry, forming a trend for interdisciplinary research. This chapter offers a brief overview of the merits of some of the most widely used neural monitoring technologies in the brain and at the periphery, identifying the need for alternative approaches and it examines the merits of ionic sensing as a possible method for obtaining alternative manifestations of neuronal activity. Existing chemical sensing technologies are then assessed in light of the requirements of such an application.

56.1 Central and Peripheral Nervous System Monitoring

56.1.1 Monitoring the Brain Function

Single neurons or groups of neurons firing together in the brain generate chemical and electrical signals, of which the latter have been studied extensively through electrical measurements. Although it is possible to monitor single neuron signals by penetrating the cell membrane (intracellular monitoring through patch-clamping) [56.3], such techniques damage the cell and are only applied in basic neuroscience research that focuses on the underlying mechanisms of ion channels (Fig. 56.1a). Methods for monitoring naturally occurring or stimulation induced neural activity for clinical applications usually include measurements in the extra-neural medium, through electrodes that do not penetrate the neural membrane. Because neural conduction involves the extracellular medium, it is possible to make such measurements to detect both localized and more *global* variations in neuronal activity [56.4, 5]. Such methods are extensively used in medical diagnostics especially for the detection of epileptic activity prior to and during seizures. Neural recordings can be conducted in vivo on living tissue and in vitro on excised, appropriately prepared and maintained brain tissue [56.3]. In vivo extracellular potential recording involves monitoring potentials that relate to trains of pulses of collective activity of groups of neurons. Such techniques date back to the beginning of the 20th century. To assess large scale potential fluctuations, electro-encephalography (EEG) is used in clinical diagnostics, to record brain activity using electrodes placed across the scalp [56.6] (Fig. 56.1b). The technique has relatively low spatial sensitivity due to the tissue and bone layers located between the electrodes and the brain. In order to overcome this limitation, electrodes can be placed intracranially to gain a higher quality signal and a better definition of the location of its source. The so called electrocorticogram (ECoG) is equivalent to the EEG but with direct electrode contact on the brain surface [56.7].

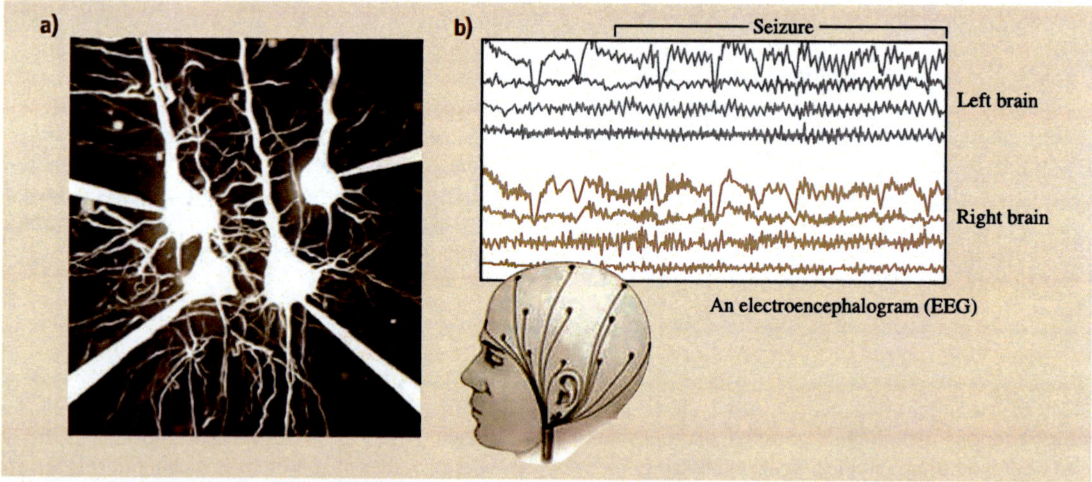

Fig. 56.1 **(a)** Simultaneous patch clamping of multiple pyramidal neurons in a cortical brain slice (after [56.1]). **(b)** Typical EEG setup and seizure-related recordings (after [56.2])

Microelectrodes for Cortical Interfacing

Conventional cortical monitoring is carried out with the use of electrode arrays placed on the brain surface (epicortical electrodes, Fig. 56.2a). Recent advances in microfabrication technology have allowed the development of implantable biomedical microsystems that include thin and flexible substrates with monolithically integrated electrode arrays and circuitry, forming hybrid electronic assemblies. Bare integrated circuits (ICs) can be placed on flexible structures, which present a significant advantage in terms of biocompatibility compared to rigid structures, as they reduce scar tissue development [56.8]. To protect implantable electrodes from the influence of the biological environment, and to generate an appropriate tissue response at the implantation site, encapsulation processes can be used with materials such as parylene and silicone.

To extract more localized neural activity than that monitored through epicortical interfaces, needle microelectrodes (Fig. 56.2b) can be used to penetrate layers of the brain (but without penetrating actual neural cells) [56.2]. These provide high resolution *intracortical field potential* and access to deeper brain structures such as the hippocampus. Contrary to the flexible substrates used for epicortical interfaces, intracortical (or intracerebral) electrodes are penetrating probes usually consisting of post-processed silicon wafers with platinum-coated metal pads in vector array arrangements. Different electrode geometries built to detect bioelectric signals include three dimensional arrays [56.9], two dimensional laminar electrodes [56.10], and subcortical silicon probes [56.2]. The probes must be robust enough to penetrate both the dura and the pia mater and at the same time they must be as small as possible to minimize brain tissue damage.

In both *epicortical* and *intracortical* methods, pairs of electrodes pick up potential differences across the extracellular fluid due to the nerve-generated ionic currents flowing through it, much like measuring the potential at various points along the length of an electrical resistor while current flows through it. The materials and physical dimensions of the electrodes are extremely important for the quality of the signal because of the capacitance formed at the electrode–tissue interface. The differential signals between electrode pairs, or between each electrode and a common one acting as the physiological ground, are amplified by specifically designed neural amplifiers, described later in this section [56.11, 12].

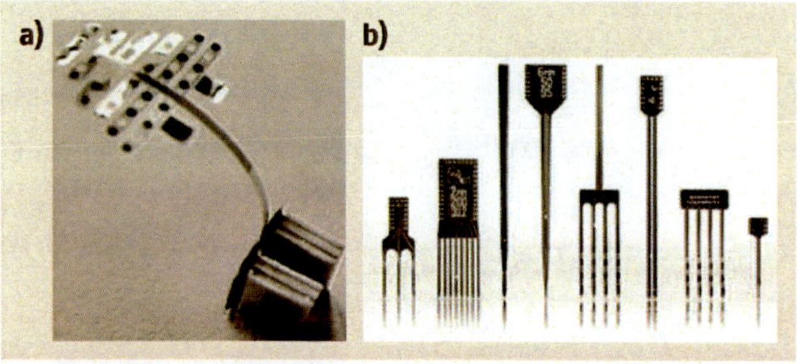

Fig. 56.2 **(a)** Polyimide microelectrodes by the IBMT group at Fraunhofer Institute used for epicortical interfacing in mice (after [56.6]). **(b)** Intracortical probes by NeuroNexus (after [56.2])

The Need for Different Brain Monitoring Modalities

Despite the immense technological progress in the field of brain function monitoring, the EEG/ECoG methods have limitations in the insight they can offer [56.13, 14]. The main limitations of the cortical interfaces are:

- A lack of adequate localization in the presence of activity from distant brain structures when epicortical interfaces are used
- Tissue damage and high risk when intracortical electrodes are used
- Dislocation of electrodes during brain inflammation
- Severe issues with wire fragility when electronics are not embedded on the actual electrode.

In epilepsy, the so-called seizure precursor detection problem currently remains unsolved [56.13, 14]. The development of alternative monitoring methods, such as functional magnetic resonance imaging (fMRI), as well as their widespread appeal in medical diagnostics [56.15], makes a very strong case for the necessity of research into other modalities of brain activity. This will be revisited later in this chapter.

56.1.2 Peripheral Nerve Monitoring Methods

Although *action potentials* across the neural membrane are of mV order, the potentials monitored outside the nerve are a result of the ionic currents propagating along the nerves (Fig. 56.3a) scaled by the impedance of extracellular fluid. In the PNS nerve fibers located proximally to the brain or the spinal cord are bundled into groups called fascicles, that later branch out to individual organs or muscle groups. Fascicles are also grouped together initially, forming nerves. Nerves can be penetrated by needle electrodes for assessing the activity

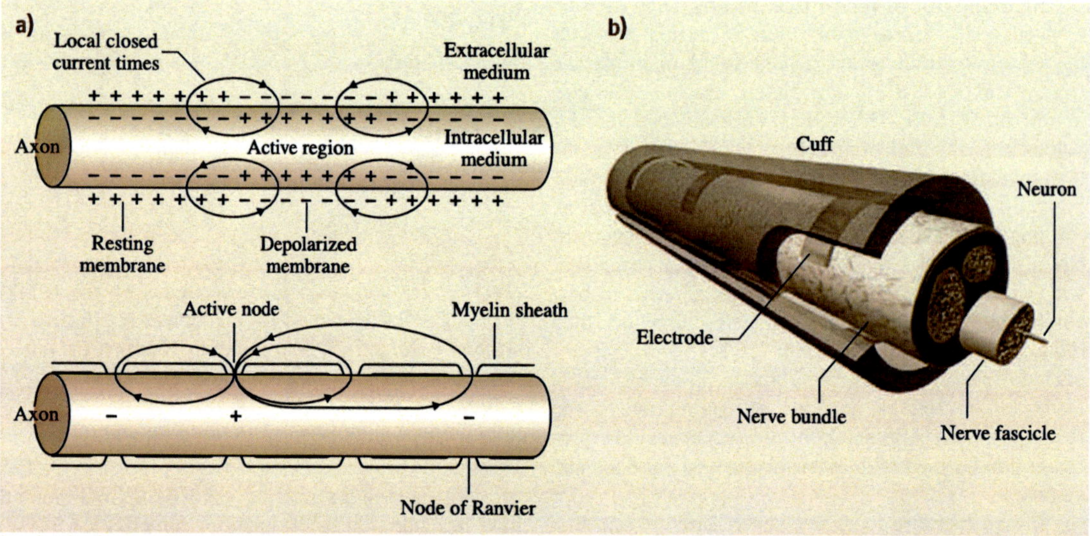

Fig. 56.3a,b Neuron conduction. **(a)** *Top*: unmyelinated fiber; *bottom*: myelinated fiber. **(b)** A tripolar cuff electrode placed around a nerve (after [56.12])

Part E | 56.1

of individual fascicles without the need to access these groups of fibers close to the particular organ they innervate. In that way surgery is minimized to a single access location that has the possibility to provide information for sensory or motor activity relating to peripheral structures located distally. However, using such needles (or *intrafascicular*) electrodes damages neural tissue and is inappropriate for chronic implantation. Implanted cuff electrodes (Fig. 56.3b) are appropriate for chronic recording of naturally occurring neural activity (electroneurogram or ENG) because they are non-invasive to the nerve and assist in reducing interference from sources outside the cuff, such as the electromyogram (EMG) generated from muscles nearby. The amplitude of the signals recorded by such electrodes placed externally to the nerves is in the order of a few microvolts [56.12, 16–19].

Characteristics and Issues of Cuff Electrodes

The conventional cuff is an insulating flexible tube typically made of silicone rubber or polyimide, with metallic (usually platinum) ring electrodes attached to its inside walls, encircling the peripheral nerve of interest [56.12, 20]. Cuffs have been extensively used in neural signal monitoring research, due to their mechanical stability and non-invasiveness to the nerve tissue [56.21]. They have demonstrated the vast potential of well performing systems for diagnostics and even more for rehabilitation, in cases of e.g. epilepsy and spinal cord injury [56.22, 23]. Ideally, the ability of the cuff electrode (or *cuff*) to linearize the internal potential field generated by external sources allows interference to be eliminated. This is more effective when *tripolar* cuffs are used (i. e., cuffs with three equally spaced ring electrodes), in combination with appropriate amplifier configurations.

In the PNS, after decades of research, cuffs – and even more so all other electrodes – have not yet become part of mainstream healthcare practice due to their inherent limitations. The main limitations of the conventional cuffs are:

- Low ENG signal amplitude (microvolts), in the presence of up to three orders-of-magnitude greater interference from myoelectric activity of nearby muscles
- Electrical noise, which introduces constraints in both electrode and amplifier designs
- Constraints on cuff and electrode dimensions
- Poor *selectivity*, in terms of which fascicle is active inside a nerve bundle and what the direction of the signal is (sensory or motor).

Myoelectric interference rejection has been mainly addressed by the use of three-electrode cuffs in combination with appropriate amplifier schemes such as *quasi-tripole* (QT) and *true-tripole* (TT) (Fig. 56.4) [56.12]. However, their EMG cancelation effectiveness is highly

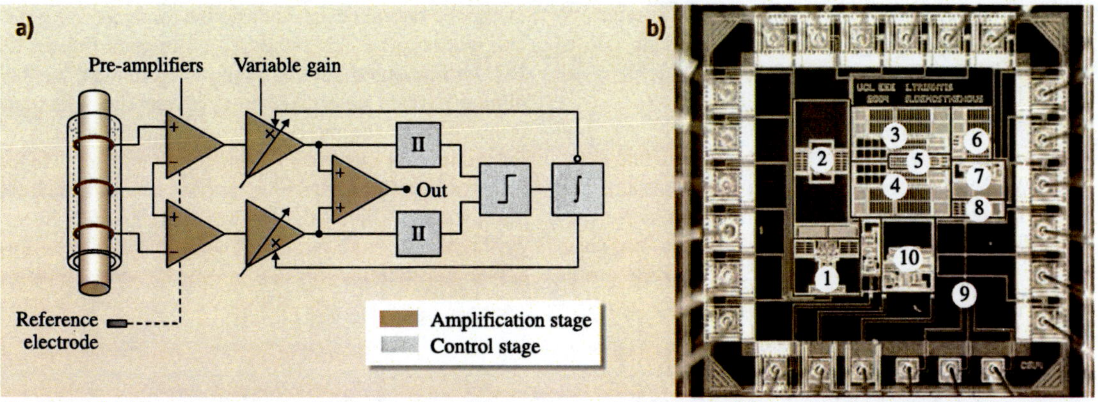

Fig. 56.4 **(a)** The adaptive tripole (AT) neural amplifier concept (after [56.12]). **(b)** AT chip microphotograph with main subcircuits numbered: (1) preamplifiers (2) band pass filters (3, 4) variable gain operational transconductance amplifiers (OTAs), (5) biasing circuitry, (6) full-wave rectifiers, (7) AT output stage, (8) feedback OTA, (9) comparator, (10) integrated integrator (after [56.12])

degraded by cuff systematic errors. These errors were given the term *cuff imbalance* by *Triantis* et al. [56.12], who developed the auto-adjustable *adaptive tripole* (AT) IC amplifier that minimized the effect of imbalance, significantly improving the quality of ENG recordings [56.12]. In the context of the same work, it was shown that there is still a need for greater immunity to myoelectric interference, as cuff imbalance also depends on the distance and orientation of the cuff relative to active muscles [56.17]. Thus, when different muscle groups become active, or in the presence of nearby stimulation causing an interfering stimulus artifact, the AT has to continuously adjust its EMG cancelation scheme.

The Need for Different Peripheral Monitoring Modalities

Even if EMG interference and noise factors are overcome, the quality of recordings strongly depends on the cuff having specific dimensions, sometimes making it too long for certain important implantation locations, for example, inside the dura close to the spine. Moreover, cuff electrodes are usually implanted in close proximity to the central nervous system, around large nerves. Therefore, employing ring electrodes does not offer any possibility of identifying which of these organs are active each time a signal is recorded. Various solutions have addressed this issue of *fascicle selectivity* by replacing ring electrodes with segmented multi-electrode rings. Although these approaches have indicated some degree of selectivity, they suffer from the aforementioned limitations: the small area of the electrode segments translates into higher electrode noise relative to full ring electrodes; and amplifier configurations used in such schemes exhibit poor electromyogram (EMG) cancelation. Finally, the unavoidable increase in cuff lengths and in the number of wires connecting the cuff with the implant greatly reduces the robustness of the systems and limits the possible implantation sites.

56.1.3 Neural Amplifier Specifications

The signals provided by the electrode contacts are amplified by custom designed neural amplifiers [56.16, 24, 25] featuring high gain to raise the signal to the appropriate range for A-D conversion. Generally, the term *neural amplifier* is used to describe devices designed to amplify neural signals, irrespective of whether these are applied to the PNS or the CNS. However, the neural signal amplitude and frequency ranges are not the same in both cases, with stimulation-induced signals also exhibiting different characteristics. Thus, a device designed for a specific type of neural signals cannot be used at a different neuroprosthetic platform, even though many of the neural amplifier characteristics are similar. Neural amplifiers mainly have to deal with three typical types of neural signals:

1. Brain cortical EEG, requiring an amplifier gain range from 50 to 100 and bandwidth from 300 to 500 Hz
2. Peripheral neural signals during stimulation which require a gain range of 500 to 1000 and bandwidth of around 3 kHz

3. Natural peripheral neural activity (ENG) requiring gains in the order of and bandwidth from 800 to 10 kHz.

The front-end noise of the amplifiers must be very low and care must be taken to choose the optimal technology [56.26]. Amplification is typically implemented in two stages, the first one providing low-noise preamplification to remove the necessity for low-noise design for the subsequent stage [56.24]. Low power integrated circuit design is desirable for a recording system that is to be integrated into an implant [56.11, 12, 27]. As implanted devices cannot be grounded the amplifiers have to be connected to a reference electrode somewhere inside the body, usually in the vicinity of the monitoring interface. Design constraints include the need for large input dynamic range to accommodate for microvolt to millivolt signals, whilst maintaining linearity, high gain, high common-mode rejection, and low power consumption [56.12].

56.2 Chemistry of Neural Activity

Although immense progress has been achieved in technology for interfacing with and for monitoring neural signals, present day nerve recording interfaces suffer from several limitations, as mentioned previously. This calls for a more careful inspection of the fundamentals of neural conduction so as to consider alternative manifestations of neural activity. Consequently, technologies that can be utilized to obtain access to this additional information can be assessed and developed. Indeed, as nerve conduction in both the brain and the periphery is based on ionic currents, rather than on electrons, chemical sensing may be promising for an in-depth approach to neural monitoring.

56.2.1 Ionic Aspect of Brain Activity

In the CNS there are two main observable chemical variations that relate to neural operation: Variations of neurotransmitters and ions (Fig. 56.5a). Neurons receive input from other neurons through neurotransmitters – comprising of amino acids, peptides, and monoamines – and when a threshold is exceeded, they generate an ionic output, whose electrical expression is the *action potential* [56.3]. At the neuronal membrane level, the input and the output are realized by ionic currents of potassium (K^+), sodium (Na^+) (Fig. 56.5b), and chloride (Cl^-) at the axons with an additional ionic current of calcium (Ca^{2+}) at the somas. Neurotransmitters do not exclusively determine the inhibitory or excitatory activity, but actually it is the ionic milieu that regulates neural response and signalling. For example, K^+, Cl^- ionic homeostasis was shown in [56.28] to affect neurotransmitter signalling defects in epileptic patients. Generally, variations in ionic concentrations in the brain have been observed in pathological states such as epilepsy, pain syndromes, movement disorders, and in certain psychiatric diseases [56.10].

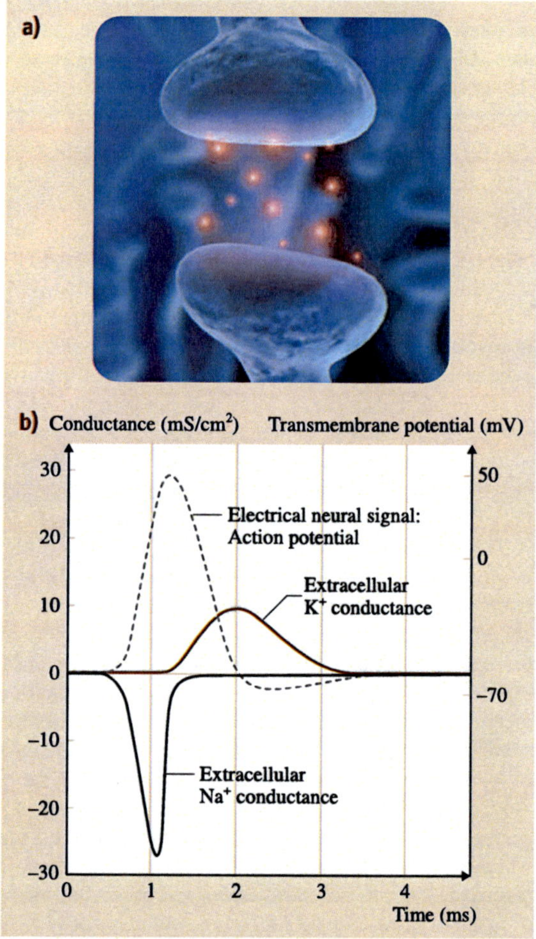

Fig. 56.5 **(a)** Depiction of a neuron's synapse exchanging neurotransmitters with another (after [56.2]). **(b)** The main ionic components of the action potential: K^+ and Na^+ conductances (after [56.3])

Thus ionic variation is of fundamental importance to neural operation. Ionic currents change the potential of both the intra- and extracellular space. This can be measured by appropriate electrical sensing methods, however without any information on its underlying ionic composition. The flow of these ionic currents – and thus the excitability of the cells – is determined by the relatively steady ionic concentrations of the intra- and extracellular space, regulated by slower transmembrane ion pumps [56.3].

56.2.2 Ionic Aspect of Peripheral Nerve Signals

Similarly to brain neurons, the fundamental mechanism of peripheral neural activity is related to the inflow and outflow of sodium (Na^+) and potassium (K^+) ions across a cell membrane (Fig. 56.5b). When an action potential is generated, sodium (Na^+) and potassium (K^+) ions are allowed to enter or to exit the cell by their respective ion channels (gates). Under resting conditions, typical mammalian extraneural concentrations of Na^+ and K^+ are approximately 150 and 5.5 mmol, respectively, while the respective intra-neural concentrations are approximately 15 and 150 mmol [56.3].

When a cell membrane is depolarized (becomes less negative) past a certain threshold, voltage-gated sodium channels open, resulting in a great and rapid influx of Na^+ ions. The influx of Na^+ ions temporarily drives the electric potential across the cell membrane away from its resting value (approximately −75 mV in a mammalian cell) towards a more positive value. This is followed by an efflux of K^+ ions through K^+ channels resulting in a rebalancing of the membrane potential. The influx of Na^+ depolarizes the adjoining part of the membrane causing it to follow the same procedure and, therefore, the action potential propagates as a wave along the axon. During an action potential, the K^+ concentration in the immediate vicinity of the active region of the nerve increases by more than 4 mmol [56.29], i. e. it approximately doubles. The duration of each of these ionic pulses, which depend on the fiber type, are of the order of a few milliseconds [56.3].

56.3 Chemical Neural Sensing Technology and Challenges

56.3.1 Requirements for a Novel Method

Based on the previous sections, the need for a monitoring method that will discriminate between the fundamental components of neural signals makes it desirable to incorporate chemical sensors on neural interfacing platforms to allow ionic neural recording. A chemical neural monitoring platform based primarily on the measurement of K^+ and Na^+ sensing should feature microsensors capable of mmol sensitivity, embedded on flexible substrates. The sensors should ideally feature millisecond-order time response and sensing should be carried out without affecting neural operation. Beyond the *proof of principle* stage, such an ionic neural sensing interface should also feature acceptable longevity, depending on the application (e.g. implantability would require a life span of the order of several years) as well as biocompatibility and robustness.

56.3.2 State of the Art in Chemical Neural Recording Platforms

Before proceeding to examine the technologies available for achieving the requirements listed above for ionic sensing of neural activity, an overview of somewhat relevant concepts that have been reported in the literature is offered below, providing a very useful perspective.

In [56.30] *Strong* et al. presented a miniaturized platform including platinum electrodes on a silicon substrate with integrated front-end circuitry for sensing electrical signals as well as neurotransmitters. Even though the approach was not presented experimentally at that stage it seems very promising in terms of miniaturization and integration. However, the sensing setup proposed used a voltammetric approach (this particular sensing approach is examined along with others in the next section) resulting in a DC biasing of up to 1.6 V across the sensing electrodes. This presents a possible disadvantage affecting the neural operations that need to be monitored. The method was proposed for sensing the neurotransmitter dopamine.

Medtronic touches upon ionic sensing for biomeasurements as it reports K^+ and Na^+ ion-selective electrodes (ISEs) in a patent [56.31], where an attempt is presented to overcome the unsuitability of reference electrodes used with such sensors. These sensors are typically used for analytical chemistry applications and the reference electrodes are bulky and made of silver or silver-chloride, which is not preferred for biocompatible interfacing, as it is polarizable. The patent focuses on ion measurements in blood demonstrating a technology that could be used for biointerfacing, although it is not directly suitable for miniaturization and, therefore, does not meet the abovementioned criteria for neural monitoring.

In [56.32] *Sudakov-Boreysha* et al. presented the use of ion-sensitive field effect transistors (ISFETs – described in detail later) on brain catheter probes for the diagnosis of ischemia. The probes penetrated the skull to monitor blood pH levels. The authors focused mainly on encapsulation issues while bioelectric recordings were not mentioned. Even though the method does not focus on neural activity per se, the setup was interesting as it interfaced with the brain using a kind of sensor that can potentially have a robust miniaturized front end.

The European project, NeuroProbes [56.33], aims at recording either electrical information in the brain (mainly) or chemical information in the form of neurotransmitters. Similarly, the *Golden Brain* EU project [56.34] senses neurotransmitters using microprobes. The respective sensing methods used require DC signals between electrodes, with possible effects on neuronal operation, and the response time is in the order of tens of seconds. These projects highlight the importance of researching alternative neural monitoring forms and they proposed miniaturized interfaces. Moreover, although they do not include ionic measurements, they underscore the importance of microfabrication and portability in the relevant technologies.

56.3.3 Ionic Sensing – State of the Art and Challenges

Ion Sensors

There are three main electro-analytical methods for biasing chemical sensors. The first two, *coulometry* and *voltammetry*, involve current measurements and they require DC signals (voltage or current) between – or across – a pair of electrodes. The *potentiometric* configuration is the only passive method where there is practically no DC current flowing between the electrodes [56.35, 36], making it more appropriate for biointerfacing and in vivo sensing.

Conventional apparatus used for ion sensing in analytical chemistry experiments cannot be used for a miniaturized neural sensing application as it is too cumbersome and inappropriate for biosignal recording [56.37]. However, chemical sensors based on ion-sensitive field effect transistors (ISFETs) are much more able to meet the above requirements. ISFETs are miniaturized, solid-state potentiometric transducers used predominantly for pH measurements [56.37].

Since their introduction in 1970 by *Bergveld* [56.37], ion-sensitive field effect transistors (ISFET) are today used in a wide range of sensing applications in the fields of clinical medicine, biotechnology, and the environment [56.38]. Their attractive features, such as small size, fast response time, the prospect of monolithic integration [56.39], low cost, as well as the possibility to combine them with various kinds of biologically active materials (e.g., enzymes, antibodies, DNA, and cells) make them well suited for *micro total analytical systems* (μTAS) or *Lab-on-Chip* devices [56.40, 41]. Overall they offer cheap and portable devices for real time monitoring that could be potentially used for various biomedical and point of care diagnostic applications.

ISFETs features meet most of the requirements listed in the previous section. Even though they are mostly used for pH measurements, they can be made sensitive to specific ions. This is accomplished by deposition of appropriate ionophores [56.42], effectively replacing the transistor gate contact with a chemical membrane permeable to specific ions only [56.37, 43–46]. In that case, they are called *chemFETs* [56.37]. The response of the modified ISFET (chemFET) stems from the voltage that develops across the interface formed between the membrane and the aqueous solution, to the reversible binding of the analyte with the ion carrier. An intermediate layer is often applied to establish a reversible electron transfer pair at the ion-selective membrane–ISFET interface. Even though other types of ion sensors exist, chemFET sensors are the most suitable choice for neural monitoring because a) most of the other ion sensor types are unsuitable for this application due to their size and to their inability to be incorporated on flexible substrates, and b) ISFETs are robust [56.39], and several sensing units can be stacked on a single chip (Fig. 56.6a) [56.47].

Although ISFETs feature fast temporal response [56.37] depositing a chemical membrane on their gate to turn them into chemFETs makes them much slower because ions have now to permeate the membrane before reaching the ISFET gate. This drawback may be minimized by reducing the membrane thickness; however, this impacts on sensor robustness and lifetime. The titration diagram in Fig. 56.6b indicates how a fast and a slow ISFET-based sensor respond to step concentration changes. For reasons of cost it is desirable to fabricate ISFETs using the well established CMOS process, with silicon nitride as the sensing layer. Their power consumption can be very low when they are designed and biased in *weak inversion* oper-

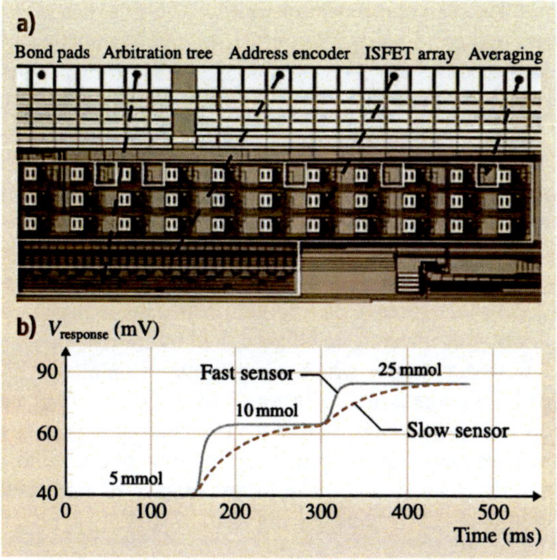

Fig. 56.6 **(a)** CMOS ISFET pixels incorporated in an array and monolithically integrated to processing circuitry (after [56.55]). **(b)** Diagram of ISFET output voltage ($V_{response}$) corresponding to ion concentration (log) step changes vs. response time for a fast and a slow sensor

ation [56.47–53]. Several studies have described their use for detecting substances like pH, urea, alcohol, and glucose [56.54].

ISFET Fabrication in CMOS Technology

ISFETs are very closely related to the transistors found in CMOS-based processes and industrial production of stand-alone ion-sensitive probes is realized through the modification of such processes [56.2]. However, it makes perfect sense to try and accommodate the realization of ISFETs in unmodified conventional CMOS foundry processes. On the one hand, this would introduce certain drawbacks, given, for example, that an optimized dielectric layer would not be possible to achieve as all devices have to rely on the gate-oxide layer provided by the technology. Furthermore, this dielectric layer would not be able to serve as the sensing layer itself, as an inherent feature of standard CMOS technologies is that the gate-oxide is automatically covered by the polysilicon gate. On the other hand, if accommodation of ISFETs in a CMOS process could somehow be achieved, the benefits would outweigh any disadvantages. That would not only provide access to a relatively cheap and widely available method of mass manufacturing but, equally importantly, it would provide access to the monolithic implementation of all those capabilities

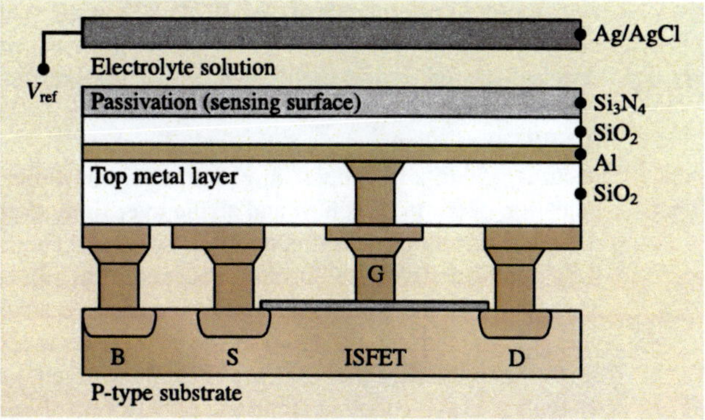

Fig. 56.7 Layers of an ISFET fabricated in CMOS process (after [56.55])

that come with silicon chips, including memory, logic and analog and digital signal processing, and conditioning. Indeed, *Bausells* [56.56] first demonstrated the feasibility of this idea by getting the chip's passivation layer, covering an otherwise floating CMOS-MOSFET gate, to act as the ion-sensing membrane (Fig. 56.7). Since then, this concept has found commercial applications and today a number of companies are understood to make use of this technology [56.2].

Sensor Technology Challenges

Although ISFET-based chemical detection of neural activity is a promising prospective technology, there are still several non-trivial challenges that need to be addressed, including drift and response time. ISFETs suffer from temperature dependence and long term temporal drift [56.57]. The temperature dependence is not expected to be an important issue inside the body, where temperature variation is minimal. Moreover, any temperature dependence can be eliminated using standard electronic circuit techniques, i. e. PTAT or bandgap reference [56.58]. Temporal drift is related to the capacitance at the front end of the ISFET sensing terminal, as the device is in effect a floating gate MOS transistor, with variations on the gate charge sensed with respect to an independent reference electrode. Similar issues affect all sensors that have an insulator-solution interface due to hydration effects of the insulating layer causing capacitance variations.

Post-processing steps to minimize drift involve briefly applying hydrofluoric acid on the sensing area before use or replacing the insulating layer of the sensor with one that has better properties, although this is not a choice if a standard CMOS fabrication process is used. The use of appropriate dedicated circuitry can also improve performance of CMOS ISFETs as they can be monolithically integrated on the devices to perform drift cancelation processing. Devices that are indifferent to the target ions are often used as reference FETs (REFETs) to eliminate drift through differential connectivity [56.59]. In general, any platform for chemical detection needs to be robust and stable over time, even though this is much more important in analytical chemistry applications where absolute concentration values are needed. In bio-signal monitoring the main focus is variation monitoring, therefore, if the drift rate is much slower than the expected changes it can be filtered out by the use of AC coupling front-end circuitry involving filtering or chopping.

Irrespective to the choice of sensing technique, one of the most dominant factors in a chemical biosignal monitoring setup will be diffusion. The detection of a particular ion will greatly depend on the distance between the sensor and the source, which in this case would be the membrane of the firing neurons. Although this raises the demands for sensor sensitivity and calls for platforms that will allow very high proximity to the source, the authors expect that diffusion will also prove to be one of the beneficial factors related to electrical interfacing. Since the dominant extraneural ions are sodium and chloride, potassium or calcium specific sensors placed adequately closely to the targeted neurons will possibly benefit from diffusion, as distant sources that usually interfere with the desired measurements will mostly cause shifting of the dominant ions, while their native potassium and calcium will not reach the sensing site. Electrically, any ionic current will be picked up by the electrodes, while ion-specific measurements will only be affected by the chosen species.

Perhaps the greatest issue in ionic neural sensing will be the improvement of the sensor's response time. Bare ISFETs for pH sensing have been reported to exhibit millisecond-order response [56.37], however, with the application of ionophores on the sensing area to turn them into ion-specific chemFETs, this increases up to several minutes. Thus, reducing response time presents a major challenge. Chemistry and micro-fabrication experts that are members of the authoring group have managed to drop chemFET response times from several minutes down to several hundreds of milliseconds. Depending on the application, this could prove to be satisfactory for the envelope detection of trains of pulses or for the detection of slow global variations of e.g. potassium in the brain, which has been reported to relate to pre-epileptic states and exhibits a variation rate of hundreds of milliseconds [56.60]. Further challenges are life span and biocompatibility. The life span is often related to encapsulation issues and degradation of ion-sensitive membranes. Commercially available devices have been reported to last several months [56.59]. Again, this may prove to be limiting for a chronic, implantable platform, however, the technology may still prove to be invaluable for intra-operative applications or for monitoring hospitalized patients. The information obtained may also be useful for an initial post-operative period for fine-tuning and calibration of electrical monitoring interfaces, assisting with adjustments that have to do with filtering, imbalance, and interference in the few weeks during which scarring tissue stabilizes. Finally, biocompatibility in terms of ill-effects to tissue due to the materials used at the front end of the sensors has been extensively researched by some of the authors of this chapter in the context of sensors of stem cell cultures, with very promising results [56.61].

Sensor Assessment

While determining the sensitivity of the sensors is quite straightforward, their response time is not so easy to assess. Shifting a sensor from a beaker to another with a different concentration can hardly assist in determining response time down to millisecond accuracy, while using a pipette to change the concentration of the solution in which the sensor is immersed involves too many parameters that mainly relate to diffusion (volume of solution, distance between pipette and sensor, stirring, or static solution – to name a few) rather than the inherent speed of the sensing membrane. As diffusion is so dominant the setup used for testing the sensing speed has to be somewhat more sophisticated, including software for modeling and, ultimately, the use of microfluidic platforms.

Due to mass limiting phenomena and ion perfusion, all sensors coated with membranes experience a lag in response time. Out of a number of approaches used to assess the influence of the membrane on the response time, incorporating a sensor into a microfluidic platform seems to be the most robust and accurate technique as it allows sensor response measurements in a highly controlled environment.

The first attempt towards integrating microfluidics and field effect transistors was reported in 1999, referred to as flowFETs. These devices were used as controlling and switching elements in microfluidic networks [56.62]. *Sharma* et al. have reported a silicon on insulator-based microfluidic device with monolithically integrated FETs for microTAS applications [56.63]. An extended gate field effect transistor (EGFET)-based bio-

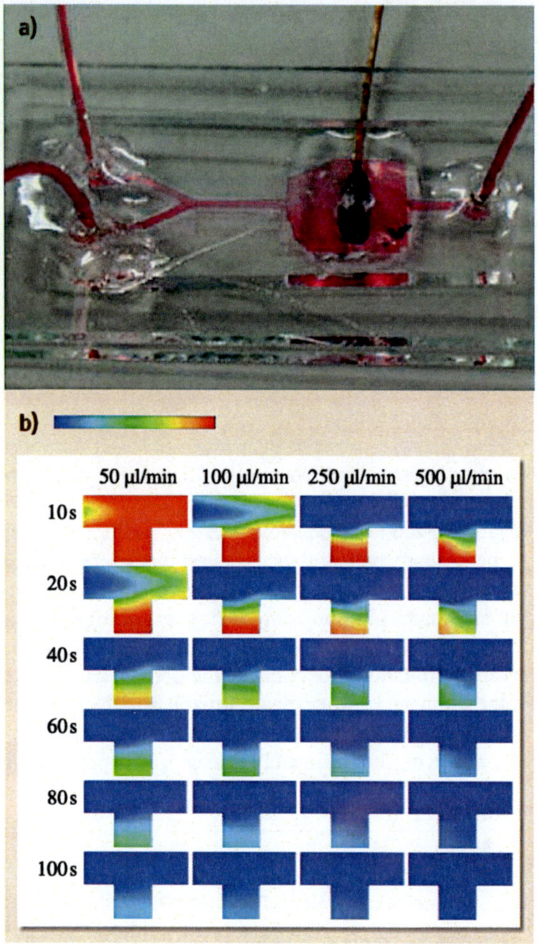

Fig. 56.8 (a) Microfluidic platform developed by some of the authors, to assess single chemFET response time. (b) The flow variations inside the microfluidic channel were simulated to assess its performance (after [56.69]) ▶

sensor integrated with a silicon microfluidic channel for the electronic detection of streptavidin-biotin protein complexes has been reported by Kim et al. In that instance, gold was used as the extended gate coated with self-assembled monolayers of thiols, onto which the biotin avidin complex was immobilized [56.64]. *Truman* et al. reported the use of silicon-based ISFETs for monitoring transport and chemical properties of liquids in microfluidic systems [56.65]. Reliable time-resolved ion and molecular transport sensing in chemoreceptive neurons in microfluidic channels using transistors has been demonstrated by *Jacquot* et al. [56.66]. Tetraphenylborate derivatives have been used to modify ISFETs on polymeric microfluidic devices for sensing cationic surfactants in dental rinses [56.67]. In another instance, a microfluidic chip was used for cell culture and a large transistor-based sensor array chip for direct extracellular imaging [56.68]. A literature survey does not reveal any solid state sensors integrated microfluidic device that could be used for simultaneous determination of cations. Members of the group of authors of this chapter have, therefore, proceeded to the fabrication and testing of a microfluidic platform for assessing chemFETs for biosignal sensing; the initial results of this work are very encouraging (Fig. 56.8a). The work was complemented by finite element method simulations to establish the validity of the microfluidic platform (Fig. 56.8b).

56.4 Conclusion

This chapter has presented an overview of neural sensing methods examining the prospects of an alternative method: ion sensing. The concept of ionic instead of electrical neural monitoring is theoretically very attractive, as it may serve to offer a greater insight into the underlying processes of neural activity. It may also offer increased selectivity and interference immunity due to better localization. Furthermore, the Nerstian nature of the chemical interface will offer a natural logarithmic signal companding to the monitored variations. In effect this will reduce the required input dynamic range of the front-end amplifiers and it will greatly minimize the effects of interference. For example, the three orders of magnitude between the microvolt neural signal and the millivolt muscle interference in ENG recording platforms will translate into a factor of three between the two signals, if these were generated by the same ionic species at the same distance from the sensing surface. The benefits of such a technology are therefore evident, as the issues of signal localization and interference reduction are two of the most dominant bottlenecks in conventional neural monitoring arrangements. It is, however, equally evident that such a technology would require the sensors to be taken out of the context of analytical chemistry, which demands accurate and stable measurements of

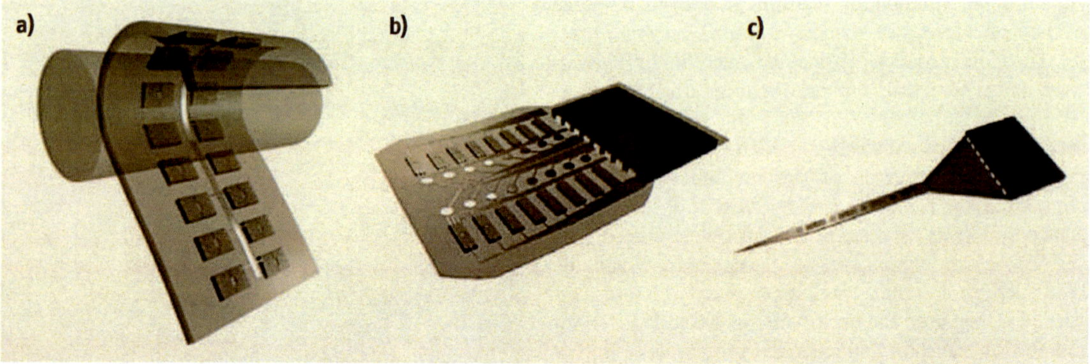

Fig. 56.9 **(a)** Concept of arrays of ISFETs attached to a cuff for PNS sensing, as detailed in 56.5; **(b,c)** similar concepts for cortical interfaces

absolute concentration values, to a new focus where response time and miniaturization are key factors. With regards to the targeted biology itself, the brain seems to be a more realistic focus, even though peripheral nerve preparations are generally easier to use for testing and developing new interfacing technologies. This is mainly due to the ion flow reducing effects of the layers surrounding fibers inside the nerves, namely the epineurium and the perineurium [56.70]. Nerves can be treated in vitro to reduce the effects of these layers, however, this ultimately makes a non-penetrating ion-sensing technology inappropriate for PNS interfacing, especially in vivo. Ultimately, the process of developing neural sensing modified ISFETs has huge potential in offering the enabling technology for all sorts of vital signal monitoring, mainly intraoperatively. The authors are developing relevant platforms (the concepts are shown in Fig. 56.9); they have formed an interdisciplinary group in the process of pursuing this target and are hopeful that they will create a trend of alternative monitoring concepts in the field of neuroprostheses.

References

56.1 P.J. Sjöström, M. Häusser: A Cooperative Switch Determines the Sign of Synaptic Plasticity in Distal Dendrites of Neocortical Pyramidal Neurons, Neuron **51**(2), 227–238 (2006)

56.2 http://www.primalbody-primalmind.com/blog/?p=366

56.3 J. Malmivuo, R. Plonsey: *Bioelectromagnetism: Principles and Applications of Bioelectric and Biomagnetic Fields* (Oxford Univ. Press, Oxford 1995)

56.4 I.F. Triantis, V. Woods, A. Eftekhar, P. Georgiou, T.G. Constandinou, E.M. Drakakis, C. Toumazou: Advances in neural interfacing, IEEE Circuits Syst. Soc. Newsl. **1**(1) (2007)

56.5 I. F. Triantis, C. Toumazou: Method and Apparatus for Measuring Activity in the Peripheral Nervous System, Patent WO/2008/007065 (2008)

56.6 J. Choi, K.P. Koch, W. Poppendieck, M. Lee, T. Doerge, H.S. Shin: A flexible microelectrode for mouse EEG, Conf. Proc. IEEE Eng. Med. Biol. Soc. 2009 (IEEE, 2009) pp. 1600–1603

56.7 N. Nakasato, M.F. Levesque, D.S. Barth, C. Baumgartner, R.L. Rogers, W.W. Sutherling: Selective electrical interfaces with the nervous system, Electroencephalogr. Clin. Neurophys. **91**(3), 171–178 (1994)

56.8 K.P. Koch, A. Ramachandran, W. Poppendieck, D. Feili, K.P. Hoffmann: Polymer-based implantable electrodes: state of the art and future prospects, Proc. Mater. Res. Soc. Spring Meeting (San Francisco, USA 2006)

56.9 W.L.C. Rutten: Selective electrical interfaces with the nervous system, Annu. Rev. Biomed. Eng. **4**, 407–452 (2002)

56.10 D. Fabó, Z. Maglóczky, L. Wittner, A. Pék, L. Eross, S. Czirják, J. Vajda, A. Sólyom, G. Rásonyi, A. Szucs, A. Kelemen, V. Juhos, L. Grand, B. Dombovári, P. Halász, T.F. Freund, E. Halgren, G. Karmos, I. Ulbert: Properties of in vivo interictal spike generation in the human subiculum, Brain **131**(Pt 2), 485–499 (2008)

56.11 R.R. Harrison, C. Charles: A low-power low-noise CMOS amplifier for neural recording applications, IEEE J. Solid-State Circuits **38**, 958–965 (2003)

56.12 I.F. Triantis: An Adaptive Amplifier for Cuff Imbalance Correction and Interference Reduction in

Nerve Signal Recording. Ph.D. Thesis (University College London, London 2005)
56.13 M. Todd: EEGs, EEG processing, and the bispectral index, Anesthesiology **89**, 815–817 (1998)
56.14 E. Niedermeyer, F.L. da Silva: *Electroencephalography: Basic Principles, Clinical Applications, and Related Fields*, 5th edn. (Lippincott Williams & Wilkins, Philadelphia 2004)
56.15 N.K. Logothetis, J. Pauls, M. Augath, T. Trinath, A. Oeltermann: Neurophysiological investigation of the basis of the fMRI signal, Nature **412**, 150–157 (2001)
56.16 I.F. Triantis, A. Demosthenous, N. Donaldson: On cuff imbalance and tripolar ENG amplifier configurations, IEEE Trans. Biomed. Eng. **52**, 314–320 (2005)
56.17 I.F. Triantis, A. Demosthenous: The effect of interference source proximity on cuff imbalance, IEEE Trans. Biomed. Eng. **53**, 354–357 (2006)
56.18 I.F. Triantis, A. Demosthenous: Tripolar-cuff deviation from ideal model: Assessment by bioelectric field simulations and saline-bath experiments, Med. Eng. Phys. **30**(5), 550–562 (2008)
56.19 Z.M. Nicolic, B.D. Popovic, R. Stein, Z. Kenwell: Instrumentation for ENG and EMG recordings in FES systems, IEEE Trans. Biomed. Eng. **41**, 703–706 (1994)
56.20 L. Andreasen, J. Struijk: On the importance of configuration and closure of nerve cuff electrodes for recording, Proc. 20th Annu. Int. Conf. IEEE Eng. Med. Biol. Soc., Vol. 20 (1998) pp. 3004–3007
56.21 J.J. Struijk, M. Thomsen, J.O. Larsen, T. Sinkjaer: Cuff electrodes for long-term recording of natural sensory information, IEEE Eng. Med. Biol. **18**, 91–98 (1999)
56.22 I.F. Triantis, A. Demosthenous, N. Donaldson: An ENG amplifier for EMG cancellation and cuff imbalance removal, EPSRC/IEEE Postgraduate Res. Conf. Electron. Photon. Commun. Softw. (PREP'03) (Exeter, UK, 2003) pp. 105–106
56.23 J. Rozman, B. Cetina: Electroneurograms recorded from the left vagus nerve of a dog, Proc. 24th IASTED Int. Conf. Biomed. Eng. (Innsbruck 2006) pp. 191–196
56.24 I.F. Triantis, A. Demosthenous: A high-performance adaptive ENG amplifier, IEEE BioCAS (Singapore 2004)
56.25 I.F. Triantis, A. Demosthenous: A BiCMOS ENG amplifier with high SIR output, IEEE Int. Symp. Circuits Syst. – ISCAS '05 (2005)
56.26 D.B. Popovic, R.B. Stein, K.L. Jonanovic, D. Rongching, A. Kostov, W.W. Armstrong: Sensory nerve recording for closed-loop control to restore motor functions, IEEE Trans. Biomed. Eng. **40**, 1024–1031 (1993)
56.27 J. Georgiou: Micropower Electronics for Neural Prosthetics. Ph.D. Thesis (Imperial College London, London 2002)
56.28 G. Huberfeld, L. Wittner, S. Clemenceau, M. Baulac, K. Kaila, R. Miles, C. Rivera: Perturbed chloride homeostasis and GABAergic signaling in human temporal lobe epilepsy, J. Neurosci. **27**(37), 9866–9873 (2007)
56.29 D.L. Eng, J.D. Kocsis: Activity-dependent changes in extracellular potassium and excitability in turtle olfactory nerve, J. Neurophysiol. **57**, 740–754 (1987)
56.30 T.D. Strong, S.M. Martin, R.F. Franklin, R.B. Brown: Integrated electrochemical neurosensors, ISCAS 2006 IEEE Int. Symp. Circuits Syst. (2006), 4 pp.
56.31 Y. Qingshan: Ion Sensor for Long Term Use in Complex Medium, Patent WO/2006/014758 (2006)
56.32 L. Sudakov-Boreysha, U. Dinnar, Y. Nemirovsky: New ISFET catheters encapsulation techniques for brain pH in-vivo monitoring, 11th IEEE Int. Conf. Electron. Circuits Syst., ICECS 2004 (2004) pp. 424–426
56.33 NeuroProbes Project: ftp://ftp.cordis.europa.eu/pub/ist/docs/mnd/neuroprobes_en.pdf
56.34 Golden Brain Project: ftp://ftp.cordis.europa.eu/pub/ist/docs/fet/openf-18.pdf
56.35 A. Radomska, S. Singhal, T. Cass: Protein engineering for biosensors. In: *Body Sensor Networks*, ed. by G.-Z. Yang (Springer, London 2006) pp. 89–116
56.36 B.A. Patel, C.A. Anastassiou, D. O'Hare: Biosensor design and interfacing. In: *Body Sensor Networks*, ed. by G.-Z. Yang (Springer, London 2006) pp. 41–88
56.37 P. Bergveld: Thirty years of ISFETOLOGY – What happened in the past 30 years and what may happen in the next 30 years, Sens. Actuators B: Chem. **88**, 1–20 (2003)
56.38 M.J. Schöning, Poghossian, A.M. Schning: Recent advances in biologically sensitive field-effect transistors (BioFETs), Analyst **127**, 1137–1151 (2002)
56.39 P.A. Hammond, D. Ali, D.R.S. Cumming: Design of a single-chip pH sensor using a conventional 0.6-μm CMOS process, IEEE Sens. J. **4**, 706–712 (2004)
56.40 D.R. Reyes, D. Iossifidis, P.A. Auroux, A. Manz: Micro total analysis systems. 1. Introduction, theory, and technology, Anal. Chem. **74**(12), 2623–2636 (2002)
56.41 P.A. Auroux, D. Iossifidis, D.R. Reyes, A. Manz: Micro total analysis systems. 2. Analytical standard operations and applications, Anal. Chem. **74**(12), 2637–2652 (2002)
56.42 Y. Alifragis, A. Volosirakis, N.A. Chaniotakis, G. Konstantinidis, A. Adikimenakis, A. Georgakilas: Potassium selective chemically modified field effect transistors based on AlGaN/GaN two-dimensional electron gas heterostructures, Biosens. Bioelectron. **22**, 2796–2801 (2007)
56.43 X. Yue, E.M. Drakakis, H. Ye, M. Lim, A. Mantalaris, N. Panoskaltsis, A. Radomska, C. Toumazou, T. Cass: An on-line, multi-parametric, multi-channel physicochemical monitoring platform for stem cell culture bioprocessing, Circuits Syst., ISCAS 2007, IEEE Int. Symp. (2007) pp. 1215–1218

56.44 M. Lim, H. Ye, N. Panoskaltsis, E.M. Drakakis, X. Yue, A.E.G. Cass, A. Radomska, A. Mantalaris: Intelligent bioprocessing for haemotopoietic cell cultures using monitoring and design of experiments, Biotechnol. Adv. **25**, 353–368 (2007)

56.45 A. Radomska, R. Koncki, K. Pyrzynska, S. Glab: Bioanalytical system for control of hemodialysis treatment based on potentiometric biosensors for urea and creatinine, Anal. Chim. Acta **523**, 193–200 (2004)

56.46 A. Radomska, E. Bodenszac, S. Glab, R. Koncki: Creatinine biosensor based on ammonium ion selective electrode and its application in flow-injection analysis, Talanta **64**, 603–608 (2004)

56.47 P. Georgiou, I.F. Triantis, T. Constandinou, C. Toumazou: Spiking Chemical Sensor (SCS): A new platform for neuro-chemical sensing, 3rd Int. IEEE/EMBS CNE (Hawaii 2007) pp. 126–129

56.48 L. Shepherd, P. Georgiou, C. Toumazou: A novel voltage-clamped CMOS ISFET sensor interface, ISCAS 2007, IEEE Int. Symp. Circuits Syst. 2007 (2007) pp. 3331–3334

56.49 L.M. Shepherd, C. Toumazou: A biochemical translinear principle with weak inversion ISFETs, IEEE Transact. Circuits Syst. I: Fundamental Theory and Applications, Regular Papers, Vol. 52 (2005) pp. 2614–2619

56.50 B. Premanode, W.P. Chan, C. Toumazou: Ultra-low power precision ISFET readout using global current feedback, Electron. Lett. **42**, 1264–1265 (2006)

56.51 L. Shepherd, C. Toumazou: Towards an implantable ultra-low power biochemical signal processor for blood and tissue monitoring, ISCAS 2005, IEEE Int. Symp. Circuits Syst., Vol. 5 (2005) pp. 5226–5229

56.52 L. Shepherd, T.G. Constandinou, C. Toumazou: Towards ultra-low power bio-inspired processing. In: *Body Sensor Networks*, ed. by G.-Z. Yang (Springer, London 2006) pp. 219–238

56.53 P. Georgiou, T.G. Constandinou, C. Toumazou: Low-power spiking chemical pixel sensor, Electron. Lett. **42**, 1331–1332 (2006)

56.54 B. Sonnleitner: Instrumentation of Biotechnological Processes. In: *Bioanalysis and Biosensors for Bioprocess Monitoring*, Vol. 66, ed. by B. Sonnleitner (Springer, Berlin Heidelberg 2000)

56.55 P. Georgiou, C. Toumazou: An adaptive ISFET chemical imager chip, ISCAS 2008, IEEE Int. Symp. Circuits Syst. (2008) pp. 2078–2081

56.56 J. Bausells, J. Carrabina, A. Errachid, A. Merlos: Ion-sensitive field-effect transistors fabricated in a commercial CMOS technology, Sens. Actuators B: Chem. **57**, 56–62 (1999)

56.57 S. Martinoia, G. Massobrio, L. Lorenzelli: Modeling ISFET microsensor and ISFET-based microsystems: a review, Sens. Actuators **105**, 14–27 (2005)

56.58 B. Razavi: *Design of Analog CMOS Integrated Circuits* (McGraw-Hill, Boston 2000)

56.59 P. Georgiou: *Chemical Bionics – A Novel Design Approach Using Ion Sensitive Field Effect Transistors* (Imperial College University of London, London 2008)

56.60 S.F. Traynelis, R. Dingledine: Role of extracellular space in hyperosmotic suppression of potassium-induced electrographic seizures, J. Neurophysiol. **61**, 927–938 (1989)

56.61 A. Radomska, S. Singhal, H. Ye, M. Lim, A. Mantalaris, X. Yue, E.M. Drakakis, A.E.G. Cass: Biocompatible ion-selective electrode for monitoring metabolic activity during the growth and cultivation of human cells, Biosens. Bioelectron. **24**, 435–441 (2008)

56.62 R.B.M. Schasfoort, S. Schlautmann, L. Hendrikse, A. van den Berg: Field-effect flow control for microfabricated fluidic networks, Science **286**, 942–945 (1999)

56.63 S. Sharma, K. Buchholz, S.M. Luber, U. Rant, M. Tornow, G. Abstreiter: Silicon-on-insulator microfluidic device with monolithic sensor integration for μTAS applications, J. Microelectromech. Syst. **15**(2), 308–313 (2006)

56.64 D.S. Kim, J.E. Park, J.K. Shin, P.K. Kim, G. Lim, S. Shoji: An extended gate FET-based biosensor integrated with a Si microfluidic channel for detection of protein complexes, Sens. Actuators B: Chem. **117**(2), 488–494 (2006)

56.65 P. Truman, P. Uhlmann, M. Stamm: Monitoring liquid transport and chemical composition in lab on a chip systems using ion sensitive FET devices, Lab Chip **6**(9), 1220–1228 (2006)

56.66 B.C. Jacquot, C. Lee, Y.N. Shen, E.C. Kan: Time-Resolved Charge Transport Sensing by Chemoreceptive Neuron MOS Transistors (CνMOS) With Microfluidic Channels, IEEE Sensors **7**(10), 1429–1434 (2007)

56.67 T. Masadome, K. Yada, S.I. Wakida: Microfluidic polymer chip integrated with an ISFET detector for cationic surfactant assay in dental rinses, Anal. Sci. **22**(8), 1065–1069 (2006)

56.68 M.J. Milgrew, M.O. Riehle, D.R.S. Cumming: A large transistor-based sensor array chip for direct extracellular imaging, Sens. Actuators B: Chem. **111**, 347–353 (2005)

56.69 S. Sharma, A. Radomska-Botelho Moniz, I. Triantis, K. Michelakis, J. Trzebinski, A. Azarbadegan, B. Field, C. Toumazou, I. Eames, A. Cass: An integrated silicon sensor with microfluidic chip for monitoring potassium and pH, Microfluid. Nanofluid. **10**(5), 1119–1125 (2011)

56.70 N.J. Abbott, G. Mitchell, K.J. Ward, F. Abdullah, I.C.H. Smith: An electrophysiological method for measuring the potassium permeability of the nerve perineurium, Brain Res. **776**, 204–213 (1997)

Part F

Medical I

Part F Medical Information Processing and Communication

57 Fusing Medical Engineering and Information Technology – Structure, Integration and Process Optimization
Hajo Tanck, Berlin, Germany
Michael Heinlein, Berlin, Germany

58 Communicating Medical Systems and Networks
Armin Gärtner, Erkrath, Germany

59 Hospital Information Systems
Peter Haas, Dortmund, Germany
Klaus A. Kuhn, München, Germany

60 Telemedicine in Germany
Rüdiger Klar, Freiburg, Germany
Ernst Pelikan, Freiburg, Germany

61 Telemedicine Using Active Implants
Klaus P. Koch, Schneidershof, Trier, Germany
Oliver Scholz, Saarbrücken, Germany

62 Fundamentals of Medical Image Processing
Thomas M. Deserno, Aachen, Germany

63 Virtual Reality in Medicine
Wolfgang Müller-Wittig, Singapore, Singapore

64 Computer-Supported Teaching and Learning Systems in Medicine
Martin Haag, Heilbronn, Germany
Martin R. Fischer, Witten, Germany

65 PACS and RIS
Christian Zapf, Erlangen, Germany
Andreas Bermann, Erlangen, Germany
Dirk Sunderbrink, Forchheim, Germany

66 3-D Postprocessing in Virtual Endoscopy
Georg-Friedemann Rust, Gauting, Germany

67 e-Health – Ambient Assisted Living and Personal Health Systems
Natasha Avila, Paris, France
Christina Sampogna, Newmarket, UK

68 Electrical Stimulation of the Nervous System
Stanisa Raspopovic, Pisa, Italy
Marco Capogrosso, Pisa, Italy
Silvestro Micera, Zürich, Switzerland

57. Fusing Medical Engineering and Information Technology – Structure, Integration and Process Optimization

Hajo Tanck, Michael Heinlein

Integrating medical engineering and IT in a consequently and especially standardized way enables hospitals to significantly increase their efficiencies. This evolves process optimizations for medical documentation and organizational advantages within the administration of the interface used.

57.1 **Standards of Interfaces** 1076
57.1.1 Health Level 7 1076
57.1.2 Digital Imaging and Communications in Medicine 1076
57.1.3 GDT 1076
57.1.4 xDT 1077
57.1.5 XML 1077

57.2 **Data Structure** 1077
57.2.1 HL7 CDA 1077
57.2.2 SNOMED 1077
57.2.3 LOINC 1077
57.2.4 Alpha ID 1077
57.2.5 Object Identifiers 1077
57.2.6 Unified Code for Units of Measure 1078

57.3 **Integrating the Healthcare Enterprise** 1078

57.4 **Integration of Medical Devices** 1078
57.4.1 Interfaces 1079
57.4.2 Connecting a Medical Device 1079
57.4.3 Organizational Requirements for the Connection 1080
57.4.4 Responsibilities 1081

57.5 **Sample Integration – From Findings to Medical Documentation** 1082

57.6 **Résumé** 1082

The combination of information technology (IT) and medical engineering (ME) is gaining more and more importance today. Years ago, only personal computers (PCs) were connected with medical devices, but nowadays systems, which consist of medical device and PC, medical devices with a complex IT infrastructure and/or medical devices with an integrated network interface, are directly integrated into IT networks. Besides creating medical findings, many systems support the medical and nursing staff in the process of medical documentation as well as in the treatment, therapy, and nursing planning. For these special fields of application, the data recorded are available for all sorts of specialized evaluations as well as scientific issues.

The integration of medical engineering into information technology of hospitals is consistent with the task of German hospitals of providing a growing quality of the treatment itself and the documentation of the treatment process along with steadily increasing cost pressure at the same time. This is guaranteed by combining medical and administrative matters.

In times of lack of budget and staff, existing resources have to be used in the most efficient way. By integrating IT and medical engineering not only workflows are designed to be used more efficiently, but also saving potential in terms of time and money can be realized.

A complete integration is based on the semantic interoperability of the different systems; for this purpose various standards are used.

The standards for a complete integration are divided into two parts:

1. Communication between two systems (interface standards) and
2. Standards for the structure of the data that have to be transferred (data structure).

57.1 Standards of Interfaces

According to the definition, an interface is described with the help of a set of rules: standardized interfaces are mutually compatible, i. e. components or modules that support the same interface can exchange standardized data among each other. Below you will find the most important standards used in health care.

57.1.1 Health Level 7

Health Level 7 (HL7) is an international health care standard for the exchange of data. HL7 was developed by the eponymous organization; currently versions 2.x and version 3 are supported. The number 7 of the name HL7 refers to the layer 7 of the ISO/OSI reference model for communication (ISO 7498-1) and expresses that the communication is described on an application level.

In the HL7 version 2.x the following message types are supported, which are divided into segments and fields:

- ADT (admission, discharge, and transfer): patient's master data and data of their stay in hospital
- ORM (order message): request of an examination
- ORU (observation results unsolicited): transfer of findings
- DFT (detail financial transactions): transfer of performance data for accounting
- BAR (billing and accounting request): transfer of performance data according to the international classification of procedures and operations in medicine (OPS Standard)
- MDM (manage document message): document management messages

57.1.2 Digital Imaging and Communications in Medicine

Digital imaging and communications in medicine (DICOM) is an open standard for the exchange of digital images and their additional information.

The DICOM standard guarantees to standardize the format for data storage as well as the communication protocol for the exchange of data. Several working groups of the DICOM Standard Committee continue to permanently develop the DICOM standard.

Increasing numbers of manufacturers of imaging systems such as digital x-ray, magnetic resonance imaging, computed tomography, endoscopy or sonography, implement the DICOM standard in their products. This results in a high interoperability between the imaging systems and the systems for image processing and the digital image archiving as well as picture archiving and communication system (PACS).

Important DICOM services are:

- Verify – the verification, if a network node supports DICOM
- Storage – the storage of data objects
- Query/retrieve – the search of a DICOM device for objects (query) and the transmission of another DICOM device (retrieve)
- Procedure step – information about the status of examination
- Storage commitment – query, if the data transmitted have been stored
- Worklist management – data transfer between the planning system and the DICOM device, where the examination have to be performed
- Presentation state storage – transfer of information on how the image material is displayed or should be displayed
- Structured reporting storage – coded transmission of medical findings
- Hanging protocols storage – storing the presentation of image series and studies.

57.1.3 GDT

The GDT (interface for the exchange of device data) has been worked out by the Qualitätsring Medizinische Software (QMS) and is implemented as a standardized interface between IT systems and medical devices. The GDT interface is part of the standard delivery package. The data that have to be exchanged are filed in a predefined list. The file name serves to clearly identify the communication partners and reads as follows:

`⟨recipient-token⟩⟨sender-token⟩.⟨consecutive number⟩.`

The consecutive number will be incremented for every new message. Thus, it is prevented that older news will be overwritten by the reading device (client) before processing. After orderly processing the message by the client, the file in the client's exchange list is deleted by the client.

57.1.4 xDT

The xDT interface was created by order of the German Association of Statutory Health Insurance Physicians (kassenärztliche Bundesvereinigung) and contains a group of data exchange formats, which are predominantly used in private practices. The formats have a common, text-oriented syntax, where every field is written as a line in the file and a common field directory based on XML (extensible markup language).

57.1.5 XML

XML was published by the World Wide Web Consortium (W3C) and the current version is the fifth edition. With the help of XML, a meta-language is defined. This language is the basis for creating custom-designed languages by means of structured and content restrictions, such as xDT.

A XML document consists of text characters, in the simplest case ASCII, and is therefore readable by humans. The definition does not contain any binary data that include unreadable texts.

57.2 Data Structure

The second important aspect in terms of integrating medical engineering and IT forms the issue about the content structure of medical document.

57.2.1 HL7 CDA

The HL7 clinical document architecture (CDA), based on XML, was worked out by the Health Level 7 group for the exchange and storage of clinical contents. A CDA document complies with a clinical document (e.g. a physician's letter, a findings report). CDA documents guarantee the data transfer between different systems.

57.2.2 SNOMED

The Systematized Nomenclature of Medicine (SNOMED) aims to index medical statements in a way that all content-related elements of the statement are recorded completely.

SNOMED CT contains 18 axes with 800 000 terms, which describe approximately 300 000 concepts (several terms per concept).

The German translation of SNOMED CT dates back to 2003 and has no longer been maintained.

57.2.3 LOINC

LOINC (logical observation identifiers names and codes) is an internationally acknowledged system for clearly en- and decoding examinations. The Regenstrief Institute (Indianapolis, USA) is responsible for maintaining and documenting the LOINC database. In Germany, the DIMDI (German Institute of Medical Documentation and Information) actively promotes the implementation of LOINC and assumes the responsibility of the central data management and the information exchange with national and international institutes, project groups and industry.

LOINC is a composition of general names and identifiers for the description of examination and test results from laboratories and clinics. LOINC is not only recommended by HL7 but also by DICOM for the structured exchange of medical findings and findings data.

57.2.4 Alpha ID

The identification number for diagnoses is a continuous number identifying entries of the alphabetical list of the ICD 10-GM (International Classification of Diseases, 10th revision). With the help of the Alpha ID medical terms can be further processed electronically; the fact that the code cannot be changed due to being maintained by the DIMDI guarantees its interoperability. Due to extending the Alpha ID by terms beyond the scope of diagnoses in the future, this terminology can be used not only for encoding diagnoses.

57.2.5 Object Identifiers

For standardized exchange of medical information, an efficient software communication in health telematics requires data objects, i.e. objects and messages have to be described clearly. Object identifiers (OID) are chains of number for identifying these objects and messages. In this case, objects are information units like institutions, classifications, information, documents or tables. If OID is used for standardized data exchange between software systems, the integration of the data between the systems is ensured.

57.2.6 Unified Code for Units of Measure

With the encoding system Unified Code for Units of Measure (UCUM), measuring units are shown in standardized form. If the applications for medical documentation show the SI standard unit system (international system of units), the transfer of e.g. the medication of a treatment is possible with clear measurements.

57.3 Integrating the Healthcare Enterprise

Last but not least, the Integrating the Healthcare Enterprise (IHE) serves as kind of a bridge between different standards and creates the technical framework for the implementation and verification of the standards' feasibility.

IHE reformulates requirements from experience into so-called Use Cases, identifies relevant standards, and develops technical guidelines in the so-called technical framework that enables a manufacturer to create their interfaces compliant to IHE. On the international *Connectathon* the manufacturers mutually test the interoperability of their systems and use the gained experiences to put it into practice.

57.4 Integration of Medical Devices

What does the integration of medical devices nowadays look like?

In many hospitals, medical devices of laboratories and radiology as well as single ultrasonic or endo-

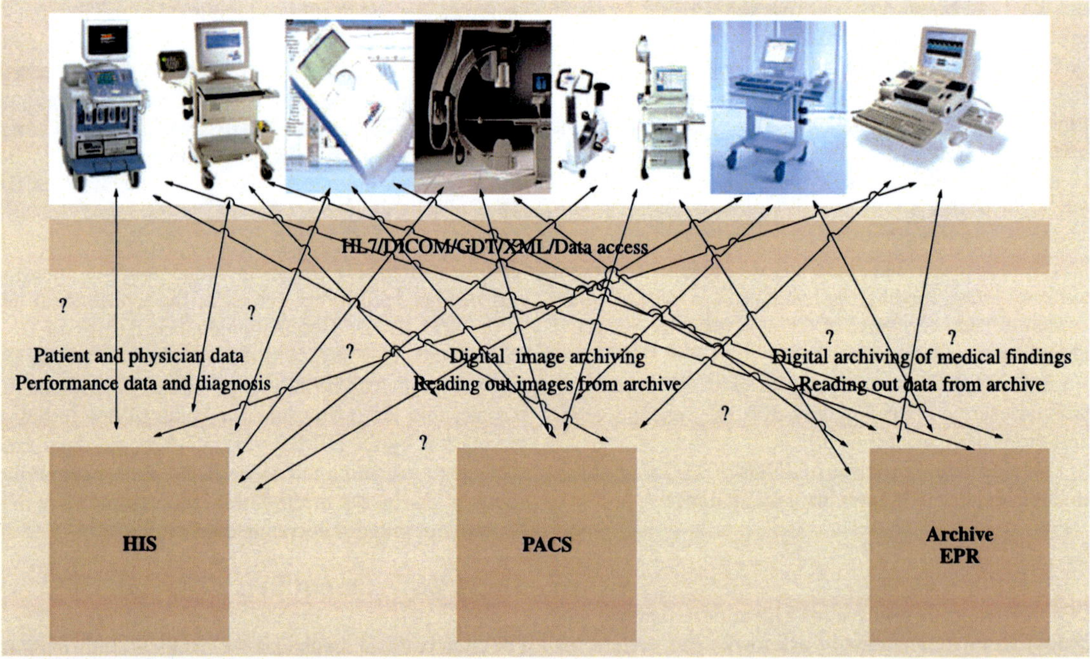

Fig. 57.1 Integration of medical devices – as of today

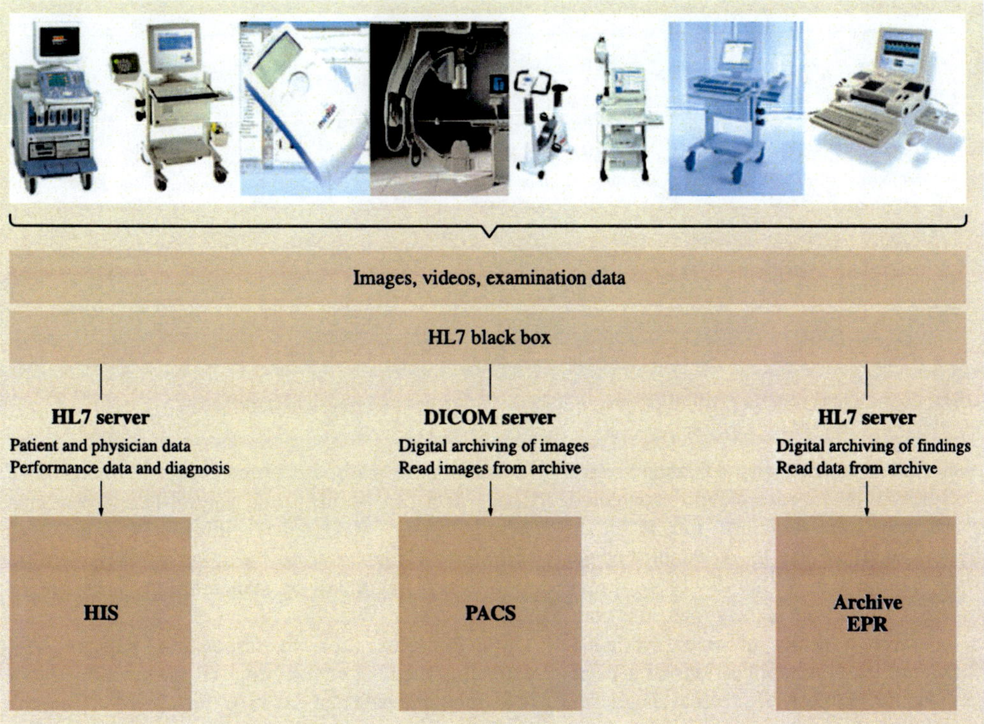

Fig. 57.2 Integration of medical devices into interface server

scopic devices are connected to the hospital information system (HIS). They are often equipped with a device interface and an interface for the data-receiving information system, which means a separate interface for every single supplier of medical devices is installed. Mostly, this is due to the fact that the devices used only have a GDT interface. The diagram in Fig. 57.1 clarifies the predominant interface structures in hospitals nowadays.

57.4.1 Interfaces

When technically implementing an interface of the medical device and IT network, both, IT and medical engineering are involved; however, from the organizational point of view the content-related result influences the hospital's treatment process and it is in the context of the organizational function that the content result along with the treatment process and affects the entire hospital structure.

Excursion to HL7 Communication Server

The communication of subsystems in a hospital can be optimized by the HL7 communication server. These components assure that not every single subsystem requires an individual interface. Using a communication server benefits in reducing the number of interfaces, as not every single system has to be connected via an individual interface. This in turn, results in a more standardized and simpler communication flow as well as cost-saving, because every separate interface demands a monetary equivalent.

The data format or the HL7 version, which the individual systems use to send their messages, is absolutely random. With the help of mapping tables, the HL7 server automatically converts these formats into the format which is required by the receiving system requires. The tables will be created once when the communication server is installed. Afterwards, the interfaces are maintained directly on the communication server. If you project the HL7 communication server technique onto the integration of medical devices, Fig. 57.2 shows the resulting system environment.

57.4.2 Connecting a Medical Device

In contrast to integrating two software systems based on a HL7 communication, the connection of medical de-

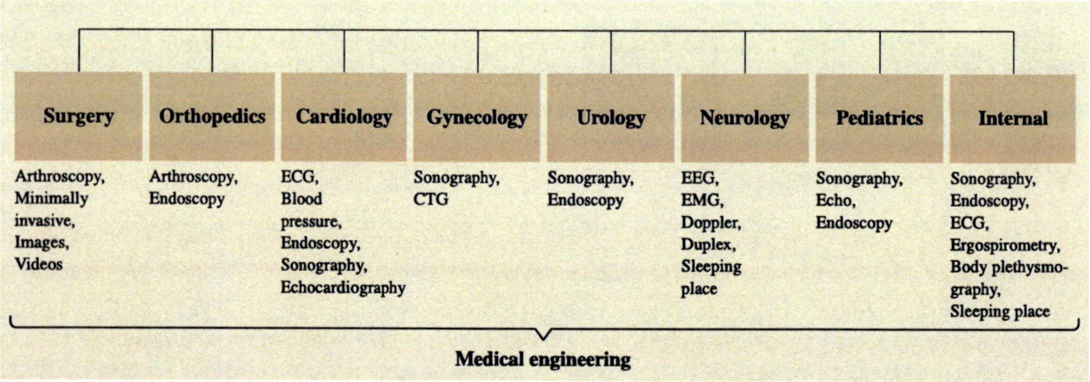

Fig. 57.3 Medical departments together with their possible medical devices and physical interfaces

vices additionally requires that each medical device and examination result will be connected and documented. For systems that only store their data locally on the device, a physical interface has to be established for each device and workstation. If the device's supplier additionally provides different applications, the complexity of the interface is further increased. Normally, the recorded data are data and applications of specialist departments for the purpose of medical diagnoses. Figure 57.3 points out the scope of physical interfaces required within in the context of individual integrations.

The connection of a device can be realized by means of proprietary interfaces or the standards (e.g. HL7/DICOM) as described in Sect. 57.2.

Attention should be paid to the facts that:

- There is no dependency created by the use of proprietary interfaces.
- IT and medical engineering have to mutually combine operating and change processes.
- The integration of medical products is effected according to the guidelines and
- Criteria of the MPG (MedizinProdukteGesetz – Medical Devices Law) and
- Standard interfaces for integration into existing systems are agreed upon.

Apart from that, it is important that it is not compulsory to use medical devices of in-house production and that you pay attention to the implementation of the IT safety concept when integrating medical devices; whereas the completing process of approval resulting in a risk analysis has also to be taken into consideration.

Secondly, the data sent from the medical devices will be forwarded to the different information systems in accordance with the purpose agreed in advance. These data can be measured values, textual findings, structured medical documents, images, PDFs etc. They can also be transferred to the information system via proprietary interfaces or standard interfaces as described in Sect. 57.2.

A joint task for all responsible persons from medicine, medical engineering, IT, and administration is defining the data that has to be transferred. Often, this results in further tasks for different specialized departments managed by the respective responsible person. The following list is an example for the IT department:

1. Adaptations to the network topology (IT/ME)
2. Need for adaption to the individual IT/ME infrastructure
3. Hard-/software requirements (operating systems etc.) that have to be considered
4. Adaptation of the storage concept/long-term archiving
5. Extending the system monitoring
6. Presentation of data consistency and workflow processes (IHE).

57.4.3 Organizational Requirements for the Connection

If we consider the integration of medical devices, it is obvious that, in terms of organizational aspects, several professional groups of the hospital with overlapping responsibilities have to be involved in this process.

While the health service's has to focus on possibilities of establishing medical diagnoses, the medical engineering will pay special attention to guarantee that the device functions without interference. The clinical

IT department is focused on the possibilities to integrate the examination results into the hospital information system, and the clinical administrative department is interested in evaluating the overall process regarding to the acquisition and consequential costs as well as the entire treatment documentation. But also the importance of the documentation possibilities will rise in the future to evaluate and justify treatment processes to cost units.

As this overall view points out, it is very important that the professional groups mentioned above work closely together. Before purchasing a device, the following sample questions have to be considered in order to satisfy the radiologic principle of *the unity of image and findings*:

- Who is responsible for process of integration with regards to the content?
- Who is responsible for technical process of integration?
- Which data should be transmitted from the medical point of view?
- Which data should be transmitted from the administrative point of view?
- Who is in charge of coordinating the specialized departments?
- Which interfaces for integration are provided by the device?
- Which interfaces support the administrative information systems?
- Do the systems that will be connected support the standards?

In addition to that, the responsible parties of a hospital have to deal with the workflow of the medical department. The convergence resulting from the integration of medical devices induces also during the use of the medical device an interference of the following aspects:

- Responsibility
- Engineering
- Organization
- Workflow/operation
- Security

57.4.4 Responsibilities

When integrating medical devices, individual responsibilities overlap during the entire cycle of operation. Especially during the operation of the medical device, it depends on the cooperation between physicians,

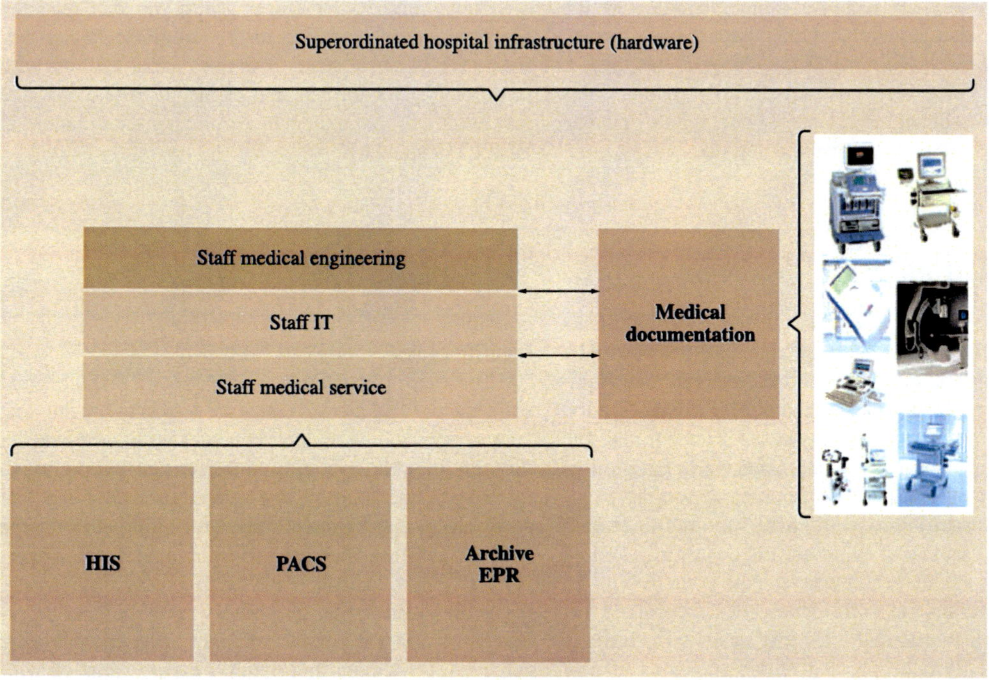

Fig. 57.4 Administrative areas of responsibility in hospitals

medical engineering and IT, whether the integration is successful or not (Fig. 57.4). The management is responsible for a secure operation by giving clear instructions and differentiating the areas of authority each person involved has to be informed about.

57.5 Sample Integration – From Findings to Medical Documentation

The information and document structures, which have hitherto been proprietarily defined between the systems with random names for measured data, document structures, features, and/or values, are normally not interpreted by all systems to be integrated. For exchanging data the classifications and terminologies mentioned in Sect. 57.3 should be applied. Using these standards guarantees that the content transmitted is readable for all systems.

Moving from the structured medical documentation to a European- and world-wide standardized medical documentation requires this compliance and will be necessary in the near future, e.g. with regard to telemedicine. These requirements will result in a gradual removal of free-textual documentation in favor of structured documentation in hospitals. Tendencies for achieving this are the nomenclatures for structured medical documentation announced by the medical associations. To avoid that a data set is transferred randomly for technical implementation of the structured documentation, another standardized format, called HL7 CDA (Clinical Document Architecture), is already available and used successfully in numerous projects for exchanging structured documents between different institutions at home and abroad.

This success is essentially based on the following aspects:

- For mechanical processing the structured content of documents are represented in the common XML format. With the help of style sheets formats readable for humans are created.
- The XML format, i.e. that the above-mentioned structure of the features used with their relations and data types are derived from a standardized information model, the HL7 RIM (reference information model). This ensures that the content is integratively represented and processed.
- Classifications and terminologies both, internal and external of HL7, will consistently be used for all types of content such as document sections, features and ranges.

Contacting the HL7 user group you can read the implementation guideline for electronic physician letters in HL7 CDA format. With this format systems with proprietary data formats can exchange structured data processible for computers by displaying their content in HL7 CDA standard.

57.6 Résumé

To sum up it can be stated that the requirements for integrating medical engineering into the hospital's information technology do not just concern the technology and organization of the hospital. If the integration is tackled correctly, this is future-oriented and the medium-term step for a comprehensive integration of IT and medical engineering in order to use the existing resources in the most efficient way and to remain competitive. Data that so far has not or just to a limited degree been exchangeable between the systems, will no longer be available in a proprietary but standardized format. Organizational and technical issues will result in adapting safety concepts and failure scenarios to new conditions for identifying and minimizing the risk of security gaps.

Connecting medical devices causes a displacement of resources and a cost reduction. Reasons for this are, for example, more efficient processes by avoiding media discontinuity, the reduction of the process period due to faster data transmission, the synergetic use of medical engineering and extended fields of application on the basis of existing medical products (Fig. 57.5). Therefore, an improvement in the treatment quality by e.g. faster diagnostics, shorter processing times, and a permanent availability of the treatment documentation is achieved. Physicians and patients highly satisfied with

	Is (current)			Demand value					Actual-theoretical	
Endoscopy and sonography	Frequentness/number/frequency	Process time (1–n employee)	Price in €	Frequentness/number/frequency	Reference parameter (per requirement, case, day ...)	Process time (1–n employee)	Qualification/role	Price in €	Deviation process times (in hours p.a.)	Deviation process costs
Arrangement of dates/priorities discussion station with endo/sono	1.895	4	0.41	632	EXA	4	NURS	0.41	84	2072
Running coordination patients transport	6.316	2	0.41	6.316	EXA	1	NURS	0.41	105	2590
Print of images (material costs)	6.701	1	0.40	6.701	EXA	0	PCE	0.40		2680
Print/filling of stationary finding	6.316	3	0.41	6.316	EXA	1.5	NURS	0.41	158	3884
Print/filling ambulatory finding	385	3	0.41	385	EXA	0	NURS	0.41	19	474
Signature stationary finding	6.316	1	0.80	6.316	EXA	0.5	ASS	0.80	53	2526
Signature ambulatory finding	385	1	0.80	385	EXA	0.5	ASS	0.80	3	154
Generation ambulatory physician's letter (cover letter)	385	5	0.80	385	EXA	1	ASS	0.80	26	1232
Communication finding (paper/HIS)	250	20	0.28	250	DAY	40	AP	0.28	–83	–1400
Sum of process advantage										14212

Fig. 57.5 Process advantages

the treatment, improved patient safety by e.g. clearly assigning patient and patient-related data, avoiding errors when transferring data and improved information of the patient and environment are further aspects.

After the fusion of medical engineering and IT, a cost-benefit analysis pointed out that there are savings of 14 000 € p.a. for an endoscopy with three workstations.

This analysis evaluates the introductory status, the increase in efficiency achieved by the users along the treatment process and the improvement potential of the entire endoscopy process. The following diagram (Fig. 57.5) describes how the process is optimized, the capacities are displaced and saved by the integration of medical engineering, the software for structured medical documentation and the hospital information system.

The optimal integration of medical engineering and IT enables to create overall concepts instead of numerous single solutions. These concepts efficiently as well as effectively support the daily work in clinics and turn future visions of data availability into reality.

Further Reading

- C.J. Evans, R.L. Wilson, F.G. Deporter: *Integrated Community Healthcare: Next Generation Strategies for Developing Provider Networks* (McGraw-Hill Professional, Dubuque 1997)
- T.R. Kucklick (Ed.): *The Medical Device R&D Handbook* (CRC, Boca Raton 2006)

58. Communicating Medical Systems and Networks

Armin Gärtner

In the last 10 years, medical devices and information technology (IT) have become inextricably connected technologically and functionally. Medical devices are no longer just connected to a personal computer (PC) in order to evaluate data but have an integrated network interface card, meaning that they can be connected directly to an IT network in order to be able to send data via the network to a server for archiving and/or further processing. The user can access this medical data for a patient at any time, in any place, and in any form. This means that the IT network in the hospital is becoming increasingly important for patient treatment and that the requirements with regard to availability, safety, and efficiency of the network are thus constantly increasing.

58.1 **Medical Networks** 1085

58.2 **Requirements for Medical Networks** 1086
58.2.1 Image Data in Radiology 1086
58.2.2 Intensive Care Network 1086
58.2.3 Transmission of Alert Data 1087

58.3 **Interconnected Medical Networks** 1087

58.4 **Risk Management, DIN EN ISO 14971** 1088

58.5 **Shared Networks** 1089

58.6 **Security Aspects of Medical Networks from a Regulatory Viewpoint** 1092

58.7 **Future Standard IEC 80001-1** 1093

References 1093

58.1 Medical Networks

Integration of a medical product into an IT network (Fig. 58.1) does not convert the IT network into a medical product.

IEC 80001-1 defines the term *medical networks* to include both active medical devices (modalities, programmable electrical medical systems (PEMS) and also network components and PCs. A medical network integrates medical devices.

The combination or integration of medical devices in IT networks as per Fig. 58.1 in the hospital

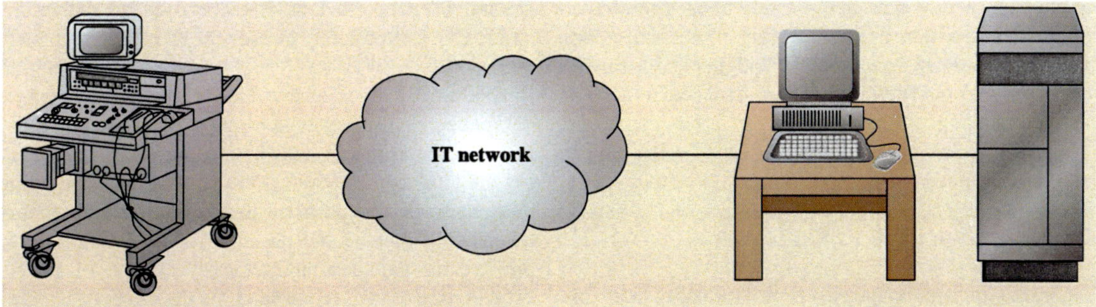

Fig. 58.1 Medical devices with a network connection

(and also other areas of the healthcare system) leads to IT networks becoming a more integral and thus more important, and also always a more highly critical, element of medical diagnostics and therapy. They therefore represent an integral part of modern patient care.

A medical network undertakes the task of relaying:

- Records of services provided, and in the private health sector, billing information,
- Image data and clinical findings, and
- Vital signs and alert data from medical devices.

For data transmission, medical devices place various demands on a network, which can be divided into three classes:

1. Transmission of a large volume of data [e.g., computed tomography (CT) series, video loops], requiring transmission rate (bandwidth).
2. Transmission of time-critical vital signs and alert data (intensive care medicine), requiring reliability.
3. Transmission of sensitive patient data, requiring safety and security.

Medical data are subject to particular requirements of

- Integrity (authenticity)
- Protection (data protection)
- Security (data transfer)
- Completeness.

A medical network must therefore ensure that these requirements are satisfied.

58.2 Requirements for Medical Networks

Medical diagnostics and therapy are today reliant on the prompt availability at any time of images, clinical findings, and other data.

Considerable problems and risks for the diagnosis and treatment of patients can therefore arise if:

- Data are not available or are incomplete.
- Data are not available promptly.
- Data such as images during an operation are suddenly no longer available.
- The transfer of vital parameters and/or alerts is not possible or is interrupted, etc.

Availability is a vital aspect of medical networks for users/operators. According to [58.1], availability is understood as meaning the fraction of time during which the application being considered or the data are available. When the availability of a service or of data is known, this value therefore gives the probability of an application and/or data being available at a particular point in time within a period of time. In hospitals, ever more applications are critical for patients and for management. A critical application or critical data transfer is defined as being an application whose failure can be associated with far-reaching consequences for the patient and the hospital.

58.2.1 Image Data in Radiology

Failure to transmit image data from a modality to an image documentation system such as a picture archiving and communication system (PACS) for storage normally does not have any critical consequences for patient care because the modalities, such as CT equipment and ultrasound equipment, etc., can buffer the images. If, for example, the switch on a radiological IT subnetwork [local area network (LAN)] fails, then the images can usually be buffered on the modality and re-sent to the PACS once the network connection has been restored. If, however, the internal web image processing is also affected by failure or disruption of the hospital-wide network, in emergency cases such as treatment of trauma patients a alternative plan should come into effect, so that any x-ray images can be printed out using a film printer and made available to the orthopedic surgeon.

58.2.2 Intensive Care Network

Intensive care networks are traditionally set up as proprietary networks with a defined junction with the general IT network. DIN EN 60601-1 3rd defines in annex H.7.3 three safety classes for IT networks in hospitals.

Failure of an intensive care network or network components is limited in its extent to one ward and must therefore be compensated for by the staff by displaying the vital signs on the system control center monitor until network functionality is restored. Patient monitors should include the principle of first-fault safety; this means that they automatically change to an entirely acoustic and visual alert in the event of failure of the network connection to the central monitoring station.

58.2.3 Transmission of Alert Data

For various reasons, technologically orientated solutions are meanwhile being offered, which can be used to perform patient monitoring via the IT network in a hospital by connecting monitoring equipment via the network to so-called alert servers or special alert software on servers. These servers or alert software enable alert functions and parameters to be configured, with which vital signs and/or alerts (bradycardia, etc.) can be passed on to remote display monitors or to mobile communications equipment.

The (partial) failure or disruption of an interconnected medical system of this type can have critical consequences for a patient (e.g., in the case of a bradycardia alert).

58.3 Interconnected Medical Networks

IT networks are increasingly used in hospitals to transmit also time-critical vital signs and alerts from patient monitors and other equipment. An IT network then becomes a central component of an interconnected medical system. Systems of this kind consist of a number of heterogeneous, autonomous single devices (medical and nonmedical devices) which are only capable of performing their task as a medical system as a result of networking. Within a medical network, it is possible to identify interconnected medical systems which are to be regarded as subsets of a medical network.

The functionality of such interconnected medical systems along the same lines as the medical electrical systems, in accordance with DIN EN 60601-1 3rd chapter 16, can be described and defined in the medical network as a subset.

The operator must define, evaluate, and ensure certain requirements for the reliability and the security of the system against disruptions and other influences. Software solutions which are marketed as medical products and implemented as server solutions in hospitals are a central component of such systems.

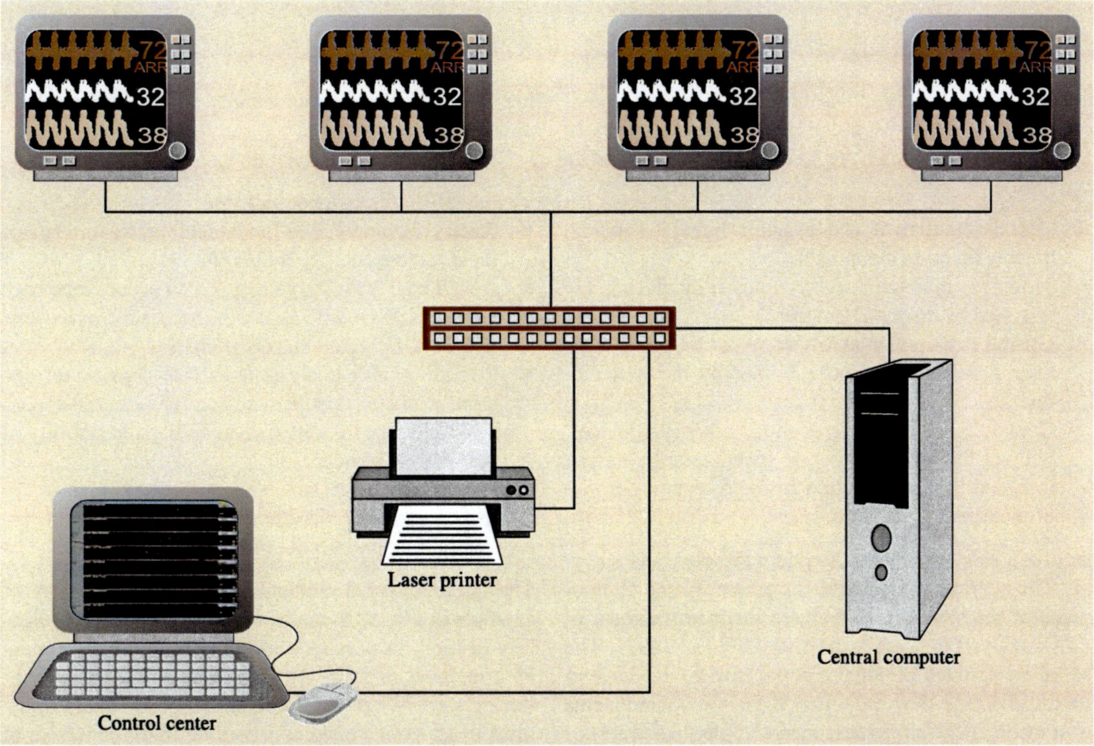

Fig. 58.2 Interconnected medical system for supplementary monitoring in the field of cardiology

The servers or alert software generally enable alert functions and parameters to be configured, with which vital signs and/or alerts (bradycardia, etc.) can be passed on to remote display monitors or else mobile communications equipment.

If electrocardiogram (ECG) alerts such as a bradycardia alert are passed on via an interconnected medical system such as this, disruption/failure can have substantial consequences for the patient if disruption/failure cannot be detected and responded to sufficiently early.

Figure 58.2 shows an example of an interconnected medical system in the field of cardiology to supplement monitoring of ambulant patients following surgery. Following surgery, the patients should be monitored using patient monitors which send vital sign alerts to a central alert server via a wireless LAN (WLAN) connection. This server or special software forwards the configurable alert messages to a telephone system, which in the event of an alert sends a message to the pager, mobile phone or digital enhanced cordless telecommunications (DECT) phone of the doctors or nursing staff.

The patient monitor and the software are marketed as individual medical devices in accordance with the medical devices directive (MDD), and the other components such as the WLAN, the hardware of the alert server, the communications system, and the mobile terminals are marketed as nonmedical devices in accordance with other directives [the low-voltage directive and the electromagnetic compatibility (EMC) and radio and telecommunications terminal equipment (R&TTE) directives].

The combination of the various components – the patient monitor as a medical device, the WLAN as transmission channel, and the PC as a nonmedical device with the alert software as a medical product and the communications system – constitutes an interconnected medical electrical system. The combination corresponds with the legal definition of a medical device system in accordance with §3 of the German Medical Devices Act [*Medizinproduktegesetz* (MPG)], because it transmits diagnostic alert data. None of the suppliers assumes responsibility for the assembly of the system and none declares conformity with the MDD 93/42/EWG.

58.4 Risk Management, DIN EN ISO 14971

Since the operator assembles this combination of a medical device system, he must carry out the process of in-house production as per §12 MPG. Every time an operator:

- Creates equipment and systems himself, without placing them on the market,
- Modifies equipment and systems in a manner not intended by the manufacturer,
- Operates equipment and systems for purposes other than the assigned purpose defined by the manufacturer, or
- Combines medical devices and nonmedical devices, without the manufacturer or manufacturers having intended this combination in the assigned purpose of the device or devices,

he produces equipment or a system in-house.

The operator must define the assembly and components of the system and evaluate them in the form of a risk analysis in accordance with DIN EN 14971. The suppliers of the individual components do not have to do this, because they only market their subcomponents as a medical device (patient monitor, alert software) or as a technical product and are neither able nor willing to assume any responsibility for the signal chain created.

What the operator must do is:

- Establish the assigned purpose
- Satisfy the applicable fundamental requirements of the directive MDD 93/42/EWG
- Develop a risk management file in accordance with DIN EN ISO 14971 and if necessary even take IEC 80001-1 into consideration
- Perform a risk analysis (establish assessment criteria, analyze risks, reduce and reassess risks, e.g., availability of the server, availability and stability of the WLAN standard, etc.)
- Assess residual risks
- Incorporate experiences and observations from using the system into the risk management file.

The interconnected medical system shown by way of example in Fig. 58.2 can be used both in a complementary fashion, in parallel with monitoring of the patient by a member of staff, but also to save on on-site monitoring staff. Whatever the intention of the operator, he must in all events ensure, according to the objective of the directive MDD 93/42/EWG, that no patient comes to any harm as a result of failure, disruption or other events during signal transmission. The risk analysis is used for this purpose and can be used to determine whether or not the risk associated with monitoring using an interconnected medical system such as this is acceptable. The risk analysis according to DIN EN 14971 therefore helps to detect dangers and risks which are not immediately recognizable and as far as possible to prevent or at least reduce these dangers and risks before they arise.

The operator must document the in-house production for every interconnected medical system, since from a legal point of view he creates a new system. This does not merely involve the legal aspects but also involves an understanding of a complex interconnected medical system by all those involved, such as the users (doctor, midwife), medical technicians, and IT staff. Risks and potential risks can be recognized and reduced prior to use by using the risk management approach according to DIN EN ISO 14971.

The operator (management) must ensure that this process is established and continued jointly by medical staff, medical technicians, and IT staff. In the event of a claim, in which a patient comes to harm as a result of a bradycardia alert not being transmitted or not being received, the operator must always demonstrate what action he took to protect the patient from risks. If he is able to account for this in the form of documentation of the in-house production, then he will at least be able to exonerate himself from the charge of gross negligence.

The operator must also decide with the users which risks exist or should be accepted with an interconnected medical system such as this. An interruption in the signal transmission or the inability to access or nonreceipt of an alert via the mobile device of the user can signify, for example, that there is no response to a bradycardia alert or that a response is not possible. The consequence of this can be the death of a patient if the user is not on site and is not notified for the reasons stated. The question is, for example, whether operators/users can and wish to accept a residual risk of one death a year and how the risk of such an event can be significantly/drastically reduced as a result of risk analysis and measures.

It is crucial that the medical staff involved help to shape this process and accept the remaining residual risk. This can, of course, also signify that a hospital arrives at the conclusion not to perform an exemplary installation such as that of an interconnected medical system, because the residual risk is considered from the point of view of those involved to be too high or unacceptable.

When carrying out the risk analysis, the individuals involved must ask themselves which risks and potential risks exist and how these can be assessed:

- Can the signal transmission be interrupted, and where and how?
- How and when is an interruption in the signal chain detected? What do the necessary responses look like?
- Will it come down to propagation delay in the signal transmission? If so, at what level and are these acceptable or not?
- Is unidirectionality sufficient?
- Is it guaranteed that an alert verifiably reaches the receiver?
- Does the IT department have command of the correct installation and administration of the WLAN?
- How is it ensured that alert data is transmitted via the network with priority (quality of service)?
- Can excessive data traffic in peak periods impair the transmission of alert data? How is this checked and verifiably prevented?
- Can malware impair the functionality of the alert server?
- What happens during patching of the alert server at the operating system level or in cases where there is antivirus protection?
- How can the system be regularly and verifiably tested for functionality and following changes?
- Have the employees who must respond to the transmission of the alert signal been, or will they be, thoroughly and comprehensively instructed about the transmission and the complexity of the system? What happens when cover is required (holiday, illness)?
- Are the batteries in the mobile terminals always charged? Is this guaranteed and how?

58.5 Shared Networks

Traditional intensive care patient monitoring with a network and control center/central server has until now been installed as a closed, proprietary network (Fig. 58.3). Manufacturers increasingly offer solutions

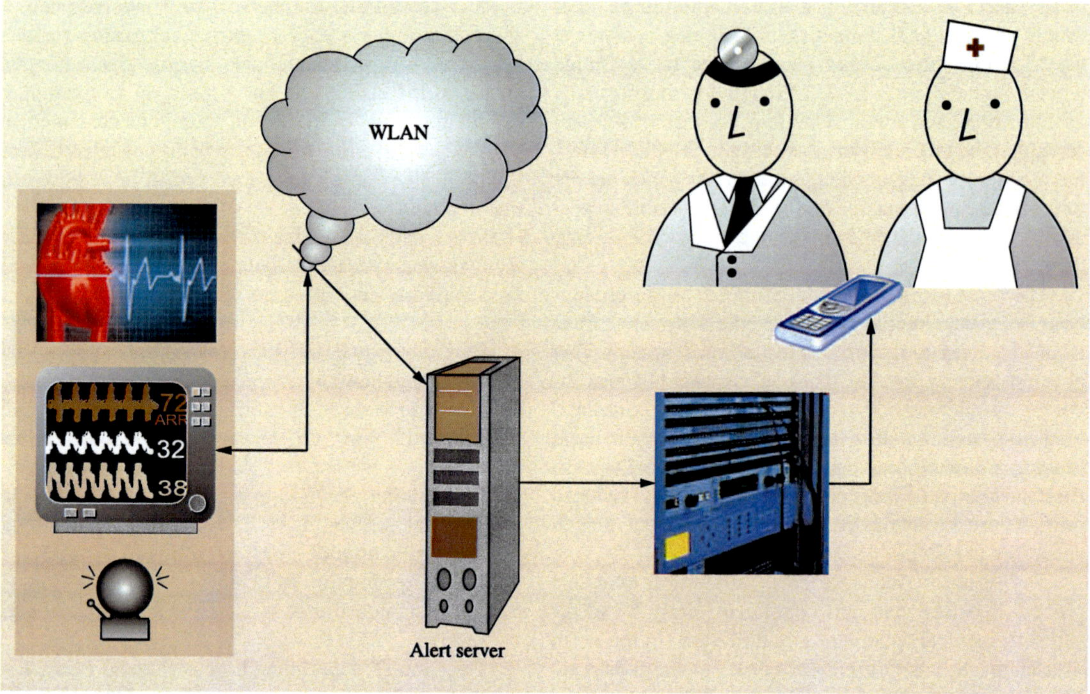

Fig. 58.3 Proprietary network in intensive care medicine

for using the existing IT network (wired or wireless) in the hospital for patient monitoring and for transferring data to the central monitoring station, under certain preconditions. Solutions of this type are described and implemented as virtualized systems or as a *shared network* form of operation. An intensive care monitoring unit can therefore use the operator-specific IT network, in other words it can share the IT network with other applications, under certain preconditions. In this context we refer to a customer support clinical network (CSCN), which is provided by the operator for the application.

Figure 58.3 shows a closed intensive care network with wired networking of the bedside monitors and the control center in the form of a so-called proprietary system, because it is frequently installed by manufacturers as a stand-alone system with its own networking.

This common solution complies with the closed network/data sharing of class C of DIN EN 60601-1 3rd (annex H7) for vital, time-critical patient data. The standard recommends using this network for all time-critical applications and processes. A network such as this is not connected to any other network because, from the viewpoint of the standard, connection to another network could bring about uncontrollable risks. Furthermore, the standard requires that the availability must be 100% and that interruptions must be avoided and must only last for a few minutes per year.

The standard in particular defines that the responsibility must be assigned to just one individual manufacturer/system contractor, which means that a manufacturer declares conformity of the intensive care system, bears the technical responsibility for installation, and, where there is a maintenance contract, also undertakes the maintenance work.

A distinction can be made between three forms of shared network operation, using the example of intensive care networks:

- Open system with wired networking of the bedside monitors and the control center via an operator-specific network
- Open system with WLAN networking of the bedside monitors and the control center via an operator-specific WLAN

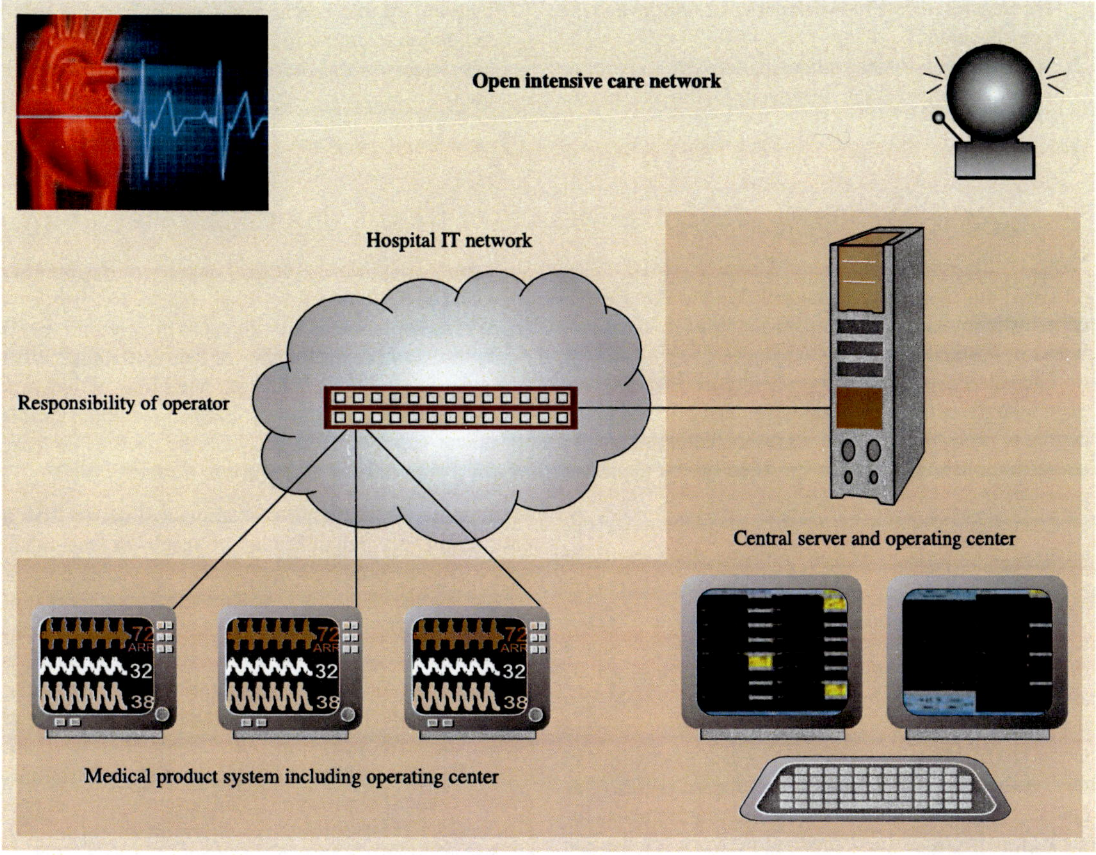

Fig. 58.4 Intensive care network in the shared network form of operation

- Open system with wired and WLAN networking of the bedside monitors and the control center via an operator-specific network (shared network).

Figure 58.4 shows the form in which *shared network* operation can be realized, using the example of an intensive care monitoring unit.

Figure 58.4 shows in principal how patient monitoring can be integrated into an existing IT network in a hospital. This means that the operator installs the network, the cabling, and the infrastructure components such as switches, or has the network, cabling, and components installed, in accordance with the specifications of the manufacturer of the patient monitoring system. For this purpose, the monitors and the control center are operated in a virtual local area network (VLAN). The intensive care network shares (shared network) the general network with other applications in order to send its patient data to the control center or to a server.

A shared network means that all users/subscribers share an installed IT infrastructure with an available, normally fixed, bandwidth (data transfer rate) in a wired or wireless network. This is usually the operator network.

This form of operation therefore results in the operator of the IT network, according to Fig. 58.4, assuming responsibility for the reliability and transmission security of the medical data and thus also for time-critical vital signs and alert data.

Manufacturers of intensive care patient monitoring do not include the operator's IT networks in their conformity assessment procedures; they only specify requirements for the network. Conformity for integrated monitoring (shared network) is only declared by a manufacturer if:

1. The manufacturer has checked the existing IT network and has validated it for its requirements for

the transfer of data between patient monitors and the control center,
2. Any changes to the validated network have been discussed and agreed in advance with the manufacturer, so that he can again perform a check and validation to determine that the changes do not have any effect on the secure and real-time transfer of data.

58.6 Security Aspects of Medical Networks from a Regulatory Viewpoint

Neither the medical devices directive (MDD) 93/42/EWG nor the German Medical Devices Act (MPG) contains requirements for the properties of medical networks in hospitals.

The directive does, however, require that medical devices and medical device systems must provide a high degree of protection for patients, users, and third parties and accomplish the services specified by the manufacturer.

Paragraph 2 of the Medical Devices Operator Ordinance obliges the operator to ensure the security of interconnected medical systems and thus of medical networks.

Medical networks are now a reality in the healthcare system. The vast majority of IT networks in hospitals are implemented as wired infrastructures due to historical factors. Parts of networks or entire networks are increasingly being implemented as radio networks in the form of a WLAN based on standard 811.3. Operators of a network must define the requirements for their IT network or their medical network for the use of interconnected medical systems and assess them for risks.

The following questions from the perspective of medical technology should help with the planning and extension of medical networks.

1. How should a medical network ideally be set up and why?
2. Which medical applications run over a network? How can these applications be divided according to priority criteria?
   - Unsecured use and transmission, no authentication.
   - Secured use and transmission, not time critical.
   - Time-critical use and transmission of vital signs and alerts, etc.
3. How does the topology behave in the event of extension or upgrading of a network?
4. How is the load (average load – peak load) or the data transfer of a network calculated? Can a propagation delay of time-critical data occur during peak load?
5. Which applications are subject to high requirements in terms of the availability of data exchange? What technology (redundancy, safeguarding of network components with battery back-up, uninterruptible power supply (UPS), etc.) can be used to increase the availability of a network or a subnetwork?

To increase the availability of a medical network with data exchange, reliability as a result of redundant hardware and software components and greater consideration for the reliability requirements from the point of view of the user can be taken into account when setting up a medical network, for example. Redundancy is generally considered to be an essential measure in order to increase the availability of networks. In the medical sector, the term *maximum reliability* is therefore also used. The operator should therefore define, for example, which services, applications, and network components are considered critical in terms of availability in order to be able to use them redundantly in the event of a failure or malfunction.

6. What security requirements are there in terms of protection from malware?

Since medical devices and medical device systems are also infected with malware in the network, the operator, represented by the medical technicians and IT staff, must operate medical devices using antimalware software (also known as antivirus protection) and must install patch management in order to permanently update both the antimalware software and the operating system.

Manufacturers of medical devices which are integrated in a medical network must at the same time provide assistance by evaluating the protection of their devices against malware within the scope of their conformity assessment procedures and by providing the operator with details about antimalware software and regular upgrades to be used.

58.7 Future Standard IEC 80001-1

IEC 80001-1 concerns the entire medical network and the medical devices integrated therein. The future standard describes a form of risk management using technology from DIN EN ISO 14971 in order to analyze, evaluate, and if necessary reduce the risks and potential risks of integrating medical devices into an IT network and of exchanging medical data over this network.

The future standard has the objectives:

- Security in the network
- An effective network
- Data and system security
- Confidentiality
- Data integrity
- Availability of the data
- Secure interconnection of medical devices and non-medical equipment within the same network.

Irrespective of the date on which this future standard will come into force, the operator is today already obliged and well advised to analyze interconnected medical systems and the medical network for risks and potential risks (e.g., malware in the network) and to take preventive measures to increase protection against malware and reduce the dangers which arise as a result. Malware in the form of computer viruses and worms such as the Conficker worm do not spare hospital networks and the PCs and medical devices within the network which have a network connection.

Further Reading

- A. Gärtner: Medizinische Netzwerke und vernetzte medizinische Systeme, Medizintechn. **2009**(6), 213–218 (2009)
- DIN EN 60601-1 (VDE 0750-1:2007-07) Medizinische elektrische Geräte – Teil 1: Allgemeine Festlegungen für die Sicherheit einschließlich der wesentlichen Leistungsmerkmale (2007)
- IEC 60601-1-SER ed1.0 Medical electrical equipment – 2010-04-29 (IEC, Geneva 2010)
- DIN EN ISO 14971:2009:10 Medical devices. Application of risk management to medical devices (Medizinprodukte – Anwendung des Risikomanagements auf Medizinprodukte) (2009)
- E DIN EN 80001-1(VDE 0756-1):2009-10 Application of risk management for IT-networks incorporating medical devices – Part 1: Roles, responsibilities and activities (IEC 62A/668/CDV:2009); (German version FprEN 80001-1:2009, Anwendung des Risikomanagements für IT-Netzwerke mit Medizinprodukten – Teil 1: Aufgaben, Verantwortlichkeiten und Aktivitäten) (Beuth, Berlin 2009)

References

58.1 M. Kappes: *Netzwerk- und Datensicherheit – Eine praktische Einführung* (Teubner, Munich 2007)

59. Hospital Information Systems

Peter Haas, Klaus A. Kuhn

In this chapter, we present an overview of health information systems with a specific focus on hospitals. We start by illustrating the different dimensions of IT support from basic documentation tasks to organizational support and medical decisions. Architectural models are described, and pros and cons of existing options are discussed. Integration challenges and the role of standards are presented. Finally, the process of selecting and implementing a hospital information system is described in detail.

59.1 **Background** 1095

59.2 **Necessity, Objectives, and Benefits of Comprehensive HIS** 1096
- 59.2.1 Necessity 1096
- 59.2.2 Objectives of Using Information Technology in Hospitals 1097
- 59.2.3 Potential Benefits 1097

59.3 **Dimensions of IT Support** 1098
- 59.3.1 Data Processing Support............. 1098
- 59.3.2 Documentation Support 1098
- 59.3.3 Organizational Support 1098
- 59.3.4 Communication Support............. 1098
- 59.3.5 Decision Support 1099

59.4 **Case Study** 1099

59.5 **Architecture and Components of HIS** 1103
- 59.5.1 Logical Architecture Model........... 1103
- 59.5.2 Implementation Alternative: Holistic Versus Heterogeneous 1105
- 59.5.3 Integration Aspects of Heterogeneous HIS 1107

59.6 **Current Trends and Prospects** 1108
- 59.6.1 Medical Documentation 1108
- 59.6.2 Integration of Medical Technology 1108
- 59.6.3 Health Information Networks 1108
- 59.6.4 Decision Support for Health Care Professionals 1109
- 59.6.5 Support for Patient Safety 1110
- 59.6.6 Genomic Data 1110

59.7 **Selection and Implementation of HIS** 1110
- 59.7.1 Preliminary Remarks 1110
- 59.7.2 Project Phases and Critical Factors 1111

59.8 **Conclusion** 1117

References 1118

59.1 Background

Health care systems in many developed countries are facing significant challenges. A rise in the average age, together with an increase in chronic illnesses and thus increasing costs, losses in insurance premiums, financial problems caused by the labor market, and the universal demand for high-quality modern medical care for all citizens – irrespective of their financial situation – are resulting in fundamental questions being asked:

- How can a balance be found between the limited funds available and the demand for health care?
- How can medical progress and the latest diagnostic tests and treatments be implemented quickly and efficiently across the board?
- How can the organization of care services be made more effective and better coordinated?
- Which instruments of health reporting are needed by government in order to be able to act quickly and appropriately?

Approaches to these problems include, for example, case and disease management, managed care, and, associated with this, improved dovetailing of the different care sectors such as in-patient and out-patient rehabilitative care. Hospitals are especially affected by these approaches and will need to be involved to a major degree in their implementation, as they are responsible for a significant proportion of complex treatments and thus also contribute essential information to the life-long treatment process of a patient and his/her continuing care.

Politicians, health insurers, medical experts, and also the general public are increasingly aware that these challenges can only be overcome with the help of health information technology. The goal is to strengthen intra and interinstitutional cooperation and to integrate the care sectors. Better access to data and knowledge is necessary to improve processes and to make decisions. Health information technology will also support the practice of evidence based medicine, characterized by *Sackett* et al. [59.1] as the conscientious, explicit, and judicious use of current best evidence in making decisions about the care of individual patients.

Boundaries between hospitals and regional and national health information networks are beginning to fade, and hospital information systems (HIS) are becoming part of health information networks, leading to the replacement of the term *hospital information systems* by *health information systems* [59.2]. Likewise, institutional medical records are being integrated into interinstitutional electronic health records.

A functioning and effective network only makes sense, however, if the nodes in this network (i. e. the participating operational information systems in the health care institutions) are available and are suitable for actually setting up this network and bringing it to life. This makes it clear that a critical success factor for the further development of the health care system is to provide health care institutions with appropriate institution-based information systems (operational information systems such as hospital information systems, medical surgery systems, emergency services information systems, etc.) and with staff with the necessary expertise.

The majority of medical information about an individual is gathered and documented in hospitals. This applies in particular to those groups mentioned initially, the chronically ill and those in old age. Essential medical procedures both of a diagnostic and therapeutic nature – which are significant also in later life and thus for further care – are performed in hospitals. The information that finds today its way into the out-patient sector is generally a brief case summary, often produced too late, which certainly helps to provide a quick overview but which is unable to answer detailed questions transparently.

It is evident that health information networks are inconceivable without well-designed and well-functioning hospital information systems, and it can even be argued that hospital information systems are the backbone of health information networks. This is also true since many problems and solutions from the microcosm of the hospital are of relevance in the macrocosm of the health IT networks. Moreover, with their EDP departments hospitals can take on the running of active network nodes for entire subregions.

In conclusion, it can thus far be determined that hospital information systems are not only a critical success factor for the respective hospital management but also collectively play a significant role in national health systems.

59.2 Necessity, Objectives, and Benefits of Comprehensive HIS

59.2.1 Necessity

The need to use comprehensive hospital information systems (HIS) is the result of four fundamental aspects:

1. The hospitals are placed under a high level of pressure with regard to their effectiveness and efficiency. This situation is manifested, among other things, in the obligations to provide proof and to communicate data, but also in complex compensation systems, for example by means of the widely used diagnosis related groups. (DRGs). Effective operational management is still only possible on the basis of a direct costing with an orientation towards medical economics and with all the necessary upstream and downstream components. Detailed information about individual treatments are necessary for this, which can only be gathered and used via a HIS that is used across the board.
2. The translation of medical knowledge – which is made available faster than previously known by means of electronic media and the Internet – into everyday clinical work can barely be achieved without appropriate support in terms of information technology. Context-sensitive tools are required, which – ideally used during individual treatment – provide the physician, in a direct and uncomplicated form, with relevant accessible information that is present in the HIS or in corresponding medical databases or knowledge bases for the particular current patient-based problem.

3. The implementation of guidelines and the organizational coordination and streamlining by means of clinical pathways cannot be achieved effectively without corresponding supporting IT functions.
4. It is clear from the statements made at the outset that hospital information systems play an essential role in the development of a health information network and in the care of patients as a whole. No efficient and networked health care system is feasible without high-performance hospital information systems.

Against this background, the question of high strategic importance is how hospitals – which ultimately produce intangible commodities such as health, recovery and relief – handle the most important production resource for the production of these commodities, information. A critical success factor for efficient medical practice – and therefore for quality, efficiency and cost effectiveness – is the fast and comprehensive availability of up-to-date information about examinations and tests. In this respect, the availability of a high-performance hospital information system for every hospital is a fundamentally crucial factor for quality, business success, and competitiveness.

59.2.2 Objectives of Using Information Technology in Hospitals

The objectives of using IT in hospitals must – as in all fields of application – always be subordinate to and derived from the business objectives. In this respect, every hospital must establish these objectives in detail. For an overview of information management and information systems architecture, we refer to [59.3–5] and for health information systems we refer to [59.6, 7].

It is generally accepted that the following *strategic objectives* can be specified:

The use of IT in hospitals must:

- Provide extensive support for the work of the management
- Enable optimization of the revenue situation
- Create cost and performance transparency
- Enable administrative procedures to be rationalized
- Contribute towards making medical organization and decision-making processes effective
- Reduce patient throughput times (examinations, operations, in-patient hospital stays, etc.)
- Make medical organizational and decision-making processes transparent
- Ensure continuous quality monitoring and improvement
- Enable information to be provided to patients, staff, and the general public
- Improve coordination/cooperation with internal and external partners
- Provide an electronic medical record.

Fundamental *operational objectives* – which follow these strategic objectives – are:

- Safeguarding of the current modes of payment and obligations to provide proof, improve liquidity → *revenue transparency*
- Introduction of (extended) basic documentation, uniform coding software: → *transparency of patient case mix*
- Operational systems in materials management, personnel management, technology, catering, etc. → *cost transparency*
- Comprehensive optimum activity recording → performance transparency, operational transparency
- Enable direct costing/activity-based costing → *transparency in the use of resources* (personnel, equipment, etc. for case groups)
- Medical organizational and documentation systems for specialist departments → *organizational transparency, transparency of documentation*
- Documentation of treatment processes, their outcomes, and their adherence to guidelines → *transparency of health care processes needed for the improvement of quality*
- The Intranet/Internet as an information medium for the different target groups → *transparency of the hospital* (e.g. for employees, patients, and the general public).

59.2.3 Potential Benefits

The potential benefits of data processing in hospitals are wide-ranging. If we draw the analogy that a hospital information system represents the brain (in this sense the memory) and nervous system (information transmission, control, monitoring, status information) of the hospital, then the enormous significance and the potential benefit of a HIS become clear.

A hospital information system:

- Enables an overall view of patient treatment
- Contributes towards integrating the different professional groups

- Relieves the medical personnel from having to duplicate work and from administrative overhead
- Enables quick access to documentation of earlier treatment
- Enables quick access to current medical knowledge
- Enables better coordination and reconciliation, e.g. using a scheduling module, and thus prompt control and regulation of the operational processes
- Enables continuous quality monitoring
- Contributes towards an increase in the quality of treatment
- Helps to avoid unnecessary examinations and tests
- Provides information about the costs accrued and their origin
- Creates operational transparency
- Helps to save costs
- Contributes towards patient satisfaction
- Increases the attractiveness of the hospital to referring doctors, patients and the general public
- Gives a competitive advantage as a result of having a suitable range of services and a faster response to market changes
- Creates a database for research in the field of clinical epidemiology, but also for health economics.

It has been shown that information technology can reduce the rate of errors by preventing errors and adverse events, by facilitating a more rapid response after an adverse event has occurred, and by tracking and providing feedback about adverse events [59.8].

However, information processing – with its nature of being an interdisciplinary technology – only provides these benefits if implemented in a balanced and coordinated manner in all operational areas, and this, therefore, requires a complete operational IT strategy. In situations where there are limited finances, strategically controlled and coordinated gradual expansion of the HIS in a hospital, therefore, becomes particularly important.

59.3 Dimensions of IT Support

The aforementioned objectives and potential benefits can by and large only be achieved if – irrespective of the degree of distribution and the integrated component systems – all aspects of IT support in the hospital are taken into consideration within the hospital information system. Major dimensions of the support provided by IT systems in hospitals are as follows.

59.3.1 Data Processing Support

Support of the processing of data, in the sense of performing complex calculations, transformations, and data conversions. For example, ascertaining the billable DRG (diagnosis related group) from available documented clinical case details. A more clinically orientated example is the calculation of dosages, for example of infusion rates. Monitoring systems also ultimately fall into this category.

59.3.2 Documentation Support

Support of documentation by providing the relevant electronic forms and text systems in order to create documents and archiving them in electronic folders. In the area of documentation relevant for billing relating to diagnoses and procedures, more complex IT modules with thesaurus functions and checks are also used. If terminologies enabling detailed descriptions – such as SNOMED – are not used, documentation support is limited. This limitation also affects decision support, e.g. when reminders can or cannot be generated from collected data. Semantic interoperability between information systems is likewise made more difficult.

59.3.3 Organizational Support

This includes support of organizational tasks and resource allocation planning (appointment schedules in diagnostic departments, bed management in wards), with process handling using a workflow management system, and support of treatment using clinical pathways.

59.3.4 Communication Support

This includes support of the internal communication in the form of communication of results, and also support of handover/shift changes, increasingly also communication with external partners. Communication support is closely linked to efficient documentation support and also to improved access to information. The integration of results and diagnoses in doctors' letters can, therefore, speed up the process of generating medical reports and improve the important communication between the in-patient and out-patient sectors. The communication of referrals is increasingly gaining

Table 59.1 Dimensions of IT support and examples of HIS functions

Dimension	Example
Data processing support	– Determining the billable DRG – Determining case-based process costs – Calculating profit margins – Calculating diagnosis-group related cost distributions – (Continually) calculating essential basic statistical measures, such as average length of stay, capacity utilization, etc. – Generating statistics and evidence (required by law)
Documentation support	– Billing-orientated documentation of diagnoses and procedures – Clinical diagnosis documentation – Documentation of procedures performed along with the corresponding results, e.g., the documentation relating to surgery, anaesthesia, test results and much more – Documentation of symptoms – Documentation of important incidents and events – Documentation of medication – Doctors' reports, documentation of case summaries integrated in an electronic medical record
Organizational support	– Appointment schedules/resource allocation planning in out-patient clinics and hospital departments – Bed management on wards – Workflow management to support the execution of examinations and generation of documents/results – Use of clinical pathways and standards of care for treatment – Monitoring of the completeness and promptness of billing documentation
Communication support	– Associated operational communication in the form of a service request or results confirmation (service communication, order entry/result reporting) – Transmission of case data to health insurers – Forwarding of discharge letters to the referring doctors – Internal operational email communication
Decision support	– Context-sensitive access to fact and knowledge bases – Clinical reminders and warnings, e.g., during order entry – Context-sensitive monitoring of test results

in importance in hospitals. When issuing referrals, reminders and warnings can be generated by checking the data.

59.3.5 Decision Support

This includes support of decision-making processes by means of operational knowledge management for context-sensitive provision of current knowledge (e.g. access to guidelines) based on given facts (e.g. alerts in case of drug interactions, etc.). Reminders and warnings are highly relevant tools for decision support [59.8, 9].

The most important functions of HIS with respect to these dimensions of support in hospitals are shown in Table 59.1. Examples of successful HIS implementations can be found in [59.10–12].

59.4 Case Study

The intention here is to illustrate, with reference to an example of a short, fictional course of treatment in the first two days of a patient's hospital stay – and thus of an in-patient case – the fundamental aspects of support provided by HIS in terms of hospital management.

Mrs Maier – an elderly lady – is to receive an artificial hip. A date has been agreed with Mrs Maier and her GP for her to be admitted to hospital on 21 February 2011. The IT support for the organization of this starts even before the effective hospital stay, due to anticipatory bed management and resource allocation planning – in the case study, a bed has therefore already been booked for Mrs Maier for the expected length of stay of 9 days.

If Mrs Maier then appears at the admission desk on the planned admission date, after briefly supplementing the information already present through central admissions – where the additional details necessary for billing can be found – she can then go directly to the ward, where a bed will already have been allocated to her, both

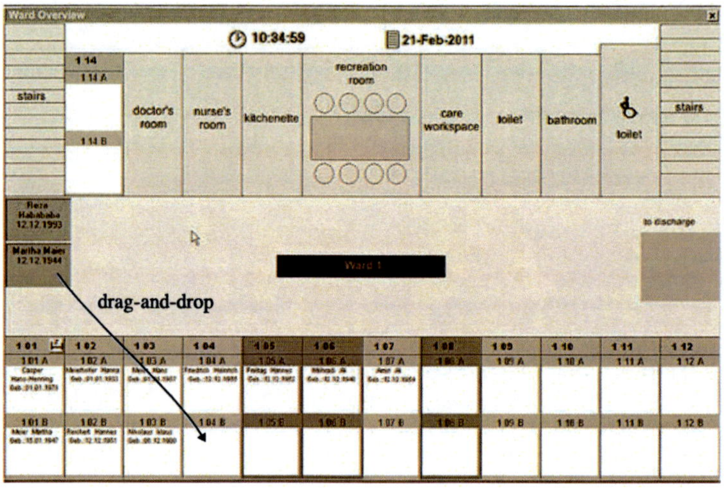

Fig. 59.1 Ward overview, including bed occupancy

in reality and also within the hospital information system (Fig. 59.1) via drag and drop. The corresponding functions with a graphical overview of bed occupancy are today realized in practically all HISs. Furthermore, double-clicking on the *bed*, demographic data, the patient's next of kin, the key caregivers, and staff also currently dealing with the patient, as well as already known critical information, are now added on the ward (Fig. 59.2).

In order to further support the treatment process, many hospitals are working on the introduction of *clinical pathways*, which determine – on a diagnosis-specific and treatment-specific basis – which procedures should be carried out in the time scale. In our case study, the corresponding pathway for a total hip replacement is as accessed for *process initialization* and uploaded into the patient's patient record. In this electronic patient record, after the allocation of the bed all the necessary procedures that are to be performed are therefore already entered automatically over time (Fig. 59.3). The fact that these are still *undone* – i. e., yet to be completed – can be seen by the status abbreviation ("r" for requested, "pl" for planned, "a" for appointment). For all special procedures in the HIS parameterization it is defined as to which types of documents/information objects (form, image, video, audio note, or report) should be recorded over all, when this procedure is finally performed. So the view of treatment process not only shows the past and future procedures over time, but also the information objects and their state (white = object still not recorded; red = object recorded, but not accessible for others than the author; green = object is entered and can be viewed by authorized users), so every user gets an overview about state of procedures and state of information objects that have to be recorded. With a double-click on a *green* = existing information object, authorized users can take a look inside the object in a simple way. Admission to the ward is likewise already documented, and direct access to the admission data and the referral form for authorized users is possible via the "*green* form" icon in first row of the line with the procedure *patient admission*.

In the next step, the physician on duty now performs the medical history check and the clinical examination and documents the results electronically. During or after the completion of this procedure, he completes the diagnosis documentation (Fig. 59.4), which is also important for the later DRG billing and electronic transfer of billing data to the health insurance fund. Because Mrs Maier is also complaining of difficulty breathing, the doctor uses the electronic order function (Fig. 59.5) to order an additional chest x-ray, additionally to the clinical pathway. On the same afternoon, in addition to the other planned procedures in the laboratory and in the x-ray department, a corresponding chest x-ray will, therefore, also be performed. The processing in the x-ray department is done using electronic job lists, which show the planned examinations for each x-ray workstation and which are then used for activity recording. Performance data, x-ray images are, therefore, available on the ward straight after the examination has taken place and, shortly afterwards, the electronic results i. e., the radiologist's report, are also available.

The next morning, the electronic medical record appears as shown in Fig. 59.6. All of the planned procedures of the previous day have been carried out (status

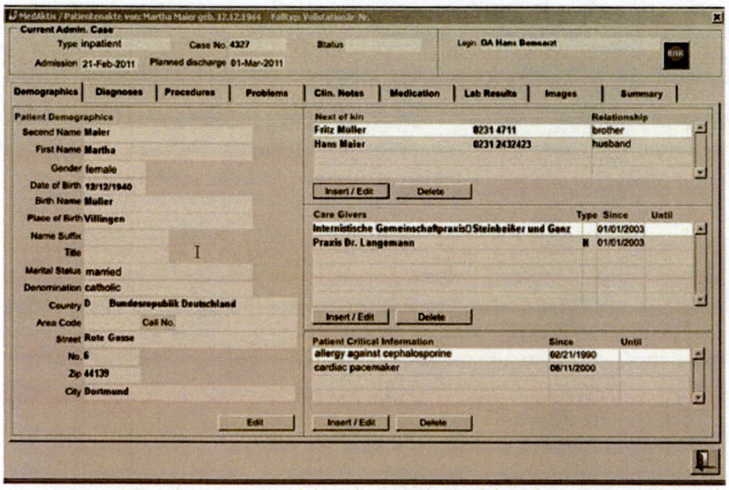

Fig. 59.2 Demographic data, the patient's next of kin, the key caregivers and staff currently dealing with the patient, as well as already known critical information added on the ward

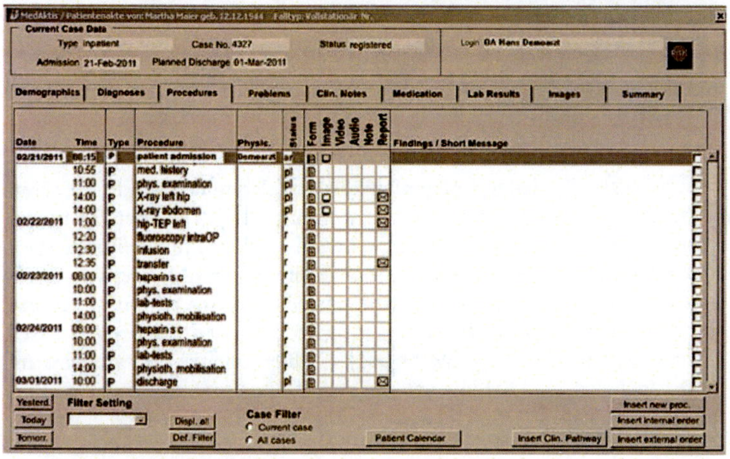

Fig. 59.3 Electronic patient record. After the allocation of the bed all necessary procedures to be performed are entered automatically over time

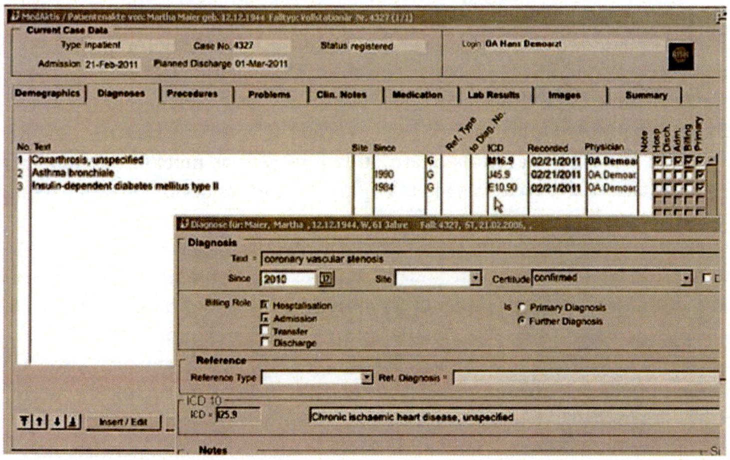

Fig. 59.4 Electronic patient record. Diagnosis documentation

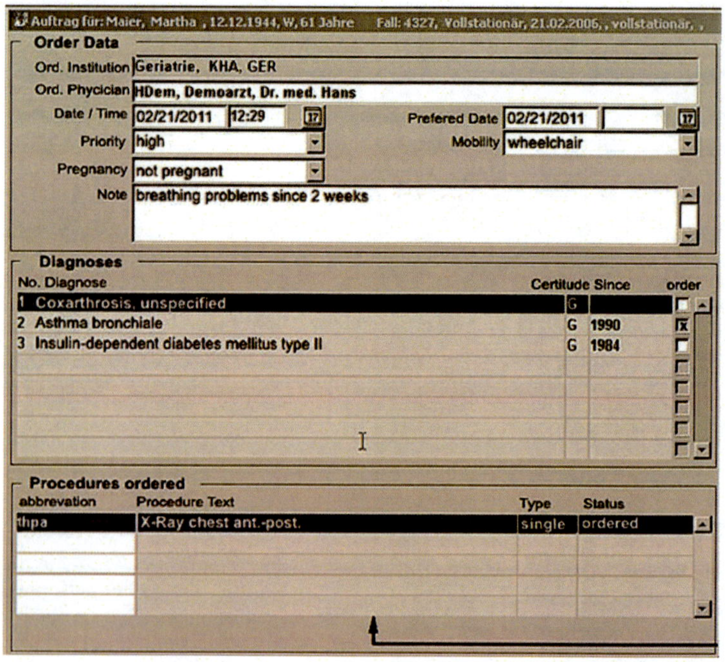

Fig. 59.5 Using the electronic order function to order an additional chest x-ray, additionally to the clinical pathway

"ar" = activity recorded or "rw" = report written), and the reports and forms can now be viewed directly via a double-click on the green icons (in the figures shown in gray). As also becomes clear from the process-orientated electronic patient record, on the previous day the OT administrator scheduled the operation for 11 o'clock ("a" = appointment given). It is also evident that, in addition to the pathway procedures included, a chest x-ray was carried out at 12:38 on the previous day.

A further account of the course of treatment and the activity-based control thereof and detailed documentation by means of the HIS should be dispensed with at this point. Overall, it has become clear that process documentation is generated within a short space of time to accompany the treatment using a modern HIS, and this process documentation creates a high level of organizational transparency and transparency of documentation – everybody involved in the treatment process has access to the important information necessary for completing his or her task and is informed in real time of the status of the course of treatment. It is also important that, at the end of the in-patient treatment, all of the information is included in the case summary – the summarizing assessment – and that the details for this can be (semi)automatically taken from the electronic medical record. This is used both to promptly send the discharge letter – potentially also in electronic form to the referring doctor – and to electronically transfer the billing information to the health insurance fund within a short space of time. For internal purposes, activity-based costing can also be carried out on the basis of the course of treatment stored in the HIS, and this activity-based costing can be used to determine the cost amounts for individual patients and diagnosis related patient groups.

On the whole, the example shows how an HIS makes the processes in hospitals more effective by supporting data processing, documentation, organization, and communication and significantly increases the clinical and administrative transparency for all those involved. Time-consuming telephone calls such as "Has Mrs Maier already been x-rayed? Where is the OT report, has it been written yet? Which services can be billed for?" are, therefore, just as much a thing of the past as the lack of economic transparency in hospital activities. HIS, therefore, supports all of the operational matters on a functional level and also supports the economic processes and the decisions on a management level. An important extension would be the transfer of de-identified data into a data warehouse.

We want to note that structured documentation of medication is needed if warnings or reminders are to be generated in the case of drug–drug interactions or contraindications. There are European countries, how-

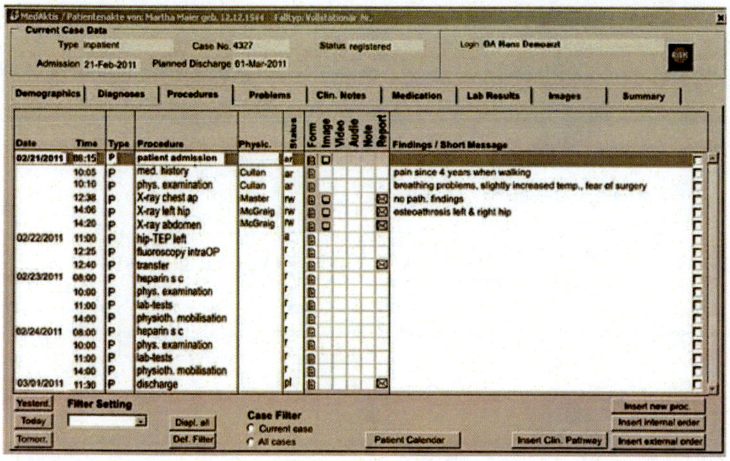

Fig. 59.6 Completed medical record 1st day of stay. All of the planned procedures have been carried out (status "ar" = activity recorded or "rw" = report written), and the reports and forms can now be viewed directly via double-click on the marked icons

ever, where such a structured documentation is not in widespread use. Likewise, the implementation of decision support tools varies significantly between European countries.

59.5 Architecture and Components of HIS

59.5.1 Logical Architecture Model

Architecture models act as a reference system and thus provide a general basis for discussion and understanding for all those involved in a design and discussion process. Essentially, a distinction must be made between models from the view of the user (owner's representation) and from the view of the computer scientist (designer's representation). Figure 59.7 shows a logical architecture model of a hospital information system – i. e. not orientated towards technical networks, protocols and computer levels, but orientated instead towards organizational units and specific application solutions and thus towards the view of the user (owner's representation).

The following component information systems can be identified.

The Administrative Information System

All the applications for *administration* and *logistics* are subsumed under this subsystem. If this also previously always included the patient data management system, then it becomes increasingly clear against the background of the new billing legislation that this should be regarded as an independent component, since it includes functions that are vital for all other information systems – even the medical ones.

The Patient Data Management System

This includes all of the *functions for the management of patient data that are necessary for accounting and for meeting the legal obligations to provide proof*. This includes the relevant and basic functions for the admission, discharge, and transfer of patients (ADT). Due to the current development in the legal context and the associated obligations to provide proof, the volume of previously simple administrative data has been expanded significantly with medical information (diagnoses, explanations, categories of care, diagnostic/therapeutic procedures). In this hybrid function – i. e. containing administrative data but also important medical data – the patient data management system is of particular importance. A bed management planning module can also be subsumed under this subsystem.

The Medical Information System

The medical information system includes all of the applications/information systems to support the documentation and organization of medical organizational units. On account of the various different orientations and the necessary specific functions of the individual modules for each application area, the medical information system itself can be subdivided into:

- Specialist department systems – e.g., surgical information systems, anaesthesia information systems,

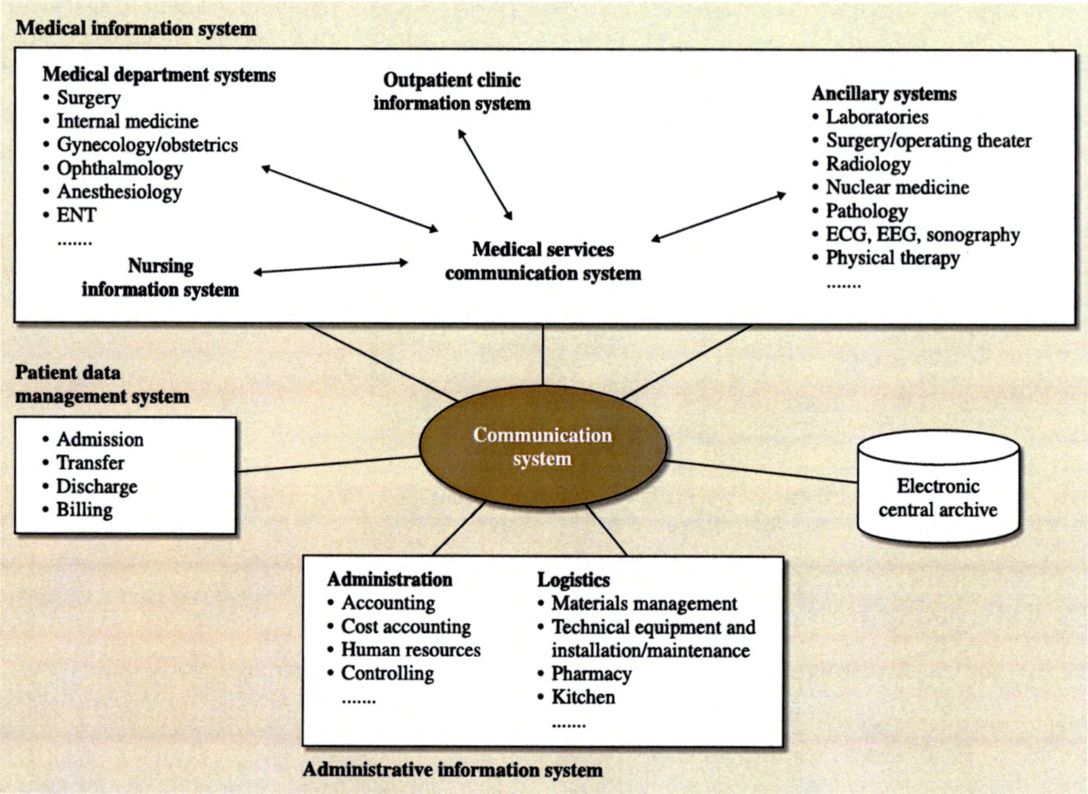

Fig. 59.7 Logical architecture model of a HIS

gynaecological information systems, etc., support the documentation of specialist medical information and also the other more subordinate purposes of documentation, such as quality management, obligations to provide evidence, billing, etc., in a specialized way within the specialist department. In addition to supporting documentation, these information systems support the medical decision-making process not only by quickly making available the most recent results by means of the service communication system, but by also allowing access to knowledge bases and electronic textbooks linked to the HIS. By means of a treatment planning module, they allow problem-based or diagnosis-based standards of care for treatment to be stored on a department-specific basis and used as a basis for planning in the case of specific treatments.

- Systems for the support of out-patient clinics – Tasks requiring specific support in out-patient clinics can be considered in particular to be the effective support of scheduling and appointment management, the effective support of out-patient treatment processes in terms of an interdisciplinary workflow, and also the documentation of specific services and out-patient billing, which is particularly sophisticated and complex in Germany.
- Diagnostic department information systems, systems in ancillary departments such as laboratory information systems, radiology information systems, pathology systems, operative documentation systems, etc., provide integrated specialist support for organization, documentation, and communication for specific diagnostic departments (function-testing departments). In addition to the very specific documentation and special workflows, whose effective execution increases throughput, special medical technology components are also incorporated here such as the online connection of laboratory analysis equipment to the laboratory information system or of the imaging modalities to the radiology information system, etc. As in out-patient clinics, diagnostic departments in which patients are ex-

amined directly also require effective appointment management.

- The clinical information system to access and manage clinical data and to communicate with other systems. This central module manages documents such as discharge letters and coded diagnoses and procedures. It collects reports and results from ancillary systems, communicates with the ADT system, and requests services from other systems. It is often based on a central repository. It provides the typical order entry/result reporting functionality. An order entry module replaces the traditional request form with online requests on the monitor and enables orders to be booked in directly on systems of ancillary departments. The individual placing the request has the option of accessing the status of his order at any time and receives the test result quickly in electronic form at the earliest possible moment.
- Nursing information system to support the planning of nursing care, nursing records including making graphs, and nursing quality management.

Ideally, a complete electronic patient record – that is to say a paperless digital collection of all treatment documents – will be possible as a result of the interaction between these various medical information systems and their components. Often, however, electronic data and paper documents coexist. Digital archives with scanned paper documents are about to change this situation.

The Communications System

From the point of view of the user, this is understood as being the service communication system described above. On a technical level and on the basis of the topology model in Fig. 59.7, however, the communications system is to be understood as an *information processor* to link, by means of software technology, a wide range of different application systems in order to enable data communication between these application systems (e.g. between specialist department systems and laboratory and radiology systems, etc.). Technically, this component is mostly based on asynchronous messaging with an interface engine or *integration engine*. Service architectures are beginning to be used more often (e.g. enterprise service bus).

Interdisciplinary Applications

This includes *applications that are of interest for many departments/users*, such as report writing and administration, diagnosis coding, knowledge servers with various databases such as the Red List, Medline, etc., office communication, spreadsheets, etc.; in the broadest sense, a nursing records system and a scheduling module are also interdisciplinary in character.

59.5.2 Implementation Alternative: Holistic Versus Heterogeneous

Besides the logical architecture, another important feature of hospital-wide information systems is the implementation in terms of software and hardware technology and the potential – which may also be determined along with the implementation – to distribute the solution between different components of the hardware infrastructure [59.2, 6, 7].

Holistic Information System

The software for all the system components to be found in the logical architecture model comes from one manufacturer and is based on a data model with an overall concept, meaning that the entire HIS is conceptually holistic, while the software can also be segmented in a modularized manner into components that can be operated individually in order to enable it to be better maintained and distributed (Fig. 59.8).

Heterogeneous or Best-of-Breed Information System

The software for the components listed in the logical architecture model comes from various different manufacturers who all work with their own data models and data storage. Communication and data synchronization between these systems is carried out using

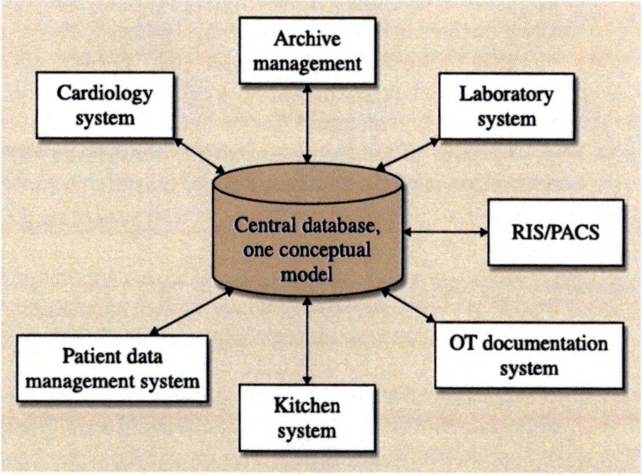

Fig. 59.8 Example of a holistic HIS

Part F | 59.5

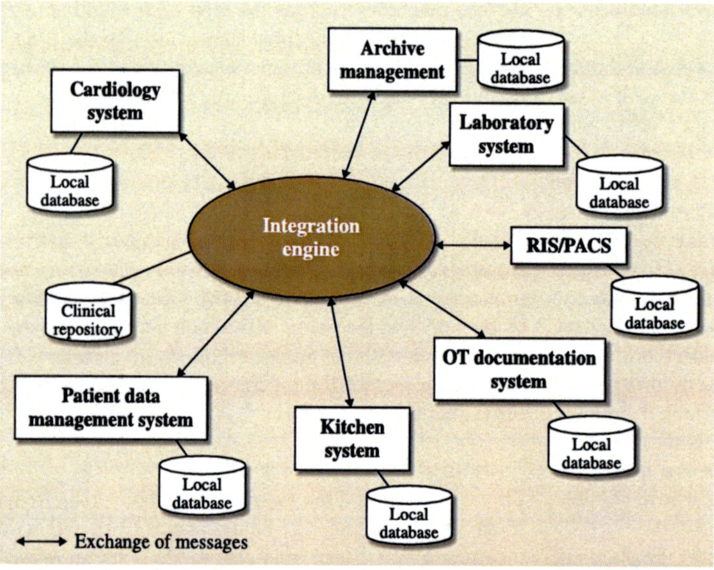

Fig. 59.9 Example of a heterogeneous HIS

Part F | 59.5

Table 59.2 Advantages and disadvantages of the two approaches of a holistic and heterogeneous information system

Holistic = all applications from a single manufacturer	Best-of-breed = applications from different manufacturers
+ If necessary, one conceptual model	– Various conceptual models
+ One vendor, one contracting party	– Interface problems
+ Less maintenance required	– Various user interfaces
+ Consistent user interface	– Various data storage systems
+ No duplicated data storage	– High level of maintenance required
– Few single medical functions – Dependent on one manufacturer	+ High level of adaptation of the individual systems to the terminology, semantics, and workflow of the field of use

appropriate linking software, e.g., an interface engine or integration engine (Fig. 59.9). This approach is known as best-of-breed. Data from different systems are usually replicated in a common physical repository. The use of virtual repositories, i. e., of federated approaches is a less common alternative. Service oriented architecture (SOA) will play an important role in the future.

Selection of the technical specifications of the HIS determines in particular the necessary effort in terms of maintenance; the more inhomogeneous a system is, the more maintenance work is required.

Both the holistic and the best-of-breed approaches to the problem have advantages and disadvantages, which are the reverse of one another. Table 59.2 shows a comparison.

In accordance with the complex and diverse nature of hospital information systems, various different types of suppliers have developed, who for the most part can be divided into the following three categories:

1. Holistic vendors, who provide a full spectrum of solutions (e.g. a clinical information system, RIS, PACS, nursing information system, etc.) in a single, integrated system. In principle, it is not necessary to integrate further systems from other manufacturers. These solutions are generally very broad in scope but sometimes go into little depth in the specific functions. Attempts are increasingly being made to implement individualization using form generators for medical data entry and management. The basic marketing point here is the integrated solution – referred to as *holistic HIS* in Fig. 59.8.
   Behind the scenes it may become clear that many overall solutions do not consist of holistic software with a uniform underlying company data model, but are assembled from systems bought from different makers who communicate with one another via coupling mechanisms. Thus, the similarity to a heterogeneous approach may be higher than expected.

2. Holistic vendors who often offer a comprehensive administrative system (ERP), sometimes connected to a central clinical repository. Contrary to 1, they do not cover the full functional spectrum, and, therefore, connect any desired medical subsystems thereto in a more or less complex manner (leading to a *heterogeneous HIS* (Fig. 59.9).
   The shared contents such as patient data, test results, etc., must be held in multiple locations and synchronized by exchanging data records (known as communication records) between the systems involved, which is often done using a communications server.
3. Specialist suppliers, who may be called niche vendors, who – usually – provide highly competent and self-contained solutions for subareas (laboratory, radiology, hygiene, etc.) and can be integrated into heterogeneous HISs by means of data communication.

Practice has shown that the larger a hospital, the less the use of a holistic information system is able to meet the demands. This can be illustrated using an example. Whereas in a 200-bed hospital the radiology department has 2 x-ray machines and this diagnostic department can be supported using the interdisciplinary service communication module of the holistic HIS, a radiology department in a 1200-bed hospital has 16 x-ray machines and further test machines, and has a complex organization. In this case, the IT support is expediently provided using a special radiology information system with a connected picture archiving system (PACS) and an online link to the individual imaging modalities.

59.5.3 Integration Aspects of Heterogeneous HIS

With a heterogeneous approach, communication between the systems is often carried out by exchanging messages. The individual systems integrated into the HIS must, therefore, have the following functionalities:

- An import module for receiving data records
- An export module for transmitting data records
- Suitable internal database structures for storing the received data (often one central repository is used)
- Dedicated functions (programs) for displaying/processing the data received from other systems.

The resulting situation is complex, as many systems have to communicate with many other systems. Therefore, an interface engine or *integration engine* is often used, which assumes the coordination and processing of communication between the various systems and also has the function of converting messages from a transmitting system into a format that can be processed by the receiving system.

On account of the complexity of the process of coupling heterogeneous systems in medicine and on account of the need for a solution of this kind in virtually all health care systems, development of the communication standard HL7 (health level seven) began in the 1980s. This communication standard includes a standardization of message types, i. e., a definition of syntax and semantics for the data records to be transmitted. If medical systems support this standard (both for import and export), they can then be coupled with one another with reduced effort and reduced costs. Figure 59.10 shows an example of the composition of a message type.

The current HL7 version is 3 [59.13]. It uses XML and has introduced the reference information model (RIM) and the clinical document architecture (CDA) [59.14]. The CDA Release 2.0 provides an exchange model for clinical documents (such as discharge summaries).

Because of these efforts it is conceivable that, against the background of international standards, the

ADT	ADT message	Chapter
MSH	Message header	2
EVN	Event type	3
PID	Patient identification	3
[PD1]	Additional demographics	3
[{NK1}]	Next of kin/associated parties	3
PV1	Patient visit	3
[PV2]	Patient visit – additional info.	3
[{DB1}]	Disability information	3
[{OBX}]	Observation/result	7
[{AL1}]	Allergy information	3
[{DG1}]	Diagnosis information	6
[DRG]	Diagnosis related group	6
[{PR1	Procedures	6
[{ROL}]	Role	12
}]		
[{GT1}]	Guarantor	6
[		
{IN1	Insurance	6
[IN2]	Insurance additional info.	6
[IN3]	Insurance add'l info – cert.	6
}		
]		
[ACC]	Accident information	6
[UB1]	Universal bill information	6
[UB2]	Universal bill 92 information	

Fig. 59.10 Example of the HL7 message type, ADT message (admission-discharge-transfer)

integration of medical application systems from different manufacturers will become increasingly easy, even if this cannot compensate for the fact that a central entity within the HIS is necessary for the management, and above all for the long-term archiving, of the electronic medical records.

59.6 Current Trends and Prospects

As the preceding illustrations have shown, hospital information systems are now comprehensive operational information systems for hospitals that support all aspects of the operational, tactical, and strategic dealings in hospital management. It is particularly important that the functionalities are adapted to a given degree and further developed in accordance with the changing general set-up of hospitals. This is accompanied by changes in the technical basis. We already have seen a development from 2-tier to 3-tier and n-tier client server implementations, and we see a development of the middleware layer with increasing use of SOA.

From the application perspective, we foresee changes in:

- Medical documentation
- Integration of medical technology
- Networking with external institutions
- Support for patient safety
- Decision support for doctors
- Integration of genetic data.

59.6.1 Medical Documentation

Medical documentation has become more fine-grained and process oriented, and it will become more and more paperless.

Order entry and result communication have been implemented in many hospitals, and this functionality will be developed further. The documentation of specialist departments has taken center stage as part of the drive towards a paperless hospital. This documentation for the specialist departments differs considerably between the different departments and must, therefore, be edited on the basis of the existing forms and converted into an electronic format. Interdisciplinary collaboration will be a focus, such as the interdisciplinary documentation of tumors. The implementation of fine-grained follow-up documentation is generally carried out by means of a tool – usually a forms generator – supplied by the manufacturer of the hospital information system, which enables the hospitals to implement specific documentation functions beyond the standard range of functions of the system supplied.

Standard terminologies are still a central challenge, and it is difficult to predict if more countries will introduce standards like SNOMED.

A relevant development towards paperless hospitals is electronic archiving. Paper-based archives will be replaced by electronic systems, comprising documents generated by computers or scanned paper records.

59.6.2 Integration of Medical Technology

The development of comprehensive paperless documentation encourages the demand for integration of information that is actually acquired by means of medical technology systems. Whereas, in the field of radiology for example, this development is largely complete as a result of the integration of modalities with PACS and radiology information systems, there are a large number of medical technology devices – such as ECG machines, EEG machines, pulmonary function testing devices, cardiac catheter measuring stations, and many more – which for the most part these days likewise record, process, and store information in digital form, but which are simply still not equipped to be interoperable such that they can communicate with other systems without any problems via standard interfaces. Considerable individual developments are still in part necessary here, but a convergence between medical technology and information technology has begun. With this integration, it is conceivable that the discontinuity between media that exists today and the distributed storage of patient medical information will increasingly be a thing of the past and that a paperless, integrated, electronic patient record, including data and documents from medical technology devices, will become available centrally in the hospital information system.

59.6.3 Health Information Networks

Hospital information systems have in the past largely been focused on assisting all aspects of the dimensions of support within a hospital which were mentioned at the beginning of this chapter. However, the ever more intensive integration of the care sectors and the accompanying need for effective cooperation and coordination between hospitals, referring doctors, medical care centers, nursing facilities, and out-patient care services also results in a considerable requirement for electronic networking between the individual information systems of these different care providers.

In many countries, regional and national health information networks have been built or are about to be built. International standards on the terminology level (SNOMED, ICD) and on the level of structure and communication (HL7, DICOM, IHE) are of relevance in these efforts [59.6, 15].

Many hospitals have, together with their HIS-vendors, tried and pursued methods of setting up information technology networks with other institutions. So-called hospital portals are used to give referring physicians the option to access, via a secure Internet connection, selected, specially provided medical information concerning the patients referred by them and to not only view this information – usually information in the form of documents, such as reports, results and discharge summaries – but also to download this information onto their local information system at the doctor's practice and to manually integrate it into their electronic index. This is done on the basis of a sophisticated access rights system, so that every referring physician only has access to the information relating to his patients and also only with the consent of every patient.

These portals have the disadvantage, however, that doctors in private practice who cooperate with multiple hospitals may have to use portals that operate differently. As a result of this development, it therefore seems advisable that automated communication of such information be implemented on the basis of transparent interoperability between hospital information systems and systems in doctors' practices. First specifications for electronic doctors' letters and projects for the exchange of such electronic doctors' letters have since been initiated. As part of two relatively large cooperative German project initiatives (eFA, electronic case record and EPA 2015, electronic patient record), current standards for technical and semantic interoperability between hospital information systems and other information systems of care providers are being developed with a physically or purely logically central patient record infrastructure. The aim is for all care providers to be able to upload any important documents generated as part of a patient's care and also detailed information, such as diagnoses, symptoms, and procedures, into a cross-institution electronic patient record such as this without any great difficulty, and thus to generate cross-institution transparency in the treatment process and the patients' specific situation, which will provide every member of staff involved in the patient's medical care with the important information in an effective and complete way. The informational self-determination of the patient and also the best conceivable security mechanisms must also be taken into consideration here.

For collaboration with the external systems, the hospital information systems require, on the one hand, an interoperability module that undertakes the technical communication with the external systems and health information network components and, on the other hand, require an extension of the internal function of integrated electronic mail inboxes and outboxes, by means of which users can access received documents and information and manually or automatically file them in the corresponding internal electronic patient records. Furthermore, it needs to be clear in the patient documentation which documents and information are of external origin or have been sent externally.

59.6.4 Decision Support for Health Care Professionals

Since the hospital information systems contain an increasing amount of clinical data, this clinical data provides the informational basis for decision support that has to be combined with knowledge, which may be represented in guidelines and pathways. A highly relevant example is the safety of drug therapy. If prescriptions and medication are documented in the HIS, then stored information can be used as the basis for checking interactions, and in the case of prescriptions contraindications can be checked and referred to on the basis of the documented diagnoses and other medical data. If the documentation is based on vocabularies and ontologies, advanced support can also be provided. It is then possible, for example, to specify possible treatment methods for certain diagnoses or to point towards the existence of clinical pathways or guidelines.

It has been shown that health information technology can reduce medical errors [59.8]. A review of medication-related clinical decision support in order entry systems has been provided in [59.16]. An example of a pragmatic approach to implementing best practices for clinical decision support systems can be found in [59.17].

Health information systems need, however, to be designed or implemented properly, as they may also induce medical errors [59.18]. We expect a growing focus on system design aspects [59.19].

59.6.5 Support for Patient Safety

Development in the health care system and in particular in hospitals has allowed productivity to increase more and more for doctors, too. There is often only a little time for communication about the patients when there is a change of shift, etc. More than ever, doctors are reliant on receiving optimum support in their patient-related collaboration through the hospital information system. On the whole, it is about maintaining and increasing patient safety within the framework of hospital treatments.

Patient safety means that a patient does not come to any harm during treatment and is not exposed to any potential health risks. Patient safety is taken very seriously worldwide. Reports on medical errors have occurred since 1991, leading to a better understanding of causes and the role of information technology in error reduction [59.8, 9, 20–22]. It can be assumed that, with the increasing penetration of IT into the medical areas of hospitals, the demand for patient safety to be supported by (pro)active, intelligent mechanisms in the HIS will also become ever more intense.

Risk management, together with incident reporting, will be of growing importance [59.23]. Various different facets and aspects can be presented here, such as:

- Structured risk assessment and presentation, and also constant and context-sensitive reference to existing risks for prescriptions and planning of procedures
- Documentation overview to give a brief patient-based overview of all the important and new information
- Support when performing procedures or when responding to unplanned incidents by displaying check lists
- Integrated messaging system, including active messaging using the HIS to send messages to the doctor on duty or the assigned doctor when new results arrive, if desired, depending on the type of results and with the ability to set different parameters depending on the patient
- Support with medication, right up to intelligent functions such as alternative medical treatment sites (AMTS), etc.
- Computer *integrated* radiology system (CIRS)
- Generation of clear transition reports.

59.6.6 Genomic Data

Health information technology plays an increasing role in translational medicine, which aims at transferring knowledge and techniques from basic research (from bench) to the development of new treatments (to bedside) [59.24]. One of the most significant developments will be the use of genetic and genomic data in clinical practice. This means that substantial security measures must be implemented and ethical considerations must be made in relation to the use of supporting information systems, because genetic data and dispositions concern not only the patient but also his/her relatives. The increasing use of genetic testing and the identification and use of biomarkers will open a completely new dimension for medical documentation, which requires special privacy protection and thus special protective mechanisms within hospital information systems.

59.7 Selection and Implementation of HIS

59.7.1 Preliminary Remarks

The implementation of information systems in the clinical areas of a hospital results in highly complex and sensitive operational sociotechnical systems, whose functionality only achieves the desired operational and strategic results if, once the system is implemented, all those involved find themselves in a working situation that is at the least consistent but preferably improved. It is, therefore, very important that, when preparing, selecting and implementing clinical information systems, the various design dimensions of operational information systems – such as the organizational structure, process organization, documentation, technical infrastructure, and user acceptance, etc. – are taken into consideration at an early stage.

The greater the extent to which an information system is implemented in human spheres of activity, the more all of these dimensions of design need to be taken into consideration. No procurement processes should, therefore, be entered into and no implementation of information systems should be carried out without precise knowledge of the given macroscopic and microscopic organization, without reflecting the aims of an information system in the aims and tasks of an organization and without any knowledge and consideration of users and the people affected.

The selection of a hospital information system must be conceived as a distinct project. Only in this way is it guaranteed that a reliable and safe decision will be reached. For this purpose, up to 10% of the total investment volume must be used to finance the selection process.

In order to be able to methodically come to a reliable purchase or development decision based on the particular requirements and the current state-of-the-art technology, the following phases must be gone through:

- Preparation phase (general and project-specific preliminary work)
- Project planning
- System analysis
- Drawing up an invitation to tender
- Selection
- Drafting a contract
- Acceptance and implementation
- Early operating phase.

These phases are necessary both for relatively large projects for the implementation of entire hospital information systems and also for smaller projects, e.g., the implementation of department information systems such as radiology or laboratory information systems.

Important milestones in the narrower sense are the completion of the system analysis, sending out the invitation to tender, conclusion of the contract, and putting the system into operation.

Those activities and circumstances whose implementation are seen as critical success factors in the overall process – and omission of which jeopardizes the success of DP projects or results in less than perfect solutions – are listed below in the form of check lists.

59.7.2 Project Phases and Critical Factors

Preparation Phase

A DP project generally begins with initiation of this project: recognized weak points should be remedied, statutory changes necessitate the use of data processing, resources for increasing effectiveness should be used, an outdated system can no longer continue to be operated, etc.

The aims drawn up here are often too short-term if the selection and use of a system is only orientated towards this initial motive, and the chance to remodel the operational organization with appropriate information technology is squandered! In contrast, the use of information technology should be geared towards the operational aims – on both an organization-wide and departmental level.

It is generally assumed that guidelines for the use and procurement of DP processes already exist within the hospital. If this is not the case, these guidelines should be defined immediately. In addition to technical standards and stipulations, they can also include specifications with respect to operating and ergonomics.

Important activities in this phase are, therefore:

- If necessary, define/amend business objectives
- Fix the objectives for the use of IT (derived from the business objectives)
- If necessary, define/amend DP guidelines
- If necessary, use the DP guidelines as a basis for general standards and stipulations
- Broadly define the project idea and project objective.

Whilst, with the exception of the last activity, these involve general guidelines and standards that ideally are already defined, the last stage that is necessary is to at least describe the specifically planned project and the initial reasons behind it.

Project Planning

Project planning involves laying all the necessary foundations in order to take on the operational project work. Orientated towards the basic set-up established by the general preliminary work, *at the end of this phase clear decision-making, reporting, and documentation structures must be defined, as well as the project responsibilities and authorities.*

Irrespective of the size of the project, there should be a separation between the strategic/tactical planning/monitoring (by a project steering committee) and the operational implementation (by a project working group). As a result, an appropriate compromise can be reached between decision-makers, representatives of the various different people affected, and the individual user groups. The project working group should consist of selected representatives of the individual occupational groups, and the amount of time that they spend working in this group pro-rata should be clarified.

The following activities should be considered during project planning:

- Forming and implementing the project steering committee with:
  - Inclusion of the management
  - Inclusion of management functions from the organizational units involved
  - Involvement of the staff committee.
- Forming and implementing the project working group with:
  - Involvement of selected representatives from the project steering committee
  - Involvement of representatives of all subsequent user groups
  - Involvement of the data processing and organization department
  - Involvement of the data protection representative.
- Definition of the project objectives:
  - Strategic objectives by the project steering committee
  - Operational objectives by the project working group.

Once these two essential groups have been installed and the objectives of the project are also set, the project-based authorities and responsibilities must be defined and all those involved and affected must be informed. It is, therefore, necessary to take into account the:

- Definition of the project-based authorities and responsibilities
- Clarification of the basic setup (staff, rooms, equipment, etc.)
- Clarification of the financing framework
- Clarification of internal resources
- Clarification of external resources/inviting offers
- Initialization and setting up of project documentation
- Rough project scheduling
- Provision to those affected with initial information and information of the project-based authorities and responsibilities
- Drawing up/update of the operational DP framework.

Once this phase has been completed, all of the decisions necessary to implement the project should then be made. It is particularly important here to provide all those involved with initial information and to keep them constantly informed. If possible, the new media should itself be used for this purpose (email with distribution lists, Intranet with corresponding project pages, etc.).

System Analysis

The selection and implementation of information systems require a high degree of operational transparency. Ultimately, for all projects for the implementation of clinical DP systems, this involves intensive migration projects that are focused on complex and sophisticated medical documentation. One of the most fundamental critical success factors for DP projects, particularly in medicine, is, therefore, the performance of a system analysis that is sufficiently detailed in terms of the knowledge and objectives of interest. It is only the results of this that generally form the basis for a definition of requirements and corresponding organizational considerations. In the past, however, it has become apparent that many hospitals at most economize on this phase – essentially on the ground work of a project – and often even entirely omit a system analysis and a definition of requirements based on the particular needs!

When carrying out the system analysis, various approaches can be used, depending on the knowledge of interest and the project objectives:

- Horizontal procedure –
  all of the affected operational units are examined with respect to a certain aspect (e.g. forms)
- Vertical procedure –
  some of the operational units of interest/all of the operational units affected are examined in detail as a whole
- Process-orientated procedure –
  all aspects are surveyed along defined operational (main) processes – also across organizational units.

The following activities should be performed:

- Factorial analysis
- Equipment analysis (staff, DP systems and data storage systems used, hardware and operating systems, other equipment)
- Structural analysis (rooms, network infrastructure, etc.)
- Quantity structures (numbers of cases, performance figures, etc.)
- Communication analysis
- Documentation analysis
- Organizational analysis
- Accounting analysis
- Weak-point analysis.

On account of the sophisticated and complex medical documentation and the organization, which is based heavily on the division of labor, the focus and the main costs are for the most part on:

- Documentation analysis:
  - Medical/accounting- and financial-controlling-based/legal
- Organizational analysis, including analysis of main processes:
  - Admission, discharge, transfer, emergency admission
  - Communication of orders (organizational means?, coordination?)
  - Communication of results
  - Report writing, including results
  - Internal interdisciplinary procedures (medical – administrative).

With the exception of interdepartmental process analyses, the analyses listed are generally to be performed with reference to specialist departments or based on diagnostic departments to be considered in isolation.

A weak-point analysis is often omitted for reasons of time or finance, or for political reasons. It is precisely this weak-point analysis that can be used with particular advantage to identify the possible potentials for improvement and increased effectiveness by means of a DP solution, and also to identify the requirements for a system that is to be procured. However, weak-point analysis means openly and honestly compiling, identifying, and communicating these weak points – which essentially means creating an organization quality circle and is, therefore, often bypassed.

At the end of the system analysis there is a correspondingly structured analysis documentation, which is used as the basis for all subsequent phases.

Drawing up the Functional Specifications

This phase can be split into the subphases:

- Creating the target concept
- Drawing up the functional definition of requirements
- Compiling the functional specifications.

The *target concept* (Fig. 59.11) serves the purpose of developing and defining the organizational, technical, and DP-procedural target on the basis of the results of the system analysis. This should as far as possible be done taking into consideration the design dimensions of operational information systems. This must in particular involve developing a clear projection of the final state of the project (that is to say which operational actions and process chains should be supported and which information objects are to be managed, etc.), orientated towards the operational objectives and the weak points found.

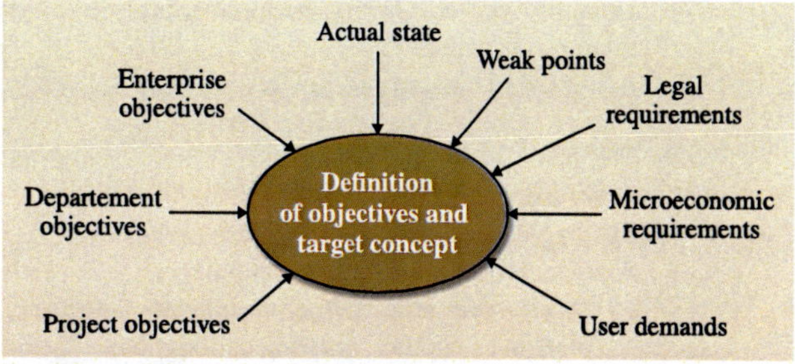

Fig. 59.11 Influencing factors on and sources of the EDP target concept

Furthermore, at this point in time it is possible, taking account of the various information sources available, both to check the strategic objectives and also to carry out a sophisticated derivation of the operation objectives of using the DP system.

The development of a target concept taking account of the factors cited above can ideally be done by moderated workshops. These workshops can be conducted both at a management level and at an operational level, and the moderator has the results of the system analysis as a background and can thus control the interactive development. The result is a target concept that has been jointly developed and accepted by all those involved, which contributes considerably towards the acceptance of the further project progression.

This target concept is then expanded further to give a *definition of requirements*, which contains all of the functional, operational, and organizational requirements for the system and is a central part of the functional specifications. The formalization and operationalization of the requirements and the expansion to include general principles for information systems are the focus of this. All of the requirements that are important for the decision and the sales agreement should be included in a structured and unambiguously worded form. The necessary additional information for sending replies must accordingly be enclosed.

The general section of the invitation to tender (introduction, terms, framework conditions) should include:

- Clear guidelines for filling out and responding to requirements

- Unambiguous eligibility requirements (e.g. minimum requirements)
- A stipulated closing date
- Names of contact people for queries
- The offer of a time-limited discussion (or alternatively a consultation)
- Envisaged time scale for the entire process
- A presentation of important parameters:
  - *Qualitative*: e.g. DP environment, interfaces, physical structure, existing cabling, etc.
  - *Quantitative*: quantity structures for numbers of cases, frequencies of examinations, number of users, etc.

Important individual activities in this phase are:

- Expand the definition of the objectives, including the business objectives and DP guidelines
- Draw up a target concept, e.g. based on questions such as:
  - Which recognized weak points should be resolved?
  - Where are the potentials to increase effectiveness?
  - Where and how can the quality of the medical work be improved by providing support?
- Define a step-by-step plan for achieving the objectives
- Set down decisive framework/boundary conditions
- Conduct a feasibility and financial viability analysis
- Form a representative definition and selection group
- Conduct market research
- Draw up a functional definition of requirements:
  - With broad inclusion of the users
  - Taking account of standards regarding structure, DP standards, etc.
  - Taking account of what is feasible and financially viable
  - Taking account of the necessary variability of the solution as a result of the target concept and definition of requirements
  - Taking account of the definition of the possible stages and timetables.
- Update the definition of the requirements for the functional specifications
- Draw up the invitation to tender on the basis of the functional specification
- Conduct the formal invitation to tender.

It is important here to ensure that the functional specifications are not too approximate but on the other hand that they are also not set out in too much detail. The requirement, for example, for specific field lengths appears to be of little help, if not in fact counterproductive.

The catalog of requirements section is further subdivided according to various functional, technical, and strategical viewpoints. Figure 59.12 shows an example of a categorization system to which further details can be added.

The biggest weak point is often a catalog of requirements with questions/requirements that have not been formulated precisely enough or that in contrast presents too many opportunities for responses.

The following possible responses are in essence conceivable:

- Free text – Questions can be answered with comprehensive explanations (example: describe the procedure for admitting patients.). Questions of this type are strictly advised against! A comparison of the tenders and the responses is possible only with a great deal of effort, if at all, and is always dependent on interpretation.
- Although categories and explanations force a definitive answer (e.g. Is health insurance billing provided? Yes – No – Explanation: ...), the categorized answer can often be qualified, hidden in the explanation. These types of questions become critical points in disputes within the scope of performance of a contract, if not before.
- Categories – A response is only possible on the basis of defined categories (provided – provided in part – not provided or yes/no, etc.). As a result, the assessment of the tenders and their comparison with the formal document can always be followed (no gray areas of interpretation). Moreover, it is sometimes useful and also acceptable to allow a measured value as an answer (e.g. maximum number of machines that can be connected, minimum storage space, frequency of mask changes for certain work, etc.).

Selection

Once the tenders have been received, they must be evaluated. The *evaluation of tenders* concerns both the formal part of the tender and also the parameters stated beyond the formal part of the tender. The formal part of the requirement should be weighted before the tenders are received. For each level in the requirements hierarchy, 100 points are given, for example (Table 59.1). Based on the responses by the tenderers, a total number

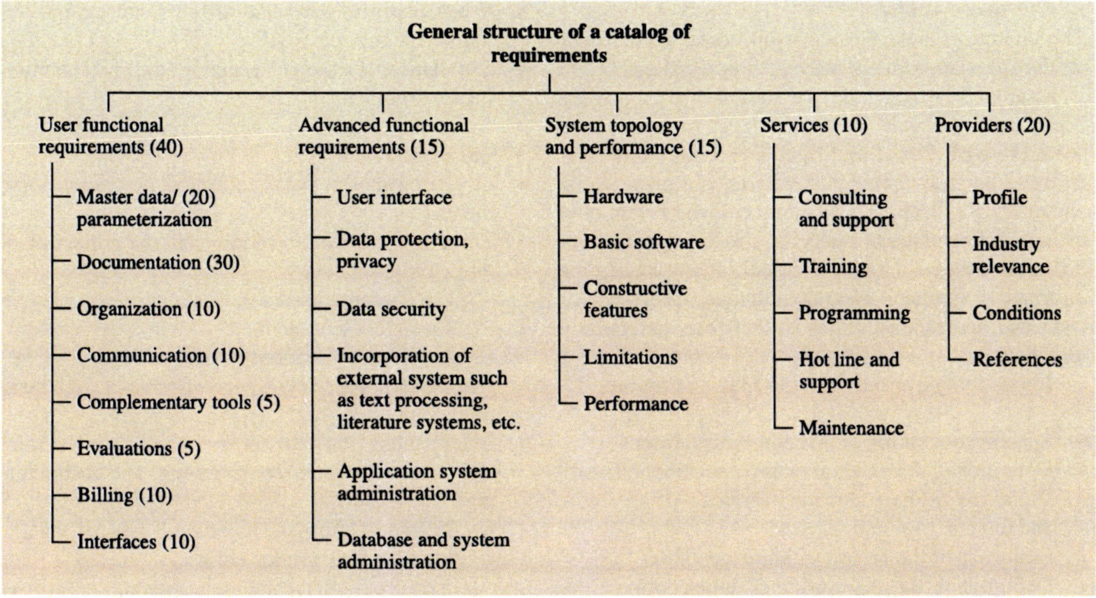

Fig. 59.12 Example of a categorization system for the catalogue of requirements with some example weightings

of points can then be calculated for the whole requirement part, which is a measure of the fulfilment of requirements. Reference should be made to the common process of cost–benefit analysis.

An important step is, after receipt of the tenders, first of all to filter these tenders so that the incomplete tenders or those that do not meet the exclusion criteria are not included in the further – more labor-intensive – selection process.

It is also necessary to go through the main stages:

- Evaluation according to the specified weightings
- Establishment of the total costs (one-off costs/running costs)
- Establishment of the costs for each point
- Establishment of the ranking (according to the absolute number of points/according to the costs per point).

The two rankings can be truly illuminating. Whereas the first ranking indicates who is offering the most in terms of function, the second indicates who will provide the most cost-effective product – irrespective of the degree to which it achieves the objectives. Depending on whether the two results correspond or differ to a vast degree, a strategic evaluation must be made with regard to available functionality and available means.

At the end of the selection process, there should be two to a maximum of three tenderers remaining, whose solutions then undergo *on-site reviews* (i. e. use in live operation) or who beforehand give in-house presentations, with all members of the project group participating. It is also helpful, for presentations at the tenderer's premises (i. e. not for on-site reviews), to specify particular case studies or cases and to allow these to be used as part of the presentation.

Important individual activities in this phase are:

- Continuation and intensification of the informal market research
- Definition of typical operational business transactions, in other words chains of medical treatment as case studies for appraisal of the solutions
- Evaluation of tenders, if necessary involving:
  - Querying, defining more precisely
  - Formal assessment and prioritization
  - Appraisal of the solutions
  - Visiting reference customers
- Informal research/validation
- Decision-making process and decision
- Putting down in writing the necessary product-related additions/amendments:
  - Clarifying ambiguities with existing modules
  - Defining the customizations/additions necessary according to the functional specifications.

Drafting a Contract

The contract should be drawn up based on the invitation to tender with the help of a professional. Faith in assurances – which in the past has all too often been tested – should be contractually translated into services owed. Consulting legal experts in the field of software law can clarify this early on and prevent disappointments. Drafting of a contract involves not only the acquisition of usage rights but also the agreement of necessary services in order to install and implement the information system. A clear definition of the division of tasks and the responsibilities is, therefore, particularly important.

Important individual activities in this phase are:

- Specification/fixing of amendments/additions
- Definition of the external range of services offered
- Definition of the contractual services: usage rights and maintenance and services
- Definition of the project of implementation
- Definition of the segregation of duties between the supplier/customer
- Definition of the responsibilities.

Acceptance and Implementation

The acceptance phase involves handover of the entire product to the customer, who then checks the product to determine whether it performs the contractually guaranteed services by means of suitable functional and quantity-orientated acceptance tests. A whole series of preliminary work is necessary in the hospital, which amounts to preparing the system for productive operation. In this respect, the work necessary for implementation of the system is necessary even before the acceptance phase, and acceptance can often only take place following a test phase in live operation because of the impracticality of testing the real operating situation.

The implementation itself can be divided into arrangement, system preparation, system adaptation and implementation, training, preparation for start-up, and start-up.

Important individual activities are as follows:

- Setting the stage technically in good time:
  - Locations and space requirements for terminals clarified?
  - Cabling sorted out from a technical and time-planning perspective?
  - Installation site for the server settled?
- Procurement of hardware, if this is not an integral part of the service owed
- Installation of a project-planning (testing) and training environment
- Definition of a training and support plan (e.g. multiplier principle)
- Gathering master data, collecting necessary documents
- Training for system operation and the project-planning group for:
  - The operating system
  - The database system
  - Application functions
- Acquiring master data (range of services, organization, employees, etc.)
- If applicable, data transfer from the existing system into a test environment, cleansing, and restructuring
  - Parametrization of functional aspects (workflow, dynamic masks, etc.)
  - Interface implementation/testing
  - Training plan for up-to-date training of all employees
  - Informing all the employees affected
  - Organization of system operation
  - Organization of first-/second-level support, fault-clearing service, call-out service
  - Installation of the entire application system (live system)
  - Quantity and load tests
  - Interface implementation/integration test
  - Setting up the production environment
  - Transfer of historic data into the production environment
- Checking all the prerequisites:
  - Have all of the necessary interfaces been implemented?
  - Are all of the interfaces functioning?
  - Has all of the master data been gathered/historic data been imported, etc.?
- Distribute user documentation.

Early Operating Phase

In the early operating phase, it is particularly important not to leave the user alone and leave the operation to run. The intended effects must now be checked with an orientation towards the target concept and, if necessary, further operational or technical system optimizations must be performed. Furthermore, user acceptance must be continuously checked in the first few weeks in order to draw conclusions from this concerning any necessary organizational changes or the need for further training:

- Check/optimize organization
- Performance evaluation and technical system optimization
- Check correct usage
- Evaluation of user acceptance
- If necessary, conduct further training
- Checking the system functionality (functional specifications versus current status), if necessary further specification and further requirements for subsequent improvements.

Further Influencing Factors that Are not Phase-Based

- Continually informing and identifying the management
- Continually involving and informing the employees
- Solutions provider/manufacturer
- Application system:
  - User interface/suitability of the tasks
  - Level of maintenance required
  - Flexibility/adaptability
  - User documentation
  - System documentation
  - Design of the hardware
- Plans by the manufacturer.

59.8 Conclusion

Hospital information systems are a crucial success factor for the successful management of hospitals. The potential benefits are great, but these benefits only become accessible when the information system is used across the board. With respect to the technical architecture, the choice is between using a holistic overall system or a heterogeneous overall system that is composed of several systems.

The implementation of information systems in hospitals requires a much more cautious approach than is the case in many other sectors. This is firstly because the core tasks – namely the direct treatment and supervision of sick people, the care given to them, catering to their needs correctly, the personal contact and connection – must not be disrupted by the use of a DP system. The system as a tool must remain in the background. This is also to remove the fears of staff and patients in this sensitive environment and to prevent these fears arising. This is a serious difference from the use of data processing in most other sectors, where core tasks (e.g. dealing with accounting records, management of materials, issuing materials, etc.) are modified and changed directly by using DP systems. However, adequate involvement of the users affected in the entire process of system selection is, therefore, of particular importance.

Critical success factors for the implementation of hospital information systems are varied, and all of the activities of the project phases described above can in this respect certainly be regarded as critical success factors.

Essentially, however, the following questions must be regarded as particularly critical:

- Is the project objective sufficiently clearly defined (not DP for the sake of DP)?
- Have the framework conditions (financial/organizational/internal political) been settled on?
- Has a clear project organization been determined?
- Have all of the responsibilities and consequences been sufficiently defined, so that the objective can also be achieved?
- Is the management behind the project?
- Is the project adequately staffed?
- Is there a clear time schedule for the project?
- Have the employees been provided with sufficient information at an early stage?
- Does a system analysis form the basis for a target concept, which itself in turn serves as the basis for the definition of requirements?
- Is sufficient employee involvement guaranteed in the definition of requirements and the selection process?
- Are the functional specifications laid out in a sufficiently refined and realistic manner?
- Is the selection process transparent?
- Has the agreement been sufficiently worked out, and are there provisions in the event of (partial) nonperformance of guaranteed services?
- Is there a clear and realistic implementation plan?
- Has sufficient training been provided?
- Has the implementation of the system in everyday operation been thought through sufficiently?
- Have all the interfaces necessary for operation been sufficiently tested?
- Is there a suitable strategy and tools for user support?

- Will checks be continuously performed to ensure correct use following implementation?
- Does the solution comply with the standards with regard to the criteria of ergonomics?

If these aspects have been given sufficient consideration, a hospital information system, which has thus been successfully implemented, can make a substantial contribution to the business success of a hospital.

References

59.1 D.L. Sackett, W.M. Rosenberg, J.A. Gray, R.B. Haynes, W.S. Richardson: Evidence based medicine: What it is and what it isn't, Br. Med. J. **312**(7023), 71–72 (1996)

59.2 K.A. Kuhn, D.A. Giuse: From hospital information systems to health information systems. Problems, challenges, perspectives, Methods Inf. Med. **40**(4), 275–287 (2001)

59.3 The Open Group: *TOGAF 9 The Open Group Architecture Framework (TOGAF)*, 9th edn. (Van Haren, Zaltbommel 2008)

59.4 J.A. Zachman: A framework for information systems architecture, IBM Syst. J. **26**(3), 276–292 (1987)

59.5 H. Krcmar: *Informationsmanagement*, 4th edn. (Springer, Berlin Heidelberg 2005)

59.6 E.H. Shortliffe, J.J. Cimino (Eds.): *Biomedical Infor matics*, 3rd edn. (Springer, Berlin Heidelberg 2006)

59.7 J. van Bemmel, M.A. Musen (Eds.): *Handbook of Medical Informatics*, 2nd edn. (Springer, Berlin Heidelberg 2002)

59.8 D.W. Bates, A.A. Gawande: Improving safety with information technology, N. Engl. J. Med. **348**, 2526–2534 (2003)

59.9 Committee on Quality of Health Care in America: *Crossing the Quality Chasm: A New Health System for the 21st Century* (National Academies Press, Washington 2001)

59.10 P.D. Clayton, J.J. Nobel: Costs and cost justification for integrated information systems in medicine. In: *Hospital Information Systems*, ed. by A.R. Bakker, C.T. Ehlers, J.R. Bryant, W.E. Hammond (North Holland, Amsterdam 1992) pp. 133–140

59.11 C.J. McDonald, J.M. Overhage, W.M. Tierney, P.R. Dexter, D.K. Martin, J.G. Suico, A. Zafar, G. Schadow, L. Blevins, T. Glazener: The Regenstrief medical record system: A quarter century experience, Int. J. Med. Inform. **54**(3), 225–253 (1999)

59.12 J.M. Teich, J.P. Glaser, R.F. Beckley, M. Aranow, D.W. Bates, G.J. Kuperman, M.E. Ward, C.D. Spurr: The Brigham integrated computing system (BICS): Advanced clinical systems in an academic hospital environment, Int. J. Med. Inform. **54**(3), 197–208 (1999)

59.13 HL7 Int.: *HL7 Standards* (Health Level Seven International, Ann Arbor 2010), available at http://www.hl7.org/implement/standards/, last access 11.13.2010

59.14 R.H. Dolin, L. Alschuler, C. Beebe, P.V. Biron, S.L. Boyer, D. Essin, E. Kimber, T. Lincoln, J.E. Mattison: The HL7 clinical document architecture, JAMIA **8**(6), 552–569 (2001)

59.15 R. Lenz, M. Beyer, K.A. Kuhn: Semantic integration in healthcare networks, Int. J. Med. Inform. **76**(2–3), 201–207 (2007)

59.16 G.J. Kuperman, A. Bobb, T.H. Payne, A.J. Avery, T.K. Gandhi, G. Burns, D.C. Classen, D.W. Bates: Medication-related clinical decision support in computerized provider order entry systems: A review, JAMIA **14**(1), 29–40 (2007)

59.17 P.A. Gross, D.W. Bates: A pragmatic approach to implementing best practices for clinical decision support systems in computerized provider order entry systems, JAMIA **14**(1), 25–28 (2007)

59.18 E. Borycki, E. Keay: Methods to assess the safety of health information systems, Healthc. Q. **13**, 47–52 (2010)

59.19 D.F. Sittig, H. Singh: A new sociotechnical model for studying health information technology in complex adaptive healthcare systems, Qual. Saf. Health Care **19**(Suppl.), 368–374 (2010)

59.20 L.L. Leape: A systems analysis approach to medical error, J. Eval. Clin. Pract. **3**(3), 213–322 (1997)

59.21 L.T. Kohn, J.M. Corrigan, M.S. Donaldson (Eds.): *To Err Is Human: Building a Safer Health System* (National Academy Press, Washington 2000), Institute of Medicine, Committee on Quality of Health Care in America

59.22 WHO: *Patient Safety* (WHO, Geneva 2009), available at http://whqlibdoc.who.int/hq/2009/WHO_IER_PSP_2009.10_eng.pdf, last access 11.07.2010

59.23 O. Levtzion-Korach, A. Frankel, H. Alcalai, C. Keohane, J. Orav, E. Graydon-Baker, J. Barnes, K. Gordon, A.L. Puopulo, E.I. Tomov, L. Sato, D.W. Bates: Integrating incident data from five reporting systems to assess patient safety: Making sense of the elephant, Jt. Comm. J. Qual. Patient Saf. **36**(9), 402–410 (2010)

59.24 I.N. Sarkar: Biomedical informatics and translational medicine, J. Transl. Med. **8**, 22 (2010)

60. Telemedicine in Germany

Rüdiger Klar, Ernst Pelikan

In Germany there are now many successful special applications of telemedicine. However, these projects are mostly not extendable to the whole nation because of the highly complex German healthcare system, limited funding, heterogeneous information technology (IT) standards in ambulatory and hospital care, the insufficient official electronic health card, different data protection and privacy regulations of federal and state governments, doubts of physicians and patients, as well as unequal costs and benefits for the different actors in telemedicine. The solutions of these problems are: better legal regulations for seamless interaction of ambulatory and hospital care, improved IT standards and support of the interoperability of telemedical and other healthcare IT, adaption of healthcare organization, workflow, and reimbursement for telemedicine services, and better information and education of all actors about the necessity and limits of telemedicine. There are no doubts about the benefits of telemedicine, but to achieve these advantages for the whole healthcare system in Germany will require much work, time, and good will.

60.1 **The Peculiar Features of German Telemedicine** 1120
60.1.1 Supporting Individual Projects 1120
60.1.2 Legal Measures to Overcome the Disadvantages of the Fragmented Healthcare System 1121
60.1.3 Electronic Health Cards, Health Professional Cards, and Official Health Telematics 1121
60.1.4 Medical Confidentiality, Data Protection, and Personal Rights 1122
60.1.5 Technical Standards 1123
60.1.6 Acceptance and Benefits for Various Players in Telemedicine 1124
60.2 **Consequences of the Peculiarities of the German System for Telemedicine** .. 1125
References 1126

Telemedicine is understood as meaning the use of telecommunications and informatics for medical applications. It is intended to improve the quality, economic efficiency, and transparency of medical care. Telemedicine uses the instruments of telematics to support communication and interaction between doctor and patient and between doctors within the framework of medical care over physical distances. Countless individual telemedical projects with often impressive solutions are increasingly being described around the world, for example ten times a year in the e-journal *Telemedicine and e-Health* [60.1] or for Germany in the annual anthology *Telemedizinführer Deutschland* (*Telemedicine Guide, Germany*) [60.2]. Important telemedical concepts and solutions are being developed by various scientific or practice-orientated professional societies and institutions or by industry, such as by the Deutsche Gesellschaft für Medizinische Informatik, Biometrie und Epidemiologie (the German Society for Medical Informatics, Biometry, and Epidemiology) [60.3], the Deutsche Gesellschaft für Gesundheitstelematik (the German Society for Health Telematics) [60.4] or the Deutsche Gesellschaft für Telemedizin (the German Society for Telemedicine) [60.5], as well as by many IT firms collectively in the Bundesverband Gesundheits-IT (the Association of IT Manufacturers in Health Care) [60.6] and by various institutions such as the Zentrum für Telematik im Gesundheitswesen (the Center for Telematics in Healthcare) [60.7], the Gesellschaft für Telematikanwendungen der Gesundheitskarte (gematik – the Society for Telematics Applications of Health Cards) [60.8], and the Deutsche Institut für Medizinische Dokumentation und Information (DIMDI – the German Society for Medical Documentation and Information) [60.9] (particular mention should be made here of its endeavors to achieve semantic interoperability). European institutions are also involved, such as the European Medicine Agency (EMEA) with its functions for the regulatory communication of data for pharmaceuticals [60.10]. Attempts are also being made to evaluate the benefits of telemedicine for patients and citizens, for the doctors involved, and for other staff in the healthcare system.

Even though the qualitative benefits are sometimes difficult to measure, they are still the focus of the objectives of telemedicine. In the case of common disorders such as diabetes mellitus (German Electromechanical Society (VDE): "Telemedicine saves millions in the treatment of diabetes") [60.11] or cardiac insufficiency [60.12, 13], particularly high goals are set but are still barely reached. However, the benefits of telemedicine are already clearly discernible for a great number of other, less prevalent diseases.

The quantitative benefits of telemedicine are likewise often cited and are said, for example, for the health system in the USA in the case of electronic prescriptions, to be worth an incredible US $ 27 billion per year and to avoid an additional 130 000 serious medication errors annually [60.14]. It is typical of such estimations of success, however, that the outlay for the electronic prescription system in the USA and the problems in terms of its acceptance are not mentioned. However, in Germany, at least, there are also thoroughly critical portrayals of the electronic prescription system which discuss these problems [60.15].

In the case of the many individual telemedical projects, it is striking that they are generally – at least in Germany – isolated from one another, and remain projects rather than products. Large-scale projects for the healthcare system are necessary to facilitate interaction between a telemedicine project and the conventional information systems as well as between different telemedicine projects. Discrepancies and system differences between a wide range of different forms of documentation must be avoided, harmonized interoperability (technical, content-related, organizational, and process-orientated cooperation) between various communication and storage technologies is necessary, the operational procedures in medical and administrative healthcare processes must be taken into consideration, and, above all, the boundaries in particular between ambulant and stationary care and the other sectors of the healthcare system must be overcome.

The introduction of the electronic health card (e-HC) began in 2006, and the associated legislation on health telematics may make an essential contribution to solving these problems for telemedicine. However, the advances made in this regard are considerably behind the original planning, and for outsiders, decisions which are barely comprehensible make this project more difficult. As a result, *increasing fatigue* is generally being observed, even among the leading protagonists of telemedicine [60.16].

The aim of this chapter is therefore to provide a critical account of the advantages of telemedicine for Germany and of its current status, but also of its limitations. Experiences with telemedicine projects at the University Medical Center in Freiburg will also be briefly discussed to illustrate our assessments by means of examples. Greater detail on the incorporation of telemedicine into health telematics in Germany as a whole can be found elsewhere [60.17].

60.1 The Peculiar Features of German Telemedicine

German telemedicine differs considerably in some aspects from telemedicine in other countries and has different – unfortunately mostly less sufficient – characteristics, at least where cross-sector projects are concerned, such as the transfer of diagnosis codes, for example, from the hospital to the Kassenärztliche Vereinigung (KV; the local state association of statutory health insurance physicians)/health insurance company, sometimes with and sometimes without additional identifiers and naturally in a different syntax. The intention here is therefore to present the most important peculiarities of German telemedicine and its parameters so that the reasons for these differences also become clear.

60.1.1 Supporting Individual Projects

The introduction of new technologies into the healthcare system begins, as is completely sensible and practical, with small straightforward pilot projects. Even in the preliminary stages of a German telemedicine project back in the 1980s, for example, a radiological picture archiving and communication system (PACS) [60.18] within a hospital information system (HIS) [60.19] was successfully tested. A lot of experience could therefore be gained with the technology and IT organization, standards of image compression and networking, and transmission of images and text, etc., and this experience is also extremely useful for expansion to a proper telemedicine system, in this case teleradiology, which extends beyond the individual care unit. Projects such as this have been and still continue to be relatively generously supported, and the specialist literature [60.2] every year describes dozens of such usually impressive projects. These projects regularly suffer from two basic problems, however.

1. They are usually only suitable for small and straightforward special applications, since for mass application in routine care they lack easy-to-handle interfaces to existing IT systems for data acquisition and data transfer. It is only in the case of special medical tasks that the additional work of redundant data processing and data control and the additional procedures are tolerated. In countries with a largely uniform IT structure for the healthcare system (as in the UK), this problem can be solved more easily, as the range of different interfaces is considerably smaller.
2. The individual projects in telemedicine rarely outgrow the care unit. Unlike the more state-controlled or centrally organized forms of healthcare provision in countries such as The Netherlands, Denmark, Sweden or Norway, the German healthcare system is fragmented, highly complex, and constructed with various heterogeneous areas of responsibility. For Germany's telemedicine system, this means that the great successes of such individual projects cannot easily be extended to the entire healthcare system. A quite considerable level of additional support is necessary, especially to expand beyond boundaries between different sectors and to integrate individual projects.

60.1.2 Legal Measures to Overcome the Disadvantages of the Fragmented Healthcare System

Even if such generous support for the integration of individual and isolated projects in telemedicine were to be awarded, this alone cannot overcome the barriers between the different sectors of the healthcare system in Germany. In order to do this, statutory and other legal activities are necessary to facilitate and improve telematic communication in the healthcare system, and in the interests of the patient, good will is required from all participants to contribute with commitment and patience to the common goal of improved and more efficient care via telemedicine.

The first important steps have indeed already been made and must be continued: In order to overcome the biggest barriers in the healthcare system, namely those between the ambulant and stationary care sectors, in recent years medical care centers and integrated care have been made possible with the GKV-Modernisierungsgesetz (GKV-GMG; statutory health insurance modernization act), the GKV-Wettbewerbsstärkungsgesetz (GKV-WSG; statutory health insurance competition strengthening act), and the Vertragsarztrechtsänderungsgesetz (VÄndG; amendment to the law governing the activities of physicians in free practice) [60.20]. The medical services of doctors, nursing staff, and rehabilitation centers can now be provided and invoiced across the different sectors, centralized at specialist facilities. This makes it easier to use telemedical services across the different sectors. Initial successes show that, despite some problems and technical deficiencies, this process has been begun correctly.

Another peculiar feature in Germany is the complex and unclear cooperation between the self-administration system in the healthcare system and the governments at national level and federal state level. The intention is often for the self-administration system initially to develop collaborative plans with its – admittedly very heterogeneous – associations of the medical profession, health insurance funds, hospitals, pharmacies, rehabilitation centers, etc., in compliance with the law, but if this is not achieved within the prescribed time limit, the governmental departments must in the short term step in by means of substitute action. As a result, a great deal of time is lost, jobs are rushed leading to errors, and harmonious cooperation suffers.

60.1.3 Electronic Health Cards, Health Professional Cards, and Official Health Telematics

Applications of telemedicine focus on medical care, that is to say on the support of diagnosis and treatment over physical distances. In this respect, telemedicine should be seen as a subarea of health telematics. The planned telematics infrastructure will provide basic services which promote the interoperability of telemedical processes. From this point of view it represents a harmonizing infrastructure which could blossom to become an integrating platform for further e-health applications. The official electronic health card (e-HC), the electronic health professional card, and the associated regulations for health telematics therefore play a crucial role in the peculiarities of the German healthcare system with respect to the legal framework conditions for telemedicine [60.8]. This involves one of the largest IT projects in the world, and cost estimate are in the region of several billions of Euros (by comparison, for the USA the costs of introducing an e-HC are estimated at US $ 150 billion); 120 000 doctors in private practices, 65 000 dentists, 21 000 pharmacists, 2200 hospitals, and many other employees in the healthcare system are involved, as are naturally all of the roughly 81 million citizens with statutory health insurance and private insurance. The laws relating to this, which were passed in 2003, provided for the introduction of the e-HC as a chip card throughout Germany by 2006, which was of course impossible to achieve. As one of the first applications of telemedicine, the e-HC is intended to implement the electronic transmission of medical referrals and, in particular, prescriptions. These electronic prescriptions are intended to improve drug safety for patients (monitoring of indications and contraindications, drug interactions, questions concerning dosage, compliance, etc.), and the administration of medication should be more efficient for all those involved (doctors, doctors' assistants, pharmacists, accounting staff, etc.). Further telemedical uses of the e-HC are in electronic patient records. Significant problems and long delays in virtually all aspects of health telematics and the e-HC have by now unfortunately arisen in Germany, and it appears that an *intolerable story of the health card in Germany* [60.15] is developing. Modest approaches – such as that of Austria, for example – appear to be more successful [60.21]. Unfortunately, the basic concept for the German e-HC has also not been synchronized in advance with neighboring EU countries or brought into agreement with EU standards [60.21, 22] and also the experiences of other countries, for example, in relation to error-prone storage of the electronic prescriptions on the e-HC (one of the *blind alleys* of patients according to *Waegemann*) [60.23] instead of on servers in the health telematics network.

It must, however, be recognized that the e-HC for the first time meets a criterion that is crucial for telemedicine, namely that of unambiguous patient identification across institutions, including the possibility for authentication and signature. This makes it possible to combine on a patient-orientated basis all of a patient's medically relevant data which accumulates at different times and in different places as part of the process of providing care and to make this data accessible to authorized individuals very quickly and in an adequately edited form. Incidentally, in the USA the absence of a *universal patient identifier* is regarded by an internationally renowned clinical journal as one of the five most important hurdles to introducing an electronic patient record and thus to telemedicine [60.24]. Without the e-HC, even if it has to be altered and expanded in some technical and functional aspects, it is not possible to successfully develop a comprehensive telemedicine system in Germany.

60.1.4 Medical Confidentiality, Data Protection, and Personal Rights

The observance of legal regulations and ethical principles relating to confidentiality in medicine, to the prevention of misuse of personal data, and in general to safeguarding of informational self-determination are essential concerns of telemedicine which necessarily are subject to an increased potential risk, as patient data from the traditional doctor–patient relationship also become potentially available elsewhere [60.25, 26].

This topic of data protection has become highly explosive in Germany. In 2007 the Ärztetag, the general meeting of the German Medical Association, rejected the e-HC with a significant majority as a *sociopolitical atom bomb* [60.27], mainly because of reservations regarding data protection, although the operating company responsible for introducing the e-HC, gematik [60.8], has developed a highly sophisticated technical plan for it. However, perhaps the complexity of this plan and even the easily comprehensible white paper constructed from it are a major reason for its rejection [60.28]. Educational work and risk analyses must therefore be performed in pilot projects, particularly from the point of view of the enormously important target group of the medical profession.

However, for patients too, dealing with the security technology which already has to be used to safeguard data protection sometimes seems to be difficult, inconvenient, and not very convincing. In addition, the majority of patients have not grown up with computers and many are very old, have dementia or are so seriously ill that they are hardly able to operate an e-HC with a PIN, an operating PIN or PUK, for example, or that it is barely possible to add prescription-free medication to the e-HC. Here, too, the process for access to the telemedicine system must be designed to be user-friendly, training must be carried out, a campaign must be launched to convince users about it, and a great deal of patience is required.

Some statutory provisions and their interpretation must naturally also be considered. For example, according to the state official for data protection in North Rhine-Westphalia, emergency data on the e-HC must not be used when admitting patients to hospital. It must be reserved solely for treatment in acute emergencies. The storage of emergency data, which was planned as the heart of the e-HC, to a large extent thus loses its relevance [60.29].

60.1.5 Technical Standards

One prerequisite for successful use of telemedicine is the interoperability of the underlying components based on data and communications standards. Medical documents and structured medical data must be represented in a standardized way. In conjunction with terminology standards, semantic indexing also becomes possible. The extensible markup language (XML) represents a general technical basis for the exchange of data between computer systems, which is also suitable for standardized description of medical documents. Various specifications of standards in the healthcare system already rely on it. This includes the HL7-based clinical document architecture (CDA) [60.30], which is used as the basis for creating standards for the representation of patient-related documents in many international projects. In Germany, the AG Sciphox/HL7 has developed definitions for the doctor's letter, the letter of referral, and the hospital admission form. The preliminary work on the specification of the electronic prescription was also done on this basis. Earlier, sector-specific exchange formats for medical communication were developed in Germany. In the area of healthcare provided by statutory health insurance physicians, the xDT data storage exchange formats have become a de facto standard. Over the course of time, a full range of standards have developed which had been initiated by different organizational units. Interoperability across the boundaries between different sectors is only possible through harmonization of the previous standardization initiatives, however. Current development in Germany is also running in this direction. The local state associations of statutory health insurance physicians (KVs) in Baden-Württemberg, Bavaria, and North Rhine-Westphalia provide their doctor-to-doctor (D2D) users with the assistance of communication of doctors' letters based on HL7 CDA release 2. The technical basis for this goes back to the VHitG initiative on *intersectoral communication*, in which companies from the software industry in the healthcare system described a guideline for implementation. The initiative is a further development of the existing specification according to HL7 CDA release 1, which is already supported in D2D.

Medical documentation as part of treatment processes is today increasingly carried out electronically in the form of electronic patient records (ePR) or electronic case records (eCR). To support communication between sectors in telemedical scenarios, a standardized architecture plays a key role for interinstitution electronic patient records. In addition to HL7 CDA as an electronic document standard, there are also a number of standards initiatives such as, for example, CEN/ISO 13606 Electronic Health Record Communication [60.31], the Standard Specification for Continuity of Care Record (CCR) from the American Society for Testing and Materials (ASTM), and the integration profile cross-enterprise document sharing (XDS) from the Integrating the Healthcare Enterprise (IHE) initiative [60.32]. An overview of these and other approaches to electronic health record (EHR) architectures can be found in *Blobel* [60.33]. It is necessary here to decide which framework architecture should be used as a basis, and in making this decision the prospect of interoperability with neighboring European countries must at least be taken into consideration in addition to national concerns. With its project to introduce the electronic health record, Austria has opted for the XDS profile and HL7 CDA as an electronic document standard for the transmission of medical results [60.21].

Because efforts towards standardization for specific implementations are at different stages of development, pragmatic initiatives are also very helpful in the interim. To accomplish harmonization for cross-project interoperable implementations, for example, for teleradiology applications, the Arbeitsgemeinschaft Informationstechnik (AGIT; German Information Technology Association) within the German x-ray society has developed a specification (http://www.tele-x-standard.de) for secure DICOM (digital imaging and communication in medicine) email communication, with which some manufacturers have affiliated themselves.

Project solutions existing today for telemedicine applications are barely integrated into the existing infrastructure of the primary systems of the relevant healthcare providers, but predominantly run alongside them. This has disadvantageous consequences for integration into the standard organizational processes. It is thus very desirable if a software architecture is chosen, abstracting from the specifics of primary systems via the provision of standardized services. For this purpose, an intermediate layer (middleware) must be used which provides web services for the typical requirements of telemedicine in terms of a service-orientated architecture (SOA). In this way, a standardized kit for telemedicine application can develop, which can be used to achieve service-orientated interoperability and which enables integration into the primary systems, without their specific implementations having to be known. This concept is exactly supported in the technological approach to telematics infrastructure. It would be desirable in this regard if appropriate services were provided in a defined and coordinated manner.

60.1.6 Acceptance and Benefits for Various Players in Telemedicine

As is always the case in highly complex, large-scale IT projects with extremely heterogeneous user groups and other involved players, the advantages of an IT system can hardly ever be felt equally by all those involved. For example, in the case of electronic prescriptions within the scope of the telematics project involving the e-HC and health professional card (HPC), significantly more additional effort is required by doctors than when writing conventional prescriptions (according to *Schweim* it takes 24 s instead of 2 s) [60.15]. Barely any immediate economical benefits or advantages of any other kind for the doctor can be recognized, but important advantages – in particular concerning drug safety – are expected for patients. Further improvements will certainly be found as a result of the test phases for introduction of the e-HC in different regions.

In telepathology, too, for example, 22 min are required to obtain a second opinion by means of a digital imaging and telepathology request, as opposed to 10 min for a conventional consultation request [60.34]. However, the German professional association of pathologists in 2007 described the issue of accounting as one of the barriers to telemedicine becoming established [60.35].

Radiologists have progressed slightly further in this regard, and teleradiological services can in principle be performed for emergency care without falling under the ban on remote medical treatment, but they have still not been included in the statutory health insurance catalog of standard benefits [60.36]. Our own experiences with a teleradiology project [60.37] demonstrate that organizational integration, technical stability, and simple handling are important criteria for acceptance. However, sustainable use requires further agreements which regulate the framework conditions for the provision of services and payment for them. It would also be desirable, however, if it were no longer necessary for upgrades in the security architecture – right up to the introduction of electronic signatures for questions and treatment recommendations as part of a teleconsultation – to be implemented on a project-specific basis and therefore in an expensive way, but rather if these upgrades could be accessed as basic services of the telematics infrastructure. The health professional card, which has still not been implemented, or certain forms of electronic signature make practical use more difficult [60.38].

In this context, the fundamental problem must also be addressed, namely that telemedicine potentially compromises the doctor–patient relationship. There is the risk that, in this very personal and generally strongly desirable relationship, the computer has a distancing effect and that the face-to-face interaction, the reassuring handshake, and in general the psychosocial component of medical care suffer as a result. *Haas* [60.38] sees this as being one of the two major reasons why *in some cases the desire to use modern IT has given way to frustration*. Technically good solutions and very large future markets for IT are not enough to introduce telemedicine with lasting success and with beneficial results. From the point of view of the doctor, too, the doctor–patient relationship can be destroyed, as the patient may conceal medical data stored on the e-HC, for example. Will the patient tell the doctor about this? The doctor will therefore have to continue to keep a parallel medical history, involving extra effort and additional cost, in order to fulfill his duty of documentation (model code of medical ethics). However, he is significantly better able to research and use the information kept in the electronic medical history and, for example, to provide the patient with copies, too.

It is clear from all of the individual telemedicine projects in Germany that appreciable advantages are already being achieved, particularly for the patient. This applies in particular to teleconsultation, including obtaining a second opinion, care cooperation including transfers and referrals, telemonitoring by means of biosignal processing, treatment management for complex widespread diseases, and information for care provision for rare and high-risk diseases and emergencies. Interviews with experts and a survey within the scope of evaluation of the teleradiology projects in

Baden-Württemberg [60.39] confirm this assessment. These projects predominantly involve teleconsultations as part of stroke care and care for accident victims with head injuries [60.40]. The majority of surveyed participants affirm that teleradiology consultations increase the quality of the decision made by the attending doctor. The benefit for patients is mainly seen as being faster and more adequate care. The main aspect of this benefit is the avoidance of unnecessary transportation and the fast provision of care during the night shift. The survey confirms that less time is required for diagnosis and treatment processes.

Not all of those involved are able to profit from telemedicine to the same extent. It is therefore imperative to create a win–win situation for as many of those involved in telemedicine as possible.

60.2 Consequences of the Peculiarities of the German System for Telemedicine

In an earlier account [60.17] we have already indicated some conditions for a successful telemedicine system. Taking into consideration the peculiar features of the German healthcare system mentioned above, but also on the basis of experience with individual telemedical projects and with the health telematics system which is currently being developed and the e-HC, we are now able to draw the following conclusions for telemedicine.

1. In the first instance, the current legal framework conditions must be improved on the basis of the varied and long-standing experience with individual telemedical projects, but also with the large-scale health telematics projects which are beginning. This involves in particular
   - extensive removal of boundaries between sectors in the healthcare system,
   - the option for healthcare providers to bill for telemedical services,
   - reliable financing of telemedicine as a standard benefit,
   - clarification of various concerns, particularly those of the medical profession regarding data protection,
   - harmonization with EU provisions and other international standardization authorities.
2. The technical integration of the e-HC and health telematics must be developed for telemedicine on the basis of an adequate and reliable technology. Basic telemedical care in Germany is not useful without the e-HC and health telematics system, but it still requires a telematics platform which is well designed and sufficiently well equipped.
3. This telematics infrastructure must be accessible to the tried and tested IT structure, some of which has already been in existence for decades, particularly to the IT structure of hospitals and doctors in private practice. Technical integration and migration aids must therefore be used for this purpose, such as middleware, SOA, EHR, and CDA.
4. Organizational changes, e.g., batch processing for electronic prescriptions, are already planned. More orientation towards processes rather than institution-orientated thinking, training courses about how to use telemedicine, educational work for all those involved concerning the benefits and the risks of telemedicine, and scientific monitoring to evaluate the telemedicine system – in particular in basic care – are necessary.
5. Clarification of the benefits of telemedicine must be conveyed convincingly, and the risks for patient and doctor, particularly in the psychosocial area, must be systematically reduced. Telemedicine must not have a negative impact on the personal patient–doctor relationship but must contribute towards reducing the distance between doctor and patient.

Taken as a whole, the benefits of telemedicine can no longer be disputed, even if its introduction into basic care in Germany still requires considerable expense, time, and patience. However, since – under the conditions listed here – at least five more years are still necessary to create the most important telemedical applications and introduce them across the board, work must now be continued with particular commitment, too.

References

60.1 http://www.liebertpub.com/tmj (last accessed May 31, 2011)

60.2 A. Jäckel (Ed.): *Telemedizinführer Deutschland* (Deutsches Medizinform, Bad Nauheim 2008)

60.3 http://www.gmds.de (last accessed May 31, 2011)

60.4 http://www.dgg-info.de (last accessed May 31, 2011)

60.5 http://www.dgtelemed.de (last accessed May 31, 2011)

60.6 http://www.bvitg.de (last accessed May 31, 2011)

60.7 http://www.ztg-nrw.de (last accessed May 31, 2011)

60.8 http://www.gematik.de (last accessed May 31, 2011)

60.9 http://www.dimdi.de (last accessed May 31, 2011)

60.10 http://www.ema.europa.eu (last accessed May 31, 2011)

60.11 http://www.diabeticus.com/pressemitteilungen/2008/03_07.html (last accessed May 31, 2011)

60.12 E. Wachs, D. Baden, M. Klingenberg, M. Wähner, C. Schultz, H. Korb: Telemedizinische Betreuungskonzepte bei chronischer Herzinsuffizienz. In: *Telemedizinführer Deutschland*, ed. by A. Jäckel (Deutsches Medizinform, Bad Nauheim 2008) pp. 136–138

60.13 F. Köhler, S. Lücke: Partnership for the heart. In: *Telemedizinführer Deutschland*, ed. by A. Jäckel (Deutsches Medizinform, Bad Nauheim 2008) pp. 132–135

60.14 The National Opinion Research Center (NORC): *Findings from the Evaluation of E-Prescribing Pilot Sites* (Agency for Healthcare Research and Quality, Rockville 2007), available at http://healthit.ahrq.gov/images/apr07norcerxreport/erxinterimevaluationreport.html

60.15 H.G. Schweim: Die unerträgliche Geschichte der Gesundheitskarte in Deutschland, GMS e-J. Med. Inform. Biom. Epidemiol. **3**, 1–5 (2007)

60.16 S.C. Semler: Elektronische Patientenakte – Stand, Perspektive und outcome. In: *Telemedizinführer Deutschland*, ed. by A. Jäckel (Deutsches Medizinform, Bad Nauheim 2008) pp. 3–4

60.17 M. Boeker, E. Pelikan: Gesundheits-Telematik – Anwendungsspektrum und Sicherheits-Infrastruktur. In: *Moderne Gesundheitskommunikation. Vom Aufklärungsgespräch zur E-Health*, ed. by K. Hurrelmann, A. Leppin (Huber, Bern 2001) pp. 259–276

60.18 E. Kotter, U. Schrader, E. Pelikan, K.H. Allmann, J. Sedat, R. Klar, U. Timmermann, M. Langer: The Freiburg large scale PACS. In: *Computer Assisted Radiology and Surgery* (Elsevier Science, Berlin 1997) p. 1007

60.19 R. Klar, S. Bucher, A. Roesner, M. Busch, U. Timmermann: Funktionalität der medizinischen Arbeitsplätze im KIS und PACS des Universitätsklinikums Freiburg (2002) http://www.uniklinikum-giessen.de/kis-ris-pacs/archiv/2002/di1150.pdf

60.20 M. Meyer, A. Ehlers: e-Health-Lösungen als Basis für eine integrierte Gesundheitsversorgung. In: *Telemedizinführer Deutschland*, ed. by A. Jäckel (Deutsches Medizinform, Bad Nauheim 2008) pp. 156–162

60.21 K.P. Pfeiffer, C.M. Auer: Herausforderungen bei der Umsetzung der elektronischen Patientenakte und Gesundheitskarte in Österreich, Bundesgesundheitsbl. Gesundheitsforsch. Gesundheitsschutz **52**, 324–329 (2009)

60.22 G. Heidenreich, B. Blobel: IT-Standards für telemedizinische Anwendungen. Der Weg zum effizienten Datenaustausch in der Medizin, Bundesgesundheitsbl. Gesundheitsforsch. Gesundheitsschutz **52**, 316–329 (2009)

60.23 P. Waegemann: e-Health und elektronische Patientenakten. In: *Telemedizinführer Deutschland*, ed. by A. Jäckel (Deutsches Medizinform, Bad Nauheim 2008) pp. 56–59

60.24 K.D. Mandl, I.S. Kohane: Tectonic shift in health information economy, N. Engl. J. Med. **358**, 1732–1737 (2008)

60.25 C. Groß: Gesundheitstelematik: Folgen für die Arzt-Patient-Beziehung, Dtsch. Ärztebl. **103**, A3469 (2006)

60.26 M. Boeker: Telematik im Gesundheitswesen, Datenschutz und Datensicherheit (WS2006/2007) http://www.medicip.uni-freiburg.de/material/materialien/telematik_datenschutz_informationssicherheit_skript.pdf

60.27 D. Borchers: Ärztetag lehnt elektronische Gesundheitskarte ab (2007) http://www.heise.de/newsticker/meldung/aerztetag-lehnt-elektronische-Gesundheitskarte-ab-179659.html

60.28 H. Krüger-Brand: Gesundheitstelematik: White Paper Sicherheit, Dtsch. Ärztebl. **105**(20), A1073 (2008)

60.29 C. Groß, U. Hein-Rusinek: Notfalldaten auf der Gesundheitskarte: Nutzen unklar, Rhein. Ärztebl. **1**, 16–17 (2008)

60.30 Health Level Seven Inc.: http://www.hl7.org (HL7, Ann Arbor 2010)

60.31 B. Blobel: EPA Modelle im Vergleich. In: *Telemedizinführer Deutschland*, Vol. 9, ed. by A. Jäckel (Deutsches Medizinform, Bad Nauheim 2008) pp. 17–24

60.32 Integrating the healthcare enterprise http://www.ihe.net (last accessed May 31, 2011)

60.33 B. Blobel: EHR architectures-comparision and trends. In: *e-Health*, ed. by B. Blobel, P.M. Pharow (IOS, Amsterdam 2008) pp. 59–73

60.34 T. Schrader, K. Hamann, P. Hufnagel, W. Schlacke, G. Kempny: T.Konsult Pathologie Zweite Mei-

nung im Rahmen des Mamma-Screenings. In: *Telemedizinführer Deutschland*, ed. by A. Jäckel (Deutsches Medizinform, Bad Nauheim 2008) pp. 109–115
60.35 I. Hosch: Advantages and disadvantages in telepathology, e-J. Pathol. Histol. **9**, 031–003 (2003)
60.36 H. Krüger-Brand: Telemedizin: Vom Projektstatus in die Routine, Dtsch. Ärztebl. **105**, A1319 (2008)
60.37 E. Pelikan, M. Philipps, S. Zimmermann: Teleradiologieprojekt Südbaden (2006) http://www.telemed-berlin.de/telemed2006/Beitraege/S06 - PELIKAN, ERNST - Teleradiologie-Projekt.pdf (last accessed May 31, 2011)
60.38 P. Haas: Die Lust und Last mit der MI aus verschiedenen Perspektiven, mdi Forum Med. **3**, 142–143 (2008)
60.39 S. Kirn: *Teleradiologieprojekte Baden-Württemberg, Begutachtung und Studentische Evaluation, Teilbericht Südbaden* (Universität Hohenheim, Hohenheim 2007), unveröffentlicht
60.40 A. Eytner: Teleradiologienetzwerk Baden-Württemberg. In: *Telemedizinführer Deutschland*, ed. by A. Jäckel (Deutsches Medizinform, Bad Nauheim 2006)

61. Telemedicine Using Active Implants

Klaus P. Koch, Oliver Scholz

Telemedicine is the area of telematics that allows transfer of information, i.e., diagnostic or therapeutic data, between two locations (spatial distance) or times (temporal distance). This includes the bidirectional transmission network between patient and doctor as well as the transmission network between two doctors. Moreover, information can be transferred without material transport. In technical realizations, wired as well as wireless communication channels are used. The possibility to transmit medically relevant data has opened additional fields of application. Examples include consulting external experts during surgical interventions, transmission of physiological data/signals obtained by the patient at home, as well as distribution of data within a hospital. In the opposite direction, data should also be able to be transmitted to therapeutic devices to adapt treatment or to monitor device function. The fields of application extend, in this regard, from transfer of x-rays to forwarding of temperature values. Modern active implants normally come with a wireless communication facility to the outside world. In particular, in the latter area, energy supply to each component also plays an important role and can be partially combined with the desired data transfer. In this chapter, the possibilities of telemedicine for the control of active implants are highlighted through a short overview of the application of telemedicine during operations and in domestic care.

61.1 **Telemedicine in the Operating Theater** ... 1129
61.2 **Telemedicine in Domestic Care** ... 1130
61.3 **Implant Telemetry** ... 1131
61.3.1 Implant Telemetry by Means of Inductive Coupling ... 1131
61.3.2 Implant Telemetry Employing Radio Communication ... 1133
61.3.3 Optical Transcutaneous Transmission ... 1134
61.3.4 Energy Supply to Medical Implants . 1134
61.3.5 Line Codes ... 1135
61.4 **Inclusion of Active Medical Implants in Telemedicine Systems** ... 1136
References ... 1136

Consultation of other experts during an operation is often possible only in large medical centers. With high levels of specialization, this becomes increasingly difficult, because experts are mostly active in different centers.

61.1 Telemedicine in the Operating Theater

The application of remote data transmission offers a good, multimodal possibility to consult other experts in the short term (Fig. 61.1). Such systems also make sense for teaching and scientific exchange, for example, with live connection to an auditorium or conference hall. In modern operating theater systems, various imaging instruments such as endoscopes and ultrasound are used [61.1]. In addition to such direct diagnostic devices, use of cameras to record the operative field and the operation room for meaningful representation of the operative sequence is advantageous. For control of imaging systems as well as communication and video conferencing equipment, appropriate devices must be available to the treating physician in the sterile area. This concerns in particular camera control to adjust the angle of view of the camera to

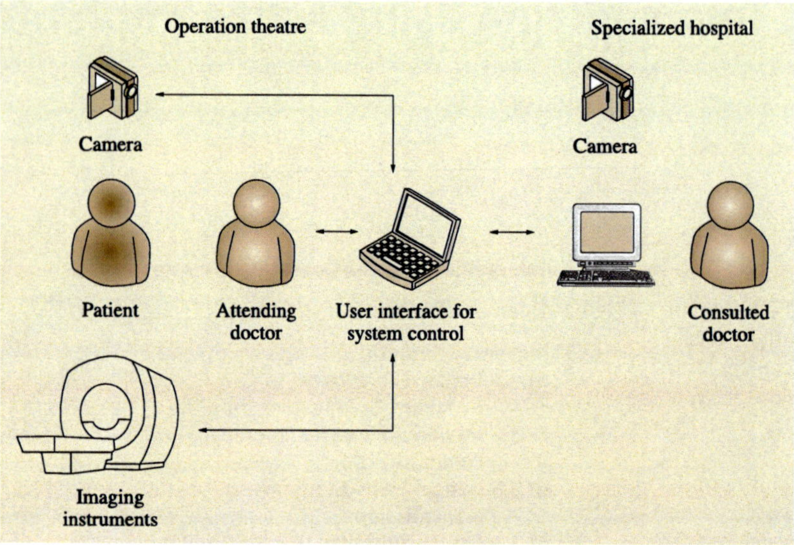

Fig. 61.1 System overview of a telemedicine-enabled operating theater

the actual operative situation. In addition, change of cameras and the link connection should be possible during surgeries from within the operating theater. Using this technology, other experts can be consulted during complicated operations without having to move the patient.

61.2 Telemedicine in Domestic Care

For recording of medically relevant data in the home environment, a wide spectrum of devices are now available on the market, extending from devices for body temperature and weight measurement to devices to record electrocardiogram (ECG) or blood pressure. Many of these devices also have interfaces for reading out the data. When coupled to a so-called house base station, such data can be transmitted to a medical service when required. Direct point-to-point connection between doctor and patient is also a possibility, although not currently implemented. Server-based systems offer the advantage of greater availability [61.2]. Also, initial reception of the data can be done by medically trained nonmedical staff (Fig. 61.2). In this way, the doctor can be relieved and thus the costs be reduced. To interface between the measuring instruments and the house base station, wired interfaces such as RS232 or universal serial bus (USB) as well as wireless interfaces such as Bluetooth are used. The house base station itself can use, in principle, a mobile phone or personal digital assistant (PDA) as a platform. Nevertheless, for many patient groups, simple operation is recommended, which is achieved rather by the development of a device [61.3], which may be as simple as an operating element that initiates only the data transfer. In addition, for wired connection of measuring instruments with mobile phones or PDAs, only an adap-

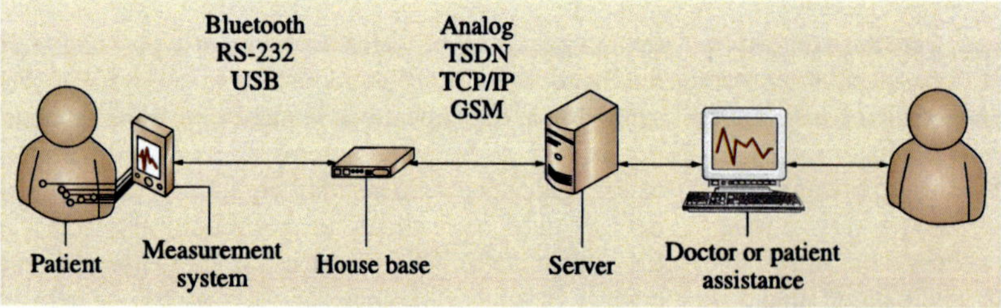

Fig. 61.2 Communication structure of domestic care systems

tor is possible. For the interface between the house base station and the server or medical staff, the possibilities differ mainly in terms of the interfaces available in the respective home environments, extending from telephone lines, integrated services digital network (ISDN) lines, asymmetric digital subscriber line (ADSL) connections, up to wireless global system for mobile communications (GSM) connections. Beside the technical problems of the supply of a huge number of interfaces and the use of communication protocols with sufficient data security, legal constraints also have to be considered with regard to data protection. Beside the aspect of cost reduction, in many cases accommodation in the home environment is often more pleasant for the patient.

61.3 Implant Telemetry

In this section the application of different telemetric approaches is described using the example of implantable systems, extending from the problems of energy supply for each component to the different possibilities for data transfer. Strictly speaking, one understands by telemetry the obtaining of measured values from distant or inaccessible measuring points, requiring the transfer of values over a distance. In medical technology usage, the term "telemetry" is often used for transmission of measured data as well as control data. This can be done by wire, but also wirelessly. A simple form of contact consists in connecting implanted systems by wire through the skin (percutaneously). Nevertheless, for chronic applications, a higher risk of complications arises because of the corresponding high risk of infection or injury. Also from a cosmetic point of view, cables that enter the skin are unsightly. This explains why most active implants are wireless, i. e., passing through the skin (transcutaneous) to link with the outside world. For special applications, the percutaneous approach remains advantageous, because there are practically no restrictions on the signal and energy transfer. Therefore, technical advancements can be applied without exposing the patient to renewed surgical intervention. This is an advantage in particular during the development of new implants [61.5, 6]. For stimulation of the visual cortex, for example, 68 electrodes for use with a camera and an electronic system for stimulation of simple visual perception [61.4] (Fig. 61.3) have been implanted. Also, temporarily implanted systems can be operated effectively using such connections. Mostly the patient is under clinical care during the application time of such systems, so expert care of the percutaneous cable feed can be guaranteed.

61.3.1 Implant Telemetry by Means of Inductive Coupling

To provide energy supply to the implant, as well as to create a bidirectional interface for data transfer, inductive interfaces are used. Here, the field created by an external transmitter coil passes through the receiving coil of the implant, transmitting energy to it based on the transformer principle. To send, in addition, data to the implant, the carrier wave of the transmitter can be modulated. For communication from the implant to the outside, usually the resistive load of the implant coil is modulated, causing a change in the transformer impedance and thereby the current in the external coil. The same principle can be observed in a classical transformer, in which the current in the primary circuit also depends on the load in the secondary circuit. This technique is used in cochlear implants because, in addition to the high data rate of several 100 kbit/s, the implant must be supplied with about 30 mW of power [61.7–9]. Currently, this approach is also used in the Freehand system, for stimulation of the arm musculature at carrier frequency of 10 MHz, for which as much as 90 mW is transferred [61.10]. In optimized cochlea implant systems (Medical Electronics), overall efficiency of up to 67% with transferred power of 250 mW has been achieved [61.11]. For implants with even higher power requirements, for example, ventricular assist devices, Dualis developed special inductive interfaces through which up to 50 W of power can be

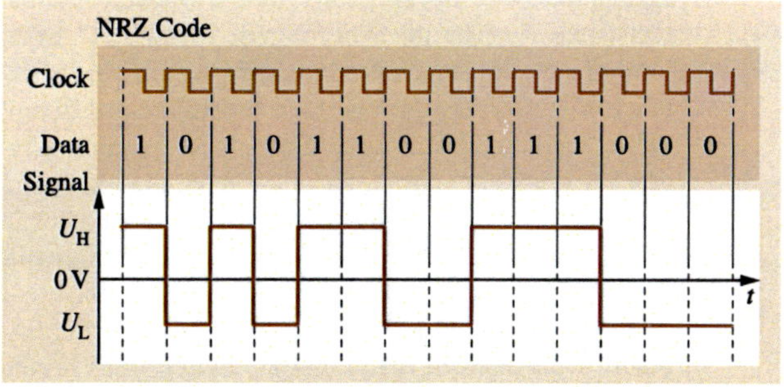

Fig. 61.3 Percutaneous cable feedthrough for stimulation of the visual cortex [61.4]

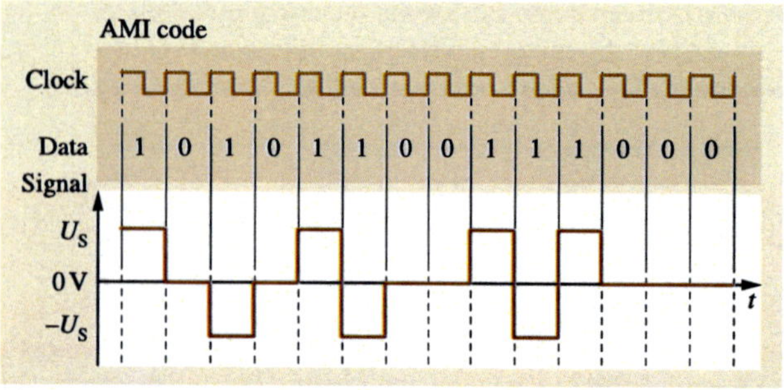

Fig. 61.4 Model of an inductive interface

transferred [61.12]. Especially in cases where a high level of miniaturization is needed and only very short distances must be bridged, systems with thin-film coils applied on their own substrates can be used [61.13]. Alternatively, attempts have been made to integrate the implant coil directly onto the communication chip, e.g. for systems for measuring intracranial pressure. However, because of the small dimensions of the receiver coil, only ranges of a few millimeters can be achieved [61.14].

As an example for combined data and energy transfer through an inductive interface, a system operating at carrier frequency of 4 MHz is described in the following [61.15]. Communication from the external unit to the implant is realized through modulation of the carrier signal amplitude. For data transmission from the implant to the extracorporeal control unit, the load of the implant is modulated. Here, the transmitter oscillating circuit is modeled as a parallel connection (C_1, L_1), and the receiver oscillating circuit as a serial connection (L_2, C_2). The simplified description of the inductive transmission is based on a simple model of a transformer in which the coupling between the coils is determined by the real coupling coefficient k (Fig. 61.4).

The transmitter can be modeled as a linear voltage source with open-circuit voltage U_0 and internal resistance R_1. The load on the receiver circuit due to the implant electronics is simulated by an ohmic resistance R_2. According to the load modulation, different values are assigned to this resistance. An important index for transfer to the implant is the voltage gain Af from the external voltage source to the load resistor in the implant [61.16, 17]. This describes the supply voltage expected at the implant U_2 as well as the transferred modulation amplitude of the carrier signal to the implant. In the applied linear model, this depends on the angular frequency ω of the carrier as follows:

$$\begin{aligned} Af &= \frac{U_2}{U_0} \\ &= \frac{\mathrm{i}\omega k\sqrt{L_1L_2}\frac{R_2}{1+\mathrm{i}\omega C_2R_2}}{\left(R_1+\mathrm{i}\omega L_1+\frac{1}{\mathrm{i}\omega C_1}\right)\left(\mathrm{i}\omega L_2+\frac{R_2}{1+\mathrm{i}\omega C_2R_2}\right)-(\omega k)^2L_1L_2}\,. \end{aligned} \tag{61.1}$$

Because the external unit should be transportable, it is battery operated, and accordingly has only a limited energy reserve. That is why the efficiency η also has to be considered during system design. The efficiency η is defined as the ratio of the active power taken up by the second side, P_2, to the active power delivered by the voltage source, P_{tot}. In the considered case, the capacitance of the transmitter circuit has no effect on the efficiency, therefore

$$\eta = \frac{P_2}{P_{tot}} = \frac{k^2\frac{L_1}{L_2}R_2}{k^2\frac{L_1}{L_2}R_2 + R_1\left(1+R_2^2\left(\frac{1}{\omega L_2}-\omega C_2\right)^2\right)}\,. \tag{61.2}$$

The final parameter to consider is the change of the primary current ΔI on modulation of the load. This primary current depends on the ratio of the open-loop voltage to U_0 and the value of the total impedance Z_{tot} through

$$\begin{aligned} \Delta I &= \frac{U_0}{|Z_{tot}(\text{with load})|} - \frac{U_0}{|Z_{tot}(\text{without load})|} \\ \text{with} \quad Z_{tot} &= R_1 + \mathrm{i}\omega L_1 + \frac{1}{\mathrm{i}\omega C_1} \\ &\quad + \frac{(k\omega)^2L_1L_2}{\mathrm{i}\omega L_2 + \frac{R_2}{\mathrm{i}\omega C_2R_2+1}}\,. \end{aligned} \tag{61.3}$$

To ensure safe operation of the transmission of energy as well as bidirectional data, optimization is necessary. This task should not be underestimated, due to the various system properties and their local extreme values and the different weighting of the properties. Another difficulty in this optimization lies in the range of variation of the parameters of the components of the transmission system. Thus, the coupling between the coils varies as a function of their relative position. Such problems can be alleviated by the use of intelligent positioning systems which indicate the effectiveness of the transfer to the patient. Further simplification of the positioning can be achieved by the use of permanent magnets integrated into the primary and secondary coils, which enables automatic positioning based on the magnetic forces and independent fixation of the external coil. With all such systems, there always remains the possibility of greater variation due to movement artifacts or changing tissue layers (fat accumulation). In addition, the components themselves are subject to manufacturing tolerances and aging. In particular, aging and changes in parasitic effects are especially critical when considering systems of implanted components in contact with body liquids.

Although inductive linkage is very well suited to combined transfer of data and energy, it is sometimes used exclusively for data transfer, in particular for implants with complete titanium housing that only allows low-frequency fields to pass. This is especially useful if, as with some simple pacemakers, only a low data rate is required.

61.3.2 Implant Telemetry Employing Radio Communication

It is remarkable how implant telemetry has changed in recent years, in particular in pacemakers. Since in the beginning only a few parameters could be read or programmed, a slow inductive link was sufficient, for which a reader coil had to be placed immediately above the implantation site. To aid the *search* for the implanted device, implant manufacturers provided a means of communication. Today, such devices incorporate many more features, allowing more than battery status and heart beat to be retrieved, for example. Modern pacemakers and implantable cardioverter–defibrillators (ICDs) even record whole intracardial electrogram (IEGM) traces and store them internally. To provide a convenient means to read the large amounts of data in an acceptable time even over longer distances (2–3 m), these devices are equipped with radio interfaces. Since the titanium casing would excessively damp electromagnetic waves, an antenna is placed outside the case. In pacemakers, the antenna is usually integrated into the polymer terminal block of the electrode wires.

The problem with radiofrequency (RF) transmission from implants is the fact that body tissues impair propagation of electromagnetic waves due to their ionic conductivity. This absorption depends not only on the type of tissue through which the transmission must pass but also on the frequency used. To a good approximation, the absorption increases with increasing frequency. However, some particular frequencies are more strongly affected, in particular those at which water molecules exhibit resonance, for instance, 2.4 GHz. Consequently, this frequency band is less suitable for implant telemetry.

Since radio waves generally propagate nearly unresisted, compared with light, strict regulations on radio traffic and channels used are necessary to avoid mutual interference or even jamming. Unfortunately, regulations differ among countries, so that it can happen that an implant of a patient cannot be read out abroad. This may occur, for instance, with implants that use the 868 MHz industrial, scientific, and medical (ISM) band, which is popular in Europe but is prohibited in the USA, where use of the 915 MHz band is instead widespread. As other devices apart from medical implants may use these unlicensed frequency bands too, definition of a RF band dedicated particularly to active medical implants and available worldwide is desirable. With the implementation of the medical implant communication service (MICS) band for wireless control and telemetry for medical implants by the International Telecommunication Union (ITU) this has been mostly achieved. Though the frequency spectrum that has been provided has to be shared with meteorological services as prime users, major problems are not expected [61.18]. In particular the overcrowded situations as experienced in the popular 2.4 GHz or 433 MHz bands (WLAN, wireless headsets, earphones, etc.) can be obviated.

The MICS frequency band lies between 402 and 405 MHz. The maximum permitted bandwidth of a MICS communication link is 300 kHz. This is sufficient to provide net transmission speeds of up to 528 kbit/s (four frequency shift keying, FSK) with Zarlink's ZL70110 transceiver chips, according to the manufacturer. An implant may initiate a communication link in emergencies only, in which case it is not required to ensure that the channel is free (listen before talk). In standard situations, only the base station may establish a connection, provided the channel is not already occupied by another link.

Presently, a number of commercially available devices already use the MICS band, e.g., implants from Medtronic and their corresponding CareLink programmer, or implants of Biotronik, St. Jude, and others. Besides pacemakers or implantable cardiac defibrillators, there exist other types of implants which make use of the MICS standard, such as implantable deep brain stimulators or blood glucose monitors. Even though not an implant in the strict sense, the camera pill from Given Imaging [61.19] transmits images taken of the gastrointestinal tract using a MICS channel.

61.3.3 Optical Transcutaneous Transmission

In cases where the transmission bandwidth of inductive links would be too low, or if one fears interference from other radio communication systems, optical transcutaneous transmission may be a viable option, provided that the skin and tissue layer to be crossed is not too thick. Examples in which this technique could be of note are, e.g., brain–computer interfaces, in which a disproportionately large amount of data [real-time multichannel electroencephalogram (EEG) recordings] has to be transmitted [61.20, 21]. Optical data transmission has also been used in fully implantable ventricular assist devices, because of the use of low-frequency magnetic fields for wireless energy transmission due to the large amount of electrical power demanded by the implant. Moreover, heavy disturbance of RF transmission channels by the operation of pumps was feared [61.22–24]. As early as the 1980s, several studies were carried out to analyze how well optical transmission suits implant telemetry [61.25]. However, transfer rates were at first rather low. Strong absorption and especially the strong dispersion due to body tissue constrained the possibilities. Nevertheless, optical transmission of data through the skin has been shown to be feasible up to 40 Mbit/s, albeit with high power demands (120 mW) and a comparatively thin skin layer (5 mm) [61.21].

One advantage of optical transcutaneous telemetry that should not be left unmentioned is the fact that the necessary components can be made very small. Indeed, manufacturers of RF transceiver chips promote their products by stating that one could implement single-chip radio solutions with them. In practice, this advantage proves to be illusionary, since additional components such as crystals, filters, and antennas are required. For optical transmission, there is no need for these devices. In place of an antenna, considerably smaller light-emitting diodes (LEDs) and photodetectors are employed. This was the main reason why, in an electrical stimulator for stimulation of the salivary glands in patients suffering from hyposalivation that had been embedded in an artificial tooth, optical transmission was used for device control [61.26]. The patient still had to open his/her mouth for successful transmission. In a more recent drug delivery system that was also embedded in a dental appliance, this was no longer necessary; here, telemetry was performed transcutaneously through the cheek [61.27]. For this, the patient was asked just to place the reader unit at the cheek, similarly to a mobile phone.

61.3.4 Energy Supply to Medical Implants

The main criteria for the choice of an implant's energy supply are the energy demands of the implant, the available space, and the possibilities to exchange the energy source. In pacemakers, primary, nonrechargeable batteries are the conventional solution. In particular, since the introduction of lithium–iodine batteries, lifespan of up to 10 years has been achieved at acceptable volume. As the active part that incorporates the battery is implanted in a subcutaneous pocket and as the distally ingrown electrode wires are attached to the active part by detachable screw locks, exchange of the active implant portion is possible with relatively low burden to the patient. The prime qualification is the very low energy consumption of pacemakers due to the low stimulation rate, the small number of electrodes, and the good coupling to the tissues. Another relevant issue is the autonomous operation of pacemakers (and deep brain stimulators, which are also powered by batteries). Both systems fulfill their role without the presence of an external control unit. Only for (de)activation or occasional adjustment of therapeutic parameters is data exchange necessary. However, battery power supply is not possible in systems with high electrode count and high stimulation rate. Examples are implants for stimulation of the upper extremities and for grasping in tetraplegics.

In implants that do not have life-sustaining functions, secondary, rechargeable batteries are already being employed; for example, Medtronic offers a neural stimulator that exhibits twice the lifespan of its primary cell-powered cousin. Recharge cycles span around 4–6 weeks. An alternative means of energy storage is utilized in implantable drug pumps. Here, energy is stored mechanically in elastic mechanical components (Medizintechnik Promedt GmbH).

To provide energy without artificial supply, there exist research activities to make use of energy prevalent in the natural environment, aiming to realize purely self-sustaining systems. These concepts can be differentiated with respect to their primary source of energy and their principle of energy conversion. Piezoelectric, capacitive, as well as electromagnetic generators belong to the class of energy converters which make use of the mechanical energy of the body. Because of the variety of locations, different possibilities of coupling exist; for example, at sinews or muscles, longitudinal forces prevail. Across joints, bending deformations can be exploited, and in body regions where pressure forces are experienced, e.g., under the sole of the foot or beneath muscles, dynamic compression conditions may be used. The major challenge for mechanical energy converters is the encapsulation of the electric components such as the piezoceramics, which needs to be deformable. Additionally, tissue reactions induced by the mechanical loading need to be carefully analyzed. Good results are expected from thermoelectric generators due to the body's continuous production of thermal energy. However, when considering the temperature gradient that prevails at possible implantation sites and the influence of environmental temperature and clothing on this gradient, little energy conversion is achievable. Research approaches of Biophan tried to optimize thermoelectric converters for temperature differentials of 1–5 °C in order to generate electric power of up to 100 μW at 4 V_{DC}. A converter area of 2.5 cm^2 is planned. However, to date, these systems have not met with success. In principle, photovoltaic cells could be a viable option if there existed enough places on the body that are not, even intermittently, covered by clothes. Furthermore, placement of such cells close beneath the skin is problematic in humans from a cosmetic point of view.

The use of fuel cells to convert metabolic products into electrical energy is another option for harvesting body energy for energy supply to active implants. The challenges here are to protect the fuel cell from catalyst poisons, to sustain a continuous fuel supply, and to remove reaction products. Partially, this may be solved through the development of suitable membranes. Although this may solve the aforementioned problems, with time, connective tissue will still cover the membrane and thus decelerate the transport of required substances. This, in turn, limits the achievable electrical power.

Due to the limited performance of implantable energy harvesters, an energy storage component, e.g., a capacitor, will be necessary in any case. This ensures that short peaks in power demand can be handled and may also serve as a backup when, at least for short periods, the harvested power is too low to sustain the implant's operation. Besides, a special energy management circuit is required to adapt the voltage levels of the generator and the electronics.

In the 1980s, even isotope batteries employing radioactive ^{238}Pu were used to power implantable pacemakers. In the meantime these have been taken off the market because of the toxicity of the material. According to the German Radiation Protection Office, 284 patients were supplied with these kind of implants, of whom 2 were still alive in 2010.

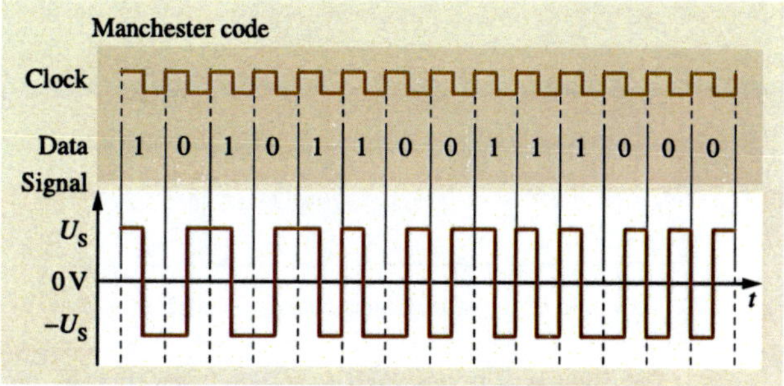

Fig. 61.5 Example of the no return to zero (NRZ) code

61.3.5 Line Codes

To transmit a serial stream of data down a transmission line or a radio link, various coding methods are employed. These differ in their transmission properties, such as the lack of a direct current (DC) component or the incorporation of a clock signal. The lack of a DC component is required if, e.g., the data stream is sent across an inductively coupled interface, to ensure galvanic insulation. Extraction of the clock signal may become necessary for successful data transmission when transmitter and receiver are excessively desynchronized.

In the following, a few codes are presented as examples.

In the no return to zero (NRZ) code the logical symbols 1 and 0 are represented by two different voltage levels (Fig. 61.5). In case of successive 0s (or 1s), a DC component appears. Moreover, it is no longer possible to extract the clock signal.

The alternate mark inversion (AMI) code is characterized by three different voltage levels (Fig. 61.6).

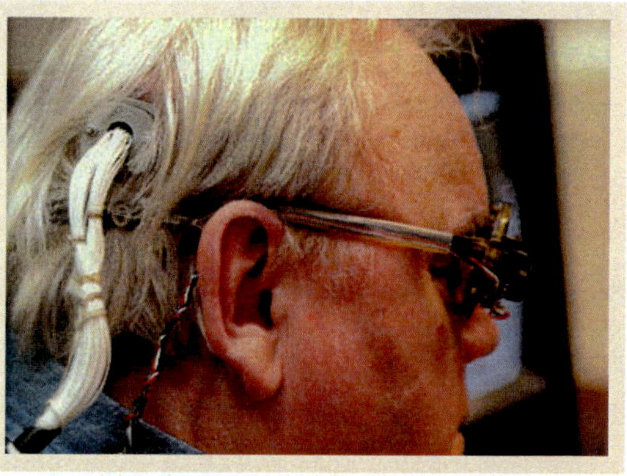

Fig. 61.6 Example of the alternate mark inversion (AMI) code

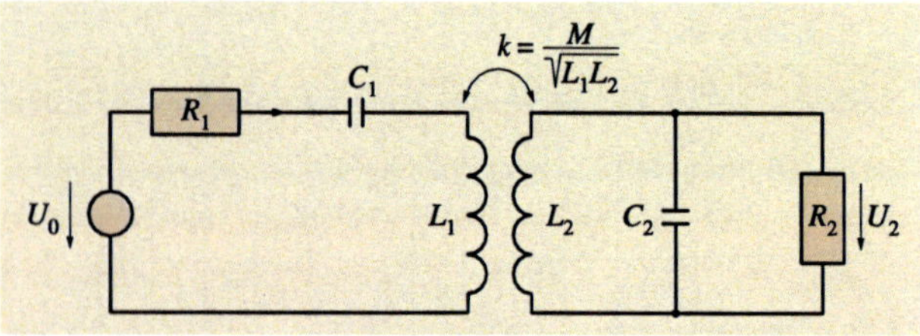

Fig. 61.7 Example of the Manchester code

The logical 0 is assigned a signal level of 0 V, and a logical 1 is marked by an applied voltage level. However, the polarity of the voltage change from one logical 1 to the next. This code allows extraction of the clock signal only during a succession of logical 1s, not from 0s.

In the Manchester code, the two logical states 1 and 0 are characterized by a polarity shift after half the symbol period (Fig. 61.7). A falling edge represents a 1, and a rising edge represents a logical 0. This ensures that the clock can always be extracted, independent of the content of the data stream. The downside of this code is that it requires a larger bandwidth, since the clock frequency is twice the bit rate.

61.4 Inclusion of Active Medical Implants in Telemedicine Systems

All major manufacturers of active implants for treatment of heart diseases now offer telemedicine services for their products, whether Merlin.net from St. Jude, Biotronik Home Monitoring, CareLink from Medtronic, etc. The Latitude system from Boston Scientific monitors not only the functions of its pacemakers but also other devices, such as a weighing scale or a blood pressure measuring device, using Bluetooth technology. The manufacturer expects that, in this way, the comprehensive health status of the patient will be available to the treating physicians, without the need for the patient to present to the practice or clinic. A study sponsored by the industry [61.28] underpins the theory that a regular flow of comprehensive pacemaker data to hospitals reduces the risk of complications.

References

61.1 H. Skupin: Telemedizin in der täglichen Routine, Kongressband 6. Würzburger Medizintechnik Kongress (2005) pp. 262–269

61.2 A. Bolz: Schlüsselkomponenten für die Integration existierender TeleHomeCare-Komponenten, Kongressband 6. Würzburger Medizintechnik Kongress (2005) pp. 262–269

61.3 S. Kiefer, M. Schäfer, F. Schera, K. Kruse: TOPCARE – A telematic homecare platform for cooperative healthcare provider networks. In: *Biomedizinische Technik, Ergänzungsband 2*, Vol. 49, ed. by U. Boenick, A. Bolz (Schiele Schön, Berlin 2004) pp. 246–247

61.4 W.H. Dobelle: Artifical vision for the blind by connecting a television camera to the visual cortex, J. Am. Soc. Artif. Int. Organs **46**, 3–9 (2000)

61.5 E. Zrenner, R. Wilke, H. Sachs, K. Bartz-Schmidt, F. Gekeler, D. Besch, U. Greppmaier, A. Harscher, T. Peters, G. Wrobel, B. Wilhelm, A. Bruckmann, A. Stett, SUBRET Study Group: Visual sensations mediated by subretinal microelectrode arrays implanted into blind retinitis pigmentosa patients, Proc. 13th Annu. Conf. IFESS (2008) pp. 218–220

61.6 R.A. Normann: Towards an artificial eye, Med. Device Technol. **8**, 14–20 (1990)

61.7 C.M. Zierhofer, E.S. Hochmair: Implementation of a telemetric monitoring system in a cochlear implant, Proc. 16th Annu. Int. Conf. IEEE Eng. Med. Biol. Soc. (1994) pp. 910–911

61.8 C.M. Zierhofer, I.J. Hochmair-Desoyer, E.S. Hochmair: Electronic design of a cochlear implant for multichannel high rate pulsatile stimulation strategies, IEEE Trans. Rehabil. Eng. **3**, 112–116 (1995)

61.9 C.M. Zierhofer, E.S. Hochmair: High-efficiency coupling-insensitive transcutaneous power and data transmission via an inductive link, IEEE Trans. Biomed. Eng. **37**(7), 716–722 (1990)

61.10 B. Smith, P.H. Peckham, M.W. Keith, D.D. Roscoe: An externally powered, multichannel, implantable stimulator for versatile control of paralyzed muscle, IEEE Trans. Biomed. Eng. **34**, 499–508 (1987)

61.11 G.A. Kendir, W. Liu, G. Wang, M. Sivaprakasam, R. Bashirullah, M.S. Humayun, J.D. Weiland: An optimal design methodology for inductive power link with class-E amplifier, IEEE Trans. Circuits Syst. **52**(5), 857–866 (2005)

61.12 B. Vodermayer, R. Gruber, T. Schmid, W. Schiller, G. Hirzinger, D. Liepsch, A. Welz: Adaptive transcoutanous energy transfer system (TET) for implantable devices, Int. J. Artif. Organs **28**(9), 885 (2005)

61.13 S. Kim, R. Bhandari, M. Klein, S. Negi, L. Rieth, P. Tathireddy, M. Toepper, H. Oppermann, F. Solzbacher: Integrated wireless neural interface based on the Utah electrode array, Biomed. Microdevices **11**(2), 453–466 (2009)

61.14 B. Flick, R. Orglmeister: A portable microsystem-based telemetric pressure and temperature measurement unit, IEEE Trans. Biomed. Eng. **47**, 12–16 (2000)

61.15 O. Scholz, J. Parramon, J.-U. Meyer, E. Valderrama: The design of an implantable telemetric device for the use in neural prostheses, Proc. 14th Int. Symp. Biotelem. (1997) pp. 265–269

61.16 E.S. Hochmair: System optimization for improved accuracy in transcutaneous signal and power transmission, IEEE Trans. Biomed. Eng. **31**, 177–186 (1984)

61.17 O. Scholz: Konzeption und Entwicklung eines Datenübertragungssystems für den Einsatz in der Neuroprothetik. Ph.D. Thesis (Universität des Saarlandes, Saarbrücken 2000)

61.18 Electronic Communications Committee (ECC) within CEPT: Coexistance between ultra-low power active medical implants (ULP-AMI) and existing radiocommunication systems and services in the frequency bands 401–402 MHz and 405–406 MHz, ECC Report **92**, 1–15 (2006)

61.19 G. Iddan, G. Meron, A. Glukhovsky, P. Swain: Wireless capsule endoscopy, Nature **405**, 417 (2000)

61.20 D.M. Ackermann, B. Smith, W. Xiao-Feng, K.L. Kilgore, P. Hunter Peckham: Designing the optical interface of a transcutaneous optical telemetry link, IEEE Trans. Biomed. Eng. **55**(4), 1365–1373 (2008)

61.21 K.S. Guillory, A.K. Misener, A. Pungor: Hybrid RF/IR transcutaneous telemetry for power and high-bandwidth data, Proc. 26th Annu. Int. Conf. IEEE EMBS (2004) pp. 4338–4340

61.22 Y. Mitamura, E. Okamoto, T. Mikami: A transcutaneous optical information transmission system for implantable motor-driven artificial hearts, Am. Soc. Artif. Int. Organs Trans. **36**(3), M278–M280 (1990)

61.23 J.M. Ahn, J.H. Lee, S.W. Choi, W.U. Kim, K.S. Omn, S.K. Park, W.G. Kim, J.R. Roh, B.G. Min: Implantable control, telemetry, and solar energy system in the moving actuator type total artificial heart, Artif. Organs **22**(3), 250–259 (1998)

61.24 E. Okamoto, Y. Yamamoto, Y. Inoue, T. Makino, Y. Mitamura: Development of a bidirectional transcutaneous optical data transmission system for artificial hearts allowing long-distance data communication with low electric power consumption, J. Artif. Organs **8**(3), 149–153 (2005)

61.25 N. Kudo, K. Shimizo, G. Matsumoto: Fundamental study on transcutaneous biotelemetry using diffused light, Front. Med. Biol. Eng. **1**(1), 19–28 (1988)

61.26 F.P. Strietzel, R. Martín-Granizo, S. Fedele, L. Lo Russo, M. Mignogna, P.A. Reichart, A. Wolff: Electrostimulating device in the management of xerostomia, Oral Dis. **13**(2), 206–213 (2006)

61.27 O. Scholz, A. Wolff, A. Schumacher, L.I. Giannola, G. Campisi, T. Ciach, T. Velten: Drug delivery from the oral cavity: Focus on a novel mechatronic delivery device, Drug Discov. Today **13**(5/6), 247–253 (2008)

61.28 J. Chen, B. Wilkoff, W. Choucair, T. Cohen, G. Crossley, W.B. Johnson, L. Mongeon, G. Serwer, L. Sherfesee: Design of the pacemaker remote follow-up evaluation and review (PREFER) trial to assess the clinical value of the remote pacemaker interrogation in the management of pacemaker patients, Trials **9**(1), 18 (2008)

62. Fundamentals of Medical Image Processing

Thomas M. Deserno

After some remarks to the background and terminology used, Sect. 62.3 deals with low-level image processing as far as necessary to understand the following chapters. Subsequently, the core steps of image analysis: feature extraction, segmentation, classification, quantitative measurements, and interpretation are presented in separate sections. Here, the focus is on segmentation of medical images, because this is of high relevance and, therefore, special methods and techniques have been developed in the medical application domain. In Sect. 62.9, certain aspects of medical data visualization are outlined. Many methods have been developed in this field specifically for clinical applications. Section 62.10 provides a brief summary of image communication. The electronic transmission and exchange of medical images will become more important in future for multimedia applications such as electronic patient records. Section 62.11 completes this chapter with an overview of past, present, and future challenges to medical image processing.

62.1 **Background** 1139
62.2 **Remarks on the Terminology** 1141
62.3 **Image Enhancement** 1142
62.3.1 Histogram Transforms 1142
62.3.2 Convolution 1143
62.3.3 Calibration 1144
62.3.4 Registration 1145
62.4 **Feature Extraction** 1147
62.5 **Segmentation** 1148
62.5.1 Pixel-Based Segmentation 1148
62.5.2 Edge-Based Segmentation 1150
62.5.3 Region-Based Segmentation 1151
62.5.4 Hybrid Segmentation Procedures 1152
62.6 **Classification** 1154
62.6.1 Statistic Classifiers 1155
62.6.2 Syntactic Classifiers 1155
62.6.3 Computational Intelligence-Based Classifiers 1155
62.7 **Quantitative Measurements** 1157
62.8 **Interpretation** 1158
62.9 **Image Data Visualization** 1158
62.9.1 The Marching Cubes Algorithm 1159
62.9.2 Surface Rendering 1160
62.9.3 Volume Rendering 1161
62.10 **Image Management** 1161
62.11 **Conclusion and Outlook** 1163
References 1165

By the increasing use of direct digital imaging systems for medical diagnostics, digital image processing is becoming more and more important in health care. In addition to originally digital methods, such as computed tomography (CT) or magnetic resonance imaging (MRI), initially analog imaging modalities such as endoscopy or radiography, have been equipped with digital sensors. Hence, medical image processing is becomming an inherent part of medical technology.

62.1 Background

Digital images are composed of individual pixels (this acronym is formed from the words picture and element), where discrete brightness or color values are assigned. They can be efficiently processed, objectively evaluated, and made available at many places at the same time by means of appropriate communication networks and protocols, such as picture archiving and communication systems (PACS) and the digital imaging and

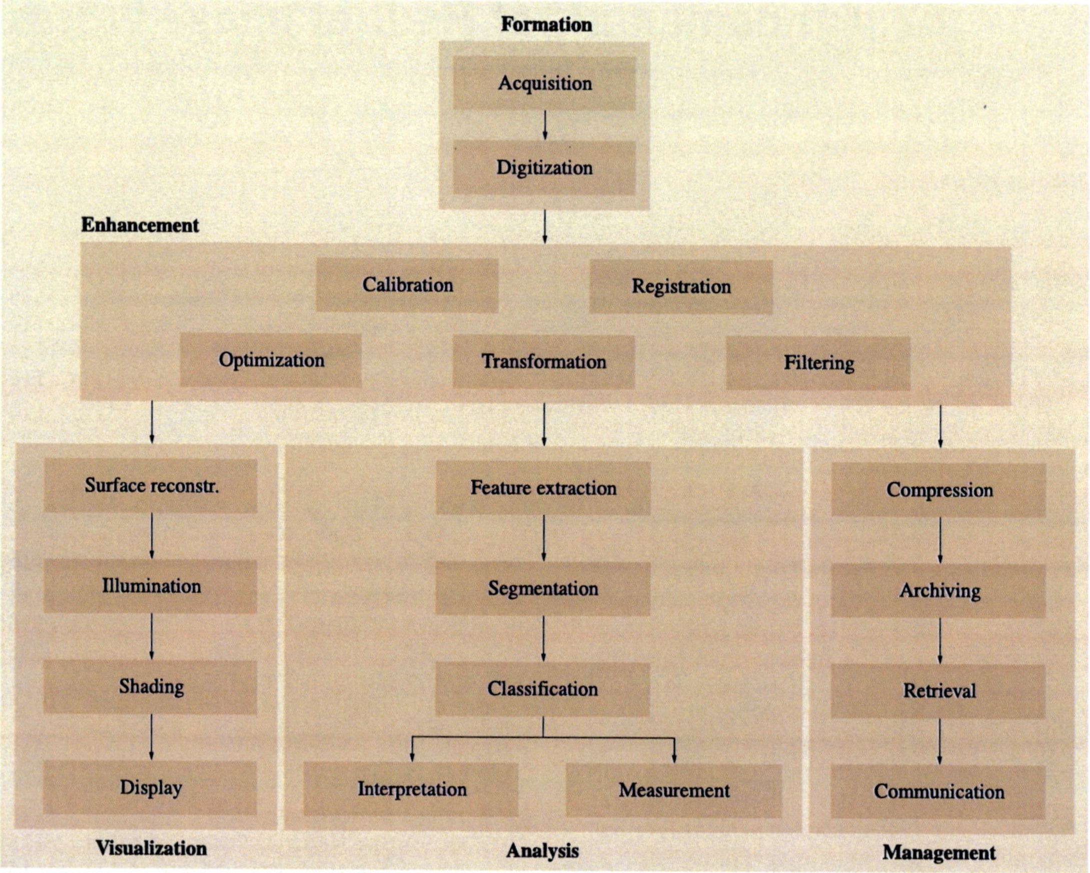

Fig. 62.1 Modules of medical image processing (after [62.1]). Medical image processing covers four main areas: image formation, analysis, visualization, and management. The algorithms of image enhancement can be assigned as pre and post in all areas

communications in medicine (DICOM) protocol, respectively. Based on digital imaging techniques, the entire spectrum of digital image processing is now applicable in medicine.

The commonly used term *medical image processing* means the provision of digital image processing for medicine. Medical image processing covers four major areas (Fig. 62.1):

1. *Image formation* includes all the steps from capturing the image to forming a digital image matrix.
2. *Image visualization* refers to all types of manipulation of this matrix, resulting in an optimized output of the image.
3. *Image analysis* includes all the steps of processing, which are used for quantitative measurements as well as abstract interpretations of medical images. These steps require a-priori knowledge on the context and content of the images, which must be integrated into the algorithms on a high level of abstraction. Thus, the process of image analysis is very specific, and developed algorithms can be transferred rarely directly into other domains of applications.
4. *Image management* sums all techniques that provide the efficient storage, communication, transmission, archiving, and access (retrieval) of image, since an uncompressed radiograph may require several megabytes of storage capacity. The methods of telemedicine are also a part of the image management.

In contrast to image analysis, which is often also referred to as high-level image processing, low-level processing denotes manual or automatic techniques, which can be implemented without a-priori knowledge on the specific content of images. This type of algorithm has similar effects regardless the content of the images. For example, histogram stretching of a radiograph improves the contrast as it does on any holiday photograph. Therefore, low-level processing methods are usually available with programs for image enhancement. On this background, this chapter gives an introduction to the methods of medical image processing.

62.2 Remarks on the Terminology

The complexity of an algorithm, the difficulty of its implementation, or the computing time required for image processing plays a secondary role for the distinction between low-level and high-level processing methods. Rather, the degree of abstraction of the a-priori knowledge is important for this meaning. Although the following definitions are not standardized in the literature, they are used consistently within this chapter:

- The *raw data* level records an image as a whole. Therefore, the totality of all pixels is regarded on this level.
- The *pixel level* refers to discrete individual pixels.
- The *edge level* represents the one-dimensional (1-D) structures, which are composed of at least two neighbored pixels.
- The *texture level* refers to two-dimensional (2-D) structures. On this level, however, the delineation of the area's contour may be unknown.
- The *region level* describes 2-D structures with a well-defined boundary.
- The *object level* associates textures or regions with a certain meaning or name, i. e. semantics is given on this level.
- The *scene level* considers the ensemble of image objects in spatial and/or temporal terms.

From an iconic (concrete) to a symbolic (abstract) description of images, information is gradually reduced. Methods of low-level image processing operate on the raw data as well as pixel, edge, or texture levels, and thus at a minimal level of abstraction. Methods of high-level image processing include the texture, region, object, and scene levels. The required abstraction can be achieved by an increased modeling of a-priori knowledge. With these definitions, a particular problem in high-level processing of medical images is immediately apparent: resulting from its complex nature, it is difficult to formulate medical a-priori knowledge such that it can be integrated directly and easily into automatic algorithms of image processing. In the literature, this is referred to as the *semantic gap*, which means the discrepancy between the cognitive interpretation of a diagnostic image by the physician (high level) and the simple structure of discrete pixels, which is used in computer programs to represent an image (low level). In the medical domain, there are three main aspects of hindering bridging this gap:

1. *Heterogeneity of images*: Medical images display organs or body parts. Even if captured with the same modality and following a standardized acquisition protocol, shape, size, and internal structures of these objects may vary remarkably not only from patient to patient (inter-subject variation) but also among different views of a patient and similar views of the same patient at different times (intra-subject variation). In other words, biological structures are subject to both, inter- and intra-individual alterability. Thus, universal formulation of a-priori knowledge is impossible.
2. *Unknown delineation of objects*: Frequently, biological structures cannot be separated from the background because the diagnostically or therapeutically relevant object is represented by the entire image.

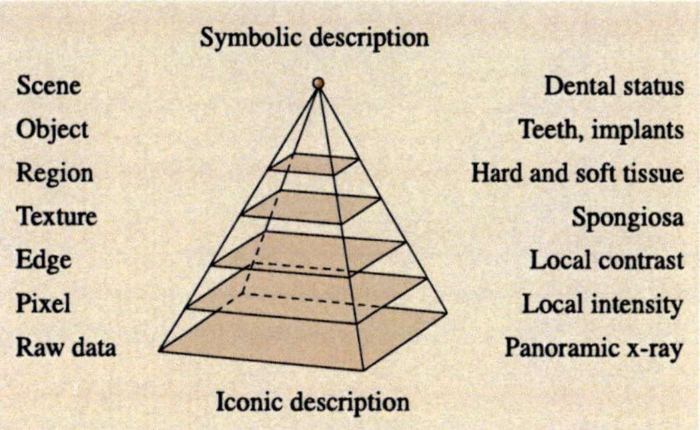

Fig. 62.2 Levels of abstraction for image processing (after [62.2]). On the pyramid's *left-hand side*, the degrees of abstraction for image processing are specified. On the *right-hand side*, the general terms are exemplified for a panoramic radiograph. At the *top*, the dental status corresponds to an abstract scene analysis, which only contains standardized information (existence and condition) on the tooth positions

Even if definable objects are observed in medical images, their segmentation is problematic because the shape or borderline itself is represented fuzzily or only partly. Hence, medically related items often can be abstracted at most on the texture level.

3. *Robustness of algorithms*: In addition to these inherent properties of medical images, which complicate their high-level processing, special requirements of reliability and robustness of medical procedures and, when applied in routine, image processing algorithms are also demanded in the medical area. As a rule, automatic analysis of images in medicine should not provide wrong measurements. This means that images that cannot be processed correctly must be automatically classified as such and rejected. Consequently, all images that are not rejected must be evaluated correctly. Furthermore, the number of rejected images must be quite small, since most medical imaging procedures are harmful and cannot be repeated just because of processing errors.

62.3 Image Enhancement

Low-level methods of imaging processing, i. e., procedures and algorithms that are performed without a-priori knowledge about the specific content of an image, are mostly applied for pre-processing of medical images. Therefore, the basic methods of histogram transforms, convolution, and (morphological) filtering are disregarded unless required for further understanding of this text. As a special preprocessing of medical images, techniques for calibration and registration are briefly introduced.

62.3.1 Histogram Transforms

Point operations (pixel transforms) are based on the histogram of the image. Modifying the pixel values, all pixels are transformed independently from their positions in the image and their immediate neighborhood. Therefore, these types of transforms are also referred to as point operations. The histogram shows the frequency distribution of pixel values (e.g., the gray scales) disregarding certain positions where the gray scales occur in the image. Simple pixel transforms can be defined using a histogram. For example, through the stretching of gray scales, the contrast of an image is improved (Fig. 62.3). After determining the histogram, upper and lower bounds are located, and a linear transform is applied that maps the lower bound to zero and the upper bound to the maximal gray scale (i. e., 255 for 8 bit images). If the histogram of the initial image does not contain all possible gray scales, the gray scale differences between neighbored pixels is enlarged, which results in an enhanced contrast.

Technically, those histogram transforms are computed by look-up tables. For all pixel values, the look-up

Part F | 62.3

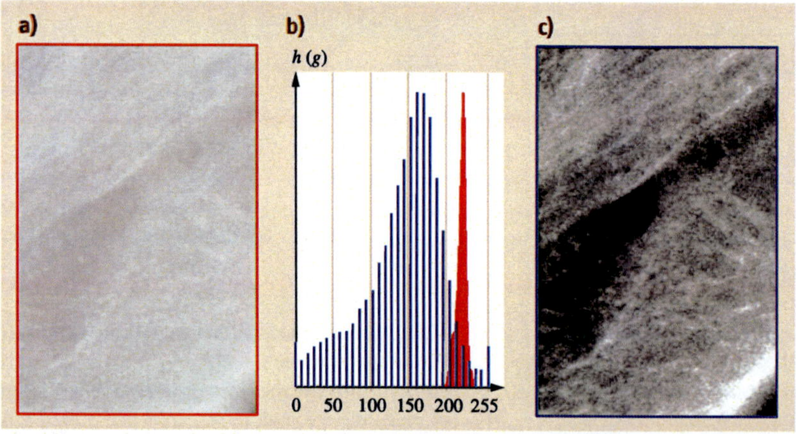

Fig. 62.3a–c Histogram stretching. A region of interest (ROI) is taken in the area of the temporomandibular joint from an intra-oral x-ray image **(a)**. Resulting from underexposure, the spongy bone structure is displayed poorly. The associated histogram **(b)** is only narrow occupied (*red*). By stretching the histogram, the bins in the histogram are linearly pulled apart (*blue*) and the contrast of resulting transformed radiograph is increased **(c)**

Table 62.1 Look-up-table for pseudo coloring [62.3]. For each value in the range of the input image, the look-up table holds a value from the range of the output image. The color palette shown here is used for pseudo coloring keeping the original brightness progression of the input image. Pseudo coloring allows presentation of data, whose range of values exceeds the length of the RGB cube's edges without reducing the information as results from windowing

Old	New pixel value			Old	New pixel value		
gray	red	gray	red	gray	red	green	blue
0	0	0	0	...	...	...	...
1	1	0	2	246	254	244	239
2	2	0	3	247	255	245	243
3	3	0	5	248	255	245	244
4	4	1	7	249	255	246	246
5	5	1	9	250	255	247	248
6	5	2	12	251	255	249	250
7	5	2	14	252	255	251	252
8	5	3	16	253	255	251	253
9	5	4	18	254	255	253	254
...	...	...	...	255	255	255	255

a)

1	1	1
1	1	1
1	1	1

b)

1	2	1
2	4	2
1	2	1

c)

−1	−2	−1
−2	12	−2
−1	−2	−1

d)

0	−1	0
−1	5	−1
0	−1	0

e)

−1	−2	−1
0	0	0
1	2	1

f)

−2	−1	0
−1	0	1
0	1	2

Fig. 62.4a–f Simple templates for discrete convolution. The sliding average (**a**) and the binomial low-pass filter (**b**) cause a smoothing of the image. The binomial high-pass filter (**c**), however, increases contrast and edges, but also the noise in the image. The templates (**a**) to (**c**) must be normalized to make sure that range domain of values is not exceeded. The contrast filter (**d**) is based on integer pixel values. The convolution with (**d**) is, therefore, very easy to calculate. The anisotropic templates (**e**) and (**f**) belong to the family of Sobel operators. Eight Sobel masks can be generated by rotation and mirroring for direction-selective edge filtering (Fig. 62.8)

table contains a new value, which can also originate from another range of values. The example in Table 62.1 assigns each gray scale with a triple for red, green, and blue (RGB). This transform is called pseudo-coloring and is frequently used in the medical domain to enhance local contrast. Computer graphic boards may limit the number of gray scales to 256 (8 bit), but offers 2563 = 16 777 216 colors. Special algorithms are recommended for pseudo-coloring in the medical context.

62.3.2 Convolution

In contrast to point operations (histogram transforms), the considered pixel are combined with the values of their neighborhood when discrete filtering is applied. The underlying mathematical operation, i. e., convolution, can be characterized with the help of so called templates (Fig. 62.4). A *template* is a mostly small, squared mask with odd lateral length. This template is mirrored along two axes (hence the name convolution is commonly used) and positioned in one corner of the input image. The image pixels under the mask are named *kernel*. Each pair of corresponding pixel values of template and kernel are multiplied and then added together. The result is registered at the position of the mask's center pixel in the output image. Then, the template is shifted row by row and column by column to the next positions on the input image, until all the positions have been visited, and the output image is thus completely calculated. The pixel values of the template determine the effect of the filter. If only positive values are used in the template, basically a (weighted) averaging is calculated in the local neighborhood of each

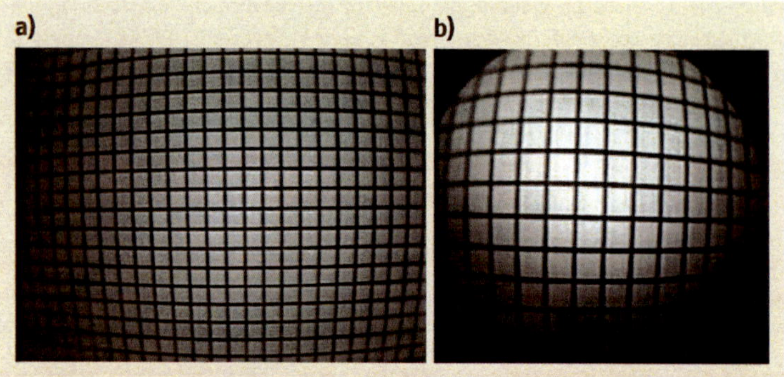

Fig. 62.5a,b Geometric distortion and brightness variations in video endoscopy (after [62.1]). With endoscopic examinations, barrel distortions are often generated, which must be corrected before the image can be analyzed quantitatively. In addition, the boundary areas in the video appear darker and blurred. Image **(a)** is generated with a rigid laryngoscope, which is used for the examination of the larynx. Image **(b)** is taken with a flexible endoscope for nasal laryngoscopy. Both endoscopes are used in clinical routine. Microscopy and other optical methods produce similar artifacts

pixel (Fig. 62.4a,b). The resulting image is smoothed and appears with reduced noise.

However, the sharpness of edges is also reduced. If the template is composed of positive and negative coefficients, the contrast in the image is intensified, and the edges are highlighted. Anisotropic (i. e., not rotationally symmetric) templates also have a preferred direction (Fig. 62.4e,f). Hereby, the contrasts can be direction-selectively strengthened.

Another approach to filtering is adapted from mathematical morphology. Although morphologic operators can also be defined to gray scale image, morphologic filtering is principally performed on binary input images, i. e., each pixel is assigned either true or false. According to a general convention, the white pixels in the binary image indicate relevant segments and the black pixels indicate the background. For printing, however, this assignment may be inverted. The binary template, which is also referred to as *structural element*, is associated to the binary image using logical operations, in particular:

- *Erosion* (based on logical AND template and binary image)
- *Dilatation* or dilation (based on logical OR template and binary image)
- *Opening* (erosion followed by dilatation using the same structuring element)
- *Closing* (dilatation followed by erosion using the same structuring element)
- *Skeleton* (e.g. by erosion with various structuring elements).

The erosion reduces the size of a segment, and the dilation leads to its enlargement. The opening removes small details on the outline of segments or the background, without affecting the total size of relevant segments. The closing is able to remove holes in the interior of a segment and smooth its contour. Here, the size of the segment is also roughly maintained. The skeleton is a path with the thickness of one pixel, which is located in the middle of the segment.

62.3.3 Calibration

If the physician intends to take quantitative measurements from an image, a careful calibration of the imaging modality is required. Both geometry (spatial domain) and brightness or color intensity (value domain) must be adapted to the modality. Calibration is device-specific but disregards the biological content captured, and thus, it is part of low-level processing methods. While reading a radiograph, calibration is made unconsciously by the radiologist. However, it must be explicitly implemented for computerized image analysis. Geometric aberration (distortions) have the consequence that relevant structures of the same size are displayed depending on the position within the image. In medicine, the positioning of the imaging device may not affect any measurements. For example, in endoscopy, resulting from the optical devices in use, so called barrel distortions originate (Fig. 62.5). Even in simple planar radiography, objects that are far away from the image plane appear larger than those located

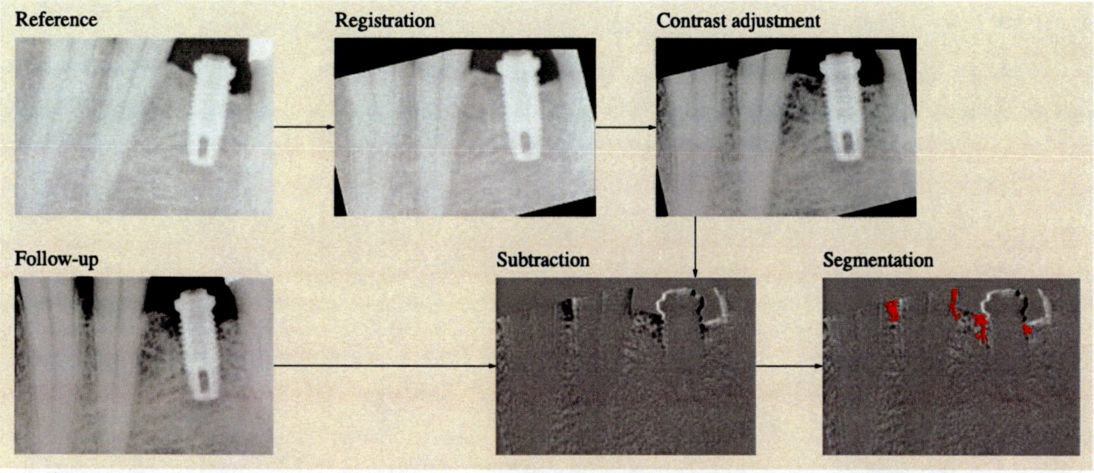

Fig. 62.6 Unimodal registration. In dental implantology, the reference image and recall recording are taken at various times. The geometric registration with subsequent contrast adjustment enables pixel-by-pixel subtraction. In the subtraction image, bone destruction is clearly emphasized and can be segmented easily on the pixel level of features (*red*)

close to the imaging device. This must be kept in mind whenever geometric measurements in digital x-rays are made and displayed to the physicians: point distances in digital images can be converted into length measurements only under the assumption of a fixed scale, which is often not fulfilled.

In the same way, is the absolute assignment of pixel values to physical measurements is usually problematic. For example, in x-ray imaging, the linear correspondence of brightness values to the accumulated absorption coefficient of the imaged structure is possible only if an aluminum (step) wedge with known x-ray absorption properties is placed beside the object. With digital video recording, a prior white balance must be performed, so that the color values correspond to reality. However, different illumination of the same scene may still alter the colors captured.

62.3.4 Registration

Often, an absolute calibration of examination procedures is not possible or only limitedly feasible. Then, registration can be used to achieve an approximation of two or more images such that at least a change in measured dimensions can be quantified. For example, an acute inflammation turns tissue into a reddish color. Under treatment, the absolute redness of the tissue is less interesting than its relative change from the findings of previous recordings. *Single-modal registration* refers to relative calibration of images that have been acquired with the same modality. For instance, images that have been taken from the same patient but at different points of time are adjusted in order to quantify the course of the disease. As in the field of calibration, we differ between geometric registration and color or contrast adjustment, if the registration is performed in the spatial domain or the value range, respectively. Figure 62.6 illustrates the diagnostic potential of registration in dental implantology. After registration, the appraisal of the status of peri-implant bone is significantly simplified by the subtraction of recall and follow-up recordings.

Multimodal registration is applied when the images to be compared are captured from different modalities. For example, a three-dimensional (3-D) rigid registration is illustrated as the movement of the hat on the head. These methods have a crucial meaning especially in neurology. Since tumor resection in the brain must be executed very carefully, in order to avoid damage of neighboring brain areas, functional and morphological brain images are registered to plan the procedure. While the morphological information in CT or MRI data can be adequately represented, function of areas in the brain is frequently localized using positron emission tomography (PET) or single photon emission computed tomography (SPECT). Thus, multimodal registration of functional and morphological datasets provides valuable additional information for diagnosis and therapy (Fig. 62.7).

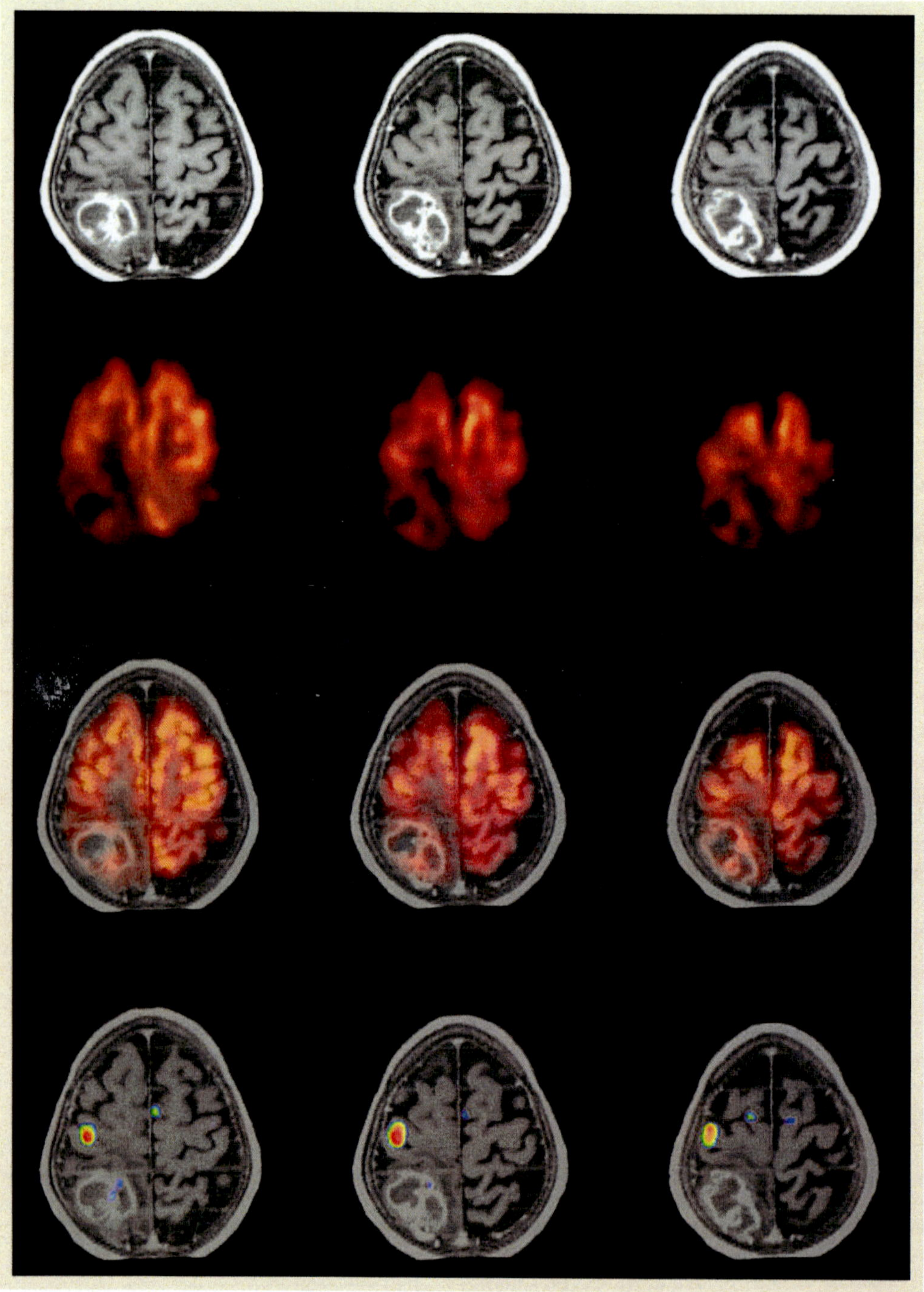

Fig. 62.7 Multimodal registration and fusion (after [62.4]). *1. Row*. T_1-weighted MRI of a 66 year old subject with right parietal glioblastoma. *2. Row*. Corresponding PET layers after multimodal registration. *3. Row*. Fusion of registered layers to support the treatment planning. *4. Row*. The fusion of MRI with PET of the sensorimotor-activated cortex area proves that the relevant area is out of focus

62.4 Feature Extraction

In Fig. 62.1, feature extraction is defined as the first stage of intelligent (high level) image analysis. It is followed by segmentation and classification, which often do not affect in the image itself, i. e. the data or pixel level, but are performed on higher abstract levels (Fig. 62.2). Therefore, the task of feature extraction is to emphasize image information on the particular level where the following algorithms operate. Consequently, information provided on other levels must be suppressed. Thus, data reduction is executed to obtain the characteristic properties.

The schema in Fig. 62.1 is greatly simplified because many connections between the modules were left out on behalf of readability. So, for example, cascades of feature extraction and segmentation at various levels of abstraction can be realized gradually, before classification is eventually performed at a high level of abstraction. Just before classification, a step of feature extraction that is based on the region level is often performed as well. *Data-based features* depend on the joint information of all pixels. Therefore, all transforms manipulating the whole matrix of an image at once can be regarded for data feature extraction.

The most famous example of a data feature transform is the Fourier transform, which describes a 2-D image in terms of frequencies, according to their amplitude and phase. Furthermore, the Hough, wavelet, or Karhunen–Loève transforms provide possibilities of data feature extraction (see Further Reading on image processing for more details). These methods are not in the focus of research in medical image processing. In fact, these procedures are rather adapted from technical areas to medical applications.

Since *pixel-based features* depend on the values of individual pixels, all point operations that have been defined in Sect. 62.3 can be regarded as feature extraction on the pixel level. Another example was already presented in Fig. 62.6, namely, the arithmetic combination of two images. The subtraction of reference and recall images after appropriate registration in both spatial and value ranges enforces local changes in the images as characteristic pixels.

Edge-based features are defined as local contrast, i. e., a strong difference of (gray scale or color) values of adjacent pixels. Thus, the discrete convolution introduced in Sect. 62.3 can be used with appropriate

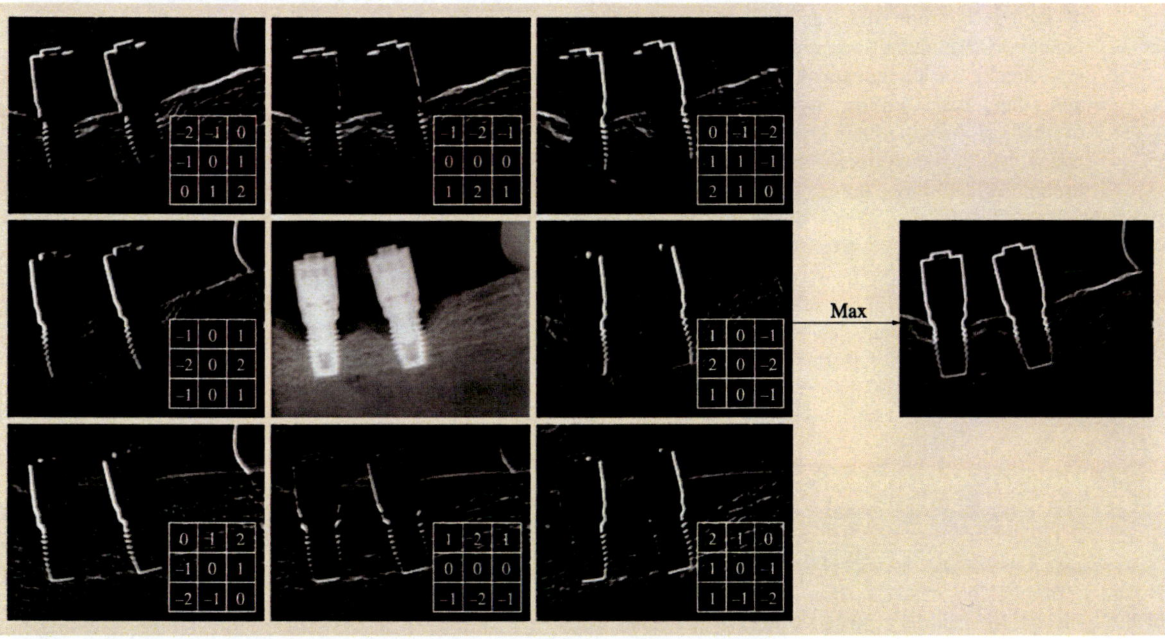

Fig. 62.8 Edge extraction with Sobel operator. The x-ray image was convolved with the eight direction-selective Sobel templates. The strong contrasts on the edges of metallic implants are further strengthened by binarization of the edge images. An isotropic edge image is obtained if, e.g., the maximum at each pixel position is chosen from the eight direction-selective subimages

templates for edge extraction. All masks for high-pass filtering amplify edges in an image. The templates of the so called Sobel operator (Fig. 62.4e,f) are particularly suited for edge extraction. Figure 62.8 exemplarily presents the result of the orientation-selective Sobel masks when applied to a dental radiograph. The edges of the metallic implants are clearly highlighted. Moreover, an isotropic Sobel-based edge image is achieved, e.g., by a linear or maximum combination of the eight subimages.

Textural features have been used in medicine for a long time. In textbooks on pathology one can read many metaphors to describe texture, such as a cobblestone-like mucosal relief, onion-like stratification of subintima, or honeycomb-like lung. As intuitive these metaphors are for people, as difficult is computational texture processing, and a variety of procedures and approaches have been developed. Texture analysis attempts to quantify objectively the homogeneity in a heterogeneous but at least subjectively periodic structure (see the spongious bone structure in Fig. 62.3c as an example). In general, we can distinguish (i) structural approaches, which are based on texture primitives (so called texel or textone) and their rules of combinations, and (ii) statistical approaches, which describe texture by a set of empirical parameters. *Regional features* are used primarily for object classification and identification. They are normally calculated for each segment after the segmentation process. The most important parameters to be mentioned here are (i) localization-descriptive measurement such as size, position, and orientation of the major axes, and (ii) delineation-descriptive measures such as shape, convexity, and length of the border. Since the degree of abstraction on the regional level is rather high as compared to the previous levels, a-priori knowledge has already been largely integrated. Therefore, universal examples cannot be specified. In fact, the definition of regional feature extraction is strongly dependent on the respective application (Sects. 62.5.3, 62.6).

62.5 Segmentation

Segmentation generally means dividing an image into connected regions. With this definition, the production of regions is emphasized as the pre-stage of classification. Other definitions emphasize the various diagnostically or therapeutically relevant image areas and, thus, focus on the most common application of medical imaging, namely, the discrimination between healthy anatomical structures in pathological tissue. Per definition, the result of segmentation is always on the regional level of abstraction (Fig. 62.2). Depending on the level of feature extraction as an input to the segmentation, we can methodically classify pixel, edge, and texture or region-oriented procedures. In addition, there are hybrid segmentation procedures, which result from the combination of single approaches.

Part F | 62.5

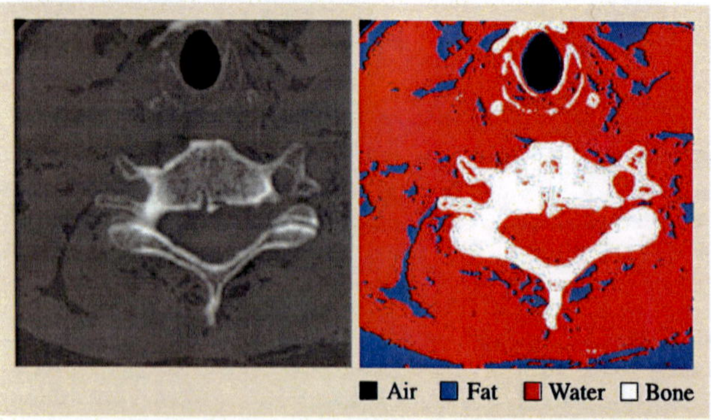

Fig. 62.9 Thresholding of a CT slice. The CT slice in the area of the spine can be statically segmented because the standardized Hounsfield units (HU) allow the definition of windows for different tissue such as bone [200 ... 3000], water [−200 ... 200], fat [−500 ... −200], and air [−1000 ... −500]

62.5.1 Pixel-Based Segmentation

Pixel-based procedures for segmentation only consider the gray scale or color value of current pixels, disregarding its surroundings. It should be pointed out that pixel-based procedures are not segmentation procedures in the strict sense of our definition. Since each pixel is considered only isolated from its neighborhood, it cannot be ensured that actually only connected components are obtained. For this reason, a post-processing is required, e.g., by morphologic filtering (Sect. 62.3).

Most pixel-based procedures use thresholds in the histogram of an image and employ more or less complex methods to determine this threshold, or statistical methods for pixel clustering are used.

Static thresholds can be applied if the assignment of pixel intensities is well known and constant for a certain type of tissue. A static threshold is independent of each image. For example, bone or soft tissue windows in the

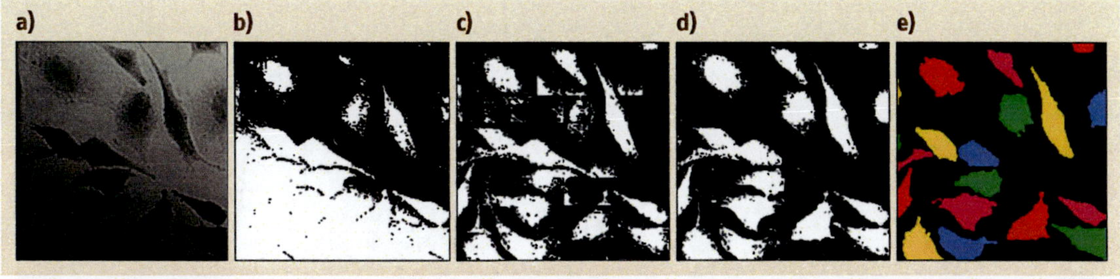

Fig. 62.10a–e Dynamic thresholding in microscopy (after [62.5]). The microscopy of a cell culture **(a)** was segmented using a global threshold **(b)**, locally adaptive **(c)** and pixel-adaptive **(d)**. According to morphological post-processing for noise reduction and a connected components analysis, the final segmentation is shown in **(e)**

CT with static thresholds on the Hounsfield units can be realized (Fig. 62.9).

Globally adaptive thresholds result from an analysis of each image. They are exclusively used in this image. The well known method of Otsu is based on a simple object/background model. The threshold in the histogram is determined such that the two resulting classes minimize the intra-class variance of gray scale values, while the inter-class variance is maximized. For example in skeletal radiography, bone, soft tissue, and background can be seen, but the actual mean gray scale of these tissue classes may vary with respect to illumination and exposure parameters. By adopting the threshold to the image, the Otsu segmentation is able to balance this variation in imaging.

Using *locally adaptive thresholds*, the threshold is computed not only for each image individually, but also for each region within an image. In the extreme case, an individual threshold is determined for every pixel position. This is particularly necessary if the simple object to background assumption is globally invalid, for instance, because of continuous brightness gradients. For example, due to the irregularity of optical illumination, the background in microscopy imaging of cell cultures (Fig. 62.10a) runs from light shades of gray (top right) to dark shades of gray (bottom left), where also the gray scale values of the cells are located. A global threshold determined with the dynamic procedure of Otsu (Fig. 62.10b) does not separate the cells from backgrounds, although the global threshold has been determined image-individually. The locally adaptive segmentation (Fig. 62.10c) leads to a significantly improved result, but isolated block artifacts appear. These artifacts can be avoided only by pixel-adaptive thresholding (Fig. 62.10d).

Pixel clustering is another way of pixel-based segmentation. This statistical method is particularly suitable if more than one value is assigned to each pixel and regarded in the segmentation process (e.g., color images). Figure 62.11 illustrates the isodata clustering algorithm (also referred to as k-means clustering) in a simple 2-D case. All pixel values are registered as data point in the 2-D feature space. Initialized by the number of segments to be obtained, the initial cluster centers are arbitrarily placed by the algorithm. Then, the following two steps are repeated iteratively until the process converges:

1. Each data point is aligned to the closest cluster center
2. Based on the current assignment, the cluster centers are re-calculated.

It can be proven mathematically that the resulting cluster centers are independent of initial positions, which may only impact the number of iterations and hence, the calculation time. However, either a fixed distance metrics (e.g., Euclidean (geometric) distance) or a data adaptive metrics (e.g. Mahalanobis distance) must be selected, which certainly impacts the clustering result. Also, the pre-defined number of cluster centers is an important parameter. If the application domain does not allow determination of the number of segments a-priori, pixel clustering can be performed for a different number of centers and the residual error of the computed model can be analyzed to determine the appropriate number of centers. Segments obtained from pixel-based analysis are usually incoherent and highly noisy (Fig. 62.9).

Therefore, post-processing is required. Noisy structures can be effectively reduced with methods of mathematical morphology. While a morphologic opening removes spread parts from the segments, holes are closed by morphologic closing (Sect. 62.3). The connected components algorithm provides each sepa-

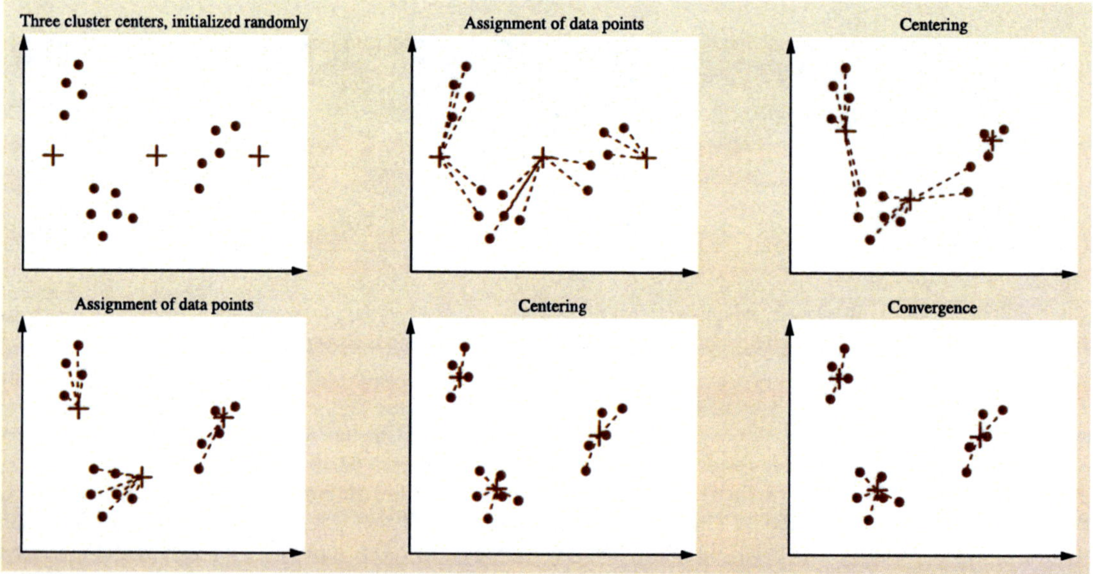

Fig. 62.11 Isodata pixel clustering. The iterative isodata algorithm for pixel clustering is exemplified in a 2-D feature space. The number of clusters is given a-priori. After arbitrary initialization, the data points are assigned to the nearest cluster center. Then, the positions of the centers are re-calculated and the assignment is updated until the process finally converges. The final location of cluster centers is not affected by their initial position. This may only have impact on the number of iterations

rated segment with a unique reference number. In the segmentation of the cell image (Fig. 62.10); clustering provides a rough cluster of cells, which is separated from the background, although many individual cells are shown separately in the image (Fig. 62.10d). After morphological post-processing and connected components analysis, cells are separated and colored (labeled) differently according to their segment number. Now, they can be further processed as independent objects (Fig. 62.10e).

62.5.2 Edge-Based Segmentation

This type of segmentation is based on the abstract level of edges and tries to capture the objects due to their closed outline in the image. Hence, edge-based segmentation procedures are only used for such problems where objects are represented as clearly defined boundaries. As described in Sect. 62.2, this occurs rather seldom when biologic tissue is imaged. One of these special cases is a metallic implant shown in a radiograph. In general, the image processing chain for edge-based segmentation is composed of edge extraction, edge completion, and object recognition. Edge extraction is usually obtained by edge-based feature extraction, as described in Sect. 62.4, such as generated with the Sobel filter (Fig. 62.8). The next steps of processing are binarization, to obtain only edge pixels and nonedge pixels, morphological filtering to reduce noise and artifacts, and, finally, a skeleton of the edge is computed. Tracing and closing of binary contours are the main tasks of the edge-based segmentation. Almost exclusively heuristical methods are used here. For example, one can search along some rays for connecting pieces of a contour of an object to skip the local gaps of the edge.

In practice, edge-based segmentation is often realized semi-automatically. With interactive live-wire segmentation, the user clicks on or near the edge of the object of interest, and the computer determines the exact edge location based on local gradients. Then, the computer calculates a cost function, which is again based on local gradients; for all paths (wire) to the current position of the cursor, the path with the least cost is displayed in real time (live) as the curser is moved manually. Therefore, the metaphor live-wire is commonly used to refer to this interactive method of segmentation. If the cursor moves far from the object, the contour is lost, but if the curser is placed near the contour again, the cost function ensures that the wire snaps back to the desired object.

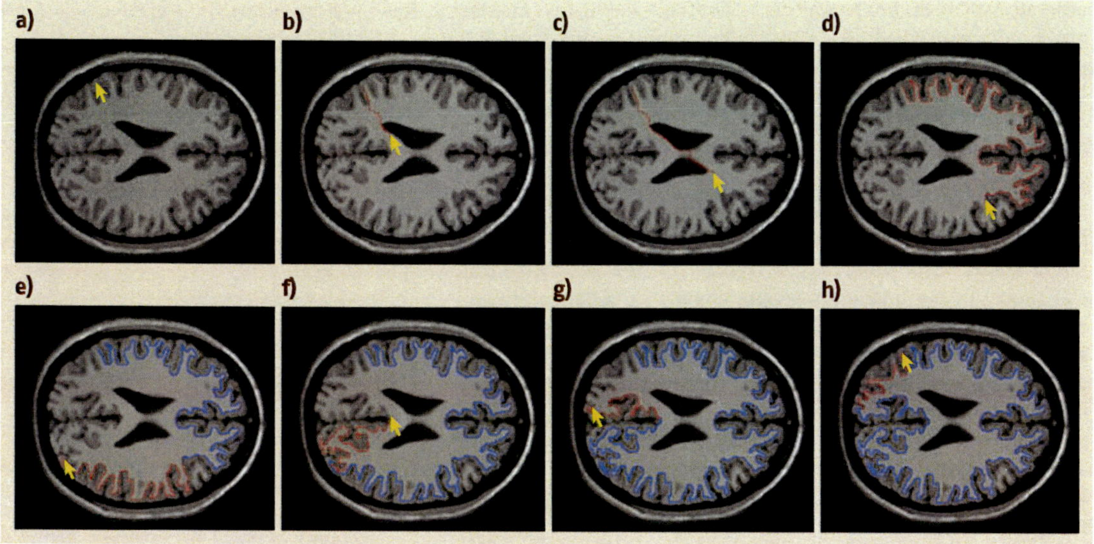

Fig. 62.12a–h Edge-based interactive live-wire segmentation in MRI (after [62.6]). The user marks a starting point with the cursor (*yellow*) on the border between white and gray matter **(a)**. The connection to the current cursor position is denoted with *red*, cf. **(b)** to **(e)**. Depending on the cursor position, the contour can also jump between very different courses **(d)**,**(e)**. So, the user can interactively place an appropriate fix point. The fixed curve segment is shown in *blue*, cf. **(e)** to **(g)**. In this example, only five points are manually marked to achieve a complete segmentation **(h)**

Finally, the user must provide only a few support points by hand and can directly verify the correctness of segmentation (Fig. 62.12). Application of such procedures can be found at computer-assisted (semiautomatic) segmentations in layers of CT data, e.g., to produce a model for surgical treatment planning.

62.5.3 Region-Based Segmentation

As an advantage of region-based segmentation, only connected segments are produced, and morphological post-processing is avoided. There are agglomerative (bottom-up), divisive (top-down), and hierarchical approaches. All approaches are based on a certain distance or similarity measure to guide the assignment of neighbored pixels or regions. Here, a variety of methods is used. The easiest is to compare the mean gray value but complex texture measures (Sect. 62.4) are also often used.

Region growing (in 3-D also referred to as volume growing) is a well known example of an agglomerative procedures. Starting from seed points, which may be placed either automatically or manually, neighboring pixels are iteratively associated to the growing areas if the distance is below a certain threshold. This process is iterated until no more merges can be carried out. From this qualitative description, the variety and sensitivity of the parameters of such procedures are already clear. The following have a special influence on the result of agglomerative segmentation:

- The number and position of seed points
- The order in which the pixels or voxels (in analogy to the term pixel, this acronym is formed from the words volume and element) are iteratively processed
- The distance or similarity measure applied
- The threshold used to guide merging.

Therefore, agglomerative algorithms for segmentation often are affected by small shifts or rotations of the input image, in particular if x and y-axes of the image matrix are transposed, which is unwanted in medical image processing. The *divisive approach* somehow inverts the agglomerative strategy. The regions are iteratively subdivided by splitting until they are sufficiently homogeneous in terms of the chosen similarity measure. As an advantage, seed points are not required, because the first split is performed throughout the whole image. As a drawback, the dividing lines are usually drawn horizontally or vertically, and this arbitrary separation may separate individual image objects. Therefore, a split is not usually performed as a self-standing segmentation procedure, but rather combined with a subsequent merging step (so called split & merge). Another drawback of divisive segmentation procedures is the resulting wedge-formed boundary of objects, which may require post-processing such as smoothing. A fundamental problem of pixel-based and region-based segmentation is the dualism between *over* and *under-segmentation*. For a definition of these terms, we rely on the general model of the image processing chain (Fig. 62.1). Here, segmentation is regarded as a pre-stage for classification, in which the extracted image segments are assigned to their semantic meaning. This can take the form of automatically assigning concrete terms for the segments (for example, the organ *heart* or the object *TPS implant* or, more abstract, a defect or an artifact). In any case, the segment should be related directly to an object, if automatic classification is to be possible. In this context, under-segmentation occurs if the resulting segments are composed from parts of several objects. Analogously, over-segmentation is obtained if a particular object is disintegrated into several segments. The big problem with the segmentation of medical images is that over and under-segmentation usually occur simultaneously.

Hierarchical procedures are one of the concepts to deal with the dualism between over and under segmentation. Starting on a lower resolution of the image, where it is represented with a small number of pixels only, the chance of splitting objects into more than one segment is decreased. Then, the exact outline of each segment is reconstructed on higher resolutions, where more details are contained.

62.5.4 Hybrid Segmentation Procedures

In the practice of medical image processing, hybrid approaches of segmentation have come to be of greatest importance. Here, one is trying to combine the advantages of individual (usually edge-based and region-based) algorithms without maintaining their disadvantages. Two widely used approaches, the watersheds transform and the active contour models, are exemplarily discussed in the following sections.

The *watershed transform* extends an agglomerative, regional segmentation procedure with edge-based aspects of segmentation. The numerical gray scales of a pixel are interpreted as the elevation level in a row of mountains, and crest lines are formed in the relief. If it is raining, the water drops fall onto the mountains, flowing downhill into the valleys, and form small lakes in the local minima. During the successive flooding of the landscape, neighboring pools may be merged. The merging of pools, however, is prevented by artificial watersheds. Frequently, the watershed segmentation is applied to a gradient image rather than the original, since the water reservoirs then directly correspond with visual image regions. For segmentation of medical images, the watershed transform especially has the following advantages:

- From the region-based idea of the flooding process, continuous segments are determined inherently.
- From the edge-based approach of the watersheds, the objects are exactly delineated.
- The problem of under-segmentation is avoided, since the merging of smaller pools is prevented by the watersheds.

However, the watershed transform results in a strong over-segmentation (for example Fig. 62.13b), which must be handled appropriately before any object recognition in the image is successful. This can be done with those methods of region merging that already have been discussed. Nevertheless, over-segmentation depicts in principle the difficulty of the watershed transform.

Active contour models are based on an edge-based segmentation in consideration of region-based aspects and an object-based model of a-priori knowledge. In the medical application domain, so called snake and balloon approaches are applied for segmentation of 2-D and 3-D image data and the tracing of contours in 2-D image and 3-D image sequences, i. e. 3-D and four-dimensional (4-D) data, respectively. The contour of the objects, which is usually closely modeled, is presented by individual nodes, which are – in the simplest case – piecewise connected with straight lines forming a closed polygon. For the nodes, a scalar quality measure (e.g., energy) is calculated and optimized in the local environment of the nodes. Alternatively, adjusted forces are determined that directly move the nodes. The iterative segmentation process completes at minimal energy or if an optimum balance of forces has been found. Thus, the potential of this approach is kept in the choice of capable quality criterions (e.g., energy) or forces. The classical *snake* approach models an internal and an external quality criterion, both as undirected energy. The internal energy results from a predefined elasticity and stiffness of the contour, which is high in places of strong bends or buckling. The external energy is calculated from an edge-filtered image. The external energy is small if the contour runs along edges. The

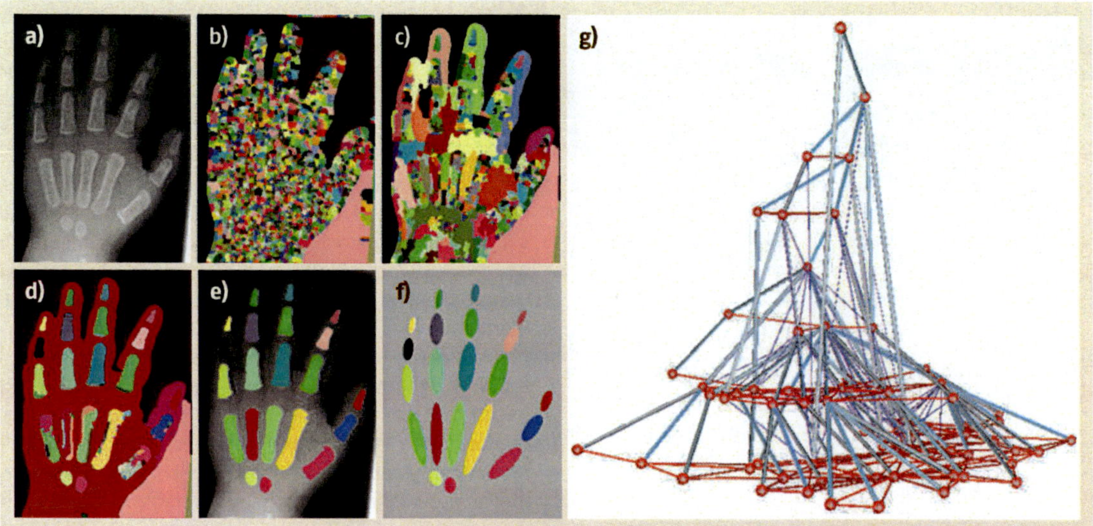

Fig. 62.13a–g Hierarchical region merging. The skeletal radiograph of the hand **(a)** has been segmented at various levels of resolution, cf. **(b)** to **(d)**. The initial step is obtained with the watershed transform (Sect. 62.5.4). Depending on the size of the objects, they can be localized in the appropriate level **(e)**, approximated by ellipses **(f)**, or visualized as nodes in a graph **(g)**

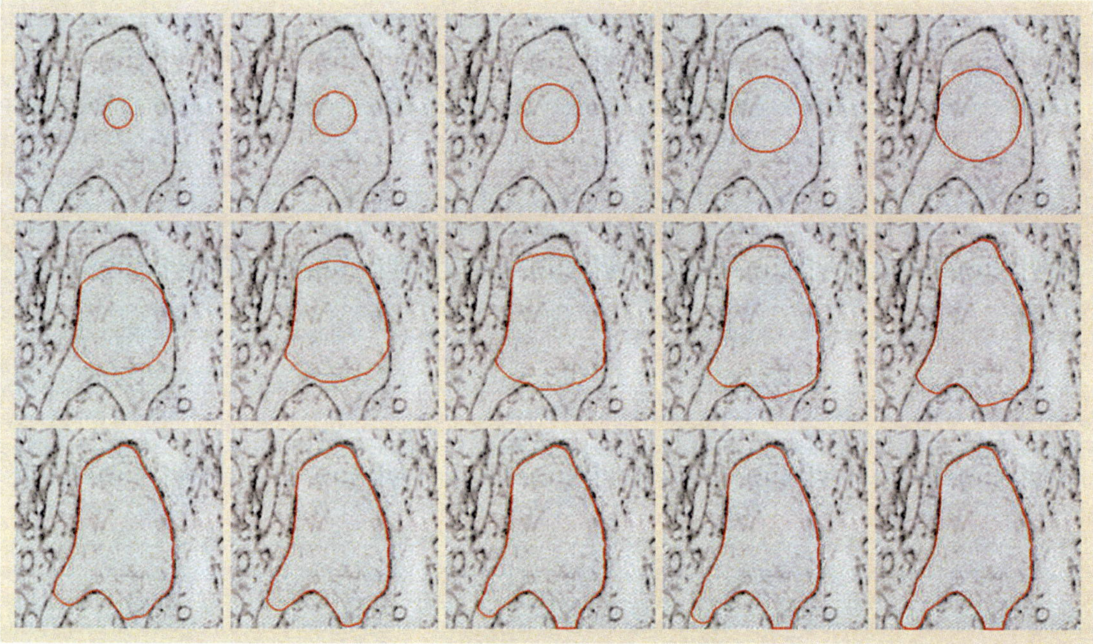

Fig. 62.14 Ballon segmentation of motoneuron cell membrane (after [62.7]). The frames show the balloon at different iterations. By touching the cell membrane, the strong image forces prevent further movement of the active contour. In this application, the internal forces correspond physically to a membrane. This is clearly recognizable at the adhesion border of the balloons reaching the dendrites (*bottom left*)

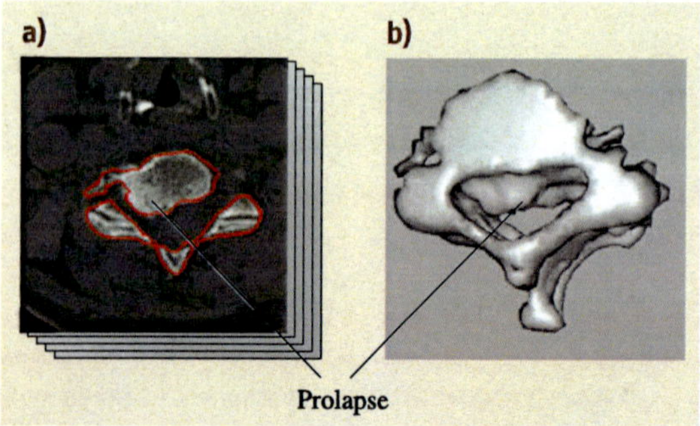

Fig. 62.15a,b Segmentation with a 3-D balloon model (after [62.8]). The CT of a spine **(a)** was segmented with a 3-D balloon. In the surface-based rendering after automatic segmentation, the prolapse is clearly visible **(b)**. The visualization is based on Phong shading (Sect. 62.9)

idea behind this approach is an edge-based segmentation combined with the a-priori knowledge that medical or biological objects rarely have sharp bending boundaries. With an optimal weighting of energy terms, the contour course is primarily determined by the information of edges in the image. However, if the object's contour is partially covered or incompletely captured, the internal energy ensures an appropriate interpolation of the region's shape.

As simply as this approach has been formulated verbally, as difficult is its implementation. During the iteration, the number of nodes must be constantly adjusted to the current size of the contour. Furthermore, crossovers and entanglements of the moving contour must be avoided. The classical snake approach also requires an already precisely positioned starting contour, which often must be defined interactively. Then, the two steps of segmentation, i. e. localization and delineation are performed by man and machine, respectively. This concept was also applied in the first publications of this segmentation method. For contour tracking of moving objects in image sequences, the segmentation of image at time t serves as an initial contour of iteration in image $t+1$. After a single initialization for the image $t=0$, the procedure runs automatically. Hence, fluoroscopy and endoscopy are suitable modalities for the application of the snake approach to track moving objects.

Balloons are based on forces rather than energies. Besides the internal and external force, an inner pressure or suction is modeled, which lets the contour continuously expand or shrink. Figure 62.14 shows the inflation movement of a balloon to segment the cell membrane, which is visualized by the synaptic boutons of contacting dendrites in a microscopy of a motoneuron. Although segmentation is done without an accurate initial contour, the balloon nestles in the course of iteration better and better to the real contour of cell membrane. Another advantage of the balloon model is that this concept is directly transferable to higher dimensions (Fig. 62.15).

In recent developments of active contour models, it has been attempted to incorporate further a-priori knowledge, e.g., in the form of anatomical models. Prototypes of the expected object shapes are integrated into the algorithm. In each iteration, the distance of the current object shape to a suitable selected prototype is modeled as an additional force on the node.

With those extensions, a break out of the active contour model is prevented also for long passages of the local object boundary without sufficient edge information. The complex and time-consuming parameterization of an active contour model for a specific application can be based on manual and also automatic reference segmentations. For the latter approach, different combinations of parameters are determined and the segmentation is performed for all cases. All resulting segmented contours are compared with the appropriate reference contour, a-priori defined as the ground truth of the training data. Then that set of parameters with the best approximation of the reference contour is selected.

62.6 Classification

According to the general processing chain (Fig. 62.1), the task of the classification step is to assign all connected regions obtained from the segmentation to particularly specified classes of objects. Usually, region-based features that capture the characteristics of the objects sufficiently abstractedly are used to guide the classification process. In this case, another feature extraction step is performed between segmentation and classification, which is not visualized in Fig. 62.1. These features must be sufficiently discriminative and suitably adopted to the application, since they fundamentally impact the resulting quality of the classifier. For all types of classifiers, we can differ supervised (trained), unsupervised (untrained) and learning classification. For example, pixel clustering, which was already introduced for pixel-based segmentation, is an unsupervised classification process. As a goal, individual objects are divided into similar groups (Fig. 62.11).

If the classification is used for identification of objects, the general principles or an exemplary reference must be available, from which the ground truth of classification can be created. The features of these samples are then used for parameterization and optimization of the classifier. Through this training, the performance of the classifier can be drastically improved. However, supervised object classification is always problematic if the patterns that are classified differ remarkably from the trained patterns. In such cases, the training set does not sufficiently reflect the real world. A learning classifier has advantages here, because it changes its parameterization with each performed classification, even after the training phase. In the following, however, we assume a suitable set of features and a sufficiently characteristic and large set of samples.

The classification itself reverts mostly to known numerical (statistical) and nonnumerical (syntactic) procedures, as well as the newer approaches of computational intelligence, such as neural networks, evolutionary algorithms, and fuzzy logic. In general, the individual features, which can be determined by different procedures, are summarized either to numerical feature vectors (also referred to as signature) or abstract strings of symbols. For example, a closed contour object can be described by its Fourier-descriptors as a feature vector, or by means of basic line items such as straight, convex, and concave forming a symbol chain.

62.6.1 Statistic Classifiers

Statistical classification regards object identification as a problem of the statistical decision theory. Parametric procedures for classification are based on the assumption of distribution functions for the feature specifications of the objects, and the parameters of the distribution functions are determined from the sample. Nonparametric methods, however, waive such model assumptions, which are not always possible in medical image processing. A classical example of such a nonparametric statistical object classifier is the nearest neighbor (NN) classifier. All features span the feature space, and each sample is represented by a point in this feature space. Based on the signature of a segment that has not been included in the training, which is assigned to its nearest neighbor in feature space, the segment is classified to the associated class of the assigned feature vector. The k nearest neighbor (KNN) classifier assigns the majority class from the k nearest neighbors in feature space (usually, $k = 3$ or 5). Often, the Euclidean distance is computed in the feature space. An example of the KNN classifier is given in Fig. 62.16.

62.6.2 Syntactic Classifiers

In symbol chains, it is neither useful nor possible to define distance measurements or metrics and to evaluate the similarity between two symbol chains, such as used for feature vectors. An exception to this statement is given with the Levenshtein distance, which is defined as the smallest number of modifications such as exchange, erase, or insert, required to transform a symbol chain into another.

The syntactic classification is, therefore, based on grammar, which can possibly generate an infinite amount of symbol chains with finite symbol formalism. A syntactic classifier can be understood as a knowledge-based classification system (*expert system*), because the classification is based on a formal heuristic symbolic representation of expert knowledge, which is transferred into image processing systems by means of facts and rules. If the expert system is able to create new rules, a learning classifier is also realizable as a knowledge-based system.

It should be noted that the terms expert system or expert knowledge, however, are not standardized in the literature. Therefore, primitive image processing systems, which use simple heuristics as implemented distinction of cases to classification or object identification, are also referred to as knowledge-based.

62.6.3 Computational Intelligence-Based Classifiers

As part of the artificial intelligence, the methods of computational intelligence include neural networks, evolutionary algorithms and fuzzy logic. These methods have their examples in biological information processing, which is particularly much more powerful for tasks of object recognition than today's computers. Therefore, they are frequently used in medical image processing for classification and object identification. Thereby all the procedures have a mathematically-based, complex background.

Artificial neural networks simulate the information processing in the human brain. They consist of many simply constructed basic elements (neurons), which are arranged and linked in several layers. Each neuron calculates the weighted sum of its input excitations,

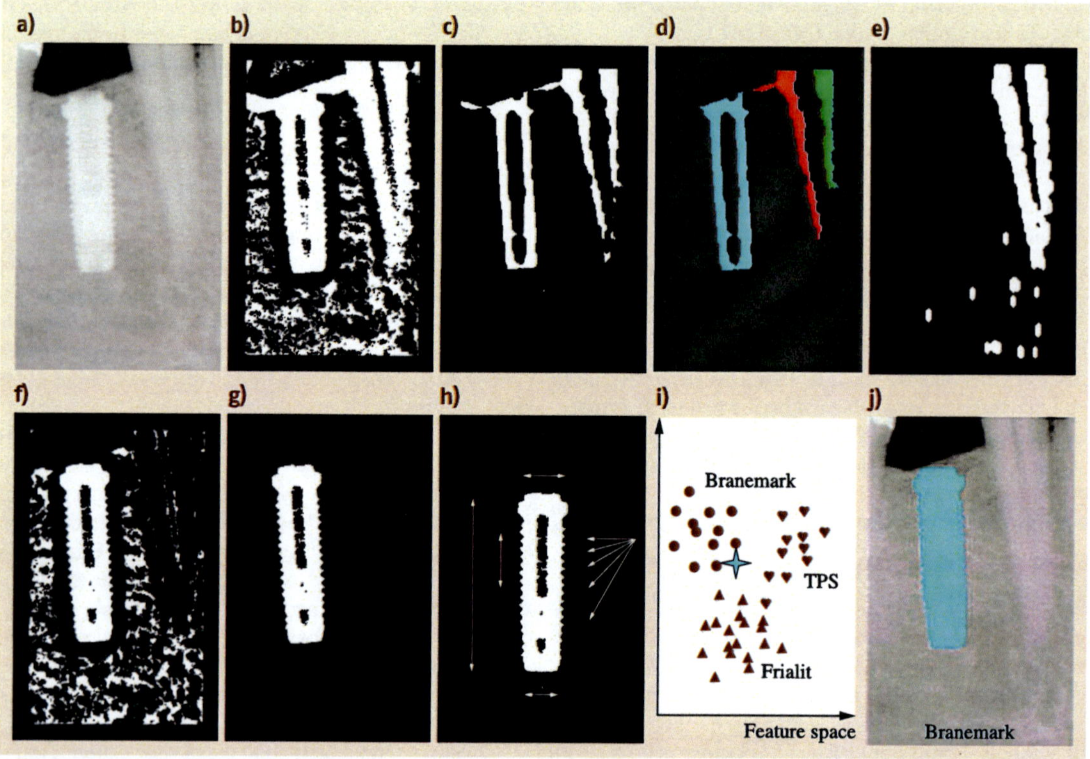

Fig. 62.16a–j Identification of dental fixtures (IDEFIX) (after [62.9]). An implant is shown in the intra-oral radiograph of the lower jaw **(a)**. For the feature extraction, the image is binarized with local-adaptive thresholding **(b)**. The morphological filtering (erosion) separates individual areas **(c)** and eliminates interference. In this example, three regions were segmented **(d)**. Further processing is shown for the blue segment. After its fade-out, the morphological erosion is compensated by a subsequent dilation **(e)**, and subtracted from the intermediate image **(b)**. Any coordinate of blue segment from **(d)** identifies the corresponding region, which can now be extracted **(g)** and aligned into a normal position using the Karhunen–Loève transform. Geometric dimensions are determined as region-based features and stored in a feature vector (signature). As part of the training, the reference measures of different implant types have been recorded in the feature space. The classification in the feature space is done with the statistical KNN classifier **(i)**, which identifies the blue segment reliably as Branemark implant screw **(j)**

which is mapped over a nonlinear function (characteristic curve) to the output. The number of layers, the number of neurons per layer, the network's topology, and the characteristic curve of the neurons are predefined within the network dimensioning step. On the one hand, heuristics are usually applied rather than methodological derivations. On the other hand, the individual weights of the excitements are identified numerically during the training of the network. Then, the network remains unchanged and can be used as a classifier. *Evolutionary algorithms* are based on the constant repetition of a cycle of mutation and selection following the Darwinian paradigm of the survival of the fittest. Genetic algorithms work on a number of individuals (the population). The crossing of two randomly selected individuals and afterwards the mutation changes the population. A fitness function evaluates the population in terms of their goodness to problem solution. Although the selections are equipped with a random component, fit individuals are frequently selected for reproduction. Evolutionary algorithms can solve complex optimization problems amazingly well, but for object classification, they are less successfully used than other methods.

The idea of *fuzzy logic* is to extend the binary (true or false) computer model with some uncertainty or blur, which also exists in the real world. Many of our sensory impressions are qualitative and imprecise and, therefore, unsuitable for accurate measurements. For example, a pixel is perceived as dark, bright, or even very bright, but not as a pixel with gray scale value 231. Fuzzy quantities are mathematically based on the fuzzy set theory, in which the belonging of an element to a set of elements is not restricted to the absolute states true (1) or false (0), but continuously defined within the entire interval [0..1]. Besides classification, applications of fuzzy logic in medical image processing can be found also for preprocessing (contrast enhancement), feature extraction (edge extraction and skeleton), and segmentation.

62.7 Quantitative Measurements

While the visual appraisal by experts is qualitative and sometimes subject to strong inter-individual as well as intra-individual fluctuations, in principle, a suitable computer-aided analysis of medical images can deliver objective and reproducible results. First of all, this requires a precise calibration of the imaging modality. Furthermore, partial (volume) effects of the imaging system and particularities of the discrete pixel topology must be taken into account and handled accordingly to ensure reliable and reproducible measurements.

The digitalization of the local area or volume of a pixel or voxel, respectively, always yields an averaging of the measured value in the appropriate field. For example in CT, a voxel containing different tissue is assigned a certain Hounsfield value that results from the proportional mean of the individual Hounsfield values of the covered tissue classes. Thus, a voxel containing only bone and air preserves the Hounsfield value of soft tissue and, thus, may distort quantitative measurements. In general, this partial (volume) effect occurs in all modalities and must be accounted for appropriately in any automatic measurement.

The common paradigms of Euclidean geometry do not apply in the discrete pixel domain. For example, the discrete representations of two straight lines may not join in a common pixel although the lines cross. Furthermore, different neighborhood concepts of discrete pixels topology have remarkable impact on the result of automatic image measurements. For example, the areas identified in region growing may be significantly larger

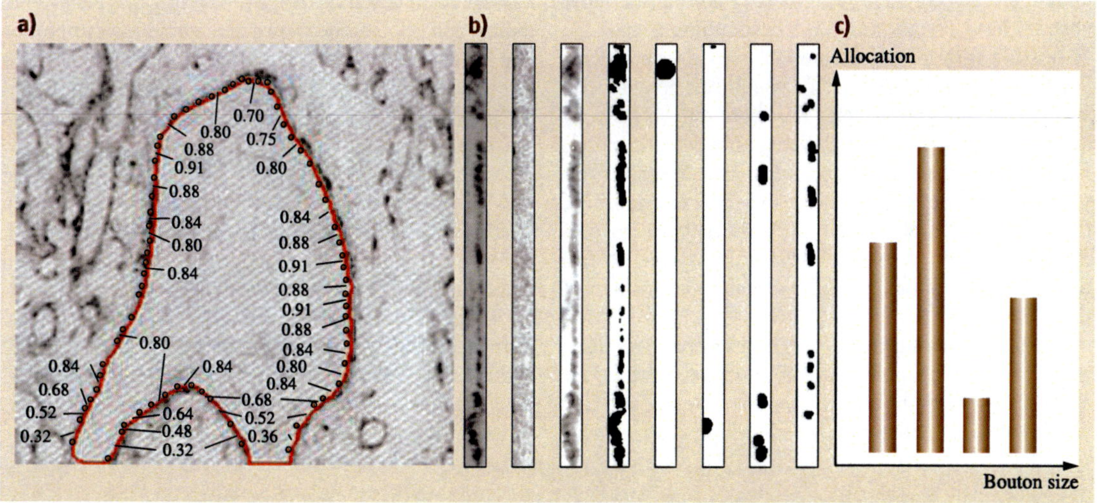

Fig. 62.17a–c Quantification of synaptic boutons on a cell membrane (after [62.10]). The cell membrane was segmented with a balloon (Fig. 62.12). Analyzing the impact of internal versus external forces at a certain vertex, local confidences can be determined to fuzzily classify the affiliation of the contour section to the actual cell membrane **(a)**. The cell contour is extracted, linearized, normalized, and binarized before the occupation of the cell membrane with synaptic boutons of different sizes are analyzed by morphological filtering **(b)**. The confidence values are considered for averaging the occupation measure along the cell membrane **(c)**

if the 8-neighborhood is applied, i. e., if eight adjacent pixels are analyzed instead of the four direct neighbors. We will now discuss some examples for image measurements. For instance, in Fig. 62.16, geometrical features were used for the automatic classification of implant systems. The feature measures were extracted on the abstract level of the region. Frequently, further measures were extracted after object identification, which use the information of the certain object detected, i. e., they operate on the level of objects. For instance, in Fig. 62.16i, we can use the knowledge that the blue segment corresponds to a Branemark implant to parameterize a special morphological filter that is adapted to the geometry of Branemark implants and measure the number of windings of the screw. Another example of object-based image measurements is given in Fig. 62.17. The result of balloon segmentation of a cell membrane (Fig. 62.14) was labeled automatically with local confidence values based on model assumptions (Fig. 62.17a).

These values indicate the contour segment belonging to a cell membrane and thus a classification via fuzzy logic. To increase the robustness and reliability of measurements, the confidence values are accounted along the contour, which is extracted, linearized, normalized, and morphologically analyzed (Fig. 62.17b), such that, finally, a reliable distribution statistics of connecting boutons according to their size is obtained (Fig. 62.17c).

62.8 Interpretation

Image interpretation may be understood in the sense of analyzing an abstract scene that corresponds to the ambitious goal of developing a *visual sense for machines*, which is similarly universal and powerful to that of humans. The previously discussed processing modules focused on automatic detection of objects as well as their properties. Now, we proceed to analyze the order of individual objects in space and/or time. Thus, the fundamental step of image interpretation is to generate a spatial-temporal scene description on the most abstract level (symbolic image description, Fig. 62.2). A suitable form of representation is the attributed relational graph (semantic web), which can be analyzed at different hierarchy levels. Therefore, the so far considered grid matrix of pixels (iconic image description, Fig. 62.2) is inappropriate for image interpretation. The primitives of the graph (node) and their relationships (edges) must be abstracted from the segmented and identified objects or object parts in the image. So far, only a few algorithms can execute this level of abstraction.

Examples for the abstraction of primitives are given by the numerous approaches to shape reconstruction: shape-from-shading, -texture, -contour, stereo, etc. Examples of the abstraction of relationships can be found at the depth reconstruction by trigonometric analysis of the projective perspective. Recently, considerable progress has been achieved in symbolic image analysis in the fields of industrial image processing and robotics. Because of the special peculiarities of medical imagery, which were discussed before, the transfer of these approaches into health care applications and medical image processing has only met with limited success so far.

Figure 62.18 displays the automatic extraction of a dental chart based on image processing of a panoramic radiograph. It clearly shows the immense difficulties that have to be overcame by the automatic interpretation of medical images. Initially, the segmentation and identification of all relevant image objects and object parts must succeed so that the semantic network can be built. This includes the instances (tooth 1, tooth 2, etc.) of the previously identified objects (e.g., tooth, crown, filling). The interpretation of the scene based on the network must be carried out in a further, not less difficult step of processing. Thus, all teeth must be named according to their position and shape. Then, crowns, bridges, fillings, and carious processes can be registered in the dental chart. However, the automation of this process, which can be accomplished by dentist in a few minutes, is not yet possible with sufficient robustness.

62.9 Image Data Visualization

Under the concept of image visualization, we summarized all the transforms that serve the optimized output of the image. In medicine, this particularly includes the realistic visualization of three-dimensional data. Such techniques have found broad applications in medical research, diagnostics, treatment planning, and therapy. In contrast to problems from the general area of computer graphics, in medical applications the displayed objects

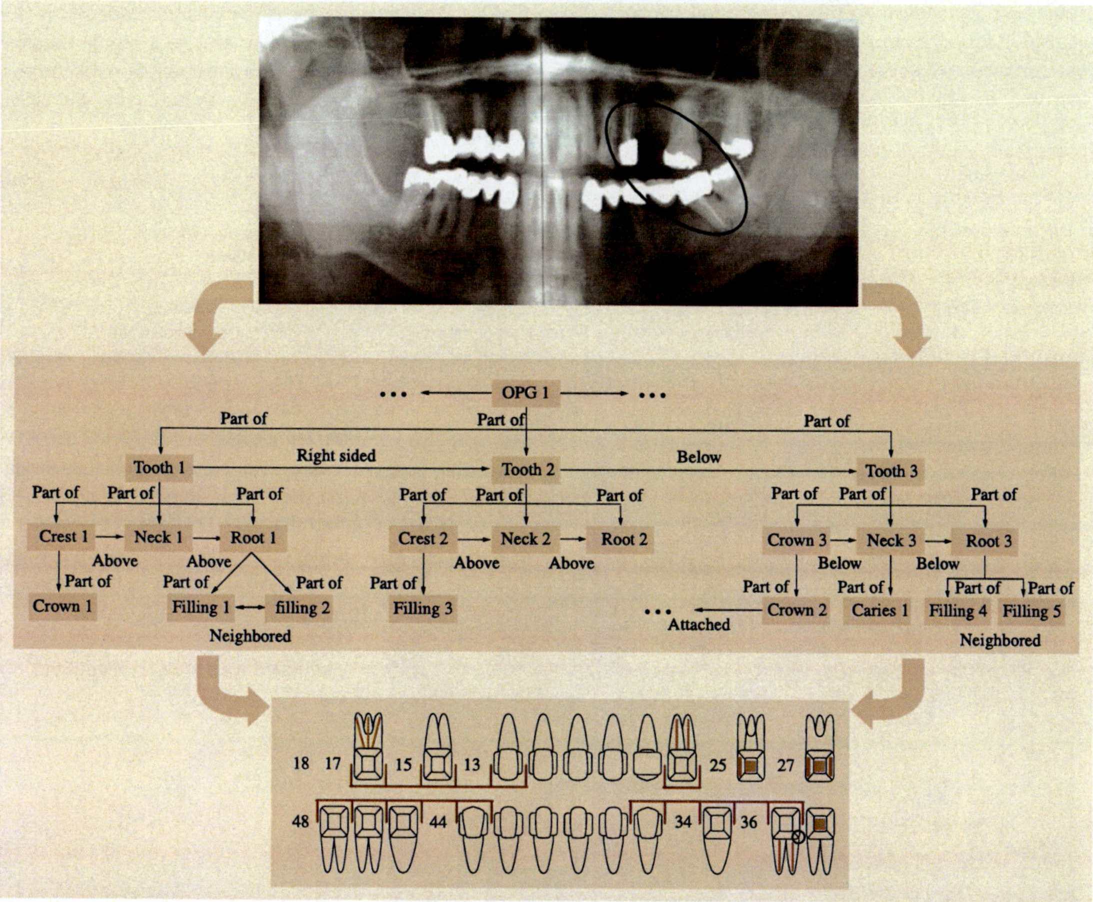

Fig. 62.18 Scheme of automatic image interpretation. The panoramic radiograph contains all relevant information of the dental chart. The symbolic description of the scene is obtained with a semantic network. Despite its already considerable complexity, the shown part of the network represents only the marked ROI. In the dental chart, information is coded differently. The teeth are named in accordance with the key of the Fédération Dentaire Internationale (FDI): the leading digit denotes the quadrant clockwise, the second digit denotes refer to the number of the tooth, counting from inside to outside. Existing teeth are represented by templates, in which dental fillings, crowns, and bridges are recorded. The *black circle* at tooth 37 (say: three, seven) indicates a carious process

are not given implicitly by formal, mathematical expressions, but as an explicit set of voxels. Consequently, specific methods have been established for medical visualization. These methods are based either on a surface reconstruction or on a direct volume visualization, and lighting and shading are also considered (Table 62.2).

62.9.1 The Marching Cubes Algorithm

The marching cubes algorithm was specifically developed for surface reconstruction from medical volumes. Here, the voxel is no longer interpreted as a cube of finite edge length but as a point. It is equivalent to a point grid for visualizing volumes. In this, a cube is considered with four corners in each of the two adjacent layers. Utilizing symmetry, the complex problem of surface production is reduced to only 15 different topologies, which can be calculated most efficiently since the polygon descriptions that belong to the basic topologies can be stored in a look-up table. Similar to the process of spatial convolution, the cube is positioned successively at all points in a volume dataset

Table 62.2 Taxonomy of 3-D visualization methods (after [62.11]). Triangulation for surface-based rendering is described in textbooks on computer graphics. The marching cubes approach is described in the text. As a simple example of a surface-based direct volume rendering methods, depth shading visualizes the length of rays passing through the volume until they hit the surface. Integral shading codes the sum of voxel values along the ray as gray scale. It is, therefore, frequently used to obtain radiograph-like images based on CT data

Taxonomy of 3-D visualization methods			
Surface reconstruction and rendering		**Direct volume rendering**	
Surface-oriented methods	Volume-oriented methods	Surface-oriented methods	Volume-oriented methods
Examples: Triangulation	Examples: Cuberille approach Marching cubes	Examples: Depth shading Depth gradient shading Gray gradient shading	Examples: Integral shading Transparent shading Maximum projection

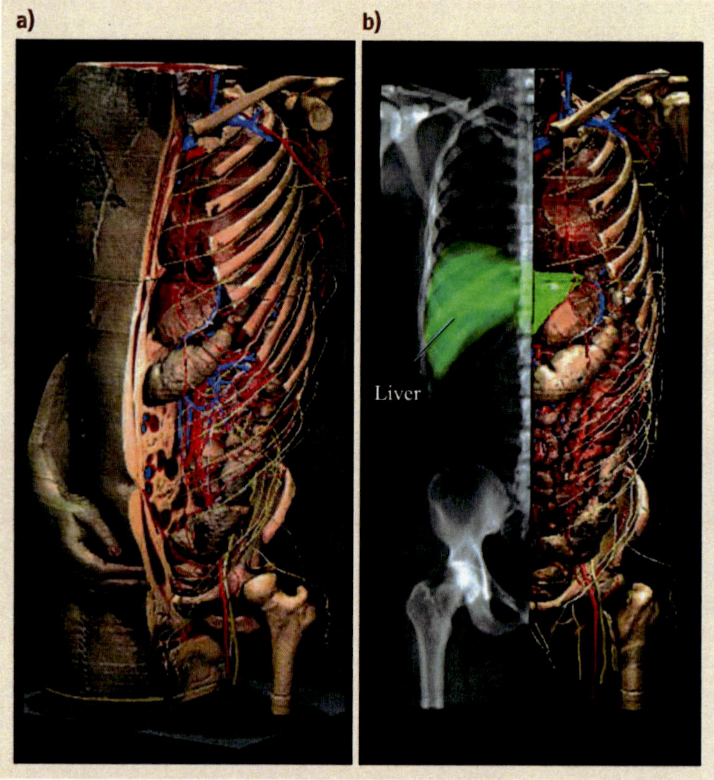

Fig. 62.19a,b 3-D visualization with Voxel-Man (after [62.12]). This 3-D model of the internal organs is based on the visible human data. The Voxel-Man 3-D Navigator provides unprecedented details and numerous interactive possibilities **(a)**. Direct volume rendering and surface-based visualization of segmented objects are combined with integral shading **(b)**

(marching). After completion of the marching cubes algorithm, a segmented volume is transformed into a triangulated surface. However, the surface is build from a very large number of triangles, which may be reduced significantly by heuristic procedures without any discernible loses of quality. Reducing the number of elements to visualize supports real-time visualization of the volume.

62.9.2 Surface Rendering

To generate photorealistic presentations of the volume surface, the lighting is simulated analogously to natural scenes. According to the lighting model by Phong, ambient light is created through overlapping of multiple reflections, diffuse scattering on nonshiny surfaces, and direct mirroring on shiny surfaces. While the intensity of the ambient light remains constant in the scene for all surface segments, the intensities of diffuse and shiny reflections depend on the orientation and characteristics of surfaces as well as their distances and directions to the light source and the observation point of viewing. Without *shading*, one can recognize the initial triangles. This is a nasty artifact in computer graphics. Therefore, various strategies for shading have been developed to improve the visual impression significantly. For instance, Gouraud shading results in smooth blunt surfaces and Phong shading also provides realistic reflections. In newer applications, transparencies are also modeled to glance at encapsulated objects. Moreover, textures or other bitmaps on the surfaces can be projected to reach a more realistic impression of the scene.

62.9.3 Volume Rendering

Direct volume visualization is obtained from preliminary calculation of the object surface. The visualization is based directly on the voxel data and, therefore, possible without any segmentation. This strategy allows visualization of medical 3-D and 4-D data by radiologists for interactive localization of pathological areas. The volume is processed either along the data layers (back-to-front or front-to-back) or along an imaginary light ray. Based on the observer position, rays will be pursued through the volume (ray-tracing). Hereby, the recursive followup of secondarily reflected rays is also possible (ray-casting). Although quite realistic visualizations can be provided to the observer, problems arising from the discrete nature of pixel topology (see above) have led to a multitude of algorithmic variants, which cannot be addressed here.

In general, parameters are extracted from voxel intensity along the rays and applied as gray or color value at the corresponding position in the viewing plane. This procedure is also referred to as shading. With the method of surface-based shading, light source and image plane are placed on the same side of the object, while volume-oriented procedures radiograph the entire object according to x-ray imaging, i. e., the object is located between light sources and the observation (Table 62.2). Combining direct volume with surface-based approaches, amazingly realistic scenes can be created (Fig. 62.19).

62.10 Image Management

Introductorily, we have summarized all image manipulation techniques that serve the effective archiving (short and long-term), transmission (communication), and the access (retrieval) of data (Fig. 62.1). For all three points, the specifics in medical applications and the health care environment have led to specific solutions, which are briefly introduced in the following section.

Already in the 1970s, the invention of computed tomography and its integration with clinical routine involved the installation of the first picture archiving and communication system (PACS), whose main task was the archiving of image data. The core problem of archiving medical images is the immensely large data volume. A simple x-ray image with 40×40 cm (e.g. a chest radiograph) with a resolution of five line pairs per millimeter and 10 bit = 1024 gray levels per pixel already requires a storage capacity of more than 10 MB. In a university hospital x-ray imaging, CT, and MRI accumulate to about 2 TB of image data each year (Table 62.3). This estimate can easily increase tenfold with the resolution-increased novel modalities such as spiral CT and whole-body MRI. For instance, in Germany, according to relevant legislations, data must be kept for at least 30 years. Therefore, efficient storage, retrieval, and communication of medical images have required effective compression and high speed networks. Due to noise in medical images, lossless compression usually has a limited effect of compression rates of two or three. It is only in recent years that feasible hybrid storage concepts have become available. So, the relevance of storage and fetching of medical image data is decreasingly important.

With increasing digitization of diagnostic imaging, the motto for medical information systems, i. e., to provide the right information at the right time and the right place, is projected to the field of medical image processing. Hence, communication is the core of today's PACS. Image data is not only transferred electronically within a department of radiology or the hospital, but also between widely separated institutions. For this task, simple bitmap formats such as the tagged image file format (TIFF) or the graphics interchange format (GIF) are inadequate, because beside the images, which might have been captured with different dimensions, medical meta information on patients (e.g., identifier (ID), name, date of birth, ...), the modality (e.g., device, parameters, ...) and organization (e.g., investigation, study, ...) must also be transferred in a standardized

Table 62.3 Data volume in 1999 at Aachen University Hospital (about 1500 beds) [62.1]. The data are taken from the Annual Report 1999 of the University Hospital of RWTH Aachen University, Aachen, Germany. The data is based on the Departments of (i) Diagnostic Radiology, (ii) Neuroradiology, (iii) Nuclear Medicine, and (iv) Dentistry, Oral, and Maxillofacial Surgery for a total of 47 199 inpatient and 116 181 outpatient images. Services (such as ultrasound, endoscopic, or photographic) from other departments have been excluded. For modalities of nuclear medicine, 20 slices per study are assumed. For comparison, the total number of analysis performed in the central laboratory of the Institute for Clinical Chemistry and Pathobiochemistry was estimated with an average of 10 measured values per analysis with a highest precision of 64 bit. However, the annual image data volume is about 10 000 times larger

Modality	Resolution (spatial domain)	Resolution (value range)	MB per image	Units in 1999	GB per year
Chest radiographs	4000×4000	10	10.73	74 056	775.91
Skeleton radiographs	2000×2000	10	4.77	82 911	386.09
CT	512×512	12	0.38	816 706	299.09
MRI	512×512	12	0.38	540 066	197.78
Other radiographs	1000×1000	10	1.19	69 011	80.34
Panoramics and skull	2000×1000	10	2.38	7599	17.69
Ultrasound	256×256	6	0.05	229 528	10.11
Dental radiographs	600×400	8	0.23	7542	1.69
PET	128×128	12	0.02	65 640	1.50
SPECT	128×128	12	0.02	34 720	0.79
For comparison					
Laboratory tests	10	64	0.00	4 898 387	0.36

way. Since 1995, the communication has been based on the digital imaging and communications in a medicine (DICOM) standard. In its current version, DICOM includes:

- *Structural information* about the contents of the data (object classes)
- *Commands* on what should happen to the data (service classes)
- *Protocols* for data transmission.

DICOM is based on the client/server paradigm and allows the coupling of PACS in radiology (RIS) or hospital information systems (HIS). DICOM incorporates existing standards for communication: the open system interconnection (ISO/OSI) model, the transmission control protocol/Internet protocol (TCP/IP), and health level seven (HL7) standard. Full DICOM compliance for imaging devices and image processing applications is achieved with only a few supported object or service classes, since other DICOM objects, which are not relevant for the current device, are simply handed over to the next system in the DICOM network. The synchronization between the client and server is regularized by conformance claims, which are also specified as part of the DICOM standard. However, the details of implementation of individual services are not specified in the standard, and so in practice, vendor-specific DICOM dialects have been developed, which can lead to incompatibilities when building PACS. In recent years, the integrating healthcare enterprises (IHE) initiative has become important. IHE aims at guiding the use of DICOM and other standards such that complete interoperability is achieved.

In today's DICOM archives, images can be retrieved systematically, only if the patient name with date of birth or the internal system identification number (ID) is known. Still, the retrieval is based on alphanumerical attributes, which are stored along the image data. It is obvious that diagnostic performance of PACS would be magnified significantly if images were to be directly available from the similar content of a given example image. To provide the query by example (QBE) paradigm is a major task of future systems for content-based image retrieval (CBIR). Again, this field of medical research requires conceptually different strategies to those demanded in commercial CBIR systems for other application areas, because of the diverse and complex structure of diagnostic information that is captured in medical images.

Figure 62.20 shows the system architecture of the image retrieval in medical applications (IRMA) framework (http://irma-project.org). This architecture reflects the chain of processing that we have discussed in this chapter, i. e., registration, feature extraction, segmen-

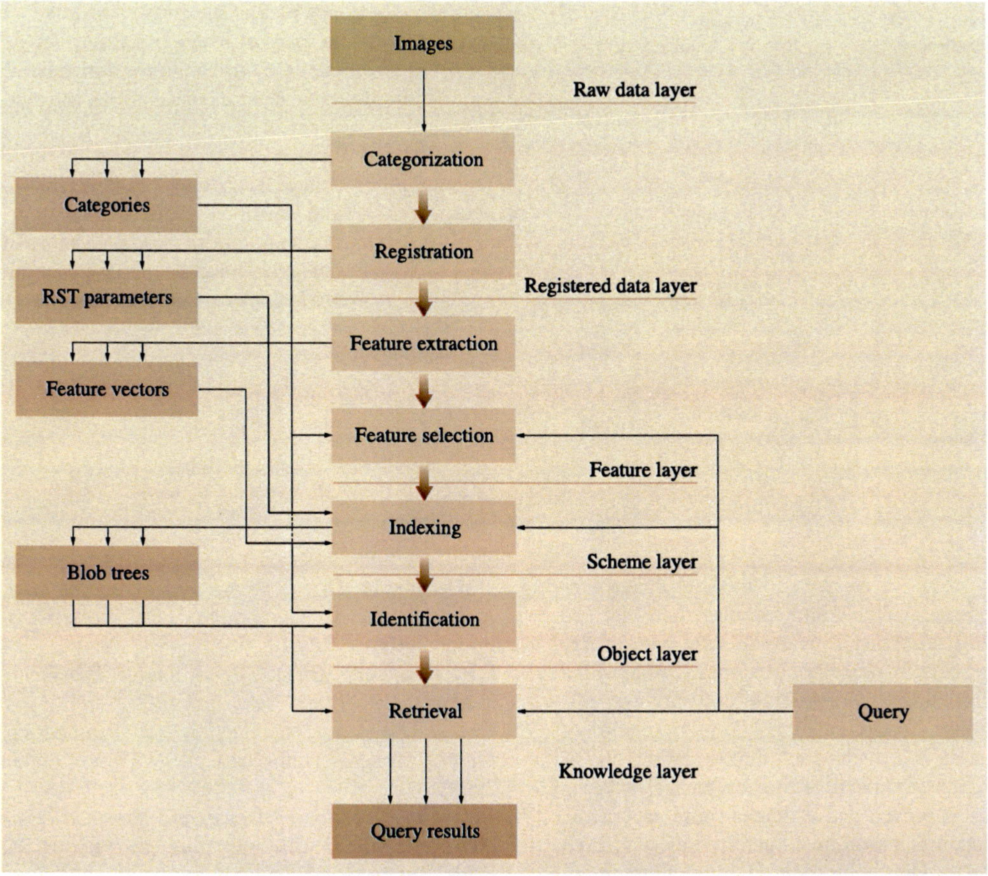

Fig. 62.20 System architecture of the IRMA framework (after [62.13]). The processing steps in IRMA are shown in the *middle column*. Categorization is based on global features and classifies images in terms of imaging modality, view direction, anatomic region, and body system. According to its category, the image geometry and contrast are registered to a reference according to rotation, scale, and translation (RST). The abstraction relies on local features, which are selected specifically to context and query. The retrieval itself is performed efficiently on abstracted and thus information-reduced levels. This architecture follows the paradigm of image analysis (Fig. 62.1). The in-between-representations on the *left side* describe the image increasingly abstract. The levels of abstraction (Fig. 62.2) are named on the *right side*

tation, classification of image objects towards the tip of the pyramid in Fig. 62.2, which is the symbolic interpretation and scene analysis. In IRMA, the image information that is relevant for retrieval is gradually condensed and abstracted. The image bitmap is symbolically represented by a semantic network (hierarchical tree structure). The nodes contain characteristic information for the represented areas (segments) of the image. Its topology describes the spatial and/or temporal condition of each object. With this technology, radiologists and doctors can be similarly supported in patient care, research, and teaching.

62.11 Conclusion and Outlook

The past, present, and future paradigms in the medical image processing are shown in Fig. 62.21. Initially (until approximately 1985), the pragmatic issues of image generation, processing, presentation, and archiving

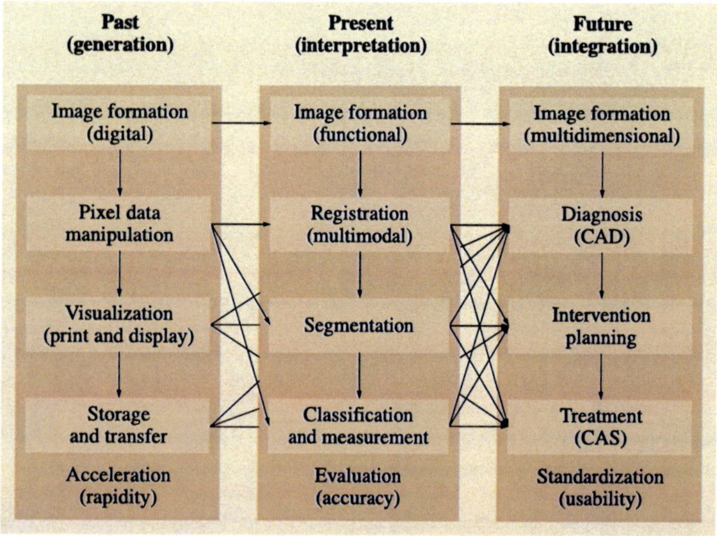

Fig. 62.21 Changing paradigms in medical image processing (after [62.14]). Until now, formation, enhancement, visualization, and management of medical images have been in the focus of research. In future, integration, standardization, and validation will be seen as major challenges for routine applications in computer-aided diagnosis (CAD), intervention planning, and therapy, such as computer-aided surgery (CAS)

stood in the focus of research in medical image processing, because available computers at that time had by far not the necessary capacity to hold and modify large image data in memory. The speed with which the image processing was possible only allowed offline calculations. Today, the automatic interpretation of medical images is a major goal. Segmentation, classification, and measurements of medical images is being continuously improved and validated more accurately, since validation is based on larger studies with high volumes of data. Hence, we focused this chapter on image analysis and the processing steps associated with it. The future development is seen in the increasing integration of algorithms and applications in the medical routine. Procedures in support of diagnosis, treatment planning, and therapy must be easily useable for physicians and, therefore, further standardized in order to ensure the necessary interoperability for clinical practice is necessary.

Further Reading

1. I.N. Bankman (Ed.): *Handbook of Medical Image Processing and Analysis*, 2nd edn. (Academic, New York 2008)
2. L. Beolchi, M.H. Kuhn (Eds.): *Medical Imaging: Analysis of Multimodality 2D/3D Images* (IOS Press, Amsterdam 1995)
3. J. Beutel, H.L. Kundel, R.L. van Metter (Eds.): *Handbook of Medical Imaging, Vol. 1: Physics and Psychophysics* (SPIE Press, Bellingham 2000)
4. D. Dougherty: *Digital Image Processing for Medical Applications* (Cambridge Univ. Press, Cambridge 2009)
5. E.R. Dougherty (Ed.): *Digital Image Processing Methods* (CRC Press, New York 1994)
6. C. Guy, D. Ffytche: *Introduction to the Principles of Medical Imaging* (Imperial College Press, London 2005)
7. J. Jan: *Medical Image Processing, Reconstruction and Restoration: Concepts and Methods* (CRC Press, Boca Raton 2005)
8. Y. Kim, S.C. Horii (Eds.): *Handbook of Medical Imaging. Vol. 3: Display and PACS* (SPIE Press, Bellingham 2000)
9. A. Meyer-Baese: *Pattern Recognition in Medical Imaging* (Academic, San Diego 2003)
10. B. Preim, D. Bartz: *Visualization in Medicine. Theory, Algorithms, and Applications* (Morgan Kaufmann, Amsterdam 2007)
11. L. Sonka, J.M. Fitzpatrik (Eds.): *Handbook of Medical Imaging. Vol. 2: Medical Image Processing and Analysis* (SPIE Press, Bellingham 2000)
12. P. Suetens: *Fundamentals of Medical Imaging*. (Cambridge Univ. Press, Cambridge 2002)
13. J.M. Tavares, R.M.N. Jorge (Eds.): *Advances in Computational Vision and Medical Image Processing: Methods and Applications* (Springer, Berlin 2008)
14. S.E. Umbaugh: *Computer Imaging: Digital Image Analysis and Processing* (CRC, Boca Raton 2005)

References

62.1 T.M. Lehmann, J. Hiltner, H. Handels: Medizinische Bildverarbeitung. In: *Handbuch der Medizinischen Informatik*, 2nd edn., ed. by T.M. Lehmann (Hanser, Munich 2005), (in German)

62.2 T.M. Lehmann, H.G. Gröndahl, D. Benn: Review: Computer-based registration for digital subtraction in dental radiology, Dentomaxillofac. Radiol. **29**(6), 323–346 (2000)

62.3 T.M. Lehmann, A. Kaser, R. Repges: A simple parametric equation for pseudocoloring grey scale images keeping their original brightness progression, Image Vis. Comput. **15**(3), 251–257 (1997)

62.4 G. Wagenknecht, H.J. Kaiser, U. Büll: Multimodale Integration, Korrelation und Fusion von Morphologie und Funktion: Methodik und erste klinische Anwendungen, RöFo Fortschr. Geb. Röntgenstr. Bildgeb. Verfahr. **170**(1), 417–426 (1999)

62.5 V. Metzler, H. Bienert, T.M. Lehmann, K. Mottaghy, K. Spitzer: A novel method for geometrical shape analysis applied to biocompatibility evaluation, ASAIO Int. J. Artif. Organs **45**(4), 264–271 (1999)

62.6 S. König, J. Hesser: Live-wires using path-graphs, Methods Inf. Med. **43**(4), 371–375 (2004)

62.7 V. Metzler, J. Bredno, T.M. Lehmann, K. Spitzer: A deformable membrane for the segmentation of cytological samples, Proc. SPIE **3338**, 1246–1257 (1998)

62.8 J. Bredno: Höherdimensionale Modelle zur Quantifizierung biologischer Strukturen. Ph.D. Thesis (RWTH, Aachen 2001), (in German)

62.9 T.M. Lehmann, W. Schmitt, H. Horn, W. Hillen: IDEFIX – Identification of dental fixtures in intraoral X-rays, Proc. SPIE **2110**, 584–595 (1996)

62.10 T.M. Lehmann, J. Bredno, V. Metzler, G. Brook, W. Nacimiento: Computer-assisted quantification of axo-somatic boutons at the cell membrane of motoneurons, IEEE Trans. Biomed. Eng. **48**(6), 706–717 (2001)

62.11 H.H. Ehricke: *Medical Imaging: Digitale Bildanalyse und -kommunikation in der Medizin* (Vieweg, Braunschweig 1997), (in German)

62.12 A. Pommert, K.H. Höhne, B. Pflesser, E. Richter, M. Riemer, T. Schiemann, U. Schumacher, U. Tiede: Ein realistisches dreidimensionales Modell der inneren Organe auf der Basis des Visible Human. In: Medizinische Bildverarbeitung. In: *Handbuch der Medizinischen Informatik*, ed. by T.M. Lehmann, 2nd edn. (Hanser, Munich 2005) pp. 72–76, (in German)

62.13 T.M. Lehmann, M.O. Güld, C. Thies, B. Fischer, K. Spitzer, D. Keysers, H. Ney, M. Kohnen, H. Schubert, B.B. Wein: Content-based image retrieval in medical applications, Methods Inf. Med. **43**(4), 354–361 (2004)

62.14 T.M. Lehmann, H.P. Meinzer, T. Tolxdorff: Advances in biomedical image analysis – Past, present and future challenges, Methods Inf. Med. **43**(4), 308–314 (2004)

63. Virtual Reality in Medicine

Wolfgang Müller-Wittig

Medicine over the past decades has undergone significant changes. In particular, surgery has become increasingly more technology based due to advances and rapid developments in device technology and computer graphics. Minimally invasive surgery (MIS) has revolutionized surgery introducing optical systems and small surgical instruments. In computer graphics Virtual Reality (VR) allows a more intuitive interaction with three-dimensional (3-D) computer generated environments. Furthermore, Augmented Reality (AR) provides a real-time overlay of reality with digital information. This transformation opens up new opportunities for diagnostics, medical education, preoperative planning and intraoperative support. The traditional *see one, do one, teach one* educational environment has now changed to simulation-based training where sophisticated surgical procedures can be planned and rehearsed in a safe virtual environment before the patient enters the operating room (OR). In the operating theatre the surgeon's abilities can be enhanced by superimposing computer generated information on the real patient resulting in better navigation and higher precision of the surgical intervention. With the increasing role of quality management and certification in the health care system, medical simulation can lead to improved quality of care and patient outcomes becoming an integral element finally.

63.1 **Virtual Reality** 1168

63.2 **Medical Applications** 1168
- 63.2.1 Anatomy Education 1170
- 63.2.2 Functional Diagnostics 1170
- 63.2.3 Virtual Endoscopy 1170
- 63.2.4 Surgical Training 1172
- 63.2.5 Preoperative Planning 1172
- 63.2.6 Intraoperative Support 1173

63.3 **VR-Based Medical Simulation** 1174

63.4 **Model Generation – Virtual Anatomy** 1175
- 63.4.1 Data Acquisition 1175
- 63.4.2 Preprocessing 1176
- 63.4.3 Segmentation 1176
- 63.4.4 Mesh Generation 1176
- 63.4.5 Decimation 1177

63.5 **Manipulations – Surgical Interventions** 1178
- 63.5.1 Modeling of Surgical Instruments 1178
- 63.5.2 Registration of Instruments 1179
- 63.5.3 Integration of Visual and Haptic Feedback 1180

63.6 **Outlook** 1182

References 1184

Rapid developments over the last 20 years have significantly influenced surgery. More changes will inevitably emerge in the coming years. Surgery will become increasingly technology based. Two essential developments which have occurred in different domains, namely computer graphics and surgery, form the basis for the topic of this chapter.

In computer graphics, virtual reality (VR) techniques introduced a new dimension in the man–machine interface. VR allows more intuitive interaction with the computer and an immersive and realistic presentation of three-dimensional (3-D) computer-generated worlds using novel input and output devices. This paradigm shift has the potential to simulate processes in such a way that the user has the feeling that he is interacting with the real world.

In surgery, the paradigm shift happened through the transition from open surgery to minimally invasive surgery (MIS), which has revolutionized surgery. MIS has developed into an indispensable diagnostic and therapeutic tool. The goal is to achieve the desired operative result with minimal trauma. The use of MIS covers all surgical fields.

Minimally invasive surgery includes surgical techniques that use optical systems to inspect cavities of the human body and small instruments to perform surgical procedures. An endoscope, with optics and light source, and other miniaturized instruments are inserted into the operative field through small incisions. With the surgeon holding the optical system with the video camera in one hand and watching the operative field on the monitor, the other hand is free to guide either a probe or a small surgical instrument.

63.1 Virtual Reality

Whereas traditional surgery allows guidance of instruments through direct visual control (one axis, eye to hand, and instrument to patient), endoscopic techniques require the coordination of two axes (eye to monitor and hand, instrument to patient). This coordination technique is called triangulation. Besides this triangulation, endoscopy requires further skills; most surgical procedures can be performed with one dominant hand performing the critical parts of the procedure, but endoscopy requires true ambidextrous activity. A further feature of minimally invasive surgical procedures is the loss of direct contact between the surgeon's hand and the actual operative field. The surgeon does not directly touch the anatomical structures. Rather, he/she manipulates these via various surgical instruments.

63.2 Medical Applications

The use of computer graphics has opened up new possibilities in medicine. At the beginning of the 1990s, *Satava* envisioned a paradigm shift in surgical training due to emerging virtual-reality-based medical simulators [63.1–3] (Fig. 63.1). Surgical education is not about training surgeons over a specific time span. Surgical

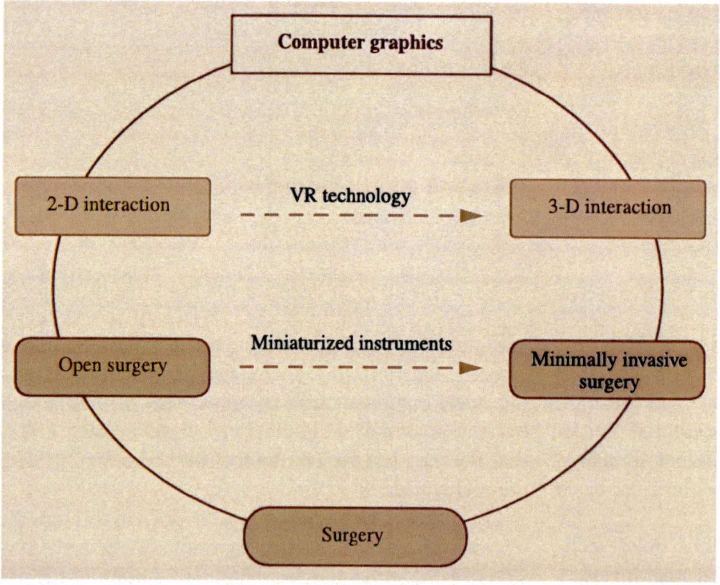

Fig. 63.1 Paradigm shift

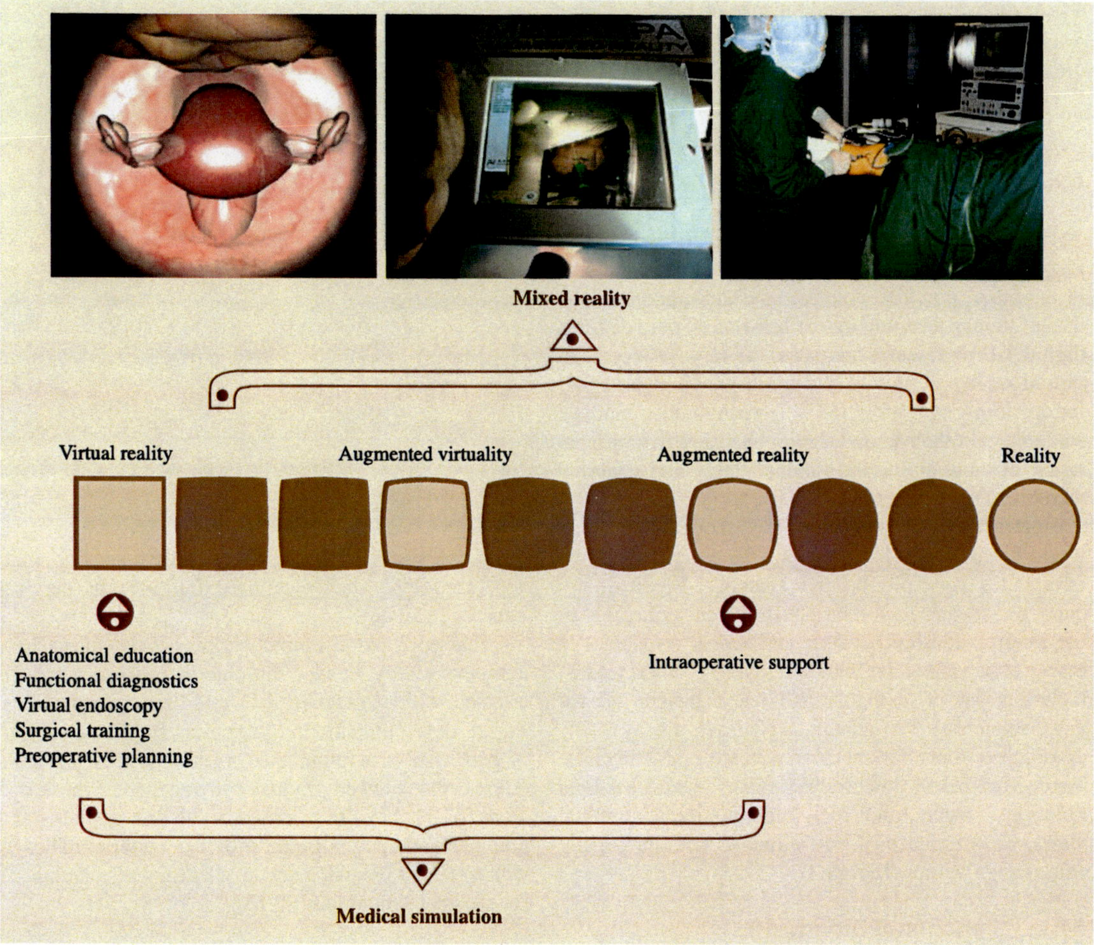

Fig. 63.2 Medical simulation. Application domains along the reality–virtuality continuum between the two poles of the real and virtual environment

training has to be qualitatively measured based on well-defined metrics. The goal was to put more stress on quality assurance in medicine. This topic gained much more attention when the Institute of Medicine (IOM) of the National Academies of the USA declared that as many as around 100 000 people die needlessly each year because of preventable medical harm [63.4]. The IOM recommended greater focus on patient safety and professional standards in healthcare, calling for periodic examinations of doctors and nurses. However, there is no evidence that there have been systematic changes to reduce preventable medical harm. Avoidable medical costs in the USA are estimated at US $500 billion each year [63.5].

Using the reality–virtuality continuum, introduced by *Paul Milgram* and *Fumio Kishino*, various technologies and their applicationq domains in medicine are described as follows (Fig. 63.2) [63.6]. Along this continuum, reality is a gradated spectrum ranging from real to virtual spaces. In virtual reality (VR), the user interacts with 3-D digital objects in real time. The term ***augmented reality*** (AR) describes the real-time overlay of reality with digital information [63.7]. A typical augmented reality system includes the following components:

- A tracking system, which captures the user's real environment

- A display (e.g., a semitransparent head-mounted display) for visualizing the digital information superimposed onto reality
- A mobile computing unit, which is responsible for 3-D registration and image generation.

63.2.1 Anatomy Education

At the beginning of the 1990s, the Visible Human Project, initiated by the National Library of Medicine in the USA, provided complex and multimodal image data of human anatomy for the first time. These detailed digital 3-D representations paved the way for the use of virtual reality to present and teach knowledge in 3-D for anatomy education [63.8]. Traditionally, medical students gain their knowledge in anatomy from preparation of cadavers or from color atlases and anatomy textbooks.

Interaction with computer-based models of the real world through new human–computer interfaces offers advantages in exploring the anatomy. Medical students understand important physiological principles or the interrelationship between anatomical structures in a way that cannot be achieved by any other means, including use of textbooks and real cadavers. Nowadays, numerous 3-D anatomical atlases are available as efficient tools for medical education. Meanwhile, Voxel-Man [63.9] and BodyViz from Visual Medical Solutions, which offer such high-quality anatomical views, are examples of cost-effective solutions using commodity laptops (Fig. 63.3).

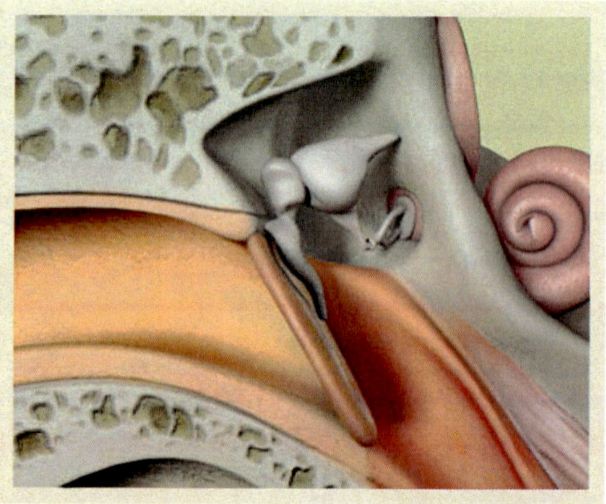

Fig. 63.3 Virtual ear (Phonak Acoustic Implants, Fraunhofer IGD)

63.2.2 Functional Diagnostics

Functional diagnostics describes simulation environments which allow analysis of the patient-specific status quo. Furthermore, the goal is to predict the surgical outcome and to validate the change in functionality that can be expected after a surgical intervention. Thus, it is necessary to simulate the planned patient-specific target state.

For instance, the project VR Whiplash has the goal of providing better diagnostics of functional disorders causing chronic pain after whiplash injuries by integrating VR and measurements of muscular activities. The problems are caused by diagnostic uncertainty and, as a result, the lack of therapy strategies. Imaging diagnostics can only display structural damage, which is, however, rare, representing only 3–5% of cases. The VR-based approach addresses the remaining large fraction of patients with functional disorders, enabling planning of patient-specific therapy, e.g., individually tailored build-up of cervical muscles [63.10].

This application domain impressively shows how computationally intense simulation processes can be coupled with interactive 3-D real-time environments. Rapid developments in computer architectures and in particular programmable graphics processing units (GPUs) has allowed gradual convergence of these two areas [63.11, 12]. For example, nVidia's computing and graphics architecture Fermi provides better graphics as well as higher precision and acceleration of calculations for medical simulation [63.13].

63.2.3 Virtual Endoscopy

Virtual endoscopy is a diagnostic technique providing computer-generated inner body views similar to actual endoscopic procedures [63.14, 15]. The goal of noninvasive 3-D simulation of endoscopic procedures using reconstructed imaging data [e.g., computer tomography (CT) and magnetic resonance imaging (MRI)] is to allow the endoscopist to simultaneously visualize virtual anatomical structures and interaction in real time (e.g., through manipulation of the viewing orientation). The potential is that many body regions which are not accessible to real endoscopy can be explored with virtual endoscopy (Fig. 63.4).

Wireless capsule endoscopy is a recently developed noninvasive imaging technique offering an alternative in diagnostics. Patients swallow a disposable capsule consisting of a pair of cameras and a wireless transmit-

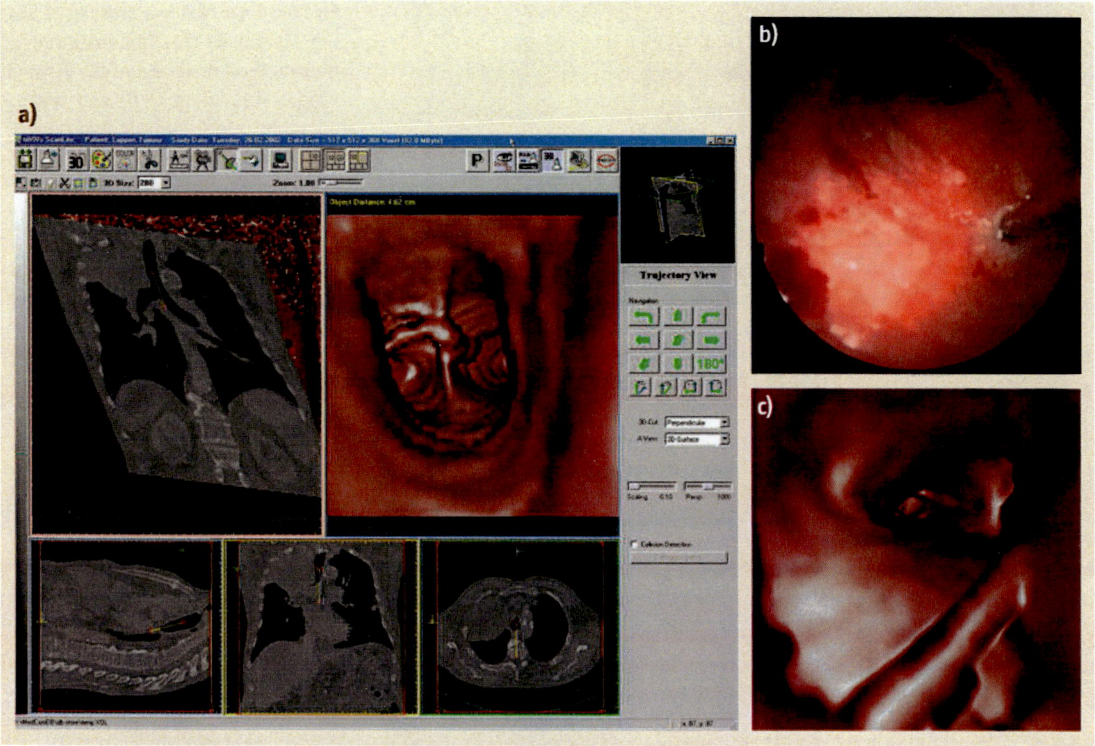

Fig. 63.4a–c Bronchoscopy simulator **(a)** based on patient-specific CT data **(b,c)**

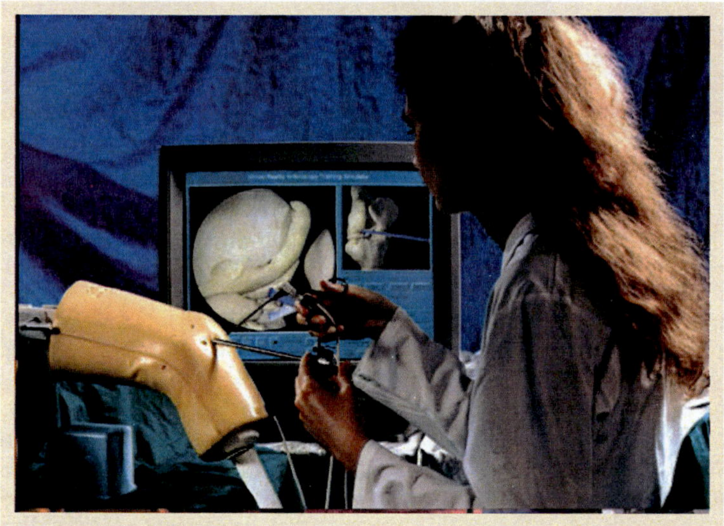

Fig. 63.5 VR arthroscopy training simulator

ter. Finally, a diagnostic video is produced based on the 50 000 images which are shot by such a PillCam during its passage through the colon. Use of this diagnostic tool is still controversial, but recent studies show that technical progress has increased the accuracy of these systems [63.16].

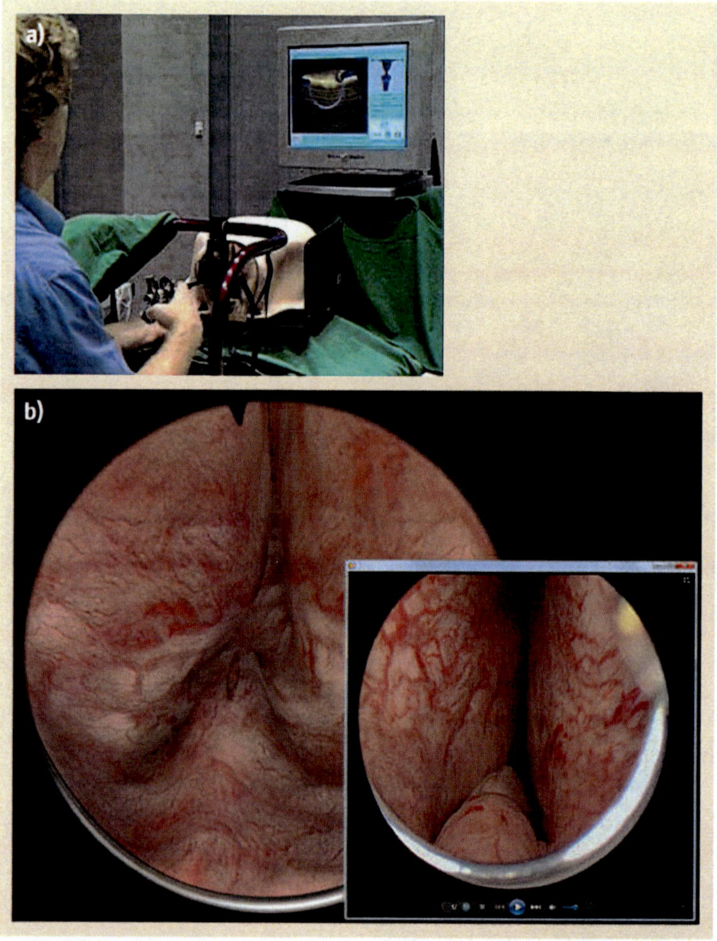

Fig. 63.6 **(a)** VR hysteroscopy training simulator. **(b)** Simulator for transurethral resection of the prostate, virtual simulation (*left*) and real operating-room (OR) movie (*right*) (VirtaMed, Zurich)

63.2.4 Surgical Training

Surgical training is one of the most promising application domains in medicine using 3-D graphics and VR technologies. Motivated by flight simulators for pilot training, a VR-based simulation system for knee arthroscopy was introduced for the first time in 1993 (Fig. 63.5) [63.17]. Since that time, various research groups have focused on the development of laparoscopy training simulators [63.18–20].

The concept of the VR-based knee arthroscopy and laparoscopy training simulators is transferable to other surgical areas (Fig. 63.5). Other minimally invasive procedures such as rhinoscopy or hysteroscopy – without direct contact with the patient – can be trained using a virtual situs (Fig. 63.6) [63.21, VirtaMed AG]. Nowadays, various medical training simulators are available commercially and successful on the market. Meanwhile, numerous medical training centers have established these VR simulators as part of their curriculum [63.22–24].

63.2.5 Preoperative Planning

Virtual reality can also support the surgeon in planning a procedure before surgery (Fig. 63.7). Before performing a surgical intervention on a real patient, the surgeon is able to practice on a virtual patient to simulate the intervention. In this manner, the safest and most effective surgical approach may be selected, requiring less time in the operating room. Here, various variants of the planned surgical procedure are simulated to identify the best path, e.g., to an injured vessel, minimizing damage to healthy structures. Furthermore, a decrease

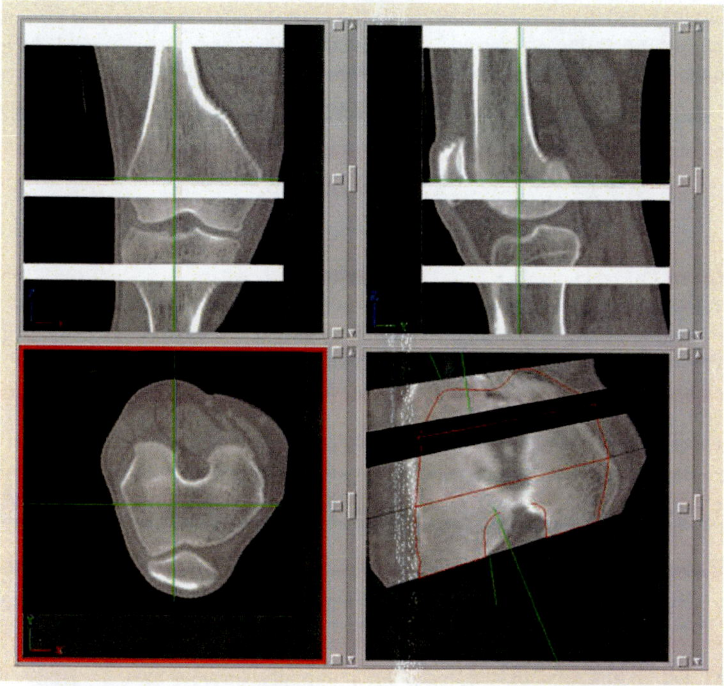

Fig. 63.7 Preoperative planning system for total knee replacement (after [63.26])

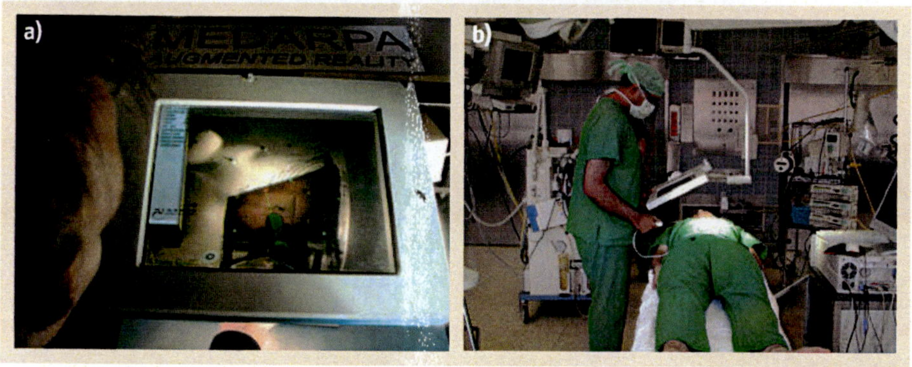

Fig. 63.8a,b *Med*ical *A*ugmented *R*eality for *P*atients (MEDARPA): augmented reality **(a)** in the operating theater **(b)**

in the rate of complications is conceivable. Postoperative treatment can be reduced, and subsequently the costs associated with this; for instance, the risk for revision is higher in case of suboptimally aligned knee implants [63.25].

63.2.6 Intraoperative Support

During surgery, the surgeon's abilities can be enhanced by fusing computer-generated data with real objects. The precision of the surgical intervention may be augmented by superimposing synthetic data on the real patient, giving the surgeon the feeling of having x-ray vision. Deeply embedded anatomic structures are visible; e.g., a tumor can be approached and extracted in the safest possible manner. In this augmented or enhanced reality (ER), surgery can be performed with greater skill. The position of surgical instruments can be monitored and displayed to the surgeon intraoperatively via monitors in the operating room.

In principle, neurosurgery has made use of augmented reality (AR) technology for many years. The microscope image is overlaid by a path selected using patient-specific CT data during the planning phase. This technology has also entered other surgical domains. Integration of medical simulation into the operating theater is becoming increasingly common [63.27–29] (Fig. 63.8).

It is obvious that, in comparison with surgical training, these AR-based intraoperative navigation systems have to fulfill very high requirements regarding accuracy. Meanwhile, more and more studies have been performed to evaluate the robustness and accuracy of AR-guided interventions under laboratory and clinical conditions in the operating theater [63.29, 30].

63.3 VR-Based Medical Simulation

In the following, challenges inherent to implementation of real-time simulations in medicine are addressed. Similar to VR applications in general, quality in medical simulation means convincing the various senses of the user in order to achieve lifelike impressions. In this context, the visual and haptic senses play essential roles. The environment and its impact on the user have to be modeled accordingly. The user should be able to interact with this environment, and beyond that should be able to modify it.

Various research in the field of medical simulation has already shown that integration of haptic displays provides realistic haptic feedback [63.31,32]. In this regard, it should be considered that tactile perception is restricted due to the surgical instruments in minimally invasive surgery. In fact, the emerging development stage in haptic display device technology can already fulfill these requirements to a certain extent.

The quality of visual feedback and the simulation of object behavior while manipulating virtual structures are the challenges of computer-based simulators. In this context, one of the great challenges is simulation and presentation of different tissue types and their elastodynamic characteristics during surgical interventions. In particular, the strong interlink between visual and haptic feedback has to be realized.

A simulation system aims at realistic modeling of a total process in its environment. Consideration of the focus on minimally invasive surgery leads to further analysis of the real environment of a surgeon. The surgeon does not directly touch anatomical structures during minimally invasive surgical proce-

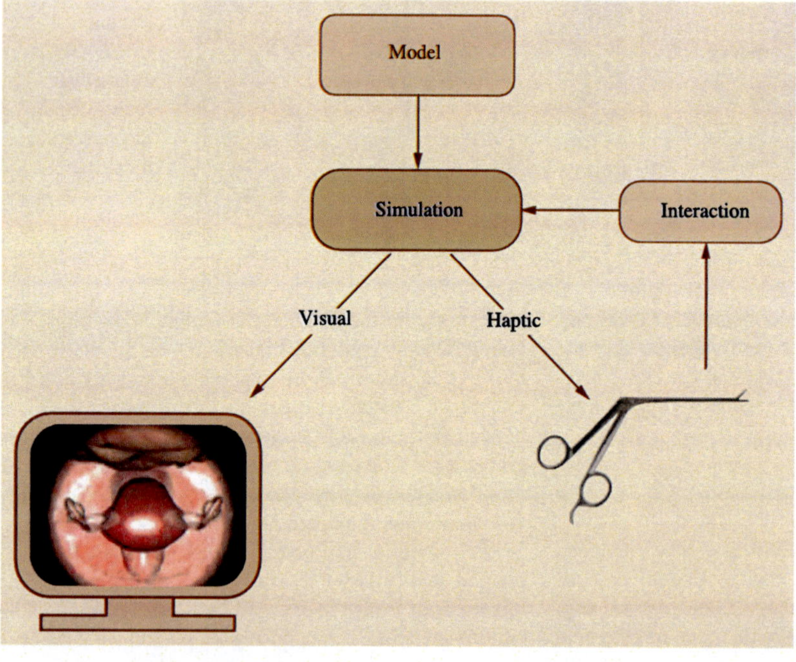

Fig. 63.9 VR-based medical simulation

dures. Rather, he manipulates these with instruments whereas he watches the operative field on a monitor – away from the patient. The surgeon guides his instruments through indirect visual control. Such an operative situation comprises a model (virtual anatomy), surgical interventions (interactions), and surgical instruments (interaction devices) as key elements. Figure 63.9 shows schematically a model of a VR-based medical simulation.

Using this model of VR-based medical simulation, the individual components are described in further detail in the following.

63.4 Model Generation – Virtual Anatomy

The generation of the anatomical region of interest is the first focus in the context of VR-based medical simulation systems. In principle, two approaches can be distinguished for the realization of 3-D models (Fig. 63.10).

On one hand, virtual anatomical structures can be generated using commercially available modeling systems which are also in use in other domains such as architecture. Anatomical models are described via surface models which consist of numerous triangles and which achieve greater realism by projecting realistic textures onto them (Fig. 63.11).

This modeling method will not be further explained in this chapter. Rather, the generation of an integrated patient-specific 3-D model based on medical image data is described in more detail. Such 3-D reconstruction opens up the opportunity to manage individual medical problems with the support of a simulation system.

Here, the anatomical model should not only contain features relevant to visual appearance but should also consider the tissue-specific characteristics required for simulation of haptic feedback. Furthermore, the concept of 3-D reconstruction should not be limited to a specific anatomical region. It should rather allow generation of 3-D representations of diverse anatomies. Finally, the generation of such a synthetic anatomical region should go beyond visualization of a static model to allow life-like simulation of a dynamic model meeting real-time conditions.

The essential steps in the 3-D reconstruction process of anatomical structures based on medical image data can be addressed as follows (Fig. 63.12).

63.4.1 Data Acquisition

Imaging modalities support both medical diagnosis and therapy planning. Multimodality medical images which have been acquired in daily clinical routine can be used for model generation. Sequences

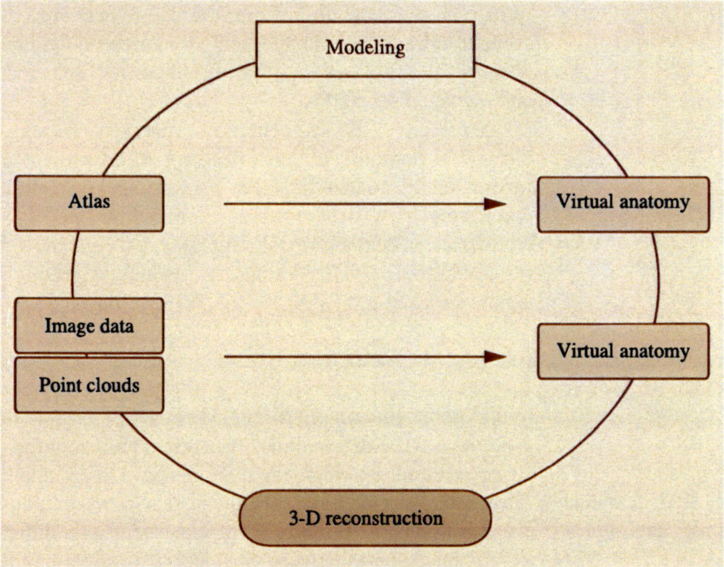

Fig. 63.10 Generation of virtual anatomy

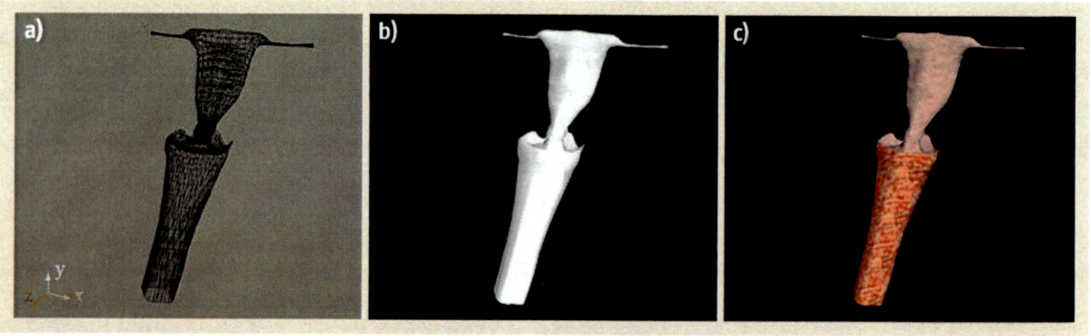

Fig. 63.11a–c Modeling of a virtual uterus. **(a)** Surface model consisting of numerous triangles. **(b)** Modeling with surface texture. **(c)** Modeling with greater realism by projecting realistic textures

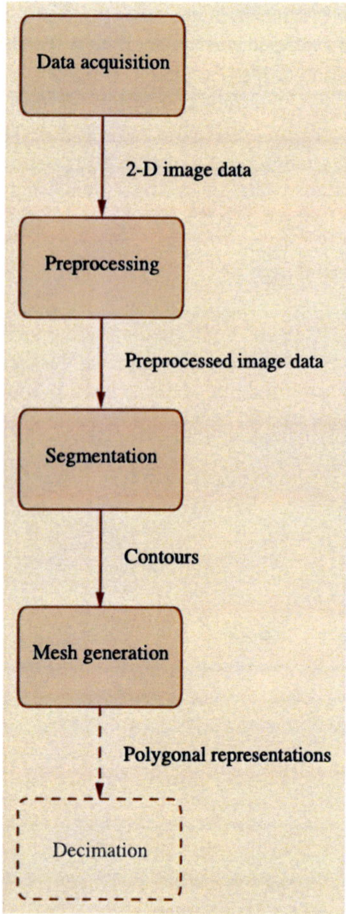

Fig. 63.12 Three-dimensional reconstruction pipeline

of image data typically from CT or MRI scanners are the basis for the reconstruction of anatomical structures.

63.4.2 Preprocessing

In this first step of the reconstruction pipeline, methods of image processing are applied to the original image data, accentuating the contrast between anatomically relevant structures and the surrounding tissue. Furthermore, signals which are redundant or noisy are removed or reduced (e.g., metallic artifacts). The goal is to improve the quality of the subsequent processing steps.

63.4.3 Segmentation

In general, segmentation describes the process of recognizing objects within an image. The result of segmentation is the classification of an image into various regions which distinguish themselves by specific features. The segmentation of medical image data aims at contour extraction of anatomical structures. Individual anatomical structures can then be identified, selected, and managed separately.

In principle, the segmentation process is challenging, since contours of an object are often presented incompletely and insufficiently. However, segmentation algorithms have been further improved so that correct detection of anatomical structures can be achieved semi-automatically. Only minimal manual intervention is now required in this regard [63.33, 34] (Fig. 63.13).

63.4.4 Mesh Generation

In this step, contours of adjacent slices are connected to generate a three-dimensional structure. Here, appropriate connections between two contours have to be found. The method has to address the fact that anatomical models might consist of highly complex structures and that significant changes between two successive slices might

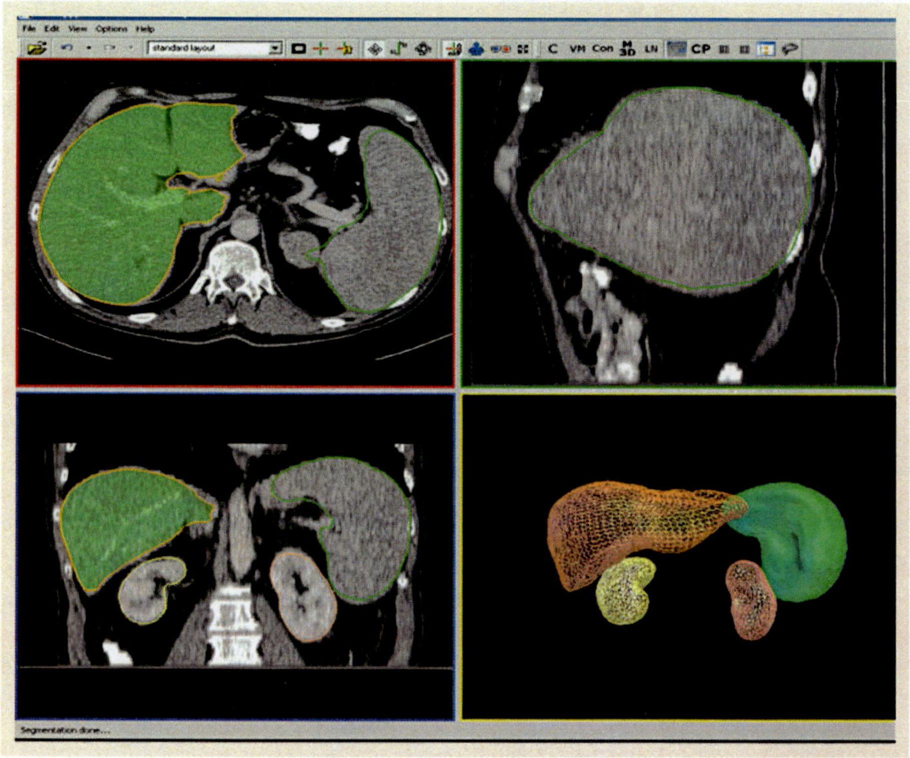

Fig. 63.13 Model-based semi-automatic segmentation of organs

occur. In case of two neighboring slices with arbitrary numbers of contours within them, it has to be guaranteed that the correct contours are connected (the correspondence problem). In case of multiple contours on a slice, the method eventually has to recognize and consider branching of an object from one slice to the next.

An algorithm which constructs meshes based on a 3-D Delaunay triangulation is able to successfully manage these problems. Further details can be found in the work of *Boissonnat* [63.35].

The result of mesh generation is shown in Fig. 63.14. Various complexities can be clearly seen, caused by the different distances between slices in the image dataset. Using this mesh, the anatomical region of interest relevant for the VR-based medical simulation (e.g., interior structures of the knee joint) can be reconstructed at higher resolution.

63.4.5 Decimation

The high resolution necessary for medical data leads to a large number of polygons. As the main objective of simulation is interactive viewing and manipulation of these 3-D objects in real time, the complex geometry of the reconstructed anatomical models may be reduced. This means that polygons with similar spatial orientation are combined during an additional decimation step. Finally, similar to the modeling process, textures are superimposed on the virtual anatomical structures, providing greater realism. These textures can be derived from an anatomical atlas or real endoscopic procedures.

Meanwhile, virtual anatomical structures can also be generated using a 3-D laser scanner based on a specialized method of stripe projection. High-resolution 3-D scans with submillimeter precision allow highly detailed 3-D digitization of real objects. A mesh generator transforms these 3-D point clouds into 3-D models. In addition, the system provides texturing of the geometric model with images of the object, resulting in a photorealistic digital model (Fig. 63.15). If needed, models at various levels of detail and complexity can be generated.

A 3-D scanner is also used in radiotherapy to guarantee beam treatment with high precision. The geometric data of the patient's surface can be used to determine the pose of the patient and to detect patient motion.

Fig. 63.14 Mesh generation of femur and tibia

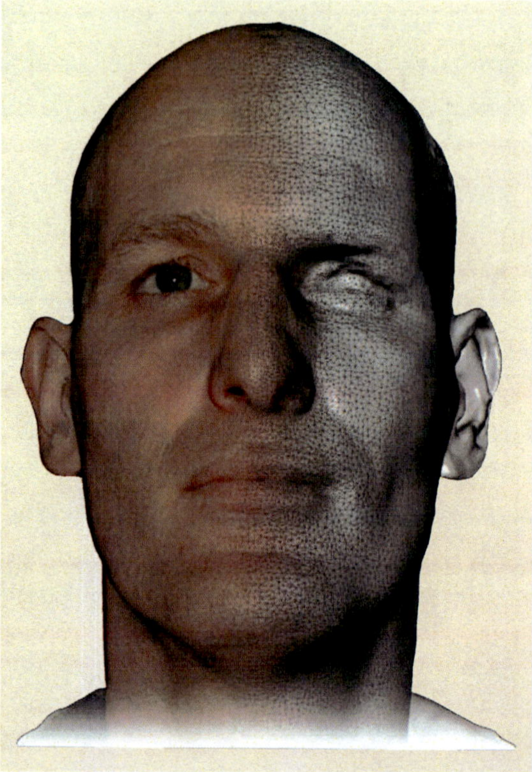

Fig. 63.15 Three-dimensional reconstruction of a head using a scanner (Polygon Technology, Darmstadt)

63.5 Manipulations – Surgical Interventions

Besides 3-D modeling, simulation of surgical instruments is a further focus in the design of VR-based medical simulators. The shape, appearance, and function of instruments must be preserved. The simulation model must allow simulation of both rigid instruments (e.g., endoscope, exploratory probe) and those with moving parts (e.g., grasping forceps, scissors). As the surgeon would do in clinical practice, the goal is to provide an environment in which he is able to maneuver instruments in the virtual anatomy derived from real patient-specific image data, as introduced earlier.

Realization of real-time 3-D interaction is one of the most important requirements. The integration and use of original surgical instruments as *input devices* is intended. VR technology has long allowed the specification of new interaction tools beyond the well-known data glove or data helmet.

63.5.1 Modeling of Surgical Instruments

Sometimes 3-D digital models of the surgical instruments cannot be provided by medical equipment manufacturers. However, virtual instruments have to be provided and integrated into the medical simulation system, where they have to be processed further in real time. Thus, the 3-D geometry of surgical instruments is generated by means of modeling systems based on technical descriptions and/or real instruments. These virtual instruments have to have the same functionality as the real ones. Furthermore, a realistic appearance of the instruments has to be produced. Here, modeling methods such as texturing can be applied to increase the realism of surgical instruments.

63.5.2 Registration of Instruments

Having introduced digital 3-D representations of instruments, these interaction devices have to be applicable in the medical simulation system. This means that they, as virtual replicas of real surgical instruments, also have to function in the same way. Movements of the instruments in the real environment have to be registered and finally transferred without delay to the virtual models which are visualized on the monitor. In this way, the surgeon gets visual feedback.

The tracking technique which is typically used for registration of movements in VR applications is also used here to realize the 3-D interaction using surgical instruments. There are various methods to determine position and registration. With reference to the physical principle used, several categories can be distinguished such as mechanical, optical, and electromagnetic tracking systems. In the following, the electromagnetic approach is briefly introduced.

A sender called a transmitter creates an electromagnetic field. Sensors (receivers) react to this electromagnetic field, delivering position and orientation information. These sensors are attached to surgical instruments, allowing registration of the instrument's movements. Meanwhile, miniature sensors are available which can even be incorporated into medical instruments with very small diameter for dynamic tracking. Cables might be inappropriate, since they limit flexibility and mobility. However, nowadays, wireless electromagnetic tracking systems are available on the market. However, a disadvantage of such tracking methods is that ferromagnetic and metallic materials interfere with the electromagnetic field, leading to distortions in the working volume.

Finally, these electromagnetic tracking systems do not provide such high accuracy with respect to position and orientation as the optical tracking systems which are primarily used in the operating theater. However, they represent a cost-effective, lightweight, and compact solution for dynamic tracking that is widely used in medical simulation systems which do not have high accuracy requirements (e.g., training simulators). Here, registration of objects can typically be achieved with accuracy of 0.8 mm in position and 0.15° in orientation. Thus, small movements of instruments can be tracked dynamically in real time. Based on the delivered sensor data, an updated view of the operating field is calculated and visualized on the monitor of the graphics personal computer (PC).

Since an endoscope has some specific optics, simulation of this should be discussed in further detail. Endoscopes can be of forward-viewing (0°) or forward-oblique direction of view (e.g., 30°, 45°, 70°). These angle optics increase the field of view by just rotating the endoscope around its axis, thus minimizing instrument motion. Further to this, some endoscopes produce viewing fields of up to 110° through the use of wide-angle optics.

As described above, the position and orientation of an instrument, here an endoscope, can be determined using sensors. Now the optics must be simulated by calculating the camera point of the instrument based on the delivered sensor data. Depending on the optics selected, the actual field of view of the endoscope is updated and visualized on the monitor. In this context, the graphical user interface contains a display window for the endoscopic view.

VR-based medical simulation systems have the advantage that endoscopes with any angle optics can be seamlessly integrated. Here, the arrow helps the surgeon determine the orientation by showing the viewing direction (Fig. 63.16).

A second window provides an overview of the operative field, which is very helpful for the determination of orientation, especially for beginners. This display window shows distinctly and visibly how the light source at the tip of the endoscope (simulated by a point source) illuminates the respective part of the virtual anatomy (Fig. 63.16, top right). In addition, other optics such as a wide-angle lens with its typical convex distortion of the endoscopic image can be simulated.

Furthermore, the simulation environment can be specified using a graphical user interface. Here, a selection of various anatomical regions and numerous surgical instruments are available. In addition, statistics are generated during simulation, allowing objective assessment of trainees' performance. A portfolio of standard training situations and tools for objective assessment are definitely the major advantages of medical training simulators in comparison with traditional training methods for learning endoscopic skills (e.g., *learning by doing* on patients).

Thus, VR technology allows the introduction of novel interaction devices. Three-dimensional real-time interaction and intuitive handling of various surgical instruments is provided in a VR-based medical simulation system. A complete set of virtual instruments such as endoscopes with various optics, exploratory probes, as

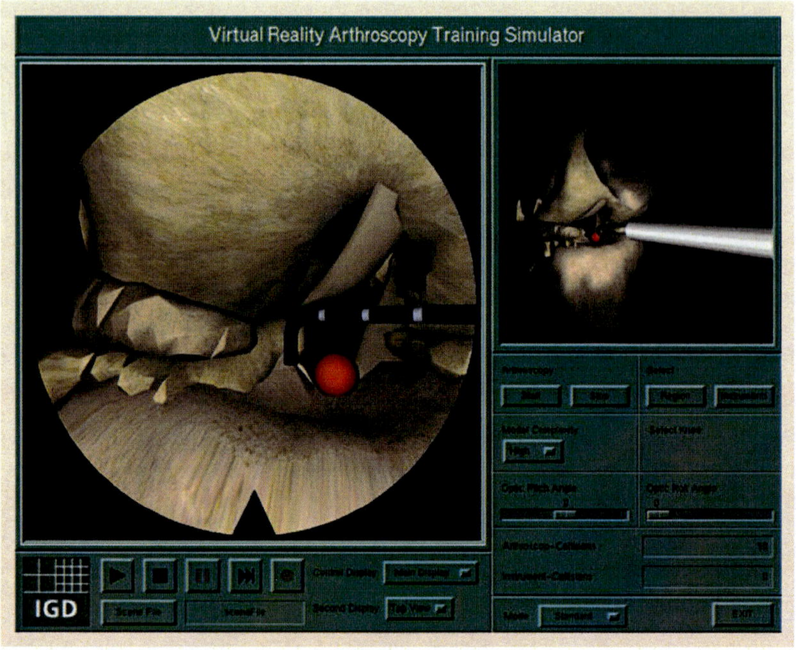

Fig. 63.16 Graphical user interface

well as grasping forceps or dissecting scissors can be simulated realistically while preserving both form and function.

Besides registration of instrument movements, fast collision detection between the various instruments and virtual anatomical structures is required. This forms the basis for simulation of tissue manipulation (e.g., deformation or cutting).

Here, deformation is exemplarily introduced as a type of interactive manipulation of anatomy to show the principles of soft tissue simulation. Two approaches for simulation of deformation are described: geometry-based and physically based methods. Deformation behavior of anatomical structures as a result of interventions with surgical instruments can be realized using *bump weighting functions* and mass–spring models or by finite-element methods (FEM), and thereby integrated into the medical simulation system.

The former method allows fast calculation of local deformations, allowing real-time simulation even on today's low-cost computers. A bump-weighting function can be specified to describe the deformation in the area close to the surgical instrument, depending on the penetration depth into the respective tissue. In this way, an individual local deformation behavior can be applied to each tissue type. This geometric approach can be realized with simple and fast transformations.

The latter approach uses methods for a physically based simulation, also addressing global deformations. Here mass–spring models allow the calculation of object deformations in real time using appropriate numerical integration with varying step size. The difficulty is to achieve a robust and stable simulation.

Finally, more and more solutions using finite-element methods are entering the medical simulation arena. Tissue-specific material properties are integrated in the FEM framework. This method achieves a high-quality simulation but unfortunately has high computational costs. Rapidly developing and recently introduced hardware architectures such as GPUs open up new opportunities to manage and speed up these complex and computationally intensive calculations. More details about this research direction can be found in *Taylor* et al. [63.12] and *Peterlik* et al. [63.36].

63.5.3 Integration of Visual and Haptic Feedback

The quality of a simulator session will be enhanced if the trainee manipulates and experiences the virtual situs through multiple sensory channels. It has been shown that visual and haptic perception play a significant role in interactive environments. Considering the specifics of these two sensory channels, the respective components

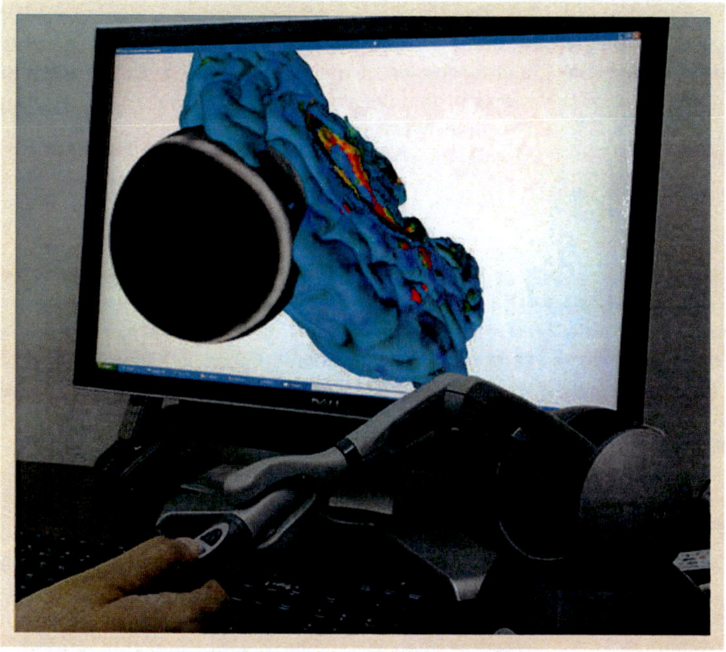

Fig. 63.17 Haptic simulation (after [63.32])

of the simulation system have to be realized with different update rates. While 20–40 Hz is sufficient for visual output to provide real-time visualization, high frequencies of up to 1000 Hz have to be generated to satisfy the human haptic sense. Finally, the visual and haptic feedback have to be seamlessly integrated and synchronized.

Two main factors have an impact on research work in the field of haptics for VR-based medical simulation systems. One cornerstone is hardware in terms of haptic device technology; the other is the software component controlling haptic displays to generate artificial resistance.

In minimally invasive surgery, the number of degrees of freedom is reduced due to fixation of the instruments at the minimal access ports. Further to this, it can be stated that neither a large working space nor high force levels are needed for the haptic interface. Certainly, in case surgical instruments come into contact with rigid objects such as bony structures, a maximal force has to be generated and the haptic display must deliver the corresponding stiffness.

As of today, six-DoF (degree of freedom) tracking systems are available and very common. Recent years have also seen an increasing number of options in the field of haptic devices. More and more six-DoF, cost-effective haptic displays have entered the market. In addition, there are haptic interfaces designed specifically for surgical interventions. This opens up new opportunities for the medical simulation domain.

The software realizes the haptic rendering by converting an internal representation of a virtual object into a representation suitable for display on a haptic device (analogously to graphics rendering) (Fig. 63.17).

Haptic simulation of minimal invasive procedures must also include the tissue-specific characteristics of anatomical structures (e.g., hard bone tissue; firm, resilient meniscus). The most relevant tissue types are classified with respect to their haptic attributes (e.g., mass, friction, elasticity) and then integrated into the VR-based simulation system. Finally, integration of different simulation models including consideration of the penetration depth of the instrument generates the corresponding object-specific haptic stimuli.

As mentioned above, haptic displays need extremely high update rates of up to 1000 Hz to avoid lag in the simulation loop which might otherwise lead to distorted haptic perception. In this context, fast collision detection is key for a believable haptic interaction. One approach is to split the virtual world into a global and a local part. The global part describes the complete virtual scene, where collision detection might be performed at less than 1000 Hz. The local part represents the region where the actual interactions between the instrument and surrounding anatomical structures take place. This region is relevant for the haptic simulation. This local region of the virtual scene is selected based on the collision detection results and then further processed by the simulation engine. Then calculations which are required for the haptic feedback are done, enabling the high update rates needed to control the haptic display.

The simulation of haptic feedback is an important research topic worldwide. A comprehensive overview and discussion of the various methods cannot be provided in this chapter. *Coles* et al. [63.37] provide an excellent overview of current developments in haptic simulation in the area of medical VR trainers.

63.6 Outlook

Rapid developments in the two areas of computer graphics and surgery will have further impacts on medical simulation in the future.

In computer graphics, the availability of increasingly powerful processor architectures opens up new possibilities for real-time simulation. In particular, GPUs, beating Moore's law by a factor of three, will allow, as well as higher visual quality and greater realism, the solution of computational power and non-graphics-related problems. Therefore, physically based methods will be integrated with greater accuracy, continuing the trend towards greater interactivity and realism.

The continuing developments will also include haptic displays, although advancements will happen at a different speed here.

The anatomical model will become more sophisticated; microstructures will be increasingly considered. In vivo measurements have been increasingly conducted, which can be the basis for mathematical modeling of the elastodynamic characteristics of organic structures. The derivation of relevant tissue-specific values is expected to lead to quality improvement in the simulation of the elastodynamic behavior of anatomical structures. A large amount of research work is focused on the collection of such empirical data for soft tissue and abdominal organs for subsequent transfer to the simulation model [63.36,38–40].

Beyond this, simulation will include further aspects of minimally invasive interventions such as coagulation, i.e., the stopping of bleeding by the use of heat. Closely linked to such simulations, physiological functions should also be mentioned, where – besides blood flow – respiration and the heart beat can be seen as the next challenges [63.41]. Realistic real-time simulation of these complex processes is one of the research challenges worldwide [63.42].

There will also be further developments in the field of minimally invasive procedures. The rapid advances in endoscopic operative techniques will continue, and entire traditional surgical domains will disappear. This will be accompanied by the increasing sophistication of instruments. Furthermore, there is a trend in surgery towards the performance of interventions via natural orifices of the body, enabling further minimization of patient trauma.

Other research work focuses on giving back the *lost feeling* to surgeons by integrating sensors to detect subtle lesions with instruments [63.43]. This includes the realistic integration of specific surgical instruments into haptic simulation. In this regard, fundamental knowledge of haptic perception is the basis for successful coupling of haptic feedback with simulators or robotic systems. One of the research challenges here is to present the forces that are applied to the patient in a realistic manner. Haptic displays do not yet accomplish this [63.44].

The introduction of new minimally invasive techniques into clinical practice has to be done via phased and structured development and validation with adequate postoperative monitoring rather than following a *trial-and-error* approach. In this context, virtual simulation of new operative procedures can make a valuable contribution to the ongoing development of minimally invasive surgery to maintain or increase its significance.

Flight simulators have been an indispensable ingredient in the training of pilots for many years. Medical simulators using VR technology and computer graphics have the potential to play a similar role in surgical training. The challenges inherent to traditional training methods of *learning by doing* or using phantoms or cadavers motivated the search for alternative training scenarios. Finally, this led to the development of VR-based simulators, which are gaining increasing levels of recognition and acceptance in minimally invasive surgery.

Furthermore, studies have provided evidence that skills acquired in a virtual training environment can be successfully transferred to the real world. Skill transfer can be improved if there is a good match between the training and the real context. VR-based training systems definitely achieve an enhancement of the learning curve without requiring patient contact. Numerous studies carried out in recent years have shown the potential for successful transfer from the simulator to the operating room [63.22, 45–50].

Today, there are already promising signs that use of VR-based training simulators in medical curricula has become a reality. Training centers such as the Queensland Health Skills Development Centre, the Minimal Invasive Surgical Centre Singapore or the Wenckebach Institute in Groningen have made medical simulators a definite part of their training programs; for instance, each year, the European Surgical Institute organizes about 80 training courses for 1000 participants, helping them to develop precision and skills with the support of VR-based simulators. Some changes in the current framework of medical education are also on the way in the USA. The general availability of medical simulators will have an impact on the way medicine is taught and practiced in the future. Thus, the American Board of Surgery already requires surgical residents to perform training sessions on simulators in the field of laparoscopy [63.51, 52].

It has to be recalled that several decades were needed to make realistic flight simulators of today's standard possible. The rapid advances in the field of computer-based training simulators and the successive improvement of their realism in recent years indicate the emergence of such systems in surgical education as well. The *permission to fail* in such simulation environments is an essential advantage over *training in the operating theater*, corresponding to a further enhancement of patient safety. Such simulation has the potential to be used to a greater extent during preoperative planning based on patient-specific data immediately before the real surgical procedure; this can be seen as a *preoperative warm-up*. In Canada, such a dry run has already been performed using the NeuroTouch simulator in the field of neurosurgery.

It is without dispute that quality management and certification are playing an increasing role in the healthcare system. This fact will finally lead to an essential reform of medical education. Here, computer-assisted training simulators with their spectrum of objective assessment will become an integral component. Is this the route to a *licence* for surgeons, as is already in place for pilots?

Furthermore, a paradigm shift is on the horizon. Not only will the surgeon receive decision support from preoperative image data, but sophisticated sensors will allow intraoperative delivery of tissue-specific characteristics. This information will be made available to the surgeon, allowing him to adapt the next surgical step; for instance, a surgical robot selects the drill channel in the bone by considering actual sensory feedback. There is no need to rely *only* on preoperative planning based on the patient-specific image dataset.

The use of robots in the operating theater has been called into question and discussed controversially. It should be considered that common industrial robots were used at that time, which can now be substituted by compact, cost-effective, modern operating-room (OR) robots (from, e.g., Mazor Surgical Technologies). Meanwhile, the benefits of robot-assisted surgery are increasingly being recognized. In the operating theater, the robot is an additional tool for the surgeon that can be selectively used during specific intraoperative steps. In this way, the intervention can be performed with greater precision and quality, resulting in a better final operative outcome. In the meantime,

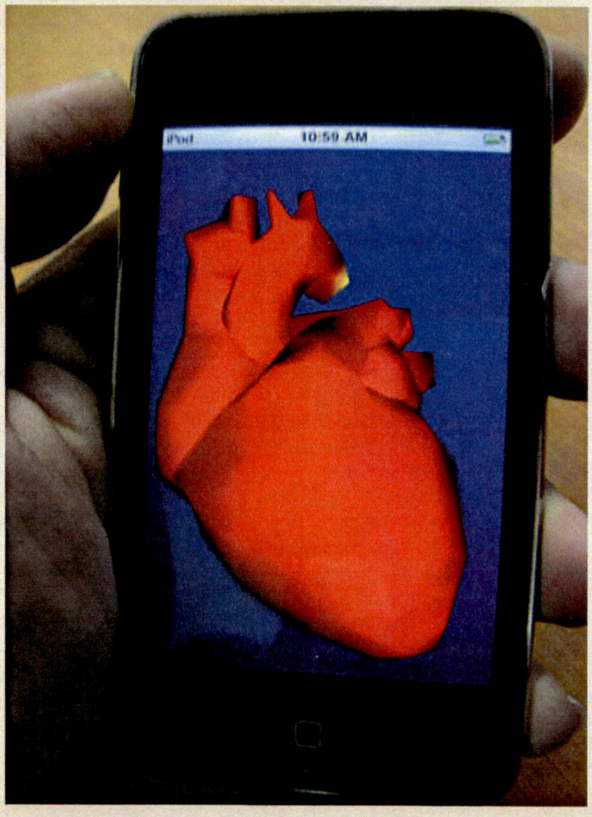

Fig. 63.18 Medical visualization on a smart phone

the global robot-assisted surgery market is expected to grow from US $ 1 billion in 2008 to US $ 14 billion by 2014.

Finally, attention should be drawn to a trend which has already entered our daily life, i.e., mobility. This will include medical simulation too; i.e., *medical simulation will go mobile.* Forecasts show that 81% of medical doctors in the USA will be equipped with – besides a stethoscope – a smart phone by 2012 [63.53]. This opens up exciting new possibilities to blend medical simulation with mobile learning environments. Considering the continuous increase of performance and functionality of mobile devices, use of 3-D visualization and AR applications *on the move* has become closer to reality (Fig. 63.18). All these developments will benefit the patient.

References

63.1 R.M. Satava: Virtual reality surgical simulator. The first steps, Surg. Endosc. **7**(3), 203–205 (1993)

63.2 R.M. Satava: Virtual reality, telesurgery, and the new world order of medicine, J. Image Guid. Surg. **1**, 12–16 (1995)

63.3 R.M. Satava: Medical virtual reality. The current status of the future, Stud. Health Technol. Inf. **29**, 100–106 (1996)

63.4 L.T. Kohn, J.M. Corrigan, M.S. Donaldson (Eds.): *To Err Is Human: Building a Safer Health System* (The National Academies, Washington 2000), Committee on Quality of Health Care in America, Institute of Medicine

63.5 R. Langreth: Fixing Hospitals, Forbes Mag. **175**(13), 68–76 (2005)

63.6 P. Milgram, F. Kishino: Taxonomy of mixed reality visual displays, IEICE Trans. Inf. Syst. E **77-D**(12), 1321–1329 (1994)

63.7 J.L. Encarnação, P. Astheimer, F. Dai, W. Felger, M. Göbel, H. Haase, S. Müller, R. Ziegler: Virtual reality technology – Enabling new dimensions in computer-supported applications, Proc. 9. Japan-Germany Forum Inf. Technol. Oita (1994)

63.8 V. Spitzer, M.J. Ackerman, A.L. Scherzinger, D. Whitlock: The visible human male: A technical report, J. Am. Med. Inf. Assoc. **3**(2), 118–130 (1996)

63.9 K.H. Höhne: *VOXEL-MAN 3-D Navigator: Brain and Skull* (Springer Electronic Media, Heidelberg 2009), Version 2.0

63.10 U. Bockholt, T. Drevensek, A. Schäffer, A. Bisler, A. Rettig, M. Kramer: Simulation zur Stimulation von Kopfbewegungen zur VR gestützten Schleudertraumatherapie, Proc. 17th Simul. Vis. Conf., Magdeburg (2006)

63.11 T. Akenine-Moller, E. Haines, N. Hoffman: *Real-Time Rendering*, 3rd edn. (Peters, Wellesley 2008)

63.12 Z.A. Taylor, O. Comas, M. Cheng, J. Passenger, D.J. Hawkes, D. Atkinson, S. Ourselin: On modelling of anisotropic viscoelasticity for soft tissue simulation: Numerical solution and GPU execution, Med. Image Anal. **13**(2), 234–244 (2009)

63.13 nVidia: NVIDIA's Next Generation CUDA Compute Architecture: Fermi (NVIDIA, Santa Clara 2009) available at http://www.nvidia.com/content/PDF/fermi_white_papers/NVIDIAFermiComputeArchitecture Whitepaper.pdf

63.14 A.E. Kaufman, S. Lakare, K. Kreeger, I. Bitter: Virtual colonoscopy, Commun. Assoc. Comput. Mach. **48**(2), 37–41 (2005)

63.15 A. Krüger, C. Kubisch, G. Strauß, B. PreimSinus: Endoscopy – Application of advanced GPU volume rendering for virtual endoscopy, IEEE Trans. Vis. Comput. Graph. **14**(6), 1491–1498 (2008)

63.16 R. Eliakim, K. Yassin, Y. Niv, Y. Metzger, J. Lachter, E. Gal, B. Sapoznikov, F. Konikoff, G. Leichtmann, Z. Fireman: Prospective multicenter performance evaluation of the second-generation colon capsule compared with colonoscopy, Endoscopy **41**, 1026–1031 (2009)

63.17 W.K. Müller, R. Ziegler, A. Bauer, E.H. Soldner: Virtual reality in surgical arthroscopic training, J. Image Guid. Surg. **1**(5), 288–294 (1995)

63.18 U.G. Kühnapfel, H.G. Krumm, C. Kuhn, M. Hübner, B. Neisius: Endosurgery simulations with KISMET, Proc. Virtual Reality World '95, Stuttgart (1995)

63.19 S. Cotin, H. Delingette, J.-M. Clément, V. Tassetti, J. Marescaux, N. Ayache: Geometric and physical representations for a simulator of hepatic surgery, Stud. Health Technol. Inf. **29**, 139–151 (1996)

63.20 A.M. Meglan, R. Raju, G.L. Merril, J.R. Merril, B.H. Nguyen, S.N. Swamy, G.A. Higgins: The Teleos virtual environment toolkit for simulation-based surgical education, Stud. Health Technol. Inf. **29**, 346–351 (1996)

63.21 U. Bockholt, Y. Jung, R.-M. Recker, M. Olbrich: Visuelles und haptisches Volumenrendering von medizinischen Bilddaten in Virtual Reality-Simulationen, 11. IFF-Wissenschaftstage 2008, ed. by M. Schenk (2008) pp. 31–37, Conf. Digest, (CD-ROM)

63.22 A. Hackethal, M. Immenroth, T. Bürger: Evaluation of target scores and benchmarks for the traversal task scenario of the minimally invasive surgical trainer-virtual reality (MIST-VR) laparoscopy simulator, Surg. Endosc. **20**(4), 645–650 (2006)

63.23 K. Maschuw, T. Osei-Agyemang, P. Weyers, R. Danila, K. Bin Dayne, M. Rothmund, I. Hassan: The impact of self-belief on laparoscopic perfor-

mance of novices and experienced surgeons, World J. Surg. **32**(9), 1911–1916 (2008)
63.24 M.W. Salkini, C.R. Doarn, N. Kiehl, T.J. Broderick, J.F. Donovan, K. Gaitonde: The role of haptic feedback in laparoscopic training using the LapMentor II, J. Endourol. **24**(1), 99–102 (2010)
63.25 E.J. Novak, M.D. Silverstein, K.J. Bozic: The Cost-Effectiveness of Computer-Assisted Navigation in Total Knee Arthroplasty, J. Bone Joint Surg. Am. **89**, 2389–2397 (2007)
63.26 W. Müller, U. Bockholt, G. Voß, A. Lahmer, M. Börner: Computer assisted preoperative planning system for total knee replacement, Proc. Fourth Annu. N. Am. Program Comput. Assist. Orthop. Surg. (CAOS), Pittsburgh USA (2000) pp. 105–107
63.27 S. Wesarg, E. Firle, B. Schwald, H. Seibert, P. Zogal, S. Röddiger: Accuracy of needle implantation in brachytherapy using a medical AR system: A phantom study, Proc. Medical Imaging 2004 (SPIE, Bellingham 2004) pp. 341–352
63.28 M. Fawad Khan, S. Dogan, A. Maataoui, J. Gurung, M. Schiemann, H. Ackermann, S. Wesarg, G. Sakas, T.J. Vogl: Accuracy of biopsy needle navigation using the Medarpa system-computed tomography reality superimposed on the site of intervention, Eur. Radiol. **15**(11), 2366–2374 (2005)
63.29 P.J. Stolka, M. Keil, G. Sakas, E. McVeigh, M.E. Allaf, R.H. Taylor, E.M. Boctor: A 3-D elastography-guided system for laparoscopic partial nephrectomies, Proc. SPIE, Medical Imaging 2010, Vol. 7625 (2010)
63.30 D. Teber, S. Guven, T. Simpfendörfer, M. Baumhauer, E.O. Güven, F. Yencilek, A.S. Gözen, J. Rassweiler: Augmented reality: A new tool to improve surgical accuracy during laparoscopic partial nephrectomy? Preliminary in vitro and in vivo results, Eur. Urol. **56**(2), 332–338 (2009)
63.31 C. Basdogan, S. De, J. Kim, M. Muniyandi, H. Kim, M.A. Srinivasan: Haptics in minimally invasive surgical simulation and training, IEEE Comput. Graph. Appl. **24**(2), 56–64 (2004)
63.32 K. Levinski, A. Sourin, V. Zagorodnov: Interactive surface-guided segmentation of brain MRI data, Comput. Biol. Med. **39**(12), 1153–1160 (2009)
63.33 K. Drechsler, C. Oyarzun Laura, Y. Chen, M. Erdt: Semi-automatic anatomical tree matching for landmark-based elastic registration of liver volumes, J. Healthc. Eng. **1**, 101–123 (2010)
63.34 M. Erdt, G. Sakas: Computer aided segmentation of kidneys using locally shape constrained deformable models on CT images, Proc. SPIE Med. Imag., Vol. 7624 (2010) pp. 762419-1–762419-8
63.35 J.D. Boissonnat: Shape reconstruction from planar cross-sections, Proc. IEEE Conf. Comput. Vis. Pattern Recognit. (1985) pp. 393–397
63.36 I. Peterlik, M. Sedef, C. Basdogan, L. Matyska: Real-time visio-haptic interaction with static soft tissue models having geometric and material nonlinearity, Comput. Graph. **34**(1), 43–54 (2010)
63.37 T. Coles, D. Meglan, N. John: The role of haptics in medical training simulators: A survey of the state of the art, IEEE Trans. Haptics **4**(1), 51–66 (2011)
63.38 M.P. Ottensmeyer: In vivo measurement of solid organ visco-elastic properties, Proc. Medicine Meets Virtual Reality 02/10 – Digital upgrades: Applying Moore's law to health, Newport Beach (IOS, Amsterdam 2002) pp. 328–333
63.39 M. Sedef, E. Samur, C. Basdogan: Real-time finite-element simulation of linear viscoelastic tissue behavior based on experimental data, IEEE Comput. Graph. Appl. **26**(5), 58–68 (2006)
63.40 B. Ahn, J. Kim: Measurement and characterization of soft tissue behavior with surface deformation and force response under large deformations, Med. Image Anal. **14**(2), 138–148 (2010)
63.41 M. Müller, S. Schirm, M. Teschner: Interactive blood simulation for virtual surgery based on smoothed particle hydrodynamics, J. Technol. Health Care **12**(1), 25–31 (2004)
63.42 S. Daenzer, K. Montgomery, R. Dillmann, R. Unterhiinninghofen: Real-time smoke and bleeding simulation in virtual surgery, Stud. Health Technol. Inf. **125**, 94–99 (2007)
63.43 E. Westebring-van der Putten, M. Hajian, R. Goossens, J. van den Dobbelsteen, J. Jakimowicz: *A Laparoscopic Grasper Handle with Integrated Augmented Tactile Feedback, Designed for Training Grasp Control*, LNCS, Vol. 6192 (Springer, Berlin, Heidelberg 2010) pp. 243–250
63.44 A.M. Okamura: Haptic feedback in robot-assisted minimally invasive surgery, Curr. Opin. Urol. **9**(1), 102–107 (2009)
63.45 N.E. Seymour, A.G. Gallagher, S.A. Roman, M.K. O'Brien, V.K. Bansal, D.K. Andersen, R.M. Satava: Virtual reality training improves operating room performance: Results of a randomized, double-blinded study, Ann. Surg. **236**(4), 458–464 (2002)
63.46 A. Hyltander, E. Liljegren, P.H. Rhodin, H. Lönroth: The transfer of basic skills learned in a laparoscopic simulator to the operating room, Surg. Endosc. **16**(9), 1324–1328 (2002)
63.47 A.G. Gallagher, A.B. Lederman, K. McGlade, R.M. Satava, C.D. Smith: Discriminative validity of the minimally invasive surgical trainer in virtual reality (MIST-VR) using criteria levels based on expert performance, Surg. Endosc. **18**, 660–665 (2004)
63.48 A.G. Gallagher, E.M. Ritter, H. Champion, G. Higgins, M.P. Fried, G. Moses, C.D. Smith, R.M. Satava: Virtual reality simulation for the operating room: Proficiency-based training as a paradigm shift in surgical skills training, Ann. Surg. **241**(2), 364–372 (2005)

63.49 K. Kahol, R.M. Satava, J. Ferrara, M.L. Smith: Effect of short-term pretrial practice on surgical proficiency in simulated environments: a randomized trial of the "preoperative warm-up" effect, J. Am. Coll. Surg. **208**(2), 255–268 (2009)

63.50 R. Lipner, J. Messenger, R. Kangilaski, D. Baim, D.R. Holmes Jr., D.O. Williams, S.B. King: A technical and cognitive skills evaluation of performance in interventional cardiology procedures using medical simulation, J. Soc. Simul. Healthc. **5**(2), 65–74 (2010)

63.51 R.M. Satava: The revolution in medical education – The role of simulation, J. Grad. Med. Edu. **1**(2), 172–175 (2009)

63.52 B. Bashankaev, S. Baido, S.D. Wexner: Augmented reality: Review of available methods of simulation training to facilitate surgical education, Surg. Endosc. **25**(1), 28–35 (2011)

63.53 M. Levy: *Physicians in 2012: The Outlook for On Demand, Mobile, and Social Digital Media* (Manhattan Research, New York 2010), Manhattan Res. Rep.

64. Computer-Supported Teaching and Learning Systems in Medicine

Martin Haag, Martin R. Fischer

In 1999, the scenario *University in the Year 2005* predicted that by 2005 half of all students would be studying at virtual universities, while there would be only a small core of traditional universities left. Obviously, this prediction did not come true, and the scenario is far from being realized, even today (Sects. 64.1–64.3). However, there is a visible trend towards using computer-supported teaching and learning systems in medical studies at traditional universities within blended learning (BL) concepts (Sect. 64.5). A meta-analysis by *Cook* et al. showed that studying supported by internet-based learning is at least equal to traditional forms of teaching [64.1]. In German-speaking countries, the term computer-based training (CBT) is frequently used for computer-supported training, but there are a plethora of other English-derived terms that are also in use, such as computer-assisted instruction (CAI), computer-assisted learning (CAL) and computer-based instruction (CBI). The term web-based training (WBT) [64.2] is used if the applications are based on web technology and accessed via the web. The term e-Learning [64.3] (electronic-learning) is somewhat broader, and – aside from CBT and WBT – also includes other forms of digital learning such as computer-supported cooperative/collaborative learning (CSCL) (Sects. 64.5 and 64.6). Against the backdrop of the historical development and moves toward reforming the medical curriculum, an overview over the relevant aspects of CBT/WBT systems in medical studies are presented in this chapter; in particular, questions regarding the integration of the curriculum and the sustainability of reforms.

64.1 **Historical Development** 1187

64.2 **Moves Towards the Reform of Medical Studies** 1189

64.3 **Developing Learning and Teaching Systems** 1189
- 64.3.1 Interaction Forms 1189
- 64.3.2 Architectures of Learning and Teaching Systems 1192
- 64.3.3 Authoring Systems 1192

64.4 **Learning Environments** 1193
- 64.4.1 Functionality of Learning Environments 1193
- 64.4.2 Interoperability and Standards 1194

64.5 **Application Scenarios for Learning and Teaching Systems** 1195

64.6 **Status of and Outlook for e-Learning in Medicine** 1195
- 64.6.1 Information Systems for CBT/WBT in Medicine 1195
- 64.6.2 Use of CBT/WBT Systems in Medicine and the Problem of Curricular Integration 1195
- 64.6.3 The Need for Research, and Outlook 1196

References 1197

64.1 Historical Development

Towards the end of the 1950s and the early 1960s, the first learning programmes were created on second-generation mainframe computers. At that time, this teaching approach was referred to as programmed teaching (PT).

Psychologists and pedagogues studied the possibilities of the new technology in great detail [64.4]. During that period, the predominant learning paradigm was behaviourism, which had a major impact on the learning programmes created. The supporters of behaviourism assumed that the human brain only had to be stimulated appropriately to achieve the desired response (the correct answer). In behaviourism, the key element is appropriate feedback for the learner in order to reinforce the correct reaction (the answer) to a stimulus.

Ultimately, the concept of behaviourism is based on rewarding the learner if they correctly answer questions and thus close in on the learning target, or punishing them if they incorrectly answer questions and thus move away from the learning target. To this end, the knowledge to be acquired is broken down into the smallest learning units, which is immediately followed by a question that must be answered correctly. If the learner is unsuccessful, the learning unit is repeated or additional help is offered. The expected success of PT for the most part did not materialize. Disenchantment soon followed the initial euphoria about computer-supported training. There were several reasons for this ([64.5, p. 39], [64.6, p. 17]): no graphical interfaces or attractive user interfaces that could be intuitively operated by people who were not specialists in computing; no authoring systems which would enable the creation of learning and teaching systems without advanced programming skills; and the hardware costs of that generation of mainframe computers were very high.

Towards the latter half of the 1970s, the first desktop personal computers became available. These were easier to handle than mainframe computers, and thus attractive to novice computer users. The available hardware and the cost-to-performance ratio continued to improve. Soon high-definition monochrome and colour monitors became available, and mice appeared as input devices. Tutorials and computer simulations were created for such personal computers (Sect. 64.3.1), but generally without too much success.

Later, however, computer-supported training staged a comeback in the form of hypermedia systems. Hypermedia refers to the merger of hypertext and multimedia. Multimedia [64.7] refers to the combination of various digital media such as text, audio clips, video clips, animations, graphics, and their interactive use on a computer. In hypertext, the individual parts of text are not arranged in a linear fashion, in contrast to book texts for instance. The individual text sections are connected using links. Due to the nonlinear arrangement of text, readers themselves are able to choose if they require extra information on a certain topic and, if so, they can choose a link to view additional information. This is an important advantage over linear text where the author controls the order in which the individual parts of a text are to be read. In conventional texts, the author also dictates the subject depth (level of detail) of each block of text.

The availability of authoring systems for the Macintosh and Windows operating systems allowed non-specialists to create hypertext with relative ease. The individual text sections created could also contain several different types of media at the same time, allowing the production of multimedia content. Without extremely expensive hardware or in-depth programming knowledge, it was then possible to create very appealing learning programmes in a relatively short space of time. Optical disks and CD-ROMs provided ways to store large amounts of data (high-resolution pictures and digital video).

Aside from the significant hardware advances that sped up the production and use of learning and teaching systems enormously, the field of teaching psychology also made progress. As behaviourism had shown itself to be a rather unsuitable paradigm on which to base teaching and learning systems, it was often replaced with a constructivist approach [64.8]. In this approach, learners are faced with complex, authentic scenarios (for example a virtual lab or patient) which require them to learn how to identify problems, connections and solutions. The aim is to integrate the newly learned material into the students' existing knowledge in such a way that they are actually able to apply it. The constructivist Cognitive Apprenticeship approach is based on assumptions of situated learning [64.9].

Modelling complex behaviour through an expert who is, figuratively speaking, looking over the learner's shoulder in an authentic situation is a central concept. The following methods are used in this approach: (1) modelling, (2) coaching, (3) scaffolding, (4) fading, (5) articulation, (6) reflection, and (7) exploration [64.10]. These approaches are strongly rooted in case-based computer-supported learning environments (virtual patients). The debate over how many instructional and how many constructivist offers should be made to the learner continues, and amongst other things depends on the prior knowledge of the learner and the educational context. A further aspect of constructivism is synchronized and asynchronized cooperative learning, meaning that problem solving should not be done in isolation but in cooperation with others.

64.2 Moves Towards the Reform of Medical Studies

The development of learning and teaching systems in medicine was, aside from developments in technology and instructional psychology, marked by moves to reform medical studies, in particular from the 1990s onwards. These reforms included (amongst others) interdisciplinary teaching that no longer distinguishes between the pre-clinical and the subsequent clinical phases of training. This was an attempt to counter the lamented deficits of non-practice-oriented inert knowledge in traditional training with problem-based learning (PBL) and case-based training with virtual patients, and to emphasize self-determined, independent learning. PBL is thought to enable students to solve clinical problems independently and to develop their capacities for independent, lifelong learning and knowledge management. In a PBL environment, a tutor presents a small group of students with a clinical case, and usually provides only some patient information and the key points of the anamnesis. The case is then typically dealt with using the following method: (1) clarification of basic comprehension issues; (2) problem definition – the group decides which questions it wants to tackle based on the case; (3) gathering ideas and approaches to a solution; (4) systematic ordering of ideas and approaches into a solution; (5) formulating the learning targets; (6) researching at home; and (7) synthesis and discussion of the learning content gathered. PBL was developed at the Canadian McMaster University towards the end of the 1960s, and has spread successfully since. In 1984, Harvard Medical School also adopted an PBL approach that uses the so-called New Pathway [64.11], which aims to transmit not only knowledge but also skills and behaviours which emphasize the relevance of life-long learning (or continuous medical education) and of skills in knowledge management. Some instances of moves toward reform in Germany include the introduction of the first pre-clinical PBL curriculum at the Private University of Witten/Herdecke in 1992 and the establishment of PBL courses in clinical studies at Munich University in 1997 in cooperation with Harvard Medical School [64.12]. 1999 saw the introduction of the subject-integrational organ- and topic-oriented Reformed Medical Studies course at the Charité in Berlin, and in 2001 the Curriculum Medicinale (HeiCuMed) in Heidelberg, which also had strong PBL components based on Harvard's New Pathway model.

Benefiting from the new Medical Licensure Act (2002) and earlier examples of reform, all departments now conduct block courses lasting several weeks. Many faculties teach practical skills at specially designed facilities (so-called skills labs). The training of communicative skills in many places is carried out using standardized patient actors who pretend to be patients. In addition, virtual patients are used in a variety of application scenarios to support self-determined, practice-related learning in various contexts.

The new Medical Licensure Act of 1 October 2003 calls for new learning and teaching methods that aim for independent learning and focus more on the acquisition of skills and abilities than was done previously. Changes to the exam system are also worth mentioning – departments must assess and mark 22 core subjects and 12 interdisciplinary subjects using alternatives to conventional multiple choice exams. Such alternatives are, for example, objective structured clinical examinations (OSCE), modified essay questions (MEQ), or key feature problems [64.13, 14].

64.3 Developing Learning and Teaching Systems

The development of CBT/WBT systems requires the collaboration of an interdisciplinary team of medical experts, programmers, curriculum developers and teaching psychologists. Prior to development, the learning targets and the planned usage scenarios must be defined [64.15, 16]. Only after this has been done can appropriate forms of interaction be defined, as well as the application architecture (Fig. 64.1).

64.3.1 Interaction Forms

Various types of learning and teaching systems can be distinguished, depending on the degree of interactivity and the topics to be taught. Actual systems themselves can potentially belong to more than one type. Presentation and browsing systems are particularly suitable for transmitting systematic knowledge.

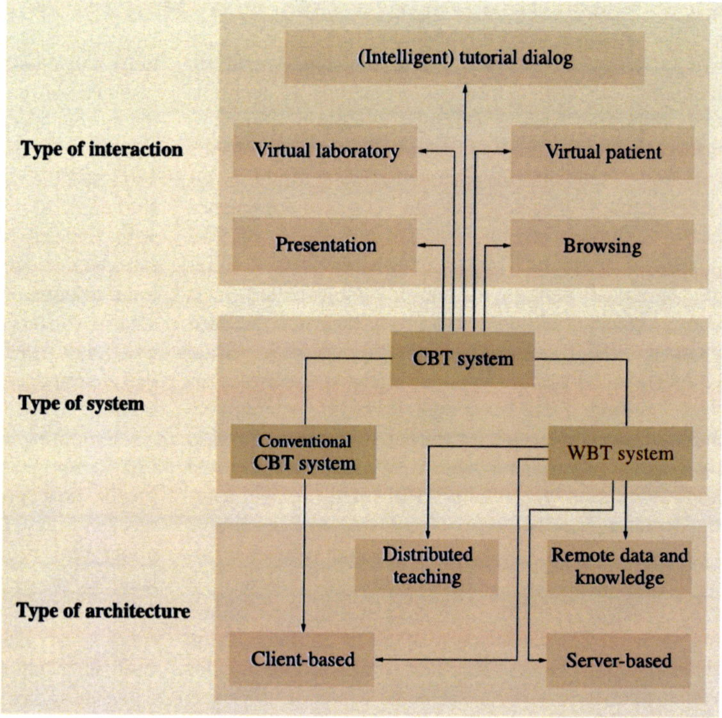

Fig. 64.1 Forms and types of interaction in CBT/WBT systems

Presentation systems are relatively quick and easy to create, as they present certain facts in a linear order. In a manner of speaking, they represent a digital form of the audio slideshow. The interaction between the learner and the computer is restricted to starting the presentation and potentially pressing the pause button or rewinding the presentation. Due to the lack of interactivity, pure presentation systems are rarely used anymore.

On the other hand, browsing systems are characterized by the inclusion of an index or a list of keywords that allow the user to jump directly to a certain chapter or page of the system. It is possible to return there at any point in the programme and at any time. Browsing systems usually contain multiple links between the individual pages (nodes). Marked text can be clicked upon by the user with a mouse, leading to a graphic, a new page, etc. These hypertext or hypermedia functions provided by the programme enable self-determined learning. In browsing systems, the computer also assumes a passive role. It is left to the learner to decide whether to use the system simply as an encyclopaedia or as a learning programme. If used as a learning programme, the learner is able to decide the order in which and how intensively they are going to work through the chapters. However, this frequently leads to uncertainty over whether all of the relevant material has been found. Many of these programmes therefore contain a question component that allows the user to detect knowledge deficits.

Most learning and teaching systems currently on the market are of the browsing type. Occasionally, virtual patients or labs (see later) also contain a browsing component that provides textbook information which can be accessed if needed.

A virtual laboratory is characterized by a mathematical model of a real process, which provides the basis for a related computer simulation. In this type of learning and teaching system, it is very important that the users are familiarized with the system, so that they understand the model, its abilities, and how to carry out actions.

In tutorial systems, the learners are accompanied through the material by the programme. The system presents the content and occasionally asks questions, which are (amongst other reasons) included to aid the discovery of the solution. If a learner cannot answer a question, they receive appropriate feedback which, for example, may contain hints about the solution or is presented along with certain information that is necessary

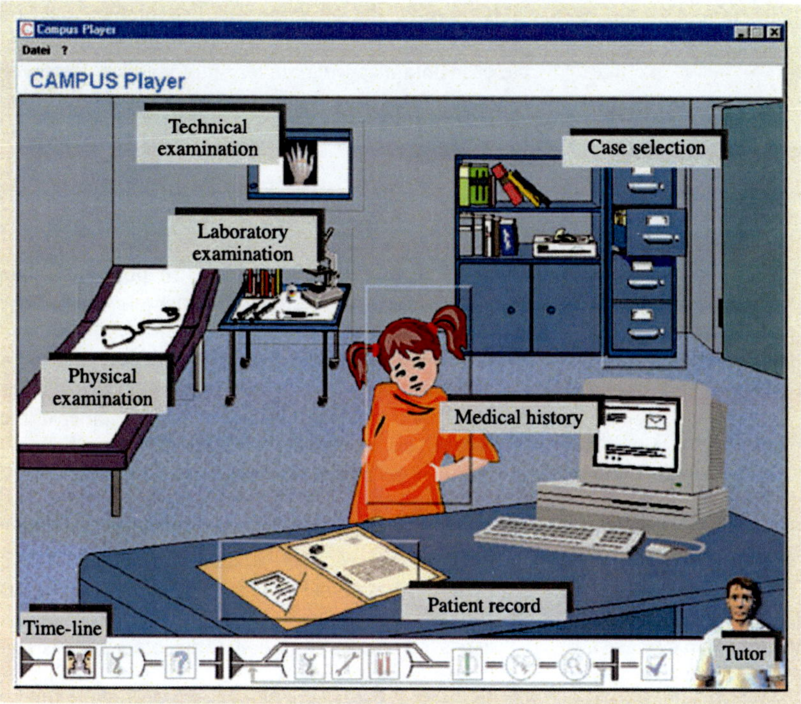

Fig. 64.2 Virtual patient in the CAMPUS system

to answer the question. In tutorial systems, the entire programme cycle is controlled by the computer. Systems that are distinctively tutorial in nature are rare on the market. Browsing systems in which questions are placed at the end of a topic, and the learner must answer these to ascertain if they have understood the material, are more common.

Intelligent tutorial systems, a distinct variety of tutorial systems, employ artificial intelligence (AI) to model user behaviour. To this end, it analyses user behaviour, prior knowledge, preferences, etc., and the AI determines the subsequent programme using this information. The difference from normal tutorial systems is that the programme sequence is not fixed within the programme. The programme usually takes on an advisory function and only intervenes if needed. One of the more difficult aspects of this approach is information modelling. The knowledge that is to be imparted must be structured and linked in such a way that the model is capable of, for example, answering questions from the user. The intelligence of a system lies in its ability to apply the information in the database in a meaningful way.

Case-based training has gained in importance over the last few years, and is increasingly found in the literature under the term "virtual patient" [64.17]. In this type of interaction, a virtual patient with a certain clinical picture is presented. Following anamnesis, physical examination, examination using technology, lab tests, diagnosis and treatment steps, students conduct the treatment of the virtual patient. There are a multitude of software systems for the development and use of virtual patients [64.18–20]. In order to better judge differences and common features, *Huwendiek* et al. [64.21] developed a typology that distinguished four main categories: general, learning targets and competences, instructional design and technology. There are particularly large variations in the instructional design of virtual patients. For example, two basic types of navigation routes through a case can be distinguished.

In the linear route type, a case consists of a predefined series of steps. At each step, the user must take decisions, answer questions or interpret information that has been presented. There are two possible ways of realising this: card-based and simulative variants. In the card-based variant, a case is represented by a stack of hypermedia cards that the user has to work through in a linear path. Here, the presentation of the virtual patient is rather abstract and not as close to reality as in the other variant. In the simulative variant, the approach to the case follows the real-life treatment pattern

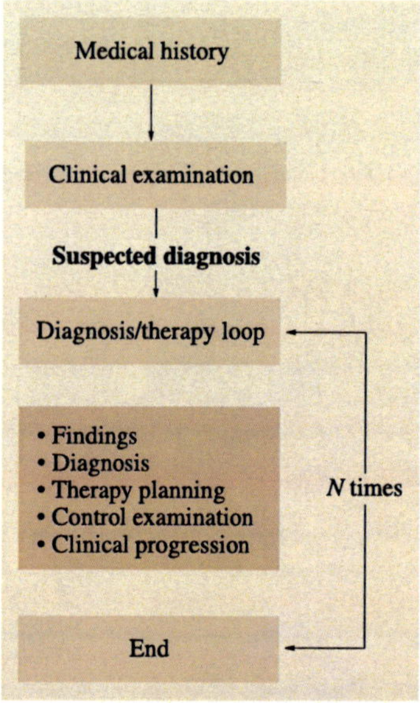

Fig. 64.3 Diagnosis–treatment loop in CAMPUS

more closely, and only presents the student with information they have asked for (knowledge on demand). In such a system, for example CAMPUS [64.22], learners take on the role of a doctor who has to examine a patient, produce a diagnosis, and then direct treatment (Fig. 64.2). Following the anamnesis and the initial physical examination, a diagnosis/treatment loop is run (or sometimes more than one), until the case can be concluded (Fig. 64.3). Treatment of the virtual patient is therefore much closer to reality. Because learners take a very active role in such systems, they are normally only suitable for students at an advanced level of study – it is very difficult to comprehend correlations and initiate appropriate actions without basic knowledge. Card-based systems such as CASUS [64.23], for example, offer the opportunity to use virtual patients at very early stages of study or for the rapid revision of material, because the case author has a great deal of freedom when designing cards using the card metaphor, and is therefore easily able to integrate basic knowledge into the case presentation.

By contrast with the linear route type, in the branched route type (as used in OpenLabyrinth [64.24] for example), several paths through a study case are possible. It is also possible to build dead-end routes. This enables case authors to realistically depict the consequences of wrong decisions, for instance. However, the effort required to create a virtual patient increases significantly in comparison with the linear route type.

Generally, the effort required to prepare medical case studies is great, especially if it is necessary to create or maintain appropriate terminology lists. However, this can be significantly reduced by using powerful authoring systems (Sect. 64.3.3).

64.3.2 Architectures of Learning and Teaching Systems

Two fundamentally different types of programme are used in learning and teaching systems:

1. *Conventional CBT systems*
   These are delivered on data storage devices such as DVDs and must be installed on the computer of the user. There is no requirement for an internet connection for it to be functional.
2. *WBT systems*
   These are based, in contrast to conventional CBT systems, on basic web technology such as HTML, JavaScript, etc., and usually have a client/server architecture. While part of the application runs on the user's computer (the client), the other part is run from a central server.
   The differences between these types result in different architectures (Fig. 64.1 [64.2]). The advantages and disadvantages of each architecture must be considered during the concept phase.

The creation of a WBT system is often a more laborious task than the creation of a conventional CBT system, although it does offer many advantages. For example, since WBT systems do not have to be installed, they are (generally speaking) platform independent and thus do not require a Windows PC to run. Such a system can therefore be used whatever the operating system, time or location, encouraging flexible and self-determined learning. Due to these advantages, new developments in this field are currently dominated by WBT systems.

64.3.3 Authoring Systems

Creating e-Learning content is a task that is made considerably easier by using powerful authoring systems, and such systems also enable members of the medical profession who do not have in-depth computing skills to perform this task. One of the most widely available commercial authoring systems is ToolBook [64.25]. This system uses a book metaphor to simplify the creation and use of learning and teaching

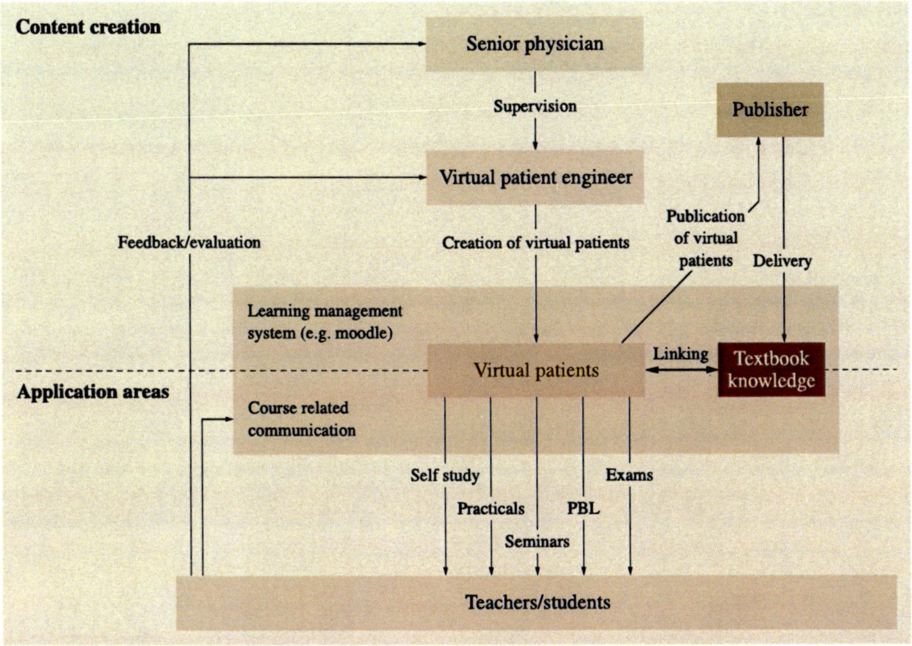

Fig. 64.4 Roles in CBT/WBT systems

systems (usually presentation and browsing systems). Other widely used authoring systems, such as Macromedia Director [64.26], use the metaphor of film – the author takes on the role of director who can control the appearance/disappearance of acting objects on/from the stage. Such authoring systems are particularly useful for the creation of animations. Authoring systems are also used for the creation of virtual patients, who are frequently produced in-house in order to cater for specific requirements. It has been shown, however, that despite the ease of use of authoring systems, it is often difficult to find authors of virtual patients. One solution is, for example, for the senior physician to assume the role of case author, who passes the work required to prepare the case to a virtual patient engineer (Fig. 64.4) before finally taking control of the prepared virtual patient. Virtual patient engineers are typically students of medicine or doctors in training.

64.4 Learning Environments

64.4.1 Functionality of Learning Environments

In the last few years, faculty and university learning environments or platforms (learning management systems (LMS)) have established themselves.

In contrast to plain collections of lecture scripts or hypertext collections on webservers, an LMS usually provides for [64.27] (Fig. 64.5):

- User administration (encrypted login)
- Course management, which allows students to book and sign up for courses
- A learning platform that includes learning objects (texts, slideshows, CBT or WBT units) and learning tools (notepad, calendar, etc.)
- An authoring tool that enables tutors to develop web documents without much knowledge of HTML and the internet
- Components for cooperative online work (computer-supported cooperative work, CSCW)
- Shared databases and repositories
- Permissions and roles allocation with differentiated permissions
- Functions that allow evaluation and exams to be conducted online.

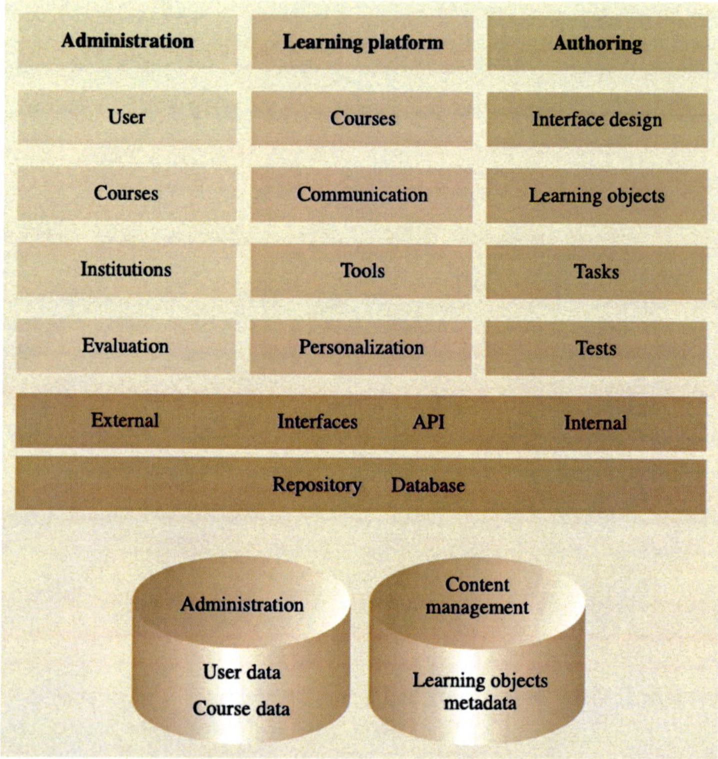

Fig. 64.5 Functionality of LMSs (after [64.27])

There are a large number of LMS systems on the market. Commercial systems aside, there are also some open source systems that are particularly attractive for cost reasons. The selection of an LMS has far-reaching consequences for a university and is not a trivial task. Care must be taken to ensure that the selected LMS possesses the necessary functionality and performance, while the licensing, running costs and cost of training tutors and students in handling the LMS should be as low as possible.

When integrating a CBT/WBT system into an LMS, it is important to make sure that users logged into the LMS system will be able to use the CBT/WBT system without the need to renew their authentication (so-called single sign-on). In addition, the LMS should be able to record the progress of each user, thus allowing them to continue their study units from appropriate points later.

64.4.2 Interoperability and Standards

Integrating a CBT/WBT system into an LMS is a significantly easier task when the LMS has standardized interfaces that support interoperability between different systems and reusability of learning objects. The SCORM (sharable courseware object reference model) standard [64.28] of the Advanced Distributed Learning Initiative (ADL) from the Department of Defense (DoD) is particularly relevant in this context. SCORM is the reference model for creating web-based exchangeable learning content with the aim of producing reusable, identifiable, interoperable and persistently stored learning objects and courses. These aims are achieved by specifying metadata that identify existing learning objects which can then be saved as SCORM packets and re-identified via the metadata. However, the downsides of standards like SCORM should also be noted; for example, there are potential bureaucracy issues, as the documentation for SCORM runs to more than 800 pages and is thus difficult to review.

In the area of virtual patients, we should mention the MedBiquitous virtual patient (MVP) specification [64.29]. The specification [64.30] currently being developed simplifies the exchange and use of virtual patients across departments significantly – an effect that has already been noticed during the early stages of working with the new standard.

64.5 Application Scenarios for Learning and Teaching Systems

Learning and teaching systems can be used in numerous ways (Fig. 64.4). At most universities, a variety of self-study programs are available in computer labs or media centres. In addition, web-based systems can often be accessed from a student's home computer via the internet. The move towards blended learning (BL) concepts – involving the compulsory integration of learning and teaching systems into lectures – has intensified recently. BL involves not only linking books, handouts, seminars, workshops, 1:1 coaching, and so on using web-based content, but also the use of virtual classrooms (for example to prepare seminars or to do follow-up work on a seminar). One of the options it provides is the use of programmes to systematically transmit knowledge and discuss and deepen learned content in face-to-face sessions. The approach in which the theoretical basics are taught in lectures and then expanded upon using virtual patients is also commonly seen; this gives students the opportunity to apply acquired information without endangering real patients. The combination of learning platforms with systematic learning materials and virtual patients is also used in the preparation of practical clinical teaching; for example, in the Klinische Fertigkeiten Online (clinical skills online) project [64.31].

They can also be used for computer-based exams. This application is of particular interest considering the workload placed on faculties to test and mark the abovementioned 34 subjects. Computer-based exams, for example those using a key feature approach, can be used to conduct valid and reliable exams using a justifiable amount of resources. The medical profession is particularly interested in such approaches that would enable formative exams using virtual patients to be taken alongside summative exams.

However, the technical and legal risks of computer-based exams should also be mentioned; for example, those associated with browser-based and client/server-based exam systems [64.32]. Special consideration must therefore be given to these aspects during development.

Lastly, learning and teaching systems are used in the area of medical CPD, as medium- to long-term residential stays in educational institutions are often neither desirable nor feasible. This is supported in Germany by the introduction of a duty to perform CPD as part of the Law on the Modernisation of Healthcare (GMG). However, there are a multitude of professional development opportunities on offer, some of which are available free of charge due to support from specialist medical publishers, medical associations or the pharmaceutical industry. This may be an obstacle to the spread of learning and teaching systems, should these be offered for a fee.

64.6 Status of and Outlook for e-Learning in Medicine

64.6.1 Information Systems for CBT/WBT in Medicine

There are a wide variety of CBT/WBT systems in medicine, as research on specialist information systems has shown. KELDAmed [64.33] at the University of Mannheim lists over 1200 learning objects, for example. The Medical Learning Resource Server of the University of Essen [64.34] currently enables users to search through over 1900 free, web-based learning objects. One problem with such information systems is ensuring that the underlying database is up-to-date. Achieving an objective evaluation of the systems on offer can also be difficult. To simplify the evaluation of systems, the working group Computer Supported Learning and Teaching Systems in Medicine [64.35] from the Society for Medical Informatics, Biometry and Epidemiology (GMDS) [64.36] has published quality criteria for digital publications in medicine [64.37]. These have been translated into various languages. The working group, which has existed for 20 years, has set itself the aim of coordinating development projects in the medical profession and improving the quality of CBT/WBT systems.

64.6.2 Use of CBT/WBT Systems in Medicine and the Problem of Curricular Integration

CBT/WBT systems are currently used by only approximately 5% of medical students, according to the experiences of various universities. In this context, the results of a study conducted at the University of Bern in 2000 are interesting. In this study, over 75% of the peo-

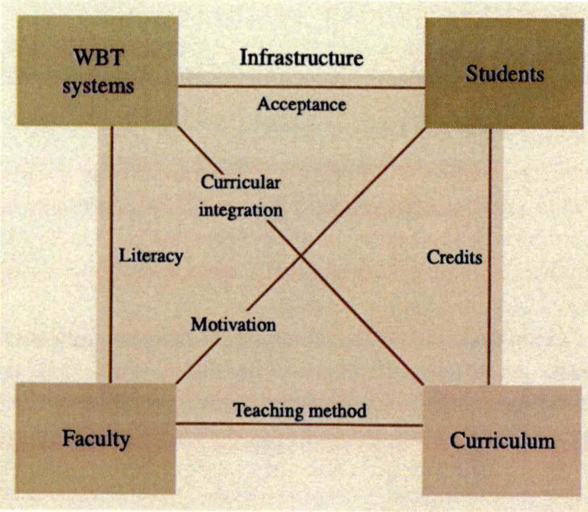

Fig. 64.6 Dependencies in the use of e-Learning

ple questioned preferred studying without a computer, and 90% of all students studied predominantly using printed media. A study from the Ludwig-Maximilians University Munich showed that almost all students completed the CASUS case studies on offer as long as they gained credits by working through the cases, so that they were relevant to the exam results of the students. When no credits were available, only about 10% of students used the case studies. This means that curricular integration of CBT/WBT systems is very important. Increasing acceptance amongst both students and tutors will continue to be a pivotal task in the future [64.38]. In particular, successful curricular integration depends on a suitable combination of CBT/WBT systems with face-to-face teaching and exams [64.39, 40]. In these circumstances, the systems are accepted and intensively used by the majority of students.

It is often argued that the reason for the lack of acceptance of these systems is that insufficient attention is paid to the risks associated with the development and application of CBT/WBT systems. The technical risks of developing a system (are we building the product right) and the application risk in particular (are we building the right product – one that is needed by users?) must be considered.

Even today, a negative attitude to computers can occasionally be encountered amongst both students and tutors. A lack of skilful handling of such systems by tutors and a lack of identification with the system hinder student motivation and student acceptance of the system. It is important to realise that tutors, especially those with active clinical duties, have very high workloads. Therefore, the question of how to get and keep doctors involved in innovative teaching is a central issue. In this context, established instances of tutor training should be mentioned, for example those that have been conducted at the University of Heidelberg for many years. Career progression towards a Master of Medical Education is also increasing in importance. Upon attaining such qualifications, tutors will possess the relevant system skills needed for the integration of CBT/WBT systems into the curriculum (Fig. 64.6).

Curricular integration only is achievable when the system contains learning incentives for the students; for example, by enabling preparation for faculty internal assessment using CBT/WBT systems [64.13].

One possible way to improve sustainability is to maintain the positive experiences gained in cross-regional cooperation to date. This includes establishing competence centres for medical education, such as those that provide e-Learning and examinations in Baden-Württemberg and Bavaria, which contribute to tutor qualification and sustainable cross-faculty networking across inner-German state borders.

Only when the problems associated with curricular integration have been solved can the vision of creating more flexible learning, reducing costs, and improving quality through the use of CBT/WBT in medical education be achieved. An obligatory national catalogue of learning targets is required to support the long-term, cross-regional establishment of CBT/WBT systems in medical training.

64.6.3 The Need for Research, and Outlook

The current state of research suggests that e-Learning and traditional methods of instruction in medicine have similar levels of effectiveness [64.1]. It has been possible to clearly show that CBT/WBT systems offer advantages in the provision of standardized teaching materials and in communication between students and tutors [64.41]. However, such studies find it difficult to perform methodological comparisons between e-Learning and (for example) face-to-face teaching, because it is difficult to control distorting factors in real-life learning situations. In addition, offers of instruction that differ in numerous ways and could be offered in a more harmonized fashion in the sense of BL are compared. In many studies, offers of e-Learning have been introduced in the last few years and compared with control groups who did not receive offers of instruction. In such cases, it is not surprising that e-Learning intervention proved to be better than no intervention. This type of study can be suspected of triviality and usually dispensed with, as one does not take an either/or approach in real teaching [64.42]. The use of the internet has been part and parcel of everyday life – including teaching – for some time now, and CBT/WBT systems are now commonly used at almost all academic institutions. In the future, studies that compare the different e-Learning integration models would be desirable, as the specific strengths and weaknesses of the e-Learning approaches offered in various educational facilities are still not clear [64.41,43]. A stronger theoretical grounding for future studies would also be desirable, as it would require more intensive cooperation between the medical profession, educationalists and psychologists. There are a plethora of research questions surrounding the foundations of case-based learning, clinical decision-making, cooperative learning in groups and inter-professional learning. In addition, the development of innovative, computer-supported formative and summative exam formats should be at the forefront of additional research efforts. Support for workplace-related evaluation methods could become increasingly important. Many technical challenges in the development of CBT/WBT systems have been surmounted in the last few years. However, there are still important topics (particularly interoperability and standards) that need further work to facilitate the exchange and technical integration of content across universities, which was created with considerable effort.

References

64.1 D.A. Cook, A.J. Levinson, S. Garside, D.M. Dupras, P.J. Erwin, V.M. Montori: Internet-based learning in the health professions, J. Am. Med. Assoc. **300**(10), 1181–1196 (2008)

64.2 M. Haag, L. Maylein, F.J. Leven, B. Tönshoff, R. Haux: Web-based training: A new paradigm in computer-assisted instruction in medicine, Int. J. Med. Inform. **53**, 79–90 (1999)

64.3 E-Learning, http://en.wikipedia.org/wiki/Elearning (last accessed July 16, 2011)

64.4 K. Schaller, G. Wodraschke (eds.): *Information und Kommunikation* (Leibnitz, Hamburg 1969)

64.5 J. Moehr: Computerunterstützter Unterricht in Kanada und den USA. In: *Computer in der Ärzteausbildung*, ed. by M.P. Baur, J. Michaelis (Oldenbourg, Munich 1990) pp. 31–50

64.6 F. Bodendorf: *Computer in der fachlichen und universitären Ausbildung* (Oldenbourg, Munich 1990)

64.7 R. Steinmetz: *Multimedia Technologie – Grundlagen, Komponenten und Systeme* (Springer, Berlin Heidelberg 2005)

64.8 R.J. Spiro, R.J. Coulson, P.J. Feltovich, D.J. Anderson: *Cognitive Flexibility Theory: Advanced Knowledge Acquisition in Ill-Structured Domains* (Erlbaum, Hillsdale 1989)

64.9 J. Brown, A. Collins, P. Duguid: Situated cognition and the culture of learning, Educ. Res. **18**(1), 32–42 (1989)

64.10 A. Collins, J. Brown, Newman: Cognitive Apprenticeship: Teaching the craft of reading, writing, and mathematics. In: *Knowing, Learning, and Instruction: Essays in Honor of Robert Glaser*, ed. by L. Resnick (Lawrence Erlbaum, Hillsdale 1989)

64.11 Harvard Medical School: http://hms.harvard.edu/admissions/default.asp?page=pathway (last accessed 15 April 2011)

64.12 R. Putz, F. Christ, H. Mandl, S. Bruckmoser, M.R. Fischer, K. Peter, G. Moore: Das Münchner Modell des Medizinstudiums (München-Harvard Educational Alliance), Med. Ausbild. **16**, 30–37 (1999)

64.13 F. Ruderich: Computerunterstützte Prüfungen in der medizinischen Ausbildung nach der neuen Approbationsordnung, Diplomarbeit im Studiengang Medizinische Informatik (Universität Heidelberg/Fachhochschule Heilbronn 2003)

64.14 M.R. Fischer, V. Kopp, M. Holzer, F. Ruderich, J. Jünger: A modified electronic key feature examination for undergraduate medical students: validation threats and opportunities, Med. Teach. **27**, 450–455 (2005)

64.15 L.H. Fall, N.B. Berman, S. Smith, C.B. White, J.C. Woodhead, A.L. Olson: Multi-institutional development and utilization of a computer-assisted learning program for the pediatrics clerkship: The CLIPP Project, Acad. Med. **80**(9), 847–855 (2005)

64.16 E.D. Kern, P.A. Thomas, M.T. Hughes (eds.): *Curriculum Development for Medical Education: A Six-Step Approach* (Johns Hopkins Univ. Press, Baltimore 2009)

64.17 D.A. Cook, M.M. Triola: Virtual patients: A critical literature review and proposed next steps, Med. Educ. **43**(4), 303–311 (2009)

64.18 M.R. Fischer: CASUS – An authoring and learing tool supporting diagnostic reasoning, Z. Hochschuldidakt. **1**, 87–98 (2000)

64.19 S. Garde, M. Bauch, M. Haag, J. Heid, S. Huwendiek, F. Ruderich, R. Singer, J.F. Leven: CAMPUS – Computer-based training in medicine as part of a problem-oriented educational strategy, Stud. Learn. Eval. Innov. Develop. **2**(1), 10–19 (2005)

64.20 M. Begg, R. Ellaway, D. Dewhurst, H. McLeod: Transforming professional healthcare narratives

into structured game-informed-learning activities, Innovate **3**(6), 1–7 (2007), Available at http://www.innovateonline.info/index.php?view article&id=419
64.21 S. Huwendiek, B. de Leng, N. Zary, M.R. Fischer, J.G. Ruiz, R. Ellaway: Towards a typology of virtual patients, Med. Teach. **31**, 743–748 (2009)
64.22 M. Haag, S. Huwendiek: http://www.virtual-patients.com (last accessed 15 April 2011)
64.23 http://www.casus.eu (last accessed 15 April 2011)
64.24 http://labyrinth.mvm.ed.ac.uk/ (last accessed 15 April 2011)
64.25 http://www.sumtotalsystems.com/products/toolbook-elearning-content.html (last accessed 15 April 2011)
64.26 http://www.adobe.com/products/director/ (last accessed 15 April 2011)
64.27 R. Schulmeister: *Lernplattformen für das virtuelle Lernen* (Oldenbourg, Munich 2003)
64.28 http://www.adlnet.gov (last accessed 15 April 2011)
64.29 MedBiquitous: http://www.medbiq.org/std_specs/standards/index.html#MVP (last accessed 15 April 2011)
64.30 V. Smothers, B. Azan, R. Ellaway: MedBiquitous Virtual Patient Specifications and Description Document Version 0.61 (MedBiquitous Consortium 2010), available at http://www.medbiq.org/working_groups/virtual_patient/VirtualPatientDataSpecification.pdf (last accessed 15 April 2011)
64.31 G. Karsten, V. Kopp, K. Brüchner, M.R. Fischer: Blended Learning zur integrierten und standardisierten Vermittlung klinischer Untersuchungstechniken: Das KliFO-Projekt, GMS Z. Med. Ausbild. **26**(1), 1–7 (2009), Doc 10
64.32 J. Heid, M. Bauch, M. Haag, F.J. Leven, I. Martsfeld, F. Ruderich, R. Singer: Computerunterstützte Prüfungen in der medizinischen Ausbildung. In: *Rechnergestützte Lehr- und Lernsysteme in der Medizin*, ed. by J. Pöppl, M. Bernauer, M. Fischer, H. Handels, R. Klar, F.J. Leven, F. Puppe, K. Spitzer (Shaker, Aachen 2004) pp. 213–218
64.33 http://www.ma.uni-heidelberg.de/apps/bibl/KELDAmed/ (last accessed 15 April 2011)
64.34 http://mmedia.medizin.uni-essen.de/portal/ (last accessed 15 April 2011)
64.35 http://www.mi.hs-heilbronn.de/gmds-cbt (last accessed 15 April 2011)
64.36 http://www.gmds.de (last accessed 15 April 2011)
64.37 S. Schulz, R. Klar, T. Auhuber, U. Schrader, A. Koop, R. Kreutz, R. Oppermann, H. Simm: Qualitätskriterien für Elektronische Publikationen in der Medizin, Inform. Biometr. Epidemiol. Med. Biol. **31**(4), 153–166 (1999), available at www.imbi.uni-freiburg.de/medinf/gmdsqc/
64.38 J. Riedel: Integration studentenzentrierter fallbasierter Lehr- und Lernsysteme in reformierten Medizinstudiengängen. Ph.D. Thesis (Universität Heidelberg, Heidelberg 2003)
64.39 I. Hege, V. Kopp, M. Adler, K. Radon, G. Mäsch, H. Lyon, M.R. Fischer: Experiences with different integration strategies of case-based e-Learning, Med. Teach. **29**(8), 791–797 (2007)
64.40 M.R. Fischer, I. Hege, A. Hörnlein, F. Puppe, B. Tönshoff, S. Huwendiek: Virtuelle Patienten in der medizinischen Ausbildung: Vergleich verschiedener Strategien zur curricularen Integration, Z. Evid. Fortb. Qualit. Gesundheitsw. **102**(10), 648–653 (2008)
64.41 M. Valcke, B. De Wever: Information and communication technologies in higher education: Evidence-based practices in medical education, Med. Teach. **28**(1), 40–48 (2006)
64.42 D. Clark: Psychological myths in e-Learning, Med. Teach. **24**, 598–604 (2002)
64.43 D.A. Cook: The failure of e-Learning research to inform educational practice, and what we can do about it, Med. Teach. **31**(2), 158–162 (2009)

65. PACS and RIS

Christian Zapf, Andreas Bermann, Dirk Sunderbrink

The digitization of radiographic imaging has been implemented for all diagnostic procedures in radiology, and digital reading has been accepted by the radiology community. Simultaneous establishment of a uniform and robust communication standard for the exchange and storage of image data (DICOM – digital imaging and communication in medicine) has long fulfilled the prerequisites for the introduction of picture archiving and communication systems (PACS) and radiology information systems (RIS). In industrialized countries, the market penetration of both systems is between 60% and 95%, depending upon market segment and type of service provider. PACS, conceived more than 20 years ago, enables image communication between individual components such as archive systems, diagnostic workstations, postprocessing workstations, and image distribution workplaces. Typically, a RIS comprises a series of software modules supporting radiology workflow such as creation of orders, scheduling, reading, reporting, medical coding, recording of services, and interfaces to a billing system.

65.1 **Radiological Workflow** 1200
65.1.1 Orders 1201
65.1.2 Admission/Scheduling 1201
65.1.3 Examination 1201
65.1.4 Reading Workflow 1202
65.1.5 Displaying and Evaluating Images 1202
65.1.6 Image Postprocessing 1202
65.1.7 Computer Aided Diagnosis (CAD) ... 1204
65.1.8 Medical Monitors 1204
65.1.9 Report Creation 1204
65.1.10 Clinical Demonstration 1205
65.1.11 Distribution of Reports/Images 1205

65.2 **Integrating PACS/RIS into the Hospital Environment** 1205
65.2.1 Integrating Information Systems 1205
65.2.2 Integrating Image Management Systems 1206
65.2.3 Teleradiology 1206
65.2.4 Mobile Devices 1206

65.3 **State-of-the-Art IT Infrastructure** 1206
65.3.1 Information Lifecycle Management 1207
65.3.2 Cloud Computing Solutions in the Hospital 1207
65.3.3 Regional IT Virtualization 1207

65.4 **Summary** 1208

References 1208

A typical RIS/PACS configuration is illustrated by Fig. 65.1. RIS and PACS are the information technology (IT) cornerstones of a radiology department. Nowadays, both systems are jointly deployed and closely integrated. Connection to a higher-level hospital information system (HIS) is typically facilitated through communication servers which link various IT systems together by means of an additional medical data standard, HL7 (Health Level 7 [65.1]). On the one hand, this enables interfacing of RIS and PACS to electronic patient records (EPR); on the other hand, it supports the medical workflow beyond the boundaries of radiology department. In the future, the rapid growth of data volumes produced by imaging systems will require integration of advanced applications for 3-D/4-D postprocessing (frequently referred to as advanced visualization) and computer-aided diagnosis (CAD) into the radiological workplace in order to provide efficient reading and reporting. The establishment of communication standards on data level is still insuffi-

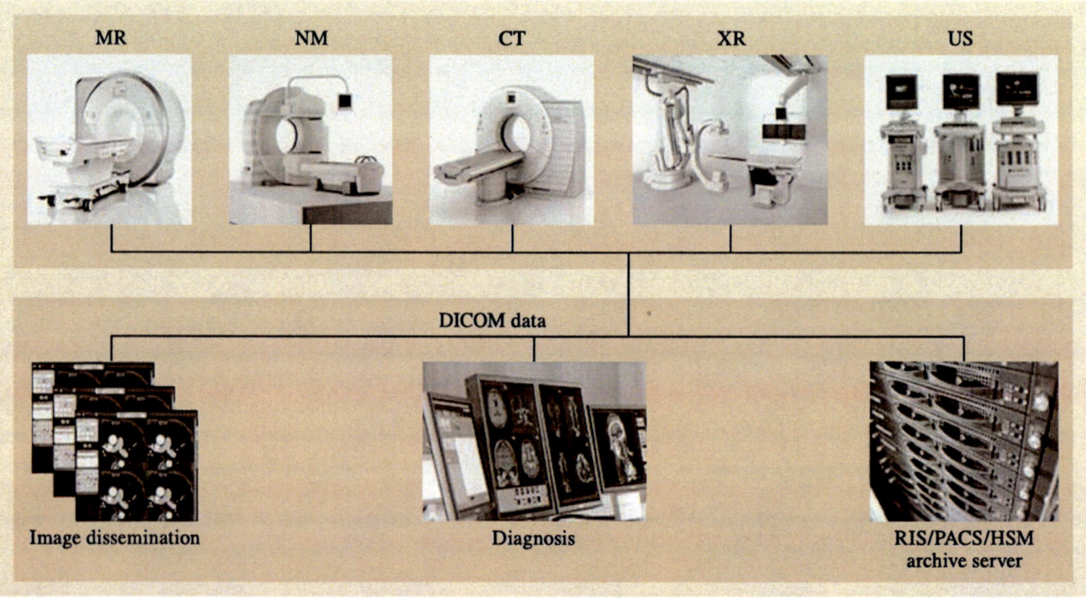

Fig. 65.1 Typical RIS/PACS configuration

cient for the optimization of a clinical workflow. As a result, the Radiological Society of North America (RSNA [65.2]) inaugurated the IHE initiative (Integrating the Healthcare Enterprise [65.3]). IHE has the objective of defining integration profiles according to standardized workflow scenarios, in an effort to validate different multi-vendor software and hardware components.

In the meantime, PACS has gone beyond the boundaries of radiology and has developed into an application for the entire healthcare enterprise. Other clinical departments such as cardiology, neurology, radiotherapy, and most recently, pathology, want to take advantage of digital image management to store their image data in a PACS archive while enjoying similar support for their respective workflows. The individual disciplines vary in the degree of IT system utilization for image data management. However, there is a trend toward standardization of the image data archive beyond departmental boundaries in order to minimize the expenditures for procurement, installation and maintenance of the systems, as well as to realize economies of scale in data management. A significant potential for optimization of patient treatment lies in the routine application of digitally supported procedures for therapy planning such as in electrophysiology, radiotherapy or computer-aided surgery (CAS), using digital image data from PACS as a planning base.

Innovations in imaging and the rapid growth of image data volumes require support of diagnostic radiology procedures by appropriate information technology such as PACS and RIS. A state-of-the-art multi-slice computed tomography scanner produces up to four thousand slices per procedure. With this amount of data, analog reading using film on a light box is impractical and there is the risk of missing diagnostically significant findings. In the following chapters, the reader will be provided with a description of the individual work steps regarding film- and paperless radiology.

65.1 Radiological Workflow

Optimization of the clinical workflow is the objective with the introduction of RIS and PACS. The focus is on improving quality and reducing cost. This can be achieved only by a seamless interaction of the individual components. A typical workflow (Fig. 65.2) will illustrate the functionality of the individual workplaces.

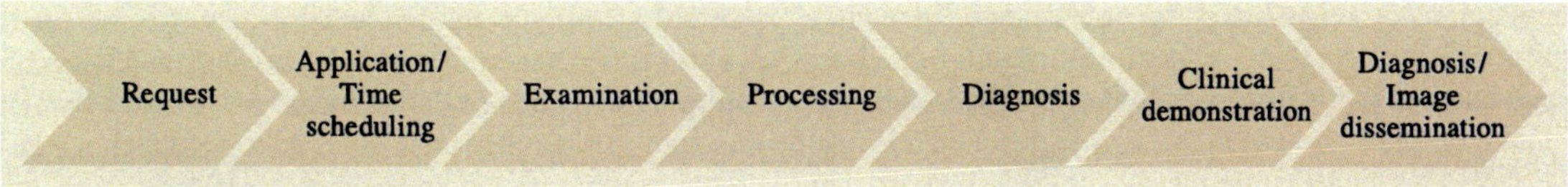

Fig. 65.2 Radiological workflow

65.1.1 Orders

A referring physician decides whether a radiology procedure is required for further treatment and orders this from the radiology department.

There are three conventional order scenarios which are to some extent used in parallel. First, orders are transmitted even today via paper, fax or telephone; second, many RIS systems provide their own electronic order modules which may also be web-based; third, many hospital information systems (HIS) offer corresponding order modules CPOE (computerized physician order entry). The advantage offered by this functionality within an HIS module is the uniform application within an enterprise. In scenario three, for example, a hospital has only one module in which a request is generated and transmitted to the connected departmental systems in HL7 format, whereas in scenario two various request modules must be used (one each per departmental information system).

If orders are not transmitted electronically, the relevant data (e.g. patient information) must be manually entered into the RIS. In each case, it is important that the clinical problems as well as the transmission of additional information such as contrast media allergies or laboratory values are as complete as possible in order to plan subsequent activities properly.

65.1.2 Admission/Scheduling

The quality of scheduling determines process efficiency in a radiology department to a large degree. It includes scheduling schemes, allocation of resources to procedures (e.g. rooms and personnel) as well as short-term adaptation to unplanned events that are common in medical care (e.g. emergencies).

Just as with orders, scheduling processes can be configured and supported in different ways by IT. Most simply, appointments can be scheduled by radiology department employees similar to managing an appointment book. Some RIS allow a person placing an order to personally choose an appointment from a selection of date and time slots. In case of further automation, a RIS can independently suggest the next convenient appointment based on preconfigured rules.

In the next step, resources are allocated to appointments depending upon the type of procedure. In almost all cases, a room with relevant equipment will be required (e.g. a MR scanner in a specific room). Often additional equipment and material (e.g. anesthesia equipment, contrast agents, consumables, etc.) and personnel are required. Therefore, multi-resource scheduling is called for, frequently requiring IT support in the RIS.

Daily planning must respond to unscheduled events such as emergencies, urgent examinations or protracted procedures (e.g. repeated respiratory artifacts in the MR). Since such events occur regularly, most RIS offer simple options for rescheduling, shifting examinations to other rooms or even the inclusion of buffers in scheduling.

Usually the ordering department receives confirmation of the scheduled appointment. Depending upon the level of HIS/RIS integration, a referring physician can track the current status of a patient in a radiology department from the workstation on his desk. A physician can see whether the patient has arrived in radiology, whether the examination has started or whether images or reports are available.

As soon as procedures are allocated to a specific imaging system (modality), the RIS generates a DICOM worklist which is queried by the modality from the RIS. As a result, the querying modality is automatically provided with the most important patient and procedure data. Depending upon the type of archive, RIS and PACS can be configured in a way that a patient's previous procedures are automatically de-archived from the long-term archive, and relevant prior procedures are available for further diagnosis (prefetching).

65.1.3 Examination

A DICOM worklist is queried by a modality from the RIS and transferred to the local worklist of the imaging modality. The patient is selected at the modality and the procedure is started. Electronic transfer guarantees the consistency of patient data. Procedure information such as type, starting time and duration, as well as additional information are registered by the modality and returned to the RIS in the DICOM format MPPS (modality performed procedure step). The MPPS data is also transmitted to the PACS, and additionally, the modality sends the study to the PACS for archiving. Transfer of images to the PACS is primarily performed in the standardized DICOM format, which is designed for manufacturer-independent data exchange between modalities and RIS/PACS. In the current 3.0 version, DICOM defines not only image formats; it also covers transfer protocols and regulates security requirements. The following services, among others, are supported:

- Archiving and transfer of images over networks (DICOM standard, parts 7 and 8)
- Archiving and exchange of images via exchangeable media (DICOM standard part 10)
- Search functions (DICOM query)
- Print functions (DICOM print)
- Workflow functions (DICOM modality worklist, modality performed procedure step)
- Image data compression (DICOM standard part 5, e.g. JPEG, JPEG2000, RLE)

DICOM conformance statements exist for most medical imaging devices. These documents describe the supported DICOM services and provide be basis for multi-vendor image data exchange. At the end of a procedure, an employee of the radiology department additionally enters on the RIS workstation which radiological services were performed and if applicable, information like the contrast agent and catheters used. The data for charge capturing and consumption of high-quality medical devices are especially relevant for medical documentation and billing.

65.1.4 Reading Workflow

Typically, the reading workflow is driven by the RIS. It prepares worklists with completed procedures and those designated for reading. The physician selects the relevant patient together with the procedure and the RIS initiates the display of the current image material and if applicable the prior procedures within the PACS viewing application.

For complex procedures requiring postprocessing, numerous specialized postprocessing applications are currently available to support and expedite the reading workflow, generating additional valuable diagnostic information. Further, a division of labor between technologist and physician is possible, in which a technologist prepares the image postprocessing from his own worklist and then upon completion, sends the case to a physician's worklist.

Once the analysis of the image material is completed, a diagnostic report is created in the RIS. Frequently the reading physicians work in a multiple-monitor environment (with up to four monitors), simultaneously displaying images and report. After completion of the report both images and report are distributed and archived.

65.1.5 Displaying and Evaluating Images

The first PACS workstations were primarily developed for working with conventional x-ray images. Reviewing images in segments on a monitor, similar to a traditional light box is still in practice. In the meantime, size and shape as well as number and arrangement of segments for image display are generally freely configurable; frequently the corresponding layout configurations can be automatically assigned by a PACS on a procedure-specific basis during loading (DICOM hanging protocols).

Operations to adjust image quality such as brightness, contrast, etc. as well as simple image manipulations such as zoom, pan, rotate, edge enhancement, smoothing, paging through slices, cine mode, etc. are now available at every PACS workstation. Since the diagnostic information of the images can be significantly affected by these adjustments (frequently the images can only be interpreted after adjustment) there is a separate DICOM standard for long-term storage of the changes in the archive (DICOM softcopy presentation state).

In order to perform objective and reproducible measurements for quantitative analysis of tissue structures based on image data, a number of basic functions are offered, such as distance, angle, and volume measurements.

65.1.6 Image Postprocessing

Image data volumes are increasing rapidly, especially in the context of CT and MR examinations. Diagnosis of such complex procedures alone using the tools described above is no longer practicable due to time and quality constraints. As a result, a number of specialized postprocessing applications have been developed, introducing a completely new paradigm of radiology reading.

The basis for most postprocessing applications is a three-dimensional reconstruction of the two-dimensional slices. Various mathematical operations interpolate the image information between the slices. Reconstruction results are dependent on the slice thickness of the original data. Thinner slices typically produce better results. The most frequent reproduction modes are MIP (maximum intensity projection), MPR (multiplanar reconstruction) and VRT (volume rendering technique) as well as their derivates (refer to Table 65.1).

MIP is frequently used to display contrast-enriched vascular structures in CT and MR angiography. MPR allows display of clip planes not acquired with the original data (e.g. coronary and sagittal in addition to the axial planes). VRTs show reconstructed objects such as the heart, skull or skeleton, as well as more complex reconstructions such as the colon.

In case of functional examinations such as cardiac ventricle function in MR and CT or functional magnetic resonance imaging (fMRI), the time dimension is critical. A number of postprocessing applications have been developed to add the fourth dimension of time to the evaluation. Image data of different modalities can be merged, resulting in additional diagnostic information. PET-CT examinations are a prime example in which the sensitivity of the PET (positron emission tomography) is combined with the morphological accuracy of CT. This allows, for example, tumors or metastases to be evaluated more precisely. In addition, special measurements, evaluation and automation algorithms have been developed for numerous diagnostic problems.

In recent years, bundling all these options into advanced visualization has inspired a number of modality and PACS vendors as well as new vendors specializing in this area to develop tailored postprocessing applications for dedicated diagnostic problems. The following list offers a selection of commonly used applications:

- CT calcium scoring
- CT coronary artery analysis
- CT and MR angiography
- CT and MR neuro perfusion
- CT cardio perfusion and ventricle function analysis
- MR cardio perfusion and ventricle function analysis, including late enhancement
- PET-CT oncology including automated compare layouts and lymph node comparisons
- CT neuro DSA
- fMRI
- MR tractography and DTI (diffusion tensor imaging).

Significant progress has been made in the development of postprocessing applications, an area currently receiving significant research and development funding. This trend has the potential to fundamentally change radiology reading, moving away from pure analysis of two-dimensional images to specialized applications tailored to specific diagnostic problems, providing anatomical, morphological, quantitative and functional results. Two essential advantages become obvious: first, this will yield a multitude of previously unavailable but relevant diagnostic data. Such data will allow more precise diagnoses and will increase the sensitivity and specificity of imaging procedures. Second, increased automation will reduce the cycle time for more extensive diagnosis such that it will eventually require less time than traditional diagnosis. The latest generation of postprocessing applications is achieving this through an extensively automated, clinically useful combination of postprocessing steps. A CT coronary artery analysis, for instance, is typically prepared in six steps in which various algorithms are applied:

1. Selection of appropriate series for reconstruction analysis (*best cardiac phase*)
2. MPR and VRT reconstruction
3. Removal of bone structures
4. Removal of lung and soft tissues
5. Removal of the blood pool (large vessels adjoining the coronary arteries and blood-filled ventricles)
6. Analysis of the center lines of the coronary arteries.

State-of-the-art postprocessing applications can perform these steps automatically, sometimes completely

Table 65.1 3-D postprocessing methods

Method	Applications
Maximum intensity projection (MIP)	Angiography
Multiplanar reconstruction (MPR)	Display of selected planes from a volume data set
Volume rendering technique (VRT)	Display of surface structures from 3-D data sets

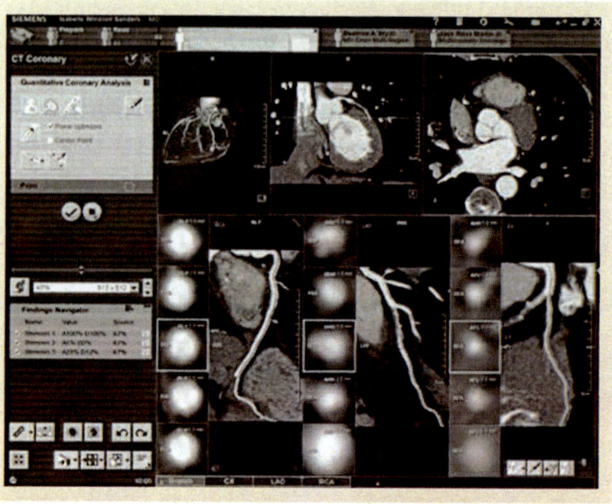

Fig. 65.3 Example of a clinical post-processing application

without manual interaction. After calling up a procedure in his worklist, the reading physician receives the procedure fully analyzed, and can start reading immediately. Also during subsequent diagnosis, the reader is presented with relevant evaluation tools specific to a procedure. Thus, with just a mouse click, a vascular structure can now be displayed in a *curved planar MPR view* for stenosis evaluation; this view shows the longitudinal slice of the blood vessel along the selected section. Automated stenosis quantification is also considered a standard instrument of vascular analysis. Figure 65.3 illustrates such an application.

Postprocessing applications have developed to a point that they are increasingly used in clinical routine. The traditional distinction between image display, evaluation and postprocessing will disappear. In the future, a radiologist will be provided with clinically useful evaluation applications tailored to answer diagnostically relevant questions.

65.1.7 Computer Aided Diagnosis (CAD)

CAD applications (computer-aided diagnosis, sometimes also called computer aided detection) go one step further. The reading physician is not only provided with customized, largely automated tools; CAD algorithms additionally examine image material autonomously with respect to a clinical question. An attempt is made to identify suspicious structures and present them to the reading physician for assessment and evaluation. Since CAD applications are significantly involved in supporting a diagnosis, they are assigned the highest risk level by regulatory authorities. As a rule, CAD applications must be validated by clinical studies following the GCP standard (good clinical practice) before regulatory approval is granted. There are currently a number of approved CAD applications commercially available, e.g. for detection of pulmonary nodules, microcalcifications in breast cancer as well as polyp detection in CT colonography. CAD applications have great potential to improve diagnostic quality, especially in the area of screening and early detection, where they might be able to significantly reduce the rate of false negative as well as possibly false positive findings.

Image postprocessing procedures and CAD make it especially clear that the potential of PACS extends well beyond image communication and the simple display of DICOM images. Ultimately, product differentiation will happen through the combination of clinical and IT know-how since the development of these applications requires not only IT knowledge, but also detailed expertise of image acquisition modalities and clinical interpretation of acquired data.

65.1.8 Medical Monitors

Images displayed on a monitor are the basis for radiological diagnosis as well as an information source for all subsequent therapeutic activities. The medical monitor is therefore the most important interface between digital information and the reading physician. For example, minimum requirements for medical monitors were established in Germany (consensus conference Halle 2001). These were subsequently more exactly specified in DIN 6868 Part 57 as well as in the Quality Assurance Directive. Röntgenverordnung § 16 requires an operator to perform acceptance testing prior to using monitors in a productive environment as well as consistency tests at regular intervals.

65.1.9 Report Creation

Once a radiologist has processed and analyzed the image data, a diagnostic report can be created directly in RIS. There are various methods for generating a report:

- The radiologist enters the diagnosis as free text into the RIS.
- The radiologist uses text modules for efficient creation of normal reports and expands the report using manually entered text.

- The radiologist dictates the findings which are later transcribed by a human operator or directly converted to text by a speech recognition system.
- The radiologist creates a structured report. With this method, the findings related to a procedure are set out in the form of values and tables. An advanced reporting system can generate clinically useful text while retaining structured information. This type of reporting is gaining importance especially with regards to more complex procedures (e.g. cardiac examinations). DICOM is providing services to standardize these reports (DICOM Structured Reporting).

Once a diagnostic report is written, it must be checked and approved by a reading physician directly or by a resident/senior physician. Only then can a report be distributed and archived as an electronic document.

65.1.10 Clinical Demonstration

During diagnosis, a radiologist can tag relevant images, identify findings using arrows, circles, etc. and correlate the cases with different clinical case demonstrations. According to this demonstration worklist, procedures can be opened on the target system in their entirety, or only with significant images; in most cases they can be displayed on a screen using a projector. During clinical demonstration, prior findings and images can be quickly accessed from the PACS. Electronic preparation of the clinical case demonstrations is efficient, saves time, and sometimes the results can only be presented in this way (especially the results of postprocessing applications).

65.1.11 Distribution of Reports/Images

The rapid distribution of results to referring physicians as the final step of a radiological workflow is an important component of the process, influencing the overall efficiency of a clinical workflow. Referring physicians must have timely access to results. In order to achieve this, image data and findings must be accessible from the HIS workstation. Technically, referring physicians can be provided, for example, with web-based access to patient or procedure data of a radiology department. Using a web browser, a referring physician can call up radiological images and view the corresponding report. The physician's access to all images or only the most relevant can be configured, as well as the image quality. In order to reduce network load as well as access times, compressed images can be displayed. It is important that a physician can retrieve the correct images in the context of the HIS workstation. Therefore integration of radiology department IT into higher-level hospital IT system is essential.

In order to display data at a workstation where it is required for a defined treatment activity, special PC systems and input devices are needed. This is relevant for example in operating theaters where sterile or OR-compatible units are required.

65.2 Integrating PACS/RIS into the Hospital Environment

A health care system without comprehensive digital information generation, storage, processing and distribution is almost not conceivable anymore. Among other things, modern information systems support the organization while enhancing cost and performance transparency both within and outside the hospital. In order to support comprehensive workflows while avoiding redundancies, it is essential that existing information and image management systems are integrated. The health care standards DICOM, HL7 and the IHE interoperability model support and foster this integration.

65.2.1 Integrating Information Systems

The intensive interdisciplinary collaboration within and outside a hospital requires access to all relevant and available patient and case data beyond departmental boundaries. Communication problems and risks can be reduced through reliable integration of different information systems and embedding them into the workflows of the respective service providers. Typically the electronic patient record (EPR) offers a comprehensive entry into related patient and case data, and is the guiding system in a hierarchically structured data model. Departmental information systems (e.g. radiology, cardiology, or laboratory) primarily provide data while assuming a secondary position with respect to data hierarchy.

This process-oriented horizontal integration of different departmental information systems with an electronic patient record would be inconceivable without the continuous update of standards such as DICOM and HL7 and interoperability models such as IHE.

In addition, the integration of data from different information systems and the related comprehensive view offer the possibility of a systematic analysis to improve business and clinical process (business intelligence).

65.2.2 Integrating Image Management Systems

In addition to radiological or cardiological images in DICOM format, state-of-the-art image management must be able to manage non-DICOM files using, for example, PDF, video and audio formats. The purpose of such a system is the acquisition, structuring and information life cycle management of heterogeneous data records as well providing these records with a very high level of accessibility and performance to all departments and physicians without respect to geographic or time-related limitations. The primary challenge for an image management system is efficient management of extensive volumes of image data.

Data management based on standards assures that image data and other documents can be stored long-term without limitation. Vendor-neutral archives (VNA) provide these functionalities and typically consist of three basic components:

1. A storage subsystem,
2. A standardized interface for data management,
3. A database indexing all storage content and changes to content.

A VNA is a common resource providing data to all IT solutions within and outside the hospital, and should meet the following requirements:

- It must be able to exchange images and diagnostic findings using the DICOM standard with clinical and PACS systems.
- It must contain interfaces that can exchange data with hospital information systems and other IT systems using the HL7 standard.
- It must be able to store all DICOM formats (including all DICOM SOP classes).
- It must also offer the option to store proprietary data formats (non-DICOM) such as pathology image data.

65.2.3 Teleradiology

Clinical examinations or diagnoses of various disciplines such as radiology, cardiology or pathology do not necessarily take place within a hospital, but can be performed by other service providers (e.g. specialists or laboratories). Generated data is transmitted to a service provider (e.g. for diagnosis, second opinion) in encrypted form. Integration of information systems and use of telematic processes allow virtual collaboration and provision of services. The physical boundaries of a hospital can be overcome taking into account any individual legal requirements for tele-applications, and new services can be offered. Authentication and authorization (who can access which data and when), as well as secure, encrypted transmission are essential for these applications.

65.2.4 Mobile Devices

Web-based software technologies combined with mobile devices offer the ability to ubiquitously access medical image and diagnostic data. Required data can be displayed, for example, in emergency situations or during patient consultation. The data formats range from simple text documents through medical images to video recordings. Thus a physician, for example, can access relevant data and supplement it, thereby significantly reducing the response time for treatment. In addition, there are no longer media discontinuities, and data does not have to be stored or copied multiple times. The exact utility and long-term significance of mobile devices to clinical daily routine has yet to be determined. The assessment and classification of mobile devices by national regulatory authorities are still in initial stages and will influence the application scenarios of these devices.

65.3 State-of-the-Art IT Infrastructure

During the definition of state-of-the-art hospital IT infrastructures, the following points require special attention:

- Scalability of the architecture from a small to a potentially regional solution
- Consideration of data security, data consistency and a high degree of data availability
- Compliance with standards for integration of heterogeneous applications and IT systems by different manufacturers for a comprehensive solution without frictional loss at the interfaces

- Modularity of the IT infrastructure to replace or update individual components without disrupting ongoing operations.

65.3.1 Information Lifecycle Management

Data management is an important component of a modern IT infrastructure. Management of image data from acquisition through archiving is a complex process addressed with various technologies to assure data availability, security, redundancy and continuity:

- Short-term storage: Depending upon configuration, it contains examination data of the least 3–12 months. Data can be accessed rapidly. Provision of data is via failsafe storage systems (e.g. RAID).
- Long-term storage and backup: Depending upon legal requirements, data must be archived up to 30 years and must be returned to accessibility within an appropriate time frame. Regular data backup is required in order to restore the original data in the event of loss (disaster recovery).

Commercially-available hierarchical storage management systems (HSM) offer the described data management functionality and additionally provide users the ability to migrate data transparently across different media.

65.3.2 Cloud Computing Solutions in the Hospital

Cloud computing (highly-scalable IT infrastructure making applications available, which can be billed after use) represents a paradigm shift for information technology in a hospital. This especially applies to information systems and image management solutions. Access to available data is via the Internet, and can be performed from any workstation world-wide. Thus clinical applications are no longer executed locally, but are made available via the cloud, and accessed there. Users only need a secure Internet connection in order to access the applications and associated data.

The achievable costs savings represent the most significant advantage for hospital IT managers. Fixed costs, previously generated by a separate IT department, can be transformed into variable costs. A hospital or individual departments such as a radiology department only need to pay for the actually required infrastructure and services, thus adapting the costs to the demand.

The possibilities offered by a cloud also play a significant role for the expansion of cooperation among service providers. This development is shaped by an immense communication requirement among the partners. In its current structure, complex networks must be maintained in order to efficiently structure collaboration. Utilization of cloud computing can cover this need more cost effectively. A complex infrastructure does not have to be created, since all users can simply access the common applications within the cloud. The flexibility inherent in the system allows adaptations due to new partners or other modifications to be performed in a timely manner and without significant financial expenditure.

65.3.3 Regional IT Virtualization

One of the challenges in current healthcare environment is the multi-site and manufacturer-independent communication and exchange of data. Regional IT virtualization plays an important role in scenarios in which several hospitals and other service providers are networked via different RIS, PACS and other IT solutions in order to provide a smooth treatment path. In this regard the following core functionalities are relevant [65.4–7]:

- Global worklist: In a regional setup, radiologists must be able to access a consolidated global worklist without knowing the actual physical location of the relevant data. The global list is thus the sum of all individual worklists of the regional participating hospitals.
- Global access to image data: During the assessment of radiological images, each radiologist accepts all cases falling under his specialty. This is a result of filtering of the global worklist by specialty. Although the images and findings-related data are physically separated, a physician has access to the data. Interoperability models such as IHE XDS (cross-document sharing) define this data exchange.
- Access control and security: granular access rights and multi-entity functionality (support of several clients/independent organizational units within an IT system) are prerequisites for this type of regional IT virtualization.
- Master Patient Index (MPI): A MPI provides functionality in order to combine information from different sources (e.g. departments) under a common identity.

65.4 Summary

Today, radiology information systems and picture archiving and communication systems are established foundations for optimizing workflows within a radiology department. Integration of postprocessing applications in the diagnostic assessment process has the potential to essentially change the reading workflow and allows the increasing data volume to be analyzed more efficiently and with higher quality.

RIS and PACS should be firmly integrated into the entire IT infrastructure in order to achieve optimal utilization. This applies internally to the hospital as well as beyond (e.g. regional associations). This integration follows existing standards such as DICOM, HL7 and IHE.

Current IT trends such as cloud computing offer the possibility of reducing costs and complexity as well as scaling RIS and PACS, dimensioning these according to demand.

References

65.1 HL7: Health Level Seven International (2011) http://www.hl7.org/

65.2 RSNA: Radiological Society of North America (RSNA, Oak Brook 2011) http://www.rsna.org/

65.3 IHE: Integrating the Healthcare Enterprise, sponsored by Healthcare Information and Management Systems Society (HIMSS) and the Radiological Society of North America (RSNA) (2011) http://www.ihe.net/

65.4 M. Benjamin, Y. Aradia, R. Shreiberab: From shared data to sharing workflow: Merging PACS and teleradiology, Eur. J. Radiol. **73**(1), 3–9 (2010)

65.5 C. Carter, B. Veale: *Digital Radiography and PACS* (Mosby-Elsevier, ST. Louis 2010)

65.6 H.K. Huang: *PACS and Imaging Informatics: Basic Principles and Applications*, 2nd edn. (Wiley-Blackwell, New York 2010)

65.7 A.A.T. Bui, R.K. Taira (Eds.): Medical Imaging Informatics. In: *PACS: A Guide to the Digital Revolution*, ed. by K.J. Dreyer, D.S. Hirschorn, J.H. Thrall, A. Mehta (Springer, New York 2006)

66. 3-D Postprocessing in Virtual Endoscopy

Georg-Friedemann Rust

Virtual endoscopy is the simulation of endoscopic interventions using methods of virtual reality and computer graphics. Usually, three-dimensional (3-D) volume data from computed tomography (CT) scans, magnetic resonance imaging (MRI), 3-D ultrasound, rotational angiography or other sources are used to generate a 3-D view of the inside of the studied structures.

In recent years, the spatial resolution of cross-sectional computed tomography imaging has improved greatly. Imaging procedures allow more detailed representation of the organs of the human body. In particular, multislice computed tomography (Chap. 16) opens new opportunities in diagnostics based on its significantly increased spatial resolution and significantly shortened examination time. This improved spatial resolution in conjunction with high-resolution rendering leads to much greater authenticity of all examined body parts. The high spatial resolution of multidetector computed tomography (MDCT) combines this with a remarkable increase of the amount of data that can be assessed.

66.1 **What Is Virtual Reality?** 1210

66.2 **Why Virtual Reality?** 1210

66.3 **Advantages of 3-D Visualization** 1210
66.3.1 Partial Volume Effect 1211
66.3.2 Two-Dimensional or Three-Dimensional Visualization 1211
66.3.3 Risks of 3-D Visualization 1213

66.4 **Conclusions** 1216

References 1216

The standard method of diagnostic imaging assessment based on analysis of axial slices is reaching its technological limit. Use of more than 1200 axial cuts is no longer a rarity with current CT scanners (2×64 detector rows or more), and the demand for alternative three-dimensional visualization techniques is increasing with the number of axial images assessed. For certain examination types, such as for hollow body organs, 3-D visualization is increasingly becoming the standard method of diagnostic assessment.

This is especially true for the intestine, and generally for hollow body organs where characterization of the cavity is becoming difficult by a significant change in physical density, such as in bronchoscopy or colonoscopy, particularly for endoluminal endoscopic views. This approach can be easily generalized for visualization of any 3-D surface. This approach can also be extended from endoluminal to extraluminal views, as in angiography, which is described as a possible future application at the end of this chapter.

The problem that often arises in 3-D visualization is the difficulty faced by the viewer in assessing the accuracy and relevance of the presented 3-D surfaces, as reality is only partially reproduced. The relevant questions are then: How authentic is the 3-D surface? What is the *best representation* of the 3-D surface?

In the reconstruction of a 3-D surface, a decision has already been made regarding which part of the dataset is considered. A correction based exclusively on the underlying 3-D surface is very limited and will not solve the inherent limitations of 3-D visualization.

The requirement is therefore to represent the *real* surface in the optimal way as a digital 3-D surface.

66.1 What Is Virtual Reality?

Reality can be represented through a corresponding physical measurement of a physical property, for instance, the physical density in computed tomography imaging (CTI). The real object is not represented in its diverse forms, but only by what we hope will be its most representative physical property. Therefore, very often the choice of this physical property will be a compromise in the corresponding visualization. It is crucial that the choice of this physical property correspond as far as possible with the needs of the application under consideration. In the case of (multidetector) computed tomography (CT), the physical property represented in the visualization is the spatial (digital) distribution of the physical density values of a given cross-sectional cut of the examined body. Based on this digitized density distribution, the (reduced) reality can be examined in three dimensions through stunningly realistic modeled images, so-called *virtual reality* (Chap. 63).

Virtual reality is a kind of simulation. Depending on the task at hand and the target of the simulation, virtual reality can be adapted to the given application. This provides an opportunity, and at the same time a risk. This is exemplified by the task of visualizing the 3-D endoluminal intestinal surface either as a stunningly realistic view as in optical endoscopy or as an authentic view. This can lead to distinct forms of visualization.

One therefore has to differentiate between the creation of images for medical assessment, where the focus is on authenticity, versus the creation of images as a kind of *modern art*.

66.2 Why Virtual Reality?

The power of virtual reality in cross-sectional imaging is based on its ability to reduce hundreds, if not thousands, of axial cuts into one 3-D volume dataset. This enables assessment to be much more intuitive. In 3-D, topographic relations can be identified more easily, quickly, and safely, which enables interventional therapies to be planned, and prepares the interventionist for the actual situation in situ.

The risks of virtual reality have been briefly mentioned: A 3-D surface is always only part of the dataset, although hopefully the *right* part, or rather the most relevant part for the current application. On the other hand, herein lies also the opportunity, namely the possibility to focus on a given region of interest of the dataset, although the question of what is the most *relevant* remains open. Even from this brief presentation, it is clear that virtual reality methods present both opportunities and risks, which also makes it difficult to provide general recommendations.

66.3 Advantages of 3-D Visualization

The key advantages of virtual endoscopic procedures are the following.

- Significantly improved spatial representation and understanding of the organ under analysis.
- Significantly longer observation time of the 3-D surface than for corresponding axial cuts. More precisely, the observation time per unit surface area is much longer than for corresponding two-dimensional (2-D) axial images. A detail of the intestinal mucosa (in virtual colonoscopy), for example, will be represented in one or two individual axial layers. Using three-dimensional methods, the observation time for such details is significantly increased compared with the two-dimensional axial slices, because during the fly-through, one is able to observe small details for longer than in corresponding 2-D images.
- Systematic investigation of the considered hollow organ in 3-D, such as the intestine, which is much more difficult and needs a longer learning curve in 2-D than in 3-D.

As an example, consider moving through the intestine, so that all structures are moving for several seconds along the observer's line of sight (Fig. 66.1). The observation time per unit surface area is much longer than if the corresponding

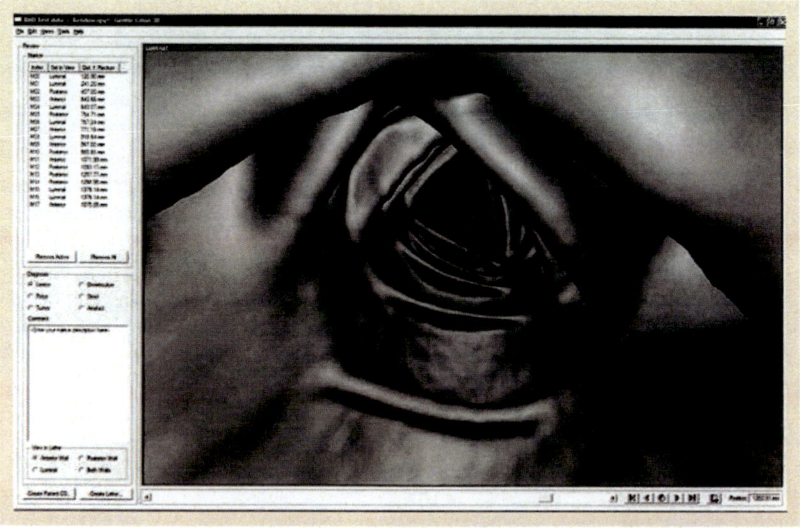

Fig. 66.1 An intraluminal virtual representation of the intestine. In a fly-through, the voxels corresponding to each detail remain in view for several seconds, which is much longer than in exclusively 2-D observation

2-D layers were considered separately. This significantly increases the physician's viewing time per unit area of mucosa surface based on the use of the 3-D surfaces available in virtual endoscopy. It is well known that the detection rate of pathological findings depends significantly on the observation time.

66.3.1 Partial Volume Effect

One of the main advantages of 3-D surfaces in comparison with 2-D surfaces is the ability to present the dataset almost free of partial volume effects. More details will be described in Sect. 66.3.2. The *partial volume effect* occurs when grey-scale values in a single axial 2-D slice of finite thickness (from CT, for example) are calculated from parts of different organs. The density value illustrated in the resulting image is then calculated not just from the density value of the given organ, but may also include density values from other organs or other parts of the organ in the same slice. This occurs when a particular organ is, along with other organs, only partially in the observed slice volume. In this case, the density value (gray value) of a 3-D pixel (voxel) in the CT slice is calculated as an average through the CT slice thickness. Therefore, another term for the *partial volume effect* is *contrast reduction*. Vice versa, visualization that is free of the partial volume effect results in high contrast. The thinner the slice thickness, the weaker the partial volume effect and the higher the resulting contrast.

66.3.2 Two-Dimensional or Three-Dimensional Visualization

The thicker the CT slice, the stronger the partial volume effect and the lower the contrast of the resulting slice image. However, 2-D images inherently suffer from the partial volume effect. In principle, 3-D surfaces can be visualized with a smaller partial volume effect than corresponding 2-D images (Fig. 66.2a), because the visualized dynamic range of the underlying images can be higher than with the standard radiological window level settings. The resulting higher contrast provides correspondingly higher detection sensitivity, particular for small structures such as flat adenomas. Figure 66.2 is an endoluminal view showing a *rough mucosa* (a combination of correlated noise from filtered back-projection and white electronic noise) that is only visible if one applies high spatial resolution in combination with high contrast. With this combination of high contrast (thin slice thickness and wide dynamic range of visualization) and high spatial resolution, the smallest lesions become much more visible than in 2-D images using the well-known *lung* window level setting. This extended image information content is correlated with higher sensitivity for artifacts, including noise. However, highly authentic 3-D visualization is required, and can be achieved through high-contrast images. Quite often, *good-looking images* are more of an issue than the authenticity of the acquired data.

The above-mentioned advantages of 3-D imaging are often negated by misunderstanding of the basis

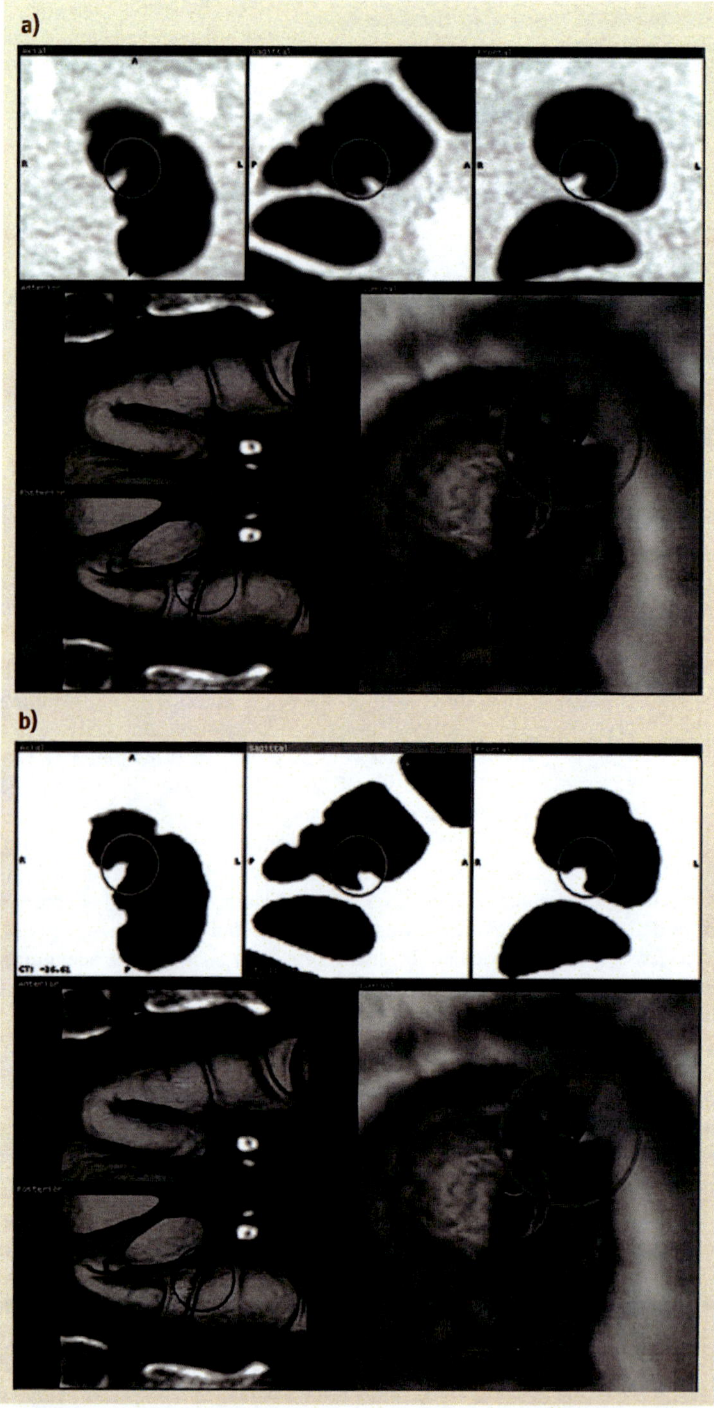

Fig. 66.2 **(a)** A 3-D image containing a small intraluminal artifact, and the corresponding location in 2-D images suffering from the partial volume effect (in the *lower left* is the intestinal intraluminal represented cut in half – split colon). **(b)** The same location as in **(a)**, but with a significant increase in the contrast of the 2-D images. The artifact in the large intraluminal representation (*bottom right*) is now seen as a small artifact in the 2-D images

for 3-D imaging: the use of 3-D imaging should not suggest that the virtual endoscopic view is a real endoscopic view. Creation of virtual views which look like real endoscopic views is only possible if either the 3-D organ surfaces are smoothed or the underlying visualization algorithm implements smoothing effects (simple volume rendering algorithms), or both.

In the case of computed tomography, the intestinal mucosa visible in a 3-D visualization is not the real intestinal mucosa but rather the CT of the intestinal mucosa, which is an important distinction from the point of view of a doctor. This means – and this also provides a simple method to verify the authenticity of the 3-D surface visualization – that all artifacts created by a CT must also be visualized in the reconstructed 3-D intestinal mucosa, including noise, beam-hardening artifacts, pitch artifacts, and edge phenomena. If such artifacts (Fig. 66.2) are not visible, the virtual mucosal surface should be interpreted with great caution.

All of these artifacts, particular memory defects resulting from the gantry, may not be visible in the 3-D surfaces, even when they are present in the original dataset, either due to the smoothing algorithm (i.e., smoothing over nearest-neighbor voxels) or due to the use of an incorrect form of volume rendering in order to create *beautiful* images that suggest that the given visualization is identical to a real view of the intraluminal intestine, which is a false approach. *Virtual reality* should not manipulate the original measured data, particular if this would prevent relevant details from being found, as this will not be *reality*. Such forms of visualization almost reduce 3-D visualization to a computer game.

66.3.3 Risks of 3-D Visualization

Figure 66.3 shows how structures can be changed even if only a simple smoothing method is applied, such as smoothing over nearest neighbors. Surfaces may appear more physiological, but at the price of loss of both detail and authenticity (Sect. 66.1). The aim of visualization for medical assessment of a 3-D surface (endoluminal or extraluminal view) must be authenticity and not manipulation of images just to produce nice images with a *realistic* appearance. It is required

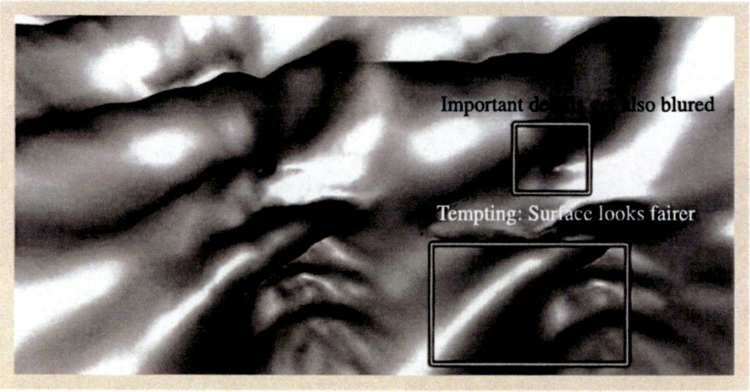

Fig. 66.3 Memory defect artifacts in an endoluminal view, resulting from memory failure of the CT gantry; the CT manufacturer and processing software cannot be displayed

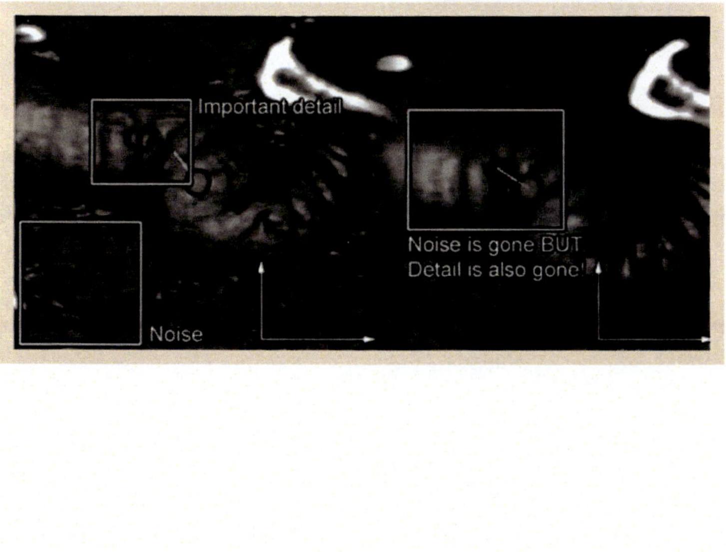

Fig. 66.4 Effect of smoothing over nearest neighbors in virtual colonoscopy

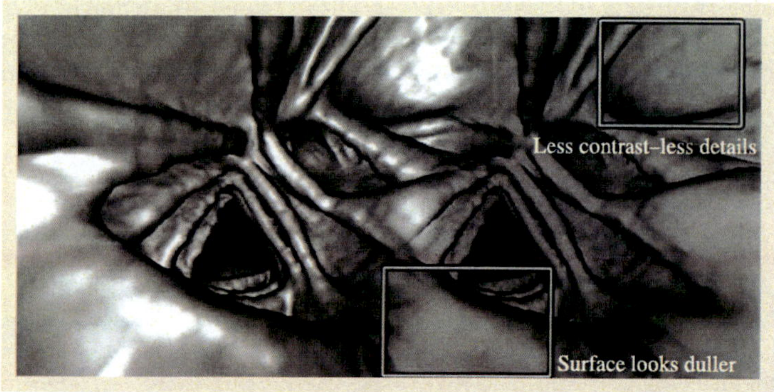

Fig. 66.5 With a nonoptimal choice of transfer function for volume rendering (ray casting), the noise increases with a corresponding loss of detail of the diverticula (*right rectangle*) and the balancing appears on the right-hand side less pronounced

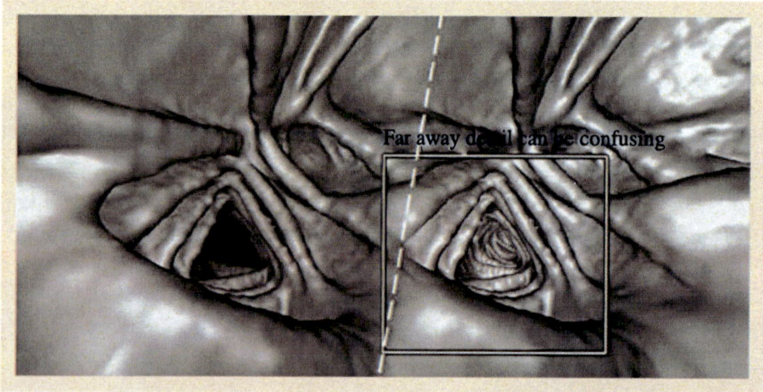

Fig. 66.6 Nonoptimal use of *diffuse lighting* can have a significant influence on the contrast of the 3-D surface under consideration

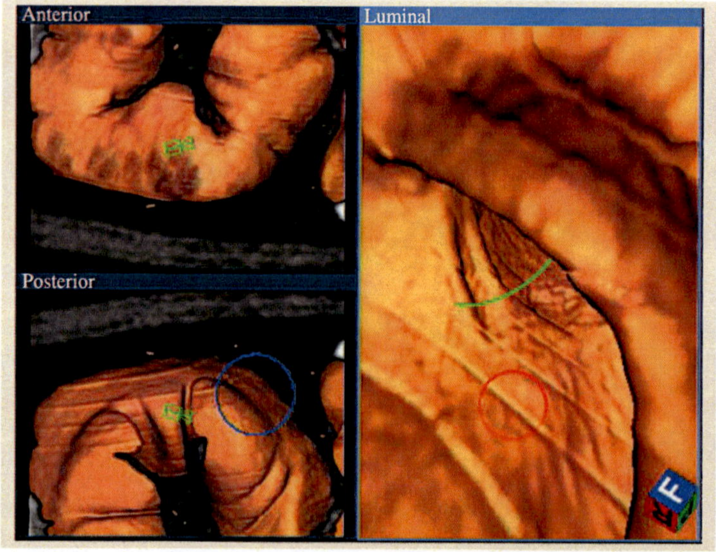

Fig. 66.7 Light absorption (*left*) focuses the attention of the viewer on mucus closer to the virtual camera

to assess the smallest artifact as well as the smallest pathology or anatomical variation. Authentic visualization is the only way to distinguish between artifacts and nonartifacts, and the only way to ensure that the diagnosis is not distorted in order to achieve false esthetics.

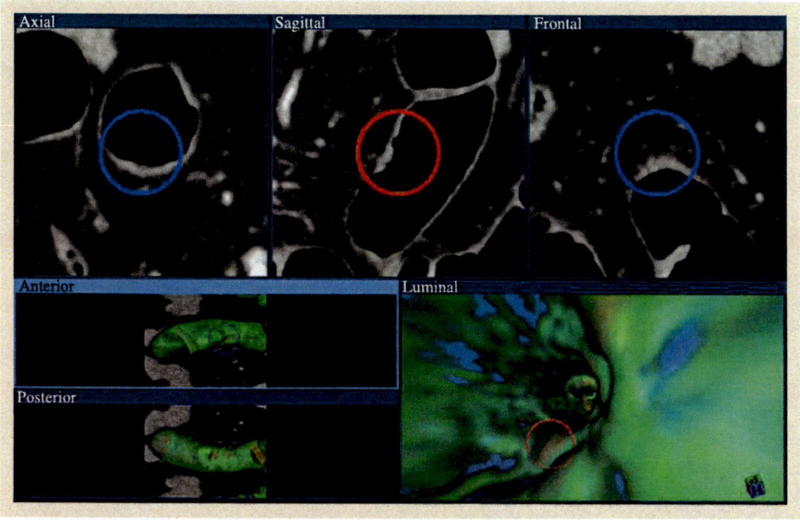

Fig. 66.8 Color-coded depth information of a flat lesion

Figure 66.4 demonstrates the effect of smoothing on image quality – in this case not by smoothing over nearest neighbors, but by too flat a transfer function used in the volume rendering. Figures 66.5, 66.6 show the options and advantages that virtual reality may have. In the virtual endoscopic view on the right of this figure, it is possible to see far into the intestine lumen. Usually, one tries to achieve undisturbed vision in order to retain focus on the region of mucosa of importance for the current investigation. The subject of endoscopic observations are the findings for the mucous membrane components that are closest to the (virtual) camera. *Distraction* of the gaze into the distance can prevent this, and in this case can be avoiding by applying absorption of light from further away, as shown in Fig. 66.6a.

Figure 66.7 shows the effect of false *diffuse lighting* in the volume rendering algorithm, resulting in loss of contrast and an associated loss of detail.

An interesting application of virtual reality is the use of a virtual view of 3-D surfaces represented not only by a single color or gray level but where depth information is presented through the use of a color map. Such coloring of 3-D surfaces reveals information beyond the surface itself, which may be even more relevant. Using this approach, color changes indicate changes in wall thickening, which is particularly relevant for virtual colonoscopy and is difficult to detect even using computer-aided diagnosis tools. The CT density value at a given depth underneath the intestinal mucosa can also be associated with a color value (i.e., a temperature scale) and projected back onto the intestinal surface. In this way, one can identify relative wall thickening and features that are very difficult to detect, such as so-called *flat lesions* and hypervascularization of the mucosa of the cavity surface, which are highly relevant for intestinal pathology. An example of a flat adenoma is shown in Fig. 66.8.

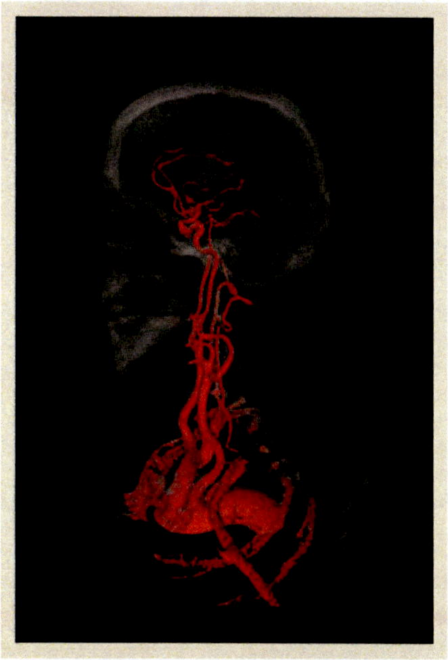

Fig. 66.9 With a single virtual high-resolution view of the vessels of the cranial base, the aneurysm in the posterior supply area of the skull can be diagnosed quickly and safely, and also visualized for the therapist in a helpful presentation in the context of other topographical anatomical structures

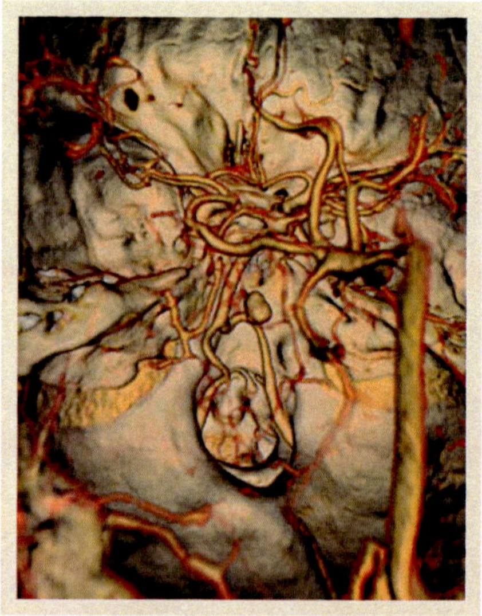

Fig. 66.10 The combination of 3-D visualization with multiple CT datasets opens up further, highly relevant forms of visualization. To date, only digital subtraction angiography (DSA) has been developed. In this application, the benefits of DSA and high-resolution subtraction CT with 3-D visualization enable high-resolution views of vascular structures in the cranial base, opening up new diagnostic and therapeutic options ◀

The topics discussed above, especially the 3-D visualization methods with the highest spatial and contrast that are technologically possible, are not limited to endoscopic issues such as endoluminal visualization, but of course apply to all 3-D surface calculations, such as for extraluminal 3-D representation of organ surfaces as in vascular imaging, e.g., enhanced intravenous angiograms. Authentic representation of the surface is of particular interest in CT angiography because even the smallest caliber variation may be of medical relevance.

66.4 Conclusions

Virtual endoscopy using multidetector CT scanners and high-performance postprocessing software enables visualization of hollow bodies. Such 3-D visualization procedures are very quick, highly accurate in terms of spatial and contrast resolution, and completely automated, offering greater accuracy than conventional 2-D imaging methods that also suffer from partial volume effects. It is of great importance to learn to differentiate between authentic visualization solutions and just *interesting* images. Manipulation of 3-D surfaces for *pretty* views of 3-D surfaces, such as the intestinal mucosa or CT angiograms, is not a good foundation for virtual reality. If anything, this only portrays a virtual (false) reality that is of no use in radiological routine. The underlying algorithms used in visualization have special importance, particularly with regard to the detectability of small lesions.

For further and deeper study we recommend [66.1–3], and practical experience based on the Visualization Tool Kit [66.4].

Further short and good examples can be found in the OpenGL Reference Books [66.5].

Not recent research but a good overview is shown by *Ecarnacao* and *Strasser* in [66.6].

References

66.1 K. H. Höhne, A. Petersik, B. Pflesser, A. Pommert, K. Priesmeyer, M. Riemer, T. Schiemann, R. Schubert, U. Tiede, M. Urban, H. Frederking, M. Lowndes, J. Morris: VOXEL-MAN 3-D Navigator: Brain and Skull. Regional, Functional and Radiological Anatomy, Version 2.0, DVD (Springer, Berlin, Heidelberg 2009)

66.2 B. Tolsdorff, A. Petersik, B. Pflesser, A. Pommert, U. Tiede, R. Leuwer, K.H. Höhne: Individual models for virtual bone drilling in mastoid surgery, Comput. Aided Surg. **14**(2), 1–7 (2009)

66.3 B. Preim, D. Bartz: *Visualization in Medicine – Theory, Algorithms, and Applications* (Morgan Kaufmann, Burlington 2007)

66.4 Kitware, Inc.: Visualization Tool Kit (Kitware, New York 2010) available at http://public.kitware.com

66.5 D. Schreiner (Ed.): *OpenGL Reference Manual*, 4th edn. (Addison-Wesley, Cranbury 2004)

66.6 J.L. Ecarnacao, W. Strasser: A VLSI implementation of the graphics standard GKS, Computergraphic **2**, 115–121 (1983)

67. e-Health – Ambient Assisted Living and Personal Health Systems

Natasha Avila, Christina Sampogna

This chapter focuses on the healthcare aspects of ambient assisted living (AAL) and the contributions that personal health systems (PHS) may bring. As part of these categories, this chapter examines the areas of self-management, tele-assistance, media and multimedia solutions, fall detection, robotics, bio-parameter monitoring, auto adaptive treatment systems, tele-therapy, tele-consultation, tele-surgical assistance, and smart homes. For each approach, the nature and purposes of the technologies will be examined and examples will be discussed. The chapter will subsequently address the benefits and the challenges ahead for AAL and PHS technologies within the broader context of e-Health. Finally, the chapter concludes with observations and trends.

67.1 **Background** 1218
67.1.1 e-Health 1219

67.2 **AAL and PHS Approaches** 1221
67.2.1 Self-Management 1222

67.3 **Benefits and Challenges Ahead** 1234
67.3.1 The Social Dimension 1234
67.3.2 Legal and Ethical Dimensions 1235
67.3.3 Societal and Global Dimensions 1237
67.3.4 Economic Dimensions 1238
67.3.5 Technological Dimensions 1240

67.4 **Conclusion** 1241
67.4.1 Observations 1241
67.4.2 Areas for Further Work 1242
67.4.3 Future Technologies 1243

References 1243

Advances in information and computer technology (ICT) over the last half-century have led to many innovations crucial for the delivery of healthcare, including the emergence of e-Health. Within the context of e-Health, this chapter examines the contribution of ICTs to the fields of ambient assisted living (AAL) and personal health systems (PHS). Albeit in different manners, both AAL and PHS include products, processes, or services that can contribute to maintaining individuals within their home environment, while providing them with the support necessary in light of their conditions or pathologies. This chapter examines the areas of self-management, teleassistance, media and multimedia solutions, fall detection, robotics, bio-parameters monitoring, auto adaptive treatment systems, teletherapy, teleconsultation, telesurgical assistance, and smart homes. For each area, the nature of the technology, the purposes for which it may be used, as well as examples of developments are outlined. Technological developments occur within a social, societal, ethical, economic, and global context. This chapter examines the benefits of these technological developments and the challenges ahead for their development, adoption, and widespread use. It concludes with some observations and proposes some trends for future developments.

Health and healthcare systems are the product of a convergence of biomedical, social, technological, economic, ethical and political factors. The delivery of effective, safe and efficient healthcare services and products to patients is a challenge facing governments around the world. The nature and scope of this challenge has become more pressing over recent decades as the global population has grown and will continue to grow, as life expectancy has increased and will continue to increase, as the burden of disease has become better understood and has shifted, and as the costs of healthcare systems and services have become unmanageable in their current state.

67.1 Background

Many healthcare systems were developed during periods where the population they served was smaller and were designed in accordance with that population. In 1960, the world population was under three billion. As of 30 March 2010, the United Nations estimated the human population of the world to have been 6 800 000 000 in 2009 [67.1].

In light of these developments, the current healthcare system model is not sustainable for much longer. The 1980s policies of cost containment are arriving at their limits (wage control and price controls plus investments postponement) [67.2]. To maintain an equivalent healthcare level and availability, the health-related expenditure would need to increase at a constant rate to cope with the geographical and nongeographical changes. As Fig. 67.1 shows, the total costs and expenditures on health have continuously increased during recent decades. Unfortunately, this increase in expenditures has not resulted in improved services. For example, extensive waiting times face patients in hospitals, in emergency rooms, and when seeing a specialist. In Canada, for instance, a recent study found that the total waiting time between referral from a general practitioner and treatment, averaged across all 12 specialties and 10 provinces surveyed, was 16.1 weeks in 2009 [67.4].

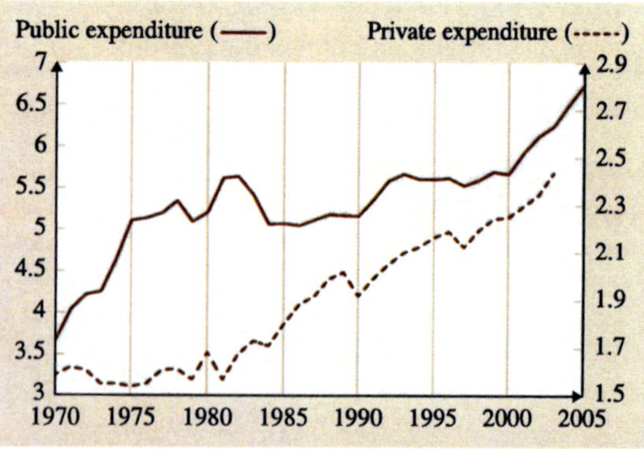

Fig. 67.1 Evolution of public and private OECD total health spending [67.1] as a percentage of GDP (after [67.2])

Moreover, other trends make the current models of healthcare systems unsustainable. The increasingly ageing population is also an important contributor to this nonsustainability. As Fig. 67.2 illustrates, by 2050 there will be significantly larger portions of the population who will be older than 65 years of age. Asia, Japan in particular, and Europe are the two regions where a significant number of countries face severe population ageing in the near future. In these regions, within 20 years many countries will face a situation where the largest population cohort will be those over 65 and the average age will be approaching 50.

This ageing of the population will result in many societal benefits. This demographical evolution also needs to inform the adaption of healthcare systems and services in order to offer effective, safe, and efficient healthcare services. As Fig. 67.3 shows, currently the

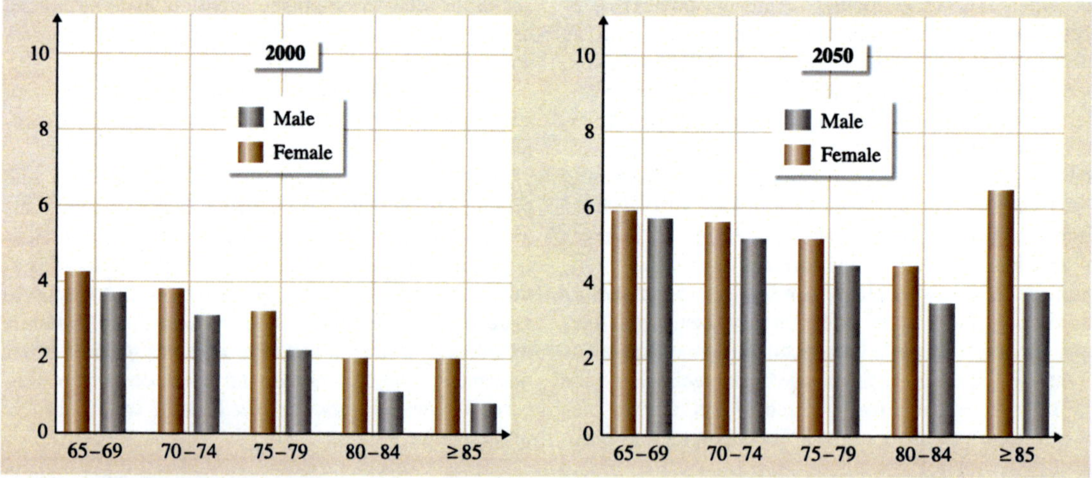

Fig. 67.2 Aging Forecast for OECD countries (after [67.3])

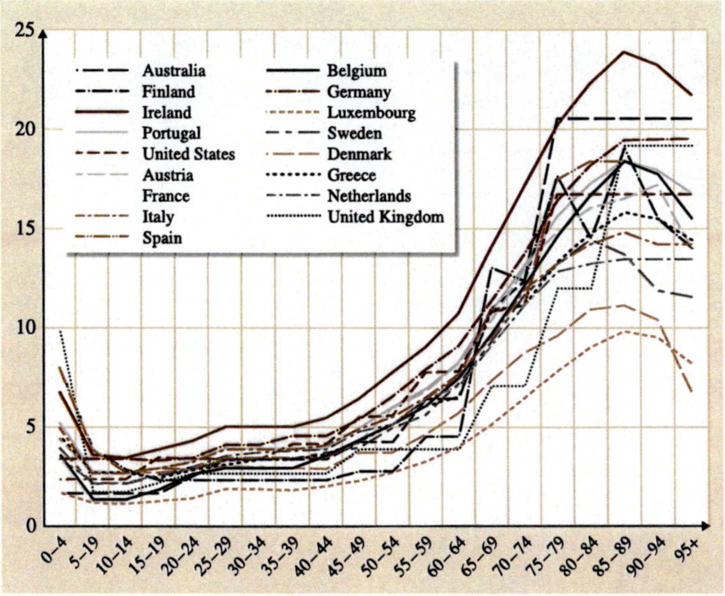

Fig. 67.3 Public healthcare expenditure by age groups as a percentage of gross domestic product (GDP) (after [67.2])

peak of the costs for healthcare services is for the portion of the population aged 75 years and older. The precariousness of the current system is compounded by the converse demographic phenomenon of the shrinking number of working adults. Currently, in Europe, there are 4 adults working to sustain each person aged over 65 years. However, by 2050, this number will decrease and there will be only 2 adults working to sustain each person aged over 65 years [67.5].

Another trend that is significantly impacting healthcare systems is the increased burden of chronic diseases. Chronic diseases are diseases of long duration and generally slow progression. Chronic diseases, such as heart disease, stroke, cancer, chronic respiratory diseases, and diabetes, are by far the leading cause of mortality in the world, representing 60% of all deaths [67.6]. In regards to those living with chronic disease, the statistics are also daunting. 860 million people worldwide suffer from one or more chronic condition. As of 2000, for example, 45% of Americans had at least one, and 23% had two or more chronic conditions. In Europe, chronic conditions account for up to 86% of all deaths.

67.1.1 e-Health

While no tool, process, or technology will address the multitude of issues currently facing healthcare systems, numerous phenomena are emerging that can contribute to finding better and different solutions. One such field is that of e-Health. e-Health can be described as an overarching term used to describe the application of information and communications technologies in the health sector. It encompasses a whole range of purposes from purely administrative through to healthcare delivery. While there is no globally-agreed upon definition or categorization of e-Health, Fig. 67.4 presents a categorization of e-Health that has been developed and used within the European Communities.

This chapter will focus on the areas of e-Health referred to as ambient assisted living (AAL) and personal health systems (PHS). The European Framework Programmes have focused considerable attention on and have set aside considerable investment for the areas of ambient assisted living and personal health systems. The Framework Programmes (FP) are programs of the European Union with the mission of funding and encouraging research and applications around the EU's key research areas. The term *personal health systems* was introduced in the 1990s (and first introduced in the Fifth Framework Programme, FP5-1998), while the term *ambient assisted living* was first widely used in 2000. Although they sometimes use different nomenclature, other economies are also allocating considerable resources in the areas of AAL and PHS.

Broadly defined, AAL is the use of information and communication technologies (ICT) in a person's daily living and working environment to enable individuals

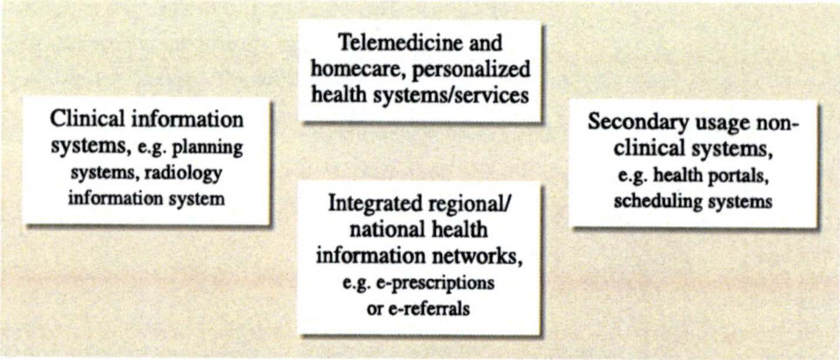

Fig. 67.4 The four components of e-Health [67.8]

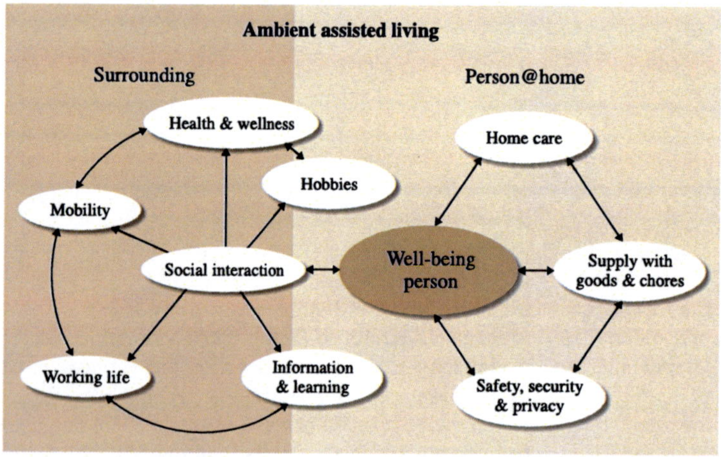

Fig. 67.5 Ambient assisted living application environments [67.9]

to stay active longer, remain socially connected, and live independently [67.7]. As Fig. 67.5 illustrates, AAL adopts a holistic and person-centric approach to ensuring the well-being of the individual. In order to achieve this objective, AAL aims to address more than healthcare aspects. As such, AAL covers many aspects such as mobility, social interaction, hobbies, chores, etc.

From a healthcare perspective, AAL includes any product, process or service that can help directly or indirectly to maintain individuals within their home environment, while providing them with the support necessary in light of their conditions or pathologies. AAL solutions aim to stabilize or even increase the autonomy of the individuals, facilitate medical treatments, and monitoring from home, and/or ease the carrying out of day-to-day tasks. AAL solutions are based on the desire of individuals to remain within their home environments as long as possible and to delay their entrance into institutionalized environments. These approaches provide solutions for the elderly, for incapacitated individuals, individuals recovering from a condition, surgery or a disease, or for individuals with cognitive, motor, or neurological needs.

Personal health systems may be considered to include wearable and/or fixed/portable systems, based on advanced sensors, microsystems, and telemedicine services for personal health monitoring [67.10, 11]. Some of the objectives that have been attributed to PHS include [67.12]:

- Detection of early signs of health status decline
- Notification of critical conditions to health professionals
- Improvement the sense of union within the family by sharing health information
- Identification of the interdependency between a person's health and their lifestyle
- Offering healthcare to remote, rural or doctor-deprived areas
- Bringing healthcare to developing countries
- Providing real-time multisourced reliable physiological data to health professionals.

The pressing need to develop and diffuse PHS technologies is their contribution to prevention, detection, diagnosis, monitoring and surveillance. For example, the European Heart Network reports that in Europe four million individuals die from cardiac diseases [67.13]. With portable heart monitoring systems capable of automatically alerting a healthcare centre capable of immediate response many lives could be saved.

67.2 AAL and PHS Approaches

This section will explore different PHS and AAL approaches. PHS or AAL approaches will generally contain one or more of the following components:

1. Data collection: The element is in charge of collecting the image, video, sound, text, or measurement of interest. The gathering may be automatic (e.g. use sensors) or manual. For example, the patient may enter text information. The PHS may be devices that are *wearable*, such as a portable heart rate chest belt, while AAL tools may be portable or fixed within their environment, such as a webcam.
2. Short-term memory: The element is responsible for ensuring the short-term storage of the data collected. There is a trend for memory chips to increase in capacity and to in reduce size and price. For example, memory cards can currently measure approximately 1.5 cm^2 and can reach 32 GB [67.14].
3. Local visualization: The element will allow the patient to visualize the fixed or continuous data collected.
4. Communication network: The element will move the data from the location of the patient to a remote location, where the server(s) is (are) located. There will generally be three parts: a transmitter (TX), the network itself, and the receiver (RX).
5. Database(s): The element will archive the information. This information can be fixed (such as patient ID or age), or incremental such as the data collected. Another type of information that is sometimes used is event tracking, such as user logins to the system (when a user accessed the system).
6. Analysis: The element will examine the data collected and the patient information to select a result and/or an action.
7. Action launcher: The element will launch an action following the analysis. The typical actions would be a flag in the database, a transmission of an SMS or email, and a local or deported sound or light alert.
8. Visualization: The element will allow the health professional, the helper, and sometimes the patient, to visualize the data collected, the patient information, the analysis rules and outcome, and the action initiated.

While not all PHS or AAL approaches will contain all of the components mentioned above, the indispensable component will be the data collection. Figure 67.6

Part F | 67.2

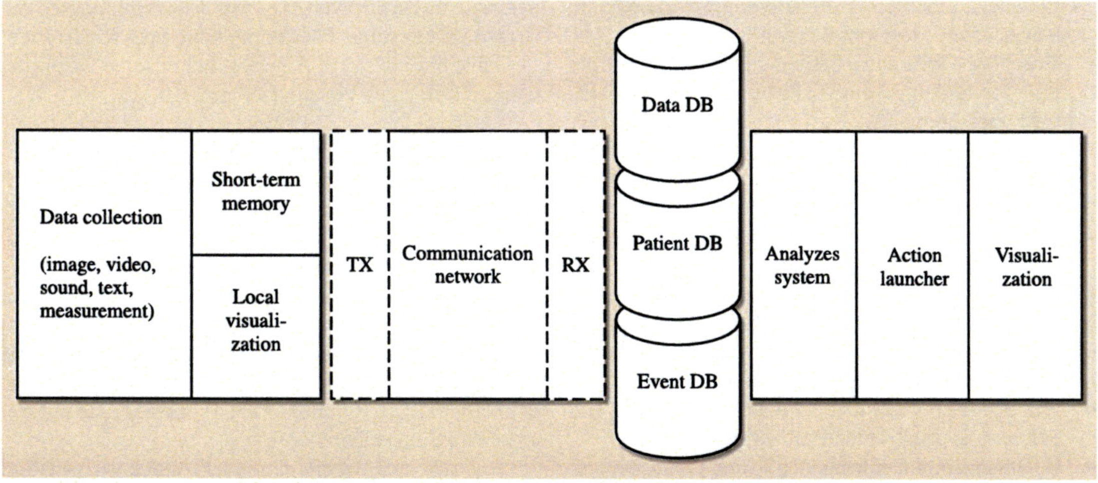

Fig. 67.6 Components of AAL and PHS

presents a visual image of the components of PHS and AAL approaches. In such approaches, the components that are not present as often include the automatic analysis and action launcher.

PHS and AAL approaches can contribute throughout the healthcare delivery process. Moreover, they can intervene in more than one stage of the healthcare delivery process. These solutions may contribute in the prevention, diagnostic, monitoring/surveillance, and treatment/therapy stages.

67.2.1 Self-Management

Self-management tools are devices, instruments, or web sites, normally available to a large number of people, that assist individuals, and sometimes their natural helpers, in healthcare prevention and monitoring.

Health Portals (Prevention – Diagnostics – Monitoring – Treatment)

Description. A health-related portal is a web site that would present information that is perceived by the person to be useful to his or her health prevention or surveillance. There is a large variety of websites, with different levels of information or service quality. Behind the websites will be government agencies, private or public hospitals, universities and research units, associations, companies, patients etc.

The health portals normally have one of the following objectives:

- Information – To facilitate the access to drug, diagnostic, or treatment information; healthcare providers, processes, comparisons; provide patient, helper of GP's testimony.
- Coaching – With some basic input information from the user, to define some risks or recommendations; track measurements.
- Exchange and community creation/reinforcement – To provide a platform where patients and their helpers can exchange on real-time or deported.
- Archiving – To provide a storage space for the person to keep health-related data.
- Reminders – To enable the person or his/her helper to plan reminders such as medication or appointments. In this case, the reminder can be directed to an email, a SMS, or a landline phone.
- Healthcare products – To facilitate the ordering and purchase of drugs or medical equipment.
- Health institution or insurance account management – To facilitate the management and understanding of the person's account situation for his or her insurance or any other organization involved in the healthcare process.

North America is the most developed region, with a strong patient/consumer demand and usage. The European Union has established guidelines for quality healthcare related websites, i. e. transparency, authority, honesty, privacy and data protection, regular information updating, accountability, responsible partnering, editorial policy, and accessibility [67.15].

Certain countries have established a certification to validate the quality of health-related websites; for example, since 2004 the Haute Autorité de Santé (HAS) in France has had the mission to deliver these certificates. This French Agency used the HONcode certification (established by the Health On the Net Foundation, Switzerland) to validate the website. However, there is no correlation between certificated sites and usage; the most-used sites are the ones that are better *referenced* in web search engines, independently of the certifications [67.16].

Examples. In 2007 Microsoft and Google entered the field with the Microsoft HealthVault and Google-Health. The features of each system are listed in Table 67.1.

At the end of 2008 and the beginning 2009, the company User Centric evaluated the feedback from 30 end-users and concluded that Google-Health was more user-friendly (navigation and data-entry) [67.17]. However, certain features such as device data collection is under development with Google Health but already available in HealthVault. The latter is more focused on B-to-B partnerships. For the moment, both services are available only in North America, but in some European countries they will be launched in 2010–2011.

Reminders (Prevention – Treatment)

Description. The outcome of a treatment will depend on the correctness of the diagnosis, the accuracy of the prescription, and finally on the proper follow up of the treatment. The patient's involvement remains very limited in the first two steps; however, the treatment is mostly his or her responsibility. The World Health Organization calculated that only half of the patients with long-term therapies followed their prescriptions correctly [67.18]. Reminders are a simple way to reduce some of the nonadherence.

Another area where reminders are important are at the beginning of neuro-degenerative diseases. Re-

Table 67.1 Comparisons between HealthVault and Google-Health

	HealthVault	Google Health
Collect health information	Manually/import	Manually/import
Archive health information	×	×
Share health records	×	×
Search health information – web	Directly/indirectly	Directly/indirectly
Search health information – associations	Indirectly	Indirectly
Push health information/health authorities	×	×
Enter medications/prescriptions	Manually/import	Manually/import
Purchase your medication	Indirectly	Indirectly
Obtain personalized coaching	Indirectly	Indirectly
Obtain drug interactions		×
Enter test results	Manually/import	Manually/import
Agenda	Indirectly	Indirectly
Reminders	Indirectly	Indirectly
Import files and images	Manually/import	Manually/import
Find a doctor	Indirectly	×
Import data from medical devices	×	
Enter weight manually	×	

Note: The features were not tested, but are declarative

minders can be useful to organize certain tasks, for example, a vocal or image reminder when it is lunch time. Unfortunately, when the disease is more advanced, reminders are not effective. Reminders can be web services that trigger an SMS, video, voice phone call, or LED; they can also be portable devices that read out a recorded or default message.

Examples. There are systems that do not require an additional device but would use the existing communication material of the user (e.g. fixed line phone, mobile, or email). The reminders can be planned in advance and delivered vocally by a call center [67.19] or delivered automatically using a synthetic voice, an SMS, or an email after scheduling from a web application [67.20].

There are other solutions that would require the acquisition of a specific-purpose device. Examples of portable reminder devices are: a watch showing the item to remember in its display (e.g. cadex watch), a pillbox that beeps and vibrates when it is time for a medication (e-pill), and a device that plays out the recorded message at specified time (e.g. memotext).

Heart Rate Monitoring (Prevention – Diagnostics – Monitoring – Treatment)

Description. A heart rate device is a portable system used to measure the heart rate (normally heartbeats per minute, or bpm). It can be visible and transmitted in real-time, or archived and transmitted once connected to a PC or communication gateway. Even though there are around ten body parts where the heart rate can be measured easily, for portability reasons most of the available devices use electrodes attached to a chest strap.

The important technical and commercial development of heart rate monitoring portable devices was done by sportsmen and women who used the measurements to improve their training. However, heart rate monitoring can be used for patients to leverage efforts, to alert in the case of a possible risky situation, or to understand the pattern during a timeframe (a day, a week, ...). It consists of a chest or finger band with an electrode strap and a connector; this band sends the heart rate information to a receiver, normally a watch or a mobile phone (Fig. 67.7). The information can be stocked in the device to be downloaded at a later stage into a PC (it can be downloaded using a cabled or a wireless solution). Once archived in the local PC or a remote server, a more detailed analysis can be performed manually or automatically. Newer models may include an accelerometer, which can help to calculate the distance completed and the mean/maximum speed of the journey.

Examples. The market leader is Polar, with a heart rate measurement using a chest strap. The reception/local archiving and visualization is done through a watch. While most other vendors follow this model for their so-

Fig. 67.7 Example of a finger heart rate measurement (after [67.21])

lutions, there are some that use different measurement points, such as the index. Personal heart rate monitors are priced from 50 to 900 USD.

Activity Monitors (Prevention)

In order to monitor the activity performed in a timeframe (a day, a week), there are activity monitoring devices that track the number of steps done, the calories burned, and the total mileage. An example is the clip by fitbit (launched at end of 2009). It has a USB connector to transmit data to a central server, which then prepares and presents the information on a webpage [67.21]. The webpage has some analysis capabilities. A competing solution tracks and provides more information to the user by using an armband with added sensors (temperature, accelerometer, and heat flux). The product is commercialized by Bodymedia [67.23].

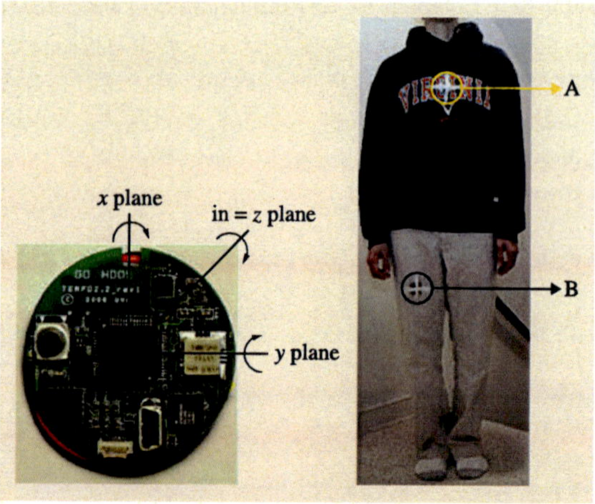

Fig. 67.8 Example of clip activity monitor (after [67.22])

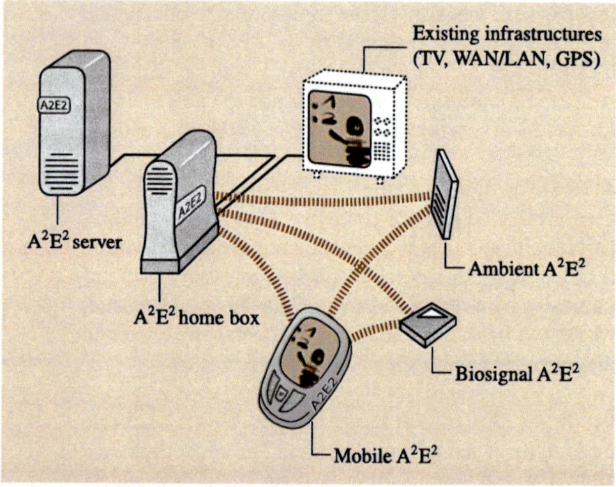

Fig. 67.9 The EU funded project A_2E_2

Motivational Exercising System (Prevention – Monitoring)

Tracking activity or calories burned can sometimes be enough motivation to regularly exercise; however, consumers above 75 years old are not the main customers of activity monitoring systems, and the benefits that can be gained from regular exercise are considerable.

The objective of an EU funded project (A2E2, Fig. 67.9) is to build a system that will induce elderly people to exercise regularly. It tracks the activity performed with ambient and personal sensors; based on the personal profile and the data of those sensors the system will then establish an adapted communication. It consists of a user interface (TV or mobile), a collection and communication system, biosensors, ambient sensors, with the archiving and analysis capabilities in a server [67.24].

Teleassistance (Monitoring)

Description. Teleassistance or home assistance is a service that consists in providing the possibility for the user to reach a person with the appropriate capacities to answer a request or to treat an emergency situation. The National Emergency phone numbers (i.e. 911 in the US and Canada, 112 in Europe) are the most widely used teleassistance service. In the 1970s, associations and companies started to propose a private or associative service that linked the customer to a call centre. It was intended to deal with isolation and emergency. The emergency is usually just a situation definition and a bridge to the National Emergency teams; there are few cases where a medical presence is available 24/7 to perform a first remote diagnostic.

Teleassistance is the most extended commercial AAL offer. In 2008 for example, there were 1.5 million subscriptions in the UK, 600 000 in Spain, and 350 000 in France (as stated by the president of the French Association of Teleassistance Companies, AFRATA, Claude Mordelet [67.25]). The variation is linear to the financial engagement of the public sector (subsidies).

Traditionally, teleassistance is limited to a home support using a communication device connected to the fixed line, with a reach of around 100 m using a portable pendant or bracelet (with a red button to declare an alert). Pushing a button sends an alert to the screen of the call center who will call back the user; the communication device answers automatically in loud-speakers. The new generation of services expands the offering from the following perspectives:

- Service (keep a double of the house key, order a taxi, do a hairdresser appointment, provide the weather forecast...).
- Intelligence (activity detection, GPS, recorded reminders...).
- Mobility (medium or long-range alert and communication...).
- Environment coverage (include fire detection, window or door breaking detection...).

Consequently, assistance and monitoring start to overlap, and this trend will continue in the coming years. An area where commercial remote assistance is just starting is the monitoring and control of medical equipment such as pumps and prostheses.

The typical system components are:

- The alert system – normally portable devices such as the bracelet or pendant, but they can also be fixed, for example a button fixed to the wall of the restrooms
- The communication gateway – which will receive the local alarm and send it to the remote server
- Storage and presentation – the alerts are stored in a database and presented to the appropriate teams for action.

In the case of a mobile alert system, the first two components can be combined in the same device.

Examples. Examples of new generation services include home safety services, which include environment monitoring, and are widely proposed by the West Lothian County [67.26]. Another example available in France is témo [67.27], which uses a mobile alert and communication system with GPS. One of the largest players, Tunstall, started offering environment monitored assistance in addition to the traditional assistance plus some basic telemonitoring services. In Bologna, Italy, a project run by the Department of the INAIL Prosthesis Centre in Vigorso di Budrio provides a good example of remote assistance to tune medical devices based on patient-located data. The project consists in remotely analyzing and controlling an upper-limb prosthesis via a microprocessor controlled arm (MCA) [67.28].

Media and Multimedia Solutions (Prevention – Diagnostics – Monitoring – Treatment)

Description. Images, audio, and videos can be exchanged in real-time or deferred time; the exchange can be patient to health professional(s), health professional to health professional(s), and patient to patient(s). The approach of sharing audio/video/image in healthcare is useful for second opinion processes, better monitoring (e.g. regular wound evolution follow-up), diagnosis (e.g. detailed images for dermatology problems), or even improve treatments (e.g. identify secondary effects that the patient is not aware of, or does not think important to share).

When the video is real-time, it is referred to as videoconference or videoteleconference (VTC). In order to make the exchange operational, it is necessary to have a large bandwidth, otherwise the images would freeze and the sound would not be synchronized to the image. The minimum recommended bandwidth is 384 kbits/s. When there is a high bandwidth, the screen(s) quality is high and the screen(s) size sufficiently large to allow showing the persons and objects in their real size, the term telepresence is used.

Image and audio sharing is more accessible than videoconferences, since it does not require such an important bandwidth, and digital cameras or microphone are accessible to a larger portion of the population in Western countries, with a trend to having a phone equipped with a digital camera, which allows a real-time dispatch from practically anywhere. The quality of the camera or image will depend on the requirement; however, a 10 mega pixels digital camera can be easily found at less than 100 USD, which would have an

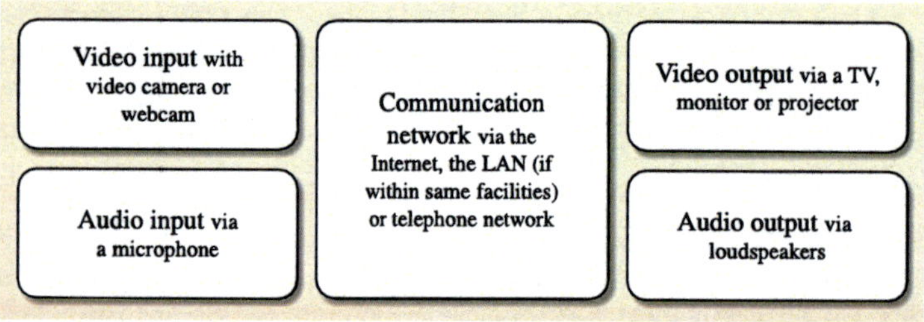

Fig. 67.10 Components of a VTC solution

enough quality for most of the applications. The components of a VTC solution are shown in Fig. 67.10. Image sharing will replace video input/output by an image input, while the audio solution will not have the video input/output.

Examples. In Sherbrooke, Canada, a new service was implemented in 2010, named TASP or *téléassistance en soins de plaies*. At the patient's home, a visiting nurse will film a wound and in parallel, at the Centre Hospitalier Universitaire de Sherbrooke, a nurse receives the video. There is a live exchange between both nurses regarding the progress of the wound healing and the treatment [67.29].

In Argentina, with the commercialized telepresence solutions from Cisco Systems, Dr. Juan P. Garrahan, in the Paediatric Hospital of Buenos Aires uses it for diagnostics, monitoring, and second opinion. Garrahan claims that 85% of consultations do not require a referral to the hospital [67.30] and in many of those cases the telepresence can be an ideal solution considering that Argentina is 3000 km long.

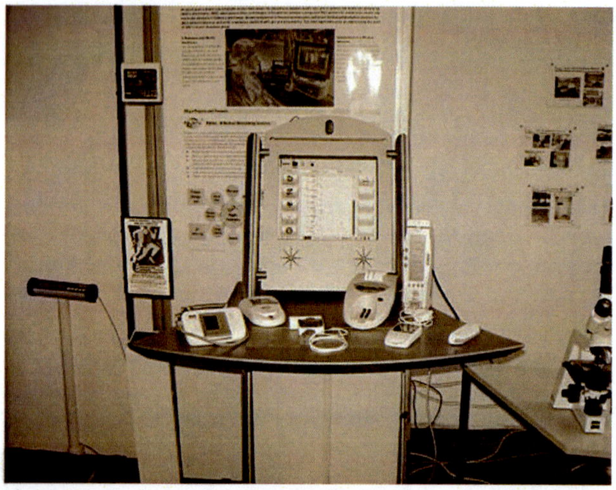

Fig. 67.11 TopCare kiosk (Fraunhofer Institute, St. Ingbert, Germany)

Telemend is a project funded by the EuropAid Cooperation Office. It has the objective of leveraging telemedicine tools to diagnose infectious diseases (malaria) in Colombia. It used TopCare Kiosk [67.31, 32] (Fig. 67.11) in remote areas of Colombia (Buenaventure and Guaviare). The kiosk is a PC equipped with a microscope, a blood analyzer, and an x-ray scan. The kiosk is manipulated by a nurse who will add the picture taken from a blood sample viewed from the microscope to the patient file. The information is then sent via a mobile network to a remote server, which allows any authenticated doctor to access the information in order to provide a diagnosis or monitor treatment.

Fall Detection

Description. Falls are an important problem in all modern societies. A report presented in 2007 in the WHO Global Report on Falls Prevention in Older Age states that 28–35% of people above 65 suffer a fall each year [67.33]; the number increases for seniors above 70 years old, since the percentage increased to 32-42% [67.34]. These falls are cause of hospital admission in 1.6-3.0 per 10 000 population [67.35] (data from Australia, Canada, and the United Kingdom).

Consequently, research and development has been trying to provide solutions to automatically detect falls in elderly people. The objective being to be able to identify premonitory symptoms of a fall before it occurs. Presently, fall detection causes numerous false-positives, and more reliable solutions are limited in their portability and/or cost.

Fall detection uses one or various portable and/or fixed sensor(s) that will identify the movement characteristics of a (several) body part(s), or the behavior pattern (e.g. electrical activity in the person's environment [67.36]). This information is then analyzed locally or remotely to determine the probability of a fall having occurred; if the outcome is a potential fall, then the notification system displays or sends an alarm, i. e. an SMS, email, or phone call.

The most common wearable sensor is a portable three-dimensional accelerometer. Possible environmental devices are video cameras, or presence sensors such as infrared or contact sensors (which will determine, for example, if a person is lying on the floor).

More accurate sensors include the monitoring of other parameters such as body temperature. Using accelerometers, it was also demonstrated that using two sensors positioned in different body areas is more reliable than using a single sensor [67.22], and it was proven that monitoring movements at the waist and head [67.37] provides more reliable information at other parts.

Examples. Vivago has chosen to identify a fall by looking into post-fall symptoms using a watch that monitors wrist movements. A reduction or absence of movement during a predefined time will launch an automatic alert. The EU funded project SIMBAD is an example of environmental solutions. In this project, fall detection is performed using wall-mounted pyroelectric infrared sensors. These sensors capture and analyze the movements without the need for invasive environmental elements such as the video [67.38].

Bioparameter Monitoring (Prevention – Diagnostics – Monitoring – Treatment)

Description. Telemonitoring consists in remotely surveying the health status of a patient who is home or institution-based. In order for the monitoring to be data-based, it requires a medical device to regularly or continuously measure the patient's physiological parameters; it is especially pertinent for people with chronic diseases.

The advantages of home vital parameter monitoring are faster data collection and analysis, avoidance of travel time and costs, integration of the monitoring in the patient's daily routine, and the possibility for the health professional to visualize data over long periods of time.

The vital parameters that can be measured remotely are numerous; the most common ones are weight, glucose levels, body temperature, hcg (human chorionic gonadotropic) for pregnancy tests, cholesterol and triglycerides levels, blood oxygen saturation, blood pressure, and heart rate (ECG). The measurements can be taken continuously, at regular periods, or at random periods depending on the pathology, the patient profile, the sensors, and the medical need.

As mentioned in previous sections, some parameters are already widely monitored in self-management tools; in this section, the interest is within a medical usage. There are parameters that can, in certain cases, be meaningful to the patient, but in most cases the interpretation belongs solely to the health professional.

A solution that monitors physiological parameters will be normally built using the following four components:

- The vital parameter measurement: the patient parameter will be measured using a medical device. The devices will follow one of the three types of procedures: noninvasive procedures (no skin breakage or penetration), minimally invasive procedures (when the device goes under the skin but without open surgical intervention such as subdermal implants), and invasive procedures devices (requiring heavy surgical intervention and from medium to long-term hospitalization).
- Local and remote data collection: the data is collected at the patient location and then transmitted to a distant server to be treated.
- Storage and presentation: while monitoring, it is essential that the data be presented in a meaningful manner to the health professional and that it is archived to obtain history and trends that are useful for the monitoring and treatment of a patient.

The vision of the European Union is a future where patients will have a combination of wearable or implantable multisensors combined with environmental sensors [67.12], but there are still some years of development necessary to achieve this.

Examples.

Glucose Monitoring. For diabetes patients, it is a requirement (it is vital for type 1 diabetes or juvenile diabetes) to measure the glucose concentration in their blood. Presently, meters can undertake a correct reading with only 0.3–1 μl of blood, and provide the reading within 3–5 s, archiving up to 2000 readings. Normally, these devices are sold with software that organizes, presents, and can perform some basic analysis. A key feature is the possibility to share the date with health professionals. An example is the Contour USB glucose meter (Fig. 67.12), which resembles a common USB key.

There is also room for niche adaptations for certain populations. An interesting variation of the product described above was FDA cleared in December 2009. It is a glucose meter for children and teenagers, called DID-

Fig. 67.12 Bayer's Contour USB glucose meter (Bayer Diabetes, Basel, Switzerland)

GET. It communicates with the Nintendo consoles DS and DS light, combining games to the readings. It includes web access to a community and the possibility to be rewarded for the results.

Many current commercial solutions have the advantages of ease-of-portability and ease-of-connection to any PC. However, they are limited to the time of reading, for example, no reading is performed during the night while the patient sleeps. Consequently, for certain patients, there is a need for continuous glucose monitoring systems (CGM). These systems continuously track the glucose level without patient intervention [67.39]. They use a sensor inserted under the skin (normally injected by the patient him or herself, refer to image below). The inserted sensor collects the measurement and communicates it to a local receiver/transmitter that then transfers the data directly or via a wireless terminal (mobile phone). The visible data can be sampled every 1–5 min.

Presently, when using a CGMs, the patient needs to replace the injected sensor every 5 to 7 days. Additionally, their accuracy has not yet reached that of traditional blood testing readers, needing calibration with a traditional blood strap reader. The CGM products are Abbott's FreeStyle Navigator, DexCom's Seven+Plus (Fig. 67.13), and Medtronic's Guardian.

Some examples of noninvasive CGM exist, e.g. a watch that uses low electric current to pull glucose through the skin (GlucoWatch G2 Biographer, by Cygnus). However, the FDA was not fully satisfied since it caused skin irritation in 50% of patients and the readings for low glucose levels were not precise. The product is no longer commercialized but developments around it continue [67.40].

Heart Rate Data Monitoring. Heart rate monitoring has been done for many years, i.e., measuring the heart rate or the more detailed ECG. Due to electrodes, microelectronics, and communications developments, the once cumbersome heavy devices that the patient carried for couple of hours while the device stored the data, are

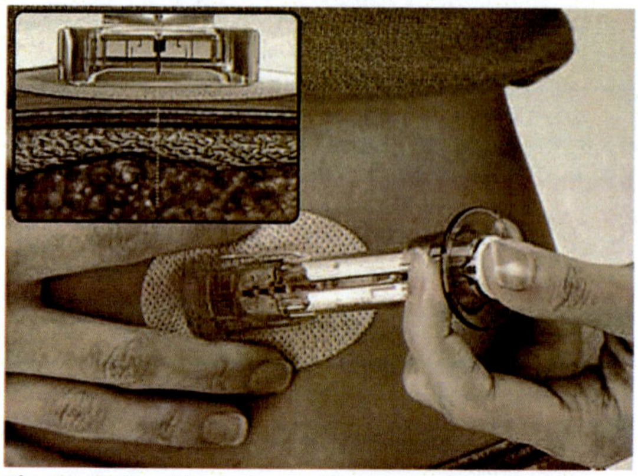

Fig. 67.13 DexCom's Seven+Plus sensor insertion (source: DexCom, San Diego, http://www.dexcom.com)

being replaced by real-time highly-portable solutions, some with basic automatic analysis. There are many advantages of monitoring real-time heart activity, such as the immediate detection of an issue, the monitoring of the outcome of treatment, and the efficient collection of information for research purposes.

The efficiency has also been clearly observed: a European trial concentrating in heart diseases was performed with patients leaving the hospital after a stay longer than 48 hours. Three groups were compared: a group receiving traditional care, a group receiving nurse phone support, and finally a group using a telemonitoring system. This system included a pressure, weight, and ECG home monitoring device. The data collected from these devices was sent via Bluetooth to a hub that used the fixed line to transmit the information to a server. The server presented the data to a nurse or technician (care manager) who escalated, if required, to a physician. This study lasted 2 years and accepted 426 patients. The study conclusion was that the nurse service group and the telemedicine group had a lower level of mortality (16–18%) among patients compared to the status-quo care [67.41].

A new stage on portability is the integration of electrodes into textiles (smart textiles) [67.42], e.g. the VitalJacket (for up to 5 d monitoring and 72 h battery autonomy) commercialized by Biodevices (a spin-off of Aveiro University) [67.43]. The Fraunhofer Institute has developed a smart textile ECG monitor using a proprietary mobile terminal and electrodes claimed to be flexible, biocompatible, and biostable [67.44].

Smart textiles facilitate the portability, however, implanted solutions are the next step. At the end of 2007 an implanted ECG monitoring system received FDA clear-

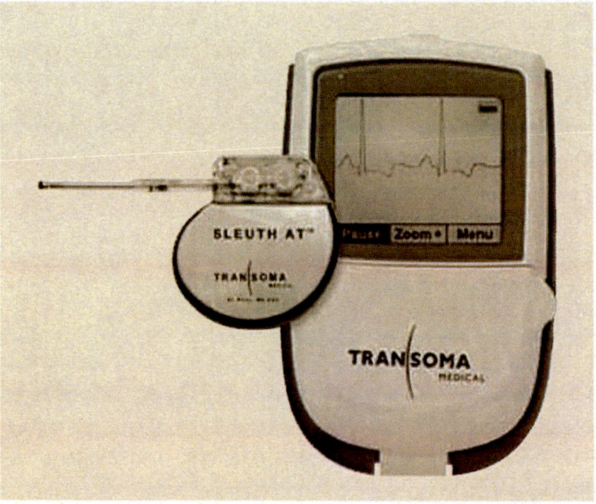

Fig. 67.14 The Sleuth implanted sensor and the handheld wireless receiver (after [67.45])

ance. The Sleuth ECG monitoring system manufactured by Transoma Medical (Fig. 67.14) was first implanted in 2008 with US patients for arrhythmia detection or monitoring; the implant done by Dr. Daoud and Dr. Hummel at The Ohio State University Medical Center in Columbus. The information was regularly collected and sent to a monitoring center where cardiac technicians filtered information before sending reports to the health professionals.

Dr. Daoud stated that [67.47]:

> *The Sleuth implantable recorder and the wireless communication system provide excellent recording fidelity and ease of transmission of ECG data that is continuously monitored and stored. Also, I can select the arrhythmia detection parameters, and the telemetry tracings are promptly sent in a user-friendly report, making it easy for me and my staff to care for our patients.*

Transoma Medical is, unfortunately, an example of the lack of synchronization between covering healthcare needs and finding a workable business model, despite the success and medical acceptance of their solution, in December 2009, Transoma ceased operations *due to the difficult market conditions* [67.45].

Despite the shown benefits, there are still a few large deployments, but some regions are more active. In Brussels, for example, 500 patients are participating in the *Belgium Heart Failure Project* [67.46] to reduce heart failure spleen mortality. Instead of using ECGs, the patients are provided with a tensiometer and their mobile phone for an SMS communication.

Sleep Apnoea Detection. HealthGear is a system tested to monitor sleep apnoea using a set of sensors communicating the data via Bluetooth to a smart mobile phone that archives, transmits, presents, and performs a basic analysis of the data received [67.48]. The sensor used for sleep apnoea detection is an oximeter for blood oxygen levels and pulse; however, the architecture could support multiple sensors. It was tested with 20 individuals, with very positive outcomes regarding usability; 100% participants declared themselves willing to use it and to recommend it, and the accuracy of the readings and analysis correctly identified cases of obstructive sleep apnoea (OSA).

Others. There are portable solutions that monitor different vital parameters; however their capabilities have not been used in full. A commercially available application is Equivital, with a two-lead ECG, temperature, and movement sensors. It monitors the heart rate, the ECG, the respiratory rate, body temperature, and movement/body orientation (Fig. 67.16). It uses a chest strap that collects information and uses bluetooth to send the data to a local PC or a gateway that then sends it to a remote server. This multisensor device is mainly used by military personnel and professional sportsmen, but its application in the medical area is still underdeveloped.

Auto Adaptive Treatment Systems (Diagnostics – Treatment)

Descriptions. Auto adaptive treatment occurs when a treatment is capable of changing itself based on real-

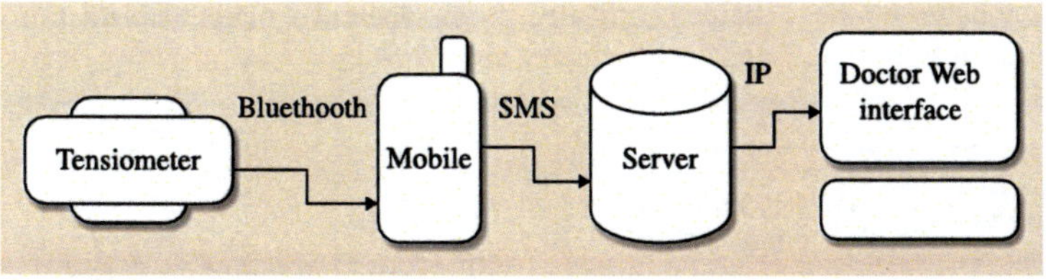

Fig. 67.15 The Belgium heart failure project (after [67.46])

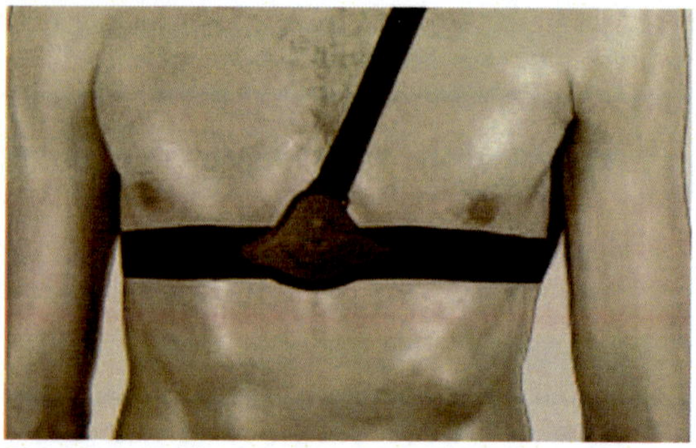

Fig. 67.16 Multiparameter band strap (after [67.49])

time or delayed information. Note: teletherapy is used to define radiation that is performed at a distance from the body.

Once a patient has been diagnosed, the health professional will determine the most appropriate treatment. The determined treatment and/or therapy may be adjusted in light of the patient response, their efficacy, and secondary and adverse effects. Monitoring of ongoing treatments or therapies and patient response are valuable for adjusting treatments and may even be crucial in cases of extreme medical condition or disease. For example, after a quadruple by-pass surgery, round the clock monitoring of the patient may be essential to ensure that there are no complications, especially when the patient is discharged from hospital.

The ideal case is a continuous monitoring feeding and modifying the treatment as needed, and the whole real-time. Typically, for example, adaptive treatment would enable patients with Type 1 diabetes to have insulin injections based on real-time continuous measurement of their blood glucose levels. Another example is having pumps for patients following cancer therapy, adapting the dose and time of medication based on the real physiological patient needs.

Figure 67.17 illustrates the common elements of auto adaptive treatment systems. Namely, a sensor(s) that collects the vital parameter(s), the analysis algorithms that will define the preferred action (reduction/increase/pause of dose or example), and the delivery treatment device that will behave based on the analysis outcome. If the analysis is done at a distant site,

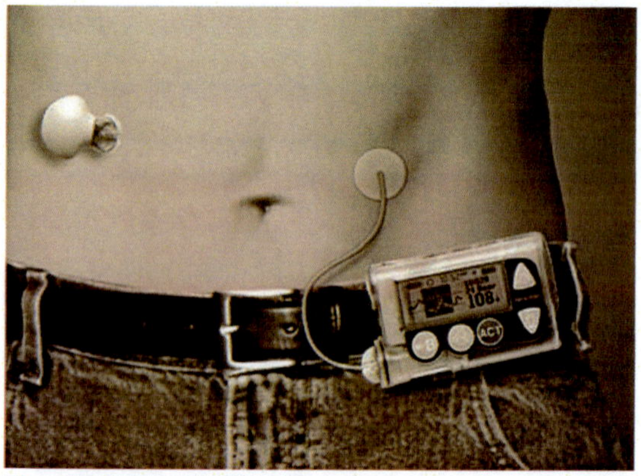

Fig. 67.18 The MiniMed Paradigm real-time system (after [67.50])

an additional component will be required: the communication channel with its transmitter/receiver both at the patient side and at the remote side.

Examples.

Diabetes. An automatic system for Type 1 diabetes patients is commercially available. This is the MiniMed Paradigm Real-Time system (Fig. 67.18), which consists of a continuous glucose meter (CGM) that monitors the glucose level. With this information the controlling device adjusts the delivery of insulin of the insulin pump [67.39].

Depression. A very different type of treatment relates to psychiatry with depressive patients. The adapted treatment was not the medication but the psychiatric sessions that were performed remotely. A comparative study was done for 6 months to evaluate the outcome of on-site treatment versus telepsychiatry (videoconference between the patient and the health professional). During 6 months, 119 depressed veterans followed either the traditional psychiatry follow-up or remote treatment. The treatment consisted of psychotropic medication, education, and short supportive counseling. The conclusion of this study was that remote treatment of depression using telepsychiatry compared with face-to-face treatment is equivalent from adherence, patient satisfaction, and costs perspectives [67.51].

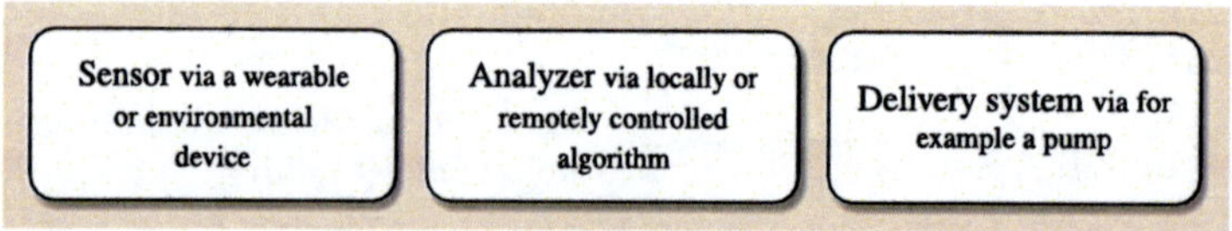

Fig. 67.17 Elements of an adaptive treatments system

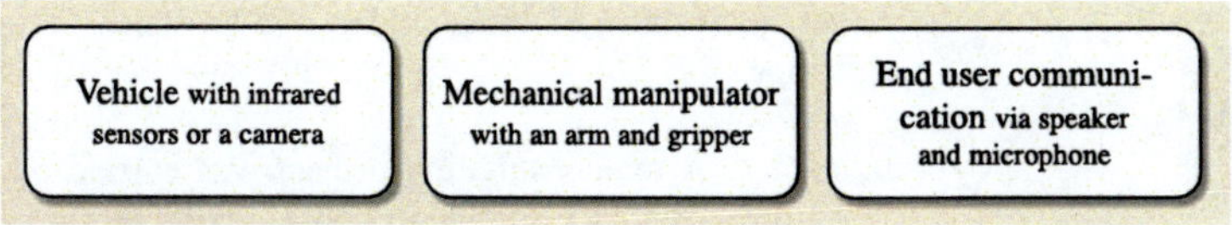

Fig. 67.19 Elements of a nurse robot

Wound Management. There are still but few examples of solutions; however different research groups are working around subjects that directly and indirectly will lead to solutions. The University of Rochester in the US is developing smart bandages with optical sensors that can monitor a change in color [67.53]. In the future, we expect to see the information collected, transmitted, and analyzed remotely, opening the possibility to automatically adapt treatments.

Robotics (Prevention – Monitoring)

Description. This approach consists of the use of robots to help perform certain basic caring actions at the patient's home, at institutions, and at hospitals. Robots are machines that operate autonomously, repetitive, programmable, or adapted tasks. Robots can do repetitive tasks (reminders, medication delivery, cleaning), specialized activities (move [67.54] or guide [67.55] a patient), interactive actions (keep a conversation), or be the communication gateway with health professionals. Most of the study outcomes show patient acceptance [67.55, 56]. In hospitals and institutions, robots help partially solve the issue of the lack of enough appropriate resources.

Nurse robots are normally mobile machines requiring a vehicle to move around the house or hospital without bumping into fixed or mobile objects (Fig. 67.19). Another key component is the mechanical manipulators, for example, an arm to deliver a drug. Finally, the end-user communication element permits us to configure or give orders to the robot, or simply, to enter information for an exchange with the robot (e.g. for a conversation).

The robot is preprogrammed with various levels of adaptation based on:

- External actions
- External inputs
- History.

The country most advanced in the area is Japan, with the technological background and one of the oldest populations.

Examples. A joint project between the University of Pittsburgh and Carnegie Mellon University has created a robot nurse for elderly people. The project's name is NurseBot, and the first robot is Florence [67.57] (in honor of Florence Nightingale). Its tasks are:

- Medication and regular life reminders (e.g. drink, go to the toilet)
- Keep company via conversation
- Provide a camera, speakers, and a microphone to allow remote health professionals to see and talk to the person
- Simple domestic appliance manipulations, such as open or close the fridge or the microwave.

Samsung has developed Romi [67.52] (Fig. 67.20), a nurse robot with the participation of the nursing school and the psychology departments of the University of Auckland (for the nurse knowledge base). Romi is polyglot (eight languages) and has a camera for face recognition, a microphone for voice recognition, arms and hands, as well as speakers.

The tasks are to:

- Deliver prescribed drugs
- Remind the patient of routines
- Motivate
- Keep a conversation.

Toyota has announced that their nurse robot will be commercialized in 2011, and Honda has spent hundreds of millions of dollars developing its human-like robot. Those are very promising news to see, in a near future, efficient and hopefully cost-efficient nurse robots in the market.

Fig. 67.20 Romi nurse robot (after [67.52])

Fig. 67.21 European project IWARD (intelligent robot swarm for attendance, recognition, cleaning, and delivery) (after [67.58])

The European project IWARD (intelligent robot swarm for attendance, recognition, cleaning, and delivery, Fig. 67.21) is targeted to providing help to institutions, clinics, and hospitals to overcome staff shortages [67.58]. The task is:

- Delivery of objects from point A to point B
- Cleaning
- Patient guidance within the establishment
- Detection of patients in unusual positions, e.g. patients sitting on the floor.

Teleconsultation (Prevention – Diagnostics –Treatment)

Description. For remote or rural areas where health professionals and specialists are not available, the possibility to perform a consultation with a remote expert is essential for the patient (he or she obtains a faster diagnosis and, if needed, treatment) and for society (to provide equal health access to all individuals while managing the health budget).

Another scenario where teleconsultation is important is during the second opinion procedure, since it normally involves two persons who are physically distant. The number of multipathologies patients is increasing with chronic diseases and ageing populations; consequently, different medical specialties require provision of an input to the diagnosis and treatment in order to treat the individual as a whole and not just an organ. The teleconsultation can consist of a phone call, an email exchange, or a chat. However, normally it involves some kind of image/video and/or sensor data exchange.

Examples. A teleconsultation project was carried out in Aarhus, Denmark (collaboration between the local hospital and university), for diabetic patients suffering from foot ulcers. The regular treatment was once a day or every other day. Moreover, twice a month the nurse organizes a remote consultation with the hospital doctor using a visio-phone. The experimentation was in 2005 and at the end of it the hospital decided to integrate the procedure to its regular operations [67.60].

A pan-European project was launched in 2005 with the name Medical Care Continuity project. The objective was to improve the follow-up of patients following the discharge from hospital. It uses a video camera that can be remotely directed by the health professional. Italy claims that these types of initiatives have reduced hospital stays by 22 days and cut hospitalization costs in half [67.61]. Maybe the Italians overestimate the benefits, but the gain for patients and hospitals is certain.

It has been seen that there are not many AAL and PHS commercial off-the-shelf products; this is changing. In 2008, Intel's Health Guide home terminal received FDA approval; it was designed for chronically ill patients to communicate with their health professional (Fig. 67.22). It has:

- Two-way video capabilities
- Interfaces with sensors (e.g. heart rate, glucose meter, oximeter) to capture the information and

Fig. 67.22 Intel's Health Guide terminal PHS6000 with tactile screen (after [67.59])

communicate to the health professional via a centralized server
- A calendar with potential reminders
- A collection of image, sound, or video education material
- A tactile screen.

It has a convenient size of 280 mm × 90 mm × 270 mm, however, it requires a broadband network connection such as ADSL, which limits the usage to home use. This terminal can be used for consultations and monitoring. A study with 200 patients above 60 started in 2010, in collaboration with General Electric and the Mayo Clinic to measure the reduction of hospitalization and emergency entries [67.59].

Telesurgical Assistance (Treatment)

Description. Telesurgical assistance is the action of performing a surgery locally with partial or total remote input. The remote involvement may consist only of a verbal opinion based on the images or sounds received, but it can go to fully controlling the surgery. Telesurgery is presently not used in AAL since it requires a setup and environment that for the moment is only possible at hospitals. However, as technology evolves, we believe that simple minimally-invasive interventions will be performed at home with the help of a trained nurse on site and experts at a remote site.

Components. An Austrian study demonstrated that remote-controlled navigation arthroscopies are reliable and accurate even for small joints (the temporomandibular joint, a facial joint) [67.62]. The same outcome was observed with a comparative local and remote intervention of 50 patients (Nissen fundoplication). The mortality was nil for both groups, and a 22 month follow-up showed comparable symptoms. However, the costs doubled for the remotely operated group with a longer time required (40 min) [67.63]. The costs and time would decrease as more remote interventions are done.

Smart Homes (Monitoring– Prevention – Diagnostics –Treatment)

The smart home relates to the home that is capable of quickly detecting a problem and even of foreseeing it; the home that will react in the most appropriate manner to the event raised; the home that palliates the physical and cognitive weakness of the elderly inhabitant using technological solutions.

The objective is an equipped home that will maintain or increase the physical and social autonomy of the individual living at home. There is no a one-size-fits-all smart home, since it needs to adapt to the specific weaknesses and environment of the person. The smart home would be equipped with a mix of the solutions presented in this chapter, plus the services and other technological solutions related to the nonhealthcare aspects such as entertainment (TV, Audio), education (TV, ...), security (closing and opening of doors/windows; intrusion detection and alert). The use of a single AAL service or tool adds a certain intelligence to the person's environment; the term smart home does not define a solution or approach but rather an attribute.

Despite the existence of standards, such as short-distance wireless communication protocols (such as zigbee, RF, or bluetooth), or standards for medical imaging (DICOM), the immediate and transparent interconnectivity of all needed components is not yet possible. There are associations such as Continua Alliance, IEEE, and mHealth Alliance working around this subject, and consequently, it can be envisaged that in the medium-term, the interconnectivity will be a pure plug-and-use.

Examples of smart-homes projects are Monami [67.64] and i2home [67.65]. The concept of the smart home was also used in a post-operative situation, where the physiotherapeutic exercises were automatically evaluated and the patient could receive feedback to ensure regularity and quality [67.66].

A subset of a smart home is the automatic adaptive environments. There are still very few environments that adapt themselves based on real-time healthcare personal parameters. There are some solutions that allow remote control of electrical or electro-mechanical devices; the control can be manually set or programmed based on the outcome of environmental sensors such as thermometers, light sensors, or fire detection (e.g. Biodom system).

The elements of an adaptive environment will be a sensor that captures health parameters, an analyzer that takes the information, determines the ideal configuration of the environmental device, and controls the device. It will, for example, adapt the heater to the body temperature of the person or the ventilation to the oximeter outcome.

67.3 Benefits and Challenges Ahead

Health and healthcare systems are complex interactions of biomedical, social, economic, ethical, and political factors. Significant advances in information communication technologies (ICTs) have made, and will continue to make, innovation in health technologies and equipment possible. ICTs have begun to positively impact the manner in which health systems provide products and services to patients, including through the development of the field of e-Health and through improved access and quality. More importantly, they hold greater promise for future improvements in health and healthcare systems. However, for e-Health approaches to provide viable and effective solutions to many of the current issues facing healthcare systems around the world, many conditions will need to fostered and brought together.

The future development and more pervasive adoption of e-Health approaches will depend on identifying where the benefits lie and the numerous challenges and issues that these technologies raise, as well as finding appropriate solutions. Some of the issues that are raised by e-Health technologies and equipment are common to other areas in the health field. For example, issues pertaining to the protection of patient privacy and confidentiality or informed consent are issues that arise within the provision of healthcare services generally. However, different facets of these issues may arise within the e-Health context. Other challenges are specific to the technologies used in and the services provided within the e-Health context and even more specific to AAL and PHS. For example, the reliability of the technologies and the consequences when they fail or the uncertainties pertaining to the legal framework are some of the issues more specific to this emerging field. Other challenges are themselves only beginning to emerge or will emerge in the future with the more widespread use of these technologies.

This section will draw attention to the numerous issues that legislators, political decision-makers, medical professionals, industry, engineering and systems design experts, patients, and civil society will need to consider and address to ensure the appropriate development, adoption, and usage of e-Health technologies and equipment.It will provide an overview of the different social, ethical, legal, societal, economic, and technological dimensions to the roll-out of e-Health solutions.

67.3.1 The Social Dimension

AAL and PHS, and e-Health more broadly, contribute to the achievement of many desired social objectives of improving healthcare services and products for patients and to improving the delivery of such healthcare [67.5]. From a patient perspective, such approaches may enable individuals to have better access to health information, to have better knowledge in regards to their health situation, to better manage their health, to have improved feedback on the impact of health services and interventions, and to make more informed choices about treatments and therapies.

From a health professional and healthcare system perspective, such approaches may enable health workers to make more effective decisions with respect to diagnosis, treatment, therapy, and monitoring. They may enable hospitals to provide higher quality and safer treatments and therapies. They may enable health professionals to contribute to addressing health problems of patients world-wide. They may improve the management of information and access to information for patient care and for administrative services. They may enable better and more timely mapping of public health threats, and they may enable improved training and sharing of knowledge along health professionals.

From a government and policymaker perspective, such approaches may enable governments to be more responsive to changing health needs. They may enable policymakers and public health authorities to be more aware of health risks, and they may foster the development of effective, safe, and efficient healthcare systems.

Stakeholders

The effective, safe, and efficient roll-out of e-Health solutions, such as those provided by AAL and PHS, will need to ensure that the appropriate people at the national, regional, and international levels are involved as early as possible. As such, it will be important for national and regional governments, all categories of health professionals, engineers, and technicians, patient and representative groups, industry, researchers and research organizations, international organizations, NGOs, and civil society to be implicated. Their diverse expertise, knowledge, technological know-how, partnerships, and investments will be essential for developing e-Health globally.

Developments and innovation within ICTs for e-Health and AAL will necessitate more effective involvement of SMEs. As agile entities they are able to make technical and business decisions more quickly and to execute them. Thus, they are better able to focus their research capacity. Well-financed SMEs are able to engage in high-risk, early-stage research and development, build strategic partnerships, and operate outside their local markets. As such, they have an important role in the development of future visions and in converting these into value propositions for patients and clients.

Participatory Involvement

Future innovation within e-Health will include engineering of context-aware and easy-to-use ICT systems that self-improve and self-adapt within their respective environments. Areas of research will include cognitive systems, interaction, robotics, and automation. For the innovations to be developed to be effective, efficient, and useful for the end-user, whether patient or caregiver, the participation of key stakeholders will be crucial. The development of technologies and the evolution of products and services will need to take into account user experience and future needs.

Carried out appropriately, stakeholder participation may be cost-effective, efficient, and eco-responsible in that the selection of technologies to be developed would reflect the needs of patients and their caregivers. Conversely, strategies that are not embedded in clear and realistic needs will be more vulnerable to failure due to lack of participation, acceptance, capacity, and other enabling factors.

There will be many challenges in carrying out stakeholder participation. Such testing will need to consider the breadth of possible users and that of possible uses. It will further need to consider the different conditions under which the technology should be used, the best manner in which to involve the testers and to simulate real world conditions and use, and the manner in which this testing respects legal and ethical frameworks. In testing AAL and PHS approaches, researchers will need to involve participants of the targeted patients. For example, elderly individuals, incapacitated individuals, or individuals with cognitive, motor, sensory, or neurological needs. As the testing will also need to assess usability, models developed involving healthy populations may not be reliable [67.67].

67.3.2 Legal and Ethical Dimensions

Existing medical and health sectors are amongst the most regulated and are covered by national, regional, and international legal instruments. Similarly, e-Health technologies are subject to the myriad of currently-existing national, regional, and international legal instruments. However, there is also a need for new policy, legislation, and regulation specific to address issues arising from e-Health technologies. Some areas where policy and legislation need to be considered include ethical considerations, privacy, confidentiality and security, and cross-border issues.

Research, testing, development, and distribution of products and services within the field of e-Health must abide by the ethical and legal frameworks generally applicable to human subjects, medical products, and technologies. While this section will review the legal and ethical issues, it is beyond the scope of this chapter to exhaustively examine such issues or to review all of the applicable legislation In addition, many of the technologies will be used for or by individuals who are in a vulnerable state. For example, the technologies may be used for elderly individuals, for individuals with a specific medical condition or disease, or for individuals with motor, cognitive, or neurological needs. These patients may be more vulnerable and thus extra care and attention is needed to ensure the protection of such patients.

Within the principles applicable to human subjects, there is the notion of minimal risk. Minimal risk has been defined to mean that the probability and magnitude of harm or discomfort anticipated in the research is not greater in and of themselves than those ordinarily encountered in daily life or during the performance of routine physical or psychological examinations or tests [67.68]. This principle entreats that patients should be exposed to minimal risk. Where it is not possible to abide by the principle of minimal risk, consideration must be given to ways in which risk can be reduced or eliminated or, alternatively, consideration must be given to employing alternative approaches.

Within the medical and health context, the cornerstone of patient protection is informed consent. This principle is recognized in international legal instruments such as the UNESCO *Universal Declaration on Bioethics and Human Rights* (2005) and the *Declaration of Helsinki on Ethical Principles for Medical Research Involving Human Subjects* of 1964 (last revised in 2008). Informed consent is a complex subject rooted in the principle that patients should be provided with pertinent information by health professionals, especially in regards to the significant risks of a given service, product or technology, as well as of the risks of particular importance to that patient. The objective is to provide useful information to enable patients to make the best decision in regards to their health and well-being. As e-Health systems become more widely implemented and are able to generate improved health and medical information specific to a patient, this will facilitate the informed consent process. Conversely, the potential use of personal health information for numerous and diverse research purposes may make the informed consent process more difficult. As personal health information may be used for broad and diverse research purposes, obtaining the type of specific consent used in most jurisdictions may not be sufficient. Moreover, the issue of secondary use of personal health information and records also challenges the current informed consent approach. Consideration has arisen of whether there may be the need to develop alternative models of consent for such broad and diverse research [67.69].

Policy decision makers and legislators should also address the issue of the possibility of discrimination and stigmatization of patients and patient populations based on medical findings and discoveries. As personal health information is increasingly centralized and completed by medical history and family medical history, a fuller picture of the current and possible future medical conditions of patients may emerge. With this fuller set of personal health information coupled with the constantly emerging medical and health findings, there may be increased risk that patients sharing disease characteristics may be the subject of discrimination or stigmatization, including from employers, insurance companies, and the general community [67.70]. In the development of e-Health strategies, addressing such issues will by key to maintaining public trust.

Privacy, Confidentiality, and Security

AAL and PHS, and e-Health, technologies will generate personal health information and data that will be stored within databases. The term database is used as a generic, catch-all term. Other terms that are used to designate such databases include *health information banks*. This collection and storage can cover quite a large definition of personal health information and data. It can include data and information generated directly by AAL and PHS technologies. It can also include information and data retrieved or derived from other health and personal records and linked or incorporated within the databases. The aggregation of significant personal health information into one or more databases increases the risks of violation of privacy, confidentiality, and raises issues of data protection and security.

This protection of privacy and confidentiality and ensuring the security of data and information will become more pressing when cross-border exchange of personal health information will be common place. Patients are increasingly seeking medical treatments and therapies outside of their country of residence. In order to provide the most appropriate treatment or therapy, health professionals will seek access to the patient's personal health information. In addition to the technological challenges, such as interoperability, numerous legal and ethical challenges will also arise. For example, when there is divergence between the laws of the different jurisdictions, issues of predominance of law and private international law/conflict of laws will arise.

In order to protect patients and maintain citizen's trust, policy-decision makers and legislators should develop the necessary legal and regulatory framework specific to the e-Health context. While some jurisdictions have already enacted legislation adapted to the specific e-Health context, thereby ensuring accrued protection of the privacy of its citizens, many others [67.5] lag behind and must rely on existing legal framework. Other European countries are in similar situations; many developing countries have yet to adopt legislation specific to e-Health. The province of British Columbia adopted the *Personal Health Information Access and Protection of Privacy* in April 2008. British Columbia is the first province in Canada to create a specific legislative framework governing access and privacy for electronic health information databases [67.71]. Some key aspects for such legislation to address include ensuring that data and information are accorded the highest level of protection, ensuring that measures are in place dealing with unforeseen and unauthorized disclosure or breach of security measures, and ensuring that unauthorized access or disclosure is significantly dissuaded and sanctioned.

Some elements for consideration in the development of principles applicable within the e-Health context include the following:

- Measures should be undertaken to ensure the respect of the individual's rights and freedom.
- Patients should be provided with detailed information of the nature of each e-Health solution/device/application prescribed for their treat-

ment and use, the inconveniences, the type of information and data that will be collected, and the diverse possible uses. The patient's consent should be obtained for the use of the solution/device/application, the collection of the data and information, and its possible uses.

- Patients should be informed and be provided with the opportunity to consent in regards to which information and data will be collected from them and which information and data will be retrieved and derived from their other health and personal records. This information should also specify the manner in which these information and data will be stored and accessed (i. e., text, audio, image, video). Patients should also be provided with the opportunity to specify which type of data and information they do not wish collected or stored.
- Patients should be informed about the uses for which their personal health information will be employed and by which entities. Patients should be provided with the opportunity to consent to the type of entities who should or should not have access to their personal health information.
- Patients should be informed about how they can access their personal health information.
- Personal health information should be available in the official languages.
- Security measures should be put in place to ensure that inappropriate and unauthorized access to patients' information and data are thwarted. Clear accountability for the storage, use, transmission, and maintenance of the information and data should be established.
- Mechanisms should be put in place for recording patients' consents and for enabling patients to subsequently modify or withdraw their previous consents.

67.3.3 Societal and Global Dimensions

The provision of effective, safe, and efficient healthcare services and products is a concern facing governments worldwide. While the specific types of concern may vary from jurisdiction to jurisdiction, many concerns cut across borders and continents. For instance, the issue of ensuring the financial sustainability of the healthcare system is a challenge that all governments face and in the coming years must address. Similarly, the promises of e-Health technologies can also benefit each jurisdiction whether developed or developing, albeit in different manners. For example, a doctor in a distant community may use teleconsultation and videoconferencing to confer with a colleague on another continent in regards to the health of a patient. The doctor and patient may be in a distant community within a developed country, such as Iqaluit in the far north of Canada, or within a developing country, such as a remote village in the Democratic Republic of Congo.

E-Health technologies and equipment can lead to a positive impact on healthcare. They can contribute to improved prevention, diagnosis, consultation, treatment, cure, and monitoring and surveillance of patients both locally, remotely, and in distant locations. Moreover, they can contribute to improving healthcare through broad and improved dissemination of public health information to citizens, facilitating the sharing of information and training among health workers. For example, the Algerian government established the *Health Algeria* network, a tool for data collection and exchange among the different actors in the health sector. It has been evaluated as one of the most effective actions for creating the environment and backbone necessary for implementing e-Health, as well as for the adoption of AAL and PHS approaches [67.72]. Improved monitoring of the incidence of public health threats and more rapid, targeted, effective, and timely responses can also be achieved through such technologies and equipment. E-Health technologies can facilitate self-study, eLearning and further professional development, as well as foster more effective and targeted research and the dissemination and access to research findings [67.73].

For e-Health to become a positive and effective force in ensuring access to healthcare around the globe, each jurisdiction must contribute to achieving common goals and shared values. Public policy must support effective, safe, efficient, and equitable e-Health systems. They should foster the use of collaborative approaches for the development of e-Health systems. They should foster the development of e-Health technologies and approaches that are supportive of local cultures and languages and that are inclusive of all communities and especially vulnerable groups.

The viability and utility of e-Health in developing countries or in least developed countries is dependent on the accessibility of ICT and health technologies and equipment. As Table 67.2 illustrates, access to needed ICT equipment and technologies is very low in developing and least developed economies. Thus, for developing and least developed economies to benefit more fully from e-Health, many more challenges will need to be addressed. Moreover, due to market distribution and market power, developing and least developed

Table 67.2 ICT infrastructure and access: core indicators, aggregate values (after [67.74])

Level of development	Number per 100 inhabitants					A6. International Internet bandwidth per inhabitant (bits)	A7. Percentage of population covered by mobile cellular telephony
	A1. Fixed telephone lines	A2. Mobile cellular telephone subscribers	A3. Computers	A4. Internet subscribers	A5. Broadband Internet subscribers		
Developed economies	51	92	62	24	19	4755	99
Transition economies	23	77	10	3	2	223	88
Developing economies	15	33	5	4	2	177	74
Least developed economies	0.9	10	0.7	0.2	0.0	7	59

economies face constraints such as insufficient telecommunications infrastructure, high telecommunications tariffs, inappropriate or weak policies, organizational inefficiency, lack of locally created content, and uneven ability to derive economic and social benefits from information-intensive activities [67.74].

Addressing these challenges should be an objective for the global community. Approaches that are available to a doctor in the Canadian far north should also be available to a rural doctor in the DR Congo. AAL and PHS technologies can bring needed healthcare services to the poorest populations located in the most remote, rural, and difficult environments. Equally, these technologies can relay news of health developments from the most remote and difficult to reach places to the rest of the world. For example, through teleassistance, health workers in remote areas in Uganda are able to immediately report incidences of communicable diseases such as cholera [67.74].

67.3.4 Economic Dimensions

The aim of e-Health is to contribute to providing effective, safe, efficient, and equitable healthcare. Within this context, AAL and PHS aim to support the decentralization of the provision of healthcare from an institutional setting to a home environment. These approaches provide some solutions for the elderly, incapacitated individuals, individuals recovering from a condition, surgery or disease, or individuals with cognitive, motor or neurological needs. These different segments will represent an increasing percentage of the population over the coming years.

Numerous pressure points militate for the rapid development of AAL and PHS technologies. The ageing of the population, the increased burden of chronic diseases, increasing longevity coupled with rapidly increasing costs all necessitate the reconsideration of the approaches used within healthcare systems. In 2005, 133 million Americans – almost 1 out of every 2 adults – had at least one chronic illness [67.75]. Each year, 7 out of 10 deaths in the United States are from chronic diseases. Heart disease, cancer and stroke account for more than 50% of all deaths each year [67.76]. In Europe, chronic conditions account for up to 86% of all deaths [67.77]. This disease burden coupled with the factor that many patients live longer means that patients will require more diverse services and support from healthcare system.

Increasingly, with a cost-containment objective, patients are being discharged from a hospital context much earlier once their condition has stabilized. Thus, the patient recovery or convalescence is shifting increasingly to a home environment. In the United States, for example, the expenditures for home healthcare rose from 12.6 billion USD in 1990 to 59 billion USD in 2007 [67.78]. As the number of patients being cared for within a home environment grows, there is an increasing need for e-Health approaches to be developed and in place to ensure the delivery of effective, safe, efficient, and equitable healthcare.

In parallel, with a system that is not able to provide all of the required services coupled with the falling number of health professionals per patient, increasingly, we are witnessing the emergence of the phenomenon of family members and other persons becoming caregivers and providing the necessary services. Such situations have led to an increase in emotional, physical, and financial stress, which in turn has often resulted in health problems for the caregivers. For example, a US National study reported that 15% of caregivers came within this category [67.79].

Primordial for the successful adoption and use of AAL and PHS approaches, and e-Health more generally, is the consideration that the benefits to the participants in the system must be greater than the investments. Benefits may be considered on a short-term scale. For example, the implementation of only one PHS approach, such as videoconferencing, may be beneficial and even cost effective. However, benefits should also be considered from a long-term and holistic approach. The participants for whom such a system should be beneficial include the patients, health professionals, public health authorities and governments, researchers and research organizations, and industry. Investment may include not only monetary contributions but also contributions in terms of expertise, knowledge, know-how, professional resources, time, and materials. Similarly, in evaluating whether the investment of resources will be effective and efficient, a holistic and long-term perspective would be adopted.

Increasing evidence is emerging that the appropriate use of e-Health technologies can lead to improved delivery of healthcare while improving efficiency. In the United States, recent studies have determined that the national implementation of fully standardized interoperability between healthcare providers and key categories of entities (e.g., specialists, laboratories, and insurance funds) may yield savings of approximately $ US 75 billion annually [67.80]. An evaluation of the e-Health approaches implemented within the province of Ontario, Canada, demonstrates the positive results in creating efficient structures while providing effective healthcare services [67.81]. They found improvements in safety and quality of services, such as 35–70% reduction in hospital admissions due to adverse drug reactions, 20–40% increase in immunization compliance, and reductions in avoidable hospital admissions and length of stay. In regards to chronic disease management, due to e-Health approaches, the province was able to increase annual eye exams for diabetics by 20–40% and decrease by 25–50% emergency room visits for asthma sufferers. The province was able to improve access to healthcare for patients in remote areas by increasing the teleconsultations by 10–20%. Through e-Health approaches, the province was able to improve efficiency, for example, by reducing by 5–15% lab tests through avoiding duplication and by increased efficiency of resource utilization (drugs and diagnostics). The e-Health system also enabled the province to improve accountability by facilitating the carrying out of evidence-based population health planning.

Patient and User Acceptance

AAL and PHS approaches, and e-Health generally, benefit patients by providing them with prompt access to their medical history, health-related and medication information, electronic communication, and decision support services at the point of care. These readily available resources help facilitate the health care delivery process, and have been found to have a positive impact on patients' health outcomes, well-being, quality of life, hospital readmission rates, and mortality rates [67.82–85].

While the potential for using such approaches to improve clinical care or patient self-management has been acknowledged by many professionals, patients may not accept these technologies for diverse reasons. The question of patient and user acceptance is a complex issue that involves human emotive and cognitive reactions. Moreover, a distinction exists between patient acceptance of a technology and patient use of the same technology. Studies have found that factors such as poor device usability, insufficient training, lack of skills, and low self-efficacy can dissuade patients from accepting and using such technologies [67.86, 87]. Motor, visual and auditory, cognitive, external, and health limitations must all be taken into account in the design of AAL and PHS approaches [67.88].

Patients' decisions to accept technology may also be related to their diseases and health conditions, attributes of the health technology itself, organizational factors such as support, the environment of use, the patients' personal characteristics, and social influences [67.67, 89–92]. Studies have shown that patients who consider themselves knowledgeable in regards to their disease and how to care for their condition tend to be more likely to accept e-Health technologies. There is a higher likelihood of acceptance of an e-Health technology by a patient if other persons they consider important thought they should use the technology [67.93]. Patients are more likely to accept and use the technology if they believe that it would be useful and that it would be easy to use. Patients are more likely to accept the technology if they believe that they are able to use the technology and that they have the resources to support its use [67.94–96].

While patient acceptance of the technology is crucial, acceptance and use of such technologies must also take into account the needs and considerations of health professionals, patients' caregiver, and public health authorities. For health professionals, reliability of the technology, utility of the measurements and data, and effectiveness of the technology are important considerations. Issues of ease of use, replacability/reparability, reliability, security, or costs are considerations for the caregiver. From the perspective of public health authorities, the elements of reliability and stability of the technology, its effectiveness and safety, the costs of the technology, training, replaceability, as well as the benefits of providing patients with more autonomy are of importance.

67.3.5 Technological Dimensions

The development of ICT technologies has made possible innovations in e-Health , such as AAL and PHS approaches. With these developments considerable benefits have emerged. However, as examined above, there remain many challenges to be addressed.

In order for such technologies to be widely accessible, challenges pertaining to connectivity remain. In order to enable connectivity, issues such as the lack of an enabling telecom policy and regulatory framework, lack of access to energy sources, such as electricity or alternative power sources, insufficient infrastructure and access, and high costs must be addressed for e-Health to be truly globally available. While these issues are more pressing for developing and least developed economies, they also need to be addressed by developed countries of considerable geographical size and with populations located in remote or distant regions.

The Issue of Scaling Up and Evidence Basis

Developing policy relies on the availability of strong, systematic, and objective evidence for and analysis of approaches that work and those that do not, and examination of the required conditions and frameworks. Within the context of ICTs and healthcare, there is currently a shortage of such systematic evidence and analysis, particularly for the application of such technologies at the population level. This is particularly apparent in regards to the scaling up of particular projects or pilot studies.

Many of the studies that have been or are being carried out are pilot studies or specific projects with a very limited scope. While such studies contribute to assessing and better evaluating the broader implementation of a specific technology, they may not be sufficient in providing a holistic view of implementing concurrently multiple approaches. While some approaches may themselves be effective and efficient, this should not lead to the conclusion that the implementation of other approaches would similarly be effective and efficient.

The large-scale roll-out of systems such as e-Health, involve significant investments and expenditures particularly of public funds. A key challenge facing governments and public health authorities is the development of a methodology that enables the systematic and holistic evaluation of how e-Health can be implemented to deliver the most effective, safe, and efficient healthcare services and products. Another key challenge is the consideration of which conditions and factors must be addressed in order for the scaling up of a particular approach within a holistic system to be effective and efficient.

Standardization, Interoperability, and Compatibility

As a fast-evolving field, many e-Health approaches and solutions face challenges in regards to standardization, interoperability, and compatibility. ICT standards are important in e-Health technologies as these approaches are heavily data and information driven and having seamless access to such information is foundational [67.97]. The lack of global standardization is a significant challenge for the effective use of e-Health technologies and approaches. For example, within Europe, it was found that there is a lack of widely used e-health standards, leading to problems with interoperability. Of the standards that do exist, the study finds that some are conflicting and many are proprietary [67.98]. The study also concluded that within a single health service provider, the standards currently used were supportive. However, the situation was unsupportive for cross-border care provision. One of the barriers identified to the adoption of international e-health standards in hospitals is hospital IT managers, who prioritize internal process functionality above commonly used standards. The respondents also agreed that managers lacked financial incentives to exchange information electronically.

Standardization/harmonization also applies to the data and information to be collected and entered into databases. For example, the lack of standards is a significant challenge in regards to ensuring the most effective and efficient use of data and information that is generated through e-Health approaches. If the identification,

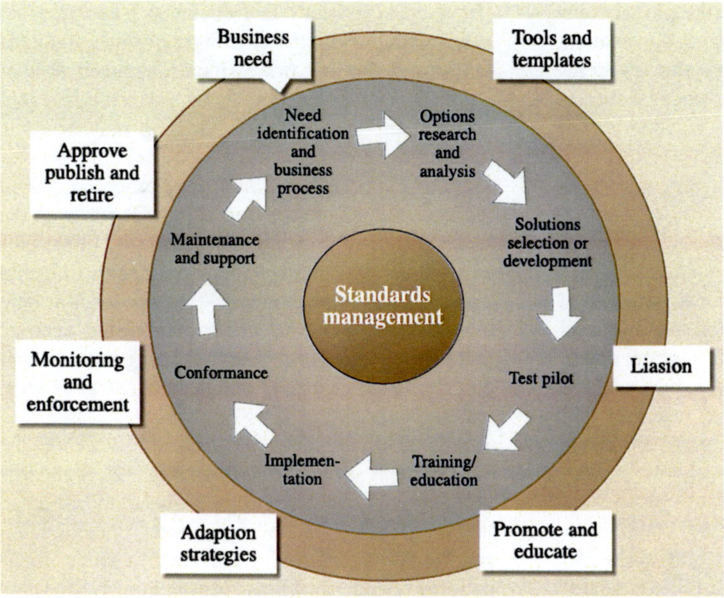

Fig. 67.23 Example of e-Health standards strategy (after [67.84])

data entry, and classification of the data and information are not standardized, the potential for secondary use of such data and information may be limited. The issue of interoperability will also be key in ensuring that resources are used efficiently. Furthermore, as more and more patients choose to undergo medical procedures outside of their country of residence, international compatibility of information and data will be another significant challenge. As Fig. 67.23 illustrates, a viable approach to developing standards and to addressing issues of interoperability and compatibility necessitates a holistic perspective that takes into account the objectives and needs all of the actors and at the different stages of use.

67.4 Conclusion

67.4.1 Observations

As governments and public health authorities search for alternative means to deliver effective, safe, efficient, and equitable healthcare services and products, the use of information and communication technologies embodied in e-Health is a promising key area. While many aspects of functioning of healthcare systems will continue to need re-vamping, e-Health will enable a patient-centric approach wherein the patient is active in the management of his or her health, while improving efficiency and effectiveness.

The increase in computer and technology literacy offers the opportunity for better integration of healthcare delivery into daily activities. Future generation of patients will be more demanding in regards to information and options for treatments. AAL and PHS approaches will provide tools for patients and their caregivers to actively monitor their health status, to learn more about their condition or disease, and to understand how to better manage it. e-Health technologies can provide important support for the prevention, detection, diagnosis, treatment, monitoring, and surveillance of a patient's condition or disease. It will permit reaction and adaptation to the changing needs and situation of the patient. Moreover, it will enable adaption within a home environment.

The full implementation of e-Health will need the involvement of all actors within the health sector, whether private or public sector, at national, regional, and international levels. As e-Health holds promise for all populations, rural or urban, remote, or central, and within developing or developed economies, the efforts of implementing this set of technologies, applications, solutions, and devices need to involve the global community. e-Health may bring many benefits; however, its full implementation will need to address and find solutions for the numerous challenges. New approaches to developing technologies, policies, and legislation will need to be considered.

67.4.2 Areas for Further Work

For legislators, policy decision-makers, medical professionals, industry, engineering, systems and design experts, patients, international organizations, and civil society to be able to ensure the appropriate development, adoption and usage of e-Health technologies and equipment, many more areas of research will need to be addressed. Some of the gaps are listed here.

1. *Internationalization*: some of the most complex issues may need to be considered and addressed at the international level. For example, standardization, interoperability, and compatibility may be issues that are best addressed at the international level in order to ensure that information, data, and technologies can be used and exchanged globally. Another area where intervention at the global level may be beneficial is the development of international instrument(s) (i. e., convention, treaty, etc.) in regards to the issues such as privacy and confidentiality, data protection, discrimination, and ethical standards.
2. *Cost – benefit analysis and evidence*: the issue of cost and expenditures in regards to healthcare systems is a significant challenge facing every government as well as the global community. While many advocates of e-Health highlight the cost containment possibility of such technologies, it is an area that requires further and more in-depth study. Complex issues of measuring impact and savings will need to be addressed; frameworks and statistical methodologies will need to be developed.
   Simplistic approaches should be avoided as they will lead to inaccurate or incomplete evaluations. For example, some robotic surgery systems cost more than $US 1 million to purchase and more than $US 100 000 a year to maintain. Hospitals can save on costs by using such robotic surgery that is less invasive and thus, decrease the length of a patient's stay due to a shorter recovery period. However, calculations taking only this aspect into account may provide an inaccurate picture. Other considerations that could be taken into account include shortened surgery times, reduced fatigue of surgical professionals, reduced risk of surgical errors, etc. With other factors taken into account, the cost–benefit evaluation may be quite different.
3. *Privacy and confidentiality*: government decision-makers will need to develop policy for ensuring the protection of privacy and confidentiality of patients and the security of their personal health information and data in support of e-Health. In addressing these issues, they will need to take into account that patients are increasingly traveling outside their country of residence to obtain medical treatments and therapies.
4. *Standards, interoperability, and compatibility*: the development of internationally-agreed standards will be a key challenge whose resolution will need to be addressed in the near future. Consideration should be given to the manner in which international standard setting organizations can play a more active role in the development of global standards necessary for the effective and efficient development of standards, interoperability, and compatibility in e-Health approaches. Consideration should also be given to the use of open sources.
5. *Technological developments*: the evolution of information and communication technologies will contribute to improving the delivery of healthcare. While it is difficult to predict the evolution of technology, the development of some key areas should be considered [67.99–102]:
   - New, pervasive, and trustworthy network and service infrastructures to replace current Internet, mobile, fixed, and audiovisual networks, including research in regards to architecture
   - ICT based on nano-scale integration, new materials, photonics, and organic electronics, which will provide new types of devices and intelligent systems
   - Context-aware ICT systems that self-improve and self-adapt within their environments, including the fields of cognitive systems, robotics, and interaction
   - New technologies and devices that respect societal goals of sustainable development. Increasingly, electronic components, devices and systems will need to focus on being smaller, more accessible, more reliable, and greener.

67.4.3 Future Technologies

Wearable and environmental sensors. Smaller, lighter, more powerful sensors equipped with wireless connectivity, permitting the measurement of single or multiparameters, including parameters that can only currently be measured with large and complex equipment.

Human–machine interaction. Devices that facilitate human–machine interaction:

- *Displays.* New lighter, thinner, faster displays. Nanoemissive display and the surface-conduction electron-emitter display, both with higher contrast levels and faster response times than LCDs.
- *Very smart textiles.* Currently smart textiles used in healthcare are mainly passive smart (sense). In the future, they will be active smart textiles (sense and react) and very smart textiles (sense, react and autoadapt).

References

67.1 United Nations: *2008 Revision of World Population Prospects* (United Nations Secretariat, New York 2009)

67.2 J. Oliveira Martins, C. de la Maisonneuve: The drivers of public expenditure on health and long-term care: An integrated approach, OECD Econ. Stud. **43**(2), 118–122 (2006), http://www.oecd.org/dataoecd/62/19/40507566.pdf (last accessed March 15, 2010)

67.3 OECD: *Society at a Glance OECD Social Indicators 2006* (OECD, Paris 2007), http://www.oecd.org/els/social/indicators/SAG (last accessed March 15, 2010)

67.4 N. Esmail: *Waiting Your Turn: Hospital Waiting Lists in Canada*, 19th edn. (Fraser Institute, Calgary 2009)

67.5 P. Lasbordes: *Télésanté 2009* (Edisante, Paris 2009)

67.6 WHO: *Chronic Diseases* (WHO, Geneva 2009)

67.7 European Commission: The Ambient Assisted Living (AAL) Joint Programme, COM 658 (EC, Luxembourg 2005)

67.8 European Commission: Accelerating the Development of the e-Health Market in Europe, e-Health, Taskforce Report 2007, Lead Market Initiative (European Communities, Luxembourg 2007)

67.9 C. Roux, M.-A. Bloch: Ambient Assisted Living an article 169 initiative, 12 November 2007 (CNSA, Paris 2007) http://www.telecom.gouv.fr/fonds_documentaire/inf/aal/ppt1107/technologieautonomie.pdf (last accessed March 15, 2010)

67.10 B. Jaffe, W.R. Cook, H. Jaffe: *Piezoelectric Ceramics* (Academic, London 1971)

67.11 European Commission: Personal Health Systems: Deployment Opportunities and ICT Research Challenges, Unit Summaries (European Parliament, Brussels 2007) http://ec.europa.eu/information_society/events/phs_2007/docs/unit-summaries-booklet.pdf (last accessed March 15, 2010)

67.12 G. van den Broek, F. Cavallo, L. Odetti, C. Wehrmann: *Ambient Assisted Living Roadmap* (IOS, Amsterdam 2010)

67.13 European Heart Network: http://www.ehnheart.org (last accessed March 15, 2010)

67.14 R. Bez, E. Camerlenghi, A. Modelli, A. Visconti: Introduction to flash memory, Proc. IEEE **91**(4), 489–502 (2003)

67.15 Commission of the European Communities: eEurope 2002: Quality criteria for health related websites, J. Med. Internet Res. **4**(3), E15 (2002)

67.16 H. Nabarette, F. Romaneix, C. Boyer, S.J. Darmoni, P.L. Rémy, E. Caniard: Certification of health-related websites in France, Presse Med. **38**(10), 1476–1483 (2009)

67.17 K. Peters, M. Niebling, T. Green, C. Slimmer, R. Schumacher: *Consumers Compare Online Personal Health Record (PHR) applications, A White Paper* (User Centric, Oakbrook Terrace 2009), http://www.usercentric.com/publications/2009/02/02/google-health-vs-microsoft-healthvault-consumers-compare-online-personal-hea (last accessed March 15, 2010)

67.18 WHO: *Adherence to Long-term Therapies, Evidence for Action* (World Health Organization, Geneva 2003)

67.19 AmeliaPlex: http://www.medicationreminders.com (last accessed March 15, 2010)

67.20 e-rappel: http://www.e-rappel.com (last accessed March 15, 2010)

67.21 Fitbit: www.fitbit.com (last accessed March 15, 2010)

67.22 Q. Li, J.A. Stankovic, M. Hanson, A. Barth, J. Lach: Accurate, Fast fall detection using gyroscopes and accelerometer-derived posture information, 6th Int. Workshop Wearable and Implantable Body Sensor Networks (Berkeley 2009)

67.23 BodyMedia: www.bodymedia.com (last accessed March 15, 2010)

67.24 http://www.a2e2.eu/ (last accessed March 15, 2010)

67.25 Gerontechnologie: Interview with Claude Mordelet, published May 20th (2008) available at http://www.gerontechnologie.net/entretien-avec-claude-mordelet-president-de-lafrata/31186 (last accessed March 15, 2010)

67.26 West Lothian Council: *Home Safety Information* (West Lothian Council, Livingston 2010), http://www.westlothian.gov.uk/media/

downloaddoc/1799563/1857743 (last accessed March 15, 2010)
67.27 Temo: www.mon-temo.com (last accessed March 15, 2010)
67.28 A. Davalli, R. Sacchetti, C. Bonivento, M. Ugolini: Tele-assistance for upper-limb myoelectric prostheses, Proc. XV World Computer Congr., Vienna (1998)
67.29 C. Pellerin, D. Strasbourg: *Téléassistance en soins de plaies: Unique au Canada* (Canada Health Infoway, Toronto 2010), (in French) available at http://www.canadahealthinfoway.com/lang-fr/about-infoway/news/news-releases/537-wound-care-teleassistance-unique-in-canada (last accessed March 15, 2010)
67.30 J. Deign: The man who is helping Argentine doctors cure at a distance, CISCO News (2010), available at http://newsroom.cisco.com/dlls/2010/ts_010410b.html (last accessed March 15, 2010)
67.31 Fraunhofer-Institut Biomedizinische Technik, St. Ingbert: http://www.ibmt.fraunhofer.de/fhg/ibmt/ (last accessed March 15, 2010)
67.32 Datamed S.A. Healthcare Information Systems, Athens: http://www.datamed.gr/index.php/home?Itemid=86 (last accessed March 15, 2010)
67.33 A.J. Blake, K. Morgan, M.J. Bendall, H. Dallosso, S.B. Ebrahim, T.H. Arie, P.H. Fentem, E.J. Bassey: Falls by elderly people at home: Prevalence and associated factors, Age Ageing **17**, 365–372 (1988)
67.34 M.E. Tinetti, M. Speechley, S.F. Ginter: Risk factors for falls among elderly persons living in the community, N. Engl. J. Med. **319**, 1701–1707 (1988)
67.35 D.J. Rose: *Fallproof! A Comprehensive Balance and Mobility Training Program* (Human Kinetics, Windsor Ontario, 2003)
67.36 G. Corte Franco, F. Gallay, M. Berenguer, C. Mourrain, P. Couturier: Non-invasive monitoring of the activities of daily living of elderly people at home – A pilot study of the usage of domestic appliances, J. Telemed. Telecare **14**(5), 231–235 (2008)
67.37 M. Kangas, A. Konttila, I. Winblad, T. Jamsa: Determination of simple thresholds for accelerometry-based parameters for fall detection, Proc. 29th Annu. Int. Conf. IEEE EMBS (IEEE, Lyon 2007) pp. 1367–1370
67.38 A. Sixsmith, N. Johnson, R. Whatmore: Pyroelectric IR sensor arrays for fall detection in the older population, J. Phys. IV (Paris) **128**, 153–160 (2005)
67.39 National Diabetes Information Clearinghouse: http://diabetes.niddk.nih.gov (last accessed March 15, 2010)
67.40 D. Mendosa: Diabetes Monitor – The Glucowatch Biographer (2010) available at http://www.diabetesmonitor.com/m91.htm (last accessed March 15, 2010)
67.41 J.G. Cleland, A.A. Louis, A.S. Rigby, U. Janssens, A.H. Balk, TEN-HMS Investigators: Noninvasive home telemonitoring for patients with heart failure at high risk of recurrent admission and death: The Trans-European network-home-care management system (TEN-HMS), J. Am. Coll. Cardiol. **45**(10), 1654–1664 (2005)
67.42 R. Paradiso: Wearable health care system for vital signs monitoring, Proc. IEEE Int. Conf. Information Technology, Applications in Biomedicine (IEEE, Bellingham 2003) pp. 283–286
67.43 VitalJacket: http://www.vitaljacket.com/ (last accessed March 15, 2010)
67.44 K.P. Hoffmann, R. Ruff: Flexible dry surface-electrodes for ECG long-term monitoring, Engineering in Medicine and Biology Society, 29th Annu. Int. Conf. IEEE (IEEE, Bellingham 2007) pp. 5740–5743
67.45 J. Vomhof Jr.: Transoma medical shuts down, Minneapolis/St. Paul Business Journal, Tuesday, December 8 (2009)
67.46 Belgium Heart Failure Project: Technologies de l'autonomie et soutien à domicile, rapport d'études, Caisse des Dépôts, Les Éditions Stratégiques, Édition juin (2009)
67.47 First US Patients Receive Transoma Medical's Sleuth Wireless Remote Implantable ECG Monitoring System, Business Wire, October 29 (2007) http://www.allbusiness.com/health-care/medical-practice-cardiology/5289811-1.html
67.48 O. Nuria, F. Flores-Mangas: HealthGear – Automatic sleep apnoea detection and monitoring with a mobile phone, J. Commun. **2**(2), 1–9 (2007)
67.49 Equivital Laboratory System, Hidalgo Limited, Swavesey: http://www.equivital.co.uk/_publicdocuments/equivital_research_brochure.pdf (last accessed March 15, 2010)
67.50 Medtronic: http://wwwp.medtronic.com/newsroom/content/1166721882051.low_resolution.jpg (last accessed March 15, 2010)
67.51 P.E. Ruskin, M. Silver-Aylaian, M.A. Kling, S.A. Reed, D.D. Bradham, J.R. Hebel, D. Barrett, F. Knowles, P. Hauser: Treatment outcomes in depression: comparison of remote treatment through telepsychiatry to in-person treatment, Am. J. Psychiatry **161**, 1471–1476 (2004)
67.52 Roboto Hosting: http://robot-hosting.com/ (last accessed March 15, 2010)
67.53 Molecular Monitoring Group, Center for Future Health: available at http://www.centerforfuturehealth.org/research/molecular_monitoring.html (last accessed March 15, 2010)
67.54 T. Hornyak: Nurse robot Riba, cnet News, January 1 (2010) http://news.cnet.com/2300-17938_105-10002074.html (last accessed March 15, 2010)
67.55 M. Montemerlo, J. Pineau, N. Roy, S. Thrun, V. Varma: Experiences with a mobile robotic guide for the elderly, National Conf. Artificial Intelligence (AAAI) (AAAI, Edmonton 2002) pp. 587–592

67.56 B. Siciliano, O. Khatib (Eds.): *Springer Handbook of Robotics* (Springer, Berlin Heidelberg 2008), Chaps. 52, 53

67.57 Nursebot project: University of Pittsburgh, Carnegie Mellon University, http://www.cs.cmu.edu/~nursebot/ (last accessed March 15, 2010)

67.58 I.A.O. Fraunhofer-Institut: http://www.iward.eu/cms/index.php (last accessed March 15, 2010)

67.59 Mayo Clinic: GE Healthcare, Intel, Mayo Clinic explore remote healthcare monitoring, Mayo Clinic News, February 28 (2010) http://www.mayoclinic.org/news2010-rst/5657.html (last accessed March 15, 2010)

67.60 Technologies de l'autonomie et soutien à domicile, rapport d'études, Caisse des Dépôts, Les Éditions Stratégiques (2009) http://www.e-alsace.net/documents/fck/file/documents_pdf/EtudesCDC.pdf

67.61 Medical Care Continuity (MCC) consortium (2005–2007) Axa Assistance, Eurogroup Alliance, France Télécom, http://fr.wikipedia.org/wiki/Medical_Care_Continuity (last accessed March 15, 2010)

67.62 A. Wagner, G. Undt, K. Schicho, F. Wanschitz, F. Watzinger, K. Murakami, C. Czerny, R. Ewers: Interactive stereotaxic teleassistance of remote experts during arthroscopic procedure, J. Arthrosc. Relat. Surg. **18**(9), 1034–1039 (2002)

67.63 M. Morino, L. Pellegrino, C. Giaccone, C. Garrone, F. Rebecchi: Randomized clinical trial of robot-assisted versus laparoscopic Nissen fundoplication, Br. J. Surg. **93**, 553–558 (2006)

67.64 MonAMI Project Office: http://www.monami.info/

67.65 J. Alexandersson: http://www.i2home.org/ (last accessed March 15, 2010)

67.66 F. Kohler, T. Schmitz-Rode, C. Disselhorst-Klug: Introducing a feedback training system for guided home rehabilitation, J. NeuroEng. Rehabil. **15**(7), 2 (2010)

67.67 C.K.L. Or, B. Karsh, D.J. Severtson, P.F. Brennan: Patient technology acceptance model (PTAM) – Exploring the determinants of consumer health information technology acceptance by homecare patients with chronic illness, Int. Conf. Healthcare Syst. Ergonomics Patient Safety, Strasburg (2008)

67.68 US Department of Health and Human Services (Ed.): Code of Federal Regulations, TITLE 45 - Public Welfare, Part 46, Protection of Human Subjects (US Department of Health and Human Services, Washington 2009)

67.69 C. Sampogna: *Creation and Governance of Human Genetic Research Database* (OECD, Paris 2006)

67.70 C. Sampogna: *Recommendation for Human Biobanks and Genetic Research Databases* (OECD, Paris 2009)

67.71 Government of British Columbia: e-Health Statute Increases Patient Access and Privacy, News Release, April 10, (2008) http://www2.news.gov.bc.ca/news_releases_2005-2009/2008HEALTH0038-000505.htm (last accessed March 15, 2010)

67.72 WHO: *Building Foundations for e-Health: Progress of Member States – Report of the WHO Global Observatory for eHealth* (World Health Organization, Geneva 2006)

67.73 A. Chetley (Ed.): *Improving Health, Connecting People: The Role of ICTs in the Health Sector of Developing Countries – A Framework Paper*, Working Paper No. 7, InfoDev (2006) available at http://www.infodev.org/en/Document.84.pdf (last accessed March 15, 2010)

67.74 S. Roberts (Ed.): *The Global Information Society: A Statistical View* (UN, New York 2008), avialable at http://www.unctad.org/en/docs/LCW190_en.pdf (last accessed March 15, 2010)

67.75 C.L. Ogden, M.D. Carroll, M.A. McDowell, K.M. Flegal: Obesity among adults in the United States – No change since 2003–2004, NCHS Data Brief No. 1 (National Center for Health Statistics, Hyattsville 2007)

67.76 H.C. Kung, D.L. Hoyert, J.Q. Xu, S.L. Murphy: Deaths: Final data for 2005, Natl Vital Statist. Rep. **56**(10), 3–10 (2008)

67.77 WHO: Largely preventable chronic diseases cause 86% of deaths in Europe Press Release EURO/05/06, Copenhagen, September (2006)

67.78 DHHS: Health United States 2009, Special Feature on Medical Technology (National Center for Health Statistics, Hyattsville, 2010) http://www.cdc.gov/nchs/data/hus/hus09.pdf#specialfeature (last accessed March 15, 2010)

67.79 MARC Tech Lab: http://marc.med.virginia.edu/projects_smarthomemonitor.html (last accessed March 15, 2010)

67.80 J. Walker, E. Pan, D. Johnston, J. Adler-Milstein, D.W. Bates, Blackford Middleton: The value of healthcare information exchange and interoperability, Health Affairs, January (2005)

67.81 A. Brown: Ontario's eHealth Strategy and Standards: Ready Set Standards!, Presentation, November (2006) available at http://sl.infoway-inforoute.ca/downloads/Stein_Brown_-_Ontario_eHealth_Strategy_and_Standards.pdf

67.82 D.H. Gustafson, R. Hawkins, E. Boberg, S. Pingree, R.E. Serlin, F. Graziano, C.L. Chan: Impact of a patient-centered, computer-based health information/support system, Am. J. Prev. Med. **16**(1), 1–9 (1999)

67.83 D. Hailey, R. Roine, A. Ohinmaa: Systematic review of evidence for the benefits of telemedicine, J. Telemed. Telecare **8**(1), 1–7 (2002)

67.84 A. Martinez, E. Everss, J.L. Rojo-Alvarez, D.P. Figal, A. Garcia-Alberola: A systematic review of the literature on home monitoring for patients with heart failure, J. Telemed. Telecare **12**(5), 234–241 (2006)

67.85 A.A. Louis, T. Turner, M. Gretton, A. Baksh, J.G.E. Cleland: A systematic review of telemonitoring for the management of heart failure, Eur. J. Heart Fail. **5**(5), 583–590 (2003)

67.86 D.R. Kaufman, V.L. Patel, C. Hilliman, P.C. Morin, J. Pevzner, R.S. Weinstock, R. Goland, S. Shea, J. Starren: Usability in the real world: Assessing medical information technologies in patients' homes, J. Biomed. Inf., **36**(1–2), 45–60 (2003)

67.87 T.L. Patterson, W.S. Shaw, D.R. Masys: Improving health through computer self-help programs: theory and practice. In: *Information Networks for Community Health*, ed. by P.F. Brennan, S.J. Schneider, E. Tornquist (Springer, New York 1997)

67.88 W.A. Rogers, B. Meyer, N. Walker, A.D. Fisk: Functional limitations to daily living tasks in the aged: A focus group analysis, Hum. Factors **40**(1), 111–125 (1998)

67.89 C.K.L. Or, R. Valdez, G. Casper, P. Carayon, L.J. Burke, P.F. Brennan, B. Karsh: Human factors and ergonomics in home care: Current concerns and future considerations for health information technology, Work **33**(2), 201–209 (2009)

67.90 T. Zayas-Cabán, P.F. Brennan: Human factors in home care. In: *Handbook of Human Factors and Ergonomics in Health Care and Patient Safety*, ed. by P. Carayon (Lawrence Erlbaum, Mahwah 2007)

67.91 G.C. Moore, I. Benbasat: Development of an instrument to measure the perceptions of adopting an information technology innovation, Inf. Syst. Res. **2**(3), 192–222 (1991)

67.92 S. Taylor, P.A. Todd: Understanding information technology usage: A test of competing models, Inform. Syst. Res. **6**(2), 144–176 (1995)

67.93 V. Venkatesh, F.D. Davis: Theoretical extension of the technology acceptance model: Four longitudinal field studies, Manag. Sci. **46**(2), 186–204 (2000)

67.94 H.P. Lu, D.H. Gustafson: An empirical study of perceived usefulness and perceived ease of use on computerized support system use over time, Int. J.Inf. Manag. **14**(5), 317–329 (1994)

67.95 E.V. Wilson, N.K. Lankton: Modeling patients' acceptance of provider-delivered e-health, J. Am. Med. Inf. Assoc. **11**(4), 241–248 (2004)

67.96 J.A. Diaz, R.A. Griffith, J.J. Ng, S.E. Reinert, P.D. Friedmann, A.W. Moulton: Patients' use of the Internet for medical information, J. Gen. Intern. Med. **17**(3), 180–185 (2002)

67.97 J. Kesan: ICT Standards and e-Health, Talk Standards, February (2010) available at http://www.talkstandards.com/ict-standardization-and-ehealth/ (last accessed March 15, 2010)

67.98 J. Hoeksma: Call for common e-health standards, e-Health Europe, November (2008), available at http://www.ehealtheurope.net/news/4293/call_for_common_e-health_standards (last accessed March 15, 2010)

67.99 K. Bonsor, J. Strickland: How robotic surgery will work (howstuffworks, Atlanta GA) available at http://science.howstuffworks.com/robotic-surgery.htm (last accessed March 15, 2010)

67.100 D. Ahmed: Hybridization of smart textiles in medical and healthcare management, AUTEX 2009 World Textile Conf. (AUTEX, Izmir 2009)

67.101 European Commission: Updated Work Programme 2009 and Work Programme 2010 – Theme 3: Information and Communication Technologies, C(2009)5893, July (2009)

67.102 J.E. Hughes Jr., M. Di Ventra, S. Evoy (Eds.): *Introduction to Nanoscale Science and Technology* (Springer, Berlin Heidelberg 2004)

68. Electrical Stimulation of the Nervous System

Stanisa Raspopovic, Marco Capogrosso, Silvestro Micera

Modern science, as we know it, is based on the concept proposed by Galileo Galilei, who designed the conceptual framework within which every scientist should act: the natural principles should be mathematically modeled and then verified in the experience.

Models are, then, the base of scientific reasoning; with accurate mathematical models (Sects. 68.2 and 68.3) we can describe reality and predict phenomena that can be verified by experimental evaluation.

One of the most exciting challenges of mankind is the modeling of the human nervous system. Mathematical models of this kind are of fundamental importance not only for neurosciences but also for applied sciences like neural prosthesis in clinical trials (for example, cochlear implants, pacemakers, and deep brain stimulators) or in new, strongly expanding fields, such as neural engineering where such models may help in optimizing state-of-the-art neuroprostheses technologies and to find new approaches for highly dimensional and difficult to observe systems, such as the brain, the spinal cord, or peripheral nerves. A very common question frequently asked by experimentally-oriented scientists is what models serve for.

For an exhaustive answer to this critical issue an entire book would be necessary, but in summary:

1. Models are a very powerful tool amenable to testing theoretical hypotheses and minimizing the use of animals in experimentations
2. If searching for optimal parameters for stimulations, or design geometry for electrodes models can minimize the high dimensional space of parameters or give hints for optimization of shape
3. Finally, they can give an insight into the basic understanding of the working principles of neuroprostheses.

68.1 Background 1247

68.2 Biophysics Models of Neuronal Response to External Fields 1248
68.2.1 Modeling of Self-Standing Neurons 1248
68.2.2 Modeling Effects of Extracellular Electrical Stimulation 1250

68.3 Finite Element (FE) Models..................... 1252
68.3.1 The Electromagnetic Problem......... 1252
68.3.2 The Finite Element Method (FEM) ... 1253
68.3.3 Models in Practice 1255

68.4 Conclusion 1256

References 1256

68.1 Background

The idea of stimulating the nervous system with the purpose of rehabilitation has its origin in Luigi Galvani's work (1737–1798). He developed the so-called *galvanic* element, by means of which nerves could be stimulated, studying the provoked muscle contractions. Later, in 1792, Alessandro Volta (1745–1827) demonstrated that the galvanic element uses electricity. So, since then nerves have been considered to be the *electric cabling* of the body. The intuition (confirmed in experiments) suggested that by means of electrical stimulation of nerves and muscles it is possible to obtain (almost) natural movements.

Neuroprostheses, which are devices meant to interact with the nervous system of disabled persons, often exploit electrical stimulation in order to develop assist devices or to deliver a therapeutic protocol.

In special cases, neuroprostheses are aimed to understand (decode) signals from the brain and to deliver signals to the tissue in order to elicit some targeted function (like deep brain stimulation [68.1], spinal epidural stimulation [68.2, 3] or peripheral functions [68.4–6]). The basic mechanism of how electric fields interact with the nervous system and how the nervous system is able to generate local fields must be investigated in order to understand how neuroprostheses work and how they can be improved.

68.2 Biophysics Models of Neuronal Response to External Fields

The principles of neural tissue excitation were deeply investigated through detailed biophysics models of neurons and fibers.

The biophysical modeling of neuronal cell response to the external electrical field evolved in two steps:

1. Significant theoretical and experimental work was carried out in order to emulate in a proper way the properties of single neural cells with different levels of detail (depending on the application field). This approach evolved until the development of neuronal and neural networks models, which are used in basic research and different everyday applications.
2. When sufficient knowledge about neuron models had been achieved, together with rapid expansion of neuroprosthesis applications, the need for models of cells in presence of artificially induced electrical fields became more and more important. Starting from analytical solutions, through special case applications, this field has now evolved into hybrid field-neuron models, accounting for both nonlinear cellular dynamics, and anisotropic electrical field distribution.

68.2.1 Modeling of Self-Standing Neurons

It is a common assumption, accepted among scientific community [68.9], that the information is transmitted through the nervous systems by means of short action potentials, traveling along the neural cells by means of complex reactions occurring across the neurons membranes.

The membrane of neural cells is a structure able to maintain a certain fixed potential difference (usually negative) between the intracellular and the extracellular space. As a complex dynamical system, if perturbed (with a sufficient amount of current injection, for example) it generates the *all-or-none* event that is the action potential.

In the first part of the 20th century the great effort in modeling the electric properties of the neural membrane gave birth to the very first electrical cir-

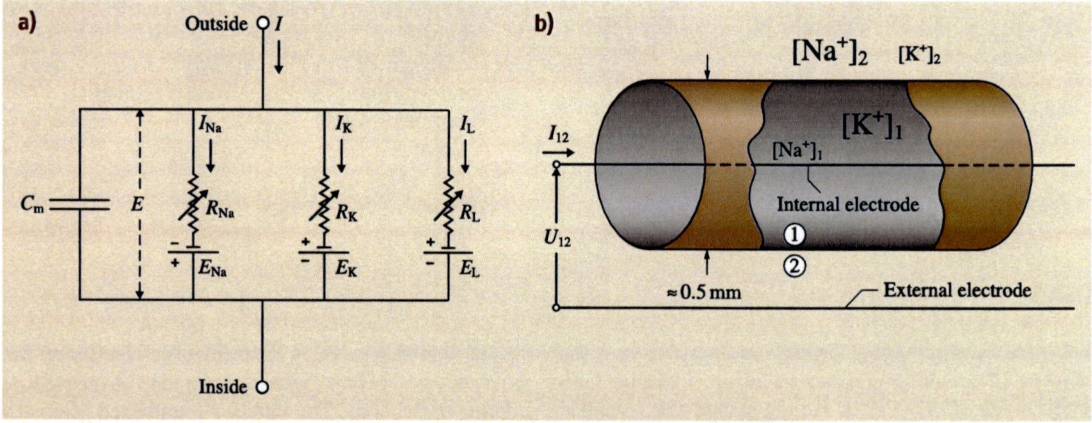

Fig. 68.1 **(a)** Equivalent electrical circuit, as used by the HH model (after [68.7]) **(b)** Typical measurement system used by HH in order to measure the effects of different currents. U12 is the potential difference across the inside (label 1) and outside (label 2) of the cell, $[Na^+]$ and $[K^+]$ are ion concentrations (after [68.8])

cuit equivalent representations of the membrane [68.9] (Fig. 68.1a). However, the breakthrough in modeling the generation of action potential was the pioneering work of *Hodgkin* and *Huxley* (HH) [68.7] who received the Nobel Prize in medicine and physiology in 1963 for this work. The HH model is still the most used, with opportune modifications, because of its clearness, intuitive representation, and many times experimentally confirmed value. The model of Hodgkin and Huxley is based on a set of four partial nonlinear differential equations reproducing the dynamics of the action potential, as related to the different ionic currents flowing through the cell's membrane. The model is based on experimental measures on the active membrane of the giant squid axon. The experiments are based on voltage clamp measurements and space clamp measurements, as can be seen schematically in Figure 68.1. For more details about experimental procedure refer to [68.7, 8].

Ion currents are used to describe the ion exchange between the inner part of the cell an the outer environment that cause depolarization of the cell body, which otherwise is kept at a constant negative potential V_{rest} with respect to the external environment. The equilibrium potential is described by the Goldman equation which relates the membrane equilibrium potential with the equilibrium potential of each ion flux flowing across the membrane [68.10, 11]

$$V_{rest} = \frac{kT}{F} \ln \frac{p_K[K^+]_e + p_{Na}[Na^+]_e + p_{Cl}[Cl^-]_i}{p_K[K^+]_i + p_{Na}[Na^+]_i + p_{Cl}[Cl^-]_e} . \tag{68.1}$$

When assuming that the membrane is a capacitor (C_m) they defined the principal equations driving the potential difference (V in (68.1)) across the membrane. The total membrane current is the sum of the ionic (I_i) currents and the capacitive ones

$$I = C_m \frac{dV}{dt} + I_i . \tag{68.2}$$

Sodium (I_{Na}), potassium (I_K), and a leakage (I_L) of ionic current were described. Since the ionic currents are actively generated to induce depolarization, the HH model describes this mechanism as an active time-varying conductance that relates in a nonlinear fashion with the voltage difference across the membrane. Conductances of different ions are driven by proper *activation* and *inactivation* variables (m, n, h), which describe the probabilistic nature of an ion gate opening and closing. The total flux of ions across the membrane causes the generation of action potentials.

Thus, the complete equation is [68.7]:

$$C_m \frac{dV}{dt} = g_{Na} m^3 h(E_{Na} - V) + g_K n^2 (E_K - V) + g_{leak}(V_{rest} - V) + I_{inj}(t) . \tag{68.3}$$

The results of the model were astonishing. They were able to replicate the action potential generation and shape perfectly with respect to the experimental measurements. Of course, (68.3) gives a very simplified insight into this very complex problem. There are several assumptions regarding the electromagnetic field theory, which hold for the model [68.12].

However, this model is able to merge physiological concepts and knowledge (like ion gates and flows across the cell membrane) with an intuitive electrical circuit model. In this way, the mathematical description is closely related in all its components to the real electrophysiology of the cell. The success of the HH model pushed the intensive studies done by mathematicians and other scientists to simplify or improve the model, depending on the nature of the field in which they exploited the neuron models.

Frankenhaeuser–Huxley [68.13] adapted the HH model for the computation in mammalian myelinated fibers, based on the voltage clamp data from *Xenopus laevis*.

A famous mathematical evolution of the HH model is the Fitzhugh–Nagumo model (FHN model), which is based on the computational work of *Fitzhugh* [68.14, 15] and practical implementation by *Nagumo* [68.16]. Fitzhugh used phase space methods and the model of the Bonhoeffer–van der Pol oscillator in order to elegantly simplify the computation of the HH equations [68.14] and introduced physiological state diagrams. He also implemented the computational model of saltatory conduction in myelinated neuronal fibers, using an algorithmic solution similar to the ones implemented in current neuronal simulators (e.g. NEURON [68.17, 18], GENESIS [68.19]), Nagumo was able to construct an active pulse transmission line using tunnel diodes, making the device able to replicate the neuronal conduction, and showing in practice the validity of modeling approach starting from the HH model.

Modeling of neuronal cells then went in two different directions: towards the simplification (in order to understand the neuronal populations' behavior and also for practical applications as neural networks), and towards very high detail (aiming to represent all possible ionic currents dynamics, molecular effects etc.) [68.20].

Models aimed to study and optimize neuroprosthesis devices are in the middle between the highly detailed and simplified models, since they have to properly describe the behavior of the tissue of interest (peripheral nerve or cerebral cortex pyramidal neurons), keeping the complexity low for computational reasons.

It is not the aim of the present text to go into the details of the evolution of further neuronal models. For the interested readers there are several excellent books on this topic such as those from [68.9] or spiking neuron models [68.22]. Nevertheless it is important to keep in mind the concepts depicted about HH and FHN model, since they are essential for the understanding of the functioning of hybrid models for practical applications of electrical stimulation of nerves.

68.2.2 Modeling Effects of Extracellular Electrical Stimulation

Electrical external stimulation requires the modeling of fibers and axons. Electric fields are, in fact, generated inside the tissues and the shape of the external electric potential along the fiber and cell membrane is responsible for the activation or the blocking of the fiber.

To study the electrical conduction properties of fibers, the cable theory [68.9] was exploited. The application of electromagnetic basic principles like the Poisson equation and the complex nonlinear model of Hodgkin and Huxley gave rise to the representation of the axon as an infinite cable interleaved by nonlinear active components (the node of Ranvier) and linear isolating ones (myelinated tracts). The modeling for extracellular electrical stimulation has its origin in experimental findings.

Louis Lapicque was a physiologist and a pioneer in the field of neural excitability by external stimulation. From Galvani's work it was known that nerves can be electrically excited. However, the stimuli used to excite nerves, derived from batteries and capacitors, were hard to control. Therefore, a central question at the time of Lapicque was how much and how long a stimulation is required to excite a nerve. A further question raised was how can the relation between the stimulus and excitability be explained by the underlying biophysics? Interestingly, after one century these questions are still challenging. In his 1907 study [68.23] *Lapicque* introduced a model of the nerve that was compared to data he obtained from frog nerve stimulation. This model, based on a simple capacitor circuit, would form the basis for later models of the membrane (as shown above in the case of the HH model). A milestone in understanding the effects of electrical stimulation of neuronal cells by external fields is proposed in the brilliant work of *McNeal* [68.24], which was the basis

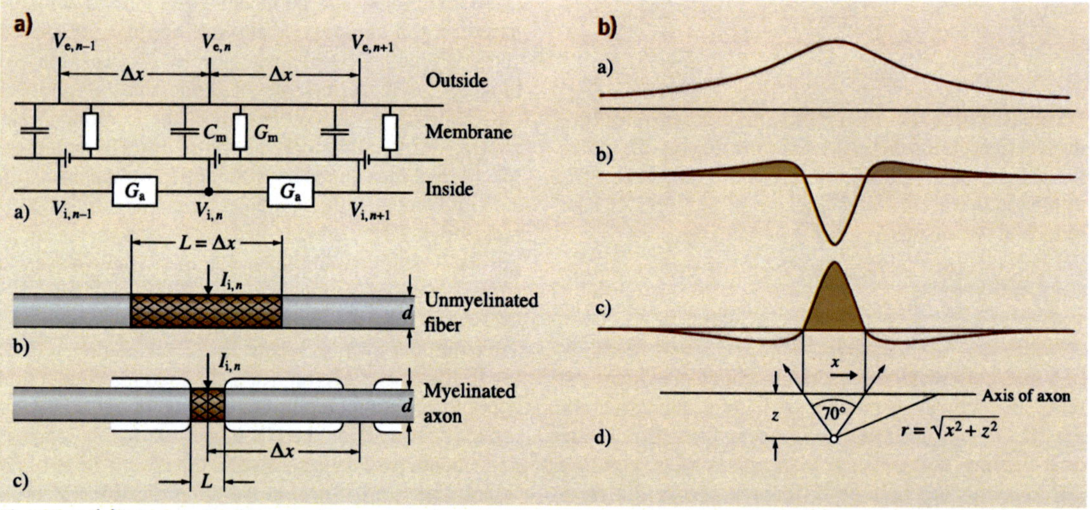

Fig. 68.2 (a) Electrical cable-circuit representation of a myelinated and an unmyelinated fiber. Making the limit for the intermodal distance (Δx) going to zero an analytic cable equation can be written, myelinated fibers are treated in an analogous way. (b) Stimulation with a monopolar electrode. a) The shape of the electric potential along the fiber, b) and c) the activating function f_n in (68.5) for anodic and cathodic current pulses. *Filled areas* represent the possible excitation zones (after [68.21] with permission)

for the so-called *activation function approach*. Furthermore, *Rattay* analytically enhanced the findings of McNeal [68.21, 25, 26] and made the calculations for the effects of monopolar, bipolar, and ring electrodes stimulation (Fig. 68.2) by using a simple circuital representation (Fig. 68.2a) and opportune approximations.

These works discovered the proportionality between the excitability of fibers and the second spatial derivative of the electric potential along the fiber (activating function), a very intuitive and useful tool for understanding the basic mechanisms of electrical stimulation effects.

For the case of myelinated fibers, the effect of external field on its excitation is, in fact, given by

$$f_n(t) = \frac{V_{e,n-1} - 2V_{e,n} + V_{e,n+1}}{\Delta x^2}\,, \tag{68.4}$$

which is the second difference quotient of the extracellular potential along the fiber. In the case of unmyelinated fibers, it simplifies to

$$f_n(t) = \frac{\partial^2 V_e(x,t)}{\partial x^2}\,. \tag{68.5}$$

Its interpretation is straightforward: An action potential can be generated at the fiber coordinate $x = n\Delta x$ if f_n is positive. In areas where f_n is negative, hyperpolarization is produced. So electrical excitability of fibers is given by the intrinsic cable-like structure of the fiber and is strongly affected by the shape of the electric potential along the direction of the fiber. This model is the point of junction between biophysics modeling and the electromagnetic models for external fields within the nerves.

Recently in analytical and computational treatment [68.28, 29], it has been shown that the activation function-based approach lead to considerable errors in prediction. Yet, it still remains the quickest and most intuitive tool for approximate estimation of the effects of neural stimulation.

With the increase of computational power, models could start to account for the realistic, point by point voltage distribution induced by current injected by means of stimulating electrode and the high nonlinearity present in axonal answers. Presently, the state-of-art models account for anisotropy of extracellular conductivity, present in real nerves, and also for the nonlinear response of cells to the extracellular stimulation. Those two aspects are solved separately: by means of the finite element method (FEM) solving for distribution of voltages generated by the injected currents and by estimating the axonal response to the electrical stimuli by calculation of neuronal dynamics. The numerical evaluation of neuronal dynamics is usually achieved using dedicated neuron computation oriented software. This kind of model could be called hybrid field-neuron models (or hybrid FEM/neuron models). A conceptual framework for these hybrid models was developed in the work on spinal cord epidural stimulations by *Coburn* [68.3, 30].

More recently, *McIntyre* and colleagues developed a model for extracellular stimulation of the central nervous system cells [68.20], which was consequently applied in different studies, with one aim being to optimize the design of cuff flat electrodes for peripheral nerve stimulation [68.5].

To couple the external electric fields with the fiber or cell, proper models to account for external stimulation must be developed. The state-of-the art, most used approach is the so-called compartmental model-

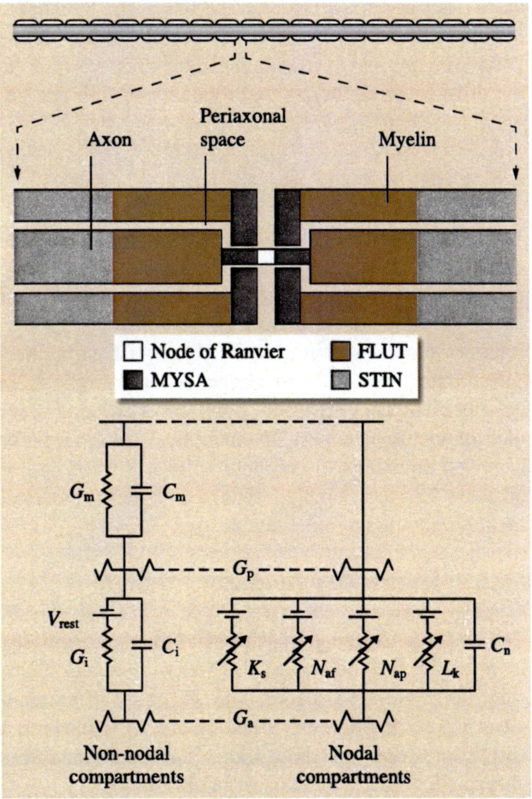

Fig. 68.3 Multicompartmental model of a myelinated fiber. The active part is represented by the Ranvier node while a realistic finite impedance representation of the myelin is exploited (after [68.27] with permission)

ing of fibers, which is based on the subdivision of fibers and cells into elementary circuit representation used to model the different parts of the cell or fiber, like axons, somas, and Ranvier nodes. Free academic software like NEURON or GENESIS can accurately reproduce the mechanism of generation of action potential of arbitrary complex cell representation. One of the most used models that implements compartmental modeling is the MRG model for the mammalian myelinated fiber (Fig. 68.3) proposed in [68.27].

This model represents the nonlinear modified Hodgkin–Huxley equations for the active compartment of the axons (Ranvier nodes) and a detailed realistic representation of the myelinated tracts. The success of this model is due to its capability to reproduce several experimental aspects of cells dynamics and to its availability; indeed this model can be found in the database available in [68.31].

The difference between state-of-the-art models resides basically in two aspects: the first is the membrane dynamics and the second is the representation of the compartments. With membrane dynamics we refer to the differential equations that relate the membrane potential to the extracellular potential; usually models differ for the choice of a particular implementation (Hodgkin–Huxley or Frankhenauser–Huxley) and on the numbers of ion channels implemented.

With compartment representation we refer to the choice of the type and number of compartments. For example, a myelinated fiber can be represented using only two compartments [68.15, 24], namely the active part, the Ranvier node, and the internodal myelinated tracts. The representation of the myelin as a perfect insulator or as a realistic resistive compartment, highly influences the outcome of the model, in particular regarding frequency effects [68.32]. Therefore, care should be taken in the choice of the model with respect to the desired outcome regarding both the biophysical representation and the compartmental representation.

However, even with the most accurate biophysical representation of the cell, or fiber, of interest, the shape of the fields generated in a stimulation process is critical for a correct prediction of nerve activation. So, accurate electromagnetic models of extracellular structures and electrodes should be implemented.

68.3 Finite Element (FE) Models

With the development of fiber models for external electrical stimulation, models of electrical sources and extracellular space became critical. Active fibers or brain, spinal cord, or peripheral nervous system neurons are placed within complex biological structures. These structures are inhomogeneous, anisotropic from an electrical and diffusion point of view, and have odd geometrical shapes. The analytical solution of the Maxwell equations to compute voltage profiles is impossible in such complex spaces and then numerical methods should be implemented

68.3.1 The Electromagnetic Problem

The electrical stimulation process of the nervous tissue is driven by the solution of the Maxwell equations inside proper solution spaces. The principles of external electrical stimulation are based on the hypothesis that current injection into tissues, of a sufficient amount, is able to elicit fiber excitation or inhibition.

Mathematically, the problem is defined by the solution of the Maxwell equations for a given space and a proper representation of the source (which can be a point process or a fully spatial representation of an electrode). In particular, the basic formulation is the Laplace equation, or alternatively the Poisson equation, for the electric potential V_e

$$\nabla \cdot \boldsymbol{\sigma} \nabla V_e = \nabla \mathbf{J}\,, \quad (68.6)$$

where $\mathbf{J}$ represents the injected current density and $\boldsymbol{\sigma}$ the conductivity tensor.

It must be stressed that this equation came from a manipulation of the fourth Maxwell equation in which we neglect the time derivative of the electric fields, using the so-called quasi-static approximation. The approximation is justified by the fact that at low frequencies (as those used for external stimulation e.g. less than MHz) wave effects can be neglected due to the specific conductivity values of tissues.

Detailed studies have been carried out to explore this basic work hypothesis and results seem to validate it, although with some limitation with respect to pulse shape and duration [68.33].

This linear partial differential equation coupled to proper boundary condition (usually of Dirichlet type) gives the solution for the stimulating electric potential. The complexity of the solution depends on the level of realism of the model, both in terms of geometrical shape and of conductivity properties.

The first models [68.24] used, for the sake of simplicity, an isotropic representation of the extracellular space, seen as an infinite space with constant and isotropic conductivity (or resistivity) σ_m. A current point source I (placed at $r = 0$) was used to represent the stimulating electrode.

In this case, the boundary conditions are

$$V_e(\infty) = 0 , \quad (68.7)$$

and thus the simple analytical solution is [68.24]

$$V_e(r) = \frac{1}{4\pi\sigma_m} \frac{I}{r} , \quad (68.8)$$

which shows the characteristic $1/r$ behavior of the monopolar solutions.

For the purpose of understanding the basic mechanisms of fiber activation and blocking, this model is a good compromise. Because, by neglecting a realistic and too complex representation of the extracellular environment, the scientist can focus on the axon properties. However, if the goal is to improve technologies or stimulating strategies of neuroprostheses a detailed representation of tissues and sources is needed.

In peripheral nerve stimulation, for example, fibers are located inside structures named fascicula, which are surrounded by a thin insulating membrane (the perinerium) and everything is embedded into the epinerium, which is the connective tissue that builds up the nerve structure (Fig. 68.4).

As demonstrated in [68.35] the shape and electrical properties of tissues can bias not only (as obvious) the solution of the Maxwell equations, but also the final prediction of the model in terms of a nerve's excitability. Since some of the essential aspects for FES are the currents necessary to elicit fiber selectivity (that is the ability to activate just the population of without stimulating the other populations), and the capacity to overcome the inverse recruitment problem [68.6] by means of more graduated activation of muscle, it is important to understand how those properties are affected by different tissues representations.

Excitation thresholds, and thus predicted selectivity and steepness of the fiber recruitment, will be affected strongly by different perinerium representations. The realistic representation of tissues inside the Maxwell equations is done by the definition of the geometrical space, and its electrical properties, by means of the tissue specific conductivity tensor. The problem is then defined in a geometrically complex anisotropic nonhomogeneous space where analytical solution is not possible.

68.3.2 The Finite Element Method (FEM)

The finite element method (FEM)is a standard numerical process used to solve partial differential equations in one or more variables in arbitrary complex spaces (for details see [68.36]).

It is based on the discretization of the geometrical space into primitive elements (mesh) where the equations are numerically integrated and then assembled into a global solution.

The subdivision of the whole space into primitive elements allows for an element-wise characterization of material properties. For the case of the electrical stimulation of tissues, the finite element method is used to solve the Maxwell equations, usually in Laplace form for the electric potential. Arbitrary complex structures such as cerebral cortex [68.1] spinal cord [68.3, 20] or peripheral nerve [68.5, 6] can be represented and

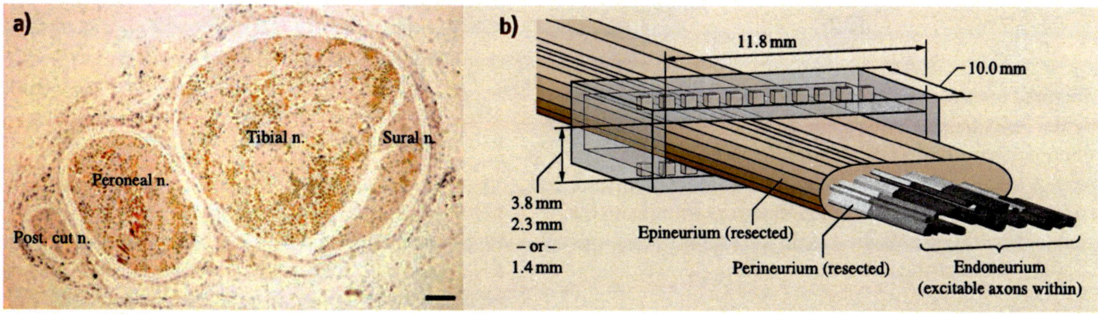

Fig. 68.4 **(a)** Histological image of a rat's sciatic nerve. Tibial, sural and, peroneal fascicles are visible (taken with permission from [68.34]). Bar = 100 μm; **(b)** a three-dimensional FE model of the nerve surrounded by a flat interface nerve electrode (FINE) (after [68.5] with permission)

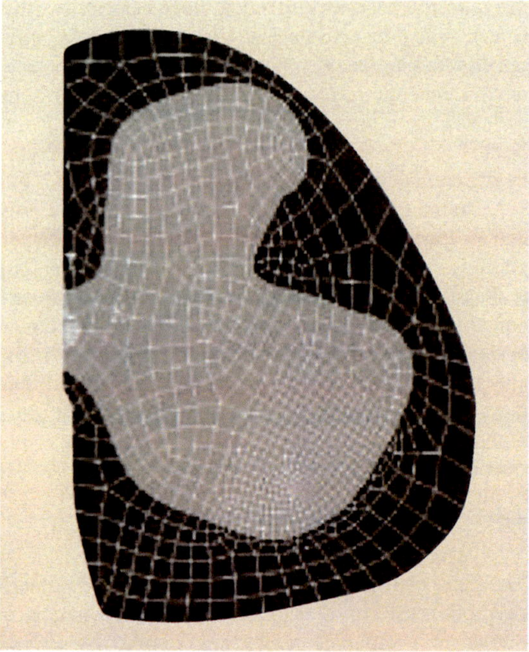

Fig. 68.5 An example of a half cat spinal cord meshed with nonuniform hexahedra (after [68.20])

electromagnetically characterized (in terms of conductivities).

The representation of structures as conductors is known as the volume conductor model. It means that structures are treated as volumes and characterized by a conductivity tensor.

The numerical process seeks for a polynomial approximation of the solution of the problem into the defined finite volume Ω. To guarantee a numerical solution, the equation system is *relaxed* into the so-called weak form

$$\int_{\Omega} v\nabla \cdot \sigma \nabla V_e = \int_{\Omega} v\nabla J \tag{68.9}$$

obtained after the multiplication of (68.6) on both sides by the function v, which can be any function with a continuous derivative, and integration on both sides. The role of the function v is to relax the mathematical constraints for the existence of the numerical solution to the problem by the use of the Green's formula. Thus, distributing the derivates over the two variables we obtain

$$\int_{\Omega} \nabla v \cdot \boldsymbol{\sigma} \nabla \boldsymbol{u} - \int_{\partial\Omega} v\hat{\boldsymbol{n}} \cdot (\boldsymbol{\sigma} \nabla \boldsymbol{u}) = - \int_{\Omega} v\nabla \cdot \mathbf{J}_p \,. \tag{68.10}$$

In this form the computation of the integral is made easier by only the presence of first-order derivates.

The subdivision of the whole volume into primitive elements allows for the solution of the weak form in each volume element with a finite set of basis functions. The basis functions serve as a local polynomial approximation of the solution in each primitive element; the degree of these functions along with the size of the elements affect the numerical solution quality. A basis function is a polynomial of given order with small support such that its value is $\psi_i = 1$ on the associated ith mesh node and $\psi_i = 0$ on every other mesh node. With such functions it is possible to represent the solution of the problem as a linear combination of the basis function and the solution at each mesh point

$$V_e(x) = \sum_{i=1}^{N} V_e(x_i)\psi_i(x) \tag{68.11}$$

and in an analogous way with the function v and the sources $\boldsymbol{J}$. Then the integrals can be solved as products between basis functions and conductivity tensor values for each volume element and assembled into a matrix $\mathbf{A}$ named stiffness matrix.

The discretized weak form is then expressed as a linear system

$$\mathbf{A} \cdot V_e = b \,, \tag{68.12}$$

which by numerical inversion gives the solution V_e for each point of the mesh.

Boundary conditions are critical to the solution of the problem, and care should be taken when defining the problem (which can have different boundary conditions with different stimulating devices). In this kind of problem sources are usually not explicitly represented by a discretized current density as in (68.1) but are expressed through continuity conditions at the boundary between media of different properties (like at the surface of an active site). In this way, the problem to be solved is expressed as a Laplace equation.

The general approach to face this problem is:

1. Choosing the proper source representation, depending on the stimulating device of interest (cuff electrodes for PNS, intraneural electrodes for PNS, microelectrodes for brain stimulation, spinal electrodes for epidural or intraspinal stimulation etc.).
2. Choosing the proper set of boundary conditions for the problem.

3. Discretization and construction of a mesh for the geometrical extracellular space in which the fiber or cell is placed.
4. Computation of the solution of the problem.

The computed solutions should then be interpolated into the fiber or cell model and used as an external stimulation process. One of the first examples of this approach is the pioneer work of *Coburn* [68.3, 30], where he computed the activation of fibers during epidural stimulation of the spinal cord. The flexibility of this approach is reflected in the number of applications applied today: peripheral nerve stimulation [68.5,6], spinal cord microstimulation [68.20], deep brain stimulation [68.1], and source localization in peripheral nerves [68.37].

The ability to model more and more complex structures and detailed biophysical characteristics of cells and fibers will lead to a better understanding of the basic principles of neurophysiology and to the optimization of actual technologies and strategies for their use. Realistic representation of anatomical structures and electrodes shapes will improve the predictions of models and help in finding new ways for stimulation and recording. Some details about the use of this model in practical applications can be found in the next section.

68.3.3 Models in Practice

The literature is rich with models of cells, fibers, and membrane dynamics, and each of these can be used in practice. So, which mode is suited for each application? The model that one should use in one's own practical or theoretical applications should not necessarily be the *last one* or that used by everybody. Rather it depends strictly on the field of interest. If the aim of a study is only to understand if some device is amenable for a particular application, then also an approximate approach, like the one proposed in the concept of the *activation function* may be an excellent tool to get an answer.

As has already been mentioned, there are many models that are amenable for electric external stimulation. However, if we focus on neuroprosthetic applications, the use of models can be summarized as follows:

1. Search for optimal parameters of stimulation in order to obtain the most effective and efficient stimuli
2. Search for the effect of geometry and placement of stimulating electrodes
3. Search for a basic understanding of therapeutical effect of a certain device, on a cellular or a systemic level.

Stimulation Parameters

Already in the very first experimental studies done in this field research tried to determine the influence of the pulse width (PW) of the current stimuli, together with frequency (f) and amplitude (a). The analysis of the strength–duration relationship for rectangular waveforms as defined by *Lapicque* [68.23] or *Weiss* [68.38] shows that a short stimulus results in greater charge efficiency, while a long stimulus results in greater power efficiency; a PW equal to the chronaxie time optimizes energy efficiency.

The effects of stimulation parameters (PW, f, a) are still the object of very intensive modeling and experimental approaches. PW effects have been studied in, among others, [68.39]. They used a genetic algorithm (GA) coupled to the compartmental model of the axonal population (MRG model) to determine the energy-optimal waveform shape for neural stimulation, making also a preliminary validation of their model in an animal model. The GA found waveforms that could substantially increase the battery life of implanted stimulators (e.g. in deep brain stimulators).

Sahin and *Tie*'s work [68.40], is an excellent example of suitable choice of models. They preferred using *Sweeny*'s model [68.41], which is similar to HH, in terms of the gate variables except for the absence of the potassium channel. They searched for the most efficient waveform that maximizes the charge injection capacity of the electrode while providing the lowest threshold charge for neural activation. Linear and exponential decrease and Gaussian waveforms were found to be the most efficient pulse shapes, giving very important hints for future experimental studies (at the moment the most conventional and almost exclusively used waveform is the rectangular one). A very common problem in FES (neural and muscular) is the so-called inverse-recruitment (e.g. the artificial activation first of the bigger diameter and then of the smaller diameter fibers, which is the opposite of the natural activation order). This problem can be addressed by studying a proper waveform to minimize inverse recruitment. There are several modeling studies trying to face this problem [68.42–45]. All are based on physical considerations that came up thanks to fiber models. In particular, they try to use two properties of fiber excitation. The first is the fact that larger diameter fibers (which in normal conditions mean also longer internodal distances) have lower excitation thresholds. This can be easily explained by the activation function concept: on larger distances the voltage quotient (68.4) is larger and then the probability of excitation is higher.

Secondly the inverse relationship of the induced voltage changes with the distance (68.8), which makes fibers closer to the electrode more sensitive to excitation. Having these properties in mind, it is possible to study proper waveform profiles that make the excitation of smaller fibers easier than larger ones; usually these waveforms have some kind of hyperpolarizing prepulse to block larger diameter fibers, and then the subsequent excitation should occur first for the smaller one, thus restoring normal recruitment [68.44, 45].

Geometrical Parameters

In searching the geometrical parameters for the optimal electrode design, or its optimal placement, many models were developed. In a recent study for pudendal afferent fibers stimulation [68.46], the FEM model has been coupled to the activation function in order to explore the parameters for different electrode geometries, finding a ring electrode to be the optimal one for the application. The results of the study were also validated in an animal (cats) model.

There are several works regarding the study of stimulation properties of different electrode geometries. Regarding extraneural stimulation, in the works [68.5,41] intensive simulation on cuff geometry showed that a flat interface can be more selective than a simple cuff and also that selectivity can be enhanced using proper stimulation paradigms on multicontact cuffs. There is not much work on intraneural stimulation, because the meshing task to model the electrode can become a critical issue. Therefore, the majority of works use point sources to stimulate nerves or the spinal cord [68.3, 47, 48]. These models, however, give important hints on which fibers can be stimulated with intraneural electrodes, and most of all they demonstrate that intraneural electrodes can reduce the inverse recruitment without particular waveforms, but just because of the fact that currents are located within the fibers.

Interpretation of Biophysical Mechanisms

A very important aspect of models is the fact that they can be used to understand the neurophysiological basis of neuroprosthetic devices. An example is the work of *McIntyre* and coworkers [68.1] on deep brain stimulation (DBS). Their models helped the understanding of the physiological mechanisms underlying DBS techniques. They were able to validate their hypothesis in experimental studies and, therefore, gave an insight into what is going on with DBS and why DBS is able to modulate brain activity.

Critical Remarks

The major problem of most of the discussed models is the general lack of rigorous experimental validations. Experimental validation is essential to understand if the hypotheses are valid and to permit us to correct the parameters of the model. Moreover, together with the model output, they can also lead towards falsification of eventual incorrect scientific assumptions. A very important aspect is that models contain only the things we implement inside, and we should be always critical of the results obtained. It would be very dangerous if a model were highly specialized to perfectly reproduce results just in one case (like a kind of overfitting), and then just slight change of parameters would destroy the stability. Therefore, whenever possible, a sensitivity analysis [68.49] should be performed to test the robustness of the model. Having a robust model, with respect to the parameters used for its construction, can also permit us to observe and understand some emerging property of the system under study, which could be difficult to do in the case of very specialized ones.

68.4 Conclusion

Models can then assess different issue in the neuroprostheses design and study. They can help during the design phase by giving hints to geometrical parameters or stimulation parameters, but they can also be used to understand if mathematical modeling of neural models is able to reproduce experimental results. This exploits the intrinsic property of science, modeling to understand reality and to improve technologies.

References

68.1 C.C. McIntyre, W.M. Grill, L.D. Sherman, N.V. Thakor: Cellular effects of deep brain stimulation: Model-based analysis of activation and inhibition, J. Neurophysiol. **91**, 1457–1469 (2004)

68.2 G. Courtine, Y. Gerasimenko, R. van den Brand, A. Yew, P. Musienko, H. Zhong, B. Song, Y. Ao, R.M. Ichiyama, I. Lavrov, R.R. Roy, M.V. Sofroniew, V.R. Edgerton: Transformation of non-functional spinal cicuits into functional states after the loss of brain input, Nat. Neurosci. **12**, 1333–1342 (2009)

68.3 B. Coburn, W.K. Sin: A theoretical study of epidural electrical stimulation of the spinal cord part I: Finite element analysis of stimulus fields, IEEE Trans. Biomed. Eng. **11**, 971–977 (1985)

68.4 S. Micera, X. Navarro, J. Carpaneto, L. Citi, O. Tonet, P.M. Rossini, M.C. Carrozza, K.P. Hoffmann, M. Vivó, K. Yoshida, P. Dario: On the use of longitudinal intrafascicular peripheral interfaces for the control of cybernetic hand prostheses in amputees, IEEE Trans. Neural Syst. Rehabil. Eng. **16**(5), 453–472 (2008)

68.5 M.A. Schiefer, R.J. Triolo, D.J. Tyler: A model of selective activation of the femoral nerve with a flat interface nerve electrode for a lower extremity neuroprosthesis, IEEE Trans. Neural Syst. Rehabil. Eng. **16**(2), 195–204 (2008)

68.6 Z. Lertmanorat, D.M. Durand: Extracellular voltage profile for reversing the recruitment order of peripheral nerve stimulation: A simulation study, Neural Eng. **1**, 202–211 (2004)

68.7 A.L. Hodgkin, A.F. Huxley: A quantitative description of membrane current and its application to conduction and excitation in nerve, J. Physiol. **117**(4), 500–544 (1952)

68.8 A.L. Hodgkin, A.F. Huxley, B. Katz: Measurement of current–voltage relations in the membrane of the giant axon of Loligo, J. Physiol. **116**, 424–448 (1952)

68.9 C. Koch: *Biophysics of Computation: Information Processing in Single Neurons* (Oxford University Press, New York 1999)

68.10 D.E. Goldman: Potential, impedance, and rectification in membranes, J. Gen. Physiol. **27**(1), 37–60 (1943)

68.11 A.L. Hodgkin, B. Katz: The effect of sodium ions on the electrical activity of the giant axon of the squid, J. Physiol. **108**, 37–77 (1949)

68.12 A.C. Smith: The electrophysics of a nerve fiber, Rev. Mod. Phys. **47**, 487–533 (1975)

68.13 B. Frankenhaeuser, A.F. Huxley: The action potential in the myelinated nerve fiber of *Xenopus laevis* as computed on the basis of voltage clamp data, J. Physiol. **171**, 302–315 (1964)

68.14 R. Fitzhugh: Thresholds and plateaus in the Hodgkin–Huxley nerve equations, J. Gen. Physiol. **43**, 867–896 (1960)

68.15 R. Fitzhugh: Computation of impulse initiation and saltatory conduction in a myelinated nerve fiber, Biophys J. **2**(1), 11–21 (1962)

68.16 J. Nagumo, S. Arimoto, S. Yoshizawa: An active pulse transmission line simulating nerve axon, Proc. IRE, **50**, 2061–2070 (1962)

68.17 M.L. Hines, N.T. Carnevale: NEURON: A tool for neuroscientists, Neuroscientist **7**, 123–135 (2001)

68.18 M.L. Hines, N.T. Carnevale: The neuron stimulation environment, Neural Comput. **9**(2), 195–204 (1997)

68.19 http://genesis-sim.org/ (last accessed April 20, 2011)

68.20 C.C. McIntyre, W.M. Grill: Extracellular stimulation of central neurons: Influence of stimulus waveform and frequency on neuronal output, J. Neurophysiol. **88**, 1592–1604 (2002)

68.21 F. Rattay: Analysis of models for extracellular fiber stimulation, IEEE Trans. Biomed. Eng. **36**(7), 676–682 (1989)

68.22 W. Gerstner, W. Kistler: *Spiking Neuron Models* (Cambridge University Press, Cambridge 2002)

68.23 L. Lapicque: Recherches quantitatives sur l'excitation electrique des nerfs traitee comme une polarisation, J. Physiol. **9**, 622–635 (1907), (in French)

68.24 D.R. McNeal: Analysis of a model for excitation of myelinated nerve, IEEE Trans. Biomed. Eng. **23**(4), 329–337 (1976)

68.25 F. Rattay: Analysis of models for external stimulation of axons, IEEE Trans. Biomed. Eng. **33**, 974–977 (1986)

68.26 F. Rattay: Ways to approximate current-distance relations for electrically stimulated fibers, J. Theor. Biol. **125**, 339–349 (1987)

68.27 C.C. McIntyre, A.G. Richardson, W.M. Grill: Modeling the excitability of mammalian nerve fibers: Influence of afterpotentials on the recovery cycle, J. Neurophysiol. **87**, 995–1006 (2002)

68.28 C.M. Zierhofer: Analysis of a linear model for electrical stimulation of axons-critical remarks on the activating function concept, IEEE Trans. Biomed. Eng. **48**(2), 173–184 (2001)

68.29 M.A. Moffitt, C.C. McIntyre, W.M. Grill: Prediction of myelinated nerve fiber stimulation thresholds: limitations of linear models, IEEE Trans. Biomed. Eng. **51**(2), 229–236 (2004)

68.30 B. Coburn: A theoretical study of epidural electrical stimulation of the spinal cord – part II: Effects on long myelinated fibers, IEEE Trans. Biomed. Eng. **11**, 978–986 (1985)

68.31 http://senselab.med.yale.edu/ (last accessed April 20, 2011)

68.32 A.G. Richardson, C.C. McIntyre, W.M. Grill: Modeling the effects of electric fields on nerve fibres: Influence of the myelin sheath, Med. Biol. Eng. Comput. **38**, 438–446 (2000)

68.33 C.A. Bossetti, M.J. Birdno, W.M. Grill: Analysis of the quasi-static approximation for calculating potentials generated by neural stimulation, J. Neural Eng. **5**(1), 44–53 (2008)

68.34 J. Badia: Topographical distribution of motor fascicles in the sciatic-tibial nerve of the rat, Muscle Nerve **42**(2), 192–201 (2010)

68.35 Y. Grinberg, M.A. Schiefer, D.J. Tyler, K.J. Gustafson: Fascicular perineurium thickness, size, and posi-

tion affect model predictions of neural excitation, IEEE Trans. Neural Syst. Rehabil. Eng. **16**(6), 572–581 (2008)
68.36 O.C. Zienkiewicz, R.L. Taylor: *Finite Element Method: Its Basis and Fundamentals* (Elsevier, Milan 2000)
68.37 J. Zariffa, M.R. Popovic: Solution space reduction in the peripheral nerve source localization problem using forward field similarities, J. Neural Eng. **5**, 191–202 (2008)
68.38 G. Weiss: Sur la possibilite de rendre comparables entre eux les appareils servant a l'excitation electrique, Arch. Ital. Biol. **35**, 413–446 (1901), (in French)
68.39 A. Wongsarnpigoon, W.M. Grill: Energy-efficient waveform shapes for neural stimulation revealed with a genetic algorithm, J. Neural Eng. **7**, 1–11 (2010)
68.40 M. Sahin, Y. Tie: Non-rectangular waveforms for neural stimulation with practical electrodes, J. Neural Eng. **4**, 227–233 (2007)
68.41 J.D. Sweeney, J.T. Mortimer, D. Durand: Modeling of mammalian myelinated nerve for functional neuromuscular electrostimulation, IEEE 9th Annu. Conf. Eng. Med. Biol. Soc. (IEEE, 1987) pp. 1577–1578
68.42 K.E. Deurloo, J. Holsheimer, P. Bergveld: The effect of subthreshold prepulses on the recruitment order in a nerve trunk analyzed in a simple and a realistic volume conductor model, Biol. Cybern. **85**, 281–291 (2001)
68.43 Z.P. Fang, J.T. Mortimer: Selective activation of small motor axons by quasitrapezoidal current pulses, IEEE Trans. Biomed. Eng. **38**(2), 168–174 (1991)
68.44 W.M. Grill, J.T. Mortimer: Inversion of the current–distance relationship by transient depolarization, IEEE Trans. Biomed. Eng. **44**, 1–9 (1997)
68.45 A. Vuckovic, N.J.M. Rijkhoff, J.J. Struijk: Different pulse shapes to obtain small fiber selective activation by anodal blocking – A simulation study, IEEE Trans. Biomed. Eng. **51**(5), 698–706 (2004)
68.46 J.P. Woock, P.B. Yoo, W.M. Grill: Finite element modeling and in vivo analysis of electrode configurations for selective stimulation of pudendal afferent fibers, BMC Urology **10**(11), 1–12 (2010)
68.47 P.H. Veltink, J.A. Van Alsté, H.B.K. Boom: Simulation of intrafascicular and extraneural nerve stimulation, IEEE Trans. Biomed. Eng. **35**(1), 69–75 (1988)
68.48 J.J. Struijk: Recruitment of dorsal column fibers in spinal cord stimulation: Influence of collateral branching, IEEE Trans. Biomed. Eng. **39**(9), 903–912 (1992)
68.49 C.C. McIntyre, W.M. Grill: Sensitivity analysis of a model of mammalian neural membrane, Biol. Cybern. **79**, 29–37 (1998)

Part G

Equipme

Part G Equipment and Tools

69 **Operating Tables – the Surgeon's Workplace**
Bernhard Kulik, Rastatt, Germany

70 **Medical Robotics**
Harald Fischer, Weingarten, Germany
Udo Voges, Eggenstein-Leopoldshafen, Germany

71 **Incubators**
Guenther Braun, Freiburg, Germany
Roland Hentschel, Freiburg, Germany

72 **Surgical Scissors**
Reiner Haag, Fridingen, Germany
Wilfried Storz, Tuttlingen, Germany

73 **Intelligent Textiles and Trends**
Christian Rotsch, Dresden, Germany
Sibylle Hanus, Greiz, Germany
Danny Schwabe, Greiz, Germany
Heike Oschatz, Greiz, Germany
Andreas Neudeck, Greiz, Germany
Uwe Möhring, Greiz, Germany

74 **Electronics in Medicine**
Paulo M. Mendes, Guimarães, Portugal
Celso P. Figueiredo, Guimarães, Portugal
Mariana Fernandes, Guimarães, Portugal
Óscar S. Gama, Braga, Portugal

69. Operating Tables – the Surgeon's Workplace

Bernhard Kulik

The central element of any operating theater is without a doubt the operating table. Wherever surgical interventions are carried out, operating tables are essential. Accordingly, the variety of tables available is very wide, ranging from simple, mobile operating room (OR) tables, right up to OR table systems with various, special OR table tops.

The type, design and functional properties of the OR table depend on the surgical discipline, the ergonomic requirements of the surgical team and, last but not least, on the financial resources available.

69.1 The History of the Operating Table 1262

69.2 The OR Table System 1263

69.3 Technology of Operating Room Table Systems 1264
69.3.1 Construction of an OR Table System 1264
69.3.2 Mobility and Flexibility of OR Table Systems 1265

69.4 Safe Patient Positioning........................ 1268
69.4.1 Decubitus Injuries 1269
69.4.2 Long-Term Position Injuries and Legal Responsibility 1270
69.4.3 Patient Positioning When Deploying Single-Pole HF Devices. 1271

69.5 Preparation: Care, Maintenance, and Hygiene 1271
69.5.1 Manual Cleaning and Disinfection 1271
69.5.2 Automatic OR Table System Cleaning and Hygiene 1271

Simple OR table designs are usually deployed in hospitals with smaller surgical departments as well as in larger hospitals with decentralized operating theaters like, for example, those with emergency department operating rooms (Fig. 69.1).

They are also often used for out-patient surgery, in so-called day clinics, or out-patient operating cen-

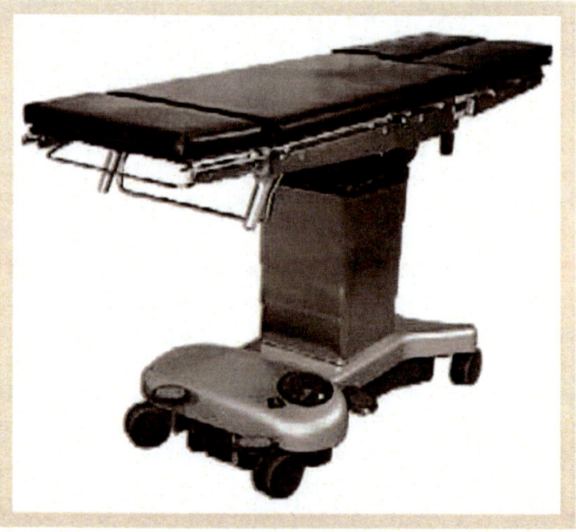

Fig. 69.1 Operating table for outpatient surgery

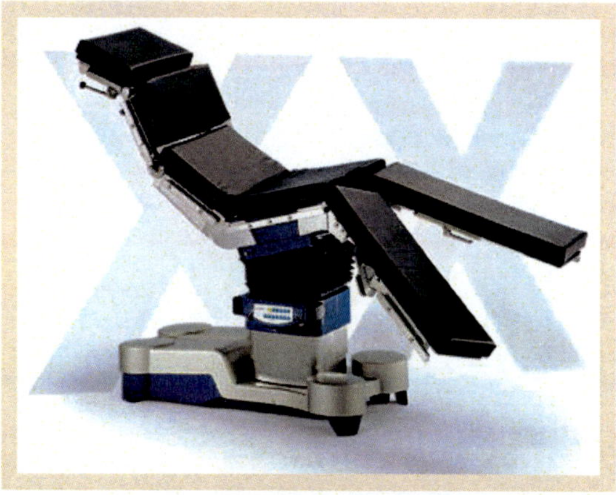

Fig. 69.2 Mobile operating table with electrical autodrive function

ters. These OR tables can be upgraded and modified by adding various accessories and, depending on the model, can be adjusted manually or electrically. An electrical autodrive function may also be implemented, assuming patient logistics calls for it (Fig. 69.2).

Modern OR table systems are characterized by the fact that the intervention-centric table tops mean they are optimized, on the one hand, to meet the special requirements of the individual surgical disciplines regarding the best patient positioning and best access to the surgical field and, as such, are a significant factor contributing towards the success of the surgical intervention. On the other hand, however, they also meet technical requirements in terms of stability and hygiene and the use of x-ray equipment.

69.1 The History of the Operating Table

Some 150 years ago, when asepsis was not even on the agenda, operations were normally carried out in the patient's bed. However, because of low bed heights, surgeons started to place their patients on higher tables, in order to gain better access to the surgical field and to give themselves a more ergonomic working position (Fig. 69.3).

The development of OR tables has kept pace with the development of surgery itself. It has occurred more or less in parallel with the continually expanding knowledge and ability of the surgeons and, over time, has been significantly influenced by specialization in the individual surgical disciplines.

Medical developments have called for improved access to the surgical field and, therefore, brought about improved patient positioning. This resulted in the emergence of *OR furniture* with table tops divided up into head, back, seat, and leg plates (Fig. 69.4).

Improvements in operating techniques made it necessary to be able to raise certain parts of the body (at the point of incision), to provide better access to the interior of the body and then to return them again to the flat position, after closing up.

Fig. 69.4 Segmented operating table top

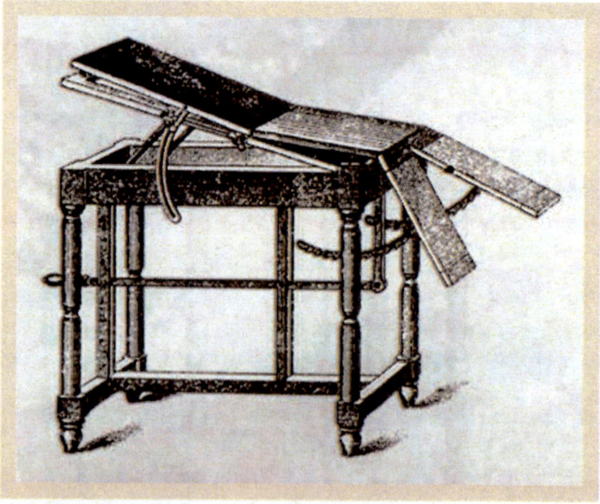

Fig. 69.3 Historic operating room furniture

Over the course of the last 70 years, general surgery has divided into individual surgical disciplines. The result of surgical specialization has been the development of special OR tables that differ in terms of the table top geometry and the layout of the various operating elements.

In the case of head operations, for example, no operating elements may be installed in this area of the OR table, as they may not be handled by nonsterile personnel during the operation, which could impact on overall sterility. As a result, special OR tables were developed for head operations.

The development of OR tables, with ever more complex table top adjustment options for other specialist surgical disciplines, followed at the same time.

A further significant factor that influenced the development of OR tables were the achievements of intraoperative imaging. It is impossible to imagine an OR these days without a mobile x-ray amplifier (Fig. 69.5).

The call for the combining of further imaging processes such as CT and MR with OR tables, has led to an increase in the use of materials other than steel, such

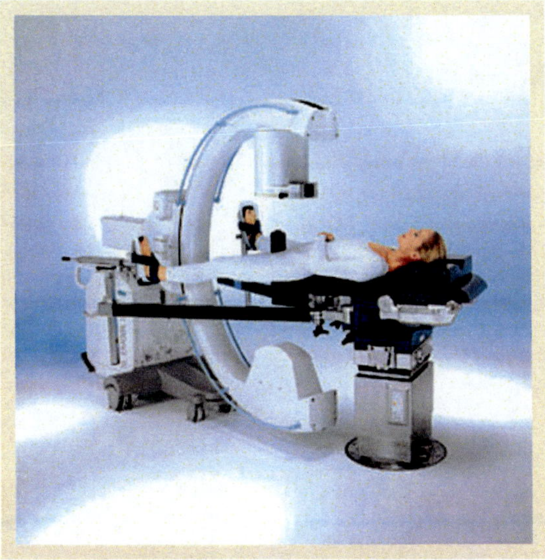

Fig. 69.5 OR table for use with mobile x-ray amplifier

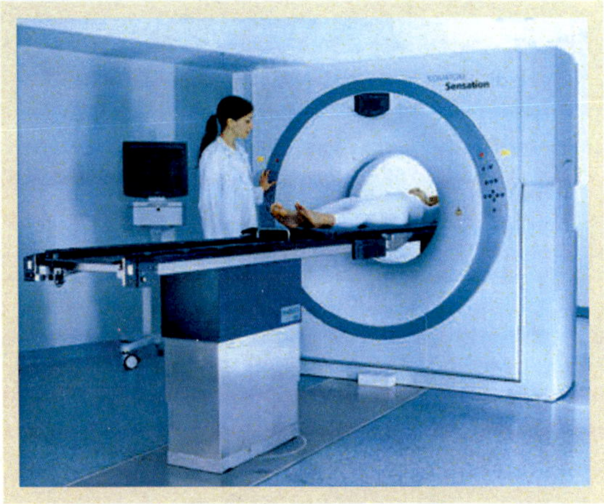

Fig. 69.6 Advanced workplace for image guided surgery (AWIGS)

as plastics reinforced with carbon fiber, as these ensure low-artifact radiolucence (Fig. 69.6).

69.2 The OR Table System

As surgeons required ever more specialized OR tables, tailored to particular operations, the development of mobile OR tables was extended to include, not just the division of the table tops but also to the table columns and/or their separation from the actual table tops, which brought with it further advantages and simplifications. The idea of an OR table system consisting of a column, a removable table top, as well as a transporter for the table top was born (Fig. 69.7a–c).

The predecessors of the current modern OR table systems had a mobile table column mounted on castors, so that the OR table could be moved and was not tied to one place, but the table top was fixed to the column. This had the disadvantage that the compact unit was very heavy for a single person to move, which had an adverse effect on mobility. As a result, the column and the table top were separated.

Modern OR table systems are available both as stationary, i. e. fixed versions, as well as mobile ones. With the stationary version, the table column is anchored to the floor and the electrics are fed through ducts in the flooring. The OR table top can be removed and is positioned over the table column using a transporter, (also known as a Lafette) and then transferred from the transporter so that it can then be wheeled away again.

Fig. 69.7a–c OR table system. **(a)** Column for stationary operating table. **(b)** Removable table top. **(c)** Transporter for the table top ▲▶

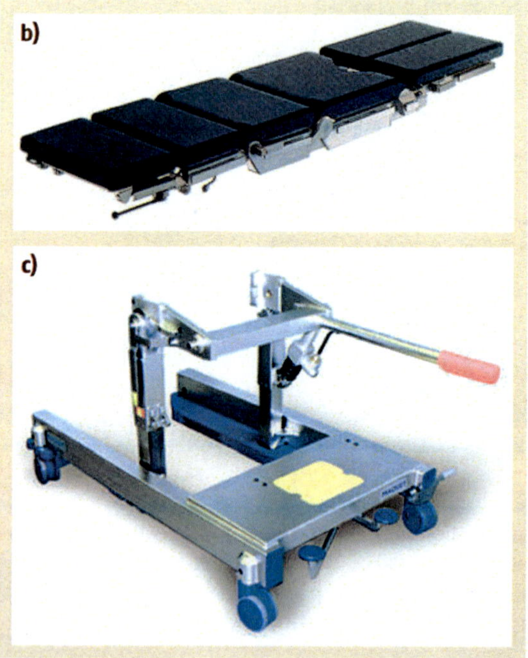

Fig. 69.7 (continued)

Fig. 69.8 Modular assembly

As the table column is bolted to the floor, there is no need for the deployment of a mobile OR table base. This gives the surgeons much more leg-room. A further advantage is that additional medical devices that are required, such as a mobile image amplifier or C-arms can be deployed and positioned at the surgical field without difficulties. The significant advantage of a mobile OR table system, on the other hand, is that this system can be moved freely within the operating theater or department. There is no need for electrical installations, as this OR system has rechargeable batteries to provide the required power, with sufficient capacity to be able to carry out some 100 operations under normal OR conditions.

OR tables or table systems are controlled using a hand-held device connected either wirelessly using infrared signals or via cable. In this way, the electrically adjustable functions can be activated from a distance, without impacting on the sterile OR environment. Individual segments of the table top, such as leg or head plates, can be removed (Fig. 69.8).

Technological advances in the area of medical technology have led to the development of high quality OR table systems with multifunctional system properties that fulfil today's high medical, hygiene and technical requirements.

69.3 Technology of Operating Room Table Systems

69.3.1 Construction of an OR Table System

The individual elements that make up the construction of an operating room table system can be described as follows:

Table Columns

An OR table system is available with a stationary column, i. e. one that is fixed to the floor, and therefore fixed to a particular location or with a mobile column, i. e. one that can be moved to different locations. Both types have their own, equally important advantages (Fig. 69.9a–c).

OR Table Tops

OR table columns can be equipped with many different, exchangeable table tops that meet the special requirements of the individual surgical disciplines and whose

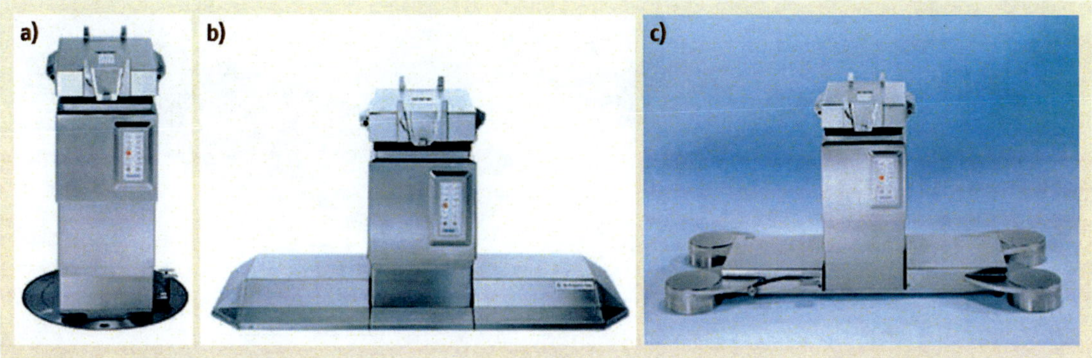

Fig. 69.9a–c Operating room table system columns. **(a)** Stationary column, **(b)** mobile column, **(c)** wheeled column

segments can be adjusted independently of each other, to achieve the required position. This means that the OR table top can be divided up in different ways, e.g. 4-, 5-, 6- or 8-way, in terms of:

- Head and foot sections, which can be swiveled up and down and removed if required.
- Middle or back sections that normally have multiple divisions and can be adjusted to achieve special slope angles.
- Leg sections that can consist of a pair of leg plates that are separated laterally or longitudinally and that can be adjusted automatically, with the option of moving or spreading both sections in a particular direction, either at the same time or independently of each other.

The OR table top is designed in such a way that adding or removing individual segments enables it to be lengthened or shortened as required. It should also have a longitudinal shift in order to facilitate whole body movement using the C-arms (Fig. 69.10a–c).

Operating Room Table Pads

The pads need to fulfill various functions. To start with, the patient must be positioned safely and carefully. The comfort of the patient is of less importance here, as the surgical procedure is done under anesthesia; it is more important to provide the greatest possible protection against nerve damage and pressure necrosis. Furthermore, the pads must be such that the surgical team has the best possible access to the area of the operation. As a result, the operation can be completed in the shortest possible time, which is particularly advantageous to the patient. Finally, the pads must comply with the strictest hygiene requirements, meaning that they must be easy to clean and disinfect.

Controllers

The table columns and the respective table tops are controlled wirelessly through an infrared or radio remote control. Alternatively, they can also be controlled using a cabled controller. In special application areas, control is also possible through a foot-operated switch, operated by the surgeon. Additional control options include the wall-mounted operating panel or an integrated OR control system, which controls the most important OR functions such as lights, OR table, endoscopy equipment, etc., either via a touchscreen monitor or though voice control (Fig. 69.11a,b).

Transporter

Using the transporter (Lafette), which is easy both to move and to manoeuvre, the OR table top is transported to the mobile or stationary table column, which then takes over the support of the table top for the duration of the operation. Lafettes are also used for transportation over longer distances and the exchange of table tops in the logistic workflow between bed transfer, pre-operation, operation, post-operation, and the return to bed transfer (Fig. 69.7c).

Accessories. Standard accessories include arm positioning equipment, lateral supports, infusion stands, leg supports, head rings, back and pelvic supports, and hand and body straps, etc.

69.3.2 Mobility and Flexibility of OR Table Systems

Modern OR table systems are characterized by their mobility, flexibility, and compatibility. It is not just the fact that a mobile OR table column allows you to select the location in the operating theater as required, it is also the fact that there is compatibility between the var-

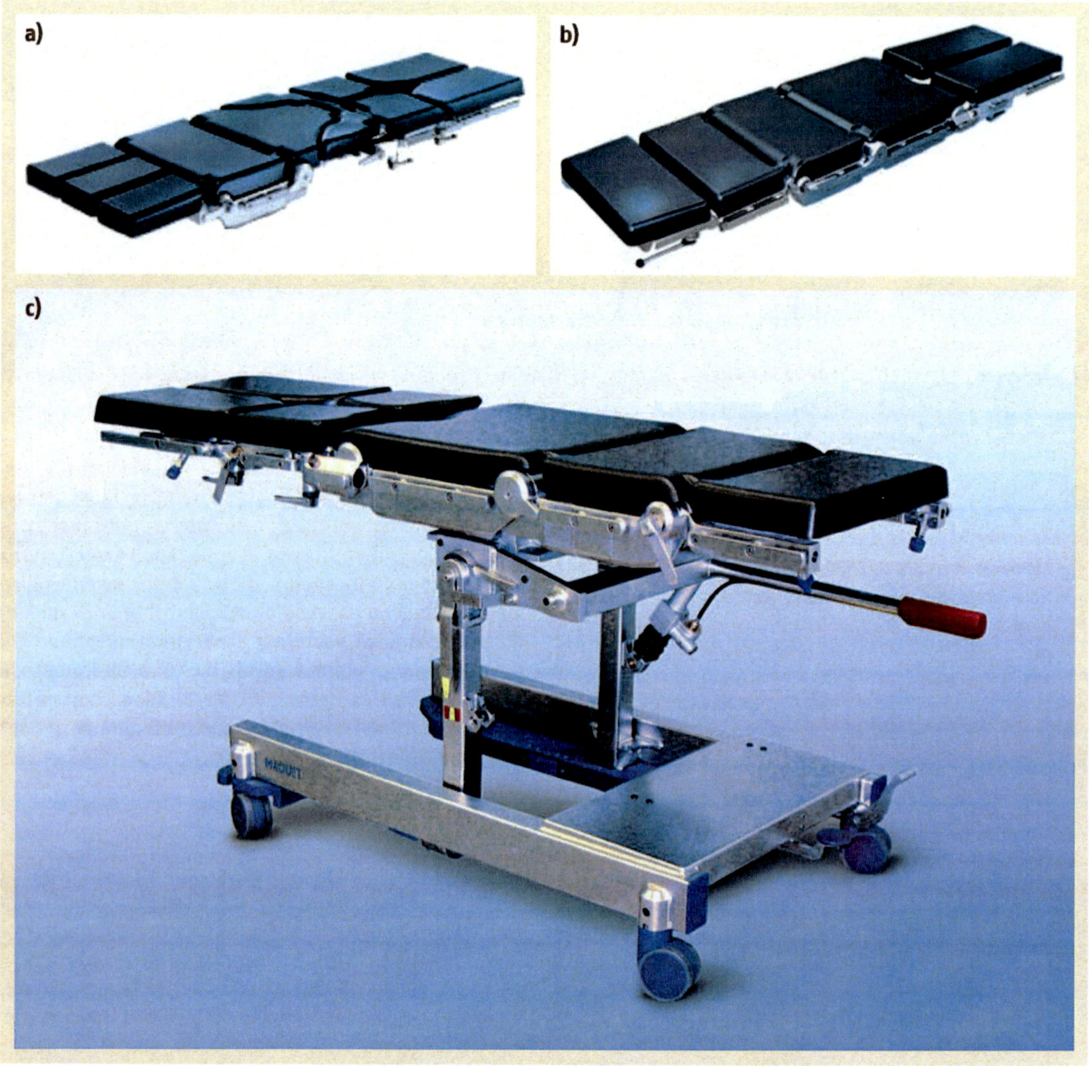

Fig. 69.10a–c Various operating room table tops. **(a)** Extensible, **(b)** universal, **(c)** on transporter

ious, differently-priced OR table systems, both mobile and stationary.

In order to guarantee flexibility, today's OR table systems are designed in such a way, or can be modified, to ensure that they meet the positioning requirements of different surgical disciplines. If you have deployed an OR table system and, should operating and positioning techniques change in the future, you only need to order a suitably modified table top that is compatible with the basic OR table column element. In this case, in an OR department equipped with mobile operating tables, it would be necessary to order a completely new OR table, which would entail considerable investment.

Today, more than ever, time savings are a significant criterion for the efficient usage of operating theaters. The use of an OR table system, i.e. a table column, two transporters as well as two OR table tops, facilitates a so-called *roundabout system*. While one surgical procedure is being finished up in the operating theater, it is possible to bring the next patient from the bed transfer room to the pre-operation room on the second

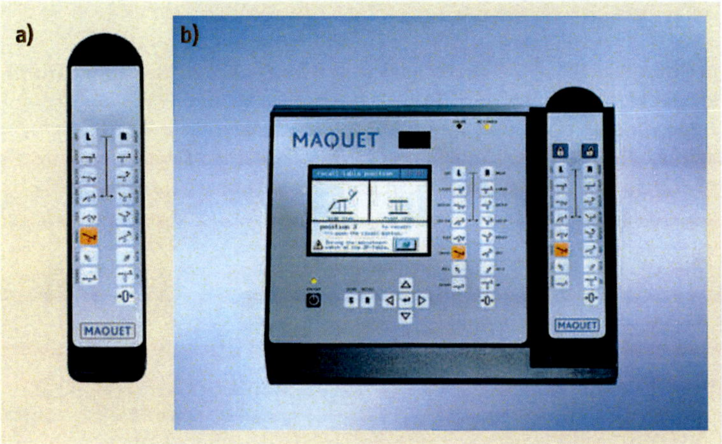

Fig. 69.11a,b Control units for table tops. **(a)** Cable control unit, **(b)** wall-mounted control unit

table top, in order to administer the anaesthetic. Once the previous operation is complete, the patient has been taken out, and the operating theater has been cleaned, the already anaesthetized patient can be moved into the theater, where he/she can be transferred, while still lying on the table top and potentially already positioned, onto the OR table column. This *roundabout system* offers the advantage that operations can continue without any great time delay factor, whilst also taking the time periods required for the effects of local anaesthetics into account.

Of course, such overlapping working practices can also be achieved using two mobile OR tables. There are, however, both advantages and disadvantages. One disadvantage is that a mobile OR table weighs between 200 and 300 kg, which means that together with the weight of the patient, in a worst case scenario, around half a metric tonne would need to be moved. One advantage is that because all functions are *on board* with a mobile OR table, the patient can be positioned and prepared ahead of time almost 100%, for the surgical intervention. If well organized, this can represent time savings (Fig. 69.12).

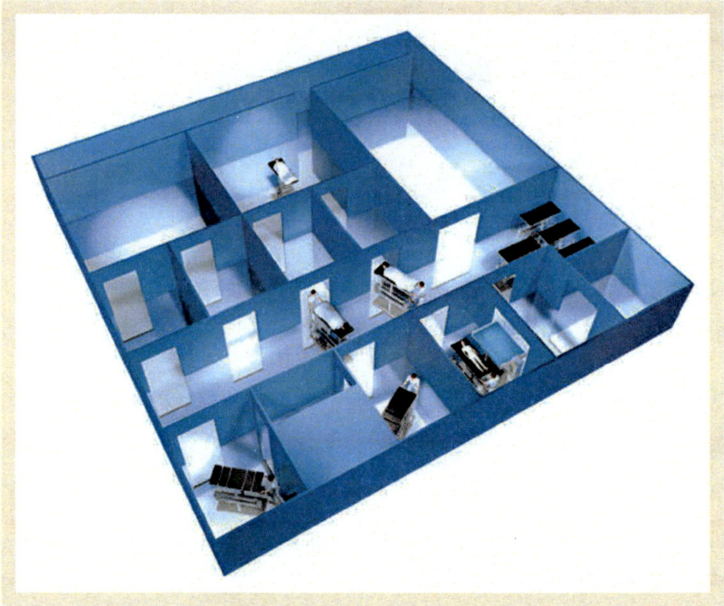

Fig. 69.12 Patient workflow

69.4 Safe Patient Positioning

Operation positioning designates the position that the patient's body is placed in, in order to guarantee the surgeon the best possible access to the operation area. Additionally, every effort is made to optimize the presentation of the anatomical structure of the operation field.

The following *standard positions* have proved themselves (Fig. 69.13a–d):

- Supine position/special supine position
- Prone position/exposed stomach position
- Lateral position

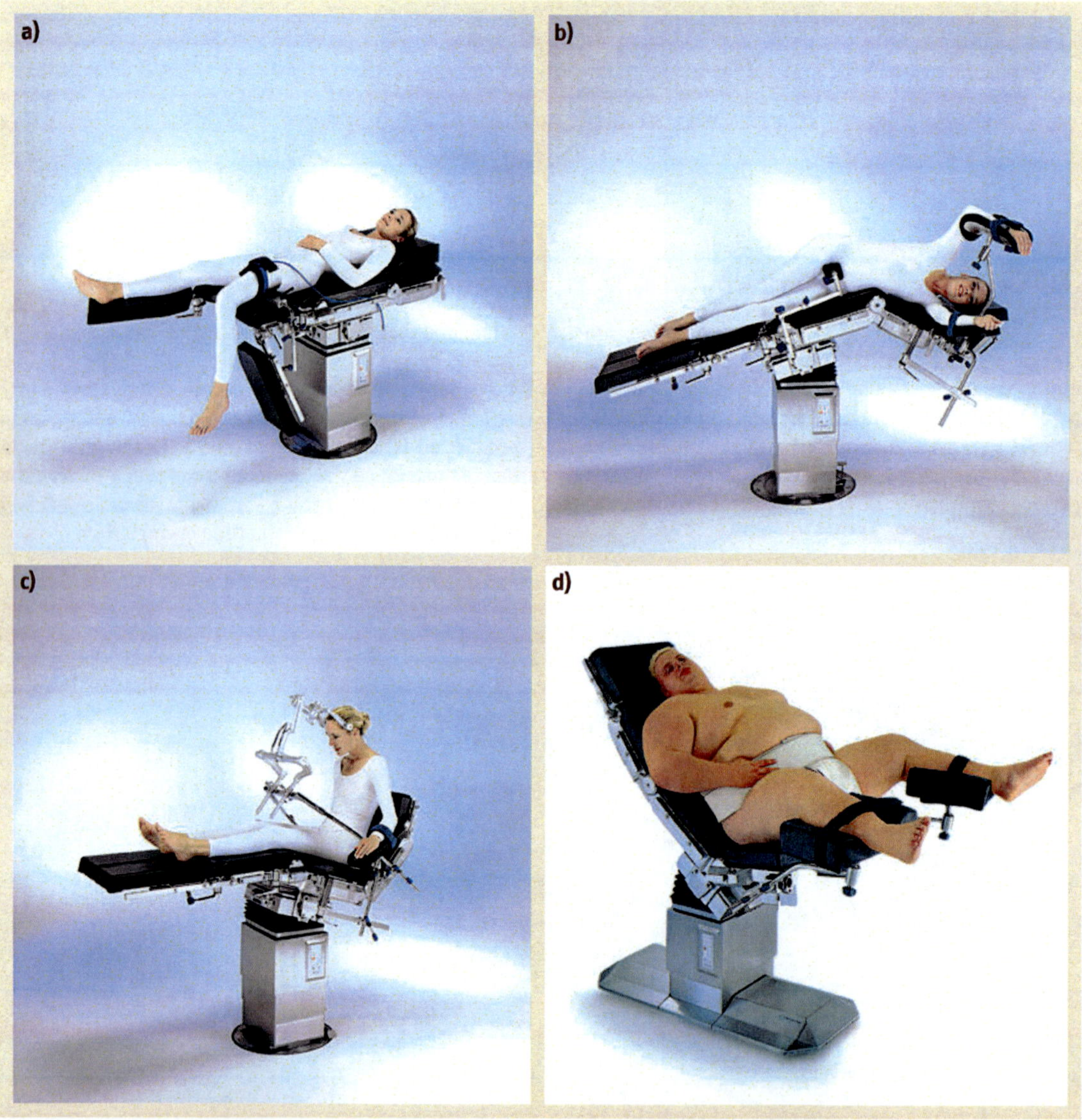

Fig. 69.13a–d Standard positions operating room tables for various surgeries. **(a)** Supine position for knee arthroscopy, **(b)** thorax surgery, **(c)** sitting position for neurosurgery, **(d)** adiposity surgery

- Dorsosacral position
- Genucubital position
- Sitting position/half seated position.

The patient ought to be positioned or be placed in the optimum operating position by the anaesthetist and the surgeon in conjunction with the OR care team for administering the anaesthetic and for the operation. Before this, the responsible specialists must decide, based on the general condition of the patient, which stresses or strains the patient can be exposed to, as a result of the positioning on the OR table. The factors that need to be considered here are the patient's age, weight and constitution, as well as their general state of health in respect of heart, lungs, circulation, metabolism, nervous system, muscle tissue, and skin tissue damage that may have been caused by metabolic dysfunctions, obesity, rheumatoid arthritis, heart and vascular weakness, or circulatory disorders, etc. The factors listed here will significantly affect the strains that a patient can withstand and must be taken into account when deciding on the positioning, as every type of patient positioning represents an additional strain.

An additional factor is that the anaesthetic and muscle relaxants actually increase the strains as they affect breathing, blood supply, and nerves in particular. Pain sensitivity is deactivated while under anaesthesia so that the patient is not aware of pain caused by pressure or strain trauma and cannot react because of interrupted adverse effects reflexes and reduced muscle tone.

This means that the patient can suffer injury even before the actual surgical intervention has started. Extreme care must be taken when making even the simplest positional change during the operation.

Endotracheal anaesthesia is administered when patients are in the normal supine position. Only once the deep anaesthesia state with relaxed muscles has been reached, can the actual operation positioning be done, taking the patient-specific factors into account (see above).

It should be ensured that the patient's arm that is going to be used for the anaesthetic and the infusion is lying evenly and extended, on a well padded arm positioning device. If necessary, the arm positioning device must be lengthened with a padded Cramer rail. If the positioning is wrong this can lead, for example, to irritation or paralysis caused by damage to the *N. radialis* or *N. ulnaris*, in spite of soft OR table padding. Overstretching the arm past the 90° angle, both abduction and supination, can cause paralysis of the plexus.

When it comes to leg positioning, the foldable leg plates on modern OR tables can be adjusted to suit legs very well, so that pressure is distributed over a large enough area and the best possible operation position can be achieved. The use of leg holders on the other hand, e.g., during gynaecological or urological interventions, carries a certain risk of causing pressure injuries and strains through incorrect positioning or deployment, by, for instance, overstretching legs that have been relaxed by anaesthetic, by the pressure of wrongly adjusted knee crutches as well as by resting the lower extremities on the leg holder bars.

The same care must be exercised when positioning the trunk. Not changing the position of the patient of a longer period of time, can lead to a further problem: decubitus (bedsores or pressure sores).

69.4.1 Decubitus Injuries

As anaesthetics and muscle relaxants can relax the skin tissue such that the arterial pressure is weaker than the external pressure, which is influenced by body weight, the blood supply can be impaired and, as a result, the skin tissue receives insufficient nutrients. Thus, for patients who are kept in the same position for a longer period time, there is and increased risk of skin and skin tissue damage.

A decubitus caused in this way can develop into necrosis (localized death of skin tissue cells), in particular when there is only a thin layer of skin covering the bone. Locations particularly at risk are:

- In supine position: heels, sacrum, elbows, shoulder blades, back of the head
- In prone position: pelvis, hips, knees, points of the toes
- In sitting position: heels, crucial ligaments, elbows, head
- In lateral position: hips, toes, knees.

At the same time, care must be taken that there is no trapping of the skin, which can lead to necrosis as a result of reduced blood flow. This risk is particularly prevalent in longer operations.

Following heart operations during which the patient's body was cooled down and single-pole high frequency devices were used, large areas of tissue necrosis were identified, which principally were diagnosed as burns. Initial tests regarding for the causes of this necrosis points that were identified as burns, by the operation team, the hospital's technical staff, health and safety and the manufacturer of the HF surgical equipment were unable come up with any explanation based on a physical cause. Only a differential diagnosis investigation of the suspicion of pressure necrosis, could definitively exclude the possibility of exogenic causes of burning.

A first indication of pressure damage is given by a reddening of the skin that does not recede immediately following a change of position.

Essentially the risk of decubitus is not greater for a person who is overweight than it is for one who is underweight. The difference is that with overweight patients, the area of damaged skin is larger, but in most cases the damage is less pronounced, while with underweight patients, the damaged area is smaller but the skin damage is more pronounced.

Reasons for the Development of Decubitus Caused by OR Positioning

The following list gives possible causes for the development of decubitus caused by OR positioning:

- OR table padding that is hard or worn out.
- Longer operations with increasingly older patients.
- Increased self-weight, particularly with obesity. On the other hand, however, also with cachectic patients where the skin is very close to the bones because of missing or reduced layers of fat under the skin. Areas of the body particularly affected here include the sacrum and the heels.
- Medicinal influences (anaesthetics), which reduce muscle and tissue tone.
- Punctiform loads resulting from positioning necessary during the operation.

Decubitus Prophylactics

Decubitus can be avoided by implementing the following countermeasures:

- Aiming for shorter operation times, as experience shows that skin tissue damage is likely to occur, at the latest, after 2 h.
- Regular and timely replacement of older or worn OR table pads with padding that is thick and soft enough.
- Careful attention to the individual segments of the OR table top and the body of the patient.
- Pressure relief brought about by the application of additional padding at the predisposed locations.
- Avoidance of skin trapping and the formation of folds or creases both in patients' skin and in the OR table padding during positioning and intra-operative patient positioning.
- Avoidance of the intra-operative distortion of patient positions.
- Carrying out positional changes during longer operations, whereby small adjustments of the motorized joints of the OR table can be used to relieve pressure on tissue and enable reperfusion.

69.4.2 Long-Term Position Injuries and Legal Responsibility

Optimized patient positioning is the best prophylactic against decubitus! As analyses of damage claims by authorized experts show operation position injuries are reported regularly in many damage claims, which represents a serious problem for the patients and increases the costs of the post-operative care.

Improper and incorrect patient positioning on the OR table may cause problems, ranging from temporary impairment up to serious and irreversible injuries. The areas most affected are:

- Nerves that are traumatized, in particular the plexus brachialis
- Eyes
- Skin
- Muscle tissue
- Tendons and ligaments.

There are comparable rules in force in different countries when it comes to clarifying the questions of legal responsibility for sustained position injuries. They indicate that while the task of positioning patients on the OR table before the operation is down to the anaesthetist, during the operation it is the job of the surgeon, taking any anaesthetic-related risks into account. However, essentially it is the joint responsibility of both the surgeon and the anaesthetist.

For the surgeon, this means that he bears the medical and legal responsibility and that any increase in risk in relation to anaesthesia, resulting from a necessary change in position, is justified. The legal responsibility rests with the anaesthetist, as part of his intra-operative duties, to make provisions for the specific risks that result from the positioning and/or to mitigate against them through the implementation of particular preventive measures. Special attention should be paid to the important issue of the risk of burning, in connection with patient positioning, during operations where single-pole high frequency devices are used.

69.4.3 Patient Positioning When Deploying Single-Pole HF Devices

High-frequency (HF) surgery equipment is used to separate specific areas of tissue using thermal energy and at the same time to cause coagulation. For this purpose, single-pole high frequency surgical devices with an active and a passive electrode are used. Because of its shape, a high current density occurs at the active electrode, the so-called cutting or coagulation electrode, as opposed to the large-surface, passive electrode (neutral electrode), which conducts the current away and, therefore, has a low current density.

Possible causes for complications when carrying out single-pole high frequency surgery are, on the one hand, that the patient has not been placed correctly in a position that will ensure that the body is not earthed and, on the other hand, the faulty application of the neutral electrode. In order to avoid localized burning caused by HF surgical equipment, the patient must be positioned in such a way that he/she is completely insulated from the OR table and its accessories, as well as being safeguarded, through a correct installation of the neutral electrode. This means:

- The patient must be positioned in such a way that he/she is not in contact with electrically conductive components such as metal parts of the OR table, holders, damp cloths, etc. with particular attention being paid to the extremities. There must be an electrically-insulating, dry, thick underlay between the patient and the OR table and holder, which must not become damp during the HF surgery, e.g. with blood or liquids used to rinse out the area of the operation.
  As dry and nonconductive fabric must be laid between patient and table padding, a minimum conductivity for the padding is laid down, in order to avoid electrostatic discharge. If this were not the case, discharge sparks caused by frictional electricity could be produced, which would represent a dangerous source of ignition energy for flammable anaesthetic gases or alcohol vapors.
- The whole surface of the neutral electrode must be applied well to the patient's body, such that it cannot become detached if the patient moves or is moved. Preferred application points are the upper and lower extremities. This prevents the occurrence of too high a transfer resistance, which interrupts the current backflow through the neutral electrode.

69.5 Preparation: Care, Maintenance, and Hygiene

69.5.1 Manual Cleaning and Disinfection

Following each operation, special attention must be paid to the preparation of the OR table tops and the OR table system transporters that were used for the operation, i.e. they must be carefully cleaned and disinfected. Particularly in smaller hospitals, cleaning and disinfection is often done manually by support staff in the operation area or by the nursing staff themselves. Because of the multiple divisions of the complete OR table system and the high hygiene demands, manual cleaning cannot always be seen to be the most reliable approach. It is also very time and staff-intensive.

Hygiene issues, maintenance work, and any repairs that may be necessary are already allowed for during design and construction. For instance, the upper part of the OR table, as well as the covers of the column and base, consist principally of smooth-surfaced, individual components made of chrome nickel steel that can be removed without any problem. Also, the electrically conductive roller bearings that are used to move and manoeuvre the mobile OR table are easily accessed from above for inspection and cleaning. Here, there are significant advantages with the OR table system since table tops and transporters can be moved effortlessly into the corresponding cleaning rooms. Any repairs can also be carried out well away from the operation area, without disturbing the OR procedures.

69.5.2 Automatic OR Table System Cleaning and Hygiene

A second option is offered by so-called decontamination machines, which automatically carry out the cleaning, disinfection and drying of suitable OR table tops, Lafettes, and OR accessories. This alternative is most frequently used in larger operating centers, as it is neither staff nor time-intensive, unlike manual cleaning and disinfection procedures. A further, not insignificant, advantage is that an automated procedure using a machine, assuming that operating instructions are correctly followed, offers maximum hygiene levels. With manual

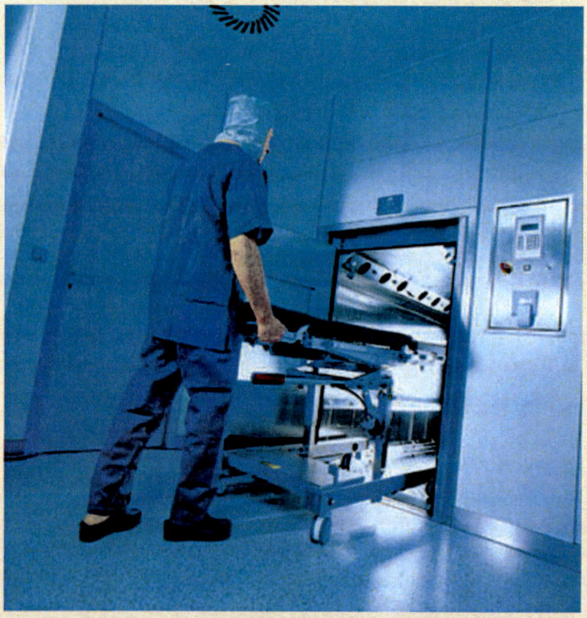

Fig. 69.14 Operating room table cleaning and decontamination machine

cleaning, a lot rests on the care practiced by the staff as well as the time invested.

On average, the cycle time of a decontamination machine, including all the individual processes such as cleaning, disinfection, intermediate drying, rinsing, and final drying is 10 min. It should also be said here, that the cycle time is dependent on the level of contamination and is, therefore, variable. All data generated during this process is recorded, in order to ensure traceability.

From the standpoint of the care staff, this represents not only a simplification of and a significant improvement in working conditions, but it also contributes to the optimizing of safe working processes in the operating area (Fig. 69.14).

Maintenance

As well as the ongoing cleaning and disinfection of the OR table as a prerequisite for aseptic working, it is essential, because of the many electrical, hydraulic, and electro-hydraulic control elements, that the OR table system is regularly maintained, in order to prevent downtimes and to ensure the safety of both patients and OR staff.

Further Reading

- C. Krettek, D. Aschemann (Eds.): *Lagerungstechniken im Operationsbereich* (Springer, Berlin Heidelberg 2005)
- P.M. Lauven, H. Stoeckel: Damage caused by patient positioning – a still unsolved problem? AINS **27**, 391–392 (1992)
- H. von der Mosel: *Medizintechnik für Pflegekräfte* (Bibliomed, Melsungen 1992)
- H. Oehmig: OP-Tisch-Technik heute – Trennung von Säule und Platte, KrankenhausTechnik **7**, 18–21 (1994)
- E. Roos: Sinn und Zweck der erdschlussfreien Patientenlagerung auf OP-Tischen, MTDialog **111**, 38–40 (1992)
- H. Schindler: (1985) *Arbeitsgebiet Operationssaal* (Enke, Stuttgart 1885)

70. Medical Robotics

Harald Fischer, Udo Voges

In 1990, robot systems for use in medicine were still the subject of research. Then, first research results were implemented in practice, and in 2000, several types of robot systems for various surgical disciplines were in clinical use. Other systems are in preparation. While some systems are applied nearly routinely, others are still in the phase of testing. Research is dedicated to solving a number of open problems and to further developing the systems, for instance, by combining robots with imaging methods or their use in special environments, e.g., in computed tomography or magnetic resonance imaging. Other objectives are further miniaturization and simplification.

After a short introduction to the fundamentals of medical robotics and the developments in this field, an overview of existing systems and their medical applications will be given. Some technical aspects of the use of such systems will be explained. The chapter is completed by an outlook.

70.1 **Fundamentals** 1273

70.2 **Development of Medical Robots** 1274

70.3 **Overview of Systems** 1275
70.3.1 Assistance Systems 1275
70.3.2 Active Holding Systems 1275
70.3.3 Master–Slave Manipulators 1276
70.3.4 Biopsy Robots 1276
70.3.5 Commercial Active Holding and Guiding Systems 1277
70.3.6 Commercial Telemanipulators 1278
70.3.7 Commercial Surgery/Biopsy Robots 1278

70.4 **Medical Applications** 1280

70.5 **Technical Aspects** 1281

70.6 **Outlook** 1282

References 1283

70.1 Fundamentals

Usually, the term "robot" is understood to refer not only to robot systems proper but also to telemanipulation systems. Strictly speaking, the only robots are the autonomously working systems known from industry, e.g., automotive manufacturing. The robot always carries out a previously defined task, in the same way. This activity is executed without any further human interference and with high repetition accuracy. The operator can only stop the system (emergency shutdown), but not make an interactive change. A change or modification is possible only by redefining the task, new programming, or new teaching of the robot.

If, however, man acts to control the system, it is referred to as a telemanipulation or remote handling system. Only the movement directly given by the operator is reproduced – if possible, simultaneously – by the working system. The input device (operation unit) moved by the operator is referred to as the master. The manipulator executing the action is called the slave (working unit). Consequently, such systems are also called master–slave manipulator systems.

Interactive systems represent a mixed type of telemanipulation and robot systems. The master is not separated from the slave, and the instrument is often controlled directly by the operator. A target path, working range, or permissible forces are defined in advance. If the operator wishes to leave this target path, he is prevented from doing so by the corresponding forces. The user-controlled instrument can leave the permissible working range only after having confirmed a warning

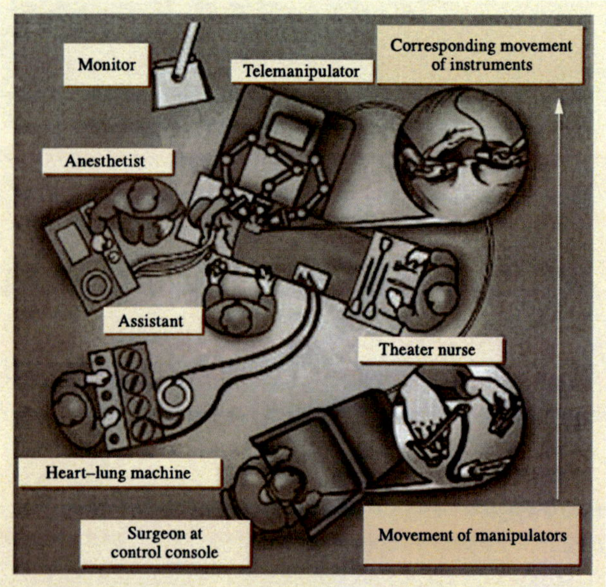

Fig. 70.1 Principle of a telemanipulator system for minimally invasive interventions (Intuitive Surgical, Sunnyvale)

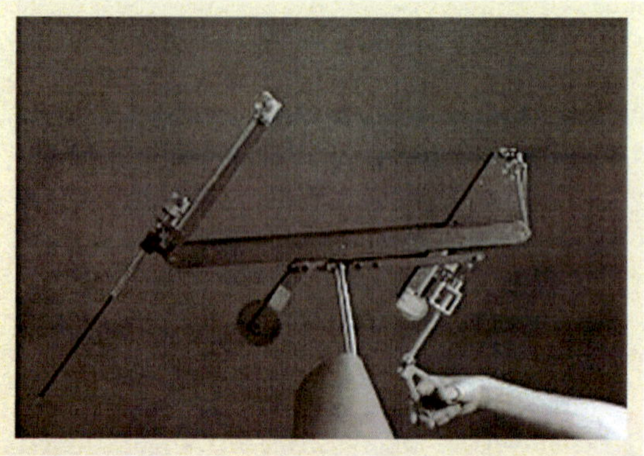

Fig. 70.2 Purely mechanical coupling between master and slave unit

message. In this way, critical operation ranges can be avoided and an operation planned, e.g., on the basis of x-ray examinations can be executed exactly.

Figure 70.1 shows the principle of a master–slave manipulator. This operation telemanipulator is used for cardiac surgery. The surgeon sits in front of the panel with the input systems (master manipulators for each hand), a viewing system, and various foot switches for operation of the instrument functions (e.g., sucking and rinsing). Movement of the master is converted into corresponding movement of the instruments, and the telemanipulator executes the corresponding actions. The assistant waits until the instruments have to be exchanged manually.

The distance between the master and the slave can be bridged in different ways. In case of direct, purely mechanical coupling of the two devices, the distance is relatively small (several meters at most). Due to the rods, tension ropes, etc. needed, however, this solution can hardly be used in medical robotics.

Figure 70.2 shows such a direct mechanical coupling between the master and slave unit. Frequently, computer systems are used instead of this direct mechanical coupling between master and slave. These systems control the coupling of master and slave. If a single computer only is applied to convert the input information of the master unit into the control information for the slave unit, a distance of several meters is possible by a cable connection. If the master and the slave are controlled by separate computers that are connected via a local area network (LAN), for instance, the distance may be as large as desired. Master and slave can even be connected across continents for telesurgery. In case of larger distance, however, delays between the action of the master and the reaction of the slave and detection of the returned video signal may occur depending on the transmission method used. This means that work has to proceed more slowly and more carefully.

70.2 Development of Medical Robots

The development of medical robotics shall be presented briefly. The systems are described in more detail in Sect. 70.4.

Use of robotics or computer-controlled systems in medicine is characterized by a stepwise process. First, endoscope guiding systems were developed (e.g., Robox [70.1], FIPS [70.2], Felix [70.3], Aesop [70.4], EndoAssist [70.5]). These systems are less safety relevant, but they facilitate the surgeon's work.

By means of endoscope guiding systems, the endoscope can be held and guided over a longer period without a human assistant. Such systems are operated by joysticks, keyboards, language, head movement, etc. Instead of explicit operation, automatic instrument tracking is possible. Major advantages of an endoscope guiding system lie in the facts that no additional assistant is required to act as a cameraman and a stable picture is obtained. The endoscopes applied for this purpose are standard systems.

Endoscope guiding systems have been developed further to become instrument guiding systems. As a rule, these are not suited for holding and guiding conventional instruments, but only specially developed, often not only rigid, but also flexible instruments. These instruments are operated via a master unit in case of the telemanipulation system Tiska [70.6]. In the case of Robodoc robots ([70.7], Caspar [70.8]), by contrast, the instrument guiding systems execute predefined tasks and have a very limited interactive input system, which mostly serves for correcting an already given track along which the system is to be moved. As use of an individual instrument guiding system is only reasonable for a robot system, and two-handed work as in case of a normal operation should be possible when using a telemanipulator, the step from the instrument guiding system to a system consisting of two instrument guiding systems and one endoscope guiding system was not long in coming (ARTEMIS [70.9], da Vinci [70.10], ZEUS [70.11]).

Using such a system, the surgeon can execute operations alone and does not require any further assistants. For the first time, solo surgery became possible.

70.3 Overview of Systems

This section gives an overview of assistance systems, telemanipulators, and robotic systems.

70.3.1 Assistance Systems

One generally distinguishes between industrial or commercially available systems, and systems for research and university purposes. Generally, these systems serve to hold instruments and cameras. Passive holding systems are positioned by hand and do not contain any actuators (motors) or active systems that are driven to the desired position. Here, only active systems will be presented.

70.3.2 Active Holding Systems

Various active holding and guiding systems were developed by the former Forschungszentrum Karlsruhe. One of these systems is presented in Fig. 70.3. It is applied to guide an endoscopic camera. As the system was designed for minimally invasive surgery (in particular, abdominal surgery), it was provided with a mechanical forced guiding system around the invariant puncture point. Electrically, the system can be moved around this point. The invariant puncture point is the point of penetration of the abdominal wall.

The mechanical construction of the device prevents mechanical stress or loading of the puncture point (invariant point).

Input takes place via a type of joystick or by language. For research purposes, a camera tracking system is available. Hence, the surgeon can move the camera intuitively to the desired position. The assistant physi-

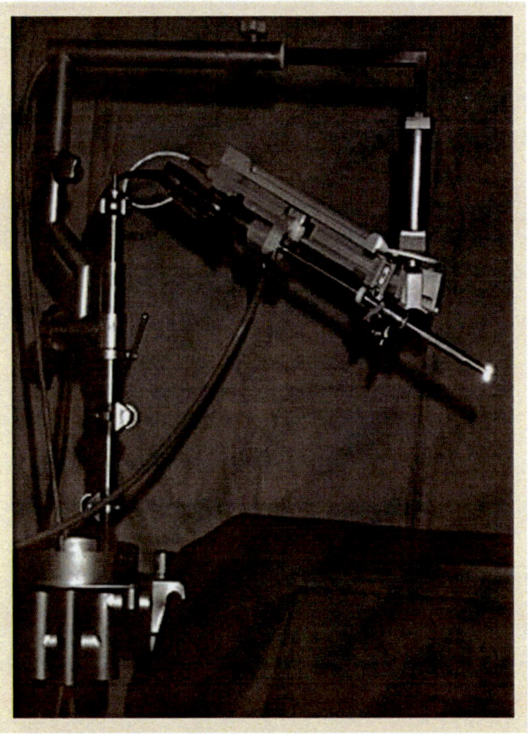

Fig. 70.3 FIPS endoscope guiding system (Forschungszentrum Karlsruhe, after [70.2])

cian no longer has to guide and position the endoscopic camera and can take over other tasks.

Such a simple camera guiding system has the great advantage of absence of tremors. The electronic assistant holds the camera without any tremors over longer periods, which facilitates operation by the surgeon, as the operative field is not blurred on the screen.

70.3.3 Master–Slave Manipulators

The first scientific telemanipulator was built as a demonstrator and tested by Forschungszentrum Karlsruhe in 1995 (Fig. 70.4). Complicated scenarios were executed successfully and precisely. The flexible distal joints of the instruments for the first time allowed bypassing of structures inside the human body. Thus, complicated interventions (surgical sutures around organs) could be performed endoscopically.

Integration of force and momentum sensors provided for force feedback. While holding his operation unit, the surgeon was able to feel the forces he exercised with the actuators in the operative area. These systems turned out to be much more complicated than initially assumed. Due to the important restrictions, integration of sensors is very difficult. Hence, the wish of the operator to use low-cost single-use tools is not satisfiable.

The master–slave manipulator of the University of Berkeley, USA, also includes force and acceleration sensors in the slave unit, i. e., in the instrument tip. These sensors feed the forces occurring in the slave unit back to the master unit. Exerted forces are fed back to the fingers of the surgeon. Such force feedback is required when the operator (master unit) and manipulator (slave unit) are no longer located in the same operation theater but are decoupled locally. If the end effectors are driven electrically, the forces acting, e.g., between the branches of endoscopic forceps are not fed back.

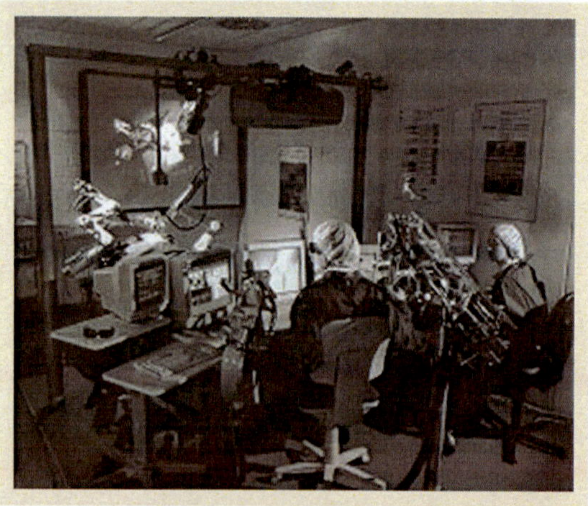

Fig. 70.4 ARTEMIS telemanipulator system (Forschungszentrum Karlsruhe, after [70.9])

If gripping on the master unit is too strong, forces on the slave side may be too high and tissue may be damaged. In case of spatial and, hence, mechanical decoupling, force feedback is an advantage, as a type of tactile feedback is regained. The surgeon obtains information about the interaction with various types of tissues. This prevents mechanical damage due to too high gripping forces.

In case of standard endoscopic operations, this feedback is of optical character only and supported by direct mechanical coupling of the endoscopic instruments. This means that the handle is connected directly (possibly via various joints) to the distal end and, hence, to the fingers of the operator. Feedback is then given by a change of color of the tissue while gripping. If the tissue becomes white, the gripping force is too high (the tissue is no longer supplied with blood and discolored).

70.3.4 Biopsy Robots

In the past years, diagnostic imaging methods and the corresponding evaluation processes have improved considerably. With increasing computer speed, it became possible to represent movement scenarios by these new tomographs in real time. Now, so-called surgical robots can be applied for simple tasks, e.g., for the biopsy of mamma carcinomas directly in the magnetic resonance tomograph. Due to the ambient conditions in such a tomograph, however, this process is very difficult. Since 1998, Forschungszentrum Karlsruhe has developed a robot for taking a tissue sample (biopsy) directly in the magnetic field of a tomograph with image control (Fig. 70.5). In November 1999, this manipulator was used for the first time for clinical evaluation [70.12].

MrBot (Fig. 70.6) also is an MR-compatible manipulator for minimally invasive access to the prostate directly in the MR. Various needle drives can be combined, such that biopsies and thermal ablations can be carried out, and even radioactive pellets can be placed precisely.

MrBot is made of nonmagnetic and dielectric materials (Fig. 70.7). Preferably, plastics, ceramics, and rubber are applied. Pneumatic stepper motors have been specially developed to drive the system. Optical sensors send the position signals to the control unit that is located outside of the MRI device.

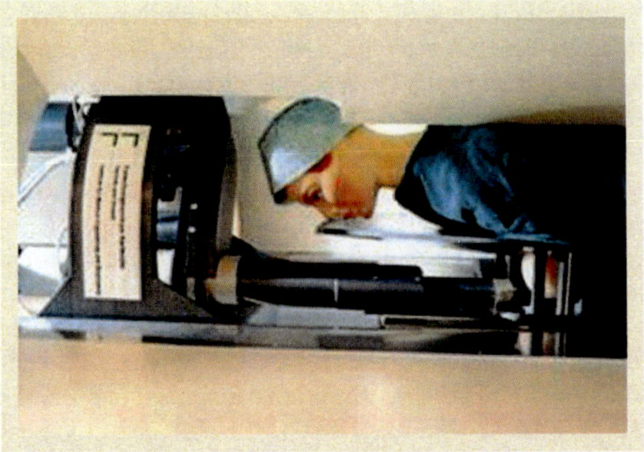

Fig. 70.5 Robitom, MRI-compatible biopsy robot (Forschungszentrum Karlsruhe, after [70.12])

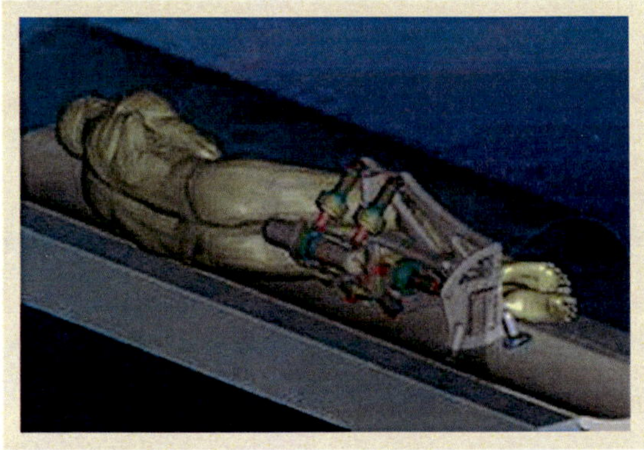

Fig. 70.6 MrBot, MR-compatible robot for biopsies and therapies (Urobotics Lab, Johns Hopkins University, Baltimore)

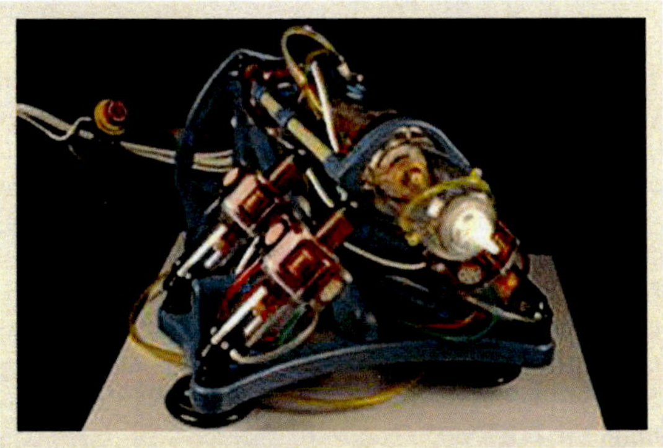

Fig. 70.7 MrBot. Motor design for controllable, precise, and safe pneumatic actuation therapies (Urobotics Lab, Johns Hopkins University, Baltimore)

Positioning accuracies in the submillimeter range and magnetic field compatibility of up to 7 T have been achieved. The tip of a needle can be positioned with accuracy ≤ 1 mm.

70.3.5 Commercial Active Holding and Guiding Systems

Presently, two approved active holding and guiding systems are still commercially available, namely FreeHand (formerly EndoAssist) by Prosurgics (formerly Armstrong Healthcare, UK) and LapMan by MedSys s.a., Belgium. In FreeHand, the camera is controlled by the head; i. e., after activating head control by a foot switch, the camera follows the movements of the head (Fig. 70.8).

LapMan is controlled via an input unit on the instrument of the surgeon. Both systems are not provided with a mechanical forced guiding system, i. e., they can be positioned arbitrarily in space. The invariant point around the puncture point in the abdominal wall is maintained by software control. In principle, the kinematically specified mechanical invariant point and the software-controlled invariant point are equivalent.

Figure 70.9 shows a camera guiding system from Medsys s.a., Belgium. Here, the invariant point is adjusted and maintained by software control. The system

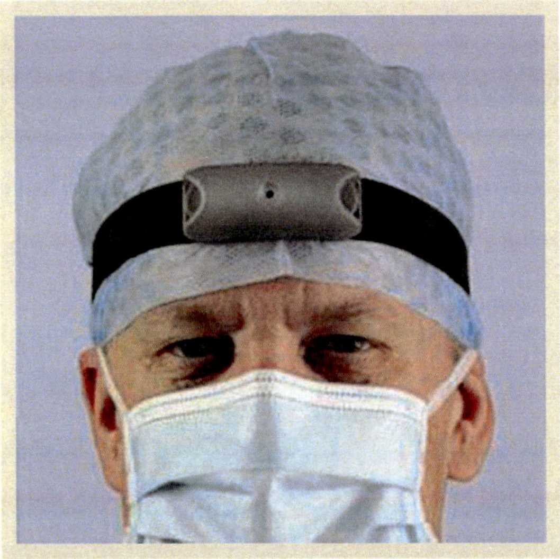

Fig. 70.8 Control unit of the FreeHand system, attached to the head of a surgeon by a rubber band (Prosurgics Ltd., Bracknell)

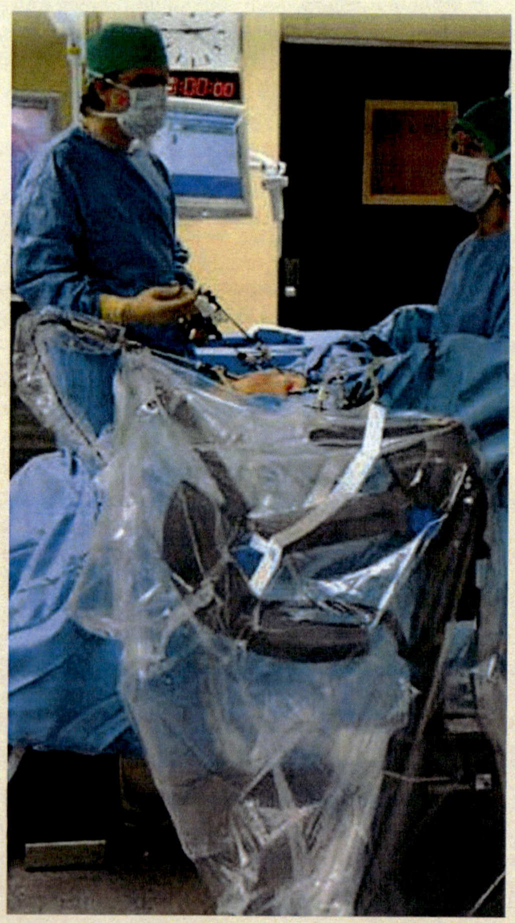

Fig. 70.9 LapMan camera guiding system (Medsys s.a., Belgium)

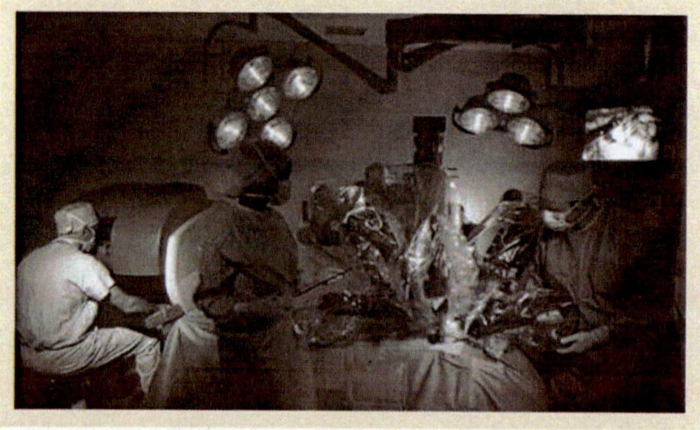

Fig. 70.10 The da Vinci telemanipulation system (Intuitive Surgical, Sunnyvale, CA)

is controlled manually via a special operation unit that is located directly on the instrument handle. The system is controlled by the index finger directly on the instrument handle.

70.3.6 Commercial Telemanipulators

The only telemanipulator system still available on the market is presented in Fig. 70.10.

This system was developed specially for cardiac surgery. In the meantime, it has been applied as a telemanipulation system in various fields of endoscopic surgery (e.g., abdomen, urology, head). Development work started in 1995. The system has been used clinically in Germany since as early as 1999. Worldwide, more than 1000 systems have been installed. In Germany, the system is used, e.g., at Leipzig, Berlin, Dresden, Frankfurt/M., Hamburg, and Munich. First, bypass operations as well as cardiac valve operations were executed in a minimally invasive manner using this system. In the meantime, operations have been performed in the fields of urology (prostate resection), gynecology, and in the head and neck area. The surgeon sits in front of a panel not more than 8 m away from the field of operation. Via screens, he is shown a three-dimensional (3-D) image of the operation area and can operate up to three flexible instruments and the endoscope via two input devices. da Vinci is a closed, integrated system that can only be used as a whole.

70.3.7 Commercial Surgery/Biopsy Robots

These systems are understood to be classical robots applied for surgical precision work, for example, for milling hip shafts. The robots move along trajectories given prior to the operation and execute a program step by step similar to computer numerically controlled (CNC) machines.

Figure 70.11 shows ROBODOC, which was presented for the first time in 1992. It is applied to precisely position hip prostheses. Apart from the robot, this system also includes a planning system by means of which the prosthesis is selected and the optimum position of the prosthesis is calculated. Since 1994, more than 12 000 operations have been carried out with ROBODOC, most of them in Germany. In the beginning, the thigh bone had to be marked with screws in a first operation for exact calculation and online monitoring of the robot milling work. The new system does without marking, which reduces strain for the patient as no previous operation is required.

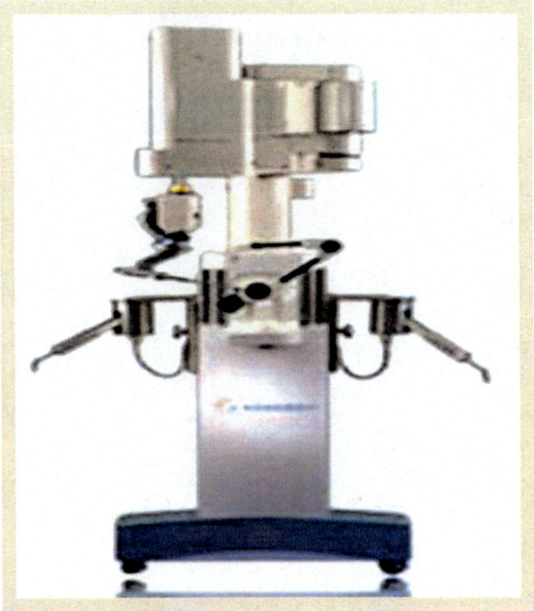

Fig. 70.11 ROBODOC operation robot (Curexo Technology Corporation, Fremont, CA)

Since 1996, CASPAR from ortoMaquet has been used in Germany. In terms of tasks and functionality, this system largely corresponded to the ROBODOC system. In addition to hip joints, knee joints could be operated with CASPAR. In the meantime, production of CASPAR has been stopped.

In Germany, SurgiScope from Jojumarie was mainly used at university hospitals and as an auxiliary system for precision work in the ear–nose–throat area [70.13]. The system was tested clinically for such work. In particular, it was applied for microscope guiding, for guiding the biopsy needle or endoscope, and as a drilling and screwing assistant. In addition, a novel type of radiation treatment was tested using this system. It was used for precise positioning of so-called radioactive pellets, e.g., in prostate resection, using which the existing tumor should be damaged specifically by radiation. In the meantime, the system has been removed from the market and is mainly used for research purposes.

Figure 70.12 shows the Innomotion system, which was the first commercially available assistance system for use directly in the magnetic resonance tomograph (MRT). The system was developed by Forschungszentrum Karlsruhe in cooperation with Professor Melzer and the industrial partner Innomedic GmbH [70.14–16].

The system was the first CE-approved MRT-compatible system worldwide in 2005. It was mainly applied for percutaneous, image-supported pain therapy of various organs such as the liver as well as of the spinal column, etc.

This precision system interacts directly with the planning software and the tomograph. It is worked on operating the system online with the MRT, i. e., the image sequences can be evaluated and implemented directly by the assistance system by the direct exchange of information. This allows for operation under direct image navigation. Precision of the intervention is increased. All materials applied are MRT compatible. Special optical sensors and pneumatic drives have been developed specifically for this system. The pneumatic actuators (pressurized air-controlled cylinders) are the first of this type worldwide and can be positioned exactly along their trajectory with an accuracy of 0.1 mm. Due to the use of MRT-compatible materials, interference with the assistance system during image acquisition is minimized considerably. In 2008, Innomedic was acquired by Synthes, USA, and Innomotion is now offered for research purposes only.

Another system of this type is B-Rob-II from the Austrian Center for Medical Innovation and Technology (ACMIT), which is presently available as a prototype only.

Under ultrasonic or computed tomography (CT) control, this system can execute biopsies (Fig. 70.13). The needle positioning unit is controlled on the basis of images and achieves precision ±0.9 mm. The needle

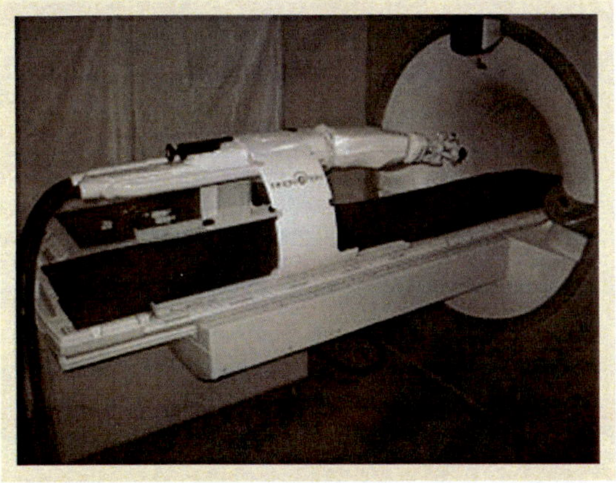

Fig. 70.12 Innomotion, MRT-compatible assistance system (Innomedic GmbH, Philippsburg-Rheinsheim, Germany)

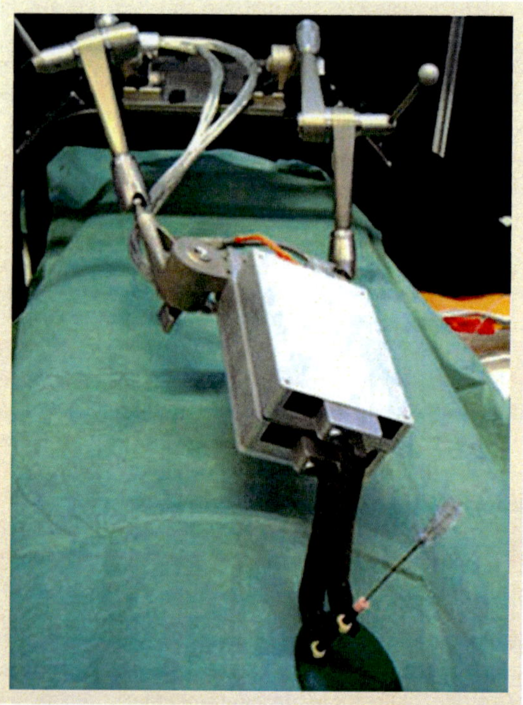

Fig. 70.13 B-Rob-II (Austrian Center for Medical Innovation and Technology, Austria) ◀

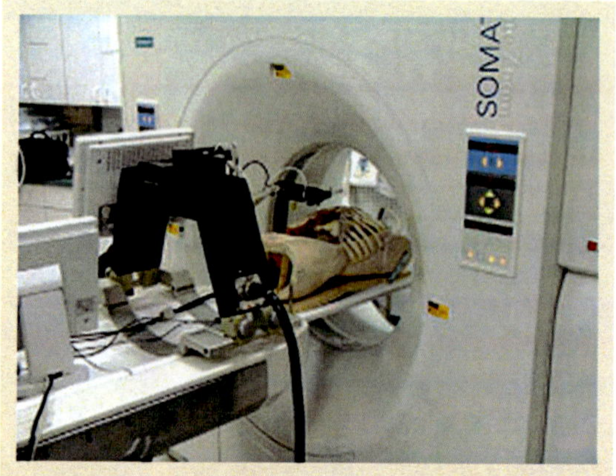

Fig. 70.14 AcuBot (Johns Hopkins Medical Institution, Baltimore)

unit can be sterilized and used several times. The system is designed in a modular manner and planned to be used in various clinical fields.

Another system to position needles under direct imaging is AcuBot from Johns Hopkins Medical Institution, USA (Fig. 70.14) [70.17].

The user interface is a 15-inch touchscreen monitor in connection with a two-axle joystick and another operation panel. The accuracy of needle positioning is >1 mm. The needle application unit is permeable to x-rays, so that it can be used directly in the CT.

70.4 Medical Applications

In abdominal surgery (e.g., cholecystectomy, prostatectomy) and in general endoscopic surgery, simple systems, such as camera guiding systems, are certainly applied most frequently. These systems supply a calm, tremor-free image. The camera is guided directly by the surgeon, so an additional assistant is not required as a cameraman.

If instrument holding systems to hold simple instruments such as forceps in a fixed position and to adjust them are used in addition to the endoscope guiding system, the surgeon can execute the operation largely alone, with the assistance of a surgical nurse only. So-called solo surgery is feasible [70.18].

Compared with manual surgery, robot systems allow for more exact surgical operations in, e.g., orthopedics (hip prostheses, knee operations), but also in otolaryngology (jaw and face surgery). The hip robot can mill out the thigh bone much more exactly than the surgeon. As a result, the prosthesis has a fitting accuracy of more than 90% compared with the manual method with about 30% contact area. Use of bone cement to support the connection between the implant and the bones can be reduced.

In neurosurgery, these exact machines are mainly used for precise guiding and holding of highly filigree operative instruments and cameras. The very high accuracy of the robot systems is combined with simple tasks, such as holding functions. The neurosurgeon profits from the robot interface to the planning software, such that exact positioning is accomplished on the basis of image data. This is of particular interest in neurosurgery. For example, Innomotion might be modified to such a system by Innomedic GmbH, Germany. Apart from endoscopic cameras, also instruments might be held and guided precisely. Accuracies of better than 1 mm are needed and feasible.

Use of telemanipulation systems enables the surgeon to perform much more precisely finer and more exact operation steps than conventional endoscopic interventions. In this connection, bypass operations in cardiac surgery and prostatectomies in urology have to be mentioned [70.19]. The highly traumatic classical open operative methods used to date can now be performed endoscopically using these precise manipulators. This is of great advantage to patients.

Generally, telemanipulation systems have the following features:

- Movement sequences can be scaled; for example, a movement of 10 cm given by the surgeon can be converted into an instrument movement of, e.g., 1 cm. More exact actions are possible.
- Movement tremors are filtered out. Tremors are eliminated, safety is increased.
- Indexing is possible. The surgeon can choose an optimum working position and does not have to operate in a tense posture according to the position of the working instrument. Optimal ergonomy for the surgeon is achieved, and fatigue-free operation becomes possible.
- It is possible to choose among various coordinate systems. It is possible to control the instruments in, e.g., world coordinates, instrument coordinates or screen coordinates. Depending on the application and the preference of the surgeon, the most favorable control method is selected.
- Forces can be measured at the instrument tip and transmitted to the surgeon at the operating unit.

Frequently, endoscopic instruments were developed for operations that had been carried out conventionally until that time. These so-called spin-offs allow for simple endoscopic interventions with novel minimally invasive surgery techniques.

70.5 Technical Aspects

Use of teletechnologies requires rethinking and familiarization [70.20]. The physician has to acquire technical understanding. Not only medical but also technical staff is required in the operation theater. An acceptance threshold has to be overcome, if the physician is no longer in direct contact with the patient, but only via telecommunication systems. In addition, a psychological inhibition threshold may result from a robot or a telemanipulator replacing the physician at the operation table and executing the intervention. It is necessary to communicate to the patient the advantages of the use of such technology. On the other hand, the patient has to trust in a physician being also present at the operating table to monitor the operation.

Such a system already exists in patient monitoring. Figure 70.15 shows the RP-7 from InTouch Technologies, USA. It is used to bring the physician to the patient via telecommunication lines (mainly the Internet). Via camera and screen, the patient can communicate with the physician and the physician sees the status of the patient. Direct presence is no longer required. The control and examination of wound healing can take place via such telenetworks all over the world.

Figure 70.16 shows that not only image and sound are transmitted to the physician but also vital parameters. The simplest control unit of the RP-7 platform is a normal laptop. Via the integrated web camera, the patient can see the physician during the ward round.

Robot and manipulator systems for use in surgery are associated with special requirements on integration in static and dynamic structures of operation theaters at hospitals. A number of concepts have been developed for retrofitting at existing installations and fitting at new constructions specially tailored to the needs of robots. Some of these requirements shall be outlined in detail below.

Construction requirements include sufficient space for rapid and easy coupling of the manipulator system to the operation table, possibly a vibration-free floor, and/or suspension from the ceiling. The operation unit must be installed such that direct view of the patient and the manipulator system as well as of other information (anesthetic data, etc.) is given, if required.

The operation staff has to undergo special training for handling of the devices and elimination of problems possibly occurring during operation. In addition, a technician has to be present for proper installation and operation of the system and correct detection and elimination of problems in due time. The interdisciplinarity of the operation staff is increased. The surgeon is required to have a certain technical un-

Fig. 70.15 RP-7 mobile robotic platform (InTouch Technologies, Santa Barbara, CA)

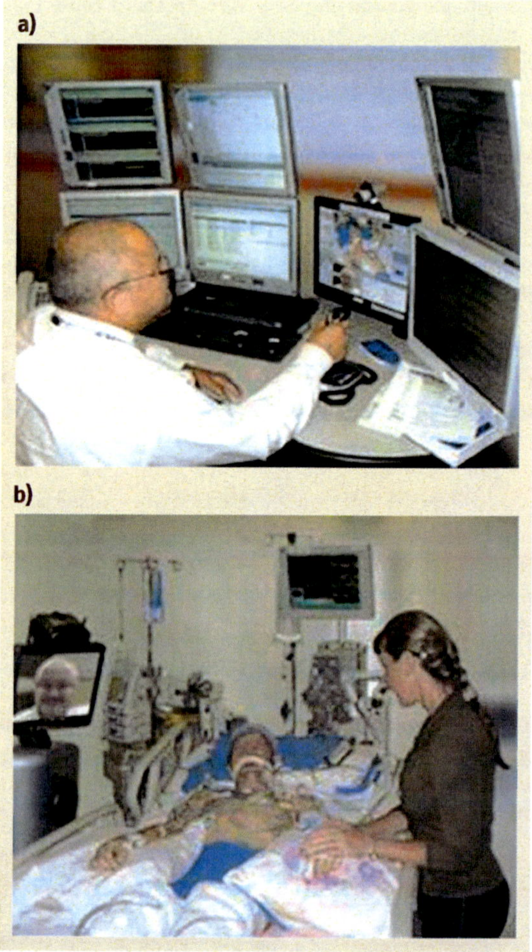

Fig. 70.16a,b Remote presence with mobile robotic platform. **(a)** Doctor's office, **(b)** remote patient (InTouch Technologies, Santa Barbara, CA)

derstanding. He has to undergo extensive training to learn how to operate the master unit and the complete telemanipulation system, and to familiarize himself with the additional functionality that differs considerably from conventional operations. The operation unit should be operable in an intuitive manner, but the surgeon has to familiarize himself with various settings that are not relevant in other cases (e.g., selection of the coordinate system, scaling, device assignment). This will ensure correct, safe, and rapid action in critical situations.

Hence, simulators and trainers are needed. Similar to flight simulators, use of training and simulation systems for telesurgery seems to be reasonable for continuous training of surgeons [70.21].

70.6 Outlook

It was demonstrated by the development and long-term testing of ARTEMIS as well as by the commercial surgical telemanipulators and robots applied meanwhile that use of teletechnologies in the operation theater results in a number of advantages. They include:

- Improved image quality by the use of endoscope guiding systems
- Improved operation quality by the more exact guiding of the instrument with the help of telemanipulation systems
- Improved operation quality by a fatigue-free, more ergonomic posture
- Improved operation quality by the integration of a 3-D viewing system.

Nevertheless, some major objectives remain to be achieved:

- Modular and flexible design of the system
- Flexible and easy-to-exchange instruments or multifunctional instruments
- Adaptation to telesurgery needs
- Various application options in all relevant surgical disciplines
- Wider usability in special environments, e.g., CT and MRI.

System modularity will not only refer to mechanical components, but also to the software. Modularity of the software system can be achieved by a distributed, object-oriented, real-time architecture. Multi-agent systems are designed in terms of both software and hardware. An open system structure will allow for integration of various master and slave systems.

Apart from telemanipulation proper, other teletechnologies will be integrated in the telepresence system to support teleconsulting and teleplanning. Applicability in various disciplines will continue to be a prerequisite for economic acceptance of such systems.

Use of teletechnologies in surgery will increase. However, they will not be applied in all areas tested. Often, only one operation will be carried out for the press to report that

> *surgeon A was the first to perform operation B with the help of robot system C. However, there will not be any successors, as the execution of such an operation B with a robot system has more drawbacks than advantages...*,

or it will take a long time until such operations will be performed widely. During the coming years, some hospitals will have to act as pioneers to identify the potential but also the limits of telesurgery. It will probably take several years until such systems will be used throughout Germany. The costs (often more than one million euros) as well as limited functionality are obstacles.

The successes achieved in the use of telemanipulation and robot systems in surgery demonstrate that such systems have a justified position in medicine and can be used reasonably. Nevertheless, there is need for further research and development. Applications can be extended. Modifications and further developments of the systems are required. If necessary, medical procedures have to be adapted to the possibilities of robotics. Entirely new operative techniques that are not feasible with purely manual operation will have to be tested. Close cooperation between engineers and physicians is required.

References

70.1 R. Oberle, U. Voges: A telerobotic approach towards endoscope guidance, 1st Eur. Congr. Eur. Assoc. Endosc. Surg. (Cologne 1993)

70.2 L. Gumb, A. Schäf, R. Trapp, G. Bueß, M. Schurr: Vorrichtung zur Führung chirurgischer Instrumente für die endoskopische Chirurgie, German Patent DE-OS 19609034 (12.9.1996), DE-PS 19609034 (1.7.1997), EP-PS 59600577 (16.9.1998)

70.3 M. Selig, H. Fischer, L. Gumb, A. Schäf, R. Cichon, U. Kappert, S. Schueler: Voice controlled camera guiding system FELIX for endoscopic cardiac surgery, 11th Annu. Sci. Meet. Soc. Minim. Invasive Ther. (SMIT) (Boston 1999)

70.4 D.R. Uecker, C. Lee, Y.F. Wang, Y. Wang: A speech-directed multimodal man-machine interface for robotically enhanced surgery, Proc. 1st Int. Symp. Med. Robot. Comput. Assist. Surg. (Pittsburgh 1994) pp. 176–183

70.5 J.M. Gilbert: The EndoAssist robotic camera holder as an aid to the introduction of laparoscopic colorectal surgery, Ann. R. Coll. Surg. Engl. **91**(5), 389–393 (2009)

70.6 B. Neisius: Konzeption und Realisierung eines experimentellen Telemanipulators für die Laparoskopie. Ph.D. Thesis (Univ. Karlsruhe, Karlsruhe 1997)

70.7 A. Bauer, M. Börner, A. Lahmer: Experiences with a computer-assisted robot (RobodocTM) for cementless hip replacement. In: *2nd IARP Workshop on Medical Robotics*, ed. by R. Dillmann, E. Holler,

Part G | 70

H.P. Meinzer (Forschungszentrum Karlsruhe, Karlsruhe 1997) pp. 133–134
70.8 J. Petermann, M. Schierl, P.F. Heeckt, L. Gotzen: The CASPAR-system (computer assisted surgery planning and robotics) in the reconstruction of the ACL, First follow-up results, 5th Int. Symp. CAOS (Davos 2000)
70.9 U. Voges, H. Breitwieser, P. Dautzenberg, M. Dickerhof, E. Holler, M. Hübner, H. Knüppel, H.G. Krumm, U. Kühnapfel, C. Kuhn, K. Leinemann, B. Neisius, R. Oberle, M. Schmitt, R. Trapp, T. Vollmer: Experimenteller Telemanipulator für die minimal invasive Chirurgie, Wiss. Ber. FZKA-5670 (1995) pp. 106–111
70.10 H. Shennib, A. Bastawisy, M.J. Mack, F.H. Moll: Computer-assisted telemanipulation: an enabling technology for endoscopic coronary artery bypass, Ann. Thorac. Surg. **66**, 1060–1063 (1998)
70.11 E.R. Stephenson, S. Snakholkar, C.T. Ducko, R.J. Damiano: Robotically assisted microsurgery for endoscopic coronary artery bypass grafting, Ann. Thorac. Surg. **66**, 1064–1067 (1998)
70.12 H. Fischer, M. Selig, J. Vagner: Manipulator für die Therapie von Mammakarzinomen im Magnetresonanztomographen (MRT), Nachr. Forschungszent. Karlsr. **32**(1–2), 33–38 (2000)
70.13 V. Urban, M. Wapler, T. Weisener, R. Schonmayr: A tactile feedback hexapod operating robot for endoscopic procedures, Neurol. Res. **21**(1), 28–30 (1999)
70.14 B. Gutmann, A. Lukoschek, H. Fischer, A. Melzer: Development of an assistant robotic system for interventional radiology inside CT and MR scanners, 2. Jahrestag. Dtsch. Ges. Comput. Roboterassist. Chir. (Nürnberg, 2003), C-II-6
70.15 A. Melzer, B. Gutmann, A. Lukoschek, M. Mark, W. Zylka, H. Fischer: Experimental evaluation of an MRI compatible telerobotic system for CT MRI guided interventions, Radiol. Suppl. **226**, 409 (2003)
70.16 O. Wendt, T. Remmele, B. Gutmann, A. Melzer, H. Fischer: Assistant manipulator system for image guided percutaneous interventions, 3. Jahrestag. Dtsch. Ges. Comput. Roboterassist. Chir. (Munich, 2004) p. s33
70.17 D. Stoianovici, K. Cleary, A. Patriciu, D. Mazilu, A. Stanimir, N. Craciunoiu, V. Watson, L. Kavoussi: AcuBot: A robot for radiological interventions, IEEE Trans. Robot. Automat. **19**(5), 927–930 (2003)
70.18 M.O. Schurr, G. Buess, B. Neisius, U. Voges: Robotics and allied technologies in endoscopic surgery, Surgical Technol. Int. VII (Universal Medical, San Francisco 1998) pp. 83–88
70.19 M. Selig, H. Fischer, L. Gumb, A. Schäf, R. Ullrich, B. Vogel, R. Cichon, M. Cornelius, U. Kappert, S. Schüler: Minimal invasive Herzchirurgie, Nachr. Forschungszent. Karlsr. **32**, 55–60 (2000)
70.20 U. Voges, E. Holler: Teletechniken und chirurgische Robotik, Nachr. Forschungszent. Karlsr. **32**(1–2), 39–54 (2000)
70.21 H. Cakmak, H. Maass, U. Kühnapfel: VSOne, a virtual reality simulator for laparoscopic surgery, Minim. Invasive Ther. Allied Technol. **14**, 134–144 (2005)

71. Incubators

Guenther Braun, Roland Hentschel

This article describes fundamental information about neonatal incubators which are typically used on intensive care unites (NICU). In a short abstract the historical development is described. State of technology is specified and also unsolved questions identified. Depending on the intended use different incubator types are shown. In addition to the functions of a neonatal incubator also risks of incubator therapy will be presented.

71.1 **Historical Background** ... 1285
71.2 **Construction and Function of an Incubator** ... 1286
71.2.1 Temperature Regulation ... 1286
71.2.2 Regulation of Humidity ... 1286
71.2.3 Regulation of Oxygen ... 1287
71.2.4 Scales ... 1287
71.2.5 X-ray Drawer ... 1287
71.3 **Incubator Models** ... 1287
71.3.1 Stationary Incubators ... 1287
71.3.2 Transport Incubators ... 1287
71.3.3 Special Incubators ... 1288
71.4 **Risks of Incubator Therapy** ... 1289
71.4.1 Temperature ... 1289
71.4.2 Oxygen Therapy ... 1289
71.5 **Hygiene** ... 1289
71.6 **Unsolved Problems** ... 1290
References ... 1290

71.1 Historical Background

Incubators are used to stabilize and maintain the thermal balance in premature and newborn babies. In the year 1907 the pediatrician Pierre Budin described in his book *The Nurseling* the connection between mortality and rectal temperature. Newborns whose rectal temperature was warmed to 36–37 °C reached a survival rate of 77%, however in babies with rectal temperature between 32.5 °C and 33.5 °C the survival rate was only 10%.

It is assumed that incubators were used by the Egyptians as early as 200 AD. In the year 1947, a specially designed incubator was launched onto the market by the American physician Chapple. This incubator was the forerunner of present-day incubators, including a transparent acrylic glass cover, a bacteria filter for air intake, an integrated air circulation system, air humidification, and an alarm device against overheating, which was first put into practice in this model [71.1].

Premature babies are not capable of maintaining their thermal balance independently [71.2]. The ratio of body surface to body volume of a newborn is 2.7 times higher than that of an adult. In neonates with birth weight of 1000 g the ratio is 4 times higher.

Loss of warmth can occur basically in four ways:

1. Heat conduction by release of heat to the mattress
2. Convection (cooling by airflow)
3. Evaporation from the skin
4. Thermal radiation from babies to the surrounding area.

Loss of warmth is reduced by vasoconstriction. By increasing vessel resistance, the extremities are cooled first, before the body temperature falls. Comparison between body temperature and peripheral parts of the body enables an early indication of thermal imbalance [71.3].

Only newborn children have brown fat tissue, which lies between the shoulder blades, behind the heart and the large vessels. These energy reserves have limited ability to maintain normal body temperature. Newborns are also not capable of producing warmth by increasing muscle activity (shivering). Due to their thin skin, more liquid is lost by evaporation from premature babies than from mature newborns.

Cold stress must absolutely be avoided for the following reasons:

- Less intake of oxygen, e.g., with a negative influence on the development of the lungs due to insufficient surfactant production. Hence, breathing problems could develop, which would have to be compensated by artificial respiration.
- Metabolism will be influenced negatively, e.g., hypoglycemia or metabolic acidosis (too low pH). The risk of pathological jaundice (icterus gravis) is increased.
- Increased danger of infection.
- Negative effect on growth.

It is therefore necessary to keep the core temperature constant and dehydration through perspiration at a minimum.

71.2 Construction and Function of an Incubator

Incubators create a microclimate, which can differ considerably from the surrounding air. In this microclimate, temperature, humidity, and oxygen content can, up to a certain limit, be regulated individually. Furthermore, the incubator isolates the patient in terms of protection from pathogens which could be transferred via the air.

As a rule, incubators are designed for body weight up to 5 kg. The user can adjust the height as well as the inclination of the mattress. This allows good *ergonomics* and easy access to the patient by nursing staff. The incubator can also be lowered to allow for optimal mother–child contact.

71.2.1 Temperature Regulation

The incubator sucks in room air via an exchangeable air filter. The room air is heated by a heating element, which has a fan in the middle. The warm air is conducted via a ventilation shaft along the long sides of the incubator to the patient. As a result, a shield of warm air is formed. This shield reduces the cooling of the incubator when it is open. As a rule, the air is released at the front end. The air temperature is measured by several NTC (negative temperature coefficient thermistor) sensors. The challenge for the manufacturer is to create a draught-free, homogeneous atmospheric environment. The fan (aerator) has to meet special requirements, e.g., it has to run as silently as possible, to avoid stress and the danger of hearing damage to the patient. Incubators run at operating noise of less than 50 dB(A). In future developments, the goal is to lower this noise level even further.

The user has two possibilities to control the temperature of the incubator:

1. Air temperature regulation: The user chooses a suitable air temperature. The temperature is measured by means of a temperature sensor in the incubator and compared with the *set level (point)*.
   The set temperature in the incubator is thus kept constant. If the actual temperature differs from the set temperature, the user will be informed by an alarm signal. As a rule, the alarm limit is ±1.5 °C from the set level (point).
2. Skin temperature regulation: The user places a skin temperature sensor on the thorax, and a second one can be attached to the extremities. The desired temperature on the thorax is set by the user.
   The incubator regulates the air temperature depending on the skin temperature. If the actual temperature differs from the set temperature, the user will be informed by an alarm signal. As a rule, the alarm limit is ±0.5 °C from the set point.

Incubators are not capable of cooling. At high outside temperatures and with large babies, the set temperature cannot be maintained. The temperature can, depending on the type of incubator, be set from 20 °C to 39 °C at 0.1 °C intervals.

71.2.2 Regulation of Humidity

Two construction principles are used to create moisture: either water is evaporated over a heater and then, after cooling, passed into the incubator, or it is created over a heated washbasin in the incubator cell. Humidity is measured using capacitive components.

Moisture is necessary not only to maintain the thermal balance of the baby but also to moisten the mouth, nose, and throat. The relative humidity can usually be set between 30% and 99%.

The user has, depending on the model, two possibilities to control the humidity:

1. Manual humidity regulation: The user chooses a suitable relative humidity. The relative humidity is measured by means of a humidity sensor and compared with the set level.
2. Automatic humidity regulation: In the automatic mode the relative humidity will be increased at higher air temperature.

Depending on the temperature, air can hold different levels of humidity, so that when dropping below a critical temperature accumulation of condensation occurs, whereas at higher temperatures, moisture can be absorbed.

As a result of the ratio of the outside temperature to the high inside temperature, there is the danger of condensation accumulating on the incubator *glass*. This may limit the possibility to monitor the baby. The effect is increased when humidity > 70% is chosen. Numerous manufacturers offer optional interior glass for better isolation. In practice, these have not become well established, especially because condensation between the two glass panels cannot be wiped away by nursing staff without great effort.

Incubators are not capable of drying air. If a humidity set level is chosen which is lower than the surrounding air humidity, the actual value achieved will never be lower than the surrounding humidity.

71.2.3 Regulation of Oxygen

As an option, with all equipment available on the market, the air can be oxygenated. The air, depending on the construction of the incubator, can be oxygenated up to 70 vol. %.

With stationary incubators the oxygen is supplied via the central gas supply. With transportable incubators for ambulance or helicopter transport, operation is possible via the central gas supply as well as from gas bottles. The oxygen in the incubator is measured by fuel cells, which are applied also in long-term respirators. Due to the potential danger of a wrong dose of O_2, O_2 testing is designed to use two measuring devices. The signals from two fuel cells are compared with each other. If the values differ considerably from each other, the fuel cells can be automatically or manually calibrated by the user.

71.2.4 Scales

Not only body temperature but also body weight is an important diagnostic parameter in neonatology. Today's technology implements scales into the incubator. In this way, *cold stress* is avoided for the newborn when it is weighed in compartment air in the incubator instead of on conventional scales in room air.

71.2.5 X-ray Drawer

To make handling of necessary x-rays easier, in newer incubator models a drawer is built in under the mattress. Common x-ray cassettes can be inserted into this drawer. In this way, the newborn is not laid down directly onto a cold, hard x-ray plate.

71.3 Incubator Models

Incubators are basically divided into three type of models.

71.3.1 Stationary Incubators

These are used for long-term care of premature and newborn babies in recovery rooms and intensive care units (Fig. 71.1).

71.3.2 Transport Incubators

These are used for *in-house transport* from the delivery room to the pediatric intensive care unit or for transfer to other clinics by vehicle or helicopter. The latter incubators have their own electricity supply via accumulators as well as a gas supply via gas bottles. In the hospital or in the vehicle the equipment can be supplied with energy via the 220 V network and via the central gas supply, in order to save the mobile energy supply. Incubators mounted on a chassis with vibration absorbers are adapted to a standard retractable carriage (gurney). Due to limited space during transport, e.g., in helicopters, as well as the short transport times, the humidification option is not available in these devices. Intensive care transport incubators are equipped with a respirator, a patient control monitor, a syringe pump, and a suction device (Fig. 71.2).

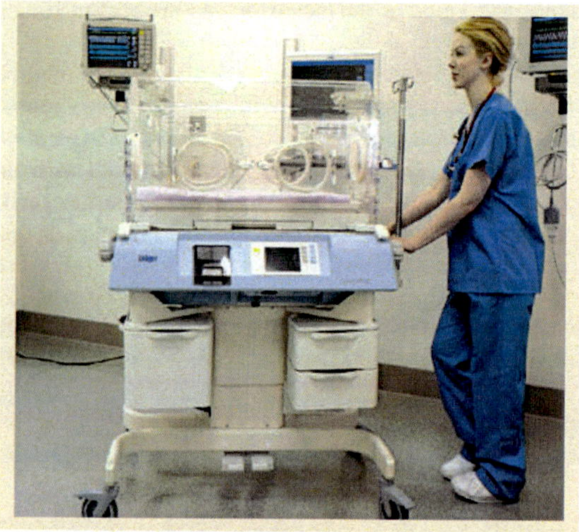

Fig. 71.1 Stationary incubator (courtesy of Draeger)

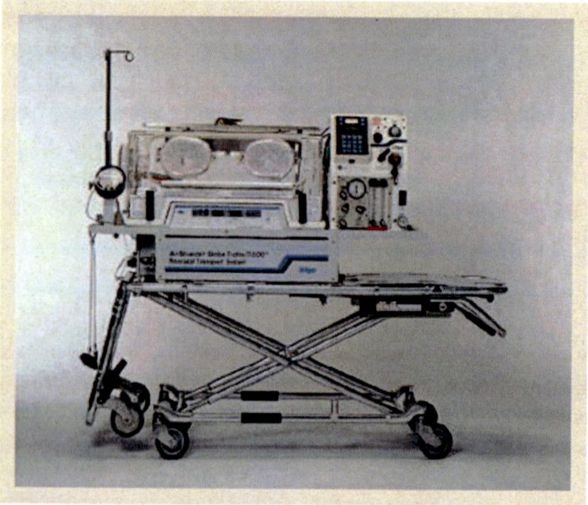

Fig. 71.2 Transport incubator with gurney (courtesy of Draeger)

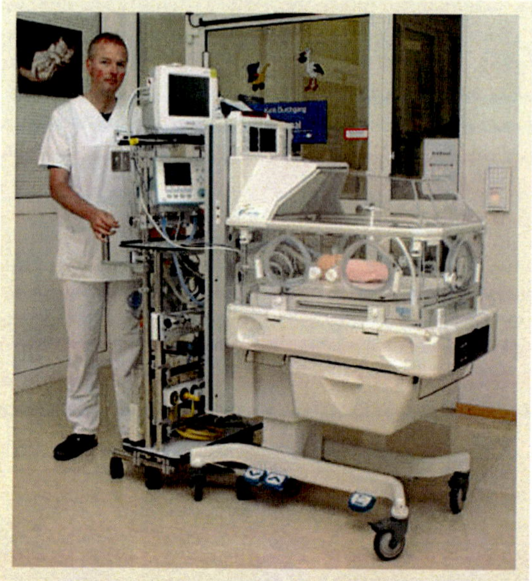

Fig. 71.3 Incubator for in-house transportation with docking carriage (courtesy of University Hospital Freiburg and General Electric Company)

71.3.3 Special Incubators

For examinations in magnetic resonance imaging (MRI), there are nonmagnetic transportable incubators, which can be inserted into the MRT. With this apparatus, air humidification and artificial ventilation are also possible as well as being fitted with a pulse oximeter for monitoring the patient. The apparatus is also suitable for thorax and brain examinations.

The problem of *in-house transportation* can be considerably simplified by a novel transport concept. Until now, for therapy each baby had to be looked after in a reanimation unit, then transferred in a transport incubator from the delivery room or the *sectio-op* to the children's ward to the place of treatment. This means repeated relocation (shifting) and therefore cold stress for the newborn, as well as the danger of unintended manipulation of infusion and respiration tubes (Fig. 71.3).

Usually, newborns are initially treated in reanimation units, i.e., a *warming bed* with a heated gel mattress and an integrated radiant heater (radiator). After stabilization in the delivery room, patients are often transferred to intensive care within a few minutes. Babies are not transferred to stationary incubators until they are in the intensive care unit. So, patients are relocated twice at room temperature and experience cold stress. The reanimation unit and the transport incubator must go through a time-consuming upgrade process. To improve this situation, a stationary incubator with an integrated heat radiator is connected to a *docking carriage (trolley)*. Thus, this incubator can be used as an initial reanimation unit. On the docking carriage (trolley), all the necessary relevant equipment can be found for an intensive care unit: respiration equipment, aspiration (suction, extraction), patient monitor, syringe pumps, and a hand anesthesia bag. The docking carriage (trolley) is equipped with O_2 and pressurized air bottles to ensure breathing during transport. The trolley is connected to the incubator without using tools. After arrival at the neonatal intensive care unit, the trolley is then connected directly in the intensive care room to carry on with therapy using the ward's equipment. Using this concept, cold and shifting (relocation) stress can be minimized (reduced). Resources are saved, and handling by staff is facilitated.

71.4 Risks of Incubator Therapy

71.4.1 Temperature

As described above, stabilization of body temperature of newborns is of great importance. Therefore, tightly controlled temperature monitoring is indispensable. For newborns, the rectal temperature is measured every 2 h. If the baby is *thermally stable*, the cycle can be expanded to 4 h. If the baby cools, it is called hypothermia. If the child is warmed by the incubator to over 37 °C, it is called hyperthermia.

As a result of increased liquid loss due to hyperthermia, electrolyte dysfunction can result. Further outcomes are hyperventilation and tachycardia. In thermostable children, skin temperature correlates very well with body temperature. Therefore, use of the skin temperature regulation operating mode simplifies the suitable choice of temperature by the user.

On no account should skin temperature regulation be used for children who are in a state of shock, because in this situation unintentional hyperthermia could occur. Also for babies with fever, skin temperature regulation is only to be used with great care, because in this case the skin temperature is higher than the body temperature. Very early premature babies experience extreme perspiration, resulting in high fluid loss, in the first days of life. A further exception is the patient experiencing shock, or skin – surface temperature due to infection, which will cause unreliable readings on the incubators skin temperature monitoring.

71.4.2 Oxygen Therapy

Oxygen treatment can pose a high potential risk for incorrect dosage. Therefore, when administrating O_2, arterially measured O_2 partial pressure must be determined. With O_2 therapy it is indispensable to monitor the patient continuously with a *pulse oximeter* or a transcutaneous O_2 *probe*. If there is undersupply (*hypoxemia*), then there is the danger of respiratory insufficiency resulting in apnoea. Continuous undersupply of O_2 can cause brain damage.

Overdose of O_2 (*hyperoxemia*) in premature babies before the 39th week of gestation can lead to serious eye damage. It can even cause retinal detachment, a very serious retinopathy.

Oxygen is associated with the danger of explosion and fire hazard. Therefore, for safety reasons, after disinfecting the hands, one should wait until the disinfectant has dried. For this reason no disinfectant or inflammable liquids such as alcohol or *benzine* should be put in or on the incubator.

71.5 Hygiene

Due to the warm climate and high humidity, an incubator is an ideal breeding ground for germs, leading to an additional risk of infection for the newborn. Therefore, after terminating the treatment of a patient, or after 7 days at the latest, a new incubator must be provided and the used incubator carefully cleaned and wiped with disinfectant.

In particular, excessive build-up of condensation in the interior of the chamber is a risk for early fungal or bacterial contamination.

71.6 Unsolved Problems

The following points should be given special consideration in a newborn's environment:

- Total reduction of noise, and the possibility to regulate light entering the chamber.
- It should be noted that, although no limits on electromagnetic fields in patient healthcare environments have been reported, their effect on heart rate and physiological functions will become an issue [71.4].
- A problem noticed in the field is international standardization of holders for patient gurneys. To ease international patient transport it is desirable to implement international standardization in the future.

References

71.1 H. Frankenberger, A. Güthe: *Inkubatoren* (TÜV Rheinland, Cologne 1991)

71.2 K. Bauer, C.P. Speer: *Pädiatrie* (Springer, Berlin, Heidelberg 2001)

71.3 H. Brück, P.A. Püschner: *Thermomonitoring* (Drägerwerk, Lübeck 1996)

71.4 R. Antonucci, A. Annalisa, V. Fanos: The infant incubator in the neonatal intensive care unit: Unresolved issues and future developments, J. Perinat. Med. **37**, 587–598 (2009)

72. Surgical Scissors

Reiner Haag, Wilfried Storz

Scissors still represent an indispensable tool for any surgical discipline, in medical practices no less than in clinical departments. It is probably safe to say that among all surgical instruments, scissors are used the most frequently – prior, during, and after the operation. Using scissors is extremely easy, with the benefits being not only excellent, but manifold and almost unlimited. In fact, there is hardly any better universal alternative when it comes to transecting or dissecting tissue or cutting sutures or any other kind of auxiliary materials. Yet in spite of the easy handling, the demands made on the surgeon or nurses are very high and complex as well – and they differ according to the task at hand. This has led to the development of a wide range of scissors that includes no less than 2000 different versions of the product.

72.1 **The History of Scissors** ... 1292
- 72.1.1 Paired Knives ... 1292
- 72.1.2 End-Jointed Scissors ... 1293
- 72.1.3 Bow Scissors ... 1293
- 72.1.4 Pivoted Scissors ... 1293

72.2 **The Function and Design of Scissors** ... 1293

72.3 **Materials** ... 1294
- 72.3.1 Soldering a Hard-Metal Inset into the Cutting Edge ... 1294
- 72.3.2 Welding Hard Metal in Place ... 1294
- 72.3.3 Special Coatings ... 1294
- 72.3.4 Titanium Scissors ... 1294

72.4 **Manufacture of Surgical Scissors** ... 1294
- 72.4.1 Steps in the Production Cycle ... 1295

72.5 **Diversification Overview** ... 1297
- 72.5.1 Surgical Standard Scissors ... 1297
- 72.5.2 Surgical Scissors – Dissecting Scissors ... 1297
- 72.5.3 Suture or Ligature Scissors ... 1299
- 72.5.4 Wire Cutting Scissors ... 1301
- 72.5.5 Microsurgical Scissors ... 1302
- 72.5.6 Vascular Scissors ... 1305
- 72.5.7 Gynecological Scissors ... 1306
- 72.5.8 Gynecological Scissors for Obstetrical Use ... 1308

72.6 **Handling and Care** ... 1310
- 72.6.1 General Instructions ... 1311
- 72.6.2 Materials Used in Scissor Manufacturing ... 1311
- 72.6.3 Water Qualities ... 1311
- 72.6.4 Preparation for Cleaning and Disinfection ... 1312
- 72.6.5 Manual Cleaning and Machine Cleaning ... 1312
- 72.6.6 Machine Cleaning and Thermal Disinfection ... 1312

72.7 **Inspection, Testing, and Care** ... 1314

72.8 **Packaging** ... 1314

72.9 **Current Terminology** ... 1315

72.10 **Steam Sterilization with Saturated Steam** 1315
- 72.10.1 Steam Quality ... 1315
- 72.10.2 Release and Storage ... 1316

72.11 **Quality Characteristics** ... 1316
- 72.11.1 Material ... 1316
- 72.11.2 Surface ... 1316
- 72.11.3 Form ... 1316
- 72.11.4 Action ... 1316
- 72.11.5 Eye Rings ... 1316
- 72.11.6 Quality of Cut ... 1317

72.12 **Future Developments** ... 1317
- 72.12.1 Users ... 1317
- 72.12.2 Industry ... 1317

72.13 **Bipolar Scissors** ... 1318

References ... 1319

Part G | 72

72.1 The History of Scissors

The question of where and when scissors were exactly invented has not been finally solved to date, but there are justified reasons to assume that the first scissor-like devices came into use between 1300 and 600 BC.

72.1.1 Paired Knives

In their most primitive form, scissors consisted of two matching knives whose cutting edges were sim-

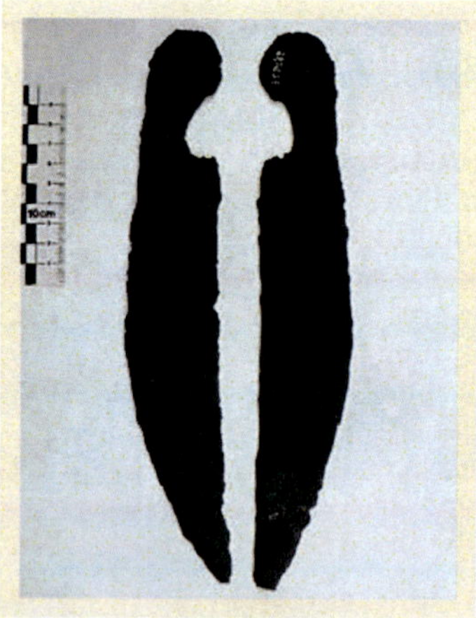

Fig. 72.1 Paired blades

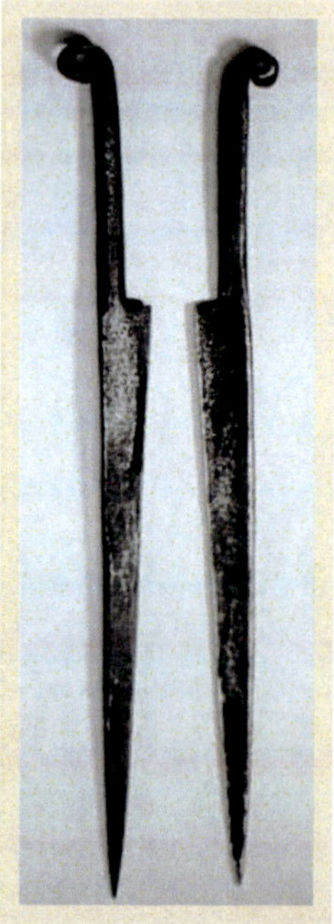

Fig. 72.2 Paired blades

Fig. 72.3 Spring bow scissors

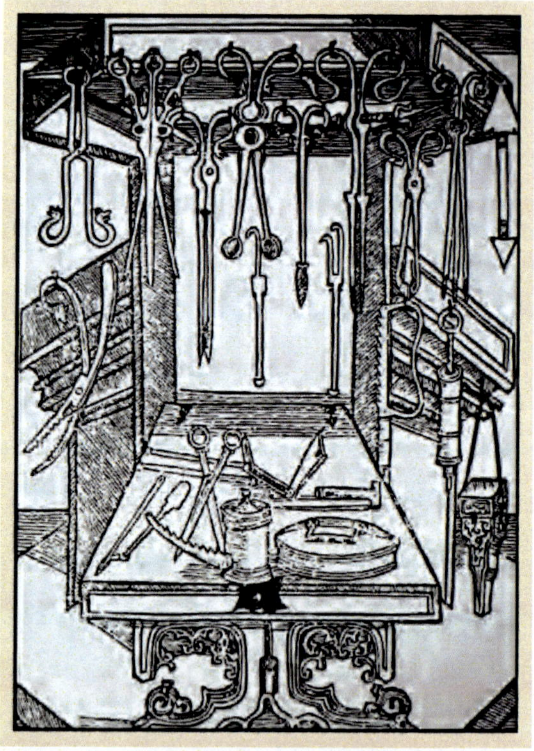

Fig. 72.4 The chamber with surgical instruments of Brunschwig

ply moved towards each other. As these paired blades were not connected to each other, the user had to work with both hands. The left hand held and supported one of the knives from below while the right hand guided the other, upper knife in such a way that the cutting edges glided closely past each other in a shearing or cutting motion while the user applied moderate pressure (Figs. 72.1, 72.2).

72.1.2 End-Jointed Scissors

End-jointed scissors represented a further advancement. They consisted of two blades that were hinged together at their back ends by way of a bolt. Their use was rather cumbersome, as the two blades had to be pressed upon each other for cutting and subsequently had to be pulled apart manually.

72.1.3 Bow Scissors

Compared with end-jointed scissors, spring bow scissors (omega-shaped back end) represent a significant improvement because they can be operated single-handedly and open by themselves due to the mechanical tension in the material. The first products of this sort probably date back to 300 BC (Fig. 72.3).

72.1.4 Pivoted Scissors

As more and more experience was gained in steel processing and user demands grew, pivoted scissors were increasingly used from the 13th century. This, in turn, led to new fields of use, not least in surgery. However, it was not until 1497 when H. Brunschwig, in his surgery book, presented a chamber full of surgical instruments, among them a pair of scissors (Fig. 72.4).

72.2 The Function and Design of Scissors

Depending on their design and field of use, surgical scissors are intended for transecting tissue, bones, organs, dressing materials, and other medical supplies. Moreover, scissors are quite frequently used for dissecting and manipulating certain tissue structures and organs, due to the fact that they are extremely easy to handle and give the surgeon a *feel* that provides valuable feedback information, thus allowing vital conclusions to be drawn.

The fields of use and practical application of surgical scissors are highly varied and extremely complex. This explains why there is such a multitude of different types and variants to choose from. However, the basic design is identical for all the different types of scissors (Fig. 72.5). At the front or distal end there are the blades, the shanks with the eye rings constitute the handle or back end, and somewhere in the middle there is the pivot or box lock.

When cutting with a pair of scissors, the tissue to be severed is grasped with the blades. As the instrument is closed, the cutting point travels along the blades from the box lock to the tips. This is termed the *action* of a pair of scissors. It is essential here that the mechanical tension between the two blades remains constant from the middle of the blades all the way to the tips. A satisfactory cut performance and quality cannot be achieved if this requirement is not met.

The length of the cut depends on the length of the shear blades and/or the length of the tissue seized by the blades. It is important here that the cutting edges of the blades are geometrically arranged in such a way that a very acute crossing angle is obtained when closing the scissors. This ensures that the tissue is cleanly transected at each cutting point – and only there.

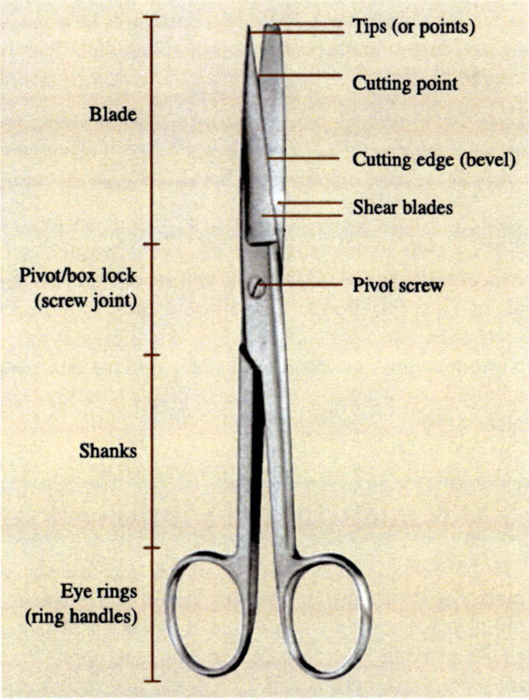

Fig. 72.5 General main parts of a pair of scissors

72.3 Materials

The most fundamental prerequisite for a high-quality pair of surgical scissors is the use of high-quality, forged steel satisfying EN ISO 7153-1:2000 requirements.

Commonly used steels include:

- 1.4117/X38CrMoV15 with a carbon content of 0.38%, a chromium content of 15% and a molybdenum content of 0.5%
- 1.4034/X46Cr13 with a carbon content of 0.42–0.50% and a chromium content of 12.5–14.5%
- 1.4021/X20Cr13 with a carbon content of 0.16–0.25% and a chromium content of 12–14%.

The following general rule applies: the higher the carbon content, the higher the hardness; and the higher the chromium content, the higher the corrosion resistance. The hallmark of hard-metal scissors is their gilded handles. There are two ways of fabrication.

72.3.1 Soldering a Hard-Metal Inset into the Cutting Edge

A hard-metal plate is soldered into the milled-out part of the scissor blade. This plate is extremely hard. However, due to the three different metals involved that occupy different positions in the electrochemical series, potential differences can lead to corrosion as a result of instrument reprocessing. Scissors with significantly curved blades are difficult or even too costly to manufacture using this method.

72.3.2 Welding Hard Metal in Place

Stellite, a hard metal alloy with a carbon content of 2.4%, a chromium content of 33%, and a tungsten content of 13% is welded into the milled-out part of the blade. Since this produces a hard weld, the scissor blade must subsequently be ground in a complex and time-consuming process. However, as only two materials are involved in this case, it is easier to create strongly bent blades.

72.3.3 Special Coatings

Hard-metal scissors featuring a special, super-hard coating are increasingly offered. Their surface hardness (Vickers hardness, HV) ranges between 2000 HV and 5000 HV.

Commonly used materials for coatings are:

- AlTiN Aluminum titanium nitrite, anthracite-colored
- TiN Titanium nitrite, gold-colored
- ZrO_2 Zirconium oxide, grayish.

User benefits include:

- Increased life due to the extremely high surface hardness
- Due to their darker color, AlTiN scissors minimize light reflections
- Less friction due to superior surface reticulation (cross-linkage); this results in smoother action and facilitates cleaning.

72.3.4 Titanium Scissors

For special applications, e.g. in an open MRI environment where antimagnetic materials are required, titanium scissors are an excellent choice. As titanium is an amorphous material, however, the manufacturing processes are significantly more complex and time-consuming, so they have to be completely separated from conventional machining processes. To take an example: The use of grinding belts previously used for conventional metal scissors would pose a risk of metal particles being ground into the titanium surface, thus leading to artifacts in the MR image in open MRI procedures and, consequently, to poorer image quality. In the worst case, the image would even be useless. Besides, the brittle material shortens the product's life, reduces cutting performance and makes the opening and closing action less smooth, compared with conventional scissors.

72.4 Manufacture of Surgical Scissors

The ease with which scissors can be used stands in marked contrast to the complexities of the manufacturing process. The whole cycle is made up of many single working steps. While machining is possible in some respects, there is still a lot of manual craftsmanship involved in the process. Even today, the production of high-quality scissors requires the skill of experienced surgical technicians or specialized *scissor makers*. Of course, their work is supported by advanced manufacturing technology. Nonetheless, precision working is a must in all production stages in order to achieve optimum results. The following table provides an overview of just the most important steps in the process, distinguishing between conventional manufacturing and an advanced process based on robotics and CNC technology (Figs. 72.6, 72.7).

72.4.1 Steps in the Production Cycle

Steps 1–6 of the preproduction cycle and step 1 of the final production cycle are *automated and standardized parts* of advanced manufacturing processes using robots and CNC technology.

Optional Technical Features of Surgical Scissors

Preproduction:

1. Drill core hole
2. Mill joint area projection (box lock corner)
3. Mill shanks

Table 72.1 Technical variants of surgical scissors

Surgical scissors	Description
Standard, without TC (hard-metal inset) (Fig. 72.8)	Low cost
With knife-type grinding	Sharp, knife-type cut
With knife-type grinding and bevel	Sharp and effortless, knife-type cut Minimized traumatization
With knife-type grinding, bevel and microtoothing	Sharp and effortless, knife-type cut Minimized traumatization Prevents tissue slippage
With TC, soldered or welded (Fig. 72.9)	Excellent edge retention
TC, with knife-type grinding	Sharp, knife-type cut Increased edge retention
TC, with knife-type grinding and bevel	Sharp and effortless, knife-type cut Minimized traumatization Increased edge retention
TC, with knife-type grinding, bevel and microtoothing	Sharp and effortless, knife-type cut Minimized traumatization Prevents tissue slippage Increased edge retention
TC, with serrated edge	Excellent edge retention Prevents tissue slippage Increased edge retention as standard
With ceramic coating (Fig. 72.10)	Long-term edge retention Excellent gliding property Glare-free surface Easy to clean
Titanium (Fig. 72.11)	Antimagnetic, thus MRI-compatible Only half the weight of steel scissors Effortless use, prevents fatigue

Fig. 72.6 Manufacturing of scissors in 1980

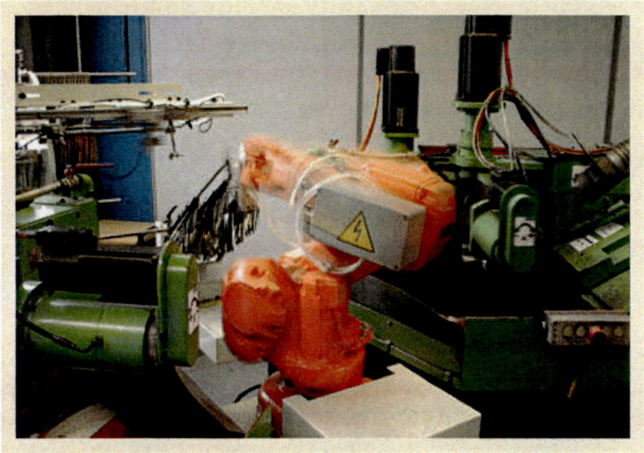

Fig. 72.7 Manufacturing of scissors in 2010

4. Prepare locations for hard-metal insets
5. Bend upper and lower blades
6. Tap lower shear blade and counterbore upper shear blade
7. Align lower/upper blade
8. Weld lower/upper blade
9. Grind blades on outside and inside

Final production:

1. Grind eye rings on outside and inside
2. Mount/rough-grind
3. Check prior to tempering (hardening)
4. Wash (remove lubricants, etc.)
5. Tempering and annealing
6. Check hardness
7. Mount/prepare for polishing

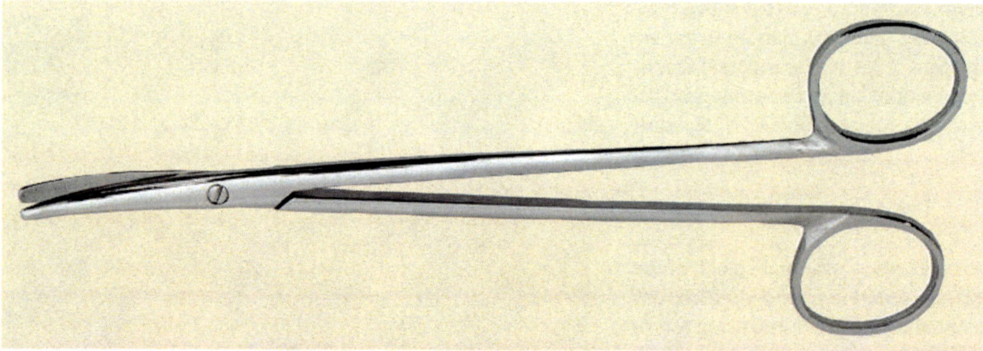

Fig. 72.8
Standard scissors

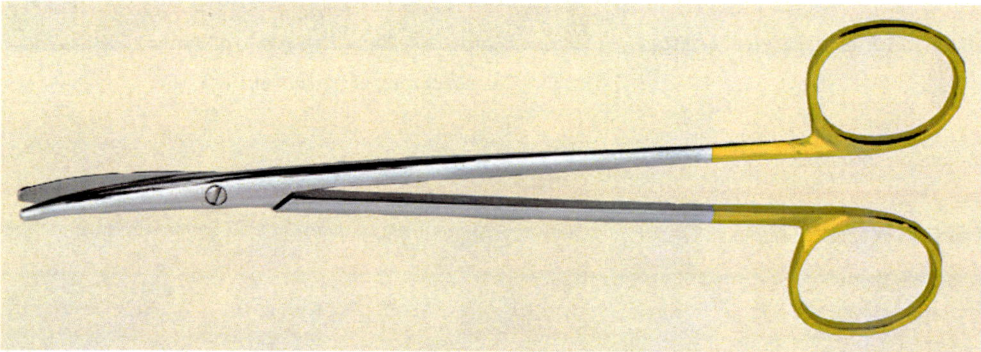

Fig. 72.9
Standard scissors with TC

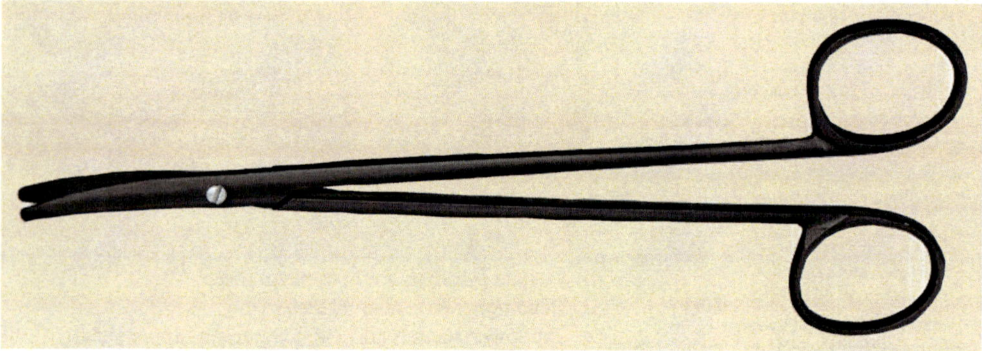

Fig. 72.10
Standard scissors with ceramic coating

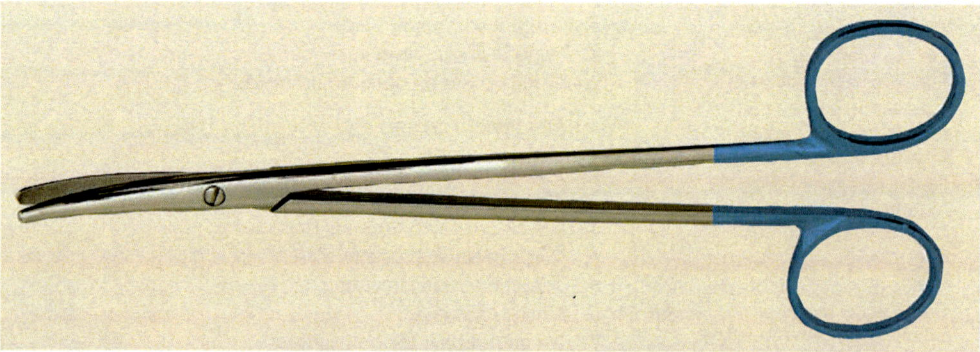

Fig. 72.11
Standard titanium scissors

8. Polish
9. Check
10. Wash/electropolish
11. Stamp, break edges, brush box lock
12. Depolish, brighten handles
13. Wash
14. Finalize, sharpen bevel of lower blade
15. Check
16. Wash
17. Gold-plate eye rings
18. Wash
19. Letter
20. Passivate
21. Final inspection

72.5 Diversification Overview

72.5.1 Surgical Standard Scissors

This type of scissors is usually used for cutting auxiliary materials needed for operations, e.g. dressing materials, flexible drain tubes, and – mostly – sutures. However, these scissors are also used for tissue transection in many surgical fields. Standard scissors are routinely included in almost all surgical instrument sets.

Types of Scissors

Standard scissors are medium strong to sturdy scissors, in straight and bent-up design.

Basis Sturdy Models. Blade tip designs are:

- Pointed/pointed (Fig. 72.12)
- Blunt/blunt (Fig. 72.13)
- Pointed/blunt (Fig. 72.14)
- The classic example blunt/blunt and bent-up, so-called Cooper scissors (Fig. 72.15a,b).

Scissor Lengths

Various lengths of scissors are commonly at the disposal of the surgeon. They cover lengths from 11 to 20 cm.

72.5.2 Surgical Scissors – Dissecting Scissors

Field of Use

Dissecting scissors (Fig. 72.20) are frequently used in surgical interventions – specifically where tissue needs to be transected and dissected in the deeper regions of the surgical site. Tissue cutting is done simply by closing the scissors effortlessly, whereas dissection is

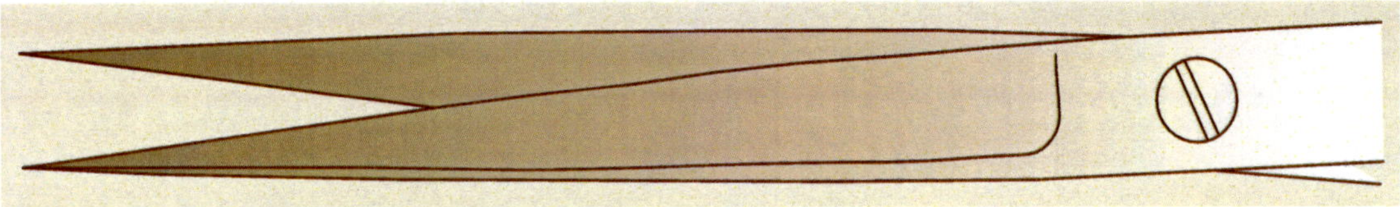

Fig. 72.12 Pointed/pointed blade tip design

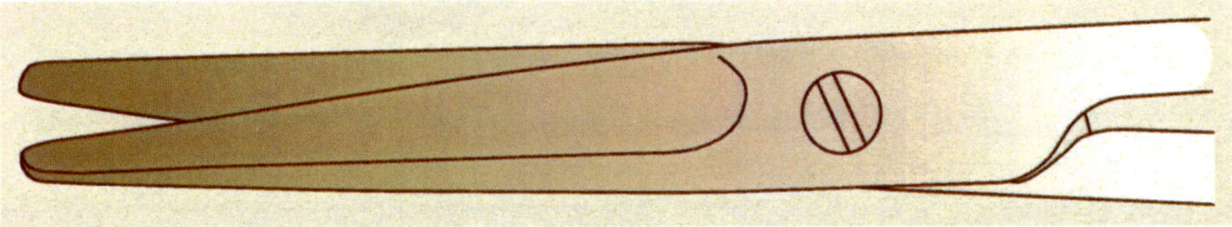

Fig. 72.13 Blunt/blunt blade tip design

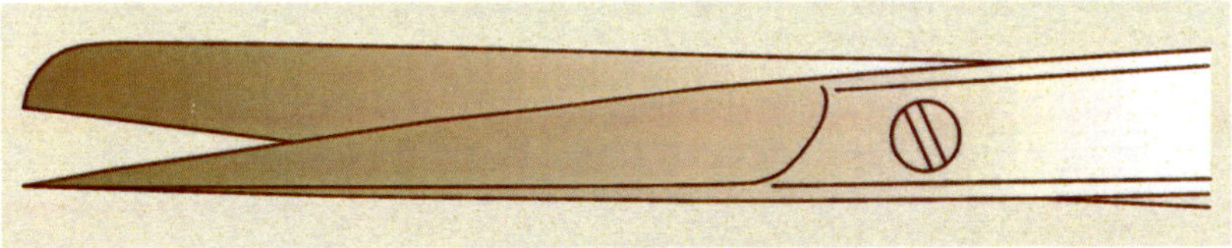

Fig. 72.14 Blade tip design pointed/blunt

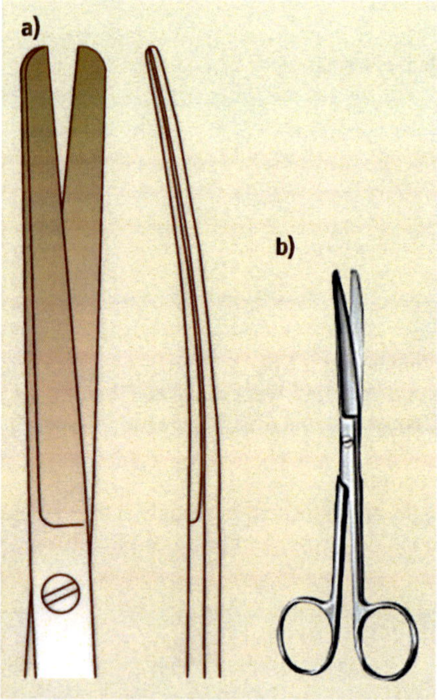

Fig. 72.15a,b Blunt/blunt and bent-up scissors, so-called Cooper scissors

performed by spreading the blades repeatedly in a *snapping* action. Blunt dissection (enucleation) is done with the scissors closed.

Wrongly though, dissecting scissors are frequently used also for cutting sutures during ligations. While this is understandable from the point of view of convenience (no need to exchange scissors), such use clearly reduces edge life. Special ligature scissors are available for cutting suture material.

Most Frequent Use

Mainly in soft tissue, abdominal and visceral surgery.

Types of Scissors

Medium-strong to slim, elegant, bent-up scissors, straight dissecting scissors are seldom used.

Most Importantly Used

By Lexer and Mayo (Figs. 72.21, 72.22), Metzenbaum, and Mayo-Stille. The best-known examples are the Metzenbaum scissors (Figs. 72.23, 72.24).

Typical Features

The hallmark of dissecting scissors are their rounded tips and the round back of the shear blades. When closed, these scissors must form smooth, round dissect-

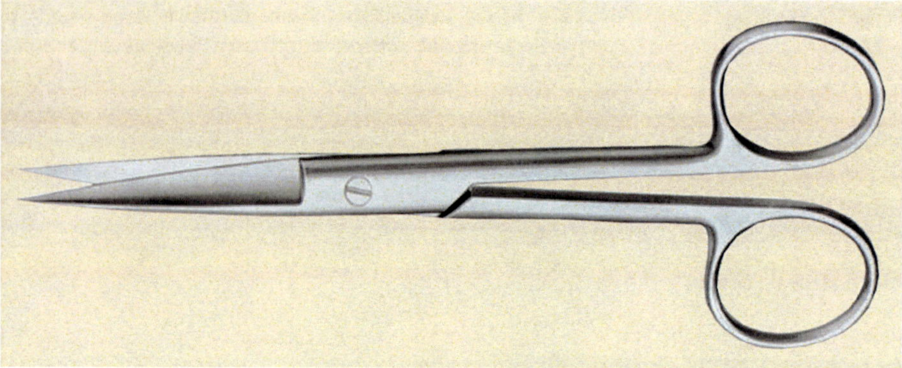

Fig. 72.16 Surgical scissors, length 11 cm

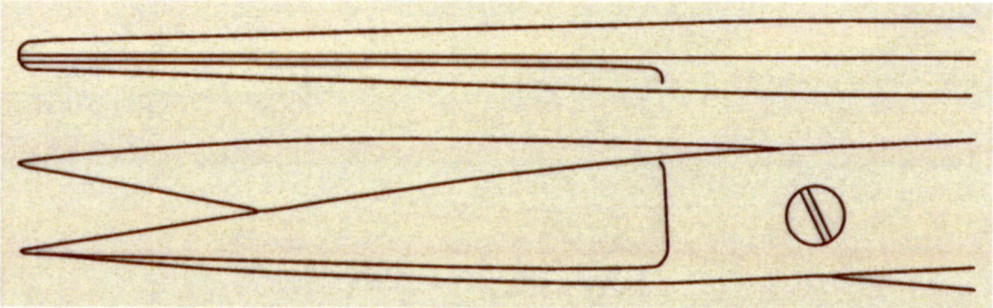

Fig. 72.17 Surgical scissors, length 16 cm

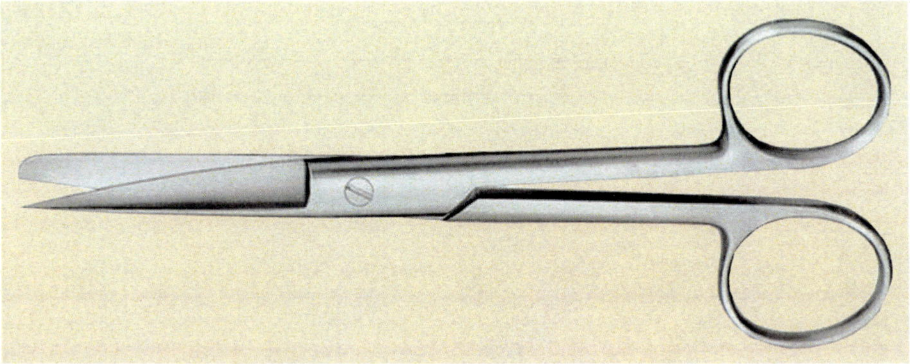

Fig. 72.18 Surgical scissors, length 18 cm

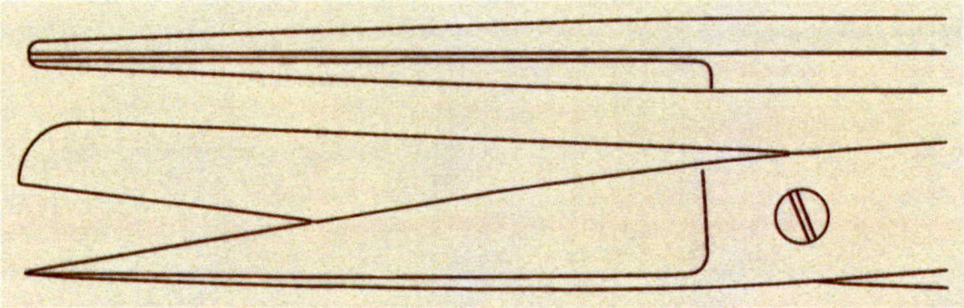

Fig. 72.19 Surgical scissors, length 20 cm

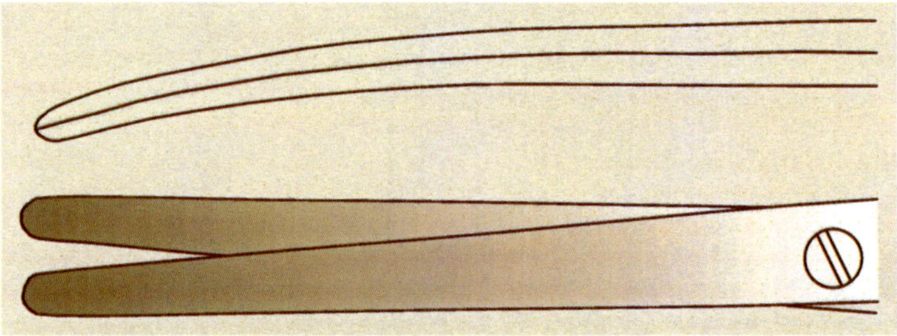

Fig. 72.20 Dissecting scissors

ing instruments without any edges or protrusions on the blades. Another important requirement is their smooth action ensuring an even cutting performance to give the surgeon a perfect *feel* for the tissue to be divided.

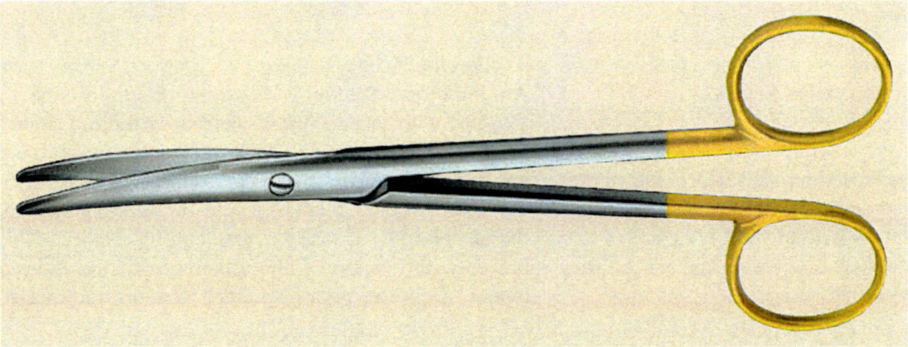

Fig. 72.21 Mayo and Lexer scissors

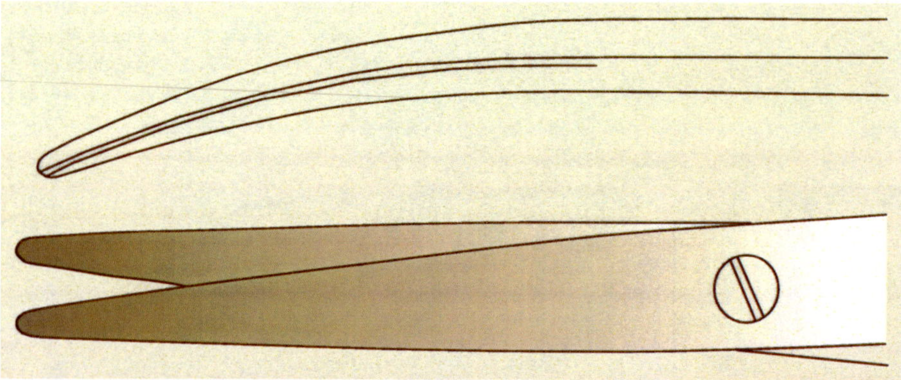

Fig. 72.22 Mayo and Lexer scissors

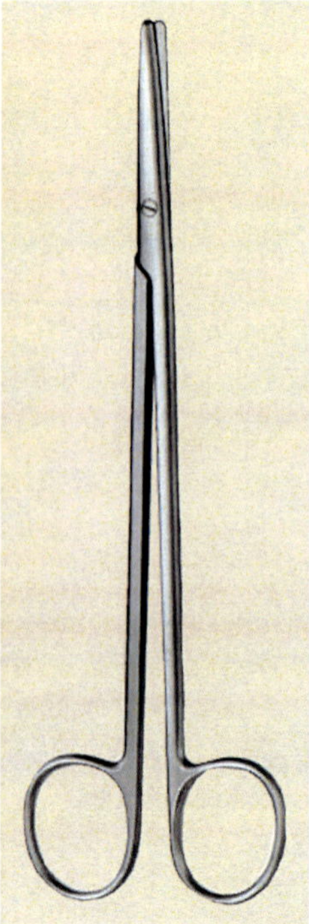

Fig. 72.23 Metzenbaum scissors

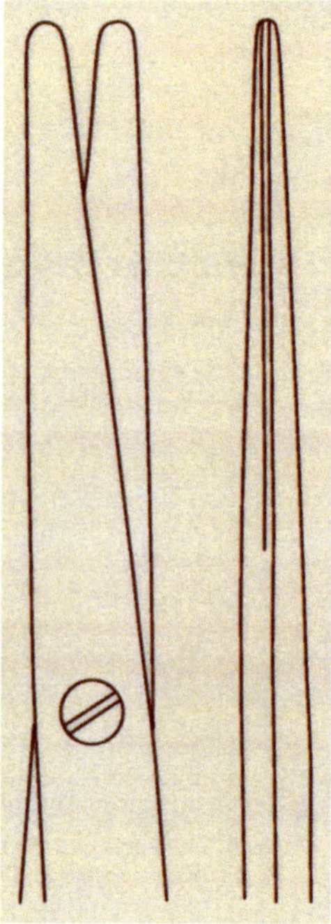

Fig. 72.24 Metzenbaum scissors

72.5.3 Suture or Ligature Scissors

Field of Use

Once the wound has healed on the surface, the sutures are removed using suture or ligature scissors. To this end, the surface ligation is simply cut with such a pair of scissors and anatomical or dressing forceps are then used to pull out the sutures.

Types of Scissors

An important feature is the slim design of one of the blades to ensure that it can be easily slid underneath the ligation to cut the suture without applying any pressure to the postoperative scar or tension to the suture. Some models feature a *suture groove* at the distal end of one of the cutting blades. This prevents the suture from slipping off the scissor blades. Delicate, pointed-pointed scissors (similar to

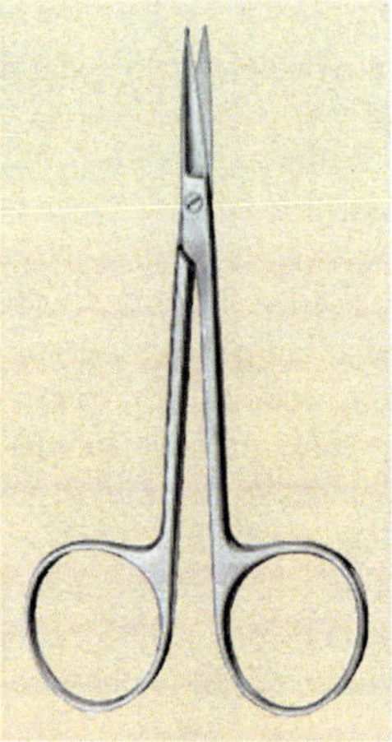

Fig. 72.25 Iris scissors

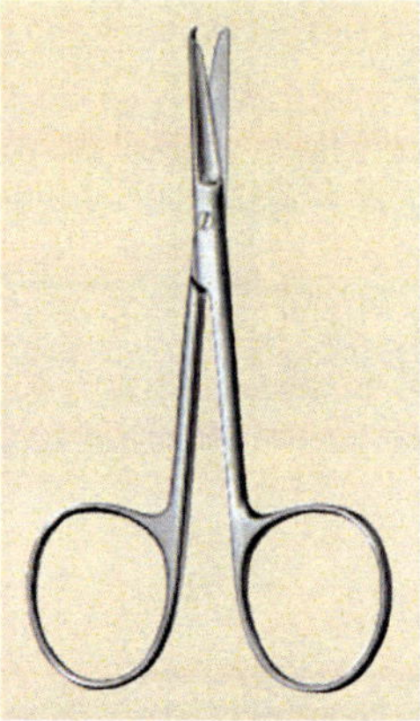

Fig. 72.26 Spencer scissors

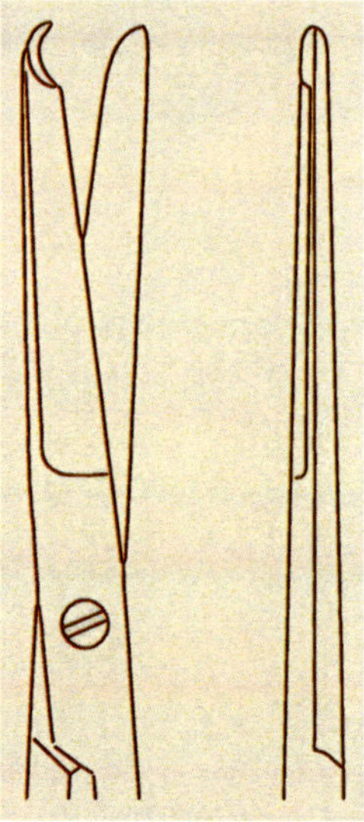

Fig. 72.27 Spencer scissors

iris scissors) are also used sometimes for this purpose.

Scissor Lengths

Various lengths of scissors are available from 9 to 15 cm.

72.5.4 Wire Cutting Scissors

Field of Use

For certain interventions (e.g. cerclages), flexible metal wires are used as suture or fixation materials. To

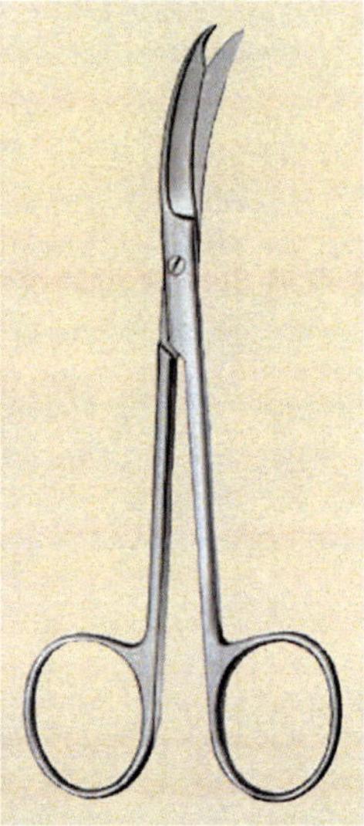

Fig. 72.28 Northben scissors

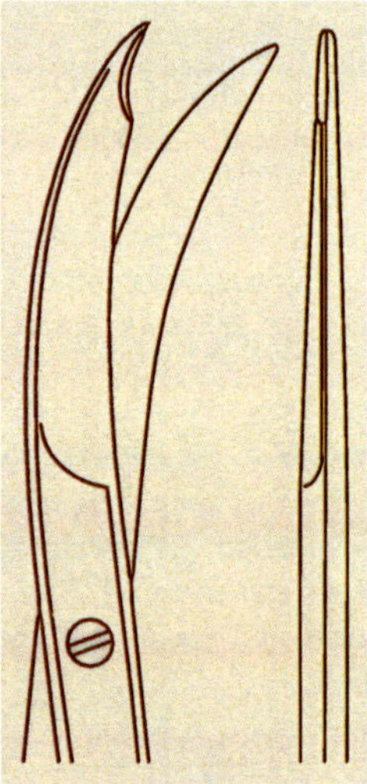

Fig. 72.29 Northben scissors

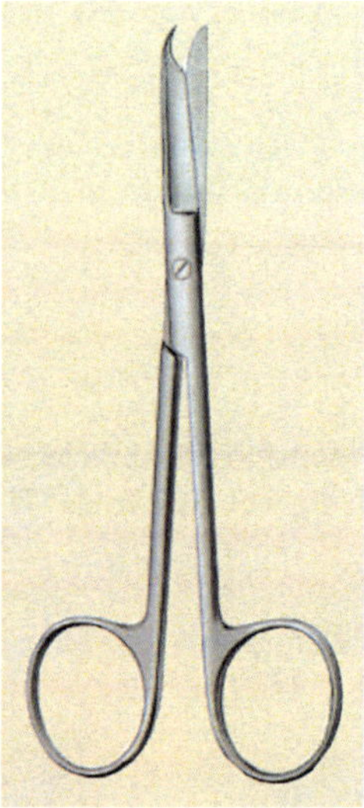

Fig. 72.30 Spencer ligature sicssors

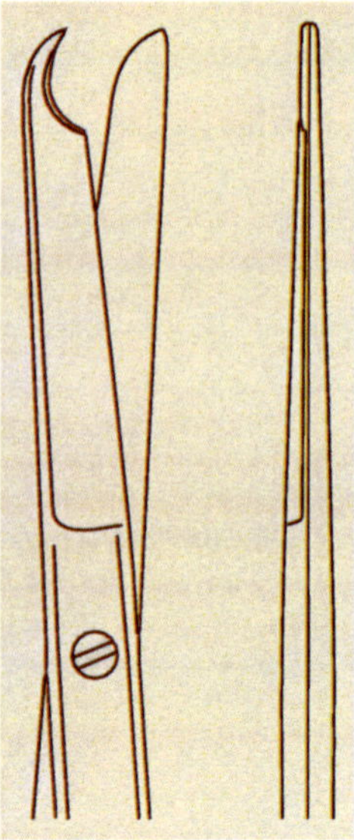

Fig. 72.31 Spencer ligature scissors

cut such materials, short and stout scissors are required.

Typical Features
The shear blades are extremely short as they are exclusively intended for cutting wires and not for providing a clean cut through tissue or dressing material. Some models include a *wire cutter* for secure grasping of the wire and have angled cutting blades.

Scissor Lengths
Various lengths of scissors are available with lengths from 10 to 16 cm.

72.5.5 Microsurgical Scissors

Microsurgical or microscissors are used across a wide range of fields nowadays, including almost all special disciplines (periodontology, endodontics, plastic surgery, hand surgery, orthopedics, cardiac surgery, vascular surgery, pediatric and neonatal surgery, gynecology, urology, neurosurgery, and ophthalmology).

Every microsurgical operation – including tendon, nerve and vascular suturing usually performed under the microscope – requires precise, light-weight mi-

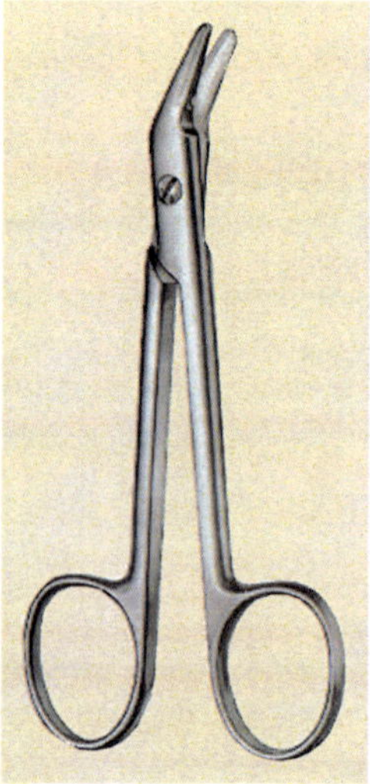

Fig. 72.32 Universal scissors

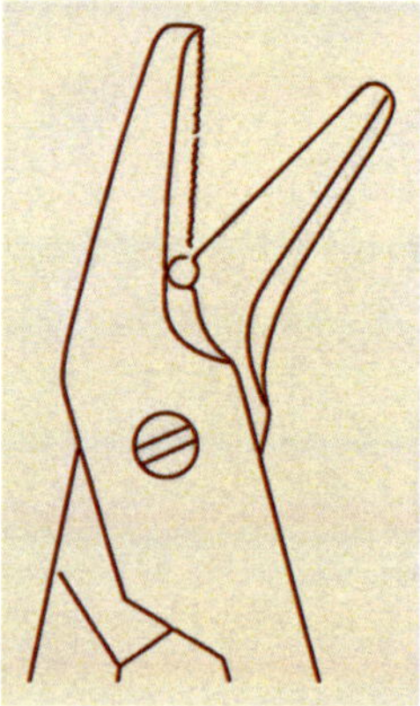

Fig. 72.33 Universal scissors

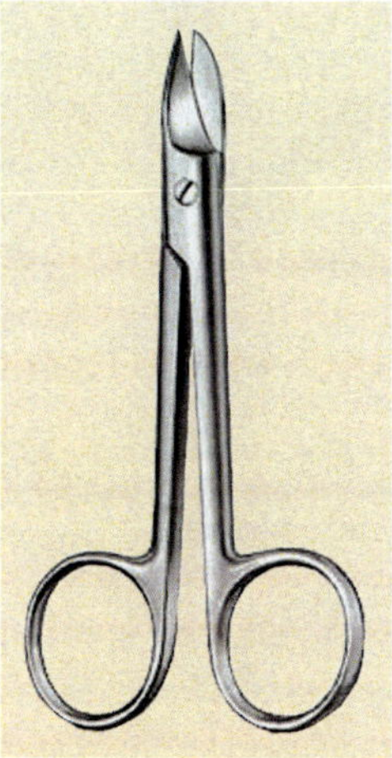

Fig. 72.34 Beebee scissors

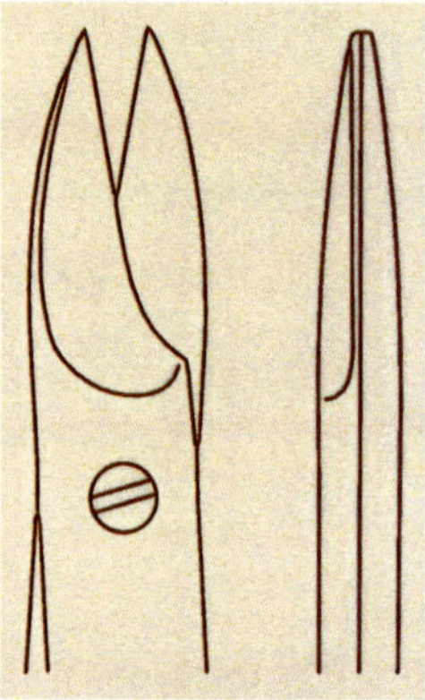

Fig. 72.35 Beebee scissors

croinstruments perfectly fitting the surgeon's hand for ultrafine interventions.

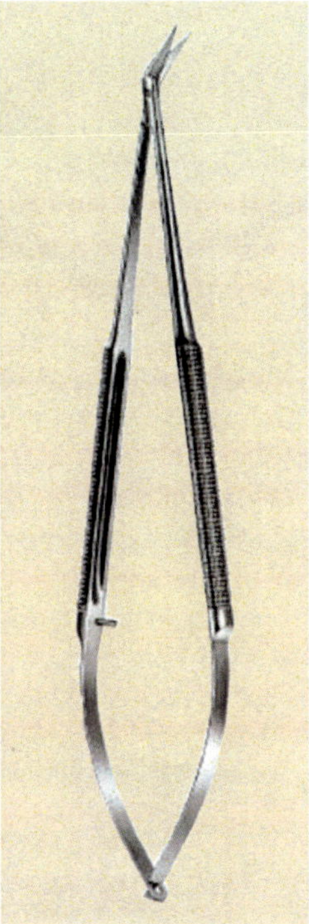

Fig. 72.36 Microsurgical scissors

Types of Scissors

Adventitia scissors have sharply pointed tips and straight blades. They are primarily used for freshening up and trimming structures that have already been prepared by blunt dissection. Besides, they are used for cutting fine sutures (>8/0). Dissecting scissors have rounded tips (designed for blunt dissection) and curved blades. They are used for severing nerves and vessels

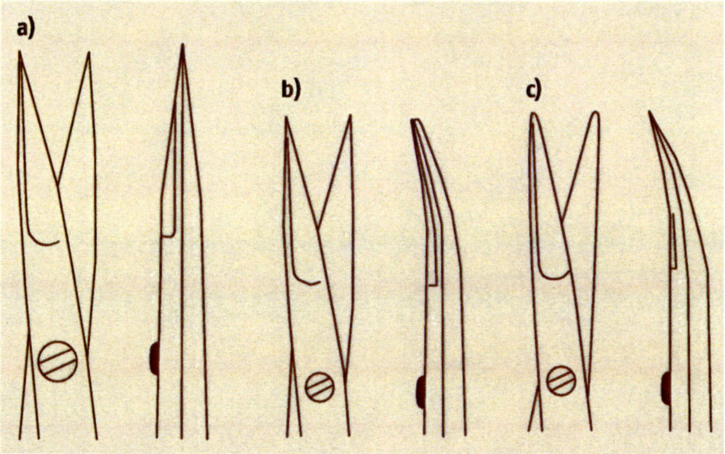

Fig. 72.37 **(a)** Adventitia scissors,**(b)** dissecting scissors,**(c)** toothed scissors

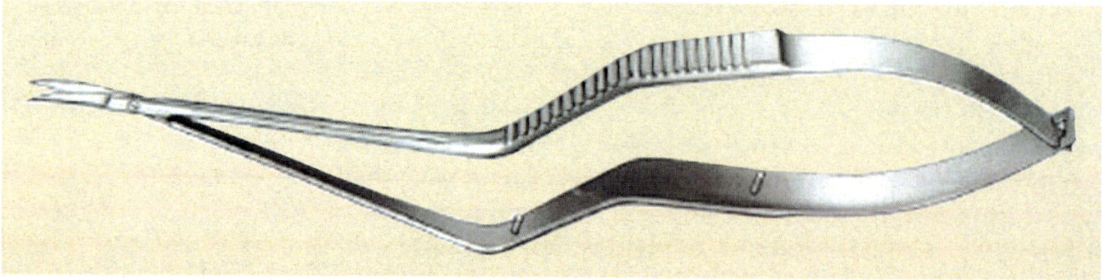

Fig. 72.38 Microscissors

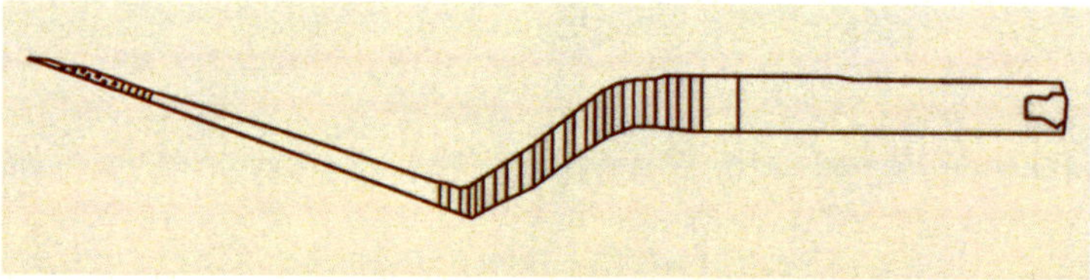

Fig. 72.39 Microscissors

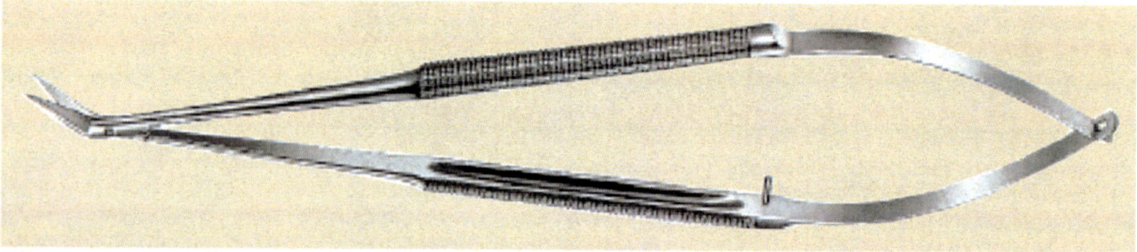

Fig. 72.40 Microscissors

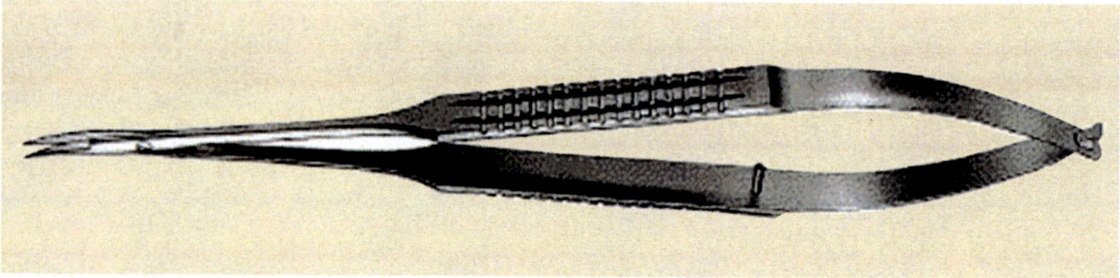

Fig. 72.41 Coronary scissors

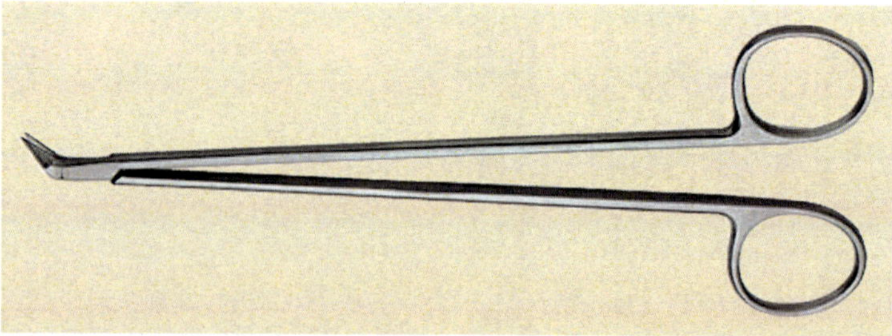

Fig. 72.42 Micro-Hegemann

from surrounding structures (soft tissue, lymph nodes) or from each other while leaving them fully intact. Toothed scissors offer the best controlled cut including the dissection of nerve tissue.

An important differentiation must be made between microscissors with round handles and those with flat handles. Round handles support the rotational movement required when passing the needle through the tissue. Nowadays, round-handle instruments are predominantly used.

Microscissors can be made of steel or titanium. Titanium scissors frequently have diamond-coated cutting edges for increased edge life and significantly lower weight.

Designs

Straight, bent-up, angled (with different angulation).

Important Users

Castroviejo, Micro-Hegemann, Barraquer, Vannas, Yasargil, Jacobson.

Scissor Lengths

Various lengths of scissors are available from 9 to 24 cm.

72.5.6 Vascular Scissors

Field of Use

Vascular scissors are used for cutting and dissecting anywhere in the arterial and venous system of the human circulation (peripheral, central, and coronary vessels). In addition, they are suitable for cutting sutures. Depending on vessel diameter, different models and designs are used, including microscissors. Particularly long vascular scissors (Potts–Smith) are also suitable for opening other ducts of the human body (such as the biliary duct for removing gallstones).

Types of Scissors

A basic differentiation is made between adventitia and dissecting vascular scissors. Adventitia scissors have

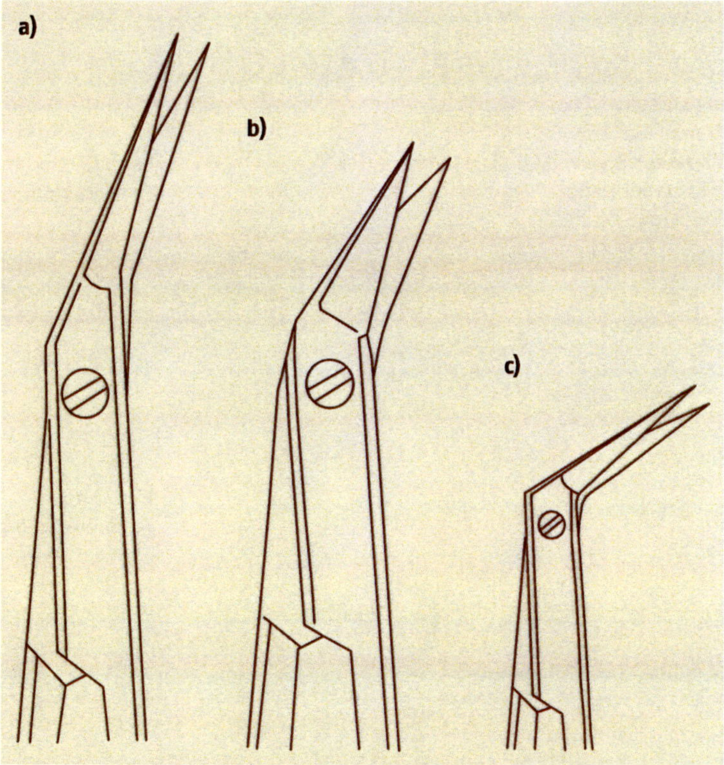

Fig. 72.43 (a) 25° angled coronary scissors, (b) 45° angled coronary scissors, (c) 60° angled coronary scissors

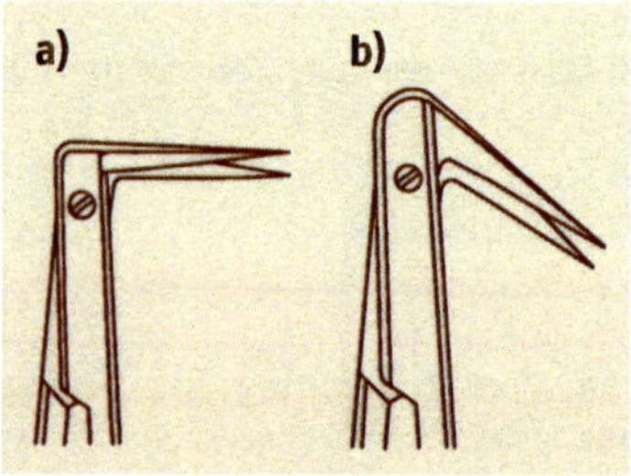

Fig. 72.44 (a) 90° angled coronary scissors, (b) 125° angled coronary scissors

two pointed tips, while dissecting scissors have two blunt tips. There are also models with a finely toothed blade to offer an even better, more controlled cut in certain situations.

Designs
Straight, bent-up, angled (with different angulation).

Important Users
Potts, Potts-Smith, De Bakey, Diethrich, Hegemann, Favoloro.

Typical Features
Adventitia scissors have straight blades for an optimal cut offering the smallest possible cutting surfaces. The extremely fine tips are intended for high-precision trimming without obstructing the surgeon's field of view. Dissecting scissors have ergonomically curved blades to ensure that the surgeon can fully concentrate on his task instead of the position of his hand. Safe dissection is guaranteed by the rounded tips.

Scissor Lengths
From 13 cm to 24 cm.

72.5.7 Gynecological Scissors

Field of Use
Exclusive use on the internal female genital organs.

Most Frequent Intervention
Hysterectomy (vaginal/abdominal total extirpation of the uterus with or without adnexectomy).

Types of Scissors
Uterine scissors, parametrium scissors (also called hysterectomy scissors).

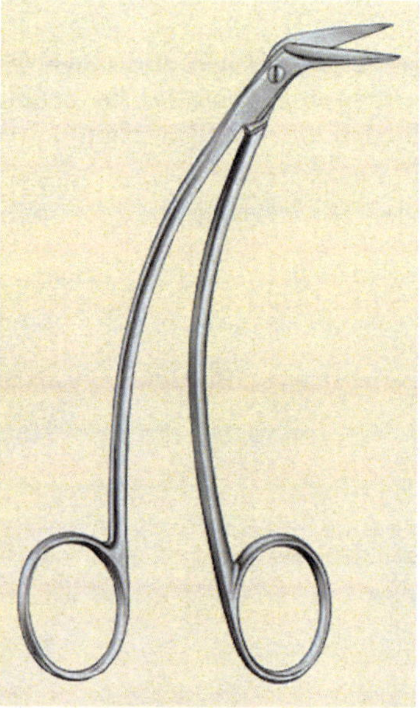

Fig. 72.45 Favaloro scissors

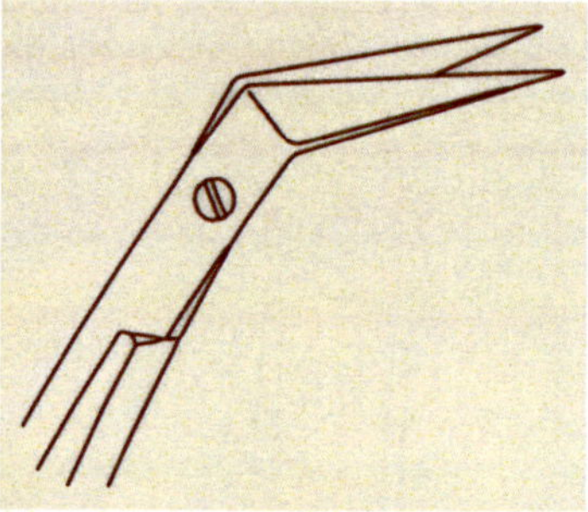

Fig. 72.46 Favaloro scissors

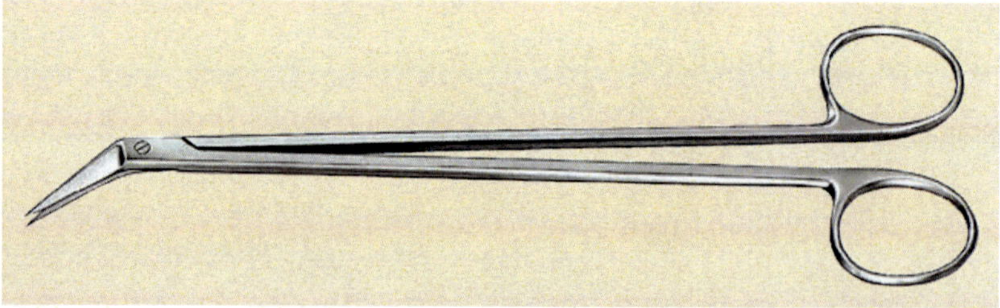

Fig. 72.47 Potts-Smith scissors

Most Important Users

Sims, Siebold, Wertheim.

Typical Features

All in all these are very stout scissors with short, bent-up blades. Both tips are always blunt. When closed they offer a round tip for dissecting and spreading.

The blades must be very short in relation to the shanks because this provides the leverage required for putting adequate pressure on the cutting edges. The blades as such must be sufficiently thick to prevent their

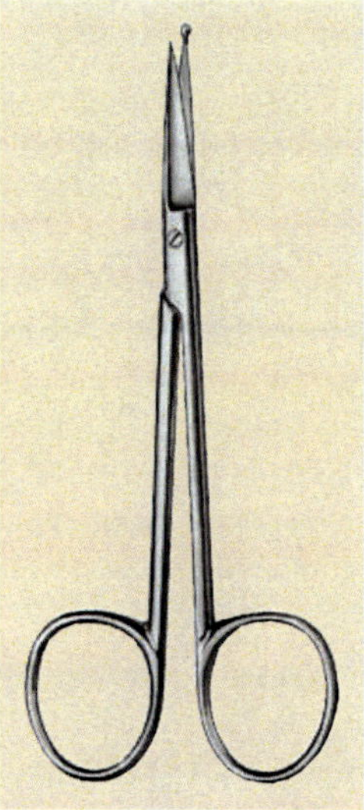

Fig. 72.50 Vascular scissors

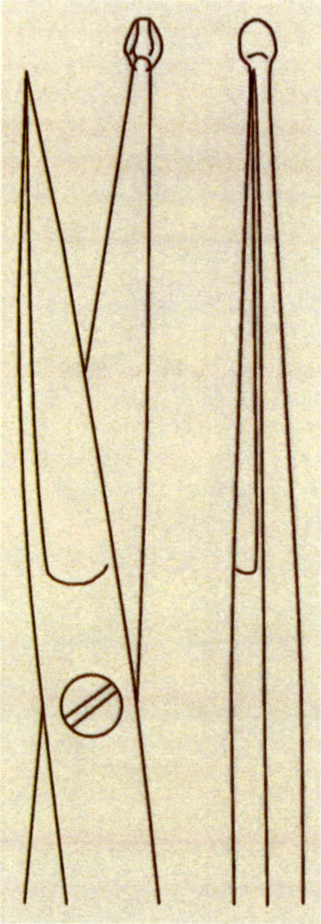

Fig. 72.51 Vascular scissors

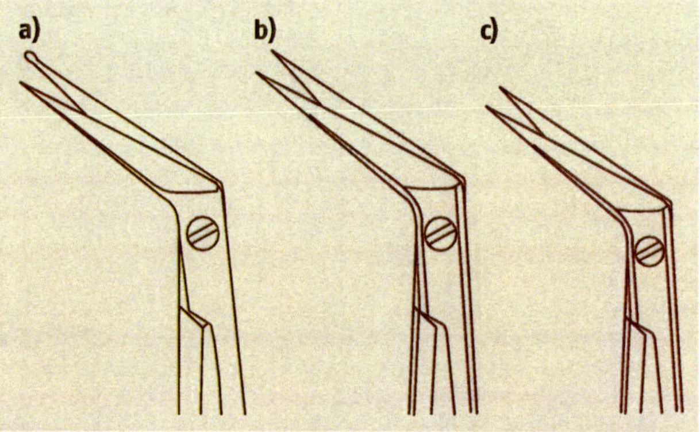

Fig. 72.48 (a) 60° angled Potts–Smith scissors button ended, (b) 60° angled Potts–Smith scissors, (c) 40° angled Potts–Smith scissors

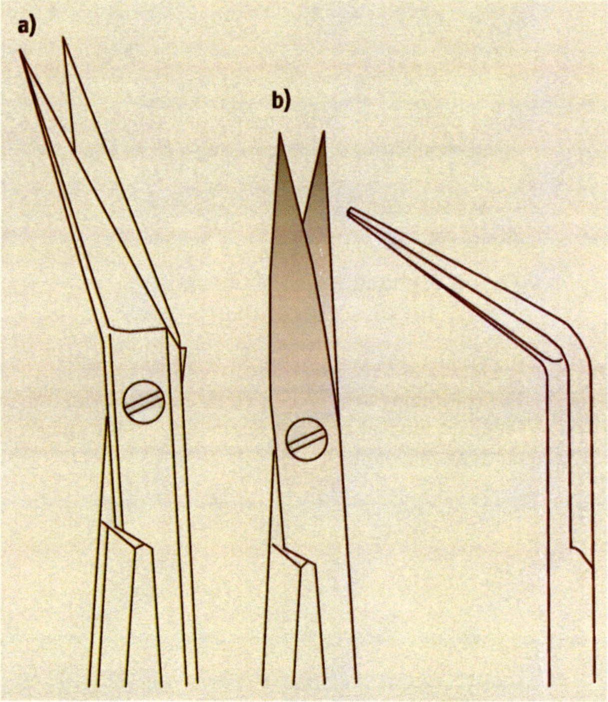

Fig. 72.49 (a) 25° angled Potts–Smith scissors, (b) Bent-up Potts–Smith scissors

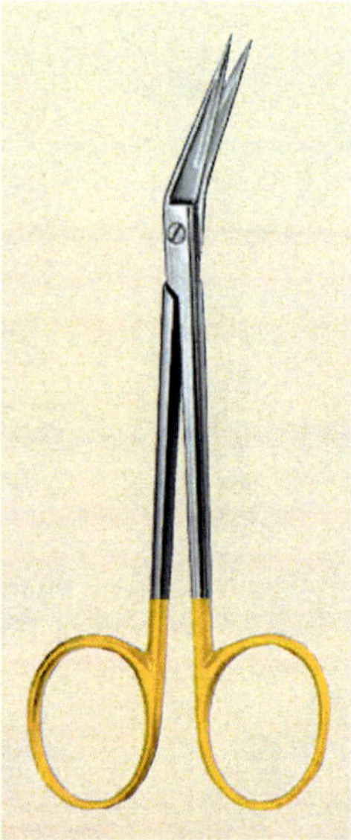

Fig. 72.52 Angled scissors

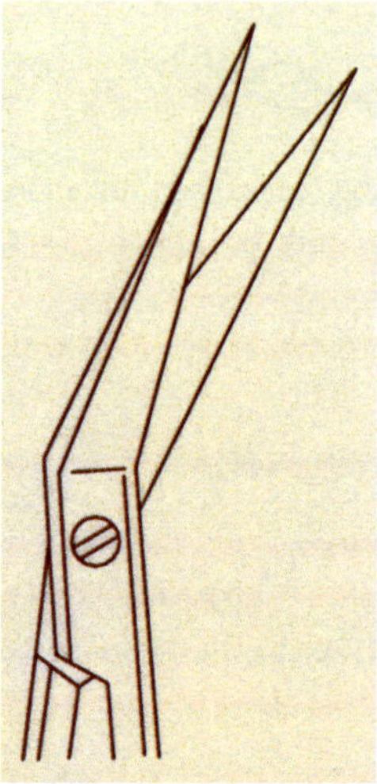

Fig. 72.53 Angled scissors

being pushed apart when transecting rough tissue (e.g. in the parametrium), thus ensuring a clean cut.

Scissor Lengths

20–28 cm.

72.5.8 Gynecological Scissors for Obstetrical Use

Fields of Use

For episiotomy to prevent perineal laceration/rupture or facilitate surgical delivery. For cutting off the umbilical cord (omphalotomy).

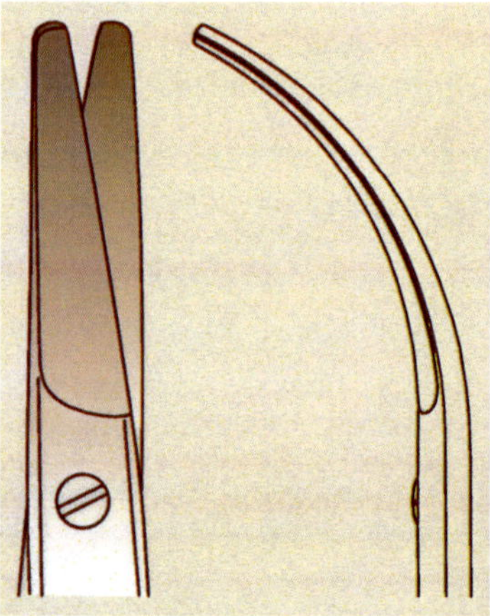

Fig. 72.54 Gynecological scissors

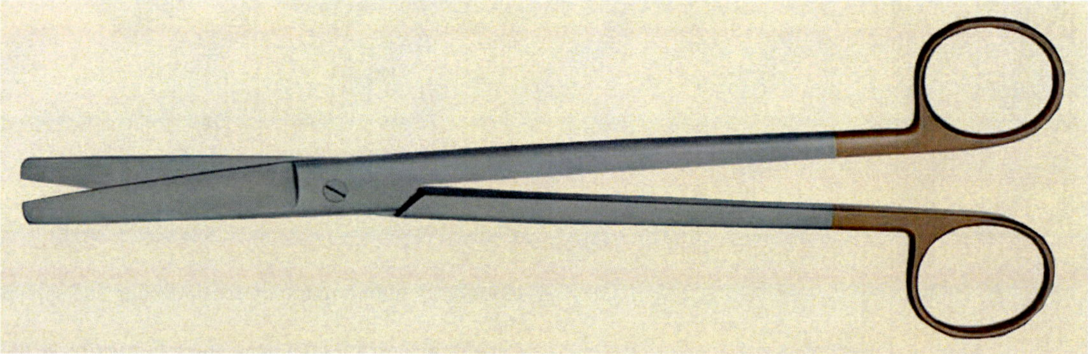

Fig. 72.55 Sims uterine scissors

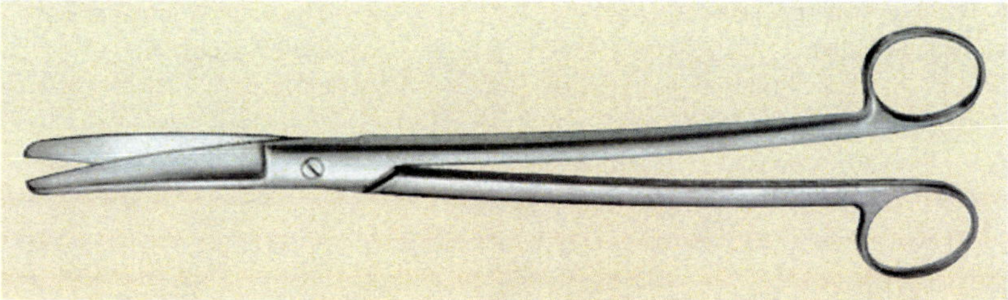

Fig. 72.56 Sims–Siebold scissors

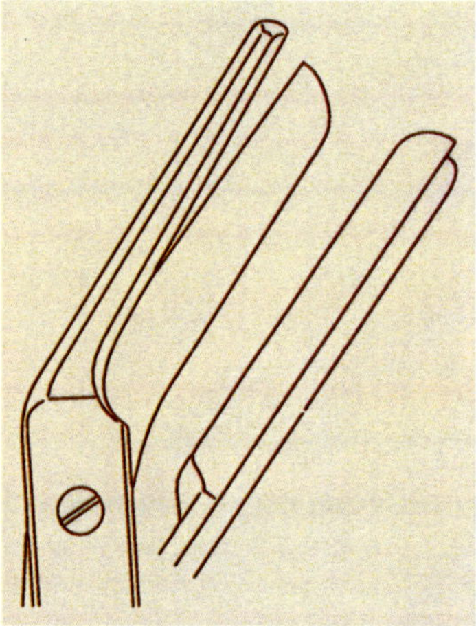

Fig. 72.57 Episiotomy scissors, Braun-Stadler

Types of Scissors

Episiotomy scissors, umbilical scissors.

Most Important Users

Braun-Stadler, Waldmann, Busch, Schuhmacher.

Typical Features

Episiotomy scissors are sturdy scissors with bent (elbowed or lunated) blades to protect both the mother and the baby against injury. The crescent-shaped blades are designed for best adaptation to the skull of the baby when inserting the scissors into the obstetric canal. A clean and even cut in such rough tissue must be guaranteed as well.

Umbilical scissors are short, stout scissors which are available in various designs all of which guar-

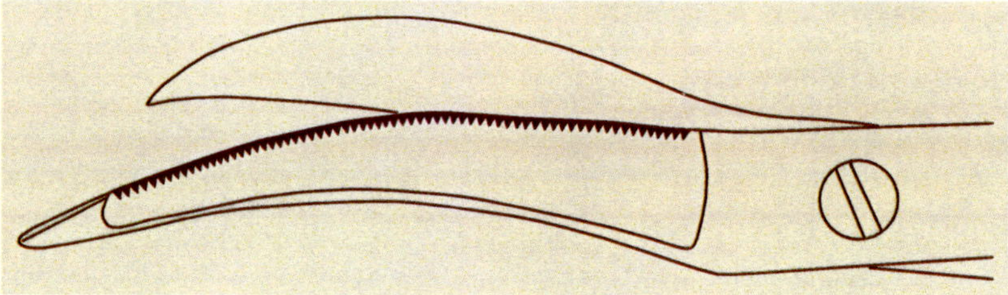

Fig. 72.58 Episiotomy scissors, Waldmann

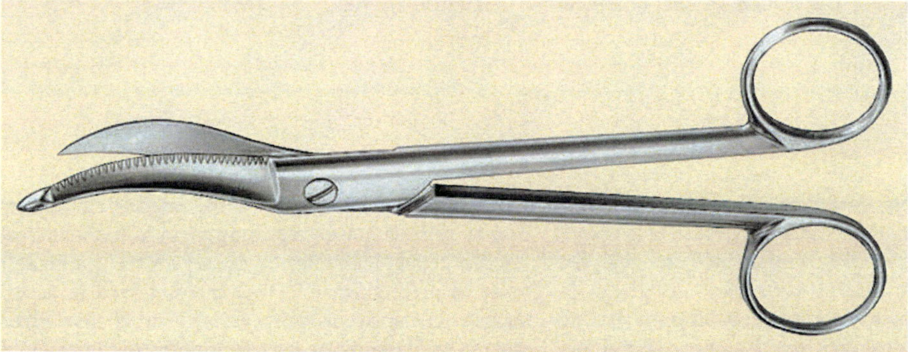

Fig. 72.59 Episiotomy scissors, Waldmann

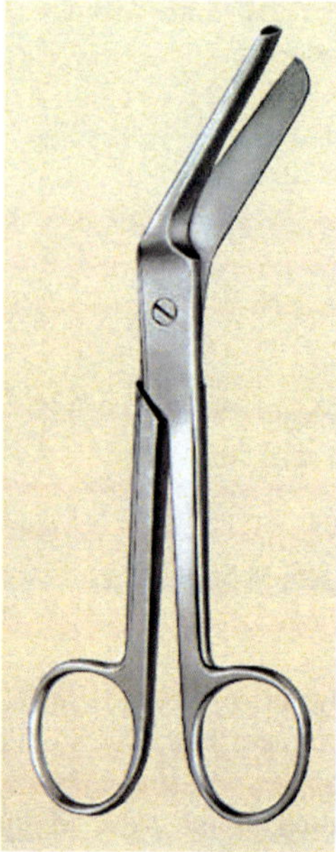

Fig. 72.60 Episiotomy scissors, Braun-Stadler

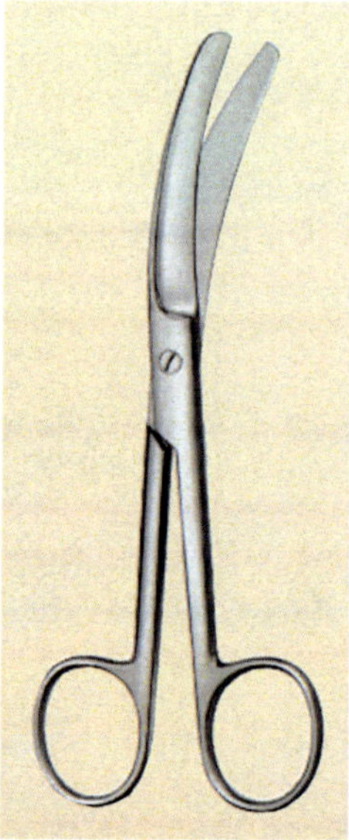

Fig. 72.61 Umbilical scissors, Busch

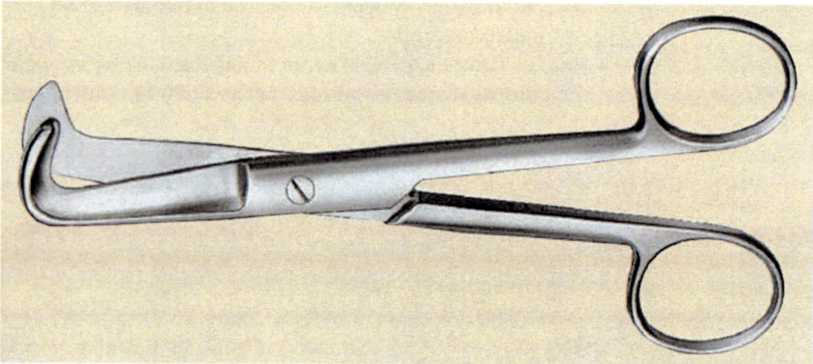

Fig. 72.62 Umbilical scissors, Schuhmacher

antee a clean and safe transection of the umbilical cord.

Scissor Lengths

14–22 cm.

72.6 Handling and Care

Surgical instruments, including all surgical scissors, may be used only for their intended purpose in the specified medical fields by adequately trained and qualified personnel. It is the treating physician's or user's responsibility to select the appropriate models for specific applications or surgical uses, provide adequate training and information on the proper handling of the instruments, and ensure sufficient experience in using them.

Prior to their first use and before each subsequent use, as well as before returning them to the manufacturer for repair, maintenance, or service, the instruments must be cleaned, disinfected, and sterilized in accordance with standardized processing instructions derived from a whole body of legal and normative provisions (laws, regulations, standards, guidelines, and recommendations).

It should be noted that proper processing and care are essential for retaining the function and value of these high-quality products – which constitute important material assets for any hospital – over many years.

Important standards, guidelines and recommendations include:

- The joint recommendation issued by the Robert Koch Institute (RKI) and the German Federal Institute for Drugs and Medical Devices (BfArM) on *Hygiene requirements for processing medical devices (Anforderungen an die Hygiene bei der Aufbereitung von Medizinprodukten)*. It requires a quality management and validated procedures for processing reusable medical devices.
- DIN EN ISO 15883 standard specifies and defines the requirements for cleaning and disinfecting apparatus (washer-disinfectors) and for validating the processes used.
- DIN EN ISO 17664 standard specifies the information to be provided by the manufacturer with regard to proper reprocessing of medical devices.

72.6.1 General Instructions

The processing cycle for medical devices generally includes:

- Preparatory work (pretreatment, collecting, precleaning, and dismantling where applicable)
- Cleaning, disinfecting, final rinse, drying
- Visual inspection for cleanness and perfect condition of the material
- Care and repair
- Functional tests
- Labeling
- Packaging, sterilization, release, and storage.

Table 72.2 Highest value for contaminations in feed water

Substance/property	Feed water
Evaporation residue	≤ 10 mg/l
Silicates (SiO_2)	≤ 1 mg/l
Iron	≤ 0.2 mg/l
Cadmium	≤ 0.005 mg/l
Lead	≤ 0.05 mg/l
Rest of heavy metal residues (except iron, cadmium, lead)	≤ 0.1 mg/l
Chlorides (Cl^-)	≤ 2 mg/l
Phosphates (P_2O_5)	≤ 0.5 mg/l
Conductivity (at 25 °C/77 °F)	≤ 5 μS/cm
pH-value (degree of acidity)	5–7.5
Appearance	colorless, clear, no sediments
Hardness (total of alkaline earth ions)	≤ 0.02 mmol/l
Note: Compliance with limits should be checked using recognized analytical methods	

72.6.2 Materials Used in Scissor Manufacturing

Apart from the standardized materials (DIN EN ISO 7153-1) used, which are described in a separate chapter, there are a number of other parameters that should also be observed during processing because they are important factors for retaining the value of the instruments in the long term.

72.6.3 Water Qualities

Due to the high quantities required in the processing cycle, water is an essential factor for achieving a good cleaning result in any machine cleaning process. Depending on the type of items to be processed, the water quality can adversely affect their long-term value retention, specifically as the overall salt content (evaporation residue) of the water can lead to unwelcome deposits on the processed goods, thus causing damage to the material.

An unfavorable water composition, therefore, can adversely affect both the process used and the appearance and materials of the instruments. Consequently, the water quality should be assessed and taken into consideration from the start, i.e. when planning the sanitary installations. For example, excessive chloride concentrations can lead to pitting on the instruments, while other substances – such as silicates/silicic acids – can cause discoloration (staining, spotting). In addition to all the natural substances contained in the water, rust may also be found in tap water. Rust is invariably a result of corroded piping, forming deposits on the instruments that subsequently lead to stains (extraneous rust) and corrosion. This can be prevented by using demineralized water for the final rinse!

The DIN EN ISO 15883-1 standard (item 6.4.2.2) provides a list of parameters that should be checked and assessed under any circumstances. DIN EN 285 (Appendix B, Table B1) specifies limit values for the boiler feed water of a steam sterilizers. Such water quality can be recommended for the final rinse of machine-based instrument processing cycles. The values in Table 72.2 can be recommended for orientation.

Fig. 72.63 Formation of rust due to multihour immersion in physiological saline

72.6.4 Preparation for Cleaning and Disinfection

The first steps are already taken in the operating room (OR). Wherever possible, residues of hemostatic, skin disinfecting or lubricating agents, or acid medicines should be removed before storing the instruments away for reprocessing after use.

Stainless steel instruments should never be immersed in physiological saline (NaCl solution) because extended contact with this medium will lead to pitting and stress-crack corrosion.

Formation of rust is due to multihour immersion in physiological saline. The improper practice of *throwing off* instruments after use is a source of damage, especially for scissors. Hard-metal scissors are particularly critical instruments since their insets are very brittle, in spite or because of their hardness. Improper handling (e.g. allowing them to drop on the tips) can lead to loss of the hard-metal insets (tip break-off or cutting edge nicks). However, nicking cannot occur in scissors with welded-on hard-metal edges.

It is important to store the instruments properly on instrument trays suitable for machine cleaning. Effective cleaning requires jointed instruments (such as forceps, pliers, scissors) to be processed in open conditions in order to minimize overlapping surfaces. Microsurgical instruments such as microscissors must be stored on special racks or using suitable supports for proper fixation (Fig. 72.64).

72.6.5 Manual Cleaning and Machine Cleaning

Machine cleaning should be given preference over manual cleaning. In any case, effective cleaning is a prerequisite for effective disinfection and subsequent sterilization. Thermal disinfection should be preferred over chemothermal disinfection methods. Standardized cleaning and disinfection can best be achieved with machine cleaning. The international standard EN ISO 15883, as well as national guidelines and recommendations require the exclusive use of validated machine cleaning processes.

72.6.6 Machine Cleaning and Thermal Disinfection

For thermal processes, the disinfecting effect is usually determined parametrically. To this end, the F value used for sterilization with moist heat was translated into the *A_0-value concept* for thermal cleaning and disinfecting processes using moist heat and as such integrated into the DIN EN ISO 15883 standard. A disinfecting method using moist heat is expected to guarantee that a defined temperature maintained over a defined period of time has a predictable lethal effect on vegetative microorganisms. The appropriate lower temperature limit has been set to 65 °C (149 °F).

The decisive criterion for defining the required temperature impact is the heat resistance (thermoresistance) of the vegetative microorganisms to be killed. This parameter is quantitatively expressed in the D value.

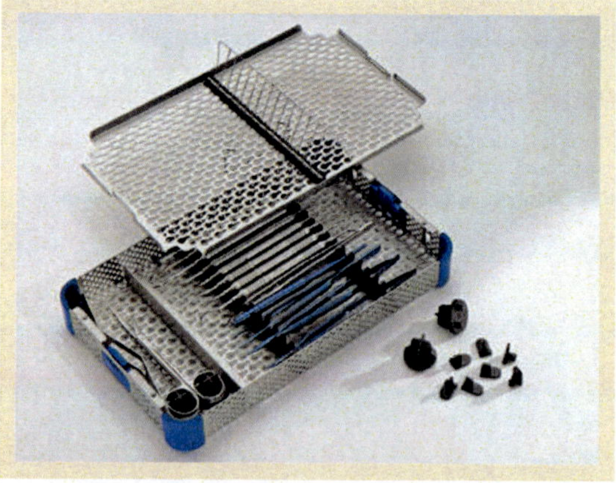

Fig. 72.64 Instrument tray for storing and fixing delicate instruments

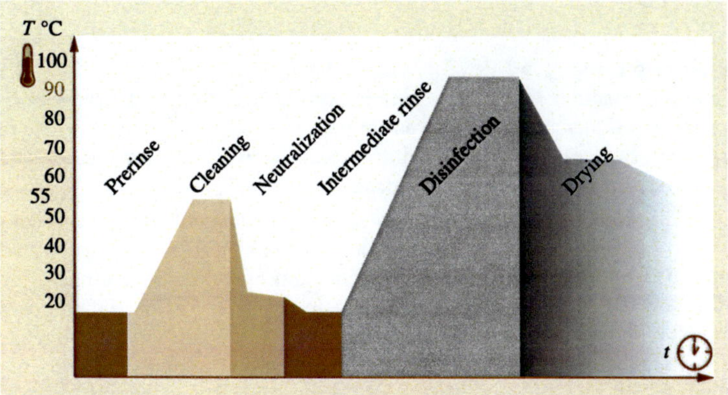

Fig. 72.65 Machine processing program with thermal disinfection typically includes the steps shown (after [72.1])

A_0, the time equivalent (in s) given by the disinfecting process at 80 °C (176 °F) in relation to a microorganism with a z value of 10 ° (50 °F).

Terms/Parameters

A	Time equivalent (in s) at 80 °C, by which a defined disinfecting effect can be achieved.
A_0 value	Killing power expressed as time equivalent (in seconds) at a temperature of 80 °C that is process-transferred to the product, for microorganisms with $z = 10$ °C.
Z value	Temperature change (in K) required for changing the D value by a factor of 10.
D value	Decimal reduction value, the time (in min) required at a defined temperature for killing 90% of a population of microorganisms.

The structure of the program to be used depends on the (e.g. hygienic) performance requirements and the goods to be processed (Fig. 72.65).

The following steps are processed with thermal disinfection:

1. *Prerinse*: Pure cold water (demineralized if available) to remove coarse soiling and foaming substances.
2. *Cleaning and process chemicals*: Warm or cold water (demineralized if available); cleaning is normally done at 40–60 °C (104–140 °F) for at least 5 min.
   A differentiation is made between alkaline and neutral cleaners. Neutral cleaning agents usually contain nonionic, low-foam tensides, and products with or without enzymes are used for this purpose.
   If the recommendations of the vCJD Task Force (see corresponding RKI guideline) are followed, a cleaner solution with a pH-value higher than 10 must be used.
   Should the water used contain excessive chloride concentrations, this can cause pitting and stress-crack corrosion on the instruments. Using alkaline cleaners or demineralized water minimizes such corrosion.
3. *First intermediate rinse – warm or cold water*: Adding an acid-based neutralizer facilitates the removal of alkaline cleaner residues. Even when using neutral cleaners, it is advisable to use a neutralizer if the water quality is poor (e.g. if the salt content is high). This prevents the formation of deposits.
4. *Second intermediate rinse*: Pure warm or cold water (demineralized if available), no additions.
5. *Thermal disinfection/final rinse*: Demineralized water, with thermal disinfection carried out at temperatures of 80–95 °C (176–203 °F) using an appropriate exposure time according to the A_0 concept (DIN EN ISO 15883). Using demineralized water prevents the formation of stains/spotting, deposits and corrosion. If a final rinse agent is added to shorten the drying time, make sure that it is both biocompatible and compatible with the instruments to be processed.
6. *Drying*: Sufficient drying must always be ensured, either as part of the automatic cleaning program or by taking adequate measures. As concerns the process chemicals used, it is mandatory to observe the product manufacturer's instructions regarding concentration, temperature, and exposure time. This is important for obtaining best results while treating the instruments as gently as possible. Automatic liquid dosing devices must be controllable. Other methods will not be discussed here.

72.7 Inspection, Testing, and Care

Sufficient cleanness is a basic prerequisite for successful sterilization. The instruments must be macroscopically clean, i. e., free from any visible residues (to be verified by inspection). Critical areas such as the handle structures, joints or jaw grooves, atraumatic toothing in particular, need to be checked with special care.

Care measures must be taken prior to performing the functional checks. This requires the targeted application of care agents to joints, box locks, threads, and friction surfaces (e.g. of scissors, forceps, punches) following careful cleaning and disinfection. As this prevents metal-on-metal friction, it helps to prevent fretting corrosion.

Requirements for care agents suitable for surgical instruments are:

- Paraffin/white-oil basis
- Biocompatible according to the valid European or US pharmacopoeia
- Approved for steam sterilization procedures and vapor-permeable.

Instruments may never be treated with care agents containing silicone, as this could lead to stiff action and jeopardize the effectiveness of steam sterilization. As the various instruments are designed for specific purposes (e.g. scissors), inspections must be such that instruments which no longer serve their purpose will be sorted out reliably. If in doubt, appropriate inspection and test measures should be agreed upon with the manufacturer of the product. Instruments that are returned to the manufacturer for repair must first be sent through the entire cleaning and sterilization cycle for hygienic reasons.

72.8 Packaging

As a rule, medical devices must be sterilized in suitable packaging (Figs. 72.66, 72.67). Instruments for use in the OR are assembled as complete, intervention-specific sets/trays and packaged as such. For thermostable instruments, steam sterilization should be preferred.

The purpose of packaging systems for terminally sterilized medical devices is to enable sterilization, provide physical protection, maintain their sterility up to the time of use, and allow for their aseptic placement for use in the OR. The general requirements applying to packaging for terminally sterilized medical devices, including validation requirements for forming, sealing, and assembly processes, have been defined in the DIN EN ISO 11607 series. In addition, parts 2 to 10 of the DIN EN 868 series contain test methods and reference values for specific materials used for preformed sterile barrier systems and for packaging systems.

Fig. 72.66 Instrument tray according to DIN 58952-3, loaded

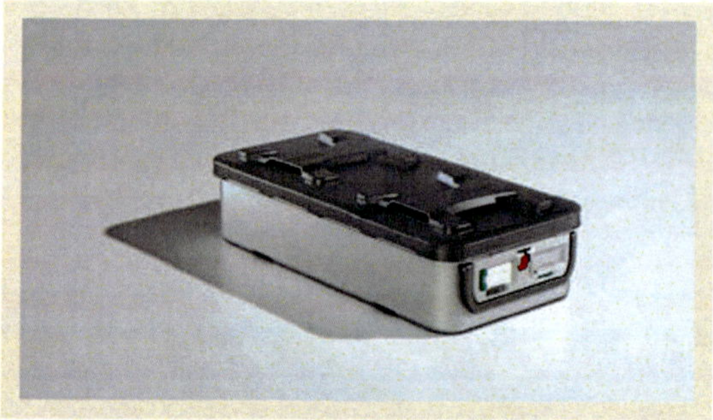

Fig. 72.67 Reusable sterilization container according to ISO 11607 and DIN EN 868-8

72.9 Current Terminology

Sterile Barrier System

Minimum packaging that prevents the ingress of microorganisms and allows the product to be placed ready for use in aseptic condition at the point of use.

Protective Packaging

Material configuration designed to prevent damage to the sterile barrier system and its contents from the time of assembly to the time of use.

Packaging System

Combination of sterile barrier system and protective packaging

Sterile Barrier System + Protective Packaging = Packaging System

Materials and (preformed) sterile barrier systems considered suitable (depending on the sterilization method used) are:

- Paper bags, sealable transparent pouches, and reels
- Plastic film constructions (composite foils)
- Sterilization paper (smooth, creped)
- Nonwoven wrapping materials
- Form/fill/seal (ffs) processes
- Packaging processes (e.g. four-side closure)
- Reusable sterilization containers.

72.10 Steam Sterilization with Saturated Steam

Reference standards: DIN EN 285:2006+A2:2009 and DIN EN ISO 17665-1 1:2006. The current standard is a validated method of steam sterilization, usually at 134 °C (273.2 °F).

The sterilization and holding times are subject to national regulations and guidelines and, therefore, cannot be defined globally. It is the operator's responsibility to ensure that the reprocessing/sterilization performed with given CSSD equipment, materials, and personnel achieves the desired results. This requires validation and routine monitoring of the process.

72.10.1 Steam Quality

The steam used for sterilization must be free from impurities and must not adversely affect the sterilization process or damage the sterilizer or the goods to be sterilized. To ensure this, the reference values specified in Tables B.1 + B.2 of EN 285 (Table 72.3) for the quality of the boiler feed water and the condensate must not be exceeded. Noncompliance can lead to corrosion due to rust particles from the piping system, for example, or to instrument

Table 72.3 Suggested maximum values of steam impurities in steam sterilizers (after [DIN EN 285 (edition 2009-08)])

	Table B.2 Steam condensate feed line	Steam condensate after contact with sterilized products	Table B.1 Feeding water for steam generation
Evaporation residue	– mg/l	≤ 30 mg/l	≤ 10 mg/l
Silicon dioxide	≤ 0.1 mg/l	≤ 0.1 mg/l	≤ 1 mg/l
Iron	≤ 0.1 mg/l	– mg/l	≤ 0.2 mg/l
Cadmium	≤ 0.005 mg/l	– mg/l	≤ 0.005 mg/l
Lead	≤ 0.05 mg/l	– mg/l	≤ 0.05 mg/l
Rest of heavy metals	≤ 0.1 mg/l*	≤ 0.1 mg/l	≤ 0.1 mg/l*
Chloride	≤ 0.1 mg/l	≤ 0.5 mg/l	≤ 2 mg/l
Phosphate	≤ 0.1 mg/l	≤ 0.1 mg/l	≤ 0.5 mg/l
Conductivity (at 25 °C)	≤ 3 μS/cm	≤ 35 μS/cm	≤ 5 μS/cm
pH-value	5–7	–	5–7.5
Appearance	Colorless, clean, without sediment	Clean, colorless	Colorless, clean, without sediment
Hardness	≤ 0.02 mmol/l	mmol/l	≤ 0.02 mmol/l

*Except iron, cadmium, lead

staining (discoloration) due to excessive silicic acid content.

If humidity is retained inside packaging systems, this can also lead to rusty instruments. Adequate measures to prevent residual humidity can be agreed upon with the manufacturer of the sterilizer.

72.10.2 Release and Storage

Follow-up treatment after sterilization includes:

- Intermediate storage of the hot sterile items immediately after removal from the sterilizer, to allow them to cool down (half an hour is usually sufficient)
- Sorting of the sterile goods with inspection for cleanness, dryness, and integrity (including packaging and labeling)
- Checking/assessing the sterilization result based on the quality records of the lot documentation
- Sorting out and blocking sterilized items that fail to meet the requirements
- Release for use.

These tasks may only be carried out by authorized personnel. Transport should be effected according to commissioning requirements, but always with adequate protection. It is recommended to use containers and transport them in closed trolleys.

Proper Storage

A dust-free and dry environment is an essential requirement for protected storage of sterile goods and prevention of corrosion damage. For the same reason, temperature fluctuations should be avoided. The maximum permitted storage period depends on the type of packaging and the storage conditions. This means up to six months for sterile supplies packaged in compliance with standard requirements and stored in a dust-free room. Storage beyond the expiration date – which has to be determined by the user in each case – is not permitted.

72.11 Quality Characteristics

72.11.1 Material

A basic prerequisite for best quality is the use of high-quality, forged steel satisfying ISO 7153-1-2000 requirements. The raw materials of reproducible high quality should preferably be of German origin. Commonly used steels are 1.4117, 1.4034 and 1.4021.

72.11.2 Surface

An optimal surface treatment comprises several working steps:

- First, at least 0.3 mm of the surface of the blank must be ground off to remove slag inclusions and surface defects. This creates the basic conditions needed for reliable resistance to corrosion.
- This is followed by mechanical polishing and, preferably, additional electropolishing.
- Finally, the product receives its surface appearance (mirror, matte, or glass bead-blasted finish).

To consistently meet the growing demands in the field of instrument reprocessing (e.g. due to the use of high-alkaline cleaners), a passivation step is increasingly added, especially for high-quality instruments. This optimizes the chromium–iron ratio on the surface of the instrument. The objective is to enrich the surface with as many of the *nobler* chromium molecules as possible in order to improve the instrument's resistance to the chemical attacks typically encountered during reprocessing, thus enhancing its resistance to corrosion.

72.11.3 Form

The closed scissor tips must fit perfectly so cutting with the tips is also possible. The fit of the two blades is good in closed scissors if one covers the other exactly.

72.11.4 Action

Scissors should be easy to open and close, with the blades gliding smoothly in place over each other. Moreover, their operation must be balanced to offer an even and controlled cut over the entire length of the blades (i. e. from the start of the cut, with the instrument wide open, to complete closure). Blades with adequate hollow grinding on the inside support smooth action.

72.11.5 Eye Rings

The eye rings are the working surfaces for the fingers. Their ergonomic shape should ensure pain-free and nonfatiguing handling even for extended periods. The inside of the eye rings, in particular, must provide a smooth, burr-free surface finish.

72.11.6 Quality of Cut

The material to be cut must be transected cleanly and smoothly. No *slippage* may occur and the material must not get squeezed between the blades, either, so they must be sharp. The cutting performance is supported or enhanced by the so-called *bevel* – i. e. additional grinding (chamfering) of the cutting edge of the lower blade. Ideally, the bevel should form an angle of 75° relative to the cutting edge. An adequate mechanical tension between the shear blades – ideally as low as possible and as high as necessary – ensures a good cutting performance. Test methods differ greatly, given the complex and diverse demands resulting from the different materials to be cut – ranging from human tissues to sutures. For example, coarse surgical scissors are performance-tested using multilayer cotton material, whereas the cutting efficiency of delicate microsurgical scissors is verified with ultra-fine latex.

72.12 Future Developments

72.12.1 Users

The high-quality scissors currently available on the market offer very high quality standards. Nonetheless, the requirements on surgical instruments continue to grow as a result of the new and improved surgical techniques being used. This trend is additionally fostered by the cost and time pressure that also affects the operating room, translating into a growing need for ever better surgical instruments – more durable, precise, and functional.

72.12.2 Industry

In the industrial sector, an enormous range of new opportunities for improving certain functions has been developed in recent years. Innovations from the fields of nano and coating technology, along with new materials such as ceramics, nitinol or carbon fibers, are bound to gradually replace existing functions of surgical instruments over time – and most likely improve them as well. This trend is enhanced by the potential use of new manufacturing technologies.

A relatively recent example from the history of scissors is the development of bipolar scissors. Apart from the conventional cutting and dissecting functions, bipolar scissors additionally offer a hemostatic function due to the simultaneous use of HF current. The clinical advantage derives from the fact that no instrument exchange is required, as the cutting, dissecting and coagulating functions are integrated into a single instrument. An additional benefit comes into play because users are already familiar with the instrument, so there is no real need to change the accustomed surgical technique. This saves the surgeon a lot of time and besides, he can fully concentrate on his task.

In terms of development and manufacture, completely new steps and methods were required to enable the use of two current potentials at the distal end of a pair of scissors as required for bipolar current application. With the given constraints, this necessitated the use of new insulation materials such as ceramics because only with a ceramic cutting edge (coated or pure) was it possible to fit the instrument with an insulating edge. Its production, in turn, called for a completely new manufacturing technique termed *ceramic injection molding* (CIM) that requires relatively complex molds, but then the resulting parts are always fully identical in size and shape.

Consequently, the metal base body of the scissors must always be fully identical as well. A perfect form fit is particularly crucial where the ceramic element is to be bonded to the metal body, as otherwise a poor fit would lead to fracture of the element during mounting. In addition, the whole process must be well matched to the opposite scissor blade to ensure the smooth action of the instrument under slight mechanical tension. In terms of controlled production conditions, this can only be achieved if the scissors or metal blades consistently have the same size and shape – which also means that manual production steps must be completely eliminated and all parts must be machined, ideally by robot-assisted CNC manufacturing. Only this approach allows the meaningful production of reproducible parts. While on the surface, the design of bipolar scissors needed just a few changes, consistent implementation actually led to enormous production process modifications requiring extremely high investments.

72.13 Bipolar Scissors

For some time now, bipolar scissors (Fig. 72.68) have been used in almost all surgical disciplines. Preferred fields of application in clinical use are (among others):

- Intestinal surgery (e.g. hemicolectomy, sigmoidectomy)
- Urology (e.g. prostatectomy)
- ENT (e.g. neck dissection)
- General surgery (adhesions)
- Gynecology (e.g. hysterectomy).

Outlook

Continued development will be of great importance, with focus on specialized, indication-specific multifunctional instruments. It will certainly be a challenge to enhance the solidity and durability of the scissors while

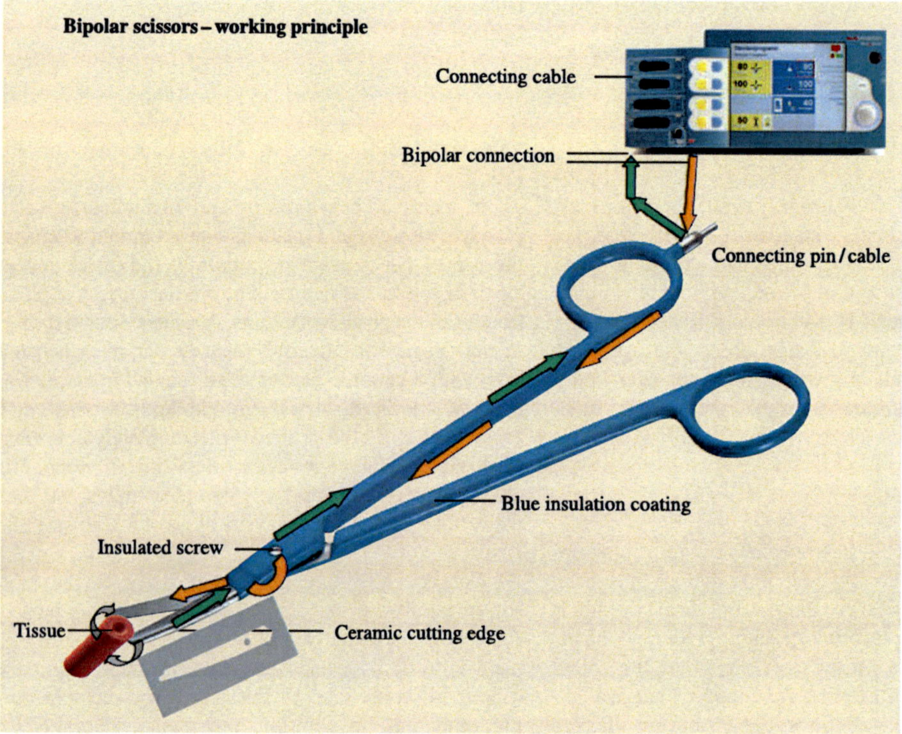

Fig. 72.68 Bipolar scissors – working principle

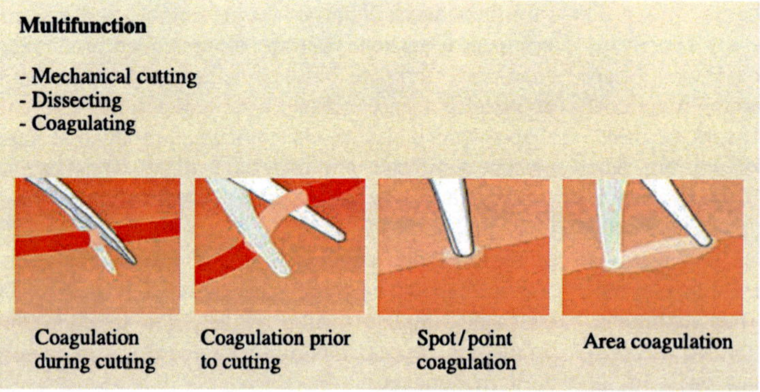

Fig. 72.69 Multifunction of bipolar scissors

making their use still easier in daily practice. Moreover, to ensure the greatest possible safety for the patients, it will also be important for users, surgeons, and surgical staff to keep themselves well-informed about the current and future developments taking place in the industry.

Further Reading

- DIN EN 285:2009-08 Sterilization – Steam sterilizers – Large sterilizers (Beuth, Berlin 2009)
- C.J.S. Thompson: *The History and Evolution of Surgical Instruments* (Martino, Eastford 2000)
- KLS Martin, Umkirch, Germany, Product Catalogue Instruments
- Lawton GmbH & KG, Fridingen, Germany: Product Catalogue Instruments
- J.A. Nideffer: *Learning Surgical Instruments* (CreateSpace, New York 2009)
- Association of Surgical Technologists (AST): *Surgical Technology for the Surgical Technologist* (Delmar, Albany 2001)
- J.K. Fuller: *Surgical Technology*, 4th edn. (Saunders Elsevier, New York 2005)

References

72.1 Instrument Processing Working Group: *Proper Maintenance of Instruments*, 8th edn. (IPWG, Mörfelden-Walldorf 2004)

73. Intelligent Textiles and Trends

Christian Rotsch, Sibylle Hanus, Danny Schwabe, Heike Oschatz, Andreas Neudeck, Uwe Möhring

Textiles are common materials for many medical applications. They are used, for example, as bandages, medical stockings, or scrubs. Developments during the last 10 years in the areas of wearable electronics, smart textiles and material research offer new possibilities to create medical textiles with a higher level of functionality and allow the development of completely new active medical textiles. This trend was made possible by the interdisciplinary cooperation of engineers and scientists from textile research, electronics, informatics, and mechanical engineering, together with medical experts.

The integration of electronic devices into textile base materials enables new possibilities for personal monitoring and therapeutical systems for sports and medical applications. Regarding the demographic development in industrial countries, such wearable monitoring devices may be very interesting in addition to common systems.

These new intelligent textiles are a result of the combination of textile and nontextile technologies. In addition to common textile properties, new functions will be realized by the integration of conductive leads. Besides the integration of common electronic components (e.g. sensors, amplifiers) in different clothes, the research activities focus on the development and application of textile and textile-based sensors and actuators.

73.1 **Textile Manufacturing Technologies and Applications** 1321
73.1.1 Textiles for Medical Application 1321
73.1.2 Weaving 1322
73.1.3 Embroidery 1322
73.1.4 Warp Knitting 1323
73.1.5 Braiding 1324
73.1.6 Conductive Thread Materials 1325
73.2 **Sensory Applications of Textiles** 1326
73.2.1 EMG and ECG Monitoring 1326
73.2.2 Respiratory Monitoring 1329
73.3 **Active Textiles – Therapeutical Applications** 1330
73.3.1 Textile Electrodes for Electrical Muscle Stimulation 1330
73.3.2 Textile Interactive Medical Agent Depots and Disposing Systems 1331
73.3.3 Heatable Textiles for Therapeutic Treatments 1331
73.4 **Passive Medical Textiles for Therapy** 1333
73.4.1 Reusable 3-D Knitted Elastic Short Traction Bandages 1333
73.4.2 Embroidered Implants for Tissue Engineering 1335
References 1335

73.1 Textile Manufacturing Technologies and Applications

73.1.1 Textiles for Medical Application

Novel functional special textiles, resulting from the combination of textile and nontextile technologies, are an interesting alternative to the usual systems for medicine and medical technology. For example, by using different thread materials, textile and nontextile manufacturing technologies or coatings, common textile structures obtain a new level of functionality.

In addition to known applications of functional medical textiles like compression textiles, scrubs, or antidecubitus systems, active medical textiles offer new applications like the identification of hospital clothing by textile RFID tags for industrial laundry or

Fig. 73.1 Woven LED structure (Philips, Lumalive; TITV Greiz)

the monitoring of vital parameters by textile-based sensors. These so-called smart textiles are made of a textile base structure and, for example, electronic devices.

In addition to the common textile properties, new functions will be realized by the integration of conductive leads. Besides the integration of common electronic components (e.g. sensors, amplifiers) in different clothes, the activities of different research institutes are concentrated in the development and the application of textile and textile-based sensors and actuators. The base of such developments is, beside other materials, a conductive thread material. One example for a commercial product, which was developed at TITV Greiz, is the so-called ELITEX material. By the combination of electronic and textile typical properties, it can be used in all textile manufacturing technologies, like weaving, embroidery, or warp knitting. This is the first step for different product developments. Other attempts were made, for example, by using metal-based wires in combination with textile thread materials.

The goal is to integrate and not only to attach a non-textile technology into the textile structure to combine the advantages of textiles, like breathability, washability, stretchability, or a good wearing comfort with technical functions, like sensing or heating [73.1].

Textiles for medical applications can be separated into

- Functional medical textiles
  - Wound treatment
  - Compression textiles
  - Antidecubitus systems
  - Implants
  - Filtration
  - Hygienic textiles
  - Electromagnetic shielding
- Active medical textiles
  - Medical agent depot and disposing systems
  - Monitoring textiles, textile-based sensors
  - Therapeutical electrodes, e.g. functional muscle stimulation
  - Electrode systems
  - RFID tags
  - Heating and cooling textiles for thermotherapeutic applications.

A short overview of textile manufacturing technologies is given in the following sections.

73.1.2 Weaving

Woven fabrics are produced by right-angled crossing of at least two-thread systems, named warp and weft. The warp threads move in a lengthwise direction through the loom. The thread are lifted or lowered in a defined order (pattern) by a shedding machine (Fig. 73.2).

Then the weft thread is inserted. This can be done by a projectile, a rapier or by air. After that the weft thread is fixed at the woven fabric end. The high and low position of the warp threads now changes for the next weft insertion. This change is carried out by a dobby machine, where several threads on one frame make the same pattern. By using a jacquard dobby it is possible to control each warp thread individually.

Typical applications of weaving technology for the realization of smart textiles are multilayer structures for energy and data lines, antennas, textile electrodes, or photonic/light emitting textiles. These photonic textiles can be realized by the integration of LEDs in the textile structure (Fig. 73.1).

With a high density of threads up to 120 per cm multilayer matrices e.g. conductive boards can be created.

73.1.3 Embroidery

In the traditional definition, embroidery is a technique of decorative needlework in which designs are created by stitching strands of some material onto a layer of another material.

Up until now, most embroidery used textiles threaded stitches onto a woven or nonwoven fabric, but recently embroidery has become more and more important in technical applications ranging from fiber reinforcement up to electronic circuits. Stitches can

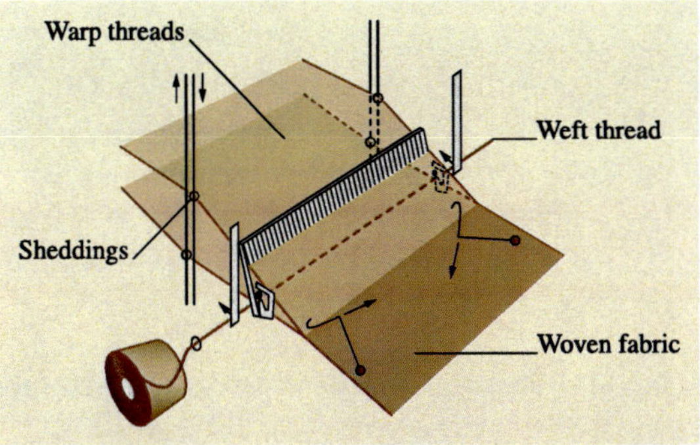

Fig. 73.2 Weaving machine with sheddings, schematic (TITV Greiz)

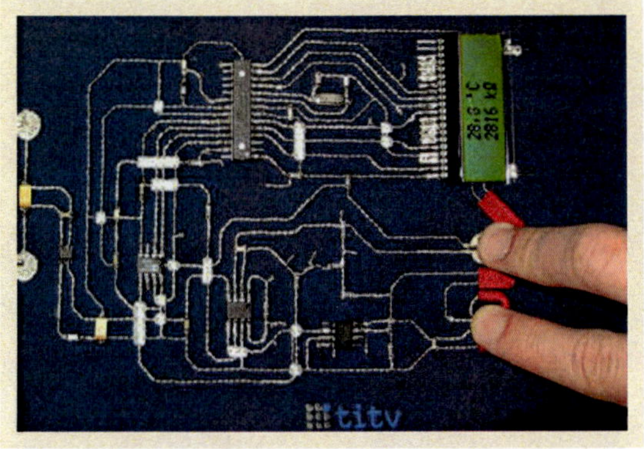

Fig. 73.3 Embroidered conductive circuit board (after [73.3], TITV Greiz)

be executed in wires and embroidery can be worked onto nontraditional materials such as plastic foils. One unique feature of embroidery is the possibility of placing stitches in any desired direction forward, backwards, and sideways. Very complex multilayer patterns, like conductive circuit boards (Fig. 73.3) can be produced in this way [73.2].

For the realization of embroidered structures two main principles exist. On the one hand, a thread is stitched onto a base material and fixed by a second thread from the lower side of the textile material (two-thread system). On on the other hand, the functional thread can be fixed on the base structure (soutage technology) by two separate threads, compared to the two-thread system. Tubes, wires, carbon, or glass fibers can be fixed onto textiles with the soutage technology, for example.

Besides the fixation of fibers and threads it is also possible to fix and to interconnect electronic components (Fig. 73.4). With embroidery technology the components can be fixed mechanically and the electronic interconnection can also be achieved.

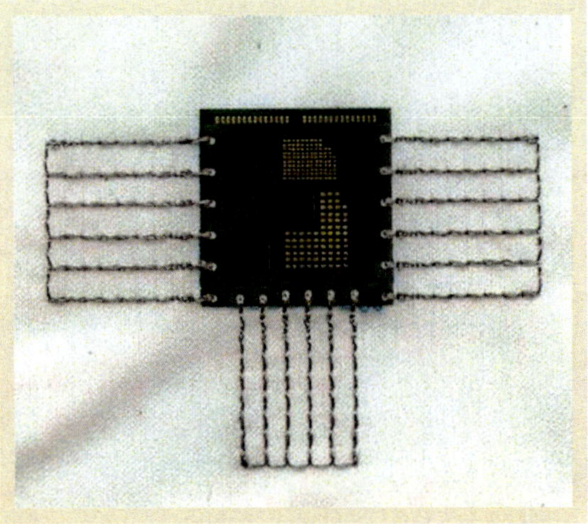

Fig. 73.4 Flex-foil substrate fixed and connected by embroidering technology (after [73.4], TITV Greiz)

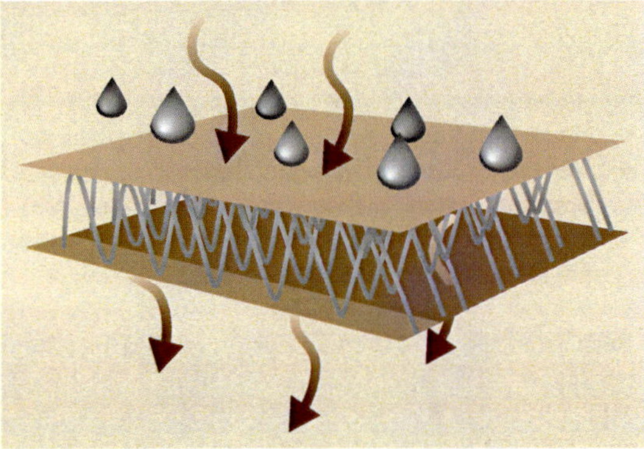

Fig. 73.5 Warp-knitted 3-D spacer fabric (after [73.4], TITV Greiz)

73.1.4 Warp Knitting

Warp knitted structures are special mesh structures. The fabric is created by the interloping of the yarns that are being knitted parallel to the length of the textile. A special type of warp knitted textile is the so-called three-dimensional spacer fabric (3-D spacer fabric). The 3-D spacer consists of two warp knitted surfaces that are connected and kept at a distance by third thread system – the pole yarn (Fig. 73.5). The two textile surfaces can be modified individually with respect to the binding or the materials used. Thus, the spacer fabric can be indi-

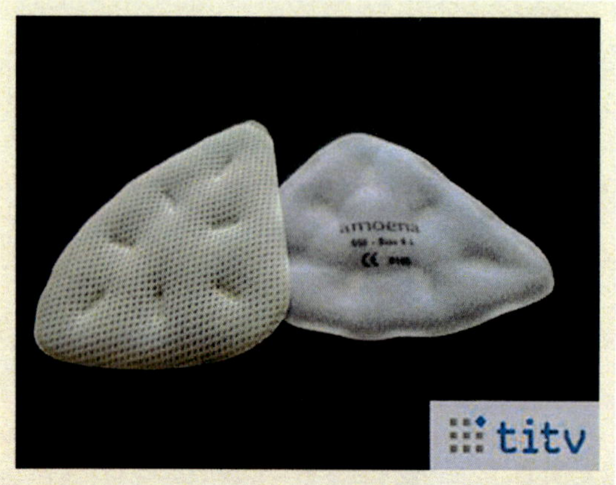

Fig. 73.6 Prosthesis pad in bras, cover made of spacer fabrics (after [73.4], TITV Greiz)

Fig. 73.7 Manufacturing of a braid (example) (TITV Greiz)

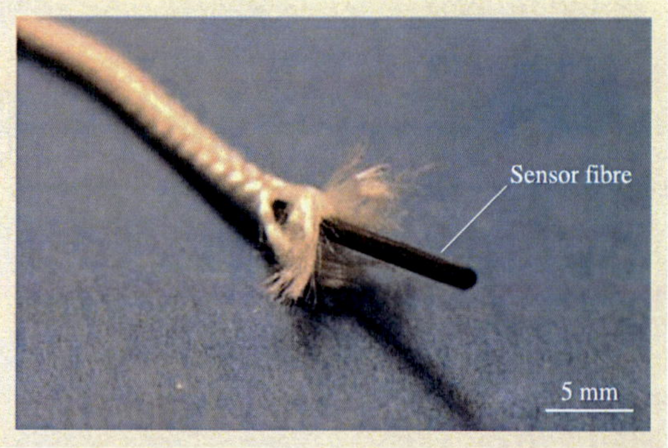

Fig. 73.8 Braid with sensor fibre as core material (TITV Greiz; ITA, Aachen)

vidually modified regarding the respective applications. For example, the thickness can be varied from 1.5 mm up to 60 mm.

The typical properties of the spacer fabrics are:

- Pressure stability and good pressure distribution
- Moisture transport by special thread materials in the upper and lower surface
- Good thermal & air circulation and good thermophysiological properties.

Because of the good thermophysiological properties of the spacer fabric, the material is used in several medical applications, e.g. the lining of orthesis or prosthesis pads (Fig. 73.6) in bras. Further applications are compression bandages, antidecubitus mattresses, insoles, or hip protectors.

73.1.5 Braiding

A braid is manufactured by interlacing three or more threads forming a flat or tubular narrow fabric. The yarn is directly fed from the rotating bobbins (Fig. 73.7) to the braiding point. Defined by the movement of the clappers to one another the binding, and thus the structure, of the braids can be varied. The dimension of a braid depends only on the count of the yarn, the number of threads, and the type of binding [73.2]. Braids can be used, for example, for energy transfer by using metal wires or litz wires.

Currently, two main variants of tubular narrow fabrics exist. A thread or another material can be placed inside the braid during the manufacturing process. The reinforcement of tubes or the shielding of a conductive thread or wire are typical applications. It is also possible to manufacture a braid without inner core material. Artificial vessels are typical medical applications for these types of braids. Other medical applications of braids are tendon replacements or surgical sutures. For example, PVDF braids are combined with titanium screws for the fixation of artificial tendons [73.5].

With the integration of nontextile materials in the braiding process new applications are possible. Figure 73.8 shows a bending angle sensor. Sensor fibers are integrated in the braid during the manufacturing process.

Besides the described structures braiding technology can also be used to create complex three-dimensional structures e.g. for fiber composites. Complete fiber reinforced components, for example for the aircraft industry, can be produced with 3-D braiding manufacturing machines [73.6].

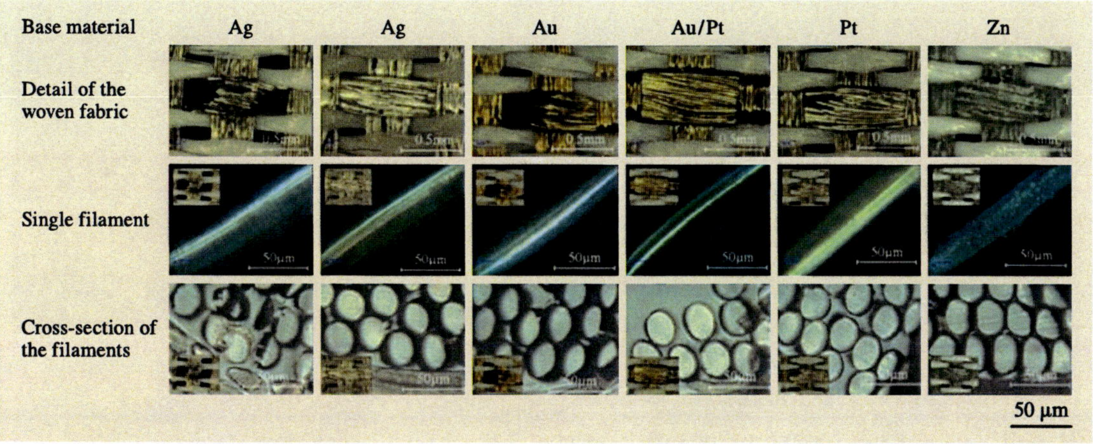

Fig. 73.9 Overview of different ELITEX variants, base material (Ag, Statex) is galvanized with silver (Ag), gold (Au), platinum (Pt) and zinc (Zn) (TITV Greiz)

73.1.6 Conductive Thread Materials

The basis for the integration of sensor and electronic devices in textiles is a conductive thread material. Actually, there are different strategies and materials to realize the transfer of data and energy in textiles. On the one hand, there are different possibilities to use common copper or steel fibers to integrate electronic parts in textiles. Some of them can be used in a regular textile process, e.g. weaving. Often there are some problems with the long-term stability of these metallic materials. If these threads are bent many times, for example, during the wearing or washing process they can break. On the other hand there have been many attempts to create conductive textile threads. They have the advantage of a very high stability against bending and low weight. However, normally it is not possible to use these structures as data lines for low-volt signals. The resistance of these materials is too high. Typically, these threads are used as energy lines, e.g. for textile heating systems. Different strategies are pursued to reduce the resistance without losing the advantage of low weight and flexibility.

One possibility is the new galvanization process, developed by the TITV Greiz, Greiz, Germany, to handle textiles. By different electrochemical processes a silver-coated polyamide fiber is again galvanized with a silver coating. So it is possible to reduce the resistance of the base material from about 700–800 Ω/m to 15–20 Ω/m (base yarn count of 235 dtex). The final value depends on the material applied (Fig. 73.9) and the type of finishing process of the yarns. For example, a different number of threads can be twisted together so the resistance can be minimized. Woven structures with a resistance of about 1 Ω/m can be realized with these yarns [73.2, 7].

By additional coating processes these conductive threads can be insulated e.g. by a TPU or PVC coating (Fig. 73.10). Another possibility is the coating of the manufactured textiles, e.g. narrow fabrics in which the conductive threads have been integrated (Fig. 73.11).

To realize stable electronic interconnections, different strategies have been used. For example the material can be soldered (Fig. 73.12) or glued under special conditions and it can be crimped with adapted tools. This is an important factor for the industrial implementation of smart textiles.

Fig. 73.10 Insulated ELITEX thread (TITV Greiz)

Fig. 73.11 Photonic textile, conductive narrow fabric, water resistant insulation by coating (TITV Greiz)

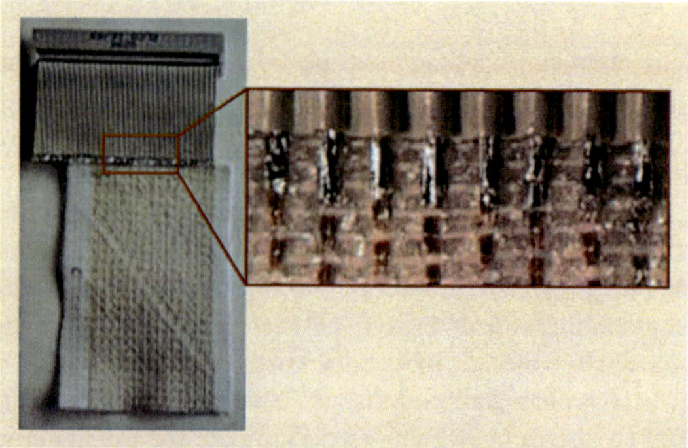

Fig. 73.12 Soldered interconnection between commercial IDE cable and woven bus structure (TITV Greiz)

Fig. 73.13 Core mantel thread construction of textile and metallic components (TITV Greiz)

Besides the use of galvanized thread materials, other possibilities to create conductive yarns or fibers also exist.

Examples are:

- Metal Fiber yarns
- Metal wire/textiles yarn construction (Fig. 73.13)
- Metalized yarn, e.g. by chemical or PVD processes
- Carbon fibers
- Fibers filled with conductive particles, e.g. silver, carbon, ICP or CNT particles
- Fibers with a conductive preparation
- ICP (intrinsic conductive polymers) fibers
- CNT (carbon nanotube) fibers.

It depends on the application and the available manufacturing technology which material should be used.

73.2 Sensory Applications of Textiles

73.2.1 EMG and ECG Monitoring

The monitoring of parameters like EMG or ECG is very important for different medical and sports medical applications from the preventive or therapeutical points of view. For surface EMG and ECG electric body signs are recorded by glued, plastic, or metal permanent electrodes separately fixed on the skin. The wearing comfort of these electrodes can be estimated as unsatisfactory for patients, especially during long-term applications. The displacement of the electrodes caused by physical activity and sweat development between skin and electrode is a common reason for wrong measurements. In some cases, the electrolytic gel causes skin irritations.

Reliable monitoring systems are required without negative influence to the wearing comfort especially in the fields of medical care and therapy, sports, and occupational medicine.

The combination of textiles as a flexible and light material with foreign technologies opens new ways to create new intelligent products for body-closed applications for medical and medical technique applications. This includes the integration of conventional sensors and sensor systems in different pieces of clothing, but also the development of textile and textile-based sensors [73.8].

Currently there are several approaches to integrate a monitoring system in vests or shirts. Research insti-

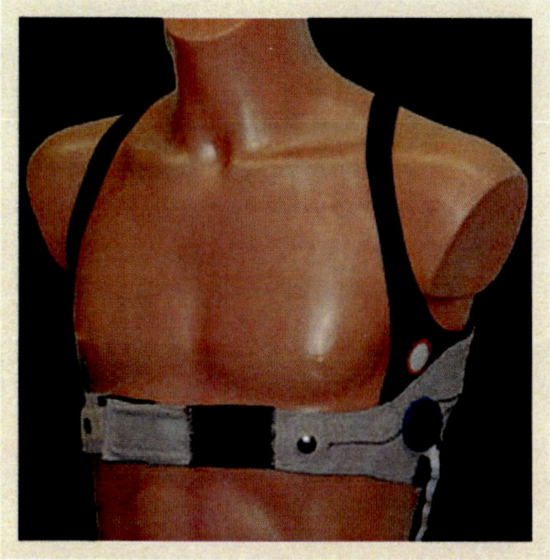

Fig. 73.14 MμGuard monitoring vest monitoring vest with silicone based dry contact electrode (after [73.10])

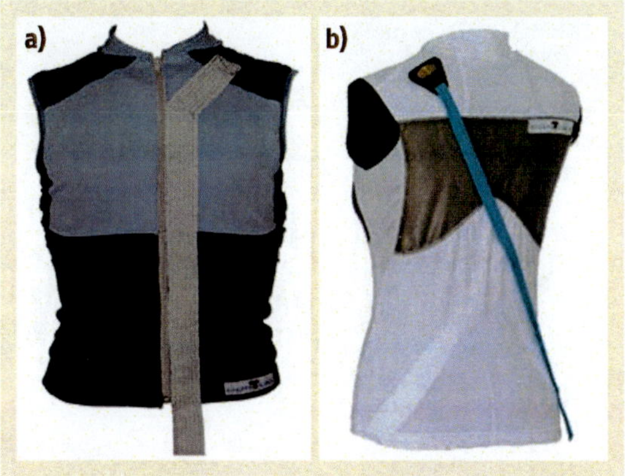

Fig. 73.15a,b ConText vest with textile-based contactless electrodes. **(a)** Prototype II, **(b)** final prototype (after [73.8])

tutes and industrial companies are working on these topics in several national and international research projects. The systems Life-Guard, SmartShirt, VitalJacket, LifeShirt and MμGuard are only a few examples of these aspirations [73.9, 10].

These monitoring shirts are designed for long-term monitoring. One aspect to increase the comfort of the wearer is the use of flexible dry electrodes. Conductive silicone-based electrodes are a promising strategy [73.13].

Textile electrode systems can work by direct skin contact or alternatively without contact by capacitive coupling.

There are currently two main strategies to monitor EMG and ECG signals. The electrical current can be measured by contact electrodes and noncontact electrodes. In the first case, a conductive textile structure (Fig. 73.16) can be used to measure the electric signals. The quality of the signals depends on different aspects, e.g. the contact area, the structure of the electrode and the surface, and inner conductivity. Normally the textile electrodes need to be moistened by sweat or special contact gels. Currently, there exist different strategies to coat the conductive thread materials by conductive polymers or ionic liquids. First results are promising [73.11].

In the second case, the idea is to monitor the EMG and ECG signals by noncontact electrodes. In different research projects the idea of a noncontact ECG monitoring has been investigated. Besides the design of the electrodes there are special requirements referring to the electronic components, the shielding, and the analyzing process [73.14, 15].

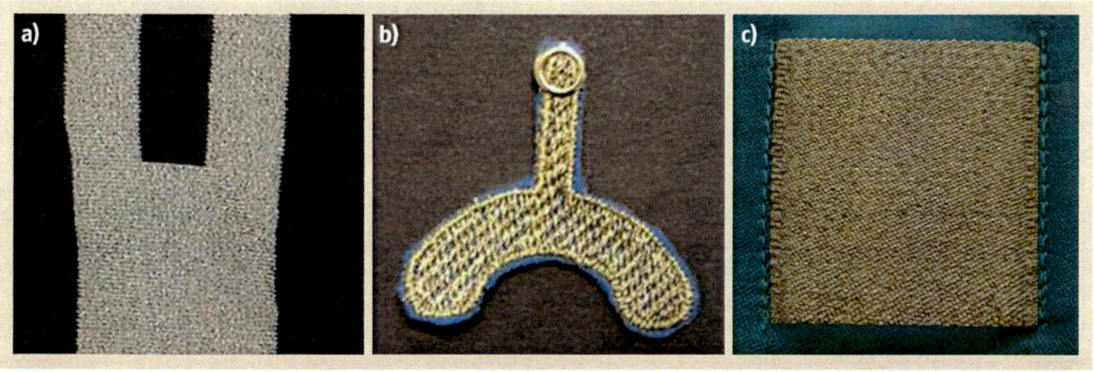

Fig. 73.16a–c Textile electrodes. Examples **(a)** knitted, **(b)** embroidered, **(c)** woven structure (after [73.11, 12], TITV Greiz)

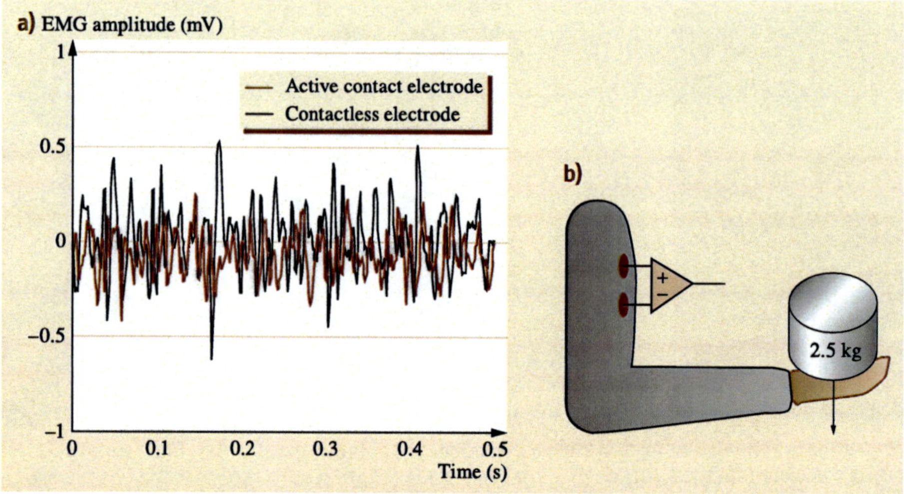

Fig. 73.17 (**a**) Recorded EMG signals with both an active and (**b**) contactless electrode (after [73.17])

In the EU research project ConText (IST-027291), TITV Greiz worked together with the project partners Philips Electronics Nederland BV (NL), The Catholic University of Leuven (B), Technische Universität Berlin (D), Clothing Plus Oy (FIN), and the Netherlands Organization for Applied Scientific Research, TNO (NL) on the contactless recording of EMG and ECG signals using only several textile and textile-based capacitive sensors with the relating electronic devices [73.16]. The textile sensors in the ConText project were manufactured by weaving, embroidery, printing, and lamination technology. The final prototype of the sensor consists of a laminated structure and woven fabrics for energy and data transfer.

Another possibility for the realization of such a sensor fabric lies in different woven multilayer structures. So it is, for example, possible to create a textile capacitor that can be used as the sensor area. It consists of two conductive layers of ELITEX material that are insulated by two textile layers in the middle. The capacitance of the fabric can be adapted by changing the conductive thread material or the linkage. By a special type of weaving it is also possible to integrate the connectors for the data transfer.

The measuring principle is similar to the contact electrode of a pair of electrodes. These electrodes are built by two conductive layers that are separated by an insulating layer and work as a capacitor. Such a configuration can be used to connect the biopotential signal capacitively to an amplifier. In practice, two electrodes are used and connected to a differential amplifier. The differential signal is then sent to the measurement system to be sampled, processed, and analyzed. Figure 73.17 shows some results of the comparison of common contact and the first prototypes of the contactless electrodes.

For a real integration of various electronic devices in a textile structure, it is necessary to use textile or textile-like bus structures for the transfer of energy or data. These structures should have mechanical qualities that are comparable to those of base textile, e.g. a shirt or a vest. The woven structures are made of a special multilayer design, which allows the creation of conductive lines of insulated layers on the upper and lower sides. These narrow fabrics can be used for analog and digital data transfer. For sensible measuring applications it is also possible to create shielded bus structures. By means of a special weave it is also possible to create flexible textile narrow fabrics with integrated conductive threads. A very high strain rate is realized without a change of the electric resistance in contrast to other commercially available materials. The resistance of these normally used materials depends on the stress rate. Because of the changing resistance it can be difficult to use these materials for data lines, especially if it is necessary to transfer low-level or low-voltage signals. There are no problems with these new elastic conductive structures. They can be used in elastic clothes, for example. So the stress–strain behavior of the clothes will not be influenced by nonelastic structures, and the stress will not influence the resistance of the bus structures.

Besides the monitoring of EMG and ECG, these sensors also provide information about the physical and

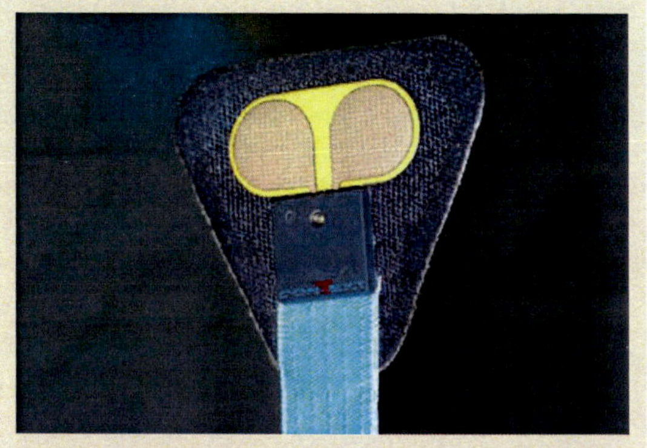

Fig. 73.18 Textile-based capacitive EMG electrode, ConText project (after [73.11])

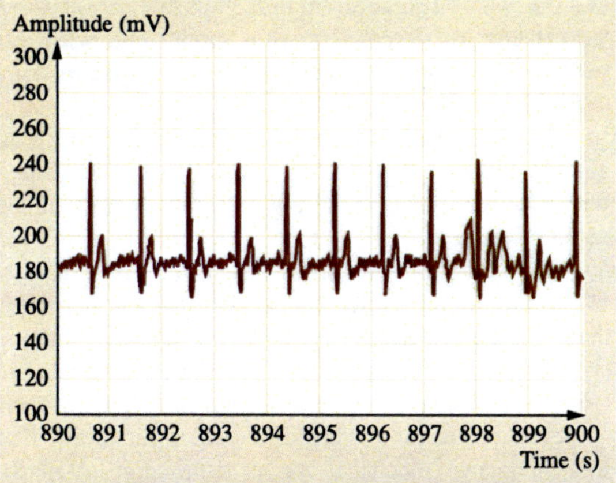

Fig. 73.19 ECG – Signal measured with contactless capacitve electrode (after [73.11], Context project)

mental stress level in addition to the general muscle potential by special algorithms [73.16, 17].

73.2.2 Respiratory Monitoring

Reliable respiratory monitoring is very important for different sport and medical applications. Besides the respiration rate, additional parameters exist, e.g. the breath volume, which deliver important information about the training condition or the sleep behavior of the sportsman or patient.

Different technological possibilities can be used to monitor respiration. The spectrum ranges from simple elongation measurement to monitoring the respiration rate up to the 3-D bioimpedance spectroscopy of the thorax to obtain detailed information about the breath volume and the health status of the lungs e.g. fluid re-

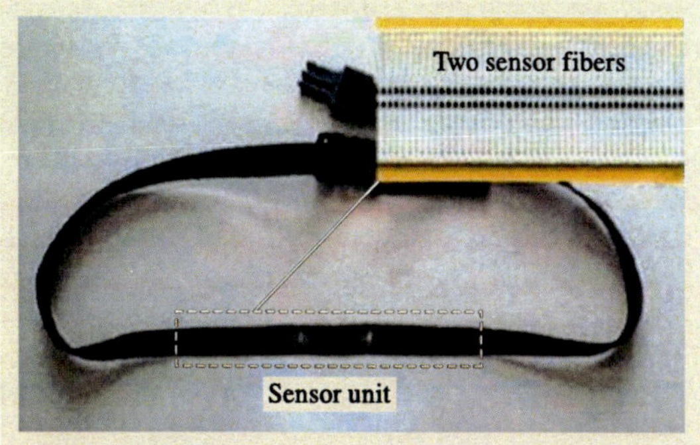

Fig. 73.20 Respiratory monitoring belt (after [73.18], TITV Greiz)

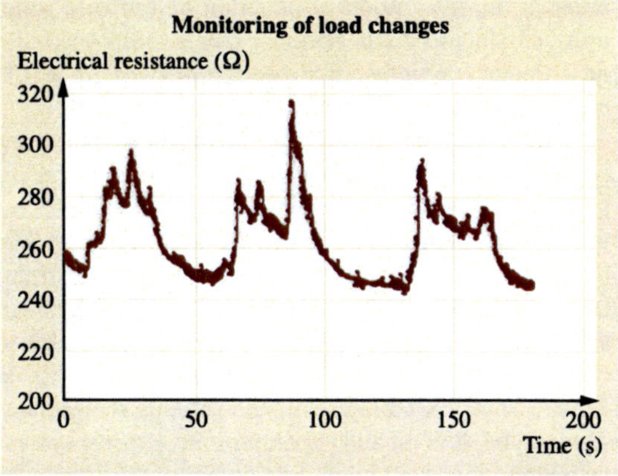

Fig. 73.21 Respiratory monitoring, results (examples) (TITV Greiz)

tention in the lungs. The type of the measuring principle depends on the medical indication.

A very easy and comfortable way to monitor only the respiration rate is the integration of an elongation sensor in an elastic belt or a shirt. The sensor system is limited to only a few electronic components that can be integrated into textiles without influencing the wearing comfort.

The respiratory monitoring belt shown in Fig. 73.20 is composed of an inelastic belt, which can be easily fixed on the thorax, and a sensor unit. The sensor unit is an elastic narrow fabric with integrated sensor fibers. A carbon filled rubber is used as sensor fibers. In the form used here, this material has a nearly linear elongation specific resistance. By inhaling, the change in the resistance can be measured through the sensor's elongation. The peaks in Fig. 73.21 show the moments of the maximum elongations. With an additional electronic device it is very easy to measure the respiration rate by measuring the change of resistance. A calibration modus at the beginning of the measuring process can be done at maximal and minimal thorax status.

For reliable monitoring it is very important to use a weave that enables a stable fixation of the sensor fibers in the narrow fabric. The narrow fabric and the sensor fibers should be configured in such a way that they always work in the elastic elongation range in order to obtain long-term stable results.

Such an easy monitoring system may be integrated in different clothes as well as heart rate monitoring belts, e.g. the Polar WearLink (by Polar Electro Oy, Finland) and may deliver additional results during sports, stress tests, or during sleep.

73.3 Active Textiles – Therapeutical Applications

73.3.1 Textile Electrodes for Electrical Muscle Stimulation

Besides the integration or adaption of sensoric functions, conductive textiles provide the possibility to build up actuatory systems, like heating devices or textile electrodes for electromyostimulation.

The effect of electricity for medical purposes, specific muscle formation among sportsmen, or for rehabilitation aspects is known today. Electrotherapies use the different effects to the human organism. The main applications are pain therapy and muscle stimulation for training effects. For electrostimulation, metal electrodes or adhesive Ag/AgCl electrodes, which can only be used a few times, are fixed onto the skin at defined spots by therapists or the patients themselves. This can be difficult and uncomfortable, especially near the extremities due to perspiration under the surface of the electrodes and leads to reduced wearing comfort and effect. To improve the wearing comfort and effect, TITV Greiz developed a new system of electrodes on the basis of conductive yarns. It was integrated in several textile elastic garments like track pants and shirts and improved wearing comfort was achieved.

To reduce the contact impedance between the skin and the electrodes, different finishings and preparations were tested on a technical skin model and with test persons by electrochemical impedance spectroscopy (EIS) [73.1, 11].

Common textile tests like wash stability and tests against abrasion and sanitizing ability were combined with electronic tests regarding the homogeneity of the conductive areas.

In cooperation with the Institute of Physiotherapy of the Friedrich Schiller University of Jena, Germany, these textile electrode systems were tested for the stimulation of femoral muscles. Textile and classic (glued) electrodes were compared with respect to their therapeutical effect during different applications, like transcutaneous electrical nerve stimulation (TENS), electrical muscle stimulation (EMS), or electromyogram (EMG) as biofeedback. Besides the parameters of

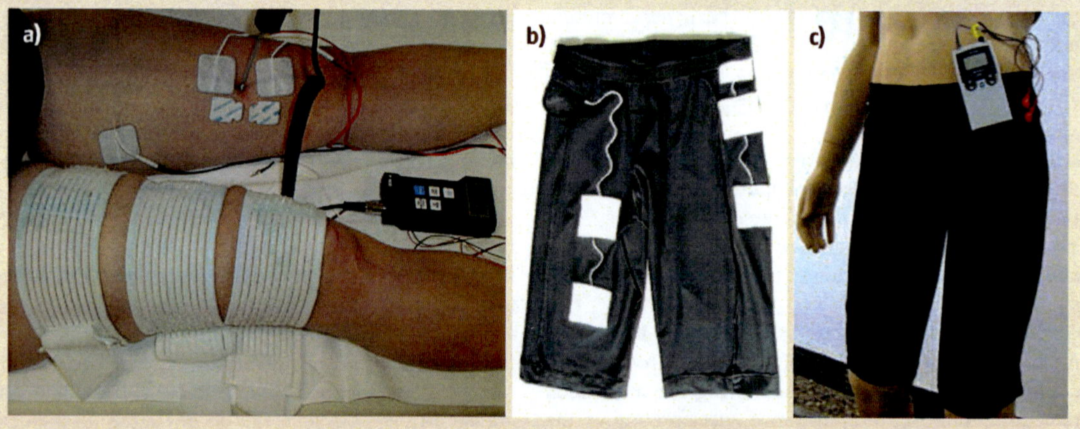

Fig. 73.22 (a) Test person with textile and glued electrodes [73.12]. **(b,c)** Prototype of a stimulation system for the femoral muscles (after [73.8], TITV Greiz)

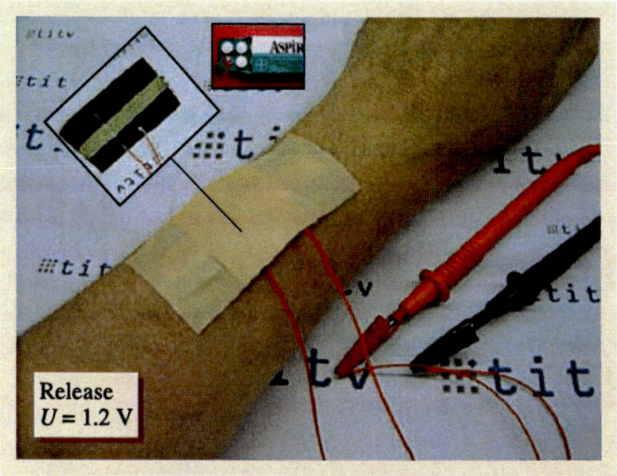

Fig. 73.23 Basic implementation of the interactive fabric structure in an interactive textile TTS (TITV Greiz)

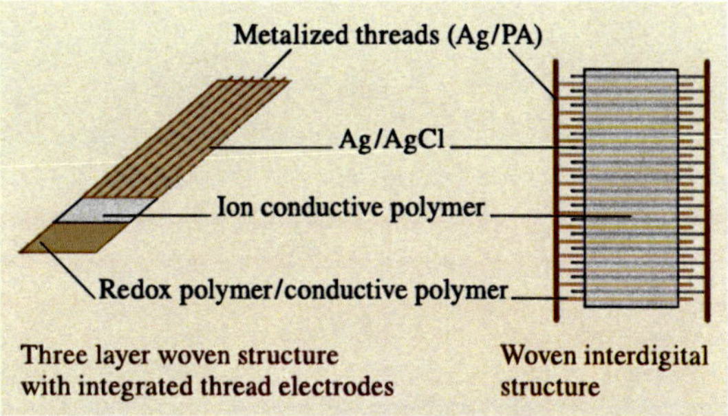

Fig. 73.24 Electrochemical modified thread electrodes as an interactive drug store and delivery systems (principle) (TITV Greiz)

the electronic therapy devices, different parameters like skin resistance, skin temperature, and the subjective impression of the test persons were registered. The results of the first study with 25 test persons show a comparable behavior of the common and the new textile electrodes. A target of further developments is the certification of the textile electrodes as a medical product class 2a [73.11, 12].

73.3.2 Textile Interactive Medical Agent Depots and Disposing Systems

The storage and controlled release of pharmaceuticals in the form of transdermal therapeutical systems (TTS) offers interesting alternatives to conventional pharmaceutical delivery in the form of tablets or injections. In addition to bypassing the digestive tract by the direct transition into the blood vessels and the associated lower dose, transdermal systems are easily applicable.

Furthermore, the patient's acceptance increases because the distances between the application forms are much higher in comparison to other pharmaceutical types. The dose is only determined by the type of storage of the pharmaceuticals in the support material. So, the danger of misuse is reduced to a minimum. Typical applications are analgesics (e.g. morphine in long-term applications), or nicotine.

In addition to the benefits of TTS some disadvantages also exist. Insufficient breathability and low moisture transport are two aspects. This can lead to occlusion of the covered skin areas, thereby favoring the occurrence of skin irritations. Furthermore, there are problems with the adhesive strength of the plaster, which are reduced during the application time. The plaster can also become detached during washing or during physical activity.

Another disadvantage during the application is the start of the release process by removing the protection layer (release liner) of the system. Once the system has fixed on the skin, the release process starts and will not end until the storage is empty or the whole system is removed.

Textile multilayer woven fabrics with different thread preparations e.g. conductive, redox, and ion conductive polymers have been investigated as one possibility of creating interactive textile base structures (Fig. 73.23) that can be used as an agent depot and delivery system. During the many experiments it was found that redox polymer modified ELITEX materials can store agents as anions that are chemically bounded on the polymer chain. The anions can be interactively electrochemically released again by applying a voltage or an appropriate reducing agent. By a suitable textile construction it was possible to create first prototypes of a textile pharmaceutical store and delivery system (Fig. 73.24). It is possible to store and release between 50 mg and up to 500 mg acetylsalicylic acid with such a structure [73.4, 7]. These interactive textile structures can be considered as first steps to textile-based interactive transdermal therapeutical systems. The developed prototypes could be combined with additional iontophoretic electrodes for a defined transfer of the released active ingredients through the skin.

73.3.3 Heatable Textiles for Therapeutic Treatments

Textile-based heating systems were one of the first developments in the area of active textiles. Several com-

Fig. 73.25 Thermal treatment bandage with controller, lumbar bandage (after [73.19], TITV Greiz)

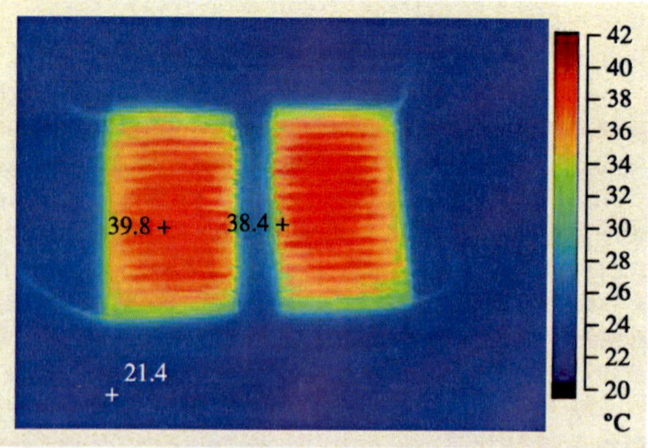

Fig. 73.26 Thermal treatment bandage, temperature distribution (after [73.19], TITV Greiz)

mercial products are already available. Mostly, these heatable textiles are used in sports applications e.g. snow boots, gloves etc. Usually combinations of metal wires, carbon fibers, and metalized fabrics are used in these applications.

Heatable Bandages for Therapeutic Treatment

Besides these commercial or sport applications, textile heating systems can also be used for therapeutical aspects. Among them, thermal treatment methods are the oldest and most commonly used therapies. The attainable effects range from a general improvement of well-being and mental relaxation to the locally defined influence of muscle tone, circulation, inflammation, pain, and the influence of the hormonal and the immune system. Thermal applications show a great diversity, which is multiplied by various application techniques. The whole body or only certain parts of the body are heated up. The heat can be applied in such a way that only the body surface or the deeper tissues are heated up. The application time can vary between 10 and 60 min.

Recent medical research results show that thermotherapeutical applications with temperatures between 37 and 45 °C and an extended period of use are very effective. Many of the applications in this area are not mobile or transportable. Therapeutical applications where the patient can move freely often have disadvantages in the dosage of the heat and the wearing comfort.

Figures 73.25 and 73.26 show a lumbar bandage for thermal treatment. It consists of an elastic bandage, two textile heating modules, and the controller device.

The heating modules are woven narrow fabrics with integrated electrical insulated heating threads. In contrast to other production technologies, the integration of the heating fibers is done directly during the manufacturing process of the narrow fabric. Thus, the textile properties e.g. flexibility and drapability, can still be retained.

The modified manufacturing process allows the integration of the heating threads regarding the defined requirements, e.g. distribution and quantity of heat, power supply, and electronic connection [73.19].

The result is a mobile bandage for thermal treatment that meets the medical, electronic, and textile requirements. It is characterized by:

- A high wearing comfort
- Optical attractiveness
- Easy handling
- Nonstationary applications
- Enhancement of patient compliance = treatment results.

Heating System for Long-Term Operations

To keep the patient's body heated during an operation, different kinds of active principles are implemented: active and passive textile passive systems, as well as convective and conductive systems that are based on nontextiles.

An important trend is the combination of heating and pressure reduction. Especially during long-term operations local pressure strain in combination with moisture accumulation results in bedsores that not only bring consequential costs but are also an additional stress for the patient. Systems that include both functional principles can have many advantages in this case, not only economical ones. Three-dimensional structures (3-D spacer fabrics) have many advantages with respect to the prevention of bedsores, because especially three-dimensional knitted textiles are not only able to allow

Fig. 73.27 Spacer fabric heating structure, heating threads integrated in the pole area (after [73.12], TITV Greiz)

Fig. 73.28 Textile heating system based on a compound of 3-D-spacer fabric and a flat knitted heating device (after [73.4])

air permeability but also moisture transport due to their mesh structure. The proven pressure reducing properties of knitted spacer fabrics are, in this case, a sort of acceptable additional feature.

A wide range of electrical conductive materials for the integration in the spacer fabric's pole area have been investigated, such as multifilament wires, carbon wires, and metalized polymer yarns. Furthermore, possibilities for electrical contact have been investigated, whereupon the focus has been put on economical processes and reversibility.

The research work of TITV Greiz showed that the integration of heating wires into the spacer fabric's pole area without intermeshing must be stated as an optimum (Fig. 73.27). The adaption of the heating performance to the operation purpose, the free choice of the contact, and economical processing are convincing advantages [73.12].

By measuring the allocation of heat and temperature on the surfaces and determining the clothing-physiological properties, the samples proved their adequacy for applications in the field of medicine (hypothermia prevention) and also in the field of rescue missions (saving victims from cold areas). Along with a battery supported power supply in combination with a controller, mobile systems with the advantages of low weight, high flexibility, and reliability can be developed.

73.4 Passive Medical Textiles for Therapy

73.4.1 Reusable 3-D Knitted Elastic Short Traction Bandages

Three-dimensional (3-D) spacer fabrics are already being used as therapeutic means for decubitus prophylaxis in hospitals and care or as bedding material for orthopedic shoes. The goal of the project was the development of 3-D knitted elastic bandages with adjusted widths for medical applications, especially for the therapy of lymph edema. Particular attention was paid to reusability, economical handling of resources, and their recirculation to the value added chain.

Lymph edema diseases have an increasing tendency. An early and comprehensive treatment is important. Compression therapy is the only possibility to stop the disease from further diffusion. The method of choice here is the complex anticongestion therapy, which is characterized by an anticongestive first step (manual lymph drainage + special compression bandaging as flanking action) and a conserving second step to keep and optimize the success of treatment [73.20].

With the development of elastic 3-D short traction bandages costs and time can be saved. Better thermo-physiological conditions, the prevention of thermal congestion and optimal breathability help the patients to improve their quality of life. Economical as well as ecological effects are achievable. The new elastic 3-D bandages (Fig. 73.29) are totally different from the products and materials known so far. Two textile outer surfaces and an aired zone between make these ban-

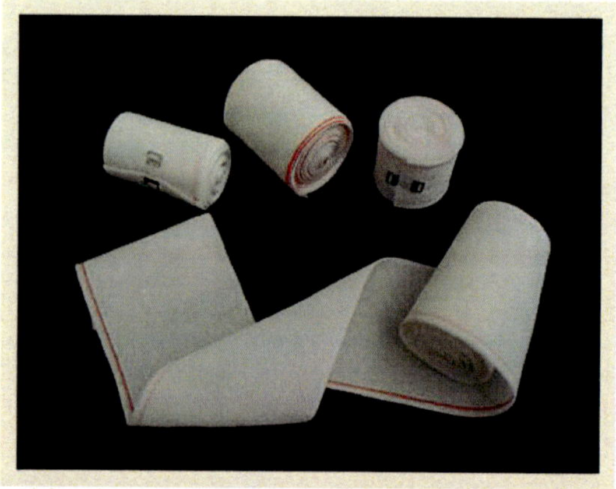

Fig. 73.29 3-D-knitted elastic short traction bandage (after [73.4], TITV Greiz)

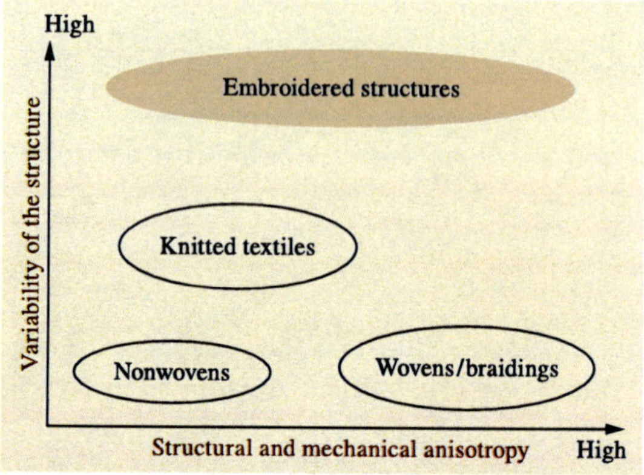

Fig. 73.31 Variability of embroidery technology compared to other textile manufacturing technologies (after [73.21])

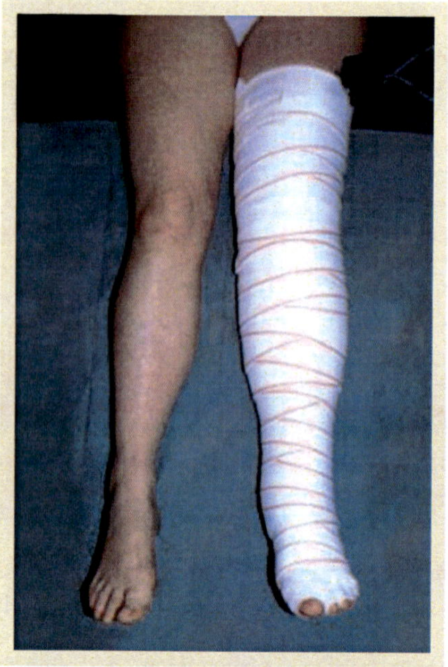

Fig. 73.30 Patient with 3-D-knitted elastic short traction bandage and fixed pressure sensor (TITV Greiz)

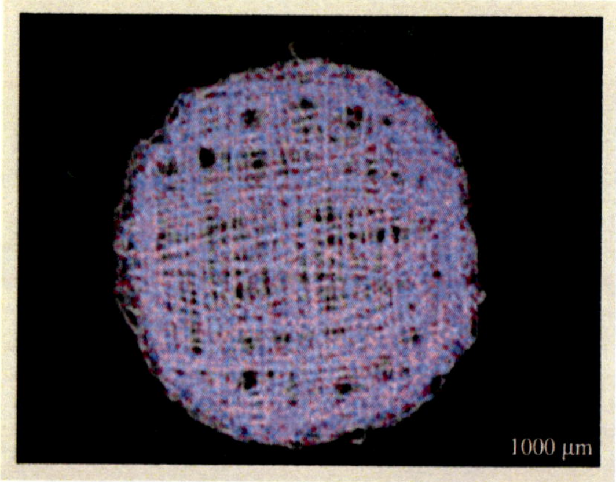

Fig. 73.32 Embroidered scaffold construction (after [73.1], TITV Greiz)

dages not only compressible but also cushioning. The bandage of edema patients can be realized with a simplified wrapping technique and without any cushioning with comparable efficiency.

The manufacturing technology with a double Raschel technique allows a number of bandages to be produced at the same time with different widths, specific materials and defined elasticity. So, favorable conditions for the following processes (relaxation, washing, fixing) can be created. The test series included tests with different materials, weaves, mesh densities, and combinations of material. Elongation values as known from standard short traction bandages were achieved by the use of elastane and partial elastic bicomponent yarns.

Clothing-physiological and skin-sensory tests were made by BPI Hohenstein, Germany, with several 3-D bandages in comparison to standard bandages. It was shown that knitted materials made of spacer fabrics achieved a significantly better comfort note than the standard materials.

Application studies with medical staff showed that 3-D bandages have better handling and wearing comfort in comparison to standard bandage materials

(Fig. 73.30). Although the bandages are more voluminous, an easier wrapping technique was pointed out.

This development is an interesting alternative for bandages, especially for patients with lymph oedema during the complex anticongestion therapy [73.22].

73.4.2 Embroidered Implants for Tissue Engineering

Embroidery is an interesting alternative for the development of three-dimensional textile-based implant constructions like scaffolds and patch grafts for tissue engineering. According to the principle of chemical embroidery, it is possible to create individual and application related scaffolds and patch grafts.

In comparison to conventionally used nonwoven, gel, foam, and foil constructions, embroidered implants offer some advantages:

- Improved mechanical stability
- Function and force flow appropriate construction
- Individual and local defined configuration – pore size, shaping etc. in terms of structure compatibility
- Selective material combination/scaffold constructions (absorbable, nonabsorbable fibers) [73.1].

Embroidered solutions for textile implant constructions offer comparable solutions to woven, braided, knitted or nonwoven materials with the advantage of an individual modification of the mechanical and structural properties by local variation of thread material and embroidered structure. So, embroidered constructions can be realized individually relaying the needed specifications [73.3, 21].

Since each implant causes a reaction with the tissue of the patient through the interface between implant and human tissues, the use of biocompatible materials in the embroidery thread technology is essential.

The base material e.g. nonwoven, which serves as a carrier of the embroidered structures, is composed of polyvinyl alcohol and is removed after the embroidery process. A cytotoxic effect caused by the base material must be excluded.

Figure 73.32 exemplifies an embroidered scaffold construction in a 10-layer construction with biocompatible yarns.

References

73.1 C. Rotsch, S. Hanus, D. Schwabe, A. Neudeck, U. Möhring, D. Zschenderlein: Textile solutions for diagnostic and therapeutic applications, Proc. Int. Conf. Latest Advances in High Tech Textiles and Textile-Based Materials, 80 Years Department of Textiles (2009) pp. 41–47

73.2 U. Möhring, A. Neudeck, W. Scheibner: Textile micro system technology. In: *Intelligent Textiles and Clothing*, Woodhead Textiles Ser., Vol. 54, ed. by H.R. Mattila (Woodhead, Cambridge 2006) pp. 342–354

73.3 W. Scheibner, C. Rotsch, S. Gimpel, U. Möhring: Textilien für sensorische und aktuatorische Applikationen, Proc. Energieautarke Sensorik, 5. GMM-Workshop: Mikrosensorik mit autarker Energieversorgung und drahtloser Signalübertragung (VDE, Berlin 2008), (in German)

73.4 C. Rotsch, S. Hanus, A. Neudeck, D. Schwabe, H. Oschatz, U. Möhring: News from textile research – special textiles for medicine and medical techniques, Orthop. Tech. **12**, 918–923 (2006)

73.5 S. Schneemelcher: Textile Implantate in der Medizin. In: *Technische Textilien*, ed. by P. Knecht (Deutscher Fachverlag, Frankfurt 2006) pp. 285–299, (in German)

73.6 B. Wulfhorst: *Textile Fertigungsverfahren* (Hanser, München 1998) pp. 187–202, (in German)

73.7 U. Möhring: Galvanische Oberflächenveredlung von textilen Polyamidstrukturen, Thüringer Werkstofftag 2006. In: *Polymere in Thüringen, Vorträge und Poster*, ed. by T. Keller, K. Jandt (Köster, Berlin 2006), (in German)

73.8 C. Rotsch, H. Oschatz, S. Hanus, K. Gnewuch, D. Zschenderlein, A. Neudeck, U. Möhring: Textile electrode – systems for sensory and actuatory applications, Proc. Jakob Müller Institute of Narrow Fabrics, 9th Narrow Fabrics Conference Frick, Switzerland (2009)

73.9 C. Rotsch, A. Neudeck, U. Möhring, H.-J. Holland, A. Heinig, H.G. Despang: Textilbasiertes Sensorsystem zum Langzeitmonitoring von Pflegebedürftigen und Risikopatienten, Proc. VDE 1. Deutscher Kongress Technologien – Anwendungen – Management (VDE, Berlin 2008) pp. 199–202, (in German)

73.10 S. Lamparth, S. Fuhrhop, M. Kirst, G. v. Wagner, J. Ottenbacher: A mobile device for textile-integrated long-term ECG monitoring, Telemedicine and e-Health, IFMBE Proceedings Vol. 25(5) (Springer, Berlin Heidelberg 2009) pp. 278–281

73.11 C. Rotsch, H. Oschatz, S. Hanus, A. Neudeck, K. Gnewuch, U. Möhring: Textile und textilbasierte Elektrodensysteme – Neue Möglichkeiten für sensorische und aktuatorische Applikationen.

In: *Dresdner Beiträge zur Medizintechnik*, ed. by R. Poll, J. Füssel, E. Koch (TUDpress, Dresden 2008) pp. 26–31, (in German)

73.12 C. Rotsch, S. Hanus, D. Schwabe, H. Oschatz, U. Möhring: Solutions for medicine from the textile research, Orthop. Tech. **8**, 634–638 (2008), (in German)

73.13 K.P. Hoffmann, R. Ruff: Flexible dry surface-electrodes for ECG long-term monitoring, Proc. 29th Annu. Int. Conf. IEEE-EMBS, Lyon (2007) pp. 5739–5742

73.14 A. Aleksandrowicz, M. Walter, S. Leonhardt: Wireless ECG measurement system with capacitive coupling, Biomed. Tech. **52**, 185–192 (2007)

73.15 M. Oehler: Kapazitive Elektroden zur Messung bioelektrischer Signale, Dissertation, Braunschweig 2009, Ber. Inst. Elektr. Messtech. Grundl. Elektrotechn. **32**, 1–167 (Tech. Univ. Carolo-Wilhelmina Braunschweig 2009), (in German)

73.16 T. Linz, L. Gourmelon, G. Langereis: Contactless EMG sensors embroidered onto textile, 4th Int. Workshop on Wearable and Implantable Body Sensor Networks, IFMBE Proceedings, Vol. 13 (Springer, Berlin Heidelberg 2007) pp. 29–34

73.17 G.R. Langereis, L. de Voogd-Claessen, A. Sipliä, C. Rotsch, A. Spaepen, T. Linz: ConText contactless sensors for body monitoring incorporated in textiles, Proc. IEEE Conf. Portable Information Devices, Orlando (2007) pp. 11–15

73.18 H. Ahlers, W. Scheibner: Resistiver Dehnungssensor, Patent 10360309 (2005), DE 10360309 A1 2005.07.28, (in German)

73.19 H. Oschatz, S. Hanus, F. Siegl, U. Möhring: Entwicklung von Schmaltextilien mit thermoregulierenden Eigenschaften, Forw. Textile Technol. **08**, 42 (2010), (in German)

73.20 P.D. Asmussen, B. Söllner: *Kompressionstherapie – Prinzipien und Praxis* (Urban Fischer, München 2004), (in German)

73.21 E. Karamuk, J. Mayer, E. Wintermantel: Sticktechnologie für medizinische Textilien und Tissue Engineering. In: *Medizintechnik – Life Science Engineering*, ed. by E. Wintermantel, S.-W. Ha (Springer, Berlin Heidelberg 2008) pp. 951–960, (in German)

73.22 M. Heide, U. Möhring, D. Schwabe: 3-D-gewirkte elastische Kurzzugbinden auf dem Prüfstand "Haut", Kettenwirk-Praxis **2**, 22–24 (2006), (in German)

74. Electronics in Medicine

Paulo M. Mendes, Celso P. Figueiredo, Mariana Fernandes, Óscar S. Gama

Electronics has long made an undeniable and valuable contribution in the field of medicine. New medical electronic devices are based on available medical knowledge combined with technologies available in the electronics field. Electronics in medicine has a wide range of applications, from diagnostics to therapy, always aiming to provide new tools to improve the well-being of the population.

Nowadays, miniaturization and integration technologies are driving the development of new electronic medical devices that should be portable, wearable, and unobtrusive, but with controlled costs. Regardless of the technologies involved, usually the main goal of such devices is the measurement of physiological variables for use in diagnostics, monitoring, and therapy.

Therefore, this chapter starts with the description of several signals that are commonly recorded, the corresponding electronics and sensing mechanisms used, and problems associated with such acquisition systems, focusing on wireless and wearable devices. It ends with examples of electronic devices in development essentially for remote health monitoring.

This chapter is not about electronics nor on their design, since there are many good books on this topic already, written either specifically for the (bio)medical field or for the general field of electronics. The aim of this chapter is rather to present to the reader those aspects that are common and specific to many medical electronic devices, and to give an overview of what is required and the challenges faced when designing electronics in this field as well as a few such examples under development.

74.1 **Basics** ... 1338
74.1.1 Fields of Application ... 1338
74.1.2 Designing Electronics in Medicine 1339
74.1.3 Medical Electronics Specific Requirements ... 1339

74.2 **Electronic Sensing** ... 1341
74.2.1 Health Condition Monitoring ... 1341
74.2.2 Biosignals and Transducers ... 1342

74.3 **Electronics for Wireless Health Monitoring** 1344
74.3.1 Requirements for Wireless Devices ... 1345
74.3.2 Data Acquisition ... 1346
74.3.3 System Integration ... 1347
74.3.4 Wireless Communications ... 1348

74.4 **Power Supply** ... 1348
74.4.1 Battery Power Budget Considerations ... 1349
74.4.2 Wireless Power ... 1350
74.4.3 Wireless Power Link Analysis ... 1350
74.4.4 Wireless Power Charger ... 1351
74.4.5 Energy Harvesting ... 1352

74.5 **Wearable Medical Electronics** ... 1353
74.5.1 Wearable Systems ... 1355
74.5.2 Categories of Wearable Systems ... 1355
74.5.3 Design Requirements ... 1356
74.5.4 Sensors for Wearable Systems ... 1357
74.5.5 Sensing Methodologies ... 1357

74.6 **Electronics in Medicine at Work** ... 1358
74.6.1 Wireless and Wearable Low-Power Health Monitoring Systems ... 1359
74.6.2 Sensors ... 1359
74.6.3 Wireless Link ... 1360
74.6.4 Optical Biopotential Recording ... 1360
74.6.5 Optical Electrodes ... 1360
74.6.6 Optical Signal Acquisition ... 1361
74.6.7 Localization Solutions ... 1361
74.6.8 Ambient Assisted Living Applications ... 1363
74.6.9 Wireless Link Design for Biomedical Applications ... 1366

References ... 1373

The definition of electronics in medicine has many facets and is hard to define. We may therefore start with the question: What is *electronics in medicine*? Is electronics not always electronics? We use the following definition [74.1,2]:

Medical electronics – A branch of electronics in which electronic instruments and equipment are used for such medical applications as diagnosis, therapy, research, anesthesia control, cardiac control, and surgery.

74.1 Basics

From our point of view, electronics in medicine encompasses all aspects of electronics that deal with the problems found when electronics are used to understand or solve a medical problem, not necessarily restricted to humans. Electronics in medicine may include the study of the electrical behavior of the human body and how it responds to electrical stimuli. Those stimuli may be in the form of voltages, electric currents, electric fields or magnetic fields. In this case, electronics is used as a model to represent such behavior.

Electronics may be used as a solution to describe biological activities, as happens with the widely used Hodgkin–Huxley model for a neuron cell [74.3], as a solution to help restore lost electrical activity, as is the case with pacemakers [74.4–6] or Parkinson disease [74.7, 8], to help in the replacement of full human body parts [74.9–11], for diagnostics [74.12–15], for therapy [74.16], or for monitoring [74.17–19].

In this sense, it is difficult to write a chapter that includes all the aspects, difficulties, and aspects associated with electronics in medicine. Moreover, there are different perspectives and different views when looking at medical electronics, corresponding to electrical engineering, chemical engineering, ethical aspects, security and privacy issues, physics, and medicine, to mention just a few. However, it is possible to identify two different sides of the same coin: electronics and medicine. On the electronics side, no one is very interested in knowing what is going on behind the electronic system under development. Rather, the system is purely developed to fulfill a set of requirements. From the medical side, what is important is how the system can help to solve an existing problem and how the system can be used, e.g., *please give me a system that enervates the bladder, when placed inside the spinal cord, close to C4–C5*. Fortunately, these two worlds are getting closer, and many great things are expected from electronics in medicine.

Since this field includes many applications, from magnetic resonance imaging (MRI) to electrocardiogram (ECG) recording, through biosensors and DNA chips, this is a very broad field, and it is impossible to include and describe all the electronics used in medical applications. Therefore, this chapter introduces the reader to several issues that must be addressed when developing electronic devices for application in medicine. There are many problems that must be tackled and solved when using electronics in medicine. Of course, some of these problems, fortunately, are also shared by other application fields.

74.1.1 Fields of Application

The most obvious application of medical electronics is their use for medical or clinical applications (diagnostics, treatment, and monitoring). However, electronics developed for these applications may have other important fields of applications. One application that is closely related to medical applications is human augmentation, which may apply at different levels, of which two can be highlighted: sensorial [74.20–24] and physical capabilities augmentation [74.23, 25, 26]. For the former, we can give the example of brain–computer interface (BCI) systems that can be used to detect what is happening in the field of vision of a soldier, before he is able to become aware of it [74.27–30]. For the latter, we can give the example of exoskeletons, which can be used to help soldiers in transportation of heavy loads (as in the movie *Avatar*) during long walking periods, or to help an old person with problems in locomotion.

Another very important field of electronics in medicine is its use for ambient assisted living (AAL) applications (Chap. 67, [74.31]). Electronics are used to record, understand, and control those environments. The main idea is to create smart environments that can sense several physical variables to allow implementation of control and decision loops, making that environment *smart*. Such smart environments may be used, e.g., for environmental control, or, focusing on the medical field, to help people with their daily activities.

One major problem they try to tackle is to deal with the ageing of the world population [74.32], for which new innovative technological solutions are expected from the field of medical electronics.

Electronics in medicine may be also used for performance improvement [74.33, 34] and performance monitoring [74.34, 35, 35–37]. These last two applications are closely related, since performance monitoring is a way to understand what can be done to improve it. Monitoring of performance can be done by measuring the physiological signals of the user, or by measuring the outcome of the human action.

Last but not least, many advanced electronics developed for application in medicine may be used for entertainment and interaction purposes [74.23, 38, 39]. All devices developed to monitor posture, position, or muscle signals, or brain activity, can be used as new paths to control new interfaces, leading to a new concept of human–machine interfaces [74.21, 23, 33].

74.1.2 Designing Electronics in Medicine

The design of electronics for medical applications implies the fulfillment of a set of requirements due to the nature of their operation. Such electronics must be designed to be small and capable of being integrated into the environment where they will operate. Another important aspect is that these systems may be required to operate in very harsh environments (inside the stomach, where lots of substances are prepared to digest almost everything), using only a single battery, with no possibility of replacement during tests, and many times only possible using surgical procedures. This requires the use of low-power design techniques, requiring biocompatibility and long-term usage capability.

Since in many applications, the electronics are required to interface with the human body, in order to record signals or to deliver signals or drugs, or to monitor their effect on the human body, one feature of such electronic devices is the ability to register very small variations of the variables under measurement, and many times it is necessary to record many signals at the same time, leading to challenges in terms of analog-to-digital conversion. Most electronics that are used in medicine are designed to sense and transmit a variable, or to control an actuator. In this way, most of the developed solutions have their roots in instrumentation or communications electronics.

Electronic devices in the medical field are usually based on state-of-the-art technologies in order to maximize benefits. They make use of the latest developments in terms of micro/nanoelectronics, micro/nanosensors, smart material technologies, device integration, device packaging, smart systems, etc. However, beyond all the electronics, most intelligent devices will only succeed if they are also designed to comply with the requirements of the software agents that are used to monitor, detect, and provide alerts based on the measured health condition.

Another important requirement is the ability to operate as a clinical tool. Many electronic devices being developed that claim potential use in the medical field are far from this goal, because there is a big difference between developing a device to monitor the heart rate while jogging versus use in a clinical application. Before clinical use, a long process of certification has to be completed [74.2, 4, 40]. Simultaneously, many such electronics must also be reliable for operation inside a clinical environment, such as a hospital. As is well known, mobile phones are certified electronics devices, but they are not allowed inside many clinical facilities. This is also the situation for many wireless technologies.

Beyond the certification requirements, which are more demanding than in other fields of electronics, the development of electronics for medicine poses its own technical challenges, due to the various operating constraints. Also, legal and cultural aspects, that differ from country to country, have to be dealt with. Moreover, since the developed solutions are, usually, for use by medical doctors, it is also necessary to break the acceptance barrier for some new technologies, despite the certification. Finally, and not less important for many applications, the proposed electronic solutions must gain user acceptance, in particular if they are for use outside the clinical environment.

74.1.3 Medical Electronics Specific Requirements

At first glance, the field of medical electronics may not seem as challenging as it really is. Signal amplitudes and frequency ranges are not that demanding, the operating range is not that large, the operating voltages and currents are, in general, normal, etc. This leads to the use of standard amplifiers, filters, analog-to-digital converters (ADCs), digital-to-analog converters (DACs), microprocessors, microcontrollers, etc. So, what are the problems when designing medical electronics? The problem lies, as in all other electronic fields, in the combination of the requirements specific to such devices.

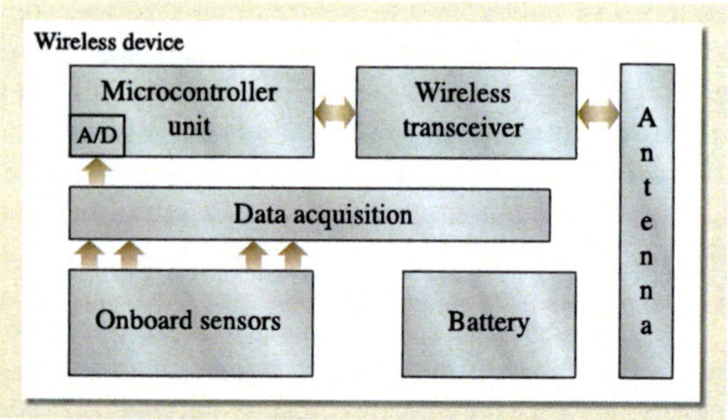

Fig. 74.1 Main electronic components required for advanced clinical data acquisition

Figure 74.1 shows a schematic of the electronics involved in a typical solution for a biomedical device. This represents a solution for an advanced device, meaning that only the most sophisticated devices require all the modules represented. In other cases, only a subset of such modules will be required; for example, not all electronic modules used in medicine require use of a microcontroller or a bidirectional wireless link.

Bearing in mind the discussion above, a set of challenges may be identified when developing electronics for medicine. These challenges may be found in the areas of data acquisition, filtering, power autonomy, and communications. Furthermore, for AAL applications, it is desirable that devices may be easily worn by the user, leading to miniaturization and wearability requirements. This has also implications in the fashion field, because many people do not like to use electronic devices because they are not fashionable.

The challenge at the data acquisition level results from the combination of a set of requirements particular to this type of systems. The frequency range of interest includes the 50/60 Hz frequency used by power lines. This is a significant problem at various levels. When designing power supplies for operation with these devices, special care is required since the direct-current (DC) voltages are obtained from a 50/60 Hz alternating-current (AC) voltage that will introduce a very strong oscillating component into the supplied power. This interfering component may easily become larger than the signal to be measured, and the minimum signal amplitude that can be measured will be increased. To avoid this, either very careful design of the power supply is required, or a battery is used as the power supply. Since many systems are designed for use on-body and in-body, the battery solution solves many of these problems. However, power lines may also be coupled to the human body, and when measuring biopotentials, interference is still a problem even when using batteries as the power supply. It is very easy to have a few millivolts superimposed on the measured biopotential due to the power lines, which is of the same order of magnitude as the biopotential itself. One possible solution is to use a notch filter to remove this component. However, this is a challenge because such filters must have a very high quality factor in order to remove only the 50 /60 Hz component, because the signal components of interest are around this frequency, and they must have also a high attenuation factor. Attenuation of the order of 30–40 dB is sufficient for most applications, although some may benefit from values up to 70–80 dB. However, this high attenuation introduces a new challenge, namely on how to obtain such devices with very low noise factor. Most designs with off-the-shelf components only enable devices with noise on the order of a millivolt.

Another problem during signal acquisition is associated with subject activity, which has frequency components inside the frequency band of interest, introducing so-called movement artifacts. Since these slow components are inside the band of interest, they cannot simply be removed by standard low- or high-pass filtering. More complex filters, such as adaptive filtering, must be used to remove these components.

At the acquisition level, we have another important aspect, especially for biopotentials, which is the well-known amplifier common mode rejection ratio (CMRR). Since the subject under measurement is at a different potential with respect to the power supply, often the output signal may be a few millivolts, even when both inputs are at the same potential (common potential). The standard solution is to connect the subject to a reference potential, thereby lowering the common mode voltage and reducing its effect on the recorded signal. However, this solution requires the use of an extra electrode, requiring also extra room to accommodate it, which is not desirable when designing miniaturized devices.

Power autonomy is also a very important issue and is easy to understand when we think about other power-limited devices, such as mobile phones. When dealing with medical electronics, this issue may become more complicated, since some devices are required to operate inside the human body for a long period, as happens with pacemakers. In terms of power autonomy, we may divide the systems into three categories: systems to go inside the body (such as endoscopic capsules), systems to operate inside the human body (such as pacemakers), and systems to operate on (outside) the body (such as wearable devices for AAL). Usually, when recording a single channel [electroencephalogram (EEG), ECG, etc.], it is not difficult to find a solution that operates on a small battery for a few days. The problem is when it is required to record many channels (such as for EEG in epilepsy applications), or when it is also required to actuate another device to obtain the required reading (e.g., a sphygmomanometer). When readings from many channels are required, one option is to design all the electronics based on low-power approaches. The other option, which is also valid for situations where actuators are also present, is to integrate energy harvesting/scavenging technologies. Despite some promising proposals, such approaches are not yet available as solutions to fully power a complete system. However, they may be a potential solution to extend the autonomy of existing systems, and in the near future it is expected that they may replace batteries in some applications.

We have also to consider communications, in which regard the desirable solution is usually to use wireless communications. However, this introduces new challenges, because this subsystem is typically the most power hungry. To reduce the power consumption of this subsystem, an as efficient as possible power amplifier design must be used, and antennas should also be selected as efficient as possible [74.41,42]. The challenge regarding antenna design results from the operation of wireless systems in low frequency bands in order to decrease the power required for transmission, but which degrades antenna efficiency because they must be electrically small to fit inside the space available in the device.

74.2 Electronic Sensing

Electronics in medicine may include many applications, as mentioned above; for example, ultrasound and MRI imaging systems are often used in diagnostics, as are ECG recording systems, all of which look very different at a high level. However, looking in more detail, all of them are designed to sense physical variables to aid clinical diagnosis. In this way, in most medical electronic devices, the requirement is to design adequate instrumentation electronics to interface with a transducer that is used to convert a physical variable of interest into an electrical signal, which is finally used to help in the diagnostics.

The success of electronics in medicine relies on their ability to help medical doctors in the determination of a set of parameters that are compared against a range of values considered as healthy. After beating the challenge of determining which parameter is important to measure, and in which range it must be, it is up to the medical electronics to provide solutions to perform these measurements reliably and in a solution that can be easily deployed. In this way, instrumentation is a very important field when dealing with medical electronics. Many medical devices deal with the measurement of specific variables, which may be chemical, physical or electrical. These measurements are supported by appropriate electronics, as well by a transducing mechanism to convert, e.g., an ionic concentration to a voltage, proportional to it.

Some recordings may be obtained by direct measurements, as for measuring body temperature, where a temperature sensor is used. Alternatively, they may be obtained by indirect measurements, such as measuring breathing rate using a temperature sensor, where temperature changes of air flow are measured to determine the breathing rate. In this way, sensing techniques and sensors are also very important fields when dealing with medical electronics. They are essential, because without sensors there is no medical information to process, analyze, transmit or store.

74.2.1 Health Condition Monitoring

Health condition monitoring requires the use of a set of sensors to monitor a wide variety of signals, either clinical or environmental, forming a sensing component, for example, surface electrodes to measure bioelectric signals, together with the acquisition electronics.

The measured signals can be classified according to their nature and/or type of monitoring. Figure 74.2 presents a few different categories of signals that are commonly measured by sensors, as well as the main parameters assessed.

Physiological signals include bioelectric events and other biochemical and physical parameters that are crucial for assessment of the user's state of health. They can be detected by means of biosensors, which are a particular category of wearable sensors that have the ability to monitor biological properties and processes.

Each signal is important for capturing the entire status of the subject and the surrounding environment. On

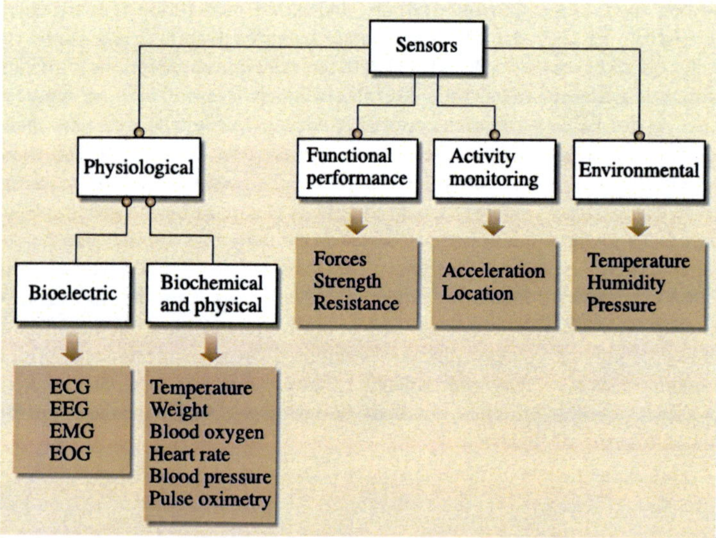

Fig. 74.2 Categories and types of possible biosignals measured to infer an individual health condition

top of such acquisition systems, use of *intelligence* is usually required so that the system can automatically detect any relevant health condition and, based on that, trigger alarms or send warnings to healthcare providers or the patient being monitored [74.19,43].

74.2.2 Biosignals and Transducers

Biosignals offer an interface between the human body and electronic systems, and they are directly related to the physiology and function of organs and organ systems, hence allowing evaluation of the health condition of a subject through their acquisition and monitoring.

An electronic system is used to measure physiological signals. Those signals may be electrical, or if not, they are translated to electrical signals using one of various transducer or translation methods.

Table 74.1 presents examples of the most commonly monitored biosignals for the human body, and their nominal values [74.44–46].

Table 74.1 allows a few conclusions to be drawn regarding the signal and transducer characteristics involved in the process of health monitoring. A few that are important to highlight are: signal amplitude, frequency, power, and the signal translation mechanism.

Signal Amplitude

Bioelectric signals are not difficult to measure from this point of view. From Table 74.1 we can see that the most demanding signals are in the μV range, which, when compared with the radiofrequency (RF) signal at an antenna input (in the range of pV), is very high. When looking for nonelectrical signals, it may also be observed that both signal amplitudes and their dynamic ranges are not very challenging.

One main problem related with signal amplitudes lies in the fact that different signals are in the same range of amplitudes and frequencies. This is a problem since some signals tends to act as interference to others, thus appearing as so-called artifacts. As an example, electrocardiography (ECG) or electrooculography (EOG) signals usually appear overlapped on electroencephalography (EEG) signals. Even simple patient movement, which occurs on the order of a few Hertz, interferes with signals such as ECG, respiration rate, and blood pressure.

Signal Frequency

From this point of view, physiological signals are also not difficult to measure. The maximum frequency is on the order of a few kilohertz. However, since many of them are in the same frequency band, as mentioned above, they tend to interfere with one another. Another common problem is electromagnetic fields coming from power lines (operating in the frequency range 50–60 Hz), which are easily coupled through the power source (that may also introduce a 100 Hz component due to the use of full-wave rectifiers), or because the human body acts as an antenna for those fields. Sometimes the coupled signal may have higher amplitude than the biosignal being measured. One solution is to use batteries to power some electronic devices, avoid-

Part G | 74.2

Table 74.1 Description and properties of commonly monitored biosignals

Biosignal	Range	Frequency	Sensor type
Blood flow	1–300 ml/s	0–20 Hz	Electromagnetic or ultrasonic
Blood pressure	0–400 mmHg	0–50 Hz	Cuff or strain gage
Blood oxygen saturation	95–99%	0–50 Hz	Paired infrared LED and photodiode
Cardiac output	4–25 l/min	0–20 Hz	Fick, dye dilution
Electrocardiogram (ECG)	0.5–4 mV	0.05–150 Hz	Skin electrodes
Electroencephalogram (EEG)	5–300 μV	0.5–150 Hz	Scalp electrodes
Electrooculogram (EOG)	10–100 μV	0–10 Hz	Skin electrodes
Electromyogram (EMG)	0.1–5 mV	0–10 kHz	Needle electrodes
Electroretinogram (ERT)	0–900 μV	0–50 Hz	Contact lens electrodes
Action potentials	−80–80 mV	10–10 kHz	Neuronal electrodes
Local field potentials	0.1–1 mV	0.1–500 Hz	Neuronal electrodes
pH	3–13 pH units	0–1 Hz	pH electrode
pCO_2	40–100 mmHg	0–2 Hz	pCO_2 electrode
pO_2	30–100 mmHg	0–2 Hz	pO_2 electrode
Pneumotachography	0–600 l/min	0–40 Hz	Pneumotachometer
Respiration rate	2–50	0.1–10 Hz	Impedance, electroactive
Temperature	32–40 °C	0–0.1 Hz	Thermistor
Activity		0–100 Hz	Accelerometer, GPS, Wi-Fi
Body weight	0–150 kg	1 time/day	Weighing machine
Glucose	70–180 mg/dL 3–8 mmol/L	1–5 times/day	Glucose meter

ing interference from the power supply. To remove the effect of interference picked up by electromagnetic coupling, the solution is to use an instrument amplifier not with an input impedance as a high as possible but rather designed to *kill* the 50/60 Hz component. However, the bioelectric signal has limited power. Another problem, due to the low frequencies involved, is the requirement for high capacity values, leading to large capacitors for filtering blocks. This has implications in terms of device dimensions and response time for the transitions of such systems.

Signal Power

Many of the important signals that need to be measured are biosignals that originate from body electrical activity. One characteristic of such signals is their small amplitude. However, that is not the main problem with their acquisition; the main problem is their low power. Figure 74.3 shows a typical acquisition system used for bioelectric signals. Figure 74.3 shows the bioelectric signal source, the electrodes (which are responsible for the interface between the ionic human body system and the electronic system), and a high-input-impedance amplifier, usually an instrumentation amplifier. Figure 74.4 shows the equivalent electric circuit.

Since the source current is on the order of pA, the available power is also very small. This is a problem when implementing the proposed solution for power line noise cancelation. The interference is canceled, but the biosignal is also attenuated. This means that any small current flow to the instrumentation amplifier will lead to a voltage drop on the electrodes, which will make the available voltage even smaller and more difficult to measure.

Signal Translation

When the biosignal to be measured is not an electrical signal, it is necessary to use a transducer to obtain an electric signal proportional to the physical variable being measured. Oximetry provides a signal obtained

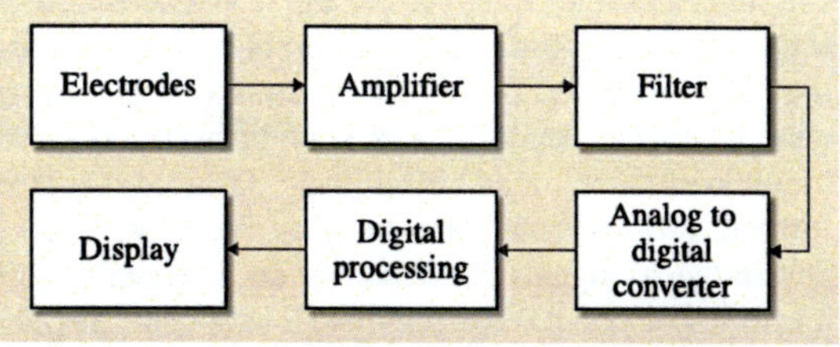

Fig. 74.3 Typical acquisition system for bioelectric signals

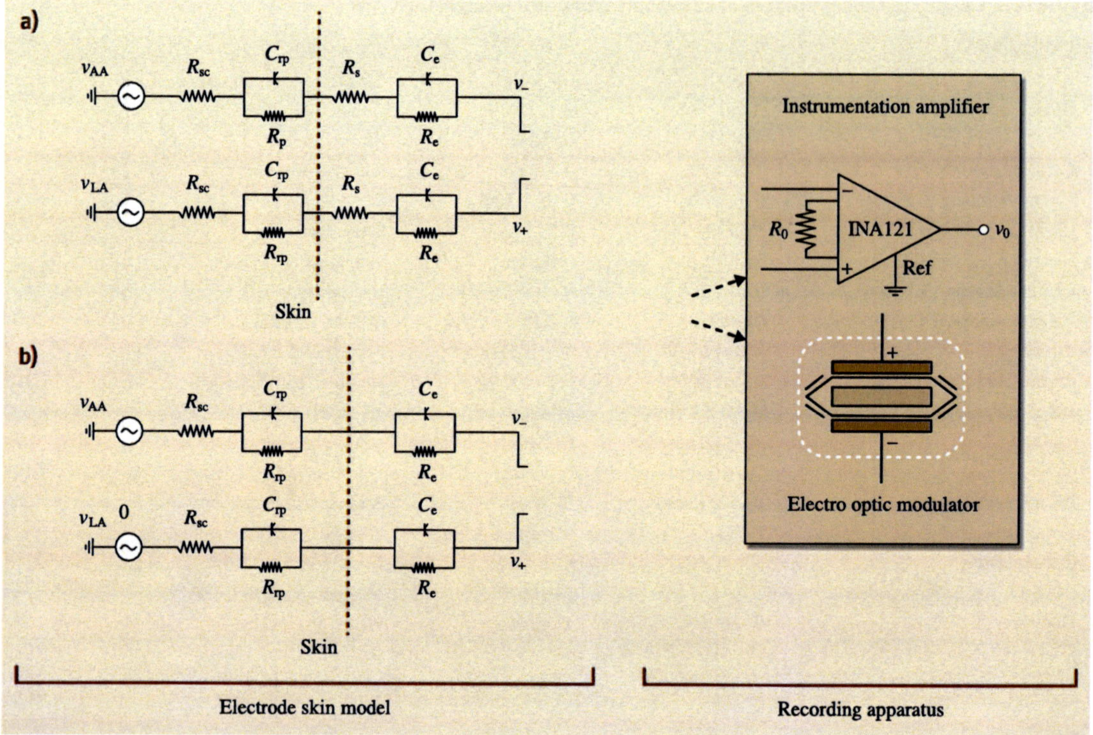

Fig. 74.4a,b Equivalent circuit for biopotential transduction using **(a)** wet/dry electrodes and **(b)** capacitive coupled electrodes

from measurements of light, spirometry results in a signal obtained from air flow, and the respiration rate may be obtained from temperature variation or by abdominal volume variation. These are only a few examples, but one key aspect may be highlighted: most of these devices require also an actuator and they tend to be bulky. In this way, measurements become more difficult when portable devices are required, and even more difficult when wearable devices are considered. When portable or wearable devices are required, often it is difficult to find a solution that is reliable enough and at the same time acceptable for use by the patient.

74.3 Electronics for Wireless Health Monitoring

Wireless health monitoring systems (WHMS) are among the most promising applications of wireless communication technology, and they have been a very active research topic with considerable advances during the last decade. These systems consist of groups of wireless-enabled sensors attached to the human body in a noninvasive or minimally invasive fashion, together forming a wireless body-area network to acquire diverse physiological signals (also known as biosignals) and transmit them via radio to a remote location, for display or monitoring to enable assessment of the user's health condition. Thorough reviews and surveys on WHMS can be found in [74.47–49].

The development of WHMS envisions the improvement of the quality and efficiency of healthcare, which is currently facing increasing demand and, consequently, increasing global costs. The tendency for increasing healthcare demand is mainly motivated by the ageing of the world population, which is expected to cause an increase in the incidence of chronic diseases and physical disabilities [74.47]. According to the United Nations, the world population is currently enduring an unprecedented and irreversible ageing process, which will cause a growth in the proportion of elderly people (over 60 years old) to a third of the population in developed countries, and to 22% of the total world population by the year 2050 [74.50]. Therefore, there is a growing need for health monitoring solutions to provide more efficient use of healthcare resources to caregivers and better quality of life to the elderly or diseased, relieving excessive pressure on healthcare demand and costs.

Use of wireless technology in healthcare provides new means for close observation of the health condition of patients. Typically, observation of a patient takes place when he feels some sign of illness and reactively seeks healthcare. Alternatively, his health condition may be followed by a caregiver, requiring the patient either to visit a healthcare facility or to stay there in case of prolonged examination or intervention. This leads to time and location constraints for the patient and can only provide limited periods of observation of his health condition, which may mean that diagnosis of health complications or pathologies may be late or overlooked. Moreover, conventional health monitoring equipment uses wired connections to the patient, hindering the activity of monitored patients.

On the other hand, WHMS inherently provide mobility to the patients, allowing them to be monitored, virtually, anytime and anywhere within the range of the wireless network. Hence, WHMS gives patients the possibility to be monitored in a convenient location, typically at home, with reduced disturbance to their daily activities, helping them to adopt a healthier lifestyle and improving their quality of life and independence. In addition, WHMS also offer the possibility of continuous health monitoring. Following the current trend towards miniaturized and low-power-consumption electronics development, WHMS can be used for long periods of time (e.g., 24 h), providing approximately continuous evaluation of the health condition of a subject. Furthermore, continuous health monitoring is a very useful feature for enhancement of preventive care and quality of treatment. By monitoring a subject continuously, any slight changes in health condition in relation to the long-term circadian pattern can be detected at an early stage by the caregiver, allowing for better decision-making and anticipating the occurrence of severe health complications. Thus, continuous health monitoring offers better treatment possibilities for patients suffering from chronic diseases or enduring rehabilitation processes. Subjects without illness may also be continuously monitored to receive feedback on their activities and modify their behavior to prevent illness by the promotion of a healthier lifestyle.

74.3.1 Requirements for Wireless Devices

Even though WHMS have the potential to provide great benefit in healthcare, they have not yet matured to a point of general use, as their application to real-life situations and acceptance by caregivers and patients is hindered by a considerable number of application and user requirement issues, including the following:

Ergonomics and Wearability

A WHMS needs to be as small and lightweight as possible so that the user feels comfortable when using it over long periods of time, and it should not affect the patient's daily activities. Ideally, the WHMS should go unnoticed by both the user and others, as the users of WHMS may become stigmatized if it is perceivable. An appropriate solution is integration of the WHMS into textile garments, resulting in wearable or smart-textiles monitoring systems [74.49, 51].

Low Power Consumption

WHMS, as a mobile application, are required to include their own power source, which is typically in the form of a battery, although energy harvesting solutions appear as a possible alternative in the future [74.52]. The power sources of these devices can only offer limited operation time, and therefore it is essential to reduce the power consumption to minimal levels, to ensure appropriate monitoring periods.

Quality of Service (QoS)

The wireless communication protocol implemented on the WHMS must include a proof mechanism to ensure that the biosignal data are reliably delivered to the base station, especially for continuous monitoring and if an emergency situation is detected.

User Friendliness

WHMS operation should be autonomous but perceivable by the user of the system, either the patient or healthcare personnel, and should require as little maintenance as possible, in order to achieve user acceptance.

Privacy and Security

The health condition of a patient must be regarded as private information, therefore data encryption methods and protection against unauthorized access should be implemented.

WHMS are expected to be a valuable tool to address the effects of population ageing on healthcare. They provide benefit to both patients and caregivers, enabling improved quality of life and independent living for patients and improved efficiency and better use of healthcare resources to caregivers. However, the design of such systems is a challenging task, and solutions able to reach wide acceptance by patients and caregivers are yet to be implemented.

74.3.2 Data Acquisition

Acquisition of biosignals from the human body requires similar instrumentation and methods as for any other signal which requires transducers. However, biosignals, namely those which are electrical, typically have small amplitudes and bandwidths and hence are very susceptible to interference or artifacts, which can easily be superimposed on the biosignal to be monitored. Moreover, part of the information contained in the biosignal may not be readily available from the raw signal, requiring further analog or digital processing for feature extraction [74.53]. Therefore, the design of acquisition electronics for biosignals must contain suitable instrumentation to ensure the integrity of the information contained in the biosignal, while any forms of interference or artifacts must be minimized, so that the acquired biosignal is useful for health condition monitoring or diagnostics, since a corrupted signal could provide misleading or meaningless information. Figure 74.5 shows a generic block diagram of instrumentation for biosignal acquisition.

Sensors, represented in the first block, are responsible for the conversion of clinically relevant physical quantities into an electrical output that can be processed by electronic instrumentation. Different sensor types are used according to the physical quantity associated with the biosignal. For electrical biosignals, it might be presumed that sensors would not be required since an electrical property, namely current flow on the human body, is measured. However, transducers are indeed required for acquisition of electrical biosignals, since the current flow on the human body is due to ion transport whereas current flow in the acquisition instrumentation is due to electron movement. This conversion is provided by biopotential electrodes [74.53], which are placed on the surface of the human body, as close as possible to the organ that generates the biosignal, such as the heart (ECG) or brain (EEG). For nonelectrical biosignals, various types of sensors are used depending on the nature of the biosignal, such as mechanical (e.g., sphygmomanometer for blood pressure monitoring) or thermal (e.g., thermocouple for temperature measurement) devices, and the locations of sensor placement vary according to where the signal is strongest or the measurement most accurate.

Following the sensor stage, the signals obtained are usually weak in amplitude and may have considerable levels of added interference, as mentioned above. Thus, analog processing, featuring low-noise amplification and filtering, is required to increase the amplitude of the acquired signal, match the input range of the analog-to-digital converter, and limit its bandwidth to the range of interest of the monitored biosignal. Care must be taken when designing or choosing amplifiers for analog processing of biosignals. Desirable features for a biosignal amplifier include [74.45, 54] high input impedance (about 10 MΩ to minimize the current drawn from the monitored signal and hence signal distortion), low output impedance (to provide as much power as possible to the load), high differential gain (10^3 or greater), high common mode rejection ratio (to minimize electromagnetic interference and electrical noise), and frequency response temperature stability. Filtering of the biosignal is usually required, since biosignals are susceptible to interference and artifacts. Filtering of biosignals can be performed in either analog or digital fashion. Analog filters are implemented using modified amplifier stages that attenuate unwanted frequency components from the acquired biosignal while provid-

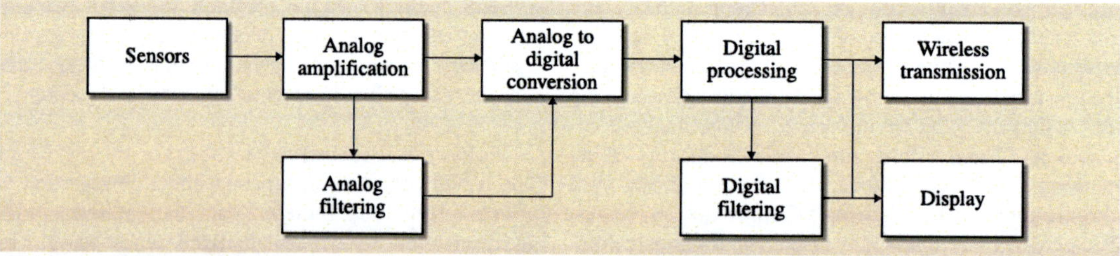

Fig. 74.5 Typical block diagram of the necessary instrumentation for biosignal acquisition

ing gain to those which are relevant, thus contributing to removal of low- and high-frequency noise. If necessary, electromagnetic interference from power lines may be further attenuated using notch filters centered at 50 or 60 Hz. Besides, analog filtering is also used to limit the bandwidth of the monitored signal according to the sampling frequency of the analog-to-digital converter, thus avoiding aliasing. Although analog filters are essentially used to remove undesirable signal features, their design must ensure that the filter itself does not cause distortion of the relevant frequency components of the monitored signal. To minimize signal distortion, analog filters should be designed to provide a constant modulus and linear phase frequency response over the pass band [74.45, 53].

After the analog filtering stage, the monitored signal is converted to digital format using standard sampling and quantization methods such as sigma–delta or successive approximations [74.55]. In digital format, biosignals can be handled by digital processors, such as microcontrollers or microprocessors, and be further processed using digital filtering techniques. Digital filters offer advantages over analog filters, as they can be easily modified by changes in programming code, as opposed to changing components in analog designs, and they do not suffer from component tolerances or environmental influences such as temperature. Still, digital filtering and processing techniques require that the processor has enough available computational power, depending on the complexity of the algorithms to be executed by the processor. This may not be the case when the processor is already performing other tasks, such as control of acquisition or communications.

Lastly, the biosignal can be displayed to physicians and users, as occurs in common hospital instrumentation (e.g., intensive care monitors) or, in wireless health monitoring systems, sent wirelessly to a remote location either in real time or offline, to allow observation and monitoring.

As described above, signal acquisition appears to be a straightforward procedure. However, many problems can be encountered when performing data acquisition from biomedical devices for use in clinical applications. Many of these problems are due to the fact that the subject under observation is moving, particularly when considering wireless devices. This causes changes in the measurement conditions. These measurement conditions include the registering positioning of the transducer and changes in body impedance due to movement, chest movement due to arm movement, as well as breathing. For clinical applications, data acquisition may often be conditioned by the requirement that the measurement device must be applied by a caregiver, and not by someone without medical training. Otherwise the signals have no clinical significance.

Signal acquisition may also be conditioned by requirements from the medical side and not from the signal side. In this way, it is a practice to show the acquired signal to a medical doctor and check if the signal *is good enough for diagnostics* and not look at the time and spectral content of the signal and decide all the details based on that. It is difficult to find specifications for the desired characteristics of signal acquisition. For many biosignals, it is therefore hard to specify the sampling frequency, number of bits for digitalization, resolution, signal latency, delay between signals, etc.

74.3.3 System Integration

When developing electronic devices to give mobility to patients, it may be attempted to achieve different degrees of mobility. A system may allow the patient to move inside a clinical facility. In this case, the device must be designed to be portable, so the patient can move around the facility. Alternatively, the system may allow the patient to be monitored at home. This is similar to the previous case, but with additional requirements. Such a system must have a set of alarms to inform the patient that something is not right (not only in terms of their medical condition, but also that something is wrong with the equipment itself or its use). Moreover, since the user may be required to place the equipment himself, it should be planned in such a way that the patient is able to place it without the need for caregiver assistance, otherwise the equipment will not be used or will be used improperly. Also, since the equipment may be required to stay at the patient's home for some time, desirably, it must be designed so that it can be cleaned at his home. Also, it must be more resistant to potential crash accidents. Another approach is to design the electronic device so that the patient can wear it as if it were just another piece of clothing.

Regarding wearable devices, a few different approaches may be adopted. One solution is based on discrete modules that are placed at different locations where signals need to be measured. However, for some applications, this may not be desirable; for example, to record an ECG, it is necessary to have physical connections between the different leads and the instrument's amplifier, and this solution implies the use of a few wires on the patient body. Moreover, with this solution, it is difficult to clean and control correct placement of the different measuring devices. The other solution is to design the sensors to be integrated with clothes, with the electronics to be attached to those sensors at a conveniently location. This implies integration of wires with textiles and the development of interconnection technologies to interface wires and electronic devices on textiles. Another approach is to embed all the electronics, wires, and sensors on a polymeric foil, leading to a smart sheet of material with the ability to record the desired signals. Still using the integration approach, instead of electric methods, optical fibers with optical sensors may be integrated, thereby reducing the problem of integrating wires and electronics [74.56].

For wearable devices, the system should be developed taking into account that the user must be able to apply it himself. No help will be provided from any health professional. In this way, it must be developed in such a way that, after putting it on as any regular piece of clothing, the system should only require intuitive adjustments to achieve correct operation. Moreover, since it is something to wear, it must be able to undergo the same cleaning processes as regular clothing.

74.3.4 Wireless Communications

As in many other scenarios, wireless communication is a desirable solution for sensors used for data acquisition in biomedical applications. It is possible to distinguish three main situations: communication between devices inside the human body (intrabody), communication between devices on the body, and communications between devices on the body and the outside world (body to air).

A few aspects may be considered when implementing wireless communication for these different situations. For devices inside the human body, the available power of those devices is very limited, it is very difficult to recharge them from outside, and RF losses are very high since the wireless medium is highly conductive. This implies that the link must be made at low frequencies (kHz to MHz). This is a problem because, at these frequencies, antennas tend to be very large or, if their size is reduced to fit inside the devices, antenna efficiency drops.

For devices on the body, one main characteristic is operation of the device very close to a highly conductive medium. This has implications for antenna design and signal propagation. Antennas must be designed so that the signal propagates close to the human body, otherwise power is wasted, and it may be required to operate in the antenna near field, as opposed to the standard operation in the far field. It is difficult to predict where devices must be placed to obtain an efficient wireless link.

In the latter two scenarios, a module with limited power may be used if easily rechargeable. The main challenge is to hide the antenna on the user environment while maintaining device operation. Solutions such as placing monopole antennas on the user's hat, which may work, are not desired since they are not fashionable. Many small modules are already sold with small antennas on them and seem to be very convenient for direct use. However, these antennas are designed to operate in free air, and when they are placed close to the human body, their operating frequency and bandwidth, as well their radiation diagram, will be modified.

When dealing with wireless communication, one main problem is to guarantee that data are delivered with sufficient quality of service (QoS) for the desired purpose. It is already possible to find many devices that transmit the measured data wirelessly, but it is very hard to find one that can be used for clinical applications. One problem is the lack of solutions to guarantee QoS for biomedical wireless links. Since data are to be used for clinical decisions, they must be obtained through solutions that guarantee that acquired data are correct and complete, otherwise patient health may be at risk.

74.4 Power Supply

All electronics and sensors require a power supply for operation. Electronic devices in medicine must have a very small form factor, including the power supply. In this regard, it is necessary to use batteries with very high charge density. However, despite all the solutions available, there is always a need to recharge the batteries. Batteries for wireless wearable devices may operate for 1 day or for 1 week, but, as for cellular phones, recharging will always be required for system operation.

Several wireless AAL solutions for monitoring have been proposed in the past few years, e.g., [74.48, 49]. Most of them are based on standards for communications on wireless local-area and sensor networks, such as Wi-Fi (802.11), Bluetooth (802.15.1) and ZigBee (802.15.4), because of the availability of commercial modules that can be customized for any wireless application. However, a few solutions that make use of other unlicensed industrial, scientific, and medical (ISM) bands exist. Thorough reviews on wireless sensor networks and related standards can be found in [74.17,57].

Since wireless monitoring is required, those systems must operate on battery power, or they must use an energy harvesting scheme [74.58]. Either way, system power consumption must be kept as low as possible. This leads to the design of systems where power consumption must be optimized. This is done by selecting the most power-efficient components available. If energy is consumed carelessly, the wireless acquisition system may rapidly become completely useless due to lack of available power. To prevent such failure, energy should be carefully saved using different approaches; for example, if the patient is in the normal state, then the sampling rate of sensors can be reduced to save power, or if the battery charge becomes low then its energy should be reserved for more vital tasks; i.e., the monitoring activity should adapt in accordance with the patient's clinical state to save energy. It should be realized that, to save further energy, communication protocols should be simple, and data should be aggregated, compressed, and transmitted in loaded packets, since computation demands much less energy than transmission. However, attention must be paid to the delay, which tends to increase linearly with packet length.

74.4.1 Battery Power Budget Considerations

Since the battery is the main power source for wireless devices, it should be selected to fit the requirements of the electronics which it will supply. The main requirement is its ability to power the electronic device for as long as possible, always considering a minimum period that is acceptable for operation. In many applications this minimum period should not be less than 1 day. After battery autonomy, and many times as important, come the issues of battery dimensions, format, and conformability.

However, other issues must also be taken into account when performing this selection. One is the recharging time, which must be appropriate for the application, since many devices use embedded batteries, which are difficult to replace. This time is usually acceptable if it is below 3 h, the time that the device may not be used. Also, due to battery lifetime, often it is convenient to use a battery with a protection circuit, which prevents the voltage from dropping below a safe value. In this way, the recharging is faster, at lower currents, and the device will have increased lifetime. Another parameter that is used for battery selection is its nominal current. However, its ability for peak currents also must be taken into account. Higher values of currents may be required, e.g., when a module has wireless communications. When the wireless link is in transmission mode, peak currents may be required. Finally, the operating temperature also interferes with battery lifetime.

Notwithstanding attempts to follow the previous guidelines, battery selection may be a hard task to accomplish, first because often it may not be easy to know beforehand the real power consumption of the device. Power consumption depends, e.g., on the data rate, operating mode (sleep, standby, transmitting, receiving), error rate, antenna efficiency due to antenna positioning, and power level of transmission [74.59]. This implies that tests have to be carried out under real scenarios, or that the battery has to be overspecified.

Another problem is that it is only possible to determine the power consumption accurately when the system is already assembled and ready for operation, where predicted values of system autonomy may diverge from those expected [74.59]. This may be due to several factors, including incomplete manufacturer specifications, only standalone power consumption being taken into account, and power consumption dependence on sampling rates, clock frequency, etc.

Figure 74.6 shows three modules (Wi-Fi, ZigBee, and proprietary) designed with off-the-shelf components that were used for power autonomy testing purposes. The measured current value for the Wi-Fi module, during transmission, exceeds that expected by about 60 mA. Also, in sleep mode, the measured current was 1.5 mA, far above the expected value of 6 μA.

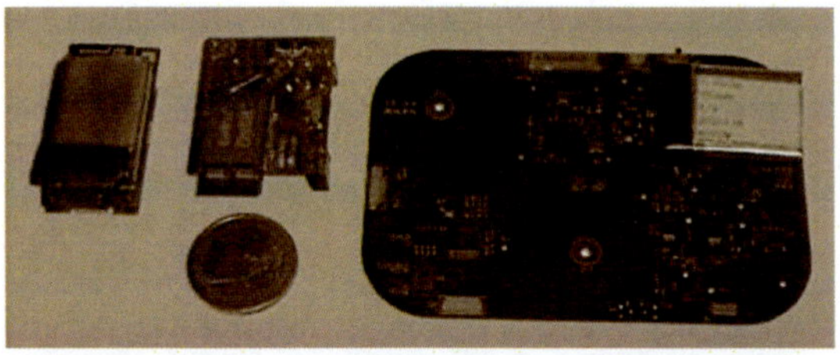

Fig. 74.6 Wireless modules used for autonomy testing: Wi-Fi (*left*), ZigBee (center), and proprietary (*right*)

The measured values of the ZigBee module were similar to the values specified in the datasheet. The ZigBee module presented higher consumption than the proprietary module but lower than the Wi-Fi module. The measured current consumption of the proprietary module was slightly higher than the value specified in the product information. The proprietary module had the lowest power consumption, and the measured power consumption was close to the predicted value.

From the power measurements, it is also evident that it may be worthless to design extremely low-power acquisition electronics, sometimes sacrificing acquisition performance, when the main power consumption may be due to the wireless subsystem that, even for the lower-power solution, is one order of magnitude higher.

Finally, it was verified that all the devices were able to work on a very small battery for periods of 1–3 days, which we believe is acceptable for AAL solutions. The Wi-Fi device, as expected, was more power hungry; however, it transmits at much higher power levels, and the data rate is also much higher. Since all microcontroler units have similar power consumption, it is expected that, if the RF power consumption is reduced, the overall Wi-Fi system consumption of this device may be significantly reduced.

74.4.2 Wireless Power

Most medical electronic devices requiring the use of batteries will need to be recharged. The most desirable solution would be one where the user does not need to pay too much attention to its use. The easiest approach is to use a wireless power link, and even more interesting would be that the device may be recharged without taking it out. This is desirable because it is uncomfortable to remove and replace the device, or because continuous monitoring is required. Another benefit of such solutions is power cable removal, allowing for easy fulfillment of requirements such as devices without connectors or that are completely sealed. Moreover, it allows the development of innovative solutions, such as recharging devices that are inside the body, or to recharge devices as we sleep. For short distances, two solutions may be used for wireless recharging: capacitive or inductive. It is also possible to use microwave links or optical links, or other forms of energy, but these options lie outside the scope of this chapter.

From basic transformer theory, we know that the voltage and current available at the receiver depend on the voltage and current at the transmitter. The inductive solution for wireless power transmission uses two inductors: the primary and the secondary. On the transmission side, a power amplifier is used to drive the primary inductor. This power amplifier must be matched to the inductor in order for the available power to be transmitted to the inductor and the maximum power to be obtained in the secondary. The inductor acting as a receiver at the secondary side is connected to a power rectifier, after which comes a circuit to control the battery charge. To obtain maximum efficiency, the receiving inductor should also be matched to the load. However, as the battery charges, the current usually decreases and the load presented to the inductor also changes. This implies that, for continuous matching, the matching circuit should change during the charging period.

74.4.3 Wireless Power Link Analysis

To understand the losses expected from this antenna solution, a wireless link was implemented and analyzed. An external (transmitter) subsystem consisting of a Collpitts oscillator was used to generate an AC signal to feed the power amplifier. The implanted (receiver) subsystem has an AC–DC converter for power recharging purposes, and the data link path is preceded by a notch filter for power link rejection [74.60]. In this way, it is possible to power the receiver as well as to receive data transmitted wirelessly. The wireless link was implemented using two coils in parallel with the following dimensions: width, 2 cm; length, 4 cm; thickness, 0.254 mm; number of turns, 18; distance between each turn, 0.254 mm (Fig. 74.7).

As the matching network for the inductors, four capacitors were used, two in series (4.7 nF) and the other two in parallel (2.2 nF), in order to avoid losses due to power mismatch in the wireless link. The match-

Fig. 74.7 Setup based on the inductive link antenna

ing network was trimmed with the help of simulation to optimize the efficiency of the wireless link. The wireless link was designed to operate at a distance of 3 mm between the two inductive coils. To investigate the link performance, measurements, simulations, and analytical models were used. Measurements were made for the inductive wireless link and compared with simulation results. The link efficiency was computed and then compared with the efficiency obtained from analytical models for other antenna solutions. Return loss measurements were performed using a vector network analyzer, enabling efficiency measurements. On the other hand, all simulations were done using Ansoft Designer SV software.

First, the inductive wireless link was characterized without any matching network and compared with the results obtained from simulations. The results showed high losses, but were made with the purpose of providing a way to design the matching network as well as to validate the simulations performed previously. Using the measured results as an input for the simulator, it was possible to design the corresponding matching network and compute the capacitors required for matching. Figure 74.8 shows the result of that matching.

It can be observed that the value of S11 was below −10 dB, meaning that the power is being injected, and S21 was −3.90 dB, meaning that 65% of the energy is not being transmitted.

After designing the matching network, the wireless link was measured again. The measured *S*-parameters are shown in Fig. 74.9. The efficiency measured for the inductive wireless link with real capacitors was not as high as that obtained theoretically, because of capacitor tolerances and quality factors. So, for 1 MHz, the value of S11 was −19.25 dB, which means that energy is entering port 1 (the transmitter), although the value of S21 is below the theoretical value (−5.02 dB), meaning that more energy is being lost (83.7%).

To test the effect that neural tissue may have on the performance of the system when used to recharge neuronal biodevices wirelessly, the validated model was used for testing the case when a slice of human tissue with thickness of 3 mm was placed between the two coils. Figure 74.10 shows the simulation results for such an inductive wireless link.

These results were not as good as the results shown in Fig. 74.8 for air, due to signal attenuation caused by the tissue. In this case, the dielectric properties were selected to match the neurological tissue behavior. The results of this analysis show that energy was being injected, because the value of S11 was simulated and tuned to be below −10 dB (−13.33 dB). On the other hand, as illustrated in Fig. 74.10, S21 was −4.51 dB, which means that 75% of the energy was being lost, which includes antenna and tissue losses.

74.4.4 Wireless Power Charger

Despite the losses expected when using wireless power transmission, there are situations where implementation of this option is required [74.61]. Figure 74.11 shows a prototype of a system developed to recharge

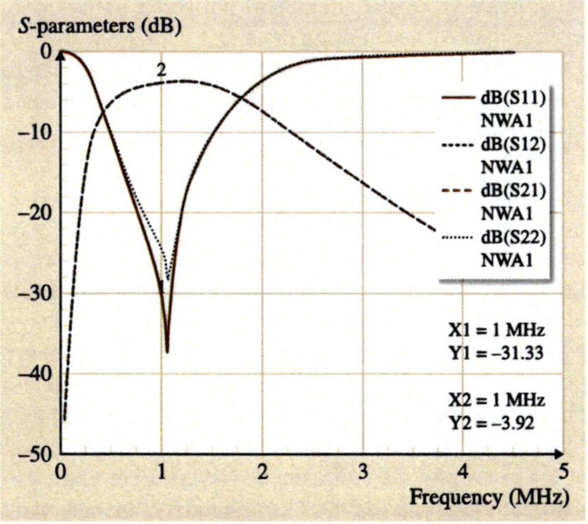

Fig. 74.8 Measured *S*-parameters and matching with ideal capacitors

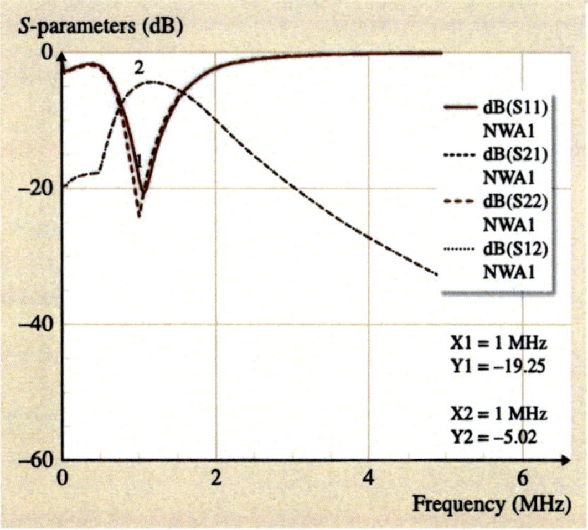

Fig. 74.9 Measured *S*-parameters for the full wireless link

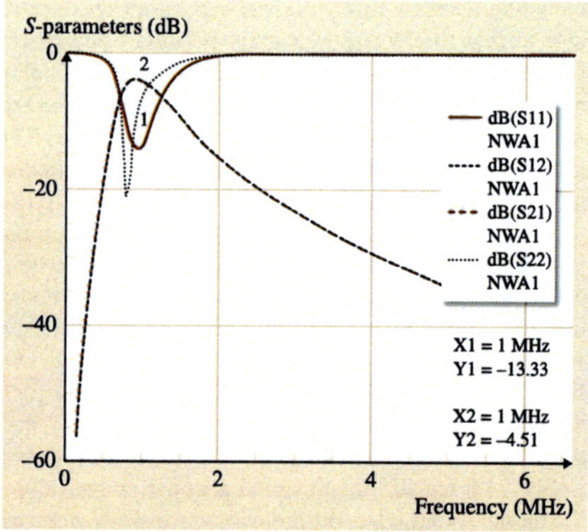

Fig. 74.10 Simulated *S*-parameters, with neurological tissue's properties

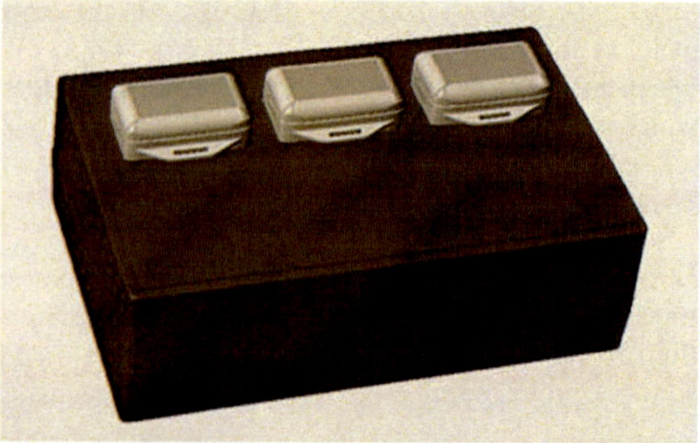

Fig. 74.11 Wireless charger device

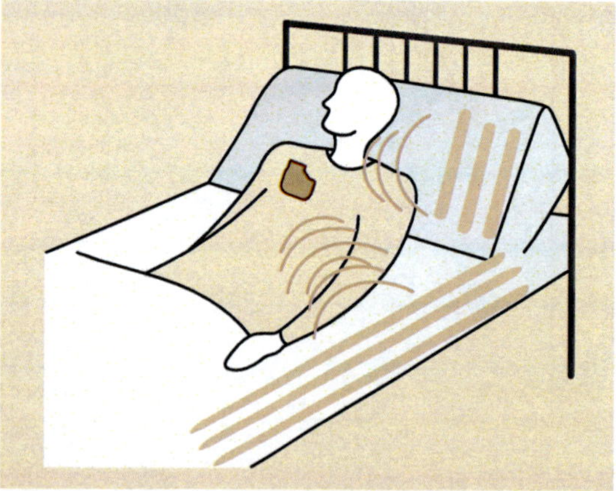

Fig. 74.12 Wireless power (power mattress and power pillow)

three wireless modules used for AAL applications. With the developed wireless link, it is possible to recharge the three modules in about 2.5 h. The modules have autonomy of approximately 20 h for their planned application.

The solution presented in Fig. 74.11, despite being effective, requires removal of the wearable device for recharging. A desirable solution would have the ability to recharge such devices when the user is performing some activity, close to a location where it is possible to place the charger. One solution that may be adopted is the use of a charger format that may be transformed into a *power pillow* or a *power mattress* (Fig. 74.12).

Another solution is to place chargers on equipment at the gymnasium, or on a desk or desk chair.

The idea behind these concepts is to provide a solution to recharge the wearable devices without removing them. Using this concept, the power is transmitted to the system's battery while the user is lying in bed, either to sleep or when at the hospital for diagnostics or treatment. However, these kinds of solutions, beyond the technical problems that must be solved before their implementation, require also deeper knowledge of the effect that the recharging electromagnetic field may have on user health.

Despite the many benefits envisaged by this technology, it still has a long way to go before it can be widely deployed. This technology suffers from a few problems, which are hard to solve, with link efficiency being one of the most important [74.62, 63]. To achieve efficiencies on the order of 70–80%, very special conditions must be met. Regular links will not reach above 50%, even for short transmission distances. However, this solution may be very interesting for applications where the charging device and wireless module may be placed close together.

74.4.5 Energy Harvesting

Another option available to power wireless devices is the use of some form of energy harvesting. Any form is valid, since it is possible to use that energy to charge a supercapacitor or a battery. Many forms of energy conversion may be used to benefit from the available energy sources: thermal, electric, mechanical or chemical [74.58]. Despite the many proposed solutions, they still cannot provide sufficient power to drive complete medical electronic devices continuously. In fact, it is already possible to find devices operating on this kind of power sources. However, they tend to work in very specific conditions and for very specific situations. The amount of power that can be obtained from these solutions is still below the requirements [74.58]. Moreover, they require very advanced electronic solutions. All the required electronics must be based on very low-power designs, and wireless transmission must occur at very low-power levels.

However, and notwithstanding these limitations, this may be a potential solution as an extra power source to help increase battery lifetime.

74.5 Wearable Medical Electronics

Electronics in medicine may be used on fixed devices, portable devices or wearable devices. In this section, electronics in medicine will be addressed from the perspective of wearable devices.

Wearable technology represents a new, emerging field with increasing potential in several aspects of modern healthcare, with greater importance at the delivery point of care services. As a result, remote and ambulatory monitoring inside or outside of the healthcare environment, and support for disabled, rehabilitating or chronically ill patients, is now a reality. The growing demand for wearable devices is being driven by the considerable need for preventive instead of reactive medicine, the global increase of health awareness, and the need for proactive personal healthcare on a daily basis.

Figure 74.13 show examples of the most commonly monitored biosignals on the human body [74.44–46].

One aspect that may be highlighted is that many important biosignals are electric signals, therefore they can be measured with similar instrumentation. On the other hand, different technologies must be available to monitor all the required signals, since they have different origins. Another aspect is that, to measure all the required signals, sensing devices must be placed in different body locations, which requires the development of specific solutions.

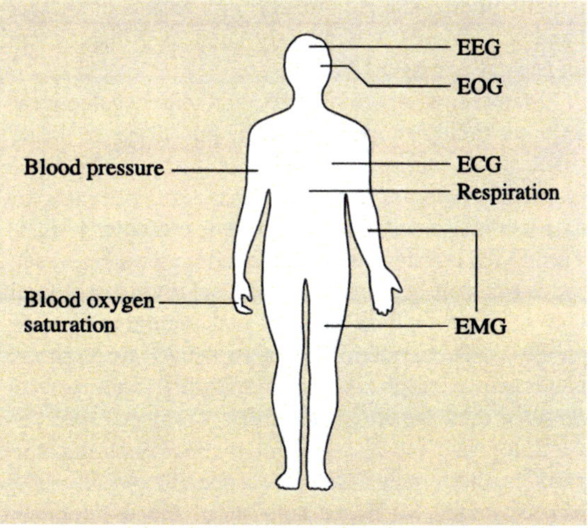

Fig. 74.13 Examples of noninvasively measurable biosignals on the surface of the human body

A wearable medical device can be described as an unobtrusive, autonomous, and ubiquitous system that supports continuous, multiparameter monitoring and treatment, and telemetric abilities [74.64, 65]. This contributes to a shift of health services from conventional hospital-centered towards individual-centered healthcare, which together with wireless technologies allows continuous feedback of relevant information to the user and/or clinical professionals. In addition, this approach improves early detection of and timely response to possible health threats [74.64, 66].

Since they are wearable, these devices are truly portable, being supported either directly on the human body or in a piece of clothing. This results in a new set of requirements, including minimal size, light weight, portability, unobtrusiveness, high levels of flexibility and integration, ease of use and comfortable wearing, increased lifecycle, and power autonomy. Such requirements can now be fulfilled, given the trends towards miniaturization of electrical and electronic equipment, as well as novel integration and sensing techniques. In particular, advances in sensing technologies allow monitoring and processing of two types of signals: vital signs such as cardiac activity, blood pressure, and temperature; and other important parameters that are more related to the patient's condition, such as body kinematics, sensorial and cognitive reactivity, and others [74.64, 65].

Due to the overall results of the advances in both the technology and healthcare sectors, it has become possible to establish a new healthcare paradigm: personalized health systems. These will enable decentralization of healthcare towards a system that will give the user a more proactive role in their care, providing better monitoring and feedback through a comfortable and unobtrusive solution. This not only contributes to improved quality of life, but also allows for preventive care by effective diagnosis at the earliest signs of disease. Likely to be of benefit to the chronically ill and disabled, wearable health devices are an attractive solution for patients undergoing rehabilitation, providing them with independent living, since they enable recording and collection of relevant data in the different situations of individual daily life [74.65, 67].

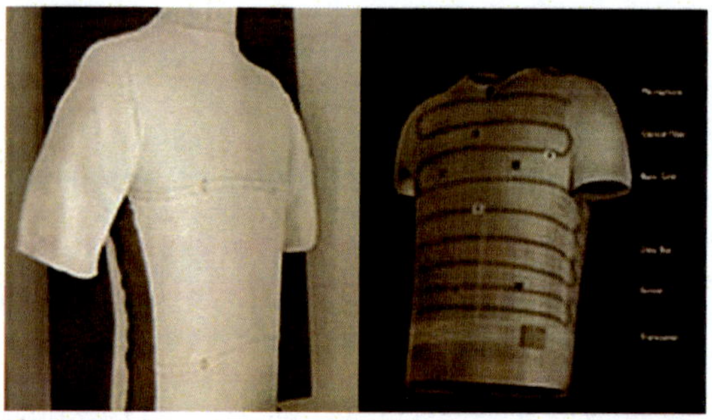

Fig. 74.14 SensatexSmart Shirt (Sensatex, Inc., Bethesda, MD)

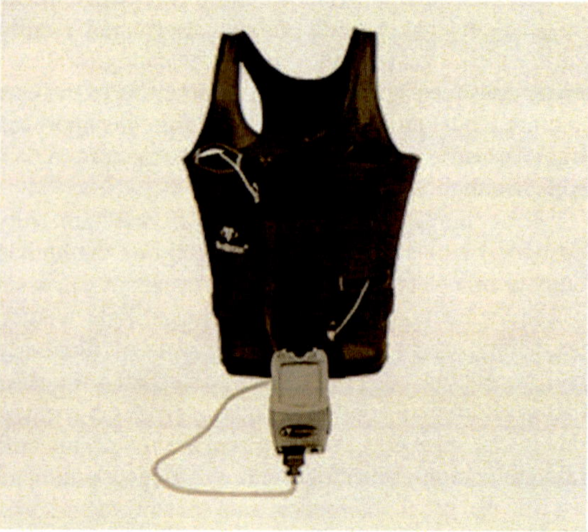

Fig. 74.15 Lifeshirt from Vivometrics (Vivometrics, Ventura, CA)

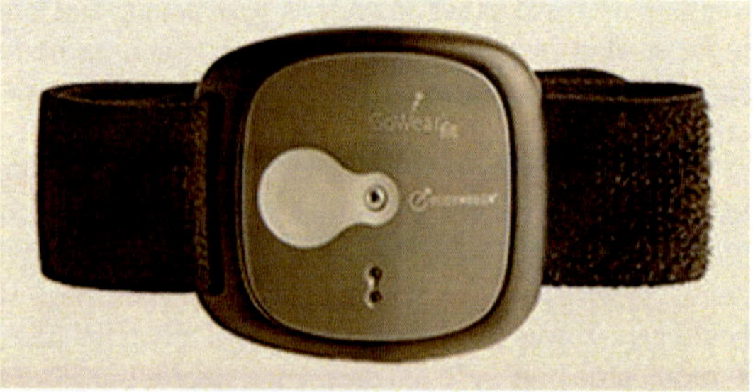

Fig. 74.16 BodyMedia arm band (BodyMedia, Pittsburgh, PA)

A number of wearable devices in the healthcare sector have emerged in the past few years, ranging from simple monitoring of daily routine, to miniaturization and integration of sensors to enhance the overall performance of wearable systems. The Georgia Institute of Technology (Atlanta) jointly with the US Navy proposed one of the first wearable solutions, consisting of a wearable vest embedded with optical fibers and sensors, working also as a data bus – the Georgia Tech Wearable Motherboard (GTWM) [74.68]. All the components are integrated into the fabric, creating a flexible device, which was manufactured essentially for use under combat conditions. The GTWM has the ability to identify the exact location of injuries or physical problems, helping to determine which of them needs immediate attention. This device was placed on the market by Sensatex, Inc., as a product named the Sensatex Smart Shirt (Fig. 74.14).

Companies such as Vivometrics and Bodymedia have also entered the market with wearable monitoring systems. The Lifeshirt (Fig. 74.15) is a product by Vivometrics and consists of a wearable physiological monitor in the form of a chest and shoulder strap, providing noninvasive ambulatory monitoring of pulmonary cardiac function and posture [74.69].

Bodymedia introduced a wearable and wireless body monitor that continuously collects physiological and lifestyle data such as heart rate, heart flux, body and ambient temperature, and galvanic skin response, among others. The Bodymedia Fit consists of an armband (Fig. 74.16), a wirelessly connected display, and an activity manager [74.70].

Many research groups have started to develop wearable technologies with main applications in health science. In the work entitled *AMON: A Wearable Medical Computer for High Risk Patients*, a wearable medical monitoring computer is presented [74.71]. The AMON system was developed under a European-sponsored consortium and consists of a wrist-worn unit with monitoring, data analysis, and communication capabilities. This system is mainly intended for high-risk patients in need of constant monitoring. The paper by *Paradiso* and *De Rossi* describes a project funded by the European Commission jointly with Philips and other partners, both industrial and academic [74.72]. This project, called MyHeart, consists of the development of functional clothes with on-body sensors and electronics to acquire, process, and evaluate physiological data. *Choi* and *Jiang* have developed a wearable sensor device in the form of a belt-type sensor head, which is composed of conductive fabric and polyvinylidene fluoride (PVDF) film, for monitoring cardiorespiratory signals during sleep [74.73]. *Pandian* and coworkers [74.74] have presented a wearable vest – the Smart vest – with integrated sensors for monitoring of ECG, photoplethysmogram (PPG), body temperature, blood pressure, galvanic skin response (GSR), and heart rate. In a more recent work, *Kanellos* and his coworkers developed a flexible optical-fiber-based pressure sensor comprising fiber-embedded Bragg grating elements [74.75].

The aim of this section is to introduce the enabling technologies behind the design of wearable devices in healthcare, as well as the basic requirements for their design.

74.5.1 Wearable Systems

The ongoing miniaturization of computing hardware and electronics as well as new smart materials and integration techniques have led to the development of noninvasive sensors and actuators for monitoring physiological and kinesiological parameters. These systems can include various types of small physiological sensors, transmission modules, and processing capabilities, facilitating low-cost wearable unobtrusive solutions for continuous all-day, any-place health, mental, and activity status monitoring.

A wearable system can be described as a smart/functional material with embedded sensors, processing and communication units, which may or may not contain attachable peripherals or other appliances. One can refer to this type of system as wearable electronics.

To design a wearable system, three areas of work need to be properly covered. First, it is important to develop unobtrusive wearable sensors to record physiological and kinesiological data reliably. Secondly, these sensors need to be implemented into a substrate material that allows multisensor integration. Finally, it is important to provide the system with acquisition and communication infrastructures, to enhance mobility of individuals, and some degree of intelligence to improve system performance at a clinical level [74.64, 65].

Figure 74.17 depicts an architectural layer for an ideal wearable health system, which is composed of a functional/smart substrate, embedded electronics, and attachable peripherals/appliances.

Wearable system components may be divided into three main categories: clothing, embedded electronics, and attachable components. The first includes all the substrate materials that act as a functional or smart structure by allowing embedding of sensors, signal processing units, communication infrastructures, actuating mechanism, and power generation, among others. The most common substrate materials used in these systems are textiles or flexible polymeric materials that will enable the design of garments with multifunctional nature [74.76]. The embedded components include all the necessary electronics, optics, or other devices that will provide sensing, actuating, signal processing, communication infrastructures, power generation, and other necessary functions. If necessary and desirable, attachable peripherals and other appliances can be included into the wearable system. Examples of these types of components are personal digital assistants (PDAs), displays, keyboard, and control knobs, among others.

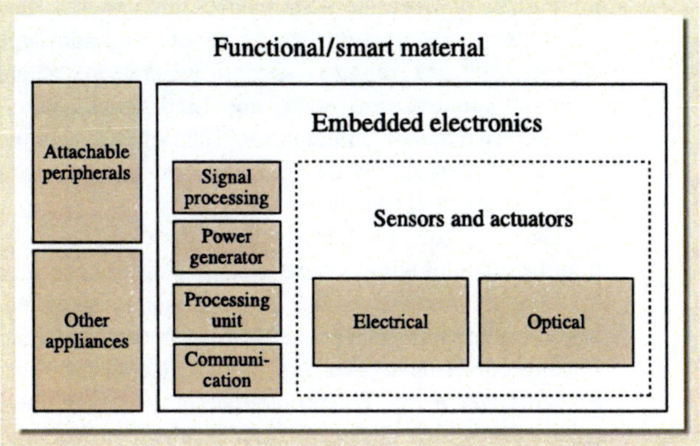

Fig. 74.17 Architectural layer of an ideal wearable health system

Focusing on the sensing technologies, new approaches are required, especially to provide noncontact methodologies to improve embedding of these components as well as to contribute to the design of totally wearable and highly comfortable functional garments. Sensors are used to monitor all the necessary physiological parameters and the physical environment surrounding the user, allowing the user's health condition to be monitored. They can either be embedded in the material, or the material itself works as a sensing element. These components will be further discussed below.

74.5.2 Categories of Wearable Systems

Wearable systems can be classified according to the level of integration of the components into the smart/functional material, i. e., the substrate. There are three types of wearable systems according to this classification: first generation, based on attached hardware components and sensors; second generation, where these components are embedded; and third generation, where innovative integration techniques during the substrate material production allow for the design of multisensory clothing and/or accessories.

The first category refers to all devices based on plug-in/out methodologies, where a supporting mechanism for attaching the necessary components is provided. These can include ECG monitoring wristwatches, and sensing components that can be attached to a t-shirt, a vest or even a cap. The problem associated with these types of wearable devices is their lack of comfort, and they are not yet a practical solution from the user perspective. In addition, since they are based on attached components that usually are not completely hidden, these systems generally do not allow for a completely unnoticeable solution.

The drawbacks of the first generation of wearable systems led to the creation of a new generation based on partially integrated/embedded architectures, where all the necessary components are fixed to the substrate material. This not only eliminates the need for qualified personnel or for the user to place the components, running the risk of misplacement, but also allows for a more practical and discrete solution. However, there is still a considerable difference from a normal garment or accessories, meaning that the components do not have a sufficient level of integration into the substrate, providing relatively comfortable solutions but still remaining perceptible.

Research and progress in integration techniques during the fabrication process allowed for the design of a third generation of wearable health devices. This type of system represents the cutting edge of wearable technology, allowing the design of smart, functional, and multisensing materials, which are apparently normal, due to this high level of integration. A very popular example of a third-generation wearable system is the electronic textile e-Textile, which consists of high-knowledge-content garments made from multifunctional fabrics. By blending the components into the user's ordinary clothing, it is possible to achieve an ideal wearable system, minimizing the hassle of wearing the device.

Despite the existence of these three generations of wearable devices, with their drawbacks and advantages, they all provide enhancement of user quality of life, allowing for real-time multiparameter monitoring under various conditions.

74.5.3 Design Requirements

The design of wearable systems requires an additional set of requirements, especially when compared with stationary equipment. If the final goal is to provide a customized multisensing wearable system that can be worn during daily life as normally as before, it is imperative that the wearable system cover the following set of key points [74.64, 65]:

Interaction
The device should be able to interact with the environment through a network of sensors placed in different parts of the clothing or accessories, allowing the creation of a certain awareness of the physiological and emotional state of the user, as well as of the surrounding environment. This is extremely important when designing smart and functional devices that are required to monitor, control, manipulate, interact, and actuate on a real-time basis. Data handling, decision support, and feedback are also crucial to establish a good interaction between the device, the components, and the user himself.

Autonomy, Safety, and Reliability
This aspect deals with the power and operational requirements which have to be met, without which autonomous, safe, and reliable operation of the device would be impossible.

Easy User Interface
It is important to maintain an easy interface between the user and the wearable device, minimizing the user's cognitive effort and involvement in the process.

Integration
It is crucial to improve integration of the components into the substrate in order to enable normal clothing and accessories with multifunctional capabilities. In addition, it is also important to interfere as little as possible with the user's physical activity, which is achieved by full integration of components into the substrate material.

Along with these key points come the user and design requirements for wearable systems. From the user's perspective, the most important features are minimal size and light weight, unobtrusive interface, user-friendliness, comfortable wear, ergonomics, portability, and increased lifecycle. When considering other perspectives in terms of design and performance, in addition to taking into account the user's requirements, there are a few key requirements to consider. Power autonomy, housing conditions, wireless communication, continuous time monitoring, and high levels of flexibility and integration are the main requirements from the design and performance perspectives.

By fulfilling the above requirements it is possible to design a wearable system that allows the user to benefit from its multifunctional and multisensing capabilities, without compromising the user's daily life activities.

74.5.4 Sensors for Wearable Systems

In wearable devices, a wide range of sensors are used to measure physiological and environmental conditions. The first type of sensors – biosensors – are used to monitor a clinical condition or process, therefore measuring biopotentials and other physiological parameters. On the other hand, the second type of sensors – peripheral sensors – are responsible for sensing the surrounding environmental conditions, enhancing the awareness of the system.

Not every sensor can be used in a wearable context, especially from the user's perspective. Not only do their physical attributes, such as size and weight, have to be taken into account, but also their noninvasive character and ease of placement must be considered. In addition, these sensors must ideally produce an electrical output in order to be digitally processed, being durable and reliable and with low power consumption [74.64, 67].

Usually, sensors used in wearable devices are based on conventional sensing elements, such as piezoelectric materials, electrodes, and others, with integrated communication infrastructure (e.g., wireless communication). This eliminates the problems associated with cable management and allows sensors to be positioned at various locations without any complications. In an ideal system, sensors should not require any contact with the user, since they could benefit from integration into clothing, which from the user's perspective contributes to the design of more practical, comfortable, and easy-to-use devices.

74.5.5 Sensing Methodologies

A number of technologies can be applied as sensing methodologies, such as electrical, mechanical, optical, piezoelectric, among others (Table 74.2) [74.12, 77, 78]. However, it is important to consider the demands associated with sensor integration into textiles or other flexible materials.

Sensors are composed of materials that respond to a physical stimulus, such as a change in temperature, pressure, electric field or illumination. They can range from skin surface electrodes to conductive yarns, providing comfortable and practical solutions for the user. Several materials can be used as the sensing component, and they can be divided into different categories, each with a specific effect.

The general acquisition system of a typical sensor used in wearable devices is composed of several functional blocks, from signal acquisition to output of the final result. Figure 74.18 shows a generic acquisition block diagram for a wearable sensor.

The transducer converts the desired signal or variable into another form of energy, which is then amplified and filtered to remove undesired signal components. The obtained analog signal is converted to digital form by an ADC and further processed using many different algorithms. If necessary, the processed signals can be converted back to analog form to drive

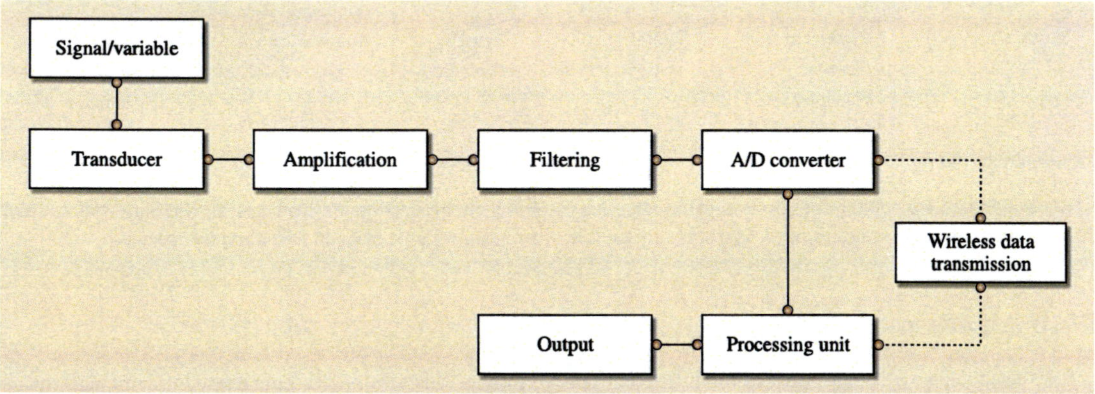

Fig. 74.18 Signal acquisition block diagram of a generic wearable sensor

Table 74.2 Different transducer effects applied in the sensing mechanism of wearable devices

Transducer effect	Sensing devices	Stimulus	Response	Examples of materials	Examples of detected signals
Electrochemical	Bioelectrodes, electrochemical cells	Biochemical activity	Electrical current	Copper, aluminum, Ag/AgCl, stainless steel, conductive yarns, electroactive polymers (EAPs)	EEG, ECG, EMG, EOG, skin electrode impedance
Electrooptic	Electrooptic modulators	Electric field	Birefringence	Lithium niobate ($LiNbO_3$), lithium tantalite ($LiTaO_3$), deuterated potassium dihydrogen phosphate (KD_2PO_4), electrooptic polymers	EEG, ECG, EMG, EOG
Electrostriction	Electrostrictor transducers Example: pressure sensor	**Direct effect (nonlinear)** Mechanical stress	Electric polarization	Lead magnesium niobate, lead lanthanum zirconate titanate (PLZT), EAPs	Carotid pulse, heart apex pulse, ECG, movement
		Reverse effect (nonlinear) Electric field	Dimensional change		
Electroluminescence	Light emission devices	Electric field	Light emission	EAPs	ECG, EMG
Photoluminescence	Photoluminescense sensors Example: UV radiation sensor	Incident light	Light emission	EAPs	UV radiation
Piezoelectricity	Piezoelectric transducers Example: strain gages	**Direct effect (linear)** Mechanical stress	Electric polarization	Polyvinylidene fluoride (PVDF), barium titanate ($BaTiO_3$), lithium niobate ($LiNbO_3$), lead zirconate titanate (PZT)-based compounds, EAPs	Carotid pulse, heart apex pulse, ECG [Samjin Choi]
		Reverse effect (linear) Electric field	Dimensional change		
Pyroelectricity	Pyroelectric sensors Example: infrared sensors	Temperature change	Electric polarization	All pyroelectrics are piezoelectrics	Body temperature, ambient temperature

specific devices. For better design, it is often necessary to include wireless communication systems that enable the sensing component to transmit the data to a control processing unit or even to a database service.

74.6 Electronics in Medicine at Work

After introducing several aspects of medical electronics and several signals that may be recorded by such systems, this section presents a few implementations of electronic systems designed for medical applications.

74.6.1 Wireless and Wearable Low-Power Health Monitoring Systems

A prototype wearable, low-power acquisition module for a WHMS is shown in Fig. 74.19. The acquisition module provides continuous monitoring of single-lead ECG, temperature, and activity [74.79]. The module is used on the chest of the monitored subject and can be attached using either a chest band or by integration into a textile garment. In addition, the wearability of the acquisition module is supported by its small size and lightweight design, and by the assembly of all of its electronic components on a flexible polyimide printed circuit board (PCB), which in turn houses dry and flexible electrodes [74.80], directly attached to the PCB without use of wired connections. Unlike conventional silver/silver chloride electrodes, these dry and flexible electrodes are able to acquire a biosignal without the need for skin preparation with electrolyte gel, which is suitable for use in long-term and continuous health monitoring systems.

74.6.2 Sensors

The sensors contained in the acquisition module were chosen considering its continuous monitoring and low-power operation features. The acquisition module includes three flexible, dry ECG electrodes with a low-power analog front-end for single-lead ECG monitoring, a negative-temperature-coefficient (NTC) thermistor for surface temperature monitoring, and a low-power three-axis accelerometer for activity monitoring.

The analog front-end designed for ECG monitoring is shown in Fig. 74.20. A pair of the flexible dry electrodes at a short distance from each other (65 mm) is used to acquire a single-lead ECG, when contacting the skin on the chest, close to the heart. The remaining electrode is used as a reference driver connection to reduce common mode interference.

The performance of the analog front-end was designed to achieve a trade-off between low power consumption and suitable common mode rejection ratio (CMRR) by appropriate choice of the instrumentation amplifier (current consumption of 93 μA and CMRR of 115 dB). A notch filter was included in the front-end to attenuate electromagnetic interference from surrounding power sources, since the acquisition module should be able to monitor the subject even in electromagnetically polluted environments. Furthermore, a bandpass filter is used to limit the bandwidth of the acquired signal to the ECG frequency range and to attenuate the DC offset. The complete analog front-end achieved current consumption of about 220 μA.

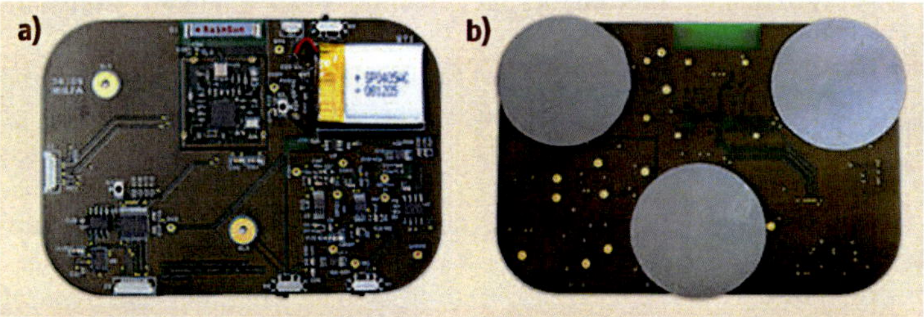

Fig. 74.19 (a) Wearable low-power acquisition module for a WHMS, showing electronics on the top layer and (b) flexible dry electrodes for ECG monitoring on the bottom layer

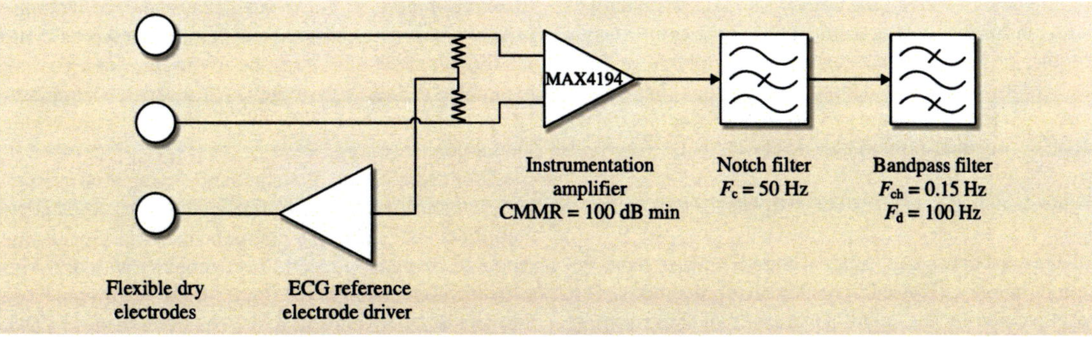

Fig. 74.20 Low-power ECG analog front-end included in the WHMS prototype

The activity of the monitored subject is estimated by the LIS302DL digital accelerometer which has a configurable range of ± 2 g or ± 8 g, featuring low current consumption (300 μA). The local temperature at the acquisition module can be estimated within the range of 20 and 40 °C, by the change in resistance in the NTC thermistor.

74.6.3 Wireless Link

The wireless communication interface between the acquisition module and a base station [e.g., personal computer (PC) or PDA] is provided by the Toumaz Sensium (Toumaz Technology Ltd., Abingdon [74.81]). The Sensium is a system-on-chip solution containing a low-power transceiver platform, 8051 microcontroller, and reconfigurable sensor interface. The Sensium allows the creation of wireless body-area networks composed of a base station and up to eight target nodes which communicate using a proprietary protocol aimed at minimization of power consumption. The wireless link between Sensium devices operates in the 868/915 MHz ISM band, providing a 50 kbps data rate.

For the design of the presented acquisition module, the use of the Toumaz Sensium solution with a proprietary wireless communication protocol was preferred over the use of standard protocols for wireless sensor networks or wireless local-area networks, such as Bluetooth (802.15.1), ZigBee (802.15.4) or Wi-Fi (802.11). The use of the proprietary protocol has the advantage of low power consumption, achieving longer operation times suitable for continuous monitoring, and avoiding the crowded 2.4 GHz ISM band, thus aiming to achieve better resilience to interference and improved quality of service. On the other hand, use of the proprietary protocol and transceiver reduced transmission power and data rate to achieve low power performance, thus only allowing transmission distances of 10 m and a data rate of 50 kbps. However, this transmission distance may be suitable for monitoring a subject at home or in specific rooms, or if the subject carries a smartphone or PDA as a base station. In addition, the 50 kbps data rate is sufficient for continuous monitoring of most biosignals, considering their limited bandwidth.

74.6.4 Optical Biopotential Recording

In the past few years, optical-based sensors have become increasingly used in healthcare applications. This is explained by the many advantages of these sensors that include immunity to electromagnetic interference, resistance to harsh environments, freedom from electrical wires, small size, light weight, and multiplexing capabilities [74.82]. Moreover, many approaches for optical fiber integration have been developed, with particular interest for wearable health devices, leading to easier optical fiber integration into textiles and other wearable materials [74.56, 83, 84].

74.6.5 Optical Electrodes

When dealing with optical electrodes, several modifications must be made to the acquisition system shown in Fig. 74.18. A light source and detector must be included, since the sensing methodology is based on alteration of light properties. Figure 74.21 shows a typical acquisition system for an optical sensor.

In this case, the acquisition system must include a light source that will pass through an optical transducer. In the presence of a particular signal, the optical transducer will suffer alterations, for example, a dimensional change. This alteration will produce a shift in light properties, whether intensity, phase, polarization or other. Afterwards, a photodetector is responsible for collecting this modulated light and converting this modification into a desired result. The latter is dependent on the type of photodetector applied in this stage; for example, a photodiode can be used to convert light intensity into current or voltage. The signal is then processed according to the remaining block diagrams, which are similar to those in Fig. 74.18.

An important feature of optical-based sensors is the ability to enable easier contactless measurements of physiological events, particularly electrophysiological signals. This can be achieved using specific transducer effects by which a material exhibits a particular response in the presence of a stimulus such as an external electric field. Materials exhibiting this type of stimulus–response mechanism are classified as electroactive materials; some such materials are included in Table 74.2. The electroactive component works as the sensing element, which can be in the form of a coating material, such as a hydrogel or a piezoelectric material, or even used as a device such as an electrooptic modulator.

For several years, the standard biopotential recording systems consisted of bioelectrodes, instrumentation amplifiers, and filtering components. The signal was recorded by placing electrodes in contact with the skin, allowing conversion of the ionic current within the body into an electronic current in the electrode leads [74.54]. The coupling between the electrode and the skin was

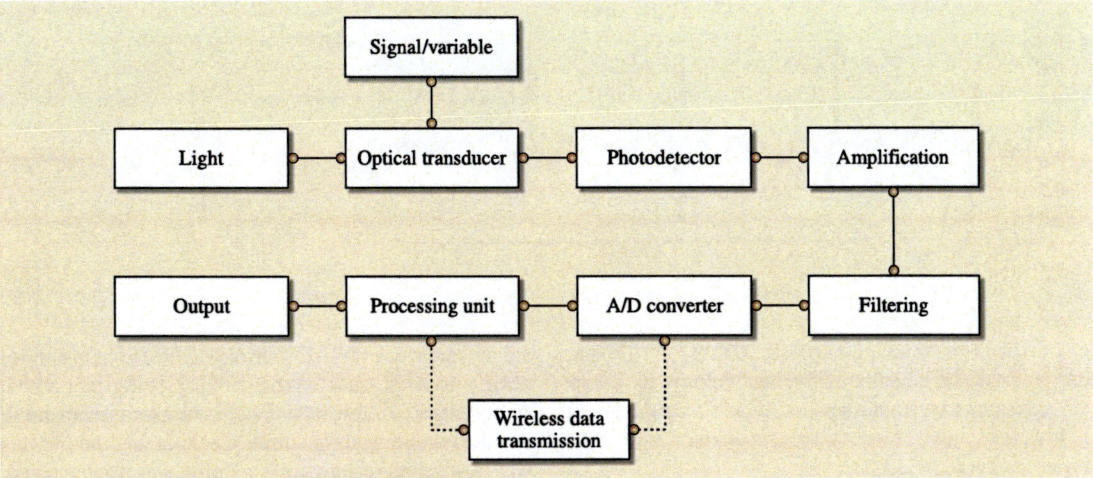

Fig. 74.21 Optical sensor acquisition block diagram

usually made by an electrolyte that could be either a gel/solution for wet electrodes or simply sweat for dry electrodes. This standard recording system, despite its good performance in terms of sensitivity, is not very suitable for wearable applications, since it requires good electrical and/or physical contact with the skin [74.85].

Novel dry and textile-based wearable electrodes have recently been proposed to overcome the limitations associated with such electrodes. These include, for example, conductive rubber electrodes [74.86], Cu-sputtered textile electrodes [74.87], conductive fabric sheets and polyvinyl difluoride (PVDF) film electrodes [74.73], and polymeric dry electrodes [74.88]. However, these still require physical contact with the skin, making optical sensors the strongest candidate for contactless measurements in wearable devices.

In recent work, an electroactive hydrogel was suggested as a transducer for optical-based biopotential measurements for applications in wearable systems [74.89]. This electroactive material, polyacrylamide hydrogel (PAAM), when submitted to an external electric field, undergoes a bending process, altering its mass and volume properties. Likewise, the amount of light passing through the hydrogel and reaching the photodetection stage will change. Although many configurations can be applied in the sensor design, the desirable solution could be a fiber pigtail, with the PAAM hydrogel placed at the tip of the fiber. This material showed an adequate sensitivity and frequency response, as well as the ability to perform better as sample thickness decreases, making it an eligible sensing component for biopotential recording applications.

74.6.6 Optical Signal Acquisition

An example of an optical-based sensor for biopotential acquisition is the electrooptic modulator, which can be used to modulate light (e.g., intensity, polarization or phase) in response to an electric field. The observed modification of the light property is proportional to the biopotential, making it possible to translate this result into a biopotential recording. The electrooptic (EO) sensor proposed is a photonic fiber optic based device, whose main functional stages are optical signal generation, electrooptic modulation, and detection. These EO sensors may be developed and assembled so as to be easily integrated into fabrics or other wearable materials. Figure 74.22 shows an ECG signal obtained using this method as the recording technique.

As shown in this figure, this EO technique has the required frequency response and sensitivity to acquire an ECG signal with its different signal components, such as the QRS complex.

74.6.7 Localization Solutions

One well-known requirement in AAL solutions is the ability to track someone inside an environment using the already deployed wireless infrastructure. Several wireless AAL solutions for monitoring have been proposed in the past few years, e.g., [74.48, 49]. One solution is to use the Wi-Fi infrastructure and deploy small modules, worn by the subject being monitored. Then, using the Wi-Fi network infrastructure, the module

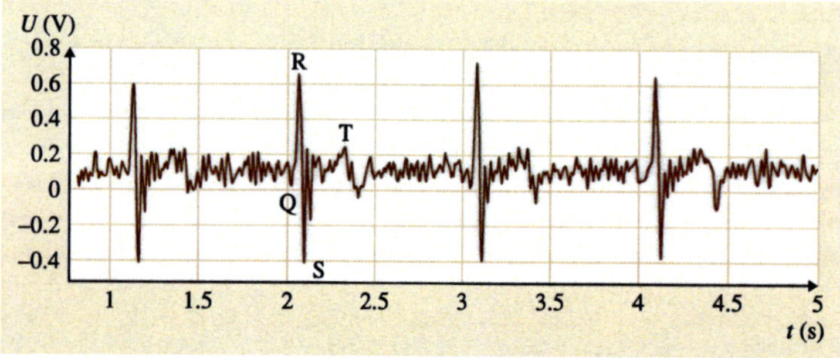

Fig. 74.22 ECG signal recorded with an electrooptic modulator

can be used for tracking purposes [74.48, 49]. Despite the well-known benefits of this technology, its acceptance requires wearable devices, which must be small, lightweight, and with enough power autonomy for comfortable operation [74.61].

Figure 74.23 shows a system that enables localization of a walking user inside a building, using a device that was developed to be used as a wristwatch [74.61]. This is the smallest localization device known for use in localization applications, based on a Wi-Fi infrastructure. This module was designed to fulfill clinical environment compliance, which requires a hermetically sealed module made from clinically compatible materials. Since it is to be worn as a wristwatch, wearability and fashionable design were taken into account. Moreover, for ease of use and cleaning, the module uses wireless recharging technology.

Since the localization device had to be designed to be worn on the subject's wrist, the module should not greatly exceed the dimensions of a wristwatch. Also, and to reduce deployment costs, the localization technology is based on the installed Wi-Fi network, so the localization device must use a Wi-Fi module. The designed module uses an off-the-shelf low-power Wi-Fi tag. The tag communicates with existing wireless local area network (WLAN) infrastructure to help control centers to keep track of the subjects being monitored. A key feature of this device is its waterproof capability and the requirement for simple cleaning methods. In this way, a wireless recharging capability was implemented, which enables full recharging of the battery when the module is placed on a base station for around 2 h. Moreover, as onboard sensors, it has a motion detector and a photonic wrist removal detector.

Fig. 74.23 Magtag, side-by-side with a wristwatch

Localization Ability

Figure 74.24 shows the software front-end used in combination with the developed module.

Initial tests showed a resolution of 2–5 m (90% of the time) for an indoor environment. Those tests showed that the tag meets the project requirements. The performance obtained makes the tag viable for all the expected project scenarios, namely in hospitals and healthcare markets.

Module Autonomy

Since the module was expected to be as small as possible, the platforms under test were powered by a 150 mAh, 3.7 V, $20\times30\times4.0\,\mathrm{mm}^3$ Li-ion battery. However, the indicated supply battery may change, due to different autonomy constraints.

All the tests were carried out using a small resistor in series with the battery, and the voltage drop was measured. From that, the current consumption was extracted. Table 74.3 shows the results obtained when using this procedure.

At the moment, this device is used only for localization purposes. In normal operating mode, the device sleeps for 20 s, then awakes, detects its position, and sends it to a nearby wireless access point.

From Table 74.3, it can be observed that, during transmission, the measured current value exceeds the expected value by about 60 mA. Also, in sleep mode, the current measured was 1.5 mA, far more than the ex-

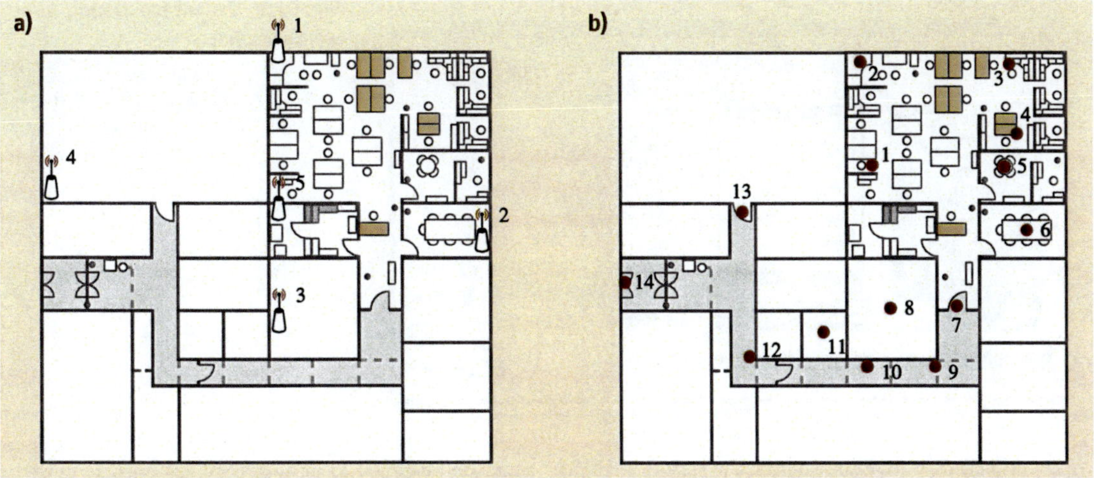

Fig. 74.24a,b Tracking operation. **(a)** Testbed access point location. **(b)** Tag positions used for localization error assessment

Table 74.3 Wi-Fi module power consumption

Operating mode	Current consumption Measured ($V_{cc} = 3.3$ V)	Expected
Tx mode	270 mA @ 18 dBm	212 mA @ 18 dBm
Rx mode	50 mA	40 mA
Sleep mode	1.5 mA	6 μA

pected value of 6 μA. To test battery autonomy, a 6 s transmission interval was used, for which the battery life was 20 h. All these tests were carried out with all the active sensors turned off.

74.6.8 Ambient Assisted Living Applications

Ambient assisted living (AAL) (Chap. 67) is an emergent technology that envisions offering new solutions for healthcare [74.31]. Those solutions will improve the quality of life of the elderly population and reduce costs associated with healthcare. However, before that happens, new solutions, including both hardware and software, must be available in order to acquire and store the required signals, to process and extract information from those signals, and to detect a set of features required to produce alarms and/or electronic assistance.

Despite the availability of several platforms for signal acquisition [74.49], a few technological issues must be solved before they can be used. The platforms must offer quality of service, be wearable, and operate for a comfortable period of time. Quality of service is required, otherwise no company will be able to offer such a highly critical service. When hiring a service for health monitoring, it must not fail, or the people involved must be aware of how safe the system is. Also, nobody will be willing to wear a device that reveals that he is being monitored, and it is also unpleasant to wear a device that needs to be recharged too often. At least, is must be able to operate for one full day. In this way, the system may be replaced when the person goes to bed or gets up.

Once the relevant data are available, an AAL system will allow remote monitoring of one, or a set of subjects, and their health and activity. Based on this and a set of information associated with each subject, it is possible to offer different levels of services. One will be self-reminders that are preset by the subject; the other will be reminders by some family member or caregiver; and finally, a health service based on clinical data may also be offered, if good enough health engines are available to extract the health condition.

Intelligence at Application Level

Figure 74.25 shows a snapshot of an application developed to collect data from several sensors, both biometric data (monitoring health and activity) and data from home automation [74.43]. Based on home automation, it is possible to infer the daily activity of a subject and help him in their routines.

This interface includes three main blocks. One block acts as a pill dispenser with the possibility to send text messages through the short message ser-

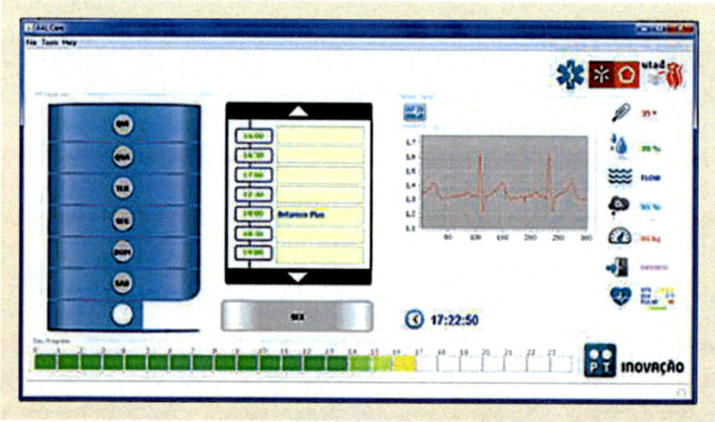

Fig. 74.25 Patient monitoring software snapshot

vice (SMS) and/or multimedia message service (MMS) with previously recorded familiar voices to provide advice regarding prescriptions. The second block provides a day (timeline) progress bar with emergency status (colors from green to red) linked to the continuous health monitoring system. The third block is a sensor panel showing data from several sensors for ECG, temperature, humidity, water flux, weight, door opening/closing, and arterial pressure with voice messages corresponding to high or very low arterial pressure readings.

Acquisition Hardware

The software operates based on the data acquired from the various sensors. Figure 74.26 shows the main blocks of the acquisition hardware. It is presented as worn by a person, but it is equally valid for data acquisition from home automation.

Since a few of those sensors must be mobile and must be placed on the person, the acquisition system must be designed to have special features. It must be wireless, offer quality of service, have low power consumption, be wearable, and operate for at least 1 day on batteries. This led to the use of a modified version of a ZigBee-based platform.

For data acquisition, a small and power-efficient module was designed. Moreover, since the set of signals that may be relevant for this system are highly heterogeneous, the module was designed so that it can be configured for different applications. The most demanding signal in this scenario is the electrocardiogram, corresponding to measurement of the electrical activity of the heart muscle, which may range from about 0.5 mV to a maximum of 10 mV. It has a spectral range of 0.05–150 Hz for diagnostic ECG and 0.5–40 Hz for monitoring ECG.

The requirements of the front-end of an ECG sensor circuit include the capability to sense low-amplitude signals in the range of 0.05–10 mV, high common mode rejection ratio (CMRR), high input impedance,

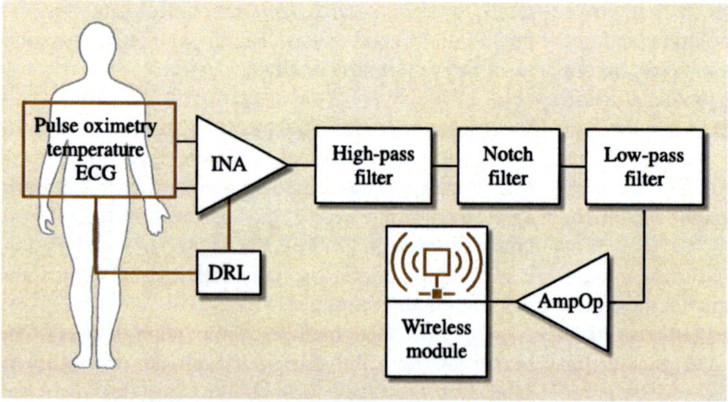

Fig. 74.26 Block diagram used for the data acquisition system (DRL – driven right leg; INA – instrumentation amplifier)

low input leakage current, flat response for 0.5–40 Hz, low weight, low power (<1 mA), and signal amplification over a range specified by the analog-to-digital converter (ADC).

The first stage of amplification is one of the most important components of the system. This amplifier presents excellent features for acquisition of biopotentials such as high input impedance, high differential gain, and high common mode rejection ratio. The integrated circuit (IC) used was the INA333 from Texas Instruments. The ECG is often contaminated with undesired spectral components. To attenuate these, the circuit must include a high-pass filter, notch filter, and a low-pass filter. All the implemented filters are active filters, so operational amplifiers were used. The ICs chosen to build the filters were the OPA2369 from Texas Instruments. The OPA2369 was chosen mainly because of its low quiescent current (0.7 μA/channel), and low offset voltage (250 μV). The high-pass filter must attenuate frequencies from zero to the cutoff frequency, thus it blocks the undesired DC components. The proposed filter is a second-order Sallen–Key high-pass filter with a Butterworth response. The filter's main features are its quality factor (Q) of 0.707, its unity gain, its cutoff frequency f_c of 0.45 Hz, and its slope of −40 dB/decade. The ECG is frequently contaminated with 50/60 Hz power-line interference, or electrostatic charge. The strategy to attenuate this kind of interference involved a notch filter or band stop filter centered at 50 Hz. The implemented notch filter is a second-order Sallen–Key with gain of 1.5, Q of 1, f_c of 50 Hz, and slope of −40 dB/decade. The low-pass filter is used to attenuate frequencies beyond the desired frequency for ECG measurement. These frequencies can come from sources such as radiofrequency and communication equipment. A second-order low-pass Sallen–Key filter with Butterworth response was implemented. The filter has unit gain, Q of 0.707, and cutoff frequency of 40 Hz. The filtered signals have to be amplified so that the ADC from the ZigBit module can measure it. The gain of this amplification stage is about 60, so the total gain of the system is 300. Finally, a driven right leg (DRL) circuit was also implemented. This circuit is used to amplify, invert, and feedback the common mode signal to the body, preventing saturation of the amplifiers, and reducing the output voltage due to CMRR. Figure 74.27 shows the frequency response for the described system.

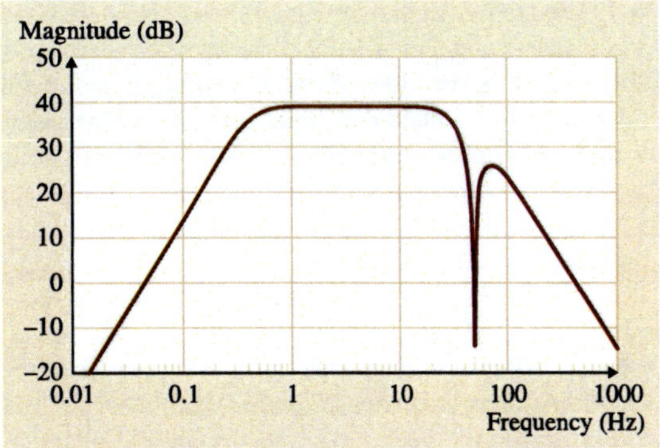

Fig. 74.27 Frequency response of the front-end electronic acquisition system

Despite the low-power design, the system was designed to achieve high rejection of the 50 Hz component. This allows for a higher degree of system miniaturization, since when registering ECG signals with very close electrodes, interference may become a problem.

Since this is intended to be a flexible platform to acquire data and feed an AAL system, it was designed to be small and modular. Figure 74.28 shows the device dimensions.

The main achievement was the low current consumption for ECG acquisition (90 μA) and the small size ($33\times30\,\mathrm{mm}^2$). This module, combined with a transceiver module, allows acquisition of almost every signal produced by any transducers that may be found useful for use in AAL systems.

The developed platform was designed to acquire signals from a range of sensors, and to transmit them

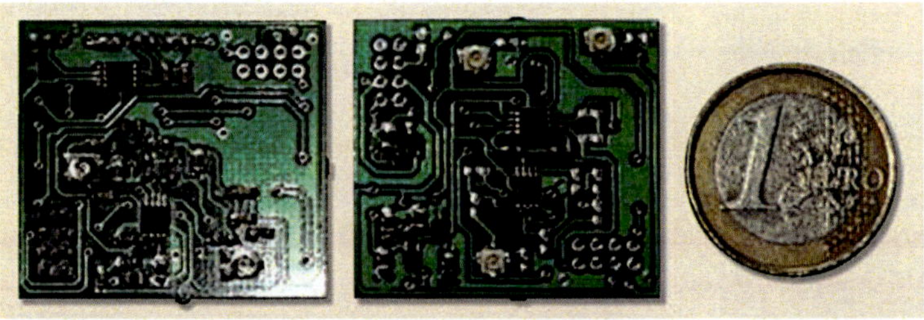

Fig. 74.28 Photograph of developed PCB layout

to a base station where the signals were registered and processed to monitor a subject during their daily life. The module is very low power, allowing operation for 1 day, and the wireless data were delivered with quality of service, making it suitable for AAL applications in controlled environments.

74.6.9 Wireless Link Design for Biomedical Applications

When a patient's clinical state turns from a noncritical situation into a critical one, a context change occurs, and consequently, the healthcare network should adapt its performance requirements to this new situation [74.90]; for instance, higher monitoring activity and lower transmission delay of vital signals might be required when the clinical situation of a patient changes from noncritical to critical. Hence, healthcare networks should provide QoS facilities for e-emergency services, since these clearly demand high reliability, guaranteed bandwidth, and short delays.

A personal healthcare system consists of a group of sensors attached noninvasively to a patient for sensing of physiological parameters. These body sensor networks (BSN) have been used in hospitals during recent decades using conventional wired equipment, hence not allowing the patient to move around freely. However, recent advances in wireless sensor technology are changing this scenario by allowing mobile and permanent monitoring of patients, even during their normal daily activities. In such healthcare systems, the information sensed at the patient's body is transmitted to a wireless base station, located no more than a few tens of meters away, and then delivered to a remote diagnosis center through a communication infrastructure.

Healthcare systems should be able to accomplish at least one crucial aim: to monitor a patient and, when an emergency occurs, immediately trigger an event to alert the patient and/or warn a remote caregiver. In this way, both the patient and the caregiver can ensure timely implementation of the appropriate procedure in accordance with the clinical episode. In addition, the system should be able to trigger an alert anticipating the case where the patient is unaware of the seriousness of his/her health situation.

Wireless sensor networks (WSN) are expected to be used in emergency healthcare applications in the near future. Since e-emergency systems should be totally reliable and efficient in order to provide valuable assistance to patients, quality of service (QoS) must be present at distinct protocol levels, including the media access control (MAC) layer.

Vital Signals Requirements

Emergency medical care requires monitoring of several vital signals simultaneously; for example, a patient in an ambulance is monitored for blood pressure, heart and respiration rates, and temperature. Besides these primary signals, other information may be captured to help diagnosis and medical decision, such as electrocardiogram (ECG), blood glucose level, blood oxygen saturation, heart and breathing sounds, or even imaging in cases of trauma. All these data should be promptly accessible for comparison, computer-aided analysis, and decision-making. However, to achieve a more encompassing picture of the clinical situation, information about the patient's personal characteristics (risk factors, degree of disease, age, etc.) and environmental context (e.g., in bed or mobile, alone or not, at work or at home, etc.) should also be provided [74.91].

Table 74.4 presents the electrical characteristics of the vital signals usually monitored in emergency medical care [74.92, 93] and Table 74.5 shows the range of typically monitored values.

If some signal exceeds its threshold, the local supervisor node should trigger an alarm to inform a caregiver or the patient himself. Table 74.5 presents typical

Table 74.4 Vital signal electrical characteristics

Vital signal	Frequency range (Hz)	Sampling rate (Hz)	Resolution (bits)
ECG (per lead)	0.01...60–125	120–250	16
Temperature	0...0.1–1	0.2–2	12
Oximetry	0...30	60	12
Arterial pressure	0...60	120	12
Respiration rate	0.1...10	20	12
Cardiac rate	0.4...5	10	12

Table 74.5 Alert detection parameters

Alert type	Detection parameter
Low SpO_2	$SpO_2 < 90\%$
Bradycardia	HR < 40 bpm
Tachycardia	HR > 150 bpm
HR change	Δ HR / 5 min > 19%
HR stability	Max HR variability from past 4 reads > 10%
BP change	Systolic or diastolic change > ±11%

thresholds for SpO_2, heart rate, and blood pressure signals for alert detection [74.94].

QoS Considerations for e-Health

Some authors argue that differentiation based on data priority is inherent to wireless sensor networks (WSN), since it is normal to have sensors to monitor distinct physical parameters simultaneously, just as happens in BSNs. Here, the importance of the collected information is necessarily distinct, and therefore the network must prioritize the transmission of critical data when a sudden clinical change in the patient occurs; for example, in patients with cardiac diseases, heart activity information is more important than body temperature data. Depending on the patient's clinical condition, the priority assigned to a vital signal can change dynamically.

Most current BSNs only offer best-effort service [74.95], which is not the most adequate for e-emergency. The given examples indicate that those networks require QoS support in critical cases. This would ensure adequate bandwidth for higher-priority streams for efficient data delivery, even in case of interference or fading.

QoS mechanisms are usually deployed in networks to guarantee consistent service levels concerning certain parameters, such as packet loss ratio, transmission delay, jitter, and available bandwidth. These are the traditional end-to-end parameters used to characterize the performance of any communication infrastructure, including BSNs; for instance, the total delay to an ECG signal being displayed in the monitor should be less than 3 s for useful real-time analysis by a cardiologist [74.95]. To guarantee that jitter does not affect the estimation of the R-wave fiducial point, which considerably modifies the spectrum, ECG signals require a minimum sampling rate of 250 Hz. It should be noted that no significant difference between ECG traces are detected when sampling the signal at rates between 250 and 500 Hz, but significant reduction in peak amplitude values and inaccurate interval measurements are obtained at 125 samples/s [74.96].

Approaches over than prioritization may additionally be used to provide QoS; for example, for efficiency reasons, a large packet length is chosen for noncritical situations. However, as soon as an emergency occurs, the packet size is reduced to meet the low-delay QoS requirement, and signals considered irrelevant to this emergency episode are sampled at a lesser rate, or not at all. Moreover, the computation power may be lowered to a minimum, since all data must be forwarded, as opposed to regular operation where, to save energy, the cardiorespiratory rhythm can be computed onboard before sending it. Alternatively, an ECG signal could be processed in the sensor itself to extract its relevant features. In this way, only information about an event is transmitted (e.g., QRS features and the corresponding timestamp of the R-peak), hence reducing the traffic load and saving energy.

The vital signals captured from the patient's body must be delivered to a remote diagnosis and supporting center, through some available communication infrastructure, such as the Internet or a mobile cellular network. As a result, the delivered QoS depends necessarily on the network infrastructure chosen for the delivery of the mobile health services. Therefore, to meet the required QoS, the mobile health services platform needs to be able to acquire and use contextual information about the QoS offered by communication network infrastructures available at the patient's current location and time. High availability and reliability are the most desirable features that these network infrastructures should offer, as well as QoS guarantees for bandwidth, end-to-end delay, jitter, and loss.

e-Health Wireless Systems with QoS

Despite the number of healthcare systems already developed, few encompass QoS requirements. To assess how QoS support is being deployed in healthcare sensor networks, some projects were analyzed, as well as the QoS requirements that the respective authors considered important to incorporate in their implementations. Based on the related literature, QoS projects for e-health can be grouped according to the topics: (a) frameworks with QoS, (b) QoS through reconfiguration, and (c) algorithms for QoS enhancement [74.90].

The target deployment of each project is diverse, such as remote patient monitoring using standard protocols and mobile communication networks, mobile telemedicine under heavy traffic conditions, signal reconstruction, and BSN lifetime improvement. Thus, the QoS support approach is also distinct in each project: transmission priority scheduling along with data compression, self-reconfiguration of the BSN by data feedback, restoration algorithms to recover missing packets, and power management model for energy saving. Table 74.6 presents an overview of diverse QoS aspects that the authors have considered important to benefit their projects with, as well as the target environments where the projects are intended to be deployed. The target deployment of each project is diverse, such as remote patient monitoring using standard protocols and mobile communication networks,

Table 74.6 Examples of QoS solutions for healthcare systems

Topic	Description	QoS goals	Target deployment
Frameworks with QoS	DiffServ based on transmission priority scheduling, data compression	Low delay and loss rate for vital signals, better bandwidth utilization	Remote medical applications using zigbee and mobile cellular networks
	Wireless diffserv infrastructure	Low loss rate, guaranteed bandwidth	Reliable network for mobile telemedicine under extreme traffic conditions
QoS through reconfiguration	Send feedback information into BSN to self-reconfiguration	Application level improvement	Remote and local patient monitoring
	Middleware system for reconfiguring bsns	Application level improvement	Increase BSN lifetime while respecting the qos needs
Algorithms for QoS enhancement	Restoration algorithms	Recover missing packets	Reconstruct a more functional signal for the doctor
	Power management model based on application states	Energy saving	Reserve power to the patient's more vital tasks

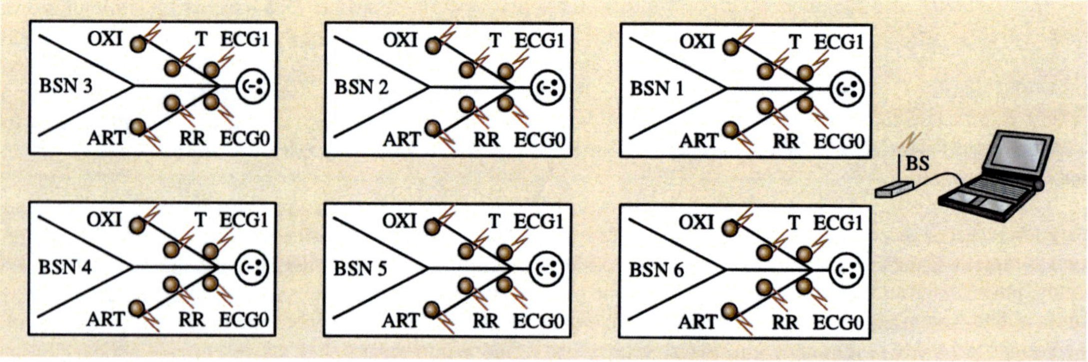

Fig. 74.29 Layout of the hospital room, with a patient being monitored in each bed

mobile telemedicine under heavy traffic conditions, signal reconstruction, and BSN lifetime improvement. Thus, the QoS support approach is also distinct in each project: transmission priority scheduling along with data compression, self-reconfiguration of the BSN by data feedback, restoration algorithms to recover missing packets, and power management model for energy saving.

Case Study

Aware of the diverse QoS approaches presented above, Fig. 74.29 shows an experimental testbed to deploy QoS solutions based on a real clinical scenario [74.97]. The scenario is based on a hospital room containing six beds with one patient per bed. Each patient is monitored by a personal BSN, and one BS collects the vital signals of all patients. The vital signals being monitored are temperature (T, Temp), oximetry (OXI), arterial pressure (ART), respiration rate (RR), and ECG data. All monitored signals must be transmitted through a low-rate channel to the BS with appropriated QoS, as specified next.

According to the IEEE 1073 group, a wireless ECG electrode should generate 4 kbps of data, and the latency introduced by framing the data samples and the transmission delay should be below 500 ms. Since ECG signals are the most demanding in terms of QoS, we take this value as the maximum delay that any vital signal should have in this study. Continuous healthcare monitoring normally uses a three-electrode ECG device, therefore this situation is assumed in our study. According to Table 74.1, each BSN produces a maximum aggregated rate of 10 424 kbps, hence resulting in a maximum total traffic inside the hospital room of 62 544 kbps. Besides guaranteeing this minimum goodput and latency below 500 ms, the e-health system must also guarantee null packet loss for every vital signal, low energy consumption, and balanced energy drainage in every BSN.

ZigBee is a short-range, low-power, low-data-rate standard for wireless sensor networks that supports a maximum data rate of 250 kbps in the 2.4 GHz band. Therefore, a ZigBee WSN is able to handle the whole traffic generated inside the hospital room without congestion. Nevertheless, other factors which may significantly affect QoS have to be considered, such as access to the wireless channel and errors which often occur during transmission.

Simulation results have shown that ZigBee is not adequate for several sensors to transmit ECG signals to a BS with full efficiency. The packet delivery efficiency starts to drop when three or more ECG devices belong to the same network channel. This is because ZigBee relies on carrier sense multiple access with collision avoidance (CSMA-CA), a contention-based MAC protocol that is vulnerable to collisions. Lost packets may be retransmitted, but there is a maximum number of five retries allowed by ZigBee before declaring channel access failure. ZigBee seems to be more adequate for BSNs that do not have large amounts of data to transfer with only several small data packets per hour, such as implanted medical sensors.

Diverse MAC protocols have been developed for wireless networks using contention-based or scheduled access techniques [74.98]. Traditional contention-based protocols assume that traffic is distributed stochastically. However, traffic in a WSN tends to be highly correlated and/or dominantly periodic, and so such protocols may not perform well in these conditions. Some CSMA-based protocols have been developed (e.g., Sensor-MAC (S-MAC) [74.98]) to help save energy in sensor network applications where nodes remain idle for long periods until detecting an event (e.g., surveillance), hence normally generating very little traffic in the network. Hence, it is inadequate to apply such MAC protocols in networks requiring high throughput and low latency, such as emergency biomedical applications. Time-division multiple access (TDMA) protocols divide time into slots that sensors can use to transmit data without needing to contend for the medium. Since quality of service is easier to assure in a collision-free environment than in a contentious-prone medium, and as a base station is available to keep the WSN scheduled and synchronized, a solution may be to use a TDMA-based MAC protocol to satisfy the QoS, such as the low-power, real-time protocol (LPRT) [74.99].

LPRT is a simple, beacon-based, low-power, real-time protocol that uses the available bandwidth dynamically and efficiently. Its highly grained superframe starts with the transmission of a beacon frame (B) by the BS, followed by the contention access period (CAP), also called the contention period (CP) [74.99]. The CAP may be used for the (dis)association or configuration of a body sensor network (BSN). The contention-free period (CFP) follows the CAP. The CFP is composed of the normal transmission period (NTP) and the retransmission period (RP). The NTP is used for sensor nodes to transmit new data. Lost data are retransmitted in the RP. Data packets are sent in contiguous slots of the CFP. The attribution of slots in the CFP is announced through a list of allocation fields carried in the payload of each beacon. Each allocation field contains the association identification and the initial transmission slot (ITS) for every sensor node in the WSN. Data frames transmitted to the BS during the NTP are acknowledged by the ACK bitmap present in the beacon of the next superframe.

The beacon size may become relatively large, since it is directly dependent on the number of sensor nodes associated with the BS. A sensor node transmits data in the superframe only if the corresponding beacon carrying the list of allocation fields is received, and a single retransmission procedure is used in case of transmission failure. These characteristics may lead to a significant packet loss ratio if communications occur in a wireless channel with an appreciable bit error ratio (BER). To improve its robustness against bit errors, a solution based on short-size beacons has been proposed.

Proposals to Improve LPRT

To define a MAC protocol more robust to channel errors, solutions based on short-size beacons, autonomous transmissions, and multiple retransmissions have been added to LPRT. RP is defined after the CAP so that a packet is retransmitted away from an eventual burst error condition responsible for the transmission failure that occurred during the last NTP. RP is not placed after the NTP to avoid a sensor node transmitting in the NTP with a variable packet size.

Short-Size Beacons. To assure good performance of the e-emergency WSN, the percentage of lost beacons should be very low. A strategy to accomplish this goal is to send beacons with a convenient transmission power, since the bit error ratio (BER) of the channel decreases as the signal-to-noise ratio (SNR) increases. In addition, the beacon frame length should be as small as possible. So, whenever possible, the beacon payload contains only the ACK bitmap to acknowledge the frames correctly received during the NTP of the last superframe, and the CAP size of the current superframe (assuming that the start slot of the CAP is known). In this case, sensor nodes must run an algorithm to compute which slots should be used to (re)transmit data without interfering with each other, in accordance with a predefined order schema. Using this strategy, the energy consumption in each BSN improves too, since smaller-size beacons are received by the sensor nodes.

According to the received ACK bitmap, each sensor node must calculate the corresponding superframe slots to transmit its data. If a sensor node does not receive a beacon or a short sequence of beacons, it may continue to send its new data in the NTP (i. e., autonomous transmissions), since a sensor node clock drift of the order of microseconds should permit the WSN to continue to be synchronized during a few consecutive beacon intervals. However, the sensor node cannot retransmit any data in the RP, because the ACK bitmap is not available and so it does not know how the RP slots are being allocated to the sensor nodes.

To save energy, it might be tempting to retransmit the lost data aggregated to the new data sent in the NTP, instead of retransmitting it in the RP. However such a strategy should be avoided, because the number of slots allocated in the NTP to each sensor node becomes variable. Consequently, if a sensor node does not receive the beacon, it has no way to know a priori which slots to use for transmission. Indeed, if a sensor node does not receive the ACK bitmap, it does not know which sensor nodes are going to transmit aggregated data in the current superframe, making it impossible to compute the new slot allocation schema. This situation does not occur if the NTP is used only to transmit new data packets. Hence, aggregated retransmissions in the NTP are not recommended, except in cases where aggregation does not imply taking more slots, such as for packets carrying temperature data.

Besides the ACK bitmap and the CAP size, a beacon may need to send reconfiguration instructions if a new clinical situation is detected in some BSN; for instance, higher monitoring activity and lower transmission delay of the vital signals might be required when a patient's clinical situation changes from noncritical to critical. In this case, the BS should inform all sensor nodes about the new situation and eventually reconfigure the WSN.

Retransmission in the RP. The retransmission order in the RP depends on the ACK bitmap received from the BS. Using an increasing slot sequence, firstly data of all sensor nodes having the highest priority and the bit false in the ACK bitmap are retransmitted successively. Then, data of all sensor nodes having the second highest priority and the bit false in the ACK bitmap are retransmitted successively, and so on. If there are not enough slots to permit all required retransmissions, then schedule truncation is done to guarantee that no retransmission occurs in the NTP. It should be noted that retransmissions may not be the appropriate error recovery mechanism if losses are due to fading.

Retransmission in the CAP. If a sensor node does not receive the beacon, then it fails to receive the ACK bitmap, and therefore it does not know if the BS correctly received the packet sent in the NTP of the last superframe. Retransmitting that packet properly in the RP is impossible, because such a sensor node cannot compute the slot allocation schema of the RP. A solution to overcome this problem is to retransmit it during the CAP using the slotted CSMA-CA. This procedure should improve the packet loss ratio at the expense of some energy consumption and CAP slot wastage, since the sensor node may be transmitting a duplicated packet. Indeed, a packet already received by the BS may be retransmitted again in the CAP of the next superframe if the sensor node does not receive the beacon.

Simulation Testbed

To test the solutions proposed in the previous section and to compare them with LPRT, the Castalia simulator was used. For this purpose, a case study and distinct MAC protocol operation modes were programmed in Castalia. For each test run, a simulation time of 1 h was defined. Extending the simulation time would not significantly affect the results. During that time period, the BS sends around 16 363 beacons, carrying in the payload only the ACK bitmap (4B). It is assumed that the BER is equal in both communication directions.

In some graphics the probability P of a fully loaded packet being received at destination is used instead of the BER. The two parameters are related by

$$\mathrm{BER} = 1 - P^{1/(8\times\mathrm{MPS})}\,, \qquad (74.1)$$

where MPS is the maximum physical frame size. Considering MPS = 133 B, the BER changes between 0 and around 6.5×10^{-4} as P decreases from 1 to 0.5 through the simulation runs. Typical values for BER in a good real wireless channel are of the order of 10^{-5} and in a bad channel may be less than 10^{-3}.

Operation Modes

Five distinct operation modes were defined for the simulation framework: 0, 1, 2, 3, and 4 [74.100]. In operation mode 0 there are no retransmissions in case of lost frames. This operation mode serves basically to define the limiting lines in the plots by considering the worst case, corresponding to no error control mechanism in the system.

In operation mode 1 lost packets are retransmitted once at RP, and these retransmitted packets are not acknowledged. This operation mode implements basically the LPRT protocol.

In operation mode 2, packets with payload size above a certain threshold may be retransmitted during RP at most twice. The threshold was set at 40 bytes, so ECG and ART sensors have two chances to retransmit lost data. The second retransmission occurs only if the frame is not correctly received by the BS during the first retransmission. For this reason, the first retransmission must be acknowledged, although the second retransmission does not need to be. Packets with payload size below the defined threshold may be retransmitted once in the RP and are not acknowledged.

In operation mode 3, packets with payload size above a certain threshold may be retransmitted at most thrice in RP.

Finally, operation mode 4 is similar to operation mode 2, but if a sensor does not receive a beacon, it tries to send in the CAP the packet transmitted in the NTP of the last superframe. In all operation modes, frames transmitted in the NTP are acknowledged through the ACK table sent in the payload of the next beacon. No operation mode allows aggregated transmissions in the NTP or RP.

Compared with mode 1, the improvement achieved when the WSN operates in modes 2, 3, and 4 is clear from Fig. 74.30 and Fig. 74.31. These operation modes present ECG packet loss below 1% while the probability P is a little above 0.8. Such improvement is due to the multiple retransmission process occurring in the superframe. When the WSN operates in mode 3 instead of mode 4, the improvement in ECG traffic is significant for $P < 0.75$. However, for ART traffic it is almost irrelevant whether mode 3 or 4 is used. This is because the probability of losing an ART packet is lower than the probability of losing an ECG packet, as the packet size is smaller in the former.

Since the oximetry, respiration rate, and temperature sensors may only retransmit once in RP, no significant difference is noted in such traffic when operating in modes 1, 2 or 3 (Fig. 74.32). However, an improvement is noticed in oximetry and respiration rate traffic when operating in mode 4. Temperature traffic does not exhibit such improvement because it cannot make use of the CAP.

To better evaluate the cost of the various operating modes in terms of consumed energy, we firstly consider that samplings performed by the sensing devices consume no power. Figure 74.33 presents for each operation mode the average lifetime per initial energy of the battery for each sensor type, relative to the simulations run with probability $P = 0.75$.

To determine how many minutes a type of sensor would live under such conditions, the values on the y-axis must be multiplied by the energy available in the battery when the sensor starts working. The available

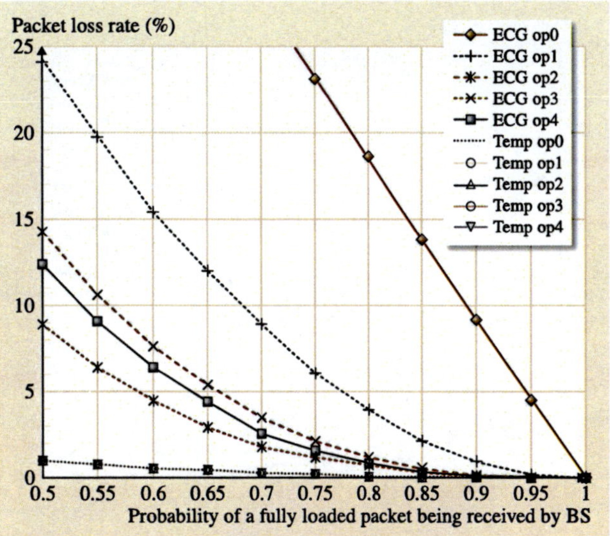

Fig. 74.30 Loss rate for ECG and temperature traffic

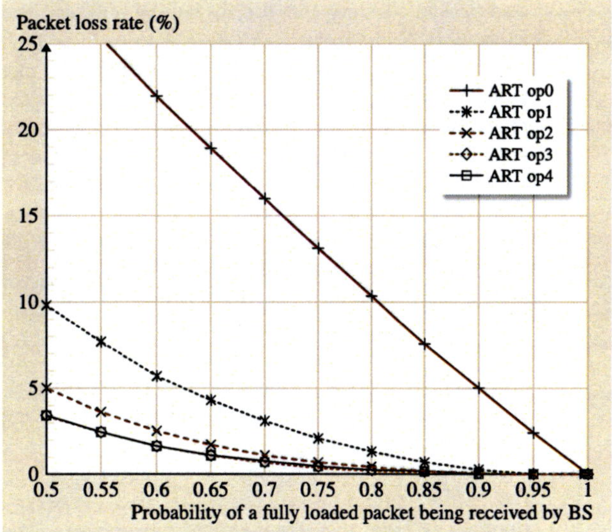

Fig. 74.31 Loss rate for ART traffic

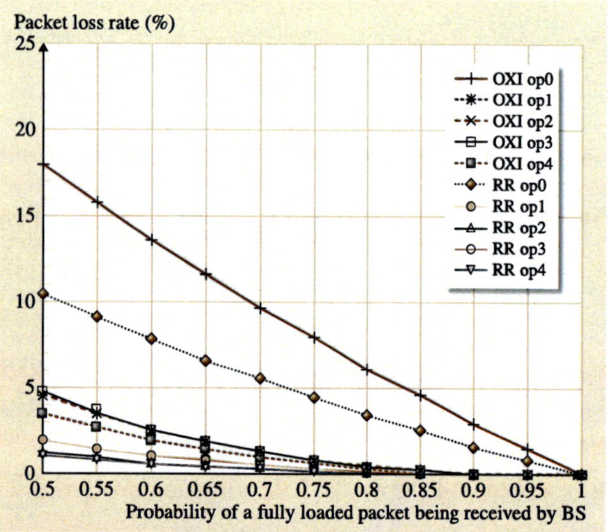

Fig. 74.32 Loss rate for OXI, and RR traffic

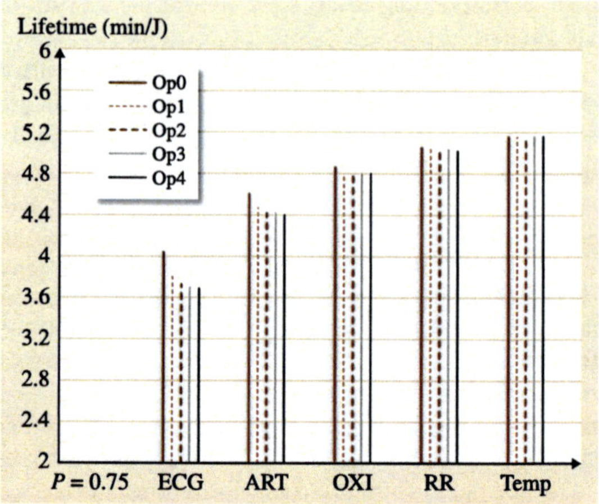

Fig. 74.33 Lifetime for sensing consumption of 0 mJ

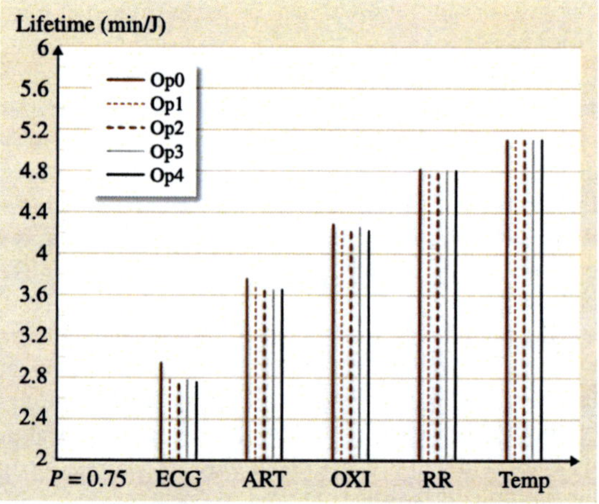

Fig. 74.34 Lifetime for sensing consumption of 0.01 mJ

energy E in a battery with n cells, having U (volts) and capacity of C (mAh) for each cell, is $E = n\,U\,C \times 3.6$ J; for instance, the 3 V, 40 mAh rechargeable Li polymer battery used by an ECG sensor node contains 432 J of initial energy before starting to power a sensor. So, according to Fig. 74.33, an ECG sensor operated in mode 4 and powered by that battery would live for 3.68×432 min ≈ 26.5 h. For every sensor type, the energy consumption is not significantly increased if the probability $P > 0.75$ and the operation mode is 1, 2, 3 or 4.

Next, it was considered that each sampling performed by the sensing devices consumes 0.01 mJ of energy. Figure 74.34 presents the results for this situation. The effect of sampling rate on the lifetimes of the sensors is clear.

As ECG sensors have both the highest sampling rate and the largest transmitted packet size, the energy drainage is faster than for the other sensor types. The differences observed in each sensor type for the various operating modes are almost indistinguishable.

Conclusions

Since wireless e-emergency networks should be reliable and efficient, they need to support QoS at distinct protocol levels, including the MAC layer. The deployment of LPRT in an e-health system leads to low power consumption, controlled latency, and throughput efficiency. However, as simulations have proved, this approach lacks reliability if the wireless channel presents a given bit error rate. To define a MAC protocol that is more robust to bit error conditions than LPRT, some solutions based on short-size beacons and multiple retransmissions were tested. The results showed that such strategies lead to meaningful improvements in packet loss rate, without significantly increasing energy consumption.

Despite presenting the best performance regarding the loss rate parameter, we foresee that a network operating in mode 3 may face scalability problems, since considerable bandwidth wastage occurs in the RP. Since operation mode 4 offers better scalability and a good compromise in terms of packet loss rate, we believe that an e-health network operating in this mode and using short-size beacons may present interesting performance.

References

74.1 D. Lapedes: *McGraw-Hill Dictionary of Scientific and Technical Terms* (McGraw-Hill, New York 2000)

74.2 J.C. Schuh: Medical device regulations and testing for toxicologic pathologists, Toxicol. Pathol. **36**, 63–69 (2008)

74.3 A. Hodgkin, A. Huxley: A quantitative description of membrane current and its application to conduction and excitation in nerve, Bull. Math. Biol. **52**, 25–71 (1990)

74.4 J. Webster: *Medical Instrumentation: Application and Design* (Wiley-India, New Delhi 2009)

74.5 F. Silveira, D. Flandre: *Low Power Analog CMOS for Cardiac Pacemakers: Design and Optimization in Bulk and SOI Technologies* (Kluwer Academic, Dordrecht 2004)

74.6 M. Schoenfeld: Pacemaker programmers: An updated synopsis, Card. Electrophysiol. Rev. **3**, 20–23 (1999)

74.7 S. Majerus: A Low-Power Wireless Transceiver for Deeply Implanted Biomedical Devices. Ph.D. Thesis (Case Western Reserve University, Cleveland 2008)

74.8 G. Qin, Y. Liu, H. Liu, Y. Ding, X. Qi, R. Du: Fabrication of bio-microelectrodes for deep-brain stimulation using microfabrication and electroplating process, Microsyst. Technol. **15**, 933–939 (2009)

74.9 P. Chappell, A. Cranny, D. Cotton, N. White, S. Beeby: Sensory motor systems of artificial and natural hands, Int. J. Surg. **5**, 436–440 (2007)

74.10 C.C. Santina: The past, present and future of bionic devices restoring cranial nerve function, research in vestibular, Science **8**, 81–82 (2009)

74.11 R. Hendriks: Bionic eyes retinal implants, still in their infancy, provide a new vision of the future, Nerac Insight, March (2008)

74.12 J. Luprano, J. i Caros, A. Ridolfi, S. Pasche, B. Gros: New generation of smart sensors for biochemical and bioelectrical applications, Personalized Health 2007 Conf. (2007)

74.13 J. El-Ali, P. Sorger, K. Jensen: Cells on chips, Nature **442**, 403–411 (2006)

74.14 D. Psaltis, S. Quake, C. Yang: Developing optofluidic technology through the fusion of microfluidics and optics, Nature **442**, 381–386 (2006)

74.15 H. Craighead: Future lab-on-a-chip technologies for interrogating individual molecules, Nature **442**, 387–393 (2006)

74.16 I. Chilibon: Ultrasound transducer for medical therapy, Sens. Actuators A **142**, 124–129 (2008)

74.17 M. Healy, T. Newe, E. Lewis: Wireless sensor node hardware: A review, IEEE Sensors (2008) pp. 621–624

74.18 A. Pantelopoulos, N. Bourbakis, A. Center: A survey on wearable biosensor systems for health monitoring, IEEE EMBS 2008, 30th Annu. Int. Conf. (2008) pp. 4887–4890

74.19 M. Chan, D. Estève, C. Escriba, E. Campo: A review of smart homes-Present state and future challenges, Comput. Meth. Progr. Biomed. **91**, 55–81 (2008)

74.20 B.A. Parviz: Augmented reality in a contact lens, IEEE Spectrum **9**, 1–4 (2009)

74.21 R. Brooks: I, Rodney Brooks, am a robot, IEEE Spectrum **45**, 71–75 (2008), http://ieeexplore.ieee.org/stamp/stamp.jsp?tp=&arnumber=4531466

74.22 K. Warwick: Robot-human interaction – Practical experiments with a cyborg. In: *Informatics in Control, Automation and Robotics I*, ed. by J. Braz, H. Araújo, A. Vieira, B. Encarnação (Springer, Dordrecht 2006) pp. 3–10

74.23 M.R.R. Powell: *Future Cyborgs: Human-Machine Interface for Virtual Reality Applications* (Air University Stroming Media, Washington 2007)

74.24 S. Buerger, R. Olsson III, K. Wojciechowski, E. Yepez III, D. Novick, K. Peterson, T. Turner, J. Wheeler, B. Rohrer, D. Kholwadwala: Portable, chronic neural interface system design for sensory augmentation, 3rd Int. IEEE/EMBS Conf. Neural Eng. CNE'07 (2007) pp. 23–27

74.25 R. Bogue: Exoskeletons and robotic prosthetics: A review of recent developments, Ind. Robot **36**, 421–427 (2009)

74.26 T. Onishi, T. Arai, K. Inoue, Y. Mae: Development of the basic structure for an exoskeleton cyborg system, Artif. Life Robot. **7**, 95–101 (2003)

74.27 W. Karlen: Adaptive Wake and Sleep Detection for Wearable Systems. Ph.D. Thesis (Ecole Polytechnique Fédérale de Lausanne, Lausanne 2009)

74.28 M. Falkenstein, J. Hoormann, S. Christ, J. Hohnsbein: ERP components on reaction errors and their functional significance: A tutorial, Biol. Psychol. **51**, 87–107 (2000)

74.29 C. Berka, D. Levendowski, M. Cvetinovic, M. Petrovic, G. Davis, M. Lumicao, V. Zivkovic, M. Popovic, R. Olmstead: Real-time analysis of EEG indexes of alertness, cognition, and memory acquired with a wireless EEG headset, Int. J. Human-Computer Interact. **17**, 151–170 (2004)

74.30 A. Buttfield, P. Ferrez, J. Millan: Towards a robust BCI: Error potentials and online learning, IEEE Trans. Neural Syst. Rehabil. Eng. **14**, 164–168 (2006)

74.31 N. Vodjdani: The ambient assisted living joint programme, Technol. Conf. 2008 ESTC (2008) pp. 1–2

74.32 A. Milenkovic, C. Otto, E. Jovanov: Wireless sensor networks for personal health monitoring: Issues and an implementation, Comput. Commun. **29**, 2521–2533 (2006)

74.33 L. Summerer, D. Izzo, L. Rossini: *Brain Machine Interfaces for Space Applications* (Academic, New York 2009)

74.34 M. Bächlin, K. Förster, G. Tröster: SwimMaster: A wearable assistant for swimmer, Proc. 11th Int. Conf. Ubiquitous Comput. – Ubicomp '09, New York (ACM, New York 2009) p. 215

74.35 A.J. Wixted, D.V. Thiel, A.G. Hahn, C.J. Gore, D.B. Pyne, D.A. James: Measurement of energy expenditure in elite athletes using MEMS-based triaxial accelerometers, IEEE Sens. J. **7**, 481–488 (2007)

74.36 S. Korsakas, A. Vainoras, V. Miskinis, R. Ruseckas: Electrocardiosignals and motion signals telemonitoring and analysis system, Comput. Cardiol. **32**, 363–366 (2005)

74.37 F. Fruett: Development of an electromechanical sensor system to monitor sports activities, Measurement **4**, 6–10 (2009)

74.38 A. Nijholt, J. Van Erp, D. Heylen: BrainGain: BCI for HCI and games, Proc. AISB Symp. Brain Comput. Interfaces Human Comput. Interact.: Convergence of Ideas (Aberdeen 2008) pp. 32–35

74.39 B. Shim, S. Lee, J. Shin: Implementation of a 3-dimensional game for developing balanced brainwave, software engineering research, management and applications, 5th IEEE ACIS Int. Conf. SERA (2007) pp. 751–758

74.40 M. Cheng: *Medical Device Regulations: Global Overview and Guiding Principles* (World Health Organization, Geneva 2003)

74.41 P. Mendes, J. Correia: *MEMS Micro-Antennas for Wireless Biomedical Systems* (Nova Publisher, Hauppauge 2007)

74.42 C.P. Figueiredo, P.M. Mendes: *Antennas for Wireless Biomedical Devices, Antennas* (Nova Publisher, Hauppauge 2009)

74.43 O.S. Gama, H. Martins, C. Pereira, S. Soares, A. Valente, V.S. Ribeiro, P. Carvalho, P.M. Mendes: A platform with combined environmental and physiological wireless data acquisition for AAL applications, ISAmI 2010 Conf. Int. Symp. Ambient Intell. (Guimarães 2010)

74.44 U. Varshney: *Pervasive Healthcare Computing: EMR/HER, Wireless and Health Monitoring* (Springer, New York 2009)

74.45 N. Thakor: *Biopotentials and Electrophysiology Measurement* (Johns Hopkins School of Medicine, Baltimore 1999)

74.46 A. Cohen: Biomedical signals. In: *The Biomedical Engineering Handbook*, ed. by J.D. Bronzino (CRC, Boca Raton 1999), Chap. 52-1

74.47 Y. Hao, R. Foster: Wireless body sensor networks for health-monitoring applications, Physiol. Meas. **29**(11), 27–56 (2008)

74.48 U. Varshney: Pervasive healthcare and wireless health monitoring, Mob. Netw. Appl. **12**, 113–127 (2007)

74.49 A. Pantelopoulos, N. Bourbakis: A survey on wearable sensor-based systems for health monitoring and prognosis, IEEE Trans. Syst. Man Cybern. **40**, 1–12 (2010)

74.50 United Nations: *World Population Ageing, Population* (2009)

74.51 F. Axisa, P.M. Schmitt, C. Gehin, G. Delhomme, E. Mcadams, A. Dittmar: Flexible technologies and smart clothing for citizen medicine, home healthcare, and disease prevention, IEEE Trans. Inform. Technol. Biomed. **9**, 325–336 (2005)

74.52 Q.A. Khan, S.J. Bang: *Energy Harvesting for Self Powered Wearable Health Monitoring System* (Health, San Francisco 2009) pp. 1–5

74.53 L. Mainardi, A. Bianchi, S. Cerutti: Digital biomedical signal acquisition and processing. In: *The Biomedical Engineering Handbook*, ed. by J.D. Bronzino (CRC, Boca Raton 1995)

74.54 M. Neuman: *Biopotential Electrodes, Medical Instrumentation* (Wiley, New York 1998)

74.55 B. Le, T. Rondeau, J. Reed, C. Bostian: Analog-to-digital converters, IEEE Signal Process. Mag. **22**(6), 69–77 (2005)

74.56 A.F. Silva, F. Goncalves, L.A. Ferreira, F.M. Araújo, P.M. Mendes, J.H. Correia: PVC smart sensing foil for advanced strain measurements, IEEE Sensors J. **10**, 1149–1155 (2010)

74.57 P. Baronti, P. Pillai, V. Chook, S. Chessa, A. Gotta, Y. Hu: Wireless sensor networks: A survey on the state of the art and the 802.15.4 and ZigBee standards, Comput. Commun. **30**, 1655–1695 (2007)

74.58 B.P. Mitcheson, M. Ieee, E.M. Yeatman, S.M. Ieee, G.K. Rao, S.M. Ieee, A.S. Holmes, T.C. Green: Energy harvesting from human and machine motion for wireless electronic devices, Proc. IEEE **96**, 1457–1486 (2008)

74.59 C.P. Figueiredo, O.S. Gama, C.M. Pereira, S. Silva, L. Domingues, K. Hoffmann, P.M. Mendes: Autonomy suitability of wireless modules for ambient assisted living applications, SENSORCOMM 2010, Fourth Int. Conf. Sensor Technol. Appl. (2010)

74.60 D.P. Ferreira, B.N. Lourenço, F.J. Rodrigues, P.M. Mendes: Design of efficient wireless links for invasive biomedical microdevices, BioMed 2010, Seventh IASTED Int. Conf. Biomed. Eng. (2010)

74.61 S. Silva, L. Domingues, P.M. Mendes: MagTag – A wearable wrist device for localization applications, 4th Int. ICST Conf. Pervasive Comput. Technol. Healthcare 2010 (Munique 2010)

74.62 Y. Kim, S. Kang, M. Lee, B. Yu, T. Zyung: Optimization of wireless power transmission through resonant coupling, 2009 Compat. Power Electron. (2009) pp. 426–431

74.63 A. Kurs, A. Karalis, R. Moffatt, J. Joannopoulos, P. Fisher, M. Soljacic: Wireless power transfer via

strongly coupled magnetic resonances, Science **317**, 83 (2007)
74.64 C. Glaros, D.I. Fotiadis: Wearable devices in healthcare, Intell. Paradig. Healthc. Enterp. **264**, 237–264 (2005)
74.65 X. Teng, Y. Zhang, C. Poon, P. Bonato: Wearable medical systems for p-health, IEEE Rev. Biomed. Eng. **1**, 62–74 (2008)
74.66 C. Gopalsamy, S. Park, R. Rajamanickam, S. Jayaraman: The wearable motherboard: The first generation of adaptive and responsive textile structures (ARTS) for medical applications, Virtual Real. **4**, 152–168 (1999)
74.67 J. Winters, Y. Wang: Wearable sensors and telerehabilitation, IEEE Eng. Med. Biol. Mag. **22**, 56–65 (2003)
74.68 S. Park, C. Gopalsamy, R. Rajasanickam, S. Jayaraman: The wearable motherboard: A flexible information infrastructure or sensate liner for medical applications, Med. Meets Virtual Real. **62**, 252–258 (1999), IOS, Amsterdam
74.69 M. Sackner, D. Keenan, D. Inman: Method and system for extracting cardiac parameters from plethysmographic signals, Patent 7604603 (2009)
74.70 Bodymedia Corp. Pittsburgh, PA
74.71 P. Lukowicz, U. Anliker, J. Ward, G. Troster, E. Hirt, C. Neufelt: AMON: A wearable medical computer for high risk patients, Proc. 6th Int. Symp. Wearable Comput. (2002) pp. 133–134
74.72 R. Paradiso, D. De Rossi: Advances in textile technologies for unobtrusive monitoring of vital parameters and movements, Eng. Med. Biol. Soc. 28th Annu. Int. Conf. IEEE (2006) pp. 392–395
74.73 S. Choi, Z. Jiang: A novel wearable sensor device with conductive fabric and PVDF film for monitoring cardiorespiratory signals, Sens. Actuators A **128**, 317–326 (2006)
74.74 P. Pandian, K. Mohanavelu, K. Safeer: TM, Smart Vest: Wearable multi-parameter remote physiological monitoring system, Med. Eng. Phys. **30**, 466–477 (2008)
74.75 G. Kanellos, G. Papaioannou, D. Tsiokos, C. Mitrogiannis, G. Nianios, N. Pleros: Two dimensional polymer-embedded quasi-distributed FBG pressure sensor for biomedical applications, Opt. Express **18**, 179–186 (2010)
74.76 F. Carpi, D. De Rossi: Electroactive polymer-based devices for e-textiles in biomedicine, IEEE Trans. Inform. Technol. Biomed. **9**, 295–318 (2005)
74.77 G. Knopf, A. Bassi: *Smart Biosensor Technology* (CRC, Boca Raton 2007)
74.78 R. Lane, B. Craig: Materials that sense and respond – An introduction to smart materials, AMPTIAC Newslett. **7**, 9–14 (2003)
74.79 C. Figueiredo, K. Becher, K.-P. Hoffmann, P. Mendes: Low power wireless acquisition module for wearable health monitoring systems, IEEE Int. Eng. Med. Biol. Soc. 32nd Annu. Conf. (IEEE, 2010) pp. 704–707
74.80 K. Hoffmann, R. Ruff: Flexible dry surface-electrodes for ECG long-term monitoring, Eng. Med. Biol. Soc. (2007) pp. 5739–5742
74.81 A.C. Wong, G. Kathiresan, C.K. Chan, O. Eljamaly, A.J. Burdett: A 1 V wireless transceiver for an ultra low power SoC for biotelemetry applications, ESSCIRC 2007 – 33rd Eur. Solid-State Circuits Conf. IEEE (2007) pp. 127–130
74.82 M. El-Sherif, J. Yuan, A. Macdiarmid: Fiber optic sensors and smart fabrics, J. Intell. Mater. Struct. **11**, 407–414 (2000)
74.83 F. Berghmans, H. Thienpont, P.V. Daele, P. Dubruel, W. Urbanczyk, J. Rayss, J.D. Webb, J. Vlekken, K. Kalli, K. Sugden: Photonic skins for optical sensing: Highlights of the PHOSFOS Project, 20th Int. Conf. Opt. Fibre Sens. 7503 (2009)
74.84 A. Lymberis, D.D. Rossi: Smart Fabrics: Integrating fiber optic sensors and information networks, Stud. Health Technol. Inform. (2004) pp. 317–323
74.85 A. Searle, L. Kirkup: A direct comparison of wet, dry and insulating bioelectric recording electrodes, Physiol. Meas. **21**, 271–283 (2000)
74.86 C. Ryu, S. Nam, S. Kim: Conductive rubber electrode for wearable health monitoring, Eng. Med. Biol. Soc. (2005)
74.87 S. Jang, J. Cho, K. Jeong, G. Cho: Exploring possibilities of ECG electrodes for bio-monitoring smartwear with Cu sputtered fabrics, Human-Comput. Interact. Platf. Tech. (2007) pp. 1130–1137
74.88 J. Baek, J. An, J. Choi, K. Park, S. Lee: Flexible polymeric dry electrodes for the long-term monitoring of ECG, Sens. Actuators A **143**, 423–429 (2008)
74.89 M. Fernandes, N. Dias, J. Nunes, M. El Tahchi, S. Lanceros-Méndez, J. Correia, P. Mendes: Wearable brain cap with contactless electroencephalogram measurement for brain-computer interface applications, Proc. 4th Int. IEEE EMBS Conf. Neural Eng. IEEE (2009) pp. 387–390
74.90 O. Gama, P. Carvalho, J. Afonso, P. Mendes: Quality of service support in wireless sensor networks for emergency healthcare services, Eng. Med. Biol. Soc. 30th Annu. Int. Conf. IEEE (2008) pp. 1296–1299
74.91 I. Gondal, M. Iqbal, M. Woods, S. Sehgal: Integrated sensing and diagnosis – the next step in real time patient health care, 6th IEEE/ACIS Int. Conf. Comput. Inf. Sci. (2007) pp. 581–586
74.92 M. Paksuniemi, H. Sorvoja, E. Alasaarela, R. Myllylä: Wireless sensor and data transmission needs and technologies for patient monitoring in the operating room and intensive care unit, IEEE Eng. Med. Biol. 27th Annu. Conf. (2005) pp. 5182–5185
74.93 S. Arnon, D. Bhastekar, D. Kedar, A. Tauber: A comparative study of wireless communication network configurations, IEEE Wireless Commun. (2003)
74.94 T. Gao, D. Greenspan, M. Welsh, R. Juang, A. Alm: Vital signs monitoring and patient tracking over

a wireless network, 27th Annu. Int. Conf. IEEE EMBS (2005)

74.95 A. Iglesias, R. Istepanian, J. García Moros: Performance study of real-time ECG transmission in wireless networks, Int. Special Topic Conf. Inf. (2006)

74.96 I. Lamprinos, A. Prentza, D. Koutsouris: Communication protocol requirements of patient personal area networks for telemonitoring, Technol. Health Care **14**, 171–187 (2006)

74.97 O.S. Gama, P. Carvalho, J.A. Afonso, P.M. Mendes: Wireless sensor networks with QoS for e-health and e-emergency applications, Proc. 2nd Int. Workshop on e-Health Serv. Technol., Porto (2008) pp. 60–69

74.98 I. Ye, J. Heidemann, D. Estrin: MAC with coordinated adaptive sleeping for wireless sensor networks, Int. Conf. Sensor Technol. Appl. **12**(3), 493–506 (2004)

74.99 J. Afonso, L. Rocha, H. Silva, J. Correia: MAC protocol for low-power real-time wireless sensing and actuation, Circuits Sys. (2006)

74.100 Ó. Gama, P. Carvalho, J. Afonso, P. Mendes: Trade-off analysis of a MAC protocol for wireless e-emergency systems, Sens. Sys. Softw. (2009) pp. 222–235

Appendix

Table A.1 Electromagnetic spectrum

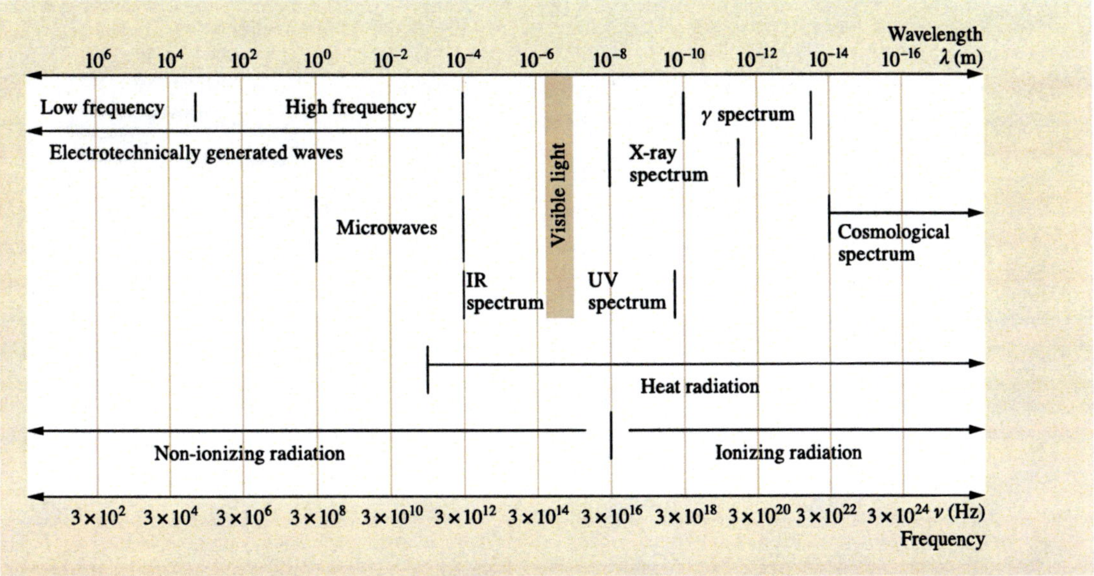

Table A.2 Frequency- and signal voltage range of different bioelectrical signals (after J. Eichmeier: *Medizinische Elektronik*, 3rd edn. (Springer, Berlin, Heidelberg 1997) p. 26)

Bioelectrical signal	Abbreviation	Frequency range (Hz)	Signal voltage range U (mV)
Lead, intracellular	L_{ic}	$0-10^{4}$	50–130
Action potential	AP or E	5×10^{4}	–
Electroencephalography	EEG	1–100	0.005–0.1
Electrodermography	EDG	0–1	0.1–5
Electrogastrography	EGG	0.02–0.2	0.2–1
Electrohysterography	EHG	0–200	0.1–8
Electrocardiography	ECG	0.2–200	0.1–3
Electrocorticography	ECoG	$10-10^{4}$	0.015–0.3
Electromyography	EMG		
surface electrode		$10-10^{3}$	0.1–5
depth electrode		$10-10^{4}$	0.05–5
Electroretinography	ERG	0.1–100	0.02–0.3
Energy bioelectrical signals: $3\times10^{-12}-7\times10^{-15}$ W s			

Radionuclides (Selection) and Dosimetric Base Items

Table A.3 Radionuclides (selection) (after W. Kauffmann, E. Moser, R. Sauer: *Radiologie*, 2nd edn. (Urban Fischer, München 2001) p. 14)

Radionuclide	Half-life ($T_{1/2}$)	Radiator	β energy (MeV)	γ energy (MeV)	Application
^{3}H	12.26 a	β	0.018		D, M
^{14}C	5730 a	β	0.156		D, M
^{32}P	14.28 d	β	1.710		RT
^{51}Cr	27.8 d	γ		0.323	D, M
^{57}Co	270 d	γ		0.122; 0.136	D, M
^{49}Fe	45.6 d	β; γ	1.570	1.095; 1.292	D, M
^{60}Co	5.26 a	β; γ	0.33; 0.09	1.17; 1.33	RT
^{67}Ga	78 h	γ	0.3; 0.018–0.9		D, M
^{75}Se	120.4 d	γ		0.14; 0.28	D, M
^{87}Sr	2.8 h	γ		0.388	D, M
^{90}Sr	27.7 a	β	0.546; 0.17		RT
^{90}Y	64 h/2.66 d	γ	0.92; 2.27		RT
^{99m}Tc	6 h	γ		0.14	D, M
$^{106}Ru + ^{106}Rh$	368 d	β	1	0.04–2.4	RT
^{111}In	2.8 d	γ		0.172; 0.247	D, M
^{113m}In	1.66 h	γ		0.393	D, M
^{123}J	13.6 h	γ		0.159	D, M, RT
^{125}J	60.2 d	γ		0.035	D, M, RT
^{131}J	8.05 d	β; γ	0.806	0.364	D, M, RT
^{133}Xe	5.27 d	β; γ	0.346	0.081	D, M
^{137}Cs	30 a	β; γ	0.51; 1.176	0.662	D, M, RT
^{192}Ir	74.2 d	β; γ	0.24–0.67; 0.17	0.317	RT
^{197}Hg	2.71 d	γ		0.077	D, M
^{198}Au	2.7 d	β	0.962; 0.31	0.412	D, M, RT

m = metastable, D = diagnostic, M = marker, RT = radiation therapy

Table A.4 Dosimetric base items (after G.H. Hartmann: Konzepte in Strahlenphysik und Dosimetrie. In: *Medizinische Physik*, Vol. 2, ed. by W. Schlegel, J. Bille (Springer, Berlin, Heidelberg 2002) p. 73)

Base item	Derivation	SI-unit	Stochastic
Energy imparted	$R_{in} - R_{out} + \sum Q$	J	•
Linear energy	ϵ_1 / l	$J\,m^{-1}$	•
Specific energy	ϵ / m	$J\,kg^{-1} = Gy$	•
Absorbed dose	dW/dm	$J\,kg^{-1} = Gy$	
Absorbed dose rate	dD/dt	$J\,kg^{-1}\,s^{-1} = Gy\,s^{-1}$	
Kerma	dE_{tr}/dm	$J\,kg^{-1} = Gy$	
Kerma rate	dK/dt	$J\,kg^{-1}\,s^{-1} = Gy\,s^{-1}$	
Specific ionization	dQ/dm	$C\,kg^{-1} = R$	
Ion dose rate	dJ/dt	$A\,kg^{-1}$	
Exposure	dQ/dm	$C\,kg^{-1}$	
Dose equivalent	$q \cdot D$	$J\,kg^{-1} = Sv$	
Equivalent dose rate	dH/dt	$Sv\,s^{-1}$	

International System of Units – SI, CGS and Prefixes

Table A.5 International systems of units (SI)

Quantity	Name	Symbol	Expression in terms of other SI units or SI base units
SI base units			
Length	Meter	m	
Mass	Kilogram	kg	
Time	Second	s	
Electric current	Ampere	A	
Thermodynamic temperature	Kelvin	K	
Amount of substance	Mole	mol	
Luminous intensity	Candela	cd	
SI derived units			
Area	Square meter	m^2	
Volume	Cubic meter	m^3	
Speed, velocity	Meter per second	m/s	
Acceleration	Meter per second squared	m/s^2	
Wave number	Reciprocal meter	m^{-1}	
Mass density	Kilogram per cubic meter	kg/m^3	
Specific volume	Cubic meter per kilogram	m^3/kg	
Current density	Ampere per square meter	A/m^2	
Magnetic field strength	Ampere per meter	A/m	
Amount-of-substance concentration	Mole per cubic meter	mol/m^3	
Luminance	Candela per square meter	cd/m^2	
Mass fraction	Kilogram per kilogram, which may be represented by the number 1	$kg/kg = 1$	
Plane angle	Radian	rad	$m/m = 1$
Solid angle	Steradian	sr	$m^2/m^2 = 1$
Frequency	Hertz	Hz	1/s
Force	Newton	N	$m\,kg/s^2$
Pressure stress	Pascal	Pa	$N/m^2 = m^{-1}\,kg\,s^{-2}$
Energy work, quantity of heat	Joule	J	$N\,m = m^2\,kg\,s^{-2}$
Power radiant flux	Watt	W	$J\,s = m^2\,kg\,s^{-3}$
Electric charge, quantity of electricity	Coulomb	C	A s
Electric potential difference, electromotive force	Volt	V	$W/A = m^2\,kg\,s^{-3}\,A^{-1}$
Capacitance	Farad	F	$C/V = m^{-2}\,kg^{-1}\,s^4\,A^2$
Electric resistance	Ohm	Ω	$V/A = m^{-2}\,kg\,s^{-3}\,A^{-2}$
Electric conductance	Siemens	S	$A/V = m^{-2}\,kg^{-1}\,s^3\,A^2$
Magnetic flux	Weber	Wb	$V\,s = m^2\,kg\,s^{-2}\,A^{-1}$
Magnetic flux density	Tesla	T	$Wb/m^2 = kg\,s^{-2}\,A^{-1}$
Inductance	Henry	H	$Wb/A = m^2\,kg\,s^{-2}\,A^{-2}$
Celsius temperature	Degree Celsius	°C	K
Luminous flux	Lumen	lm	$cd\,sr = m^2\,m^{-2}\,cd = cd$
Illuminance	Lux	lx	$lm/m^2 = m^2\,m^{-4}\,cd = m^{-2}\,cd$
Activity (of a radionuclide)	Becquerel	Bq	s^{-1}

Table A.5 (continued)

Quantity	Name	Symbol	Expression in terms of other SI units or SI base units
Absorbed dose, specific energy (imparted), kerma	Gray	Gy	$J/kg = m^2/s^2$
Dose equivalent	Sievert	Sv	$J/kg = m^2/s^2$
Dynamic viscosity	Pascal second	Pa s	
Moment of force	Newton meter	N m	
Surface tension	Newton per meter	N/m	kg/s^2
Angular velocity	Radian per second	rad/s	
Angular acceleration	Radian per second squared	rad/s^2	
Heat flux density, irradiance	Watt per square meter	W/m^2	kg/s^3
Heat capacity, entropy	Joule per Kelvin	J/K	$m^2\,kg/s^2\,K$
Specific heat capacity, specific entropy	Joule per kilogram Kelvin	J/(kg K)	$m^2/s^2\,K$
Specific energy	Joule per kilogram	J/kg	m^2/s^2
Thermal conductivity	Watt per meter Kelvin	W/(m K)	$m\,kg/s^3\,K$
Energy density	Joule per cubic meter	J/m^3	$kg/m\,s^2$
Electric field strength	Volt per meter	V/m	$m\,kg/s^3\,A$
Electric charge density	Coulomb per cubic meter	C/m^3	$A\,s/m^3$
Electric flux density	Coulomb per square meter	C/m^2	$A\,s/m^2$
Permittivity	Farad per meter	F/m	$s^4\,A^2/m^3\,kg$
Permeability	Henry per meter	H/m	$m\,kg/s^2\,A^2$
Molar entropy, molar heat capacity	Joule per mole Kelvin	J/(mol K)	$m^2\,kg/s^2\,K\,mol$
Exposure (x- and γ-rays)	Coulomb per kilogram	C/kg	A s/kg
Absorbed dose rate	Gray per second	Gy/s	m^2/s^3
Radiant intensity	Watt per steradian	W/sr	
Radiance	Watt per square meter steradian	$W/(m^2\,sr)$	

Table A.6 Centimetre-gram-second system (CGS units)

Quantity	CGS unit abbreviation	Equivalent in SI units
Length	cm	10^{-2} m
Time	s	
Mass	g	10^{-3} kg
Velocity	cm/s	10^{-2} m/s
Force	dyn	10^{-5} N
Energy	erg	10^{-7} J
Power	erg/s	10^{-7} W
Pressure	Ba	10^{-1} Pa
Viscosity	P	10^{-1} Pa s
Wavenumber	cm^{-1}	$100\,m^{-1}$
Stokes	cm^2/s	$10^{-4}\,m^2/s$
Magnetic field	G	10^{-4} T
Magnetic field strength or intensity	Oe	$1000/(4\pi)$ A/m
Luminance	cd/cm^2	$10^4\,cd/m^2$

Table A.7 SI prefixes

Prefix-symbol	Exponent E *n* of decimal numbers	Factor 10^n	Prefix-symbol	Exponent E *n* of decimal numbers	Factor 10^n
yotta- (Y)	E 24	10^{24}	deci- (d)	E −1	10^{-1}
zetta- (Z)	E 21	10^{21}	centi- (c)	E −2	10^{-2}
exa- (E)	E 18	10^{18}	milli- (m)	E −3	10^{-3}
peta- (P)	E 15	10^{15}	micro- (μ)	E −6	10^{-6}
tera- (T)	E 12	10^{12}	nano- (n)	E −9	10^{-9}
giga- (G)	E 9	10^{9}	pico- (p)	E −12	10^{-12}
mega- (M)	E 6	10^{6}	femto- (f)	E −15	10^{-15}
kilo- (k)	E 3	10^{3}	atto- (a)	E −18	10^{-18}
hecto- (h)	E 2	10^{2}	zepto- (z)	E −21	10^{-21}
deca- (da)	E 1	10^{1}	yocto- (y)	E −24	10^{-24}
	E	10			

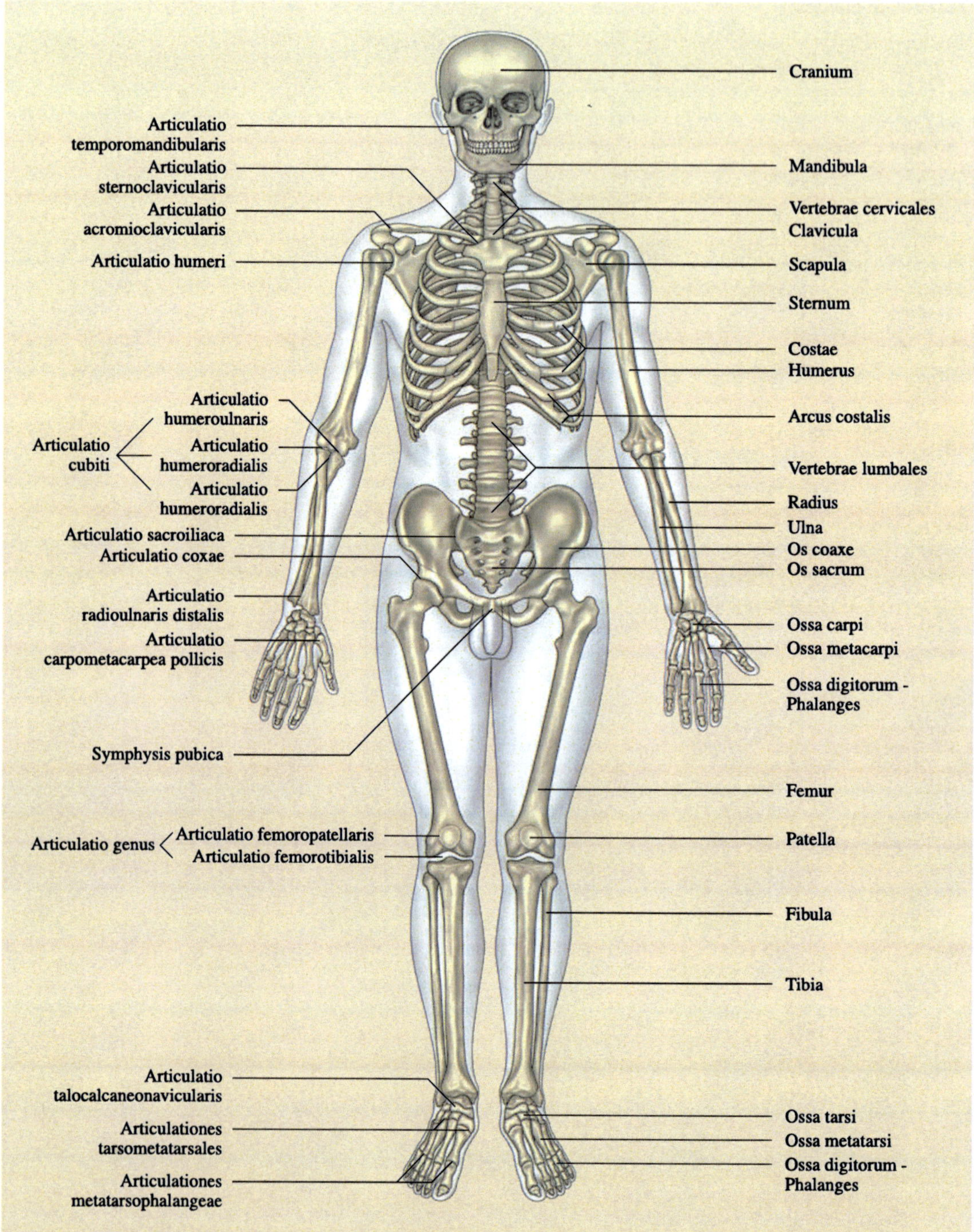

Fig. A.1 Human skeleton (ventral) (from K. Zilles, B. Tillmann: *Anatomie* (Springer, Berlin, Heidelberg 2010))

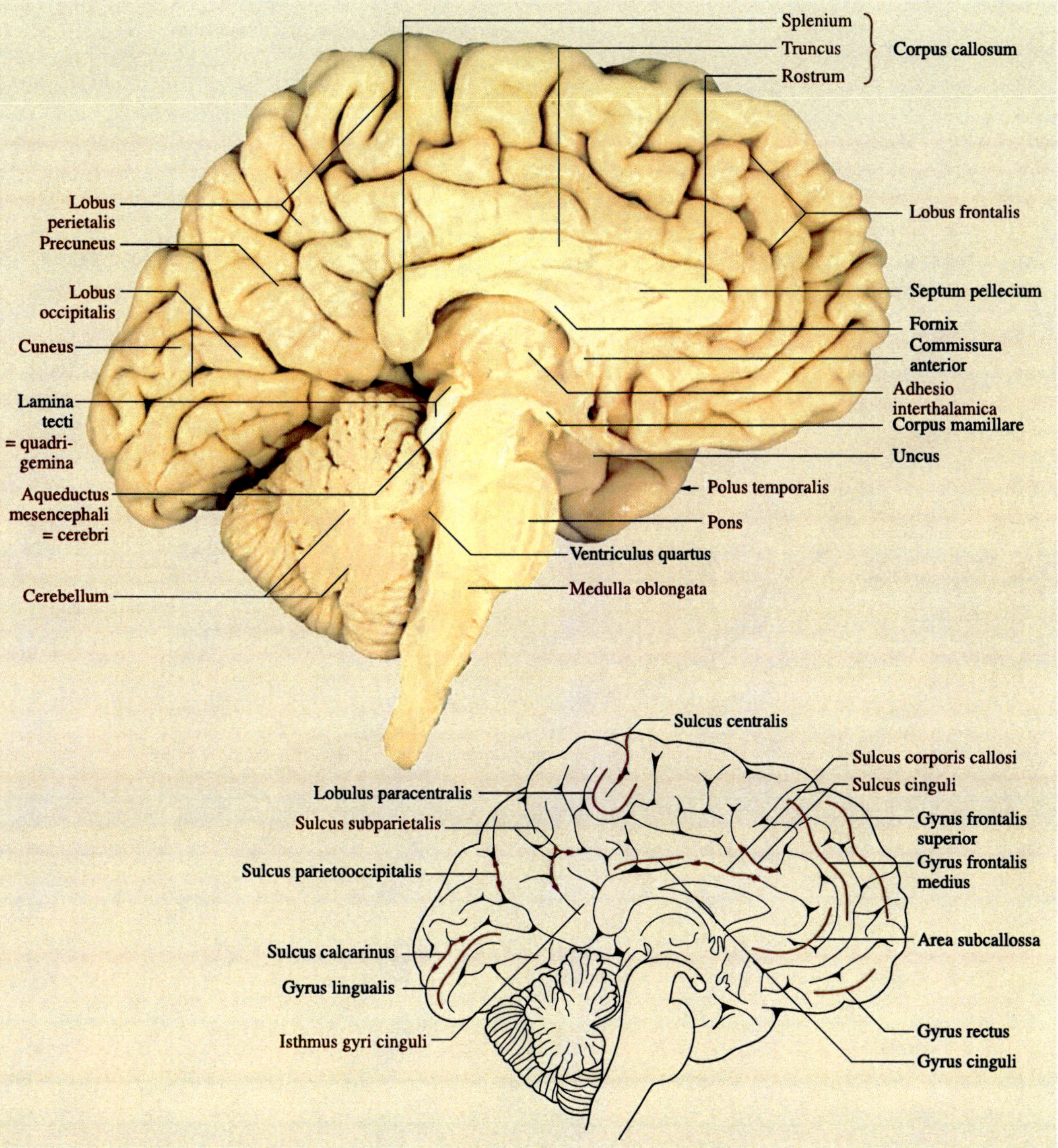

Fig. A.2 Human brain, median sagittal transaction (from B. Tillmann: Atlas der Anatomie, 2nd edn. (Springer, Berlin, Heidelberg 2010))

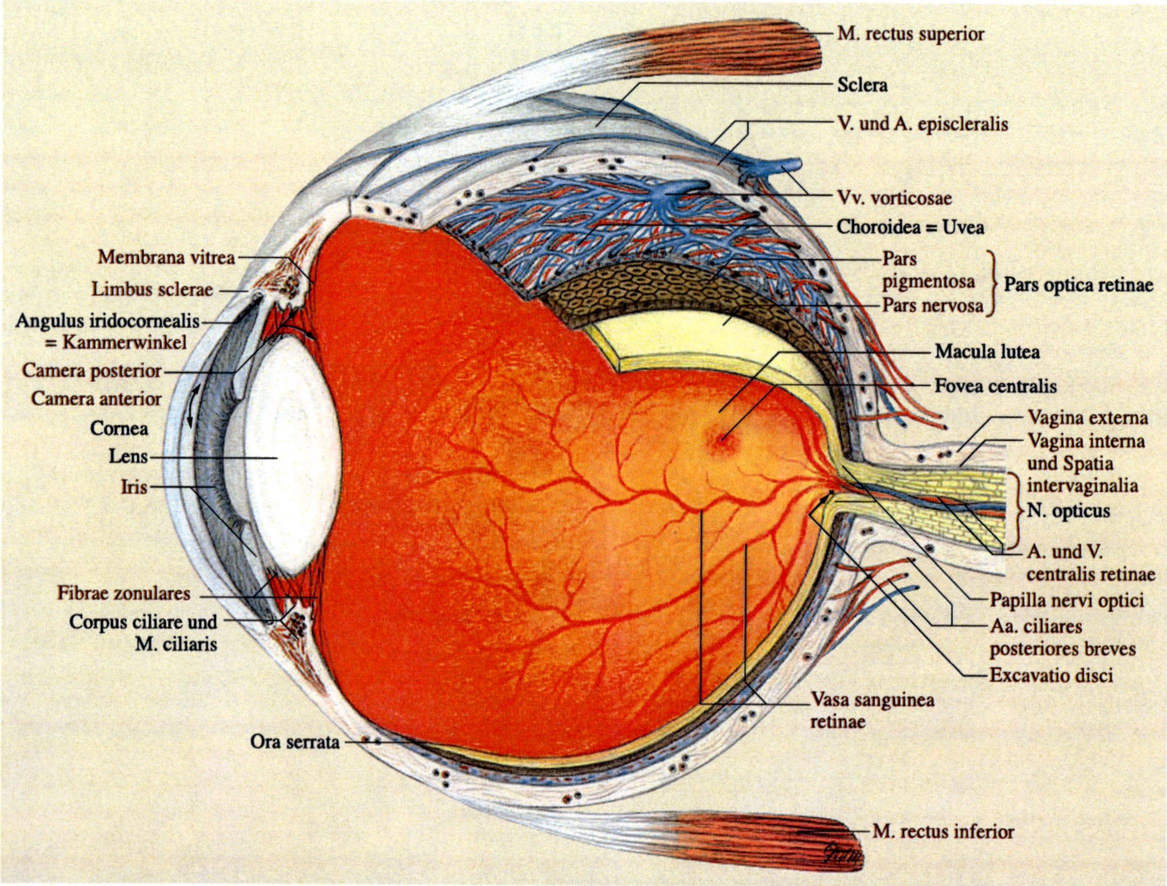

Fig. A.3 Eye ball, view from left lateral (from B. Tillmann: Atlas der Anatomie, 2nd edn. (Springer, Berlin, Heidelberg 2010))

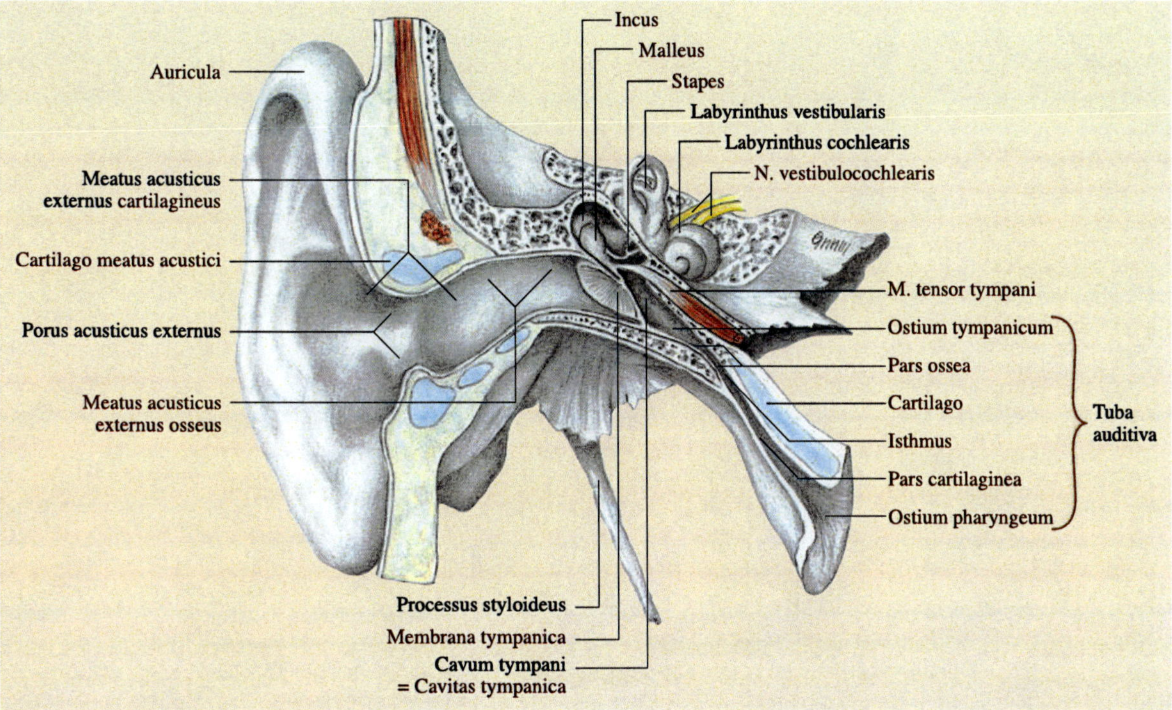

Fig. A.4 Outer and inner ear, right-hand side (from B. Tillmann: Atlas der Anatomie, 2nd edn. (Springer, Berlin, Heidelberg 2010))

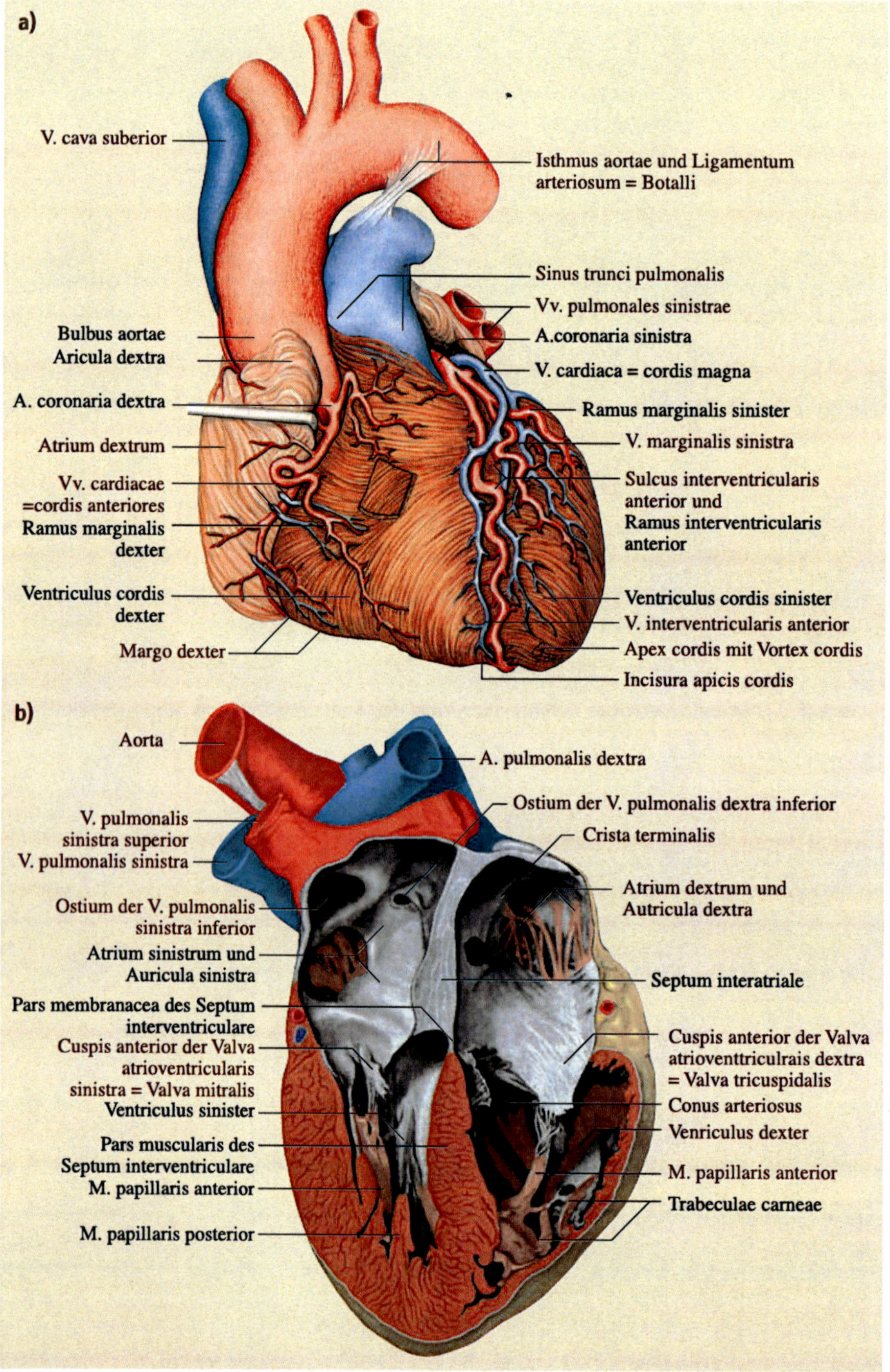

Fig. A.5a,b Human heart. **(a)** Front, **(b)** frontal section through the middle part (from B. Tillmann: Atlas der Anatomie, 2nd edn. (Springer, Berlin, Heidelberg 2010))

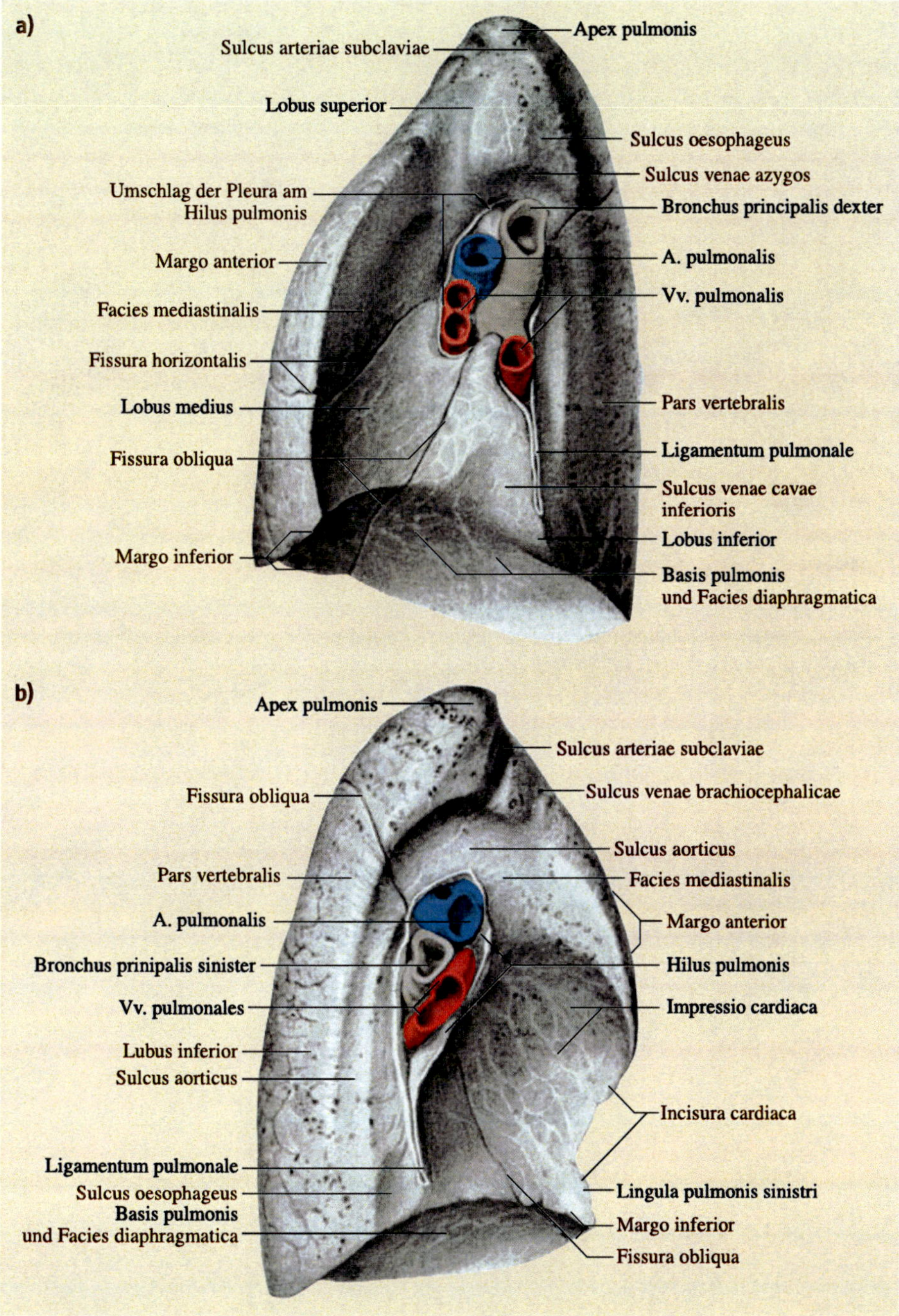

Fig. A.6 Human lung. (**a**) Right lung, medial. (**b**) Left lung, medial (from B. Tillmann: Atlas der Anatomie, 2nd edn. (Springer, Berlin, Heidelberg 2010))

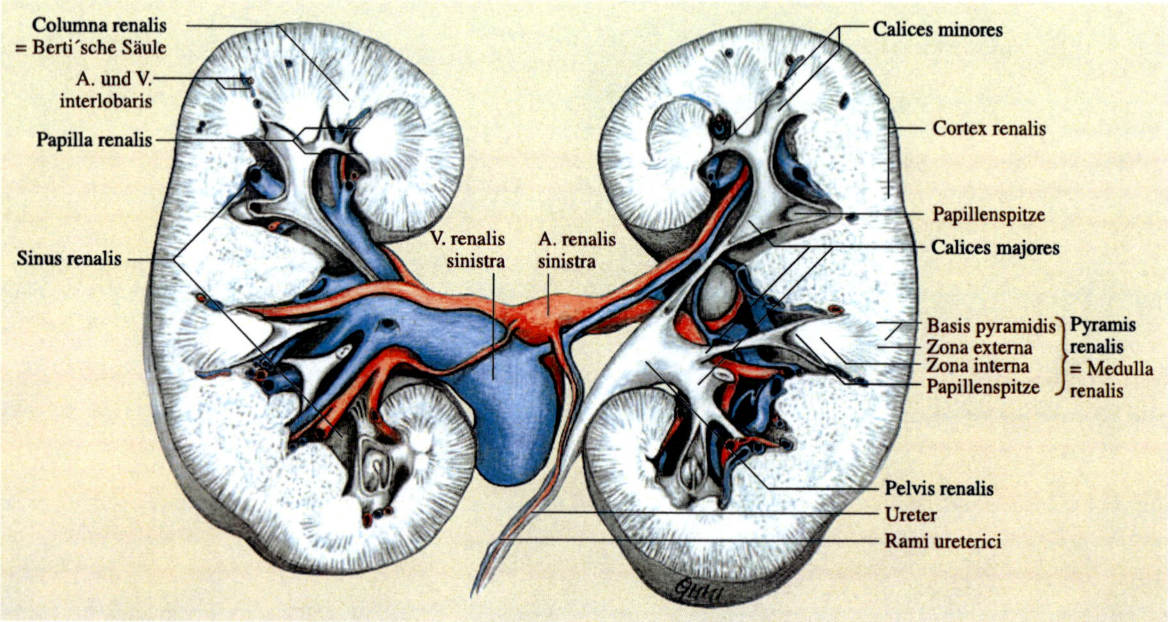

Fig. A.7 Median longitudinal section through a left kidney. View of the front (*left*) and posterior cut surface (*right*) (from B. Tillmann: Atlas der Anatomie, 2nd edn. (Springer, Berlin, Heidelberg 2010))

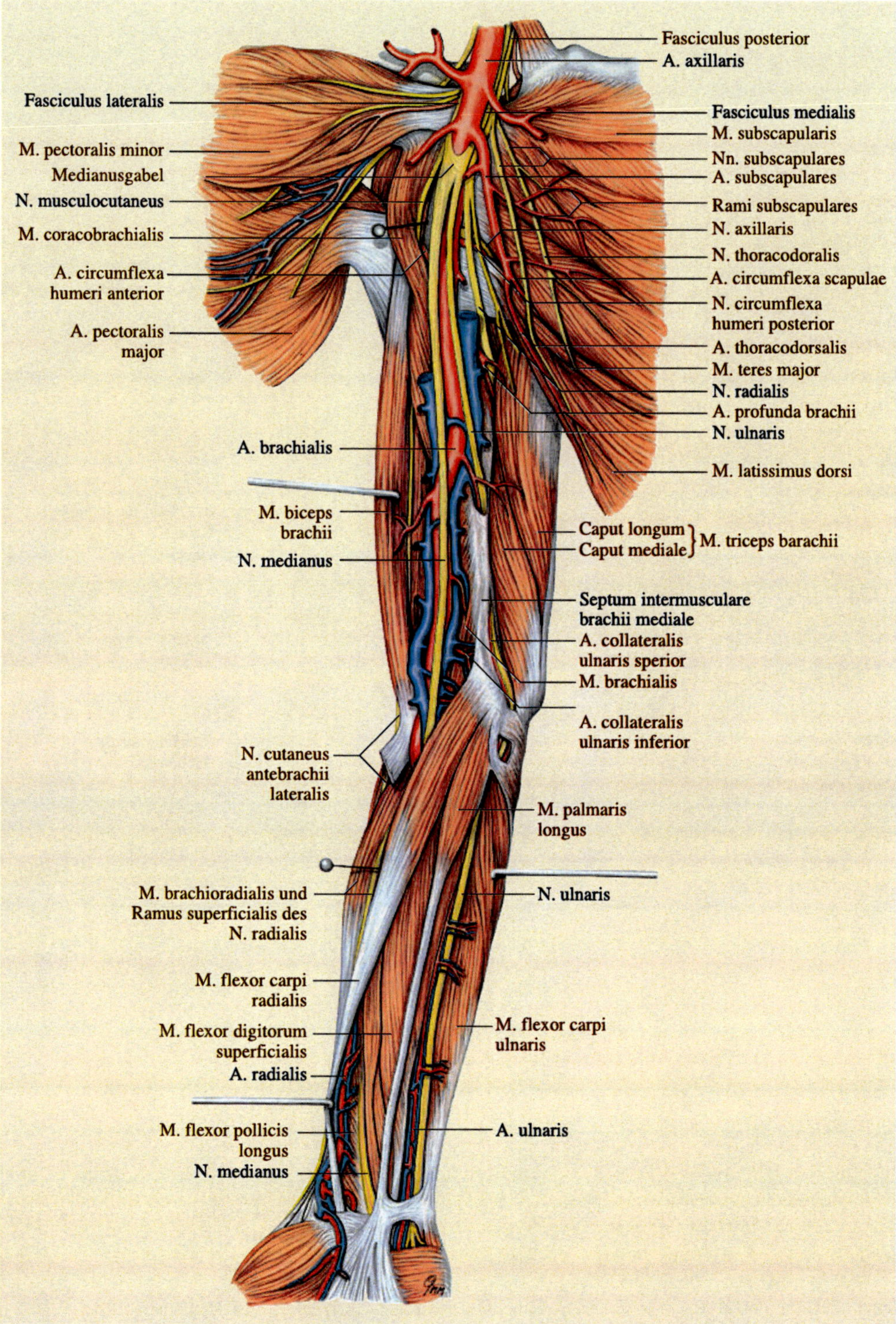

Fig. A.8 Muscels and nerves of a right arm (from B. Tillmann: Atlas der Anatomie, 2nd edn. (Springer, Berlin, Heidelberg 2010))

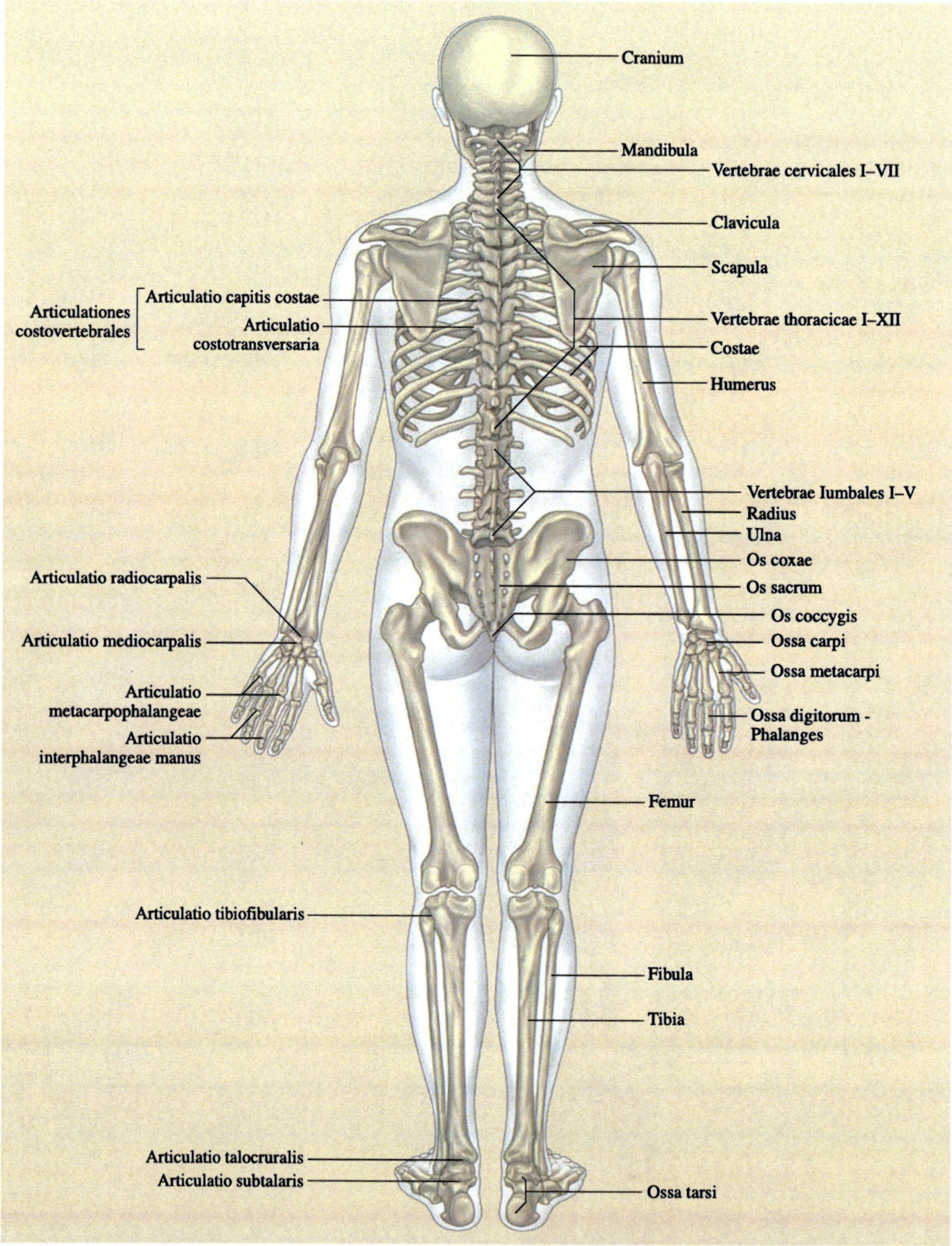

Fig. A.9 Human skeleton (dorsal) (from K. Zilles, B. Tillmann: Anatomie (Springer, Berlin, Heidelberg 2010))

Acknowledgements

C.16 Computed Tomography

by Thorsten M. Buzug

The author would like to acknowledge the great support from Philips, Siemens, and General Electric with photographs of their products. Special thanks go to Annette Halstrick and Dr. Hans-Dieter Nagel, Philips Medical Systems, Hamburg as well as Leon de Vries, Philips Medical Systems, Best. Furthermore, I have to thank Doris Pischitz and Jürgen Greim, Siemens Medical Solutions, Forchheim as well as Dieter Keidel, General Electric, Solingen. I also have to acknowledge the help of Tobias Knopp for copyediting the text of this chapter.

C.18 Medical Infrared Imaging

by Gerald C. Holst, Thorsten M. Buzug

T.M.B. would like to thank Arcangelo Merla, Functional Infrared Imaging Laboratory, University G. d'Annunzio, Italy, for supporting this chapter with Fig. 18.7. Additionally, T.M.B. would like to thank IASTED and ACTA Press for permission to reuse Fig. 18.6, and IEEE for permission to reuse Fig. 18.8.

C.19 Endoscopy

by Martin Leonhard, Klaus-Martin Irion

The authors would like to thank Melanie Schwitkowski for her considerable assistance during the writing of the chapter and Horst Christoph Weiss for his contribution on the reprocessing of endoscopes and instruments.

C.22 Near-Infrared Spectroscopy (NIRS)

by John McNulty, Michael Born, Robert S. Pozos

The authors would like to extend their thanks to Eddie Kwon, Jon Lopez, and C. Draskovich for their assistance in many experiments.

C.24 Magnetic Particle Imaging

by Jörn Borgert, Bernhard Gleich, Thorsten M. Buzug

The authors would like to thank Sven Biederer, Maren Bobek, Claas Bontus, Marlitt Erbe, Jürgen Kanzenbach, Tobias Knopp, Michael Kuhn, Kerstin Lüdtke-Buzug, Jürgen Rahmer, Timo Sattel, Ingo Schmale, Joachim D. Schmidt, Jürgen Weizenecker, and Oliver Woywode for valuable discussions and comments.

J.B. and B.G. acknowledge the support of the Federal Ministry of Education and Research, Germany (BMBF), grant number 13N9079.

C.25 MR-Guided Interventions and Surgery

by Andreas Melzer, Erwin Immel, Rachel Toomey, Fabiola Fernandez-Gutierrez

Andreas Melzer gratefully acknowledges the funding of the Scottish Office for the Northern Research Partnership and support of the MR endoscopy project by Olympus and J. Lewin, J. Duerk, M. Wendt, Dept. of MRI at Case Western Reserve University, Cleveland, OH, USA. A. Buecker, E. Spüntrup, N. Krämer and Dept. of Radiology University Aachen, Germany, are recognized with gratitude for their work on Vena Cave Filter animal study. The authors acknowledge the funding of the European Framework 7 for the Marie Curie Initial training Network www.IIIOS.eu.

C.26 Devices and Materials in MRI

by Gregor Schaefers, Andreas Melzer

We would like to express our gratitude to Dr. Hans Engels for reviewing and commenting on this chapter, Thomas Bertsch for performing MR testing of stents and preparation of Fig. 26.8a,b, Dr. Bernd Guttmann for conception and preparation of Fig. 26.10, Daniel Sachtler for testing and images prepared for Fig. 26.1a–d, and Sentürk Konak for MR testing and images for Fig. 26.11a,b.

Andreas Melzer gratefully acknowledges the funding of the Scottish Office for the Northern Research Partnership.

D.33 Application of Shock Waves and Pressure Pulses in Medicine

by Friedrich Ueberle

I would like to dedicate this publication to my very esteemed colleague Dr. Wolfgang Hepp, who unfortunately passed away. He was dedicated to the development of ESWL since its pioneer days. I would like to thank all colleagues from the ESWL community for their always friendly and cooperative support.

D.36 Mechanical Circulatory Support Systems

by Roland Hetzer, Ewald Hennig

We thank Dr. med. Dipl. Psych. W. Albert for contributing Sect. 36.7. We also thank Anne M. Gale of the Deutsches Herzzentrum for editorial assistance with the chapter.

D.37 Neural Interfaces for Implanted Stimulators
by Xiao Liu, Andreas Demosthenous, Nick Donaldson

The authors would like to acknowledge the financial support from the UK Engineering and Physical Science Research Council under Grant EP/F009593/1. They would also like to thank Mr. Peter Langlois and Dr. Anne Vanhoest for useful discussion on this work.

D.42 Rehabilitation and Therapeutic Robotics
by Loredana Zollo, Dino Accoto, Silvia Sterzi, Eugenio Guglielmelli

The authors would like to warmly thank Antonino Salerno and Massimo Vespignani for their fruitful collaboration to the implementation and the experimental validation ofthe CBM-motus control system.

E.55 Intraoperative Neuromonitoring
by Werner Kneist, Daniel W. Kauff

The authors are grateful to Prof. Dr. med. Th.J. Musholt for the Figs. 55.4, 55.5 and 55.6a. Many thanks to Prof. Dr. med. A. Keilmann for Fig. 55.6b and to Prof. Dr. med. M. Herrmann from the Institute of Anatomy of the University of Ulm for hand drawing, the photograph of Fig. 55.8, helpful discussion, and demonstration of the morphologic conditions for nerve-sparing surgery of pelvic organs.

About the Authors

Dino Accoto

Università Campus Bio-Medico di Roma
CIR – Center for Integrated Research
Rome, Italy
d.accoto@unicampus.it

Chapter D.42

Dino Accoto is an Assistant Professor in Industrial Bioengineering at the Università Campus Bio-Medico di Roma. He received the PhD in biomedical robotics from the Scuola Superiore Sant'Anna in 2002. Since 2004, he has been with the Laboratory of Biomedical Robotics and Biomicrosystems at Università Campus Bio-Medico di Roma. His main research interests are in wearable robotics, microengineering and biomechatronics.

Albert J. Augustin

Städtisches Klinikum Karlsruhe
Augenklinik
Karlsruhe, Germany
albertjaugustin@googlemail.com

Chapter B.13

Albert J. Augustin is currently a Professor of Ophthalmology and Chairman of the Department of Ophthalmology at the Klinikum Karlsruhe, Germany. Prior to this, he was Professor of Ophthalmology at the University of Mainz where he was Head of the Laboratory for Pathophysiology of the Eye and Assistant Medical Director at Bonn University Eye Hospital. Dr. Augustin is the recipient of numerous awards and has published a number of books and over 140 papers. His main research activities relate to pathobiochemistry and pathophysiology of oxidative damage to the eye in age-related macular degeneration, diabetic retinopathy, proliferative vitreoretinopathy and retinal ischemia. He also established a combination therapy for wet AMD and initiated trials investigating the efficacy and safety of Triple Therapy.

Natasha Avila

e-medicis
Paris, France
natasha.avila@emedicis.com

Chapter F.67

Natasha Avila is the Founder and Managing Director of e-medicis. Prior to this she held positions at General Electric Healthcare where she was responsible for the European Operation Services of the IS product line around medical imaging and managed Lucent's professional services unit in France. She worked as a Consultnat in the US, Germany and UK for the telecommunications and banking industry. She has a Master of Engineering from Cornell University and a MBA degree from Cambridge University. She is member of the French Society of Geriatrics.

Michael Bateman

University of Minnesota
Department of Biomedical Engineering
and Surgery
Minneapolis, MN, USA
batem034@umn.edu

Chapter D.43

Michael Bateman obtained his Masters degree in Mechanical Engineering from the University of Bristol, UK, in 2004. He went on to work for CG Versatile Fittings, UK, as a product design engineer for a year before moving to Minnesota to pursue a career in biomedical engineering. He is currently in the fourth year of his graduate studies at the University of Minnesota in Biomedical Engineering and is striving for his PhD degree in 2011.

Andreas Bermann

Siemens Healthcare
Imaging and Therapy Division, SYNGO
Erlangen, Germany
andreas.bermann@siemens.com

Chapter F.65

Andreas Bermann received his degree as medical doctor from the University Medical School of Düsseldorf. He is now responsible for the product definition of the medical imaging software syngoTM at Siemens Healthcare. With his team of highly specialized software engineers, physicists and medical doctors he is developing software for imaging modalities, advanced visualization applications and PACS. In 1998 he achieved his Masters Degree in Business Administration from the University of Bayreuth.

Ulrich Böckler

Sorin Group Deutschland GmbH
München, Germany
ulrich.boeckler@sorin.com

Chapter D.32

Ulrich Böckler is holder of the European Certificate in Cardiovascular Perfusion (ECCP). For several years he was Chief of the Perfusion Department at the German Heart Centre Munich. He is currently working as Product Manager heart lung machines (HLM) for Sorin Group Deutschland GmbH.

Armin Bolz

Karlsruhe Institute of Technology
Institute for Biomedical Engineering
Karlsruhe, Germany
armin.bolz@kit.edu

Chapter D.38

Armin Bolz received his PhD in Physics from the University of Erlangen. He spent some years with Biotronik as head of R&D for pacemakers and defibrillators. Since 1999 Armin Bolz is a Professor for Biomedical Engineering at the Karlsruhe Institute of Technology (KIT). His current research interests are cardiovascular diagnostics, electrostimulation and defibrillation. He also actively supports young start-ups to transfer latest research results into products.

Jörn Borgert

Philips Technologie GmbH
Forschungslaboratorien
Tomographic Imaging Group
Hamburg, Germany
joern.borgert@philips.com

Chapter C.24

Jörn Borgert received his PhD degree in Theoretical Physics from the University Kiel, Germany. He is now Senior Scientist at Philips Research. His current research interests are tomographic imaging and, in particular, magnetic particle imaging (MPI). He received the innovators award of the Federal Ministry of Education and Research of Germany for the application of MPI in breast cancer treatment in 2008.

Michael Born

San Diego State University
Department of Biology
San Diego, CA, USA
Michael.j.born@gmail.com

Chapter C.22

Michael Born received his Bachelor of Science degree in Bioengineering from San Diego State University. He is currently attending graduate school at Northwestern University in Chicago, Illinois where he is working on his Masters Degree, also in Bioengineering.

Guenther Braun

University Medical Center Freiburg
Paediatrics and Adolescent Medicine
Freiburg, Germany
guenther.braun@uniklinik-freiburg.de

Chapter G.71

Guenther Braun received the Masters degree 1996 in Biomedical Engineering from the University of Applied Science in Ulm. Currently he is responsible for the biomedical department of the University Childrens Hospital in Freiburg.

Dirk Büchel

Medical Faculty of Tübingen
Tübingen, Germany
buechel.dirk@t-online.de

Chapter A.6

Dr. sc. hum. Dipl.-Ing. Dirk Büchel is a usability expert who developed the standard procedure for the usability engineering process and final usability testing of medical products within his dissertation.

Thorsten M. Buzug

University of Lübeck
Institute of Medical Engineering
Lübeck, Germany
buzug@imt.uni-luebeck.de

Chapters C.16, C.18, C.24

Professor Thorsten M. Buzug received his PhD in 1993 from the University of Kiel. After a postdoctoral position at the German Federal Armed Forces Underwater Acoustics and Marine Geophysics Research Institute he joined the Philips Research Laboratories Hamburg. As leader of the Philips research cluster Medical Image Processing he was responsible for several. In 2006 he has been appointed as Director of the Institute of Medical Engineering at the University of Lübeck. His research is focused on medical imaging and image computing.

Marco Capogrosso

Scuola Superiore Sant Anna
ARTS lab
Pisa, Italy
m.capogrosso@sssup.it

Chapter F.68

Marco Capogrosso received the MS degree in Applied Physics from the Faculty of Mathematical, Physical and Natural Sciences at the University of Pisa in 2009. He is currently working towards a PhD degree in biomedical robotics at the Scuola Superiore Sant Anna, Pisa. His research interests are in neural modelling, finite element models, functional imaging, and neuroscience.

Peter H. Cossmann

Med Tech Consulting
Wettingen, Switzerland
peter.cossmann@medtechconsulting.ch

Chapter D.35

Peter H. Cossmann received his Master degree in Medical Physics from the ETH Zurich and his PhD from the University of Bern. He is currently Head Consultant at Med Tech Consulting in Wettingen. He was former Head of Medical Physics of the Institutes for Radiotherapy at Hirslanden Klinik Aarau and Zurich. Dr. Cossmann holds Lectureships in Biomedical Engineering and Physics and his current areas of research include on-board imaging, cone-beam computed tomography, gating and dosimetry.

Andreas Demosthenous

University College London
Electronic and Electrical Engineering
London, UK
a.demosthenous@ee.ucl.ac.uk

Chapter D.37

Andreas Demosthenous is a Professor of Analogue and Biomedical Electronics in the Department of Electrical and Electronic Engineering, University College London, UK. His main area of research is analog and mixed-signal integrated circuits for biomedical, communication, sensor and signal processing applications. He has numerous collaborations for interdisciplinary research and has published over 150 articles in journals and international conference proceedings.

Thomas M. Deserno

RWTH Aachen University
Aachen, Germany
Deserno@ieee.org

Chapter F.62

Thomas M. Deserno (born as Lehmann) is Professor of Medical Informatics at RWTH Aachen University, where he heads the Medical Image Processing Group within the Department of Medical Informatics. His research interests include discrete realizations of continuous image transforms, applications to quantitative measurements for computer-assisted diagnoses, and content-based image retrieval from large medical databases.

Nick Donaldson

University College London
Medical Physics & Bioengineering
London, UK
nickd@medphys.ucl.ac.uk

Chapter D.37

Professor Donaldson studied Engineering at Cambridge University and did a PhD at University College London. He directs the Implanted Devices Group and has been Principal Investigator for many projects related to implants and functional electrical stimulation. His research interests include implant technology, neuroprosthesis that use nerve signals as inputs; stimulators of nerve roots; and electrical stimulation for recreational exercise of paralyzed legs.

William K. Durfee

University of Minnesota
Department of Mechanical Engineering
Minneapolis, MN, USA
wkdurfee@umn.edu

Chapter B.14

William Durfee is with the Department of Mechanical Engineering at the University of Minnesota. He received the AB degree in Engineering and Applied Physics from Harvard University and the MS and PhD degrees in Mechanical Engineering from MIT. His professional interests include design of medical devices, rehabilitation engineering, advanced orthotics, biomechanics and physiology of human muscle including electrical stimulation of muscle, product design and design education.

Günter Edlinger

g.tec medical engineering GmbH
Graz, Austria
edlinger@gtec.at

Chapter E.52

Günter Edlinger studied control engineering and received his PhD in biomedical engineering from the University of Technology Graz where he was an Assistant Professor from 1993 to 2001. He is co-founder and CEO of g.tec medical engineering GmbH and has been working there since 1999. His research interests comprise invasive and noninvasive brain computer interfaces, neurophysiological instrumentation aspects and inverse problems.

Amir Eftekhar

Imperial College London, South Kensington Campus
Centre for Bio-Inspired Technology
London, UK
a.eftekhar@imperial.ac.uk

Chapter E.56

Amir Eftekhar is a Research Associate in the Centre for Bio-Inspired Technology, one of the founders of the Institute of Biomedical Engineering, and now a part of the EEE department. His research involves applications of interfaces for neurological studies including electrode technology for interfacing with the central and peripheral nervous system, front-end electronics (analog/digital) for implantable or portable systems to interface with these electrodes and advanced signal processing methods.

Mariana Fernandes

University of Minho
Department of Industrial Electronics, DEI – Campus de Azurém
Guimarães, Portugal
mfernandes@dei.uminho.pt

Chapter G.74

Mariana Fernandes received the Integrated Master degree in Biomedical Engineering from the University of Minho (2002-2007). She is currently pursuing the PhD degree within the MIT|Portugal Program in bioengineering at the Department of Industrial Electronics, University of Minho. Her research area is the development of wearable and contactless technologies to monitor bioelectric signals.

Eduardo Fernandez

Universidad Miguel Hernández
Bioengineering Institute
Elche, Spain
e.fernandez@umh.es

Chapters B.11, D.41

Eduardo Fernandez is a Professor of Cellular Biology and Director of the Visual Rehabilitation and Neuroprosthesis Unit in the Bioengineering Institute of the University Miguel Hernández (Spain). He received a MD degree from the University of Alicante (1986) and a PhD in Neuroscience in 1990. He is actively working on the development of visual neuroprostheses and brain machine interfaces. He is also working on brain plasticity and cortical reorganization in severe vision loss.

Fabiola Fernandez-Gutierrez

University of Dundee
Institute for Medical Science and Technology
Dundee, UK
f.fernandezgutierrez@dundee.ac.uk

Chapter C.25

Fabiola Fernandez-Gutierrez received a degree in Telecommunication Engineering and a Master degree in Automatic Control, Robotics and Telematics from the University of Sevila, Spain. She is now working on her PhD at the Institute for Medical Science and Technology on the European Project Integrated Interventional Imaging Operating System (IIIOS) focusing on the design and improvement of ergonomics workflow for multimodality image-guided procedures.

Celso P. Figueiredo

University of Minho, Campus de Azurém
Department of Industrial Electronics
Guimarães, Portugal
celso.figueiredo@dei.uminho.pt

Chapter G.74

Celso Pitães Figueiredo received his MSc degree in Biomedical Engineering – Medical Electronics from the University of Minho in 2007, Guimarães, Portugal. He is currently working towards his PhD in Biomedical Engineering at the University of Minho, in collaboration with the Fraunhofer Institute for Biomedical Engineering, St. Ingbert, Germany. His thesis involves the research and development of wearable and wireless systems for health monitoring.

Authors

Harald Fischer

Creamedix GmbH
Weingarten, Germany
hafi@creamedix.eu

Chapter G.70

Harald Fischer received his Diploma and PhD degree in Biomedical Engineering from the Technical University of Karlsruhe, Germany. He worked for several years as the Head of the Institute of Biomedical Engineering and Biophysics at the Karlsruhe Institute of Technology (formerly Research Center Karlsruhe) in the field of robotic surgery and new materials for surgical interventions. He is now the CEO of Creamedix GmbH producing medical implants out of shape memory alloys. His research interests are new materials for medical applications, medical coatings and cell growing applications.

Martin R. Fischer

Private University Witten/Herdecke gGmbH
Fakultät für Gesundheit
Witten, Germany
martin.fischer@uni-wh.de

Chapter F.64

Dr. Fischer is a General internist and Endocrinologist and is Professor, Dean for Education, and Director of the Institute for Teaching and Educational Research in Health Sciences at the Faculty for Medicine of Witten/Herdecke University, Germany. He received the Master of Medical Education (MME) from the University of Berne, Switzerland, and is also certified in medical informatics.

Óscar S. Gama

Universidade do Minho,
Campus de Gualtar
Departamento de Informática
Braga, Portugal
osg@di.uminho.pt

Chapter G.74

Óscar Gama finished his education in Electronic Engineering at the University of Oporto in 1989 and received a Master degree in Informatics from the University of Minho (Braga) in 2003. He is currently a PhD student in the Electronics Department of the University of Minho. His research focuses on self-reconfiguring e-health wireless sensor networks with quality of service.

Stephan Garbe

University Hospital of Bonn
Radiology Department
Bonn, Germany
garbe@uni-bonn.de

Chapter D.45

Stephan Garbe received his PhD in Physics from the University of Münster in 1995. After a postdoctoral position at the Risø National Lab, he worked as Medical Physicist first at the Department for Radiotherapy at the University Hospital of Essen and later at the Department for Radiology at the University Hospital of Bonn. Since 2007 he is Head of the Medical Physics Department. His current research interests are in-vivo dosimetry, dosimetric verification in IMRT and quality assurance in radiotherapy.

Belinda Garner

Imperial College London, South Kensington Campus
Institute of Biomedical Engineering
London, UK
b.garner@imperial.ac.uk

Chapter E.56

Belinda Garner received her Bachelor of Science (Honours) degree in Neuroscience/ Anatomy and Cell Biology and her PhD degree in Behavioural Neuroscience from the University of Melbourne, Australia. Her research interests include understanding the role of stress in the development and maintenance of psychiatric disorders and in the development of novel technology for monitoring brain and peripheral nerve function.

Armin Gärtner

Ingenieurbüro für Medizintechnik
Erkrath, Germany
armin.gaertner@t-online.de

Chapter F.58

Armin Gärtner successfully completed his engineering studies at the University of Applied Sciences in Gießen, Germany and has been working in the hospital sector for many years. He is an Official Expert for medical technology at the Chamber of Industry and Commerce – District Düsseldorf and owner of an engineering office for medical technology. He is Member of the Standards Developing Organization 80001 and author of numerous publications and specialized books regarding subjects of medical engineering and information technology.

Bernhard Gleich

Philips Technologie GmbH
Tomographic Imaging Group
Hamburg, Germany
bernhard.gleich@philips.com

Chapter C.24

Bernhard Gleich received his Diploma in Physics from the Institute of Surface Chemistry and Catalysis, University Ulm, Germany. He is now Senior Scientist at Philips Research. His current research interests are tomographic imaging and magnetic particle imaging. He is one of the inventors of magnetic particle imaging (MPI).

Rudolf Götz

Aloka GmbH
Willich, Germany
R.Goetz@aloka-europe.com

Chapter C.17

Rudolf Götz received the Dipl.-Ing. in Electronic Engeneering from the FH Karlsruhe in 1972 and the Dipl.-Ing. in Medical Engineering and Biological Cybernetics from the University of Karlsruhe in 1979. Currently he is product manager for ultrasound at ALOKA GmbH, Germany.

Christoph Guger

g.tec medical engineering GmbH, Guger Technologies OG
Schiedlberg, Austria
guger@gtec.at

Chapter E.52

Dr. Christoph Guger studied Biomedical Engineering at the University of Technology Graz and Johns Hopkins University in Baltimore, USA. Then he did research work at the Department of Medical Informatics at the University of Technology Graz and received his PhD degree in 1999 with the design of the first real-time brain computer interface (BCI) system with continuous feedback. He also developed the real-time analysis with common spatial patterns which is still the fastest and most accurate approach for oscillatory BCI. He is co-founder of g.tec where he works since 1999.

Eugenio Guglielmelli

Universita' Campus Bio-Medico
CIR – Center For Integrated Research
Rome, Italy
E.Guglielmelli@Unicampus.It

Chapter D.42

Eugenio Guglielmelli (MSc 1991, PhD 1995 both in Electronical Engineering from the University of Pisa, Italy) is Full Professor of Bioengineering at Università Campus Bio-Medico (Rome, Italy) where he serves as the Director of the Laboratory of Biomedical Robotics and Biomicrosystems. His main research interests are in human-centred robot design, biomechatronics, biomorphic robot control, neurobotics, rehabilitation and assistive robotics.

Martin Haag

Heilbronn University
Medical Informatics
Heilbronn, Germany
martin.haag@hs-heilbronn.de

Chapter F.64

Martin Haag is a Professor for Software Engineering and Head of the Lab e-Learning in Medicine of the Centre for e-Learning Technology at Heilbronn University, Germany and head of the Centre for Virtual Patients at the University of Heidelberg, Germany. His research group focuses on development of virtual patients and secure electronic examinations.

Reiner Haag

Lawton GmbH & Co. KG
Fridingen, Germany
reiner.haag@lawton.de

Chapters D.34, G.72

Reiner Haag received the Diplom Ingenieur degree in Electronics and Mechanical Engineering from the University of Applied Science Furtwangen, Germany, in 1985. After 6 years in R&D and 10 years International Product Management in electrosurgery he had gained 2 years of experience in sales of endoscopic instruments. He has eight publications and 3 patents. Since 2002 he is General Manager at Lawton Medizintechnik, a Company producing special high-end surgical instruments.

Peter Haas

University of Applied Sciences and Arts Dortmund
Medical Informatics
Dortmund, Germany
haas@fh-dortmund.de

Chapter F.59

Academic Studies Medical Informatics 1977–1982, Project Manager in a large-scale hospital for the implementation of a radiological information system from 1983 to 1984, Project Manager and later Head of Department of Medical Informatics at the University of Heidelberg 1985–1989, Marketing Manager Public Authorities Software AG 1989–1990, Director Sales, Marketing, training hospital information systems vendor 1991–1993. Since 1994 Professor for Medical Informatics. His research acivities concern information systems in healthcare and electronic patient records, health telematics, knowledge management based systems in healthcare.

Andreas Hahn

Sorin Group Deutschland GmbH
General Management
München, Germany
andreas.hahn@sorin.com

Chapter D.32

Andreas received the MS degree in Electronics and Telecommunications from the TH Karlsruhe and the Doctors degree from the Technical University of Breslau. He is more than 35 years active in medical industry. Starting as R&D Engineer, Head of R&D for medical lasers, VP operations of MBB Medical Technology, he is now General Manager of Sorin Group Deutschland GmbH within CP BU. He is Member of the Operations Council for the Sorin Group. Board member of Laser- und Medizin-Technologie LMTB GmbH. He holds more than 30 patents for medical lasers and heart-lung machines.

Sibylle Hanus

Textile Research Institute Thuringia-Vogtland
Textile Structures and Materials
Greiz, Germany
s.hanus@titv-greiz.de

Chapter G.73

Sibylle Hanus received the diplomas in Physical Technics and Engineering Economics from the University of Applied Sciences Zwickau, Germany. Since 1999 she is working at the Textile Research Institute Thuringia-Vogtland, Greiz, Germany as scientist in the development of medical textiles, especially in the field of invasive medical devices. Since 2008 she is team leader of the group Textile Structures and Materials.

Michael Heinlein

Mednovo Medical Software Solutions GmbH
Berlin, Germany
info@mednovo.de

Chapter F.57

Michael Heinlein is co-founder and co-owner of Mednovo Medical Software Solutions GmbH, founded in 2002. As Managing Director, he focuses on marketing, sales, support, business development and organization. Mednovo is providing software solutions in healthcare and offers an overall IT concept, optimizing the process of a daily routine in hospitals, clinics and medical engineering companies.

Ewald Hennig

Deutsches Herzzentrum Berlin
Berlin, Germany
Hennig@dhzb.de

Chapter D.36

Ewald Hennig received a PhD in Aeronautical Engineering from the Technical University Berlin in 1965.There he was with the Department of Biomedical Research until 1971. In 1972 he joined the Research Group "Mechanical Heart Assist and Total Heart Replacement" of. E. S. Bücherl at the Free University of Berlin as Technical Director of Experimental Surgery. In 1985 he was Visiting Professor at the Second Medical University of Shanghai. Since 1988 he has been a member of the DHZB-MCSS Research Group (Prof. Hetzer) as Senior VAD Coordinator. Research activities concern BiVAD and TAH with rotary blood pumps.

Roland Hentschel

Universitätsklinikum Freiburg
Neonatologie/Intensivmedizin, Zentrum für Kinder- und Jugendmedizin
Freiburg, Germany
roland.hentschel@uniklinik-freiburg.de

Chapters E.54, G.71

Roland Hentschel is Head of the Department of Neonatology and Intensive Care Medicine, University Children's Hospital Freiburg, since 1996. He is specialized in Neonatology, Intensive Care Medicine and Pediatric Pulmonology. In 2003 he was appointed University Lecturer for neonatology and pediatrics. His main research and clinical work focuses on newborn lung function (grantee of the German Research Council (DFG) from 1990–1996), mechanical ventilation, transfusion medicine, and ethics in neonatology.

Roland Hetzer

Deutsches Herzzentrum Berlin
Berlin, Germany
hetzer@dhzb.de

Chapter D.36

Roland Hetzer, MD, PhD, is Chairman of the Deutsches Herzzentrum Berlin since its foundation in 1986. He received his training in surgery at the Hannover Medical School and specialized in Cardiothoracic and Vascular Surgery after completing his fellowship training at the Pacific Medical Center and Stanford University. He became staff member (1978) in this field at the Hannover Medical School and Associate Professor (1983) when he started heart transplantation. At Berlin within the complete scope of cardiothoracic surgery he is heading the largest programme of mechanical hearts worldwide.

Lothar Heuser

Klinikum der Ruhr-Universität Bochum
Institut für Diagnostische & Interventionelle Radiologie, Knappschaftskrankenhaus
Bochum, Germany
lothar.heuser@rub.de

Chapter C.15

Lothar J. Heuser was fellow at the Departments of Diagnostic Radiology und Radiation Therapy at the University of Cologne where he received his venia legendi and became an Associate Professor. Since 1986 he is Professor of Radiology and Chaiman at the Ruhr-Universität Bochum and Chairman of the Department of Diagnostic and Interventional Radiology, Neuroradiology and Nuclear Medicine. Current research interests are digital image processing and interventional procedures.

Ullrich Hieronymi

Dräger Medical Deutschland GmbH
Lübeck, Germany
ullrich.hieronymi@draeger.com

Chapters E.47, E.48, E.49, E.50

Ullrich Hieronymi received his biomedical engineering graduate and Doctor of Engineering Science from the Ilmenau College of Advanced Technology in 1971 and 1975, respectively. For 17 years he took the biomedical engineering part in a research department for intensive care medicine of a Berlin hospital. Since 1992 until 2011 he worked as a monitoring application specialist in the medical device industry. After his semi-retirement he now proofreads texts on monitoring.

Robert Hitchcock

University of Utah
Bioengineering
Salt Lake City, UT, USA
r.hitchcock@utah.edu

Chapter D.40

Robert Hitchcock is an Assistant Professor in the Department of Bioengineering at The University of Utah and Director of the University's bioDesign program. He has over 26 years of experience bringing new products to the marketplace and has been involved as an engineer, manager, consultant and executive with a number of leading medical device companies.

Gerhard Hoffmann

Orangedental GmbH & Co KG
Biberach, Germany
gh@orangedental.de

Chapter C.20

Gerhard Hoffmann is a Ing.(grad) in Hospital Engineering at the University of Applied Sciences Gießen. He has many years of an experience in medical technology industry. As Director of Sales his focus is on dental diagnostic imaging systems where he specializes in digital inclusion and cone-beam computed tomography (CBCT).

Klaus-Peter Hoffmann

Fraunhofer Institute for Biomedical Engineering
Medical Engineering and Neuroprosthetics
St. Ingbert, Germany
klaus.hoffmann@ibmt.fraunhofer.de

Chapters B.9, B.10, B.11, D.39, D.40, D.41, E.46

Dr. Hoffmann is currently a Professor of Biomedical Engineering at the Hochschule für Technik und Wirtschaft des Saarlandes. He is acting as the Head of the Department Medical Engineering and Neuroprosthetics at the Fraunhofer Institute for Biomedical Engineering in St. Ingbert. His main research interests include smart interfaces in Clinical Neurophysiology as well as implantable microsystems especially sensors and actuators for neuroprosthetics.

Gerald C. Holst

JCD Publishing Company
Winter Park, FL, USA
Jerry@JCDPublishing.com

Chapter C.18

Dr. Gerald C. Holst is a lecturer and consultant for thermal imaging system design, applications, and testing. He developed software that simulates imaging system performance. He authored 6 books which are published by JCD Publishing Company. Dr. Holst is a SPIE Fellow and chairs the SPIE conference on Infrared Imaging Systems since 1990.

Sebastian Hoth

University of Heidelberg
ENT Hospital – Audiology
Heidelberg, Germany
sebastian.hoth@med.uni-heidelberg.de

Chapter B.12

Professor Sebastian Hoth, Head of the Audiology Unit at the University ENT Hospital in Heidelberg, studied Physics at the University of Erlangen and was awarded his PhD in 1982. He completed his post-doctoral thesis in 1996 and was appointed Professor in 2004. His scientific activity is dedicated to objective audiometry, cochlear implant and other implantable hearing systems. Parallel to this, he is deeply involved in audiology education, the training of students and the formation of professionals.

Stephen A. Howard

University of Minnesota
Department of Biomedical Engineering and Surgery
Minneapolis, MN, USA
howa0255@umn.edu

Chapter D.43

Stephen Howard is a PhD graduate student in Biomedical Engineering at the University of Minnesota. He earned his Masters degree in Biochemistry and Chemistry at Gustavus Adolphus College in St. Peter, MN. He is interested in the effects of fatty acids on the membrane proteins and ion channels, particularly looking at signal modification and performance pre and post ischemia.

Bernhard Hug

KLS Martin GmbH + Co. KG
Umkirch, Germany
Bernhard.Hug@klsmartin.com

Chapter D.34

Bernhard Hug studied Biomedical Engineering in Gießen and Berlin and received his PhD from the Faculty of Mathematics and Natural Sciences of the Humboldt-Universität zu Berlin, Germany. He is Head of Development for Electrosurgical Accessories at KLS Martin, a company of the KLS Martin Group. A main focus of his research activities is the clinical application of electrosurgical technology in different medical specialties.

Paul A. Iaizzo

University of Minnesota
Department of Surgery
Minneapolis, MN, USA
iaizz001@umn.edu

Chapters B.14, D.43

Paul Iaizzo earned the PhD degree in Physiology, and is Professor of Surgery, Integrative Biology and Physiology, Anesthesiology, and Carlson School of Management at the University of Minnesota. He heads the Visible Heart® Laboratory, where his main research focus is translational systems physiology and his research group has developed a unique isolated working large mammalian heart model.

Erwin Immel

University of Dundee
Institute for Medical Science and Technology
Dundee, UK
e.z.immel@dundee.ac.uk

Chapter C.25

Erwin Immel graduated 2006 from a Five-Year Degree in Medical Engineering at the University of Applied Science in Gelsenkirchen, where he received a Master of Science degree. Before starting his PhD, Erwin worked as Research and Development Engineer at Biophan Europe. The research topic of his PhD projects (University of Dundee) is the improvement of the visualization of vascular implants such as vena cava filters, stents, and prosthetic heart valves in MRI.

Klaus-Martin Irion

Karl Storz GmbH & Co. KG
Tuttlingen, Germany
K.Irion@karlstorz.de

Chapter C.19

Klaus-Martin Irion is the Global Vice President of Research and Technology at Karl Storz, Tuttlingen, Germany. He received his engineering diploma and his doctorate from the Faculty of Electronic Engineering of the University of Stuttgart. His scientific work focused on high-frequency ultrasonic image processing and tissue characterization at the Institute of Biomedical Engineering, Stuttgart. Dr. Irion is over 20 years with Karl Storz. Before taking over his position as Global Vice President of Research and Technology he was Head of R&D/Electronic Engineering.

Heinz-Michael Just

Klinikum Nürnberg
Institute of Hospital Epidemiology, Medical Microbiology and Infectious Diseases
Nürnberg, Germany
just@klinikum-nuernberg.de

Chapter A.3

Heinz-Michael Just, MD, is a specialist for microbiology, virology and hospital epidemiology, for hygiene and environmental medicine and a specialist for infectious diseases. He is Head of the Institute of Hospital Epidemiology, Medical Microbiology and Infectious Diseases, Klinikum Nürnberg, Gemany. He is also a Member of the Commission for Hospital Hygiene and Infectious Disease Prevention (KRINKO) appointed by the Ministry of Health as well as a Member of the Committee for Biological Agents (ABAS) appointed by the Ministry of Labour.

Daniel W. Kauff

University Medicine of the Johannes Gutenberg University Mainz
Department of General and Abdominal Surgery
Mainz, Germany
Kauff@ach.klinik.uni-mainz.de

Chapter E.55

Daniel W. Kauff received his medical doctorate from the University Medicine of the Johannes- Gutenberg University of Mainz, Germany. Since 2008, he is working as a physician and research associate at the Department of General and Abdominal Surgery, University Medicine of Mainz, Germany. His current scientific interests are colorectal surgery and intraoperative pelvic neuromonitoring.

Rüdiger Klar

Universitätsklinik Freiburg
Institut für Medizinische Biometrie und Medizinische Informatik
Freiburg, Germany
ruediger.klar@web.de

Chapter F.60

Prof. em. Dr. rer. nat. Rüdiger Klar studied in Bonn and Hamburg and worked two years in a software company at Hannover. From 1986 till 2007 he was Head of the Department for Medical Informatics at the University Hospital Freiburg. His scientific and practical work deals with patient care systems, telemedicine, medical classifications and IT standards, health economy and medical decision making. He is member of several international scientific societies and editorial boards of scientific journals. He was President of the German Society of Medical Informatics, Biometry and Epidemiology and is still active in this field.

Werner Kneist

University Medicine of the Johannes Gutenberg University Mainz
Department of General and Abdominal Surgery
Mainz, Germany
werner.kneist@unimedizin-mainz.de

Chapter E.55

Werner Kneist is currently an Associate Professor in the Department of General and Abdominal Surgery of the University Medicine of the Johannes Gutenberg University Mainz. His research focuses on colorectal surgery with a special interest in the oncological and functional outcome. Over the past years he and his interdisciplinary group have developed a series of innovative techniques for intraoperative neuromonitoring of pelvic autonomic nerves.

Klaus P. Koch

University of Applied Sciences Trier
Department of Engineering
Schneidershof, Trier, Germany
koch@fh-trier.de

Chapter F.61

Klaus Peter Koch received his PhD in 2003. From 1998 to 2007 he worked at the Fraunhofer Institute for Biomedical Engineering. Currently he is Full Professor at the University of Applied Sciences Trier, Germany. His research interest is system integration for neural implants, design and simulation of electrodes for recording and stimulation, active implantable electronics and artefact models.

Heike Kramme

Titisee, Germany
kramme-titisee@t-online.de

Chapter A.1

Heike Kramme is a Graduate in Administration Sciences after studies in Law and Management Sciences focusing on health care at the Universities of Freiburg and Constance, Germany. Heike Kramme has many years of professional experience in human resources management, among others in publishing. She has various publications about health policy and health economics.

Rüdiger Kramme

Titisee, Germany
kramme-titisee@t-online.de

Chapters A.1, A.4, B.7, B.8, D.28, E.47, E.48, E.49, E.50

Rüdiger Kramme is a Graduate Engineer and studied Biomedical and Hospital Engineering as well as National Economy in Gießen und Freiburg, Germany. He has many years of professional experience in sales, marketing and human resources development in medical device industry for consumer and investment capital goods. He also is a Lecturer for Medical Engineering at the University of Applied Sciences in Gießen. Rüdiger Kramme has authored numerous scientific publications in magazines and books. He is the founder and Editor of the standard work Medizintechnik – Verfahren, Systeme, Informationsverarbeitung and the Dictionary Technische Medizin published by Springer.

Ursula Krechel

University Clinic Freiburg
IT-Department
Freiburg, Germany
Ursula.krechel@uniklinik-freiburg.de

Chapter B.9

Uschi Krechel has studied Electronic Engineering at the Technical University of Aachen, Germany and worked as an Application Engineer in German companies building electrophysiological equipment such as EMG, EEG, Electronystagmographs for 13 years. Among these were Toennies and Jaeger. Currently she works as an IT Specialist at the University Clinic of Freiburg, Germany.

Harald Kronberg

Michelstadt, Germany
haraldkronberg@freenet.de

Chapters E.49, E.51

Harald Kronberg received the Master degree and the PhD in Physics both from Göttingen University. He then spent seven years in applied research of the Göttingen Max-Planck-Institute of Experimental Medicine. In 1984, he joined the Applied Research Department of Hellige in Freiburg, now part of General Electric Healthcare. Until his retirement end of 2009, he was Marketing Manager of GE's patient monitoring platforms.

Authors

Klaus A. Kuhn

Technische Universität München (TUM)
Institute of Medical Statistics and Epidemiology
München, Germany
klaus.kuhn@tum.de

Chapter F.59

Klaus Kuhn is a Full Professor of Medical Informatics at the Technical University Munich and a member of both the Medical and the Informatics Faculty. He holds degrees in Medicine, Mathematics, and Computer Science. He chaired the Editorial Board of Medinfo 2007, the World Congress of Medical Informatics. He is past President of the German Association of Medical Informatics, Biometry, and Epidemiology, and he is a Fellow of ACMI. His research interests are information systems with a special focus on translational research.

Bernhard Kulik

Maquet GmbH & Co. KG
Rastatt, Germany
b.kulik@maquet.de

Chapter G.69

Bernhard Kulik is Director of Product Management Operating Room tables at Maquet GmbH & Co. KG in Rastatt, Germany. He has co-authored some chapters in medical textbooks published by Springer Medizin Verlag, Heidelberg.

Timothy G. Laske

Medtronic, Inc.
Mounds View, MN, USA
tim.g.laske@medtronic.com

Chapter D.43

Tim Laske has a BS degree in Biological Sciences and Mechanical Engineering from Michigan Technological University. He received his Masters degree in Mechanical Engineering from the University of Michigan, Ann Arbor, and his PhD in Biomedical Engineering from the University of Minnesota where he serves as an Adjunct Associate Professor. Currently, Tim is a Senior Product Development Director in the Structural Heart Business at Medtronic and a Bakken Fellow. He is responsible for research and development for current and future heart valve repair and replacement products.

Martin Leonhard

Karl Storz GmbH & Co. KG
Research and Technology
Tuttlingen, Germany
m.leonhard@karlstorz.de

Chapter C.19

Martin Leonhard is Head of Technology Management at Karl Storz, Tuttlingen, Germany. He received his physics diploma and doctorate both from the University of Ulm. He worked on time-resolved spectroscopy at the Max-Planck-Institute in Martinsried and on medical laser applications at the Munich-Großhadern University Hospital. He was Deputy Head of Marketing Europe before taking his current position to identify and evaluate future technologies for Karl Storz.

Xiao Liu

University College London
Department of Electronic and Electrical Engineering
London, UK
x.liu@ee.ucl.ac.uk

Chapter D.37

Xiao Liu received the B.Eng. degree from Xi'an Jiaotong University in 2003, the MSc degree from University of Southampton in 2004, and the PhD degree from the University College London (UCL) in 2009. He is currently a Research Associate at UCL. His research interests include analogue and mixed-signal integrated circuit design for biomedical applications, neuroprosthesis and microelectronic sensor design.

Martin Maier

Philips Healthcare
Böblingen, Germany
martin.maier@philips.com

Chapter E.53

Martin Maier holds an engineering degree in Telecommunications from the University of Applied Science in Esslingen and an MBA in International Marketing from the European School of Business in Reutlingen. During his career Martin has gathered experience in the area of production, software quality engineering, software development, product support and product marketing. In his current role as Global Senior Product Manager he is responsible for the lifecycle of a complete product family in healthcare industry.

Giovanni Maio

University of Freiburg
Institute of Bioethics and History of Medicine
Freiburg, Germany
maio@ethik.uni-freiburg.de

Chapter A.2

Giovanni Maio has the Chair of Bioethics at the University of Freiburg. He owns a Master degree in Philosophy, a PhD in Medicine and a habilitation thesis in Medical Ethics. He is Director of the Interdisciplinary Center for Ethics and Director of the Institute of Bioethics and History of Medicine. He is member of ethics committees of the German Parliament and of the Medical Association. He has written over 200 publications on biomedical ethics and takes a critical look on the mere technological self-conception of modern medicine.

Albrecht Malkmus

GE Medical Systems Information Technologies
Freiburg, Germany
albrecht.malkmus@ge.com

Chapter A.5

Albrecht Malkmus received the Masters degree in Electronic Engineering and Information Technology from the University in Frankfurt, Germany. He is currently the Regulatory Affairs Leader and Member of the Quality Management Department at GE Medical Systems Information Technologies in Freiburg, Germany. He is member of the German National Standardization Committee and the Regulatory Affairs Working Group of the ZVEI.

Authors

Kei Masani

Toronto Rehabilitation Institute
Lyndhurst Centre
Toronto, Ontario, Canada
k.masani@utoronto.ca

Chapter D.44

Kei Masani received the PhD degree in Physical and Health Education from the University of Tokyo, Tokyo, Japan. He is now a Scientist at Toronto Rehab, Toronto, Canada. He received the Young Investigator Award from the Japanese Society of Biomechanics in 2000. His research interests are in understanding human movement and movement variability, from the view of neuro-mechanical interaction and sensory-motor integration.

Ulrich Matern

wwH-c GmbH
Tübingen, Germany
ulrich.matern@wwH-c.com

Chapter A.6

Ulrich Matern is a surgeon (Associated Professor, Medical Faculty of Univ. Tübingen, Germany) and CEO and owner of wwH-c GmbH Tübingen, Germany. wwH-c operates the Experimental-OR as a lab for research, development and testing of medical and hospital technologies as well as a teaching and training centre for hospital engineers, physicians and nursing staff. Dr. Matern represents the German Society of Surgeons within the German Standardization Committee for Usability. He is working as a consultant for hospitals worldwide regarding planning of new facilities and also auditing hospitals using the Best Hospital Practice standard that was developed by wwH-c together with TÜV.

John McNulty

San Diego State University
Department of Biology
San Diego, CA, USA
jmcnulty24@gmail.com

Chapter C.22

John McNulty received his Bachelor of Science degree in Biology from San Diego State University. He is currently working as a Research Assistant in the lab of Dr. Robert Pozos in San Diego, California. John is in the process of applying to medical school where he wants to pursue a career in surgery.

Andreas Melzer

University Dundee
IMSaT
Dundee, Scotland, UK
a.melzer@dundee.ac.uk

Chapters C.25, C.26

Andreas Melzer, MD DDS, is Professor of Medical Technology and Founding Director of the Institute for Medical Science and Technology IMSaT at the University Dundee. Funded by Northern Research Partnership it is hosting GE's first European Centre of Excellence for MRI-guided interventions and surgery. 60 scientist are working in the field of multimodality image-guided diagnosis and therapy, biophotonics, ultrasound, MRI, interventional and surgical technologies and nanobiotechnology.

Paulo M. Mendes

University of Minho, Campus de Azurém
Department of Industrial Electronics
Guimarães, Portugal
paulo.mendes@dei.uminho.pt

Chapter G.74

Paulo Mendes received the PhD degree in Industrial Electronics from theUniversity of Minho, Protugal, in 2005. Since then he has been an Assistant Professor at the University of Minho and a researcher in the micro/nanotechnologies and biomedical applications research group (Algoritmi). He has published more than 90 journal and conference papers. His main research interests are in wearable devices, wireless sensor networks, and antenna miniaturization for biomedical applications.

Silvestro Micera

ETH Zürich, ETL K 10.1
Institut für Automatik
Zürich, Switzerland
micera@control.ee.ethz.ch

Chapters D.39, F.68

Silvestro Micera received the University degree in Electrical Engineering from the University of Pisa in 1996 and the PhD degree in Bio-Engineering from Scuola Superiore Sant'Anna (SSSA) in 2000. From 2000 to 2009 he has been Assistant Professor at SSSA where he is now the Head of the Neuroprosthesis Group. He is currently Adjunct Assistant Professor at the ETH Zurich.

Kostis Michelakis

Imperial College London, South Kensington Campus
Institute of Biomedical Engineering
London, UK
k.michelakis@imperial.ac.uk

Chapter E.56

Kostis Michelakis is Research Fellow and Bionanotechnology Facility Manager in the Institute of Biomedical Engineering at Imperial College London. His current research is concerned with micro/nano-fabricated devices and systems at the interface between biochemistry and microelectronics. He has worked in the area of micro/nano fabrication for over 20 years. He received his PhD from the University of Athens and has previously held posts at FORTH (Greece) and the Department of Electrical and Electronic Engineering of Imperial College London.

Uwe Möhring

Textile Research Institute
Thuringia-Vogtland
Greiz, Germany
u.moehring@titv-greiz.de

Chapter G.73

Dr. Uwe Möhring received the doctorate for metal organic catalysis from the Julius-Maximilians-University Würzburg in 1995. Prior to his appointment as Managing Director of TITV Greiz in 2001, he worked for Degussa Corporation. At the TITV Greiz he brought together a team of 60 interdisciplinary scientists for textile microsystems technology and special surface technologies of textiles. The focus is on medical applications of textiles and the development of textile electrodes and sensor structures as well as special intelligent textiles for health and disease management.

Wolfgang Müller-Wittig

Nanyang Technological University (NTU)
Fraunhofer IDM@NTU
Singapore
wolfgang.mueller-wittig@fraunhofer.sg

Chapter F.63

Wolfgang Müller-Wittig is Director of the Project Centre for Interactive Digital Media at Nanyang Technological University. He is also an Adjunct Associate Professor at the NTU School of Computer Engineering. He received his Dr.-Ing. in Computer Science from the Technische Universität Darmstad, Germanyt. His research interests include highly interactive 3-D computer graphics for manufacturing, engineering, edutainment, cultural heritage, and biomedical sciences.

Andreas Neudeck

Textile Research Institute
Thuringia-Vogtland
Greiz, Germany
a.neudeck@titv-greiz.de

Chapter G.73

Andreas Neudeck received his Master of Education from the Applied Pedagogical University of Halle, Germany and a PhD in Physical Chemistry from the same University. After his PhD he joined the Leibniz Institute for Solid State and Materials Research Dresden (IFW) for a Habilitation Fellowship of the DFG. During his DFG Fellowship he spent one year at the Université Paris 7. He obtained his Habilitation (DSc) in Physical Chemistry from Dresden University of Technology in 1999. Since 2000 he works at the Textile Research Institute Thuringia-Vogtland (TITV), Germany and is now responsible for the Advanced Research as Representative Head of Research and Development.

Wolfgang R. Nitz

Siemens AG Healthcare Sector
Erlangen, Germany
wolfgang.nitz@siemens.com

Chapter C.23

Wolfgang R. Nitz received his Master degree in Power Electronics from the University of Siegen and holds a Master degree in Experimental Nuclear Physics from the University of Karlsruhe. He received his PhD from the University of Würzburg and is an Assistant Professor for Experimental Radiology at the University of Regensburg. He has been awarded the first Wilhelm-Conrad-Röntgen-Preis in 2001 for his work on vascular interventions under MR guidance. He is currently heading the patent and licensing management of the MR business of the Siemens AG.

Heike Oschatz

Textile Research Institute
Thuringia-Vogtland
Textile Structures and Materials
Greiz, Germany
h.oschatz@titv-greiz.de

Chapter G.73

Heike Oschatz received the diploma in Textile Technology from the University of Applied Sciences Reichenbach, Germany. Since 1990 she has been working at the Textile Research Institute Thuringia-Vogtland (TITV Greiz), Germany, as a scientist in the field of textile materials. She is working as project manager in several national and European projects with focus on textile technology and the development of special woven or knitted textiles for clothing, medicine and technology.

Ernst Pelikan

University Medical Center Freiburg
Hospital Computer Department
Freiburg, Germany
ernst.pelikan@uniklinik-freiburg.de

Chapter F.60

Ernst Pelikan received his Diploma in Physics from the Albert-Ludwigs-University of Freiburg, Germany in 1983. He joined the Hospital Computer Center at the University Medical Center Freiburg in 1984, where he is the Head of the Department of Communications and Networks since 1989. His research interests are in network design, network security, standards-based medical data exchange and telemedicine.

Thomas Peyn

Dräger Medical AG & Co. KG
Department for Respiratory Care
Lübeck, Germany
thomas.peyn@draeger.com

Chapter D.27

Thomas Peyn, electrical engineering technician and training manager, has been with Dräger since February 1996. Firmly convinced that clinical application of sophisticated technology is the basis for enhanced therapeutic success his focus and passion is education and training especially within the field of applied health care technology.

Doris Pommi

Siemens Healthcare
Forchheim, Germany
doris.pommi@siemens.com

Chapter C.21

Doris Pommi is product manager for interventional radiology at Siemens Healthcare in Germany. She has a background as Medical Laboratory Scientist (MTRA) and marketing consultant.

Milos R. Popovic

University of Toronto
Rehabilitation Engineering Laboratory,
Institute of Biomaterials and Biomedical Engineering
Toronto, Ontario, Canada
milos.popovic@utoronto.ca

Chapter D.44

Milos R. Popovic received his PhD in Mechanical Engineering from the University of Toronto, Ontario, Canada in 1996, and the Dipl. Electrical Engineer degree from the University of Belgrade, Yugoslavia in 1990. Dr. Popovic is the Toronto Rehab Chair in Spinal Cord Injury Research. He is also a Professor in the Institute of Biomaterials and Biomedical Engineering at the University of Toronto. His fields of expertise are functional electrical stimulation. His interests are in the areas of neuro-rehabilitation, physiological control systems, assistive technology, and brain machine interfaces.

Robert S. Pozos

San Diego State University
Department of Biology
San Diego, CA, USA
bpozos@gmail.com

Chapters B.10, C.22

Dr. Robert Pozos, Professor of Biology at San Diego State University, has had extensive experience studying physiological signals in humans in extreme environments.His expertise is in the area of thermal regulation in hot and cold environments. The importance of the role of peripheral circulation invarious environments has been addressed using various technologies specifically near infrared spectroscopy. Utilizing this technology, he has demonstrated that the peripheral circulation has complex controls that respond differentially in different anatomical areas.

Anna Radomska-Botelho Moniz

Imperial College London, South Kensington Campus
Institute of Biomedical Engineering
London, UK
a.radomska@imperial.ac.uk

Chapter E.56

Anna Radomska is a Research Associate at the Institute of Biomedical Engineering in the Imperial College London. The main fields of her research cover development and applications of chemical sensors and biosensors in clinical and biomedical analysis. She received her PhD with distinction in Analytical Chemistry from Warsaw University, Department of Chemistry in 2004. During her PhD she developed several potentiometric and optical biosensors which were successfully used for monitoring, control and assessment of haemodialysis therapy.

Stanisa Raspopovic

Scuola Superiore Sant Anna
ARTS lab
Pisa, Italy
s.raspopovic@sssup.it

Chapter F.68

Stanisa Raspopovic received his MS degree from the Department of Engineering at the University of Pisa, Italy. He is currently working towards the PhD in Biomedical Robotics at the Scuola Superiore Sant Anna, Pisa. His research interests are in neuroprostheses, neural interfaces in animal and human models, neural modelling, and processing of PNS and CNS signals for understanding neural control.

Annette Reinhardt

Klinikum Nürnberg
Institute of Hospital Epidemiology
Nürnberg, Germany
annette.reinhardt@klinikum-nuernberg.de

Chapter A.3

Dr. Annette Reinhardt studied Medicine in Regensburg, Erlangen (Germany) and Winterthur (Switzerland). Since 1993 she has been employed as Doctor of Medicine at the Department of Hospital Hygiene, Clinical Microbiology and Infectious Diseases of the Klinikum Nürnberg, Germany. During her career, Dr. Reinhardt has gathered experience in medical microbiology and healthcare-associated infection surveillance. Her main activity is hospital hygiene with a focus on adapting existing recommendations.

Cristiano Rizzo

Micromed S.p.A.
Mogliano Veneto (TV), Italy
cristiano.rizzo@micromed-it.com

Chapter E.52

Cristiano Rizzo received his Master Degree in Electronic Engineering with Biomedical Specialization from the University of Padova, Italy in 1996. Working for Micromed S.p.A since 1987 as a programmer, he moved to the Micromed French office in 1994 for 2 years. Back in 1996 as Project Development Manager he is actually in Marketing and Sales.

Eckhard Roggenkamp

Klinikum Nuremberg
Institute of Hospital Epidemiology
Nürnberg, Germany
eckhard.roggenkamp@klinikum-nuernberg.de

Chapter A.3

Eckhard Roggenkamp earned his Diploma as Engineering at the University of Applied Sciences Lübeck, Germany. Since 1977 he is at the Institute of Hygiene, Microbiology and Infectiology at the Klinikum Nürnberg, Germany. His major tasks are controlling of sterilization processes for medical devices, validation of washer-disinfectors, inspection of ventilation in buildings and rooms of health care, advising the clinic in all technical concerns of hygiene. He is Vice-chairman of the DIN Commission Disinfection Apparatus.

Christopher Rolfes

University of Minnesota
Department of Biomedical Engineering and Surgery
Minneapolis, MN, USA
rolf0053@umn.edu

Chapter D.43

Christopher Rolfes graduated from St. Olaf College in 2006 with a double major in Physics and Mathematics. After this, he worked for a year as a post-baccalaureate researcher at the National Heart, Lung and Blood Institute in Bethesda, MD. While at the NHLBI, he helped to develop programming for an automated microtome, which is capable of producing three-dimensional reconstructions of mouse hearts with micrometer scale resolution. He is currently a PhD graduate candidate in Biomedical Engineering at the University of Minnesota.

Christian Rotsch

Fraunhofer Institute for Machine Tools and Forming Technology IWU
Department Adaptronics and Acoustics
Dresden, Germany
christian.rotsch@iwu.fraunhofer.de

Chapter G.73

Christian Rotsch received his Diploma in Physical Techniques with the specialization in Biomedical Engineering from the University of Applied Sciences Zwickau, Germany. Between 2004 and 2009 he worked at the Textile Research Institute Thuringia-Vogtland, Germany, as scientist, project and group leader of the smart textiles group in several national and European projects. Since 2009 he works at the Fraunhofer Institute for Machine Tools and Forming Technology. His field of work is the development of active and adaptronic devices for medical applications.

Georg-Friedemann Rust

Rendoscopy AG
Gauting, Germany
gfr@rendoscopy.de

Chapter F.66

Georg-Friedemann Rust studied Physics and Medicine. His PhD was performed at the DESY, German Electron Synchrotron Radiation Labatory in Hamburg together with the IMDM, University Hospital Hamburg, Eppendorf. Dr. Rust changed to the Ludwig-Maximilians University, Department of Clinical and Interventional Radiology. During the Research activities he worked in different fields, such as 3-D Visualization and x-ray dosage issues. During his further research activities and established scientific cooperations with the International Atomic and Energy Commission (IAEA) of the UN. He also founded together of the Bavarian Government a spin-off company, named Rendoscopy AG. Dr. Rust is currently working as a radiologist in the General Hospital Bogenhausen, Munich.

Christina Sampogna

University of Cambridge
Newmarket, Suffolk, UK
Christina.Sampogna@cantab.net

Chapter F.67

Christina Sampogna is an international lawyer, policy advisor, and economist whose experience covers government, industry, public sector, non-profit sector, and international organizations. Having completed studies in Economics (BA) and in Law (LL. L, LL.B.), she received her LL.M. from the University of Cambridge and her D.E. A. in European Community Law from the College of Europe, Bruges. Professionally qualified in diverse jurisdictions, Christina has developed expertise in numerous legal domains and industrial sectors, including emerging technologies (life sciences; information and communication technologies), health, international trade, and environment/ sustainable development.

Gregor Schaefers

MR:comp GmbH
Gelsenkirchen, Germany
schaefers@mrcomp.com

Chapter C.26

Gregor Schaefers obtained his Dipl.-Ing. (FH)degree in Medical Engineering from the University of Applied Sciences Gelsenkirchen, Germanyin 2001. He is co-founder, shareholderand Managing Directorof MR:comp GmbH, testing laboratory for MR safety and compatibility. He is a memberofDIN, ISO, IEC,ASTMstandardization committees workingin developmentand optimization ofMR testing methods.

Rolf M. Schlegelmilch

SMT medical GmbH & Co.
Würzburg, Germany
r.schlegelmilch@smt-medical.com

Chapter B.8

Rolf Schlegelmilch has more than 30 years of experience of in the medical equipment industry . With a focus on diagnostic applications he specializes in pulmonary function and vascular testing, life style and activity monitoring, as well as anti-aging diagnostics. He earned a degree in Mathematics and Economies from the University of Würzburg and worked in several management functions. He is a founder and Managing Director of SMT medical.

Oliver Scholz

HTW Saarland,
University of Applied Sciences
Saarbrücken, Germany
oliver.scholz@htw-saarland.de

Chapter F.61

Oliver Scholz is currently heading the Bio-telemetry Working Group of Fraunhofer IBMT. His main field of research is biomedical telemetry and bio-monitoring incorporating wireless links for medical implant control plus wireless powering and physiological data acquisition. In recent years, this expertise directed him more and more into communication and powering systems for micro robots.

Frank Schön

Aloka Holding Europe AG
Zug, Switzerland
F.Schoen@aloka-europe.com

Chapter C.17

Frank Schön is Marketing Director of Aloka Holding Europe AG. He studied Medicine, Business and Computer Sciences at the Universities of Mainz and Hagen, Germany. He was educated in Internal Medicine and Cardiology at the University Clinic of Essen. After several years of practicing medicine he started in 1999 with Aloka Ultrasound in Germany. After positions in R&D of Aloka Tokyo, Managing Director of Aloka Germany from 2005 to 2010, he is now responsible for International Marketing in Europe.

Arthur Schultz

Hannover Medical School
EEG Monitoring Research Group
Hannover, Germany
ab.schultz@t-online.de

Chapter E.51

PD Dr. Dr. Arthur Schultz, Anaesthesiologist and Medical Computer Scientists (GMDS), is University Lecturer at Hannover Medical School, Hannover, Germany. He is also Head of the Interdisciplinary EEG Monitoring Research Group at Hannover Medical School/Clinic Region Hannover focusing on the automatic classification of electroencephalograms (EEGs) recorded during anaesthesia and intensive care.

Barbara Schultz

Hannover Medical School
Informatik/Biometrie
Hannover, Germany
ab.schultz@t-online.de

Chapter E.51

PD Dr. Barbara Schultz is a University Teacher at Hannover Medical School, Hannover, Germany, and a Member of the Interdisciplinary EEG Monitoring Research Group at Hannover Medical School. Her main research area is electroencephalography (EEG) in patients in the operating theatre and in the intensive care unit with a focus on automatic classification of the EEG.

Dirk Schulze

DDZ Breisgau
Freiburg, Germany
dirk.schulze@ddz-breisgau.de

Chapter C.20

Dirk Schulze received the PhD degree in Dentistry from Hamburg University in 2007. He has been with University Freiburg until the end of 2009 and is currently working in a private dental imaging center. His research activities are in further processing dental CBCT data, cephalometric analysis and evaluation of radiation exposure due to dental imaging devices.

Wilhelm Schütte

Kreiskrankenhaus Gummersbach,
Academic Teaching Hospital of the
University of Cologne
Department of Medical Radiation Physics
Gummersbach, Germany
wilhelm.schuette@kkh-gummersbach.de

Chapter D.45

After his graduation in Physics (Dipl. Phys.) in 1986, Wilhelm Schütte has been working as a Medical Physicist at the Department of Medical Radiation Physics of Gummersbach Hospital (Academic Teaching Hospital of the University of Cologne). He has published numerous articles in national and international scientific journals. His interest is in dosimetry of high-energy radiation, radiation treatment planning and IMRT.

Danny Schwabe

Textile Research Institute
Thuringia-Vogtland
Greiz, Germany
d.schwabe@titv-greiz.de

Chapter G.73

Danny Schwabe received the Diploma in Textile and Leather-Technology at the University of Applied Science Zwickau (FH Reichenbach), Germany. Since 2002, he is working at the TITV Greiz as a scientific employee in the field of 3-D warp knitting. He investigated several projects dealing with medical textiles such as compression bandages, operating -table covers, heating textiles and hip protection pads.

Sanjiv Sharma

Imperial College London, South
Kensington Campus
Institute of Biomedical Engineering
London, UK
sanjiv.sharma@imperial.ac.uk

Chapter E.56

Sanjiv Sharma is a Research Officer in the Institute of Biomedical Engineering at Imperial College London. His main research areas are micro and submicron fluidic systems, noninvasive and label free detection systems. He did his PhD in Chemistry at the Regional Research Laboratory, Bhopal, India. He has held research positions in France (Institut de Physique Nucléaire d' Orsay and Hospital St. Antoine, Paris) and Germany (Walter Schottky Institut, Technische Universität München).

Erich Siegel

Dräger Medical AG & Co. KG
Lübeck, Germany
erich.siegel@draeger.com

Chapter D.30

Erich Siegel received the Master degree and the PhD degree in Physics both from the University of Karlsruhe, Germany. He was assistant and lecturer at the department of applied physics at the University of Bonn. More than 20 years he is engaged in the field of medical technology at Dräger Medical in Lübeck, Germany, where he now is Training Manager of anaesthesiology.

Florian Solzbacher

University of Utah
Electrical and Computer Engineering
Salt Lake City, UT, USA
florian.solzbacher@utah.edu

Chapters D.40, E.46

Dr. Solzbacher is Director of the Microsystems Laboratory at the University of Utah, Co-Director of the Utah Nanotechnology Institute, President of Blackrock Microsystems and an Associate Professor in Electrical and Computer Engineering with adjunct appointments in Materials Science and Bioengineering. His research focuses on harsh environment microsystems and materials, including implantable, wireless microsystems for biomedical and healthcare applications. Professor Solzbacher received his MSc EE from the Technical University Berlin in 1997 and his PhD from the Technical University Ilmenau in 2003.

Silvia Sterzi

Università Campus Bio-Medico
Physical Medicine and Rehabilitation
Rome, Italy
s.sterzi@unicampus.it

Chapter D.42

Silvia Sterzi, MD, is Associate Professor of Physical Medicine and Rehabilitation at THE Campus Bio-Medico University in Rome, Italy, where she serves as the Head of the Department of Rehabilitation and Director of Postgraduate School of Physical Medicine and Rehabilitation. Dr. Sterzi is a Member of the European Board of Physical Medicine and Rehabilitation.

Wilfried Storz

Gebrüder Martin GmbH & Co. KG
Tuttlingen, Germany
wilfried.storz@klsmartin.com

Chapter G.72

Wilfried Storz holds the position of Officer with general commercial power of attorney in the Central Marketing Department of Gebrüder Martin GmbH. His special areas of responsibility are surgical instruments for various medical disciplines and sterilization containers. Besides he has been an active member of the DIN NA 027 Precision Mechanics and Optics Standards Committee (NAFuO), which is part of the Surgical Instruments Working Committee (NA 027-02-01 AA).

Dirk Sunderbrink

Siemens AG – Healthcare Sector
Imaging and Therapy Department
Forchheim, Germany
dirk.sunderbrink@siemens.com

Chapter F.65

Dirk Sunderbrink received a degree in Medical Health Engineering from the University of Applied Sciences Hamburg and in Industrial Engineering from the University of Applied Sciences Bielefeld, Germany. He joined Siemens in 1995 and served for more than 10 years in sales and business management for radiological information systems. Currently he holds the position of Global Marketing Manager for cardiology in angiography business unit. Prior to joining Siemens he was product manager for MRI and CT of Toshiba Medical Systems Germany.

Hajo Tanck

Mednovo Medical Software Solutions GmbH
Berlin, Germany
info@mednovo.de

Chapter F.57

Hajo Tanck earned his degree in Business Informatics from the University of Applied Sciences Flensburg, Germany. His professional career is closely connected to the healthcare sector. Mr. Tanck works for Mednovo Medical Software Solutions GmbH in the sales department and project management and is responsible for the support of complex projects.

Prashant Tathireddy

University of Utah
Electrical and Computer Engineering
Salt Lake City, UT, USA
p.tathireddy@utah.edu

Chapter D.40

Prashant Tathireddy received the PhD degree in Chemical Engineering from the University of Utah and has been with the Department of Electrical and Computer since September 2005. He received Fraunhofer's 'Outstanding Research Scientist Fellowship' Award in 2007 and was posted as a guest scientist for over a year at the Fraunhofer Institute for BiomedicalEngineering (IBMT), St. Ingbert, Germany. Professor Tathireddy's current research focuses on design, fabrication process development and testing of chronically implantable medical microdevices.

Rachel Toomey

University of Dundee
Institute for Medical Science and Technology
Dundee, UK
r.toomey@dundee.ac.uk

Chapter C.25

Rachel holds a BSc (Hons.) in Radiography and a PhD in the area of medical image display from University College Dublin. She is currently employed at the Institute for Medical Science and Technology, University of Dundee. Her research interests include image perception, development of training models and integration of imaging modalities in the operating theatre.

Iasonas F. Triantis

Sensors Systems and Circuits Research Group
Department of Electronic and Electrical Engineering
London, UK
i.triantis@ee.ucl.ac.uk

Chapter E.56

Iasonas Triantis is a Senior Researcher in microelectronics for electrical impedance tomography for cancer imaging at the Electrical and Electronic Engineering Department University College London. During his PhD at the same department, and later as a Research Associate at Imperial College London, he designed neural sensing and stimulation chips and researched alternative neural interfacing methods. He also managed a neural interfacing group and a laboratory.

Jakub Trzebinski

Imperial College London, South Kensington Campus
Institute of Biomedical Engineering
London, UK
jakub.trzebinski09@imperial.ac.uk

Chapter E.56

Jakub Trzebinski received his Bachelor degree (Honours) from the National University of Ireland, Galway and currently is a PhD student at Brunel University and Imperial College London working on electrochemical biosensors, in particular glucose sensors. His research area includes understanding and favourably modifying biosensor-tissue interactions.

Friedrich Ueberle

Hamburg University of Applied Sciences
Department of Life Sciences/Biomedical Technology
Hamburg, Germany
friedrich.ueberle@haw-hamburg.de

Chapter D.33

Friedrich Ueberle received his Diploma in Electrical Engineering from the University of Karlsruhe, Germany. During his PhD he was involved in research for piezoelectric lithotripters. Afterwards, he continued shock wave research and application for 14 years in industry. Since 2002, he is a Professor for Biomedical Device Technology at the Hamburg University of Applied Sciences. His research focuses on ultrasound, shockwaves, medical imaging, and sensors.

Hans-Peter Uhlig

Dresden, Germany
cu-uhlig@t-online.de

Chapter A.4

Hans-Peter Uhlig works as a Consulting Engineer in Dresden, Germany specializing in electrical installations for medical facilities. He is member of the Saxon Chamber of Engineers. Since more than 10 years he contributes to standardization of electrical installations for medical facilities within the German Commission for Electrical (DKE), Electronic and Information Technologies of DIN and VDE. He has authored and co-authored one specialist book and several other publications.

Jörg Vienken

Fresenius Medical Care
Bad Homburg, Germany
joerg.vienken@fmc-ag.com

Chapter D.31

Joerg Vienken has a degree in Chemical Engineering from the Technical University of Darmstadt and a Doctoral Degree in Biophysics and Engineering from the Technical University of Aachen (RWTH), Germany. Since 25 years he is working in leading positions in the medical device industry. He currently serves as Vice President BioSciences at Fresenius Medical Care in Bad Homburg, Germany. He is member of the Editorial Board of several scientific journals dedicated to biomaterials, artificial organs and medical device technology.

Udo Voges

Karlsruher Institut für Technologie
Institut für Angewandte Informatik
Eggenstein-Leopoldshafen, Germany
udo.voges@kit.edu

Chapter G.70

Udo Voges received his Diploma in Mathematics from the Ludwig-Maximilians-Universität in Munich, Germany and his PhD degree in Computer Science from the Universität Karlsruhe, Germany. He is Senior Scientist at the Karlsruher Institut für Technologie (KIT), Institut für Angewandte Informatik. His research interests are in medical engineering, systems engineering and system safety.

Birgit Wacker

Philips Healthcare
Böblingen, Germany
birgit.wacker@philips.com

Chapter E.53

Birgit Wacker received the degree in Informatics from Hochschule Furtwangen University, Germany in 1988. She started her career at Hewlett Packard as software development engineer for clinical information systems and is now working for Philips Healthcare in international marketing of healthcare IT solutions.

Golam Abu Zakaria

Hospital of the University of Cologne
Department of Medical Radiation Physics
Gummersbach, Germany
zakaria@kkh-gummersbach.de

Chapter D.45

Golam Abu Zakaria received his PhD in Medical Physics from Heidelberg University. He currently is chairman of the Department of Medial Radiation Physics at Gummersbach Hospital (University of Cologne) and Professor of Medical Physics/Biomedical Engineering at Gono University, Bangladesh. He has published more than 65 journal and conference papers.

Christian Zapf

Siemens AG – Healthcare Sector
Imaging and Therapy – SYNGO
Erlangen, Germany
christian.zapf@siemens.com

Chapter F.65

Christian Zapf received his Master degree in Computer Science from Clemson University, SC, USA. He joined Siemens Healthcare in 1993 and his primary focus has been on radiology and cardiology IT (RIS, CVIS), image management solutions (PACS), and the magnetic resonance business. As Vice President Business Strategy for the SYNGO business unit of Siemens Healthcare, he is currently responsible for the strategy of the radiology and cardiology IT business of Siemens Healthcare.

Frank Zgoda

Laser- und Medizin-Technologie GmbH
Berlin, Germany
f.zgoda@LMTB.de

Chapter D.29

After studying chemistry and physics at the Freie Universität Berlin, Germany, Frank Zgoda joined the Laser- und Medizin-Technologie GmbH Berlin in 1989. He is consulting manufacturers of innovative medical devices in the fields of laser safety and medical device safety and develops and delivers laser safety courses for physicians, LSOs, and supervisors. He is a member of the German standards committee on medical lasers.

Loredana Zollo

Università Campus Bio-Medico di Roma
Laboratory of Biomedical Robotics and Biomicrosystems
Rome, Italy
l.zollo@unicampus.it

Chapter D.42

Loredana Zollo received her PhD in Bioengineering in 2004. In 2004, she entered as post-doc fellow at Università Campus Bio-Medico di Roma and joined the Laboratory of Biomedical Robotics and Biomicrosystems. In 2008 she became Assistant Professor in Bioengineering. Her research interests are in the fields of rehabilitation and neuro-robotics. She is co-chair of the IEEE-RAS TC on Rehabilitation and Assistive Robotics.

Detailed Contents

List of Abbreviations XXXVII

Part A Medical Technology Basics

1 **Technology in Medicine: Its Role and Significance in Terms of Health Policy**
Rüdiger Kramme, Heike Kramme 3
1.1 A Short History 3
1.2 Early Breakthroughs of Medical Technology 3
1.3 Analog to Digital 4
1.4 Health Policy 5
1.5 New Key Areas 5
1.6 Innovation Versus Financial Resources 6

2 **Medicine Is More Than Applied Technology for Human Beings**
Giovanni Maio 7
2.1 Technology Suggests Feasibility and Controllability 7
2.2 Technology Knows No Bounds 8
2.3 Technology Is Unable to Answer the Question of Meaning 9
2.4 Technology Alone Does Not Make Medicine Humane 9
References 10

3 **Hygiene in Medical Technology**
Heinz-Michael Just, Eckhard Roggenkamp, Annette Reinhardt 11
3.1 Background 12
3.1.1 Employee Protection 12
3.1.2 Patient Protection 12
3.2 Causes of Infection 13
3.3 Vaccinations 13
3.4 Disinfection Methods 14
3.4.1 Basics of Disinfection 14
3.4.2 Disinfection Processes 14
3.4.3 Chemical Disinfecting Agents 15
3.4.4 Carrying out Manual Disinfection 15
3.4.5 Physical Disinfection Processes 16
3.4.6 Application Times and Ranges of Action 19
3.4.7 Comparison of Chemical and Physical Disinfection Processes 20

3.5 Sterilization Methods 21
3.5.1 Sterilization Processes 21
3.6 Hygiene of Noninvasive Technology Equipment 25
3.6.1 Equipment Used on the Patient 25
3.6.2 Equipment Not Used on the Patient 26
3.6.3 Repair and Maintenance 26
3.7 Hygiene of Invasive Technology Equipment 26
3.8 Practical Examples 26
3.8.1 Postoperative Wound Infections 28
3.8.2 Ventilator-Associated Pneumonias (VAP) 28
3.8.3 Catheter-Related Septicemia 29
3.8.4 Catheter-Related Urinary Tract Infection 29
3.8.5 Dialysis 29
3.8.6 Creutzfeldt–Jakob Disease 31
3.9 Regulations 31
3.9.1 Technical Regulations for Hazardous Substances 31
3.9.2 Standards 32
References 33

4 **Technical Safety of Electrical Medical Technology Equipment and Systems**
Rüdiger Kramme, Hans-Peter Uhlig 35
4.1 General Information Regarding the Safety of Technical Systems 36
4.2 Attaining Safety in Medical Institutions 36
4.3 Minimum Requirements for ME Equipment 37
4.4 Areas Used for Medical Purposes 40
4.5 Electrical Systems According to the Nature of the Connection to Earth 42
4.6 Protection Against Shock Currents 42
4.7 Power Supply 44
4.8 Power Sources for Safety Purposes with Accumulators 44
4.9 Final Circuits and Plug Sockets 45
4.10 Static Electricity 45
4.11 Electromagnetic Compatibility 46
4.12 Conclusions 47
References 47

5 **Quality Management in Medical Technology**
Albrecht Malkmus 49
5.1 Objectives of a Quality Management System 49
5.1.1 Concepts 49
5.1.2 Objectives and Clientele of a QMS 50
5.1.3 Regulatory Requirements 50
5.2 Elements of a Quality Management System 54
5.3 Organization of a Quality Management System 54

5.4 Implementation of a QMS ... 56
5.4.1 The QMS and the (End) Customers ... 56
5.4.2 The QMS and the Management ... 56
5.4.3 The QMS and the Employees ... 56
5.4.4 The QMS and Satisfaction of the Regulatory Requirements 56
5.5 Product Quality ... 57
5.6 Concluding Remarks ... 57
References ... 58

6 **Usability of Medical Devices**
Ulrich Matern, Dirk Büchel ... 59
6.1 What Is Usability? ... 59
6.2 Usability in Medical Technology – Obligation or Opportunity? ... 60
6.3 Usability in Medical Technology – Why? ... 61
6.4 Development of Usable Devices – How Is this Done? ... 62
6.5 Testing of Usable Devices – How Is this Done? ... 64
6.6 Assessment of Usability ... 67
6.7 Usability Development, Testing, and Assessment – An Example ... 67
References ... 70

Part B Functional Diagnostics Devices

7 **Basic Diagnostics in Cardiology**
Rüdiger Kramme ... 75
7.1 Electrocardiography ... 75
7.2 Electrocardiograph Equipment Technology and PC ECG ... 76
7.2.1 Physical and Technological Principles ... 76
7.2.2 Equipment Classification ... 77
7.2.3 Recording Systems ... 78
7.2.4 Electrode Technology ... 78
7.2.5 System Properties ... 79
7.2.6 Operating Modes ... 79
7.3 ECG Methods ... 79
7.4 Lead Systems ... 80
7.4.1 Measurements from the Surface of the Body ... 80
7.4.2 Standard Leads ... 80
7.4.3 Augmented and Reduced Leads from the Surface of the Body ... 82
7.4.4 Invasive Leads ... 83
7.5 Methodological Notes ... 83
7.6 The Diagnostic Value of the ECG ... 83
7.7 Complications ... 84
7.8 Technical Safety Aspects of ECG Systems ... 84
7.9 Long-Term ECG ... 84
7.9.1 Leads ... 84

7.10 Long-Term ECG Systems ... 85
7.11 Computer-Based Assessment ... 85
7.12 Heart Rate Variability and Heart Rate Turbulence ... 87
7.13 Indications for Long-Term Electrocardiography ... 87
7.14 The Significance of the Long-Term ECG ... 87
7.15 The Exercise ECG ... 88
7.16 Equipment Technology ... 88
7.16.1 Physical and Technological Principles ... 88
7.16.2 Ergometry Measuring Station ... 88
7.16.3 Types of Ergometers ... 88
7.17 Reduced Exercise ECG Leads ... 89
7.18 Automatic ST Measuring Programs ... 90
7.19 Exercise Test ... 90
7.19.1 Stress Intensity ... 90
7.20 Methodological Notes ... 93
7.21 The Diagnostic Value of Ergometry ... 93
7.22 Indications ... 93
7.23 Abort Criteria and Safety Measures ... 93
7.24 Technical Safety Aspects ... 94
7.25 Notes on Planning ... 94
Further Reading ... 94

8 **Pulmonary Function Testing**
Rolf M. Schlegelmilch, Rüdiger Kramme ... 95
8.1 Spirometry ... 95
8.1.1 Flow Recording ... 96
8.1.2 Technology ... 96
8.1.3 Methodology of Spirometry ... 101
8.1.4 Cooperation-Free Pulmonary Function Tests ... 104
8.2 Advanced Cardiopulmonary Function Testing ... 105
8.2.1 Overview ... 105
8.2.2 Body Plethysmography ... 106
8.2.3 Diffusion Capacity ... 109
8.2.4 Nitrogen-Washout Test for Determination of FRC ... 111
8.2.5 Ergospirometry ... 111
8.2.6 Noninvasive Determination of Cardiac Output ... 114
8.2.7 Metabolic Activity Monitoring ... 115
8.2.8 Planning and Laboratory Space Considerations ... 116
References ... 116

9 **Devices and Methods in Clinical Neurophysiology**
Klaus-Peter Hoffmann, Ursula Krechel ... 119
9.1 Basics ... 119
9.1.1 Neurophysiological Basics ... 119
9.1.2 Technical Basics ... 121

9.2 Electroencephalograph ... 129
9.2.1 Signals ... 129
9.2.2 Device Technology ... 130
9.2.3 Methodology ... 132
9.2.4 EEG Recording Methods ... 134
9.2.5 Evaluation and Signal Analysis ... 136
9.2.6 Special Methods ... 138
9.3 Electromyograph ... 138
9.3.1 Signals ... 139
9.3.2 Device Technology ... 139
9.3.3 Electromyography ... 142
9.3.4 Electroneurography ... 145
9.3.5 Evoked Potentials ... 148
9.3.6 Event Related Potentials ... 156
Further Reading ... 157

10 **Sleep Diagnostic Systems**
Klaus-Peter Hoffmann, Robert S. Pozos ... 159
10.1 Function and Application ... 159
10.2 Sleep Diagnostics, Sleep Laboratories, and Sleep Apneas ... 160
10.2.1 Sleep Diagnostics ... 160
10.2.2 Sleep Laboratory ... 161
10.2.3 Sleep Apneas ... 162
10.2.4 Sudden Infant Death Syndrome ... 162
10.3 Technology ... 162
10.3.1 Devices for Monitoring of Selected Parameters for Special Applications ... 162
10.3.2 Ambulant Devices for Diagnostics ... 162
10.3.3 Devices for Diagnostics in Sleep Laboratories ... 163
10.3.4 Therapy Devices ... 163
10.4 Sleep Diagnostic Procedures ... 165
10.4.1 Ambulatory Polygraphy ... 165
10.4.2 Small Polysomnography ... 165
10.4.3 Extensive Polysomnography ... 165
10.4.4 Multiple Sleep Latency Test ... 165
10.4.5 Maintenance of Wakefulness Test (MWT) ... 165
10.4.6 Therapeutic Monitoring ... 165
10.4.7 Intensive Monitoring (Epilepsy) ... 166
10.4.8 EEG After Sleep Deprivation ... 166
10.5 Signal Recording and Signal Processing ... 166
10.5.1 Signal Recording ... 166
10.5.2 Visual Evaluation ... 169
10.5.3 Computer-Aided Evaluation ... 169
10.6 Fields of Application ... 171
10.7 Methodical Instructions ... 172
10.7.1 Handling/Application ... 172
10.7.2 Artifacts ... 173

10.8 Medical Significance of Sleep Diagnostics 173
10.9 Therapy 175
10.9.1 Behavioral Medical Psychotherapeutic Therapy 175
10.9.2 Chronobiological Therapy 175
10.9.3 Medicinal Therapy 175
10.9.4 Machine-Aided Therapy 175
10.9.5 Operational Therapy 175
10.10 Safety Aspects 176
10.10.1 Mobile Devices 176
10.10.2 Devices in the Sleep Laboratory 176
10.11 Planning Advice 176
10.11.1 Required Space 176
10.11.2 Required Time 176
10.11.3 Staff 176
Further Reading 176

11 **Nystagmography**
Klaus-Peter Hoffmann, Eduardo Fernandez 179
11.1 Application 179
11.2 Eye Movements 180
11.2.1 Saccades 180
11.2.2 Smooth Pursuit Movements 181
11.2.3 Nystagmuses 181
11.2.4 Convergence Movements 181
11.2.5 Torsional Movements 181
11.3 Technology and Methods 181
11.3.1 Frenzel Glasses 182
11.3.2 Electronystagmography (ENG) and Electrooculography (EOG) 182
11.3.3 Photoelectronystagmography (PENG) and Infrared Oculography (IROG) 182
11.3.4 Magnetooculography 183
11.3.5 Electromagnetic Technology (Search-Coil System) 183
11.3.6 Videooculography (VOG) 184
11.4 Methods 184
11.4.1 Saccades 185
11.4.2 Smooth Pursuit Movements 185
11.4.3 Optokinetic Nystagmus 186
11.4.4 Vestibular Nystagmus 186
11.5 Signal Recording and Signal Processing 187
11.6 Medical Significance 187
11.7 Safety Aspects 188
11.8 Spatial Planning 188
References 189

12 **Audiometry**
Sebastian Hoth 191
12.1 Physical, Technical and Physiological Bases of Audiometry 191
12.1.1 The Physical Description of Sound 192
12.1.2 Anatomy and Physiology of the Auditory System 194
12.1.3 Hearing Disorders 198
12.2 Behavioral Audiometric Assessment 200
12.2.1 Stimulus and Perception 200
12.2.2 Temporal Aspects of Hearing 201
12.2.3 Pure Tone Audiometry 202
12.2.4 Recruitment 204
12.2.5 Speech Audiometry 206
12.2.6 Binaural Hearing 210
12.3 Objective Audiometric Assessment 212
12.3.1 Impedance Audiometry 212
12.3.2 Otoacoustic Emissions (OAE) 215
12.3.3 Evoked Potentials of the Auditory System 221
12.4 Technical Hearing Devices 228
12.4.1 Conventional Hearing Aids 228
12.4.2 Implantable Hearing Systems 236
12.4.3 Cochlear Implants 237
References 242

13 **Measurement Techniques in Ophthalmology**
Albert J. Augustin 245
13.1 Measurement of Intraocular Pressure 246
13.1.1 Goldmann's Applanation Tonometry 246
13.1.2 Schiötz's Impression Tonometry 246
13.1.3 Noncontact Tonometry (Air-Puff Tonometer) 246
13.1.4 Mackay–Marg Tonometry (Tonopen) 247
13.1.5 The Pascal Dynamic Contour Tonometer (DCT) 247
13.2 Optical Coherence Tomography (OCT) 247
13.2.1 OCT Macula 247
13.2.2 OCT of the Optic Nerve Head (ONH) 248
13.2.3 OCT with Visante 249
13.3 Laser-Scanning Tomography with the Heidelberg Retina Tomograph (HRT) 249
13.4 Nerve Fiber Polarimetry with GDx 251
13.5 The Rostock Cornea Module (Confocal Laser Microscope) 252
13.6 Automatic Refractometry 252
13.7 Visually Evoked Potential (VEP) 253
13.7.1 Flash VEP 253
13.7.2 Pattern VEP 254
13.8 The Ganzfeld ERG (Ganzfeld Electroretinogram) 254
13.9 Pattern Electroretinography (Pattern ERG, PERG) 257
13.10 Multifocal ERG (mfERG) 259
13.11 Electrooculograms (EOG) 260

13.12 Adaptometry ... 261
13.13 Aberrometry (Wavefront Analysis) ... 261
13.14 Keratometry ... 263
13.15 Retinoscopy or Skiascopy ... 263
13.16 Ultrasound ... 263
13.16.1 A-Scan Ultrasound, Ultrasound Biometry ... 264
13.16.2 Ultrasound Pachymetry A-Scan ... 264
13.16.3 Ultrasound B-Scan ... 264
13.17 Corneal Topography ... 265
13.18 The Orbscan ... 265
13.19 Scheimpflug Examination ... 267
13.20 Fluorescence Angiography of the Retina (Sodium-Fluorescein) ... 267
13.21 Fluorescence Angiography of the Retina (Indocyanine Green) ... 269
13.22 Visual Field Measurement (Perimetry) ... 269
13.23 Exophthalmometry ... 271
References ... 271

14 **Functional Force Assessment of Skeletal Muscles**
Paul A. Iaizzo, William K. Durfee ... 273
14.1 The Need for Skeletal Muscle Force Assessment ... 273
14.2 Manual Muscle Strength Testing ... 275
14.2.1 Apparatus ... 276
14.3 Advanced Muscle Assessment Methods ... 278
14.3.1 Measuring Muscle Dynamics ... 278
14.3.2 Electromyogram ... 279
14.4 Stimulated Muscle Force Assessment ... 279
14.5 Stimulated Muscle Force Assessment in Animal Models ... 283
14.6 Conclusion ... 285
References ... 285

Part C Medical Imaging

15 **Digital Radiography**
Lothar Heuser ... 291
15.1 Historical Background ... 291
15.2 From Analog to Digital Image ... 292
15.2.1 Spatial Resolution, Dynamic Range, and Dose Quantum Efficiency ... 292
15.2.2 Differences Between Analog and Digital Images ... 294
15.3 Digital Imaging Systems in Radiology ... 296
15.3.1 Digital Image Intensifier Radiography and Conventional DSA ... 296
15.3.2 Digital Storage Screens ... 297
15.3.3 Digital Flat-Panel Detectors (Planar Detectors) ... 300

15.4 Digital Image Processing 305
15.4.1 Frequency Filtering 306
15.4.2 Multifrequency Filtering 307
15.4.3 Dual-Spectrum Radiography (Energy Subtraction) and Temporal Subtraction 307
15.5 Image Communication and Archiving 307
15.5.1 Example 1 – Hospital Outpatient or Patient at a Doctor's Surgery 308
15.5.2 Example 2 – Hospital Inpatient 308
References 309

16 **Computed Tomography**
Thorsten M. Buzug 311
16.1 Background 311
16.2 Milestones of Computed Tomography 313
16.2.1 Analog Geometric Tomography and Tomosynthesis 314
16.2.2 Generations of CT Systems 314
16.2.3 Spiral CT 317
16.3 Computed Tomography Technology 318
16.3.1 X-Ray Generation 318
16.3.2 Detector Systems 321
16.4 Image Reconstruction 323
16.4.1 Fourier-Slice Theorem 323
16.4.2 Filtered Backprojection 324
16.4.3 Raw-Data Interpolation for Spiral CT 324
16.4.4 Artifacts 326
16.5 Scan Planning and Applications 329
16.5.1 Scan Planning 329
16.5.2 Hounsfield Units and Gray-Value Mapping 329
16.5.3 Three-Dimensional Data Representation 331
16.5.4 Clinical Applications 335
16.6 Dose 335
16.7 Special System Designs 338
16.7.1 Electron-Beam CT 339
16.7.2 Volume CT 340
16.7.3 Micro-CT 340
16.7.4 PET-CT 340
16.7.5 Dual-Source CT 341
References 341

17 **Ultrasound Diagnostics**
Rudolf Götz, Frank Schön 343
17.1 Basic Physical Principles 344
17.1.1 Principle 344
17.1.2 Generation of Sound Waves 344
17.1.3 Reflection 344
17.1.4 Spatial Mapping – Transit Time 345

17.1.5 Penetration Depth, Axial Resolution and Frequency Ranges 346
17.1.6 Influencing Factors: Pressure and Temperature 346
17.1.7 Second Harmonic 347
17.1.8 Broadband Harmonics 348
17.2 Visualization of the Blood Flow and Vascular System 348
17.2.1 Doppler 348
17.2.2 B-Mode 349
17.2.3 Ultrasound Contrast Medium 349
17.3 Equipment Technology 350
17.3.1 The Basic Design of an Ultrasound Unit 350
17.3.2 Doppler Ultrasonography 352
17.3.3 Types of Probes 356
17.3.4 Focusing 358
17.4 Three-Dimensional Ultrasound (3-D, Real-Time 3-D) 360
17.4.1 Acquisition Techniques 360
17.4.2 3-D Reconstruction 362
17.4.3 New and Additional Technologies 363
17.5 Operation of an Ultrasound Unit 367
17.5.1 General Conditions 367
Further Reading 367

18 **Medical Infrared Imaging**
Gerald C. Holst, Thorsten M. Buzug 369
18.1 Background 369
18.2 Infrared System Design 370
18.2.1 Detectors 371
18.2.2 Performance Metrics 371
18.2.3 Image Processing 372
18.3 Infrared Physics 372
18.3.1 Planck's Blackbody Law 372
18.3.2 Camera Formula 372
18.3.3 Emissivity and Reflectivity 373
18.4 IR Imaging in Medical Applications 373
18.5 Specific Applications 374
18.5.1 Circulatory Disturbance 374
18.5.2 Sports Medicine 374
18.5.3 Skin Cancer 375
18.6 Limitations of IR Imaging in Medical Applications 377
18.7 Summary 377
References 378

19 **Endoscopy**
Martin Leonhard, Klaus-Martin Irion 379
19.1 Basics 380
19.1.1 Terminology 380
19.1.2 Endoscopy and the Strengths of Optical Perception 380

19.1.3 Challenges 381
19.1.4 History 382
19.2 Endoscopes and Endoscopic Accessories 382
19.2.1 Light Sources and Illumination 382
19.2.2 Imaging: Endoscopes and Image Sensors 383
19.2.3 Video Chain 386
19.2.4 Monitors 386
19.2.5 Image and Video Documentation 386
19.2.6 Telemedicine in Endoscopy 387
19.2.7 Endoscopic Instruments 387
19.2.8 Preparation of Endoscopes and Instruments 388
19.2.9 Peripheral Units 389
19.2.10 Endoscopy Workstations 390
19.3 Integrated Operating Theaters 391
19.4 Medical Applications 393
19.4.1 Most Common Procedures 393
19.4.2 Special Procedures 393
19.4.3 Miniaturization 394
19.4.4 Controlled Access 394
19.4.5 Alternatives to Surgical Microscopes 394
19.4.6 Gastroenterological Endoscopy 394
19.5 Tissue Differentiation 394
19.5.1 Molecular Imaging/Fluorescence Endoscopy 395
19.5.2 Infrared Endoscopy 396
19.5.3 Zoom Endoscopes, Endomicroscopes 396
19.5.4 Ultrasound Endoscopes 396
19.5.5 Optical Coherence Tomography 397
19.5.6 Virtual Chromoendoscopy 397
19.6 Further Future Developments 398
19.6.1 Natural Orifice Translumenal Endoscopic Surgery – NOTES . 398
19.6.2 Single Port 399
19.6.3 Simulation 399
19.6.4 Endorobotics 400
19.6.5 Capsule Endoscopy 401
19.6.6 Endonavigation 401
References 402

20 **Cone-Beam Computed Tomography and Navigation**
Dirk Schulze, Gerhard Hoffmann 405
20.1 Technical Background of Dental Digital Volume Tomography 405
20.1.1 Data Acquisition 405
20.1.2 Data Reconstruction 406
20.1.3 Data Evaluation 407
20.2 Areas of Application of Dental CBCT 408
20.2.1 Preservative Dentistry 408
20.2.2 Prosthodontics 409
20.2.3 Functional Diagnostics and Therapy 409

20.2.4 Surgical Dentistry ... 410
20.2.5 Implantology ... 410
20.2.6 Cranio-Maxillofacial Surgery ... 411
20.2.7 Orthodontics ... 411
20.2.8 Technical Restrictions of the Procedure ... 412
20.2.9 Dose ... 412
20.2.10 Navigation ... 412
20.2.11 Prerequisites for Navigated Procedures ... 412
20.2.12 Further Processing ... 412
20.2.13 Registration ... 412
References ... 413

21 **Interventional Radiology – Angiography**
Doris Pommi ... 417
21.1 Definition of Digital Subtraction Angiography ... 417
21.1.1 Components Required for DSA ... 418
21.2 Application Range for Angiography ... 419
21.3 Advantages of Interventional Radiology Procedures ... 420
21.4 Trends of Development ... 420
21.4.1 Enhanced Computer Performance ... 420
21.4.2 Special Visualizations ... 420
21.4.3 Hybrid Applications ... 420
Further Reading ... 421

22 **Near-Infrared Spectroscopy (NIRS)**
John McNulty, Michael Born, Robert S. Pozos ... 423
22.1 NIRS – Technical ... 423
22.2 NIRS Technology: Engineering Aspects ... 425
22.2.1 Theory of NIRS Penetration ... 425
22.3 Instrumentation and Equipment ... 427
22.4 New Developments: Multidepth Differential Approach ... 430
22.5 Clinical Application and Study of NIRS ... 431
22.5.1 Overview ... 431
22.5.2 Thenar ... 431
22.5.3 Forearm ... 431
22.5.4 Brain ... 431
22.5.5 Lower Limb ... 432
22.5.6 Breast Tumors and Kidneys ... 432
22.6 Does Skin Blood Flow Affect NIRS Measurements? ... 433
22.6.1 Forearm Skin Blood Flow Study Versus StO_2 ... 434
22.6.2 Thenar Skin Blood Flow Study Versus StO_2 ... 435
22.7 Future of NIRS ... 436
References ... 437

23 **Magnetic Resonance Imaging**
Wolfgang R. Nitz 439
23.1 History of MRI 439
23.2 MRI – System Components 441
23.2.1 The Magnet – The Magnetic Field Strength B_0 441
23.2.2 The System for the Magnetic Field Gradient 442
23.2.3 The Radiofrequency System 444
23.2.4 Measurement Control, Acquisition, and Image Reconstruction Systems 445
23.3 MRI – Basic Principles and Applications 446
23.3.1 Slice Selection and Spatial Encoding 446
23.3.2 The Spin-Echo Sequence 447
23.3.3 The Multi-Echo Spin-Echo Sequence 449
23.3.4 The Gradient Echo Sequence 450
23.3.5 The Sequence Family 451
23.3.6 MRI Spectroscopy 452
23.4 MRI – Safety-Relevant Aspects 453
23.4.1 Attraction and Torque Due to Strong Magnetic Fields 453
23.4.2 RF Interaction with the Patient's Body 454
23.4.3 Interaction with Active and Passive Implants 454
23.4.4 Interaction Based on Changes in Magnetic Field Gradients 454
23.4.5 Safety Issues in Conjunction with Loss of Superconductivity (Quenching) 455
23.4.6 Gadolinium-Containing Contrast Agents and NSF 455
23.5 MRI – Pictures of the Future 456
23.5.1 Magnetic Field Strength 456
23.5.2 RF Technology 456
23.5.3 Application Development 456
23.5.4 Hybrid Systems 457
23.5.5 Theranostics – Therapy Under Image Guidance 458
References 458

24 **Magnetic Particle Imaging**
Jörn Borgert, Bernhard Gleich, Thorsten M. Buzug 461
24.1 Introduction 461
24.2 A Brief History of Magnetic Particle Imaging 462
24.3 How Magnetic Particle Imaging Works 463
24.3.1 Signal Generation and Acquisition 463
24.3.2 Spatial Coding: Selection Field 464
24.3.3 Performance Upgrade I: Drive Field 467
24.3.4 Performance Upgrade II: Focus Field 467
24.4 From Data to Images – Reconstruction 468
24.5 Beyond General Purpose Systems – Special Geometry 470
24.5.1 Unlimited Access – Single Sided Magnetic Particle Imaging 470
24.5.2 Zero-Dimensional Imaging – Magnetic Particle Spectroscopy 471
24.5.3 MPI/MRI Hybrid Systems 471

24.6 Putting MPI to Use – Applications 472
24.6.1 Cardiovascular 472
24.6.2 Oncology, Sentinel Lymph Node Imaging and Hyperthermia 473
24.6.3 Cell Labeling and Tracking 473
24.6.4 Gastrointestinal and Lung Imaging 474
References 474

25 **MR-Guided Interventions and Surgery**
Andreas Melzer, Erwin Immel, Rachel Toomey, Fabiola Fernandez-Gutierrez 477
25.1 MRI Basics 478
25.2 MRI Image Guidance for Interventions and Surgery in Comparison with CT and Ultrasound 479
25.3 MR Systems Design and Setup for Interventions and Surgery 479
25.4 Instruments for Interventional and Intraoperative MRI 482
25.5 MR-Applicable Endoscopic Instrument Systems 482
25.6 Instrument Representation and Tracking in MRI 483
25.7 MR-Guided Robotics and Navigation 485
25.7.1 Technology of MR-Compatible Robotic Instrument Guiding Systems 486
25.7.2 Technique for Robotic-Assisted MR-Guided Interventions. 487
25.7.3 Evaluation of Targeting Precision 488
25.7.4 MRI-Compatible Robot MrBot for Prostate Biopsy 489
25.7.5 Technical Issues of MR Robotics 489
25.8 Hybrid Multimodal Imaging for MR-Guided Diagnosis and Therapy. 491
25.9 Therapeutic MR-Guided Imaging 493
25.10 MR-Guided Delivery of Implants 494
25.10.1 MR-Guided Delivery of Stents and Stent Grafts 495
25.10.2 MR-Guided Heart Valve Prosthesis 496
25.10.3 MR-Guided Delivery of Vena Cava Filter 496
25.10.4 MR-Guided Delivery of Closure Devices for Cardiac Septal Defects 497
25.11 Conclusions 498
References 498

26 **Devices and Materials in MRI**
Gregor Schaefers, Andreas Melzer 503
26.1 MR Safety 504
26.2 Interactions in the MR Environment 504
26.2.1 Magnetically Induced Displacement Force (Static, Dynamic) 505
26.2.2 Magnetically Induced Torque (Static, Dynamic) 505
26.2.3 Radiofrequency-Induced Heating 505
26.2.4 Gradient-Induced Interaction 506
26.2.5 Safe Operation of Devices Within the MR Environment (Dependent on Individual Demands) 506

Detailed Cont.

26.2.6 Image Quality ... 507
26.2.7 MR Image Artifacts Due to the Medical Device ... 507
26.3 Examples of MR Artifacts Caused by Medical Devices ... 510
26.3.1 Stent Artifacts Summary ... 511
26.4 Evaluation of MRI Artifacts of Implants ... 512
26.5 MR Safety Labeling ... 514
26.5.1 Terms, Definitions, Icons, and Marking ... 515
26.6 Interpretation of MR Labeling ... 517
26.7 Discussion ... 518
References ... 519

Part D Therapeutic Devices

27 **Long-Term Ventilators for Intensive Therapy**
Thomas Peyn ... 525
27.1 Tasks of the Ventilator ... 525
27.2 Function and Components of a Long-Term Ventilator ... 526
27.2.1 Power Supply ... 527
27.2.2 Gas Mixer ... 527
27.2.3 Pressure Generator ... 527
27.2.4 Breathing System ... 528
27.2.5 Gas Humidifier ... 528
27.2.6 Expiratory Valve ... 528
27.2.7 Operating and Display Unit ... 528
27.2.8 Alarm System ... 529
27.2.9 Patient Monitoring ... 529
27.3 Technical Implementation ... 529
27.3.1 Continuous Flow Systems ... 529
27.3.2 Demand Flow Systems ... 530
27.3.3 Combined Flow Systems ... 530
27.4 Controlling the Ventilator ... 530
27.4.1 Start of Inspiration ... 530
27.4.2 Switchover Behavior ... 531
27.5 Ventilation Procedures ... 531
27.5.1 Controlled Ventilation (Total Respiratory Substitution) ... 531
27.5.2 Supported Spontaneous Breathing ... 534
27.5.3 Spontaneous Breathing (Without Respiratory Support) ... 536
27.5.4 Mixed Ventilation ... 536
27.6 Ventilation Extras and Special Functions ... 540
27.6.1 Inverse Ratio Ventilation ... 540
27.6.2 Sigh ... 540
27.6.3 AutoFlow ... 540
27.6.4 Automatic Tube Compensation ... 542
27.6.5 Noninvasive Ventilation ... 542
27.6.6 Apnea Ventilation ... 543

27.7 Patient Monitoring and Alarm Limits 543
27.8 Weaning Strategy and SmartCare/PS 543

28 **Defibrillators and ICD Systems**
Rüdiger Kramme 545
28.1 Defibrillator Technology 546
28.1.1 Physical Principles 546
28.1.2 System Properties 548
28.2 Therapeutic Intervention 549
28.2.1 Defibrillation/Cardioversion 549
28.2.2 External and Internal Defibrillation 550
28.3 Methodological Notes 550
28.3.1 Electrodes and Contact Agents 550
28.4 Complications 551
28.5 Technical Safety Aspects 551
28.5.1 Use 551
28.5.2 Device 551
28.6 Implantable Cardioverter-Defibrillators 551
28.6.1 ICD Development 552
28.6.2 System Technology 553
28.6.3 Algorithms 554
28.6.4 Electrodes 554
28.6.5 Complications 555
28.6.6 Function Checking 555
References 556

29 **Laser Systems**
Frank Zgoda 557
29.1 History of the Laser 558
29.2 Physics and Technology 558
29.2.1 Laser Medium 559
29.2.2 Laser Radiation 560
29.2.3 Laser Systems 562
29.2.4 Beam Delivery Systems 563
29.3 Application Methods 564
29.3.1 Noncontact Mode 564
29.3.2 Contact Mode 564
29.3.3 Gas or Liquid Flushing During Laser Cutting 564
29.3.4 Special Applicators 564
29.4 Biophysical Effects on Tissue 564
29.4.1 Absorption and Penetration Depth 564
29.4.2 Photochemical Effects 566
29.4.3 Photothermal Effects 566
29.4.4 Photodecomposition Effects 567

29.5 Laser Types in Medicine 567
29.5.1 The Ruby Laser 569
29.5.2 The Neodymium:YAG Laser 569
29.5.3 The Frequency-Doubled Nd:YAG Laser (KTP Laser) 569
29.5.4 The Erbium:YAG Laser 569
29.5.5 The Holmium:YAG Laser 569
29.5.6 The Alexandrite Laser 569
29.5.7 The Diode Laser 570
29.5.8 The CO_2 Laser 570
29.5.9 Argon-Ion and Krypton-Ion Lasers 570
29.5.10 The Excimer Laser 570
29.5.11 The HeNe Laser 571
29.5.12 The Dye Laser 571
29.5.13 The Free-Electron Laser 571
29.6 Fields of Use 571
29.6.1 The Eye 571
29.6.2 The Body Surface 571
29.6.3 The Vascular System 572
29.6.4 Open Surgery 572
29.6.5 Endoscopy 573
29.6.6 Photodynamic Therapy 573
29.6.7 Lithotripsy 573
29.6.8 Laser-Induced Thermotherapy 573
29.7 Safety Aspects 574
29.7.1 Permissible Exposure and Laser Classification 574
29.7.2 Safety Requirements for Laser Devices 576
29.7.3 Protective Measures and Application Safety 576
29.8 Future Prospects 577
References 577

30 **Inhalational Anesthesia Devices**
Erich Siegel 579
30.1 Anesthesia Devices in General Anesthesia 579
30.2 Functional Principle, Medical Aspects 580
30.3 Functional Principle of the Main Components 582
30.3.1 Drug Dosing 582
30.3.2 Ventilator and Breathing System 585
30.3.3 Monitoring 590
30.3.4 Sensor Principles for Device Monitoring 592
30.4 Safe Operation Prerequisites 594
30.4.1 Connection Prerequisites 594
30.4.2 Training and Continuing Medical Education 594
30.4.3 Cleaning 594
30.4.4 Maintenance and Pre-use Check 595
Further Reading 595

31 Extracorporeal Blood Purification Systems
Jörg Vienken 597
31.1 Historical Perspective 598
31.1.1 Prologue 598
31.1.2 Blood Purification: The Search for Principles 598
31.1.3 Membranes, Transport Processes, and Dialyzers 598
31.1.4 Preparatory Work for Extracorporeal Blood Circulation for Blood Purification 599
31.1.5 Removal of Nephrotoxic Substances Through Dialysis as an Experimental Therapy 600
31.2 Blood Purification for the Therapy of Renal Failure 600
31.2.1 Principles, Control Variables, and Conditions for Hemodialysis 600
31.2.2 Dialysis Membranes 601
31.2.3 Dialyzer Construction 604
31.2.4 Performance Parameters 605
31.2.5 Low-Flux, High-Flux, and High-Performance Dialyzers 608
31.3 Dialysis Machines and Additional Equipment: Use and Conditions 611
31.3.1 Determination of Fluid Status with Whole-Body Bioimpedance Spectroscopy 613
31.3.2 Description of the Technology and Methodology 613
31.3.3 Validation of Clinical Data 614
31.4 Blood Purification in Liver Replacement Therapy 614
31.4.1 Detoxification 615
31.4.2 Synthetic Liver Function 616
References 617

32 Heart-Lung Machines
Ulrich Böckler, Andreas Hahn 621
32.1 Historical Development of Extracorporeal Circulation 621
32.2 Extracorporeal Circulation 623
32.3 Structure and Function of the Heart-Lung Machine 623
32.3.1 Blood Pumps and Their Function 624
32.3.2 Oxygenator and Gas Exchange Function 624
32.3.3 Tubing Systems for Extracorporeal Circulation 625
32.3.4 Blood Filters 625
32.3.5 Cardiotomy Reservoir 625
32.3.6 Cannulae 626
32.4 Components of the Heart-Lung Machine 627
32.4.1 Basic Components 627
32.4.2 Additional HLM Components 629
32.5 Extracorporeal Circulation 629
32.5.1 Preparing the ECC 630
32.5.2 Carrying Out the Perfusion 630
32.5.3 Patient Monitoring and Therapy During ECC 631
32.5.4 Complications During Extracorporeal Circulation 633

32.6 Differentiation of Heart–Lung Machines 635
32.6.1 Modular Systems 635
32.6.2 Nonmodular Systems 636
32.7 Aspects of Technical Safety 636
32.8 Prospects for Further Development 637
References 638

33 **Application of Shock Waves and Pressure Pulses in Medicine**
Friedrich Ueberle 641
33.1 Introduction – Historical Development 642
33.2 Definitions of Physical Terms: Acoustics – Sound Waves – Pressure Pulses – Shock Waves 643
33.2.1 Wave Forms Used in Acoustic Therapies 643
33.2.2 Sound Waves – Mechanical Waves in Media 644
33.2.3 Definition of Shock Wave and Pressure Pulse 645
33.2.4 Mechanical Effects of Sound Waves at Interfaces 645
33.2.5 Mechanical Effects of Sound Waves Caused by Nonlinear Properties of the Media 647
33.2.6 Thermal Effects Caused by Pressure Pulses 649
33.3 The Acoustic Field of a Lithotripter – Basics of Measurement Technology 649
33.3.1 Sensors and Measurement Technology 649
33.3.2 Measurement Parameters of Lithotripter Sound Fields 651
33.3.3 Energy Parameters of the Sound Field 653
33.3.4 Significance of the Pressure Pulse Parameters in Lithotripsy 655
33.3.5 Which Physical Effects Play a Significant Role in Stone Comminution? 657
33.3.6 Effect of the Pressure Pulses on Tissue 658
33.4 Generation of Pressure Pulses for Extracorporeal Lithotripsy (ESWL) and Extracorporeal Shock Wave Therapy (ESWT) 660
33.4.1 Point Sources and Extended Pressure Pulse Sources 660
33.4.2 Generation of Spherical Pressure Pulses and Shock Waves 660
33.4.3 Generation of Planar Sound Waves 662
33.5 Extracorporeal Lithotripsy (ESWL) and Extracorporeal Shock Wave Therapy (ESWT) in Practice 664
33.5.1 System Components of a Lithotripter 664
33.5.2 Approaches to Online Disintegration Tracking During Treatment 667
33.6 The Patient 667
33.6.1 Patient Bedding for ESWL and ESWT Treatment 667
33.6.2 Stone- and Patient-Related Influences 667
33.7 Assessment of the Clinical Efficiency of Lithotripters 668
33.7.1 The Controversy About the Right Focal Size 669
33.7.2 Side Effects and Safety 669

33.8 Associations and Societies for Lithotripsy and Pressure Pulse Therapy 670
References 671

34 **High-Frequency Surgery**
Bernhard Hug, Reiner Haag 677
34.1 Development of High-Frequency Surgery 678
34.2 Physical and Technical Principles 681
34.2.1 Bioelectrical and Biothermal Effects on Tissue 681
34.2.2 Coagulation 683
34.2.3 Electrotomy (Cutting) 683
34.3 Technology and Techniques 684
34.3.1 HF Generator Technology 684
34.3.2 HF Application Technique 685
34.3.3 Leakage Currents 686
34.4 Types of Current and Their Application 687
34.4.1 Cutting Currents 687
34.4.2 Coagulation Currents 691
39.5 Methodical Instructions for Application and Safety 697
34.5.1 Patient and User Safety 697
34.5.2 Neutral Electrode 698
34.5.3 HF Instruments and Cables 699
34.5.4 HF Surgical Unit 699
34.5.5 Incidences During Application of HF Surgery 699
34.6 Outlook 700
References 701

35 **Medical Radiation Therapy**
Peter H. Cossmann 703
35.1 X-Radiation 704
35.2 Historical Development of Radiation Therapy 704
35.3 Physical and Technical Principles of Radiation Physics 705
35.3.1 Types of Radiation 705
35.3.2 Radiation Sources 706
35.3.3 Interactions with Matter 706
35.3.4 Dosimetry 707
35.4 Forms of Therapy 708
35.4.1 Percutaneous Therapy 709
35.4.2 Brachytherapy 709
35.4.3 Radionuclide Therapy 709
35.5 Equipment Technology for the Generation of Radiation 710
35.5.1 X-ray Equipment 710
35.5.2 Cobalt Unit 711
35.5.3 Linear Accelerators 711
35.5.4 Miniature Accelerators 712
35.5.5 Brachytherapy Sources and Brachytherapy Equipment 713
35.5.6 Radioactive Implants 714

35.6 Special Techniques and Newer Developments in Teletherapy 716
35.6.1 Intensity-Modulated Radiotherapy 716
35.6.2 Breathing-Adapted Radiotherapy (BART) 716
35.6.3 Image-Guided Radiotherapy 717
35.6.4 Dynamic Adaptive Radiotherapy 718
35.6.5 Stereotactic Radiosurgery/Radiotherapy (SRS/SRT) 718
35.6.6 Tomotherapy 719
35.6.7 Intensity-Modulated Arc Therapy 719
35.6.8 Current Developments in Linear Accelerator Technology ... 720
References 721

36 **Mechanical Circulatory Support Systems**
Roland Hetzer, Ewald Hennig 723
36.1 Introduction – History 724
36.2 Indications for Application of MCSS 725
36.2.1 Bridge to Transplant 726
36.2.2 Bridge to Recovery 726
36.2.3 Bridge to Bridge = Bridge to Decision 727
36.2.4 Permanent Support or Replacement of the Cardiac Pump Function (Destination Therapy) 727
36.3 Classification of MCSS 727
36.3.1 Physical Pumping Principles and Concepts 727
36.3.2 Pulsatile Systems 727
36.3.3 Nonpulsatile Systems – Rotary Blood Pumps 727
36.4 Today's Systems 728
36.4.1 Pulsatile Systems 728
36.4.2 Nonpulsatile Systems (Axial and Centrifugal Rotary Blood Pumps) 732
36.4.3 MCSS for Pediatric Patients 736
36.4.4 Summary 737
36.5 Complications 739
36.5.1 Device Malfunction 739
36.5.2 Hemolysis 740
36.5.3 Thromboembolism 740
36.5.4 Bleeding 740
36.5.5 Infection 740
36.5.6 Problems at Anastomoses 741
36.5.7 Right Ventricular Insufficiency 741
36.5.8 Anatomic Obstruction of Flow 741
36.5.9 Arrhythmias 741
36.6 Technical Follow-Up and Care 742
36.7 Psychosomatic Syndromes and Quality of Life During Treatment with MCS 742
36.7.1 Psychosomatic Syndromes During MCS 742
36.7.2 Profound Psychological Processes 743
36.7.3 Coping Skills 743
36.7.4 Quality of Life 743

36.7.5 Therapeutic Consequences ... 743
36.7.6 Summary ... 743
36.8 Overview and Outlook ... 744
References ... 747

37 **Neural Interfaces for Implanted Stimulators**
Xiao Liu, Andreas Demosthenous, Nick Donaldson ... 749
37.1 Stimulating Electrodes ... 751
37.1.1 Electrode–Electrolyte Interface and Circuit Model ... 751
37.1.2 Safety Issues ... 753
37.2 Implantable Cable Management ... 753
37.2.1 Current Limitations ... 753
37.2.2 Solutions ... 755
37.3 Design of Stimulator Output Stage ... 756
37.3.1 Stimulation Pulses ... 757
37.3.2 Current Generator Circuits ... 757
37.3.3 Safe Output Stage Circuits ... 759
37.4 Conclusions ... 763
References ... 764

38 **Cardiac Pacemaker Systems**
Armin Bolz ... 767
38.1 Structure of a Pacemaker System ... 768
38.1.1 The Programming Device ... 768
38.1.2 The Pacemaker ... 769
38.1.3 The Pacemaker Electrode ... 769
38.2 The Functionality of a Cardiac Pacemaker ... 771
38.2.1 The Stimulation Function ... 772
38.3 Stimulation Modes ... 774
38.3.1 International (NBG) Pacemaker Code ... 775
38.3.2 VVI Pacemaker ... 775
38.3.3 Two-Chamber Pacemaker ... 776
38.3.4 Frequency-Adaptive Pacemakers ... 780
38.3.5 Antitachyarrhythmia Pacemakers ... 781
38.3.6 Biatrial Stimulation Systems ... 781
38.3.7 Biventricular Stimulation Systems ... 782
38.3.8 Selection Criteria ... 782
References ... 782

39 **Introduction to Neuroprosthetics**
Klaus-Peter Hoffmann, Silvestro Micera ... 785
39.1 Neuroprostheses ... 785
39.2 Application of Neural Prostheses ... 787
39.2.1 Cardiac Pacemaker ... 787
39.2.2 Cochlear Implant ... 787

39.2.3 Deep Brain Stimulation 787
39.2.4 Bladder Stimulation 788
39.2.5 Human–Computer Interfaces 788
39.2.6 Extremity Prostheses 788
39.3 Specific Technological Features 788
39.3.1 Biocompatibility 788
39.3.2 Encapsulation 789
39.3.3 Energy Supply 789
39.3.4 Electronics 789
39.3.5 Signal Processing 790
39.3.6 Signal Transmission 790
39.4 Biological–Technical Interface 790
39.4.1 Electrodes 790
39.4.2 Interfaces for a Bionic Hand Prosthesis 792
39.4.3 Thin-Film Longitudinal Intrafascicular Electrode (tf-LIFE) . 793
39.5 Future Developments 798
References 799

40 **Implantable Microsystems**
Prashant Tathireddy, Florian Solzbacher, Robert Hitchcock, Klaus-Peter Hoffmann 801
40.1 Market, Applications, and Common Requirements 801
40.2 Sensors 805
40.2.1 Pressure Sensors 805
40.2.2 Voltage Sensors 805
40.2.3 Biochemical Measurement Sensors 810
40.3 In vitro and in vivo Testing 814
References 816

41 **Visual Prostheses**
Eduardo Fernandez, Klaus-Peter Hoffmann 821
41.1 The Case for Artificial Vision 822
41.2 Visual Pathways: From Real Vision to Visual Neuroprostheses 822
41.3 Current Approaches to Visual Prostheses 824
41.3.1 Retinal Stimulation 824
41.3.2 Optic Nerve Stimulation 826
41.3.3 LGN Stimulation 826
41.3.4 Intracortical Stimulation 826
41.4 Engineering Visual Neuroprostheses 827
41.5 Safe and Effective Stimulation of Visual Pathways Through Multiple Microelectrodes 829
41.6 Selection of Suitable Subjects for a Visual Prosthesis 830
41.7 Challenges and Future Perspectives 831
References 832

42 **Rehabilitation and Therapeutic Robotics**
Loredana Zollo, Dino Accoto, Silvia Sterzi, Eugenio Guglielmelli 835
42.1 Background 835
42.2 Human-Centered Approach to Rehabilitation Robot Design 837
42.2.1 CBM-motus Mechanical Design 839
42.2.2 CBM-motus Control System Design 840
42.2.3 Experimental Validation 841
42.3 Robot-Based Measure of Patient's Performance 844
42.3.1 Kinematic and Dynamic Performance Indicators 845
42.3.2 Case Study 846
42.4 Conclusions and Further Readings 850
References 851

43 **Cardiac Devices and Testing**
Michael Bateman, Stephen A. Howard, Christopher Rolfes, Timothy G. Laske, Paul A. Iaizzo 855
43.1 Background 856
43.2 Selected Landmark Events in Cardiac Devices and Surgery 856
43.3 Market Released Cardiac Devices 857
43.4 Device Development 857
43.4.1 The Six Phases of Device Development 858
43.4.2 Intellectual Property 859
43.4.3 Device Prototyping 860
43.4.4 Device Testing 861
43.4.5 The Visible Heart 864
43.4.6 Animal Testing 864
43.4.7 Clinical Testing and Regulatory Approval 866
43.5 The Anatomy of a Device 867
43.5.1 Functionality 867
43.5.2 Biocompatibility 868
43.5.3 Durability 870
43.5.4 Design for Manufacture 871
43.6 Emerging Cardiac Device Technology 871
43.6.1 Atrial Fibrillation 871
43.6.2 Heart Failure 872
43.6.3 Left Ventricular Assist Devices and Artificial Hearts 872
43.6.4 Heart Transplant 872
43.6.5 Valves 872
43.6.6 Vascular Disease 873
43.6.7 Hybrid Operating Rooms 873
43.6.8 Novel Devices 873
43.7 Conclusions 874
References 874

44 **Functional Electrical Stimulation in Rehabilitation and Neurorehabilitation**
Kei Masani, Milos R. Popovic 877

44.1 The Basis of Electrical Stimulation 878
44.1.1 History – The Use of Electrical Stimulation in Medicine 878
44.1.2 Physiology 878
44.1.3 Technology 879
44.1.4 Muscle Fatigue 882
44.2 Neuroprosthetic Use of FES 883
44.2.1 Neuroprosthesis for Drop Foot 883
44.2.2 Neuroprosthesis for Walking 883
44.2.3 Neuroprostheses for Standing 885
44.2.4 Neuroprostheses for Reaching and Grasping 885
44.2.5 Neuroprostheses for Bladder Function 887
44.2.6 Neuroprostheses for Respiration 887
44.2.7 Neuroprostheses for Hearing 888
44.2.8 Therapeutic FES Besides FES Therapy 888
44.3 FES Therapy 889
44.4 Other Uses of Electrical Stimulation 890
44.4.1 Pressure Ulcers 890
44.4.2 Diabetes 890
44.5 Concluding Remarks 890
References 890

45 **Treatment Planning and Patient Treatment**
Golam Abu Zakaria, Wilhelm Schütte, Stephan Garbe 897

45.1 Principles of Radiotherapy and Treatment Planning 898
45.1.1 Principles of Radiotherapy 898
45.1.2 Principles of Treatment Planning and Treatment 899
45.2 Imaging in Treatment Planning 901
45.2.1 Computed Tomography 901
45.2.2 Magnetic Resonance Imaging 902
45.2.3 Positron Emission Tomography 902
45.3 Basic Techniques in External Beam Therapy 902
45.3.1 Simple Treatment Planning 902
45.3.2 Three-Dimensional Conformal Treatment Planning (3-D CRT) 903
45.3.3 Inverse Planning 903
45.3.4 Intensity Modulated Radiation Therapy 903
45.3.5 Dynamic Delivery Techniques 904
45.3.6 Tomotherapy 904
45.3.7 Image Guided Radiotherapy (IGRT) 906
45.3.8 Respiratory Guided Radiotherapy 906
45.3.9 Intraoperative Electron Radiation Therapy 906

45.3.10 Stereotactic Irradiation 907
45.3.11 Total Body Irradiation with Photons 907
45.3.12 Total Skin Electron Beam Therapy 908
45.4 Target Volumes and Organ at Risk 909
45.5 Modern Treatment Planning Systems 910
45.5.1 Basic Components of a Treatment Planning System 911
45.5.2 Physics Module 911
45.5.3 Import Module 912
45.5.4 Target Definition or Contouring Module 912
45.5.5 Beam Modeling Module 913
45.5.6 Dose Calculation Module 913
45.5.7 Evaluation and Optimization Module 913
45.5.8 Plan Analysis Module 914
45.5.9 Export Module 914
45.6 Simulation of the Patient and the First Treatment 915
45.6.1 Conventional Simulation 916
45.6.2 Virtual Simulation 916
45.7 Quality Control in Radiation Therapy 917
45.7.1 Personnel 917
45.7.2 Planning system 917
45.7.3 Medical Linear Accelerator and Simulator 917
References 919

Part E Monitoring

46 **Recording and Processing of Biosignals**
Klaus-Peter Hoffmann, Florian Solzbacher 923
46.1 Measuring in Medicine 923
46.1.1 Biosignals 924
46.1.2 Biological Measuring Chain 924
46.1.3 Artifacts 933
46.2 Registration of Biological Signals 934
46.2.1 Bioacoustic Signals 934
46.2.2 Biochemical Signals 934
46.2.3 Bioelectric and Biomagnetic Signals 935
46.2.4 Biomechanical Signals 937
46.2.5 Biooptical Signals 940
46.2.6 Biothermal Signals 940
46.3 Measurement and Signal Analysis from a Metrological Point of View 940
46.3.1 Biostatistical Methods 941
46.3.2 Biosignal Analysis 941
46.4 Test Planning and Clinical Studies 944
Further Reading 945

47 **Monitoring Systems**
Ullrich Hieronymi, Rüdiger Kramme 947
47.1 Fields of Use for Patient Monitoring Systems 948
47.2 Types of Monitors 949
47.2.1 Stand-Alone Monitors 949
47.2.2 Transport Monitors 949
47.2.3 Telemetry Devices 949
47.2.4 Preconfigured or Compact Monitors 949
47.2.5 Modular Monitors 950
47.2.6 Information Monitors with Integrated Transport 950
47.2.7 Integrated Acute Medical System 951
47.3 Monitor Screen Content 951
47.3.1 Display 951
47.3.2 Channels 951
47.3.3 Waveform Representation 951
47.3.4 Screen Configuration 951
47.4 Handling 952
47.4.1 Controls 952
47.4.2 User Guidance and Operating Philosophy 952
47.5 Alarms and Events 952
47.6 Trend Display 952
47.7 Automatic Calculations 953
47.8 Advanced System Properties 953
47.8.1 Data Integration at the Workstation 953
47.8.2 Monitor Network 953
47.9 Central Monitoring and Documentation 953
Further Reading 954

48 **Cardiovascular Monitoring**
Ullrich Hieronymi, Rüdiger Kramme 955
48.1 Monitoring the Cardiac Function 955
48.1.1 Electrocardiogram (ECG) and Heart Rate (HR) 955
48.1.2 Arrhythmia Monitoring 957
48.2 Monitoring the Circulatory Function (Hemodynamic Monitoring) 957
48.2.1 Pulse 958
48.2.2 Discontinuous Noninvasive Blood Pressure (NIBP) 959
48.2.3 Continuous Noninvasive Arterial Blood Pressure Measurement (CNAP) 961
48.2.4 Invasive Pressure Measurement in the High-Pressure and the Low-Pressure Systems 962
48.2.5 The Ideal Pressure Measurement System – Requirements . 963
48.2.6 Determining the Cardiac Output (CO) 966
48.2.7 Calculation of Hemodynamic Variables 969
References 969

49 **Respiratory Monitoring and Pulse Oximetry**
Ullrich Hieronymi, Rüdiger Kramme, Harald Kronberg 971
49.1 Respiratory Mechanics 971
49.1.1 Monitoring of the Respiration Rate 971
49.1.2 Pressure and Flow 972
49.2 Gas Exchange 973
49.2.1 Gas Monitoring: Preliminary Remarks 973
49.2.2 Pulse Oximetry 976
49.2.3 Transcutaneous Blood Gas Measurement 979
49.2.4 Measuring the Respiratory Gases 981
49.2.5 Capnography 983
References 985

50 **Temperature Monitoring**
Rüdiger Kramme, Ullrich Hieronymi 987
50.1 Hyperthermia and Hypothermia 987
50.2 Measuring Sites for Temperature Measurement 987
50.3 Temperature Sensors and Probes 987
50.4 Methodological Notes 989
Further Reading 990

51 **Cerebral Monitoring**
Barbara Schultz, Arthur Schultz, Harald Kronberg 991
51.1 EEG Monitoring 991
51.1.1 Kinds of EEG Activity 991
51.1.2 Spectral Analysis 992
51.1.3 Other Parameters 992
51.1.4 EEG Stages of Anesthesia and Sedation 992
51.1.5 Course of EEG Stages During Anesthesia 992
51.1.6 Technical Requirements 994
51.1.7 Benefit of EEG Monitoring in the Operating Room 994
51.1.8 Benefit of EEG Monitoring in the Intensive Care Unit 995
51.2 Intracranial Pressure 995
51.2.1 The Physiology of Intracranial Pressure 996
51.2.2 Intracranial Pressure Measurement to Monitor Treatment . 997
References 1000

52 **Brain Computer Interface**
Günter Edlinger, Cristiano Rizzo, Christoph Guger 1003
52.1 Introduction to BCI 1003
52.2 Measuring Brain Activity 1004
52.2.1 Invasive BCIs 1004
52.2.2 Partially Invasive BCIs 1004
52.2.3 Noninvasive BCIs 1004
52.2.4 Other Functional Imaging BCIs 1005

52.3 BCI System Structure 1005
52.3.1 Noninvasive (EEG) BCI Systems 1006
52.3.2 SMR-based BCIs 1007
52.3.3 P300-based BCI 1010
52.4 Conclusions 1014
References 1015

53 **Fetal Monitoring**
Birgit Wacker, Martin Maier 1019
53.1 Cardiotocography (CTG) 1019
53.1.1 Areas of Application 1019
53.1.2 Measuring Procedures 1019
53.1.3 Direct ECG 1020
53.1.4 Toco Procedure 1020
53.1.5 Intrauterine Pressure Measurement 1021
53.1.6 Additional Functions and Options 1021
53.1.7 CTG Selection Criteria 1022
53.2 Obstetric Monitoring Systems 1022
53.2.1 Areas of Application 1022
53.2.2 System Setup 1022
53.2.3 Collection and Processing of Data from the Connected CTGs 1023
53.2.4 Assessment of the CTG Data and Raising the Alarm 1024
53.2.5 Presentation of the CTG Data 1026
53.2.6 Patient Data Management 1026
53.2.7 Archiving and Retrieving Data via Long-Term Storage Media 1028
53.2.8 Computer Communication Within Obstetrics 1029
53.2.9 Computer Communication with Other Hospitals 1029
53.2.10 Communication with the Hospital Administration Computer 1029
53.2.11 Exporting Patient Data to Other Hospital Systems 1029
53.2.12 Connection to Lab Computer 1030
53.2.13 Access to Internal Hospital Computer Systems 1030
Further Reading 1030

54 **Neonatal Monitoring**
Roland Hentschel 1031
54.1 Electrocardiogram 1032
54.2 Impedance Pneumography 1033
54.3 Combined Cardiorespiratory Analysis 1034
54.4 Pulse Oximetry 1035
54.5 Transcutaneous Measurement of the Partial Pressure 1037
54.6 Measurement of the $PtcCO_2$ (Transcapnode) 1038
54.7 Measurement of the $PtcO_2$ (Transoxode) 1038
54.8 Monitoring the Oxygenation – Which Method? 1039
54.9 Setting Alert Limits and Limit Values 1040
References 1041

55 **Intraoperative Neuromonitoring**
Werner Kneist, Daniel W. Kauff 1043
55.1 General Principles 1043
55.1.1 Intraoperative Nerve Mapping 1045
55.1.2 Intermittent Intraoperative Neuromonitoring 1046
55.1.3 Continuous Intraoperative Neuromonitoring 1046
55.2 Neuromonitoring Signals 1046
55.2.1 Signal Recording and Processing 1046
55.2.2 Signal Analysis 1047
55.3 Scope of Application 1048
55.3.1 Thyroid Surgery 1048
55.3.2 Neurosurgery 1050
55.3.3 Orthopedic Surgery 1052
55.3.4 Otolaryngology and Head and Neck Surgery 1053
55.3.5 Pelvic Surgery 1054
55.4 Quality Management 1056
55.5 Guidelines and Legal Aspects 1056
Further Reading 1057

56 **Ionic Neural Sensing**
Iasonas F. Triantis, Anna Radomska-Botelho Moniz, Kostis Michelakis, Sanjiv Sharma, Jakub Trzebinski, Belinda Garner, Amir Eftekhar 1059
56.1 Central and Peripheral Nervous System Monitoring 1060
56.1.1 Monitoring the Brain Function 1060
56.1.2 Peripheral Nerve Monitoring Methods 1061
56.1.3 Neural Amplifier Specifications 1063
56.2 Chemistry of Neural Activity 1064
56.2.1 Ionic Aspect of Brain Activity 1064
56.2.2 Ionic Aspect of Peripheral Nerve Signals 1065
56.3 Chemical Neural Sensing Technology and Challenges 1065
56.3.1 Requirements for a Novel Method 1065
56.3.2 State of the Art in Chemical Neural Recording Platforms ... 1065
56.3.3 Ionic Sensing – State of the Art and Challenges 1066
56.4 Conclusion 1069
References 1070

Part F Medical Information Processing and Communication

57 **Fusing Medical Engineering and Information Technology – Structure, Integration and Process Optimization**
Hajo Tanck, Michael Heinlein 1075
57.1 Standards of Interfaces 1076
57.1.1 Health Level 7 1076
57.1.2 Digital Imaging and Communications in Medicine 1076
57.1.3 GDT 1076
57.1.4 xDT 1077
57.1.5 XML 1077

Detailed Cont.

57.2 Data Structure 1077
57.2.1 HL7 CDA 1077
57.2.2 SNOMED 1077
57.2.3 LOINC 1077
57.2.4 Alpha ID 1077
57.2.5 Object Identifiers 1077
57.2.6 Unified Code for Units of Measure 1078
57.3 Integrating the Healthcare Enterprise 1078
57.4 Integration of Medical Devices 1078
57.4.1 Interfaces 1079
57.4.2 Connecting a Medical Device 1079
57.4.3 Organizational Requirements for the Connection 1080
57.4.4 Responsibilities 1081
57.5 Sample Integration – From Findings to Medical Documentation 1082
57.6 Résumé 1082
Further Reading 1083

58 **Communicating Medical Systems and Networks**
Armin Gärtner 1085
58.1 Medical Networks 1085
58.2 Requirements for Medical Networks 1086
58.2.1 Image Data in Radiology 1086
58.2.2 Intensive Care Network 1086
58.2.3 Transmission of Alert Data 1087
58.3 Interconnected Medical Networks 1087
58.4 Risk Management, DIN EN ISO 14971 1088
58.5 Shared Networks 1089
58.6 Security Aspects of Medical Networks from a Regulatory Viewpoint . 1092
58.7 Future Standard IEC 80001-1 1093
References 1093

59 **Hospital Information Systems**
Peter Haas, Klaus A. Kuhn 1095
59.1 Background 1095
59.2 Necessity, Objectives, and Benefits of Comprehensive HIS 1096
59.2.1 Necessity 1096
59.2.2 Objectives of Using Information Technology in Hospitals .. 1097
59.2.3 Potential Benefits 1097
59.3 Dimensions of IT Support 1098
59.3.1 Data Processing Support 1098
59.3.2 Documentation Support 1098
59.3.3 Organizational Support 1098
59.3.4 Communication Support 1098
59.3.5 Decision Support 1099

59.4 Case Study ... 1099
59.5 Architecture and Components of HIS ... 1103
59.5.1 Logical Architecture Model ... 1103
59.5.2 Implementation Alternative: Holistic Versus Heterogeneous ... 1105
59.5.3 Integration Aspects of Heterogeneous HIS ... 1107
59.6 Current Trends and Prospects ... 1108
59.6.1 Medical Documentation ... 1108
59.6.2 Integration of Medical Technology ... 1108
59.6.3 Health Information Networks ... 1108
59.6.4 Decision Support for Health Care Professionals ... 1109
59.6.5 Support for Patient Safety ... 1110
59.6.6 Genomic Data ... 1110
59.7 Selection and Implementation of HIS ... 1110
59.7.1 Preliminary Remarks ... 1110
59.7.2 Project Phases and Critical Factors ... 1111
59.8 Conclusion ... 1117
References ... 1118

60 **Telemedicine in Germany**
Rüdiger Klar, Ernst Pelikan ... 1119
60.1 The Peculiar Features of German Telemedicine ... 1120
60.1.1 Supporting Individual Projects ... 1120
60.1.2 Legal Measures to Overcome the Disadvantages of the Fragmented Healthcare System ... 1121
60.1.3 Electronic Health Cards, Health Professional Cards, and Official Health Telematics ... 1121
60.1.4 Medical Confidentiality, Data Protection, and Personal Rights ... 1122
60.1.5 Technical Standards ... 1123
60.1.6 Acceptance and Benefits for Various Players in Telemedicine ... 1124
60.2 Consequences of the Peculiarities of the German System for Telemedicine ... 1125
References ... 1126

61 **Telemedicine Using Active Implants**
Klaus P. Koch, Oliver Scholz ... 1129
61.1 Telemedicine in the Operating Theater ... 1129
61.2 Telemedicine in Domestic Care ... 1130
61.3 Implant Telemetry ... 1131
61.3.1 Implant Telemetry by Means of Inductive Coupling ... 1131
61.3.2 Implant Telemetry Employing Radio Communication ... 1133
61.3.3 Optical Transcutaneous Transmission ... 1134
61.3.4 Energy Supply to Medical Implants ... 1134
61.3.5 Line Codes ... 1135

61.4 Inclusion of Active Medical Implants in Telemedicine Systems 1136
References 1136

62 **Fundamentals of Medical Image Processing**
Thomas M. Deserno 1139
62.1 Background 1139
62.2 Remarks on the Terminology 1141
62.3 Image Enhancement 1142
62.3.1 Histogram Transforms 1142
62.3.2 Convolution 1143
62.3.3 Calibration 1144
62.3.4 Registration 1145
62.4 Feature Extraction 1147
62.5 Segmentation 1148
62.5.1 Pixel-Based Segmentation 1148
62.5.2 Edge-Based Segmentation 1150
62.5.3 Region-Based Segmentation 1151
62.5.4 Hybrid Segmentation Procedures 1152
62.6 Classification 1154
62.6.1 Statistic Classifiers 1155
62.6.2 Syntactic Classifiers 1155
62.6.3 Computational Intelligence-Based Classifiers 1155
62.7 Quantitative Measurements 1157
62.8 Interpretation 1158
62.9 Image Data Visualization 1158
62.9.1 The Marching Cubes Algorithm 1159
62.9.2 Surface Rendering 1160
62.9.3 Volume Rendering 1161
62.10 Image Management 1161
62.11 Conclusion and Outlook 1163
References 1165

63 **Virtual Reality in Medicine**
Wolfgang Müller-Wittig 1167
63.1 Virtual Reality 1168
63.2 Medical Applications 1168
63.2.1 Anatomy Education 1170
63.2.2 Functional Diagnostics 1170
63.2.3 Virtual Endoscopy 1170
63.2.4 Surgical Training 1172
63.2.5 Preoperative Planning 1172
63.2.6 Intraoperative Support 1173
63.3 VR-Based Medical Simulation 1174
63.4 Model Generation – Virtual Anatomy 1175
63.4.1 Data Acquisition 1175
63.4.2 Preprocessing 1176
63.4.3 Segmentation 1176

63.4.4 Mesh Generation ... 1176
63.4.5 Decimation ... 1177
63.5 Manipulations – Surgical Interventions ... 1178
63.5.1 Modeling of Surgical Instruments ... 1178
63.5.2 Registration of Instruments ... 1179
63.5.3 Integration of Visual and Haptic Feedback ... 1180
63.6 Outlook ... 1182
References ... 1184

64 Computer-Supported Teaching and Learning Systems in Medicine
Martin Haag, Martin R. Fischer ... 1187
64.1 Historical Development ... 1187
64.2 Moves Towards the Reform of Medical Studies ... 1189
64.3 Developing Learning and Teaching Systems ... 1189
64.3.1 Interaction Forms ... 1189
64.3.2 Architectures of Learning and Teaching Systems ... 1192
64.3.3 Authoring Systems ... 1192
64.4 Learning Environments ... 1193
64.4.1 Functionality of Learning Environments ... 1193
64.4.2 Interoperability and Standards ... 1194
64.5 Application Scenarios for Learning and Teaching Systems ... 1195
64.6 Status of and Outlook for e-Learning in Medicine ... 1195
64.6.1 Information Systems for CBT/WBT in Medicine ... 1195
64.6.2 Use of CBT/WBT Systems in Medicine and the Problem of Curricular Integration ... 1195
64.6.3 The Need for Research, and Outlook ... 1196
References ... 1197

65 PACS and RIS
Christian Zapf, Andreas Bermann, Dirk Sunderbrink ... 1199
65.1 Radiological Workflow ... 1200
65.1.1 Orders ... 1201
65.1.2 Admission/Scheduling ... 1201
65.1.3 Examination ... 1201
65.1.4 Reading Workflow ... 1202
65.1.5 Displaying and Evaluating Images ... 1202
65.1.6 Image Postprocessing ... 1202
65.1.7 Computer Aided Diagnosis (CAD) ... 1204
65.1.8 Medical Monitors ... 1204
65.1.9 Report Creation ... 1204
65.1.10 Clinical Demonstration ... 1205
65.1.11 Distribution of Reports/Images ... 1205
65.2 Integrating PACS/RIS into the Hospital Environment ... 1205
65.2.1 Integrating Information Systems ... 1205
65.2.2 Integrating Image Management Systems ... 1206
65.2.3 Teleradiology ... 1206
65.2.4 Mobile Devices ... 1206

65.3 State-of-the-Art IT Infrastructure 1206
65.3.1 Information Lifecycle Management 1207
65.3.2 Cloud Computing Solutions in the Hospital 1207
65.3.3 Regional IT Virtualization 1207
65.4 Summary 1208
References 1208

66 **3-D Postprocessing in Virtual Endoscopy**
Georg-Friedemann Rust 1209
66.1 What Is Virtual Reality? 1210
66.2 Why Virtual Reality? 1210
66.3 Advantages of 3-D Visualization 1210
66.3.1 Partial Volume Effect 1211
66.3.2 Two-Dimensional or Three-Dimensional Visualization 1211
66.3.3 Risks of 3-D Visualization 1213
66.4 Conclusions 1216
References 1216

67 **e-Health – Ambient Assisted Living and Personal Health Systems**
Natasha Avila, Christina Sampogna 1217
67.1 Background 1218
67.1.1 e-Health 1219
67.2 AAL and PHS Approaches 1221
67.2.1 Self-Management 1222
67.3 Benefits and Challenges Ahead 1234
67.3.1 The Social Dimension 1234
67.3.2 Legal and Ethical Dimensions 1235
67.3.3 Societal and Global Dimensions 1237
67.3.4 Economic Dimensions 1238
67.3.5 Technological Dimensions 1240
67.4 Conclusion 1241
67.4.1 Observations 1241
67.4.2 Areas for Further Work 1242
67.4.3 Future Technologies 1243
References 1243

68 **Electrical Stimulation of the Nervous System**
Stanisa Raspopovic, Marco Capogrosso, Silvestro Micera 1247
68.1 Background 1247
68.2 Biophysics Models of Neuronal Response to External Fields 1248
68.2.1 Modeling of Self-Standing Neurons 1248
68.2.2 Modeling Effects of Extracellular Electrical Stimulation 1250
68.3 Finite Element (FE) Models 1252
68.3.1 The Electromagnetic Problem 1252
68.3.2 The Finite Element Method (FEM) 1253
68.3.3 Models in Practice 1255

68.4 Conclusion 1256
References 1256

Part G Equipment and Tools

69 **Operating Tables – the Surgeon's Workplace**
Bernhard Kulik 1261
69.1 The History of the Operating Table 1262
69.2 The OR Table System 1263
69.3 Technology of Operating Room Table Systems 1264
69.3.1 Construction of an OR Table System 1264
69.3.2 Mobility and Flexibility of OR Table Systems 1265
69.4 Safe Patient Positioning 1268
69.4.1 Decubitus Injuries 1269
69.4.2 Long-Term Position Injuries and Legal Responsibility 1270
69.4.3 Patient Positioning When Deploying Single-Pole HF Devices 1271
69.5 Preparation: Care, Maintenance, and Hygiene 1271
69.5.1 Manual Cleaning and Disinfection 1271
69.5.2 Automatic OR Table System Cleaning and Hygiene 1271
Further Reading 1272

70 **Medical Robotics**
Harald Fischer, Udo Voges 1273
70.1 Fundamentals 1273
70.2 Development of Medical Robots 1274
70.3 Overview of Systems 1275
70.3.1 Assistance Systems 1275
70.3.2 Active Holding Systems 1275
70.3.3 Master–Slave Manipulators 1276
70.3.4 Biopsy Robots 1276
70.3.5 Commercial Active Holding and Guiding Systems 1277
70.3.6 Commercial Telemanipulators 1278
70.3.7 Commercial Surgery/Biopsy Robots 1278
70.4 Medical Applications 1280
70.5 Technical Aspects 1281
70.6 Outlook 1282
References 1283

71 **Incubators**
Guenther Braun, Roland Hentschel 1285
71.1 Historical Background 1285
71.2 Construction and Function of an Incubator 1286
71.2.1 Temperature Regulation 1286
71.2.2 Regulation of Humidity 1286
71.2.3 Regulation of Oxygen 1287

71.2.4 Scales 1287
71.2.5 X-ray Drawer 1287
71.3 Incubator Models 1287
71.3.1 Stationary Incubators 1287
71.3.2 Transport Incubators 1287
71.3.3 Special Incubators 1288
71.4 Risks of Incubator Therapy 1289
71.4.1 Temperature 1289
71.4.2 Oxygen Therapy 1289
71.5 Hygiene 1289
71.6 Unsolved Problems 1290
References 1290

72 **Surgical Scissors**
Reiner Haag, Wilfried Storz 1291
72.1 The History of Scissors 1292
72.1.1 Paired Knives 1292
72.1.2 End-Jointed Scissors 1293
72.1.3 Bow Scissors 1293
72.1.4 Pivoted Scissors 1293
72.2 The Function and Design of Scissors 1293
72.3 Materials 1294
72.3.1 Soldering a Hard-Metal Inset into the Cutting Edge 1294
72.3.2 Welding Hard Metal in Place 1294
72.3.3 Special Coatings 1294
72.3.4 Titanium Scissors 1294
72.4 Manufacture of Surgical Scissors 1294
72.4.1 Steps in the Production Cycle 1295
72.5 Diversification Overview 1297
72.5.1 Surgical Standard Scissors 1297
72.5.2 Surgical Scissors – Dissecting Scissors 1297
72.5.3 Suture or Ligature Scissors 1299
72.5.4 Wire Cutting Scissors 1301
72.5.5 Microsurgical Scissors 1302
72.5.6 Vascular Scissors 1305
72.5.7 Gynecological Scissors 1306
72.5.8 Gynecological Scissors for Obstetrical Use 1308
72.6 Handling and Care 1310
72.6.1 General Instructions 1311
72.6.2 Materials Used in Scissor Manufacturing 1311
72.6.3 Water Qualities 1311
72.6.4 Preparation for Cleaning and Disinfection 1312
72.6.5 Manual Cleaning and Machine Cleaning 1312
72.6.6 Machine Cleaning and Thermal Disinfection 1312
72.7 Inspection, Testing, and Care 1314
72.8 Packaging 1314

72.9 Current Terminology ... 1315
72.10 Steam Sterilization with Saturated Steam ... 1315
72.10.1 Steam Quality ... 1315
72.10.2 Release and Storage ... 1316
72.11 Quality Characteristics ... 1316
72.11.1 Material ... 1316
72.11.2 Surface ... 1316
72.11.3 Form ... 1316
72.11.4 Action ... 1316
72.11.5 Eye Rings ... 1316
72.11.6 Quality of Cut ... 1317
72.12 Future Developments ... 1317
72.12.1 Users ... 1317
72.12.2 Industry ... 1317
72.13 Bipolar Scissors ... 1318
References ... 1319

73 **Intelligent Textiles and Trends**
Christian Rotsch, Sibylle Hanus, Danny Schwabe, Heike Oschatz, Andreas Neudeck, Uwe Möhring ... 1321
73.1 Textile Manufacturing Technologies and Applications ... 1321
73.1.1 Textiles for Medical Application ... 1321
73.1.2 Weaving ... 1322
73.1.3 Embroidery ... 1322
73.1.4 Warp Knitting ... 1323
73.1.5 Braiding ... 1324
73.1.6 Conductive Thread Materials ... 1325
73.2 Sensory Applications of Textiles ... 1326
73.2.1 EMG and ECG Monitoring ... 1326
73.2.2 Respiratory Monitoring ... 1329
73.3 Active Textiles – Therapeutical Applications ... 1330
73.3.1 Textile Electrodes for Electrical Muscle Stimulation ... 1330
73.3.2 Textile Interactive Medical Agent Depots and Disposing Systems ... 1331
73.3.3 Heatable Textiles for Therapeutic Treatments ... 1331
73.4 Passive Medical Textiles for Therapy ... 1333
73.4.1 Reusable 3-D Knitted Elastic Short Traction Bandages ... 1333
73.4.2 Embroidered Implants for Tissue Engineering ... 1335
References ... 1335

74 **Electronics in Medicine**
Paulo M. Mendes, Celso P. Figueiredo, Mariana Fernandes, Óscar S. Gama 1337
74.1 Basics ... 1338
74.1.1 Fields of Application ... 1338
74.1.2 Designing Electronics in Medicine ... 1339
74.1.3 Medical Electronics Specific Requirements ... 1339

74.2 Electronic Sensing 1341
74.2.1 Health Condition Monitoring 1341
74.2.2 Biosignals and Transducers 1342
74.3 Electronics for Wireless Health Monitoring 1344
74.3.1 Requirements for Wireless Devices 1345
74.3.2 Data Acquisition 1346
74.3.3 System Integration 1347
74.3.4 Wireless Communications 1348
74.4 Power Supply 1348
74.4.1 Battery Power Budget Considerations 1349
74.4.2 Wireless Power 1350
74.4.3 Wireless Power Link Analysis 1350
74.4.4 Wireless Power Charger 1351
74.4.5 Energy Harvesting 1352
74.5 Wearable Medical Electronics 1353
74.5.1 Wearable Systems 1355
74.5.2 Categories of Wearable Systems 1355
74.5.3 Design Requirements 1356
74.5.4 Sensors for Wearable Systems 1357
74.5.5 Sensing Methodologies 1357
74.6 Electronics in Medicine at Work 1358
74.6.1 Wireless and Wearable Low-Power Health Monitoring Systems 1359
74.6.2 Sensors 1359
74.6.3 Wireless Link 1360
74.6.4 Optical Biopotential Recording 1360
74.6.5 Optical Electrodes 1360
74.6.6 Optical Signal Acquisition 1361
74.6.7 Localization Solutions 1361
74.6.8 Ambient Assisted Living Applications 1363
74.6.9 Wireless Link Design for Biomedical Applications 1366
References 1373

Appendix 1377
Acknowledgements 1391
About the Authors 1393
Detailed Contents 1415
Subject Index 1455

Subject Index

χ^2 test 940
O_2 saturation 940
β_2-microglobulin 606
1-chip camera 384
2-(fluorine-18)-fluoro-2-desoxy-D-glucose (^{18}F-FDG) 341
2.3-DPG 1039
3-chip
– camera 384
– endocamera image 385

A

AAL components 1221
AAL solution 1350, 1361
abdominal aortic aneurysm (AAA) 494
abdominal pressure 805
aberration-free refraction 262
aberrometry 261
ablation of tumor 493
abort criterium 93
absolute
– dosimetry 918
– muscle force 276
absolute claudication distance (ACD) 432
absorption 373, 425
– coefficient 427, 564
– infrared 593
– of sound waves 647
acceleration voltage 318
accelerator tube 712
acceptance phase 1116
access instrument 387
access resistance 752
accessory decontamination 18
accident and emergency (A&E) 948
achromatic processing module 827
achromatopsia 261
acid-base status 632
acidosis 1039
acoustic
– coupling 479
– cross-talk 204
– feedback 237
– lens 662
– noise 506
– Ohm 193, 212
– pressure 350
– property of media 644
– stimulation 217
– stimulator 142
– therapy 643
acoustic evoked potential (AEP) 149, 150
acquisition time 316, 336
action potential 121, 878, 1249
– generation 1249
activated clotting time (ACT) 616, 632
activation function 1255
– approach 1251
active
– electrode 239, 682, 687
– holding system 1275
– medical implant 1136
– medical textile 1322, 1330
– messaging 1110
– shielding 505
active contour 1152
– model 1154
activity
– detection 1225
– manager 1354
– matrix 828
– monitor 1224
– monitoring 115
– of physicians 1121
actual cycle time 537
AcuBot 1280
acute sinoatrial (SA) 768
adaptive
– filtering 902
– planning 905
– system 1005
– treatments system element 1230
adaptive probe off detection (APOD) 978
adaptive tripole (AT) 1063
adaptometry 261
adhesive electrode 548
adipose tissue mass 614
adiposity surgery 1268
administrative procedure 1097
admission, discharge, and transfer (ADT) 1103
adsorption of blood components 610
advanced combination encoders (ACE) 241
advanced electronics 1339
advanced multimodality image-guided OR (AMIGO) 492
advanced neurofeedback tool 1004
advanced visualization 1203
advanced workplace for image guided surgery (AWIGS) 1263
Adventitia scissors 1303
adverse event 1098
aerobic endurance exercise 890
afterloading technology 715
AgCl electrode 123
agglomerative segmentation 1151
aiming beam 562
air
– arterial embolism 634
– bone gap 203
– puff tonometer 246
– temperature regulation 1286
air conduction (AC) 202
– testing 204
airborne transmission 27
airway
– adapter 981
– pressure 972
– resistance 105, 107, 109
airway pressure release ventilation (APRV) 539
alarm 948
– audible 952
– automatic 1022
– limit 543
– priority 952
alarm system 529
– principle 1025
– rule-based algorithm 1025
albumin-bound toxin 597, 615
alert
– limit 1040
– software 1087
– system 1032, 1225
alexandrite laser 569
algebraic reconstruction (ART) 407, 470
aliasing 353
alkalosis 1039
allocation planning 1098
all-or-nothing law 767
alpha-wave 129
alternate binaural loudness balance (ABLB) 204

alternate mark inversion (AMI) 1135
alternative medical treatment sites (AMTS) 1110
alveolar pressure 106
alveolitis 110
ambient
– intelligence 1004
– light 1161
ambient assisted living (AAL) 1217, 1219, 1338, 1363
– application environment 1220
ambient temperature and pressure, dry (ATPD) 973
ambient temperature and pressure, saturated (ATPS) 973
ambulatory polygraphy 165
American Society for Testing and Materials (ASTM) 1123
American Society of Cardiology (AHA) 551
American Speech and Hearing Association (ASHA) 200
American Thoracic Society (ATS) 97
amperometric glucose electrode 935
amperometric sensor 927
amplifier
– concept 1063
– noise 128
amplitude mode (A-mode) 351
amplitude modulation following response (AMFR) 226
amyotrophic lateral sclerosis (ALS) 889, 1003
anaerobic threshold 113
anaesthetic gas 973
– measurement error 982
– vaporizer 629
anaesthetic ventilation 973
analog
– geometric tomography 313
– image 292
analog blurring 314
analog-to-digital converter (ADC) 294, 297, 1365
anastomosis 741
anatomic obstruction of flow 741
anatomical
– field marker 407
– model 1182
– structure 1176
anatomy
– of a device 867
– virtual 1175, 1178
anchorable probe head 771
anesthesia 588
– balanced 579, 580
– intravenous 993
– ventilation modes 588
anesthesia device 579, 594, 595
– cleaning 594
– maintenance 595
anesthesia staff 579
anesthetic
– agent 579, 583
concentration 584
– gas 580
aneurysm 305
aneurysma clip 156
angiographic image acquisition 333
angiography 417
– application range 419
– catheter 417
– flat-panel detector 303
– fluorescence 267
– procedure 419
– with CO_2 298
angiography system 418, 420
– monoplane 418, 419
angioplasty 420
– balloon 419
angled scissors 1308
angular dose modulation 336
animal
– model 283
– testing 864
anisotropic template 1143
ankle dorsiflexion 279
antenna efficiency 1348
anterior myocardial infarction 768
anterior–anterior method 550
anterior–posterior 330
– method 550
anthrax 19
anti-arrhythmic therapy 87
anticongestion therapy 1335
antifaradization capacitor 685
antimicrobial action of disinfectants 14
antisaccade 180
antiscatter grid 321
antitachyarrhythmia pacemaker 781
antitachycardic pacing (ATP) 554
aortic dissection 633
apical cannula 741
apnea 160, 1033, 1035
– alarm 543
– monitoring 972
– ventilation 543
apnea index (AI) 168
apoptosis 899
apparent diffusion coefficient (ADC) 452
applanation tonometry 246, 999
application
– neutral electrode 698
– of shock waves 643
– of stimuli 933
– of telemedicine 1121
– rule 697
application-specific integrated circuits (ASICS) 801
applied part 684
approval process 866
approximal caries diagnostics 408
Arbeitsgemeinschaft Informationstechnik (AGIT) 1123
architecture model 1103
archiving and retrieval 1023
area index 845
area of induction 505
argon-assisted
– coagulation 683, 691, 693
advantages 694
– cutting 690
argon-ion laser 558, 570
arm flexion 279
array electrode 555
array of ISFETs 1070
arrhythmia 741
– monitoring 950, 957
– ventricular 87
ARTEMIS 1276, 1282
arterial
– air embolism 634
– cannulation 626
– pressure 963
arterial blood
– gas 980
– pump 624
– temperature 377
arterial function (FMD) 366
arterial O_2 content (aO_2) 631
arterial saturation (SaO_2) 632
arteriovenous O_2 concentration difference ($avDO_2$) 631
arthroscopy
– knee 1172
– training simulator 1171
articulated arm 563
artifact 305, 408, 933
– measurement 933
– motion 328, 338
– suppression 257
– testing 507

artificial
– immunization 13
– intelligence 944, 1155
– kidney 599, 600
– neural network 1155
– opsin 825
– respiration 12, 1032
– retina 829
– urine 599
– ventilatory support 887
– vision 822
assessment matrix 66
assessment of the CTG data 1023, 1024
assist device, adaptation 742
assisted spontaneous breathing (ASB) 534
Association for the Advancement of Medical Instrumentation (AAMI) 550, 606
astigmatism 263
– inversus 265
– rectus 265
ASTM standard 503
asynchronous cardioversion 549
atmospheric transmittance 370
atrial
– electrode 555
– fibrillation 871
– probe 770
– stimulus 778
atrial septal defect (ASD) 724
atrioventricular (AV)
– block 768
– defibrillator 552
atrium-controlled DDD pacemaker
– flow diagram 777
– functional diagram 778
atrium-controlled ventricular pacemaker 782
attention-deficit/hyperactivity disorder (ADHD) 1004
attenuation
– of sound waves 647
– profile 324
– value 332
– value of water 329
attributed relational graph 1158
audible alarm 952
audiometric assessment 200
audiometric testing 226
audiometry 212
– normal 151
– objective 151
auditory
– canal 989
– fatigue 201
– system 194
auditory brainstem (ABI) 199
auditory brainstem response (ABR) 222, 223
auditory evoked potential (AEP) 212, 221
auditory neuropathy spectrum disorder (ANSD) 200
auditory processing disorder (APD) 200, 206
auditory steady state response (ASSR) 226
augmented lead 82
augmented reality (AR) 1169, 1173
auscultatory method, functional principle 959
authentic view 1210
authenticity of the acquired data 1211
authoring system 1192
auto adaptive treatment system 1230
auto immune reaction 198
auto start 695
auto stop 695
autocorrelation function 942
autologous chromophores 565
automated external defibrillator (AED) 547, 549
automated refractometer (AR) 263
automatic
– adaptive environment 1233
– alarm 1022
– image interpretation 1159
– OR table system cleaning 1271
– refractometry 252
– segmentation 1154
– static perimetry 269
– treadmill 836
automatic exposure control (AEC) 337
automatic gain control (AGC) 229
automatic tube compensation (ATC) 542
automatic volume control (AVC) 229
autonomous active implant 799
autonomous nervous system 120
autoregressive parameter 992
average noise level 220
average reference 124
average value 941
averaged potential 145
a-wave 255
axial flow MCSS 733
axial image 1210
axial resolution 346

B

baby
– newborn 1285
– premature 1285
backdiffusion 609
backfiltration 608, 609
background radiation 373
backward telemetry 239
bacterial colonization 12, 13
balanced anesthesia 579, 580
ballon segmentation 1153
balloon
– angioplasty 419
– catheter 966
– dilatation 420, 715
– model 1154
– segmentation 1158
bandage
– 3-D 1334
– compression 1324
– knitted elastic 1333
band-pass noise 209
baroflex control diagram 780
barrel distortion 1144
basal-cell carcinoma 376
basalioma 377
baseline shift 83
basic bicarbonate 31
basic cardiology diagnostics 75
basilar membrane 197
batch washer system 16
battery lifetime 1349
battery-operated ME equipment 37
battery-supported power supply (BPS) 42, 44
beam
– delivery system 563
– hardening artifact 327
– quality 563
– steering device 712
– transmission system 558
bed occupancy 1100
bed-to-bed communication 952
Beer–Lambert law 425
behaviourism 1187, 1188
behind the ear (BTE) 228
Belgium heart failure project 1229
Belgrade grasping-reaching system 885
bellows unit 586
bench top test 859
benign prostate tissue (BPH) 393

Berger effect 132
Berlin Heart
– EXCOR 724, 728, 736
– INCOR 734
beta-wave 129
biatrial stimulation system 781
bicycle ergometer 88, 111, 374
bidirectional interface 792, 1131
bidirectional wireless link 1340
bilateral phrenic nerve stimulation 887
bileaflet valve 869
binary digit (bit) 293
binaural
– hearing 210
– loudness summation 210
binaural intelligibility level difference (BILD) 211, 236
binaural masking level difference (BMLD) 211
binocular representation 823
bio impedance spectroscopy 1329
bioacoustic signal 934
bioactive
– implant 799
– substance 798
biochemical
– sensor 810
– signal 934
– voltage 455
biocompatibility 788, 868, 1339
biocompatible material 787, 814
biodegradable coating 873
bioelectric
– generator 927
bioelectric potential 926, 935
– cortical 943
bioelectric signal 790, 935
– acquisition system 1343
– amplitude 1342
– frequency 1342
– power 1343
bioelectrical effect 681
bioelectrical potential
– difference 123
bioelectrode 1358, 1360
bioheat equation 377
bioimpedance, whole-body 613
bioinspired retina-like encoder 824
bioinspired visual encoder 828
biological
– artifact 924, 934
– characterization 797
– indicator 18
– measurement 924
– model 944
– pressure limit 659
– response 870
– signal registration 934
– system simulation 944
– tissue 448
– tissue electrical properties 940
biomagnetic examination 927
biomagnetic field 928
biomechanical signal 936
biometry 263
bionic glove 885
bionic hand 798
– prosthesis 792
bionic neuron method 755
biooptical signal 940
bioparameter monitoring 1227
biophotonic imaging 491
biophysical mechanism 1256
biophysics modeling 1248
biopotential
– acquisition 1361
– difference 80
– electrode 1346
biopsy robot 1277
bioreactor 617
biosensor 925
biosignal 139, 172, 923, 924
– acquisition 923, 1346
– analysis time domain 941
– artifact 173
– commonly monitored 1343
– digital transmission 933
– physical properties 924
– recording 163
– storage and registry 932
– telemetric transmission 932
biosignal amplifier 930
– system 1006
– wireless 1012
biostimulation 566
biothermal
– effect 681
– signal 940
biphasic defibrillation 546
biphasic pulse 757
– form 547
biphasic shock 546
biplane angiography system 418
bipolar
– cell 255, 825
– circuit 686
– clamp forceps 696
– coagulation 682, 691, 695
– cutting 690
– electric stimulation 1048, 1054
– lead 80
– recording 123, 182
– scissors, working principle 1318
– stimulation 770
– vessel sealing 691, 696
bipolar transurethral resection (bipol-TUR) 690
bis acrylamide nitrogen gelatin (BANG) 708
bit error ratio (BER) 1369
biventricular assist device (BVAD) 726
biventricular defibrillator system (CRT-ICD) 553
biventricular stimulation 782
blackbody radiation law 372
bladder
– control implant 754
– function 887
– stimulation 788
blend cut 688
blended learning (BL) 1187, 1195
blind defibrillation 548
blind test 944
blink reflex 147, 575
blitz VEP 253
Bloch equation 440
blocking capacitor 761
blood
– component adsorption 610
– oxygenated 1039
blood clot imaging 497
blood filter 624, 625
blood flow
– Doppler 356
– localized muscle 429
– rate 608
– velocity 936
– velocity measurement 938
– visualization 348, 349
blood gas 973
– exercise-related 113
blood gas analysis (BGA) 627, 973
blood glucose
– concentration control 810
– monitor 1133
blood oxygenation 580
blood oxygenation-dependent imaging 451
blood pressure 162, 580, 936, 1032, 1038
– invasive 580
– invasive measurement 938
blood pump 623, 624, 745
– pneumatic 724
blood purification 597, 600
– principle 598

blood volume 936
blood-level concentration 585
blood–retina barrier 267
Bluetooth technology 1136
bluetooth technology 77
blue-yellow perimetry 269
blunt dissection 1298
body array 445
body coil 444
body composition monitor (BCM) 613
body floating (BF) 684
body plethysmography 105
– clinical value 108
– principle 106
body sensor network (BSN) 1366, 1369
body temperature 591, 929, 987
– device 802
body temperature and pressure, saturated (BTPS) 973
body temperature pressure, saturated (BTPS) 98
body weight 1343
bodymedia arm band 1354
bone
– anchored hearing aid 237
– marrow transplantation 908
– pathology 411
– removal 333
– window 331, 332
bone conduction (BC) 197, 202, 203
borderline risk 36
bottom-up segmentation 1151
bovine serum 811
bow scissors 1293
box calibration 107
Boyle's law 107
brachytherapy 489, 708, 709
– equipment 713
bradycardia 1031, 1032
bradycardias 776
braid with sensor fibre 1324
brain 159, 821
– imaging-compatible robotic system 851
– induced modification 851
– monitoring modality 1061
– perfusion 335
brain activity 1004
– ionic aspect 1064
brain–computer interface (BCI) 1003
– enabled control 1014
– illiteracy 1014
– system structure 1005
brain–machine interfaces 890
brainstem
– AEP 150
– pathway 823
– surgery 1049
brainstem electric response audiometry (BERA) 222
brainstem evoked response audiometry (BERA) 151
braking principle 89
branch–pod–pea structure 756
Braunwald–Cutter valve 869
breath-by-breath method 113
breathing
– control system 717
– disruption 526
– equivalent 113
– filter 98
– resistance 99
– system 528, 580, 582, 585
breathing disorder
– sleep-related 160
– therapy device 163
breathing-adapted radiotherapy (BART) 716
bridge to bridge (BTB) 727, 739
bridge to decision (BTD) 727, 739
bridge to recovery (BTR) 725, 726
bridge to transplantation (BTT) 724, 725, 732, 872
brightness 294
– mode (B-mode) 351
broadband harmonics 348
B-Rob-II 1279
bronchopulmonary dysplasia 1034
bronchoscopy 1209
browsing system 1189
bubble monitor 628
building law 45
bump-weighting function 1180
Burian–Allen electrode 255
business-to-business sector (B2B) 50
Buysse 696
b-wave 255
bypass
– operation 1278, 1281
– principle 584

C

Ca^{2+}-troponin kinetics 882
cable management 754
Cabrera's circle 82
calcium scoring 335
calculation algorithm
– collapse cone 913
– Monte Carlo method 913
– pencil beam 913
calibration 331, 1145
– measurement 464
– process 427
calorimetry 708
camera
– formula 372
– guiding system 1277, 1278, 1280
– pill 1133
– system 381
cancer fulguration 679
capacitive
– charging 752
– coupling 686
– sensor 805
capillary membrane 600
– Moiré structure 605
capnography 548, 971, 983
– benefits 985
– indications 985
– limitations and sources of error 985
capsule endoscopy 401
capture efficiency 321
carbon dioxide output 113
carbon fiber 1326
carbon nanotube (CNT) 1326
carbonization 567, 683, 688
carboxyhemoglobin (COHb) 425, 974
carboxyhemoglobin content (SpCO) 978
cardiac
– arrhythmia 85, 957
– electrical action potential 75
– pressure 805
– pump function 727
– valve operation 1278
cardiac device 870
– design 856
– development 857
– durability 870
– market released 858
– surgery 856
– technology 871
– testing 865
cardiac floating (CF) 684
cardiac output (CO) 936, 958, 966
cardiac pacemaker 772, 786, 787, 933
– block diagram 772
– electrode 771
– functionality 771

cardiac pumping function (EF) 551
cardiac resynchronization therapy (CRT) 552
cardiac surgery 631, 855
– equipment 12
– off-pump 623
cardiacassist PTVA 733
cardio CT 335
cardiogenic shock 726
cardiology 1088
cardioplegia 626
– control 629
– pump 624
cardiopulmonary
– bypass 634
– function testing 105
– stress testing 105
cardiorespiratory
– analysis 1034
– event 168
– function 162
– monitoring 1034
cardiotocography (CTG) 1019
cardiotomy reservoir 624, 625
cardiovascular
– function 580
– marker 810
– monitoring 955
cardiovascular disease (CVD) 472
cardioversion 554
– complications 551
– method overview 549
care 1314
caries diagnostics 408
C-arm 1264
– angiography 304
CARMAT artificial heart 744
CARMEDA method 734
case-based training 1191
case-specific protocol 420
CASPAR 1279
catalytic reaction 23
categorization system 1115
catheter 419
– direct measurement 996
– intra-arterial 962
– multifiber 564
– Swan–Ganz 965
catheter-related septicemia 29
cathlab 482
cautery 677
cavitation
– generation of 648
– water jet 657
CBM-motus 837
CCD sensor 184
CCD technology 302
cell labeling 473
cell membrane permeability 494
cell proliferation 899
cell tracking 473
cellular dynamics 1248
cellulose membrane 604
cellulose nitrate 599
Center for Disease Control and Prevention (CDC) 28
central apnoea 1031
central auditory processing disorder (CAPD) 200
central conduction time 224
central display and control module (CDM) 635
central monitoring station 1090
central motor conduction time (CMCT) 156
central nervous system (CNS) 119, 807, 890, 996, 1059
central station 952, 953
central venous catheter (CVC) 964
central venous oxygen saturation 979
central venous pressure (CVP) 964, 969
central wake station 164
centrifugal pump 624
ceramics 789
cerebral blood flow (CBF) 335
cerebral blood flow measurements using NIRS (CBF_{NIRS}) 432
cerebral blood volume (CBV) 335, 997
cerebral hemorrhage 305
cerebral immaturity 1031
cerebral perfusion pressure (CPP) 995, 996
cerebrogram 993
cerebrospinal fluid (CSF) 448, 996
cervical muscle 1170
charge
– accumulation 757
– balance 752, 757
– mode 758
– per phase 753
– storage capacity 796
charge-coupled device (CCD) 296, 340, 383
– sensor 406
chemFET 1066
chemical disinfection
– agent 15
– disadvantage 20
chemical neural recording 1065
chemical process 14
chemical shift 451
chemoembolization 419
chemotherapy 898
– agent 494
chest x-ray workstation 302
China compulsory certificate (CCC) 52
choroidal neovascularization (CNV) 269
chromophore 425
chronic cardiomyopathy 725
chronic implantation 865
chronic pain 1170
chronically ill patient 743
chronobiological disorder 168
chronobiological therapy 175
chronotropic insufficiency 780
CiCa device 616
circulatores 598
circulatory
– disturbance 374
– function monitoring 955, 957
circumneural electrode 790
CIS strategy 240
citrate anticoagulation 616
Clarion CI 888
class AP equipment 37
class APG equipment 37
cleaning and disinfection 1312
clearance 604, 609
client/server paradigm 1162
clinical case demonstration 1205
clinical complication 869
clinical decision support 1109
clinical document architecture (CDA) 1107, 1123
clinical efficiency of lithotripters 668
– focal size 669
– side effects and safety 669
clinical information system 1105, 1110
clinical measurement 998
clinical neurophysiology 120
– devices and methods 119
– safety aspects 129
clinical outpatient department 391
clinical pathway 1097, 1100
clinical study 944
clinical target volume (CTV) 909, 910
clinical testing 61, 857
clip activity monitor 1224
closed detector ring 315

closed loop
– setup 1005
– system 799, 1003
closed-bore high-field MRI 481
closing 1144
clothing sensor system 1326
cloud computing
– in the hospital 1207
– reducing costs 1208
cluster analysis 941
clustering 1149, 1150
CO_2 laser 558
CO_2 laser 570
CO_2 absorber 585
CO_2 laser 563
CO_2 partial pressure 582
coagulation 566
– argon-assisted 683, 691, 693
– bipolar 682
– contact 691
– forced 691, 693
– interstitial 564
– monopolar 682
– time 682
coated electrode contact 796
coating
– biodegradable 873
– process 795, 1325
cobalt unit 711, 718
cochlear damage 205
cochlear implant (CI) 237, 786, 787, 878, 888, 1131
– components 238
– market 804
cocktail party effect 211
coded filling device 584
cognitive apprenticeship 1188
coherence 561
coil
– arrangement 445
– sensitivity 466
cold
– cardioplegic solution 626
– light 382
– stress 1287
cold cathode tube 705
collapsed cone algorithm 911
collector electrode 80
collimation 561
collimator grid 321
collodium 599
colloquial speech 207
Collpitts oscillator 1350
colonoscopy 1209
color Doppler 352, 354
color lookup table (CLT) 408
color temperature 383
color vision 381
color-coded depth information 1215
combined cleaning and disinfection 18
combined flow system 530
comfortable stimulation level (CL) 241
Comité Consultatif International Téléphonique et Télégraphique (CCITT) 209
commissioning 917
common disorders 1120
common mode rejection ratio (CMRR) 126, 1340, 1359, 1364
communication
– bed-to-bed 952
– electronics 1339
– gateway 1225
– protocol 1131
– support 1098
– unit 1355
compact monitor 949
comparative analysis 836
compatibility datasheet 518
compensating setup 720
compex motion 885, 887
complementary metal–oxide–semiconductor (CMOS) 383
complete
– deafness 237
– failure mode 861
completely in the canal (CIC) 228
complex impedance 212
components of AAL 1221
compressed air 527
compressed analog stimulation strategy (CAS) 239
compression
– bandage 1324
– textile 1321
computational intelligence-based classifier 1155
computed force 844
computer communication 1023
computer EEG 134
computer graphics 1168
computer integrated radiology system (CIRS) 1110
computer performance 420
computer tomography (CT) 4, 129, 241, 291, 311, 329, 405, 420, 477, 478, 493, 708, 901, 1139, 1210
– angiography 334, 472, 1216
– applied dose 335
– artifact 327
– based positioning 486
– cone beam 901
– guidance 479
– slice 915
– subtraction 1216
– value 329
– x-ray tube
heat problem 319
computer tomography imaging (CTI) 1210
computer-aided diagnosis (CAD) 1164, 1199, 1204
computer-aided surgery (CAS) 1164, 1200
computer-assisted documentation 391
computer-assisted instruction (CAI) 1187
computer-assisted learning (CAL) 1187
computer-based
– exam
legal risks 1195
– training system 1189
computer-based instruction (CBI) 1187
computer-based training (CBT) 1187
computer-controlled system 1274
computerized 106
– evaluation program 84
computerized axial tomography (CAT) 311
computerized physician order entry (CPOE) 1201
concentration measurement 593
– infrared absorption 593
concentric
– contraction 274
– needle electrode 927
concept
– development 859
– of intraoperative monitoring 1045
condition for hemodialysis 600
conductive
– circuit board 1323
– embroidered circuit 1323
– footwear 46
– hearing 198
– housing 43
– loss 204
– silicone-based electrode 1327
– thread material 1325
– yarn 1326
conductometric sensor 927

condyles 410
cone of confusion 210
cone-beam computer tomography (CBCT) 339, 405, 717, 906
– system 407
– technical restrictions 412
cone-beam computer tomography (CT) 901
cone-beam detector 322
cone-rod dystrophy 256
confidence value 1157
confocal endoscopy 396
confocal laser microscope 252
– Rostock cornea module 252
conformance claim 1162
congenital Leber's amaurosis (LCA) 256
congenital vascular disorder (CVD) 573
congestive heart failure 872
connector configuration 770
constant power algorithm 689
constrained motion 841
constructivism 1188
consultation request 1124
contact
– coagulation 691
– endoscopy 396
– gel 550
– transmission 27
– voltage 42
contact quality monitors (CQM) 698
contact-lens
– adjustment 263
– electrode 259
contactless capacitive electrode 1329
contaminated equipment 12
contamination 12
content-based image retrieval (CBIR) 1162
contention access period (CAP) 1369
contention-free period (CFP) 1369
context of use 63
ConText vest 1327
contingent negative variation (CNV) 223
continued fever 987
continuing medical education 594
continuity of care record (CCR) 1123
continuous flow system 529
continuous glucose monitoring systems (CGM) 1228
continuous interleaved sampling (CIS) 239
continuous intraoperative neuromonitoring (cIONM) 1046
continuous noninvasive arterial blood pressure (CNAP) 961
continuous positive airway pressure (CPAP) 530
continuous wave (CW) 352, 353, 425, 642
contour
– model 1152
– tracking 1154
contractile element (CE) 278
contralateral routing of signals (CROS) 231
contrast 1211
– agent 455, 478, 487
– enhanced angiography 340
– enhancement 1157
– filter 1143
– harmonic 350
– medium 298, 417, 419
– registration 1145
contrast medium
– nephrotoxic 498
control loop 841
control unit 585, 1277
controlled automatic gain control (AGC) 230
controlled ventilation 530, 531, 588
convective forces 602
conventional
– CBT system 1192
– detector 292
convergence movement 181
conversion layer 299
– scanning electromicrograph 301
convex array ultrasound probe 364
convex probe 350, 358
convolution 1143
cooled detector 371
cooler
– mechanical 371
– thermoelectric 371
Coolidge tube 705
cooner wire 795
Cooper cable 754
cooperative/collaborative learning (CSCL) 1187
cooximetry 1036
coping strategy 743
coregistration 490
Coriolis torque 841
cornea
– applanation 246
– radius of curvature 263
– thickness 249, 265
– topography 265, 266
coronal reformatting 333
coronary
– heart disease 715
– scissors 1304
– sinus electrode 555
– stent 334
coronary artery analysis (CT) 1203
coronary artery bypass grafting (CABG) 873
coronary artery disease (CAD) 472
corrosion
– of electrode 789
– stability 793
Corti's organ 196
cortical
– bioelectric potential 943
– monitoring 1061
– potential distribution 943
– stimulation response 155
cortical electric response audiometry (CERA) 222
cortico-muscular connectivity 890
corticospinal tract lesion 279
cost savings in healthcare 446
cost transparency 1097
cost-effective haptic display 1181
cost-effective OR robot 1183
coulometry 1066
counter torque 505
coupling
– acoustic 479
– capacitor 761, 772
course of EEG stage 992
cranial nerve 195
cranial pressure measurement 997
craniofacial traumatology 411
craniomandibular dysfunction 409
cranio-maxillofacial surgery 411
crank ergometry 888
creatine phosphokinease (CPK) 472
crest factor 687, 693
Creutzfeldt–Jakob disease 31
critical
– data transfer 1086
– operation range 1274
cross-channel signal comparison 1024
cross-circulation 856
cross-contamination 97
cross-enterprise document sharing (XDS) 1123
cryogenic tissue ablation 483
cryoprobes 390

CsI fiber structure 323
CT system
– 64 row 320
– classification 314
CTG
– connection system 1024
– graph 1026
– monitoring 1022
– selection criteria 1022
Cuberille approach 1160
cuff
– electrode 790, 1062
– geometry 1256
cuff size 961
CUNY (The City University of New York) 209
cup area 249
cup disk area ratio 249
cuprophan 599, 604
curative radiotherapy 898
current density 685
current mode 758
current-based impedance control 843
curricular integration 1195
customizable display 420
cut-off frequency 354
cut-off point 611
cutting
– argon-assisted 690
– depth 688
– performance 1299
– speed 688
– tissue 677
– voltage 684
cutting current 687, 688
– fractionated 689
CV (consonant–vowel) 209
CyberHand 798
Cyberknife 490, 907

D

d'Arsonval 679
da Vinci telemanipulation system 1278
damage zone 567
damp heat 16
Darwinian paradigm 1156
data
– in clinical practice 1110
– integration 953
– local network 307
– management 392, 1022
– packet 756
– reconstruction 406, 407
– registration 412
– representation three-dimensional 331
– security 1131
– transmission 1086
data acquisition 324, 328, 1364
– biosignal 1346
data processing in hospitals 1097
– potential benefit 1097
data transfer
– critical 1086
– rate 1005
day clinic 1261
DDD pacemaker 769, 776, 779
– atrium-controlled 777
– clinical assessment 779
DDDR pacemaker 554
de Keating-Hart 679
dead zone 99
deafness 203
– complete 237
decimeter device 680
decision support 1099, 1108
decision-making process 1097
decontamination
– system 18
decontamination of accessories 18
decubitus
– injury 1269
– locations particularly at risk 1269
– mattress 1324
– prophylactics 1270
deep brain stimulation (DBS) 787, 802, 1133, 1134, 1256
deep hypnotic stage 993
default cycle time 537
defibrillation
– complications 551
– electrode 550
– method overview 549
– monophasic 546
– proof 684
– safety aspects 551
defibrillation threshold (DFT) 554, 555
defibrillator
– effectiveness 547
– implant 547
– physical principle 546
– portable, system components 546
– semiautomated 548
– semi-automated 547
defibrillator device 548
definition and icon, historical definition 515
definition of shock wave 645
deflecting magnet 712
degree of controllability 7
degree of freedom (DoF) 487, 1181
degree of interactivity 1189
degree of painlessness 580
DEGRO 917
Delaunay triangulation 1177
delivery of implants, MR guided 494
delivery of liquid 624
delivery system 564
delta-wave 129
demand
– flow system 530
– pacemaker 775
demineralized water 17, 31
demographic patient data 1101
dental planning 335
dental radiology 314
dentist 408
deoxygenated blood 1039
department information system 1111, 1205
depression 1230
depth dose curve 898, 899
depth filter 625
depth of field 394
depth shading 1160
dermatosis 12
descending neurogenic evoked potential (DNEP) 1053
desflurane 583
desiccation 691, 692
design
– draft 69
– freeze 859
– of electronics 1339
– of scissors 1293
design for manufacture (DFM) 871
desired speech level (DSL) 234
destination therapy (DT) 725, 727
detection
– limit 465, 467
– of flutter 957
– rate 1211
detector
– conventional 292
– dynamics 323
– flat-panel 303, 322
– GaAs 371
– gas 321
– solid-state 322, 916
– type 427
detector angular subtense (DAS) 371

detector array 315, 326
– gas-based 321
determination of physical activity 115
detoxification 615
Deutsche und Internationale Gesellschaft für Stosswellentherapie (DIGEST) 670
Deutsches Herzzentrum Berlin (DHZB) 724
Deutsches Institut für Medizinische Dokumentation und Information (DIMDI) 1119
development 597
– error 63
device
– development process 858
– dimension 505
– effectiveness 59
– efficiency 59
– implanted 870
– malfunction 739
– material 505, 506
– size dimension 1365
– switching 44
device components 591
– monitoring 591
device monitoring 580, 592
– paramagnetic sensor 592
DFT wire 770
diabetes 889, 890
– macular edema 248
– retinopathy 257
diagnosis 1164
– documentation 1101
– treatment loop 1192
diagnosis related group (DRG) 1096
diagnostic 510, 1158
– functional 119, 409, 1170
– imaging assessment 1209
– procedure 121
– ultrasound frequency range 346
– value of ergometry 93
– x-ray 491
diagnostic methods 944
dialysance 609
dialysis
– complications 613
– dose 605
– efficiency 605
– hemodialysis 30
– machine 12
– machine equipment 611
– membrane 601, 606
– peritoneal 29
– reimbursement 605
– treatment performance parameter 605
– water 30
dialysis fluid 31, 600, 601
– contamination 602
dialyzer
– construction 604
– high-performance 608
– performance 605, 606
diamagnetic material 505
diaphragm pneumotachometer 100
diastolic dysfunction 356
diathermy 677
differential infrared thermography (DIT) 373
differential nonlinearity (DNL) 760
differential path length 426
differential-diagnostic assessment 411
diffraction of sound wave 647
diffuse lighting 1215
diffuse reflection 1161
diffusion
– assessment 111
– capacity 109
– testing 105
diffusion tensor imaging (DTI) 452, 457
diffusion weighted imaging (DWI) 452
diffusive
– clearance 610
– permeability 602
digital
– archive 307
– archiving 386
– cassette 299
– detector system 296
– encoding, fundamental advantage 352
– flat-panel detector 300
– fluoroscopy 293, 296
– imaging 291, 295
– information generation 1205
– luminescent screen, readout process 299
– object 1169
– phosphor storage 297
– thermal array 78
digital image
– formation 1164
– parameter 294
– plate 292
– processing 305
– technology 307
digital imaging and communications in medicine (DICOM) 307, 392, 407, 487, 1139, 1162, 1199
– hanging protocol 1202
– RT 912
– worklist 1201
digital radiography 293
– advantage in exposure 294
digital reconstructed radiograph (DRR) 906
digital signal processor (DSP) 790
digital subtraction angiography (DSA) 292, 417, 1216
– blood vessel 418
digitally reconstructed radiograph (DRR) 916
digital-to-analog converter (DAC) 297
dilatation 419, 1144
– balloon 420, 715
dilution method 967
dimensions of IT-support 1099
DIN 14155 61
DIN 58921 22
DIN 58946-6 22
DIN 58948 24
DIN 58955 section 1–7 19
DIN 6800-2 918
DIN 6827-1 911, 914
DIN 6847-5 918
DIN 6874-5 918
DIN EN 1422 24
DIN EN 285 19
DIN EN 554 22
DIN EN IEC 62366 61
DIN EN ISO 13407 62
DIN EN ISO 15883 16, 17
DIN EN ISO 17665 22
DIN EN ISO 17665-1 22
DIN EN ISO 9000:2005 (Concepts) 54
DIN EN ISO 9004:2000 54
DIN EN ISO 15883-1 17
diode laser 567, 570
direct
– catheter measurement 996
– ECG 1021
– energy conversion detector 301
– radiography 304
direct acoustic cochlear stimulation (DACS) 236
direct current (DC) 546, 762, 1135
discharge 1027
– phase 762

discontinuous system-based ECG 85
discrete convolution 1143
discriminant analysis 941
disease management 1095
disinfectant 12
– active substance 14
– cost-saving 15
– disposal 16
– exposure time 15
– pretreatment 12
– procedure 15
– selection 15
– sequence 15
– solution 15
disinfection 14, 389
– instrument 15
– manual 15
– method 14
– of bedding 19
– of instruments 15
– of surfaces 14
– physical 16
– steam 19
– thermal 16
disinfection process 14
– application time 19
– ranges of action 19
disintegrate kidney stone 642
disintegration
– diameter 657
– efficacy 653, 657
– threshold 658
– tracking 667
disk area 249
displacement force 505
display calculation 363
display unit 528, 590
disposable 699
dissecting scissors 1297, 1299, 1303
distorted haptic perception 1181
distortion product otoacoustic emissions (DPOAE) 216
distribution of results 1205
distribution system operator (DSO) 44
divisive segmentation 1151
docking station 950
doctors' letters based on HL7 CDA 1123
doctor-to-doctor (D2D) 1123
documentation 1098
– analysis 1113
– of a defibrillation 546
domestic appliance manipulation 1231
Doppler
– effect 1019
– shift 348, 425, 426
– technique 349
– ultrasonography 352
– ultrasound 381
dose
– calculation 911
– constraint 903
– distribution 716, 904, 915
– efficiency 302
– exposure 294
– modulation 337
– parameter 613
– quantum efficiency 292
– reduction 337
dose quantum efficiency (DQE) 292
dose volume histogram (DHV) 915
dose volume histogram (DVH) 903, 912
dosimetry 707
– absolute 918
– Fricke 708
double blind method 944
double prismatic joint 840
double-insulated equipment 37
DPF measurement 426
drafting of a contract 1116
drilling template 409
drive field 463, 467
drop foot stimulator 883
droplet transmission 27
drug
– dosing 582
– dosing unit 580
– efficacy 162
– testing 945
dry ECG electrode 1359
dry heat 16
DTL (Dawson–Trick–Litzkow) 255
dual scanning 300
dual-chamber ICD 553
dual-exposure single-detector technique 307
dual-plane x-ray 303
dual-sensor system 780
dual-source CT 339, 341
Duchenne muscular dystrophy (DMD) 285
duplex mode 352
DuraHeart LVAS blood pump 735
dye laser 567, 571
dynamic adaptive radiotherapy (DART) 718
dynamic collimator rotation 720
dynamic compression 230
dynamic contour tonometry (DCT) 247
dynamic flow 355
dynamic focusing 358
dynamic friction 841
dynamic performance indicator 845
dynamic range 304, 1211
dynamic range compression (DRC) 306
dynamic range reduction (DRR) 306
dynamic robot-based evaluation 848
dynamic thresholding 1149
dynamic torque 505
dynamometer 276, 277
dyspnea 104

E

ear canal probe 213
ear mold 232
early auditory evoked potentials (EAEP) 222
early detection 991
early receptor potential (ERP) 255
early tumor detection 395
eccentric contraction 274
echo time (TE) 447
echo train length (ETL) 449
eddy current 505, 511
– brake 89
– induction 510
edge enhancement 299
edge extraction 1148, 1150, 1157
edge level 1141
edge-based segmentation 1150
EDP target concept 1113
effective acoustic amplification 235
effective dose 412
effective energy 653
effective membrane surface 607
effective power density 561
effective QMS 50
efferent pathway 120
efficacy of a therapy 280
efficient defibrillation 550
eFlow technology 355
e-Health 1119
– confidentiality 1236
– cross-border issue 1235

– economic dimension 1238
– internationalization 1242
– privacy 1236
– security 1236
– social dimension 1234
– societal and global dimension 1237
– technological dimension 1240
e-health 1217, 1219
e-Health wireless system 1367
Einthoven's bipolar limb lead 81
Einthoven's triangle 81
e-Learning 1187
– current state 1196
– status 1195
e-Learning intervention 1196
e-Learning research 1196
electric
– field distribution 770
– field strength 693
– spark 687
– transducer 927
electric and acoustic stimulation (EAS) 241
electric response audiometry (ERA) 222
electrical
– accident 35
– autodrive 1262
– biosignal 1346
– cardioversion 549
– circuit model 1249
– energy 654
– equipment 35
– excitation process 75
– interference field 46
– nerve stimulus 681
– potential 79, 121
– safety 749
– stimulation 1250
– stimulation process 1252
– stimulator 142
– stimulus 749
electrical fast transient (EFT) 803
electrical muscle stimulation (EMS) 1330
electrical system
– planning and installation 36
– protective measure 42
electrically evoked auditory brainstem responses (E-ABR) 242
electrocardiogram (ECG) 75, 472, 591, 955, 1032, 1343
– acquisition 1365
– data management 79
– database 87
– diagnostic hint 84
– electrode 1033
– episode 85
– exercise 88
– exercise indication 93
– fetal 1020
– interferences 956
– leads and electrodes 956
– loss rate 1371
– measurement 77
– monitoring electrode positions 956
– monitoring system 1228
– neonatology 1032
– nomenclature 76
– preamplifier 76
– recording 936
– recording equipment 77
– recording method 79
– signal detection algorithm 554
– test procedure 91
electrocautery 677
electrochemical cell 1358
electrochemical characterization 796
electrochemical contact 79
electrochemical contact potential 77
electrochemical impedance 793
electrochemical impedance spectroscopy (EIS) 1330
electrochemical potential 927
electrochemical reaction 592
electrochemical sensor 811
electrocoagulation 679
electrocortical grid array (ECoG) 807
electrocorticography (ECoG) 138
electrode 122, 1004
– active 239, 682, 687
– anchoring 771
– atrial 555
– biological characterization 797
– characterization 795
– conductor 770
– connecting scheme 131
– contactless capacitive 1329
– coronary sinus 555
– corrosion of 789
– cuff 790, 1062
– dry ECG 1359
– epimyocardial 555
– equivalent circuit diagram 927
– esophageal pulse 550
– extracochlear 238
– finger 555
– flexible 790
– flexible dry 1359
– flexible polyimide 790
– ground 152
– implanted stimulating 750
– indifferent 82
– iontophoretic 1331
– iridium 753
– metal 123
– monopole needle 927
– multichannel array 888
– needle 927, 1062
– nerve cuff 826
– nonpolarizable 79
– optical 1360
– overview 141
– passive fixation 771
– *p*CO2 926
– percutaneous 880
– performance 753
– placement 128, 132
– polarizing 123
– positioning 552
– positions for ECG monitoring 956
– recording 77, 122, 151, 152
– resistance 122
– screw 771
– shaft 790
– sieve 791
– stimulation 888, 1251
– suction unit 90
– technology 78
– thin-film 793, 810
– tip 770
– transition impedance 127
– transvenous 554
– type 78
– ventricular 775
– working 794
electrode material 793
– electrochemical impedance 793
electrode–electrolyte interface 751, 757
– electrical circuit model 751
electrodesiccation 679
electroencephalography (EEG) 4, 119, 129, 991, 1003, 1343
– age-related change 994
– alteration 991
– amplifier 1005
– analysis 992
– assessment 994
– based BCI 1006
– blink artifact 137

– change stage 992
– characteristics 132
– classification 992
– computer-based method 992
– conventional visual 992
– device standard component 130
– instrumentation real-time 1003
– monitor Narcotrend 992
– monitoring 992
intensive care unit 995
operating room 994
– muscle activity 137
– recording cap 1005
– sensor 805
– spectral analysis 992
– stages A–F 993
electroencephalography (EEG) signal 223
– analysis 136
– quality 994
electroluminescent screen (EL) 951
electrolyte
– balance 613
– imbalance 1031
electrolytic effect 681
electromagnetic (EM) 803
– biocompatibility 789
– fields 445
– generator 1134
– interference 1346
– model 1251
– tracking navigation 485
electromagnetic compatibility (EMC) 46, 504
electromagnetic interference (EMI) 504
electromagnetic shock wave emitter (EMSE) 662
electromyography (EMG) 119, 138, 142, 162, 279, 881, 1053, 1062, 1330, 1343
– circuit diagram 139
– examination 140, 927
– index 993
– macro needle 144
– potential
amplitude analysis 145
– sensor 805
– signal 933
– stimulator 142
electron
– density conversion 905
– gun 712
– radiation 898
electron-beam computed tomography (EBCT) 339
electroneurography (ENG) 119, 145, 1062
electronic
– artifact 328
– CTG archive 1028
– dosing 582
– image converter 386
– integration 790
– interconnection 1323
– module 1340
– order form 1102
– sensing 1341
– x-ray detector 314
electronic case record (eFA) 1109
electronic device
– frequency response 127
– on textile 1348
– system integration 1347
electronic health card (e-HC) 1122
electronic health record (EHR) 1096, 1123
electronic patient record (EPR) 1101, 1109, 1199, 1205
electronic probe 357
– convex 358
– focal constriction 358
– focal zone 358
– linear 357
– sector 358
electronics 789
– in medicine 1338
electronystagmography (ENG) 182
– technology 182
electrooculography (EOG) 182, 260, 1343
– technology 182
electroophthalmology 151
electrooptic (EO) 1361
– modulator 1358, 1362
electropneumatic driving unit 729
electroretinogram (ERG) 253, 830, 1343
electrostrictor transducer 1358
electrosurgery 677, 1318
electrotherapy 550
electrotomy 677, 683
elimination of aneurysms 420
ellipsoidal reflector 661
elongation sensor 1329
embedded sensor 1355
embolization 419
embroidered
– conductive circuit 1323
– scaffold construction 1335
embroidery 1322
– technology 1334
EMC environment class 46
emergency medicine 391
emergency route 37
emergency supply of equipment 93
emergency treatment 995
emissivity 373
emittance 373
emotional distress 742
emphysema 110
EN ISO 13485:2009 52
EN ISO 14971:2008 52
EN 207 577
encapsulation 794
– material 789
end-effector trajectory 843
endocapsule 401
endoluminal endoscopic view 1209
endomicroscope 396
endonavigation 401
endorobot 400
endoscope
– core components 382
– guiding system 1274, 1275
– image sensor 383
– remote control 400
endoscopic
– accessories 382
– cystoscopy 395
– image 381
– image chain 380
– image monitor 386
– instrument 387
– instrument system 482
– polypectomy 689
– procedure 380
– surgery 482
– tumor treatment 390
endoscopy 379, 573
– capsule 401
– challenges 381
– confocal 396
– contact 396
– data management 392
– flexible 383
– fluorescence 395
– HF electrosurgery 390
– infrared 396
– laser light 390
– optical 1210
– rules 698
– standard light source 383
– system integration 392
– virtual 333, 381, 1170, 1211
– workstation 390
endothermy 677
endotoxin passage 602

endotracheal tube 595
energy
– conversion 1352
– focal 653
– generation 789
– management circuit 1135
– shock waveform 546
– subtraction 307
energy flux density (ED) 652
– threshold 659
enhanced cornea compensation (ECC) 252
enhanced reality (ER) 1173
EPID 916
epidural ICP measurement, prerequisite 999
epileptic activity 166
epileptic pattern 131
epimyocardial electrode 555
epiretinal device 825
epiretinal implant 826
episiotomy scissors
– Braun-Stadler 1309, 1310
– Waldmann 1309
equipment
– double-insulated 37
– not used on the patient 26
– repair and maintenance 26
– trolley, multiple sockets 45
– used on the patient 25
equivalent circuit diagram 122
equivalent rectangular bandwidth (ERB) 221
erase bar mode 951
erbium:YAG laser 569
erectile dysfunction 172
ERG potential 254
ergometry 88
– measuring station 89
ergonomic property 59
ergonomics and wearability 1345
ergospirometry 105, 106, 111, 112
– clinical value 113
– parameters 114
erosion 1144
esophageal pulse electrode 550
esophagus stimulation 768
ESTRO 917
etCO_2 measurement 983
ethylene oxide (EO) 23, 604
– adsorption 23
– desorption 23
– exhaust air 23
– sterilization, substance 22
ETHZ-ParaCare neuroprosthesis 885
eTracking 366
Euclidean distance 1149
Euclidean geometry 1157
EUROMACS VAD registry 746
European law 50
European Medicine Agency (EMEA) 1120
European Register for Mechanical Circulatory Support 746
European Respiratory Society (ERS) 97
European Society of Cardiology (ESC) 551
European Union (EU) 754
European Uremic Toxin Working Group (EUTox) 606
Eustachian tube 194, 213
evaluation of tenders 1114
event recording 1033
event related potential (ERP) 156, 223
event storage 952
event-related desynchronization (ERD) 1007
event-related synchronization (ERS) 1007
event-triggered report 554
evoked compound action potentials (TECAP) 242
evoked magnetic field 937
evoked potential (EP) 119, 937
– recording parameter 149
evoked torque 280
evolutionary algorithm 1155, 1156
exact surgical operation 1280
examination of patients 923
excimer laser 570
excitation of cavitation 648
excitatory postsynaptic potential (EPSP) 121, 129
excitatory synapse 121
excited nuclear configuration 426
execution time 845
exercise
– aerobic endurance 890
– electrocardiogram (ECG) 88
– related blood gas 113
– testing 374
exophthalmometry 271
expert system 700, 1155
expiratory CO_2 104
expiratory gas flow 972
expiratory reserve volume (ERV) 96, 101
expiratory valve 528, 533, 538
explosive atmosphere 45
exporting patient data 1029
exposure 300
– dose 294
– duration 566
expulsion dynamics 933
extended field of view (EFOV) 363
extended gate field effect transistor (EGFET) 1068
extensible markup language (XML) 1123
external
– beam therapy 902, 907
– cardiotocograph, measuring procedures 1020
– defibrillator 546, 549
– earpiece 888
– pressure sensor 998
– stimulation 750
external heart–lung machine (HLM) 621
exteroceptive sensor 792
extinction coefficient 425
extracellular
– electrical stimulation 1250
– space 613
extracellular water (ECW) 614
extracochlear electrode 238
extracorporeal circulation (ECC) 621, 623, 629
– historical development 621, 622
– performance 630, 632–634
– references 638
– structure and function 623–626
– technical aspects 635–637
extracorporeal pulsatile MCSS 728
– long-term application 728
extracorporeal shock wave lithotripsy (ESWL) 573, 641, 664
– historical development 642
extracorporeal shock wave treatment (ESWT) 641, 642, 664
– long-term side effects 659
extraluminal view 1209
extraneural stimulation 1256
extremity prosthesis 788
eye applicator 714
eye movement (EOG) 933
– control 180

F

fabrics with integrated electrical insulated heating 1332
failure modes and effects analysis (FMEA) 862
fan-beam geometry 315

far infrared (FIR) 370
Faradaic
– process 752
– resistance 752
Faraday cage effect 507, 681, 685
fasciculation 143
fashionable electronic device 1340
fast field echo (FFE) 450
fast Fourier transformation (FFT) 128, 447
fast low-angle shot (FLASH) 450
fast spin echo (FSE) 449
Fastl noise 209
fault current 42
Favaloro scissors 1306
FDA approval 866
FDRC 231
feature extraction 1006, 1147, 1150, 1156, 1157
feature vector 1156
Federal Communications Commission (FCC) 803
Federal Drug Administration (FDA) 486
Fédération Dentaire Internationale (FDI) 1159
feedback
– acoustic 237
– control system 836
Feldkamp algorithm 305, 406
FEM model 1256
femoral skin 374
Fermi resonance 426
ferromagnetic material 505
ferrous sulfate dosimetry 708
ferrum candens 678
fetal
– electrocardiogram (ECG) 1020
– movement pattern 1019
– outcome 1027
fetal heart rate 1019, 1024
– ultrasound 1019
fiber
– activation 1253
– bundle 605
– optic hydrophone 650, 651
fibrillation 143, 549
– atrial 871
– scoring algorithm 548
fibrous encapsulation 793
Fick's principle 966
field
– anatomical marker 407
– effect transistor 927
– hydrophone 651
field free point (FFP) 464
field of view (FoV) 371, 405, 446, 463
fill factor 321
film emulsion 293
film–screen system 295
filter fouling 608
filtered backprojection (FBP) 324, 406
financial resource 6
fine structure processing (FSP) 239
Finetech–Brindley stimulator 887
finger
– electrode 555
– flexor 886
– perimetry 269
– sensor 976
finite element (FE) model 1252
finite element method (FEM) 1180, 1251, 1253
first generation of CT 314
first stage of labor 1027
first treatment 915
first-fault philosophy 36
Fitzhugh–Nagumo model (FHN model) 1249
fixed-rate ventricular stimulation 775
fixed-tube detector position 330
flash simulation 68
flat interface nerve electrode (FINE) 1253
flat-panel detector 303, 322
Fleisch pitot tube 99
Fleisch pneumotachometer 99
flex-foil substrate 1323
flexible
– dry electrode 1359
– electrode 790
– endoscopy 383
– implantable microelectrode 791
– microimplant 789
– polyimide electrode 790
flicker point 263
flip angle (FA) 478, 485
floating mass transducer (FMT) 236
floating microelectrode 827
flooring 45
flow
– artifact 510
– condition 740
– cycled mode 531
– directed catheter 965
– generator 527
– rate 624
– sensor 97, 98
– system, continuous 529
– trigger system 531
– volume loop 96, 972
flowmeter 98, 104
fluid-attenuated inversion recovery (FLAIR) 452
fluorescence
– angiography 267
– decay 321
– endoscopy 395
fluoroscopy 720
– flat-panel detector 303
focal
– constriction 358
– energy 653
– length 561
– volume 652
focus field 463, 467
focusing of spherical wave 660
follow-up documentation 1108
Food and Drug Administration (FDA) 453, 866
force
– feedback 412, 1276
– level 276
– sensor 841
– transducer 280
forced
– coagulation 691, 693
– expiration 101
– oscillation technique 104
– spirogram parameter 102
– spirometry 96, 101, 102
forced expired volume in one second (FEV_1) 96
forced vital capacity (FVC) 95
force–velocity property 278
fore-vacuum 21
form of respiration 532
formaldehyde (FO) 24
formaldehyde (FO) gas sterilization
– adsorption and desorption 24
– operating staff 24
– process sequence 24
– requirements 24
– substance 24
formant 206
forward planning 911, 913
Foucault's method 253
four-beam box 914
four-dimensional (4-D) 1152
four-dimensional CT (4-D CT) 906
Fourier
– analysis 348
– descriptor 1155
– ellipsometer 251
– transform 1147

Fourier transform infrared spectroscopy (FTIR) 796
Fourier transformation (FT) 440, 790
Fourier-slice theorem 324
four-level laser 560
fourth generation of CT 317
FPSA-system 615
fractional saturation 976
fractionated cutting current 689
fractionated vacuum 21
– process 19, 20
fractionation 907
Frank–Condon principle 425
Franz Volhard 597, 600
free motion 841
free-electron laser (FEL) 567, 571
FreeHand system 885
– control unit 1277
freezing technique 483
Freiburg speech intelligibility test 207
French Association of Teleassistance Companies (AFRATA) 1225
Frenzel glasses 182
frequency 192
– adaptive pacemaker 780
– filtering 306
– harmonic 347
– of apnoeas 1033
– response 930
– spectrum 220
– train 882
– transfer function 234
– variation 149
frequency shift keying (FSK) 1133
fresh gas decoupling 587
fresh gas supply 586
– continuous 586
– discontinuous 587
Fricke dosimetry 708
front-end electronic acquisition system 1365
fuel cell 592, 1135
– electrochemical reaction 592
fulguration 679, 693
– current 688
– device 679
full dynamic range compression 231
full scale clinical trial 866
full width at half maximum (FWHM) 465, 652, 653
full wireless link 1351
fully automated defibrillator 547
functional
– area 55
– diagnostics 119, 409, 1170
– disorder 172
– image formation 1164
– medical textile 1322
– recovery 836
– saturation 976
– specification 63
– textile 1321
functional electrical stimulation (FES) 749, 877
– induced muscle fatigue 882
– stimulation 759
– technology 877
functional electrical stimulation therapy (FET) 889
functional hemoglobin 974
– absorption spectra 976
functional magnetic resonance imaging (fMRI) 830, 1005, 1061, 1203
functional MRI (fMRI) 451
functional residual capacity (FRC) 101, 105, 107
fundamental wavelength 347
fundus camera 267
future developments 1317
future standard 1093
fuzzy logic 1155, 1156
fuzzy set theory 1157
F test 940

G

GaAs detector 371
Galenus of Pergamon 598
gall duct stone lithotripsy 643
galvanic
– potential 77
– process 789
galvanized thread material 1326
gamma knife 719
gantry 330
– based imaging 717
Ganzfeld ERG (Ganzfeld electroretinogram) 253, 254
gap detection threshold 201
gas
– blender 629
– detector 321
– dilution method 105
– exchange 624
– humidifier 528
– ion tube 704
– laser 567
– monitoring 973
– sterilization 795
– wet scrubbing 23
gastroscopy 394
gateway 953
gating spring theory 197
general supply (GS) 44
generalized autocalibrating partially parallel acquisitions (GRAPPA) 445
generation
– of cavitation 648
– of pressure pulse 660
– of radiation 710
– of shockwave 660
generator circuit 757, 760
generic wearable sensor signal acquisition 1357
genetic algorithm (GA) 1255
genetic data 1108
geometric
– aberration 1144
– analog tomography 313
– distortion 1144
geometrical parameter 1256
geometry 918
Georg Haas 599, 600
George Frédérique Parrot 598
Georgia Tech Wearable Motherboard (GTWM) 1354
German Federal Institute for Drugs and Medical Devices (BfArM) 61, 1311
German Medical Devices Act (MPG) 551, 1092
German Sleep Society (DGSM) 161
German system for telemedicine 1125
German telemedicine 1120
glass 789
– fiber bundles 383
glaucoma 249, 257
– damage 252
– diagnostics 247
glaucomatous damage 249
global ejection fraction (GEF) 968
global end-diastolic volume (GEDV) 968
glucose
– concentration 811
– meter 1228
– monitoring 1227
glucose oxidase (GOx) 811
goal-directed robotic therapy 848
goal-oriented control 1010
Goldberger's lead 81

Goldman equation 1249
Göttingen sentence test 209
gradient
– field strength 467
– magnetic field 506
– slew rate 506
gradient echo (GE) 450, 478, 488, 510, 512
– sequence 488
gradient echo sequence (GE) 485
gradient-induced interaction 506
gradiometer 937
grammar 1155
graphical image processing 363
graphical user interface (GUI) 487, 1180
graphics interchange format (GIF) 1161
graphics processing unit (GPU) 407, 1170
Graseby capsule 1035
grasping in tetraplegics 1134
Grass stimulator 798
gravitational torque 841
gravity compensation 839
gray level 293
gray matter (GM) 448
grayscale
– dynamic 292
– interval 332
– value 1157
gripping force 1276
gross domestic product (GDP) 1219
gross tumor volume (GTV) 909, 910
ground
– electrode 152
– truth 1154
growth function 221
guidelines of a QMS 56
gynecological scissors 1306, 1308
gyromagnetic ratio 478

H

H_2O_2 sterilization 24
– process sequence 24
haematocrit 1038
hair cell
– loss 198
– sensory 195
Hall sensor 929
– schematic setup 929
Hall voltage 929
hand 14
– diagnostic test 277
handling system 400
hands-free technology 361
haptic
– display 1174
– feedback 381, 1174, 1180
– perception, distorted 1181
– simulation 1181
hard x-ray 327
hard-disc drive (HDD) 294
hardening artifact 410
hard-metal scissors 1294
harmonic frequency 347
Hartmann–Shack sensor 262
hazard 35
– specific 40
head coil 445
head mounted display (HMD) 1013, 1170
head operation 1262
health
– condition monitoring 1341
– information network 1096, 1108
– insurer 1096
– policy 5
– portal 1222
– telematics 1119, 1121
health level 7 (HL7) 1107, 1162
– interface 308, 1029
– message type 1107
health monitoring 1345
– system 1359
health professional card (HPC) 1124
healthcare delivery 804
hearing
– assessment 227
– binaural 210
– conductive 198
– device 228
– disorder 198
– impairment 191, 206
– nerve 194
– nerve malformation 199
– normal 203
– spatial 193
– system 236
– threshold 201
hearing aid 228, 229, 233, 236
– implantable 236
– multichannel 231
– nonlinear 233
– parameter 232
hearing in noise test (HINT) 209
hearing level (HL) 201
hearing loss (HL) 198, 234
– for speech 208
heart
– block 856
– defect 1037
– replacement system 724
– signal 771
– surgery, human 856
heart rate (HR) 167, 955, 956, 1035
– data monitoring 1228
– measurement 1224
– monitoring 1223
– standard values 1033
– turbulence 87
– variability 87
heart transplant (HTX) 724, 725, 872
heart valve
– design 862
– prosthesis 486, 496
heart–lung machine
– components 627–629
– modular 635
– semi-modular 635
– structure and function 623
heart–lung procedure 856
HeartMate X 747
heat and moisture exchanger (HME) 528
heat quantity 682
heat transfer capacity 695
heatable
– bandage 1332
– textile 1332
heating testing for implant 505
heat-resistant item 21
heat-resistant virus 18
Hebb's law 890
heel effect 321
helical rise 326
helical tomotherapy 905
hematoporphyrin 558
hemodiafiltration (HDF) 600, 602, 608
– online 601
hemodialysis 608
– condition 600
– equipment 30
hemodynamic 631
– monitoring 957
hemofiltration (HF) 608, 611
hemoglobin (Hb) 425, 565, 631, 973
– excitation coefficient 425
hemolysis 740
hemoperfusion 611
hemophan 604
hemopump 727

hemostasis 677
HeNe laser 571
hepatitis B 18
hepatitis C virus (HCV) 614
hermeticity of the sealing 754
heterogeneous HIS 1105, 1107
heterogeneous information system 1106
HgCdTe 371
HH model, equivalent electrical circuit 1248
hierarchical region merging 1153
high level disinfection (HLD) 389
high-contrast structure 408
high-definition (HD) 381
high-dose rate (HDR) 709, 713
highest quantum efficiency 301
high-flux membrane 607
high-frequency (HF) 478, 1271
– current 683
– current type 687
– electrosurgery 390
– generator 684
– knife 684
– monopolar technique 685
– noise 942
– OR table 1271
– sound wave 344
– surgery 677
– surgical unit, application and safety 697
– train 763
high-frequency current-switching (HFCS) 762
high-frequency-induced interstitial tumor therapy (HFITT) 692
high-performance dialyzer 608
high-precision radiation treatment 718
high-resolution imaging 419
hip replacement 1278
Hippocrates 3
histogram stretching 1142
Hittorf tube 705
HL (hearing level) 151
HL7 (health level 7) 1199
HLM
– semimodular configuration 636
– technical safety 636
Hodgkin and Huxley model (HH) 1249, 1338
holistic HIS 1105, 1106
hollow body organs 1209
hollow fiber membrane 600
holmium:YAG laser 569
Holter monitoring 84
home 1027
horizontal eye movement 187
horizontal open MRI 480
horizontal procedure 1112
hospital information system (HIS) 106, 308, 392, 1162, 1199
– bundle electrocardiography 83
– function 1099
– vendor types 1106
hospital information systems (HIS) 1096
hospital IT infrastructure 1206
hospital room 1368
hot air sterilization 22
hot-wire
– anemometer 100
– sensor 593
Hounsfield unit (HU) 329, 1148
Hounsfield value 1157
hour-glass figure 265
house base station 1130
human
– heart surgery 856
– interface 585
– motion parameter 837
– skin regulation 428
human–computer interface 788
humane medicine 9
hybrid
– BCI 1014
– electronic assembly 1061
– field-neuron model 1248
– operating room 873
– segmentation 1152
hydraulic dynamometer 940
hydrogel osmolality 814
hydrostatic pressure 961
hydrostatic suction 626
hygiene
– guideline 12, 30
– in medical technology 11
– measure 12
– of spirometer 98
– plan 14
hypercapnia 1039
hypermedia 1188
hyperoxia 1037, 1039
hyperpolarization 196, 1251
– prepulse 1256
hypertext 1190
hyperthermia 473, 987
hyperthyreosis 987
hyperventilation 133
hypnotic drug effect 991
hypofluorescence 267
hypoglycaemias 1031
hypoperfusion 429, 980, 983
hypopnea 170
hypopnea index (HI) 168
hypothalamus 119
hypothermia 632
– induced 991
hypothermic ECC 632
hypovolaemia 979
hypoxia 1039
hysteroscopy 394
– training simulator 1172

I

ICG angiography 269
iconic image description 1158
ideal wearable health system 1355
identification number (ID) 1162
identification of dental fixtures (IDEFIX) 1156
identification of objects 1155
IEC 60601-2-22 576
IEC 60825-1 576
IEC EN DIN 60601-1-6 60, 64
IEC EN DIN 62366 60
Igaki-Tamai stent 873
image
– 2-D 1210
– 3-D 1278
– 3-chip endocamera 385
– analog 292
– analysis 1140
– artifact, instrument 510
– brightness 295
– communication 1161
– converter, electronic 386
– data compression 1202
– documentation 386
– enhancement 1140
– generation 1170
– information 365
– intensifier 294, 312, 418
– interpretation 1158
– iterative warping 307
– management 1140, 1161
– management system 1206
– manipulation 1161
– matrix 294
– of organ, frequency 346
– optimization 365
– postprocessing 1202
– quality 304, 507
– reconstruction 311, 317, 323, 445
– retrieval 1161
– sensor 383
– spatial 420

– transmission 307, 383
– visualization 1140, 1158
image acquisition 468
– ECG-triggered 336
image formation 1140, 1164
– functional 1164
image processing 295, 365, 372, 944
– high-level 1141
– low-level 1141
– system 1155
image retrieval in medical applications (IRMA) 1162
(image) latency 386
image-guided 719
– modality 899
– surgery 335
image-guided radiotherapy (IGRT) 717, 902, 906
imaging
– 3-D 421
– artifact 510
– diagnostics 1170
– high-resolution 419
– in treatment planning 901
– interventional 335
– intraoperative 478
– lung 474
– method, overview 343
– narrowband 397
– processing 1142
– second harmonic 347
– system 296
– two-dimensional (2-D) 467
immersive feedback 1004
immobilization 900
– support 899
immune response 804
impedance
– audiometry 212
– complex 212
– interelectrode 751, 752, 758
– of sound wave 193
– pneumography 971, 972, 1033, 1034
– spectroscopy 795
– transducer 194
impedance cardiogram (ICG) 968
impedance control 841
– current-based 843
implant
– bioactive 799
– electronics 789
– for deep brain stimulation 786
– heating testing for 505
– manufacturer 504
– passive 454
– radioactive 714
– subretinal 825
– technology 768
implant telemetry 1131
– transfer rate 1134
implantable
– blood pump 745
– cable 753
– cardiac defibrillator 1133
– device market 804
– energy harvester 1135
– hearing aid 236
– hearing system 237
– infusion pump market 804
– loop recorder 85
– neural interface 807
– pressure sensor 805
– pulsatile MCSS 730
– sensor 805, 806
– signal processor chip 807
– stimulation system 750
– stimulator 749, 760
– system 885
implantable cardioverter/defibrillator (ICD) 551, 741, 802
– development 552
– device technology, function checking 555
– system technology 553
implantable microsystem 801
– design and testing 802
– medical application 801
– testing 815
implantation procedure 754
implanted
– device 870
– electronics 827
– epimysial stimulation electrode 885
– ICD system 552
– material 804
– smart stimulator 799
– stimulating electrode 750
– unit 888
implantology 410
implementation 1111
– of HIS 1110
importing lab results 1030
improved patient positioning 1262
improvement of quality 1097
improvement to diagnostic 5
impulse
– generation 545
– oscillometry 105
incidence 699
incubator 1285, 1286
– regulation of humidity 1286
– regulation of oxygen 1287
indexing 1281
indicator dilution curve 939
indices for unperturbed motion 847
indifferent electrode 82
indirect registration 931
individual activity 1115
indocyanine green (ICG) 269, 396
induced
– emission 560
– hypothermia 991
inductive coupling 1131
inductive interface 790
inductive link antenna 1350
inductive transmission 1132
industrial, scientific, and medical (ISM) 1349
ineffective respiratory work 1034
infection 12
– cause 13
infection transmission 27
– protective measures 27
infectious disease 12
information
– access 950
– lifecycle management 1207
– management 1097
– monitor 950
– network, health 1096, 1108
– processing 1155
information and computer technology (ICT) 1217
information system 1110
– architecture 1097
– implementation 1111, 1117
– investment volume 1111
– selection process 1111
information technology (IT) 4, 6, 393, 1085, 1199
– hospital infrastructure 1206
– in hospitals 1097
– networks in healthcare 1085
– portal 950
infrared (IR)
– absorption 593
– coagulator 683
– endoscopy 396
– imaging band 370
– imaging, medical applications 377
– measuring cell 982
– sensitive semiconductor 929
– sensor 1358
– spectrometer 935

infrared (IR) imaging system 369
– detector 371
infrared functional imaging (IRFI) 369
infrared oculography (IROG) 182
infusion
– apparatus 12
– pump market 804
inhalation anesthesia machine 580
inhaled air 974
inhibiting mode 776
inhibitory postsynaptic potential (IPSP) 121, 129
initial claudication distance (ICD) 432
initial transmission slot (ITS) 1369
inner corneal layer 252
inner ear 194
inner hair cell (IHC) 195, 196
innervation pattern 143
Innsbruck sentence test 209
input
– data 912
– impedance 126
– sensitivity of a pacemaker 771
InSb 371
insertion activity 143
insertion gain (IG) 235
in-situ diagnostics 577
in-situ measurement 235
insomnia diagnostics 161
inspection 1314
inspiration line 586
inspiration/expiration (I/E) 972
inspiratory
– O_2 concentration 529, 580, 581
– O_2 measurement 583
– phase 530
– ventilation pressure 529
inspiratory capacity (IC) 96, 101
inspiratory reserve volume (IRV) 96, 101
installation qualification (IQ) 17, 22
instant thrombosis 495
instantaneous field of view (IFoV) 371
Institute of Medicine (IOM) 61, 1169
instruction, traditional methods 1196
instructional design 1191
instrument
– disinfection 15
– displacement 484
– guiding system 1275
– holding system 1280
– recognition 700
– tracking 1275
instrumental diagnostic procedure 409
insufflator 389
insulation fault 37
integral nonlinearity (INL) 760
integral shading 1160
integrated
– acute medical system 951
– operating theater 391, 392
integrated circuit (IC) 789, 795, 1061
integrating healthcare enterprise (IHE) 1123, 1162, 1200
integration engine 1105
Integration of Interventional Imaging in the Operating System (IIIOS) 493
integration of medical technology 1108
integration of modalities 1108
intellectual property (IP) 857, 859
intelligent robot swarm for attendance, recognition, cleaning and delivery (IWARD) 1232
intelligent tutorial system 1191
intensity modulated arc therapy (IMAT) 719
intensity modulated radiation therapy (IMRT) 716, 719, 902, 903
intensive care
– medicine network 1090
– network 1086, 1090
intensive care unit (ICU) 948
intensive care ventilator 526, 585
– component 527
intensive therapy 525
interaction concept 63, 69
interactive data exchange 78
interactive fabric structure 1331
interactive system 1273
interconnected medical system 1087
interconnection technology 1348
interdisciplinary collaboration 1205
interelectrode impedance 751, 752, 758
interface for data transmission 351
interface server 1079
interface with air 345
interference
– coupling 780
– suppression 47
– voltage 76
intermittent fever 987
intermittent intraoperative neuromonitoring (iIONM) 1046
intermittent mandatory ventilation 539
intermittent positive pressure ventilation (IPPV) 588
internal
– backfiltration 610
– cardiotocograph, measuring procedures 1020
– filtration 610
– hospital computer system 1030
– paddle 548
– power source 44
– web image processing 1086
internal defibrillation 550
– maximum energy pulse 550
international (NBG) pacemaker code 775
international classification of sleep disorders (ICSD) 160
International Electrotechnical Commission (IEC) 861
International Organization for Standardization (ISO) 861
international pacemaker code (NBG) 553
International Society for Musculoskeletal Shockwave Therapy (ISMST) 670
internationally recognized regulation 93
inter-peak latency 151
interpolated surrogate data 327
interruption in the signal chain 1089
interstitial coagulation 564
interstitial fluid (ISF) 811
inter-subject variation 1141
intervention planning 1164
interventional
– imaging 335
– procedure 477
– radiology 419
– scanning 504
– therapy 1210
– tools 516
intestine lumen 1215
intra-aortic axial flow pump 727
intra-arterial catheter 962
intracardiac suction tube 624, 627
intracardial
– ECG 932
– stimulation using implants 768
intracardial electrogram (IEGM) 1133

Subject Index

intracellular space 613
intracellular water (ICW) 614
intracochlear electrode array 238
intracortical
– field potential 1061
– penetrating electrode 827
– probe 1061
– stimulation 826
intracranial blood volume 998
intracranial pressure (ICP) 805, 995
– dynamic component 997
– measurement system, bandwidth 998
– monitoring 995
– pathological values 997
– pulse curve, fine structure 997
– range 997
– technical option 998
– volume characteristic 996
intraocular lens 263
intraocular pressure (IOP) 246, 247, 805, 936
intraoperational (IOP) 805
intraoperative
– assessment 1046
– examination 396
– imaging 478
– measurement 556
– MR imaging 477, 480
– navigation 410
– nerve mapping 1043
– radiotherapy 902
– verification of nerve damage 1043
intraoperative electron radiation therapy (IOERT) 906
intraoperative monitoring 140
– concept of 1045
intraoperative neuromonitoring (IONM) 1043, 1045, 1056
– equipment 1050
– legal aspects 1056
– methods and scopes 1044
– signal interpretation 1047
intraoral scanner 409
intraoral x-ray image 1142
intra-subject variation 1141
intrathoracic blood volume (ITBV) 968
intrathoracic gas volume 105, 107, 109
intrauterine pressure measurement 1021
intravenous anesthesia 993
intravenous anesthetic agent 584
intrinsic conductive polymer (ICP) 1326
intubated patient 542
invariant point 1277
invasive
– blood pressure 580
– lead 83
– measurement of blood pressure 938
– technology equipment, hygiene 26
invasive pressure measurement 962
– requirements 963
inverse planning 903
investigational device exemption (IDE) 866
invitation to tender 1113
in-vivo targeting 486
ion concentration 1067
ion gate opening 1249
ionic
– component action potential 1064
– composition 1065
– conduction 79
– sensor 1066
ionization 707
– chamber 321
ion-selective electrode (ISE) 1065
ion-sensitive field effect transistor (ISFET) 1066
iontophoretic electrode 1331
iPhone for consultation 1029
iridium electrode 753
Iris scissors 1301
iris surface 265
iron oxide based nanoparticle 463
– magnetization 463
irradiation 933
– tissue 566
– volume 910
irreversible cell damage 683
ISFET for pH sensing 1068
ISNT norm 252
ISO 14937 25
ISO 14971 60
ISO/TR 14969 (Guidelines) 54
isocenter 909, 914, 918
isocentrical-beams technique 904
isodata clustering 1149, 1150
isodose distribution 914
isoflurane 583
isokinetic contraction 274
isolating transformer 40
isometric contraction 274
isophone 200
isotonic contraction 274
isotropic voxel 407
ISO 13485:2009 52
iterative
– design process 69
– image warping 307
ITU (International Telecommunication Union) 1133

J

Jarvik Heart 2000 Flow Maker 735
Jean Antoine Nollet 598
J-electrode 771
jerk
– index 845
– magnitude 845
– nystagmus 181
Johan Jacob Abel 599, 600
Joule's law 682
just noticeable difference (jnd) 200

K

Karhunen–Loève transform 1147, 1156
Karlsruhe Institute of Technology (KIT) 486
Kassenärztliche Vereinigung (KV); local state association of statutory health insurance physicians 1120
keratoconus 265
keratometry 263
keyhole surgery 380
kidney
– artificial 599, 600
– disease 597
– lithotripter 660
– patient 600
kidney stone 642
– disintegrate 642
kinematic architecture 840
kinematic module 839
kinetic perimetry 269
Kirchhoff's law 373
k-means clustering 1149
k-nearest neighbor (KNN) classifier 1155
knee
– arthroplasty 889
– arthroscopy 1172
– joint 1177
knitted elastic bandage 1333
knowledge-based classification 1155
Korotkoff sounds 959
krypton-ion laser 570
KTP laser 569
KUKA robot arm 490

L

labor and delivery (hospital/birth center) 1027
labor progress chart 1026
lactate dehydrogenase (LDH) 740
laminar flow 98
Langevin theory 465
laparoscopy 1183
– intervention 400
large-caliber duodenoscope 394
Larmor frequency 442, 478, 496
laryngeal surgery 394
laser 390
– 3-D scanner 1177
– angioplasty 564
– application fields 568
– classification 574
– contact mode 564
– cutting, liquid flushing 564
– diode 567, 570
– Doppler 426
– efficiency 562
– erbium:YAG 569
– excimer 570
– eye-protectors 576
– four-level 560
– gas 567
– generated shock waves 661
– HeNe 571
– holmium:YAG 569
– induced process 559
– interference biometry 264
– krypton-ion 570
– KTP 569
– medical use 558
– medium 559
– Nd:YAG 558, 569
– noncontact mode 564
– profilometry 796
– pulse duration 567
– ruby 558, 569
– scanning tomography 249
– solid-state 567
– three-level 560
– types in medicine 567
– wavelengths 559
laser lithotripsy (LLT) 573
laser radiation 561
– fields of use 571
– penetration depth 564
laser system 562
– safety aspect 574
laser-induced thermotherapy (LITT) 573
– MRT control 574
lateral and palmar grasp 886
lateral geniculate nucleus (LGN) 824, 826
layered piezoelectric source 663
lead system 80
leakage current 42, 684, 686
leakage resistance flooring 45
lean tissue mass 614
learning
– algorithm category 428
– classification 1154
– environments 1193
learning management systems (LMS) 1193
learning system 1187
– application scenarios 1195
– architecture 1192
Leber's disease 259
left atrial pressure (LAP) 965
left ventricular assist device (LVAD) 724, 872
leg ergometry 889
length ratio 845
Lenz's law 505
leukemia
– acute lymphatic leukemia (ALL) 908
– acute myeloic leukaemia (AML) 908
– chronic myeloic leukemia (CML) 908
– myelodysplastic syndrome (MDS) 908
level
– monitor 628
– of safety 42
Levenshtein distance 1155
LGN stimulation 826
life cycle phase 57
life reminder 1231
lifeguard 1327
lifeshirt 1327, 1354
life-supporting medical equipment and systems 46
life-sustaining VAD 739
– blood pump 739
ligature scissors 1300
light
– detector 425
– emission device 1358
– guidance effect 323
– source 382, 425
– stimulus 255
light amplification by stimulated emission of radiation (LASER) 558
light spot hydrophone (LSHD) 651
light-emitting diode (LED) 383, 483, 562, 1134
lighting 1159, 1160
– model 1160
Lilly pneumotachometer 99
limb lead 80
limbic system 119
limit of neural damage 753
limits to technical progress 6
linear accelerator 711, 905, 906
– based plan verification 916
– output calibration 912
– technology 720
linear correlation coefficient 941
linear discriminant analyzer (LDA) 1010
linear probe 350, 357, 364
liposome 494
liquid- crystal display (LCD) 951
liquid-crystal display (LCD) 483
lithium ion dilution (LiDCO) 968
lithotripsy 655
– ureter stone 643
lithotripter
– acoustic field 649
– coupling 664
– system components 664
live wire segmentation 1151
liver
– cells 617
– perfusion 335
– replacement therapy 614
– support systems 615
live-wire 1150
local
– adaptive thresholding 1156
– contrast enhancement 306
– data network 307
local area network (LAN) 953, 1086
local field potentials (LFP) 1004
localization solution 1361
localized muscle blood flow 429
logical architecture model 1104
logistics 1103
long range alert 1225
longitudinal compression 644
longitudinal dose modulation 336

longitudinal nuclear magnetization 441
long-term
– archive 1201
– biocompatibility 809, 864
– circulatory support device 726
– ECG 84
– EEG 134, 136, 162
– MCS system 725
– monitoring 932, 947, 1327
– position injury 1270
– recording 163
– stability 1325
– storage 292
– storage and backup 1207
– storage media 1022, 1028
– test 64
– usage capability 1339
long-term ventilator 525, 526
– component 526
– task 526
long-wavelength infrared (LWIR) 370
look-up table (LUT) 299, 1143, 1159
loop-closing contact 505
loss of superconductivity 455
loss of vision 821
loudness 201
– binaural summation 210
– growth function 235
– intensity function 205
– scaling 205
low-dose rate (LDR) 709
low-energy x-ray 319
low-field MR system 442
low-flow system 581
low-flux membrane 606
low-level laser therapy (LLLT) 571
low-noise differential amplifier 76
low-power acquisition electronic 1350
low-power acquisition module 1359
low-power design 1339, 1353
low-power ECG analog 1359
low-power, real-time protocol (LPRT) 1369
low-voltage application 789
lumbal point 153
lumbal vertebra 153
luminescent layer 299
lung
– edema 110
– emphysema 334
– imaging 474
– measurement 335
– volume determination 111
– volume subdivision 102, 105
– window 1211

M

MμGuard 1327
– monitoring vest 1327
machine cleaning 1312
– procedure 17
– quality assurance 17
machine learning 428
machine loading 17
machine processing program 1313
machine-aided therapy 175
Mackay–Marg tonometry 247
macroelectrode 927
magnetic
– flux density 505
– interference field 46
– nanoparticle, core diameter 466
– stimulator 142
– transducer 927
magnetic field 478
– gradient 442
– sensor 484
– strength 504
magnetic material 461
– spatial distribution 464
magnetic particle imaging (MPI) 461, 462, 472
– applications 472
– cardiovascular 472
– cardiovascular system 472
– continuous mode 468
– cuboid 468
– detection limit 466
– FFP path 468
– field-of-view 467
– focus field strength 467
– fundamental frequency 463
– gastro-intestinal application 474
– history 462
– hybrid imaging 471
– iron concentration 466
– lymph node imaging 473
– magentic resonance imaging (MRI) hybrid system 471
– medical applications 472
– multistation mode 468
– neuro vascular 474
– oncology 473
– particle core size 465
– particle optimization 462, 471
– peripheral nerve stimulation (PNS) 467
– physiological compatibility 462
– regularization 470
– scanner design 462
– sentinel lymph node imaging 473
– signal 463
– signal generation 463
– signal strength 466
– single sided 470
– single sided coil assembly 470
– single-sided 462
– spatial coding 464
– spatial encoding 463
– spatial resolution 464, 465
– special geometry 470
– specific absorption rate (SAR) 467
– stroke 474
– system function 469
– temperature measurement 462
– temporal resolution 464
– therapy response monitoring 474
– tracer material 463
– tracer optimization 471
– volume-of-interest 467
magnetic particle spectrometer (MPS) 462, 471
magnetic particle spectroscopy (MPS) 470
magnetic resonance (MR) 440, 493, 503
– angiography 444, 1203
– compatible 516
– compatible manipulator 1276
– contrast agent 510
– electromagnetic field characteristic 506
– environment 506
– examination noise exposure 455
– guided biopsy 482
– guided endoscopic surgery 483
– guided intervention 498
– guided robotic 485, 486
– guided stent implantation 482
– image artifact 507, 516
– implant 494, 514
– interaction 504
– labeling 517
– moving components 505
– PET 457
– pulse sequence 507
– sequence 505
– system safe operation 506
– testing 503
– thermometry 478
magnetic resonance (MR) safety
– categories and icons 515

– icon 516
– labeling 514
– requirement 504
magnetic resonance (MR) scanner
– environment 505
– for safety testing 515
– induced voltage 506
– interventional application 504
magnetic resonance (MR) scanning 504
– risks and pitfalls 517
magnetic resonance (MR) signal 441
– fading 447
magnetic resonance guided focused ultrasound system (MRgFUS) 478, 490, 493
magnetic resonance imaging (MRI) 242, 311, 420, 439, 503, 708, 803, 902, 1139
magnetic resonance imaginge (MRI)
– basic principles 446
– component 441
– future 456
– guided cannula intervention 489
– guided laparoscopic surgery 483
– guided valve repair 496
– interaction with device 515
– resonance frequency 511
– safety 453
– scanner 487
– spectroscopy 452
– target precision 488
magnetic resonance spectroscopy (MRS) 452
magnetic resonance tomograph (MRT) 1279
magnetic resonance tomography (MRT) 292
magnetically induced
– displacement force 505
– torque 505
magnetization 461, 468, 478
– curve 465
– imaging 461
magnetoencephalography (MEG) 138, 1005
magnetooculography (MOG) 183
magnification 380
Mahalanobis distance 1149
main electronic component 1340
main static magnetic field 504
mains-operated ME equipment 37
maintenance 12, 37, 55, 595
– anesthesia device 595
– costs 45
maintenance of wakefulness test (MWT) 165
malpositions of the condyle 410
mammalian visual system 822
Manchester code 1136
mandatory minute volume 537
mandatory phase 537
mandatory ventilation 532
– mode 531
manipulation instrument 387
manipulator system 486
man–machine interface 60
– interaction concept 63
manual
– disinfection 15
– ventilation 588
manual breathing bag 580
manual muscle test (MMT) 275
manufacturer-independent communication 1207
manufacturing technology 787
map of the retina 248
marching cubes 1160
– algorithm 1159
market approval 866
marketing instrument 59
MARS 615, 616
Masimo rainbow 978
mask image 417
mask technology, advantage 298
mass flowmeter 100
mass screening 5
mass transfer area coefficient 610
master–slave
– coupling 1274
– manipulator system 1273
matching network 1351
material
– diamagnetic 505
– erosion 871
– ferromagnetic 505
– galvanized thread 1326
– implanted 804
– nonconducting 486
– paramagnetic 505
– property 346
– tracer 472
maternal
– ECG 1020
– heart rate 1024
matrix array 360
– ultrasound probe 362
matrix inversion 469
maxillary sinusitis 411
maximum intensity projection (MIP) 1203
maximum permissible exposure (MPE) 574, 576
maximum physical frame (MPS) 1370
maximum projection 1160
maximum-likelihood expectation maximization (MLEM) 328
Maxwell configuration 464
Mayo and Lexer scissors 1299, 1300
mean arterial pressure (APm, also MAP) 958
mean transit time (MTT) 335
measurement 933
– assessment technique 273
– body temperature 939
– control 445
– error 931
– flow 937
– force 940
– mass 939
– medical process 923
– of movement 937
– shape 939
– size 939
– statistical evaluation 940
– velocity 938
– volume 939
measuring chain
– invasive pressure measurement 963
– transmission behavior 931
measuring unit 590
mechanical
– cooler 371
– decoupling 1276
– dosing 582
– ventilation 28, 532
– wave 644
mechanical circulatory support system (MCSS) 724
– classification 727
– clinically proven 728
– for pediatric patients 736
mechanical index (Mi) 350
mechanics and control 837
mechanoelectric transducer 927
mechatronic design 838
MedBiquitous virtual patient (MVP) specification 1194
media access control (MAC) 1366
median filter 306
median scalp point 153
medical
– active implant 1136
– application of informatics 1119

– communication, exchange format 1123
– diagnosis 1175
– diagnostics 1086
– documentation 1108, 1123
– grade computer 953
– information system 1103, 1104
– institution safety 36
– linear accelerator 917
– monitors 386
– record 1096
– research 1158
– robotic 1274
– simulation 1174
– simulation system 1179
– training simulator 1172
– treatment method 566
– visualization on smart phone 1183
– VR trainer 1182
medical device 867, 1085
– functionality 867
– grade 67
– lifecycle phase 57
– operator ordinance 1092
– routine clinical operation 65
– safety 60
– standards 32
– technical requirements 62
– usability 60
– validation of usability before 66
medical device directive (MDD) 52, 861, 1088
medical electrical (ME) 214
– device 1337
– mains-operated 35
– specific requirement 1339
medical electrical (ME) equipment
– battery-operated 37
– first fault 40
– handheld 40
– insulation class 37
– requirements 40
medical image 1206
– application 1148
– automatic interpretation 1158
– standards 1233
medical image processing 1140, 1164
– fuzzy logic 1157
Medical Implant Communication Service (MICS) 803, 1133
medical IT system 43
– power socket 44
medical laser 558
– components 559, 562
– smart systems 577
medical network 1085
– availability 1086, 1092
– requirements 1086
medical technology 3
– electromagnetic compatibility 47
– hygiene 11
– international regulatory requirements 52
– regulatory guideline 49
– standard 52
– usability 60
medical textile 1321
– active 1322, 1330
– functional 1322
medicinal therapy 175
medium-dose rate (MDR) 709
Medizinproduktegesetz (MPG) 1088
Medtronic Interstim stimulator 887
megavoltage CT (MVCT) 905
melanin 565
membrane
– diffusion 599
– dynamics 1252
– hydrophone 651
– mucous 14
– polymer 602
– surface 607
– tympanic 194
memory-guided saccade 180
mental exertion 59
menu structure 63
mesh
– filter 625
– generation 1176
metabolic
– heat production 377
– marker 810
– monitoring 105
– parameter 926
– pathway 424
metal
– artifact 327, 412
– electrode 123
– fiber yarn 1326
– shadow 327
– shape-memory 482
– wire/textile yarn 1326
metalized polymer yarn 1333
metalloproteins in hemoglobin 425
metering valve 529
methemoglobin (MetHb) 425, 974, 1036
methemoglobin content (SpMet) 978
metrological
– check 989
– registration 182
Metzenbaum scissors 1300
mHealth alliance 1233
Michigan electrode array 808
micro flex interconnection (MFI) 795
micro total analytical system (μTAS) 1066
microbiological quality of water 31
microbolometer 371
microcatheter 805
microcirculatory blood perfusion 426
microclimate 1286
microcoil tracking 490
microcontroller (MC) 138
micro-CT 339, 340
microelectrode 794, 927
– array 808, 1005
– processing 794
microelectromechanical systems (MEMS) 801
microelectronic prosthesis 821
microfluidic platform 1069
micro-Hegemann 1304, 1305
microphotodiode array (MPDA) 825
microprocessor 1339
– controlled HF surgical unit 680
microscissors 1304
microstimulator 755
microstream technology 984
microstructured platinum 795
microsurgical scissors 1302
microsystem 801
– testing 814
microwave 680
midbrain implants (AMI) 199
middle ear 214
– dysfunction 204
– function 213
– infection 198
– reflex 214
middle latency auditory evoked potential (MAEP) 226
middle latency response audiometry (MLRA) 222
midsized molecule 606

mid-wavelength infrared (MWIR) 370
MIMiC collimator 905
miniature accelerator 712
minimally invasive
– instrument 387
– procedure 5
minimally invasive surgery (MIS) 380, 391, 399, 686, 1167
minimization of risks 49
minimum pulse voltage 774
minimum resolvable temperature (MRT) 371
Minnesota code 4
minute volume (MV) 537, 586
MIS simulator 400
mismatch negativity (MMN) 223
mixed apnea 162, 1031
mixed ventilation 533, 588
mixing bag method 113
mobile
– device 1029, 1206
– MR 482
– operating table 1261
– recording 164
– robotic platform 1282
– x-ray amplifier 1263
modality performed procedure step (MPPS) 1202
mode of activation 882
model
– 3-D 1175
– anatomical 1182
– animal 283
– balloon 1154
– contour 1152
model of axonal population (MRG model) 1255
modeling the biological system 838
modern medical care 1095
modern monitoring system 949
modern patient care 1086
modular
– concept of HLM system 635
– HLM 635
– monitor 949
modularity 1283
modulation field 463
modulation transfer function (MTF) 319
molar extinction coefficient 982
molecular genetics 821
molecular imaging 4, 395
monitor
– armband 115
– level 628
– modular 949
– network 947, 953
– unit 903
monitored biosignal 1353
monitoring 590, 937, 1087, 1224
– belt 1329
– brain function 1060
– cardiovascular 955
– cortical 1061
– device 36
– drug dosing 591
– gas 973
– health condition 1341
– hemodynamic 957
– intraoperative 140
– long-term 932, 947, 1327
– metabolic 105
– neural activity 1059
– of pelvic autonomic nerves 1048
– of respiration 161
– parameter 579
– peripheral 1063
– pulse 959
– sensor 591
– shirt 1327
– short-term 947
– SIDS 1035
– technique 762
– temperature 987
– the oxygen status, limit values 1037
– triplet 1024
– twins 1024
– unit 580, 590
monitoring system 163
– in vests or shirts 1326
monochromaticity 561
monolithic circuit 789
monophasic
– defibrillation 546
– pulse 757
– pulse form 547
monoplane angiography system 418, 419
monopolar
– circuit 685
– coagulation 682
monopole needle electrode 927
monoterminal circuit 686
Monte Carlo method 911
Moorfield's regression analysis 249, 250
moral boundary of technology 8
morbus Refsum 256
morphologic
– closing 1144, 1149
– dilatation 1144
– erosion 1144
– filtering 1148
– opening 1144
Mößbauer effect 197
most frequent use 1298
most important users 1307
most importantly used 1298
motion
– artifact 328, 338
– capability 848
– compensation 376
– dynamics 849
– sequence 937
motion mode (M-mode) 352
motivational exercising system 1224
motor
– disability 888
– exercise 835
– imagery 1008
– nerve 146
– nerve conduction 147
– recovery 889
motor evoked potential (MEP) 149, 155, 1053
motor unit action potential (MUAP) 145
motor unit potential (MUP) 145
motorized instrument 390
move during scan (MDS) 457
MPR in dentistry 408
MrBot 489
MRT-compatible system 1279
MTF restoration 306
mucosectomy 689
mucous membrane 14
multichannel
– afterloading device 713
– EEG recording 1134
– electrode array 888
– hearing aid 231
multidetector computer tomography (MDCT) 1209
– scanner 1216
multidimensional image formation 1164
multi-echo concept 451
multielectrode array (MEA) 826
multifiber catheter 564
multifilament wire 1333
multifocal ERG (mfERG) 258, 259
multifrequency filtering 307
multifrequency tympanometry (MFT) 214
multifunction 1318

multileaf collimator (MLC) 716, 903
multimodal
– analysis 850
– imaging 491
– registration 1145
– registration and fusion 1146
multimodality image-guided diagnosis and therapy setup (MITOS) 492
multiparameter monitoring 1353
multiplanar display 363, 479
multiplanar reconstruction (MPR) 407, 1203
multiplanar reformatting (MPR) 333
multiple imaging modality 420
multiple microphones 242
multiple pyramidal neurons 1060
multiple sclerosis (MS) 150, 151, 199
multiple sleep latency test (MSLT) 165
multiple sockets 40
multiplexer 789
multi-resource scheduling 1201
multislice CT (MSCT) 322, 339
multislice detector 320
– array 339
multivariate
– calibration 427
– statistical method 941
multiwavelength algorithm 427
muscle
– absolute force 276
– contraction 455
– dynamic property 274
– dynamics 278
– external testing 275
– potential 143
– saturation 433
– strength, manual testing 275
– work 936
muscle function 273
– assessment technique 273
– measurement 273
muscular system 120, 121
MWIR detector 371
myelinated nerve fiber 121
myelinated neuronal fiber 1249
– multicompartmental model 1251
myocardial arrest 626
myocardial fiber 545
myocardial infarction 356
myocarditis 725
myoglobin 425

N

n of *m* strategy 239
N_2 washout 105
– technical characteristics 111
N_2O dosing 583
N_2O uptake 581
N-acetyl-aspartate (NAA) 452
narrow fabric 1326
narrowband
– imaging 397
– noise 235
nasal continuous positive airway pressure (nCPAP) 160, 163
nasal flow 1035
nasopharynx 194
NASPE/BPEG generic pacemaker code (NBG) 775
National Acoustics Laboratories (NAL) 234
National Center for Image Guided Therapy (NCIGT) 493
natural atrial excitation 778
natural orifice translumenal endoscopic surgery (NOTES) 398
navigated procedure 412
navigation 412
– intraoperative 410
Nd:YAG laser 558, 569
– frequency-doubled 569
– vascular system 572
near infrared (NIR) 370, 396
– light spectrum 427
– overtone band 426
– signal, qualitative analysis 428
near infrared radiation (NIR) 424
near infrared spectroscopy (NIRS) 423, 424, 1005
– clinical application 431
– device 427
– signal quantification 428
near infrared spectroscopy (NIRS) monitoring
– breast 432
– kidney 432
near infrared spectroscopy (NIRS) technology 425
– future 436
nearest neighbor (NN) classifier 1155
neck
– coil 445
– flexion 279
– surgery 1053
need of amplification 235
needle
– crystal screen 302
– hydrophone 651
– thermometer 988
needle electrode 927, 1062
– concentric 927
needle image plate (NIP) 300
negative temperature coefficient (NTC) 1359
neighborhood 1158
neonatal
– EEG 125
– monitoring 1032
neonatology 1031, 1036
– adhesive sensors 1036
– monitoring the oxygen status 1037
nephrogenic systemic fibrosis (NSF) 453, 455
nephrotoxic
– contrast medium 498
– iodine 478
Nernst 679
– law 681
nerve conduction 936
nerve conduction velocity (NCV) 146
nerve cuff electrode 826
nerve fiber indicator (NFI) 252
nerve fiber polarimetry 251
nerve structure 1253
nervous system 119
– disorder 807
– function 807
nervous tissue, electrical stimulation process 1252
NESS Handmaster 885, 886
network
– neuronal 129
– shared 1090
network interface card 1085
networked health care system 1097
networking 1108
neural
– activity 1064
– amplifier 1063
– excitability 1250
– sensing technology 1065
– spike signal 1003
– stimulator 749, 1134
– tissue excitation 1248
neural interface 749
– implantable 807
neural network 944, 1155
– artificial 1155
neural pathways (NP) 1045

neural prosthesis 785, 789
– biocompatibility 788
– closed-loop 798
– encapsulation 789
– energy supply 789
– schematic 786
neuroelectrode array 802
neurogenesis 835
neuromodulation 887
neuromonitoring
– methods 1053
– signals 1046
neuromuscular disorder 280
neuromuscular electrical stimulation (NMES) 877
neuromuscular transition pulse 148
neuron's synapse 1064
neuronal 1155, 1248
– conduction 148, 1062
– network 129
– signal 792
neuronal cell 1338
– modeling 1249
neuronal response 1248
– biophysics model 1248
neurophysiology
– diagnostic procedure 121
– method 121
neuroplasticity 890
neuroprosthesis 877, 878, 1248
– for grasping 886
neuroprosthetic intervention 807
neuropsychological dysfunction 742
neuroradiology 418
neurorehabilitation 850, 877
neurostimulation 1046
neurosurgeon 907
neurosurgery 394, 1050, 1173, 1280
neurotransmitter 1064
neutral conductor 37
neutral electrode (NE) 685, 698
New York Heart Association (NYH) 725
newborn
– nursery 1027
– pathophysiological disorder 1031
– respiration 1031
newborn infant 1033
– electrodes 1032
– impedance pneumography 1033
– respiratory disorder 1033
– single-channel ECG 1032
NIBP measurement 961
NICU 1027
night blindness 261
nitinol
– guide wire 481
– susceptibility 494
– wire 510
nitrogen-washout test 111
nitrous oxide 580
– cutoff 583
no return to zero (NRZ) 1135
node of Ranvier 1251
noise
– acoustic 506
– band-pass 209
– Fastl 209
– narrowband 235
– voltage 466
noise equivalent differential temperature (NEDT) 371
noise equivalent temperature (NET) 371
noise equivalent temperature difference (NETD) 371
noise-induced hearing loss (NIHL) 199
noisy image 507
nonconducting material 486
noncontact application 694
noncontact tonometry 246
nondestructive testing (NDT) 323, 370
nondispersive infrared measurement or spectroscopy (NDIR) 982
noninvasive
– measurable biosignal 1353
– surgery 5
– technology equipment 25
noninvasive blood pressure (NIBP) 591, 959
– meter 94
noninvasive ventilation (NIV) 542
nonlinear
– hearing aid 233
– sound velocity 347
– stimulus block 217
– transfer function 931
– transmission 931
nonlinearity coefficient 648
nonmodular system 636
nonpenetrating electrode 809
nonphysiological effect 780
nonpolarizable electrode 79
nonpulsatile system 727, 732
nonretinal input 823
nonstick technology 695
nonsupervised method 428
nonvisual information 823
normal
– audiometry 151
– basic interval 776
– hearing 203
– hearing ear 219
normal transmission period (NTP) 1369
normalization point 914
normalized mean speed 845
Northben scissors 1301
notch noise 223
NTC sensor 1286
nuclear magnetic moment 439
nuclear magnetic resonance (NMR) 902
nuclear medicine 462
Nucleus 22 888
nuclide 716
nurse robot 1231
nursing information system 1105
Nyquist limit 354
nystagmography 179
– application 187

O

O_2 saturation 161, 1032
O_2 uptake 581
obesity epidemic 810
object identification 1155
object level 1141
objective audiometry 151
obstetric monitoring
– processing of data 1023
– system 1022
– system features 1022
– working area 1023
obstructive apnoea 1031
obstructive patients 109
obstructive sleep apnoea (OSA) 1229
obturator nerve stimulation 691
occupational
– disease 12
– health 15
– medicine 1326
OCT of the macula 247
ocular hypertension 257
oculomotor system 179
OECD total health spending 1218
oesophageal temperature measurement 989
Oguchi's disease 261
Ohm's law of acoustics 193
Oldenburg sentence test 209
onboard storage 803

one-dimensional (1-D) 1141
– treatment 903
one-shock strategy 550
ONH margin damage 252
online diagnosis 85
online hemodiafiltration (OL-HDF) 612
open ear gain (OEG) 235
open surgery 572, 687
open system interconnection (ISO/OSI) 1162
opening 1144
operating medical device 61
operating mode for emergencies 740
operating phase 1116
operating qualification 17
operating room (OR) 420, 1261
– hybrid 873
– management system 393
– positioning, development of decubitus 1270
operating room (OR) table 1263, 1271
– base 1264
– column 1267
– head operations 1262
– maintenance 1271
– manual cleaning 1271
– mobile x-ray amplifier 1263
– pad 1265
– top 1264
operating room (OR) table system 1263, 1264
– cleaning 1271
– hygiene 1271
– mobility 1265
– modular assembly 1264
– technology 1264
operating table 1261
– mobile 1261
– system 46
– top, segmented 1262
operating theater (OT) 948
– integrated 391, 392
– lighting 45
operation robot 1279
operational knowledge management 1099
operational logic 63
operational objectives 1097
operational qualification (OQ) 22
operational sociotechnical system 1110
operational therapy 175
operational transparency 1112
operator interface 67
ophthalmology 571
– measurement techniques 246
opioid 579
opposing beams 914
optic nerve 821, 823
– stimulation 826
– transduction velocity 253
optic nerve head (ONH) 247–249
optic neuritis 257
optical
– biopotential recording 1360
– electrode 1360
– endoscopy 1210
– fiber 563
– penetration depth 567
– perception 380
– sensor acquisition 1361
– transcutaneous transmission 1134
optical coherence tomography (OCT) 247, 397
optimal electrode design 1256
optimized image 483
optoelectronic tracking 484
optokinetic nystagmus (OKN) 181, 186
orbscan topography 265
order scenario 1201
organ at risk (OAR) 909, 912
organ motion 910
organ system 1342
organizational analysis 1113
organizational requirements 62
organizational transparency 1097
organizational unit 1112
Orssengo–Pye formula 246
orthodontics 411
orthopantomography 314
orthopedics 1280
oscillating potential 256
oscillometric measurement
– double-tube system 959
– functional principle 959
osmolality sensor 811
ossicular chain 196
osteomyelitis 411
otitis media with effusion (OME) 213
otoacoustic emission (OAE) 212, 215, 216
– evoked (EOAE) 215
– spontaneous (SOAE) 215
otolaryngology 1280
otosclerosis 204
otoscope 186
ototoxic drug 198
Otsu algorithm 1149
outer ear 194
outer hair cell (OHC) 195, 196, 215
out-patient surgery 1261
output stage circuit 759
– monitoring 760
output-controlled (AGC/o) 229
overfolding artifact 445
over-segmentation 1152
overtone band 426
overview display 1026
overview scan 330
oxygen
– deficiency signal 583
– hemoglobin dissociation 1039
– hemoglobin dissociation curve 975
– measurement 424
– monitoring 1040
– partial pressure 1036
– ratio controller 583
– transfer 424
– uptake 113, 114
oxygen concentration 527
– measurement 592
oxygen content (SpOC) 974, 978
oxygen desaturation index (ODI) 168
oxygen saturation (SpO_2) 167, 939, 1035
oxygen saturation (SpO_2) 591, 974
oxygen transport 974
– terms 975
oxygenated
– blood 1039
– tissue hemoglobin 436
oxygenation 526, 579
– monitoring 1039
oxygenator 623, 624
oxyhemoglobin (O_2Hb) 974

P

P300 evoked potentials 1007
pacemaker 691, 767, 768, 802, 957
– antenna 1133
– battery 769
– DDD 769, 776, 779
– DDDR 554
– demand 775
– electrode 769
– energy consumption 1134
– frequency-adaptive 780
– input sensitivity of a 771
– manual 769
– parameter 769

– programming device 768
– selection criteria 782
– single-chamber 775
– syndrome 776, 779
– system 768
– two-chamber 776
packaging material 22
paediatric urology 393
pain diagnostics 370
paired *t*-test 428
palliative radiotherapy 898
pancreatic duct stones lithotripsy 643
paper mode 951
paracorporeal ventricular assist device (PVAD) 730
paradoxical respiration 1034
parallel acquisition techniques (PAT) 445
parallel sectional image 361
parallel transmit (pTX) 445
paramagnetic
– coating 482
– material 505
– nanoparticle 482
– O_2 measuring cell 981
– sensor 592
parameter pitch 326
parameterization 1154
parametrium 1308
parasitic resonant frequency 510
parastep electrical stimulation 884
paroxysm change 136
partial
– pressure 974
– volume effect 1157
partial least-squares (PLS) 428
partially implantable hearing system 236
partial-volume artifact 327
particle core diameter 465
parts per million (ppm) 452
Pascal dynamic contour tonometer (DCT) 247
passive
– discharging 757
– electrode fixation 771
– finger flexion 886
– holding system 1275
– implant 454
– motion system 836
– navigation 412
– telemetry system 803
patch clamping 1060
patent ductus arteriosus (PDA) 432
path of infection 13
pathologic adaptation 206
pathologic fatigue 206
pathologic nystagmus 181
pathological loudness perception 206
patient
– active role 837
– administration 1026
– bedding for ESWL 667
– care 1163
– device interaction 869
– flow for obstetrics 1027
– flowsheet-based report 1028
– global status 850
– motion 847
– movement compensation 490
– noise limit 465
– performance 844, 850
– poststroke 844
– record 1026
– safety 61, 507, 1108, 1110, 1169
– specific CT data 1171
– transfer 481
– virtual 1172, 1191
patient data
– demographic 1101
– exporting 1029
– management 1023, 1026, 1103
patient monitoring 529, 543, 591, 947, 1090
– parameter 993
– sensor 591
– sensor principles 594
patient monitoring system
– classification 948
– fields of use 948
– patient transport 948
patient positioning 419, 1270, 1271
– high-frequency (HF) 1271
– improved 1262
pattern electroretinography (PERG) 257
pattern recognition 145
pattern VEP 253
patterned electrical stimulation 822
patterned phosphene 822
PCB layout 1365
*p*CO2 electrode 926
PDCA cycle 54
peak clipping (PC) 230
peak expiratory flow (PEF) 101
peak flowmeter 101
PediaFlow VAD 737
pediatric cardiology 304, 418
PediPump design 737
pelvic fracture treatment 486
pelvic surgery 1047, 1049, 1054
– neuromonitoring setup 1055
Penaz method 961
pencil beam 315
– algorithm 911
– principle 316
penetrating 807
– electrode array 807
people-orientated medical care 5
perception threshold (PT) 241, 772
percutaneous
– cable feed 1131
– electrode 880
– hand neuroprosthesis 880
– intervention 477
– radiation therapy 708
– therapy 709
percutaneous transluminal angioplasty (PTA) 419
percutaneous transluminal coronary angioplasty (PTCA) 715
performance
– index 850
– management 40
performance qualification (PQ) 17, 22
perfusion
– disorder 269
– pressure range 997
perfusion index (PI) 977
perimetry 269
– automatic static 269
– finger 269
periodic breathing 1031, 1033
periodic checks 918
periodic leg movements (PMS) 168
peripheral
– microcirculation 1036
– monitoring 1063
– signal processing 239
peripheral arterial disease (PAD) 432
peripheral nerve stimulation (PNS) 444, 453, 454, 467, 504
peripheral nervous system (PNS) 120, 1059
permanent artificial heart 745
permanent germ 13
permanent threshold shift (PTS) 199
permeability, hydraulic 606
perrotary nystagmus 186
personal computer (PC) 138, 1022, 1179, 1188
personal digital assistant (PDA) 1130

personal health systems (PHS) 1217, 1219
personal healthcare system 1366
personal protection 16
personal vaccine protection 14
personalized health system 1353
personnel
– medical physicist 917
– MTRA 917
– radiotherapist 917
PET-CT 457
PET-CT scanner 339
PFC syndrome 1037
phantom measurement 708, 902
pharmaceutical affairs law (PAL) 52
pharmaco EEG 134
pharmacokinetics 945
pharmacological intervention 821
phase encoding 446
phased array principle 358
Phong model 1160
Phong shading 1154, 1161
phosphene 822
phosphor plate 294
phosphorus spectroscopy 453
photoablation 564, 567
photoactivation 566
photochemical effect 566
photodecomposition effect 566
photodetector 1360
photodisruption 567
photodynamic diagnosis (PDD) 395
photodynamic therapy 558, 564
photodynamic therapy (PDT) 390, 566, 573
photoelectric interaction 321
photoelectric transducer 929
photoelectron 321
photoelectronystagmography (PENG) 182
photoinduction 566
photoluminescense sensor 1358
photon detector 371
photonic bandgap fiber 563
photonic textile 1326
photonic/light emitting textile 1322
photon-only machine 711
photoreceptor 823
photosensitization 566
photosensitizer 573
photostimulation 133
photostimulation approach 825
photothermal effect 566
phototransduction 255
phrenic nerve pacing 887
physical activity 106
physical compensator 716
physical disinfection 16
– advantages 20
physical exertion 59
physical interface 1080
physical medicine 835
physical pumping principle 727
physiological
– artifact 136, 933
– calculation 953
– parameter 947, 1357
– sensor 1355
– signal 1342
– state diagram 1249
PiCCO technology 967, 968
pick-and-go principle 951
picture archiving and communication system (PACS) 295, 308, 392, 487, 914, 1086, 1120, 1139, 1161, 1199
picture measurement 253
piezoceramic sensor 650
piezoelectric effect 344
piezoelectric transducer 928, 1358
piezoresistive pressure sensor 247, 814
piezoresistive sensor 593
pigment anomaly 571
PillCam 1171
pilot scan 329
pinch grip device 277
pixel
– based procedure 1148
– based segmentation 1154
– by-pixel subtraction 1145
– clustering 1150
– level 1141
– transform 1142
placebo effect 944
Placido plate imaging 265
planar
– sound wave 662
– x-ray 313
planar detector
– CT 304
– working principle 303
Planck's blackbody radiation law 372
plan–do–check–act cycle 57
planning
– adaptive 905
– of radiotherapy 335
– process 900
– RT treatment 900
– system 410
planning system 917
– 3-D 898
plate electrode 550
plausibility check 1026
pleth variability index (PVI) 978
plethysmogram 1035
plug socket, safety requirements 45
pneumatic
– actuation therapy 1277
– actuator 1279
– blood pump 724
pneumotachograph 937, 972
pneumotachography 1343
point operation 1143
point-to-point motion 845
point-to-point movement 846
polarizing electrode 123
polyacrylamide hydrogel (PAAM) 1361
polyacrylonitrile (PAN) 602, 603
polyamide (PA) 602, 603
polychromatic
– radiation 331
– x-ray 319
polygraphic signal 162
polyimide microelectrode 1061
polyimide printed circuit board (PCB) 1359
polymer-based flexible device 809
polyneuropathy 152
polysomnographic signal 166
polysomnography 163, 165
polysulfone (PSu) 602, 603
polyurethane (PUR) 604
polyvinyl difluoride (PVDF) 1361
porcine model 488
portable EEG 135
portable FES system 880
position
– long-term injury 1270
– sensor 360
– tracking 1012
positioning guide 412
positioning of implant 410
positive end-expiratory pressure (PEEP) 528, 972
positive pressure pulse 645
positron emission tomography (PET) 4, 311, 340, 472, 491, 493, 902, 1005, 1145, 1203
post delivery 1027
post interventional MR imaging 510
postdefibrillation arrhythmia 551
postmarket assessment 859
postoperative complication 572

postoperative dysfunctions 1049
postoperative therapy 1043
postoperative wound infection 28
postprocessing application 1203
postprocessing methods 1203
postprocessing software 1216
poststroke patient 844
post-ventricular atrial refractory period (PVARP) 778
potential
– bioelectric 926, 935
– cortical distribution 943
– equalization 42
– galvanic 77
potentiometric
– configuration 1066
– sensor 927
potting material 605
Potts-Smith scissors 1306
power amplifier 1350
power consumption 1341
power density 566
power Doppler 355
power Doppler technique 355
power for coagulation 682
power link rejection 1350
power mattress 1352
power spectral density 1008
power supply (PS) 44
– safe operation 44
power wireless device 1352
preauricular point 125
precise dosage control 559
precision radiotherapy 906
preclinical structural evaluation 863
preclinical study, goals 865
precordial lead 84
predicted values 103
pre-fetching 1201
pregnancy 1027
pre-interventional dataset 412
premature infant 1032, 1037
– pulse oximetry 1036
preoperative data 412
preoperative planning system 1173
preoperative tomographic diagnostics 410
preparation
– for cleaning 1312
– of endoscope, cleaning 388
– of instrument 388
preplanning 900
prepolarized MRI 471
pre-pregnancy 1027
prescription of gain and output (POGO) 234
presentation of the CTG data 1023
presentation system 1190
pressure
– acoustic 350
– generator 527
– monitor 628
– profile 609
– transducer 964
– ulcer 890
pressure control, biphasic positive airway pressure (PC-BIPAP) 533
pressure control-airway pressure release ventilation (PC-APRV) 539
pressure controlled (PC) 531
pressure measurement 593, 937
– intrauterine 1021
– piezoresistive sensor 593
pressure pulse
– duration 652
– effect 658
– medicin 641
– on tissue 658
– parameter 655
– rise time 652
– thermal effects 649
pressure pulse source 660
– electro hydraulic 661
– field parameters 656
– focal parameters 656
– parameters 657
– piezoelectric 663
– spark-gap source 660
pressure sensor 802, 963, 1358
– external 998
– implantable 805
pressure support (PS) 534, 588, 589
– level 534
pressure-controlled mixed ventilation 538
pressure-controlled ventilation (PCV) 533, 589
pressure-cycled mode 531
pressure-volume loop 972
pre-use check 595
prevention costs 50
preventive measure 55
primary acoustic energy 654
primary prevention 1043
priming volume 604
principal component regression (PCR) 428
principle of windowing 332
print functions 1202
probe 350
– atrial 770
problem-based learning 1189
process control 1010
process oriented record 1101
processing on visual information 823
process-orientated procedure 1112
product
– constraint 859
– development 856
– quality 57
– realization 50
professional standards in healthcare 1169
programmable electrical medical system (PEMS) 948, 1085
programmed teaching (PT) 1187
project planning 1111
projection integral 324
projection profile 324
projection space 327
projects in telemedicine 1121
prometh adsorber 616
Prometheus system 615, 616
proper storage 1316
proportional assist ventilation (PAV) 534
proportional pressure support (PPS) 534
proportional–integral–derivative (PID) 839
proprietary module 1350
proprietary network 1090
proprioceptive sensor 792
prospects of life 8
prostate
– biopsy 489
– carcinoma 715
– intervention 489
prostatectomy 1281
prosthesis
– extremity 788
– microelectronic 821
prosthesis-pad 1324
prosthetic cardiac valve, specification 868
prosthetic heart valve 871
– ISO guidelines 871
prosthetic robotics 890
prosthodontics 409
protection
– class 37
– objective 37
– of employees 12
– of patients 12, 37
protective clothing 16
protective device 35

protective earth (PE) 37, 42
protective measure 43
protective packaging 1315
protein molecule 478
protocol 308
proton density (PD) 446
proton resonance frequency (PRF) 478
proton spectroscopy 453
prototype device 860
prototype testing 864
provocation method 133
pseudarthrosis 643
pseudo coloring 371, 1143
psychiatric disorder 5
psychoacoustic methods 212
psychological disorder 743
psychological shock reaction 742
psychosomatic syndrome 742
psychotherapeutic therapy 175
public healthcare expenditure 1219
pulley–string system 842
pulmonary
– artery catheter 967
– blood flow 432
– dysfunction 726
– fibrosis 110
– gas distribution 104
– gas transport 101
– pressure 805
pulmonary artery pressure (PAP) 958, 963, 965
pulmonary capillary wedge pressure (PCWP) 965
pulmonary function testing (PF) 95, 971
– cooperation-free 104
– parameters 105
– quality 102
pulmonary vascular resistance (PVR) 969
pulsatile and nonpulsatile absorption 977
pulsatile flow control 628
pulsatile IR signal 978
pulsatile system 727
pulse
– CO-oximetry 978
– generator 772
– monitoring 959
– oximeter 939, 1289
– oximetry 971, 976, 977, 1035, 1037, 1039
– stimulation 546
– working capacity 91
pulse intensity integral (PII) 652
pulse transit time (PTT) 171
pulse wave velocity (PWV) 366
pulse width (PW) 1255
pulsed wave (PW) 353
pulsed-dose rate (PDR) 709
pulseless support 739
pulses per second (pps) 238
pupil localization and diameter 265
pure cut 688
pure tone 193
– audiogram 207
– audiometer 202
PVDF membrane 650
– contact-free 650
pyroelectric sensor 1358

Q

QRS
– class memory 87
– complex 545
– morphology 86
– trigger signal 77
– triggering 547
qualitative imaging 373
quality 50, 1283
– assurance 17
– assurance costs 50
– assurance in medicine 1169
– audit 55
– beam 563
– characteristics 1316
– control 917
– factor 65
– illumination 382
– of cut 1317
– of life 743, 1004
– of product 49
– policy 53
quality management (QM)
– concept 49
– element, ISO 13485 53
– officer 53
– organization 55
quality management system (QMS) 49
– element 54
– implementation 56
– market requirements 50
– objective 50
– organization 54
quality of service (QoS) 1345, 1366
– considerations, e-Health 1367
– solutions for healthcare system 1368
quantitative imaging 461
quantum detection 406
quantum efficiency 319, 321, 371
quantum-well infrared photodetector (QWIP) 371
quartz microbalance 939
quasi-tripole (QT) 1062
query by example (QBE) 1162
question of meaning 9

R

R peak 171
radial flow MCSS 735
radiation
– biophysical effect 564
– constant 372
– dose 292
– generation of 710
– physics 705
– source 706
radiation therapy (RT) 707, 719, 898
– backward process 716
– high-precision 718
– historical development 704
– percutaneous 708
radio communication 1133
radioactive
– decay 706
– implant 714
– isotope 716
– seed 507
– stent 716
radiochemotherapy (RCTX) 898
radiofrequency (RF) 440, 483, 494, 1342
– artifact 495, 507
– induced heating 505, 516
– interaction 454
– power 1350
– surgery 46, 677
– system 444
– transmission 1133
– wireless neural recording 813
radiographic slice 313
Radiological Society of North America (RSNA) 324, 1200
radiological window level setting 1211
radiological workflow 1200
radiology 312, 1199
– department, order 1201
– hybrid OR 420
– interventional 419
radiology information system (RIS) 308

radiology information systems (RIS) 1199
radiolucency 490
radionuclide 711
– therapy 708, 709
radiosurgery 677
radiotherapy
– intraoperative 902
– licensure 917
– treatment, indication 899
radon space 324
railway nystagmus 181
random saccade 180
random stimulation train 882
range of motion (ROM) 275
rapid acquisition with relaxation enhancement (RARE) 449
RapidArc 904
rarefaction pressure 654
– spatial distribution 654
Raschel technique 1334
raw data level 1141
raw EEG 992, 993
– autoregressive parameter 992
ray casting 1161, 1214
ray-tracing 1161
– aberrometer 262
– sensor 262
reading workflow 1202
reading workplace 419
readout process 300
real ear to coupler difference (RECD) 234
real environment 1169
realistic operator interface 69
reality–virtuality continuum 1169
real-time 3-D interaction 1178
real-time imaging 462
real-time MRI guidance 488
real-time tracking 490
reason for using ventilator 526
rebreathing system 581
receiver coil 463
receiver electronics, noise 463
recharge cycle 1134
Rechtschaffen and Kales 169
reconstructed imaging data 1170
reconstruction 468
– 3-D 304
– axis 408
– of anatomical structure 1176
– pipeline 1176
record and verification system 914
recording
– coil 465
– electrode 77, 122, 151, 152
– program 127
– system 78
recording of bioelectrical potential 122
– surface electrode 928
recovery phase 92
recurrent laryngeal nerve (RLN) 1048, 1051
red alarm 952
red blood cell (RBC) 473
– labeling 473
redox polymer modified ELITEX 1331
reduced lead 83
reduction in blood perfusion 374
redundant array of independent discs (RAID) 294
redundant data processing 1121
refelected pressure 646
reference electrode 152, 237
reference FET (REFET) 1068
reference frequency response 232
reference information model (RIM) 1107
reference point 914
refertilization 572
reflected sound energy 646
reflection coefficient 345
reflection of ultrasound 933
reflectory excitation process 148
reform medical studies 1189
refraction of sound waves 646
refractometry 252
– automatic 252
region growing 1151, 1157
region level 1141
region merging 1152
region of interest (ROI) 372, 1142
regional IT virtualization 1207
regional lymph nodes 910
region-based segmentation 1151
registration 412, 1145
– 3-D 1170
– indirect 931
regression line 941
regularization 470
regulatory compliance 802
regulatory pathway 804
rehabilitation 105, 242, 749, 822, 835, 877, 1247
– and therapeutic robotics 835
– robot design 837, 839
– robotic machine 850
– robotics 890
rehabilitation therapy
– collaboration 839
– robot-aided 839
ReinHeart 745
Reissner's membrane 196
relative data 912
relative dose distribution 918
relaxation time 448, 478
release and storage 1316
reliability 1086
reliable design 5
reliable MR safety 518
reliable technology 37
REM sleep 168
remittent fever 987
remote consultation 1029
remote data transmission 1129
remote handling system 1273
remote presence 1282
remote rehabilitation 839
removable table top 1263
René Dutrochet 598
repetition frequency pulse (PRF) 353
repetition time (TR) 447
replacement valve 872
report creation 1204
reprocessing 12
– of intravascular catheters 29
reproducibility 924
– of the patient setup 900
reproducible recording 125
reproductive medicine 7
research 1163
– protocol 865
resectoscope for tissue ablation 393
reservoir bag 580
reservoir capacitor 772
residual current protective device (RCD) 43
residual risk 36
residual vision 830
residual volume (RV) 101
resistance of germs 21
resistive motion 848, 849
resolution 371, 380
resonance frequency 840
resonance tomographs (MRT) 129
resonant
– marker 485
– stent 495
– stent structure 510
respiration 135, 1032
– artificial 12, 1032
– flow 929, 937

respiration rate (RR) 971, 1034, 1343
– standard values 1033
respiratory
– amplitude 168
– deficit 1033
– depression 526
– distress 1031
– effort 167
– flow 163
– flow velocity 936
– mechanics 971
– minute volume 581
– monitoring 1329
– monitoring belt 1329
– movement 484
– pattern 1033, 1034
– pause 1031
– rate setting RR 533
– ratio 113
– sensor 780
– volume 936
– work 526
respiratory disturbance index (RDI) 168
respiratory gas 973, 981
– absorption spectra 982
respiratory guided radiotherapy (RGRT) 902
restenosis 715
resting potential 121
resuscitation quality 549
retina 821
– artificial 829
– cross section 823
retinal
– ganglion cell 828
– implant 787
– message 823
– stimulation 824
– thickness 248
– visual prosthesis 825
retinis pigmentosa 261
retinopathy 257
retinoscopy 263
retinotopic gradient 827
retransmission 1370
retrieval of data 1028
retrocochlear hearing disorder 199
reusable 699
reusable sterilization container, instrument tray 1314
revenue transparency 1097
reverse osmosis 30
reverse recruitment 882
reversible cell damage 683
rheobase 774
rhizotomy 887
right ventricular insufficiency 741
rigid endoscopes 383
Rinne test 202
RIS/PACS configuration 1199
risk
– analysis 1089
– based infection control measures 13
– estimation 861
– management 40, 49, 1110
– management DIN EN ISO 14971 1088
Robert Koch Institute (RKI) 13, 1311
ROBODOC 1278
robot 1183, 1231
– aided motor therapy 844
– assisted surgery 1183
– for biopsy 1277
– performance 838
– psychological inhibition 1281
– surgical 1183, 1276
– system 1273
robotic
– arm 486
– assistance system 486
– assisted MR 487
– control 489
– technology 419, 835
– therapy, goal-directed 848
rod lens
– endoscope 483
– telescope 384
roller pump 624
room-based
– imaging 718
– laser system 916
root mean square (RMS) 218, 262
Rostock cornea module 252
rotary blood pump 727
rotating anode disk 320
rotation, scale, translation (RST) 1163
rotational speed 88
routine activation method 133
routine EEG 134
RPE atrophy 267
rubber electrode 1361
ruby laser 558, 569
rules
– and argon beamer 698
– in endoscopy 698

S

saccade 180, 185
saccadic eye movement 943
sacro-lumbar anterior root stimulator implant (SLARSI) 754
safe operation 594
– connection prerequisites 594
– interaction 506
– power supply 44
safe patient positioning 1268
– standard positions 1268
– various surgeries 1268
safe usability 60
safety 59, 749, 754
– class 36
– glass 576
– issue 453
– margin 902
– measure 35, 93, 576
– of drug therapy 1109
– of escape 37
– of ME equipment 37
– requirement 504, 576
– standard 36
safety engineering
– direct 36
– indicative 36
– indirect 36
safety testing
– MR scanner 515
– standard 515
salivary stone lithotripsy 643
saltatory conduction 1249
sampling artifact 328
sarcoidosis 110
sarcomere 274
sarcoplasmic reticulum 882
saturated steam 1315
saturation magnetization 465
saturation of tissue oxygen (StO_2) 424
scala tympani 195
scaling factor 426
scalp point 153
scan planning 329
scanning electron microscopy (SEM) 293, 796
scanning frequency 353
scanning laser ophthalmoscope 251
scanogram 329
scatter artifact 328
scattering applicator 564
scene level 1141
scheduling scheme 1201

Scheimpflug examination 267
Scheiner's principle 253
Schiötz's impression tonometry 246
scientific exchange 1129
scientific properness 9
scintillation camera 930
scintillation detector 316, 708
scintillator detector 321
scissors
– coronary 1304
– Favaloro 1306
– gynecological 1306, 1308
– hard-metal 1294
– history 1292
– Iris 1301
– ligature 1300
– main parts 1293
– manufacturing 1295
– Mayo and Lexer 1299, 1300
– Metzenbaum 1300
– microsurgical 1302
– Northben 1301
– Potts-Smith 1306
– Sims–Siebold 1309
– Spencer 1301
– Spencer ligature 1302
– spring bow 1292
– surgical 1291, 1295
– titanium 1294, 1296
– toothed 1303
– types 1297
– wire cutting 1302
sclerosis 410
SCORM (sharable courseware object reference model) 1194
scout view 329
screen configuration 951
screen printing 795
screw electrode 771
search functions 1202
search-coil system 183
second harmonic imaging 347
second stage of labor 1027
secondary digitization 292
secondary layer membrane 608
secondary reconstruction 331
second-generation CT 316
second-order gradiometer 937
sector phased array 362
sector probe 350, 358
seed implantation 714
see-saw breathing 1031
segmentation 412, 1148, 1164
– automatic 1154
– balloon 1158
– of organs 1177
segmented
– dataset 412
– operating table top 1262
selection criteria 782
selection field 464
– gradient 465
selection gradient (GS) 447
selective stimulation 186
self-administration system 1121
self-expanding stent 510
self-standing neuron 1248
semiautomated defibrillator 548
send coil 463
sensing
– electronic 1341
– function 771
– methodology 1357
– technique 1068
sensitive patient data 1086
sensitivity 293
– coil 466
– of sensor 1068
sensitivity encoding (SENSE) 445
sensor 1341, 1357
– biochemical 810
– capacitive 805
– conductometric 927
– dynamic properties 931
– elongation 1329
– embedded 1355
– exteroceptive 792
– finger 976
– flow 97, 98
– force 841
– Hartmann–Shack 262
– hot-wire 593
– implantable 805, 806
– infrared 1358
– ionic 1066
– network 1349
– paramagnetic 592
– piezoceramic 650
– position 360
– potentiometric 927
– principles 594
– pyroelectric 1358
– steel ball 650
– subcutaneous glucose 802
– technology 1067
– temperature 987
– UV radiation 1358
– zero-point stability 998
sensorimotor coordination system 836
sensorimotor rhythm 1007
sensorineural hearing disorder 198
sensory
– conduction disturbance 145
– feedback 1183
– hair cell 195
– neural dysfunction 785
sentence test 209
sentinel lymph node imaging 473
sequential stimulation 883
series element (SE) 278
series-blocking-capacitor 763
service-oriented architecture (SOA) 1124
service-oriented architecture (SOA) 1106
sevoflurane 583
shading 1159, 1161
shaft electrode 790
shape reconstruction 1158
shape-from-contour 1158
shape-from-shading 1158
shape-from-stereo 1158
shape-from-texture 1158
shape-memory metal 482
shared network 1090
shiny reflection 1161
shock
– configuration 546
– front 645
– tube 662
shock current 42
– protection 42, 43
shockwave
– biotechnological application 643
– effect of tumors 643
– generation of 660
– measurement technology 649
– medicin 641
– pulse repetition rate 649
– sensor 649
shockwave lithotripsy
– influences 667
– number of pulses 668
short increment sensitivity index (SISI) 205
short therapy time 713
short-range telemetric equipment 769
short-tau inversion recovery (STIR) 452
short-term
– monitoring 947
– storage 1207
shortwave 680
short-wavelength infrared (SWIR) 370
shutter glasses 1012

sick
– newborn infant 1037
– sinus syndrome 555
SIDS monitoring 1035
sieve electrode 791
sieving coefficient (SC) 610
sigmoid psychometric function 208
signal
– acquisition 1347
– adaptation 790
– amplitude 1342
– biochemical 934
– biothermal 940
– classification 790
– evaluation 992
– generation 469
– IQ 978
– recording 122, 166
– strength 466
– transmission 790
signal analysis 135, 940
– frequency domain 943
signal extraction technology (SET) 977
signal processing 166, 187
– artifact 128
– clinical neurophysiology 128
signal-to-noise ratio (SNR) 208, 294, 327, 374, 441, 507
signature 1155, 1156
silicon detector 427
silver bromide 293
silver-coated polyamide fiber 1325
simple sampling recorder 85
Sims–Siebold scissors 1309
simulation 916, 1210
– environment 65, 1170
– haptic 1181
– of surgical instruments 1178
– testbed 1370
simulator 914, 917, 1282
single
– cell recording 1005
– failure safety 773
– port surgery 399
single fiber electromyography (EMG) 144
single photon emission computed tomography (SPECT) 472, 1145
single-bed display 1026
single-breath method 105, 109
single-breath test 110
– clinical value 110
single-chamber
– ICD 553
– pacemaker 775
single-exposure dual-detector technique 307
single-fiber electrode 927
single-sided MPI scanner 470
single-use adhesive electrode 90
sinogram of attenuation 324
sinusoidal oscillation 348
sitting position 1268
six-DoF tracking system 1181
size of artifact 510
skeletal muscle 273
skeleton 1150
skeletonization 1144, 1157
skin 14
– cancer 375
– emissivity 373
– marker 900
– surface cooling 374
– surface measurements 989
– temperature 369
– temperature measurement 373
– temperature regulation 1286, 1289
– tumors 571
– vasoconstriction 374
skin blood flow (SkBF) 431, 433
– brain 431
– forearm 431
– thenar 431
skin tissue damage 1269
– locations particularly at risk 1269
skull base surgery 1047, 1049
sleep
– apnea 162
– apnea detection 1229
– architecture 169
– EEG 135
– laboratory 159, 160
– phase classification 168
sleep diagnostic 160
– medical significance 173
– procedure 165
sleep disorder 159
– diagnosis 160
– international classification 174
– multistage 163
– signal recording 166
– therapy 175
slice interpolation 326
slice selection 446
slice-thickness focusing 358
sliding-window approach 904
sliding-window technique 716
slip-ring technology 317, 326
slot-scan method 302
slow AEP (SAEP) 222
slow cortical potentials (SCP) 1007
slow spirometry 96
slow vital capacity 102
smart
– home 1233
– homes project 1233
– textile 1228, 1322, 1325, 1345
SmartShirt 1327
smooth pursuit movement 185
snake algorithm 1152
snoring microphone 167
Sobel operator 1143, 1148
social lifestyle 5
soda lime 585
soft
– tissue window 331
– x-ray 327
solid-state
– detector 322, 916
– laser 567
solo surgery 1275
solvent drag 600, 602, 611
somatosensory evoked potential (SEP) 140, 149, 152, 1053
sone 201
sonoporation 494
sorbitol/mannitol 690
Sorin tilting disc valve 729
sound
– conduction 345
– frequency 196
– particle velocity 192
– propagation velocity 192
– velocity, nonlinear 347
sound field
– energy parameter 653
– geometry 356
– parameters 657
sound pressure level (SPL) 193
sound wave 192
– diffraction of 647
– generation 344
– impedance of 193
– penetration depth 346
– transmission 346
– travel speed in tissue 345
sound wave at interfaces 645
– transmission and reflection 645
source recording 124
soutage technology 1323
space clamp measurement 1249
spacer fabric 1324, 1333
– 3-D 1323
spacer yarns 605
spark intensity 689
spark-gap source 660

spatial
– (digital) distribution of physical density 1210
– distribution 464
– domain 1144
– encoding 446
– gradient of magnetic field 505
– hearing 193
– image 420
– information 410, 446
– mapping 345
– representation 1210
– resolution 292, 317, 371, 466, 479, 1209
spatially resolved refractometer 262
specific absorption rate (SAR) 467, 505, 518
specific MR environment 505
specific thermal capacity 682
speckle reduction 365
spectral analysis 992
spectral entropy 992
spectral radiant exitance 372
spectral response 372
speech audiometry 206
– reliability 209
speech in noise test (SPIN) 209
speech perception threshold (SRT) 206
speech processor 888
speech recognition system, report creation 1205
speech signal 206
speech test 209
speech transformation 207
speech understanding 231
speed index 845
Spencer
– ligature scissors 1302
– scissors 1301
spherical sound source 663
sphygmomanometer 959, 1341
spike-wave complex 130
spiking neuron model 1250
spin 478
spin echo (SE) 478, 510
spinal cord injury (SCI) 749, 877
spinal reflex circuit 119
spine surgery 394, 1047, 1049, 1052
spin-echo sequence 447, 448
– multi-echo 449
spiral CT 317, 1161
– technology 326
spirometry 95, 96
– calibration 97
– forced 96, 101, 102
– hygiene 97
– measuring range 97
spirometry sensor 98
spoiled gradient recalled acquired steady state (SPGR) 450
spontaneous activity 143
spontaneous breathing (SB) 534, 536
– activity 531
– support 588
spontaneous control 537
spontaneous emission 559
spontaneous eye movement 179
spontaneous minute volume 537
spontaneous proportional pressure support (SPN-PPS) 535
spontaneous respiration 983
spontaneous-continuous positive airway pressure (SPN-CPAP) 536
sports medical application 1326
sports medicine 1326
spray coagulation 683, 691, 693
spring bow scissors 1292
sputtered platinum surface 796
ST measuring program 90
ST segment analysis 1024
ST waveform analysis (STAN) 1020
staff shortage 1232
stages of the QMS 52
stainless-steel medical alloy 494
stand-alone monitor 949
standard deviation 941
standard deviation (SD) 1033
standard EEG recording, recording method 134
standard flash (SF) 255
standard lead 80
standard position 1268
standard scissors with ceramic coating 1296
standard scissors with TC 1296
standard temperature and pressure, dry (STPD) 973
standard titanium scissors 1296
standardization 1164
– commission for ergometry 91
– of ergometry application 90
standardized investigation program 92
standing wave 193
stapedius reflex 214
star shot at different gantry angles 918
Stargardt's disease 261
static friction 841
static magnetic field 505
stationary incubator 1287
stationary operating table column 1263
statistical classification 1155
statistical KNN classifier 1156
statutory health insurance competition strengthening act (GKV-WSG) 1121
statutory health insurance modernization act (GKV-GMG) 1121
steady-state PERG 257
steady-state visually evoked potentials (SSVEP) 1007
steam
– disinfection 19
– flow process 19
– quality 1315
– sterilization 21, 1314
steel ball sensor 650
steepening sound wave 647
steered compound imaging (SCI) 364
stem cell labeling 474
stenosis treatment 420
stent 715
– coronary 334
– graft 495
– implantation 419
– radioactive 716
– resonant 495
– self-expanding 510
stereocystoscope 385
stereoendoscope 385
stereotactic body radiation therapy (SBRT) 907
stereotactic irradiation 902, 907
stereotactic radiosurgery (SRS) 907
stereotactic radiosurgery/radiotherapy (SRS/SRT) 718
sterility 14
sterilization 389
– ETO 605
– gas 795
– hot air 22
– method 21
– overpressure process 23
– physicochemical proces 22
– process 21
– requirements 24
– steam 21, 1314
– technology standards 32
– vacuum process 23

Subject Index

stimulated muscle force
– animal model 283
– assessment 279
stimulation 179
– external 750
– extracellular electrical 1250
– fixed-rate ventricular 775
– implant for deep brain 786
– in the atrium 778
– interface 749
– LGN 826
– mode 758, 759
– monitor 455
– parameter 882, 1047, 1255
– parameter modulation 882
– pulse 757
– selective 186
– threshold 774
– threshold curve 774
– transient intracardial 768
stimulation system 772
– equivalent circuit 772
– implantable 750
stimulator
– acoustic 142
– Finetech–Brindley 887
– Grass 798
– implantable 749, 760
stimulator output stage 756
– active control 762
– charge imbalance 762
– monitoring 762
stimulus
– atrial 778
– generation 187
stimulus frequency otoacoustic emissions (SFOAE) 215
stone comminution 657
stone material data 644
storage
– device 295
– long-term 292
strategic objectives 1097
stray current 46
Strehl number 262
stress intensity 90
stress limit 91
striate cortex 823
stripe projection 1177
stroke
– patient 334
– rehabilitation 1004
stroke volume (SV) 957
strong magnetic field 453
– attraction and torque 453
structural element 1144
structure
– anatomical 1176
– of domestic care system 1130
structured risk assessment 1110
subcutaneous
– defibrillation electrode 554
– glucose sensor 802
subjective loudness 200
subretinal implant 825
sudden deafness 198
sudden infant death syndrome (SIDS) 162, 1034
sunlight 383
super blend cut 688
superconductive quantum interference device (SQUID) 138, 937
superparamagnetic iron oxide (SPIO) 461
superposition-algorithm 911
super-pulse mode 570
supervised learning 1006
supervised method 428
supervised training 1154
supine position 1268
supporting ventilation mode 531
supra-aural headphone 204
supraventricular extrasystole 551, 1032
surface
– 3-D 363, 1210, 1211
– based rendering 1154, 1160
– based visualization 1160
– charge 928
– coil 444
– compatibility 788
– display 363
– ECG 1326
– electrode 807, 878, 936
– microscope 375
– rendering 333, 1160
– temperature 940
surface electromyography (EMG) 135, 1326
surgery
– keyhole 380
– laryngeal 394
– neck 1053
– noninvasive 5
– open 572, 687
– solo 1275
– spine 394, 1047, 1049, 1052
– thyroid 1047, 1048
– traditional 5
surgical
– dentistry 410
– field 1262
– instrument 1292
– intervention 420
– microscope 394
– operation, exact 1280
– outcome 1170
– planning 411
– robot 1183, 1276
– standard scissors 1297
– steel 482
surgical scissors 1291, 1295
– handling and care 1310
– manufacture 1294
surgically implanted device 867
SurgiScope 1279
surrounding tissue conductivity 506
survival test 64
susceptibility artifact 507, 510
susceptibility-weighted imaging (SWI) 451
Swan–Ganz catheter 965
switching device 44
switchover behavior 531
symbol chain 1155
symbolic image description 1158
sympathetic tone 119
SynCardia CardioWest TAH 732
synchronized defibrillation 549
synchronized intermittent mandatory ventilation (SIMV) 590
synchronous operating mode 547
syntactic classification 1155
synthetic liver function 616
system
– adaptive 1005
– integration 392
– matrix 469
systemic arterial pressure (SAP) 995
systemic heparinization 632
systemic vascular resistance (SVR) 969
system-on-a-chip (SOC) 750

T

table column 1264
table top
– conductive 46
– removable 1263
tachycardia 1032
tactile feedback 1276
tagged image file format (TIFF) 1161
TAHROT study 745
TandemHeart 733

target concept 1113
target volume 909, 910
targeted hygiene measure 12
targeting precision 488
teaching 1163
teaching system
– application scenarios 1195
– architecture 1192
technical
– artifact 136, 934
– component 579
– development in ECG 75
– feature 1295
– imperative 8
– safety 35
technicians' responsibility 8
technology
– bounds 8
– feasibility and controllability 7
– moral boundary of 8
technophobia 6
teleassistance 1224
téléassistance en soins de plaies (TASP) 1226
telecommunication 1029
– medical application 1119
teleconsultation 1124, 1232
telemanipulation system 400, 1273
telemanipulator system 1276
– minimally invasive intervention 1274
– psychological inhibition 1281
telematics 5, 1119, 1129
– infrastructure 1124
telemedical concept 1119
telemedicine 387, 392, 1119, 1123, 1140
– control of active implants 1129
– enabled operating theater 1130
– guide, Germany 1119
– home environment 1130
– modern operating theater system 1129
– projects in Germany 1124
telemetric
– approach 1131
– transmission 933
telemetry 803, 932
– device 949
– passive system 803
telenetwork 1281
telepathology 1124
telepresence 1283
teleradiological service 1124
teleradiology 1123, 1206
telerehabilitation 836
telesurgery 1274, 1283
– assistance 1233
– simulation system 1282
teletherapy 716, 898
temperature 1286
– calibration 371
– conduction 567
– monitor 628
– monitoring 987
– signature 369, 374
temperature sensor 987
– overview 988
template 1143
temporal behavior 233
temporal difference limen 201
temporal dose modulation 336
temporal subtraction technique 307
temporary germ 13
temporary threshold shift (TTS) 199, 201
temporomandibular disease 411
tenodesis grasp 886
Tesla 679
test instrument 18
test of Weber 202
test planning 944
testing 1314
– animal 864
– disinfection 18
– equipment 61
– muscle 274
– the cleaning 18
tetanic contraction 279
textile
– base materials 1321
– based capacitive EMG-electrode 1329
– based contactless electrode 1327
– based heating system 1331
– based implant construction 1335
– based sensor 1321
– capacitor 1328
– compression 1321
– electrode 1322, 1330, 1361
– electrode for electromyostimulation 1330
– electrode system 1327
– functional 1321
– garment 1345
– heatable 1332
– heating system 1333
– medical applications 1321
– sensor 1328
– smart 1228, 1322, 1325, 1345
textile heating module 1332
– manufacturing process 1332
textural feature 1148
texture 1151
– level 1141
tf-LIFE 798
– electrode assignment 793
– fields of application 798
thenar 431
– skin blood flow 435
theranostic 458
therapeutic electron radiation 711
therapeutic hyperthermia 898
therapeutic monitoring 84, 165
therapeutic MR-guided imaging 493
therapeutic tool 1167
therapy 1158
– acoustic 643
– anticongestion 1335
– external beam 902, 907
– interventional 1210
– machine-aided 175
– of renal failure 600
– percutaneous 709
– planning 715, 1175
thermal
– ablation 419
– comb 932
– damage 683
– disinfection 14, 1313
– effect 681
– imaging band 370
– imbalance 1285
– process 14
– radiation sensor 988
– treatment bandage 1332
thermistor 929, 988
thermocouple 929, 930
thermodilution curve 967
thermodynamic inflammatory response 566
thermoelectric
– converter 1135
– transducer 929
– voltage 929
thermoelectric cooler (TEC) 371
thermographic measurement 940
thermoluminescent dosimeters (TLD) 708
thermoregulation 374
– sequence 376
theta activity 170
theta-wave 129
thickness of the cornea 265
thin-film electrode 793, 810
thin-film longitudinal intrafascicular electrode (tf-LIFE) 792, 793, 797

thin-film microelectrode 794
– micromachining 794
thin-film transistor (TFT) 300, 322, 483, 951
– weight advantage 386
third stage of labor and delivery 1027
third thread system 1323
third-generation scanner 317, 323
Thomas Graham 598
thoracic electrical bioimpedance (TEB) 968
thoracic impedance 546, 547
thoracoabdominal asynchrony 1034
thorax surgery 1268
thread electrode 1331
thread material 1324
– conductive 1325
three-dimensional (3-D) 1145
– cephalometry 411
– dataset 1210
– digitization 1177
– imaging 360, 467
– reconstruction 408
– representation of instruments 1179
– spacer fabric 1323
three-dimensional conformal treatment planning (3-D CRT) 903
three-dimensional rotation angiography (3-D DSA) 304
three-level laser 560
threshold
– breach alarm 1025
– method 412
– of hearing 233
threshold tone decay test (TTDT) 205
thresholding 1148, 1149
thromboembolism 740
thrombolysis 419
thumb adduction 279
thyroid surgery 1047, 1048
thyroidectomy 1051
tibial nerve 153
tidal
– breathing 105
– volume 96, 101, 590
Tikhonov regularization 470
time-based electromyographic signal 1048
time-critical vital signs 1086
time-cycled mode 531
time-division multiple access (TDMA) 1369
time-of-flight MR angiography (ToF-MRA) 451
times of echo (TE) 478
times of repetition (TR) 478
time-varying conductance 1249
tinnitus 228
tip tracking 507
tissue
– characterization 398
– class 1157
– cryogenic ablation 483
– engineering 1335
– harmonic imaging 348
– oxygen perfusion 424
– oxygenation device 428
– perfusion 429
– resistance 682, 683
– specific relaxation time 440
– vitality 335
tissue Doppler imaging (TDI) 356
tissue hemoglobin index (THI) 424
tissue hemoglobin measurement (THC) 425
tissue oxygen saturation (StO_2) 436
tissue oxygenation index (TOI) 429
titanium
– housing 789
– scissors 1294
toco procedure 1020
tomogram
– 2-D 397
– 3-D 397
tomographic image reconstruction 324
tomosynthesis 313
tomotherapy 904
tonometry 805
– Mackay–Marg 247
tonotopic stimulation 790
toothed scissors 1303
TopCare Kiosk 1226
top-down
– biomechatronic design 838
– segmentation 1151
topogram 329
topography of larynx 1051
topology model 1105
torque control 844
torsional movement 181
total artificial heart (TAH) 724, 732
total body irradiation (TBI) 902, 907
– parallel-opposing-beams technique 908
– translational technique 908
total functional hemoglobin 976
total imaging matrix 457
total intravenous anesthesia (TIVA) 579, 1046
total lung capacity (TLC) 101
total perfusion time 629
total peripheral resistance (TPR) 969
total skin electron beam therapy (TSEBT)
– continuous rotation 908
– several field arrangement 908
total skin irradiation with electrons 902
total work 846
tracer material 472
tracking operation 1363
tracking saccade 180
tracking technique 1179
trademark 859
traditional surgery 5
trainer 1282
training 594
– computer-based system 1189
– data 1154
– scenario 1004
transcapnode 980, 1038
transcatheter delivered therapy 864
transcranial attenuation 204
transcranial magnetic stimulation (TMS) 830
transcutaneous
– blood gas measurement 976
– oxygen partial pressure $ptcO_2$ 979
– partial pressure 1037
– stimulation 767
transcutaneous electrical nerve stimulation (TENS) 1330
transcutaneous energy transmission (TET) 741, 745, 747
transdermal therapeutical systems (TTS) 1331
transducer 1019
– effect, sensing mechanism 1358
– force 280
– impedance 194
– pressure sensor 29
transduction velocity 253
transesophageal echocardiography (TEE) 741
transfer of images 1202
transformer theory 1350
transgastric operation 398
transient
– intracardial stimulation 768
– phenomenon 346

transient evoked otoacoustic emission (TEOAE) 215, 216, 219
translational medicine 1110
transmembrane pressure (TMP) 606, 607
transmission
– line coding 1135
– module 1355
– photometry 977, 1035
– protocol 933
transmission control protocol/Internet protocol (TCP/IP) 1162
transmitted
– ECG signal 85
– sound energy 646
– sound wave 344
transmitter 1179
transmitting coil 478
transoxode 1038
transparency of patient data 1097
transparent shading 1160
transport incubator 1288
transthermy 677
transthoracic pacemaker 548
transurethral resection (TUR) 682
transvenous electrode 554
transverse excitation 444
transverse nuclear magnetization 440, 441
trapezoidal technique 364
trauma 335
– patient 431
traveling wave 196
treadmill ergometer 92, 111
treated volume 910
treatment delivery 900
treatment of disc herniation 394
treatment of epilepsy 822
treatment planning 898, 901, 1158
– principles 899
treatment planning system (TPS) 905, 910
– basic component 911
– beam modelling module 913
– contouring module 912
– dose calculation module 913
– evaluation and optimization module 913
– export module 914
– import module 912
– physics module 911
– plan analysis module 914
trend display 952
triceps brachii muscle 886
triple-chamber ICD 553
triplex mode 355
trocar 387
troponin 472
true-tripole (TT) 1062
TruST algorithm 956
tube compensation (TC) 542
tube operated camera 296
tube system 12
tubing system 623
– for ECC 625
tubular narrow fabric 1324
tumor
– ablation 494
– ablation of 493
– localization 915
– perfusion 335
– volume 909, 910
tungsten anode, heat capacity 319
TUR syndrome 690
turbine flowmeter 100
turbo spin echo (TSE) 449
tutorial system 1190
two-chamber cardiac pacemaker 776
two-dimensional (2-D) 1141, 1210
– imaging 467
– reconstruction 326
two-dimensional brightness mode (2-D B-mode) 352
two-stage cannula 626
tympanic membrane 194
tympanometry 213
type B protection 684
types of scissors 1297

U

ultrafiltration 602, 607
– rate 606, 610
ultrafiltration coefficient (UF) 606, 608
ultrahard photon radiation 711
ultrahard x-radiation 707
ultrasound (US) 263, 346, 477–479
– biometry 264
– contrast medium 349
– diagnostic 343
– digital encoding 352
– Doppler method 937
– endoscope 396
– flowmeter 100
– guided ablation 494
– image generation 344
– mediated targeted drug delivery 494
– pachymetry 264
– wave 933
ultrasound (US) probe 344
– design, type of probe 350
– integrated 3-D/4-D 362
ultrasound (US) targeting 666
– inline 666
– online 666
ultrasound (US) unit
– basic design 350
– component 350, 351
– operation 367
ultrasound pneumography (UPG) 104
ultrasound-induced targeted drug delivery (UITDD) 493
ultraviolet (UV) 370
uncooled detector 371
undersaturation of blood 1031
under-segmentation 1152
undulant fever 987
unimodal registration 1145
unipolar lead 80
unipolar recording 123
unipolar stimulation 770
United States Renal Data System (USRDS) 612
univariate
– biostatistical method 941
– statistics 941
unperturbed motion 847, 849
unplanned incident 1110
unrecognized RLN palsy 1050
unsupervised training 1154
urea clearance 605
ureter stone lithotripsy 643
urine
– artificial 599
– production 633
US Assist Device Registry INTERMACS level 726
US Food and Drug Administration (FDA) 754
usability
– assessment matrix 67
– criterium 64
– incident 61
– international standardization 60
– laboratory 64
– orientated development 64
– problem 65
– test 64
usability engineering 59
– stages of development 62
usable device
– development 62
– testing 64
use of resources 1097

useful signal 77
useful work 846
user error 61
user guidance 947, 952
user safety 697
user–product interface 60
Utah electrode array (UEA) 790, 808
Utah wireless neural interface 809
uterine activity 1019
uterine fibroid 493
UV radiation sensor 1358
UV-laser radiation 567
U test 940

V

vaccination 13
VAD system 728
VAD-induced hemolysis 740
validated processes 17
validation 17, 87
value domain 1144
value-adding process 49
vaporization 567, 683
vaporizer principle 584
variability 209
variable pressure support 535
variant Creutzfeldt–Jakob disease (vCJD) 29
vascular
– intervention 417
– narrowing 374
– scissors 1305, 1307
– system 572
vascular occlusion test (VOT) 431
vasoconstriction 435, 1285
vasodilation 435
VC (vowel–consonant) 209
vegetative nervous system 120
velocity
– blood flow 936
– of hemoglobin 426
vena cava filter (VCF) 496
vendor-neutral archive (VNA) 1206
venous
– cannulation 626
– oxygen saturation 979
– pressure 963
venous saturation (SvO_2) 632
ventilation
– controlled 530, 531, 588
– frequency 539
– mandatory 532
– manual 588
– measurement 112
– mechanical 28, 532
– mode 536
– parameter 529
– procedure 531
– reason 526
ventilator 525, 580, 581, 585, 587, 973
– associated 28
– associated pneumonia 28
– breathing circuit 28
– breathing system 585
– gas flow behavior 529
– monitoring 587
– parameter settings 581
– reason for using 526
– subcomponents 585
ventricle-controlled DDD pacemaker, flow diagram 779
ventricular
– arrhythmia 87
– electrode 775
– fibrillation 548, 549
– flutter 549, 551
– probe 770
ventricular assist device (VAD) 724, 1131
ventricular extrasystole (VES) 777, 957, 1032
Venturi tube 100
verification system
– LANTIS (Siemens) 915
– MOSAIQ (IMPAC, Elekta) 915
– VARiS (Varian) 915
vertical eye movement 182
vertical open-configuration MR 480
vertical procedure 1112
very long-wave infrared (VLWIR) 370
vessel
– analysis 335
– sealing 683, 696
– wall 715
vestibular system 179, 186
vestibulo-ocular reflex (VOR) 181
vibrational energy 426
video chain 386
video EEG 135
videocolonoscope 395
videocolonoscopy 401
videocystoscope 385
videoendoscope 384
videogastroscopy 401
videooculography (VOG) 184
– examination method 184
videoteleconference (VTC) 1225
viewing angle 951
virtual
– anatomical structure 1170
– anatomy 1175, 1178
– chromoendoscopy 397
– colonoscopy 1213
– endoscopy 333, 381, 1170, 1211
– laboratory 1190
– mucosal surface 1213
– patient 1172, 1191
– simulation 901, 916
– uterus 1176
virtual LAN (VLAN) 953
virtual reality (VR) 1004, 1013, 1167, 1169, 1210
– based medical simulator 1168
– method 1210
– smart home control 1013
Visante OCT 249
viscous friction 841
visible heart apparatus 864
visible human Project 1170
visual
– alarm 952
– center 823
– control 394
– cortex 822, 824
– evaluation 169
– feedback 1180
– field 270
– field defect 269
– impairment 821
– information processing 824
– sensation 823
– stimulator 142
– system, functional segregation 824
visual evoked potential (VEP) 149, 151, 253, 830
visual neuroprosthesis 830
– criteria 830
– functional blocks 829
visual pathway 823
– safe stimulation 829
visual prosthesis 822
– future perspectives 831
visualization 1160, 1164
– 3-D 419, 1160, 1209, 1213
– of the blood vessel 417
visually handicapped 821
vita maxima 113
vital capacity (VC) 101
vital function 120
vital signal requirement 1366
VitalJacket 1327
vividiffusion apparatus 599
volatile anaesthetics 45